New Pattern

Mathematics

for JEE (Advanced)

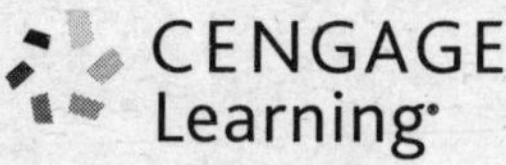

Andover • Melbourne • Mexico City • Stamford, CT • Toronto • Hong Kong • New Delhi • Seoul • Singapore • Tokyo

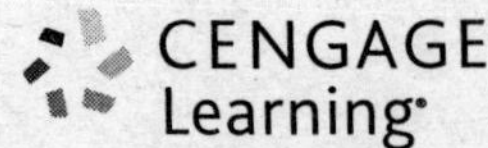

New Pattern
Mathematics
for JEE (Advanced)

ISBN-13: 978-81-315-1960-8
ISBN-10: 81-315-1960-0

Cengage Learning India Pvt. Ltd.
418, F.I.E., Patparganj
Delhi 110092

Cengage Learning is a leading provider of customized learning solutions with office locations around the globe, including Andover, Melbourne, Mexico City, Stamford (CT), Toronto, Hong Kong, New Delhi, Seoul, Singapore, and Tokyo. Locate your local office at:
www.cengage.com/global

Cengage Learning products are represented in Canada by Nelson Education, Ltd.

For product information, visit **www.cengage.co.in**

Printed in India
First Impression 2013

Contents

Section A: Trigonometry

Section B: Algebra

Section C: Coordinate Geometry

Section D: Calculus

Section E: Vectors & 3D Geometry

Preface

Over the years, the pattern of the Joint Entrance Examination, conducted by the Indian Institutes of Technology, has witnessed a vast change. We know it well that merely cramming formulae and theory does not ensure success in the examination. It takes a thorough understanding of concepts, guided practice, and hard work to crack one of the world's most prestigious engineering entrance examinations. Keeping this in mind, this book, *New Pattern Mathematics for IIT-JEE*, presents a comprehensive collection of various types of questions based on the latest pattern and trends in the IIT-JEE. This book is extremely useful to the aspirants who, having studied theory thoroughly, are looking for a diverse pool of questions to practice. They can also hone their problem-solving skills by comparing their approaches with that adopted in the solutions to the questions provided in this book.

This book has been divided into five sections: Trigonometry, Algebra, Coordinate Geometry, Calculus, and Vectors & 3D Geometry. The sections have been further divided into chapters. Each chapter features all types of problems asked in the examination—Objective Type, Multiple Correct Answers Type, Reasoning Type, Comprehension Type, Matrix-Match Type, and Integer Type. A mix of questions helps stimulate and strengthen multi-dimensional problem-solving skills of an aspirant. Solutions to most of the questions have been provided in a separate section at the end of each chapter. We believe that this book will immensely help IIT aspirants strengthen their grasp of concepts and achieve success in the examination.

Section A

Trigonometry

1. Trigonometric Ratios and Identities
2. Trigonometric Equations
3. Inverse Trigonometric Functions
4. Solutions and Properties of Triangle

CHAPTER 1

Trigonometric Ratios and Identities

Objective Type

Each question has four choices a, b, c, and d, out of which *only* *one* answer is correct. Find the correct answer.

1. Which of the following is correct?

   a. $\sin 1^\circ > \sin 1$ **b.** $\sin 1^\circ < \sin 1$

   c. $\sin 1^\circ = \sin 1$ **d.** $\sin 1^\circ = \dfrac{\pi}{180} \sin 1$

2. The equation $\sin^2 \theta = \dfrac{x^2 + y^2}{2xy}$ is possible if

   a. $x = y$ **b.** $x = -y$

   c. $2x = y$ **d.** None of these

3. If $1 + \sin x + \sin^2 x + \sin^3 x + \cdots \infty$ is equal to $4 + 2\sqrt{3}$, $0 < x < \pi$, then x is equel to

   a. $\dfrac{\pi}{6}$

   b. $\dfrac{\pi}{4}$

   c. $\dfrac{\pi}{3}$ or $\dfrac{\pi}{6}$

   d. $\dfrac{\pi}{3}$ or $\dfrac{2\pi}{3}$

4. If $\dfrac{\sin x}{a} = \dfrac{\cos x}{b} = \dfrac{\tan x}{c} = k$, then $bc + \dfrac{1}{ck} + \dfrac{ak}{1 + bk}$ is equal to

   a. $k\left(a + \dfrac{1}{a}\right)$ **b.** $\dfrac{1}{k}\left(a + \dfrac{1}{a}\right)$

   c. $\dfrac{1}{k^2}$ **d.** $\dfrac{a}{k}$

5. If A, B, C are angles of a triangle, then

   $2\sin \dfrac{A}{2} \operatorname{cosec} \dfrac{B}{2} \sin \dfrac{C}{2} - \sin A \cot \dfrac{B}{2} - \cos A$ is

   a. independent of A, B, C

   b. function of A, B

   c. function of C

   d. none of these

6. The least value of $6 \tan^2 \phi + 54 \cot^2 \phi + 18$ is

   I: 54 when A.M. ≥ G.M. is applicable for $6 \tan^2 \phi$, $54 \cot^2 \phi$, 18.

   II: 54 when A.M. ≥ G.M is applicable for $6 \tan^2 \phi$, $54 \cot^2 \phi$ and 18 added further.

   III: 78 when $\tan^2 \phi = \cot^2 \phi$.

   a. I is correct

   b. I and II are correct

   c. III is correct

   d. None of the above is correct

7. If $5 \tan \theta = 4$, then $\dfrac{5 \sin \theta - 3 \cos \theta}{5 \sin \theta + 2 \cos \theta}$ is equal to

   a. 0 **b.** 1

   c. 1/6 **d.** 6

8. If $2 \sec 2\theta = \tan \phi + \cot \phi$ then one of the values of $\theta + \phi$ is

   a. $\pi/2$ **b.** $\pi/4$

   c. $\pi/3$ **d.** None of these

9. If $\sin x + \operatorname{cosec} x = 2$, then $\sin^n x + \operatorname{cosec}^n x$ is equal to

   a. 2 **b.** 2^n

   c. 2^{n-1} **d.** 2^{n-2}

10. A quadratic equation whose roots are $\operatorname{cosec}^2\theta$ and $\sec^2\theta$ can be

a. $x^2-2x+2=0$ b. $x^2-3x+3=0$
c. $x^2-5x+5=0$ d. None of these

11. If $\pi<\alpha<\dfrac{3\pi}{2}$, then $\sqrt{\dfrac{1-\cos\alpha}{1+\cos\alpha}}+\sqrt{\dfrac{1+\cos\alpha}{1-\cos\alpha}}$ is equal to

a. $2/\sin\alpha$ b. $-2/\sin\alpha$
c. $1/\sin\alpha$ d. $-1/\sin\alpha$

12. The value of $\cos\dfrac{\pi}{7}+\cos\dfrac{2\pi}{7}+\cos\dfrac{3\pi}{7}+\cos\dfrac{4\pi}{7}+\cos\dfrac{5\pi}{7}+\cos\dfrac{6\pi}{7}+\cos\dfrac{7\pi}{7}$ is

a. 1 b. –1
c. 0 d. None of these

13. The least value of $2\sin^2\theta+3\cos^2\theta$ is

a. 1 b. 2
c. 3 d. 5

14. The greatest value of $\sin^4\theta+\cos^4\theta$ is

a. 1/2 b. 1
c. 2 d. 3

15. If $f(x)=\cos^2\theta+\sec^2\theta$, then

a. $f(x)<1$ b. $f(x)=1$
c. $2>f(x)>1$ d. $f(x)\geq 2$

16. If $f(x)=\sin^6 x+\cos^6 x$, then range of $f(x)$ is

a. [1/4, 1] b. [1/4, 3/4]
c. [3/4, 1] d. None of these

17. If $a\leq 3\cos x+5\sin(x-\pi/5)\leq b$ for all x, then (a, b) is

a. $(-\sqrt{19}, \sqrt{19})$ b. $(-17, 17)$
c. $(-\sqrt{21}, \sqrt{21})$ d. None of these

18. The equation $\sin x(\sin x+\cos x)=k$ has real solutions if and only if k is a real number such that

a. $0\leq k\leq\dfrac{1+\sqrt{2}}{2}$ b. $2-\sqrt{3}\leq k\leq 2+\sqrt{3}$
c. $0\leq k\leq 2-\sqrt{3}$ d. $\dfrac{1-\sqrt{2}}{2}\leq k\leq\dfrac{1+\sqrt{2}}{2}$

19. If $\cot(\alpha+\beta)=0$, then $\sin(\alpha+2\beta)$ is equal to

a. $-\sin a$ b. $\sin b$
c. $\cos\alpha$ d. $\cos\beta$

20. If $\dfrac{x}{\cos\theta}=\dfrac{y}{\cos\left(\theta-\dfrac{2\pi}{3}\right)}=\dfrac{z}{\cos\left(\theta+\dfrac{2\pi}{3}\right)}$, then $x+y+z$ is equal to

a. 1 b. 0
c. –1 d. None of these

21. $\sin^{2n}x+\cos^{2n}x$ lies between

a. –1 and 1 b. 0 and 1
c. 1 and 2 d. None of these

22. The roots of the equation $4x^2-2\sqrt{5}\,x+1=0$ are

a. sin 36°, sin 18° b. sin 18°, cos 36°
c. sin 36°, cos 18° d. cos 18°, cos 36°

23. If $\dfrac{3\pi}{4}<\alpha<\pi$, then $\sqrt{2\cot\alpha+\dfrac{1}{\sin^2\alpha}}$ is equal to

a. $1+\cot\alpha$ b. $-1-\cot\alpha$
c. $1-\cot\alpha$ d. $-1+\cot\alpha$

24. If $f(\theta)=5\cos\theta+3\cos\left(\theta+\dfrac{\pi}{3}\right)+3$, then range of $f(\theta)$ is

a. [–5, 11] b. [–3, 9]
c. [–2, 10] d. [–4, 10]

25. If α, β, γ, δ are the smallest positive angles in ascending order of magnitude which have their sines equal to the positive quantity k, then the value of $4\sin\dfrac{\alpha}{2}+3\sin\dfrac{\beta}{2}+2\sin\dfrac{\gamma}{2}+\sin\dfrac{\delta}{2}$ is equal to

a. $2\sqrt{1-k}$ b. $2\sqrt{1+k}$
c. $\dfrac{\sqrt{1+k}}{2}$ d. None of these

26. Let $A_0A_1A_2A_3A_4A_5$ be a regular hexagon inscribed in a circle of unit radius. Then the product of the lengths of the line segments A_0A_1, A_0A_2 and A_0A_4 is

a. $\dfrac{3}{4}$ b. $3\sqrt{3}$
c. 3 d. $\dfrac{3\sqrt{3}}{2}$

27. If $\sin\theta_1+\sin\theta_2+\sin\theta_3=3$, then $\cos\theta_1+\cos\theta_2+\cos\theta_3$ is equal to

a. 3 b. 2
c. 1 d. 0

28. If $\sin^2\theta=\dfrac{x^2+y^2+1}{2x}$, then x must be

a. –3 b. –2
c. 1 d. None of these

29. If $\sin(\alpha+\beta)=1$, $\sin(\alpha-\beta)=\dfrac{1}{2}$, then $\tan(\alpha+2\beta)\tan(2\alpha+\beta)$ is equal to

a. 1 b. – 1
c. 0 d. None of these

30. The expression $3\left[\sin^4\left(\frac{3}{2}\pi-\alpha\right)+\sin^4(3\pi+\alpha)\right]$ $\times 2\left[\sin^6\left(\frac{3\pi}{2}-\alpha\right)+\cos^6\left(\frac{5\pi}{2}+\alpha\right)\right]$ is equal to

a. 0 b. 1
c. 3 d. None of these

31. If cosec θ – cot $\theta = q$, then the value of cosec θ is

a. $\theta+\frac{1}{q}$ b. $\theta-\frac{1}{q}$
c. $\frac{1}{2}\left(q+\frac{1}{q}\right)$ d. None of these

32. If $x = \sec\theta - \tan\theta$ and $y = \operatorname{cosec}\theta + \cot\theta$, then

a. $xy+1=x-y$ b. $xy+1=y-x$
c. $xy+1=x+y$ d. None of these

33. If $x=\frac{2\sin\theta}{1+\cos\theta+\sin\theta}$, then $\frac{1-\cos\theta+\sin\theta}{1+\sin\theta}$ is equal to

a. $1+x$ b. $1-x$
c. x d. $1/x$

34. If $\theta = \pi/4n$, then the value of $\tan\theta\tan 2\theta \cdots \tan(2n-2)\theta\tan(2n-1)\theta$ is

a. –1 b. 1
c. 0 d. 2

35. If A, B, C, D are the smallest positive angles in ascending order of magnitude which have their sines equal to the positive quantity k, then the value of $4\sin\frac{A}{2}+3\sin\frac{B}{2}+2\sin\frac{C}{2}+\sin\frac{D}{2}$ is

a. $2\sqrt{1-k}$ b. $2\sqrt{1+k}$
c. $2\sqrt{k}$ d. None of these

36. If sec α and cosec α are the roots of $x^2-px+q+0$, then

a. $p^2=q(q-2)$ b. $p^2=q(q+2)$
c. $p^2+q^2=2q$ d. None of these

37. If $\sin x+\sin^2 x=1$, then the value of $\cos^{12} x + 3\cos^{10} x + 3\cos^8 x + \cos^6 x - 2$ is equal to

a. 0 b. 1
c. –1 d. 2

38. If $\cos(A-B)=3/5$ and $\tan A\tan B=2$, then

a. $\cos A\cos B=\frac{1}{5}$ b. $\sin A\sin B=-\frac{2}{5}$
c. $\cos A\cos B=-\frac{1}{5}$ d. $\sin A\sin B=-\frac{1}{5}$

39. If $(1+\tan\alpha)(1+\tan 4\alpha)=2$, $\alpha\in(0,\pi/16)$ then α is equal to

a. $\frac{\pi}{20}$ b. $\frac{\pi}{30}$
c. $\frac{\pi}{40}$ d. $\frac{\pi}{60}$

40. If $A=\sin 45°+\cos 45°$ and $B=\sin 44°+\cos 44°$, then

a. $A>B$ b. $A<B$
c. $A=B$ d. None of these

41. $\frac{1}{4}\left[\sqrt{3}\cos 23°-\sin 23°\right]$ is equal to

a. cos 43° b. cos 7°
c. cos 53° d. None of these

42. If $\cos\theta_1=2\cos\theta_2$, then $\tan\frac{\theta_1-\theta_2}{2}\tan\frac{\theta_1+\theta_2}{2}$ is equal to

a. $\frac{1}{3}$ b. $-\frac{1}{3}$
c. 1 d. –1

43. The value of $\frac{3+\cot 80°\cot 20°}{\cot 80°+\cot 20°}$ is equal to

a. cot 20° b. tan 50°
c. cot 50° d. cot $\sqrt{20°}$

44. If $\tan\beta=2\sin\alpha\sin\gamma\operatorname{cosec}(\alpha+\gamma)$, then cot α, cot β, cot γ are in

a. A.P. b. G.P.
c. H.P. d. None of these

45. In a triangle ABC, if $\sin A\cos B=1/4$ and $3\tan A=\tan B$, then $\cot^2 A$ is equal to

a. 2 b. 3
c. 4 d. 5

46. tan 100° + tan 125° + tan 100° tan 125° is equal to

a. 0 b. 1/2
c. –1 d. 1

47. tan 20° + tan 40° + $\sqrt{3}$ tan 20° tan 40° is equal to

a. $\frac{1}{\sqrt{3}}$ b. $\sqrt{3}$
c. $-\frac{1}{\sqrt{3}}$ d. $-\sqrt{3}$

48. $\frac{\sqrt{2}-\sin\alpha-\cos\alpha}{\sin\alpha-\cos\alpha}$ is equal to

a. $\sec\left(\frac{\alpha}{2}-\frac{\pi}{8}\right)$ b. $\cos\left(\frac{\pi}{8}-\frac{\alpha}{2}\right)$
c. $\tan\left(\frac{\alpha}{2}-\frac{\pi}{8}\right)$ d. $\cot\left(\frac{\alpha}{2}-\frac{\pi}{2}\right)$

49. If $\sin\theta_1 - \sin\theta_2 = a$ and $\cos\theta_1 + \cos\theta_2 = b$, then

a. $a^2 + b^2 \geq 4$ **b.** $a^2 + b^2 \leq 4$

c. $a^2 + b^2 \geq 3$ **d.** $a^2 + b^2 \leq 2$

50. If $\dfrac{1+\sin 2x}{1-\sin 2x} = \cot^2(a+x), \forall x \in R \sim \left(n\pi + \dfrac{\pi}{4}\right), n \in N$, then a is equal to

a. $\pi/4$ **b.** $\pi/2$

c. $3\pi/4$ **d.** None of these

51. If $\tan\alpha$ is equal to the integral solution of the inequality $4x^2 - 16x + 15 < 0$ and $\cos\beta$ is equal to the slope of the bisector of the first quadrant, then $\sin(\alpha+\beta)\sin(\alpha-\beta)$ is equal to

a. 3/5 **b.** 3/5

c. $2/\sqrt{5}$ **d.** 4/5

52. If $\dfrac{\cos(x-y)}{\cos(x+y)} + \dfrac{\cos(z+t)}{\cos(z-t)} = 0$, then the value of $\tan x \tan y \tan z \tan t$ is equal to

a. 1 **b.** –1

c. 2 **d.** –2

53. Let $f(n) = 2\cos nx, \forall n \in N$, then $f(1)f(n+1) - f(n)$ is equal to

a. $f(n+3)$ **b.** $f(n+2)$

c. $f(n+1)f(2)$ **d.** $f(n+2)f(2)$

54. If in a triangle ABC, $\sin A \cos B = 1/4$ and $3\tan A = \tan B$, then the triangle is

a. right angled **b.** equilateral

c. isosceles **d.** none of these

55. If A and B are acute positive angles which satisfy the equalities $3\sin^2 A + 2\sin^2 B = 1$ and $3\sin 2A - 2\sin 2B = 0$, then $A + 2B$ is equal to

a. π **b.** $\pi/2$

c. $\pi/4$ **d.** $\pi/6$

56. Let $f(\theta) = \dfrac{\cot\theta}{1+\cot\theta}$ and $\alpha + \beta = \dfrac{5\pi}{4}$, then the value $f(\alpha)f(\beta)$ is

a. 1/2 **b.** 1/2

c. 2 **d.** None of these

57. If $y = (1 + \tan A)(1 - \tan B)$ where $A - B = \pi/4$, then $(y+1)^{y+1}$ is equal to

a. 9 **b.** 4

c. 27 **d.** 81

58. If $\sin(y+z-x)$, $\sin(z+x-y)$, $\sin(x+y-z)$ are in A.P., then $\tan x$, $\tan y$, $\tan z$ are in

a. A.P. **b.** G.P.

c. H.P. **d.** None of these

59. If $\cos\alpha + \cos\beta = 0 = \sin\alpha + \sin\beta$, then $\cos 2\alpha + \cos 2\beta$ is equal to

a. $-2\sin(\alpha+\beta)$ **b.** $-2\cos(\alpha+\beta)$

c. $2\sin(\alpha+\beta)$ **d.** $2\cos(\alpha+\beta)$

60. If $x_1, x_2, x_3, \ldots, x_n$ are in A.P. whose common difference is α, then the value of $\sin\alpha(\sec x_1 \sec x_2 + \sec x_2 \sec x_3 + \cdots + \sec x_{n-1}\sec x_n)$ is

a. $\dfrac{\sin(n-1)\alpha}{\cos x_1 \cos x_n}$

b. $\dfrac{\sin n\alpha}{\cos x_1 \cos x_n}$

c. $\sin(n-1)\alpha \cos x_1 \cos x_n$

d. $\sin n\alpha \cos x_1 \cos x_n$

61. If $\tan\dfrac{\pi}{9}$, x, and $\tan\dfrac{5\pi}{18}$ are in A.P. and $\tan\dfrac{\pi}{9}$, y, and $\tan\dfrac{7\pi}{18}$ are also in A.P., then

a. $2x = y$ **b.** $x > 2$

c. $x = y$ **d.** None of these

62. If $\tan\alpha = \dfrac{m}{m+1}$ and $\tan\beta = \dfrac{1}{2m+1}$, then $\alpha + \beta$ is equal to

a. $\dfrac{\pi}{3}$ **b.** $\dfrac{\pi}{4}$

c. $\dfrac{\pi}{6}$ **d.** None of these

63. Let α and, β be such that $\pi < \alpha - \beta < 3\pi$. If $\sin\alpha + \sin\beta = -21/65$ and $\cos\alpha + \cos\beta = -17/65$, then the value of $\cos(\alpha-\beta)/2$ is

a. $-3/\sqrt{130}$ **b.** $3/\sqrt{130}$

c. 6/65 **d.** –6/65

64. If $\dfrac{\sin(x+y)}{\sin(x-y)} = \dfrac{a+b}{a-b}$, then $\dfrac{\tan x}{\tan y}$ is equal to

a. b/a **b.** a/b

c. ab **d.** None of these

65. $\dfrac{\sin 3\theta + \sin 5\theta + \sin 7\theta + \sin 9\theta}{\cos 3\theta + \cos 5\theta + \cos 7\theta + \cos 9\theta}$ is equal to

a. $\tan 3\theta$ **b.** $\cot 3\theta$

c. $\tan 6\theta$ **d.** $\cot 6\theta$

66. If x, y, z are in A.P, then $\dfrac{\sin x - \sin z}{\cos z - \cos x}$ is equal to

a. $\tan y$ **b.** $\cot y$

c. $\sin y$ **d.** $\cos y$

67. If $\cos 25° + \sin 25° = p$, then $\cos 50°$ is

a. $\sqrt{2-p^2}$ **b.** $-\sqrt{2-p^2}$

c. $p\sqrt{2-p^2}$ **d.** $-p\sqrt{2-p^2}$

68. $\dfrac{\sin^2 A-\sin^2 B}{\sin A\cos A-\sin B\cos B}$ is equal to

a. $\tan(A-B)$ b. $\tan(A+B)$

c. $\cot(A-B)$ d. $\cot(A+B)$

69. If $\tan A=\dfrac{1-\cos B}{\sin B}$, then $\tan 2A$ is

a. $\tan 2A=\tan B$

b. $\tan 2A=\tan^2 B$

c. $\tan 2A=\tan^2 B+2\tan B$

d. none the these

70. If $a+b=3-\cos 4\theta$ and $a-b=4\sin 2\theta$, then ab is always less than or equal to

a. $\dfrac{1}{2}$ b. 1

c. $\dfrac{2}{3}$ d. $\dfrac{3}{4}$

71. The value of $\cos^2 10°-\cos 10°\cos 50°+\cos^2 50°$ is equal to

a. $\dfrac{4}{3}$ b. $\dfrac{1}{3}$

c. $\dfrac{3}{4}$ d. 3

72. The numerical value of $\tan 20°\tan 80°\cot 50°$ is equal to

a. $\sqrt{3}$ b. $\dfrac{1}{\sqrt{3}}$

c. $2\sqrt{3}$ d. $\dfrac{1}{2\sqrt{3}}$

73. If $\tan^2\theta=2\tan^2\phi+1$, then $\cos 2\theta+\sin^2\phi$ equals

a. -1 b. 0

c. 1 d. None of these

74. The value of $\cot 70°+4\cos 70°$ is

a. $\dfrac{1}{\sqrt{3}}$ b. $\sqrt{3}$

c. $2\sqrt{3}$ d. $\dfrac{1}{2}$

75. If x_1 and x_2 are two distinct roots of the equation $a\cos x+b\sin x=c$, then $\tan\dfrac{x_1+x_2}{2}$ is equal to

a. $\dfrac{a}{b}$ b. $\dfrac{b}{a}$

c. $\dfrac{c}{a}$ d. $\dfrac{a}{c}$

76. Given that $(1+\sqrt{1+x})\tan y=1+\sqrt{1-x}$. Then $\sin 4y$ is equal to

a. $4x$ b. $2x$

c. x d. None of these

77. If $\cos x=\tan y$, $\cos y=\tan z$, and $\cos z=\tan x$, then the value of $\sin x$ is

a. $2\cos 18°$ b. $\cos 18°$

c. $\sin 18°$ d. $2\sin 18°$

78. If $\sin 2\theta=\cos 3\theta$ and θ is an acute angle, then $\sin\theta$ equals

a. $\dfrac{\sqrt{5}-1}{4}$ b. $-\left(\dfrac{\sqrt{5}-1}{4}\right)$

c. $\dfrac{\sqrt{5}+1}{4}$ d. $\dfrac{-\sqrt{5}-1}{4}$

79. If θ_1 and θ_2 are two values lying in $[0, 2\pi]$ for which $\tan\theta=\lambda$, then $\tan\dfrac{\theta_1}{2}\tan\dfrac{\theta_2}{2}$ is equal to

a. 0 b. -1

c. 2 d. 1

80. If $\tan\theta=\sqrt{n}$, where $n\in N, \geq 2$, then $\sec 2\theta$ is always

a. a rational number

b. an irrational number

c. a positive integer

d. a negative integer

81. If $\sin x+\cos x=\dfrac{\sqrt{7}}{2}$ where $x\in A$, then $\tan\dfrac{x}{2}$ is equal to

a. $\dfrac{3-\sqrt{7}}{3}$ b. $\dfrac{\sqrt{7}-2}{3}$

c. $\dfrac{4-\sqrt{7}}{4}$ d. None of these

82. The values of $\sin^2\dfrac{\pi}{8}+\sin^2\dfrac{3\pi}{8}+\sin^2\dfrac{5\pi}{8}+\sin^2\dfrac{7\pi}{8}$ is

a. 1 b. 2

c. $1\dfrac{1}{8}$ d. $2\dfrac{1}{8}$

83. If $x\in\left(\pi, \dfrac{3\pi}{2}\right)$, then $4\cos^2\left(\dfrac{\pi}{4}-\dfrac{x}{2}\right)+\sqrt{4\sin^4 x+\sin^2 2x}$ is always equal to

a. 1 b. 2

c. -2 d. None of these

84. $\cos^3 x \sin 2x = \sum_{x=0}^{n} a_r \sin(rx),\ \forall x \in R$ then

a. $n=5, a_1=\frac{1}{2}$ b. $n=5, a_1=\frac{1}{4}$

c. $n=5, a_2=\frac{1}{8}$ d. $n=5, a_2=\frac{1}{4}$

85. The value of $\cos 2(\theta+\phi)+4\cos(\theta+\phi)\sin\theta\sin\phi + 2\sin^2\phi$ is

a. independent of θ only

b. independent of ϕ only

c. independent of both θ and ϕ

d. dependent on θ and ϕ

86. If $\cos 2B = \frac{\cos(A+C)}{\cos(A-C)}$, then $\tan A, \tan B, \tan C$ are in

a. A.P. b. G.P.

c. H.P. d. None of these

87. If $\cos x = \frac{2\cos y - 1}{2-\cos y}$, where $x, y \in (0, \pi)$, then $\tan\frac{x}{2}\cot\frac{y}{2}$ is equal to

a. $\sqrt{2}$ b. $\sqrt{3}$

c. $\frac{1}{\sqrt{2}}$ d. $\frac{1}{\sqrt{3}}$

88. If $\tan x = b/a$, then $\sqrt{(a+b)/(a-b)} + \sqrt{(a-b)/(a+b)}$ is equal to

a. $2\sin x/\sqrt{\sin 2x}$ b. $2\cos x/\sqrt{\cos 2x}$

c. $2\cos x/\sqrt{\sin 2x}$ d. $2\sin x/\sqrt{\cos 2x}$

89. If α is a root of $25\cos^2\theta + 5\cos\theta - 12 = 0$, $\frac{\pi}{2} < \alpha < \pi$, then $\sin 2\alpha$ is equal to

a. $\frac{24}{25}$ b. $-\frac{24}{25}$

c. $\frac{13}{18}$ d. $-\frac{13}{18}$

90. The value of $\tan 9° - \tan 27° - \tan 63° + \tan 81°$ is

a. 2 b. 3

c. 4 d. None of these

91. If $\sin^{-1} a + \sin^{-1} b + \sin^{-1} c = \pi$, then $a\sqrt{1-a^2} + b\sqrt{1-b^2} + c\sqrt{1-c^2}$ is equal to

a. $a+b+c$ b. $a^2b^2c^2$

c. $2abc$ d. $4abc$

92. If $A+B+C=\frac{3\pi}{2}$, then $\cos 2A + \cos 2B + \cos 2C$ is equal to

a. $1-4\cos A\cos B\cos C$

b. $4\sin A\sin B\sin C$

c. $1+2\cos A\cos B\cos C$

d. $1-4\sin A\sin B\sin C$

93. In triangle ABC, $\tan\frac{A}{2}$, $\tan\frac{B}{2}$, $\tan\frac{C}{2}$ are in H.P., then the value of $\cot\frac{A}{2}\cot\frac{C}{2}$ is equal to

a. 1 b. 2

c. 3 d. 4

94. If any triangle ABC, $\sin^2 A - \sin^2 B + \sin^2 C$ is always equal to

a. $2\sin A\sin B\cos C$ b. $2\sin A\cos B\sin C$

c. $2\sin A\cos B\cos C$ d. $2\sin A\sin B\sin C$

95. If $\tan^2\alpha\tan^2\beta + \tan^2\beta\tan^2\gamma + \tan^2\gamma\tan^2\alpha + 2\tan^2\alpha\tan^2\beta\tan^2\gamma = 1$, then the value of $\sin^2\alpha + \sin^2\beta + \sin^2\gamma$ is

a. 0 b. 1

c. 1 d. None of these

96. In a triangle ABC, $\frac{\sin A + \sin B + \sin C}{\sin A + \sin B - \sin C}$ is equal to

a. $\tan\frac{A}{2}\cot\frac{B}{2}$ b. $\cot\frac{A}{2}\tan\frac{B}{2}$

c. $\cot\frac{A}{2}\cot\frac{B}{2}$ d. $\tan\frac{A}{2}\tan\frac{B}{2}$

97. $\frac{\sin 2A + \sin 2B + \sin 2C}{\sin A + \sin B + \sin C}$ is equal to

a. $8\sin\frac{A}{2}\sin\frac{B}{2}\sin\frac{C}{2}$ b. $8\cos\frac{A}{2}\cos\frac{B}{2}\cos\frac{C}{2}$

c. $8\tan\frac{A}{2}\tan\frac{B}{2}\tan\frac{C}{2}$ d. $8\cot\frac{A}{2}\cot\frac{B}{2}\cot\frac{C}{2}$

98. If $\cos^2 A + \cos^2 B + \cos^2 C = 1$, then ΔABC is

a. equilateral b. isosceles

c. right angled d. None of these

99. In a triangle ABC, $\tan A + \tan B + \tan C = 6$ and $\tan A\tan B = 2$, then the values of $\tan A$, $\tan B$, and $\tan C$ are

a. 1,2,3 b. 3, 2/3, 7/3

c. 4, 1/2, 3/2 d. None of these

100. The value of $\tan 6°\tan 42°\tan 66°\tan 78°$ is

a. 1 b. $\frac{1}{2}$

c. $\frac{1}{4}$ d. $\frac{1}{8}$

101. If $0<\alpha<\frac{\pi}{6}$, then $\alpha(\text{cosec}\,\alpha)$ is

a. less than $\frac{\pi}{6}$ b. greater than $\frac{\pi}{6}$

c. less than $\frac{\pi}{3}$ d. greater than $\frac{\pi}{3}$

102. The number of integral values of a for which there is no solution of the equation $a\cos x+\cot x+1=\text{cosec}\,x$ is

a. 1 b. 2

c. 3 d. 4

103. If $\left|\cos\theta\left\{\sin\theta+\sqrt{\sin^2\theta+\sin^2\alpha}\right\}\right|\le k$, then the value of k is

a. $\sqrt{1+\cos^2\alpha}$ b. $\sqrt{1+\sin^2\alpha}$

c. $\sqrt{2+\sin^2\alpha}$ d. $\sqrt{2+\cos^2\alpha}$

104. If $\sin\theta_1\sin\theta_2-\cos\theta_1\cos\theta_2+1=0$, then the value of $\tan\left(\frac{\theta_1}{2}\right)\cot\left(\frac{\theta_2}{2}\right)$ is always equal to

a. -1 b. 1

c. 2 d. -2

105. The numerical value of $\tan\frac{\pi}{3}+2\tan\frac{2\pi}{3}+4\tan\frac{4\pi}{3}+8\tan\frac{8\pi}{3}$ is equal to

a. $-5\sqrt{3}$ b. $-\frac{5}{\sqrt{3}}$

c. $5\sqrt{3}$ d. $\frac{5}{\sqrt{3}}$

106. $\tan^6\frac{\pi}{9}-33\tan^4\frac{\pi}{9}+27\tan^2\frac{\pi}{9}$ is equal to

a. 0 b. $\sqrt{3}$

c. 3 d. 9

107. If $\cos x+\cos y-\cos(x+y)=3/2$, then

a. $x+y=0$ b. $x=2y$

c. $x=y$ d. $2x=y$

108. If $a\sin x+b\cos(x+\theta)+b\cos(x-\theta)=d$, then the minimum value of $|\cos\theta|$ is equal to

a. $\frac{1}{2|b|}\sqrt{d^2-a^2}$ b. $\frac{1}{2|a|}\sqrt{d^2-a^2}$

c. $\frac{1}{2|d|}\sqrt{d^2-a^2}$ d. None of these

109. If $\frac{\sin x}{\sin y}=\frac{1}{2}$, $\frac{\cos x}{\cos y}=\frac{3}{2}$ where $x, y\in\left(0,\frac{\pi}{2}\right)$, then the value of $\tan(x+y)$ is equal to

a. $\sqrt{13}$ b. $\sqrt{14}$

c. $\sqrt{17}$ d. $\sqrt{15}$

110. If $x\in(\pi, 2\pi)$ and $\frac{\sqrt{1+\cos x}+\sqrt{1-\cos x}}{\sqrt{1+\cos x}-\sqrt{1-\cos x}}=\cot\left(a+\frac{x}{2}\right)$, then a is equal to

a. $\frac{\pi}{4}$ b. $\frac{\pi}{2}$

c. $\frac{\pi}{3}$ d. None of these

111. If $\tan x=n\tan y$, $n\in R^+$, then maximum value of $\sec^2(x-y)$ is equal to

a. $\frac{(n+1)^2}{2n}$ b. $\frac{(n+1)^2}{n}$

c. $\frac{(n+1)^2}{2}$ d. $\frac{(n+1)^2}{4n}$

112. If $\cot^2 x=\cot(x-y)\cot(x-z)$, then $\cot 2x$ is equal to $(x\neq\pm\pi/4)$

a. $\frac{1}{2}(\tan y+\tan z)$ b. $\frac{1}{2}(\cot y+\cot z)$

c. $\frac{1}{2}(\sin y+\sin z)$ d. None of these

113. If A, B, C are acute positive angles such that $A+B+C=\pi$ and $\cot A\cot B\cot C=k$, then

a. $K\le\frac{1}{3\sqrt{3}}$ b. $K\ge\frac{1}{3\sqrt{3}}$

c. $K<\frac{1}{9}$ d. $K>\frac{1}{3}$

114. If $u=\sqrt{a^2\cos^2\theta+b^2\sin^2\theta}+\sqrt{a^2\sin^2\theta+b^2\cos^2\theta}$, then the difference between the maximum and minimum values of u^2 is given by

a. $2(a^2+b^2)$ b. $2\sqrt{a^2+b^2}$

c. $(a+b)^2$ d. $(a-b)^2$

Multiple Correct Answers Type

Each question has four choices a, b, c, and d, out of which one or more answers are correct.

1. If $\cos\beta$ is the geometric mean between $\sin\alpha$ and $\cos\alpha$, where $0<\alpha, \beta<\frac{\pi}{2}$, then $\cos 2\beta$ is equal to

a. $-2\sin^2\left(\frac{\pi}{4}-\alpha\right)$ b. $-2\cos^2\left(\frac{\pi}{4}+\alpha\right)$

c. $2\sin^2\left(\frac{\pi}{4}+\alpha\right)$ d. $2\cos^2\left(\frac{\pi}{4}-\alpha\right)$

2. If $0 \le \theta \le \pi$ and $81^{\sin^2\theta} + 81^{\cos^2\theta} = 30$, then θ is

a. 30° b. 60°

c. 120° d. 150°

3. Suppose *ABCD* (in order) is a quadrilateral inscribed in a circle. Which of the following is/are always true?

a. $\sec B = \sec D$ b. $\cot A + \cot C = 0$

c. $\operatorname{cosec} A = \operatorname{cosec} C$ d. $\tan B + \tan D = 0$

4. Which of the following statements are always correct (where Q denotes the set of rationals)?

a. $\cos 2\theta \in Q$ and $\sin 2\theta \in Q \Rightarrow \tan\theta \in Q$ (if defined)

b. $\tan\theta \in Q \Rightarrow \sin 2\theta$, $\cos 2\theta$ and $\tan 2\theta \in Q$ (if defined)

c. if $\sin\theta \in Q$ and $\cos\theta \in Q \Rightarrow \tan 3\theta \in Q$ (if defined)

d. if $\sin\theta \in Q \Rightarrow \cos 3\theta \in Q$

5. Which of the following quantities are rational?

a. $\sin\left(\frac{11\pi}{12}\right)\sin\left(\frac{5\pi}{12}\right)$

b. $\operatorname{cosec}\left(\frac{9\pi}{10}\right)\sec\left(\frac{4\pi}{5}\right)$

c. $\sin^4\left(\frac{\pi}{8}\right) + \cos^4\left(\frac{\pi}{8}\right)$

d. $\left(1+\cos\frac{2\pi}{9}\right)\left(1+\cos\frac{4\pi}{9}\right)\left(1+\cos\frac{8\pi}{9}\right)$

6. In which of the following sets the inequality $\sin^6 x + \cos^6 x > \frac{5}{8}$ holds good?

a. $\left(-\frac{\pi}{8}, \frac{\pi}{8}\right)$ b. $\left(\frac{3\pi}{8}, \frac{5\pi}{8}\right)$

c. $\left(\frac{\pi}{4}, \frac{3\pi}{4}\right)$ d. $\left(\frac{7\pi}{8}, \frac{9\pi}{8}\right)$

7. Which of the following inequalities hold true in any triangle *ABC*?

a. $\sin\frac{A}{2}\sin\frac{B}{2}\sin\frac{C}{2} \le \frac{1}{8}$

b. $\cos\frac{A}{2}\cos\frac{B}{2}\cos\frac{C}{2} \le \frac{3\sqrt{3}}{8}$

c. $\sin^2\frac{A}{2} + \sin^2\frac{B}{2} + \sin^2\frac{C}{2} < \frac{3}{4}$

d. $\cos^2\frac{A}{2} + \cos^2\frac{B}{2} + \cos^2\frac{C}{2} \le \frac{9}{4}$

8. For $\alpha = \pi/7$ which of the following hold(s) good?

a. $\tan\alpha\tan 2\alpha\tan 3\alpha = \tan 3\alpha - \tan 2\alpha - \tan\alpha$

b. $\operatorname{cosec}\alpha = \operatorname{cosec} 2\alpha + \operatorname{cosec} 4\alpha$

c. $\cos\alpha - \cos 2\alpha + \cos 3\alpha = 1/2$

d. $8\cos\alpha\cos 2\alpha\cos 4\alpha = 1$

9. Which of the following is/are correct?

a. $(\tan x)^{\ln(\sin x)} > (\cot x)^{\ln(\sin x)}, \forall x \in \left(0, \frac{\pi}{4}\right)$

b. $4^{\ln \operatorname{cosec} x} < 5^{\ln \operatorname{cosec} x}, \forall x \in \left(0, \frac{\pi}{2}\right)$

c. $\left(\frac{1}{2}\right)^{\ln(\cos x)} < \left(\frac{1}{3}\right)^{\ln(\cos x)}, \forall x \in \left(0, \frac{\pi}{2}\right)$

d. $2^{\ln(\tan x)} > 2^{\ln(\sin x)}, \forall x \in \left(0, \frac{\pi}{2}\right)$

10. Which of the following do/does not reduce to unity?

a. $\dfrac{\sin(180° + A)}{\tan(180° + A)}\dfrac{\cot(90° + A)}{\tan(90° + A)}\dfrac{\cos(360° - A)\operatorname{cosec} A}{\sin(-A)}$

b. $\dfrac{\sin(-A)}{\sin(180° + A)} - \dfrac{\tan(90° + A)}{\cot A} + \dfrac{\cos A}{\sin(90° + A)}$

c. $\dfrac{\sin 24°\cos 6° - \sin 6°\ 66°}{\sin 21°\cos 39° - \cos 51°\sin 69°}$

d. $\dfrac{\cos(90° + A)\sec(-A)\tan(180° - A)}{\sec(360° + A)\sin(180° + A)\cot(90° - A)}$

11. Which of the following identities wherever defined hold(s) good?

a. $\cot\alpha - \tan\alpha = 2\cot 2\alpha$

b. $\tan(45° + \alpha) - \tan(45° - \alpha) = 2\operatorname{cosec} 2\alpha$

c. $\tan(45° + \alpha) + \tan(45° - \alpha) = 2\sec 2\alpha$

d. $\tan\alpha + \cot\alpha = 2\tan 2\alpha$

12. A circle centred at O has radius 1 and contains the point A. Segment AB is tangent to the circle at A and $\angle AOB = \theta$. If point C lies on OA and BC bisects the angle ABO then OC equals

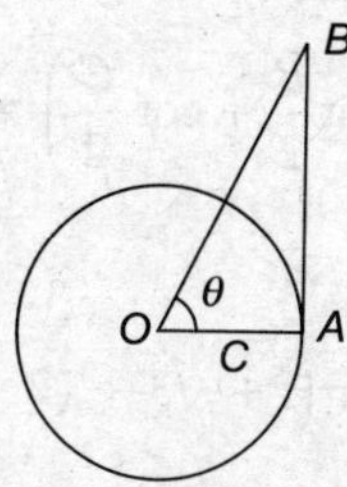

Fig. 1.1

a. $\sec\theta(\sec\theta - \tan\theta)$ b. $\frac{\cos^2\theta}{1+\sin\theta}$

c. $\frac{1}{1+\sin\theta}$ d. $\frac{1-\sin\theta}{\cos^2\theta}$

13. The expression $(\tan^4 x + 2\tan^2 x + 1)\cos^2 x$ when $x = \pi/12$ can be equal to

a. $4(2-\sqrt{3})$ b. $4(\sqrt{2}+1)$

c. $16\cos^2\frac{\pi}{12}$ d. $16\sin^2\frac{\pi}{12}$

14. Let α, β, and γ be some angles in the first quadrant satisfying $\tan(\alpha+\beta) = 15/8$ and $\operatorname{cosec}\gamma = 17/8$ then which of the following holds good?

a. $\alpha+\beta+\gamma=\pi$

b. $\cot\alpha\cot\beta\cot\gamma = \cot\alpha + \cot\beta + \cot\gamma$

c. $\tan\alpha + \tan\beta + \tan\gamma = \tan\alpha\tan\beta\tan\gamma$

d. $\tan\alpha\tan\beta + \tan\beta\tan\gamma + \tan\gamma\tan\alpha = 1$

15. $(a+2)\sin\alpha + (2a-1)\cos\alpha = (2a+1)$ if $\tan\alpha$ is

a. $\frac{3}{4}$ b. $\frac{4}{3}$

c. $\frac{2a}{a^2+1}$ d. $\frac{2a}{a^2-1}\alpha$

16. If $\sin\beta$ is the geometric mean between $\sin\alpha$ and $\cos\alpha$, then $\cos 2\beta$ is equal to

a. $2\sin^2\left(\frac{\pi}{4}-\alpha\right)$

b. $2\cos^2\left(\frac{\pi}{4}-\alpha\right)$

c. $2\cos^2\left(\frac{\pi}{4}-\alpha\right)$

d. $2\sin^2\left(\frac{\pi}{4}-\alpha\right)$

Reasoning Type

Each question has four choices a, b, c, and d, out of which *only one* is correct. Each question contains Statement 1 and Statement 2.

a. Both the statements are True and Statement 2 is the correct explanation of Statement 1.

b. Both the statements are True but Statement 2 is Not the correct explanation of Statement 1.

c. Statement 1 is True and Statement 2 is False.

d. Statement 1 is False and Statement 2 is True.

1. **Statement 1:** If $x+y+z=xyz$, then at most one of the numbers can be negative.

   Statement 2: In a triangle ABC, $\tan A + \tan B + \tan C = \tan A \tan B \tan C$ and there can be at most one obtuse angle in a triangle.

2. **Statement 1:** $\cos 1 < \cos 7$.

   Statement 2: $1 < 7$.

3. **Statement 1:** $\tan 4 < \tan 7.5$.

   Statement 2: $\tan x$ is always an increasing function.

4. **Statement 1:** $\cos 1 < \sin 1$.

   Statement 2: In the first quadrant cosine decreases but sine increases.

5. **Statement 1:** If $f(\theta) = (\sin\theta + \operatorname{cosec}\theta)^2 + (\cos\theta + \sec\theta)^2$, then the minimum value of $f(\theta)$ is 9.

   Statement 2: Maximum value of $\sin 2\theta$ is 1.

6. **Statement 1:** If $\sin^2\theta_1 + \sin^2\theta_2 + \cdots + \sin^2\theta_n = 0$, then the different sets of values of $(\theta_1, \theta_2, \cdots, \theta_n)$ for which $\cos\theta_1 + \cos\theta_2 + \cdots + \cos\theta_n = n-4$ is $n(n-1)$.

   Statement 2: If $\sin^2\theta_1 + \sin^2\theta_2 + \cdots + \sin^2\theta_n = 0$, then $\cos\theta_1, \cos\theta_2, \cdots, \cos\theta_n = \pm 1$.

7. **Statement 1:** The minimum value of $27^{\cos 2x}81^{\sin 2x}$ is $\frac{1}{243}$.

   Statement 2: The minimum value of $a\cos\theta + b\sin\theta$ is $-\sqrt{a^2+b^2}$.

8. **Statement 1:** If A, B, C are the angles of a triangle such that angle A is obtuse then $\tan B \tan C > 1$.

   Statement 2: In any triangle, $\tan A = \frac{\tan B + \tan C}{\tan B \tan C - 1}$.

9. **Statement 1:** $\tan 5°$ is an irrational number.

   Statement 2: $\tan 15°$ is an irrational number.

10. **Statement 1:** $\sin\frac{\pi}{18}$ is a root of $8x^3 - 6x + 1 = 0$.

    Statement 2: For any $\theta \in R$, $\sin 3\theta = 3\sin\theta - 4\sin^3\theta$.

11. Let f be any one of the six trigonometric functions. Let $A, B \in R$ satisfying $f(2A) = f(2B)$.

Statement 1: $A = n\pi + B$, for some $n \in Z$.

Statement 2: 2π is one of the period of f.

12. Statement 1: $\sin 3 < \sin 1 < \sin 2$.

Statement 2: $\sin x$ is positive in first and second quadrants.

13. Statement 1: The maximum and minimum values of the function $f(x) = \dfrac{1}{3\sin x + 4\cos x - 2}$ do not exist.

Statement 2: The given function is an unbounded function.

14. Statement 1: The minimum value of the expression $\sin\alpha + \sin\beta + \sin\gamma$ is negative, where α, β, γ are real numbers such that $\alpha + \beta + \gamma = \pi$.

Statement 2: If α, β, and γ are angles of a triangle then $\sin\alpha + \sin\beta + \sin\gamma = \cos\dfrac{\alpha}{2}\cos\dfrac{\beta}{2}\cos\dfrac{\gamma}{2}$.

15. Statement 1: If in a triangle $\sin^2 A + \sin^2 B + \sin^2 C = 2$ then one of the angles must be 90°.

Statement 2: In any triangle $\sin^2 A + \sin^2 B + \sin^2 C = 2 + 2\cos A\cos B\cos C$.

16. Statement 1: In a triangle, the least value of the sum of cosines of its angles is unity.

Statement 2: $\cos A + \cos B + \cos C = 1 + 4\sin\dfrac{A}{2}\sin\dfrac{B}{2}\sin\dfrac{C}{2}$, if A, B, C are the angles of a triangle.

17. Let α, β, and γ satisfy $0 < \alpha < \beta < \gamma < 2\pi$ and $\cos(x+\alpha) + \cos(x+\beta) + \cos(x+\gamma) = 0$, $\forall x \in R$.

Statement 1: $\gamma - \alpha = \dfrac{2\pi}{3}$.

Statement 2: $\cos\alpha + \cos\beta + \cos\gamma = 0$ and $\sin\alpha + \sin\beta + \sin\gamma = 0$.

18. If $A + B + C = \pi$, then

Statement 1: $\cos^2 A + \cos^2 B + \cos^2 C$ has its minimum value $\dfrac{3}{4}$.

Statement 2: Maximum value of $\cos A\cos B\cos C = \dfrac{1}{8}$.

19. Statement 1: If $xy + yz + zx = 1$ where $x, y, z \in R^+$, then $\dfrac{x}{1+x^2} + \dfrac{y}{1+y^2} + \dfrac{z}{1+z^2} = \dfrac{2}{\sqrt{(1+x^2)(1+y^2)(1+z^2)}}$.

Statement 2: In a triangle ABC, $\sin 2A + \sin 2B + \sin 2C = 4\sin A\sin B\sin C$.

20. Statement 1: In any triangle ABC,

$\ln\left(\cot\dfrac{A}{2} + \cot\dfrac{B}{2} + \cot\dfrac{C}{2}\right) = \ln\cot\dfrac{A}{2} + \ln\cot\dfrac{B}{2} + \ln\cot\dfrac{C}{2}$.

Statement 2: $\ln\left(1 + \sqrt{3} + (2 + \sqrt{3})\right) = \ln 1 + \ln\sqrt{3} + \ln(2 + \sqrt{3})$.

Linked Comprehension Type

Based upon each paragraph, three multiple choice questions have to be answered. Each question has 4 choices a, b, c, and d, out of which *only one* is correct.

For Problems 1–3

If $\sin\alpha = A\sin(\alpha + \beta)$, $A \neq 0$, then

1. The value of $\tan\alpha$ is

a. $\dfrac{A\sin\beta}{1 - A\cos\beta}$ **b.** $\dfrac{A\sin\beta}{1 + A\cos\beta}$

c. $\dfrac{A\cos\beta}{1 - A\sin\beta}$ **d.** $\dfrac{A\sin\beta}{1 + A\cos\beta}$

2. The value of $\tan\beta$ is

a. $\dfrac{\sin\alpha(1 + A\cos\beta)}{A\cos\alpha\cos\beta}$ **b.** $\dfrac{\sin\alpha(1 - A\cos\beta)}{A\cos\alpha\cos\beta}$

c. $\dfrac{\cos\alpha(1 - A\sin\beta)}{A\cos\alpha\cos\beta}$ **d.** $\dfrac{\cos\alpha(1 + A\sin\beta)}{A\cos\alpha\cos\beta}$

3. Which of the following is not the value of $\tan(\alpha + \beta)$?

a. $\dfrac{\sin\beta}{\cos\beta - A}$ **b.** $\dfrac{\sin\alpha\cos\alpha}{A\cos\beta - \sin^2\alpha}$

c. $\dfrac{\sin\alpha\cos\alpha}{A\cos\beta + \sin^2\alpha}$ **d.** None of these

For Problems 4–6

If α, β, γ, and δ are the solutions of the equation $\tan\left(\theta + \dfrac{\pi}{4}\right) = 3\tan 3\theta$, no two of which have equal tangents, then

4. The value of $\tan\alpha + \tan\beta + \tan\gamma + \tan\delta$ is

a. 1/3 **b.** 8/3

c. –8/3 **d.** 0

5. The value of $\tan\alpha\tan\beta\tan\gamma\tan\delta$ is

a. –1/3 **b.** – 2

c. 0 **d.** None of these

6. The value of $\dfrac{1}{\tan\alpha}+\dfrac{1}{\tan\beta}+\dfrac{1}{\tan\gamma}+\dfrac{1}{\tan\delta}$ is

a. -8 b. 8
c. 2/3 d. 1/3

For Problems 7–9

$\sin\alpha+\sin\beta=3$ and $\cos\alpha+\cos\beta=4$

7. The value of $\sin(\alpha+\beta)$ is

a. $\dfrac{24}{25}$ b. $\dfrac{13}{25}$
c. $\dfrac{12}{13}$ d. none of these

8. The value of $\cos(\alpha+\beta)$ is

a. $\dfrac{12}{25}$ b. $\dfrac{7}{25}$
c. $\dfrac{12}{13}$ d. none of these

9. The value of $\tan(\alpha+\beta)$ is

a. $\dfrac{25}{7}$ b. $\dfrac{25}{12}$
c. $\dfrac{25}{13}$ d. $\dfrac{24}{7}$

For Problems 10–12

To find the sum $\sin^2\dfrac{2\pi}{7}+\sin^2\dfrac{4\pi}{7}+\sin^2\dfrac{8\pi}{7}$ we follow the following method.

Put $7\theta=2n\pi$, where n is any integer.

Then $\sin 4\theta=\sin(2n\pi-3\theta)=-\sin 3\theta$. (1)

This means that $\sin\theta$ takes the values 0, $\pm\sin(2\pi/7)$, $\pm\sin(4\pi/7)$ and $\pm\sin(8\pi/7)$.

Since $\sin(6\pi/7)=\sin(8\pi/7)$, from equation (1), we now get

$2\sin 2\theta\cos 2\theta=4\sin^3\theta-3\sin\theta$

$\Rightarrow 4\sin\theta\cos\theta(1-2\sin^2\theta)=\sin\theta(4\sin^2\theta-3)$

Rejecting the value $\sin\theta=0$,

$\Rightarrow 4\cos\theta(1-2\sin^2\theta)=4\sin^2\theta-3$

$\Rightarrow 16\cos^2\theta(1-2\sin^2\theta)^2=(4\sin^2\theta-3)^2$

$\Rightarrow 16(1-\sin^2\theta)(1-4\sin^2\theta+4\sin^4\theta)=16\sin^4\theta-24\sin^2\theta+9$

$\Rightarrow 64\sin^6\theta-112\sin^4\theta-56\sin^2\theta-7=0$

This is a cubic in $\sin^2\theta$ with the roots $\sin^2(2\pi/7)$, $\sin^2(4\pi/7)$, and $\sin^2(8\pi/7)$.

The sum of these roots is

$$\sin^2\frac{2\pi}{7}+\sin^2\frac{4\pi}{7}+\sin^2\frac{8\pi}{7}=\frac{112}{64}=\frac{7}{4}.$$

Now answer the following questions.

10. The value of $\left(\tan^2\dfrac{\pi}{7}+\tan^2\dfrac{2\pi}{7}+\tan^2\dfrac{3\pi}{7}\right)\times\left(\cot^2\dfrac{\pi}{7}+\cot^2\dfrac{2\pi}{7}+\cot^2\dfrac{3\pi}{7}\right)$ is

a. 105 b. 35
c. 210 d. None of these

11. The value of $\dfrac{\tan^2\frac{\pi}{7}+\tan^2\frac{2\pi}{7}+\tan^2\frac{3\pi}{7}}{\cot^2\frac{\pi}{7}+\cot^2\frac{2\pi}{7}+\cot^2\frac{3\pi}{7}}$ is

a. 7 b. 35/3
c. 21/5 d. None of these

12. The value of $\tan^2\dfrac{\pi}{7}\tan^2\dfrac{2\pi}{7}\tan^2\dfrac{3\pi}{7}$ is

a. -3 b. -7
c. -5 d. None of these

For Problems 13–15

An altitude BD and a bisector BE are drawn in the triangle ABC from the vertex B. It is known that the length of side $AC=1$, and the magnitudes of the angles BEC, ABD, ABE, and BAC form an A.P.

13. The area of circle circumscribing ΔABC is

a. $\dfrac{\pi}{8}$ b. $\dfrac{\pi}{4}$
c. $\dfrac{\pi}{2}$ d. π

14. Let O be the circumcentre of ΔABC, the radius of the circle inscribed in ΔBOC is

a. $\dfrac{1}{8\sqrt{3}}$ b. $\dfrac{1}{4\sqrt{3}}$
c. $\dfrac{1}{2\sqrt{3}}$ d. $\dfrac{1}{2}$

15. Let B' be the image of point B with respect to side AC of ΔABC, then the length BB' is equal to

a. $\dfrac{\sqrt{3}}{4}$ b. $\dfrac{\sqrt{2}}{4}$
c. $\dfrac{1}{\sqrt{2}}$ d. $\dfrac{\sqrt{3}}{2}$

Matrix-Match Type

Each question contains statements given in two columns which have to be matched.

Statements (a, b, c, d) in column I have to be matched with statements (p, q, r, s) in column II. If the correct matches

are a → p, a → s, b → q, b → r, c → p, c → q and d → s, then the correctly bubbled 4 × 4 matrix should be as follows:

	p	q	r	s
a	(p) filled	(q)	(r)	(s) filled
b	(p)	(q) filled	(r) filled	(s)
c	(p) filled	(q) filled	(r)	(s)
d	(p)	(q)	(r)	(s) filled

1. If $\cos\theta - \sin\theta = \frac{1}{5}$ where $0 < \theta < \frac{\pi}{2}$

Column I	Column II
a. $(\cos\theta + \sin\theta)/2$	**p.** $\frac{4}{5}$
b. $\sin 2\theta$	**q.** $\frac{7}{10}$
c. $\cos 2\theta$	**r.** $\frac{24}{25}$
d. $\cos\theta$	**s.** $\frac{7}{25}$

2. For all real values of θ

Column I	Column II
a. $A = \sin^2\theta + \cos^4\theta$	**p.** $A \in [-1, 1]$
b. $A = 3\cos^2\theta + \sin^4\theta$	**q.** $A \in \left[\frac{3}{4}, 1\right]$
c. $A = \sin^2\theta - \cos^4\theta$	**r.** $A \in [2\sqrt{2}, \infty)$
d. $A = \tan^2\theta + 2\cot^2\theta$	**s.** $A \in [1, 3]$

3. If $\cos\alpha + \cos\beta = 1/2$ and $\sin\alpha + \sin\beta = 1/3$

Column I	Column II
a. $\cos\left(\frac{\alpha+\beta}{2}\right)$	**p.** $\pm\frac{\sqrt{13}}{12}$
b. $\cos\left(\frac{\alpha-\beta}{2}\right)$	**q.** $\frac{2}{3}$
c. $\tan\left(\frac{\alpha-\beta}{2}\right)$	**r.** $\pm\frac{3}{\sqrt{13}}$
d. $\tan\left(\frac{\alpha-\beta}{2}\right)$	**s.** $\pm\sqrt{\frac{131}{13}}$

4.

Column I	Column II
a. $\sin(410° - A)\cos(400° + A) + \cos(410° - A)\sin(400° + A)$ is equal to	**p.** -1
b. $\frac{\cos^2 1° - \cos^2 2°}{2\sin 3°\sin 1°}$ is equal to	**q.** 0
c. $\sin(-870°) + \operatorname{cosec}(-660°) + \tan(-855°) + 2\cot(840°) + \cos(480°) + \sec(900°)$	**r.** $\frac{1}{2}$
d. If $\cos\theta = \frac{4}{5}$ where $\theta \in \left(\frac{3\pi}{2}, 2\pi\right)$ and $\cos\phi = \frac{3}{5}$ where $\phi \in \left(0, \frac{\pi}{2}\right)$ then $\cos(\theta - \phi)$ has the value equal to	**s.** 1

5.

Column I	Column II
a. The maximum value of $\{\cos(2A + \theta) + \cos(2B + \theta)\}$, where A, B are constants, is	**p.** $2\sin(A+B)$
b. The maximum value of $\{\cos 2A + \cos 2B\}$, where $(A+B)$ is constant and $A, B \in (0, \pi/2)$, is	**q.** $2\sec(A+B)$
c. The minimum value of $\{\sec 2A + \sec 2B\}$, where $(A+B)$ is constant and $A, B \in (0, \pi/4)$, is	**r.** $2\cos(A+B)$
d. The minimum value of $\sqrt{\{\tan\theta + \cot\theta - 2\cos 2(A+B)\}}$ where A, B are constants and $\theta \in (0, \pi/2)$, is	**s.** $2\cos(A-B)$

6.

Column I	Column II
a. $\dfrac{\cos 20° + \cos 80° - \sqrt{3}}{\cos 50°}$	**p.** -1
b. $\cos 0° + \cos\dfrac{\pi}{7} + \cos\dfrac{2\pi}{7} + \cos\dfrac{3\pi}{7} + \cos\dfrac{4\pi}{7} + \cos\dfrac{5\pi}{7} + \cos\dfrac{6\pi}{7}$	**q.** $-\dfrac{3}{4}$
c. $\cos 20° + \cos 40° + \cos 60° - 4\cos 10° \cos 20° \cos 30°$	**r.** 1
d. $\cos 20° \cos 100° + \cos 100° \cos 140° - \cos 140° \cos 200°$	**s.** 0

Integer Type

1. In a triangle ABC, $BC = 8$, $CA = 6$, and $AB = 10$. A line dividing the triangle ABC into two regions of equal areas is perpendicular to AB at the point X. Then the value of $BX/\sqrt{2}$ is ______.
2. The exact value of cosec 10° + cosec 50° – cosec 70° is ______.
3. Let ABCDEFGHIJKL be a regular dodecagon. The value of $(AB/AF) + (AF/AB)$ is ______.
4. ABC is an equilateral triangle of side 4 cm. If R, r, h are the circumradius, inradius, and altitude, respectively, then $(R + r)/h$ is equal to ______.
5. If $x = (4l)/(1 + l^2)$ and $y = (2 - 2l^2)/(1 + l^2)$, where l is a real parameter, then $x^2 - xy + y^2$ lies between $[a, b]$. The value of $a + b$ is ______.
6. Let x, y, z, t be real numbers such that $x^2 + y^2 = 9$, $z^2 + t^2 = 4$, and $xt - yz = 6$. Then the greatest value of $P = xz$ is ______.
7. For each natural number k, let C_k denote a circle with radius k centimeter and centre at the origin O. On the circle C_k, a particle moves k cm in the counter-clockwise direction. After completing its motion on C_k, the particle moves to C_{k+1} in the radial direction. The motion of the particle continues in this manner. The particle starts at (1, 0). If the particle crosses the positive direction of x-axis for the first time on the circle C_n, then the value of n is ______.
8. If $3\sin\theta + 5\cos\theta = 5$, then $|5\sin\theta - 3\cos\theta|$ is equal to ______.
9. If $(\sec A + \tan A)(\sec B + \tan B)(\sec C + \tan C) = (\sec A - \tan A)(\sec B - \tan B)(\sec C - \tan C)$, then the value of each side is ______.
10. If $(\cos^4 A/\cos^2 B) + (\sin^4 A/\sin^2 B) = 1$, then the value of $(\cos^4 B/\cos^2 A) + (\sin^4 B/\sin^2 A)$ is ____.
11. The value of $\cos^2(\pi/16) + \cos^2(3\pi/16) + \cos^2(5\pi/16) + \cos^2(7\pi/16)$ is ______.
12. Find the maximum value of $[1 + \sin((\pi/4) + \theta) + 2\sin((\pi/4) - \theta)]$, (where [.] represents the greatest integer function) is ______.
13. If x is the A.M. of $\tan(\pi/9)$ and $\tan(5\pi/18)$ and y is the A.M. of $\tan(\pi/9)$ and $\tan(7\pi/18)$, then the value of y/x is ______.
14. If $p(x) = \sin x(\sin^3 x + 3) + \cos x(\cos^3 x + 4) + (1/2)\sin^2 2x + 4\cos x + 5$, then the least value of $p(x)$ is ______.
15. The value of $\sqrt{3}$ cosec 20° – sec 20° is ______.
16. The value of $2\left[\cos^4\left(\dfrac{\pi}{8}\right) + \cos^4\left(\dfrac{3\pi}{8}\right) + \cos^4\left(\dfrac{5\pi}{8}\right)\right.$ is ______.
17. The value of $64\cos(2\pi/15)\cos(4\pi/15)\cos(8\pi/15)\cos(14\pi/15)$ is ______.
18. If ABC is a triangle and $\tan(A/2)$, $\tan(B/2)$, $\tan(C/2)$ are in H.P., then the square of the minimum value of $(\cot(B/2))^2$ is ______.
19. The least integral value of $(\tan x)/(\tan 3x)$ is ______.
20. Suppose $\sin^3 x \sin 3x = \sum_{m=0}^{n} C_m \cos mx$ is an identity in x, where $C_0, C_1, \ldots, C_n$ are constants then the value of n is ______.
21. If $\cos(x - y)$, $\cos x$, and $\cos(x + y)$ are in H.P., then $[\cos x \sec(y/2)]^2$ is equal to ______.
22. The value of $3(\sin x - \cos x)^4 + 6(\sin x + \cos x)^2 + 4(\sin^6 x + \cos^6 x) - 7$ is ______.
23. The expression $3\{\sin^4((3/2)\pi - \alpha) + \sin^4(3\pi + \alpha)\} - 2\{\sin^6((1/2)\pi + \alpha) + \sin^6(5\pi - \alpha)\}$ is equal to ______.
24. For a positive integer n, let $f_n(\theta) = (\tan(\theta/2)) \times (1 + \sec\theta)(1 + \sec 2\theta)(1 + \sec 4\theta)\ldots(1 + \sec 2^n\theta)$, then the value of $f_2(\pi/16)$ is ______.
25. If $\dfrac{\sin\theta \sin 2\theta + \sin 3\theta \sin 6\theta + \sin 4\theta \sin 13\theta}{\sin\theta \cos 2\theta + \sin 3\theta \cos 6\theta + \sin 4\theta \cos 13\theta} = \tan k\theta$ then value of k is ______.
26. The value of $\sum_{r=1}^{2009} \cos\left(\dfrac{r\pi}{6}\right)$ is ________.
27. If $\sin\theta + \sin^2\theta + \sin^3\theta = 1$, then $\cos^6\theta - 4\cos^4\theta + 8\cos^2\theta = ?$
28. If $\sin^2\theta_1 + \sin^2\theta_2 + \sin^2\theta_3 = 0$, then number of possible values of $\cos\theta_1 + \cos\theta_2 + \cos\theta_3$ is ________.
29. The least value of $18\sin^2\theta + 2\,\text{cosec}^2\theta - 3$ is ________.

30. If the value of tan (37.5°) can be expressed as $\sqrt{a} - \sqrt{b} + \sqrt{c} - \sqrt{d}$, where $a, b, c, d \in N$ and $(a > b > c > d)$, then the value of ad/bc is equal to__________.

31. The least value of $\text{cosec}^2 x + 4 \sec^2 x$ is__________.

32. If $\theta = 3\alpha$ and $\sin \theta = \dfrac{3}{5}$, then the value of the expression $\dfrac{3\,\text{cosec}\,\alpha - 4 \sec \alpha}{2}$ is__________.

33. In a triangle ABC, $3 \sin A + 4 \cos B = 6$ and $3 \cos A + 4 \sin B = 1$, then the value of $\text{cosec}^2 C$ is__________.

34. Suppose ABC is a triangle with three acute angles A, B and C. Then the sum of the all possible quadrant numbers, where the point whose coordinates are $(\cos B - \sin A, \sin B - \cos A)$ lies is__________.

ANSWERS AND SOLUTIONS

Objective Type

1. b. Since $f(x) = \sin x$ is an increasing function for $0 < x < \frac{\pi}{2}$ and 1 radian is approximately 57°.

Therefore,

$1° < 1^c \Rightarrow \sin 1° < \sin 1$.

2. a. We have $\frac{x^2+y^2}{2xy} = \frac{1}{2}\left(\frac{x}{y}+\frac{y}{x}\right)$

Now, $\sin^2\theta = \frac{x^2+y^2}{2xy} \Rightarrow \frac{x^2+y^2}{2xy} \geq 0$

$[\because \sin^2\theta \geq 0]$

$\Rightarrow$ x and y have the same sign

Now, $\frac{x^2+y^2}{2xy} = \frac{1}{2}\left(\frac{x}{y}+\frac{y}{x}\right) \Rightarrow \frac{x^2+y^2}{2xy} \geq 1$

But $\sin^2\theta \leq 1$. Therefore, $\frac{x^2+y^2}{2xy} = 1 \Rightarrow x = y$.

3. d. Since $0 < x < \pi$. Therefore, $\sin x > 0$

We have $1 + \sin x + \sin^2 x + \cdots \infty = 4 + 2\sqrt{3}$

$$\Rightarrow \quad \frac{1}{1-\sin x} = 4 + 2\sqrt{3}$$

$$\Rightarrow \quad \sin x = 1 - \frac{1}{4+2\sqrt{3}}$$

$$\Rightarrow \quad \sin x = \frac{3+2\sqrt{3}}{4+2\sqrt{3}} = \frac{\sqrt{3}}{2}$$

$$\Rightarrow \quad x = \frac{\pi}{3} \text{ or } \frac{2\pi}{3}$$

4. b. $\frac{\cos x \tan x}{k^2} + \frac{1}{\tan x} + \frac{\sin x}{1+\cos x}$

$$= \frac{\sin x}{k^2} + \frac{\cos x\,(1+\cos x) + \sin^2 x}{\sin x\,(1+\cos x)}$$

$$= \frac{a}{k} + \frac{1}{\sin x} = \frac{a}{k} + \frac{1}{ak}$$

5. a. $2\sin\frac{A}{2}\operatorname{cosec}\frac{B}{2}\left(\sin\frac{C}{2} - \cos\frac{A}{2}\cos\frac{B}{2}\right) - \cos A$

$$= 2\sin\frac{A}{2}\operatorname{cosec}\frac{B}{2}\left(-\sin\frac{A}{2}\sin\frac{B}{2}\right) - \cos A$$

$$= -2\sin^2\frac{A}{2} - \cos A = -1.$$

6. b. Applying A.M. ≥ G.M. in $6\tan^2\phi$, $54\cot^2\phi$, 18, we get

$$\frac{6\tan^2\phi + 54\cot^2\phi + 18}{3} \geq (6 \times 54 \times 18)^{1/3} \geq 18$$

This is true if $6\tan^2\phi = 54\cot^2\phi = 18$

$$\Rightarrow \quad \tan^2\phi = 3 \text{ and } \cot^2\phi = \frac{1}{3}$$

$\Rightarrow$ I and II are correct.

7. c. $5\tan\theta = 4 \Rightarrow \tan\theta = \frac{4}{5}$

Now $\frac{5\sin\theta - 3\cos\theta}{5\sin\theta + 2\cos\theta} = \frac{5\frac{\sin\theta}{\cos\theta} - 3}{5\frac{\sin\theta}{\cos\theta} + 2} = \frac{5\tan\theta - 3}{5\tan\theta + 2}$

$$= \frac{5 \times \frac{4}{5} - 3}{5 \times \frac{4}{5} + 2} = \frac{1}{6}$$

8. b. $2\sec 2\theta = \tan\phi + \cot\phi$

$$\Rightarrow \quad \frac{2}{\cos 2\theta} = \frac{\sin^2\phi + \cos^2\phi}{\sin\phi\cos\phi}$$

$$\Rightarrow \quad \frac{2}{\cos 2\theta} = \frac{1}{\sin\phi\cos\phi}$$

$$\Rightarrow \cos 2\theta = \sin 2\phi$$

$$\Rightarrow 2\theta = 90° - 2\phi$$

$$\Rightarrow \theta + \phi = \frac{\pi}{4}$$

9. a. $\sin x + \operatorname{cosec} x = 2$

$\Rightarrow (\sin x - 1)^2 = 0$

$\Rightarrow \sin x = 1$

$\Rightarrow \sin^n x + \operatorname{cosec}^n x = 1 + 1 = 2$

10. c. $\sec^2\theta + \operatorname{cosec}^2\theta = \frac{1}{\cos^2\theta} + \frac{1}{\sin^2\theta} = \frac{4}{\sin^2 2\theta} \geq 4$

Also, $\sec^2\theta\operatorname{cosec}^2\theta = \frac{4}{\sin^2 2\theta} \geq 4$

Hence, the only equation which can have roots $\operatorname{cosec}^2\theta$ and $\sec^2\theta$ is $x^2 - 5x + 5 = 0$.

11. b. $\sqrt{\frac{1-\cos\alpha}{1+\cos\alpha}} + \sqrt{\frac{1+\cos\alpha}{1-\cos\alpha}}$

$$= \frac{1-\cos\alpha + 1 + \cos\alpha}{\sqrt{1-\cos^2\alpha}} = \frac{2}{\pm\sin\alpha} = \frac{2}{-\sin\alpha}$$

$$\left(\text{since } \pi < \alpha < \frac{3\pi}{2}\right)$$

12. b. We have $\cos\frac{\pi}{7}+\cos\frac{2\pi}{7}+\cos\frac{3\pi}{7}+\cos\frac{4\pi}{7}+\cos\frac{5\pi}{7}+\cos\frac{6\pi}{7}+\cos\frac{7\pi}{7}$

$$=\left(\cos\frac{\pi}{7}+\cos\frac{6\pi}{7}\right)+\left(\cos\frac{2\pi}{7}+\cos\frac{5\pi}{7}\right)+\left(\cos\frac{3\pi}{7}+\cos\frac{4\pi}{7}\right)+\cos\pi$$

$$=\left(\cos\frac{\pi}{7}-\cos\frac{\pi}{7}\right)+\left(\cos\frac{2\pi}{7}-\cos\frac{2\pi}{7}\right)+\left(\cos\frac{3\pi}{7}-\cos\frac{3\pi}{7}\right)+\cos\pi$$

$$=\cos\pi=-1$$

13. b. $2\sin^2\theta+3\cos^2\theta=2(\sin^2\theta+\cos^2\theta)+\cos^2\theta$

$=2+\cos^2\theta\geq 2\quad[\because \cos^2\theta>0]$

14. b. $\sin^4\theta+\cos^4\theta=(\sin^2\theta+\cos^2\theta)^2-2\sin^2\theta\cos^2\theta$

$=1-2\sin^2\theta\cos^2\theta\leq 1$

15. d. We have

$f(x)=\cos^2\theta+\sec^2\theta$

$=(\cos\theta-\sec\theta)^2+2\cos\theta\sec\theta$

$=2+(\cos\theta-\sec\theta)^2\geq 2$

16. a. $f(x)=\cos^6 x+\sin^6 x$

$=(\cos^2 x+\sin^2 x)(\sin^4 x+\cos^4 x-\cos^2 x\sin^2 x)$

$=[(\sin^2 x+\cos^2 x)^2-3\sin^2 x\cos^2 x]$

$=1-\frac{3}{4}\sin^2 2x$

$\Rightarrow f(x)\in\left[\frac{1}{4},1\right]$

17. a. $f(x)=3\cos x+5\sin(x-\pi/5)$

$=-\frac{1}{2}\cos x+5\times\frac{\sqrt{3}}{2}\sin x$

Then,

$$-\sqrt{\left(\frac{1}{2}\right)^2+\left(\frac{5\sqrt{3}}{2}\right)^2}\leq f(x)\leq\sqrt{\left(\frac{1}{2}\right)^2+\left(\frac{5\sqrt{3}}{2}\right)^2}$$

$\Rightarrow -\sqrt{19}\leq f(x)\leq\sqrt{19}$

18. d. Given expression is $1-\cos 2x+\sin 2x=2k$

$\Rightarrow \sin 2x-\cos 2x=2k-1$

$\Rightarrow \sin(2x-\alpha)=\frac{2k-1}{\sqrt{2}}$

$\Rightarrow -1\leq\frac{2k-1}{\sqrt{2}}\leq 1$

$\Rightarrow \frac{1-\sqrt{2}}{2}\leq k\leq\frac{1+\sqrt{2}}{2}$

19. d. $\cot(\alpha+\beta)=0$

$\Rightarrow \tan(\alpha+\beta)=\infty\Rightarrow \alpha+\beta=\pi/2$

$\Rightarrow \sin(\alpha+2\beta)=\sin(90°+\beta)=\cos\beta$

20. b. We have

$$\frac{x}{\cos\theta}=\frac{y}{\cos\left(\theta-\frac{2\pi}{3}\right)}=\frac{z}{\cos\left(\theta+\frac{2\pi}{3}\right)}$$

Therefore, each ratio is equal to

$$\frac{x+y+z}{\cos\theta+\cos\left(\theta-\frac{2\pi}{3}\right)+\cos\left(\theta+\frac{2\pi}{3}\right)}=\frac{x+y+z}{0}$$

$\Rightarrow\ x+y+z=0.$

21. b. Since $0\leq\sin^{2n}x\leq\sin^2 x$

$0\leq\cos^{2n}x\leq\cos^2 x$

$[\because \sin^4 x=\sin^2 x\sin^2 x\leq\sin^2 x.\ \therefore\ \sin^4 x\leq\sin^2 x$ and so on$]$

$\Rightarrow 0\leq\sin^{2n}x+\cos^{2n}x\leq\sin^2 x+\cos^2 x=1$

$\Rightarrow 0\leq\sin^{2n}x+\cos^{2n}x\leq 1$

22. b. $4x^2-2\sqrt{5}x+1=0$

Let α and β be the roots. Therefore,

$\alpha+\beta=\frac{2\sqrt{5}}{4}=\frac{\sqrt{5}}{2},\ \alpha\beta=\frac{1}{4}$

Since $\sin 18°=\frac{\sqrt{5}-1}{4}$, $\cos 36°=\frac{\sqrt{5}+1}{4}$

$\sin 18°+\cos 36°=\frac{2\sqrt{5}}{4}=\frac{\sqrt{5}}{2}\quad\sin 18°\cos 36°$

$=\frac{5-1}{16}=\frac{4}{16}=\frac{1}{4}$

Here the required roots are sin 18° and cos 36°.

23. b. $\sqrt{2\cot\alpha+\frac{1}{\sin^2\alpha}}=\sqrt{2\cot\alpha+\operatorname{cosec}^2\alpha}$

$=\sqrt{2\cot\alpha+1+\cot^2\alpha}=|1+\cot\alpha|$

$=-1-\cot\alpha$

[since $\cot a<-1$ when $\frac{3\pi}{4}<\alpha<\pi$,

$|1+\cot\alpha|=-1-\cot\alpha$]

24. d. $f(\theta)=5\cos\theta+3\cos\left(\theta+\frac{\pi}{3}\right)+3$

$=5\cos\theta+\frac{3}{2}\cos\theta-\frac{3\sqrt{3}}{2}\sin\theta+3$

$=\frac{13}{2}\cos\theta-\frac{3\sqrt{3}}{2}\sin\theta+3$

$$= \sqrt{\left(\frac{169}{4}+\frac{27}{4}\right)} \sin(\theta-\alpha)+3.$$

Thus, range of $f(\theta)$ is $[-4, 10]$.

25. b. Since $\alpha < \beta < \gamma < \delta$ and

$\sin \alpha = \sin \beta = \sin \gamma = \sin \delta = K$, therefore $\beta = \pi - \alpha$, $\gamma = 2\pi + \alpha$, $\delta = 3\pi - \alpha$

$$\Rightarrow 4 \sin \frac{\alpha}{2} + 3 \sin \frac{\beta}{2} + 2 \sin \frac{\gamma}{2} + \sin \frac{\delta}{2}$$

$$= 4 \sin \frac{\alpha}{2} + 3 \cos \frac{\alpha}{2} - 2 \sin \frac{\alpha}{2} - \cos \frac{\alpha}{2}$$

$$= 2 \sin \frac{\alpha}{2} + 2 \cos \frac{\alpha}{2}$$

$$= 2\sqrt{1+\sin\alpha} = 2\sqrt{1+K}$$

26. c. Let O be the centre of the circle

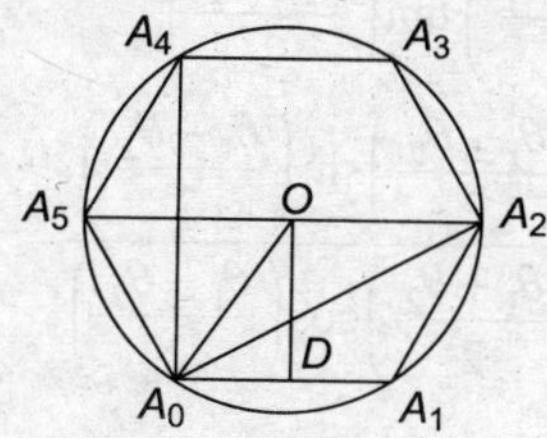

Fig. 1.2

Since $\angle A_0OA_1 = \frac{360°}{6} = 60°$,

A_0OA_1 is an equilateral triangle

$\therefore A_0A_1 = 1$ [radius of circle = 1]

Also $A_0A_2 = A_0A_4 = 2OD = 2[OA_0] \sin 60°$

$$= 2(1)\frac{\sqrt{3}}{2} = \sqrt{3}$$

Hence, $(A_0A_1)(A_0A_2)(A_0A_4) = (1)(\sqrt{3})(\sqrt{3}) = 3$

27. d. The given relation is satisfied only when $\sin \theta_1 = \sin \theta_2 = \sin \theta_3 = 1$

$\Rightarrow \cos \theta_1 = \cos \theta_2 = \cos \theta_3 = 0$

$\Rightarrow \cos \theta_1 + \cos \theta_2 + \cos \theta_3 = 0$

28. c. $\sin^2 \theta \le 1$

$$\Rightarrow \frac{x^2+y^2+1}{2x} \le 1 \Rightarrow x^2+y^2-2x+1 \le 0 \quad [\text{as } x > 0]$$

$\Rightarrow (x-1)^2 + y^2 \le 0$

It is possible, iff $x - 1$ and $y = 0$.

29. a. $\sin(\alpha+\beta) = 1$

$$\Rightarrow \alpha+\beta = \frac{\pi}{2} \quad \sin(\alpha-\beta)$$

$$= \frac{1}{2} \Rightarrow \alpha - \beta = \frac{\pi}{6}$$

Solving, we get $\alpha = \frac{\pi}{3}$ and $\beta = \frac{\pi}{6}$

Now $\tan(\alpha+2\beta)\tan(2\alpha+\beta)$

$$= \tan\left(\frac{\pi}{3}+\frac{\pi}{3}\right)\tan\left(\frac{2\pi}{3}+\frac{\pi}{6}\right)$$

$$= \tan\frac{2\pi}{3}\tan\frac{5\pi}{6} = \left(-\cot\frac{\pi}{3}\right)\left(-\cot\frac{\pi}{6}\right)$$

$$= \left(-\frac{1}{\sqrt{3}}\right)(-\sqrt{3}) = 1$$

30. b. L.H.S. $= 3(\cos^4\alpha + \sin^4\alpha) - 2(\cos^6\alpha + \sin^6\alpha)$

$= 3(1 - 2\sin^2\alpha\cos^2\alpha) - 2[1^3 - 3\sin^2\alpha\cos^2\alpha 1] = 1$

$a^3 + b^3 = (a+b)^3 - 3ab(a+b)$

Here $a + b = \sin^2\alpha + \cos^2\alpha = 1$.

31. c. $\operatorname{cosec}\theta - \cot\theta = q$ (1)

$$\Rightarrow \operatorname{cosec}\theta + \cot\theta = \frac{1}{q}$$

$$\Rightarrow \operatorname{cosec}\theta = \frac{1}{2}[q + (1/q)], \text{ on addition.}$$

32. b. $$xy + 1 = \frac{1-\sin\theta}{\cos\theta}\frac{1+\cos\theta}{\sin\theta} + 1 = \frac{1-\sin\theta+\cos\theta}{\sin\theta\cos\theta}$$

$$= \frac{(\sin^2\theta+\cos^2\theta)}{\sin\theta\cos\theta} - \frac{(\sin\theta-\cos\theta)}{\sin\theta\cos\theta}$$

$= (\tan\theta + \cot\theta) - (\sec\theta - \operatorname{cosec}\theta)$

$= (\operatorname{cosec}\theta + \cot\theta) - (\sec\theta - \tan\theta) = y - x$

33. c. Multiplying x above and below by $1 - \cos\theta + \sin\theta$, we get,

$$x = \frac{2\sin\theta(1-\cos\theta+\sin\theta)}{(1+\sin\theta)^2 - \cos^2\theta}$$

$$= \frac{2\sin\theta(1-\cos\theta+\sin\theta)}{(1+\sin\theta)^2 - (1-\sin^2\theta)}$$

Putting $1 - \sin^2\theta = (1 + \sin\theta)(1 - \sin\theta)$, we get

$$\frac{2\sin\theta}{2\sin\theta}\frac{1-\cos\theta+\sin\theta}{1+\sin\theta}$$

34. b. $2n\theta = \pi/2$

$\therefore \theta, (2n-1)\theta = (\pi/2) - \theta; 2\theta, (2n-2)\theta = (\pi/2) - 2\theta, \ldots$

They form complementary angles A and B so that $\tan A \tan B = \tan A \cot A = 1$ for each pair.

35. b. $A < B < C < D$ is in ascending order (1)

$\sin A = \sin B = \sin C = \sin D = k$ (positive) (2)

We have chosen the values to satisfy the conditions of equations (1) and (2)

$$E = 4\sin\frac{A}{2} + 3\sin\frac{\pi - A}{2} + 2\sin\frac{2\pi + A}{2} + \sin\frac{3\pi - A}{2}$$

$$= 4\sin\frac{A}{2} + 3\cos\frac{A}{2} - 2\sin\frac{A}{2} - \cos\frac{A}{2}$$

$$= 2\left(\sin\frac{A}{2} + \cos\frac{A}{2}\right)$$

$$= 2\sqrt{\sin^2\frac{A}{2} + \cos^2\frac{A}{2} + 2\sin\frac{A}{2}\cos\frac{A}{2}}$$

$$= 2\sqrt{1 + \sin A} = 2\sqrt{1+k}$$

Hence, (b) is the correct answer.

36. b. $\sec\alpha + \operatorname{cosec}\alpha = p$, $\sec\alpha \operatorname{cosec}\alpha = q$

$$\Rightarrow \frac{\sin\alpha + \cos\alpha}{\sin\alpha\cos\alpha} = p \text{ and } \frac{1}{\sin\alpha\cos\alpha} = q$$

$$\Rightarrow \frac{1 + 2\sin\alpha\cos\alpha}{\sin^2\alpha\cos^2\alpha} = p^2$$

$$\Rightarrow \frac{1 + \frac{2}{q}}{\frac{1}{q^2}} = p^2$$

$$\Rightarrow q^2\left(1 + \frac{2}{q}\right) = p^2 \Rightarrow q(q + 2) = p^2$$

37. c. We have $\sin x + \sin^2 x = 1$

$\Rightarrow \sin x = 1 - \sin^2 x \Rightarrow \sin x = \cos^2 x$

Now $\cos^{12} x + 3\cos^{10} x + 3\cos^8 x + \cos^6 x - 2$

$= \sin^6 x + 3\sin^5 x + 3\sin^4 x + \sin^3 x - 2$

$= (\sin^2 x)^3 + 3(\sin^2 x)^2 \sin x + 3(\sin^2 x)(\sin x)^2 + (\sin x)^3 - 2$

$= (\sin^2 x + \sin x)^3 - 2 = (1)^3 - 2 = -1$

38. a. $\cos(A - B) = \frac{3}{5}$

$\Rightarrow 5\cos A\cos B + 5\sin A\sin B = 3$ (1)

From second relation,

$\sin A\sin B = 2\cos A\cos B$ (2)

$\Rightarrow \cos A\cos B = \frac{1}{5}$ and $\sin A\sin B = \frac{2}{5}$

39. a. We know that $(1 + \tan A)(1 + \tan B) = 2$

$\Rightarrow \tan A + \tan B = 1 - \tan A\tan B$

$\Rightarrow \tan(A + B) = 1$. Therefore

$$A + B = \frac{\pi}{4}$$

or $\alpha + 4\alpha = \frac{\pi}{4}$.

Hence $\alpha = \frac{\pi}{20}$

40. a. $A = \frac{1}{\sqrt{2}} + \frac{1}{\sqrt{2}} = \sqrt{2}$

$$B = \sqrt{2}\left[\frac{1}{\sqrt{2}}\sin 44^\circ + \frac{1}{\sqrt{2}}\cos 44^\circ\right]$$

$= \sqrt{2}[\sin(45^\circ + 44^\circ)]$

$B = \sqrt{2}\sin 89^\circ < \sqrt{2}\sin 90^\circ = \sqrt{2}$

Hence $A > B \Rightarrow$ (a)

41. d. $\frac{1}{4}(\sqrt{3}\cos 23^\circ - \sin 23^\circ)$

$= \frac{1}{2}(\cos 30^\circ\cos 23^\circ - \sin 30^\circ\sin 23^\circ)$

$= \frac{1}{2}\cos(30^\circ + 23^\circ) = \frac{1}{2}\cos 53^\circ$

42. b. $\tan\left(\frac{\theta_1 - \theta_2}{2}\right)\tan\left(\frac{\theta_1 + \theta_2}{2}\right)$

$$= \frac{\sin\left(\frac{\theta_1 + \theta_2}{2}\right)\sin\left(\frac{\theta_1 - \theta_2}{2}\right)}{\cos\left(\frac{\theta_1 + \theta_2}{2}\right)\cos\left(\frac{\theta_1 - \theta_2}{2}\right)}$$

$$= \frac{\sin^2\frac{\theta_1}{2} - \sin^2\frac{\theta_2}{2}}{\cos^2\frac{\theta_1}{2} - \sin^2\frac{\theta_2}{2}}$$

$$= \frac{1 - \cos\theta_1 - (1 - \cos\theta_2)}{1 + \cos\theta_1 - (1 - \cos\theta_2)} = -\frac{1}{3}$$

43. b. $$\frac{3 + \frac{\cos 80^\circ\cos 20^\circ}{\sin 80^\circ\sin 20^\circ}}{\frac{\cos 80^\circ}{\sin 80^\circ} + \frac{\cos 20^\circ}{\sin 20^\circ}}$$

$$= \frac{3\sin 80^\circ\sin 20^\circ + (\cos 80^\circ\cos 20^\circ + \sin 80^\circ\sin 20^\circ)}{\sin 20^\circ\cos 80^\circ + \cos 20^\circ\sin 80^\circ}$$

$$= \frac{-\cos 100^\circ + \cos 60^\circ + \cos 60^\circ}{\sin 100^\circ}$$

$$= \frac{1 - \cos 100^\circ}{\sin 100} = \tan 50^\circ.$$

44. a. $\tan\beta = 2\sin\alpha\sin\gamma\operatorname{cosec}(\alpha + \gamma) = \frac{2\sin\alpha\sin\gamma}{\sin(\alpha + \gamma)}$

$$\Rightarrow \cot\beta = \frac{\sin(\alpha + \gamma)}{2\sin\alpha\sin\gamma}$$

$$\Rightarrow 2\cot\beta = \frac{\sin\alpha\cos\gamma + \cos\alpha\sin\gamma}{\sin\alpha\sin\gamma} = \cot\alpha + \cot\gamma$$

$\Rightarrow \cot\alpha, \cot\beta, \cot\gamma$ are in A.P.

45. b. $3\sin A\cos B = \sin B\cos A$

$\Rightarrow \cos A \sin B = \frac{3}{4}$

$\Rightarrow \sin(A+B) = 1$

$\Rightarrow C = \frac{\pi}{2},\ B = \frac{\pi}{2} - A$

$\Rightarrow 3 \tan A = \tan\left(\frac{\pi}{2} - A\right)$

$\Rightarrow 3 = \cot^2 A$

46. d. $\tan(100° + 125°) = \frac{\tan 100° + \tan 125°}{1 - \tan 100° \tan 125°}$. Therefore,

$\tan 225° = \frac{\tan 100° + \tan 125°}{1 - \tan 100° \tan 125°}$,

i.e., $1 = \frac{\tan 100° + \tan 125°}{1 - \tan 100° \tan 125°}$

i.e., $\tan 100° + \tan 125° + \tan 100° \tan 125° = 1$.

47. b. We know that $\tan(20° + 40°) = \frac{\tan 20° + \tan 40°}{1 - \tan 20° \tan 40°}$

$\Rightarrow \sqrt{3} = \frac{\tan 20° + \tan 40°}{1 - \tan 20° \tan 40°}$

$\Rightarrow \sqrt{3} - \sqrt{3} \tan 20° \tan 40° = \tan 20° + \tan 40°$

$\Rightarrow \tan 20° + \tan 40° + \sqrt{3} \tan 20° \tan 40° = \sqrt{3}$

48. c. $\frac{\sqrt{2} - \sin\alpha - \cos\alpha}{\sin\alpha - \cos\alpha}$

$= \frac{\sqrt{2} - \sqrt{2}\left(\frac{1}{\sqrt{2}}\sin\alpha + \frac{1}{\sqrt{2}}\cos\alpha\right)}{\sqrt{2}\left(\frac{1}{\sqrt{2}}\sin\alpha - \frac{1}{\sqrt{2}}\cos\alpha\right)}$

$= \frac{\sqrt{2} - \sqrt{2}\cos\left(\alpha - \frac{\pi}{4}\right)}{\sqrt{2}\sin\left(\alpha - \frac{\pi}{4}\right)}$

$= \frac{\sqrt{2}(1 - \cos\theta)}{\sqrt{2}\sin\theta}$, where $\theta = \alpha - \frac{\pi}{4}$

$= \frac{2\sin^2(\theta/2)}{2\sin(\theta/2)\cos(\theta/2)}$

$= \tan\frac{\theta}{2} = \tan\left(\frac{\alpha}{2} - \frac{\pi}{8}\right)$

49. b. $\sin\theta_1 - \sin\theta_2 = a,\ \cos\theta_1 + \cos\theta_2 = b$

$\Rightarrow a^2 + b^2 = 2 + 2\cos(\theta_1 + \theta_2)$

$\Rightarrow 0 \le a^2 + b^2 \le 4$

50. c. $\frac{1 + \sin 2x}{1 - \sin 2x} = \frac{(\sin x + \cos x)^2}{(\sin x - \cos x)^2}$

$= \left(\frac{1 + \tan x}{1 - \tan x}\right)^2$

$= \left(\tan\left(\frac{\pi}{4} + x\right)\right)^2 = \tan^2\left(\frac{\pi}{4} + x\right)$

$= \cot^2\left(\frac{\pi}{2} + \frac{\pi}{4} + x\right) = \cot^2\left(\frac{3\pi}{4} + x\right)$

$\Rightarrow a = \frac{3\pi}{4}$

51. d. We have $4x^2 - 16x + 15 < 0 \Rightarrow \frac{3}{2} < x < \frac{5}{2}$. So, integral solution of $4x^2 - 16x + 15 < 0$ is $x = 2$. Thus $\tan\alpha = 2$.
It is given that $\cos\beta = \tan 45° = 1$. Therefore, $\sin(\alpha + \beta)\sin(\alpha - \beta)$

$= \sin^2\alpha - \sin^2\beta$

$= \frac{1}{1 + \cot^2\alpha} - (1 - \cos^2\beta)$

$= \frac{1}{1 + \frac{1}{4}} - 0 = \frac{4}{5}$

52. b. $\frac{\cos(x-y)}{\cos(x+y)} + \frac{\cos(z+t)}{\cos(z-t)} = 0$

$\Rightarrow \frac{1 + \tan x \tan y}{1 - \tan x \tan y} + \frac{1 - \tan z \tan t}{1 + \tan z \tan t} = 0$

$\Rightarrow 1 + \tan z \tan t + \tan x \tan y + \tan x \tan y \tan z \tan t + 1 - \tan z \tan t - \tan x \tan y + \tan x \tan y \tan z \tan t = 0$

$\Rightarrow \tan x \tan y \tan z \tan t = -1$.

53. b. $f(n) = 2\cos nx$

$\Rightarrow f(1) f(n+1) - f(n)$

$= 4\cos x \cos(n+1)x - 2\cos nx$

$= 2[2\cos(n+1)x \cos x - \cos nx]$

$= 2[\cos(n+2)x + \cos nx - \cos nx$

$= 2\cos(n+2)x = f(n+2)$

54. a. $\frac{\tan A}{\tan B} = \frac{1}{3} \Rightarrow \frac{\sin A \cos B}{\cos A \sin B} = \frac{1}{3}$

Put $\sin A \cos B = \frac{1}{4}$

$\Rightarrow \cos A \sin B = \frac{3}{4}$

$\Rightarrow \sin(A+B) = \frac{1}{4} + \frac{3}{4} = 1$

$\Rightarrow \sin C = 1 = \sin\frac{\pi}{2}$

$\Rightarrow C = \frac{\pi}{2}$. Hence, the triangle is right angled.

55. b. $3\sin^2 A + 2\sin^2 B = 1$

$\Rightarrow 3\sin^2 A = \cos 2B$

Also $3\sin 2A - 2\sin 2B = 0$

$\Rightarrow \sin 2B = \frac{3}{2}\sin 2A$ (2)

Now, $\cos(A+2B) = \cos A\cos 2B - \sin A\sin 2B$

$= \cos A\, 3\sin^2 A - \sin A\,\frac{3}{2}\sin 2A$

$= 3\sin^2 A\cos A - 3\sin^2 A\cos A = 0$

$\therefore A + 2B = \frac{\pi}{2}$

56. a. $f(\beta) = f\left(\frac{5\pi}{4} - \alpha\right) = \dfrac{\cot\left(\frac{5\pi}{4} - \alpha\right)}{1 + \cot\left(\frac{5\pi}{4} - \alpha\right)}$

$= \dfrac{1}{1 + \tan\left(\frac{5\pi}{4} - \alpha\right)}$

$= \dfrac{1}{1 + \dfrac{1 - \tan\alpha}{1 + \tan\alpha}} = \dfrac{1 + \tan\alpha}{2}$

As $f(\alpha) = \dfrac{\cot\alpha}{1 + \cot\alpha} = \dfrac{1}{1 + \tan\alpha}$

Hence, $f(\alpha)f(\beta) = \frac{1}{2}$

57. c. $A - B = \frac{\pi}{4} \Rightarrow \tan(A - B) = \tan\frac{\pi}{4}$

$\Rightarrow \dfrac{\tan A - \tan B}{1 + \tan A\tan B}$

$\Rightarrow \tan A - \tan B - \tan A\tan B = 1$

$\Rightarrow \tan A - \tan B - \tan A\tan B + 1 = 2$

$\Rightarrow (1 + \tan A)(1 - \tan B) = 2 \Rightarrow y = 2$

Hence, $(y+1)^{y+1} = (2+1)^{2+1} = (3)^3 = 27.$

58. a. Apply $b - a = c - b$ for A.P.

$2\cos z\sin(x - y) = 2\cos x\sin(y - z)$

Divide by $2\cos x\cos y\cos z$, etc.

$\tan x - \tan y = \tan y - \tan z$

59. b. $(\cos\alpha + \cos\beta)^2 - (\sin\alpha + \sin\beta)^2 = 0$

$\Rightarrow (\cos^2\alpha + \cos^2\beta + 2\cos\alpha\cos\beta) - (\sin^2\alpha + \sin^2\beta + 2\sin\alpha\sin\beta) = 0$

$\Rightarrow \cos 2\alpha + \cos 2\beta = -2(\cos\alpha\cos\beta - \sin\alpha\sin\beta)$

$\Rightarrow \cos 2\alpha + \cos 2\beta = -2\cos(\alpha + \beta).$

60. a. We have

$\sin\alpha(\sec x_1\sec x_2 + \sin\alpha\sec x_2\sec x_3 + \cdots + \sin\alpha\sec x_{n-1}\sec x_n)$

$= \dfrac{\sin(x_2 - x_1)}{\cos x_1\cos x_2} + \dfrac{\sin(x_3 - x_2)}{\cos x_2\cos x_3} + \cdots + \dfrac{\sin(x_n - x_{n-1})}{\cos x_{n-1}\cos x_n}$

$= \tan x_2 - \tan x_1 + \tan x_3 - \tan x_2 + \cdots + \tan x_n - \tan x_{n-1}$

$= \tan x_n - \tan x_1 = \dfrac{\sin(x_n - x_1)}{\cos x_n\cos x_1} = \dfrac{\sin(n-1)\alpha}{\cos x_n\cos x_1}.$

$[\because x_n = x_1 + (n-1)\alpha]$

61. a. By the given conditions

$\tan\frac{\pi}{9} + \tan\frac{5\pi}{18} = 2x\tan\frac{\pi}{9} + \tan\frac{7\pi}{18} = 2y$

$\Rightarrow 2x = \tan 20° + \tan 50°$

$= \dfrac{\sin 20°}{\cos 20°} + \dfrac{\sin 50°}{\cos 50°}$

$= \dfrac{\sin 20°\cos 50° + \cos 20°\sin 50°}{\cos 20°\cos 50°}$

$= \dfrac{\sin 70°}{\cos 20°\cos 50°}$

$= \dfrac{\cos 20°}{\cos 20°\cos 50°} = \dfrac{1}{\cos 50°} = \dfrac{1}{\sin 40°}$

$= \operatorname{cosec} 40°$

$2y = \tan 20° + \tan 70°$

$= \dfrac{\sin 20°}{\cos 20°} + \dfrac{\sin 70°}{\cos 70°}$

$= \dfrac{\sin 90°}{\cos 20°\cos 70°}$

$= \dfrac{1}{\cos 20°\cos 70°} = \dfrac{1}{\cos 20°\sin 20°}$

$= \dfrac{2}{2\sin 20°\cos 20°} = \dfrac{2}{\sin 40°}$

$= 2\operatorname{cosec} 40°$

Hence, $2y = 2(2x) \Rightarrow y = 2x$

62. b. We have $\tan\alpha = \dfrac{m}{m+1}$ and $\tan\beta = \dfrac{1}{2m+1}$

Now $\tan(\alpha + \beta) = \dfrac{\tan\alpha + \tan\beta}{1 - \tan\alpha\tan\beta}$

$= \dfrac{\dfrac{m}{m+1} + \dfrac{1}{2m+1}}{1 - \dfrac{m}{(m+1)}\dfrac{1}{(2m+1)}}$

$= \dfrac{2m^2 + m + m + 1}{2m^2 + m + 2m + 1 - m}$

$= \dfrac{2m^2 + 2m + 1}{2m^2 + 2m + 1} = 1$

$\Rightarrow \tan(a+b) = \tan\dfrac{\pi}{4}$

Hence, $\alpha+\beta = \dfrac{\pi}{4}$.

Trick: As $\alpha+\beta$ is independent of m, put $m = 1$, then

$\tan\alpha = \dfrac{1}{2}$ and $\tan\beta = \dfrac{1}{3}$.

Therefore, $\tan(\alpha+\beta) = \dfrac{(1/2)+(1/3)}{1-(1/6)} = 1.$

Hence, $\alpha+\beta = \dfrac{\pi}{4}$.

(Also check for other values of m)

63. a. We have $\sin\alpha + \sin\beta = -\dfrac{21}{65}$ (1)

$\cos\alpha + \cos\beta = -\dfrac{17}{65}$ (2)

Squaring equation (1), we get

$\sin^2\alpha + \sin^2\beta + 2\sin\alpha\sin\beta = \left(\dfrac{21}{65}\right)^2$ (3)

Squaring equation (2), we get

$\cos^2\alpha + \cos^2\beta + 2\cos\alpha\cos\beta = \left(\dfrac{27}{65}\right)^2$ (4)

Adding equations (3) and (4), we get

$$2 + 2\cos(\alpha-\beta) = \frac{1}{(65)^2}\left[(27)^2 + (21)^2\right]$$

$$= \frac{1}{(65)^2}(729 + 441)$$

$$\Rightarrow 2 + 2\cos(\alpha-\beta) = \frac{1}{(65)^2}(1170) = \frac{18}{65}$$

$$\Rightarrow 1 + \cos(\alpha-\beta) = \frac{9}{65}$$

$$\Rightarrow 2\cos^2\frac{\alpha-\beta}{2} = \frac{9}{65}$$

$$\Rightarrow \cos\frac{\alpha-\beta}{2} = -\frac{3}{\sqrt{130}}$$

$[\because \pi < \alpha-\beta < 3\pi \Rightarrow \dfrac{\pi}{2} < \dfrac{\alpha-\beta}{2} < \dfrac{3\pi}{2}$

$\Rightarrow \cos\left(\dfrac{\alpha-\beta}{2}\right) < 0]$

64. b. $\dfrac{\sin(x+y)}{\sin(x-y)} = \dfrac{a+b}{a-b}$

$$\Rightarrow \frac{\sin(x+y)+\sin(x-y)}{\sin(x+y)-\sin(x-y)} = \frac{(a+b)+(a-b)}{(a+b)-(a-b)}$$

$$\Rightarrow \frac{2\sin x\cos y}{2\cos x\sin y} = \frac{2a}{2b}$$

$$\Rightarrow \frac{\tan x}{\tan y} = \frac{a}{b}$$

65. c. $\dfrac{\sin 3\theta + \sin 5\theta + \sin 7\theta + \sin 9\theta}{\cos 3\theta + \cos 5\theta + \cos 7\theta + \cos 9\theta}$

$$= \frac{(\sin 3\theta + \sin 9\theta) + (\sin 5\theta + \sin 7\theta)}{(\cos 3\theta + \cos 9\theta) + (\cos 5\theta + \cos 7\theta)}$$

$$= \frac{2\sin 6\theta\cos 3\theta + 2\sin 6\theta\cos\theta}{2\cos 6\theta\cos 3\theta + 2\cos 6\theta\cos\theta}$$

$$= \frac{2\sin 6\theta(\cos 3\theta + \cos\theta)}{2\cos 6\theta(\cos 3\theta + \cos\theta)} = \tan 6\theta$$

66. b. $\dfrac{\sin x - \sin z}{\cos z - \cos x} = \dfrac{2\cos\left(\dfrac{x+z}{2}\right)\sin\left(\dfrac{x-z}{z}\right)}{2\sin\left(\dfrac{x+z}{z}\right)\sin\left(\dfrac{x-z}{z}\right)}$

$$= \cot\left(\frac{x+z}{2}\right) = \cot(y)$$

67. c. $\cos 50° = \cos^2 25° - \sin^2 25°$

$= (\cos 25° + \sin 25°)(\cos 25° - \sin 25°)$

$= p(\cos 25° - \sin 25°)$ (1)

Now $(\cos 25° - \sin 25°)^2 + (\cos 25° + \sin 25°)^2$

$= 1 + 1$

$\Rightarrow \cos 25° - \sin 25° = \sqrt{2-p^2}$ (2)

We have taken positve sign as $\cos 25° > \sin 25°$. Therefore,

$\cos 50° = p\sqrt{2-p^2}$, by equations (1) and (2).

68. b. $\dfrac{\sin^2 A - \sin^2 B}{\sin A\cos A - \sin B\cos B} = \dfrac{2\sin(A+B)\sin(A-B)}{\sin 2A - \sin 2B}$

$$= \frac{2\sin(A+B)\sin(A-B)}{2\sin(A-B)\cos(A+B)}$$

$$= \tan(A+B)$$

69. a. $\tan A = \dfrac{1-\cos B}{\sin B} = \dfrac{2\sin^2(B/2)}{2\sin(B/2)\cos(B/2)}$

$$= \tan\frac{B}{2} \Rightarrow \tan 2A = \tan B$$

70. b. On adding, we get $a = \dfrac{3 - \cos 4\theta + 4\sin 2\theta}{2}$

$$= (1 + \sin 2\theta)^2$$

On subtracting, we get

$b = (1 - \sin 2\theta)^2$

$\Rightarrow ab = \cos^4 2\theta \le 1$

71. c. $\cos^2 10° - \cos 10°\cos 50° + \cos^2 50°$

$$=\frac{1}{2}[1+\cos 20° -(\cos 60° +\cos 40°)+(1+\cos 100°)]$$

$$=\frac{1}{2}[1+\cos 20° - \frac{1}{2} - \cos 40° + 1 - \cos 80°]$$

$$=\frac{1}{2}\left[\frac{3}{2} + \cos 20° - (2\cos 60° \cos 20°)\right] = \frac{3}{4}$$

72. a. $\tan 20° \tan 80° \cot 50°$

$$= \tan(50° - 30°)\tan(50° + 30°)\frac{1}{\tan 50°}$$

$$= \frac{\tan 50° - \frac{1}{\sqrt{3}}}{1 + \frac{\tan 50°}{\sqrt{3}}} \times \frac{\tan 50° + \frac{1}{\sqrt{3}}}{1 - \frac{\tan 50°}{\sqrt{3}}} \times \frac{1}{50°}$$

$$= \frac{(\sqrt{3}\tan 50° - 1)(\sqrt{3}\tan 50° + 1)}{(\sqrt{3} + \tan 50°)(\sqrt{3} - \tan 50°)\tan 50°}$$

$$= \frac{3\tan^2 50° - 1}{3\tan 50° - \tan^3 50°} = -\cot(150°).$$

73. b. $\tan^2\theta = 2\tan^2\phi + 1$

$\Rightarrow 1 + \tan^2\theta = 2(1 + \tan^2\phi)$

$\Rightarrow \sec^2\theta = 2\sec^2\phi$

$\Rightarrow \cos^2\phi = 2\cos^2\theta$

$\Rightarrow \cos^2\phi = 1 + \cos 2\theta$

$\Rightarrow \cos 2\theta = \cos^2\phi - 1$

$\Rightarrow \cos 2\theta = -\sin^2\phi$

$\Rightarrow \sin^2\phi + \cos 2\theta = 0$

74. b. $\cot 70° + 4\cos 70° = \frac{\cos 70° + 4\sin 70°\cos 70°}{\sin 70°}$

$$= \frac{\cos 70° + 2\sin 140°}{\sin 70°}$$

$$= \frac{\cos 70° + 2\sin(180° - 40°)}{\sin 70°}$$

$$= \frac{\sin 20° + \sin 40° + \sin 40°}{\sin 70°}$$

$$= \frac{2\sin 30°\cos 10° + \sin 40°}{\sin 70°}$$

$$= \frac{\sin 80° + \sin 40°}{\sin 70°}$$

$$= \frac{2\sin 60°\cos 20°}{\sin 70°} = \sqrt{3}$$

75. b. $a\cos x + b\sin x = c$

$$\Rightarrow \frac{a\left(1 - \tan^2\frac{x}{2}\right)}{1 + \tan^2\frac{x}{2}} + \frac{2b\tan\frac{x}{2}}{1 + \tan^2\frac{x}{2}} = c$$

$$\Rightarrow (c + a)\tan^2\frac{x}{2} - 2b\tan\frac{x}{2} + c - a = 0$$

$$\Rightarrow \tan\frac{x_1}{2} + \tan\frac{x_2}{2} = \frac{2b}{c + a}$$

and $\tan\frac{x_1}{2}\tan\frac{x_2}{2} = \frac{c - a}{c + a}$

$$\Rightarrow \tan\left(\frac{x_1 + x_2}{2}\right) = \frac{\frac{2b}{c + a}}{1 - \frac{c - a}{c + a}} = \frac{2b}{2a} = \frac{b}{a}$$

76. c. $\tan y = \frac{1 + \sqrt{1 - x}}{1 + \sqrt{1 + x}}$

If $x = \cos\theta$, then

$\sqrt{1 - x} = \sqrt{2}\sin(\theta/2)$, $\sqrt{1 + x}$

$= \sqrt{2}\cos(\theta/2)$

$$\Rightarrow \tan y = \frac{\sqrt{2}\left[\frac{1}{\sqrt{2}} + \sin\frac{\theta}{2}\right]}{\sqrt{2}\left[\frac{1}{\sqrt{2}} + \cos\frac{\theta}{2}\right]} = \frac{\sin\frac{\pi}{4} + \sin\frac{\theta}{2}}{\cos\frac{\pi}{4} + \cos\frac{\theta}{2}}$$

$$\Rightarrow \tan y = \frac{2\sin\left(\frac{\pi}{8} + \frac{\theta}{4}\right)\cos\left(\frac{\pi}{8} - \frac{\theta}{4}\right)}{2\cos\left(\frac{\pi}{8} + \frac{\theta}{4}\right)\cos\left(\frac{\pi}{8} - \frac{\theta}{4}\right)}$$

$$\Rightarrow \tan y = \tan\left(\frac{\pi}{8} + \frac{\theta}{4}\right)$$

$$\Rightarrow 4y = \frac{\pi}{2} + \theta$$

$\Rightarrow \sin 4y = \cos\theta = x$

77. d. We have $\cos x = \tan y$

$\Rightarrow \cos^2 x = \tan^2 y$

$\Rightarrow \cos^2 x = \sec^2 y - 1$

$\Rightarrow \cos^2 x = \cot^2 z - 1$

$\Rightarrow \cos^2 x = \cot^2 z - 1$ [$\because \cos y = \tan z \therefore \sec y = \cot z$]

$\Rightarrow 1 + \cos^2 x = \cot^2 z$

$\Rightarrow 1 + \cos^2 x = \frac{\tan^2 x}{1 - \tan^2 x}$ [$\because \cos z = \tan x$]

$$\Rightarrow 1 + \cos^2 x = \frac{\sin^2 x}{\cos^2 x - \sin^2 x}$$

$\Rightarrow 2\sin^4 x - 6\sin^2 x + 2 = 0$

$$\Rightarrow \sin^2 x = \frac{3 - \sqrt{5}}{2}$$

$$\Rightarrow \sin^2 x = \left(\frac{\sqrt{5}-1}{2}\right)^2$$

$$\Rightarrow \sin x = \frac{\sqrt{5}-1}{2} = 2\sin 18°$$

78. a. $\sin 2\theta = \cos 3\theta$

$\Rightarrow 2\sin\theta\cos\theta = 4\cos^3\theta - 3\cos\theta$

$\Rightarrow 2\sin\theta = 4(1-\sin^2\theta) - 3 = 4\sin^2\theta + 2\sin\theta - 1 = 0$

$$\Rightarrow \sin\theta = \frac{\sqrt{5}-1}{4}$$

79. b. $\tan\theta = \lambda$

$$\Rightarrow \frac{2\tan\theta/2}{1-\tan^2\theta/2} = \lambda$$

$$\Rightarrow \lambda\tan^2\frac{\theta}{2} + 2\tan\frac{\theta}{2} - \lambda = 0$$

$$\Rightarrow \tan\frac{\theta_1}{2}\tan\frac{\theta_2}{2} = -1$$

80. a. $\tan\theta = \sqrt{n}$

$$\Rightarrow \cos 2\theta = \frac{1-\tan^2\theta}{1+\tan^2\theta} = \frac{1-n}{1+n} = \text{rational.}$$

81. b. $\sin x + \cos x = \dfrac{\sqrt{7}}{2}$

$$\Rightarrow \frac{2\tan\frac{x}{2}}{\left(1+\tan^2\frac{x}{2}\right)} + \frac{1-\tan^2\frac{x}{2}}{1+\tan^2\frac{x}{2}} = \frac{\sqrt{7}}{2}$$

$$\Rightarrow (\sqrt{7}+2)\tan^2\frac{x}{2} - 4\tan\frac{x}{2} + (\sqrt{7}-2) = 0$$

$$\Rightarrow \tan\frac{x}{2} = \frac{4 \pm \sqrt{16-4(7-4)}}{2(\sqrt{7}+2)}$$

$$= \frac{1}{(\sqrt{7}+2)} \text{ as } \frac{x}{2} < \frac{\pi}{8}$$

$$\Rightarrow \tan\frac{x}{2} = \frac{\sqrt{7}-2}{3}$$

82. b. $\sin\dfrac{7\pi}{8} = \sin\left(\pi - \dfrac{\pi}{8}\right) = \sin\dfrac{\pi}{8}$ and

$$\sin\frac{5\pi}{8} = \sin\left(\pi - \frac{3\pi}{8}\right) = \sin\frac{3\pi}{8}$$

$$\text{The given value} = 2\left[\sin^2\frac{\pi}{8} + \sin^2\frac{3\pi}{8}\right]$$

$$= 2\left[\sin^2\frac{\pi}{8} + \cos^2\frac{\pi}{8}\right]$$

$$= 2(1) = 2$$

$$\left[\because \sin\frac{3\pi}{8} = \sin\left(\frac{\pi}{2} - \frac{\pi}{8}\right) = \cos\frac{\pi}{8}\right]$$

83. b. $4\cos^2\left(\dfrac{\pi}{4} - \dfrac{x}{2}\right) + \sqrt{4\sin^4 x + \sin^2 2x}$

$$= 4\cos^2\left(\frac{\pi}{4} - \frac{x}{2}\right) + \sqrt{4\sin^2 x(\cos^2 x + \sin^2 x)}$$

$$= 2\left(1 + \cos\left(\frac{\pi}{2} - x\right)\right) + 2|\sin x|$$

$$= 2 + 2\sin x - 2\sin x, \text{ as } x \in \left(\pi, \frac{3\pi}{2}\right) = 2$$

84. b. $\cos^3 x \sin 2x = \cos^2 x \cos x \sin 2x$

$$= \left(\frac{1-\cos 2x}{2}\right)\left(\frac{2\sin 2x\cos x}{2}\right)$$

$$= \frac{1}{4}(1-\cos 2x)(\sin 3x + \sin x)$$

$$= \frac{1}{4}\left[\sin 3x + \sin x - \frac{1}{2}(2\sin 3x \cos 2x) - \frac{1}{2}(2\cos 2x \sin x)\right]$$

$$= \frac{1}{4}\left[\sin 3x + \sin x - \frac{1}{2}(\sin 5x + \sin x) - \frac{1}{2}(\sin 3x - \sin x)\right]$$

$$= \frac{1}{4}\left[\sin x + \frac{1}{2}\sin 3x - \frac{1}{2}\sin 5x\right]$$

$$\Rightarrow a_1 = \frac{1}{4}, a_3 = \frac{1}{8}, n = 5$$

85. b. Given expression is

$2\sin^2\phi + 4\cos(\theta+\phi)\sin\theta\sin\phi + \cos 2(\theta+\phi)$

$= (1-\cos 2\phi) + 4\cos(\theta+\phi)\sin\theta\sin\phi + 2\cos^2(\theta+\phi) - 1$

$= -\cos 2\phi + 4\cos(\theta+\phi)\sin\theta\sin\phi + 2\cos^2(\theta+\phi)$

$= -\cos 2\phi + 2\cos(\theta+\phi)[\cos(\theta+\phi) + 2\sin\theta\sin\phi]$

$= -\cos 2\phi + 2\cos(\theta+\phi)[\cos\theta\cos\phi + \sin\theta\sin\phi]$

$= -\cos 2\phi + 2\cos(\theta+\phi)\cos(\theta-\phi)$

$= -\cos 2\phi + \cos 2\theta + \cos 2\phi = \cos 2\theta$

86. b. $\dfrac{\cos 2B}{1} = \dfrac{\cos(A+C)}{\cos(A-C)}$

Applying componendo and dividendo, we get

$$\Rightarrow \frac{1-\cos 2B}{1+\cos 2B} = \frac{\cos(A-C) - \cos(A+C)}{\cos(A-C) + \cos(A+C)}$$

$$\Rightarrow \frac{2\sin^2 B}{2\cos^2 B} = \frac{2\sin A\sin C}{2\cos A\cos C}$$

$\Rightarrow \tan^2 B = \tan A \tan C$

$\Rightarrow \tan A, \tan B, \tan C$ are in G.P.

87. b. $\cos x = \dfrac{2\cos y - 1}{2 - \cos y}$

$$\Rightarrow \frac{1-\tan^2\frac{x}{2}}{1+\tan^2\frac{x}{2}} = \frac{\frac{2(1-\tan^2 y/2)}{1+\tan^2 y/2} - 1}{2 - \frac{1-\tan^2 y/2}{1+\tan^2 y/2}}$$

$$\Rightarrow 6\tan^2 y/2 = 2\tan^2\frac{x}{2}$$

$$\Rightarrow \tan\frac{x}{2}\cot\frac{y}{2} = \sqrt{3}$$

88. b. We have $\sqrt{\left(\dfrac{a+b}{a-b}\right)} + \sqrt{\left(\dfrac{a-b}{a+b}\right)}$

$$= \frac{a+b+a-b}{\sqrt{(a^2-b^2)}}$$

$$= \frac{2a}{\sqrt{(a^2-b^2)}} = \frac{2}{\sqrt{[1-(b/a)^2]}}$$

$$= \frac{2}{\sqrt{(1-\tan^2 x)}} = \frac{2\cos x}{\sqrt{(\cos^2 x - \sin^2 x)}}$$

$$= \frac{2\cos x}{\sqrt{(\cos 2x)}}$$

89. b. Since α is a root of $25\cos^2\theta + 5\cos\theta - 12 = 0$,

$$25\cos^2\alpha + 5\cos\alpha - 12 = 0$$

$$\Rightarrow \cos\alpha = \frac{-5 \pm \sqrt{25+1200}}{50} = -\frac{4}{5} \quad \left[\because \frac{\pi}{2} < \alpha < \pi\right]$$

and $\sin\alpha = \sqrt{1 - \dfrac{16}{25}} = \dfrac{3}{5}$ $\quad [\because \sin 2\alpha = 2\sin\alpha\cos]$

$$= 2\left(\frac{3}{5}\right)\left(\frac{-4}{5}\right) = \frac{-24}{25}$$

90. c. We have $\tan 9° - \tan 27° - \tan 63° + \tan 81°$

$$= (\tan 9° + \tan 81°) - (\tan 27° + \tan 63°)$$

$$= \frac{1}{\sin 9° \cos 9°} - \frac{1}{\sin 27° \cos 27°}$$

$$= \frac{2}{\sin 18°} - \frac{2}{\sin 54°}$$

$$= 2\left[\frac{\sin 54° - \sin 18°}{\sin 54° \sin 18°}\right]$$

$$= 2\left[\frac{2\cos 36° \sin 18°}{\sin 18° \cos 36°}\right] = 4$$

91. c. Let $A = \sin^{-1}a$, $B = \sin^{-1}b$, and $C = \sin^{-1}c$.
So $A + B + C = \pi$

$$a\sqrt{1-a^2} + b\sqrt{1-b^2} + c\sqrt{1-c^2}$$

$$= \frac{1}{2}(\sin 2A + \sin 2B + \sin 2C)$$

$$= \frac{1}{2}[4\sin A \sin B \sin C] = 2abc.$$

92. d. $\cos 2A + \cos 2B + \cos 2C$

$$= 2\cos(A+B)\cos(A-B) + \cos 2C$$

$$= 2\cos\left(\frac{3\pi}{2} - C\right)\cos(A-B) + \cos 2C$$

$$= -2\sin C\cos(A-B) + 1 - 2\sin^2 C$$

$$= 1 - 2\sin C(\cos(A-B) + \sin C)$$

$$= 1 - 2\sin C\{\cos(A-B) + \sin[3\pi/2 - (A+B)]\}$$

$$= 1 - 2\sin C[\cos(A-B) - \cos(A+B)]$$

$$= 1 - 4\sin A \sin B \sin C$$

Trick: Consider $A = B = C = \dfrac{\pi}{2}$,

then $\cos 2A + \cos 2B + \cos 2C = 3\cos\pi = -3$

Also $1 - 4\sin A \sin B \sin C = 1 - \dfrac{\pi}{2} = -3$.

93. c. $\dfrac{2}{\tan\frac{B}{2}} = \dfrac{1}{\tan\frac{A}{2}} + \dfrac{1}{\tan\frac{C}{2}}$

$$\Rightarrow 2\tan\frac{A}{2}\tan\frac{C}{2} = \tan\frac{B}{2}\tan\frac{C}{2} + \tan\frac{B}{2}\tan\frac{A}{2}$$

$$= 1 - \tan\frac{A}{2}\tan\frac{C}{2}$$

$$\Rightarrow \tan\frac{A}{2}\tan\frac{C}{2} = \frac{1}{3}$$

$$\Rightarrow \cot\frac{A}{2}\cot\frac{C}{2} = 3$$

94. b. $\sin^2 A - \sin^2 B + \sin^2 C = \sin(A+B)\sin(A-B) + \sin^2 C$

$$= \sin C(\sin(A-B) + \sin C)$$

$$= \sin C(\sin(A-B) + \sin(A+B))$$

$$= 2\sin A\cos B\sin C.$$

95. c. $\sin^2\alpha + \sin^2\beta + \sin^2\gamma$

$$= \frac{\tan^2\alpha}{1+\tan^2\alpha} + \frac{\tan^2\beta}{1+\tan^2\beta} + \frac{\tan^2\gamma}{1+\tan^2\gamma}$$

$$= \frac{x}{1+x} + \frac{y}{1+y} + \frac{z}{1+z}$$

[where $x = \tan^2\alpha, y = \tan^2\beta, z = \tan^2\gamma$]

$$= \frac{(x+y+z) + (xy+yz+zx+2xyz) + xy + yz + zx + xyz}{(1+x)(1+y)(1+z)}$$

$$= \frac{1+x+y+z+xy+yz+zx}{(1+x)(1+y)(1+z)} = 1.$$

$[\because xy + yz + zx + 2xyz = \gamma]$

96. c. $D^r = \sin A + \sin B - \sin C$

$$= 2\sin\frac{A+B}{2}\cos\frac{A-B}{2} - 2\sin\frac{C}{2}\cos\frac{C}{2}$$

$$= 2\cos\frac{C}{2}\left[\cos\left(\frac{A-B}{2}\right) - \sin\frac{C}{2}\right]$$

$$= 2\cos\frac{C}{2}\left[\cos\frac{A-B}{2} - \cos\frac{A+B}{2}\right]$$

$$= 2\cos\frac{C}{2}\left[2\sin\frac{A}{2}\sin\frac{B}{2}\right]$$

$$= 4\sin\frac{A}{2}\sin\frac{B}{2}\cos\frac{C}{2}.$$

Also $\sin A + \sin B + \sin C = 4\cos\frac{A}{2}\cos\frac{B}{2}\cos\frac{C}{2}$

$$\Rightarrow \frac{\sin A + \sin B + \sin C}{\sin A + \sin B - \sin C} = \cot\frac{A}{2}\cot\frac{B}{2}$$

97. a. $\dfrac{\sin 2A + \sin 2B + \sin 2C}{\sin A + \sin B + \sin C} = \dfrac{4\sin A \sin B \sin C}{4\cos\frac{A}{2}\cos\frac{B}{2}\cos\frac{C}{2}}$

$$= 8\sin\frac{A}{2}\sin\frac{B}{2}\sin\frac{C}{2}$$

$$\left[\because \sin A = 2\sin\frac{A}{2}\cos\frac{A}{2}\right]$$

98. c. We have $\cos^2 A + \cos^2 B - (1 - \cos^2 C) = 0$

$\Rightarrow \cos^2 A + \cos^2 B - \sin^2 C = 0$

$\Rightarrow \cos^2 A + \cos(B + C)\cos(B - C) = 0$

$\Rightarrow 2\cos A \cos B \cos C = 0$

$\Rightarrow$ Either A or B or C is 90°.

99. a. In a triangle,

$\tan A + \tan B + \tan C = \tan A \tan B \tan C$ (1)

$\Rightarrow 6 = 2\tan C$

$\Rightarrow \tan C = 3$

Also $\tan A + \tan B = 6 - 3 = 3$ (2)

$\Rightarrow$ $\tan A$ and $\tan B$ are roots $x^2 - 3x + 2 = 0$ by equations (1) and (2).

$\Rightarrow \tan A, \tan B = 2, 1$ or $1, 2$ and $\tan C = 3$.

100. a. We have $\tan 6° \tan 42° \tan 66° \tan 78°$

$= \tan 6° \tan(60° - 18°) \tan(60° + 6°) \tan(60° + 18°)$

$$= \frac{\tan 6° \tan(60° + 6°) \tan 18° \tan(60° - 18°) \tan(60° + 18°)}{\tan 18°}$$

$$= \frac{\tan 6° \tan(60° + 6°) \tan(3 \times 18°)}{\tan 18°}$$

$$= \frac{\tan 6° \tan(60° - 6) \tan(60° + 6)}{\tan 18°} = \frac{\tan 18°}{\tan 18°} = 1$$

101. c.

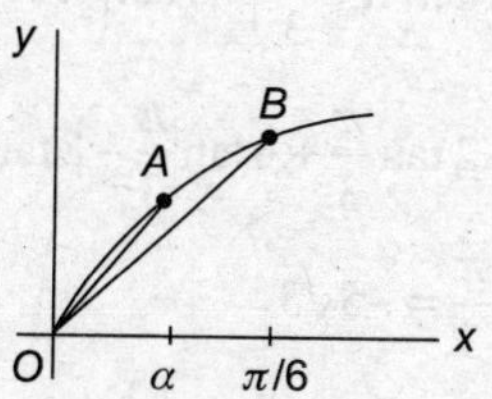

Fig. 1.3

In the graph of $y = \sin x$, let $A \equiv (\alpha, \sin\alpha)$, $B = \left(\frac{\pi}{6}, \sin\frac{\pi}{6}\right)$

Clearly, slope of $OA >$ slope of OB,

so $\dfrac{\sin\alpha}{\alpha} > \dfrac{\sin\frac{\pi}{6}}{\frac{\pi}{6}} = \dfrac{3}{\pi} \Rightarrow \dfrac{\alpha}{\sin\alpha} < \dfrac{\pi}{3}$.

102. c. $\sin x + \cos x = 1 - \alpha \sin x \cos x$

$\Rightarrow a^2 \sin^2 x \cos^2 x - 2(a + 1)\sin x \cos x = 0$

$$\Rightarrow \sin 2x\left(\frac{a^2}{2}\sin 2x - 2(a+1)\right) = 0$$

$\Rightarrow$ either $\sin 2x = 0, \forall a \in R$ or $\sin 2x = \dfrac{4(a+1)}{a^2}$

for no solution $\left|\dfrac{4(a+1)}{a^2}\right| > 1$

$\Rightarrow a \in \left(2 - \sqrt{2}, 2 + \sqrt{2}\right)$

$\Rightarrow a$ cannot be equal to 1, 2, 3.

103. b. Let $u = \cos\theta\left\{\sin\theta + \sqrt{\sin^2\theta + \sin^2\alpha}\right\}$

$\Rightarrow (u - \sin\theta\cos\theta)^2 = \cos^2\theta(\sin^2\theta + \sin^2\alpha)$

$\Rightarrow u^2\tan^2\theta - 2u\tan\theta + u^2 - \sin^2\alpha = 0$

Since $\tan\theta$ is real, $4u^2 - 4u^2(u^2 - \sin^2\alpha) \ge 0$.

$\Rightarrow u^2 - (1 + \sin^2\alpha) \ge 0$

$\Rightarrow |u| \le \sqrt{1 + \sin^2\alpha}$

104. a. $\sin\theta_1 \sin\theta_2 - \cos\theta_1 \cos\theta_2 = -1$

$\Rightarrow \cos(\theta_1 + \theta_2) = 1$

$\Rightarrow \theta_1 + \theta_2 = 2n\pi, n \in I$

$$\Rightarrow \frac{\theta_1}{2} + \frac{\theta_2}{2} = n\pi$$

Thus, $\tan\dfrac{\theta_1}{2}\cot\dfrac{\theta_2}{2} = \tan\dfrac{\theta_1}{2}\cot\left(n\pi - \dfrac{\theta_1}{2}\right)$

$$= -\tan\frac{\theta_1}{2}\cot\frac{\theta_1}{2} = -1$$

105. a. $\tan\frac{\pi}{3}+2\tan\frac{2\pi}{3}+4\tan\frac{4\pi}{3}+8\tan\frac{8\pi}{3}$

$=\tan\frac{\pi}{3}+2\tan\left(\pi-\frac{\pi}{3}\right)+4\tan\left(\pi+\frac{\pi}{3}\right)+8\tan\left(3\pi-\frac{\pi}{3}\right)$

$=\tan\frac{\pi}{3}-2\tan\frac{\pi}{3}+4\tan\frac{\pi}{3}-8\tan\frac{\pi}{3}$

$=-5\tan\frac{\pi}{3}=-5\sqrt{3}$

106. c. Since $\tan 3\theta=\dfrac{3\tan\theta-\tan^3\theta}{1-3\tan^2\theta}$

Putting $\theta=\frac{\pi}{9}$, we get

$\tan\frac{\pi}{3}=\dfrac{3\tan\frac{\pi}{9}-\tan^3\frac{\pi}{9}}{1-3\tan^2\frac{\pi}{9}} \Rightarrow 3\left(1-3\tan^2\frac{\pi}{9}\right)^2$

$=\left(3\tan\frac{\pi}{9}-\tan^3\frac{\pi}{9}\right)^2$

$\Rightarrow \tan^6\frac{\pi}{9}-33\tan^4\frac{\pi}{9}+27\tan^2\frac{\pi}{9}=3.$

107. c. $\cos x+\cos y-\cos(x+y)=\frac{3}{2}$

$\Rightarrow 2\cos\left(\frac{x+y}{2}\right)\cos\left(\frac{x-y}{2}\right)-2\cos^2\left(\frac{x+y}{2}\right)$

$+1=\frac{3}{2}$

$\Rightarrow 2\cos^2\left(\frac{x+y}{2}\right)-2\cos\left(\frac{x+y}{2}\right)\cos\left(\frac{x-y}{2}\right)$

$+\frac{1}{2}=0$

Since $\cos\left(\frac{x+y}{2}\right)$ is always real,

discriminant ≥ 0.

$\Rightarrow 4\cos^2\left(\frac{x-y}{2}\right)-4\geq 0$

$\Rightarrow \cos^2\left(\frac{x-y}{2}\right)\geq 1$

$\Rightarrow \cos^2\left(\frac{x-y}{2}\right)=1$

$\Rightarrow \frac{x-y}{2}=0 \Rightarrow x=y$

108. a. $a\sin x+b\cos(x+\theta)+b\cos(x-\theta)=d$

$\Rightarrow a\sin x+2b\cos x\cos\theta=d$

$\Rightarrow |d|\leq\sqrt{a^2+4b^2\cos^2\theta}$

$\Rightarrow \dfrac{d^2-a^2}{4b^2}\leq\cos^2\theta$

$\Rightarrow |\cos\theta|\geq\dfrac{\sqrt{d^2-a^2}}{2|b|}$

109. d. $\dfrac{\sin x}{\sin y}=\dfrac{1}{2},\ \dfrac{\cos x}{\cos y}=\dfrac{3}{2}\Rightarrow\dfrac{\tan x}{\tan y}=\dfrac{1}{3}$

$\tan(x+y)=\dfrac{\tan x+\tan y}{1-\tan x\tan y}=\dfrac{4\tan x}{1-3\tan^2 x}$

Also $\sin y=2\sin x$, $\cos y=\frac{2}{3}\cos x$

$\Rightarrow \sin^2 y+\cos^2 y=4\sin^2 x+\dfrac{4\cos^2 x}{9}=1$

$\Rightarrow 36\tan^2 x+4=9\sec^2 x=9(1+\tan^2 x)$

$\Rightarrow 27\tan^2 x=5$

$\Rightarrow \tan x=\dfrac{\sqrt{5}}{3\sqrt{3}}$

$\Rightarrow \tan(x+y)=\dfrac{\frac{4\sqrt{5}}{3\sqrt{3}}}{1-\frac{15}{27}}=\dfrac{4\sqrt{5}\times 27}{12\times 3\sqrt{3}}=\sqrt{15}$

110. a. $\sqrt{1+\cos x}=\sqrt{2\cos^2\frac{x}{2}}=\sqrt{2}\left|\cos\frac{x}{2}\right|$

and $\sqrt{1-\cos x}=\sqrt{2\sin^2\frac{x}{2}}=\sqrt{2}\left|\sin\frac{x}{2}\right|$

$\Rightarrow \dfrac{\sqrt{1+\cos x}+\sqrt{1-\cos x}}{\sqrt{1+\cos x}-\sqrt{1-\cos x}}=\dfrac{\left|\cos\frac{x}{2}\right|+\left|\sin\frac{x}{2}\right|}{\left|\cos\frac{x}{2}\right|-\left|\sin\frac{x}{2}\right|}$

$=\dfrac{-\cos\frac{x}{2}+\sin\frac{x}{2}}{-\cos\frac{x}{2}-\sin\frac{x}{2}}$

$=\dfrac{\cos\frac{x}{2}-\sin\frac{x}{2}}{\cos\frac{x}{2}+\sin\frac{x}{2}}$

$=\dfrac{1-\tan\frac{x}{2}}{1+\tan\frac{x}{2}}$

$=\tan\left(\frac{\pi}{4}-\frac{x}{2}\right)$

$=\cot\left(\frac{\pi}{2}-\left(\frac{\pi}{4}-\frac{x}{2}\right)\right)$

$=\cot\left(\frac{\pi}{4}+\frac{x}{2}\right)$

111. d. $\tan x = n\tan y$, $\cos(x-y)$

$= \cos x\cos y + \sin x\sin y$

$\Rightarrow \cos(x-y) = \cos x\cos y\,(1+\tan x\tan y)$

$= \cos x\cos y\,(1+n\tan^2 y)$

$$\Rightarrow \sec^2(x-y) = \frac{\sec^2 x\sec^2 y}{(1+n\tan^2 y)^2}$$

$$= \frac{(1+\tan^2 x)(1+\tan^2 y)}{(1+n\tan^2 y)^2}$$

$$= \frac{(1+n^2\tan^2 y)(1+\tan^2 y)}{(1+n\tan^2 y)^2}$$

$$= 1+\frac{(n-1)^2\tan^2 y}{(1+n\tan^2 y)^2}$$

Now, $\left(\dfrac{1+n\tan^2 y}{2}\right)^2 \ge n\tan^2 y$

$$\Rightarrow \frac{\tan^2 y}{(1+n\tan^2 y)^2} \le \frac{1}{4n}$$

$$\Rightarrow \sec^2(x-y) \le 1+\frac{(n-1)^2}{4n} = \frac{(n+1)^2}{4n}$$

112. b. $\cot^2 x = \cot(x-y)\cot(x-z)$

$$\Rightarrow \cot^2 x = \left(\frac{\cot x\cot y+1}{\cot y-\cot x}\right)\left(\frac{\cot x\cot z+1}{\cot z-\cot x}\right)$$

$\Rightarrow \cot^2 x\cot y\cot z - \cot^3 x\cot y - \cot^3 x\cot z + \cot^4 x$

$= \cot^2 x\cot y\cot z + \cot x\cot y + \cot x\cot z + 1$

$\Rightarrow \cot^3 x\,(\cot y + \cot z) + \cot x\,(\cot y + \cot z) + 1 - \cot^4 x = 0$

$\Rightarrow \cot x(\cot y + \cot z)(1+\cot^2 x) + (1-\cot^2 x)(1+\cot^2 x) = 0$

$\Rightarrow (\cot x(\cot y + \cot z) + (1-\cot^2 x)) = 0$

$$\Rightarrow \frac{\cot^2 x-1}{2\cot x} = \frac{1}{2}(\cot y+\cot z) = \cot 2x$$

113. a. $A+B+C=\pi$

$\Rightarrow \tan A + \tan B + \tan C = \tan A\tan B\tan C$

Now, A.M. $\ge$ G.M.

$$\Rightarrow \frac{\tan A+\tan B+\tan C}{3} \ge (\tan A\tan B\tan C)^{1/3}$$

$$\Rightarrow \frac{\tan A+\tan B+\tan C}{3} \ge (\tan A\tan B\tan C)^{1/3}$$

$\Rightarrow (\tan A\tan B\tan C)^{2/3} \ge 3$

$$\Rightarrow \left(\frac{1}{K}\right)^{2/3} \ge 3$$

$$\Rightarrow K \le \frac{1}{3\sqrt{3}}$$

114. d. $u^2 = (a^2\cos^2\theta + b^2\sin^2\theta) + (a^2\sin^2\theta + b^2\sin^2\theta)$

$$+2\sqrt{a^2\cos^2\theta+b^2\sin^2\theta}\,\sqrt{a^2\sin^2\theta+b^2\cos^2\theta}$$

$= a^2+b^2$

$$+2\sqrt{\sin^2\theta\cos^2\theta\left(a^4+b^4\right)+a^2b^2\left(\sin^4\theta+\cos^4\theta\right)}$$

$= (a^2+b^2)$

$$+2\sqrt{a^2b^2\left(1-2\sin^2\theta\cos^2\theta\right)+\left(a^4+b^4\right)\sin^2\theta\cos^2\theta}$$

$$= (a^2+b^2)+2\sqrt{a^2b^2+\left(a^2-b^2\right)^2\sin^2\theta\cos^2\theta}$$

$$= (a^2+b^2)+2\sqrt{a^2b^2+\frac{\left(a^2-b^2\right)^2}{4}\sin^2 2\theta}$$

$$\text{Max. } u^2 = (a^2+b^2) = 2\sqrt{a^2b^2+\frac{\left(a^2-b^2\right)^2}{4}}$$

Min. $u^2 = (a^2+b^2)+2ab$

$$\text{Hence, difference} = 2\sqrt{a^2b^2+\frac{\left(a^2-b^2\right)^2}{4}} - 2ab$$

$$= \sqrt{4a^2b^2+a^4+b^4-2a^2b^2} - 2ab$$

$$= \sqrt{\left(a^2+b^2\right)^2} - 2ab = a^2+b^2-2ab$$

$= (a-b)^2$

Multiple Correct Answers Type

1. a, b.

$2\sin\alpha\cos\alpha = 2\cos^2\beta$

$\sin 2\alpha = 1+\cos 2\beta$. Therefore,

$\cos 2\beta = -(1-\sin 2\alpha)$

$$= -\left(1-\cos\left(\frac{\pi}{2}-2\alpha\right)\right)$$

$$= -2\sin^2\left(\frac{\pi}{4}-\alpha\right) = -2\cos^2\left(\frac{\pi}{4}+\alpha\right)$$

2. a., b.

$\cos^2\theta = 1 - \sin^2\theta$. Let $81^{\sin^2\theta} = z$. Therefore,

$z + \frac{81}{z} = 30$ or

$z^2 - 30z + 81 = 0 \Rightarrow (z-27)(z-3) = 0$

$\therefore 81^{\sin^2\theta} = 3^{4\sin^2\theta} = 3^3$ or 3^1

$\Rightarrow \sin^2\theta = \frac{3}{4}, \frac{1}{4}$ or $\sin\theta = \pm\frac{\sqrt{3}}{2}, \pm\frac{1}{2}$

$\therefore \theta = \frac{\pi}{6}, \frac{\pi}{3}$ or $30°, 60°$. $\because 0 \le \theta \le \pi$

3. b., c., d.

Opposite angles of a cyclic quadrilateral are supplementary.

4. a., b., c.

a. $\tan\theta = \frac{1-\cos 2\theta}{\sin 2\theta} \Rightarrow$ (a) is correct.

b. $\sin 2\theta = \frac{2\tan\theta}{1+\tan^2\theta}$

$\cos 2\theta = \frac{1-\tan^2\theta}{1+\tan^2\theta}$

$\tan 2\theta = \frac{2\tan\theta}{1-\tan^2\theta}$

$\Rightarrow$ (b) is correct.

c. $\tan 3\theta = \frac{\sin 3\theta}{\cos 3\theta} \Rightarrow$ (c) is correct.

d. $\sin\theta = \frac{1}{3}$ which is rational but $\cos 3\theta = \cos\theta$ $(4\cos^2\theta - 3)$ which is irrational.

$\Rightarrow$ (d) is incorrect.

5. a., b., c., d.

a. $\sin\left(\frac{11\pi}{12}\right)\sin\left(\frac{5\pi}{12}\right) = \sin\left(\frac{\pi}{12}\right)\cos\left(\frac{\pi}{12}\right)$

$= \frac{1}{2}\sin\left(\frac{\pi}{6}\right) = \frac{1}{4} \in Q.$

b. $\text{cosec}\left(\frac{9\pi}{10}\right)\sec\left(\frac{4\pi}{5}\right) = -\text{cosec}\left(\frac{\pi}{10}\right)\sec\left(\frac{\pi}{5}\right)$

$= \frac{1}{\sin 18°\cos 36°}$

$= -\frac{16}{(\sqrt{5}-1)(\sqrt{5}+1)}$

$= -4 \in Q$

c. $\sin^4\left(\frac{\pi}{8}\right) + \cos^4\left(\frac{\pi}{8}\right)$

$= 1 - 2\sin^2\left(\frac{\pi}{8}\right)\cos^2\left(\frac{\pi}{8}\right)$

$= 1 - \frac{1}{2}\sin^2\left(\frac{\pi}{4}\right) = 1 - \frac{1}{4} = \frac{3}{4} \in Q$

d. $2\cos^2\frac{\pi}{9}\ 2\cos^2\frac{2\pi}{9}\ 2\cos^2\frac{4\pi}{9} = 8$ (cos 20° cos 40° cos 80°) $= \frac{1}{8} \in Q$

6. a., b., d.

$(\sin^2 x + \cos^2 x)^3 - 3\sin^2 x\cos^2 x(\sin^2 x + \cos^2 x) > \frac{5}{8}$

$\Rightarrow 1 - 3\sin^2 x\cos^2 x > \frac{5}{8}$

$\Rightarrow \frac{3}{8} > 3\sin^2 x\cos^2 x$

$\Rightarrow 1 - 2\sin^2 2x > 0$

$\Rightarrow \cos 4x > 0$

$\Rightarrow 4x \in \left(-\frac{\pi}{2}, \frac{\pi}{2}\right)$

$\Rightarrow 4x \in \left(2n\pi - \frac{\pi}{2}, 2n\pi + \frac{\pi}{2}\right)$, $n \in Z$, generalising now verify.

7. a., b., d.

c. $\sum\sin^2\frac{A}{2} = \frac{1}{3}[3 - (\cos A + \cos B + \cos C)]$

$= \frac{3}{2} - \frac{1}{2}(\cos A + \cos B + \cos C)$

but $[\cos A + \cos B + \cos C)]_{max} = 3/2$. Therefore,

$\sum\sin^2\frac{A}{2}\Big]_{min} = \frac{3}{2} - \frac{3}{4} = \frac{3}{4}$

Hence $\sum\sin^2\frac{A}{2} \ge \frac{3}{4}$

$\Rightarrow$ (c) is incorrect

a, b, d are correct and hold good in an equilateral triangle as their maximum values.

8. a., b., c.

a. $\tan\alpha\tan 2\alpha\tan 3\alpha = \tan 3\alpha - \tan 2\alpha - \tan\alpha$ always holds good.

b. R.H.S. $= \frac{\sin 4\alpha + \sin 2\alpha}{\sin 2\alpha\sin 4\alpha}$

$= \frac{2\sin 3\alpha\cos\alpha}{\sin 2\alpha\sin 4\alpha}$

$$= \frac{1}{\sin\alpha} = \text{cosec}\,\alpha \text{ (using } \pi/7 = \alpha)$$

$\Rightarrow$ (b) is correct

c. $\cos\alpha + \cos 3\alpha + \cos 5\alpha$

$$= \frac{\sin 3\alpha}{\sin\alpha}\cos 3\alpha = \frac{\sin 6\alpha}{2\sin\alpha} = \frac{\sin\frac{6\pi}{7}}{2\sin\frac{\pi}{7}}$$

$$= \frac{\sin\left(\pi + \frac{\pi}{7}\right)}{2\sin\frac{\pi}{7}} = \frac{1}{2}$$

Also $\cos 2\alpha = \cos\frac{2\pi}{7} = -\cos\left(\pi - \frac{5\pi}{7}\right)$

$$= -\cos\left(\frac{5\pi}{7}\right)$$

$$= -\cos 5\alpha.$$

d. $8\cos\alpha\cos 2\alpha\cos 4\alpha = \frac{\sin 8\alpha}{\sin\alpha}$

$$= \frac{\sin\frac{8\pi}{7}}{\sin\frac{\pi}{7}} = -\frac{\sin\frac{\pi}{7}}{\sin\frac{\pi}{7}} = -1$$

9. a., b., c., d.

a. For $x \in \left(0, \frac{\pi}{4}\right)$, $\tan x < \cot x$

Also $\ln(\sin x) < 0$

$\Rightarrow (\tan x)^{\ln(\sin x)} > (\cot x)^{\ln(\sin x)}$

b. For $x \in \left(0, \frac{\pi}{2}\right)$, $\text{cosec}\, x \geq 1$

$\Rightarrow \ln(\text{cosec}\, x) \geq 0$

$\Rightarrow 4^{\ln \text{cosec}\, x} < 5^{\ln(\text{cosec}\, x)}$

c. $x \in \left(0, \frac{\pi}{2}\right) \Rightarrow \cos x \in (0, 1)$

$\Rightarrow \ln(\cos x) < 0$

Also $\frac{1}{2} > \frac{1}{3}$

$\Rightarrow \left(\frac{1}{2}\right)^{\ln(\cos x)} < \left(\frac{1}{3}\right)^{\ln(\cos x)}$

d. For $x \in \left(0, \frac{\pi}{2}\right)$

$\sin x < \tan x$

$\Rightarrow \ln(\sin x) < \ln(\tan x)$

$\Rightarrow 2^{\ln(\sin x)} < 2^{\ln(\tan x)}$

10. a., c., d.

a. 1

b. 3

c. $\frac{\sin 24°\cos 6° - \cos 24°\sin 6°}{\sin 21°\cos 39° - \sin 39°\cos 21°}$

$$= \frac{\sin(18°)}{\sin(-18)} = -1$$

d. -1

11. a., c.

a. $\frac{\cos^2\alpha - \sin^2\alpha}{\sin\alpha\cos\alpha} = 2\cot 2\alpha$

b. $\frac{1+t}{1-t} - \frac{1-t}{1+t}$ where $t = \tan\alpha$

$$= \frac{(1+t)^2 - (1-t)^2}{1-t^2}$$

$$= \frac{4t}{1-t^2} = \frac{2\times 2\tan\alpha}{1-\tan^2\alpha}$$

$= 2\tan 2\alpha$

$\Rightarrow$ (b) is incorrect.

c. $\frac{1+t}{1-t} + \frac{1-t}{1+t} = \frac{(1+t)^2 + (1-t)^2}{1-t^2}$

$$= \frac{2(1+t^2)}{1-t^2} \quad \text{(where } t = \tan\alpha)$$

$$= \frac{2}{\cos 2\alpha} = 2\sec 2\alpha$$

$\Rightarrow$ (c) is correct

d. $\tan\alpha + \cot\alpha = \frac{1}{\cos\alpha\sin\alpha}$

$$= \frac{2}{\sin 2\alpha} = 2\,\text{cosec}\, 2\alpha$$

$\Rightarrow$ (d) is incorrect

12. a., c., d.

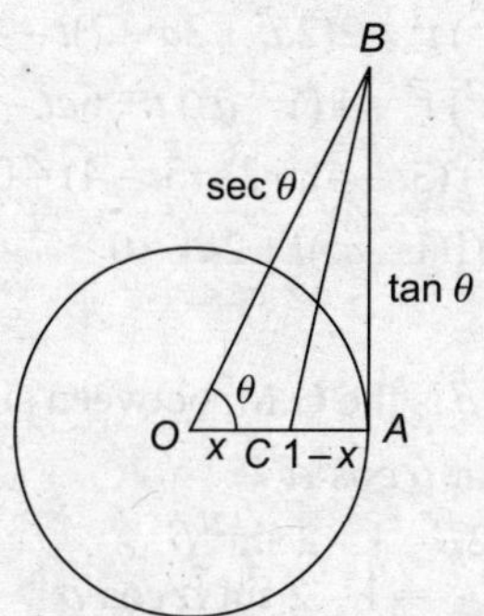

Fig. 1.4

Using property of angle bisector, we get

$$\frac{\sec\theta}{\tan\theta}=\frac{x}{1-x}$$

$$\Rightarrow x=\frac{\sec\theta}{\sec\theta+\tan\theta}=\sec\theta(\sec\theta-\tan\theta)$$

$$=\frac{1}{1+\sin\theta}.$$

13. a., d.

$$y=\frac{(1+\tan^2 x)^2}{1+\tan^2 x}=1+\tan^2 x$$

$$=1+\left(2-\sqrt{3}\right)^2$$

$$=8-4\sqrt{3}=4(2-\sqrt{3})$$

$$=4\left[\left(\sqrt{\frac{3}{2}}-\frac{1}{\sqrt{2}}\right)^2\right]$$

$$=4\left[4\left(\frac{\sqrt{3}-1}{2\sqrt{2}}\right)^2\right]$$

$$=16\sin^2\frac{\pi}{12}$$

14. b., d.

$\tan(\alpha+\beta)=\dfrac{15}{8}$ and $\operatorname{cosec}\gamma=\dfrac{17}{8}\Rightarrow\tan\gamma=\dfrac{8}{15}$

$\therefore\ \alpha+\beta+\gamma=\dfrac{\pi}{2}\Rightarrow$ (b) is true.

Also $\tan(\alpha+\beta)=\dfrac{\tan\alpha+\tan\beta}{1-\tan\alpha\tan\beta}$

$$\Rightarrow\cot\gamma=\frac{\tan\alpha+\tan\beta}{1-\tan\alpha\tan\beta}.$$

$$\Rightarrow\tan\alpha\tan\beta+\tan\beta\tan\gamma+\tan\gamma\tan\alpha=1$$

15. b., d.

Divide by $\cos\alpha$ and square both sides and let $\tan\alpha=t$ so that $\sec^2\alpha=1+t^2$

$\Rightarrow[(a+2)t+(2a-1)]^2=[(2a+1)^2(1+t^2)]$

$\Rightarrow t^2[(a+2)^2-(2a+1)]+2(a+2)(2a-1)t+[(2a-1)^2-(2a+1)^2]=0$

$\Rightarrow 3(1-a^2)t^2+2(2a^2+3a-2)t-4\times 2a=0$

$\Rightarrow 3(1-a^2)t^2-4(1-a^2)t+6at-8a=0$

$\Rightarrow t(1-a^2)(3t-4)+2a(3t-4)=0$

$\Rightarrow(3t-4)[(1-a^2)t+2a]=0$

16. a., c.

Since $\sin\beta$ is the G.M. between $\sin\alpha$ and $\cos\alpha$,

$\sin^2\beta=\sin\alpha\cos\alpha$

Now $\cos 2\beta=1-2\sin^2\beta$

$=1-2\sin\alpha\cos\alpha$

$=(\cos\alpha-\sin\alpha)^2$

$$=2\left(\frac{1}{\sqrt{2}}\cos\alpha-\frac{1}{\sqrt{2}}\sin\alpha\right)^2$$

$$=2\sin^2\left(\frac{\pi}{4}-\alpha\right)$$

Also $\cos 2\beta=2\cos^2\left\{\dfrac{\pi}{2}-\left(\dfrac{\pi}{4}-\alpha\right)\right\}$

$$=2\cos^2\left(\frac{\pi}{4}+\alpha\right),$$

$$\Rightarrow t=\tan\alpha=\frac{4}{3}\text{ or }\frac{2a}{a^2-1}$$

Reasoning Type

1. d. Statement 1 is wrong as z can be written as $\dfrac{-(x+y)}{1-xy}$.

$\Rightarrow$ for any values of xy $(xy\neq 1)$, we get a value of z and statement 2 is correct.

2. b. $\cos 7=\cos(2\pi+7-2\pi)=\cos(7-2\pi)=\cos(0.72)$

Now 1 radian and 0.72 radian angles are for first quadrant where $\cos x$ is decreasing, hence $\cos 1<\cos 0.72$ or $\cos 1<\cos 7$.

But statement 2 is not correct explanation for $\cos 1<\cos 7$.

Note that $\cos 0.5>\cos 7$.

3. b. $\tan 4=\tan(\pi+(4-\pi))=\tan(4-\pi)=\tan 0.86$

$\tan 7.5=\tan(2\pi+(7.5-2\pi))=\tan(7.5-2\pi)=\tan 1.22$

Now both angles, i.e., 0.86 radians and 1.22 radians are for first quadrant, hence $\tan 0.86<\tan 1.22$ as $\tan x$ is an increasing function. But this is not always true, as $3>1$ but $\tan 3<\tan 1$.

Hence, both statements are true but statement 2 is not the correct explanation of statement 1.

4. b. In first quadrant, $\cos\theta>\sin\theta$ for $\theta\in(\pi/4,\pi/2)$.

Hence $\cos 1<\sin 1$.

Also in first quadrant cosine is decreasing and sine is increasing, but this is not correct reason for which $\cos 1<\sin 1$. Thus correct answer is (b).

5. a. $f(\theta)=(\sin\theta+\operatorname{cosec}\theta)^2+(\cos\theta+\sec\theta)^2$

$$=\sin^2\theta+\cos^2\theta+\operatorname{cosec}^2\theta+\sec^2\theta+4$$

$$=5+\frac{1}{\sin^2\theta}+\frac{1}{\cos^2\theta}$$

$$=5+\frac{4}{\sin^2 2\theta}\geq 9.$$

Hence correct answer is (a).

6. d. $\sin^2\theta_1+\sin^2\theta_2+\cdots+\sin^2\theta_n=0$
$\Rightarrow \sin\theta_1=\sin\theta_2=\cdots=\sin\theta_n=0$
$\Rightarrow \cos^2\theta_1, \cos^2\theta_2, \cdots, \cos^2\theta_n=1$
$\Rightarrow \cos\theta_1, \cos\theta_2, \cdots, \cos\theta_n=\pm 1$
Now $\cos\theta_1+\cos\theta_2+\cdots+\cos\theta_n=n-4$ means two of $\cos\theta_1, \cos\theta_2, \cdots, \cos\theta_n$ must be -1 and the other be 1. Now two values from $\cos\theta_1, \cos\theta_2, \cdots, \cos\theta_n$ can be selected in ${}^nC_2$ ways. Hence, the number of solutions is ${}^nC_2=\dfrac{n(n-1)}{2}$.
Hence, statement 1 is false, but statement 2 is correct.

7. a. Let $y=27^{\cos 2x}\times 81^{\sin 2x}=3^{3\cos 2x+4\sin 2x}$
Now $-\sqrt{3^2+4^2}\le 3\cos 2x+4\sin 2x\le\sqrt{3^2+4^2}$
or $-5\le 3\cos 2x+4\sin 2x\le 5$
$\Rightarrow 3^{-5}\le 3^{3\cos 2x+4\sin 2x}\le 3^5$
Hence, correct answer is (a).

8. d. Obviously in triangle ABC

$\tan A=\tan(\pi-(B+C))$

$=-\tan(B+C)$

$=\dfrac{\tan B+\tan C}{\tan B\tan C-1}$

If angle A is obtuse than $\tan A<0$

$\Rightarrow \dfrac{\tan B+\tan C}{\tan B\tan C-1}<0$

$\Rightarrow \tan B\tan C<1$ (as B and C will be acute,
Hence, $\tan B+\tan C>0$).
Thus statement 1 is false and statement 2 is true.

9. a. We know that $\tan 15°=2-\sqrt{3}$ which is an irrational number. Hence, statement 2 is true.
Statement 1 is also true as if tan 5° is a rational number,

then $\tan 15°=\dfrac{3\tan 5°-\tan^3 5°}{1-3\tan^2 5°}$ should be a rational number, which is not true.
Hence, tan 5° is an irrational number.
Obviously, statement 2 is a correct reasoning for statement 1.

10. a. Statement 2 is true as it is one of the standard results of multiple angles.
Putting $A=\pi/18$ in the formula $\sin 3A=3\sin A-4\sin^3 A$, we get $8x^3-6x+1=0$, where $x=\sin\dfrac{\pi}{18}$. Hence, statement 1 is also true because of statement 2.

11. a. Statement 2 is true, because each trigonometric function has a principle period of π or 2π and hence 2π is one of the periods of every trigonometric function.
Thus $f(2A)=f(2B)$
$\Rightarrow 2A=2n\pi+2B$, for some $n\in Z$
$\Rightarrow A=n\pi+B$
$\Rightarrow$ Statement 1 is true because of statement 2.

12. b. From Fig. 1.5, clearly, $\sin 3<\sin 1<\sin 2$
But statement 2 is not sufficient to ensure this.
Hence answer is (b).

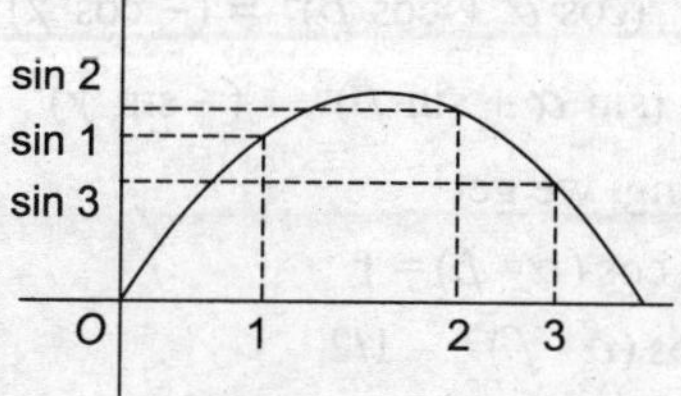

Fig. 1.5

13. a. Let $g(x)=3\sin x+4\cos x-2$

Maximum value of $g(x)=\sqrt{3^2+4^2}-2=3$

Minimum value of $g(x)=-\sqrt{3^2+4^2}-2=-7$

Therefore, the range of $f(x)=\dfrac{1}{g(x)}$ is $R-\left[-\dfrac{1}{7},\dfrac{1}{3}\right]$

$\Rightarrow$ It is an unbounded function.
$\Rightarrow f(x)$ has no maximum and no minimum values.

14. b. Statement 2 is true as it is one of the identities in triangle.
R.H.S. in statement 2 is always positive as $\alpha,\beta,\gamma\in(0,\pi/2)$
Statement 1 is true as select $\alpha=2\pi$, $\beta=-\pi/2$, $\gamma=-\pi/2$
Then $\sin\alpha+\sin\beta+\sin\gamma=0-1-1=-2$, which shows that minimum value will be negative.
But statement 2 is not the correct explanation of statement 1, as $\alpha+\beta+\gamma=\pi$ does not follow that α,β,γ are angles of a triangle.

15. a. $\sin^2 A+\sin^2 B+\sin^2 C$

$=\dfrac{1-\cos 2A}{2}+\dfrac{1-\cos 2B}{2}+\dfrac{1-\cos 2C}{2}$

$=\dfrac{3-(\cos 2A+\cos 2B+\cos 2C)}{2}$

$=\dfrac{3-(-1-4\cos A\cos B\cos C)}{2}$

$=2+2\cos A\cos B\cos C \quad (1)$

Hence, statement 2 is true.
From statement 1 using equation (1),
$\cos A\cos B\cos C=0$, then either A, B or C is 90°.

Both statement 1 and statement 2 are true and statement 2 is the correct explanation of statement 1.

16. d. Statement 2 is true as it is one of the conditional identities in the triangle. Since R.H.S. > 1 in statement 2, statement 1 is false.

17. d. Given $\cos x \sum \cos\alpha - \sin x \sum \sin\alpha = 0,\ \forall\ x \in R$

Hence, $\cos\alpha + \cos\beta + \cos\gamma = 0$

and $\sin\alpha + \sin\beta + \sin\gamma = 0$

$\Rightarrow$ Statement 2 is true.

Now $(\cos\alpha + \cos\beta)^2 = (-\cos\gamma)^2$

and $(\sin\alpha + \sin\beta)^2 = (-\sin\gamma)^2$

Adding, we get

$2 + 2\cos(\alpha - \beta) = 1$

$\Rightarrow \cos(\alpha - \beta) = -1/2$

Similarly $\cos(\beta - \gamma) = -1/2$ and $\cos(\gamma - \alpha) = -1/2$

Now $0 < \alpha < \beta < \gamma < 2\pi$

$\Rightarrow\ \beta - \alpha < \gamma - \alpha$

Hence, $\beta - \alpha = \dfrac{2\pi}{3}$ and $\gamma - \alpha = \dfrac{4\pi}{3}$

$\Rightarrow$ Statement 1 is false.

18. a. $\cos^2 A + \cos^2 B + \cos^2 C = 1 - 2\cos A\cos B\cos C$

$\therefore \sum \cos^2 A\big|_{\min} = 1 - 2 \times \dfrac{1}{8} = 1 - \dfrac{1}{4} = \dfrac{3}{4}$

19. d. Let $x = \cot A, y = \cot B$, and $z = \cot C$. So

$\sum \cot A \cot B = 1$

$\Rightarrow A + B + C = n\pi$

In statement 1

$$\text{L.H.S.} = \frac{1}{2}[\sin 2A + \sin 2B + \sin 2C]$$

$$= 2\sin A\sin B\sin C$$

$$\text{R.H.S.} = \frac{2}{|\operatorname{cosec} A \operatorname{cosec} B \operatorname{cosec} C|}$$

$$= \pm 2\sin A\sin B\sin C = \text{L.H.S.}$$

Statement 2 is true as it is one of the conditional identities in the triangle.

20. b. In triangle $ABC, A + B + C = \pi$

and $\cot\dfrac{A}{2} + \cot\dfrac{B}{2} + \cot\dfrac{C}{2} = \cot\dfrac{A}{2}\cot\dfrac{B}{2}\cot\dfrac{C}{2}$

$$\Rightarrow \ln\left(\cot\frac{A}{2} + \cot\frac{B}{2} + \cot\frac{C}{2}\right)$$

$$= \ln\cot\frac{A}{2} + \ln\cot\frac{B}{2} + \ln\cot\frac{C}{2}$$

Hence, statement 1 is true.

In statement 2, R.H.S. $= \ln 1 + \ln\sqrt{3} + \ln(2 + \sqrt{3})$

$$= \ln(1\sqrt{3}(2 + \sqrt{3}))$$

$$= \ln(2\sqrt{3} + 3) = \text{R.H.S.}$$

But statement 2 does not explain statement 1.

Linked Comprehension Type

For Problems 1–3

1. a., 2. b., 3. c.

Sol 1. a. $\sin\alpha = A\sin(\alpha + \beta) = A(\sin\alpha\cos\beta + \sin\beta\cos\alpha)$

$\Rightarrow \sin\alpha(1 - A\cos\beta) = A\sin\beta\cos\alpha$

$$\Rightarrow\quad \tan\alpha = \frac{A\sin\beta}{(1 - A\cos\beta)} \qquad (1)$$

2. b. $\tan\beta = \dfrac{\sin\beta}{\cos\beta} = \dfrac{(1 - A\cos\beta)\tan\alpha}{A\cos\beta}$

$$= \frac{(1 - A\cos\beta)\sin\alpha}{A\cos\alpha\cos\beta}$$

[From equation (1) and (2)]

3. c. $\tan(\alpha + \beta) = \dfrac{\tan\alpha + \tan\beta}{1 - \tan\alpha\tan\beta}$

$$= \frac{\dfrac{A\sin\beta}{1 - A\cos\beta} + \dfrac{\sin\beta}{\cos\beta}}{1 - \dfrac{A\sin\beta\sin\beta}{(1 - A\cos\beta)\cos\beta}}$$

$$= \frac{A\sin\beta\cos\beta + \sin\beta - A\sin\beta\cos\beta}{\cos\beta - A\cos^2\beta - A\sin^2\beta}$$

$$= \frac{\sin\beta}{\cos\beta - A}$$

Also $\tan(a + b) = \dfrac{\tan\alpha + \tan\beta}{1 - \tan\alpha\tan\beta}$

$$= \frac{\dfrac{\sin\alpha}{\cos\alpha} + \dfrac{\sin\alpha(1 - A\cos\beta)}{A\cos\alpha\cos\beta}}{1 - \dfrac{\sin^2\alpha(1 - A\cos\beta)}{A\cos^2\alpha\cos\beta}}$$

[From equation (2)]

$$= \frac{[A\sin\alpha\cos\beta + \sin\alpha - A\sin\alpha\cos\beta]\cos\alpha}{A\cos^2\alpha\cos\beta - \sin^2\alpha + A\sin^2\alpha\cos\beta}$$

$$= \frac{\sin\alpha\cos\alpha}{A\cos\beta - \sin^2\alpha}$$

For Problems 4–6

4. d., 5. a., 6. b.

Sol. We have $\tan\left(\theta + \frac{\pi}{4}\right) = 3\tan 3\theta$

$$\Rightarrow \frac{1+\tan\theta}{1-\tan\theta} = 3 \times \frac{3\tan\theta - \tan^3\theta}{1-3\tan^2\theta}$$

$$\Rightarrow \frac{1+t}{1-t} = 3\left(\frac{3t-t^3}{1-3t^2}\right) \text{ (putting } t = \tan\theta)$$

$$\Rightarrow 3t^4 - 6t^2 + 8t - 1 = 0$$

Hence,

S_1 = sum of roots = $t_1 + t_2 + t_3 + t_4 = 0$;

S_2 = sum of product of roots taken two at a time = -2

S_3 = sum of product of roots taken three at time $= -8/3$

S_4 = product of all roots = $-1/3$.

For Problems 7–9

7. a., 8. b., 9. d.

Sol. $\sin\alpha + \sin\beta = 3$ (1)

$$\Rightarrow 2\sin\left(\frac{\alpha+\beta}{2}\right)\cos\left(\frac{\alpha-\beta}{2}\right) = 3 \quad (2)$$

$\cos\alpha + \cos\beta = 4$ (3)

$$\Rightarrow 2\cos\left(\frac{\alpha+\beta}{2}\right)\cos\left(\frac{\alpha-\beta}{2}\right) = 4 \quad (4)$$

Dividing (2) by (4), we have

$$\tan\left(\frac{\alpha+\beta}{2}\right) = \frac{3}{4}$$

$$\Rightarrow \sin(\alpha+\beta) = \frac{2\tan\left(\frac{\alpha+\beta}{2}\right)}{1+\tan^2\left(\frac{\alpha+\beta}{2}\right)} = \frac{2\frac{3}{4}}{1+\left(\frac{3}{4}\right)^2} = \frac{24}{25}$$

$$\text{and } \cos(\alpha+\beta) = \frac{1-\tan^2\left(\frac{\alpha+\beta}{2}\right)}{1+\tan^2\left(\frac{\alpha+\beta}{2}\right)} = \frac{1-\left(\frac{3}{4}\right)^2}{1+\left(\frac{3}{4}\right)^2}$$

$$= \frac{7}{25}$$

For Problems 10–12

10. a., 11. c., 12. b.

Sol. Let $\theta = \frac{n\pi}{7}$ (so that $7\theta = n\pi$)

$$\Rightarrow 4\theta + 3\theta = n\pi$$

$$\Rightarrow \tan 4\theta = \tan(n\pi - 3\theta)$$

$$\Rightarrow \tan 4\theta = -\tan 3\theta$$

$$\Rightarrow \frac{4\tan\theta - 4\tan^3\theta}{1-6\tan^2\theta + \tan^4\theta} = \frac{3\tan\theta - \tan^3\theta}{1-3\tan^2\theta}$$

$$\Rightarrow \frac{4z-4z^3}{1-6z^2+z^4} = -\frac{3z-z^3}{1-3z^2}$$

[Where $\tan\theta = z$ (say)]

$$\Rightarrow (4-4z^2)(1-3z^2) = -(3-z^2)(1-6z^2+z^4)$$

$$\Rightarrow z^6 - 21z^4 + 35z^2 - 7 = 0 \quad (1)$$

This is a cubic equation in z^2, i.e. in $\tan^2\theta$.

The roots of this equation are, therefore,

$\tan^2\frac{\pi}{7}$, $\tan^2\frac{2\pi}{7}$, and $\tan^2\frac{3\pi}{7}$.

From equation (1), sum of the roots $= \frac{-(-21)}{1} = 21$

$$\Rightarrow \tan^2\frac{\pi}{7} + \tan^2\frac{2\pi}{7} + \tan^2\frac{3\pi}{7} = 21 \quad (2)$$

Putting $\frac{1}{y}$ in place of z in equation (1), we get

$$-7y^6 + 35y^4 - 21y^2 + 1 = 0$$

or $7y^6 - 35y^4 = 21y^2 - 1 = 0$ (3)

This is a cubic equation in y^2 i.e. in $\cot^2\theta$.

The roots of this equation are, therefore,

$\cot^2\frac{\pi}{7}$, $\cot^2\frac{2\pi}{7}$, and $\cot^2\frac{3\pi}{7}$.

Sum of the roots of equation (3) $= \frac{35}{7} = 5$

$$\Rightarrow \cot^2\frac{\pi}{7} + \cot^2\frac{2\pi}{7} + \cot^2\frac{3\pi}{7} = 5 \quad (4)$$

By multiplying equations (2) and (4), we get

$$\left(\tan^2\frac{\pi}{7} + \tan^2\frac{2\pi}{7} + \tan^2\frac{3\pi}{7}\right)$$

$$\times\left(\cot^2\frac{\pi}{7} + \cot^2\frac{2\pi}{7} + \cot^2\frac{3\pi}{7}\right) = 21 \times 5 = 105$$

For Problems 13–15

13. b., 14. b., 15. d.

Sol. Angles BEC, ABD, ABE, and BAC are in A.P.
Let $\angle BEC = \alpha - 3\beta, \angle ABD = \alpha - \beta, \angle ABE = \alpha + \beta$ and $\angle BAC = \alpha + 3\beta$

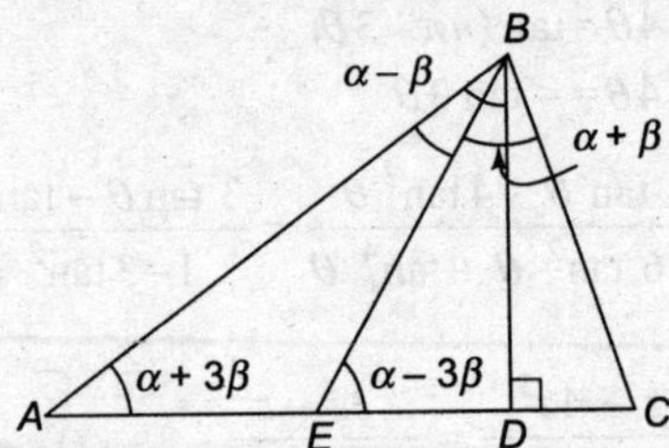

Fig. 1.6

Now, $\alpha - 3\beta = (\alpha + 3\beta) + (\alpha + \beta)$

[Using exterior angle theorem]

$\Rightarrow \alpha = -7\beta$. Therefore,

$\beta = -\dfrac{\pi}{24}, \ \alpha = \dfrac{7\pi}{24}$ and

from ΔABD, $\alpha - \beta + \alpha + 3\beta = \dfrac{\pi}{2}$

$\Rightarrow 2\beta + 2\beta = \dfrac{\pi}{2} \Rightarrow \alpha + \beta = \dfrac{\pi}{4}$

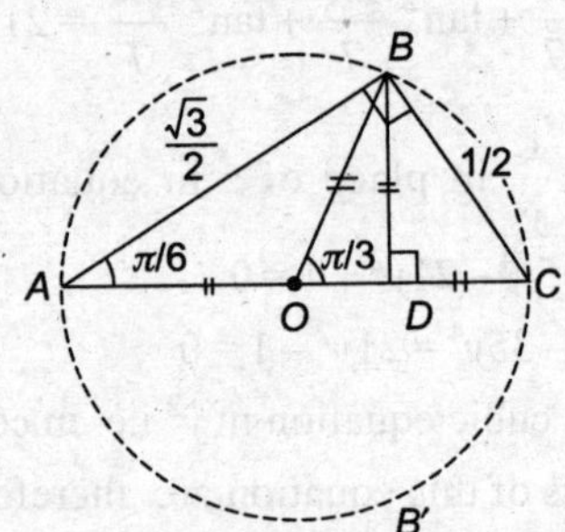

Fig. 1.7

$\therefore \angle B = 2(\alpha + \beta) = \dfrac{\pi}{2}, \angle A = \dfrac{\pi}{6}, \angle C = \dfrac{\pi}{3}$

$\Rightarrow ABC$ is a 30°–90°–60° triangle.

13. Area of the circle circumscribing ΔABC

$$= \pi \left(\frac{1}{2}\right)^2$$

$$= \frac{\pi}{4}.$$

14. ΔBOC is equilateral $\Rightarrow r = \dfrac{\Delta}{s} = \dfrac{\frac{\sqrt{3}}{2}\left(\frac{1}{2}\right)^2}{\frac{1}{2}\left(\frac{3}{2}\right)} = \dfrac{1}{4\sqrt{3}}$

15. $BD = OB \sin\dfrac{\pi}{3} = \dfrac{1}{2}\sin\dfrac{\pi}{3} = \dfrac{\sqrt{3}}{4}$

$\therefore \quad BB' = 2BD = \dfrac{\sqrt{3}}{2}$

Matrix-Match Type

1. a → q; b → r; c → s; d → p.

$\cos\theta - \sin\theta = \dfrac{1}{5}$, where $0 < \theta < \dfrac{\pi}{2}$. (1)

Squaring both sides of equation (1),

$1 - \sin 2\theta = \dfrac{1}{25}$

$\Rightarrow \sin 2\theta = \dfrac{24}{25} \quad \Rightarrow \quad \cos 2\theta = \dfrac{7}{25}$

Also,

$(\cos\theta + \sin\theta)^2 = (\cos\theta - \sin\theta)^2 + 4\cos\theta\sin\theta$

$= \dfrac{1}{25} + 2\sin 2\theta$

$= \dfrac{1}{25} + \dfrac{48}{25} = \dfrac{49}{25}$

$\Rightarrow \cos\theta + \sin\theta = \dfrac{7}{5}$ (2)

$\Rightarrow (\cos\theta + \sin\theta)/2 = \dfrac{7}{10}$

Also solving equations (1) and (2), we get

$\cos\theta = \dfrac{4}{5}$

2. a → q; b → s; c → a; d → r.

a. $A = \sin^2\theta + \cos^4\theta$

$$= \frac{1 - \cos 2\theta}{2} + \left(\frac{1 + \cos 2\theta}{2}\right)^2$$

$$= \frac{1}{2} - \frac{1}{2}\cos 2\theta + \frac{1}{4} + \frac{1}{2}\cos 2\theta + \frac{1}{4}\cos^2 2\theta$$

$$= \frac{3}{4} + \frac{1}{4}\left(\frac{\cos 4\theta + 1}{2}\right) = \frac{3}{4} + \frac{1}{8} + \frac{1}{8}\cos 4\theta$$

Now, $-1 \le \cos 4\theta \le 1$

$\Rightarrow \dfrac{-1}{8} \le \dfrac{\cos 4\theta}{8} \le \dfrac{1}{8}$

$\Rightarrow \dfrac{3}{4} + \dfrac{1}{8} - \dfrac{1}{8} \le \dfrac{3}{4} + \dfrac{1}{4}\left(\dfrac{1 + \cos 4\theta}{2}\right) \le \dfrac{3}{4} + \dfrac{1}{8} + \dfrac{1}{8}$

$\Rightarrow \dfrac{3}{4} \le A \le 1$

b. $A = 3\cos^2\theta + \sin^4\theta$

$$= 3\frac{1 + \cos 2\theta}{2} + \left(\frac{1 - \cos 2\theta}{2}\right)^2$$

$$= \frac{3 + 3\cos 2\theta}{2} + \frac{1 - 2\cos 2\theta + \cos^2 2\theta}{4}$$

$$=\frac{7+4\cos 2\theta+\cos^2 2\theta}{4}$$

$$=\frac{(\cos 2\theta+2)^2+3}{4}$$

Now, $1\le\cos 2\theta+2\le 3$

$$\Rightarrow 1\le\frac{(\cos 2\theta+2)^2+3}{4}\le 3$$

c. $A=\sin^2\theta-\cos^4\theta$

$$=\frac{1-\cos 2\theta}{2}-\left(\frac{1+\cos 2\theta}{2}\right)^2$$

$$=\frac{1}{2}-\frac{1}{2}\cos 2\theta-\frac{1}{4}-\frac{1}{2}\cos 2\theta-\frac{1}{4}\cos^2 2\theta$$

$$=\frac{1}{4}-\cos 2\theta-\frac{1}{4}\cos^2 2\theta$$

$$=-\left(\frac{1}{4}\cos^2 2\theta+\cos 2\theta-\frac{1}{4}\right)$$

$$=\frac{5}{4}-\left(\frac{1}{2}\cos 2\theta+1\right)^2$$

Now, $-\frac{1}{2}\le\frac{1}{2}\cos 2\theta\le\frac{1}{2}$

$$\Rightarrow \frac{1}{2}\le\frac{1}{2}\cos 2\theta+1\le\frac{3}{2}$$

$$\Rightarrow \frac{1}{4}\le\left(\frac{1}{2}\cos 2\theta+1\right)^2\le\frac{9}{4}$$

$$\Rightarrow -\frac{9}{4}\le-\left(\frac{1}{2}\cos 2\theta+1\right)^2\le-\frac{1}{4}$$

$$\Rightarrow -1\le\frac{5}{4}-\left(\frac{1}{2}\cos 2\theta+1\right)^2\le 1$$

d. $A=\tan^2\theta+2\cot^2\theta$

$$=(\tan\theta-\sqrt{2}\cot\theta)^2+2\sqrt{2}\ge 2\sqrt{2}$$

3. a → q; b → r; c → p; d → s.

$\cos\alpha+\cos\beta=1/2$

$$\Rightarrow 2\cos\left(\frac{\alpha+\beta}{2}\right)\cos\left(\frac{\alpha-\beta}{2}\right)=\frac{1}{2} \quad (1)$$

$\sin\alpha+\sin\beta=1/3$

$$\Rightarrow 2\sin\left(\frac{\alpha+\beta}{2}\right)\cos\left(\frac{\alpha-\beta}{2}\right)=\frac{1}{3} \quad (2)$$

Dividing equation (2) by (1),

$$\tan\left(\frac{\alpha+\beta}{2}\right)=\frac{2}{3}$$

$$\Rightarrow \cos\left(\frac{\alpha+\beta}{2}\right)=\pm\frac{3}{\sqrt{13}}$$

Squaring and adding the given results, we have

$$2+2\cos(\alpha-\beta)=\frac{13}{36}$$

$$\Rightarrow \cos(\alpha-\beta)=-\frac{59}{72}$$

Now, $2\cos^2\left(\frac{\alpha-\beta}{2}\right)-1=\cos(\alpha-\beta)$

$$\Rightarrow 2\cos^2\left(\frac{\alpha-\beta}{2}\right)=1-\frac{59}{72}=\frac{13}{72}$$

$$\Rightarrow \cos\left(\frac{\alpha-\beta}{2}\right)=\pm\frac{\sqrt{13}}{12}$$

$$\Rightarrow \tan\left(\frac{\alpha-\beta}{2}\right)=\pm\sqrt{\frac{131}{13}}$$

4. a → s; b → r; c → p; d → q.

a. $\sin(410°+400°)=\sin 810°$

$$=\sin(720°+90°)$$

$$=\sin 90°=1$$

b. $\frac{\sin^2 2°-\sin^2 1°}{2\sin 3°\sin 1°}=\frac{\sin 3°\sin 1°}{2\sin 3°\sin 1°}=\frac{1}{2}$

c. $\sin(-870°)+\operatorname{cosec}(-660°)+\tan(-855°)+2\cot(840°)+\cos(480°)+\sec(900°)$

$=-\sin(810°+60°)-\operatorname{cosec}(720°-60°)-\tan(810°+45°)+2\cot 120°+\cos 120°+\sec 180°$

$$=-\frac{1}{2}+\frac{2}{\sqrt{3}}+1-\frac{2}{\sqrt{3}}-\frac{1}{2}-1=-1$$

d. $\cos(\theta-\phi)=\cos\theta\cos\phi+\sin\theta\sin\phi$

$$=\frac{4}{5}\times\frac{3}{5}-\frac{3}{5}\times\frac{4}{5}=0$$

5. a → s; b → r; c → q; d → p

a. $\{\cos(2A+\theta)+\cos(2B+\theta)\}$

$$=2\cos(A-B)\cos(A+B+\theta)$$

Maximum value is $2\cos(A-B)$ when $\cos(A+B+\theta)=1$

b. $\{\cos 2A+\cos 2B\}$

$2\cos(A+B)\cos(A-B)$

Maximum value is $2\cos(A-B)$ when

$\cos(A+B)=1$

c. For $y=\sec x, x \in (0, \pi/2)$, tangent drawn to it any point lies completely below the graph of

$y=\sec x$, thus $\dfrac{\sec 2A+\sec 2B}{2} \geq \sec(A+B)$

$\Rightarrow \sec A+\sec B \geq \sec(A+B)$

Hence minimum value is $A+B$

d. $\sqrt{\{\tan\theta+\cot\theta-2\cos 2(A+B)\}}$

$=\sqrt{\left(\sqrt{\tan\theta}-\sqrt{\cot\theta}\right)^2+2-2\cos 2(A+B)}$

$=\sqrt{\left(\sqrt{\tan\theta}-\sqrt{\cot\theta}\right)^2+4\sin^2(A+B)}$

Minimum value occurs when $\sqrt{\tan\theta}=\sqrt{\cot\theta}$

and minimum value is $\sqrt{4\sin^2(A+B)}=2\sin(A+B)$

6. **a → s; b → r; c → p; d → q.**

a. $\cos 20° + \cos 80° - \sqrt{3}\cos 50°$

$=2\cos 30°\cos 50° - \sqrt{3}\cos 50°$

$=\sqrt{3}\cos 50° - \sqrt{3}\cos 50° = 0$

b. $\cos 0° + \cos\dfrac{\pi}{7} + \cos\dfrac{2\pi}{7} + \cos\dfrac{3\pi}{7} + \cos\dfrac{4\pi}{7} + \cos\dfrac{5\pi}{7} + \cos\dfrac{6\pi}{7}$

$=1+\left(\cos\dfrac{\pi}{7}+\cos\dfrac{6\pi}{7}\right)+\left(\cos\dfrac{2\pi}{7}+\cos\dfrac{5\pi}{7}\right)+\left(\cos\dfrac{3\pi}{7}+\cos\dfrac{4\pi}{7}\right)$

$=1+\left(\cos\dfrac{\pi}{7}+\cos\left(\pi-\dfrac{\pi}{7}\right)\right)+\left(\cos\dfrac{2\pi}{7}+\cos\left(\pi-\dfrac{2\pi}{7}\right)\right)+\left(\cos\dfrac{3\pi}{7}+\cos\left(\pi-\dfrac{3\pi}{7}\right)\right)$

$=1+0+0+0=1$

c. $\cos 20° + \cos 40° + \cos 60° - 4\cos 10°\cos 20°\cos 30°$

$=\cos 20° + \cos 40° + \cos 60° - 4\cos 10°\cos 20°\cos 30°$

$=2\cos 30°\cos 10° + 2\cos^2 30° - 1 - 4\cos 10°\cos 20°\cos 30°$

$=2\cos 30°(\cos 10° + \cos 30°) - 1 - 4\cos 10°\cos 20°\cos 30°$

$=2\cos 30°(2\cos 10°\cos 20°) - 1 - 4\cos 10°\cos 20°\cos 30° = -1$

d. $\cos 20°\cos 100° + \cos 100°\cos 140° - \cos 140°\cos 200°$

$=\dfrac{1}{2}(\cos 120° + \cos 80° + \cos 240° + \cos 40° - \cos 340° - \cos 60°)$

$=\dfrac{1}{2}\left(-\dfrac{1}{2}+\cos 80° - \dfrac{1}{2} + \cos 40° - \cos 340° - \dfrac{1}{2}\right)$

$=\dfrac{1}{2}\left(-\dfrac{3}{2}+\cos 80° + \cos 40° - \cos 20°\right)$

$=\dfrac{1}{2}\left(-\dfrac{3}{2}+2\cos 60°\cos 20° - \cos 20°\right)=\dfrac{1}{2}\left(-\dfrac{3}{2}\right)$

$=-\dfrac{3}{4}$

Integer Type

1. **(4)** $2\times\dfrac{x\times y}{2}=\dfrac{8\times 6}{2}=24$

$x(x\tan B)=24$

$\Rightarrow x^2\times\dfrac{3}{4}=24$

Fig. 1.8

$\Rightarrow x^2=32 \Rightarrow x=4\sqrt{2}$

2. **(6)** $\dfrac{1}{\sin 10°}+\dfrac{1}{\sin 50°}-\dfrac{1}{\sin 70°}$

$=\dfrac{1}{\cos 80°}+\dfrac{1}{\cos 40°}-\dfrac{1}{\cos 20°}$

$=\dfrac{\cos 40°\cos 20° + \cos 80°\cos 20° - \cos 40°\cos 80°}{\cos 20°\cos 40°\cos 80°}$

$=8[\cos 20°(\cos 40° + \cos 80°) - \cos 40°\cos 80°]$

$=8[2\cos 20°\cos 60°\cos 20° - \cos 40°\cos 80°]$

$=4[2\cos^2 20° - 2\cos 40°\cos 80°]$

$= 4\,[1 + \cos 40° - (\cos 120° + \cos 40°)]$

$= 4 \times \dfrac{3}{2} = 6$

3. (4)

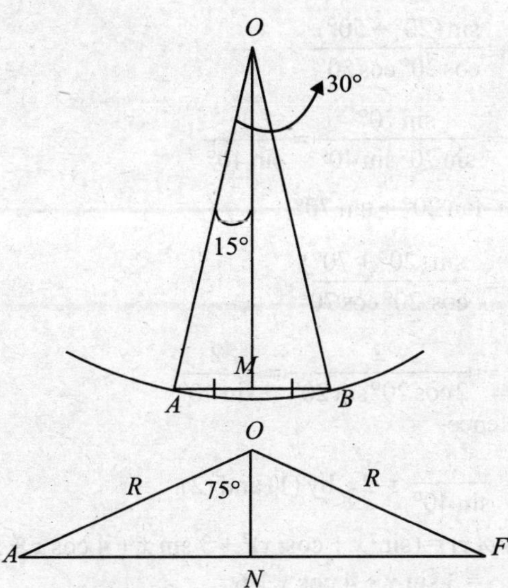

Fig. 1.9

$AM = R \sin \dfrac{\pi}{12} \Rightarrow AB = 2R \sin \dfrac{\pi}{12}$

Similarly, $AN = R \sin \dfrac{5\pi}{12} \Rightarrow AF = 2R \sin \dfrac{5\pi}{12}$.

Therefore,

$$\frac{AB}{AF} + \frac{AF}{AB} = \frac{\sin \frac{\pi}{12}}{\sin \frac{5\pi}{12}} + \frac{\sin \frac{5\pi}{12}}{\sin \frac{\pi}{12}}$$

$$= \tan \frac{\pi}{12} + \cot \frac{\pi}{12}$$

$$= (2 - \sqrt{3}) + \left(2 + \sqrt{3}\right) = 4$$

4. (1)

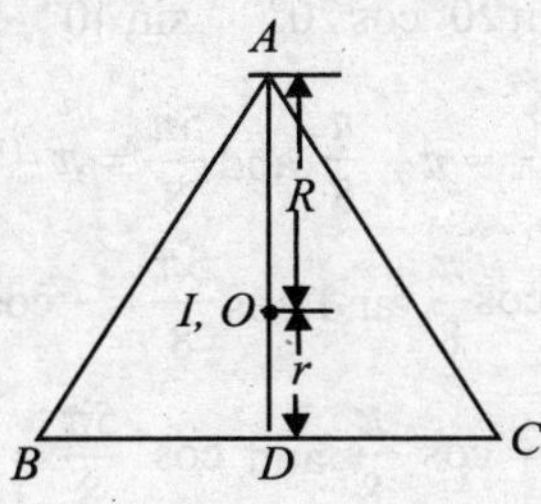

Fig. 1.10

In an equilateral triangle, the circumcentre (O) and incentre (I) coincide.

Also from the diagram, we have

$R + r = h$

$\Rightarrow \dfrac{R + r}{h} = 1$

5. (8) $l = \tan \theta$

$x = 2 \sin 2\theta$ and $y = 2 \cos 2\theta$

$E = x^2 - xy + y^2 = 4 - 4 \sin 2\theta \cos 2\theta = 4 - 2 \sin 4\theta$

$E \in [2, 6] \Rightarrow a + b = 8$

6. (3) Let $x = 3 \cos \theta$, $y = 3 \sin \theta$,

$z = 2 \cos \phi$, and $t = 2 \sin \phi$.

$\Rightarrow 6 \cos \theta \sin \phi - 6 \sin \theta \cos \phi = 6$

$\Rightarrow \sin (\phi - \theta) = 1$

$\Rightarrow \phi - \theta = 90° \Rightarrow \phi = 90 + \theta$. Therefore,

$x = 3 \cos \theta$, $y = 3 \sin \theta$

and $z = -2 \sin \theta$, $t = 2 \cos \theta$.

$P = xz = -6 \sin \theta \cos \theta = -3 \sin 2\theta$

Hence, $P_{\max} = 3$.

7. (7)

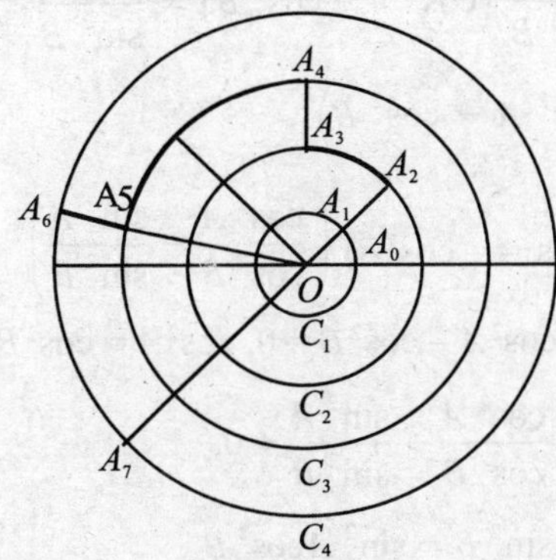

Fig. 1.11

The motion of the particle on the first four circles is shown with bold line in the figure. Note that on every circle, the particle travels just one radian. The particle crosses the positive direction of x-axis for the first time on C_n, where n is the least positive integer such that $n \ge 2\pi \Rightarrow n = 7$.

8. (3) Given $3 \sin \theta + 5 \cos \theta = 5$ (1)

Let $5 \sin \theta - 3 \cos \theta = x$ (2)

Squaring and adding, we get

$(9 \sin^2 \theta + 25 \cos^2 \theta + 30 \sin \theta \cos \theta) + (25 \sin^2 \theta + 9 \cos^2 \theta - 30 \sin \theta \cos \theta) = 25 + x^2$

$\Rightarrow 9(\sin^2 \theta + \cos^2 \theta) + 25 (\sin^2 \theta + \cos^2 \theta) = 25 + x^2$

$\Rightarrow 34 = 25 + x^2$ or, $x^2 = 9$

$\Rightarrow x = \pm 3$

9. (1) Let $(\sec A + \tan A)(\sec B + \tan B)(\sec C + \tan C) = x$

(1)

Then $(\sec A - \tan A)(\sec B - \tan B)(\sec C - \tan C) = x$ (2)

Multiplying (1) and (2), we get

$(\sec^2 A - \tan^2 A)(\sec^2 B - \tan^2 B)(\sec^2 C - \tan^2 C) = x^2$

or $x^2 = 1$ or $x = \pm 1$.

Hence, each side $= \pm 1$.

10. Given $\dfrac{\cos^4 A}{\cos^2 B}+\dfrac{\sin^4 A}{\sin^2 B}=1(\cos^2 A+\sin^2 A)$

$$\Rightarrow \frac{\cos^4 A}{\cos^2 B}-\cos^2 A=\sin^2 A-\frac{\sin^4 A}{\sin^2 B}$$

$$\Rightarrow \frac{\cos^2 A(\cos^2 A-\cos^2 B)}{\cos^2 B}=\sin^2 A\frac{(\sin^2 B-\sin^2 A)}{\sin^2 B}$$

$$\Rightarrow \frac{\cos^2 A}{\cos^2 B}(\cos^2 A-\cos^2 B)=\frac{\sin^2 A}{\sin^2 B}$$

$$[(1-\cos^2 B)-(1-\cos^2 A)]$$

$$\Rightarrow \frac{\cos^2 A}{\cos^2 B}(\cos^2 A-\cos^2 B)=\frac{\sin^2 A}{\sin^2 B}$$

$$(\cos^2 A-\cos^2 B)$$

$$\Rightarrow (\cos^2 A-\cos^2 B)\left(\frac{\cos^2 A}{\cos^2 B}-\frac{\sin^2 A}{\sin^2 B}\right)=0$$

When $\cos^2 A-\cos^2 B=0$, $\cos^2 A=\cos^2 B$ (1)

When $\dfrac{\cos^2 A}{\cos^2 B}-\dfrac{\sin^2 A}{\sin^2 B}=0$,

$\cos^2 A\sin^2 B=\sin^2 A\cos^2 B$

$\Rightarrow \cos^2 A\,(1-\cos^2 B)=(1-\cos^2 A)\cos^2 B$

$\Rightarrow \cos^2 A-\cos^2 A\cos^2 B=\cos^2 B-\cos^2 A\cos^2 B$

$\Rightarrow \cos^2 A=\cos^2 B$ (2)

Thus in both cases,

$\cos^2 A=\cos^2 B$

$\Rightarrow 1-\sin^2 A=1-\sin^2 B$ or $\sin^2 A=\sin^2 B$ (3)

Now $\dfrac{\cos^4 B}{\cos^2 A}+\dfrac{\sin^4 B}{\sin^2 A}=\dfrac{\cos^4 B}{\cos^2 B}+\dfrac{\sin^4 B}{\sin^2 B}$ [From (3)]

11. **(2)** L.H.S. =

$$\cos^2\frac{\pi}{16}+\cos^2\frac{3\pi}{16}+\cos^2\left(\frac{\pi}{2}-\frac{3\pi}{16}\right)+\cos^2\left(\frac{\pi}{2}-\frac{\pi}{16}\right)$$

$$=\cos^2\frac{\pi}{16}+\cos^2\frac{3\pi}{16}+\sin^2\frac{3\pi}{16}+\sin^2\frac{\pi}{16}$$

$$=\left(\cos^2\frac{\pi}{16}+\sin^2\frac{\pi}{16}\right)+\left(\cos^2\frac{3\pi}{16}+\sin^2\frac{3\pi}{16}\right)$$

$$=1+1=2$$

12. **(3)** $1+\sin\left(\dfrac{\pi}{4}+\theta\right)+2\sin\left(\dfrac{\pi}{4}-\theta\right)$

$$=1+\sin\left(\frac{\pi}{4}+\theta\right)+2\cos\left[\frac{\pi}{2}-\left(\frac{\pi}{4}-\theta\right)\right]$$

$$=1+\sin\left(\frac{\pi}{4}+\theta\right)+2\cos\left(\frac{\pi}{4}+\theta\right)$$

Hence, the maximum value is $1+\sqrt{5}$.

13. **(2)** $2x=\tan 20°+\tan 50°$

$$=\frac{\sin(20°+50°)}{\cos 20°\cos 50°}$$

$$=\frac{\sin 70°}{\sin 70°\sin 40°}=\frac{1}{\sin 40°} \quad (1)$$

$2y=\tan 20°+\tan 70°$

$$=\frac{\sin(20°+70°)}{\cos 20°\cos 70°}$$

$$=\frac{2}{2\cos 20°\sin 20°}=\frac{2}{\sin 40°} \quad (2)$$

Hence,

$y=\dfrac{2}{\sin 40°}=2x$, by (1) and (2).

14. **(1)** $p(x)=(\sin^2 x+\cos^2 x)^2+3\sin x+4\cos x+5$

$=3\sin x+4\cos x+6$

$\Rightarrow 1\le p(x)\le 11$

15. **(4)** L.H.S. $=\dfrac{\sqrt{3}}{\sin 20°}-\dfrac{1}{\cos 20°}=\dfrac{\sqrt{3}\cos 20°-\sin 20°}{\sin 20°\cos 20°}$

$$=\frac{2\left(\frac{\sqrt{3}}{2}\cos 20°-\frac{1}{2}\sin 20°\right)}{\sin 20°\cos 20°}$$

$$=\frac{2(\sin 60°\cos 20°-\cos 60°\sin 20°)}{\sin 20°\cos 20°}$$

$$=\frac{2\sin(60°-20°)}{\sin 20°\cos 20°}=\frac{2\sin 40°}{\sin 20°\cos 20°}$$

$$=\frac{4\sin 40°}{2\sin 20°\cos 20°}=\frac{4\sin 40°}{\sin 40°}=4$$

16. We have $\dfrac{7\pi}{8}=\pi-\dfrac{\pi}{8}$ and $\dfrac{5\pi}{8}=\pi-\dfrac{3\pi}{8}$. Therefore,

$\cos\dfrac{7\pi}{8}=-\cos\dfrac{\pi}{8}$ and $\cos\dfrac{5\pi}{8}=-\cos\dfrac{3\pi}{8}$

$\Rightarrow \cos^4\dfrac{7\pi}{8}=\cos^4\dfrac{\pi}{8}$ and $\cos^4\dfrac{5\pi}{8}=\cos^4\dfrac{3\pi}{8}$

L.H.S. $=2\cos^4\dfrac{\pi}{8}+2\cos^4\dfrac{3\pi}{8}$

$$=2\left[\left(\cos^2\frac{\pi}{8}\right)^2+\left(\cos^2\frac{3\pi}{8}\right)^2\right]$$

$$= 2\left[\frac{1+\cos(\pi/4)}{2}\right]^2 + \left[\frac{1+\cos(3\pi/4)}{2}\right]^2$$

$$\left(\because \frac{1+\cos 2\theta}{2} = \cos^2\theta\right)$$

$$= \frac{2}{4}\left[\left(1+\cos\frac{\pi}{4}\right)^2 + \left(1+\cos\frac{3\pi}{4}\right)^2\right]^2$$

$$= \frac{2}{4}\left[\left(1+\frac{1}{\sqrt{2}}\right)^2 + \left(1-\frac{1}{\sqrt{2}}\right)^2\right]$$

$$= \frac{1}{2}\left[\left(1+\frac{1}{\sqrt{2}}\right)^2 + \left(1-\frac{1}{\sqrt{2}}\right)^2\right]$$

$$= \frac{1}{2}\left[\left(1+\frac{1}{2}+\sqrt{2}\right) + \left(1+\frac{1}{2}-\sqrt{2}\right)\right]$$

$$= \frac{3}{2} = \text{R.H.S.}$$

17. (4) We have

$$\text{L.H.S.} = \cos\frac{2\pi}{15}\cos\frac{4\pi}{15}\cos\frac{8\pi}{15}\cos\left(\pi-\frac{\pi}{15}\right)$$

$$= \left(\cos\frac{2\pi}{15}\cos\frac{4\pi}{15}\cos\frac{8\pi}{15}\right)\left(-\cos\frac{\pi}{15}\right)$$

$$= -\cos\frac{\pi}{15}\cos\frac{2\pi}{15}\cos\frac{4\pi}{15}\cos\frac{8\pi}{15}$$

$= -\cos A\cos(2A)\cos(2^2A)\cos(2^3A)$,

where $A = \frac{\pi}{15}$

$$= -\left(\frac{\sin 2^4 A}{2^4\sin A}\right) = -\frac{\sin 16A}{2^4\sin A}$$

$$= -\frac{\sin(15A+A)}{16\sin A} \quad (\because 15A = \pi)$$

$$= \frac{-\sin(\pi+A)}{16\sin A}$$

$$= \frac{\sin A}{16\sin A} = \frac{1}{16} = \text{R.H.S.}$$

18. (3) $A+B+C=\pi$

$$\Rightarrow \frac{A}{2}+\frac{B}{2} = \frac{\pi}{2}-\frac{C}{2}$$

$$\Rightarrow \cot\left(\frac{A}{2}+\frac{B}{2}\right) = \cot\left(\frac{\pi}{2}-\frac{C}{2}\right)$$

$$\Rightarrow \frac{\cot\frac{A}{2}\cot\frac{B}{2}-1}{\cot\frac{A}{2}+\cot\frac{B}{2}} = \tan\frac{C}{2} = \frac{1}{\cot\frac{C}{2}}$$

$$\Rightarrow \cot\frac{A}{2}\cot\frac{B}{2}\cot\frac{C}{2} = \cot\frac{A}{2}+\cot\frac{B}{2}+\cot\frac{C}{2} \quad (1)$$

Since $\tan(A/2)$, $\tan(B/2)$, $\tan(C/2)$ are in H.P.,

$\cot\frac{A}{2}$, $\cot\frac{B}{2}$, $\cot\frac{C}{2}$ are in A.P.

$$\Rightarrow \cot\frac{A}{2}+\cot\frac{C}{2} = 2\cot\frac{B}{2}$$

Hence, equation (1) becomes

$$\cot\frac{A}{2}\cot\frac{B}{2}\cot\frac{C}{2} = 3\cot\frac{B}{2}$$

$$\Rightarrow \cot\frac{A}{2}\cot\frac{C}{2} = 3$$

$\Rightarrow$ G.M. of $\cot\frac{A}{2}$ and $\cot\frac{C}{2}$ is

$$\sqrt{\cot\frac{A}{2}\cot\frac{C}{2}} = \sqrt{3}$$

and A.M. of $\cot\frac{A}{2}$ and $\cot\frac{C}{2}$ is

$$\frac{\cot\frac{A}{2}+\cot\frac{C}{2}}{2} = \cot\frac{B}{2}$$

But A.M. $\ge$ G.M. Therefore,

$$\cot\frac{B}{2} \ge \sqrt{3}$$

$\Rightarrow$ Minimum value of $\cot\frac{B}{2}$ is $\sqrt{3}$.

19. (4) Let $y = \frac{\tan x}{\tan 3x}$

$$\Rightarrow y = \frac{\tan x\left(1-3\tan^2 x\right)}{3\tan x-\tan^3 x}$$

$$\Rightarrow y = \frac{1-3\tan^2 x}{3-\tan^2 x}$$

$$\Rightarrow 3y-(\tan^2 x)y = 1-3\tan^2 x$$

$$\Rightarrow (y-3)\tan^2 x = 3y-1$$

$$\Rightarrow \tan^2 x = \frac{3y-1}{y-3}$$

$$\Rightarrow \frac{3y-1}{y-3} > 0 \quad \text{(L.H.S. being a prefect square)}$$

+ve   – ve   + ve

$-\infty$ 1/3   3

Fig. 1.12

$\Rightarrow y < \frac{1}{3}$ *or* $y > 3$

Thus y never lies between 1/3 and 3.

20. **(6)** According to the question, we have expressed L.H.S. in the form

$c_0 + c_1 \cos x + c_2 \cos 2x + \ldots + c_n \cos nx$

Now $\sin^3 x \sin 3x = \left(\frac{3\sin x - \sin 3x}{4}\right)\sin 3x$

$= \left(\frac{3\sin x \sin 3x - \sin^2 3x}{4}\right)$

$= \left[\frac{3(\cos 2x - \cos 4x) - (1 - \cos 6x)}{8}\right]$

Hence, $n = 6$.

21. **(2)** We have

$\frac{2}{\cos x} = \frac{1}{\cos(x-y)} + \frac{1}{\cos(x+y)} = \frac{2\cos x \cos y}{\cos^2 x - \sin^2 y}$

$\Rightarrow \cos^2 x - 2\sin^2 y = \cos^2 x \cos y$

$\Rightarrow \cos^2 x(1 - \cos y) = \sin^2 y$

$\Rightarrow \cos^2 x\left(2\sin^2 \frac{y}{2}\right) = 4\sin^2 \frac{y}{2}\cos^2 \frac{y}{2}$

$\Rightarrow \cos^2 x = 2\cos^2 \frac{y}{2}$

$\Rightarrow \cos^2 x \sec^2 \frac{y}{2} = 2$

$\Rightarrow \cos x \sec \frac{y}{2} = \pm\sqrt{2}$

22. **(6)** $3(\sin x - \cos x)^4 + 6(\sin x + \cos x)^2 + 4(\sin^6 x + \cos^6 x) - 7$

$= 3(1 - \sin 2x)^2 + 6(1 + \sin 2x) + 4[(\sin^2 x + \cos^2 x)^3 - 3\sin^2 x \cos^2 x(\sin^2 x + \cos^2 x)] - 7$

$= 3(1 - 2\sin 2x + \sin^2 2x) + (6 + 6\sin 2x) + 4\left(1 - \frac{3}{4}\sin^2 2x\right) - 7$

$= 13 + 3\sin^2 2x - 3\sin^2 2x - 7$

$= 13 - 7 = 6$

23. **(1)** $3\left[\sin^4\left(\frac{3}{2}\pi - \alpha\right) + \sin^4(3\pi + \alpha)\right]$

$-2\left[\sin^6\left(\frac{1}{2}\pi + \alpha\right) + \sin^6(5\pi - \alpha)\right]$

$= 3(\cos^4 \alpha + \sin^4 \alpha) - 2(\cos^6 \alpha + \sin^6 \alpha)$

$= 3(1 - 2\sin^2 \alpha \cos^2 \alpha) - 2[(\sin^2 \alpha \cos^2 \alpha)^3 - 3\sin^2 \alpha \cos^2 \alpha(\sin^2 \alpha + \cos^2 \alpha)]$

$= 3(1 - 2\sin^2 \alpha \cos^2 \alpha) - 2(1 - 3\sin^2 \alpha \cos^2 \alpha)$

$= 1$

24. **(1)** $f_n(\theta) = \frac{\sin(\theta/2)}{\cos(\theta/2)}\left[\frac{2\cos^2(\theta/2)}{\cos\theta}\frac{2\cos^2\theta}{\cos 2\theta}\frac{2\cos^2 2\theta}{\cos 4\theta}\ldots\right]$

$= \frac{\sin\theta}{\cos\theta}\left[\frac{2\cos^2\theta}{\cos 2\theta}\frac{2\cos^2 2\theta}{\cos 4\theta}\ldots\right]$

$= \frac{\sin 2\theta}{\cos 2\theta}\left[\frac{2\cos^2 2\theta}{\cos 4\theta}\ldots\right]$

$= \tan 2^n \theta$

$f_2\left(\frac{\pi}{16}\right) = \tan\left(4 \times \frac{\pi}{16}\right) = \tan\frac{\pi}{4} = 1$

25. **(9)** Multiplying the numerator and denominator by 2, we get

$\frac{\cos\theta - \cos 3\theta + \cos 3\theta - \cos 9\theta + \cos 9\theta - \cos 17\theta}{\sin 3\theta - \sin\theta + \sin 9\theta - \sin 3\theta + \sin 17\theta - \sin 9\theta}$

$= \frac{\cos\theta - \cos 17\theta}{\sin 17\theta - \sin\theta} = \frac{2\sin 9\theta \sin 4\theta}{2\sin 4\theta \cos 9\theta}$

$= \tan 9\theta = \tan k\theta$

$\Rightarrow k = 9$

26. **(0)** $S = \cos\theta + \cos 2\theta + \ldots + \cos n\theta$ where $\theta = \pi/6$ and $n = 2009$

$S = \frac{\sin\frac{n\theta}{2}}{\sin\frac{\theta}{2}}\cos\left((n+1)\frac{\theta}{2}\right)$

Now $(n+1)\frac{\theta}{2} = \left(\frac{2010}{2}\right)\frac{\pi}{6} = (335)\frac{\pi}{2}$

Hence $\cos(n+1) = 0$

$\Rightarrow S = 0$

27. **(4)** We have $\sin\theta(1 + \sin^2\theta) = 1 - \sin^2\theta$

$\Rightarrow \sqrt{(1 - \cos^2\theta)(2 - \cos^2\theta)} = \cos^2\theta$

28. **(4)** $\sin^2\theta_1 + \sin^2\theta_2 + \sin^2\theta_3 = 0$

$\Rightarrow \cos\theta_1, \cos\theta_2, \cos\theta_3 = \pm 1$

$\Rightarrow \cos\theta_1 + \cos\theta_2 + \cos\theta_3 = 3$ (when all are + 1)

$= -3$ (when all are -1)

$= 1$ (when two are $+1$ and one -1)

$= -1$ (when two are -1 and one $+ 1$)

29. **(9)** $18\sin^2\theta + 2\operatorname{cosec}^2\theta - 3$

$= (\sqrt{18}\sin\theta - \sqrt{2}\operatorname{cosec}\theta)^2 + 9$

Hence least value is 9.

30. **(1)** $\tan 37.5° = \tan\left(\frac{75}{2}\right)^\circ = \frac{1 - \cos 75°}{\sin 75°}$

$$= \frac{1-\frac{\sqrt{3}-1}{2\sqrt{2}}}{\frac{\sqrt{3}+1}{2\sqrt{2}}} = \frac{2\sqrt{2}-\sqrt{3}+1}{\sqrt{3}+1} = \frac{(2\sqrt{2}-\sqrt{3}+1)(\sqrt{3}-1)}{2}$$

$$= 2\sqrt{6} - 4 - 2\sqrt{2}$$

$$= \frac{2(\sqrt{6}-3+\sqrt{3})-(2\sqrt{2}-\sqrt{3}+1)}{2}$$

$$= \frac{2\sqrt{6}-4+2\sqrt{3}-2\sqrt{2}}{2} = \sqrt{6}-\sqrt{4}+\sqrt{3}-\sqrt{2}$$

$\therefore a=6, b=4, c=3, d-2$

$$\Rightarrow \frac{ad}{bc} = \frac{12}{12} = 1$$

31. **(9)** $\text{cosec}^2 x + 4\sec^2 x$

$= 1 + \cot^2 x + 4 + 4\tan^2 x$

$= 5 + \cot^2 x + 4\tan^2 x$

$= 5 + (\cot x - 2\tan x)^2 + 4$

$= 9 + (\cot x - 2\tan x)^2 \geq 9$

Hence least value is 9.

32. **(5)** $3\,\text{cosec}\,\alpha - 4\sec\alpha = \dfrac{3}{\sin\alpha} - \dfrac{4}{\cos\alpha}$

$$= \frac{5}{\sin\alpha\cos\alpha}\left[\frac{3}{5}\cos\alpha - \frac{4}{5}\sin\alpha\right]$$

$$= 5\left[\frac{\sin 3\alpha\cos\alpha - \cos 3\alpha\sin\alpha}{\sin\alpha\cos\alpha}\right]$$

$= 10$

33. **(4)** Square and adding,

$9 + 16 + 24\sin(A+B) = 37$

$24\sin(A+B) = 12$

$$\sin(A+B) = \frac{1}{2} \Rightarrow \sin C = \frac{1}{2}$$

$C = 30°$ or $150°$

If $C = 150°$, then even of $B = 0$ and $A = 30°$

The quantity $3\sin A + 4\cos B$,

$$3 \times \frac{1}{2} + 4 = 5\frac{1}{2} < 6$$

Hence $C = 150°$ is not possible $\Rightarrow \angle C = 30°$ only

34. **(2)** Since A, B, C are acute angles

$\therefore A + B > \pi/2$

$A > \pi/2 - B$

$\sin A - \cos B > 0$

$\Rightarrow \cos B - \sin A < 0$ (1)

Again, $B > \pi/2 - A$

$\sin B > \cos A$

$\sin B - \cos A > 0$ (2)

From (1) and (2),

x-coordinate is – ve and y-coordinate is + ve

$\Rightarrow$ lies in 2nd quadrant only

CHAPTER 2

Trigonometric Equations

Objective Type

Each question has four choices a, b, c, and d, out of which *only one* answer is correct.

1. If $\sin\theta = \frac{1}{2}$ and $\cos\theta = -\frac{\sqrt{3}}{2}$, then the general value of q is $(n \in Z)$

   a. $2n\pi + \frac{5\pi}{6}$ b. $2n\pi + \frac{\pi}{6}$
   c. $2n\pi + \frac{7\pi}{6}$ d. $2n\pi + \frac{\pi}{4}$

2. The most general value for which $\tan\theta = -1$, $\cos\theta = \frac{1}{\sqrt{2}}$ is $(n \in Z)$

   a. $n\pi + \frac{7\pi}{4}$ b. $n\pi + (-1)^n \frac{7\pi}{4}$
   c. $2n\pi + \frac{7\pi}{4}$ d. None of these

3. If $\cos p\theta + \cos q\theta = 0$, then the different values of θ are in A.P. where the common difference is

   a. $\frac{\pi}{p+q}$ b. $\frac{\pi}{p-q}$
   c. $\frac{2\pi}{p+q}$ d. $\frac{3\pi}{p \pm q}$

4. If $\cos\theta + \cos 7\theta + \cos 3\theta + \cos 5\theta = 0$, then θ is equal to $(n \in Z)$

   a. $n\pi$ b. $n\pi/2$
   c. $n\pi/4$ d. $n\pi/8$

5. If $3\tan^2\theta - 2\sin\theta = 0$, then $\theta =$ is equal to $(n \in Z)$

   a. $2n\pi$ b. $n\pi + (-1)^n \frac{\pi}{6}$
   c. $n\pi - (-1)^n \frac{\pi}{6}$ d. $n\pi + \frac{\pi}{3}$

6. If $\sin\theta$, 1, $\cos 2\theta$ are in G.P., then θ is equal to $(n \in Z)$

   a. $n\pi + (-1)^n \frac{\pi}{2}$ b. $n\pi + (-1)^{n-1} \frac{\pi}{2}$
   c. $2n\pi$ d. None of these

7. The sum of all the solutions of the equation $\cos\theta \cos\left(\frac{\pi}{3} + \theta\right) \cos\left(\frac{\pi}{3} - \theta\right) = \frac{1}{4}$, $\theta \in [0, 6\pi]$

   a. 15π b. 30π
   c. $\frac{100\pi}{3}$ d. None of these

8. If $\sec\theta - 1 = (\sqrt{2} - 1)\tan\theta$, then θ is equal to $(n \in Z)$

   a. $(2n-1)\pi$ b. $2n\pi + \frac{\pi}{4}$
   c. $2n\pi - \frac{\pi}{4}$ d. $2n\pi + \frac{\pi}{3}$

9. The total number of solution of $\sin^4 x + \cos^4 x = \sin x \cos x$ in $[0, 2\pi]$ is equal to

   a. 2 b. 4
   c. 6 d. None of these

10. Number of solutions the equation $\sin 5x + \sin 3x + \sin x = 0$ for $0 \le x \le \pi$ is

    a. 1 b. 2
    c. 3 d. None of these

11. The sum of all the solution of the equation $\cot\theta = \sin 2\theta$, $(\theta \neq n\pi$, n integer), $0 \le \theta \le \pi$ is

a. $3\pi/2$ b. π
c. $3\pi/4$ d. 2π

12. The number of solutions of the equation $12\cos^3 x - 7\cos^2 x + 4\cos x = 9$ is

a. 0 b. 2
c. Infinite d. None of these

13. Which of the following is not the general solution of the equation $2^{\cos 2x} + 1 = 3.2^{-\sin^2 x}$?

a. $n\pi, n \in Z$ b. $\left(n+\frac{1}{2}\right)\pi, n \in Z$
c. $\left(n-\frac{1}{2}\right)\pi, n \in Z$ d. None of these

14. The general solution of the equation $\cos x \cos 6x = -1$ is

a. $x = (2n+1)\pi, n \in Z$ b. $x = 2n\pi, n \in Z$
c. $x = n\pi, n \in Z$ d. None of these

15. The equation $\cos x + \sin x = 2$ has

a. only one solution
b. two solutions
c. no solution
d. infinite number of solutions

16. If $0 \le x \le 2\pi$, then the number of solutions of equation $3(\sin x + \cos x) - 2(\sin^3 x + \cos^3 x) = 8$ is

a. 0 b. 1
c. 2 d. 4

17. If $\frac{1}{6}\sin\theta$, $\cos\theta$, $\tan\theta$ are in G.P., then θ is equal to $(n \in Z)$

a. $2n\pi \pm \frac{\pi}{3}$ b. $2n\pi \pm \frac{\pi}{6}$
c. $n\pi + (-1)^n \frac{\pi}{3}$ d. $n\pi + \frac{\pi}{3}$

18. The number of solutions of $2\sin^2 x + \sin^2 2x = 2$, $x \in [0, 2\pi]$ is

a. 4 b. 5
c. 7 d. 6

19. General solution of $\sin^2 x - 5\sin x\cos x - 6\cos^2 x = 0$ is

a. $x = n\pi - \pi/4, n \in Z$ only
b. $n\pi + \tan^{-1} 6, n \in Z$ only
c. both (a) and (b)
d. none of these

20. General solution of the equation $\tan\theta + \tan 4\theta + \tan 7\theta = \tan\theta \tan 4\theta \tan 7\theta$ is

a. $\theta = n\pi/12$, where $n \in Z$
b. $\theta = n\pi/9$, where $n \in Z$
c. $\theta = n\pi + \pi/12$, where $n \in Z$
d. none of these

21. The number of solutions of $\sec^2\theta + \operatorname{cosec}^2\theta + 2\operatorname{cosec}^2\theta = 8$, $0 \le \theta \le \pi/2$ is

a. 4 b. 3
c. 0 d. 2

22. The total number of solutions of $\tan x + \cot x = 2\operatorname{cosec} x$ in $[-2\pi, 2\pi]$ is

a. 2 b. 4
c. 6 d. 8

23. If $i = \sqrt{-1}$, which of the following is true for $z = (3 + 2i\sin\theta)/(1 - 2i\sin\theta)$.

a. z is purely real for $\theta = n\pi \pm \pi/3$
b. z is purely imaginary for $\theta = n\pi \pm \pi/2$
c. z is purely real for $\theta = n\pi$
d. none of these

24. If $\sin x + \cos x = \sqrt{\left(y + \frac{1}{y}\right)}$, $x \in [0, \pi]$, then

a. $x = \frac{\pi}{4}, y = 1$ b. $y = 0$
c. $y = 2$ d. $x = \frac{3\pi}{4}$

25. The complete solution of the equation. $7\cos^2 x + \sin x\cos x - 3 = 0$, is given by

a. $n\pi + \frac{\pi}{2}$ $(n \in Z)$
b. $n\pi - \frac{\pi}{2}$ $(n \in Z)$
c. $n\pi + \tan^{-1}\left(\frac{3}{4}\right)$ $(n \in Z)$
d. $n\pi + \frac{3\pi}{4}$, $k\pi + \tan^{-1}\left(\frac{4}{3}\right)$ $(k, n \in Z)$

26. Let $\theta \in [0, 4\pi]$ satisfying the equation $(\sin\theta + 2)(\sin\theta + 3)(\sin\theta + 4) = 6$. If the sum of all the values of θ is of the form $k\pi$, then the value of k is

a. 6 b. 5
c. 4 d. 2

27. If the inequality $\sin^2 x + a\cos x + a^2 > 1 + \cos x$ holds for any $x \in R$ then the largest negative integral value of a is

a. -4 b. -3
c. -2 d. -1

28. The number of solution of the equation $\sin^4 x - \cos^2 x\sin x + 2\sin^2 x + \sin x = 0$ in $0 \le x \le 3\pi$ is

a. 3 b. 4
c. 5 d. 6

29. If $x, y \in [0, 2\pi]$ and $\sin x + \sin y = 2$, then the value of $x + y$ is

a. π b. $\frac{\pi}{2}$
c. 3π d. None of these

30. For $n \in Z$, the general solution of the equation $(\sqrt{3}-1)\sin\theta+(\sqrt{3}+1)\cos\theta=2$ is

a. $\theta=2n\pi\pm\frac{\pi}{4}+\frac{\pi}{12}$

b. $\theta=n\pi+(-1)^n\frac{\pi}{4}+\frac{\pi}{12}$

c. $\theta=2n\pi\pm\frac{\pi}{4}$

d. $\theta=n\pi+(-1)^n\frac{\pi}{4}-\frac{\pi}{12}$

31. If $\sin 6\theta+\sin 4\theta+\sin 2\theta=0$, then θ is equal to $(n\in Z)$

a. $\frac{n\pi}{4}$ or $n\pi\pm\frac{\pi}{3}$ b. $\frac{n\pi}{4}$ or $n\pi\pm\frac{\pi}{6}$

c. $\frac{n\pi}{4}$ or $2n\pi\pm\frac{\pi}{6}$ d. None of these

32. The value of $\cos y\cos\left(\frac{\pi}{2}-x\right)-\cos\left(\frac{\pi}{2}-y\right)\cos x+\sin y\cos\left(\frac{\pi}{2}-x\right)+\cos x\sin\left(\frac{\pi}{2}-y\right)$ is zero if

a. $x=0$ b. $y=0$

c. $x=y$ d. $n\pi+y-\frac{\pi}{4}$ $(n\in Z)$

33. General solution of $\tan 5\theta=\cot 2\theta$ is $(n\in Z)$

a. $\theta=\frac{n\pi}{7}+\frac{\pi}{2}$ b. $\theta=\frac{n\pi}{7}+\frac{\pi}{3}$

c. $\frac{n\pi}{7}+\frac{\pi}{14}$ d. $\theta=\frac{n\pi}{7}-\frac{\pi}{14}$

34. One root of the equation $\cos x-x+\frac{1}{2}=0$ lies in the interval

a. $\left(0,\frac{\pi}{2}\right)$ b. $\left(-\frac{\pi}{2},0\right)$

c. $\left(\frac{\pi}{2},\pi\right)$ d. $\left(\pi,\frac{3\pi}{2}\right)$

35. $\tan\left(\frac{p\pi}{4}\right)=\cot\left(\frac{q\pi}{4}\right)$ if $(n\in Z)$

a. $p+q=0$ b. $p+q=2n+1$

c. $p+q=2n$ d. $p+q=2(2n+1)$

36. The range of y such that the equation in x, $y+\cos x=\sin x$ has a real solution is

a. $[-2,2]$ b. $[-\sqrt{2},\sqrt{2}]$

c. $[-1,1]$ d. $\left[-\frac{1}{2},\frac{1}{2}\right]$

37. One of the general solutions of the equation $4\sin^4 x+\cos^4 x=1$ is

a. $n\pi\pm\alpha/2$, $\alpha=\cos^{-1}(1/5)$, $\forall n\in Z$

b. $n\pi\pm\alpha/2$, $\alpha=\cos^{-1}(3/5)$, $\forall n\in Z$

c. $2n\pi\pm\alpha/2$, $\alpha=\cos^{-1}(1/3)$, $\forall n\in Z$

d. none of these

38. The number of roots of the equation $(1-\tan\theta)(1+\sin 2\theta)=1+\tan\theta$ for $\theta\in[0,2\pi]$ is

a. 3   b. 4

c. 5   d. None of these

39. The number of solutions of the equation $\sin x+\sin 2x+\sin 3x=\cos x+\cos 2x+\cos 3x$, $0\le x\le 2\pi$, is

a. 7   b. 5

c. 4   d. 6

40. The number of values of θ which satisfy the equation $\sin 3\theta-\sin\theta=4\cos^2\theta-2$, $\theta\in[0,2\pi]$, is

a. 4   b. 5

c. 7   d. 0

41. One of the general solutions of the equation $4\sin\theta\sin 2\theta\sin 4\theta=\sin 3\theta$, is

a. $(3n\pm1)\pi/12$, $\forall n\in Z$ b. $(4n\pm1)\pi/9$, $\forall n\in Z$

c. $(3n\pm1)\pi/9$, $\forall n\in Z$ d. $(3n\pm1)\pi/3$, $\forall n\in Z$

42. The general solutions of the equation $\tan\theta+\tan 2\theta+\tan 3\theta=0$, is

a. $\theta=n\pi/6$, $n\in Z$ only

b. $\theta=n\pi\pm\alpha$, $n\in Z$, where $\tan\alpha=1/\sqrt{2}$ only

c. Both a and b

d. None of these

43. The general solution of the equation $\sin 3\alpha=4\sin\alpha\sin(x+\alpha)\sin(x-\alpha)$ is

a. $n\pi\pm\pi/4$, $\forall n\in Z$ b. $n\pi\pm\pi/3$, $\forall n\in Z$

c. $n\pi\pm\pi/9$, $\forall n\in Z$ d. $n\pi\pm\pi/12$, $\forall n\in Z$

44. One of the general solution of the equation $\sqrt{3}\cos\theta-3\sin\theta=4\sin 2\theta\cos 3\theta$ is

a. $m\pi+\pi/18$, $m\in Z$ b. $m\pi/2+\pi/6$, $\forall m\in Z$

c. $m\pi/3+\pi/18$, $m\in Z$ d. None of these

45. The equation $\sin^4 x+\cos^4 x+\sin 2x+\alpha=0$ is solvable for

a. $-\frac{5}{2}\le\alpha\le\frac{1}{2}$ b. $-3\le a\le 1$

c. $-\frac{3}{2}\le\alpha\le\frac{1}{2}$ d. $-1\le a\le 1$

46. Consider the system of linear equations in x, y, and z,

$(\sin 3\theta)x-y+z=0$

$(\cos 2\theta)x+4y+3z=0$

$2x+7y+7z=0$,

then which of the following can be the values of θ for which the system has a non-trivial solution

a. $n\pi+(-1)^n\pi/6$, $\forall n\in Z$

b. $n\pi+(-1)^n\pi/3$, $\forall n\in Z$

c. $n\pi+(-1)^n\pi/9$, $\forall n\in Z$

d. None of these

47. The smallest positive x satisfying the equation $\log_{\cos x} \sin x + \log_{\sin x} \cos x = 2$, is
a. $\pi/2$ b. $\pi/3$
c. $\pi/4$ d. $\pi/6$

48. Number of ordered pairs which satisfy the equation $x^2 + 2x \sin(xy) + 1 = 0$, (where $y \in [0, 2\pi]$), are
a. 1 b. 2
c. 3 d. 0

49. The general solution of the equation $8 \cos x \cos 2x \cos 4x = \sin 6x/\sin x$ is
a. $x = n\pi/7 + \pi/21, \forall n \in Z$
b. $x = 2n\pi/7 + \pi/14, \forall n \in Z$
c. $x = n\pi/7 + \pi/14, \forall n \in Z$
d. $x = n\pi + \pi/14, \forall n \in Z$

50. If $\cos 3x + \sin\left(2x - \frac{7\pi}{6}\right) = -2$, then x is equal to ($k \in Z$)
a. $\frac{\pi}{3}(6k+1)$ b. $\frac{\pi}{3}(6k-1)$
c. $\frac{\pi}{3}(2k+1)$ d. None of these

51. If $(1 - \tan\theta)(1 + \tan\theta)\sec^2\theta + 2^{\tan^2\theta} = 0$, then the number of values of θ in the interval $\left(-\frac{\pi}{2}, \frac{\pi}{2}\right)$ are
a. 1 b. 2
c. 3 d. 4

52. The equation, $a \sin x + \cos 2x = 2a - 7$, possesses a solution which is
a. $a > 6$ b. $2 \le a \le 6$
c. $a > 2$ d. None of these

53. The total number of solutions of $\cos x = \sqrt{1 - \sin 2x}$ in $[0, 2\pi]$ is equal to
a. 2 b. 3
c. 5 d. None of these

54. The number of solutions of $\sum_{r=1}^{5} \cos rx = 5$ in the interval $[0, 2\pi]$ is
a. 0 b. 1
c. 5 d. 10

55. The number of values of x for which $\sin 2x + \cos 4x = 2$ is
a. 0 b. 1
c. 2 d. infinite

56. Let α, β be any two positive values of x for which $2 \cos x$, $|\cos x|$ and $1 - 3\cos^2 x$ are in G.P. The minimum value of $|\alpha - \beta|$ is
a. $\frac{\pi}{3}$ b. $\frac{\pi}{4}$
c. $\frac{\pi}{2}$ d. None of these

57. The general solution of the equation $\sin^{100} x - \cos^{100} x = 1$ is
a. $2n\pi + \frac{\pi}{3}, n \in I$ b. $n\pi + \frac{\pi}{2}, n \in I$
c. $n\pi + \frac{\pi}{4}, n \in I$ d. $2n\pi - \frac{\pi}{3}, n \in I$

58. The total number of solutions of $|\cot x| = \cot x + \frac{1}{\sin x}, x \in [0, 3\pi]$ is
a. 1 b. 2
c. 3 d. Zero

59. If $\tan(A - B) = 1$ and $\sec(A + B) = \frac{2}{\sqrt{3}}$, then the smallest positive values of A and B, respectively, are
a. $\frac{25\pi}{24}, \frac{19\pi}{24}$ b. $\frac{19\pi}{24}, \frac{25\pi}{24}$
c. $\frac{31\pi}{24}, \frac{13\pi}{24}$ d. $\frac{13\pi}{24}, \frac{31\pi}{24}$

60. If $3 \tan(\theta - 15°) = \tan(\theta + 15°)$, then θ is equal to ($n \in Z$)
a. $n\pi + \frac{\pi}{4}$ b. $n\pi + \frac{\pi}{8}$
c. $n\pi + \frac{\pi}{3}$ d. None of these

61. If $\tan 3\theta + \tan\theta = 2\tan 2\theta$, then θ is equal to ($n \in Z$)
a. $n\pi$ b. $\frac{n\pi}{4}$
c. $2n\pi$ d. None of these

62. The set of all x in $(-\pi, \pi)$ satisfying $|4 \sin x - 1| < \sqrt{5}$ is given by
a. $\left(-\frac{\pi}{10}, \frac{3\pi}{10}\right)$ b. $\left(\frac{\pi}{10}, \frac{3\pi}{10}\right)$
c. $\left(\frac{\pi}{10}, \frac{3\pi}{10}\right)$ d. None of these

63. $\sin x + \cos x = y^2 - y + a$ has no value of x for any y if a belongs to
a. $(0, \sqrt{3})$ b. $(-\sqrt{3}, 0)$
c. $(-\infty, -\sqrt{3})$ d. $(\sqrt{3}, \infty)$

64. The solution of the equation $4\sin^2 x + \tan^2 x + \operatorname{cosec}^2 x + \cot^2 x - 6 = 0$ is
a. $n\pi \pm \frac{\pi}{4}$ b. $2n\pi \pm \frac{\pi}{4}$
c. $n\pi + \frac{\pi}{3}$ d. $n\pi - \frac{\pi}{6}$

65. The number of solutions of $x \in [0, 2\pi]$ for which $[\sin x + \cos x] = 3 + [-\sin x] + [-\cos x]$ (where $[\cdot]$ denotes the greatest integer function) is
a. 0 b. 4
c. infinite d. 1

66. The equation $\cos^8 x + b\cos^4 x + 1 = 0$ will have a solution if b belongs to

a. $(-\infty, 2]$ b. $[2, \infty)$
c. $(-\infty, -2]$ d. None of these

67. The number of values of y in $[-2\pi, 2\pi]$ satisfying the equation $|\sin 2x| + |\cos 2x| = |\sin y|$ is

a. 3 b. 4
c. 5 d. 6

68. If both the distinct roots of the equation $|\sin x|^2 + |\sin x| + b = 0$ in $[0, \pi]$ are real, then the values of b are

a. $[-2, 0]$ b. $(-2, 0)$
c. $[-2, 0)$ d. $(-2, 0]$

69. If $|2\sin\theta - \operatorname{cosec}\theta| \geq 1$ and $\theta \neq \frac{n\pi}{2}$, $n \in I$, then

a. $\cos 2\theta \geq \frac{1}{2}$ b. $\cos 2\theta \geq \frac{1}{4}$
c. $\cos 2\theta \leq \frac{1}{2}$ d. $\cos 2\theta \leq \frac{1}{4}$

70. The number of solutions of the equation $\sin^3 x\cos x + \sin^2 x\cos^2 x + \sin x\cos^3 x = 1$ in the interval $[0, 2\pi]$ is

a. 4 b. 2
c. 1 d. 0

71. $e^{|\sin x|} + e^{-|\sin x|} + 4a = 0$ will have exactly four different solutions in $[0, 2\pi]$ if

a. $a \in R$ b. $a \in \left[-\frac{e}{4}, -\frac{1}{4}\right]$
c. $a \in \left[\frac{-1-e^2}{4e}, \infty\right)$ d. None of these

72. The total number of solutions of $\ln|\sin x| = -x^2 + 2x$ in $\left[-\frac{\pi}{2}, \frac{3\pi}{2}\right]$ is equal to

a. 1 b. 2
c. 4 d. None of these

73. The total number of ordered pairs (x, y) satisfying $|x| + |y| = 4$, $\sin\left(\frac{\pi x^2}{3}\right) = 1$; is equal to

a. 2 b. 3
c. 4 d. 6

74. The total number of solutions of $\sin\{x\} = \cos\{x\}$ (where {.} denotes the fractional part) in $[0, 2\pi]$ is equal to

a. 5 b. 6
c. 8 d. None of these

75. If $a, b \in [0, 2\pi]$, and the equation $x^2 + 4 + 3\sin(ax + b) - 2x = 0$ has at least one solution, then the value of $(a + b)$ can be

a. $\frac{7\pi}{2}$ b. $\frac{5\pi}{2}$
c. $\frac{9\pi}{2}$ d. None of these

76. The equation $\tan^4 x - 2\sec^2 x + a^2 = 0$ will have at least one solution if

a. $|a| \leq 4$ b. $|a| \leq 2$
c. $|a| \leq \sqrt{3}$ d. None of these

77. Complete the set of values of x in $(0, \pi)$ satisfying the equation $1 + \log_2 \sin x + \log_2 \sin 3x \geq 0$ is

a. $\left(\frac{2\pi}{3}, \frac{3\pi}{4}\right]$ b. $\left(\frac{\pi}{3}, \frac{2\pi}{3}\right)$
c. $\left(0, \frac{\pi}{2}\right) \cup \left(\frac{2\pi}{3}, \pi\right)$ d. $\left(\frac{\pi}{2}, \frac{2\pi}{3}\right)$

Multiple Correct Answers Type

Each question has four choices a, b, c, and d, out of which *one or more* answers are correct.

1. Solve $4\sin^4 x + \cos^4 x = 1$, then x is equal to $(n \in Z)$

a. $n\pi$ b. $n\pi \pm \sin^{-1}\sqrt{\frac{2}{5}}$
c. $\frac{2n\pi}{3}$ d. $2n\pi \pm \frac{\pi}{4}$

2. If $\sin^3\theta + \sin\theta\cos\theta + \cos^3\theta = 1$, then x is equal to $(n \in Z)$

a. $2n\pi$ b. $2n\pi + \frac{\pi}{2}$
c. $2n\pi - \frac{\pi}{2}$ d. $n\pi$

3. A general solution of the equation $\tan^2\theta + \cos 2\theta = 1$ is $(n \in Z)$

a. $n\pi - \frac{\pi}{4}$ b. $2n\pi + \frac{\pi}{4}$
c. $n\pi + \frac{\pi}{4}$ d. $2n\pi - \frac{\pi}{4}$

4. If $\sin x + \cos x = \sqrt{y + \frac{1}{y}}$ for $x \in [0, \pi]$, then

a. $x = \frac{\pi}{4}$ b. $y = 0$
c. $y = 1$ d. $x = \frac{3\pi}{4}$

5. $\sin\theta + \sqrt{3}\cos\theta = 6x - x^2 - 11$, $0 \leq \theta \leq 4\pi$, $x \in R$, holds for

a. no values of x and θ.
b. one value of x and two values of θ.
c. two values of x and two values of θ.
d. two point of values of (x, θ).

6. If $\sin^2 x - 2\sin x - 1 = 0$ has exactly four different solutions in $x \in [0, n\pi]$ then value/values of n are ($n \in N$)
a. 5
b. 3
c. 4
d. 6

7. For the smallest positive values of x and y the equation $2(\sin x + \sin y) - 2\cos(x - y) = 3$ has a solution then which of the following is/are true
a. $\sin\frac{x+y}{2} = 1$
b. $\cos\left(\frac{x-y}{2}\right) = \frac{1}{2}$
c. Number of ordered pairs (x, y) is 2
d. Number of ordered pairs (x, y) is 3

8. For the equation $1 - 2x - x^2 = \tan^2(x+y) + \cot^2(x+y)$,
a. exactly one value of x exists
b. exactly two values of x exists
c. $y = -1 + n\pi + \pi/4, n \in Z$
d. $y = 1 + n\pi + \pi/4, n \in Z$

9. If $x + y = \pi/4$ and $\tan x + \tan y = 1$, then ($n \in Z$)
a. $\sin x = 0$ always
b. when $x = n\pi + \pi/4$ then $y = -n\pi$
c. when $x = n\pi - \pi/4$ then $y = n\pi$
d. when $x = n\pi + \pi/4$ then $y = n\pi - \frac{\pi}{4}$

10. If $x + y = 2\pi/3$ and $\sin x/\sin y = 2$, then
a. the number of values of $x \in [0, 4\pi]$ are 4
b. number of values of $x \in [0, 4\pi]$ are 2
c. number of values of $y \in [0, 4\pi]$ are 4
d. number of values of $y \in [0, 4\pi]$ are 8

11. Let $\tan x - \tan^2 x > 0$ and $|2\sin x| < 1$. Then the intersection of which of the following two sets satisfies both the inequalities.
a. $x > n\pi + \pi/6, n \in Z$
b. $x > n\pi - \pi/6, n \in Z$
c. $x < n\pi - \pi/4, n \in Z$
d. $x < n\pi + \pi/4, n \in Z$

12. If $\cos(x + \pi/3) + \cos x = a$ has real solutions, then
a. the number of integral values of a is 3
b. the sum of number of integral values of a is 0
c. when $a = 1$, the number of solutions for $x \in [0, 2\pi]$ is 3
d. when $a = 1$, the number of solutions for $x \in [0, 2\pi]$ is 2

13. For $0 \le x \le 2\pi$, then $2^{\operatorname{cosec}^2 x}\sqrt{\frac{1}{2}y^2 - y + 1} \le \sqrt{2}$
a. is satisfied by exactly one value of y.
b. is satisfied by exactly two value of x.
c. is satisfied by x for which $\cos x = 0$.
d. is satisfied by x for which $\sin x = 0$.

14. If the equation, $\sin^2 x - a\sin x + b = 0$ has only one solution in $(0, \pi)$ then which of the following statements are correct.
a. $a \in (-\infty, 1] \cup [2, \infty)$
b. $b \in (-\infty, 0] \cup [1, \infty)$
c. $a = 1 + b$
d. None of these

15. If the equation $(\operatorname{cosec}^2\theta - 4)x^2 + (\cot\theta + \sqrt{3})x + \cos^2\frac{3\pi}{2} = 0$ holds true for all real x then the most general values of θ can be given by ($n \in Z$)
a. $2n\pi + \frac{11\pi}{6}$
b. $2n\pi + \frac{5\pi}{6}$
c. $2n\pi \pm \frac{7\pi}{6}$
d. $n\pi \pm \frac{11\pi}{6}$

16. If $(\sin\alpha)x^2 - 2x + b \ge 2$ for all the real values of $x \le 1$ and $\alpha \in (0, \pi/2) \cup (\pi/2, \pi)$ then the possible real values of b is/are
a. 2
b. 3
c. 4
d. 5

Reasoning Type

Each question has four choices a, b, c, and d, out of which *only one* is correct. Each question contains Statement 1 and Statement 2.

a. if both the statements are True and Statement 2 is the correct explanation of Statement 1
b. if both the statements are True but Statement 2 is Not the correct explanation of Statement 1
c. if Statement 1 is True and Statement 2 is False.
d. if Statement 1 is False and Statement 2 is True.

1. **Statement 1:** The value of x for which $(\sin x + \cos x)^{1+\sin 2x} = 2$, when $0 \le x \le \pi$ is $\pi/4$ only.
Statement 2: The maximum value of $\sin x + \cos x$ occurs when $x = \pi/4$.

2. **Statement 1:** The equation $\sin^2 x + \cos^2 y = 2\sec^2 z$ is solvable when only $\sin x = 1$, $\cos y = 1$, and $\sec z = 1$, where $x, y, z \in R$.
Statement 2: The maximum value of $\sin x$ and $\cos y$ is 1 and minimum value of $\sec z$ is 1.

3. **Statement 1:** Equation, $x\sin x = 1$, has four roots for $x \in (-\pi, \pi)$.
Statement 2: The graph of $y = \sin x$ and $y = 1/x$ cuts exactly two times for $x \in (0, \pi)$.

4. **Statement 1:** $\sin x = a$, where $-1 < a < 0$, then for $x \in [0, n\pi]$ has $2(n-1)$ solutions, $\forall n \in N$.
Statement 2: $\sin x$ takes the value a exactly two times when we take one complete rotation covering all the quadrant starting from $x = 0$.

5. **Statement 1:** Equation $\sqrt{1 - \sin 2x} = \sin x$ has one solution for $x \in [0, \pi/4]$.
Statement 2: $\cos x > \sin x$ when $x \in [0, \pi/4]$.

6. **Statement 1:** The number of solutions of the equation $|\sin x| = |x|$ is only one.
Statement 2: $|\sin x| \geq 0 \ \forall x \in R$.

7. **Statement 1:** General solution of $\frac{\tan 4x - \tan 2x}{1 + \tan 4x \tan 2x} =$ is 1 $x = \frac{n\pi}{2} + \frac{\pi}{8}, n \in I$.
Statement 2: General solution of $\tan \alpha = 1$ is $\alpha = n\pi + \frac{\pi}{4}, n \in I$.

8. **Statement 1:** The equation $\sin(\cos x) = \cos(\sin x)$ has no real solution.
Statement 2: $\sin x \pm \cos x$ is bounded in $\left[-\sqrt{2}, \sqrt{2}\right]$.

9. **Statement 1:** Equation $\sin x = e^x$ has infinite solutions.
Statement 2: $y = e^x$ is unbounded function.

10. **Statement 1:** Number of solution of $n|\sin x| = m|\cos x|$, (where $m, n \in Z$) in $[0, 2\pi]$ is independent of m and n.
Statement 2: Multiplying trigonometric functions by constant changes only range of the function but period remains same.

Linked Comprehension Type

Based upon each paragraph, three multiple choice questions have to be answered. Each question has four choices a, b, c, and d, out of which *only one* is correct.

For Problems 1–3

Consider the cubic equation $x^3 - (1 + \cos\theta + \sin\theta)x^2 + (\cos\theta\sin\theta + \cos\theta + \sin\theta)x - \sin\theta\cos\theta = 0$ whose roots are x_1, x_2, and x_3.

1. The value of $x_1^2 + x_2^2 + x_3^2$ equals
   a. 1 b. 2
   c. $2\cos\theta$ d. $\sin\theta(\sin\theta + \cos\theta)$

2. Number of values of θ in $[0, 2\pi]$ for which at least two roots are equal
   a. 3 b. 4
   c. 5 d. 6

3. Greatest possible difference between two of the roots if $\theta \in [0, 2\pi]$ is
   a. 2 b. 1
   c. $\sqrt{2}$ d. $2\sqrt{2}$

For Problems 4–6

Consider the equation $\sec\theta + \operatorname{cosec}\theta = a$, $\theta \in (0, 2\pi) - \{\pi/2, \pi, 3\pi/2\}$

4. If the equation has four real roots then
   a. $|a| \geq 2\sqrt{2}$ b. $|a| < 2\sqrt{2}$
   c. $a \geq -2\sqrt{2}$ d. None of these

5. If the equation has two real roots then
   a. $|a| \geq 2\sqrt{2}$ b. $a < 2\sqrt{2}$
   c. $|a| < 2\sqrt{2}$ d. None of these

6. If the equation has no real roots then
   a. $|a| \geq 2\sqrt{2}$ b. $a < 2\sqrt{2}$
   c. $|a| < 2\sqrt{2}$ d. None of these

For Problems 7–9

Consider the system of equations
$\sin x \cos 2y = (a^2 - 1)^2 + 1$,
$\cos x \sin 2y = a + 1$

7. The number of values of a for which the system has a solution is
   a. 1 b. 2
   c. 3 d. Infinite

8. The number of values of $x \in [0, 2\pi]$ when the system has solution for permissible values of a is
   a. 1 b. 2
   c. 3 d. 4

9. The number of values of $y \in [0, 2\pi]$ when the system has solution for permissible values of a
   a. 2 b. 3
   c. 4 d. 5

For Problems 10–12

Consider the equation $\int_0^x (t^2 - 8t + 13)dt = x\sin(a/x)$

10. The number of real values of x for which the equation has solution is
    a. 1 b. 2
    c. 3 d. Infinite

11. If x takes the values for which the equation has a solution then, the number of values of $a \in [0, 100]$
    a. 2 b. 1
    c. 5 d. 3

12. One of the solutions of $|y - \cos a| < x$ is where x and a are values which satisfy the given equation is
    a. $y \in [-5, 7]$ b. $y \in [-7, 5]$
    c. $y \in [5, 7]$ d. None of these

For Problems 13–15

Consider the system of equations
$x\cos^3 y + 3x\cos y\sin^2 y = 14$
$x\sin^3 y + 3x\cos^2 y\sin y = 13$

13. The value/values of x are
    a. $\pm 5\sqrt{5}$ b. $\pm\sqrt{5}$
    c. $\pm 1/\sqrt{5}$ d. None of these

14. The number of values of $y \in [0, 6\pi]$ is
    a. 5 b. 3
    c. 4 d. 6

15. The value of $\sin^2 y + 2\cos^2 y$ is
    a. 4/5 b. 9/5
    c. 2 d. None of these

Matrix-Match Type

Each question contains statements given in two columns which have to be matched. Statements (a, b, c, d) in column I have to be matched with statements (p, q, r, s) in column II. If the correct match are a → p, a → s; b → q; c → p, c → q and d → s; then the correctly bubbled 4 × 4 matrix should be as follows:

	p	q	r	s
a	(p) shaded	(q)	(r)	(s) shaded
b	(p)	(q) shaded	(r) shaded	(s)
c	(p) shaded	(q) shaded	(r)	(s)
d	(p)	(q)	(r)	(s) shaded

1.

Column I (Equation)	Column II (Solution)
a. $\cos^2 2x + \cos^2 x = 1$	**p.** $x = \left\{n\pi + \dfrac{\pi}{4}\right\} \cup \left\{n\pi + \dfrac{\pi}{6}\right\} n \in Z$
b. $\cos x + \sqrt{3}\sin x = \sqrt{3}$	**q.** $x = \dfrac{n\pi}{3}, n \in Z$
c. $1 + \sqrt{3}\tan^2 x = \left(1 + \sqrt{3}\right)\tan x$	**r.** $x = (2n-1)\dfrac{\pi}{6}, n \in Z$
d. $\tan 3x - \tan 2x - \tan x = 0$	**s.** $x = \left\{2n\pi + \dfrac{\pi}{2}\right\} \cup \left\{2n\pi + \dfrac{\pi}{6}\right\}, n \in Z$

2.

Column I (Equation)	Column II (Solutions)
a. $x^3 + x^2 + 4x + 2\sin x = 0$ in $0 \le x \le 2\pi$	**p.** 4
b. $\sin e^x \cos e^x = 2^{x-2} + 2^{-x-2}$	**q.** 1
c. $\sin 2x + \cos 4x = 2$	**r.** 2
d. $30\,|\sin x| = x$ in $0 \le x \le 2\pi$	**s.** 0

3.

Column I (Equation)	Column II (Solution)
a. $\max\limits_{\theta \in R} \{5\sin\theta + 3\sin(\theta - \alpha)\} = 7$ then the set of possible values of α is	**p.** $x = 2n\pi + 3\pi/4$, $n \in Z$
b. $x \ne \dfrac{n\pi}{2}$ and $(\cos x)^{\sin^2 x - 3\sin x + 2} = 1$	**q.** $\alpha = 2n\pi \pm \dfrac{\pi}{3}; n \in Z$
c. $\sqrt{(\sin x)} + 2^{1/4}\cos x = 0$	**r.** $2n\pi + \cos^{-1}(1/3)$, $n \in Z$
d. $\log_5 \tan x = (\log_5 4)(\log_4 (3\sin x))$	**s.** no solution

Integer Type

1. The number of solutions of the equation $16^{\sin^2 x} + 16^{\cos^2 x} = 10, 0 \le x < 2\pi$, is ______.

2. The number of solutions of the equation $e^{\sin x} - e^{-\sin x} - 4 = 0$ is ______.

3. If the equation $a\sin x + \cos 2x = 2a - 7$ possesses a solution, then the number of integral values of a is ______.

4. The total number of integral values of n so that $\sin x(\sin x + \cos x) = n$ has at least one solution is ______.

5. The number of integral values of k for which the equation $7\cos x + 5\sin x = 2k + 1$ has a solution is ______.

6. If $x, y \in [0, 2\pi]$, then the total number of ordered pairs (x, y) satisfying $\sin x \cos y = 1$ is ______.

7. The number of solutions of the equation $1 + \sin x \sin^2(x/2) = 0$ is ______.

8. The number of solutions of the equation $\sin x = x/10$ is ______.

9. The number of integral values of a for which the equation, $4\sin[x + (\pi/3)]\cos[x - (\pi/6)] = a^2 + \sqrt{3}\sin 2x - \cos 2x$ has a solution, is ______.

10. The number of solutions of the system of equations $\sin x + \sin y = \sin(x + y)$ and $|x| + |y| = 1$ is ______.

11. The smallest positive root of the equation $\sqrt{\sin(1-x)} = \sqrt{\cos x}$ lies in the quadrant number ______.

12. The number of solutions of the equation $\sin x + \sin\left[(\pi/8)\sqrt{(1-\cos 2x)^2 + \sin^2 2x}\right] = 0$, $x \in [5\pi/2, 7\pi/2]$ is ______.

13. $\cos(a - b) = 1$ and $\cos(a + b) = 1/e$, where $a, b \in [-\pi, \pi]$. The number of pairs of a, b that satisfy both the equations is ______.

14. The number of solutions of the equation $[(\sqrt{3}-1)/\sin x]+[(\sqrt{3}+1)/\cos x] = 4\sqrt{2}$ for $x \in (0, \pi/2)$ is ______.

15. If the sum of all roots of the equation $\sin[\pi \log_3 (1/x)] = 0$ in $(0, 2\pi)$ is S, then the value of $[S]$, where [.] represents the greatest integer function, is ______.

16. If the minimum value of the expression $y = (27)^{\cos x} + (81)^{\sin x}$ is m, then the value of $[1/m]$ is (where [.] represents greatest integer function).

17. If $(4^x + 4^{1-x})\ (1 + \tan^2 2y)\ (3 + \cos 3z) = 8$, then the value of cosec y is ______.

ANSWERS AND SOLUTIONS

Objective Type

1. a. $\sin\theta = \frac{1}{2}$ and $\cos\theta = -\frac{\sqrt{3}}{2}$

$\Rightarrow \theta$ lies in the second quadrant.

$\Rightarrow \sin\theta = \sin\frac{5\pi}{6}$; $\cos\theta = \cos\frac{5\pi}{6}$

Hence, $\theta = 2n\pi + \frac{5\pi}{6}$

2. c. Since $\tan\theta < 0$ and $\cos\theta > 0$, θ lies in the fourth quadrant. Then $\theta = 7\pi/4$.

Hence, the general value of θ is $2n\pi + 7\pi/4, n \in Z$.

3. c. $\cos p\theta = -\cos q\theta = \cos(\pi - q\theta)$

$\Rightarrow p\theta = 2n\pi \pm (\pi - q\theta)$

$\Rightarrow (p \mp q)\,\theta = (2n \pm 1)\,\pi$

$\Rightarrow \theta = \frac{(2n \pm 1)\pi}{(p \mp q)}, n \in Z$

$\Rightarrow \theta = \frac{r\pi}{p \pm q}$, where $r = -3, -1, 1, 3, \ldots$

$\Rightarrow \theta = \ldots, \frac{-3\pi}{p \pm q}, \frac{-\pi}{p \pm q}, \frac{\pi}{p \pm q}, \frac{3\pi}{p \pm q}, \ldots$

Shown above is an A.P. of common difference $\frac{2\pi}{p \pm q}$.

4. d. $(\cos\theta + \cos 7\theta) + (\cos 3\theta + \cos 5\theta) = 0$

$\Rightarrow 2\cos 4\theta(\cos 3\theta + \cos\theta) = 0$

$\Rightarrow 4\cos 4\theta\cos 2\theta\cos\theta = 0$

$\Rightarrow 4 \times \frac{1}{2^3\sin\theta}\left(\sin 2^3\theta\right) = 0$

$\Rightarrow \sin 8\theta = 0$ or $\theta = n\pi/8, n \in Z$.

5. b. $3\frac{\sin^2\theta}{\cos^2\theta} - 2\sin\theta = 0$, $\cos\theta \neq 0$

$\Rightarrow 3\sin^2\theta - 2\sin\theta(1 - \sin^2\theta) = 0$

$\Rightarrow 3\sin^2\theta - 2\sin\theta(1 - \sin^2\theta) = 0$

$\Rightarrow \sin\theta(2\sin^2\theta + 3\sin\theta - 2) = 0$

$\Rightarrow \sin\theta(2\sin\theta - 1)(\sin\theta + 2) = 0$

$\Rightarrow \sin\theta = 0, 1 - 2$ (rejected)

$\Rightarrow \theta = n\pi,\ n\pi + (-1)^n\frac{\pi}{6}, n \in Z.$

6. b. We have $1^2 = \sin\theta\cos 2\theta$

$\Rightarrow 1 - \sin\theta(1 - 2\sin^2\theta) = 0$

$\Rightarrow 2\sin^3\theta - \sin\theta + 1 = 0$

$\Rightarrow (\sin\theta + 1)(2\sin^2\theta - 2\sin\theta + 1) = 0$

$\Rightarrow \sin\theta = -1.$

The other factor gives imaginary roots.

$\Rightarrow \theta = n\pi + (-1)^n\left(-\frac{\pi}{2}\right) = n\pi - (-1)n\frac{\pi}{2}$

$= n\pi + (-1)n^{-1}\frac{\pi}{2}, n \in Z.$

7. b. $2\cos\theta[\cos 120° + \cos 2\theta] = 1$

$\Rightarrow 2\cos\theta\left(-\frac{1}{2} + 2\cos^2\theta - 1\right) = 1$

$\Rightarrow 4\cos^3\theta - 3\cos\theta - 1 = 0$

$\Rightarrow \cos 3\theta = 1 = \cos 0$

$\Rightarrow 3\theta = 2n\pi$ or $\theta = \frac{2n\pi}{3}, n \in Z$

Given the values so that $2n$ does not exceed 18. Therefore, $n = 0, 1, 2, 3, \ldots, 9$

Hence, the sum $= \frac{2\pi}{3}\sum_1^9 n = \frac{2\pi}{3} \times \frac{9(9+1)}{2} = 30\pi$

8. b. $\sec\theta - 1 = (\sqrt{2} - 1)\tan\theta$

$\Rightarrow \frac{1 - \cos\theta}{\cos\theta} = \frac{(\sqrt{2} - 1)\sin\theta}{\cos\theta}$

$\Rightarrow 2\sin^2(\theta/2) = (\sqrt{2} - 1)\,2\sin(\theta/2)\cos(\theta/2)$

$\Rightarrow \sin(\theta/2) = 0$ or $\tan(\theta/2) = (\sqrt{2} - 1) = \tan(\pi/8)$

$\Rightarrow \theta/2 = n\pi$ or $\frac{\theta}{2} = n\pi + \frac{\pi}{8}, n \in Z$

$\Rightarrow \theta = 2n\pi$ or $\theta = 2n\pi + \frac{\pi}{4}$

9. a. $\sin^4 x + \cos^4 x = \sin x\cos x$

$\Rightarrow (\sin^2 x + \cos^2 x)^2 - 2\sin^2 x\cos^2 x = \sin x\cos x$

$\Rightarrow 1 - \frac{\sin^2 2x}{2} = \frac{\sin 2x}{2}$

$\Rightarrow \sin^2 2x + \sin 2x - 2 = 0$

$\Rightarrow (\sin 2x + 2)(\sin 2x - 1) = 0$

$\Rightarrow \sin 2x = 1$

$\Rightarrow 2x = (4n + 1)\frac{\pi}{2}$

$\Rightarrow x = (4n + 1)\frac{\pi}{4}, n \in Z$

$\Rightarrow x = \frac{\pi}{4}, \frac{5\pi}{4}$ $[\because x \in [0, 2\pi]]$

Thus there are two solutions.

10. c. $\sin 3x + (\sin 5x + \sin x) = 0$

$\Rightarrow \sin 3x + (2\sin 3x\cos 2x) = 0$

$\Rightarrow \sin 3x = 0$ or $\cos 2x = -\frac{1}{2} = \cos\frac{2\pi}{3}$

$\Rightarrow x = n\pi/3$ or $x = n\pi \pm \frac{\pi}{3}, n \in Z$

Then $x = 0$, $\pi/3$, and $2\pi/3$. Hence, there are three solutions.

11. a. From the given relation

$\cos\theta = (2\sin\theta\cos\theta)\sin\theta$, $\sin\theta \neq 0$

$\Rightarrow \sin\theta = \pm\dfrac{1}{\sqrt{2}}$ or $\cos\theta = 0$

$\Rightarrow \theta = \dfrac{\pi}{4}, \dfrac{3\pi}{4}, \dfrac{\pi}{2}$ $[\because \theta \in [0, \pi]]$

Then the sum of roots is $\dfrac{3\pi}{2}$.

12. c. The given equation is

$(\cos x - 1)(12\cos^2 x + 5\cos x + 9) = 0$

$\Rightarrow \cos x = 1$ only as the other factor gives imaginary roots

$\Rightarrow \cos x = 1 \quad \Rightarrow x = 2n\pi, n \in Z$

Hence it has infinite solutions as $n \in Z$.

13. d. $\cos 2x = 1 - 2\sin^2 x$ and put $2^{-\sin^2 x} = t$

$\Rightarrow 2^{\cos 2x} = 2^{1-2\sin^2 x} = 2\left(2^{-\sin^2 x}\right)^2 = 2t^2$

$\Rightarrow 2t^2 - 3t + 1 = 0$

$\Rightarrow t = 1, 1/2$

$\Rightarrow 2^{-\sin^2 x} = 1 = 2^0$

$\Rightarrow \sin^2 x = 0$ or $x = n\pi, n \in Z$

from $2^{-\sin^2 x} = \dfrac{1}{2} = 2^{-1}$

$\Rightarrow \sin^2 x = 1$ or $x = n\pi \pm \dfrac{\pi}{2}, n \in Z$

14. a. $\cos x \cos 6x = -1$

$\Rightarrow 2\cos x\cos 6x = -2 \Rightarrow \cos 7x + \cos 5x = 2$

$\Rightarrow \cos 7x = -1$ and $\cos 5x = -1$

The value of x satisfying these two equations simultaneously and lying between 0 and 2π is π.

Therefore, the general solution is $x = 2n\pi + \pi, n \in Z$.

$\Rightarrow x = (2n + 1)\pi, n \in Z$

15. c. This is possible only when $\sin x = \cos x = 1$, which does not hold simultaneously.

Hence, no solution.

16. a. The given equation is $3(\sin x + \cos x) - 2(\sin x + \cos x)(1 - \sin x\cos x) = 8$

$\Rightarrow (\sin x + \cos x)[3 - 2 + 2\sin x\cos x] = 8$

$\Rightarrow (\sin x + \cos x)[\sin^2 x + \cos^2 x + 2\sin x\cos x] = 8$

$\Rightarrow (\sin x + \cos x)^3 = 8$

$\Rightarrow \sin x + \cos x = 2$

Above solution is not possible. Hence, the given equation has no solution.

17. a. $\cos^2\theta = \dfrac{1}{6}\sin\theta\tan\theta$

$\Rightarrow 6\cos^3\theta = 1 - \cos^2\theta$

$\Rightarrow 6\cos^3\theta + \cos^2\theta - 1 = 0$

$\Rightarrow (2\cos\theta - 1)(3\cos^2\theta + 2\cos\theta + 1) = 0$

$\Rightarrow \cos\theta = \dfrac{1}{2} \Rightarrow \theta = 2n\pi \pm \dfrac{\pi}{3}, n \in Z$

The other factor gives imaginary roots.

18. d. $(1 - \cos 2x) + (1 \cos^2 2x) = 2$

$\Rightarrow \cos 2x(\cos 2x + 1) = 0,$

$\Rightarrow \cos 2x = 0$ or $\cos 2x = -1$

$\Rightarrow 2x = (2n + 1)\pi/2$ or $2x = (2n \pm 1)\pi, n \in Z$

$\Rightarrow x = (2n + 1)\pi/4$ or $x = (2n \pm 1)\pi/2, n \in Z$

Hence, the solutions are $\pi/4, 3\pi/4, 5\pi/4, 7\pi/4, \pi/2, 3\pi/2$.

19. c. Dividing the given equation by $\cos^2 x$, as $\cos x = 0$ does not satisfy the equation, we have

$\tan^2 x - 5\tan x - 6 = 0$

$\Rightarrow (\tan x + 1)(\tan x - 6) = 0$

$\Rightarrow \tan x = -1$ or $\tan x = 6$

If $\tan x = -1 = \tan(-\pi/4)$, then $x = n\pi - \pi/4, \forall n \in Z$

and, if $\tan x = 6 = \tan\alpha$ (say)

$\Rightarrow \alpha = \tan^{-1} 6$, then $x = n\pi + \alpha = n\pi + \tan^{-1} 6, \forall n \in Z$

Hence, $x = n\pi - \pi/4, n\pi + \tan^{-1} 6, n \in Z$

20. d. From the given equation, we have

$\dfrac{\tan\theta + \tan 4\theta}{1 - \tan\theta\tan 4\theta} = -\tan 7\theta$

$\Rightarrow \tan(\theta + 4\theta) = -\tan 7\theta$

$\Rightarrow \tan 5\theta = \tan(-7\theta)$

$\Rightarrow 5\theta = n\pi - 7\theta$

$\Rightarrow \theta = n\pi/12$, where $n \in Z$, but $n \neq 6, 18, 30, \ldots$

21. d. We have

$\dfrac{1}{\sin^2\theta\cos^2\theta} + \dfrac{2}{\sin^2\theta} = 8$, $\sin\theta \neq 0$,

$\cos\theta \neq 0,$

$\Rightarrow 1 + 2\cos^2\theta = 8\sin^2\theta\cos^2\theta$

$\Rightarrow 1 + 2\cos^2\theta = 8\cos^2\theta(1 - \cos^2\theta)$

$\Rightarrow 8\cos^4\theta - 6\cos^2\theta + 1 = 0$

$\Rightarrow (4\cos^2\theta - 1)(2\cos^2\theta - 1) = 0$

$\Rightarrow \cos^2\theta = 1/4 = \cos^2(\pi/3)$

or $\cos^2\theta = 1/2 = \cos^2(\pi/4)$

$\Rightarrow \theta = n\pi \pm \pi/3$ or $\theta = n\pi \pm \pi/4, n \in Z$

Hence, for $0 \le \theta \le \pi/2$, $\theta = \pi/3$, $\theta = \pi/4$

22. b. $\tan x + \cot x = 2\operatorname{cosec} x$

$\Rightarrow \dfrac{\sin x}{\cos x} + \dfrac{\cos x}{\sin x} = \dfrac{2}{\sin x}$

$\Rightarrow \dfrac{1}{\sin x\cos x} = \dfrac{2}{\sin x}$

$\Rightarrow \cos x = \dfrac{1}{2}$

$\Rightarrow x = \pm\dfrac{\pi}{3}, \pm\dfrac{5\pi}{3}$

Thus there are four solutions.

23. d. Let $Z = \dfrac{3 + 2i\sin\theta}{1 - 2i\sin\theta} = \dfrac{(3 + 2i\sin\theta)(1 + 2i\sin\theta)}{(1 - 2i\sin\theta)(1 + 2i\sin\theta)}$

$$= \frac{\left(3-4\sin^2\theta\right)+8i\sin\theta}{1+4\sin^2\theta}$$

Real part of $Z = \dfrac{3-4\sin^2\theta}{1+4\sin^2\theta}$ and imaginary part of $Z = \dfrac{8\sin\theta}{1+4\sin^2\theta}$

Z is real if imaginary part $= \dfrac{8\sin\theta}{1+4\sin^2\theta} = 0$

or $\sin\theta = 0$ or $\theta = n\pi, \forall\, n \in I$.

Z is purely imaginary if the real part $(3-4\sin^2\theta)/(1+4\sin^2\theta)=0$

or $\sin^2\theta = 3/4 = \sin^2(\pi/3)$ or $\theta = n\pi \pm \pi/3, \forall\, n \in I$

24. a. $\because \dfrac{y+(1/y)}{2} \ge \sqrt{y\dfrac{1}{y}} \Rightarrow \sqrt{\left(y+\dfrac{1}{y}\right)} \ge \sqrt{2}$

[$\because$ A.M.$\ge$G.M.]

But $|\sin x + \cos x| \le \sqrt{2}$, Which is possible only when $y + \dfrac{1}{y} = 2$

$\Rightarrow y = 1$ and $x = \dfrac{\pi}{4}$

25. d. Since $7\cos^2 x + \sin x\cos x - 3 = 0$,
dividing the equation by $\cos^2 x$, we get
$7 + \tan x - 3\sec^2 x = 0$
$\Rightarrow 7 + \tan x - 3(1+\tan^2 x) = 0$
$\Rightarrow 3\tan^2 x - \tan x - 4 = 0$
$\Rightarrow (\tan x + 1)(3\tan x - 4) = 0$

$\Rightarrow \tan x = -1$ or $\tan x = \dfrac{4}{3}$

$\Rightarrow x = n\pi + \dfrac{3\pi}{4}$ or $x = k\pi + \tan^{-1}\left(\dfrac{4}{3}\right)$, where $(k, n \in Z)$

26. b. $(\sin\theta+2)(\sin\theta+3)(\sin\theta+4)=6$
L.H.S. >6 and R.H.S. $=6$

$\Rightarrow$ equality only one hold if $\sin\theta = -1$

$\Rightarrow \theta = \dfrac{3\pi}{2}, \dfrac{7\pi}{2}$ $\therefore$ sum $= 5\pi \Rightarrow k = 5$

27. b. $\sin^2 x + a\cos x + a^2 > 1 + \cos x$
put $x = 0$
$a + a^2 > 2$
$a^2 + a - 2 > 0$
$(a+2)(a-1) > 0$

Therefore

−2 1

Fig. 2.1

Therefore, largest negative integral value of $a = -3$

28. b. $\sin^4 x - \cos^2 x\sin x + 2\sin^2 x + \sin x = 0$
$\Rightarrow \sin x[\sin^3 x - \cos^2 x + 2\sin x + 1] = 0$
$\Rightarrow \sin x[\sin^3 x - 1 + \sin^2 x + 2\sin x + 1] = 0$
$\Rightarrow \sin x[\sin^3 x + \sin^2 x + 2\sin x] = 0$
$\Rightarrow \sin^2 x = 0$ or $\sin^2 x + \sin x + 2 = 0$
[Not possible for real x]
$\Rightarrow$ not possible for real x
$\Rightarrow \sin x = 0$
Hence, the solutions are $x = 0, \pi, 2\pi, 3\pi$.

29. a. Since $x \in [0, 2\pi]$ and $y \in [0, 2\pi]$,
and $\sin x + \sin y = 2$,
this is possible only when $\sin x = 1$ and $\sin y = 1$

$\Rightarrow x = \dfrac{\pi}{2}$ and $y = \dfrac{\pi}{2}$

Hence, $x + y = \pi$.

30. a. $(\sqrt{3}-1)\sin\theta + (\sqrt{3}+1)\cos\theta = 2$

$\Rightarrow \dfrac{(\sqrt{3}-1)}{2\sqrt{2}}\sin\theta + \left(\dfrac{\sqrt{3}+1}{2\sqrt{2}}\right)\cos\theta = \dfrac{1}{\sqrt{2}}$

$\Rightarrow \sin\dfrac{\pi}{12}\sin\theta + \cos\dfrac{\pi}{12}\cos\theta = \cos\dfrac{\pi}{4}$

$\Rightarrow \cos\left(\theta - \dfrac{\pi}{12}\right) = \cos\dfrac{\pi}{4}$

$\Rightarrow \theta - \dfrac{\pi}{12} = 2n\pi \pm \dfrac{\pi}{4}, n \in Z$

$\Rightarrow \theta = 2n\pi \pm \dfrac{\pi}{4} + \dfrac{\pi}{12}$

31. a. $\sin 6\theta + \sin 4\theta + \sin 2\theta = 0$

$\Rightarrow (\sin 6\theta + \sin 2\theta) + \sin 4\theta = 0$

$\Rightarrow 2\sin 4\theta\cos 2\theta + \sin 4\theta = 0$

$\Rightarrow \sin 4\theta(2\cos 2\theta + 1) = 0$

$\Rightarrow \sin 4\theta = 0$ or $\cos 2\theta = -\dfrac{1}{2} = \cos\dfrac{2\pi}{3}$

$\Rightarrow 4\theta = n\pi$ or $2\theta = 2n\pi \pm \dfrac{2\pi}{3}, n \in Z$

$\Rightarrow \theta = \dfrac{n\pi}{4}$ or $\theta = n\pi \pm \dfrac{\pi}{3}$

32. d. The given expression
$= \cos y\sin x - \sin y\cos x + \sin y\sin x + \cos x\cos y$
$= \sin(x-y) + \cos(x-y)$
$\therefore \sin(x-y) + \cos(x-y) = 0$

$\Rightarrow \sqrt{2}\left(\dfrac{1}{\sqrt{2}}\sin(x-y) + \dfrac{1}{\sqrt{2}}\cos(x-y)\right) = 0$

$\Rightarrow \sin\left(x - y + \dfrac{\pi}{4}\right) = 0$

$\Rightarrow \dfrac{\pi}{4} + x - y = n\pi$

$\Rightarrow x - y = n\pi - \dfrac{\pi}{4}$

$\Rightarrow x = n\pi - \dfrac{\pi}{4} + y$ where $n \in Z$

33. c. $\tan 5\theta = \cot 2\theta = \tan\left(\frac{\pi}{2} - 2\theta\right)$

$\Rightarrow 5\theta = n\pi + \frac{\pi}{2} - 2\theta, n \in Z$

$\Rightarrow 7\theta = n\pi + \frac{\pi}{2} \Rightarrow \theta = \frac{n\pi}{7} + \frac{\pi}{14}, n \in Z$

34. a. Let $f(x) = \cos x - x + \frac{1}{2}$

$f(0) = 1 + \frac{1}{2} > 0$

$f\left(\frac{\pi}{2}\right) = 0 - \frac{\pi}{2} + \frac{1}{2} = \frac{1-\pi}{2} < 0$

$\Rightarrow$ one root lies in the interval $\left(0, \frac{\pi}{2}\right)$

35. d. $\tan\left(\frac{p\pi}{4}\right) = \cot\left(\frac{q\pi}{4}\right) = \tan\left(\frac{\pi}{2} - \frac{q\pi}{4}\right)$

$\Rightarrow \frac{p\pi}{4} = n\pi + \frac{\pi}{2} - \frac{q\pi}{4}$

$\Rightarrow \frac{p}{4} = n + \frac{1}{2} - \frac{q}{4}$

$\Rightarrow \frac{p+q}{4} = \frac{2n+1}{2}$

$\Rightarrow p + q = 2(2n+1)$

36. b. $y = \sin x - \cos x = \sqrt{2}\left[\frac{1}{\sqrt{2}}\sin - \frac{1}{\sqrt{2}}\cos x\right]$

$= \sqrt{2}\sin\left(x - \frac{\pi}{4}\right) \Rightarrow -\sqrt{2} \le y \le \sqrt{2}$

$\Rightarrow$ range of y is $[-\sqrt{2}, \sqrt{2}]$

37. a. $4\sin^4 x + \cos^4 x = 1$

$\Rightarrow (2\sin^2 x)^2 + \frac{1}{4}(2\cos^2 x)^2 = 1$

$\Rightarrow (1 - \cos 2x)^2 + \frac{1}{4}(1 + \cos 2x)^2 = 1$

$\Rightarrow 5\cos^2 2x - 6\cos 2x + 1 = 0$

$\Rightarrow (\cos 2x - 1)(5\cos 2x - 1) = 0$

$\Rightarrow \cos 2x = 1$ or $\cos 2x = 1/5$

$\Rightarrow 2x = 2n\pi$ or $2x = 2n\pi \pm \alpha$,

where $\alpha = \cos^{-1}(1/5), \forall n \in Z$

38. c. $(1 - \tan\theta)[1 + 2\tan\theta/(1 + \tan^2\theta)] = 1 + \tan\theta$

$\Rightarrow (1 - \tan\theta)(1 + \tan\theta)^2 = (1 + \tan\theta)(1 + \tan^2\theta)$

$\Rightarrow (1 + \tan\theta)[(1 - \tan^2\theta) - (1 + \tan^2\theta)] = 0$

$\Rightarrow -2\tan^2\theta = 0, (1 + \tan\theta) = 0$

$\Rightarrow \tan\theta = 0$, or $\tan\theta = -1$

$\Rightarrow \theta = n\pi$ or $n\pi - \pi/4, \forall n \in Z$, for $\theta \in [0, 2\pi]$ $\theta = 0$,

$\pi, 2\pi, \frac{3\pi}{4}, \frac{7\pi}{4}$

39. d. We have $(\sin x + \sin 3x) + \sin 2x = (\cos x + \cos 3x) + \cos 2x$,

$\Rightarrow 2\sin 2x\cos x + \sin 2x = 2\cos 2x\cos x + \cos 2x$

$\Rightarrow \sin 2x(2\cos x + 1) = \cos 2x(2\cos x + 1)$

$\Rightarrow (2\cos x + 1)(\sin 2x - \cos 2x) = 0$

$\Rightarrow \cos x = -1/2$ or $\sin 2x - \cos 2x = 0$

$\Rightarrow x = 2n\pi \pm 2\pi/3$ or $\tan 2x = 1 = \tan(\pi/4)$

$\Rightarrow x = 2n\pi \pm 2\pi/3$ or $x = (4n+1)\pi/8, n \in Z$

But here $0 \le x \le 2\pi$

Hence $x = \pi/8, 5\pi/8, 2\pi/3, 9\pi/8, 4\pi/3, 13\pi/8$

40. b. $3\sin\theta - 4\sin^3\theta - \sin\theta = 2(2\cos^2\theta - 1)$

$\Rightarrow 2\sin\theta(1 - 2\sin^2\theta) = 2\cos 2\theta$

$\Rightarrow 2\cos 2\theta(\sin\theta - 1) = 0$

$\Rightarrow \cos 2\theta = 0$ or $\sin\theta = 1$

$\Rightarrow 2\theta = (2n+1)\pi/2$ or $\theta = 2n\pi + \pi/2, \forall n \in Z$

$\Rightarrow \theta = (2n+1)\pi/4$, or $\theta = (4n+1)\pi/2, \forall n \in Z$

Hence, $\theta = \pi/4, 3\pi/4, 5\pi/4, 7\pi/4, \pi/2$

$(\because \theta \in [0, 2\pi])$

41. c. We have $4\sin\theta\sin 2\theta\sin 4\theta = 3\sin\theta - 4\sin^3\theta$

$\Rightarrow \sin\theta[4\sin 2\theta\sin 4\theta - 3 + 4\sin^2\theta] = 0$

$\Rightarrow \sin\theta[2(\cos 2\theta - \cos 6\theta) - 3 + 2(1 - \cos 2\theta)] = 0$

$\Rightarrow \sin\theta(-2\cos 6\theta - 1) = 0$

$\Rightarrow \sin\theta = 0$ or $\cos 6\theta = -1/2$

$\Rightarrow \theta = n\pi$ or $6\theta = 2n\pi \pm 2\pi/3, \forall n \in Z$

$\Rightarrow \theta = n\pi$ or $\theta = (3n \pm 1)\pi/9, \forall n \in Z$

42. b. From the given equation, we have $\tan\theta + \tan 2\theta + \tan(\theta + 2\theta) = 0$

$\Rightarrow (\tan\theta + \tan 2\theta) + \frac{\tan\theta + \tan 2\theta}{1 - \tan\theta\tan 2\theta} = 0$

$\Rightarrow (\tan\theta + \tan 2\theta)\left[1 + \frac{1}{1 - \tan\theta\tan 2\theta}\right] = 0$

$\Rightarrow (\tan\theta + \tan 2\theta)(2 - \tan\theta\tan 2\theta) = 0$

$\Rightarrow \tan\theta = \tan(-2\theta)$ or $2 - \tan\theta[(2\tan\theta)/(1 - \tan^2\theta)] = 0$

$\Rightarrow \theta = n\pi - 2\theta$ or $1 - 2\tan^2\theta = 0$

$\Rightarrow \theta = n\pi/3$ or $\tan^2\theta = 1/2 = \tan^2\alpha$ (say),

Hence, $\theta = n\pi \pm \alpha$, where $\tan\alpha = 1/\sqrt{2}$.

43. b. We have $\sin 3\alpha = 4\sin\alpha(\sin^2 x - \sin^2\alpha)$

$\Rightarrow 3\sin\alpha - 4\sin^3\alpha = 4\sin\alpha\sin^2 x - 4\sin^3\alpha$

$\Rightarrow 3\sin\alpha = 4\sin\alpha\sin^2 x$ or $\sin\alpha = 0$

If $\sin\alpha \ne 0$, $\sin^2 x = 3/4 = (\sqrt{3}/2)^2 = \sin^2(\pi/3)$

$\therefore x = n\pi \pm \pi/3, \forall n \in Z$

If $\sin\alpha = 0$, i.e., $\alpha = n\pi$, equation becomes an identity.

44. c. We have $\sqrt{3}\cos\theta - 3\sin\theta = 2(\sin 5\theta - \sin\theta)$

$\Rightarrow (\sqrt{3}/2)\cos\theta - (1/2)\sin\theta = \sin 5\theta$

$\Rightarrow \cos(\theta + \pi/6) = \sin 5\theta = \cos(\pi/2 - 5\theta)$

$\Rightarrow \theta + \pi/6 = 2n\pi \pm (\pi/2 - 5\theta)$

$\Rightarrow \theta = n\pi/3 + \pi/18$ or $\theta = -n\pi/2 + \pi/6, \forall n \in Z$

45. c. $\sin^4 x + \cos^4 x + \sin 2x + \alpha = 0$

$\Rightarrow (\sin^2 x + \cos^2 x)^2 - 2\sin^2 x\cos^2 x + \sin 2x + \alpha = 0$

$\Rightarrow \sin^2 2x - 2\sin 2x - 2 - 2\alpha = 0$

Let $\sin 2x = y$. Then the given equation becomes $y^2 - 2y - 2(1+\alpha) = 0$ where $-1, \le y \le 1$, $(\because -1 \le \sin 2x \le 1)$
For real, discriminant ≥ 0

$\Rightarrow 3+2\alpha \ge 0 \quad \Rightarrow \alpha \ge -\frac{3}{2}$

Also $-1 \le y \le 1 \Rightarrow -1 \le 1 - \sqrt{3+2\alpha} \le 1$

$\Rightarrow 3+2\,\theta \le 4 \quad \Rightarrow \alpha \le \frac{1}{2}$. Thus $-\frac{3}{2} \le \alpha \le \frac{1}{2}$

46. a. Since the system has a non-trivial solution, the determinant of coefficients = 0

$$\Rightarrow \begin{vmatrix} \sin 3\theta & -1 & 1 \\ \cos 2\theta & 4 & 3 \\ 2 & 7 & 7 \end{vmatrix} = 0$$

$\Rightarrow \sin 3\,\theta(28-21) - \cos 2\theta(-7-7) + 2(-3-4) = 0$
$\Rightarrow \sin 3\,\theta + 2\cos 2\theta - 2 = 0$
$\Rightarrow (3\sin\theta - 4\sin^3\theta) + 2(1-2\sin^2\theta) - 2 = 0$
$\Rightarrow 4\sin^3\theta + 4\sin^2\theta - 3\sin\theta = 0$
$\Rightarrow \sin\theta(2\sin\theta - 1)(2\sin\theta + 3) = 0$
$\Rightarrow \sin\theta = 0$ or $\theta = n\pi + (-1)^n \pi/6, \forall\, n \in Z$

47. c. Let $\log_{\cos x} \sin x = t$, then the given equation is $t + \frac{1}{t} = 2$

$\Rightarrow (t-1)^2 = 0 \Rightarrow t = 1$
$\Rightarrow \log_{\cos x} \sin x = 1$ or $\sin x = \cos x \Rightarrow \tan x = 1$
$\Rightarrow x = \pi/4$

48. b. Given $x^2 + 2x\sin(xy) + 1 = 0$
$\Rightarrow [x + \sin(xy)]^2 + [1 - \sin^2(xy)] = 0$
$\Rightarrow x + \sin(xy) = 0$ and $\sin^2(xy) = 1$
$\sin^2(xy) = 1$ gives $\sin(xy) = 1$ or -1
If $\sin(xy) = 1 \Rightarrow x = -1 \Rightarrow \sin(-y) = 1 \Rightarrow \sin y = -1$,
Then the ordered pair is $(1, 3\pi/2)$
if $\sin(xy) = -1 \Rightarrow x = 1 \Rightarrow \sin y = -1 \Rightarrow (-1, 3\pi/2)$
Thus there are two ordered pairs.

49. c. The given equation is $8\sin x\cos x\cos 2x\cos 4x = \sin 6x\ (\sin x \ne 0)$
$\Rightarrow \sin 8x = \sin 6x \Rightarrow 2\cos 7x\sin x = 0$
As $\sin x \ne 0$, $\cos 7x = 0$ or $7x = n\pi + \pi/2$
i.e., $x = n\pi/7 + \pi/14, n \in Z$

50. d. We have $\cos 3x + \sin\left(2x - \frac{7\pi}{6}\right) = -2$

$\Rightarrow 1 + \cos 3x + 1 + \sin\left(2x - \frac{7\pi}{6}\right) = 0$

$\Rightarrow (1 + \cos 3x) + 1 - \cos\left(2x - \frac{2\pi}{3}\right) = 0$

$\Rightarrow 2\cos^2\frac{3x}{2} + 2\sin^2\left(x - \frac{\pi}{3}\right) = 0$

$\Rightarrow \cos\frac{3x}{2} = 0$ and $\sin\left(x - \frac{\pi}{3}\right) = 0$

$\Rightarrow \frac{3x}{2} = \frac{\pi}{2}, \frac{3\pi}{2}, \ldots$, and $x - \frac{\pi}{3} = 0, \pi, 2\pi, \ldots$

$\Rightarrow x = \frac{\pi}{3}$

Therefore, the general solution of $\cos\frac{3x}{2} = 0$ and $\sin\left(x - \frac{\pi}{3}\right) = 0$ is $x = 2k\pi + \frac{\pi}{3} = \frac{\pi}{3}(6k+1)$, where $k \in Z$.

51. b. Let $\tan^2\theta = t$
$\Rightarrow 1 - t^2 + 2^t = 0$
It is clearly satisfied by $t = 3$ by inspection
$\Rightarrow \tan^2\theta = 3$
$\Rightarrow \theta = \pm\pi/3$ in the given interval.

52. b. We can write the given equation as
$a\sin x + (1 - 2\sin^2 x) = 2a - 7$
$\Rightarrow 2\sin^2 x - a\sin x + 2a - 8 = 0$

$\Rightarrow \sin x = \frac{a \pm \sqrt{a^2 - 8(2a-8)}}{4} = \frac{a \pm (a-8)}{4}$

$\Rightarrow \sin x = (a-4)/2$
$\Rightarrow -1 \le (a-4)/2 \le 1$
$\Rightarrow -2 \le (a-4) \le 2$
$\Rightarrow 2 \le a \le 6$

53. a. $\cos x = \sqrt{1 - \sin 2x} = |\sin x - \cos x|$

(i) $\sin x < \cos x \Rightarrow x \in \left[0, \frac{\pi}{4}\right) \cup \left(\frac{5\pi}{4}, 2\pi\right]$ (1)

Then the given equation is $\cos x = \cos x - \sin x$
$\Rightarrow \sin x = 0 \Rightarrow x = \pi, 2\pi$
$\Rightarrow x = 2\pi$ [From (1)]

(ii) $\sin x \ge \cos x \Rightarrow x \in \left(\frac{\pi}{4}, \frac{5\pi}{4}\right)$

$\Rightarrow \cos x = \sin x - \cos x$
$\Rightarrow \tan x = 2$
$\Rightarrow x = \tan^{-1}$ (2)
Hence, there are two solutions.

54. b. $\sum_{r=1}^{5} \cos rx = 5$

$\Rightarrow \cos x + \cos 2x + \cos 3x + \cos 4x + \cos 5x = 5$, which is possible only when $\cos x = \cos 2x = \cos 3x = \cos 4x = \cos 5x = 1$ and is satisfied by $x = 0$ only.
Hence, there is only one solution.

55. a. $\sin 2x + \cos 4x = 2$
It is possible only when $\sin 2x = 1$ and $\cos 4x = 1$.

Therefore, $2x = 2n\pi + \frac{\pi}{2}$ and $4x = 2m\pi$

$\Rightarrow x = n\pi + \frac{\pi}{4}$ and $x = \frac{m\pi}{2}$, where $m, n \in Z$

Then the solution is

$\left(n\pi + \frac{\pi}{4}, n \in I\right) \cap \left(\frac{m\pi}{2}, m \in Z\right) = \phi$

56. d. $\cos^2 x = 2\cos x(1 - 3\cos^2 x)$

$\Rightarrow 6\cos^3 x + \cos^2 x - 2\cos x = 0$

$\Rightarrow \cos x = 0, \frac{1}{2}, -\frac{2}{3}$

$\Rightarrow x = \frac{\pi}{2}, \frac{\pi}{3}, \cos^{-1}\left(-\frac{2}{3}\right)$ [$\because \alpha, \beta$ are positive]

If $\alpha = \frac{\pi}{2}$ and $\beta = \frac{\pi}{3}, |\alpha - \beta| = \frac{\pi}{6}$

57. b. We have, $\sin^{100} x - \cos^{100} x = 1$

$\Rightarrow \sin^{100} x = 1 + \cos^{100} x$

Since, the L.H.S. never exceeds 1 while the R.H.S. exceeds 1 unless $\cos x = 0$

Then $x = n\pi + \frac{\pi}{2}, n \in I$

58. c. $|\cot x| = \cot x + \frac{1}{\sin x}$

If $\cot x > 0 \Rightarrow \cot x = \cot x + \frac{1}{\sin x} = 0$

$\Rightarrow \frac{1}{\sin x} = 0$ which is not possible.

If $\cot x \le 0 \Rightarrow -\cot x = \cot x + \frac{1}{\sin x}$

$\Rightarrow -2\cot x = \frac{1}{\sin x}$

$\Rightarrow \cos x = -\frac{1}{2}$

$\Rightarrow x = \frac{2\pi}{3}, \frac{4\pi}{3}, \frac{8\pi}{3}$ $(\because x \in [0, 3\pi])$

59. a. $\tan(A - B) = 1$

$\Rightarrow A - B = n_1\pi + \frac{\pi}{4} = \frac{\pi}{4}, \frac{3\pi}{4}, -\frac{3\pi}{4}, \ldots$

$\sec(A + B) = \frac{2}{\sqrt{3}}$

$\Rightarrow A + B = 2n_2\pi \pm \frac{\pi}{6} = \frac{\pi}{6}, \frac{11\pi}{6}, \ldots$

For the least positive values of A and B,

$A + B = \frac{11\pi}{6}, A - B = \frac{\pi}{4} \Rightarrow B = \frac{19\pi}{24}, A = \frac{25\pi}{24}$

60. a. Let $A = \theta + 15°, B = \theta - 15°$

$\Rightarrow A + B = 2\theta$ and $A - B = 30°$

Now $\frac{\tan A}{\tan B} = \frac{3}{1}$

$\Rightarrow \frac{\tan A + \tan B}{\tan A - \tan B} = \frac{3+1}{3-1}$ (applying comp.– divdn)

$\Rightarrow \frac{\sin(A+B)}{\sin(A-B)} = 2$

$\Rightarrow \sin 2\theta = 2\sin 30° = 1$

$\Rightarrow 2\theta = 2n\pi + \frac{\pi}{2}$ or $\theta = n\pi + \frac{\pi}{4}, n \in Z$

61. a. $\tan 3\theta + \tan\theta = 2\tan 2\theta$

$\Rightarrow \tan 3\theta - \tan 2\theta = \tan 2\theta - \tan\theta$

$\Rightarrow \frac{\sin(3\theta - 2\theta)}{\cos 3\theta \cos 2\theta} = \frac{\sin(2\theta - \theta)}{\cos 2\theta \cos\theta}$

$\Rightarrow \sin\theta(2\sin\theta\sin 2\theta) = 0$

$\Rightarrow \sin\theta = 0$ or $\sin 2\theta = 0$

$\Rightarrow \theta = n\pi$ or $2\theta = n\pi, n \in Z$

But $\theta = \frac{n\pi}{2}$ is rejected as when n is odd, $\tan\theta$ is not defined and when n is even $= 2r$ then $\theta = r\pi$.

Then $\theta = n\pi, n \in I$ is the only solution.

62. a. We have $|4\sin x - 1| < \sqrt{5}$

$\Rightarrow -\sqrt{5} < 4\sin x - 1 < \sqrt{5}$

$\Rightarrow -\left(\frac{\sqrt{5}-1}{4}\right) < \sin x < \left(\frac{\sqrt{5}+1}{4}\right)$

$\Rightarrow -\sin\frac{\pi}{10} < \sin x < \cos\frac{\pi}{5}$

$\Rightarrow \sin\left(-\frac{\pi}{10}\right) < \sin x < \sin\left(\frac{\pi}{2} - \frac{\pi}{5}\right)$

$\Rightarrow \sin\left(-\frac{\pi}{10}\right) < \sin x < \sin\frac{3\pi}{10}$

$\Rightarrow x \in \left(-\frac{\pi}{10}, \frac{3\pi}{10}\right)$

63. d. $y^2 - y + a = \left(y - \frac{1}{2}\right)^2 + a - \frac{1}{4}$

Since $-\sqrt{2} \le \sin x + \cos x \le \sqrt{2}$, the given equation will have no real value x for any y if $a - \frac{1}{4} > \sqrt{2}$

i.e., $a \in \left(\sqrt{2} + \frac{1}{4}, \infty\right)$

$\Rightarrow a \in (\sqrt{3}, \infty)$ (as $\sqrt{2} + 1/4 < \sqrt{3}$)

64. a. $(2\sin x - \operatorname{cosec} x)^2 + (\tan x - \cot x)^2 = 0$

$\Rightarrow \sin^2 x = \frac{1}{2}$ and $\tan^2 x = 1$

$\Rightarrow x \in n\pi \pm \frac{\pi}{4}, n \in 2$

65. a. $[\sin x + \cos x] = 3 + [-\sin x] + [-\cos x]$

Maximum value of left-hand side is 1 and minimum of right hand side is also 1

$\Rightarrow [\sin x + \cos x] = 3 + [-\sin x] + [-\cos x] = 1$

$\Rightarrow x \in \pi \pm \frac{\pi}{4}$

$\Rightarrow [\sin x + \cos x] = 1, [-\sin x] = -1, [-\cos x] = -1$

which is not possible.

66. c. $\cos^8 x + b\cos^4 x + 1 = 0$

$\Rightarrow b = -\left(\cos^4 x + \dfrac{1}{\cos^4 x}\right) \le -2, \forall x \in R$

$\Rightarrow b \in (-\infty, -2]$

67. b. Here $1 \le |\sin 2x| + |\cos 2x| \le \sqrt{2}$ and $|\sin y| \le 1$ so solution is possible only when $|\sin y| = 1$

$\Rightarrow \sin y = \pm 1 \Rightarrow y = \pm \dfrac{\pi}{2}, \pm \dfrac{3\pi}{2}$

68. b. Given that $|\sin x|^2 + |\sin x| + b = 0$

$\Rightarrow |\sin x| = \dfrac{-1 \pm \sqrt{1-4b}}{2}$

$\Rightarrow 0 \le \dfrac{-1 \pm \sqrt{1-4b}}{2} < 1 \Rightarrow -2 < b < 0$

69. a. $|2\sin\theta - \operatorname{cosec}\theta| \ge 1$

$\Rightarrow |2\sin^2\theta - 1| \ge |\sin\theta|$

$\Rightarrow |\cos 2\theta| \ge |\sin\theta|$

$\Rightarrow 2\cos^2 2\theta \ge 1 - \cos 2\theta$

$\Rightarrow 2\cos^2 2\theta + \cos 2\theta - 1 \ge 0$

$\Rightarrow (2\cos 2\theta - 1)(\cos 2\theta + 1) \ge 0$

$\Rightarrow \cos 2\theta \ge \dfrac{1}{2}$ (As $\cos\theta \ne 0$, i.e., $\cos 2\theta \ne -1$)

70. d. The given equation can be written as

$\sin x \cos x [\sin^2 x + \sin x \cos x + \cos^2 x] = 1$

$\Rightarrow \sin x \cos x [1 + \sin x \cos x] = 1$

$\Rightarrow \sin 2x [2 + \sin 2x] = 4$

$\Rightarrow \sin 2x = \dfrac{-2 \pm \sqrt{4+16}}{2} = -1 \pm \sqrt{5}$,

which is not possible.

71. d. $e^{|\sin x|} + e^{-|\sin x|} + 4a = 0$, let $t = e^{|\sin x|}$

$\Rightarrow t \in [1, e]$

$\Rightarrow t + \dfrac{1}{t} + 4a = 0$

$\Rightarrow t^2 + 4at + 1 = 0$

This quadratic expression should have two distinct roots in $[1, e]$

$\Rightarrow 16a^2 - 4 > 0, f(1) = 1 + 4a + 1 \ge 0,$

$f(e) = e^2 + 4ae + 1 \ge 0, 1 < - a < e$

$\Rightarrow |a| > \dfrac{1}{2}, a \ge -\dfrac{1}{2}, a \ge \dfrac{-1-e^2}{4e}, -\dfrac{e}{4} < a < -\dfrac{1}{4}$

Clearly, there is no value of a satisfying the above inequalities simultaneously.

72. b.

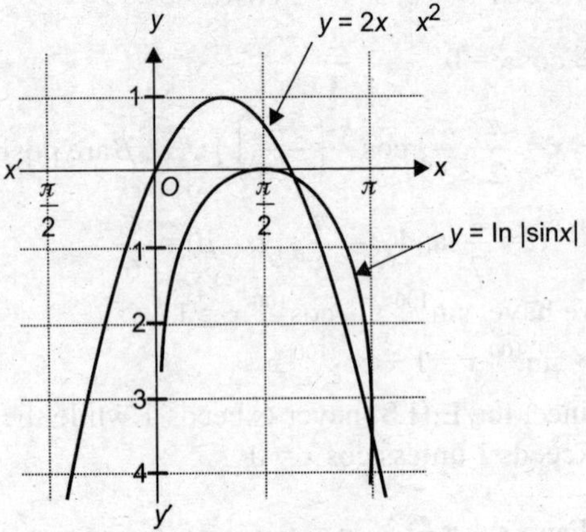

Fig. 2.2

$\ln|\sin x| = -x(x-2)$

Graphs of $y = \ln|\sin x|$ and $y = -x(x-2)$ meet exactly two times in $\left[-\dfrac{\pi}{2}, \dfrac{3\pi}{2}\right]$.

73. c. $|x| + |y| = 4$, $\sin\left(\dfrac{\pi x^2}{3}\right) = 1$

$\Rightarrow |x|, |y| \in [0, 4], \dfrac{\pi x^2}{3} = (4n+1)\dfrac{\pi}{2}$

$\Rightarrow x^2 = \dfrac{(4n+1)3\pi}{2} = \dfrac{3\pi}{2}$, as $|x| \le 4$

$\Rightarrow |x| = \sqrt{\dfrac{3\pi}{2}}, |y| = 4 - \sqrt{\dfrac{3\pi}{2}}$

Thus there are four ordered pairs.

74. b. $\sin\{x\} = \cos\{x\}$

$\tan\{x\} = 1$

$\tan(\pi/4) = 1 < \tan 1$

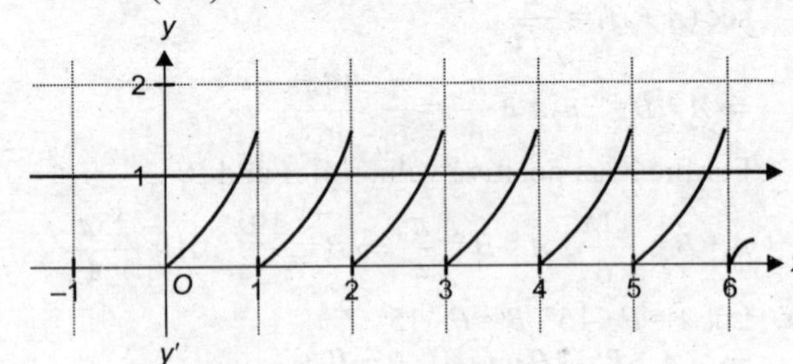

Fig. 2.3

Graphs of $y = \tan\{x\}$ and $y = 1$ meet exactly six times in $[0, 2\pi]$

75. a. $x^2 + 4 - 2x + 3\sin(ax+b) = 0$

$(x-1)^2 + 3 + 3\sin(ax+b) = 0$

$\Rightarrow x = 1, \sin(ax+b) = -1$

$\Rightarrow \sin(a+b) = -1$

$\Rightarrow a + b = (4n-1)\dfrac{\pi}{2}, n \in I \Rightarrow a + b = \dfrac{7\pi}{2}$

[From given options]

76. c. $\tan^4 x - 2\sec^2 x + a^2 = 0$

$\Rightarrow \tan^4 x - 2(1 + \tan^2 x) + a^2 = 0$

$\Rightarrow \tan^4 x - 2\tan^2 x + 1 = 3 - a^2$

$\Rightarrow (\tan^2 x - 1)^2 = 3 - a^2$

$\Rightarrow 3 - a^2 \geq 0 \Rightarrow a^2 \leq 3 \Rightarrow |a| \leq \sqrt{3}$

77. a. $1 + \log_2 \sin x + \log_2 \sin 3x \geq 0$

$\Rightarrow \sin x, \sin 3x > 0$

$\Rightarrow \log_2 (2 \sin x \sin 3x) \geq 0$

$\Rightarrow 2 \sin x \sin 3x \geq 1$

For $\sin x > 0$

$\Rightarrow x \in (0, \pi)$ (1)

$\sin 3x > 0$ (2)

$\Rightarrow 3x \in (0, \pi) \cup (2\pi, 3\pi)$

$\Rightarrow x \in \left(0, \frac{\pi}{3}\right) \cup \left(\frac{2\pi}{3}, \pi\right)$

$\Rightarrow \frac{1}{4} \leq \sin^2 x \leq \frac{1}{2}$

For $2 \sin x \sin 3x \geq 1$

$\Rightarrow 2\sin^2 x (3 - 4\sin^2 x) \geq 1$

$\Rightarrow 8\sin^4 x - 6\sin^2 + 1 \leq 0$

$\Rightarrow (2\sin^2 x - 1)(4\sin^2 x - 1) \leq 0$

$\Rightarrow \frac{1}{2} \leq \sin x \leq \frac{1}{\sqrt{2}}$

$\Rightarrow x \in \left[\frac{\pi}{3}, \frac{\pi}{4}\right] \cup \left[\frac{2\pi}{3}, \frac{3\pi}{4}\right]$ (3)

Thus, finally $x \in \left(\frac{2\pi}{3}, \frac{3\pi}{4}\right]$.

[From equations (1), (2), (3)]

Multiple Correct Answers Type

1. a., b. We have $4\sin^4 x + \cos^4 x = 1$

$\Rightarrow 4\sin^4 x = 1 - \cos^4 x$

$= (1 - \cos^2 x)(1 + \cos^2 x)$

$= \sin^2 x (2 - \sin^2 x)$

$\Rightarrow \sin^2 x [5\sin^2 x - 2] = 0$

$\Rightarrow \sin x = 0$ or $\sin x = \pm\sqrt{2/5}$

$\Rightarrow x = n\pi$ or $x = n\pi \pm \sin^{-1}\sqrt{\frac{2}{5}}, n \in Z$

2. a., b. $(\sin^3 \theta + \cos^3) - (1 - \sin\theta\cos\theta) = 0$

$\Rightarrow (\sin\theta + \cos\theta)(1 - \sin\theta\cos\theta) - (1 - \sin\theta\cos\theta) = 0$

$\Rightarrow (1 - \sin\theta\cos\theta)(\sin\theta + \cos\theta - 1) = 0$

$\Rightarrow \sin\theta + \cos\theta = 1$, $2\sin\theta\cos\theta = 2$ or $\sin 2\theta = 2$ (not possible)

$\Rightarrow \sin\theta + \cos\theta = 1$

$\Rightarrow \theta - \frac{\pi}{4} = 2n\pi \pm \frac{\pi}{4}, n \in Z$

$\Rightarrow \theta = 2n\pi$ or $2n\pi + \frac{\pi}{2}$

3. a., c. We have $\tan^2\theta = 1 - \cos 2\theta$

$= 2\sin^2\theta$ or $\operatorname{cosec}^2\theta \tan^2\theta$

$= 2$ or $(1 + \cot^2\theta)\tan^2\theta$

$= 2$ or $\tan^2\theta + 1 = 2$

$\Rightarrow \tan^2\theta = 1 \Rightarrow \tan\theta = \pm 1 \Rightarrow \theta = n\pi \pm \frac{\pi}{4}$

Moreover, $\tan^2\theta = 2\sin^2\theta \Rightarrow \sin^2\theta = 0 \Rightarrow \theta = n\pi$

4. a., c. $y + \frac{1}{y} \geq 2 \Rightarrow \sqrt{y + \frac{1}{y}} \geq 2$

But $\sin x + \cos x \leq \sqrt{2}$

$\Rightarrow y + \frac{1}{y} = 2$ and $\sin x + \cos x = \sqrt{2}$

$\Rightarrow y = 1$ and $\sin\left(x + \frac{\pi}{4}\right) = 1$ or $y = 1$ and $x = \frac{\pi}{4}$

5. b., d. $\sin\theta + \sqrt{3}\cos\theta = -2 - (x^2 - 6x + 9) = -2 - (x - 3)^2$

So $\sin\theta + \sqrt{3}\cos\theta \leq -2$

But the minimum value of $\sin\theta + \sqrt{3}\cos\theta$

$= -\sqrt{1 + 3} = -2$

We have $\sin\theta + \sqrt{3}\cos\theta = -2$ and then $x = 3$

Hence, $x = 3$ and $\cos\left(\theta - \frac{\pi}{6}\right) = -1$, i.e., $\theta - \frac{\pi}{6}$

$= \pi, 3\pi$

6. a., c. $\sin^2 x - 2\sin x - 1 = 0$

$\Rightarrow (\sin x - 1)^2 = 2 \Rightarrow \sin x - 1 = \pm\sqrt{2} \Rightarrow \sin x = 1 - \sqrt{2}$

as $\sin x \not> 1$

There are two solutions in $[0, 2\pi]$ and two more in $[2\pi, 4\pi]$.

Thus $n = 4, 5$.

7. a., b., c. The given equation is $2(\sin x + \sin y) - 2\cos(x - y) = 3$

$\Rightarrow 2 \times 2 \sin\frac{x+y}{2}\cos\frac{x-y}{2} - 2\left[2\cos^2\frac{x-y}{2} - 1\right] = 3$

$\Rightarrow 4\cos^2\left(\frac{x-y}{2}\right) - 4\sin\left(\frac{x+y}{2}\right)\cos\left(\frac{x-y}{2}\right) + 1 = 0$

$\Rightarrow \cos\left(\frac{x-y}{2}\right)$

$= \frac{4\sin\left(\frac{x+y}{2}\right) \pm \sqrt{16\sin^2\left(\frac{x-y}{2}\right) - 16}}{8}$

$\therefore \sin^2\left(\frac{x+y}{2}\right) \geq 1 \Rightarrow \sin\frac{x-y}{2} = \pm 1$

since x and y are smallest and positive,

$\sin\frac{x+y}{2} = 1$ and $\frac{x+y}{2} = \frac{\pi}{2}$

i.e., $x + y = \pi$ (1)

Also $\cos\left(\frac{x-y}{2}\right)=\frac{1}{2}$

$\Rightarrow x-y=\frac{2\pi}{3}$ or $-\frac{2\pi}{3}$ (2)

From equations (1) and (2), we get

$\left(x=\frac{5\pi}{6}, y=\frac{\pi}{6}\right)$ or $\left(x=\frac{\pi}{6}, y=\frac{5\pi}{6}\right)$

8. a., d. $1-2x-x^2=\tan^2(x+y)+\cot^2(x+y)$

$\Rightarrow -(x+1)^2=[\tan(x+y)-\cot(x+y)]^2$

Now L.H.S. ≤ 0 and R.H.S. ≥ 0

$\Rightarrow -(x+1)^2=[\tan(x+y)-\cot(x+y)]^2=0$

$\Rightarrow x=-1$ and $\tan(x+y)=\cot(x+y)$

$\Rightarrow x=-1$ and $\tan^2(-1+y)=1$

$\Rightarrow x=-1$ and $-1+y=n\pi\pm\pi/4, n\in Z$

9. b., c. From $\tan x+\tan y=1$, we have

$\frac{\sin x}{\cos x}+\frac{\sin y}{\cos y}=1$

$\Rightarrow \sin x\cos y+\sin y\cos x=\cos x\cos y$

$\Rightarrow 2\sin(x+y)=2\cos x\cos y$

$\Rightarrow 2\sin(x+y)=\cos(x+y)+\cos(x-y)$

$\Rightarrow 2\sin(\pi/4)=\cos(\pi/4)+\cos(x-y)$,

$\Rightarrow \cos(x-y)=1/\sqrt{2}=\cos(\pi/4)$

$\Rightarrow x-y=2n\pi\pm\pi/4, \forall n\in Z$ (1)

Also we have $x+y=\pi/4$ (2)

From equations (1) and (2), we have

$x=n\pi+\pi/4$ and $y=-n\pi, \forall n\in Z$

or $x=n\pi$ and $y=-n\pi+\pi/4, \forall n\in Z$

10. a., c. $x+y=2\pi/3\Rightarrow y=2\pi/3-x$. Therefore,

$\sin x=2\sin(2\pi/3-x)$

$=2[(\sqrt{3}/2)\cos x+(1/2)\sin x]$

$=\sqrt{3}\cos x+\sin x$

$\Rightarrow \cos x=0\Rightarrow x=n\pi+\pi/2$

$\Rightarrow y=\frac{2\pi}{3}-n\pi-\frac{\pi}{2}=\frac{\pi}{6}-n\pi$

Hence, for $x\in[0,4\pi]$, $x=\pi/2, 3\pi/2, 5\pi/2, 7\pi/2$ and for $x\in[0,4\pi]$, $y=\pi/6, 7\pi/6, 13\pi/6, 19\pi/6$

11. a., d. $\tan x-\tan^2 x>0$

$\Rightarrow \tan x(\tan x-1)<0$

$\Rightarrow 0<\tan x<1$

$\Rightarrow 0<x<\pi/4$

$\Rightarrow n\pi<x<n\pi+\pi/4, n\in Z$ (generalising)

$|\sin x|<\frac{1}{2}$

$\Rightarrow -\frac{1}{2}<\sin x<\frac{1}{2}\Rightarrow -\pi/6<x<\pi/6$

$\Rightarrow -\pi/6+n\pi<x<\pi/6+n\pi, n\in Z$ (generalising)

Then the common values are $n\pi+\pi/6<x<n\pi+\pi/4$.

12. a., b., d. $\cos(x+\pi/3)+\cos x=a$

$\Rightarrow \frac{1}{2}\cos x-(\sqrt{3}/2)\sin x+\cos x=a$

$\Rightarrow (3/2)\cos x-(\sqrt{3}/2)\sin x=a$

$\Rightarrow -\sqrt{\left(\frac{9}{4}+\frac{3}{4}\right)}\le a\le\sqrt{\left(\frac{9}{4}+\frac{3}{4}\right)}$

$\Rightarrow -\sqrt{3}\le a\le\sqrt{3}$ (1)

Hence, there are three integral values of $a=-1, 0, 1$ whose sum is 0.

For $a=1$, the given equation is $(\sqrt{3}/2)\cos x-(1/2)\sin x=1/\sqrt{3}$

$\Rightarrow \cos(x+\pi/6)=1/\sqrt{3}$

$\Rightarrow x+\pi/6=2n\pi\pm\alpha$, where $\alpha=\cos^{-1}(1/\sqrt{3})$

$\Rightarrow x=2n\pi-\pi/6\pm\alpha$

Hence, solutions for $a=1$ in $[0,2\pi]$ are $\cos^{-1}(1/\sqrt{3})-\pi/6$, $11/6-\cos^{-1}(1/\sqrt{3})$.

13. a., b., c. The given inequality can be written as

$2^{\operatorname{cosec}^2 x}\sqrt{(y-1)^2+1}\le\sqrt{2}$ (1)

Since $\operatorname{cosec}^2 x\ge 1$ for all real x, therefore

$2^{\operatorname{cosec}^2 x}\ge 2$ (2)

Also $(y-1)^2+1\ge 1\Rightarrow\sqrt{(y-1)^2+1}\ge 1$ (3)

From (1) and (2), we get

$2^{\operatorname{cosec}^2 x}\sqrt{(y+1)^2+1}\ge 2$ (4)

From equations (1) and (4), equality holds only when

$2^{\operatorname{cosec}^2 x}=2$ and $\sqrt{(y-1)^2+1}=1$

$\Rightarrow \operatorname{cosec}^2 x=1$ and $(y-1)^2+1=1\Rightarrow\sin x=\pm 1$ and $y=1$

$\Rightarrow x=\frac{\pi}{2}, \frac{3\pi}{2}$ and $y=1$

Hence, the solution of the given inequality is, $x=\pi/2, 3\pi/2$ and $y=1$.

14. a., b., c. $\sin^2 x-a\sin x+b=0$ has only one solution in $(0,\pi)$

$\Rightarrow \sin x=1$ gives one solution and $\sin x=\alpha$ gives other solution such that $\alpha>1$ or $\alpha\le 0$

$\Rightarrow (\sin x-1)(\sin x-\alpha)$ is the same equation as $\sin^2 x-a\sin x+b=0$

$\Rightarrow 1+\alpha=a$ and $\alpha=b$

$\Rightarrow 1+b=a$ and $b>1$ or $b\le 0$

$\Rightarrow b\in(-\infty,0]\cup[1,\infty)$, and $a\in(-\infty,1]\cup[2,\infty)$

15. a., b. Given that the quadratic equation is an identity.

Therefore, $\operatorname{cosec}^2\theta=4$ and $\cot\theta=-\sqrt{3}$

$\Rightarrow \operatorname{cosec}\theta = 2$ or -2 and $\tan\theta = -\frac{1}{\sqrt{3}}$

$\theta = \frac{5\pi}{6}$ or $\frac{11\pi}{6}$

16. c., d. Abscissa corresponding to the vertex is given by

$x = \frac{1}{\sin\alpha} > 1$ is the vertex

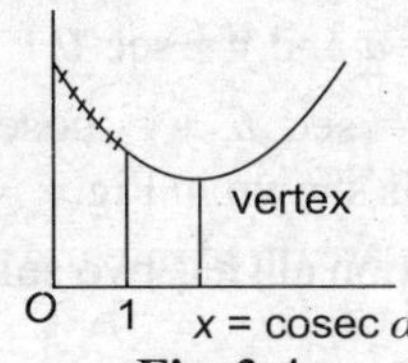

Fig. 2.4

The graph of $f(x) = (\sin\alpha)x^2 - 2x + b$, $\forall x \le 1$, is shown as

So minimum of $f(x) = (\sin\alpha)x^2 - 2x + b - 2$ must be greater than zero but minimum is at $x = 1$

i.e., $\sin\alpha - 2 + b - 2 \ge 0$

$\Rightarrow b \ge 4 - \sin\alpha$, $\alpha \in (0, \pi)$ $\Rightarrow b \ge 4$ as $\sin\alpha > 0$ in $(0, \pi)$

Reasoning Type

1. a. $(\sin x + \cos x)^{1 + \sin 2x} = 2$

$\Rightarrow (\sin x + \cos x)^{(\sin x + \cos x)^2} = 2$

Now we know that maximum value of $\sin x + \cos x$ is $\sqrt{2}$ which occurs at $x = \pi/4$

Also the given equation has roots only if $\sin x + \cos x = \sqrt{2}$

Hence there is only one solution for $0 \le x \le \pi$

Thus correct answer is *a*.

2. d. We know that $\sin^2 x \le 1$ and $\cos^2 y \le 1$, then $\sin^2 x + \cos^2 y \le 2$

Also $\sec^2 z \ge 1$, then $2\sec^2 z \ge 2$.

Hence the given equation is solvable only if $\sin^2 x + \cos^2 y = 2$ and $2\sec^2 z = 2$, for which $\sin^2 x$, $\cos^2 y$, $\sec^2 z = 1$

Then $\sin x$, $\cos y$, $\sec z = \pm 1$

Hence statement 1 is false and statement 2 is true.

3. a.

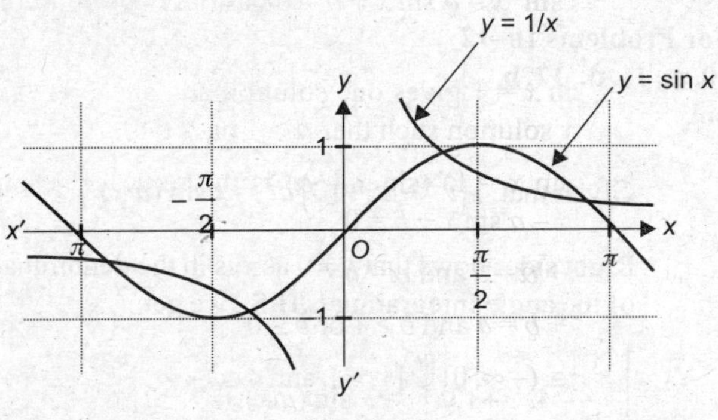

Fig. 2.5

Draw the graphs of $y = \sin x$ and $y = 1/x$ and verify.

4. d. When $n = 1$, we have interval $[0, \pi]$ which covers only first and second quadrant, in which $\sin x = -1/2$ is not possible. Hence, the number of solutions are zero. Also from $2(n-1)$ we have zero solution when $n = 1$.

For $n = 2$ we have interval $[0, 2\pi]$ which covers all the quadrant only once. Hence the number of solutions are two.

Also from $2(n-1)$ we have 2 solutions, when $n = 2$.

For $n = 3$ we have interval $[0, 3\pi]$, which covers third and fourth quadrants only once. Hence, the number of solutions are two. But from $2(n-1)$, we have four solutions which condradicts.

Hence, statement 1 is false, statement 2 is true.

5. a. $\sqrt{1 - \sin 2x} = \sin x$

$\Rightarrow \sqrt{(\sin x - \cos x)^2} = \sin x$

$\Rightarrow |\sin x - \cos x| = \sin x$

$\Rightarrow \cos x - \sin x = \sin x$

$\Rightarrow 2\sin x = \cos x$

$\Rightarrow \tan x = 2$, which is not possible for $x \in [0, \pi/4]$ as $\tan x \in [0, 1]$ for these values of x.

6. a. Draw the graphs of $y = |\sin x|$ and $y = |x|$ and verify.

7. d. Given $\tan 2x = 1$. So

$2x = n\pi + \frac{\pi}{4}$, $n \in Z$ (note that $\tan 4x$ is not defined for these values of x)

Hence, the given equation has no solution.

Statement 1 is false and statement 2 is true.

8. a. $\cos(\sin x) = \sin(\cos x)$

$\Rightarrow \cos(\sin x) = \cos(\pi/2 - \cos x)$

$\Rightarrow \sin x = 2n\pi \pm (\pi/2 - \cos x)$, $n \in Z$ (1)

Taking positive sign, $\sin x = 2n\pi + \pi/2 - \cos x$

or $(\cos x + \sin x) = \frac{1}{2}(4n + 1)\pi$

Now L.H.S. $\in [-\sqrt{2}, \sqrt{2}]$, hence

$-\sqrt{2} \le \frac{1}{2}(4n + 1)\pi\ \sqrt{2}$.

$\Rightarrow \frac{-2\sqrt{2} - \pi}{4\pi} \le n \le \frac{2\sqrt{2} - \pi}{4\pi}$, which is not satisfied by any integer n.

Similarly taking negative sign we have $\sin x - \cos x = (4n - 1)\pi/2$, which is also not satisfied for any integer n.

Hence, there is no solution.

9. b. Statement 1 is true.

Also statement 2 is true but does not explain statement 1.

Consider the equation $\sin x = x^3$.

Here $y = x^3$ is unbounded function but equation has only one real root.

10. a. Let $y = n|\sin x| = m|\cos x|$

The curves $y = n|\sin x|$ and $y = m|\cos x|$ intersect at four points in $[0, 2\pi]$.

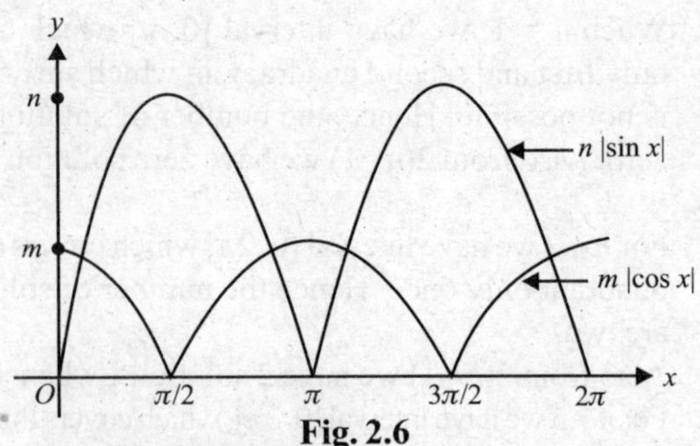

Fig. 2.6

Linked Comprehension Type

For Problems 1–3

1. b., 2. c., 3. a.

Sol.

1. b. $x^3 - (1 + \cos\theta + \sin\theta)x^2 + (\cos\theta\sin\theta + \cos\theta + \sin\theta)x - \sin\theta\cos\theta = 0$

Given cubic function is

$f(x) = (x-1)(x-\cos\theta)(x-\sin\theta)$. Therefore, roots are 1, $\sin\theta$, and $\cos\theta$. Hence,

$x_1^2 + x_2^2 + x_3^2 = 1 + \sin^2\theta + \cos^2 = 2$

2. c. Now if $1 = \sin\theta \Rightarrow \theta = \frac{\pi}{2}$

if $1 = \cos\theta \Rightarrow \theta = 0, 2\pi$

and if $\sin\theta = \cos\theta \Rightarrow \tan\theta = 1$

$\therefore \theta = \frac{\pi}{4}, \frac{5\pi}{4}$

$\therefore$ The number of values of θ in $[0, 2\pi]$ is 5.

3. a. Again the maximum possible difference between the two roots is 2

$1 - \sin\theta = 2$ when $\theta = 3\pi/2$

or $1 - \cos\theta = 2$ when $\theta = \pi$

For Problems 4–6

4. a., 5. c., 6. d.

Sol.

See Fig.2.7 for the solution

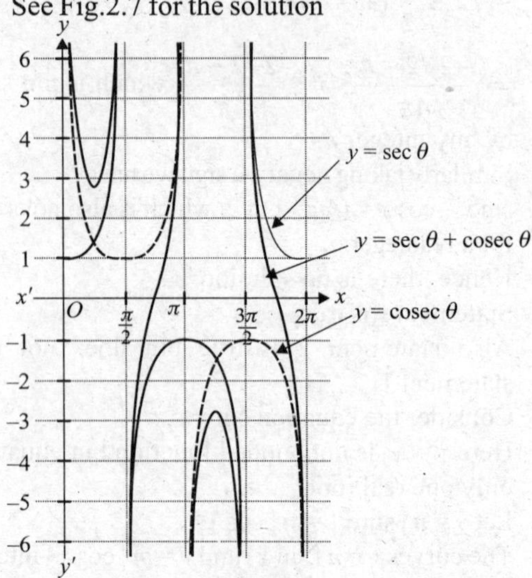

Fig. 2.7

Graph with thin solid line is of $y = \sec\theta$

Graph with dashed line is of $y = \text{cosec}\,\theta$

Graph with thick solid line is of $y = \sec\theta + \text{cosec}\,\theta$

$\sec\theta + \text{cosec}\,\theta = a$ (1)

has solution where

graphs of $y = a$ and $y = \sec\theta + \text{cosec}\,\theta$ intersect.

Graph of $y = \sec\theta$, $y = \text{cosec}\,\theta$ and $y = \sec\theta + \text{cosec}\,\theta$ is as shown in Fig. 2.5.

Clearly equation (1) has two solutions if

$-2\sqrt{2} < y < 2\sqrt{2}$.

Equation (1) has four solutions if $y \le -2\sqrt{2}$ or $y \ge 2\sqrt{2}$

In any case equation (1) has always two roots.

For Problems 7–9

7. a., 8. b., 9. d.

Sol. The given system is

$\sin x\cos 2y = (a^2-1)^2 + 1$,

$\cos x\sin 2y = a + 1$ (1)

Since the L.H.S. of the equations do not exceed 1, the given system may have solutions only for a's such that

$(a^2-1)^2 + 1 \le 1$ and $-1 \le a + 1 \le 1$ (2)

$(a^2-1)^2 + 1 \le 1 \Rightarrow (a^2-1)^2 \le 0 \Rightarrow (a^2-1)^2 = 0$

$\Rightarrow a = 1$.

But $a = 1$ does not satisfy the $\cos x\sin 2y = a + 1$

Thus $a = -1$ only and then we get

$\sin x\cos 2y = 1$

$\cos x\sin 2y = 0$ (3)

Adding and subtracting, we get

$\sin(x + 2y) = 1$

$\sin(x - 2y) = 1$ (4)

Obviously the system (4) yields

$x + 2y = 2n\pi + \pi/2$ and $x - 2y = 2m\pi + \pi/2$; $n, m \in I$

Hence, $x = (n+m)\pi + \pi/2$; $y = \frac{1}{2}(n-m)\pi$, $n, m \in I$

For Problems 10–12

10. a., 11. d., 12. b.

Sol.

Given that $\int_0^x \left(t^2 - 8t + 13\right) dt = x\sin(a/x)$ (1)

Right side shows that $x \ne 0$ as x is in the denominator of the angle Integrating L.H.S., we get

$\left[\frac{t^3}{3} - 4t^2 + 13t\right]_0^x = x\sin(a/x)$

or $(1/3)[x^3 - 12x^2 + 39x] = x\sin(a/x)$

or $\sin(a/x) = (1/3)[x^2 - 12x + 39]$ [$\because x \neq 0$]
$= (1/3)\{(x-6)^2 + 3\}$
$= (1/3)(x-6)^2 + 1$

But $\sin(a/x) \leq 1$, so $\sin(a/x) = 1$, which is possible only for $x = 6$.

Then we have $\sin(a/6) = 1$

or $a/6 = 2n\pi + \pi/2$ or $a = 12n\pi + 3\pi, n \in Z$

Hence, $x = 6, a = 12n\pi + 3\pi, n \in Z$

For $a \in [0, 100]$, there are exactly 3 value of $a = 3\pi, 15\pi, 27\pi$

$|y - \cos a| < x \Rightarrow |y + 1| < 6 \Rightarrow y \in [-7, 5]$

For Problems 13–15

13. a., 14. d., 15. b.

Sol. The given equations are

$x\cos^3 y + 3x\cos y \sin^2 y = 14$ and (1)

$x\sin^3 y + 3x\cos^2 y \sin y = 13$ (2)

Adding equations (1) and (2), we have

$x(\cos^3 y + 3\cos y \sin^2 y + 3\cos^2 y \sin y \sin^3 y) = 27$

$\Rightarrow x(\cos y + \sin y)^3 = 27$

$\Rightarrow x^{1/3}(\cos y + \sin y) = 3$ (3)

Subtracting (2) from (1), we have

$x(\cos^3 y + 3\cos y \sin^2 y - 3\cos^2 y \sin y - \sin^3 y) = 1$

$\Rightarrow x(\cos y - \sin y)^3 = 1$

$\Rightarrow x^{1/3}(\cos y - \sin y) = 1$ (4)

Dividing (3) and (4), we get

$\cos y + \sin y = 3\cos y - 3\sin y$

$\Rightarrow \tan y = 1/2$

Case I:

$\sin y = 1/\sqrt{5}$ and $\cos y = 2/\sqrt{5}$

$y = 2n\pi + \alpha$, where $0 < \alpha < \pi/2$ and $\sin\alpha = 1/\sqrt{5}$

i.e., y lies in the first quadrant

From equations (3), $x^{1/3}(3/\sqrt{5}) = 3$ or $x = 5/\sqrt{5}$

Case II:

$\sin y = -1/\sqrt{5}$ and $\cos y = -2/\sqrt{5}$

$y = 2n\pi + (\pi + \alpha)$, where $0 < \alpha < \pi/2$ and $\sin\alpha = -1/\sqrt{5}$

i.e., y lies in the 3rd quadrant

Therefore, from equation (3), $x^{1/3}(-3/\sqrt{5}) = 3$ or $x = -5\sqrt{5}$

Thus $\sin^2 y + 2\cos^2 y = 1/5 + 8/5 = 9/5$

Also there are exactly six values of $y \in [0, 6\pi]$, three in 1st quadrant and three in 3rd quadrant.

Matrix-Match Type

1. a → r; b → s; c → p; d → q.

a. $\cos^2 2x - \sin^2 x = 0$

$\Rightarrow \cos 3x \cos x = 0$

$\Rightarrow \cos 3x = 0$ or $\cos x = 0$

$\Rightarrow 3x = (2n-1)\dfrac{\pi}{2}$ or $x = (2n-1)\dfrac{\pi}{2}, n \in Z$

$\Rightarrow x = (2n-1)\dfrac{\pi}{6}$ or $x = (2n-1)\dfrac{\pi}{2}$

Hence, the general solution is $(2n-1)\dfrac{\pi}{6}$ as $(2n-1)\dfrac{\pi}{2}$ is contained in $(2n-1)\dfrac{\pi}{6}$.

b. $\cos x + \sqrt{3}\sin x = \sqrt{3}$

$\Rightarrow \dfrac{\cos x}{2} + \dfrac{\sqrt{3}}{2}\sin x = \dfrac{\sqrt{3}}{2}$

$\Rightarrow \cos\left(x - \dfrac{\pi}{3}\right) = \cos\dfrac{\pi}{6}$

$\Rightarrow x - \dfrac{\pi}{3} = 2n\pi \pm \dfrac{\pi}{6}, n \in Z$

$\Rightarrow x - \dfrac{\pi}{3} = 2n\pi + \dfrac{\pi}{6}$ or $2n\pi - \dfrac{\pi}{6}$

$\Rightarrow x = 2n\pi + \dfrac{\pi}{2}$ or $x = 2n\pi + \dfrac{\pi}{6}$

c. $\sqrt{3}\tan^2 x - (\sqrt{3}+1)\tan x + 1 = 0$

$\Rightarrow \sqrt{3}\tan x(\tan x - 1) - (\tan x - 1) = 0$

$\Rightarrow (\tan x - 1)(\sqrt{3}\tan x - 1) = 0$

$\tan x = 1 \Rightarrow x = n\pi + \dfrac{\pi}{4}, n \in Z$

or $\tan x = \dfrac{1}{\sqrt{3}} \Rightarrow x = n\pi + \dfrac{\pi}{6}, n \in Z$

d. $\tan 3x - \tan 2x - \tan x = 0$

or $\tan x \tan 2x \tan 3x = 0$

$x = n\pi$ or $\dfrac{n\pi}{2}$ (rejected) or $\dfrac{n\pi}{3}$

$\Rightarrow$ General solution $\dfrac{n\pi}{3}$ as $n\pi$ is contained in $\dfrac{n\pi}{3}$.

2. a → q; b → s; c → s; d → p.

a. Here $x^3 + (x+2)^2 + 2\sin x = 4$.

Clearly, $x = 0$ satisfies the equation.

If $0 < x \leq \pi, x^3 + (x+2)^2 + 2\sin x > 4$

If $\pi < x \leq 2\pi, x^3 + (x+2)^2 + 2\sin x > 27 + 25 - 2$

So $x = 0$ is the only solution.

b. Here $\dfrac{1}{2}\sin(2e^x) = \dfrac{1}{4}\left(2^x + 2^{-x}\right) \geq \dfrac{1}{4}2 = \dfrac{1}{2}$

[$\because$ A.M.≥G.M.]

$\Rightarrow \sin(2e^x) \geq 1 \Rightarrow \sin(2e^x) = 1$

But equality can hold when $2^x = 2^{-x} = 1$, i.e., $x = 0$.

Then $\sin(2.e^0) = 1$, which is not true.
Hence no solution.

c. $\sin 2x + \cos 4x = 2$
$\Rightarrow \sin 2x = 1, \cos 4x = 1$. Therefore,
$1 - 2\sin^2 2x = 1$ or $1 - 2 = 1$ (absurd)

d. The given solution is $|\sin x| = x/30$
The solution of this equation are the point of intersections of the curves
$y = |\sin x|$ and $y = x/30$.

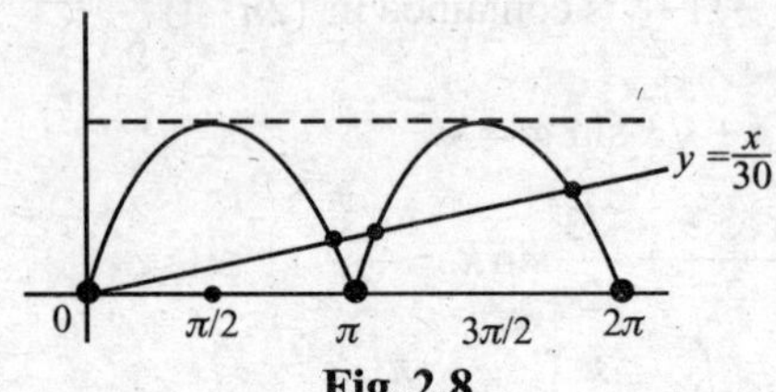

Fig. 2.8

Since there are four points of intersection in $0 \le x \le 2\pi$, the equation has four solutions

3. a → q; b → s; c → p; d → r.

a. $5\sin\theta + 3(\sin\theta\cos\alpha - \cos\theta\sin\alpha)$
$= (5 + 3\cos\alpha)\sin\theta - 3\sin\alpha\cos\theta$

$\Rightarrow \max_{\theta \in R} \{5\sin\theta + 3\sin(\theta - \alpha)\}$

$= \sqrt{(5+3\cos\alpha)^2 + 9\sin^2\alpha}$

$= \sqrt{34 + 30\cos\alpha}$

The given equation is $34 + 30\cos\alpha = 49$.

$\Rightarrow \cos\alpha = 1/2 \Rightarrow \alpha = 2n\pi \pm \frac{\pi}{3}, n \in Z$

b. $(\cos x)^{\sin^2 x - 3\sin x + 2} = 1$
$\Rightarrow \sin^2 x - 3\sin x + 2 = 0$
$\Rightarrow (\sin x - 2)(\sin x - 1) = 0$
$\Rightarrow \sin x = 1$
But this does not satisfy the equation because $0^\circ = 1$ is absurd.

c. $\sqrt{(\sin x)} + 2^{1/4}\cos x = 0$ (1)
$\Rightarrow \sqrt{(\sin x)} > 0$ and so $\cos x < 0$,
i.e., x lies in second quadrant or third quadrant.
Equation (1) can be rewritten as

$2^{1/4}\cos x = -\sqrt{(\sin x)}$.

Squaring $\sqrt{2}\cos^2 x = \sin x$
$\Rightarrow \sqrt{2}\sin^2 x + \sin x - \sqrt{2} = 0$
$\Rightarrow (\sqrt{2}\sin x - 1)(\sin x + \sqrt{2}) = 0$
$\sin x \ne -\sqrt{2}$ $\therefore$ $\sin x = 1/\sqrt{2}$ (2)
From equation (2) $\cos x < 0$ so $x = 3\pi/4 \in$ second quadrant.

and so the general value of x is given by $x = 2n\pi + 3\pi/4, n \in Z$

d. $\log_5 \tan x = (\log_5 4)(\log_4(3\sin x))$
$\Rightarrow \log_5 \tan x = \log_5(3\sin x)$ (1)
Since $\log x$ is real when $x > 0$,
We have $\tan x > 0$, $\sin x > 0$, So
x lies in first quadrant. Now equation (1) gives
$\tan x = 3\sin x$ or $\cos x = 1/3$
Hence, $x = 2n\pi + \cos^{-1}(1/3)$

Integer Type

1. (8) $16^{\sin^2 x} + 16^{1-\sin^2 x} = 10$ (1)

If $16^{\sin^2 x} = t$, then

$t + \frac{16}{t} = 10$

Thus equation (1) becomes
$t^2 - 10t + 16 = 0$
$\Rightarrow t = 2, 8$

$\Rightarrow 16^{\sin^2 x} = 16^{1/4}$ or $16^{3/4}$

$\Rightarrow \sin x = \pm\frac{1}{2}, \pm\frac{\sqrt{3}}{2}$

Now $\sin x = \frac{1}{2} \Rightarrow x = \frac{\pi}{6}, \frac{5\pi}{6}$

$\sin x = -\frac{1}{2} \Rightarrow x = \frac{7\pi}{6}$ or $\frac{11\pi}{6}$

$\sin x = \frac{\sqrt{3}}{2} \Rightarrow x = \frac{\pi}{3}, \frac{2\pi}{3}$

$\sin x = -\frac{\sqrt{3}}{2} \Rightarrow x = \frac{4\pi}{3}, \frac{5\pi}{3}$

Hence, there will be eight solutions in all.

2. (0) Putting $e^{\sin x} = t$, we get
$t^2 - 4t - 1 = 0$
$\Rightarrow t = e^{\sin x} = 2 \pm \sqrt{5}$
Now $\sin x \in [-1, 1]$
$\Rightarrow e^{\sin x} \in [e^{-1}, e^1]$ and $2 \pm \sqrt{5} \notin [e^{-1}, e^1]$
Hence, there does not exist any solution.

3. (5) We can write the given equation as
$a\sin x + (1 - 2\sin^2 x) = 2a - 7$
$\Rightarrow 2\sin^2 x - a\sin x + 2a - 8 = 0$

$\Rightarrow \sin x = \frac{a \pm \sqrt{a^2 - 8(2a-8)}}{4} = \frac{a \pm (a-8)}{4}$

$\Rightarrow \sin x = (a - 4)/2$ [$\because$ $\sin x = 2$ is not possible]

The equation has a solution if $-1 \le (a-4)/2 \le 1$

$\Rightarrow -2 \le (a-4) \le 2$

$\Rightarrow 2 \le a \le 6$

4. **(2)** $\sin x(\sin x + \cos x) = n$

$\Rightarrow \sin^2 x + \sin x \cos x = n$

$$\Rightarrow \frac{1-\cos 2x}{2} + \frac{\sin 2x}{2} = n$$

$\Rightarrow \sin 2x - \cos 2x = 2n - 1$

$\Rightarrow -\sqrt{2} \le 2n - 1 \le \sqrt{2}$

$$\Rightarrow \frac{1-\sqrt{2}}{2} \le n \le \frac{1+\sqrt{2}}{2}$$

$\Rightarrow n = 0, 1$

5. **(8)** We know that $a\cos\theta + b\sin\theta = c$ has a solution only if $|c| \le \sqrt{a^2+b^2}$.

Then for the given equation, we must have

$|2k+1| \le \sqrt{74}$

$\Rightarrow -\sqrt{74} < 2k+1 < \sqrt{74}$

$\Rightarrow -8 < 2k+1 < 8$ [For integral solutions]

$\Rightarrow k = -4, -3, -2, -1, 0, 1, 2, 3$

Thus eight values will satisfy the above inequality.

6. **(3)** $\sin x \cos y = 1$

$\Rightarrow \sin x = 1, \cos y = 1$ or $\sin x = -1, \cos y = -1$

If $\sin x = 1, \cos y = 1$, then $x = \pi/2, y = 0, 2\pi$.

If $\sin x = -1, \cos y = -1$, then $x = 3\pi/2, y = \pi$.

Thus the possible ordered pairs are $(\pi/2, 0), (\pi/2, 2\pi), (3\pi/2, \pi)$.

7. $1 + \sin x \sin^2 \frac{x}{2} = 0$

$\Rightarrow 2 + 2\sin x \sin^2 \frac{x}{2} = 0$

$\Rightarrow 2 + \sin x(1 - \cos x) = 0$

$\Rightarrow 4 + 2\sin x - \sin 2x = 0$

$\Rightarrow \sin 2x = 2\sin x + 4$, which is not possible for any value of x

As L.H.S. has the maximum value 1 and R.H.S. has the minimum value 2.

Hence, there is no solution.

8. **(7)** Let $f(x) = \sin x$ and $g(x) = x/10$.

Also we know that

$-1 \le \sin x \le 1$

$$\Rightarrow -1 \le \frac{x}{10} \le 1$$

$\Rightarrow -10 \le x \le 10$

Thus the sketch of both curves when $x \in [-10, 10]$ is as follows.

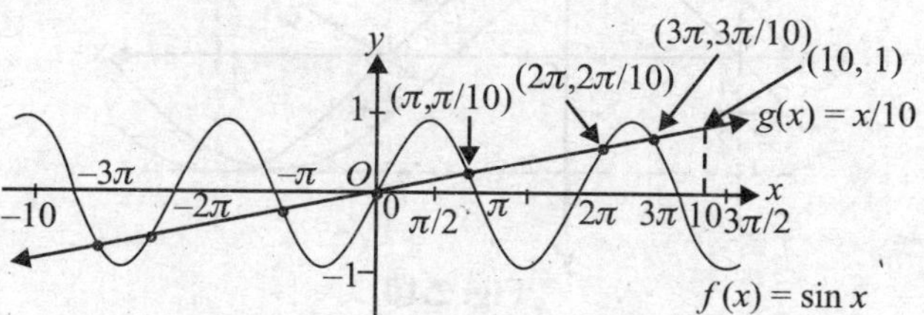

Fig. 2.9

From the above figure, it can be seen that $f(x) = \sin x$ and $g(x) = x/10$ intersect at seven points. So the number of solutions is 7.

9. **(5)** The given equation can be re-written as

$2\{\sin[2x + (\pi/6)] + \sin(\pi/2)\} = a^2 + \sqrt{3}\sin 2x - \cos 2x$

$\Rightarrow \cos 2x = (a^2 - 2)/2$

$\Rightarrow 2\cos^2 x = a^2/2$ or $\cos^2 x = (a/2)^2$ (1)

$\Rightarrow a^2 \le 4$ or $-2 \le a \le 2$ (2)

For $a = 0$, the given equation is reduced to

$\cos x = 0$, i.e., $x = n\pi + (\pi/2)$ (3)

Hence, $-2 \le a \le 2$

10. **(6)** $\sin x + \sin y = \sin(x+y)$, $|x| + |y| = 1$

$$\Rightarrow 2\sin\frac{x+y}{2}\left(\cos\frac{x-y}{2} - \cos\frac{x+y}{2}\right) = 0$$

$$\Rightarrow 4\sin\frac{x+y}{2}\sin\frac{x}{2}\sin\frac{y}{2} = 0$$

(i) $\sin\frac{x+y}{2} = 0$

$\Rightarrow x + y = 2n\pi,\ n \in Z$

$\Rightarrow x + y = 0$

(ii) $\sin\frac{x}{2} = 0$

$\Rightarrow x = 2m\pi,\ m \in Z$

$\Rightarrow x = 0$

(iii) $\sin\frac{y}{2} = 0$

$\Rightarrow y = 2p\pi,\ p \in Z$

$\Rightarrow y = 0$

Also $|x| + |y| = 1$

If $x = 0$, then $|y| = 1 \Rightarrow y = \pm 1$.

If $y = 0$, then $|x| = 1 \Rightarrow x = \pm 1$.

If $y = -x$, then $|x| + |-x| = 2 \Rightarrow x = \pm(1/2)$ and $y = \mp(1/2)$.

Hence, solutions are (0, 1), (0, –1), (1, 0), (–1, 0), (1/2, – 1/2), and (– 1/2, 1/2).

11. The given equation is possible if $\sin(1-x) \ge 0$ and $\cos x \ge 0$.

Squaring, we get $\sin(1-x) = \cos x$.

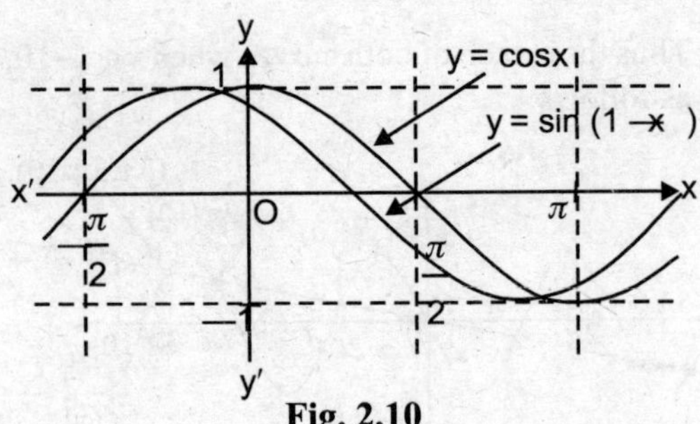

Fig. 2.10

$\Rightarrow \cos\left[\frac{\pi}{2}-(1-x)\right]=\cos x$

$\Rightarrow \frac{\pi}{2}-1+x=2n\pi \pm x, n\in Z$

$\Rightarrow x=\dfrac{2n\pi-\frac{\pi}{2}+1}{2}$

For $n=2$, $x=(7\pi/4)+(1/2)$ is the smallest positive root of the given equation.

12. (1) The given equation is

$\sin x+\sin\left(\frac{\pi}{8}\sqrt{(1-\cos 2x)^2+\sin^2 2x}\right)=0$

Now $\sin\left(\frac{\pi}{8}\sqrt{(1-\cos 2x)^2+\sin^2 2x}\right)$

$=\sin\frac{\pi}{8}\sqrt{2-2\cos 2x}$

$=\sin\left(\frac{\pi}{4}|\sin x|\right)$

So $\sin x+\sin\left(\frac{\pi}{4}|\sin x|\right)=0$ (1)

The equation has a solution only if $\sin x\le 0$.
The graph of $f(x)=\sin x\le 0$ is shown here.

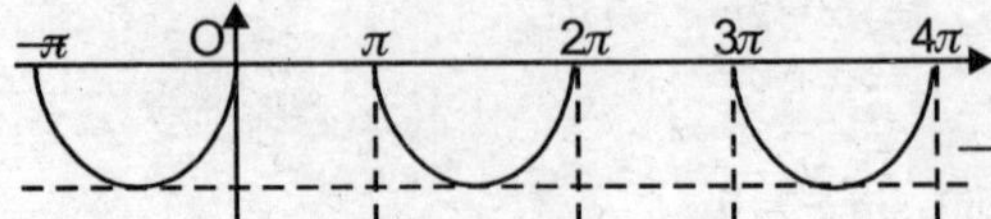

Fig. 2.11

The graph of $y=\sin[(\pi/4)|\sin x|]$ is given below.

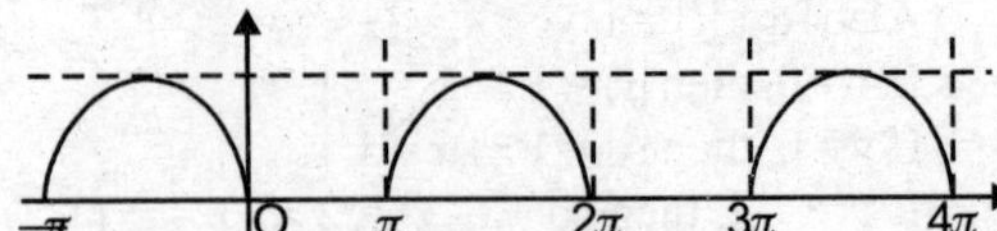

Fig. 2.12

Hence, the required solution of (1) is 3π.

13. (4) Given $\cos(\alpha-\beta)=1$ and $\cos(\alpha+\beta)=1/e$, where $\alpha,\beta\in[-\pi,\pi]$.

Now $\cos(\alpha-\beta)=1$

$\Rightarrow \alpha-\beta=0\Rightarrow \alpha=\beta$. Therefore,

$\cos(\alpha+\beta)=1/e$

$\Rightarrow \cos 2\alpha=1/e$

Since $0<1/e<1$ and $2\alpha\in[-2\pi,2\pi]$, there will be two values of 2α satisfying $\cos 2\alpha=1/e$ in $[0,2\pi]$ and two in $[-2\pi,0]$.

Therefore, there will be four values of α in $[-\pi,\pi]$ and four corresponding values of β.

Hence, there are four sets of (α,β).

14. (2) The equation is

$$\frac{\left(\frac{\sqrt{3}-1}{2\sqrt{2}}\right)}{\sin x}+\frac{\left(\frac{\sqrt{3}+1}{2\sqrt{2}}\right)}{\cos x}=2$$

$$\Rightarrow \frac{\sin\frac{\pi}{12}}{\sin x}+\frac{\cos\frac{\pi}{12}}{\cos x}=2$$

$\Rightarrow \sin\left(\frac{\pi}{12}+x\right)=\sin 2x$. Therefore,

$\frac{\pi}{12}+x=2x$ or $\frac{\pi}{12}+x=\pi-2x$

$\Rightarrow x=\frac{\pi}{12}$ or $\frac{11\pi}{36}$

15. (4) $\pi\log_3\left(\frac{1}{x}\right)=k\pi, k\in I$

$\log_3\left(\frac{1}{x}\right)=k\Rightarrow x=3^{-k}$

Possible values of k are $-1, 0, 1, 2, 3, \ldots$

$S=3+1+\frac{1}{3}+\frac{1}{3^2}+\frac{1}{3^3}+\cdots\infty$

$=\frac{3}{1-(1/3)}=\frac{9}{2}$

16. (7) $y=(3)^{3\cos x}+(3)^{4\sin x}$

Now using $AM\ge GM$

$\frac{3^{3\cos x}+3^{4\sin x}}{2}\ge\left(3^{3\cos x}\times 3^{4\sin x}\right)^{1/2}$

$\Rightarrow 3^{3\cos x}+3^{4\sin x}\ge 2\sqrt{3^{3\cos x+4\sin x}}\ge 2\sqrt{3^{-5}}$

$(-5\le 3\cos x+4\sin x\le 5)$

$\therefore 3^{3\cos x}+3^{4\sin x}\ge 2\sqrt{3^{-5}}$

$=\frac{2}{3^{5/2}}=\frac{2}{3\times 3\times\sqrt{3}}=\frac{2}{\sqrt{234}}=\sqrt{\frac{4}{243}}$

17. (1) Here $4^x+4^{1-x}\ge 4$

Also $3+\cos 3z\ge 2$

Now LHS = RHS is possible only when

$1+\tan^2 2y=1$

$\Rightarrow \tan^2 2y=0$

$\Rightarrow \tan 2y=0$

$\Rightarrow y=\pi/2$

CHAPTER 3

Inverse Trigonometric Functions

Objective Type

Each question has four choices a, b, c, and d, out of which *only one* answer is correct. Find the correct answer.

1. The principal value of $\sin^{-1}(\sin 10)$ is
   a. 10 b. $10-3\pi$
   c. $3\pi-10$ d. None of these

2. $\cos^{-1}\left(\cos\left(\frac{5\pi}{4}\right)\right)$ is given by
   a. $\frac{5\pi}{4}$ b. $\frac{3\pi}{4}$
   c. $\frac{-\pi}{4}$ d. None of these

3. The value of $\sin^{-1}(\sin 12) + \cos^{-1}(\cos 12)$ is
   a. zero b. $24-2\pi$
   c. $4\pi-24$ d. None of these

4. The value of the expression $\sin^{-1}\left(\sin\frac{22\pi}{7}\right)$ $+\cos^{-1}\left(\cos\frac{5\pi}{3}\right)+\tan^{-1}\left(\tan\frac{5\pi}{7}\right)+\sin^{-1}(\cos 2)$ is
   a. $\frac{17\pi}{42}-2$ b. -2
   c. $\frac{-\pi}{21}-2$ d. None of these

5. The value of $\sin^{-1}(\cos(\cos^{-1}(\cos x) + \sin^{-1}(\sin x)))$ where $x \in \left(\frac{\pi}{2}, \pi\right)$ is
   a. $\frac{\pi}{2}$ b. $-\pi$
   c. π d. $-\frac{\pi}{2}$

6. $\cos^{-1}(\cos(2\cot^{-1}(\sqrt{2}-1)))$ is equal to
   a. $\sqrt{2}-1$ b. $\frac{\pi}{4}$
   c. $\frac{3\pi}{4}$ d. None of these

7. The value of $\sin^{-1}\left(\cot\left(\sin^{-1}\sqrt{\frac{2-\sqrt{3}}{4}}+\cos^{-1}\frac{\sqrt{12}}{4}+\sec^{-1}\sqrt{2}\right)\right)$ is
   a. 0 b. $\frac{\pi}{2}$
   c. $\frac{\pi}{3}$ d. None of these

8. The value of $\cos^{-1}\sqrt{\frac{2}{3}}-\cos^{-1}\frac{\sqrt{6}+1}{2\sqrt{3}}$ is
   a. $\frac{\pi}{3}$ b. $\frac{\pi}{4}$
   c. $\frac{\pi}{2}$ d. $\frac{\pi}{6}$

9. The value of $\cos\left(\frac{1}{2}\cos^{-1}\frac{1}{8}\right)$ is

a. $\frac{3}{4}$ b. $-\frac{3}{4}$

c. $\frac{1}{16}$ d. $\frac{1}{4}$

10. If $\tan(x+y)=33$ and $x=\tan^{-1}3$, then y will be

a. 0.3 b. $\tan^{-1}(1.3)$

c. $\tan^{-1}(0.3)$ d. $\tan^{-1}\left(\frac{1}{18}\right)$

11. The value of $\tan\left(\frac{1}{2}\cos^{-1}\left(\frac{\sqrt{5}}{3}\right)\right)$ is

a. $\frac{3+\sqrt{5}}{2}$ b. $3+\sqrt{5}$

c. $\frac{1}{2}(3-\sqrt{5})$ d. None of these

12. $\tan^{-1}\left[\frac{\cos x}{1+\sin x}\right]$ is equal to

a. $\frac{\pi}{4}-\frac{x}{2}$, for $x\in\left(-\frac{\pi}{2},\frac{3\pi}{2}\right)$

b. $\frac{\pi}{4}-\frac{x}{2}$, for $x\in\left(-\frac{\pi}{2},\frac{\pi}{2}\right)$

c. $\frac{\pi}{4}-\frac{x}{2}$, for $x\in\left(\frac{3\pi}{2},\frac{5\pi}{2}\right)$

d. $\frac{\pi}{4}-\frac{x}{2}$, for $x\in\left(-\frac{3\pi}{2},-\frac{\pi}{2}\right)$

13. $\tan^{-1}\left[\frac{\sqrt{1+x^2}+\sqrt{1-x^2}}{\sqrt{1+x^2}-\sqrt{1-x^2}}\right]$ is equal to

a. $\frac{\pi}{4}+\frac{1}{2}\cos^{-1}x^2$ b. $\frac{\pi}{4}+\cos^{-1}x^2$

c. $\frac{\pi}{4}+\frac{1}{2}\cos^{-1}x$ d. $\frac{\pi}{4}-\frac{1}{2}\cos^{-1}x^2$

14. $4\tan^{-1}\frac{1}{5}-\tan^{-1}\frac{1}{239}$ is equal to

a. π b. $\pi/2$

c. $\pi/3$ d. $\pi/4$

15. $\sin^{-1}\left[x\sqrt{1-x}-\sqrt{x}\sqrt{1-x^2}\right]$ is equal to

a. $\sin^{-1}x+\sin^{-1}\sqrt{x}$ b. $\sin^{-1}x-\sin^{-1}\sqrt{x}$

c. $\sin^{-1}\sqrt{x}-\sin^{-1}x$ d. None of these

16. $\tan^{-1}\left(\frac{x}{y}\right)-\tan^{-1}\left(\frac{x-y}{x+y}\right)$ is

a. $\frac{\pi}{2}$ b. $\frac{\pi}{3}$

c. $\frac{\pi}{4}$ d. $\frac{\pi}{4}$ or $-\frac{3\pi}{4}$

17. If $\tan^{-1}\frac{a+x}{a}+\tan^{-1}\frac{a-x}{a}=\frac{\pi}{6}$, then $x^2=$

a. $2\sqrt{3}a$ b. $\sqrt{3}a$

c. $2\sqrt{3}a^2$ d. None of these

18. If $\cot^{-1}\frac{n}{\pi}>\frac{\pi}{6}$, $n\in N$, then the maximum value of n is

a. 6 b. 7

c. 5 d. None of these

19. $\sec^2(\tan^{-1}2)+\text{cosec}^2(\cot^{-1}3)$ is equal to

a. 5 b. 13

c. 15 d. 6

20. If $A=\tan^{-1}\left(\frac{x\sqrt{3}}{2K-x}\right)$ and $B=\tan^{-1}\left(\frac{2x-K}{K\sqrt{3}}\right)$, then the value of $A-B$ is

a. 0° b. 45°

c. 60° d. 30°

21. The value of $\sec\left[\tan^{-1}\frac{b+a}{b-a}-\tan^{-1}\frac{a}{b}\right]$ is

a. 2 b. $\sqrt{2}$

c. 4 d. 1

22. If a $\sin^{-1}x-$ b $\cos^{-1}x=c$, then $a\sin^{-1}x+b\cos^{-1}x$ is equal to

a. 0 b. $\frac{\pi ab+c(b-a)}{a+b}$

c. $\frac{\pi}{2}$ d. $\frac{\pi ab+c(a-b)}{a+b}$

23. The number of solutions of the equation

$$\cos^{-1}\left(\frac{1+x^2}{2x}\right)-\cos^{-1}x=\frac{\pi}{2}+\sin^{-1}x$$ is given by

a. 0 b. 1

c. 2 d. 3

24. The sum of the solutions of the equation

$$2\sin^{-1}\sqrt{x^2+x+1}+\cos^{-1}\sqrt{x^2+x}=\frac{3\pi}{2}$$ is

a. 0 b. -1

c. 1 d. 2

25. The number of solutions of the equation $\tan^{-1}(1+x) + \tan^{-1}(1+x) = \frac{\pi}{2}$ is

a. 2 b. 3

c. 1 d. 0

26. The number of solutions of the equation $\sin^{-1} x + \sin^{-1}(1-x) = \cos^{-1} x$ is

a. 1 b. 0

c. 2 d. None of these

27. If $\tan^{-1}\frac{1-x}{1+x} = \frac{1}{2}\tan^{-1} x$, then x is equal to

a. 1 b. $\sqrt{3}$

c. $\frac{1}{\sqrt{3}}$ d. None of these

28. For the equation $\cos^{-1} x + \cos^{-1} 2x + \pi = 0$, the number of real solutions is

a. 1 b. 2

c. 0 d. ∞

29. The value of a for which $ax^2 + \sin^{-1}(x^2 - 2x + 2) + \cos^{-1}(x^2 - 2x + 2) = 0$ has a real solution, is

a. $\frac{\pi}{2}$ b. $-\frac{\pi}{2}$

c. $\frac{2}{\pi}$ d. $-\frac{2}{\pi}$

30. The number of real solutions of the equation

$\tan^{-1}\sqrt{x^2 - 3x + 2} + \cos^{-1}\sqrt{4x - x^2 - 3} = \pi$ is

a. one b. two

c. zero d. infinite

31. If $3\tan^{-1}\left(\frac{1}{2+\sqrt{3}}\right) - \tan^{-1}\frac{1}{x} = \tan^{-1}\frac{1}{3}$, then x is equal to

a. 1 b. 2

c. 3 d. $\sqrt{2}$

32. If $\tan^{-1} x + 2\cot^{-1} x = \frac{2\pi}{3}$, then x is equal to

a. $\frac{\sqrt{3}-1}{\sqrt{3}+1}$ b. 3

c. $\sqrt{3}$ d. $\sqrt{2}$

33. If $\sin^{-1}\left(\frac{2x}{1+x^2}\right) - 4\cos^{-1}\left(\frac{1-x^2}{1+x^2}\right) + 2\tan^{-1}\left(\frac{2x}{1-x^2}\right) = \frac{\pi}{3}$, where $|x| < 1$, then x is equal to

a. $\frac{1}{\sqrt{3}}$ b. $-\frac{1}{\sqrt{3}}$

c. $\sqrt{3}$ d. $-\frac{\sqrt{3}}{4}$

34. If $\sin^{-1}\left(\frac{5}{x}\right) + \sin^{-1}\frac{12}{x} = \frac{\pi}{2}$, then x is equal to

a. $\frac{7}{13}$ b. $\frac{4}{3}$

c. 13 d. $\frac{13}{7}$

35. If $\sin^{-1}(x-1) + \cos^{-1}(x-3) + \tan^{-1}\left(\frac{x}{2-x^2}\right) = \cos^{-1} k + \pi$, then the value of k is

a. 1 b. $-\frac{1}{\sqrt{2}}$

c. $\frac{1}{\sqrt{2}}$ d. None of these

36. If $\cot^{-1} x + \cot^{-1} y + \cot^{-1} z = \frac{\pi}{2}$, $x, y, z > 0$ and $xy < 1$, then $x + y + z$ is also equal to

a. $\frac{1}{x} + \frac{1}{y} + \frac{1}{z}$ b. xyz

c. $xy + yz + zx$ d. None of these

37. If $\sin^{-1} x + \sin^{-1} y + \sin^{-1} z = \pi$, then

a. $x^2 + y^2 + z^2 + xyz = 0$ b. $x^2 + y^2 + z^2 + 2xyz = 0$

c. $x^2 + y^2 + z^2 + xyz = 1$ d. $x^2 + y^2 + z^2 + 2xyz = 1$

38. If $\tan^{-1} x + \tan^{-1} y + \tan^{-1} z = \frac{\pi}{2}$, then

a. $x + y + z - xyz = 0$ b. $x + y + z + xyz = 0$

c. $xy + yz + zx + 1 = 0$ d. $xy + yz + zx - 1 = 0$

39. If $\cos^{-1}\sqrt{p} + \cos^{-1}\sqrt{1-p} + \cos^{-1}\sqrt{1-q} = \frac{3\pi}{4}$, then the value of q is

a. 1 b. $\frac{1}{\sqrt{2}}$

c. $\frac{1}{3}$ d. $\frac{1}{2}$

40. If $\sin^{-1} a + \sin^{-1} b + \sin^{-1} c = \pi$, then the value of $a\sqrt{(1-a^2)} + b\sqrt{(1-b^2)} + c\sqrt{(1-c^2)}$ will be

a. $2abc$ b. abc

c. $\frac{1}{2}abc$ d. $\frac{1}{3}abc$

41. If $\sin^{-1} x + \sin^{-1} y + \sin^{-1} z = \pi$, then $x^4 + y^4 + z^4 + 4x^2y^2z^2 = K(x^2y^2 + y^2z^2 + z^2x^2)$, where K is equal to

a. 1 b. 2
c. 4 d. None of these

42. If $\cos^{-1} x - \cos^{-1} \frac{y}{2} = \alpha$ then $4x^2 - 4xy \cos \alpha + y^2$ is equal to
a. 4 b. $2 \sin^2 \alpha$
c. $-4 \sin^2 \alpha$ d. $4 \sin^2 \alpha$

43. The value of x which satisfies the equation
$2 \tan^{-1} 2x = \sin^{-1} \frac{4x}{1+4x^2}$ is valid in the interval
a. $\left[\frac{1}{2}, \infty\right)$ b. $\left(-\infty, -\frac{1}{2}\right]$
c. $[-1, 1]$ d. $\left[-\frac{1}{2}, \frac{1}{2}\right]$

44. If $x \in [-1, 0)$, then $\cos^{-1}(2x^2 - 1) - 2\sin^{-1} x$ is equal to
a. $-\frac{\pi}{2}$ b. π
c. $\frac{3\pi}{2}$ d. -2π

45. If $2 \sin^{-1} x = \sin^{-1}(2x\sqrt{1-x^2})$, then
a. $[-1, 1]$ b. $\left[-\frac{1}{\sqrt{2}}, 1\right]$
c. $\left[-\frac{1}{\sqrt{2}}, \frac{1}{\sqrt{2}}\right]$ d. None of these

46. If $x_1 = 2\tan^{-1}\left(\frac{1+x}{1-x}\right)$, $x_2 = \sin^{-1}\left(\frac{1-x^2}{1+x^2}\right)$ where $x \in (0, 1)$, then $x_1 + x_2$ is
a. 0 b. 2π
c. π d. None of these

47. The value of $\sin(2 \sin^{-1}(0.8))$ is
a. $\sin 1.2°$ b. $\sin 1.6°$
c. 0.48 d. 0.96

48. The value of $\tan(\sin^{-1}(\cos(\sin^{-1} x)))$ $\tan(\cos^{-1}(\sin(\cos^{-1} x)))$ where $x \in (0, \pi/2)$ is equal to
a. 0 b. 1
c. -1 d. None of these

49. If $\sin^{-1}\left(\frac{2a}{1+a^2}\right) + \sin^{-1}\left(\frac{2b}{1+b^2}\right) = 2\tan^{-1} x$, then x is equal to $[a, b \in (0, 1)]$
a. $\frac{a-b}{1+ab}$ b. $\frac{b}{1+ab}$
c. $\frac{b}{1-ab}$ d. $\frac{a+b}{1-ab}$

50. If x takes negative permissible values, then $\sin^{-1} x$ is equal to
a. $\cos^{-1}\sqrt{1-x^2}$ b. $-\cos^{-1}\sqrt{1-x^2}$
c. $\cos^{-1}\sqrt{x^2-1}$ d. $\pi - \cos^{-1}\sqrt{1-x^2}$

51. If $x^2 + y^2 + z^2 = r^2$, then $\tan^{-1}\left(\frac{xy}{zr}\right) + \tan^{-1}\left(\frac{yz}{xr}\right)$ $+ \tan^{-1}\left(\frac{xz}{yr}\right)$ is equal to
a. π b. $\frac{\pi}{2}$
c. 0 d. None of these

52. If $f(x) = \sin^{-1}\left(\frac{\sqrt{3}}{2}x - \frac{1}{2}\sqrt{1-x^2}\right)$, $-\frac{1}{2} \le x \le 1$, then $f(x)$ is equal to
a. $\sin^{-1}\left(\frac{1}{2}\right) - \sin^{-1}(x)$ b. $\sin^{-1} x - \frac{\pi}{6}$
c. $\sin^{-1} x + \frac{\pi}{6}$ d. None of these

53. If $x \in (0, 1)$, then the value of $\tan^{-1}\left(\frac{1-x^2}{2x}\right)$ $+ \cos^{-1}\left(\frac{1-x^2}{1+x^2}\right)$ is equal to
a. $-\frac{\pi}{2}$ b. zero
c. $\frac{\pi}{2}$ d. π

54. The trigonometric equation $\sin^{-1} x = 2\sin^{-1} a$ has a solution for
a. all real values b. $|a| < \frac{1}{2}$
c. $|a| \le \frac{1}{\sqrt{2}}$ d. $\frac{1}{2} < |a| < \frac{1}{\sqrt{2}}$

55. If $2\tan^{-1} x = \pi + \tan^{-1}\left(\frac{2x}{1-x^2}\right)$, then
a. $x > 1$ b. $x < 1$
c. $x > -1$ d. $-1 < x < 1$

56. $\sum_{r=1}^{n} \sin^{-1}\left(\frac{\sqrt{r} - \sqrt{r-1}}{\sqrt{r(r+1)}}\right)$ is equal to
a. $\tan^{-1}(\sqrt{n}) - \frac{\pi}{4}$ b. $\tan^{-1}(\sqrt{n+1}) - \frac{\pi}{4}$
c. $\tan^{-1}(\sqrt{n})$ d. $\tan^{-1}(\sqrt{n+1})$

57. $\sum_{r=1}^{n} \tan^{-1}\left(\frac{2^{r-1}}{1+2^{2r-1}}\right)$ is equal to

a. $\tan^{-1}(2^n)$
b. $\tan^{-1}(2^n)-\frac{\pi}{4}$
c. $\tan^{-1}(2^{n+1})$
d. $\tan^{-1}(2^{n+1})-\frac{\pi}{4}$

58. $\sum_{m=1}^{n} \tan^{-1}\left(\frac{2m}{m^4+m^2+2}\right)$ is equal to

a. $\tan^{-1}\left(\frac{n^2+n}{n^2+n+2}\right)$
b. $\tan^{-1}\left(\frac{n^2-n}{n^2-n+2}\right)$
c. $\tan^{-1}\left(\frac{n^2+n+2}{n^2+n}\right)$
d. None of these

59. Value of $\sum_{r=0}^{\infty} \tan^{-1}\left(\frac{1}{1+r+r^2}\right)$ is equal to

a. $\frac{\pi}{2}$
b. $\frac{3\pi}{4}$
c. $\frac{\pi}{4}$
d. None of these

60. If $\sin^{-1} x = \theta + \beta$ and $\sin^{-1} y = \theta - \beta$, then $1 + xy =$

a. $\sin^2\theta + \sin^2\beta$
b. $\sin^2\theta + \cos^2\beta$
c. $\cos^2\theta + \cos^2\theta$
d. $\cos^2\theta + \sin^2\beta$

61. If $u = \cot^{-1}\sqrt{\tan\alpha} - \tan^{-1}\sqrt{\tan\alpha}$, then $\tan\left(\frac{\pi}{4}-\frac{u}{2}\right)$ is equal to

a. $\sqrt{\tan\alpha}$
b. $\sqrt{\cot\alpha}$
c. $\tan\alpha$
d. $\cot\alpha$

62. $\tan\left[\frac{\pi}{4}+\frac{1}{2}\cos^{-1}\frac{a}{b}\right] + \tan\left[\frac{\pi}{4}-\frac{1}{2}\cos^{-1}\frac{a}{b}\right]$ is equal to

a. $\frac{2a}{b}$
b. $\frac{2b}{a}$
c. $\frac{a}{b}$
d. $\frac{b}{a}$

63. $2\tan^{-1}\left[\sqrt{\frac{a-b}{a+b}}\tan\frac{\theta}{2}\right]$ is equal to

a. $\cos^{-1}\left(\frac{a\cos\theta+b}{a+b\cos\theta}\right)$
b. $\cos^{-1}\left(\frac{a+b\cos\theta}{a\cos\theta+b}\right)$
c. $\cos^{-1}\left(\frac{a\cos\theta}{a+b\cos\theta}\right)$
d. $\cos^{-1}\left(\frac{b\cos\theta}{a\cos\theta+b}\right)$

64. $\cot^{-1}\left[\frac{\sqrt{1-\sin x}+\sqrt{1+\sin x}}{\sqrt{1-\sin x}-\sqrt{1+\sin x}}\right]$, $\left(\text{where } x \in \left[0, \frac{\pi}{2}\right]\right)$, is equal to

a. $\pi - x$
b. $2\pi - x$
c. $\frac{x}{2}$
d. $\pi - \frac{x}{2}$

65. The value of $\tan^{-1}\left(\frac{x\cos\theta}{1-x\sin\theta}\right) - \cot^{-1}\left(\frac{\cos\theta}{x-\sin\theta}\right)$ is

a. 2θ
b. θ
c. $\theta/2$
d. independent of θ

66. If $x \in \left(-\frac{\pi}{2}, \frac{\pi}{2}\right)$, then the value of $\tan^{-1}\left(\frac{\tan x}{4}\right) + \tan^{-1}\left(\frac{3\sin 2x}{5+3\cos 2x}\right)$ is

a. $\frac{x}{2}$
b. $2x$
c. $3x$
d. x

67. $\cot^{-1}\left(\sqrt{\cos\alpha}\right) - \tan^{-1}\left(\sqrt{\cos\alpha}\right) = x$, then $\sin x$ is equal to

a. $\tan^2\frac{\alpha}{2}$
b. $\cot^2\frac{\alpha}{2}$
c. $\tan a$
d. $\cot\frac{\alpha}{2}$

68. $\tan\left(\frac{\pi}{4}+\frac{1}{2}\cos^{-1}x\right) + \tan\left(\frac{\pi}{4}-\frac{1}{2}\cos^{-1}x\right)$, $x \neq 0$, is equal to

a. x
b. $2x$
c. $\frac{2}{x}$
d. None of these

69. The greatest and the least values of $(\sin^{-1} x)^3 + (\cos^{-1} x)^3$ are

a. $-\pi/2, \pi/2$
b. $-\pi^3/8, \pi^3/8$
c. $\pi^3/32, 7\pi^3/8$
d. None of these

70. Range of $f(x) = \sin^{-1} x + \tan^{-1} x + \sec^{-1} x$ is

a. $\left(\frac{\pi}{4}, \frac{3\pi}{4}\right)$
b. $\left[\frac{\pi}{4}, \frac{3\pi}{4}\right]$
c. $\left\{\frac{\pi}{4}, \frac{3\pi}{4}\right\}$
d. None of these

71. The range of $\tan^{-1}\left(\frac{2x}{1+x^2}\right)$ is

a. $\left[-\frac{\pi}{4}, \frac{\pi}{4}\right]$
b. $\left(-\frac{\pi}{2}, \frac{\pi}{2}\right)$
c. $\left(-\frac{\pi}{2}, \frac{\pi}{4}\right]$
d. $\left[\frac{\pi}{4}, \frac{\pi}{2}\right]$

72. If $[\cot^{-1} x] + [\cos^{-1} x] = 0$, where $[\cdot]$ denotes the greatest integer function, then the complete set of values of x is

a. $(\cos 1, 1]$
b. $(\cos 1, \cos 1)$
c. $(\cot 1, 1]$
d. None of these

73. $\sin^{-1}(\sin 5) > x^2 - 4x$ holds if

a. $x = 2 - \sqrt{9 - 2\pi}$
b. $x = 2 + \sqrt{9 - 2\pi}$
c. $x > 2 + \sqrt{9 - 2\pi}$
d. $x \in (2 - \sqrt{9 - 2\pi},\ 2 + \sqrt{9 - 2\pi})$

74. $\frac{\alpha^3}{2}\operatorname{cosec}^2\left(\frac{1}{2}\tan^{-1}\frac{\alpha}{\beta}\right) + \frac{\beta^3}{2}\sec^2\left(\frac{1}{2}\tan^{-1}\left(\frac{\beta}{\alpha}\right)\right)$ is equal to

a. $(\alpha - \beta)(\alpha^2 + \beta^2)$
b. $(\alpha + \beta)(\alpha^2 - \beta^2)$
c. $(\alpha + \beta)(\alpha^2 + \beta^2)$
d. None of these

75. The value of $\lim_{|x| \to \infty} \cos(\tan^{-1}(\sin(\tan^{-1} x)))$ is

a. -1
b. $\sqrt{2}$
c. $-\frac{1}{\sqrt{2}}$
d. $\frac{1}{\sqrt{2}}$

76. $\sin^{-1}(3x - 2 - x^2) + \cos^{-1}(x^2 - 4x + 3) = \frac{\pi}{4}$ can have a solution for $x \in$

a. $[1, 2]$
b. $\left(\frac{3+\sqrt{5}}{2}, 2+\sqrt{2}\right)$
c. $\left(\frac{3-\sqrt{5}}{2}, 2-\sqrt{2}\right)$
d. None of these

77. If $2^{2\pi/\sin^{-1} x} - 2(a+2)2^{\pi/\sin^{-1} x} + 8a < 0$ for at least one real x, then

a. $\frac{1}{8} \le a < 2$
b. $a < 2$
c. $a \in R - \{2\}$
d. $a \in \left[0, \frac{1}{8}\right) \cup (2, \infty)$

78. The number of integral values of k for which the equation $\sin^{-1} x + \tan^{-1} x = 2k + 1$ has a solution is

a. 1
b. 2
c. 3
d. 4

79. If $\tan^{-1}(\sin^2\theta - 2\sin\theta + 7) + \cot^{-1}(5^{\sec^2 y} + 1) = \frac{\pi}{2}$, then the value of $\cos^2\theta - \sin\theta$ is equal to

a. 0
b. -1
c. 1
d. None of these

80. Complete solution set of $[\cot^{-1} x] + 2[\tan^{-1} x] = 0$, where $[\cdot]$ denotes the greatest integer function, is equal to

a. $(0, \cot 1)$
b. $(0, \tan 1)$
c. $(\tan 1, \infty)$
d. $(\cot 1, \tan 1)$

Multiple Correct Answers Type

Each question has four choices a, b, c and d, out of which *one or more* answers are correct.

1. If $\alpha, \beta (\alpha < \beta)$ are the roots of the equation $6x^2 + 11x + 3 = 0$, then which of the following are real?

a. $\cos^{-1}\alpha$
b. $\sin^{-1}\beta$
c. $\operatorname{cosec}^{-1}\alpha$
d. both $\cot^{-1}\alpha$ and $\cot^{-1}\beta$

2. $2\tan^{-1}(-2)$ is equal to

a. $-\cos^{-1}\left(\frac{-3}{5}\right)$
b. $-\pi + \cos^{-1}\frac{3}{5}$
c. $-\frac{\pi}{2} + \tan^{-1}\left(-\frac{3}{4}\right)$
d. $-\pi + \cot^{-1}\left(-\frac{3}{4}\right)$

3. If α, β, γ are the roots of $\tan^{-1}(x-1) + \tan^{-1} x + \tan^{-1}(x+1) = \tan^{-1} 3x$, then

a. $\alpha + \beta + \gamma = 0$
b. $\alpha\beta + \beta\gamma + \gamma\alpha = -1/4$
c. $\alpha\beta\gamma = 1$
d. $|\alpha - \beta|_{max} = 1$

4. If $f(x) = \sin^{-1}x + \sec^{-1}x$ is defined, then which of the following value/values is/are in its range

a. $-\pi/2$
b. $\pi/2$
c. π
d. $3\pi/2$

5. If $(\sin^{-1} x + \sin^{-1} w)(\sin^{-1} y + \sin^{-1} z) = \pi^2$, then

$$D = \begin{vmatrix} x^{N_1} & y^{N_2} \\ z^{N_3} & w^{N_4} \end{vmatrix} \quad (N_1, N_2, N_3, N_4 \in N)$$

a. has a maximum value of 2
b. has a minimum value of 0
c. 16 different D are possible
d. has a minimum value of -2

6. Indicate the relation which is/are true in their respective domain

a. $\tan|\tan^{-1} x| = |x|$ b. $\cot|\cot^{-1} x| = |x|$

c. $\tan^{-1}|\tan x| = |x|$ d. $\sin|\sin^{-1} x| = |x|$

7. If α is a real number for which $f(x) = \ln \cos^{-1} x$ is defined, then a possible value of $[\alpha]$ (where $[\cdot]$ denotes the greatest integer function) is

a. 0 b. 1

c. -1 d. -2

8. Which of the following is a rational number?

a. $\sin\left(\tan^{-1} 3 + \tan^{-1} \frac{1}{3}\right)$

b. $\cos\left(\frac{\pi}{2} - \sin^{-1} \frac{3}{4}\right)$

c. $\log_2\left(\sin\left(\frac{1}{4}\sin^{-1}\frac{\sqrt{63}}{8}\right)\right)$

d. $\tan\left(\frac{1}{2}\cos^{-1}\frac{\sqrt{5}}{3}\right)$

9. If $z = \sec^{-1}\left(x + \frac{1}{x}\right) + \sec^{-1}\left(y + \frac{1}{y}\right)$ where $xy < 0$, then the possible values of z is (are)

a. $\frac{8\pi}{10}$ b. $\frac{7\pi}{10}$

c. $\frac{9\pi}{10}$ d. $\frac{21\pi}{20}$

10. If $f(x) = (\sin^{-1} x)^2 + (\cos^{-1} x)^2$, then

a. $f(x)$ has the least value of $\frac{\pi^2}{8}$

b. $f(x)$ has the greatest value of $\frac{5\pi^2}{8}$

c. $f(x)$ has the least value of $\frac{\pi^2}{16}$

d. $f(x)$ has the greatest value of $\frac{5\pi^2}{4}$

11. If $\sin^{-1} x + \sin^{-1} y = \frac{\pi}{2}$ and $\sin 2x = \cos 2y$, then

a. $x = \frac{\pi}{8} + \sqrt{\frac{1}{2} - \frac{\pi^2}{64}}$ b. $y = \sqrt{\frac{1}{2} + \frac{\pi^2}{64}} - \frac{\pi}{12}$

c. $x = \frac{\pi}{12} + \sqrt{\frac{1}{2} - \frac{\pi^2}{64}}$ d. $y = \sqrt{\frac{1}{2} - \frac{\pi^2}{64}} - \frac{\pi}{8}$

12. If $\cot^{-1}\left(\frac{n^2 - 10n + 21.6}{\pi}\right) > \frac{\pi}{6}, n \in N$, then n can be

a. 3 b. 2

c. 4 d. 8

13. If $S_n = \cot^{-1}(3) + \cot^{-1}(7) + \cot^{-1}(13) + \cot^{-1}(21) + \cdots$ n terms, then

a. $S_{10} = \tan^{-1}\frac{5}{6}$ b. $S_\infty = \frac{\pi}{4}$

c. $S_6 = \sin^{-1}\frac{4}{5}$ d. $S_{20} = \cot^{-1} 1.1$

14. The value of k $(k > 0)$ such that the length of the longest interval in which the function $f(x) = \sin^{-1}|\sin kx| + \cos^{-1}(\cos kx)$ is constant is $\frac{\pi}{4}$ is/are

a. 8 b. 4

c. 12 d. 16

15. The equation $1 + x^2 + 2x \sin(\cos^{-1} y) = 0$ is satisfied by

a. exactly one value of x b. exactly two values of x

c. exactly one value of y d. exactly two values of y

16. To the equation $2^{2\pi/\cos^{-1}x} - \left(a + \frac{1}{2}\right)2^{\pi/\cos^{-1}x} - a^2 = 0$ has only one real root, then

a. $1 \le a \le 3$ b. $a \ge 1$

c. $a \le -3$ d. $a \ge 3$

17. If $\sin^{-1}\left(a - \frac{a^2}{3} + \frac{a^3}{9} + \cdots\right) + \cos^{-1}(1 + b + b^2 + \cdots) = \frac{\pi}{2}$, then

a. $b = \frac{2a-3}{3a}$ b. $b = \frac{3a-2}{2a}$

c. $a = \frac{3}{2-3b}$ d. $a = \frac{2}{3-2b}$

18. If $2\tan^{-1} x + \sin^{-1}\frac{2x}{1+x^2}$ is independent of x, then

a. $x > 1$ b. $x < -1$

c. $0 < x < 1$ d. $-1 < x < 0$

19. $\cos^{-1} x + \cos^{-1}\left(\frac{x}{2} + \frac{1}{2}\sqrt{3 - 3x^2}\right)$ is equal to

a. $\frac{\pi}{3}$ for $x \in \left[\frac{1}{2}, 1\right]$

b. $\frac{\pi}{3}$ for $x \in \left[0, \frac{1}{2}\right]$

c. $2\cos^{-1}x - \cos^{-1}\frac{1}{2}$ for $x \in \left[\frac{1}{2}, 1\right]$

d. $2\cos^{-1}x - \cos^{-1}\frac{1}{2}$ for $x \in \left[0, \frac{1}{2}\right]$

20. Which of the following quantities is/are positive?
  a. $\cos(\tan^{-1}(\tan 4))$ b. $\sin(\cot^{-1}(\cot 4))$
  c. $\tan(\cos^{-1}(\cos 5))$ d. $\cot(\sin^{-1}(\sin 4))$

Reasoning Type

Each question has four choices a, b, c, and d, out of which *only one* is correct. Each question contains Statement 1 and Statement 2.

a. Both the statements are True and Statement 2 is the correct explanation of Statement 1.
b. Both the statements are True but Statement 2 is Not the correct explanation of Statement 1.
c. Statement 1 is True and Statement 2 is False.
d. Statement 1 is False and Statement 2 is True.

1. **Statement 1:** Number of roots of the equation $\cot^{-1}x + \cos^{-1}2x + \pi = 0$ is zero.
   Statement 2: Range of $\cot^{-1}x$ and $\cos^{-1}x$ is $(0, \pi)$ and $[0, \pi]$, respectively.
2. **Statement 1:** Range of $f(x) = \tan^{-1}x + \sin^{-1}x + \cos^{-1}x$ is $(0, \pi)$.
   Statement 2: $f(x) = \tan^{-1}x + \sin^{-1}x + \cos^{-1}x = \frac{\pi}{2} + \tan^{-1}x$, for $x \in [-1, 1]$.
3. **Statement 1:**
   $\operatorname{cosec}^{-1}\left(\frac{1}{2} + \frac{1}{\sqrt{2}}\right) > \sec^{-1}\left(\frac{1}{2} + \frac{1}{\sqrt{2}}\right)$.
   Statement 2: $\operatorname{cosec}^{-1}x < \sec^{-1}x$ if $1 \le x < \sqrt{2}$.
4. **Statement 1:** $\sin^{-1}\left(\frac{1}{\sqrt{e}}\right) > \tan^{-1}\left(\frac{1}{\sqrt{\pi}}\right)$.
   Statement 2: $\sin^{-1}x > \tan^{-1}y$ for $x > y$, $\forall x, y \in (0, 1)$.
5. **Statement 1:** Principal value of $\cos^{-1}(\cos 30)$ is $30 - 9\pi$.
   Statement 2: $30 - 9\pi \in [0, \pi]$.
6. Let $f(x) = \sin^{-1}\left(\frac{2x}{1+x^2}\right)$.
   Statement 1: $f'(2) = -\frac{2}{5}$.
   Statement 2: $\sin^{-1}\left(\frac{2x}{1+x^2}\right) = \pi - 2\tan^{-1}x, \forall x > 1$.
7. **Statement 1:** Domain of $\tan^{-1}x$ and $\cot^{-1}x$ is R.
   Statement 2: $f(x) = \tan x$ and $g(x) = \cot x$ are unbounded functions.
8. **Statement 1:** $\tan^{-1}\left(\frac{3}{4}\right) + \tan^{-1}\left(\frac{1}{7}\right) = \frac{\pi}{4}$.
   Statement 2: For $x > 0, y > 0$,
   $\tan^{-1}\left(\frac{x}{y}\right) + \tan^{-1}\left(\frac{y-x}{y+x}\right) = \frac{\pi}{4}$.
9. **Statement 1:** Principal value of $\sin^{-1}(\sin 3)$ can be 3 if we restrict the domain of $f(x) = \sin x$ to $[\pi/2, 3\pi/2]$.
   Statement 2: The restriction that the principal values of $\sin^{-1}(\sin x)$ is $[-\pi/2, -\pi/2]$ is a matter of convention. We could have allowed principal values $[\pi/2, 3\pi/2]$ without affecting the condition required for definition of inverse function.

Linked Comprehension Type

Based upon each paragraph, three multiple choice questions have to be answered. Each question has four choices a, b, c, and d, out of which *only one* is correct.

For Problems 1–3

For $x, y, z, t \in R$, $\sin^{-1}x + \cos^{-1}y + \sec^{-1}z \ge t^2 - \sqrt{2\pi}\, t + 3\pi$.

1. The value of $x + y + z$ is equal to
   a. 1 b. 0
   c. 2 d. –1
2. The principal value of $\cos^{-1}(\cos 5t^2)$ is
   a. $\frac{3\pi}{2}$ w b. $\frac{\pi}{2}$
   c. $\frac{\pi}{3}$ d. $\frac{2\pi}{3}$
3. The value of $\cos(\min\{x, y, z\})$ is
   a. 0 b. $\pi/2$
   c. π d. $\pi/3$

For Problems 4–6

$ax + b(\sec(\tan^{-1}x)) = c$ and $ay + b(\sec(\tan^{-1}y)) = c$.

4. The value of xy is

a. $\frac{2ab}{a^2-b^2}$ b. $\frac{c^2-b^2}{a^2-b^2}$

c. $\frac{c^2-b^2}{a^2+b^2}$ d. None of these

5. The value of $x+y$ is

a. $\frac{2ac}{a^2-b^2}$ b. $\frac{c^2-b^2}{a^2-b^2}$

c. $\frac{c^2-b^2}{a^2+b^2}$ d. None of these

6. The value of $\frac{x+y}{1-xy}$ is

a. $\frac{2ab}{a^2-c^2}$ b. $\frac{2ac}{a^2-c^2}$

c. $\frac{c^2-b^2}{a^2+b^2}$ d. None of these

For Problems 7–9

Consider the system of equations $\cos^{-1}x+(\sin^{-1}y)^2=\frac{p\pi^2}{4}$ and $(\cos^{-1}x)(\sin^{-1}y)^2=\frac{\pi^2}{16}, p\in Z.$

7. The value of p for which system has a solution

a. 1 b. 2

c. 0 d. −1

8. The value of x which satisfies the system of equations is

a. $\cos\frac{\pi^2}{8}$ b. $\sin\frac{\pi^2}{4}$

c. $\cos\frac{\pi^2}{2}$ d. None of these

9. Which of the following is not the value of y that satisfies the system of equations.

a. 1 b. −1

c. 1/2 d. None of these

For Problems 10–12

Let $\cos^{-1}(4x^3-3x)=a+b\cos^{-1}x.$

10. If $x\in\left[-\frac{1}{2},-1\right)$, then the value of $a+b\pi$ is

a. 2π b. 3π

c. π d. -2π

11. $x\in\left[-\frac{1}{2},\frac{1}{2}\right]$, then the principal value of $\sin^{-1}\left(\sin\frac{a}{b}\right)$ is

a. $-\frac{\pi}{3}$ b. $\frac{\pi}{3}$

c. $-\frac{\pi}{6}$ d. $\frac{\pi}{6}$

12. If $x\in\left(\frac{1}{2},1\right]$, then the value of $\lim_{y\to a} b\cos(y)$ is

a. −1/3 b. −3

c. 1/3 d. 3

Matrix-Match Type

Each question contains statements given in two columns which have to be matched. Statements a, b, c, d in column I have to be matched with statements p, q, r, s in column II. If the correct matches are a→p, a→s, b→q, b→r, c→p, c→q and d→s; then the correctly bubbled 4 × 4 matrix should be as follows:

	p	q	r	s
a	(p)	q	r	(s)
b	p	(q)	(r)	s
c	(p)	(q)	r	s
d	p	q	r	(s)

1.

Column I	Column II
a. $\cos^{-1}(4x^3-3x)=3\cos^{-1}x$, then x can take values	p. [1/2, 1]
b. $\sin^{-1}(3x-4x^3)=3\sin^{-1}x$, then x can take values	q. [−1/2, 0]
c. $\cos^{-1}(4x^3-3x)=3\sin^{-1}x$, then x can take values	r. $[0, \sqrt{3}/2]$
d. $\sin^{-1}(3x-4x^3)=3\cos^{-1}x$, then x can take values	s. [0, 1/2]

2.

Column I	Column II
a. $(\sin^{-1}x)^2+(\sin^{-1}y)^2=\frac{\pi^2}{2}$ $\Rightarrow x^3+y^3=$	p. 1
b. $(\cos^{-1}x)^2+(\cos^{-1}y)^2=2\pi^2$ $\Rightarrow x^5+y^5=$	q. −2
c. $(\sin^{-1}x)^2(\cos^{-1}y)^2=\frac{\pi^4}{4}$ $\Rightarrow \lvert x-y\rvert=$	r. 0
d. $\lvert\sin^{-1}x-\sin^{-1}y\rvert=\pi\Rightarrow x^y=$	s. 2

3.

Column I	Column II
a. $x \in [\pi, 2\pi] \Rightarrow |\tan^{-1}(\tan x)|$ can be	p. $|x-2\pi|$
b. $x \in [\pi, 2\pi] \Rightarrow |\cot^{-1}(\cot x)|$ can be	q. $|x-\pi|$
c. $x \in [-\pi, \pi] \Rightarrow |\sin^{-1}(\sin x)|$ can be	r. $|x|$
d. $x \in [-\pi, \pi] \Rightarrow |\cos^{-1}(\cos x)|$ can be	r. $|x+\pi|$

4.

Column I	Column II
a. $\sin^{-1}\frac{4}{5}+2\tan^{-1}\frac{1}{3}=$	p. $\pi/6$
b. $\sin^{-1}\frac{12}{13}+\cos^{-1}\frac{4}{5}+\tan^{-1}\frac{63}{16}=$	q. $\pi/2$
c. If $A=\tan^{-1}\frac{x\sqrt{3}}{2\lambda-x}$ and $B=\tan^{-1}\left(\frac{2x-\lambda}{\lambda\sqrt{3}}\right)$ then the value of $A-B$ is	r. $\pi/4$
d. $\tan^{-1}\frac{1}{7}+2\tan^{-1}\frac{1}{3}=$	s. π

5.

Column I	Column II
a. Range of $f(x)=\sin^{-1}x+\cos^{-1}x+\cot^{-1}x$ is	p. $\left[0,\frac{\pi}{2}\right)\cup\left(\frac{\pi}{2},\pi\right]$
b. Range of $f(x)=\cot^{-1}x+\tan^{-1}x+\operatorname{cosec}^{-1}x$ is	q. $\left[\frac{\pi}{2},\frac{3\pi}{2}\right]$
c. Range of $f(x)=\cot^{-1}x+\tan^{-1}x+\cos^{-1}x$ is	r. $\{0,\pi\}$
d. Range of $f(x)=\sec^{-1}x+\operatorname{cosec}^{-1}x+\sin^{-1}x$ is	s. $\left[\frac{3\pi}{4},\frac{5\pi}{4}\right]$

Integer Type

1. If $\cos^{-1}\lambda+\cos^{-1}\mu+\cos^{-1}\gamma=3\pi$, then the value of $\lambda\mu+\mu\gamma+\gamma\lambda$ is ______.
2. If $(\sin^{-1}x)^2+(\sin^{-1}y)^2+(\sin^{-1}z)^2=3\pi^2/4$, then the maximum value of $x+y+z$ is ______.
3. If $x>y>z>0$, then the value of
$$\left[\cot^{-1}\frac{xy+1}{x-y}+\cot^{-1}\frac{yz+1}{y-z}+\cot^{-1}\frac{zx+1}{z-x}\right],$$
where [.] represents the greatest integer function, is ______.
4. If the range of $f(x)=\sin^{-1}x+\tan^{-1}x+\cos^{-1}x$ is $[a, b]$, then the value of $2b/a$ is ______.
5. The value of $\sqrt{x}$, where x is the value for which
$$\sin^{-1}\frac{14}{|x|}+\sin^{-1}\frac{2\sqrt{15}}{|x|}=\frac{\pi}{2}$$ is ______.
6. The least integral value of x for which $\tan^{-1}x>\cot^{-1}x$ is ______.
7. If $\tan^{-1}[(1/2)\tan 2A]+\tan^{-1}(\cot A)+\tan^{-1}(\cot^3 A)=k(\pi/4)$, for $0<A<\pi/4$, then the value of k is ______.
8. The number of integral values of x for which $\sin^{-1}[(4x)/(x^2+4)]+2\tan^{-1}(-x/2)$ is independent at x is ___.
9. If $(x-1)(x^2+1)>0$, then the value of
$$\left|\operatorname{cosec}\left(\frac{1}{2}\tan^{-1}\frac{2x}{1-x^2}-\tan^{-1}x\right)\right|$$ is ______.
10. The number of integral values of x for which $f(x)=\sin^{-1}[(1+x^2)/(2x)]$ is defined is ______.
11. The number of positive integral solutions of the equation $\tan^{-1}x+\cos^{-1}(y/\sqrt{1-y^2})=\sin^{-1}(3/\sqrt{10})$ is ______.
12. The number of roots of the equation $\sin^{-1}6x+\sin^{-1}6\sqrt{3}x=-\pi/2$ is ______.
13. The number of solutions of the equation
$$\frac{(\sin^{-1}x)^3+(\cos^{-1}x)^3}{(\tan^{-1}x+\cot^{-1}x)^3}=7$$ is ______.
14. The number of integral values of $8a$ for which the equation $(\sin^{-1}x)^3+(\cos^{-1}x)^3=a\pi^3$ has a solution is ______.
15. If $\theta=\cot^{-1}2+\cot^{-1}8+\cot^{-1}18+\cdots$, then the value of $7\tan\theta$ is ______.
16. Number of solutions of the equation $\cos(\cos^{-1}x)=\operatorname{cosec}(\operatorname{cosec}^{-1}x)$ is ______.
17. Number of real solutions of the equation $\sqrt{1+\cos 2x}=\sqrt{2}\sin^{-1}(\sin x)$, $-\pi\le x\le\pi$, is ______.

18. Number of integral values of x for which $\left|\cos^{-1}\left(\frac{1-x^2}{1+x^2}\right)\right| < \frac{\pi}{3}$ is______.

19. Number of solutions of the equation $3\cos^{-1}x - \pi x - \frac{\pi}{2} = 0$ is______.

20. The number of integral values of x, for which $f(x) = \sin^{-1}(\sqrt{x^2-2x+1} - 2)$ is defined, is______.

21. If $\sin^{-1}x + \sin^{-1}y = \frac{\pi}{2}$, then the value of $\frac{1+x^4+y^4}{x^2 - x^2y^2 + y^2}$ is______.

22. α_1 and α_2 satisfy $\sin^{-1}\frac{2x}{1+x^2} = \tan^{-1}\frac{2x}{1-x^2}$, and $|\alpha_1 - \alpha_2| < K$, for all α_1 and α_2, then K is equal to______.

23. If $\sin^{-1}\left(\frac{4x}{x^2+4}\right) + 2\tan^{-1}\left(-\frac{x}{2}\right)$ is independent at x then number of integral values of x is______.

24. The ratio of maximum and minimum value of $A = (\tan^{-1}x)^3 + (\cot^{-1}x)^3$ is______.

ANSWERS AND SOLUTIONS

Objective Type

1. c. $\sin^{-1}(\sin 10) = \sin^{-1}[\sin(3\pi - 10)] = 3\pi - 10$

2. b. $\cos^{-1}\left(\cos\frac{5\pi}{4}\right) = \cos^{-1}\left(\cos\left(2\pi - \frac{5\pi}{4}\right)\right)$

$= \cos^{-1}\left(\cos\frac{3\pi}{4}\right) = \frac{3\pi}{4}$

3. a. $\sin^{-1}(\sin 12) + \cos^{-1}(\cos 12)$

$= \sin^{-1}(\sin(12 - 4\pi)) + \cos^{-1}(\cos(4\pi - 12))$

$= 12 - 4\pi + 4\pi - 12 = 0$

4. a. $\sin^{-1}\sin\left(\frac{22\pi}{7}\right) = \sin^{-1}\sin\left(3\pi + \frac{\pi}{7}\right) = -\frac{\pi}{7}$

$\cos^{-1}\cos\left(\frac{5\pi}{3}\right) = \cos^{-1}\cos\left(2\pi - \frac{\pi}{3}\right) = \frac{\pi}{3}$

$\tan^{-1}\tan\left(\frac{5\pi}{7}\right) = \tan^{-1}\tan\left(\pi - \frac{2\pi}{7}\right) = -\frac{2\pi}{7}$

$\sin^{-1}\cos(2) = \frac{\pi}{2} - \cos^{-1}\cos 2 = \frac{\pi}{2} - 2$

Required value $= -\frac{\pi}{7} + \frac{\pi}{3} - \frac{2\pi}{7} + \frac{\pi}{2} - 2,$

$= \frac{(-18+35)\pi}{42} - 2 = \frac{17\pi}{42} - 2$

5. d. $\sin^{-1}(\cos(\cos^{-1}(\cos x) + \sin^{-1}(\sin x)))$

$= \sin^{-1}(\cos(x + \pi - x))$ [as $x \in (\pi/2, \pi)$]

$= \sin^{-1}(\cos\pi) = \sin^{-1}(-1) = -\frac{\pi}{2}$

6. c. $\cos^{-1}(\cos(2\cot^{-1}(\sqrt{2}-1)))$

$= \cos^{-1}\left(\cos\left(2\left(67.5^\circ\right)\right)\right)$

$= \cos^{-1}\left(\cos\left(135^\circ\right)\right) = 135^\circ$

$= \frac{3\pi}{4}$

7. a. We have

$\sin^{-1}\left(\cot\left(\sin^{-1}\sqrt{\frac{2-\sqrt{3}}{4}} + \cos^{-1}\frac{\sqrt{12}}{4} + \sec^{-1}\sqrt{2}\right)\right)$

$= \sin^{-1}\left(\cot\left(\sin^{-1}\left(\frac{\sqrt{3}-1}{2\sqrt{2}}\right) + \cos^{-1}\frac{\sqrt{3}}{2} + \cos^{-1}\frac{1}{\sqrt{2}}\right)\right)$

$= \sin^{-1}[\cot(15^\circ + 30^\circ + 45^\circ)]$

$= \sin^{-1}(\cot(90^\circ)) = \sin^{-1}(0) = 0$

8. d. $\cos^{-1}\left(\sqrt{\frac{2}{3}}\right) - \cos^{-1}\left(\frac{\sqrt{6}+1}{2\sqrt{3}}\right)$

$= \tan^{-1}\left(\frac{1}{\sqrt{2}}\right) - \tan^{-1}\left(\frac{\sqrt{3}-\sqrt{2}}{1+\sqrt{6}}\right)$

$= \tan^{-1}\left(\frac{1}{\sqrt{2}}\right) - \left[\tan^{-1}\sqrt{3} - \tan^{-1}\sqrt{2}\right]$

$= \left(\tan^{-1}\frac{1}{\sqrt{2}} + \tan^{-1}\sqrt{2}\right) - \tan^{-1}\sqrt{3}$

$= \frac{\pi}{2} - \frac{\pi}{3}$

$= \frac{\pi}{6}$

9. a. Let $\cos^{-1}\left(\frac{1}{8}\right) = \theta$, where $0 < \theta < \pi$, then

$\frac{1}{2}\cos^{-1}\frac{1}{8} = \frac{1}{2}\theta$

$\Rightarrow \cos\left(\frac{1}{2}\cos^{-1}\frac{1}{8}\right) = \cos\frac{\theta}{2}$

Now, $\cos^{-1}\frac{1}{8} = \theta \quad \Rightarrow \cos\theta = \frac{1}{8}$

$\Rightarrow \cos^2\frac{\theta}{2} = \frac{9}{16} \quad \Rightarrow \cos\frac{\theta}{2} = \frac{3}{4}$

10. c. $x = \tan^{-1} 3 \quad \Rightarrow \tan x = 3$

$\tan(x+y) = 33$

$\Rightarrow \frac{\tan x + \tan y}{1 - \tan x \tan y} = 33$

$\Rightarrow \frac{3 + \tan y}{1 - 3\tan y} = 33$

$\Rightarrow 3 + \tan y = 33 - 99\tan y$

$\Rightarrow 100\tan y = 30$

$\Rightarrow \tan y = 0.3 \quad \Rightarrow y = \tan^{-1}(0.3)$

11. c. Let $\cos^{-1}\left(\frac{\sqrt{5}}{3}\right) = \alpha$. Then $\cos\alpha = \frac{\sqrt{5}}{3}$,

where $0 < \alpha < \frac{\pi}{2}$

Now, $\tan\frac{\alpha}{2}=\sqrt{\frac{1-\cos\alpha}{1+\cos\alpha}}=\sqrt{\frac{1-\sqrt{5}/3}{1+\sqrt{5}/3}}$

$$=\sqrt{\frac{3-\sqrt{5}}{3+\sqrt{5}}}=\sqrt{\frac{(3-\sqrt{5})^2}{9-5}}=\frac{1}{2}(3-\sqrt{5})$$

12. a. $\tan^{-1}\left[\frac{\cos x}{1+\sin x}\right]$

$$=\tan^{-1}\left[\frac{\sin(\pi/2-x)}{1+\cos(\pi/2-x)}\right]$$

$$=\tan^{-1}\left[\frac{2\sin(\pi/4-x/2)\cos(\pi/4-x/2)}{2\cos^2(\pi/4-x/2)}\right]$$

$$=\tan^{-1}\tan\left(\frac{\pi}{4}-\frac{x}{2}\right)=\frac{\pi}{4}-\frac{x}{2}$$

$$\Rightarrow -\frac{\pi}{2}<\frac{\pi}{4}-\frac{x}{2}<\frac{\pi}{2}$$

$$\Rightarrow -\frac{3\pi}{4}<-\frac{x}{2}<\frac{\pi}{4}$$

$$\Rightarrow -\frac{\pi}{4}<\frac{x}{2}<\frac{3\pi}{4}$$

$$\Rightarrow -\frac{\pi}{2}<\frac{x}{2}<\frac{3\pi}{2}$$

13. a. $\tan^{-1}\left[\frac{\sqrt{1+x^2}+\sqrt{1-x^2}}{\sqrt{1+x^2}-\sqrt{1-x^2}}\right]$

$$=\tan^{-1}\left[\frac{\sqrt{1+\cos 2\theta}+\sqrt{1-\cos 2\theta}}{\sqrt{1+\cos 2\theta}-\sqrt{1-\cos 2\theta}}\right]$$

[Putting $x^2=\cos 2\theta \Rightarrow \theta=\frac{1}{2}\cos^{-1}x^2$]

$$=\tan^{-1}\left[\frac{\sqrt{2}\cos\theta+\sqrt{2}\sin\theta}{\sqrt{2}\cos\theta-\sqrt{2}\sin\theta}\right]$$

$\left[\sqrt{2\sin^2\theta}=\sqrt{2}\sin\theta,\text{ as } x^2=\cos 2\theta\in[0,1)\right.$

$\Rightarrow \theta\in(0,\pi/4]$

$$=\tan^{-1}\left[\frac{1+\tan\theta}{1-\tan\theta}\right]=\tan^{-1}\left[\frac{\tan\frac{\pi}{4}+\tan\theta}{1-\tan\frac{\pi}{4}\tan\theta}\right]$$

$$=\tan^{-1}\tan\left(\frac{\pi}{4}+\theta\right)=\frac{\pi}{4}+\theta=\frac{\pi}{4}+\frac{1}{2}\cos^{-1}x^2$$

14. d. $4\tan^{-1}\frac{1}{5}-\tan^{-1}\frac{1}{239}$

$$=2\tan^{-1}\left(\frac{2/5}{1-1/25}\right)-\tan^{-1}\frac{1}{239}$$

$$=2\tan^{-1}\frac{5}{12}-\tan^{-1}\frac{1}{239}$$

$$=\tan^{-1}\left(\frac{2(5/12)}{1-(5/12)^2}\right)-\tan^{-1}\frac{1}{239}$$

$$=\tan^{-1}\frac{120}{119}-\tan^{-1}\frac{1}{239}$$

$$=\tan^{-1}\left(\frac{(120/119)-(1/239)}{1+(120/119)\times(1/239)}\right)$$

$$=\tan^{-1}\left(\frac{28569}{28569}\right)=\tan^{-1}(1)=\frac{\pi}{4}$$

15. b. Let $x=\sin\theta$ and $\sqrt{x}=\sin\phi$, where $x\in[0,1]$

$\Rightarrow \theta,\phi\in[0,\pi/2]$

$$\Rightarrow \theta-\phi\in\left[\frac{-\pi}{2},\frac{\pi}{2}\right]$$

Now, $\sin^{-1}\left(x\sqrt{1-x}-\sqrt{x}\sqrt{1-x^2}\right)$

$$=\sin^{-1}\left(\sin\theta\sqrt{1-\sin^2\phi}-\sin\phi\sqrt{1-\sin^2\theta}\right)$$

$$=\sin^{-1}(\sin\theta\cos\phi-\sin\phi\cos\theta)$$

$$=\sin^{-1}\sin(\theta-\phi)=\theta-\phi=\sin^{-1}(x)-\sin^{-1}(\sqrt{x})$$

16. c. $\tan^{-1}\frac{x}{y}-\tan^{-1}\left(\frac{x-y}{x+y}\right)$

$$=\tan^{-1}\frac{x}{y}-\tan^{-1}\left(\frac{1-y/x}{1+y/x}\right)$$

$$=\tan^{-1}\frac{x}{y}-\left(\tan^{-1}1-\tan^{-1}\frac{y}{x}\right)$$

$$=\tan^{-1}\frac{x}{y}+\tan^{-1}\frac{y}{x}-\frac{\pi}{4}$$

$$=\tan^{-1}\frac{x}{y}+\cot^{-1}\frac{x}{y}-\frac{\pi}{4}$$

$$=\frac{\pi}{2}-\frac{\pi}{4}=\frac{\pi}{4}$$

17. c. The given equation is $\tan^{-1}\frac{a+x}{a}+\tan^{-1}\frac{a-x}{a}=\frac{\pi}{6}$

$$\Rightarrow \tan^{-1}\left(\frac{\frac{a+x}{a}+\frac{a-x}{a}}{1-\frac{a+x}{a}\frac{a-x}{a}}\right)=\frac{\pi}{6}$$

$$\Rightarrow \frac{2a^2}{x^2}=\tan\frac{\pi}{6}=\frac{1}{\sqrt{3}} \Rightarrow x^2=2\sqrt{3}a^2$$

18. c. $\cot^{-1}\frac{n}{\pi}>\frac{\pi}{6}$

$\Rightarrow \frac{n}{\pi}<\cot\frac{\pi}{6}$

[as $\cot^{-1}x$ is a decreasing function]

$\Rightarrow \frac{n}{\pi}<\sqrt{3} \quad \Rightarrow n<\sqrt{3}\,\pi \quad \Rightarrow n<5.46$

$\Rightarrow$ Maximum value of n is 5.

19. c. Let $\tan^{-1}2=\alpha \Rightarrow \tan\alpha=2$

and $\cot^{-1}3=\beta \Rightarrow \cot\beta=3$

$\sec^2(\tan^{-1}2)+\operatorname{cosec}^2(\cot^{-1}3)$

$=\sec^2\alpha+\operatorname{cosec}^2\beta$

$=1+\tan^2\alpha+1+\cot^2\beta$

$=2+(2)^2+(3)^2=15$

20. d. $\tan(A-B)=\frac{\tan A-\tan B}{1+\tan A\tan B}$

$$=\frac{\frac{\sqrt{3}x}{2K-x}-\frac{2x-K}{\sqrt{3}K}}{1+\frac{\sqrt{3}x}{2K-x}\frac{2x-K}{\sqrt{3}K}}$$

$$=\frac{3Kx-(2x-K)(2K-x)}{(2K-x)\sqrt{3}K+\sqrt{3}x(2x-K)}$$

$$=\frac{3Kx-(4Kx-2x^2-2K^2+Kx)}{2\sqrt{3}K^2-\sqrt{3}Kx+2\sqrt{3}x^2-\sqrt{3}Kx}$$

$$=\frac{2x^2-2Kx+2K^2}{2\sqrt{3}x^2-2\sqrt{3}Kx+2\sqrt{3}K^2}=\frac{1}{\sqrt{3}}=\tan 30°$$

Hence, $A-B=30°$

21. b. $\tan^{-1}\frac{b+a}{b-a}-\tan^{-1}\frac{a}{b}=\tan^{-1}\frac{\frac{b+a}{b-a}-\frac{a}{b}}{1+\frac{b+a}{b-a}\frac{a}{b}}$

$$=\tan^{-1}\frac{b^2+ab-ab+a^2}{b^2-ab+ab+a^2}$$

$$=\tan^{-1}\frac{a^2+b^2}{a^2+b^2}=\tan^{-1}1=\frac{\pi}{4}$$

$\Rightarrow$ Required value $=\sec\left(\frac{\pi}{4}\right)=\sqrt{2}$

22. d. $a\sin^{-1}x-b\cos^{-1}x=c$

We have

$b\sin^{-1}x+b\cos^{-1}x=\frac{b\pi}{2}$

$\Rightarrow (a+b)\sin^{-1}x$

$=\frac{b\pi}{2}+c$

$\Rightarrow \sin^{-1}x=\frac{\left(\frac{b\pi}{2}\right)+c}{a+b}$

$\Rightarrow \cos^{-1}x=\frac{\pi ab+c(a-b)}{a+b}$

23. b. $\cos^{-1}\left(\frac{1+x^2}{2x}\right)=\frac{\pi}{2}+(\sin^{-1}x+\cos^{-1}x)$

$\Rightarrow \cos^{-1}\left(\frac{1+x^2}{2x}\right)=\pi$

$\Rightarrow \left(\frac{1+x^2}{2x}\right)=\cos\pi=-1$

$\Rightarrow x^2+1+2x=0 \qquad \Rightarrow x=-1$

24. b. $0\le x^2+x+1\le 1$ and $0\le x^2+x\le 1$

$\therefore x=-1, 0$

for $x=-1$

L.H.S. $=2\sin^{-1}1+\cos^{-1}0=\frac{3\pi}{2}$

$\therefore x=-1$ is a solution.

For $x=0$, L.H.S. $=2\sin^{-1}1+\cos^{-1}0=\frac{3\pi}{2}$

$\therefore x=0$ is a solution.

Hence, sum of the solutions $=-1$.

25. c. $\tan^{-1}(1+x)+\tan^{-1}(1-x)=\frac{\pi}{2}$

$\Rightarrow \tan^{-1}(1+x)=\frac{\pi}{2}-\tan^{-1}(1-x)$

$\Rightarrow \tan^{-1}(1+x)=\cot^{-1}(1-x)$

$\Rightarrow \tan^{-1}(1+x)=\tan^{-1}\left(\frac{1}{1-x}\right)$

$\Rightarrow 1+x=\frac{1}{1-x} \Rightarrow 1-x^2=1 \Rightarrow x=0$

26. c. We have $\sin^{-1}x+\sin^{-1}(1-x)=\cos^{-1}x$

$\Rightarrow \sin(\sin^{-1}x+\sin^{-1}(1-x))=\sin(\cos^{-1}x)$

$\Rightarrow x\sqrt{1-(1-x)^2}+\sqrt{1-x^2}(1-x)=\sqrt{1-x^2}$

$\Rightarrow x\sqrt{1-(1-x)^2} = x\sqrt{1-x^2}$

$\Rightarrow x = 0$ or $2x - x^2 = 1 - x^2$

$\Rightarrow x = 0$ or $x = \dfrac{1}{2}$

27. c. We have $\tan^{-1}\dfrac{1-x}{1+x} = \dfrac{1}{2}\tan^{-1}x$

$$\Rightarrow \tan^{-1}\left[\frac{1-\tan\theta}{1+\tan\theta}\right] = \frac{1}{2}\theta$$

[Putting $x = \tan\theta$]

$$\Rightarrow \tan^{-1}\left[\frac{\tan\frac{\pi}{4} - \tan\theta}{1 + \tan\frac{\pi}{4}\tan\theta}\right] = \frac{\theta}{2}$$

$$\Rightarrow \tan^{-1}\tan\left(\frac{\pi}{4} - \theta\right) = \frac{\theta}{2}$$

$$\Rightarrow \frac{\pi}{4} - \theta = \frac{\theta}{2}$$

$$\Rightarrow \theta = \frac{\pi}{6} = \tan^{-1}x$$

$$\Rightarrow x = \tan\frac{\pi}{6} = \frac{1}{\sqrt{3}}$$

28. c. We have, $\cos^{-1}x + \cos^{-1}(2x) = -\pi$, which is not possible as $\cos^{-1}x$ and $\cos^{-1}2x$ never take negative values.

29. b. The given equation is $ax^2 + \sin^{-1}((x-1)^2 + 1) + \cos^{-1}((x-1)^2 + 1) = 0$.
Now, $-1 \le (x-1)^2 + 1 \le 1 \Rightarrow x = 1$

So we have $a + \dfrac{\pi}{2} = 0 \Rightarrow a = -\dfrac{\pi}{2}$

30. c. Since $\sqrt{x^2 - 3x + 2} \ge 0$,

$$0 \le \tan^{-1}\sqrt{x^2 - 3x + 2} < \frac{\pi}{2}$$

and $\sqrt{4x - x^2 - 3} \ge 0$

$$\Rightarrow 0 < \cos^{-1}\sqrt{4x - x^2 - 3}$$

$\le \dfrac{\pi}{2}$. Adding, we have

$0 <$ L.H.S. $< \pi$

$\Rightarrow$ The given equation has no solution.

31. b. The given equation can be written as

$$3\tan^{-1}\left(2 - \sqrt{3}\right) = \tan^{-1}\left(\frac{1}{x}\right) + \tan^{-1}\left(\frac{1}{3}\right)$$

$$\Rightarrow 3(15°) = \tan^{-1}\frac{\frac{1}{x} + \frac{1}{3}}{1 - \frac{1}{x}\frac{1}{3}} \Rightarrow 1 = \frac{3+x}{3x-1} \Rightarrow x = 2$$

32. c. $\tan^{-1}x + 2\cot^{-1}x = \dfrac{2\pi}{3}$

$$\Rightarrow \tan^{-1}x = 2\left(\frac{\pi}{3} - \cot^{-1}x\right)$$

$$= 2\left(\frac{\pi}{3} - \left(\frac{\pi}{2} - \tan^{-1}x\right)\right) = 2\left(-\frac{\pi}{6} + \tan^{-1}x\right)$$

$$\Rightarrow \tan^{-1}x = \frac{\pi}{3} \Rightarrow x = \tan\frac{\pi}{3} = \sqrt{3}$$

33. a. Put $x = \tan\theta$. Therefore,

$$3\sin^{-1}\frac{2\tan\theta}{1+\tan^2\theta} - 4\cos^{-1}\frac{1-\tan^2\theta}{1+\tan^2\theta}$$

$$+ 2\tan^{-1}\frac{2\tan\theta}{1-\tan^2\theta} = \frac{\pi}{3}$$

$$\Rightarrow 3\sin^{-1}(\sin 2\theta) - 4\cos^{-1}(\cos 2\theta)$$

$$+ 2\tan^{-1}(\tan 2\theta) = \frac{\pi}{3}$$

$$\Rightarrow 3(2\theta) - 4(2\theta) + 2(2\theta) = \frac{\pi}{3} \Rightarrow 2\theta = \frac{\pi}{3}$$

$$\Rightarrow \theta = \frac{\pi}{6} \Rightarrow x = \tan\frac{\pi}{6} = \frac{1}{\sqrt{3}}$$

34. c. Put $\sin^{-1}\dfrac{5}{x} = A \Rightarrow \dfrac{5}{x} = \sin A$

$$\sin^{-1}\frac{12}{x} = B \Rightarrow \frac{12}{x} = \sin B \Rightarrow A + B = \frac{\pi}{2}$$

$$\Rightarrow \sin A = \sin\left(\frac{\pi}{2} - B\right) = \cos B = \sqrt{1 - \sin^2 B}$$

$$\Rightarrow \frac{5}{x} = \sqrt{1 - \frac{144}{x^2}} \Rightarrow \frac{169}{x^2} = 1$$

$\Rightarrow x^2 = 169 \quad \Rightarrow \quad x = 13$
($\because x = -13$ does not satisfy the given equation)

35. c. $\sin^{-1}(x-1) \Rightarrow -1 \le x - 1 \le 1 \Rightarrow 0 \le x \le 2$
$\cos^{-1}(x-3) \Rightarrow -1 \le x - 3 \le 1 \Rightarrow 2 \le x \le 4$
$\therefore x = 2$

So, $\sin^{-1}(2-1) + \cos^{-1}(2-3) + \tan^{-1}\dfrac{2}{2-4}$
$= \cos^{-1}k + \pi$

$\Rightarrow \sin^{-1}1 + \cos^{-1}(-1) + \tan^{-1}(-1) = \cos^{-1}k + \pi$

$$\Rightarrow \frac{\pi}{2} + \pi - \frac{\pi}{4} = \cos^{-1}k + \pi$$

$$\Rightarrow \cos^{-1}k = \frac{\pi}{4} \Rightarrow k = \frac{1}{\sqrt{2}}$$

36. b. $\cot^{-1}x+\cot^{-1}y+\cot^{-1}z=\frac{\pi}{2}$

$\Rightarrow \frac{\pi}{2}-\tan^{-1}x+\frac{\pi}{2}-\tan^{-1}y+\frac{\pi}{2}-\tan^{-1}z=\frac{\pi}{2}$

$\Rightarrow \tan^{-1}x+\tan^{-1}y+\tan^{-1}z=\pi$

$\Rightarrow \tan^{-1}x+\tan^{-1}y=\pi-\tan^{-1}z$

$\Rightarrow \tan(\tan^{-1}x+\tan^{-1}y)=\tan(\pi-\tan^{-1}z)$

$\Rightarrow \frac{x+y}{1-xy}=-z$

$\Rightarrow x+y+z=xyz$

37. d. Given that $\sin^{-1}x+\sin^{-1}y+\sin^{-1}z=\pi$

$\Rightarrow \sin^{-1}(x)+\sin^{-1}(y)+\sin^{-1}(z)=\sin^{-1}(-1)$

$\Rightarrow \sin^{-1}(x)+\sin^{-1}(y)=\pi-\sin^{-1}(z)$

$\Rightarrow \sin^{-1}\left(xy-\sqrt{1-x^2}\sqrt{1-y^2}\right)=\sin^{-1}(-z)$

$\Rightarrow xy-\sqrt{(1-x^2)(1-y^2)}=-z$

$\Rightarrow (xy+z)=\sqrt{(1-x^2)(1-y^2)}$

Squaring both sides, we get $x^2+y^2+z^2+2xyz=1$

Trick: Put $x=y=z=\frac{1}{2}$ so that $\cos^{-1}\frac{1}{2}+\cos^{-1}\frac{1}{2}$

$+\cos^{-1}\frac{1}{2}=\pi.$

38. d. Given that $\tan^{-1}x+\tan^{-1}y+\tan^{-1}z=\frac{\pi}{2}$

$\Rightarrow \tan^{-1}\left[\frac{x+y+z-xyz}{1-xy-yz-xz}\right]=\frac{\pi}{2}$

Hence, $xy+yz+zx-1=0$

39. d. Let $\alpha=\cos^{-1}\sqrt{p}$, $\beta=\cos^{-1}\sqrt{1-p}$, and

$\gamma=\cos^{-1}\sqrt{1-q}$

$\Rightarrow \cos\alpha=\sqrt{p}$, $\cos\beta=\sqrt{1-p}$, and

$\cos\gamma=\sqrt{1-q}$

Therefore, $\sin\alpha=\sqrt{1-p}$, $\sin\beta=\sqrt{p}$, and

$\sin\gamma=\sqrt{q}$

The given equation may be written as $\alpha+\beta+\gamma=\frac{3\pi}{4}$

$\Rightarrow \alpha+\beta=\frac{3\pi}{4}-\gamma$

$\Rightarrow \cos(\alpha+\beta)=\cos\left(\frac{3\pi}{4}-\gamma\right)$

$\Rightarrow \cos\alpha\cos\beta-\sin\alpha\sin\beta=\cos\left(\pi-\left(\frac{\pi}{4}+\gamma\right)\right)$

$=-\cos\left(\frac{\pi}{4}+\gamma\right)$

$\Rightarrow \sqrt{p}\sqrt{1-p}-\sqrt{1-p}\sqrt{p}$

$=-\left(\frac{1}{\sqrt{2}}\sqrt{1-q}-\frac{1}{\sqrt{2}}\sqrt{q}\right)$

$\Rightarrow 0=\sqrt{1-q}-\sqrt{q}\Rightarrow 1-q=q\Rightarrow q=\frac{1}{2}$

40. a. Let $\sin^{-1}a=A$, $\sin^{-1}b=B$, $\sin^{-1}c=C$

$\Rightarrow \sin A=a$, $\sin B=b$, $\sin C=c$

and $A+B+C=\pi$

$\Rightarrow \sin 2A+\sin 2B+\sin 2C=4\sin A\sin B\sin C$ (1)

$\Rightarrow \sin A\cos A+\sin B\cos B+\sin C\cos C$

$=2\sin A\sin B\sin C$

$\Rightarrow \sin A\sqrt{(1-\sin^2 A)}+\sin B\sqrt{(1-\sin^2 B)}$

$+\sin C\sqrt{1-\sin^2 C}=2\sin A\sin B\sin C$ (2)

$\Rightarrow a\sqrt{(1-a^2)}+b\sqrt{(1-b^2)}+c\sqrt{(1-c)^2}=2abc$

Trick: Let $a=\frac{1}{\sqrt{2}}$, $b=\frac{1}{\sqrt{2}}$, $c=1$.

Then $a\sqrt{1-a^2}+b\sqrt{1-b^2}+c\sqrt{1-c^2}$

$=\frac{1}{\sqrt{2}}\sqrt{1-\frac{1}{2}}+\frac{1}{\sqrt{2}}\sqrt{1-\frac{1}{2}}+1\sqrt{1-1}=1.$

41. b. Since $\sin^{-1}x+\sin^{-1}y+\sin^{-1}z=\pi$,

$\sin^{-1}x+\sin^{-1}y=\pi-\sin^{-1}z$

$\Rightarrow \sin^{-1}\left(x\sqrt{1-y^2}+y\sqrt{1-x^2}\right)=\pi-\sin^{-1}(z)$

$\Rightarrow x\sqrt{1-y^2}+y\sqrt{1-x^2}=\sin(\pi-\sin^{-1}(z))$

$=\sin(\sin^{-1}z)=z$

$\Rightarrow x^2(1-y^2)=z^2+y^2(1-x^2)-2zy\sqrt{1-x^2}$

$\Rightarrow (x^2-y^2-z^2)^2=4y^2z^2(1-x^2)$

$\Rightarrow x^4+y^4+z^4-2x^2y^2-2x^2z^2+2y^2z^2=4y^2z^2-4x^2y^2z^2$

$\Rightarrow x^4+y^4+z^4+4x^2y^2z^2=2(x^2y^2+y^2z^2+z^2x^2)$

$\Rightarrow K=2$

42. d. We have $\cos^{-1}x-\cos^{-1}\frac{y}{2}=\alpha$

$\Rightarrow x=\cos\left(\cos^{-1}\frac{y}{2}+\alpha\right)$

$=\cos\left(\cos^{-1}\frac{y}{2}\right)\cos\alpha-\sin\left(\cos^{-1}\frac{y}{2}\right)\sin\alpha$

$$= \frac{y}{2}\cos\alpha - \sqrt{1-\frac{y^2}{4}}\sin\alpha$$

$\Rightarrow 2x = y\cos\alpha - \sin\alpha\sqrt{4-y^2}$

$\Rightarrow 2x - y\cos\alpha = -\sin\alpha\sqrt{4-y^2}$

Squaring, we get

$4x^2 + y^2\cos^2\alpha - 4xy\cos\alpha = 4\sin^2\alpha - y^2\sin^2\alpha$

$\Rightarrow 4x^2 - 4xy\cos\alpha + y^2 = 4\sin^2\alpha$

43. d. $2\tan^{-1}2x = \sin^{-1}\dfrac{4x}{1+4x^2}$

$\Rightarrow -\dfrac{\pi}{2} \le 2\tan^{-1}2x \le \dfrac{\pi}{2}$

$\Rightarrow -\dfrac{\pi}{4} \le \tan^{-1}2x \le \dfrac{\pi}{4}$

$\Rightarrow -1 \le 2x \le 1$

$\Rightarrow -\dfrac{1}{2} \le x \le \dfrac{1}{2}$

44. b. $\cos^{-1}(2x^2-1) = 2\pi - 2\cos^{-1}x$ (as $x < 0$)

$\Rightarrow \cos^{-1}(2x^2-1) - 2\sin^{-1}x = 2\pi - 2\cos^{-1}x - 2\sin^{-1}x$

$= 2\pi - 2(\cos^{-1}x + \sin^{-1}x)$

$= 2\pi - 2\dfrac{\pi}{2} = \pi$

45. c. $2\sin^{-1}x = \sin^{-1}\left(2x\sqrt{1-x^2}\right)$

Range of the right hand angle is $\left[-\dfrac{\pi}{2}, \dfrac{\pi}{2}\right]$

$\Rightarrow -\dfrac{\pi}{2} \le 2\sin^{-1}x \le \dfrac{\pi}{2}$

$\Rightarrow \dfrac{-\pi}{4} \le \sin^{-1}x \le \dfrac{\pi}{4}$

$\Rightarrow x \in \left[-\dfrac{1}{\sqrt{2}}, \dfrac{1}{\sqrt{2}}\right]$

46. c. $x_1 = 2\tan^{-1}\left(\dfrac{1+x}{1-x}\right)$ and

$x_2 = \sin^{-1}\left(\dfrac{1-x^2}{1+x^2}\right) = \tan^{-1}\left(\dfrac{1-x^2}{2x}\right)$

Now $\dfrac{1+x}{1-x} > 1$

$$\Rightarrow x_1 = \pi + \tan^{-1}\left(\frac{2\left(\frac{1+x}{1-x}\right)}{1-\left(\frac{1+x}{1-x}\right)^2}\right)$$

$= \pi + \tan^{-1}\left(\dfrac{1-x^2}{-2x}\right)$

$= \pi - \tan^{-1}\left(\dfrac{1-x^2}{2x}\right)$

$\Rightarrow x_1 + x_2 = \pi$

47. d. $\sin(2\sin^{-1}(0.8))$

$= \sin(\sin^{-1}(2\times 0.8\sqrt{1-(0.8)^2}))$

$= \sin(\sin^{-1}0.96)$

$= 0.96$

48. b. $\tan(\sin^{-1}(\cos(\sin^{-1}x)))\tan(\cos^{-1}(\sin(\cos^{-1}x)))$

$= \tan(\sin^{-1}(\cos(\cos^{-1}\sqrt{1-x^2})))$

$\tan\{\cos^{-1}(\sin(\sin^{-1}\sqrt{1-x^2}))\}$

$= \tan(\sin^{-1}\sqrt{1-x^2})\tan(\cos^{-1}\sqrt{1-x^2})$

$= \tan(\cos^{-1}x)\tan(\sin^{-1}x)$

$= \tan(\cos^{-1}x)\tan(\pi/2 - \cos^{-1}x)$

$= \tan(\cos^{-1}x)\cot(\cos^{-1}x) = 1$

49. d. Since $\sin^{-1}\left(\dfrac{2x}{1+x^2}\right) = 2\tan^{-1}x$ for $x \in (-1, 1)$,

$\sin^{-1}\left(\dfrac{2a}{1+a^2}\right) + \sin^{-1}\left(\dfrac{2b}{1+b^2}\right) = 2\tan^{-1}x$

$\Rightarrow 2\tan^{-1}a + 2\tan^{-1}b = 2\tan^{-1}x$

$\Rightarrow \tan^{-1}a + \tan^{-1}b = \tan^{-1}x$

$\Rightarrow \tan^{-1}\left(\dfrac{a+b}{1-ab}\right) = \tan^{-1}x$

$\Rightarrow x = \dfrac{a+b}{1-ab}$

50. b. If $x < 0$, then $\sin^{-1}x < 0$ but $\cos^{-1}\sqrt{1-x^2}$ is always positive.

So $\sin^{-1}x = -\cos^{-1}\sqrt{1-x^2}$.

51. b. We have

$\dfrac{xy}{zr}\dfrac{yz}{xr} = \dfrac{y^2}{r^2} = \dfrac{y^2}{x^2+y^2+z^2} < 1$

$\Rightarrow \tan^{-1}\left(\dfrac{xy}{zr}\right) + \tan^{-1}\left(\dfrac{yz}{xr}\right) + \tan^{-1}\left(\dfrac{xz}{yr}\right)$

$$= \tan^{-1}\left(\frac{\frac{xy}{zr} + \frac{yz}{xr}}{1 - \frac{xy}{zr}\frac{yz}{xr}}\right) + \tan^{-1}\left(\frac{xz}{yr}\right)$$

$$= \tan^{-1}\left(\frac{\frac{y(x^2+z^2)}{xzr}}{\frac{r^2-y^2}{r^2}}\right)+\tan^{-1}\left(\frac{xz}{yr}\right)$$

$$= \tan^{-1}\left(\frac{\frac{yr(x^2+z^2)}{xz}}{(x^2+z^2)}\right)+\tan^{-1}\left(\frac{xz}{yr}\right)$$

$$= \tan^{-1}\left(\frac{yr}{xz}\right)+\tan^{-1}\left(\frac{xz}{yr}\right)=\frac{\pi}{2}$$

52. b. Let $x=\sin\theta$ where $-\frac{1}{2}\le x\le 1 \Rightarrow -\frac{\pi}{6}\le\theta\le\frac{\pi}{2}$

Then $f(x)=\sin^{-1}\left(\frac{\sqrt{3}}{2}x-\frac{1}{2}\sqrt{1-x^2}\right)$

$$=\sin^{-1}\left(\frac{\sqrt{3}}{2}\sin\theta-\frac{1}{2}\cos\theta\right)$$

$$=\sin^{-1}\left(\sin\left(\theta-\frac{\pi}{6}\right)\right)$$

$$=\theta-\frac{\pi}{6}=\sin^{-1}x-\frac{\pi}{6}$$

$$\left(\because \theta-\frac{\pi}{6}\in\left[\frac{-\pi}{3},\frac{\pi}{3}\right]\right)$$

53. c. Let $y=\tan^{-1}\left(\frac{1-x^2}{2x}\right)+\cos^{-1}\left(\frac{1-x^2}{1+x^2}\right)$.

Put $x=\tan\theta$. As $x\in(0,1)$, $\theta\in\left(0,\frac{\pi}{4}\right)$ and $\frac{\pi}{2}-2\theta\in(0,\pi/2)$. Therefore,

$y=\tan^{-1}(\cot 2\theta)+\cos^{-1}(\cos 2\theta)$

$$=\tan^{-1}\left(\tan\left(\frac{\pi}{2}-2\theta\right)\right)+\cos^{-1}(\cos 2\theta)$$

$$=\frac{\pi}{2}-2\theta+2\theta=\frac{\pi}{2}$$

54. c. $\sin^{-1}x=2\sin^{-1}a$

Now $-\frac{\pi}{2}\le\sin^{-1}x\le\frac{\pi}{2}$

$$\Rightarrow -\frac{\pi}{2}\le 2\sin^{-1}a\le\frac{\pi}{2}$$

$$\Rightarrow -\frac{\pi}{4}\le\sin^{-1}a\le\frac{\pi}{4}$$

$$\Rightarrow \frac{1}{\sqrt{2}}\le a\le\frac{1}{\sqrt{2}}\Rightarrow |a|\le\frac{1}{\sqrt{2}}$$

55. a. Let $\tan^{-1}x=\theta$,

where $-\frac{\pi}{2}<\theta<\frac{\pi}{2}\Rightarrow -\pi<2\theta<\pi$

Let $\frac{\pi}{2}<2\theta<\pi\Rightarrow\frac{\pi}{4}<\theta<\frac{\pi}{2}$

$$\Rightarrow \frac{\pi}{4}<\tan^{-1}x<\frac{\pi}{2}\Rightarrow x>1$$

$$\Rightarrow \tan^{-1}\left(\frac{2\tan\theta}{1-\tan^2\theta}\right)$$

$$=\tan^{-1}(\tan 2\theta)$$

$$=\tan^{-1}(\tan(2\theta-\pi)=2\theta-\pi$$

$$=2\tan^{-1}x-\pi$$

56. c. $\sin^{-1}\left(\frac{\sqrt{r}-\sqrt{r-1}}{\sqrt{r(r+1)}}\right)=\tan^{-1}\left(\frac{\sqrt{r}-\sqrt{r-1}}{1+\sqrt{r(r-1)}}\right)$

$$\Rightarrow\sum_{r=1}^{n}\sin^{-1}\left(\frac{\sqrt{r}-\sqrt{r-1}}{\sqrt{r(r+1)}}\right)$$

$$=\sum_{r=1}^{n}(\tan^{-1}\sqrt{r}-\tan^{-1}\sqrt{r-1})$$

$$=\tan^{-1}\sqrt{n}$$

57. b. $\sum_{r=1}^{n}\tan^{-1}\left(\frac{2^{r-1}}{1+2^{2r-1}}\right)$

$$=\sum_{r=1}^{n}\tan^{-1}\left(\frac{2^{r-1}}{1+2^r\,2^{r-1}}\right)$$

$$=\sum_{r=1}^{n}\tan^{-1}\left(\frac{2^r-2^{r-1}}{1+2^r\,2^{r-1}}\right)$$

$$=\sum_{r=1}^{n}[\tan^{-1}(2^r)-\tan^{-1}(2^{r-1})]=\tan^{-1}(2^n)-\tan^{-1}(1)$$

$$=\tan^{-1}(2^n)-\frac{\pi}{4}$$

58. a. We have $\sum_{m=1}^{n}\tan^{-1}\left(\frac{2m}{m^4+m^2+2}\right)$

$$=\sum_{m=1}^{n}\tan^{-1}\left(\frac{2m}{1+(m^2+m+1)(m^2-m+1)}\right)$$

$$=\sum_{m=1}^{n}\tan^{-1}\left(\frac{(m^2+m+1)-(m^2-m+1)}{1+(m^2+m+1)(m^2-m+1)}\right)$$

$$=\sum_{m=1}^{n}[\tan^{-1}(m^2+m+1)-\tan^{-1}(m^2-m+1)]$$

$$= (\tan^{-1} 3 - \tan^{-1} 1) + (\tan^{-1} 7 - \tan^{-1} 3) + (\tan^{-1} 13 - \tan^{-1} 7) + \cdots + [\tan^{-1}(n^2+n+1) - \tan^{-1}(n^2-n+1)]$$

$$= \tan^{-1}(n^2+n+1) - \tan^{-1} 1 = \tan^{-1}\left(\frac{n^2+n}{2+n^2+n}\right)$$

For $n \to \infty$, sum $= \tan^{-1}(1) = \frac{\pi}{4}$.

59. a. $\tan^{-1}\left(\frac{1}{1+r+r^2}\right) = \tan^{-1}\left(\frac{r+1-r}{1+r(r+1)}\right)$

$$= \tan^{-1}(r+1) - \tan^{-1}(r)$$

$$\Rightarrow \sum_{r=0}^{n}[\tan^{-1}(r+1) - \tan^{-1}(r)] = \tan^{-1}(n+1) - \tan^{-1}(0)$$

$$= \tan^{-1}(n+1)$$

$$\Rightarrow \sum_{r=0}^{\infty}\tan^{-1}\left(\frac{1}{1+r+r^2}\right) = \tan^{-1}(\infty) = \frac{\pi}{2}$$

60. b. Obviously, $x = \sin(\theta+\beta)$ and $y = \sin(\theta-\beta)$

$$\Rightarrow 1 + xy = 1 + \sin(\theta+\beta)\sin(\theta-\beta)$$

$$= 1 + \sin^2\theta - \sin^2\beta$$

$$= \sin^2\theta + \cos^2\beta$$

61. a. Let $\sqrt{\tan\alpha} = \tan x$,

Then $u = \cot^{-1}(\tan x) - \tan^{-1}(\tan x)$

$$= \frac{\pi}{2} - x - x = \frac{\pi}{2} - 2x$$

$$\Rightarrow 2x = \frac{\pi}{2} - u \Rightarrow \frac{\pi}{4} - \frac{u}{2}$$

$$\Rightarrow \tan x = \tan\left(\frac{\pi}{4} - \frac{u}{2}\right)$$

$$\Rightarrow \sqrt{\tan\alpha} = \tan\left(\frac{\pi}{4} - \frac{u}{2}\right)$$

62. b. $\tan\left[\frac{\pi}{4} + \frac{1}{2}\cos^{-1}\frac{a}{b}\right] + \tan\left[\frac{\pi}{4} - \frac{1}{2}\cos^{-1}\frac{a}{b}\right]$

Let $\frac{1}{2}\cos^{-1}\frac{a}{b} = \theta \Rightarrow \cos 2\theta = \frac{a}{b}$

Thus, $\tan\left[\frac{\pi}{4} + \theta\right] + \tan\left[\frac{\pi}{4} - \theta\right]$

$$= \frac{1+\tan\theta}{1-\tan\theta} + \frac{1-\tan\theta}{1+\tan\theta}$$

$$= \frac{1+\tan^2\theta + 2\tan\theta + 1 + \tan^2\theta - 2\tan\theta}{(1-\tan^2\theta)}$$

$$= \frac{2(1+\tan^2\theta)}{1-\tan^2\theta} = \frac{2}{\cos 2\theta} = \frac{2}{a/b} = \frac{2b}{a}$$

63. a. $2\tan^{-1}\left[\sqrt{\frac{a-b}{a+b}}\tan\frac{\theta}{2}\right]$

$$= \cos^{-1}\left[\frac{1 - \left(\frac{a-b}{a+b}\right)\tan^2\frac{\theta}{2}}{1 + \left(\frac{a-b}{a+b}\right)\tan^2\frac{\theta}{2}}\right]$$

$$\left(\because 2\tan^{-1} x = \cos^{-1}\frac{1-x^2}{1+x^2}\right)$$

$$= \cos^{-1}\left[\frac{(a+b) - (a-b)\tan^2\frac{\theta}{2}}{(a+b) + (a-b)\tan^2\frac{\theta}{2}}\right]$$

$$= \cos^{-1}\left[\frac{a\left(1 - \tan^2\frac{\theta}{2}\right) + b\left(1 + \tan^2\frac{\theta}{2}\right)}{a\left(1 + \tan^2\frac{\theta}{2}\right) + b\left(1 - \tan^2\frac{\theta}{2}\right)}\right]$$

$$= \cos^{-1}\left[\frac{\frac{a\left(1-\tan^2\frac{\theta}{2}\right)}{1+\tan^2\frac{\theta}{2}} + b}{a + b\left(\frac{1-\tan^2\frac{\theta}{2}}{1+\tan^2\frac{\theta}{2}}\right)}\right]$$

$$= \cos^{-1}\left[\frac{a\cos\theta + b}{a + b\cos\theta}\right]$$

64. d. $\cot^{-1}\left[\frac{\sqrt{1-\sin x} + \sqrt{1+\sin x}}{\sqrt{1-\sin x} - \sqrt{1+\sin x}}\right]$

$$= \cot^{-1}\left[\frac{(\sqrt{1-\sin x} + \sqrt{1+\sin x})}{(\sqrt{1-\sin x} - \sqrt{1+\sin x})}\frac{(\sqrt{1-\sin x} + \sqrt{1+\sin x})}{(\sqrt{1-\sin x} + \sqrt{1+\sin x})}\right]$$

$$= \cot^{-1}\left[\frac{(1-\sin x) + (1+\sin x) + 2\sqrt{1-\sin^2 x}}{(1-\sin x) - (1+\sin x)}\right]$$

$$= \cot^{-1}\left[\frac{2(1+\cos x)}{-2\sin x}\right]$$

$$= \cot^{-1}\left[-\frac{2\cos^2(x/2)}{2\sin(x/2)\cos(x/2)}\right]$$

$$= \cot^{-1}\left(-\cot\frac{x}{2}\right) = \cot^{-1}\left[\cot\left(\pi - \frac{x}{2}\right)\right] = \pi - \frac{x}{2}$$

65. b. $\tan^{-1}\left(\dfrac{x\cos\theta}{1-x\sin\theta}\right) - \cot^{-1}\left(\dfrac{\cos\theta}{x-\sin\theta}\right)$

$$= \tan^{-1}\left(\frac{x\cos\theta}{1-x\sin\theta}\right) - \tan^{-1}\left(\frac{x-\sin\theta}{\cos\theta}\right)$$

$$= \tan^{-1}\left(\frac{\dfrac{x\cos\theta}{1-x\sin\theta} - \dfrac{x-\sin\theta}{\cos\theta}}{1+\left(\dfrac{x\cos\theta}{1-x\sin\theta}\right)\left(\dfrac{x-\sin\theta}{\cos\theta}\right)}\right)$$

$$= \tan^{-1}\left(\frac{x\cos^2\theta - x + \sin\theta + x^2\sin\theta - x\sin^2\theta}{\cos\theta - x\cos\theta\sin\theta + x^2\cos\theta - x\cos\theta\sin\theta}\right)$$

$$= \tan^{-1}\left(\frac{-x\sin^2\theta - x + \sin\theta + x^2\sin\theta - x\sin^2\theta}{\cos\theta - 2x\cos\theta\sin\theta + x^2\cos\theta}\right)$$

$$= \tan^{-1}\left(\frac{-2x\sin^2\theta + \sin\theta + x^2\sin\theta}{\cos\theta - 2x\cos\theta\sin\theta + x^2\cos\theta}\right)$$

$$= \tan^{-1}\left(\frac{\sin\theta(-2x\sin\theta + 1 + x^2)}{\cos\theta(1-2x\sin\theta + x^2)}\right)$$

$$= \tan^{-1}(\tan\theta) = \theta$$

66. d. $\tan^{-1}\left(\dfrac{\tan x}{4}\right) + \tan^{-1}\left(\dfrac{3\sin 2x}{5+3\cos 2x}\right)$

$$= \tan^{-1}\left(\frac{\tan x}{4}\right) + \tan^{-1}\left(\frac{\dfrac{6\tan x}{1+\tan^2 x}}{5+\dfrac{3(1-\tan^2 x)}{1+\tan^2 x}}\right)$$

$$= \tan^{-1}\left(\frac{\tan x}{4}\right) + \tan^{-1}\left(\frac{6\tan x}{8+2\tan^2 x}\right)$$

$$= \tan^{-1}\left(\frac{\tan x}{4}\right) + \tan^{-1}\left(\frac{3\tan x}{4+\tan^2 x}\right)$$

$$= \tan^{-1}\left(\frac{\dfrac{\tan x}{4} + \dfrac{3\tan x}{4+\tan^2 x}}{1-\dfrac{3\tan^2 x}{4(4+\tan^2 x)}}\right) \left[\text{as } \left|\frac{\tan x}{4}\ \frac{3\tan x}{4+\tan^2 x}\right| < 1\right]$$

$$= \tan^{-1}\left(\frac{16\tan x + \tan^3 x}{16+\tan^2 x}\right)$$

$$= \tan^{-1}(\tan x) = x$$

67. a. $\cot^{-1}\left(\sqrt{\cos\alpha}\right) - \tan^{-1}\left(\sqrt{\cos\alpha}\right) = x$

$$\Rightarrow \tan^{-1}\left(\frac{1}{\sqrt{\cos\alpha}}\right) - \tan^{-1}\left(\sqrt{\cos\alpha}\right) = x$$

$$\Rightarrow \tan^{-1}\frac{\dfrac{1}{\sqrt{\cos\alpha}} - \sqrt{\cos\alpha}}{1+\dfrac{1}{\sqrt{\cos\alpha}}\sqrt{\cos\alpha}} = x$$

$$\Rightarrow \tan^{-1}\frac{1-\cos\alpha}{2\sqrt{\cos\alpha}} = x$$

$$\Rightarrow \tan x = \frac{1-\cos\alpha}{2\sqrt{\cos\alpha}}$$

$$\Rightarrow \cot x = \frac{2\sqrt{\cos\alpha}}{1-\cos\alpha}$$

$$\Rightarrow \operatorname{cosec} x = \sqrt{1+\frac{4\cos\alpha}{(1-\cos\alpha)^2}} = \frac{1+\cos\alpha}{1-\cos\alpha}$$

$$\Rightarrow \sin x = \frac{1-\cos\alpha}{1+\cos\alpha} = \frac{2\sin^2\alpha/2}{2\cos^2\alpha/2} = \tan^2\alpha/2$$

68. c. $\tan\left(\dfrac{\pi}{4} + \dfrac{1}{2}\cos^{-1}x\right) + \tan\left(\dfrac{\pi}{4} - \dfrac{1}{2}\cos^{-1}x\right)$

$$= \frac{1+\tan\left(\frac{1}{2}\cos^{-1}x\right)}{1-\tan\left(\frac{1}{2}\cos^{-1}x\right)} + \frac{1-\tan\left(\frac{1}{2}\cos^{-1}x\right)}{1+\tan\left(\frac{1}{2}\cos^{-1}x\right)}$$

$$= \frac{\left(1+\tan\left(\frac{1}{2}\cos^{-1}x\right)\right)^2 + \left(1-\tan\left(\frac{1}{2}\cos^{-1}x\right)\right)^2}{1-\tan^2\left(\frac{1}{2}\cos^{-1}x\right)}$$

$$= 2\frac{1+\tan^2\left(\frac{1}{2}\cos^{-1}x\right)}{1-\tan^2\left(\frac{1}{2}\cos^{-1}x\right)}$$

$$= \frac{2}{\cos(\cos^{-1}x)} = \frac{2}{x}$$

69. c. We have, $(\sin^{-1}x)^3 + (\cos^{-1}x)^3$

$= (\sin^{-1}x + \cos^{-1}x)^3 - 3\sin^{-1}\cos^{-1}x(\sin^{-1}x + \cos^{-1}x)$

$$= \frac{\pi^3}{8} - 3(\sin^{-1}x\cos^{-1}x)\frac{\pi}{2}$$

$$= \frac{\pi^3}{8} - \frac{3\pi}{2}\sin^{-1}x\left(\frac{\pi}{2} - \sin^{-1}x\right)$$

$$= \frac{\pi^3}{8} - \frac{3\pi^2}{4}\sin^{-1}x + \frac{3\pi}{2}(\sin^{-1}x)^2$$

$$= \frac{\pi^3}{8} + \frac{3\pi}{2}\left[(\sin^{-1}x)^2 - \frac{\pi}{2}\sin^{-1}x\right]$$

$$= \frac{\pi^3}{8} + \frac{3\pi}{2}\left[\left(\sin^{-1}x - \frac{\pi}{4}\right)^2\right] - \frac{3\pi^3}{32}$$

$$= \frac{\pi^3}{32} + \frac{3\pi}{2}\left(\sin^{-1}x - \frac{\pi}{4}\right)^2$$

So the least value is $\frac{\pi^3}{32}$ when $\left(\sin^{-1}x - \frac{\pi}{4}\right) = 0.$

And the greatest value occurs when

$$\left(\sin^{-1}x - \frac{\pi}{4}\right)^2 = \left(-\frac{\pi}{2} - \frac{\pi}{4}\right)^2 = \frac{9\pi^2}{16}.$$

Therefore, the greatest value is

$$\frac{\pi^3}{32} + \frac{9\pi^2}{16} \times \frac{3\pi}{2} = \frac{7\pi^3}{8}$$

70. c. $f(x) = \sin^{-1}x + \tan^{-1}x + \sec^{-1}x$, clearly domain of $f(x)$ is $x = \pm 1$.

Thus, the range is $\{f(1), f(-1)\}$, i.e., $\left\{\frac{\pi}{4}, \frac{3\pi}{4}\right\}$.

71. a. $1 + x^2 \ge 2|x|$

$$\Rightarrow \frac{2|x|}{1+x^2} \le 1$$

$$\Rightarrow -1 \le \frac{2x}{1+x^2} \le 1 \Rightarrow \quad \tan^{-1}\left(\frac{2x}{1+x^2}\right) \in \left[-\frac{\pi}{4}, \frac{\pi}{4}\right]$$

72. c. $[\cot^{-1}x] + [\cos^{-1}x] = 0$

As $\cos^{-1}x, \cot^{-1}x \ge 0, \quad [\cot^{-1}x] = [\cos^{-1}x] = 0$

$[\cot^{-1}x] = 0 \quad \Rightarrow x \in (\cot 1, \infty)$ (1)

$[\cos^{-1}x] = 0 \quad \Rightarrow x \in (\cos 1, 1]$ (2)

Hence, from equations (1) and (2), $x \in (\cot 1, 1]$

73. d. $\frac{3\pi}{2} < 5 < \frac{5\pi}{2}$

$\Rightarrow \sin^{-1}(\sin 5) = 5 - 2\pi$

Given $\sin^{-1}(\sin 5) > x^2 - 4x$

$\Rightarrow x^2 - 4x + 4 < 9 - 2\pi$

$\Rightarrow (x-2)^2 < 9 - 2\pi$

$\Rightarrow -\sqrt{9-2\pi} < x - 2 < \sqrt{9-2\pi}$

$\Rightarrow 2 - \sqrt{9-2\pi} < x < 2 + \sqrt{9-2\pi}$

74. c. $\frac{\alpha^3}{2}\operatorname{cosec}^2\left(\frac{1}{2}\tan^{-1}\frac{\alpha}{\beta}\right) + \frac{\beta^3}{2}\sec^2\left(\frac{1}{2}\tan^{-1}\frac{\beta}{\alpha}\right)$

$$= \alpha^3 \frac{1}{1-\cos\left(\tan^{-1}\left(\frac{\alpha}{\beta}\right)\right)} + \beta^3 \frac{1}{1+\cos\left(\tan^{-1}\frac{\beta}{\alpha}\right)}$$

$$= \alpha^3 \frac{1}{1-\cos\left(\cos^{-1}\left(\frac{\beta}{\sqrt{\alpha^2+\beta^2}}\right)\right)} + \beta^3 \frac{1}{1+\cos\left(\cos^{-1}\frac{\alpha}{\sqrt{\alpha^2+\beta^2}}\right)}$$

$$= \alpha^3 \frac{1}{1-\frac{\beta}{\sqrt{\alpha^2+\beta^2}}} + \beta^3 \frac{1}{1+\frac{\alpha}{\sqrt{\alpha^2+\beta^2}}}$$

$$= \sqrt{\alpha^2+\beta^2}\left(\frac{\alpha^3}{\sqrt{\alpha^2+\beta^2}-\beta} + \frac{\beta^3}{\sqrt{\alpha^2+\beta^2}+\alpha}\right)$$

$$= \sqrt{\alpha^2+\beta^2}\left(\alpha^3\frac{(\sqrt{\alpha^2+\beta^2}+\beta)}{\alpha^2} + \beta^3\frac{(\sqrt{\alpha^2+\beta^2}-\alpha)}{\beta^2}\right)$$

$$= \sqrt{\alpha^2+\beta^2}\left[\alpha\left(\sqrt{\alpha^2+\beta^2}+\beta\right) + \beta\left(\sqrt{\alpha^2+\beta^2}-\alpha\right)\right]$$

$$= \sqrt{\alpha^2+\beta^2}\,(\alpha+\beta)\sqrt{\alpha^2+\beta^2}$$

$$= (\alpha+\beta)(\alpha^2+\beta^2)$$

75. d. $\lim_{|x|\to\infty} \cos(\tan^{-1}(\sin(\tan^{-1}x)))$

$= \cos(\tan^{-1}(\sin(\tan^{-1}\infty)))$

$= \cos(\tan^{-1}(\sin(\pi/2)))$

$= \cos(\tan^{-1}(1)) = \cos(\pi/4) = \frac{1}{\sqrt{2}}$

76. d. $\sin^{-1}(-(x-1)(x-2)) + \cos^{-1}((x-3)(x-1)) = \frac{\pi}{4}$

For $x \in [1, 2] \Rightarrow \sin^{-1}(-(x-1)(x-2)) \in [0, \pi/2)$

and $\cos^{-1}((x-3)(x-1)) \in [\pi/2, \pi]$

$\Rightarrow$ No solution in the given range

Also, $-1 \le 3x - 2 - x^2 \le 1$ and

$-1 \le x^2 - 4x + 3 \le 1$

$\Rightarrow 2 - \sqrt{2} \le x \le \frac{3+\sqrt{5}}{2}$

77. d. $2^{2\pi/\sin^{-1}x} - 2(a+2)2^{\pi/\sin^{-1}x} + 8a < 0$

$(2^{\pi/\sin^{-1}x} - 4)(2^{\pi/\sin^{-1}x} - 2a) < 0$

Now $2^{\pi/\sin^{-1}x} \in \left(0, \frac{1}{4}\right] \cup [4, \infty)$

Now for $2^{\pi/\sin^{-1}x} \in \left(0, \frac{1}{4}\right]$

$(2^{\pi/\sin^{-1}x} - 4) < 0$

$\Rightarrow 2^{\pi/\sin^{-1}x} - 2a > 0$

$\Rightarrow 2a < 2^{\pi/\sin^{-1}x}$

$\Rightarrow 2a < \frac{1}{4}$

$\Rightarrow 0 \le a < \frac{1}{8}$

Similarly, for $2^{\pi/\sin^{-1}x} \in [4, \infty), a > 2$

$\Rightarrow a \in \left[0, \frac{1}{8}\right) \cup (2, \infty)$

78. b. Given that $\sin^{-1} x + \tan^{-1} x = 2k + 1$

The range of the function $\sin^{-1} x + \tan^{-1} x$ is

$\left[\frac{-3\pi}{4}, \frac{3\pi}{4}\right]$ (as both functions are increasing)

$\Rightarrow$ Integral values of k are -1 and 0.

79. c. From the given equation,

$\sin^2 \theta - 2 \sin \theta + 7 = 5^{\sec^2 y} + 1$

$\Rightarrow (\sin \theta - 1)^2 + 6 = 5^{\sec^2 y} + 1$

L.H.S. ≤ 6, R.H.S. ≥ 6

Possible solution is $\sin \theta = -1$ when L.H.S.

= R.H.S.

$\Rightarrow \cos^2 \theta = 0 \Rightarrow \cos^2 \theta - \sin \theta = 1$

80. d. $[\cot^{-1} x] + 2[\tan^{-1} x] = 0$

$\Rightarrow [\cot^{-1} x] = 0, [\tan^{-1} x] = 0$

or $[\cot^{-1} x] = 2, [\tan^{-1} x] = -1$

$[\cot^{-1} x] = 0 \Rightarrow x \in (\cot 1, \infty)$

$[\tan^{-1} x] = 0 \Rightarrow x \in (0, \tan 1)$

$\Rightarrow$ For $[\cot^{-1} x] = [\tan^{-1} x] = 0, x \in (\cot 1, \tan 1)$

$[\cot^{-1} x] = 2 \Rightarrow x \in (\cot 3, \cot 2]$

$[\tan^{-1} x] = -1 \Rightarrow x \in [-\tan 1, 0)$

$\Rightarrow$ No such x exists.

Thus the solution set is $(\cot 1, \tan 1)$.

Multiple Correct Answers Type

1. b., c., d.

$6x^2 + 11x + 3 = 0$

$\Rightarrow (2x + 3)(3x + 1) = 0$

$\Rightarrow x = -3/2, -1/3.$

For $x = -3/2$, $\cos^{-1} x$ is not defined as domain of $\cos^{-1} x$ is $[-1, 1]$ and for $x = -1/3$, $\text{cosec}^{-1} x$ is not defined as domain of $\text{cosec}^{-1} x$ is $R - (-1, 1)$.

Howeverz, $\cot^{-1} x$ is defined for both of these values as domain of $\cot^{-1} x$ is R.

2. a., b., c.

Let $\tan^{-1}(-2) = \theta \Rightarrow \tan \theta = -2$

$\Rightarrow \theta = (-\pi/2, 0)$

$\Rightarrow 2\theta = (-\pi, 0)$

$\cos(-2\theta) = \cos 2\theta = \frac{1 - \tan^2 \theta}{1 + \tan^2 \theta} = \frac{-3}{5}$

$\Rightarrow -2\theta = \cos^{-1}\left(\frac{-3}{5}\right) = \pi - \cos^{-1}\frac{3}{5}$

$\Rightarrow 2\theta = -\pi + \cos^{-1}\frac{3}{5} = -\pi + \tan^{-1}\frac{4}{3}$

$= -\pi + \cot^{-1}\frac{3}{4}$

$= -\pi + \frac{\pi}{2} - \tan^{-1}\frac{3}{4} = -\frac{\pi}{2} - \tan^{-1}\frac{3}{4}$

$= -\frac{\pi}{2} + \tan^{-1}\left(-\frac{3}{4}\right)$

3. a., b., d.

$\tan^{-1}(x - 1) + \tan^{-1}(x) + \tan^{-1}(x + 1) = \tan^{-1} 3x$

$\Rightarrow \tan^{-1}(x - 1) + \tan^{-1}(x) = \tan^{-1} 3x - \tan^{-1}(x + 1)$

$\Rightarrow \tan^{-1}\left[\frac{(x-1) + x}{1 - (x-1)(x)}\right] = \tan^{-1}\left[\frac{3x - (x+1)}{1 + 3x(x+1)}\right]$

$\Rightarrow \frac{2x - 1}{1 - x^2 + x} = \frac{2x - 1}{1 + 3x^2 + 3x}$

$\Rightarrow (1 - x^2 + x)(2x - 1) = (1 + 3x^2 + 3x)(2x - 1)$

$\Rightarrow x = 0, \pm\frac{1}{2}$

4. b.

We know that $\sin^{-1}x$ is defined for $x \in [-1, 1]$ and $\sec^{-1}x$ is defined for $x \in (-\infty, -1] \cup [1, \infty)$.

Hence, the given function is defined for $x \in \{-1, 1\}$.

Therefore $f(1) = \pi/2, f(-1) = \pi/2$.

5. a., c., d.

$(\sin^{-1} x + \sin^{-1} w)(\sin^{-1} y + \sin^{-1} z) = \pi^2$

$\Rightarrow \sin^{-1} x + \sin^{-1} w = \sin^{-1} y + \sin^{-1} z = \pi$

or $\sin^{-1} x + \sin^{-1} w = \sin^{-1} y + \sin^{-1} z = -\pi$

$\Rightarrow x = y = z = w = 1$ or $x = y = z = w = -1$

Hence, maximum value of $\begin{vmatrix} x^{N_1} & y^{N_2} \\ z^{N_3} & w^{N_4} \end{vmatrix} = \begin{vmatrix} 1 & -1 \\ 1 & 1 \end{vmatrix} = 2$

and minimum value $\begin{vmatrix} -1 & 1 \\ 1 & 1 \end{vmatrix} = -2$

Also there are 16 different determinants as each place value be 1 or -1.

6. a., d.

Since $|\tan^{-1} x| = \begin{cases} \tan^{-1} x, \text{ if } x \geq 0 \\ -\tan^{-1} x, \text{ if } x < 0 \end{cases}$

$\Rightarrow |\tan^{-1} x| = \tan^{-1} |x| \;\forall\, x \in R$

$\Rightarrow \tan |\tan^{-1} x| = \tan \tan^{-1} |x| = |x|$

Similarly, it can be shown that options (c) and (d) are also correct. However, (b) is not correct for if we take $x = -1$, then $\cot |\cot^{-1} x| = 1$, which is not equal to -1.

7. a., c.

Domain of $f(x) = \ln \cos^{-1} x$ is $x \in [-1, 1)$. Therefore, $[\alpha] = -1$ or 0

8. a., b., c.

a. $\sin\left(\tan^{-1} 3 + \tan^{-1} \dfrac{1}{3}\right) = \sin \dfrac{\pi}{2} = 1$

b. $\cos\left(\dfrac{\pi}{2} - \sin^{-1} \dfrac{3}{4}\right) = \cos\left(\cos^{-1} \dfrac{3}{4}\right) = \dfrac{3}{4}$

c. $\sin\left(\dfrac{1}{4} \sin^{-1} \dfrac{\sqrt{63}}{8}\right)$

Let $\sin^{-1} \dfrac{\sqrt{63}}{8} = \theta$

$\Rightarrow \sin\theta = \dfrac{\sqrt{63}}{8} \Rightarrow \cos\theta = \dfrac{1}{8}$

We have $\cos \dfrac{\theta}{2} = \sqrt{\dfrac{1+\cos\theta}{2}} = \dfrac{3}{4}$

$\Rightarrow \sin \dfrac{\theta}{4} = \sqrt{\dfrac{1-\cos\dfrac{\theta}{2}}{2}} = \dfrac{1}{2\sqrt{2}}$

Now, $\log_2 \sin\left(\dfrac{1}{4} \sin^{-1} \dfrac{\sqrt{63}}{8}\right) = \log_2 \dfrac{1}{2\sqrt{2}} = -\dfrac{3}{2}$

d. $\cos^{-1} \dfrac{\sqrt{5}}{3} = \theta \Rightarrow \cos\theta = \dfrac{\sqrt{5}}{3}$. Hence,

$\tan \dfrac{\theta}{2} = \dfrac{3-\sqrt{5}}{2}$ which is irrational.

9. c., d.

$xy < 0 \Rightarrow \quad x + \dfrac{1}{x} \geq 2, y + \dfrac{1}{y} \leq -2$

or $x + \dfrac{1}{x} \leq -2, y + \dfrac{1}{y} \geq 2$

$x + \dfrac{1}{x} \geq 2 \Rightarrow \quad \sec^{-1}\left(x + \dfrac{1}{x}\right) \in \left[\dfrac{\pi}{3}, \dfrac{\pi}{2}\right)$

$y + \dfrac{1}{y} \leq -2 \Rightarrow \sec^{-1}\left(y + \dfrac{1}{y}\right) \in \left(\dfrac{\pi}{2}, \dfrac{2\pi}{3}\right]$

$\Rightarrow z \in \left(\dfrac{5\pi}{6}, \dfrac{7\pi}{6}\right)$

10. a., d.

Let $f(x) = (\sin^{-1} x)^2 + (\cos^{-1} x)^2$

$= (\sin^{-1} x + \cos^{-1} x)^2 - 2\sin^{-1} x \cos^{-1} x$

$= \dfrac{\pi^2}{4} - 2\sin^{-1} x\left[\dfrac{\pi}{2} - \sin^{-1} x\right]$

$= \dfrac{\pi^2}{4} - \pi \sin^{-1} x + 2(\sin^{-1} x)^2$

$= 2\left[\left(\sin^{-1} x\right)^2 - \dfrac{\pi}{2}\sin^{-1} x + \dfrac{\pi^2}{8}\right]$

$= 2\left(\sin^{-1} x - \dfrac{\pi}{4}\right)^2 + 2\left[\dfrac{\pi^2}{16}\right]$

Now, $-\dfrac{\pi}{2} \leq \sin^{-1} x \leq \dfrac{\pi}{2}$

$\Rightarrow -\dfrac{3\pi}{4} \leq \sin^{-1} x - \dfrac{\pi}{4} \leq \dfrac{\pi}{4}$

$\Rightarrow 0 \leq \left(\sin^{-1} x - \dfrac{\pi}{4}\right)^2 \leq \dfrac{9\pi^2}{16}$

$\Rightarrow 0 \leq 2\left(\sin^{-1} x - \dfrac{\pi}{4}\right)^2 \leq \dfrac{9\pi^2}{8}$

$\Rightarrow \dfrac{\pi^2}{8} \leq \left(\sin^{-1} x - \dfrac{\pi}{4}\right)^2 + \dfrac{2\pi^2}{16} \leq \dfrac{5\pi^2}{4}$

Hence, range of the given function is $\left[\dfrac{\pi^2}{8}, \dfrac{5\pi^2}{4}\right]$.

11. a., d.

For the given equation $0 \leq x, y \leq 1$.

Also, $\sin^{-1} x + \sin^{-1} y = \dfrac{\pi}{2}$

$\Rightarrow \sin^{-1} x = \cos^{-1} y = \sin^{-1} \sqrt{1-y^2}$

$\Rightarrow x = \sqrt{1-y^2} \quad \Rightarrow \quad x^2 + y^2 = 1 \qquad (1)$

Again, $\sin 2x = \cos 2y$

$\Rightarrow \quad \cos\left(\dfrac{\pi}{2} - 2x\right) = \cos 2y$

$\Rightarrow \quad \frac{\pi}{2} - 2x = 2n\pi \pm 2y$, where $n \in I$

$\Rightarrow \quad x \pm y = \frac{\pi}{4} - n\pi \qquad (2)$

From equations (1) and (2), we get

$x = \frac{\pi}{8} + \sqrt{\frac{1}{2} - \frac{\pi^2}{64}}$ and $y = \sqrt{\frac{1}{2} - \frac{\pi^2}{64}} - \frac{\pi}{8}$

12. a., c.

We have,

$$\cot^{-1}\left(\frac{n^2 - 10n + 21.6}{\pi}\right) > \frac{\pi}{6}$$

$$\Rightarrow \frac{n^2 - 10n + 21.6}{\pi} < \cot\frac{\pi}{6}$$

[As $\cot x$ is decreasing for $0 < x < \pi$]

$\Rightarrow \quad n^2 - 10n + 21.6 < \pi\sqrt{3}$

$\Rightarrow n^2 - 10n + 25 + 21.6 - 25 < \pi\sqrt{3}$

$\Rightarrow -\sqrt{\sqrt{3}\pi + 3.4} < n - 5 < \sqrt{\sqrt{3}\pi + 3.4}$

$\Rightarrow 5 - \sqrt{\sqrt{3}\pi + 3.4} < n < 5 + \sqrt{\sqrt{3}\pi + 3.4} \qquad (1)$

Since $\sqrt{3}\pi = 5.5$ nearly

$\therefore \sqrt{\sqrt{3}\pi + 3.4} \sim \sqrt{8.9} \sim 2.9$

$\Rightarrow 2.1 < n < 7.9$

$\therefore \quad n = 3, 4, 5, 6, 7$

{as $n \in N$}

or the minimum value of $n = 3$.

13. a., b., d.

Let t_r denote the rth term of the series 3, 7, 13, 21, ... and

$S = 3 + 7 + 13 + 21 + \cdots + t_n$

Also, $S = 3 + 7 + 13 + \cdots + t_{n-1} + t_n$

Subtracting, we get

$0 = 3 + 4 + 6 + \cdots + 2n - t_n$

$\Rightarrow t_n = 3 + 4 + 6 + \cdots + 2n$

$= 3 + \frac{1}{2}(n-1)(4 + 2n)$

$= n^2 + n + 1$

Let $T_r = \cot^{-1}(r^2 + r + 1)$

$$= \tan^{-1}\left(\frac{1}{r^2 + r + 1}\right)$$

$$= \tan^{-1}\left(\frac{r + 1 - r}{1 + r(r+1)}\right) = \tan^{-1}(r+1) - \tan^{-1} r$$

Thus, the sum of the first n terms of the given series is

$$\sum_{r=1}^{n}\left[\tan^{-1}(r+1) - \tan^{-1} r\right]$$

$$= \tan^{-1}(n+1) - \tan^{-1}(1)$$

$$= \tan^{-1}\left[\frac{n + 1 - n}{1 + 1(n+1)}\right]$$

$$= \tan^{-1}\left(\frac{n}{n+2}\right) = \tan^{-1}\left(\frac{1}{1 + \frac{2}{n}}\right)$$

$$\Rightarrow \quad S_\infty = \lim_{n\to\infty} \tan^{-1}\left(\frac{1}{1 + \frac{2}{n}}\right) = \frac{\pi}{4},$$

$S_{10} = \tan^{-1}\frac{10}{12} = \tan^{-1}\frac{5}{6}$

$S_6 = \tan^{-1}\frac{3}{4} = \sin^{-1}\frac{3}{5}$

$S_{20} = \tan^{-1}\frac{10}{11} = \cot^{-1} 1.1$

14. b.

$f(x) = \sin^{-1}|\sin kx| + \cos^{-1}(\cos kx)$

Let $g(x) = \sin^{-1}|\sin x| + \cos^{-1}(\cos x)$

$$g(x) = \begin{cases} 2x, & 0 \le x \le \frac{\pi}{2} \\ \pi, & \frac{\pi}{2} < x \le \frac{3\pi}{2} \\ 4\pi - 2x, & \frac{3\pi}{2} < x \le 2\pi \end{cases}$$

$g(x)$ is periodic with period 2π and is constant in the continuous interval $\left[2n\pi + \frac{\pi}{2}, 2n\pi + \frac{3\pi}{2}\right]$ (where $n \in I$) and $f(x) = g(kx)$.

So, $f(x)$ is constant in the interval $\left[\frac{2n\pi}{k} + \frac{\pi}{2k}, \frac{2n\pi}{k} + \frac{3\pi}{2k}\right]$

$$\Rightarrow \quad \frac{\pi}{4} = \frac{3\pi}{2k} - \frac{\pi}{2k} \Rightarrow \frac{\pi}{k} = \frac{\pi}{4} \Rightarrow k = 4$$

15. a., c.

Given equation is $x^2 + 2x\sin(\cos^{-1} y) + 1 = 0$. Since x is real, $D \ge 0$,

$4(\sin(\cos^{-1}y))^2 - 4 \ge 0$

$\Rightarrow (\sin(\cos^{-1}y))^2 \ge 1$

$\Rightarrow \sin(\cos^{-1}y) = \pm 1$

$\Rightarrow \cos^{-1}y = \pm\frac{\pi}{2} \Rightarrow y = 0$

Putting value of y in the original equation, we have $x^2+2x+1=0 \Rightarrow x=-1$.

Hence, the equation has only one solution.

16. b., c.

$$1 \le \frac{\pi}{\cos^{-1}x} < \infty \Rightarrow 2 \le 2^{\frac{\pi}{\cos^{-1}x}} < \infty$$

Hence, 2 should lie between or on the roots of $t^2 - \left(a+\frac{1}{2}\right)t - a^2 = 0$ where $t = 2^{\pi/\cos^{-1}x}$.

$\Rightarrow f(2) \le 0 \quad \Rightarrow \quad a^2+2a-3 \ge 0 \quad \Rightarrow a \in (-\infty, -3] \cup [1, \infty)$

17. a., c.

The given relation is possible when

$$a - \frac{a^2}{3} + \frac{a^3}{9} + \cdots = 1 + b + b^2 + \cdots$$

Also, $-1 \le a - \frac{a^2}{3} + \frac{a^3}{9} + \cdots \le 1$ and

$-1 \le 1 + b + b^2 + \cdots \le 1$

$\Rightarrow |b| < 1 \Rightarrow |a| < 3$ and $\dfrac{a}{1+\frac{a}{3}} = \dfrac{1}{1-b}$

$\Rightarrow \dfrac{3a}{a+3} = \dfrac{1}{1-b}$, there are infinitely many solutions.

$\Rightarrow 3a - 3ab = a + 3 \quad \Rightarrow \quad 2a - 3ab = 3$

$\Rightarrow b = \dfrac{2a-3}{3a}$ and $a = \dfrac{3}{2-3b}$

18. a., b.

We know that if $|x| \le 1$,

then $2\tan^{-1}x = \sin^{-1}\left(\dfrac{2x}{1+x^2}\right)$

and if $|x| > 1$, then $\pi - 2\tan^{-1}x = \sin^{-1}\left(\dfrac{2x}{1+x^2}\right)$

Thus, if $|x| > 1$,

$$2\tan^{-1}x + \sin^{-1}\frac{2x}{1+x^2} = 2\tan^{-1}x + \pi - 2\tan^{-1}x = \pi$$

19. a., d.

Case 1: If $0 \le x \le \frac{1}{2}$, then

$$\cos^{-1}\left(\frac{x}{2} + \frac{1}{2}\sqrt{3-3x^2}\right) = \cos^{-1}\left(x\frac{1}{2} + \sqrt{1-x^2}\,\frac{\sqrt{3}}{2}\right)$$

$$= \cos^{-1}x - \cos^{-1}\frac{1}{2}$$

Case 2: If $\frac{1}{2} \le x \le 1$, then

$$\cos^{-1}\left(\frac{x}{2} + \frac{1}{2}\sqrt{3-3x^2}\right) = \cos^{-1}\frac{1}{2} - \cos^{-1}x$$

20. a., b., c.

a. $\cos(\tan^{-1}(\tan(4-\pi))) = \cos(4-\pi) = \cos(\pi-4)$
$= -\cos 4 > 0$

b. $\sin(\cot^{-1}(\cot(4-\pi))) = \sin(4-\pi)$
$= -\sin 4 > 0$ (as $\sin 4 < 0$)

c. $\tan(\cos^{-1}(\cos(2\pi-5))) = \tan(2\pi-5)$
$= -\tan 5 > 0$ (as $\tan 5 < 0$)

d. $\cot(\sin^{-1}(\sin(\pi-4)) = \cot(\pi-4) = -\cot 4 < 0$

Reasoning Type

1. a. Statement 2 is correct, from which we can say $\cot^{-1}x + \cos^{-1}2x = -\pi$ is not possible. Hence, both the statements are correct and statement 2 is the correct explanation of statement 1.

2. d. Obviously, statement 2 is correct but when $x \in [-1, 1]$ we have $\tan^{-1}x \in [-\pi/4, \pi/4]$

$\Rightarrow \frac{\pi}{2} + \tan^{-1}x \in [\pi/4, 3\pi/4]$

Hence, statement 2 is true but statement 1 is false.

3. c. $\operatorname{cosec}^{-1}x > \sec^{-1}x$

$\Rightarrow \operatorname{cosec}^{-1}x > \frac{\pi}{2} - \operatorname{cosec}^{-1}x$

$\Rightarrow \operatorname{cosec}^{-1}x > \frac{\pi}{4}$

$\Rightarrow 1 \le x < \sqrt{2}$ and $\left(\frac{1}{2} + \frac{1}{\sqrt{2}}\right) \in [1, \sqrt{2})$

But statement 2 is false.

4. a. $\sin^{-1}x = \tan^{-1}\dfrac{x}{\sqrt{1-x^2}} > \tan^{-1}x > \tan^{-1}y$

$$\left(\because x > y, \frac{x}{\sqrt{1-x^2}} > x\right)$$

Therefore, Statement 2 is true.

Now, $e < \pi \Rightarrow \dfrac{1}{\sqrt{e}} > \dfrac{1}{\sqrt{\pi}}$

By statement 2,

$$\sin^{-1}\left(\frac{1}{\sqrt{e}}\right) > \tan^{-1}\left(\frac{1}{\sqrt{e}}\right) > \tan^{-1}\left(\frac{1}{\sqrt{\pi}}\right)$$

Therefore, statement 1 is true.

5. d. $30 - 9\pi \in [0, \pi]$ is true but it is not principal value of $\cos^{-1}(\cos 30)$ as $\cos^{-1}(\cos 30)$

$= \cos^{-1}(\cos(9\pi + (30 - 9\pi)))$
$= \cos^{-1}(-\cos(30 - 9\pi)) = \pi - (30 - 9\pi)$
$= 10\pi - 30.$

Hence, statement 2 is true but statement 1 is false.

6. a. $f(x) = \sin^{-1}\left(\dfrac{2x}{1+x^2}\right) = \pi - 2\tan^{-1}x, x \geq 1$

$\Rightarrow f'(x) = -\dfrac{2}{1+x^2} \Rightarrow f'(2) = -\dfrac{2}{5}$

Thus statement 1 is true, statement 2 is true and statement 2 is a correct explanation for statement 1.

7. b. We know that $\tan^{-1}x$ and $\cot^{-1}x$ has domain R, also $\tan x$ and $\cot x$ are unbounded functions. On the other hand, $\sec x$ is an unbounded function, but its range is $R - (-1, 1)$, not R,

Hence, correct answer is (b).

8. a. For $x > 0, y > 0$,

$$\tan^{-1}\left(\frac{x}{y}\right) + \tan^{-1}\left(\frac{y-x}{y+x}\right) \quad (1)$$

$$= \tan^{-1}\left(\frac{x}{y}\right) + \tan^{-1}\left(\frac{1-\frac{x}{y}}{1+\frac{x}{y}}\right)$$

$$= \tan^{-1}\left(\frac{x}{y}\right) + \tan^{-1}1 - \tan^{-1}\frac{x}{y} = \frac{\pi}{4}$$

Now, in formula (1), putting $\dfrac{x}{y} = \dfrac{3}{4}$, we get

$$\tan^{-1}\left(\frac{3}{4}\right) + \tan^{-1}\left(\frac{1}{7}\right) = \frac{\pi}{4}$$

Hence, both statements are correct and statement 2 is the correct explanation of statement 1.

9. a.

Linked Comprehension Type

For Problems 1–3

1. d., 2. b., 3. c.

Sol.

$\sin^{-1}x \in \left[-\dfrac{\pi}{2}, \dfrac{\pi}{2}\right]$

$\cos^{-1}y \in [0, \pi]$

$\sec^{-1}z \in \left[0, \dfrac{\pi}{2}\right) \cup \left(\dfrac{\pi}{2}, \pi\right]$

$\Rightarrow \sin^{-1}x + \cos^{-1}y + \sec^{-1}z \leq \dfrac{\pi}{2} + \pi + \pi = \dfrac{5\pi}{2}$

Also, $t^2 - \sqrt{2\pi}\, t + 3\pi = t^2 - 2\sqrt{\dfrac{\pi}{2}}\, t + \dfrac{\pi}{2} - \dfrac{\pi}{2} + 3\pi$

$= \left(t - \sqrt{\dfrac{\pi}{2}}\right)^2 + \dfrac{5\pi}{2} \geq \dfrac{5\pi}{2}$

The given inequation exists if equality holds, i.e.,

L.H.S. = R.H.S. $= \dfrac{5\pi}{2}$

$\Rightarrow x = 1, y = -1, z = -1$ and $t = \sqrt{\dfrac{\pi}{2}}$

$\Rightarrow \cos^{-1}(\cos 5t^2) = \cos^{-1}\left(\cos\left(\dfrac{5\pi}{2}\right)\right)$

$= \dfrac{\pi}{2} \cos(\min\{x, y, z\})$

$= \cos(-1) = \pi$

For Problems 4–6

4. b., 5. a., 6. b.

Sol.

Given $ax + b(\sec(\tan^{-1}x)) = c$ and $ay + b(\sec(\tan^{-1}y)) = c$.

Let $\tan^{-1}x = \alpha$ and $\tan^{-1}y = \beta$, then the given relations are

$a\tan\alpha + b\sec\alpha = c$ and $a\tan\beta + b\sec\beta = c$

From these two relations, we can conclude that equation $a\tan\theta + b\sec\theta = c$ has roots α and β.

$a\tan\theta + b\sec\theta = c$

$\Rightarrow b\sec\theta = c - a\tan\theta$

$\Rightarrow b^2\sec^2\theta = c^2 - 2ac\tan\theta + a^2\tan^2\theta$

$\Rightarrow b^2 + b^2\tan^2\theta = c^2 - 2ac\tan\theta + a^2\tan^2\theta$

$\Rightarrow (a^2 - b^2)\tan^2\theta - 2ac\tan\theta + c^2 - b^2 = 0$

$\Rightarrow$ Sum of the roots, $\tan\alpha + \tan\beta = x + y = \dfrac{2ac}{a^2 - b^2}$

and the product of roots, $\tan\alpha\tan\beta = xy = \dfrac{c^2 - b^2}{a^2 - b^2}$

and $\dfrac{x+y}{1-xy} = \dfrac{\frac{2ac}{a^2-b^2}}{1 - \frac{c^2-b^2}{a^2-b^2}} = \dfrac{2ac}{a^2 - c^2}$

For Problems 7–9

7. b., 8. d., 9. c.

Sol.

Let $\cos^{-1}x = a \Rightarrow a \in [0, \pi]$

and $\sin^{-1}y = b \Rightarrow b \in [-\pi/2, \pi/2]$

We have $a + b^2 = \dfrac{p\pi^2}{4}$ (1)

and $ab^2 = \frac{\pi^2}{16}$ (2)

Since $b^2 \in [0, \pi^2/4] \Rightarrow a + b^2 \in [0, \pi + \pi^2/4]$

So, equation (1) $\Rightarrow 0 \le \frac{p\pi^2}{4} \le \pi + \frac{\pi^2}{4}$

i.e., $0 \le p \le \frac{4}{\pi} + 1$

Since $p \in Z$, so $p = 0$, 1 or 2.

Substituting the value of b^2 from equation (1) in equation (2), we get

$$a\left(\frac{p\pi^2}{4} - a\right) = \frac{\pi^2}{16} \Rightarrow 16a^2 - 4p\,\pi^2 a + \pi^4 = 0 \quad (3)$$

since $a \in R \Rightarrow D \ge 0$

i.e., $16p^2\pi^4 - 64\pi^4 \ge 0 \Rightarrow p^2 \ge 4 \Rightarrow p \ge 2 \Rightarrow p = 2$

Substituting $p = 2$ in equation (3), we get

$16a^2 - 8\pi^2 a + \pi^4 = 0$

$\Rightarrow (4a - \pi^2)^2 = 0 \Rightarrow a = \frac{\pi^2}{4} = \cos^{-1} x$

$\Rightarrow x = \cos\frac{\pi^2}{4}$

Form equation (2), we get $\frac{\pi^4}{4}\ b^2 = \frac{\pi^2}{16}$

$\Rightarrow \quad b = \pm\frac{\pi}{2} = \sin^{-1} y \quad \Rightarrow y = \pm 1$

For Problems 10–12

10. c., 11. a., 12. d.

Sol.

Let $\cos^{-1} x = \theta$

$\Rightarrow x = \cos\theta$, where $\theta \in [0, \pi]$

$\cos^{-1}(4x^3 - 3x) = \cos^{-1}(4\cos^3\theta - 3\cos\theta)$

$= \cos^{-1}(\cos 3\theta) = \cos^{-1}(\cos\alpha)$

where $\alpha = 3\theta \in [0, 3\pi]$

Refer the graph of $y = \cos^{-1}(\cos\alpha)$, $\alpha \in [0, 3\pi]$.

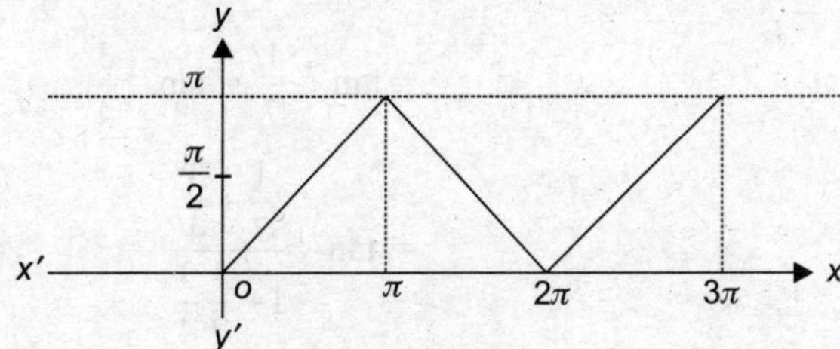

Fig. 3. 1

From the graph,

$\cos^{-1}(4x^3 - 3x) = \cos^{-1}(\cos\alpha)$

$$= \begin{cases} \alpha, & 0 \le \alpha < \pi \\ 2\pi - \alpha, & \pi \le \alpha \le 2\pi \\ \alpha - 2\pi, & 2\pi < \alpha \le 3\pi \end{cases}$$

$$= \begin{cases} 3\cos^{-1} x, & 0 \le 3\cos^{-1} x < \pi \\ 2\pi - 3\cos^{-1} x, & \pi \le 3\cos^{-1} x \le 2\pi \\ 3\cos^{-1} x - 2\pi, & 2\pi < 3\cos^{-1} x \le 3\pi \end{cases}$$

$$= \begin{cases} 3\cos^{-1} x, & 0 \le \cos^{-1} x < \pi/3 \\ 2\pi - 3\cos^{-1} x, & \pi/3 \le \cos^{-1} x \le 2\pi/3 \\ 3\cos^{-1} x - 2\pi, & 2\pi/3 < 3\cos^{-1} x \le \pi \end{cases}$$

$$= \begin{cases} 3\cos^{-1} x, & 1/2 < x \le 1 \\ 2\pi - 3\cos^{-1} x, & -1/2 \le x \le 1/2 \\ 3\cos^{-1} x - 2\pi, & -1 \le x < -1/2 \end{cases}$$

$$= \begin{cases} 3\cos^{-1} x - 2\pi, & -1 \le x < -1/2 \\ 2\pi - 3\cos^{-1} x, & -1/2 \le x \le 1/2 \\ 3\cos^{-1} x, & 1/2 < x \le 1 \end{cases}$$

For $x \in \left[-\frac{1}{2}, -1\right)$, $\cos^{-1}(4x^3 - 3x)$

$= 3\cos^{-1}x - 2\pi$

$\Rightarrow a = -2\pi$ and $b = 3 \Rightarrow a + b\pi = \pi$

For $x \in \left[-\frac{1}{2}, \frac{1}{2}\right]$, $\cos^{-1}(4x^3 - 3x) = 2\pi - 3\cos^{-1}x$

$\Rightarrow a = 2\pi$ and $b = -3$

$\Rightarrow \sin^{-1}\left(\sin\frac{a}{b}\right) = \sin^{-1}\left(\sin\frac{2\pi}{-3}\right)$

$= \sin^{-1}\left(-\frac{\sqrt{3}}{2}\right) = -\frac{\pi}{3}$

For $x \in \left(\frac{1}{2}, 1\right]$, $\cos^{-1}(4x^3 - 3x) = 3\cos^{-1}x$

$\Rightarrow \quad a = 0$ and $b = 3$

Hence, $\lim_{y \to a} b\cos(y) = \lim_{y \to 0} 3\cos(y) = 3$

Matrix-Match Type

1. a → p; b → q, s; c → r, s; d → r, s.

a. $\cos^{-1}(4x^3 - 3x) = 3\cos^{-1}x$

$0 \le \cos^{-1}(4x^3 - 3x) \le \pi$

$\Rightarrow 0 \le 3\cos^{-1}x \le \pi$

$\Rightarrow 0 \le \cos^{-1}x \le \pi/3 \quad \Rightarrow \quad 1/2 \le x \le 1$

b. $\sin^{-1}(3x - 4x^3) = 3\sin^{-1}x$

$-\pi/2 \le \sin^{-1}(3x - 4x^3) \le \pi/2$

$\Rightarrow -\pi/2 \le 3\sin^{-1}x \le \pi/2$

$\Rightarrow -\pi/6 \le \sin^{-1}x \le \pi/6$

$\Rightarrow -1/2 \le x \le 1/2$

c. $\cos^{-1}(4x^3 - 3x) = 3\sin^{-1}x$

$0 \le \cos^{-1}(3x - 4x^3) \le \pi$

$\Rightarrow 0 \le 3\sin^{-1}x \le \pi$

$\Rightarrow 0 \le \sin^{-1}x \le \pi/3$

$\Rightarrow 0 \le x \le \sqrt{3}/2$

d. $\sin^{-1}(3x - 4x^3) = 3\cos^{-1}x$

$-\pi/2 \le \sin^{-1}(3x - 4x^3) \le \pi/2$

$\Rightarrow -\pi/2 \le 3\cos^{-1}x \le \pi/2$

$\Rightarrow -\pi/6 \le \cos^{-1}x \le \pi/6$

$\Rightarrow 0 \le \cos^{-1}x \le \pi/6$

$\Rightarrow 0 \le x \le \sqrt{3}/2$

2. a. → q, r, s; b → q; c → r, s; d → p.

a. $(\sin^{-1}x)^2 + (\sin^{-1}y)^2 = \dfrac{\pi^2}{2}$

$\Rightarrow (\sin^{-1}x)^2 = (\sin^{-1}y)^2 = \dfrac{\pi^2}{4}$

$\Rightarrow \sin^{-1}x = \pm\dfrac{\pi}{2}, \sin^{-1}y = \pm\dfrac{\pi}{2}$

$\Rightarrow x = \pm 1$ and $y = \pm 1$

$\Rightarrow \quad x^3 + y^3 = -2, 0, 2$

b. $(\cos^{-1}x)^2 + (\cos^{-1}y)^2 = 2\pi^2$

$\Rightarrow (\cos^{-1}x)^2 = (\cos^{-1}y)^2 = \pi$

$\Rightarrow x = y = -1$

$\Rightarrow x^5 + y^5 = -2$

c. $(\sin^{-1}x)^2(\cos^{-1}y)^2 = \dfrac{\pi^4}{4}$

$\Rightarrow (\sin^{-1}x)^2 = \dfrac{\pi^2}{4}$ and $(\cos^{-1}y)^2 = \pi^2$

$\Rightarrow (\sin^{-1}x) = \pm\dfrac{\pi}{2}$ and $(\cos^{-1}y) = \pi$

$\Rightarrow x = \pm 1$ and $y = -1$

$\Rightarrow -|x - y| = 0, 2$

d. $|\sin^{-1}x - \sin^{-1}y| = \pi$

$\Rightarrow \sin^{-1}x = -\dfrac{\pi}{2}$ and $\sin^{-1}y = \dfrac{\pi}{2}$

or $\sin^{-1}x = \dfrac{\pi}{2}$ and $\sin^{-1}y = -\dfrac{\pi}{2}$

$\Rightarrow x^y = 1^{(-1)}$ or $(-1)^1 = 1$ or -1

3. a → p, q; b → q; c → q, r, s; d → r.

Refer the graphs of $y = \sin^{-1}(\sin x)$, $y = \cos^{-1}(\cos x)$ and $y = \tan^{-1}(\tan x)$.

4. a → q; b → s; c → p; d → r.

a. $\sin^{-1}\dfrac{4}{5} = \tan^{-1}\dfrac{4}{3}$

$2\tan^{-1}\dfrac{1}{3} = \tan^{-1}\dfrac{3}{4} = \cot^{-1}\dfrac{4}{3}$

and $\tan^{-1}x + \cot^{-1}x = \dfrac{\pi}{2}$

b. $\tan^{-1}\dfrac{12}{5} + \tan^{-1}\dfrac{3}{4} + \tan^{-1}\dfrac{63}{16}$

$= \pi + \tan^{-1}\dfrac{48 + 15}{20 - 36} + \tan^{-1}\dfrac{63}{16}$

$= \pi - \tan^{-1}\dfrac{63}{16} + \tan^{-1}\dfrac{63}{16} = \pi$

c. $A = \tan^{-1}\dfrac{x\sqrt{3}}{2\lambda - x}$ and $B = \tan^{-1}\left(\dfrac{2x - \lambda}{\lambda\sqrt{3}}\right)$

Now, $\tan(A - B) = \dfrac{\tan A - \tan B}{1 + \tan A\tan B}$

$= \dfrac{\dfrac{x\sqrt{3}}{2\lambda - x} - \dfrac{2x - \lambda}{\lambda\sqrt{3}}}{1 + \dfrac{x\sqrt{3}}{2\lambda - x}\cdot\dfrac{2x - \lambda}{\lambda\sqrt{3}}}$

$= \dfrac{3x\lambda + (2x - \lambda)(x - 2\lambda)}{\sqrt{3}\left[\lambda(2\lambda - x) + x(2x - \lambda)\right]}$

$= \dfrac{1}{\sqrt{3}}\left[\dfrac{2x^2 - 2\lambda x + 2\lambda^2}{2x^2 - 2\lambda x + 2\lambda^2}\right] = \dfrac{1}{\sqrt{3}} = \tan 30°$

$\therefore A - B = 30°$

d. $\tan^{-1}\dfrac{1}{7} + 2\tan^{-1}\dfrac{1}{3} = \tan^{-1}\dfrac{1}{7} + \tan^{-1}\dfrac{2\frac{1}{3}}{1 - \frac{1}{9}}$

$= \tan^{-1}\dfrac{1}{7} + \tan^{-1}\dfrac{3}{4}$

$= \tan^{-1}\dfrac{\frac{1}{7} + \frac{3}{4}}{1 - \frac{1}{7}\frac{3}{4}}$

$= \tan^{-1}1 = \pi/4$

5. a → s; b → p; c → q; d → r.

a. $f(x) = \sin^{-1}x + \cos^{-1}x + \cot^{-1}x$

$= \dfrac{\pi}{2} + \cot^{-1}x, x \in [-1, 1]$

For $x \in [-1, 1]$, $\cot^{-1} x \in \left[\frac{\pi}{4}, \frac{3\pi}{4}\right]$

$\Rightarrow \frac{\pi}{2} + \cot^{-1} x \in \left[\frac{3\pi}{4}, \frac{5\pi}{4}\right]$

b. $f(x) = \cot^{-1} x + \tan^{-1} x + \text{cosec}^{-1} x$

$= \frac{\pi}{2} + \text{cosec}^{-1} x$, where $x \in (-\infty, -1] \cup [1, \infty)$

Now $\text{cosec}^{-1} x \in \left[-\frac{\pi}{2}, 0\right) \cup \left(0, \frac{\pi}{2}\right]$

$\Rightarrow \frac{\pi}{2} + \text{cosec}^{-1} x \in \left[0, \frac{\pi}{2}\right) \cup \left(\frac{\pi}{2}, \pi\right]$

c. $f(x) = \cot^{-1} x + \tan^{-1} x + \cos^{-1} x$

$= \frac{\pi}{2} + \cos^{-1} x$, where $x \in [-1, 1]$

$\Rightarrow \frac{\pi}{2} + \cos^{-1} x \in \left[\frac{\pi}{2}, \frac{3\pi}{2}\right]$

d. $\sec^{-1} x + \text{cosec}^{-1} x + \sin^{-1} x$, where $x \in \{-1, 1\}$

$= \frac{\pi}{2} + \sin^{-1} x$, where $x \in \{-1, 1\}$

Hence, $f(x) \in \{0, \pi\}$

Integer Type

1. **(3)** We know that $0 \le \cos^{-1} x \le \pi$.

   Hence, from the question

   $\cos^{-1} \lambda = \pi, \cos^{-1} \mu = \pi, \cos^{-1} \gamma = \pi$

   ($\because \cos^{-1} \lambda + \cos^{-1} \mu + \cos^{-1} \gamma = 3\pi$ only when each value attains its maximum)

   $\Rightarrow \lambda = \mu = \gamma = -1$

   $\Rightarrow \lambda\mu + \mu\gamma + \gamma\lambda = 3$

2. **(3)** $(\sin^{-1} x)^2 + (\sin^{-1} y)^2 + (\sin^{-1} z)^2 = \frac{3\pi^2}{4}$

   $\Rightarrow \sin^{-1} x = \sin^{-1} y = \sin^{-1} z = \pm\frac{\pi}{2}$

   For maximum of $x + y + z$,

   $x = y = z = 1$

   Hence, $(x + y + z)_{\max} = 3$.

3. **(3)** $\cot^{-1} \frac{xy+1}{x-y} + \cot^{-1} \frac{yz+1}{y-z} + \cot^{-1} \frac{zx+1}{z-x}$

   $= \tan^{-1} \frac{x-y}{1+xy} + \tan^{-1} \frac{y-z}{1+yz} + \pi + \tan^{-1} \frac{z-x}{1+zx}$

   $\left[\because \tan^{-1}\left(\frac{1}{x}\right) = \begin{cases} \cot^{-1} x, & \text{for } x > 0 \\ -\pi + \cot^{-1} x, & \text{for } x < 0 \end{cases}\right]$

   $= \tan^{-1} x - \tan^{-1} y + \tan^{-1} y - \tan^{-1} z + \pi + \tan^{-1} z - \tan^{-1} x$

   $= \pi$

4. **(6)** Clearly, the domain of the function is $[-1, 1]$.

   Also $\tan^{-1} x \in \left[-\frac{\pi}{4}, \frac{\pi}{4}\right]$ for $x \in [-1, 1]$.

   Now $\sin^{-1} x + \cos^{-1} x = \frac{\pi}{2}$ for $x \in [-1, 1]$.

   Thus $f(x) = \tan^{-1} x + \frac{\pi}{2}$, where $x \in [-1, 1]$.

   Hence, the range is

   $\left[-\frac{\pi}{4} + \frac{\pi}{2}, \frac{\pi}{4} + \frac{\pi}{2}\right] = \left[\frac{\pi}{4}, \frac{3\pi}{4}\right]$.

5. **(4)** $\sin^{-1} \frac{14}{|x|} + \sin^{-1} \frac{2\sqrt{15}}{|x|} = \frac{\pi}{2}$

   $\Rightarrow \sin^{-1} \frac{14}{|x|} = \frac{\pi}{2} - \sin^{-1} \frac{2\sqrt{15}}{|x|}$

   $= \cos^{-1} \frac{2\sqrt{15}}{|x|}$

   $= \sin^{-1} \sqrt{1 - \left(\frac{2\sqrt{15}}{|x|}\right)^2}$

   for $0 \le \frac{2\sqrt{15}}{|x|} \le 1$ or $|x| \ge 2\sqrt{15}$.

   $\Rightarrow \left(\frac{14}{|x|}\right)^2 = 1 - \left(\frac{2\sqrt{15}}{|x|}\right)^2$

   $\Rightarrow |x| = 16$

   $\Rightarrow x = \pm 16$, which satisfy $|x| \ge 2\sqrt{15}$.

6. **(2)** $\tan^{-1} x > \cot^{-1} x$

   $\Rightarrow \tan^{-1} x > \pi/2 - \tan^{-1} x$

   $\Rightarrow \tan^{-1} x > \pi/4$

   $\Rightarrow x > 1$

7. **(4)** For $0 < A < \pi/4$, $\cot A > 1$

   $\Rightarrow \cot A \cot^3 A > 1$

   $\Rightarrow \tan^{-1}\left(\frac{1}{2} \tan 2A\right) + \tan^{-1}(\cot A) + \tan^{-1}(\cot^3 A)$

   $= \tan^{-1}\left(\frac{\tan A}{1 - \tan^2 A}\right) + \pi + \tan^{-1}\left(\frac{\cot A + \cot^3 A}{1 - \cot^4 A}\right)$

   $= \tan^{-1}\left(\frac{\tan A}{1 - \tan^2 A}\right) + \pi + \tan^{-1}\left(\frac{\cot A}{1 - \cot^2 A}\right)$

   $= \tan^{-1}\left(\frac{\tan A}{1 - \tan^2 A}\right) + \pi + \tan^{-1}\left(\frac{\tan A}{\tan^2 A - 1}\right)$

   $= \pi = 4(\pi/4) = 4 \tan^{-1}(1)$

8. **(5)** $\sin^{-1}\left(\dfrac{4x}{x^2+4}\right) + 2\tan^{-1}\left(-\dfrac{x}{2}\right)$

$$= \sin^{-1}\left(\dfrac{2\times\dfrac{x}{2}}{\left(\dfrac{x}{2}\right)^2+1}\right) - 2\tan^{-1}\dfrac{x}{2}$$

$$= 2\tan^{-1}\dfrac{x}{2} - 2\tan^{-1}\dfrac{x}{2} = 0$$

Here $\left|\dfrac{x}{2}\right| \le 1$

$|x| \le 2 \Rightarrow -2 \le x \le 2$

9. **(1)** $(x-1)(x^2+1) > 0$

$\Rightarrow x > 1$

$$\sin\left[\dfrac{1}{2}\tan^{-1}\left(\dfrac{2x}{1-x^2}\right) - \tan^{-1}x\right]$$

$$= \sin\left[\dfrac{1}{2}(-\pi + 2\tan^{-1}x) - \tan^{-1}x\right]$$

$$= \sin\left(-\dfrac{\pi}{2}\right) = -1$$

10. **(2)** $f(x) = \sin^{-1}\left(\dfrac{1+x^2}{2x}\right)$ is defined for

$-1 \le \dfrac{1+x^2}{2x} \le 1$ or $\left|\dfrac{1+x^2}{2x}\right| \le 1$

$\Rightarrow |1+x^2| \le |2x|$, for all x

$\Rightarrow 1+x^2 \le |2x|$, for all x [As $1+x^2 > 0$]

$\Rightarrow x^2 - 2|x| + 1 \le 0$

$\Rightarrow |x|^2 - 2|x| + 1 \le 0$ [As $x^2 = |x^2|$]

$\Rightarrow (|x|-1)^2 \le 0$

But $(|x|-1)^2$ is either always positive or zero.

Thus $(|x|-1)^2 = 0$

$\Rightarrow |x| = 1 \Rightarrow x = \pm 1$

Thus the domain of $f(x)$ is $\{-1, 1\}$.

11. **(2)** $\tan^{-1}x + \cos^{-1}\dfrac{y}{\sqrt{1-y^2}} = \sin^{-1}\dfrac{3}{\sqrt{10}}$

$\Rightarrow \tan^{-1}x + \tan^{-1}\left(\dfrac{1}{y}\right) = \tan^{-1}(3)$

$\Rightarrow \tan^{-1}\left(\dfrac{1}{y}\right) = \tan^{-1}3 - \tan^{-1}(x)$

$\Rightarrow \tan^{-1}\left(\dfrac{1}{y}\right) = \tan^{-1}\left(\dfrac{3-x}{1+3x}\right)$

$\Rightarrow y = \dfrac{1+3x}{3-x}$

As x, y are positive integers, $x = 1, 2$ and corresponding $y = 2, 7$.

Hence, the solutions are $(x, y) = (1, 2)\ (2, 7)$, i.e., two solutions.

12. **(1)** Let us transfer $\sin^{-1}6\sqrt{3}x$ to the right-hand side of the equation and calculate the sine of both sides of the resulting equation.

$\sin(\sin^{-1}6x) = \sin(-\sin^{-1}6\sqrt{3}\,x - \pi/2)$

$\Rightarrow 6x = -\sin(\sin^{-1}6\sqrt{3}x + \sin^{-1}1)$ [using $\sin^{-1}(-x) = -\sin^{-1}x$]

$\Rightarrow 6x = -\sin(\sin^{-1}\sqrt{1-108x^2})$ (1)

Squaring both sides, we get

$36x^2 = 1 - 108x^2$

$\Rightarrow 144x^2 = 1$

$\Rightarrow x = 1/12$ and $x = -1/12$

Let us verify.

Substituting $x = 1/2$ in the given equation, we get

$$\sin^{-1}\left(-\dfrac{1}{2}\right) + \sin^{-1}\left(-\dfrac{\sqrt{3}}{2}\right) = -\dfrac{\pi}{6} - \dfrac{\pi}{3} = -\dfrac{\pi}{2}$$

Thus $x = 1/12$ is a root of the given equation.

But substituting in (1), we get

L.H.S. $= 6x = \dfrac{1}{2}$

R.H.S. $= -\sqrt{1-108x^2} = -1/2$

Thus L.H.S. $\ne$ R.H.S. of equation (1)

Thus $x = -1/12$ is a root of the given equation as it satisfies the given equation and equation (1).

13. We know that $\tan^{-1}x + \cot^{-1}x = \pi/2, \forall x \in R$.

So the given equation can be written as

$$(\sin^{-1}x)^3 + (\cos^{-1}x)^3 = 7\left(\dfrac{\pi^3}{8}\right)$$

$$\Rightarrow (\sin^{-1}x + \cos^{-1}x)^3 - 3(\sin^{-1}x)(\cos^{-1}x)(\sin^{-1}x + \cos^{-1}x) = 7\left(\dfrac{\pi}{2}\right)^3$$

$$\Rightarrow \left(\dfrac{\pi}{2}\right)^3 - 3(\sin^{-1}x)(\cos^{-1}x)\left(\dfrac{\pi}{2}\right) = 7\left(\dfrac{\pi}{2}\right)^3$$

$$\Rightarrow (\sin^{-1}x)(\cos^{-1}x) = -\dfrac{\pi^2}{2}$$

As the maximum value of $\cos^{-1}x$ is π ($\cos^{-1}x$ is always ≥ 0), and the minimum value of $\sin^{-1}x$ is $-(\pi/2)$, and this happens at the same x, i.e., at $x = -1$, so the minimum value of $(\sin^{-1}x)(\cos^{-1}x)$ is

$\left(-\frac{\pi}{2}\right)(\pi) = -\frac{\pi^2}{2}$.

So $(\sin^{-1} x)(\cos^{-1} x) = -\frac{\pi^2}{2}$

$\Rightarrow x = -1$ only.

14. **(7)** $(\sin^{-1} x)^3 + (\cos^{-1} x)^3 = a\pi^3$

$\Rightarrow (\sin^{-1} x + \cos^{-1} x)[(\sin^{-1} x + \cos^{-1} x)^2 - 3\sin^{-1} x \cos^{-1} x] = a\pi^3$

$$\Rightarrow \frac{\pi^2}{4} - 3\sin^{-1} x \cos^{-1} x = 2a\pi^2$$

$$\Rightarrow \sin^{-1} x\left(\frac{\pi}{2} - \sin^{-1} x\right) = \frac{\pi^2}{12}(1-8a)$$

$$\Rightarrow (\sin^{-1} x)^2 - \frac{\pi}{2}\sin^{-1} x = -\frac{\pi^2}{12}(1-8a)$$

$$\Rightarrow \left(\sin^{-1} x - \frac{\pi}{4}\right)^2 = \frac{\pi^2}{12}(8a-1) + \frac{\pi^2}{16}$$

$$\Rightarrow \left(\sin^{-1} x - \frac{\pi}{4}\right)^2 = \frac{\pi^2}{48}(32a-1)$$

Now $\sin^{-1} x \in \left[-\frac{\pi}{2}, \frac{\pi}{2}\right]$

$$\Rightarrow -\frac{3\pi}{4} \le \sin^{-1} x - \frac{\pi}{4} \le \frac{\pi}{4}$$

$$\Rightarrow 0 \le \left(\sin^{-1} x - \frac{\pi}{4}\right)^2 \le \frac{9\pi^2}{16}$$

$$\Rightarrow 0 \le \frac{\pi^2}{48}(32a-1) \le \frac{9\pi^2}{16}$$

$$\Rightarrow 0 \le 32a - 1 \le 27$$

$$\Rightarrow \frac{1}{32} \le a \le \frac{7}{8}$$

Thus the required set of values of a is $[1/32, 7/8]$.

15. **(7)** Let t_n denote the nth term of the series. Then $t_n = \cot^{-1} 2n^2$

$$= \tan^{-1}\frac{1}{2n^2}$$

$$= \tan^{-1}\frac{(2n+1)-(2n-1)}{1+(4n^2-1)}$$

$$= \tan^{-1}(2n+1) - \tan^{-1}(2n-1)$$

Putting $n = 1, 2, 3, \cdots$, etc., in (1), we get

$t_1 = \tan^{-1} 3 - \tan^{-1} 1$

$t_2 = \tan^{-1} 5 - \tan^{-1} 3$

$t_3 = \tan^{-1} 7 -- \tan^{-1} 5$

$t_n = \tan^{-1}(2n+1) - \tan^{-1}(2n-1)$

Adding, we get

$S_n = \tan^{-1}(2n+1) - \tan^{-1} 1$.

As $n \to \infty$, $\tan^{-1}(2n-1) \to \pi/2$.

Hence, the required sum is $\pi/4$.

16. **(2)** $\cos(\cos^{-1} x) = x$ for $x \in [-1, 1]$

$\text{cosec}(\text{cosec}^{-1} x) = x$ for $x \in (-\infty, -1] \cup [1, \infty)$

$\Rightarrow \cos(\cos^{-1} x) = \text{cosec}(\text{cosec}^{-1} x)$ for $x = \pm 1$ only.

Hence two roots only.

17. **(2)** Here $|\cos x| = \sin^{-1}(\sin x)$

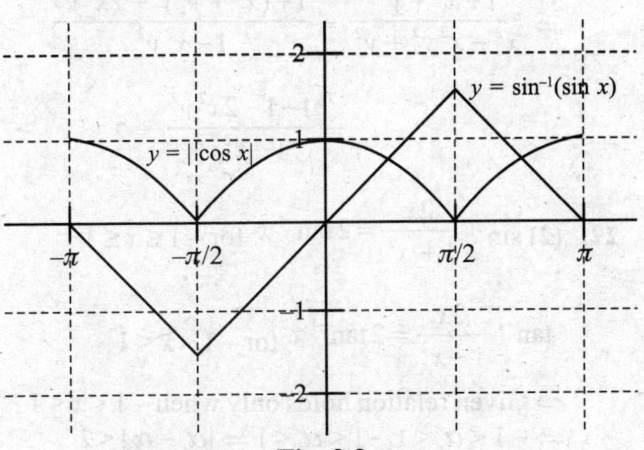

Fig. 3.2

From the graph number of solutions are 2

18. **(1)** We have, $\left|\cos^{-1}\left(\frac{1-x^2}{1+x^2}\right)\right| < \frac{\pi}{3}$

$$\Rightarrow -\frac{\pi}{3} < \cos^{-1}\left(\frac{1-x^2}{1+x^2}\right) < \frac{\pi}{3}$$

$$\Rightarrow 0 \le \cos^{-1}\frac{1-x^2}{1+x^2} < \frac{\pi}{3} \Rightarrow \frac{1}{2} < \frac{1-x^2}{1+x^2} \le 1$$

$$\Rightarrow 1 + x^2 < 2(1-x^2) \le 2(1+x^2) \Rightarrow 0 \le x^2 < \frac{1}{3}$$

$$\Rightarrow -\frac{1}{\sqrt{3}} < x < \frac{1}{\sqrt{3}}$$

19. **(1)** $3\cos^{-1} x - \pi x - \frac{\pi}{2} = 0 \Rightarrow \cos^{-1} x = \frac{\pi x}{3} + \frac{\pi}{6}$

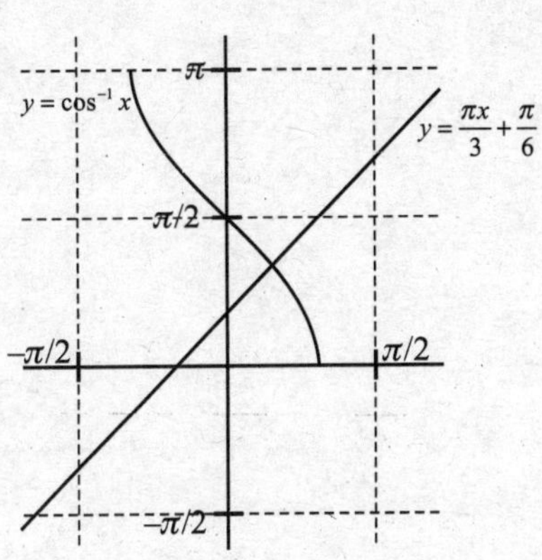

Fig. 3.3

20. **(6)** $f(x)$ is defined for $-1 \le |x-1| - 2 \le 1$

$\Rightarrow 1 \le |x-1| \le 3$

$\Rightarrow 1 \le (x-1) \le 3$ or $-3 \le x-1 \le -1$

$\Rightarrow 2 \le x \le 4$ or $-2 \le x \le 0 \Rightarrow x \in [2, 4] \cup [-2, 0]$

21. **(2)** $\sin^{-1} x + \sin^{-1} y = \dfrac{\pi}{2} \Rightarrow \sin^{-1} x = \dfrac{\pi}{2} - \sin^{-1} y$

$\Rightarrow \sin^{-1} x = \sin^{-1}\sqrt{1-y^2} \Rightarrow x^2 + y^2 = 1$

$$\Rightarrow \frac{1+x^4+y^4}{x^2 - x^2y^2 + y^2} = \frac{1+(x^2+y^2)^2 - 2x^2y^2}{1-x^2y^2}$$

$$= \frac{1+1-2x^2y^2}{1-x^2y^2} = 2$$

22. **(2)** $\sin^{-1}\dfrac{2x}{1+x^2} = 2\tan^{-1} x$ for $-1 \le x \le 1$

$\tan^{-1}\dfrac{2x}{1-x^2} = 2\tan^{-1} x$ for $-1 < x < 1$

$\Rightarrow$ Given relation holds only when $-1 < x < 1$

$\Rightarrow -1 < \alpha_1 < 1, -1 < \alpha_2 < 1 \Rightarrow |\alpha_1 - \alpha_2| < 2$

23. **(5)** $\sin^{-1}\left(\dfrac{4x}{x^2+4}\right) + 2\tan^{-1}\left(-\dfrac{x}{2}\right)$

$$= \sin^{-1}\left(\frac{2 \times \frac{x}{2}}{\left(\frac{x}{2}\right)^2 + 1}\right) - 2\tan^{-1}\frac{x}{2}$$

$$= 2\tan^{-1}\frac{x}{2} - 2\tan^{-1}\frac{x}{2} = 0$$

Here $\left|\dfrac{x}{2}\right| \le 1 \Rightarrow |x| \le 2 \Rightarrow -2 \le x \le 2$

24. **(4)** $A = (\tan^{-1} x)^3 + (\cot^{-1} x)^3$

$A = (\tan^{-1}x + \cot^{-1}x)^3 - 3\tan^{-1} x \cot^{-1} x (\tan^{-1} x + \cot^{-1} x)$

$$\Rightarrow A = \left(\frac{\pi}{2}\right)^3 - 3\tan^{-1} x \cot^{-1} x \times \frac{\pi}{2}$$

$$\Rightarrow A = \frac{\pi^3}{8} - \frac{3\pi}{2}\tan^{-1} x\left(\frac{\pi}{2} - \tan^{-1} x\right)$$

$$\Rightarrow A = \frac{\pi^3}{32} + \frac{3\pi}{2}\left(\tan^{-1} x - \frac{\pi}{4}\right)^2 \text{ as } x > 0$$

$$\Rightarrow \frac{\pi^2}{32} \le A < \frac{\pi^3}{8}$$

CHAPTER 4

Solutions and Properties of Triangle

Objective Type

Each question has four choices a, b, c, and d, out of which *only one* answer is correct.

1. If in a ΔABC, $\sin A \cos B = \frac{\sqrt{2}-1}{\sqrt{2}}$ and $\sin B \cos A = \frac{1}{\sqrt{2}}$, then the triangle is
   a. equilateral　b. isosceles
   c. right angled　d. righ tangled isosceles

2. ABC is an equilateral triangle of side 4 cm. If R, r, h are the circumradius, inradius and altitude, respectively, then $\frac{R+r}{h}$ is equal to
   a. 4　b. 2
   c. 1　d. 3

3. If in a ΔABC, $a^2 + b^2 > c^2$, then the triangle is
   a. acute angled at C
   b. obtuse angled at C
   c. an acute angled triangle
   d. None of these

4. A piece of paper is in the shape of a square of side 1 m long. It is cut at the four corners to make a regular polygon of eight sides (octagon). The area of the polygon is
   a. $2(\sqrt{2}-1)\,\text{m}^2$　b. $(\sqrt{2}-1)\,\text{m}^2$
   c. $\frac{1}{\sqrt{2}}\,\text{m}^2$　d. None of these

5. If A, B and C are angles of a triangle such that angle A is obtuse, then $\tan B \tan C$ will be less than
   a. $\frac{1}{\sqrt{3}}$　b. $\frac{\sqrt{3}}{2}$
   c. 1　d. None of these

6. In a ΔABC, $\angle B = \frac{\pi}{3}$. The range of values of x, where $x = \sin A \sin C$, is the interval
   a. $\left[\frac{1}{4}, \frac{3}{4}\right]$　b. $\left(0, \frac{3}{4}\right)$
   c. $\left(0, \frac{3}{4}\right]$　d. $\left[\frac{1}{4}, \frac{3}{4}\right]$

7. If in ΔABC, AC is double of AB, then the value of $\cot \frac{A}{2} \cot \frac{B-C}{2}$ is equal to
   a. $\frac{1}{3}$　b. $-\frac{1}{3}$
   c. 3　d. $\frac{1}{2}$

8. In a right angled isosceles triangle the ratio of the circumradius and inradius is
   a. $2(\sqrt{2}+1):1$　b. $(\sqrt{2}+1):1$
   c. 2:1　d. $\sqrt{2}:1$

9. In triangle ABC, $a = 5$, $b = 3$ and $c = 7$, the value of $3 \cos C + 7 \cos B$ is equal to
   a. 5　b. 10
   c. 7　d. 3

10. If in triangle ABC, $\angle B = 90°$ then $\tan^2(A/2)$ is
    a. $\frac{b-c}{b+c}$　b. $\frac{b+c}{b-c}$
    c. $\frac{2bc}{b-c}$　d. None of these

11. In Δ ABC, if $b^2 + c^2 = 2a^2$, then value of $\frac{\cot A}{\cot B + \cot C}$ is

a. $\frac{1}{2}$ b. $\frac{3}{2}$

c. $\frac{5}{2}$ d. $\frac{5}{3}$

12. If $\sin\theta$ and $-\cos\theta$ are the roots of the equation $ax^2 - bx - c = 0$, where a, b, c are the sides of a triangle ABC, then $\cos B$ is equal to

a. $1-\frac{c}{2a}$ b. $1-\frac{c}{a}$

c. $1+\frac{c}{2a}$ d. $1+\frac{c}{3a}$

13. In ΔABC, $a^2+b^2+c^2=ac+ab\sqrt{3}$, then the triangle is

a. equilateral b. isosceles

c. right angled d. None of these

14. In a ΔABC, $(a+b+c)(b+c-a)=kbc$ if

a. $k<0$ b. $k>0$

c. $0<k<4$ d. $k>4$

15. If one side of a triangle is double the other, and the angles on opposite sides differ by 60°, then the triangle is

a. equilateral b. obtuse angled

c. right angled d. acute angled

16. In a triangle ABC, if $r_1=2r_2=3r_3$, then $a:b$ is equal to

a. $\frac{5}{4}$ b. $\frac{4}{5}$

c. $\frac{7}{4}$ d. $\frac{4}{7}$

17. If in a triangle $\left(1-\frac{r_1}{r_2}\right)\left(1-\frac{r_1}{r_3}\right)=2$, then the triangle is

a. right angled b. isosceles

c. equilateral d. None of these

18. In an equilateral triangle, the inradius, circumradius and one of the ex-radii are in the ratio

a. 2:3:5 b. 1:2:3

c. 1:3:7 d. 3:7:9

19. In any ΔABC, if $\cot\frac{A}{2}$, $\cot\frac{B}{2}$, $\cot\frac{C}{2}$ are in A.P., then a, b, c are in

a. A.P. b. G.P.

c. H.P. d. None of these

20. In a ΔABC, if $A=30°$, $b=2$, $c=\sqrt{3}+1$, then $\frac{C-B}{2}$ is equal to

a. 15° b. 30°

c. 45° d. None of these

21. In a triangle ABC, if $a:b:c=7:8:9$ then $\cos A:\cos B$ is equal to

a. $\frac{11}{63}$ b. $\frac{22}{63}$

c. $\frac{2}{9}$ d. None of these

22. In a triangle ABC, if $\cos A+\cos B+\cos C=\frac{7}{4}$, then $\frac{R}{r}$ is equal to

a. $\frac{3}{4}$ b. $\frac{4}{3}$

c. $\frac{2}{3}$ d. $\frac{3}{2}$

23. In a ΔABC, $\cot\frac{A}{2}+\cot\frac{B}{2}+\cot\frac{C}{2}$ is equal to

a. $\frac{\Delta}{r^2}$ b. $\frac{(a+b+c)^2}{abc}2R$

c. $\frac{\Delta}{r}$ d. $\frac{\Delta}{Rr}$

24. In triangle ABC, $a^2+c^2=2002b^2$, then $\frac{\cot A+\cot C}{\cot B}$ is equal to

a. $\frac{1}{2001}$ b. $\frac{2}{2001}$

c. $\frac{3}{2001}$ d. $\frac{4}{2001}$

25. If the hypotenuse of a right-angled triangle is four times the length of the perpendicular drawn from the opposite vertex to it, then the difference of the two acute angles will be

a. 60° b. 15°

c. 75° d. 30°

26. In ΔABC, if $\sin^2\frac{A}{2}$, $\sin^2\frac{B}{2}$, $\sin^2\frac{C}{2}$ are in H.P., then a, b, c will be in

a. A.P. b. G.P.

c. H.P. d. None of these

27. Given $b=2$, $c=\sqrt{3}$, $\angle A=30°$, then inradius of ΔABC is

a. $\frac{\sqrt{3}-1}{2}$ b. $\frac{\sqrt{3}+1}{2}$

c. $\frac{\sqrt{3}-1}{4}$ d. None of these

28. If P is a point on the altitude AD of the triangle ABC such that $\angle CBP = B/3$, then AP is equal to

a. $2a\sin\frac{C}{3}$ b. $2b\sin\frac{C}{3}$

c. $2c\sin\frac{B}{3}$ d. $2c\sin\frac{C}{3}$

29. In triangle ABC, $\angle A = \pi/2$, then $\tan\frac{C}{2}$ is equal to

a. $\frac{a-c}{2b}$ b. $\frac{a-b}{2c}$

c. $\frac{a-c}{b}$ d. $\frac{a-b}{c}$

30. If the angles A, B, C of a triangle are in A.P. and sides a, b, c are in G.P., then a^2, b^2, c^2 are in

a. A.P. b. H.P.

c. G.P. d. None of these

31. In a Δ ABC, $\sin A + \sin B + \sin C = 1 + \sqrt{2}$ and $\cos A + \cos + \cos C = \sqrt{2}$ if the triangle is

a. equilateral b. isosceles

c. right angled d. right angled isosceles

32. If $\frac{r}{r_1} = \frac{r_2}{r_3}$, then

a. $A = 90°$ b. $B = 90°$

c. $C = 90°$ d. None of these

33. If in a ΔABC, $\cos 3A + \cos 3B + \cos 3C = 1$, then one angle must be exactly equal to

a. 90° b. 45°

c. 120° d. None of these

34. If $\cos B \cos C + \sin B \sin C \sin^2 A = 1$, then the triangle ABC is

a. isosceles and right angled

b. equilateral

c. isosceles whose equal angles are greater than $\pi/4$

d. none

35. In triangle ABC, internal angle bisector $\angle A$ makes an angle θ with side BC. Value of $\sin\theta$ is equal to

a. $\left|\sin\left(\frac{B-C}{2}\right)\right|$ b. $\left|\sin\left(\frac{B}{2}-C\right)\right|$

c. $\cos\left(\frac{B-C}{2}\right)$ d. $\cos\left(\frac{B}{2}-C\right)$

36. In an acute angled triangle ABC, $r + r_1 = r_2 + r_3$ and $\angle B > \frac{\pi}{3}$ then

a. $b+2c<2a<2b+2c$ b. $b+4c<4a<2b+4c$

c. $b+4c<4a<4b+4c$ d. $b+3c<3a<3b+3c$

37. In a triangle ABC, $\angle A = 30°$, $BC = 2 + \sqrt{5}$, then the distance of the vertex A from the orthocentre of the triangle is

a. 1 b. $(2+\sqrt{5})\sqrt{3}$

c. $\frac{\sqrt{3}+1}{2\sqrt{2}}$ d. $\frac{1}{2}$

38. If I is the incentre of a triangle ABC, then the ratio $IA:IB:IC$ is equal to

a. $\text{cosec}\frac{A}{2}:\text{coses}\frac{B}{2}:\text{cosec}\frac{C}{2}$

b. $\sin\frac{A}{2}:\sin\frac{B}{2}:\sin\frac{C}{2}$

c. $\sec\frac{A}{2}:\sec\frac{B}{2}:\sec\frac{C}{2}$

d. None of these

39. In a Δ ABC, the value of $\frac{a\cos A + b\cos B + c\cos C}{a+b+c}$ is

a. $\frac{R}{r}$ b. $\frac{R}{2r}$

c. $\frac{r}{R}$ d. $\frac{2r}{R}$

40. If D is the mid point of the side BC of a triangle ABC and AD is perpendicular to AC, then

a. $3b^2 = a^2 - c^2$ b. $3a^2 = b^2 - 3c^2$

c. $b^2 = a^2 - c^2$ d. $a^2 + b^2 = 5c^2$

41. Two medians drawn from the acute angles of a right angled triangle intersect at an angle $\pi/6$. If the length of the hypotenuse of the triangle is 3 units, then the area of the triangle (in sq. units) is

a. $\sqrt{3}$ b. 3

c. $\sqrt{2}$ d. 9

42. For a triangle ABC, $R = \frac{5}{2}$ and $r = 1$. Let I be the incentre of the triangle and D, E and F be the feet of the perpendiculars from I to BC, CA and AB, respectively. The value of $\frac{ID \times IE \times IF}{IA \times IB \times IC}$ is equal to

a. $\frac{5}{2}$ b. $\frac{5}{4}$

c. $\frac{1}{10}$ d. $\frac{1}{5}$

43. If the median of ΔABC through A is perpendicular to AB, then

a. $\tan A + \tan B = 0$ b. $2\tan A + \tan B = 0$

c. $\tan A + 2\tan B = 0$ d. None of these

44. In Δ ABC, the median AD divides $\angle BAC$ such that $\angle BAD : \angle CAD = 2:1$. Then $\cos\frac{A}{3}$ is equal to

a. $\frac{\sin B}{2\sin C}$ b. $\frac{\sin C}{2\sin B}$

c. $\frac{2\sin B}{\sin C}$ d. None of these

45. If in a ΔABC, $b = 3$ cm, $c = 4$ cm and the length of the perpendicular from A to the side BC is 2 cm, then the number of solutions of the triangle is

a. 1 b. 0

c. 3 d. 2

46. In triangle ABC, $\sum \sin\frac{A}{2} = \frac{6}{5}$ and $\sum II_1 = 9$ (where I_1, I_2, I_3 are ex-centres and I is in-centre, then circumradius R is equal to

a. $\frac{15}{8}$ b. $\frac{15}{4}$

c. $\frac{15}{2}$ d. $\frac{4}{12}$

47. In a triangle ABC, medians AD and CE are drawn. If $AD = 5$, $\angle DAC = \pi/8$, and $\angle ACE = \pi/4$, then the area of the triangle ABC is equal to

a. $\frac{25}{9}$ b. $\frac{25}{3}$

c. $\frac{25}{18}$ d. $\frac{10}{3}$

48. In a triangle ABC, if $\tan(A/2) = 5/6$ and $\tan(B/2) = 20/37$, the sides a, b and c are in

a. A.P. b. G.P.

c. H.P. d. None of these

49. If H is the orthocentre of a acute angled triangle ABC whose circumcircle is $x^2 + y^2 = 16$, then circumdiameter of the triangle HBC is

a. 1 b. 2

c. 4 d. 8

50. In a triangle ABC, $a = 5$, $b = 4$, and $c = 3$. G is the centroid of the triangle. Circumradius of triangle GAB is equal to

a. $2\sqrt{13}$ b. $\frac{5}{12}\sqrt{13}$

c. $\frac{5}{3}\sqrt{13}$ d. $\frac{3}{2}\sqrt{13}$

51. In triangle ABC, line joining circumcentre and incentre is parallel to side AC, then $\cos A + \cos C$ is equal to

a. –1 b. 1

c. –2 d. 2

52. In triangle ABC, line joining the circumcentre and orthocentre is parallel to side AC, then value of $\tan A \tan C$ is equal to

a. $\sqrt{3}$ b. 3

c. $3\sqrt{3}$ d. None of these

53. If in a ΔABC, $8R^2 = a^2 + b^2 + c^2$, then the triangle ABC is

a. right angled b. isosceles

c. equilateral d. None of these

54. In a triangle ABC, $\frac{a}{b} = \frac{2}{3}$ and $\sec^2 A = \frac{8}{5}$. Then number of triangles satisfying these conditions are

a. 0 b. 1

c. 2 d. 3

55. We are given b, c and $\sin B$ such that B is acute and $b < c \sin B$. Then

a. no triangle is possible

b. one triangle is possible

c. two triangles are possible

d. a rightangled triangle is possible

56. If a, b and A are given in a triangle and c_1, c_2 are the possible values of the third side, then $c_1^2 + c_2^2 - \cos A$ is equal to

a. $4a^2 \sin 2A$ b. $4a^2 \sin^2 A$

c. $4a^2 \cos 2A$ d. $4a^2 \cos^2 A$

57. In a ΔABC, a, b, A are given and c_1, c_2 are two values of the third side c. The sum of the areas of the two triangles with sides a, b, c_1 and a, b, c_2 is

a. $(1/2)\, b^2 \sin 2A$ b. $(1/2)\, a^2 \sin 2A$

c. $b^2 \sin 2A$ d. None of these

58. The area of the circle and the area of a regular polygon of n sides and of perimeter equal to that of the circle are in the ratio of

a. $\tan\left(\frac{\pi}{n}\right) : \frac{\pi}{n}$ b. $\cos\left(\frac{\pi}{n}\right) : \frac{\pi}{n}$

c. $\sin\frac{\pi}{n} : \frac{\pi}{n}$ d. $\cot\left(\frac{\pi}{n}\right) : \frac{\pi}{n}$

59. The ration of the area of a regular polygon of n sides inscribed in a circle to that of the polygon of same number of sides circumscribing the same circle is 3:4. Then the value of n is

a. 6 b. 4

c. 8 d. 12

60. In a triangle ABC, if P, Q, R divides sides AB, BC, and CA, respectively, in the raito $k1$ (in order). If the ratio $\left(\frac{\text{area } PQR}{\text{area } ABC}\right)$ is $\frac{1}{3}$, then k is equal to

a. 1 b. 2

c. 1/2 d. None of these

61. The side of a triangle ABC are in A.P. (order being a, b, c) and satisfy $\frac{2!}{1!9!} + \frac{2!}{3!7!} + \frac{1}{5!5!} = \frac{8^a}{(2b)!}$, then the value of $\cos A + \cos B$ is

a. $\frac{12}{7}$ b. $\frac{13}{7}$

c. $\frac{11}{7}$ d. $\frac{10}{7}$

62. If point P lies on sides of a right angled triangle ABC, then $PA + PB + PC$ is minimum when P is the

a. orthocentre

b. circumcentre

c. mid point of the smallest side

d. None of these

63. In any triangle ABC, $\dfrac{a^2+b^2+c^2}{R^2}$ has the maximum value of

a. 3 b. 6
c. 9 d. None of these

64. If a, b, c are the sides of a triangle, then the minimum value of $\dfrac{2a}{b+c-a}+\dfrac{2b}{c+a-b}+\dfrac{2c}{a+b-c}$ is

a. 3 b. 9
c. 6 d. 1

65. In any triangle, minimum value of $\dfrac{r_1r_2r_3}{r^3}$ is equal to

a. 1 b. 9
c. 27 d. None of these

66. In a convex quadrilateral $ABCD$, $AB = a$, $BC = b$, $CD = c$, $DA = d$. This quadrilateral is such that a circle can be inscribed in it and a circle can be also circumscribed about it, then $\tan^2 A/2$ is equal to

a. $\dfrac{ad}{bc}$ b. $\dfrac{ab}{cd}$
c. $\dfrac{cd}{ab}$ d. $\dfrac{bc}{ad}$

67. In triangle ABC, $\sin A$, $\sin B$ and $\sin C$ are in A.P, then

a. the altitudes are in H.P.
b. the altitudes are in A.P.
c. the altitudes are in G.P.
d. None of these

68. In triangle ABC, $\angle C = 2\pi/3$ and CD is the internal angle bisector of $\angle C$, meeting the side AB at D. Length CD is equal to

a. $\dfrac{ab}{2(a+b)}$

b. $\dfrac{2ab}{a+b}$

c. $\dfrac{2ab}{\sqrt{3(a+b)}}$

d. $\dfrac{ab}{a+b}$

69. In ΔABC, let R = circumradius, r = inradius, if r be the distance between the circumcentre and the incentre, then ratio R/r is equal to

a. $\sqrt{2}-1$ b. $\sqrt{3}-1$
c. $\sqrt{2}+1$ d. $\sqrt{3}+1$

70. In the adjacent figure AB is the diameter of the circle, centered at 'O'. If $\angle COA = 60°$, $AB = 2R$, $AC = d$ and $CD = 1$, then l is equal to

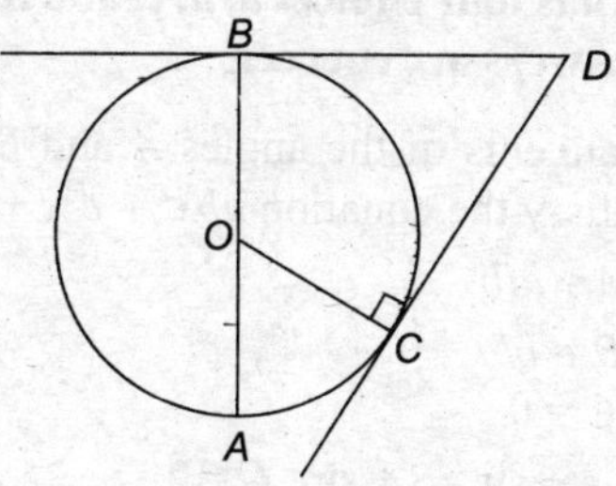

Fig. 4.1

a. $d\sqrt{3}$ b. $d/\sqrt{3}$
c. $3d$ d. $\sqrt{3}d/2$

71. In triangle ABC, base BC and area of triangle Δ are fixed. Locus of the centroid of triangle ABC is a straight line that is

a. parallel to side BC
b. right bisector of side BC
c. right angle of BC
d. inclined at an angle $\sin^{-1}\left(\dfrac{\sqrt{\Delta}}{BC}\right)$ to side BC

72. Let AD be a median of the ΔABC. If AE and AF are medians of the triangle ABC and ADC, respectively, and $AD = m_1$, $AE = m_2$, $AF = m_3$, then $a^2/8$ is equal to

a. $m_2^2+m_3^2-2m_1^2$ b. $m_1^2+m_2^2-2m_3^2$
c. $m_1^2+m_3^2-2m_2^2$ d. None of these

73. A variable triangle ABC is circumscribed about a fixed circle of unit radius. Side BC always touches the circle at D and has fixed direction. If B and C vary in such a way that $(BD)(CD) = 2$, then locus of vertex A will be a straight line

a. parallel to side BC
b. perpendicular to side BC
c. making an angle $\pi/6$ with BC
d. making an angle $\sin^{-1}(2/3)$ with BC

74. In a triangle ABC, $\angle A = \pi/3$ and its incircle is of unit radius. If the radius of the circle touching the sides AB, AC internally and incircle externally is x, then the value of x is

a. 1/2 b. 1/4
c. 1/3 d. None of these

75. The radii r_1, r_2, r_3 of the described circles of the triangle ABC are in H.P. If the area of the triangle is 24 cm^2 and its perimeter is 24 cm, then the length of its largest side is

a. 10 b. 9
c. 8 d. None of these

Multiple Correct Answers Type

Each question has four choices a, b, c, and d, out of which *one or more* answers are correct.

1. If the tangents of the angles A and B of a triangle ABC satisfy the equation $abx^2 - c^2x + ab = 0$, then
   a. $\tan A = a/b$
   b. $\tan B = b/a$
   c. $\cos C = 0$
   d. $\sin^2 A + \sin^2 B + \sin^2 C = 2$

2. In a triangle, the angles are in A.P. and the lengths of the two larger sides are 10 and 9, respectively, then the length of the third side can be
   a. $5 + \sqrt{6}$ b. 0.7
   c. $5 - \sqrt{6}$ d. None of these

3. In a triangle ABC if $2a^2b^2 + 2b^2c^2 = a^4 + b^4 + c^4$, then angle B is equal to
   a. 45° b. 135°
   c. 120° d. 60°

4. If in a triangle ABC, a, b, c and angle A are given and $c \sin A < a < c$, then
   a. $b_1 + b_2 = 2c \cos A$ b. $b_1 + b_2 = c \cos A$
   c. $b_1 b_2 = c^2 - a^2$ d. $b_1 b_2 = c^2 + a^2$

5. There exists a triangle ABC satisfying
   a. $\tan A + \tan B + \tan C = 0$
   b. $\frac{\sin A}{2} = \frac{\sin B}{3} = \frac{\sin C}{7}$
   c. $(a + b)^2 = c^2 + ab$ and $\sqrt{2}(\sin A + \cos A) = \sqrt{3}$
   d. $\sin A + \sin B = \frac{\sqrt{3}+1}{2}$, $\cos A \cos B = \frac{\sqrt{3}}{4} = \sin A \sin B$

6. CF is the internal bisector of angle C of ΔABC, then CF is equal to
   a. $\frac{2ab}{a+b}\cos\frac{C}{2}$ b. $\frac{a+b}{2ab}\cos\frac{C}{2}$
   c. $\frac{b \sin A}{\sin\left(B + \frac{C}{2}\right)}$ d. None of these

7. The sides of a ΔABC satisfy the equation $2a^2 + 4b^2 + c^2 = 4ab + 2ac$. Then
   a. the triangle is isosceles
   b. the triangle is obtuse
   c. $B = \cos^{-1}\frac{7}{8}$
   d. $A = \cos^{-1}\frac{1}{4}$

8. Let ABC be an isosceles triangle with base BC. If r is the radius of the circle inscribed in the ΔABC and r_1 be the radius of the circle escribed opposite to the angle A, then the product $r_1 r$ can be equal to
   a. $R^2 \sin^2 A$ b. $R^2 \sin^2 2B$
   c. $\frac{1}{2}a^2$ d. $\frac{a^2}{4}$

   where R is the radius of the circumcircle of the ΔABC.

9. If in a triangle $\sin^4 A + \sin^4 B + \sin^4 C = \sin^2 B \sin^2 C + 2 \sin^2 C \sin^2 A + 2 \sin^2 A \sin^2 B$, then its angle A is equal to
   a. 30° b. 120°
   c. 150° d. 60°

10. The area of a regular polygon of n sides is (where r is inradius, R is circumradius and a is side of the triangle)
    a. $\frac{nR^2}{2}\sin\left(\frac{2\pi}{n}\right)$ b. $nr^2 \tan\left(\frac{\pi}{n}\right)$
    c. $\frac{na^2}{4}\cot\frac{\pi}{n}$ d. $nR^2 \tan\left(\frac{\pi}{n}\right)$

11. In an acute angled triangle ABC, AD is the altitude. Circle drawn with AD as its diameter cuts the AB and AC at P and Q, respectively. Length PQ is equal to
    a. $\frac{\Delta}{2R}$ b. $\frac{abc}{4R^2}$
    c. $2R \sin A \sin B \sin C$ d. $\frac{\Delta}{R}$

12. If A is the area and $2s$ the sum of the sides of a triangle, then
    a. $A \leq \frac{s^2}{4}$ b. $A \leq \frac{s^2}{3\sqrt{3}}$
    c. $A < \frac{s^2}{\sqrt{3}}$ d. None of these

13. If the angles of a triangle are 30° and 45°, and the included side is $(\sqrt{3}+1)$ cm, then
    a. Area of the triangle is $\frac{1}{2}(\sqrt{3}+1)$ sq. units
    b. Area of the triangle is $\frac{1}{2}(\sqrt{3}-1)$ sq. units
    c. Ratio of greater side to smaller side is $\frac{\sqrt{3}+1}{\sqrt{2}}$
    d. Ratio of greater side to smaller side is $\frac{1}{4\sqrt{3}}$

14. Sides of a ΔABC are in A.P. If $a < \min\{b, c\}$, then $\cos A$ may be equal to

a. $\frac{4b-3c}{2b}$ b. $\frac{3c-4b}{2c}$

c. $\frac{4c-3b}{2b}$ d. $\frac{4c-3b}{2c}$

15. Length of the tangents from A, B, C to the incircle are in A.P., then

a. r_1, r_2, r_3 are in H.P. b. r_1, r_2, r_3 are in A.P.

c. a, b, c are in A.P. d. $\cos A = \frac{4c-3b}{2b}$

16. If sides of triangle ABC are a, b and c such that $2b = a + c$, then

a. $\frac{b}{c} > \frac{2}{3}$ b. $\frac{b}{c} > \frac{1}{3}$

c. $\frac{b}{c} < 2$ d. $\frac{b}{c} < \frac{3}{2}$

17. In ΔABC, if $\cos\frac{A}{2} = \sqrt{\frac{b+c}{2c}}$, then

a. Area of triangle is $\frac{1}{2}ab$

b. Circumradius is equal to $\frac{1}{2}c$

c. Area of triangle is $\frac{1}{2}bc$

d. Circumradius is equal to $\frac{1}{2}a$

18. If the sides of a right angled triangle are in G.P., then the cosine of the acute angle of the triangle is

a. $\frac{\sqrt{5}-1}{2}$ b. $\frac{\sqrt{5}+1}{2}$

c. $\sqrt{\frac{\sqrt{5}-1}{2}}$ d. $\frac{\sqrt{\sqrt{5}+1}}{2}$

Reasoning Type

Each question has four choices a, b, c and d, out of which *only one* is correct. Each question contains statement 1 and statement 2.

a. Both the statements are True and Statement 2 is the correct explanation of Statement 1.

b. Both the statements are True but Statement 2 is Not the correct explanation of Statement 1.

c. Statement 1 is True and Statement 2 is False.

d. Statement 1 is False and Statement 2 is True.

1. Statement 1: If side BC and ratio of r_2 and r_3 of an acute angled triangle is given, then the locus of A is a hyperbola.

Statement 2: If base of triangle is given and difference of two variable sides is constant (less than the base), then locus of variable vertex is a hyperbola.

2. Statement 1: In ΔABC, $\sin A + \sin B > \sin C$.

Statement 2: In ΔABC, sum of two sides is always greater than

3. In acute angled ΔABC, $a > b > c$

Statement 1: $r_1 > r_2 > r_3$.

Statement 2: $\cos A < \cos B < \cos C$.

4. Statement 1: The incentre of the triangle formed by the feet of altitudes from the vertices of triangle ABC to the opposite sides is the orthocentre of the triangle ABC.

Statement 2: The inentre of triangle ABC is orthocentre of the triangle $I_1I_2I_3$, where I_1, I_2, I_3 are excentres of triangle ABC.

5. Statement 1: If I is incentre of ΔABC and I_1 excentre opposite to A and P is the intersection of II_1 and BC, then $IP \cdot I_1P = BP \cdot PC$.

Statement 2: In a ΔABC, I is incentre and I_1 is excentre opposite to A, then IBI_1C must be square.

6. Statement 1: If the quadrilateral Q_1 formed by joining mid-points of sides of another quadrilateral Q_2 is cyclic, then Q_1 is necessarily a rectangle.

Statement 2: The quadrilateral Q_1 formed by joining midpoints of sides of another quadrilateral Q_2 is always a parallelogram.

7. Statement 1: If a triangle DEF is constructed by joining the mid-points of sides of triangle ABC, inscribed in a circle of radius 5 units and having one of the sides of 8 units. Then maximum area of triangle DEF is 16 sq. units.

Statement 2: $\Delta DEF = (1/4)\,\Delta ABC$

8. Statement 1: In ΔABC, the centroid (G) divides line joining orthocenter (H) and circumcenter in ratio 2:1.

Statement 2: The centroid (G) divides the median AD in ratio 2:1.

9. Statement 1: Circumradius of $\Delta I_1I_2I_3$ is $2R$.

Statement 2: Circumradius of the triangle formed by feet of altitudes of ΔABC is $R/2$.

10. Statement 1: If the incircle of the triangle ABC passes through its circumcenter, then

$$\sin\frac{A}{2}\sin\frac{B}{2}\sin\frac{C}{2} = \sqrt{2}.$$

Statement 2: Distance between the circumcenter and incenter is $\sqrt{R^2 - 2rR}$.

11. **Statement 1:** In a triangle ABC, D is a point on the side AB such that $CD^2 = AD \cdot DB$, then the greatest value of $\sin A \sin B$ is $\sin^2 C/2$.
Statement 2: Greatest value of $\sin A \sin B$ occurs when CD is the angle bisector of angle C.

12. **Statement 1:** If a, b, c are the sides of a triangle, then the minimum value of
$$\frac{2a}{b+c-a}+\frac{2b}{c+a-b}+\frac{2c}{a+b-c} \text{ is } 9.$$
Statement 2: A.M. $\geq$ G.M. $\geq$ H.M.

13. **Statement 1:** If $C=45°$, $B=15°$, then the line joining A and circumcenter (O) divides BC in ratio 2.
Statement 2: Line joining A and circumcenter (O) divides BC in ratio $\frac{\sin 2C}{\sin 2B}$.

14. **Statement 1:** If $a=3, b=7, c=8$, and internal angle bisector AI meets BC at D (where I is incentre), then $AI/ID = 11/2$.
Statement 2: Internal angle bisector of angle A divides the side BC in ratio AB/AC.

Linked Comprehension Type

Based upon each paragraph, three multiple choice questions have to be answered. Each question has four choices a, b, c, and d, out of which *only one* is correct.

For Problems 1–3

Given that $\Delta = 6, r_1 = 2, r_2 = 3, r_3 = 6$.

1. Circumradius R is equal to
a. 2.5 b. 3.5
c. 1.5 d. None of these

2. Inradius is equal to
a. 2 b. 1
c. 1.5 d. 2.5

3. Difference between the greatest and the least angle is
a. $\cos^{-1}\frac{4}{5}$ b. $\tan^{-1}\frac{3}{4}$
c. $\cos^{-1}\frac{3}{5}$ d. None of these

For Problems 4–6

Let $a=6, b=3$ and $\cos(A-B)=\frac{4}{5}$.

4. Area of the triangle is equal to
a. 9 b. 12
c. 11 d. 10

5. Angle C is equal to
a. $\frac{3\pi}{4}$ b. $\frac{\pi}{4}$
c. $\frac{\pi}{2}$ d. None of these

6. Value of $\sin A$ is equal to
a. $\frac{1}{2\sqrt{5}}$ b. $\frac{1}{\sqrt{3}}$
c. $\frac{1}{\sqrt{5}}$ d. $\frac{2}{\sqrt{5}}$

For Problems 7–9

p_1, p_2, p_3 are altitudes of a triangle ABC from the vertices A, B, C and Δ is the area of the triangle.

7. The value of $p_1^{-2}+p_2^{-2}+p_3^{-2}$ is equal to
a. $\frac{a+b+c}{\Delta}$ b. $\frac{a^2+b^2+c^2}{4\Delta^2}$
c. $\frac{a^2+b^2+c^2}{\Delta^2}$ d. None of these

8. The value of $\frac{1}{p_1}+\frac{1}{p_2}+\frac{1}{p_3}$ is equal to
a. $\frac{2}{r}$ b. $\frac{2s}{\Delta}$
c. $\frac{8R}{abc}$ d. None of these

9. The value of $\frac{\cos A}{p_1}+\frac{\cos B}{p_2}+\frac{\cos C}{p_3}$ is equal to
a. $\frac{1}{R}$ b. $\frac{a^2+b^2+c^2}{2R}$
c. $\frac{\Delta}{2R}$ d. None of these

For Problems 10–12

Let O be a point inside a ΔABC such that $\angle OAB = \angle OBC = \angle OCA = \theta$.

10. $\cot A + \cot B + \cot C$ is equal to
a. $\tan^2\theta$ b. $\cot^2\theta$
c. $\tan\theta$ d. $\cot\theta$

11. $\operatorname{cosec}^2 A + \operatorname{cosec}^2 B + \operatorname{cosec}^2 C$ is equal to
a. $\cot^2\theta$ b. $\operatorname{cosec}^2\theta$
c. $\tan^2\theta$ d. $\sec^2\theta$

12. Area of ΔABC is equal to
a. $\left(\frac{a^2+b^2+c^2}{4}\right)\tan\theta$ b. $\left(\frac{a^2+b^2+c^2}{4}\right)\cot\theta$
c. $\left(\frac{a^2+b^2+c^2}{2}\right)\tan\theta$ d. $\left(\frac{a^2+b^2+c^2}{2}\right)\cot\theta$

For Problems 13–15

Let D, E, F be the feet of altitudes from the vertices of acute-angled triangle ABC to the sides BC, AC and AB, respectively. Triangle DEF is defined as pedal triangle of triangle ABC. (R and r are circumradius and inradius of triangle ABC, respectively.)

13. Consider the following statements:
 i. orthocentre of the triangle ABC is the incentre of the triangle DEF
 ii. A, B, C are excentres of triangle DEF
 a. only (i) is true
 b. only (ii) is true
 c. both (i) and (ii) are true
 d. both (i) and (ii) are false

14. Circumradius of the pedal triangle of triangle ABC is
 a. $R/2$ **b.** $r/2$
 c. $R/4$ **d.** $r/4$

15. If x, y, z are the sides of pedal triangle, then $x+y+z$ is equal to
 a. $\Delta R/2$ **b.** $\Delta/2R$
 c. ΔR **d.** None of these

For Problems 16–18

Incircle of ΔABC touches the sides BC, AC, AB at D, E, F, respectively. Then answer the following questions.

16. $\angle DEF$ is equal to
 a. $\dfrac{\pi - B}{2}$ **b.** $\pi - 2B$
 c. $A - C$ **d.** None of these

17. Area of ΔDEF is
 a. $2r^2 \sin(2A)\sin(2B)\sin(2C)$
 b. $2r^2 \cos\dfrac{A}{2}\cos\dfrac{B}{2}\cos\dfrac{C}{2}$
 c. $2r^2 \sin(A-B)\sin(B-C)\sin(C-A)$
 d. None of these

18. The length of side EF is
 a. $r\sin\dfrac{A}{2}$ **b.** $2r\sin\dfrac{A}{2}$
 c. $r\cos\dfrac{A}{2}$ **d.** $2r\cos\dfrac{A}{2}$

For Problems 19–21

Internal bisectors of ΔABC meets the circumcircle at points D, E, F,

19. The length of side EF is
 a. $2R\cos\left(\dfrac{A}{2}\right)$ **b.** $2R\sin\left(\dfrac{A}{2}\right)$
 c. $R\cos\left(\dfrac{A}{2}\right)$ **d.** $2R\cos\left(\dfrac{B}{2}\right)\cos\left(\dfrac{C}{2}\right)$

20. Area of ΔDEF is
 a. $2R^2\cos^2\left(\dfrac{A}{2}\right)\cos^2\left(\dfrac{B}{2}\right)\cos^2\left(\dfrac{C}{2}\right)$
 b. $2R^2\sin\left(\dfrac{A}{2}\right)\sin\left(\dfrac{B}{2}\right)\sin\left(\dfrac{C}{2}\right)$
 c. $2R^2\sin^2\left(\dfrac{A}{2}\right)\sin^2\left(\dfrac{B}{2}\right)\sin^2\left(\dfrac{C}{2}\right)$
 d. $2R^2\cos\left(\dfrac{A}{2}\right)\cos\left(\dfrac{B}{2}\right)\cos\left(\dfrac{C}{2}\right)$

21. Ratio of area of triangle ABC and triangle DEF is
 a. ≥ 1 **b.** ≤ 1
 c. $\geq 1/2$ **d.** $\leq 1/2$

Matrix-Match Type

Each question contains statements given in two columns which have to be matched.
Statements (a, b, c, d) in column I have to be matched with statements (p, q, r, s) in column II. If the correct match are a→p, a→s; b→q, b→r, c→p, c→q and d→s; then the correctly bubbled 4 × 4 matrix should be as follows:

	p	q	r	s
a	(p) ●	q	r	(s) ●
b	p	(q) ●	(r) ●	s
c	(p) ●	(q) ●	r	s
d	p	q	r	(s) ●

1.

Column I	Column II
a. $b > c\sin B, b < c$ and B is acute angle	**p.** 0
b. $b > c\sin B, c < b$, and B is acute angle	**q.** 2
c. $b > c\sin B, c < b$ and B is obtuse angle	**r.** data insufficient
d. $b > c\sin B, c > b$ and B is an obtuse angle	**s.** 1

2. In acute angled triangle ABC

Column I	Column II
a. $\cos A, \cos B, \cos C$ are in A.P.	**p.** Distances of orthocentre from vertices of triangle are in A.P.
b. $\sin(A/2), \sin(B/2), \sin(C/2)$ are in A.P.	**q.** Distances of orthocentre from sides of triangle are in H.P.
c. Distances of circumcentre from the vertices of the triangle ABC are in A.P.	**r.** Distances of incentre from vertices of in H.P.

d. Circumradii of triangles OBC, OAC, and OAB are in H.P. (where O is circumcentre of triangle ABC)	s. Distances of incentre from excentres of triangle are in A.P.

3.

Column I	Column II
a. If the sines of the angles A and B of a triangle ABC satisfy the equation $c^2x^2 - c(a+b)x + ab = 0$ the triangle can be	p. right angled
b. If one angle of a triangle is 30° and the lengths of the sides adjacent to it are 40 and $40\sqrt{3}$, the triangle can be	q. isosceles
c. If the angles of a triangle ABC satisfy the equation $81^{\sin^2 x} + 81^{\cos^2 x} = 30$, then the triangle can be	r. equilateral
d. In a triangle ABC, $\cos A \cos B + \sin A \sin B \sin C = 1$, then the triangle can be	s. obtuse angled

4. Let O be the circumcentre, H be the orthocentre, I be the incentre and I_1, I_2, I_3 be the excentres.

Column I	Column II
a. Angle subtended by OI at vertex A	p. $\lvert B-C \rvert$
b. Angle subtended by HI at vertex A	q. $\frac{\lvert B-C \rvert}{2}$
c. Angle subtended by OH at vertex A	r. $\frac{B+C}{2}$
d. Angle subtended by I_2I_3 at I_1	s. $\frac{B}{2}-C$

5.

Column I	Column II
a. $\cot\frac{A}{2} = \frac{b+c}{a}$	p. always right angled
b. $a \tan A + b \tan B = (a+b)\tan\left(\frac{A+B}{2}\right)$	q. always isosceles
c. $a \cos A = b \cos B$	r. may be right angled
d. $\cos A = \frac{\sin B}{2 \sin C}$	s. may be right angled isosceles

Integer Type

1. The value of $(a \cos A + b \cos B + c \cos C)/(R \sin A \sin B \sin C)$ is ______.
2. In a triangle, the angles A, B, C are in A.P. Then the value of $\frac{a+c}{\cos[(A-C)/2]\sqrt{a^2 - ac + c^2}}$ is ______.
3. If $b + c = 3a$, then the value of $\cot(B/2)\cot(C/2)$ is______.
4. The radius of the incircle of a triangle with sides are 18, 24, and 30 cm is ______.
5. In any triangle ABC, the least value of $(r_1 + r_2 + r_3)/r$ is ______.
6. The two adjacent sides of a cyclic quadrilateral are 2 and 5 and the angle between them is 60°. If the area of the quadrilateral is $4\sqrt{3}$, then the length of larger of the remaining two sides is ______.
7. Perpendiculars are drawn from the angles A, B, C of an acute-angled triangle on opposite sides and produced to meet the circumscribing circle. If these produced parts are a, b, c, respectively, then the value of $\frac{(a/\alpha)+(b/\beta)+(c/\gamma)}{\tan A + \tan B + \tan C}$ is ______.
8. Let $A_0A_1A_2A_3A_4A_5$ be a regular hexagon inscribed in a circle of unit radius. Then the product of the lengths of the line segments A_0A_1, A_0A_2, in line and A_0A_4 is ______.
9. Three circles touch each other externally. The tangent at their point of contact meets at a point whose distance from the point of contact is 4. Then the square root of the ratio of the product of the radii to the sum of the radii of the circles is ______.
10. The sides of a triangle are three consecutive natural numbers and its largest angle is twice the smallest one. Then the length of the largest side of the triangle is ______.

11. In an equilateral triangle, three coins of radii 1 unit each are kept so that they touch each other and also the sides of the triangle. If L is the side length of triangle, then the value of $[L]$, where [.] represents the greatest integer function, is _____.

12. Let $ABCD$ be a quadrilateral with area 18, with side AB parallel to side CD and $AB = 2CD$. Let AD be a perpendicular to AB and CD. If a circle is drawn inside the quadrilateral $ABCD$ touching all the sides, then its radius is _____.

13. A circle is inscribed in an equilateral triangle of side length 6 units. The area of any square inscribed in this circle is _____.

14. The number of positive integral values of $a \in [1, 10]$ for which $a^2 + 2a$, $2a + 3$, and $a^2 + 3a + 8$ are the sides of a triangle is _____.

15. The area of a triangle ABC is equal to $(a^2 + b^2 - c^2)$, where a, b and c are the side of the triangle. The value of $\tan C$ equals _____.

16. In a triangle ABC, if $A - B = 120°$ and $R = 8r$ where R and r have their usual meanings, then $8\cos C$ equals to _____.

17. If the sides of a triangle are in $A.P.$ and its largest and the smallest angle are θ and ϕ, then $\dfrac{\cos\theta + \cos\phi}{(1-\cos\theta)(1-\cos\phi)}$ equals _____.

18. If the side of a regular pentagon is 'a' and its diagonal is of length 'b', then $\dfrac{a^2}{b^2} + \dfrac{b^2}{a^2} =$ _____.

19. ABC is a triangle. h_a is the length of the altitude from A. h_b has a similar meaning. Given $a \le h_a$ and $b \le h_b$. If $c = 4$ then the maximum area of the triangle is _____.

ANSWERS AND SOLUTIONS

Objective Type

1. c. Adding, $\sin(A+B)=1$

and subtracting, $\sin(A-B)=\dfrac{\sqrt{2}-2}{\sqrt{2}}=1-\sqrt{2}\neq 0$

$\therefore A+B=90°,\ A\neq B$

2. c.

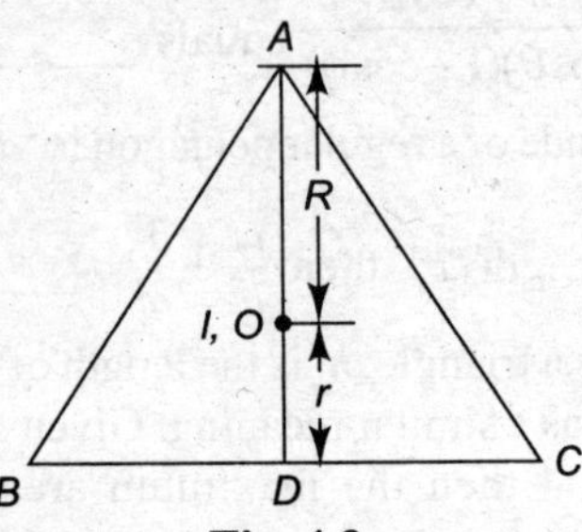

Fig. 4.2

In equilateral triangle circumcentre (O) and incentre (I) coincide.

Also from the diagram $R+r=h \Rightarrow \dfrac{R+r}{h}=1$

3. a. $\cos C=\dfrac{a^2+b^2-c^2}{2ab}>0$

$\Rightarrow 0°<C<90°$

4. a.

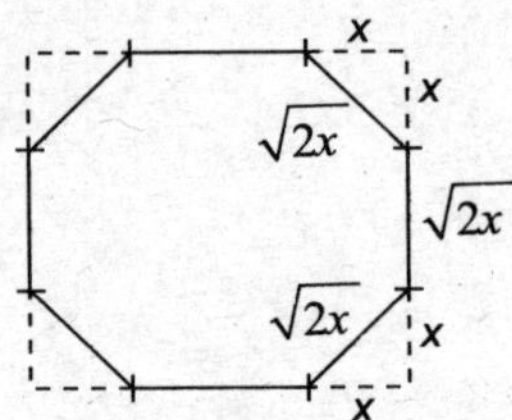

Fig. 4.3

Clearly, $x+\sqrt{2}x+x=1$. So, $x=\dfrac{1}{2+\sqrt{2}}$.

The required area $=\left(1^2-4\times\dfrac{1}{2}x^2\right)\text{ m}^2$

$=\left\{1-2\dfrac{1}{(2+\sqrt{2})^2}\right\}\text{ m}^2$

$=\left\{1-\dfrac{1}{(\sqrt{2}+1)^2}\right\}\text{ m}^2$

$=\left[1-(\sqrt{2}-1)^2\right]\text{ m}^2$

$=2(\sqrt{2}-1)\text{ m}^2$

5. c. As A is an obtuse angle,

$90°<A<180°$

$\Rightarrow 90°<180-(B+C)<180°$

$\Rightarrow 0<B+C<90°$

$\Rightarrow B+C<90° \Rightarrow B<90°-C$

$\therefore \tan B<\tan(90°-C)$

$\tan B<\cot C \Rightarrow \tan B\tan C<1$

6. d. $x=\dfrac{1}{2}[\cos(A-C)-\cos(A+C)]$

$=\dfrac{1}{2}\left[\cos(A-C)+\dfrac{1}{2}\right]$

But $0\le\cos(A-C)\le 1$

$\Rightarrow \dfrac{1}{2}\left(0+\dfrac{1}{2}\right)\le x\le\dfrac{1}{2}\left(1+\dfrac{1}{2}\right)$

7. c. Here $b=2c$.

Now, $\tan\dfrac{B-C}{2}=\dfrac{b-c}{b+c}\cot\dfrac{A}{2}$

$\cot\dfrac{A}{2}\cot\dfrac{B-C}{2}=\dfrac{b+c}{b-c}=\dfrac{3c}{c}=3$

8. b.

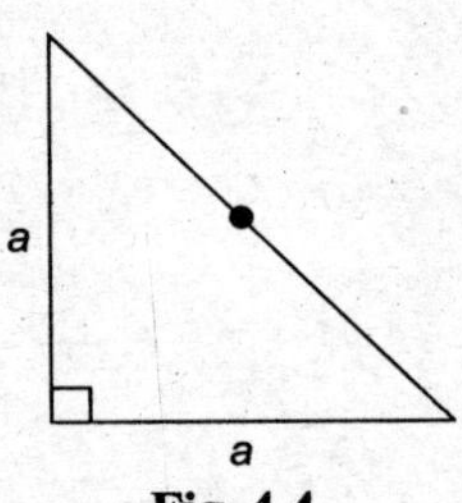

Fig. 4.4

Here, $R=\dfrac{\sqrt{2}a}{2}=\dfrac{a}{\sqrt{2}}$

$r=\dfrac{\Delta}{s}=\dfrac{\frac{1}{2}a^2}{\frac{1}{2}\left(a+a+\sqrt{2}a\right)}=\dfrac{a}{2+\sqrt{2}}$

$\therefore \dfrac{R}{r}=\dfrac{a}{\sqrt{2}}\times\dfrac{2+\sqrt{2}}{a}=\sqrt{2}+1$

9. a. $\cos C=\dfrac{a^2+b^2-c^2}{2ab}=\dfrac{25+9-49}{2\times5\times3}=-\dfrac{1}{2}$

$\cos B=\dfrac{a^2+c^2-b^2}{2ac}=\dfrac{25+49-9}{2\times5\times7}=\dfrac{13}{14}$

$\Rightarrow 3\cos C+7\cos B=-\dfrac{3}{2}+\dfrac{13}{2}=5$

10. a. $\angle B=90° \Rightarrow \cos A=\dfrac{c}{b}$

$\Rightarrow \dfrac{1-\tan^2\frac{A}{2}}{1+\tan^2\frac{A}{2}}=\dfrac{c}{b}$

$$\Rightarrow \frac{\left(1+\tan^2\frac{A}{2}\right)-\left(1-\tan^2\frac{A}{2}\right)}{\left(1+\tan^2\frac{A}{2}\right)+\left(1-\tan^2\frac{A}{2}\right)}=\frac{(b-c)}{(b+c)}$$

$$\Rightarrow \tan^2\frac{A}{2}=\frac{b-c}{b+c}$$

11. a. $\frac{\cot A}{\cot B+\cot C}$

$$=\frac{\frac{R(b^2+c^2-a^2)}{abc}}{\frac{R(a^2+c^2-b^2)}{abc}+\frac{R(a^2+b^2-c^2)}{abc}}$$

$$=\frac{b^2+c^2-a^2}{2a^2}=\frac{2a^2-a^2}{2a^2}=\frac{1}{2}$$

12. c. Here, $\sin\theta-\cos\theta=\frac{b}{a}$ and $\sin\theta\cos\theta=\frac{c}{a}$

$$\Rightarrow 1-2\sin\theta\cos\theta=\frac{b^2}{a^2}$$

$$\Rightarrow 1-\frac{2c}{a}=\frac{b^2}{a^2}$$

$$\Rightarrow a^2-b^2=2ac$$

Hence, $\cos B=\frac{a^2+c^2-b^2}{2ac}=\frac{2ac+c^2}{2ac}=1+\frac{c}{2a}$

13. c. We have, $a^2+b^2+c^2-ac-ab\sqrt{3}=0$

$$\Rightarrow \frac{a^2}{4}-ac+c^2+\frac{3a^2}{4}+b^2-ab\sqrt{3}=0$$

$$\Rightarrow \left[\frac{a}{2}-c\right]^2+\left[\frac{\sqrt{3}a}{2}-b\right]^2=0$$

$\Rightarrow a=2c$ and $2b=\sqrt{3}a \Rightarrow b^2+c^2=a^2$

Hence, the triangle is right angled.

14. c. $(a+b+c)(b+c-a)=kbc$

$\Rightarrow (b+c)^2-a^2=kbc$

$\Rightarrow b^2+c^2-a^2=(k-2)bc$

$\Rightarrow 2bc\cos A=(k-2)bc$

$$\Rightarrow \cos A=\frac{k-2}{2}$$

Now, A being the angle of a triangle,

$-1<\cos A<1 \Rightarrow -2<k-2<2$

$\Rightarrow 0<k<4$

15. c. $a=2b$ and $A-B=60°$

We know that $\tan\frac{A-B}{2}=\frac{a-b}{a+b}\cot\frac{C}{2}$

$$\Rightarrow \tan 30°=\frac{1}{3}\cot\frac{C}{2}\Rightarrow \tan\frac{C}{2}=\frac{1}{\sqrt{3}}$$

$\Rightarrow C=60°$

Hence, $A+B=120° \Rightarrow 2A=180°$

$\Rightarrow A=90°, B=30°, C=60°$

16. a. $r_1=2r_2=3r_3$

$$\Rightarrow \frac{\Delta}{s-a}=2\frac{\Delta}{s-b}=3\frac{\Delta}{s-c}$$

$$\Rightarrow \frac{1}{s-a}=\frac{2}{s-b}=\frac{3}{s-c}=k \text{ (say)}$$

$\Rightarrow s-a=\frac{1}{k}$, $s-b=\frac{2}{k}$, and $s-c=\frac{3}{k}$

Adding, $3s-(a+b+c)=\frac{6}{k}$

$\Rightarrow s=\frac{6}{k}\Rightarrow a=\frac{5}{k}$ and $b=\frac{4}{k}\Rightarrow \frac{a}{b}=\frac{5}{4}$

17. a. We have, $\left(1-\frac{s-b}{s-a}\right)\left(1-\frac{s-c}{s-a}\right)=2$

$\Rightarrow 2(b-a)(c-a)=4(s-a)^2$

$\Rightarrow 2(bc-ac-ab+a^2)=(2s-2a)^2$

$\Rightarrow 2(bc-ac-ab+a^2)=(b+c-a)^2$

$\Rightarrow a^2=b^2+c^2$

$\Rightarrow$ triangle is right angled

18. b. We have, $\Delta=\frac{\sqrt{3}}{4}a^2$, $s=\frac{3a}{2}$. Therefore,

$$r=\frac{\Delta}{s}=\frac{a}{2\sqrt{3}},\ R=\frac{abc}{4\Delta}=\frac{a^3}{\sqrt{3}a^2}=\frac{a}{\sqrt{3}}$$

and $r_1=\frac{\Delta}{s-a}=\frac{\sqrt{3}/4a^2}{a/2}=\frac{\sqrt{3}}{2}a$

Hence, $r:R:r_1=\frac{a}{2\sqrt{3}}:\frac{a}{\sqrt{3}}:\frac{\sqrt{3}}{2}a=1:2:3$

19. a. $\cot\frac{A}{2}$, $\cot\frac{B}{2}$, $\cot\frac{C}{2}$ are in A.P.

$$\Rightarrow 2\cot\frac{B}{2}=\cot\frac{A}{2}+\cot\frac{C}{2}$$

$$\Rightarrow 2\sqrt{\frac{s(s-b)}{(s-a)(s-c)}}=\sqrt{\frac{s(s-a)}{(s-b)(s-c)}}+\sqrt{\frac{s(s-c)}{(s-a)(s-b)}}$$

$\Rightarrow 2(s-b)=s-a+s-c$

$\Rightarrow 2b=a+c$

$\Rightarrow a, b, c$ are in A.P.

20. b. We have, $\tan\frac{C-B}{2}=\frac{c-b}{c+b}\cot\frac{A}{2}$

$$\tan\left(\frac{C-B}{2}\right) = \frac{\sqrt{3}+1-2}{\sqrt{3}+1+2}\cot 15°$$

$$= \frac{\sqrt{3}-1}{\sqrt{3}+3}\frac{1}{\tan(45-30°)}$$

$$= \frac{\sqrt{3}-1}{\sqrt{3}+3}\cdot\frac{\sqrt{3}+1}{\sqrt{3}-1} = \frac{1}{\sqrt{3}}$$

$$= \tan 30°$$

$$\Rightarrow \frac{C-B}{2} = 30°.$$

21. d. $\cos A = \dfrac{b^2+c^2-a^2}{2bc} = \dfrac{64+81-49}{2\times 8\times 9}$

$$= \frac{145-49}{144} = \frac{96}{144}$$

$$\cos B = \frac{a^2+c^2-b^2}{2ac} = \frac{49+81-64}{2\times 7\times 9} = \frac{66}{126} = \frac{11}{21}$$

22. b. $\cos A + \cos B + \cos C = \dfrac{7}{4}$

$$\Rightarrow 1+4\sin\frac{A}{2}\sin\frac{B}{2}\sin\frac{C}{2} = \frac{7}{4}$$

$$\Rightarrow 4\sin\frac{A}{2}\sin\frac{B}{2}\sin\frac{C}{2} = \frac{3}{4}$$

$(\because r = 4R\sin A/2 \sin B/2 \sin C/2)$

$$\Rightarrow \frac{r}{R} = \frac{3}{4} \Rightarrow \frac{R}{r} = \frac{4}{3}$$

23. a. $\cot\dfrac{A}{2}+\cot\dfrac{B}{2}+\cot\dfrac{C}{2}$

$$= \frac{s(s-a)}{\Delta}+\frac{s(s-b)}{\Delta}+\frac{s(s-c)}{\Delta}$$

$$= \frac{s}{\Delta}[3s-(a+b+c)]$$

$$= \left(\frac{a+b+c}{2}\right)^2 \times \frac{4R}{abc} = s^2\times\frac{1}{\Delta}$$

$$= \frac{\Delta^2}{r^2}\times\frac{1}{\Delta} = \frac{\Delta}{r^2}$$

24. b. $\dfrac{\cot A+\cot C}{\cot B} = \dfrac{\sin(A+C)\sin B}{\sin A\sin C\cos B}$

$$= \frac{\sin^2 B}{\sin A\sin C\cos B}$$

$$= \frac{4R^2 b^2}{4R^2\, ac\cos B}$$

$$= \frac{2b^2}{2ac\cos B} = \frac{2b^2}{a^2+c^2-b^2}$$

$$= \frac{2b^2}{2002b^2-b^2} = \frac{2}{2001}$$

25. a.

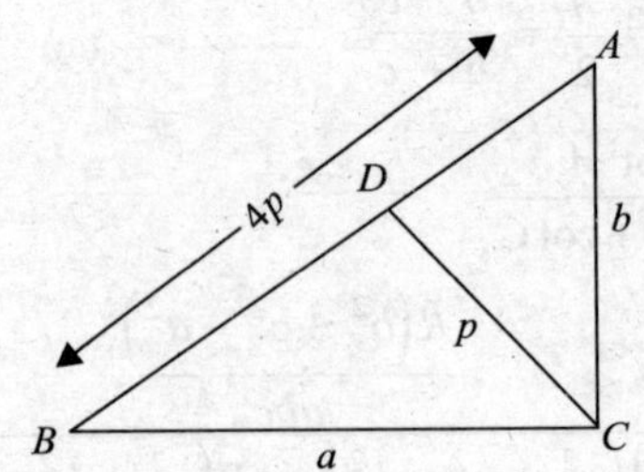

Fig. 4.5

$$\Delta = \frac{1}{2}ab = \frac{1}{2}p\,4p \Rightarrow ab = 4p^2$$

Also, $a^2+b^2 = c^2 = 16p^2$

$\therefore (a-b)^2 = a^2+b^2-2ab = 8p^2$

Also, $(a+b)^2 = a^2+b^2+2ab = 24p^2$

$$\tan\frac{A-B}{2} = \frac{a-b}{a+b}\cot\frac{C}{2} = \frac{1}{\sqrt{3}}1$$

$$\Rightarrow \frac{A-B}{2} = 30° \Rightarrow A-B = 60°$$

26. c. $\dfrac{1}{\sin^2\frac{A}{2}}, \dfrac{1}{\sin^2\frac{B}{2}}, \dfrac{1}{\sin^2\frac{C}{2}}$ are in A.P.

$$\Rightarrow \frac{1}{\sin^2\frac{C}{2}}-\frac{1}{\sin^2\frac{B}{2}} = \frac{1}{\sin^2\frac{B}{2}}-\frac{1}{\sin^2\frac{A}{2}}$$

$$\Rightarrow \frac{ab}{(s-a)(s-b)}-\frac{ac}{(s-a)(s-c)}$$

$$= \frac{ac}{(s-a)(s-c)}-\frac{bc}{(s-b)(s-c)}$$

$$\Rightarrow \left(\frac{a}{s-a}\right)\left(\frac{b(s-c)-c(s-b)}{(s-b)(s-c)}\right)$$

$$= \left(\frac{c}{s-c}\right)\left(\frac{a(s-b)-b(s-a)}{(s-a)(s-b)}\right)$$

$$\Rightarrow ab-ac = ac-bc$$

$$\Rightarrow ab+bc = 2ac$$

$$\Rightarrow \frac{1}{c}+\frac{1}{a} = \frac{2}{b}, \text{ i.e., } a, b, c \text{ are in H.P.}$$

27. a. $b=2, c=\sqrt{3}, \angle A=30°$

$$a=\sqrt{b^2+c^2-2bc\cos A}$$

$$=\sqrt{4+3-2\times 2\sqrt{3}\,\frac{\sqrt{3}}{2}}=1$$

$$\Rightarrow r=(s-a)\tan\frac{A}{2}=\frac{b+c-a}{2}\tan\frac{A}{2}=\frac{\sqrt{3}+1}{2}$$

$$\tan 15°=\frac{\sqrt{3}+1}{2}\cdot\frac{\sqrt{3}-1}{\sqrt{3}+1}=\frac{\sqrt{3}-1}{2}$$

28. c.

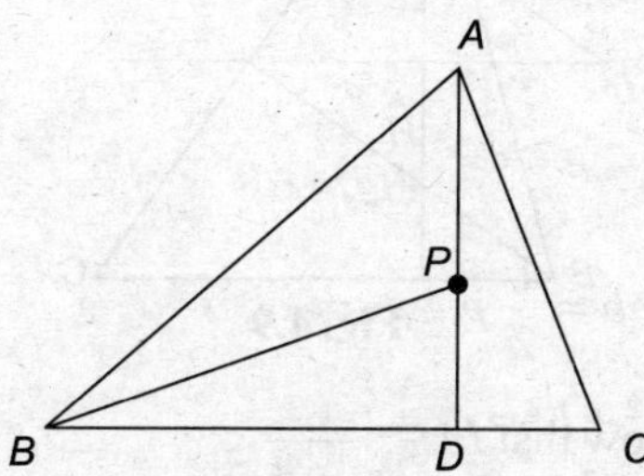

Fig. 4.6

$\angle BPA=90°+B/3, \angle ABP=2B/3$

In ΔABP

$$\frac{AP}{\sin(2B/3)}=\frac{c}{\sin(90°+B/3)}=\frac{c}{\cos(B/3)}$$

[by sine rule]

$$\Rightarrow AP=\frac{c\sin(2B/3)}{\cos(B/3)}=\frac{2c\sin(B/3)\cos(B/3)}{\cos(B/3)}$$

$$\Rightarrow AP=2c\sin(B/3)$$

29. d. $\angle A=\frac{\pi}{2}\Rightarrow a^2=b^2+c^2$

$$\sin C=\frac{c}{a}=\frac{2\tan\frac{C}{2}}{1+\tan^2\frac{C}{2}}$$

Fig. 4.7

$$\Rightarrow c\tan^2\frac{C}{2}-2a\tan\frac{C}{2}+c=0$$

$$\Rightarrow \tan\frac{C}{2}=\frac{2a\pm\sqrt{4a^2-4c^2}}{2c}=\frac{2a\pm 2b}{2c}=\frac{a\pm b}{c}$$

$$=\frac{a-b}{c}$$

Because if $\tan\frac{C}{2}=\frac{a+b}{c}$, then $\tan\frac{C}{2}>1$

$\Rightarrow \frac{C}{2}>\frac{\pi}{4}\Rightarrow C>\frac{\pi}{2}$ which is not possible.

30. a. Since the angles are in A.P.,

$2B=A+C$

$\Rightarrow 3B=A+B+C$

$\Rightarrow 3B=180°$

$\Rightarrow B=60°\Rightarrow \cos B=\frac{1}{2}$

$$\Rightarrow \frac{c^2+a^2-b^2}{2ac}=\frac{1}{2}$$

$\Rightarrow c^2+a^2-b^2=ac$

$\Rightarrow c^2+a^2-b^2=b^2$ [$\because a, b, c$ are in G.P. $\therefore b^2=ac$]

$\Rightarrow c^2+a^2=2b^2$

$\Rightarrow a^2, b^2, c^2$ are in A.P.

31. d. If the triangle is equilateral,

$$\sin A+\sin B+\sin C=\frac{3\sqrt{3}}{2}$$

If the triangle is isosceles,
let $A=30°, B=30°, C=120°$. Then,

$$\sin A+\sin B+\sin C=1+\frac{\sqrt{3}}{2}.$$

If the triangle is right angled,
let $A=90°, B=30°, C=60°$.

Then, $\sin A+\sin B+\sin C=\frac{3+\sqrt{3}}{2}$

If the triangle is right angled isosceles, then one of the angles is 90° and the remaining two are 45° each, so that

$\sin A+\sin B+\sin C=1+\sqrt{2}$

and $\cos A+\cos B+\cos C=\sqrt{2}$

32. c. $\frac{r}{r_1}=\frac{r_2}{r_3}$

$\Rightarrow rr_3=r_1r_2$

$$\Rightarrow \frac{\Delta}{s}\frac{\Delta}{s-c}=\frac{\Delta}{s-a}\frac{\Delta}{s-b}$$

$$\Rightarrow \frac{(s-a)(s-b)}{s(s-c)}=1$$

$$\Rightarrow \tan^2\frac{C}{2}=1$$

$$\Rightarrow \tan\frac{C}{2}=1$$

$$\Rightarrow \frac{C}{2}=45°\Rightarrow C=90°$$

33. c. Since $\cos 3A + \cos 3B + \cos 3C = 1$

$\Rightarrow 4\sin\frac{3A}{2}\sin\frac{3B}{2}\sin\frac{3C}{2} = 0$

Either $\frac{3A}{2} = 180°$ or $\frac{3B}{2} = 180°$ or $\frac{3C}{2} = 180°$

Either $A = 120°$ or $B = 120°$ or $C = 120°$

34. a. Given, $\cos B\cos C + \sin B\sin C\sin^2 A = 1$ (1)

Now, we know that $\sin^2 A \le 1$ (2)

Also, $\sin B$ and $\sin C$ are positive.

$\Rightarrow \sin B\sin C\sin^2 A \le \sin B\sin C$ (3)

$\Rightarrow 1 - \cos B\cos C \le \sin B\sin C,$

[by using equation (1)]

$\Rightarrow \cos(B - C) \ge 1 \Rightarrow \cos(B - C) = 1 \Rightarrow B - C = 0$

$\Rightarrow B = C$

Also, $\sin^2 A = 1$, i.e., $A = \pi/2$. Hence, the triangle is right-angled isosceles.

35. c.

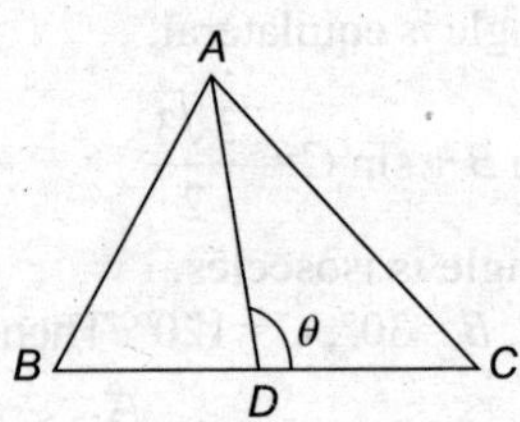

Fig. 4.8

$\theta = \pi - \left(C + \frac{A}{2}\right)$

$\Rightarrow \sin\theta = \sin\left(C + \frac{A}{2}\right)$

$= \sin\left(C + \frac{\pi}{2} - \frac{B + C}{2}\right)$

$= \cos\left(\frac{B - C}{2}\right)$

36. d. $r - s\, r_2 = r_3 - r_1$

$\Rightarrow \frac{\Delta}{s} - \frac{\Delta}{s - b} = \frac{\Delta}{s - c} - \frac{\Delta}{s - a}$

$\Rightarrow \frac{-b}{s(s - b)} = \frac{c - a}{(s - a)(s - c)}$

$\Rightarrow \frac{(s - a)(s - c)}{s(s - b)} = \frac{a - c}{b}$

$\Rightarrow \tan^2\frac{B}{2} = \frac{a - c}{b}$

But $\frac{B}{2} \in \left(\frac{\pi}{6}, \frac{\pi}{4}\right)$

$\Rightarrow \tan^2\frac{B}{2} \in \left(\frac{1}{3}, 1\right)$

$\Rightarrow \frac{1}{3} < \frac{a - c}{b} < 1$

$\Rightarrow b < 3a - 3c < 3b$

$\Rightarrow b + 3c < 3a < 3b + 3c$

37. b. $R = \frac{a}{2\sin A} = \frac{2 + \sqrt{5}}{2\sin 30°} = \frac{2 + \sqrt{5}}{2 \times \frac{1}{2}} = (2 + \sqrt{5})$

Now, $AH = 2R\cos A = 2(2 + \sqrt{5})\cos 30°$

$= (2 + \sqrt{5})\sqrt{3}$

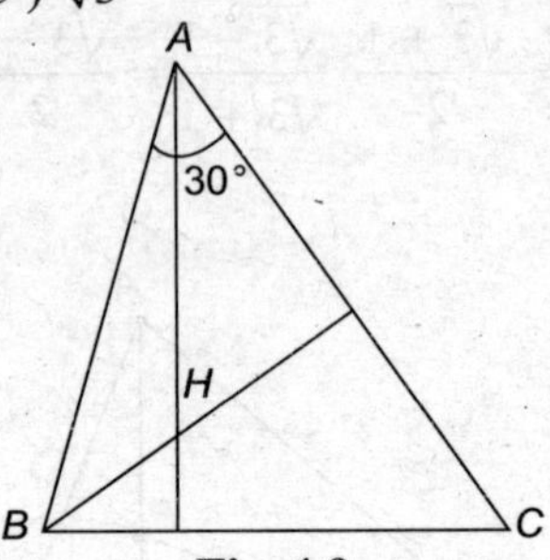

Fig. 4.9

38. a. We know that $IA = \frac{r}{\sin\frac{A}{2}}$

$\Rightarrow IA : IB : IC = \operatorname{cosec}\frac{A}{2} : \operatorname{cosec}\frac{B}{2} : \operatorname{cosec}\frac{C}{2}$

39. c. $\frac{a\cos A + b\sin B + c\cos C}{a + b + c}$

$= \frac{2R\sin A\cos A + 2R\sin B\cos B + 2R\sin C\cos C}{2s}$

$= \frac{R}{2s}(\sin 2A + \sin 2B + \sin 2C)$

$= \frac{R}{2s} 4\sin A\sin B\sin C = \frac{4R}{2s}\frac{abc}{8R^3} = \frac{abc}{4sR^2}$

But $R = \frac{abc}{4\Delta}$, $r = \frac{\Delta}{s}$

So the value $= \frac{4\Delta R}{4\frac{\Delta}{r}R^2} = \frac{r}{R}$

40. a. From the right-angled ΔCAD, we have

$\cos C = \frac{b}{a/2}$

$\Rightarrow \frac{2b}{a} = \frac{a^2 + b^2 - c^2}{2ab}$

$\Rightarrow a^2 + b^2 - c^2 = 4b^2 \Rightarrow a^2 - c^2 = 3b^2$

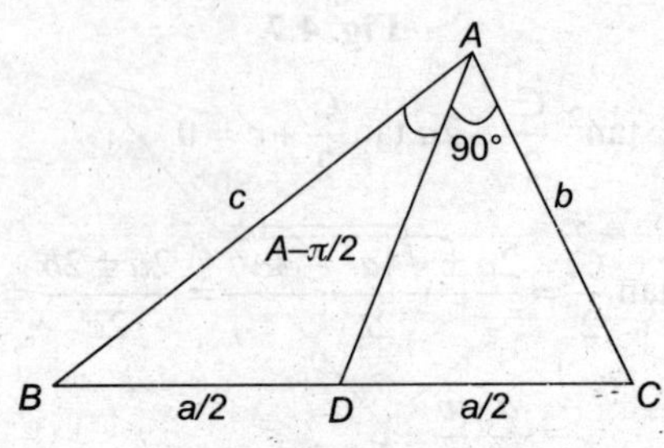

Fig. 4.10

41. a.

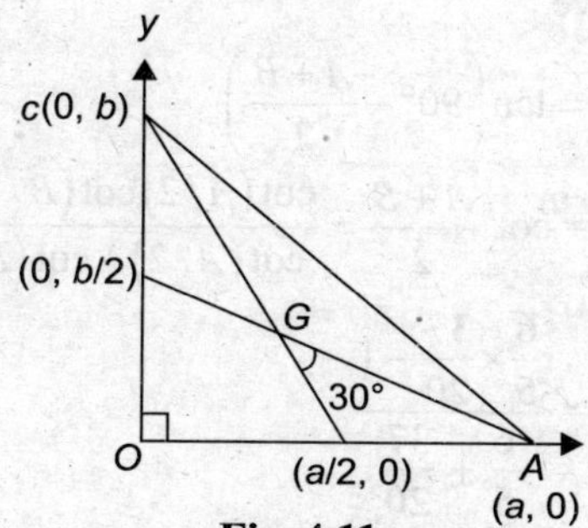

Fig. 4.11

Slope of $GC = \frac{-2b}{a}$, Slope of $AG = \frac{-b}{2a}$,

$$\tan 30° = \frac{\frac{3b}{2a}}{1+\frac{b^2}{a^2}} \text{ and } a^2+b^2 = 9$$

$$\Rightarrow \frac{1}{\sqrt{3}} = \frac{3ba}{2(a^2+b^2)}$$

$$\Rightarrow \frac{1}{2}ab = \left(\frac{a^2+b^2}{3\sqrt{3}}\right) = \frac{9}{3\sqrt{3}} = \sqrt{3}$$

42. c.

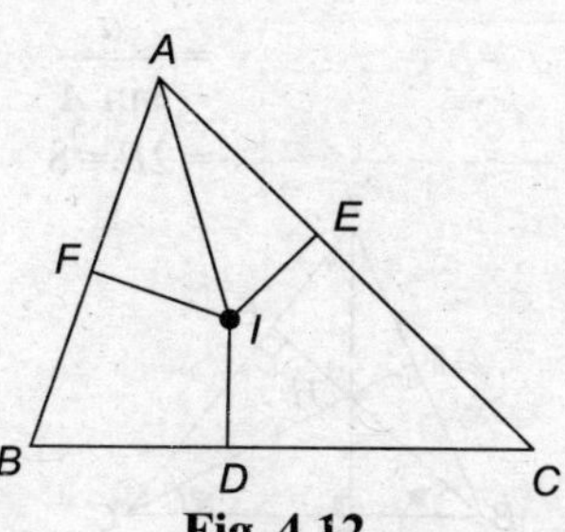

Fig. 4.12

In triangles AIF and AIE,

$$\frac{IF}{\sin A/2} = AI = \frac{IE}{\sin A/2} \Rightarrow AI^2 = \frac{IE \cdot IF}{\sin^2 A/2}$$

$$\Rightarrow \frac{ID \times IE \times IF}{IA \times IB \times IC} = \sin\frac{A}{2}\sin\frac{B}{2}\sin\frac{C}{2} = \frac{r}{4R} = \frac{1}{10}$$

43. c. We have $BD = DC$ and $\angle DAB = 90°$. Draw CN perpendicular to BA produced, then in $\Delta\, BCN$, we have $DA = \frac{1}{2}CN$ and $AB = AN$

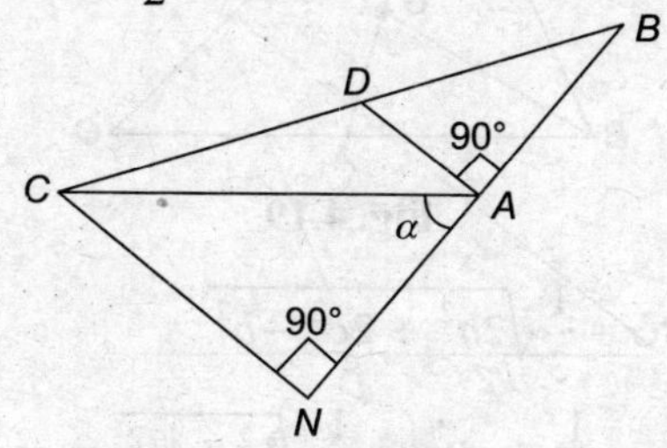

Fig. 4.13

Let $\angle CAN = \alpha$, Therefore,

$$\tan A = \tan(\pi - \alpha) = -\tan\alpha = -\frac{CN}{NA} = -2\frac{AD}{AB}$$

$$= -2\tan B$$

$$\Rightarrow \tan A + 2\tan B = 0$$

44. a.

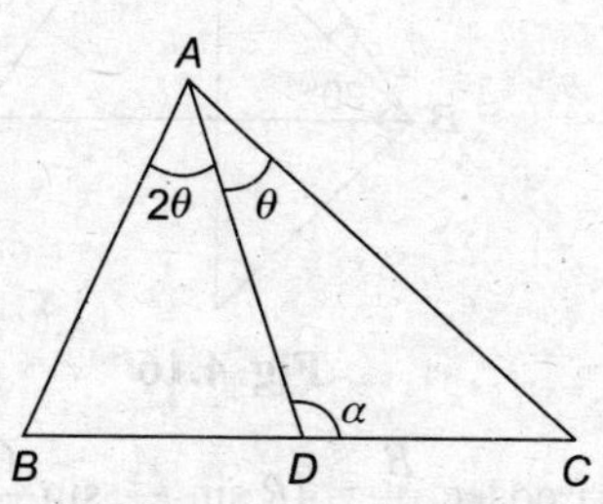

Fig. 4.14

Let $\frac{A}{3} = \angle CAD = \theta$

Now, by m–n theorem,

$$(1+1)\cot\alpha = 1\cot 2\theta - 1\cot\theta$$

$$\Rightarrow 2\cot(B+2\theta) = \cot 2\theta - \cot\theta$$

$$\Rightarrow \cot(B+2\theta) + \cot\theta = \cot 2\theta - \cot(B+2\theta)$$

$$\Rightarrow \frac{\sin(B+3\theta)}{\sin(B+2\theta)\ \sin\theta} = \frac{\sin B}{\sin(B+2\theta)\ \sin 2\theta}$$

$$\Rightarrow \frac{\sin(B+A)}{\sin\theta} = \frac{\sin B}{\sin 2\theta}$$

$$\Rightarrow \sin C = \frac{\sin B}{2\cos\theta}$$

$$\Rightarrow \cos\frac{A}{3} = \frac{\sin B}{2\sin C}$$

45. d.

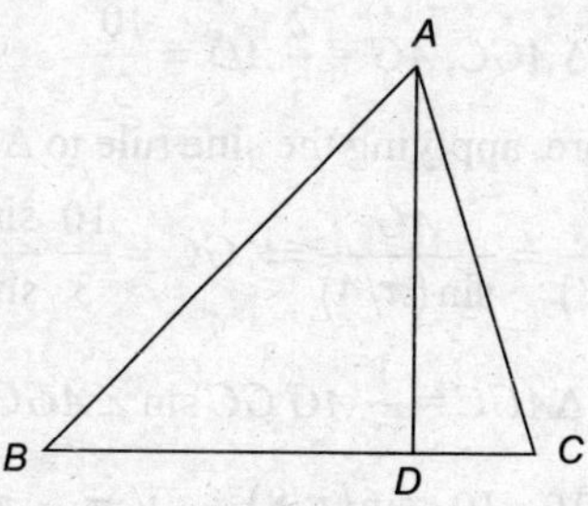

Fig. 4.15

$$CD = \sqrt{3^2 - 2^2} = \sqrt{5}$$

$$BD = \sqrt{4^2 - 2^2} = 2\sqrt{3}$$

Now, $ar\,\Delta ABC = \frac{1}{2} \times 2 \times (\sqrt{5} + 2\sqrt{3}) = \frac{1}{2} \times 4 \times 3 \sin A$

$$\Rightarrow \sin A = \frac{\sqrt{5}+2\sqrt{3}}{6} < 1.$$

Also, $a > b, a > c$
Hence, A can have two values.

46. d.

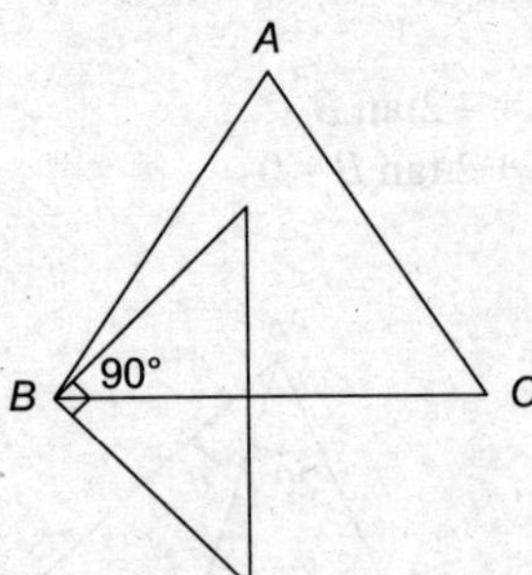

Fig. 4.16

$$BI = r \operatorname{cosec}\frac{B}{2} = 4R\sin\frac{A}{2}\sin\frac{C}{2}$$

$$BI_1 = \sec\frac{B}{2} = 4R\sin\frac{A}{2}\cos\frac{C}{2}$$

$$\Rightarrow II_1 = \sqrt{(BI)^2 + (BI')^2} = 4R\sin\frac{A}{2}$$

$$\Rightarrow \sum II_1\ 4R\sum\sin\frac{A}{2} \Rightarrow R = \frac{15}{8}$$

47. b.

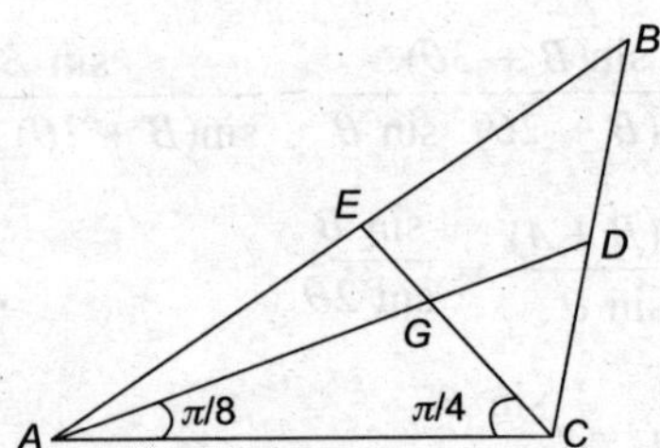

Fig. 4.17

Let G be the point of intersection of the medians of triangle ABC. Then the area of ΔABC is three times that of ΔAGC.

Now in ΔAGC, $AG = \frac{2}{3}AD = \frac{10}{3}$.

Therefore, applying the sine rule to ΔAGC, we get

$$\frac{GC}{\sin(\pi/8)} = \frac{AG}{\sin(\pi/4)} \Rightarrow GC = \frac{10}{3}\frac{\sin(\pi/8)}{\sin(\pi/4)}$$

$$\text{Area of } \Delta AGC = \frac{1}{2}AG\, GC \sin\angle AGC$$

$$= \frac{1}{2}\times\frac{10}{3}\times\frac{10}{3}\frac{\sin(\pi/8)}{\sin(\pi/4)}\sin\left(\frac{\pi}{2}+\frac{\pi}{8}\right)$$

$$= \frac{50}{9}\frac{\sin(\pi/8)\cos(\pi/8)}{\sin(\pi/4)}$$

$$= \frac{50}{18} = \frac{25}{9}$$

$$\text{Area of } \Delta ABC = 3\times\frac{25}{9} = \frac{25}{3}$$

48. a. We have

$$\tan\frac{C}{2} = \tan\left(90° - \frac{A+B}{2}\right)$$

$$= \cot\frac{A+B}{2} = \frac{\cot(A/2)\cot(B/2) - 1}{\cot(A/2) + \cot(B/2)}$$

$$= \frac{\frac{6}{5}\times\frac{37}{20} - 1}{\frac{6}{5} + \frac{37}{20}}$$

$$= \frac{222 - 100}{120 + 185} = \frac{122}{305} = \frac{2}{5}$$

Also,

$$\tan\frac{A}{2}\tan\frac{C}{2} = \sqrt{\frac{(s-b)(s-c)}{s(s-a)}}\sqrt{\frac{(s-a)(s-b)}{s(s-c)}}$$

$$\Rightarrow \frac{5}{6}\times\frac{2}{5} = \frac{s-b}{s}$$

$\Rightarrow 3(s-b) = s$

$\Rightarrow 2s = 3b \quad \Rightarrow \quad a + b + c = 3b$

$\Rightarrow a + c = 2b$

$\Rightarrow a, b$ and c are in A.P.

49. d. Circumdiameter of $\Delta HBC = \frac{a}{\sin\angle BHC}$

$$= \frac{a}{\sin A}$$

$$= 2R = 8$$

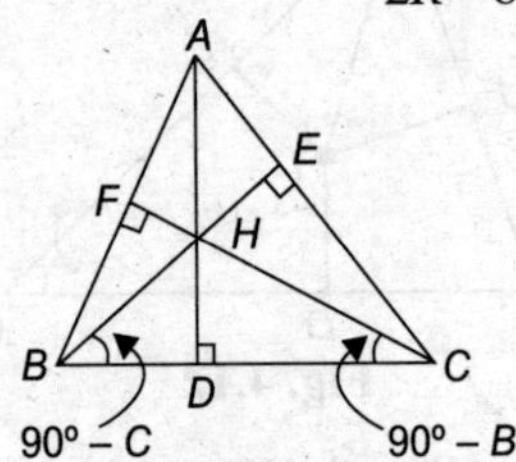

Fig. 4.18

50. b. $AG = \frac{2}{3}AA_1,\ BG = \frac{2}{3}BB_1$

$$\Rightarrow AG = \frac{1}{3}\sqrt{2b^2 + 2c^2 - a^2}$$

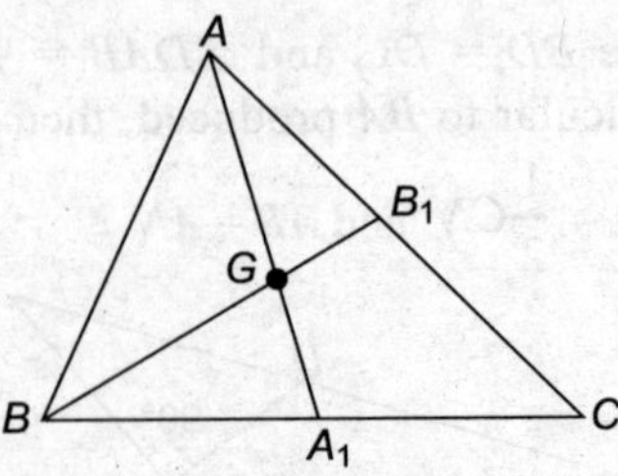

Fig. 4.19

and $BG = \frac{1}{3}\sqrt{2a^2 + 2c^2 - b^2}$

$\Rightarrow AG = \frac{1}{3}a,\ BG = \frac{1}{3}\sqrt{b^2 + 4c^2}$ as $a^2 = b^2 + c^2$

$\Rightarrow AG = \frac{5}{3},\ BG = \frac{1}{3}\sqrt{16+36} = \frac{2}{3}\sqrt{13}$

Also, $AB = c = 3$ and $\Delta_{GAB} = \frac{1}{3}\Delta_{ABC} = 2$

If R_1 be the circumradius of triangle GAB then

$R_1 = \frac{(AG)(BG)(AB)}{4\Delta_{GAB}} = \frac{5}{3}\times\frac{2}{3}\sqrt{13}\times 3\times\frac{1}{4\times 2}$

$= \frac{5\sqrt{13}}{12}$ units

51. b. Distance of circumcentre from side AC is $R\cos B$ and distance of incentre from side AC is r

$\Rightarrow R\cos B = r$

$\Rightarrow \frac{r}{R} = \cos B$

$\Rightarrow 4\sin\frac{A}{2}\sin\frac{B}{2}\sin\frac{C}{2} = \cos B$

$\Rightarrow \cos A + \cos B + \cos C - 1 = \cos B$

$\Rightarrow \cos A + \cos C = 1$

52. b. Distance of circumcentre from side AC is $R\cos B$ and distance of orthocentre from side AC is $2R\cos A\cos C$

$\Rightarrow R\cos B = 2R\cos A\cos C$

$\Rightarrow -\cos(A+C) = 2\cos A\cos C$

$\Rightarrow \sin A\sin C = 3\cos A\cos C$

$\Rightarrow \tan A\tan C = 3$

53. a. We have, $8R^2 = a^2 + b^2 + c^2$

$\Rightarrow 8R^2 = 4R^2(\sin^2 A + \sin^2 B + \sin^2 C)$

$\Rightarrow \sin^2 A + \sin^2 B + \sin^2 C = 2$

$\Rightarrow 1 - \cos^2 A + 1 - \cos^2 B + \sin^2 C = 2$

$\Rightarrow (\cos^2 A - \sin^2 C) + \cos^2 B = 0$

$\Rightarrow \cos(A+C)\cos(A-C) + \cos^2 B = 0$

$\Rightarrow -\cos B[\cos(A-C) - \cos B] = 0$

$\Rightarrow -\cos B[\cos(A-C) + \cos(A+C)] = 0$

$\Rightarrow -2\cos A\cos B\cos C = 0$

$\Rightarrow -2\cos A\cos B\cos C = 0$

$\Rightarrow \cos A = 0$ or $\cos B = 0$ or $\cos C = 0$

$\Rightarrow A = \frac{\pi}{2}$ or $B = \frac{\pi}{2}$ or $C = \frac{\pi}{2}$

54. c. We have $\frac{a}{2} = \frac{b}{3} = k$ and $\sec^2 A = \frac{8}{5}$

$\Rightarrow \cos^2 A = \frac{5}{8}$

$\Rightarrow \frac{5}{8} = \left(\frac{9k^2 + c^2 - 4k^2}{6kc}\right)^2 = \left(\frac{5k^2 + c^2}{6kc}\right)^2$

$\Rightarrow 45k^2c^2 = 50k^4 + 20k^2c^2 + 2c^4$

$\Rightarrow 2c^4 - 25k^2c^2 + 50k^4 = 0$

$\Rightarrow c^2 = \frac{25k^2 \pm \sqrt{625k^4 - 400k^4}}{4}$

$\Rightarrow \frac{25k^2 \pm 15k^2}{4} = 10k^2,\ \frac{5}{2}k^2$

There are two possible valid values of c^2. Hence, there exist two triangles satisfying the given conditions.

55. a. $\frac{\sin B}{b} = \frac{\sin C}{c}$

$\Rightarrow \sin C = \frac{c}{b}\sin B > 1$ $[\because b < \sin B \text{ (Given)}]$

which is impossible.

Hence, no triangle is possible.

56. d. We have $\cos A = \frac{b^2 + c^2 - a^2}{2bc}$

$\Rightarrow c^2 - 2bc\cos A + b^2 - a^2 = 0$, which is quadratic in c

$\therefore c_1 + c_2 = 2b\cos A$ and $c_1c_2 = b^2 - a^2$ (1)

$\therefore c_1^2 + c_2^2 - 2c_1c_2\cos 2A$

$= (c_1+c_2)^2 - 2c_1c_2 - 2c_1c_2\cos 2A$

[Using equation (1)]

$= (c_1+c_2)^2 - 2c_1c_2(1+\cos 2A)$

$= 4b^2\cos^2 A - 2(b^2 - a^2)\,2\cos^2 A = 4a^2\cos^2 A$

57. a. We have,

$\cos A = \frac{c^2 + b^2 - a^2}{2bc}$

$\Rightarrow c^2 - 2bc\cos A + b^2 - a^2 = 0$

It is given that c_1 and c_2 are roots of this equation.

Therefore, $c_1 + c_2 = 2b\cos A$ and $c_1c_2 = b^2 - a^2$

$\Rightarrow 2R(\sin C_1 + \sin C_2) = 2R\sin B\cos A$

$\Rightarrow \sin C_1 + \sin C_2 = 2\sin B\cos A$

Now, sum of the areas of two triangles

$= \frac{1}{2}ab\sin C_1 + \frac{1}{2}ab\sin C_2$

$= \frac{1}{2}ab(\sin C_1 + \sin C_2)$

$= \frac{1}{2}ab(2\sin B\cos A)$

$= ab\sin B\cos A = b(b\sin A)\cos A$

$= \frac{1}{2}b^2\sin 2A.$

58. a. Let r be the radius of the circle and A_1 be its area. Then $A_1 = \pi r^2$.

Since the perimeter of the circle is same as the perimeter of a regular polygon of n sides, therefore $2\pi r = na$, where a is the length of one side of the regular polygon.

$\Rightarrow a = \frac{2\pi r}{n}$

Let A_2 be the area of the polygon.

Then, $A_2 = \frac{1}{4}\pi a^2 \cot\left(\frac{\pi}{n}\right) = \frac{\pi^2 r^2}{n}\cot\left(\frac{\pi}{n}\right)$

$\Rightarrow A_1 : A_2 = \pi r^2 : \frac{\pi^2 r^2}{n}\cot\left(\frac{\pi}{n}\right) = \tan\left(\frac{\pi}{n}\right) : \frac{\pi}{n}$

59. a. Let a be the radius of the circle, then the ratio of the area of regular polygons of n sides inscribed and circumscribing the same circle is

$$\frac{S_1}{S_2} = \frac{\frac{1}{2}na^2 \sin(2\pi/n)}{na^2 \tan(\pi/n)} = \frac{3}{4}$$

$\Rightarrow \cos^2(\pi/n) = 3/4$

$\Rightarrow \cos(\pi/n) = \sqrt{3}/2$

$\Rightarrow \pi/n = \pi/6 \Rightarrow n = 6$

60. b.

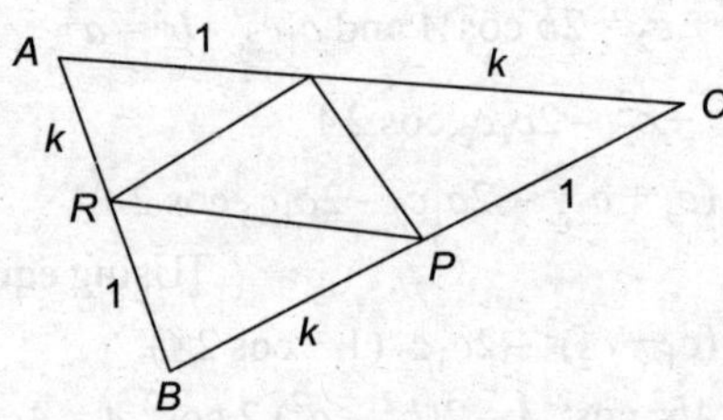

Fig. 4.20

Area of triangle $ARQ = \frac{b}{(k+1)} \times \frac{kc}{(K+1)} \times \frac{1}{2}\sin A$

$\Rightarrow \frac{k\Delta}{(k+1)^2}$

Similarly, area $(\Delta BPR) = (\Delta PCQ) = \frac{k\Delta}{(k+1)^2}$

$\Rightarrow$ area $= \left(\frac{k^2 - k + 1}{k^2 + 2k - 1}\right)\Delta R$

61. a. $\frac{2!}{1!9!} + \frac{2}{3!7!} + \frac{1}{5!5!}$

$= \frac{2}{10!}({}^{10}C_1) + \frac{2}{10!}({}^{10}C_3) + \frac{1}{10!}{}^{10}C_5$

$= \frac{1}{10!}({}^{10}C_1 + {}^{10}C_9 + {}^{10}C_3 + {}^{10}C_7 + {}^{10}C_5)$

$= \frac{1}{10!}(2^{10-1}) = \frac{2^9}{10!} = \frac{8^a}{(2b)!}$ (given)

$\Rightarrow 3a = 9, 2b = 10 \Rightarrow a = 3, b = 5$

Also, a, b, c are in A.P. $\Rightarrow c = 7$

$\cos A = \frac{b^2 + c^2 - a^2}{2bc} = \frac{25 + 49 - 9}{2\times5\times7} = \frac{65}{2\times5\times7} = \frac{13}{14}$

$\cos B = \frac{a^2 + c^2 - b^2}{2ac} = \frac{9 + 49 - 25}{2\times3\times7} = \frac{11}{14}$

$\Rightarrow \cos A + \cos B = \frac{13 + 11}{14} = \frac{24}{14} = \frac{12}{7}$

62. a.

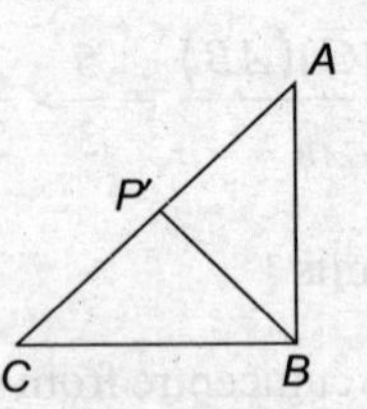

Fig. 4.21

Let point P is on side BC, then $(PA + PB + PC)_{\min}$

$\Rightarrow P$ is foot of perpendicular from vertex A

$\Rightarrow B$ and P' are critical points as shown in Fig. 4.21.

When P coincides with B, $PA + PB + PC = a + b$

Similarly, for P', $PA + PB + PC = c + a\sin B$

But $c + a\sin B > a + b \Rightarrow B$ is the required point, i.e., the orthocentre.

63. c. In triangle ABC,

$\frac{a^2 + b^2 + c^2}{R^2}$

$= 4(\sin^2 A + \sin^2 B + \sin^2 C)$

$= 8(1 + \cos A\cos B\cos C) \le 8\left(1 + \frac{1}{8}\right) \le 9$

64. c. Let $a + b + c = 2s$, then we have to find the minimum value of

$$\frac{a}{s-a} + \frac{b}{s-b} + \frac{c}{s-c} = -3 + \frac{s}{s-a} + \frac{s}{s-b} + \frac{s}{s-c}$$

Also, $\frac{\frac{s}{s-a} + \frac{s}{s-b} + \frac{s}{s-c}}{3} \ge \frac{3}{\frac{s-a}{s} + \frac{s-b}{s} + \frac{s-c}{s}}$

$\because \frac{s-a}{s} + \frac{s-b}{s} + \frac{s-c}{s} = 1,$

$\frac{s}{s-a} + \frac{s}{s-b} + \frac{s}{s-c} \ge 9$

Thus minimum value of the expression is 6.

65. c. $\frac{r_1 r_2 r_3}{r^3} = \left(\frac{r_1}{r}\frac{r_2}{r}\frac{r_3}{r}\right)^{1/3} \ge \frac{3}{\frac{r}{r_1} + \frac{r}{r_2} + \frac{r}{r_3}}$

$\ge \frac{3}{\frac{s-a}{s} + \frac{s-b}{s} + \frac{s-c}{s}} \ge \frac{3s}{s-a+s-b+s-c} \ge \frac{3s}{s}$

$\Rightarrow \frac{r_1}{r}\frac{r_2}{r}\frac{r_3}{r} \ge 27$

66. d. Since a circle can be inscribed in the quadrilateral, thus $a + c = b + d$.

Further it is also concyclic, thus $\angle A + \angle C = \pi$

$BD^2 = a^2 + d^2 - 2ad \cos A$

Also, $BD^2 = b^2 + c^2 - 2bc \cos C = b^2 + c^2 + 2bc \cos A$

$\Rightarrow 2 \cos A (bc + ad) = a^2 + d^2 - b^2 - c^2$

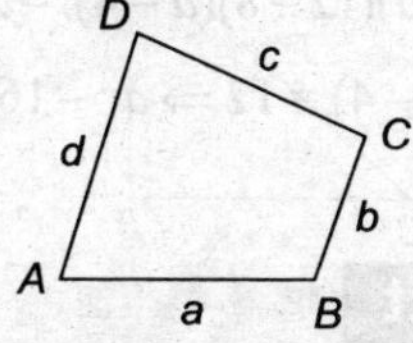

Fig. 4.22

or $2 \cos A = \dfrac{a^2 + d^2 - b^2 - c^2}{bc + ad}$

Since $a + c = b + d \Rightarrow a - d = b - c$

$\Rightarrow \quad sa^2 + d^2 - 2ad = b^2 + c^2 - 2bc$

$\Rightarrow \quad a^2 + d^2 - b^2 - c^2 = 2(ad - bc)$

$$\Rightarrow \quad \cos A = \frac{ad - bc}{bc + ad} = \frac{1 - \tan^2 \frac{A}{2}}{1 + \tan^2 \frac{A}{2}}$$

$$\Rightarrow \quad \tan^2 \frac{A}{2} = \frac{bc}{ad}$$

67. a.

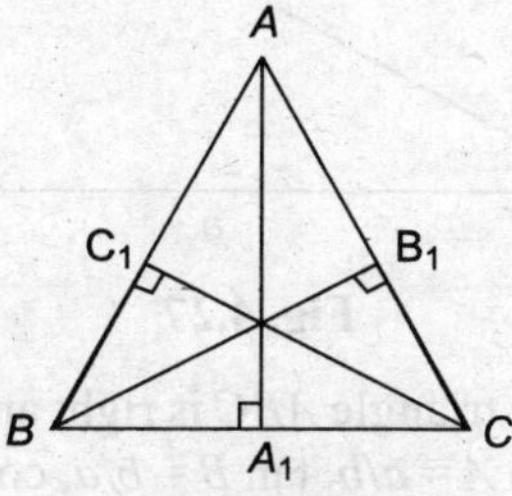

Fig. 4.23

Altitudes are AA_1, BB_1 and CC_1.

$\Delta = \dfrac{1}{2}(AA').\, a = \dfrac{1}{2}(BB_1)b = \dfrac{1}{2}(CC_1)c$

$\Rightarrow \quad AA_1 = \dfrac{2\Delta}{a}, BB_1 = \dfrac{2\Delta}{b}, CC_1 = \dfrac{2\Delta}{c}$

$\sin A, \sin B, \sin C$ are in A.P.

$\Rightarrow \quad a, b, c$ are in A.P.

$\Rightarrow \quad \dfrac{1}{a}, \dfrac{1}{b}, \dfrac{1}{c}$ are in H.P.

$\Rightarrow \quad \dfrac{2\Delta}{a}, \dfrac{2\Delta}{b}, \dfrac{2\Delta}{c}$ are in H.P.

$\Rightarrow \quad AA_1, BB_1, CC_1$ are in H.P.

68. d. $CD = \dfrac{2ab}{a+b} \cos \dfrac{C}{2} = \dfrac{2ab}{a+b} \cos \dfrac{\pi}{3} = \dfrac{ab}{a+b}$

69. c. We know that $OI = \sqrt{R^2 - 2Rr} = r$

$\Rightarrow R^2 - 2Rr - r^2 = 0$

$\Rightarrow R = r \pm \sqrt{2}r = (\sqrt{2} + 1)r \Rightarrow \dfrac{R}{r} = \sqrt{2} + 1$

70. a. $AC = d, OA = OB = r, CD = BD = 1, \angle COA = \dfrac{\pi}{3}$

$\therefore AC^2 = OA^2 + OC^2 - 2AO\, OC \cos \dfrac{\pi}{3}$

$\Rightarrow d^2 = 2r^2 - 2r^2 \dfrac{1}{2} = r^2$

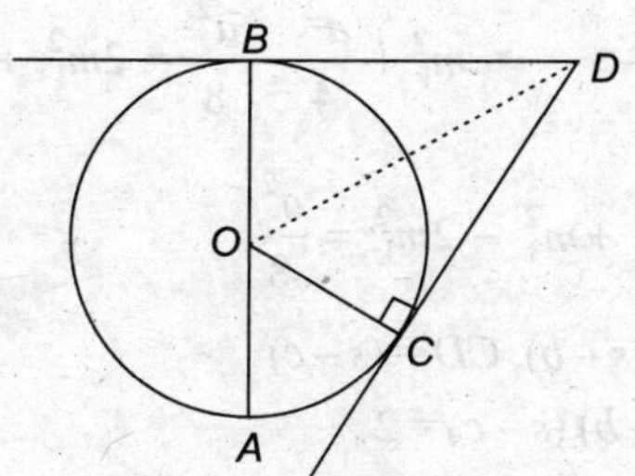

Fig. 4.24

Also, $\angle BOD = \angle COD = \dfrac{2\pi}{3 \times 2} = \dfrac{\pi}{3}$

$\Rightarrow \tan \dfrac{\pi}{3} = \dfrac{BD}{OB} = \dfrac{l}{r} \Rightarrow l = r\sqrt{3} = d\sqrt{3}$

71.a. $\Delta = \dfrac{1}{2}(BC)\,h$, where h is the distance of vertex A from side BC

$\Delta_{GBC} = \dfrac{\Delta}{3} = \dfrac{(BC)h}{6}$, where G is the centroid

$\Rightarrow h = \dfrac{2\Delta}{BC} =$ constant

Thus distance of vertex A from side is fixed. This in turn implies that distance of centroid from side BC will be fixed, hence locus of G will be a line parallel to BC.

72. c. In $\Delta ABC, AD^2 = m_1{}^2 = \dfrac{c^2 + b^2}{2} - \dfrac{a^2}{4}$

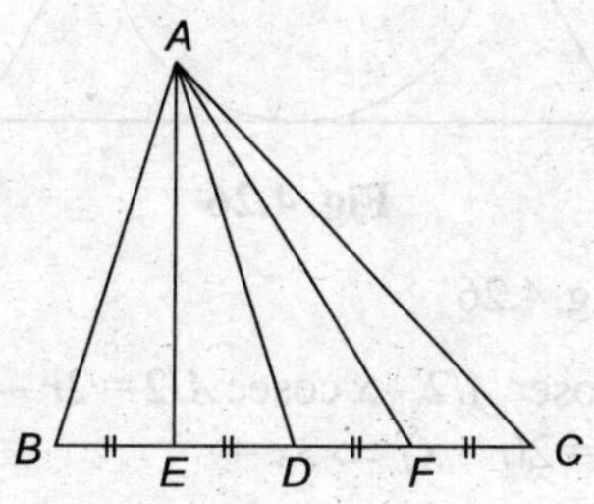

Fig. 4.25

In ΔABD, $AE^2 = m_2^2 = \dfrac{AD^2 + c^2}{2} - \dfrac{\left(\frac{a}{2}\right)^2}{4}$

$AF^2 = m_3^2 = \dfrac{AD^2 + b^2}{2} - \dfrac{\left(\frac{a}{2}\right)^2}{4}$

$\therefore\ m_2^2 + m_3^2 = AD^2 + \dfrac{b^2 + c^2}{2} - \dfrac{a^2}{8}$

$= m_1^2 + \dfrac{a^2}{4} - \dfrac{a^2}{8} = 2m_1^2 + \dfrac{a^2}{8}$

$\Rightarrow m_2^2 + m_3^2 - 2m_1^2 = \dfrac{a^2}{8}$

73. a. $BD = (s-b)$, $CD = (s-c)$

$\Rightarrow (s-b)(s-c) = 2$

$\Rightarrow s(s-a)(s-b)(s-c) = 2s(s-a)$

$\Rightarrow \Delta^2 = 2s(s-a)$

$\Rightarrow \dfrac{\Delta^2}{s^2} = \dfrac{2(s-a)}{s} = 1$

(radius of incircle of triangle ABC)

$\Rightarrow \dfrac{a}{s} = \text{constant}$

Now, $\Delta = \dfrac{1}{2}aH_a$, where H_a is the distance of A from BC

$\Rightarrow \dfrac{\Delta}{s} = \dfrac{1}{2}\dfrac{aH_a}{s} = 1 \Rightarrow H_a = \dfrac{2s}{a} = \text{constant}$

$\Rightarrow$ Locus of A will be a straight line parallel to side BC.

74. c.

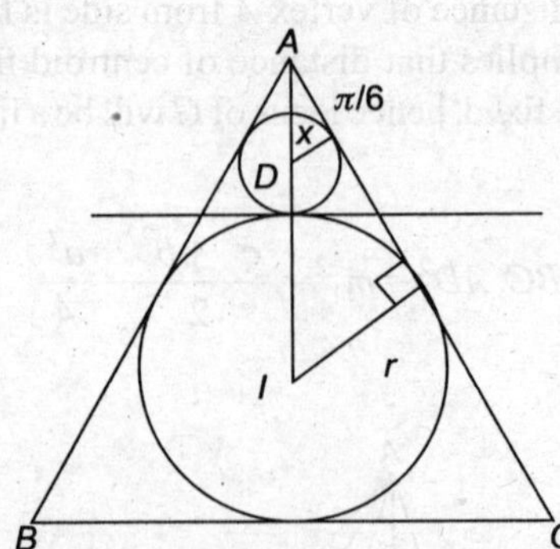

Fig. 4.26

From Fig. 4.26

$DI = r\operatorname{cosec} A/2 - x\operatorname{cosec} A/2 = 2r - 2x$

$\Rightarrow r + x = 2(r - x) \Rightarrow 3x = r$

$\Rightarrow x = \dfrac{r}{3} = \dfrac{1}{3}$

75. a. Given r_1, r_2, r_3 are in H.P.

$\Rightarrow a, b, c$ are in A.P.

and $a + b + c = 24 = 2s$

$\Rightarrow 2b + b = 24 \Rightarrow b = 8 \Rightarrow c = 16 - a$

Also given $\sqrt{s(s-a)(s-b)(s-c)} = 24$

$\Rightarrow \sqrt{12(12-a)(12-8)(a-4)} = 24$

$\Rightarrow (12-a)(a-4) = 12 \Rightarrow a^2 - 16a + 60 = 0$

$\Rightarrow a = 6, 10$

Multiple Correct Answers Type

1. a., b., c., d.

From the given equation, we get

$\tan A + \tan B = c^2/ab$ and $\tan A \tan B = 1$

Since $\tan(A+B) = \dfrac{\tan A + \tan B}{1 - \tan A \tan B}$, we have

$A + B = \dfrac{\pi}{2}$ and hence $C = \dfrac{\pi}{2}$

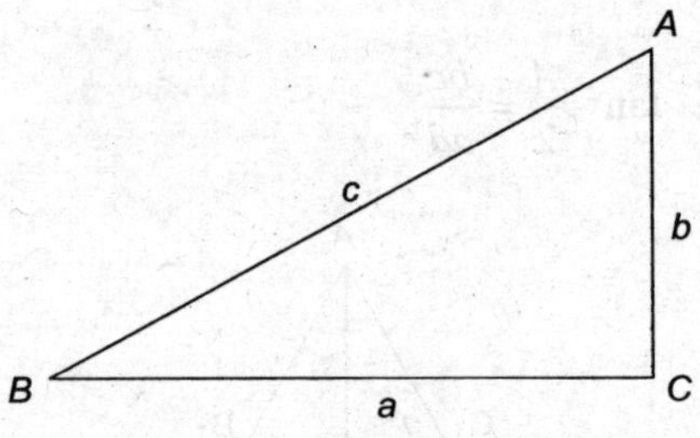

Fig. 4.27

Therefore, triangle ABC is right angled at C.

Hence, $\tan A = a/b$, $\tan B = b/a$, $\cos C = 0$, $\sin A = a/c$, $\sin B = b/c$ and $\sin C = 1$.

So, $\sin^2 A + \sin^2 B + \sin^2 C = \dfrac{a^2}{c^2} + \dfrac{b^2}{c^2} + 1$

$= \dfrac{a^2 + b^2}{c^2} + 1 = 1 + 1 = 2$

$[\because a^2 + b^2 = c^2]$

Hence, all options are correct.

2. a., c.

Let the angle be $A = x - d$, $B = x$, $C = x + d$.

Then $x - d + x + x + d = 180° \Rightarrow x = 60°$.

Therefore, two larger angles are $B = 60°$ and C.

Here $b = 9$ and $c = 10$

Now, $\cos B = \dfrac{c^2 + a^2 - b^2}{2ac}$

$$\Rightarrow \frac{1}{2} = \frac{100 + a^2 - 81}{20a}$$

$\Rightarrow a^2 - 10a + 19 = 0$

$\Rightarrow a = 5 \pm \sqrt{6}$

3. a., b.

$2a^2b^2 + 2b^2c^2 = a^4 + b^4 + c^4$

Also, $(a^2 + b^2 + c^2)^2 = a^4 + b^4 + c^4 - 2(a^2b^2 + b^2c^2 - c^2a^2)$

$\Rightarrow (a^2 - b^2 + c^2)^2 = 2c^2a^2$

$$\Rightarrow \frac{a^2 - b^2 + c^2}{2ca} = \pm\frac{1}{\sqrt{2}} = \cos B$$

$\Rightarrow B = 45°$ or $135°$

4. a., c.

From the cosine formula, $\cos A = \dfrac{b^2 + c^2 - a^2}{2bc}$

$\Rightarrow b^2 - (2c\cos A)\,b + (c^2 - a^2) = 0$, which is a quadratic equation in b

since $c \sin A < a < c$

Therefore, two triangles will be obtained, but this is possible when two values of the third side are also obtained. Clearly, two values of sides b will be b_1 and b_2. Let these are roots of the above equation.

$\therefore$ sum of roots $= b_1 + b_2 = 2c\cos A$, hand $b_1 b_2 = c^2 - a^2$

5. c., d.

a. $\tan A + \tan B + \tan C = \tan A \tan B \tan C = 0$

$\Rightarrow$ either A or B or $C = 0$

Not possible in a triangle.

b. By sine rule $\dfrac{a}{2} = \dfrac{b}{3} = \dfrac{c}{7} = \lambda$ (say)

$a + b = 5\lambda$, $c = 7\lambda$, i.e., $a + b < c$

Not possible in a triangle as sum of two sides is greater than the third.

c. Given that $(a + b)^2 = c^2 + ab$

$\Rightarrow a^2 + b^2 - c^2 + ab = 0$

$\Rightarrow 2ab\cos C + ab = 0$

$\Rightarrow \cos C = -\dfrac{1}{2}$ or $\angle C = 120°$

Also given $\sqrt{2}(\sin A + \cos A) = \sqrt{3}$

$$\Rightarrow \sin(A + 45°) = \frac{\sqrt{3}}{2}$$

$\Rightarrow A + 45° = 60°$

$\Rightarrow A = 15°$ and hence $B = 45°$

This is possible in a triangle.

d. Given that $\cos A \cos B = \dfrac{\sqrt{3}}{4} = \sin A \sin B$

$$\Rightarrow \cos(A - B) = \frac{\sqrt{3}}{2} \Rightarrow A - B = \frac{\pi}{6}$$

and $\cos(A + B) = 0 \Rightarrow A + B = \dfrac{\pi}{2} \Rightarrow A = 60°, B = 30°$

Hence, $\sin A + \sin B = \dfrac{\sqrt{3}+1}{2}$. This is possible in a triangle.

6. a., c.

$\Delta = \Delta_1 + \Delta_2$

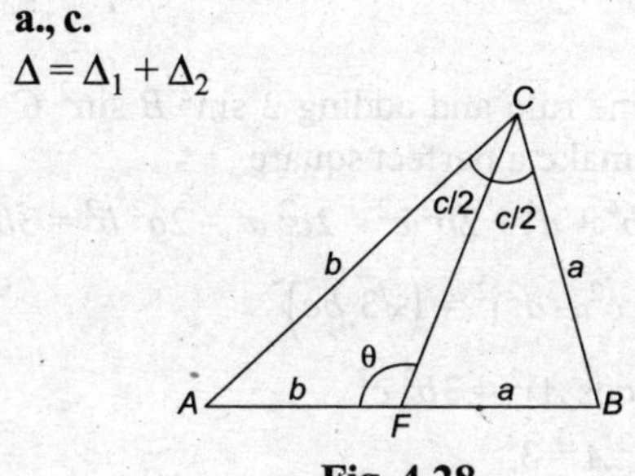

Fig. 4.28

$$\frac{1}{2}ab\sin C = \frac{1}{2}b(CF)\sin\frac{C}{2} + \frac{1}{2}a(CF)\sin\frac{C}{2}$$

$$\Rightarrow CF = \frac{ab\sin C}{(a+b)\sin\frac{C}{2}} = \frac{2ab\cos\frac{C}{2}}{a+b}$$

Again in ΔCFB by sine rule

$$\frac{CF}{\sin B} = \frac{a}{\sin(\pi - \theta)} = \frac{a}{\sin\theta} = \frac{a}{\sin\left(B + \frac{C}{2}\right)}$$

$$\because \theta + B + \frac{C}{2} = \pi$$

$$\Rightarrow CF = \frac{a\sin B}{\sin\left(B + \frac{C}{2}\right)} = \frac{b\sin A}{\sin\left(B + \frac{C}{2}\right)} \text{ (by sine rule)}$$

7. a., c., d.

$(a^2 - 2ac + c^2) + (a^2 - 4ab + 4b^2) = 0$

$\Rightarrow (a - c)^2 + (a - 2b)^2 = 0$

$\Rightarrow a = c$ and $a = 2b$

$\Rightarrow$ Triangle is isosceles.

Also $\cos B = \dfrac{a^2 + c^2 - b^2}{2ac} = \dfrac{7b^2}{8b^2} = \dfrac{7}{8}$

$$\cos A = \frac{b^2 + c^2 - a^2}{2bc} = \frac{1}{4}$$

8. a., b., d.

$$r = \frac{\Delta}{s},\ r_1 = \frac{\Delta}{s - a}$$

$$r_1 r = \frac{\Delta^2}{s(s - a)} = \frac{s(s-a)(s-b)(s-c)}{s(s-a)}$$

$$= (s-b)(s-c) = (s-b)^2 \qquad [\because b = c]$$

$$= \frac{(2s-2b)^2}{4}$$

$$= \frac{(a+b+c-2b)^2}{4} \qquad [\because b = c]$$

$$= \frac{a^2}{4} = \frac{4R^2 \sin^2 A}{4} = R^2 \sin^2 A$$

Also if $\angle B = \theta \Rightarrow \angle A = \pi - 2\theta$

$r_1 r = R^2 \sin^2(\pi - 2\theta) = R^2 \sin^2 2\theta = R^2 \sin^2 2B$

9. a., c.

Using sine rule and adding $2\sin^2 B \sin^2 C$ to both sides to make a perfect square

$\Rightarrow a^4 + b^4 + c^4 + 2b^2c^2 - 2c^2a^2 - 2a^2b^2 = 3b^2c^2$

$\Rightarrow (b^2 + c^2 - a^2)^2 = \left(\sqrt{3}\, bc\right)^2$

$\Rightarrow (2bc \cos A)^2 = 3b^2c^2$

$\Rightarrow 4\cos^2 A = 3$

$\Rightarrow \cos A = \dfrac{\sqrt{3}}{2} \Rightarrow A = 30° \text{ or } 150°$

10. a, b, c.

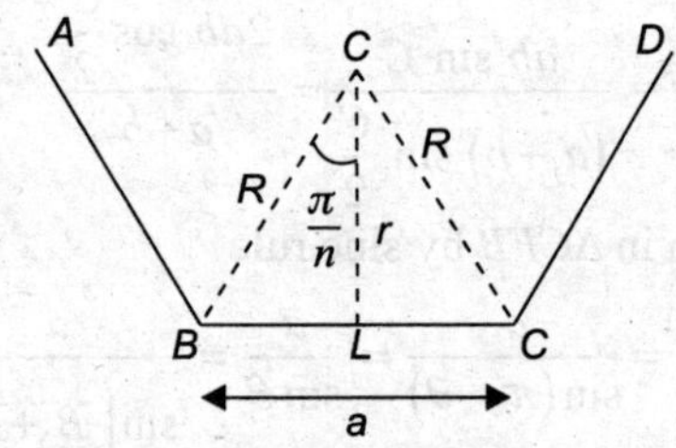

Fig. 4.29

Area of polygon

$\Delta = n$ times the area of the ΔOBC

$$= \frac{na^2}{4} \cot\frac{\pi}{n} \qquad (1)$$

Now, $a = 2r \tan \dfrac{\pi}{n}$

$\Rightarrow \Delta = nr^2 \tan\left(\dfrac{\pi}{n}\right)$ [From equation (1)]

Also, $a = 2R \sin \dfrac{\pi}{n}$

$\Rightarrow \Delta = \dfrac{nR^2}{2} \sin\left(\dfrac{2\pi}{n}\right)$ [From equation (1)]

11. c., d.

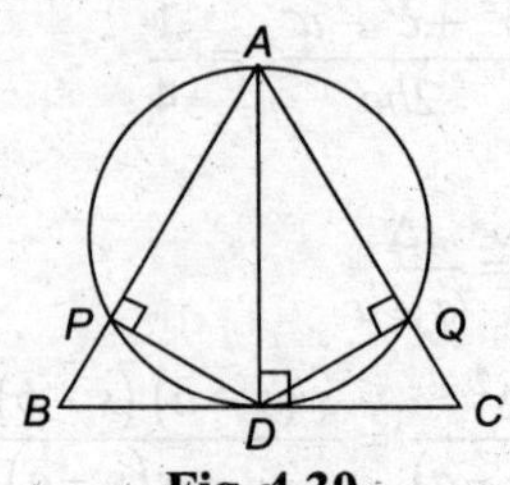

Fig. 4.30

$AD = c \sin B \qquad \left(\text{as } \angle ADP = \angle AEP = \dfrac{\pi}{2}\right)$

Using sine rule in triangle APQ, we get

$\dfrac{PQ}{\sin A} = 2R' = AD = c \sin B$

$\Rightarrow PQ = c \sin B \sin A = c \dfrac{b}{2R} \sin A = \dfrac{2\Delta}{2R} = \dfrac{\Delta}{R}$

12. a., b.

We have $2s = a + b + c$, $A^2 = s(s-a)(s-b)(s-c)$

Now, A.M. $\geq$ G.M.

$\Rightarrow \dfrac{s + (s-a) + (s-b) + (s-c)}{4}$

$\geq [s(s-a)(s-b)(s-c)]^{1/4}$

$\Rightarrow \dfrac{4s - 2s}{4} \geq [A^2]^{1/4} \Rightarrow s/2 \geq A^{1/2} \Rightarrow A \leq s^2/4$

Also,

$\dfrac{(s-a) + (s-b) + (s-c)}{3} \geq [(s-a)(s-b)(s-c)]^{1/3}$

or $\dfrac{s}{3} \geq \left[\dfrac{A^2}{s}\right]^{1/3}$ or $\dfrac{A^2}{s} \leq \dfrac{s^3}{27} \Rightarrow A \leq \dfrac{s^2}{3\sqrt{3}}$

13. a.

We have $\dfrac{\sin A}{a} = \dfrac{\sin B}{b} = \dfrac{\sin C}{c}$

$\Rightarrow \dfrac{\sin 105°}{\sqrt{3}+1} = \dfrac{\sin 30°}{b} = \dfrac{\sin 45°}{c}$

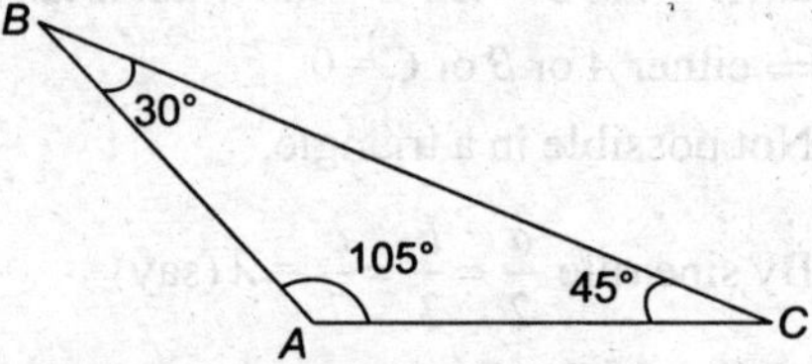

Fig. 4.31

$\Rightarrow b = \dfrac{(\sqrt{3}+1)}{2\sin 105°},\ c = \dfrac{\sqrt{3}+1}{\sqrt{2}\sin 105°}$

Area of $\Delta ABC = \dfrac{1}{2} bc \sin A = \dfrac{1}{2} bc \sin 105°$

$$= \frac{1}{2}\frac{(\sqrt{3}+1)^2}{2\sqrt{2}\sin(60° + 45°)}$$

$$= \frac{(\sqrt{3}+1)^2}{4\sqrt{2}\left[\frac{\sqrt{3}}{2}\frac{1}{\sqrt{2}} + \frac{1}{\sqrt{2}}\frac{1}{2}\right]}$$

$$= \frac{(\sqrt{3}+1)^2}{2(\sqrt{3}+1)} = \frac{1}{2}(\sqrt{3}+1)$$

14. a., c.

Sides are in A.P. and $a < \min\{b, c\}$

$\Rightarrow$ order of A.P. can be b, c, a or c, b, a

Case I

If $2c = a + b$

$$\cos A = \frac{b^2 + c^2 - a^2}{2bc} = \frac{b^2 + c^2 - (2c - b)^2}{2bc}$$

$$= \frac{4b - 3c}{2b}$$

Case II

If $2b = a + c$

$$\cos A = \frac{b^2 + c^2 - a^2}{2bc} = \frac{b^2 + c^2 - (2b - c)^2}{2bc}$$

$$= \frac{4c - 3b}{2b}$$

15. a., c., d.

Given that $s - a, s - b, s - c$ are in A.P.

$\Rightarrow a, b, c$ are in A.P.

$$\Rightarrow \frac{\Delta}{s-a}, \frac{\Delta}{s-b}, \frac{\Delta}{s-c} \text{ are in H.P.}$$

$\Rightarrow r_1, r_2, r_3$ are in A.P.

$$\text{Also, } \cos A = \frac{b^2 + c^2 - a^2}{2bc} = \frac{b^2 + c^2 - (2b - c)^2}{2bc}$$

$$= \frac{4c - 3b}{2b}$$

16. a., c., d.

$$a + b > c \Rightarrow 2b - c + b > c \Rightarrow 3b > 2c \Rightarrow \frac{b}{c} > \frac{2}{3}$$

Also, $b + c > a \Rightarrow b + c > 2b - c \Rightarrow 2c > b$

$$\Rightarrow \frac{b}{c} < 2$$

17. a., b.

$$\cos\frac{A}{2} = \sqrt{\frac{b+c}{2c}} \Rightarrow \sqrt{\frac{b+c}{2c}} = \sqrt{\frac{s(s-a)}{bc}}$$

$\Rightarrow b^2 + bc = 2s(s - a)$

$\Rightarrow b^2 + bc = 2s(s - a)$

$$\Rightarrow b^2 + bc = (a + b + c)\left(\frac{b + c - a}{2}\right)$$

$\Rightarrow a^2 + b^2 = c^2 \Rightarrow \angle \pi/2$

18. a., c.

a, b, c are in G.P. $\Rightarrow b^2 = ac$

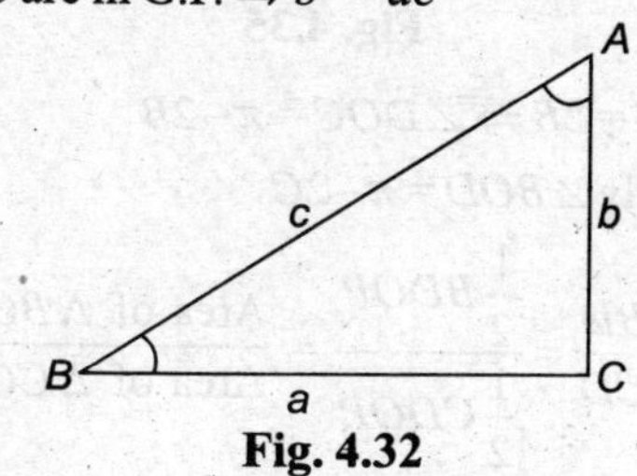

Fig. 4.32

Also, triangle is right angled

$\Rightarrow c^2 = a^2 + b^2$

$\Rightarrow c^2 = a^2 + ac$

$$\Rightarrow \left(\frac{a}{c}\right)^2 + \left(\frac{a}{c}\right) - 1 = 0$$

$$\Rightarrow \frac{a}{c} = \cos B = \frac{-1 \pm \sqrt{5}}{2}$$

$$\Rightarrow \cos B = \frac{\sqrt{5} - 1}{2} \qquad [B \text{ is acute}]$$

$$\cos A = \sin B = \sqrt{1 - \cos^2 B}$$

$$= \sqrt{1 - \left(\frac{\sqrt{5} - 1}{2}\right)^2}$$

$$= \sqrt{\frac{\sqrt{5} - 1}{2}}$$

Reasoning Type

1. a. Given a is constant and $\frac{r_2}{r_3} = k$ (where k is constant)

$$\text{Now, } \frac{r_2}{r_3} = k \Rightarrow \frac{\frac{\Delta}{s-b}}{\frac{\Delta}{s-c}} = k$$

$$\Rightarrow \frac{s-c}{s-b} = k$$

$$\Rightarrow \frac{(s-c) - (s-b)}{s - c + s - b} = \frac{k-1}{k+1}$$

$$\Rightarrow \frac{b - c}{2s - (b + c)} = \frac{k-1}{k+1} \Rightarrow b - c = a\frac{k-1}{k+1} = \text{constant}$$

$\Rightarrow$ locus of vertex A is hyperbola.

2. a. In ΔABC, $a + b > c$

$\Rightarrow 2R \sin A + 2R \sin B > 2R \sin C$

$\Rightarrow \sin A + \sin B > \sin C$

3. b. $a > b > c$

$\Rightarrow s - a > s - b > s - c$

$$\Rightarrow \frac{\Delta}{s-a} > \frac{\Delta}{s-b} > \frac{\Delta}{s-c} \Rightarrow r_1 > r_2 > r_3$$

Also, cosine is a decreasing function for $x \in (0, \pi/2)$

For $a > b > c \Rightarrow A > B > C \Rightarrow \cos A < \cos B < \cos C$

Hence, both statements are true, but statement 2 does not explain statement 1.

4. a.

5. c.

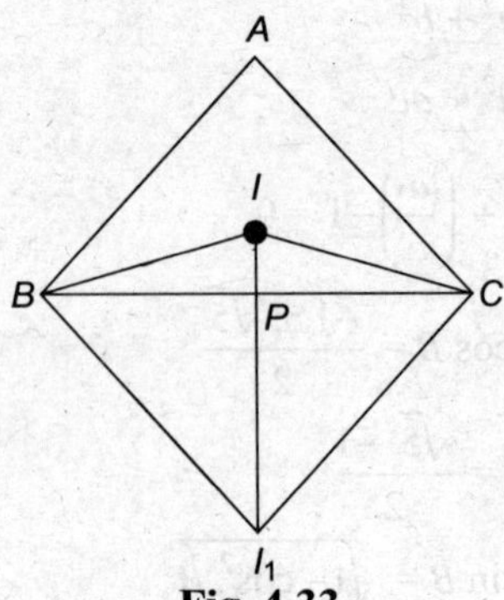

Fig. 4.33

$\angle ICI_1 = \dfrac{\pi}{2}$

$\angle IBI_1 = \dfrac{\pi}{2}$,

$\therefore BICI_1$ is a cyclic quadrilateral.

$\Rightarrow BP \cdot PC = IP \cdot I_1P$

6. a. We know that equadilateral θ_1 is a parallealogram and θ_1 is also cyclic so it becomes a rectangle.

7. d. Statement 2 is obviously true.

$$\Rightarrow \Delta_{DEF\ \max} = \frac{1}{4}\Delta_{ABC\ \max} = \frac{1}{4}\times\frac{8\times 8}{2} = 8 \text{ units}$$

8. b.

9. a. Circumradius of the pedal triangle (triangle formed by the feet of altitudes) of ΔABC is $R/2$.
Also, ΔABC is pedal triangle of $\Delta I_1I_2I_3$. Hence, answer is (b).

10. d. Statement 2 is true
Distance between circumcentre (O) and incentre (I) is $\sqrt{R^2 - 2rR}$.
If incircle of passes through its circumcentre then

$\sqrt{R^2 - 2rR} = r$

$$\Rightarrow \left(\frac{r}{R}\right)^2 + 2\left(\frac{r}{R}\right) - 1 = 0$$

$$\Rightarrow \frac{r}{R} = \frac{-2\pm\sqrt{4+4}}{2} = \sqrt{2}-1 \text{ as } \frac{r}{R} > 0$$

$$\Rightarrow \sin\frac{A}{2}\sin\frac{B}{2}\sin\frac{C}{2} = \frac{\sqrt{2}-1}{4}$$

11. a.

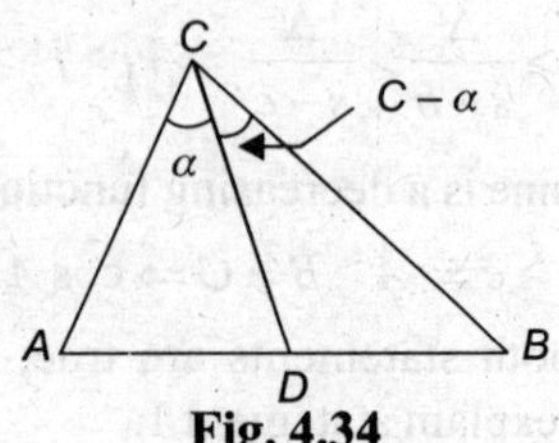

Fig. 4.34

Let $\angle ACD = \alpha \Rightarrow \angle DCB = (C-\alpha)$.
Applying the sine rule in $\Delta\ ACD$ and in $\Delta\ DCB$ respectively, we get

$$\frac{AD}{\sin\alpha} = \frac{CD}{\sin A} \text{ and } \frac{BD}{\sin(C-\alpha)} = \frac{CD}{\sin B}$$

$$\Rightarrow \frac{AD\ BD}{\sin\alpha\ \sin(C-\alpha)} = \frac{CD^2}{\sin A\ \sin B}$$

$$\Rightarrow \sin A \sin B = \sin\alpha \sin(C-\alpha)\ (CD^2 = AD\ BD)$$

$$\Rightarrow \frac{1}{2}[\cos(2\alpha - C) - \cos C]$$

$$= \frac{1}{2}\left[\cos(2\alpha - C) - 1 + 2\sin^2\frac{C}{2}\right]$$

$$= \sin^2\frac{C}{2} - \frac{1}{2}[1-\cos(2\alpha - C)]$$

Now, $1 - \cos(2\alpha - C) \ge 0$

$\Rightarrow \sin A\sin B \le \sin^2\dfrac{C}{2}$ and equality sign holds if

$1 - \cos(2\alpha - C) = 0$

$\Rightarrow \alpha = \dfrac{C}{2}$

It means that the equality sign holds if CD is the internal bisector of the angle C.

12. d. Let $a+b+c = 2s$, then we have to find the minimum value of

$$\frac{a}{s-a} + \frac{b}{s-b} + \frac{c}{s-c} = -3 + \frac{s}{s-a} + \frac{s}{s-b} + \frac{s}{s-c}$$

Also, $\dfrac{\dfrac{s}{s-a} + \dfrac{s}{s-b} + \dfrac{s}{s-c}}{3} \ge \dfrac{3}{\dfrac{s-a}{s} + \dfrac{s-b}{s} + \dfrac{s-c}{s}}$

$\because \dfrac{s-a}{s} + \dfrac{s-b}{s} + \dfrac{s-c}{s} = 1,$

$$\frac{s}{s-a} + \frac{s}{s-b} + \frac{s}{s-c} \ge 9$$

Thus the minimum value of the required expression is 6.

13. a.

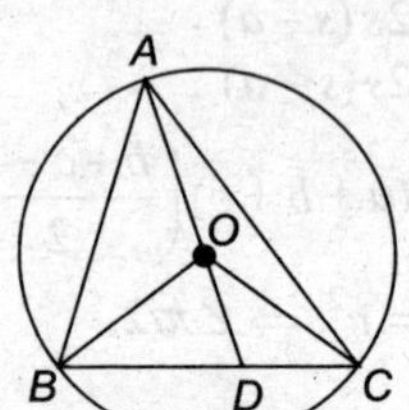

Fig. 4.35

$\angle AOC = 2B \Rightarrow \angle DOC = \pi - 2B$

Similarly $\angle BOD = \pi - 2C$

$$\text{Now, } \frac{BD}{CD} = \frac{\frac{1}{2}BDOP}{\frac{1}{2}CDOP} = \frac{\text{Area of } \Delta BOD}{\text{Area of } \Delta COD}$$

$$=\frac{\frac{1}{2}OBOD\sin\angle BOD}{\frac{1}{2}OCOD\sin\angle COD}=\frac{\sin 2C}{\sin 2B}=\frac{\sin 90°}{\sin 30°}$$
$$=2$$

14. d. Statement 2 is a geometrical fact.

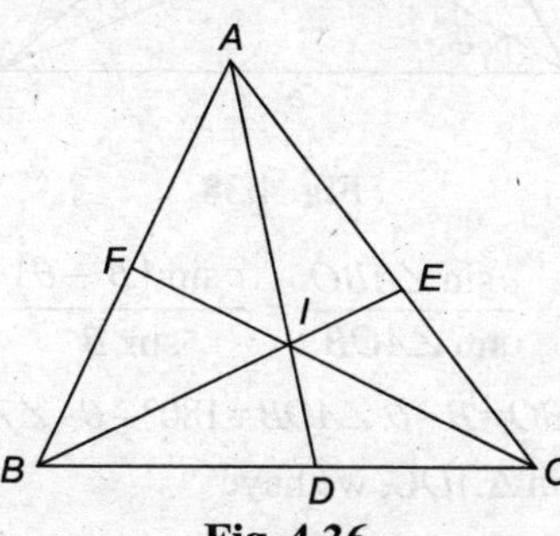

Fig. 4.36

Let AD, BE, and CF are, respectively, the internal bisectors of the angles A, B and C.

As AD bisects $\angle BAC$ internally,

$$\therefore \frac{BD}{DC}=\frac{BA}{AC}=\frac{c}{b} \quad (1)$$
$$\Rightarrow \frac{DC}{BD}=\frac{b}{c}$$
$$\Rightarrow \frac{DC}{BD}+1=\frac{b}{c}+1$$
$$\Rightarrow \frac{DC+BD}{BD}=\frac{b+c}{c}$$
$$\Rightarrow \frac{BC}{BD}=\frac{b+c}{c}$$
$$\Rightarrow \frac{a}{BD}=\frac{b+c}{c}$$
$$\Rightarrow BD=\frac{ac}{b+c} \quad (2)$$

Since BI is the bisector of $\angle B$, so it divides AD in the ratio $AI : ID$.

$$\therefore \frac{AI}{ID}=\frac{AB}{BD}=\frac{c}{\frac{ac}{b+c}}=\frac{b+c}{a} \quad \text{[Using equation 2)]}$$
$$\Rightarrow AI : ID=(7+8):3=5:1 \quad (3)$$

Linked Comprehension Type

For Problems 1–3

1. a., 2. b., 3. c.

Sol. We have,

$$r_1=\frac{\Delta}{s-a}=2, r_2=\frac{\Delta}{s-b}=3, r_3=\frac{\Delta}{s-c}=6$$

Given $\Delta=6$,

$$\therefore s-a=3 \quad (1)$$
$$s-b=2 \quad (2)$$
$$s-c=1 \quad (3)$$

Adding equations (1) and (2), $2s-a-b=5$

$\Rightarrow a+b+c-a-b=5$ or $c=5$

Adding equations (1) and (3), $2s-a-c=4$, $\therefore b=4$ and adding equations (2) and (3), $2s-b-c=3$,

$\therefore a=3$

Hence, the sides of the triangle are $a=3, b=4, c=5$.

Since the triangle is right angled, greatest angle is 90°.

Also, the least angle is opposite to side a, which is $\sin^{-1}\frac{3}{5}$.

$$\Rightarrow 90°-\sin^{-1}\frac{3}{5}=\cos^{-1}\frac{3}{5}$$

Also, $R=\frac{abc}{4\Delta}=\frac{60}{24}=2.5$

$$r=\frac{\Delta}{s}=\frac{6}{6}=1$$

For Problems 4–6

4. a., 5. c., 6. d.

Sol. $\cos(A-B)=\frac{4}{5}$

$$\Rightarrow \frac{1-\tan^2\frac{A-B}{2}}{1+\tan^2\frac{A-B}{2}}=\frac{4}{5}$$
$$\Rightarrow \tan^2\frac{A-B}{2}=\frac{1}{9}$$
$$\Rightarrow \tan\frac{A-B}{2}=\frac{1}{3}$$

Now, $\tan\frac{A-B}{2}=\frac{a-b}{a+b}\cot\frac{C}{2}$

$$\Rightarrow \frac{1}{3}=\frac{6-3}{6+3}\cot\frac{C}{2}$$
$$\Rightarrow \cot\frac{C}{2}=1 \Rightarrow C=\frac{\pi}{2}$$

Area of triangle $=\frac{1}{2}ab\sin C=\frac{1}{2}\times 6\times 3\times 1=9.$

$$\frac{a}{\sin A}=\frac{\sqrt{a^2+b^2}}{1}$$
$$\Rightarrow \frac{6}{\sin A}=\sqrt{45}$$
$$\Rightarrow \sin A=\frac{2}{\sqrt{5}}$$

For Problems 7–9

7. b., 8. d., 9. a.

Sol.

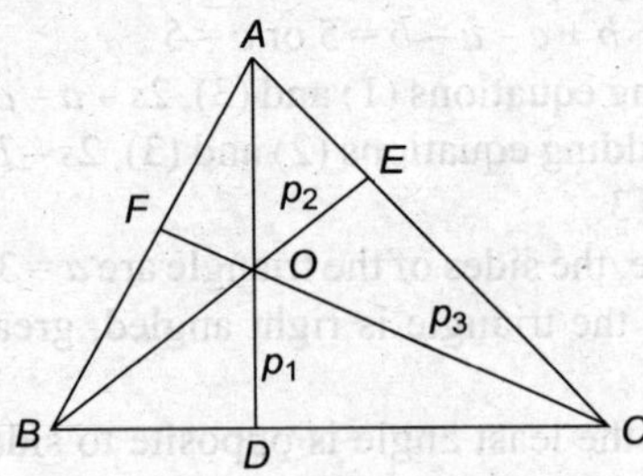

Fig. 4.37

7. We have $\frac{1}{2}ap_1 = \Delta, \frac{1}{2}bp_2 = \Delta, \frac{1}{2}cp_3 = \Delta$

$$\Rightarrow p_1 = \frac{2\Delta}{a}, p_2 = \frac{2\Delta}{b}, p_3 = \frac{2\Delta}{c}$$

$$\therefore \frac{1}{p_1^2} + \frac{1}{p_2^2} + \frac{1}{p_3^2} = \frac{a^2 + b^2 + c^2}{4\Delta^2}$$

8. $\frac{1}{p_1} + \frac{1}{p_2} + \frac{1}{p_3} = \frac{a}{2\Delta} + \frac{b}{2\Delta} + \frac{c}{2\Delta}$

$$= \frac{a+b+c}{2\Delta} = \frac{2s}{2\Delta} = \frac{s}{\Delta} = \frac{1}{r}$$

9. L.H.S. $= \cos A \frac{a}{2\Delta} + \cos B \frac{b}{2\Delta} + \cos C \frac{c}{2\Delta}$

$$= \frac{1}{2\Delta}(2R\sin A\cos A + 2R\sin B\cos B + 2R\sin C\cos C)$$

$$= \frac{R}{2\Delta}(\sin 2A + \sin 2B + \sin 2C)$$

$$= \frac{R}{2\Delta}(4\sin A\sin B\sin C)$$

$$= \frac{4R}{2\Delta}\frac{a}{2R}\frac{b}{2R}\sin C$$

$$= \frac{1}{\Delta R}\left(\frac{1}{2}ab\sin C\right)$$

$$= \frac{1}{\Delta R}\Delta = \frac{1}{R}$$

For Problems 10–12

10. d., 11. b., 12. a.

Sol. Applying sine formula in ΔAOB, we have

$$\frac{OA}{\sin\angle ABO} = \frac{AB}{\sin\angle AOB}$$

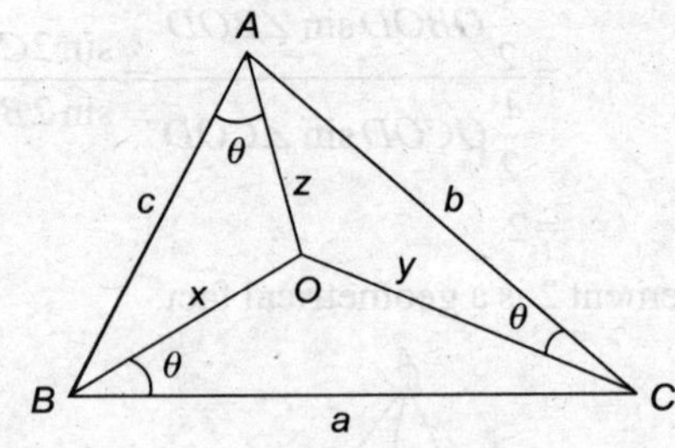

Fig. 4.38

$$\Rightarrow OA = \frac{c\sin\angle ABO}{\sin\angle AOB} = \frac{c\sin(B-\theta)}{\sin B} \quad (1)$$

$(\because \angle ABO = B-\theta, \angle AOB = 180° - \theta - \angle ABO = 180° - B)$

Again in ΔAOC, we have

$$\frac{OA}{\sin\angle ACO} = \frac{AC}{\sin\angle AOC} \quad (2)$$

$$\Rightarrow OA = \frac{b\sin\angle ACO}{\sin\angle AOC} = \frac{b\sin\theta}{\sin A}$$

$[\because \angle ACO = \theta, \angle OAC = A - \theta, \angle AOC = 180° - \theta - \angle OAC = 180° - A]$

From equations (1) and (2), we have

$$\frac{c\sin(B-\theta)}{\sin B} = \frac{b\sin\theta}{\sin A}$$

$$\Rightarrow c\sin A\sin(B-\theta) = b\sin\theta\sin B$$

$$\Rightarrow c\sin A\sin(B-\theta) = b\sin B\sin\theta\sin(A+C)$$

$$\Rightarrow 2R\sin C\sin A(\sin B\cos\theta - \cos B\sin\theta)$$
$$= 2R\sin B\sin\theta(\sin A\cos C + \cos A\sin C)$$

Dividing both sides by $2R\sin\theta\sin A\sin B\sin C$, we get

$$\cot\theta - \cot B = \cot C + \cot A$$

$$\Rightarrow \cot\theta = \cot A + \cot B + \cot C \quad (3)$$

Squaring both sides, we have

$$\cot^2\theta = \cot^2 A + \cot^2 B + \cot^2 C + 2(\cot A\cot B + \cot\theta B\cot C + \cot C\cot A)$$

$$\Rightarrow \text{cosec}^2\theta - 1 = (\text{cosec}^2 A - 1) + (\text{cosec}^2 B - 1) + (\text{cosec}^2 C - 1) + 2$$

(Since in ΔABC, $\cot A\cot B + \cot B\cot C + \cot C\cot A = 1$)

$$\Rightarrow \text{cosec}^2\theta = \text{cosec}^2 A + \text{cosec}^2 B + \text{cosec}^2 C$$

Area of triangle ABC,

$$\Delta = \Delta_1 + \Delta_2 + \Delta_3 = \frac{1}{2}[aOB + bOC + cOA]\sin\theta$$

$$= \frac{1}{4}\tan\theta[2aOB\cos\theta + 2bOC\cos\theta + 2cOA\cos\theta]$$

$$= \frac{1}{4}\tan\theta[a^2 + b^2 + c^2]$$

For Problems 13 – 15

13. c., 14. a., 15. d.

Sol.

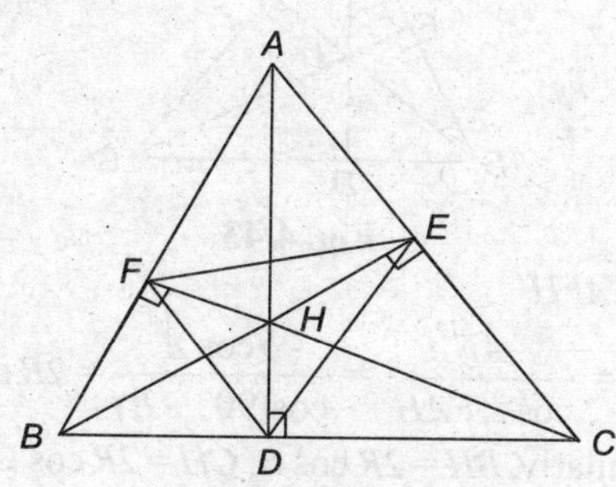

Fig. 4.39

Points B, F, H, D are concyclic

$\Rightarrow \angle HDF = \angle HBF = 90° - A\ (= \angle EBA)$

Points E, H, D, C are concyclic

$\Rightarrow \angle HDE = \angle HCE = 90° - A\ (= \angle FCA)$

Hence, line AD is an angle bisector of $\angle FDE$

$\Rightarrow H$ is incentre of triangle DEF.

Also, $AB \perp CF$, then AB is the external angle bisector of angle F in $\Delta\, DEF$ (as two bisectors of line are always perpendicular).

Similarly, AC and BC are other external bisectors, then A, B, C are excentres of $\Delta\, DEF$.

To find the side EF of pedal triangle:

We must consider a triangle which contains EF, one of which is ΔBEF.

From ΔBFC, $BF = a \cos B$.

Also, $\angle FBE = \angle ABE = 90° - A$

Points B, F, E, C are concyclic, so

$\angle BEF = \angle BCF = 90° - B$.

Then in ΔBEF, by sine rule

$$\frac{EF}{\sin(90° - A)} = \frac{a \cos B}{\sin(90° - B)} \Rightarrow EF = a \cos A$$

Similarly, $DE = c \cos C$ and $FD = b \cos B$

$$\Rightarrow x + y + z = a \cos A + b \cos B + c \cos C$$
$$= R\,(2 \sin A \cos A + \sin B \cos B + \sin C \cos C)$$
$$= R\,(\sin 2A + \sin 2B + \sin 2C)$$
$$= 4R \sin A \sin B \sin C$$
$$= 4R \frac{abc}{8R^3} = \frac{2\Delta}{R}$$

For circumradius R' of triangle DEF, by sine rule,

$$2R' = \frac{EF}{\sin \angle EDF} = \frac{a \cos A}{\sin(180° - 2A)} = \frac{a \cos A}{2 \sin A \cos A} = R$$

$\Rightarrow R' = R/2$

For Problems 16 – 18

16. a., 17. b., 18. d.

Sol.

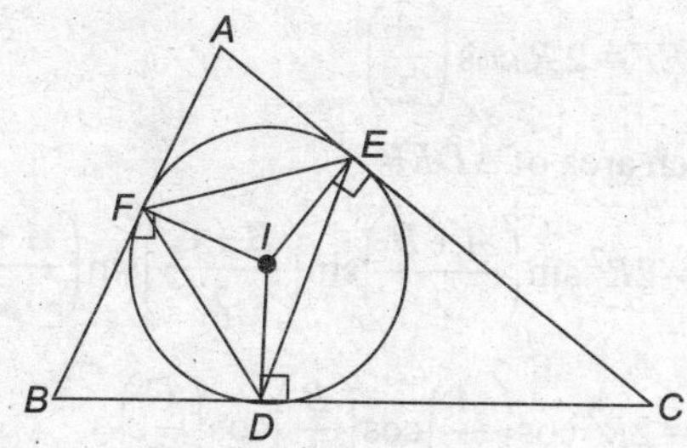

Fig. 4.40

Points I, D, C, E are concyclic

$\Rightarrow \angle EID = \pi - C$

Also, I is circumcentre of ΔDEF,

$$\Rightarrow \angle DFE = \frac{1}{2}\angle DIE = \frac{\pi - C}{2}$$

Similarly, $\angle FDE = \dfrac{\pi - A}{2}$, $\angle FED = \dfrac{\pi - B}{2}$

$\Rightarrow$ area of $\Delta DEF = 2r^2 \sin(\angle D) \sin(\angle E) \sin(\angle F)$

$$= 2r^2 \cos\frac{A}{2} \cos\frac{B}{2} \cos\frac{C}{2}$$

Also by sine rule, in ΔDEF

$$\frac{EF}{\sin \angle D} = 2r \Rightarrow EF = 2r \cos\frac{A}{2}$$

For Problems 19 – 21

19. a., 20. d., 21. b.

Sol.

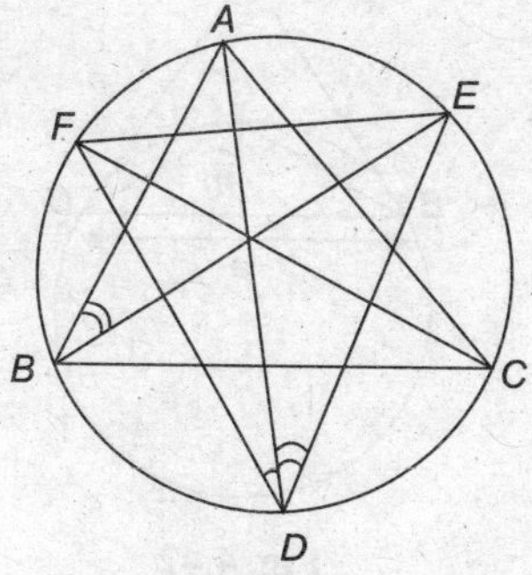

Fig. 4.41

$$\text{Area of } \Delta ABC = \frac{abc}{4R} = 2R^2 \sin A \sin B \sin C \quad (1)$$

$\angle ADE = \angle ABE = B/2$

Similarly, $\angle FDA = \angle FCA = C/2 \Rightarrow \angle FDE = \dfrac{B + C}{2}$

and $\angle DEF = \dfrac{A + C}{2}$ and $\angle DFE = \dfrac{A + B}{2}$

In ΔDEF, by sine rule, $\dfrac{EF}{\sin(\angle D)} = 2R$

$\Rightarrow EF = 2R\cos\left(\dfrac{A}{2}\right)$

Then area of ΔDEF

$$= 2R^2 \sin\left(\frac{A+B}{2}\right)\sin\left(\frac{B+C}{2}\right)\sin\left(\frac{A+C}{2}\right)$$

$$= 2R^2 \cos\left(\frac{A}{2}\right)\cos\left(\frac{B}{2}\right)\cos\left(\frac{C}{2}\right)$$

Now $\dfrac{\Delta_{ABC}}{\Delta_{DEF}} = \dfrac{2R^2 \sin A \sin B \sin C}{2R^2\cos\left(\frac{A}{2}\right)\cos\left(\frac{B}{2}\right)\cos\left(\frac{C}{2}\right)}$

$$= 8\sin\left(\frac{A}{2}\right)\sin\left(\frac{B}{2}\right)\sin\left(\frac{C}{2}\right) \le 1$$

$\Rightarrow \Delta_{ABC} \le \Delta_{DEF}$

Matrix-Match Type

1. a → q; b → s; c → p; d → p.

2. a → p, q; b → r, s; c → p, q; d → p, q.

O is circumcentre, then in ΔOMB, $OM = R\cos A$ (distance of circumcentre from BC).
Similarly, distances of circumcentre from AC and AB are $R\cos B$ and $R\cos C$.

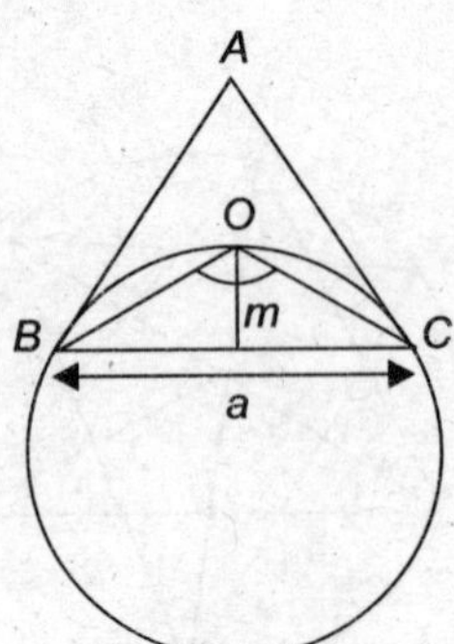

Fig. 4.42

Applying sine rule in triangle OBC,

$$2R_1 = \frac{a}{\sin 2A} = \frac{a}{2\sin A\cos A} = \frac{R}{\cos A}$$

$$\Rightarrow R_1 = \frac{R}{2\cos A}$$

Similarly, $R_2 = \dfrac{R}{\cos B}$ and $R_3 = \dfrac{R}{\cos C}$ (where R_1, R_2, R_3 are circumradius of ΔOBC, ΔOCA, and ΔOAB, respectively)

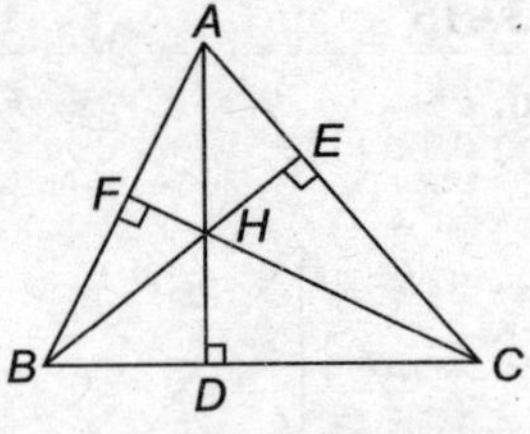

Fig. 4.43

In ΔAFH,

$$AH = \frac{AF}{\cos\angle FAH} = \frac{b\cos A}{\cos(90° - B)} = 2R\cos A$$

Similarly, $BH = 2R\cos B$, $CH = 2R\cos C$.

Also, $HF = AF\tan\angle FAH = b\cos A\cot B = 2R\cos A\cos B$.

Similarly, $HE = 2R\cos B\cos C$ and $HF = 2R\cos A\cos C$.

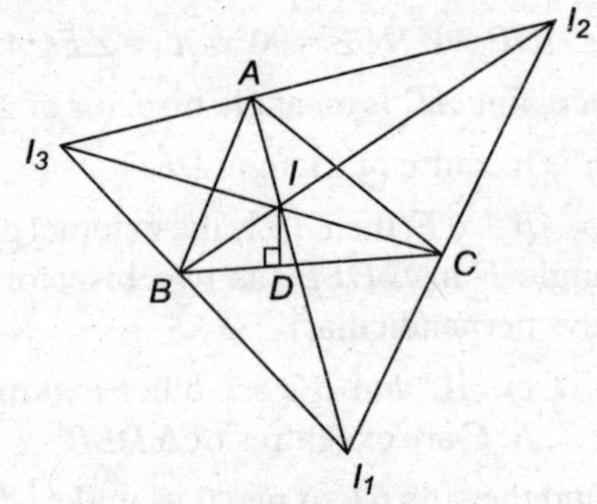

Fig. 4.44

In ΔIDB, $BI = \dfrac{ID}{\sin\angle IBD} = \dfrac{r}{\sin(B/2)}$

Similarly, $CI = \dfrac{r}{\sin(C/2)}$ and $AI = \dfrac{r}{\sin(A/2)}$

Also in ΔIBI_1, $II_1 = \dfrac{BI}{\cos\angle BII_1} = \dfrac{r/\sin(B/2)}{\cos\left(\frac{\pi}{2} - \frac{C}{2}\right)}$

$$= \frac{r}{\sin(B/2)\sin(C/2)}$$

Similarly, $II_2 = \dfrac{r}{\sin(A/2)\sin(C/2)}$

and $II_3 = \dfrac{r}{\sin(A/2)\sin(B/2)}$

3. a → p; b → q, r; c → q, r, s; d → p, q.

a. $\sin A$, $\sin B$ are roots of $c^2x^2 - c(a+b)x + ab = 0$

$$\therefore \sin A + \sin B = \frac{c(a+b)}{c^2}$$

$$\Rightarrow 2R(a+b) = \frac{c(a+b)}{c^2}$$

$$\Rightarrow 2R = \frac{1}{c} \quad (1)$$

From sine formula,

$$\frac{\sin A}{a} = \frac{\sin B}{b} = \frac{\sin C}{c} = 2R = \frac{1}{c}$$

[using equation (1)]

$\Rightarrow \sin C = 1$

$\Rightarrow C = 90°$

b.

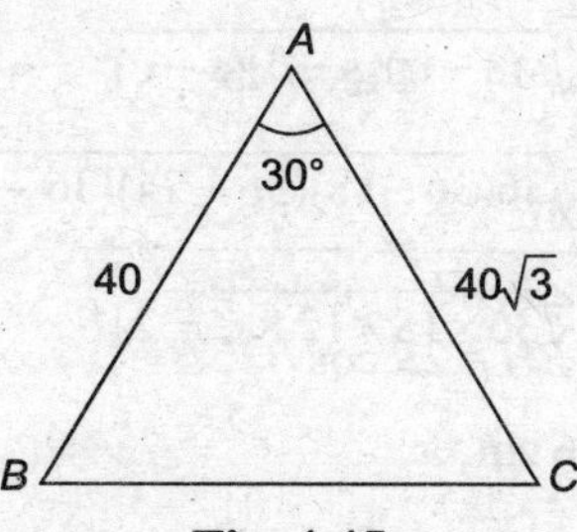

Fig. 4.45

Using cosine formula,

$$\cos 30° = \frac{40^2 + (40\sqrt{3})^2 - a^2}{2 \times 40 \times 40\sqrt{3}}$$

$\Rightarrow a = 40$

$\therefore AB = BC = 40$

Therefore, ΔABC is isosceles

Also, $\angle A = \angle C = 30°$

$\Rightarrow \angle B = 120°$

Therefore, ΔABC is an obtuse angled Δ.

c. $81^{\sin^2 x} + 81^{\cos^2 x} = 30 \Rightarrow 81^{\sin^2 x} + \dfrac{81}{81^{\sin^2 x}} = 30$

Put $81^{\sin 2}x = t$

$$t + \frac{81}{t} = 30 \Rightarrow t = 27, t = 3$$

When $t = 3$, $81^{\sin 2 x} = 3 \Rightarrow 3^{4\sin^2 x} = 3$

$\Rightarrow x = 30°$ or $150°$

When $t = 27$, $3^{4\sin^2 x} = 3^3 \Rightarrow x = 60°$ or $120°$

Therefore the triangle can be isosceles, equilateral or obtuse angled but not right angled as $x = 90°$ does not satisfy the given equation.

d. In ΔABC,

$\cos A \cos B + \sin A \sin B \sin C = 1$

$$\Rightarrow \sin C = \frac{1 - \cos A \cos B}{\sin A \sin B} \quad (1)$$

Since $0 < \sin C \leq 1$

[$\because \angle C$ is an angle of the triangle]

$$\Rightarrow 0 < \frac{1 - \cos A \cos B}{\sin A \sin B} \leq 1$$

$\Rightarrow 1 - \cos A \cos B \leq \sin A \sin B$

$\Rightarrow 1 \leq \sin A \sin B + \cos A \cos B$

$\Rightarrow 1 \leq \cos(A - B)$ or $\cos(A - B) \geq 1$

$\Rightarrow A - B = 0 \Rightarrow A = B \quad (2)$

Also from (1) and (2), $\sin C = \dfrac{1 - \cos^2 A}{\sin^2 A}$

$\sin C = 1 \Rightarrow C = 90°$

From equations (2) and (3), ΔABC is a right-angled isosceles triangle.

4. **a → q; b → q; c → p; d → r.**

5. **a → p, r; b → q, r; c → r, s; d → q, r. s.**

a. $\cot \dfrac{A}{2} = \dfrac{b + c}{a}$

$$\Rightarrow \frac{\cos(A/2)}{\sin(A/2)} = \frac{\sin B + \sin C}{\sin A}$$

$$\Rightarrow \frac{\cos(A/2)}{\sin(A/2)} = \frac{2\sin\frac{B+C}{2}\cos\frac{B-C}{2}}{2\sin\frac{A}{2}\cos\frac{A}{2}}$$

$$\Rightarrow \cos\frac{A}{2} = \cos\frac{B-C}{2}$$

$\Rightarrow A = B - C$

$\Rightarrow A + C = B$

$\Rightarrow A + B + C = 2B = \pi \Rightarrow B = \pi/2$

b. If in ΔABC, $a \tan A + b \tan B = (a + b) \tan \frac{1}{2}(A + B)$

Rewrite the given relation as

$$a\left[\tan A - \frac{1}{2}(A + B)\right] = b\left[\tan \frac{1}{2}(A + B) - \tan B\right]$$

$$\Rightarrow \frac{a \sin\left(A - \frac{A+B}{2}\right)}{\cos A \cos\frac{A+B}{2}} = \frac{b \sin\left(\frac{A+B}{2} - B\right)}{\cos\frac{A+B}{2}\cos B}$$

$$\Rightarrow \frac{2R \sin A \sin\frac{1}{2}(A - B)}{\cos A}$$

$$= \frac{2R \sin B \sin\frac{1}{2}(A - B)}{\cos B}$$

$$\Rightarrow \sin\frac{1}{2}(A - B)\,(\tan A - \tan B) = 0$$

$\Rightarrow A = B$ from either factor

c. $a \cos A = b \cos B$

$2R \sin A \cos A = 2R \sin B \cos B$

$\Rightarrow \sin 2A = \sin 2B$

$\Rightarrow 2A = 2B$ or $\pi - 2B$

$\Rightarrow A = B$ or $A + B = \pi/2$

$\Rightarrow \Delta ABC$ is isosceles or $C = \pi/2$,

i.e., ΔABC is right angled

d. $\cos A = \dfrac{\sin B}{2\sin C}$

$$\Rightarrow \frac{b^2 + c^2 - a^2}{2bc} = \frac{b}{2c}$$

$\Rightarrow b^2 + c^2 - a^2 = b^2$

$\Rightarrow c^2 = a^2 \Rightarrow c = a$

$\Rightarrow$ The triangle is isosceles

Integer Type

1. **(4)** $a \cos A = 2R \sin A \cos A = R \sin 2A$

$\Rightarrow a \cos A + b \cos B + c \cos C$

$= R (\sin 2A + \sin 2B + \sin 2C)$

$= 4R \sin A \sin B \sin C$

2. Since the angles A, B, C are in A.P.,

$A + C = 2B$.

But $A + B + C = 180°$

$\Rightarrow 3B = 180° \Rightarrow B = 60°$

Now $\cos B = \dfrac{1}{2} = \dfrac{a^2 + c^2 - b^2}{2ac}$

$\Rightarrow a^2 + c^2 - b^2 = ac$

$\Rightarrow a^2 - ac + c^2 = b^2$

$$\Rightarrow \frac{a + c}{\sqrt{a^2 - ac + c^2}} = \frac{a + c}{b}$$

$$= \frac{k(\sin A + \sin C)}{k \sin B}$$

$$= \frac{2\sin\left(\frac{A + C}{2}\right)\cos\left(\frac{A - C}{2}\right)}{\sin B}$$

$$= \frac{2\sin 60°}{\sin 60°}\cos\frac{1}{2}\left(\frac{A - C}{2}\right)$$

$$= \cos\frac{1}{2}\left(\frac{A - C}{2}\right)$$

3. $$\cot\frac{B}{2}\cot\frac{C}{2} = \sqrt{\frac{s(s - b)}{(s - a)(s - c)}}\sqrt{\frac{s(s - c)}{(s - a)(s - b)}}$$

$$= \frac{s}{s - a}$$

$$= \frac{2s}{2s - 2a}$$

$$= \frac{a + b + c}{b + c - a} = \frac{4a}{2a} = 2,$$

4. **(6)** $r = \dfrac{\Delta}{s}$

$$s = \frac{a + b + c}{2} = \frac{18 + 24 + 30}{2} = 36$$

$\Rightarrow \Delta = \sqrt{s(s - a)(s - b)(s - c)}$

$= \sqrt{36(36 - 18)(36 - 24)(36 - 30)}$

$= \sqrt{36 \times 18 \times 12 \times 6} = 216$

$\dfrac{216}{36} = 6$ cm

5. **(9)** We have already proved that

$$\frac{1}{r_1} + \frac{1}{r_2} + \frac{1}{r_3} = \frac{1}{r}$$

Now using A.M. $\geq$ H.M., we get

$$\frac{r_1 + r_2 + r_3}{3} \geq \frac{3}{\frac{1}{r_1} + \frac{1}{r_2} + \frac{1}{r_3}} = 3r$$

$$\Rightarrow \frac{r_1 + r_2 + r_3}{r} \geq 9$$

6. **(3)** Let $ABCD$ be the cyclic quadrilateral in which $AB = 2$ and $BC = 5$, $\angle ABC = 60°$.

$\angle ADC = 180° - 60° = 120°$

Area of $ABCD$ = Area of ΔABC + Area of ΔACD

$$\Rightarrow 4\sqrt{3} = \frac{1}{2} AB \times BC \sin 60° + \frac{1}{2} CD \times DA \sin 120°$$

$= \dfrac{1}{2} \times 2 \times 5 \times (\sqrt{3}/2) + \dfrac{1}{2} xy(\sqrt{3}/2)$, where $CD = x$, $AD = y$

$\Rightarrow xy = 6$ (1)

Form ΔABC, we get

$AC^2 = AB^2 + BC^2 - 2AB \times BC \cos 60°$

$= 4 + 25 - 20 \times (1/2) = 19$ (2)

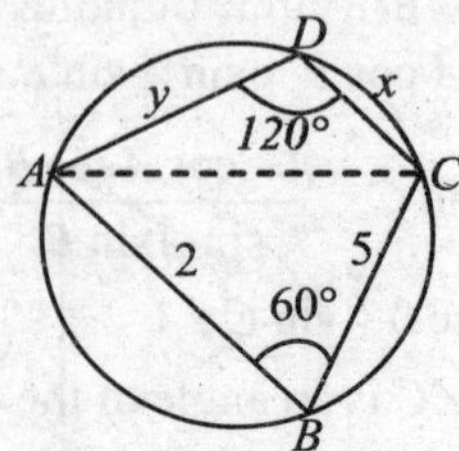

Fig. 4.46

Also from ΔACD, we get

$AC^2 = CD^2 + DA^2 - 2CD \times DA \cos 120°$

$= x^2 + y^2 + xy$

$= x^2 + y^2 + 6$ [Using (1)] (3)

Now from (2) and (3), we have

$x^2+y^2+6=19$ or $x^2+y^2=13$ (4)
$\Rightarrow x^4-13x^2+36=0$
$\Rightarrow x^2=4, 9$
$\Rightarrow x=2, 3 \Rightarrow y=3, 2$
Hence, the other two sides of the cyclic quadrilateral are 3 and 2.

7. **(7)** Let AD be the perpendicular from A on BC. When AD is produced, it meets the circumscribing circle at E. From the question, $DE=\alpha$.

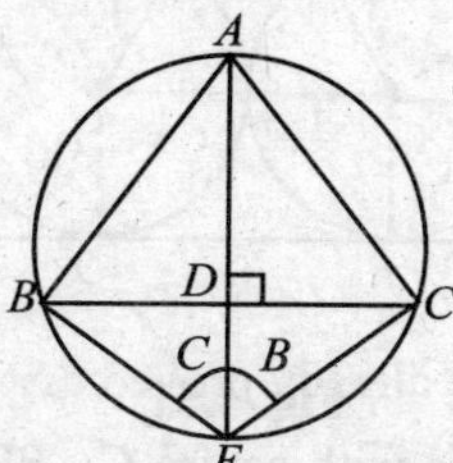

Fig. 4.47

Since angles in the same segment are equal, $\angle AEB=\angle ACB=\angle C$ and $\angle AEC=\angle ABC=\angle B$.
From the right-angled triangle BDE, we get

$$\tan C=\frac{BD}{DE} \quad (1)$$

From the right-angled triangle CDE, we get

$$\tan B=\frac{CD}{DE} \quad (2)$$

Adding (1) and (2), we get

$$\tan B+\tan C=\frac{BD+CD}{DE}=\frac{BC}{DE}=\frac{a}{\alpha} \quad (3)$$

Similarly, $\tan C+\tan A=\dfrac{b}{\beta}$ (4)

and $\tan A+\tan B=\dfrac{c}{\gamma}$ (5)

Adding (3), (4), and (5), we get

$$\frac{a}{\alpha}+\frac{b}{\beta}+\frac{c}{\gamma}=2(\tan A+\tan B+\tan C)$$

8. **(3)** Given that $A_0A_1A_2A_3A_4A_5$ is a regular hexagon inscribed in a circle of radius 1.

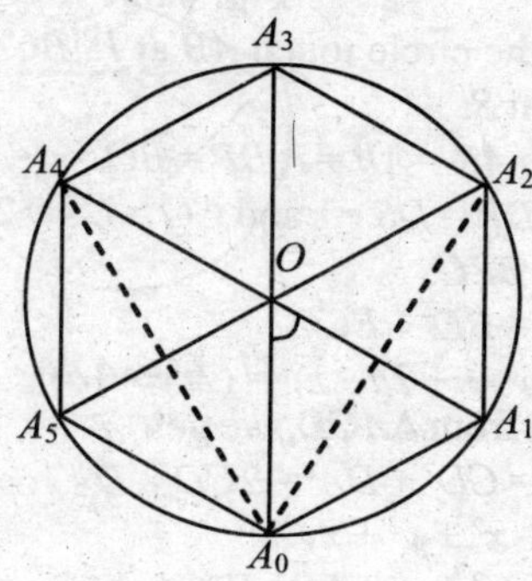

Fig. 4.48

$$\angle A_0OA_1=\frac{360°}{6}=60°$$

But in ΔOA_0A_1, $OA_0=OA_1=1$. Therefore,
$\angle OA_0A_1=\angle OA_1A_0=60°$
$\Rightarrow \Delta OA_0A_1$ is equilateral.
$\Rightarrow A_0A_1=1=A_1A_2=A_2A_3=A_3A_4=A_4A_5=A_5A_0$ and $\angle A_0A_1A_2=120°$.
Using cosine rule, we get

$$\cos 120°=\frac{(A_0A_1)^2+(A_1A_2)^2-(A_0A_2)^2}{2\,(A_0A_1)\,(A_1A_2)}$$

$$\Rightarrow -\frac{1}{2}=\frac{1+1-(A_0A_2)^2}{2\times 1\times 1}$$

$$\Rightarrow A_0A_2=\sqrt{3}$$

Thus by symmetry, $A_0A_4=\sqrt{3}$

$$\Rightarrow A_0A_1\times A_0A_2\times A_0A_4=1\times\sqrt{3}\times\sqrt{3}=3$$

9. **(4)** Let us consider three circles with centres at A, B, C and radii r_1, r_2, r_3, respectively. The circles touch each other externally at P, Q, R.
Let the common tangents at P, Q, and R meet each other at O.
Then $OP=OQ=OR=4$ (given) (Lengths of tangents from a point to a circle are equal.)
Also $OP\perp AB$, $OQ\perp AC$, $OR\perp BC$.

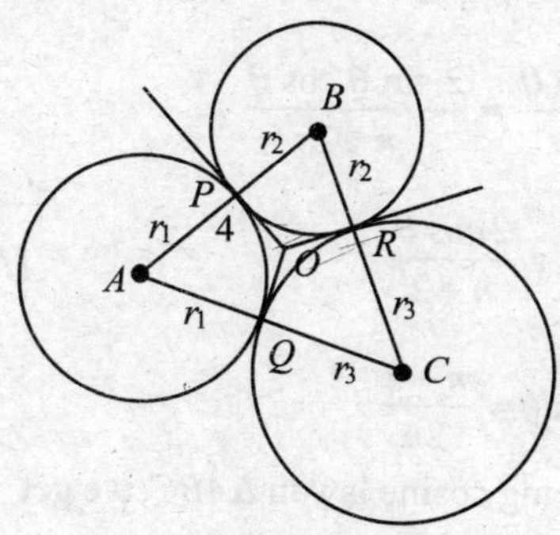

Fig. 4.49

$\Rightarrow O$ is the incentre of ΔABC.
Thus for ΔABC,

$$s=\frac{(r_1+r_2)+(r_2+r_3)+(r_3+r_1)}{2}$$

$$\Rightarrow s=r_1+r_2+r_3$$

$$\Rightarrow \Delta=\sqrt{(r_1+r_2+r_3)\,r_1r_2r_3} \text{ (Hero's formula)}$$

Now $r=\dfrac{\Delta}{s}$

$$\Rightarrow 4=\frac{\sqrt{(r_1+r_2+r_3)r_1r_2r_3}}{r_1+r_2+r_3}$$

$$\Rightarrow 4 = \frac{\sqrt{r_1 r_2 r_3}}{\sqrt{r_1 + r_2 + r_3}}$$

$$\Rightarrow \frac{r_1 r_2 r_3}{r_1 + r_2 + r_3} = \frac{16}{1}$$

$$\Rightarrow r_1 r_2 r_3 : r_1 + r_2 + r_3 = 16 : 1$$

10. **(6)** Let the sides of the triangle be n, $n + 1$, $n + 2$, where $n \in N$.

Let $a = n, b = n + 1, c = n + 2$

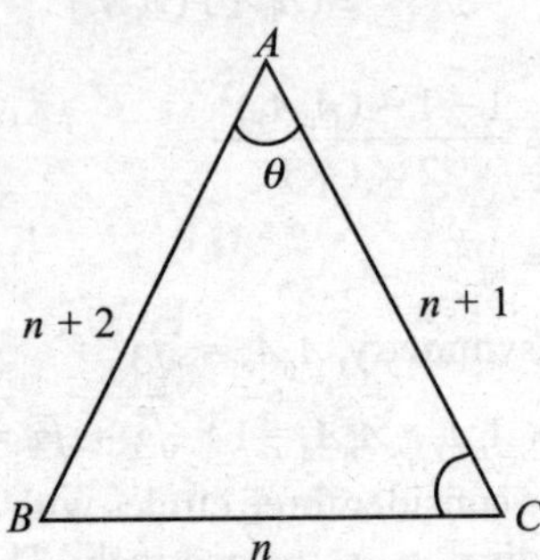

Fig. 4.50

Let the smallest angle $\angle A = \theta$.

Then the greatest $\angle C = 2\theta$.

Applying sine law in ΔABC, we get

$$\frac{\sin\theta}{n} = \frac{\sin 2\theta}{n+2}$$

$$\Rightarrow \frac{\sin\theta}{n} = \frac{2\sin\theta\cos\theta}{n+2}$$

$$\Rightarrow \frac{1}{n} = \frac{2\cos\theta}{n+2} \qquad [\text{As } \sin\theta \neq 0)]$$

$$\Rightarrow \cos\theta = \frac{n+2}{2n} \qquad (1)$$

Applying cosine law in ΔABC, we get

$$\cos\theta = \frac{(n+1)^2 + (n+2)^2 - n^2}{2(n+1)(n+2)} \qquad (2)$$

Comparing the values of $\cos\theta$ from (1) and (2), we get

$$\frac{(n+1)^2 + (n+2)^2 - n^2}{2(n+1)(n+2)} = \frac{n+2}{2n}$$

$\Rightarrow (n+2)^2(n+1) = n(n+2)^2 + n(n+1)^2 - n^3$

$\Rightarrow n(n+2)^2 + (n+2)^2 = n(n+2)^2 + n(n+1)^2 - n^3$

$\Rightarrow n^2 + 4n + 4 = n^3 + 2n^2 + n - n^3$

$\Rightarrow n^2 - 3n - 4 = 0$

$\Rightarrow (n+1)(n-4) = 0$

$\Rightarrow n = 4 \qquad [\text{As } n \neq -1]$

Hence, the sides of the triangle are 4, 4 + 1, 4 + 2, i.e., 4, 5, 6.

11. The situation is as shown in the figure.

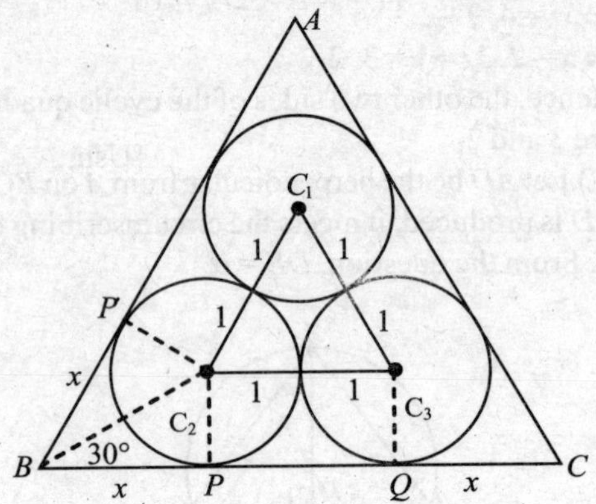

Fig. 4.51

For the circle with centre C_2, BP and BP' are two tangents. Therefore, BC_2 must be the bisector of $\angle B$. But $\angle B = 60°$ ($\because \Delta ABC$ is equilateral.)

So $\angle C_2BP = 30°$

$$\Rightarrow \tan 30° = \frac{1}{x}$$

$\Rightarrow x = \sqrt{3}$

Hence, $BC = BP + PQ + QC$

$= x + 2 + x = 2 + 2\sqrt{3}$

12. **(4)** Given $AB \parallel CD$, $CD = 2AB$.

Let $AB = a$, then $CD = 2a$.

Let the radius of the circle be r.

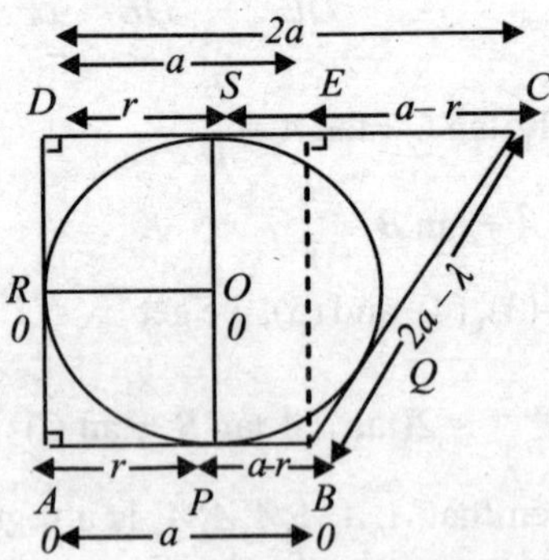

Fig. 4.52

Let the circle touch AB at P, BC at Q, AD at R, and CD at S.

Then $AR = AP = r$, $BP = BQ = a - r$

and $DR = DS = r$ and $CQ = CS = 2a - r$.

In ΔBEC,

$BC^2 = BE^2 + EC^2$

$\Rightarrow (a - r + 2a - r)^2 = (2r)^2 + (a)^2$

$\Rightarrow 9a^2 + 4r^2 - 12ar = 4r^2 + a^2$

$$\Rightarrow a = \frac{3}{2}r \qquad (1)$$

Also area of quadrilateral $ABCD$ is 18

$\Rightarrow$ Area of quadrilateral $ABED$ + area of $\Delta BCE = 18$

$\Rightarrow (a \times 2r) + (\frac{1}{2} \times a \times 2r) = 18$

$\Rightarrow ar = 6$

$\Rightarrow \frac{3r^2}{2} = 6$ [Using equation (1)]

$\Rightarrow r^2 = 4$

$\Rightarrow r = 2$

13. (6)

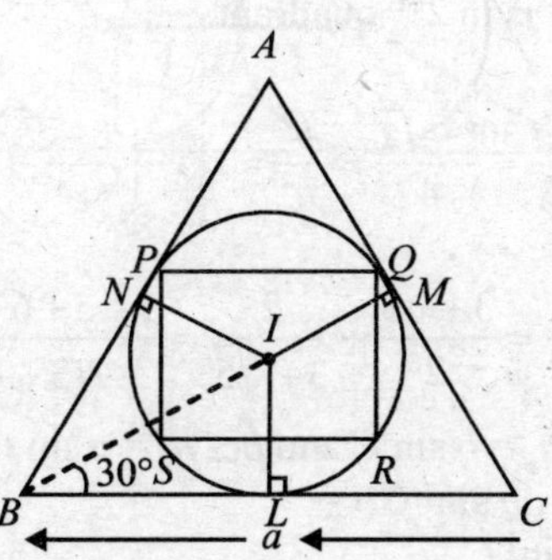

Fig. 4.53

Given that ABC is an equilateral triangle of side length 6 units. The radius of a circle inscribed in it is r.
So in ΔIBL,

$\tan 30° = \frac{r}{3} \Rightarrow r = \sqrt{3}$.

If $PQRS$ is the square inscribed in the circle of radius r, side of square $= 2(r \sin 45)$

$= \frac{2r}{\sqrt{2}}$

$= \frac{2}{\sqrt{2}} \times \sqrt{3}$

$= \sqrt{6}$

Hence, area of square is 6 sq. units.

14. (5) If $a^2 + 2a$, $2a + 3$, $a^2 + 3a + 8$ are the sides of a triangle, then the sum of any two sides is greater than the third side.

Let $x = a^2 + 2a$, $y = 2a + 3$, $z = a^2 + 3a + 8$.

Then $x + y > z$

$\Rightarrow a^2 + 4a + 3 > a^2 + 3a + 8$

$\Rightarrow a > 5$ (1)

Also $y + z > x$

$\Rightarrow a^2 + 5a + 11 > a^2 + 2a$

$\Rightarrow 3a > -11$

$\Rightarrow a > -11/3$ (2)

And $z + x > y$

$\Rightarrow 2a^2 + 5a + 8 > 2a + 3$

$\Rightarrow 2a^2 + 3a + 5 > 0$

Here coefficient of $a^2 > 0$ and $D = 9 - 40$, which is negative.

It is true for all values of a.

Combining (1) and (2), we get $a > 5$.

15. (4) $\cos C = \frac{a^2 + b^2 - c^2}{2ab}$

now $\Delta = a^2 + b^2 - c^2$

hence $\cos C = \frac{\Delta}{2ab}$ (1)

also $\Delta = \frac{1}{2} ab \sin C \Rightarrow \frac{2\Delta}{\sin C} = ab \Rightarrow \sin C = \frac{2\Delta}{ab}$ (2)

From (1) and (2),

$\tan C = \frac{2\Delta}{ab} \times \frac{2ab}{\Delta} = 4$

16. (7) $R = 8r = 8\left(4R \sin\frac{A}{2} \sin\frac{B}{2} \sin\frac{C}{2}\right)$

$\therefore 2 \sin\frac{A}{2} \sin\frac{B}{2} \sin\frac{C}{2} = \frac{1}{16}$

$\left(\cos\frac{A-B}{2} - \cos\frac{A+B}{2}\right) \sin\frac{C}{2} = \frac{1}{16}$

$\sin\frac{C}{2} \times \left(\frac{1}{2} - \sin\frac{C}{2}\right) = \frac{1}{16}$; $\sin^2\frac{C}{2} = \frac{1}{4}$

$\cos C = 1 - 2\sin^2\frac{C}{2} = 1 - \frac{1}{8} = \frac{7}{8}$

17. (4) Let a, b, c be the sides of the triangle so that $(a + c)/2 = b$. If a is the smallest side, then angle opposite to this side is smallest. If a is the smallest, then c is the largest side and angle opposite to this side is greatest. Hence,

$a + c = 2b \Rightarrow \sin\phi + \sin\theta = 2\sin(\pi - (\theta + \phi))$

$= 2\sin(\theta + \phi)$

or $2\sin\frac{\theta+\phi}{2}\cos\frac{\theta-\phi}{2} = 4\sin\frac{\theta+\phi}{2}\cos\frac{\theta+\phi}{2}$

$\Rightarrow \cos\frac{\theta-\phi}{2} = 2\cos\frac{\theta+\phi}{2}$ (1)

Multiply both sides of (1) by $2\cos\frac{\theta-\phi}{2}$

$2\cos^2\frac{\theta-\phi}{2} = 4\cos\frac{\theta-\phi}{2}\cos\frac{\theta+\phi}{2}$

$\Rightarrow 1 + \cos(\theta - \phi) = 2(\cos\theta + \cos\phi)$ (2)

Now, multiply both sides of (1) by $2\cos\frac{\theta+\phi}{2}$

$2\cos\frac{\theta+\phi}{2}\cos\frac{\theta-\phi}{2} = 4\cos^2\frac{\theta+\phi}{2}$

$\Rightarrow \cos\theta + \cos\phi = 2\,(1 + \cos(\theta + \phi))$ (3)

From (2) and (3), we have

$2\,[1 + \cos(\theta - \phi)] + 2\,[1 + \cos(\theta + \phi)] = 5\,(\cos\theta + \cos\phi)$

$\Rightarrow 4\,(1 + \cos\theta\cos\phi) = 5\,(\cos\theta + \cos\phi)$

$\Rightarrow 4\,[1 + \cos\theta\cos\phi - \cos\phi - \cos\theta - \cos\phi] = \cos\theta + \cos\phi$

Hence $4\,(1 - \cos\theta)\,(1 - \cos\phi) = \cos\theta + \cos\phi$

Hence $4\,(1 - \cos\theta)\,(1 - \cos\phi) = \cos\theta + \cos\phi$

18. (3) Use cosine rule in ΔABE

$$\frac{2a^2 - b^2}{2a^2} = \cos\left(\frac{3\pi}{5}\right)$$

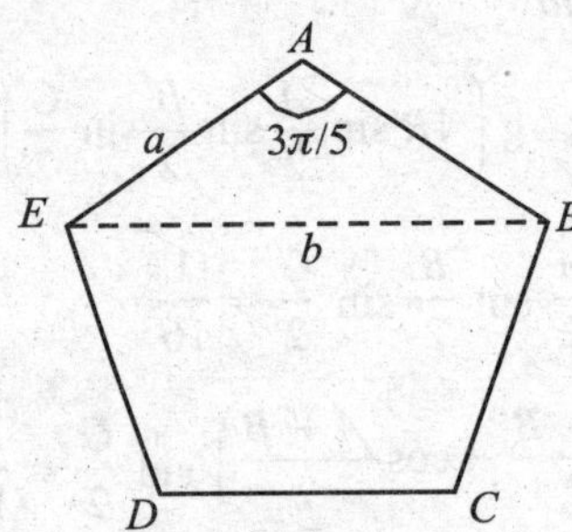

Fig. 4.54

$$1 - \frac{b^2}{2a^2} = \cos\frac{3\pi}{5}$$

$$\Rightarrow 2\left(1 - \cos\frac{3\pi}{5}\right) = \frac{b^2}{a^2}$$

Now let $\frac{3\pi}{5} = \theta \Rightarrow 2\theta = 3\pi - 3\theta$

$\Rightarrow \sin 2\theta = \sin 3\theta \Rightarrow 2\cos\theta = 4\cos^2\theta - 1$

$$\Rightarrow \cos\theta = \frac{2 \pm \sqrt{4 + 16}}{8} = \frac{1 \pm \sqrt{5}}{4}$$

But $\frac{3\pi}{5}$ is in 2nd quadrant.

$$\Rightarrow \cos\frac{3\pi}{5} = \frac{1 - \sqrt{5}}{4} \Rightarrow \frac{b^2}{a^2} = 2\left[1 - \left(\frac{1 - \sqrt{5}}{4}\right)\right] = \frac{3 + \sqrt{5}}{2}$$

$$\frac{a^2}{b^2} + \frac{b^2}{a^2} = \frac{3 + \sqrt{5}}{2} + \frac{2}{3 + \sqrt{5}} = \frac{9 + 5 + 6\sqrt{5} + 4}{2(3 + \sqrt{5})} = 3$$

19. (4) $a \le h_a = b \sin C$ and $b \le h_b = a \sin C$

$\Rightarrow ab \le ab \sin^2 C$

$\Rightarrow 1 \le \sin^2 C$

Hence $\sin C = 1$ (*i.e.*, $C = \frac{\pi}{2}$)

Hence maximum area of the triangle is $\frac{1}{2}a^2$

$$= \frac{1}{2}\left(\frac{4}{\sqrt{2}}\right)^2 = 4$$

Section B

Algebra

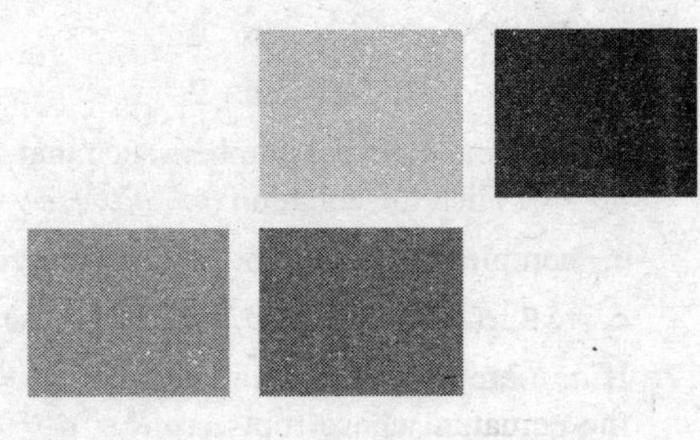

CHAPTER 1

Quadratic Equations and Expressions

Objective Type

Each question has four choices a, b, c, and d, out of which *only one* is correct. Find the correct answer.

1. If $x = 2 + 2^{2/3} + 2^{1/3}$, then the value of $x^3 - 6x^2 + 6x$ is
   a. 3 **b.** 2 **c.** 1 **d.** −2
2. The least value of $x^2 + 4y^2 + 3z^2 - 2x - 12y - 6z + 14$ is
   a. 1 **b.** no least value
   c. 0 **d.** none of these
3. Number of positive integers n for which $n^2 + 96$ is a perfect square is
   a. 8 **b.** 12 **c.** 4 **d.** infinite
4. If $x, y \in R$ satisfy the equation $x^2 + y^2 - 4x - 2y + 5 = 0$, then the value of the expression $[(\sqrt{x}-\sqrt{y})^2+4\sqrt{xy}]/(x+\sqrt{xy})$ is
   a. $\sqrt{2}+1$ **b.** $\frac{\sqrt{2}+1}{2}$
   c. $\frac{\sqrt{2}-1}{2}$ **d.** $\frac{\sqrt{2}+1}{\sqrt{2}}$
5. The number of real roots of the equation $|x|^2 - 3|x| + 2 = 0$ is
   a. 2 **b.** 1 **c.** 4 **d.** 3
6. If $x = 1 + i$ is a root of the equation $x^3 - ix + 1 - i = 0$, then the other real root is
   a. 0 **b.** 1
   c. −1 **d.** none of these
7. The number of roots of the equation $\sqrt{x-2}(x^2-4x+3)=0$ is
   a. three **b.** four
   c. one **d.** two
8. The curve $y = (\lambda + 1)x^2 + 2$ intersects the curve $y = \lambda x + 3$ in exactly one point, if λ equals
   a. $\{-2, 2\}$ **b.** $\{1\}$
   c. $\{-2\}$ **d.** $\{2\}$
9. If the expression $x^2 + 2(a + b + c)x + 3(bc + ca + ab)$ is a perfect square, then
   a. $a = b = c$ **b.** $a = \pm b = \pm c$
   c. $a = b \neq c$ **d.** none of these
10. If one root of the equation $ax^2 + bx + c = 0$ is square of the other, then $a(c - b)^3 = cX$, where X is
    a. $a^3 - b^3$ **b.** $a^3 + b^3$
    c. $(a - b)^3$ **d.** none of these
11. If $x^2 + px + 1$ is a factor of the expression $ax^3 + bx + c$, then
    a. $a^2 - c^2 = ab$ **b.** $a^2 + c^2 = -ab$
    c. $a^2 - c^2 = -ab$ **d.** none of these
12. Sum of the non-real roots of $(x^2 + x - 2)(x^2 + x - 3) = 12$ is
    a. −1 **b.** 1
    c. −6 **d.** 6
13. If $(ax^2 + c)y + (a'x^2 + c') = 0$ and x is a rational function of y and ac is negative, then
    a. $ac' + a'c = 0$ **b.** $a/a' = c/c'$
    c. $a^2 + c^2 = a'^2 + c'^2$ **d.** $aa' + cc' = 1$
14. Let p and q be roots of the equation $x^2 - 2x + A = 0$ and let r and s be the roots of the equation $x^2 - 18x + B = 0$. Also, $p < q < r < s$ are in arithmetic progression, then the values of A and B are
    a. 3, −77 **b.** 3, 77
    c. −3, −77 **d.** −3, 77
15. The number of irrational roots of the equation $4x(x^2+x+3)+5x(x^2-5x+3)=-3/2$ is

a. 4 b. 0
c. 1 d. 2

16. Let a, b and c be real numbers such that $4a + 2b + c = 0$ and $ab > 0$. Then the equation $ax^2 + bx + c = 0$ has
a. complex roots b. exactly one root
c. real roots d. none of these

17. If α, β are the roots of the equation $x^2 - 2x + 3 = 0$. Then the equation whose roots are $P = \alpha^3 - 3\alpha^2 + 5\alpha - 2$ and $Q = \beta^3 - \beta^2 + \beta + 5$ is
a. $x^2 + 3x + 2 = 0$ b. $x^2 - 3x - 2 = 0$
c. $x^2 - 3x + 2 = 0$ d. none of these

18. If α, β be the roots of the equation $2x^2 - 35x + 2 = 0$, then the value of $(2\alpha - 35)^3 (2\beta - 35)^3$ is equal to
a. 8 b. 1
c. 64 d. none of these

19. If a, b, c are three distinct positive real numbers then the real roots of $ax^2 + 2b|x| - c = 0$ is
a. 0 b. 4
c. 2 d. none of these

20. If $p(q - r)x^2 + q(r - p)x + r(p - q) = 0$ has equal roots, then $2/q =$
a. $\frac{1}{p}+\frac{1}{r}$ b. $p + r$
c. $\frac{1}{p}+r$ d. $p+\frac{1}{r}$

21. Let $a \neq 0$ and $p(x)$ be a polynomial of degree greater than 2. If $p(x)$ leaves remainders a and $-a$ when divided, respectively, by $x + a$ and $x - a$, the remainder when $p(x)$ is divided by $x^2 - a^2$ is
a. $2x$ b. $-2x$
c. x d. $-x$

22. The quadratic $x^2 + ax + b + 1 = 0$ has roots which are positive integers, then $(a^2 + b^2)$ can be equal to
a. 50 b. 37
c. 61 d. 19

23. The sum of values of x satisfying the equation
$(31+8\sqrt{15})^{x^2-3}+1=(32+8\sqrt{15})^{x^2-3}$ is
a. 3 b. 0
c. 2 d. none of these

24. If α, β be the roots of the equation $ax^2 + bx + c = 0$, then value of $(a\alpha^2+c)(a\alpha+b)+(a\beta^2+c)(a\beta+b)$ is
a. $\frac{b(b^2-2ac)}{4a}$ b. $\frac{b^2-4ac}{2a}$
c. $\frac{b(b^2-2ac)}{a^2c}$ d. none of these

25. A quadratic equation whose product of roots x_1 and x_2 is equal to 4 and satisfying the relation $x_1(x_1-1)+x_2(x_2-1)=2$ is
a. $x^2 - 2x + 4 = 0$ b. $x^2 + 2x + 4 = 0$
c. $x^2 + 4x + 4 = 0$ d. $x^2 - 4x + 4 = 0$

26. If $a, b, c, d \in R$, then the equation $(x^2 + ax - 3b)(x^2 - cx + b)(x^2 - dx + 2b) = 0$ has
a. 6 real roots b. at least 2 real roots
c. 4 real roots d. 3 real roots

27. If α and β are the roots of the equation $x^2 + px + q = 0$, and α^4 and β^4 are the roots of $x^2 - rx + q = 0$, the roots of $x^2 - 4qx + 2q^2 - r = 0$ are always
a. both non-real b. both positive
c. both negative d. opposite in sign

28. If the roots of the equation $(a - 1)(x^2 + x + 1)^2 = (a + 1)(x^4 + x^2 + 1)$ are real and distinct then the value of $a \in$
a. $(-\infty, 3]$ b. $(-\infty,-2) \cup (2, \infty)$
c. $[-2, 2]$ d. $[-3, \infty)$

29. If $b_1b_2 = 2(c_1 + c_2)$, then at least one of the equations $x^2 + b_1x + c_1 = 0$ and $x^2 + b_2x + c_2 = 0$ has
a. maginary roots
b. real roots
c. purely imaginary roots
d. none of these

30. The integral values of m for which the roots of the equation $mx^2 + (2m - 1)x + (m - 2) = 0$ are rational are given by the expression [where n is an integer]
a. n^2 b. $n(n + 2)$
c. $n(n + 1)$ d. none of these

31. Suppose A, B, C are defined as $A = a^2b + ab^2 - a^2c - ac^2$, $B = b^2c + bc^2 - a^2b - ab^2$, and $C = a^2c + ac^2 - b^2c - bc^2$, where $a > b > c > 0$ and the equation $Ax^2 + Bx = C + 0$ has equal roots, then a, b, c are in
a. A.P. b. G.P.
c. H.P. d. A.G.P.

32. The coefficient of x in the equation $x^2 + px + q = 0$ was wrongly written as 17 in place of 13 and the roots thus found was -2 and -15. Then the roots of the correct equation are
a. $-3, 10$ b. $-3, -10$
c. $3, -10$ d. none of these

33. If $a(p + q)^2 + 2bpq + c = 0$ and $a(p + r)^2 + 2bpr + c = 0$ $(a \neq 0)$, then
a. $qr = p^2$ b. $qr = p^2 + \frac{c}{a}$
c. $qr = -p^2$ d. none of these

34. If the roots of the equation $ax^2 - bx + c = 0$ are α, β then the roots of the equation $b^2cx^2 - ab^2x + a^3 = 0$ are
a. $\frac{1}{\alpha^3+\alpha\beta}, \frac{1}{\beta^3+\alpha\beta}$ b. $\frac{1}{\alpha^2+\alpha\beta}, \frac{1}{\beta^2+\alpha\beta}$
c. $\frac{1}{\alpha^4+\alpha\beta}, \frac{1}{\beta^4+\alpha\beta}$ d. none of these

35. If α and β, α and γ, α and δ are the roots of the equations $ax^2 + 2bx + c = 0$, $2bx^2 + cx + a = 0$, and $cx^2 + ax + 2b = 0$, respectively, where a, b and c are positive real numbers, then $\alpha + \alpha^2 =$

a. abc b. $a+2b+c$
c. -1 d. 0

36. $x^2 - xy + y^2 - 4x - 4y + 16 = 0$ represents
a. a point
b. a circle
c. a pair of straight lines
d. none of these

37. If α, β be the non-zero roots of $ax^2 + bx + c = 0$ and α^2, β^2 be the roots of $a^2x^2 + b^2x + c^2 = 0$, then a, b, c are in
a. G.P. b. H.P.
c. A.P. d. none of these

38. If the roots of the equation $ax^2 + bx + c = 0$ are of the form $(k+1)/k$ and $(k+2)/k+1$, then $(a + b + c)^2$ is equal to
a. $2b^2 - ac$ b. Σa^2
c. $b^2 - 4ac$ d. $b^2 - 2ac$

39. If α, β are the roots of $ax^2 + bx + c = 0$ and $\alpha + h, \beta + h$ are the roots of $px^2 + qx + r = 0$, then $h =$
a. $-\frac{1}{2}\left(\frac{a}{b}-\frac{p}{q}\right)$ b. $\left(\frac{b}{a}-\frac{q}{p}\right)$
c. $\frac{1}{2}\left(\frac{b}{a}-\frac{q}{p}\right)$ d. none of these

40. If α, β be the roots of the equation $(x - a)(x - b) + c = 0$ $(c \neq 0)$, then the roots of the equation $(x - c - \alpha)(x - c - \beta) = c$ are
a. $a + c$ and $b + c$ b. $a - c$ and $b - c$
c. a and $b + c$ d. $a + c$ and b

41. If α, β are the roots of $ax^2 + c = bx$, then the equation $(a + cy)^2 = b^2y$ in y has the roots
a. $\alpha\beta^{-1}, \alpha^{-1}\beta$ b. α^{-2}, β^{-2}
c. α^{-1}, β^{-1} d. α^2, β^2

42. If the roots of the equation $x^2 + 2ax + b = 0$ are real and distinct and they differ by at most $2m$, then b lies in the interval
a. $(a^2, a^2 + m^2)$ b. $(a^2 - m^2, a^2)$
c. $[a^2 - m^2, a^2)$ d. none of these

43. If the ratio of the roots of $ax^2 + 2bx + c = 0$ is same as the ratio of the $px^2 + 2qx + r = 0$, then
a. $\frac{2b}{ac} = \frac{q^2}{pr}$ b. $\frac{b}{ac} = \frac{q}{pr}$
c. $\frac{b^2}{ac} = \frac{q^2}{pr}$ d. none of these

44. If one root of $x^2 - x - k = 0$ is square of the other, then $k =$
a. $2 \pm \sqrt{5}$ b. $2 \pm \sqrt{3}$
c. $3 \pm \sqrt{2}$ d. $5 \pm \sqrt{2}$

45. If α and β be the roots of the equation $x^2 + px - 1/(2p^2) = 0$, where $p \in R$, then the minimum value of $\alpha^4 + \beta^4$ is
a. $2\sqrt{2}$ b. $2 - \sqrt{2}$
c. 2 d. $2 + \sqrt{2}$

46. If α, β are real and α^2, β^2 are the roots of the equation $a^2x^2 + x + 1 - a^2 = 0$ $(a > 1)$, then $\beta^2 =$
a. a^2 b. $1 - \frac{1}{a^2}$
c. $1 - a^2$ d. $1 + a^2$

47. If α and β are the roots of the equation $x^2 - ax + b = 0$ and $A_n = \alpha^n + \beta^n$, then which of the following is true?
a. $A_{n+1} = aA_n + bA_{n-1}$
b. $A_{n+1} = bA_n + aA_{n-1}$
c. $A_{n+1} = aA_n - bA_{n-1}$
d. $A_{n+1} = bA_n - aA_{n-1}$

48. The value of m for which one of the roots of $x^2 - 3x + 2m = 0$ is double of one of the roots of $x^2 - x + m = 0$ is
a. -2 b. 1
c. 2 d. none of these

49. If the equations $ax^2 + bx + c = 0$ and $x^3 + 3x^2 + 3x + 2 = 0$ have two common roots, then
a. $a = b = c$ b. $a = b \neq c$
c. $a = -b = c$ d. none of these

50. Number of values of a for which equations $x^3 + ax + 1 = 0$ and $x^4 + ax^2 + 1 = 0$ have a common root
a. 0 b. 1
c. 2 d. infinite

51. Let $p(x) = 0$ be a polynomial equation of the least possible degree, with rational coefficients, having $\sqrt[3]{7} + \sqrt[3]{49}$ as one of its roots. Then the product of all the roots of $p(x) = 0$ is
a. 56 b. 63
c. 7 d. 49

52. If $\alpha, \beta, \gamma, \sigma$ are the roots of the equation $x^4 + 4x^3 - 6x^2 + 7x - 9 = 0$, then the value of $(1 + \alpha^2)(1 + \beta^2)(1 + \gamma^2)(1 + \sigma^2)$ is
a. 9 b. 11
c. 13 d. 5

53. If $(m_r, 1/m_r)$, $r = 1, 2, 3, 4$ be four pairs of values of x and y that satisfy the equation $x^2 + y^2 + 2gx + 2fy + c = 0$, then value of $m_1 m_2 m_3 m_4$ is
a. 0 b. 1
c. -1 d. none of these

54. If roots of an equation $x^n - 1 = 0$ are $1, a_1, a_2, \ldots, a_{n-1}$, then the value of $(1 - a_1)(1 - a_2)(1 - a_3)(1 - a_{n-1})$ will be
a. n b. n^2
c. n^n d. 0

55. If $\tan\theta_1, \tan\theta_2, \tan\theta_3$ are the real roots of the $x^3 - (a + 1)x^2 + (b - a)x - b = 0$, where $\theta_1 + \theta_2 + \theta_3 \in (0, \pi)$, then $\theta_1 + \theta_2 + \theta_3$ is equal to
a. $\pi/2$ b. $\pi/4$
c. $3\pi/4$ d. π

56. If α, β, γ are the roots of $x^3 - x^2 - 1 = 0$ then the value of $[(1 + \alpha)/(1 - \alpha)] + [(1 + \beta)/(1 - \beta)] + [(1 + \gamma)/(1 - \gamma)]$ is equal to

a. −5 b. −6
c. −7 d. −2

57. Let r, s, and t be the roots of the equation, $8x^3 + 1001x + 2008 = 0$. The value of $(r + s)^3 + (s + t)^3 + (t + r)^3$ is

a. 251 b. 751
c. 735 d. 753

58. If x be real, then $x/(x^2-5x+9)$ lies between

a. −1 and −1/11 b. 1 and −1/11
c. 1 and 1/11 d. none of these

59. If x is real, then the maximum value of $(3x^2 + 9x + 17)/(3x^2+9x+7)$ is

a. 1/4 b. 41
c. 1 d. 17/7

60. If $a, b \in R$, $a \neq 0$ and the quadratic equation $ax^2 - bx + 1 = 0$ has imaginary roots then $(a + b + 1)$ is

a. positive b. negative
c. zero d. dependent on the sign of b

61. If the expression $[mx-1+(1/x)]$ is non-negative for all positive real x, then the minimum value of m must be

a. −1/2 b. 0
c. 1/4 d. 1/2

62. Suppose that $f(x)$ is a quadratic expression positive for all real x. If $g(x) = f(x) + f'(x) + f''(x)$, then for any real x

a. $g(x) < 0$ b. $g(x) > 0$
c. $g(x) = 0$ d. $g(x) \geq 0$

63. Let $f(x) = ax^2 - bx + c^2$, $b \neq 0$ and $f(x) \neq 0$ for all $x \in R$. Then

a. $a + c^2 < b$ b. $4a + c^2 > 2b$
c. $9a - 3b + c^2 < 0$ d. none of these

64. x_1 and x_2 are the roots of $ax^2 + bx + c = 0$ and $x_1x_2 < 0$. Roots of $x_1(x - x_2)^2 + x_2(x - x_1)^2 = 0$ are

a. real and of opposite signs
b. negative
c. positive
d. non-real

65. If a, b, c, d are four consecutive terms of an increasing A.P. then the roots of the equation $(x - a)(x - c) + 2(x - b)(x - d) = 0$ are

a. non-real complex b. real and equal
c. integers d. real and distinct

66. If roots of $x^2 - (a - 3)x + a = 0$ are such that at least one of them is greater than 2, then

a. $a \in [7, 9]$ b. $a \in [7, \infty)$
c. $a \in [9, \infty)$ d. $a \in [7, 9)$

67. Let $f(x) = ax^2 + bx + c$, $a, b, c \in r$. If $f(x)$ takes real values for real values of x and non-real values for non-real values of x, then

a. $a = 0$
b. $b = 0$
c. $c = 0$
d. nothing can be said about a, b, c.

68. All the values of m for which both the roots of the equation $x^2 - 2mx + m^2 - 1 = 0$ are greater than −2 but less than 4, lie in the interval

a. $-2 < m < 0$ b. $m > 3$
c. $-1 < m < 3$ d. $1 < m < 4$

69. If the roots of the quadratic equation $(4p - p^2 - 5)x^2 - (2p - 1)x + 3p = 0$ lie on either side of unity, then the number of integral values of p is

a. 1 b. 2
c. 3 d. 4

70. The interval of a for which $\tan^2 x - (a - 4)\tan x + 4 - 2a = 0$ has at least one solution $\forall x \in [0, \pi/4]$

a. $a \in (2, 3)$ b. $a \in [2, 3]$
c. $a \in (1, 4)$ d. $a \in [1, 4]$

71. The range of a for which the equation $x^2 + ax - 4 = 0$ has its smaller root in the interval $(-1, 2)$ is

a. $(-\infty, -3)$ b. $(0, 3)$
c. $(0, \infty)$ d. $(-\infty, -3) \cup (0, \infty)$

72. If both roots of the equation $ax^2 + x + c - a = 0$ are imaginary and $c > -1$, then

a. $3a > 2 + 4c$ b. $3a < 2 + 4c$
c. $c < a$ d. none of these

73. The set of all possible real values of a such that the inequality $(x - (a - 1))(x - (a^2 + 2)) < 0$ holds for all $x \in (-1, 3)$ is

a. $(0, 1)$ b. $(\infty, -1]$
c. $(-\infty, -1)$ d. $(1, \infty)$

74. Consider the equation $x^2 + 2x - n = 0$, where $n \in N$ and $n \in [5, 100]$. Total number of different values of 'n' so that the given equation has integral roots is

a. 8 b. 3
c. 6 d. 4

75. Total number of values of a so that $x^2 - x - a = 0$ has integral roots, where $a \in N$ and $6 \leq a \leq 100$, is equal to

a. 2 b. 4
c. 6 d. 8

76. Total number of integral values of a so that $x^2 - (a + 1)x + a - 1 = 0$ has integral roots is equal to

a. 1 b. 2
c. 4 d. none of these

77. The number of values of k for which $[x^2 - (k - 2)x + k^2] \times [x^2 + kx + (2k - 1)]$ is a perfect square is

a. 2 b. 1
c. 0 d. none of these

78. If α, β are the roots of $x^2 + px + q = 0$ and $x^{2n} + p^nx^n + q^n = 0$ and if (α/β), (β/α) are the roots of $x^n + 1 + (x + 1)^n = 0$, then $n(\in N)$

a. must be an odd integer
b. may be any integer
c. must be an even integer
d. cannot say anything

79. The number of positive integral solutions of $x^4 - y^4 = 3789108$ is

a. 0 **b.** 1
c. 2 **d.** 4

80. If $xy = 2(x + y)$, $x \le y$ and $x, y \in N$, then the number of solutions of the equation are

a. two **b.** three
c. no solution **d.** infinitely many solutions

81. If α, β, γ are such that $\alpha + \beta + \gamma = 2$, $\alpha^2 + \beta^2 + \gamma^2 = 6$, $\alpha^3 + \beta^3 + \gamma^3 = 8$, then $\alpha^4 + \beta^4 + \gamma^4$ is

a. 18 **b.** 10
c. 15 **d.** 36

82. The number of integral values of a for which the quadratic equation $(x + a)(x + 1991) + 1 = 0$ has integral roots are

a. 3 **b.** 0
c. 1 **d.** 2

83. The number of real solutions of the equation $(9/10)^x = -3 + x - x^2$ is

a. 2 **b.** 0
c. 1 **d.** none of these

84. If a, b and c are real numbers such that $a^2 + b^2 + c^2 = 1$, then $ab + bc + ca$ lies in the interval

a. $[1/2, 2]$ **b.** $[-1, 2]$
c. $[-1/2, 1]$ **d.** $[-1, 1/2]$

85. If the equation $\cot^4 x - 2 \operatorname{cosec}^2 x + a^2 = 0$ has at least one solution then, sum of all possible integral values of a is equal to

a. 4 **b.** 3
c. 2 **d.** 0

86. Let x, y, z, t be real numbers $x^2 + y^2 = 9$, $z^2 + t^2 = 4$ and $xt - yz = 6$. Then the greatest value of $P = xz$ is

a. 2 **b.** 3
c. 4 **d.** 6

87. If a, b, c be distinct positive numbers, then the nature of roots of the equation $1/(x - a) + 1/(x - b) + 1/(x - c) = 1/x$ is

a. all real and distinct
b. all real and at least two are distinct
c. at least two real
d. all non-real

88. If $(b^2 - 4ac)^2 (1 + 4a^2) < 64a^2$, $a < 0$, then maximum value of quadratic expression $ax^2 + bx + c$ is always less than

a. 0 **b.** 2
c. -1 **d.** -2

89. For $x^2 - (a + 3)|x| + 4 = 0$ to have real solutions, the range of a is

a. $(-\infty, -7] \cup [1, \infty)$ **b.** $(-3, \infty)$
c. $(-\infty, -7]$ **d.** $[1, \infty)$

90. If the quadratic equation $4x^2 - 2(a + c - 1)x + ac - b = 0$ $(a > b > c)$

a. both roots are greater than a
b. both roots are less than c
c. both roots lie between $c/2$ and $a/2$
d. exactly one of the roots lies between $c/2$ and $a/2$

91. If the equation $x^2 + ax + b = 0$ has distinct real roots and $x^2 + a|x| + b = 0$ has only one real root, then which of the following is true

a. $b = 0, a > 0$ **b.** $b = 0, a < 0$
c. $b > 0, a < 0$ **d.** $b < 0, a > 0$

92. The equation $2^{2x} + (a - 1)2^{x+1} + a = 0$ has roots of opposite signs then exhaustive set of values of a is

a. $a \in (-1, 0)$ **b.** $a < 0$
c. $a \in (-\infty, 1/3)$ **d.** $a \in (0, 1/3)$

93. If the equation $|x^2 + bx + c| = k$ has four real roots, then

a. $b^2 - 4c > 0$ and $0 < k < \dfrac{4c - b^2}{4}$

b. $b^2 - 4c < 0$ and $0 < k < \dfrac{4c - b^2}{4}$

c. $b^2 - 4c > 0$ and $k > \dfrac{4c - b^2}{4}$

d. none of these

94. $P(x)$ is a polynomial with integer coefficients such that for four distinct integers a, b, c, d; $P(a) = P(b) = P(c) = P(d) = 3$. If $P(e) = 5$ (e is an integer), then

a. $e = 1$ **b.** $e = 3$
c. $e = 4$ **d.** no real value of e

95. The number of integral values of x satisfying $\sqrt{-x^2 + 10x - 16} < x - 2$ is

a. 0 **b.** 1
c. 2 **d.** 3

96. If $x^2 + ax - 3x - (a + 2) = 0$ has real and distinct roots, then minimum value of $(a^2 + 1)/(a^2 + 2)$ is

a. 1 **b.** 0
c. $\dfrac{1}{2}$ **d.** $\dfrac{1}{4}$

97. The set of values of a for which $(a - 1)x^2 - (a + 1)x + a - 1 \ge 0$ is true for all $x \ge 2$

a. $(-\infty, 1)$ **b.** $\left(1, \dfrac{7}{3}\right)$
c. $\left(\dfrac{7}{3}, \infty\right)$ **d.** none of these

98. The value of the expression $x^4 - 8x^3 + 18x^2 - 8x + 2$ when $x = 2 + \sqrt{3}$

a. 2 **b.** 1
c. 0 **d.** 3

Multiple Correct Answers Type

Each question has four choices a, b, c and d, out of which *one or more* answers are correct.

1. If $x, y \in R$ and $2x^2 + 6xy + 5y^2 = 1$, then

a. $|x| \leq \sqrt{5}$ b. $|x| \geq \sqrt{5}$

c. $y^2 \leq 2$ d. $y^2 \leq 4$

2. If the equation whose roots are the squares of the roots of the cubic $x^3 - ax^2 + bx - 1 = 0$ is identical with the given cubic equation, then

a. $a = 0, b = 3$

b. $= b = 0$

c. $a = b = 3$

d. a, b are roots of $x^2 + x + 2 = 0$

3. If the equation $ax^2 + bx + c = 0$, $a, b, c \in R$ have non-real roots, then

a. $c(a - b + c) > 0$ b. $c(a + b + c) > 0$

c. $c(4a - 2b + c) > 0$ d. none of these

4. If $c \neq 0$ and the equation $p(2x) = a(x + c) + b(x - c)$ has two equal roots, then p can be

a. $(\sqrt{a} - \sqrt{b})^2$ b. $(\sqrt{a} + \sqrt{b})^2$

c. $a + b$ d. $a - b$

5. If the equations $4x^2 - x - 1 = 0$ and $3x^2 + (\lambda + \mu)x + \lambda - \mu = 0$ have a root common then the rational values of λ and μ are

a. $\lambda = \frac{-3}{4}$ b. $\lambda = 0$

c. $\mu = \frac{3}{4}$ d. $\mu = 0$

6. If the equation $ax^2 + bx + c = 0$ $(a > 0)$ has two real roots α and β such that $\alpha < -2$ and $\beta > 2$, then which of the following statements is/are true?

a. $a - |b| + c < 0$ b. $c < 0, b^2 - 4ac > 0$

c. $4a - 2|b| + c < 0$ d. $9a - 3|b| + c < 0$

7. If the following figure shows the graph of $f(x) = ax^2 + bx + c$, then

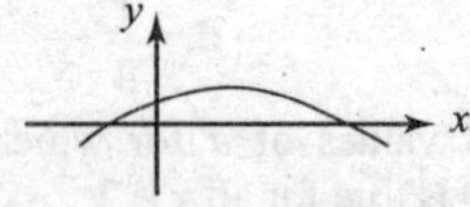

Fig. 1.1

a. $ac < 0$ b. $bc > 0$

c. $ab > 0$ d. $abc < 0$

8. If $\cos x - y^2 - \sqrt{y - x^2 - 1} \geq 0$, then

a. $y \geq 1$ b. $x \in R$

c. $y = 1$ d. $= 0$

9. The value of x satisfying the equation $2^{2x} - 8 \times 2^x = -12$ is

a. $1 + \frac{\log 3}{\log 2}$ b. $\frac{1}{2}\log 6$

c. $1 + \log\frac{3}{2}$ d. 1

10. If α, β are the roots of the quadratic equation $ax^2 + bx + c = 0$ then which of the following expression will be the symmetric function of roots

a. $\left|\log\frac{\alpha}{\beta}\right|$ b. $\alpha^2\beta^5 + \beta^2\alpha^5$

c. $\tan(\alpha - \beta)$ d. $\left(\log\frac{1}{\alpha}\right)^2 + (\log\beta)^2$

11. If the quadratic equation $ax^2 + bx + c = 0$ $(a > 0)$ has $\sec^2\theta$ and $\operatorname{cosec}^2\theta$ as its roots, then which of the following must hold good?

a. $b + c = 0$ b. $b^2 - 4ac \geq 0$

c. $c \geq 4a$ d. $4a + b \geq 0$

12. Let $a, b, c \in Q^+$ satisfying $a > b > c$. Which of the following statement(s) hold true for the quadratic polynomial $f(x) = (a + b - 2c)x^2 + (b + c - 2a)x + (c + a - 2b)$?

a. The mouth of the parabola $y = f(x)$ opens upwards

b. Both roots of the equation $f(x) = 0$ are rational

c. x-coordinate of vertex of the graph is positive

d. Product of the roots is always negative

13. The graph of the quadratic trinomial $y = ax^2 + bx + c$ has its vertex at $(4, -5)$ and two x-intercepts one positive and one negative. Which of the following holds good?

a. $a > 0$ b. $b < 0$

c. $c < 0$ d. $8a = b$

14. If the roots of the equation, $x^3 + px^2 + qx - 1 = 0$ form an increasing G.P., where p and q are real, then

a. $p + q = 0$

b. $p \in (-3, \infty)$

c. one of the root is unity

d. one root is smaller than 1 and one root is greater than 1

15. If $(\sin\alpha)x^2 - 2x + b \geq 2$, for all real values of $x \leq 1$ and $\alpha \in (0, \pi/2) \cup (\pi/2, \pi)$, then possible real values of b is/are

a. 2 b. 3

c. 4 d. 5

16. If every pair from among the equations $x^2 + ax + bc = 0$, $x^2 + bx + ca = 0$, and $x^2 + cx + ab = 0$ has a common root, then

a. the sum of the three common roots is $(-1)/2\,(a + b + c)$

b. the sum of the three common roots is $2(a + b + c)$

c. the product of the three common roots is abc

d. the product of the three common roots is $a^2b^2c^2$

17. If a, b, c are in G.P., then the roots of the equation $ax^2 + bx + c = 0$ are in the ratio

a. $\frac{1}{2}(-1 + i\sqrt{3})$ b. $\frac{1}{2}(1 - i\sqrt{3})$

c. $\frac{1}{2}(-1 - i\sqrt{3})$ d. $\frac{1}{2}(1 + i\sqrt{3})$

18. If $ax^2 + (b - c)x + a - b - c = 0$ has unequal real roots for all $c \in R$, then

a. $b < 0 < a$ b. $a < 0 < b$

c. $b < a < 0$ d. $b > a > 0$

19. Given that α, γ are roots of the equation $Ax^2 - 4x + 1 = 0$, and β, δ the roots of the equation of $Bx^2 - 6x + 1 = 0$, such that α, β, γ, and δ are in H.P., then

a. $a = 3$ **b.** $a = 4$
c. $b = 2$ **d.** $b = 8$

20. If the equations $x^2 + px + q = 0$ and $x^2 + p'x + q' = 0$ have a common root, then show that it must be equal to

a. $\frac{pq' - p'q}{q - q'}$ **b.** $\frac{q - q'}{p' - p}$
c. $\frac{p' - p}{q - q'}$ **d.** $\frac{pq' - p'q}{p - p'}$

21. If $x^3 + 3x^2 - 9x + c$ is of the form $(x - \alpha)^2 (x - \beta)$, then c is equal to

a. 27 **b.** −27
c. 5 **d.** −5

22. If the equations $x^2 + bx - a = 0$ and $x^2 - ax + b = 0$ have a common root, then

a. $a + b = 0$ **b.** $a = b$
c. $a - b = 1$ **d.** $a + b = 1$

23. If $(x^2 + ax + 3)/(x^2 + x + a)$ takes all real values for possible real values of x, then

a. $4a^3 + 39 < 0$ **b.** $4a^3 + 39 \geq 0$
c. $a \geq \frac{1}{4}$ **d.** $a < \frac{1}{4}$

24. If $\cos^4 \theta + \alpha$, $\sin^4 \theta + \alpha$ are the roots of the equation $x^2 + 2bx + b = 0$ and $\cos^2 \theta + \beta$, $\sin^2 \theta + \beta$ are the roots of the equation $x^2 + 4x + 2 = 0$, then the values of b are

a. 2 **b.** −1
c. −2 **d.** 1

25. If the roots of the equation $x^2 + ax + b = 0$ are c and d, then roots of the equation $x^2 + (2c + a)x + c^2 + ac + b = 0$ are

a. c **b.** $d - c$
c. $2c$ **d.** 0

26. If a, b, $c \in R$ and $abc < 0$, then the equation $bcx^2 + 2(b + c - a)x + a = 0$, has

a. both positive roots
b. both negative roots
c. real roots
d. one positive and one negative root

27. For the quadratic equation $x^2 + 2(a + 1)x + 9a - 5 = 0$, which of the following is are true?

a. If $2 < a < 5$, then roots are of opposite sign.
b. If $a < 0$, then roots are of opposite sign.
c. If $a > 7$, then both roots are negative.
d. If $2 \leq a \leq 5$, then roots are unreal.

28. Let $P(x) = x^2 + bx + c$, where b and c are integers. If $P(x)$ is a factor of both $x^4 + 6x^2 + 25$ and $3x^4 + 4x^2 + 28x + 5$, then

a. $P(x) = 0$ has imaginary roots
b. $P(x) = 0$ has roots of opposite sign
c. $P(1) = 4$
d. $P(1) = 6$

30. If $|ax^2 + bx + c| \leq 1$ for all x in [0, 1], then

a. $|a| \leq 8$ **b.** $|b| > 8$
c. $|c| \leq 1$ **d.** $|a| + |b| + |c| \leq 17$

31. Let $f(x) = ax^2 + bx + c$. Consider the adjacent diagram. Then

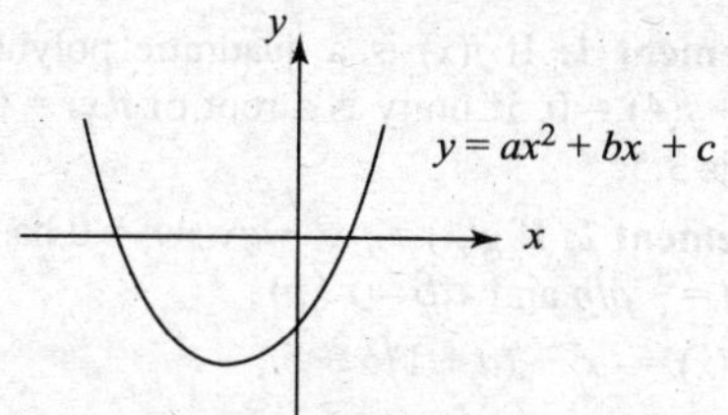

Fig. 1.2

a. $c < 0$ **b.** $b > 0$
c. $a + b - c > 0$ **d.** $abc < 0$

Reasoning Type

Each question has four choices a, b, c and d, out of which *only one* is correct. Each question contains Statement 1 and Statement 2.

a. Both the statements are True and Statement 2 is the correct explanation of Statement 1.
b. Both the statements are True but Statement 2 is NOT the correct explanation of Statement 1.
c. Statement 1 is True and Statement 2 is False.
d. Statement 1 is False and Statement 2 is True.

1. Consider the function $f(x) = \ln (ax^3 + (a + b) x^2 + (b + c) x + c)$.

Statement 1: Domain of the functions is $(-1, \infty) \sim \{-(b/*2a)\}$, where $a > 0$, $b^2 - 4ac = 0$.

Statement 2: $ax^2 + bx + c = 0$ has equal roots when $b^2 - 4ac = 0$.

2. Statement 1: If $a > 0$ and $b^2 - ac < 0$, then domain of the function $f(x) = \sqrt{ax^2 + 2bx + c}$ is R.

Statement 2: If $b^2 - ac < 0$, then $ax^2 + 2bx + c = 0$ has imaginary roots.

3. Statement 1: If the equations $ax^2 + bx + c = 0$ and $x^2 - 3x + 4 = 0$ have exactly one root common, then at least one of a, b, c is imaginary.

Statement 2: If a, b, c are not all real, then $ax^2 + bx + c = 0$ can have one root real and one root imaginary.

4. Statement 1: If $\cos^2 \pi/8$ is a root of the equation $x^2 + ax + b = 0$ where $a, b \in$, then ordered pair (a, b) is $[-1, (1/8)]$.

Statement 2: If $a + mb = 0$ and m is irrational, then $a, b = 0$.

5. **Statement 1:** If $a^2 + b^2 + c^2 < 0$, then if roots of the equation $ax^2 + bx + c = 0$ are imaginary, then they are not complex conjugates.

   Statement 2: Equation $ax^2 + bx + c = 0$ has complex conjugate roots when a, b, c are real.

6. **Statement 1:** Equation $ix^2+(i-1)x-(1/4)-i=0$ has imaginary roots.

   Statement 2: If $a = i$, $b = i - 1$ and $c=-(1/5)-i$, then $b^2 - 4ac < 0$.

7. **Statement 1:** If $f(x)$ is a quadratic polynomial satisfying $f(2) + f(4) = 0$. If unity is a root of $f(x) = 0$, then the other root is 3.5.

   Statement 2: If $g(x) = px^2 + qx + r = 0$ has roots α, β, then $\alpha+\beta=-q/p$ and $\alpha\beta=(r/p)$.

8. Let $f(x) = -x^2 + (a + 1)x + 5$.

   Statement 1: $f(x)$ is positive for some $\alpha<x<\beta$ and for all $a \in R$.

   Statement 2: $f(x)$ is always positive for all $x \in R$ and for same real a.

9. Let a, b, c be real such that $ax^2 + bx + c = 0$ and $x^2 + x + 1 = 0$ have a common root.

   Statement 1: $a = b = c$

   Statement 2: Two quadratic equations with real coefficients cannot have only one imaginary root common.

10. **Statement 1:** The equation $(x-p)(x-r) + \lambda(x-q)(x-s) = 0$, where $p<q<r<s$, has non-real roots.

    Statement 2: The equation $px^2 + qx + r = 0$ $(p, q, r \in R)$ has non-real roots if $q^2 - 4pr < 0$.

11. **Statement 1:** If $px^2 + qx + r = 0$ is a quadratic equation $(p, q, r \in R)$ such that its roots are α, β and $p + q + r < 0$, $p - q + r < 0$ and $r > 0$, then $[\alpha] + [\beta] = -1$, where $[\cdot]$ denotes greatest integer function.

    Statement 2: If for any two real numbers a and b, function $f(x)$ is such that $f(a) f(b) < 0 \Rightarrow f(x)$ has at least one real root lying in (a, b).

12. **Statement 1:** If $0<\alpha<(\pi/4)$, then the equation $(x - \sin\alpha)(x - \cos\alpha) - 2 = 0$ has both roots in $(\sin\alpha, \cos\alpha)$.

    Statement 2: If $f(a)$ and $f(b)$ possess opposite signs, then there exist at least one solution of the equation $f(x) = 0$ in open interval (a, b).

13. **Statement 1:** If all real values of x obtained from the equation $4^x - (a-3)2^x + (a-4) = 0$ are non-positive, then $a \in (4, 5]$.

    Statement 2: If $ax^2 + bx + c$ is non-positive for all real values of x, then $b^2 - 4ac$ must be negative or zero and a must be negative.

14. **Statement 1:** If $(a^2-4)x^2 + (a^2-3a+2)x + (a^2-7a+10) = 0$ is an identity, then the value of a is 2.

    Statement 2: If $a - b = 0$, then $ax^2 + bx + c = 0$ is an identity.

15. **Statement 1:** If the roots of $x^5 - 40x^4 + Px^3 + Qx^2 + Rx + S = 0$ are in G.P. and sum of their reciprocal is 10, then $|S| = 64$.

    Statement 2: $x_1 x_2 x_3 x_4 x_5 = -S$, where x_1, x_2, x_3, x_4, x_5 are the roots of the given equation.

16. **Statement 1:** If a, b, c, a_1, b_1, c_1 are rational and equations $ax^2 + 2bx + c = 0$ and $a_1x^2 + 2b_1x + c_1 = 0$ have one and only one root in common, then both $b^2 - ac$ and $b_1^2 - a_1c_1$ must be perfect squares.

    Statement 2: If two quadratic equations with rational coefficients have a common irrational root $p+\sqrt{q}$, then both roots will be common.

17. **Statement 1:** If $a, b, c \in Z$ and $ax^2 + bx + c = 0$ has an irrational root, then $|f(\lambda)|\geq 1/q^2$, where $\lambda\in Q=p/q$ and $f(x) = ax^2 + bx + c$.

    Statement 2: If $a, b, c \in Q$ and $b^2 - 4ac$ is positive but not a perfect square, then roots of equation $ax^2 + bx + c = 0$ are irrational and always occur in conjugate pair like $2+\sqrt{3}$ and $2-\sqrt{3}$.

18. **Statement 1:** The number of values of a for which $(a^2-3a+2)x^2 + (a^2-5a+6)x + a^2 - 4 = 0$ is an identity in x is 2.

    Statement 2: If $a = b = c = 0$, then equation $ax^2 + bx + c = 0$ is an identity in x.

19. **Statement 1:** If roots of the equation $x^2 - bx + c = 0$ are two consecutive integers, then $b^2 - 4c = 1$.

    Statement 2: If a, b, c are odd integer, then the roots of the equation $4abc\,x^2 + (b^2 - 4ac)x - b = 0$ are real and distinct.

20. Let $ax^2 + bx + c = 0$, $a \neq 0$ $(a, b, c \in R)$ has no real roots and $a + b + 2c = 2$.

    Statement 1: $ax^2 + bx + c > 0, \forall x \in R$.

    Statement 2: $a + b$ is positive.

21. Consider a general expression of degree 2 in two variables as $f(x, y) = 5x^2 + 2y^2 - 2xy - 6x - 6y + 9$.

    Statement 1: $f(x, y)$ can be resolved into two linear factors over real coefficients.

    Statement 2: If we compare $f(x, y)$ with $ax^2 + by^2 + 2hxy + 2gx + 2fy + c = 0$, discriminant of $f(x, y)$, i.e., $abc + 2fgh - af^2 - bg^2 - ch^2 = 0$.

22. **Statement 1:** The equation $x^2 + (2m+1)x + (2n+1) = 0$, where m and n are integers, cannot have any rational roots.

    Statement 2: The quantity $(2m+1)^2 - 4(2n+1)$, where $m, n \in I$ can never be a perfect square.

Linked Comprehension Type

Based upon each paragraph, some multiple choice questions have to be answered. Each question has four choices a, b, c and d, out of which *only one* is correct.

For Problems 1–3

Consider an unknown polynomial which when divided by $(x-3)$ and by $(x-4)$ leaves remainders as 2 and 1, respectively. Let $R(x)$ be the remainder when this polynomial is divided by $(x-3)(x-4)$.

1. If equation $R(x) = x^2 + ax + 1$ has two distinct real roots, then values of a are

a. $(-2, 2)$ b. $(-\infty, -2) \cup (2, \infty)$
c. $(-2, \infty)$ d. all real numbers

2. If $R(x) = px^2 + (q - 1)x + 6$ has no distinct real roots and $p > 0$, then least value of $3p + q$ is
a. -2 b. $2/3$
c. $-1/3$ d. none of these

3. Range of $f(x) = [R(x)]/(x^2 - 3x + 2)$ is
a. $[-2, 2]$
b. $(-\infty, -2-\sqrt{3}] \cup [-2+\sqrt{3}, \infty)$
c. $(-\infty, -7-4\sqrt{3}] \cup [-7+4\sqrt{3}, \infty)$
d. none of these

For Problems 4–6

Consider the quadratic equation $ax^2 - bx + c = 0$, $a, b, c \in N$, which has two distinct real roots belonging to the interval $(1, 2)$.

4. The least value of a is
a. 4 b. 6
c. 7 d. 5

5. The least value of b is
a. 10 b. 11
c. 13 d. 15

6. The least value of c is
a. 4 b. 6
c. 7 d. 5

For Problems 7–9

Consider the equation $x^4 + 2ax^3 + x^2 + 2ax + 1 = 0$, where $a \in R$. Also range of function $f(x) = x + (1/x)$ is $(-\infty, -2] \cup [2, \infty)$.

7. If equation has at least two distinct positive real roots then all possible values of a are
a. $(-\infty, -1/4)$ b. $(5/4, \infty)$
c. $(-\infty, -3/4)$ d. none of these

8. If equation has at least two distinct negative real roots, then all possible values of a are
a. $(3/4, \infty)$ b. $(-5/4, \infty)$
c. $(-\infty, 1/4)$ d. none of these

9. If exactly two roots are positive and two roots are negative, then number of integral values of a is
a. 2 b. 1
c. 0 d. 3

For Problems 10–12

Let $f(x) = x^2 + b_1x + c_1$, $g(x) = x^2 + b_2x + c_2$. Let the real roots of $f(x) = 0$ be α, β and real roots of $g(x) = 0$ be $\alpha + h, \beta + h$. The least value of $f(x)$ is $-1/4$. The least value of $g(x)$ occurs at $x = 7/2$

10. The least value of $g(x)$ is
a. $-\frac{1}{4}$ b. -1
c. $-\frac{1}{3}$ d. $-\frac{1}{2}$

11. The value of b_2 is
a. -5 b. 9
c. -8 d. -7

12. The roots of $f(x) = 0$ are
a. $3, -4$ b. $-3, 4$
c. $3, 4$ d. $-3, -4$

For Problems 13–15

In the given figure, vertices of ΔABC lie on $y = f(x) = ax^2 + bx + c$. The ΔABC is rightangled isosceles triangle whose hypotenuse $AC = 4\sqrt{2}$ units.

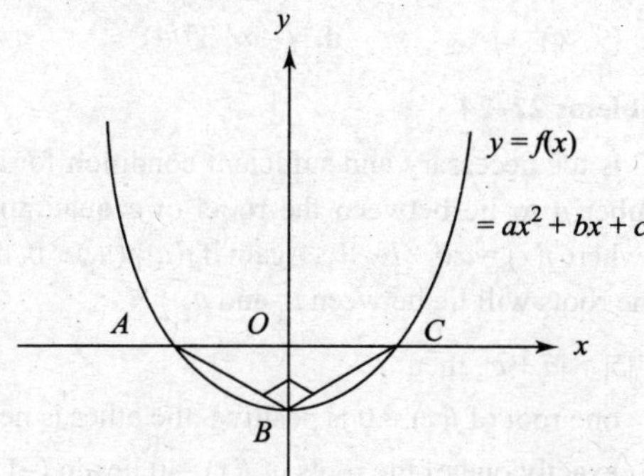

Fig. 1.3

13. $y = f(x)$ is given by
a. $y = x^2 - 2\sqrt{2}$ b. $y = x^2 - 12$
c. $y = \frac{x^2}{2} - 2$ d. $y = \frac{x^2}{2\sqrt{2}} - 2\sqrt{2}$

14. Minimum value of $y = f(x)$ is
a. -4 b. -2
c. $-2\sqrt{2}$ d. none of these

15. Number of integral values of k for which one root of $f(x) = 0$ is more than k and other less than k
a. 6 b. 4
c. 5 d. 7

For Problems 16–18

Consider the inequality $9^x - a3^x - a + 3 \le 0$, where a is a real parameter.

16. The given inequality has at least one negative solution for $a \in$
a. $(-\infty, 2)$ b. $(3, \infty)$
c. $(-2, \infty)$ d. $(2, 3)$

17. The given inequality has at least one positive solution for $a \in$
a. $(-\infty, -2)$ b. $[3, \infty)$
c. $(2, \infty)$ d. $[-2, \infty)$

18. The given inequality has at least one real solution for $a \in$
a. $(-\infty, 3)$ b. $[2, \infty)$
c. $(3, \infty)$ d. $[-2, \infty)$

For Problems 19–21

Consider the inequation $x^2 + x + a - 9 < 0$.

19. The values of the real parameter a so that the given inequation has at least one positive solution:

a. $(-\infty, 37/4)$ b. $(-\infty, \infty)$
c. $(3, \infty)$ d. $(-\infty, 9)$

20. The values of the real parameter a so that the given inequation has at least one negative solution.

a. $(-\infty, 9)$ b. $(37/4, \infty)$
c. $[9, 37/4)$ d. none of these

21. The values of the real parameter a so that the given inequation is true, $\forall\, x \in (-1, 3)$:

a. $(-\infty, -3)$ b. $(-3, \infty)$
c. $[9, \infty)$ d. $(-\infty, 37/4)$

For Problems 22–24

$af(\mu) < 0$ is the necessary and sufficient condition for a particular real number μ to lie between the roots of a quadratic equation $f(x) = 0$, where $f(x) = ax^2 + bx + c$. Again if $f(\mu_1)f(\mu_2) < 0$, then exactly one of the roots will lie between μ_1 and μ_2.

22. If $|b| > |a + c|$, then

a. one root of $f(x) = 0$ is positive, the other is negative
b. exactly one of the roots of $f(x) = 0$ lies in $(-1, 1)$
c. 1 lies between the roots of $f(x) = 0$
d. both the roots of $f(x) = 0$ are less than 1

23. If $a(a + b + c) < 0 < (a + b + c)c$, then

a. one root is less than 0, the other is greater than 1
b. exactly one of the roots lies in $(0, 1)$
c. both the roots lie in $(0, 1)$
d. at least one of the roots lies in $(0, 1)$

24. If $(a + b + c)c < 0 < a(a + b + c)$, then

a. one root is less than 0, the other is greater than 1
b. one root lies in $(-\infty, 0)$ and other in $(0, 1)$
c. both roots lie in $(0, 1)$
d. one root lies in $(0, 1)$ and other in $(1, \infty)$

For Problems 25–26

The real numbers x_1, x_2, x_3 satisfying the equation $x^3 - x^2 + \beta x + \gamma = 0$ are in A.P.

25. All possible values of β are

a. $\left(-\infty, \frac{1}{3}\right)$ b. $\left(-\infty, -\frac{1}{3}\right)$
c. $\left(\frac{1}{3}, \infty\right)$ d. $\left(-\frac{1}{3}, \infty\right)$

26. All possible values of γ are

a. $\left(-\frac{1}{9}, \infty\right)$ b. $\left(-\frac{1}{27}, +\infty\right)$
c. $\left(\frac{2}{9}, +\infty\right)$ d. none of these

Matrix-Match Type

Each question contains statements given in two columns, which have to be matched. Statements a, b, c, d in column I have to be matched with statements p, q, r, s in column II. If the correct matches are a → p, a → s, b → q, b → r, c → p, c → q and d → s, then the correctly bubbled 4 × 4 matrix should be as follows:

	p	q	r	s
a	(p) ●	(q)	(r)	(s) ●
b	(p)	(q) ●	(r) ●	(s)
c	(p) ●	(q) ●	(r)	(s)
d	(p)	(q)	(r)	(s) ●

1. Match the following for the equation $x^2 + a|x| + 1 = 0$, where a is a parameter.

Column I	Column II
a. No real roots	p. $a < -2$
b. Two real roots	q. ϕ
c. Three real roots	r. $a = -2$
d. Four distinct real roots	s. $a \geq 0$

2. Column I contains rational algebraic expression and column II contains possible integers, which lie in their range.

Column I	Column II
a. $y = \dfrac{x^2 - 2x + 4}{x^2 + 2x + 4}, x \in R$	p. 1
b. $y = \dfrac{x^2 - 3x - 2}{2x - 3}, x \in R$	q. 4
c. $y = \dfrac{2x^2 - 2x + 4}{x^2 - 4x + 3}, x \in R$	r. −3
d. $x^2 - (a - 3)x + 2 < 0, \forall\, x \in (-2, 3)$	s. −10

3.

Column I	Column II
a. If a, b, c and d are four zero real number such that $(d + a - b)^2 + (d + b - c)^2 = 0$ and the roots of the equation $a(b - c)x^2 + b(c - a)x + c(a - b) = 0$ are real and equal then	p. $a + b + c = 0$
b. If the roots of the equation $(a^2 + b^2)x^2 - 2b(a + c)x + (b^2 + c^2) = 0$ are real and equal, then	q. a, b, c are in A.P.

c. If the equation $ax^2 + bx + c = 0$ and $x^3 - 3x^2 + 3x - 1 = 0$ have a common real root, then	**r.** a, b, c are in G.P.
d. Let a, b, c be positive real numbers such that the expression $bx^2 + (\sqrt{(a+c)^2 + 4b^2})x + (a + c + (a + c)$ is non-negative $\forall\, x \in R$, then	**s.** a, b, c are in H.P.

4.

Column I (Number of positive integers for which)	Column II
a. One root is positive and the other is negative for the equation $(m - 2)x^2 - (8 - 2m)x - (8 - 3m) = 0$	**p.** 0
b. Exactly one root of equation $x^2 - m(2x - 8) - 15 = 0$ lies in interval (0, 1)	**q.** infinite
c. The equation $x^2 + 2(m + 1)x + 9m - 5 = 0$ has both roots negative	**r.** 1
d. The equation $x^2 + 2(m - 1)x + m + 5 = 0$ has both roots lying on either sides of 1	**s.** 2

5.

Column I	Column II
a. If $x^2 + ax + b = 0$ has roots α, β and $x^2 + px + q = 0$ has roots $-\alpha, \gamma$, then	**p.** $(1 - bq)^2 = (a - pb)(p - aq)$
b. If $x^2 + ax + b = 0$ has roots α, β and $x^2 + px + q = 0$ has roots $1/\alpha$, γ, then	**q.** $(4 - bq)^2 = (4a + 2pb)(-2p - aq)$
c. If $x^2 + ax + b = 0$ has roots α, β and $x^2 + px + q = 0$ has roots $-2/\alpha, \gamma$, then	**r.** $(1 - 4bq)^2 = (a + 2bp)(-2p - 4aq)$
d. If $x^2 + ax + b = 0$ has roots α, β and $x^2 + px + q = 0$ has roots $-1/2\alpha, \gamma$, then	**s.** $(q - b)^2 = (aq + bp)(p - a)$

Integer Type

1. The value of $\sqrt{6 + \sqrt{6 + \sqrt{6 + \ldots \infty}}}$ is ___.
2. If the expression $ax^4 + bx^3 - x^2 + 2x + 3$ has the remainder $4x + 3$ when divided by $x^2 + x - 2$, then the value of $a + 4b$ is ___.
3. Given that $x^2 + x - 6$ is a factor of $2x^4 + x^3 - ax + bx + a + b - 1$, then the value of $a - 3b$ is ___.
4. If $f(x) = ax^2 + bx + c$, $g(x) = -ax^2 + bx + c$, where $ac \neq 0$, then the minimum number of real roots of the equation $f(x)\, g(x) = 0$ is ___.
5. If α, β are the roots of $375\, x^2 - 25x - 2 = 0$ and $s_n = \alpha^n + \beta^n$, then the value of $1/\,(3 \lim\limits_{n\to\infty} \sum_{r=1}^{n} s_r)$ is ___.
6. If $x^2 + 3x + 5 = 0$ and $ax^2 + bx + c = 0$ have a common root and $a, b, c \in N$, then the minimum value of $a + b + c$ is ___.
7. If $ax^2 + bx + c = 0$ and $bx^2 + cx + a = 0$ have a common root and a, b, c are non-zero real numbers, then the value of $(a^3 + b^3 + c^3)/(abc)$ is ___.
8. If the equations $x^2 + ax + 12 = 0$, $x^2 + bx + 15 = 0$, and $x^2 + (a + b)x + 36 = 0$ have a common positive root, then the value of $|a - b|$ is ___.
9. The equations $ax^2 + bx + a = 0$ and $x^3 - 2x^2 + 2x - 1 = 0$ have two roots common. Then the value of $a + b$ is ___.
10. If α, β, γ are the roots of the equation $x^3 + 4x + 1 = 0$, then the value of $(\alpha + \beta)^{-1} + (\beta + \gamma)^{-1} + (\gamma + \alpha)^{-1}$ is ___.
11. The positive value of m for which the expression $2x^2 + mxy + 3y^2 - 5y - 2$ can be resolved into two rational linear factors is ___.
12. The least value of $(6x^2 - 22x + 21)/(5x^2 - 18x + 17)$ is ___.
13. The least value of n such that $(n - 2)\, x^2 + 8x + n + 4 > 0$, $\forall\, x \in R$, where $n \in N$ is ___.
14. If the inequality $(mx^2 + 3x + 4)/(x^2 + 2x + 2) < 5$ is satisfied for all $x \in R$, then the greatest value of $[m]$, where [.] represents the greatest integer function is ___.
15. If $ax^2 + bx - 18 = 0$ does not have distinct real roots, then the least value of $3a + b$. is ___.
16. If the expression $(ax^2 + 3x - 4)/(3x - 4x^2 + a)$ can take any real value, then the greatest value of a is ___.
17. Let a, b, c be real numbers such that $a + 2b + c = 4$. Then the maximum value of $(ab + bc + ca)$ is ___.
18. If $(x^2 + x^2 + 2)^2 - (a - 3)\,(x^2 + x + 1)\,(x^2 + x + 2) + (a - 4)\,(x^2 + x + 1)^2 = 0$ has at least one root, then the least integer just greater than the maximum value of a is ___.
19. The integral value of a so that the roots of the equation $x^2 - 2x - (a^2 - 1) = 0$ lie between the roots of the equation $x^2 - 2(a + 1)x + a(a - 1) = 0$ is ___.
20. The number of integral pairs (x, y) for which $2x^2 - 3xy - 2y^2 = 7$ is ___.
21. If k is the integral part of the greatest root of the equation $x^3 - 10x^2 - 11x - 100 = 0$, then the value of $k - 6$ is ___.
22. If x, y, z are distinct positive numbers such that $x + (1/y) = y + (1/z) = z + (1/x)$, then the value of xyz is ___.
23. The number of integers n such that the equation $nx^2 + (n + 1)\, x + (n + 2) = 0$ has rational roots only is ___.
24. If $x^2 + \frac{1}{x^2} = 7$, then the value of $\frac{1}{2}\left|x^3 + \frac{1}{x^3}\right|$ $(x > 0)$ is ___.
25. If a, b, c are real numbers such that $a^2 + 2b = 7$, $b^2 + 4c = -7$, and $c^2 + 6a = -14$, then the value of $(a^2 + b^2 + c^2)/2$ is ___.
26. Let a be a real number. Then the minimum number of real roots of the equation $(x^2 + ax + 1)\,(3x^2 + ax - 3) = 0$ is ___.
27. The least integral value of a for which all roots of the equation $x^4 - 4x^3 - 8x^2 + a = 0$ are real is ___.
28. Let S be a square of unit area. Consider any quadrilateral which has one vertex on each side of S. If a, b, c, and d denote the lengths of the sides of the quadrilateral, then the greatest value of $a^2 + b^2 + c^2 + d^2$ is ___.

29. If $x^2 + (a - b)x + (1 - a - b) = 0$, where $a, b \in R$, then the least integral value of a for which the equation has unequal real roots for all values of b is _______.

30. The sum of all real roots of the equation $|x - 2|^2 + |x - 2| - 2 = 0$ is _______.

31. For the equation $3x^2 + px + 3 = 0, p > 0$, if one of the roots is the square of the other, then the value of p is ___.

32. Let α, β be the roots of $x^2 - x + p = 0$ and γ, δ be the roots of $x^2 - 4x + q = 0$. If $\alpha, \beta, \gamma, \delta$ are in G.P., then the value of $a/(2p)$ is _______.

33. Let a, b be the roots of the equation $x^2 - 10cx - 11d = 0$ and c, d be those of $x^2 - 10ax - 11b = 0$.
If $a + b + c + d = p^2q$, where p is a prime number, then the value of $p - q$ $(a \neq b \neq c \neq d)$ is _______.

34. The complex numbers $1 + i$ and $1 + 2i$ are both roots of the equation $x^5 - 6x^4 + Ax^3 + Bx^2 + Cx + D = 0$, where $A, B, C, D \in R$. The value of $|D/4|$ is _______.

35. p, q, r are integers such that $f(x) = x^4 + 4x^3 + 6px^2 + 4qx + r$, is divisible by $(x^3 + 3x^2 + 9x + 3)$. The value of $(p + q + r)$ is equal to _______.

36. Let $a > 2, a \in N$ be a constant. If there are just 18 positive integers satisfying the inequality $(x - a)(x - 2a)(x - a^2) < 0$, then the value of a is _______.

37. Let x, y, z be non-negative integers such $x + y + z = 12$. Then the maximum value of $(xyz + xy + yz + zx)/16$ is _______.

38. The possible value of a for which $\dfrac{x^3 - 6x^2 + 11x - 6}{x^3 + x^2 - 10x + 8} + \dfrac{a}{30} = 0$ does not have a real solution is _______.

39. Suppose a cubic polynomial $f(x) = x^3 + px^2 + qx + 72$ is divisible by both $x^2 + ax + b$ and $x^2 + bx + a$ (where a, b, p, q are constants and $a \neq b$), then the value of p is _______.

ANSWERS AND SOLUTIONS

Objective Type

1. b. Given,

$x - 2 = 2^{2/3} + 2^{1/3}$

Cubing both sides, we get

$(x-2)^3 = 2^2 + 2 + 3 \times 2^{2/3} \times 2^{1/3}(x-2) = 6 + 6(x-2)$

or

$x^3 - 6x^2 + 12x - 8 = -6 + 6x$

$\therefore x^3 - 6x^2 + 6x = 2$

2. a. Let, $f(x, y, z) = x^2 + 4y^2 + 3z^2 - 2x - 12y - 6z + 14$

$= (x-1)^2 + (2y-3)^2 + 3(z-1)^2 + 1$

For the least value of $f(x, y, z)$,

$x - 1 = 0$. $2y - 3 = 0$ and $z - 1 = 0$

$\therefore x = 1, y = 3/2, z = 1$

Hence the least value of $f(x, y, z)$ is $f(1, 3/2, 1) = 1$.

3. c. Let m be a positive integer for which

$n^2 + 96 = m^2$

$\Rightarrow m^2 - n^2 = 96 \Rightarrow (m+n)(m-n) = 96$

$\Rightarrow (m+n)\{(m+n) - 2n\} = 96$

$\Rightarrow m + n$ and $m - n$ must be both even

As $96 = 2 \times 48$ or 4×24 or 6×16 or 8×12, number of solutions is 4.

4. d. $f(x, y) = (x-2)^2 + (y-1)^2 = 0$

$\Rightarrow x = 2$ and $y = 1$

$$\therefore E = \frac{(\sqrt{2}-1)^2 + 4\sqrt{2}}{2+\sqrt{2}} = \frac{(\sqrt{2}+1)^2}{\sqrt{2}(\sqrt{2}+1)} = \frac{\sqrt{2}+1}{\sqrt{2}}$$

5. c. We have,

$|x|^2 - 3|x| + 2 = 0$

$\Rightarrow (|x|-1)(|x|-2) = 0$

$\Rightarrow |x| = 1, 2 \Rightarrow x = \pm 1, \pm 2$

6. c. Clearly, $x = -1$ satisfies the equation.

7. d. Given equation is satisfied by $x = 1, 2, 3$. But for $x = 1$, $\sqrt{x-2}$ is not defined. Hence, number of roots is 2 and the roots are $x = 2$ and 3.

8. c. As $(\lambda + 1)x^2 + 2 = \lambda x + 3$ has only one solution, so

$D = 0$

$\Rightarrow \lambda^2 - 4(\lambda+1)(-1) = 0$

$\Rightarrow \lambda^2 + 4\lambda + 4 = 0$

$\Rightarrow (\lambda + 2)^2 = 0$

$\therefore \lambda = -2$

9. a. Given quadratic expression is

$x^2 + 2(a+b+c)x + 3(bc+ca+ab)$.

This quadratic expression will be a perfect square if the discriminant of its corresponding equation is zero. Hence,

$4(a+b+c)^2 - 4 \times 3(bc+ca+ab) = 0$

$\Rightarrow (a+b+c)^2 - 3(bc+ca+ab) = 0$

$\Rightarrow a^2 + b^2 + c^2 + 2ab + 2bc + 2ca - 3(bc+ca+ab) = 0$

$\Rightarrow a^2 + b^2 + c^2 - ab\ bc - ca = 0$

$$\Rightarrow \frac{1}{2}[2a^2 + 2b^2 + 2c^2 - 2ab - 2bc - 2ca] = 0$$

$$\Rightarrow \frac{1}{2}[(a^2+b^2-2ab)+(b^2+c^2-2bc)+(c^2+a^2-2ca)]=0$$

$$\Rightarrow \frac{1}{2}[(a-b)^2+(b-c)^2+(c-a)^2]=0,$$

which is possible only when $(a-b)^2 = 0$, $(b-c)^2 = 0$ and $(c-a)^2 = 0$, i.e., $a = b = c$.

10. c. If one root is square of the other root of the equation $ax^2 + bx + c = 0$, then

$\beta = \alpha^2 \Rightarrow \alpha^2 + \alpha = -b/a$ and $\alpha^2 a = c/a$

By eliminating α, we get

$b^3 + ac^2 + a^2c = 3abc$,

which can be written in the form $a(c-b)^3 = c(a-b)^3$.

Alternative solution:

Let the roots be 2 and 4. Then the equation is

$x^2 - 6x + 8 = 0$.

Here obviously,

$$X = \frac{a(c-b)^3}{c} = \frac{1(14)^3}{8} = \frac{14}{2} \times \frac{14}{2} \times \frac{14}{2} = 7^3,$$

which is given by $(a-b)^3 = 7^3$.

11. a. Given that $x^2 + px + 1$ is a factor of $ax^3 + bx + c = 0$. Then let $ax^3 + bx + c = (x^2 + px + 1)(ax + \lambda)$, where λ is a constant. Then equating the coefficients of like powers of x on both sides, we get

$0 = ap + \lambda$, $b = p\lambda + a$, $c = \lambda$

$$\Rightarrow p = -\frac{\lambda}{a} = -\frac{c}{a}$$

Hence,

$$b = \left(-\frac{c}{a}\right)c + a$$

or

$ab = a^2 - c^2$

12. a. Put $x^2 + x = y$, so that Eq. (1) becomes

$(y-2)(y-3) = 12$

$\Rightarrow y^2 - 5y - 6 = 0$

$\Rightarrow (y-6)(y+1) = 0 \Rightarrow y = 6, -1$

When $y = 6$, we get

$x^2 + x - 6 = 0$

$\Rightarrow (x+3)(x-2) = 0$ or $x = -3, 2$

When $y = -1$, we get

$x^2 + x + 1 = 0$

$\Rightarrow x = \omega, \omega^2$

and their sum is -1.

13. b. Given, $(ax^2 + c)y + (a'x^2 + c') = 0$

or $x^2(ay + a') + (cy + c') = 0$.

If x is rational, then the discriminant of the above equation must be a perfect square. Hence,

$0 - 4(ay + a')(cy + c')$ must be a perfect square

$\Rightarrow -acy^2 - (ac' + a'c)y - a'c'$ must be a perfect square

$\Rightarrow (ac' + a'c)^2 - 4ac\,a'c' = 0$ $\quad [\because D = 0]$

$\Rightarrow (ac' - a'c)^2 = 0$

$\Rightarrow ac' = a'c$

$\Rightarrow \dfrac{a}{a'} = \dfrac{c}{c'}$

14. d. Let the four numbers in A.P. be $p = a - 3d$, $q = a - d$, $r = a + d$, $s = a + 3d$. Therefore,

$p + q = 2$, $r + s = 18$

Given that $pq = A$, $rs = B$.

$\therefore p + q + r + s = 4a = 20$

$\Rightarrow a = 5$

Now,

$p + q = 2 \Rightarrow 10 - 4d = 2$

$r + s = 18 \Rightarrow 10 + 4d = 18$

$\therefore \quad d = 2$

Hence, the numbers are $-1, 3, 7, 11$.

$pq = A = -3$, $rs = B = 77$

15. d. Here, $x = 0$ is not a root. Divide both the numerator and denominator by x and put $x + (3/x) = y$ to obtain

$$\frac{4}{y+1} + \frac{5}{y-5} = -\frac{3}{2} \Rightarrow y = -5, 3$$

$x + (3/x) = -5$ has two irrational roots and $x + (3/x) = 3$ has imaginary roots.

16. c. Clearly, $x = 2$ is a root of the equation and imaginary roots always occur in pairs. Therefore, the other root is also real.

17. c. Given, α, β are roots of equation

$x^2 - 2x + 3 = 0$

$\Rightarrow \alpha^2 - 2\alpha + 3 = 0 \qquad (1)$

and

$\beta^2 - 2\beta + 3 = 0 \qquad (2)$

$\Rightarrow \alpha^2 = 2\alpha - 3 \Rightarrow \alpha^3 = 2\alpha^2 - 3\alpha$

$\Rightarrow P = (2\alpha^2 - 3\alpha) - 3\alpha^2 + 5\alpha - 2$

$= -\alpha^2 + 2\alpha - 2 = 3 - 2 = 1, \qquad$ [Using (1)]

Similarly, we can have $Q = 2$.

Now, sum of roots is 3 and product of roots is 2. Hence, the equired equation is $x^2 - 3x + 2 = 0$.

18. c. Since α, β are the roots of the equation $2x^2 - 35x + 2 = 0$,

$2\alpha^2 - 35\alpha = -2$ or $2\alpha - 35 = \dfrac{-2}{\alpha}$

and

$2\beta^2 - 35\beta = -2$ or $2\beta - 35 = \dfrac{-2}{\beta}$

Now,

$(2\alpha - 35)^3(2\beta - 35)^3$

$$= \left(\frac{-2}{\alpha}\right)^3\left(\frac{-2}{\beta}\right)^3 = \frac{8\times 8}{\alpha^3\beta^3} = \frac{64}{1} = 64 \qquad [\because \alpha\beta = 1]$$

19. c. Here a, b, c are positive. So,

$|x| = -b + \sqrt{b^2 + ac}$

Hence, x has two real values, neglecting $|x| = -b - \sqrt{b^2 + ac}$, as $|x| \geq 0$.

20. a. Since $p(q - r) + q(r - p) + r(p - q) = 0$, so one root is 1 and the other root is $r(p - q)/[p(q - r)]$. Since both the roots are equal, we have

$$\frac{rp - rq}{pq - pr} = 1$$

$\Rightarrow rp - rq = pq - pr$

$\Rightarrow 2rp = q(p + r)$

$\Rightarrow \dfrac{2}{q} = \dfrac{p + r}{pr} = \dfrac{1}{p} + \dfrac{1}{r}$

21. d. We are given that $p(-a) = q$ and $p(a) = -a$

[since when a polynomial $f(x)$ is divided by $x - a$, remainder is $f(a)$]. Let the remainder, when $p(x)$ is divided by $x^2 - a^2$, be $Ax + B$. Then,

$p(x) = Q(x)(x^2 - a^2) + Ax + B \qquad (1)$

where $Q(x)$ is the quotient. Putting $x = a$ and $-a$ in (1), we get

$p(a) = 0 + Aa + B \Rightarrow -a = Aa + B \qquad (2)$

and

$p(-a) = 0 - aA + B \Rightarrow a = -aA + B \qquad (3)$

Solving (2) and (3), we get

$B = 0$ and $A = -1$

Hence, the required remainder is $-x$.

22. a. $x^2 + ax + b + 1 = 0$ has positive integral roots α and β. Hence,

$(\alpha + \beta) = -a$ and $\alpha\beta = b + 1$

$\Rightarrow (\alpha + \beta)^2 + (\alpha\beta - 1)^2 = a^2 + b^2$

$\Rightarrow a^2 + b^2 = (\alpha^2 + 1)(\beta^2 + 1)$.

$\Rightarrow a^2 + b^2$ can be equal to 50 (since other options have prime numbers)

23. b. $(31 + 8\sqrt{15})^{x^2-3}\, 1 = (32 + 8\sqrt{15})^{x^2-3}$

$\Rightarrow (31 + 8\sqrt{15})^{x^2-3} + 1^{x^2-3} = (32 + 8\sqrt{15})^{x^2-3}$

$\Rightarrow x^2 - 3 = 1$ or $x = \pm 2$ [$\because a^n + b^n = (a+b)^n$]

24. c. $a\alpha^2 + c = -b\alpha,\ a\alpha + b = -\dfrac{c}{\alpha}$

Hence, the given expression is

$$\frac{b}{c}(\alpha^2+\beta^2) = \frac{b(b^2-2ac)}{a^2c}$$

25. a. We have,

$x_1 x_2 = 4$

$$\Rightarrow x_2 = \frac{4}{x_1}$$

$$\therefore \frac{x_1}{x_1-1} + \frac{x_2}{x_2-1} = 2$$

$$\Rightarrow \frac{x_1}{x_1-1} + \frac{\frac{4}{x_1}}{\frac{4}{x_1}-1} = 2$$

$$\Rightarrow \frac{x_1}{x_1-1} + \frac{4}{4-x_1} = 2$$

$\Rightarrow 4x_1 - x_1^2 + 4x_1 - 4 = 2(x_1-1)(4-x_1)$

$\Rightarrow x_1^2 - 2x_1 + 4 = 0$

$\Rightarrow x^2 - 2x + 4 = 0$

26. b. The discriminants of the given equations are $D_1 = a^2 + 12b$, $D_2 = c^2 - 4b$ and $D_3 = d^2 - 8b$.

$\therefore D_1 + D_2 + D_3 = a^2 + c^2 + d^2 \geq 0$

Hence, at least one of D_1, D_2, D_3 is non-negative. Therefore, the equation has at least two real roots.

27. d. α, β are roots of $x^2 + px + q = 0$. Hence $\alpha + \beta = -p$ and $\alpha\beta = q$.

Now, α^4, β^4 are roots $x^2 - px + q = 0$. Hence, $\alpha^4 + \beta^4 = r, \alpha^4\beta^4 = q$.

Now, for equation $x^2 - 4qx + 2q^2 - r = 0$, product of roots is $2q^2 - r = 2(\alpha\beta)^2 - (\alpha^4 + \beta^4)$

$= -(\alpha^2 - \beta^2)^2$

< 0

As product of roots is negative, so the roots must be real.

28. b. $x^4 + x^2 + 1 = (x^2+1)^2 - x^2$

$= (x^2 + x + 1)(x^2 - x + 1)$

$$x^2 + x + 1 = \left(x + \frac{1}{2}\right)^2 + \frac{3}{4} \neq 0,\ \forall x$$

Therefore, we can cancel this factor and we get

$(a-1)(x^2 - x + 1) = (a+1)(x^2 - x + 1)$

or

$x^2 - ax + 1 = 0$

It has real and distinct roots if $D = a^2 - 4 > 0$.

29. b. Let D_1 and D_2 be discriminants of $x^2 + b_1x + c_1 = 0$ and $x^2 + b_2x + c_2 = 0$, respectively. Then,

$D_1 + D_2 = b_1^2 - 4c_1 + b_2^2 - 4c_2$

$= (b_1^2 + b_2^2) - 4(c_1 + c_2)$

$= b_1^2 + b_2^2 - 2b_1b_2$ [$\because b_1 b_2 = 2(c_1 + c_2)$]

$= (b_1 - b_2)^2 \geq 0$

$\Rightarrow D_1 \geq 0$ or $D_2 \geq 0$ or D_1 and D_2 both are positive

Hence, at least one of the equations has real roots.

30.c. Discriminant $D = (2m-1)^2 - 4(m-2)m = 4m + 1$ must be perfect square. Hence,

$4m + 1 = k^2$, say for some $k \in I$

$$\Rightarrow m = \frac{(k-1)(k+1)}{4}$$

Clearly, k must be odd. Let $k = 2n + 1$.

$$\therefore m = \frac{2n(2n+2)}{4} = = n(n+1),\ n \in I$$

31. c. $A = a(b-c)(a+b+c)$

$B = b(c-a)(a+b+c)$

$C = c(a-b)(a+b+c)$

Now,

$Ax^2 + Bx + C = 0$

$\Rightarrow (a+b+c)\{a(b-c)x^2 + b(c-a)x + c(a-b)\} = 0$

Given that roots are equal. Hence,

$D = 0$

$\Rightarrow b^2(c-a)^2 - 4ac(b-c)(a-b) = 0$

$\Rightarrow b^2c^2 - 2ab^2c + b^2a^2 - 4a^2bc + 4acb^2 + 4a^2c^2 - 4abc^2 = 0$

$\Rightarrow b^2c^2 + b^2a^2 + 4a^2c^2 + 2ab^2c - 4a^2bc - 4abc^2 = 0$

$\Rightarrow (bc + ab - 2ac)^2 = 0$

$\Rightarrow bc + ab = 2ac$

$$\Rightarrow \frac{1}{a} + \frac{1}{c} = \frac{2}{b}$$

$\Rightarrow a, b, c$ are in H.P.

32. b. Correct equation is

$x^2 + 13x + q = 0$ (1)

Incorrect equation is

$x^2 + 17x + q = 0$ (2)

Given that roots of Eq. (1) are -2 and -15. Therefore, product of the roots of incorrect equation is

$q = (-2)(-15) = 30$. From (1), the correct equation is

$x^2 + 13x + 30 = 0$

$\Rightarrow x = -3, -10$

33. b. Given,

$a(p+q)^2 + 2bpq + c = 0$ and $a(p+r)^2 + 2bpr + c = 0$

$\Rightarrow q$ and r satisfy the equation $a(p+x)^2 + 2bpx + c = 0$

$\Rightarrow q$ and r are the roots of $a(p+x)^2 + 2bpx + c = 0$

or

$ax^2 + 2(ap + bp)x + c + ap^2 = 0$

$$\Rightarrow qr = \text{product of roots} = \frac{c + ap^2}{a} = p^2 + \frac{c}{a}$$

34. b. Multiplying the given equation by c/a^3, we get

$$\frac{b^2c^2}{a^3}x^2 - \frac{b^2c}{a^2}x + c = 0$$

$$\Rightarrow a\left(\frac{bc}{a^2}x\right)^2 - b\left(\frac{bc}{a^2}\right)x + c = 0$$

$$\Rightarrow \frac{bc}{a^2}x = \alpha, \beta$$

$$\Rightarrow (\alpha+\beta)\alpha\beta\, x = \alpha, \beta$$

$$\Rightarrow x = \frac{1}{(\alpha+\beta)\alpha}, \frac{1}{(\alpha+\beta)\beta}$$

35. d. Here,

$$\alpha + \beta = -\frac{2b}{a}$$

$$\gamma + \alpha = -\frac{c}{2b}, \alpha + \delta = -\frac{a}{c}$$

$$\alpha\beta = \frac{c}{a}, \alpha\gamma = \frac{a}{2b}, \alpha\delta = \frac{2b}{c}$$

$$\Rightarrow \alpha + \delta = \frac{1}{\alpha\beta} \Rightarrow \alpha^2\beta + \alpha\beta\delta = -1 \quad (1)$$

$$\alpha + \beta = \frac{1}{\alpha\gamma} \Rightarrow \alpha^2\gamma + \alpha\beta\gamma = -1 \quad (2)$$

$$\alpha + \gamma = \frac{1}{\alpha\delta} \Rightarrow \alpha^2\delta + \alpha\beta\gamma = -1 \quad (3)$$

Solving Eqs. (1), (2) and (3), we get $\alpha = -1$.

$\therefore \alpha + \alpha^2 = (-1) + (-1)^2 = -1 + 1 = 0$

36. a. Given equation is

$x^2 - (y + 4)x + y^2 - 4y + 16 = 0$

Since x is real, so,

$D \geq 0$

$\Rightarrow (y + 4)^2 - 4(y^2 - 4y + 16) \geq 0$

$\Rightarrow -3y^2 + 24y - 48 \geq 0$

$\Rightarrow y^2 - 8y + 16 \leq 0$

$\Rightarrow (y - 4)^2 \leq 0$

$\Rightarrow y - 4 = 0$

$\Rightarrow y = 4$

Since the equation is symmetric in x and y, therefore $x = 4$ only.

37. a. α, β are roots of $ax^2 + bx + c = 0$. Hence,

$$\alpha + \beta = -\beta/\alpha \quad (1)$$

$$\alpha\beta = \frac{c}{a} \quad (2)$$

α^2, β^2 are roots of $a^2x^2 + b^2x + c^2 = 0$. Hence,

$$\alpha^2 + \beta^2 = -\frac{b^2}{a^2} \quad (3)$$

$$\alpha^2\beta^2 = \frac{c^2}{a^2} \quad (4)$$

Now, from (3),

$$(\alpha + \beta)^2 - 2\alpha\beta = -\frac{b^2}{a^2}$$

$$\Rightarrow \left(\frac{-b}{a}\right)^2 - 2\frac{c}{a} = \frac{-b^2}{a^2}$$

$$\Rightarrow 2\frac{b^2}{a^2} = \frac{2c}{a}$$

$\Rightarrow b^2 = ac \Rightarrow a, b, c$ are in G.P.

38. c. We have,

$$\frac{k+1}{k} + \frac{k+2}{k+1} = -\frac{b}{a} \quad (1)$$

and

$$\frac{k+1}{k}\,\frac{k+2}{k+1} = \frac{c}{a}$$

$$\Rightarrow \frac{k+2}{k} = \frac{c}{a} \text{ or } \frac{2}{k} = \frac{c}{a} - 1 = \frac{c-a}{a} \text{ or } k = \frac{2a}{c-a} \quad (2)$$

Now, eliminate k. Putting the value of k in Eq. (1), we get

$$\frac{c+a}{2a} + \frac{2c}{c+a} = -\frac{b}{a}$$

$\Rightarrow (c + a)^2 + 4ac = -2b\,(a + c)$

$\Rightarrow (a + c)^2 + 2b\,(a + c) = -4ac$

Adding b^2 to both sides, we have

$(a + b + c)^2 = b^2 - 4ac$

39. c. We have,

$$\alpha + \beta = -\frac{b}{a}, \alpha\beta = \frac{c}{a}$$

$$\alpha + h + \beta + h = -\frac{q}{p}, (\alpha + h)(\beta + h) = \frac{r}{p}$$

$$\Rightarrow \alpha + \beta + 2h = -\frac{q}{p}$$

$$\Rightarrow -\frac{b}{a} + 2h = \frac{-q}{p} \qquad [\because a + b = -\frac{b}{a}]$$

$$\Rightarrow h = \frac{1}{2}\left(\frac{b}{a} - \frac{q}{p}\right)$$

40. a. Given, α, β are roots of the equation

$(x - a)(x - b) + c = 0.$

Then, by factor theorem,

$(x - a)(x - b) + c = (x - \alpha)(x - \beta)$

Replacing x by $x - c$, we get

$(x - c - a)(x - c - b) + c = (x - c - \alpha)(x - c - \beta)$

$\Rightarrow (x - c - \alpha)(x - c - \beta) - c = [x - (c + a)]\,[x - (c + b)]$

Then, again by factor theorem roots of the equation $(x - c - \alpha)(x - c - \beta)\, c = 0$ are $a + c$ and $b + c$.

41. b. $ax^2 - bx + c = 0$

$$\alpha + \beta = \frac{b}{a}, \alpha\beta = \frac{c}{a}$$

Also,

$(a + cy)^2 = b^2y$

$\Rightarrow c^2 y^2 - (b^2 - 2ac)y + a^2 = 0$

$$\Rightarrow \left(\frac{c}{a}\right)^2 y^2 - \left(\left(\frac{b}{a}\right)^2 - 2\left(\frac{c}{a}\right)\right)y + 1 = 0.$$

$\Rightarrow (\alpha\beta)^2 y^2 - (\alpha^2 + \beta^2)y + 1 = 0$

$\Rightarrow y^2 - (\alpha^{-2} + \beta^{-2})y + \alpha^{-2}\beta^{-2} = 0$

$\Rightarrow (y - \alpha^{-2})(y - \beta^{-2}) = 0$

Hence the roots are α^{-2}, β^{-2}.

42. c. Let the roots be α, β. Therefore,

$\alpha + \beta = -2a$ and $\alpha\beta = b$

Given,

$|\alpha - \beta| \le 2m$

$\Rightarrow |\alpha - \beta|^2 \le (2m)^2$

$\Rightarrow (\alpha + \beta)^2 - 4ab \le 4m^2$

$\Rightarrow 4a^2 - 4b \le 4m^2$

$\Rightarrow a^2 - m^2 \le b$ and discriminant $D > 0$ or $4a^2 - 4b > 0$

$\Rightarrow a^2 - m^2 \le b$ and $b < a^2$

Hence, $b \in [a^2 - m^2, a^2)$.

43. c. Let the roots of the equation $ax^2 + 2bx + c = 0$ be α and β and roots of the equation $px^2 + 2qx + r = 0$ be γ and δ. Given,

$$\frac{\alpha}{\beta} = \frac{\gamma}{\delta} \Rightarrow \frac{\alpha}{\gamma} = \frac{\beta}{\delta}$$

$$\Rightarrow \frac{\alpha + \beta}{\gamma + \delta} = \sqrt{\frac{\alpha\beta}{\gamma\delta}}$$

$$\Rightarrow \frac{-\frac{2b}{a}}{-\frac{2q}{p}} = \sqrt{\frac{\frac{c}{a}}{\frac{r}{p}}}$$

$$\Rightarrow \frac{b^2}{ac} = \frac{q^2}{pr}$$

44. a. Let α and α^2 be the roots of $x^2 - x - k = 0$. Then,

$\alpha + \alpha^2 = 1$ and $a^3 = -k$

$\Rightarrow (-k)^{1/3} + (-k)^{2/3} = 1$

$\Rightarrow -k^{1/3} + k^{2/3} = 1$

$\Rightarrow (k^{2/3} - k^{1/3})^3 = 1$

$\Rightarrow k^2 - k - 3k(k^{2/3} - k^{1/3}) = 1$

$\Rightarrow k^2 - k - 3k(1) = 1$

$\Rightarrow k^2 - 4k - 1 = 0$

$\Rightarrow k = 2 \pm \sqrt{5}$

45. d. Here,

$\alpha^4 + \beta^4 = (\alpha^2 + \beta^2)^2 - 2\alpha^2\beta^2$

$= \{(\alpha + \beta)^2 - 2\alpha\beta\}^2 - 2(\alpha\beta)^2$

$$= \left(p^2 + \frac{1}{p^2}\right)^2 - \frac{1}{2p^4} = p^4 + \frac{1}{2p^4} + 2$$

$$= \left(p^2 - \frac{1}{\sqrt{2}p^2}\right)^2 + 2 + \sqrt{2} \ge 2 + \sqrt{2}$$

Thus, the minimum value of $\alpha^4 + \beta^4$ is $2 + \sqrt{2}$.

46. b. We have,

$$\alpha^2 + \beta^2 = -\frac{1}{a^2}, \alpha^2\beta^2 = \frac{1 - a^2}{a^2}$$

$$\Rightarrow \beta^2\left[-\frac{1}{a^2} - \beta^2\right] = \frac{1 - a^2}{a^2}$$

$\Rightarrow \beta^2[-1 - a^2\beta^2] = 1 - a^2$

$\Rightarrow a^2\beta^4 + \beta^2 + 1 - a^2 = 0$

$$\Rightarrow \beta^2 = \frac{-1 \pm \sqrt{1 - 4a^2(1 - a^2)}}{2a^2}$$

$$\Rightarrow \frac{-1 \pm \sqrt{4a^4 - 4a^2 + 1}}{2a^2} = \frac{-1 \pm \sqrt{(2a^2 - 1)^2}}{2a^2}$$

$$= \frac{-1 \pm (2a^2 - 1)}{2a^2} \qquad [\because a > 1, \therefore 2a^2 - 1 > 0]$$

$$\therefore \beta^2 = \frac{-1 + 2a^2 - 1}{2a^2} \qquad [\because \beta^2 > 0]$$

$$= \frac{2(a^2 - 1)}{2a^2} = \frac{a^2 - 1}{a^2} = 1 - \frac{1}{a^2}$$

47. c. $A_{n+1} = \alpha^{n+1} + \beta^{n+1}$

$= \alpha^{n+1} + \alpha^n\beta + \beta^{n+1} + \alpha\beta^n - \alpha^n\beta - \alpha\beta^n$

$= \alpha^n(\alpha + \beta) + \beta^n(\beta + \alpha) - \alpha\beta(\alpha^{n-1} + \beta^{n-1})$

$= \alpha^n(\alpha + \beta) + \beta^n(\beta + \alpha) - \alpha\beta(\alpha^{n-1} + \beta^{n-1})$

$= (\alpha + \beta)(\alpha^n + \beta^n) - \alpha\beta(\alpha^{n-1} + \beta^{n-1})$

$= aA_n - bA_{n-1}$

48. a. Let α be the root of $x^2 - x + m = 0$ and 2α be the root of $x^2 - 3x + 2m = 0$. Then,

$\alpha^2 - \alpha + m = 0$ and $4\alpha^2 - 6\alpha + 2m = 0$

Eliminating α, $m^2 = -2m \Rightarrow m = 0, m = -2$

49. a. We have,

$x^3 + 3x^2 + 3x + 2 = 0$

$\Rightarrow (x^3 - 1) + 3(x^2 + x + 1) = 0$

$\Rightarrow (x^2 + x + 1)(x - 1 + 3) = 0$

$\Rightarrow (x + 2)(x^2 + x + 1) = 0 \Rightarrow x = -2, w, w^2$

Equations $x^3 + 3x^2 + 3x + 2 = 0$ and $ax^2 + bx + c = 0$ have two common roots. Therefore, w and w^2 are common roots of the two equations. Hence, $a = b = c = 1$.

50. b. Given equations are

$x^3 + ax + 1 = 0$

or

$x^4 + ax^2 + x = 0$ (1)

and

$x^4 + ax^2 + 1 = 0$ (2)

From (1) – (2), we get $x = 1$. Thus, $x = 1$ is the common roots. Hence,

$1 + a + 1 = 0 \Rightarrow a = -2$

51. a. $x = \sqrt[3]{7} + \sqrt[3]{49}$

$\Rightarrow x^3 = 7 + 49 + 3\sqrt[3]{7}\ \sqrt[3]{49}(\sqrt[3]{7} + \sqrt[3]{49}) = 56 + 21x$

$\Rightarrow x^3 - 21x - 56 = 0$

Therefore, the product of roots is 56.

52. c. Since $\alpha, \beta, \gamma, \sigma$ are the roots of the given equation,

$x^4 + 4x^3 - 6x^2 + 7x - 9 = (x - \alpha)(x - \beta)(x - \gamma)(x - \sigma)$

Putting $x = i$ and then $x = -i$, we get

$1 - 4i + 6 + 7i - 9 = (i - \alpha)(i - \beta)(i - \gamma)(i - \sigma)$

and

$1 + 4i + 6 - 7i - 9 = (-i - \alpha)(-i - \beta)(-i - \gamma)(-i - \sigma)$

Multiplying these two equations, we get

$(-2 + 3i)(-2 - 3i) = (1 + \alpha^2)(1 + \beta^2)(1 + \gamma^2)(1 + \sigma^2)$

$\Rightarrow 13 = (1 + \alpha^2)(1 + \beta^2)(1 + \gamma^2)(1 + \sigma^2)$

53. b. If $[m_r, (1/m_r)]$ satisfy the given equation $x^2 + y^2 + 2gx + 2fy + c = 0$, then

$$m_r^2 + \frac{1}{m_r^2} + 2gm_r + \frac{2f}{m_r} + c = 0$$

$\Rightarrow m_r^4 + 2gm_r^3 + cm_r^2 + 2fm_r + 1 = 0$

Now, roots of given equation are m_1, m_2, m_3, m_4. The product of roots

$$m_1 m_2 m_3 m_4 = \frac{\text{constant term}}{\text{coefficient of } m_r^4} = \frac{1}{1} = 1$$

54. a. Clearly,

$x^n - 1 = (x - 1)(x - a_1)(x - a_2) \cdots (x - a_{n-1})$

$$\Rightarrow \frac{x^n - 1}{x - 1} = (x - a_1)(x - a_2) \cdots (x - a_{n-1})$$

$\Rightarrow 1 + x + x^2 + \cdots + x^n = (x - a_1)(x - a_2) \cdots (x - a_{n-1})$

$\Rightarrow n = (1 - a_1)(1 - a_2) \cdots (1 - a_{n-1})$ [putting $x = 1$]

55. b. $\tan\theta_1 + \tan\theta_2 + \tan\theta_3 = (a + 1)$

$\Sigma \tan\theta_1 \tan\theta_2 = (b - a)$

$\tan\theta_1 \tan\theta_2 \tan\theta_3 = b$

$\therefore \tan(\theta_1 + \theta_2 + \theta_3)$

$$= \frac{\Sigma \tan\theta_1 - \Pi \tan\theta_1}{1 - \Sigma \tan\theta_1 \tan\theta_2}$$

$$= \frac{a + 1 - b}{1 - (b - a)} = 1$$

$$\Rightarrow \theta_1 + \theta_2 + \theta_3 = \frac{\pi}{4}$$

56. a. $\Sigma\alpha = 1, \Sigma\alpha\beta = 0, \alpha\beta\gamma = 1$

$$\Sigma\frac{1+\alpha}{1-\alpha} = -\Sigma\frac{-\alpha+1-2}{1-\alpha} = \Sigma\left(\frac{2}{1-\alpha} - 1\right)$$

$$= 2\Sigma\frac{1}{1-\alpha} - 3$$

Now,

$$\frac{1}{(x-\alpha)} + \frac{1}{(x-\beta)} + \frac{1}{(x-\gamma)} = \frac{3x^2 - 2x}{x^3 - x^2 - 1}$$

$$\Rightarrow \frac{1}{1-\alpha} + \frac{1}{1-\beta} + \frac{1}{1-\gamma} = \frac{3-2}{1-1-1} = -1$$

$$\Rightarrow \frac{1+\alpha}{1-\alpha} = -5$$

57. d. Equation $8x^3 + 1001x + 2008 = 0$ has roots r, s, and t.

$$r + s + r = 0,\ rst = -\frac{2008}{8} = -251$$

Now, let $r + s = A$, $s + t = B$, $t + r = C$.

$\therefore A + B + C = 2(r + s + t) = 0$

Hence,

$A^3 + B^3 + C^3 = 3ABC$

$\therefore (r + s)^3 + (s + t)^3 + (t + r)^3$

$= 3(r + s)(s + t)(t + r)$

$= 3(r + s + t - t)(s + t + r - r)(t + r + s - s)$

$= -3rst$ (as $r + s + t = 0$)

$= 3(251) = 753$

58. b. Let,

$$\frac{x}{x^2 - 5x + 9} = y$$

$\Rightarrow yx^2 - 5yx + 9y = x$

$\Rightarrow yx^2 - (5y + 1)x + 9y = 0$

Now, x is real, so

$D \geq 0$

$\Rightarrow (-(5y + 1))^2 - 4y(9y) \geq 0$

$\Rightarrow -11y^2 + 10y + 1 \geq 0$

$\Rightarrow 11y^2 - 10y - 1 \leq 0$

$\Rightarrow (11y + 1)(y - 1) \leq 0$

$$\Rightarrow -\frac{1}{11} \leq y \leq 1$$

59. b. Let,

$$y = \frac{3x^2 + 9x + 17}{3x^2 + 9x + 7}$$

$\Rightarrow 3x^2(y - 1) + 9x(y - 1) + 7y - 17 = 0$

Since x is real, so,

$D \geq 0$

$\Rightarrow 81(y - 1)^2 - 4 \times 3(y - 1)(7y - 17) \geq 0$

$\Rightarrow (y - 1)(y - 41) \leq 0 \Rightarrow 1 \leq y \leq 41$

Therefore, the maximum value of y is 41.

60. a. $D = b^2 - 4a < 0 \Rightarrow a > 0$

Therefore, the graph is concave upwards.

$f(x) > 0, \forall x \in R$

$\Rightarrow f(-1) > 0$

$\Rightarrow a + b + 1 > 0$

61. c. We know that $ax^2 + bx + c \geq 0$, $\forall x \in R$, if $a > 0$ and $b^2 - 4ac \leq 0$. So,

$$mx - 1 + \frac{1}{x} \geq 0 \Rightarrow \frac{mx^2 - x + 1}{x} \geq 0$$

$\Rightarrow mx^2 - x + 1 \geq 0$ as $x > 0$.

Now,

$mx^2 - x + 1 \geq 0$ if $m > 0$ and $1 - 4m \leq 0$

$\Rightarrow m > 0$ and $m \geq 1/4$

Thus, the minimum value of m is 1/4.

62. b. Let $f(x) = ax^2 + bx + c$ be a quadratic expression such that $f(x) > 0$ for all $x \in R$. Then, $a > 0$ and $b^2 - 4ac < 0$. Now,

$g(x) = f(x) + f'(x) + f''(x)$

$\Rightarrow g(x) = ax^2 + x\,(b + 2a) + (b + 2a + c)$

Discriminant of $g(x)$ is

$D = (b + 2a)^2 - 4a\,(b + 2a + c)$

$= b^2 - 4a^2 - 4ac$

$= (b^2 - 4ac) - 4a^2$

$< 0 \quad (\because b^2 - 4ac < 0)$

Therefore, $g(x) > 0$ for all $x \in R$.

63. b. Here, $ax^2 - bx + c^2 = 0$ does not have real roots. So,

$D < 0 \Rightarrow b^2 - 4ac^2 < 0 \Rightarrow a > 0$

Therefore, $f(x)$ is always positive. So,

$f(2) > 0 \Rightarrow 4a - 2b + c^2 > 0$

64. a. $x_1(x - x_2)^2 + x_2(x - x_1)^2 = 0$

$\Rightarrow x^2(x_1 + x_2) - 4x\,x_1x_2 + x_1x_2\,(x_1 + x_2) = 0$

$D = 16(x_1x_2)^2 - 4x_1x_2(x_1 + x_2)^2 > 0\ (\because x_1x_2 < 0)$

The product of roots is $x_1x_2 < 0$. Thus, the roots are real and of opposite signs.

65. d. Given that $a < b < c < d$. Let

$f(x) = (x - a)\,(x - c) + 2(x - b)(x - d)$

$\Rightarrow f(b) = (b - a)\,(b - c) < 0$

and

$f(d) = (d - a)\,(d - c) > 0$

Hence, $f(x) = 0$ has one root in (b, d). Also, $f(a)f(c) < 0$. So the other root lies in (a, c). Hence, roots of the equation are real and distinct.

66. c. $x^2 - (a - 3)\,x + a = 0$

$\Rightarrow D = (a - 3)^2 - 4a$

$= a^2 - 10a + 9$

$= (a - 1)\,(a - 9)$

Case I:

Both the roots are greater than 2.

$$D \geq 0, f(2) > 0, -\frac{B}{2A} > 2$$

$$\Rightarrow (a - 1)\,(a - 9) \geq 0; 4 - (a - 3)2 + a > 0; \frac{a - 3}{2} > 2$$

$\Rightarrow a \in (-\infty, 1] \cup [9, \infty); a < 10; a > 7$

$\Rightarrow a \geq [9, 10]$ (1)

Case II:

One root is greater than 2 and the other root is less than or equal to 2. Hence,

$f(2) \leq 0$

$\Rightarrow 4 - (a - 3)2 + a \leq 0$

$\Rightarrow a \geq 10$ (2)

From (1) and (2),

$a \in [9, 10) \cup [10, \infty) \Rightarrow a \in [9, \infty)$

67. a. Suppose $a \neq 0$. We rewrite $f(x)$ as follows:

$$f(x) = a\left\{x^2 + \frac{b}{a}x + \frac{c}{a}\right\}$$

$$= a\left\{\left(x + \frac{b}{2a}\right)^2 + \frac{4ac - b^2}{4a^2}\right\}$$

$$f\left(-\frac{b}{2a} + i\right) = a\left\{\left(-\frac{b}{2a} + i + \frac{b}{2a}\right)^2 + \frac{4ac - b^2}{4a^2}\right\}$$

$$= a\left\{-1 + \frac{4ac - b^2}{4a^2}\right\}, \text{ which is a real number}$$

This is against the hypothesis. Therefore, $a = 0$.

68. c. The given equation is

$x^2 - 2mx + m^2 - 1 = 0$

$\Rightarrow (x - m)^2 - 1 = 0$

$\Rightarrow (x - m + 1)\,(x - m - 1) = 0$

$\Rightarrow x = m - 1, m + 1$

From given condition,

$m - 1 > -2$ and $m + 1 < 4$

$\Rightarrow m > -1$ and $m > 3$

Hence, $-1 < m < 3$.

69. b. Note that coefficient of x^2 is $(4p - p^2 - 5) < 0$. Therefore, the graph is concave downward. According to the question, 1 must lie between the roots. Hence,

$f(1) > 0$

$\Rightarrow 4p - p^2 - 5 - 2p + 1 + 3p > 0$

$\Rightarrow -p^2 + 5p - 4 > 0$

$\Rightarrow p^2 - 5p + 4 < 0$

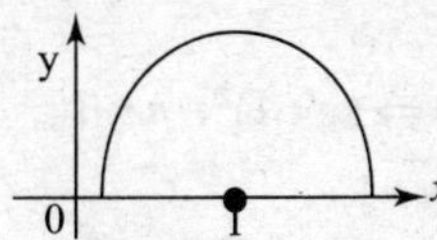

Fig. 1.4

$\Rightarrow (p - 4)\,(p - 1) < 0$

$\Rightarrow 1 < p < 4$

$\Rightarrow p \in \{2, 3\}$

70. b. $$\tan x = \frac{a - 4 \pm \sqrt{(a - 4)^2 - 4(4 - 2a)}}{2}$$

$$= \frac{a - 4 \pm a}{2} = a - 2, -2$$

$\therefore \tan x = a - 2$ $\quad (\because \tan x \neq -2)$

$\because x \in \left[0, \frac{\pi}{4}\right]$

$\therefore 0 \leq a - 2 \leq 1$

$\Rightarrow 2 \leq a \leq 3$

71. a. Clearly, $f(-1) > 0, f(2) < 0$. Now,

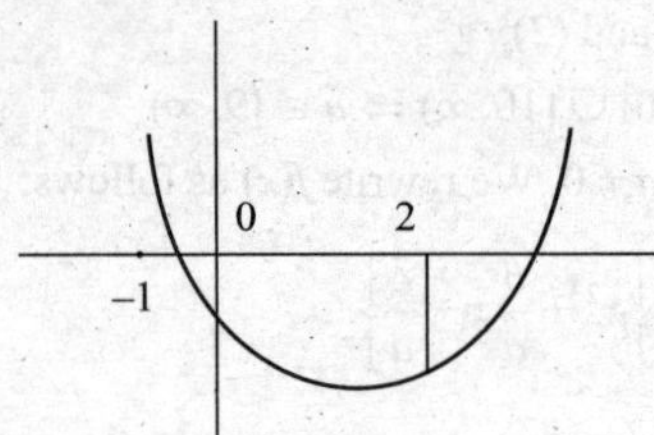

Fig. 1.5

$f(0) = -4 < 0$

$\Rightarrow f(-1) = 1 - a - 4 > 0$ and $f(2) = 4 + 2a - 4 < 0$

$\Rightarrow a < -3$ and $a < 0$

$\Rightarrow a \in (-\infty, -3)$

72. b. Let,

$f(x) = ax^2 + x + c - a$

$f(1) = c + 1 > 0 \quad [\because c > -1]$

Therefore, given expression is positive $\forall x \in R$. So,

$f\left(\frac{1}{2}\right) > 0$

$\Rightarrow \frac{a}{4} + \frac{1}{2} + c - a > 0$

$\Rightarrow 4c - 3a + 2 > 0$

$\Rightarrow 4c + 2 > 3a$

73. b.

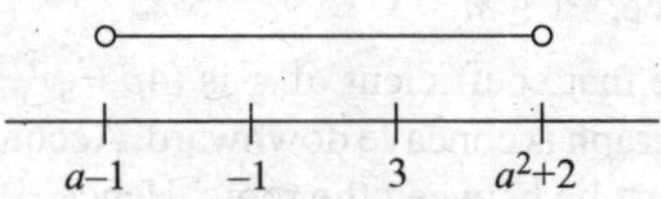

Fig. 1.6

We have,

$a - 1 \leq -1$ and $a^2 + 2 \geq 3$

$a \leq 0$ and $a^2 \geq 1$

Hence, $a \leq -1$.

74. a. $x^2 + 2x - n = 0 \Rightarrow (x + 1)^2 = n + 1$

$\Rightarrow x = -1 \pm \sqrt{n+1}$

Thus, $n + 1$ should be a perfect square. Now,

$n \in [5, 100] \Rightarrow n + 1 \in [6, 101]$

Perfect square values of $n + 1$ are 9, 16, 25, 36, 49, 64, 81, 100. Hence, number of values is 8.

75. d. $x^2 - x - a = 0, D = 1 + 4a =$ odd

D must be perfect square of some odd integer. Let

$D = (2\lambda + 1)^2$

$\Rightarrow 1 + 4a = 1 + 4\lambda^2 + 4\lambda$

$\Rightarrow a = \lambda(\lambda + 1)$.

Now,

$a \in [6, 100]$

$\Rightarrow a = 6, 12, 20, 30, 42, 56, 72, 90$

Thus a can attain eight different values.

76. a. $x^2 - (a + 1)x + a - 1 = 0$

$\Rightarrow (x - a)(x - 1) = 1$

Now, $a \in I$ and we want x to be an integer. Hence,

$x - a = 1, x - 1 = 1$ or $x - a = -1, x - 1 = -1$

$\Rightarrow a = 1$ in both cases

77. b. For given situation, $x^2 - (k - 2)x + k^2 = 0$ and $x^2 + kx + 2k - 1 = 0$ should have both roots common or each should have equal roots. If both roots are common, then

$\frac{1}{1} = \frac{-(k-2)}{k} = \frac{k^2}{2k-1}$

$\Rightarrow k = -k + 2$ and $2k - 1 = k^2 \Rightarrow k = 1$

If both the equations have equal roots, then

$(k - 2)^2 - 4k^2 = 0$ and $k^2 - 4(2k - 1) = 0$

$\Rightarrow (3k - 2)(-k - 2) = 0$ and $k^2 - 8k + 4 = 0$

(no common value)

Therefore, $k = 1$ is the only possible value.

78. c. We have,

$\alpha + \beta = -p$ and $\alpha\beta = q$ (1)

Also, since α, β are the roots of $x^{2n} + p^n x^n + q^n = 0$, we have

$\alpha^{2n} + p^n\alpha^n + q^n = 0$ and $\beta^{2n} + p^n\beta^n + q^n = 0$

Subtracting the above relations, we get

$(\alpha^{2n} - \beta^{2n}) + p^n(\alpha^n - \beta^n) = 0$, Therefore,

$\alpha^n + \beta^n = -p^n$ (2)

Given, α/β or β/α is a root of $x^n + 1 + (x + 1)^n = 0$. So,

$(\alpha/\beta)^n + 1 + [(\alpha/\beta) + 1]^n = 0$

$\Rightarrow (\alpha^n + \beta^n) + (\alpha + \beta)^n = 0$

$\Rightarrow -p^n + (-p)^n = 0$ [Using (1) and (2)]

It is possible only when n is even.

79. a. Since R.H.S. is an even integer. so L.H.S. is also an even integer. So, either both x and y are even integers, or both of them are odd integers. Now,

$x^4 - y^4 = (x - y)(x + y)(x^2 + y^2)$

$\Rightarrow x - y, x + y, x^2 + y^2$ must be an even integer

Therefore, $(x - y)(x + y)(x^2 + y^2)$ must be divisible by 8. But R.H.S. is not divisible by 8. Hence, the given equation has no solution.

80. a. $xy = 2(x + y) \Rightarrow y(x - 2) = 2x$

$\therefore y = \frac{2x}{x-2}$ but $x, y \in N$

By trial, $x = 3, 4, 6$. Then $y = 6, 4, 3$. But $x \leq y$. Therefore, $x = 3, 4$ and $y = 6, 4$ are two solutions.

81. a. We have,

$(\alpha+\beta+\gamma)^2=\alpha^2+\beta^2+\gamma^2+2(\beta\gamma+\gamma\alpha+\alpha\beta)$

$\Rightarrow 4=6+2(\beta\gamma+\gamma\alpha+\alpha\beta)$

$\Rightarrow \beta\gamma+\gamma\alpha+\alpha\beta=-1$

Also,

$\alpha^3+\beta^3+\gamma^3-3\alpha\beta\gamma=(\alpha+\beta+\gamma)(\alpha^2+\beta^2+\gamma^2-\beta\gamma-\gamma\alpha-\alpha\beta)$

$\Rightarrow 8-3\alpha\beta\gamma=2(6+1)$

$\Rightarrow 3\alpha\beta\gamma=8-14=-6$ or $\alpha\beta\gamma=-2$

Now,

$(\alpha^2+\beta^2+\gamma^2)^2=\Sigma\alpha^4+2\Sigma\alpha^2\beta^2$

$=\Sigma\alpha^4+2[(\Sigma\alpha\beta)^2-2\alpha\beta\gamma(\Sigma\alpha)]$

$\Rightarrow \Sigma\alpha^4=36-2[(-1)^2-2(-2)(2)]=18$

82. d. $(x+a)(x+1991)+1=0$

$\Rightarrow (x+a)(x+1991)=-1$

$\Rightarrow (x+a)=1$ and $x+1991=-1$

$\Rightarrow a=1993$ or

$x+a=-1$ and $x+1991=1$

83. b. Let $f(x)=-3+x-x^2$. Then $f(x)<0$ for all x, because coefficient of x^2 is less than 0 and $D<0$. Thus, L.H.S. of the given equation is always positive whereas the R.H.S. is always less than zero. Hence, there is no solution.

84. c. Put $ab+bc+ca=t$. Now,

$(a+b+c)^2=a^2+b^2+c^2+2t$

$\Rightarrow (a+b+c)^2=1+2t$

$\Rightarrow 1+2t\ge 0$

$\Rightarrow -\frac{1}{2}\le t$

Again,

$(a-b)^2+(b-c)^2+(c-a)^2=2-2t$

$\Rightarrow 2-2t\ge 0$

$\Rightarrow t\le 1$

$\Rightarrow -\frac{1}{2}\le t\le t\le 1$

85. d. $\cot^4 x-2(1+\cot^2 x)+a^2=0$

$\Rightarrow \cot^4 x-2\cot^2 x+a^2-2=0$

$\Rightarrow (\cot^2 x-1)^2=3-a^2$

Now, for at least one solution

$3-a^2\ge 0$

$\Rightarrow a^2-3\le 0$

$\therefore a\in[-\sqrt{3},\sqrt{3}]$

Integral values are $-1, 0, 1$. Therefore,

sum = 0

86. b. $x=3\cos\theta;\ y=3\sin\theta$

$z=2\cos\phi;\ t=2\sin\phi$

$\therefore 6\cos\theta\sin\phi-6\sin\theta\cos\phi=6$

$\Rightarrow \sin(\phi-\theta)=1$

$\Rightarrow \phi=90°+\theta$

$\Rightarrow P=xz=-6\sin\theta\cos\theta=-3\sin 2\theta$

$\Rightarrow P_{\max}=3$

87. a. The equation on simplifying gives

$x(x-b)(x-c)+x(x-c)(x-a)+x(x-a)(x-b)-(x-a)(x-b)(x-c)=0$ (1)

Let,

$f(x)=x(x-b)(x-c)+x(x-c)(x-a)+x(x-a)(x-b)-(x-a)(x-b)(x-c)$

We can assume without loss of generality that $a<b<c$.

Now,

$f(a)=a(a-b)(a-c)>0$

$f(b)=b(b-c)(b-a)<0$

$f(c)=c(c-a)(c-b)>0$

So, one root of (1) lies in (a, b) and second root in (b, c). Obviously the third root must also be real.

88. b. $\frac{(b^2-4ac)^2}{16a^2}<\frac{4}{1+4a^2}$ (1)

Now,

$\max(ax^2+bx+c)=-\frac{b^2-4ac}{4a}$

Also,

$\frac{-2}{\sqrt{1+4a^2}}<-\frac{b^2-4ac}{4a}<\frac{2}{\sqrt{1+4a^2}}$ [From (1)]

So, maximum value is always less than 2 (when $a\to 0$).

89. d. $a=\frac{x^2+4}{|x|}-3$

$=|x|+\frac{4}{|x|}-3$

$\ge 2\sqrt{|x|\frac{4}{|x|}}-3$ ($\because$ A.M. $\ge$ G.M.)

$=4-3=1$

$\Rightarrow a\ge 1$

90. d. Here, $f(x)=(2x-a)(2x-c)+(2x-b)$. So,

$f\left(\frac{a}{2}\right)=a-b,\ f\left(\frac{c}{2}\right)=c-b$

Now,

$f\left(\frac{a}{2}\right)f\left(\frac{c}{2}\right)=(a-b)(c-b)<0\ (a>b>c)$

Hence, exactly one of the roots lies between $c/2$ and $a/2$.

91. a. Since the equation $x^2+ax+b=0$ has distinct real roots and $x^2+a|x|+b=0$ has only one real root, so one root of the equation $x^2+ax+b=0$ will be zero and other root will be negative. Hence, $b=0$ and $a>0$.

Graph of $y=ax^2+bx+c$ according to conditions given in question

Graph of $y=ax^2+b|x|+c$

Fig. 1.7

92. c. The given equation is

$2^{2x} + (a-1)2^{x+1} + a = 0$

or

$t^2 + 2(a-1)t + a = 0$, where $2^x = t$

Now, $t = 1$ should lie between the roots of this equation.

$\therefore 1 + 2(a-1) + a < 0 \Rightarrow a < \frac{1}{3}$

93. a.

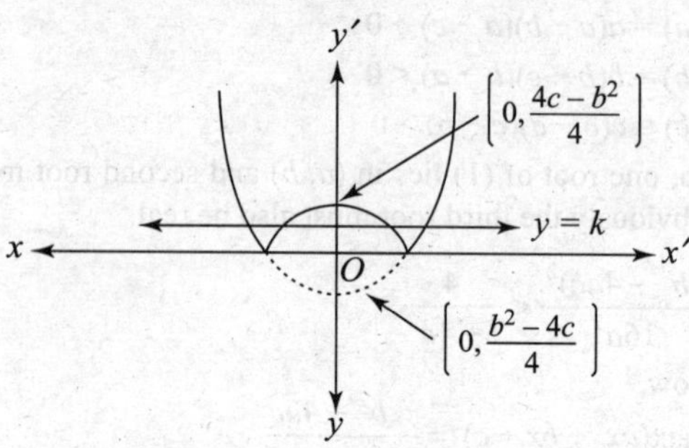

Fig. 1.8

For the equation to have four real roots, the line $y = k$ must intersect $y = |x^2 + bx + c|$ at four points.

$\therefore D > 0$ and $k \in \left(0, \frac{-D}{4}\right)$

94. d. $P(a) = P(b) = P(c) = P(d) = 3$

$\Rightarrow P(x) = 3$ has a, b, c, d as its roots

$\Rightarrow P(x) - 3 = (x-a)(x-b)(x-c)(x-d)\,Q(x)$

[$\because Q(x)$ has integral coefficient]

If $P(e) = 5$, then

$(e-a)(e-b)(e-c)(e-d)\,Q(e) = 2$

This is possible only when at least three of the five integers $(e-a)$, $(e-b)$, $(e-c)$, $(e-d)$, $Q(e)$ are equal to 1 or −1. Hence, two of them will be equal, which is not possible. Since a, b, c, d are distinct integers, therefore $P(e) = 5$ is not possible.

95. d. $\sqrt{-x^2 + 10x - 16} < x - 2$

We must have

$-x^2 + 10x - 16 \ge 0$

$\Rightarrow x^2 - 10x + 16 \le 0$

$\Rightarrow 2 \le x \le 8$ (1)

Also,

$-x^2 + 10x - 16 < x^2 - 4x + 4$

$\Rightarrow 2x^2 - 14x + 20 > 0$

$\Rightarrow x^2 - 7x + 10 > 0$

$\Rightarrow x > 5$ or $x < 2$ (2)

From (1) and (2),

$5 < x \le 8 \Rightarrow x = 6, 7, 8$

96. c. $D > 0 \Rightarrow (a-3)^2 + 4(a+2) > 0$

$\Rightarrow a^2 - 6a + 9 + 4a + 8 > 0$

$\Rightarrow a^2 + 2a + 17 > 0$

$\Rightarrow a \in R$

$\therefore \frac{a^2+1}{a^2+2} = 1 - \frac{1}{a^2+2} \ge \frac{1}{2}$

97.c. Given,

$(a-1)x^2 - (a+1)x + a - 1 \ge 0$

$\Rightarrow a(x^2 - x + 1) - (x^2 + x + 1) \ge 0$

$\Rightarrow a \ge \frac{x^2+x+1}{x^2-x+1} = 1 + \frac{2x}{x^2-x+1}$

$= 1 + \frac{2}{x + \frac{1}{x} - 1}$ (1)

Let $y = x + (1/x)$. Now, y is increasing in $[2, \infty)$. Hence,

$1 + \frac{2}{x + \frac{1}{x} - 1} \in \left(1, \frac{7}{3}\right]$

For all $x \ge 2$, Eq. (1) should be true. Hence, $a > 7/3$.

98. b. $x = 2 + \sqrt{3}$

$\Rightarrow (x-2)^2 = 3$

$\Rightarrow x^2 - 4x + 1 = 0$ (1)

$\Rightarrow (x-2)^4 = 9$

$\Rightarrow x^4 - 8x^3 + 24x^2 - 32x + 16 = 9$

$\Rightarrow x^4 - 8x^3 + 18x^2 - 8x + 2 + 6(x^2 - 4x + 1) - 1 = 0$

Using (1), we get

$x^4 - 8x^3 + 18x^2 - 8x + 2 = 1$

Multiple Correct Answers Type

1. a, c. $2x^2 + 6xy + 5y^2 = 1$ (1)

Eq. (1) can be rewritten as

$2x^2 + (6y)x + 5y^2 - 1 = 0$

Since x is real,

$36y^2 - 8(5y^2 - 1) \ge 0$

$\Rightarrow y^2 \le 2$

$\Rightarrow -\sqrt{2} \le y \le \sqrt{2}$

Eq (1) can also be rewritten as

$5y^2 + (6x)y + 2x^2 - 1 = 0$

Since y is real,

$36x^2 - 20(2x^2 - 1) \ge 0$

$\Rightarrow 36x^2 - 40x^2 + 20 \ge 0$

$\Rightarrow -4x^2 \ge -20$

$\Rightarrow x^2 \le 5$

$\Rightarrow -\sqrt{5} \le x \le \sqrt{5}$

2. b, c, d.

Given equation is $x^3 - ax^2 + bx - 1 = 0$. If roots of the equation be α, β, γ, then

$$\alpha^2 + \beta^2 + \gamma^2 = (\alpha + \beta + \gamma)^2 - 2(\alpha\beta + \beta\gamma + \gamma\alpha)$$
$$= a^2 - 2b$$
$$\alpha^2\beta^2 + \beta^2\gamma^2 + \gamma^2\alpha^2 = (\alpha\beta + \beta\gamma + \gamma\alpha)^2 - 2\alpha\beta\gamma(\alpha + \beta + \gamma)$$
$$= b^2 - 2a$$
$$\alpha^2\beta^2\gamma^2 = 1$$

So the equation whose roots are $\alpha^2, \beta^2, \gamma^2$ is given by

$$x^3 - (a^2 - 2b)x^2 + (b^2 - 2a)x - 1 = 0$$

It is identical to

$$x^3 - ax^2 + bx - 1 = 0$$
$$\Rightarrow a^2 - 2b = a \text{ and } b^2 - 2a = b$$

Eliminating b, we get

$$\frac{(a^2 - a)^2}{4} - 2a = \frac{a^2 - a}{2}$$
$$\Rightarrow a\{a(a-1)^2 - 8 - 2(a-1)\} = 0$$
$$\Rightarrow a(a^3 - 2a^2 - a - 6) = 0$$
$$\Rightarrow a(a-3)(a^2 + a + 2) = 0$$
$$\Rightarrow a = 0 \text{ or } a = 3 \text{ or } a^2 + a + 2 = 0,$$

which gives $b = 0$ or $b = 3$ or $b^2 + b + 2 = 0$. So, $a = b = 0$ or $a = b = 3$ or a, b are roots of $x^2 + x + 2 = 0$.

3. a, b, c.

Since the roots of $ax^2 + bx + c = 0$ are non-real, so $f(x) = ax^2 + bx + c$ will have same sign for every value of x. Hence,

$$f(0) = c, f(1) = a + b + c, f(-1) = a - b + c$$
$$f(2) = 4a - 2b + c$$
$$\Rightarrow c(a + b - c) > 0, c(a - b + c) > 0, c(4a - 2b + c) > 0$$

4. a, b.

We can write the given equation as

$$\frac{p}{2x} = \frac{(a+b)x + c(b-a)}{x^2 - c^2}$$
$$\Rightarrow p(x^2 - c^2) = 2(a+b)x^2 - 2c(a-b)x$$
$$\Rightarrow (2a + 2b - p)x^2 - 2c(a-b)x + pc^2 = 0$$

For this equation to have equal roots,

$$c^2(a-b)^2 - pc^2(2a + 2b - p) = 0$$
$$\Rightarrow (a-b)^2 - 2p(a+b) + p^2 = 0 \quad [\because c^2 \neq 0]$$
$$\Rightarrow [p - (a+b)]^2 = (a+b)^2 - (a-b)^2 = 4ab$$
$$\Rightarrow p - (a+b) = \pm 2\sqrt{ab}$$
$$\Rightarrow p = a + b \pm 2\sqrt{ab} = (\sqrt{a} \pm \sqrt{b})^2$$

5. a, d.

Roots of $4x^2 - x - 1 = 0$ are irrational. So one root common implies both roots are common. Therefore,

$$\frac{4}{3} = \frac{-1}{\lambda + \mu} = \frac{-1}{\lambda - \mu}$$
$$\Rightarrow \lambda = \frac{-3}{4}, \ \mu = 0$$

6. a, b, c.

$$f(x) = ax^2 + bx + c$$
$$f(0) = c < 0, D > 0 \Rightarrow b^2 - 4ac > 0$$
$$f(1) < 0 \text{ and } f(-1) < 0$$
$$\Rightarrow a - |b| + c < 0$$

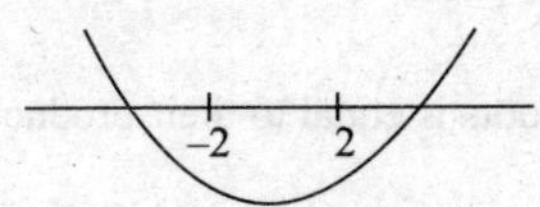

Fig. 1.9

$$f(2) < 0 \text{ and } f(-2) < 0$$
$$\Rightarrow 4a - 2|b| + c < 0$$

Nothing can be said about $f(3)$ or $f(-3)$, whether it is positive or negative.

7. a, b, d.

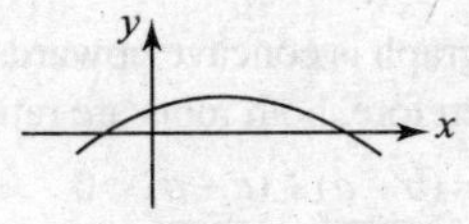

Fig. 1.10

From the graph,

$$f(0) = c > 0 \quad (1)$$

Also, the graph is concave downward. Hence,

$$a < 0 \quad (2)$$

Further, abscissa of the vertex,

$$-\frac{b}{2a} > 0 \Rightarrow b > 0 \quad (3)$$

From (1), (2), (3),

$$ac < 0, ab < 0 \text{ and } bc > 0$$

8. c, d.

$$\cos x - y^2 - \sqrt{y - x^2 - 1} \geq 0 \quad (1)$$

Now, $\sqrt{y - x^2 - 1}$ is defined when $y - x^2 - 1 \geq 0$ or $y \geq x^2 + 1$. So minimum value of y is 1. From (1),

$$\cos x - y^2 \geq \sqrt{y - x^2 - 1},$$

where $\cos x - y^2 \leq 0$ [as when $\cos x$ is maximum (=1) and y^2 is minimum (=1), so $\cos x - y^2$ is maximum]. Also,

$$\sqrt{y - x^2 - 1} \geq 0$$

Hence,

$$\cos x - y^2 = \sqrt{y - x^2 - 1} = 0$$
$$\Rightarrow y = 1 \text{ and } \cos x = 1, y = x^2 + 1$$
$$\Rightarrow x = 0, y = 1$$

9. a, d.

$$2^x = t$$
$$t^2 - 8t + 12 = 0$$
$$(t - 6)(t - 2) = 0$$

$2^x = 6 \Rightarrow x = \log_2 6 = 1 + \dfrac{\log 3}{\log 2}$

$2^x = 2 \Rightarrow x = 1$

10. a, b, d.

Symmetric functions are those which do not change by interchanging α and β.

11. a, b, c.

Sum of the roots is equal to their product and the roots are real. Hence,

$-\dfrac{b}{a} = \dfrac{c}{a}$

$\Rightarrow b + c = 0$

$\Rightarrow b^2 - 4ac \geq 0$

12. a, b, c.

$f(x) = Ax^2 + Bx + C$

$A = a + b - 2c = (a - c) + (b - c) > 0$

$\Rightarrow A > 0$

Hence, the graph is concave upwards. Also, $x = 1$ is obvious solution; therefore, both roots are rational.

$b + c - 2a = \underbrace{(b - a)}_{-\text{ve}} + \underbrace{(c - a)}_{-\text{ve}} < 0$

$\Rightarrow B < 0$

$\therefore$ vertex $= -\dfrac{B}{2A} > 0$

Hence, abscissa of the vertex $a > 0$. Option (d) need not be correct as with $a = 5$, $b = 4$, $c = 2$, $P < 0$ and with $a = 6$, $b = 3$, $c = 2$, $P > 0$.

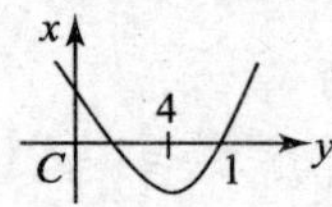

Fig. 1.11

13. a, b, c.

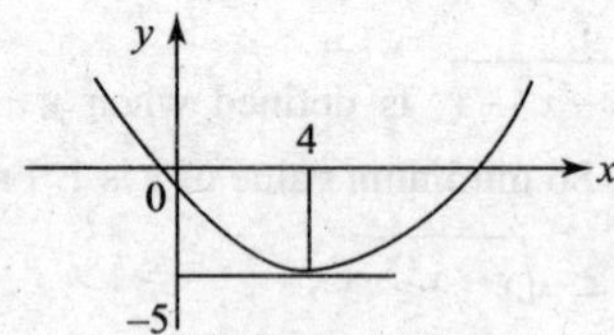

Fig. 1.12

From figure,

$a > 0$

$-\dfrac{b}{2a} = 4 \Rightarrow -\dfrac{b}{2a} > 0$

$\therefore b < 0$

$f(0) = c < 0$

Also,

$-\dfrac{b}{2a} = 4 \Rightarrow 8a + b = 0$

14. a, c, d.

Let the roots be a/r, a, ar, where $a > 0$, $r > 1$. Now,

$(a/r) + a + ar = -p$ (1)

$a(a/r) + a(ar) + (ar)(a/r) = q$ (2)

$(a/r)(a)(ar) = 1$ (3)

$\Rightarrow a^3 = 1$

$\Rightarrow a = 1$

Hence, (c) is correct. From (1), putting $a = 1$, we get

$-p - 3 > 0 \quad \left(\because r + \dfrac{1}{r} > 2\right)$

$\Rightarrow -p > 3 \Rightarrow p < -3$

Hence, (b) is not correct. Also,

$(1/r) + 1 + r = -p$ (4)

From (2), putting $a = 1$, we get

$(1/r) + r + 1 = q$ (5)

From (4) and (5), we have

$-p = q \Rightarrow p + q = 0$

Hence, (a) is correct. Now, as $r > 1$

$a/r = 1/r < 1$

and

$ar = r > 1$

Hence, (d) is correct.

15. a, b.

Given, $(\sin \alpha)\, x^2 - 2x + b \geq 2$.

Let $f(x) = (\sin \alpha)x^2 - 2x + b - 2$.

Abscissa corresponding to the vertex is given by

$x = \dfrac{1}{\sin \alpha} > 1$

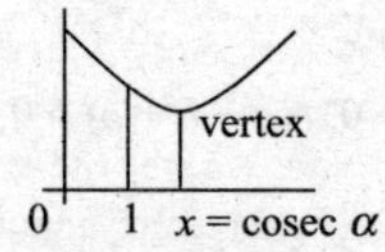

Fig. 1.13

The graph of $f(x) = (\sin \alpha)x^2 - 2x + b - 2$, $\forall\, x \leq 1$, is shown in the figure. Therefore, minimum of $f(x) = (\sin \alpha)x^2 - 2x + b - 2$ must be greater than zero but minimum is at $x = 1$. That is,

$\sin \alpha - 2 + b - 2 \geq 0$, $b \geq 4 - \sin \alpha$, $\alpha \in (0, \pi)$

16. a, c.

Since each pair has common root, let the roots be α, β for Eq. (1); β, γ for Eq. (2) and γ, α for Eq. (3). Therefore,

$\alpha + \beta = -\alpha$, $\alpha\beta = bc$

$\beta + \gamma = -b$, $\beta\gamma = ca$

$\gamma + \alpha = -c$, $\gamma\alpha = ab$

Adding, we get

$2(\alpha + \beta + \gamma) = -(a + b + c)$

$\Rightarrow \alpha + \beta + \gamma = -\dfrac{1}{2}(a + b + c)$

Also by multiplying product of roots, we have

$\alpha^2 \beta^2 \gamma^2 = a^2 b^2 c^2 \Rightarrow \alpha\beta\gamma = abc$

17. a, c.

Given,

$b^2 = ac$

$\Rightarrow \left(\frac{b}{a}\right)^2 = \frac{c}{a}$

$\Rightarrow \left(-\frac{b}{a}\right)^2 = \frac{c}{a}$

$\Rightarrow (\alpha+\beta)^2 = \alpha\beta$

$\Rightarrow \alpha^2+\beta^2+\alpha\beta = 0$

$\Rightarrow \left(\frac{\alpha}{\beta}\right)^2 + \left(\frac{\alpha}{\beta}\right) + 1 = ($

$\Rightarrow \frac{\alpha}{\beta} = \omega, \omega^2$

18. c, d.

We have,

$D = (b-c)^2 - 4a(a-b-c) > 0$

$\Rightarrow b^2 + c^2 - 2bc - 4a^2 + 4ab + 4ac > 0$

$\Rightarrow c^2 + (4a - 2b)c - 4a^2 + 4ab + b^2 > 0$ for all $c \in R$

Discriminant of the above expression in c must be negative. Hence,

$(4a - 2b)^2 - 4(-4a^2 + 4ab + b^2) < 0$

$\Rightarrow 4a^2 - 4ab + b^2 + 4a^2 - 4ab - b^2 < 0$

$\Rightarrow a(a-b) < 0$

$\Rightarrow a < 0$ and $a - b > 0$ or $a > 0$ and $a - b < 0$

$\Rightarrow b < a < 0$ or $b > a > 0$

19. a, d.

Since $\alpha, \beta, \gamma, \delta$ are in H.P., hence $1/\alpha, 1/\beta, 1/\gamma, 1/\delta$ are in A.P. and they may be taken as $a-3d, a-d, a+d, a+3d$. Replacing x by $1/x$, we get the equation whose roots are $1/\alpha, 1/\beta, 1/\gamma, 1/\delta$. Therefore, equation $x^2 - 4x + A = 0$ has roots $a-3d, a+d$ and equation $x^2 - 6x + B = 0$ has roots $a-d, a+3d$. Sum of the roots is

$2(a-d) = 4, 2(a+d) = 6$

$\therefore a = 5/2, d = 1/2$

Product of the roots is

$(a-3d)(a+d) = A = 3$

$(a-d)(a+3d) = B = 8$

20. a, b.

Equations $x^2 + px + q = 0$ and $x^2 + p'x + q' = 0$ have a common root. Therefore,

$(q - q')^2 = (pq' - p'q)(p' - p)$ (1)

Subtracting two equations, we have

$$x = \frac{q-q'}{p'-p}$$

Also using (1),

$$x = \frac{q-q'}{p'-p} = \frac{pq'-p'q}{q-q'}$$

21. b, c.

$f(x) = x^3 + 3x^2 - 9x + c$ is of the form $(x-\alpha)^2(x-\beta)$, showing that α is a double root so that $f'(x) = 0$ has also one root α, i.e., $3x^2 + 6x - 9 = 0$ has one root α. Hence, $x^2 + 2x - 3 = 0$ or $(x+3)(x-1) = 0$ has the root α which can be either -3 or 1. If $\alpha = 1$, then $f(x) = 0$ gives $c - 5 = 0$ or $c = 5$. If $\alpha = -3$, then $f(x) = 0$ gives

$-27 + 27 + 27 + c = 0$

$\therefore c = -27$

22. a, c.

If α be the common root, then

$\alpha^2 + b\alpha - a = 0$ and $\alpha^2 - a\alpha + b = 0$

Subtracting,

$\alpha(b + a) - (a + b) = 0$

$\Rightarrow (a + b)(\alpha - 1) = 0$

$\Rightarrow a + b = 0$ or $\alpha = 1$

When $\alpha = 1$, then from any equation we have $a - b = 1$.

23. a, d.

Let,

$$\frac{x^2+ax+3}{x^2+x+a} = y$$

$\Rightarrow x^2(1-y) - x(y-a) + 3 - ay = 0$

$\because x \in R$

$(y-a)^2 - 4(1-y)(3-ay) \ge 0$

$\Rightarrow (1-4a)y^2 + (2a+12)y + a^2 - 12 \ge 0$ (1)

Now, (1) is true for all $y \in R$, if $1 - 4a > 0$ and $D \le 0$. Hence,

$a < \frac{1}{4}$ and $4(a+6)^2 - 4(a^2-12)(1-4a) \le 0$

$\Rightarrow a < \frac{1}{4}$ and $4a^3 - 36a + 48 \le 0$

$\Rightarrow a < \frac{1}{4}$ and $4a^3 \le 36a - 48$

$\Rightarrow 4a^3 < 36\left(\frac{1}{4}\right) - 48$

$\Rightarrow 4a^3 + 39 < 0 \quad \left[\because a < \frac{1}{4}\right]$

24. a, b.

Here,

$\cos^2\theta - \sin^2\theta = \cos 2\theta$

$\Rightarrow \cos^4\theta - \sin^4\theta = \cos 2\theta$

$\Rightarrow (-2b)^2 - 4b = (-4)^2 - 4 \times 2$

(since L.H.S. is difference of roots of first equation and R.H.S. is difference of roots of second equation)

$\Rightarrow 4b^2 - 4b = 16 - 8 = 8$

$\Rightarrow 4b^2 - 4b - 8 = 0$

$\Rightarrow b^2 - b - 2 = 0$

$\Rightarrow (b+1)(b-2) = 0$

$\Rightarrow b = 2, -1$

25. b, d.

Let $f(x) = x^2 + ax + b$. Then,

$x^2 + (2c + a)x + c^2 + ac + b = f(x + c)$

Thus, the roots of $f(x + c) = 0$ will be $0, d - c$.

26. c, d.

Product of roots is

$\frac{a}{bc} < 0 \quad [\because abc < 0]$

Hence, roots are real and of opposite sign.

27. c, d.

Given equation is

$x^2 + 2(a + 1)x + 9a - 5 = 0$

$D = 4(a + 1)^2 - 4(9a - 5) = 4(a - 1)(a - 6)$

$\therefore D \geq 0 \Rightarrow a \leq 1$ or $a \geq 6 \Rightarrow$ roots are real

If $a < 0$, then $9a - 5 < 0$. Hence, the products of roots is less than 0. So, the roots are of opposite sign. If $a > 7$, then sum of roots is $-2(a + 1) < 0$. Product of roots is greater than 0.

28. a, c.

Since $P(x)$ divides both of them, hence $P(x)$ also divides

$(3x^4 + 4x^2 + 28x + 5) - 3(x^4 + 6x^2 + 25)$

$= -14x^2 + 28x - 70$

$= -14(x^2 - 2x + 5)$

which is a quadratic. Hence,

$P(x) = x^2 - 2x + 5 \Rightarrow P(1) = 4$

30. a., d.

On putting $x = 0$, 1 and 1/2, we get

$-1 \leq c \leq 1$ (1)

$-1 \leq a + b + c \leq 1$ (2)

$-4 \leq a + 2b + 4c \leq 4$ (3)

From (1), (2), (3), we get

$|b| \leq 8$ and $|a| \leq 8$

$\Rightarrow |a| + |b| + |c| \leq 17$

31. a, b, c, d.

Let $f(x) = ax^2 + bx + c$.

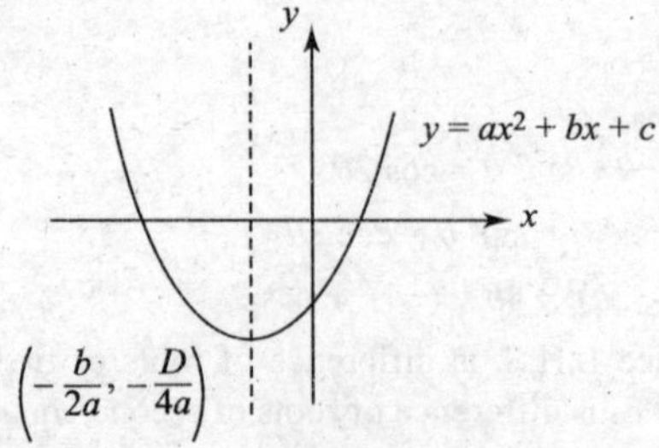

Fig. 1.14

From the diagram, we can see that $a > 0$, $c < 0$ and $-[b(2a)] < 0$. Hence, $b > 0$. Therefore, $a + b - c > 0$

Reasoning Type

1. a. We must have

$ax^3 + (a + b)x^2 + (b + c)x + c > 0$

$\Rightarrow ax^2(x + 1) + bx(x + 1) + c(x + 1) > 0$

$\Rightarrow (x + 1)(ax^2 + bx + c) > 0$

$\Rightarrow a(x + 1)\left(x + \frac{b}{2a}\right)^2 > 0$ as $b^2 = 4ac$

$\Rightarrow x > -1$ and $x \neq -\frac{b}{2a}$

2. b. If $a > 0$, then graph of $y = ax^2 + 2bx + c$ is concave upward. Also if $b^2 - ac < 0$, then the graph always lies above the x-axis; hence, $ax^2 + 2bx + c > 0$ for all real values of x. Thus, domain of function $f(x) = \sqrt{ax^2 + 2bx + c}$ is R.

If $b^2 - ac < 0$, then $ax^2 + 2bx + c = 0$ has imaginary roots. Then the graph of $y = ax^2 + 2bx + c$ never cuts x-axis, or y is either always positive or always negative. Hence, both the statements are correct but statement 2 is not correct explanation of statement 1.

3. a. $ax^2 + bx + c = 0$ has two complex conjugate roots only if all the coefficients are real. If all the coefficients are not real then it is not necessary that both the roots are imaginary. Hence, statement 1 is true.

Now, equation $x^2 - 3x + 4 = 0$ has two complex conjugate roots. If $ax^2 + bx + c = 0$ has all coefficients real, then there will be two common roots. But if there is only one root common, then at least one of a, b, c must be non-real.

Thus both the statements are true and statement 2 is correct explanation of statement 1.

4. a. $\cos\frac{\pi}{4} = 2\cos^2\frac{\pi}{8} - 1$

$\Rightarrow \cos^2\frac{\pi}{8} = \left(\frac{1}{\sqrt{2}} + 1\right)\frac{1}{2}$

$\Rightarrow \cos^4\frac{\pi}{8} = \frac{1}{4}\left(\frac{1}{2} + 1 + \frac{2}{\sqrt{2}}\right) = \left(\frac{3}{2} + \sqrt{2}\right)\frac{1}{4}$

$\Rightarrow \frac{1}{4}\left(\frac{3}{2} + \sqrt{2}\right) + \frac{a}{2}\left(\frac{1}{\sqrt{2}} + 1\right) + b = 0$

($\because \cos^2 \pi/8$ a root of equation)

$\Rightarrow \left(\frac{3}{8} + \frac{a}{2} + b\right) + \sqrt{2}\left(\frac{1}{4} + \frac{a}{a}\right) = 0$

Since a and b are rational, so

$\frac{1}{4} + \frac{a}{4} = 0, \frac{3}{8} + \frac{a}{2} + b = 0$

$\Rightarrow a = -1, b = \frac{1}{8}$

Thus, both the statements are correct and statement 2 is correct explanation of statement 1.

5. a. If $a^2 + b^2 + c^2 < 0$, then all a, b, c are not real or at least one of a, b, c is imaginary number. Hence roots of equation $ax^2 + bx + c = 0$ has no complex conjugate roots, even though the roots are complex. Hence statement 1 is true. Statement 2 is obviously true. Also, statement 2 is correct explanation of statement 1.

6. b. $ix^2 + (i-1)x - \frac{1}{2} - i = 0$

$$\Rightarrow x = \frac{-(i-1) \pm \sqrt{(i-1)^2 - 4(i)\left(-\frac{i}{2} - i\right)}}{2i} = \frac{-(i-1) \pm \sqrt{-4}}{2i}$$

Thus, roots are imaginary. Also, we have $b^2 - 4ac = -4 < 0$, but this is not the correct reason for which roots are imaginary as coefficients of the equation are imaginary.

Hence, both the statements are correct but statement 2 is not correct explanation of statement 1.

7. a. $f(x) = (x - 1)(ax + b)$

$f(2) = 2a + b$

$f(4) = 3(4a + b) = 12a + 3b$

$f(2) + f(4) = 14a + 4b = 0$

$\Rightarrow \frac{-b}{a} = 3.5$

Now, sum of roots is. Hence, the other root is 3.5.

8. c. Here, $f(x)$ is a downward parabola.

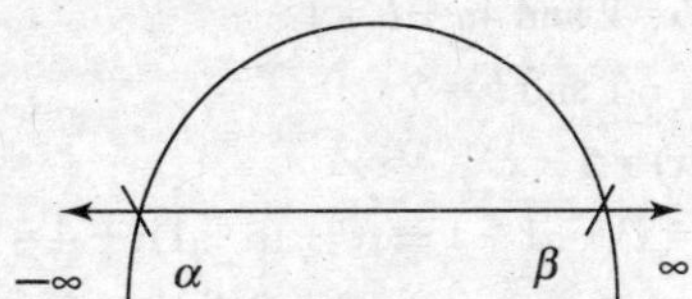

Fig. 1.15

$D = (a + 1)^2 + 20 > 0$

From the graph, clearly, statement 1 is true but statement 2 is false.

9. a. $x^2 + x + 1 = 0$

$D = -3 < 0$

Therefore, $x^2 + x + 1 = 0$ and $ax^2 + bx + c = 0$ have both the roots common. Hence,

$a = b = c$

10. d. Statement 2 is obviously true. Let,

$f(x) = (x - p)(x - r) + \lambda (x - q)(x - s) = 0$

Then,

$f(p) = \lambda (p - q)(p - s)$

$f(r) = \lambda (r - q)(r - s)$

$\Rightarrow f(p) f(r) < 0$

Hence, there is a root between p and r. Thus, statement 1 is false.

11. a. Given equation is

$px^2 + qx + r = 0$

Let,

$f(x) = px^2 + qx + r$

$f(0) = r > 0$

$f(1) = p + q + r < 0$

$f(-1) = p - q + r < 0$

Hence, one root lies in $(-1, 0)$ and the other in $(0, 1)$.

$\therefore [\alpha] = -1$ and $[\beta] = 0$

$\Rightarrow [\alpha] + [\beta] = -1$

Therefore, statement 2 is true and is correct explanation of statement 1.

12. d. Let $f(x) = (x - \sin \alpha)(x - \cos \alpha) - 2$. Then,

$f(\sin \alpha) = -2 < 0, f(\cos \alpha) = -2 < 0$

Also, as $0 < a < \pi/4$, hence, in $\alpha < \cos \alpha$. Therefore, equation $f(x) = 0$ has one root in $(-\infty, \sin \alpha)$ and other in $(\cos \alpha, \infty)$.

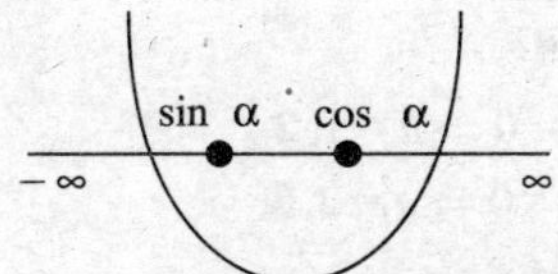

Fig. 1.16

13. b. The equation can be written as

$(2^x)^2 - (a - 3) 2^x + (a - 4) = 0$

$\Rightarrow 2^x = 1$ and $2^x = a - 4$

We have,

$x \leq 0$ and $2^x = a - 4$ $\quad [\because x$ is non-positive$]$

$\therefore 0 < a - 4 \leq 1 \Rightarrow 4 < a \leq 5$

$\therefore a \in (4, 5]$

14. c. Clearly, Statement 1 is true but Statement 2 is false, since, $ax^2 + bx + c = 0$ is an identity when $a = b = c = 0$.

15. d. Roots of the equation $x^5 - 40x^4 + px^3 + qx^2 + rx + s = 0$ are in G.P. Let the roots be a, ar, ar^2, ar^3, ar^4. Therefore,

$$a + ar + ar^2 + ar^3 + ar^4 = 40 \quad (1)$$

and

$$\frac{1}{a} + \frac{1}{ar} + \frac{1}{ar^2} + \frac{1}{ar^3} + \frac{1}{ar^4} = 10 \quad (2)$$

From (1) and (2),

$$ar^2 = \pm 2 \quad (3)$$

Now, the product of roots is $a^5r^{10} = (ar^2)^5 = \pm 32$.

$\therefore |S| = 32$

16. a. Given equations are

$$ax^2 + 2bx + c = 0 \quad (1)$$

$$a_1x^2 + 2b_1x + c_1 = 0 \quad (2)$$

Since Eqs. (1) and (2) have only one common root and a, b, c, a_1, b_1, c_1 are rational, therefore, common root cannot be imaginary or irrational (as irrational roots occur in conjugate pair when coefficients are rational, and complex roots always occur in conjugate pair).

Hence, the common root must be rational. Therefore, both the roots of Eqs. (1) and (2) will be rational. Therefore, $4(b^2 - ac)$ and $4(b_1^2 - a_1c_1)$ must be perfect squares (squares of rational numbers). Hence, $b^2 - ac$ and $b_1^2 - a_1c_1$ must be perfect squares.

17. a. Let $f(x) = ax^2 + bx + c$. Since coefficient are integers and one root is irrational, so both the roots are irrational. Hence, for any $\lambda \in Q$,

$f(\lambda) \neq 0 \Rightarrow |f(\lambda)| > 0$

$$\Rightarrow \left|\frac{ap^2}{q^2} + \frac{bp}{q} + c\right| > 0, \quad \text{where } \lambda = \frac{p}{q},\ p, q \in Z$$

$$\Rightarrow \frac{1}{q^2}\left|ap^2 + bpq + cq^2\right| > 0$$

Now, $a, b, c, p, q \in I$. Hence,

$|ap^2 + bpq + cq^2| \geq 1$

$$\Rightarrow |f(\lambda)| \geq \frac{1}{q^2}$$

18. d. $a^2 - 3a + 2 = 0 \Rightarrow a = 1, 2$

$a^2 - 5a + 6 = 0 \Rightarrow a = 2, 3,$

$a^2 - 4 = 0 \Rightarrow a = \pm 2$

Therefore, $a = 2$ is the only solution.

Hence, statement 1 is false. Statement 2 is true by definition.

19. b. According to statement 1, given equation is

$x^2 - bx + c = 0$

Let α, β be two roots such that

$|\alpha - \beta| = 1$

$\Rightarrow (\alpha + \beta)^2 - 4\alpha\beta = 1$

$\Rightarrow b^2 - 4c = 1$

According to statement 2, given equation is $4abc\ x^2 + (b^2 - 4ac)\ x - b = 0$. Hence,

$D = (b^2 - 4ac)^2 + 16ab^2c$

$= (b^2 + 4ac)^2 > 0$

Hence, roots are real and unequal.

20. c. $f(x) = ax^2 + bx + c$

Given,

$f(0) + f(1) = 2$

$\Rightarrow f(x) > 0\ \forall\ x \in R$

Hence, statement 1 is true. Let,

$f(x) = x^2 - x + 1$

$a + b = 0$

Hence, statement 2 is false.

21. d. $f(x, y) = (2x - y)^2 + (x + y - 3)^2$ (1)

Therefore, statement 1 is false as it represents a point (1, 2).

22. a. $D = \underbrace{\underbrace{(2m+1)^2}_{\text{odd}} - \underbrace{4(2n+1)}_{\text{even}}}_{\text{odd}}$

For rational root, D must be a perfect square. As D is odd, let D be perfect square of $2l + 1$, where $l \in Z$.

$(2m + 1)^2 - 4(2n + 1) = (2l + 1)^2$

$\Rightarrow (2m + 1)^2 - (2l + 1)^2 = 4(2n + 1)$

$\Rightarrow [(2m + 1) + (2l + 1)]\ [2(m - l)] = 4(2n + 1)$

$\Rightarrow (m + l + 1)\ (m - l) = (2n + 1)$ (1)

R.H.S. of (1) is always odd but L.H.S. is always even. Hence, D cannot be a perfect square. So, the roots cannot be rational.

Hence, statement 1 is true, statement 2 is true and statement 2 is correct explanation for statement 1.

Linked Comprehension Type

For Problems 1–3

1. d., 2. c., 3. c.

Sol. Let unknown polynomial be $P(x)$. Let $Q(x)$ and $R(x)$ be the quotient and remainder, respectively, when it is divided by $(x - 3)(x - 4)$. Then,

$P(x) = (x - 3)(x - 4)Q(x) + R(x)$

Then, we have

$R(x) = ax + b$

$\Rightarrow P(x) = (x - 3)(x - 4)Q(x) + ax + b$

Given that $P(3) = 2$ and $P(4) = 1$. Hence,

$3a + b = 2$ and $4a + b = 1$

$\Rightarrow a = -1$ and $b = 5$

$\Rightarrow R(x) = 5 - x$

1. $x + 5 = x^2 + ax + 1 \Rightarrow x^2 + (a + 1)x - 4 = 0$

Given that roots are real and distinct.

$\therefore D > 0 \Rightarrow (a + 1)^2 + 16 > 0,$

which true for all real x.

2. $-x + 5 = px^2 + (q - 1)x + 6 \Rightarrow px^2 + qx + 1 = 0$

Now, $p > 0$ and equation has no distinct real roots or equation has real and equal or imaginary roots. Then,

$px^2 + qx + 1 \geq 0,\ \forall\ x \in R$

$\Rightarrow f(3) \geq 0 \Rightarrow 9p + 3q + 1 \geq 0 \Rightarrow 3p + q \geq -1/3$

Hence, the least value of $3p + q$ is $-1/3$.

3. c. $f(x) = y = \dfrac{-x + 5}{x^2 - 3x + 2}$

$\Rightarrow yx^2 + (1 - 3y)x + 2y - 5 = 0$

Now, if x is real, then

$D \geq 0$

$\Rightarrow (1 - 3y)^2 - 4y(2y - 5) \geq 0$

$\Rightarrow y^2 + 14y + 1 \geq 0$

$$\Rightarrow y \in \left(-\infty, \frac{-14 - \sqrt{192}}{2}\right] \cup \left[\frac{-14 + \sqrt{192}}{2}, \infty\right)$$

$(-\infty, -7 - 4\sqrt{3}] \cup [-7 + 4\sqrt{3}, \infty)$

For Problems 4–6

4. d., 5. b., 6. b.

Sol. $ax^2 - bx + c = 0$

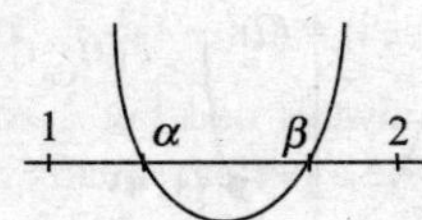

Fig. 1.17

Let $f(x) = ax^2 - bx + c$ be the corresponding quadratic expression and α, β be the roots of $f(x) = 0$. Then,

$f(x) = a(x - \alpha)(x - \beta)$

Now,

$$af(1) > 0,\ af(2) > 0,\ 1 < -\frac{b}{2a} < 2,\ b^2 - 4ac > 0$$

$\Rightarrow a(1 - \alpha)(1 - \beta) > 0,\ a(2 - \alpha)(2 - \beta) > 0,\ 2a < b < 4a,$

$b^2 - 4ac > 0$

$\Rightarrow a^2(1 - \alpha)(1 - \beta)(2 - \alpha)(2 - \beta) > 0$

$\Rightarrow a^2(\alpha - 1)(2 - \alpha)(\beta - 1)(2 - \beta) > 0$

As $f(1)$ and $f(2)$ both are integers and $f(1) > 0$, and $f(2) > 0$, so

$f(1)\ f(2) > 0$

$\Rightarrow f(1)\ f(2) \geq 1$

$\Rightarrow 1 \leq a^2(\alpha - 1)(2 - \alpha)(\beta - 1)(2 - \beta)$

Now,

$$\frac{(\alpha - 1) + (2 - \alpha)}{2} \geq ((\alpha - 1)(2 - \alpha))^{1/2}$$

$$\Rightarrow (\alpha - 1)(2 - \alpha) \leq \frac{1}{4}$$

Similarly,

$$(\beta - 1)(2 - \beta) \leq \frac{1}{4}.$$

$$\Rightarrow (\alpha - 1)(2 - \alpha)(\beta - 1)(2 - \beta) < \frac{1}{16}$$

As $\alpha \neq \beta$, so

$a^2 > 16 \Rightarrow a \geq 5$

$\Rightarrow b^2 > 20c$ and $b > 10 \Rightarrow b \geq 11$

Also,

$b^2 > 100 \Rightarrow c > 5 \Rightarrow c \geq 6$

For Problems 7–9

7. c., 8. a., 9. c.

Sol. Given equation is

$$x^4 + 2ax^3 + x^2 + 2ax + 1 = 0 \qquad (1)$$

or

$$\left(x^2 + \frac{1}{x^2}\right) + 2a\left(x + \frac{1}{x}\right) + 1 = 0$$

$$\Rightarrow \left(x + \frac{1}{x}\right)^2 + 2a\left(x + \frac{1}{x}\right) - 1 = 0$$

$$\Rightarrow t^2 + 2at - 1 = 0 \qquad (2)$$

where $t = x + (1/x)$. Now,

$$\left(x + \frac{1}{x}\right) \geq 2$$

or

$$\left(x + \frac{1}{x}\right) \leq -2$$

$\therefore t \geq 2$ or $t \leq -2$

7. Now, Eq. (1) will have at least two positive roots, when at least one root of Eq. (2) will be greater than 2. From Eq. (2),

$$D = 4a^2 - 4(-1) = 4(1 + a^2) > 0,\ \forall\ a \in R \qquad (3)$$

Let the roots of Eq. (2) be α, β. If $\alpha, \beta \leq 2$, then

$$\Rightarrow f(2) \geq 0 \text{ and } \frac{-B}{2A} < 2$$

$$\Rightarrow 4 + 4a - 1 \geq 0 \text{ and } -\frac{2a}{2} < 2$$

$$\Rightarrow a \geq -\frac{3}{4} \text{ and } a > -2$$

$$\Rightarrow a \geq -\frac{3}{4}$$

Therefore, at least one root will be greater than 2. Then,

$$a < -\frac{3}{4} \qquad (4)$$

Combining (3) and (4), we get

$$a < -\frac{3}{4}$$

Hence, at least one root will be positive if $a \in [-\infty, -(3/4)]$.

8. Now, Eq. (1) will have at least two roots negative, when at least one root of Eq. (2) will be less than −2. If $\alpha, \beta \geq -2$, then

$$f(-2) \geq 0 \text{ and } -\frac{B}{2A} > -2$$

$$\therefore 4 - 4a - 1 \geq 0 \text{ and } -\frac{2a}{2} > -2$$

$$\therefore a \leq \frac{3}{4} \text{ and } a < 2$$

$$\therefore a \leq \frac{3}{4} \qquad (5)$$

Combining (3) and (5), at least one root will be less than −2 for Eq. (2) if

$$a > \frac{3}{4}$$

$$\therefore a \in \left(\frac{3}{4},\ \infty\right)$$

9. If exactly two roots are positive, then other two roots are negative. Then – 2 and 2 must lie between the roots. So,

$f(-2) < 0$ and $f(2) < 0$

$\Rightarrow a > 3/4$ and $a < -3/4$

Hence, no such values of a exist.

For Problems 10–12

10. a., 11. d., 12. c.

Sol. $(\beta - \alpha) = ((\beta + h) - (\alpha + h))$

$(\beta + \alpha)^2 - 4\alpha\beta = [(\beta + h) + (\alpha + h)]^2 - 4(\beta + h)(\alpha + h)$

$(-b_1)^2 - 4c_1 = (-b_2)^2 - 4c_2$

$D_1 = D_2$ (1)

The least value of $f(x)$ is

$$-\frac{D_1}{4} = -\frac{1}{4} \Rightarrow D_1 = 1 \text{ and } D_2 = 1$$

Therefore, the least value of

$$g(x) \text{ is } -\frac{D_2}{4} = -\frac{1}{4}$$

The least value of $g(x)$ occurs at

$$\frac{b_2}{2} = \frac{7}{2} \Rightarrow b_2 = -7$$

$$b_2^2 - 4c_2 = D_2$$

$$49 - 4c_2 = 1 \Rightarrow \frac{48}{4} = c_2 \Rightarrow c_2 = 12$$

$$x^2 - 7x + 12 = 0 \Rightarrow x = 3, 4$$

For Problems 13–15

13. d., 14. c., 15. d.

Sol.

13. $AC = 4\sqrt{2}$

$$\therefore AB = BC = \frac{4\sqrt{2}}{\sqrt{2}} = 4 \text{ units}$$

$$OB = \sqrt{4^2 - (2\sqrt{2})^2} = 2\sqrt{2}$$

$$\therefore A(-2\sqrt{2}, 0), B(2\sqrt{2}, 0), C(0, -2\sqrt{2})$$

Since $y = ax^2 + bx + c$ passes through A, B and C, we get

$$y = \frac{x^2}{2\sqrt{2}} - 2\sqrt{2}$$

14. Minimum value of $y = x^2/(2\sqrt{2}) - 2\sqrt{2}$ is $2\sqrt{2}$ at $x = 0$.

15. $f(x) = 0$

$\Rightarrow x^2/(2\sqrt{2}) - 2\sqrt{2} = 0$ or $x = \pm\, 2\sqrt{2}$

Therefore, number of integral values of k for which k lies in $(-2\sqrt{2}, 2\sqrt{2})$ is 5.

For Problems 16–18

16. d., 17. c., 18. b.

Sol. Given that $9^x - a3^x - a + 3 \le 0$

Let $t = 3^x$. Then,

$t^2 - at - a + 3 \le 0$

or

$t^2 + 3 \le a(t + 1)$

where $t \in R^+$, $\forall\, x \in R$

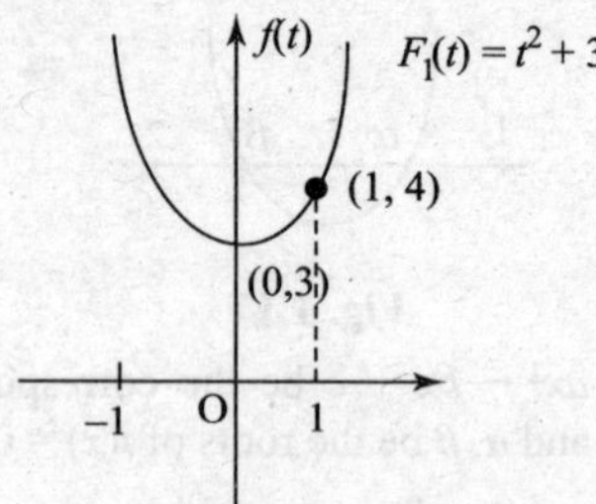

Fig. 1.18

Let $f_1(t) = t^2 + 3$ and $f_2(t) = a(t + 1)$.

16. For $x < 0$, $t \in (0, 1)$. That means (1) should have at least one solution in $t \in (0, 1)$. From (1), it is obvious that $a \in R^+$. Now $f_2(t) = a(t + 1)$ represents a straight line. It should meet the curve $f_1(t) = t^2 + 3$, at least once in $t \in (0, 1)$.

$f_1(0) = 3, f_1(1) = 4, f_2(0) = a, f_2(1) = 2a$

If $f_1(0) = f_2(0)$, Then $a = 3$; if $f_1(1) = f_2(1)$, then $a = 2$. Hence, the required range is $a \in (2, 3)$.

17. For at least one positive solution, $t \in (1, \infty)$. That means graphs of $f_1(t) = t^2 + 3$ and $f_2(t) = a(t + 1)$ should meet at least once in $t \in (1, \infty)$. If $a = 2$, both the curves touch each other at (1, 4). Hence, the required range is $a \in (2, \infty)$.

18. In this case both graphs should meet at least once in $t \in (0, \infty)$. For $a = 2$ both the curves touch, hence, the required range is $a \in [2, \infty)$.

For Problems 19–21

19. d., 20. c., 21. a.

Sol. Let $f(x) = x^2 + x + a - 9$.

$x^2 + x + a - 9 < 0$ has at least one positive solution, then either both the roots of equation $x^2 + x + a - 9 = 0$ are non-negative or 0 lies between the roots.

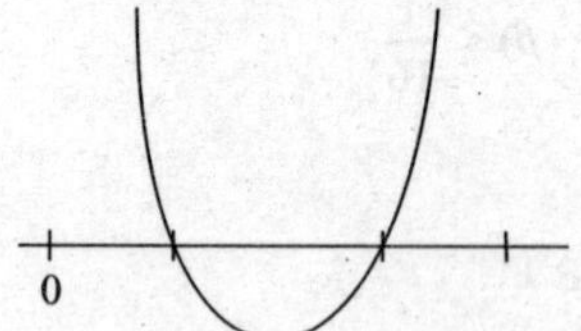

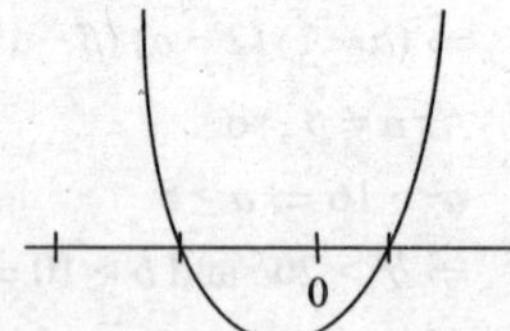

Fig. 1.19

For case I, sum or roots is $-1/2 > 0$, which is not possible.

For case II,

$f(0) < 0 \Rightarrow a - 9 < 0 \Rightarrow a < 9$

20.

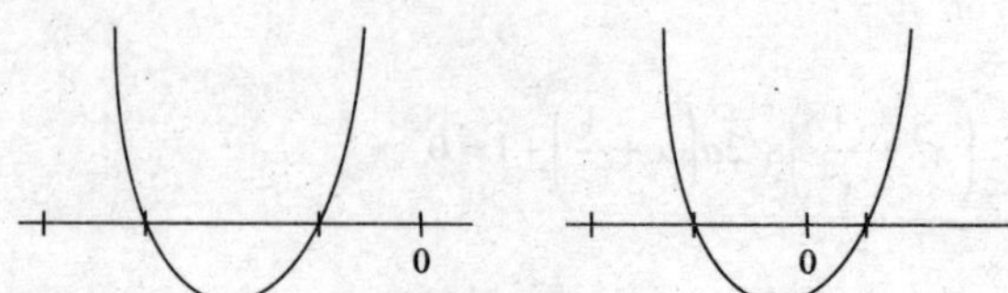

Fig. 1.20

If $x^2 + x + a - 9 < 0$ has at least one negative solution, then either both the roots of equation $x^2 + x + a - 9 = 0$ are non-positive or 0 lies between the roots.

For case I, sum of roots is $-1/2 < 0$. Product of roots is $a - 9 > 0 \Rightarrow a \geq 9$ and

$D > 0 \Rightarrow 1 - 4(a - 9) > 0 \Rightarrow a < \frac{37}{4}$

Hence, $9 \leq a < 37/4$.

21. If $x^2 + x + a - 9 < 0$ is true $\forall x \in (-1, 3)$, then $f(-1) < 0$ and $f(3) < 0$.

$\therefore 1 - 1 + a - 9 < 0$ and $9 + 3 + a - 9 < 0$

$\Rightarrow a < 9$ and $a < -3$

$\Rightarrow a < -3$

For Problems 22–24

22. b., 23. a., 24. b.

Sol.

22. $b^2 > (a + c)^2$

$\Rightarrow (a + c - b)(a + c + b) < 0$

$\Rightarrow f(-1) f(1) < 0$

So, there is exactly one root in $(-1, 1)$.

23. $af(1) < 0$ and $f(0) f(1) > 0$

$\Rightarrow af(1) < 0$ and $af(0) < 0$

Hence, both the numbers 0 and 1 lie between the roots.

24. $f(0) f(1) < 0$ and $af(1) > 0$

$\Rightarrow f(0) f(1) < 0$ and $af(0) < 0$

Hence, exactly one root lies in (0, 1) and 0 lies between the roots.

For Problems 25–26

25. a., 26. b.

Sol. From the question, the real roots of $x^3 - x^2 + \beta x + \gamma = 0$ are x_1, x_2, x_3 and they are in A.P. As x_1, x_2, x_3 are in A.P., let $x_1 = a - d, x_2 = a, x_3 = a + d$. Now,

$$x_1 + x_2 + x_3 = \frac{-1}{1} = 1$$

$$\Rightarrow a - d + a + a + d = 1$$

$$\Rightarrow a = \frac{1}{3} \quad (1)$$

$$x_1x_2 + x_2x_3 + x_3x_1 = \frac{\beta}{1} = \beta$$

$$\Rightarrow (a - d)\, a + a\,(a + d) + (a + d)(a - d) = \beta \quad (2)$$

$$x_1x_2x_3 = -\frac{\gamma}{1} = -\gamma$$

$$\Rightarrow (a - d)\, a\,(a + d) = -\gamma \quad (3)$$

From (1) and (2), we get

$3a^2 - d^2 = b$

$$\Rightarrow 3\frac{1}{9} - a^2 = \beta, \text{ so } \beta = \frac{1}{3} - d^2 < \frac{1}{3}$$

From (1) and (3), we get

$$\frac{1}{3}\left(\frac{1}{9} - d^2\right) = -\gamma$$

$$\Rightarrow \gamma = \frac{1}{3}\left(d^2 - \frac{1}{9}\right) > \frac{1}{3}\left(-\frac{1}{9}\right) = -\frac{1}{27}$$

$$\therefore \beta \in \left(-\infty, \frac{1}{3}\right) \text{ and } \gamma \in \left(-\frac{1}{27}, +\infty\right)$$

Matrix-Match Type

1. a → s; b → r; c → q; d → p.

Obviously when $a \geq 0$, we have no roots as all the terms are followed by +ve sign. Also for $a = -2$, we have

$x^2 - 2|x| + 1 = 0$

or

$|x| - 1 = 0 \Rightarrow x = \pm 1$

Hence, the equation has two roots.

Also when $a < -2$, for given equation

$$|x| = \frac{-a \pm \sqrt{a^2 - 4}}{2} > 0$$

Hence, the equation has four roots as $|-a| > \sqrt{a^2 - 4}$. Obviously, the equation has no three real roots for any value of a.

2. a → p; b → p, q, r, s; c → p, q, s; d → r, s.

a. $y = \dfrac{x^2 - 2x + 4}{x^2 + 2x + 4}$

$\Rightarrow x^2y + 2xy + 4y = x^2 - 2x + 4$

$\Rightarrow (y - 1)x^2 + 2(y + 1)x + 4(y - 1) = 0$

$D \geq 0$

$\Rightarrow 4(y + 1)^2 - 16(y - 1)^2 \geq 0$

$\Rightarrow (y + 1)^2 - (2y - 2)^2 \geq 0$

$\Rightarrow (3y - 1)(3 - y) \geq 0$

$\Rightarrow (3y - 1)(y - 3) \leq 0 \Rightarrow y \in \left[\frac{1}{3}, 3\right]$

$\Rightarrow \{1\} \Rightarrow P$

b. $y = \dfrac{x^2 - 3x - 2}{2x - 3}$

$\Rightarrow x^2 - 3x - 2 = 2xy - 3y$

$\Rightarrow x^2 - (3 + 2y)x + (3y - 2) = 0$

$D \geq 0$

$\Rightarrow (3 + 2y)^2 - 4(3y - 2) \geq 0$

$\Rightarrow 9 + 4y^2 + 12y - 12y + 8 \geq 0$

$\Rightarrow 4y^2 + 17 \geq 0$

which is always true. Hence,

$y \in R \Rightarrow \{1, 4, -3, -10\} \Rightarrow p, q, r, s$

c. $y = \dfrac{2x^2 - 2x + 4}{x^2 - 4x + 3}$

$\Rightarrow x^2y - 4xy + 3y = 2x^2 - 2x + 4$

$(y-2)x^2 + 2(1-2y)x + 3y - 4 = 0$

$D \geq 0$

$4(1-2y)^2 - 4(y-2)(3y-4) \geq 0$

$\Rightarrow 1 + 4y^2 - 4y - (3y^2 - 10y + 8) \geq 0$

$\Rightarrow y^2 + 6y - 7 \geq 0$

$\Rightarrow (y+7)(y-1) \geq 0$

$\Rightarrow y \geq 1$ or $y \leq -7$

$\Rightarrow \{1, 4, -10\} \Rightarrow p, q, s$

d. $f(x) = x^2 - (a-3)x + 2 < 0, \forall x \in [-2, -1]$

$\Rightarrow f(-2) < 0$ and $f(-1) < 0$

$\Rightarrow 4 + 2(a-3) + 2 < 0$ and $1 + (a-2) + 2 < 0$

$\Rightarrow a < 0$ and $a < -1$

$\Rightarrow a < -1$

$\Rightarrow a \in \{-10, -3\}$

3. a → q, r, s; b → r; c → p; d → q.

a. $d + a - b = 0$ and $d + b - c = 0$

$d = b - a$ and $d = c - b$

$\therefore b - a = c - b \Rightarrow 2b = a + b \Rightarrow a, b, c$ are in A.P.

Also $x = 1$ satisfies the second equation. Therefore, the other root is also 1. Product of roots is 1.

$\therefore c(a-b) = a(b-c) \Rightarrow b = \dfrac{2ac}{a+c} \Rightarrow a, b, c$ are in H.P.

Therefore, a, b, c are in A.P. and a, b, c in H.P. Hence, a, b, c are in G.P.

b. $(a^2 + b^2)x^2 - 2b(a+c)x + (b^2 + c^2) = 0$

The roots are real and equal. Hence,

$4b^2(a+c)^2 - 4(a^2+b^2)(b^2+c^2) = 0$

$\Rightarrow b^2(a^2 + c^2 + 2ac) - (a^2b^2 + a^2c^2 + b^4 + b^2c^2) = 0$

$\Rightarrow b^2a^2 + b^2c^2 + 2ab^2c - a^2b^2 - a^2c^2 - b^4 - b^2c^2 = 0$

$\Rightarrow 2ab^2c - a^2c^2 - b^4 = 0 \Rightarrow (b^2 - ac)^2 = 0$

Hence, $b^2 = ac$. Thus a, b, c are in G.P.

c. $(x-1)^3 = 0 \Rightarrow x = 1$ is the common root. Hence, $a + b + c = 0$.

d. $(a+c)^2 + 4b^2 - 4b(a+c) \leq 0, \forall x \in R$

$\Rightarrow ((a+c) - 2b)^2 \leq 0$

$\Rightarrow a + c = 2b$

$\Rightarrow a, b, c$ in A.P.

4. a → r; b → r; c → q; d → p.

a. $(m-2)x^2 - (8-2m)x - (8-3m) = 0$ has roots of opposite signs. The product of roots is

$$-\frac{8-3m}{m-2} < 0$$

$$\Rightarrow \frac{3m-8}{m-2} < 0$$

$\Rightarrow 2 < m < 8/3$

b. Exactly one root of equation $x^2 - m(2x-8) - 15 = 0$ lies in interval (0, 1).

$f(0)f(1) < 0$

$\Rightarrow (0 - m(-8) - 15)(1 - m(-6) - 15) < 0$

$\Rightarrow (8m - 15)(6m - 15) < 0$

$\Rightarrow 15/8 < m < 15/6$

c. $x^2 + 2(m+1)x + 9m - 5 = 0$ has both roots negative. Hence, sum of roots is

$-2(m+1) < 0$ or $m > -1$ (1)

Product of roots is

$9m - 5 > 0 \Rightarrow m > 5/9$ (2)

Discriminant,

$D \geq 0 \Rightarrow 4(m+1)^2 - 4(9m-5) \geq 0$

$\Rightarrow m^2 - 7m + 6 \geq 0$

$\Rightarrow m \leq 1$ or $m \geq 6$ (3)

Hence, for (1), (2), and (3), we get

$$m \in \left(\frac{5}{9}, 1\right] \cup [6, \infty)$$

d. $f(x) = x^2 + 2(m-1)x + m + 5 = 0$ has one root less than 1 and the other root greater than 1. Hence,

$f(1) < 0$

$\Rightarrow 1 + 2(m-1) + m + 5 < 0$

$\Rightarrow m < -4/3$

5. a → s; b → p; c → q; d → r.

a. $x^2 + ax + b = 0$ has root α. Hence,

$\alpha^2 + a\alpha + b = 0$ (1)

$x^2 + px + q = 0$ has roots $-\alpha, \gamma$. Hence,

$\alpha^2 - p\alpha + q = 0$ (2)

Eliminating α from (1) and (2), we get

$(q-b)^2 = (aq + bp)(-p - a)$

$\Rightarrow (q-b)^2 = (aq + bp)(p - a)$

b. $x^2 + ax + b = 0$ has root is α, β. Hence,

$\alpha^2 + a\alpha + b = 0$ (1)

$x^2 + px + q = 0$ has root $1/\alpha$. Hence,

$q\alpha^2 + p\alpha + 1 = 0$ (2)

Eliminating α from (1) and (2), we get

$(1 - bq)^2 = (a - pb)(p - aq)$

c. $x^2 + ax + b = 0$ has roots α, β. Hence,

$\alpha^2 + a\alpha + b = 0$ (1)

$x^2 + px + q = 0$ has roots $-2/\alpha, \gamma$. Hence,

$q\alpha^2 - 2p\alpha + 4 = 0$ (2)

Eliminating α from (1) and (2), we get

$(4 - bq)^2 = (4a + 2pb)(-2p - aq)$

d. $x^2 + ax + b = 0$ has roots α, β. Hence,

$\alpha^2 + a\alpha + b = 0$ (1)

$x^2 + px + q = 0$ has roots $-1/2\alpha, \gamma$. Hence,

$4q\alpha^2 - 2p\alpha + 1 = 0$ (2)

Eliminating α from (1) and (2), we get

$(1 - 4bq)^2 = (a + 2bp)(-2p - 4aq)$

Integer Type

1. **(3)** Let $x = \sqrt{6+\sqrt{6+\sqrt{6+\ldots\infty}}}$

Then, $x = \sqrt{6+x}$

$\Rightarrow x^2 = 6 + x$

$\Rightarrow x^2 - x - 6 = 0$

$\Rightarrow (x-3)(x+2) = 0$

$\Rightarrow x = 3$ or $x = -2$

But the given expression is positive. So, $x = 3$.

Hence, the value of the given expression is 3.

2. **(9)** Let $f(x) = ax^4 + bx^3 - x^2 + 2x + 3$.

Now $x^2 + x - 2 = (x+2)(x-1)$

Given $f(-2) = a(-2)^4 + b(-2)^3\ (-2)^2 + 2(-2) + 3$

$= 4(-2) + 3$

$\Rightarrow 16a\ 8b - 4 - 4 + 3 = -5$

$\Rightarrow 2a - b = 0$ (1)

Also $f(1) = a + b - 1 + 2 + 3 = 4(1) + 3$

$\Rightarrow a + b = 3$ (2)

From (1) and (2) we get $a = 1$, $b = 2$.

3. **(7)** $x^2 + x - 6 = (x+3)(x+2)$

Let $f(x) = 2x^4 + x^3 - ax^2 + bx + a + b - 1$. Therefore,

$f(-3) = 2(-3)^4 + (-3)^3 - a(-3)^2 - 3b + a + b - 1 = 0$

$\Rightarrow 134 - 8a - 2b = 0$

$\Rightarrow 4a + b = 67$ (1)

and $f(2) = 2(2)^4 + 2^3 - a(2)^2 + 2b + a + b - 1 = 0$

$\Rightarrow 39 - 3a + 3b = 0$

$\Rightarrow a - b = 13$ (2)

From (1) and (2) we get a = 16, b = 3.

4. **(2)** Let D_1 and D_2 be the discriminates of $ax^2 + bx + c = 0$ and $-ax^2 + bx + c = 0$ respectively.

Then $D_1 = b^2 - 4ac$ and $D_2 = b^2 + 4ac$.

Now $ac \neq 0 \Rightarrow$ either $ac > 0$ or $ac < 0$.

If $ac > 0$, then $D_2 > 0$. Therefore, the roots of $-ax^2 + bx + c = 0$ are real.

If $ac < 0$, then $D_1 > 0$. Therefore, the roots of $ax^2 + bx + c = 0$ are real.

Thus, $f(x)\, g(x)$ has at least two real roots.

5. **(4)** Since α, β are the roots of $375x^2 - 25x - 2 = 0$,

$\alpha + \beta = \frac{25}{375} = \frac{1}{15}$ and $\alpha\beta = -\frac{2}{375}$

$\Rightarrow \lim_{n\to\infty} \sum_{r=1}^{n} S_r = \lim_{n\to\infty} \sum_{r=1}^{n} (\alpha^r + \beta^r)$

$= (\alpha + \alpha^2 + \alpha^3 + \ldots \infty) + (\beta + \beta^2 + \beta^3 + \ldots \infty)$

$= \frac{\alpha}{1-\alpha} + \frac{\beta}{1-\beta}$

$= \frac{\alpha - \alpha\beta + \beta - \alpha\beta}{(1-\alpha)(1-\beta)}$

$= \frac{\alpha + \beta - 2\alpha\beta}{1 - (\alpha+\beta) + \alpha\beta}$

$= \frac{\frac{1}{15} + \frac{4}{375}}{1 - \frac{1}{15} - \frac{2}{375}}$

$= \frac{25+4}{375-25-2} = \frac{29}{348} = \frac{1}{12}$

6. **(9)** The roots of $x^2 + 3x + 5 = 0$ are non-real.

Thus the given equations will have two common roots. Therefore,

$\frac{a}{1} = \frac{b}{3} = \frac{c}{5} = \lambda$

$\Rightarrow a + b + c = 9\lambda$

Thus the minimum value of $a + b + c = 9$.

7. **(3)** $ax^2 + bx + c = 0$ and $bx^2 + cx + a = 0$ have a common root. Therefore,

$(bc - a^2)^2 = (ab - c^2)(ac - b^2)$

$\Rightarrow b^2c^2 + a^4 - 2a^2bc = a^2bc - ab^3 - ac^3 + b^2c^2$

$\Rightarrow a^4 + ab^3 + ac^3 = 3a^2bc$

$\Rightarrow \frac{a^3 + b^3 + c^3}{abc} = 3$

8. **(1)** We have

$x^2 + ax + 12 = 0$ and

$x^2 + bx + 15 = 0$

Adding these two equations, we get

$2x^2 + (a+b)x + 27 = 0$.

Now subtracting it from the last equation, we get

$x^2 - 9 = 0 \Rightarrow x = 3, -3$.

Thus the common positive root is 3. Therefore,

$9 + 12 + 3a = 0 \Rightarrow a = -7$ and

$9 + 3b + 15 = 0 \Rightarrow b = -8$.

9. **(0)** We have $x^3 - 2x^2 + 2x - 1 = 0$

$\Rightarrow (x-1)(x^2 - x + 1) = 0$

$\Rightarrow x = 1$ or $x = -\omega, -\omega^2$

Since $ax^2 + bx + a = 0$ and $x^3 - 2x^2 + 2x - 1 = 0$ have two roots in common, $-\omega$ and $-\omega^2$ are the common roots (as both roots of a quadratic equation are either real or non-real).

Now $-\omega$ is a root of $ax^2 + bx + a = 0$. Therefore,

$a(1 + \omega^2) - b\omega = 0$

$\Rightarrow a(-\omega) - b\omega = 0$ (as $1 + \omega + \omega^2 = 0$)

$\Rightarrow a + b = 0$

10. **(4)** For the given equation

$\alpha + \beta + \gamma = 0$

$\alpha\beta + \beta\gamma + \alpha\gamma = 4$ and $\alpha\beta\gamma = -1$.

Now

$(\alpha+\beta)^{-1} + (\beta+\gamma)^{-1} + (\gamma+\alpha)^{-1}$

$= (-\gamma)^{-1} + (-\beta)^{-1} + (-\alpha)^{-1}$

$= -\dfrac{\alpha\beta + \beta\gamma + \alpha\gamma}{\alpha\beta\gamma}$

$= -\dfrac{(-4)}{1}$

$= 4$

11. **(7)** We know that $ax^2 + 2hxy + by^2 + 2gx + 2fy + c$ can be resolved into two linear factors if and only if

$abc + 2fgh - af^2 - bg^2 - ch^2 = 0$.

The given expression is $2x^2 + mxy + 3y^2 - 5y - 2$ (1)

Here $a = 2, h = \frac{m}{2}, b = 3, g = 0, f = \frac{-5}{2}, c = -2$.

The expression $2x^2 + mxy + 3y^2 - 5y - 2$ will have two linear factors if and only if

$abc + 2fgh - af^2 - bg^2 - ch^2 = 0$. Therefore,

$2\times 3(-2) + 2\left(\frac{-5}{2}\right)(0)\left(\frac{m}{2}\right) - 2\left(\frac{-5}{2}\right)^2 - 3\times 0^2 - (-2)\left(\frac{m}{2}\right)^2 = 0$

$\Rightarrow -12 - \frac{25}{2} + \frac{m^2}{2} = 0$

$\Rightarrow m^2 = 49$

$\Rightarrow m = \pm 7$

12. **(1)** Let $\dfrac{6x^2 - 22x + 21}{5x^2 - 18x + 17} = y$. Therefore,

$(6 - 5y)x^2 - 2x(11 - 9y) + 21 - 17y = 0$

Since x is real,

$4(11 - 9y)^2 - 4(6 - 5y)(21 - 17y) \geq 0$

$\Rightarrow -4y^2 + 9y - 5 \geq 0$

$\Rightarrow 4y^2 - 9y + 5 \leq 0$

$\Rightarrow 4(y - 1)(y - 5/4) \leq 0$

$\Rightarrow 1 \leq y \leq 5/4$

Hence, the least value of the given expression is 1.

13. **(5)** $(n - 2)x^2 + 8x + n + 4 > 0$, σ $\forall x \in R$

$\Rightarrow 64 - 4(n - 2)(n + 4) < 0$

$\Rightarrow 16 - (n^2 + 2n - 8) < 0$

$\Rightarrow n^2 + 2n - 24 > 0$

$\Rightarrow (n + 6)(n - 4) > 0$

$\Rightarrow n > 4$ as $n \in N$

$\Rightarrow n \geq 5$

$\Rightarrow$ The least value of n is 5.

14. **(3)** Since $x^2 + 2x + 2 = (x + 1)^2 + 1 > 0, \forall x \in R$

$\dfrac{mx^2 + 3x + 4}{x^2 + 2x + 2} < 5$

$\Rightarrow (m - 5)x^2 - 7x - 6 < 0, \forall x \in R$

$\Rightarrow m - 5 < 0$ and $D < 0$

$\Rightarrow m < 5$ and $49 + 24(m - 5) < 0$

$\Rightarrow m < \dfrac{71}{21}$

15. **(6)** $ax^2 + bx - 18 = 0$ does not have distinct real roots. Therefore,

$f(x) = ax^2 + bx - 18 \leq 0, \forall x \in R$, if $a < 0$

or $f(x) = ax^2 + bx - 18 \geq 0, \forall x \in R$, if $a > 0$.

But $f(0) = 6 > 0$. Therefore,

$f(x) = ax^2 + bx - 18 \geq 0, \forall x \in R$

$\Rightarrow f(3) = 9a + 3b - 18 \geq 0$

$\Rightarrow 3a + b \geq 6$

$\Rightarrow$ The least value of $3a + b$ is 6.

16. **(7)** Let $y = \dfrac{ax^2 + 3x - 4}{3x - 4x^2 + a}$

$\Rightarrow (a + 4y)x^2 + (3 - 3y)x - 4 - ay = 0$

Since x is real,

$D \geq 0$

$\Rightarrow 9(1 - y)^2 + 4(a + 4y)(4 + ay) \geq 0$

$\Rightarrow (9 + 16a)y^2 + (-18 + 4a^2 + 64)y + (9 + 16a) \geq 0, \forall y \in R$

[$\because y$ takes any real value]

$\Rightarrow 9 + 16a > 0$ and $(4a^2 + 46)^2 - 4(9 + 16a)^2 \leq 0$

$\Rightarrow a > -\dfrac{9}{16}$ and $(4a^2 + 46 - 18 - 32)(4a^2 + 46 + 18 + 32a) \leq 0$

$\Rightarrow a > -\dfrac{9}{16}$ and $(a^2 - 8a + 7)(a^2 + 8a + 16) \leq 0$

$\Rightarrow a > -\dfrac{9}{16}$ and $1 \leq a \leq 7$

$\Rightarrow 1 \leq a \leq 7$

17. **(4)** Given $a + 2b + c = 4 \Rightarrow a = 4 - 2b - c$

Let $ab + bc + ca = x$

or $a(b + c) + bc = x$

$\Rightarrow (4 - 2b - c)(b + c) + bc = x$

$\Rightarrow 4b + 4c - 2b^2 - 2bc - bc - c^2 + bc = x$

$\Rightarrow 2b^2 - 4b + 2bc - 4c + c^2 + x = 0$

$\Rightarrow 2b^2 + 2(c - 2)b - 4c + c^2 + x = 0$

Since $b \in R$,

$4(c - 2)^2 - 4\times 2(-4c + c^2 + x) \geq 0$

$\Rightarrow c^2 - 4c + 4 + 8c - 2c^2 - 2x \geq 0$

$\Rightarrow c^2 - 4c + 2x - 4 \leq 0$

Since $c \in R$,

$16 - 4(2x - 4) \geq 0$

$\Rightarrow x \leq 4$

Therefore, the maximum value of $(ab + bc + ac)$ is 4.

18. **(7)** Let $t = x^2 + x + 1$. Therefore,

$\in \left[\dfrac{3}{4}, \infty\right)$

$\Rightarrow (t + 1)^2 - (a - 3)t(t + 1) + (a - 4)t^2 = 0$

$\Rightarrow t^2 + 2t + 1 - (a - 3)(t^2 + t) + (a - 4)t^2 = 0$

$\Rightarrow t(2 - a + 3) + 1 = 0$

$\Rightarrow t = \dfrac{1}{(a - 5)}$

$\Rightarrow \dfrac{1}{a - 5} \geq \dfrac{3}{4} \Rightarrow \dfrac{19 - 3a}{(a - 5)} \geq 0$

$\Rightarrow a \in \left(5, \dfrac{19}{3}\right]$

19. **(0)** $x^2 - 2x - (a^2 - 1) = 0$ (1)

$x^2 - 2(a+1)x + a(a-1)$ (2)

For equation (1), $x = \dfrac{2 \pm \sqrt{4 + 4(a^2-1)}}{2} = 1 \pm a$

Now the roots of equation (1) lie between the roots of equation (2).

Hence, the graphs of the expressions of equations (1) and (2) are

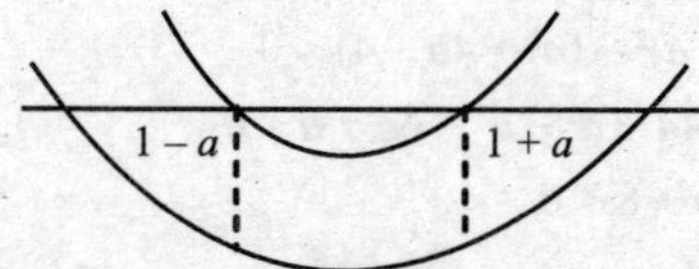

Fig. 1.21

Let $f_1(x) = x^2 - 2x - (a^2 - 1)$ and $f_2(x) = x^2 - 2(a+1)x + a(a-1)$.

If the roots of $f_1(x) = 0$ are real, then

$D \geq 0$

$\Rightarrow 4 + 4(a^2 - 1)\ 0$

$\Rightarrow a^2 \geq 0$, which is always true.

Also $f_2(1-a) < 0$ and $f_2(1+a) < 0$

$\Rightarrow (1-a)^2 - 2(a+1)(1-a) + a(a-1) < 0$

$\Rightarrow (1-a)[(1-a) - 2a - 2 - a] < 0$

$\Rightarrow (1-a)(-4a-1) < 0$

$\Rightarrow (a-1)(4a+1) < 0$

$\Rightarrow -\frac{1}{4} < a < 1$ (3)

And

$(1+a)^2 - 2(a+1)(a+1) + a(a-1) < 0$

$\Rightarrow -(a+1)^2 + a(a-1) < 0$

$\Rightarrow -a^2 - 2a - 1 + a^2 - a < 0$

$\Rightarrow 3a + 1 > 0$

$\Rightarrow a > -\dfrac{1}{3}$ (4)

From (3) and (4), the common values of $-\frac{1}{4} < a < 1$

20. We are given that $(2x + y)(x - 2y) = 7$.

Since x and y are integers, L.H.S. is the product of two integers and R.H.S. is also the product of two the integers 7 and 1, or 1 and 7, or −7 and −1, or −1 and −7.

Hence, we can choose

$2x + y = 7$ and $x - 2y = 1$ (1)

$2x + y = 1$ and $x - 2y = 7$ (2)

$2x + y = -7$ and $x - 2y = -1$ (3)

$2x + y = -1$ and $x - 2y = -7$ (4)

Solve them as usual and only (1) and (3) give integral solutions as 3, 1 for (1) and −3, −1 for (3). Equations (2) and (4), when solved, do not give integral values of x and y.

21. **(5)** $x^3 - 10x^2 - 11x - 100 = 0$

Let $f(x) = x^3 - 10x^2 - 11x - 100$. Therefore,

$f'(x) = 3x^2 - 20x - 11$

For $3x^2 - 20x - 11 = 0$

$x = \dfrac{20 \pm \sqrt{400 + 132}}{6} = \dfrac{10 \pm \sqrt{133}}{3}$

Hence, the graph of $y = f(x)$ is

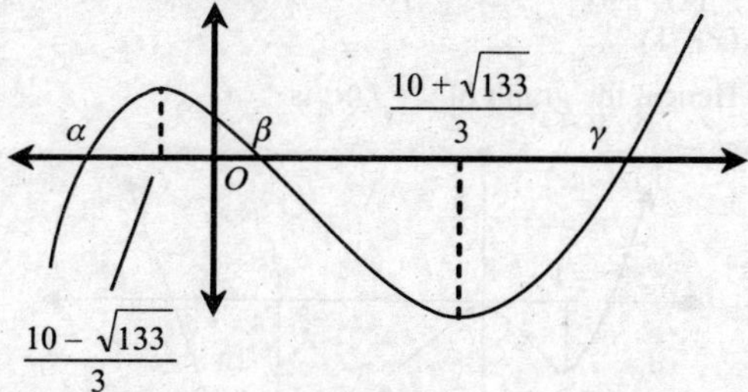

Fig. 1.22

Now $\dfrac{10 + \sqrt{133}}{3} \cong 7.16$

Now $f(8) = 8^3 - 10(8)^2 - 11(8) - 100 < 0$

$f(9) = 9^3 - 10(9)^2 - 11(9) - 100 < 0$

$f(10) = 10^3 - 10(10)^2 - 11(10) - 100 < 0$

$f(11) = 11^3 - 10(11)^2 - 11(11) - 100 < 0$

$f(12) = 12^3 - 10(12)^2 - 11(12) - 100 > 0$

$\Rightarrow \gamma \in (11, 12)$

$\Rightarrow [r] = 11$

22. **(1)** $x - y = \dfrac{y - z}{yz}, y - z = \dfrac{z - x}{zx}, z - x = \dfrac{x - y}{xy}$

Therefore, $(x - y)(y - z)(z - x) = \dfrac{(x-y)(y-z)(z-x)}{(xyz)^2}$

$\Rightarrow xyz = 1$

23. **(3)** Rational root $\Rightarrow D$ must be a perfect square and $D > 0$

$D = (n+1)^2 - 4n(n+2)$

$D = -3n^2 - 6n + 1 = 4 - 3(n+1)^2$

$\alpha, \beta = \dfrac{-6 \pm \sqrt{36 + 12}}{6} = \dfrac{-3 \pm 2\sqrt{3}}{6} = -1 + \dfrac{2\sqrt{3}}{3}$

or $-1 - \dfrac{2\sqrt{3}}{3}$

Hence, the possible integral values are − 2, − 1, 0, and all of them make D a perfect square.

24. **(9)** $\left(x + \frac{1}{x}\right)^2 - 2 = 7$

$\Rightarrow \left(x + \frac{1}{x}\right)^2 = 9$

$\Rightarrow x + \frac{1}{x} = 3, \neq -3\ (x > 0)$

$\Rightarrow \left(x + \frac{1}{x}\right)^3 = 27$

$\Rightarrow x^3 + \frac{1}{x^3} + 3\left(x + \frac{1}{x}\right) = 27$

$\Rightarrow x^3 + \frac{1}{x^3} = 27 - 9 = 18$

25. **(7)** $a^2 + 2b = 7$

$b^2 + 4c = -7$

$c^2 + 6a = -14$

$a^2 + b^2 + c^2 + 2b + 4c - 6a = -14$

$(a+3)^2 + (b+1)^2 + (c+2)^2 - 14 = -14$

$\Rightarrow a = -3, b = -1$, and $c = -2$

$\Rightarrow a^2 + b^2 + c^2 = 14$

26. **(2)** $D_1 = a^2 - 4, D_2 = a^2 + 36 > 0$

27. **(2)** Given $x^4 - 4x^3 - 8x^2 + a = 0$.

Let $f(x) = x^4 - 4x^3 - 8x^2$. Therefore,

$f'(x) = 4x^3 - 12x^2 - 16x = 4x(x^2 - 3x - 4) = 4x(x-4)(x+1)$

Hence, the graph of $y = f(x)$ is

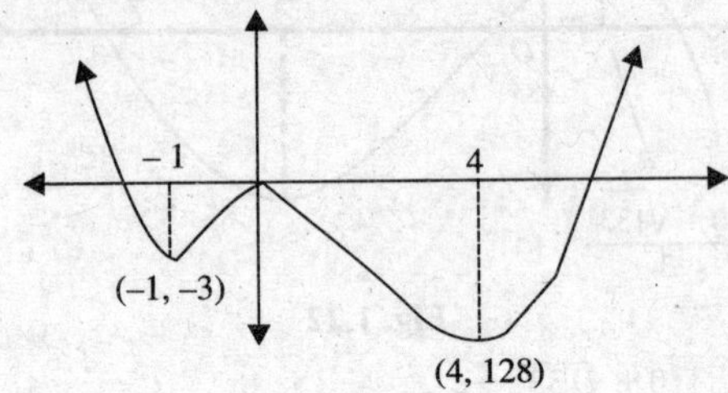

Fig. 1.23

Now the graph of $y = f(x) + a$ moves up or down depending on the values of a.

It is clear that if the equation $f(x) + a = 0$ has four real roots, then $0 \le a < 3$.

28. **(4)**

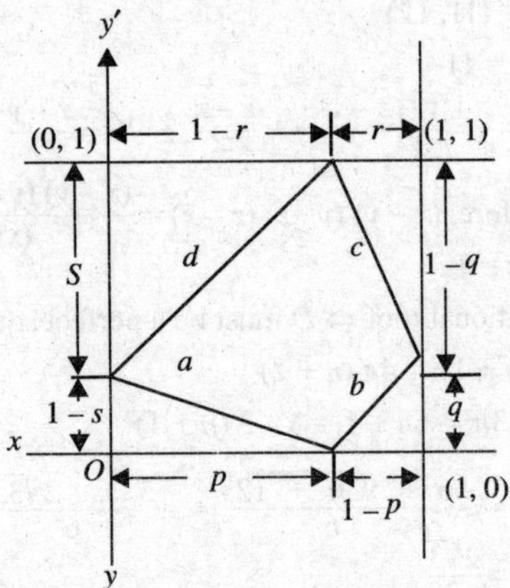

Fig. 1.24

$a^2 = p^2 + (1-s)^2$

$b^2 = (1-p)^2 + q^2$

$c^2 = (1-q)^2 + r^2$

$d^2 = (1-r)^2 + s^2$

Therefore, $a^2 + b^2 + c^2 + d^2 = [p^2 + (1-p)^2] + [q^2 + (1-q)^2] + [r^2 + (1-r)^2] + [s^2 + (1-s)^2]$,

where $p, q, r, s \in [0, 1]$.

Now consider the function

$y = x^2 + (1-x)^2, 0 \le x \le 1$

$\Rightarrow y = 2x^2 - 2x + 1$, which has vertex $\left(-\frac{b}{2a}, -\frac{D}{4a}\right)$

$\equiv \left(\frac{1}{2}, \frac{1}{2}\right)$.

Hence, the minimum value is $\frac{1}{2}$ when $x = \frac{1}{2}$ and the maximum value is 1 when $x = 1$.

Therefore, the minimum value of $a^2 + b^2 + c^2 + d^2 \frac{1}{2} + \frac{1}{2} + \frac{1}{2} + \frac{1}{2} = 2$ is and the maximum value is $1 + 1 + 1 + 1 = 4$.

29. **(2)** The given equation is

$x^2 + (a-b)x + (1-a-b) = 0, a, b \in R$.

For this equation to have unequal real roots $\forall\, b$,

$D > 0$

$\Rightarrow (a-b)^2 - 4(1-a-b) > 0$

$\Rightarrow a^2 + b^2 - 2ab - 4 + 4a + 4b > 0$

$\Rightarrow b^2 + b(4-2a) + a^2 + 4a - 4 > 0$ (1)

This is a quadratic expression in b, and it will be true $\forall\, b \in R$. Therefore, its discriminant < 0

$\Rightarrow (4-2a)^2 - 4(a^2 + 4a - 4) < 0$

$\Rightarrow (2-a)^2 - (a^2 + 4a - 4) < 0$

$\Rightarrow 4 - 4a + a^2 - a^2 - 4a + a < 0$

$\Rightarrow -8a + 8 < 0$

$\Rightarrow a > 1$

30. **(4)** $|x-2|^2 + |x-2| - 2 = 0$

$\Rightarrow (|x-2| + 2)(|x-2| - 1) = 0$

$\Rightarrow |x-2| - 1 = 0$

$\Rightarrow x - 2 = \pm 1$

$\Rightarrow x = 1, 3$

Therefore, the sum of the roots is $3 + 1 = 4$.

31. **(3)** Let α, α^2 be the roots of $3x^2 + px + 3 = 0$.

Now $S = \alpha + \alpha^2 = -p/3, p = \alpha^3 = 1$

$\Rightarrow \alpha = 1, \omega, \omega^2$

$\alpha + \alpha^2 = -p/3 \Rightarrow \omega + \omega^2 = -p/3$

$\Rightarrow -1 = -p/3 \Rightarrow p = 3$

32. **(8)** Clearly, $\alpha + \beta = 1, \alpha\beta = p, \gamma + \delta = 4$, and $\gamma\delta = q, (p, q \in I)$

Since $\alpha, \beta, \gamma, \delta$ are in G.P. (with common ratio r), $\alpha + \alpha r = 1, \alpha(r^2 + r^3) = 4$

$\Rightarrow \alpha(1 + r) = 1, \alpha r^2(1 + r) = 4$

$\Rightarrow r^2 \times 1 = 4 \Rightarrow r^2 = 4 \Rightarrow r = 2, -2$

If $r = 2, \alpha + 2\alpha = 1 \Rightarrow \alpha = \frac{1}{3}$.

If $r = -2, \alpha - 2\alpha = 1 \Rightarrow \alpha = -1$

But $p = \alpha\beta \in I$. Therefore, $r = -2$ and $\alpha = -1$.

For $p = -2, q = \alpha^2 r^5 = 1(-2)^5 = -32$.

33. **(1)** The roots of $x^2 - 10cx - 11d = 0$ are a and b. Therefore, $a + b = 10c$ and $ab = 11d$.

c and d are the roots of $x^2 - 10ax - 11b = 0$. Therefore, $c + d = 10a$ and $cd = -11b$.

$a + b + c + d = 10(a + c)$ and $abcd = 121bd$

$\Rightarrow b + d = 9(a + c)$ and $ac = 121$

Also we have $a^2 - 10ac - 11d = 0$ and $c^2 - 10ac - 11b = 0$

$\Rightarrow a^2 + c^2 - 20ac - 11(b + d) = 0$

$\Rightarrow (a + c)^2 - 22 \times 121 - 99(a + c) = 0$

$\Rightarrow a + c = 121$ or -22

For $a + c = -22$, we get $a = c$.

Therefore, rejecting these values we have $a + c = 121$.

Therefore, $a + b + c + d = 10(a + c) = 1210$.

34. **(5)** Obviously roots are

$1+i$, $1-i$, $1+2i$ and $1-2i$ and r(say), where r is real

Now Sum of the roots = 6 (from the given equation)

$\therefore 4+r=6 \Rightarrow r=2$

Product of the roots = $(1-i^2)(1-4i^2).\ 2$

$\therefore -D = 2\times 5\times 2$

$\Rightarrow D=-20$

35. **(8)** $(x^4+4x^3+6px^2+4qx+r)=(x+a)(x^3+3x^2+9x+3)$

Comparing coefficients

$a=1$; $6p=12 \Rightarrow p=2$

$4q=12 \Rightarrow q=3$ and $r=3$

hence $p+q+r=8$

36. **(5)** As $a>2$, hence $a^2>2a>a>2$

Now $(x-a)(x-2a)(x-a^2)<0 \Rightarrow$ the solution set is as

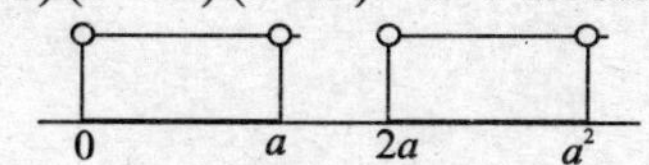

Fig. 1.25

Between $(0, a)$ there are $(a-1)$ positive integers

Between $(2a, a^2)$ there are (a^2-2a-1) positive integers

$\therefore a^2-2a-1+a-1=18 \Rightarrow a^2-a-20=0$

$(a-5)(a+4)=0$

$\therefore a=5$ and $a=-4$ but $a>2$ $\therefore a=5$

37. **(7)** $xyz+xy+yz+zx$

$=(x+1)(y+1)(z+1)-(x+y+z)-1$

$abc-13$

where $a=x+1$, $b=y+1$, $c=z+1$ are positive integers whose sum is 15.

abc is largest when $a=b=c=5$

Thus the answer is

$5\times 5\times 5-13=112$

$$\therefore \frac{xyz+xy+yz+zx}{16}=\frac{112}{16}=7$$

38. **(5)** Let $f(x)+\dfrac{a}{30}=0$

where $f(x)=\dfrac{x^3-6x^2+11x-6}{x^3+x^2-10x+8}=\dfrac{(x-1)(x-2)(x-3)}{(x-1)(x-2)(x+4)}$

$=\dfrac{(x-3)}{x+4}, x\neq 1, 2, -4.$

Range of $f(x)=R-\left\{1,-\frac{2}{5},-\frac{1}{6}\right\}$

So the equation does not have solution if

$\dfrac{a}{30}=-1, \dfrac{2}{5}, \dfrac{1}{6}$

$a=-30, 12, 5$

39. **(0)** Since cubic is divisible by both x^2+ax+b and x^2+bx+a

Therefore, x^2+ax+b and x^2+bx+a must have a common roots

$x^2+ax+b=0$

$x^2+bx+a=0$

Subtract

$\Rightarrow x(a-b)=(a-b)$

$\Rightarrow x=1$

Therefore, common roots

for $x^2+ax+b=0$ are 1 and α $\Rightarrow 1\,\alpha=b \Rightarrow \alpha=b$

for $x^2+bx+a=0$ are 1 and β $\Rightarrow 1\,\beta=a \Rightarrow \beta=a$

$\Rightarrow$ roots of cubic be 1, a, b

$\Rightarrow 1\times a\times b=-72$ (1)

and $a+b+1=0$ (2)

(from $x^2+ax+b=0$ put $x=1$)

$\therefore a-\dfrac{72}{a}=-1$

$\Rightarrow a^2+a-72=0$

$(a+9)(a-8)=0$

$a=-9, 8$

Therefore, roots are $1, -9, 8$

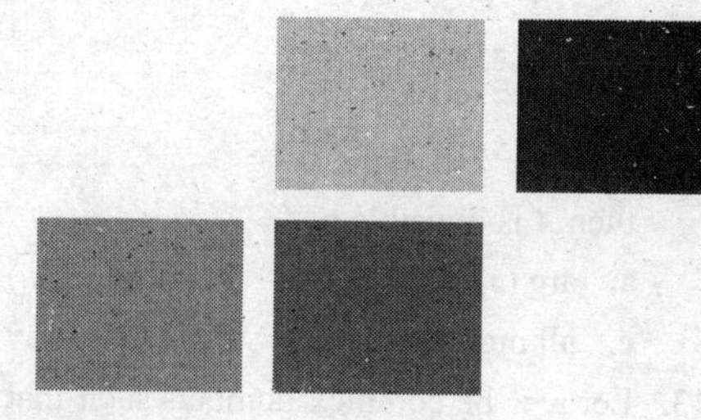

CHAPTER 2

Complex Numbers

Objective Type

Each question has four choices a, b, c and d, out of which *only one* answer is correct.

1. If $a < 0, b > 0$ then $\sqrt{a}\sqrt{b}$ is equal to

   a. $-\sqrt{|a|b}$ **b.** $\sqrt{|a|b}\,i$

   c. $\sqrt{|a|b}$ **d.** none of these

2. If $x = 9^{1/3}\, 9^{1/9}\, 9^{1/27} \cdots \infty$, $y = 4^{1/3}\, 4^{-1/9}\, 4^{1/27} \cdots \infty$, and $z = \sum_{r=1}^{\infty} (1+i)^{-r}$, then $\arg(x + yz)$ is equal to

   a. 0 **b.** $\pi - \tan^{-1}\left(\frac{\sqrt{2}}{3}\right)$

   c. $-\tan^{-1}\left(\frac{\sqrt{2}}{3}\right)$ **d.** $-\tan^{-1}\left(\frac{2}{\sqrt{3}}\right)$

3. Consider the equation $10z^2 - 3iz - k = 0$, where z is a complex variable and $i^2 = -1$. Which of the following statements is true?

   a. For real positive numbers k, both roots are purely imaginary.

   b. For all complex numbers k, neither root is real.

   c. For all purely imaginary numbers k, both roots are real and irrational.

   d. For real negative numbers k, both roots are purely imaginary.

4. The number of solutions of the equation $z^2 + \bar{z} = 0$ is

   a. 1 **b.** 2 **c.** 3 **d.** 4

5. If $a^2 + b^2 = 1$, then $(1 + b + ia)/(1 + b - ia) =$

   a. 1 **b.** 2 **c.** $b + ia$ **d.** $a + ib$

6. The expression $\left[\dfrac{1+\sin\frac{\pi}{8} + i\cos\frac{\pi}{8}}{1+\sin\frac{\pi}{8} - i\cos\frac{\pi}{8}}\right]^8 =$

   a. 1 **b.** -1 **c.** i **d.** $-i$

7. If z_1 and z_2 be complex numbers such that $z_1 \neq z_2$ and $|z_1| = |z_2|$. If z_1 has positive real part and z_2 has negative imaginary part, then $[(z_1 + z_2)/(z_1 - z_2)]$ may be

   a. Purely imaginary **b.** Real and positive

   c. Real and negative **d.** None of these

8. If $|z_1| = |z_2|$ and $\arg((z_1/z_2) = \pi$, then $z_1 + z_2$ is equal to

   a. 0 **b.** purely imaginary

   c. purely real **d.** none of these

9. If $\cos\alpha + 2\cos\beta + 3\cos\gamma = \sin\alpha + 2\sin\beta + 3\sin\gamma = 0$, then the value of $\sin 3\alpha + 8\sin 3\beta + 27\sin 3\gamma$ is

   a. $\sin(a + b + \gamma)$ **b.** $3\sin(\alpha + \beta + \gamma)$

   c. $18\sin(\alpha + \beta + \gamma)$ **d.** $\sin(\alpha + 2\beta + 3)$

10. If centre of a regular hexagon is at origin and one of the vertices on Argand diagram is $1 + 2i$, then its perimeter is

    a. $2\sqrt{5}$ **b.** $6\sqrt{2}$ **c.** $4\sqrt{5}$ **d.** $6\sqrt{5}$

11. If $z(1 + a) = b + ic$ and $a^2 + b^2 + c^2 = 1$, then $[(1 + iz)/(1 - iz)] =$

    a. $\dfrac{a+ib}{1+c}$ **b.** $\dfrac{b-ic}{1+a}$

    c. $\dfrac{a+ic}{1+b}$ **d.** none of these

12. If z_1, z_2, z_3 are three complex numbers and

$$A = \begin{vmatrix} \arg z_1 & \arg z_2 & \arg z_3 \\ \arg z_2 & \arg z_3 & \arg z_1 \\ \arg z_3 & \arg z_1 & \arg z_2 \end{vmatrix}$$

then A is divisible by

a. $\arg(z_1 + z_2 + z_3)$ **b.** $\arg(z_1 z_2 z_3)$
c. all numbers **d.** cannot say

13. Let z, w be complex numbers such that $\bar{z} + i\bar{w} = 0$ and $\arg zw = \pi$. Then $\arg z$ equals

a. $\frac{\pi}{4}$ **b.** $\frac{\pi}{2}$ **c.** $\frac{3\pi}{4}$ **d.** $\frac{5\pi}{4}$

14. If for complex numbers z_1 and z_2, $\arg(z_1) - \arg(z_2) = 0$, then $|z_1 - z_2|$ is equal to

a. $|z_1| + |z_2|$ **b.** $|z_1| - |z_2|$
c. $\left||z_1| - |z_2|\right|$ **d.** 0

15. If $k > 0$, $|z| = |w| = k$ and $\alpha = \frac{z - \bar{w}}{k^2 + z\bar{w}}$, then $\text{Re}(\alpha)$ equals

a. 0 **b.** $k/2$
c. k **d.** none of these

16. If $z = x + iy$ and $x^2 + y^2 = 16$, then the range of $||x| - |y||$ is

a. $[0, 4]$ **b.** $[0, 2]$
c. $[2, 4]$ **d.** none of these

17. If $k + |k + z^2| = |z|^2$ $(k \in R^-)$, then possible argument of z is

a. 0 **b.** π
c. $\pi/2$ **d.** none of these

18. If $z = x + iy$ $(x, y \in R, x \neq -1/2)$, the number of values of z satisfying $|z|^n = z^2\,|z|^{n-2} + z|z|^{n-2} + 1 \cdot (n \in N, n > 1)$ is

a. 0 **b.** 1 **c.** 2 **d.** 3

19. If x and y are complex numbers, then the system of equations $(1 + i)x + (1 - i)y = 1$, $2ix + 2y = 1 + i$ has

a. unique solution
b. no solution
c. infinite number of solutions
d. none of these

20 Number of solutions of the equation $z^3 + [3(\bar{z})^2]/|z| = 0$ where z is a complex number is

a. 2 **b.** 3 **c.** 6 **d.** 5

21. The principal argument of the complex number $[(1+i)^5(1+\sqrt{3}i)^2]/[-2i(-\sqrt{3}+i)]$ is

a. $\frac{19\pi}{12}$ **b.** $-\frac{7\pi}{12}$ **c.** $-\frac{5\pi}{12}$ **d.** $\frac{5\pi}{12}$

22. The polynomial $x^6 + 4x^5 + 3x^4 + 2x^3 + x + 1$ is divisible by

a. $x + \omega$ **b.** $x + \omega^2$
c. $(x + \omega)(x + \omega^2)$ **d.** $(x - \omega)(x - \omega^2)$

where ω is one of the imaginary cube roots of unity.

23. If $(\cos\theta + i\sin\theta)(\cos 2\theta + i\sin 2\theta) \cdots (\cos n\theta + i\sin n\theta) = 1$, then the value of θ is

a. $4m\pi$ **b.** $\frac{2m\pi}{n(n+1)}$
c. $\frac{4m\pi}{n(n+1)}$ **d.** $\frac{m\pi}{n(n+1)}$

24. Given $z = (1 + i\sqrt{3})^{100}$, then [RE(z)/IM(z)] equals

a. 2^{100} **b.** 2^{50} **c.** $\frac{1}{\sqrt{3}}$ **d.** $\sqrt{3}$

25. If $z = (i)^{(i)^{(i)}}$ where $i = \sqrt{-1}$, then $|z|$ is equal to

a. 1 **b.** $e^{-\pi/2}$
c. $e^{-\pi}$ **d.** none of these

26. If $z = i\log(2 - \sqrt{-3})$, then $\cos z =$

a. i **b.** $2i$ **c.** 1 **d.** 2

27. If the equation $z^4 + a_1z^3 + a_2z^2 + a_3z + a_4 = 0$, where a_1, a_2, a_3, a_4 are real coefficients different from zero, has a purely imaginary root, then $[a_3/(a_1a_2)] + (a_1a_4)(a_2a_3)$ has the value equal to

a. 0 **b.** 1 **c.** -2 **d.** 2

28. Suppose A is a complex number and n Œ N, such that $A^n = (A + 1)^n = 1$, then the least value of n is

a. 3 **b.** 6 **c.** 9 **d.** 12

29. Number of complex numbers z such that $|z| = 1$ and $|z/\bar{z} + \bar{z}/z| = 1$ is $[\arg(z) \in (0, 2\pi)]$

a. 4 **b.** 6
c. 8 **d.** more than 8

30. If α, β be the roots of the equation $u^2 - 2u + 2 = 0$ and if $\cot\theta = x + 1$, then $[(x+\alpha)^n - (x+\beta)^n]/[\alpha - \beta]$ is equal to

a. $\frac{\sin n\theta}{\sin^n\theta}$ **b.** $\frac{\cos n\theta}{\cos^n\theta}$
c. $\frac{\sin n\theta}{\cos^n\theta}$ **d.** $\frac{\cos n\theta}{\sin^n\theta}$

31. Dividing $f(z)$ by $z - i$, we obtain the remainder i and dividing it by $z + i$, we get the remainder $1 + i$, then remainder upon the division of $f(z)$ by $z^2 + 1$ is

a. $\frac{1}{2}(z+1)+i$ **b.** $\frac{1}{2}(iz+1)+i$
c. $\frac{1}{2}(iz-1)+i$ **d.** $\frac{1}{2}(z+i)+1$

32. If $z_1 z_2 \in C$, $z_1^2 + z_2^2 \in R$, $z_1(z_1^2 - 3z_2^2) = 2$ and $z_2(3z_1^2 - z_2^2) = 11$, then the value of $z_1^2 + z_2^2$ is

a. 10 **b.** 12 **c.** 5 **d.** 8

33. z_1 and z_2 are two distinct points in an Argand plane. If $a|z_1| = b|z_2|$ (where $a, b \in R$), then the point $(az_1/bz_2) + (bz_2/az_1)$ is a point on the

a. line segment $[-2, 2]$ of the real axis
b. line segment $[-2, 2]$ of the imaginary axis
c. unit circle $|z| = 1$
d. the line with $\arg z = \tan^{-1} 2$

34. If $x^2 + x + 1 = 0$, then the value of $(x + 1/x)^2 + (x^2 + 1/x^2)^2 + \cdots + (x^{27} + 1/x^{27})^2$ is

a. 27 **b.** 72 **c.** 45 **d.** 54

35. If ω be a complex n^{th} root of unity, then $\sum_{r=1}^{n}(ar+b)\,\omega^{r-1}$ is equal to

a. $\frac{n(n+1)a}{2}$ **b.** $\frac{nb}{1-n}$
c. $\frac{na}{\omega-1}$ **d.** none of these

36. The roots of the cubic equation $(z+ab)^3=a^3$, such that $a\neq 0$, represent the vertices of a triangle of sides of length

a. $\frac{1}{\sqrt{3}}|ab|$ **b.** $\sqrt{3}\,|a|$ **c.** $\sqrt{3}\,|b|$ **d.** $|a|$

37. Sum of common roots of the equations $z^3+2z^2+2z+1=0$ and $z^{1985}+z^{100}+1=0$ is

a. -1 **b.** 1 **c.** 0 **d.** 1

38. If $\left|\frac{z_1}{z_2}\right|=1$ and $\arg(z_1z_2)=0$, then

a. $z_1=z_2$ **b.** $|z_2|^2=z_1z_2$
c. $z_1z_2=1$ **d.** none of these

39. If z_1 and z_2 are the complex roots of the equation $(x-3)^3+1=0$, then z_1+z_2 equals to

a. 1 **b.** 3 **c.** 5 **d.** 7

40. Which of the following is equal to $\sqrt[3]{-1}$?

a. $\frac{\sqrt{3}+\sqrt{-1}}{2}$ **b.** $\frac{-\sqrt{3}+\sqrt{-1}}{\sqrt{-4}}$
c. $\frac{\sqrt{3}-\sqrt{-1}}{\sqrt{-4}}$ **d.** $-\sqrt{-1}$

41. If $|z-1|\leq 2$ and $|\omega z-1-\omega^2|=a$ (where ω is a cube root of unity) then complete set of values of a is

a. $0\leq a\leq 2$
b. $\frac{1}{2}\leq a\leq\frac{\sqrt{3}}{2}$
c. $\frac{\sqrt{3}}{2}-\frac{1}{2}\leq a\leq\frac{1}{2}+\frac{\sqrt{3}}{2}$
d. $0\leq a\leq 4$

42. If $|z^2-3|=3|z|$ then the maximum value of $|z|$ is

a. 1 **b.** $\frac{3+\sqrt{21}}{2}$
c. $\frac{\sqrt{21}-3}{2}$ **d.** none of these

43. If $|2z-1|=|z-2|$ and z_1, z_2, z_3 are complex numbers such that $|z_1-\alpha|<\alpha, |z_2-\beta|<\beta$, then $|z_1+x_2/\alpha+\beta)|$

a. $<|z|$ **b.** $<2|z|$ **c.** $>|z|$ **d.** $>2|z|$

44. If z_1 is a root of the equation $a_0z^n+a_1z^{n-1}+\cdots+a_{n-1}z+a_n=3$, where $|a_i|<2$ for $i=0,1,\ldots,n$. Then

a. $|z_1|>\frac{1}{3}$ **b.** $|z_1|<\frac{1}{4}$ **c.** $|z_1|>\frac{1}{4}$ **d.** $|z|<\frac{1}{3}$

45. If $8iz^3+12z^2-18z+27i=0$, then

a. $|z|=\frac{3}{2}$ **b.** $|z|=\frac{2}{3}$ **c.** $|z|=1$ **d.** $|z|=\frac{3}{4}$

46. If $|z|<\sqrt{2}-1$, then $|z^2+2z\cos\alpha|$ is

a. less than 1 **b.** $\sqrt{2}+1$
c. $\sqrt{2}-1$ **d.** none of these

47. If the complex number z satisfies the condition $|z|\geq 3$, then the least value of $[x+(1/z)]$ is equal to

a. 5/3 **b.** 8/3
c. 11/3 **d.** none of these

48. Let $|z_r-r|\leq r, \forall r=1,2,3,\ldots,n$. Then $\left|\sum_{r=1}^{n}z_r\right|$ is less than

a. n **b.** $2n$
c. $n(n+1)$ **d.** $\frac{n(n+1)}{2}$

49. If $|z^2-1|=|z|^2+1$, then z lies on

a. a circle **b.** a parabola
c. an ellipse **d.** none of these

50. If $|z|=1$ then the point representing the complex number $-1+3z$ will lie on

a. a circle **b.** a straight line
c. a parabola **d.** a hyperbola

51. If $z=(\lambda+3)-i\sqrt{5-\lambda^2}$, then the locus of z is

a. ellipse **b.** semicircle
c. parabola **d.** straight line

52. If $A(z_1), B(z_2), C(z_3)$ are the vertices of the triangle ABC such that $(z_1-z_2)/(z_3-z_2)=(1/\sqrt{2})-(i/1/\sqrt{2})$, the triangle ABC is

a. equilateral **b.** right angled
c. isosceles **d.** obtuse angled

53. If z_1, z_2, z_3 are the vertices of an equilateral triangle ABC such that $|z_1-i|=|z_2-i|=|z_3-i|$, then $|z_1+z_2+z_3|$ equals to

a. $3\sqrt{3}$ **b.** $\sqrt{3}$ **c.** 3 **d.** $\frac{1}{3\sqrt{3}}$

54. The greatest positive argument of complex number satisfying $|z-4|=\text{Re}(z)$ is

a. $\frac{\pi}{3}$ **b.** $\frac{2\pi}{3}$ **c.** $\frac{\pi}{2}$ **d.** $\frac{\pi}{4}$

55. The complex number associated with the vertices A, B, C of ΔABC are $e^{i\theta}, \omega, \bar{\omega}$, respectively [where $\omega, \bar{\omega}$ are the complex cube roots of unity and $\cos\theta>\text{Re}(\omega)$], then the complex number of the point where angle bisector of A meets the circumcircle of the triangle, is

a. $e^{i\theta}$ **b.** $e^{-i\theta}$ **c.** $\omega\bar{\omega}$ **d.** $\omega+\bar{\omega}$

56. The maximum area of the triangle formed by the complex coordinates z, z_1, z_2 which satisfy the relations $|z-z_1|=|z-z_2|$ and $|z-[(z_1+z_2/2|\leq r$, where $r>|z_1-z_2|$ is

a. $\frac{1}{2}|z_1-z_2|^2$ **b.** $\frac{1}{2}|z_1-z_2|r$
c. $\frac{1}{2}|z_1-z_2|^2r^2$ **d.** $\frac{1}{2}|z_1-z_2|r^2$

57. Locus of z if $\arg[z-(1+i)] = \begin{cases} \frac{3\pi}{4} & \text{when } |z| \le |z-2| \\ \frac{-\pi}{4} & \text{when } |z| > |z-2| \end{cases}$ is

a. straight lines passing through (2, 0)

b. straight lines passing through (2, 0), (1, 1)

c. a line segment

d. a set of two rays

58. If z is a complex number such that $-\pi/2 \le \arg z \le \pi/2$, then which of the following inequality is true?

a. $|z-\bar{z}| \le |z|(\arg z - \arg \bar{z})$

b. $|z-\bar{z}| \ge |z|(\arg z - \arg \bar{z})$

c. $|z-\bar{z}| < (\arg z - \arg \bar{z})$

d. none of these

59. If z is a complex number lying in the fourth quadrant of Argand plane and $|[kz/k+1]+2i| > \sqrt{2}$ for all real value of k $(k \ne -1)$, then range of $\arg(z)$ is

a. $\left(-\frac{\pi}{8}, 0\right)$ b. $\left(-\frac{\pi}{6}, 0\right)$

c. $\left(-\frac{\pi}{4}, 0\right)$ d. none of these

60 If z is a complex number then the locus of z satisfying the condition $|2z-1| = |z-1|$ is

a. perpendicular bisector of line segment joining 1/2 and 1

b. circle

c. parabola

d. none of the above curves

61. If $|z_2 + iz_1| = |z_1| + |z_2|$ and $|z_1| = 3$ and $|z_2| = 4$, then area of ΔABC, if affixes of A, B and C are z_1, z_2 and $[(z_2 - iz_1)/(1-i)]$ respectively, is,

a. $\frac{5}{2}$ b. 0 c. $\frac{25}{2}$ d. $\frac{25}{4}$

62. If a complex number z satisfies $|2z+10+10i| \le 5\sqrt{3}-5$, then the least principal argument of z is

a. $-\frac{5\pi}{6}$ b. $-\frac{11\pi}{12}$ c. $-\frac{3\pi}{4}$ d. $-\frac{2\pi}{3}$

63. If t and c are two complex numbers such that $|t| \ne |c|$, $|t| = 1$ and $z = (at+b)/(t-c)$, $z = x+iy$. Locus of z is (where a, b are complex numbers)

a. line segment b. straight line

c. circle d. none of these

64. The number of complex numbers z satisfying $|z-3-i| = |z-9-i|$ and $|z-3+3i| = 3$ are

a. one b. two

c. four d. none of these

65. Let a be a complex number such that $|a| < 1$ and $z_1, z_2, z_3, \ldots$ be the vertices of a polygon such that $z_k = 1 + a + a^2 + \cdots + a^{k-1}$ for all $k = 1, 2, 3, \ldots$ then $z_1, z_2, \ldots$ lie within the circle

a. $\left|z - \frac{1}{1-a}\right| = \frac{1}{|a-1|}$ b. $\left|z + \frac{1}{1-a}\right| = \frac{1}{|a-1|}$

c. $\left|z + \frac{1}{1-a}\right| = |a-1|$ d. $\left|z + \frac{1}{1-a}\right| = |a+1|$

66. Let $\lambda \in R$, the origin and the non-real roots of $2z^2 + 2z + \lambda = 0$ form the three vertices of an equilateral triangle in the Argand plane then λ is

a. 1 b. $\frac{2}{3}$ c. 2 d. -1

67. Let $z = 1 - t + i\sqrt{t^2+t+2}$, where t is a real parameter. The locus of z in the Argand plane is

a. a hyperbola b. an ellipse

c. a straight line d. none of these

68. If $z^2 + z|z| + |z^2| = 0$, then the locus of z is

a. a circle b. a straight line

c. a pair of straight lines d. none of these

69. The roots of the equation $t^3 + 3at^2 + 3bt + c = 0$ are z_1, z_2, z_3 which represent the vertices of an equilateral triangle, then

a. $a^2 = 3b$ b. $b^2 = a$

c. $a^2 = b$ d. $b^2 = 3a$

70. If z lies on the circle $|z-2i| = 2\sqrt{2}$ then the value of $\arg[(z-2)/(x+2)]$ is equal to

a. $\frac{\pi}{3}$ b. $\frac{\pi}{4}$ c. $\frac{\pi}{6}$ d. $\frac{\pi}{2}$

71. $P(z)$ be a variable point in the Argand plane such that $|z| = \text{minimum } \{|z-1|, |z+1|\}$ then $z + \bar{z}$ will be equal to

a. -1 or 1 b. 1 but not equal to -1

c. -1 but not equal to 1 d. none of these

72. The locus of point z satisfying $\text{Re}\left(\frac{1}{z}\right) = k$, where k is a non-zero real number, is

a. a straight line b. a circle

c. an ellipse d. a hyperbola

73. z_1 and z_2 lie on a circle with centre at the origin. The point of intersection z_3 of the tangents at z_1 and z_2 is given by

a. $\frac{1}{2}(\bar{z}_1 + \bar{z}_2)$ b. $\frac{2z_1z_2}{z_1+z_2}$

c. $\frac{1}{2}\left(\frac{1}{\bar{z}_1} + \frac{1}{\bar{z}_2}\right)$ d. $\frac{z_1+z_2}{\bar{z}_1\bar{z}_2}$

74. If $|z_1| = |z_2| = |z_3| = 1$ and $z_1 + z_2 + z_3 = 0$, then area of the triangle whose vertices are z_1, z_2, z_3 is

a. $3\sqrt{3}/4$ b. $\sqrt{3}/4$ c. 1 d. 2

75. z_1, z_2, z_3, z_4 are distinct complex numbers representing the vertices of a quadrilateral $ABCD$ taken in order. If $z_1 - z_4 = z_2$

$-z_3$ and $\arg(z_4 - z_1)/(z_2 - z_1) = \pi/2$, then the quadrilateral is

a. rectangle b. rhombus

c. square d. trapezium

76. If $\arg\left(\dfrac{z_1 - \dfrac{z}{|z|}}{\dfrac{z}{|z|}}\right) = \dfrac{\pi}{2}$ and $\left|\dfrac{z}{|z|} - z_1\right| = 3$ then $|z_1|$ equals to

a. $\sqrt{26}$ b. $\sqrt{10}$ c. $\sqrt{3}$ d. $2\sqrt{2}$

77. The points $z_1 = 3 + \sqrt{3}i$ and $z_2 = 2\sqrt{3} + 6i$ are given on a complex plane. The complex number lying on the bisector of the angle formed by the vectors z_1 and z_2 is

a. $z = \dfrac{(3+2\sqrt{3})}{2} + \dfrac{\sqrt{3}+2}{2}i$

b. $z = 5 + 5i$

c. $z = -1 - i$

d. none of these

78. Let C_1 and C_2 are concentric circles of radius 1 and 8/3, respectively, having centre at (3, 0) on the Argand plane. If the complex number z satisfies the inequality

$\log_{1/3}\left(\dfrac{|z-3|^2 + 2}{11|z-3|-2}\right) > 1$ then

a. z lies outside C_1 but inside C_2

b. z lies inside of both C_1 and C_2

c. z lies outside both of C_1 and C_2

d. none of these

79. If $|z - 2 - i| = |z|\left|\sin\left(\dfrac{\pi}{4} - \arg z\right)\right|$, then locus of z is

a. a pair of straight lines

b. circle

c. parabola

d. ellipse

80. If z is a complex number having least absolute value and $|z - 2 + 2i| = 1$, then $z =$

a. $(2 - 1/\sqrt{2})(1 - i)$ b. $(2 - 1/\sqrt{2})(1 + i)$

c. $(2 + 1/\sqrt{2})(1 - i)$ d. $(2 + 1/\sqrt{2})(1 + i)$

81. If $z = 3/(2 + \cos\theta + i\sin\theta)$, then locus of z is

a. a straight line

b. a circle having centre on y-axis

c. a parabola

d. a circle having centre on x-axis

82. If p and q are distinct prime numbers, then the number of distinct imaginary numbers which are p^{th} as well as q^{th} roots of unity are

a. min (p, q) b. max (p, q)

c. 1 d. zero

83. If α is the nth root of unity, then $1 + 2\alpha + 3\alpha^2 + \cdots$ to n terms equal to

a. $\dfrac{-n}{(1-\alpha)^2}$ b. $\dfrac{-n}{1-\alpha}$

c. $\dfrac{-2n}{1-\alpha}$ d. $\dfrac{-2n}{(1-\alpha)^2}$

84. Given z is a complex number with modulus 1. Then the equation $[(1 + ia/(1 - ia)]^4 = z$ has

a. all roots real and distinct

b. two real and two imaginary

c. three roots real and one imaginary

d. one root real and three imaginary

85. Roots of the equations are $(z + 1)^5 = (z - 1)^5$ are

a. $\pm i\tan\left(\dfrac{\pi}{5}\right), \pm i\tan\left(\dfrac{2\pi}{5}\right)$

b. $\pm i\cot\left(\dfrac{\pi}{5}\right), \pm i\cot\left(\dfrac{2\pi}{5}\right)$

c. $\pm i\cot\left(\dfrac{\pi}{5}\right), \pm i\tan\left(\dfrac{2\pi}{5}\right)$

d. none of these

86. The value of z satisfying the equation $\log z + \log z^2 + \cdots + \log z^n = 0$ is

a. $\cos\dfrac{4m\pi}{n(n+1)} + i\sin\dfrac{4m\pi}{n(n+1)}$, $m = 1, 2, \ldots$

b. $\cos\dfrac{4m\pi}{n(n+1)} - i\sin\dfrac{4m\pi}{n(n+1)}$, $m = 1, 2, \ldots$

c. $\sin\dfrac{4m\pi}{n} + i\cos\dfrac{4m\pi}{n}$, $m = 1, 2, \ldots$

d. 0

87. If $n \in N > 1$ then sum of real part of roots of $z^n = (z + 1)^n$ is equal to

a. $\dfrac{n}{2}$ b. $\dfrac{(n-1)}{2}$ c. $-\dfrac{n}{2}$ d. $\dfrac{(1-n)}{2}$

88. Which of the following represents a point in an Argand plane, equidistant from the roots of the equation $(z + 1)^4 = 16z^4$?

a. (0, 0) b. $\left(-\dfrac{1}{3}, 0\right)$

c. $\left(\dfrac{1}{3}, 0\right)$ d. $\left(0, \dfrac{2}{\sqrt{5}}\right)$

89. $1, z_1, z_2, z_3, \ldots, z_{n-1}$ are the nth roots of unity, then the value of $[1/3(3 - z_1)] + [1/(3 - z_2)] + \cdots + [1/(3 - z_{n-1})]$ is equal to

a. $\dfrac{n\,3^{n-1}}{3^n - 1} + \dfrac{1}{2}$ b. $\dfrac{n\,3^{n-1}}{3^n - 1} - 1$

c. $\dfrac{n\,3^{n-1}}{3^n - 1} + 1$ d. none of these

Multiple Correct Answers Type

Each question has four choices a, b, c and d, out of which *one or more* answers are correct.

1. If $z = \omega, \omega^2$, where ω is a non-real complex cube root of unity, are two vertices of an equilateral triangle in the Argand plane then the third vertex may be represented by

 a. $z = 1$ b. $z = 0$ c. $z = -2$ d. $z = -1$

2. $P(z_1)$, $Q(z_2)$, $R(z_3)$, and $S(z_4)$ are four complex numbers representing the vertices of a rhombus taken in order on the complex plane, then which one of the following is/are correct?

 a. $\dfrac{z_1 - z_4}{z_2 - z_3}$ is purely real

 b. $\text{amp}\dfrac{z_1 - z_4}{z_2 - z_4} \neq \text{amp}\dfrac{z_2 - z_4}{z_3 - z_4}$

 c. $\dfrac{z_1 - z_3}{z_2 - z_4}$ is purely imaginary

 d. $|z_1 - z_3| \neq |z_2 - z_4|$

3. If $z^3 + (3 + 2i)z + (-1 + ia) = 0$ has one real root, then the value of a lies in the interval ($a \in R$)

 a. $(-2, 1)$ b. $(-1, 0)$ c. $(0, 1)$ d. $(-2, 3)$

4. A rectangle of maximum area is inscribed in the circle $|z - 3 - 4i| = 1$. If one vertex of the rectangle is $4 + 4i$, then another adjacent vertex of this rectangle can be

 a. $3 + 4i$ b. $3 + 5i$ c. $3 + 3i$ d. $3 - 3i$

5. If $|z_1| = 15$ and $|z_2 - 3 - 4i| = 5$, then

 a. $|z_1 - z_2|_{min} = 5$ b. $|z_1 - z_2|_{min} = 10$

 c. $|z_1 - z_2|_{max} = 20$ d. $|z_1 - z_2|_{max} = 25$

6. If the points $A(z)$, $B(-z)$, and $C(1 - z)$ are the vertices of an equilateral triangle ABC, then

 a. sum of possible z is 1/2

 b. sum of possible z is 1

 c. product of possible z is 1/4

 d. product of possible z is 1/2

7. If $|(z - z_1)/(z - z_2)| = 3$, where z_1 and z_2 are fixed complex numbers and z is a variable complex number, then z lies on a

 a. Circle with z_1 as its interior point

 b. Circle with z_2 as its interior point

 c. Circle with z_1 as its exterior point

 d. Circle with z_2 as its exterior point

8. If $\arg(z + a) = \pi/6$ and $\arg(z - a) = 2\pi/3$ ($a \in R^+$), then

 a. $|z| = a$ b. $|z| = 2a$

 c. $\arg(z) = \dfrac{\pi}{2}$ d. $\arg(z) = \dfrac{\pi}{3}$

9. The value(s) of $(-i)^{1/3}$ is/are

 a. $\dfrac{\sqrt{3} - i}{2}$ b. $\dfrac{\sqrt{3} + i}{2}$

 c. $\dfrac{-\sqrt{3} - i}{2}$ d. $\dfrac{-\sqrt{3} + i}{2}$

10. If $1, z_1, z_2, z_3, \ldots, z_{n-1}$ be the nth roots of unity and ω be a non-real complex cube root of unity, then the product $\prod_{r=1}^{n-1} (\omega - z_r)$ can be equal to

 a. 0 b. 1 c. -1 d. $1 + \omega$

11. If the equation, $z^3 + (3 + i)z^2 - 3z - (m + i) = 0$, where $m \in R$, has at least one real root, then m can have the value equal to

 a. 1 b. 2 c. 3 d. 5

12. Let $P(x)$ and $Q(x)$ be two polynomials. Suppose that $f(x) = P(x^3) + xQ(x^3)$ is divisible by $x^2 + x + 1$, then

 a. $P(x)$ is divisible by $(x - 1)$ but $Q(x)$ is not divisible by $x - 1$

 b. $Q(x)$ is divisible by $(x - 1)$ but $P(x)$ is not divisible by $x - 1$

 c. Both $P(x)$ and $Q(x)$ are divisible by $x - 1$

 d. $f(x)$ is divisible by $x - 1$

13. If $\text{amp}(z_1 z_2) = 0$ and $|z_1| = |z_2| = 1$, then

 a. $z_1 + z_2 = 0$ b. $z_1 z_2 = 1$

 c. $z_1 = \bar{z}_2$ d. none of these

14. If $|z - (1/z)| = 1$ then

 a. $|z|_{max} = \dfrac{1 + \sqrt{5}}{2}$ b. $|z|_{min} = \dfrac{\sqrt{5} - 1}{2}$

 c. $|z|_{max} = \dfrac{\sqrt{5} - 2}{2}$ d. $|z|_{min} = \dfrac{\sqrt{5} - 1}{\sqrt{2}}$

15. z_0 is a root of the equation $z^n \cos\theta_0 + z^{n-1}\cos\theta_1 + \cdots + z\cos\theta_{n-1} + \cos\theta_n = 2$, where $\theta_i \in R$, then

 a. $|z_0| > 1$ b. $|z_0| > \dfrac{1}{2}$

 c. $|z_0| > \dfrac{1}{4}$ d. $|z_0| > \dfrac{3}{2}$

16. If from a point P representing the complex number z_1 on the curve $|z| = 2$, two tangents are drawn to the curve $|z| = 1$, meeting at points $Q(z_2)$ and $R(z_3)$, then

 a. complex number $(z_1 + z_2 + z_3)/3$ will be on the curve $|z| = 1$

 b. $\left(\dfrac{4}{\bar{z}_1} + \dfrac{1}{\bar{z}_2} + \dfrac{1}{\bar{z}_3}\right)\left(\dfrac{4}{z_1} + \dfrac{1}{z_2} + \dfrac{1}{z_3}\right) = 9$

 c. $\arg\left(\dfrac{z_2}{z_3}\right) = \dfrac{2\pi}{3}$

 d. orthocentre and circumcnetre of ΔPQR will coincide

17. If z_1, z_2 are two complex numbers ($z_1 \neq z_2$) satisfying $|z_1^2 - z_2^2| = |\bar{z}_1^2 + \bar{z}_2^2 - 2\bar{z}_1\bar{z}_2|$, then

a. $\frac{z_1}{z_2}$ is purely imaginary

b. $\frac{z_1}{z_2}$ is purely real

c. $|\arg z_1 - \arg z_2| = \pi$

d. $|\arg z_1 - \arg z_2| = \frac{\pi}{2}$

18. A complex number z is rotated in anticlockwise direction by an angle α and we get z' and if the same complex number z is rotated by an angle α in clockwise direction and we get z'' then

a. z', z, z'' are in G.P. b. z', z, z'' are in H.P.

c. $z' + z'' = 2z \cos \alpha$ d. $z'^2 + z''^2 = 2z^2 \cos 2\alpha$

19. Let z be a complex number satisfying equation $z^p = \bar{z}^q$, where $p, q \in N$, then

a. if $p = q$, then number of solutions of equation will be infinite

b. if $p = q$, then number of solutions of equation will be finite

c. if $p \neq q$, then number of solutions of equation will be $p + q + 1$.

d. if $p \neq q$, then number of solutions of equation will be $p + q$

20. If $z_1 = 5 + 12i$ and $|z_2| = 4$ then

a. maximum $(|z_1 + iz_2|) = 17$

b. minimum $(|z_1 + (1 + i)z_2|) = 13 - 4\sqrt{2}$

c. minimum $\left|\frac{z_1}{z_2 + \frac{4}{z_2}}\right| = \frac{13}{4}$

d. maximum $\left|\frac{z_1}{z_2 + \frac{4}{z_2}}\right| = \frac{13}{3}$

21. If $p = a + b\omega + c\omega^2$, $q = b + c\omega + a\omega^2$, and $r = c + a\omega + b\omega^2$ where $a, b, c \neq 0$ and ω is the complex cube root of unity, then

a. If p, q, r lie on the circle $|z| = 2$, the triangle formed by these points is equilateral.

b. $p^2 + q^2 + r^2 = a^2 + b^2 + c^2$

c. $p^2 + q^2 + r^2 = 2(pq + qr + rp)$

d. none of these

22. z_1, z_2, z_3 and z'_1, z'_2, z'_3 are non-zero complex numbers such that $z_3 = (1 - \lambda)z_1 + \lambda z_2$ and $z'_3 = (1 - \mu)z'_1 + \mu z'_2$ then which of the following statements is/are true?

a. If $\lambda, \mu \in R - \{0\}$, then z_1, z_2 and z_3 are collinear and z'_1, z'_2, z'_3 are collinear separately.

b. If λ, μ are complex numbers, where $\lambda = \mu$ then triangles formed by points z_1, z_2, z_3 and z'_1, z'_2, z'_3 are similar.

c. If λ, μ are distinct complex numbers, then points z_1, z_2, z_3 and z'_1, z'_2, z'_3 are not connected by any well defined geometry.

d. If $0 < \lambda < 1$, then z_3 divides the line joining z_1 and z_2 internally and if $\mu > 1$ then z'_3 divides the line joining of z'_1, z'_2 externally.

23. If $|z - 3| = \min\{|z - 1|, |z - 5|\}$, then Re($z$) equals to

a. 2 b. $\frac{5}{2}$ c. $\frac{7}{2}$ d. 4

24. If n is a natural number ≥ 2, such that $z^n = (z + 1)^n$, then

a. roots of equation lie on a straight line parallel to y-axis

b. roots of equation lie on a straight line parallel to x-axis

c. sum of the real parts of the roots is $-[(n-1)/2]$

d. none of these

25. If $|z - 1| = 1$, then

a. $\arg(z - 1 - i)/z$ can be equal to $-\pi/4$

b. $(z - 2)/z$ is purely imaginary number

c. $(z - 2)/z$ is purely real number

d. if $\arg(z) = \theta$, where $z \neq 0$ and θ is acute, then $1 - 2/z = i \tan \theta$

26. If $z = x + iy$, then the equation $|(2z - i)/(z + 1)| = m$ represents a circle when m can be

a. 1/2 b. 1

c. 2 d. $3 < r < 2\sqrt{3}$

27. Given that the two curves $\arg(z) = \pi/6$ and $|z - 2\sqrt{3}i| = r$ intersect in two distinct points, then

a. $[r] \neq 2$ b. $0 < r < 3$ c. $r = 6$ d. $r > 3$

($[r]$ represents integral part of r)

28. If P and Q are represented by the complex numbers z_1 and z_2, such that $|1/z_2 + 1/z_1| = |1/z_2 - 1/z_1|$, then

a. ΔOPQ (where O is the origin) is equilateral

b. ΔOPQ is right angled

c. the circumcentre of ΔOPQ is $\frac{1}{2}(z_1 + z_2)$

d. the circiucentre of ΔOPQ is $\frac{1}{3}(z_1 + z_2)$

29. Given $z = f(x) + i\,g(x)$ where f, g: $(0, 1) \to (0, 1)$ are real valued functions. Then, which of the following does not hold good?

a. $z = \frac{1}{1 - ix} + i\left(\frac{1}{1 + ix}\right)$

b. $z = \frac{1}{1 + ix} + i\left(\frac{1}{1 - ix}\right)$

c. $z = \frac{1}{1 + ix} + i\left(\frac{1}{1 + ix}\right)$

d. $z = \frac{1}{1 - ix} + i\left(\frac{1}{1 - ix}\right)$

30. Given that the complex numbers which satisfy the equation $z\bar{z}^3 + \bar{z}z^3 = 350$ form a rectangle in the Argand plane with the length of its diagonal having an integral number of units, then

a. area of rectangle is 48 sq. units

b. if z_1, z_2, z_3, z_4 are vertices of rectangle then $z_1+z_2+z_3+z_4 = 0$

c. rectangle is symmetrical about real axis

d. $\arg(z_1 - z_3) = \frac{\pi}{4}$ or $\frac{3\pi}{4}$

31. Equation of tangent drawn to circle $|z| = r$ at the point $A(z_0)$ is

a. $\text{Re}\left(\frac{z}{z_0}\right) = 1$ b. $z\bar{z}_0 + z_0\bar{z} = 2r^2$

c. $\text{Im}\left(\frac{z}{z_0}\right) = 1$ d. $\text{Im}\left(\frac{z_0}{z}\right) = 1$

32. z_1 and z_2 are the roots of the equation $z^2 - az + b = 0$, where $|z_1| = |z_2| = 1$ and a, b are non-zero complex numbers, then

a. $|a| \le 1$ b. $|a| \le 2$

c. $\arg(a^2) = \arg(b)$ d. $\arg a = \arg(b^2)$

33. Let z_1, z_2, z_3 be the three non-zero complex numbers such that $z_2 \ne 1$, $a = |z_1|$, $b = |z_2|$ and $c = |z_3|$. Let,

$$\begin{vmatrix} a & b & c \\ b & c & a \\ c & a & b \end{vmatrix} = 0$$

Then

a. $\arg\left(\frac{z_3}{z_2}\right) = \arg\left(\frac{z_3 - z_1}{z_2 - z_1}\right)^2$

b. orthocentre of triangle formed by z_1, z_2, z_3 is $z_1 + z_2 + z_3$

c. if triangle formed by z_1, z_2, z_3 is equilateral, then its area is $\frac{3\sqrt{3}}{2}|z_1|^2$

d. if triangle formed by z_1, z_2, z_3 is equilateral then $z_1+z_2+z_3 = 0$

34. Locus of complex number satisfying $\arg[(z - 5 + 4i)/(z + 3 - 2i)] = -\pi/4$ is the arc of a circle

a. whose radius is $5\sqrt{2}$

b. whose radius is 5

c. whose lengh (of arc) is $(15\pi)/\sqrt{2}$

d. whose centre is $-2 - 5i$

35. If α is a complex constant such that $\alpha z^2 + z + \bar{\alpha} = 0$ has a real root, then

a. $\alpha + \bar{\alpha} = 1$

b. $\alpha + \bar{\alpha} = 0$

c. $\alpha + \bar{\alpha} = -1$

d. the absolute value of the real root is 1

36. If $\sqrt{5 - 12i} + \sqrt{-5 - 12i} = z$, then principal value of arg z can be

a. $-\frac{\pi}{4}$ b. $\frac{\pi}{4}$ c. $\frac{3\pi}{4}$ d. $-\frac{3\pi}{4}$

Reasoning Type

Each question has four choices a, b, c, and d, out of which *only one* is correct. Each question contains Statement 1 and Statement 2.

a. Both the statements are True and Statement 2 is the correct explanation of Statement 1.

b. Both the statements are True but Statement 2 is Not the correct explanation of Statement 1.

c. Statement 1 is True and Statement 2 is False.

d. Statement 1 is False and Statement 2 is True.

1. **Statement 1:** If $\arg(z_1z_2) = 2\pi$, then both z_1 and z_2 are purely real (z_1 and z_2 have principal arguments).

Statement 2: Principal argument of complex number and between $-\pi$ and π.

2. **Statement 1:** If n is an odd integer greater than 3 but not a multiple of 3, prove that $(x + 1)^n - x^n - 1$ is divisible by $x^3 + x^2 + x$.

Statement 2: If n is an odd integer greater than 3 but not a multiple of 3, we have $1 + \omega^n + \omega^{2n} = 3$.

3. **Statement 1:** If $z_1 + z_2 = a$ and $z_1 z_2 = b$, where $a = \bar{a}$ and $b = \bar{b}$, then $\arg(z_1z_2) = 0$.

Statement 2: The sum and product of two complex numbers are real if and only if they are conjugate of each other.

4. **Statement 1:** If $x + (1/x) = 1$ and $p = x^{4000} + (1/x^{4000})$ and q is the digit at unit place in the number $2^{2^n} + 1$, $n \in N$ and $n > 1$, then the value of $p + q$ is 8.

Statement 2: If ω, ω^2 are the roots of $x + 1/x = -1$, then $x^2 + 1/x^2 = -1$, $x^3 + 1.x^3 = 2$.

5. **Statement 1:** Let z_1 and z_2 are two complex numbers such that $|z_1 - z_2| = |z_1 + z_2|$ then the orthocentre of ΔAOB is $[(z_1 + z_2)/2]$.

(where O is origin).

Statement 2: In case of right angled triangle, orthocentre is that point at which the triangle is right angled.

6. **Statement 1:** Locus of z satisfying $|z - 1| + |z - 8| = 5$ is an ellipse.

Statement 2: Sum of focal distances of any point on ellipse is constant for an ellipse.

7. **Statement 1:** $|z_1 - a| < a$, $|z_2 - b| < b$, $|z_3 - c| < c$, where a, b, c are positive real numbers, then $|z_1 + z_2 + z_3|$ is greater than $2|a + b + c|$.

Statement 2: $|z_1 \pm z_2| \le |z_1| + |z_2|$.

8. Let fourth roots of unity be z_1, z_2, z_3 and z_4 respectively.

Statement 1: $z_1^2 + z_2^2 + z_3^2 + z_4^2 = 0$.

Statement 2: $z_1 + z_2 + z_3 + z_4 = 0$.

9. **Statement 1:** If the equation $ax^2 + bx + c = 0$, $0 < a < b < c$, has non-real complex roots z_1 and z_2, then $|z_1| > 1$, $|z_2| > 1$.

Statement 2: Complex roots always occur in conjugate pairs.

10. **Statement 1:** If z_1, z_2 are the roots of the quadratic equation $az^2 + bz + c = 0$ such that $\text{Im}(z_1 z_2) \ne 0$, then at least one of a, b, c is imaginary.

Statement 2: If quadratic equation having real coefficients has complex roots, then roots are always conjugate to each other.

11. Statement 1: The product of all values of $(\cos\alpha + i\sin\alpha)^{3/5}$ is $\cos 3\alpha + i\sin 3\alpha$.

Statement 2: The product of fifth roots of unity is 1.

12. Statement 1: If $|z_1| = |z_2| = |z_3|$, $z_1 + z_2 + z_3 = 0$ and (z_1), $B(z_2)$, $C(z_3)$ are the vertices of ΔABC, then one of the values of $\arg[(z_2 + z_3 - 2z_1)/(z_3 - z_2)]$ is $\pi/2$.

Statement 2: In equilateral triangle orthocentre coincides with centroid.

13. Statement 1: Let z be a complex number, then the equation $z^4 + z + 2 = 0$ cannot have a root, such that $|z| < 1$.

Statement 2: $|z_1 + z_2| \le |z_1| + |z_2|$

14. If $z_1 \ne -z_2$ and $|z_1 + z_2| = \left|(1/z_1) + (1/z_2)\right|$ then

Statement 1: z_1z_2 is unimodular.

Statement 2: z_1 and z_2 both are unimodular.

Linked Comprehension Type

Based upon each paragraph, the relevant multiple choice questions have to be answered. Each question has four choices a, b, c, and d, out of which *only one* is correct.

For Problems 1–4

Consider the complex numbers z_1 and z_2 satisfying the relation $|z_1 + z_2|^2 = |z_1|^2 + |z_2|^2$.

1. Complex number $z_1\bar{z}_2$ is

a. purely real **b.** purely imaginary
c. zero **d.** none of these

2. Complex number z_1/z_2 is

a. purely real **b.** purely imaginary
c. zero **d.** none of these

3. One of the possible argument of complex number $i(z_1/z_2)$

a. $\dfrac{\pi}{2}$ **b.** $-\dfrac{\pi}{2}$
c. 0 **d.** none of these.

4. Possible difference between the argument of z_1 and z_2 is

a. 0 **b.** π
c. $-\dfrac{\pi}{2}$ **d.** none of these

For Problems 5–8

Consider the complex numbers $z = (1 - i\sin\theta)/(1 + i\sin\theta)$.

5. The value of θ for which z is purely real is

a. $n\pi - \dfrac{\pi}{4}, n \in I$ **b.** $n\pi + \dfrac{\pi}{4}, n \in I$
c. $n\pi, n \in I$ **d.** none of these

6. The value of θ for which z is purely imaginary is

a. $n\pi - \dfrac{\pi}{4}, n \in I$ **b.** $n\pi - \dfrac{\pi}{4}, n \in I$
c. $n\pi, n \in I$ **d.** no real values of θ

7. The value of θ for which z is unimodular is

a. $n\pi \pm \dfrac{\pi}{6}, n \in I$ **b.** $n\pi \pm \dfrac{\pi}{3}, n \in I$
c. $n\pi \pm \dfrac{\pi}{4}, n \in I$ **d.** no real values of θ

8. If argument of z is $\pi/4$, then

a. $\theta = n\pi$, $n \in I$ only
b. $\theta = (2n + 1)$, $n \in I$ only
c. both $\theta = n\pi$ and $\theta = (2n + 1)\dfrac{\pi}{2}$, $n \in$ I
d. none of these

For Problems 9–11

Consider a quadratic equation $az^2 + bz + c = 0$ where a, b, c are complex numbers.

9. The condition that the equation has one purely imaginary root is

a. $(c\bar{a} - a\bar{c})^2 = -(b\bar{c} + c\bar{b})(a\bar{b} + \bar{a}b)$
b. $(c\bar{a} + a\bar{c})^2 = (b\bar{c} + c\bar{b})(a\bar{b} + \bar{a}b)$
c. $(c\bar{a} - a\bar{c})^2 = (b\bar{c} - c\bar{b})(a\bar{b} - \bar{a}b)$
d. none of these

10. If equation has two purely imaginary roots, then which of the following is not true

a. $a\bar{b}$ is purely imaginary
b. $b\bar{c}$ is purely imaginary
c. $c\bar{a}$ is purely real
d. none of these

11. The condition that the equation has one purely real root is

a. $(c\bar{a} - a\bar{c})^2 = (b\bar{c} + c\bar{b})(a\bar{b} - \bar{a}b)$
b. $(c\bar{a} - a\bar{c})^2 = (b\bar{c} - c\bar{b})(a\bar{b} + \bar{a}b)$
c. $(c\bar{a} - a\bar{c})^2 = (b\bar{c} + c\bar{b})(a\bar{b} + \bar{a}b)$
d. $(c\bar{a} - a\bar{c})^2 = (b\bar{c} - c\bar{b})(a\bar{b} - \bar{a}b)$

For Problems 12–14

Consider the equation $az + b\bar{z} + c = 0$, where $a, b, c \in Z$.

12. If $|a| \ne |b|$, then z represents

a. circle **b.** straight line
c. one point **d.** ellipse

13. If $|a| = |b|$ and $\bar{a}c \ne b\bar{c}$, then z has

a. infinite solutions **b.** no solutions
c. finite solutions **d.** cannot say anything

14. If $|a| = |b| \ne 0$ and $\bar{a}c = b\bar{c}$, then $az + b\bar{z} + c = 0$ represents

a. an ellipse **b.** a circle
c. a point **d.** a straight line

For Problems 15–17

Let z be a complex number satisfying $z^2 + 2z\lambda + 1 = 0$, where λ is a parameter which can take any real value.

15. The roots of this equation lie on a certain circle if

a. $-1 < \lambda < 1$ **b.** $\lambda > 1$
c. $\lambda < 1$ **d.** none of these

16. One root lies inside the unit circle and one outside if

a. $-1 < \lambda < 1$ **b.** $\lambda > 1$
c. $\lambda < 1$ **d.** none of these

17. For every large value of λ, the roots are approximately

a. $-2\lambda, 1/\lambda$ b. $-\lambda, -1/\lambda$

c. $-2\lambda, -1/2\lambda$ d. none of these

For Problems 18–20

Consider the equation $az^2 + z + 1 = 0$ having purely imaginary root where $a = \cos\theta + i\sin\theta$, $i = \sqrt{-1}$ and function $f(x) = x^3 - 3x^2 + 3(1 + \cos\theta)x + 5$, then answer the following questions.

18. Which of the following is true about $f(x)$?

a. $f(x)$ decreases for $x \in [2n\pi, (2n+1)\pi]$, $n \in Z$

b. $f(x)$ decreases for $x \in \left[(2n-1)\frac{\pi}{2}, (2n+1)\frac{\pi}{2}\right]$, $n \in Z$

c. $f(x)$ is non-monotonic function

d. $f(x)$ increases for $x \in R$.

19. Which of the following is true?

a. $f(x) = 0$ has three real distinct roots

b. $f(x) = 0$ has one positive real root

c. $f(x) = 0$ has one negative real root

d. $f(x) = 0$ has three but not distinct roots

20. Number of roots of the equation $\cos 2\theta = \cos\theta$ in $[0, 4\pi]$ is

a. 2 b. 3

c. 4 d. 6

For Problems 21–23

Complex numbers z satisfy the equation $|z - 4/z)| = 2$.

21. The difference between the least and the greatest moduli of complex numbers is

a. 2 b. 4

c. 1 d. 3

22. The value of arg (z_1/z_2), where z_1 and z_2 are complex numbers with the greatest and the least moduli can be

a. 2π b. π

c. $\pi/2$ d. none of these

23. Locus of z if $|z - z_1| = |z - z_2|$, where z_1 and z_2 are complex numbers with the greatest and the least moduli is

a. line parallel to real axis

b. line parallel to imaginary axis

c. line having positive slope

d. line having negative slope

For Problems 24–26

Consider ΔABC in Argand plane. Let $A(0)$, $B(1)$, and $C(1 + i)$ be its vertices and M be the mid-point of CA. Let z be a variable complex number on the line BM. Let u be another variable complex number defined as $u = z^2 + 1$.

24. Locus of u is

a. parabola b. ellipse

c. hyperbola d. none of these

25. Axis of locus of u is

a. imaginary axis b. real axis

c. $z + \bar{z} = 2$ d. none of these

26. Directrix of locus of u is

a. imaginary axis b. $z - \bar{z} = 2i$

c. real axis d. none of these

For Problems 27–29

In an Argand plane z_1, z_2 and z_3 are, respectively, the vertices of an isosceles triangle ABC with $AC = BC$ and $\angle CAB = \theta$. z_4 is the centre of triangle

27. The value of $AB \times AC/(IA)^2$ is

a. $\frac{(z_2 - z_1)(z_3 - z_1)}{(z_4 - z_1)^2}$ b. $\frac{(z_2 - z_1)(z_1 - z_3)}{(z_4 - z_1)^2}$

c. $\frac{(z_4 - z_1)}{(z_2 - z_1)(z_3 - z_1)}$ d. none of these

28. The value of $(z_4 - z_1)^2 (\cos\theta + 1)\sec\theta$ is

a. $\frac{(z_2 - z_1)(z_3 - z_1)}{(z_4 - z_1)}$ b. $(z_2 - z_1)(z_3 - z_1)$

c. $(z_2 - z_1)(z_3 - z_1)^2$ d. $\frac{(z_2 - z_1)(z_3 - z_1)}{(z_4 - z_1)^2}$

29. The value of $(z_2 - z_1)^2 \tan\theta \tan\theta/2$ is

a. $(z_1 + z_2 - 2z_3)$

b. $(z_1 + z_2 - z_3)(z_1 + z_2 - z_4)$

c. $-(z_1 + z_2 - 2z_3)(z_1 + z_2 - 2z_4)$

d. none of these

For Problems 30–32

$A(z_1)$, $B(z_2)$, $C(z_3)$ are the vertices of a triangle ABC inscribed in the circle $|z| = 2$. Internal angle bisector of the angle A meets the circumcircle again at $D(z_4)$.

30. Complex number representing point D is

a. $z_4 = \frac{z_1 z_2}{z_3}$ b. $z_4 = \frac{z_2 z_3}{z_1}$

c. $z_4 = \frac{z_1 z_2}{z_3}$ d. none of these

31. arg $[z_4/(z_2 - z_3)]$ is equal to

a. $\frac{\pi}{4}$ b. $\frac{\pi}{3}$

c. $\frac{\pi}{2}$ d. $\frac{2\pi}{3}$

32. For fixed positions of $B(z_2)$ and $C(z_3)$ all the bisectors (internal) of $\angle A$ will pass through a fixed point which is

a. H.M. of z_2 and z_3 b. A.M. of z_2 and z_3

c. G.M. of z_2 and z_3 d. none of these

Matrix-Match Type

Each question contains statements given in two columns which have to be matched. Statements a, b, c, d in column I have to be matched with statements p, q, r, s in column II. If the correct match is a→p, a→s, b→r, c→p, c→q and d→s, then the correctly bubbled 4 × 4 matrix should be as follows:

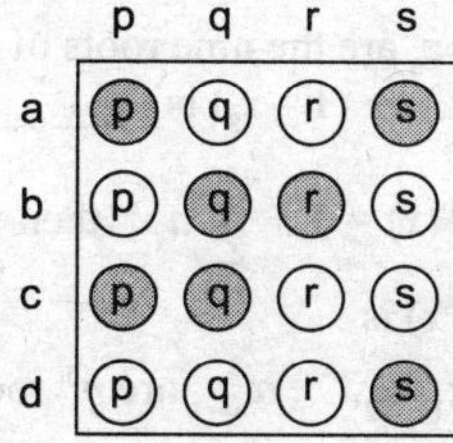

1.

Column I	Column II: argument of $z = a + ib$
a. $ab > 0$	**p.** $-\tan^{-1}\left\|\frac{b}{a}\right\|$
b. $ab < 0$	**q.** $\pi - \tan^{-1}\left\|\frac{b}{a}\right\|$
c. $a^2 + b^2 = 0$	**r.** $\tan^{-1}\frac{b}{a}$
d. $ab = 0$	**s.** $-\pi + \tan^{-1}\frac{b}{a}$
	t. not defined
	u. 0 or $\frac{\pi}{2}$

2.

Column I	Column II
a. $z^4 - 1 = 0$	**p.** $z = \cos\frac{\pi}{8} + i\sin\frac{\pi}{8}$
b. $z^4 + 1 = 0$	**q.** $z = \cos\frac{\pi}{8} - i\sin\frac{\pi}{8}$
c. $iz^4 + 1 = 0$	**r.** $z = \cos\frac{\pi}{4} + i\sin\frac{\pi}{4}$
d. $iz^4 - 1 = 0$	**s.** $z = \cos 0 + i\sin 0$

3.

Column I	Column II (Locus)
a. $\|z - 1\| = \|z - i\|$	**p.** pair of straight lines
b. $\|z + \bar{z}\| + \|z - \bar{z}\| = 2$	**q.** a line through the origin
c. $\|z + \bar{z}\| = \|z - \bar{z}\|$	**r.** circle
d. If z lies on $\|z\| = 1$, then $2/z$ lies on	**s.** square

4. Which of the condition/conditions in column II are satisfied by the quadrilateral formed by z_1, z_2, z_3, z_4 in order given in column I?

Column I	Column II
a. parallelogram	**p.** $z_1 - z_4 = z_2 - z_3$
b. rectangle	**q.** $\|z_1 - z_3\| = \|z_2 - z_4\|$
c. rhombus	**r.** $\frac{z_1 - z_2}{z_3 - z_4}$ is purely real
d. square	**s.** $\frac{z_1 - z_3}{z_2 - z_4}$ is purely imaginary
	t. $\frac{z_1 - z_2}{z_3 - z_2}$ is purely imaginary

5.

Column I	Column II
a. If $\|z - 2i\| + \|z - 7i\| = k$, then locus of z is an ellipse if $k =$	**p.** 7
b. If $\|(2z - 3)/(3z - 2)\| = k$, then locus of z is a circle if 2/3 is a point inside circle and 3/2 is outside the circle if $k =$	**q.** 8
c. If $\|z - 3\| - \|z - 4i\| = k$, then locus of z is a hyperbola if k is	**r.** 2
d. If $\|z - (3 + 4i)\| = k/50\|a\bar{z} + \bar{a}z + b\|$, where $a = 3 + 4i$, then locus of z is a hyperbola with $k =$	**s.** 4
	t. 5

6.

Column I	Column II
a. The value of $\sum_{n=1}^{5}(x^n + 1/x^n)^2$ when $x^2 - x + 1 = 0$ is	**p.** 2
b. If $\left[\frac{1 + \cos\theta + i\sin\theta}{\sin\theta + i(1 + \cos\theta)}\right]^4 = \cos n\theta + i\sin n\theta$, then $n =$	**q.** 4
c. The adjacent vertices of a regular polygon of n sides having centre at origin are the points z and $\bar{z}$. If $\text{Im}(z)/\text{Re}(z) = \sqrt{2} - 1$, then the value of $n/4$ is	**r.** 9
d. $1/50\left\{\sum_{r=1}^{10}(r - \omega)(r - \omega)^2\right\}$ = (where ω is cube root of unity)	**s.** 8

Integer Type

1. If $z = x + iy$, $z^{1/3} = a - ib$, and $\frac{x}{a} - \frac{y}{b} = k(a^2 - b^2)$, then the value of k is ____.

2. The value of $e^{2mi\cot^{-1}p}\left(\frac{pi + 1}{pi - 1}\right)^m$ is____.

3. The number of complex numbers z satisfying $Re(z^2) = 0$, $|z| = \sqrt{3}$ is____.

4. If α and β are different complex numbers with $|\beta| = 1$, then the value of $\left|\frac{\beta - \alpha}{1 - \bar{\alpha}\beta}\right|$ is ____.

5. If $|z_1| = 1$, $|z_2| = 2$, $|z_3| = 3$ and $|9z_1z_2 + 4z_1z_3 + z_2z_3| = 12$, then the value of $|z_1 + z_2 + z_3|$ is_____.

6. Let $\left|\dfrac{\bar{z}_1 - 2\bar{z}_2}{2 - z_1\bar{z}_2}\right| = 1$ and $|z_2| \neq 1$, where z_1 and z_2 are complex numbers, then the value of $3|z_1|$ is

7. The number of ordered pair(s) (z, ω) of the complex numbers z and ω satisfying the system of equations $z^3 + \bar{\omega}^7 = 0$ and $z^5 \times \omega^{11} = 1$ is_____.

8. If $|z_1 - 1| \leq 1$, $|z_2 - 2| \leq 2$, $|z_3 - 3| \leq 3$, then the greatest value of $(|z_1 + z_2 + z_3|)/2$ is_____.

9. If z is any complex number such that $|z + 4| \leq 3$, then the greatest value of $|z + 1|$ is_____.

10. The difference between the least and the greatest values of $|z_1 + z_2|$ if $z_1 = 24 + 7i$ and $|z_2| = 6$ is_____.

11. The number of points of intersection of the curves $\arg(z - 3i) = \dfrac{3\pi}{4}$ and $\arg(2z + 1 - 2i) = \dfrac{\pi}{4}$ is_____.

12. Let z be a complex number satisfying the equation $(z^3 + 3)^2 = -16$, then the value of $|z|^3$ is_____.

13. Let z be not a real number such that $\dfrac{1 + z + z^2}{1 - z + z^2} \in R'$ then the value of $7|z|$ is_____.

14. If a, b, c are non zero complex numbers of equal modulus and satisfy $az^2 + bz + c = 0$, then the difference between the greatest and the least values of $|z|$ is.

15. If $\dfrac{2z_1}{3z_2}$ is a purely imaginary number, then the value of $\left|\left|\dfrac{z_1 - z_2}{z_1 + z_2}\right|\right|$ is_____.

16. The absolute value of the expression $\left(\cos\dfrac{\pi}{2} + i\sin\dfrac{\pi}{2}\right)\left(\cos\dfrac{\pi}{2^2} + i\sin\dfrac{\pi}{2^2}\right)\cdots$ to ∞ is_____.

17. The number of common roots of $x^{12} - 1 = 0$, $x^4 + x^2 + 1 = 0$ is_____.

18. z_1 and z_2 are the roots of $3z^2 + 3z + b = 0$. If $O(0)$, $A(z_1)$, $B(z_2)$ is an equilateral triangle, then the value of $5b$ is_____.

19. Let z be *a* complex number having the argument θ, $0 < \theta < \pi/2$ and satisfying the equation $|z - 3i| = 3$. Then the value of $3\left|\cot\theta - \dfrac{6}{z}\right|$ is_____.

20. If $|z - 2 - 3i|^2 + |z - 4 - 3i|^2 = \lambda$ represents the equation of a circle with the least radius, then the value of λ is_____.

21. The number of complex numbers which satisfy both the equations $|z - 1 - i| = \sqrt{2}$ and $|z + 1 + i| = 2$ is_____.

22. If $|z - 1| + |z + 3| \leq 8$, then the greatest value of $|z - 4|$ is_____.

23. If ω is an imaginary fifth root of unity, then the value of $\log_2 \log_{1/8}\left|1 + \omega + \omega^2 + \omega^3 - \dfrac{1}{\omega}\right|$ is_____.

24. If $1, z_1, z_2, z_3, \ldots z_8$ are the nine roots of unity, then the value of $(1 - z_1)(1 - z_2)\cdots(1 - z_8)$ is_____.

25. If $\alpha = e^{\frac{i2\pi}{7}}$ and $f(x) = 1 + \sum_{k=1}^{20} a_k x^k$ then the value of $f(x) + f(\alpha x) + \cdots + f(\alpha^6 x)$ is_____

26. If $n \geq 3$ and $1, \alpha_1, \alpha_2, \ldots \alpha_{n-1}$ are n^{th} roots of unity, then the value of the sum $\sum_{1 \leq i < j \leq n-1} \alpha i \alpha j$ is_____.

27. Given α, β are the 5th and 4th non–real roots of unity respectively, then the value of $(1 + \alpha)(1 + \beta)(1 + \alpha^2)(1 + \beta^2)(1 + \beta^3)(1 + \alpha^4)$ is_____.

28. If the six solutions of $x^6 = -64$ are written in the form $a + bi$, where a and b are real, then the product of those solutions with $a > 0$ is_____.

29. If z_r $(r = 1, 2, 3, \ldots, 50)$ are the roots of the equation $\sum_{r=0}^{50} (z)^r = 0$, then the value of $\left|\sum_{r=1}^{50} \dfrac{1}{z_r - 1}\right|$ is the square of_____.

30. Consider a triangle having vertices at the points $A\left(\dfrac{2}{\sqrt{3}}e^{i\frac{\pi}{2}}\right)$, $B\left(\dfrac{2}{\sqrt{3}}e^{-i\frac{\pi}{6}}\right)$, $C\left(\dfrac{2}{\sqrt{3}}e^{-i\frac{5\pi}{2}}\right)$. Let P be any point on its incircle, then the value of $AP^2 + BP^2 + CP^2$ is_____.

31. If $|z| = 4$, then the area of the triangle on the Argand diagram formed by the complex numbers z, iz, and $z + iz$ is_____.

32. Let $z_1 = 10 + 6i$ and $z_2 = 4 + 6i$. If z is any complex number such that the argument of $\dfrac{(z - z_1)}{(z - z_2)}$ is $\dfrac{\pi}{4}$, then the integral value of $|z - 7 - 9i|$ i is_____.

33. Complex numbers z_1, z_2, z_3 are the vertices A, B, C, respectively, of an isosceles right-angled triangle with right angle at C. The value of $\dfrac{(z_1 - z_2)^2}{(z_1 - z_3)(z_3 - z_2)}$ is_____.

34. If $iz^3 + z^2 - z + i = 0$, then the value of $7|z|$ is_____.

35. For all complex numbers z_1, z_2 satisfying $|z_1| = 12$ and $|z_2 - 3 - 4i| = 5$, the minimum value of $|z_1 - z_2|$ is_____.

36. If ω $(\neq 1)$ be a cube root of unity and $(1 + \omega^2)^n = (1 + \omega^4)^n$, then the least positive value of n is_____

37. a, b, c be integers, not all simultaneously equal and ω be the cube root of unity $(\omega \neq 1)$, then the minimum value of $|a + b\omega + c\omega^2|$ is_____.

38. Let $z = x + iy$ be a complex number where x and y are integers. Then the greatest composite value less than or equal to 10 by which the area of the rectangle is divisible whose vertices are the roots of the equation $z\bar{z}^3 + \bar{z}z^3 = 350$ is_____.

39. If ω and ω^2 are the non-real cube roots of unity and $\dfrac{1}{a + \omega} + \dfrac{1}{b + \omega} + \dfrac{1}{c + \omega} = 2\omega^2$ and $\dfrac{1}{a + \omega^2} + \dfrac{1}{b + \omega^2} + \dfrac{1}{c + \omega^2} = 2\omega$, then the value of $\dfrac{1}{a + 1} + \dfrac{1}{b + 1} + \dfrac{1}{c + 1}$ is_____.

40. When the polynomial $5x^3 + Mx + N$ is divided by $x^2 + x + 1$, the remainder is 0. The value of $|M + N|$ is_____.

41. Let $Z_1 = (8 + i)\sin\theta + (7 + 4i)\cos\theta$ and $Z_2 = (1 + 8i)\sin\theta + (4 + 7i)\cos\theta$ are two complex numbers. If $Z_1 Z_2 = a + ib$, where $a, b \in R$, then the largest value of $(a + b)^{1/3}\ \forall\ \theta \in R$, is________.

42. Modulus of non-zero complex number z, satisfying $\bar{z} + z = 0$ and $|z|^2 - 4zi = z^2$ is ________.

43. Let Z_1, Z_2, Z_3 be three complex numbers such that $|z_1| = |z_2| = |z_3| = 1$ and $\frac{z_1^2}{z_2 z_3} + \frac{z_2^2}{z_1 z_3} + \frac{z_3^2}{z_1 z_2} + 1 = 0$, then sum of possible values of $|z_1 + z_2 + z_3|$ is________.

44. Number of ordered pair(s) (a, b) of real numbers such that $(a + ib)^7 = a - ib$ holds good, is________.

45. Let Z_1 and Z_2 be complex numbers such that $|z_1| = |z_2| = r (r > 0)$. then the minimum value of $\left(\frac{r(z_1 + z_2)}{r^2 + z_1 z_2}\right)^2 + \left(\frac{r(z_1 - z_2)}{r^2 - z_1 z_2}\right)^2$ is ________.

ANSWERS AND SOLUTIONS

Objective Type

1. b. Verify by selecting particular values of a and b.
Let $a = -9$ and $b = 4$. Then,

$$\sqrt{a}\sqrt{b} = \sqrt{-9}\sqrt{4} = (3i)(2) = 6i$$

From option (a), we have

$$-\sqrt{|a|b} = -\sqrt{|-9|\times 4} = -\sqrt{36} = -6$$

From option (b), we have

$$\sqrt{|a|b}\,i = \sqrt{|-9|\times 4}\;i = 6i$$

2. c. $$x = 9^{\frac{1}{3}+\frac{1}{9}+\frac{1}{27}+\cdots} = 9^{\frac{\frac{1}{3}}{1-\frac{1}{3}}} = 9^{\frac{1}{2}} = 3$$

$$y = 4^{\frac{1}{3}-\frac{1}{9}+\frac{1}{27}+\cdots} = 4^{\frac{\frac{1}{3}}{1+\frac{1}{3}}} = 4^{\frac{1}{4}} = \sqrt{2}$$

$$z = \sum_{r=1}^{\infty}(1+i)^{-r} = \frac{1}{1+i} + \frac{1}{(1+i)^2} + \frac{1}{(1+i)^3} + \cdots$$

$$= \frac{\frac{1}{1+i}}{1-\frac{1}{1+i}} = \frac{1}{i} = -i$$

Let $a = x + yz = 3 - \sqrt{2}\,i$ (fourth quadrant). Then,

$$\arg a = -\tan^{-1}\left(\frac{\sqrt{2}}{3}\right)$$

3. d. $10z^2 - 3iz - k = 0$

$$\Rightarrow z = \frac{3i \pm \sqrt{-9+40k}}{20}$$

Now, $D = -9 + 40k$. If $k = 1$, then $D = 31$. So (a) is false.

If k is a negative real number, then D is a negative real number. So (d) is true.

If $k = i$, then $D = -9 + 40i = 16 + 40i - 25 = (4 + 5i)^2$, and the roots are $(1/5) + (2/5)i$ and $-(1/5 - 1/10)i$. So (c) is false.

If $k = 0$ (which is a complex number), then the roots are 0 and $(3/10)i$. So (b) is false.

4. d. Let $z = x + iy$, so that $\bar{z} = x + iy$.

$\therefore z^2 + \bar{z} = 0$

$\Rightarrow (x^2 - y^2 + x) + i(2xy - y) = 0$

Equating real and imaginary parts, we get

$$x^2 - y^2 + x = 0 \qquad (1)$$

and

$$2xy - y = 0 \Rightarrow y = 0 \text{ or } x = \frac{1}{2}$$

If $y = 0$, then (1) gives $x^2 + x = 0 \Rightarrow x = 0$ or $x = -1$

If $x = 1/2$, then from (1),

$$y^2 = \frac{1}{4} + \frac{1}{2} = \frac{3}{4} \Rightarrow y = \pm\frac{\sqrt{3}}{2}$$

Hence, there are four solutions in all.

5. c. Given that $a^2 + b^2 = 1$. Therefore,

$$\frac{1+b+ia}{1+b-ia} = \frac{(1+b+ia)(1+b+ia)}{(1+b-ia)(1+b+ia)}$$

$$= \frac{(1+b)^2 - a^2 + 2ia(1+b)}{1+b^2+2b+a^2}$$

$$= \frac{(1-a^2) + 2b + b^2 + 2ia(1+b)}{2(1+b)}$$

$$= \frac{2b^2 + 2b + 2ia(1+b)}{2(1+b)}$$

$$= b + ia$$

6. b. Let,

$$\sin\frac{\pi}{8} + i\cos\frac{\pi}{8} = z$$

$$\Rightarrow \left[\frac{1+\sin\frac{\pi}{8} + i\cos\frac{\pi}{8}}{1+\sin\frac{\pi}{8} - i\cos\frac{\pi}{8}}\right]^8$$

$$= \left(\frac{1+z}{1+\frac{1}{z}}\right)^8$$

$$= z^8$$

$$= \left(\sin\frac{\pi}{8} + i\cos\frac{\pi}{8}\right)^8$$

$$= \left(\cos\left(\frac{\pi}{2}-\frac{\pi}{8}\right) + i\sin\left(\frac{\pi}{2}-\frac{\pi}{8}\right)\right)^8$$

$$= \left(\cos\frac{3\pi}{8} + i\sin\frac{3\pi}{8}\right)^8$$

$$= \cos 3\pi = -1$$

7. a. Let $z_1 = a + ib$ and $z_2 = c - id$, where $a > 0$ and $d > 0$. Then,

$$|z_1| = |z_2| \Rightarrow a^2 + b^2 = c^2 + d^2 \qquad (1)$$

Now,

$$\frac{z_1+z_2}{z_1-z_2} = \frac{(a+ib)+(c-id)}{(a+ib)-(c-id)}$$

$$= \frac{[(a+c)+i(b-d)][(a-c)-i(b+d)]}{[(a-c)+i(b+d)][(a-c)-i(b+d)]}$$

$$= \frac{(a^2+b^2)-(c^2+d^2)-2(ad+bc)i}{a^2+c^2-2ac+b^2+d^2+2bd}$$

$$= \frac{-(ad+bc)i}{a^2+b^2 \quad ac+bd} \qquad \text{[Using (1)]}$$

Hence, $(z_1+z_2)/(z_1-z_2)$ is purely imaginary. However, if $ad + bc = 0$, then $(z_1+z_2)/(z_1-z_2)$ will be equal to zero. According to the conditions of the equation, we can have $ad + bc = 0$.

8. a. We have,

$$\arg\left(\frac{z_1}{z_2}\right)=\pi$$

$\Rightarrow \arg(z_1) - \arg(z_2) = \pi$

$\Rightarrow \arg(z_1) = \arg(z_2) + \pi$

Let $\arg(z_2) = \theta$. Then $\arg(z_1) = \pi + \theta$.

$\therefore z_1 = |z_1|[\cos(\pi + \theta) + i\sin(\pi + \theta)]$

$= |z_1|(-\cos\theta - i\sin\theta)$

and

$z_2 = |z_2|(\cos\theta + i\sin\theta)$

$= |z_1|(\cos\theta + i\sin\theta)\ (\because |z_1| = |z_2|)$

$= -z_1$

$\Rightarrow z_1 + z_2 = 0$

9. c. Let

$a = \cos\alpha + i\sin\alpha$

$b = \cos\beta + i\sin\beta$

$c = \cos\gamma + i\sin\gamma$

Then,

$a + 2b + 3c = (\cos\alpha + 2\cos\beta + 3\cos\gamma) + i(\sin\alpha + 2\sin\beta + 3\sin\gamma)$

$= 0$

$\Rightarrow a^3 + 8b^3 + 27c^3 = 18\,abc$

$\Rightarrow \cos 3\alpha + 8\cos 3\beta + 27\cos 3\gamma = 18\cos(\alpha + \beta + \gamma)$

and

$\sin 3\alpha + 8\sin 3\beta + 27\sin 3\gamma = 18\sin(\alpha + \beta + \gamma)$

10. d.

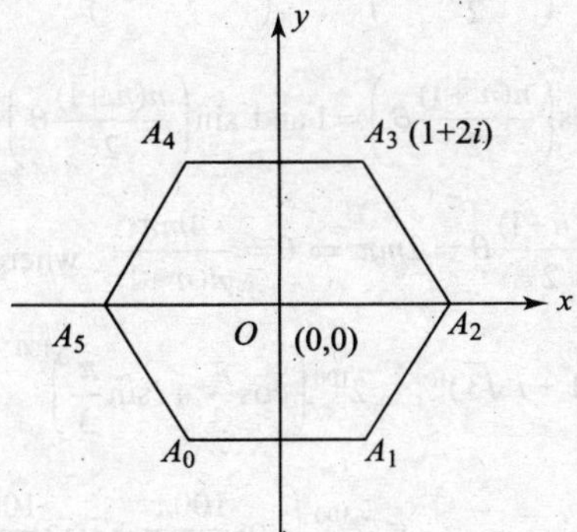

Fig. 2.1

Let the vertices be $z_0, z_1, \ldots, z_5$ w.r.t. centre O as origin and $|z_0| = \sqrt{5}$.

$\therefore A_0A_1 = |z_1 - z_0| = |z_0e^{i\theta} - z_0|$

$= |z_0|\,|\cos\theta + i\sin\theta - 1|$

$= \sqrt{5}\sqrt{(\cos\theta - 1)^2 + \sin^2\theta}$

$= \sqrt{5}\sqrt{2(1-\cos\theta)}$

$= \sqrt{5}\,2\sin(\theta/2)$

$\Rightarrow A_0A_1 = \sqrt{5}\,2\sin\left(\frac{\pi}{6}\right) = \sqrt{5}\left(\because \theta = \frac{2\pi}{6} = \frac{\pi}{3}\right)$ (1)

Hence, the perimeter of regular polygon is $6\sqrt{5}$.

11. a. $\frac{1+iz}{1-iz} = \frac{1+i[b+ic)/(1+a)]}{1-i[b+ic)/(1+a)]}$

$= \frac{1+a-c+ib}{1+a+c-ib}$

$= \frac{(1+a-c+ib)(1+a+c+ib)}{(1+a+c)^2+b^2}$ $[\because a^2+b^2+c^2=1]$

$= \frac{1+2a+a^2-b^2-c^2+2ib+2iab}{1+a^2+c^2+b^2+2ac+2(a+c)}$

$= \frac{2a+2a^2+2ib+2iab}{2+2ac+2(a+c)}$

$= \frac{a+a^2+ib+iab}{1+ac+(a+c)}$

$= \frac{a(a+1)+ib(a+1)}{(a+1)(c+1)}$

$= \frac{a+ib}{c+1}$

12. b. If z_1, z_2, z_3 are three complex numbers, then

$$A = \begin{vmatrix} \arg z_1 & \arg z_2 & \arg z_3 \\ \arg z_2 & \arg z_3 & \arg z_1 \\ \arg z_3 & \arg z_1 & \arg z_2 \end{vmatrix}$$

$$\Rightarrow A = (\arg z_1 + \arg z_2 + \arg z_3)\begin{vmatrix} 1 & \arg z_2 & \arg z_3 \\ 1 & \arg z_3 & \arg z_1 \\ 1 & \arg z_1 & \arg z_2 \end{vmatrix}$$

(Using $C_1 \to C_1 + C_2 + C_3$)

$$\Rightarrow A = \arg(z_1z_2z_3)\begin{vmatrix} 1 & \arg z_2 & \arg z_3 \\ 1 & \arg z_3 & \arg z_1 \\ 1 & \arg z_1 & \arg z_2 \end{vmatrix}$$

Hence, A is divisible by $\arg(z_1z_2z_3)$.

13. c. $\bar{z} + i\bar{w} = 0$

$\Rightarrow z - iw = 0$

$\Rightarrow z = iw$ (1)

$\arg zw = \pi$

$\Rightarrow \arg z + \arg w = \pi$

$\Rightarrow \arg z + \arg\frac{z}{i} = \pi$ [Using (1)]

$\Rightarrow \arg z + \arg z - \arg i = \pi$

$\Rightarrow 2\arg z - \frac{\pi}{2} = \pi$

$\Rightarrow 2\arg z = \frac{3\pi}{2}$

$\Rightarrow \arg z = \frac{3\pi}{4}$

14. c. We have,

$|z_1 - z_2|^2 = |z_1|^2 + |z_2|^2 - 2|z_1|\,|z_2| \cos(\theta_1 - \theta_2)$

where $\theta_1 = \arg(z_1)$ and $\theta_2 = \arg(z_2)$. Given,

$\arg(z_1 - z_2) = 0$

$\Rightarrow |z_1 - z_2|^2 = |z_1|^2 + |z_2|^2 - 2|z_1|\,|z_2|$

$= (|z_1| - |z_2|)^2$

$\Rightarrow |z_1 - z_2| = ||z_1| - |z_2||$

15. a. $\alpha = \dfrac{z - \bar{w}}{k^2 + z\bar{w}} \Rightarrow \bar{\alpha} = \dfrac{\bar{z} - w}{k^2 + \bar{z}w}$

But $z\bar{z} = w\bar{w} = k^2$. Hence,

$$\Rightarrow \bar{\alpha} = \frac{\dfrac{k^2}{z} - \dfrac{k^2}{\bar{w}}}{k^2 + \dfrac{k^2}{z}\dfrac{k^2}{\bar{w}}} = \frac{\bar{w} - z}{z\bar{w} + k^2} = -\alpha$$

$\Rightarrow \alpha + \bar{\alpha} = 0$

$\Rightarrow \mathrm{Re}(\alpha) = 0$

16. a. Here $x = 4\cos\theta$, $y = 4\sin\theta$.

$\therefore ||x| - |y|| = |4|\cos\theta| - 4|\sin\theta||$

$= 4||\cos\theta| - |\sin\theta||$

$= 4\sqrt{1 - 2|\cos\theta||\sin\theta|}$

$= 4\sqrt{1 - |\sin 2\theta|}$

Hence, the range is [0, 4].

17. c. $|k + z^2| = |z^2| - k = |z^2| + |k|$

$\Rightarrow k, z^2$, and $0 + i0$ are collinear

$\Rightarrow \arg(z^2) = \arg(k)$

$\Rightarrow 2\arg(z) = \pi$

$\Rightarrow \arg(z) = \dfrac{\pi}{2}$

18.b. The given equation is

$|z|^n = (z^2 + z)\,|z|^{n-2} + 1$

$\Rightarrow z^2 + z$ is real

$\Rightarrow z^2 + z = \bar{z}^2 + \bar{z}$

$\Rightarrow (z - \bar{z})(z + \bar{z} + 1) = 0$

$\Rightarrow z = \bar{z} = x$ as $z + \bar{z} + 1 \neq 0$ $(x \neq -1/2)$

Hence, the given equation reduces to

$x^n = x^n + x|x|^{n-2} + 1$

$\Rightarrow x|x|^{n-2} = -1$

$\Rightarrow x = -1$

So number of solutions is 1.

19. c. Observing carefully the system of equations, we find

$\dfrac{1+i}{2i} = \dfrac{1-i}{2} = \dfrac{1}{1+i}$

Hence, there are infinite number of solutions.

20. d. Given,

$$z^3 + \frac{3(\bar{z})^2}{|z|} = 0$$

Let,

$z = re^{i\theta}$

$\Rightarrow r^3 e^{i3\theta} + 3re^{-i2\theta} = 0$

Since r cannot be zero, so $re^{i5\theta} = -3$,

which will hold for $r = 3$ and five distinct values of θ. Thus there are five solutions.

21. c.
$$\frac{(1+i)^5\left(1+\sqrt{3}i\right)^2}{-2i\left(-\sqrt{3}+i\right)} = \frac{(\sqrt{2})^5\left(\dfrac{1}{\sqrt{2}} + \dfrac{i}{\sqrt{2}}\right)^5 2^2\left(\dfrac{1}{2} + \dfrac{\sqrt{3}}{2}i\right)^2}{2i2\left(\dfrac{\sqrt{3}}{2} - \dfrac{i}{2}\right)}$$

$\Rightarrow$ Argument $= \dfrac{5\pi}{4} + \dfrac{2\pi}{3} - \dfrac{\pi}{2} + \dfrac{\pi}{6} = \dfrac{19\pi}{12}$

Therefore, the principal argument is $-5\pi/12$.

22. d. Let $f(x) = x^6 + 4x^5 + 3x^4 + 2x^3 + x + 1$. Hence,

$f(\omega) = \omega^6 + 4\omega^5 + 3\omega^4 + 2\omega^3 + \omega + 1$

$= 1 + 4\omega^2 + 3\omega + 2 + \omega + 1$

$= 4(\omega^2 + \omega + 1)$

$= 0$

Hence $f(x)$ is divisible by $x - \omega$. Then $f(x)$ is also divisible by $x - \omega^2$ (as complex roots occur in conjugate pairs).

$f(-\omega) = (-\omega)^6 + 4(-\omega)^5 + 3(-\omega)^4 + 2(-\omega)^3 + (-\omega) + 1$

$= \omega^6 - 4\omega^5 + 3\omega^4 - 2\omega^3 - \omega + 1$

$= 1 - 4\omega^2 + 3\omega - 2 - \omega + 1$

$\neq 0$

23. c. We have,

$(\cos\theta + i\sin\theta)(\cos 2\theta + i\sin 2\theta)\cdots$

$\times(\cos n\theta + i\sin n\theta) = 1$

$\Rightarrow \cos(\theta + 2\theta + 3\theta + \cdots + n\theta) + i\sin(\theta + 2\theta + \cdots + n\theta) = 1$

$\Rightarrow \cos\left(\dfrac{n(n+1)}{2}\theta\right) + i\sin\left(\dfrac{n(n+1)}{2}\theta\right) = 1$

$\Rightarrow \cos\left(\dfrac{n(n+1)}{2}\theta\right) = 1$ and $\sin\left(\dfrac{n(n+1)}{2}\theta\right) = 0$

$\Rightarrow \dfrac{n(n+1)}{2}\theta = 2m\pi \Rightarrow \theta = \dfrac{4m\pi}{n(n+1)}$, where $m \in Z$

24. c. $z = (1 + i\sqrt{3})^{100} = 2^{100}\left(\cos\dfrac{\pi}{3} + i\sin\dfrac{\pi}{3}\right)^{100}$

$= 2^{100}\left(\cos\dfrac{100\pi}{3} + i\sin\dfrac{100\pi}{3}\right)$

$= 2^{100}\left(-\cos\dfrac{\pi}{3} - i\sin\dfrac{\pi}{3}\right)$

$= 2^{100}\left(-\dfrac{1}{2} - \dfrac{i\sqrt{3}}{2}\right)$

$\Rightarrow \dfrac{\mathrm{Re}(z)}{\mathrm{Im}(z)} = \dfrac{-1/2}{-\sqrt{3}/2} = \dfrac{1}{\sqrt{3}}$

25. a. $i = \cos\frac{\pi}{2} + i\sin\frac{\pi}{2} = e^{\frac{i\pi}{2}}$

$\Rightarrow i^i = \left(e^{\frac{i\pi}{2}}\right)^i = e^{-\frac{\pi}{2}}$

$\Rightarrow z = (i)^{(i)^i} = i^{e^{-\frac{\pi}{2}}}$

$\Rightarrow |z| = 1$

26. d. $z = i\log(2-\sqrt{3})$

$\Rightarrow e^{iz} = e^{i^2\log(2-\sqrt{3})} = e^{-\log(2-\sqrt{3})}$

$\Rightarrow e^{iz} = e^{\log(2-\sqrt{3})^{-1}} = e^{\log(2+\sqrt{3})} = (2+\sqrt{3})$

$\Rightarrow \cos z = \frac{e^{iz}+e^{-iz}}{2} = \frac{(2+\sqrt{3})+(2-\sqrt{3})}{2} = 2$

27. b. Let xi be the root where $x \neq 0$ and $x \in R$

$x^4 - a_1x^3i - a_2x^2 + a_3xi + a_4 = 0$

$\Rightarrow x^4 - a_2x^2 + a_4 = 0$ (1)

and

$a_1x^3 - a_3x = 0$ (2)

From Eq. (2),

$a_1x^2 - a_3 = 0$

$\Rightarrow x^2 = a_3/a_1$ (as $x \neq 0$)

Putting the value of x^2 in Eq. (1), we get

$\frac{a_3^2}{a_1^2} - \frac{a_2a_3}{a_1} + a_4 = 0$

$\Rightarrow a_3^2 + a_4a_1^2 = a_1a_2a_3$

$\Rightarrow \frac{a_3}{a_1a_2} + \frac{a_1a_4}{a_2a_3} = 1$ (dividing by $a_1a_2a_3$)

28. b. Let $A = x + iy$. Given,

$|A| = 1 \Rightarrow x^2 + y^2 = 1$

and

$|A+1| = 1 \Rightarrow (x+1)^2 + y^2 = 1$

$\Rightarrow x = -\frac{1}{2}$ and $y = \pm\frac{\sqrt{3}}{2}$

$\Rightarrow A = \omega$ or ω^2

$\Rightarrow (\omega)^n = (1+\omega)^n = (-\omega^2)^n$

Therefore, n must be even and divisible by 3.

29. c. Let $z = \cos x + i\sin x$, $x \in [0, 2\pi)$. Then,

$1 = \left|\frac{z}{\bar{z}} + \frac{\bar{z}}{z}\right|$

$= \frac{|z^2+\bar{z}^2|}{|z|^2}$

$= |\cos 2x + i\sin 2x + \cos 2x - i\sin 2x| = 2|\cos 2x|$

$\Rightarrow \cos 2x = \pm 1/2$

Now,

$\cos 2x = 1/2$

$\Rightarrow x_1 = \frac{\pi}{6}, x_2 = \frac{5\pi}{6}, x_3 = \frac{7\pi}{6}, x_4 = \frac{11\pi}{6}$

$\cos 2x = -\frac{1}{2}$

$\Rightarrow x_5 = \frac{\pi}{3}, x_6 = \frac{2\pi}{3}, x_7 = \frac{4\pi}{3}, x_8 = \frac{5\pi}{3}$

30. a. $u^2 - 2u + 2 = 0 \Rightarrow u = 1 \pm i$

$\Rightarrow \frac{(x+\alpha)^n - (x+\beta)^n}{\alpha - \beta}$

$= \frac{[(\cot\theta - 1) + (1+i)]^n - [(\cot\theta - 1) + (1-i)]^n}{2i}$

$[\because \cot\theta - 1 = x]$

$= \frac{(\cos\theta + i\sin\theta)^n - (\cos\theta - i\sin\theta)^n}{\sin^n\theta\, 2i}$

$= \frac{2i\sin n\theta}{\sin^n\theta\, 2i}$

$= \frac{\sin n\theta}{\sin^n\theta}$

31. b. $f(z) = g(z)(z-i)(z+i) + az + b$; $a, b \in C$

Given,

$f(i) = i \Rightarrow ai + b = i$ (1)

and

$f(-i) = 1 + i$

$\Rightarrow a(-i) + b = 1 + i$ (2)

From (1) and (2), we have

$a = \frac{i}{2}, b = \frac{1}{2} + i$

Hence, the required remainder is

$az + b = (1/2)iz + (1/2) + i.$

32. c. We have,

$z_1(z_1^2 - 3z_2^2) = 2$ (1)

$z_2(3z_1^2 - z_2^2) = 11$ (2)

Multiplying (2) by i and adding it to (1), we get

$z_1^3 - 3z_2^2z_1 + i(3z_1^2z_2 - z_2^3) = 2 + 11i$

$\Rightarrow (z_1 + iz_2)^3 = 2 + 11i$ (3)

Multiplying (2) by i and subtracting it from (1), we get

$z_1^3 - 3z_2^2z_1 - i(3z_1^2z_2 - z_2^3) = 2 - 11i$

$\Rightarrow (z_1 - iz_2)^3 = 2 - 11i$ (4)

Multiplying (3) and (4), we get

$(z_1^2 + z_2^2)^3 = 2^2 - 121i^2 = 4 + 121 = 125$

$\Rightarrow z_1^2 + z_2^2 = 5$

33. a. Assuming $\arg z_1 = \theta$ and $\arg z_2 = \theta + \alpha$,

$\frac{az_1}{bz_2} + \frac{bz_2}{az_1} = \frac{a|z_1|e^{i\theta}}{b|z_2|e^{i(\theta+\alpha)}} + \frac{b|z_2|e^{i(\theta+\alpha)}}{a|z_1|e^{i\theta}}$

$= e^{-i\alpha} + e^{i\alpha} = 2\cos\alpha$

Hence, the point lies on the line segment $[-2, 2]$ of the real axis.

34. d. $x^2 + x + 1 = 0 \Rightarrow x = \omega$ or ω^2

Let $x = \omega$. Then,

$x + \frac{1}{x} = \omega + \frac{1}{\omega} = \omega + \omega^2 = -1$

$x^2 + \frac{1}{x^2} = \omega^2 + \frac{1}{\omega^2} = \omega^2 + \omega = -1$

$$x^3+\frac{1}{x^3}=\omega^3+\frac{1}{\omega^3}=2$$

$$x^4+\frac{1}{x^4}=\omega^4+\frac{1}{\omega^4}=\omega+\omega^2=-1, \text{ etc.}$$

$$\therefore \left(x+\frac{1}{x}\right)^2+\left(x^2+\frac{1}{x^2}\right)^2+\left(x^3+\frac{1}{x^3}\right)^2+\cdots+\left(x^{27}+\frac{1}{x^{27}}\right)^2$$

$$=\left[\left(x+\frac{1}{x}\right)^2+\left(x^2+\frac{1}{x^2}\right)^2+\left(x^4+\frac{1}{x^4}\right)^2+\cdots+\left(x^{26}+\frac{1}{x^{26}}\right)^2\right]$$

$$+\left[\left(x^3+\frac{1}{x^3}\right)^2+\left(x^6+\frac{1}{x^6}\right)^2+\left(x^9+\frac{1}{x^9}\right)\right.$$

$$\left.\cdots+\left(x^{27}+\frac{1}{x^{27}}\right)^2\right]$$

$$=18+9(2^2)=54$$

35. c. $E=\sum_{r=1}^{n}(ar+b)\omega^{r-1}$

$$=(a+b)+(2a+b)\omega+(3a+b)\omega^2+\cdots+(na+b)\omega^{n-1}$$

$$=b\underbrace{(1+\omega+\omega^2+\cdots+\omega^{n-1})}_{\text{zero}}+a(1+2\omega+3\omega^2+\cdots+n\omega^{n-1})$$

Now,

$$S=1+2\omega+3\omega^2+\cdots+n\omega^{n-1}$$

$$S\omega=\omega+2\omega^2+\cdots+(n-1)\omega^{n-1}+n\omega^n$$

$$\therefore S(1-\omega)=\underbrace{1+\omega+\omega^2+\cdots+\omega^{n-1}}_{\text{zero}}-n\omega^n=-n\ (\because \omega^n=1)$$

$$\Rightarrow S=\frac{n}{\omega-1}$$

$$\Rightarrow E=\frac{na}{\omega-1}$$

36. b. Taking cube roots of both sides, we get

$$z+ab=a(1)^{1/3}=a,\ a\omega,\ a\omega^2$$

where

$$\omega=-\frac{1}{2}+i\frac{\sqrt{3}}{2},\ \omega^2=-\frac{1}{2}-i\frac{\sqrt{3}}{2}$$

$$\therefore z_1=a-ab,\ z_2=a\omega-ab,\ z_3=a\omega^2-ab$$

$$|z_1-z_2|=|a(1-\omega)|$$

$$=|a|\left|1-\left(-\frac{1}{2}+i\frac{\sqrt{3}}{2}\right)\right|$$

$$=|a|\left|\frac{3}{2}-i\frac{\sqrt{3}}{2}\right|$$

$$=|a|\left(\frac{9}{4}+\frac{3}{4}\right)^{1/2}=\sqrt{3}|a|$$

Similarly,

$$|z_2-z_3|=|z_3-z_1|=\sqrt{3}|a|$$

37. a. We have,

$$z^3+2z^2+2z+1=0$$

$$\Rightarrow (z^3+1)+2z(z+1)=0$$

$$\Rightarrow (z+1)(z^2+z+1)=0$$

$$\Rightarrow z=-1,\ \omega,\ \omega^2$$

Since $z=-1$ does not satisfy $z^{1985}+z^{100}+1=0$ while $z=\omega$, ω^2 satisfy it, hence sum is $\omega+\omega^2=-1$.

38. b. Let $z_1=|z_1|(\cos\theta_1+i\sin\theta_1)$. Now,

$$\left|\frac{z_1}{z_2}\right|=1\Rightarrow |z_1|=|z_2|$$

Also,

$$\arg(z_1z_2)=0\Rightarrow \arg(z_1)+\arg(z_2)=0$$

$$\Rightarrow \arg(z_2)=-\theta_1$$

$$\Rightarrow z_2=|z_2|(\cos(-\theta_1)+i\sin(-\theta_1))$$

$$=|z_1|(\cos\theta_1-i\sin\theta_1)=\bar{z}_1$$

$$\Rightarrow \bar{z}_2=\overline{(\bar{z}_1)}=z_1$$

$$\Rightarrow |z_2|=z_1z_2$$

39. d. $(x-3)^3+1=0$

$$\Rightarrow \left(\frac{x-3}{-1}\right)^3=1$$

$$\Rightarrow \frac{x-3}{-1}=1,\ \omega,\ \omega^2$$

$$\Rightarrow x=2,\ 3-\omega,\ 3-\omega^2$$

Hence, the um of complex roots is $6-(\omega+\omega)^2=6+1=7$.

40. b. $x=\sqrt[3]{-1}$

$$\Rightarrow x^3=-1$$

$$\Rightarrow (-x)^3=1$$

$$\Rightarrow -x=1,\ \omega,\ \omega^2$$

$$\Rightarrow x=-1,\ -\omega,\ -\omega^2$$

$$=-1,\ \frac{1+\sqrt{3}i}{2},\ \frac{1-\sqrt{3}i}{2}$$

$$=-1,\ \frac{-\sqrt{3}+i}{2i},\ \frac{\sqrt{3}+i}{2i}$$

$$=-1,\ \frac{-\sqrt{3}+\sqrt{-1}}{\sqrt{-4}},\ \frac{\sqrt{3}+\sqrt{-1}}{\sqrt{-4}}$$

41. d. $|\omega z-1-\omega^2|=a$

$$\Rightarrow |z+1|=a\Rightarrow |z-1+2|=a$$

$$\Rightarrow |z-1|+2\ge a\Rightarrow 0\le a\le 4$$

42. b. $|z^2-3|\ge |z|^2-3$

$$\Rightarrow 3|z|\ge |z|^2-3$$

$$\Rightarrow |z|^2-3|z|-3\le 0$$

$$\Rightarrow 0<|z|\le \frac{3+\sqrt{21}}{2}$$

43. b. $|2z-1|=|z-2|$

$$\Rightarrow |2z-1|^2=|z-2|^2$$

$$\Rightarrow (2z-1)(2\bar{z}-1)=(z-2)(\bar{z}-2)$$

$$\Rightarrow 4z\bar{z}-2\bar{z}-2z+1=z\bar{z}-2\bar{z}-2z+4$$

$\Rightarrow 3|z|^2 = 3$

$\Rightarrow |z| = 1$

Again,

$|z_1 + z_2| = |z_1 - \alpha + z_2 - \beta + \alpha + \beta|$

$\leq |z_1 - \alpha| + |z_2 - \beta| + |\alpha + \beta|$

$< \alpha + \beta + |\alpha + \beta| = 2|\alpha + \beta|$ [$\because \alpha, \beta > 0$]

$\therefore \left|\dfrac{z_1 + z_2}{\alpha + \beta}\right| < 2$

$\Rightarrow \left|\dfrac{z_1 + z_2}{\alpha + \beta}\right| < 2\,|z|$

44. a. $a_0 z^n + a_1 z^{n-1} + a_2 z^{n-2} + \cdots + a_{n-1} z + a_n = 3$

$\Rightarrow |3| = |a_0 z^n + a_1 z^{n-1} + \cdots + a_{n-1} z + a_n|$

$\Rightarrow 3 \leq |a_0|\,|z|^n + |a_1|\,|z|^{n-1} + \cdots + |a_{n-1}|\,|z| + |a_n|$

$\Rightarrow 3 < 2(|z|^n + |z|^{n-1} + \cdots + |z| + 1)$

$\Rightarrow 1 + |z| + |z|^2 + \cdots + |z|^n > \dfrac{3}{2}$

If $|z| \geq 1$, the inequality is clearly satisfied. For $|z| < 1$, we must have,

$\dfrac{1 - |z|^{n+1}}{1 - |z|} > \dfrac{3}{2}$

$\Rightarrow 2 - 2|z|^{n+1} > 3 - 3|z|$

$\Rightarrow 2|z|^{n+1} < 3|z| - 1$

$\Rightarrow 3|z| - 1 > 0$

$\Rightarrow |z| > \dfrac{1}{3}$

45. a. $8iz^3 + 12z^2 - 18z + 27i = 0$

$\Rightarrow 4iz^2(2z - 3i) - 9(3z - 2i) = 0$

$\Rightarrow (2z - 3)(4iz^2 - 9) = 0$

$\Rightarrow \quad z = \dfrac{3i}{2}$ and $z^2 = \dfrac{9}{4i}$

$\Rightarrow \quad |z| = \dfrac{3}{2}$ and $|z^2| = \dfrac{9}{4}$

$\Rightarrow \quad |z| = \dfrac{3}{2}$

46. a. $|z^2 + 2z\cos\alpha| \leq |z^2| + |2z\cos\alpha|$

$= |z|^2 + 2|z|\,|\cos\alpha| \leq |z|^2 + 2|z|$

$< (\sqrt{2} - 1)^2 + 2(\sqrt{2} - 1) = 1$

47. b. $\left|z + \dfrac{1}{z}\right| \geq \left|\,|z| - \dfrac{1}{|z|}\right|$

Hence, the least value occurs when $|z| = 3$.

$\therefore \left|z + \dfrac{1}{z}\right|_{\text{least}} = 3 - \dfrac{1}{3} = \dfrac{8}{3}$

48. c. $\left|\sum_{r=1}^{n} z_r\right| \leq \sum_{r=1}^{n} |z_r| \leq \sum_{r=1}^{n} |z_r - r| + \sum_{r=1}^{n} r \leq 2\sum_{r=1}^{n} r$

49. d. Let $z = x + iy$. Then,

$|z^2 - 1| = |z|^2 + 1$

$\Rightarrow |(x^2 - y^2 - 1) + 2ixy| = x^2 + y^2 + 1$

$\Rightarrow (x^2 - y^2 - 1)^2 + 4x^2y^2 = (x^2 + y^2 + 1)^2$

$\Rightarrow x = 0$

Hence, z lies on imaginary axis.

50. a. $|z| = 1$, let $\alpha = -1 + 3z$

$\Rightarrow \alpha + 1 = 3z$

$\Rightarrow |\alpha + 1| = 3|z| = 1$

Hence, 'α' lies on a circle centred at -1 and radius equal to 1.

51. b. Let $z = x + iy$. Then,

$x = \lambda + 3$ and $y = -\sqrt{5 - \lambda^2}$

$\Rightarrow (x - 3)^2 = \lambda^2$ (1)

and

$y^2 = 5 - \lambda^2$ (2)

From (1) and (2),

$(x - 3)^2 = 5 - y^2 \Rightarrow (x - 3)^2 + y^2 = 5$

Obviously it is a semicircle as $y < 0$. Hence part of the circle lies below the x-axis.

52. c.

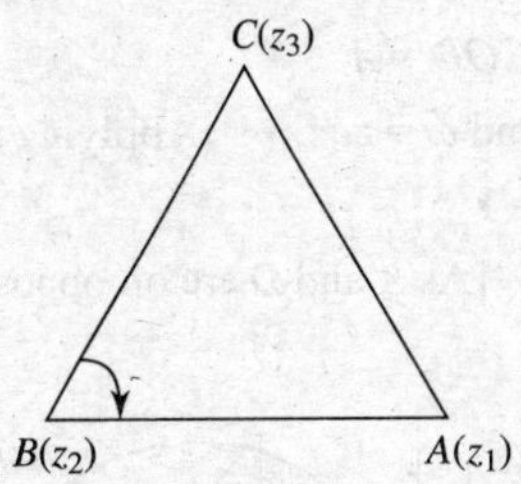

Fig. 2.2

$\dfrac{z_1 - z_2}{z_3 - z_2} = \dfrac{1}{\sqrt{2}} + \dfrac{i}{\sqrt{2}} = e^{i\pi/4}$

$\angle CBA = \dfrac{\pi}{4}$

Also,

$|z_1 - z_2| = |z_3 - z_2|$

Hence, ΔABC is isosceles.

53. c. Given that

$|z_1 - i| = |z_2 - i| = |z_3 - i|$

Hence, z_1, z_2, z_3 lie on the circle whose centre is i. Also given that the triangle is equilateral. Hence centroid and cirucmcentre coincides.

$\therefore \dfrac{z_1 + z_2 + z_3}{3} = i$

$\Rightarrow |z_1 + z_2 + z_3| = 3$

54. d.

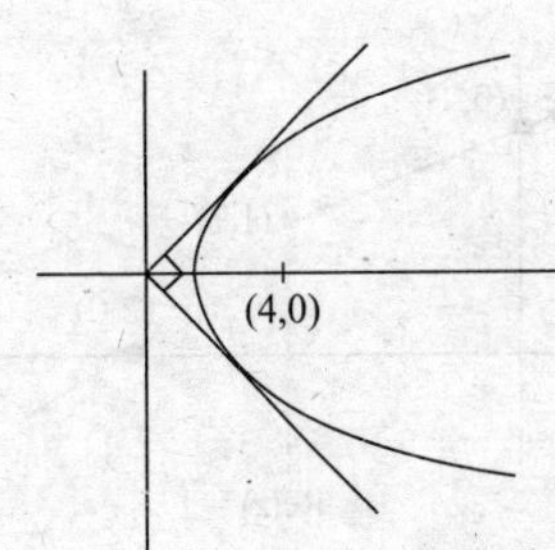

Fig. 2.3

$|z - 4| = \text{Re}(z)$

$\Rightarrow \sqrt{(x-4)^2 + y^2} = x$

$\Rightarrow x^2 - 8x + 16 + y^2 = x^2$

$\Rightarrow y^2 = 8(x - 2)$

The given relation represents the part of the parabola with focus (4, 0) lying above x-axis and the imaginary axis as the directirx. The two tangents from directrix are at right angle. Hence, the greatest positive argument of z is $\pi/4$.

55. d.

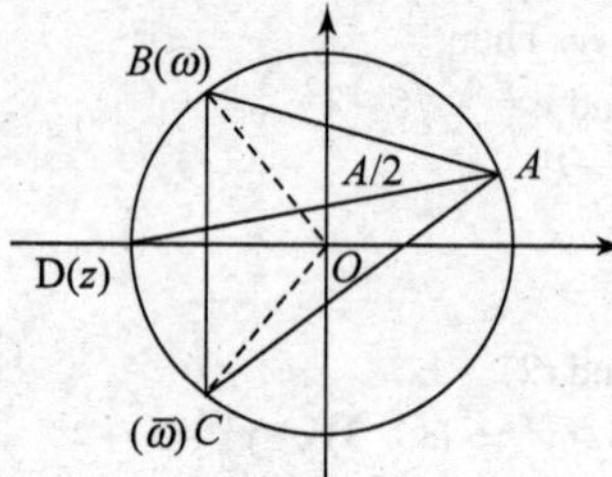

Fig. 2.4

Clearly,

$\angle DOB = \angle COD = A$

$\Rightarrow z = \omega e^{iA}$ and $\bar{\omega} = ze^{iA}$ [Applying rotation about O]

$\Rightarrow z^2 = \omega\bar{\omega} = 1$

$\Rightarrow z = -1$ [As A and D are on opposite sides of BC]

56. b.

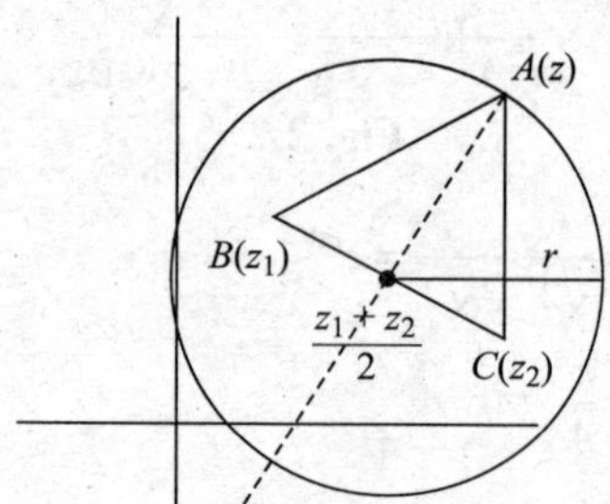

Fig. 2.5

By the given conditions, the area of the triangle ABC is given by $(1/2)|z_1 - z_2|r$.

57. d. The given equation is written as

$$\arg[z - (1 + i)] = \begin{cases} \dfrac{3\pi}{4}, & \text{when } x \le 2 \\ \dfrac{-\pi}{4}, & \text{when } x > 2 \end{cases}$$

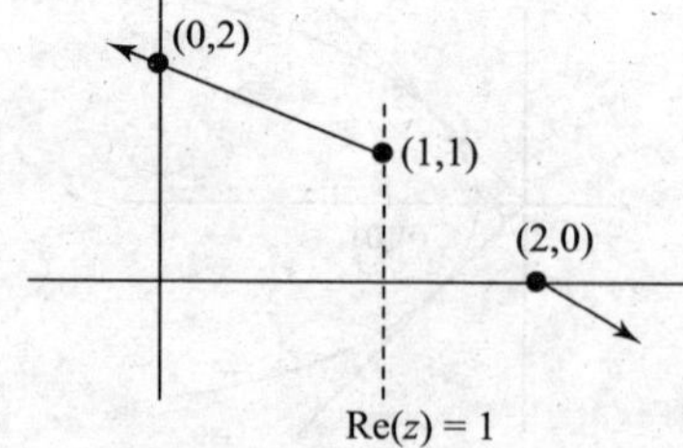

Fig. 2.6

Therefore, the locus is a set of two rays.

58. a.

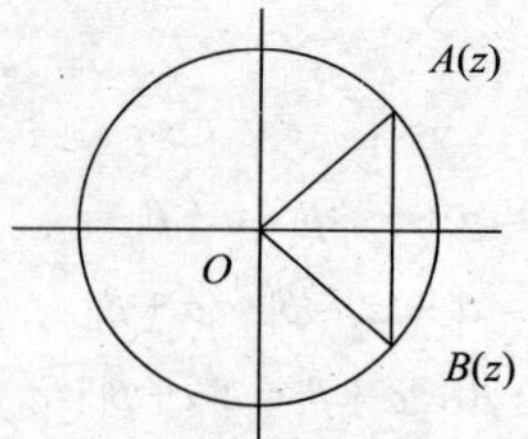

Fig. 2.7

$|z - \bar{z}|$ is the length AB while $|z|(\arg z - \arg\bar{z})$ is arc length AB. Therefore,

$|z - \bar{z}| \le |z| (\arg z - \arg\bar{z})$

59. c.

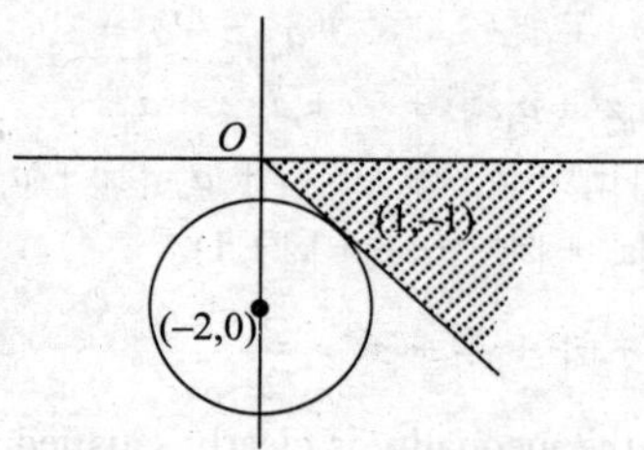

Fig. 2.8

$kz/(k + 1)$ represents any point lying on the line joining origin and z.

Given,

$$\left|\frac{kz}{k+1} + 2i\right| > \sqrt{2}$$

Hence, $kz/(k + 1)$ should lie outside the circle $|z + 2| > \sqrt{2}$. So, z should lie in the shaded region.

$$\therefore -\frac{\pi}{4} < \arg(z) < 0$$

60. b. $2\left|z - \dfrac{1}{2}\right| = |z - 1|$

$$\therefore \frac{|z-1|}{\left|z - \dfrac{1}{2}\right|} = 2$$

So, locus of z is a circle.

61. d. $|z_2 + iz_1| = |z_1| + |z_2|$

$\Rightarrow iz_1$, $0 + i0$, and z_2 are collinear

$\Rightarrow \arg(iz_1) = \arg(z_2)$

$\Rightarrow \arg(z_2) - \arg(z_1) = \dfrac{\pi}{2}$

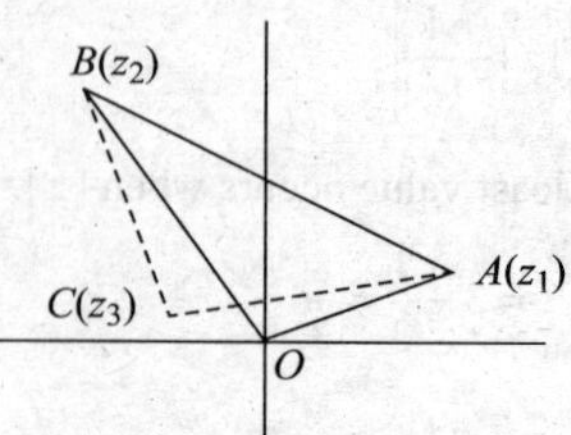

Fig. 2.9

Let,

$$z_3 = \frac{z_2 - iz_1}{1-i}$$

$\Rightarrow (1-i)z_3 = z_2 - iz_1$

$\Rightarrow z_2 - z_3 = i(z_1 - z_3)$

$\therefore \angle ACB = \frac{\pi}{2}$

and

$|z_1 - z_3| = |z_2 - z_3|$

$\Rightarrow AC = BC$

$\because AB^2 = AC^2 + BC^2$

$\Rightarrow AC = \frac{5}{\sqrt{2}}$ $(\because AB = 5)$

Therefore, area of ΔABC is $(1/2)AC \times BC = AC^2/2 = 25/4$ sq. units

62. a.

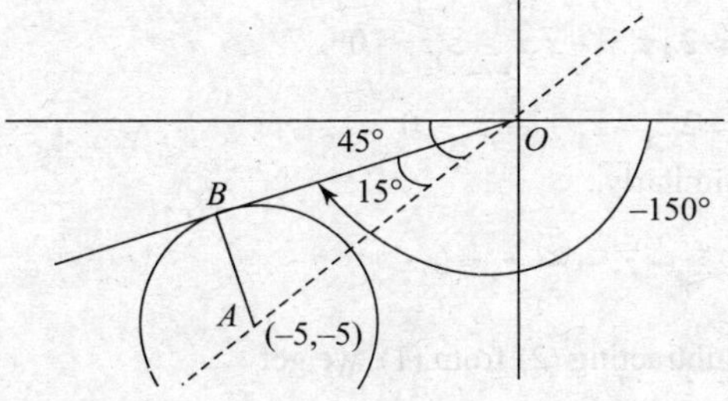

Fig. 2.10

$|2z + 10 + 10i| \le 5\sqrt{3} - 5$

$\Rightarrow |z + 5 + 5i| \le \frac{5(\sqrt{3}-1)}{2}$

Point B has least principal argument. Now,

$AB = \frac{5(\sqrt{3}-1)}{2}$

$OA = 5\sqrt{2}$

$\angle AOB = \frac{\pi}{12}$

$\therefore \arg(z) = -\frac{5\pi}{6}$

63. c. $z = \frac{at+b}{t-c} \Rightarrow t = \frac{b+cz}{z-a}$

Now,

$|t| = 1$

$\Rightarrow \left|\frac{b+cz}{z-a}\right| = 1$

$\Rightarrow \left|\frac{z+\frac{b}{c}}{z-a}\right| = \frac{1}{|c|}$ $(\neq 1 \text{ as } |c| \neq |t|)$

$\Rightarrow$ locus of z is a circle

64. a. Let $z = x + iy$. Then,

$|z - 3 - i| = |z - 9 - i|$

$\Rightarrow \sqrt{(x-3)^2 + (y-1)^2} = \sqrt{(x-9)^2 + (y-1)^2}$

$\Rightarrow x = 6$

$|z - 3 + 3i| = 3$

$\Rightarrow \sqrt{(x-3)^2 + (y+3)^2} = 3$

For $x = 6$, $y = -3$.

$\therefore z = 6 - 3i$

65. a. Given,

$z_k = 1 + a + a^2 + \cdots + a^{k-1} = \frac{1-a^k}{1-a}$

$\Rightarrow z_k - \frac{1}{1-a} = -\frac{a^k}{1-a}$

$\Rightarrow \left|z_k - \frac{1}{1-a}\right| = \frac{|a|^k}{|1-a|} < \frac{1}{|1-a|}$ $[\because |a| < 1]$

Hence, z_k lies within the circle.

$\therefore \left|z - \frac{1}{1-a}\right| = \frac{1}{|1-a|}$

66. b. $2z^2 + 2z + \lambda = 0$

Let the roots be z_1, z_2. Then,

$z_1 + z_2 = -1$ and $z_1 z_2 = \frac{\lambda}{2}$

$0, z_1, z_2$ form an equilateral triangle.

$\therefore z_1^2 + z_2^2 = z_1 z_2$

$\Rightarrow (z_1 + z_2)^2 = 3z_1 z_2$

$\Rightarrow 1 = 3\frac{\lambda}{2}$

$\Rightarrow \lambda = \frac{2}{3}$

67. a. $x + iy = 1 - t + i\sqrt{t^2 + t + 2}$

$\Rightarrow x = 1 - t,\ y = \sqrt{t^2 + t + 2}$

Eliminating t,

$y^2 = t^2 + t + 2 = (1-x)^2 + 1 - x + 2 = \left(x - \frac{3}{2}\right)^2 + \frac{7}{4}$

$\Rightarrow y^2 - \left(x - \frac{3}{2}\right)^2 = \frac{7}{4}$, which is a hyperbola

68. c. $z^2 + z|z| + |z|^2 = 0 \Rightarrow \left(\frac{z}{|z|}\right)^2 + \frac{z}{|z|} + 1 = 0$

$\Rightarrow \frac{z}{|z|} = \omega, \omega^2 \Rightarrow z = \omega|z|$ or $z = \omega^2|z|$

$\Rightarrow x + iy = |z|\left(\frac{-1}{2} + \frac{i\sqrt{3}}{2}\right)$ or $x + iy = |z|\left(\frac{-1}{2} - \frac{i\sqrt{3}}{2}\right)$

$\Rightarrow x = -\frac{1}{2}|z|,\ y = |z|\frac{\sqrt{3}}{2}$ or $x = -\frac{|z|}{2},\ y = -\frac{|z|\sqrt{3}}{2}$

$\Rightarrow y + \sqrt{3}x = 0$ or $y - \sqrt{3}x = 0 \Rightarrow y^2 - 3x^2 = 0$

69. c. $S_1 = \Sigma z_1 = -3a,\ S_2 = \Sigma z_1 z_2 = 3b$

Since the triangle is equilateral, we have

$\Sigma z_1^2 = \Sigma z_1 z_2$

$\Rightarrow (\Sigma z_1)^2 - 2\Sigma z_1 z_2 = \Sigma z_1 z_2$

$\Rightarrow (\Sigma z_1)^2 = 3\Sigma z_1 z_2$

$\Rightarrow (-3a)^2 = 3(3b)$

$\Rightarrow 9a^2 = 9b$

$\Rightarrow a^2 = b$

70. b.

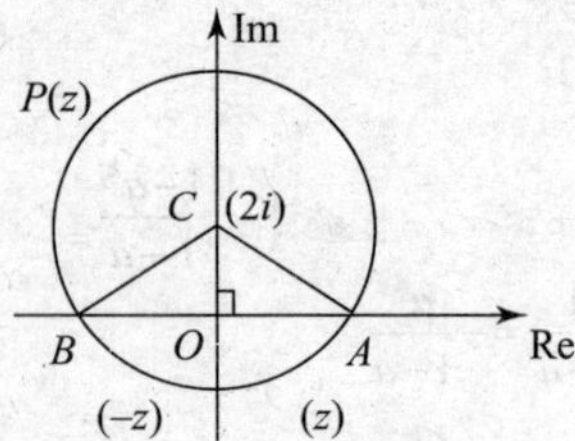

Fig. 2.11

$CA = CB = 2\sqrt{2}, OC = 2$

$\Rightarrow OA = OB = 2$

$\Rightarrow A \equiv 2 + 0i, B = -2 + 0i$

Clearly,

$\angle BCA = \pi/2$

$\Rightarrow \angle BPA = \pi/4$

$$\Rightarrow \arg\left(\frac{z-2}{z+2}\right) = \frac{\pi}{4}$$

71. a.

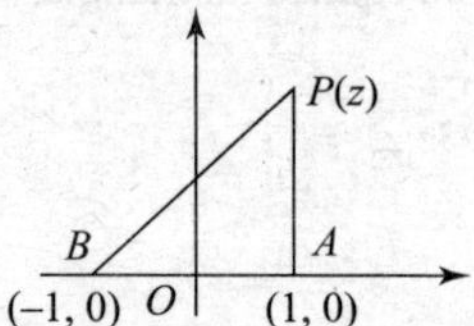

Fig. 2.12

When $|z - 1| < |z + 1|$ (or $x > 0$)

$|z| = |z - 1|$

$\Rightarrow x^2 + y^2 = (x - 1)^2 + y^2$

$\Rightarrow x = 1/2$

$\Rightarrow z + \bar{z} = 1$

When $|z - 1| > |z + 1|$ (or $x < 0$),

$|z| = |z + 1|$

$\Rightarrow x^2 + y^2 = (x + 1)^2 + y^2$

$\Rightarrow x = -1/2$

$\Rightarrow z + \bar{z} = -1$

72. b. Let $z = x + iy$. Then,

$$\text{Re}\left(\frac{1}{z}\right) = k$$

$$\Rightarrow \text{Re}\left(\frac{1}{x+iy}\right) = k$$

$$\Rightarrow \text{Re}\left(\frac{x}{x^2+y^2} - \frac{iy}{x^2+y^2}\right) = k$$

$$\Rightarrow \frac{x}{x^2+y^2} = k$$

$$\Rightarrow x^2 + y^2 - \frac{1}{k}x = 0,$$

which is a circle.

73. b.

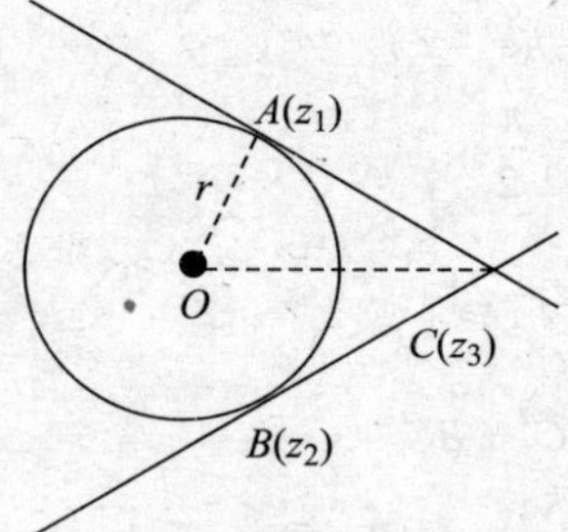

Fig. 2.13

As ΔOAC is a right-angled triangle with right angle at A, so

$$|z_1|^2 + |z_3 - z_1|^2 = |z_3|^2$$

$$\Rightarrow 2|z_1|^2 - \bar{z}_3 z_1 - \bar{z}_1 z_3 = 0$$

$$\Rightarrow 2\bar{z}_1 - \bar{z}_3 - \frac{\bar{z}_1}{z_1} z_3 = 0 \quad (1)$$

Similarly,

$$2\bar{z}_2 - \bar{z}_3 - \frac{\bar{z}_2}{z_2} z_3 = 0 \quad (2)$$

Subtracting (2) from (1), we get

$$2(\bar{z}_2 - \bar{z}_1) = z_3\left(\frac{\bar{z}_1}{z_1} - \frac{\bar{z}_2}{z_2}\right)$$

$$\Rightarrow \frac{2r^2(z_1 - z_2)}{z_1 z_2} = z_3 r^2\left(\frac{z_2^2 - z_1^2}{z_1^2 z_2^2}\right) \quad [\because |z_1|^2 = |z_2|^2 = r^2]$$

$$\Rightarrow z_3 = \frac{2z_1 z_2}{z_2 + z_1}$$

74. a. $|z_1| = |z_2| = |z_3| = 1$

Hence, the circumcentre of triangle is origin. Also, centroid $(z_1 + z_2 + z_3)/3 = 0$, which coincides with the circumcentre. So the triangle is equilateral. Since radius is 1, length of side is $a = \sqrt{3}$. Therefore, the area of the triangle is $(\sqrt{3}/4)a^2 = (3\sqrt{3}/4)$.

75. a.

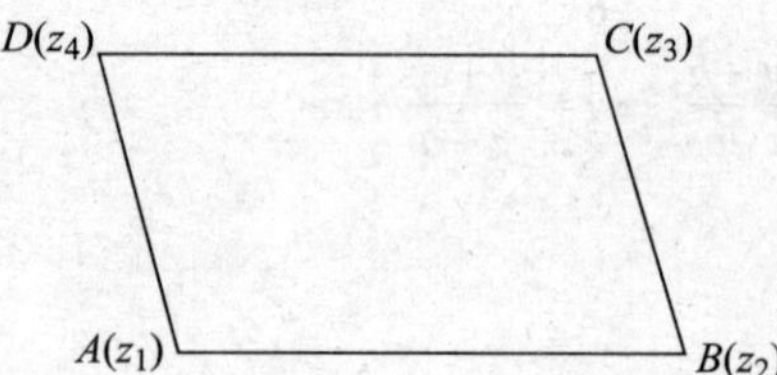

Fig. 2.14

The first condition implies that $[(z_1 + z_3)/2] + [(z_2 + z_4)/2]$, i.e., diagonals AC and BD bisect each other. Hence, quadrilateral is a parallelogram. The second condition implies that the angle between AD and AB is 90°. Hence the parallelogram is a rectangle.

76. b. Given that

$$\arg\left(\frac{z_1 - \frac{z}{|z|}}{\frac{z}{|z|}}\right) = \frac{\pi}{2}$$

and

$$\left|\frac{z}{|z|} - z_1\right| = 3,$$

from which we can establish the following geometry.

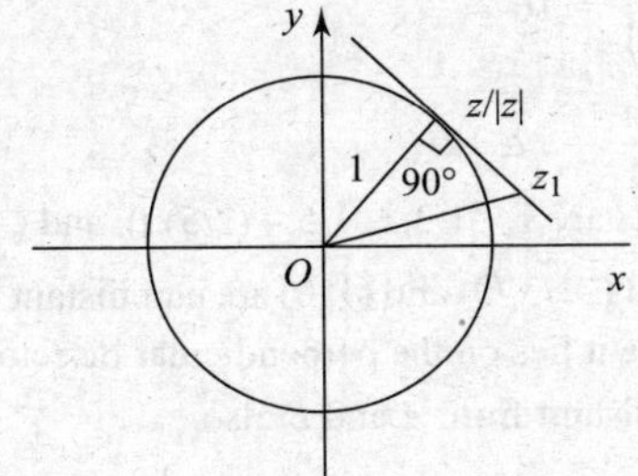

Fig. 2.15

From the diagram,

$$\left|\frac{z}{|z|} - z_1\right| = 3, |z_1| = \sqrt{9+1} = \sqrt{10}$$

77. b. Note that $z_1 = 3 + \sqrt{3}i$ lies on the line $y = (1/\sqrt{3})^x$ and $z_2 = 2\sqrt{3} + 6i$ lies on the line $y = \sqrt{3}x$.

Hence $z = 5 + 5i$ will only lie on the bisector of z_1 and z_2, i.e., $y = x$.

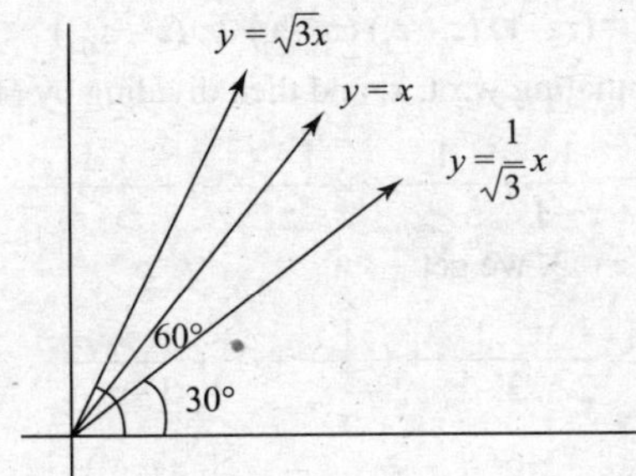

Fig. 2.16

78. d. $\log_{1/3}\left(\frac{|z-3|^2+2}{11|z-3|-2}\right) > 1$

$$\Rightarrow \frac{|z-3|^2+2}{11|z-3|-2} < \frac{1}{3}$$

$\Rightarrow (3t-8)(t-1) < 0$ (where $|z-3| = t$)

$\Rightarrow 1 < |z-3| < 8/3$

Hence, z lies between the two concentric circles.

79. c. We have,

$$|(x-2) + i(y-1)| = |z| \left|\frac{1}{\sqrt{2}}\cos\theta - \frac{1}{\sqrt{2}}\sin\theta\right|$$

where $\theta = \arg z$.

$$\sqrt{(x-2)^2 + (y-1)^2} = \frac{1}{\sqrt{2}}|x-y|,$$

which is a parabola

80. a. We have,

$|z - 2 + 2i| = 1$

$\Rightarrow |z - (2 - 2i)| = 1$

Hence, z lies on a circle having centre at $(2, -2)$ and radius 1. It is evident from the figure that the required complex number z is given by the point P. We find that OP makes an angle $\pi/4$ with OX and

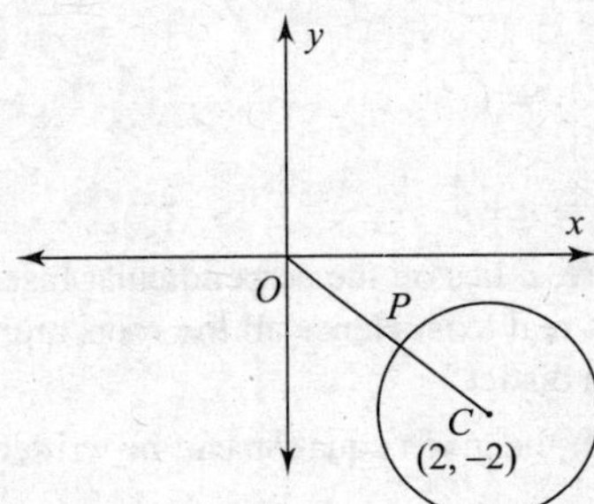

Fig. 2.17

$$OP = OC - CP = \sqrt{2^2+2^2} - 1 = 2\sqrt{2} - 1$$

So coordinates of P are $[(2\sqrt{2}-1)\cos(\pi/4), -(2\sqrt{2}-1)\sin(\pi/4)]$, i.e., $((2-1/\sqrt{2}), -(2-1/\sqrt{2}))$.

Hence,

$$z = \left(2 - \frac{1}{\sqrt{2}}\right) + \left\{-\left(2 - \frac{1}{\sqrt{2}}\right)\right\}i = \left(2 - \frac{1}{\sqrt{2}}\right)(1-i)$$

81. d. Given,

$$z = \frac{3}{2 + \cos\theta + i\sin\theta}$$

$$\Rightarrow \cos\theta + i\sin\theta = \frac{3}{z} - 2 = \frac{3-2z}{z}$$

$$\Rightarrow 1 = \frac{|3-2z|}{|z|} \quad \text{[taking modulus]}$$

$$\Rightarrow \frac{\left|z - \frac{3}{2}\right|}{|z|} = \frac{1}{2}$$

Hence, locus of z is a circle.

82. d. We have,

$$e^{\frac{2\pi ri}{p}} = e^{\frac{2\pi m}{q}}$$

$r = 0, 1, \ldots, p - 1$

$m = 0, 1, \ldots, q - 1$

This is possible iff $r = m = 0$ but for $r = m = 0$ we get 1 which is not an imaginary number.

83. b. Let,

$S = 1 + 2a + 3a^2 + \cdots + n\alpha^{n-1}$

$\Rightarrow \alpha S = \alpha + 2\alpha^2 + 3\alpha^3 + \cdots + (n-1)\alpha^{n-1} + n\alpha^n$

On subtracting, we get

$S(1-\alpha) = 1 + [\alpha + \alpha^2 + \alpha^{n-1}] - n\alpha^n$

$$= 1 + \frac{\alpha(1-\alpha^{n-1})}{1-\alpha} - n\alpha^n$$

$$\Rightarrow S = \frac{1}{1-\alpha} + \frac{\alpha - \alpha^n}{(1-\alpha)^2} - \frac{n\alpha^n}{1-\alpha} \qquad [\because \alpha^n = 1]$$

$$= \frac{1}{1-\alpha} + \frac{\alpha - 1}{(1-\alpha)^2} - \frac{n}{1-\alpha} = -\frac{n}{1-\alpha}$$

84. a. $\left(\frac{1+ia}{1-ia}\right)^4 = z$

$$\Rightarrow \left|\frac{1+ia}{1-ia}\right|^4 = |z|$$

$$\Rightarrow \left|\frac{a-i}{a+i}\right|^4 = 1$$

$$\Rightarrow |a-i| = |a+i|$$

Therefore, a lies on the perpendicular bisector of i and $-i$, which is real axis. Hence all the roots are real. Obviously roots are distict.

85. b. For $z \neq 1$, the given equation can be written as

$$\left(\frac{z+1}{z-1}\right)^5 = 1$$

$$\Rightarrow \frac{z+1}{z-1} = e^{(2k\pi i)/5}$$

where $k = -2, -1, 1, 2$.

If we denote this value of z by z_k, then

$$z_k = \frac{e^{(2k\pi i)/5}+1}{e^{(1k\pi i)/5}-1}$$

$$= \frac{e^{(k\pi i)/5}+e^{(-k\pi i)/5}}{e^{(k\pi i)/5}-e^{(-k\pi i)/5}}$$

$$= -i \cot\left(\frac{k\pi}{5}\right), k = -2, -1, 1, 2$$

Therefore, roots of the equation are

$\pm i \cot(\pi/5), \pm i \cot(2\pi/5)$.

86. a. We have,

$$\log z + \log z^2 + \log z^3 + \cdots + \log z^n = 0$$

$$\Rightarrow \log(z\, z^2\, z^3 \cdots z^n) = 0$$

$$\Rightarrow \log\left(z^{\frac{n(n+1)}{2}}\right) = 0$$

$$\Rightarrow z^{\frac{n(n+1)}{2}} = 1$$

$$\Rightarrow z = 1^{\frac{2}{n(n+1)}}$$

$$= (\cos 0° + i \sin 0°)^{\frac{2}{n(n+1)}}$$

$$= (\cos 2m\pi + i \sin 2m\pi)^{\frac{2}{n(n+1)}}, m = 0, 1, 2, 3, \ldots$$

$$= \cos\frac{4m\pi}{n(n+1)} + i \sin\frac{4m\pi}{n(n+1)}, m = 0, 1, 2, \ldots$$

87. d. The equation $z^n = (z+1)^n$ will have exactly $n-1$ roots. We have,

$$\left(\frac{z+1}{z}\right)^n = 1$$

$$\Rightarrow \left|\frac{z+1}{z}\right| = 1$$

$$\Rightarrow |z+1| = |z|$$

Therefore, z lies on the right bisector of the segment connecting the points (0, 0) and (–1, 0). Thus Re(z) = –1/2.

Hence, roots are collinear and will have their real parts equal to –1/2.

Hence sum of the real parts of roots is $-1/2(n-1)$.

88. c. $\left(\frac{z+1}{z}\right)^4 = 16$

$$\Rightarrow \frac{z+1}{z} = \pm 2, \pm 2i$$

The roots are 1, –1/3, (–1/5 – (2/5) i), and (–1/5 + (2/5) i).

Note that (–1/3,/0) and (1, 0) are equidistant from (–1/3, /0) and since it lies on the perpendicular bisector of AB, it will be equidistant from A and B also.

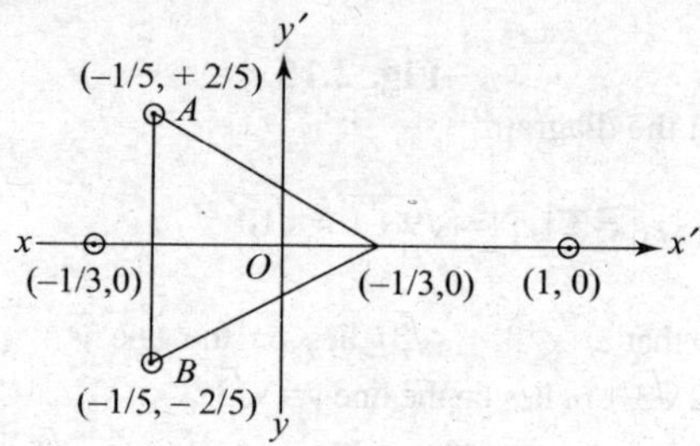

Fig. 2.18

89. d. $(z^n - 1) = (z-1)(z-z_1)(z-z_2)\ldots(z-z_{n-1})$ (1)

Differentiating w.r.t. x, and then dividing by (1), we have

$$\frac{nz^{n-1}}{z^n-1} = \frac{1}{z-1} + \frac{1}{z-z_1} + \frac{1}{z-z_2} + \cdots + \frac{1}{z-z_{n-1}}$$

Putting $z = 3$, we get

$$\frac{n3^{n-1}}{3^n-1} = \frac{1}{2} + \frac{1}{3-z_1} + \frac{1}{3-z_2} + \cdots + \frac{1}{3-z_{n-1}}$$

$$\Rightarrow \frac{1}{3-z_1} + \frac{1}{3-z_2} + \cdots + \frac{1}{3-z_{n-1}} = \frac{n3^{n-1}}{3^n-1} - \frac{1}{2}$$

Multiple Correct Answers Type

1. a., c.

Clearly, we have to find it for real z. Let $z = x$. Then,

$$|x-w| = |x-w^2| = |w-w^2|$$

$$\Rightarrow \left(x+\frac{1}{2}\right)^2 + \frac{3}{4} = \left|\frac{-1+\sqrt{3}i}{2} - \frac{-1-\sqrt{3}i}{2}\right|^2 = 3 \Rightarrow x + \frac{1}{2} = \pm\frac{3}{2}$$

$$\Rightarrow x = 1, -2$$

2. a., b., c., d.

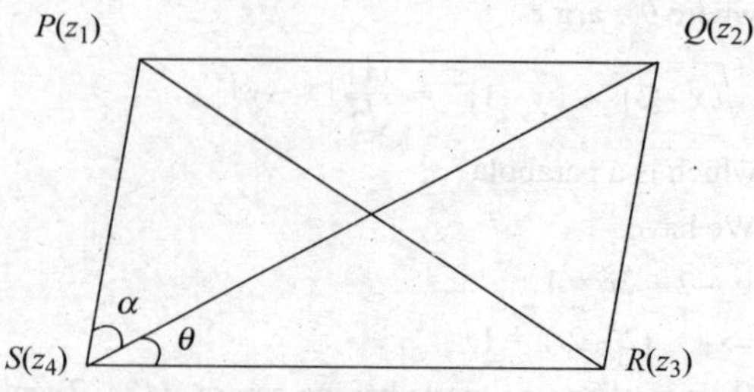

Fig. 2.19

a. $PS \parallel QR \Rightarrow \arg\left(\frac{z_1 - z_4}{z_2 - z_3}\right) = 0 \Rightarrow \frac{z_1 - z_4}{z_2 - z_3}$ is purely real

b. $\theta \neq \alpha \Rightarrow \text{amp}\left(\frac{z_1 - z_4}{z_2 - z_4}\right) \neq \text{amp}\left(\frac{z_2 - z_4}{z_3 - z_4}\right)$

c. Diagonals of rhombus are perpendicular. Hence, $(z_1 - z_3)/(z_2 - z_4)$ is purely imaginary.

d. Diagonals of rhombus are not equal. Hence, $|z_1 - z_3| \neq |z_2 - z_4|$.

3. a., b., d.

Let $z = \alpha$ be a real root. Then,

$\alpha^3 + (3 + 2i)\alpha + (-1 + i\alpha) = 0$

$\Rightarrow (\alpha^3 + 3\alpha - 1) + i(\alpha + 2\alpha) = 0$

$\Rightarrow \alpha^3 + 3\alpha - 1 = 0$ and $\alpha = -\alpha/2$

$\Rightarrow -\frac{a^3}{8} - \frac{3a}{2} - 1 = 0$

$\Rightarrow a^3 + 12a + 8 = 0$

Let $f(a) = a^3 + 12a + 8$.

$\therefore f(-1) < 0, f(0) > 0, f(-2) < 0, f(1) > 0$, and $f(3) > 0$

Hence, $a \in (-1, 0)$ or $a \in (-2, 1)$ or $a \in (-2, 3)$.

4. b., c.

Clearly, the inscribed rectangle is a square. Let the adjacent vertex be z. Then,

$$\frac{3 + 4i - (z)}{(3 + 4i) - (4 + 4i)} = e^{\pm i\pi/2}$$

$\Rightarrow 3 + 4i - z = \pm i(-1)$

$\Rightarrow z = 3 + 4i \pm i = 3 + 5i$ or $3 + 3i$

5. a., d.

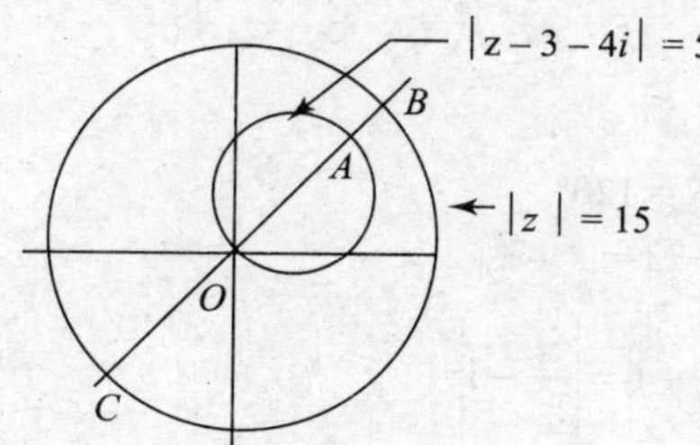

Fig. 2.20

We have

$|z_1| = 15, |z_2 - 3 - 4i| = 5$

Minimum value of $|z_1 - z_2|$ is $AB = OB - OA = 15 - 10 = 5$. Maximum value of $|z_1 - z_2|$ is $CA = OA + OC = 10 + 15 = 25$.

6. a., c.

Triangle ABC is equilateral. Hence,

$z^2 + (-z)^2 + (1 - z)^2 = z(-z) + z(1 - z) + (-z)(1 - z)$

$\Rightarrow 3z^2 - 2z + 1 = -z^2$

$\Rightarrow 4z^2 - 2z + 1 = 0$

$\Rightarrow z = \frac{2 \pm \sqrt{-12}}{8} = \frac{1 \pm \sqrt{3}i}{4}$

$\Rightarrow \text{Re}(z) = \frac{1}{4}$

7. b., c.

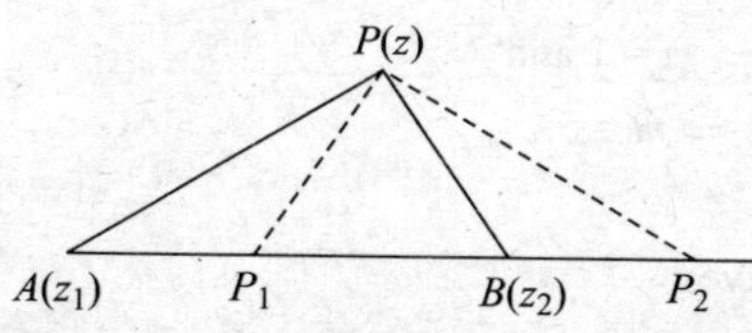

Fig. 2.21

Let internal and external bisectors of $\angle APB$ meet the line joining A and B at P_1 and P_2, respectively. Hence,

$AP_1:P_1B \equiv PA:PB \equiv 3:1$ (internal division)

$AP_2:P_2B \equiv PA:PB \equiv 3:1$ (external division)

Thus P_1 and P_2 are fixed points. Also,

$\angle P_1PP_2 = \frac{\pi}{2}$

Thus P lies on a circle having P_1P_2 as its diameter. Clearly, $B(z_2)$ lies inside this circle.

8. a., d.

Refer the figure, z lies on the point of intersection of the rays from A and B. $\angle ACB$ is a right angle and OBC is an equilateral triangle. Hence,

$$OC = a \Rightarrow z = a\left(\cos\frac{\pi}{3} + i\sin\frac{\pi}{3}\right)$$

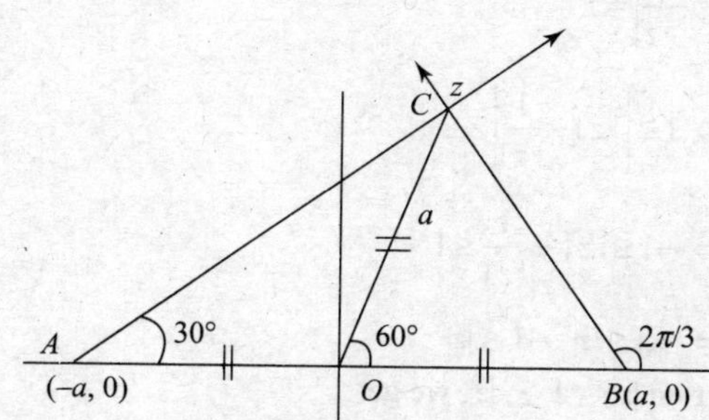

Fig. 2.22

9. a., c.

$(-i)^{1/3} = (i^3)^{1/3} = i, i\omega, i\omega^2$

where

$\omega = \frac{-1 + \sqrt{3}i}{2}$

Hence roots are $i, (-\sqrt{3} - i)/3, (-\sqrt{3} - i)/2$.

10. a., b., d.

$x^n - 1 = (x - 1)(x - z_1)(x - z_2) \cdots (x - z_{n-1})$

$\Rightarrow \frac{x^n - 1}{x - 1} = (x - z_1)(x - z_2) \cdots (x - z_{n-1})$

Putting $x = \omega$, we have

$$\prod_{r=1}^{n-1} (\omega - z_r) = \frac{\omega^n - 1}{\omega - 1} = \begin{cases} 0, & \text{if } n = 3k, k \in Z \\ 1, & \text{if } n = 3k + 1, k \in Z \\ 1 + \omega, & \text{if } n = 3k + 2, k \in Z \end{cases}$$

11. a., d.

If α is a real root, then

$\alpha^3 + (3 + i)\alpha^2 - 3\alpha - (m + i) = 0$

$\therefore \alpha^3 + 3\alpha^2 - 3\alpha - m = 0$ and $\alpha^2 - 1 = 0$

$\Rightarrow \alpha = 1$ or -1. So

$\alpha = 1 \Rightarrow m = 1$ and

$\alpha = -1 \Rightarrow m = 5$.

12. c., d.

We have,

$x^2 + x + 1 = (x - \omega)(x - \omega^2)$

Since $f(x)$ is divisible by $x^2 + x + 1$, $f(\omega) = 0$, $f(\omega^2) = 0$, so

$P(\omega^3) + \omega Q(\omega^3) = 0 \Rightarrow P(1) + \omega Q(1) = 0$ (1)

$P(\omega^6) + \omega^2 Q(\omega^6) = 0 \Rightarrow P(1) + \omega^2 Q(1) = 0$ (2)

Solving (1) and (2), we obtain

$P(1) = 0$ and $Q(1) = 0$

Therefore, both $P(x)$ and $Q(x)$ are divisible by $x - 1$. Hence, $P(x^3)$ and $Q(x^3)$ are divisible by $x^3 - 1$ and so by $x - 1$ Since $f(x) = P(x^3) + xQ(x^3)$, we get $f(x)$ is divisible by $x - 1$.

13. b., c.

$\text{amp}(z_1 z_2) = 0 \Rightarrow \text{amp } z_1 + \text{amp } z_2 = 0$

$\therefore \text{amp } z_1 = -\text{amp } z_2 = \text{amp } \bar{z}_2$

Since $|z_1| = |z_2|$, we get $|z_1| = |\bar{z}_2|$. So, $z_1 = \bar{z}_2$. Also, $z_1 z_2 = \bar{z}_2 z_2 = |z_2|^2 = 1$ because $|z_2| = 1$.

14. a., b.

$$\left|z - \frac{1}{z}\right| = 1$$

$$\Rightarrow 1 \ge \left||z| - \frac{1}{|z|}\right|$$

$$\Rightarrow -1 \le |z| - \frac{1}{|z|} \le 1$$

$$\Rightarrow -|z| \le |z|^2 - 1 \le |z|$$

From $|z|^2 - 1 \ge -|z|$, we get

$$|z|^2 + |z| - 1 \ge 0$$

$$\Rightarrow |z| \ge \frac{-1 + \sqrt{5}}{2} \quad (1)$$

From $|z|^2 - 1 \le |z|$, we get

$$|z|^2 - |z| - 1 \le 0$$

$$\Rightarrow \frac{1 - \sqrt{5}}{2} \le |z| \le \frac{1 + \sqrt{5}}{2} \quad (2)$$

From (1) and (2), we get

$$\Rightarrow \frac{-1 + \sqrt{5}}{2} \le |z| \le \frac{1 + \sqrt{5}}{2}$$

$$\Rightarrow |z|_{\min} = \frac{\sqrt{5} - 1}{2}, \ |z|_{\max} = \frac{1 + \sqrt{5}}{2}$$

15. a., b., c.

$$z^n \cos\theta_0 + z^{n-1}\cos\theta_1 + \cdots + z\cos\theta_{n-1} + \cos\theta_n = 2$$

$$\Rightarrow 2 = |z_0^n \cos\theta_0 + z_0^{n-1}\cos\theta_1 + \cdots + z_0\cos\theta_{n-1} + \cos\theta_n|$$

$$\Rightarrow 2 \le |z_0|^n |\cos\theta_0| + |z|^{n-1}|\cos\theta_1| + \cdots + |z_0||\cos\theta_{n-1}| + |\cos\theta_n|$$

$$\Rightarrow 2 \le |z_0|^n + |z_0|^{n-1} + |z_0|^{n-2} + \cdots + |z_0| + 1$$

which is clearly satisfied for $|z_0| \ge 1$. If $|z_0| < 1$, then

$$2 < 1 + |z_0| + |z_0|^2 + \cdots + |z|^n + \cdots \infty$$

$$\Rightarrow 2 < \frac{1}{1 - |z_0|}$$

$$\Rightarrow |z_0| > \frac{1}{2}$$

16. a., b., c., d.

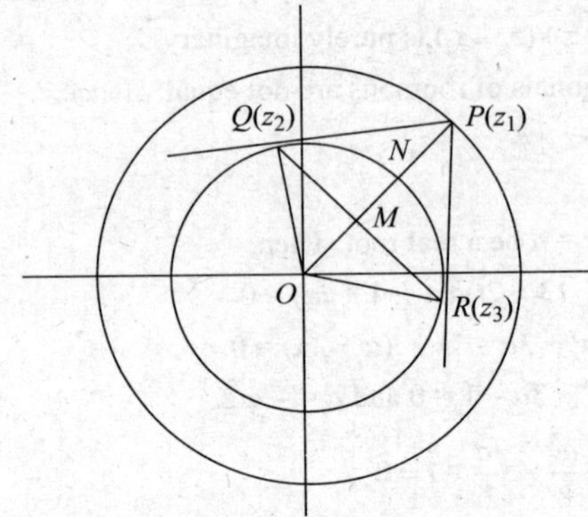

Fig. 2.23

Since $OQ = 1$ and $OP = 2$, so $\sin(\angle OPQ) = 1/2$ and hence $\angle QPR = \pi/3$. Then ΔPQR is equilateral. Also, $OM \perp QR$. Then from ΔOMQ, $OM = 1/2$. Hence $MN = 1/2$. Then centroid of ΔPQR lies on $|z| = 1$.

As PQR is an equilateral triangle, so orthocentre, circumcentre and centroid will coincide. Now,

$$\left|\frac{z_1 + z_2 + z_3}{3}\right| = 1$$

$$\Rightarrow |z_1 + z_2 + z_3|^2 = 9$$

$$\Rightarrow (z_1 + z_2 + z_3)(\bar{z}_1 + \bar{z}_2 + \bar{z}_3) = 9$$

$$\Rightarrow \left(\frac{4}{z_1} + \frac{1}{z_2} + \frac{1}{z_3}\right)\left(\frac{4}{z_1} + \frac{1}{z_2} + \frac{1}{z_3}\right) = 9$$

and

$$\angle QOR = 120°$$

17. a., d. $|z_1 + z_2| = |z_1 - z_2|$

$$\Rightarrow \left|\frac{z_1}{z_2} + 1\right| = \left|\frac{z_1}{z_2} - 1\right|$$

$\Rightarrow \dfrac{z_1}{z_2}$ lies on perpendicular bisector of 1 and -1

$\Rightarrow \dfrac{z_1}{z_2}$ lies on imaginary axis

$\Rightarrow \dfrac{z_1}{z_2}$ is purely imaginary

$$\Rightarrow \arg\left(\frac{z_1}{z_2}\right) = \pm\frac{\pi}{2}$$

$$|\arg(z_1) - \arg(z_2)| = \frac{\pi}{2}$$

18. a., c., d.

$$z' = ze^{i\alpha} \quad (1)$$

$$z'' = ze^{-i\alpha} \quad (2)$$

$$\therefore z'z'' = z^2$$

$\Rightarrow z', z, z''$ are in G.P.

Also,

$$\left(\frac{z'}{z}\right)^2 + \left(\frac{z''}{z}\right)^2 = 2\cos 2\alpha$$

$$\Rightarrow z'^2 + z''^2 = 2z^2 \cos 2\alpha$$

$$\Rightarrow z'^2 + z''^2 = 2z^2 (2\cos^2\alpha - 1)$$

$$\Rightarrow z'^2 + z''^2 + 2z^2 = 4z^2 \cos^2\alpha$$

$$\Rightarrow z'^2 + z''^2 + 2z'z'' = 4z^2 \cos^2\alpha$$

$$\Rightarrow (z' + z'')^2 = 4z^2 \cos^2\alpha$$

$$\Rightarrow z' + z'' = 2z \cos \alpha$$

19. a., c. If $p = q$, then equation becomes $z^p = \bar{z}^p$ and it has infinite number of solutions because any $z \in R$ will satisfy it. If $p \neq q$, let $p > q$, then $z^p = \bar{z}^q$.

$$\therefore |z|^p = |z|^q$$

$$\Rightarrow |z|^p (|z|^{p-q} - 1) = 0$$

$$\Rightarrow |z| = 0 \text{ or } |z| = 1$$

$$|z| = 0 \Rightarrow z = 0 + i0$$

$$|z| = 1 \Rightarrow z = e^{i\theta}$$

$$\Rightarrow e^{(p+q)\theta_i} = 1$$

$$\Rightarrow z = 1^{1/(p+q)}$$

Therefore, the number of solutions is $p + q + 1$.

20. a., b., d.

$$z_1 = 5 + 12i, |z_2| = 4$$

$$|z_1 + iz_2| \le |z_1| + |z_2| = 13 + 4 = 17$$

$$\therefore |z_1 + (1+i)z_2| \ge ||z_1| - |1+i||z_2||$$

$$= 13 - 4\sqrt{2}$$

$$\therefore \min(|z_1 + (1+i)z_2|) = 13 - 4\sqrt{2}$$

$$\left|z_2 + \frac{4}{z_2}\right| \le |z_2| + \frac{4}{|z_2|} = 4 + 1 = 5$$

$$\left|z_2 + \frac{4}{z_2}\right| \ge |z_2| - \frac{4}{|z_2|} = 4 - 1 = 3$$

$$\therefore \max\left|\frac{z_1}{z_2 + \frac{4}{z_2}}\right| = \frac{13}{3} \text{ and } \min\left|\frac{z_1}{z_2 + \frac{4}{z_2}}\right| = \frac{13}{5}$$

21. a., c.

$$p + q + r = a + b\omega + c\omega^2$$
$$+ b + c\omega + a\omega^2$$
$$+ c + a\omega + b\omega^2$$

$$\therefore p + q + r = (a + b + c)(1 + \omega + \omega^2) = 0 \quad (1)$$

p, q, r lie on the circle $|z| = 2$, whose circumcentre is origin. Also $(p + q + r)/3 = 0$. Hence the cenroid coincides with circumcentre. So, the triangle is equilateral. Now, $(p + q + r)^2 = 0$.

$$\Rightarrow p^2 + q^2 + r^2 = -2pqr\left[\frac{1}{p} + \frac{1}{q} + \frac{1}{r}\right]$$

$$= -2pqr\left[\frac{1}{a+b\omega+c\omega^2} + \frac{1}{b+c\omega+a\omega^2} + \frac{1}{c+a\omega+b\omega^2}\right]$$

$$= -2pqr\left[\frac{1}{\omega^2(a\omega+b\omega^2+c)} + \frac{1}{\omega(b\omega^2+c+a\omega)} + \frac{1}{c+a\omega+b\omega^2}\right]$$

$$= \frac{-2pqr}{a\omega + b\omega^2 + c}\left[\frac{1}{\omega^2} + \frac{1}{\omega} + \frac{1}{1}\right] = 0 \quad (2)$$

Hence,

$$p^2 + q^2 + r^2 = 2(pq + qr + rp)$$

22. a., b., c., d.

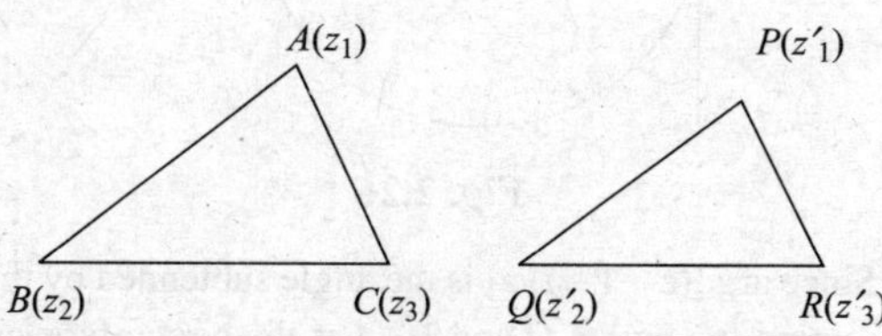

Fig. 2.24

$$z_3 = (1-\lambda)z_1 + z_2 = \frac{(1-\lambda)z_1 + \lambda z_2}{1 - \lambda + \lambda}$$

Hence, z_3 divides the line joining $A(z_1)$ and $B(z_2)$ in the ratio $(\lambda : 1) - \lambda$. That means the given points are collinear. Also, the ratio $\lambda/(1-\lambda) > 0$ (or $0 < \lambda < 1$) if z_3 divides the line joining z_1 and z_2 internally and $\mu/(1-\mu) < 0$ (or $\mu < 0$ or $\mu > 1$) if z_3' divides the line joining z_1', z_2' externally.

When λ, μ are complex numbers, where $\lambda = \mu$, we have $z_3 = (1-\lambda)z_1 + \lambda z_2$ and $z_3' = (1-\lambda)z_1' + \lambda z_2'$. Comparing the value of λ, we have

$$\frac{z_3 - z_1}{z_2 - z_1} = \frac{z_3' - z_1'}{z_2' - z_1'}$$

$$\Rightarrow \left|\frac{z_3 - z_1}{z_2 - z_1}\right| = \left|\frac{z_3' - z_1'}{z_2' - z_1'}\right| \text{ and } \arg\left(\frac{z_3 - z_1}{z_2 - z_1}\right) = \arg\left(\frac{z_3' - z_1'}{z_2' - z_1'}\right)$$

$$\Rightarrow \frac{AC}{AB} = \frac{PQ}{PR} \text{ and } \angle BAC = \angle QPR$$

Hence, triangles ABC and PQR are similar.

23. a., d.

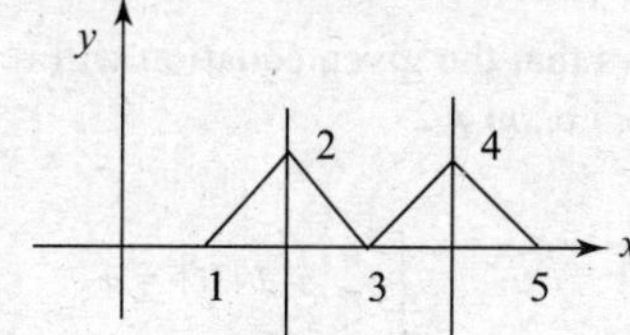

Fig. 2.25

24. a., c.

Given,

$$z^n = (z+1)^n \Rightarrow |z^n| = |(z+1)^n|$$

$$\therefore |z|^n = |z+1|^n \Rightarrow |z| = |z+1|$$

$$\Rightarrow |z|^2 = |z+1|^2$$

$$\Rightarrow x^2 + y^2 = (x+1)^2 + y^2, \text{ where } z = x + iy$$

$$\Rightarrow x = -\frac{1}{2}$$

Hence, z lies on the line $x = -1/2$. Hence sum of real parts of the roots is $-(n-1)/2$ (since equation has $n - 1$ roots).

25. a., b., d.

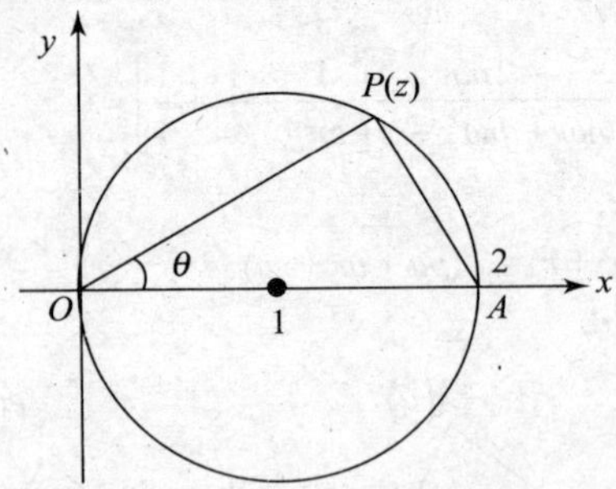

Fig. 2.26

Since arg $[(z - 1 - i)/z]$ is the angle subtended by the chord joining the points O and $1 + i$ at the circumference of the circle $|z - 1| = 1$, so it is equal to $-\pi/4$. The line joining the points $z = 0$ and $z = 2 + 0i$ is the diameter.

$$\arg\frac{z-2}{z} = \pm\frac{\pi}{2}$$

$$\Rightarrow \frac{z-2}{z-0} \text{ is purely imaginary}$$

We have,

$$\angle OPA = \frac{\pi}{2}$$

$$\Rightarrow \arg\left(\frac{2-z}{0-z}\right) = \frac{\pi}{2} \Rightarrow \frac{z-2}{z} = \frac{AP}{OP}i$$

Now in ΔOAP,

$$\tan\theta = \frac{AP}{OP}$$

Thus.

$$\frac{z-2}{z} = i\tan\theta$$

26. a., b., d.

$$\left|\frac{2z-i}{z+1}\right| = m \Rightarrow \left|z - \frac{i}{2}\right| = \frac{m}{2}|z+1|$$

This shows that the given equation will represent a circle, if $m/2 \neq 1$, i.e., $m \neq 2$.

27. a., d.

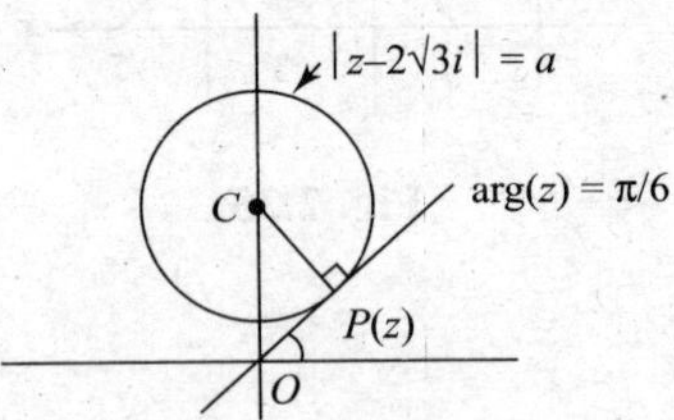

Fig. 2.27

$CP = r$, $OC = 2\sqrt{3}$, $\angle COP = \pi/3$

$$\Rightarrow CP = OC\sin\frac{\pi}{3} = 2\sqrt{3}\,\frac{\sqrt{3}}{2} = 3$$

Thus, when $r = 3$, the circle touches the line. Hence, for two distinct points of intersection $3 < r < 2\sqrt{3}$.

28. b., c.

We have,

$$\left|\frac{1}{z_2} + \frac{1}{z_1}\right| = \left|\frac{1}{z_2} - \frac{1}{z_1}\right| \Rightarrow |z_1 + z_2| = |z_1 - z_2|$$

Squaring both sides, we have

$$|z_1|^2 + |z_2|^2 + 2(z_1\bar{z}_2 + \bar{z}_1 z_2) = |z_1|^2 + |z_2|^2 - 2(z_1\bar{z}_2 + \bar{z}_1 z_2)$$

$$\Rightarrow 4(z_1\bar{z}_2 + \bar{z}_1 z_2) = 0$$

$$\Rightarrow \left(\frac{z_1}{z_2}\right) = \frac{\pi}{2} = \arg\left(\frac{z_1 - 0}{z_2 - 0}\right)$$

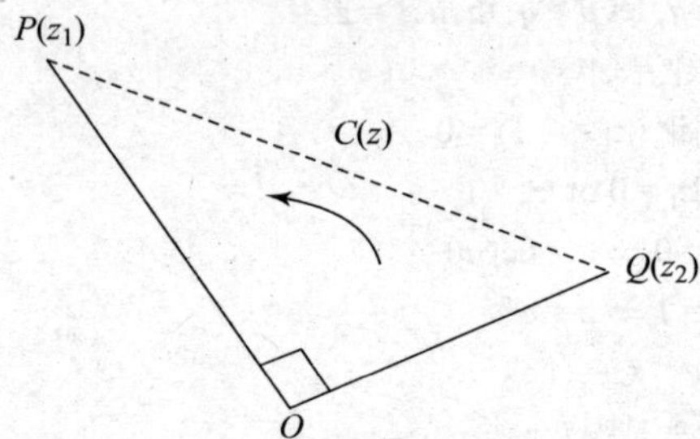

Fig. 2.28

That is angle between z_2, O and z_1 is a right angle, taken in order, as shown in the above diagram. Now, the circumcentre of the above diagram will lie on the line PQ as diameter and is represented by C which is the centre of PQ, such that $z = (z_1 + z_2)/2$, where z is the affix of circumcentre.

29. a., c., d.

Choice (a) on simplification gives

$$z = \frac{1+x}{1+x^2} + i\frac{1+x}{1+x^2}$$

For $x = 0.5$, $f(0.5) > 1$ which is out of range, Hence, (a) is not correct. From choice (b),

$$z = \frac{1-x}{1+x^2} + i\frac{1-x}{1+x^2}$$

$f(x)$ and $g(x) \in (0, 1)$ if $x \in (0, 1)$. Hence, (b) is correct. From choice (c),

$$z = \frac{1+x}{1+x^2} + \frac{1-x}{1+x^2}i$$

Hence, (c) is not correct. From choice (d),

$$z = \frac{1-x}{1+x^2} + \frac{1+x}{1+x^2}i$$

Hence, (d) is not correct.

30. a., b., c.

Let $z = x + iy$ where x, y satisfy the given equation. Hence,

$(x^2 + y^2)(x^2 - y^2) = 175$

$\Rightarrow x^2 + y^2 = 25$ and $x^2 - y^2 = 7$ (as all other possibilities will give non-integral solutions)

Hence, possible values of z will be $4 + 3i$, $4 - 3i$, $-4 + 3i$ and $-4 - 3i$. Clearly, it will form a rectangle having length of the diagonal 10.

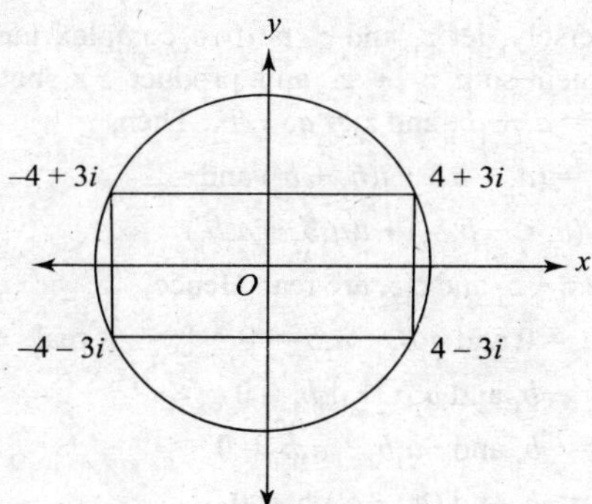

Fig. 2.29

From the diagram, options (a), (b), (c) are correct.

31. a., b.

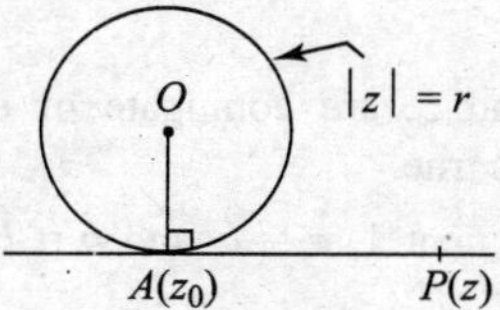

Fig. 2.30

$$\angle OAP = \frac{\pi}{2}$$

$\Rightarrow \frac{z - z_0}{z_0}$ is purely imaginary

$$\Rightarrow \frac{z - z_0}{z_0} + \frac{\bar{z} - \bar{z}_0}{\bar{z}_0} = 0 \Rightarrow \frac{z}{z_0} + \frac{\bar{z}}{\bar{z}_0} = 2 \quad (1)$$

$$\Rightarrow \text{Re}\left(\frac{z}{z_0}\right) = 1$$

From (1),

$$z\bar{z}_0 + z_0\bar{z} = 2|z_0|^2 = 2r^2$$

32. b., c.

z_1 and z_2 are the roots of the equation $z^2 - az + b = 0$. Hence,

$z_1 + z_2 = a, z_1z_2 = b$

Now,

$|z_1 + z_2| \le |z_1| + |z_2|$

$\Rightarrow |z_1 + z_2| = |a| \le 1 + 1 = 2 \quad (\because |z_1| = |z_2| = 1)$

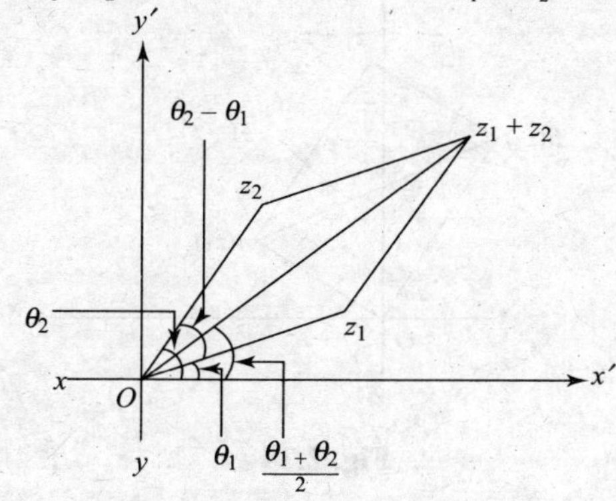

Fig. 2.31

$$\Rightarrow \arg(a) = \frac{1}{2}[\arg(z_2) + \arg(z_1)]$$

Also,

$\arg(b) = \arg(z_1z_2) = \arg(z_1) + \arg(z_2)$

$\Rightarrow 2\arg(a) = \arg(b)$

33. a., b., d. Given,

$$\begin{vmatrix} a & b & c \\ b & c & a \\ c & a & b \end{vmatrix} = 0$$

$\Rightarrow a^3 + b^3 + c^3 - 3abc = 0$

$\Rightarrow (a + b + c)(a^2 + b^2 + c^2 - ab - bc - ca) = 0$

$\Rightarrow \frac{1}{2}(a+b+c)[(a-b)^2 + (b-c)^2 + (c-a)^2] = 0$

$\Rightarrow (a - b)^2 + (b - c)^2 + (c - a)^2 = 0$

$\Rightarrow a = b = c$ [$\because a + b + c \ne 0, \because z_1 \ne 0, \therefore |z_1| = a \ne 0$ etc.]

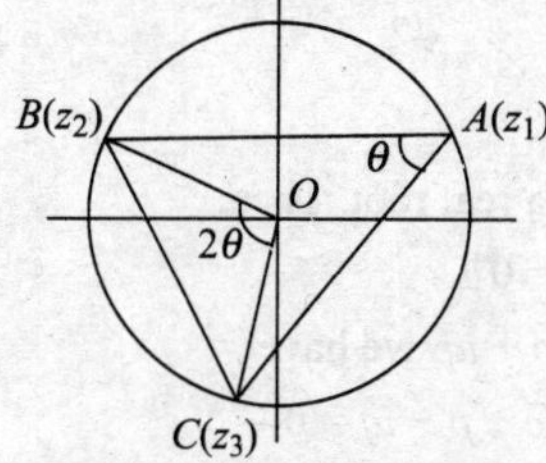

Fig. 2.32

Hence, $OA = OB = OC$, where O is the origin and A, B, C are the points representing z_1, z_2 and z_3, respectively. Therefore, O is circumcentre of ΔABC. Now,

$$\arg\left(\frac{z_3}{z_2}\right) = \angle BOC \quad (1)$$

$$= 2\angle BAC = 2\arg\left(\frac{z_3 - z_1}{z_2 - z_1}\right) \quad (2)$$

$$= \arg\left(\frac{z_3 - z_1}{z_3 - z_1}\right)^2 \quad [\because \angle BOC = 2\angle BAC]$$

Hence,

$$\arg\left(\frac{z_3}{z_2}\right) = \arg\left(\frac{z_3 - z_1}{z_2 - z_1}\right)^2$$

Also, centroid is $(z_1 + z_2 + z_3)/3$. Since $HG:GO \equiv 2:1$ (where H is orthocentre and G is centroid), then orthocentre is $z_1 + z_2 + z_3$ (by section formula). When triangle is equilateral centroid coincides with circumcentre. Hence, $z_1 + z_2 + z_3 = 0$.

Also, the area for equilateral triangle is $(\sqrt{3}/3)L^2$, where L is length of side. Since radius is $|z_1|$, $L = \sqrt{3}|z_1|$, hence area is $(3\sqrt{3}/4)|z_1|^2$.

34. a., c., d.

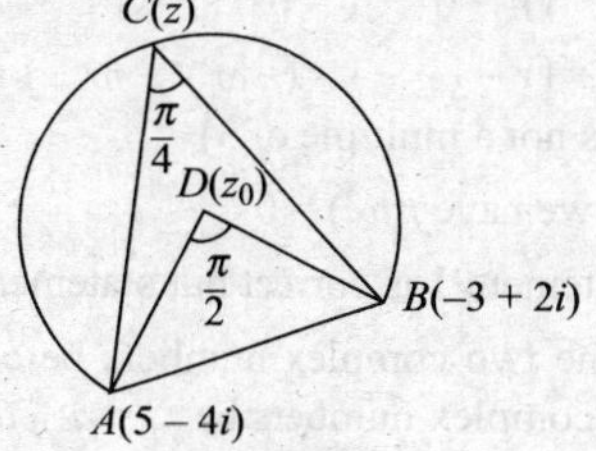

Fig. 2.33

$$\frac{z_0-(-3+2i)}{z_0-(5-4i)}=\frac{BD}{AD}e^{i\pi/2}=i$$

$\Rightarrow z_0+3-2i=iz_0-5i-4$

$\Rightarrow z_0=-2-5i$

$\Rightarrow$ Radius $AD=|5-4i-(-2-5i)|$

$=|7+i|$

$=\sqrt{50}=5\sqrt{2}$

Length of arc $=\frac{3}{4}$(perimeter of circle)

$=\frac{3}{4}(2\pi\times 5\sqrt{2})$

$=\frac{15\pi}{\sqrt{2}}$

35. a, d.

Let $z=c$ be a real root. Then,

$ac^2+c+\bar{a}=0$ (1)

Putting $a=p+iq$, we have

$(p+iq)c^2+c+p-iq=0$

$\Rightarrow pc^2+c+p=0$ and $qc^2-q=0 \Rightarrow c=\pm 1$ $(\because q\neq 0)$

$\therefore$ (1) $\Rightarrow a\pm 1+\bar{a}=0$

Also,

$|c|=1$

36. a, b., c., d.

$\sqrt{5-12i}=\sqrt{(3-2i)^2}=\pm(3-2i)$

$\sqrt{-5-12i}=\sqrt{(2-3i)^2}=\pm(2-3i)$

$\Rightarrow z=\sqrt{5-12i}+\sqrt{-5-12i}$

$=-1-i,-5+5i,5-5i,1+i$

Therefore, principal values of arg z are $-2\pi/4$, $3\pi/4$, $-\pi/4$, $2\pi/4$.

Reasoning Type

1. a. $\arg(z_1z_2)=2\pi \Rightarrow \arg(z_1)+\arg(z_2)=2\pi \Rightarrow \arg(z_1)=\arg(z_2)=\pi$, as principal arguments are from $-\pi$ to π.

Hence both the complex numbers are purely real. Hence both the statements are true and statement 2 is correct explanation of statement 1.

2. c. $x^3+x^2+x=x(x^2+x+1)=x(x-\omega)(x-\omega^2)$

Now $f(x)=(x+1)^n-x^n-1$ is divisible by x^3+x^2+x. Then $f(0)=0, f(\omega)=0\, f(\omega^2)=0$. Now,

$f(0)=(0+1)^n-0^n-1=0$

$f(\omega)=(\omega+1)^n-\omega^n-1=(-\omega^2)^n-\omega^n-1=-(\omega^{2n}+\omega^n+1)$
$=0$ (as n is not a multiple of 3)

Similarly, we have $f(\omega^2)=0$.

Hence, statement 1 is correct but statement 2 is false.

3. a. First, let the two complex numbers be conjugate of each other. Let complex numbers be $z_1=a+ib$ and $z_2=a-ib$. Then, $z_1+z_2=(a+ib)+(a-ib)=2a$, which is real and $z_1z_2=(a+ib)(a-ib)=a^2-i^2b^2=a^2+b^2$, which is real.

Conversely, let z_1 and z_2 be two complex numbers such that their sum z_1+z_2 and product z_1z_2 both are real. Let $z_1=a_1+ib_1$ and $z_2=a_2+ib_2$. Then,

$z_1+z_2=(a_1+a_2)+i(b_1+b_2)$ and

$z_1z_2=(a_1a_2-b_1b_2)+i(a_1b_2+a_2b_1)$

Now, z_1+z_2 and z_1z_2 are real. Hence,

$b_1+b_2=0$ and $a_1b_2+a_2b_1=0$ $[\because z$ is real $\Rightarrow \text{Im}(z)=0]$

$\Rightarrow b_2=-b_1$ and $a_1b_2+a_2b_1=0$

$=-b_1$ and $-a_1b_1+a_2b_1=0$

$=-b_1$ and $(a_2-a_1)\,b_1=0$

$=-b_1$ and $a_2-a_1=0$

$=-b_1$ and $a_2=a_1$

$\Rightarrow z_2=a_2+ib_2=a_1-ib_1$

$=\bar{z}_1$

Hence, z_1 and z_2 are conjugate of each other. Hence, statement 2 is true.

Also in statement 1, $a=\bar{a}$ and $b=\bar{b}$, then a and b are real.

Thus, z_1+z_2 and z_1z_2 are real. So,

$z_2=\bar{z}_2$

$\Rightarrow \arg(z_1z_2)=\arg(z_1\bar{z}_1)=\arg(|z_1|^2)=0$

Hence, statement 1 is correct and statement 2 is correct explanation of statement 1.

4 d. $x+\frac{1}{x}=1$

$\Rightarrow x^2-x+1=0$

$\therefore x=-\omega,-\omega^2$

Now for $x=-\omega$,

$P=\omega^{4000}+\frac{1}{\omega^{4000}}=\omega+\frac{1}{\omega}=-1$

Similarly for $x=-\omega^2$, $P=-1$. For $n>1$,

$2^n=4k$

$\therefore\ 2^{2^n}=2^{4k}=(16)^k=$ a number with last digit 6

$\Rightarrow q=6+1=7$

Hence, $p+q=-1+7=6$.

5. a.

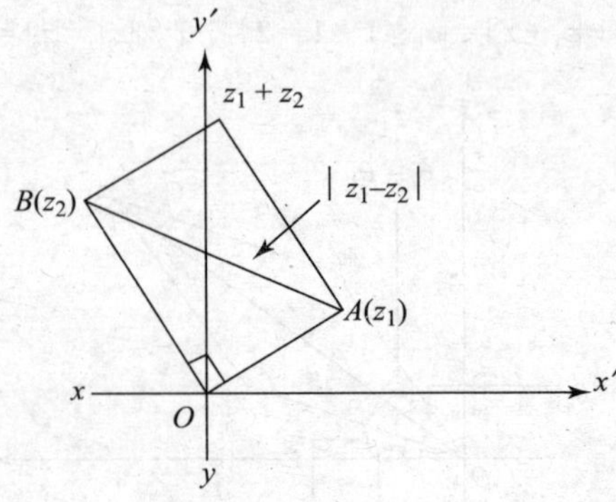

Fig. 2.34

From the diagram when $|z_1-z_2|=|z_1+z_2|$, OAB is right-angled triangle. Hence orthocentre is mid-point of AB, i.e., $(z_1+z_2)/2$.

6. d. Statement 2 is true as it is the definition of an ellipse. Statement 1 is false as distance between 1 and 8 is 7 but $|z-1|+|z-8| = 5 < 7$. Hence, no such z exists.

7. d. $|z_1+z_2+z_3| = |z_1-a+z_2-b+z_3-c+(a+b+c)|$

$\le |z_1-a|+|z_2-b|+|z_3-c|+|a+b+c|$

$\le 2|a+b+c|$

Hence, $|z_1+z_2+z_3|$ is less than $2|a+b+c|$.

8. b. Fourth roots of unity are $-1, 1, -i$ and i.

$\therefore z_1^2+z_2^2+z_3^2+z_4^2 = 0$ and $z_1+z_2+z_3+z_4 = 0$

9. a. If roots of $ax^2+bx+c=0$, $0<a<b<c$, are non-real, then they will be the conjugate of each other. Hence,

$z_2 = \bar{z}_1 \Rightarrow |z_1| = |z_2|$

Now,

$z_1z_2 = \dfrac{c}{a} > 1 \Rightarrow |z_1|^2 > 1$

$\Rightarrow |z_1| > 1$

$\Rightarrow |z_2| > 1$

10. a. We have,

$az^2+bz+c=0$ (1)

and z_1, z_2 [roots of (1)] are such that $\text{Im}(z_1z_2) \ne 0$. So z_1 and z_2 are not conjugate of each other. That is complex roots of (1) are not conjugate of each other, which implies that coefficients a, b, c cannot all be real. Hence, at least one of a, b, c is imaginary.

11. b. We have,

$(\cos\theta + i\sin\theta)^{3/5} = (\cos 3\theta + i\sin 3\theta)^{1/5}$

$= [\cos(2r\pi+3\theta) + i\sin(2r\pi+3\theta)]^{1/5}$ where $r = 0, 1, 2, 3, 4$

$= e^{i\left(\frac{2r\pi+3\theta}{5}\right)}$, $r = 0, 1, 2, 3, 4$

Hence, product of all values of $(\cos\theta+i\sin\theta)^{3/5}$ is

$e^{i\frac{3\theta}{5}}\, e^{i\left(\frac{2\pi+3\theta}{5}\right)}\, e^{i\left(\frac{4\pi+3\theta}{2}\right)}\, e^{i\left(\frac{6\pi+3\theta}{5}\right)}\, e^{i\left(\frac{8\pi+3\theta}{5}\right)}$

$= e^{i3\theta+4\pi i}$

$= e^{i4\pi}\, e^{i3\theta}$

$= \cos 3\theta + i\sin 3\theta$

Also product of roots of the equation $x^5 - 1 = 0$ is 1. Hence statement 2 is true. But it is not correct explanation of statement 1.

12. b.

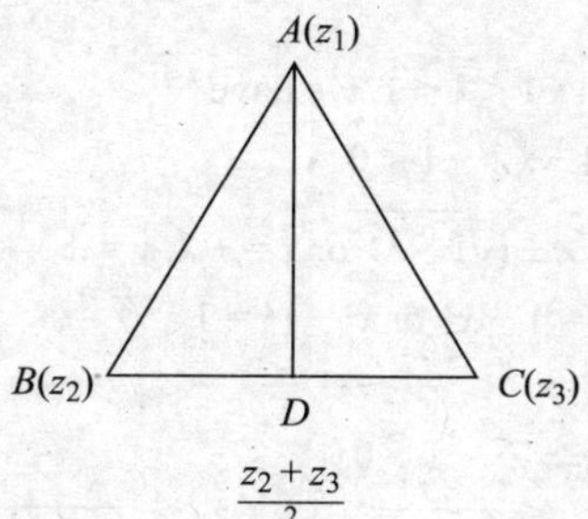

Fig. 2.35

$\arg\left(\dfrac{z_2+z_3-2z_1}{z_3-z_2}\right) = \arg\left(2\left(\dfrac{\frac{z_2+z_3}{2}-z_1}{z_3-z_2}\right)\right)$

$= \arg\left(\dfrac{\frac{z_2+z_3}{2}-z_1}{z_3-z_2}\right)$

$(z_2+z_3)/2$ is the mid-point of side BC. Clearly, line joining A and mid-point of BC will be perpendicular to side BC. Thus,

$\arg\left(\dfrac{\frac{z_2+z_3}{2}-z_1}{z_3-z_2}\right) = \dfrac{\pi}{2}$

Hence, statement 2 is also true. However, it does not explain statement 1.

13. a. Suppose there exists a complex number z which satisfies the given equation and is such that $|z| < 1$. Then,

$z^4+z+2 = 0 \Rightarrow -2 = z^4+z \Rightarrow |-2| = |z^4+z|$

$\Rightarrow 2 \le |z^4| + |z| \Rightarrow 2 < 2$, because $|z| < 1$

But $2 < 2$ is not possible. Hence, given equation cannot have a root z such that $|z| < 1$.

14. c. $|z_1+z_2| = \left|\dfrac{z_1+z_2}{z_1z_2}\right|$

$\Rightarrow |z_1+z_2|\left(1-\dfrac{1}{|z_1z_2|}\right) = 0$

$\Rightarrow |z_1z_2| = 1$

Hence, statement 1 is unimodular. However, it is not necessary that $|z_1| = |z_2| = 1$. Hence, statement 2 is false.

Linked Comprehension Type

For Problems 1–4

1. b., 2., b., 3., c., 4., c.

Sol. Given that

$|z_1+z_2|^2 = |z_1|^2+|z_2|^2$

$\Rightarrow |z_1|^2+|z_2|^2+z_1\bar{z}_2+\bar{z}_1z_2 = |z_1|^2+|z_2|^2$

$\Rightarrow z_1\bar{z}_2+\bar{z}_1z_2 = 0$ (1)

$\Rightarrow \dfrac{z_1}{z_2}+\dfrac{\bar{z}_1}{\bar{z}_2} = 0$ (dividing by $z_2\bar{z}_2$)

$\Rightarrow \dfrac{z_1}{z_2}+\overline{\left(\dfrac{z_1}{z_2}\right)} = 0$ (2)

From (1), $z_1\bar{z}_2$ is purely imaginary. From (2), z_1/z_2 is purely imaginary. Hence,

$\arg\left(\dfrac{z_1}{z_2}\right) = \pm\dfrac{\pi}{2} \Rightarrow \arg(z_1) - \arg(z_2) = \pm\dfrac{\pi}{2}$

Also, $i(z_1/z_2)$ is purely real. Hence its possible arguments are 0 and π.

For Problems 5–8

5. a, 6. d, 7. c, 8. d.

Sol.

$$z = \frac{1 - i\sin\theta}{1 + i\cos\theta} = \frac{(1 - i\sin\theta)(1 - i\cos\theta)}{(1 + i\cos\theta)(1 - i\cos\theta)}$$

$$= \frac{(1 - \sin\theta\cos\theta) - i(\cos\theta + \sin\theta)}{(1 + \cos^2\theta)}$$

If z is purely real, then

$\cos\theta + \sin\theta = 0$ or

$\tan\theta = -1$

$\Rightarrow \theta = n\pi - \frac{\pi}{4}, n \in I$

If z is purely imaginary, $1 - \sin\theta\cos\theta = 0$ or $\sin\theta\cos\theta = 1$, which is not possible. Hence,

$$|z| = \left|\frac{1 - i\sin\theta}{1 + i\cos\theta}\right| = \frac{\sqrt{1 + \sin^2\theta}}{\sqrt{1 + \cos^2\theta}}$$

If $|z| = 1$, then

$\cos^2\theta = \sin^2\theta \Rightarrow \tan^2\theta = 1 \Rightarrow \theta = n\pi \pm \frac{\pi}{4}, n \in I$

We have,

$$\arg(z) = \frac{-(\cos\theta + \sin\theta)}{(1 - \sin\theta\cos\theta)}$$

Now,

$\arg(z) = 1$

$$\Rightarrow \frac{-(\cos\theta + \sin\theta)}{(1 - \sin\theta\cos\theta)} = 1$$

$\Rightarrow \cos^2\theta + \sin^2\theta + 2\sin\theta\cos\theta = 1 + \sin^2\theta\cos^2\theta - 2\sin\theta\cos\theta$

$\Rightarrow 1 + 4\sin\theta\cos\theta = 1 + \sin^2\theta\cos^2\theta$

$\Rightarrow \sin^2\theta\cos^2\theta - 4\sin\theta\cos\theta = 0$

$\Rightarrow \sin\theta\cos\theta(\sin\theta\cos\theta - 4) = 0$

$\Rightarrow \sin\theta\cos\theta = 0 \quad (\because \sin\theta\cos\theta = 4$ is not possible)

$\Rightarrow \theta = (2n + 1)\pi$ or $\theta = (4n - 1)\pi/2, n \in I$

$(\because -\cos\theta - \sin\theta > 0)$

For Problems 9–11

9. a., 10., d., 11., d.

Sol.

9. Let z_1 (purely imaginary) be a root of the given equation. Then,

$z_1 = -\bar{z}_1$

and

$az_1^2 + bz_1 + c = 0 \quad (1)$

$\Rightarrow \overline{az_1^2 + bz_1 + c} = \bar{0}$

$\Rightarrow \bar{a}\bar{z}_1^2 + \bar{b}\bar{z}_1 + \bar{c} = 0$

$\Rightarrow \bar{a}z_1^2 - \bar{b}z_1 + \bar{c} = 0 \quad [\text{as } \bar{z}_1 = -z_1] \quad (2)$

Now Eqs. (1) and (2) must have one common root. So

$(c\bar{a} - a\bar{c})^2 = (b\bar{c} + c\bar{b})(-a\bar{b} - \bar{a}b)$

Let z_1 and z_2 be two purely imaginary roots. Then,

$\bar{z}_1 = -z_1, \bar{z}_2 = -z_2$

Now,

$az^2 + bz + c = 0 \quad (1)$

$\Rightarrow \overline{az^2 + bz + c} = \bar{0}$

$\Rightarrow \bar{a}\bar{z}^2 + \bar{b}\bar{z} + \bar{c} = 0$

$\Rightarrow \bar{a}z^2 - \bar{b}z + \bar{c} = 0 \quad (2)$

Eq. (1) and (2) must be identical as their roots are same.

$\therefore \frac{a}{\bar{a}} = -\frac{b}{\bar{b}} = \frac{c}{\bar{c}}$

$\Rightarrow a\bar{c} = \bar{a}c, a\bar{b} + \bar{a}b = 0$ and $b\bar{c} + \bar{b}c = 0$

Hence, $a\bar{c}$ is purely real and $a\bar{b}$ and $b\bar{c}$ are purely imaginary.

Let z_1 (purely real) be a root of the given equation. Then,

$z_1 = \bar{z}_1$

and

$az_1^2 + bz_1 + c = 0 \quad (1)$

$\Rightarrow \overline{az_1^2 + bz_1 + c} = \bar{0}$

$\Rightarrow \bar{a}\bar{z}_1^2 + \bar{b}\bar{z}_1 + \bar{c} = 0$

$\Rightarrow \bar{a}z_1^2 + \bar{b}z_1 + \bar{c} = 0 \quad (2)$

Now (1) and (2) must have one common root. Hence,

$(c\bar{a} - a\bar{c})^2 = (b\bar{c} - c\bar{b})(a\bar{b} - \bar{a}b)$

For Problems 12–14

12. c., 13., b., 14., d.

Sol. $az + b\bar{z} + c = 0 \quad (1)$

$\Rightarrow \bar{a}\bar{z} + \bar{b}z + \bar{c} = 0 \quad (2)$

Eliminating $\bar{z}$ from (1) and (2), we get

$$z = \frac{c\bar{a} - b\bar{c}}{|b|^2 - |a|^2}$$

If $|a| \neq |b|$, then z represents one point on the Argand plane. If $|a| = |b|$ and $\bar{a}c \neq b\bar{c}$, then no such z exists. Adding (1) and (2),

$(\bar{a} + b)\bar{z} + (a + \bar{b})z + (c + \bar{c}) = 0$

This is of the form $A\bar{z} + \bar{A}z + B = 0$, where $B = c + \bar{c}$ is real.

Hence locus of z is a straight line.

For Problems 15–17

15. a., 16., b., 17., c.

$$z = -\lambda \pm \sqrt{\lambda^2 - 1}$$

Case I:

When $-1 < \lambda < 1$, we have

$\lambda^2 < 1 \Rightarrow \lambda^2 - 1 < 0$

$z = -\lambda \pm i\sqrt{1 - \lambda^2}$ or $x = -\lambda, y = \pm\sqrt{1 - \lambda^2}$

$\Rightarrow y^2 = 1 - x^2$ or $x^2 + y^2 = 1$

Case II:

$\lambda > 1 \Rightarrow \lambda^2 - 1 > 0$

$z = -\lambda \pm \sqrt{\lambda^2 - 1}$ or $x = -\lambda \pm \sqrt{\lambda^2 - 1}, y = 0$

Roots are $(-\lambda + \sqrt{\lambda^2 - 1}, 0)$, $(-\lambda - \sqrt{\lambda^2 - 1}, 0)$. One root lies inside the unit circle and the other root lies outside the unit circle.

Case III:

When λ is very large,

$$z=-\lambda-\sqrt{\lambda^2-1}\approx-2\lambda$$

$$z=-\lambda+\sqrt{\lambda^2-1}=\frac{(-\lambda+\sqrt{\lambda^2-1})(-\lambda-\sqrt{\lambda^2-1})}{(-\lambda-\sqrt{\lambda^2-1})}$$

$$=\frac{1}{-\lambda-\sqrt{\lambda^2-1}}=-\frac{1}{2\lambda}$$

For Problems 18–20

18. d., 19. d., 20. c.

Sol. We have

$$az^2+z+1=0 \quad (1)$$

$$\Rightarrow \overline{az^2+z+1}=0 \quad \text{[Taking conjugate of both sides]}$$

$$\Rightarrow \bar{a}z^2-z+1=0 \quad (2)$$

[Since z is purely imaginary $\bar{z}=-z$]

Eliminating z from (1) and (2) by cross-multiplication rule,

$$(\bar{a}-a)^2+2(a+\bar{a})=0\Rightarrow\left(\frac{\bar{a}-a}{2}\right)^2+\frac{a+\bar{a}}{2}=0$$

$$\Rightarrow-\left(\frac{a-\bar{a}}{2i}\right)^2+\left(\frac{a+\bar{a}}{2}\right)=0\Rightarrow-\sin^2\theta+\cos\theta=0$$

$$\Rightarrow\cos\theta=\sin^2\theta \quad (3)$$

Now,

$f(x)=x^3-3x^2+3(1+\cos\theta)x+5$

$f'(x)=3x^2-6x+3(1+\cos\theta)$

Its discriminant is

$36-36(1+\cos\theta)=-36\cos\theta=-36\sin^2\theta<0$

Hence, $f(x)$ is increasing $\forall\ x\in R$. Also, $f(0)=5$, then $f(x)=0$ has one negative root. Now,

$\cos 2\theta=\cos\theta\Rightarrow 1-2\sin^2\theta=\cos\theta$

$\Rightarrow 1-2\cos\theta=\cos\theta$

$\Rightarrow \cos\theta=1/3,$

which has four roots for $\theta\in[0,4\pi]$.

For Problems 21–23

21. a., 22. b., 23. b.

Sol.

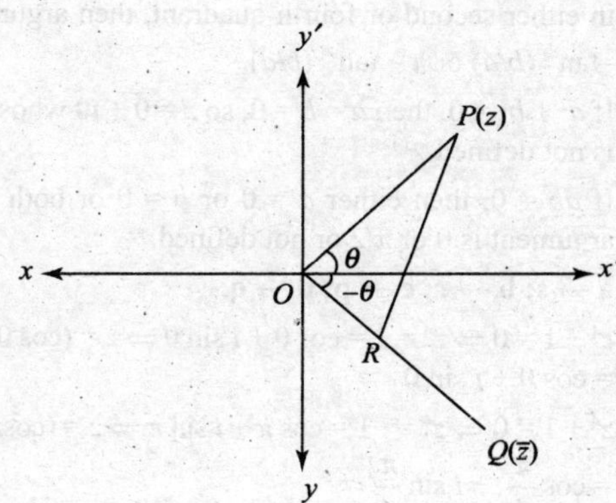

Fig. 2.36

We have

$$\left|\,|z|-\left|\frac{4}{z}\right|\,\right|\le\left|z-\frac{4}{z}\right|=2$$

$$\Rightarrow-2\le|z|-\frac{4}{|z|}\le 2$$

$\Rightarrow |z|^2+2|z|-4\ge 0$ and $|z|^2-2|z|-4\le 0$

$\Rightarrow (|z|+1)^2-5\ge 0$ and $(|z|-1)^2\le 5$

$\Rightarrow (|z|+1+\sqrt{5})(|z|+1-\sqrt{5})\ge 0$ and $(|z|-1+\sqrt{5})\times(|z|-1-\sqrt{5})\le 0$

$\Rightarrow |z|\le-\sqrt{5}-1$ or $|z|\ge\sqrt{5}-1$ and $\sqrt{5}-1\le|z|\le\sqrt{5}+1$

$\Rightarrow \sqrt{5}-1\le|z|\le\sqrt{5}+1$

Hence, the least modulus is $\sqrt{5}-1$ and the greatest modulus is $\sqrt{5}+1$. Also,

$$|z|=\sqrt{5}+1\Rightarrow\frac{4}{|z|}=\sqrt{5}-1$$

Now,

$$\frac{4}{z}=\frac{4\bar{z}}{|z|^2}$$

Hence, $4/z$ lies in the direction of $\bar{z}$.

$$\left|z-\frac{4}{z}\right|=PR=2\text{ (given)}$$

We have,

$OP=\sqrt{5}+1$ and $OR=\sqrt{5}-1$

$$\Rightarrow\cos 2\theta=\frac{OP^2+OR^2+PR^2}{2\,OP\cdot OR}$$

$$=\frac{(\sqrt{5}+1)^2+(\sqrt{5}-1)^2-4}{2(5-1)}=1$$

$\Rightarrow 2\theta=0, 2\pi$

$\Rightarrow \theta=0,\pi$

$\Rightarrow z$ is purely real

$\Rightarrow z=\pm(\sqrt{5}+1)$

Similarly for $|z|=\sqrt{5}-1$, we have $z=\pm(\sqrt{5}-1)$.

For Problems 24–26

24. a., 25. c., 26. b.

Sol. $BM\equiv y-0=-1(x-1)$

$x+y=1$

$\therefore\ \sqrt{u-1}=t+i(1-t)$

$u=2t+2it(1-t)$

$x=2t$ and $y=2t(1-t)$

$y=x(1-x/2)$

$2y=2x-x^2$

$$\Rightarrow(x-1)^2=-2\left(y-\frac{1}{2}\right),$$

which is a parabola. Its axis is $x=1$, i.e., $z+\bar{z}=2$ and directrix is $y=1$, i.e., $z-\bar{z}=2i$.

For Problems 27–29

27. a., 28. b., 29. c.

Sol. 27.

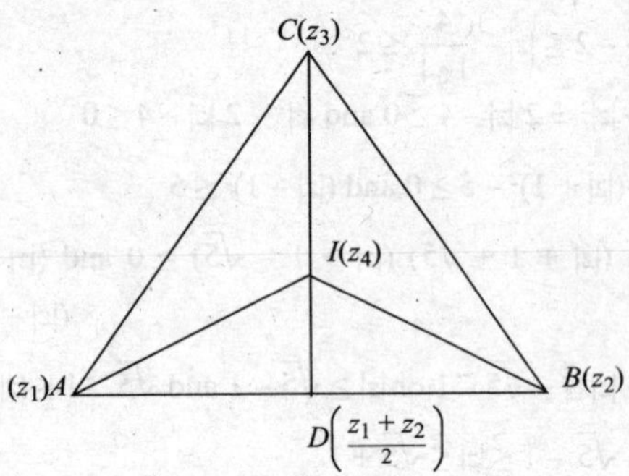

Fig. 2.37

$$\frac{AB\times AC}{(IA)^2} = \frac{AB}{IA}\times\frac{AC}{IA}$$

$$\angle IAB = \frac{\theta}{2}, \angle IAC = \frac{\theta}{2}$$

$$\frac{z_2 - z_1}{z_4 - z_1} = \frac{|z_2 - z_1|}{|z_4 - z_1|} e^{-\frac{i\theta}{2}}$$

and

$$\frac{z_3 - z_1}{z_4 - z_1} = \frac{|z_3 - z_1|}{|z_4 - z_1|} e^{\frac{i\theta}{2}}$$

Multiplying,

$$\frac{z_2 - z_1}{z_4 - z_1}\frac{z_3 - z_1}{z_4 - z_1} = \frac{|z_2 - z_1|}{|z_4 - z_1|}\frac{|z_3 - z_1|}{|z_4 - z_1|}$$

$$\Rightarrow \frac{(z_2 - z_1)(z_3 - z_1)}{(z_4 - z_1)^2} = \frac{AB\times AC}{IA^2} \quad (1)$$

28. From (1),

$$\frac{(z_2 - z_1)(z_3 - z_1)}{(z_4 - z_1)^2} = 2\left(\frac{AD}{IA}\right)^2\left(\frac{AC}{AD}\right) \quad (\because AB = 2\,AD)$$

$$\Rightarrow (z_2 - z_1)(z_3 - z_1) = (z_4 - z_1)^2\, 2\cos^2\frac{\theta}{2}\sec\theta$$

$$= (z_4 - z_1)^2 (\cos\theta + 1)\sec\theta$$

29. Keeping in mind that $\tan\theta = CD/AD$ and $\tan(\theta/2) = ID/BD$, We have,

$$\frac{z_3 - \frac{z_1 + z_2}{2}}{z_1 - \frac{z_1 + z_2}{2}} = \frac{\left|z_3 - \frac{z_1 + z_2}{2}\right|}{\left|z_1 - \frac{z_1 + z_2}{2}\right|} e^{-i\frac{\pi}{2}}$$

$$\Rightarrow \frac{2z_3 - z_1 - z_2}{z_1 - z_2} = \frac{CD}{AD} e^{-i\frac{\pi}{2}} \quad (1)$$

and

$$\frac{z_4 - \frac{z_1 + z_2}{2}}{z_2 - \frac{z_1 + z_2}{2}} = \frac{\left|z_4 - \frac{z_1 + z_2}{2}\right|}{\left|z_2 - \frac{z_1 + z_2}{2}\right|} e^{i\frac{\pi}{2}}$$

$$\Rightarrow \frac{2z_4 - z_1 - z_2}{z_2 - z_1} = \frac{ID}{BD} e^{i\frac{\pi}{2}} \quad (2)$$

Multiplying (1) and (2), we have

$$\frac{2z_3 - z_1 - z_2}{z_1 - z_2}\frac{2z_4 - z_1 - z_2}{z_2 - z_1} = \frac{CD}{AD}\frac{ID}{BD} = \tan\theta\tan\frac{\theta}{2}$$

$$\Rightarrow (z_2 - z_1)^2 \tan\theta\tan\frac{\theta}{2} = -(z_1 + z_2 - 2z_3)(z_1 + z_2 - 2z_4)$$

For Problems 30–32

30. d., 31. c., 32. c.

Sol.

30. $\angle BOD = 2\angle BAD = A$

$\angle COD = 2\angle CAD = A$

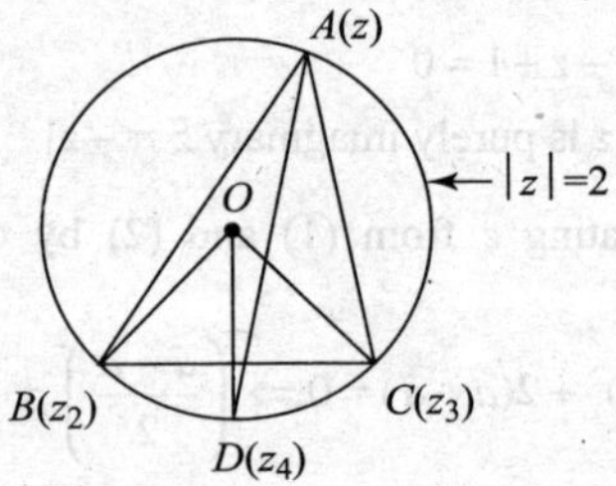

Fig. 2.38

$\frac{z_4}{z_2} = e^{iA}, \frac{z_3}{z_4} = e^{iA}$ [From rotation about the point O]

$\Rightarrow z_4^2 = z_2 z_3$

31. Clearly, OD bisects $\angle BAC$ of isosceles triangle BOC. Thus angle between segments OD and BC is $\pi/2$.

$$\therefore \arg\left(\frac{z_4}{z_2 - z_3}\right) = \frac{\pi}{2}$$

Matrix-Match Type

1. a → r, s; b → q, p; c → t; d → u, t.

a. If $ab > 0$, then either a, b both are positive or both a, b are negative. Hence, $z = a + ib$ lies in either first or third quadrant, then argument of z is $\tan^{-1} b/a$ or $-\pi + \tan^{-1} b/a$.

b. If $ab < 0$, then a and b have opposite signs, then z lies in either second or fourth quadrant, then argument of z is $-\tan^{-1}(b/a)$ or $\pi - \tan^{-1}(b/a)$.

c. If $a^2 + b^2 = 0$, then $a = b = 0$, so $z = 0 + i0$ whose argument is not defined.

d. If $ab = 0$, then either $a = 0$ or $b = 0$ or both are 0, then argument is 0 or $\pi/2$ or not defined.

2. a → s; b → r; c → p; d → q.

a. $z^4 - 1 = 0 \Rightarrow z^4 = 1 = \cos 0 + i\sin 0 \Rightarrow z = (\cos 0 + i\sin 0)^{1/4}$
$= \cos 0 + i\sin 0$

b. $z^4 + 1 = 0 \Rightarrow z^4 = -1 = \cos\pi + i\sin\pi \Rightarrow z = (\cos\pi + i\sin\pi)^{1/4}$
$= \cos\frac{\pi}{4} + i\sin\frac{\pi}{4}$

c. $iz^4 + 1 = 0 \Rightarrow z^4 = i = \cos\frac{\pi}{2} + i\sin\frac{\pi}{2}$

$$\Rightarrow z = \left(\cos\frac{\pi}{2} + i\sin\frac{\pi}{2}\right)^{1/4} = \cos\frac{\pi}{8} + i\sin\frac{\pi}{8}$$

d. $iz^4 - 1 = 0 \Rightarrow z^4 = -i = \cos\frac{\pi}{2} - i\sin\frac{\pi}{2}$

$$\Rightarrow z = \left(\cos\frac{\pi}{2} + i\sin\frac{\pi}{2}\right)^{1/4} = \cos\frac{\pi}{8} - i\sin\frac{\pi}{8}$$

3. a → q; b → s; c → p; d → r.

a. $|z - 1| = |z - i|$

Hence it lies on the perpendicular bisector of the line joining (1, 0) and (0, 1) which is a straight line passing through the origin.

b. $|z + \bar{z}| + |z - \bar{z}| = 2$

$\Rightarrow |x| + |y| = 1$

Hence, z lies on a square.

c. Let $z = x + iy$. Then,

$|z + \bar{z}| = |z - \bar{z}|$

$\Rightarrow |2x| = |2iy|$

$\Rightarrow |x| = |y|$

$\Rightarrow x = \pm y$

Hence, the locus of z is a pair of straight lines.

d. Let $Z = 2/z$. Then,

$$|Z| = \left|\frac{2}{z}\right| = \frac{2}{|z|} = \frac{2}{1} = 2$$

This shows that Z lies on a circle with centre at the origin and radius 2 units.

4. a → p, r; b → p, q, r, t; c → p, r, s; d → p, q, r, s, t.

Fig. 2.39

In a parallelogram, the midpoints of diagonals coincide

$$\frac{z_1 + z_3}{2} = \frac{z_2 + z_4}{2}$$

$\Rightarrow z_1 - z_4 = z_2 - z_3$

Also in a parallelogram, $AB \| CD$. Hence,

$$\arg\left(\frac{z_1 - z_2}{z_3 - z_4}\right) = 0$$

$\Rightarrow \frac{z_1 - z_2}{z_3 - z_4}$ is purely real

In a rectangle, adjacent sides are perpendicular. Hence,

$$\arg\left(\frac{z_1 - z_2}{z_3 - z_2}\right) = \frac{\pi}{2}$$

$\Rightarrow \frac{z_1 - z_2}{z_3 - z_2}$ is purely imaginary

Also in a rectangle,

$AC = BD \Rightarrow |z_1 - z_3| = |z_2 - z_4|$

In a rhombus,

$AC \perp BD \Rightarrow \frac{z_1 - z_3}{z_2 - z_4}$ is purely imaginary

5. a → p, q; b → p, q, r, s, t; c → r, s; d → p, q.

a. $|z - 2i| + |z - 7i| = k$ is ellipse if $k > |7i - 2i|$ or $k > 5$

b. $\left|\frac{2z-3}{3z-2}\right| = k \Rightarrow \left|\frac{z - \frac{3}{2}}{z - \frac{2}{3}}\right| = \frac{3k}{2} \Rightarrow 3k/2 > 1 \Rightarrow k > 2/3$

c. $|z - 3| - |z - 4i| = k$ is hyperbola, if $k < |3 - 4i| \Rightarrow 0 < k < 5$

d. $|z - (3 + 4i)| = \frac{k}{50}|a\bar{z} + \bar{a}z + b|$

$$\Rightarrow |z - (3 + 4i)| = \frac{k}{5}\frac{|a\bar{z} + \bar{a}z + b|}{2|3 + 4i|}$$

This is hyperbola if $k/5 > 1 \Rightarrow k > 5$.

6. a → s; b → q; c → p; d → r.

a. $x^2 - x + 1 = 0$

$$\Rightarrow x = \frac{1 \pm i\sqrt{3}}{2}$$

$= -\omega, -\omega^2$

$$\Rightarrow \left(x^n + \frac{1}{x^n}\right)^2 = (-1)^{2n}\left(\omega^n + \frac{1}{\omega^n}\right)^2$$

$$= (\omega^n + \omega^{2n})^2 \quad (1)$$

$$\because \frac{1}{\omega^n} = \frac{\omega^{2n}}{\omega^{3n}} = \omega^{2n}$$

Now,

$$1 + \omega^n + \omega^{2n} = \frac{1 - \omega^{3n}}{1 - \omega^n} = 0 \text{ for } n \neq 3p$$

$\therefore \omega^n + \omega^{2n} = -1$ for $n \neq 3p$

$= 2$ for $n = 3p$

$$\therefore \sum_{n=1}^{5}\left(x^n + \frac{1}{x^n}\right)^2 = 8$$

b. In the expression,

$$\left[\frac{1 + \cos\theta + i\sin\theta}{\sin\theta + i(1 + \cos\theta)}\right]^4$$

numerator is

$$1 + \cos\theta + i\sin\theta = 2\cos\frac{\theta}{2}\left[\cos\frac{\theta}{2} + i\sin\frac{\theta}{2}\right]$$

$$= 2\cos\frac{\theta}{2}\, e^{i\theta/2}$$

and denominator is

$-i^2\sin\theta + i(1 + \cos\theta) = i$ [conjugate of numerator]

$$= i\,2\cos\frac{\theta}{2}\, e^{-\theta/2}$$

$\therefore E = \left(\frac{N^r}{D^r}\right) = \left[\frac{1}{i}\frac{e^{i\theta/2}}{e^{-i\theta/2}}\right]^4 = \frac{1}{i^4}e^{4i\theta}$

$= \cos 4\theta + i \sin 4\theta$

$\therefore n = 4$

c. We know that if $z = re^{i\theta}$, then $\bar{z} = re^{-i\theta}$.

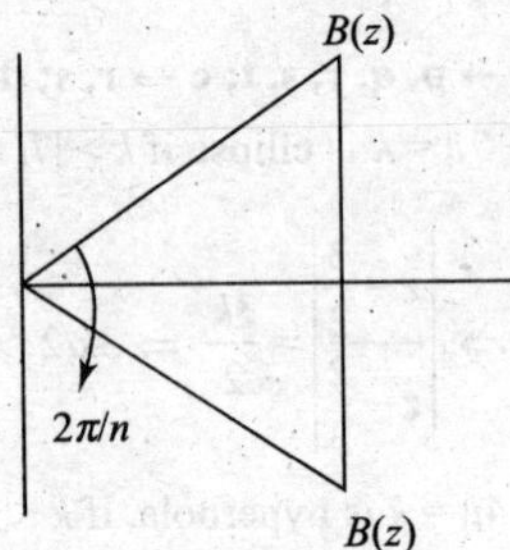

Fig. 2.40

$\therefore \frac{\text{Im}(z)}{\text{Re}(z)} = \frac{r\sin\theta}{r\cos\theta} = \tan\theta = \tan\frac{\pi}{n} = \sqrt{2}-1$

$\Rightarrow \tan\frac{\pi}{n} = \tan\frac{\pi}{8} \Rightarrow n = 8$

d. $\sum_{r=1}^{10}(r-\omega)(r-\omega^2) = \sum_{r=1}^{10}(r^2+r+1)$

$= \Sigma r^2 + \Sigma r + 10$

$= \frac{10\times11\times21}{6} + \frac{10\times11}{2} + 10$

$= 450$

$\Rightarrow \frac{1}{50}\left\{\sum_{r=1}^{10}(r-\omega)(r-\omega)^2\right\} = 9$

Integer Type

1. (4) $(x+iy)^{1/3} = a - ib$

$\Rightarrow x + iy = (a-ib)^3$

$= (a^3 - 3ab^2) + i(b^3 - 3a^2b)$

$\Rightarrow x = a^3 - 3ab^2, y = b^3 - 3a^2b$

$\Rightarrow \frac{x}{a} = a^2 - 3b^2$ and $\frac{y}{b} = b^2 - 3a^2$

$\Rightarrow \frac{x}{a} - \frac{y}{b} = a^2 - 3b^2 - b^2 + 3a^2$

$\Rightarrow \frac{x}{a} - \frac{y}{b} = 4(a^2-b^2) = k(a^2-b^2)$. Therefore,

$k = 4$

2. (1) Let $\cot^{-1} p = \theta$, then $\cot\theta = p$

Now L.H.S. $= e^{2mi\theta}\left(\frac{i\cot\theta+1}{i\cot\theta-1}\right)^m$

$= e^{2mi\theta}\left[\frac{i(\cot\theta - i)}{i(\cot\theta+i)}\right]^m$

$= e^{2mi\theta}\left[\frac{(\cot\theta-i)}{(\cot\theta+i)}\right]^m$

$= e^{2mi\theta}\left(\frac{\cos\theta - i\sin\theta}{\cos\theta+i\sin\theta}\right)^m$

$= e^{2mi\theta}\left(\frac{e^{-\theta}}{e^{i\theta}}\right)^m$

$= e^{2mi\theta}(e^{-2i\theta})^m$

$= e^{2mi\theta}e^{-2mi\theta} = e^0 = 1$

3. (4) $z = x + iy$

$\Rightarrow z^2 = x^2 - y^2 + 2ixy$

$\Rightarrow \text{Re}(z^2) = x^2 - y^2$

Also $|z| = \sqrt{x^2+y^2}$

$\Rightarrow x^2 - y^2 = 0, x^2 + y^2 = 3$

$\Rightarrow x^2 = y^2 = \frac{3}{2}$

$\Rightarrow x = \pm\sqrt{\frac{3}{2}}, y = \pm\sqrt{\frac{3}{2}} \Rightarrow z = \pm\sqrt{\frac{3}{2}} \pm \sqrt{\frac{3}{2}}i$

Thus there are four such complex numbers.

4. (1) Given $|\beta| = 1 \Rightarrow \beta\bar{\beta} = 1 \Rightarrow \beta = \frac{1}{\bar{\beta}}$ (1)

Now $\left|\frac{\beta-\alpha}{1-\bar{\alpha}\beta}\right| = \left|\frac{\frac{1}{\bar{\beta}}-\alpha}{1-\bar{\alpha}\beta}\right|$

$= \left|\frac{1}{\bar{\beta}}\right|\left|\frac{1-\alpha\bar{\beta}}{1-\bar{\alpha}\beta}\right|$

$= \frac{1}{|\bar{\beta}|}\left|\frac{1-\alpha\bar{\beta}}{1-\bar{\alpha}\beta}\right|$

$= \left|\frac{1-\bar{\alpha}\beta}{1-\bar{\alpha}\beta}\right| = 1$

5. (2) $|z_1| = 1 \Rightarrow z_1\bar{z}_1 = 1$; $|z_2| = 2 \Rightarrow z_2\bar{z}_2 = 4$; $|z_3| = 3$

$\Rightarrow z_3\bar{z}_3 = 9$

Also, $|9z_1z_2 + 4z_1z_3 + z_2z_3| = 12$

$\Rightarrow |z_1z_2z_3\bar{z}_3 + z_1z_2z_3\bar{z}_2 + z_1\bar{z}_1z_2z_3| = 12$

$\Rightarrow |z_1z_2z_3|\,|\bar{z}_1 + \bar{z}_2 + \bar{z}_3| = 12$

$\Rightarrow |z_1|\,|z_2|\,|z_3|\,|\bar{z}_1 + \bar{z}_2 + \bar{z}_3| = 12$

$\Rightarrow 6|\bar{z}_1 + \bar{z}_2 + \bar{z}_3| = 12$

$\Rightarrow |\bar{z}_1 + \bar{z}_2 + \bar{z}_3| = 2$

$\Rightarrow |z_1 + z_2 + z_3| = 2$

6. (6) $\left|\frac{\bar{z}_1 - 2\bar{z}_2}{2 - z_1\bar{z}_2}\right| = 1$

$\Rightarrow |\bar{z}_1 - 2\bar{z}_2|^2 = |2 - z_1\bar{z}_2|^2$

$\Rightarrow (\bar{z}_1 - 2\bar{z}_2)(\bar{\bar{z}}_1 - 2\bar{\bar{z}}_2) = (2 - z_1\bar{z}_2)(\overline{2 - z_1\bar{z}_2})$

$\Rightarrow (\bar{z}_1 - 2\bar{z}_2)(z_1 - 2z_2) = (2 - z_1\bar{z}_2)(2 - \bar{z}_1z_2)$

$\Rightarrow z_1\bar{z}_1 - 2\bar{z}_1z_2 - 2z_1\bar{z}_2 + 4z_2\bar{z}_2 = 4 - 2\bar{z}_1z_2 - 2z_1\bar{z}_2 + z_1\bar{z}_1 z_2\bar{z}_2$

$\Rightarrow |z_1|^2 + 4|z_2|^2 = 4 + |z_1|^2|z_2|^2$

$\Rightarrow |z_1|^2 - |z_1|^2|z_2|^2 + 4|z_2|^2 - 4 = 0$

$\Rightarrow |z_1|^2(1 - |z_2|^2) + 4(|z_2|^2 - 1) = 0$

$\Rightarrow (|z_2|^2 - 1)(|z_1|^2 - 4) = 0$

$\Rightarrow |z_1| = 2$ (as $|z_2| \neq 1$)

7. $z^3 + \bar{\omega}^7 = 0$

$\Rightarrow z^3 = -\bar{\omega}^7$

$\Rightarrow |z|^3 = |\bar{\omega}|^7 = |\omega|^7$

$\Rightarrow |z|^{15} = |\omega|^{35}$ (1)

Again $z^5 \times \omega^{11} = 1$

$\Rightarrow |z|^5 \times |\omega|^{11} = 1$

$\Rightarrow |z|^{15} \times |\omega|^{33} = 1$ (2)

From (1) and (2) $\Rightarrow |z| = |\omega| = 1$

Again $-(\bar{\omega})^{35} = \dfrac{1}{\omega^{33}}$

$\Rightarrow -(\bar{\omega})^2 = 1$

$\Rightarrow (\omega)^2 = -1 = i^2$

$\Rightarrow \omega = -i$ or i

8. **(6)** $|z_1 + z_2 + z_3| = |(z_1 - 1) + (z_2 - 2) + (z_3 - 3) + 6| \le |z_1 - 1| + |z_2 - 2| + |z_3 - 3| + 6 \le 1 + 2 + 3 + 6 = 12$

9. **(6)** $|z + 1| = |z + 4 - 3|$

$= |(z + 4) + (-3)|$

$\le |z + 4| + |-3|$

$= |z + 4| + 3$

$\le 3 + 3 = 6$ $[\because |z + 4| \le 3]$

$\Rightarrow$ The greatest value of $|z + 1| = 6$

10. **(6)** $|z_1 + z_2| \le |z_1| + |z_2|$

$= |24 + 7i| + 6$

$= 25 + 6$

$= 31$

Also,

$|z_1 + z_2| = |z_1 - (-z_2)| \ge ||z_1| - |z_2||$

$\Rightarrow |z_1 + z_2| \ge |25 - 6|$

$= 19$

Hence, the least value of $|z_1 + z_2|$ is 19 and greatest value is 25

11. **(0)** Let $z = x + iy$, then

$\arg(z - 3i) = \arg(x + iy - 3i) = \dfrac{3\pi}{4}$

$\Rightarrow x < 0, y - 3 > 0$ $\left(\because \dfrac{3\pi}{4} \text{ is in II quadrant}\right)$

and $\dfrac{y-3}{x} = \tan\dfrac{3\pi}{4} = -1$

$\Rightarrow y = -x + 3 \;\forall\, x < 0$ and $y > 3$ (1)

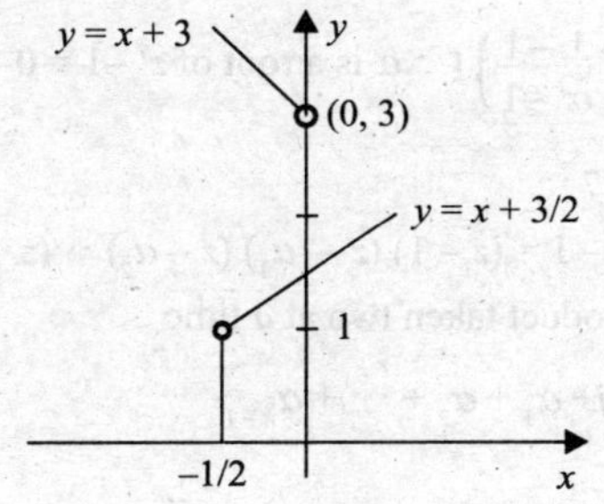

Fig. 2.41

and $\arg(2z + 1 - 2i) = \arg[(2x + 1) + i(2y - 2)] = \dfrac{\pi}{4}$

$\Rightarrow 2x + 1 > 0, 2y - 2 > 0$ $\left(\because \dfrac{\pi}{4} \text{ is in I quadrant}\right)$

and $\dfrac{2y-2}{2x+1} = \tan\dfrac{\pi}{4} = 1$

$\Rightarrow 2y - 2 = 2x + 1$

$\Rightarrow y = x + \dfrac{3}{2} \;\forall\, x > -\dfrac{1}{2}, y > 1$ (2)

From equations (1) and (2), we get the graph.

It is clear from the graph that the two lines do not intersect.

Therefore, there is no point of intersection

12. **(5)** $(z^3 + 3)^2 = 16i^2$

$z^3 + 3 = 4i$ or $-4i$

$\Rightarrow z^3 = -3 + 4i$ or $-3 - 4i$

$\Rightarrow |z^3| = |-3 + 4i| = 5$

$\Rightarrow |z| = 5^{1/3}$

13. **(7)** $\dfrac{1 + z + z^2}{1 - z + z^2}$

$= \dfrac{1 - z + z^2 + 2z}{1 - z + z^2}$

$= 1 + \dfrac{2}{z - 1\,(1/z)}$

Hence, $z + \dfrac{1}{z} \in R$;

$\Rightarrow z + \dfrac{1}{z} = \bar{z} + \dfrac{1}{\bar{z}}$

$\Rightarrow (z - \bar{z}) = \dfrac{1}{\bar{z}} - \dfrac{1}{z} = \dfrac{z - \bar{z}}{|z|^2}$

$\Rightarrow (z - \bar{z})(|z|^2 - 1) = 0$

$\Rightarrow |z| = 1$; ($z = \bar{z}$ is not possible as z is not real)

14. **(1)** $z\bar{z} = 1 \Rightarrow z = \dfrac{1}{\bar{z}}$

Now $z = (z_1 + z_2 + z_3 + \cdots + z_n)\left(\dfrac{1}{z_1} + \dfrac{1}{z_2} + \cdots + \dfrac{1}{z_n}\right)$

$= (z_1 + z_2 + \cdots + z_n)(\bar{z}_1 + \bar{z}_2 + \cdots + \bar{z}_3)$

$= (z_1 + z_2 + \cdots + z_n)\overline{(z_1 + z_2 + \cdots + z_n)}$

$= |z_1 + z_2 + \cdots + z_n|^2$, which is real.

$|z_1 + z_2 + \cdots + z_n|^2 \le |z_1| + |z_2| + \cdots + |z_n| = n^2$

Therefore $0 < z \le n^2$

$\dfrac{\sqrt{5} - 1}{2} \le |z| \le \dfrac{\sqrt{5} + 1}{2}$

15. **(1)** As given, let $\dfrac{2z_1}{3z_2} = iy$ or $\dfrac{z_1}{z_2} = \dfrac{3}{2}iy$, so that

$$\left|\frac{z_1 - z_2}{z_1 z_2}\right| = \left|\frac{\frac{z_1}{z_2} - 1}{\frac{z_1}{z_2} + 1}\right| = \left|\frac{\frac{3}{2}iy - 1}{\frac{3}{2}iy + 1}\right| = \left|\frac{1 - \frac{3}{2}iy}{1 + \frac{3}{2}iy}\right| = 1$$

$(\because |z| = |\bar{z}|)$

16. **(1)** $\left(\cos\dfrac{\pi}{2} + i\sin\dfrac{\pi}{2}\right)\left(\cos\dfrac{\pi}{2^2} + i\sin\dfrac{\pi}{2^2}\right) \cdots$ to ∞

$= \cos\left(\dfrac{\pi}{2} + \dfrac{\pi}{2^2} + \cdots\right) + i\sin\left(\dfrac{\pi}{2} + \dfrac{\pi}{2^2} + \cdots\right)$

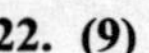

$$= \cos\left[\frac{\pi}{2}\left(1+\frac{1}{2}+\frac{1}{2^2}+\cdots\right)\right] + i\sin\left[\frac{\pi}{2}\left(1+\frac{1}{2}+\frac{1}{2^2}+\cdots\right)\right]$$

$$= \cos\left[\frac{\pi}{2}\left(\frac{1}{1-\frac{1}{2}}\right)\right] + i\sin\left[\frac{\pi}{2}\left(\frac{1}{1-\frac{1}{2}}\right)\right]$$

$$= \cos\pi + i\sin\pi = -1$$

17. (4) $x^{12} - 1 = (x^6+1)(x^6-1) = (x^6+1)(x^2-1)(x^4+x^2+1)$
Common roots are given by $x^4+x^2+1=0$. Therefore,

$$x^2 = \frac{-1 \pm i\sqrt{3}}{2} = \omega, \omega^2 \text{ or } \omega^4, \omega^2 \qquad (\because \omega^3 = 1)$$

or $x = \pm\omega^2, \pm\omega$

18. (5) $z_1 + z_2 = -1, z_1 z_2 = \frac{b}{3}$
Triangle OAB is equilateral. so we have

$0^2 + z_1^2 + z_2^2 = 0 \times z_1 + 0 \times z_2 + z_1 z_2$

$\Rightarrow (z_1+z_2)^2 - 2z_1z_2 = z_1z_2$

$\Rightarrow 1 = 3z_1z_2 = 3\frac{b}{3}$

$\Rightarrow b = 1$

19. (3) Let $z = r(\cos\theta + i\sin\theta)$.
Now $r = \text{OA}\sin\theta = 6\sin\theta$.

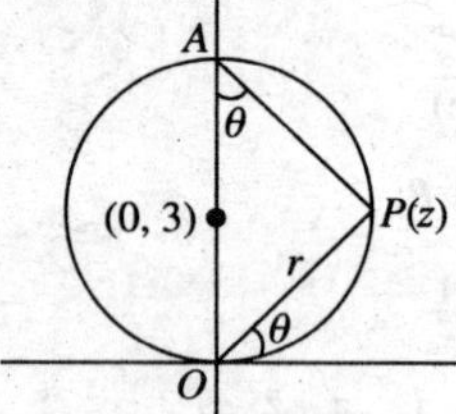

Fig. 2.42

$z = 6\sin\theta(\cos\theta + i\sin\theta)$

$$\Rightarrow \frac{6}{z} = \frac{1}{\sin\theta(\cos\theta + i\sin\theta)}$$

$$= \frac{\cos\theta - i\sin\theta}{\sin\theta}$$

$= -i + \cot\theta$

$\Rightarrow \cot\theta - \frac{6}{z} = \text{i}$

20. (4)

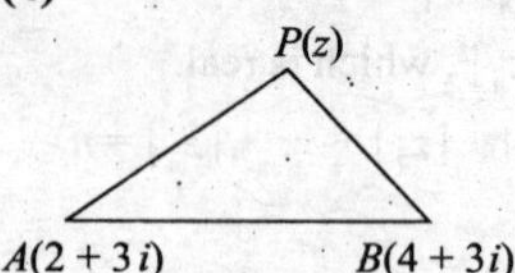

Fig. 2.43

If P(z) lies on a circle and $PB^2 + PA^2 = \lambda$ (constant) (for least radius, A and B are end points of diameter), then $\lambda = AB^2 \Rightarrow \lambda = 4$

21. (2) The given equations represent circles.

$(x-1)^2 + (y-1)^2 = 2$ and $(x+1)^2 + (y+1)^2 = 4$

$\Rightarrow C_1(1, 1), r_1 = \sqrt{2}, C_2(-1, -1), r_2 = 2$

$C_1C_2 = \sqrt{8} = 2\sqrt{2} = 2.8, r_1 + r_2 = 2 + \sqrt{2} = 3.41$

since $C_1C_2 < r_1 + r_2$ and $C_1C_2 > r_2 - r_1$, the two circles intersect at two points. The two common points will satisfy both.

22. (9)

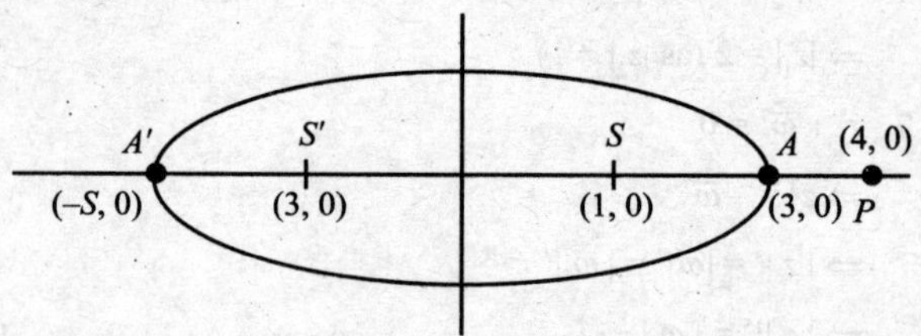

Fig. 2.44

Given $|z-1| + |z+3| \le 8$. Then z lies inside or on the ellipse whose foci an (1, 0) and (–3, 0) and vertices are (–5, 0) and (3, 0). Clearly the minimum and maximum values of $|z-4|$ are 1 and 9, respectively, representing the distances PA and PA.

Thus $1 \le |z-4| \le 9$.

23. (3) Here $\omega^5 = 1$ Therefore, $\omega^{-1} = \omega^4$
Also $1 + \omega + \omega^2 + \omega^3 + \omega^4 = 0$, Therefore,

$\log_{1/8}\left|1 + \omega + \omega^2 + \omega^3 - \frac{1}{\omega}\right| = \log_{1/8}|1 + \omega + \omega^2 + \omega^3 - \omega^4|$
$= \log_{1/8}|-2\omega^4| = \log_{1/8} 2 = 3$

24. (9) $z^9 - 1 = (z-1)(z-z_1)(z-z_2)\cdots(z-z_8)$

$\Rightarrow \frac{z^9-1}{z-1} = (z-z_1)(z-z_2)\cdots(z-z_8)$

$\Rightarrow 1 + z + z^2 + \cdots z^8 = (z-z_1)(z-z_2)\cdots(z-z_8)$

Putting $z = 1$, we get

$(1-z_1)(1-z_2)\cdots(1-z_8) = 1 + 1 \cdots + 1 = 9$

25. (7) $\alpha = e^{\frac{i2\pi}{7}}$ = seventh root of unity or root of the equation $z^7 - 1 = 0$

Its other roots are 1, $e^{\frac{i2\pi}{7}} = \alpha^2, \alpha^3, \alpha^4, \alpha^5, \alpha^6$

Now $f(x) + f(\alpha x) + \cdots + f(\alpha^6 x)$

$= f(x) + f(\alpha x) + \cdots + f(\alpha^6 x)$

$$= 7 + \sum_{k=1}^{20} a_k x^k + \sum_{k=1}^{20} a_k(\alpha x)^k + \cdots + \sum_{k=1}^{20} a_k(\alpha^6 x)^k$$

$$= 7 + \sum_{k=1}^{20} a_k(x^k + \alpha^k x^k + \alpha^{2k}x^k + \cdots + \alpha^6 x^k)$$

$$= 7 + \sum_{k=1}^{20} a_k(x^k + \alpha^k x^k + a^{2k}x^k + \cdots + a^{6k}x^k)$$

$$= 7 + \sum_{k=1}^{20} a_k\left(x^k\frac{(\alpha^k)^7 - 1}{\alpha^k - 1}\right)$$

$$= 7 + \sum_{k=1}^{20} a_k\left(x^k\frac{(\alpha^7)^k - 1}{\alpha^k - 1}\right)$$

$$= 7 + \sum_{k=1}^{20} a_k\left(x^k\frac{1-1}{\alpha^k - 1}\right) (\because \alpha \text{ is a root of } z^7 - 1 = 0$$

$\Rightarrow \alpha^7 - 1 = 0) = 7$

26. (1) We have $z^n - 1 = (z-1)(z-\alpha_1)(z-\alpha_2)\cdots(z-\alpha_{n-1})$
Now sum of product taken two at a time

$$S = \sum_{1 \le i \le j \le n-1} \alpha i \alpha j + \alpha_1 + \alpha_2 + \cdots + \alpha_{n-1}$$

Now $S = 0$ as the coefficient of z^{n-1} is zero and $\alpha_1 + \alpha_2 + \cdots + \alpha_{n-1} = -1$

$$\Rightarrow 0 = \sum_{1 \le i \le j \le n-1} \alpha i \alpha j - 1$$

$$\Rightarrow \sum_{1 \le i \le j \le n-1} \alpha i \alpha j = 1$$

27. (0) As α is the 5th non–real root of unity,

$\alpha^4 + \alpha^3 + \alpha^2 + \alpha + 1 = 0$

β is the 4th non-real root of unity,

$\beta^3 + \beta^2 + \beta + 1 = 0$

Now

$(1 + \alpha)(1 + \alpha^2)(1 + \alpha^4)(1 + \beta)(1 + \beta^2)(1 + \beta^3)$

$= (1 + \alpha + \alpha^2 + \alpha^3)(1 + \alpha^4)(1 + \beta + \beta^2 + \beta^3)(1 + \beta^3) = 0$

28. (4) $x^6 = -64$

$\Rightarrow \left(\frac{x}{2i}\right)^6 = 1$

$\Rightarrow \frac{x}{2i} = \cos\left(\frac{2n\pi}{6}\right) + i \sin\left(\frac{2n\pi}{6}\right)$, where $n = 0, 1, 2, 3, 4, 5$

$\Rightarrow x = 2i \cos\left(\frac{n\pi}{3}\right) - 2 \sin\left(\frac{n\pi}{3}\right)$, where $n = 0, 1, 2, 3, 4, 5$

For $n = 4$ and 5, we have positive real part

$\Rightarrow$ required product

$$= \left[2i \cos\left(\frac{4\pi}{3}\right) - 2 \sin\left(\frac{4\pi}{3}\right)\right] \left[2i \cos\left(\frac{5\pi}{3}\right) - 2 \sin\left(\frac{5\pi}{3}\right)\right]$$

$$= 4\left[-i\frac{1}{2} + \frac{\sqrt{3}}{2}\right]\left[i\frac{1}{2} + \frac{\sqrt{3}}{2}\right] = 4$$

29. (5) $E = \frac{1}{z_1 - 1} + \frac{1}{z_2 - 1} + \cdots + \frac{1}{z_{50} - 1}$, where $z_1, z_2, \cdots, z_{50}$ are the roots of the equation $z^{51} - 1 = 0$ other than 1.

Let $1 + z + z^2 + \cdots + z^{50} = (z - z_1)(z - z_2)(z - z_{50})$

$\Rightarrow \log(1 + z + z^2 + \cdots + z^{50}) = \log[(z - z_1)(z - z_2)(z - z_{50})]$

Differentiate both sides w.r.t. z and put $z = 1$

$$\Rightarrow \frac{1 + 2z + 3z^2 + \cdots + 50z^{49}}{1 + z + z^2 + \cdots + z^{50}}$$

$$= \frac{1}{z - z_1} + \frac{1}{z - z_2} + \cdots + \frac{1}{z - z_{50}}$$

$$\Rightarrow \frac{50 \times 51}{2 \times 51} = -\left[\frac{1}{z_1 - 1} + \frac{1}{z_2 - 1} + \cdots + \frac{1}{z_{50} - 1}\right].$$ Therefore,

$$\therefore \sum \frac{1}{z_r - 1} = -25$$

30. (5)

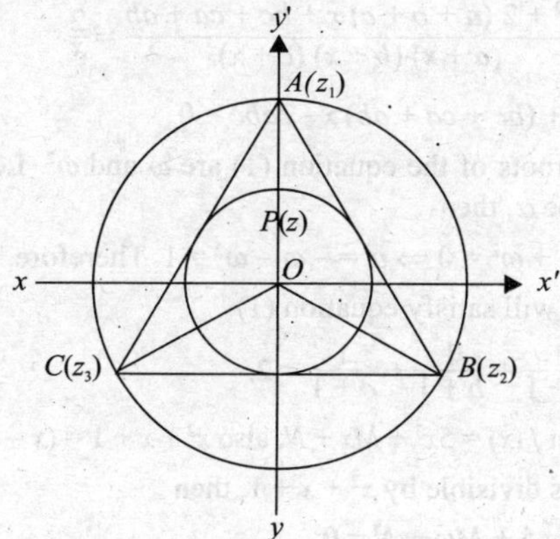

Fig. 2.45

Let $z_1 = \frac{2}{\sqrt{3}} e^{-i\frac{\pi}{2}}, z_2 = \frac{2}{\sqrt{3}} e^{-i\frac{\pi}{6}}, z_3 = \frac{2}{\sqrt{3}} e^{-i\frac{5}{6}\pi}.$

Clearly, the points lie on the circle $|z| = \frac{2}{\sqrt{3}}$

And Δ ABC is equilateral.

$\Rightarrow$ Centroid coincides with circumcentre

$\Rightarrow z_1 + z_2 + z_3 = 0$ and $\bar{z}_1 + \bar{z}_2 + \bar{z}_3 = 0$

Since the triangle is equilateral, inradius $r = ID = \frac{1}{\sqrt{3}}$

$\Rightarrow$ Equation of the in circle is $|z| = \frac{1}{\sqrt{3}}$

Let $P(z)$ be any point on the incircle

Now, $AP^2 = |z - z_1|^2 = |z|^2 + |z_1|^2 - (\bar{z}z_1 + \bar{z}z_1)$

Similarly, $BP^2 = |z|^2 + |z_2|^2 - (z\bar{z}_2 + \bar{z}z_2)$

and $CP^2 = |z|^2 + |z_3|^2 - (z\bar{z}_3 + \bar{z}z_3)$

$\Rightarrow AP^2 + BP^2 + CP^2$

$= 3|z|^2 + |z_1|^2 + |z_2|^2 + |z_3|^2 - z(\bar{z}_1 + \bar{z}_2 + \bar{z}_3) - \bar{z}(z_1 + z_2 + z_3)$

$= 3 \times \frac{4}{3} + \frac{1}{3} + \frac{1}{3} + \frac{1}{3} - z(0) - \bar{z}(0)$

$= 5$

31. (8)

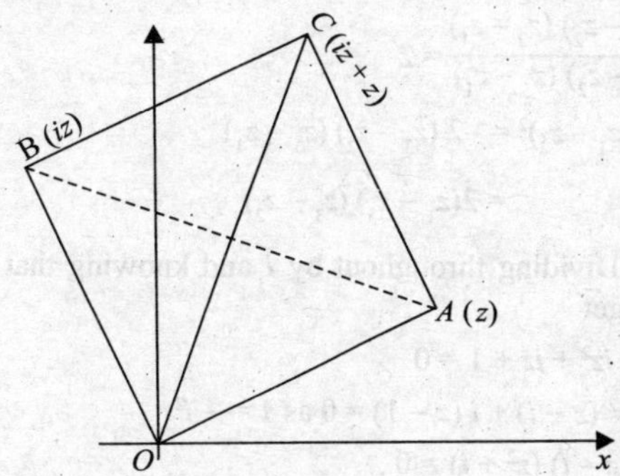

Fig. 2.46

Let $A(z)$, $B(iz)$, $C(z + iz)$

We know that iz is obtained by rotating OA through an angle of 90°. Also point $z + iz$ can be obtained by completing the parallelogram, two of whose adjacent sides are OA and OB.

From the Argand diagram, it is clear that

$Ar(\Delta ABC)$

$= Ar(\Delta OAB)$

$= \frac{1}{2} \times OA \times OB$ [$\because$ *rt* $\angle$*ed* at point *O*]

$= \frac{1}{2} |z| \times |iz|$

$= \frac{1}{2} |z|^2$

32. $\arg\left(\frac{z - z_1}{z - z_2}\right) = \frac{\pi}{4}$

$\Rightarrow$ Locus of z is the major arc whose centre is z_0, say.

Applying rotation at z_0, we have

$$\frac{z_0 - (10 + 6i)}{z_0 - (4 + 6i)} = \frac{|z_0 - (10 + 6i)|}{|z_0 - (4 + 6i)|} e^{i\frac{\pi}{2}}$$

$$\Rightarrow \frac{z_0 - (10 + 6i)}{z_0 - (4 + 6i)} = i$$

$\Rightarrow z_0 - 10 - 6i = iz_0 - 4i + 6$

$\Rightarrow z_0 = 7 + 9i$

Thus the centre is $7 + 9i$ and z is any point on the arc.

Hence, $|z - (7 + 9i)| = |10 + 6i - (7 + 9i)| = 3\sqrt{2}$.

33. (3)

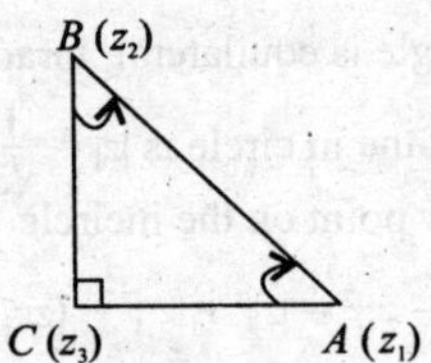

Fig. 2.47

Applying rotation about point C, $\frac{z_2 - z_3}{z_1 - z_3} e^{i\pi/2}$ (1)

Applying rotation about point B, $\frac{z_1 - z_2}{z_3 - z_2} = \sqrt{2}e^{i\pi/4}$ (2)

Applying rotation about point A, $\frac{z_2 - z_1}{z_3 - z_1} = \sqrt{2}e^{-i\pi/4}$ (3)

Multiplying (2) and (3), we get

$$\frac{(z_1 - z_2)(z_2 - z_1)}{(z_3 - z_2)(z_3 - z_1)} = 2$$

$$\Rightarrow (z_1 - z_2)^2 = -2(z_3 - z_2)(z_3 - z_1)$$

$$= 2(z_1 - z_3)(z_3 - z_2)$$

34. (7) Dividing throughout by i and knowing that $\frac{1}{i} = -i$, we get

$z^3 - iz^2 + iz + 1 = 0$

$\Rightarrow z^2(z - i) + i(z - 1) = 0$ as $1 = -i^2$

$\Rightarrow (z - i)(z^2 + i) = 0$

$\Rightarrow z = i$ or $z^2 = -i$

$\Rightarrow |z| = |i| = 1$ or $|z^2| = |z|^2 = |-i| = 1$

$\Rightarrow |z| = 1$

Hence in either case $|z| = 1$

35. (2) $|z_1| = 12$

$\Rightarrow z_1$ lies on a circle with centre (0, 0) and radius 12 units and $|z_2 - 3 - 4i| = 5$

$\Rightarrow z_2$ lies on a circle with centre (3, 4) and radius 5 units.

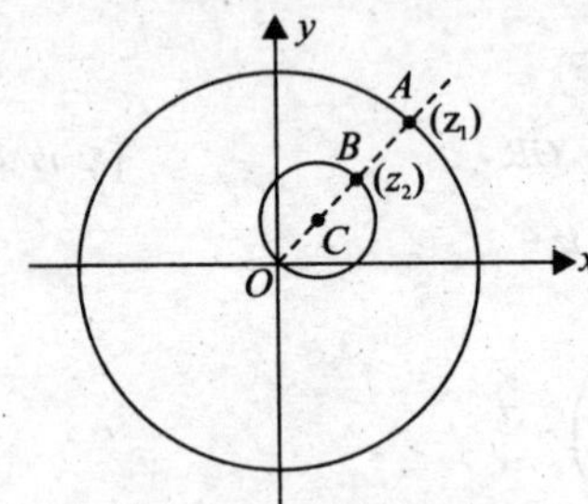

Fig. 2.48

From the figure it is clear that $|z_1 - z_2|$, i.e., distance between z_1 and z_2 will be minimum when they lie at A and B respectively, i.e., on the diagram as shown.

Then $|z_1 - z_2| = AB = OA - OB = 12 - 2(5) = 2$.

As above is the minimum value, we must have $|z_1 - z_2| \geq 2$

36. (3) $(1 + \omega^2)^n = (1 + \omega^4)^n$

$\Rightarrow (-\omega)^n = (1 + \omega)^n = (-\omega^2)^n$

$\Rightarrow \omega^n = 1$

$\Rightarrow$ least positive value of n is 3

37. (8) Given that a, b, c are integers not all equal, ω is cube root of unity $\neq 1$, then

$|a + b\omega + c\omega^2|$

$= a + b\left(\frac{-1 + i\sqrt{3}}{2}\right) + c\left(\frac{-1 - i\sqrt{3}}{2}\right)$

$= \left(\frac{2a - b - c}{2}\right) + i\left(\frac{b\sqrt{3} - c\sqrt{3}}{2}\right)$

$= \frac{1}{2}\sqrt{(2a - b - c)^2 + 3(b - c)^2}$

$= \frac{1}{2}\sqrt{4a^2 + b^2 + c^2 - 4ab + 2ab - 4ac + 3b^2 + 3c^2 - 6bc}$

$= \sqrt{a^2 + b^2 + c^2 - ab - bc - ca}$

$= \sqrt{\frac{1}{2}[(a - b)^2 + (b - c)^2 + (c - a)^2]}$

R.H.S. will be minimum when $a = b = c$, but we cannot take $a = b = c$ as per question.

Therefore the minimum value is obtained when any two are zero and the third is a minimum magnitude integer, i.e., 1.

Thus $b = c = 0$, $a = 1$ gives us the minimum value 1.

38. (8) $z\bar{z}(\bar{z}^2 + z^2) = 350$

Put $z = x + iy$

$(x^2 + y^2)(x^2 - y^2) = 175$

$(x^2 + y^2)(x^2 - y^2) = 5 \times 5 \times 7$

$x^2 + y^2 = 25$

$x^2 - y^2 = 7$

$x = \pm 4, y = \pm 3$

$x, y \in$ I

Area $= 8 \times 6 = 48$ sq. unit.

39. (2) The given relations can be rewritten as $\frac{1}{a + \omega} + \frac{1}{b + \omega} + \frac{1}{c + \omega} = \frac{2}{\omega}$ and $\frac{1}{a + \omega^2} + \frac{1}{b + \omega^2} + \frac{1}{c + \omega^2} = \frac{2}{\omega^2}$.

Therefore,

ω and ω^2 are the roots of $\frac{1}{a + x} + \frac{1}{b + x} + \frac{1}{c + x} = \frac{2}{x}$

$$\Rightarrow \frac{3x^2 + 2(a + b + c)x + bc + ca + ab}{(a + x)(b + x)(c + x)} = \frac{2}{x}$$

$\Rightarrow x^3 + (bc + ca + ab)x - 2abc = 0$ (1)

Two roots of the equation (1) are ω and ω^2. Let the third root be α, then

$\alpha + \omega + \omega^2 = 0 \Rightarrow \alpha = -\omega - \omega^2 = 1$. Therefore,

$\alpha = 1$ will satisfy equation (1)

$\Rightarrow \frac{1}{a + 1} + \frac{1}{b + 1} + \frac{1}{c + 1} = 2.$

40. (5) Let $f(x) = 5x^3 + Mx + N$, also $x^2 + x + 1 = (x - \omega)(x - \omega^2)$

$f(x)$ is divisible by $x^2 + x + 1$, then

$f(\omega) = 5 + M\omega + N = 0$

and $f(\omega^2) = 5 + M\omega^2 + N = 0$

$\Rightarrow M = 0; N = -5$

$\Rightarrow M + N = -5$

41. (5) $Z_1 = (8 \sin\theta + 7 \cos\theta) + i(\sin\theta + 4\cos\theta)$

$Z_2 = (\sin\theta + 4\cos\theta) + i(8\sin\theta + 4\cos\theta)$

Hence $\left.\begin{array}{l} Z_1 = x + iy \\ Z_2 = y + ix \end{array}\right]$, where $x = (8 \sin\theta + 7\cos\theta)$ and $y = (\sin\theta + 4\cos\theta)$

$Z_1Z_2 = (xy - xy) + i(x^2 + y^2) = 0 + i(x^2 + y^2) = a + ib$
$\Rightarrow a = 0; b = x^2 + y^2$

Now, $x^2 + y^2 = (8\sin\theta + 7\cos\theta)^2 + (\sin\theta + 4\cos\theta)^2$

$= 65\sin^2\theta + 65\cos^2\theta + 120\sin\theta\cos\theta$

$= 65 + 60\sin 2\theta$

Hence $Z_1\,Z_2|_{\max} = 125$

42. (2) $\bar{z} = -z$

$z\bar{z} - 4zi = z^2$

$\Rightarrow \bar{z} - 4i = z \Rightarrow 2z = -4i$

$z = -2i$

Hence, $|z| = 2$

43. (3) Given $|z_1| = |z_2| = |z_3| = 1 \Rightarrow z_1 = \frac{1}{\bar{z}_1}$

Also, $\frac{z_1^2}{z_2z_3} + \frac{z_2^2}{z_1z_3} + \frac{z_3^2}{z_1z_2} + 1 = 0$

$\Rightarrow (z_1)^3 + (z_2)^3 + (z_3)^3 - 3z_1z_2z_3 = -4z_1z_2z_3$

$(z_1 + z_2 + z_3)[(z_1)^2 + (z_2)^2 + (z_3)^2 - \Sigma z_1z_2] = -4z_1z_2z_3$

$\Sigma z_1[(\Sigma z_1)^2 - \Sigma z_1z_2]$

$= -4z_1z_2z_3$

Let $z_1 + z_2 + z_3 = z$

$\Rightarrow z[z^2 - 3\Sigma z_1z_2] = -4z_1z_2z_3$

$z^3 = 3z\Sigma z_1z_2 - 4z_1z_2z_3$

$z^3 = z_1z_2z_3\left[3z\left(\frac{1}{z_1} + \frac{1}{z_2} + \frac{1}{z_3}\right)^4\right]$

$= z_1z_2z_3\,[3z(\bar{z}_1 + \bar{z}_2 + \bar{z}_3) - 4]$

$\Rightarrow z^3 = z_1z_2z^3\,[3\,|z|^2 - 4]$ (1)

Now, if $|z| \geq 2/\sqrt{3}$, then

$|z|^3 = 3\,|z^2| - 4 \Rightarrow |z|^3 - 3\,|z^2| + 4 = 0$

$\Rightarrow (|z| - 2)(|z|^2 - |z| - 2) = 0$

$\Rightarrow (|z| - 2)(|z| - 2)(|z| + 1) = 0 \Rightarrow |z| = 2$ or $|z| = -1$ (rejected)

Now, if $0 < |z| < 2/\sqrt{3}$, then Eq. (1) becomes

$|z|^3 = 4 - 3\,|z|^2 \Rightarrow |z|^3 + 3\,|z|^2 - 4 = 0$

$\Rightarrow (|z| - 1)(|z| + 2)^2 = 0 \Rightarrow |z| = +1$ or $|z| = -2$ (rejected)

Hence $|z| = \{1, 2\}$, where $|z| = z_1 + z_2 + z_3$

44. (9) Let $z = a + ib \Rightarrow \bar{z} = a - ib$

Hence, we have $z^7 = \bar{z}$

$\therefore |z|^7 = |\bar{z}| = |z|$

$\Rightarrow |z|\,[|z|^6 - 1] = 0$

$\Rightarrow |z| = 0$ or $|z| = 1$

If $|z| = 0, z = 0$

If $|z| = 1; z^8 = z\bar{z} = |z|^2 = 1$

$\Rightarrow$ 8 values of z Total number of ordered pairs = 9

45. (1) Let $z_1 = re^{i2x}$

$z_2 = re^{i2y}$

$\frac{r(z_1 + z_2)}{r^2 + z_1z_2} = \frac{\cos(x - y)}{\cos(x + y)}$

$\frac{r(z_1 + z_2)}{r^2 + z_1z_2} = \frac{\sin(y - x)}{\sin(y + x)}\left(\frac{r(z_1 + z_2)}{r^2 + z_1z_2}\right)^2 + \left(\frac{r(z_1 - z_2)}{r^2 - z_1z_2}\right)^2$

$= \frac{\cos^2(x - y)}{\cos^2(x + y)} + \frac{\sin^2(x - y)}{\sin^2(x + y)} \geq \cos^2(x - y) + \sin^2(x - y) = 1$

CHAPTER 3

Progression and Series

Objective Type

Each question has four choices a, b, c and d, out of which *only one* answer is correct.

1. If $\log_2(5 \times 2^x + 1)$, $\log_4(2^{1-x} + 1)$, and 1 are in A.P., then x equals
   a. $\log_2 5$ **b.** $1 - \log_5 2$
   c. $\log_5 2$ **d.** none of these

2. If three positive real numbers a, b, c are in A.P. such that $abc = 4$, then the minimum value of b is
   a. $2^{1/3}$ **b.** $2^{2/3}$
   c. $2^{1/2}$ **d.** $2^{3/2}$

3. The maximum sum of the series $20 + 19\frac{1}{3} + 18\frac{2}{3} + \cdots$ is
   a. 310 **b.** 300
   c. 320 **d.** none of these

4. The largest term common to the sequences 1, 11, 21, 31, … to 100 terms and 31, 36, 41, 46, … to 100 terms is
   a. 381 **b.** 471
   c. 281 **d.** none of these

5. If the sum of m terms of an A.P. is the same as the sum of its n terms, then the sum of its $(m + n)$ terms is
   a. mn **b.** $-mn$
   c. $1/mn$ **d.** 0

6. The ratio of the sum to n terms of two A.P.'s is $(7n + 1)$: $(4n + 27)$. Then the ratio of their m^{th} terms is
   a. $(14m + 6):(8m - 23)$ **b.** $(14m - 6):(8m - 23)$
   c. $(14m - 6):(8m + 23)$ **d.** $(14m + 6):(8m + 23)$

7. If the ratio of the sum to n terms of two A.P.'s is $(5n + 3)$: $(3n + 4)$, then the ratio of their 17th terms is
   a. 172:99 **b.** 168:103
   c. 175:99 **d.** 171:103

8. 150 workers were engaged to finish a piece of work in a certain number of days. Four workers stopped working on the second day, four more workers stopped their work on the third day and so on. It took 8 more days to finish the work. Then the number of days in which the work was completed is
   a. 29 days **b.** 24 days
   c. 25 days **d.** none of these

9. In an A.P. of which a is the first term, if the sum of the first p terms is zero, then the sum of the next q terms is
   a. $-\frac{a(p+q)p}{q+1}$ **b.** $\frac{a(p+q)p}{p+1}$
   c. $-\frac{a(p+q)q}{p-1}$ **d.** none of these

10. If S_n denotes the sum of first n terms of an A.P. and $\frac{S_{3n} - S_{n-1}}{S_{2n} - S_{2n-1}} = 31$, then the value of n is
   a. 21 **b.** 15
   c. 16 **d.** 19

11. If a, b, and c are in A.P. then $a^3 + c^3 - 8b^3$ is equal to
   a. 2 abc **b.** 6 abc
   c. 4 abc **d.** none of these

12. The number of terms of an A.P. is even; the sum of the odd terms is 24 and of the even terms is 30; and the last term exceeds the first by 10/2, then the number of terms in the series is

a. 3 b. 4

c. 6 d. 5

13. If a, $\frac{1}{b}$, c and $\frac{1}{p}$, q, $\frac{1}{r}$ form two A.P.'s of the same common difference, then a, q, c are in A.P. if

a. p, b, r are in A.P. b. $\frac{1}{p}, \frac{1}{b}, \frac{1}{r}$ are in A.P.

c. p, b, r are in G.P. d. none of these

14. Suppose that $F(n + 1) = \frac{2F(n)+1}{2}$ for $n = 1, 2, 3, \ldots$ and $F(1) = 2$. Then, $F(101)$ equals

a. 50 b. 52

c. 54 d. none of these

15. If the sum of n terms of an A.P. is $cn\ (n - 1)$, where $c \neq 0$, then sum of the squares of these terms is

a. $c^2n^2\ (n + 1)^2$ b. $\frac{2}{3}c^2\ n\ (n - 1)\ (2n - 1)$

c. $\frac{2c^2}{3}\ n\ (n + 1)\ (2n + 1)$ d. none of these

16. Consider an A.P. $a_1, a_2, a_3, \ldots$ such that $a_3 + a_5 + a_8 = 11$ and $a_4 + a_2 = -2$, then the value of $a_1 + a_6 + a_7$ is

a. -8 b. 5

c. 7 d. 9

17. If $a_1, a_2, a_3, \ldots$ are in A.P., then a_p, a_q, a_r are in A.P. if p, q, r are in

a. A.P. b. G.P.

c. H.P. d. none of these

18. Let $a_1, a_2, a_3 \ldots$ be terms of an A.P. If $\frac{a_1 + a_2 + \cdots + a_p}{a_1 + a_2 + \cdots + a_q} = \frac{p^2}{q^2}$, $p \neq q$, then $\frac{a_6}{a_{21}}$ equals

a. 41/11 b. 7/2

c. 2/7 d. 11/41

19. If S_n denotes the sum of first n terms of an A.P. whose first term is a and $\frac{S_{nx}}{S_x}$ is independent of x, then $S_p =$

a. p^3 b. p^2a

c. pa^2 d. a^3

20. Three numbers form an increasing G.P. If the middle number is doubled, then the new numbers are in A.P. The common ratio of the G.P. is

a. $2-\sqrt{3}$ b. $2+\sqrt{3}$

c. $\sqrt{3}-2$ d. $3+\sqrt{2}$

21. If $a_1, a_2, a_3\ (a_1 > 0)$ are three successive terms of a G.P. with common ratio r, the value of r for which $a_3 > 4a_2 - 3a_1$ holds is given by

a. $1 < r < 3$ b. $-3 < r < -1$

c. $r > 3$ or $r < 1$ d. none of these

22. Let $S \subset (0, \pi)$ denote the set of values of x satisfying the equation $8^{1+|\cos x|+\cos^2 x+|\cos^3 x|+\cdots \text{ to } \infty} = 4^3$. Then $S =$

a. $\{\pi/3\}$ b. $\{\pi/3, -2\pi/3\}$

c. $\{-\pi/3, 2\pi/3\}$ d. $\{\pi/3, 2\pi/3\}$

23. If $|a| < 1$ and $|b| < 1$, then the sum of the series $1 + (1 + a)b + (1 + a + a^2)b^2 + (1 + a + a^2 + a^3)b^3 + \cdots$ is

a. $\frac{1}{(1-a)(1-b)}$ b. $\frac{1}{(1-a)(1-ab)}$

c. $\frac{1}{(1-b)(1-ab)}$ d. $\frac{1}{(1-a)(1-b)(1-ab)}$

24. If $(p + q)^{\text{th}}$ term of a G.P. is a and its $(p - q)^{\text{th}}$ term is b where $a, b \in R^+$, then its pth term is

a. $\sqrt{\frac{a^3}{b}}$ b. $\sqrt{\frac{b^3}{a}}$

c. $\sqrt{ab}$ d. none of these

25. If the sides of a triangle are in G.P., and its larger angle is twice the smallest, then the common ratio r satisfies the inequality

a. $0 < r < \sqrt{2}$ b. $1 < r < \sqrt{2}$

c. $1 < r < 2$ d. none of these

26. The value of $0.2^{\log_{\sqrt{5}}\left(\frac{1}{4}+\frac{1}{8}+\frac{1}{16}+\cdots\right)}$ is

a. 4 b. log 4

c. log 2 d. none of these

27. If $(1 + x)(1 + x^2)(1 + x^4) \cdots (1 + x^{128}) = \sum_{r=0}^{n} x^r$, then n is equal to

a. 256 b. 255

c. 254 d. none of these

28. If x, y, z are in G.P. and $a^x = b^y = c^z$, then

a. $\log_b a = \log_a c$ b. $\log_c b = \log_a c$

c. $\log_b a = \log_c b$ d. none of these

29. The geometric mean between -9 and -16 is

a. 12 b. -12

c. -13 d. none of these

30. If S denotes the sum to infinity and S_n the sum of n terms of the series $1 + \frac{1}{2} + \frac{1}{4} + \frac{1}{8} + \cdots$, such that $S - S_n < \frac{1}{1000}$, then the least value of n is

a. 8 b. 9

c. 10 d. 11

31. Value of $y = (0.64)^{\log_{0.25}\left(\frac{1}{3}+\frac{1}{3^2}+\frac{1}{3^3}+\cdots \text{ up to } \infty\right)}$ is

a. 0.9 b. 0.8

c. 0.6 d. 0.25

32. The number of terms common between the series $1 + 2 + 4 + 8 + \ldots$ to 100 terms and $1 + 4 + 7 + 10 + \ldots$ to 100 terms is

a. 6 b. 4
c. 5 d. none of these

33. After striking the floor, a certain ball rebounds $(4/5)^{th}$ of height from which it has fallen. Then the total distance that it travels before coming to rest, if it is gently dropped from a height of 120 m is
a. 1260 m b. 600 m
c. 1080 m d. none of these

34. The sum of an infinite G.P. is 57 and the sum of their cubes is 9747, then common ratio of the G.P. is
a. 1/3 b. 2/3
c. 1/6 d. none of these

35. If $a^2 + b^2$, $ab + bc$ and $b^2 + c^2$ are in G.P., then a, b, c are in
a. A.P. b. G.P.
c. H.P. d. none of these

36. If $(r)_n$ denotes the number $rrr \ldots$ (n digits), where $r = 1, 2, 3, \ldots, 9$ and $a = (6)_n$, $b = (8)_n$, $c = (4)_{2n}$, then
a. $a^2 + b + c = 0$ b. $a^2 + b - c = 0$
c. $a^2 + b - 2c = 0$ d. $a^2 + b - 9c = 0$

37. Let $a = 1\,1\,1 \ldots 1$ (55 digits), $b = 1 + 10 + 10^2 + \cdots + 10^4$, $c = 1 + 10^5 + 10^{10} + 10^{15} + \cdots + 10^{50}$, then
a. $a = b + c$ b. $a = bc$
c. $b = ac$ d. $c = ab$

38. Let a_n be the nth term of a G.P. of positive numbers. Let $\sum_{n=1}^{100} a_{2n} = \alpha$ and $\sum_{n=1}^{100} a_{2n-1} = \beta$, such that $\alpha \neq \beta$, then the common ratio is
a. α/β b. β/α
c. $\sqrt{\alpha/\beta}$ d. $\sqrt{\beta/\alpha}$

39. The sum of 20 terms of a series of which every even term is 2 times the term before it, and every odd term is 3 times the term before it, the first term being unity is
a. $\left(\frac{2}{7}\right)(6^{10} - 1)$ b. $\left(\frac{3}{7}\right)(6^{10} - 1)$
c. $\left(\frac{3}{5}\right)(6^{10} - 1)$ d. none of these

40. In a G.P. the first, third and fifth terms may be considered as the first, fourth and sixteenth terms of an A.P. Then the fourth term of the A.P., knowing that its first term is 5 is
a. 10 b. 12
c. 16 d. 20

41. If the pth, q^{th}, and r^{th} terms of an A.P. are in G.P., then common ratio of the G.P. is
a. $\frac{pr}{q^2}$ b. $\frac{r}{p}$
c. $\frac{q+r}{p+q}$ d. $\frac{q-r}{p-q}$

42. If the pth, q^{th}, r^{th}, and s^{th} terms of an A.P. are in G.P., then $p - q, q - r, r - s$ are in
a. A.P. b. G.P.
c. H.P. d. none of these

43. If a, b, c, d are in G.P., then $(b - c)^2 + (c - a)^2 + (d - b)^2$ is equal to
a. $(a - d)^2$ b. $(ad)^2$
c. $(a + d)^2$ d. $(a/d)^2$

44. If a, b, c are digits, then the rational number represented by $0.cababab \ldots$ is
a. $cab/990$ b. $(99c + ba)/990$
c. $(99c + 10a + b)/99$ d. $(99c + 10a + b)/990$

45. The sum of an infinite geometric series is 162 and the sum of its first n terms is 160. If the inverse of its common ratio is an integer, then which of the following is not a possible first term?
a. 108 b. 144
c. 160 d. none of these

46. Let $f(x) = 2x + 1$. Then the number of real number of values of x for which the three unequal numbers $f(x)$, $f(2x)$, $f(4x)$ are in G.P. is
a. 1 b. 2
c. 0 d. none of these

47. Concentric circles of radii 1, 2, 3, …, 100 cm are drawn. The interior of the smallest circle is coloured red and the angular regions are coloured alternately green and red, so that no two adjacent regions are of the same colour. Then, the total area of the green regions in sq. cm is equal to
a. 1000π b. 5050π
c. 4950π d. 5151π

48. Let $\{t_n\}$ be a sequence of integers in G.P. in which $t_4:t_6 = 1:4$ and $t_2 + t_5 = 216$. Then t_1 is
a. 12 b. 14
c. 16 d. none of these

49. If $x, 2y, 3z$ are in A.P., where the distinct numbers x, y, z are in G.P., then the common ratio of the G.P. is
a. 3 b. $\frac{1}{3}$
c. 2 d. $\frac{1}{2}$

50. If S_p denotes the sum of the series $1 + r^p + r^{2p} + \cdots$ to ∞ and s_p the sum of the series $1 - r^p + r^{2p} - r^{3p} + \cdots$ to ∞, $|r| < 1$, then $S_p + s_p$ in terms of S_{2p} is
a. $2S_{2p}$ b. 0
c. $\frac{1}{2}S_{2p}$ d. $-\frac{1}{2}S_{2p}$

51. If x, y, z are real and $4x^2 + 9y^2 + 16z^2 - 6xy - 12yz - 8zx = 0$, then x, y, z are in
a. A.P. b. G.P.
c. H.P. d. none of these

52. If $a_1, a_2, \ldots, a_n$ are in H.P., then $\frac{a_1}{a_2 + a_3 + \cdots + a_n}, \frac{a_2}{a_1 + a_3 + \cdots + a_n}, \ldots, \frac{a_n}{a_1 + a_2 + \cdots + a_{n-1}}$ are in

a. A.P. b. G.P.

c. H.P. d. none of these

53. If $H_1, H_2, \ldots, H_{20}$ be 20 harmonic means between 2 and 3, then $\frac{H_1+2}{H_1-2}+\frac{H_{20}+3}{H_{20}-3}=$

a. 20 b. 21

c. 40 d. 38

54. Let a_1, a_2, a_3, a_4, and a_5 be such that a_1, a_2, and a_3 are in A.P., a_2, a_3 and a_4 are in G.P., and a_3, a_4 and a_5 are in H.P. Then $\log_e a_1, \log_e a_3$, and $\log_e a_5$ are in

a. G.P. b. A.P.

c. H.P. d. none of these

55. If a, b, c are in A.P., then $\frac{a}{bc}, \frac{1}{c}, \frac{2}{b}$ and will be in

a. A.P. b. G.P.

c. H.P. d. none of these

56. If x, $2x + 2$, and $3x + 3$ are first three terms of a G.P., then the fourth term is

a. 27 b. −27

c. 13.5 d. −13.5

57. Sum of three numbers in G.P. be 14. If one is added to first and second and 1 is subtracted from the third, the new numbers are in A.P. The smallest of them is

a. 2 b. 4

c. 6 d. 10

58. If a, b, and c are in A.P., p, q, and r are in H.P. and ap, bq, and cr are in G.P., then $\frac{p}{r}+\frac{r}{p}$ is equal to

a. $\frac{a}{c}-\frac{c}{a}$ b. $\frac{a}{c}+\frac{c}{a}$

c. $\frac{b}{q}+\frac{q}{b}$ d. $\frac{b}{q}-\frac{q}{b}$

59. If a, b, and c are in A.P. and $b-a, c-b$, and a are in G.P., then $a:b:c$ is

a. 1:2:3 b. 1:3:5

c. 2:3:4 d. 1:2:4

60. Let a, b, and c be distinct real numbers which are in G.P. If $x \in R$ is such that $a + x, b + x$, and $c + x$ are in H.P., then x equals

a. a b. b

c. c d. $(a + b + c)/3$

61. In a sequence of 21 terms, the first 11 terms are in A.P. with common difference 2 and the last 11 terms are in G.P. with common ratio 2. If the middle term of the A.P. is equal to the middle term of the G.P., then the middle term of the entire sequence is

a. −10/31 b. 10/31

c. 32/31 d. −31/32

62. $a, b, c, d \in R^+$ such that a, b, and c are in A.P. and b, c and, d are in H.P., then

a. $ab = cd$ b. $ac = bd$

c. $bc = ad$ d. none of these

63. If in a progression $a_1, a_2, a_3, \ldots$, etc., $(a_r - a_{r+1})$ bears a constant ratio with $a_r \times a_{r+1}$, then the terms of the progression are in

a. A.P. b. G.P.

c. H.P. d. none of these

64. If a, b, and c are in G.P., then $a + b$, $2b$, and $b + c$ are in

a. A.P. b. G.P.

c. H.P. d. none of these

65. If a, x, and b are in A.P., a, y, and b are in G.P. and a, z, b are in H.P. such that $x = 9z$ and $a > 0, b > 0$, then

a. $|y| = 3z$ b. $x = 3|y|$

c. $2y = x + z$ d. none of these

66. Let $n \in N, n > 25$. Let A, G, H denote the arithmetic mean, geometric mean and harmonic mean of 25 and n. The least value of n for which $A, G, H \in \{25, 26, \ldots, n\}$ is

a. 49 b. 81

c. 169 d. 225

67. The 15th term of the series $2\frac{1}{2}+1\frac{7}{13}+1\frac{1}{9}+\frac{20}{23}+\cdots$ is

a. $\frac{5}{12}$ b. $\frac{10}{21}$

c. $\frac{10}{23}$ d. none of these

68. If a, b, and c are in G.P. and x, y, respectively, be A.M.'s between a, b and b, c, then the value of $\frac{a}{x}+\frac{c}{y}$ is

a. 1 b. 2

c. 1/2 d. none of these

69. If a, b, and c are in A.P. and p, p' are, respectively, A.M. and G.M. between a and b while q, q' are, respectively, the A.M. and G.M. between b and c, then

a. $p^2 + q^2 = p'^2 + q'^2$ b. $pq = p'q'$

c. $p^2 - q^2 = p'^2 - q'^2$ d. none of these

70. If $a_1, a_2, \ldots, a_n$ are in A.P. with common difference $d \neq 0$, then sum of the series $\sin d[\sec a_1 \sec a_2 + \sec a_2 \sec a_3 + \cdots + \sec a_{n-1} \sec a_n]$ is

a. $\operatorname{cosec} a_n - \operatorname{cosec} a$ b. $\cot a_n - \cot a$

c. $\sec a_n - \sec a_1$ d. $\tan a_n - \tan a_1$

71. The sum of the series $a - (a + d) + (a + 2d) - (a + 3d) + \cdots$ up to $(2n + 1)$ terms is

a. $-nd$ b. $a + 2nd$

c. $a + nd$ d. $2nd$

72. The sum to 50 terms of the series

$$1+2\left(1+\frac{1}{50}\right)+3\left(1+\frac{1}{50}\right)^2+\cdots \text{ is given by}$$

a. 2500 b. 2550

c. 2450 d. none of these

73. The sum to 50 terms of the series $\frac{3}{1^2}+\frac{5}{1^2+2^2}+\frac{7}{1^2+2^2+3^2}+\cdots$ is

a. $\frac{100}{17}$ **b.** $\frac{150}{17}$

c. $\frac{200}{51}$ **d.** $\frac{50}{17}$

74. If $\frac{1}{1^2}+\frac{1}{2^2}+\frac{1}{3^2}+\cdots$ to $\infty=\frac{\pi^2}{6}$, then $\frac{1}{1^2}+\frac{1}{3^2}+\frac{1}{5^2}+\cdots$ equals

a. $\pi^2/8$ **b.** $\pi^2/12$

c. $\pi^2/3$ **d.** $\pi^2/2$

75. Coefficient of x^{18} in $(1+x+2x^2+3x^3+\cdots+18x^{18})^2$ is equal to

a. 995 **b.** 1005

c. 1235 **d.** none of these

76. $\lim_{n\to\infty}\sum_{r=1}^{n}\frac{r}{1\times3\times5\times7\times9\times\cdots\times(2r+1)}$ is equal to

a. $\frac{1}{3}$ **b.** $\frac{3}{2}$

c. $\frac{1}{2}$ **d.** none of these

77. If $1+\sum_{r=1}^{30} r\times r!$ is divisible by the product of n consecutive integers, then n is

a. composite number **b.** odd number

c. divisible by 3 **d.** none of these

78. If $\sum_{r=1}^{n} r^4 = I(n)$, then $\sum_{r=1}^{n}(2r-1)^4$ is equal to

a. $I(2n)-I(n)$ **b.** $I(2n)-16I(n)$

c. $I(2n)-8I(n)$ **d.** $I(2n)-4I(n)$

79. Value of $\left(1+\frac{1}{3}\right)\left(1+\frac{1}{3^2}\right)\left(1+\frac{1}{3^4}\right)\left(1+\frac{1}{3^8}\right)\cdots\infty$ is equal to

a. 3 **b.** $\frac{6}{5}$

c. $\frac{3}{2}$ **d.** none of these

80. If $x_1, x_2, \ldots, x_{20}$ are in H.P. and x_1, 2, x_{20} are in G.P., then $\sum_{r=1}^{19} x_r x_{r+1} =$

a. 76 **b.** 80

c. 84 **d.** none of these

81. The value of $\sum_{r=0}^{n}(a+r+ar)(-a)^r$ is equal to

a. $(-1)^n[(n+1)a^{n+1}-a]$ **b.** $(-1)^n(n+1)a^{n+1}$

c. $(-1)^n\frac{(n+2)a^{n+1}}{2}$ **d.** $(-1)^n\frac{na^n}{2}$

82. If $b_i = 1-a_i$, $na=\sum_{i=1}^{n}a_i$, $nb=\sum_{i=1}^{n}b_i$, then $\sum_{i=1}^{n}a_ib_i+\sum_{i=1}^{n}(a_i-a)^2=$

a. ab **b.** $-nab$

c. $(n+1)ab$ **d.** nab

83. The sum of the series $\frac{x}{1-x^2}+\frac{x^2}{1-x^4}+\frac{x^4}{1-x^8}+\cdots$ to infinite terms, if $|x|<1$, is

a. $\frac{x}{1-x}$ **b.** $\frac{1}{1-x}$

c. $\frac{1+x}{1-x}$ **d.** 1

84. If $a_1, a_2, a_3, \ldots, a_{2n+1}$ are in A.P., then $\frac{a_{2n+1}-a_1}{a_{2n+1}+a_1}+\frac{a_{2n}-a_2}{a_{2n}+a_2}+\cdots+\frac{a_{n+2}-a_n}{a_{n+2}+a_n}$ is equal to

a. $\frac{n(n+1)}{2}\times\frac{a_2-a_1}{a_{n+1}}$ **b.** $\frac{n(n+1)}{2}$

c. $(n+1)(a_2-a_1)$ **d.** none of these

85. The sum of $i-2-3i+4\cdots$ up to 100 terms, where $i=\sqrt{-1}$, is

a. $50(1-i)$ **b.** $25i$

c. $25(1+i)$ **d.** $100(1-i)$

86. Let $S=\frac{4}{19}+\frac{44}{19^2}+\frac{444}{19^3}+\cdots$ up to ∞. Then S is equal to

a. 40/9 **b.** 38/81

c. 36/171 **d.** none of these

87. If $H_n = 1+\frac{1}{2}+\cdots+\frac{1}{n}$, then value of $S_n = 1+\frac{3}{2}+\frac{5}{3}+\cdots+\frac{99}{50}$ is

a. $H_{50}+50$ **b.** $100-H_{50}$

c. $49+H_{50}$ **d.** $H_{50}+100$

88. If the sum to infinity of the series $1+2r+3r^2+4r^3+\cdots$ is 9/4, then value of r is

a. 1/2 **b.** 1/3

c. 1/4 **d.** none of these

89. The sum of series $1+\frac{4}{5}+\frac{7}{5^2}+\frac{10}{5^3}+\cdots\infty$ is

a. 7/16 **b.** 5/16

d. 105/64 **d.** 35/16

90. The sum $\frac{1^3}{1}+\frac{1^3+2^3}{1+3}+\frac{1^3+2^3+3^3}{1+3+5}+\cdots$ to 16 terms is

a. 246 **b.** 646

c. 446 **d.** 746

91. The sum $1+3+7+15+31+\cdots$ to 100 terms is

a. $2^{100}-102$ **b.** $2^{99}-101$

c. $2^{101}-102$ **d.** none of these

92. If $a_1, a_2, a_3, \ldots a_n$ are in H.P. then $a_1a_2+a_2a_3+\cdots+a_{n-1}a_n$ is

a. $(n-1)a_1a_n$ **b.** $n\,a_1a_n$

c. $(n-1)a_1/a_n$ **d.** none of these

93. The coefficient of x^{49} in the product $(x-1)(x-3)\cdots(x-99)$ is

a. -99^2 **b.** 1
c. -2500 **d.** none of these

94. The sum of 20 terms of the series whose r^{th} term is given by $T(n) = (-1)^n \dfrac{n^2+n+1}{n!}$ is

a. $\dfrac{20}{19!}-2$ **b.** $\dfrac{21}{20!}-1$
c. $\dfrac{21}{20!}$ **d.** none of these

95. Consider the sequence 1, 22, 4444, 88888888, ... Then 1025th term will be

a. 2^9 **b.** 2^{11}
c. 2^{10} **d.** 2^{12}

96. If $t_n = \dfrac{1}{4}(n+2)(n+3)$ for $n = 1, 2, 3, \ldots$, then $\dfrac{1}{t_1}+\dfrac{1}{t_2}+\dfrac{1}{t_3}+\cdots+\dfrac{1}{t_{2003}} =$

a. $\dfrac{4006}{3006}$ **b.** $\dfrac{4003}{3007}$
c. $\dfrac{4006}{3008}$ **d.** $\dfrac{4006}{3009}$

97. The sum of $0.2 + 0.004 + 0.00006 + 0.0000008 + \cdots$ to ∞ is

a. $\dfrac{200}{891}$ **b.** $\dfrac{2000}{9801}$
c. $\dfrac{1000}{9801}$ **d.** none of these

98. The 100th term of the series $2\dfrac{1}{2}+1\dfrac{7}{13}+1\dfrac{1}{9}+\dfrac{20}{23}+\cdots$ is

a. $\dfrac{20}{503}$ **b.** $\dfrac{20}{497}$
c. 20(503) **d.** none of these

99. If $1^2 + 2^2 + 3^2 + \cdots + 2003^2 = (2003)(4007)(334)$ and $(1)(2003) + (2)(2002) + (3)(2001) + \cdots + (2003)(1) = (2003)(334)(x)$, then x equals

a. 2005 **b.** 2004
c. 2003 **d.** 2001

100. If t_n denotes the nth term of the series $2 + 3 + 6 + 11 + 18 + \cdots$, then t_{50} is

a. $49^2 - 1$ **b.** 49^2
c. $50^2 + 1$ **d.** $49^2 + 2$

101. The positive integer n for which $2\times 2^2 + 3\times 2^3 + 4\times 2^4 + \cdots + n\times 2^n = 2^{n+10}$ is

a. 510 **b.** 511
c. 512 **d.** 513

102. If $1-\dfrac{1}{3}+\dfrac{1}{5}-\dfrac{1}{7}+\dfrac{1}{9}-\dfrac{1}{11}+\cdots=\dfrac{\pi}{4}$, then value of $\dfrac{1}{1\times 3}+\dfrac{1}{5\times 7}+\dfrac{1}{9\times 11}+\cdots$ is

a. $\pi/8$ **b.** $\pi/6$
c. $\pi/4$ **d.** $\pi/36$

103. The coefficient of x^{19} in the polynomial $(x-1)(x-2)(x-2^2)\cdots(x-2^{19})$ is

a. $2^{20} - 2^{19}$ **b.** $2^{20} - 1$
c. 2^{20} **d.** none of these

104. If $b_{n+1} = \dfrac{1}{1-b_n}$ for $n \ge 1$ and $b_1 = b_3$, then $\sum_{r=1}^{2001} b_r^{2001}$ is equal to

a. 2001 **b.** -2001
c. 0 **d.** none of these

105. If $(1^2 - t_1) + (2^2 - t_2) + \cdots + (n^2 - t_n) = 1/3n(n^2 - 1)$, then t_n is equal to

a. n^2 **b.** $2n$
c. $n^2 - 2n$ **d.** none of these

106. If $ax^3 + bx^2 + cx + d$ is divisible by $ax^2 + c$, then a, b, c, d are in

a. A.P. **b.** G.P.
c. H.P. **d.** none of these

107. If $(1 + 3 + 5 + \cdots + p) + (1 + 3 + 5 + \cdots + q) = (1 + 3 + 5 + \cdots + r)$ where each set of parentheses contains the sum of consecutive odd integers as shown, the smallest possible value of $p + q + r$ (where $p > 6$) is

a. 12 **b.** 21
c. 45 **d.** 54

108. In a geometric series, the first term is a and common ratio is r. If S_n denotes the sum of the n terms and $U_n = \sum_{n=1}^{n} S_n$ then $rS_n + (1-r)U_n$ equals

a. 0 **b.** n
c. na **d.** nar

109. The line $x + y = 1$ meets x-axis at A and y-axis at B, P is the mid-point of AB;
P_1 is the foot of the perpendicular from P to OA;
M_1 is that of P_1 from OP;
P_2 is that of M_1 from OA;
M_2 is that of P_2 from OP;
P_3 is that of M_2 from OA; and so on.
If P_n denotes the nth foot of the perpendicular on OA, then OP_n is

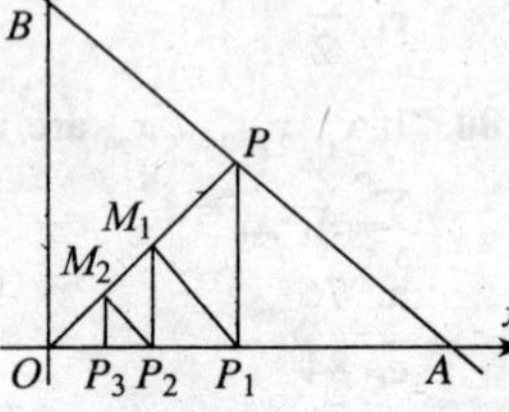

Fig. 3.1

a. $\left(\dfrac{1}{2}\right)^{n-1}$ **b.** $\left(\dfrac{1}{2}\right)^{n}$
c. $\left(\dfrac{1}{2}\right)^{n+1}$ **d.** none of these

110. If $(1-p)(1+3x+9x^2+27x^3+81x^4+243x^5)=1-p^6$, $p \neq 1$, then the value of $\frac{p}{x}$ is

a. $\frac{1}{3}$ **b.** 3

c. $\frac{1}{2}$ **d.** 2

111. ABC is a right-angled triangle in which $\angle B = 90°$ and $BC = a$. If n points $L_1, L_2, \ldots, L_n$ on AB is divided in $n+1$ equal parts and $L_1M_1, L_2M_2, \ldots, L_nM_n$ are line segments parallel to BC and $M_1, M_2, \ldots, M_n$ are on AC, then the sum of the lengths of $L_1M_1, L_2M_2, \ldots, L_nM_n$ is

a. $\frac{a(n+1)}{2}$ **b.** $\frac{a(n-1)}{2}$

c. $\frac{an}{2}$ **d.** none of these

112. Let T_r and S_r be the rth term and sum up to r^{th} term of a series respectively. If for an odd number n, $S_n = n$ and $T_n = \frac{T_{n-1}}{n^2}$, then T_m (m being even) is

a. $\frac{2}{1+m^2}$ **b.** $\frac{2m^2}{1+m^2}$

c. $\frac{(m+1)^2}{2+(m+1)^2}$ **d.** $\frac{2(m+1)^2}{1+(m+1)^2}$

113. $ABCD$ is a square of length a, $a \in N$, $a > 1$. Let $L_1, L_2, L_3, \ldots$ be points on BC such that $BL_1 = L_1L_2 = L_2L_3 = \cdots = 1$ and $M_1, M_2, M_3, \ldots$ be points on CD such that $CM_1 = M_1M_2 = M_2M_3 = \cdots = 1$. Then $\sum_{n=1}^{a-1}(AL_n^2 + L_nM_n^2)$ is equal to

a. $\frac{1}{2}a(a-1)^2$ **b.** $\frac{1}{2}(a-1)(2a-1)(4a-1)$

c. $\frac{1}{2}a(a-1)(4a-1)$ **d.** none of these

114. If x, y, and z are in G.P. and $x+3$, $y+3$, and $z+3$ are in H.P., then

a. $y=2$ **b.** $y=3$

c. $y=1$ **d.** $y=0$

115. If x, y, and z are distinct prime numbers, then

a. x, y, and z may be in A.P. but not in G.P.

b. x, y, and z may be in G.P. but not in A.P.

c. x, y, and z can neither be in A.P. nor in G.P.

d. none of these

Multiple Correct Answers Type

Each question has four choices a, b, c and d, out of which *one or more* answers are correct.

1. For the series, $S = 1 + \frac{1}{(1+3)}(1+2)^2 + \frac{1}{(1+3+5)}(1+2+3)^2 + \frac{1}{(1+3+5+7)}(1+2+3+4)^2 + \cdots$

a. 7th term is 16

b. 7th term is 18

c. sum of first 10th terms is $\frac{505}{4}$

d. sum of first 10th terms is $\frac{405}{4}$

2. If sum of an infinite G.P. $p, 1, 1/p, 1/p^2, \ldots$ is 9/2, then value of p is

a. 2 **b.** 3/2

c. 3 **d.** 9/2

3. If $\sum_{r=1}^{n} r(r+1)(2r+3) = an^4 + bn^3 + cn^2 + dn + e$, then

a. $a - b = d - c$

b. $e = 0$

c. a, $b - 2/3$, $c - 1$ are in A.P.

d. $(b+d)/a$ is an integer

4. The terms of an infinitely decreasing G.P. in which all the terms are positive, the first term is 4, and the difference between the third and fifth term is 32/81, then

a. $r = 1/3$ **b.** $r = 2\sqrt{2}/3$

c. $S_\infty = 6$ **d.** none of these

5. The consecutive digits of a three digit number are in G.P. If the middle digit be increased by 2, then they form an A.P. If 792 is subtracted from this, then we get the number constituting of same three digits but in reverse order. Then number is divisible by

a. 7 **b.** 49

c. 19 **d.** none of these

6. If $S_n = 1^2 - 2^2 + 3^2 - 4^2 + 5^2 - 6^2 + \cdots$, then

a. $S_{40} = -820$ **b.** $S_{2n} > S_{2n+2}$

c. $S_{51} = 1275$ **d.** $S_{2n+1} > S_{2n-1}$

7. Given that $x + y + z = 15$ when a, x, y, z, b are in A.P. and $\frac{1}{x} + \frac{1}{y} + \frac{1}{z} = \frac{5}{3}$ when a, x, y, z, b are in H.P. Then

a. G.M. of a and b is 3

b. one possible value of $a + 2b$ is 11

c. A.M. of a and b is 6

d. greatest value of $a - b$ is 8

8. Let $a_1, a_2, a_3, \ldots, a_n$ be in G.P. such that $3a_1 + 7a_2 + 3a_3 - 4a_5 = 0$. Then common ratio of G.P. can be

a. 2 **b.** $\frac{3}{2}$

c. $\frac{5}{2}$ **d.** $-\frac{1}{2}$

9. If $S = \frac{1}{\sqrt{2}+\sqrt{5}} + \frac{1}{\sqrt{5}+\sqrt{8}} + \frac{1}{\sqrt{8}+\sqrt{11}} + \cdots$ n terms, then the value of the above expression is

a. $\frac{\sqrt{3n+2}-\sqrt{2}}{3}$ **b.** $\frac{n}{\sqrt{2+3n}+\sqrt{2}}$

c. less than n **d.** less than $\sqrt{\frac{n}{3}}$

10. If a, b, and c are in H.P. then the value of $\dfrac{(ac+ab-bc)(ab+bc-ac)}{(abc)^2}$ is

a. $\dfrac{(a+c)(3a-c)}{4a^2c^2}$ **b.** $\dfrac{2}{bc}-\dfrac{1}{b^2}$

c. $\dfrac{2}{bc}-\dfrac{1}{a^2}$ **d.** $\dfrac{(a-c)(3a+c)}{4a^2c^2}$

11. If $p(x)=\dfrac{1+x^2+x^4+\cdots+x^{2n-2}}{1+x+x^2+\cdots+x^{n-1}}$ is a polynomial in x, then n can be

a. 5 **b.** 10

c. 20 **d.** 17

12. For an increasing A.P. $a_1, a_2, \ldots, a_n$ if $a_1+a_3+a_5=-12$ and $a_1a_3a_5=80$, then which of the following is/are true?

a. $a_1=-10$ **b.** $a_2=-1$

c. $a_3=-4$ **d.** $a_5=-2$

13. If $n>1$, the values of the positive integer m for which n^m+1 divides $a=1+n+n^2+\cdots+n^{63}$ is/are

a. 8 **b.** 16

c. 32 **d.** 64

14. If p, q, and r are in A.P., then which of the following is/are true?

a. p^{th}, q^{th}, and r^{th} terms of A.P, are in A.P.

b. p^{th}, q^{th}, and r^{th} terms of G.P. are in G.P.

c. p^{th}, q^{th}, and r^{th} terms of H.P, are in H.P.

d. none of these

15. If $1+2x+3x^2+4x^3+\cdots\infty\geq 4$, then

a. least value of x is 1/2

b. greatest value of x is 4/3

c. least value of x is 2/3

d. greatest value of x is 3/2

16. Let $E=\dfrac{1}{1^2}+\dfrac{1}{2^2}+\dfrac{1}{3^2}+\cdots$. Then,

a. $E<3$ **b.** $E>3/2$ **c.** $E>2$ **d.** $E<2$

17. If the sum of n terms of an A.P. is given by $S_n=a+bn+cn^2$ where a, b, c are independent of n, then

a. $a=0$

b. common difference of A.P. must be $2b$

c. common difference of A.P. must be $2c$

d. first term of A.P. is $b+c$

18. If $x^2+9y^2+25z^2=xyz\left(\dfrac{15}{x}+\dfrac{5}{y}+\dfrac{3}{z}\right)$, then

a. x, y, and z are in H.P.

b. $\dfrac{1}{x}, \dfrac{1}{y}, \dfrac{1}{z}$ are in A.P.

c. x, y, z are in G.P.

d. $\dfrac{1}{x}, \dfrac{1}{y}, \dfrac{1}{z}$ are in G.P.

19. If a, b, c, and d are four unequal positive numbers which are in A.P., then

a. $\dfrac{1}{a}+\dfrac{1}{d}>\dfrac{1}{b}+\dfrac{1}{c}$

b. $\dfrac{1}{a}+\dfrac{1}{d}<\dfrac{1}{b}+\dfrac{1}{c}$

c. $\dfrac{1}{b}+\dfrac{1}{c}>\dfrac{4}{a+d}$

d. $\dfrac{1}{a}+\dfrac{1}{d}=\dfrac{1}{b}+\dfrac{1}{c}$

20. The next term of the G.P. x, x^2+2, and x^3+10 is

a. $\dfrac{729}{16}$ **b.** 6

c. 0 **d.** 54

21. In the 20th row of the triangle

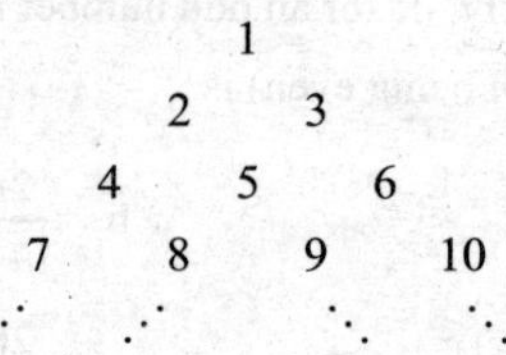

a. last term = 210

b. first term = 191

c. sum = 4010

d. sum = 4200

22. If A_1, A_2; G_1, G_2; and H_1, H_2 are two A.M.'s and H.M.'s, respectively, between two quantities a and b, then ab is equal to

a. A_1H_2 **b.** A_2H_1

c. G_1G_2 **d.** none of these

23. Let $S_1, S_2, \ldots$ be squares such that for each $n\geq 1$, the length of a side of S_n equals the length of a diagonal of S_{n+1}. If the length of a side of S_1 is 10 cm, then for which of the following values of n is the area of S_n less than 1 sq. cm?

a. 7 **b.** 8

c. 9 **d.** 10

24. If $\dfrac{1}{b-a}+\dfrac{1}{b-c}=\dfrac{1}{a}+\dfrac{1}{c}$, then

a. a, b, and c are in H.P.

b. a, b, and c are in A.P.

c. $b=a+c$

d. $3a=b+c$

25. If a, b, and c are in G.P. and x and y, respectively, be arithmetic means between a, b and b, c, then

a. $\dfrac{a}{x}+\dfrac{c}{y}=2$ **b.** $\dfrac{a}{x}+\dfrac{c}{y}=\dfrac{c}{a}$

c. $\dfrac{1}{x}+\dfrac{1}{y}=\dfrac{2}{b}$ **d.** $\dfrac{1}{x}+\dfrac{1}{y}=\dfrac{2}{ac}$

Reasoning Type

Each question has four choices a, b, c, and d, out of which *only one* is correct. Each question contains Statement 1 and Statement 2.

a. Both the Statements are True and Statement 2 is the correct explanation of Statement 1.

b. Both the Statements are True but Statement 2 is Not the correct explanation of Statement 1.

c. Statement 1 is True and Statement 2 is False.

d. Statement 1 is False and Statement 2 is True.

1. Statement 1: The numbers $\sqrt{2}$, $\sqrt{3}$, $\sqrt{5}$ cannot be the terms of a single A.P. with non-zero common difference.

Statement 2: If p, q, r $(p \neq q)$ are terms (not necessarily consecutive) of an A.P., then there exists a rational number k such that $(r-q)/(q-p) = k$.

2. Statement 1: In a G.P. if the $(m+n)^{\text{th}}$ term be p and $(m-n)^{\text{th}}$ term be q, then its mth term is $\sqrt{pq}$.

Statement 2: In a G.P. of terms $a_1, a_2, a_3, \ldots, a_n$, terms at same distance are in G.P.

3. Statement 1: There are infinite geometric progressions for which 27, 8, and 12 are three of its terms (not necessarily consecutive).

Statement 2: Given terms are integers.

4. Statement 1: If $x^2 + 9y^2 + 25z^2 = xyz\left(\dfrac{15}{x} + \dfrac{5}{y} + \dfrac{3}{z}\right)$, then x, y, z are in H.P.

Statement 2: If $a_1^{\,2} + a_2^{\,2} + \cdots + a_n^{\,2} = 0$, then $a_1 = a_2 = a_3 = \cdots = a_n = 0$.

5. Statement 1: Coefficient of x^{14} in $(1 + 2x + 3x^2 + \cdots + 16x^{15})^2$ is 560.

Statement 2: $\sum_{r=1}^{n} r(n-r) = \dfrac{n(n^2-1)}{6}$.

6. Statement 1: $x = 1111\cdots91$ times is composite number.

Statement 2: 91 is composite number.

7. Let $a, r \in R - \{0, 1, -1\}$ and n be an even number.

Statement 1: $a \times ar \times ar^2 \cdots ar^{n-1} = (a^2 r^{n-1})^{n/2}$.

Statement 2: Product of ith term from the beginning and from the end in a G.P. is independent of i.

8. Statement 1: Sum of the series $1^3 - 2^3 + 3^3 - 4^3 + \cdots + 11^3 = 378$.

Statement 2: For any odd integer $n \geq 1$, $n^3 - (n-1)^3 + \cdots + (-1)^{n-1}\, 1^3 = \dfrac{1}{4}(2n-1)(n+1)^2$.

9. Statement 1: If an infinite G.P. has second term x and its sum is 4, then x belongs to $(-8, 1)$.

Statement 2: Sum of an infinite G.P. is finite if for its common ratio r, $0 < |r| < 1$.

10. Statement 1: Let $p_1, p_2, \ldots, p_n$ and x be distinct real number such that $\left(\sum_{r=1}^{n-1} p_r^2\right) x^2 + 2\left(\sum_{r=1}^{n-1} p_r p_{r+1}\right) x + \sum_{r=2}^{n} p_r^2 \leq 0$, then $p_1, p_2, \ldots, p_n$ are in G.P.

Statement 2: If $\dfrac{p_2}{p_1} = \dfrac{p_3}{p_2} = \cdots = \dfrac{p_n}{p_{n-1}}$, then $p_1, p_2, \ldots, p_n$ are in G.P.

11. Statement 1: $1^{99} + 2^{99} + \cdots + 100^{99}$ is divisible by 10100.

Statement 2: $a^n + b^n$ is divisible by $a + b$ if n is odd.

12. Statement 1: If the A.M. of two numbers is 5/2, G.M. of the numbers is 2, then the H.M. will be 8/5.

Statement 2: For a group of positive numbers (G.M.)^2 = (A.M.) × (H.M.).

Linked Comprehension Type

Based upon each paragraph, three multiple choice questions have to be answered. Each question has four choices a, b, c and d, out of which *only one* is correct.

For Problems 1–3

Sum of certain consecutive odd positive integers is $57^2 - 13^2$.

1. Number of integers are

a. 40 **b.** 37 **c.** 44 **d.** 51

2. The least value of an integer is

a. 22 **b.** 27 **c.** 31 **d.** 43

3. The greatest integer is

a. divisible by 7 **b.** divisible by 11

c. divisible by 9 **d.** none of these

For Problems 4–6

Consider three distinct real numbers a, b, c in a G.P. with $a^2 + b^2 + c^2 = t^2$ and $a + b + c = t$. Sum of the common ratio and its reciprocal is denoted by S.

4. Complete set of α^2 is

a. $\left(\dfrac{1}{3}, 3\right)$ **b.** $\left[\dfrac{1}{3}, 3\right]$

c. $\left(\dfrac{1}{3}, 3\right) - \{1\}$ **d.** $\left(-\infty, \dfrac{1}{3}\right) \cup (3, \infty)$

5. Complete set of S is

a. $(-2, 2)$ **b.** $(-\infty, -2) \cup (2, \infty)$

c. $(-1, 1)$ **d.** $(-\infty, -1) \cup (1, \infty)$

6. If a, b, and c also represent the sides of a triangle, then the complete set of α^2 is

a. $\left(\dfrac{1}{3}, 3\right)$ **b.** $(2, 3)$

c. $\left[\dfrac{1}{3}, 2\right]$ **d.** $\left(\dfrac{\sqrt{5}+3}{2}, 3\right)$

For Problems 7–9

In a G.P., the sum of the first and last term is 66, the product of the second and the last but one is 128 and the sum of the terms is 126.

7. If an increasing G.P. is considered, then the number of terms in G.P. is

a. 9 **b.** 8

c. 12 **d.** 6

8. If the decreasing G.P. is considered, then the sum of infinite terms is

a. 64 **b.** 128

c. 256 **d.** 729

9. In any case, the difference of the least and greatest term is

a. 78

b. 126

c. 126

d. none of these

For Problems 10–12

Four different integers form an increasing A.P. One of these numbers is equal to the sum of the squares of the other three numbers. Then

10. The product of all numbers is

a. −2 **b.** 1

c. 0 **d.** 2

11. The sum of all the four numbers is

a. 3 **b.** 0

c. 4 d 2

12. The common difference of the four numbers is

a. 1 **b.** 3

c. 2 **d.** − 2

For Problems 13–15

Consider the sequence in the form of groups (1), (2, 2), (3, 3, 3), (4, 4, 4, 4), (5, 5, 5, 5, 5), ...

13. The 2000th term of the sequence is not divisible by

a. 3

b. 9

c. 7

d. none of these

14. The sum of first 2000 terms is

a. 84336

b. 96324

c. 78466

d. none of these

15. The sum of the remaining terms in the group after 2000th term in which 2000th term lies is

a. 1088

b. 1008

c. 1040

d. none of these

For Problems 16–18

There are two sets A and B each of which consists of three numbers in A.P. whose sum is 15 and where D and d are the common differences such that $D - d = 1$. If $\frac{p}{q} = \frac{7}{8}$, where p and q are the product of the numbers, respectively, and $d > 0$ in the two sets.

16. Sum of the product of the numbers in set A taken two at a time is

a. 51 **b.** 71

c. 74 **d.** 86

17. Sum of the product of the numbers in set B taken two at a time is

a. 54 **b.** 64

c. 73 **d.** 81

18. Value of $q - p$ is

a. 20 **b.** 30

c. 15 **d.** 25

For Problems 19–21

Let $A_1, A_2, A_3, ..., A_m$ be the A.M.'s between −2 and 1027 and $G_1, G_2, G_3, ..., G_n$ be the G.M's between 1 and 1024. The product of G.M.'s is 2^{45} and sum of A.M.'s is 1025×171.

19. The value of $\sum_{r=1}^{n} G_r$ is

a. 512 **b.** 2046

c. 1022 **d.** none of these

20. The number of A.M.'s is

a. 1243 **b.** 1542

c. 1330 **d.** none of these

21. The numbers $2A_{171}, G_5^2 + 1, 2A_{172}$ are in

a. A. P. **b.** G.P.

c. H.P. **d.** none of these

For Problems 22–24

Two consecutive numbers from 1, 2, 3, ..., n are removed. The A.M. of the remaining numbers is 105/4.

22. The value of n lies in

a. [45, 55] **b.** [52, 60]

c. [41, 49] **d.** none of these

23. The removed numbers

a. lie between 10 and 20

b. are greater than 10

c. are less than 15

d. none of these

24. Sum of all numbers

a. exceeds 1600

b. is less than 1500

c. lies between 1300 and 1500

d. none of these

For Problems 25–27

Two A.P.'s have the same numbers. The ratio of the last term of the first progression to first term of the second progression is equal to the ratio of the last term of the second progression to the first term of the first progression and is equal to 4, the ratio of the sum of the n terms of the first progression to the sum of the n terms of the second progression is equal to 2.

25. The ratio of their common difference is

a. 12 **b.** 24

c. 26 **d.** 9

26. The ratio of their nth term is

a. 6/5

b. 7/2

c. 9/5

d. none of these

27. Ratio of their first term is

a. 2/7 **b.** 3/5

c. 4/7 **d.** 2/5

For Problems 28–30

The numbers a, b, and c are between 2 and 18, such that

(i) their sum is 25

(ii) the numbers 2, a, and b are consecutive terms of an A.P.

(iii) the numbers b, c, 18 are consecutive terms of a G.P.

28. The value of abc is

a. 500

b. 450

c. 720

d. none of these

29. Roots of the equation $ax^2 + bx + c = 0$ are

a. real and positive

b. real and negative

c. imaginary

d. real and of opposite sign

30. If a, b, and c are roots of the equation $x^3 + qx^2 + rx + s = 0$, then the value of r is

a. 184 **b.** 196

c. 224 **d.** none of these

For Problems 31–33

Let $T_1, T_2, T_3, \ldots, T_n$ be the terms of a sequence and let $(T_2 - T_1) = T'_1, (T_3 - T_2) = T'_2, \ldots, (T_n - T_{n-1}) = T'_{n-1}$.

Case I:

If $T'_1, T'_2, \ldots, T'_{n-1}$ are in A.P., then T_n is quadratic in n. If $T'_1 - T'_2$, $T'_2 - T'_3, \ldots$, are in A.P., then T_n is cubic in n.

Case II:

If $T'_1, T'_2, \ldots, T'_{n-1}$ are not in A.P., but in G.P., then $T_n = ar^n + b$, where r is the common ratio of the G.P. $T'_1, T'_2, T'_3, \ldots$ and $a, b \in R$. Again, if $T'_1, T'_2, \ldots, T'_{n-1}$ are not in G.P. but $T'_2 - T'_1, T'_3 - T'_2, \ldots, T'_{n-2}$ are in G.P., then T_n is of the form $ar^n + bn + c$ and is the C.R. of the G.P. $T'_2 - T'_1, T'_3 - T'_2 T'_4 - T'_3, \ldots$ and $a, b, c \in R$.

31. The sum of 20 terms of the series $3 + 7 + 14 + 24 + 37 + \ldots$ is

a. 4010

b. 3860

c. 4240

d. none of these

32. The 100th term of the series $3 + 8 + 22 + 72 + 226 + 1036 + \cdots$ is divisible by 2^n, then maximum value of n is

a. 4 **b.** 2

c. 3 **d.** 5

33. For the series $2 + 12 + 36 + 80 + 150 + 252 + \cdots$, the value of $\lim_{n \to \infty} T_n/n^3$ is (where T_n is n^{th} term)

a. 2

b. 1/2

c. 1

d. none of these

Matrix-Match Type

Each question contains statements given in two columns which have to be matched. Statements a, b, c, d in column I have to be matched with statements p, q, r, s in column II. If the correct matches are a→p, a→s, a→q, b→r, c→p, c→q and d→s, then the correctly bubbled 4 × 4 matrix should be as follows:

	p	q	r	s
a	**p**	q	r	**s**
b	p	**q**	**r**	s
c	**p**	**q**	r	s
d	p	q	r	**s**

1.

Column I	Column II
a. If a, b, c are in G.P., then $\log_a 10$, $\log_b 10$, $\log_c 10$ are in	**p.** A.P.
b. If $\dfrac{a + be^x}{a - be^x} = \dfrac{b + ce^x}{b - ce^x} = \dfrac{c + de^x}{c - de^x}$, then a, b, c, d are in	**q.** H.P.
c. If a, b, c are in A.P.; a, x, b are in G.P. and b, y, c are in G.P., then x^2, b^2, y^2 are in	**r.** G.P.
d. If x, y, z are in G.P., $a^x = b^y = c^z$, then $\log a$, $\log b$, $\log c$ are in	**s.** none of these

2.

Column I	Column II
a. If $\Sigma n = 210$, then Σn^2 is divisible by the greatest prime number which is greater than	**p.** 16
b. Between 4 and 2916 is inserted odd number $(2n + 1)$ G.M's. Then the $(n + 1)$th G.M. is divisible by greatest odd integer which is less than	**q.** 10
c. In a certain progression, three consecutive terms are 40, 30, 24, 20. Then the integral part of the next term of the progression is more than	**r.** 34
d. $1+\frac{4}{5}+\frac{7}{5^2}+\frac{10}{5^3}+\cdots$ to $\infty = \frac{a}{b}$, where H.C.F.$(a, b) = 1$, then $a - b$ is less than	**s.** 30

Integer Type

1. If the sum of the first $3n$ terms is equal to the next n terms of an A.P. whose common difference is non-zero, then the reciprocal of the ratio of the sum of the first $2n$ terms to the next $2n$ terms is ___.

2. If 11 A.M.'s are inserted between 28 and 10, then the number of integral A.M.'s. is ___.

3. Between 1 and 31, m arithmetic means are inserted so that the ratio of the seventh and $(m-1)^{\text{th}}$ means is 5 : 9. Then the value of $m/2$ is ___.

4. If $a_1 = 3$, $a_n = 96$, and $S_n = 189$, the number of terms in the G.P. is ___.

5. Three non-zero numbers a, b, and c are in A.P. If we increase a by 1 or increase c by 2, the numbers become in G.P. Then the value of $b - a$ is ___.

6. If the continued product of three numbers in G.P. is 216 and the sum of their products in pairs is 156, then the common ratio can be ___.

7. If nine arithmetic means and nine harmonic means are inserted between 2 and 3 alternatively, then the value of $A+\frac{6}{H}$ (where A is any of the A.M.'s and H the corresponding H.M.) is ___.

8. If the first two terms of a H.P. are 2/5 and 12/13, respectively, then which number of term is the largest term?

9. If the sum to infinity of the series $3 + (3 + d)(1/4) + (3 + 2d)(1/4^2) + \ldots\infty$ is 44/9, then the value of d is ___.

10. If $S = 1/(1 \times 3 \times 5) + 1/(3 \times 5 \times 7) + 1/(5 \times 7 \times 9) + \ldots\infty$, then the value of $[36S]$, (where [.] represents the greatest integer function) is ___.

11. If the terms of the A.P. $\sqrt{a-x}, \sqrt{x}, \sqrt{a+x}, \ldots$ are all integers, where $a, x > 0$, and a is the least composite value, then $a/5$ is ___.

12. Along a road lie an odd number of stones placed at intervals of 10 m. These stones have to be assembled around the middle stone. A person can carry only one stone at a time. A man carried the job with one of the end stones by carrying them in succession. In carrying all the stones, he covered a distance of 3 km. Then the number of stones is the square of ___.

13. If the sum of n terms of the series $[(2n + 1)/(2n - 1)] + 3[(2n + 1)(2n - 1)]^2 + 5[(2n + 1)(2n - 1)] + \ldots\infty$ is 36, then the value of n is ___.

14. If $\log_3 2$, $\log_3(2^x - 5)$, and $\log_3[2^x - (7/2)]$ are in A.P., then the value of x is ___.

15. Let $a_1, a_2, \ldots, a_{10}$ be in A. P. and $h_1, h_2, \ldots, h_{10}$ be in H.P. If $a_1 = h_1 = 2$ and $a_{10} = h_{10} = 3$, then a_4h_7 is ___.

16. $\sum_{k=1}^{360}\{1/[k\sqrt{k+1} + (k + 1)\sqrt{k}]\}$ is the ratio of two relative prime positive integers m and n. The value of $|m - n|$ is___.

17. If $S=\frac{1}{3^2+1}+\frac{1}{4^2+2}+\frac{1}{5^2+3}+\frac{1}{6^2+4}+\ldots\infty$, then the value of $\left[\frac{1}{S}\right]$, (where [·] represents the greatest integer function) is ___.

18. Consider the 10 numbers $ar, ar^2, ar^3, \ldots ar^{10}$. Their sum is 18 and the sum of their reciprocals is 6. If the product of these 10 numbers is a^b (where a and b are prime), then the value of $a + b$ is ___.

19. The value of the ratio $\left(1+\frac{1}{2^2}+\frac{1}{3^2}+\frac{1}{4^2}+\cdots\right)/\left(1-\frac{1}{2^2}+\frac{1}{3^2}-\frac{1}{4^2}+\cdots\right)$ is ___.

20. Let $a_n = 16, 4, 1, \ldots$ be a geometric sequence. Define P_n as the product of the first n terms. Then the value of $\frac{\sum_{n=1}^{\infty}\sqrt[n]{P_n}}{4}$ is ___.

21. If $a_n = (3/4) - (3/4)^2 + (3/4)^3 + \ldots (-1)^{n-1}(3/4)^n$ and $b_n = 1 - a_n$, then the least natural number n_0 such that $b_n > a_n\ \forall\, n \geq n_0$ is ___.

22. Let S_k, $k = 1, 2, \ldots, 100$, denote the sum of an infinite geometric series whose first term is $(k-1)/k!$ and the common ratio is $1/k$, then the value of $(100^2/100!) + \sum_{k=1}^{100}|(k^2 - 3k + 1)S_k|$ is ___.

23. Let $a_1, a_2, a_3, \ldots a_{11}$ be real numbers satisfying $a_1 = 15$, $27 - 2a_2 > 0$ and $a_k = 2a_{k-1} - a_{k-2}$ for $k = 3, 4, \ldots 11$. If $\frac{a_1^2+a_2^2+\cdots+a_{11}^2}{11} = 90$, then the value of $\frac{a_1+a_2+\cdots+a_{11}}{11}$ is ___.

24. The value of n, where n is a positive integer satisfying the equation $2 + (6 \times 2^2 - 4.\ 2) + (6 \times 3^2 - 4.3) + \cdots + (6 \times n^2 - 4n) = 140$, is __________.

25. The sum of the first n terms of the sequence 1, $(1 + 2)$, $(1 + 2 + 2^2)$,…, $(1 + 2 + 2^2 + \cdots + 2^{k-1})$ is of the form $2^{n+R} + Sn^2 + nT + U$ for all $n \in N$. Then the value of $|R + S + T + U|$ is __________.

26. Let $x_1, x_2, x_3,\ldots, x_n$ be a sequence of integers such that
a. $-1 \le x_i \le 2$ for $i = 1, 2, 3,\ldots, n$
b. $x_1 + x_2 + x_3 + \cdots + x_n = 19$
c. $x_1^2 + x_2^2 + \cdots + x_n^2 = 99$
Let m and M be the minimum and maximum possible values of $x_1^3 + x_2^3 + \cdots + x_n^3$, respectively. Then the value of M/m is__________.

27. The sequence $a_1, a_2, a_3,\ldots$ satisfies $a_1 = 19$, $a_9 = 99$ and for all $n \ge 3$, a_n is the arithmetic mean of the first $(n - 1)$ terms, then the value of $[a_2/20]$ is (where [.] represents greatest integer function) __________.

28. N is the set of all natural numbers and R is the set of all real numbers. A function $f: N \to R$ is given by

$$f(n) = \frac{4n + \sqrt{4n^2 - 1}}{\sqrt{2n+1} + \sqrt{2n-1}}, \text{ then the value of } \left[\frac{\sum_{r=1}^{40} f(r)}{40}\right]$$

is (where [.] represents greatest integer function) _____.

ANSWERS AND SOLUTIONS

Objective Type

1. d. The given numbers are in A.P. Therefore,

$2\log_4(2^{1-x}+1) = \log_2(5\times 2^x+1)+1$

$\Rightarrow 2\log_{2^2}\left(\frac{2}{2^x}+1\right) = \log_2(5\times 2^x+1)+\log_2 2$

$\Rightarrow \frac{2}{2}\log_2\left(\frac{2}{2^x}+1\right) = \log_2(5\times 2^x+1)2$

$\Rightarrow \log_2\left(\frac{2}{2^x}+1\right) = \log_2(10\times 2^x+2)$

$\Rightarrow \frac{2}{2^x}+1 = 10\times 2^x+2$

$\Rightarrow \frac{2}{y}+1 = 10y+2$, where $2^x = y$

$\Rightarrow 10y^2+y-2=0$

$\Rightarrow (5y-2)(2y+1)=0$

$\Rightarrow y = 2/5$ or $y=-1/2$

$\Rightarrow 2^x = 2/5$ or $2^x = -1/2$

$\Rightarrow x = \log_2(2/5)$ [$\because$ 2^x cannot be negative]

$\Rightarrow x = \log_2 2 - \log_2 5$

$\Rightarrow x = 1-\log_2 5$

2. b. Since a, b, c are in A.P., $b-a=d$ and $c-b=d$, where d is the common difference of the A.P.

$\therefore a = b-d$ and $c = b+d$

Now, $abc = 4$

$\Rightarrow (b-d)\,b\,(b+d) = 4$

$\Rightarrow b(b^2-d^2)=4$

But, $b(b^2-d^2) < b\times b^2$

$\Rightarrow b(b^2-d^2) < b^3$

$\Rightarrow 4 < b^3$

$\Rightarrow b^3 > 4$

$\Rightarrow b > 2^{2/3}$

Hence, the minimum value of b is $2^{2/3}$.

3. a. n^{th} term of the series is $20+(n-1)(-2/3)$.

For the sum to be maximum,

n^{th} term ≥ 0

$\Rightarrow 20+(n-1)\left(-\frac{2}{3}\right)\geq 0$

$\Rightarrow n \leq 31$

Thus, the sum of 31 terms is maximum and is

$\frac{31}{2}\left[40+30\times\left(-\frac{2}{3}\right)\right] = 310$

4. d. 100th term of 1, 11, 21, 31, … is $1+(100-1)10 = 991$.

100th term of 31, 36, 41, 46, … is $31+(100-1)5 = 526$.

Let the largest common term be 526. Then,

$526 = 31+(n-1)10$

$\Rightarrow n = 50.5$

But n is an integer; hence, $n = 50$. Hence, the largest common term is $31+(50-1)10 = 521$.

5. d. Let a be the first term and d be the common difference of the given A.P. Then,

$S_m = S_n \Rightarrow \frac{m}{2}[2a+(m-1)d] = \frac{n}{2}[2a+(n-1)d]$

$\Rightarrow 2a(m-n)+\{m(m-1)-n(n-1)\}\,d=0$

$\Rightarrow 2a(m-n)+\{(m^2-n^2)-(m-n)\}d=0$

$\Rightarrow (m-n)\,[2a+(m+n-1)d]=0$

$\Rightarrow 2a+(m+n-1)d = 0$ [$\because m-n\neq 0$] (1)

Now, $S_{m+n} = \frac{m+n}{2}[2a+(m+n-1)d] = \frac{m+n}{2}\times 0 = 0$

[Using (1)]

6. c. Let a_1, a_2 be the first terms and d_1, d_2 the common differences of the two given A.P.'s. Then the sums of their n terms are given by

$S_n = \frac{n}{2}[2a_1+(n-1)d_1]$ and $S_n' = \frac{n}{2}[2a_2+(n-1)d_2]$

$\therefore \frac{S_n}{S_n'} = \frac{\frac{n}{2}[2a_1+(n-1)d_1]}{\frac{n}{2}[2a_2+(n-1)d_2]} = \frac{2a_1+(n-1)d_1}{2a_2+(n-1)d_2}$ (1)

It is given that

$\frac{S_n}{S_n'} = \frac{7n+1}{4n+27}$

$\Rightarrow \frac{2a_1+(n-1)d_1}{2a_2+(n-1)d_2} = \frac{7n+1}{4n+27}$

To find the ratio of the mth terms of the two given A.P.'s, we replace n by $(2m-1)$ in (1). We get,

$\frac{2a_1+(2m-2)d_1}{2a_2+(2m-2)d_2} = \frac{7(2m-1)+1}{4(2m-1)+27}$

$\Rightarrow \frac{a_1+(m-1)d_1}{a_2+(m-1)d_2} = \frac{14m-6}{8m+23}$

Hence, the ratio of mth terms of the two A.P.'s is $(14m-6):(8m+23)$.

7. b. $\frac{\frac{n}{2}(2a+(n-1)d)}{\frac{n}{2}(2a'+(n-1)d')} = \frac{5n+3}{3n+4}$

$\Rightarrow \frac{(2a+(2n-2)d)}{(2a'+(2n-2)d')} = \frac{5(2n-1)+3}{3(2n-1)+4}$ (replace n by $2n-1$)

$$\Rightarrow \frac{(a+(n-1)d)}{(a'+(n-1)d')} = \frac{10n-2}{6n+1}$$

$$\Rightarrow \frac{(a+(17-1)d)}{(a'+(17-1)d')} = \frac{168}{103}$$

8. c. Suppose the work is completed in n days when the workers stopped working. Since four workers stopped working every day except the first day. Therefore, the total number of workers who worked all the n days is the sum of n terms of an A.P. with first term 150 and common difference –4, i.e.,

$$\frac{n}{2}[2\times150+(n-1)\times-4] = n(152-2n)$$

Had the workers not stopped working, then the work would have finished in $(n - 8)$ days with 150 workers working on each day. Therefore, the total number of workers who would have worked all the n days is $150(n - 8)$.

$\therefore n(152 - 2n) = 150(n - 8)$

$\Rightarrow n^2 - n - 600 = 0$

$\Rightarrow (n - 25)(n + 24) = 0$

$\Rightarrow n = 25$

Thus, the work is completed in 25 days.

9. c. Given, $S_p = 0$. Therefore,

$$\frac{p}{2}[2a+(p-1)d] = 0 \Rightarrow d = \frac{-2a}{p-1} \quad (1)$$

Sum of next q terms is sum of an A.P. whose first term will be

$T_{p+1} = a + pd.$

$$\therefore S = \frac{q}{2}[2(a+pd)+(q-1)d]$$

$$= \frac{q}{2}[2a+(p-1)d+(p+q)d]$$

$$= \frac{q}{2}\left[0-(p+q)\frac{2a}{p-1}\right]$$

$$= -a\frac{(p+q)q}{p-1} \quad \text{[Using (1)]}$$

10. b. $S_{3n} = \frac{3n}{2}[2a+(3n-1)d]$

$S_{n-1} = \frac{n-1}{2}[2a+(n-2)d]$

$\Rightarrow S_{3n} - S_{n-1}$

$$= \frac{1}{2}[2a(3n-n+1)] + \frac{d}{2}[3n(3n-1)-(n-1)(n-2)]$$

$$= \frac{1}{2}[2a(2n+1)+d(8n^2-2)]$$

$= a(2n + 1) + d(4n^2 - 1)$

$= (2n + 1)[a + (2n - 1)d]$

$S_{2n} - S_{2n-1} = T_{2n} = a + (2n - 1)d$

$$\Rightarrow \frac{S_{3n}-S_{n-1}}{S_{2n}-S_{2n-1}} = (2n+1)$$

Given,

$$\frac{S_{3n}-S_{n-1}}{S_{2n}-S_{2n-1}} = 31 \Rightarrow n = 15$$

11. d. $2b = a + c$

$\Rightarrow 8b^3 = (a + c)^3 = a^3 + c^3 + 3ac(a + c)$

$\Rightarrow 8b^3 = a^3 + c^3 + 3ac(2b)$

$\Rightarrow a^3 + c^3 - 8b^3 = -6abc$

12. b. Let the series have $2n$ terms and the series is $a, a + d, a + 2d, \ldots, a + (2n - 1)d$.

According to the given conditions, we have

$[a + (a + 2d) + (a + 4d) + \cdots + (a + (2n - 2)d)] = 24$

$\Rightarrow \frac{n}{2}[2a+(n-1)2d] = 24$

$\Rightarrow n[a + (n - 1)d] = 24 \quad (1)$

Also,

$[(a + d) + (a + 3d) + \cdots + (a + (2n - 1)d)] = 30$

$\Rightarrow \frac{n}{2}[2(a+d)+(n-1)2d] = 30$

$\Rightarrow n[(a + d) + (n - 1)d] = 30 \quad (2)$

Also, the last term exceeds the first by 21/2. Therefore,

$a + (2n - 1)d - a = 21/2$

$\Rightarrow (2n - 1)d = 21/2 \quad (3)$

Now, subtracting (1) from (2),

$nd = 6 \quad (4)$

Dividing (3) by (4), we get

$$\frac{2n-1}{n} = \frac{21}{12}$$

$\Rightarrow n = 4$

13. b. Since a, q, and c are in A.P., so

$2q = a + c$

$$\Rightarrow \frac{1}{p}+\frac{1}{r} = \frac{2}{b}$$

$$\Rightarrow \frac{1}{p}, \frac{1}{b}, \frac{1}{r} \text{ are in A.P.}$$

14. b. Given,

$$F(n+1) = \frac{2F(n)+1}{2}$$

$\Rightarrow F(n + 1) - F(n) = 1/2$

Hence, the given series is an A.P. with common difference 1/2 and first term being 2. $F(101)$ is 101st term of A.P. given by $2 + (101-1)(1/2) = 52$.

15. b. If t_r be the rth term of the A.P., then

$t_r = S_r - S_{r-1}$

$= cr(r - 1) - c(r - 1)(r - 2)$

$= c(r - 1)(r - r + 2) = 2c(r - 1)$

We have,

$t_1^2 + t_2^2 + \cdots + t_n^2 = 4c^2(0^2 + 1^2 + 2^2 + \cdots + (n - 1)^2)$

$$= 4c^2\,\frac{(n-1)n(2n-1)}{6}$$

$$= \frac{2}{3}\,c^2n\,(n-1)\,(2n-1)$$

16. c. Given that

$a_3 + a_5 + a_8 = 11$

$\Rightarrow\ a + 2d + a + 4d + a + 7d = 11$

$\Rightarrow 3a + 13d = 11$ (1)

Given,

$a_4 + a_2 = -2$

$\Rightarrow a + 3d + a + d = -2$

$\Rightarrow a = -1 - 2d$ (2)

Putting the value of a from (2) in (1), we get

$3(-1-2d) + 13d = 11 \Rightarrow 7d = 14 \Rightarrow d = 2$ and $a = -5$

$\Rightarrow a_1 + a_6 + a_7 = 7$

17. a. If p, q, r are in A.P., then in an A.P. or G.P. or an H.P. $a_1, a_2, a_3, \ldots,$ the terms a_p, a_q, a_r are in A.P., G.P. or H.P., respectively.

18. d. Given, $a_1, a_2, a_3, \ldots$ are terms of A.P.

$$\therefore\ \frac{a_1 + a_2 + \cdots + a_p}{a_1 + a_2 + \cdots + a_q} = \frac{p^2}{q^2}$$

$$\Rightarrow \frac{\frac{p}{2}[2a_1 + (p-1)d]}{\frac{q}{2}[2a_1 + (q-1)d]} = \frac{p^2}{q^2}$$

$$\Rightarrow \frac{2a_1 + (p-1)d}{2a_1 + (q-1)d} = \frac{p}{q}$$

$\Rightarrow [2a_1 + (p-1)d]\,q = [2a_1 + (q-1)d]\,p$

$\Rightarrow 2a_1\,(q-p) = d[(q-1)\,p - (p-1)q]$

$\Rightarrow 2a_1\,(q-p) = d(q-p)$

$\Rightarrow 2a_1 = d$

$$\therefore\ \frac{a_6}{a_{21}} = \frac{a_1 + 5d}{a_1 + 20d} = \frac{a_1 + 10a_1}{a_1 + 40a_1} = \frac{11}{41}$$

19. b.
$$\frac{S_{nx}}{S_x} = \frac{\frac{nx}{2}[2a + (nx-1)d]}{\frac{x}{2}[2a + (x-1)d]}$$

$$= \frac{n[(2a-d) + nxd]}{(2a-d) + xd}$$

For $\frac{S_{xn}}{S_x}$ to be independent of x,

$2a - d = 0 \Rightarrow 2a = d$

Now,

$$S_p = \frac{p}{2}[2a + (p-1)d] = p^2\,a$$

20. b. Let the three numbers be a/r, a, ar. As the numbers form an increasing G.P., so, $r > 1$. It is given that a/r, $2a$, ar are in A.P. Hence,

$$4a = \frac{a}{r} + ar$$

$\Rightarrow r^2 - 4r + 1 = 0$

$\Rightarrow r = 2 \pm \sqrt{3}$

$= 2 + \sqrt{3}\quad [\because r > 1]$

21. c. Let a_1, a_2, and a_3 be first three consecutive terms of G.P. with common ratio r. Then,

$a_2 = a_1\,r$ and $a_3 = a_1\,r^2$

Now,

$a_3 > 4a_2 - 3a_1$

$\Rightarrow a_1r^2 > 4a_1r - 3a_1$

$\Rightarrow r^2 > 4r - 3$

$\Rightarrow r^2 - 4r + 3 > 0$

$\Rightarrow (r-1)\,(r-3) > 0$

$\Rightarrow r < 1$ or $r > 3$

22. d. We know that

$-1 \le \cos x \le 1$

$\Rightarrow |\cos x| \le 1$

But,

$x \in S \Rightarrow x \in (0, \pi) \Rightarrow |\cos x| < 1$

Now,

$$8^{1 + |\cos x| + \cos^2 x + |\cos^3 x| + \cdots \text{ to } \infty} = 4^3$$

$\Rightarrow 8^{1/1-|\cos x|} = 8^2$

$$\Rightarrow \frac{1}{1 - |\cos x|} = 2$$

$$\Rightarrow |\cos x| = \frac{1}{2}$$

$$\Rightarrow \cos x = \pm\frac{1}{2}$$

$\Rightarrow x = \pi/3, 2\pi/3$

$\Rightarrow S = \{\pi/3, 2\pi/3\}$

23. c. We have,

$1 + (1+a)b + (1+a+a^2)b^2 + (1+a+a^2+a^3)b^3 + \cdots \infty$

$$= \sum_{n=1}^{\infty}(1 + a + a^2 + \cdots + a^{n-1})b^{n-1}$$

$$= \sum_{n=1}^{\infty}\left(\frac{1-a^n}{1-a}\right)b^{n-1}$$

$$= \sum_{n=1}^{\infty}\frac{b^{n-1}}{1-a} - \sum_{n=1}^{\infty}\frac{a^nb^{n-1}}{1-a}$$

$$= \frac{1}{1-a}\sum_{n=1}^{\infty}b^{n-1} - \frac{a}{1-a}\sum_{n=1}^{\infty}(ab)^{n-1}$$

$$= \frac{1}{1-a}[1 + b + b^2 + \cdots \infty] - \frac{a}{1-a}[1 + ab + (ab)^2 + \cdots \infty]$$

$$= \frac{1}{1-a} \times \frac{1}{1-b} - \frac{a}{(1-a)(1-ab)}$$

$$= \frac{1}{(1-ab)(1-b)}$$

24. c. Let A be first term and r be the common ratio. We have,

$a = Ar^{p+q-1},\ b = Ar^{p-q-1}$

$\Rightarrow ab = A^2 \times r^{2p-2}$

$\Rightarrow \sqrt{ab} = Ar^{p-1} = p^{\text{th}}$ term

25. b. Let the sides of the triangle be a/r, a and ar, with $a > 0$ and $r > 1$. Let α be the smallest angle, so that the largest angle is 2α. Then α is opposite to the side a/r, and 2α is positive to the side ar. Applying sine rule, we get

$$\frac{a/r}{\sin\alpha} = \frac{ar}{\sin 2\alpha}$$

$$\Rightarrow \frac{\sin 2\alpha}{\sin\alpha} = r^2$$

$$\Rightarrow 2\cos\alpha = r^2$$

$$\Rightarrow r^2 < 2$$

$$\Rightarrow r < \sqrt{2}$$

Hence, $1 < r < \sqrt{2}$.

26. a. We have,

$$\frac{1}{4} + \frac{1}{8} + \frac{1}{16} + \cdots = \frac{1/4}{1-1/2} = \frac{1}{2}$$

Hence,

$$0.2^{\log_{\sqrt{5}}\left(\frac{1}{4}+\frac{1}{8}+\frac{1}{16}+\cdots\right)} = 0.2^{\log_{\sqrt{5}}\frac{1}{2}}$$

$$= \left(\frac{1}{5}\right)^{\log_{\sqrt{5}}\frac{1}{2}}$$

$$= (5^{-1})^{2\log_5\frac{1}{2}}$$

$$= (5)^{-2\log_5\frac{1}{2}}$$

$$= (5)^{\log_5 4}$$

$$= 4$$

27. b. Degree of x on L.H.S. is

$$1 + 2 + 4 + \cdots + 128 = 1 + 2 + 2^2 + \cdots + 2^7$$

$$= \frac{2^8 - 1}{2 - 1}$$

$$= 255$$

28. c. x, y, and z are in G.P. Hence,

$$y^2 = xz \qquad (1)$$

We have,

$a^x = b^y = c^z = \lambda$ (say)

$\Rightarrow x \log a = y \log b = z \log c = \log \lambda$

$$\Rightarrow x = \frac{\log\lambda}{\log a}, y = \frac{\log\lambda}{\log b}, z = \frac{\log\lambda}{\log c}$$

Putting the values of x, y, and z in (1), we get

$$\left(\frac{\log\lambda}{\log b}\right)^2 = \frac{\log\lambda}{\log a}\frac{\log\lambda}{\log c}$$

$\Rightarrow (\log b)^2 = \log a \log c$

$\Rightarrow \log_b a = \log_c b$

29. b. Required G.M. is $-\sqrt{-9 \times -16} = -12$.

30. d. We have,

$$S = \frac{1}{1-\frac{1}{2}} = 2$$

$$S_n = \frac{(1-1/2^n)}{(1-1/2)} = 2\left(1-\frac{1}{2^n}\right) = 2 - \frac{1}{2^{n-1}}$$

$$\therefore S - S_n < \frac{1}{1000} \Rightarrow \frac{1}{2^{n-1}} < \frac{1}{1000}$$

$\Rightarrow 2^{n-1} > 1000$

$\Rightarrow n - 1 \geq 10$

$\Rightarrow n \geq 11$

Hence, the least value of n is 11.

31. b. We have,

$$S = \frac{1}{3} + \frac{1}{3^2} + \frac{1}{3^3} + \cdots = \frac{1/3}{1-1/3} = \frac{1}{2}$$

$$\therefore\ y = (0.64)^{\log_{0.25}(1/2)}$$

$$\Rightarrow \log y = \frac{\log(1/2)}{\log(0.25)}\log(0.64)$$

$$= \frac{-\log 2}{-\log 4}\log(0.64)$$

$$= \frac{1}{2}\log(0.64) = \log 0.8$$

$\Rightarrow y = 0.8$

32. c. For G.P., $t_n = 2^{n-1}$; for A.P., $T_m = 1 + (m-1)3 = 3m - 2$.

They are common if $2^{n-1} = 3m - 2$. For G.P. 100th term is 2^{99}. For A.P. 100th term is $1 + (100 - 1)3 = 298$. Now we must choose value of m such that $3m - 2$ is of type 2^{n-1}. Hence, $3m - 2 = 1, 2, 4, 8, 16, 32, 64, 128, 256$ for which $m = 1, 4/3, 2, 10/3, 6, 34/2, 22, 130/3, 86$. Hence, possible values of m are 1, 2, 6, 22, 86. Hence, there are five common terms.

33. c. Initially the ball falls from a height of 120 m. After striking the floor, it rebounds and goes to a height of $\frac{4}{5} \times (120)$ m. Now, it falls from a height of $\frac{4}{5} \times (120)$ m and after rebounding goes to a height of $\frac{4}{5}\left(\frac{4}{5}(120)\right)$ m. This process is continued till the ball comes to rest.

Hence, the total distance travelled is

$$120 + 2\left[\frac{4}{5}(120) + \left(\frac{4}{5}\right)^2(120) + \cdots \infty\right]$$

$$= 120 + 2\left[\frac{\frac{4}{5}(120)}{1-\frac{4}{5}}\right] = 1080 \text{ m}$$

34. b. Let a be the first term and r the common ratio of the G.P. Then, the sum is given by

$$\frac{a}{1-r} = 57 \quad (1)$$

Sum of the cubes is 9747. Hence,

$a^3 + a^3 r^3 + a^3 r^6 + \cdots = 9747$

$$\Rightarrow \frac{a^3}{1-r^3} = 9747 \quad (2)$$

Dividing the cube of (1) by (2), we get

$$\frac{a^3}{(1-r)^3} \frac{(1-r^3)}{a^3} = \frac{(57)^3}{9747}$$

$$\Rightarrow \frac{1+r^3}{(1-r)^3} = 19$$

$$\Rightarrow \frac{1+r+r^2}{(1-r)^2} = 19$$

$\Rightarrow 18r^2 - 39r + 18 = 0$

$\Rightarrow (3r-2)(6r-9) = 0$

$\Rightarrow r = 2/3$ or $r = 3/2$

$= 2/3$ [$\because r \neq 3/2$, because $-1 < r < 1$ for an infinite G.P.]

35. b. $a^2 + b^2, ab + bc, b^2 + c^2$ are in G.P.

$\Rightarrow (ab + bc)^2 = (a^2 + b^2)(b^2 + c^2)$

$\Rightarrow a^2b^2 + b^2c^2 + 2ab^2c = a^2b^2 + a^2c^2 + b^2c^2 + b^4$

$\Rightarrow b^4 + a^2c^2 - 2ab^2c = 0$

$\Rightarrow (b^2 - ac)^2 = 0$

$\Rightarrow b^2 = ac$

$\Rightarrow a, b,$ and c are in G.P.

36. b. $a = (6)_n = 6\,6\,6 \cdots 6$ (n digits)

$= 6 \times 1 + 6 \times 10 + 6 \times 10^2 + \cdots + 6 \times 10^{n-1}$

$$= \frac{6}{9}(10^n - 1) = \frac{2}{3}(10^n - 1)$$

$$b = \frac{8}{9}(10^n - 1),\ c = \frac{4}{9}(10^{2n} - 1)$$

Now,

$$a^2 + b = \frac{4}{9}(10^n - 1)^2 + \frac{8}{9}(10^n - 1)$$

$$= \frac{4}{9}(10^n - 1)(10^n - 1 + 2)$$

$$= \frac{4}{9}(10^{2n} - 1) = c$$

37. b. $a = 1 + 10 + 10^2 + \cdots + 10^{54}$

$$= \frac{10^{55} - 1}{10 - 1} = \frac{10^{55} - 1}{10^5 - 1} \times \frac{10^5 - 1}{10 - 1} = bc$$

38. a. Let a be the first term and r be the common ratio of the given G.P. Then,

$$\alpha = \sum_{n=1}^{100} a_{2n} \Rightarrow \alpha = a_2 + a_4 + \cdots + a_{200}$$

$= ar + ar^3 + \cdots + ar^{199}$

$= ar(1 + r^2 + r^4 + \cdots + r^{198})$

$$\beta = \sum_{n=1}^{100} a_{2n-1} \Rightarrow \beta = a_1 + a_3 + \cdots + a_{199}$$

$= a + ar^2 + \cdots + ar^{198}$

$= a(1 + r^2 + \cdots + r^{198})$

Clearly, $\alpha/\beta = r$.

39. c. The series is

$1 + 2 + 2 \times 3 + 2^2 \times 3 + 2^2 \times 3^2 + 2^3 \times 3^2 + \cdots$ to 20 terms

$= (1 + 2 \times 3 + 2^2 \times 3^2 + \cdots$ to 10 terms)

$+ (2 + 2^2 \times 3 + 2^3 \times 3^2 + \cdots$ to 10 terms)

$$= \frac{1(2^{10}3^{10} - 1)}{6 - 1} + \frac{2(2^{10}3^{10} - 1)}{6 - 1}$$

$$= \left(\frac{3}{5}\right)(6^{10} - 1)$$

40. d. $a = 5,\ ar^2 = a + 3d,\ ar^4 = a + 15d$

$\therefore 5r^2 = 5 + 3d,\ 5r^4 = 5 + 15d$

$\Rightarrow r^4 = 1 + 3d$

$\Rightarrow 25r^4 = 25 + 75d$

$\Rightarrow (5 + 3d)^2 = 25 + 75d$

$\Rightarrow 25 + 30d + 9d^2 = 25 + 75d$

$\Rightarrow 9d^2 - 45d = 0$

$\Rightarrow d = 5, 0$

$\Rightarrow T_4 = a + 3d = 5 + 15 = 20$

41. d. p^{th}, q^{th}, r^{th} terms of A.P. are

$a + (p-1)d = x \quad (1)$

$a + (q-1)d = xR \quad (2)$

$a + (r-1)d = xR^2, \quad (3)$

where R is common ratio of G.P.

Subtracting (2) from (3) and (1) from (2) and then dividing the former by the later, we have

$$\frac{q-r}{p-q} = \frac{xR^2 - xR}{xR - x} = R$$

42. b. Given that

$a + (p-1)d = A$

$a + (q-1)d = AR$

$a + (r-1)d = AR^2$

$a + (s-1)d = AR^3,$

where R is common ratio of G.P. Now,

$$p - q = \frac{A - AR}{d},\ q - r = R\left(\frac{A - AR}{d}\right),$$

$$r - s = R^2\left(\frac{A - AR}{d}\right)$$

Clearly, $p - q, q - r, r - s$ are in G.P.

43. a. Let r be the common ratio of the G.P., a, b, c, d. Then,

$b = ar,\ c = ar^2$ and $d = ar^3$

$\therefore (b - c)^2 + (c - a)^2 + (d - b)^2$

$= (ar - ar^2)^2 + (ar^2 - a)^2 + (ar^3 - ar)^2$

$= a^2r^2(1 - r)^2 + a^2(r^2 - 1)^2 + a^2r^2(r^2 - 1)^2$

$= a^2(r^6 - 2r^3 + 1)$

$= a^2(1 - r^3)^2$

$= (a - ar^3)^2$

$= (a - d)^2$

44. d. Let $P = 0.cababab \ldots$

$\Rightarrow 10P = c.ababab\ldots \quad (1)$

and

$1000P = cab.ababab\ldots \quad (2)$

Subtracting Eq. (1) from Eq. (2), we have

$990P = cab - c$

or

$$P = \frac{100c + 10a + b - c}{990} = \frac{99c + 10a + b}{990}$$

45. d. $S_\infty = \frac{a}{1-r} = 162$

$S_n = \frac{a(1-r^n)}{1-r} = 160$

Dividing,

$1 - r^n = \frac{160}{162} = \frac{80}{81}$

$\Rightarrow 1 - \frac{80}{81} = r^n$

$\Rightarrow r^n = \frac{1}{81}$ or $\left(\frac{1}{r}\right)^n = 81$ (1)

Now, it is given that $1/r$ is an integer and n is also an integer.

Hence, the relation (1) implies that $1/r = 3, 9$ or 81 so that $n = 4, 2$ or 1.

$\therefore a = 162\left(1-\frac{1}{3}\right)$ or $162\left(1-\frac{1}{9}\right)$ or $162\left(1-\frac{1}{81}\right)$

$= 108$ or 144 or 160

46. d. $f(x) = 2x + 1$

$\Rightarrow f(2x) = 2(2x) + 1 = 4x + 1$ and

$f(4x) = 2(4x) + 1 = 8x + 1$

Now, $f(x), f(2x), f(4x)$ are in G.P. Hence,

$(4x + 1)^2 = (2x + 1)(8x + 1)$

$\Rightarrow 2x = 0$

Hence, $f(x), f(2x)$, and $f(4x)$ is equal to 1 which contradicts the given condition. Therefore, a, b, and c are in H.P.

47. b.

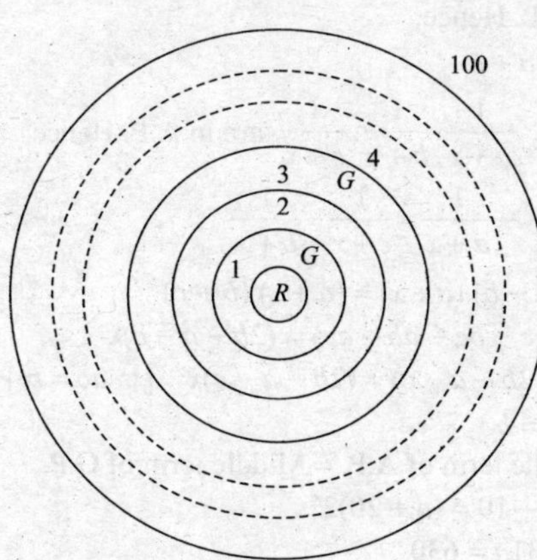

Fig. 3.2

$\pi\left[(r_2^2 - r_1^2) + (r_4^2 - r_3^2) + \cdots + (r_{100}^2 - r_{99}^2)\right]$

$= \pi\,[r_1 + r_2 + r_3 + r_4 + \cdots + r_{100}]$

$(\because r_2 - r_1 = r_4 - r_3 = \cdots = r_{100} - r_{99} = 1)$

$= \pi\,[1 + 2 + 3 + \cdots + r_{100}] = 5050\pi$ sq. cm

48. a. $\frac{t_4}{t_6} = \frac{1}{4} \Rightarrow \frac{ar^3}{ar^5} = \frac{1}{4} \Rightarrow r^2 = 4 \Rightarrow r = \pm 2$

Also,

$t_2 + t_5 = 216$

$\Rightarrow ar + ar^4 = 216$

$\Rightarrow a + 8a = 108$

$\Rightarrow a = 12$ (when $r = 2$)

49. b. x, y, and z are in G.P. Hence,

$y = xr, z = xr^2$

Also, x, $2y$, and $3z$ are in A.P. Hence,

$4y = x + 3z$

$\Rightarrow 4xr = x + 3xr^2$

$\Rightarrow 3r^2 - 4r + 1 = 0$

$\Rightarrow (3r - 1)(r - 1) = 0$

$\Rightarrow r = 1/3$ [$r \neq 1$ is not possible as x, y, z are distinct]

50. a. $S_p = \frac{1}{1-r^p}$, $s_p = \frac{1}{1+r^p}$, $S_{2p} = \frac{1}{1-r^{2p}}$

Clearly,

$S_p + s_p = \frac{2}{1-r^{2p}} = 2S_{2p}$

51. c. Multiplying the given expression by 2 and rewriting it, we have

$(2x - 3y)^2 + (3y - 4z)^2 + (4z - 2x)^2 = 0$

$\Rightarrow 2x = 3y = 4z$

$\Rightarrow \frac{1}{x}, \frac{1}{y}, \frac{1}{z}$ are in A.P.

$\Rightarrow x, y, z$ are in H.P.

52. c. $a_1, a_2, \ldots, a_n$ are in H.P.

$\Rightarrow \frac{1}{a_1}, \frac{1}{a_2}, \ldots, \frac{1}{a_n}$ are in A.P.

$\Rightarrow \frac{a_1 + a_2 + a_3 + \cdots + a_n}{a_1}, \frac{a_1 + a_2 + a_3 + \cdots + a_n}{a_2}, \ldots, \frac{a_1 + a_2 + a_3 + \cdots + a_n}{a_n}$ are in A.P.

$\Rightarrow 1 + \frac{a_2 + a_3 + \cdots + a_n}{a_1}, 1 + \frac{a_1 + a_3 + \cdots + a_n}{a_2}, \ldots, 1 + \frac{a_1 + a_2 + \cdots + a_{n-1}}{a_n}$ are in A.P.

$\Rightarrow \frac{a_2 + a_3 + \cdots + a_n}{a_1}, \frac{a_1 + a_3 + \cdots + a_n}{a_2}, \ldots, \frac{a_1 + a_2 + \cdots + a_{n-1}}{a_n}$ are in A.P.

$\Rightarrow \frac{a_1}{a_2 + a_3 + \cdots + a_n}, \frac{a_2}{a_1 + a_3 + \cdots + a_n}, \ldots, \frac{a_n}{a_1 + a_2 + \cdots a_{n-1}}$ are in H.P.

53. c. $\frac{H_1 + 2}{H_1 - 2} + \frac{H_{20} + 3}{H_{20} - 3} = \frac{\frac{1}{2} + \frac{1}{H_1}}{\frac{1}{2} - \frac{1}{H_1}} + \frac{\frac{1}{3} + \frac{1}{H_{20}}}{\frac{1}{3} - \frac{1}{H_{20}}}$

$$= \frac{\frac{1}{2}+\frac{1}{2}+d}{\frac{1}{2}-d-\frac{1}{2}} + \frac{\frac{1}{3}+\frac{1}{3}-d}{\frac{1}{3}+d-\frac{1}{3}}$$

$$= \frac{\frac{2}{2}+d}{-d} + \frac{\frac{2}{3}-d}{d}$$

$$= \frac{\frac{2}{3}-1}{d} - 2$$

$= 2 \times 21 - 2$ $\left[\text{as also, } \frac{1}{3} = \frac{1}{2} + 21d\right]$

$= 40$

54. b. We have,

a_1, a_2, a_3 are in A.P. $\Rightarrow 2a_2 = a_1 + a_3$ (1)

a_2, a_3, a_4 are in G.P. $\Rightarrow a_3^2 = a_2a_4$ (2)

a_3, a_4, a_5 are in H.P. $\Rightarrow a_4 = \frac{2a_3a_5}{a_3 + a_5}$ (3)

Putting $a_2 = \frac{a_1 + a_3}{2}$ and $a_4 = \frac{2a_3a_5}{a_3 + a_5}$ in (2), we get

$a_3^2 = \frac{a_1 + a_3}{2} \times \frac{2a_3a_5}{a_3 + a_5}$

$\Rightarrow a_3^2 = a_1a_5$

Hence, a_1, a_3, and a_5 are in G.P. So, $\log_e a_1$, $\log_e a_3$ and $\log_e a_5$ are in A.P.

55. d. a, b, and c are in A.P. Hence,

$2b = a + c$ (1)

$\frac{a}{bc} + \frac{2}{b} = \frac{a+2c}{bc} \neq \frac{2}{c}$

$\Rightarrow \frac{a}{bc}, \frac{1}{c}, \frac{2}{b}$ are not in A.P.

$\frac{bc}{a} + \frac{b}{2} = \frac{2bc+ab}{2a} \neq c$

Hence, the given numbers are not in H.P. Again,

$\frac{a}{bc}\frac{2}{b} = \frac{2a}{b^2c} \neq \frac{1}{c^2}$

Therefore, the given numbers are not in G.P.

56. d. x, $2x + 2$, $3x + 3$ are in G.P. Hence,

$(2x + 2)^2 = x(3x + 3)$

$\Rightarrow 4x^2 + 8x + 4 = 3x^2 + 3x$

$\Rightarrow x^2 + 5x + 4 = 0$

$\Rightarrow x = -1, -4$

So, the G.P. is $-4, -6, -9, \ldots$ (considering $x = -4$, as for $x = -1$, $2x + 2 = 0$). Hence, the fourth term is $-9 \times 1.5 = -13.5$.

57. a. Let the numbers be a, ar, ar^2. Then,

$a + ar + ar^2 = 14$ (given) (1)

Now,

$a + 1$, $ar + 1$, $ar^2 - 1$ are in A.P.

$\Rightarrow 2(ar + 1) = a + 1 + ar^2 - 1$

$\Rightarrow 2ar + 2 = a + ar^2$ (2)

From (1) and (2),

$2ar + 2 = 14 - ar$

$\Rightarrow 3ar = 12$

$\Rightarrow ar = 4$ (3)

From (1),

$a + 4 + 4r = 14$

$\Rightarrow a + 4r = 10$ (4)

From (2) and (4),

$a + \frac{16}{a} = 10 \Rightarrow a = 2, 8$

Hence, the smallest number is 2.

58. b. $\frac{p}{r} + \frac{r}{p} = \frac{p^2 + r^2}{pr} = \frac{(p+r)^2 - 2pr}{pr}$

$= \frac{\frac{4p^2r^2}{q^2} - 2pr}{pr}$ $\left[\because p, q, r \text{ are in H.P. } \therefore q = \frac{2pr}{p+r}\right]$

$= \frac{4pr}{q^2} - 2 = \frac{4b^2}{ac} - 2$

$[\because ap, bq, cr \text{ are in A.P.} \Rightarrow b^2q^2 = acpr]$

$= \frac{(a+c)^2}{ac} - 2$ $[a, b, c \text{ are in A.P.} \Rightarrow 2b = a + c]$

$= \frac{a}{c} + \frac{c}{a}$

59. a. We have,

$2b = a + c$

$(c - b)^2 = (b - a)a$

$\Rightarrow (b - a)^2 = (b - a)a$ $[2b = a + c \Rightarrow b - a = c - b]$

$\Rightarrow b = 2a$

$\Rightarrow c = 3a$ [Using $2b = a + c$]

$\Rightarrow a:b:c = 1:2:3$

60. b. As a, b, c are distinct and are in G.P., so a, b, c cannot be in A.P. Hence,

$2b \neq a + c$

Now, $\frac{1}{a+x}, \frac{1}{b+x}, \frac{1}{c+x}$ are in A.P. Hence,

$\frac{1}{b+x} - \frac{1}{a+x} = \frac{1}{c+x} - \frac{1}{b+x}$

$\Rightarrow (a - b)(c + x) = (a + x)(b - c)$

$\Rightarrow (ac - bc - ab + ac) = (2b - a - c)x$

$\Rightarrow b(2b - a - c) = (2b - a - c)x$ $[\because ac = b^2]$

$\Rightarrow x = b$

61. a. Middle term of A.P. = Middle term of G.P.

$\Rightarrow a + 10 = (a + 20)2^5$

$\Rightarrow -31a = 630$

$\Rightarrow a = -630/31$

Now, middle term of entire sequence is

$a + 20 = -\frac{630}{31} + 20 = -\frac{10}{31}$

62. c. $2b = a + c$, $c = \frac{2bd}{b+d}$

$\Rightarrow 2bd = c(b + d)$

$\Rightarrow (a+c)d = c(b+d)$ [as $2b = a+c$]

$\Rightarrow ad + cd = bc + cd$

$\Rightarrow bc = ad$

63. c. $\dfrac{a_r - a_{r+1}}{a_r\, a_{r+1}} = k$ (constant)

$\Rightarrow \dfrac{1}{a_{r+1}} - \dfrac{1}{a_r} = k$

$\Rightarrow \dfrac{1}{a_1}, \dfrac{1}{a_2}, \ldots, \dfrac{1}{a_n}$ are in A.P.

$\Rightarrow a_1, a_2, a_3, \ldots,$ are in H.P.

64. c. Let $a = 1, b = 2, c = 4$ Then,

$a + b = 3, 2b = 4, b + c = 6$

$\Rightarrow \dfrac{1}{4} - \dfrac{1}{3} = -\dfrac{1}{12}$ and $\dfrac{1}{6} - \dfrac{1}{4} = -\dfrac{1}{12}$

Hence, $a + b, 2b, b + c$ are in H.P.

65. b. x is A.M. of a and b, y is G.M. of a and b, z is H.M. of a and b.

$y^2 = xz$

Also given,

$x = 9z$

$\Rightarrow x = 9y^2/x \Rightarrow 9y^2 = x^2 \Rightarrow x = 3|y|$

66. d. $A = \dfrac{25+n}{2}, G = 5\sqrt{n}, H = \dfrac{50n}{25+n}$

As A, G, H are natural numbers, n must be odd perfect square. Now, H will be a natural number, if we take $n = 225$.

67. a. Reciprocals of the terms of the series are 2/5, 13/20, 9/10, 23/20, … or 8/20, 13/20, 18/20, 23/20, …. Its nth term is

$\dfrac{8+(n-1)5}{20} = \dfrac{5n+3}{20}$

Therefore, the 15th term is 20/48 = 5/12.

68. b. Given, $b^2 = ac$ and $x = \dfrac{a+b}{2}, y = \dfrac{b+c}{2}$. Therefore,

$\dfrac{a}{x} + \dfrac{c}{y} = \dfrac{2a}{a+b} + \dfrac{2c}{b+c}$

$= \dfrac{2a(b+c) + 2c(a+b)}{(a+b)(b+c)}$

$= 2\dfrac{2ac + ab + bc}{ab + ac + b^2 + bc}$

$= 2\dfrac{2ac + ab + bc}{2ac + ab + bc}$

$= 2$

69. c. $2b = a + c$

a, p, b, q, c are in A.P. Hence,

$p = \dfrac{a+b}{2}$ and $q = \dfrac{b+c}{2}$

Again, a, p', b, q', and c are in G.P. Hence,

$p' = \sqrt{ab}$ and $q' = \sqrt{bc}$

$\Rightarrow p^2 - q^2 = \dfrac{(a-c)(a+c+2b)}{4}$

$= (a-c)b$

$= ab - bc$

$= p'^2 - q'^2$

70. d. As $a_1, a_2, a_3, \ldots, a_{n-1}, a_n$ are in A.P., hence

$d = a_2 - a_1 = a_3 - a_2 = \cdots = a_n - a_{n-1}$

$\sin d[\sec a_1 \sec a_2 + \sec a_2 \sec a_3 + \cdots + \sec a_{n-1} \sec a_n]$

$= \dfrac{\sin(a_2 - a_1)}{\cos a_1 \cos a_2} + \dfrac{\sin(a_3 - a_2)}{\cos a_2 \cos a_3} + \cdots + \dfrac{\sin(a_n - a_{n-1})}{\cos a_{n-1} \cos a_n}$

$= (\tan a_2 - \tan a_1) + (\tan a_3 - \tan a_2) + \cdots +$

$(\tan a_n - \tan a_{n-1})$

$= \tan a_n - \tan a_1$

71. c. $S = [a - (a+d)] + [(a+2d) - (a+3d)] + \cdots + [(a+(2n-2)d)] - a + (2n-1)d] + (a + 2nd)$

$= [(-d) + (-d) + \cdots + n \text{ times}] + a + 2nd$

$= -nd + a + 2nd$

$= a + nd$

72. a. Let $1 + (1/50) = x$. Let S be the sum of 50 terms of the given series. Then,

$S = 1 + 2x + 3x^2 + 4x^3 + \cdots + 49x^{48} + 50x^{49}$ (1)

$xS = x + 2x^2 + 3x^3 + 4x^3 + \cdots + 49x^{48} + 50x^{49}$ (2)

$(1-x)S = 1 + x + x^2 + x^3 + \cdots + x^{48} - 50x^{49}$

[Subtracting (2) from (1)]

$\Rightarrow S(1-x) = \dfrac{1-x^{50}}{1-x} - 50x^{50}$

$\Rightarrow S(-1/50) = -50(1 - x^{50}) - 50x^n$

$\Rightarrow \dfrac{1}{50} S = 50$

$\Rightarrow S = 2500$

73. a. Let T_r be the rth term of the given series. Then,

$T_r = \dfrac{2r+1}{1^2 + 2^2 + \cdots + r^2}$

$= \dfrac{6(2r+1)}{(r)(r+1)(2r+1)}$

$= 6\left(\dfrac{1}{r} - \dfrac{1}{r+1}\right)$

So, sum is given by

$\sum_{r=1}^{50} T_r = 6\sum_{r=1}^{50}\left(\dfrac{1}{r} - \dfrac{1}{r+1}\right)$

$= 6\left[\left(1 - \dfrac{1}{2}\right) + \left(\dfrac{1}{2} - \dfrac{1}{3}\right) + \left(\dfrac{1}{3} - \dfrac{1}{4}\right) + \cdots + \left(\dfrac{1}{50} - \dfrac{1}{51}\right)\right]$

$= 6\left[1 - \dfrac{1}{51}\right]$

$= \dfrac{100}{17}$

74. a. $\frac{1}{1^2}+\frac{1}{3^2}+\frac{1}{5^2}+\frac{1}{7^2}+\cdots$

$$=\left(\frac{1}{1^2}+\frac{1}{2^2}+\frac{1}{3^2}+\frac{1}{4^2}+\frac{1}{5^2}+\frac{1}{6^2}+\frac{1}{7^2}+\cdots\right)-\left(\frac{1}{2^2}+\frac{1}{4^2}+\frac{1}{6^2}+\cdots\right)$$

$$=\frac{\pi^2}{6}-\frac{1}{4}\left(\frac{1}{1^2}+\frac{1}{2^2}+\frac{1}{3^2}+\cdots\right)$$

$$=\frac{\pi^2}{6}-\frac{1}{4}\left(\frac{\pi^2}{6}\right)$$

$$=\frac{\pi^2}{8}$$

75. b. Coefficient of x^{18} in $(1+x+2x^2+3x^3+\cdots+18x^{18})^2$

= Coefficient of x^{18} in $(1+x+2x^2+3x^3+\cdots+18x^{18})\times(1+x+2x^2+3x^3+\cdots+18x^{18})$

$$=1\times 18+1\times 17+2\times 16+\cdots+17\times 1+18\times 1$$

$$=36+(1\times 17+2\times 16+\cdots+17\times 1)$$

$$=36+\sum_{r=1}^{17}r(18-r)$$

$$=36+18\sum_{r=1}^{17}r-\sum_{r=1}^{17}r^2$$

$$=36+18\frac{18(17)}{2}-\frac{17(18)(35)}{6}$$

$$=1005$$

76. c. $T(r)=\frac{r}{1\times 3\times 5\times\cdots\times(2r+1)}$

$$=\frac{2r+1-1}{2(1\times 3\times 5\cdots(2r+1))}$$

$$=\frac{1}{2}\left(\frac{1}{1\times 3\times 5\cdots(2r-1)}-\frac{1}{1\times 3\times 5\cdots(2r+1)}\right)$$

$$=-\frac{1}{2}[V(r)-V(r-1)]$$

$$\Rightarrow\sum_{r=1}^{n}T(r)=-\frac{1}{2}(V(n)-V(0))$$

$$=\frac{1}{2}\left(1-\frac{1}{1\times 3\times 5\times\cdots\times(2n+1)}\right)$$

$$\Rightarrow\lim_{n\to\infty}\sum_{r=1}^{n}\frac{r}{1\times 3\times 5\times 7\times 9\times\cdots\times(2r+1)}$$

$$=\lim_{n\to\infty}\frac{1}{2}\left(1-\frac{1}{1\times 3\times 5\times\cdots\times(2n+1)}\right)$$

$$=\frac{1}{2}$$

77. d. $r\times r!=(r+1-1)\times r!$

$$=(r+1)!-r!$$

$$=V(r)-V(r-1)$$

$$\Rightarrow\sum_{r=1}^{30}r(r!)=V(31)-V(0)$$

$$=(31)!-1$$

$$\Rightarrow 1+\sum_{r=1}^{30}r(r!)=31!,$$

which is divisible by 31 consecutive integers which is a prime number.

78. b. $I(2n)=1^4+2^4+3^4+\cdots+(2n-1)^4+(2n)^4$

$$=[(1^4+3^4+5^4+\cdots+(2n-1)^4]+2^4(1^4+2^4+3^4+4^4+\cdots+n^4)$$

$$=\sum_{r=1}^{n}(2r-1)^4+16\times I(n)$$

$$\Rightarrow\sum_{r=1}^{n}(2r-1)^4=I(2n)-16I(n)$$

79. c. Consider the first product,

$$P=\left(1+\frac{1}{3}\right)\left(1+\frac{1}{3^2}\right)\left(1+\frac{1}{3^4}\right)\left(1+\frac{1}{3^8}\right)\cdots\left(1+\frac{1}{3^{2^n}}\right)$$

$$=\frac{\left(1-\frac{1}{3}\right)\left(1+\frac{1}{3}\right)\left(1+\frac{1}{3^2}\right)\left(1+\frac{1}{3^4}\right)\left(1+\frac{1}{3^8}\right)\cdots\left(1+\frac{1}{3^{2^n}}\right)}{\left(1-\frac{1}{3}\right)}$$

$$=\frac{\left(1-\frac{1}{3^2}\right)\left(1+\frac{1}{3^2}\right)\left(1+\frac{1}{3^4}\right)\left(1+\frac{1}{3^8}\right)\cdots\left(1+\frac{1}{3^{2^n}}\right)}{\left(1-\frac{1}{3}\right)}$$

$$=\frac{\left(1-\frac{1}{3^4}\right)\left(1+\frac{1}{3^4}\right)\left(1+\frac{1}{3^8}\right)\cdots\left(1+\frac{1}{3^{2^n}}\right)}{\left(1-\frac{1}{3}\right)}$$

$$=\frac{1}{\left(1-\frac{1}{3}\right)}\left(1-\left(\frac{1}{3}\right)^{2^{n+1}}\right)$$

$$=\frac{3}{2}\left(1-\left(\frac{1}{3}\right)^{2^{n+1}}\right)$$

$$\Rightarrow\left(1+\frac{1}{3}\right)\left(1+\frac{1}{3^2}\right)\left(1+\frac{1}{3^4}\right)\left(1+\frac{1}{3^8}\right)\ldots\text{ infinity}$$

$$=\lim_{n\to\infty}\frac{3}{2}\left(1-\left(\frac{1}{3}\right)^{2^{n+1}}\right)$$

$$=\frac{3}{2}$$

80. a. Clearly, $\frac{1}{x_1},\frac{1}{x_2},\ldots,\frac{1}{x_{20}}$ will be in A.P. Hence,

$$\frac{1}{x_2}-\frac{1}{x_1}=\frac{1}{x_3}-\frac{1}{x_2}=\cdots=\frac{1}{x_{r+1}}-\frac{1}{x_r}=\cdots=\lambda\text{ (say)}$$

$$\Rightarrow\frac{x_r-x_{r+1}}{x_rx_{r+1}}=\lambda$$

$$\Rightarrow x_rx_{r+1}=-\frac{1}{\lambda}(x_{r+1}-x_r)$$

$$\Rightarrow \sum_{r=1}^{19} x_r x_{r+1} = -\frac{1}{\lambda}\sum_{r=1}^{19}(x_{r+1}-x_r)$$

$$= -\frac{1}{\lambda}(x_{20}-x_1)$$

Now,

$$\frac{1}{x_{20}} = \frac{1}{x_1} + 19\lambda$$

$$\Rightarrow \frac{x_1 - x_{20}}{x_1 x_{20}} = 19\lambda$$

$$\Rightarrow \sum_{r=1}^{19} x_r x_{r+1} = 19x_1x_{20} = 19 \times 4 = 76$$

(∵ x_1, 2, x_{20} are in G.P., then $x_1x_{20} = 4$)

81. b. $T_r = r(-a)^r + (r+1)\,a(-a)^r$

$= r(-a)^r - (r+1)(-a)^{r+1}$

$= v_r - v_{r+1}$ (say)

So,

$$\sum_{r=0}^{n} T_r = \sum_{r=0}^{n}(v_r - v_{r+1})$$

$= v_0 - v_{n+1}$

$= -(n+1)(-a)^{n+1}$

82. d. $\Sigma a_i b_i = \Sigma a_i(1 - a_i)$

$= na - \Sigma a_i^2$

$= na - \Sigma(a_i - a + a)^2$

$= na - \Sigma[(a_i - a)^2 + a^2 + 2a(a_i - a)]$

$= na - \Sigma(a_i - a)^2 - \Sigma a^2 - 2a\,\Sigma(a_i - a)$

$\Rightarrow \Sigma a_i b_i + \Sigma(a_i - a)^2 = na - na^2 - 2a(na - na)$

$= na(1 - a) = nab$

$$\left[\begin{array}{l} \because \Sigma b_i = \Sigma 1 - \Sigma a_i \\ \therefore nb = n - na \\ \text{or } a + b = 1 \end{array}\right]$$

83. a. The general term of the given series is

$$t_n = \frac{x^{2^{n-1}}}{1 - x^{2^n}} = \frac{1 + x^{2^{n-1}} - 1}{\left(1 + x^{2^{n-1}}\right)\left(1 - x^{2^{n-1}}\right)}$$

$$\Rightarrow t_n = \frac{1}{1 - x^{2^{n-1}}} - \frac{1}{1 - x^{2^n}}$$

Now,

$$S_n = \sum_{n=1}^{n} t_n = \left[\left\{\frac{1}{1-x} - \frac{1}{1-x^2}\right\} + \left\{\frac{1}{1-x^2} - \frac{1}{1-x^4}\right\} + \cdots + \left\{\frac{1}{1-x^{2^{n-1}}} - \frac{1}{1-x^{2^n}}\right\}\right]$$

$$= \frac{1}{1-x} - \frac{1}{1-x^{2^n}}$$

Therefore, the sum to infinite terms is

$$\lim_{n\to\infty} S_n = \frac{1}{1-x} - 1$$

$$= \frac{x}{1-x} \quad \left[\because \lim_{n\to\infty} x^{2n} = 0, \text{ as } |x| < 1\right]$$

84. a. The general term can be given by

$$t_{r+1} = \frac{a_{2n+1-r} - a_{r+1}}{a_{2n+1-r} + a_{r+1}},\ r = 0, 1, 2, \ldots, n-1$$

$$= \frac{a_1 + (2n - r)\,d - \{a_1 + rd\}}{a_1 + (2n - r)\,d + \{a_1 + rd\}}$$

$$= \frac{(n - r)d}{a_1 + nd}$$

Therefore, the required sum is

$$S_n = \sum_{r=0}^{n-1} t_{r+1}$$

$$= \sum_{r=0}^{n-1} \frac{(n-r)\,d}{a_1 + nd}$$

$$= \left[\frac{n + (n-1) + (n-2) + \cdots + 1}{a_1 + nd}\right] d$$

$$= \frac{n(n+1)\,d}{2a_{n+1}}$$

$$= \frac{n(n+1)}{2}\ \frac{a_2 - a_1}{a_{n+1}} \quad [\because d = a_2 - a_1]$$

85. a. Let,

$S = i - 2 - 3i + 4 + 5i + \cdots + 100i^{100}$

$= i + 2i^2 + 3i^3 + 4i^4 + 5i^5 + \cdots + 100i^{100}$

$\Rightarrow iS = i^2 + 2i^3 + 3i^4 + \cdots + 99i^{100} + 100i^{101}$

$\Rightarrow S - iS = i + [i^2 + i^3 + i^4 + \cdots + i^{100}] - 100i^{101}$

$$\Rightarrow S(1 - i) = i + i^2\,\frac{(1 - i^{99})}{(1 - i)} - 100i^{101}$$

$$= i\,\frac{(1 + i)}{(1 - i)} - 100i$$

$$= i + \frac{(1+i)^2}{2} - 100i = -100i$$

$$\Rightarrow S = \frac{-100i}{1 - i} = -50i(1 + i) = -50(i - 1) = 50(1 - i)$$

86. b. $$S = \frac{4}{19} + \frac{44}{19^2} + \frac{444}{19^3} + \cdots \quad (1)$$

$$\Rightarrow \frac{1}{19}S = \frac{4}{19^2} + \frac{44}{19^3} + \cdots \quad (2)$$

Subtracting (2) from (1), we get

$$\frac{18}{19}S = \frac{4}{19} + \frac{40}{19^2} + \frac{400}{19^3} + \cdots$$

$$= \frac{\frac{4}{19}}{1 - \frac{10}{19}} = 4/9$$

$\Rightarrow S = 38/81$

87. b. $$S_n = 1 + \frac{3}{2} + \frac{5}{3} + \cdots + \frac{99}{50}$$

$$= (2 - 1) + \left(2 - \frac{1}{2}\right) + \left(2 - \frac{1}{3}\right) + \cdots + \left(2 - \frac{1}{50}\right)$$

$= 100 - H_{50}$

88. b. $S = 1 + 2r + 3r^2 + 4r^3 + \cdots$

$rS = r + 2r^2 + 3r^3 + 4r^4 + \cdots$

$(1-r)S = 1 + r + r^2 + r^3 + \cdots$

$= \frac{1}{1-r}$

$\Rightarrow S = \frac{1}{(1-r)^2}$

Given, $S = 9/4 \Rightarrow \frac{1}{(1-r)^2} = 9/4$

$\Rightarrow 1 - r = \pm\frac{2}{3}$

$\Rightarrow r = 1/3$ or $5/3$

Hence, $r = 1/3$ as $-1 < r < 1$.

89. d. Let,

$$S = 1 + \frac{4}{5} + \frac{7}{5^2} + \frac{10}{5^3} + \cdots\infty$$

Then,

$$\frac{1}{5}S = \frac{1}{5} + \frac{4}{5^2} + \frac{7}{5^3} + \cdots\infty$$

$$\Rightarrow S\left(1-\frac{1}{5}\right) = 1 + 3\left[\frac{1}{5} + \frac{1}{5^2} + \frac{1}{5^3} + \cdots\infty\right]$$

$$\Rightarrow \frac{4}{5}S = 1 + 3\times\frac{1/5}{1-(1/5)} = 1 + \frac{3}{4} = \frac{7}{4}$$

$$\Rightarrow S = \frac{35}{16}$$

90. c. $T_n = \frac{1^3+2^3+3^3+\cdots+n^3}{1+3+5+\cdots n \text{ terms}} = \frac{\Sigma n^3}{\frac{n}{2}[2\times 1 + (n-1)2]}$

$$= \frac{1}{4}\times\frac{n^2(n+1)^2}{n^2} = \frac{1}{4}(n^2+2n+1) \quad (1)$$

Now,

$$S_n = \frac{1}{4}(\Sigma n^2 + 2\Sigma n + n)$$

$$= \frac{1}{4}\left[\frac{n(n+1)(2n+1)}{6} + 2\times\frac{n(n+1)}{2} + n\right]$$

$$= \frac{n}{24}[2n^2 + 3n + 1 + 6n + 6 + 6]$$

$$= \frac{n}{24}[2n^2 + 9n + 13]$$

Putting $n = 16$, we get

$$S_{16} = \frac{16}{24}[2(256) + 144 + 13]$$

$$= \frac{2}{3}(669) = 446$$

91. c. Here the successive differences are 2, 4, 8, 16, … which are in G.P.

$S = 1 + 3 + 7 + 15 + 31 + \cdots + T_{100}$

$S = (2^1 - 1) + (2^2 - 1) + (2^3 - 1) + \cdots + (2^{100} - 1)$

$= (2 + 2^2 + 2^3 + \cdots + 2^{100}) - 100$

$$= 2\left(\frac{2^{100}-1}{2-1}\right) - 100$$

$= 2^{101} - 102$

92. a. $a_1, a_2, a_3, \ldots, a_n$ are in H.P. Hence, $1/a_1, 1/a_2, 1/a_3, \ldots 1/a_n$ are in A.P. Also,

$$\frac{1}{a_2}-\frac{1}{a_1}, \frac{1}{a_3}-\frac{1}{a_2}, \ldots, \frac{1}{a_n}-\frac{1}{a_{n-1}} = d$$

$$\Rightarrow a_1a_2 + a_2a_3 + \cdots + a_{n-1}a_n$$

$$= \frac{a_1-a_2}{d} + \frac{a_2-a_3}{d} + \cdots + \frac{a_{n-1}-a_n}{d}$$

$$= \frac{a_1-a_n}{d}$$

$$= (n-1)a_1a_n \left[\text{as } \frac{1}{a_n} = \frac{1}{a_1} + (n-1)d\right]$$

93. c. Here, number of factors is 50. Therefore, the coefficient of x^{49} is

$$-1-3-5-\cdots-99 = -\frac{50}{2}(1+99) = -2500$$

94. b. $T_r = (-1)^r\frac{r^2+r+1}{r!}$

$$= (-1)^r\left[\frac{r}{(r-1)!} + \frac{1}{(r-1)!} + \frac{1}{r!}\right]$$

$$= (-1)^r\left[\frac{1}{(r-2)!} + \frac{r}{(r-1)!} + \frac{1}{(r-1)!} + \frac{1}{r!}\right]$$

$$= \left[\frac{(-1)^r}{r!} + \frac{(-1)^r}{(r-1)!}\right] + \left[\frac{(-1)^r}{(r-1)!} + \frac{(-1)^r}{(r-2)!}\right]$$

$$= \left[\frac{(-1)^r}{r!} - \frac{(-1)^{r-1}}{(r-1)!}\right] - \left[\frac{(-1)^{r-1}}{(r-1)!} - \frac{(-1)^{r-2}}{(r-2)!}\right]$$

$$= V(r) - V(r-1)$$

$$\therefore \sum_{r=1}^{n} T_r = V(n) - V(0) = \left[\frac{(-1)^n}{n!} - \frac{(-1)^{n-1}}{(n-1)!}\right] - 1$$

Therefore, the sum of 20 terms is

$$\left[\frac{1}{20!} - \frac{-1}{19!}\right] - 1 = \frac{21}{20!} - 1$$

95. c. Let the 1025th term fall in the nth group. Then

$1 + 2 + 4 + \cdots + 2^{n-1} < 1025 \le 1 + 2 + 4 + \cdots + 2^n$

$\Rightarrow 2^{n-1} < 1026 \le 2^{n+1}$

$\Rightarrow n = 10$

$\Rightarrow$ 1025th term is 2^{10}

96. d. Let $t_n = (1/4)(n+2)(n+3)$. Then

$$\frac{1}{t_1} + \frac{1}{t_2} + \frac{1}{t_3} + \cdots + \frac{1}{t_{2003}}$$

$$= 4\left[\frac{1}{3\times 4} + \frac{1}{4\times 5} + \frac{1}{5\times 6} + \cdots + \frac{1}{2005\times 2006}\right]$$

$$= 4\left[\frac{1}{3} - \frac{1}{2006}\right]$$

$$= 4\times\frac{2003}{3(2006)} = \frac{4006}{3009}$$

97. b. $S = \frac{2}{10} + \frac{4}{10^3} + \frac{6}{10^5} + \frac{8}{10^7} + \cdots$ to ∞ (1)

$$= \frac{\frac{2}{10}}{1-\frac{1}{10^2}} + \frac{2\times\left(\frac{1}{10^2}\right)}{\left(1-\frac{1}{10^2}\right)^2}$$

$$= \frac{20}{99} + \frac{200}{9801}$$

$$= \frac{2000}{9801}$$

98. a. Reciprocals of the terms of the series are 2/5, 13/20, 9/10, 23/20, … or 8/10, 13/20, 18/20, 23/20, …. Its nth term is $\frac{8+(n-1)5}{20} = \frac{5n+3}{20}$. Therefore, the rth term of the required series is $20/(5n+3)$.

99. a. $S = (1)(2003) + (2)(2002) + (3)(2001) + \cdots + (2003)(1)$

$$= \sum_{r=1}^{2003} r(2003-(r-1))$$

$$= \sum_{r=1}^{2003} r(2004-r)$$

$$= \sum_{r=1}^{2003} 2004r - \sum_{r=1}^{2003} r^2$$

$$= \frac{2004\times 2003\times 2004}{2} - 2003\times 4007\times 334$$

$= 2003 \times 334 \times (6012 - 4007)$

$= 2003 \times 334 \times 2005$

Hence, $x = 2005$.

100. d. $2 + 3 + 6 + 11 + 18 + \cdots$

$= (0^2 + 2) + (1^2 + 2) + (2^2 + 2) + (3^2 + 2) + \cdots$

Hence, $t_{50} = 49^2 + 2$.

101. d. We have,

$2^{n+10} = 2 \times 2^2 + 3 \times 2^3 + 4 \times 2^4 + \cdots + n \times 2^n$

$\Rightarrow 2(2^{n+10}) = 2 \times 2^3 + 3 \times 2^4 + \cdots + (n-1) \times 2^n + n \times 2^{n+1}$

$\Rightarrow$ Subtracting, we get

$-2^{n+10} = 2 \times 2^2 + 2^3 + 2^4 + \cdots + 2^n - n \times 2^{n+1}$

$$= 8 + \frac{8(2^{n-2}-1)}{2-1} - n\times 2^{n+1}$$

$= 8 + 2^{n+1} - 8 - n \times 2^{n+1} = 2^{n+1} - (n)\, 2^{n+1}$

$\Rightarrow 2^{10} = 2n - 2 \Rightarrow n = 513$

102. a. We have,

$$\frac{\pi}{4} = \left(1-\frac{1}{3}\right) + \left(\frac{1}{5}-\frac{1}{7}\right) + \left(\frac{1}{9}-\frac{1}{11}\right) + \cdots$$

$$= \frac{2}{1\times 3} + \frac{2}{5\times 7} + \frac{2}{9\times 1} + \cdots$$

$$\Rightarrow \frac{1}{1\times 3} + \frac{1}{5\times 7} + \frac{1}{9\times 11} + \cdots = \frac{\pi}{8}$$

103. b. The coefficient of x^{19} in the polynomial $(x-1)(x-2)(x-2^2) \cdots (x-2^{19})$ is

$$1 + 2 + 2^2 + \cdots + 2^{19} = 1\left(\frac{2^{20}-1}{2-1}\right)$$

$= 2^{20} - 1$

104. b. $b_2 = \frac{1}{1-b_1}$

$$b_3 = \frac{1}{1-b_2} = \frac{1}{1-\frac{1}{1-b_1}} = \frac{1-b_1}{-b_1} = \frac{b_1-1}{b_1}$$

$b_1 = b_3 \Rightarrow b_1^2 - b_1 + 1 = 0$

$\Rightarrow b_1 = -\omega$ or $\omega^2 \Rightarrow b_2 = \frac{1}{1+\omega} = -\omega$ or ω^2

$$\sum_{r=1}^{2001} b_r^{2001} = 2001\,(-1)^{2001} = -2001$$

105. d. $(1^2 - t_1) + (2^2 - t_2) + \cdots + (n^2 - t_n) = \frac{1}{3}n(n^2-1)$

$\Rightarrow 1^2 + 2^2 + 3^2 + \cdots + n^2 - \{t_1 + t_2 + \cdots + t_n\} = \frac{1}{3}n(n^2-1)$

$$\Rightarrow \frac{n(n+1)(2n+1)}{6} - S_n = \frac{1}{3}n(n^2-1)$$

$$\Rightarrow S_n = \frac{n(n+1)}{6}[2n+1-2(n-1)]$$

$$= \frac{n(n+1)}{6}[2n+1-2n+2]$$

$$= \frac{n(n+1)}{2}$$

$$\Rightarrow S_{n-1} = \frac{n(n-1)}{2}$$

$\Rightarrow T_n = S_n - S_{n-1} = n$

106. d. Since $ax^3 + bx^2 + cx + d$ is divisible by $ax^2 + c$, when $ax^3 + bx^2 + cx + d$ is divided by $ax^2 + c$ the remainder should be zero. If $ax^3 + bx^2 + cx + d$ is divided by $ax^2 + c$, then the remainder is $(bc/a) - d$. Therefore,

$$\frac{bc}{a} - d = 0$$

$\Rightarrow bc = ad$

$$\Rightarrow \frac{b}{a} = \frac{d}{c}$$

Hence, from this, a, b, c, d are not necessarily in G.P.

107. b. We know that $1 + 3 + 5 + \cdots + (2k-1) = k^2$. Thus, the given equation can be written as

$$\left(\frac{p+1}{2}\right)^2 + \left(\frac{q+1}{2}\right)^2 = \left(\frac{r+1}{2}\right)^2$$

$\Rightarrow (p+1)^2 + (q+1)^2 = (r+1)^2$

As $p > 6$, $p + 1 > 7$, we may take $p + 1 = 8$, $q + 1 = 6$, $r + 1 = 10$. Hence,

$p + q + r = 21$

108. c. $S_n = \dfrac{a(r^n - 1)}{r - 1}$

$$\Rightarrow U_n = \sum_{n=1}^{n} \frac{a(r^n - 1)}{r-1} = \frac{a}{r-1}\sum_{n=1}^{n}(r^n - 1)$$

$$\Rightarrow U_n = \frac{a}{r-1}\{r + r^2 + \cdots + r^n - n\}$$

$$= \frac{a}{r-1}\left\{\frac{r(r^n - 1)}{r-1} - n\right\}$$

$$\Rightarrow (r-1)\,U_n = \frac{ar(r^n - 1)}{r-1} - an$$

$$\Rightarrow (r-1)\,U_n = rS_n - an$$

$$\Rightarrow rS_n + (1-r)\,U_n = an$$

109. b. We have,

$$(OM_{n-1})^2 = (OP_n)^2 + (P_nM_{n-1})^2$$
$$= 2(OP_n)^2$$
$$= 2\alpha_n^2 \text{ (say)}$$

Also,

$$(OP_{n-1})^2 = (OM_{n-1})^2 + (P_{n-1}M_{n-1})^2$$

$$\Rightarrow \alpha_{n-1}{}^2 = 2\alpha_n^2 + \frac{1}{2}\alpha_{n-1}{}^2$$

$$\Rightarrow \alpha_n = \frac{1}{2}\alpha_{n-1}$$

$$\Rightarrow OP_n = \alpha_n = \frac{1}{2}\alpha_{n-1} = \frac{1}{2^2}\alpha_{n-1} = \cdots = \frac{1}{2^n} = \left(\frac{1}{2}\right)^n$$

110. b. $(1 - p)(1 + 3x + 9x^2 + 27x^3 + 81x^4 + 243x^5) = 1 - p^6$

$$\Rightarrow 1 + 3x + 9x^2 + 27x^3 + 81x^4 + 243x^5 = \frac{1-p^6}{1-p}$$

$$\Rightarrow 1 + 3x + 9x^2 + 27x^3 + 81x^4 + 243x^5$$
$$= 1 + p + p^2 + p^3 + p^4 + p^5$$

Comparing, we get $p = 3x$ or $p/x = 3$.

111. c.

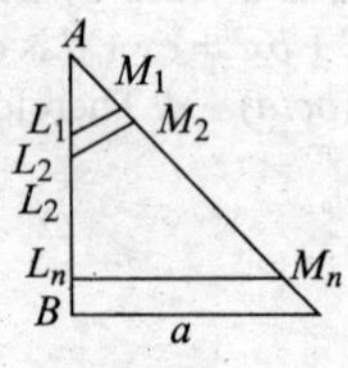

Fig. 3.3

$$\frac{AL_1}{AB} = \frac{L_1M_1}{BC}$$

$$\Rightarrow \frac{1}{n+1} = \frac{L_1M_1}{a}$$

$$\Rightarrow L_1M_1 = \frac{a}{n+1}$$

$$\frac{AL_2}{AB} = \frac{L_2M_2}{BC}$$

$$\Rightarrow \frac{2}{n+1} = \frac{L_2M_2}{a} \Rightarrow L_2M_2 = \frac{2a}{n+1}, \text{ etc.}$$

Hence, the required sum is

$$\frac{a}{n+1} + \frac{2a}{n+1} + \frac{3a}{n+1} + \cdots + \frac{na}{n+1} = \frac{a}{n+1}\,\frac{n(n+1)}{2}$$

$$= \frac{an}{2}$$

112. d. $S_n - S_{n-2} = 2$

$$\Rightarrow T_n + T_{n-1} = 2$$

Also, $T_n + T_{n-1} = \left(\dfrac{1}{n^2} + 1\right)T_{n-1} = 2$

$$\Rightarrow T_{n-1} = \frac{2}{1 + \dfrac{1}{n^2}} = \frac{2n^2}{1+n^2}$$

So, $T_m = \dfrac{2(m+1)^2}{1 + (m+1)^2}$

113. c.

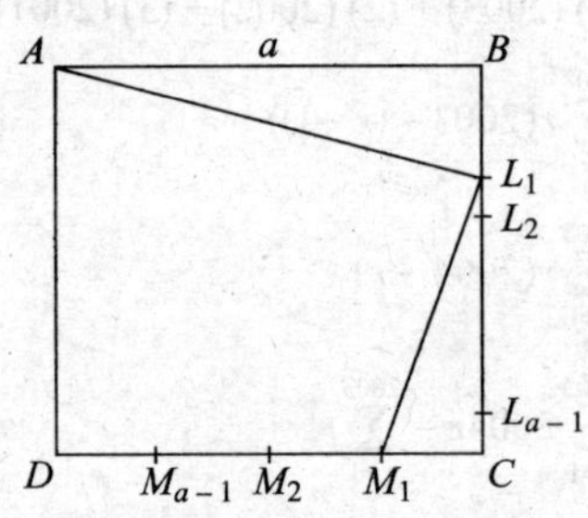

Fig. 3.4

$(AL_1)^2 + (L_1M_1)^2 = (a^2 + 1^2) + \{(a-1)^2 + 1^2\}$

$(AL_2)^2 + (L_2M_2)^2 = (a^2 + 2^2) + \{(a-2)^2 + 2^2\}$

$\vdots$

$(AL_{a-1})^2 + (L_{a-1}M_{a-1})^2 = a^2 + (a - 1^2) + \{1^2 + (a-1)^2\}$

Therefore, the required sum is

$$(a-1)a^2 + \{1^2 + 2^2 + \cdots + (a-1)^2\} + 2\{1^2 + 2^2 + \cdots + (a-1)^2\}$$

$$= (a-1)a^2 + 3\,\frac{(a-1)a(2a-1)}{6}$$

$$= a(a-1)\left(a + \frac{2a-1}{2}\right)$$

$$= \frac{1}{2}(a-1)(4a-1)$$

114. b. x, y, z are in G.P. Hence,

$$y^2 = xz$$

Now, $x + 3, y + 3, z + 3$ are in H.P. Hence,

$$y + 3 = \frac{2(x+3)(z+3)}{(x+3)+(z+3)}$$

$$= \frac{2[xz + 3(x+z) + 9]}{[(x+z)+6]}$$

$$= \frac{2[y^2 + 3(x+z) + 9]}{[x+z+6]}$$

Obviously, $y = 3$ satisfies it.

115. a. x, y, z are in G.P.

$\Leftrightarrow y^2 = xz$

$\Leftrightarrow x$ is a factor of y (not possible)

Taking $x = 3, y = 5, z = 7$, we have x, y, z are in A.P. Thus x, y, z may be in A.P. but not in G.P.

Multiple Correct Answers Type

1. a., c.

$$S = 1 + \frac{1}{(1+3)}(1+2)^2 + \frac{1}{(1+3+5)}(1+2+3)^2 + \frac{1}{(1+3+5+7)}(1+2+3+4)^2 + \cdots$$

The rth term is given by

$$T_r = \frac{1}{r^2}(1+2+\cdots+r)^2$$

$$= \frac{1}{r^2}\left\{\frac{r(r+1)}{2}\right\}^2$$

$$= \frac{r^2+2r+1}{4}$$

$\therefore T_7 = 16$ and

$$S_{10} = \sum_{r=1}^{10} T_r = \frac{1}{4}\left\{\frac{(10)(10+1)(20+1)}{6} + (10)(10+1) + 10\right\}$$

$$= \frac{505}{4}$$

2. b., c.

We have

$$\frac{p}{1-1/p} = \frac{9}{2}$$

$\Rightarrow 2p^2 - 9p + 9 = 0$

$\Rightarrow p = 3/2, 3$

3. a., b., c., d.

$an^4 + bn^3 + cn^2 + dn + e$

$$= 2\sum_{r=1}^{n} r(r+1)(r+2) - \sum_{r=1}^{n} r(r+1)$$

$$= \frac{2}{4}n(n+1)(n+2)(n+3) - \frac{1}{3}n(n+1)(n+2)$$

$$= \frac{1}{6}(3n^4 + 16n^3 + 27n^2 + 14n)$$

4. a., b., c.

Given that $a = 4$, $T_3 - T_5 = 32/81$. Hence,

$a(r^2 - r^4) = 32/81$

or $r^4 - r^2 + 8/81 = 0$

or $81r^4 - 81r^2 + 8 = 0$

or $(9r^2 - 8)(9r^2 - 1) = 0$

$\therefore r^2 = 8/9, 1/9$

Therefore, the value of r is to be positive since all the terms are positive.

For $r = 1/3$,

$$S_\infty = \frac{a}{1-r} = \frac{4}{1-\frac{1}{3}} = \frac{4\times 3}{2} = 6$$

Similarly, we can find S_∞ when $r = 2\sqrt{2}/3$.

5. a., b., c.

Let the three-digit number be xyz. According to given condition, we have

$y^2 = xz$ (1)

$2(y + 2) = x + z$ (2)

$100x + 10y + z - 792 = 100z + 10y + x$

$\Rightarrow x - z = 8$ (3)

Squaring (2) and (3) and subtracting, we have

$4xz = 4(y + 2)^2 - 64$ (4)

$\Rightarrow y^2 = (y + 2)^2 - 16$ [Using (1)]

$\Rightarrow y = 3$

$\Rightarrow x + z = 10$ [Using (2)]

$\Rightarrow x = 9, z = 1$

Hence, the number is $931 = 7^2 \times 19$.

6. a., b., c.

Clearly, n^{th} term of the given series is negative or positive accordingly as n is even or odd, respectively.

Case I: When n is even: In this case, the given series is

$S_n = 1^2 - 2^2 + 3^2 - 4^2 + \cdots + (n-1)^2 - n^2$

$= (1^2 - 2^2) + (3^2 - 4^2) + \cdots + ((n-1)^2 - n^2)$

$= (1-2)(1+2) + (3-4)(3+4) + \cdots + ((n-1) - (n))(n-1+n)$

$= -(1 + 2 + 3 + 4 + \cdots + (n-1) + n)$

$$= -\frac{n(n+1)}{2} \quad (1)$$

Case II: When n is odd: In this case, the given series is

$S_n = (1^2 - 2^2) + (3^2 - 4^2) + \cdots + \{(n-2)^2 - (n-1)^2\} + n^2$

$= (1-2)(1+2) + (3-4)(3+4) + \cdots + ((n-2) - (n-1)) + ((n-2) + (n-1)) + n^2$

$= -(1 + 2 + 3 + 4 + \cdots + (n-2) + (n-1)) + n^2$

$$= -\frac{(n-1)(n-1+1)}{2} + n^2 = \frac{n(n+1)}{2} \quad (2)$$

$\Rightarrow S_{40} = -820$ [Using (1)]

$S_{51} = 1275$ [Using (2)]

Also,

$S_{2n} > S_{2n+2}$ [From (1)]

$S_{2n+1} < S_{2n-1}$ [From (2)]

7. a., b., d.

$$x + y + z = 3\left(\frac{a+b}{2}\right)$$

$$\Rightarrow 15 = 3\frac{(a+b)}{2}$$

$\Rightarrow a + b = 10$ (1)

$$\frac{1}{x} + \frac{1}{y} + \frac{1}{z} = \frac{3\left(\frac{1}{a} + \frac{1}{b}\right)}{2}$$

$$\Rightarrow \frac{5}{3} = \frac{3(a+b)}{2ab} = \frac{3\times 10}{2ab}$$

$\Rightarrow ab = 9$ (2)

From (1) and (2), $a = 9$, $b = 1$ or $a = 1$ and $b = 9$. Hence, G.M. $= \sqrt{ab} = 3$, $a + 2b = 11$ or 19.

8. b., d.

Given,

$3a_1 + 7a_2 + 3a_3 - 4a_5 = 0$

$\Rightarrow 7(a_1 + a_2 + a_3) = 4(a_1 + a_3 + a_5)$

$\Rightarrow 7(1 + r + r^2) = 4(1 + r^2 + r^4)$

$\Rightarrow 7 = 4(r^2 - r + 1)$

$\Rightarrow 4r^2 - 4r + 1 = 4$

$\Rightarrow (2r - 1)^2 = 4$

$\Rightarrow 2r - 1 = \pm 2$

$\Rightarrow r = 3/2, -1/2$

9. b., c.

$$\frac{1}{\sqrt{2}+\sqrt{5}} + \frac{1}{\sqrt{5}+\sqrt{8}} + \frac{1}{\sqrt{8}+\sqrt{11}} + \cdots n \text{ terms}$$

$$= \frac{\sqrt{5}-\sqrt{2}}{3} + \frac{\sqrt{8}-\sqrt{5}}{3} + \cdots + \frac{\sqrt{5+(n-1)3} - \sqrt{2+(n-1)3}}{3}$$

$$= \frac{\sqrt{3n+2}-\sqrt{2}}{3}$$

$$= \frac{3n+2-2}{3(\sqrt{3n+2}+\sqrt{2})}$$

$$= \frac{n}{\sqrt{3n+2}+\sqrt{2}}$$

$$= \frac{n}{\sqrt{2+3n}+\sqrt{2}} < \frac{n}{\sqrt{3n}} < n$$

10. a., b.

$$\left(\frac{1}{b}+\frac{1}{c}-\frac{1}{a}\right)\left(\frac{1}{c}+\frac{1}{a}-\frac{1}{b}\right)$$

$$= \left(\frac{1}{b}+\frac{1}{c}-\frac{2}{b}+\frac{1}{c}\right)\left(\frac{1}{c}+\frac{1}{b}-\frac{1}{c}\right)$$

$$= \left(\frac{2}{c}-\frac{1}{b}\right)\frac{1}{b} = \frac{2}{bc}-\frac{1}{b^2}$$

Also by eliminating b, we get the given expression $\frac{(a+c)(3a-c)}{4a^2c^2}$.

11. a., d.

$$p(x) = \left(\frac{1-x^{2n}}{1-x^2}\right)\left(\frac{1-x}{1-x^n}\right) = \frac{1+x^n}{1+x}$$

As $p(x)$ is a polynomial, $x = -1$ must be a zero of $1 + x^n$. Hence, $1 + (-1)^n = 0$. So, n must be odd.

12. a., c., d.

$a_1 + a_3 + a_5 = -12$

$a + a + 2d + a + 4d = -12 \quad (d > 0)$

$a + 2d = -4$ (1)

$a_1 a_3 a_5 = 80$

$a(a + 2d)(a + 4d) = 80$

or $a(-4)(-4 - 2d + 4d) = 20$ [From (1)]

or $(-4 - 2d)(-4 + 2d) = -20 \Rightarrow d = \pm 3$

Since A.P. is increasing, so $d = +3$; $a = -10$. Hence,

$$\left.\begin{array}{l} a_1 = -10;\ a_2 = -7 \\ a_3 = a + 2d = -10 + 6 = -4 \\ a_5 = a + 4d = -10 + 8 = -2 \end{array}\right\}$$

13. a., b., c.

$$a = \frac{n^{64}-1}{n-1}$$

$$= (n + 1)(n^2 + 1)(n^4 + 1)(n^8 + 1)(n^{16} + 1)(n^{32} + 1)$$

14. a., b., c.

If p, q, r are in A.P., then p^{th}, q^{th} and r^{th} terms are equal distant terms which are always in the same series of which they are terms.

15. a., d.

$S = 1 + 2x + 3x^2 + 4x^3 + \cdots \infty$

$\Rightarrow xS = x + 2x^2 + 3x^3 + 4x^4 + \cdots \infty$

$\Rightarrow (1 - x)S = 1 + x + x^2 + \cdots \infty = \dfrac{1}{1-x}$

$\Rightarrow S = \dfrac{1}{(1-x)^2}$

Now,

$$S \ge 4 \Rightarrow \frac{1}{(1-x)^2} > 4$$

$$\Rightarrow (x-1)^2 \le \frac{1}{4}$$

$$\Rightarrow -\frac{1}{2} \le x - 1 \le \frac{1}{2}$$

$$\Rightarrow \frac{1}{2} \le x \le \frac{3}{2} \Rightarrow \frac{1}{2} \le x < 1$$

16. a., b., d.

$$E < 1 + \frac{1}{(1)(2)} + \frac{1}{(2)(3)} + \cdots$$

$$= 1 + \left(1 - \frac{1}{2}\right) + \left(\frac{1}{2} - \frac{1}{3}\right) + \cdots = 2$$

$$E > 1 + \frac{1}{(2)(3)} + \frac{1}{(3)(4)} + \cdots$$

$$= 1 + \left(\frac{1}{2} - \frac{1}{3}\right) + \left(\frac{1}{3} - \frac{1}{4}\right) + \cdots = \frac{3}{2}$$

17. a., c., d.

$$S_n = \frac{n}{2}\left[2a' + (n-1)d\right] = a + bn + cn^2$$

$$\Rightarrow na' + \frac{n(n-1)}{2}d = a + bn + cn^2$$

$$\Rightarrow \left(a' - \frac{d}{2}\right)n + \frac{n^2 d}{2} = a + bn + cn^2$$

On comparing,

$$a=0,\ b=a'-\frac{d}{2},\ c=\frac{d}{2} \Rightarrow d=2c$$

18. a., b.

$x^2+9y^2+25z^2=15yz+5zx+3xy$

$\Rightarrow (x)^2+(3y)^2+(5z)^2-(x)(3y)-(3y)(5z)-(x)(5z)=0$

$\Rightarrow \frac{1}{2}[(x-3y)^2+(3y-5z)^2+(x-5z)^2]=0$

$\Rightarrow x-3y=0,\ 3y-5z=0,\ x-5z=0$

$$\Rightarrow \frac{1}{x}=\frac{1}{3y} \text{ and } \frac{5}{3y}=\frac{1}{z} \text{ and } \frac{1}{5z}=\frac{1}{x}$$

$$\Rightarrow \frac{1}{x}+\frac{1}{z}=\frac{1}{3y}+\frac{5}{3y}=\frac{2}{y}$$

Therefore, $1/x$, $1/y$, and $1/z$ are in A.P. and x, y, and z are in H.P.

19. a., c.

Let $b=a+p$, $c=a+2p$, $d=a+3p$ (where p is common difference). Then,

$$\frac{\frac{1}{a}+\frac{1}{d}}{\frac{1}{b}+\frac{1}{c}}=\frac{\frac{1}{a}+\frac{1}{a+3p}}{\frac{1}{a+p}+\frac{1}{a+2p}}$$

$$=\frac{(a+p)(a+2p)}{a(a+3p)}$$

$$=\frac{a^2+3ap+2p^2}{a^2+3ap}>1$$

$$\therefore \frac{1}{a}+\frac{1}{d}>\frac{1}{b}+\frac{1}{c}$$

$$\left(\frac{1}{b}+\frac{1}{c}\right)(a+d)=\left(\frac{1}{a+p}+\frac{1}{a+2p}\right)(a+a+3p)$$

$$=\frac{(2a+3p)^2}{a^2+3ap+2p^2}$$

$$=4+\frac{p^2}{a^2+3ap+2p^2}>4$$

20. a., d.

x, x^2+2, x^3+10 are in G.P. Hence,

$x(x^3+10)=(x^2+2)^2=x^4+4x^2+4$

$\Rightarrow 4x^2-10x+4=0$

$\Rightarrow 2x^2-5x+2=0$

$\Rightarrow x=2,\ \frac{1}{2}$

The fourth term of G.P. is

$$(x^3+10)r=(x^3+10)\left(\frac{x^2+2}{x}\right)$$

$$=\begin{cases}54 \text{ when } x=2\\ \frac{729}{16} \text{ when } x=\frac{1}{2}\end{cases}$$

21. a., b., c.

Last term in n^{th} row is

$$1+2+3+\cdots+n=\frac{1}{2}n(n+1) \quad (1)$$

As terms in the nth row forms an A.P. with common difference 1, so

First term = Last term – $(n-1)$ (1)

$$=\frac{1}{2}n(n+1)-n+1$$

$$=\frac{1}{2}(n^2-n+2) \quad (2)$$

$$\text{Sum of terms}=\frac{1}{2}n\left[\frac{1}{2}(n^2-n+2)+\frac{1}{2}(n^2+n)\right]$$

$$=\frac{1}{2}n(n^2+1) \quad (3)$$

Now, put $n=20$ in (1), (2), (3) to get required answers.

22. a., b., c.

Since A_1, A_2 are two arithmetic means between a and b, therefore, a, A_1, A_2 b are in A.P. with common difference d given by

$$d=\frac{b-a}{2+1}=\frac{b-a}{3} \quad \left[\text{using } d=\frac{b-a}{n+1}\right]$$

Now,

$$A_1=a+d=a+\frac{b-a}{3}=\frac{2a+b}{3}$$

and

$$A_2=a+2d=a+2\left(\frac{b-a}{3}\right)=\frac{a+2b}{3}$$

It is given that G_1G_2 are two G.M.'s between a and b. Therefore, a, G_1, G_2, b are in G.P. with common ratio r given by

$$r=\left(\frac{b}{a}\right)^{\frac{1}{2+1}}=\left(\frac{b}{a}\right)^{1/3} \quad \left[\because r=\left(\frac{b}{a}\right)^{1/n+1}\right]$$

Now,

$$G_1=ar=a\left(\frac{b}{a}\right)^{1/3}=a^{2/3}\,b^{1/3}$$

and

$$G_2=ar^2=a\left(\frac{b}{a}\right)^{2/3}=a^{1/3}\,b^{2/3}$$

It is also given that H_1, H_2 are two H.M's between a and b, therefore, a, H_1, H_2, b are in H.P. Hence, $1/a$, $1/H_1$, $1/H_2$, $1/b$, are in A.P. with common difference D given by

$$D=\frac{a-b}{(2+1)ab}=\frac{a-b}{3ab} \quad \left[\because D=\frac{a-b}{(n+1)ab}\right]$$

Now,

$$\frac{1}{H_1}=\frac{1}{a}+D=\frac{1}{a}+\frac{a-b}{3ab}=\frac{a+2b}{3ab}$$

$$\Rightarrow H_1 = \frac{3ab}{a+2b}$$

$$\frac{1}{H_2} = \frac{1}{a} + 2D$$

$$= \frac{1}{a} + 2D$$

$$= \frac{1}{a} + \frac{2(a-b)}{3ab}$$

$$= \frac{2a+b}{3ab}$$

$$\Rightarrow H_2 = \frac{3ab}{2a+b}$$

We have,

$$A_1H_2 = \frac{2a+b}{3} \times \frac{3ab}{2a+b} = ab,$$

$$A_2H_1 = \frac{a+2b}{3} \times \frac{3ab}{a+2b} = ab,$$

$$G_1G_2 = (a^{2/3} b^{1/3})(a^{1/3} b^{2/3}) = ab$$

$$\therefore A_1H_2 = A_2H_1 = G_1G_2 = ab$$

23. b., c., d.

We have length of a side of S_n is equal to the length of a diagonal of S_{n+1}. Hence,

Length of a side of $S_n = \sqrt{2}$(Length of a side of S_{n+1})

$$\Rightarrow \frac{\text{Length of a side of } S_{n+1}}{\text{Length of side of } S_n} = \frac{1}{\sqrt{2}}, \text{ for all } n \geq 1$$

Hence, sides of $S_1, S_2, ..., S_n$ form a G.P. with common ratio $1/\sqrt{2}$ and first term 10.

$$\therefore \text{Side of } S_n = 10\left(\frac{1}{\sqrt{2}}\right)^{n-1} = \frac{10}{2^{\frac{n-1}{2}}}$$

$$\Rightarrow \text{Area of } S_n = (\text{side})^2 = \left(\frac{10}{2^{\frac{n-1}{2}}}\right)^2 = \frac{100}{2^{n-1}}$$

Now, area of $S_n < 1 \Rightarrow n = b, c, d$.

24. a., c.

$$\frac{1}{b-a} + \frac{1}{b-c} = \frac{1}{a} + \frac{1}{c}$$

$$\Rightarrow \frac{1}{b-a} - \frac{1}{c} = \frac{1}{a} - \frac{1}{b-c}$$

$$\Rightarrow \frac{c-b+a}{c(b-a)} = \frac{b-c-a}{a(b-c)}$$

$$\Rightarrow c-b+a = 0 \text{ or } \frac{1}{c(b-a)} = \frac{1}{a(c-b)}$$

$$\Rightarrow b = a+c \quad \text{or} \quad ac - bc = ab - ac$$

$$\Rightarrow b = a+c \quad \text{or} \quad b = \frac{2ac}{a+c}$$

25. a, c.

a, b, c are in G.P. Hence,

$b^2 = ac$ (1)

x is A.M. of a and b. Hence,

$2x = a + b$ (2)

y is A.M. of b and c. Hence,

$2y = b + c$ (3)

$$\therefore \frac{a}{x} + \frac{c}{y} = a \times \frac{2}{a+b} + c \times \frac{2}{b+c} \quad \text{[Using (2) and (3)]}$$

$$= 2\left[\frac{ab+ac+ac+bc}{ab+ac+b^2+bc}\right]$$

$$= 2 \quad \text{[Using (1)]}$$

Again,

$$\frac{1}{x} + \frac{1}{y} = \frac{2}{a+b} + \frac{2}{b+c}$$

$$= \frac{2(a+c+2b)}{ab+ac+b^2+bc}$$

$$= \frac{2(a+c+2b)}{ab+2b^2+bc} \quad (\because b^2 = ac)$$

$$= \frac{2(a+c+2b)}{b(a+c+2b)}$$

$$= \frac{2}{b}$$

Reasoning Type

1. a. Let p, q, r be the lth, m^{th}, and n^{th} terms of an A.P. Then $p = (a + (l-1)d$, $q = a + (m-1)d$ and $r = a + (n-1)d$

Hence, $r - q = (n-m)d$ and $q - p = (m-l)d$, so that

$$\frac{r-q}{p-q} = \frac{(n-m)d}{(m-l)d} = \frac{n-m}{m-l} \quad (\because d \neq 0) \tag{1}$$

Since l, m, n are positive integers and $m \neq l$, $(n-m)/(m-l)$ is a rational number. From (1), using $p = \sqrt{2}, q = \sqrt{3}, r = \sqrt{5}$, we have

$$\frac{\sqrt{5}-\sqrt{3}}{\sqrt{2}-\sqrt{3}} = \frac{n-m}{m-l} \text{ (which is not possible.)}$$

Hence, $\sqrt{2}$, $\sqrt{3}$, $\sqrt{5}$ cannot be the terms of an A.P.

2. a. Statement 2 is true as it is a property of sequence in G.P.

Now T_{m-n}, T_m, and T_{m+n} are in G.P. ($\because T_m$ from T_{m-n} and T_{m+n} from T_m are at same distance)

$$\therefore T_m^2 = T_{m-n} T_{m+n}$$

$$\Rightarrow T_m = \sqrt{pq}$$

3. b. Let, if possible, 8 be the first term and 12 and 27 be m^{th} and n^{th} terms, respectively. Then,

$12 = ar^{m-1} = 8r^{m-1}$, $27 = 8r^{n-1}$

$$\Rightarrow \frac{3}{2} = r^{m-1}, \left(\frac{3}{2}\right)^3 = r^{n-1} = r^{3(m-1)}$$

$\Rightarrow n-1=3m-3$ or $3m=n+2$

$\Rightarrow \frac{m}{1}=\frac{n+2}{3}=k$ (say)

$\therefore m=k, n=3k-2$

By giving k different values, we get the integral values of m and n.

Hence there can be **infinite** number of G.P.'s whose any three terms will be 8, 12, 27 (not consecutive). Obviously, statement 2 is not a correct explanation of statement 1.

4. a. $x^2+9y^2+25z^2=xyz\left(\frac{15}{x}+\frac{5}{y}+\frac{3}{z}\right)$

$\Rightarrow x^2+9y^2+25z^2-15yz-5xz-3xy=0$

$\Rightarrow 2x^2+18y^2+50z^2-30yz-10xz-6xy=0$

$\Rightarrow (x-3y)^2+(3y-5z)^2+(5z-x)^2=0$

$\Rightarrow x-3y=0, 3y-5z=0, 5z-x=0$

$\Rightarrow x=3y=5z=k$ (say)

$\Rightarrow x=k, y=k/3, z=k/5$

Hence, x, y, z are in H.P. Hence, option (a) is correct.

5. a. Coefficient of x^{14} in $(1+2x+3x^2+\cdots+16x^{15})^2$

$=$ Coefficient of x^{14} in $(1+2x+3x^2+\cdots+16x^{15})(1+2x+3x^2+\cdots+16x^{15})$

$=1\times 15+2\times 14+\cdots+15\times 1$

$=\sum_{r=1}^{15} r(16-r)$

Also,

$$\sum_{r=1}^{n} r(n-r)=\sum_{r=1}^{n} nr-\sum_{r=1}^{n} r^2$$

$$=n\frac{n(n+1)}{2}-\frac{n(n+1)(2n+1)}{6}$$

$$=\frac{n(n+1)}{6}(3n-(2n+1))$$

$$=\frac{n(n^2-1)}{6}$$

$$\Rightarrow \sum_{r=1}^{15} r(16-r)=\frac{15(15^2-1)}{6}=560$$

Hence, option (a) is correct.

6. b. $x=1111\cdots 91$ times

$=1+10+10^2+10^3+\cdots+10^{90}$

$$=\frac{1(10^{91}-1)}{10-1}$$

$$=\frac{(10^{13\times 7}-1)}{10-1}$$

$$=\frac{((10^{13})^7-1)}{10^{13}-1}\times\frac{(10^{13}-1)}{10-1}$$

$=(1+10^{13}+10^{26}+\cdots+10^{78})\times(1+10+10^2+\cdots+10^{12})$

$=$ composite numbers

But statement 2 is not a correct explanation of statement 1 as 111 has 1 digit 3 times, and 3 is a prime number but $111=3\times 37$ is a composite number. Hence, (b) is the correct option.

7. a. We have,

$$a\times ar\times\ldots\times ar^{n-1}=a^n\times r^{1+2+\cdots+(n-1)}=a^n r^{\frac{n(n-1)}{2}}$$

$$=(a^2r^{n-1})^{n/2}$$

Hence, statement 1 is true.

Also, $(a\times r^{i-1})(a\times r^{n-ik})=a^2\times r^{n-1}$, which is independent of k. Hence, statement 2 is a correct explanation for statement 1, as in the product of $a, ar, ar^2, \ldots ar^{n-1}$, there are $n/2$ groups of numbers, whose product is a^2r^{n-1}. Hence, (a) is the correct option.

8. d. For odd integer n, we have

$S_n=n^3-(n-1)^3+\cdots+(-1)^{n-1}1^3$

$=1^3-2^3+3^3-4^3+\cdots+n^3$

$=[1^3+2^3+3^3+\cdots+(n-1)^3+n^3]-2[2^3+4^3+6^3+\cdots+(n-1)^3]$

$$=\frac{n^2(n+1)^2}{4}-2\times 2^3\left[1^3+2^3+\cdots+\left(\frac{n-1}{2}\right)^3\right]$$

$$=\frac{n^2(n+1)^2}{4}-2^4\frac{\left(\frac{n-1}{2}\right)^2\left(\frac{n-1}{2}+1\right)^2}{4}$$

$$=\frac{n^2(n+1)^2}{4}-\frac{(n-1)^2(n+1)^2}{4}$$

$$=\frac{(n+1)^2}{4}\left[n^2-(n-1)^2\right]$$

$$=\frac{1}{4}(2n-1)(n+1)^2$$

Now, putting $n=11$ in above formula, $S_{11}=756$. Hence statement 1 is false and statement 2 is correct.

9. d. Sum $=\frac{x/r}{1-r}=4$ (where r is common ratio)

$x=4r(1-r)=4(r-r^2)$

For $r\in(-1,1)-\{0\}$

$r-r^2\in\left(-2,\frac{1}{4}\right)-\{0\}$

$\Rightarrow x\in(-8,1)-\{0\}$

10. b. The **given** inequality is

$(p_1^2+p_2^2+\cdots+p_{n-1}^2)x^2+2(p_1p_2+p_2p_3+\cdots+p_{n-1}p_n)x+(p_2^2+\cdots+p_n^2)\le 0$

$\Rightarrow (p_1x+p_2)^2+(p_2x+p_3)^2+\cdots+(p_{n-1}x+p_n)^2\le 0$ (1)

But each one of the terms on the L.H.S. is a perfect square and hence is positive or zero.

Therefore, (1) holds only if

$p_1x+p_2=0=p_2x+p_3=p_3x+p_4=\cdots=p_{n-1}x+p_n$

$$\Rightarrow -x=\frac{p_2}{p_1}=\frac{p_3}{p_2}=\cdots=\frac{p_n}{p_{n-1}}$$

Hence, $p_1, p_2, \ldots, p_n$ are in G.P.

11. a. Statement 2 is true as

$$\frac{a^n+b^n}{a+b}=\frac{a^n-(-b)^n}{a-(-b)}$$

$$=a^{n-1}-a^{n-2}b+a^{n-3}b^2-\cdots(-1)^{n-1}b^{n-1}$$

Now,

$1^{99}+2^{99}+\cdots+100^{99}=(1^{99}+100^{99})+(2^{99}+99^{99})+\cdots+(50^{99}+51^{99})$

Each bracket is divisible by 101. Hence, the sum is divisible by 101. Also,

$1^{99}+2^{99}+L\cdots 100^{99}=(1^{99}+99^{99})+(2^{99}+98^{99})+\cdots+(49^{99}+51^{99})+50^{99}+100^{99}$

Here, each bracket and 50^{99} and 100^{99} are divisible by 100. Hence sum is divisible by 100. Hence sum is divisible by $101\times 100=10100$.

12. c. For two positive numbers $(GM)^2=(AM)\times(HM)$. This is not true for numbers greater than 2. Hence statement 1 is true but statement 2 is false.

Linked Comprehension Type

For Problems 1–3

1. c., 2.. b., 3. d.

Sol.

Let the odd integers be $2m+1, 2m+3, 2m+5, \ldots$ and let their number be n. Then,

$$57^2-13^2=(n/2)[2(2m+1)+(n-1)\times 2]$$
$$=n(2m+n)$$
$$=2mn+n^2$$

$\Rightarrow 57^2-13^2=(n+m)^2-m^2$

$\Rightarrow m=13$ and $n+m=57$

$\Rightarrow n=57-13=44$

Hence, the required odd integers are 27, 29, 31, …, 113.

For Problems 4–6

4. c., 5. c., 6. d.

Sol.

4. a, b, c are in G.P. Hence, a, ar, ar^2 are in G.P. So,

$$\frac{a^2+a^2r^2+a^2r^4}{(a+ar+ar^2)^2}=\frac{t^2}{\alpha^2t^2}=\frac{1}{\alpha^2}$$

$$\alpha^2=\frac{r^2+r+1}{r^2-r+1}$$

Let $\alpha^2=y$.

$$\therefore y=\frac{r^2+r+1}{r^2-r+1}$$

$(y-1)r^2-r(y+1)+(y-1)=0$

For real r,

$(y+1)^2-4(y-1)^2\geq 0$

$\Rightarrow \frac{1}{3}\leq y\leq 3$

But $y\neq 1/3, 1, 3$ $(\because r\neq 1,-1,0)$

$\therefore \frac{1}{3}<y<3$ and $y\neq 1$

$$\alpha^2\in\left(\frac{1}{3},3\right)-\{1\}$$

5. $s=r+\frac{1}{r}$

$s\in(-\infty,-2)\cup(2,\infty)$

6. Let $b=ar$, $c=ar^2$ and $r>0$.

As sum of two sides is more than the third side, we have,

$$r\in\left(\frac{\sqrt{5}-1}{2},\frac{\sqrt{5}+1}{2}\right)-\{1\}$$

$$\Rightarrow r+\frac{1}{r}-1\in(1,\sqrt{5}-1)$$

$$\text{As } \alpha^2=\frac{r^2+r+1}{r^2-r+1}=1+\frac{2}{r+\frac{1}{r}-1}$$

$$\therefore \alpha^2\in\left(\frac{\sqrt{5}+3}{2},3\right)$$

For Problems 7–9

7. d., 8. b., 9. d.

Sol.

Let a be the first term and r the common ratio of the given G.P.

Further, let there be n terms in the given G.P. Then,

$a_1+a_n=66\Rightarrow a+ar^{n-1}=66$ (1)

$a_2\times a_{n-1}=128$

$\Rightarrow ar\times ar^{n-2}=128$

$\Rightarrow a^2r^{n-1}=128$

$\Rightarrow a\times(ar^{n-2})=128\Rightarrow ar^{n-1}=\frac{128}{a}$

Putting this value of ar^{n-1} in (1), we get

$a+\frac{128}{a}=66$

$\Rightarrow a^2-66a+128=0$

$\Rightarrow (a-2)(a-64)=0$

$\Rightarrow a=2, 64$

Putting $a=2$ in (1), we get

$2+2\times r^{n-1}=66\Rightarrow r^{n-1}=32$

Putting $a=64$ in (1), we get

$64+64r^{n-1}=66\Rightarrow r^{n-1}=\frac{1}{32}$

For an increasing G.P., $r>1$. Now,

$S_n=126$

$$\Rightarrow 2\left(\frac{r^n-1}{r-1}\right)=126$$

$$\Rightarrow \frac{r^n-1}{r-1}=63$$

$$\Rightarrow \frac{r^{n-1}\times r-1}{r-1}=63$$

$$\Rightarrow \frac{32r-1}{r-1} = 63$$

$\Rightarrow r = 2$

$\therefore r^{n-1} = 32 \Rightarrow 2^{n-1} = 32 = 2^5 \Rightarrow n - 1 = 5 \Rightarrow n = 6$

For decreasing G.P., $a = 64$ and $r = 1/2$. Hence, the sum of infinite terms is $64/[1 - (1/2)] = 128$.

For Problems 10–12

10. c., 11. d., 12. a.

Sol.

Let the four integers be $a - d$, a, $a + d$ and $a + 2d$, where a and d are integers and $d > 0$. Now,

$a + 2d = (a - d)^2 + a^2 + (a + d)^2$

$\Rightarrow 2d^2 - 2d + 3a^2 - a = 0$ (1)

$\therefore d = \frac{1}{2}\left[1 \pm \sqrt{1 + 2a - 6a^2}\right]$ (2)

Since d is a positive integer, so

$1 + 2a - 6a^2 > 0$

$\Rightarrow 6a^2 - 2a - 1 < 0$

$$\Rightarrow \frac{1-\sqrt{7}}{6} < a < \frac{1+\sqrt{7}}{6}$$

$\Rightarrow a = 0$ [$\because$ a is an integer]

Hence, from (2),

$d = 1$ or 0

But since $d > 0$,

$\therefore d = 1$

Hence, the four numbers are $-1, 0, 1, 2$.

For Problems 13–15

13. d., 14. a., 15. b.

Sol.

1, 2, 2, 3, 3, 3, 4, 4, 4, 4, …

Let us write the terms in the groups as follows: 1, (2, 2), (3, 3, 3), (4, 4, 4, 4), … consisting of 1, 2, 3, 4, … terms. Let 2000th term fall in nth group. Then,

$$\frac{(n-1)n}{2} < 2000 \le \frac{n(n+1)}{2}$$

$\Rightarrow n(n - 1) < 4000 \le n(n + 1)$

Let us consider,

$n(n - 1) < 4000$

$\Rightarrow n^2 - n - 4000 < 0$

$$\Rightarrow n < \frac{1+\sqrt{16001}}{2} \Rightarrow n < 64$$

We have

$n(n + 1) \ge 4000 \Rightarrow n^2 + n - 4000 \ge 0 \Rightarrow n \ge 63$

That means 2000th term falls in 63rd group. That also means that the 2000th term is 63. Now, total number of terms up to 62nd group is $(62 \times 63)/2 = 1953$. Hence, sum of first 2000 terms is

$1^2 + 2^2 + \cdots + 62^2 + 63(2000 - 1953)$

$$= \frac{62(63)125}{6} + 63 \times 47 = 84336$$

Sum of the remaining terms is $63 \times 16 = 1008$.

For Problems 16–18

16. b, 17. a, 18. c.

Sol. Let numbers in set A be $a - D$, a, $a + D$ and these in set B be $b - d$, b, $b + d$. Now,

$3a = 3b = 15$

$\Rightarrow a = b = 5$

Set $A = \{5 - D, 5, 5, + D\}$

Set $B = \{5 - d, 5, 5 + d\}$

where $D = d + 1$

Also,

$$\frac{p}{q} = \frac{5(25-D^2)}{5(25-d^2)} = \frac{7}{8}$$

$\Rightarrow 25(8 - 7) = 8(d + 1)^2 - 7d^2$

$\Rightarrow d = -17, 1$ but $d > 0 \Rightarrow d = 1$

So, the numbers in set A are 3, 5, 7 and the numbers in set B are 3, 4, 6.

Now, sum of product of numbers in set A taken two at a time is $3 \times 5 + 3 \times 7 + 5 \times 7 = 71$. The sum of product of numbers in set A taken two at a time is $3 \times 4 + 4 \times 6 + 3 \times 6 = 54$. Also,

$p = 3 \times 5 \times 7 = 105$ and $q = 4 \times 5 \times 6 = 120$

$\Rightarrow q - p = 15$

For Problems 19–21

19. c., 20. b., 21. a.

Sol.

19. $G_1 G_2 \cdots G_n = (\sqrt{1 \times 1024})^n = 2^{5n}$

Given,

$2^{5n} = 2^{45} \Rightarrow n = 9$

Hence,

$r = (1024)^{\frac{1}{9+1}} = 2$

$\Rightarrow G_1 = 2, r = 2$

$$\Rightarrow G_1 + G_2 + \cdots + G_n = \frac{2 \times (2^9 - 1)}{2-1} = 1024 - 2 = 1022$$

20. b. $A_1 + A_2 + A_3 + \cdots + A_{m-1} + A_m = 1025 \times 771$

$$\Rightarrow m\left(\frac{-2 + 1027}{2}\right) = 1025 \times 771$$

$\Rightarrow m = 1542$

21. a. We have,

$A_{171} + A_{172} = -2 + 1027 = 1025$

$$\Rightarrow \frac{2A_{171} + 2A_{172}}{2} = 1025$$

Also,

$G_5 = 1 \times 2^5 = 32$

$\Rightarrow G_5^2 = 1024$

$\Rightarrow G_5^2 + 1 = 1025$

$\Rightarrow 2A_{171}, G_5^2 + 1, 2A_{172}$ are in A.P.

For Problems 22–24

22. a., 23. c., 24. b.

Sol.

Let m and $(m + 1)$ be the removed numbers from 1, 2, ..., n. Then, sum of the remaining numbers is $[n(n + 1)/2] - (2m + 1)$. From given condition,

$$\frac{105}{4} = \frac{\frac{n(n+1)}{2} - (2m+1)}{(n-2)}$$

$\Rightarrow 2n^2 - 103n - 8m + 206 = 0$

Since n and m are integers, so n must be even. Let $n = 2k$. Then,

$$m = \frac{4k^2 + 103(1 - k)}{4}$$

Since m is an integer, then $1 - k$ must be divisible by 4. Let $k = 1 + 4t$. Then we get $n = 8t + 2$ and $m = 16t^2 - 95t + 1$. Now,

$1 \le m < n$

$\Rightarrow 1 \le 16t^2 - 95t + 1 < 8t + 2$

Solving, we get $t = 6$. Hence,

$n = 50$ and $m = 7$

Hence, the removed numbers are 7 and 8. Also, sum of all numbers is $50(50 + 1)/2 = 1275$.

For Problems 25–27

25. c., 26. b., 27. a.

Sol.

Let the first term a and common difference d of the first A.P. and the first term b and common difference e of the second A.P. and let the number of terms be n. Then,

$$\frac{a + (n-1)d}{b} = \frac{b + (n-1)e}{a} = 4 \quad (1)$$

$$\frac{\frac{n}{2}[2a + (n-1)d]}{\frac{n}{2}[2b + (n-1)e]} = 2 \quad (2)$$

From (1) and (2), we get

$a - 4b + (n - 1)d = 0$ (3)

$b - 4a + (n - 1)e = 0$ (4)

$2a - 4b + (n - 1)d - 2(n - 1)e = 0$ (5)

$4 \times (3) + (4)$ gives

$-15b + 4(n - 1)d + (n - 1)e = 0$ (6)

$(4) + 2 \times (5)$ gives

$-7b + 2(n - 1)d - 3(n - 1)e = 0$ (7)

Further, $15 \times (7) - 7 \times (6)$ gives

$2(n - 1)d - 52(n - 1)e = 0$

or $d = 26e$ $(\because n > 1)$

$\therefore d/e = 26$

Putting $d = 26e$ in (3) and solving it with (4), we get $a = 2(n - 1)e, b = 7(n - 1)e$

Then, the ratio of their n^{th} terms is

$$\frac{2(n-1)e + (n-1)26e}{7(n-1)e + (n-1)e} = \frac{7}{2}$$

For Problems 28–30

28. d., 29. c., 30. b.

Sol.

We have,

$a + b + c = 25$ (1)

$2a = b + 2$ (2)

$c^2 = 18b$ (3)

Eliminating a from (1) and (2), we have

$$b = 16 - \frac{2c}{3}$$

Then from (3),

$$c^2 = 18\left(16 - \frac{2c}{3}\right)$$

$\Rightarrow c^2 + 12c - 18 \times 16 = 0$

$\Rightarrow (c - 12)(c + 24) = 0$

Now, $c = -24$ is not possible since it does not lie between 2 and 18. Hence, $c = 12$. Then from (3), $b = 8$ and finally from (2), $a = 5$.

Thus, $a = 5$, $b = 8$ and $c = 12$. Hence, $abc = 5 \times 8 \times 12 = 480$.

Also, equation $ax^2 + bx + c = 0$ is $5x^2 + 8x + 12 = 0$, which has imaginary roots.

If a, b, c are roots of the equation $x^3 + qx^2 + rx + s = 0$, then sum of product of roots taken two at a time is $r = 5 \times 8 + 5 \times 12 + 8 \times 12 = 196$.

For Problems 31–33

31. c., 32. c., 33. c.

Sol.

31. Clearly, here the differences between the successive terms are $7 - 3$, $14 - 7$, $24 - 14$, ..., i.e., 4, 7, 10, ... which are in A.P.

$\therefore T_n = an^2 + bn + c$

Thus, we have

$3 = a + b + c$

$7 = 4a + 2b + c$

$14 = 9a + 3b + c$

Solving, we get $a = 3/2$, $b = -1/2$, $c = 2$. Hence,

$$T_n = \frac{1}{2}(3n^2 - n + 4)$$

$$\therefore S_n = \frac{1}{2}\left[3\Sigma n^2 - \Sigma n + 4n\right]$$

$$= \frac{1}{2}\left[3\frac{n(n+1)(2n+1)}{6} - \frac{n(n+1)}{2} + 4n\right]$$

$= \frac{n}{2}(n^2 + n + 4)$

$\Rightarrow S_{20} = 4240$

32. The first differences are 5, 14, 50 194, 770, ….

The second differences are 9, 36, 144, 576, ….

They are in G.P. whose n^{th} term is $ar^{n-1} = a4^{n-1}$.

Therefore, T_n of the given series will be of the form

$T_n = a4^{n-1} + bn + c$

$T_1 = a + b + c = 3$

$T_2 = 4a + 2b + c = 8$

$T_3 = 16a + 3b + c = 22$

Solving, we have $a = 1, b = 2, c = 0$.

$\therefore T_n = 4^{n-1} + 2n$

$\Rightarrow T_{100} = 4^{99} + 200$

$= 2^{198} + 2^3 \times 25$

$= 8(2^{195} + 25)$ (which is divisible by 8)

33. Given series is $2 + 12 + 36 + 80 + 150 + 252 + \cdots$

The first differences are 10, 24, 44, 70, 102, …

The second differences are 14, 20, 26, 32, … which are in A.P.

Hence, general term of the series is

$T_n = an^3 + bn^2 + cn + d$

$\Rightarrow 2 = a + b + c + d$

$12 = 8a + 4b + 2c + d$

$36 = 27a + 9b + 3c + d$

$80 = 64a + 16b + 4c + d$

Solving for a, we get $a = 1$.

$$\therefore \lim_{n\to\infty} \frac{T_n}{n^3} = \lim_{n\to\infty}\left(1 + \frac{b}{n} + \frac{c}{n^2} + \frac{d}{n^3}\right) = 1$$

Matrix-Match Type

1. a → q; b → r; c → p; d → r.

a. a, b, c are in G.P. Hence,

$b^2 = ac$

$\Rightarrow 2\log_{10} b = \log_{10} a + \log_{10} c$

$$\Rightarrow \frac{2}{\log_b 10} = \frac{1}{\log_a 10} + \frac{1}{\log_c 10}$$

$$\Rightarrow \frac{2}{y} = \frac{1}{x} + \frac{1}{z}$$

Hence, x, y, z are in H.P.

b. $$\frac{a + be^x}{a - be^x} = \frac{b + ce^x}{b - ce^x} = \frac{c + de^x}{c - de^x}$$

$$\Rightarrow \frac{2a}{a - be^x} - 1 = \frac{2b}{b - ce^x} - 1 = \frac{2c}{c - de^x} - 1$$

$$\Rightarrow \frac{a - be^x}{a} = \frac{b - ce^x}{b} = \frac{c - de^x}{c}$$

$$\Rightarrow 1 - \frac{b}{a}e^x = 1 - \frac{c}{b}e^x = 1 - \frac{d}{e}e^x$$

$$\Rightarrow \frac{b}{c} = \frac{c}{d} = \frac{d}{c} = r \text{ (say)}$$

Hence, a, b, c, d are in G.P.

c. Given, $2b = a + c, x^2 = ab, y^2 = bc$. Now,

$x^2 + y^2 = b(a + c) = b\, 2b = 2b^2$

$\Rightarrow x^2 + y^2 = 2b^2$

Hence, x^2, b^2, y^2 are in A.P.

d. $x \log a = y \log b = z \log c = k$ (say)

Also,

$y^2 = xz$

$$\Rightarrow \frac{k^2}{(\log b)^2} = \frac{k^2}{\log a \log c}$$

Hence, $\log a, \log b, \log c$ are in G.P.

2. a → p, q, r, s; b → r, s; c → p, q; d → r, s.

a. $\Sigma n = 210$

$\Rightarrow n(n + 1) = 420$

$\Rightarrow (n - 20)(n + 21) = 0$

$\Rightarrow n = 20$

Hence,

$$\Sigma n^2 = \frac{n}{6}(n + 1)(2n + 1)$$

$$= \frac{20}{6}(21)(41)$$

$= (10)(7)(41)$

Hence, the greatest prime number by which Σn^2 is divisible is 41.

b. $4, G_1, G_2, \ldots, G_{n+1}, \ldots, G_{2n}, G_{2n+1}, 2916$

G_{n+1} will be the middle mean of $(2n + 1)$ odd means and it will be equidistant from the first and last terms. Hence, $4, G_{n+1}, 2916$ will also be in G.P. So,

$\Rightarrow G_{n+1}^2 = 4 \times 2916$

$= 4 \times 9 \times 324$

$= 4 \times 9 \times 4 \times 81$

$\Rightarrow G_{n+1} = 2 \times 3 \times 2 \times 9 = 108$

Hence, the greatest odd number by which G_{n+1} is divisible is 27.

c. Terms are 40, 30, 24, 20. Now,

$$\frac{1}{30} - \frac{1}{40} = \frac{1}{120}$$

$$\Rightarrow \frac{1}{24} - \frac{1}{30} = \frac{6}{24 \times 30} = \frac{1}{120}$$

and $$\frac{1}{20} - \frac{1}{24} = \frac{4}{20 \times 24} = \frac{1}{120}$$

Hence, 1/30, 1/24, 1/20 are in A.P. with common difference $d = 1/120$. Hence, the next term is $1/20 + 1/120 = 7/120$.

Therefore, the next term of given series is $\frac{120}{7} = 17\frac{1}{7}$.

Hence, the integral part of $17\frac{1}{7}$ is 17.

d. $S = 1 + \frac{4}{5} + \frac{7}{5^2} + \frac{10}{5^3} + \cdots \infty$

$$\Rightarrow \frac{1}{5}S = \frac{1}{5} + \frac{4}{5^2} + \frac{7}{5^3} + \cdots$$

$$\Rightarrow S\left(1 - \frac{1}{5}\right) = 1 + 3\left[\frac{1}{5} + \frac{1}{5^2} + \frac{1}{5^3} + \cdots \infty\right]$$

$$\Rightarrow \frac{4}{5}S = 1 + 3\left[\frac{1/5}{1 - 1/5}\right] = 1 + \frac{3}{4} = \frac{7}{4}$$

$$\Rightarrow S = \frac{35}{16}$$

$\Rightarrow a = 35$ and $b = 16$

$\Rightarrow a - b = 19$

Integer Type

1. (5)

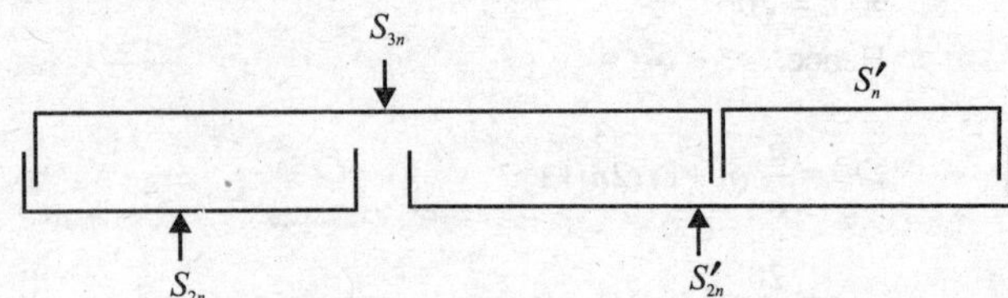

Fig. 3.5

Given $S_{3n} = S'_n = S_{4n} - S_{3n}$

$\Rightarrow 2S_{3n} = S_{4n}$

$$\Rightarrow 2 \times \frac{3n}{2}[2a + (3n - 1)d) = \frac{4n}{2}(2a + (4n - 1)d]$$

$\Rightarrow 12a + (18n - 6)d = 8a + (16n - 4)d$

$\Rightarrow 4a = (-n + 2)d$

$\Rightarrow 2a = (1 - n)d$ (1)

Now we have to find $\frac{S_{2n}}{S'_{2n}}$,

$$\frac{S_{2n}}{S'_{2n}} = \frac{S_{2n}}{S_{4n} - S_{2n}}$$

$$= \frac{\frac{2n}{2}[2a + (2n - 1)d]}{\frac{4n}{2}[2a + (4n - 1)d] - \frac{2n}{2}[2a + (2n - 1)d]}$$

$$= \frac{2((1 - n)d + (2n - 1)d)}{4(1 - n)d + (4n - 1)d) - 2((1 - n)d + (2n - 1)d)}$$

$$= \frac{2nd}{10nd} = \frac{1}{5}$$

2. (5) Let $A_1, A_2, A_3, \ldots, A_{11}$ be 11 A.M.'s between 28 and 10. Therefore,

$28, A_1, A_2, \ldots, A_{11}$, 10 are in A.P.

Let d be the common difference of the A.P.

Number of terms = 13

$10 = T_{13} = T_1 + 12d = 28 + 12d$

$$\Rightarrow d = \frac{10 - 28}{12} = -\frac{18}{12} = -\frac{3}{2}$$

$\Rightarrow$ Number of integral A.M.'s is 5.

3. (7) Let $A_1, A_2, \ldots, A_m$ be m arithmetic means between 1 and 31.

Then $1, A_1, A_2, \ldots, A_m$, 31 is an A.P. with common difference

$$d = \frac{31 - 1}{m - 1} = \frac{30}{m + 1}\left[\text{Using} : d = \frac{b - a}{n + 1}\right]$$

Now $A_7 = 1 + 7d$

$$\Rightarrow A_7 = 1 + \frac{7 \times 30}{m + 1} = \frac{m + 211}{m + 1}$$

and $A_{m-1} = 1 + (m - 1)d$

$$= 1 + \frac{30}{m + 1}(m - 1)$$

$$= \frac{31m - 29}{m + 1}$$

It is given that $\frac{A_7}{A_{m-1}} = \frac{5}{9}$

$$\Rightarrow \frac{m + 211}{31m - 29} = \frac{5}{9}$$

$\Rightarrow 9m + 1899 = 155m - 145$

$\Rightarrow 146m = 2044$

$\Rightarrow m = 14$

4. (6) Let r be the common ratio of the given G.P. Then $a_n = 96$. Therefore,

$a_1 r^{n-1} = 96$

$\Rightarrow 3r^{n-1} = 96$

$\Rightarrow r^{n-1} = 32$ (1)

Now $S_n = 189$

$$\Rightarrow a_1\left(\frac{r^n - 1}{r - 1}\right) = 189$$

$$\Rightarrow 3\left(\frac{(r^{n-1})r - 1}{r - 1}\right) = 189$$

$$\Rightarrow 3\left(\frac{32r - 1}{r - 1}\right) = 189 \quad \text{[Using (1)]}$$

$\Rightarrow 32r - 1 = 63r - 63$

$\Rightarrow 31r = 62$

$\Rightarrow r = 2$

Putting $r = 2$ in (1), we get

$2^{n-1} = 32$

$\Rightarrow 2^{n-1} = 2^5$

$\Rightarrow n - 1 = 5$

$\Rightarrow n = 6$

5. **(4)** Since a, b, c are in A.P,

$2b = a + c$ (1)

Again by the given condition,

$a + 1, b, c$ are in G.P. and $a, b, c + 2$ are in G.P.Therefore,

$b^2 = (a + 1)c$ (2)

and $b^2 = a(c + 2)$ (3)

By (2) and (3),

$(a + 1)c = a(c + 2)$

$\Rightarrow ac + c = ac + 2a$

$\Rightarrow c = 2a$

$\Rightarrow$ (2) gives $b^2 = (a + 1)2a$.

Also (1) gives $2b = a + 2a = 3a$.

$\Rightarrow b = \frac{3a}{2}$

$\Rightarrow \frac{9a^2}{4} = (a + 1)2a$

$\Rightarrow \frac{9a}{8} = a + 1$

$\Rightarrow a = 8$

$\Rightarrow c = 2 \times 8 = 16$

$\Rightarrow 2b = 8 + 16 = 24$

$\Rightarrow b = 12$

6. **(3)** Let the three numbers be $a/r, a, ar$.

Since their product is 216,

$(a/r) \times a \times ar = 216$

$\Rightarrow a^3 = 216$

$\Rightarrow a = 6$

Since the sum of their products in pairs is 156,

$\frac{a}{r} \times a + a \times ar + \frac{a}{r} \times ar = 156$

$\Rightarrow a^2\left(\frac{1}{r} + r + 1\right) = 156$

$\Rightarrow 36\left(\frac{1 + r^2 + r}{r}\right) = 156$

$\Rightarrow 3(r^2 + r + 1) = 13r$

$\Rightarrow 3r^2 - 10r + 3 = 0$

$\Rightarrow (3r - 1)(r - 3) = 0$

$\Rightarrow r = \frac{1}{3}$ or 3

Hence, putting the values of a and r, we get the required numbers as 18, 6, 2 or 2, 6, 18.

7. **(5)** Let $H_1, H_2, \cdots, H_9$ be nine harmonic means between 2 and 3.

$2, H_1, H_2, \cdots, H_9, 3$ are in HP. Therefore,

$\frac{1}{2}, \frac{1}{H_1}, \frac{1}{H_2}, \cdots, \frac{1}{H_9}, \frac{1}{3}$ are in A.P.

With common difference $D = \frac{2 - 3}{(9 + 1) \times 2 \times 3} = -\frac{1}{60}$

$\left[\because D = \frac{a - b}{(n - 1)ab}\right]$

$\frac{1}{H_i} = \frac{1}{2} + iD$, where $i = 1, 2, 3, \cdots, 9$

$\Rightarrow \frac{1}{H_i} = \frac{1}{2} + -\frac{i}{60}$

$\Rightarrow \frac{6}{H_i} = 3 - \frac{i}{10}$ (1)

Let $A_1, A_2, \cdots, A_9$ be nine A.M.'s between 2 and 3.

Then $2, A_1, A_2, \cdots, A_9, 3$ are in A.P.

With common difference $d = \frac{3 - 2}{9 + 1} = \frac{1}{10}$ $\left[\because d = \frac{b - a}{n + 1}\right]$

$A_i = 2 + id$, where $i = 1, 2, \cdots, 9$

$\Rightarrow A_i = 2 + \frac{i}{10}$ (2)

From (1) and (2), we have

$A_i + \frac{6}{H_i} = 3 - \frac{i}{10} + 2 + \frac{i}{10} = 5$ for $i = 1, 2, \cdots, 9$.

8. **(2)** Let the H.P. be $\frac{1}{a}, \frac{1}{a + d}, \frac{1}{a + 2d}, \frac{1}{a + 3d}, \cdots$

Then $\frac{1}{a} = \frac{2}{5}$ and $\frac{1}{a + d} = \frac{12}{13}$

$\Rightarrow a = \frac{5}{2}$ and $d = -\frac{17}{12}$

Now, n^{th} term of the H.P. is $\frac{1}{a + (n - 1)d} = \frac{12}{47 - 17n}$.

So the nth term is largest when $47 - 17n$ has the least value.

Clearly, $\frac{12}{47 - 17n}$ is least for $n = 2$.

Hence, 2nd term is the largest term.

9. **(2)** $S = 3 + (3 + d)\frac{1}{4} + (3 + 2d)\frac{1}{4^2} + \cdots \infty$ (1)

$\Rightarrow \frac{1}{4}S = (3)\frac{1}{4} + (3 + d)\frac{1}{4^2} + \cdots \infty$ (2)

Subtracting (2) from (1), we get

$\frac{3}{4}S = 3 + (d)\frac{1}{4} + (d)\frac{1}{4^2} + \cdots \infty$

$= 3 + \frac{\frac{d}{4}}{1 - \frac{1}{4}}$

$= 3 + \frac{d}{3}$

$\Rightarrow S = 4 + \frac{4d}{9}$

Given $4 + \frac{4d}{9} = \frac{4}{9}$

$\Rightarrow \frac{4d}{9} = \frac{8}{9}$

$\Rightarrow d = 2$

10. **(2)** $T(r) = \frac{1}{(2r - 1)(2r + 1)(2r + 3)}$

$= \frac{(2r + 3) - (2r - 1)}{4(2r - 1)(2r + 1)(2r + 3)}$

$= \frac{1}{4}\left[\frac{1}{(2r - 1)(2r + 1)} - \frac{1}{(2r + 1)(2r + 3)}\right]$

$= -\frac{1}{4}\left[\frac{1}{(2r + 1)(2r + 3)} - \frac{1}{(2r - 1)(2r + 1)}\right]$

$= -\frac{1}{4}[V(r) = -V(r - 1)]$

$\Rightarrow$ Sum of n terms

$$S_n = \sum_{r=1}^{n} T(r) = -\frac{1}{4}[V(n) - V(0)]$$

$$= -\frac{1}{4}\left[\frac{1}{(2n+1)(2n+3)} - \frac{1}{3}\right]$$

$$= \frac{1}{4}\left[\frac{1}{3} - \frac{1}{(2n+1(2n+3)}\right]$$

$$\Rightarrow S = \lim_{n\to\infty} S_n$$

$$= \lim_{n\to\infty} \frac{1}{4}\left[\frac{1}{3} - \frac{1}{(2n+1)(2n+3)}\right]$$

$$= \frac{1}{12}$$

$\Rightarrow 36S = 3$

$\Rightarrow [36S] = 2$ (as $36S$ is not exactly 3, it is in fact slightly lesser than 3).

11. **(4)** $\sqrt{a-x}, \sqrt{x}, \sqrt{a+x}$ are in A.P.

$\Rightarrow 2\sqrt{x} = \sqrt{a-x} + \sqrt{a+x}$

$\Rightarrow 4x = a - x + a + x + 2\sqrt{a^2 - x^2}$ (squaring both sides)

$\Rightarrow 2x - a = \sqrt{a^2 - x^2}$

$\Rightarrow 4x^2 - 4ax + a^2 = a^2 - x^2$

$\Rightarrow 5x^2 = 4ax$

$\Rightarrow a = \frac{5x}{4}$ (as $a, x > 0$)

Now x must be a perfect square as $\sqrt{x}$ is an integer.

$\Rightarrow x = 1, 4, 9, 16, \ldots$, etc.

For $x = 1$, $a = \frac{5}{4}$ (rational number)

For $x = 4$, $a = 5$ (prime number)

For $x = 9$, $a = \frac{45}{4}$ (rational number)

For $x = 16$, $a = 20$ (composite number)

Hence, the least composite number is 20.

12. **(5)** Let there be $(2n + 1z)$ stones. Clearly, one stone lies in the middle and n stones on each side of it in a row.

Let P be the mid-stone and let A and B be the end stones on the left and right of P, respectively.

Clearly, there are n intervals each of length 10 m on both sides of P.

Now, suppose that the man starts from A. He picks up the end stone on the left of the mid-stone and goes to the mid-stone, drops it and goes to $(n-1)^{\text{th}}$ stone on the left, picks it up, goes to the mid-stone and drops it.

This process is repeated till he collects all stones on the left of the mid-stone at the mid-stone.

So the distance covered in collecting stones on the left of the mid-stone

$= 10 \times n + 2[10 \times (n-1) + 10 \times (n-2) + \cdots + 10 \times 2 + 10 \times 1]$

After collecting all stones on the left of the mid-stone, the man goes to the stone B on the right side of the mid-stone, picks it up, goes to the mid-stone and drops it.

Then he goes to $(n-1)^{\text{th}}$ stone on the right and the process is repeated till he collects all stones at the mid-stone.

The distance covered in collecting the stones on the right side of the mid-stone

$= 2[10 \times n + 10 \times (n-1) + 10 \times (n-2) + \cdots + 10 \times 2 + 10 \times 1]$

Therefore, the total distance covered

$= 10 \times n + 2[10 \times (n-1) + 10 \times (n-2) + \cdots + 10 \times 2 + 10 \times 1] + 2[10 \times n + 10 \times (n-1) + \cdots + 10 \times 2 + 10 \times 1]$

$= 4[10 \times n + 10 \times (n-1) + \cdots + 10 \times 2 + 10 \times 1] - 10 \times n$

$= 40[1 + 2 + 3 + \cdots + n] - 10n$

$= \left[\frac{n}{2}(1+n)\right] - 10n$

$= 20n(n+1) - 10n$

$= 20n^2 + 10n$

But the total distance covered is 3 km = 3000 m. Therefore,

$20n^2 + 10n = 3000$

$\Rightarrow 2n^2 + n - 300 = 0$

$\Rightarrow (n-12)(2n+25) = 0$

$\Rightarrow n = 12$

Hence, the number of stones $= 2n + 1 = 25$

13. **(4)** Let $\frac{2n+1}{2n-1} = r$.

Then the given series is

$$S = r + 3r^2 + 5r^3 + 7r^4 + \cdots + (2n-1)r^n$$

$$rS = r^2 + 3r^3 + 5r^4 + \cdots + (2n-3)r^n + (2n-1)r^{n+1}$$

$$(1-r)S = r + 2r^2 + 2r^3 + \cdots + 2r^n - (2n-1)r^{n+1}$$

$$\Rightarrow 36(1-r) = r + \frac{2r^2(1-r^{n-1})}{1-r} - (2n-1)r^{n+1}$$

$\Rightarrow 36(1-r)^2 = r - r^2 + 2r^2 - 2r^{n+1} - (2n-1)r^{n+1} + (2n-1)r^{n+2}$

$$\Rightarrow 36(1-r)^2 = r + r^2 - (2n+1)r^{n+1} + (2n-1)r^{n+2}$$

$$\Rightarrow 36(1-r)^2 = r + r^2 - (2n-1)\left[\frac{2n+1}{2n-1}r^{n+1} - r^{n+2}\right]$$

$$\Rightarrow 36(1-r)^2 = r + r^2 - (2n-1)[r \times r^{n+1} - r^{n+2}]$$

$$\Rightarrow 36(1-r)^2 = r(1+r)$$

$$\Rightarrow 36\left[1 - \frac{2n+1}{2n-1}\right]^2 = \frac{2n+1}{2n-1}\left[1 + \frac{2n+1}{2n-1}\right]$$

$$\Rightarrow 36\left[\frac{-2}{2n-1}\right]^2 = \frac{2n+1}{2n-1}\left[\frac{4n}{2n-1}\right]$$

$$\Rightarrow 36 = n(2n+1)$$

$$\Rightarrow n = 4$$

14. **(3)** Given that $\log_3 2$, $\log_3(2^x - 5)$, $\log_3[2^x - (7/2)]$ are in A.P. Therefore,

$2\log_3(2^x - 5) = \log_3[2^x - (7/2)] + \log_3 2$

$$\Rightarrow (2^x - 5)^2 = 2\left(2^x - \frac{7}{2}\right)$$

$$\Rightarrow (2^x)^2 - 10 \times 2^x + 25 - 2 \times 2 + 7 = 0$$

$$\Rightarrow (2^x)^2 - 12 \times 2^x + 32 = 0$$

Let $2^x = y$, then we get

$y^2 - 12y + 32 = 0$

$\Rightarrow (y-4)(y-8) = 0$

$\Rightarrow y = 4$ or 8

$\Rightarrow 2^x = 2^2$ or 2^3

$\Rightarrow x = 2$ or 3

But for $\log_3(2^x - 5)$ and $\log_3[2^x - (7/2)]$ to be defined

$2^x - 5 > 0$ and $2^x - (7/2) > 0$

$\Rightarrow 2^x > 5$ and $2^x > 7/2$

$\Rightarrow 2^x > 5$

$\Rightarrow x \neq 2$. Therefore, $x = 3$.

15. (6) $a_1 = h_1 = 2,\ a_{10} = h_{10} = 3$

$3 = a_{10} = 2 + 9d \Rightarrow d = 1/9$

Therefore, $a_4 = 2 + 3d = 7/3$.

Also $3 = h_{10} \Rightarrow \frac{1}{3} = \frac{1}{h_{10}} = \frac{1}{2} + 9D$

$\Rightarrow D = -\frac{1}{54}$

$\Rightarrow \frac{1}{h_7} = \frac{1}{2} + 6D = \frac{1}{2} - \frac{1}{9} = \frac{7}{18}$. Therefore,

$a_4 h_7 = \frac{7}{3} \times \frac{18}{7} = 6.$

16. (1) $T_k = \frac{1}{\sqrt{k}\sqrt{k+1}\,[\sqrt{k} + \sqrt{k+1}]}$

$= \frac{\sqrt{k+1} - \sqrt{k}}{\sqrt{k}\,\sqrt{k+1}}$

$= \frac{1}{\sqrt{k}} - \frac{1}{\sqrt{k+1}}$

$S = \sum_{k=1}^{360}\left(\frac{1}{\sqrt{k}} - \frac{1}{\sqrt{k+1}}\right)$

$= 1 - \frac{1}{\sqrt{361}} = 1 - \frac{1}{19} = \frac{18}{19}$

Therefore, $n - m = 1$.

17. (2) $T_n = \frac{1}{n^2 + (n-2)}$

$= \frac{1}{(n+2)(n-1)},\ n = 3, 4, 5, \ldots$

$= \frac{1}{3}\left[\frac{1}{n-1} - \frac{1}{n+2}\right]$

$S = \sum_{n=3}^{\infty} T_n$

$= \frac{1}{3}\left(\frac{1}{2} - \frac{1}{5}\right) + \frac{1}{3}\left(\frac{1}{3} - \frac{1}{6}\right) + \frac{1}{3}\left(\frac{1}{4} - \frac{1}{7}\right) + \frac{1}{3}\left(\frac{1}{5} - \frac{1}{8}\right)$

$= \frac{1}{3}\left[\frac{1}{2} + \frac{1}{3} + \frac{1}{4}\right] = \frac{1}{3}\left[\frac{6+4+3}{12}\right] = \frac{13}{36}$

18. (8) Given $\frac{ar(r^{10} - 1)}{r - 1} = 18$ (1)

Also $\frac{\frac{1}{ar}\left(1 - \frac{1}{r^{10}}\right)}{1 - \frac{1}{r}} = 6$

$\Rightarrow \frac{1}{ar^{11}}\,\frac{(r^{10} - 1)r}{r - 1} = 6$

$\Rightarrow \frac{1}{a^2r^{11}}\,\frac{ar(r^{10} - 1)}{r - 1} = 6$ (2)

From (1) and (2), we get

$\frac{1}{a^2r^{11}} \times 18 = 6 \Rightarrow a^2r^{11} = 3.$

So $P = a^{10}r^{55} = (a^2r^{11})^5 = 3^5 = 243$.

19. (2) Let $t = 1 - \frac{1}{2^2} + \frac{1}{3^2} - \frac{1}{4^2} + \ldots$ and $S = 1 + \frac{1}{2^2} + \frac{1}{3^2} + \cdots$

$s - t = \left(1 + \frac{1}{2^2} + \frac{1}{3^2} + \frac{1}{4^2} + \cdots\right) - \left(1 - \frac{1}{2^2} + \frac{1}{3^2} - \frac{1}{4^2} + \cdots\right)$

$= \frac{2}{2^2} + \frac{2}{4^2} + \frac{2}{6^2} + \ldots = \frac{2}{2^2}\left(\frac{1}{1^2} + \frac{1}{2^2} + \frac{1}{3^2} + \ldots\right) = \frac{s}{2}$

20. (8) For the G.P. $a, ar, ar^2, \ldots$

$P_n = a\,(ar)\,(ar^2) \ldots (ar^{n-1}) = a^n r^{n(n-1)/2}$

$S = \sum_{n=1}^{\infty} \sqrt[n]{P_n} = \sum_{n=1}^{\infty} ar^{(n-1)/2}$

Now $\sum_{n=1}^{\infty} ar^{(n-1)/2} = a\,[1 + \sqrt{r} + r + r\sqrt{r} + \cdots \infty] = \frac{a}{1 - \sqrt{r}}$

Given $a = 16$ and $r = 1/4$. Therefore,

$S = \frac{16}{1 - (1/2)} = 32.$

21. (6) $a_n = \frac{3}{4} - \left(\frac{3}{4}\right)^2 + \left(\frac{3}{4}\right)^3 + \cdots + (-1)^{n-1}\left(\frac{3}{4}\right)^n$

$= \frac{\frac{3}{4}\left(1 - \left(-\frac{3}{4}\right)^n\right)}{1 - \left(-\frac{3}{4}\right)}$

$= \frac{3}{4}\left(1 - \left(-\frac{3}{4}\right)^n\right)$

We have $b_n = 1 - a_n$ and $b_n > a_n$ for $n \geq n_0$. Therefore,

$1 - a_n > a_n \quad \Rightarrow 2a_n < 1$

$\Rightarrow \frac{6}{7}\left[1 - \left(-\frac{3}{4}\right)^n\right] < 1$

$\Rightarrow -\left(-\frac{3}{4}\right)^n < \frac{1}{6}$

$\Rightarrow (-3)^{n+1} < 2^{2n-1}$

For n to be even, the inequality always holds. For n to be odd, it holds for $n \geq 7$.

Therefore, the least natural number for which it holds is 6.

22. (3) $S_k = \frac{\frac{k-1}{k!}}{1 - \frac{1}{k}} = \frac{1}{(k-1)!}$

$\sum_{k=2}^{100}\left|(k^2 - 3k + 1)\frac{1}{(k-1)!}\right|$

$\sum_{k=2}^{100}\left|\frac{(k-1)^2 - k}{(k-1)!}\right|$

$$= \Sigma \frac{k-1}{(k-2)!} - \frac{k}{(k-1)!}$$

$$= \left|\frac{2}{1!} - \frac{3}{2!}\right| + \left|\frac{3}{2!} - \frac{4}{3!}\right| + \cdots$$

$$= \frac{2}{1!} - \frac{1}{0!} + \frac{2}{1!} - \frac{3}{2!} + \frac{3}{2!} - \frac{4}{3!} + \ldots + \frac{99}{98!} - \frac{100}{99!}$$

$$= 3 - \frac{100}{99!}$$

23. **(0)** $a_k = 2a_{k-1} - a_{k-2} \Rightarrow a_1, a_2, \ldots, a_{11}$ are in A.P. Therefore,

$$a_1^2 + a_2^2 + \cdots + a_{11}^2 = \frac{11a^2 + 35 \times 11d^2 + 10ad}{11} = 90$$

$$225 + 35d^2 + 150d = 90$$

$$35d^2 + 150d + 135 = 0$$

$$d = -3, -9/7$$

Given $a_2 < 27/2$. Therefore, $d = -3$ and $d - 9/7$. Hence,

$$\frac{a_1 + a_2 + \cdots + a_{11}}{11} = \frac{11}{2}[30 - 10 \times 3] = 0.$$

24. (4) $S = 1 + 6(2^2 + 3^2 + 4^{23} + \cdots + n^2) + 1 - 4(2 + 3 + 4 + \cdots + n)$

$$6\sum_{r=1}^{n} r^2 - 4\sum_{r=1}^{n} r = 140$$

$$\Rightarrow n(n+1)(2n+1) - 2n(n+1) = 140$$

$$\Rightarrow n(n+1)(2n-1) = 4 \times 5 \times 7$$

$$\Rightarrow n = 4$$

25. (2) $T_n = 1 + 2 + 2^2 + \cdots + 2^{n-1}$

(G.P with $a = 1$ and $r = 2$)

$= (2^n - 1)$

Hence sum $\sum_{n=1}^{n} (2^n - 1) = [(2 + 2^2 + 2^3 + \ldots + 2^n) - n]$

$= 2(2^n - 1) - n = 2^{n+1} - 2 - n$

$= 2^{n+1} + (0)n^2 + (-1)n + (-2)$

Hence $R = 1$; $S = 0$; $T = -1$; $U = -2$

$R + S + T + u = -2$

26. **(7)** Let a, b, c denote the numbers of –1's, 1's, and 2's in the sequence respectively. We need not consider the zeroes. Then a, b, c are non-negative integers satisfying $-a + b + 2c = 19$, $a + b + 4c = 99$

$\therefore a = 40 - c$, $b = 59 - 3c$

$0 \le c \le 19$, since $b \ge 0$

So, $x_1^3 + x_2^3 + \cdots + x_n^3 = -a + b + 8c$

$= 19 + 6c$

The minimum is achieved when

$c = 0$ $(a = 40, b = 59)$

The maximum is achieved when

$a = 21$ and $b = 2$.

Thus $m = 19$ and $M = 133$

So, $\frac{M}{m} = \frac{133}{19} = 7$

27. **(8)** For $n \ge 3$, $a_n = \frac{a_1 + a_2 + a_3 + \ldots + a_{n-1}}{n-1}$

$$\Rightarrow (n-1)a_n = a_1 + a_2 + \ldots + a_{n-1}$$

$$a_{n+1} = \frac{a_1 + a_2 + \ldots + a_n}{n}$$

$$= \frac{(n-1)a_n + a_n}{n}$$

$$= \frac{na_n}{n} = a_n$$

$$a_{n+1} = a_n \Rightarrow a_9 = a_3 = 99$$

$$a_3 = \frac{a_1 + a_2}{2} \Rightarrow 99 + \frac{19 + a_2}{2}$$

$$a_2 = 179$$

28. **(9)** To simplify elegantly, put $a = \sqrt{n+1}$ and $b = \sqrt{2n-1}$. This leads to the expression

$$f(n) = \frac{(2n+1)^{\frac{3}{2}} - (2n-1)^{\frac{3}{2}}}{2}$$

Hence $f(1) + \cdots + f(40) = \frac{(81)^{3/2} - (1)^{3/2}}{2} = 364$

CHAPTER 4

Inequalities Involving Means

Objective Type

Each question has four choices a, b, c and d, out of which *only one* is correct.

1. The minimum value of $\frac{x^4+y^4+z^2}{xyz}$ for positive real numbers x, y, z is
   a. $\sqrt{2}$ **b.** $2\sqrt{2}$
   c. $4\sqrt{2}$ **d.** $8\sqrt{2}$
2. A rod of fixed length k slides along the coordinate axes. If it meets the axes at $A(a, 0)$ and $B(0, b)$, then the minimum value of $\left(a+\frac{1}{a}\right)^2+\left(b+\frac{1}{b}\right)^2$ is
   a. 0 **b.** 8
   c. $k^2-4+\frac{4}{k^2}$ **d.** $k^2+4+\frac{4}{k^2}$
3. The least value of $6\tan^2\phi + 54\cot^2\phi + 18$ is
   (I) 54 when A.M. ≥ G.M. is applicable for $6\tan^2\phi$, $54\cot^2\phi$, 18
   (II) 54 when A.M. ≥ G.M. is applicable for $6\tan^2\phi$, $54\cot^2\phi$ and 18 is added further (III) 78 when $\tan^2\phi = \cot^2\phi$
   a. (I) is correct, (II) is false
   b. (I) and (II) are correct
   c. (III) is correct
   d. none of the above are correct
4. If ab^2c^3, $a^2b^3c^4$, $a^3b^4c^5$ are in A.P. $(a, b, c > 0)$, then the minimum value of $a+b+c$ is
   a. 1 **b.** 3
   c. 5 **d.** 9
5. If $y = 3^{x-1} + 3^{-x-1}$, then the least value of y is
   a. 2 **b.** 6
   c. 2/3 **d.** 3/2
6. Minimum value of $(b+c)/a+(c+a)/b+(a+b)/c$ (for real positive numbers a, b, c) is
   a. 1 **b.** 2
   c. 4 **d.** 6
7. If the product of n positive numbers is n^n, then their sum is
   a. a positive integer **b.** divisible by n
   c. equal to $n + (1/n)$ **d.** never less than n^2
8. The minimum value of $P = bcx + cay + abz$, when $xyz = abc$, is
   a. $3abc$ **b.** abc
   c. abc **d.** $4abc$
9. If l, m, n be the three positive roots of the equation $x^3 - ax^2 + bx + 48 = 0$, then the minimum value of $(1/l) + (2/m) + (3/n)$equals
   a. 1 **b.** 2
   c. 3/2 **d.** 5/2
10. If positive numbers a, b, c be in H.P., then equation $x^2 - kx + 2b^{101} - a^{101} - c^{101} = 0$ $(k \in R)$ has
    a. both roots positive
    b. both roots negative
    c. one positive and one negative root
    d. both roots imaginary
11. For $x^2 - (a+3)|x| + 4 = 0$ to have real solutions, the range of a is
    a. $(-\infty, -7] \cup [1, \infty)$ **b.** $(-3, \infty)$
    c. $(-\infty, -7]$ **d.** $[1, \infty)$
12. If a, b, c are the sides of a triangle, then the minimum value of $\frac{a}{b+c-a}+\frac{b}{c+a-b}+\frac{c}{a+b-c}$ is equal to

a. 3 b. 6
c. 9 d. 12

13. If $a, b, c, d \in R^+ - \{1\}$, then the minimum value of $\log_d a + \log_b d + \log_a c + \log_c b$ is

a. 4 b. 2
c. 1 d. none of these

14. If $a, b, c \in R^+$, then $\frac{bc}{b+c} + \frac{ac}{a+c} + \frac{ab}{a+b}$ is always

a. $\le \frac{1}{2}(a+b+c)$ b. $\ge \frac{1}{3}\sqrt{abc}$
c. $\le \frac{1}{3}(a+b+c)$ d. $\ge \frac{1}{2}\sqrt{abc}$

15. If $a, b, c \in R^+$, then $(a+b+c)\left(\frac{1}{a}+\frac{1}{b}+\frac{1}{c}\right)$ is always

a. ≥ 12 b. ≥ 9
c. ≤ 12 d. none of these

16. If $a, b, c \in R^+$, then the minimum value of $a(b^2+c^2)+b(c^2+a^2) + c(a^2+b^2)$ is equal to

a. abc b. $2abc$
c. $3abc$ d. $6abc$

17. If $a, b, c, d \in R^+$ and a, b, c, d are in H.P., then

a. $a+d > b+c$ b. $a+b > c+d$
c. $a+c > b+d$ d. none of these

18. If $a, b, c \in R^+$ such that $a+b+c=18$, then the maximum value of $a^2 b^3 c^4$ is equal to

a. $2^{18} \times 3^2$ b. $2^{18} \times 3^3$
c. $2^{19} \times 3^2$ d. $2^{19} \times 3^3$

19. $f(x) = \frac{(x-2)(x-1)}{(x-3)}, \forall x > 3$. The minimum value of $f(x)$ is equal to

a. $3+2\sqrt{2}$ b. $3+2\sqrt{3}$
c. $3\sqrt{2}+2$ d. $3\sqrt{2}-2$

20. If $a > 0$, then least value of $(a^3+a^2+a+1)^2$ is

a. $64a^2$ b. $16a^4$
c. $16a^3$ d. none of these

Multiple Correct Answers Type

Each question has four choices a, b, c and d, out of which *one or more* answers are correct.

1. If A is the area and $2s$ the sum of the sides of a triangle, then

a. $A \le \frac{s^2}{4}$ b. $A \le \frac{s^2}{3\sqrt{3}}$
c. $A < \frac{s^2}{\sqrt{3}}$ d. none of these

2. If x, y, z are positive numbers in A.P., then

a. $y^2 \ge xz$ b. $xy + yz \ge 2xz$
c. $\frac{x+y}{2y-x} + \frac{y+z}{2y-z} \ge 4$ d. none of these

3. For positive real numbers a, b, c such that $a+b+c=p$, which one holds?

a. $(p-a)(p-b)(p-c) \le \frac{8}{27}p^3$
b. $(p-a)(p-b)(p-c) \ge 8abc$
c. $\frac{bc}{a} + \frac{ca}{b} + \frac{ab}{c} \le p$
d. none of these

4. If first and $(2n-1)^{\text{th}}$ terms of an A.P., G.P. and H.P. are equal and their n^{th} terms are a, b, c, respectively, then

a. $a=b=c$ b. $a+c=b$
c. $a>b>c$ d. $ac-b^2=0$

Linked Comprehension Type

Based upon each paragraph, three multiple choice questions have to be answered. Each question has four choices a, b, c and d, out of which *only one* is correct.

For Problems 1–3

Roots of the equation $f(x) = x^6 - 12x^5 + bx^4 + cx^3 + dx^2 + ex + 64 = 0$ are positive.

1. Which of the following has the greatest absolute value?

a. b b. c c. d d. e

2. Which of the following has the least absolute value?

a. b b. c c. d d. e

3. Remainder when $f(x)$ is divided by $x-1$ is

a. 2 b. 1 c. 3 d. 10

For Problems 4–6

Equation $x^4 + ax^3 + bx^2 + cx + 1 = 0$ has real roots (a, b, c are non-negative).

4. Minimum non-negative real value of a is

a. 10 b. 9 c. 6 d. 4

5. Minimum non-negative real value of b is

a. 12 b. 15 c. 6 d. 10

6. Minimum non-negative real value of c is

a. 10 b. 9 c. 6 d. 4

Integer Type

1. The minimum value of $4^{\sin^2 x} + 4^{\cos^2 x}$ is ______.

2. The least value of sec A + sec B + sec C in an acute angled triangle is ______.

3. If $yz + zx + xy = 12$, where x, y, z are positive values, then the greatest value of xyz is ______.

4. If $a, b > 0$ such that $a^3 + b^3 = 2$, then the maximum value of $a + b$ is ___.

5. If a, b, c are real numbers such that $0 < a < 1$, $0 < b < 1$, $0 < c < 1$, and $a+b+c=2$, then the least value of $\frac{a}{1-a} \times \frac{b}{1-a} \times \frac{c}{1-c}$ is ______.

6. In ΔABC, the least value of cosec $(A/2)$ + cosec $(B/2)$ + cosec $(C/2)$ is ______.

7. The least value of the expression $2\log_{10} x - \log_x (0.01)$, for $x > 1$, is ______.

8. If a, b, c, d are positive real numbers such that $a+b+c+d = 2$, then the maximum value of $M = (a+b)(c+d)$ is ___.

ANSWERS AND SOLUTIONS

Objective Type

1. b. Using A.M. ≥ G.M., we have

$x^4 + y^4 \geq 2x^2y^2$ and $2x^2y^2 + z^2 \geq \sqrt{8}xyz$

$$\Rightarrow \frac{x^4 + y^4 + z^2}{xyz} \geq \sqrt{8}$$

2. d. $a^2 + b^2 = k^2$

$$\left(a+\frac{1}{a}\right)^2 + \left(b+\frac{1}{b}\right)^2 = a^2 + b^2 + 4 + \frac{1}{a^2} + \frac{1}{b^2}$$

$$= k^2 + 4 + \frac{k^2}{a^2b^2}$$

$$\geq k^2 + 4 + \frac{k^2}{\left(\frac{a^2+b^2}{2}\right)^2}$$

$$= k^2 + 4 + \frac{4}{k^2}$$

3. b. Applying A.M. ≥ G.M. in 6 $\tan^2 \phi$, 54 $\cot^2 \phi$, 18, we get

$$\frac{6\tan^2\phi + 54\cot^2\phi + 18}{3} \geq (6 \times 54 \times 18)^{1/3} \geq 18$$

This is true if

$6 \tan^2 \phi = 54 \cot^2 \phi = 18$

$\Rightarrow \tan^2 \phi = 3$ and $\cot^2 \phi = \frac{1}{3}$

Hence, the statements I and II are correct.

4. b. A.M. ≥ G.M.

$$\Rightarrow \frac{a+b+c}{3} \geq (abc)^{1/3}$$

$$\Rightarrow a + b + c \geq 3\,(abc)^{1/3} \quad (1)$$

But given ab^2c^3, $a^2b^3c^4$, $a^3b^4c^5$ are in A.P. ($\because abc \neq 0$). Hence,

$2abc = 1 + a^2b^2c^2$

$\Rightarrow (abc - 1)^2 = 0$

$\therefore abc = 1$

Now from Eq. (1), we get

$a + b + c \geq 3(1)^{1/3}$

$\Rightarrow (a + b + c) \geq 3$

Hence, minimum value of $a + b + c$ is 3.

5. c. $y = \frac{3^x}{3} + \frac{1}{3^x\,3}$

Now,

A.M. ≥ G.M.

$$\Rightarrow \frac{\frac{3^x}{3} + \frac{1}{3^x 3}}{2} \geq \sqrt{\frac{3^x}{3}\frac{1}{3^x 3}}$$

$$\Rightarrow 3^{x-1} + 3^{-x-1} \geq \frac{2}{3}$$

6. d. We have A.M. ≥ G.M. Therefore,

$$\frac{a}{b} + \frac{b}{a} + \frac{b}{c} + \frac{c}{b} + \frac{c}{a} + \frac{a}{c} \geq 6$$

$$\Rightarrow \frac{b+c}{a} + \frac{c+a}{b} + \frac{a+b}{c} \geq 6$$

Hence, the least value is 6.

7. d. Let $a_1, a_2, \cdots, a_n$ be n positive integers such that $a_1a_2 \cdots a_n = n^n$. Since A.M. ≥ G.M., hence,

$$\frac{a_1 + a_2 + \cdots + a_n}{n} \geq (a_1a_2 \cdots a_n)^{1/n}$$

$$\Rightarrow \frac{a_1 + a_2 + \cdots + a_n}{n} \geq n$$

$$\Rightarrow a_1 + a_2 + \cdots + a_n \geq n^2$$

8. a. A.M. ≥ G.M.

$$\Rightarrow \frac{bcx + cay + abz}{3} \geq (a^2b^2c^2xyz)^{1/3}$$

$\Rightarrow bcx + cay + abz \geq 3xyz$

$\Rightarrow bcx + acy + abz \geq 3abc$

9. c. Consider $1/l$, $2/m$, $3/n$ and use A.M. ≥ G.M. Then,

$$\frac{1}{3}\left(\frac{1}{l} + \frac{2}{m} + \frac{3}{n}\right) \geq \left(\frac{1}{l}\frac{2}{m}\frac{3}{n}\right)^{1/3} = \left(\frac{6}{lmn}\right)^{1/3}$$

But $lmn = 48$. Therefore,

$$\frac{1}{3}\left(\frac{1}{l} + \frac{2}{m} + \frac{3}{n}\right) \geq \left(\frac{6}{48}\right)^{1/3} = \frac{1}{2}$$

$$\therefore \left(\frac{1}{l} + \frac{2}{m} + \frac{3}{n}\right)_{\min} = \frac{3}{2}$$

10. c. a, b, c are in H.P. Hence, H.M. of a and c is b.

$\therefore \sqrt{ac} > b \quad (\because$ G.M. > H.M.)

Since A.M. > G.M., so

$$\frac{a^{101} + c^{101}}{2} > (\sqrt{ac})^{101} > b^{101} \quad [\because \sqrt{ac} > b]$$

$\Rightarrow 2b^{101} - a^{101} - c^{101} < 0$

Let,

$f(x) = x^2 - kx + 2b^{101} - a^{101} - c^{101}$

$\therefore f(0) = 2b^{101} - a^{101} - c^{101} < 0$

Hence, equation $f(x) = 0$ has one root in $(-\infty, 0)$ and other in $(0, \infty)$.

11. d. $a = \dfrac{x^2+4}{|x|} - 3$

$= |x| + \dfrac{4}{|x|} - 3$

$\geq 2\sqrt{|x| \times \dfrac{4}{|x|}} - 3 \quad (\because \text{A.M.} \geq \text{G.M.})$

$\Rightarrow a \geq 1$

12. a. Given expression is

$\dfrac{1}{2}\sum \dfrac{2a}{b+c-a} = \dfrac{1}{2}\sum\left(\dfrac{2a}{b+c-a}+1\right) - \dfrac{3}{2}$

$= \dfrac{1}{2}(a+b+c)\sum\left(\dfrac{1}{b+c-a}\right) - \dfrac{3}{2}$

Now,

$(a+b+c) = \sum(b+c-a)$

Applying A.M. $\geq$ H.M., minimum value of the expression is

$(1/2 \times 9) - 3/2 = 3.$

13. a. A.M. $\geq$ G.M.

$\Rightarrow \dfrac{\log_d a + \log_c b + \log_a c + \log_b d}{4}$

$\geq \sqrt[4]{\dfrac{\log a}{\log d} \times \dfrac{\log b}{\log c} \times \dfrac{\log c}{\log a} \times \dfrac{\log c}{\log b}}$

$\Rightarrow \log_d a + \log_b d + \log_a c + \log_c b \geq 4$

14. a. Using A.M. and H.M. inequality, we get

$\dfrac{2bc}{b+c} \leq \dfrac{b+c}{2}, \dfrac{2ac}{a+c} \leq \dfrac{a+c}{2}, \dfrac{2ab}{a+b} \leq \dfrac{a+b}{2}$

$\Rightarrow \dfrac{bc}{b+c} + \dfrac{ac}{a+c} + \dfrac{ab}{a+b} \leq \dfrac{1}{2}(a+b+c)$

15. b. $\dfrac{a+b+c}{3} \geq \dfrac{3}{\dfrac{1}{a}+\dfrac{1}{b}+\dfrac{1}{c}} \quad (\because \text{A.M.} \geq \text{H.M.})$

$\Rightarrow (a+b+c)\left(\dfrac{1}{a}+\dfrac{1}{b}+\dfrac{1}{c}\right) \geq 9$

16. d. $a(b^2+c^2) + b(c^2+a^2) + c(a^2+b^2)$

$= ab^2 + ac^2 + bc^2 + ba^2 + ca^2 + cb^2$

Using A.M. $\geq$ G.M., we get

$\dfrac{a(b^2+c^2)}{6} + \dfrac{b(c^2+a^2)}{6} + \dfrac{c(a^2+b^2)}{6} \geq (a^6b^6c^6)^{1/6}$

$\Rightarrow a(b^2+c^2) + b(c^2+a^2) + c(a^2+b^2) \geq 6abc$

17. a. a, b, c, d are in H.P. Hence, b is H.M. of a and c; c is H.M. of b and d. Using A.M. $\geq$ H.M., we get

$\dfrac{a+c}{2} > b \Rightarrow a+c > 2b$

and

$\dfrac{b+d}{2} > c \Rightarrow b+d > 2c$

$\Rightarrow a+c+b+d > 2b+2c$

$\Rightarrow a+d > b+c$

18. d. $a+b+c = 18$

$\Rightarrow 2 \times \dfrac{a}{2} + 3 \times \dfrac{b}{3} + 4 \times \dfrac{c}{4} = 18$

Using weighted A.M. and G.M. inequality, we get

$\dfrac{2 \times \dfrac{a}{2} + 3 \times \dfrac{b}{3} + 4 \times \dfrac{c}{4}}{9} \geq \left(\left(\dfrac{a}{2}\right)^2\left(\dfrac{b}{3}\right)^3\left(\dfrac{c}{4}\right)^4\right)^{1/9}$

$\Rightarrow 2^9 \geq \dfrac{a^2}{2^2} \times \dfrac{b^3}{3^3} \times \dfrac{c^4}{4^4}$

$\Rightarrow a^2b^3c^4 \leq 3^3 \times 2^{19}$

19. a. Let,

$x - 3 = t$

$\Rightarrow x - 2 = (t+1)$ and $x - 1 = t + 2$

$\Rightarrow f(x) = \dfrac{(x-2)(x-1)}{(x-3)}$

$= \dfrac{(t+1)(t+2)}{t}$

$= \dfrac{(t^2+3t+2)}{t}$

$= t + \dfrac{2}{t} + 3 \geq 3 + 2\sqrt{2}$ (using A.M. $\geq$ G.M., as $t > 0$)

20. c. We have,

$\dfrac{a^3+1}{2} \geq \sqrt{a^3 \times 1}$

and

$\dfrac{a^2+a^1}{2} \geq \sqrt{a^2a^1}$

Adding, we get

$\dfrac{a^3+a^2+a+1}{2} \geq 2\sqrt{a^3}$

Multiple Correct Answers Type

1. a, b.

We have,

$2s = a+b+c$

$A^2 = s(s-a)(s-b)(s-c)$

Now, A.M. ≥ G.M.

$$\Rightarrow \frac{s+(s-a)+(s-b)+(s-c)}{4} \geq [s\,(s-a)\,(s-b)\,(s-c)]^{1/4}$$

$$\Rightarrow \frac{4s-2s}{4} \geq [A^2]^{1/4}$$

$\Rightarrow s/2 \geq A^{1/2} \Rightarrow A \leq s^2/4$

Also,

$$\frac{(s-a)+(s-b)+(s-c)}{3} \geq [(s-a)\,(s-b)\,(s-c)]^{1/3}$$

$$\Rightarrow \frac{s}{3} \geq \left[\frac{A^2}{s}\right]^{1/3}$$

$$\Rightarrow \frac{A^2}{s} \leq \frac{s^3}{27}$$

$$\Rightarrow A \leq \frac{s^2}{3\sqrt{3}}$$

2. a, c.

A.M. of x and z is y. G.M. of x and z is $\sqrt{xz}$.

Now, A.M. ≥ G.M. $\Rightarrow y^2 \geq xz$

Also, A.M ≥ H.M. $\Rightarrow y \geq \dfrac{2xz}{x+z}$

$$\frac{x+y}{2y-x} = \frac{x+y}{x+z-x} = \frac{x+y}{z} \text{ and } \frac{y+z}{2y-z} = \frac{y+z}{x}$$

$$\therefore \frac{\frac{x+y}{2y-x}+\frac{y+z}{2y-z}}{2} \geq \sqrt{\frac{x+y}{z}\frac{y+z}{x}}$$

$$= \sqrt{1+\frac{y(x+y+z)}{xz}} \quad [\because x+z=2y]$$

$$= \sqrt{1+\frac{3y^2}{xz}}$$

$$\therefore \frac{x+y}{2y-x}+\frac{y+z}{2y-z} \geq 2\sqrt{1+3\frac{y^2}{xz}} \geq 4 \quad [\because y^2 \geq xz]$$

3. a, b.

Using A.M. ≥ G.M. one can show that

$(b+c)\,(c+a)\,(a+b) \geq 8abc$

$\Rightarrow (p-a)\,(p-b)\,(p-c) \geq 8abc$

Therefore, (b) holds. Also,

$$\frac{(p-a)+(p-b)+(p-c)}{3} \geq [(p-a)(p-b)(p-c)]^{1/3}$$

$$\Rightarrow \frac{3p-(a+b+c)}{3} \geq [(p-a)(p-b)(p-c)]^{1/3}$$

$$\Rightarrow \frac{2p}{3} \geq [(p-a)(p-b)(p-c)]^{1/3}$$

$$\Rightarrow (p-a)\,(p-b)\,(p-c) \leq \frac{8p^3}{27}$$

Therefore, (a) holds. Again,

$$\frac{1}{2}\left(\frac{bc}{a}+\frac{ca}{b}\right) \geq \sqrt{\left(\frac{bc}{a}\frac{ca}{b}\right)}$$

and so on. Adding the inequalities, we get

$$\frac{bc}{a}+\frac{ca}{b}+\frac{ab}{c} \geq a+b+c = p$$

Therefore, (c) does not hold.

4. c., d.

Consider the A.P. Since a is equidistant from the first term α and the last term β of the A.P., α, a, β are in A.P. Hence, a is the A.M. of α and β. So,

$$a = \frac{\alpha+\beta}{2}$$

Similarly, b and c are the G.M. and H.M. i.e., $b = \sqrt{\alpha\beta}$ and $c = \dfrac{2\alpha\beta}{\alpha+\beta}$

Since A.M., G.M. and H.M. are in G.P. and A.M. ≥ G.M. ≥ H.M., a, b, c are in G.P. and $a \geq b \geq c$.

Linked Comprehension Type

For Problems 1–3

1. c., 2. a., 3. b.

Sol. Let roots of equation $x^6 - 12x^5 + bx^4 + cx^3 + dx^2 + ex + 64 = 0$ be x_i, $i = 1, 2, \ldots, 6$. Now,

$$x_1+x_2+x_3+x_4+x_5+x_6 = 12$$

and

$$x_1\,x_2\,x_3\,x_4\,x_5\,x_6 = 64$$

Thus,

$$\frac{x_1+x_2+\cdots+x_6}{6} = 2 \text{ and } (x_1\,x_2\,x_3\,x_4\,x_5\,x_6)^{1/6} = 2$$

$\Rightarrow$ A.M. = G.M.

$\Rightarrow x_1 = x_2 = x_3 = x_4 = x_5 = x_6 = 2$

Hence, the given equation is equivalent to

$(x-2)^6 = 0$

or

$x^6 - 12x^5 + 60x^4 - 160x^3 + 240x^2 - 192x - 64 = 0$

$\therefore f(1) = 1 - 12 + 60 - 160 + 240 - 192 + 64 = 1$

For Problems 4–6

4. d., 5. c., 6. d.

Sol. For non-negative values of a, roots must be negative. Let the roots be x_1, x_2, x_3, x_4 (< 0). Then,

$$x_1 + x_2 + x_3 + x_4 = -a$$

$$x_1x_2 + x_1x_3 + x_1x_4 + x_2x_3 + x_2x_4 + x_3x_4 = b$$

$$x_1x_2x_3 + x_2x_3x_4 + x_3x_4x_1 + x_4x_1x_2 = -c$$

$$x_1x_2x_3x_4 = 1$$

Now,

$$\frac{(-x_1)+(-x_2)+(-x_3)+(-x_4)}{4} \ge [(-x_1)(-x_2)(-x_3)(-x_4)]^{1/4}$$

[$\because$ A.M. $\ge$ G.M.]

$$\Rightarrow \frac{a}{4} \ge 1$$

$$\Rightarrow a \ge 4$$

Hence, the minimum value of a is 4. Similarly,

$$\frac{x_1x_2 + x_1x_3 + \cdots + x_3x_4}{6} \ge \left[x_1^{\,3}x_2^{\,3}x_3^{\,3}x_4^{\,3}\right]^{1/4}$$

$$\Rightarrow \frac{b}{6} \ge 1$$

$$\Rightarrow b \ge 6$$

Hence, the minimum value of b is 6. Finally,

$$\frac{-x_1x_2x_3 - x_2x_3x_4 - x_3x_4x_1 - x_4x_1x_2}{4} \ge \left[x_1^{\,3}x_2^{\,3}x_3^{\,3}x_4^{\,3}\right]^{1/4}$$

$$\Rightarrow \frac{c}{4} \ge 1$$

$$\Rightarrow c \ge 4$$

Hence, the minimum value of c is 4.

Integer Type

1. (4) Using A.M. $\ge$ G.M., we have

$$4^{\sin^2 x} + 4^{\cos^2 x} \ge 2\sqrt{4^{\sin^2 x} \times 4^{\cos^2 x}}$$

$$= 2\sqrt{4^{\sin^2 x + \cos^2 x}}$$

$$= 2\sqrt{4}$$

$$= 4$$

2. (6) In an acute angled triangle, sec A, sec B, sec C are positive.

Now A.M. $\ge$ H.M.

$$\Rightarrow \frac{\sec A + \sec B + \sec C}{3} \ge \frac{3}{\cos A + \cos B + \cos C}$$

But in ΔABC, $\cos A + \cos B + \cos C \le 3/2$

$$\Rightarrow \frac{\sec A + \sec B + \sec C}{3} \ge 2$$

$$\Rightarrow \sec A + \sec B + \sec C \ge 6$$

3. (8) Let $A = xyz$

$$\Rightarrow A^2 = x^2y^2z^2 = (xy)(yz)(zx) \qquad (1)$$

So A is maximum if A^2 is maximum.

Now $xy + yz + zx = 12$. Therefore,

$(xy)(yz)(zx)$ will be maximum if all the factors are equal.

$\Rightarrow xy = yz = zx = 4$, when A^2 is maximum.

The maximum value of $A^2 = 4 \times 4 \times 4$

$\Rightarrow$ The maximum value of $A = xyz$ is 8

4. (2) A.M. of $(1/3)^{\text{th}}$ power $\le$ $(1/3)^{\text{th}}$ power of A.M. Therefore,

$$\frac{(a^3)^{1/3} + (b^3)^{1/3}}{2} \le \left(\frac{a^3 + b^3}{2}\right)^{1/3}$$

$$\Rightarrow \frac{a+b}{2} \le 1$$

$$\Rightarrow a + b \le 2$$

5. (8) Put $a = y + z$, $b = z + x$, $c = x + y$ so that $x + y + z = 1$. Since $x = 1 - a$, $y = 1 - b$, $z = 1 - c$, and $0 < a < 1$, $0 < b < 1$, $0 < c < 1$,

if follows that $x, y, z > 0$

$$\text{Now } \frac{a}{1-a} \times \frac{b}{1-b} \times \frac{c}{1-c} = \frac{[(y+z)(z+x)(x+y)]}{xyz} \qquad (1)$$

A.M. $\ge$ G.M.

$$\Rightarrow \frac{y+z}{2} \ge \sqrt{yx},\ \frac{z+x}{2} \ge \sqrt{zx},\ \frac{x+y}{2} \ge \sqrt{xy}$$

$$\text{Multiplying we get } \frac{[(y+z)(z+x)(x+y)]}{(xyz)} \ge 8 \qquad (2)$$

From (1) and (2), we get

$$\frac{a}{1-a} \times \frac{b}{1-b} \times \frac{c}{1-c} \ge 8.$$

6. (6) In ΔABC, we know that $\sin(A/2)\sin(B/2)\sin(C/2)$ $d(1/8)$.

Now A.M. $\ge$ G.M.

$$\Rightarrow \frac{\operatorname{cosec}\frac{A}{2} + \operatorname{cosec}\frac{B}{2} + \operatorname{cosec}\frac{C}{2}}{3} \ge \left(\operatorname{cosec}\frac{A}{2}\operatorname{cosec}\frac{B}{2}\operatorname{cosec}\frac{C}{2}\right)^{1/3}$$

$$\Rightarrow \frac{\operatorname{cosec}\frac{A}{2} + \operatorname{cosec}\frac{B}{2} + \operatorname{cosec}\frac{C}{2}}{3} \ge \left(\frac{1}{\sin\frac{A}{2}\sin\frac{B}{2}\sin\frac{C}{2}}\right)^{1/3}$$

$$\Rightarrow \frac{\operatorname{cosec}\frac{A}{2} + \operatorname{cosec}\frac{B}{2} + \operatorname{cosec}\frac{C}{2}}{3} \ge (8)^{1/3}$$

$$\Rightarrow \operatorname{cosec}\frac{A}{2} + \operatorname{cosec}\frac{B}{2} + \operatorname{cosec}\frac{C}{2} \ge 6$$

7. (2)

$$2\log_{10} x - \log_x (0.01) = 2\log_{10} x - \frac{\log_{10}(0.01)}{\log_{10} x}$$

$$= 2\log_{10} + \frac{2}{\log_{10} x}$$

$$= 2\left(\log_{10} x + \frac{2}{\log_{10} x}\right)$$

(Here $x > 1 \Rightarrow \log_{10} x > 0$)

Now A.M. $\ge$ G.M.

$$\Rightarrow \frac{\log_{10} x + \frac{2}{\log_{10} x}}{2} \ge \left(\log_{10} x \frac{2}{\log_{10} x}\right)^{1/2}$$

$$\Rightarrow \log_{10} x + \frac{2}{\log_{10} x} \ge 2$$

8. (1) As A.M. $\ge$ G.M. for positive real numbers, we get

$$\frac{(a+b)+(c+d)}{2} \ge \sqrt{(a+b)(c+d)}$$

$$\Rightarrow M \le 1$$

Also $(a+b)(c+d) > 0$ $\qquad$ ($\because a, b, c, d > 0$)

Therefore, $0 \le M \le 1$.

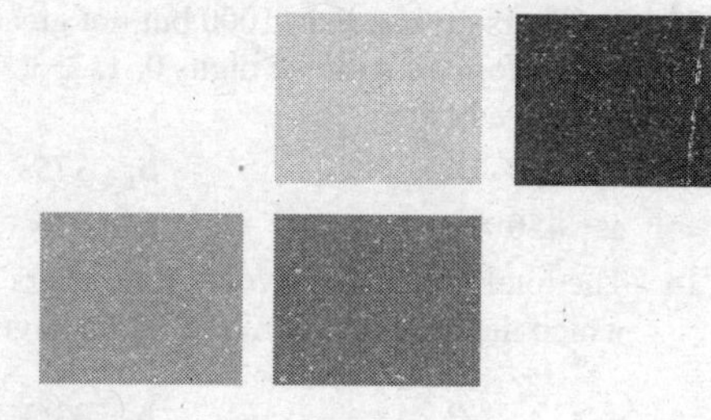

CHAPTER 5

Permutation and Combination

Objective Type

Each question has four choices a, b, c and d, out of which *only one* answer is correct.

1. If $\alpha = {}^mC_2$, then ${}^\alpha C_2$ is equal to
   a. ${}^{m+1}C_4$ b. ${}^{m-1}C_4$
   c. $3\ {}^{m+2}C_4$ d. $3\ {}^{m+1}C_4$

2. If ${}^nC_3 + {}^nC_4 > {}^{n+1}C_3$, then
   a. $n > 6$ b. $n > 7$
   c. $n < 6$ d. none of these

3. The value of $\sum_{r=0}^{n-1} {}^nC_r/({}^nC_r + {}^nC_{r+1})$ equals
   a. $n + 1$ b. $n/2$
   c. $n + 2$ d. none of these

4. In a city no two persons have identical set of teeth and there is no person without a tooth. Also no person has more than 32 teeth. If we disregard the shape and size of tooth and consider only the positioning of the teeth, the maximum population of the city is
   a. 2^{32} b. $(32)^2 - 1$
   c. $2^{32} - 1$ d. 2^{32-1}

5. In a room, there are 12 bulbs of the same wattage, each having a separate switch. The number of ways to light the room with different amount of illumination is
   a. $12^2 - 1$ b. 2^{12}
   c. $2^{12} - 1$ d. 12^2

6. The number of possible outcomes in a throw of n ordinary dice in which at least one of the dice shows an odd number is
   a. $6^n - 1$ b. $3^n - 1$
   c. $6^n - 3^n$ d. none of these

7. Let A be a set of n (≥ 3) distinct elements. The number of triplets (x, y, z) of the A elements in which at least two coordinates is equal to
   a. ${}^nP_3$ b. $n^3 - {}^nP_3$
   c. $3n^2 - 2n$ d. $3n^2(n - 1)$

8. The total number of flags with three horizontal strips in order, which can be formed using two identical red, tow identical green and two identical white strips is equal to
   a. $4!$ b. $3 \times (4!)$
   c. $2 \times (4!)$ d. none of these

9. The number of five-digit numbers that contain 7 exactly once is
   a. $(41)(9^3)$ b. $(37)(9^3)$
   c. $(7)(9^4)$ d. $(41)(9^4)$

10. A variable name in certain computer language must be either an alphabet or an alphabet followed by a decimal digit. The total number of different variable names that can exist in that language is equal to
    a. 280 b. 290
    c. 286 d. 296

11. The number less than 1000 that can be formed using the digits 0, 1, 2, 3, 4, 5 and is repeated in the formed number is equal to
    a. 130 b. 131
    c. 156 d. 155

12. Total number of six-digit numbers that can be formed, having the property that every succeeding digit is greater than the preceding digit, is equal to
    a. ${}^9C_3$ b. ${}^{10}C_3$
    c. ${}^9P_3$ d. ${}^{10}P_3$

13. Numbers greater than 1000 but not greater than 4000, which can be formed with the digits 0, 1, 2, 3, 4 (repetition of digits is allowed), are

a. 350 **b.** 375
c. 450 **d.** 576

14. The total number of five-digit numbers of different digits in which the digit in the middle is the largest is

a. $\sum_{n=4}^{9} {}^nP_4$ **b.** 33(3!)
c. 30(3!) **d.** none of these

15. The number of four-digit numbers that can be made with the digits 1, 2, 3, 4 and 5 in which at least two digits are identical is

a. $4^5 - 5!$ **b.** 505
c. 600 **d.** none of these

16. The number of nine non-zero digits such that all the digits in the first four places are less than the digit in the middle and all the digits in the last four places are greater than that in the middle is

a. 2(4!) **b.** 3(7!)/2
c. 2(7!) **d.** ${}^4P_4 \times {}^4P_4$

17. Total number of words that can be formed using all letters of the word BRIJESH that neither begins with *I* nor ends with *B* is equal to

a. 3720 **b.** 4920
c. 3600 **d.** 4800

18. The total number of six-digit natural numbers that can be made with the digits 1, 2, 3, 4, if all digits are to appear in the same number at least once is

a. 1560 **b.** 840
c. 1080 **d.** 480

19. Total number of six-digit numbers in which all and only odd digits appear is

a. $\frac{5}{2}(6!)$ **b.** 6!
c. $\frac{1}{2}(6!)$ **d.** none of these

20. Total number less than 3×10^8 and can be formed using the digits 1, 2, 3 is equal to

a. $\frac{1}{2}(3^9 + 4 \times 3^8)$ **b.** $\frac{1}{2}(3^9 - 3)$
c. $\frac{1}{2}(7 \times 3^8 - 3)$ **d.** $\frac{1}{2}(3^9 - 3 + 3^8)$

21. If all the permutations of the letters in the word OBJECT are arranged (and numbered serially) in alphabetical order as in a dictionary, then the 717th word is

a. TOJECB **b.** TOJECB
c. TOCJEB **d.** TOJCBE

22. In a three-storey building, there are four rooms on the ground floor, two on the first and two on the second floor. If the rooms are to be allotted to six persons, one person occupying one room only, the number of ways in which this can be done so that no floor remains empty is

a. ${}^8P_6 - 2(6!)$ **b.** ${}^8P_6$
c. ${}^8P_5\,(6!)$ **d.** none of these

23. The total number not more than 20 digits that are formed by using the digits 0, 1, 2, 3 and 4 is

a. 5^{20} **b.** $5^{20} - 1$
c. $5^{20} + 1$ **d.** none of these

24. The number of ways in which we can select four numbers from 1 to 30 so as to exclude every selection of four consecutive numbers is

a. 27378 **b.** 27405
c. 27399 **d.** none of these

25. The number of words of four letters containing equal number of vowels and consonants, where repetition is allowed, is

a. 105^2 **b.** 210×243
c. 105×243 **d.** 150×21^2

26. The number of different seven digit numbers that can be written using only the three digits 1, 2 and 3 with the condition that the digit 2 occurs twice in each number is

a. ${}^7P_2\, 2^5$ **b.** ${}^7C_2\, 2^5$
c. ${}^7C_2\, 5^2$ **d.** none of these

27. The total number of three-letter words that can be formed from the letter of the word SAHARANPUR is equal to

a. 210 **b.** 237
c. 247 **d.** 227

28. The number of ways to fill each of the four cells of the table with a distinct natural number such that the sum of the numbers is 10 and the sums of the numbers placed diagonally are equal is

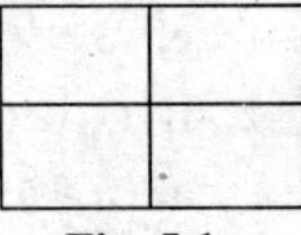

Fig. 5.1

a. 4 **b.** 8
c. 24 **d.** 6

29. The number of ways of arranging *m* positive and *n* ($< m + 1$) negative signs in a row so that no two negative signs are together is

a. ${}^{m+1}P_n$ **b.** ${}^{n+1}P_m$
c. ${}^{m+1}C_n$ **d.** ${}^{n+1}C_m$

30. Three boys of class X, four boys of class XI and five boys of class XII sit in a row. The total number of ways in which these boys can sit so that all the boys of same class sit together is equal to

a. $(3!)^2\,(4!)\,(5!)$ **b.** $(3!)\,(4!)^2(5!)$
c. $(3!)\,(4!)\,(5!)$ **d.** $(3!)\,(4!)\,(5!)^2$

31. A library has *a* copies of one book, *b* copies each of two books, *c* copies each of three books, an single copy of *d* books. The total number of ways in which these books can be arranged in a shelf is equal to

a. $\frac{(a + 2b + 3c + d)!}{a!\,(b!)^2\,(c!)^3}$ **b.** $\frac{(a + 2b + 3c + d)!}{a!(2b!)(c!)^3}$

c. $\frac{(a + b + 3c + d)!}{(c!)^3}$ **d.** $\frac{(a + 2b + 3c + d)!}{a!(2b)!(3c)!}$

32. The sum of the digits in the unit's place of all numbers formed with the help of 3, 4, 5, 6 taken all at a time is
a. 18 b. 432
c. 108 d. 144

33. The sum of all the numbers of four different digits that can be made by using the digits 0, 1, 2 and 3 is
a. 26664 b. 39996
c. 38664 d. none of these

34. The sum of all four-digit numbers that can be formed by using the digits 2, 4, 6, 8 (when repetition of digits is not allowed) is
a. 133320 b. 533280
c. 53328 d. none of these

35. The number of ordered pairs of integers (x, y) satisfying the equation $x^2 + 6x + y^2 = 4$ is
a. 2 b. 8
c. 6 d. none of these

36. The number of five-digit telephone numbers having at least one of their digits repeated is
a. 90000 b. 100000
c. 30240 d. 69760

37. How many numbers can be made with the digits 3, 4, 5, 6, 7, 8 lying between 3000 and 4000, which are divisible by 5 while repetition of any digit is not allowed in any number?
a. 60 b. 12
c. 120 d. 24

38. The number of ways in which 10 candidates $A_1, A_2, \ldots, A_{10}$ can be ranked such that A_1 is always above A_{10} is
a. 5! b. 2(5!)
c. 10! d. $\frac{1}{2}(10!)$

39. In the decimal system of numeration of six-digit numbers in which the sum of the digits is divisible by 5 is
a. 180000 b. 540000
c. 5×10^5 d. none of these

40. To fill 12 vacancies there are 25 candidates of which five are from scheduled caste. If three of the vacancies are reserved for scheduled caste candidates while the rest are open to all, the number of ways in which the selection can be made is
a. ${}^5C_3 \times {}^{22}C_9$ b. ${}^{22}C_9 - {}^5C_3$
c. ${}^{22}C_3 + {}^5C_3$ d. none of these

41. If the difference of the number of arrangements of three things from a certain number of dissimilar things and the number of selections of the same number of things from them exceeds 100, then the least number of dissimilar things is
a. 8 b. 6
c. 5 d. 7

42. Among 10 persons, A, B, C are to speak at a function. The number of ways in which it can be done if A wants to speak before B and B wants to speak before C is
a. 10!/24 b. 9!/6
c. 10!/6 d. none of these

43. In how many ways can a team of 11 players be formed out of 25 players, if six out of them are always to be included and five always to be excluded
a. 2020 b. 2002
c. 2008 d. 8002

44. In how many ways can a team of six horses be selected out of a stud of 16, so that there shall always be three out of A B C A' B' C', but never A A', B B' or C C' together
a. 840 b. 1260
c. 960 d. 720

45. There are two bags each containing m balls. If a man has to select equals number of balls from both the bags the number of ways in which he can do so if he must choose at least one ball from each bag is
a. m^2 b. ${}^{2m}C_m$
c. ${}^{2m}C_m - 1$ d. none of these

46. The number of ways in which the letters of the word PERSON can be placed in the squares of the given figure so that no row remains empty is

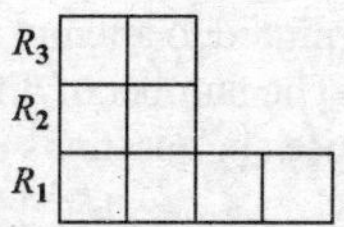

Fig. 5.2

a. 24 × 6! b. 26 × 6!
c. 26 × 7! d. 27 × 6!

47. The number of words of four letters that can be formed from the letters of the word EXAMINATION is
a. 1464 b. 2454
c. 1678 d. none of these

48. The letters of word ZENITH are written in all possible ways. If all these words are written in the order of a dictionary, then the rank of the word ZENITH is
a. 716 b. 692
c. 698 d. 616

49. A class contains three girls and four boys. Every Saturday, five go on a picnic (a different group of students is sent every week). During the picnic, each girl in the group is given a doll by the accompanying teacher. If all possible groups of five have gone for picnic once, the total number of dolls that the girls have got is
a. 21 b. 45
c. 27 d. 24

50. A teacher takes three children from her class to the zoo at a time, but she does not take the same three children to the zoo more than once. She finds that she went to the zoo 84 times more than a particular child has gone to the zoo. The number of children in her class is
a. 12 b. 10
c. 60 d. none of these

51. Number of ways in which a lawn-tennis mixed double be made from seven married couples if no husband and wife play in the same set is
a. 240 b. 420
c. 720 d. none of these

52. In a class tournament, all participants were to play different games with one another. Two players fell ill after having played three games each. If the total number of games played in the

tournament is equal to 84, the total number of participants in the beginning was equal to

a. 10 b. 15
c. 12 d. 14

53. A person always prefers to eat 'parantha' and 'vegetable dish' in his meal. How many ways can he make his platter in a marriage party if there are three types of paranthas, four types of 'vegetable dish', three types of 'salads' and two types of 'sauces'?

a. 3360 b. 4096
c. 3000 d. none of these

54. The number of even divisors of the number $N = 12600 = 2^3 3^2 5^2 7$ is

a. 72 b. 54
c. 18 d. none of these

55. A candidate is required to answer six out of 10 questions, which are divided into two groups, each containing five questions. He is not permitted to attempt more than four questions from either group. The number of different ways in which the candidate can choose six questions is

a. 50 b. 150
c. 200 d. 250

56. There were two women participating in a chess tournament. Every participant played two games with the other participates. The number of games that the men played among themselves proved to exceed by 66 number of games that the man played with the women. The number of paticipants is

a. 6 b. 11
c. 3 d. none of these

57. Two teams are to play a series of five matches between them. A match ends in a win, loss or draw for a team. A number of people forecast the result of each match and no two people make the same forecast for the series of matches. The smallest group of people in which one person forecasts correctly for all the matches will contain n people, where n is

a. 81 b. 243
c. 486 d. none of these

58. Ten IIT and 2 DCE students sit in a row. The number of ways in which exactly 3 IIT students sit between 2 DCE students is

a. ${}^{10}C_3 \times 2! \times 3! \times 8!$ b. $10! \times 2! \times 3! \times 8!$
c. $5! \times 2! \times 9! \times 8!$ d. none of these

59. A team of four students is to be selected from a total of 12 students. The total number of ways in which the team can be selected such that two particular students refuse to be together and other two particular students wish to be together only is equal to

a. 220 b. 182
c. 226 d. none of these

60. In an election, the number of candidates is one greater than the persons to be elected. If a voter can vote in 254 ways, the number of candidates is

a. 7 b. 10
c. 8 d. 6

61. Two players P_1 and P_2 play a series of $2n$ games. Each game can result in either a win or a loss for P_1. The total number of ways in which P_1 can win the series of these games is equal to

a. $\frac{1}{2}(2^{2n} - {}^{2n}C_n)$ b. $\frac{1}{2}(2^{2n} - 2 \times {}^{2n}C_n)$
c. $\frac{1}{2}(2^n - {}^{2n}C_n)$ d. $\frac{1}{2}(2^n - 2 \times {}^{2n}C_n)$

62. In an examination of nine papers, a candidate has to pass in more papers than the number of papers in which he fails in order to be successful. The number of ways in which he can be unsuccessful is

a. 255 b. 256
c. 193 d. 319

63. A student is allowed to select at most n books from a collection of $(2n + 1)$ books. If the total number of ways in which he can select at least one book is 63, then the value of n is

a. 2 b. 3
c. 4 d. 5

64. In a group of 13 cricket players, four are bowlers. Find out in how many ways can they form a cricket team of 11 players in which at least 2 bowlers are included.

a. 55 b. 72
c. 78 d. none of these

65. A person predicts the outcome of 20 cricket matches of his home team. Each match can result in a either win, loss or tie for the home team. Total number of ways in which he can make the predictions so that exactly 10 predictions are correct is equal to

a. ${}^{20}C_{10} \times 2^{10}$ b. ${}^{20}C_{10} \times 3^{20}$
c. ${}^{20}C_{10} \times 3^{10}$ d. ${}^{20}C_{10} \times 2^{20}$

66. The number of different ways in which five 'alike dashes' and eight 'alike dots' can be arranged using only seven of these 'dashes' and 'dots' is

a. 350 b. 120
c. 1287 d. none of these

67. Let there be $n \geq 3$ circles in a plane. The value of n for which the number of radical centres is equal to the number of radical axes is (assume that all radical axes and radical centre exist and are different)

a. 7 b. 6
c. 5 d. none of these

68. The number of ways of choosing a committee of two women and three men from five women and six men, if Mr. A refuses to serve on the committee if Mr. B is a member and Mr. B can only serve, if Miss C is the member of the committee is

a. 60 b. 84
c. 124 d. none of these

69. The last digit of $(1! + 2! + \cdots + 2005!)^{500}$ is

a. 9 b. 2
c. 7 d. 1

70. $ABCD$ is a convex quadrilateral and 3, 4, 5 and 6 points are marked on the sides AB, BC, CD and DA, respectively. The number of triangles with vertices on different sides is

a. 270 b. 220
c. 282 d. 342

71. There are 10 points in a plane of which no three points are collinear and four points are concyclic. The number of different circles that can be drawn through at least three points of these points is

a. 116 b. 120
c. 117 d. none of these

72. n lines are drawn in a plane such that no two of them are parallel and no three of them are concurrent. The number of different points at which these lines will cut is

a. $\sum_{k=1}^{n-1} k$ b. $n(n-1)$
c. n^2 d. none of these

73. The number of triangles that can be formed with 10 points as vertices, n of them being collinear, is 110. Then n is

a. 3 b. 4
c. 5 d. 6

74. There are three coplanar parallel lines. If any p points are taken on each of the lines, the maximum number of triangles with vertices on these points is

a. $3p^2(p-1)+1$ b. $3p^2(p-1)$
c. $p^2(4p-3)$ d. none of these

75. The maximum number of points of intersection of five lines and four circles is

a. 60 b. 72
c. 62 d. none of these

76. If m parallel lines in a plane are intersected by a family of n parallel lines, the number of parallelograms that can be formed is

a. $\frac{1}{4}mn(m-1)(n-1)$ b. $\frac{1}{2}mn(m-1)(n-1)$
c. $\frac{1}{4}m^2n^2$ d. none of these

77. The number of integral solutions of $x+y+z=0$ with $x \geq -5$, $y \geq -5$, $z \geq -5$ is

a. 134 b. 136
c. 138 d. 140

78. The number of ways in which we can get a score of 11 by throwing three dice is

a. 18 b. 27
c. 45 d. 56

79. In how many different ways can the first 12 natural numbers be divided into three different groups such that numbers in each group are in A.P.?

a. 1 b. 5
c. 6 d. 4

80. Fifteen identical balls have to be put in five different boxes. Each box can contain any number of balls. The total number of ways of putting the balls into the boxes so that each box contains at least two balls is equal to

a. ${}^9C_5$ b. ${}^{10}C_5$
c. ${}^6C_5$ d. ${}^{10}C_6$

81. If n objects are arranged in a row, then the number of ways of selecting three of these objects so that no two of them are next to each other is

a. ${}^{n-2}C_3$ b. ${}^{n-3}C_2$
c. ${}^{n-3}C_3$ d. none of these

82. The number of ways to give 16 different things to three persons A, B, C so that B gets one more than A and C gets two more than B, is

a. $\frac{16!}{4!5!7!}$ b. $4!\,5!\,7!$
c. $\frac{16!}{3!5!8!}$ d. none of these

83. The number of ways in which we can distribute mn students equally among m sections is given by

a. $\frac{(mn)!}{n!}$ b. $\frac{(mn)!}{(n!)^m}$
c. $\frac{(mn)!}{m!n!}$ d. $(mn)^m$

84. $2m$ white counters and $2n$ red counters are arranged in a straight line with $(m+n)$ counters on each side of a central mark. The number of ways of arranging the counters, so that the arrangements are symmetrical with respect to the central mark, is

a. ${}^{m+n}C_m$ b. ${}^{2m+2n}C_{2m}$
c. $\frac{1}{2}\frac{(m+n)!}{m!n!}$ d. none of these

85. A person buys eight packets of TIDE detergent. Each packet contains one coupon, which bears one of the letters of the word TIDE. If he shows all the letters of the word TIDE, he gets one free packet. If he gets exactly one free packet, then the number of different possible combinations of the coupons is

a. ${}^7C_3$ b. ${}^8C_4$
c. ${}^8C_3$ d. 4^4

86. There are three copies each of four different books. The number of ways in which they can be arranged in a shelf is

a. $\frac{12!}{(3!)^4}$ b. $\frac{12!}{(4!)^3}$
c. $\frac{21!}{(3!)^4 4!}$ d. $\frac{12!}{(4!)^3 3!}$

87. The number of ways in which 12 books can be put in three shelves with four on each shelf is

a. $\frac{12!}{(4!)^3}$ b. $\frac{12!}{(3!)(4!)^3}$
c. $\frac{12!}{(3!)^3 4!}$ d. none of these

88. The total number of ways in which $2n$ persons can be divided into n couples is

a. $\frac{2n!}{n!n!}$ b. $\frac{2n!}{(2!)^n}$
c. $\frac{2n!}{n!(2!)^n}$ d. none of these

89. Let $x_1, x_2, \ldots, x_k$ be the divisors of positive integer n (including 1 and n). If $x_1 + x_2 + \cdots + x_k = 75$, then $\sum_{i=1} 1/x_i$ is equal to

a. $\frac{75}{n^2}$ b. $\frac{75}{n}$

c. $\frac{75}{k}$ d. none of these

90. Let $A = \{x_1, x_2, x_3, \ldots, x_7\}$, $B = \{y_1, y_2, y_3\}$. The total number of functions $f: A \to B$ that are on to and there are exactly three element x in A such that $f(x) = y_2$ is equal to

a. 490 b. 510

c. 630 d. none of these

91. The total number of ways in which n^2 number of identical balls can be put in n numbered boxes (1, 2, 3, ..., n) such that i^{th} box contains at least i number of balls is

a. ${}^{n^2}C_{n-1}$ b. ${}^{n^2-1}C_{n-1}$

c. ${}^{\frac{n^2+n-2}{2}}C_{n-1}$ d. none of these

92. The total number of ways in which 15 identical blankets can be distributed among four persons so that each of them gets at least two blankets is equal to

a. ${}^{10}C_3$ b. ${}^{9}C_3$

c. ${}^{11}C_3$ d. none of these

93. Number of ways in which 25 identical things be distributed among five persons if each gets odd number of things is

a. ${}^{25}C_4$ b. ${}^{12}C_8$

c. ${}^{14}C_{10}$ d. ${}^{13}C_3$

94. Number of ways in which Rs. 18 can be distributed among four persons such that nobody receives less than Rs. 4 is

a. 4^2 b. 2^4

c. 4! d. none of these

95. In how many ways can 17 persons depart from railway station in 2 cars and 3 autos, given that 2 particular persons depart by same car (4 persons can sit in a car and 3 persons can sit in an auto)?

a. $\frac{15!}{2!4!(3!)^3}$ b. $\frac{16!}{(2!)^2 4!(3!)^3}$

c. $\frac{17!}{2!4!(3!)^3}$ d. $\frac{15!}{4!(3!)^3}$

96. The total number of ways of selecting six coins out of 20 one-rupee coins, 10 fifty-paise coins and 7 twenty-five paise coins is

a. 28 b. 56

c. ${}^{37}C_6$ d. none of these

97. Let $f(n, k)$ denote the number of ways in which k identical balls can be coloured with n colours so that there is at least one ball of each colour. Then $f(2n, n)$ must be equal to

a. ${}^{2n}C_n$ b. ${}^{2n-1}C_{n+1}$

c. ${}^{2n-1}C_n$ d. none of these

98. The total number of ways in which three distinct numbers in A.P. can be selected from the set {1, 2, 3, ..., 24} is equal to

a. 66 b. 132

c. 198 d. none of these

99. The total number of ways of selecting two number from the set {1, 2, 3, 4, ..., $3n$} so that their sum is divisible by 3 is equal to

a. $\frac{2n^2-n}{2}$ b. $\frac{3n^2-n}{2}$

c. $2n^2 - n$ d. $3n^2 - n$

100. Among the 8! permutations of the digits 1, 2, 3, ..., 8, consider those arrangements which have the following property. If we take any five consecutive positions, the product of the digits in these positions is divisible by 5. The number of such arrangements is equal to

a. 7! b. 2. (7!)

c. ${}^7C_4$ d. none of these

101. The total number of divisors of 480, that are of the form $4n + 2$, $n \geq 0$, is equal to

a. 2 b. 3

c. 4 d. none of these

102. The total number of times, the digit 3 will be written, when the integers having less than 4 digits are listed is equal to

a. 300 b. 310

c. 302 d. 306

103. Straight lines are drawn by joining m points on a straight line to n points on another line. Then excluding the given points, the number of point of intersections of the lines drawn is (no two lines drawn are parallel and no three lines are concurrent)

a. $\frac{1}{4}mn(m-1)(n-1)$ b. $\frac{1}{2}mn(m-1)(n-1)$

c. $\frac{1}{2}m^2n^2$ d. $\frac{1}{4}m^2n^2$

104. In a polygon, no three diagonals are concurrent. If the total number of points of intersection of diagonals interior to the polygon is 70, then the number of diagonals of the polygon is

a. 20 b. 28

c. 8 d. none of these

105. Two packs of 52 cards are shuffled together. The number of ways in which a man can be dealt 26 cards so that he does not get two cards of the same suit and same denomination is

a. ${}^{52}C_{26} \times 2^{26}$ b. ${}^{104}C_{26}$

c. $2 \times {}^{52}C_{26}$ d. none of these

106. There are $(n + 1)$ white and $(n + 1)$ black balls each set numbered 1 to $n + 1$. The number of ways in which the balls can be arranged in a row so that the adjacent balls are of different colours is

a. $(2n + 2)!$ b. $(2n + 2)! \times 2$

c. $(n + 1)! \times 2$ d. $2\{(n + 1)!\}^2$

107. The number of three-digit numbers of the form xyz such that $x < y$ and $z \leq y$ is

a. 276 b. 285

c. 240 d. 244

108. A is a set containing n different elements. A subset P of A is chosen. The set A is reconstructed by replacing the elements of P. A subset Q of A is again chosen. The number of ways of choosing P and Q so that $P \cap Q$ contains exactly two elements is

a. ${}^nC_3 \times 2^n$ **b.** ${}^nC_2 \times 3^{n-2}$
c. 3^{n-2} **d.** none of these

109. Messages are conveyed by arranging four white, one blue and three red flags on a pole. Flags of the same colour are alike. If a message is transmitted by the order in which the colours are arranged, the total number of messages that can be transmitted if exactly six flags are used is

a. 45 **b.** 65
c. 125 **d.** 185

110. Twenty persons are sitting in a particular arrangement around a circular table. Three persons are to be selected for leaders. The number of ways of selection of three persons such that no two were sitting adjacent to each other is

a. 600 **b.** 900
c. 800 **d.** none of these

111. A seven-digit number without repetition and divisible by 9 is to be formed by using seven digits out of 1, 2, 3, 4, 5, 6, 7, 8, 9. The number of ways in which this can be done is

a. 9! **b.** 2(7!)
c. 4 (7!) **d.** none of these

112. n is selected from the set {1, 2, 3, …, 10} and the number $2^n + 3^n + 5^n$ is formed. Total number of ways of selecting n so that the formed number is divisible by 4 is equal to

a. 50 **b.** 49
c. 48 **d.** none of these

113. The number of distinct natural numbers up to a maximum of four digits and divisible by 5, which can be formed with the digits 0, 1, 2, 3, 4, 5, 6, 7, 8, 9 each digit not occurring more than once in each number is

a. 1246 **b.** 952
c. 1106 **d.** none of these

114. A man has three friends. The number of ways he can invite one friend everyday for dinner on six successive nights so that no friend is invited more than three times is

a. 640 **b.** 320
c. 420 **d.** 510

115. There are four letters and four directed envelopes. The number of ways in which all the letters can be put in the wrong envelope is

a. 8 **b.** 9
c. 16 **d.** none of these

116. A bag contains four one-rupee coins, two twenty-five paisa coins and five ten-paisa coins. In how many ways can an amount, not less than Re 1 be taken out from the bag? (Consider coins of the same denominations to be identical.)

a. 71 **b.** 72
c. 73 **d.** 80

117. In a certain test, there are n questions. In the test 2^{n-i} students gave wrong answers to at least i questions, where $i = 1, 2, \ldots, n$. If the total number of wrong answers given is 2047, then n is equal to

a. 10 **b.** 11
c. 12 **d.** 13

118. Rajdhani Express going from Bombay to Delhi stops at five intermediate stations, 10 passengers enter the train during the journey with 10 different ticket of two classes. The number of different sets of tickets they may have is

a. ${}^{15}C_{10}$ **b.** ${}^{20}C_{10}$
c. ${}^{30}C_{10}$ **d.** none of these

119. A train timetable must be compiled for various days of the week so that two trains twice a day depart for three days, one train daily for two days and three trains once a day for two days. How many different timetables can be compiled?

a. 140 **b.** 210
c. 133 **d.** 72

120. The total number of positive integral solution of $15 < x_1 + x_2 + x_3 \le 20$ is equal to

a. 685 **b.** 785
c. 1125 **d.** none of these

Multiple Correct Answers Type

Each question has four choices a, b, c, and d, out of which *one or more* answers are correct.

1. Number of ways in which three numbers in A.P. can be selected from 1, 2, 3, …, n is

a. $\left(\frac{n-1}{2}\right)^2$ if n is even **b.** $\frac{n(n-2)}{4}$ if n is even
c. $\frac{(n-1)^2}{4}$ if n is odd **d.** none of these

2. Kanchan has 10 friends among whom two are married to each other. She wishes to invite five of them for a party. If the married couples refuse to attend separately, then the number of different ways in which she can invite five friends is

a. ${}^8C_5$ **b.** $2 \times {}^8C_3$
c. ${}^{10}C_5 - 2 \times {}^8C_4$ **d.** none of these

3. A forecast is to be made of the results of five cricket matches, each of which can be a win or a draw or a loss for Indian team. Let,

p = number of forecasts with exactly one error
q = number of forecasts with exactly three errors and
r = number of forecasts with all five errors

Then the correct statement(s) is/are

a. $2q = 5r$ **b.** $8p = q$
c. $8p = 5r$ **d.** $2(p + r) > q$

4. Ten persons numbered 1, 2, …, 10 play a chess tournament, each player playing against every other player exactly one game. It is known that no game ends in a draw. If $w_1, w_2, \ldots, w_{10}$ are the number of games won by players 1, 2, 3, …, 10, respectively, and $l_1, l_2, \ldots, l_{10}$ are the number of games lost by the players 1, 2, …, 10, respectively, then,

a. $\sum w_i = \sum l_i = 45$ b. $w_i + l_i = 9$

c. $\sum w l_i^2 = 81 + \sum l_i^2$ d. $\sum w_i^2 = \sum l_i^2$

5. The number of ways of choosing triplet (x, y, z) such that $z \geq \max\{x, y\}$ and $x, y, z \in \{1, 2, \ldots, n, n+1\}$ is

a. ${}^{n+1}C_3 + {}^{n+2}C_3$ b. $n(n+1)(2n+1)/6$

c. $1^2 + 2^2 + \cdots + n^2$ d. $2({}^{n+2}C_3) - {}^{n+1}C_2$

6. Number of ways in which 200 people can be divided in 100 couples is

a. $\dfrac{(200)!}{2^{100}(100)!}$ b. $1 \times 3 \times 5 \cdots 199$

c. $\left(\dfrac{101}{2}\right)\left(\dfrac{102}{2}\right)\cdots\left(\dfrac{200}{2}\right)$ d. $\dfrac{(200)!}{(100)!}$

7. If a seven-digit number made up of all distinct digits 8, 7, 6, 4, 3, x and y is divisible by 3, then

a. maximum value of $x - y$ is 9

b. maximum value of $x + y$ is 12

c. minimum value of xy is 0

d. minimum value of $x + y$ is 3

8. If n is number of necklaces which can be formed using 17 identical pearls and two identical diamonds and similarly m is number of necklaces which can be formed using 17 identical pearls and different diamonds, then

a. $n = 9$

b. $m = 18$

c. $n = 18$

d. $m = 9$

9. Let $f(n)$ be the number of regions in which n coplanar circles can divide the plane. If it is known that each pair of circles intersect in two different point and no three of them have common point of intersection, then

a. $f(20) = 382$

b. $f(n)$ is always an even number

c. $f^{-1}(92) = 10$

d. $f(n)$ can be odd

10. Given that the divisors of $n = 3^p \times 5^q \times 7^r$ are of the form $4\lambda + 1$, $\lambda \geq 0$. Then

a. $p + r$ is always even

b. $p + q + r$ is always odd

c. q can be any integer

d. if p is odd then r is even

11. Number of ways of selecting three integers from $\{1, 2, 3, \ldots, n\}$ if their sum is divisible by 3 is

a. $3({}^{n/3}C_3) + (n/3)^3$ if $n = 3k$, $k \in N$

b. $2({}^{(n-1)/3}C_3) + ({}^{(n+2)/3}C_3) + ((n-1)/3)^2(n+2)$, if $n = 3k+1$, $k \in N$

c. $2({}^{(n-1)/3}C_3) + ({}^{(n+2)/3}C_3) + ((n-1)/3)^2(n+2)$, if $n = 3k+2$, $k \in N$

d. independent of n

12. Number of points of intersection of n straight lines if n satisfies ${}^{n+5}P_{n+1} = \dfrac{11(n-1)}{2} \times {}^{n+3}P_n$ is

a. 15 b. 28

c. 21 d. 10

13. Number of shortest ways in which we can reach from the point (0, 0, 0) to point (3, 7, 11) in space where the movement is possible only along the x-axis, y-axis and z-axis or parallel to them and change of axes is permitted only at integral points (an integral point is one which has its coordinate as integer) is

a. equivalent to number of ways of dividing 21 different objects in three groups of size 3, 7, 11

b. equivalent to coefficient of y^3z^7 in the expansion of $(1 + y + z)^{21}$

c. equivalent to number of ways of distributing 21 different objects in three boxes of size 3, 7, 11

d. equivalent to number of ways of arranging 21 objects of which 3 are alike of one kind, 7 are alike of second type and 11 are alike of third type

14. Number of ways in which 30 identical things are distributed among six persons is

a. ${}^{17}C_5$ if each gets odd number of things

b. ${}^{16}C_{11}$ if each gets odd number of things

c. ${}^{14}C_5$ if each gets even number of things (excluding 0)

d. ${}^{15}C_{10}$ if each gets even number of things (excluding 0)

15. If N denotes the number of ways of selecting r objects out of n distinct objects $(r \geq n)$ with unlimited repetition but with each object included at least once in selection, then N is equal to

a. ${}^{r-1}C_{r-n}$ b. ${}^{r-1}C_n$

c. ${}^{r-1}C_{n-1}$ d. none of these

16. A is a set containing n elements. A subset P_1 of A is chosen. The set A is reconstructed by replacing the elements of P_1. Next, a subset P_2 of A is chosen and again the set is reconstructed by replacing the elements of P_2. In this way, m (>1) subsets $P_1, P_2, \ldots, P_m$ of A are chosen. The number of ways of choosing $P_1, P_2, \ldots, P_m$ is

a. $(2^m - 1)^n$ if $P_1 \cap P_2 \cap \cdots \cap P_m = \phi$

b. 2^{mn} if $P_1 \cup P_2 \cup \cdots \cup P_m = A$

c. 2^{mn} if $P_1 \cap P_2 \cap \cdots \cap P_m = \phi$

d. $(2^m - 1)^n$ if $P_1 \cup P_2 \cup \cdots \cup P_m = A$

17. If $10! = 2^p \times 3^q \times 5^r \times 7^s$, then

a. $2q = p$

b. $pqrs = 64$

c. number of divisors of 10! is 280

d. number of ways of putting 10! as a product of two natural numbers is 135

18. If $P = 21(21^2 - 1^2)(21^2 - 2^2)(21^2 - 3^2) \cdots (21^2 - 10^2)$, then P is divisible by

a. 22! b. 21!

c. 19! d. 20!

19. Let n is of four-digit integer in which all the digits are different. If x is number of odd integers and y is number of even integers, then

Reasoning Type

Each question has four choices a, b, c and d, out of which *only one* is correct. Each question contains Statement 1 and Statement 2.

a. Both the statements are True and Statement 2 is the correct explanation of Statement 1.

b. Both the statements are True but Statement 2 is Not the correct explanation of Statement 1.

c. Statement 1 is True and Statement 2 is False.

d. Statement 1 is False and Statement 2 is True.

1. **Statement 1:** The number of positive integral solutions of $abc = 30$ is 27.

   Statement 2: Number of ways in which three prizes can be distributed among three persons is 3^3.

2. **Statement 1:** Number of ways in which 10 identical toys can be distributed among three students if each receives at least two toys is ${}^9C_2$.

   Statement 2: Number of positive integral solutions of $x + y + z + w = 7$ is ${}^6C_3$

3. **Statement 1:** $(n^2)!/(n!)^n$ is a natural number for all $n \in N$.
   Statement 2: Number of ways in which n^2 objects can be distributed among n persons equally is $(n^2)!/(n!)^n$.

4. **Statement 1:** The number of ways of writing 1400 as a product of two positive integers is 12.
   Statement 2: 1400 is divisible by exactly three prime factors.

5. **Statement 1:** Let $E = \{1, 2, 3, 4\}$ and $F = \{a, b\}$. Then the number of onto functions from E to F is 14.
   Statement 2: Number of ways in which four distinct objects can be distributed into two different boxes is 14 if no box remains empty.

6. **Statement 1:** Number of ways in which India can win the series of 11 matches is 2^{10}. (if no match is drawn).
   Statement 2: For each match there are two possibilities, either India wins or loses.

7. **Statement 1:** Number of ways in which Indian team (11 players) can bat, if Yuvraj wants to bat before Dhoni and Pathan wants to bat after Dhoni is 11!/3!.
   Statement 2: Yuvraj, Dhoni and Pathan can be arranged in batting order in 3! ways.

8. **Statement 1:** When number of ways of arranging 21 objects of which r objects are identical of one type and remaining are identical of second type is maximum, then maximum value of ${}^{13}C_r$ is 78.

   Statement 2: ${}^{2n+1}C_r$ is maximum when $r = n$.

9. **Statement 1:** Total number of five-digit numbers having all different digits and divisible by 4 can be formed using the digits {1, 3, 2, 6, 8, 9} is 192.

   Statement 2: A number is divisible by 4, if the last two digits of the number are divisible by 4.

10. **Statement 1:** Number of ways in which 30 can be partitioned into three unequal parts, each part being a natural number is 61.

    Statement 2: Number of ways of distributing 30 identical objects in three different boxes is ${}^{30}C_2$.

11. **Statement 1:** If $p, q < r$, the number of different selections of $p + q$ things taking r at a time, where p things are identical and q other things are identical, is $p + q - r + 1$.

    Statement 2: If $p, q > r$, the number of different selections of $p + q$ things taking r at a time, where p things are identical and q other things are identical, is $r - 1$.

12. **Statement 1:** The number of ways in which three distinct numbers can be selected from the set $\{3^1, 3^2, 3^3, ..., 3^{100}, 3^{101}\}$ so that they form a G.P. is 2500.
    Statement 2: If a, b, c are in A.P., then $3^a, 3^b, 3^c$ are in G.P.

13. **Statement 1:** Number of ways in which two persons A and B select objects from two different groups each having 20 different objects such that B selects always more objects than A (including the case when A selects no object) is $(2^{40} - {}^{40}C_{20})/2$.

    Statement 2: The sum $\sum_{0 \le i < j \le n} {}^nC_i{}^nC_j = (2^{2n} - {}^{2n}C_n)/2$.

14. **Statement 1:** Number of ways of selecting 10 objects from 42 objects of which, 21 objects are identical and remaining objects are distinct is 2^{20}.

    Statement 2: ${}^{42}C_0 + {}^{42}C_1 + {}^{42}C_2 + \cdots + {}^{42}C_{21} = 2^{41}$.

15. **Statement 1:** Number of terms in the expansion of $(x + y + z + w)^{50}$ is ${}^{53}C_3$.

    Statement 2: Number of non-negative solution of the equation $p + q + r + s = 50$ is ${}^{53}C_3$.

16. **Statement 1:** The number of ways in which n persons can be seated at a round table, so that all shall not have the same neighbours in any two arrangements is $(n - 1)!/2$.

    Statement 2: Number of ways of arranging n different beads in circles in which is $(n - 1)!/2$.

17. **Statement 1:** Number of zeros at the end of 50! is equal to 12.

    Statement 2: Exponent of 2 in 50! is 47.

Linked Comprehension Type

Based upon each paragraph, three multiple choice questions have to be answered. Each question has four choices a, b, c and d, out of which *is only one* is correct.

For Problems 1–3

We have to choose 11 players for a cricket team from eight batsmen, six bowlers, four all rounders and two wicket keepers in the following conditions.

1. The number of selections when almost one all rounder and one wicket keeper will play

a. ${}^{4}C_{1} \times {}^{14}C_{10} + {}^{2}C_{1} \times {}^{14}C_{10} + {}^{4}C_{1} \times {}^{2}C_{1} \times {}^{14}C_{9} + {}^{14}C_{11}$
b. ${}^{4}C_{1} \times {}^{15}C_{11} + {}^{15}C_{11}$
c. ${}^{4}C_{1} \times {}^{15}C_{10} + {}^{15}C_{11}$
d. none of these

2. Number of selections when two particular batsmen do not want to play when a particular bowler will play

a. ${}^{17}C_{10} + {}^{19}C_{11}$ b. ${}^{17}C_{10} + {}^{19}C_{11} + {}^{17}C_{11}$
c. ${}^{17}C_{10} + {}^{20}C_{11}$ d. ${}^{19}C_{10} + {}^{19}C_{11}$

3. Number of selections when a particular batsman and a particular wicket keeper do not want to play together

a. $2\,{}^{18}C_{10}$ b. ${}^{19}C_{11} + {}^{18}C_{10}$
c. ${}^{19}C_{10} + {}^{19}C_{11}$ d. none of these

For Problems 4–6

Twelve persons are to be arranged around two round tables such that one table can accommodate seven persons and another five persons only. Answer the following questions.

4. Number of ways in which these 12 persons can be arranged is

a. ${}^{12}C_{5}6!4!$ b. $6!4!$
c. ${}^{12}C_{5}6!4!$ d. none of these

5. Number of ways of arrangements if two particular persons A and B do not want to be on the same table is

a. ${}^{10}C_{4}\,6!\,4!$ b. $2\,{}^{10}C_{6}\,6!\,4!$
c. ${}^{11}C_{6}\,6!\,4!$ d. none of these

6. Number of ways of arrangement if two particular persons A and B want to be together and consecutive is

a. ${}^{10}C_{7}\,6!\,3!2! + {}^{10}C_{5}\,4!5!2!$ b. ${}^{10}C_{5}\,6!\,3! + {}^{10}C_{7}\,4!5!$
c. ${}^{10}C_{7}\,6!\,2! + {}^{10}C_{5}\,5!2!$ d. none of these

For Problems 7–9

Five balls are to be placed in three boxes. Each box should hold all the five balls so that no box remains empty.

7. Number of ways if balls are different but boxes are identical is

a. 30 b. 25
c. 21 d. 35

8. Number of ways if balls and boxes are identical is

a. 3 b. 1
c. 2 d. none of these

9. Number of ways if balls as well as boxes are identical but boxes are kept in a row is

a. 10 b. 15
c. 20 d. 6

For Problems 10 – 12

Let $f(n)$ denote the number of different ways in which the positive integer n can be expressed as the sum of 1s and 2s. For example, $f(4) = 5$, since $4 = 2 + 2 = 2 + 1 + 1 = 1 + 2 + 1 = 1 + 1 + 2 = 1 + 1 + 1 + 1$. Note that order of 1s and 2s is important.

10. The value of $f(6)$ is

a. 12 b. 13
c. 18 d. 21

11. The value of $f(f(6))$ is

a. 400 b. 350
c. 377 d. none of these

12. $f: N \to N$ is

a. one-one and onto
b. one-one and into
c. many-one and onto
d. many-one and into

For Problems 13–15

There are m seats in the first row of a theatre, of which n are to be occupied.

13. The number of ways of arranging n persons if no two persons sit side by side is

a. $\dfrac{(m-n+1)!}{(m-3n+1)!}$ b. $\dfrac{(m-n+1)!}{(m-2n)!}$
c. $\dfrac{(m-n+1)!}{(m-2n+1)!}$ d. $\dfrac{(m-n+2)!}{(m-2n-1)!}$

14. If n is even, the number of ways of arranging n persons if each person has exactly one neighbour is

a. $({}^{n}P_{n/2})({}^{m-n+1}P_{n/2})$ b. $({}^{n}P_{n})({}^{m-n+1}P_{n/2})$
c. $({}^{n}P_{n/2})({}^{m-n+1}P_{n})$ d. none of these

15. The number of ways of arranging n persons, if out of any two seats located symmetrically in the middle of the row at least one is empty is

a. $({}^{m/2}C_{n})(2^{n}) - 1$ b. ${}^{m/2}P_{n}$
c. $({}^{m/2}P_{n})(2^{n}-1)$ d. $({}^{m/2}P_{n})(2^{n})$

For Problems 16–18

Consider the letters of the word MATHEMATICS.

16. Possible number of words taking all letters at a time such that at least one repeating letter is at odd position in each word is

a. $\dfrac{11!}{2!2!2!} - \dfrac{9!}{2!2!}$ b. $\dfrac{9!}{2!2!2!}$
c. $\dfrac{9!}{2!\,2!}$ d. $\dfrac{11!}{2!\,2!\,2!}$

17. Possible number of words taking all letters at a time such that in each word both M's are together and both T's are together but both A's are not together is

a. $\dfrac{11!}{2!\,2!\,2!} - \dfrac{10!}{2!\,2!}$ b. $7!\,{}^{8}C_{2}$
c. $\dfrac{6!\,4!}{2!\,2!}$ d. $\dfrac{9!}{2!\,2!\,2!}$

18. Possible number of words in which no two vowels are together is

a. $7!\,{}^{8}C_{4}\dfrac{4!}{2!}$ b. $\dfrac{7!}{2!}\,{}^{8}C_{4}\dfrac{4!}{2!}$
c. $\dfrac{7!}{2!\,2!}\,{}^{8}C_{4}\dfrac{4!}{2!}$ d. $\dfrac{7!}{2!\,2!\,2!}\,{}^{8}C_{4}\dfrac{4!}{2!}$

Matrix-Match Type

Each question contains statements given in two columns, which have to be matched. Statements a, b, c, d in column I have to be matched with statements p, q, r, s in column II. If the correct match are a→p, a→s, , b→q, b→r, c→p, c→q and d→s, then the correctly bubbled 4 × 4 matrix should be as follows:

	p	q	r	s
a	(p) ●	(q)	(r)	(s) ●
b	(p)	(q) ●	(r) ●	(s)
c	(p) ●	(q) ●	(r)	(s)
d	(p)	(q)	(r)	(s) ●

1.

Column I	Column II
a. Number of straight lines joining any two of 10 points of which four points are collinear	**p.** 30
b. Maximum number of points of intersection of 10 straight lines in the plane	**q.** 60
c. Maximum number of points of intersection of six circles in the plane	**r.** 40
d. Maximum number of points of intersection of six parabolas	**s.** 45

2. Consider a 6 × 6 chessboard. Then match the following columns.

Column I	Column II
a. Number of rectangles	**p.** $^{10}C_5$
b. Number of squares	**q.** 441
c. Number of ways three squares can be selected if they are not in same row or column	**r.** 91
d. In how many ways eleven '+' signs can be arranged in the squares if no row remains empty	**s.** 2400

3. Consider convex polygon having 35 diagonals. Then match the following column.

Column I	Column II
a. Number of triangles joining the vertices of the polygon	**p.** 210
b. Number of points of intersections of diagonal which lies inside the polygon	**q.** 120
c. Number of triangles in which exactly one side is common with that of polygon	**r.** 10
d. Number of triangles in which exactly two sides are common with that of polygon	**s.** 60

4. A function is defined as $f : \{a_1, a_2, a_3, a_4, a_5, a_6\} \rightarrow \{b_1, b_2, b_3\}$.

Column I	Column II
a. Number of subjective functions	**p.** is divisible by 9
b. Number of functions in which $f(ai) \neq bi$	**q.** is divisible by 5
c. Number of invertible functions	**r.** is divisible by 4
d. Number of many one functions	**s.** is divisible by 3
	t. not possible

5.

Column I	Column II
a. Total number of function f: {1, 2, 3, 4, 5} → {1, 2, 3, 4, 5} that are on to and $f(i) \neq i$ is equal to	**p.** divisible by 11
b. If $x_1x_2x_3 = 2 \times 5 \times 7^2$, then the number of solution set for (x_1, x_2, x_3) where $x_i \in N, x_i > 1$ is	**q.** divisible by 7
c. Number of factors of 3780 are divisible by either 3 or 2 or both is	**r.** divisible by 3
d. Total number of divisors of $n = 2^5 \times 3^4 \times 5^{10}$ that are of the form $4\lambda + 2$, $\lambda \geq 1$ is	**s.** divisible by 4

6.

Column I	Column II
a. Four dice (six faced) are rolled. The number of possible outcomes in which at least one dice shows 2 is	**p.** 210
b. Let A be the set of 4-digit number $a_1a_2a_3a_4$ where $a_1 > a_2 > a_3 > a_4$. Then $n(A)$ is equal to	**q.** 480
c. The total number of three-digit numbers, the sum of whose digits is even, is equal to	**r.** 671
d. The number of four-digit numbers that can be formed from the digits 0, 1, 2, 3, 4, 5, 6, 7 so that each number contains digit 1 is	**s.** 450

7.

Column I	Column II
a. The number of five-digit numbers having the product of digits 20 is	**p.** > 70
b. A closest has five pairs of shoes. The number of ways in which four shoes can be drawn from it such that there will be no complete pair is	**q.** < 60

c. Three ladies have each brought their one child for admission to a school. The principal wants to interview the six persons one by one subject to the condition that no mother is interviewed before her child. The number of ways in which interviews can be arranged is	r. ∈ (50, 110)
d. The figures 4, 5, 6, 7, 8 are written in every possible order. The number of numbers greater than 56000 is	s. ∈ (40, 70)

8.

Column I	Column II
a. If a denotes the number of permutations of $x+2$ things taken all at a time, b the number of permutations of x things taken 11 at a time and c the number of permutations of $x-11$ things taken all at a time such that $a = 182\ bc$, then the value of x is product of	p. 6
b. The number of six-digit numbers that can be made with the digits 0, 1, 2, 3, 4 and 5 so that even digits occupy odd places is product of	q. 5
c. The number of five-digit numbers that can be made using the digits 1 and 2 and in which at least one digit is different is product of	r. 4
d. In a polygon the number of diagonals is 54. The number of sides of the polygon is product of	s. 3

Integer Type

1. The digit at the unit place of the sum $(1!)^2 + (2!)^2 + (3!)^2 + \cdots + (2008!)^2$ is________.
2. If n is the number of distinct rational numbers x such that $0 < x < 1$ and $x = \frac{p}{q}$, where $p, q \in \{1, 2, 3, 4, 5, 6\}$, then the value of $n - 5$ is________.
3. The number of polynomials of the form $x^3 + ax^2 + bx + c$ that are divisible by $x^2 + 1$, where $a, b, c \in \{1, 2, 3, \ldots\ldots, 9\}$ is________.
4. The total number of 'n' digit numbers ($n > 1$), having the property that no two consecutive digits are same is 729, then the value of n is________.
5. If $a, b \in \{1, 2, 3, 4, 5, 6\}$, Then the number of ordered pairs (a, b) if $\operatorname{Lt}_{x\to 0} \left(\frac{a^x + b^x}{2}\right)^{\frac{2}{x}} = 6$ is________.
6. If $a, b, c \in \{1, 2, 3, 4, 5, 6\}$. find the number of ways b can be selected if $f(x) = x^3 + ax^2 + bx + c$ is an increasing function.
7. If the number of zeros at the end of 130! is 2^n, then the value of n is________.
8. The remainder when $1! + 2! + 3! + 4! + \cdots + n!$ is divided by 15, if $n \geq 5$, is________.
9. If the ratio ${}^{2n}C_3 : {}^{n}C_3$ is equal to 11 : 1, then the value of n is________.
10. A person wishes to make up as many different parties as he can out of 16 friends. Each party consists of the same number. Then number of friends he should invite at a time is________.
11. If ${}^{15}C_{3r} : {}^{15}C_{r+1} = 11 : 3$, then the value of '$r$' is________.
12. The greatest composite number by which the total number of ways of selecting five letters from the letters of the word INDEPENDENT is divisible is________.
13. A person is permitted to select at least one and at most n coins from a collection of $(2n + 1)$ distinct coins. If the total number of ways in which he can select coins is 255, then the value of n is________.
14. If n is number of ways in which 15 identical blankets can be distributed among six beggars such that everyone gets at least one blanket and two particular beggars get equal blankets and another three particular beggars get equal blankets, then the value of $n/2$ is________.
15. Sohan has x children by his first wife. Geeta has $(x + 1)$ children by her first husband. They marry and have children of their own. The whole family has 24 children. Assuming that two children of the same parents do not fight, then the maximum value of x is________.
16. An n-digit number is a positive number with exactly n digits. Nine hundred distinct n-digit numbers are to be formed using only the three digits 2, 5, and 7. The smallest value of n for which this is possible is________.
17. Let T_n denote the number of triangles which can be formed using the vertices of a regular polygon of n sides. If $T_{n+1} - T_n = 21$, then n equals________.
18. If r, s, t are prime numbers and p, q are positive integers such that the L.C.M of p, q is $r^2t^4s^2$, then the greatest composite number by which the number of ordered pair (p, q) is divisible by is________.
19. $(9!)!$ is divisible by $(9!)^{p\,!}$, then the greatest value of p is________.
20. The members of a chess club took part in a round robin competition in which each plays everyone else once. All members scored the same number of points, except four juniors whose total score was 17.5. If there are n members, then the value of $n/3$ is________.
21. A student is allowed to select atmost n books from a collection of $(2n + 1)$ books. If the total number of ways in which he can select at least one book is 63, then the value of n is________.
22. There are four balls of different colours and four boxes of colours, same as those of the balls. The number of ways in which the balls, one each in a box, could be placed such that a ball does not go to a box of its own colour is________.
23. Six persons A,B,C,D,E, and F are to be seated at a circular table. If n is the number of ways in which this can be done such that A always has either B or C on his right and B always has either C or D on his right, then $n/2$ is________.

24. If number of function defined from f: {1, 2, 3, 4, 5, 6} $\rightarrow$ {7, 8, 9, 10} such that the sum $f(1) + f(2) + f(3) + f(4) + f(5) + f(6)$ is odd is 2^n, then the value of $n - 4$ is ______.

25. If number of ways in which a person can walk up a stairway which has 7 steps, if he can take 1 or 2 steps up the stairs at a time, is N, then the value of $N/3$ is ______.

26. There are unique integers a_2, a_3, a_4, a_5, a_6, a_7 such that $\frac{5}{7} = \frac{a_2}{2!} + \frac{a_3}{3!} + \frac{a_4}{4!} + \frac{a_5}{5!} + \frac{a_6}{6!} + \frac{a_7}{7!}$, where $0 \le a_i < i$ for $i = 2, 3, \ldots, 7$. Then find the value of $a_2 + a_3 + a_4 + a_5 + a_6 + a_7$.

ANSWERS AND SOLUTIONS

Objective Type

1. d. $\alpha = {}^mC_2 \Rightarrow \alpha = \dfrac{m(m-1)}{2}$

$$\therefore {}^{\alpha}C_2 = \frac{\alpha(\alpha-1)}{2} = \frac{1}{2}\frac{m(m-1)}{2}\left\{\frac{m(m-1)}{2}-1\right\}$$

$$= \frac{1}{8}m(m-1)(m-2)(m+1)$$

$$= \frac{1}{8}(m+1)m(m-1)(m-2) = 3\,{}^{m+1}C_4$$

2. a. ${}^nC_3 + {}^nC_4 > {}^{n+1}C_3$

$\Rightarrow {}^{n+1}C_4 > {}^{n+1}C_3 \qquad [\because {}^nC_r + {}^nC_{r+1} = {}^{n+1}C_{r+1}]$

$$\Rightarrow \frac{{}^{n+1}C_4}{{}^{n+1}C_3} > 1$$

$$\Rightarrow \frac{n-2}{4} > 1$$

$\Rightarrow n > 6$

3. b. $$\sum_{r=0}^{n-1}\frac{{}^nC_r}{{}^nC_r + {}^nC_{r+1}} = \sum_{r=0}^{n-1}\frac{1}{1+\frac{{}^nC_{r+1}}{{}^nC_r}}$$

$$= \sum_{r=0}^{n-1}\frac{1}{1+\frac{n-r}{r+1}}$$

$$= \sum_{r=0}^{n-1}\frac{r+1}{n+1} = \frac{1}{n+1}\sum_{r=0}^{n-1}(r+1)$$

$$= \frac{1}{(n+1)}[1+2+\cdots+n] = \frac{n}{2}$$

4. c. We have 32 places for teeth. For each place, we have two choices either there is a tooth or there is no tooth. Therefore, the number of ways to fill up these places is 2^{32}. As there is no person without a tooth, the maximum population is $2^{32} - 1$.

5. c. For each bulb there are two possibilities. It will be switched either on or off. Hence, total number of ways in which the room can be illuminated is $2^{12} - 1$.

6. c. The total number of ways is $6 \times 6 \times$ L to n times $= 6^n$. The total number of ways to show only even numbers is $3 \times 3 \times \ldots$ to n times $= 3^n$. Therefore, the required number of ways is $6^n - 3^n$.

7. c. Total number of triplets without restriction is $n \times n \times n$.

The number of triplets with all different coordinates is ${}^nP_3$.

Therefore, the required number of triplets is $n^3 - n(n-1)(n-2)$.

8. a. All strips are of different colours, then number of flags is $= 3! = 5$. When two strips are of same colour, then number of flags is ${}^3C_1 \times (3!/2) \times {}^2C_1 = 18$. Total number of flags is $6 + 18 = 24 = 4!$

9. a. If 7 is used at first place, the number of numbers is 9^4 and for any other four places it is 8×9^3.

10. c. Total number of variables if only alphabet is used is 286.

Total number of variables if alphabets and digits both are used is 26×10. Hence, the total number of variables is $26(1 + 10) = 286$.

11. b. The number of one-digit numbers is 6.

The number of two-digit numbers is $5 \times 5 = 25$.

The number of thee-digit numbers is $5 \times 5 \times 4 = 100$.

Hence, the total number are is 131.

12. a. $x_1 < x_2 < x_3 < x_4 < x_5 < x_6$, when the number is $x_1x_2x_3x_4x_5x_6$

Clearly no digit can be zero. Also, all the digits are distinct. So, let us first select six digits from the list of digits 1, 2, 3, 4, 5, 6, 7, 8, 9 which can be done in ${}^9C_6$ ways.

After selecting these digits they can be put only in one order.

Thus, total number of such numbers is ${}^9C_6 \times 1 = {}^9C_3$.

13. b. Numbers greater than 1000 and less than or equal to 4000 will be of 4 digits and will have either 1 (except 1000) or 2 or 3 in the first place with 0 in each of remaining places. After fixing first place, the second place can be filled by any of the five digits. Similarly the third place can be filled up in 5 ways and fourth place can be filled up in 5 ways. Thus, there will be $5 \times 5 \times 5 = 125$ ways in which 1 will be in first place but this also includes 1000. Hence, there will be 124 numbers having 1 in the first place. Similarly, 125 for each 2 or 3. One number will be there in which 4 will be in the first place, i.e., 4000. Hence, the required number of ways is $124 + 125 + 125 + 1 = 375$.

14. d.

Middle digit	Digits available for remaining four places	Pattern	Number of ways filling remaining four places
4	0, 1, 2, 3	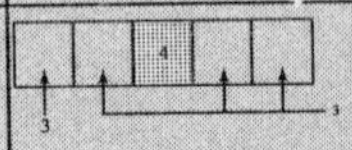 [diagram]	$3 \times {}^3P_3$
5	0, 1, …, 4	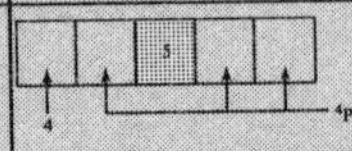 [diagram]	$4 \times {}^4P_3$
6	0, 1, …, 5	…	$5 \times {}^5P_3$
7	0, 1, …, 6	…	$6 \times {}^6P_3$
8	0, 1, …, 7	…	$7 \times {}^7P_3$
9	0, 1, …, 8	…	$8 \times {}^8P_3$

15. b. The number of numbers when repetition is allowed is 5^4.

The number of numbers when digits cannot be repeated is $^5P_5$.

Therefore, the required number of numbers is $5^4 - 5!$.

16. d. According to given conditions, numbers can be formed by the following format:

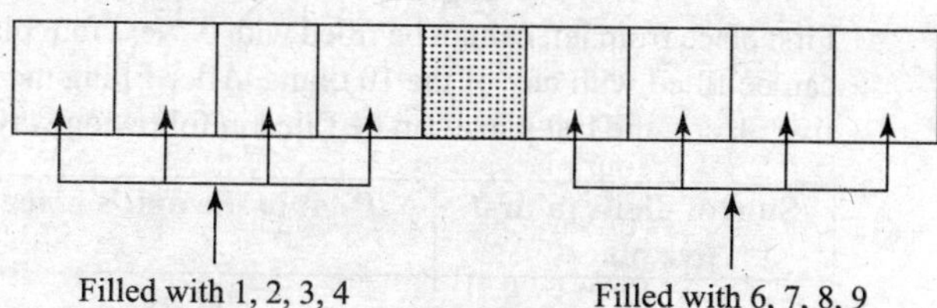

Fig. 5.3

The required number of numbers is $^4P_4 \times {}^4P_4$.

17. a. Total number of words without any restriction is 7!

Total number of words beginning with I is 6!

Total number of words ending with *B* is 6!

Total number of words beginning with I and ending with *B* is 5!

Thus the total number of required words is $7! - 6! - 6! + 5! = 7! - 2(6!) + 5!$.

18. a. There can be two types of numbers.

i. Any one of the digits 1, 2, 3, 4 appears thrice and the remaining digits only once, i.e., of the type 1, 2, 3, 4, 4, 4, etc. Number of ways of selection of digit which appears thrice is $^4C_1$.

Then the number of numbers of this type is $(6!/3!) \times {}^4C_1 = 480$.

ii. Any two of the digits 1, 2, 3, 4 appears twice and the remaining two only once, i.e., of the type 1, 2, 3, 3, 4, 4, etc. The number of ways of selection of two digits which appear twice is $^4C_2$. Then the number of numbers of this type is $[6!/(2!2!) \times {}^4C_2]$. Therefore, the required number of numbers is $480 + 1080 = 1560$.

19. a. Clearly, one of the odd digits 1, 3, 5, 7, 9 will be repeated. The number of selections of the sixth digit is $^5C_1 = 5$. Then the required number of numbers is $5 \times (6!/2!)$.

20. c. Formed number can be utmost of nine digits. Total number of such numbers is

$$3 + 3^2 + 3^3 + \cdots + 3^8 + 2 \times 3^8$$

$$= \frac{3(3^8-1)}{3-1} + 2\times 3^8 = \frac{3^9 - 3 + 4\times 3^8}{2} = \frac{7\times 3^8 - 3}{2}$$

21. d. The order of letters of the word OBJECT is B C E J O T.

Words starting with B can be formed in 5! ways.

Words starting with C can be formed in 5! ways.

Words starting with E can be formed in 5! ways.

Words starting with J can be formed in 5! ways.

Words starting with O can be formed in 5! ways.

Words starting with TB can be formed in 4! ways.

Words starting with TC can be formed in 4! ways.

Words starting with TE can be formed in 4! ways.

Words starting with TJ can be formed in 4! ways.

Words starting with TOB can be formed in 3! ways.

Words starting with TOC can be formed in 3! ways.

Words starting with TOE can be formed in 3! ways.

Words starting with TOJB can be formed in 2! ways.

Words starting with TOJC can be formed in 2! ways.

Therefore, the total number of words is 718 words.

Hence 717th word is TOJCBE.

22. a. The number of ways of allotment without any restriction is $^8P_6$. Now, it is possible that all rooms of second floor or third floor are not occupied. Thus, there are two ways in which one floor remains unoccupied. Hence, the number of ways of allotment in which a floor is unoccupied is $2 \times 6!$ Hence, number of ways in which none of the floor remains unoccupied is $^8P_6 - 2(6!)$.

23. a. Each position can be filled in 5 ways. Hence, the total number of numbers is 5^{20}.

24. a. The number of ways of selecting four numbers from 1 to 30 without any restriction is $^{30}C_4$. The number of ways of selecting four consecutive [i.e. (1, 2, 3, 4), (2, 3, 4, 5), …, (27, 28, 29, 30)] number is 27. Hence, the number of ways of selecting four integers which excludes consecutive four selections is

$$^{30}C_4 - 27 = \frac{30\times 29\times 28\times 27}{24} - 27 = 27378$$

25. d. Let us first select two places for vowel, which can be selected from 4 places in $^4C_2$ ways. Now this places can be filled by vowels in $5 \times 5 = 25$ ways as repetition is allowed. The remaining two places can be filled by consonants in 21×21 ways. Then the total number of words is $^4C_2 \times 25 \times 21^2 = 150 \times 21^2$.

26. b. Other than 2, remaining five places can be filled by 1 and 3 for each place. The number of ways for five places is $2 \times 2 \times 2 \times 2 \times 2 = 2^5$. For 2, selecting 2 places out of 7 is $^7C_2$. Hence, the required number of ways is $^7C_2 \times 2^5$.

27. c. 1 S, 3A, 1 H, 2 R, 1 N, 1 P, 1 U

When all letters are different corresponding ways is $^7C_3 \times 3! = {}^7C_3 = 210$. When two letters are of one kind and other is different, corresponding number of ways is $^2C_1 \times {}^6C_1 \times (3!/2!) = 36$. When all letters are alike, corresponding number of ways is 1. Thus, total number of words that can be formed is $210 + 36 + 1 = 247$.

28. b. The natural numbers are 1, 2, 3, 4. Clearly, in one diagonal we have to place 1, 4 and in the other 2, 3.

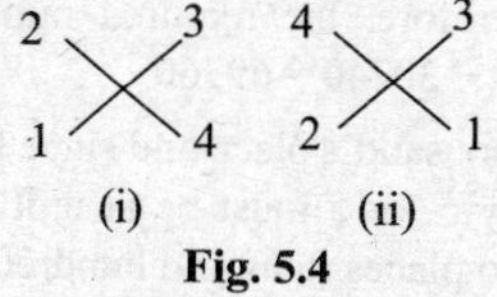

Fig. 5.4

The number of ways in (i) is $2! \times 2! = 4$.

The number of ways in (ii) is $2! \times 2! = 4$.

Therefore, the total number of ways is 8.

29. c. First arrange m positive signs. The number of ways is just 1 (as all + signs are identical). Now, $m + 1$ gaps are created of which n are to be selected for placing '–' signs. Then the total number of ways of doing so is ${}^{m+1}C_n$. After selecting the gaps '–' signs can be arranged in one way only.

30. a. We can think of three packets. One consisting of three boys of class X, other consisting of four boys of class XI and last one consisting of five boys of class XII. These packets can be arranged in 3! ways and contents of these packets can be further arranged in 3! 4! and 5! ways, respectively. Hence, the total number of ways is 3!× 3! × 4! × 5!

31. a. The total number of books is $a + 2b + 3c + d$. The total number of ways in which these books can be arranged in a shelf (in same row) is

$$\frac{(a+2b+3c+d)!}{a!(b!)^2(c!)^3}$$

32. c. Required sum is $3!(3 + 4 + 5 + 6) = 6 \times 18 = 108$.

[If we fix 3 in the unit place, other three digits can be arranged in 3! ways. Similarly for 4, 5, 6.]

33. c. The number of numbers with 0 in the unit's place is 3! = 6.

The number of numbers with 1 or 2 or 3 in the unit's place is $3! - 2! = 4$. Therefore the sum of the digits in the unit's place is $6 \times 0 + 4 \times 1 + 4 \times 2 + 4 \times 3 = 24$.

Similarly, for the ten's and hundred's places, the number of numbers with 1 or 2 in the thousand's place is 3! Therefore, the sum of the digits in the thousand's place is $6 \times 1 + 6 \times 2 + 6 \times 3 = 36$.

Hence, the required sum is $36 \times 1000 + 24 \times 100 + 24 \times 10 + 24$.

34. a. Total numbers ending with 2 is 3! as after fixing 2 in the unit's place other three places can be filled by 3! ways. Thus, 2 appears in the unit's place 3! times.

Similarly, all other digits 4, 6 and 8 also appear 3! times. Then sum of the digits in the unit's place is $6(2 + 4 + 6 + 8) = 120$ units. Similarly, sum of digits in ten's place is 120 tens and that in hundred's place is 120 hundreds, etc. Hence, sum of all the 24 numbers is $120(1 + 10 + 10^2 + 10^3) = 120 \times 1111 = 133320$.

35. b. $(x + 3)^2 + y^2 = 13$

$\Rightarrow x + 3 = \pm 2, y = \pm 3$ or $x + 3 = \pm 3, y = \pm 2$

36. d. Using the digits 0, 1, 2, ..., 9 the number of five digit telephone numbers which can be formed is 10^5 (since repetition is allowed). The number of five digit telephone, numbers which have none of the digits repeated is ${}^{10}P_5 = 30240$. Therefore, the required number of telephone numbers is $10^5 - 30240 = 69760$.

37. b. 3 must be at thousand's place and since the number should be divisible by 5, or 5 must be at unit's place. Now, we have to fill two places (tens and hundreds), i.e., ${}^4P_2 = 12$.

38. d. Two positions for A_1 and A_{10} can be selected in ${}^{10}C_2$ ways. Rest eight students can be ranked in 8! ways. Hence total number of ways is ${}^{10}C_2 \times 8! = (1/2)(10!)$.

39. a.

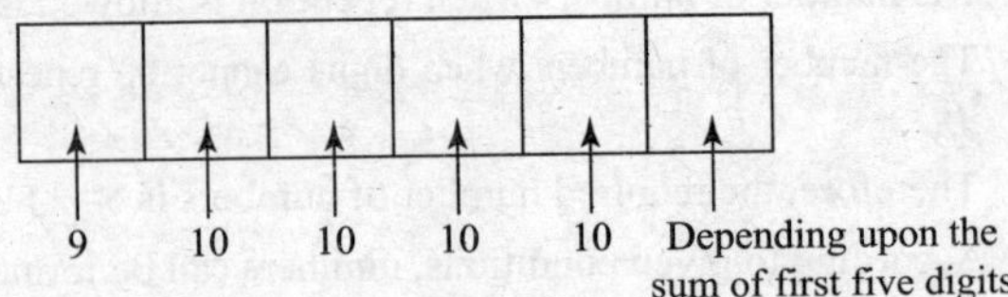

Fig. 5.5

First place from left cannot be filled with 0. Next four places can be filled with any of the 10 digits. After filling the first five places, the last place can be filled in following ways:

Sum of digits in first five places	Digit in the unit's place
$5k$	0 or 5
$5k + 1$	4 or 9
$5k + 2$	3 or 8
$5k + 3$	2 or 7
$5k + 4$	1 or 6

Thus, in any case the last place can be filled in two ways. Hence, the required number of numbers is $9 \times 10^4 \times 2$.

40. a. The selection can be made in ${}^5C_3 \times {}^{22}C_9$ ways.

(Since third vacancies are filled from five candidates in ${}^5C_3$ ways and now remaining candidates are 22 and remaining seats are 9.)

41. d. Here,

${}^nP_3 - {}^nC_3 > 100$

$\Rightarrow \frac{n!}{(n-3)!} - \frac{n!}{3!(n-3)!} > 100$

$\Rightarrow \frac{5}{6}n(n-1)(n-2) > 100$

$\Rightarrow n(n-1)(n-2) > 120$

$\Rightarrow n(n-1)(n-2) > 6 \times 5 \times 4$

$\Rightarrow n = 7, 8, ...$

42. c. Places for A, B, C can be chosen in ${}^{10}C_3$ ways. Remaining seven persons can speak in 7! ways. Hence, the number of ways in which they can speak is ${}^{10}C_3 \times 7! = 10!/6$.

43. b. Since five players are always to be excluded and six players always to be included, therefore five players are to be chosen from 14. Hence, required number of ways is ${}^{14}C_5 = 2002$.

44. c.

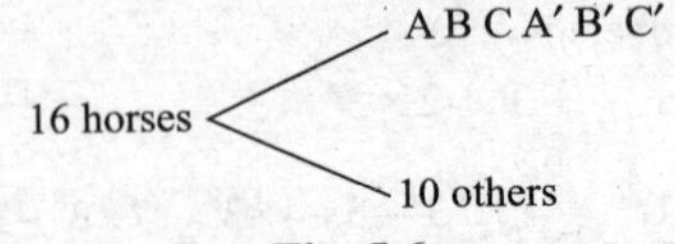

Fig. 5.6

The number of ways is ${}^{10}C_3 \times$ number of ways of choosing out of $ABC\ A'B'C'$, so that AA', BB' or CC' are not together

$= {}^{10}C_3$ (one from each of pairs AA', BB', CC')

$= {}^{10}C_3 \times 8$

$= \frac{10 \times 9 \times 8}{1 \times 2 \times 3} 8$

$= 960$

45. b. The number of ways of selecting r ($0 \le r \le m$) balls out of m is ${}^mC_r$. Therefore, the number of ways if selecting r balls from each of the bag is $({}^mC_r)^2$. Further the number of ways of selecting equal number of balls from each of the two bags, choosing at least one from each bag, is

$({}^mC_1)^2 + ({}^mC_2)^2 + \cdots + ({}^mC_m)^2 = {}^{2m}C_m - 1$

$[\because ({}^mC_0)^2 + ({}^mC_1)^2 + \cdots + ({}^mC_m)^2 = {}^{2m}C_m]$

46. b. There are six different letters. We have to select six squares, taking at least one from each row and then arranging in each selection. Let us first select places in each row such that no row remains empty.

R_1	R_2	R_3	Number of selections
1	1	4	${}^2C_1\,{}^2C_1\,{}^4C_4 = 4$
1	2	3	${}^2C_1\,{}^2C_2\,{}^4C_3 = 8$
2	1	3	${}^2C_2\,{}^2C_1\,{}^4C_3 = 8$
2	2	2	${}^2C_2\,{}^2C_2\,{}^4C_2 = 6$

Therefore, the total number of selections of six squares is $4 + 8 + 8 + 6 = 26$. For each selection of six squares, the number of arrangements of 6 letters is $6! = 720$. Hence, the required number of ways is $26 \times 720 = 18720$.

47. b. There are 11 letters A, A; I, I; N, N; E, X, M, T, O. For the selection of four letters we have the following possibilities:

a. 2 alike, 2 alike

b. 2 alike, 2 different

c. All four different

a. There are three pairs of two letters. So the number of ways of selection of two pairs is ${}^3C_2$ and permutation of these four letters is $4! / 2!\,2!$ Therefore, the number of words in this case is ${}^3C_2 \times 4! / 2!\,2! = 18$.

b. We have to select one pair from three pairs and two distinct letters from remaining seven distinct letters. For illustration, let us select both A, A; then we have I, N, E, X, M, T, O, i.e., 7 as remaining distinct letters. Hence, the number of selections is ${}^3C_1 \times {}^7C_2$ and these 4 (2 same, 2 distinct) can be permuted in $4!/2!$ ways. Therefore, number of words is ${}^3C_1 \times {}^7C_2 \times 4! / 2! = 3 \times 21 \times 12 = 756$.

c. There are eight distinct letters so number of words of four letters is ${}^8C_4 \times 4! = 1680$. By sum rule, the total number of words is $18 + 756 + 1680 = 2454$.

48. d. The total number of words is $6! = 720$. Let us write the letters of word ZENITH alphabetically, i.e., EHINTZ.

For ZENITH word start with	Word starting with	Number of words
Z	E	5!
	H	5!
	I	5!
	N	5!
	T	5!
ZEN	ZEH	3!
	ZEI	3!
ZENI	ZENH	2!
ZENIT	ZENIH	1
	Total number of words before ZENITH	615

Hence, there are 615 words before ZENITH, so the rank of ZENITH is 616.

49. b.

Number of girls	Number of boys	Number of groups going to picnic	Total number of dolls
1	4	${}^3C_1\,{}^4C_4$	$1({}^3C_1\,{}^4C_4) = 3$
2	3	${}^3C_2\,{}^4C_3$	$2({}^3C_2\,{}^4C_3) = 24$
3	2	${}^3C_3\,{}^4C_2$	$3({}^3C_3\,{}^4C_2) = 18$
		Total	45

50. b. The number of times the teacher goes to the zoo is ${}^nC_3$. The number of times a particular child goes to the zoo is equal to number of ways two other children can be selected who accompany a particular child, i.e., ${}^{n-1}C_2$. From the question,

${}^nC_3 - {}^{n-1}C_2 = 84$

or $(n - 1)(n - 2)(n - 3) = 6 \times 84 = 9 \times 8 \times 7 \Rightarrow n - 1 = 9$

51. b. We first select two men out of seven in ${}^7C_2$ ways. Now we exclude the wives of these two selected men and so select two ladies from remaining five ladies in ${}^5C_2$ ways. Let A, B be two men and X, Y be the ladies playing in one set. Then we can have

i. A and X plying against B and Y.

ii. A and Y playing against B and X.

Then the total number of ways is ${}^7C_2 \times {}^5C_2 \times 2 = 21 \times 10 \times 2 = 420$.

52. b. Suppose there are n players in the beginning. The total number of games to be played was equal to ${}^nC_2$ and each player would have played $n - 1$ games.

Let us assume that A and B fell ill. Now the total number of games of A and B is $(n - 1) + (n - 1) - 1 = 2n - 3$. But they have played three games each. Then their remaining number of games is $2n - 3 - 6 = 2n - 9$. Given,

${}^nC_2 - (2n - 9) = 84$

$\Rightarrow n^2 - 5n - 150 = 0$

$\Rightarrow n = 15$

Alternative solution:

The number of games excluding A and B is ${}^{n-2}C_2$. But before leaving A and B played 3 games each. Then,

${}^{n-2}C_2 + 6 = 84$

Solving this equation, we get $n = 15$.

53. a. The number of ways he can select at least one parantha is $2^3 - 1 = 7$. The number of ways he can select at least one vegetable dish is $2^4 - 1 = 15$. The number of ways he can select zero or more items from salads and sauces is 2^5. Hence, the total number of ways is $7 \times 15 \times 32 = 3360$.

54. b. Number of even divisors is equal to number of ways in which one or more '2', zero or more '3', zero or more '5' and zero or more '7' can be selected, and is given by (3)(2 + 1)(2 + 1)(1 + 1) = 54.

55. c. The number of ways the candidate can choose questions under the given conditions is enumerated below.

Group 1	Group 2	Number of ways
4	2	$({}^5C_4)({}^5C_2) = 50$
3	3	$({}^5C_3)({}^5C_3) = 100$
2	4	$({}^5C_2)({}^5C_4) = 50$
	Total number of ways	200

56. c. Let there be n men participants. Then the number of games that the men play between themselves is $2 \times {}^nC_2$ and the number of games that the men played with the women is $2 \times (2n)$.

$\therefore 2 \times {}^nC_2 - 2 \times 2n = 66$ (by hypothesis)

$\Rightarrow n^2 - 5n - 66 = 0 \Rightarrow n = 11$

Hence, the number of participants is 11 men + 2 women = 13.

57. b. The smallest number of people

= total number of possible forecasts

= total number of possible results

$= 3 \times 3 \times 3 \times 3 \times 3$

58. a. Three IIT students who will be between the IIT students can be selected in ${}^{10}C_3$ ways. Now, two DCE students having three IIT students between them can be arranged in 2!×3! ways. Finally, a group of above five students and the remaining seven students together can be arranged in 8! ways. Hence, total number of ways is ${}^{10}C_3 \times 2! \times 3! \times 8!$

59. c. Let S_1 and S_2 refuse to be together and S_3 and S_4 want to be together only. The total number of ways when S_3 and S_4 are selected is $({}^8C_2 + {}^2C_1 \times {}^8C_1) = 44$. The total ways when S_3 and S_4 are not selected is $({}^8C_4 + {}^2C_1 \times {}^8C_3) = 182$. Thus, the total number of ways is 44 + 182 = 226.

60. c. Let there be n candidates. Then,

${}^nC_1 + {}^nC_2 + \cdots + {}^nC_{n-1} = 254$

$\Rightarrow 2^n - 2 = 254$

$\Rightarrow 2^n = 2^8 \Rightarrow n = 8$

61. b. P_1 must win at least $n + 1$ games. Let P_1 win $n + r$ games ($r = 1, 2, \ldots, n$). Therefore, corresponding number of ways is ${}^{2n}C_{n+r}$. The total number of ways is

$$\sum_{r=1}^{n} {}^{2n}C_{n+r} = {}^{2n}C_{n+1} + {}^{2n}C_{n+2} + \cdots + {}^{2n}C_{2n}$$

$$= \frac{2^{2n}}{2} - {}^{2n}C_n$$

$$= \frac{1}{2}(2^{2n} - 2 \times {}^{2n}C_n)$$

62. b. The candidate is unsuccessful if he fails in 9 or 8 or 7 or 6 or 5 papers. Therefore, the number of ways to be unsuccessful is

$${}^9C_9 + {}^9C_8 + {}^9C_7 + {}^9C_6 + {}^9C_5 = {}^9C_0 + {}^9C_1 + {}^9C_2 + {}^9C_3 + {}^9C_4$$

(recall the concept of half series)

$$= \frac{1}{2}({}^9C_0 + {}^9C_1 + \cdots + {}^9C_9)$$

$$= \frac{1}{2} \times 2^9 = 2^8$$

63. b. Since the student is allowed to select at most n books out of $(2n + 1)$ books, in order to select one book he has the choice to select one, two, three, …, n books. Thus, if T is the total number of ways of selecting one book, then

$$T = {}^{2n+1}C_1 + {}^{2n+1}C_2 + \ldots + {}^{2n+1}C_n = 63 \qquad (1)$$

Again the sum of binomial coefficients is

$${}^{2n+1}C_0 + {}^{2n+1}C_1 + {}^{2n+1}C_2 + \cdots + {}^{2n+1}C_n + {}^{2n+1}C_{n+1} + {}^{2n+1}C_{n+2} + \ldots + {}^{2n+1}C_{2n+1} = (1+1)^{2n+1} = 2^{2n+1}$$

or

$${}^{2n+1}C_0 + 2({}^{2n+1}C_1 + {}^{2n+1}C_2 + \cdots + {}^{2n+1}C_n) + {}^{2n+1}C_{2n+1} = 2^{2n+1}$$

$\Rightarrow 1 + 2(T) + 1 = 2^{2n+1}$

$\Rightarrow 1 + T = \dfrac{2^{2n+1}}{2} = 2^{2n}$

$\Rightarrow 1 + 63 = 2^{2n}$

$\Rightarrow 2^6 = 2^{2n} \Rightarrow n = 3$

64. c. The number of ways can be given as follows:

2 bowlers and 9 other players: ${}^4C_2 \times {}^9C_9$

3 bowlers and 8 other players: ${}^4C_3 \times {}^9C_8$

4 bowlers and 7 other players: ${}^4C_4 \times {}^9C_7$

Hence, required number of ways is

$6 \times 1 + 4 \times 9 + 1 \times 36 = 78$.

65. a. Matches whose predictions are correct can be selected in ${}^{20}C_{10}$ ways. Now each wrong prediction can be made in 2 ways. Thus, the total number of ways is ${}^{20}C_{10} \times 2^{10}$.

66. b.

Dashes	Dots	Arrangements
5	2	${}^7C_2$
4	3	${}^7C_3$
3	4	${}^7C_4$
2	5	${}^7C_5$
1	6	${}^7C_6$
0	7	${}^7C_7$

The total number of ways is ${}^7C_2 + {}^7C_3 + {}^7C_4 + {}^7C_5 + {}^7C_6 = 2^7 - 8 = 120$.

67. c. For a radical centre, three circles are required. The total number of radical centres is ${}^nC_3$. The total number of radical axis is ${}^nC_2$. Now,

${}^nC_2 = {}^nC_3 \Rightarrow n = 5$

68. c. i. Miss C is taken

(A) B included $\Rightarrow$ A excluded $\Rightarrow {}^4C_1 \times {}^4C_2 = 24$

(B) B excluded $\Rightarrow {}^4C_1 \times {}^5C_3 = 40$

ii. Miss C is not taken

$\Rightarrow$ B does not come; $\Rightarrow {}^4C_2 \times {}^5C_3 = 60$

$\Rightarrow$ Total = 124

Alternative method:

Case I:

Mr. B is present

$\Rightarrow A$ is excluded and C included

Hence, the number of ways is ${}^4C_2\,{}^4C_1 = 24$.

Case II:

Mr. B is absent

$\Rightarrow$ No constraint

Hence, the number of ways is ${}^5C_3\,{}^5C_2 = 100$.

$\therefore$ Total = 124.

69. d. $N = 1! + 2! + \cdots + 2005!$

$= (1! + 2! + 3! + 4!) + (5! + \cdots + 2005!)$

$= 33 +$ an integer having 0 in its unit's place

$=$ an integer having 3 in its unit's place

Hence, N^{500} is an integer having 1 in its unit's place.

70. d.

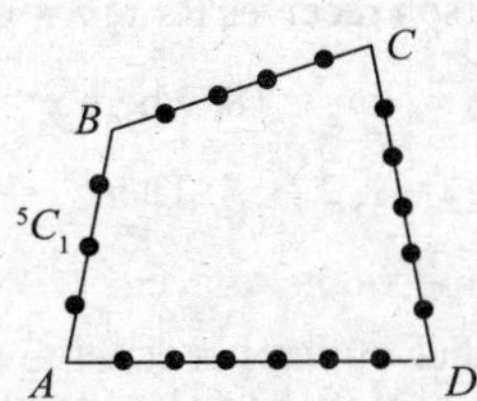

Fig. 5.7

The number of triangles with vertices on sides AB, BC, CD is ${}^3C_1 \times {}^4C_1 \times {}^5C_1$.

Similarly, for other cases, the total number of triangles is ${}^3C_1 \times {}^4C_1 \times {}^5C_1 + {}^3C_1 \times {}^4C_1 \times {}^6C_1 + {}^3C_1 \times {}^5C_1 \times {}^6C_1 + {}^4C_1 \times {}^5C_1 \times {}^6C_1 = 342$.

71. c. Number of points required for the fixed circle is 3. So, first select any three points from the 10 points in ${}^{10}C_3$ ways.

In these ways, circle with four concyclic points is selected in ${}^4C_3$ ways. But it should be taken once then total number of circles is $({}^{10}C_3 - {}^4C_3) + 1$.

72. a. The number of points of intersection is equal to the number of ways two lines are selected, which is given by

$${}^nC_2 = \frac{n(n-1)}{2} = \sum_{k=1}^{n-1} K$$

73. c. Out of 10 points let n points are collinear. Then the number of triangles is

${}^{10}C_3 - {}^nC_3 = 110$

$$\Rightarrow \frac{10\times9\times8}{6} - \frac{n(n-1)(n-2)}{6} = 110$$

$\Rightarrow n(n-1)(n-2) = 60$

$\Rightarrow n = 5$

74. c. Select any three points from total $3p$ points, which can be done ${}^{3p}C_3$ ways. But this also includes selection of three collinear points. Now three collinear points from each straight line can be selected in ${}^pC_3$ ways. Then the number of triangles is ${}^{3p}C_3 - 3{}^pC_3 = p^2(4p-3)$.

75. c. Two circles intersect at two distinct points. Two straight lines intersect at one point. One circle and one straight line intersect at two distinct points. Then the total numbers of points of intersections are as follows:

Number of ways of selection	Points of intersection
Two straight lines: ${}^5C_2$	${}^5C_2 \times 1 = 10$
Two circles: ${}^4C_2$	${}^4C_2 \times 2 = 12$
One line and one circle: ${}^5C_1 \times {}^4C_1$	${}^5C_1 \times {}^4C_1 \times 2 = 40$
Total	62

76. a. The number of selection of two parallel lines from m lines is ${}^mC_2$.

The number of selection of two parallel lines from n lines is ${}^nC_2$.

Hence, the number of parallelograms lines is

$${}^mC_2 \times {}^nC_2 = \frac{1}{4}mn(m-1)(n-1)$$

77. b. Let $x = p - 5$, $y = q - 5$ and $z = r - 5$, where $p, q, r \geq 0$.

Then the given equation reduces to

$p + q + r = 15$ (1)

Now, we have to find non-negative integral solution of Eq. (1). The total number of such solutions is ${}^{15+3-1}C_{3-1} = {}^{17}C_2 = 136$.

78. b. Dice is marked with numbers 1, 2, 3, 4, 5, 6. If the sum of dice in three throws is 11, then observations must be 1, 4, 6; … 1, 5, 5; … 2, 3, 6; … 2, 4, 5; … 3, 3, 5; … 3, 4, 4.

We can get this observation in $3! + 3!/2! + 3! + 3! + 3!/2! + 3!/2! = 27$ ways.

79. d. No group of four numbers from the first 12 natural numbers can have the common difference 4.

If one group including 1 is selected with the common difference 1, then the other two group can have the common difference 1 or 2.

If one group including 1 is selected with the common difference 2, then one of the other two groups can have the common difference 2 and the remaining group will have common difference 1.

If one group including 1 is selected with the common difference 3, then the other two groups can have the common difference 3.

Therefore, the required number of ways is $2 + 1 + 1 = 4$.

80. a. Let the balls put in the box are x_1, x_2, x_3, x_4 and x_5. We have,

$x_1 + x_2 + x_3 + x_4 + x_5 = 15, x_i \geq 2$

$\Rightarrow (x_1 - 2) + (x_2 - 2) + (x_3 - 2) + (x_4 - 2) + (x_5 - 2) = 5$

$\Rightarrow y_1 + y_2 + y_3 + y_4 + y_5 = 5, y_i = x_i - 2 \geq 0$

The total number of ways is equal to number of non-negative integral solutions of the last equation, which is equal to ${}^{5+5-1}C_5 = {}^9C_5$.

81. a. Let x be the number of objects to the left of the first object chosen, y the number of objects between the first and the second, z the number of objects between the second and the third and u the number of objects to the right of the third objects. Then, $x, u \geq 0$; $y, z \geq 1$ and $x + y + z + u = n - 3$. Let $y_1 = y - 1$ and $z_1 = z - 1$. Then, $y_1 \geq 0, z_1 \geq 0$ such that $x + y_1 + z_1 + u = n - 5$.

The total number of non-negative integral solutions of this equation is ${}^{n-5+4-1}C_{4-1} = {}^{n-2}C_3$.

82. a. Obviously, A, B and C get 4, 5 and 7 objects, respectively. Then, number of distribution ways is equal to number of division of ways, which is given by 16!/(4!5!7!).

83. b. The number of ways is

$$\frac{(mn)!}{(n!)^m m!} m! = \frac{(mn)!}{(n!)^m}$$

84. a. $m + n$ counters on one side can be arranged in $\frac{(m+n)!}{m!n!}$ ways.

For each arrangement on one side, corresponding arrangement on the other side is just one as arrangements are symmetrical. Hence, the total number of arrangements is

$$\frac{(m+n)!}{m!n!} = {}^{m+n}C_m$$

85. a. Let x_1, x_2, x_3, x_4 be the number of times T, I, D, E appears on the coupon. Then we must have $x_1 + x_2 + x_3 + x_4 = 8$, where $1 \leq x_1, x_2, x_3, x_4 \leq 8$ (as each letter must appear once). Then the required number of combinations of coupons is equivalent to number of positive integral solutions of the above equation, which is further equivalent to number of ways of eight identical objects distributed among four persons when each gets at least one objects, and is given by ${}^{8-1}C_{4-1} = {}^7C_3$.

86. a. Here, we have to divide 12 books into four sets of three books each. Therefore, the required number of ways is

$$\frac{12!}{(3!)^4 4!} \times 4!$$

87. a. Since the shelves which are to receive the books are different, therefore the required number of ways is $12!/(4!)^3$.

88. c. Here, we are dividing $2n$ people in n groups of 2 each, and we are concerned with mere grouping. Hence, the required number of ways is $\frac{2n!}{n!(2!)^n}$.

89. b. $\sum_{i=1}^{k} \frac{1}{x_i} = \frac{1}{x_1} + \frac{1}{x_2} + \cdots + \frac{1}{x_k} = \frac{\Sigma x_i}{n} = \frac{75}{n}$

(as L.C.M. of $x_1, x_2, \ldots, x_n$ is n)

90. a. Three elements from set A can be selected in ${}^7C_3$ ways. Their image has to be y_2. Remaining two images can be assigned to remaining four pre-images in 2^4 ways. But the function is onto, hence the number of ways is $2^4 - 2$. Then the total number of functions is ${}^7C_3 \times 14 = 490$.

91. c. If we put minimum number of balls required in each box, balls left are $n(n - 1)/2$ which can be put in ${}^{(n^2+n-1)/2}C_{n-1}$ ways without restriction.

92. a. Let the blankets received by the persons are x_1, x_2, x_3, and x_4. We have

$x_1 + x_2 + x_3 + x_4 = 15$ and $x_i \geq 2$

$\Rightarrow (x_1 - 2) + (x_2 - 2) + (x_3 - 2) + (x_4 - 2) = 7$

$\Rightarrow y_1 + y_2 + y_3 + y_4 = 7$ (where $y_i = x_i - 2 \geq 0$)

The required number is equal to the number of non-negative integral solutions of this equation which is equal to ${}^{4+7-1}C_7$, i.e., ${}^{10}C_7 = {}^{10}C_3$.

93. c. Let person P_i gets x_i number of things such that

$x_1 + x_2 + x_3 + x_4 + x_5 = 25$

Let $x_i = 2\lambda_i + 1$, where $\lambda_i \geq 0$. Then,

$2(\lambda_1 + \lambda_2 + \lambda_3 + \lambda_4 + \lambda_5) + 5 = 25$

$\Rightarrow \lambda_1 + \lambda_2 + \lambda_3 + \lambda_4 + \lambda_5 = 10$

We have to simply obtain the number of non-negative integral solutions of this equation, which is equal to ${}^{14}C_4$.

94. d. Suppose i^{th} person receives Rs x_i; $i = 1, 2, 3, 4$

Then, $x_1 + x_2 + x_3 + x_4 = 18$, where $x_i \geq 4$

Let $y_i = x_i - 4$, $i = 1, 2, 3, 4$. Then,

$y_1 + y_2 + y_3 + y_4 = 6$

The total number of ways is equal to number of solutions of the above equation, which is given by ${}^{6+4-1}C_{4-1} = {}^9C_3 = 84$.

95. d. Make one group of two persons, one group of four persons and three groups of three persons among 15 persons (except two particular persons). Hence the number of ways by grouping method is

$$\frac{15!}{2!4!(3!)^3 3!}$$

Now we add that two persons in the group of two persons and thus number of arrangements of these groups into cars and autos is

$$\frac{15!}{2!4!(3!)^3 3!} \times 2! \times 3! = \frac{15!}{4!(3!)^3}$$

96. a. Since the total number of selections of r things from n things where each thing can be repeated as many times as one can is ${}^{n+r-1}C_r$. Therefore the required number is ${}^{3+6-1}C_6 = 28$.

97. c. $f(2n, n)$ must be equal to number of positive integer solutions of $x_1 + x_2 + \cdots + x_n = 2n$, which must be equal to ${}^{2n-1}C_{n-1} = {}^{2n-1}C_n$.

98. b. Let the numbers selected be x_1, x_2, x_3. We must have

$2x_2 = x_1 + x_3$

$\Rightarrow x_1 + x_3 =$ even.

Therefore, x_1, x_3 both are odd or both are even.

If x_1 and x_3 both are even, we can select them in ${}^{12}C_2$ ways.

Similarly, if x_1 and x_3 both are odd, we can again select them in ${}^{12}C_2$ ways. Thus, the total number of ways is $2 \times {}^{12}C_2 = 132$.

99. b. Given number can be rearranged as

$1, 4, 7, \ldots, 3n-2 \rightarrow 3\lambda - 2$

$2, 5, 8, \ldots, 3n-1 \rightarrow 3\lambda - 1$

$3, 6, 9, \ldots, 3n \rightarrow 3\lambda$

That means, we must take two numbers from last row or one number each from first and second rows. Therefore, the total number of ways is

$${}^nC_2 + {}^nC_1 \times {}^nC_1 = \frac{n(n-1)}{2} + n^2 = \frac{3n^2 - n}{2}$$

100. b. Let the arrangement be $x_1 x_2 x_3 x_4 x_5 x_6 x_7 x_8$ clearly 5 should occupy the position x_4 or x_5. Thus required number is 2(7!).

101. c. $480 = 2^5 \times 3 \times 5$

Now, $4n + 2 = 2(2n + 1)$ = odd multiple of 2. Thus, the total number of such divisors is $1 \times 2 \times 2 = 4$.

102. a.

Number of times 3 used	Pattern	Numbers of type	Number of times 3 appears
1	– – 3 – 3 – 3 – –	3 × 9 × 9	1(3 × 9 × 9)
2	– 3 3 33 – 3 – 3	3 × 9	2(3 × 9)
3	333	1	3
		Total	300

Any place other than 3 is filled by 9 ways as 0 can appear anywhere which gives all types of numbers like single digit, two digits, etc.

Alternative solution:

A three-digit block from 000 to 999 means 1000 numbers, each number constituting three digits. Hence, the total numbers of digits which we have to write is 3000.

Since the total number of digits is 10 (0 to 9) no digit is filled preferentially. Therefore, number of times we write 3 is 3000/10 = 300.

103. b.

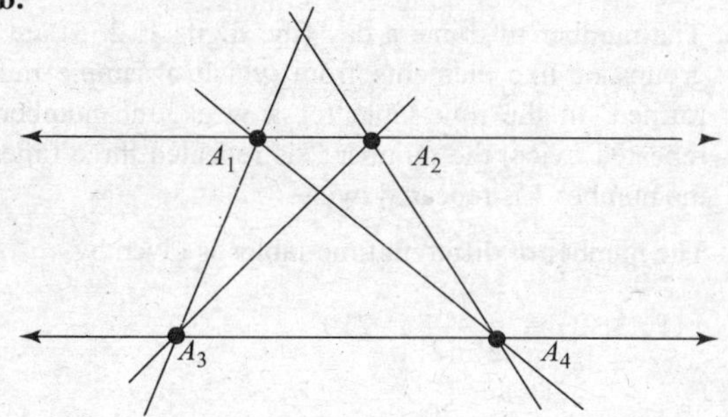

Fig. 5.8

For intersection point we must have two straight lines, for which two points from each straight line must be selected. Now selection of these points can be done in ${}^mC_2 \times {}^nC_2$ ways. Now as shown in diagram these four points can give two different sets of straight lines, which generate two distinct points of intersection.

Then total number of points of intersection is ${}^mC_2 \times {}^nC_2 \times 2$.

104. a.

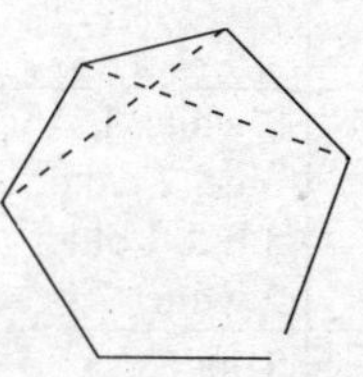

Fig. 5.9

Let the number of sides be n. A selection of four vertices of the polygon gives an interior intersection.

$\therefore {}^nC_4 = 70$

$\Rightarrow n(n-1)(n-2)(n-3) = 24 \times 70 = 8 \times 7 \times 6 \times 5$

$\Rightarrow n = 8$

105. a. 26 cards can be chosen out of 52 cards in ${}^{52}C_{26}$ ways. There are two ways in which each card can be dealt, because a card can be either from the first pack or from the second. Hence the total number of ways is ${}^{52}C_{26} \times 2^{26}$.

106. d. Since the balls are to be arranged in a row so that the adjacent balls are of different colours, we can therefore begin with a white ball or a black ball. If we begin with a white ball, we find that $n+1$ white balls numbered 1 to $n+1$ can be arranged in a row in $(n+1)!$ ways. Now $n+2$ places are created among $n+1$ white balls which can be filled by $n+1$ black balls in $(n+1)!$ ways.

So, the total number of arrangements in which adjacent balls are of different colours and first ball is a white ball is $(n+1)! \times (n+1)! = [(n+1)!]^2$. But we can begin with a black ball also. Hence, the required number of arrangements is $2[(n+1)!]^2$.

107. a. If zero is included it will be at $z \Rightarrow {}^9C_2$

$$\text{If zero is excluded} \begin{cases} x, y, z \text{ all diff.} & \Rightarrow {}^9C_3 \times 2! \\ x = z < y & \Rightarrow {}^9C_2 \\ x < y = z & \Rightarrow {}^9C_2 \end{cases}$$

The total number of ways is 276.

Alternative method:

y can be from 2 to 9; so total number of ways is

$$\sum_{r=2}^{9} (r^2 - 1) = 276$$

108. b. The two common elements can be selected in ${}^nC_2$ ways. Remaining $n-2$ elements, each can be chosen in three ways, i.e., $a \in P$ and $a \notin Q$ or $a \in Q$ and $a \notin P$ or a is neither in P nor in Q. Therefore, the total number of ways is ${}^nC_2 \times 3^{n-2}$.

109. d. We will consider the following cases

Case	Flags	No. of signals
4 alike and 2 others alike	4 white and 2 red	$\frac{6!}{4!\,2!} = 15$
4 alike and 2 others different	4 white, 1 red and 1 blue	$\frac{6!}{4!} = 30$
3 alike and 3 others alike	3 white, 3 red	$\frac{6!}{3!\,3!} = 20$
3 alike and 2 other alike and 1 different	3 white, 1 blue, 2 red or 3 red, 1 blue, 2 white	${}^2C_1 \times \frac{6!}{3!2!} = 120$
	Total	185

110. c. The total number of ways of selection without restriction is ${}^{20}C_3$. The number of ways of selection when two are adjacent is $20 \times {}^{16}C_1$. The number of ways of selection when all the three are adjacent is 20. The required number of ways is

$${}^{20}C_3 - 20 \times 16 - 20 = \frac{20\times19\times18}{6} - 20\times16 - 20$$

$$= 20[57-16-1]$$

$$= 20\times40 = 800$$

111. c. Sum of seven digits is a multiple of 9. Sum of numbers 1, 2, 3, 4, 5, 6, 7, 8, 9 is 45; so two left digits should also have sum of 9. The pairs of left numbers are (1, 8), (2, 7), (3, 6), (4, 5). With each pair left number of 7-digit number is 7! So with all 4 pairs, total number is 4 × 7!.

112. b. If n is odd,

$3^n = 4\lambda_1 - 1$, $5^n = 4\lambda_2 + 1$

$\Rightarrow 2^n + 3^n + 5^n$ is divisible by 4 if $n \geq 2$

Thus, $n = 3, 5, 7, 9, \ldots, 99$, i.e., n can take 49 different values. If n is even, $3^n = 4\lambda_1 + 1$, $5^n = 4\lambda_2 + 1$

$\Rightarrow 2^n + 3^n + 5^n$ is not divisible by 4

as $2^n + 3^n + 5^n$ will be in the form of $4\lambda + 2$.

Thus, the total number of ways of selecting n is equal to 49.

113. c.

Number of digits	Numbers ending with 0	Numbers ending with 5	Total
×	0	1	1
× ×	8	9	17
× × ×	9·8 = 72	8·8 = 64	136
× × × ×	9·8·7 = 504	8·8·7 = 448	952
		Total	1106

114. d. Let x, y, z be the friends and a, b, c denote the case when x is invited a times, y is invited b times and z is invited c times.

Now, we have the following possibilities:

$(a, b, c) = (1, 2, 3)$ or $(3\ 3\ 0)$ or $(2\ 2\ 2)$

[grouping of 6 days of week]

Hence, the total number of ways is

$$\frac{6!}{1!2!3!}3! + \frac{6!}{3!3!2!}3! + \frac{6!}{(2!2!2!)3!}3!$$

$$= 360 + 60 + 90 = 510$$

115. b. There is concept of derangement. The required number is

$$4!\left[1 - \frac{1}{1!} + \frac{1}{2!} - \frac{1}{3!} + \frac{1}{4!}\right] = 9$$

116. c. When at least one one-rupee coin is selected we can select any number of twenty-five paisa coins and ten paisa coins. Then number of ways of such selection is 4 (2 + 1)(5 + 1) = 72 as we can select zero or more twenty-five paisa and ten-paisa coins to ensure that amount selected is Re. 1 or more.

But when none of one-rupee coins is selected we have to select all twenty-five paisa coins and ten-paisa coins to ensure sum of Re. 1, which can be done only in one way. Then the total number of ways is 73.

117. b. Since the number of students giving wrong answers to at least i questions ($i = 1, 2, \ldots, n$) is 2^{n-i}.

The number of students answering exactly i ($1 \leq i \leq -1$) questions wrongly = {the number of students answering at least i questions wrongly, $i = 1, 2, \ldots$)} – {the number of students answering at least ($i+1$) questions wrongly ($2 \leq i+1 \leq n$)} $= 2^{n-i} - 2^{n-(i+1)}$ ($1 \leq i \leq n-1$).

Now, the number of students answering all the n questions wrongly is $2^{n-2} = 2^0$. Thus, the total number of wrong answers is

$$1(2^{n-1} - 2^{n-2}) + 2(2^{n-2} - 2^{n-3}) + 3(2^{n-3} - 2^{n-4}) + \cdots$$
$$+ (n-1)(2^1 - 2^0) + n(2^0) = 2^{n-1} + 2^{n-2} + 2^{n-3} + \cdots + 2^0$$
$$= 2^n - 1 \qquad (\because \text{ it is a G.P.})$$

Therefore, as given,

$$2^n - 1 = 2047 \Rightarrow 2^n = 2048 = 2^{11} \Rightarrow n = 11$$

118. c. For a particular class the total number of different tickets from first intermediate station is 5. Similarly, number of different tickets from second intermediate station is 4. So the total number of different tickets is 5 + 4 + 3 + 2 + 1 = 15. And same number of tickets for another class is equal to total number of different tickets, which is equal to 30 and number of selection is ${}^{30}C_{10}$.

119. b. The number of trains a day (the digits 1, 2, 3) are three groups of like elements from which a sample must be formed. In the time-table for a week, the number 1 is repeated twice, the number 2 is repeated three times and the number 3 is repeated twice.

The number of different time-tables is given by

$$p(2, 3, 2) = \frac{7!}{2!3!2!} = 210$$

120. a. $15 < x_1 + x_2 + x_3 \leq 20$

$\Rightarrow x_1 + x_2 + x_3 = 16 + r, r = 0, 1, 2, 3, 4$

Now, the number of positive integral solution of $x_1 + x_2 + x_3 = 16 + r$ is ${}^{13+r+3-1}C_{13+r}$, i.e., ${}^{15+r}C_{13+r} = {}^{15+r}C_2$.

Thus, the total number of solutions is

$$\sum_{r=0}^{4} {}^{15+r}C_2 = {}^{15}C_2 + {}^{16}C_2 + {}^{17}C_2 + {}^{18}C_2 + {}^{19}C_2$$

$$= \frac{1}{2}(15\times14+16\times15+17\times16+18\times17+19\times18)$$

$$= 685$$

Multiple Correct Answers Type

1. b., c.

If a, b, c are in A.P., then a and c both are odd or both are even.

Case I: n is even.

The number of ways of selection of two even numbers a and c is ${}^{n/2}C_2$. Number of ways of selection of two odd numbers is ${}^{n/2}C_2$. Hence the number of A.P.'s is

$$2\,{}^{n/2}C_2 = 2\frac{\frac{n}{2}\left(\frac{n}{2}-1\right)}{2} = \frac{n(n-2)}{4}$$

Case II: n is odd.

The number of ways of selection of two odd numbers a and c is ${}^{(n+1)/2}C_2$. The number of ways of selection of two even numbers a and c is ${}^{(n-1)/2}C_2$. Hence the number of A.P.'s is

${}^{(n+1)/2}C_2 + {}^{(n-1)/2}C_2$

$$= \frac{\left(\frac{n+1}{2}\right)\left(\frac{n+1}{2}-1\right)}{2} + \frac{\left(\frac{n-1}{2}\right)\left(\frac{n-1}{2}-1\right)}{2}$$

$$= \frac{1}{8}(n-1)((n+1)+(n-3))$$

$$= \frac{(n-1)^2}{4}$$

2. b., c.

The number of ways of inviting, with the couple not included, is ${}^8C_5$. The number of ways of inviting with the couple included, is ${}^8C_3$. Therefore the required number of ways is

${}^8C_5 + {}^8C_3 = {}^8C_3 + {}^8C_3 \quad (\because {}^8C_5 = {}^8C_3)$

Also,

$${}^{10}C_5 - 2\times{}^8C_4 = \frac{10!}{5!5!} - 2\times\frac{8!}{4!4!}$$

$$= \frac{10\times9\times8\times7\times6}{120} - 2\times\frac{8\times7\times6\times5}{24}$$

$$= 9\times4\times7 - 140$$

$$= 112 = 2\times\frac{8!}{3!5!}$$

3. a., b., d.

$p = {}^5C_4 \times {}^2C_1 = 10$

$q = {}^5C_2({}^2C_1)^3 = 80$

$r = {}^5C_0({}^2C_1)^5 = 32$

$\Rightarrow 2q = 5r,\ 8p = q$ and $2(p + r) > q$

4. a., b., d.

Clearly, each player will play nine games. And total number of games is ${}^{10}C_2 = 45$. Clearly,

$w_i + l_i = 9$ and $\Sigma w_i = \Sigma l_i = 45$

$\Rightarrow w_i = 9 - l_i \Rightarrow w_i^2 = 81 + l_i^2 - 18l_i$

$\Rightarrow \Sigma w_i^2 = 81\times10 + \Sigma l_i^2 - 180\,\Sigma l_i = 810 + \Sigma l_i^2 - 18\times45$

$= \Sigma l_i^2$

5. b., c., d.

When $z = n + 1$, we can choose x, y from $\{1, 2, \ldots, n\}$.

When $z = n + 1$, x, y can be chosen in n^2 ways and when $z = n$, x, y can be chosen in $(n - 1)^2$ ways and so on. Therefore, the number of ways of choosing triplets is

$$n^2 + (n-1)^2 + \cdots + 1^2 = \frac{1}{6}n(n+1)(2n+1)$$

Alternatively triplets with $x = y < z$, $x < y < z$, $y < z < x$ can be chosen in ${}^{n-1}C_2$, ${}^{n+1}C_3$, ${}^{n+1}C_3$ ways. Therefore,

${}^{n+1}C_2 + 2({}^{n+1}C_3) = {}^{n+2}C_2 + {}^{n+1}C_3 = 2({}^{n+2}C_3) - {}^{n+1}C_2$

6. a., b., c.

$$\frac{(200)!}{\underbrace{2!\,2!\ldots2!}_{100\text{ times}}(100)!} = \frac{(200)!}{100!\,2^{100}}$$

$$= 1\times3\times5\cdots199$$

Also,

$$\frac{(200)!}{100!\,2^{100}} = \binom{101}{2}\binom{102}{2}\ldots\binom{200}{2}$$

7. a., b., c., d.

8, 7, 6, 4, 2, x and y

Any number is divisible by 3 if sum of digits is divisible by 3, i.e., $x + y + 27$ is divisible by 3. x and y can take values from 0, 1, 3, 5, 9. Possible pairs are (5, 1) (3, 0) (9, 0) (9, 3) and (1, 5), (0, 3) (0, 9) (3, 9).

8. a., d.

Problem is same as dividing 17 identical things in two groups.

$$\therefore n = \frac{17+1}{2} = 9$$

There is no effect if two diamonds are different as necklace can be flipped over. Hence, $n = m = 9$.

9. a., b., c.

The number of regions for n circles be $f(n)$. Clearly, $f(1) = 2$. Now,

$f(n) = f(n-1) + 2(n-1), \forall n \geq 2$

$\Rightarrow f(n) - f(n-1) = 2(n-1)$

Putting $n = 2, 3, \ldots, n$, we get

$f(n) - f(1) = 2(1 + 2 + 3 + \cdots n - 1) = (n-1)n$

$\Rightarrow f(n) = n(n-1) + 2 = (n^2 - n + 2)$ [which is always even]

$\Rightarrow f(20) = 20^2 - 20 + 2 = 382$

Also,

$n^2 - n + 2 = 92$

$\Rightarrow n^2 - n - 90 = 0 \Rightarrow n = 10$

10. a., c.

$3^p = (4-1)^p = 4\lambda_1 + (-1)^p$

$5^q = (4+1)^q = 4\lambda_2 + 1$

$7^r = (8-1)^r = 4\lambda_3 + (-1)^r$

Hence, any positive integer power of 5 will be in the form of $4\lambda_2 + 1$. Even power of 3 and 7 will be in the form of $4\lambda + 1$ and odd power of 3 and 7 will be in the form of $4\lambda - 1$. Hence, both p and r must be odd or both must be even. Thus $p + r$ is always even. Also, $p + q + r$ can be odd or even.

11. a., b., c.

When $n = 3k$, there are exactly $n/3$ integers of each type $3p$, $3p + 1$, $3p + 2$.

Now, sum of three selected integers is divisible by 3. Then either all the integers of the same type $3p$, $3p + 1$ or $3p + 2$ or one-one integer from each type. Then number of selection ways is ${}^{n/3}C_3 + {}^{n/3}C_3 + {}^{n/3}C_3 + ({}^{n/3}C_1)({}^{n/3}C_1)({}^{n/3}C_1) = 3({}^{n/3}C_3) + (n/3)^3$.

If $n = 3k + 1$, then there are $(n-1)/3$ integers of the type $3p$, $3p + 2$, and $(n + 2)/3$ integers of the type $3p + 1$. Then number of selection ways is $2({}^{(n-1)/3}C_3) + ({}^{(n+2)/3}C_3) + ((n-1)/3)^2(n + 2)$.

When $n = 3k + 2$, the number of selection ways are same as in the case of $n = 3k + 1$.

12. a., c.

$${}^{n+5}P_{n+1} = \frac{11(n-1)}{2} \times {}^{n+3}P_n$$

$$\Rightarrow {}^{n+5}P_{n+1} = \frac{(n+5)!}{4!} = \frac{11(n-1)}{2}\frac{(n+3)!}{3!}$$

$\Rightarrow (n+5)(n+4) = 22(n-1)$

After solving, we get $n = 6$ or $n = 7$.

The number of points of intersection of lines is ${}^6C_2$ or ${}^7C_2 \equiv 15$ or 21.

13. a., b., d.

Total number of units to be covered is $3 + 7 + 11 = 21$. A person can choose three units in ${}^{21}C_3$ ways. A person can choose seven units in ${}^{18}C_7$ ways. The rest 11 units can be chosen in one way. Therefore, total number of ways is ${}^{21}C_3 \times {}^{18}C_7 \times 1 = 21!/(3!7!11!)$.

14. a., c.

Let person P_i get x_i number of things such that

$x_1 + x_2 + x_3 + x_4 + x_5 + x_6 = 30$

If x_i is odd or $x_i = 2\lambda_i + 1$, where $\lambda_i \geq 0$, then

$2(\lambda_i + \lambda_2 + \lambda_3 + \lambda_4 + \lambda_5 + \lambda_6) + 6 = 30$

$\Rightarrow \lambda_1 + \lambda_2 + \lambda_3 + \lambda_4 + \lambda_5 + \lambda_6 = 12$

Then number of solutions is ${}^{12+6-1}C_{6-1} = {}^{17}C_5$. If x_i is even or $x_i = 2\lambda_p$, where $\lambda_i \geq 1$, then

$2(\lambda_i + \lambda_2 + \lambda_3 + \lambda_4 + \lambda_5 + \lambda_6) = 30$

$\Rightarrow \lambda_i + \lambda_2 + \lambda_3 + \lambda_4 + \lambda_5 + \lambda_6 = 15$

Therefore, required number of ways is ${}^{15-1}C_{6-1} = {}^{14}C_5$.

15. a., c.

Let x_i $(1 \leq i \leq n)$ be the number of objects selected of the ith type. Since each object is to be selected at least once, we must have $x_i \geq 1$ and $x_1 + x_2 + \cdots + x_n = r$. We have to find number of positive integral solutions of the above equation. Total number of such solutions is ${}^{r-1}C_{n-1} = {}^{r-1}C_{r-n}$.

16. a., d.

Let $A = \{a_1, a_2, \ldots, a_n\}$. For each $a_i (1 \leq i \leq n)$, we have either $a_i \in P_j$ or $a_i \notin P_j$ $(1 \leq j \leq m)$. That is, there are 2^m choices in which a_i $(1 \leq i \leq n)$ may belong to the P_j's. One of these, there is only one choice, in which $a_i \in P_j$ for all $j = 1, 2, \ldots, m$ which is not favourable for $P_1 \cap P_2 \cap \cdots \cap P_m$ to be ϕ. Thus, $a_i \notin P_1 \cap P_2 \cap \cdots \cap P_m$ in $2^m - 1$ ways.

Since there are n elements in set A, the total number of choices is $(2^m - 1)^n$.

Also, there is exactly one choice, in which, $a_i \notin P_j$ for all $j = 1, 2, \cdots, m$ which is not favourable for $P_1 \cup P_2 \cup \cdots \cup P_m$ to be equal to A.

Thus, a_i can belong to $P_1 \cup P_2 \cup \cdots \cup P_m$ in $(2^m - 1)$ ways.

Since there are n elements in set A, the number of ways in which $P_1 \cup P_2 \cup \cdots \cup P_m$ can be equal to A is $(2^m - 1)^n$.

17. b., c., d.

Exponent of 2 is

$$\left[\frac{10}{2}\right] + \left[\frac{10}{2^2}\right] + \left[\frac{10}{2^3}\right] = 5 + 2 + 1 = 8$$

Exponent of 3 is

$$\left[\frac{10}{3}\right] + \left[\frac{10}{3^2}\right] = 3 + 1 = 4$$

Exponent of 5 is

$$\left[\frac{10}{5}\right] = 2$$

Exponent of 7 is

$$\left[\frac{10}{7}\right] = 1$$

The number of divisors of 10! is $(8 + 1)(4 + 1)(2 + 1)(1 + 1) = 270$. The number of ways of putting N as a product of two natural numbers is $270/2 = 135$.

18. b., c., d.

$$P = 21(21+1)(21-1)(21+2)(21-2)\cdots(21+10)(21-10)$$

$$= (21-10)(21-9)\cdots(21-1)\,21\,(21+1)(21+10)\cdots(21+10)$$

$$= 41 \times 40 \cdots 11,$$

which is divisible by 21!, and hence by 20! and 19!

19. a., d.

When n is odd:

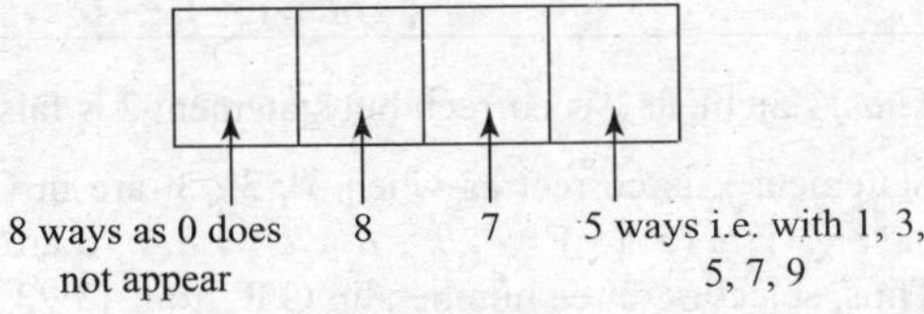

Fig. 5.10

The number of such numbers is $8 \times 8 \times 7 \times 5 = 2240$.

When n is even:

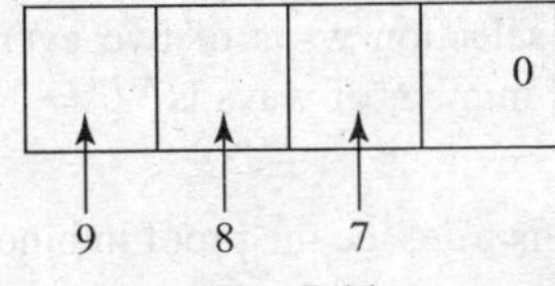

Fig. 5.11

If unit's place is filled with 0, then the total number is $9 \times 8 \times 7 = 504$.

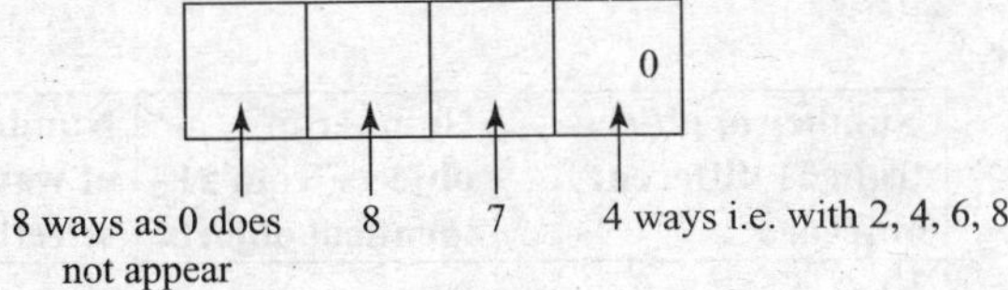

Fig. 5.12

If unit's place is not filled with 0, then the total number is $8 \times 8 \times 7 \times 4 = 1792$.

Hence, the total number of even numbers is $y = 504 + 1792 = 2296$.

Reasoning Type

1. a. We have, $30 = 2 \times 3 \times 5$. So, 2 can be assigned to either a or b or c, i.e. 2 can be assigned in three ways. Similarly, each of 3 and 5 can be assigned in three ways. Thus, the number of solutions is $3 \times 3 \times 3 = 27$.

2. d. Each student receives at least two toys. Let us first give each student one toy. Now, we are left with seven toys, which can be distributed among three students such that each receives at least one toy, which is equivalent to number of positive integral solutions of the equation $x + y + z + w = 7$, which is given by ${}^{7-1}C_{4-1} = {}^{6}C_3$.

Hence, statement 1 is false and statement 2 is correct.

3. a. Number of ways of dividing n^2 objects into n groups of same size is $\dfrac{(n^2)!}{(n!)^n n!}$.

Now number of ways of distributing these n groups among n persons is $\left[\dfrac{(n^2)!}{(n!)^n n!}\right] n! = \dfrac{(n^2)!}{(n!)^n}$ which is always an integer.

Also we know that product of r is divisible by $r!$ Now, $(n^2)! = 1 \times 2 \times 3 \times 4 \cdots n^2 = 1 \times 2 \times 3 \cdots n \times (n+1)(n+2) \cdots 2n \times (2n+1)(2n+2) \cdots 3n \times (n^2 - (n^2-1))(n^2 - (n^2-1)) \cdots n^2$

Thus, in $n^2!$ there are n rows each consisting product of n integers. Each row is divisible by $n!$

Hence, $(n^2)!$ is divisible by $(n!)^n$ or $\dfrac{(n^2)!}{(n!)^n}$ is a natural number.

Hence, both statements are correct and statement 2 is correct explanation of statement 1.

4. b. $1400 = 2^3 5^2 7$

The number of ways in which 1400 can be expressed as a product of two positive integers is

$$\frac{(3+1)(2+1)(1+1)}{2} = 12$$

Statement 2 is correct but does not explain statement 1 as it just gives the information about the prime factor about which 1400 is divisible.

5. a. In onto functions each image must be assigned at least one pre-image. Now if we consider the images a and b as two different boxes, then four distinct objects 1, 2, 3, and 4 (pre-images) can be distributed in $2^4 - {}^{2}C_1(2-1)^4 = 14$ ways.

Hence, both statements are correct and statement 2 is correct explanation of statement 1.

6. b. India must win at least six matches of 11 matches. Then number of ways in which India can win the series is ${}^{11}C_6 + {}^{11}C_7 + \cdots + {}^{11}C_{11} = 2^{10}$

Thus, both the statements are true, but statement 2 is not correct explanation of statement 1.

7. a. The batting order of 11 players can be decided in 11! ways. Now Yuvraj, Dhoni and Pathan can be arranged in 3! ways.

But the order of these three players is fixed, i.e., Yuvraj–Dhoni–Pathan. Now, 11! answer is 3! times more, hence the required answer is 11!/3!.

8. d. Number of ways of arranging 21 identical objects when r is identical of one type and remaining are identical of second type is $\dfrac{21!}{r!(21-r)!} = {}^{21}C_r$ which is maximum when $r = 10$ or 11.

Therefore, ${}^{13}C_r = {}^{13}C_{10}$ or ${}^{13}C_{11}$, hence maximum value of ${}^{13}C_r$ is ${}^{13}C_{10} = 286$.

Hence, statement 1 is false. Obviously statement 2 is true.

9. a. A number is divisible by 4, if the last two digits are divisible by 4. Last two digits can be 12, 16, 28, 32, 36, 68, 92, 96. Thus, last two places can be filled in eight ways. The remaining three places can be filled with remaining four digits in ${}^4C_3\ 3!$ ways. Total number of such numbers is $8 \times ({}^4C_3\ 3!) = 192$.

10. b. We have $a + b + c = 30$, and $a \neq b \neq c$. Let $a < b < c$.

Now, relative values of a, b, c are tabulated as follows.

a	b	c	Number of triplets (a, b, c)
1	2	27	
	3	26	
	4	25	
	⋮	⋮	
	14	15	13
2	3	25	
	4	24	
	⋮	⋮	
	13	15	11
3	4	23	
	5	22	
	⋮	⋮	
	13	14	10
4	5	21	
	6	20	
	⋮	⋮	
	12	14	8
5	6	19	
	7	18	
	⋮	⋮	
	12	13	7
6	7	17	
	⋮	⋮	
	11	13	5
7	8	15	
	⋮	⋮	
	11	12	4
8	9	13	
	10	12	2
9	10	11	1
		Total	61

Statement 2 is correct but it does not explain statement 1.

11. c. When $p, q < r$, we have selection procedure as follows :

From p identical things	From q identical things
p	$r - p$
$p - 1$	$r - (p - 1)$
$p - 2$	$r - (p - 2)$
⋮	⋮
⋮	⋮
$r - q$	q
	Total: $p + q - r + 1$

When $p, q > r$, we have selection procedure as follows:

From p identical things	From q identical things
r	0
$r - 1$	1
$r - 2$	2
⋮	⋮
⋮	⋮
0	r
	Total: $r + 1$

Thus, statement 1 is correct, but statement 2 is false.

12. a. Statement 2 is correct as when 3^a, 3^b, 3^c are in G.P., we have $(3^b)^2 = (3^a)(3^c) \Rightarrow 2b = a + c \Rightarrow a, b, c$ are in A.P. Thus, selecting three numbers in G.P. from $\{3^1, 3^2, 3^3, \ldots, 3^{100}, 3^{101}\}$ is equivalent to selecting 3 numbers from $\{1, 2, 3, \ldots, 101\}$ which are in A.P. Now, a, b, c are in A.P. if either a and c are odd or a and c are even.

Number of selection ways of two odd numbers is ${}^{51}C_2$. Number of selection ways of two even numbers is ${}^{50}C_2$. Hence, total number of ways is ${}^{51}C_2 + {}^{50}C_2 = 1275 + 1225 = 2500$.

13. a. Statement 2 is true, see the proof in binomial theorem.

Also in statement 1, if A selects i objects and B selects j objects then $i < j$. Hence number of ways is $\sum_{0 \le i < j \le 20} {}^{20}C_i\, {}^{20}C_j$.

14. c.

Number of objects from 21 different objects	Number of objects from 21 identical objects	Number of ways of selections
10	0	${}^{21}C_{10} \times 1$
9	1	${}^{21}C_9 \times 1$
⋮	⋮	⋮
0	10	${}^{21}C_0 \times 1$

Thus, total number of ways of selection is ${}^{21}C_0 + {}^{21}C_1 + {}^{21}C_2 + \cdots + {}^{21}C_{10} = 2^{20}$.

Statement 2 is false, as given series is not exact half series.

15. a. General in the expansion of $(x + y + z + w)^{50}$ is

$$\frac{50!}{p!q!r!s!} x^p y^q z^r w^s$$

where $p + q + r + s = 50$, $0 \le p, q, r, s \le 50$.

Now number of terms is equal to number of ways in which we can adjust powers of x, y, z and w such that their sum is 50, i.e., equal to the non-negative solutions of $p+q+r+s = 50$, which is given by ${}^{50+4-1}C_{4-1}$.

16. a. When n persons be seated at a round table, so that all shall not have the same neighbours in any two arrangements, clockwise and anticlockwise arrangements are considered to be the same, which is the case when n different beads are arranged in the circle. Hence, number of ways is $(n-1)!/2$.

17. b. Exponent of 2 in 50! is

$$\left[\frac{50}{2}\right]+\left[\frac{50}{4}\right]+\left[\frac{50}{8}\right]+\left[\frac{50}{16}\right]+\left[\frac{50}{32}\right]=25+12+6+3+1=47.$$

And exponent of 5 in 50! is

$$\left[\frac{50}{5}\right]+\left[\frac{50}{25}\right]=12$$

Now number of zeros in the end of 50! is equal to exponent of 10 in 50! which is equal to exponent of 5 in 50! Therefore, number of zeros in the end depends on exponent of 5, but not on the exponent of 2.

Hence, both statements 1 and 2 are true; but statement 2 is not a correct explanation for statement 1.

Linked Comprehension Type

For Problems 1–3

1. a., 2. b., 3. b.

Sol.

1. When one all rounder and 10 players from bowlers and batsmen play, number of ways is ${}^4C_1\ {}^{14}C_{10}$.
When one wicketkeeper and 10 players from bowlers and batsmen play, number of ways is ${}^2C_1\ {}^{14}C_{10}$.
When one all rounder, one wicketkeeper and nine from batsmen and bowlers play, number of ways is ${}^4C_1\ {}^2C_1\ {}^{14}C_9$.
When all elven players play from bowlers and batsmen then, number of ways is ${}^{14}C_{11}$.
Total number of selections is ${}^4C_1\ {}^{14}C_{10} + {}^2C_1\ {}^{14}C_{10} + {}^4C_1\ {}^2C_1\ {}^{14}C_9 + {}^{14}C_{11}$.

2. If the particular bowler plays then two batsmen will not play. So, rest of 10 players can be selected from 17 other players. Number of such selections is ${}^{17}C_{10}$.
If the particular bowler does not play, then number of selections is ${}^{19}C_{11}$.
If all the three players do not play, then number of selections is ${}^{17}C_{11}$.
Total number of selections is ${}^{17}C_{10} + {}^{19}C_{11} + {}^{17}C_{11}$.

3. If the particular batsman is selected, then rest of 10 players can be selected in ${}^{18}C_{10}$ ways.

If particular wicketkeeper is selected, then rest of 10 players can be selected in ${}^{18}C_{10}$ ways.

If both are not selected, then number of ways is ${}^{18}C_{11}$.

Hence, total number of ways is $2\ {}^{18}C_{10} + {}^{18}C_{11}$

$= {}^{19}C_{11} + {}^{18}C_{10}$.

For Problems 4–6

4. c., 5. b., 6. a.

Sol. 4. Seven persons can be selected for first table in ${}^{12}C_7$ ways. Now these seven persons can be arranged in 6! ways. The remaining five persons can be arranged on the second table in 4! ways.Hence, total number of ways is ${}^{12}C_5 6!4!$.

5. Here, A can sit on first table and B on the second or A on second table and B on the second table.

If A is on the first table, then remaining six for first table can be selected in ${}^{10}C_6$ ways. Now these seven persons can be arranged in 6! ways. Remaining five can be arranged on the other table in 4! ways.

Hence, total number of ways is $2\ {}^{10}C_6\ 6!4!$.

6. If A, B are on the first table, then remaining five can be selected in ${}^{10}C_5$ ways.

Now seven persons including A and B can be arranged on the first table in which A and B are together in 2!5! ways. Remaining five can be arranged on the second table in 4! ways. Total number of ways is ${}^{10}C_5\ 4!5!2!$

If A, B are on the second table, then remaining three can be selected in ${}^{10}C_3$ ways.

Now five persons including A and B can be arranged on the second table in which A and B are together 2!3! ways. Remaining seven can be arranged on the first table in 6! ways. Hence, number of ways for first table is ${}^{10}C_7\ 6!3!2!$.

For Problems 7–9

7. b., 8. c., 9. d.

Sol. If no box remains empty, then we can have (1, 1, 3) or (1, 2, 2) distribution pattern.

7. When balls are different and boxes are identical, number of distributions is equal to number of divisions in (1, 2, 3) or (1, 2, 2) ways. Hence, total number of ways is

$$\frac{5!}{1!\cdot 2!\cdot 3!}+\frac{5!}{(2!)^2\,1!\cdot 2!}=25$$

8. When balls as well as boxes are identical, we have only two ways (1, 1, 3) and (1, 2, 2). Hence, number of ways is 2.

9. When boxes are kept in a row, they will be treated as different. In this case the number of ways will be ${}^{5-1}C_{3-1} = {}^4C_2 = 6$.

For Problems 10–12

10. b., 11. c., 12. b.

Sol. $6 = 0(2) + 6(1) = 1(2) + 4(1) = 2(2) + 2(1) = 3(2) + 0(1)$

Number of 2s	Number of 1s	Number of permutations
0	6	1
1	4	$\frac{5!}{4!}=5$
2	2	$\frac{4!}{2!\,2!}=6$
3	0	$\frac{3!}{3!}=1$
		Total = 13

$\therefore f(6) = 13$

Now, $f(f(6)) = f(13)$

Number of 1s	Number of 2s	Number of permutations
13	0	1
11	1	$\frac{12!}{11!}=12$
9	2	$\frac{11!}{9!2!}=55$
7	3	$\frac{10!}{7!\,3!}=120$
5	4	$\frac{9!}{5!4!}=126$
3	5	$\frac{8!}{3!5!}=56$
1	6	$\frac{7!}{6!}=7$
		Total = 377

$\therefore f(f(6)) = f(13) = 377$

$f(1) = 1$ (1)

$f(2) = 2$ (1, 1 or 2)

$f(3) = 3$ (1, 1, 1 or 2, 1 or 1, 2)

$f(4) = 5$ (explained in the paragraph)

By taking higher value of n in $f(n)$, we always get more value of $f(n)$. Hence, $f(x)$ is one-one. Clearly, $f(x)$ is into.

For Problems 13–15

13. c., 14. a., 15. d.

Sol.

13. Let x_0 denote the number of empty seats to the left of the first person, x_i $(1 \le i \le n-1)$ the number of empty seats between the ith and $(i+1)^{th}$ persons of the nth person. Then $x_0, x_n \ge 0$ and $x_i \ge 1$ for $1 \le i \le n-1$.

$$x_0 + x_1 + \cdots + x_n = (m-n) \qquad (1)$$

Putting $x_i = y_i + 1$, where $y_i \ge 0$, we have

$$x_0 + y_1 + \cdots + y_{n-1} + x_n + (1 + 1 + 1 + \cdots + (n-1) \text{ times}) = (m-n)$$

$$\Rightarrow x_0 + y_1 + \cdots + y_{n-1} + x_n = m - n - (n-1)$$

$$\Rightarrow x_0 + y_1 + \cdots + y_{n-1} + x_n = m - 2n + 1$$

Now number of non-negative integral solutions is ${}^{n+1+(m-2n+1)-1}C_{n+1-1} = {}^{m-n+1}C_n$. Since we can permute n persons in $n!$ ways, the required number of ways is

$$\left({}^{m-n+1}C_n\right)(n!) = \frac{(m-n+1)!}{n!(m-2n+1)!}\,n! = \frac{(m-n+1)!}{(m-2n+1)!}$$

14. Let $n = 2k$, where k is some positive integer. Let x_0 denote the number of empty seats to the left of the first pair, x_i $(1 \le i \le k-1)$ the number of empty seats between i^{th} and $(i+1)^{th}$ pair and x_k the number of empty seats to the right of the kth pair. Note that $x_0, x_k \ge 0$, $x_i \ge 1$ $(1 \le i \le k-1)$ and

$$x_0 + x_1 + \cdots + x_k = (m - 2k) \qquad (2)$$

The number of integral solutions of Eq. (2) is ${}^{m-2k+1}C_k$ (see the solution to problem 13).

Since we can permute n persons in $n!$ ways, the required number of ways is

$$\left({}^{m-2k+1}C_k\right)(2k)! = \frac{(m-2k+1)!}{k!(m-3k+1)!}(2k)!$$

$$= \frac{(2k)!}{(k)}\,\frac{(m-2k+1)!}{(m-3k+1)!}$$

$$= ({}^{2k}P_k)\,({}^{m-2k+1}P_k)$$

$$= ({}^{n}P_{n/2})\,({}^{m-n+1}P_{n/2})$$

15. m is even. Let $m = 2k$, where k is some positive integer. We can choose n seats out of the k seats to the left of the middle seat in ${}^kC_n$ ways. Each chosen seat can be either empty or occupied. Thus, the number of ways of choosing seats for n persons is equal to $({}^kC_n)(2^n)$. We can arrange n persons at these seats in ${}^nP_n$ ways. Hence, the required number of arrangements is given by

$$(n!)({}^kC_n)(2^n) = ({}^kP_n)(2^n) = ({}^{m/2}P_n)(2^n)$$

For Problems 16–18

16. d., 17. b., 18. c.

Sol.

16. Since there are five even places and three pairs of repeated letters, therefore at least one of these must be at an odd place. Therefore, the number of ways is 11!/(2!2!2!).

17. Make a group of both M's and another group of T's. Then except A's we have five letters remaining. So M's, T's and the letters except A's can be arranged in 7! ways. Therefore, total number of arrangements is $7! \times {}^8C_2$.

18. Consonants can be placed in 7!/(2!2!) ways. Then there are eight places and four vowels. Therefore, number of ways is

$$\frac{7!}{2!2!}\,{}^8C_4\,\frac{4!}{2!}$$

Matrix-Match Type

1. a → r; b → s; c → p; d → q.

a. ${}^{10}C_2 - {}^4C_2 + 1 = 45 - 6 + 1 = 40$

b. $1 \times {}^{10}C_2 = 45$

c. $2 \times {}^6C_2 = 30$

d. ${}^6C_2 \times 4 = 60$

2. a → q; b → r; c → s; d → p.

a. Number of rectangles is equal to number of ways we can select two vertical lines and two horizontal lines. Total number of ways is ${}^7C_2 \times {}^7C_2 = 441$.

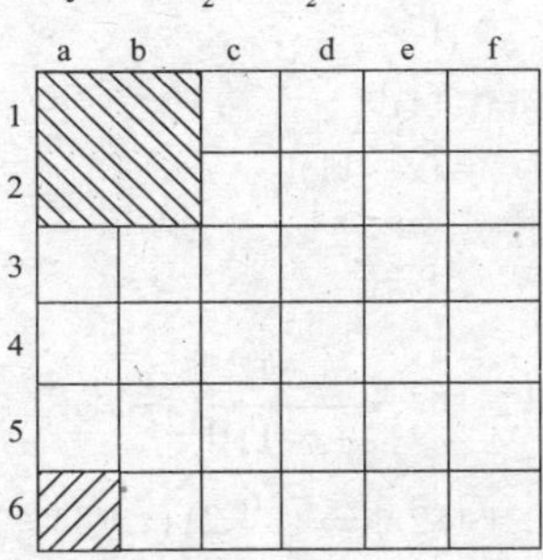

Fig. 5.13

If the square is of 1 sq. units like *a*6, then we have such 6 × 6 = 36 squares.

If the square is of 4 sq. units like shaded region of the squares *a*1, *a*2, *b*1, *b*2, then we have such five squares in the belt formed by rows 1 and 2. Similarly we have four more belts 23, 34, 45 and 56. Hence, there are 5 × 5 = 25 such squares.

Similarly we have 4 × 4, 3 × 3, 2 × 2, 1 × 1 squares of increasing sizes.

Hence, total number of squares is 1 + 4 + 9 + 16 + 25 + 36 = 91.

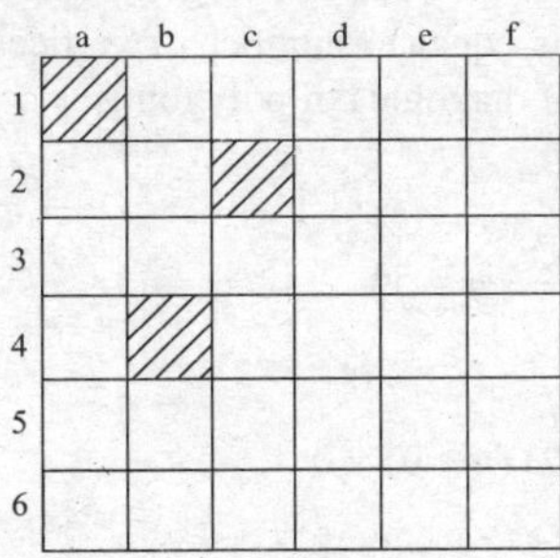

Fig. 5.14

The first square can be selected in 36 ways. If one such square *a*1 is selected, we are left with 25 squares; second square cannot be selected from row 1 and column *a*. If second square is *c*2, we are left with 16 squares, from which third square can be selected, e.g., *b*4.

Hence, number of ways of selections is 36 × 25 × 16. But in this one-by-one type of selection order of selection is also considered. Hence, actual number of ways is (36 × 25 × 16)/3! = 2400.

d. Given number of ways is equivalent to selecting 11 squares from 36 squares if no row remains empty.

Suppose $x_1, x_2, x_3, x_4, x_5, x_6$ be the number of squares selected from the 1st, 2nd, 3rd, 4th, 5th and 6th row.

Then we must have $x_1 + x_2 + x_3 + x_4 + x_5 + x_6 = 11$ (where $1 \le x_i \le 6$).

The number of positive integral solutions of the above equation is ${}^{11-1}C_{6-1} = {}^{10}C_5$.

3. a → q; b → p; c → s; d → r.

a. If polygon has *n* sides, then number of diagonals is ${}^nC_2 - n = 35$ (given). Solving we get $n = 10$. Thus, there are 10 vertices, from which ${}^{10}C_3$ (= 120) triangles can be formed.

b. Four vertices can be selected in ${}^{10}C_4$ (= 210) ways. Using these four vertices two diagonals can be formed, which has exactly one point of intersection lying inside the polygon.

Hence, number of points of intersections of diagonal which lies inside the polygon is ${}^{10}C_4 \times 1 = 210$.

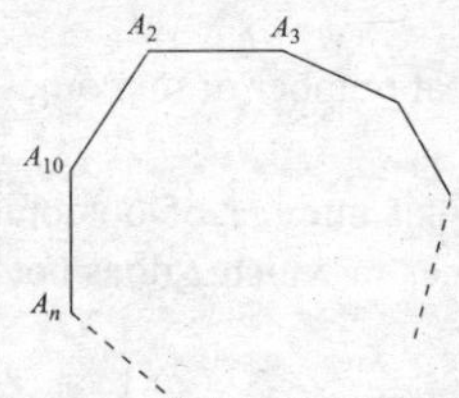

Fig. 5.15

Suppose one of the sides of the triangle is A_1A_2. Then third vertex cannot be A_3 or A_{10}. Thus, for the third vertex six vertices are left. There are six triangles in which side A_1A_2 is common with that of polygon. Similarly, for each of the sides A_2A_3, A_3A_4, ⋯, A_9A_{10} there are six triangles. Then total number of triangles is 60.

d. Triangles $A_1A_2A_3$, $A_2A_3A_4$,…,$A_8A_9A_{10}$ have two sides common with that of polygon. Hence, there are 10 such triangles.

4. a → p, q, r, s; b → p, r, s; c → q, r, s; d → s.

a. Number of subjective functions is

$3^6 - {}^3C_1(3-1)^6 + {}^3C_2(3-2)^6 = 729 - 192 + 3 = 540$

b. If $f(a_i) \ne b_i$, then pre-image a_1, a_2, a_3 cannot be assigned images b_1, b_2, b_3, respectively.

Hence, each of a_1, a_2, a_3 can be assigned images in two ways.

a_4, a_5, a_6 can be assigned images in three ways each.

Hence, number of functions is $2^3 3^3 = 216$.

c. One-one functions are not possible ar pre-images are more than images.

d. Number of many-one functions is

Total number of functions – number of one-one functions

$= 3^6 - {}^6P_3 = 729 - 120 = 609$

5. a → p, s; b → q, r; c → p, s; d → r.

a. Total number of required functions is equal to number of derangement of five objects, which is given by

$$5!\left(\frac{1}{2!} - \frac{1}{3!} + \frac{1}{4!} - \frac{1}{5!}\right) = 44$$

b. $x_1x_2x_3 = 2 \times 35 \times 7 = 2 \times 49 \times 5 = 10 \times 7 \times 7 = 14 \times 7 \times 5$

So total number of solution set is $3 \times 3! + (3!/2!) = 21$.

c. $3780 = 2^2 \times 3^3 \times 5 \times 7$

Number of divisors which are divisible by 2 but not by 3 is $2 \times 2 \times 2 = 8$.

Number of divisors which are divisible by 3 but not by 2 is $3 \times 2 \times 2 = 12$.

Number of divisors which are divisible by 2 as well as 3 is $2 \times 3 \times 2 \times 2 = 24$.

Hence total number of divisors is 44.

d. $4\lambda + 2 = 2(2\lambda + 1) =$ odd multiple of 2

Thus, total number of divisors is $1 \times 5 \times 11 - 1 = 54$. (1 is subtracted and powers of three and five are zero each and this will make $\lambda = 0$.)

6. a → r; b → p; c → s; d → q.

a. The number of possible outcomes with 2 on at least one dice

= The total number of outcomes with 2 on at least one dice

= (The total number of outcomes) – (The number of outcomes in which 2 does not appear on any dice)

= $6^4 - 5^4$

= $1296 - 625 = 671$

b. Any selection of four digits from the 10 digits 0, 1, 2, 3, ..., 9 gives one number. So, the required number of numbers is ${}^{10}C_4$.

c. Let the number be $n = pqr$. Since $p + q + r$ is even, p can be filled in 9 ways and q can be filled in 10 ways.

r can be filled in number of ways depending upon what is the sum of p and q.

If $p + q$ is odd, then r can be filled with any one of five odd digits.

If $p + q$ is even, then r can be filled with any one of five even digits.

In any case, r can be filled in five ways.

Hence, total number of numbers is $9 \times 10 \times 5 = 450$.

d. After fixing 1 at one position out of 4 places 3 places can be filled by ${}^7P_3$ ways. But for some numbers whose fourth digit is zero, such type of ways is ${}^6P_2$. Therefore, total number ways is ${}^7P_3 - {}^6P_2 = 480$.

7. a → q, s; b. → p, r; c → p, r; d → p, r.

a. There are two cases.

i. 5, 4, 1, 1, 1

Number of ways of selection is $5!/3! = 20$.

ii. 5, 2, 2, 1, 1

Number of ways of selection is $5!/(2!2!) = 30$.

Hence, total number is $20 + 30 = 50$.

b. Select 4 pairs in ${}^5C_4 = 5$ ways. Now select exactly one shoe from each of the pairs selected in $({}^2C_1)^4$ ways. This will fulfill the condition. Hence, required answer is $5 \times 16 = 80$.

c. The first child C_1 can be chosen in 3 ways; his/her mother can be interviewed in 5 ways; the second child C_2 can be chosen in 2 ways, and his/her mother can be interviewed in 3 ways.

Hence total number of ways is $3 \times 5 \times 2 \times 3 = 90$.

d. Required number of ways is $5! - 4! - 3! = 120 - 64 - 6 = 90$.

(Number will be less than 56000 only if either 4 occurs on the first place or 5, 4 occurs on the first two places.)

8. a → r, s; b → p, r; c → p, q; d → r, s

a. We have,

$a = {}^{x+2}P_{x+2} = (x + 2)!$,

$b = {}^xP_{11} = \dfrac{x!}{(x-11)!}$

$c = {}^{x-11}P_{x-11} = (x - 11)!$

Now,

$a = 182\, bc$

$\Rightarrow (x + 2)! = 182 \times \dfrac{x!}{(x-11)!}(x-11)!$

$\Rightarrow (x + 2)! = 182x! \Rightarrow (x + 2)(x + 1) = 182$

$\Rightarrow x = 12$

b. ×|×|×|

Even digits occupy odd places shown by crosses. Crosses can be filled in $2 \times 2 \times 1$ ways (Since 0 cannot go in the first place from the left). The remaining places can be filled in 3! ways.

Therefore, the required number of numbers is $2 \times 2 \times 1 \times 3! = 24$.

c. Total number of numbers without restriction is 2^5. Two numbers have all the digits equal. So, the required number of numbers is $2^5 - 2$.

d. Let number of sides of polygon be n. Number of sides of polygon is equal to number of vertices of polygon. Now number of diagonals of polygon is

${}^nC_2 - n = 54$

$\Rightarrow \dfrac{n(n-1)}{2} - n = 54$

$\Rightarrow n^2 - 3n - 162 = 0$

$\Rightarrow (n - 12)(n + 9) = 0$

$\Rightarrow n = 12$

Integer Type

1. **(7)** $S = 1 + 4 + 36 + 576 + \cdots + (2008!)^2 = 617 +$ all other terns end in zero. Hence, digit at the unit place is 7.

2. **(7)** As $0 < x < 1$, we have $p < q$

p	q
1	2, 3, 4, 5, 6
2	3, 4, 5, 6

3	4, 5, 6
4	5, 6
5	6

Thus the number of rational numbers is equal to 5 + 4 + 3 + 2 + 1 = 15

When p, q have a common factor, we get some rational numbers which are not different from those already counted.

There are four such numbers, viz, $\frac{2}{4}, \frac{2}{6}, \frac{3}{6}, \frac{4}{6}$. Therefore, is equal to the required number of rational numbers = 15 – 4 = 11.

3. **(9)** $$\begin{array}{r} x+a \\ x^2+1\,\overline{)\,x^3+ax^2+bx+c} \\ x^3+x \\ \hline ax^2+(b \quad 1)\,x+c \\ ax^2+a \\ \hline (b \quad 1)\,x+c \quad a \end{array}$$

Now the remainder $(b–1)\,x + c – a$ must be zero for any x.

Then $b – 1 = 0$ and $c – a = 0$

$\Rightarrow b = 1$ and $c = a$

c or a can be selected in 9 ways.

Hence the number of polynomials is 9.

4. **(3)**

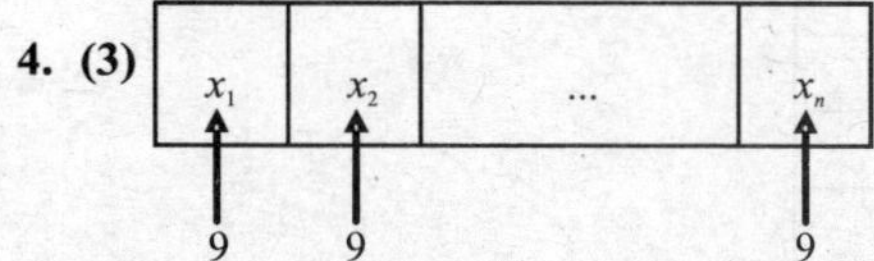

The digit 'x_1' can be selected in 9 ways as 0 cannot be selected.

The digit 'x_2' can be selected in 9 ways as 0 can be selected but the digit in position x_1 cannot be selected.

Similarly, all the remaining digits can be selected in 9 ways each.

Thus the total number of such numbers is equal to 9^n.

5. **(4)** $\lim\limits_{x\to 0}\left(\frac{1+a^x+b^x-2}{2}\right)^{\frac{2}{a^x+b^x-2}\lim\limits_{x\to 0}\left(\frac{a^x-1+b^x-1}{x}\right)} = e^{\ln ab} = ab$

$= 6$

Total number of possible ways in which a, b can take values = 6 × 6 = 36.

Total possible ways are (1, 6), (6, 1), (2, 3), (3, 2).

Total number of possible ways = 4.

6. **(6)** $f'(x) = 3x^2 + 2ax + 9$ and
$y = f(x)$ is increasing
$\Rightarrow f'(x) \geq 0,\ \forall x$ and for $f'(x) = 0$ should not form an interval

$\Rightarrow (2a)^2 - 4 \times 3 \times b \leq 0 \Rightarrow a^2 - 3b \leq 0$

this is true for exactly 16 ordered pairs (a, b), $1 \leq a, b \leq 6$.

7. **(5)** The number of zeros at the end of 130! is equal to the exponent of 10 in 130!.

Now the exponent of 10 is equal to the exponent of 5 as the exponent of 2 is higher than the exponent of 5.

Now the exponent of 5 is $\left[\frac{130}{5}\right]+\left[\frac{130}{5^2}\right]+\left[\frac{130}{5^3}\right] = 26 + 5 + 1 = 32$.

Hence, the exponent of 10 is also 32 and hence there are 32 zeros in the end of 130!

It should be noted that the exponent of 2 is equal to

$\left[\frac{130}{2}\right]+\left[\frac{130}{2^2}\right]+\left[\frac{130}{2^3}\right]+\left[\frac{130}{2^4}\right]+\left[\frac{130}{2^5}\right]+\left[\frac{130}{2^6}\right]+\left[\frac{130}{2^7}\right]$

$= 65 + 32 + 16 + 8 + 4 + 2 + 1 = 128$

Hence, the exponent of 10 is equal to exponent of 5.

8. **(3)** Let $N = 1! + 2! + 3! + 4! + 5! + 6! \cdots + n!$

Now $\frac{N}{15} = \frac{1! + 2! + 3! + 4! + 5! + \cdots + n!}{15}$

$= \frac{1! + 2! + 3! + 4!}{15} + \frac{5! + 6! \cdots n!}{15}$

$= \frac{33}{15}$ + integer (as 5!, 6! ⋯ are divisible by 15)

Hence, remainder = 3.

9. **(6)** We have,

${}^{2n}C_3 : {}^{n}C_3 = 11{:}1$

$\Rightarrow \frac{{}^{2n}C_3}{{}^{n}C_3} = \frac{1}{1}$

$\Rightarrow \frac{\frac{(2n)!}{(2n-3)!\,(3!)}}{\frac{n!}{(n-3)!\,(3!)}} = \frac{1}{1}$

$\Rightarrow \frac{(2n)!}{(2n-3)!} \times \frac{(n-3)!}{n!} = \frac{1}{1}$

$\Rightarrow \frac{(2n)\,(2n-1)\,(2n-2)}{n\,(n-1)\,(n-2)} = \frac{1}{1}$

$\Rightarrow \frac{4\,(2n-1)}{n-2} = \frac{1}{1}$

$\Rightarrow 8n - 4 = 11n - 22$

$\Rightarrow 3n = 18 \Rightarrow n = 6$

10. **(8)** The number of parties if he invites r friends at a time is ${}^{16}C_r$. Obviously, for the maximum number of parties, he must invite 8 friends at time.

11. **(3)** ${}^{15}C_{3r} : {}^{15}C_{r+1} = 11 : 3$

Clearly 'r' can be 0, 1, 2, 3, 4, and 5 But the possibilities of $r = 0$ or 5 are clearly ruled out (as ${}^{15}C_0 = {}^{15}C_{15} = 1$).

For r = 1,

${}^{15}C_{3r} = {}^{15}C_3 = \frac{15 \times 14 \times 13}{6}$ and ${}^{15}C_{r+1} = \frac{15 \times 14}{2}$.

Therefore,

${}^{15}C_{3r} : {}^{15}C_{r+1} = 13 : 3$

For r = 2,

${}^{15}C_{3r} = {}^{15}C_6 = \frac{15 \times 14 \times 13 \times 12 \times 11 \times 10}{6 \times 5 \times 4 \times 3 \times 2 \times 1}$

and ${}^{15}C_{r+1} = {}^{15}C_3 = \frac{15 \times 14 \times 13}{6}$

$\Rightarrow {}^{15}C_{3r} : {}^{15}C_{r+1} \neq 11 : 3$

For r = 3,

$${}^{15}C_{3r} = {}^{15}C_9 = \frac{15 \times 14 \times 13 \times 12 \times 11 \times 10}{6 \times 5 \times 4 \times 3 \times 2 \times 1}$$

and $${}^{15}C_{r+1} = {}^{15}C_4 = \frac{15 \times 14 \times 13 \times 12}{4 \times 3 \times 2 \times 1}$$

$\Rightarrow {}^{15}C_{3r} : {}^{15}C_{r+1} = 11 : 3$

For r = 4,

$${}^{15}C_{3r} = {}^{15}C_{12} = {}^{15}C_3 = \frac{15 \times 14 \times 13}{3 \times 2 \times 1}$$

and $${}^{15}C_{r+1} = {}^{15}C_5 = \frac{15 \times 14 \times 13 \times 12 \times 11}{5 \times 4 \times 3 \times 2 \times 1}$$

$\Rightarrow {}^{15}C_{3r} : {}^{15}C_{r+1} = 5 : 33$

Thus $r = 3$

12. (9) Given letters are I, (N, N, N), (D, D), (E, E, E), P, T.

The choice are as follows:

Choice	Ways
All the letters are distinct [∵ different letters are I, N, D E, P, T]	${}^6C_5 = 6$
3 distinct, 2 alike	${}^3C_1 \times {}^5C_3 = 30$
2 distinct, 3 alike	${}^2C_1 \times {}^5C_2 = 20$
2 alike, 2 alike, 1 distinct	${}^3C_2 \times {}^4C_1 = 12$
3 alike, 2 alike	${}^2C_1 \times {}^2C_1 = 4$
	Total = 72

13. (4) We have ${}^{2n+1}C_1 + {}^{2n+1}C_2 + \cdots + {}^{2n+1}C_n = 255$ (1)

Also the sum of binomial coefficients is equal to $= {}^{2n+1}C_0 + {}^{2n+1}C_1 + \cdots + {}^{2n+1}C_n + {}^{2n+1}C_{n+1} + \cdots + {}^{2n+1}C_{2n+1} = (1+1)^{2n+1} = 2^{2n+1}$

$\Rightarrow {}^{2n+1}C_0 + 2({}^{2n+1}C_1 + {}^{2n+1}C_2 + \cdots + {}^{2n+1}C_n) + {}^{2n+1}C_{2n+1} = 2^{2n+1}$

$\Rightarrow 1 + 2(255) + 1 = 2^{2n+1}$

$\Rightarrow 1 + 255 = 2^{2n}$

$\Rightarrow 2^{2n} = 2^8 \Rightarrow n = 4$

14. (6) The number of ways of distributing blankets is equal to the number of solutions of the equation $3a + 2b + c = 15$ $a, b, c \geq 1$, which is equal to the coefficient of t^{15} in

$(t^3 + t^6 + t^9 + t^{12} \cdots)(t^2 + t^4 + \cdots)(t + t^2 + \cdots)$

= coefficient of t^9 in $(1 + t^3 + t^6 + t^9)(1 + t^2 + t^4 + t^6 + t^8)(1 + t + t^2 + \cdots + t^9)$ (neglecting higher powers)

= coefficient of t^9 in $(1 + t^2 + t^4 + t^5 + 2t^6 + t^7 + 2t^8 + 2t^9)(1 + t + t^2 + \cdots + t^9)$

$= 1 + 1 + 1 + 1 + 1 + 1 + 2 + 1 + 2 + 2 = 12.$

15. (8) The whole family has 24 children.

Children of Geeta and Sohan are $24 - x - (x + 1) = 23 - 2x$ in number.

When a Sohan-child fights with a Geeta-child there are $x(x + 1)$ fights. When a Sohan-child fights with Sohan-Geeta child, there are $x(23 - 2x)$ fights. When a Geeta-child fights with a Sohan-Geeta child, there are

$(x + 1)(23 - 2x)$ fights. Therefore, total number of fights

$F(x) = x(x + 1) + x(23 - 2x) + (x + 1)(23 - 2x)$

$= 23 + 45x - 3x^2$

$$= 3\left[\frac{23}{3} + \frac{225}{4} - \left(x - \frac{15}{2}\right)^2\right]$$

For $F(x)$ to be maximum $x - \frac{15}{2} = 0$, i.e., $x = 7.5$

Since x is an integer, $x = 7$ or 8

for both $x = 7$ or $x = 8$.

16. (7) Distinct n digit numbers which can be formed using 2, 5, and 8 are 3^n.

We have to find n so that $3^n \geq 900$

$\Rightarrow 3^{n-2} \geq 100$

$\Rightarrow n - 2 \geq 5$

$\Rightarrow n \geq 7.$

So the least value of n is 7.

17. (7) A regular polygon of n sides has n vertices, no two of which are collinear. Out of these n points, ${}^nC_3$ triangles can be formed. Therefore,

$T_n = {}^nC_3; T_{n+1} = {}^{n+1}C_3.$

Given $T_{n+1} - T_n = 21$. Therefore,

$\Rightarrow {}^{n+1}C_3 - {}^nC_3 = 21$

$$\frac{(n+1)n(n-1)}{3 \times 2 \times 1} - \frac{n(n-1)(n-2)}{3 \times 2 \times 1} = 21$$

$\Rightarrow n(n-1)(n+1-n+2) = 126$

$\Rightarrow n(n-1) = 42$

$\Rightarrow n(n-1) = 7 \times 6$

$\Rightarrow n = 7$

18. (9) If the L.C.M. of p and q is $r^2t^4s^2$ then distribution of factors r is

p	q
r^0	r^2
r^1	r^2
r^2	r^2
r^2	r^0
r^2	r^1

Thus factor r can be distributed in $2 \times 3 - 1$ ways.

Similarly, factor t and s can be distributed in $2 \times 5 - 1$ and $2 \times 3 - 1$ ways, respectively.

Hence, the number of ordered pairs is $(2 \times 3 - 1) \times (2 \times 5 - 1) \times (2 \times 3 - 1) = 225$.

19. (8) Now, $(n!)!$ is the product of the positive integers from 1 to $n!$. We write the integers from 1 to $n!$ in $(n - 1)!$ rows as follows;

$1.\ 2.\ 3. \cdots n$

$(n + 1)(n + 2)(n + 3) \cdots (2n)$

$(2n + 1)(2n + 2) \cdots (3n)$

$(3n + 1)(3n + 2) \cdots (4n)$

$\cdots \quad \cdots \quad \cdots$

$\cdots \quad \cdots \quad \cdots$

$(n! - n + 1)(n! - n + 2)\ (n(n - 1)!)$

Each of these $(n - 1)!$ rows contains n consecutive positive integers.

The product of consecutive integers is each row in divisible by $n!$.

Thus the product of all the integers from 1 to $n!$ is divisible by $(n!)^{(n-1)!}$

Hence, the value of p is 8

20. **(9)** Let the number of members be n.

Total number of points ${}^nC_2$

${}^nC_2 - 17.5 = (n-4)\,x$ (where x is the number of point scored by each player

$\Rightarrow n(n-1) - 35 = 2(n-4)\,x$

$\Rightarrow 2x = \dfrac{n(n-1)-35}{n-4}$ (where x takes the values 0.5, 1, 1.5 etc)

$= \dfrac{n^2-n-35}{n-4}$

$= \dfrac{n(n-4)+3(n-4)-23}{n-4}$

$= (n+3) - \dfrac{23}{n-4}$

$\Rightarrow \dfrac{23}{n-4}$ must be an integer

$\Rightarrow n = 27$

21. **(3)** Number of ways in which a student can select at least one and almost n books out of $(2n+1)$ books is equal to $= {}^{2n+1}C_1 + {}^{2n+1}C_2 + {}^{2n+1}C_3 + \cdots + {}^{2n+1}C_n$

$= \frac{1}{2}[2 \times {}^{2n+1}C_1 + 2 \times {}^{2n+1}C_2 + 2 \times {}^{2n+1}C_3 + \cdots + 2 \times {}^{2n+1}C_n]$

$= \frac{1}{2}[({}^{2n+1}C_1 + {}^{2n+1}C_{2n}) + ({}^{2n+1}C_2 + {}^{2n+1}C_{2n-1}) + ({}^{2n+1}C_3 + {}^{2n+1}C_{2n-2}) + \ldots + ({}^{2n+1}C_n + {}^{2n+1}C_{n+1})]$ [Using ${}^nc_r = {}^nc_{n-r}$]

$= \frac{1}{2}[{}^{2n+1}C_1 + {}^{2n+1}C_2 + {}^{2n+1}C_3 + \cdots + {}^{2n+1}C_n + {}^{2n+1}C_{n+1} + {}^{2n+1}C_{n+2} + \cdots + {}^{2n+1}C_{2n}]$

$= \frac{1}{2}[{}^{2n+1}C_0 + {}^{2n+1}C_1 + {}^{2n+1}C_2 + \ldots + {}^{2n+1}C_{2n+1} - 1 - 1]$

$= \frac{1}{2}[2^{2n+1} - 2] = 2^{2n} - 1$

now given $2^{2n} - 1 = 63$

$\Rightarrow 2^{2n} = 64 = 2^6$

$\Rightarrow 2n = 6$

$\Rightarrow n = 3$

22. **(9)** We know that the number of derangements of n objects is equal to

$= n!\left[1 - \frac{1}{1!} + \frac{1}{2!} - \frac{1}{3!} + \frac{1}{4!} - \cdots \frac{1}{n!}\right].$

Therefore number of ways of putting all the 4 balls into boxes of different colour

$= 4!\left[1 - \frac{1}{1!} + \frac{1}{2!} - \frac{1}{3!} + \frac{1}{4!}\right]$

$= 4!\left(\frac{1}{2} - \frac{1}{6} + \frac{1}{24}\right)$

$= 24\left(\frac{12-4+1}{24}\right)$

$= 9$

23. **(9)** Let the seat occupied by A be numbered as 1 and the remaining five seats be numbered as 2, 3, 4, 5, and 6 anticlockwise. Then there arises two cases.

Case I. B is on the right of A, i.e. at no. 2.

Then seat number 3 can be occupied by C or D in ${}^2C_1$ ways and the remaining three persons can have the remaining three seats in 3! ways.

Therefore number of arrangements in this case $= 2 \times 6 = 12$ (1)

Case II. C is on the right of A, i.e., at no. 2.

Then B can occupy any seat from number 3 or 4 or 5. Then D must be on the right of B. So we are left with two persons and two seats which can be occupied in 2! ways.

Therefore number of arrangements in this case $= {}^3C_1 \times 2! = 6$ (2)

These cases are exclusive. So by sum rule, total number of arrangements $= 12 + 6 = 18$.

24. **(7)** Giving any random configuration for $f(1)$ to $f(5)$, number of functions $= 4^5$ ways (each element from 1 to 5 can be mapped in 4 ways)

Now the sum $\sum_{i=1}^{5} f(1)$ is either odd or even. If it is odd then we have only two choices for $f(6)$, i.e., element 8 or 10.

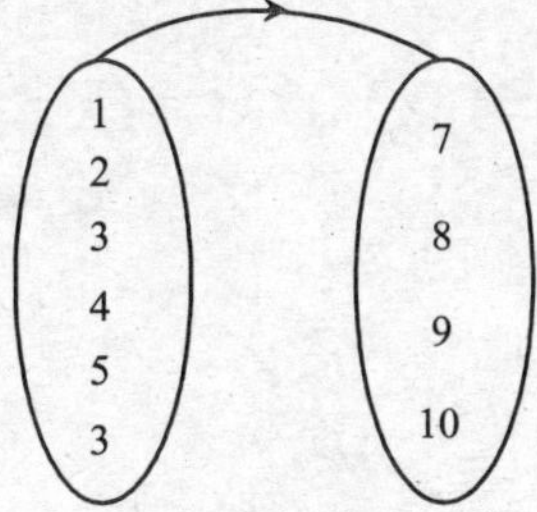

Fig. 5.16

(as odd + even = odd).

If the sum is even then also we have only choices for $f(6)$, i.e,. 7 or 9.

Hence the total functions $= 4^5 \times 2 = 2^{11}$.

25. **(7)** x denotes the number of times he can take unit step and y denotes the number of times he can take 2 steps. Therefore, $x + y = 7$

Put $x = 1, 2, 5, 7$

If $x = 1$, 1 2 2 2 $\Rightarrow \frac{4!}{3!} = 4$

$x = 3$, 1 1 1 2 2 $\Rightarrow \frac{5!}{2! \times 3!} = 10$

$x = 5$, 1 1 1 1 1 2 $\Rightarrow {}^6C_1 = 6$

$x = 7$, 1 1 1 1 1 1 1 $\Rightarrow {}^7C_0 = 1$

26. **(9)** Multiplying by 7 On both the sides, we get

$3600 = 2520a_2 + 840a_3 + 210a_4 + 42a_5 + 7a_6 + a_7$

$3600 - a_7$ is a multiple of 7!, which implies that $a_7 = 2$

Thus $\frac{3598}{7} = 514 = 360a_2 + 120a_3 + 30a_4 + 6a_5 + a_6$

$\Rightarrow 514 - a_6$ is a multiple of 6,

$\Rightarrow a_6 = 4$

Thus $\frac{510}{6} = 85 = 60a_2 + 20a_3 + 5a_4 + a_5$

Then it follows that $85 - a_5$ is a multiple of 5, where $a_5 = 0$, Continuing in this process, we obtain $a_4 = 1$, $a_3 = 1$, $a_2 = 1$. Thus the desired sum is

$1 + 1 + 1 + 0 + 4 + 2 = 9$

CHAPTER 6

Binomial Theorem

Objective Type

Each question has four choices a, b, c and d, out of which *only* *one* is correct.

1. If the coefficients of 5th, 6th, and 7th terms in the expansion of $(1 + x)^n$ be in A.P., then $n =$

a. 7 only **b.** 14 only

c. 7 or 14 **d.** none of these

2. If x^m occurs in the expansion of $(x + 1/x^2)^{2n}$, then the coefficient of x^m is

a. $\dfrac{(2n)!}{(m)!(2n-m)!}$

b. $\dfrac{(2n)!3!3!}{(2n-m)!}$

c. $\dfrac{(2n)!}{\left(\dfrac{2n-m}{3}\right)!\left(\dfrac{4n+m}{3}\right)!}$

d. none of these

3. The coefficient of x^5 in the expansion of $(1+x)^{21} + (1+x)^{22} + \cdots + (1+x)^{30}$ is

a. ${}^{51}C_5$ **b.** ${}^{9}C_5$

c. ${}^{31}C_6 - {}^{21}C_6$ **d.** ${}^{30}C_5 + {}^{20}C_5$

4. The coefficient of $1/x$ in the expansion of $(1 + x)^n (x + 1/x)^n$ is

a. $\dfrac{n!}{(n-1)!(n+1)!}$ **b.** $\dfrac{(2n)!}{(n-1)!(n+1)!}$

c. $\dfrac{(2n)!}{(2n-1)!(2n+1)!}$ **d.** none of these

5. In the expansion of $(1 + 3x + 2x^2)^6$, the coefficient of x^{11} is

a. 144 **b.** 288

c. 216 **d.** 576

6. If the coefficients of rth and $(r + 1)^{\text{th}}$ terms in the expansion of $(3 + 7x)^{29}$ are equal, then r equals

a. 15 **b.** 21

c. 14 **d.** none of these

7. If the coefficients of three consecutive terms in the expansion of $(1 + x)^n$ are in the ratio 1:7:42, then the value of n is

a. 60 **b.** 70

c. 55 **d.** none of these

8. If $(1 + 2x + x^2)^n = \sum_{r=0}^{2n} a_r x^r$, then $a_r =$

a. $({}^nC_r)^2$ **b.** ${}^nC_r \times {}^nC_{r+1}$

c. ${}^{2n}C_r$ **d.** ${}^{2n}C_{r+1}$

9. If in the expansion of $(1 + x)^n$, a, b, c are three consecutive coefficients, then $n =$

a. $\dfrac{ac+ab+bc}{b^2+ac}$ **b.** $\dfrac{2ac+ab+bc}{b^2-ac}$

c. $\dfrac{ab+ac}{b^2-ac}$ **d.** none of these

10. The coefficient of the middle term in the binomial expansion in powers of x of $(1 + \alpha x)^4$ and of $(1 - \alpha x)^6$ is the same, if α equals

a. $-\dfrac{5}{3}$ **b.** $\dfrac{10}{3}$

c. $-\dfrac{3}{10}$ **d.** $\dfrac{3}{5}$

11. The coefficient of x^n in the expansion of $(1 - x)(1 - x)^n$ is

a. $n-1$ b. $(-1)^n(1+n)$
c. $(-1)^{n-1}(n-1)^2$ d. $(-1)^{n-1}n$

12. If the coefficients of rth, $(r+1)^{th}$, and $(r+2)^{th}$ terms in the binomial expansion of $(1+y)^m$ are in A.P., then m and r satisfy the equation
a. $m^2 - m(4r+1) + 4r^2 + 2 = 0$
b. $m^2 - m(4r-1) + 4r^2 - 2 = 0$
c. $m^2 - m(4r-1) + 4r^2 + 2 = 0$
d. $m^2 - m(4r+1) + 4r^2 - 2 = 0$

13. If the coefficient of x^7 in $\left[ax^2 + \left(\frac{1}{bx}\right)\right]^{11}$ equals the coefficient of x^{-7} in $\left[ax - \left(\frac{1}{bx^2}\right)\right]^{11}$, then a and b satisfy the relation
a. $a+b=1$ b. $a-b=1$
c. $ab=1$ d. $\frac{a}{b}=1$

14. The coefficient of $a^8b^4c^9d^9$ in $(abc+abd+acd+bcd)^{10}$ is
a. 10! b. $\frac{10!}{8!4!9!9!}$
c. 2520 d. none of these

15. The coefficient of x^5 in the expansion of $(x^2-x-2)^5$ is
a. –83 b. –82
c. –86 d. –81

16. The coefficient of x^2y^3 in the expansion of $(1-x+y)^{20}$ is
a. $\frac{20!}{2!3!}$ b. $-\frac{20!}{2!3!}$
c. $\frac{20!}{5!2!3!}$ d. none of these

17. If the term independent of x in the $\left(\sqrt{x} - \frac{k}{x^2}\right)^{10}$ is 405, then k equals
a. 2, –2 b. 3, –3
c. 4, –4 d. 1, –1

18. The coefficient of x^{10} in the expansion of $(1+x^2-x^3)^8$ is
a. 476 b. 496
c. 506 d. 528

19. If the coefficient of x^n in $(1+x)^{101}(1-x+x^2)^{100}$ is non-zero, then n cannot be of the form
a. $3r+1$ b. $3r$
c. $3r+2$ d. none of these

20. The coefficient of x^{28} in the expansion of $(1+x^3-x^6)^{30}$ is
a. 1 b. 0
c. ${}^{30}C_6$ d. ${}^{30}C_3$

21. The term independent of a in the expansion of $\left(1+\sqrt{a}+\frac{1}{\sqrt{a}-1}\right)^{-30}$ is
a. ${}^{30}C_{20}$ b. 0
c. ${}^{30}C_{10}$ d. none of these

22. The coefficient of x^{53} in the expansion $\sum_{m=0}^{100} {}^{100}C_m (x-3)^{100-m} 2^m$ is
a. ${}^{100}C_{47}$ b. ${}^{100}C_{53}$
c. $-{}^{100}C_{53}$ d. $-{}^{100}C_{100}$

23. The coefficient of the term independent of x in the expansion of $\left(\frac{x+1}{x^{2/3}-x^{1/3}+1} - \frac{x-1}{x-x^{1/2}}\right)^{10}$ is
a. 210 b. 105
c. 70 d. 112

24. In the expansion of $(1+x+x^3+x^4)^{10}$, the coefficient of x^4 is
a. ${}^{40}C_4$ b. ${}^{10}C_4$
c. 210 d. 310

25. The approximate value of $(1.0002)^{3000}$ is
a. 1.6 b. 1.4
c. 1.8 d. 1.2

26. Last three digits of the number $N = 7^{100} - 3^{100}$ are
a. 100 b. 300
c. 500 d. 000

27. The expression
$$\left(\sqrt{2x^2+1}+\sqrt{2x^2-1}\right)^6 + \left(\frac{2}{\sqrt{2x^2+1}+\sqrt{2x^2-1}}\right)^6$$ is a polynomial of degree
a. 6 b. 8
c. 10 d. 12

28. The coefficient of $x^r[0 \le r \le (n-1)]$ in the expansion of $(x+3)^{n-1} + (x+3)^{n-2}(x+2) + (x+3)^{n-3}(x+2)^2 + \cdots + (x+2)^{n-1}$ are
a. ${}^nC_r(3^r - 2^n)$ b. ${}^nC_r(3^{n-r} - 2^{n-r})$
c. ${}^nC_r(3^r + 2^{n-r})$ d. none of these

29. If $(1+2x+3x^2)^{10} = a_0 + a_1x + a_2x^2 + \cdots + a_{20}x^{20}$, then a_1 equals
a. 10 b. 20
c. 210 d. none of these

30. In the expansion of $(5^{1/2}+7^{1/8})^{1024}$, the number of integral terms is
a. 128 b. 129
c. 130 d. 131

31. The total number of terms which are dependent on the value of x, in the expansion of $\left(x^2 - 2 + \frac{1}{x^2}\right)^n$ is equal to
a. $2n+1$ b. $2n$
c. n d. $n+1$

32. If the 6th term in the expansion of $\left(\frac{1}{x^{8/3}} + x^2\log_{10}x\right)^8$ is 5600, then x equals

a. 1 b. $\log_e 10$

c. 10 d. x does not exist

33. If n is an integer between 0 and 21, then the minimum value of $n!(21-n)!$ is attained for $n =$

a. 1 b. 10

c. 12 d. 20

34. In the expansion of $(3^{-x/4} + 3^{5x/4})^n$ the sum of binomial coefficient is 64 and term with the greatest binomial coefficient exceeds the third by $(n-1)$, the value of x must be

a. 0 b. 1

c. 2 d. 3

35. If the last term in the binomial expansion of $\left(2^{1/3} - \frac{1}{\sqrt{2}}\right)^n$ is $\left(\frac{1}{3^{5/3}}\right)^{\log_3 8}$, then the 5th term from the beginning is

a. 210 b. 420

c. 105 d. none of these

36. If ${}^{n+1}C_{r+1} : {}^{n}C_{r} : {}^{n-1}C_{r-1} = 11:6:3$, then $nr =$

a. 20 b. 30

c. 40 d. 50

37. The value of x for which the sixth term in the expansion of

$$\left[2^{\log_2 \sqrt{9^{x-1}+7}} + \frac{1}{2^{\frac{1}{5}\log_2(3^{x-1}+1)}}\right]^7$$ is 84 is

a. 4 b. 1 or 2

c. 0 or 1 d. 3

38. The number of integral terms in the expansion of $(\sqrt{3} + \sqrt[8]{5})^{256}$ is

a. 33 b. 34

c. 35 d. none of these

39. The number of real negative terms in the binomial expansion of $(1 + ix)^{4n-2}$, $n \in N$, $x > 0$ is

a. n b. $n + 1$

c. $n - 1$ d. $2n$

40. If in the expansion of $(a - 2b)^n$, the sum of 5th and 6th terms is 0, then the values of $a/b =$

a. $\frac{n-4}{5}$ b. $\frac{2(n-4)}{5}$

c. $\frac{5}{n-4}$ d. $\frac{5}{2(n-4)}$

41. The number of distinct terms in the expansion of $\left(x + \frac{1}{x} + x^2 + \frac{1}{x^2}\right)^{15}$ is/are

a. 255 b. 61

c. 127 d. none of these

42. The sum of the coefficients of even powers of x in the expansion of $(1 + x + x^2 + x^3)^5$ is

a. 256 b. 128

c. 512 d. 64

43. Maximum sum of coefficient in the expansion of $(1 - x\sin\theta + x^2)^n$ is

a. 1 b. 2^n

c. 3^n d. 0

44. If the sum of the coefficients in the expansion of $(a + b)^n$ is 4096, then the greatest coefficient in the expansion is

a. 924 b. 792

c. 1594 d. none of these

45. If the sum of the coefficients in the expansion of $(1 - 3x + 10x^2)^n$ is a and if the sum of the coefficients in the expansion of $(1 + x^2)^n$ is b, then

a. $a = 3b$ b. $a = b^3$

c. $b = a^3$ d. none of these

46. If $(1 + x - 2x^2)^6 = 1 + a_1x + a_2x^2 + a_3x^3 + \cdots$, then the value of $a_2 + a_4 + a_6 + \cdots + a_{12}$ will be

a. 32 b. 31

c. 64 d. 1024

47. The fractional part of $2^{4n}/15$ is $(n \in N)$

a. $\frac{1}{15}$ b. $\frac{2}{15}$

c. $\frac{4}{15}$ d. none of these

48. The value of ${}^{15}C_0^2 - {}^{15}C_1^2 + {}^{15}C_2^2 - \cdots - {}^{15}C_{15}^2$ is

a. 15 b. -15

c. 0 d. 51

49. The value of

$$\binom{30}{0}\binom{30}{10} - \binom{30}{1}\binom{30}{11} + \binom{30}{2}\binom{30}{12} + \cdots + \binom{30}{20}\binom{30}{30} =$$

a. ${}^{60}C_{20}$ b. ${}^{30}C_{10}$

c. ${}^{60}C_{30}$ d. ${}^{40}C_{30}$

50. If $f(x) = x^n$, then the value of $f(1) + \frac{f^1(1)}{1} + \frac{f^2(1)}{2!} + \cdots + \frac{f^n(1)}{n!}$, where $f^r(x)$ denotes the rth order derivative of $f(x)$ with respect to x, is

a. n b. 2^n

c. 2^{n-1} d. none of these

51. The value of ${}^{20}C_0 + {}^{20}C_1 + {}^{20}C_2 + {}^{20}C_3 + {}^{20}C_4 + {}^{20}C_{12} + {}^{20}C_{13} + {}^{20}C_{14} + {}^{20}C_{15}$ equals to

a. $2^{19} - \frac{({}^{20}C_{10} + {}^{20}C_9)}{2}$

b. $2^{19} - \frac{({}^{20}C_{10} + 2 \times {}^{20}C_9)}{2}$

c. $2^{19} - \frac{{}^{20}C_{10}}{2}$

d. none of these

52. The value of $\frac{{}^nC_0}{n}+\frac{{}^nC_1}{n+1}+\frac{{}^nC_2}{n+2}+\cdots+\frac{{}^nC_n}{2n}$ is not equal to

a. $\int_0^1 x^{n-1}(1-x)^n dx$ b. $\int_1^2 x^n(x-1)^{n-1} dx$

c. $\int_1^2 x^{n-1}(1+x)^n dx$ d. $\int_1^2 x^n(x+1)^{n-1} dx$

53. If $C_0, C_1, C_2, \ldots, C_n$ are the binomial coefficients, then $2\times C_1+2^3\times C^3+2^{5\times}C_5+\cdots$ equals

a. $\frac{3^n+(-1)^n}{2}$ b. $\frac{3^n-(-1)^n}{2}$

c. $\frac{3^n+1}{2}$ d. $\frac{3^n-1}{2}$

54. The value of $\sum_{r=1}^{n}(-1)^{r+1}\frac{{}^nC_r}{r+1}$ is equal to

a. $-\frac{1}{n+1}$ b. $-\frac{1}{n}$

c. $\frac{1}{n+1}$ d. $\frac{n}{n+1}$

55. $(1+x)^n = C_0+C_1x+C_2x^2+\cdots+C_nx^n$ then $C_0C_2+C_1C_3+C_2C_4+\cdots+C_{n-2}C_n=$

a. $\frac{(2n)!}{(n!)^2}$ b. $\frac{(2n)!}{(n-1)!(n+1)!}$

c. $\frac{(2n)!}{(n-2)!(n+2)!}$ d. none of these

56. The sum of series ${}^{20}C_0-{}^{20}C_1+{}^{20}C_2-{}^{20}C_3+\cdots+{}^{20}C_{10}$ is

a. $\frac{1}{2}\,{}^{20}C_{10}$ b. 0

c. ${}^{20}C_{10}$ d. $-{}^{20}C_{10}$

57. ${}^{404}C_4-{}^4C_1{}^{303}C_4+{}^4C_2{}^{202}C_4-{}^4C_3{}^{101}C_4$ is equal to

a. $(401)^4$ b. $(101)^4$

c. 0 d. $(201)^4$

58. If $(3+x^{2008}+x^{2009})^{2010}=a_0+a_1x+a_2x^2+\cdots+a_nx^n$, then the value of $a_0-\frac{1}{2}a_1-\frac{1}{2}a_2+a_3-\frac{1}{2}a_4-\frac{1}{2}a_5+a_6-\cdots$ is

a. 3^{2010} b. 1

c. 2^{2010} d. none of these

59. The value of $\sum_{r=0}^{10}{}^{10}C_r\,3_r\,(-2)^{10-r}$ is

a. 20 b. 10

c. 300 d. 30

60. The value of $\sum_{r=0}^{40}{}^{40}C_r\,{}^{30}C_r$ is

a. $40\,{}^{69}C_{29}$ b. $40\,{}^{70}C_{30}$

c. ${}^{69}C_{29}$ d. ${}^{70}C_{30}$

61. The value of $\sum_{r=1}^{15}\frac{r2^r}{(r+2)!}$ is equal to

a. $\frac{(17)!-12^{16}}{(17)!}$ b. $\frac{(18)!-2^{17}}{(18)!}$

c. $\frac{(16)!-2^{15}}{(16)!}$ d. $\frac{(15)!-2^{14}}{(15)!}$

62. $(n+2)\,{}^nC_0\,2^{n+1}-(n+1)\,{}^nC_1\,2^n+n\,{}^nC_2\,2^{n-1}-\cdots$ is equal to

a. 4 b. $4n$

c. $4(n+1)$ d. $2(n+2)$

63. The value of $\sum_{r=0}^{50}(-1)^r\frac{{}^{50}C_r}{r+2}$ is equal to

a. $\frac{1}{50\times 51}$ b. $\frac{1}{52\times 50}$

c. $\frac{1}{52\times 51}$ d. none of these

64. In the expansion of $[(1+x)/(1-x)]^2$, the coefficient of x^n will be

a. $4n$ b. $4n-3$

c. $4n+1$ d. one of these

65. The sum of $1+n\left(1-\frac{1}{x}\right)+\frac{n(n+1)}{2!}\left(1-\frac{1}{x}\right)^2+\cdots\infty$ will be

a. x^n b. x^{-n}

c. $\left(1-\frac{1}{x}\right)^n$ d. none of these

66. $\sum_{k=1}^{\infty}k\left(1+\frac{1}{n}\right)^{k-1}=$

a. $n(n-1)$ b. $n(n+1)$

c. n^2 d. $(n+1)^2$

67. The coefficient of x^4 in the expansion of $\{\sqrt{1+x^2}-x\}^{-1}$ in ascending powers of x, when $|x|<1$, is

a. 0 b. $\frac{1}{2}$

c. $-\frac{1}{2}$ d. $-\frac{1}{8}$

68. $1+\frac{1}{3}x+\frac{1\times 4}{3\times 6}x^2+\frac{1\times 4\times 7}{3\times 6\times 9}x^3+\cdots$ is equal to

a. x b. $(1+x)^{1/3}$

c. $(1-x)^{1/3}$ d. $(1-x)^{-1/3}$

69. $1+\frac{1}{4}+\frac{1\times 3}{4\times 8}+\frac{1\times 3\times 5}{4\times 8\times 12}+\cdots=$

a. $\sqrt{2}$ b. $\frac{1}{\sqrt{2}}$

c. $\sqrt{3}$ d. $\frac{1}{\sqrt{3}}$

70. If $|x| < 1$, then $1 + n\left(\frac{2x}{1+x}\right) + \frac{n(n+1)}{2!}\left(\frac{2x}{1+x}\right)^2 + \cdots$ is equal to

a. $\left(\frac{2x}{1+x}\right)^n$ b. $\left(\frac{1+x}{2x}\right)^n$

c. $\left(\frac{1-x}{1+x}\right)^n$ d. $\left(\frac{1+x}{1-x}\right)^n$

71. The coefficient of x^5 in $(1 + 2x + 3x^2 + \cdots)^{-3/2}$ is

a. 21 b. 25

c. 26 d. none of these

72. If $|x| < 1$, then the coefficient of x^n in expansion of $(1 + x + x^2 + x^3 + \cdots)^2$ is

a. n b. $n - 1$

c. $n + 2$ d. $n + 1$

73. If x is positive, the first negative term in the expansion of $(1 + x)^{27/5}$ is

a. 5th term b. 8th term

c. 6th term d. 7th term

74. If x is so small that x^3 and higher powers of x may be neglected, then $\frac{(1+x)^{3/2} - \left(1+\frac{1}{2}x\right)^3}{(1-x)^{1/2}}$ may be approximated as

a. $3x + \frac{3}{8}x^2$ b. $1 - \frac{3}{8}x^2$

c. $\frac{x}{2} - \frac{3}{x}x^2$ d. $-\frac{3}{8}x^2$

75. If the expansion in powers of x of the function $1/(1 - ax)(1 - bx)$ is $a_0 + a_1x + a_2x^2 + a_3x^3 + \cdots$, then a_n is

a. $\frac{b^n - a^n}{b - a}$ b. $\frac{a^n - b^n}{b - a}$

c. $\frac{a^{n+1} - b^{n+1}}{b - a}$ d. $\frac{b^{n+1} - a^{n+1}}{b - a}$

76. Value of $\sum_{k=1}^{\infty}\sum_{r=0}^{k}\frac{1}{3^k}({}^kC_r)$ is

a. $\frac{2}{3}$ b. $\frac{4}{3}$

c. 2 d. 1

77. The sum of rational term in $(\sqrt{2} + \sqrt[3]{3} + \sqrt[6]{5})^{10}$ is equal to

a. 12632 b. 1260

c. 126 d. none of these

78. If $f(x) = 1 - x + x^2 - x^3 + \cdots - x^{15} + x^{16} - x^{17}$, then the coefficient of x^2 in $f(x - 1)$ is

a. 826 b. 816

c. 822 d. none of these

79. If $p = (8 + 3\sqrt{7})^n$ and $f = p - [p]$, where $[\cdot]$ denotes the greatest integer function, then the value of $p(1 - f)$ is equal to

a. 1 b. 2

c. 2^n d. 2^{2n}

80. The value of $\sum_{r=0}^{10}(r)^{20}C_r$ is equal to

a. $20(2^{18} + {}^{19}C_{10})$ b. $10(2^{18} + {}^{19}C_{10})$

c. $20(2^{18} + {}^{19}C_{11})$ d. $10(2^{18} + {}^{19}C_{11})$

81. The last two digits of the number $(23)^{14}$ are

a. 01 b. 03

c. 09 d. none of these

82. Let $f(x) = a_0 + a_1x + a_2x^2 + \cdots + a_nx^n + \cdots$ and $\frac{f(x)}{1-x} = b_0 + b_1x + b_2x^2 + \cdots + b_nx^n + \cdots$, then

a. $b_n + b_{n-1} = a_n$ b. $b_n - b_{n-1} = a_n$

c. $b_n / b_{n-1} = a_n$ d. none of these

83. If $(1 - x^2)^n = \sum_{r=0}^{n} a_r x^r (1 - x)^{2n-r}$, then a_r is equal to

a. ${}^nC_r$ b. ${}^nC_r 3^r$

c. ${}^{2n}C_r$ d. ${}^nC_r 2^r$

84. $[({}^nC_0 + {}^nC_3 + \cdots) - (1/2)({}^nC_1 + {}^nC_2 + {}^nC_4 + {}^nC_5 + \cdots)]^2 + (3/4)({}^nC_1 - {}^nC_2 + {}^nC_4 - {}^nC_5 + \cdots)^2 =$

a. 3 b. 4

c. 2 d. 1

85. If $\frac{x^2 + x + 1}{1 - x} = a_0 + a_1x + a_2x^2 + \cdots$, then $\sum_{r=1}^{50} a_r$ is equal to

a. 148 b. 146

c. 149 d. none of these

86. 'p' is a prime number and $n < p < 2n$. If $N = {}^{2n}C_n$, then

a. p divides N b. p^2 divides N

c. p cannot divide N d. none of these

87. $\sum_{r=0}^{300} a_r x^r = (1 + x + x^2 + x^3)^{100}$. If $a = \sum_{r=0}^{300} a_r$, then $\sum_{r=0}^{300} ra_r$ is equal to

a. $300a$ b. $100a$

c. $150a$ d. $75a$

88. The value of $\sum_{r=1}^{n+1}\left(\sum_{k=1}^{n} {}^kC_{r-1}\right)$ (where $r, k, n \in N$) is equal to

a. $2^{n+1} - 2$ b. $2^{n+1} - 1$

c. 2^{n+1} d. none of these

89. If $(1 - x)^{-n} = a_0 + a_1x + a_2x^2 + \cdots + a_rx^r + \cdots$, then $a_0 + a_1 + a_2 + \cdots + a_r$ is equal to

a. $\frac{n(n+1)(n+2)\cdots(n+r)}{r!}$

b. $\frac{(n+1)(n+2)\cdots(n+r)}{r!}$

c. $\frac{n(n+1)(n+2)\cdots(n+r-1)}{r!}$

d. none of these

90. The value $\sum_{r=0}^{20} r(20-r)\,({}^{20}C_r)^2$ is equal to

a. $400\ {}^{39}C_{20}$
b. $400\ {}^{40}C_{19}$
c. $400\ {}^{39}C_{19}$
d. $400\ {}^{38}C_{20}$

Multiple Correct Answers Type

Each question has four choices a, b, c and d, out of which *one or more* answers are correct.

1. The value/values of x in the expression $\left(x + x^{\log_{10} x}\right)^5$ if the third term in the expansion is 10,00,000 is/are

a. 10
b. 100
c. $10^{-5/2}$
d. $10^{-3/2}$

2. If the coefficients of rth, $(r+1)^{\text{th}}$ and $(r+2)^{\text{th}}$ terms in the expansion of $(1+x)^{14}$ are in A.P., then r is/are

a. 5
b. 12
c. 10
d. 9

3. In the expansion of $(x+a)^n$ if the sum of odd terms be P and sum of even terms be Q, then

a. $P^2 - Q^2 = (x^2 - a^2)^n$
b. $4PQ = (x+a)^{2n} - (x-a)^{2n}$
c. $2(P^2 + Q^2) = (x+a)^{2n} + (x-a)^{2n}$
d. none of these

4. If $\left(4+\sqrt{15}\right)^n = I + f$, where n is an odd natural number, I is an integer and $0 < f < 1$, then

a. I is a natural number
b. I is an even integer
c. $(I+f)(I-f) = 1$
d. $1 - f = \left(4-\sqrt{5}\right)^n$

5. The number of values of r satisfying the equation ${}^{69}C_{3r-1} - {}^{69}C_{r^2} = {}^{69}C_{r^2-1} - {}^{69}C_{3r}$ is

a. 1
b. 2
c. 3
d. 7

6. If the 4th term in the expansion of $(ax + 1/x)^n$ is 5/2, then

a. $a = \dfrac{1}{2}$
b. $n = 8$
c. $a = \dfrac{2}{3}$
d. $n = 6$

7. The sum of the coefficient in the expansion of $(1 + ax - 2x^2)^n$ is

a. positive, when $a < 1$ and $n = 2k$, $k \in N$
b. negative, when $a < 1$ and $n = 2k+1$, $k \in N$
c. positive, when $a > 1$ and $n \in N$
d. zero, when $a = 1$

8. If $f(m) = \sum_{i=0}^{m} \binom{30}{30-i}\binom{20}{m-i}$ where $\binom{p}{q} = {}^{p}C_q$, then

a. maximum value of $f(m)$ is ${}^{50}C_{25}$
b. $f(0) + f(1) + \cdots + f(50) = 2^{50}$
c. $f(m)$ is always divisible by 50
d. The value of $\sum_{m=0}^{50} (f(m))^2 = {}^{100}C_{50}$

9. If for z as real or complex, $(1 + z^2 + z^4)^8 = C_0 + C_1 z^2 + C_2 z^4 + \cdots + C_{16} z^{32}$, then

a. $C_0 - C_1 + C_2 - C_3 + \cdots + C_{16} = 1$
b. $C_0 + C_3 + C_6 + C_9 + C_{12} + C_{15} = 3^7$
c. $C_2 + C_5 + C_8 + C_{11} + C_{14} = 3^6$
d. $C_1 + C_4 + C_7 + C_{10} + C_{13} + C_{16} = 3^7$

10. In the expansion of $(7^{1/3} + 11^{1/9})^{6561}$,

a. there are exactly 730 rational terms
b. there are exactly 5831 irrational terms
c. the term which involves greatest binomial coefficients is irrational
d. the term which involves greatest binomial coefficients is rational

11. If $(1+x)^n = C_0 + C_1x + C_2x^2 + \cdots + C_nx^n$, then $C_0 - (C_0 + C_1) + (C_0 + C_1 + C_2) - (C_0 + C_1 + C_2 + C_3) + \cdots (-1)^{n-1}(C_0 + C_1 + \cdots + C_{n-1})$, where n is even integer is

a. a positive value
b. a negative value
c. divisible by 2^{n-1}
d. divisible by 2^n

12. Let $(\sqrt{2} + 1)^6 = I + f$, where $0 < f < 1$ and I is an integer. Then

a. I is an odd integer
b. I is an even integer
c. $[(\sqrt{2} + 1)^6] = 197$, where $[\cdot]$ represents the greatest integer function
d. the value of $(I+f)(1-f) = 1$

13. The value of ${}^{n}C_1 + {}^{n+1}C_2 + {}^{n+2}C_3 + \cdots + {}^{n+m-1}C_m$ is equal to

a. ${}^{m+n}C_n$
b. ${}^{m+n}C_{n-1}$
c. ${}^{m}C_1 + {}^{m+1}C_2 + {}^{m+2}C_3 + \cdots + {}^{m+n-1}C_n$
d. ${}^{m+n}C_m$

14. If $(1+x)^n = C_0 + C_1x + C_2x^2 + \cdots + C_nx^n$, $n \in N$, then $C_0 - C_1 + C_2 - \cdots + (-1)^{m-1} C_{m-1}$ is equal to $(m < n)$

a. $\dfrac{(n-1)(n-2)\cdots(n-m+1)}{(m-1)!}(-1)^{m-1}$
b. ${}^{n-1}C_{m-1}\,(-1)^{m-1}$
c. $\dfrac{(n-1)(n-2)\cdots(n-m)}{(m-1)!}(-1)^{m-1}$
d. ${}^{n-1}C_{n-m}\,(-1)^{m-1}$

15. 10th term of $\left(3-\sqrt{\frac{17}{4}+3\sqrt{2}}\right)^{20}$

a. an irrational number
b. a rational number
c. a positive integer
d. a negative integer

16. For the expansion $(x \sin p + x^{-1} \cos p)^{10}$, $(p \in R)$,

a. the greatest value of the term independent of x is $10!/2^5(5!)^2$
b. the least value of sum of coefficient is zero
c. the greatest value of sum of coefficient is 32
d. the least value of the term independent of x occurs when $p=(2n+1)\frac{\pi}{4}$, $n \in Z$

17. Let $(1+x^2)^2(1+x)^n = \sum_{K=0}^{n+4} a_k x^k$. If a_1, a_2 and a_3 are in A.P., then the possible value/values of n is/are

a. 5 b. 4
c. 3 d. 2

18. The middle term in the expansion of $(x/2+2)^8$ is 1120; then $x \in R$ is equal to

a. -2 b. 3
c. -3 d. 2

19. For which of the following values of x, 5th term is the numerically greatest term in the expansion of $(1+x/3)^{10}$:

a. -2 b. 1.8
c. 2 d. -1.9

20. For natural numbers m, n, if $(1-y)^m(1+y)^n = 1 + a_1y + a_2y^2 + \cdots$, and $a_1 = a_2 = 10$, then

a. $m < n$ b. $m > n$
c. $m + n = 80$ d. $m - n = 20$

Reasoning Type

Each question has four choices a, b, c and d, out of which *only one* is correct. Each question contains Statement 1 and Statement 2.

a. Both the statements are True and Statement 2 is the correct explanation of Statement 1.
b. Both the statements are True but Statement 2 is Not the correct explanation of Statement 1.
c. Statement 1 is True and Statement 2 is False.
d. Statement 1 is False and Statement 2 is True.

1. **Statement 1:** The value of $({}^{10}C_0) + ({}^{10}C_0 + {}^{10}C_1) + ({}^{10}C_0 + {}^{10}C_1 + {}^{10}C_2) + \cdots + ({}^{10}C_0 + {}^{10}C_1 + {}^{10}C_2 + \cdots + {}^{10}C_9)$ is $10 \cdots 2^9$.

Statement 2: ${}^nC_1 + 2\,{}^nC_2 + 3\,{}^nC_3 + \cdots + n\,{}^nC_n = n2^{n-1}$.

2. **Statement 1:** Greatest term in the expansion of $(1+x)^{12}$, when $x = 11/10$ is 7th.

Statement 2: Seventh term in the expansion of $(1+x)^{12}$ has the factor ${}^{12}C_6$ which is greatest value of ${}^{12}C_r$.

3. **Statement 1:** Remainder when 3456^{2222} is divided by 7 is 4.

Statement 2: Remainder when 25 is divided 7 is 4.

4. **Statement 1:** Three consecutive binomial coefficients are always in A.P.

Statement 2: Three consecutive binomial coefficients are not in H.P. or G.P.

5. **Statement 1:** If $n \in N$ and n is not a multiple of 3 and $(1+x+x^2)^n = \sum_{r=0}^{2n} a_r x^r$, then the value of $\sum_{r=0}^{n} (-1)^r a_r\,{}^nC_r$ is zero.

Statement 2: The coefficient of x^n in the expansion of $(1-x^3)$ is zero, if $n = 3k+1$ or $n = 3k+2$.

6. **Statement 1:** $3^{2n+2} - 8n - 9$ is divisible by 64, $\forall\, n \in N$.

Statement 2: $(1+x)^n - nx - 1$ is divisible by x^2, $\forall\, n \in N$.

7. **Statement 1:** The number of distinct terms in $(1+x+x^2+x^3+x^4)^{1000}$ is 4001.

Statement 2: The number of distinct terms in the expansion $(a_1 + a_2 + \cdots + a_m)^n$ is ${}^{n+m-1}C_{m-1}$.

8. **Statement 1:** The coefficient of x^n in $\left(1+x+\frac{x^2}{2!}+\frac{x^3}{3!}+\cdots+\frac{x^n}{n!}\right)^3$ is $\frac{3^n}{n!}$.

Statement 2: The coefficient of x^n in e^{3x} is $\frac{3^n}{n!}$.

9. **Statement 1:** In the expansion of $(1+x)^{41}(1-x+x^2)^{40}$, the coefficient of x^{85} is zero.

Statement 2: In the expansion of $(1+x)^{41}$ and $(1-x+x^2)^{40}$, x^{85} term does not occur.

10. **Statement 1:** The total number of dissimilar terms in the expansion of $(x_1 + x_2 + \cdots + x_n)^3$ is $\frac{n(n+1)(n+2)}{6}$.

Statement 2: The total number of dissimilar terms in the expansion of $(x_1 + x_2 + x_3)^n$ is $\frac{n(n+1)(n+2)}{6}$.

11. **Statement 1:** If p is a prime number $(p \neq 2)$, then $\left[(2+\sqrt{5})^p\right] - 2^{p+1}$ is always divisible by p (where $[\cdot]$ denotes the greatest integer function).

Statement 2: If n is prime, then ${}^nC_1, {}^nC_2, {}^nC_3, \ldots, {}^nC_{n-1}$ must be divisible by n.

12. Let n be a positive integer and k be a whole number, $k \le 2n$.

Statement 1: The maximum value of ${}^{2n}C_k$ is ${}^{2n}C_n$.

Statement 2: $\frac{{}^{2n}C_{k+1}}{{}^{2n}C_k} < 1$, for $k = 0, 1, 2, \ldots, n-1$ and $\frac{{}^{2n}C_k}{{}^{2n}C_{k-1}} > 1$ for $k = n+1, n+2, \ldots, 2n$

13. **Statement 1:** The sum of coefficients in the expansion of $(3^{-x/4} + 3^{5x/4})^n$ is 2^n.

Statement 2: The sum of coefficients in the expansion of $(x+y)^n$ is 2^n when we put $x = y = 1$.

14. **Statement 1:** ${}^mC_r + {}^mC_{r-1}\,{}^nC_1 + {}^mC_{r-2}\,{}^nC_2 + \cdots + {}^nC_r = 0$, if $m + n < r$.

Statement 2: $^nC_r = 0$ if $n < r$.

15. **Statement 1:** $\sum\sum_{0\le i<j\le n}\left(\frac{i}{^nC_i}+\frac{j}{^nC_j}\right)$ is equal to $\frac{n^2}{2}\ a,$

where $a = \sum_{r=0}^{n}\frac{1}{^nC_r}$.

Statement 2: $\sum_{r=0}^{n}\frac{r}{^nC_r} = \sum_{r=0}^{n}\frac{n-r}{^nC_r}$.

Linked Comprehension Type

Based upon each paragraph, three multiple choice questions have to be answered. Each question has four choices a, b, c and d, out of which *only one* is correct.

For Problems 1–3

The 6th term in the expansion of $\left[\sqrt{\{2^{\log(10-3^x)}\}} + \sqrt[5]{\{2^{(x-2)\log 3}\}}\right]^m$ is equal to 21, if it is known that the binomial coefficient of the 2nd, 3rd and 4th terms in the expansion represents, respectively, the 1st, 3rd and 5th terms of an A.P. (the symbol log stands for logarithm to the base 10).

1. The value of m is

 a. 6 **b.** 7
 c. 8 **d.** 9

2. The sum of possible values of x is

 a. 1 **b.** 3
 c. 4 **d.** none of these

3. The minimum value of expression is

 a. 64 **b.** 32
 c. 128 **d.** none of these

For Problems 4–6

The 2nd, 3rd and 4th terms in the expansion of $(x + a)^n$ are 240, 720 and 1080, respectively.

4. The value of $(x - a)^n$ can be

 a. 64 **b.** –1
 c. –32 **d.** none of these

5. The value of least term in the expansion is

 a. 16 **b.** 160
 c. 32 **d.** 81

6. The sum of odd-numbered terms is

 a. 1664 **b.** 2376
 c. 1562 **d.** 1486

For Problems 7–9

If $(1 + x + x^2)^{20} = a_0 + a_1x + a_2x^2 \cdots + a_{40}x^{40}$, then answer the following questions.

7. The value of $a_0 + a_1 + a_2 + \cdots + a_{19}$ is

 a. $\frac{1}{2}(9^{10} + a_{20})$ **b.** $\frac{1}{2}(9^{10} - a_{20})$
 c. $\frac{9^{10}}{2}$ **d.** none of these

8. The value of $a_0^2 - a_1^2 + a_2^2 - \cdots - a_{19}^2$ is

 a. $\frac{1}{2}a_{20}(1-a_{20})$ **b.** $\frac{1}{2}a_{20}(1+a_{20})$
 c. $\frac{1}{2}a_{20}^2$ **d.** none of these

9. The value of $a_0 + 3a_1 + 5a_2 + \cdots + 81a_{40}$ is

 a. 161×3^{20} **b.** 41×3^{40}
 c. 41×3^{20} **d.** none of these

For Problems 10–12

An equation $a_0 + a_1x + a_2x^2 + \cdots + a_{99}x^{99} + x^{100} = 0$ has roots $^{99}C_0$, $^{99}C_1$, $^{99}C_2$, …, $^{99}C_{99}$.

10. The value of a_{99} is equal to

 a. 2^{98} **b.** 2^{99}
 c. -2^{99} **d.** none of these

11. The value of a_{98} is

 a. $\frac{2^{198} - {}^{198}C_{99}}{2}$ **b.** $\frac{2^{198} + {}^{198}C_{99}}{2}$
 c. $2^{99} - {}^{99}C_{49}$ **d.** none of these

12. The value of $(^{99}C_0)^2 + (^{99}C_1)^2 + \cdots + (^{99}C_{99})^2$ is equal to

 a. $2a_{98} - a_{99}^2$ **b.** $a_{99}^2 - a_{98}$
 c. $a_{99}^2 - 2a_{98}$ **d.** none of these

For Problems 13–15

Any complex number in polar form can be an expression in Euler's form as $\cos\theta + i\sin\theta = e^{i\theta}$. This form of the complex number is useful in finding the sum of series $\sum_{r=0}^{n} {}^nC_r(\cos\theta + i\sin\theta)^r$.

$$\sum_{r=0}^{n} {}^nC_r(\cos r\theta + i\sin r\theta) = \sum_{r=0}^{n} {}^nC_r e^{ir\theta}$$

$$= \sum_{r=0}^{n} {}^nC_r\left(e^{i\theta}\right)^r$$

$$= \left(1+e^{i\theta}\right)^n$$

Also, we know that the sum of binomial series does not change if r is replaced by $n - r$.

Using these facts, answer the following questions.

13. The value of $\sum_{r=0}^{100} {}^{100}C_r(\sin rx)$ is equal to

 a. $2^{100}\cos^{100}\frac{x}{2}\sin 50x$ **b.** $2^{100}\sin(50x)\cos\frac{x}{2}$
 c. $2^{101}\cos^{100}(50x)\sin\frac{x}{2}$ **d.** $2^{101}\sin^{100}(50x)\cos(50x)$

14. In triangle ABC, the value $\sum_{r=0}^{50} {}^{50}C_r a^r b^{n-r} \cos(rB-(50-r)A)$ is equal to (where a, b, c are sides of triangle opposite to angle A, B, C, respectively, and s is semi-perimeter)

a. c^{49}

b. $(a+b)^{50}$

c. $(2s-a-b)^{50}$

d. none of these

15. If

$$f(x)=\frac{\sum_{r=0}^{50} {}^{50}C_r \sin 2rx}{\sum_{r=0}^{50} {}^{50}C_r \cos 2rx},$$

then the value of $f(\pi/8)$ is equal to

a. 1 **b.** –1

c. irrational value **d.** none of these

For Problems 16–18

Let

$$P=\sum_{r=1}^{50}\frac{{}^{50+r}C_r(2r-1)}{{}^{50}C_r(50+r)},$$

$$Q=\sum_{r=0}^{50}\left({}^{50}C_r\right)^2,\ R=\sum_{r=0}^{100}(-1)^r\left({}^{100}C_r\right)^2$$

16. The value of $P-Q$ is equal to

a. 1 **b.** –1

c. 2^{50} **d.** 2^{100}

17. The value of $P-R$ is equal to

a. 1 **b.** –1

c. 2^{50} **d.** 2^{100}

18. $Q+R$ is equal to

a. $2P+1$

b. $2P-1$

c. $2P+2$

d. $2P-2$

For Problems 19–21

P is a set containing n elements. A subset A of P is chosen and the set P is reconstructed by replacing the elements of A. A subset B of P is chosen again.

19. The number of ways of choosing A and B such that A and B have no common elements is

a. 3^n

b. 2^n

c. 4^n

d. none of these

20. The number of ways of choosing A and B such that B contains just one element more than A is

a. 2^n **b.** ${}^{2n}C_{n-1}$

c. ${}^{2n}C_n$ **d.** $(3^n)^2$

21. The number of ways of choosing A and B such that B is a subset of A is

a. ${}^{2n}C_n$

b. 4^n

c. 3^n

d. none of these

Matrix-Match Type

Each question contains statements given in two columns which have to be matched. Statements a, b, c, d in column I have to be matched with statements p, q, r, s in column II. If the correct matches are a→p, a→s, b→r, c→p, c→q and d→s, then the correctly bubbled 4 × 4 matrix should be as follows:

	p	q	r	s
a	(p) ●	(q)	(r)	(s) ●
b	(p)	(q) ●	(r) ●	(s)
c	(p) ●	(q) ●	(r)	(s)
d	(p)	(q)	(r)	(s) ●

1.

Column I	Column II
a. If ${}^{(n+1)}C_4 + {}^{(n+1)}C_3 + {}^{(n+2)}C_3 > {}^{(n+3)}C_3$, then possible value/values of n is/are	**p.** 4
b. The remainder when $(3053)^{456}-(2417)^{333}$ is divided by 9 is less than	**q.** 5
c. The digit in the unit place of the number $183! + 3^{183}$ is greater than	**r.** 6
d. If sum of the coefficients of the first, second and third terms of the expansion of $\left(x^2+\frac{1}{x}\right)^m$ is 46, then the index of the term that does not contain x is greater than	**s.** 7

2.

Column I	Column II
a. ${}^{32}C_0^2 - {}^{32}C_1^2 + {}^{32}C_2^2 - \cdots + {}^{32}C_{32}^2 =$	**p.** ${}^{63}C_{32}$
b. ${}^{32}C_0^2 + {}^{32}C_1^2 + {}^{32}C_2^2 - \cdots + {}^{32}C_{32}^2 =$	**q.** ${}^{32}C_{16}$
c. $\frac{1}{32}(1\times {}^{32}C_1^2 + 2\times {}^{32}C_2^2 + \cdots + 32\times {}^{32}C_{32}) =$	**r.** 0
d. ${}^{31}C_0^2 - {}^{31}C_1^2 + {}^{31}C_2^2 - \cdots - {}^{31}C_{31}^2 =$	**s.** ${}^{64}C_{32}$

3.

Column I	Column II
a. $\sum\sum_{i\neq j} {}^{10}C_i\,{}^{10}C_j$	**p.** $\dfrac{2^{20}-{}^{20}C_{10}}{2}$
b. $\sum\sum_{0\le i\le j\le n} {}^{10}C_i\,{}^{10}C_j$	**q.** $2^{20}-{}^{20}C_{10}$
c. $\sum\sum_{0\le i<j\le n} {}^{10}C_i\,{}^{10}C_j$	**r.** 2^{20}
d. $\sum_{i=0}^{10}\sum_{j=0}^{10} {}^{10}C_i\,{}^{10}C_j$	**s.** $\dfrac{2^{20}+{}^{20}C_{10}}{2}$

4.

Column I	Column II
a. The sum of binomial coefficients of terms containing power of x more than x^{20} in $(1+x)^{41}$ is divisible by	**p.** 2^{39}
b. The sum of binomial coefficients of rational terms in the expansion of $(1+\sqrt{2})^{42}$ is divisible by	**q.** 2^{40}
c. If $\left(x+\dfrac{1}{x}+x^2+\dfrac{1}{x^2}\right)^{21}=a_0x^{-42}+a_1x^{-41}+a_2x^{-40}+\cdots+a_{82}x^{40}$, then $a_0+a_2+\cdots+a_{82}$ is divisible by	**r.** 2^{41}
d. The sum of binomial coefficients of positive real terms in the expansion of $(1+ix)^{42}$ $(x>0)$ is divisible by	**s.** 2^{38}

5.

Column I	Column II
a. The coefficient of two consecutive terms in the expansion of $(1+x)^n$ will be equal, then n can be	**p.** 9
b. If 15^n+23^n is divided by 19, then n can be	**q.** 10
c. ${}^{10}C_0\,{}^{20}C_{10}-{}^{10}C_1\,{}^{18}C_{10}+{}^{10}C_2\,{}^{16}C_{10}-\cdots$ is divisible by 2^n, then n can be	**r.** 11
d. If the coefficients of T_r, T_{r+1}, T_{r+2} terms of $(1+x)^{14}$ are in A.P., then r is less than	**s.** 12

Integer Type

1. The value of
$$\frac{(18^3+73+3\times 18\times 7\times 25)}{3^6+6\times 243\times 2+81\times\times 4+20\times 27\times 8+15\times 9\times 16+6\times 3\times 32+64}$$
is______.

2. The term in $\left(\sqrt[3]{\dfrac{a}{\sqrt{b}}}+\sqrt{\dfrac{b}{\sqrt[3]{a}}}\right)^{21}$ which has the same power of a and b is______.

3. The coefficient of x^{50} in $(1+x)^{101}(1-x+x^2)^{100}$ is______.

4. The number of terms which are free from radical signs in the expansion of $(y^{1/5}+x^{1/10})^{55}$ is______.

5. The prime number by which the coefficient of $a^3b^4c^5$ in the expansion of $(bc+ca+ab)^6$ is not divisible is______.

6. If the sum of the coefficients in the expansion of $(\alpha^2x^2-2\alpha x+1)^{51}$ vanishes, then the value of α is______.

7. The digit at the 100th place of the number 17^{256} is______.

8. The greatest power of 10 by which $101^{100}-1$ is divisible is______.

9. The positive integer just greater than $(1+0.0001)^{10000}$ is______.

10. The remainder when 5^{99} is divided by 13 is______.

11. The remainder when $1690^{2608}+2608^{16900}$ is divided by 7 is______.

12. The remainder when $x=5^{5^{5^{\cdot^{\cdot^{5}}}}}$ (24 times 5) is divided by 24 is______.

13. The degree of the polynomial $\dfrac{1}{\sqrt{4x+1}}\left\{\left(\dfrac{1+\sqrt{4x+1}}{2}\right)^7-\left(\dfrac{1-\sqrt{4x+1}}{2}\right)^7\right\}$ is______.

14. If $(2+\sqrt{3})^n=I+f$, where I and n are + positive integers and $0<f<1$, then the value of $(1-f)(I+f)$ is

15. Given that the fourth term in the expansion of $\left(2+\frac{3}{8}x\right)^{10}$ has the maximum numerical value, then the least positive value is______.

16. The value of ${}^nC_0\times{}^{2n}C_n-{}^nC_1\times{}^{2n-1}C_n+{}^nC_2\times{}^{2n-2}C_n+\cdots+(-1)^n\times{}^nC_n\times{}^nC_n$ is______.

17. Let n be an odd natural number greater than 1. Then the number of zeros at the end of the sum 99^n+1 is______.

18. The value of $\lim\limits_{n\to\infty}\dfrac{1}{3^n}\sum_{p=1}^{n}\left(\sum_{m=p}^{n}{}^nC_m\times{}^mC_p\right)$ is______.

19. If the $(r+1)$th term is the first negative term in the expansion of $(1+x)^{7/2}$, then the value of r is______.

20. If $|x|<1$, then the coefficient of x^n in the expansion of $(1+2x+3x^2+4x^3+\cdots)^{1/2}$ is______.

21. The value of ${}^{20}C_0-\dfrac{{}^{20}C_1}{2}+\dfrac{{}^{20}C_2}{3}-\dfrac{{}^{20}C_3}{4}+\cdots$ is______.

22. The sum $\sum_{i=0}^{m}\binom{9}{i}\binom{7}{m-i}$, where $\binom{p}{q}=0$ if $(p<q)$ is maximum when m is______.

23. If ${}^{n-1}C_r=(k^2-3)\,{}^nC_{r+1}$, then the maximum value of k is______.

24. If the constant term in the binomial expansion of $\left(x^2-\frac{1}{x}\right)^n$, $n\in N$ is 15 then the value of n is equal to______.

25. $\sum_{i=0}^{n}\sum_{j=0}^{n}\sum_{k=0}^{n}\binom{n}{i}\binom{n}{j}\binom{n}{k},\binom{n}{r}={}^nC_r$ is less than 5000, if n is equal to______.

ANSWERS AND SOLUTIONS

Objective Type

1. c. Coefficient of T_5 is ${}^nC_4$, that of T_6 is ${}^nC_5$ and that of T_7 is ${}^nC_6$. According to the condition, $2\,{}^nC_5 = {}^nC_4 + {}^nC_6$. Hence,

$$2\left[\frac{n!}{(n-5)!5!}\right] = \left[\frac{n!}{(n-4)!4!} + \frac{n!}{(n-6)!6!}\right]$$

$$\Rightarrow 2\left[\frac{1}{(n-5)5}\right] = \left[\frac{1}{(n-4)(n-5)} + \frac{1}{6\times 5}\right]$$

After solving, we get $n = 7$ or 14.

2. c. $T_{r+1} = {}^{2n}C_r x^{2n-r}\left(\frac{1}{x^2}\right)^r = {}^{2n}C_r x^{2n-3r}$

This contains x^m. If $2n - 3r = m$, then

$$r = \frac{2n-m}{3}$$

$$\Rightarrow \text{Coefficient of } x^m = {}^{2n}C_r,\ r = \frac{2n-m}{3}$$

$$= \frac{2n!}{(2n-r)!r!} = \frac{2n!}{\left(2n - \frac{2n-m}{3}\right)!\left(\frac{2n-m}{3}\right)!}$$

$$= \frac{2n!}{\left(\frac{4n+m}{3}\right)!\left(\frac{2n-m}{3}\right)!}$$

3. c. $(1+x)^{21} + (1+x)^{22} + \cdots + (1+x)^{30}$

$$= (1+x)^{21}\left[\frac{(1+x)^{10}-1}{(1+x)-1}\right] = \frac{1}{x}\left[(1+x)^{31} - (1+x)^{21}\right]$$

$\Rightarrow$ Coefficient of x^5 in the given expression

$= \text{Coefficient of } x^5 \text{ in } \left\{\frac{1}{x}\left[(1+x)^{31} - (1+x)^{21}\right]\right\}$

$= \text{Coefficient of } x^6 \text{ in } [(1+x)^{31} - (1+x)^{21}]$

$= {}^{31}C_6 - {}^{21}C_6$

4. b. $(1+x)^n = {}^nC_0 + {}^nC_1x + {}^nC_2x^2 + \cdots + {}^nC_nx^n$

$$\left(1+\frac{1}{x}\right)^n = {}^nC_0 + {}^nC_1\frac{1}{x} + {}^nC_2\frac{1}{x^2} + \cdots + {}^nC_n\left(\frac{1}{x}\right)^n$$

Obviously, the required coefficient of $1/x$ can be given by

${}^nC_0\,{}^nC_1 + {}^nC_1\,{}^nC_2 + \cdots + {}^nC_{n-1}\,{}^nC_n$

$= {}^nC_0\,{}^nC_{n-1} + {}^nC_1\,{}^nC_{n-2} + \cdots + {}^nC_{n-1}\,{}^nC_0$

$$= {}^{2n}C_{n-1} = \frac{(2n)!}{(n-1)!(n+1)!}$$

5. d. $(1+3x+2x^2)^6 = [1 + x(3+2x)]^6$

$= 1 + {}^6C_1\,x(3+2x) + {}^6C_2x^2(3+2x)^2 + {}^6C_3\,x^3(3+2x)^3$

$+ {}^6C_4x^4(3+2x)^4 + {}^6C_5x^5(3+2x)^5 + {}^6C_6x^6(3+2x)^6$

We get x^{11} only from ${}^6C_6x^6(3+2x)^6$. Hence, coefficient of x^{11} is ${}^6C_5 \times 3 \times 2^5 = 576$.

6. b. We have $T_{r+1} = {}^{29}C_r \times 3^{29-r} \times (7x)^r = ({}^{29}C_r \times 3^{29-r} \times 7^r)\,x^r$

Coefficient of $(r+1)^{\text{th}}$ term is ${}^{29}C_r \times 3^{29-r} \times 7^r$

and coefficient of rth term is ${}^{29}C_{r-1} \times 3^{30-r} \times 7^{r-1}$

From given condition,

${}^{29}C_r \times 3^{29-r} \times 7^r = {}^{29}C_{r-1} \times 3^{30-r} \times 7^{r-1}$

$$\Rightarrow \frac{{}^{29}C_r}{{}^{29}C_{r-1}} = \frac{3}{7} \Rightarrow \frac{30-r}{r} = \frac{3}{7} \Rightarrow r = 21$$

7. c. Let $(r+1)^{\text{th}}$, $(r+2)^{\text{th}}$, and $(r+3)^{\text{th}}$ be three consecutive terms.

Then,

${}^nC_r : {}^nC_{r+1} : {}^nC_{r+2} = 1:7:42$

Now,

$$\frac{{}^nC_r}{{}^nC_{r+1}} = \frac{1}{7} \Rightarrow \frac{r+1}{n-r} = \frac{1}{7} \Rightarrow n - 8r = 7 \quad (1)$$

$$\frac{{}^nC_{r+1}}{{}^nC_{r+2}} = \frac{7}{42} \Rightarrow \frac{r+2}{n-r-1} = \frac{1}{6} \Rightarrow n - 7r = 13 \quad (2)$$

Solving (1) and (2), we get $n = 55$.

8. c. $(1+2x+x^2)^n = \sum_{r=0}^{2n} a_rx^r \Rightarrow [(1+x)^2]^n = \sum_{r=0}^{2n} a_rx^r$

$\Rightarrow (1+x)^{2n} = \sum_{r=0}^{2n} a_rx^r$

$\Rightarrow \sum_{r=0}^{2n} {}^{2n}C_rx^r = \sum_{r=0}^{2n} a_rx^r$

$\Rightarrow a_r = {}^{2n}C_r$

9. b. Here $a = {}^nC_r$, $b = {}^nC_{r+1}$ and $c = {}^nC_{r+2}$.

Put $n = 2$, $r = 0$, then option (b) holds the condition, i.e.,

$$n = \frac{2ac + ab + bc}{b^2 - ac}$$

10. c. Middle term of $(1+\alpha x)^4$ is T_3.

Its coefficient is ${}^4C_2\,(\alpha)^2 = 6\alpha^2$.

Middle term of $(1-\alpha x)^6$ is T_4.

Its coefficient is ${}^6C_3\,(-\alpha)^3 = -20\alpha^3$.

According to question,

$6\alpha^2 = -20\alpha^3$

$\Rightarrow 3\alpha^2 + 10\alpha^3 = 0$

$\Rightarrow \alpha^2(3 + 10\alpha) = 0$

$$\Rightarrow \alpha = -\frac{3}{10}$$

11. b. $(1-x)(1-x)^n$

$= (1-x)\,[1 + n(-x) + \cdots + {}^nC_{n-1}(-x)^{n-1} + {}^nC_n(-x)^n]$

Therefore, coefficient of x^n is

${}^nC_n(-1)^n - {}^nC_{n-1}(-1)^{n-1} = (-1)^n + (-1)^n\,n$

$= (-1)^n(1+n)$

12. d. Here, the coefficients of T_r, T_{r+1} and T_{r+2} in $(1+y)^m$ are in A.P.

$\Rightarrow {}^mC_{r-1}$, ${}^mC_r$, and ${}^mC_{r+1}$ are in A.P.

$\Rightarrow 2\,{}^mC_r = {}^mC_{r-1} + {}^mC_{r+1}$

$$\Rightarrow 2\frac{m!}{r!(m-r)!}$$

$$= \frac{m!}{(r-1)!(m-r+1)!} + \frac{m!}{(r+1)!(m-r-1)!}$$

$$\Rightarrow \frac{2}{r(m-r)} = \frac{1}{(m-r+1)(m-r)} + \frac{1}{(r+1)r}$$

$$\Rightarrow m^2 - m(4r+1) + 4r^2 - 2 = 0$$

13. c. For $\left(ax^2 + \left(\frac{1}{bx}\right)\right)^{11}$,

$$T_{r+1} = {}^{11}C_r\,(ax^2)^{11-r}\left(\frac{1}{bx}\right)^r$$

$$= {}^{11}C_r\,a^{11-r}\,\frac{1}{b^r}\,x^{22-3r}$$

For x^7,

$22 - 3r = 7$

$\Rightarrow 3r = 15$

$\Rightarrow r = 5$

$$\Rightarrow T_6 = {}^{11}C_5\,a^6\,\frac{1}{b^5}\,x^7$$

$$\Rightarrow \text{Coefficient of } x^7 \text{ is } {}^{11}C_5\,\frac{a^6}{b^5}$$

Similarly, coefficient of x^{-7} in $\left(ax - \left(\frac{1}{bx^2}\right)\right)^{11}$ is ${}^{11}C_6\,\frac{a^5}{b^6}$.

Given that

$${}^{11}C_5\,\frac{a^6}{b^5} = {}^{11}C_6\,\frac{a^5}{b_6}$$

$$\Rightarrow a = \frac{1}{b}$$

$\Rightarrow ab = 1$

14. c. $a^{10}b^{10}c^{10}d^{10}\left(\frac{1}{a}+\frac{1}{b}+\frac{1}{c}+\frac{1}{d}\right)^{10}$

Therefore, the required coefficient is equal to the coefficient of $a^{-2}b^{-6}c^{-1}d^{-1}$ in $\left(\frac{1}{a}+\frac{1}{b}+\frac{1}{c}+\frac{1}{d}\right)^{10}$, which is given by

$$\frac{10!}{2!6!1!1!} = \frac{10\times 9\times 8\times 7}{2} = 2520$$

15. d. $(x-2)^5\,(x+1)^5$

$$= [{}^5C_0\,x^5 - {}^5C_1 x^4 \times 2 + \cdots]\,[{}^5C_0 + {}^5C_1 x + \cdots]$$

$\Rightarrow$ Coefficient of x^5

$$= {}^5C_0{}^5C_5 - {}^5C_1 \times 2 \times {}^5C_4 + {}^5C_2 \times 2^2 \times {}^5C_3 - {}^5C_3 \times 2^3 \times {}^5C_2 + {}^5C_4 \times 2^4 \times {}^5C_1 - {}^5C_5 \times 2^5 \times {}^5C_0$$

$$= 1 - 5\times 5\times 2 + 10\times 10\times 4 - 10\times 10\times 8 + 5\times 5\times 16 - 32$$

$$= -81$$

16. d. The general term in the expansion of $(1-x+y)^{20}$ is

$$\frac{20!}{r!s!t!}1^r(-x)^s(y)^t, \text{ where } r+s+t=20$$

For x^2y^3, we have the term

$$\frac{20!}{15!2!3!}1^{15}(-x)^2(y)^3$$

Hence, the coefficient of x^2y^3 is

$$\frac{20!}{15!2!3!}$$

17. b. $t_{r+1} = {}^{10}C_r\,(\sqrt{x})^{10-r}\left(\frac{-k}{x^2}\right)^r = {}^{10}C_r x^{5-5r/2}(-k)^r$

For this to be independent of x, r must be 2, so that

${}^{10}C_2\,k^2 = 405 \Rightarrow k = \pm 3$

18. a. We rewrite the given expression as $[1 + x^2(1-x)]^8$ and expand by using the binomial theorem. We have,

$$[1+x^2(1-x)]^8$$

$$= {}^8C_0 + {}^8C_1\,x^2\,(1-x) + {}^8C_2\,x^4\,(1-x)^2 + {}^8C_3\,x^6\,(1-x)^3 + {}^8C_4\,x^8\,(1-x)^4 + {}^8C_5\,x^{10}\,(1-x)^5 + \cdots$$

The two terms which contain x^{10} are ${}^8C_4\,x^8\,(1-x)^8$ and ${}^8C_5\,x^{10}\,(1-x)^5$. Thus, the coefficient of x^{10} in the given expression is given by

${}^8C_4$[coefficient of x^2 in the expansion of $(1-x)^4$] + ${}^8C_5$

$$= {}^8C_4(6) + {}^8C_5 = \frac{8!}{4!\,4!}(6) + \frac{8!}{3!\,5!}$$

$$= (70)(6) + 56 = 476$$

19. c. We have,

$$(1+x)^{101}\,(1-x+x^2)^{100} = (1+x)\,((1+x)\,(1-x+x^2))^{100}$$

$$= (1+x)(1+x^3)^{100} = (1+x)\,\{C_0 + C_1x^3 + C_2x^6 + \cdots + C_{100}x^{300}\}$$

$$= (1+x)\sum_{r=0}^{n}{}^nC_r x^{3r} = \sum_{r=0}^{n}{}^nC_r x^{3r} + \sum_{r=0}^{n}{}^nC_r x^{3r+1}$$

Hence, there will be no term containing $3r+2$.

20. b. $(1+x^3-x^6)^{30}$

$$= \{1 + x^3\,(1-x^3)\}^{30}$$

$$= {}^{30}C_0 + {}^{30}C_1\,x^3\,(1-x^3) + {}^{30}C_2\,x^6\,(1-x^3)^2 + \cdots$$

Obviously, each term will contain x^{3m}, $m \in N$. But 28 is not divisible by 3. Therefore, there will be no term containing x^{28}.

21. b. $\left(1 + \sqrt{a} + \frac{1}{\sqrt{a}-1}\right)^{-30}$

$$= \left(\frac{\sqrt{a}-1}{a}\right)^{30}$$

$$= \frac{1}{a^{30}}(1-\sqrt{a})^{30}$$

$$= \frac{1}{a^{30}}\{{}^{30}C_0 - {}^{30}C_1\sqrt{a} + \cdots + {}^{30}C_{30}(\sqrt{a})^{30}\}$$

There is no term independent of a.

22. c. The given sigma is the expansion of

$$[(x-3)+2]^{100} = (x-1)^{100} = (1-x)^{100}$$

Therefore, x^{53} will occur in T_{54}.

$$T_{54} = {}^{100}C_{53}\,(-x)^{53}$$

Therefore, the coefficient is $-{}^{100}C_{53}$.

23. a. We have,

$$\frac{x+1}{x^{2/3}-x^{1/3}+1}-\frac{x-1}{x-x^{1/2}}$$

$$=\frac{(x^{1/3})^3+1^3}{x^{2/3}-x^{1/3}+1}-\frac{x-1}{x^{1/2}(x^{1/2}-1)}$$

$$=\frac{(x^{1/3}+1)(x^{2/3}-x^{1/3}+1)}{x^{2/3}-x^{1/3}+1}-\frac{x^{1/2}+1}{x^{1/2}}$$

$= x^{1/3} + 1 - 1 - x^{-1/2} = x^{1/3} - x^{-1/2}$

$$\therefore \left(\frac{x+1}{x^{2/3}-x^{1/3}+1}-\frac{x-1}{x-x^{1/2}}\right)^{10}=(x^{1/3}-x^{-1/2})^{10}$$

Let T_{r+1} be the general term in $(x^{1/3} - x^{-1/2})^{10}$. Then,

$T_{r+1} = {}^{10}C_r\,(x^{1/3})^{10-r}\,(-1)^r\,(x^{-1/2})^r$

For this term to be independent of x, we must have

$\frac{10-r}{3}-\frac{r}{2}=0 \Rightarrow 20-2r-3r=0 \Rightarrow r=4$

So, the required coefficient is ${}^{10}C_4\,(-1)^4 = 210$.

24. d. $(1+x+x^3+x^4)^{10} = (1+x)^{10}\,(1+x^3)^{10}$

$=(1+{}^{10}C_1x+{}^{10}C_2x^2+{}^{10}C_3x^3+{}^{10}C_2x^4\cdots)(1+{}^{10}C_1x^3+{}^{10}C_2x^6+\cdots)$

Therefore, coefficient of x^4 is ${}^{10}C_1\,{}^{10}C_1 + {}^{10}C_4 = 310$.

25. a. $(1.0002)^{3000} = (1+0.0002)^{3000}$

$$=1+(3000)(0.0002)+\frac{(3000)(2999)}{1.2}(0.0002)^2+\cdots$$

$= 1 + (3000)\,(0.0002) = 1.6$

26. d. $(5+2)^{100} - (5-2)^{100}$

$= 2\,[{}^{100}C_1\,5^{99}\times 2 + {}^{100}C_3\,5^{97}\times 2^3 + \cdots + {}^{100}C_{99}5\times 2^{99}]$

$= 2\,[1000\times 5^{98} + {}^{100}C_3\,5^{94} + \cdots + 1000\times 2_{98}]$

$\Rightarrow$ Minimum 000 as last three digits

27. a. We have,

$$\frac{2}{\sqrt{2x^2+1}+\sqrt{2x^2-1}}=\frac{2\left(\sqrt{2x^2+1}-\sqrt{2x^2-1}\right)}{(2x^2+1)-(2x^2-1)}$$

$$=\sqrt{2x^2+1}-\sqrt{2x^2-1}$$

Thus, the given expression can be written as

$$\left(\sqrt{2x^2+1}+\sqrt{2x^2-1}\right)^6+\left(\sqrt{2x^2+1}-\sqrt{2x^2-1}\right)^6$$

But $(a+b)^6 + (a-b)^6 = 2[a^6 + {}^6C_2a^4b^2 + {}^6C_4a^2b^4 + b^6]$

Therefore,

$$\left(\sqrt{2x^2+1}+\sqrt{2x^2-1}\right)^6+\left(\sqrt{2x^2+1}-\sqrt{2x^2-1}\right)^6$$

$$= 2[(2x^2+1)^3 + 15(2x^2+1)^2\,(2x^2-1) + 15(2x^2+1)\times(2x^2-1)^2 + (2x^2-1)^3],$$

which is a polynomial of degree 6.

28. b. We have,

$(x+3)^{n-1}+(x+3)^{n-2}(x+2)+(x+3)^{n-3}(x+2)^2+\cdots+(x+2)^{n-1}$

$$=\frac{(x+3)^n-(x+2)^n}{(x+3)-(x+2)}=(x+3)^n-(x+2)^n$$

$$\left(\because \frac{x^n-a^n}{x-a}=x^{n-1}+x^{n-2}a^1+x^{n-3}a^2+\cdots+a^{n-1}\right)$$

Therefore, coefficient of x^r in the given expression is equal to coefficient of x^r in $[(x+3)^n - (x+2)^n]$, which is given by

${}^nC_r3^{n-r} - {}^nC_r2^{n-r} = {}^nC_r\,(3^{n-r} - 2^{n-r})$.

29. b. The general term in the expansion of $(1+2x+3x^2)^{10}$ is

$$\frac{10!}{r!s!t!}1^r(2x)^s(3x^2)^t=\frac{10!}{r!s!t!}2^s\times 3^t\times x^{s+2t},$$

where $r+s+t=10$

We have to find a_1, i.e., the coefficient of x.

For the coefficient of x^1, we must have $s+2t=1$.

But $r+s+t=10$

$\therefore s = 1-2t$ and $r = 9+t$, where $0 \le r, s, t \le 10$

Now, $t=0 \Rightarrow s=1, r=9$

For other values of t, we get netative values of s. so, there is only one term containing x and its coefficient is 10!/(9! 1! 0!) $2^1\times 3^0 = 20$. Hence, $a_1 = 20$.

30. b. $T_{r+1} = {}^{1024}C_r(5^{1/2})^{1024-r}(7^{1/8})^r$

Now this term is an integer if $1024-r$ is an even integer, for which $r = 0, 2, 4, 6,\ldots, 1022, 1024$ of which $r = 0, 8, 16, 2424, \ldots, 1024$ are divisible by 8 which maks $r/8$ an integer.

For A.P., $r = 0.\ 8, 16, 24, \ldots, 1024$,

$1024 = 0 + (n-1)8 \Rightarrow n = 129$

31. c. $$\left(x^2-2+\frac{1}{x^2}\right)^n=\frac{1}{x^{2n}}(x^4-2x^2+1)^n=\frac{(x^2-1)^n}{x^{2n}}$$

Total number of terms that are dependent on x is equal to number of terms in the expansion of $(x^2-1)n$ that have degree of x different from $2n$, which is given by $(n+1)-1 = n$.

32. b. It is given that 6th term in the expansion of

$\left(\frac{1}{x^{8/3}}+x^2\log_{10}x\right)^8$ is 5600, therefore

$${}^8C_5\left(x^2\log_{10}x\right)^5\left(\frac{1}{x^{8/3}}\right)^3=5600$$

$$\Rightarrow 56x^{10}(\log_{10}x)^5\frac{1}{x^8}=5600$$

$\Rightarrow x^2(\log_{10}x)^5 = 100$

$\Rightarrow x^2\,(\log_{10}x)^5 = 10^2\,(\log_{10}10)^5$

$\Rightarrow x = 10m$ which occurs when $n = 10$

33. b. $n!\,(21-n)! = 21!\dfrac{n!\;(21-n)!}{21!}=\dfrac{21!}{{}^{21}C_n}$ which is minimum when ${}^{21}C_n$ is maximum which occurs when $n = 10$

34. a. To get sum of coefficients put $x = 0$. Given that sum of coefficients is

$2^n = 64$

$\Rightarrow n = 6$

The greatest binomial coefficient is ${}^6C_3$.

Now given that

$T_4 - T_3 = 6 - 1 = 5$

$\Rightarrow {}^6C_3(3^{-x/4})^3(3^{5x/4})^3 - {}^6C_2(3^{-x/4})^2(3^{5x/4})^4 = 5$

which is satisfied by $x = 0$.

35. a. Last term of $\left(2^{1/3} - \frac{1}{\sqrt{2}}\right)^n$ is

$$T_{n+1} = {}^nC_n(2^{1/3})^{n-n}\left(-\frac{1}{\sqrt{2}}\right)^n = {}^nC_n(-1)^n\frac{1}{2^{n/2}} = \frac{(-1)^n}{2^{n/2}}$$

Also, we have

$$\left(\frac{1}{3^{5/8}}\right)^{\log_3 8} = \frac{1}{(3^{5/3})^{3\log_3 2}} = 3^{-(5/3)\log_3 2^3} = 2^{-5}$$

Thus,

$$\frac{(-1)^n}{2^{n/2}} = 2^{-5}$$

$$\Rightarrow \frac{(-1)^n}{2^{n/2}} = \frac{(-1)^{10}}{2^5}$$

$\Rightarrow n/2 = 5$

$\Rightarrow n = 10$

Now,

$$T_5 = T_{4+1} = {}^{10}C_4(2^{1/3})^{10-4}\left(-\frac{1}{\sqrt{2}}\right)^4$$

$$= \frac{10!}{4!6!}(2^{1/3})^6(-1)^4(2^{-1/2})^4$$

$= 210\,(2^2)\,(1)\,(2^{-2}) = 210$

36. d. Given,

$$\frac{{}^{n+1}C_{r+1}}{{}^nC_r} = \frac{11}{6} \Rightarrow \frac{\frac{n+1}{r+1}\times{}^nC_r}{{}^nC_r} = \frac{11}{6}$$

$\Rightarrow 6n + 6 = 11r + 11 \Rightarrow 6n - 11r = 5$ (1)

Also,

$$\frac{{}^nC_r}{{}^{n-1}C_{r-1}} = \frac{6}{3} \Rightarrow \frac{\frac{n}{r}\times{}^{n-1}C_{r-1}}{{}^{n-1}C_{r-1}} = \frac{6}{3} \Rightarrow n = 2r \quad (2)$$

From (1) and (2), $r = 5$ and $n = 10$,

$\therefore nr = 50$

37. b. By the given condition,

$84 = T_6 = T_{5+1}$

$$= {}^7C_5\left(2^{\log_2\sqrt{9^{x-1}+7}}\right)^2\left(\frac{1}{2^{\frac{1}{5}\log_2(3^{x-1}+1)}}\right)^5$$

$$= 21\; 2^{\log_2(9^{x-1}+7)}\,2^{-\log_2(3^{x-1}+1)}$$

$$\Rightarrow 4 = 2^{\log_2\frac{9^{x-1}+7}{3^{x-1}+1}} = \frac{9^{x-1}+7}{3^{x-1}+1}$$

$\Rightarrow (3^{x-1})^2 - 4 \times 3^{x-1} + 3 = 0$

$\Rightarrow (3^{x-1} - 1)(3^{x-1} - 3) = 0$

$\Rightarrow 3^{x-1} = 1$ or 3

$\Rightarrow 3^{x-1} = 3^0$ or 3^1

$\Rightarrow x - 1 = 0$ or 1

$\Rightarrow x = 1, 2$

38. a. General term,

$T_{r+1} = {}^{256}C_r(\sqrt{3})^{256-r}(\sqrt[8]{5})^r$

$= {}^{256}C_r\, 3^{\frac{256-r}{2}}\, 5^{\frac{r}{8}}$

The terms are integral if $(256-r)/2$ and $r/8$ are both positive integers.

$\therefore r = 0, 8, 16, 24, ..., 256$

Hence, there are 33 integral terms.

39. a. $T_{r+1} = {}^{4n-2}C_r(ix)^r$ (since $x > 0$)

T_{r+1} is negative, if i^r is negative and real.

$i^r = -1 \Rightarrow r = 2, 6, 10, \ldots$ which form an A.P.

$0 \le r \le 4n - 2$

$4n - 2 = 2 + (r - 1)\,4 \Rightarrow r = n$

The required number of terms is n.

40. b. $T_5 = {}^nC_4\, a^{n-4}(-2b)^4$

and $T_6 = {}^nC_5\, a^{n-5}(-2b)^5$

As $T_5 + T_6 = 0$, we get

${}^nC_4\, 2^4 a^{n-4} b^4 = {}^nC_5\, 2^5 a^{n-5} b^5$

$$\Rightarrow \frac{a^{n-4}b^4}{a^{n-5}\;b^5} = \frac{n!2^5}{5!(n-5)!}\cdot\frac{4!\,(n-4)}{n!2^4}$$

$$\Rightarrow \frac{a}{b} = \frac{2(n-4)}{5}$$

41. b. $$\left(x + \frac{1}{x} + x^2 + \frac{1}{x^2}\right)^{15} = \left(\frac{x^3 + x + x^4 + 1}{x^2}\right)^{15}$$

$$= \frac{a_0 + a_1x + a_2x^2 + \cdots + a_{60}x^{60}}{x^{30}}$$

Hence, the total number of terms is 61.

42. c. $(1 + x + x^2 + x^3)^5 = a_0 + a_1x + a_2x^2 + a_3x^3 + \cdots + a_{15}x^{15}$

Putting $x = 1$ and $x = -1$ alternatively, we have

$a_0 + a_1 + a_2 + a_3 + \cdots + a_{15} = 4^5$ (1)

$a_0 - a_1 + a_2 - a_3 + \cdots - a_{15} = 0$ (2)

Adding (1) and (2), we have

$2(a_0 + a_2 + a_4 + \cdots + a_{14}) = 4^5$

$\Rightarrow a_0 + a_2 + a_4 + \cdots + a_{14} = 2^9 = 512$

43. c. Sum of coefficients in $(1 - x\sin\theta + x^2)^n$ is $(1 - \sin\theta + 1)^n$ (putting $x = 1$)

This sum is greatest when $\sin\theta = -1$, then maximum sum is $3n$.

44. a. We know that the sum of the coefficients in a binomial expansion is obtained by replacing each variable by unit in the given expression. Therefore, sum of the coefficients in $(a + b)^n$ is given by $(1 + 1)^n$.

$\therefore\ 4096 = 2^n \Rightarrow 2^n = 2^{12} \Rightarrow n = 12$

Hence, n is even. So, the greatest coefficient is ${}^nC_{n/2}$, i.e., ${}^{12}C_6 = 924$.

45. b. We have, a = sum of the coefficients in the expansion of $(1 - 3x + 10x^2)^n = (1 - 3 + 10)^n = (8)^n$

$\Rightarrow (1 - 3x + 10x^2)^n = (2)^{3n}$ (putting $x = 1$)

Now, b = sum of the coefficients in the expansion of $(1+x^2)^n = (1 + 1)^n = 2^n$. Clearly, $a = b^3$.

46. b. $(1 + x - 2x^2)^6 = 1 + a_1x + a_2x^2 + \cdots$

Putting $x = 1$, we get

$0 = 1 + a_1 + a_2 + a_3 + \cdots + a_{12}$ (1)

Putting $x = -1$, we get

$64 = 1 - a_1 + a_2 - a_3 + \cdots + a_{12}$ (2)

(1) + (2) gives

$64 = 2[1 + a_2 + a_4 + \cdots + a_{12}]$

$\Rightarrow 1 + a_2 + a_4 + \cdots + a_{12} = 32$

$\Rightarrow a_2 + a_4 + \cdots + a_{32} = 31$

47. a. $\dfrac{2^{4n}}{15} = \dfrac{(15+1)^n}{15}$

$= \dfrac{({}^nC_0 15^n + {}^nC_1 15^{n-1} + \cdots + {}^nC_{n-1}15 + {}^nC_n)^n}{15}$

$= \text{Integer} + \dfrac{1}{15}$.

Hence, the fractional part of $\dfrac{2^{4n}}{15}$ is $\dfrac{1}{15}$.

48. c. As we know that ${}^nC_0 - {}^nC_1^2 + {}^nC_2^2 - {}^nC_3^2 + \cdots + (-1)^n\, {}^nC_n^2 = 0$ (if n is odd) and in the question $n = 15$ (odd). Hence, sum of given series is 0.

49. b. $(1 - x)^{30} = {}^{30}C_0x^0 - {}^{30}C_1x^1 + {}^{30}C_2x^2 + \cdots + (-1)^{30}\, {}^{30}C_{30}x^{30}$ (1)

$(x + 1)^{30} = {}^{30}C_0x^{30} + {}^{30}C_1x^{29} + {}^{30}C_2x^{28} + \cdots + {}^{30}C_{10}x^{20}$

$+ \cdots + {}^{30}C_{30}x^0$ (2)

Multiplying (1) and (2) and equating the coefficient of x^{20} on both sides, we get required sum is equal to coefficient of x^{20} in $(1 - x^2)^{30}$, which is given by ${}^{30}C_{10}$.

50. b. We have, $f(x) = x^n$. So,

$f^1(x) = nx^{n-1} \Rightarrow f'(1) = n$

$f^2(x) = n(n-1)x^{n-2} \Rightarrow f^2(1) = n(n-1)$

$f^3(x) = n(n-1)(n-2)x^{n-3} \Rightarrow f^3(1) = n(n-1)(n-2)$

$\vdots$

$f^n(x) = n(n-1)(n-3) \cdots 1 \Rightarrow f^n(1) = n(n-1)(n-3) \cdots 1$

$\Rightarrow f(1) + \dfrac{f^1(1)}{1} + \dfrac{f^2(1)}{2!} + \cdots \dfrac{f^n(1)}{n!}$

$= 1 + \dfrac{n}{1} + \dfrac{n(n-1)}{2!} + \dfrac{n(n-1)(n-2)}{3!} \cdots + \dfrac{n(n-1)n(n-2)\cdots 1}{n!}$

$= {}^nC_0 + {}^nC_1 + {}^nC_2 + \cdots + {}^nC_n$

$= 2^n$

51. b. Given series is ${}^{20}C_0 + {}^{20}C_1 + {}^{20}C_2 + \cdots + {}^{20}C_8$

$= \dfrac{(2^{20} - {}^{20}C_{10})}{2} - {}^{20}C_9 = 2^{19} - \dfrac{({}^{20}C_{10} + 2 \times {}^{20}C_9)}{2}$

52. c. Let,

$S = \dfrac{{}^nC_0}{n} + \dfrac{{}^nC_1}{n+1} + \dfrac{{}^nC_2}{n+2} + \cdots + \dfrac{{}^nC_n}{2n}$

$= {}^nC_0 \int_0^1 x^{n-1}dx + {}^nC_1 \int_0^1 x^n dx + \cdots + {}^nC_n \int_0^1 x^{2n-1}dx$

$= \int_0^1 [{}^nC_0x^{n-1} + {}^nC_1x^n + \cdots + {}^nC_nx^{2n-1}]dx$

$= \int_0^1 x^{n-1}(1 + x)^n\, dx$

$= \int_1^2 x^n(x - 1)^{n-1}dx$

53. b. $(1 + x)^n = C_0 + C_1x + C_2x + C_2x^2 + C_3x^3 + \cdots + C_nx^n$

$(1 - x)^n = C_0 - C_1x + C_2x^2 - C_3x^3 + \cdots + (-1)^nC_nx^n$

$\Rightarrow [(1 + x)^n - (1 - x)^n] = 2\,[C_1x + C_3x^3 + C_5x^5 + \cdots]$

$\Rightarrow \dfrac{1}{2}[(1 + x)^n - (1 - x)^n] = C_1x + C_3x^3 + C_5x^5 + \cdots$

Putting $x = 2$, we have

$2\,C_1 + 2^3\,C_3 + 2^5\,C_5 + \cdots = \dfrac{3^n - (-1)^n}{2}$

54. d. $\displaystyle\sum_{r=1}^{n}(-1)^{r+1}\frac{{}^nC_r}{(r+1)} = \frac{1}{n+1}\sum_{r=1}^{n}(-1)^{r+1}\,{}^{n+1}C_{r+1}$

$= \dfrac{1}{n+1}(0 - 1 + (n + 1)) = \dfrac{n}{n+1}$

55. c. $(1 + x)^n = C_0 + C_1x + C_2x^2 + C_3x^3 + \cdots + C_{n-1}x^{n-1} + C_nx^n$ (1)

$(x + 1)^n = C_0x^n + C_1x^{n-1} + C_2x^{n-2} + \cdots + C_{n-1}x + C_n$ (2)

Multiplying Eqs. (1) and (2) and equating the coefficient of x^{n-2}, we get

$C_0C_2 + C_1C_3 + C_2C_4 + \cdots + C_{n-2}C_n$

$=$ Coefficient of x^{n-2} in $(1 + x)^{2n}$

$= {}^{2n}C_{n-2}$

$= \dfrac{(2n)!}{(n-2)!(n+2)!}$

56. a. We know that

$(1 - 1)^{20} = {}^{20}C_0 - {}^{20}C_1 + {}^{20}C_2 - {}^{20}C_3 + \cdots$

$+ {}^{20}C_{10} - {}^{20}C_{11} + {}^{20}C_{12} - \cdots + {}^{20}C_{20} = 0$

$\Rightarrow 2\,({}^{20}C_0 - {}^{20}C_1 + {}^{20}C_2 - {}^{20}C_3 + \cdots - {}^{20}C_9) + {}^{20}C_{10} = 0$

$[\because\ {}^{20}C_{20} = {}^{20}C_0,\ {}^{20}C_{19} = {}^{20}C_1$, etc.$]$

$\Rightarrow {}^{20}C_0 - {}^{20}C_1 + {}^{20}C_2 - {}^{20}C_3 + \cdots - {}^{20}C_9 + {}^{20}C_{10}$

$= -\dfrac{1}{2}\,{}^{20}C_{10} + {}^{20}C_{10} = \dfrac{1}{2}\,{}^{20}C_{10}$

57. b. The given expression is the coefficient of x^4 in

$${}^4C_0(1+x)^{404} - {}^4C_1(1+x)^{303} + {}^4C_2(1+x)^{202} - {}^4C_3(1+x)^{101} + {}^4C_4$$

$= $ Coefficient of x^4 in $[(1+x)^{101} - 1]^4$

$= $ Coefficient of x^4 in $({}^{101}C_1\, x + {}^{101}C_2 x^2 + \cdots)^4$

$= (101)^4$

58. c. Put $x = \omega, \omega^2$

$(3 + \omega + \omega^2)^{2010} = a_0 + a_1\omega + a_2\omega^2 + \cdots$

$\Rightarrow 2^{2010} = a_0 + a_1\omega^2 + a_2\omega + a_3 + a_4\omega + \cdots \quad (1)$

and

$2^{2010} = a_0 + a_1\omega^2 + a_2\omega + a_3 + a_4\omega + \cdots \quad (2)$

Adding (1) and (2), we have

$2 \times 2^{2010} = 2a_0 - a_1 - a_2 + 2a_3 - a_4 - a_5 + 2a_6 - \cdots$

$\Rightarrow 2^{2010} = a_0 - \frac{1}{2}a_1 - \frac{1}{2}a_2 + a_3 - \frac{1}{2}a_4 - \frac{1}{2}a_5 + a_6 \cdots$

59. d. $\sum_{r=0}^{10} r\,{}^{10}C_r 3^r(-2)^{10-r} = 10\sum_{r=0}^{10} {}^9C_{r-1}3^r(-2)^{10-r}$

$= 10 \times 3\sum_{r=0}^{10} {}^9C_{r-1}3^{r-1}(-2)^{10-r}$

$= 30(3-2)^{10}$

$= 30$

60. a. $\sum_{r=0}^{40} r\,{}^{40}C_r\,{}^{30}C_r = 40\sum_{r=0}^{40} {}^{39}C_{r-1}\,{}^{30}C_r$

$= 40\sum_{r=0}^{40} {}^{39}C_{r-1}\,{}^{30}C_{30-r}$

$= 40\,{}^{39+30}C_{r-1+30-r}$

$= 40\,{}^{69}C_{29}$

61. a. $\dfrac{r \times 2^r}{(r+2)!} = \dfrac{(r+2-2)2^r}{(r+2)!}$

$= \dfrac{2^r}{(r+1)!} - \dfrac{2^{r+1}}{(r+2)!}$

$= -\left(\dfrac{2^{r+1}}{(r+2)!} - \dfrac{2^r}{(r+1)!}\right)$

$= -(V(r) - V(r-1))$

$\Rightarrow \sum_{r=1}^{15} \dfrac{r \times 2^r}{(r+2)!} = -(V(15) - V(0))$

$= -\left(\dfrac{2^{16}}{17!} - \dfrac{2}{2!}\right)$

$= 1 - \dfrac{2^{16}}{(17)!}$

62. c. $t_{r+1} = (-1)^r(n - r + 2)\,{}^nC_r\,2^{n-r+1}$

$= (n+2)\,2^{n+1}(-1)^r\,{}^nC_r\left(\dfrac{1}{2}\right)^r - 2^{n+1}(-1)^r r\,{}^nC_r\left(\dfrac{1}{2}\right)^r$

$= (n+2)\,2^{n+1}\,{}^nC_r\left(-\dfrac{1}{2}\right)^r + 2^n\, n\,{}^{n-1}C_{r-1}\left(-\dfrac{1}{2}\right)^{r-1}$

$\therefore \text{Sum} = (n+2)\,2^{n+1}\left\{{}^nC_0 - {}^nC_1 \times \dfrac{1}{2} + {}^nC_2 \times \left(\dfrac{1}{2}\right)^2 - \cdots\right\}$

$+ n\,2^n\left\{{}^{n-1}C_0 - {}^{n-1}C_1 \times \dfrac{1}{2} + {}^{n-1}C_2 \times \left(\dfrac{1}{2}\right)^2 + \cdots\right\}$

$= (n+2)\,2^{n+1}\left(1 - \dfrac{1}{2}\right)^n + n\,2^n\left(1 - \dfrac{1}{2}\right)^{n-1}$

$= 2(n+2) + 2n$

$= 4n + 4$

63. c. Here,

$T_r = (-1)^r\dfrac{{}^{50}C_r}{r+2}$

$= (-1)^r(r+1)\dfrac{{}^{50}C_r}{(r+1)(r+2)}$

$= (-1)^r(r+1)\dfrac{{}^{52}C_{r+2}}{51 \times 52}$

$= (-1)^r\dfrac{[(r+2)-1]\,{}^{52}C_{r+2}}{51 \times 52}$

$= (-1)^r\dfrac{[52\,{}^{51}C_{r+1} - {}^{52}C_{r+2}]}{51 \times 52}$

$= \dfrac{[-52\,{}^{51}C_{r+1}(-1)^r - {}^{52}C_{r+2}(-1)^{r+2}]}{51 \times 52}$

$\sum_{r=0}^{50}(-1)^r\dfrac{{}^{50}C_r}{r+2}$

$= \sum_{r=0}^{50}\dfrac{[-52\,{}^{51}C_{r+1}(-1)^{r+1} - {}^{52}C_{r+2}(-1)^{r+2}]}{51 \times 52}$

$= -52\,\dfrac{(1-1)^{51} - {}^{51}C_0}{51 \times 52} - \dfrac{(1-1)^{52} - {}^{52}C_0 + {}^{52}C_1}{51 \times 52}$

$= \dfrac{1}{51} - \dfrac{1}{52}$

$= \dfrac{1}{51 \times 52}$

Alternative solution:

$(1-x)^n = \sum_{r=0}^{n} {}^nC_r(-1)^r x^r$

$\Rightarrow x(1-x)^n = \sum_{r=0}^{n}(-1)^r\,{}^nC_r x^{r+1}$

Integrating both sides within the limits 0 to 1, we get

$\int_0^1 x(1-x)^n\,dx = \sum_{r=0}^{n}(-1)^r\dfrac{{}^nC_r}{r+2}$

$\Rightarrow \sum_{r=0}^{n}(-1)^r\dfrac{{}^nC_r}{r+2} = \int_0^1 x(1-x)^n\,dx$

$$= \int_0^1 (1-x)x^n dx$$

$$= \left.\frac{x^{n+1}}{x+1} - \frac{x^{n+2}}{n+2}\right|_0^1$$

$$= \frac{1}{n+1} - \frac{1}{n+2}$$

$$= \frac{1}{(n+1)(n+2)}$$

Now put $n = 50$.

64. a. Given term can be written as

$$(1+x)^2(1-x)^{-2} = (1+2x+x^2)[1+2x+3x^2+\cdots+(n-1) \times x^{n-2} + nx^{n-1} + (n+1)x^n + \cdots]$$

Coefficient of x^n is $(n+1+2n+n-1) = 4n$.

65. a. $1 + n\left(1-\frac{1}{x}\right) + \frac{n(n+1)}{2!}\left(1-\frac{1}{x}\right)^2 + \cdots \infty$

$$= 1 - n\left[-\left(1-\frac{1}{x}\right)\right] + \frac{-n(-n-1)}{2!}\left[-\left(1-\frac{1}{x}\right)\right]^2 + \cdots \infty$$

$$= \left[1-\left(1-\frac{1}{x}\right)\right]^{-n}$$

$$= x^n$$

66. c. $\sum_{k=1}^{n} k\left(1+\frac{1}{n}\right)^{k-1} = 1 + 2\left(1+\frac{1}{n}\right)^1 + 3\left(1+\frac{1}{n}\right)^2 + \cdots$

$$= 1 + 2t + 3t^2 + \cdots$$

$$= (1-t)^{-2}$$

$$\left[1-\left(1+\frac{1}{n}\right)\right]^{-2} = \left(\frac{1}{n}\right)^{-2} = n^2$$

67. d. $\left[\sqrt{1+x^2} - x\right]^{-1} = \frac{1}{\sqrt{1+x^2} - x} \times \frac{\left(\sqrt{1+x^2}+x\right)}{\left(\sqrt{1+x^2}+x\right)}$

$$= \frac{\sqrt{1+x^2}+x}{1+x^2-x^2} = x + \sqrt{1+x^2}$$

$$= x + (1+x^2)^{1/2}$$

$$= x + 1 + \frac{1}{2}x^2 + \frac{1}{2}\left(-\frac{1}{2}\right)\frac{x^4}{2!} + \cdots$$

Obviously, the coefficient of x^4 is $-1/8$.

68. d. Let,

$$(1+y)^n = 1 + \frac{1}{3}x + \frac{1\times 4}{3\times 6}x^2 + \frac{1\times 4\times 7}{3\times 6\times 9}x^3 + \cdots$$

$$= 1 + ny + \frac{n(n-1)}{2!}y^2 + \cdots$$

Comparing the terms, we get

$$ny = \frac{1}{3}x, \quad \frac{n(n-1)}{2!}y^2 = \frac{1\times 4}{3\times 6}x^2$$

Solving, $n = -1/3, y = -x$. Hence, the given series is $(1-x)^{-1/3}$.

69. a. Let the given series be identical with

$$(1+x)^n = 1 + nx + \frac{n(n-1)}{1\times 2}x^2 + \cdots \infty$$

$$\Rightarrow nx = \frac{1}{4} \Rightarrow n^2x^2 = \frac{1}{16}$$

Also,

$$\frac{n(n-1)}{2}x^2 = \frac{3}{32} \Rightarrow \frac{2n}{n-1} = \frac{\frac{1}{16}}{\frac{3}{32}} = \frac{2}{3}$$

$$\Rightarrow 3n = n - 1$$

$$\Rightarrow 2n = -1$$

$$\Rightarrow n = -\frac{1}{2} \quad \Rightarrow x = -\frac{1}{2}$$

$$\Rightarrow \text{Required sum} = \left(1-\frac{1}{2}\right)^{-\frac{1}{2}} = \left(\frac{1}{2}\right)^{-\frac{1}{2}} = (2)^{\frac{1}{2}} = \sqrt{2}$$

70. d. Required value is

$$\left(1-\frac{2x}{1+x}\right)^{-n} = \left(\frac{1+x-2x}{1+x}\right)^{-n} = \left(\frac{1-x}{1+x}\right)^{-n} = \left(\frac{1+x}{1-x}\right)^{n}$$

71. d. $(1+2x+3x^2+\cdots)^{-3/2} = [(1-x)^{-2}]^{-3/2}$

$$= (1-x)^3 = 1 - 3x + 3x^2 - x^3$$

Therefore, coefficient of x^5 is 0.

72. d. $(1+x+x^2+\cdots)^2 = ((1-x)^{-1})^2 = (1-x)^{-2}$

$$= 1 + 2x + 3x^2 + \cdots$$

Therefore, coefficient of x^n is $n+1$.

73. b. T_{r+1} in $(1+x)^n$ is

$$\frac{n(n-1)(n-2)\cdots(n-r+1)}{r!}x^r$$

For first negative term,

$$n - r + 1 < 0$$

$$\Rightarrow \frac{27}{5} - r + 1 < 0$$

$$\Rightarrow r > \frac{32}{5}$$

Thus, first negative term occurs when $r = 7$.

74. d. $\frac{(1+x)^{3/2} - \left(1+\frac{1}{2}x\right)^3}{(1-x)^{1/2}}$

$$= \frac{\left(1+\frac{3}{2}x+\frac{3}{8}x^2\right) - \left(1+\frac{3}{2}x+3\frac{x^2}{4}\right)}{(1-x)^{1/2}}$$

$= \frac{-3}{8} x^2 (1-x)^{-1/2}$

$= -\frac{3}{8} x^2 \left(1+\frac{x}{2}\right)$

$= -\frac{3}{8} x^2$

75. d. $\frac{1}{(1-ax)(1-bx)} = a_0 + a_1x + a_2x^2 + \cdots + a_nx^n + \cdots$

But $(1-ax)^{-1}(1-bx)^{-1} = (1 + ax + a^2x^2 + \cdots)(1 + bx + b^2x^2 + \cdots)$

$\Rightarrow$ Coefficient of x^n is $b^n + ab^{n-1} + a^2b^{n-2} + \cdots + a^{n-1}b + a^n$

$= \frac{b^{n+1} - a^{n+1}}{b-a}$

$\Rightarrow a_n = \frac{b^{n+1} - a^{n+1}}{b-a}$

76. c. $\sum_{r=0}^{k} 3^{\frac{1}{k}}({}^kC_r) = 3^{\frac{1}{k}} \sum_{k=0}^{r} {}^kC_r = \frac{2^k}{3^k}$

Thus,

$S = \frac{2/3}{1-2/3} = 2$

77. d. General term in the expansion of $(\sqrt{2}+\sqrt[3]{3}+\sqrt[6]{5})^{10}$ is

$\frac{10!}{a!b!c!}(\sqrt{2})^a(\sqrt[3]{3})^b(\sqrt[6]{5})^c$ where $a+b+c=10$.

For rational term, we have the following:

Value of *a*, *b*, *c*	Value of term
$a=4, b=0, c=6$	$\frac{10!}{4!0!6!}(\sqrt{2})^4(\sqrt[3]{3})^0(\sqrt[6]{5})^6 = 4200$
$a=10, b=0, c=0$	$\frac{10!}{10!0!0!}(\sqrt{2})^{10}(\sqrt[3]{3})^0(\sqrt[6]{5})^0 = 32$
$a=4, b=6, c=0$	$\frac{10!}{4!6!0!}(\sqrt{2})^4(\sqrt[3]{3})^6(\sqrt[6]{5})^0 = 7560$

78. b. $f(x) = 1 - x + x^2 - x^3 + \cdots - x^{15} + x^{16} - x^{17} = \frac{1-x^{18}}{1+x}$

$\Rightarrow f(x-1) = \frac{1-(x-1)^{18}}{x}$

Therefore, required coefficient of x^2 is equal to coefficient of x^3 in $1-(x-1)^{18}$, which is given by ${}^{18}C_3 = 816$.

79. a. $p = (8+3\sqrt{7})^n = {}^nC_0 8^n + {}^nC_1 8^{n-1}(3\sqrt{7}) + \cdots$

Let, $p_1 = (8-3\sqrt{7})^n = {}^nC_0 8^n - {}^7C_1 8^{n-1}(3\sqrt{7}) + \cdots$

$p_1 + p_2 = 2({}^nC_0 8^n + {}^nC_2 8^{n-2}(3\sqrt{7})^2 + \cdots) =$ even integer

p_1 clearly belongs to $(0, 1)$

$\Rightarrow [p] + f + p_1 =$ even integer

$\Rightarrow f + p_1 =$ integer

$f \in (0, 1), p_1 \in (0, 1)$

$\Rightarrow f + p \in (0, 2)$

$\Rightarrow f + p_1 = 1$

$\Rightarrow p_1 = 1 - f$

Now, $p(1-f) = pp_1 = [(8+3\sqrt{7})^n (8-3\sqrt{7})]^n = 1$

80. a. $\sum_{r=0}^{10} = \sum_{r=0}^{10} 20 \times {}^{19}C_{r-1}$

$= 20({}^{19}C_{-1} + {}^{19}C_0 + {}^{19}C_1 + \cdots + {}^{19}C_{10})$

$= 20({}^{19}C_0 + {}^{19}C_1 + \cdots + {}^{19}C_{10})$

$= 20\left(\frac{1}{2} \times 2^{19} + {}^{19}C_{10}\right)$

$= 20(2^{18} + {}^{19}C_{10})$

81. c. $(23)^{14} = (529)^7 = (530-1)^7$

$= {}^7C_0(530)^7 - {}^7C_1(530)^6 + \cdots - {}^7C_5(530)^2 + {}^7C_6\, 530 - 1$

$= {}^7C_0(530)^7 - {}^7C_1(530)^6 + \cdots - 3710 - 1 = 100m + 3709$

Therefore, last two digits are 09.

82. b. $\frac{f(x)}{1-x} = b_0 + b_1x + b_2x^2 + \cdots + b_nx^n + \cdots$

$\Rightarrow a_0 + a_1x + a_2x^2 + \cdots + a_nx^n + \cdots$

$= (1-x)(b_0 + b_1x + b_2x^2 + \cdots + b_nx^n + \cdots)$

Comparing the coefficient of x^n on both the sides,

$a_n = b_n - b_{n-1}$.

83. d. $(1-x)^n(1+x)^n = \sum_{r=0}^{n} a_r x^r (1-x)^n (1-x)^{n-r}$

$\Rightarrow (1-x+2x)^n = \sum_{r=0}^{n} a_r x^r (1-x)^{n-r}$

$\Rightarrow \sum_{r=0}^{n} {}^nC_r (1-x)^{n-r}(2x)^r = \sum_{r=0}^{n} a_r x^r (1-x)^{n-r}$

Comparing general term, we get $a_r = {}^nC_r 2^r$.

84. d. $(1+\omega)^n = {}^nC_0 + {}^nC_1\,\omega + \cdots$

$= ({}^nC_0 + {}^nC_3 + \cdots) + ({}^nC_1 + {}^nC_4 + \cdots)\left(\frac{-1+\sqrt{3}i}{2}\right)$

$+ ({}^nC_2 + {}^nC_5 + \cdots)\left(\frac{-1-\sqrt{3}i}{2}\right)$

$= ({}^nC_0 + {}^nC_3 + \cdots) - \frac{1}{2}({}^nC_1 + {}^nC_2 + {}^nC_4 + {}^nC_5 \ldots)$

$+ \frac{i\sqrt{3}}{2}({}^nC_1 - {}^nC_2 + {}^nC_4 - {}^nC_5 + \cdots)$

Equating the modulus, we get $|(-\omega^2)^n| = 1$.

85. c. $\frac{(x^2+x+1)(1-x)}{(1-x)^2} = (1-x^3)(1-x)^{-2}$

$= (1-x^3)(1+2x+3x^2+\cdots)$

Now, $a_r = (r+1) - (r-2) = 3$

But $a_1 = 2$

So,

$$\sum_{r=1}^{50} a_r = 2 + 49 \times 3 = 149$$

86. a. $N = {}^{2n}C_n = \dfrac{(2n)!}{(n!)^2} = \dfrac{(n+1)(n+2)\cdots(n+n)}{(n!)}$

$\Rightarrow (n!)\, N = (n+1)(n+2)\cdots(n+n)$

Since $n < p < 2n$, so p divides $(n+1)(n+2)\cdots(n+n)$.

87. c. $\displaystyle\sum_{r=0}^{300} a_r \times x^r = (1+x+x^2+x^3)^{100}$

Clearly, a_r is the coefficient of x^r in the expansion of $(1+x+x^2+x^3)^{100}$.

Replacing x by $1/x$ in the given equation, we get

$$\sum_{r=0}^{300} a_r \left(\frac{1}{x}\right)^r = \frac{1}{x^{300}}(x^3+x^2+x+1)^{100}$$

$$\Rightarrow \sum_{r=0}^{300} a_r x^{300-r} = (1+x+x^2+x^3)^{100}$$

Here, a_r represents the coefficient of x^{300-r} in $(1+x+x^2+x^3)^{100}$.

Thus, $a_r = a_{300-r}$

Let $I = \displaystyle\sum_{r=0}^{300} r \times a_r$

$$= \sum_{r=0}^{300} (300-r) a_{300-r}$$

$$= \sum_{r=0}^{300} (300-r) a_r$$

$$= 300 \sum_{r=0}^{300} a_r - \sum_{r=0}^{300} r a_r$$

$\Rightarrow 2I = 300a$

$\Rightarrow I = 150a$

88. a. $\displaystyle\sum_{r=1}^{n+1}\left(\sum_{k=1}^{n} {}^kC_{r-1}\right) = \sum_{r=1}^{n+1}\left(\sum_{k=1}^{n}\left({}^{k+1}C_r - {}^kC_r\right)\right)$

$$= \sum_{r=1}^{n+1}\left({}^{n+1}C_r - {}^1C_r\right)$$

$$= 2^{n+1} - 2$$

89. b. We have,

$(1-x)^{-n} = a_0 + a_1 x + a_2 x^2 + \cdots + a_r x^r + \cdots$

and

$(1-x)^{-1} = 1 + x + x^2 + x^3 + \cdots + x^r + \cdots$

Hence,

$a_0 + a_1 + a_2 + \cdots + a_r$

= Coefficient of x^r in the product of the two series

= Coefficient of x^r in $(1-x)^{-n}(1-x)^{-1}$

$$= \frac{(n+1)(n+2)\cdots(n+r)}{r!}$$

$$= {}^{r+n+1-1}C_{n+1-1} = {}^{n+r}C_n$$

90. d. $\displaystyle\sum_{r=0}^{20} r(20-r)\times({}^{20}C_r)^2 = \sum_{r=0}^{20} r\times{}^{20}C_r(20-r)\times{}^{20}C_{20-r}$

$\Rightarrow \displaystyle\sum_{r=0}^{n} 20\,{}^{19}C_{r-1}\times 20\times{}^{19}C_{19-r} = 400\sum_{r=0}^{n} {}^{19}C_{r-1}\times{}^{19}C_{19-r}$

$= 400 \times$ coefficient of x^{18} in $(1+x)^{19}(1+x)^{19}$

$= 400 \times {}^{38}C_{18}$

$= 400 \times {}^{38}C_{20}$

Multiple Correct Answers Type

1. a., c.

Inclusion of $\log x$ implies $x > 0$.

Now, 3rd term in the expansion is

$T_{2+1} = {}^5C_2\, x^{5-2}\,(x^{\log_{10} x})^2 = 1000000$ (given)

or $x^{3+2\log_{10} x} = 10^5$

Taking logarithm of both sides, we get

$(3 + 2\log_{10} x)\log_{10} x = 5$

or $2y^2 + 3y - 5 = 0$, where $\log_{10} x = y$

or $(y-1)(2y+5) = 0$ or $y = 1$ or $-5/2$

or $\log_{10} x = 1$ or $-5/2$

$\therefore x = 10^1 = 10$ or $10^{-5/2}$

2. a., d.

Coefficients of rth, $(r+1)^{th}$ and $(r+2)^{th}$ terms are ${}^{14}C_{r-1}$, ${}^{14}C_r$ and ${}^{14}C_{r+1}$.

If these coefficients are in A.P., then

$2({}^{14}C_r) = {}^{14}C_{r-1} + {}^{14}C_{r+1}$

$$\Rightarrow \frac{2(14)!}{r!(14-r)!} = \frac{(14)!}{(r-1)!(15-r)!} + \frac{(14)!}{(r+1)!(13-r)}$$

$$\Rightarrow \frac{2(14)!}{r!(14-r)!} = \frac{(14)![(r+1)r + (15-r)(14-r)]}{(r+1)!(15-r)!}$$

$\Rightarrow 2(15-r)(r+1) = 2r^2 - 28r + 210$

$\Rightarrow r^2 - 14r + 45 = 0$ or $(r-5)(r-9) = 0$

$\Rightarrow r = 5$ or 9

3. a., b., c.

We have,

$(x+a)^n = {}^nC_0 x^n + {}^nC_1 x^{n-1} a + {}^nC_2 x^{n-2} a^2 + \cdots + {}^nC_n a^n$

$= [{}^nC_0 x^n + {}^nC_2 x^{n-2} a^2 + \cdots] + [{}^nC_1 x^{n-1} a + {}^nC_3 x^{n-3} a^3 + \cdots]$

or $(x+a)^n = P + Q$ (1)

Similarly,

$(x-a)^n = P - Q$ (2)

i. (1) × (2) $\Rightarrow P^2 - Q^2 = (x^2 - a^2)^n$

ii. Squaring (1) and (2) and subtracting (2) from (1), we get

$4PQ = (x + a)^{2n} - (x - a)^{2n}$

iii. Squaring (1) and (2) and adding,

$2(P^2 + Q^2) = (x + a)^{2n} + (x - a)^{2n}$

4. a, c, d.

$I+f = \left(4+\sqrt{15}\right)^n$

Let $f' = \left(4-\sqrt{15}\right)^n$. Then $0 < f' < 1$

$$I+f = {}^nC_0\, 4^n + {}^nC_1\, 4^{n-1}\sqrt{15} + {}^nC_2\, 4^{n-2}\, 15 + {}^nC_3\, 4^{n-3}\left(\sqrt{15}\right)^3 + \cdots$$

$$f' = {}^nC_0 4^n - {}^nC_1 4^{n-1}\sqrt{15} + {}^nC_2 4^{n-2}\times 15 - {}^nC_3 4^{n-3}\left(\sqrt{15}\right)^3 + \cdots$$

$\therefore I+f+f' = 2\,({}^nC_0\, 4^n + {}^nC_2\, 4^{n-2}\times 15 + \cdots) =$ even integer

$\because 0 < f + f' < 2 \Rightarrow f + f' = 1 \Rightarrow 1 - f = f'$

Thus, I is an odd integer. Now,

$1 - f = f' = (4-\sqrt{15})^n$

$(I + f)(1 - f) = (I + f)f' = 1$

5. c., d.

$${}^{69}C_{3r-1} + {}^{69}C_{3r} = {}^{69}C_{r^2-1} + {}^{69}C_{r^2}$$

$\Rightarrow {}^{70}C_{3r} = {}^{70}C_{r^2}$

Thus, $r^2 = 3r$ or $70 - 3r = r^2$ so that $r = 0$, 3 or 7, −10.

Hence, $r = 3$ and 7 (as the given equation is not defined for $r = 0$ and −10).

6. a., d.

It is given that the fourth term in the expansion of $\left(ax + \frac{1}{x}\right)^n$ is $\frac{5}{2}$, therefore

$${}^nC_3\,(ax)^{n-3}\left(\frac{1}{x}\right)^3 = \frac{5}{2} \Rightarrow {}^nC_3\, a^{n-3}\, x^{n-6} = \frac{5}{2}$$

[$\because$ R.H.S. is independent of x]

Putting $n = 6$ in (1), we get ${}^6C_3\, a^3 = \frac{5}{2} \Rightarrow a^3 = \frac{1}{8} \Rightarrow a = \frac{1}{2}$

7. a., b., c., d.

We know that to get the sum of coefficients, we put $x = 1$. Then, sum of coefficients is $(1 + ax - 2x^2)^n$ is $(a - 1)^n$. Obviously, when $a > 1$, sum is positive for any n.

8. a, b, d.

$$f(m) = \sum_{i=0}^{m}\binom{30}{30-i}\binom{20}{m-i} = \sum_{i=0}^{m}\binom{30}{i}\binom{20}{m-i} = {}^{50}C_m$$

$f(m)$ is greatest when $m = 25$. Also,

$f(0) + f(1) + \cdots + f(50)$

$= {}^{50}C_0 + {}^{50}C_1 + {}^{50}C_2 + \cdots + {}^{50}C_{50} = 2^{50}$

Also, ${}^{50}C_m$ is not divisible by 50 for any m as 50 is not a prime number

$$\sum_{m=0}^{50}(f(m))^2 = ({}^{50}C_0)^2 + ({}^{50}C_1)^2 + ({}^{50}C_2)^2 + \cdots + ({}^{50}C_{50})^2 = {}^{100}C_{50}$$

9. a., b., d.

$(1 + z^2 + z^4)^8 = C_0 + C_1 z^2 + C_2 z^4 + \cdots + C_{16} z^{32}$ (1)

Putting $z = i$, where $i = \sqrt{-1}$,

$(1 - 1 + 1)^8 = C_0 - C_1 + C_2 - C_3 + \cdots + C_{16}$

$\Rightarrow C_0 - C_1 + C_2 - C_3 + \cdots + C_{16} = 1$

Also, putting $z = \omega$,

$(1 + \omega^2 + \omega^4)^8 = C_0 + C_1\,\omega^2 + C_2\,\omega^4 + \cdots + C_{16}\,\omega^{32}$

$\Rightarrow C_0 + C_1\omega^2 + C_2\omega + C_3 + \cdots + C_{16}\omega^2 = 0$ (2)

Putting $x = \omega^2$,

$(1 + \omega^4 + \omega^8)^8 = C_0 + C_1\omega^4 + C_2\,\omega^8 + \cdots + C_{16}\omega^{64}$

$\Rightarrow C_0 + C_1\omega + C_2\,\omega^2 + \cdots + C_{16}\omega = 0$ (3)

Putting $x = 1$,

$3^8 = C_0 + C_1 + C_2 + \cdots + C_{16}$ (4)

Adding (2), (3) and (4), we have

$3(C_0 + C_3 + \cdots + C_{15}) = 3^8$

$\Rightarrow C_0 + C_3 + \cdots + C_{15} = 3^7$

Similarly, first multiplying (1) by z and then putting 1, ω, ω^2 and adding, we get

$C_1 + C_4 + C_7 + C_{10} + C_{13} + C_{16} = 3^7$

Multiplying (1) by z^2 and then putting 1, ω, ω^2 and adding, we get

$C_2 + C_5 + C_8 + C_{11} + C_{14} = 3^7$

10. a.

General term is ${}^{6561}C_r\, 7^{\frac{6561-r}{3}}\, 11^{\frac{r}{9}}$.

To make the term free of radical sign, r should be a multiple of 9.

$\therefore r = 0, 9, 18, 27, \ldots, 6561$.

Hence, there are 730 terms. The greatest binomial coefficients are

${}^{6561}C_{\frac{6561-1}{2}}$ and ${}^{6561}C_{\frac{6561-3}{2}}$ or ${}^{6561}C_{3280}$ and ${}^{6561}C_{3279}$.

Now, 3280 are 3279 are not a multiple of 3; hence, both terms involving greatest binomial coefficients are irrational.

11. b., c.

For $n = 2m$, the given expression is

$$C_0 - (C_0 + C_1) + (C_0 + C_1 + C_2) - (C_0 + C_1 + C_2 + C_3) + \cdots (-1)^{n-1}(C_0 + C_1 + \cdots + C_{n-1})$$

$$= C_0 - (C_0 + C_1) + (C_0 + C_1 + C_2) - (C_0 + C_1 + C_2 + C_3) + \cdots - (C_0 + C_1 + \cdots + C_{2m-1})$$

$$= -(C_1 + C_3 + C_5 + \cdots + C_{2m-1})$$

$$= -(C_1 + C_3 + C_5 + \cdots + C_{n-1}) = -2^{n-1}$$

12. a., c., d.

$\left(\sqrt{2}+1\right)^6 = I + f$

$= {}^6C_0\left(\sqrt{2}\right)^6 + {}^6C_1\left(\sqrt{2}\right)^5 + \cdots + {}^6C_6$,

where I is an integer, $0 < f < 1$.

Let $f' = \left(\sqrt{2}-1\right)^6 = {}^6C_0\left(\sqrt{2}\right)^6 - {}^6C_1\left(\sqrt{2}\right)^5 + \cdots$, $0 < f' < 1$

$\therefore p + f + f' = 2[{}^6C_0\, 2^3 + {}^6C_2\, 2^2 + {}^6C_4\, 2 + {}^6C_6]$

= even integer

$\Rightarrow f+f'$ is an integer

But $f+f' \in (0, 2)$ as $0 < f, f' < 1$

$\Rightarrow f+f' = 1$

$\Rightarrow I$ is an odd integer

Also, $(I+f)(1-f) = (I+f)f' = (\sqrt{2}+1)^6(\sqrt{2}-1)^6 = 1$

$2[8 + 15 \times 4 + 15 \times 2 + 1] = 198$

$\therefore 197 < p + f < 198$

$\Rightarrow 197 < (\sqrt{2}+1)^6 < 198$

$\Rightarrow \left(\sqrt{2}+1\right)^6 = 197$

13. a., b., d.

$${}^nC_1 + {}^{n+1}C_2 + {}^{n+2}C_3 + \cdots + {}^{n+m-1}C_m$$

$$= {}^nC_{n-1} + {}^{n+1}C_{n-1} + {}^{n+2}C_{n-1} + \cdots + {}^{n+m-1}C_{n-1}$$

$$= \text{Coefficient of } x^{n-1} \text{ in } (1+x)^n + (1+x)^{n+1} + (1+x)^{n+2} + \cdots + (1+x)^{n+m-1}$$

$$= \text{Coefficient of } x^{n-1} \text{ in } (1+x)^n\left[\frac{(1+x)^m - 1}{(1+x)-1}\right]$$

$$= \text{Coefficient of } x^{n-1} \text{ in } \frac{(1+x)^{m+n} - (1+x)^n}{x}$$

$$= \text{Coefficient of } x^n \text{ in } \left[(1+x)^{m+n} - (1+x)^{n-1}\right]$$

$$= {}^{m+n}C_n$$

Similarly, we can prove

$${}^mC_1 + {}^{m+1}C_2 + {}^{m+2}C_3 + \cdots + {}^{m+n-1}C_n = {}^{m+n}C_m$$

14. a., b., d.

$$\frac{(n-1)(n-2)\cdots(n-m+1)}{(m-1)!}$$

$$= \frac{(n-1)(n-2)\cdots(n-m+1)(n-m)\cdots 2\times 1}{(n-m)!(m-1)!}$$

$$= {}^{n-1}C_{m-1}$$

$$= \text{Coefficient of } x^{m-1} \text{ in } (1+x)^{n-1}$$

$$= \text{Coefficient of } x^{m-1} \text{ in } (1+x)^n(1+x)^{-1}$$

Now,

$$(1+x)^n = C_0 + C_1x + C_2x^2 + \cdots + C_{m-1}x^{m-1} + \cdots + C_nx^n \quad (1)$$

$$(1+x)^{-1} = 1 - x + x^2 - x^3 + \cdots + (-1)^{m-1}x^{m-1} + \cdots \quad (2)$$

Collecting the coefficients of x^{m-1} in the product of (1) and (2), we get

$$(-1)^{m-1}C_0 + (-1)^{m-2}C_1 + \cdots + C_{m-1}$$

$$= \text{Coefficient of } x^{m-1} \text{ in } (1+x)^{n-1}$$

$$= {}^{n-1}C_{m-1}$$

$$\therefore C_0 - C_1 + C_2 - \cdots + (-1)^{m-1}C_{m-1}$$

$$= {}^{n-1}C_{m-1}(-1)^{m-1}$$

$$= \frac{(n-1)(n-2)\cdots(n-m+1)}{(m-1)!}(-1)^{m-1}$$

15. a., d.

We have,

$$\frac{17}{4} + 3\sqrt{2} = \frac{1}{4}(9 + 8 + 12\sqrt{2})$$

$$= \frac{1}{4}(3 + 2\sqrt{2})^2$$

$$\therefore \quad 3 - \sqrt{\frac{17}{4} + 3\sqrt{2}} = 3 - \frac{1}{2}(3 + 2\sqrt{2})$$

$$= \frac{3}{2} - \sqrt{2}$$

Hence, the 10th term of $\left(3 - \sqrt{\frac{17}{4} + 3\sqrt{2}}\right)^{20}$ is

$${}^{20}C_0\left(\frac{3}{2}\right)^{20-9}(-\sqrt{2})^9,$$

which is a positive integer and hence a rational number.

16. a., b., c.

$(x \sin p + x^{-1}\cos p)^{10}$

The general term in the expansion is

$$T_{r+1} = {}^{10}C_r(x \sin p)^{10-r}(x^{-1}\cos p)^r$$

For the term independent of x, we have $10 - 2r = 0$ or $r = 5$.

Hence, the independent term is

$${}^{10}C_5 \sin^5 p \cos^5 p = {}^{10}C_5 \frac{\sin^5 2p}{32},$$

which is the greatest when $\sin 2p = 1$.

The least value of ${}^{10}C_5 \frac{\sin^5 2p}{32}$ is $-\frac{10!}{2^5(5!)^2}$ when

$\sin 2p = -1$ or $p = (4n-1)\frac{\pi}{4}, n \in Z$.

Sum of coefficient is $(\sin p + \cos p)^{10}$, when $x = 1$ or $(1 + \sin 2p)^5$, which is least when $\sin 2p = -1$. Hence, least sum of coefficients is zero. Greatest sum of coefficient occurs when $\sin 2p = 1$. Hence, greatest sum is $2^5 = 32$.

17. b., c., d.

L.H.S. $= (1 + 2x^2 + x^4)(1 + C_1x + C_2x^2 + C_3x^3 + \cdots)$

R.H.S. $= a_0 + a_1x + a_2x^2 + a_3x^3 + \cdots$

Comparing the coefficients of $x, x^2, x^3, \ldots$

$a_1 = C_1, a_2 = C_2 + 2, a_3 = C_3 + 2C_1$ (1)

Now, $2a_2 = a_1 + a_3$ (A.P.)

$\Rightarrow 2({}^nC_2 + 2) = {}^nC_1 + ({}^nC_3 + 2\,{}^nC_1)$ [Using (1)]

$$\Rightarrow 2\frac{n(n-1)}{2} + 4 = 3n + \frac{n(n-1)(n-2)}{6}$$

$\Rightarrow n^3 - 9n^2 + 26n - 24 = 0$

$\Rightarrow (n-2)(n^2 - 7n + 12) = 0$

$\Rightarrow (n-2)(n-3)(n-4) = 0$

$\Rightarrow n = 2, 3, 4$

18. a., d.

Middle term is $\left(\frac{n}{2}+1\right)^{\text{th}}$ or $(4+1)^{\text{th}}$ or T_5

$$\Rightarrow T_5 = {}^8C_4\left(\frac{x}{2}\right)^4 \times 2^4 = 1120$$

$\Rightarrow \frac{8\times7\times6\times5}{1\times2\times3\times4}x^4 = 1120$

$\Rightarrow x^4 = \frac{1120}{70} = 16$

$\Rightarrow (x^2+4)(x^2-4) = 0$

$\Rightarrow x = \pm 2$ ($\because x \in R$)

19. a., b., c., d.

Let T_5 be numerically the greatest term in the expansion of $(1 + x/3)^{10}$.

Then,

$\left|\frac{T_5}{T_4}\right| \ge 1$ and $\left|\frac{T_6}{T_5}\right| \le 1$

Now,

$\frac{T_{r+1}}{T_r} = \frac{10-r+1}{r}\frac{x}{3}$

$\Rightarrow \left|\frac{7}{4}\times\frac{x}{3}\right| \ge 1$ and $\left|\frac{6}{5}\times\frac{x}{3}\right| \le 1$

$\Rightarrow |x| \ge \frac{12}{7}$ and $|x| \le \frac{5}{2}$ (1)

$\Rightarrow \frac{12}{7} \le |x| \le \frac{5}{2}$

$\Rightarrow x \in \left[-\frac{5}{2}, -\frac{12}{7}\right] \cup \left[\frac{12}{7}, \frac{5}{2}\right]$

20. a., c.

$(1-y)^m (1+y)^n$

$= (1 - {}^mC_1 y + {}^mC_2 y^2 - \cdots)(1 + {}^nC_1 y + {}^nC_2 y^2 + \cdots)$

$= 1 + (n-m) + \left\{\frac{m(m-1)}{2} + \frac{n(n-1)}{2} - mn\right\} y^2 + \cdots$

Given,

$a_1 = 10$

$\Rightarrow a_1 = n - m = 10$ (1)

$a_2 = \frac{m^2+n^2-m-n-2mn}{2} = 10$

$(m-n)^2 - (m+n) = 20$

$\Rightarrow m + n = 80$ (2)

Solving (1) and (2), we get $m = 35$, $n = 45$.

Reasoning Type

1. a. $({}^{10}C_0) + ({}^{10}C_0 + {}^{10}C_1) + ({}^{10}C_0 + {}^{10}C_1 + {}^{10}C_2)$

$+ \cdots + ({}^{10}C_0 + {}^{10}C_1 + {}^{10}C_2 + \cdots + {}^{10}C_9)$

$= 10\,{}^{10}C_0 + 9\,{}^{10}C_1 + 8\,{}^{10}C_2 + \cdots + {}^{10}C_9$

$= {}^{10}C_1 + 2\,{}^{10}C_2 + 3\,{}^{10}C_3 + \cdots + 10\,{}^{10}C_{10}$

$= \sum_{r=1}^{10} r\,{}^{10}C_r = 10\sum_{r=1}^{10} {}^9C_{r-1} = 10 \times 2^9$

2. b. $\frac{T_{r+1}}{T_r} = \frac{12-r+1}{r}\frac{11}{10}$

Let,

$T_{r+1} \ge T_r \Rightarrow 13 - r \ge 1.1x$

$\Rightarrow 13 \ge 2.1r$

$\Rightarrow r \le 6.19$

Hence, the greatest term occurs for $r = 6$.

Hence, 7th term is the greatest term. Also, the binomial coefficient of 7th term is ${}^{12}C_6$ which is the greatest binomial coefficient.

But this is not the reason for which T_7 is the greatest. Here, it is co-incident that the greatest term has the greatest binomial coefficient

Hence, statement 1 is true, statement 2 is true; but statement 2 is not a correct explanation of statement 1.

3. a. $3456^{2222} = (7 \times 493 + 5)^{2222}$

$= (7k + 5)^{2222}$

$= 7m + 5^{2222}$

Now,

$5^{2222} = 5^2(5^3)^{740}$

$= 25(125)^{740}$

$= 25(126 - 1)^{740}$

$= 25[7n + 1]$

$= 175n + 25$

Remainder when $175n + 25$ is divided by 7 is 4.

Hence, both the statements are correct and statement 2 is a correct explanation of statement 1.

4. d. Statement 2 is true as it is the property of binomial coefficients.

But statement 1 is false as three consecutive binomial coefficients may be in A.P. but not always.

5. a. $(1 + x + x^2)^n = \sum_{r=0}^{2n} a_r x^r$ (1)

We know that

$(1-x)^n = \sum_{r=0}^{n} (-1)^r\,{}^nC_r x^r = \sum_{r=0}^{n} (-1)^{n-r}\,{}^nC_r x^{n-r}$ (2)

Multiplying (1) and (2), we get

$\sum_{r=0}^{n} (-1)^{-r}\,{}^nC_r a_r =$ coefficient of x^n in $(1-x^3)^n$

Since $n \ne 3k$, therefore

$\sum_{r=0}^{n} (-1)^{-r} a_r\,{}^nC_r = 0$

$\Rightarrow \sum_{r=0}^{n} (-1)^r a_r\,{}^nC_r = 0$

Hence, both the statements are correct and statement 2 is a correct explanation of statement 1.

6. a. $(1+x)^n - nx - 1 = (1 + {}^nC_1 x + {}^nC_2 x^2 + \cdots + {}^nC_n x^n) - nx - 1$ (1)

$= {}^nC_2 x^2 + \cdots + {}^nC_n x^n$

$= x^2({}^nC_2 + {}^nC_3 x + \cdots + {}^nC_n x^{n-2})$

Hence, $(1+x)^n - nx - 1$ is divisible by x^2.

Now in (1), replace x by 8 and n by $n + 1$. Then, we have

$(1+8)^{n+1} - (n+1)8 - 1 = 8^2({}^nC_2 + {}^nC_3 8 + \cdots + {}^nC_n 8^{n-2})$

$\Rightarrow 9^{2n+2} - 8n - 9 = 8^2({}^nC_2 + {}^nC_3 8 + \cdots + {}^nC_n 8^{n-2})$,

which is divisible by 64.

Hence, both the statements are correct and statement 2 is a correct explanation of statement 1.

7. b. $(1+x+x^2+x^3+x^4)^{1000} = a_0 + a_1x + a_2x^2 + a_3x^3 + \cdots + a_{4000}x^{4000}$

Clearly, there are 4001 terms. Also, number of term in the expansion

$(a_1 + a_2 + \cdots + a_m)^n$ is ${}^{n+m-1}C_{m-1}$.

Clearly, statement 2 has nothing to do with statement 1.

8. a. Coefficient of x^n in $\left(1 + x + \frac{x^2}{2!} + \frac{x^3}{3!} + \cdots \frac{x^n}{n!}\right)^3$

$= $ Coefficient of x^n in $\left(1 + x + \frac{x^2}{2!} + \cdots \frac{x^n}{n!} + \cdots\right)^3$

as higher powers of x are not counted while calculating the coefficient of x^n)

$= $ Coefficient of x^n in e^{3x} is $\frac{3^n}{n!}$

9. b. $(1+x)^{41}(1-x+x^2)^{40}$

$= (1+x)(1+x)^{40}(1-x+x^2)^{40}$

$= (1+x)(1+x^3)^{40}$

$= (1+x^3)^{40} + x(1+x^3)^{40}$

$= (1 + {}^{40}C_1x^3 + {}^{40}C_2x^6 + \cdots + {}^{40}C_{40}x^{120}) + ({}^{40}C_0 + {}^{40}C_1x^4 + {}^{40}C_2x^7 + \cdots + {}^{40}C_{40}x^{121})$

Hence, the coefficient of x^{85} is zero as there is no term in the above expansion which has x^{85}.

Also, statement 2 is correct but it is not a correct explanation of statement 1.

10. b. We know that the total number of terms in $(x_1 + x_2 + \cdots + x_r)^n$ is ${}^{n+r-1}C_{r-1}$. So the total number of term in $(x_1 + x_2 + \cdots + x_n)^3$ is

$${}^{3+n-1}C_{n-1} = {}^{n+2}C_{n-1} = {}^{n+2}C_3 = \frac{(n+2)(n+1)n}{6}$$

and the total number of terms in $(x_1 + x_2 + x_3)^n$ is

$${}^{n+3-1}C_{3-1} = {}^{n+2}C_2 = \frac{(n+2)(n+1)n}{6}$$

11. a. We have,

$$\left(2+\sqrt{5}\right)^p + \left(2-\sqrt{5}\right)^p$$

$$= 2[2^p + {}^pC_2\, 2^{p-5}\, 5 + {}^pC_4\, 2^{p-4}\, 5^2 + \cdots + {}^pC_{p-1}\, 2 \times 5^{(p-1)/2}] \quad (1)$$

From (1), $\left(2+\sqrt{5}\right)^p + \left(2-\sqrt{5}\right)^p$ is an integer and

$-1 < \left(2-\sqrt{5}\right)^p < 0 \quad (\because p \text{ is odd})$

So,

$$\left[\left(2+\sqrt{5}\right)^p\right] = (2+\sqrt{5})^p + (2-\sqrt{5})^p$$

$$= 2^{p+1} + {}^pC_2\, 2^{p-1}\, 5 + \cdots + {}^pC_{p-1}\, 2^2\, 5^{(p-1)/2}$$

$$\therefore \left[\left(2+\sqrt{5}\right)^p\right] - 2^{p+1} = 2[{}^pC_2\, 2^{p-2}\, 5 + {}^pC_4 2^{p-4}\, 5^2 + \cdots + {}^pC_{p-1}\, 2 \times 5^{(p-1)/2}]$$

Now, all the binomial coefficients

$${}^pC_2 = \frac{p(p-1)}{1 \times 2},$$

$${}^pC_4 = \frac{p(p-1)(p-2)(p-3)}{1 \times 2 \times 3 \times 4}, \ldots, {}^pC_{p-1} = p$$

are divisible by the prime p. Thus, R.H.S. is divisible by p.

12. a. Statement 2 is true (can be checked easily) and that is why ${}^{2n}C_0 < {}^{2n}C_1 < {}^{2n}C_2 < \cdots < {}^{2n}C_{n-1} < {}^{2n}C_n > {}^{2n}C_{n+1} \cdots > {}^{2n}C_{2n}$.

13. b. Obviously, statement 2 is true. But to get the sum of coefficients in the expansion of $(3^{-x/4} + 3^{5x/4})^n$, we must put $x = 0$.

14. a. We know that

$${}^mC_r + {}^mC_{r-1}\,{}^nC_1 + {}^mC_{r-2}\,{}^nC_2 + \cdots + {}^nC_r$$

$= $ Coefficient of x^r in $(1+x)^m(1+x)^n$

$= $ Coefficient of x^r in $(1+x)^{m+n}$

$= {}^{m+n}C_r$

$= 0$ as $m + n < r$

15. a. $$S = \sum\sum_{0 \le i < j \le n}\left(\frac{i}{{}^nC_i} + \frac{j}{{}^nC_j}\right)$$

$$= \sum\sum_{0 \le i < j \le n}\left(\frac{n-i}{{}^nC_{n-i}} + \frac{n-j}{{}^nC_{n-j}}\right)$$

$$= n\sum\sum_{0 \le i < j \le n}\left(\frac{1}{{}^nC_i} + \frac{1}{{}^nC_j}\right) - S$$

$$\Rightarrow S = \frac{n}{2}\sum\sum_{0 \le i < j \le n}\left(\frac{1}{{}^nC_i} + \frac{1}{{}^nC_j}\right)$$

$$= \frac{n}{2}\left(\sum_{r=0}^{n-1}\frac{n-r}{{}^nC_r} + \sum_{r=1}^{n}\frac{r}{{}^nC_r}\right)$$

$$= \frac{n}{2}\left(\sum_{r=0}^{n}\frac{n}{{}^nC_r}\right)$$

$$= \frac{n^2}{2}a$$

Linked Comprehension Type

For Problems 1–3

1. b., 2. d., 3. c.

Sol. The coefficient of the 2nd, 3rd and 4th terms in the expansion are ${}^mC_1$, ${}^mC_2$ and ${}^mC_3$, which are given in A.P. Hence,

$$2\,{}^mC_2 = {}^mC_1 + {}^mC_3$$

$$\Rightarrow \frac{2m(m-1)}{2!} = m + \frac{m(m-1)(m-2)}{3!}$$

$$\Rightarrow m(m^2 - 9m + 14) = 0$$

$$\Rightarrow m(m-2)(m-7) = 0$$

$\Rightarrow m = 7$ ($\because m \neq 0$ or 2 as 6th term is given equal to 21)

Now, 6th term in the expansion, when $m = 7$, is

$${}^7C_5\left[\sqrt{\{2^{\log(10-3^x)}\}}\right]^{7-5} \times \left[\sqrt[5]{\{2^{(x-3)\log 3}\}}\right]^5 = 21$$

$$\Rightarrow \frac{7\times 6}{2!}\, 2^{\log(10-3^x)} \times 2^{(x-2)\log 3} = 21$$

$\Rightarrow 2^{\log(10-3^x)+(x-2)\log 3} = 1 = 2^0$

$\Rightarrow \log(10-3^x) + (x-2)\log 3 = 0$

$\Rightarrow \log(10-3^x)(3)^{(x-2)} = 0$

$\Rightarrow (10-3^x)\times 3^x \times 3^{-2} = 1$

$\Rightarrow 10\times 3^x - (3^x)^2 = 9$

$\Rightarrow (3^x)^2 - 10\times 3^x + 9 = 0$

$\Rightarrow (3^x-1)(3^x-9) = 0$

$\Rightarrow 3^x - 1 = 0 \Rightarrow 3^x = 1 = 3^0 \Rightarrow x = 0$

$\Rightarrow 3^x - 9 = 0 \Rightarrow 3^x = 3^2 \Rightarrow x = 2$

Hence, $x = 0$ or 2. When $x = 2$,

$$\left[\sqrt{\{2^{\log(10-3^x)}\}} + \sqrt[5]{\{2^{(x-2)\log 3}\}}\right]^m = [1+1]^7 = 128$$

When $x = 0$,

$$\left[\sqrt{\{2^{\log(10-3^x)}\}} + \sqrt[5]{\{2^{(x-2)\log 3}\}}\right]^m = \left[\sqrt{\{2^{\log 9}\}} + \sqrt[5]{\{2^{-2\log 3}\}}\right]^7$$

$$= \left[2^{\frac{\log 9}{2}} + \frac{1}{2^{\frac{\log 9}{5}}}\right]^7 > 2^7$$

Hence, the minimum value is 128.

For Problems 4–6

4. b., 5. c., 6. c.

Sol. 2nd term is ${}^nC_1 x^{n-1}a = 240$ (1)

3rd term is ${}^nC_2\, x^{n-2}\, a^2 = 720$ (2)

4th term is ${}^nC_3\, x^{n-3}\, a^3 = 1080$ (3)

Multiplying (1) and (3) and dividing by the square of (2), we get

$$\frac{{}^nC_1\times{}^nC_3}{({}^nC_2)^2} = \frac{240\times 1080}{(720)^2}$$

$$\Rightarrow \frac{n\times n(n-1)(n-2)(2!)^2}{n^2(n-1)^2\times 3!} = \frac{1}{2}$$

$\Rightarrow 4(n-2) = 3(n-1)$ ($\because n \neq 1$)

$\Rightarrow n = 5$

Putting $n = 5$, from (1) and (2), we get

$5x^4a = 240$ and $10x^3a^2 = 720$

$$\Rightarrow \frac{(5x^4a)^2}{10x^3a^2} = \frac{(240)^2}{720}$$

or $x^5 = 32$

$\therefore x = 2$

$$\therefore\ a = \frac{240}{5x^4} = \frac{48}{2^4} = 3$$

Hence, $x = 2$, $a = 3$ and $n = 5$.

$(x-a)^n = (2-3)^5 = -1$

Also,

$(2+3)^5 = 2^5 + {}^5C_1 2^4\times 3 + {}^5C_2 2^3\times 3^2 + {}^5C_3 2^2\times 3^3 + {}^5C_4 2\times 3^4 + {}^5C_5 3^5$

$= 32 + 240 + 720 + 1080 + 810 + 243$

Hence, least value of the term is 32.

Sum of odd-numbered terms is $32 + 720 + 810 = 1562$.

For Problems 7–9

7. b., 8. a., 9. c.

Sol. Let,

$$(1+x+x^2)^{20} = \sum_{r=0}^{40} a_r x^r \quad (1)$$

Replacing x by $1/x$, we get

$$\left(1+\frac{1}{x}+\frac{1}{x^2}\right)^{20} = \sum_{r=0}^{40} a_r\left(\frac{1}{x}\right)^r$$

$$\Rightarrow (1+x+x^2)^{20} = \sum_{r=0}^{40} a_r x^{40-r} \quad (2)$$

Since (1) and (2) are same series, coefficient of x^r in (1) = coefficient of x^r in (2)

$\Rightarrow a_r = a_{40-r}$

In (1), putting $x = 1$, we get

$3^{20} = a_0 + a_1 + a_2 + \cdots + a_{40}$

$= (a_0 + a_1 + a_2 + \cdots + a_{19}) + a_{20} + (a_{21} + a_{n+2} + \cdots + a_{40})$

$= 2(a_0 + a_1 + a_2 + \cdots + a_{19}) + a_{20}$

$$\Rightarrow a_0 + a_1 + a_2 + \cdots + a_{19} = \frac{1}{2}(3^{20} - a_{20}) = \frac{1}{2}(9^{10} - a_{20})$$

Also,

$a_0 + 3a_1 + 5a_2 + \cdots + 81a_{40}$

$= (a_0 + 81a_{40}) + (3a_1 + 79a_{39}) + \cdots + (39a_{19} + 43a_{21}) + 41a_{20}$

$= 82(a_0 + a_1 + a_2 + \cdots + a_{19}) + 41a_{20}$

$= 41(9^{10} - a_{20}) + 41a_{20}$

$= 41\times 3^{20}$

$a_0^2 - a_1^2 + a_2^2 - a_3^2 + \cdots$ suggests that we have to multiply the two expansions.

Replacing x by $-1/x$ in (1), we get

$$\left(1-\frac{1}{x}+\frac{1}{x^2}\right)^{20} = a_0 - \frac{a_1}{x} + \frac{a_2}{x^2} - \cdots + \frac{a_{40}}{x^{40}}$$

$\Rightarrow (1-x+x^2)^{20} = a_0x^{40} - a_1x^{39} + a_2x^{38} - \cdots + a_{40}$ (2)

Clearly,

$a_0^2 - a_1^2 + a_2^2 + \cdots + a_{40}^2$ is the coefficient of x^{40} in $(1+x+x^2)^{20}(1-x+x^2)^{20}$

= Coefficient of x^{40} in $(1+x^2+x^4)^{20}$

In $(1+x^2+x^4)^{20}$, replace x^2 by y, then the coefficient of y^{20} in $(1+y+y^2)^{20}$ is a_{20}. Hence,

$$a_0^2 - a_1^2 + a_2^2 - \cdots + a_{40}^2 = a_{20}$$

$$\Rightarrow (a_0^2 - a_1^2 + a_2^2 - \cdots - a_{19}^2) + a_{20}{}^2 + (-a_{21}^2 + \cdots + a_{40}^2) = a_{20}$$

$$\Rightarrow 2(a_0^2 - a_1^2 + a_2^2 - \cdots - a_{19}^2) + a_{20}^2 = a_{20}$$

$$\Rightarrow a_0^2 - a_1^2 + a_2^2 - \cdots - a_{19}^2 = \frac{a_{20}}{2}[1 - a_{20}]$$

For Problems 10–12

10. c., 11. a., 12. c.

Sol. $a_0 + a_1x + a_2x^2 + \cdots + a_{99}x^{99} + x^{100} = 0$ has roots ${}^{99}C_0$, ${}^{99}C_1$, ${}^{99}C_2$, …, ${}^{99}C_{99}$.

$$\Rightarrow a_0 + a_1x + a_2x^2 + \cdots + a_{99}x^{99} + x^{100}$$

$$= (x - {}^{99}C_0)(x - {}^{99}C_1)(x - {}^{99}C_2) \cdots (x - {}^{99}C_{99})$$

Now, sum of roots is

$${}^{99}C_0 + {}^{99}C_1 + {}^{99}C_2 + \cdots + {}^{99}C_{99} = -\frac{a_{99}}{\text{coefficient of } x^{100}}$$

$$\Rightarrow a_{99} = -2^{99}$$

Also, sum of product of roots taken two at a time is

$$\frac{a_{99}}{\text{coefficient of } x^{100}}$$

$$\therefore \sum\sum_{0 \le i < j \le 99} {}^{99}C_i\, {}^{99}C_j = \frac{\left(\sum_{i=0}^{99}\sum_{j=0}^{99} {}^{99}C_i\, {}^{99}C_j\right) - \sum_{i=0}^{99} ({}^{99}C_i)^2}{2}$$

$$= \frac{\left(\sum_{i=0}^{99} {}^{99}C_i 2^{99}\right) - \sum_{i=0}^{99} ({}^{99}C_i)^2}{2}$$

$$= \frac{2^{99}2^{99} - \sum_{i=0}^{99} ({}^{99}C_i)^2}{2}$$

$$= \frac{2^{198} - {}^{198}C_{99}}{2}$$

$$({}^{99}C_0)^2 + ({}^{99}C_1)^2 + \cdots + ({}^{99}C_{99})^2$$

$$= ({}^{99}C_0 + {}^{99}C_1 + {}^{99}C_2 \cdots + {}^{99}C_{99})^2 - 2\sum\sum_{0 \le i < j \le 99} {}^{99}C_i\, {}^{99}C_j$$

$$= (-a_{99})^2 - 2a_{98}$$

$$= a_{99}^2 - 2a_{98}$$

For Problems 13–15

13. a, 14. c, 15. a.

Sol. a. $\sum_{r=0}^{100} {}^{100}C_r \sin rx = \text{Im}\left(\sum_{r=0}^{100} {}^{100}C_r e^{irx}\right)$ (Im = inaginary part)

$$= \text{Im}\left(\sum_{r=0}^{100} {}^{100}C_r (e^{ix})^r\right) = \text{Im}((1 + e^{ix})^{100})$$

$$= \text{Im}\,(1 + \cos x + i \sin x)^{100}$$

$$= \text{Im}\left(2\cos^2\frac{x}{2} + 2i\sin\frac{x}{2} \times \cos\frac{x}{2}\right)^{100}$$

$$= \text{Im}\left(2\cos\frac{x}{2}\left(\cos\frac{x}{2} + i\sin\frac{x}{2}\right)\right)^{100}$$

$$= 2^{100}\cos^{100}\frac{x}{2}\sin(50x) \times \sum_{r=0}^{50} \times {}^{50}C_r a^r \times b^{50-r} \times \cos[rB - (50 - r)A]$$

$$= \text{Re}\left(\sum_{r=0}^{50} {}^{50}C_r a^r \times b^{50-r} \times e^{i(rB-(50-r)A)}\right)$$

$$= \text{Re}\left(\sum_{r=0}^{50} {}^{50}C_r (a \times e^{iB})^r \times (b \times e^{-iA})^{50-r}\right)$$

$$= \text{Re}(ae^{iB} + b\,e^{-iA})^{50}$$

$$= \text{Re}(a\cos B + ia\sin B + b\cos A - ib\sin A)^{50}$$

$$= \text{Re}(a\cos B + b\cos A)^{50} = c^{50}$$

$$\frac{\sum_{r=0}^{50} {}^{50}C_r \sin 2rx}{\sum_{r=0}^{50} {}^{50}C_r \cos 2rx} = \frac{\sum_{r=0}^{50} {}^{50}C_{50-r} \sin 2(50 - r)x}{\sum_{r=0}^{50} {}^{50}C_{50-r} \cos 2(50 - r)x}$$

$$= \frac{\sum_{r=0}^{50} {}^{50}C_r[\sin 2rx + \sin 2(50 - r)x]}{\sum_{r=0}^{50} {}^{50}C_r[\cos 2rx + \cos 2(50 - r)x]}$$

$$= \frac{\sum_{r=0}^{50} {}^{50}C_r 2\sin(50x)\cos(2r - 50)x}{\sum_{r=0}^{50} {}^{50}C_r 2\cos(50x)\cos(2r - 50)x}$$

$$= \tan(50x)$$

$$\Rightarrow f(\pi/8) = \tan(25\pi/4) = \tan(6\pi + \pi/4) = 1$$

For Problems 16–18

16. b., 17. b., 18. c.

Sol. General term of the series is

$$T(r) = \sum_{r=1}^{50} \frac{{}^{50+r}C_r(2r - 1)}{{}^{50}C_r(50 + r)}$$

$$= \frac{{}^{50+r}C_r}{{}^{50}C_r}\left(1 - \frac{50 - r + 1}{50 + r}\right)$$

$$= \frac{{}^{50+r}C_r}{{}^{50}C_r} - \frac{{}^{50+r}C_r}{{}^{50}C_r}\left(\frac{50 - r + 1}{50 + r}\right)$$

Now,

$$\frac{{}^{50+r}C_r}{{}^{50}C_r}\left(\frac{50 - r + 1}{50 + r}\right) = \frac{(50 - r + 1)(50 + r)!\,r!\,(50 - r)!}{r!\,50!\,(50 + r)\,50!}$$

$$= \frac{(50 - r + 1)!\,(50 + r - 1)!}{50!\,50!}$$

$$= \frac{{}^{50+r-1}C_{r-1}}{{}^{50}C_{r-1}}$$

$$\Rightarrow T(r) = \frac{{}^{50+r}C_r}{{}^{50}C_r} - \frac{{}^{50+r-1}C_{r-1}}{{}^{50}C_{r-1}} = V(r) - V(r - 1)$$

where $V(r) = \dfrac{{}^{50+r}C_r}{{}^{50}C_r}$

Now, sum of the given series

$$S = \sum_{r=1}^{50} T(r) = V(50) - V(0)$$

$$= \frac{^{100}C_{50}}{^{50}C_{50}} - \frac{^{50}C_0}{^{50}C_0} = {}^{100}C_{50} - 1$$

Also,

$$Q = \sum_{r=0}^{Q=50} (^{50}C_r)^2 = {}^{50}C_0^2 + {}^{50}C_1^2 + {}^{50}C_2^2 + \cdots + {}^{50}C_{50}^2 = {}^{100}C_{50}$$

$$\Rightarrow P - Q = -1$$

We know that

$$C_0^2 - C_1^2 + C_2^2 + \cdots + (-1)^n C_n^2$$

$$= \begin{cases} 0, & \text{if } n \text{ is odd} \\ (-1)^n\, {}^nC_{n/2}, & \text{if } n \text{ is even} \end{cases}$$

$$\Rightarrow \sum_{r=0}^{100} (-1)^r \left({}^{100}C_r\right)^2 = (-1)^{100}\, {}^{100}C_{50} = {}^{100}C_{50}$$

$$\Rightarrow P - R = -1$$

$$Q + R = 2\,{}^{100}C_{50} = 2P + 2$$

For Problems 19–21

19. a., 20. b., 21. c.

Sol. Suppose A contains r $(0 \le r \le n)$ elements.

Then, B is constructed by selecting some elements from the remaining $n - r$ elements. Here, A can be chosen in ${}^nC_r$ ways and B in ${}^{n-r}C_0 + {}^{n-r}C_1 + \cdots + {}^{n-r}C_{n-r} = 2^{n-r}$ ways. So, the total number of ways of choosing A and B is ${}^nC_r \times 2^{n-r}$. But r can vary from 0 to n. So, total number of ways is

$$\sum_{r=0}^{n} {}^nC_r \times 2^{n-r} = (1+2)^n = 3^n$$

If A contains r elements, then B contains $(r + 1)$ elements. Then, the number of ways of choosing A and B is ${}^nC_r \times {}^nC_{r+1} = C_r\, C_{r+1}$.

But r can vary from 0 to $(n - 1)$.

So, the number of ways is

$$\sum_{r=0}^{n-1} C_r C_{r+1} = C_0C_1 + C_1C_2 + \cdots + C_{n-1}C_n = {}^{2n}C_{n-1}$$

Let A contains r $(0 \le r \le n)$ elements.

Then, A can be chosen in ${}^nC_r$ ways. The subset B of A can have at most r elements, and the number of ways of choosing B is 2^r.

Therefore, the number of ways of choosing A and B is ${}^nC_r \times 2^r$.

But r can vary from 0 to n.

So, the total number of ways is

$$\sum_{r=0}^{n} {}^nC_r \times 2^r = (1+2)^n = 3^n$$

Matrix-Match Type

1. a → q, r, s; b → p, q, r, s; c → p, q, r; d → p, q, r.

a. ${}^{(n+1)}C_4 + {}^{(n+1)}C_3 + {}^{(n+2)}C_3 = {}^{(n+3)}C_4$

$$\Rightarrow {}^{(n+3)}C_4 > {}^{(n+3)}C_3 \Rightarrow \frac{^{n+3}C_4}{^{n+3}C_3} > 1$$

$$\Rightarrow n > 4 \text{ or } n \ge 5$$

b. $(3053)^{456} - (2417)^{333} = (339 \times 9 + 2)^{456} - (269 \times 9 - 4)^{333}$

Remainder of given number is same as remainder of $2^{456} + 4^{333}$ and

$$2^{456} + 4^{333} = (64)^{76} + (64)^{111}$$
$$= (1 + 63)^{76} + (1 + 63)^{111}$$
$$= (1 + 9 \times 7)^{76} + (1 + 9 \times 7)^{111}$$

Hence, the remainder is 2.

c. We know that $n!$ terminates in 0 for $n \ge 5$ and 3^{4n} terminates in 1 ($\because\ 3^4 = 81$).

Therefore, $3^{180} = (3^4)^{45}$ terminates in 1.

Also, $3^3 = 27$ terminates in 7.

Hence, $183! + 3^{183}$ terminates in 7.

That is, the digit in the unit place is 7.

d. We are given

$${}^mC_0 + {}^mC_1 + {}^mC_2 = 46$$
$$\Rightarrow 2m + m(m-1) = 90$$
$$\Rightarrow m^2 + m - 90 = 0$$
$$\Rightarrow m = 9 \text{ as } m > 0$$

Now, $(r + 1)^{\text{th}}$ term of $\left(x^2 + \frac{1}{x}\right)^m$ is

$${}^mC_r (x^2)^{m-r} \left(\frac{1}{x}\right)^r = {}^mC_r\, x^{2m-3r}$$

For this to be independent of x, $2m - 3r = 0 \Rightarrow r = 6$.

2. a → q; b → s; c → p; d → r.

We know that

$${}^nC_0^2 + {}^nC_1^2 + \cdots + {}^nC_n^2 = {}^{2n}C_n$$

and

$${}^nC_0^2 + {}^nC_1^2 + \cdots + {}^nC_n^2 = \begin{cases} 0, & \text{if } n \text{ is odd} \\ {}^nC_{n/2}(-1)^n, & \text{if } n \text{ is even} \end{cases}$$

From this, ${}^{31}C_0^2 - {}^{31}C_1^2 + {}^{31}C_2^2 - \cdots - {}^{31}C_{31}^2 = 0$

$${}^{32}C_0^2 - {}^{32}C_1^2 + {}^{32}C_2^2 - \cdots + {}^{32}C_{32}^2 = {}^{32}C_{16}$$

$${}^{32}C_0^2 + {}^{32}C_1^2 + {}^{32}C_2^2 - \cdots + {}^{32}C_{32}^2 = {}^{64}C_{32}$$

Also, $1/32(1 \times {}^{32}C_1^2 + 2 \times {}^{32}C_2^2 - \cdots + 32 \times {}^{32}C_{32}^2)$

$$= \frac{1}{32} \sum_{r=1}^{32} r\left({}^{32}C_r\right)^2$$

$$= \frac{1}{32} \sum_{r=1}^{32} r\, {}^{32}C_r\, {}^{32}C_{32-r}$$

$$= \frac{1}{32} \sum_{r=1}^{32} 32\, {}^{31}C_{r-1}\, {}^{32}C_{32-r}$$

$$= {}^{63}C_{31} = {}^{63}C_{32}$$

3. a → q; b → s; c → p; d → r.

In the sum of series

$$\sum_{i=1}^{n}\sum_{j=1}^{n} f(i)\times f(j) = \sum_{i=1}^{n}\left(f(i)\left(\sum_{j=1}^{n} f(j)\right)\right)$$

and j are independent. In this summation, three type of terms occur, for which $i<j$, $i>j$ and $i=j$. Also, the sum of terms when $i<j$ is equal to the sum of the terms when $i>j$ if $f(i)$ and $f(j)$ are symmetrical. So, in that case

$$\sum_{i=0}^{n}\sum_{j=0}^{n} f(i)\times f(j) = \sum\sum_{0\le i<j\le n} f(i)\times f(j)$$

$$+\sum\sum_{0\le j<i\le n} f(i)f(j) + \sum\sum_{i=j} f(i)f(j)$$

$$= 2\sum\sum_{0\le i<j\le n} f(i)f(j) + \sum\sum_{i=j} f(i)f(j)$$

$$\Rightarrow \sum\sum_{0\le i<j\le n} f(i)f(j) = \frac{\sum_{i=0}^{n}\sum_{j=0}^{n} f(i)\times f(j) - \sum\sum_{i=j} f(i)f(j)}{2}$$

a. $\displaystyle\sum\sum_{i\ne j} {}^{10}C_i\,{}^{10}C_j = \sum_{i=0}^{10}\sum_{j=0}^{10} {}^{10}C_i\,{}^{10}C_j - \sum_{i=0}^{10} {}^{10}C_i^2 = 2^{20} - {}^{20}C_{10}$

b. $\displaystyle\sum\sum_{0\le i\le j\le 10} {}^{10}C_i\,{}^{10}C_j = \frac{\sum_{i=0}^{10}\sum_{j=0}^{10} {}^{10}C_i\,{}^{10}C_j + \sum_{i=0}^{10} {}^{10}C_i^2}{2} = \frac{2^{20} + {}^{20}C_{10}}{2}$

c. $\displaystyle\sum\sum_{0\le i<j\le 10} {}^{10}C_i\,{}^{10}C_j = \frac{\sum_{i=0}^{10}\sum_{j=0}^{10} {}^{10}C_i\,{}^{10}C_j - \sum_{i=0}^{10} {}^{10}C_i^2}{2}$

$$= \frac{2^{20} - {}^{20}C_{10}}{2}$$

d. $\displaystyle\sum_{i=0}^{10}\sum_{j=0}^{10} {}^{10}C_i\,{}^{10}C_j = \sum_{i=0}^{10} {}^{10}C_i \sum_{j=0}^{10} {}^{10}C_j = 2^{20}$

4. a → p., s; b → p., q., r., s; c → p., q., r., s; d → p., q., s.

a. $\ln(1+x)^{41} = {}^{41}C_0 + {}^{41}C_1 x + {}^{41}C_2 x^2 + \cdots + {}^{41}C_{20} x^{20} + {}^{41}C_{21} x^{21} + \cdots + {}^{41}C_{41} x^{41}$

$\Rightarrow {}^{41}C_{21} + {}^{41}C_{22} + \cdots + {}^{41}C_{41} = 2^{39}$

b. $(1+\sqrt{2})^{42} = {}^{42}C_0 + {}^{42}(C_1\sqrt{2}) + {}^{42}C_2(\sqrt{2})^2 + {}^{42}C_3(\sqrt{2})^3 + \cdots + {}^{42}C_{42}(\sqrt{2})^{42}$

Sum of binomial coefficients of rational terms is

${}^{42}C_0 + {}^{42}C_2 + {}^{42}C_4 + \cdots + {}^{42}C_{42} = 2^{41}$

c. $\left(x + \dfrac{1}{x} + x^2 + \dfrac{1}{x^2}\right)^{21} = \left(\dfrac{x^3 + x + x^4 + 1}{x^2}\right)^{21}$

$$= \frac{a_0 + a_1 x + a_2 x^2 + \cdots + a_{82}x^{82}}{x^{42}} \quad (1)$$

Now, putting $x = 1$, we get

$4^{21} = a_0 + a_1 + a_2 + \cdots + a_{82}$

Putting $x = -1$, we get

$0 = a_0 - a_1 + a_2 - a_3 + \cdots + a_{82}$

Adding, we get

$4^{21} = 2(a_0 + a_2 + \cdots + a_{82})$

$\Rightarrow a_0 + a_2 + \cdots + a_{82} = 2^{41}$

d. We know that

$${}^nC_0 - {}^nC_2 + {}^nC_4 - {}^nC_6 + \cdots = 2^{n/2}\cos\frac{n\pi}{4} \quad (1)$$

and

$${}^nC_0 + {}^nC_2 + {}^nC_4 + {}^nC_6 + \cdots = 2^{n-1} \quad (2)$$

$$\Rightarrow {}^nC_0 + {}^nC_4 + {}^nC_8 + \cdots = \frac{1}{2}\left(2^{n/2}\times\cos\frac{n\pi}{4} + 2^{n-1}\right)$$

For $n = 42$,

$${}^{42}C_0 + {}^{42}C_4 + {}^{42}C_8 + \cdots = \frac{1}{2}\left(2^{21}\times\cos\frac{21\pi}{2} + 2^{41}\right)$$

$$= 2^{40}$$

5. a → p, r; b → p, q, r, s; c → p, q; d → q, r, s.

a. Let consecutive coefficients be ${}^nC_r$ and ${}^nC_{r+1}$. Then,

$$\frac{n!}{(n-r)!r!} = \frac{n!}{(n-r-1)!(r+1)!}$$

$$\Rightarrow \frac{1}{(n-r)(n-r-1)!r!} = \frac{1}{(n-r-1)!(r+1)r!}$$

$\Rightarrow r + 1 = n - r$

$\Rightarrow n = 2r + 1$

Hence, n is odd.

b. $E = (19-4)^{23} + (19+4)^{23}$

$= 2\,[19^{23} + {}^{23}C_2 \times 19^{21}\times 4^2 + \cdots + {}^{23}C_{22}\times 19\times 4^{22}]$

$= 2\times 19\,[19^{22} + {}^{23}C_2\times 19^{20}\times 4^2 + \cdots + {}^{23}C_{22}\times 4^{22}]$

$\Rightarrow E$ is divisible by 19

$\Rightarrow$ Remainder $= 0$

c. ${}^{10}C_0\,{}^{20}C_{10} - {}^{10}C_1\,{}^{18}C_{10} + {}^{10}C_2\,{}^{16}C_{10} - \cdots$

$=$ Coefficient of x^{10} in $[{}^{10}C_0(1+x)^{20} - {}^{10}C_1 \times (1+x)^{18} + {}^{10}C_2(1+x)^{16} - \cdots]$

$=$ Coefficient of x^{10} in $[{}^{10}C_0((1+x)^2)^{10} - {}^{10}C_1 \times ((1+x)^2)^9 + {}^{10}C_2((1+x)^2)^8 - \cdots]$

$=$ Coefficient of x^{10} in $[(1+x)^2 - 1]^{10}$

$=$ Coefficient of x^{10} in $[2x + x^2]^{10}$

$= 2^{10}$

d. $T_r = {}^{14}C_{r-1}x^{r-1}$; $T_{r+1} = {}^{14}C_r x^r$; $T_{r+2} = {}^{14}C_{r+1}x^{r+1}$

By the given condition,

$$2\,{}^{14}C_r = {}^{14}C_{r-1} + {}^{14}C_{r+1} \quad (1)$$

$$\Rightarrow 2 = \frac{{}^{14}C_{r-1}}{{}^{14}C_r} + \frac{{}^{14}C_{r+1}}{{}^{14}C_r}$$

$$\Rightarrow 2 = \frac{r}{14-r+1} + \frac{14-(r+1)+1}{r+1}$$

$$\Rightarrow 2 = \frac{r}{15-r} + \frac{14-r}{r+1}$$

$\Rightarrow r = 9$

Integer Type

1. The numerator is of the form

$a^3 + b^3 + 3ab\,(a+b) = (a+b)^3$

$\therefore N^r = (18 + 7)^3 = 25^3$

$\therefore D^r = 3^6 + {}^6C_1\, 3^5 \times 2^1 + {}^6C_2 3^4 \times 2^2 + {}^6C_3 3^3 \times 2^3 + {}^6C_4 3^2 \times 2^4 + {}^6C_5 3 \times 2^5 + {}^6C_6 2^6$

This is clearly the expansion of $(3 + 2)^6 = 5^6 = (25)^3$. Therefore,

$$\frac{N^r}{D^r} = \frac{(25)}{(25)^3} = 1$$

2. **(9)** We have $T_{r+1} = {}^{21}C_r \left(\sqrt[3]{\frac{a}{\sqrt{b}}}\right)^{21-r} \left(\sqrt{\frac{b}{\sqrt[3]{a}}}\right)^r$

$= {}^{21}C_r a^{7-(r/2)} b^{(2/3)r-(7/2)}$

Since the powers of a and b are same,

$$7 - \frac{r}{2} = \frac{2}{3}r - \frac{7}{2} \Rightarrow r = 9$$

3. **(0)** $(1+x)^{101}(1-x+x^2)^{100} = (1+x)[(1+x)^{100}(1-x+x^2)^{100}]$

$= (1+x)(1-x^3)^{100}$

$= (1-x^3)^{100} + x(1-x^3)^{100}$

Now the coefficient of x^{50} in $(1-x^3)^{100} + x(1-x^3)^{100}$ is equal to coefficient of x^{50} in $(1-x^3)^{100}$ + coefficient of x^{49} in $(1-x^3)^{100} = 0$ (as 49 and 50 are not multiples of 3)

4. **(6)** In the expansion of $(y^{1/5} + x^{1/10})^{55}$,

$T_{r+1} = {}^{55}C_r (y^{1/5})^{55-r}(x^{1/10})^r = {}^{55}C_r y^{11-r/5} x^{r/10}$

This T_{r+1} will be independent of radicals if the exponents $r/5$ and $r/10$ are integers, for $0 \le r \le 55$, which is possible only when $r = 0, 10, 20, 30, 40$ and 50

Therefore, there are six, terms viz, T_1, T_{11}, T_{21}, T_{31}, T_{41} and T_{51}, which are independent of radicals.

5. **(7)** In this case write

$a^3b^4c^5 = (ab)^x(bc)^y(ca)^z$ say

$\Rightarrow a^3b^4c^5 = a^{z+x} b^{x+y} c^{y+z}$

$\Rightarrow z + x = 3, x + y = 4, y + z = 5$

Adding all,

$2(x+y+z) = 12$

$\Rightarrow x + y + z = 6$

Then $x = 1, y = 3, z = 2$,

Therefore, the coefficient of $a^3b^4c^5$ in the expansion of $(bc + ca + ab)^6$ or the coefficient of $(ab)^1(bc)^3(ca)^2$ in the expansion of $(bc + ca + ab)^6$ is $\frac{6!}{1! \times 3! \times 2!} = 60$

6. **(1)** The sum of the coefficients of the polynomial $(\alpha^2x^2 - 2\alpha x + 1)^{51}$ is obtained by putting $x = 1$ in $(\alpha^2x^2 - 2\alpha x + 1)^{51}$.

Hence, from the given data,

$(\alpha^2 - 2\alpha + 1)^{51} = 0 \Rightarrow \alpha = 1$

7. **(6)** Since $17^{256} = (17^2)^{128} = (289)^{128} = (290-1)^{128}$,

$17^{256} = {}^{128}C_0 (290)^{128} - {}^{128}C_1 (290)^{127} + {}^{128}C_2 (290)^{126} - \cdots - {}^{128}C_{125} (290)^3 + {}^{128}C_{126} (290)^2 - {}^{128}C_{127} (290) + 1$

$= [{}^{128}C_0 (290)^{128} - {}^{128}C_1 (290)^{127} + {}^{128}C_2 (290)^{126} - \cdots - {}^{128}C_{125} (290)^3] + {}^{128}C_{126} (290)^2 - {}^{128}C_{127} (290) + 1$

$= 1000m + {}^{128}C_2 (290)^2 - {}^{128}C_1 (290) + 1 \ (m \in I_+)$

$= 1000m + \frac{(128)(127)}{2}(290)^2 - 128 \times 290 + 1$

$= 1000m + (128)(127)(290)(145) - 128 \times 290 + 1$

$= 1000m + (128)(290)(127 \times 145 - 1) + 1$

$= 1000m + (128)(290)(18414) + 1$

$= 1000m + 683527680 + 1$

$= m + 683527000 + 680 + 1$

$= (m + 683527) + 681$

Hence, last three digits of 17^{256} must be 681.

8. **(4)** $(1+100)^{100} = 1 + 100 \times 100 + \frac{100 \times 99}{1 \times 2}(100)^2 + \frac{100 \times 99 \times 98}{1 \times 2 \times 3}(100)^3 \ldots$

$\Rightarrow (101)^{100} - 1 = 100 \times 100 \left[1 + \frac{100 \times 99}{1 \times 2} + \frac{100 \times 99 \times 98}{1 \times 2 \times 3} 100 + \cdots\right]$

Form above it is clear that $(101)^{100} - 1$ is divisible by $(100)^2 = 10000$.

9. **(3)** We know that $e = \lim_{n \to \infty} \left(1 + \frac{1}{n}\right)^n$ and $2 < e < 3$. Therefore,

$(1 + 0.0001)^{10000} < 3$ (By putting $n = 10000$)

Also $(1 + 0.0001)^{10000}$

$= 1 + 10000 \times 10^{-4} + \frac{10000 \times 9999}{2!} \times 10^{-8} + \cdots$ upto 10001 terms.

$= 1 + 1 + \cdots \Rightarrow (1 + 0.0001)^{10000} > 2$.

Hence 3 is the positive integer just greater than $(1 + 0.0001)^{10000} > 2$.

10. **(8)** Here $5^2 = 25$, which is close to $26 = 13 \times 2$.

Hence,

$E = 5^{99} = 5 \times 5^{98} = 5 \times (5^2)^{49} = 5(26-1)^{49}$

$\Rightarrow E = 5[{}^{49}C_0 26^{49} - {}^{49}C_1 26^{48} + {}^{49}C_2 26^{48} - \cdots + {}^{49}C_{48} 26 - {}^{49}C_{49}]$

$= 5 \times 26k - 5$

Now $\frac{E}{13} = 10k - \frac{5}{13} = 10k - 1 + \frac{8}{13}$

Hence, the remainder is 8.

11. **(1)** Here, the bases 1690 and 2608 are too big. So first let us reduce them.

$1690 = 7 \times 241 + 3$ and $2608 = 7 \times 372 + 4$

Let $S = 1690^{2608} + 2608^{1690}$

$= (7 \times 241 + 3)^{2608} + (7 \times 372 + 4)^{1690}$

$= 7k + 3^{2608} + 4^{1690}$, where k is some positive integer.

Let $S' = 3^{2608} + 4^{1690}$

Clearly, the remainder in S and S' will be the same when divided by 7.

$S' = 3 \times 3^{3 \times 867} + 4 \times 4^{3 \times 563}$

$= 3 \times 27^{867} + 4 \times 64^{563}$

$= 3(28-1)^{867} + 4(63+1)^{563}$

$= 3[7n - 1] + 4[7m + 1]$,

where m and n are some positive integers

$= 7p + 1$, where p is some positive integer.

Hence, the remainder is 1.

12. **(5)** Here $5^{5^{5^{\cdot^{\cdot^{5}}}}}$ (23 times 5) is an odd natural number. Therefore,

$x = 5^{2m+1} = 5(25^m)$, where m is a natural number.

Thus, $x = 5(24 + 1)^m$

$= 5 +$ a multiple of 24

Hence, the remainder is 5.

13. $\frac{1}{\sqrt{4x+1}}\left\{\left(\frac{1+\sqrt{4x+1}}{2}\right)^7 - \left(\frac{1-\sqrt{4x+1}}{2}\right)^7\right\}$

$= \frac{1}{2^7\sqrt{4x+1}}\{2\,[{}^7C_1\sqrt{4x+1} + {}^7C_3(\sqrt{4x+1})^3$

$+ {}^7C_5(\sqrt{4x+1})^5 + {}^7C_7(\sqrt{4x+1})^7]\}$

$= \frac{1}{2^6}\{{}^7C_1 + {}^7C_3(4x+1) + {}^7C_5(4x+1)^2 + {}^7C_7(4x+1)^3\}$

Clearly, it is a polynomial of degree 3.

14. **(1)** $(2+\sqrt{3}) = I + f$

or $I + f = 2^n + {}^nC_1 2^{n-1}\sqrt{3} + {}^nC_2 2^{n-2}(\sqrt{3})^2 + {}^nC_3 2^{n-3}(\sqrt{3})^3 + \cdots$ (1)

Now $0 < 2 - \sqrt{3} < 1 \Rightarrow 0 < (2-\sqrt{3})^n < 1$

Let $(2-\sqrt{3})^n = f'$ where $0 < f' < 1$. Therefore,

$f' = 2^n - {}^nC_1 2^{n-1}\sqrt{3} + {}^nC_2 2^{n-2}(\sqrt{3})^2 - {}^nC_3 2^{n-3}(\sqrt{3})^3 + \cdots$ (2)

Adding (1) and (2)

$I + f + f' = 2\,[2^n + {}^nC_2 2^{n-2}3 + \ldots]$

or $I + f + f' =$ even integer (3)

Now $0 < f < 1$ and $0 < f' < 1$. Therefore,

$0 < f + f' < 2$

Hence, from (3) we conclude that $f + f'$ is an integer between 0 and 2. Therefore,

$\therefore f + f' = 1 \Rightarrow f' = 1 - f$ (4)

From (3) and (4) we get $I + 1 =$ even integer. Therefore, I is an odd integer.

Now $I + f = (2+\sqrt{3})^n$, $f' = 1 - f = (2-\sqrt{3})^n$. Therefore,

$(I+f)(1-f) = [(2+\sqrt{3})(2-\sqrt{3})]^n = (4-3)^n = 1$

15. **(2)** Let T_4 be the numerically greatest term in the expansion of $2^{10}\left(1+\left(\frac{3}{16x}\right)\right)^{10}$. Therefore,

$\left|\frac{T_4}{T_3}\right| \ge 1$ and $\left|\frac{T_4}{T_5}\right| \ge 1$

$\Rightarrow \left|\frac{T_4}{T_3}\right| \ge 1$ and $\left|\frac{T_5}{T_4}\right| \le 1.$

Now $\frac{T_{r+1}}{T_r} = \frac{n-r+1}{r}$

Taking $r = 3$ and $r = 4$ and x by $\frac{3x}{16}$ and $n = 10$ in the above two, we get

$\left|\frac{11-3}{3}\cdot\frac{3x}{16}\right| \ge 1$ and $\left|\frac{11-4}{4}\cdot\frac{3x}{16}\right| \le 1$

$\Rightarrow |x| \ge 2$ and $|x| \le \frac{64}{21}$ (1)

$\Rightarrow x^2 \ge 4$ and $x^2 \le \left(\frac{64}{21}\right)^2$

$\Rightarrow 4 \le x^2 \le \left(\frac{64}{21}\right)^2$

$\Rightarrow 2 \le x \le \frac{64}{21}$ or $-\frac{64}{21} \le x \le -2$

16. **(1)** We know that

${}^{2n}C_n =$ coefficient of x^n in $(1+x)^{2n}$

${}^{2n-1}C_n =$ coefficient of x^n in $(1+x)^{2n-1}$

${}^{2n-2}C_n =$ coefficient of x^n in $(1+x)^{2n-2}$

${}^{2n-3}C_n =$ coefficient of x^n in $(1+x)^{2n-3}$

$\vdots$

${}^nC_n =$ coefficient of x^n in $(1+x)^n$

Thus we have,

${}^nC_0 \times {}^{2n}C_n - {}^nC_1 \times {}^{2n-1}C_n + {}^nC_2 \times {}^{2n-2}C_n + \cdots + (-1)^n \times {}^nC_n \times {}^nC_n$

$=$ coefficient of x^n in $[C_0(1+x)^{2n} - C_1(1+x)^{2n-1} + C_2(1+x)^{2n-2} - C_3(1+x)^{2n-3} + \cdots + (-1)^n C_n(1+x)^n]$

$=$ coefficient of x^n in $(1+x)^n\,[C_0(1+x)^n - C_1(1+x)^{n-1} + C_2(1+x)^{n-2} - C_3(1+x)^{n-3} + \cdots + (-1)^n C_n]$

$=$ coefficient of x^n in $(1+x)^n\,[(1+x)-1]^n$

$=$ coefficient of x^n in $(1+x)^n x^n$

$= 1$

17. **(2)** $1 + 99^n = 1 = (100-1)^n$

$= 1 + [{}^nC_0 100^n - {}^nC_1 100^{n-1} + {}^nC_2 100^{n-2} \cdots - {}^nC_n]$

[$\because n$ are odd]

$= 100\,[{}^nC_0 100^{n-1} - {}^nC_1 100^{n-2} \cdots + {}^nC_{n-1}]$

$= 100 \times$ integer whose unit's apostrophe place is different from zero [n having odd digit at the unit's place]

Hence, number of zeros at the end of the sum $99^n + 1$ is 2.

18. $\sum_{p=1}^{n}\left(\sum_{m=p}^{n} {}^nC_m \times {}^mC_p\right)$

$= \sum_{p=1}^{n}\left(\sum_{m=p}^{n} \frac{n!}{(n-m)} \times \frac{m!}{(m-p)!p!}\right)$

$= \sum_{p=1}^{n}\left(\sum_{m=p}^{n} \frac{n!}{(n-m)} \times \frac{m!}{(m-p)!\,p!}\right)$

$= \sum_{p=1}^{n} \frac{n!}{(n-p)!\,p!}\left(\sum_{m=p}^{n} \frac{(n-p)!}{(n-m)!\,(m-p)!}\right)$

$= \sum_{p=1}^{n} {}^nC_p\left(\sum_{m=p}^{n} {}^{n-p}C_{m-p}\right)$

$= \sum_{p=1}^{n} {}^nC_p 2^{n-p}$

$= {}^nC_1 2^{n-1} + {}^nC_2 2^{n-2} + \cdots + {}^nC_n 2^0$

$= (2+1)^n - {}^nC_0 2^n$

$= 3^n - 2^n$

Also $\lim_{n\to\infty} \frac{1}{3^n} \sum_{p=1}^{n} \left(\sum_{m=p}^{n} {}^nC_m \times {}^mC_p \right)$

$= \lim_{n\to\infty} \frac{3^n - 2^n}{3^n} = 1$

19. **(5)** We have, $T_{r+1} = \frac{\frac{7}{2}\left(\frac{7}{2}-1\right)\left(\frac{7}{2}-2\right)\cdots\left(\frac{7}{2}-r+1\right)x^r}{r}$

This will be the first negative term when

$\frac{7}{2} - r + 1 < 0$ i.e., $r > \frac{9}{2}$

Hence, $r = 5$

20. **(1)** Since $1 + 2x + 3x^2 + 4x^3 + \cdots \infty = (1-x)^{-2}$

Therefore, we have

$(1 + 2x + 3x^2 + 4x^3 + \cdots \infty)^{1/2}$

$= \{(1-x)^{-2}\}^{1/2}$

$= (1-x)^{-1}$

$= 1 + x + x^2 + \cdots + x^n + \cdots \infty$

$\Rightarrow$ The coefficient of $x^n = 1$

21. **(1)** ${}^{20}C_0 - \frac{{}^{20}C_1}{2} + \frac{{}^{20}C_2}{3} - \frac{{}^{20}C_3}{4} + \cdots$

$= \sum_{r=0}^{20} \frac{{}^{20}C_r}{r} (-1)^r$

$= \sum_{r=0}^{20} \frac{{}^{21}C_{r+1}}{20+1} (-1)^r$

$= -\frac{1}{21} \sum_{r=0}^{20} {}^{21}C_{r+1} (-1)^{r+1}$

$= -[-{}^{21}C_1 + {}^{21}C_2 - {}^{21}C_3 + \cdots]$

$= -[({}^{21}C_0 - {}^{21}C_1 + {}^{21}C_2 - {}^{21}C_3 + \cdots) - {}^{21}C_0]$

$= -[(1-1)^{21} - 1]$

22. **(8)** $\sum_{i=0}^{m} {}^9C_i \times {}^7C_{m-i}$

$= {}^9C_0 \times {}^7C_m + {}^9C_1 \times {}^7C_{m-1} + {}^9C_2 \times {}^7C_{m-2} + \cdots + {}^9C_m \times {}^7C_0$

= coefficient of x^m in the expansion of the product $(1+x)^9 (x+1)^7$

= coefficient of x^m in the expansion of $(1+x)^{16}$

$= {}^{16}C_m$

Hence, the maximum value of ${}^{16}C_m$ is ${}^{16}C_8$

23. **(2)** ${}^{n-1}C_r = {}^nC_{r+1}\,(k^2 - 3)$

$\Rightarrow k^2 - 3 = \frac{{}^{n-1}C_r}{{}^nC_{r+1}} = \frac{r+1}{n}$

Since $0 \le r \le n-1$

$\Rightarrow 1 \le r+1 \le n$

$\Rightarrow \frac{1}{n} \le \frac{r+1}{n} \le 1$

$\Rightarrow \frac{1}{n} \le k^2 - 3 \le 1$

$\Rightarrow 3 + \frac{1}{n} \le k^2 \le 4 \Rightarrow \sqrt{3 + \frac{1}{n}} \le k \le 2$ as $n \to \infty$

$\Rightarrow \sqrt{3} < k \le 2$

$\Rightarrow k \in (\sqrt{3}, 2]$

24. **(6)** T_{r+1} in $\left(x^2 - \frac{1}{x}\right)^n$ is ${}^nC_r\,(x^2)^{n-r}\,(-1)^r\,x^{-r}$

$= {}^nC_r\,x^{2n-3r}\,(-1)^r$

Constant term $= {}^nC_r\,(-1)^r$ if $2n = 3r$, i.e., coefficient of $x = 0$

Hence ${}^nC_{2n/3}\,(-1)^{2n/3} = 15 = {}^6C_4 \Rightarrow n = 6$

25. **(4)** Number of ways to distribute n different things among three boys

$= \sum_{i=0}^{n} \sum_{j=0}^{n} \sum_{k=0}^{n} \binom{n}{i}\binom{n}{j}\binom{n}{k} = 2^{3n}$

for $n = 4$, $E = 2^{12} = 4096$

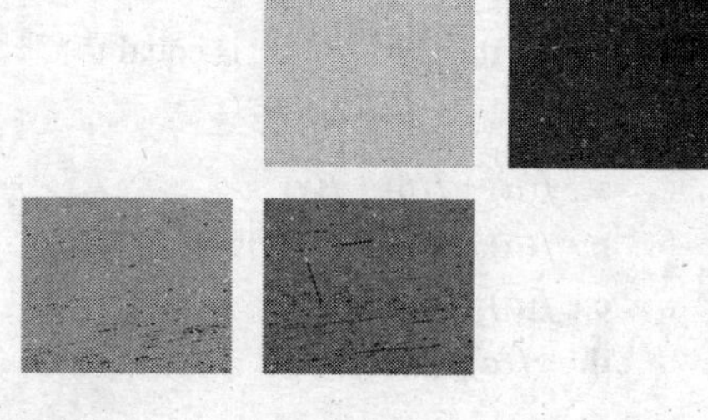

CHAPTER 7

Determinants

Objective Type

Each question has four choices a, b, c and d, out of which *only one* answer is correct. Find the correrct answer.

1. If $p+q+r=0=a+b+c$, then the value of the determinant $\begin{vmatrix} pa & qb & rc \\ qc & ra & pb \\ rb & pc & qa \end{vmatrix}$ is

   a. 0 **b.** $pa+qb+rc$
   c. 1 **d.** none of these

2. If a determinant of order 3 × 3 is formed by using the numbers 1 or −1, then the minimum value of the determinant is

   a. −2 **b.** −4
   c. 0 **d.** −8

3. If $z=\begin{vmatrix} -5 & 3+4i & 5-7i \\ 3-4i & 6 & 8+7i \\ 5+7i & 8-7i & 9 \end{vmatrix}$, then z is

   a. purely real **b.** purely imaginary
   c. $a+ib$, where $a \neq 0, b \neq 0$ **d.** $a+ib$, where $b=4$

4. If α, β, γ are the roots of $px^3+qx^2+r=0$, then the value of the determinant $\begin{vmatrix} \alpha\beta & \beta\gamma & \gamma\alpha \\ \beta\gamma & \gamma\alpha & \alpha\beta \\ \gamma\alpha & \alpha\beta & \beta\gamma \end{vmatrix}$ is

   a. p **b.** q
   c. 0 **d.** r

5. When the determinant $\begin{vmatrix} \cos 2x & \sin^2 x & \cos 4x \\ \sin^2 x & \cos 2x & \cos^2 x \\ \cos 4x & \cos^2 x & \cos 2x \end{vmatrix}$ is expanded in powers of $\sin x$, then the constant term in that expression is

   a. 1 **b.** 0
   c. −1 **d.** 2

6. If $a=\cos\theta+i\sin\theta$, $b=\cos 2\theta-i\sin 2\theta$, $c=\cos 3\theta+i\sin 3\theta$ and if $\begin{vmatrix} a & b & c \\ b & c & a \\ c & a & b \end{vmatrix}=0$, then

   a. $\theta=2k\pi, k\in Z$ **b.** $\theta=(2k+1)\pi, k\in Z$
   c. $\theta=(4k+1)\pi, k\in Z$ **d.** none of these

7. If $\begin{vmatrix} a & b-c & c+b \\ a+c & b & c-a \\ a-b & a+b & c \end{vmatrix}=0$, then the line $ax+by+c=0$ passes through the fixed point which is

   a. (1, 2) **b.** (1, 1)
   c. (−2, 1) **d.** (1, 0)

8. If $\begin{vmatrix} x^n & x^{n+2} & x^{n+3} \\ y^n & y^{n+2} & y^{n+3} \\ z^n & z^{n+2} & z^{n+3} \end{vmatrix}=(x-y)(y-z)(z-x)\left(\frac{1}{x}+\frac{1}{y}+\frac{1}{z}\right)$, then n equals

   a. 1 **b.** −1
   c. 2 **d.** −2

9. If $f(x) = a + bx + cx^2$ and α, β, γ are the roots of the equation $x^3 = 1$, then $\begin{vmatrix} a & b & c \\ b & c & a \\ c & a & b \end{vmatrix}$ is equal to

a. $f(\alpha) + f(\beta) + f(\gamma)$
b. $f(\alpha)f(\beta) + f(\beta)f(\gamma) + f(\gamma)f(\alpha)$
c. $f(\alpha)\, f(\beta) f(\gamma)$
d. $-f(\alpha)f(\beta)f(\gamma)$

10. If [ ] denotes the greatest integer less than or equal to the real number under consideration, and $-1 \le x < 0, 0 \le y < 1, 1 \le z < 2$, then the value of the determinant

$$\begin{vmatrix} [x]+1 & [y] & [z] \\ [x] & [y]+1 & [z] \\ [x] & [y] & [z]+1 \end{vmatrix} \text{ is}$$

a. $[x]$ **b.** $[y]$
c. $[z]$ **d.** none of these

11. Let $a, b, c \in R$ such that no two of them are equal and satisfy

$\begin{vmatrix} 2a & b & c \\ b & c & 2a \\ c & 2a & b \end{vmatrix} = 0$, then equation $24ax^2 + 4bx + c = 0$ has

a. at least one root in [0, 1]
b. at least one root in $\left[-\frac{1}{2}, \frac{1}{2}\right]$
c. at least one root in[−1, 0]
d. at least two roots in [0, 2]

12. If p, q, r are in A.P., then the value of determinant

$$\begin{vmatrix} a^2 + a^{2n+1} + 2p & b^2 + 2^{n+2} + 3q & c^2 + p \\ 2^n + p & 2^{n+1} + q & 2q \\ a^2 + 2^n + p & b^2 + 2^{n+1} + 2q & c^2 - r \end{vmatrix} \text{ is}$$

a. 1 **b.** 0
c. $a^2b^2c^2 - 2^n$ **d.** $(a^2 + b^2 + c^2) - 2^nq$

13. If $(x_1 - x_2)^2 + (y_1 - y_2)^2 = a^2$
$(x_2 - x_3)^2 + (y_2 - y_3)^2 = b^2$
$(x_3 - x_1)^2 + (y_3 - y_1)^2 = c^2$

and $k \begin{vmatrix} x_1 & y_1 & 1 \\ x_2 & y_2 & 1 \\ x_3 & y_3 & 1 \end{vmatrix} = (a + b + c)\,(b + c - a)\,(c + a - b)$
$\times\,(a + b - c)$, then the value of k is

a. 1 **b.** 2
c. 4 **d.** none of these

14. The value of the determinant $\begin{vmatrix} ka & k^2 + a^2 & 1 \\ kb & k^2 + b^2 & 1 \\ kc & k^2 + c^2 & 1 \end{vmatrix}$ is

a. $k(a + b)(b + c)(c + a)$
b. $k\,abc(a^2 + b^2 + c^2)$
c. $k(a - b)(b - c)(c - a)$
d. $k(a + b - c)(b + c - a)(c + a - b)$

15. If $\begin{vmatrix} 1 & 1 & 1 \\ a & b & c \\ a^3 & b^3 & c^3 \end{vmatrix} = (a - b)\,(b - c)\,(c - a)\,(a + b + c)$, where a, b, c are all different, then the determinant

$$\begin{vmatrix} 1 & 1 & 1 \\ (x-a)^2 & (x-b)^2 & (x-c)^2 \\ (x-b)(x-c) & (x-c)(x-a) & (x-a)(x-b) \end{vmatrix} \text{ vanishes}$$

when

a. $a + b + c = 0$ **b.** $x = \frac{1}{3}(a + b + c)$
c. $x = \frac{1}{2}(a + b + c)$ **d.** $x = a + b + c$

16. The determinant $\begin{vmatrix} y^2 & -xy & x^2 \\ a & b & c \\ a' & b' & c' \end{vmatrix}$ is equal to

a. $\begin{vmatrix} bx + ay & cx + by \\ b'x + a'y & c'x + b'y \end{vmatrix}$

b. $\begin{vmatrix} ax + by & bx + cy \\ a'x + b'y & b'x + c'y \end{vmatrix}$

c. $\begin{vmatrix} bx + cy & ax + by \\ b'x + c'y & a'x + b'y \end{vmatrix}$

d. $\begin{vmatrix} ax + by & bx + cy \\ a'x + b'y & b'x + c'y \end{vmatrix}$

17. If $\begin{vmatrix} b^2 + c^2 & ab & ac \\ ab & c^2 + a^2 & bc \\ ca & cb & a^2 + b^2 \end{vmatrix} = ka^2b^2c^2$, then the value of k is

a. 2 **b.** 4
c. 0 **d.** none of these

18. If a, b and c are non-zero real numbers, then

$\Delta = \begin{vmatrix} b^2c^2 & bc & b + c \\ c^2a^2 & ca & c + a \\ a^2b^2 & ab & a + b \end{vmatrix}$ is equal to

a. abc **b.** $a^2b^2c^2$
c. $bc + ca + ab$ **d.** none of these

19. The value of $\begin{vmatrix} -1 & 2 & 1 \\ 3 + 2\sqrt{2} & 2 + 2\sqrt{2} & 1 \\ 3 - 2\sqrt{2} & 2 - 2\sqrt{2} & 1 \end{vmatrix}$ is equal to

a. zero **b.** $-16\sqrt{2}$
c. $-8\sqrt{2}$ **d.** none of these

20. Let $\{D_1, D_2, D_3, ..., D_n\}$ be the set of third-order determinants that can be made with the distinct non-zero real numbers $a_1, a_2, ..., a_9$. Then

a. $\sum_{i=1}^{n} D_i = 1$ **b.** $\sum_{i=1}^{n} D_i = 0$
c. $D_i = D_j, \forall\, i, j$ **d.** None of these

21. If w is a complex cube root of unity, then value of

$$\Delta = \begin{vmatrix} a_1 + b_1 w & a_1 w^2 + b_1 & c_1 + b_1 \overline{w} \\ a_2 + b_2 w & a_2 w^2 + b_2 & c_2 + b_2 \overline{w} \\ a_3 + b_3 w & a_3 w^2 + b_3 & c_3 + b_3 \overline{w} \end{vmatrix} \text{ is}$$

a. 0 **b.** −1
c. 2 **d.** none of these

22. If x, y, z are in A.P., then the value of the determinant $\begin{vmatrix} a+2 & a+3 & a+2x \\ a+3 & a+4 & a+2y \\ a+4 & a+5 & a+2z \end{vmatrix}$ is

a. 1 **b.** 0
c. $2a$ **d.** a

23. If $a + b + c = 0$, one root of $\begin{vmatrix} a-x & c & b \\ c & b-x & a \\ b & a & c-x \end{vmatrix} = 0$ is

a. $x = 1$ **b.** $x = 2$
c. $x = a^2 + b^2 + c^2$ **d.** $x = 0$

24. In triangle ABC, if

$$\begin{vmatrix} 1 & 1 & 1 \\ \cot\frac{A}{2} & \cot\frac{B}{2} & \cot\frac{C}{2} \\ \tan\frac{B}{2} + \tan\frac{C}{2} & \tan\frac{C}{2} + \frac{A}{2} & \tan\frac{A}{2} + \tan\frac{B}{2} \end{vmatrix} = 0, \text{ then}$$

the triangle must be

a. equilateral **b.** isosceles
c. obtuse angled **d.** none of these

25. If a, b, c, d, e and f are in G.P., then the value of $\begin{vmatrix} a^2 & d^2 & x \\ b^2 & e^2 & y \\ c^2 & f^2 & z \end{vmatrix}$ depends on

a. x and y **b.** x and z
c. y and z **d.** independent of x, y and z

26. If $a^2+b^2+c^2=-2$ and $f(x)= \begin{vmatrix} 1+a^2 x & (1+b^2)\,x & (1+c^2)\,x \\ (1+a^2)\,x & 1+b^2 x & (1+c^2)\,x \\ (1+a^2)\,x & (1+b^2)\,x & 1+c^2 x \end{vmatrix}$,

then $f(x)$ is a polynomial of degree

a. 0 **b.** 1
c. 2 **d.** 3

27. The value of the determinant $\begin{vmatrix} 1 & 1 & 1 \\ {}^mC_1 & {}^{m+1}C_1 & {}^{m+2}C_1 \\ {}^mC_2 & {}^{m+1}C_2 & {}^{m+2}C_2 \end{vmatrix}$ is equal to

a. 1 **b.** −1
c. 0 **d.** none of these

28. If $x \neq y \neq z$ and $\begin{vmatrix} x & x^2 & 1+x^3 \\ y & y^2 & 1+y^3 \\ z & z^2 & 1+z^3 \end{vmatrix} = 0$, then the value of xyz is

a. 1 **b.** 2
c. −1 **d.** −2

29. If $x \neq 0, y \neq 0, z \neq 0$ and $\begin{vmatrix} 1+x & 1 & 1 \\ 1+y & 1+2y & 1 \\ 1+z & 1+z & 1+3z \end{vmatrix} = 0$, then $x^{-1} + y^{-1} + z^{-1}$ is equal to

a. −1 **b.** −2
c. −3 **d.** none of these

30. The value of $\begin{vmatrix} yz & zx & xy \\ p & 2q & 3r \\ 1 & 1 & 1 \end{vmatrix}$, where x, y, z are, respectively, p^{th}, $(2q)^{th}$ and $(3r)^{th}$ terms of an H.P., is

a. −1 **b.** 0
c. 1 **d.** none of these

31. If $a_1b_1c_1$, $a_2b_2c_2$ and $a_3b_3c_3$ are 3-digit even natural numbers and $\Delta = \begin{vmatrix} c_1 & a_1 & b_1 \\ c_2 & a_2 & b_2 \\ c_3 & a_3 & b_3 \end{vmatrix}$, then Δ is

a. divisible by 2 but not necessarily by 4
b. divisible by 4 but not necessarily by 8
c. divisible by 8
d. none of these

32. The value of the determinant of nth order, being given by $\begin{vmatrix} x & 1 & 1 & \cdots \\ 1 & x & 1 & \cdots \\ 1 & 1 & x & \cdots \\ \cdots & \cdots & \cdots & \cdots \end{vmatrix}$ is

a. $(x-1)^{n-1}(x+n-1)$ **b.** $(x-1)^n(x+n-1)$
c. $(1-x)^{-1}(x+n-1)$ **d.** none of these

33. If a, b, c are positive and are the pth, q^{th} and r^{th} terms, respectively, of a G.P., then $\Delta = \begin{vmatrix} \log a & p & 1 \\ \log b & q & 1 \\ \log c & r & 1 \end{vmatrix}$ is

a. 0 **b.** $\log(abc)$
c. $-(p+q+r)$ **d.** none of these

34. If $f'(x)=\begin{vmatrix} mx & mx-p & mx+p \\ n & n+p & n-p \\ mx+2n & mx+2n+p & mx+2n-p \end{vmatrix}$, then $y=f(x)$ represents

a. a straight line parallel to x-axis
b. a straight line parallel to y-axis
c. parabola
d. a straight line with negative slope

35. If $\begin{vmatrix} x & 3 & 6 \\ 3 & 6 & x \\ 6 & x & 3 \end{vmatrix}=\begin{vmatrix} 2 & x & 7 \\ x & 7 & 2 \\ 7 & 2 & x \end{vmatrix}=\begin{vmatrix} 4 & 5 & x \\ 5 & x & 4 \\ x & 4 & 5 \end{vmatrix}=0$, then x is equal to

a. 0
b. –9
c. 3
d. none of these

36. If $\begin{vmatrix} x^n & x^{n+2} & x^{2n} \\ 1 & x^a & a \\ x^{n+5} & x^{a+6} & x^{2n+5} \end{vmatrix}=0,\ \forall\, x\in R$, where $n\in N$, then value of a is

a. n
b. $n-1$
c. $n+1$
d. none of these

37. If $\Delta=\begin{vmatrix} 3 & 4 & 5 & x \\ 4 & 5 & 6 & y \\ 5 & 6 & 7 & z \\ x & y & z & 0 \end{vmatrix}=0$, then

a. x, y, z are in A.P.
b. x, y, z are in G.P.
c. x, y, z are in H.P.
d. none of these

38. Let $x<1$, then value of $\begin{vmatrix} x^2+2 & 2x+1 & 1 \\ 2x+1 & x+2 & 1 \\ 3 & 3 & 1 \end{vmatrix}$ is

a. non-negative
b. non-positive
c. negative
d. positive

39. Value of $\begin{vmatrix} x+y & z & z \\ x & y+z & x \\ y & y & z+x \end{vmatrix}$, where x, y, z are non-zero real numbers, is equal to

a. xyz
b. $2xyz$
c. $3xyz$
d. $4xyz$

40. Which of the following is not the root of the equation $\begin{vmatrix} x & -6 & -1 \\ 2 & -3x & x-3 \\ -3 & 2x & x+2 \end{vmatrix}=0$?

a. 2
b. 0
c. 1
d. –3

41. If $f(x)=\begin{vmatrix} x & a & a \\ a & x & a \\ a & a & x \end{vmatrix}=0$, then

a. $f'(x)=0$ and $f''(x)=0$ has one common root
b. $f(x)=0$ and $f'(x)=0$ has one common root
c. sum of roots of $f(x)=0$ is $-3a$
d. none of these

42. Roots of the equation $\begin{vmatrix} x & m & n & 1 \\ a & x & n & 1 \\ a & b & x & 1 \\ a & b & c & 1 \end{vmatrix}=0$ are

a. independent of m and n
b. independent of a, b and c
c. depend on m, n and a, b, c
d. independent of m, n and a, b, c

43. If a, b, c are different, then the value of x satisfying $\begin{vmatrix} 0 & x^2-a & x^3-b \\ x^2+a & 0 & x^2+c \\ x^4+b & x-c & 0 \end{vmatrix}=0$ is

a. c
b. c
c. b
d. 0

44. If $\begin{vmatrix} b+c & c+a & a+b \\ a+b & b+c & c+a \\ c+a & a+b & b+c \end{vmatrix}=k\begin{vmatrix} a & b & c \\ c & a & b \\ b & c & a \end{vmatrix}$, then the value of k is

a. 1
b. 2
c. 3
d. 4

45. Suppose $D=\begin{vmatrix} a_1 & b_1 & c_1 \\ a_2 & b_2 & c_2 \\ a_3 & b_3 & c_3 \end{vmatrix}$ and $D'=\begin{vmatrix} a_1+pb_1 & b_1+qc_1 & c_1+ra_1 \\ a_2+pb_2 & b_2+qc_2 & c_2+ra_2 \\ a_3+pb_3 & b_3+qc_3 & c_3+ra_3 \end{vmatrix}$. Then

a. $D'=D$
b. $D'=D(1-pqr)$
c. $D'=D(1+p+q+r)$
d. $D'=D(1+pqr)$

46. The value of the determinant $\Delta=\begin{vmatrix} 1^2 & 2^2 & 3^2 & 4^2 \\ 2^2 & 3^2 & 4^2 & 5^2 \\ 3^2 & 4^2 & 5^2 & 6^2 \\ 4^2 & 5^2 & 6^2 & 7^2 \end{vmatrix}$ is equal to

a. 1
b. 0
c. 2
d. 3

47. If $a>0$ and discriminant of $ax^2+2bx+c$ is negative, then $\Delta=\begin{vmatrix} a & b & ax+b \\ b & c & bx+c \\ ax+b & bx+c & 0 \end{vmatrix}$ is

a. Positive
b. $(ac - b)^2)(ax^2 + 2bx + c)$
c. Negative
d. 0

(AIEEE, 2002)

48. If $a_1, a_2, \ldots, a_n, \ldots$ form a G.P. and $a_i > 0$, for all $i \geq 1$,

then $\begin{vmatrix} \log a_n & \log a_{n+1} & \log a_{n+2} \\ \log a_{n+3} & \log a_{n+4} & \log a_{n+5} \\ \log a_{n+6} & \log a_{n+7} & \log a_{n+8} \end{vmatrix}$ is equal to

a. 0
b. 1
c. 2
d. 3

(AIEEE, 2005)

49. Let $\vec{a}_r = x_r\hat{i} + y_r\hat{j} + z_r\hat{k}$, $r = 1, 2, 3$ be three mutually perpendicular unit vectors, then the value of $\begin{vmatrix} x_1 & x_2 & x_3 \\ y_1 & y_2 & y_3 \\ z_1 & z_2 & z_3 \end{vmatrix}$ is equal to

a. zero
b. ± 1
c. ± 2
d. none of these

50. The number of distinct real roots of $\begin{vmatrix} \sin x & \cos x & \cos x \\ \cos x & \sin x & \cos x \\ \cos x & \cos x & \sin x \end{vmatrix}$ $= 0$ in the interval $-\pi/4 \leq x \leq \pi/4$ is

a. 0
b. 2
c. 1
d. 3

51. If $p\lambda^4 + q\lambda^3 + r\lambda^2 + s\lambda + t = \begin{vmatrix} \lambda^2+3\lambda & \lambda-1 & \lambda+3 \\ \lambda^2+1 & 2-\lambda & \lambda-3 \\ \lambda^2-3 & \lambda+4 & 3\lambda \end{vmatrix}$, then p is equal to

a. -5
b. -4
c. -3
d. -2

52. If x, y, z are different from zero and $\Delta = \begin{vmatrix} a & b-y & c-z \\ a-x & b & c-z \\ a-x & b-y & c \end{vmatrix}$ $= 0$, then the value of the expression $\frac{a}{x} + \frac{b}{y} + \frac{c}{z}$ is

a. 0
b. -1
c. 1
d. 2

53. If A, B, C are angles of a triangle, then the value of $\begin{vmatrix} e^{2iA} & e^{-iC} & e^{-iB} \\ e^{-iC} & e^{2iB} & e^{-iA} \\ e^{-iB} & e^{-iA} & e^{2iC} \end{vmatrix}$ is

a. 1
b. -1
c. -2
d. -4

54. For the equation $\begin{vmatrix} 1 & x & x^2 \\ x^2 & 1 & x \\ x & x^2 & 1 \end{vmatrix} = 0$,

a. There are exactly two distinct roots
b. There is one pair of equation real roots
c. There are three pairs of equal roots
d. Modulus of each root is 2

55. Let m be a positive integer and

$$\Delta_r = \begin{vmatrix} 2r-1 & {}^mC_r & 1 \\ m^2-1 & 2^m & m+1 \\ \sin^2(m^2) & \sin^2(m) & \sin^2(m+1) \end{vmatrix} \quad (0 \leq r \leq m).$$

Then the value of $\sum_{r=0}^{m} \Delta_r$ is given by

a. 0
b. $m^2 - 1$
c. 2^m
d. $2^m \sin^2(2^m)$

56. If $D_k = \begin{vmatrix} 1 & n & n \\ 2k & n^2+n+1 & n^2+n \\ 2k-1 & n^2 & n^2+n+1 \end{vmatrix}$ and $\sum_{k=1}^{n} D_k = 56$, then n equals

a. 4
b. 6
c. 8
d. none of these

57. The value of $\sum_{r=2}^{n} (-2)^r \begin{vmatrix} {}^{n-2}C_{r-2} & {}^{n-2}C_{r-1} & {}^{n-2}C_r \\ -3 & 1 & 1 \\ 2 & -1 & 0 \end{vmatrix}$ $(n > 2)$ is

a. $2n - 1 + (-1)^n$
b. $2n + 1 + (-1)^{n-1}$
c. $2n - 3 + (-1)^n$
d. none of these

58. The value of the determinant $\begin{vmatrix} {}^nC_{r-1} & {}^nC_r & (r+1)\,{}^{n+2}C_{r+1} \\ {}^nC_r & {}^nC_{r+1} & (r+2)\,{}^{n+2}C_{r+2} \\ {}^nC_{r+1} & {}^nC_{r+2} & (r+3)\,{}^{n+2}C_{r+3} \end{vmatrix}$ is

a. $n^2 + n - 1$
b. 0
c. ${}^{n+3}C_{r+3}$
d. ${}^nC_{r-1} + {}^nC_r + {}^nC_{r+1}$

59. $\Delta_1 = \begin{vmatrix} y^5z^6(z^3-y^3) & x^4z^6(x^3-z^3) & x^4y^5(y^3-x^3) \\ y^2z^3(y^6-z^6) & xz^3(z^6-x^6) & xy^2(x^6-y^6) \\ y^2z^3(z^3-y^3) & xz^3(x^3-z^3) & xy^2(y^3-x^3) \end{vmatrix}$ and

$\Delta_2 = \begin{vmatrix} x & y^2 & z^3 \\ x^4 & y^5 & z^6 \\ x^7 & y^8 & z^9 \end{vmatrix}$. Then $\Delta_1\Delta_2$ is equal to

a. Δ_2^3
b. Δ_2^2
c. Δ_2^4
d. None of these

60. If $l_1^2 + m_1^2 + n_1^2 = 1$, etc. and $l_1l_2 + m_1m_2 + n_1n_2 = 0$, etc. and $\Delta = \begin{vmatrix} l_1 & m_1 & n_1 \\ l_2 & m_2 & n_2 \\ l_3 & m_3 & n_3 \end{vmatrix}$, then

a. $|\Delta| = 3$
b. $|\Delta| = 2$
c. $|\Delta| = 1$
d. $\Delta = 0$

61. The value of the determinant

$$\begin{vmatrix} (a_1-b_1)^2 & (a_1-b_2)^2 & (a_1-b_3)^2 & (a_1-b_4)^2 \\ (a_2-b_1)^2 & (a_2-b_2)^2 & (a_2-b_3)^2 & (a_2-b_4)^2 \\ (a_3-b_1)^2 & (a_3-b_2)^2 & (a_3-b_3)^2 & (a_3-b_4)^2 \\ (a_4-b_1)^2 & (a_4-b_2)^2 & (a_4-b_3)^2 & (a_4-b_4)^2 \end{vmatrix}$$ is

a. dependant on a_i, $i = 1, 2, 3, 4$

b. dependant on b_i, $i = 1, 2, 3, 4$

c. dependant on a_{ij}, b_i, $i = 1, 2, 3, 4$

d. 0

62. The value of determinant $\begin{vmatrix} bc-a^2 & ac-b^2 & ab-c^2 \\ ac-b^2 & ab-c^2 & bc-a^2 \\ ab-c^2 & bc-a^2 & ac-b^2 \end{vmatrix}$ is

a. always positive
b. always negative
c. always zero
d. cannot say anything

63. Value of $\begin{vmatrix} 1+x_1 & 1+x_1x & 1+x_1x^2 \\ 1+x_2 & 1+x_2x & 1+x_2x^2 \\ 1+x_3 & 1+x_3x & 1+x_3x^2 \end{vmatrix}$ depends upon

a. x only
b. x_1 only
c. x_2 only
d. none of these

64. If $\begin{vmatrix} a^2+\lambda^2 & ab+c\lambda & ca-b\lambda \\ ab-c\lambda & b^2+\lambda^2 & bc+a\lambda \\ ca+b\lambda & bc-a\lambda & c^2+\lambda^2 \end{vmatrix}\begin{vmatrix} \lambda & c & -b \\ -c & \lambda & a \\ b & -a & \lambda \end{vmatrix}$

$= (1 + a^2 + b^2 + c^2)^3$, then the value of λ is

a. 8
b. 27
c. 1
d. −1

65. $f(x) = \begin{vmatrix} \cos x & x & 1 \\ 2\sin x & x^2 & 2x \\ \tan x & x & 1 \end{vmatrix}$. The value of $\lim\limits_{x\to 0}\dfrac{f(x)}{x}$ is equal to

a. 1
b. −1
c. zero
d. none of these

66. If the determinant $\begin{vmatrix} b-c & c-a & a-b \\ b'-c' & c'-a' & a'-b' \\ b''-c'' & c''-a'' & a''-b'' \end{vmatrix}$

$= m\begin{vmatrix} a & b & c \\ a' & b' & c' \\ a'' & b'' & c'' \end{vmatrix}$, then the value of m is

a. 0
b. 2
c. −1
d. 1

67. Let $\begin{vmatrix} x & 2 & x \\ x^2 & x & 6 \\ x & x & 6 \end{vmatrix} = Ax^4 + Bx^3 + Cx^2 + Dx + E$. Then the value of $5A + 4B + 3C + 2D + E$ is equal to

a. zero
b. −16
c. 16
d. −11

68. If $\Delta_1 = \begin{vmatrix} x & b & b \\ a & x & b \\ a & a & x \end{vmatrix}$ and $\Delta_2 = \begin{vmatrix} x & b \\ a & x \end{vmatrix}$ are the given determinants, then

a. $\Delta_1 = 3(\Delta_2)^2$
b. $\dfrac{d}{dx}(\Delta_1) = 3\Delta_2$
c. $\dfrac{d}{dx}(\Delta_1) = 3(\Delta_2)^2$
d. $\Delta_1 = 3\Delta_2^{3/2}$

69. If $y = \sin mx$, then the value of the determinant $\begin{vmatrix} y & y_1 & y_2 \\ y_3 & y_4 & y_5 \\ y_6 & y_7 & y_8 \end{vmatrix}$, where $y_n = \dfrac{d^n y}{dx^n}$, is

a. m^9
b. m^2
c. m^3
d. none of these

70. If the value of the determinant $\begin{vmatrix} a & 1 & 1 \\ 1 & b & 1 \\ 1 & 1 & c \end{vmatrix}$ is positive, then $(a, b, c > 0)$

a. $abc > 1$
b. $abc > -8$
c. $abc < -8$
d. $abc > -2$

71. If $A_1, B_1, C_1, \ldots$ are, respectively, the cofactors of the elements $a_1, b_1, c_1, \ldots$ of the determinant $\Delta = \begin{vmatrix} a_1 & b_1 & c_1 \\ a_2 & b_2 & c_2 \\ a_3 & b_3 & c_3 \end{vmatrix}$, $\Delta \neq 0$, then the value of $\begin{vmatrix} B_2 & C_2 \\ B_3 & C_3 \end{vmatrix}$ is equal to

a. $a_1^2\Delta$
b. $a_1\Delta$
c. $a_1\Delta^2$
c. $a_1^2\Delta^2$

72. Let $f(x) = \begin{vmatrix} 2\cos^2 x & \sin 2x & -\sin x \\ \sin 2x & 2\sin^2 x & \cos x \\ \sin x & -\cos x & 0 \end{vmatrix}$. Then the value of $\int_0^{\pi/2} [f(x) + f'(x)]dx$ is

a. π
b. $\pi/2$
c. 2π
d. $3\pi/2$

73. The number of positive integral solutions of the equation $\begin{vmatrix} x^3+1 & x^2y & x^2z \\ xy^2 & y^3+1 & y^2z \\ xz^2 & yz^2 & z^3+1 \end{vmatrix} = 11$ is

a. 0
b. 3
c. 6
d. 12

74. a, b, c are distinct real numbers, not equal to one. If $ax + y + z = 0$, $x + by + z = 0$ and $x + y + cz = 0$ have a non-trivial solution, then the value of $\dfrac{1}{1-a}+\dfrac{1}{1-b}+\dfrac{1}{1-c}$ is equal to

a. −1
b. 1
c. zero
d. none of these

75. If the system of linear equations $x + y + z = 6$, $x + 2y + 3z = 14$ and $2x + 5y + \lambda z = \mu$ $(\lambda, \mu \in R)$ has a unique solution, then

a. $\lambda \neq 8$ b. $\lambda = 8, \mu \neq 36$
c. $\lambda = 8, \mu = 36$ d. none of these

76. If α, β, γ are the angles of a triangle and the system of equations
$\cos(\alpha - \beta)x + \cos(\beta - \gamma)y + \cos(\gamma - \alpha)z = 0$
$\cos(\alpha + \beta)x + \cos(\beta + \gamma)y + \cos(\gamma + \alpha)z = 0$
$\sin(\alpha + \beta)x + \sin(\beta + \gamma)y + \sin(\gamma + \alpha)z = 0$
has non-trivial solutions, then triangle is necessarily
a. equilateral b. isosceles
c. right angled d. acute angled

77. Given $a = x/(y - z)$, $b = y/(z - x)$, and $c = z/(x - y)$, where x, y and z are not all zero, then the value of $ab + bc + ca$ is
a. 0 b. 1
c. −1 d. none of these

78. If $pqr \neq 0$ and the system of equations
$(p + a)x + by + cz = 0$
$ax + (q + b)y + cz = 0$
$ax + by + (r + c)z = 0$
has a non-trivial solution, then value of $\frac{a}{p} + \frac{b}{q} + \frac{c}{r}$ is
a. −1 b. 0
c. 1 d. 2

79. The system of equations
$\alpha x - y - z = \alpha - 1$
$x - \alpha y - z = \alpha - 1$
$x - y - \alpha z = \alpha - 1$
has no solution if α is
a. either −2 or 1 b. −2
c. 1 d. not −2

80. The set of equations $\lambda x - y + (\cos\theta)z = 0$, $3x + y + 2z = 0$, $(\cos\theta)x + y + 2z = 0$, $0 \le \theta < 2\pi$, has non-trivial solution(s)
a. for no value of λ and θ
b. for all values of λ and θ
c. for all values of λ and only two values of θ
d. for only one value of λ and all values of θ

81. If a, b, c are non-zeros, then the system of equations $(\alpha + a)x + \alpha y + \alpha z = 0$, $\alpha x + (\alpha + b)y + \alpha z = 0$, $\alpha x + \alpha y + (\alpha + c)z = 0$ has a non-trivial solution if
a. $\alpha^{-1} = -(a^{-1} + b^{-1} + c^{-1})$
b. $\alpha^{-1} = a + b + c$
c. $\alpha + a + b + c = 1$
d. none of these

82. If $c < 1$ and the system of equations $x + y - 1 = 0$, $2x - y - c = 0$ and $bx + 3by - c = 0$ is consistent, then the possible real values of b are
a. $b \in \left(-3, \frac{3}{4}\right)$ b. $b \in \left(-\frac{3}{2}, 4\right)$
c. $b \in \left(-\frac{3}{4}, 3\right)$ d. none of these

83. If a, b, c are in G.P. with common ratio r_1 and α, β, γ are in G.P. with common ratio r_2, and equations $ax + \alpha y + z = 0$, $bx + \beta y + z = 0$, $cx + \gamma y + z = 0$ have only zero solution, then which of the following is not true?
a. $r_1 \neq 1$ b. $r_2 \neq 1$
c. $r_1 \neq r_2$ d. none of these

84. If a, b, c are non-zero real numbers and if the equations $(a - 1)x = y + z$, $(b - 1)y = z + x$, $(c - 1)z = x + y$ have a non-trivial solution, then $ab + bc + ca$ equals
a. $a + b + c$ b. abc
c. 1 d. none of these

Multiple Correct Answers Type

Each question has four choices a, b, c and d, out of which *one or more* answers are correct. Find the correct answer.

1. Which of the following has/have value equal to zero?
a. $\begin{vmatrix} 8 & 2 & 7 \\ 12 & 3 & 5 \\ 16 & 4 & 3 \end{vmatrix}$ b. $\begin{vmatrix} 1/a & a^2 & bc \\ 1/b & b^2 & ac \\ 1/c & c^2 & ab \end{vmatrix}$
c. $\begin{vmatrix} a+b & 2a+b & 3a+b \\ 2a+b & 3a+b & 4a+b \\ 4a+b & 5a+b & 6a+b \end{vmatrix}$ d. $\begin{vmatrix} 2 & 43 & 6 \\ 7 & 35 & 4 \\ 3 & 17 & 2 \end{vmatrix}$

2. If $g(x) = \begin{vmatrix} a^{-x} & e^{x\log_e a} & x^2 \\ a^{-3x} & e^{3x\log_e a} & x^4 \\ a^{-5x} & e^{5x\log_e a} & 1 \end{vmatrix}$, then
a. graphs of $g(x)$ is symmetrical about origin
b. graphs of $g(x)$ is symmetrical about Y-axis
c. $\left.\frac{d^4 g(x)}{dx^4}\right|_{x=0} = 0$
d. $f(x) = g(x) \times \log\left(\frac{a-x}{a+x}\right)$ is an odd function

3. If $\Delta = \begin{vmatrix} -x & a & b \\ b & -x & a \\ a & b & -x \end{vmatrix}$, then a factor of Δ is
a. $a + b + x$
b. $x^2 - (a - b)x + a^2 + b^2 + ab$
c. $x^2 + (a + b)x + a^2 + b^2 - ab$
d. $a + b - x$

4. If $\Delta = \begin{vmatrix} \sin\theta\cos\phi & \sin\theta\sin\phi & \cos\theta \\ \cos\theta\cos\phi & \cos\theta\sin\phi & -\sin\theta \\ -\sin\theta\sin\phi & \sin\theta\cos\phi & 0 \end{vmatrix}$ then
a. Δ is independent of θ b. Δ is independent of ϕ
c. Δ is a constant d. $\left.\frac{d\Delta}{d\theta}\right]_{\theta=\pi/2} = 0$

5. If $\phi(\alpha, \beta) = \begin{vmatrix} \cos\alpha & -\sin\alpha & 1 \\ \sin\alpha & \cos\alpha & 1 \\ \cos(\alpha+\beta) & -\sin(\alpha+\beta) & 1 \end{vmatrix}$, then

a. $f(300, 200) = f(400, 200)$

b. $f(200, 400) = f(200, 600)$

c. $f(100, 200) = f(200, 200)$

d. none of these

6. The determinant $\Delta = \begin{vmatrix} a^2+x & ab & ac \\ ab & b^2+x & bc \\ ac & bc & c^2+x \end{vmatrix}$ is divisible by

a. x **b.** x^2

c. x^3 **d.** none of these

7. If $f(x) = \begin{vmatrix} a & -1 & 0 \\ ax & a & -1 \\ ax^2 & ax & a \end{vmatrix}$, then $f(2x) - f(x)$ is divisible by

a. x **b.** a

c. $2a + 3x$ **d.** x^2

8. $\Delta = \begin{vmatrix} 1 & 1+ac & 1+bc \\ 1 & 1+ad & 1+bd \\ 1 & 1+ae & 1+be \end{vmatrix}$ is independent of

a. a **b.** b

c. c, d, e **d.** none of these

9. If $g(x) = \dfrac{f(x)}{(x-a)(x-b)(x-c)}$, where $f(x)$ is a polynomial of degree < 3, then

a. $\int g(x)dx = \begin{vmatrix} 1 & a & f(a)\log|x-a| \\ 1 & b & f(b)\log|x-b| \\ 1 & c & f(c)\log|x-c| \end{vmatrix} \div \begin{vmatrix} 1 & a & a^2 \\ 1 & b & b^2 \\ 1 & c & c^2 \end{vmatrix} + k$

b. $\dfrac{dg(x)}{dx} = \begin{vmatrix} 1 & a & f(a)(x-a)^{-2} \\ 1 & b & f(b)(x-b)^{-2} \\ 1 & c & f(c)(x-c)^{-2} \end{vmatrix} \div \begin{vmatrix} a^2 & a & 1 \\ b^2 & b & 1 \\ c^2 & c & 1 \end{vmatrix}$

c. $\dfrac{dg(x)}{dx} = \begin{vmatrix} 1 & a & f(a)(x-a)^{-2} \\ 1 & b & f(b)(x-b)^{-2} \\ 1 & c & f(c)(x-c)^{-2} \end{vmatrix} \div \begin{vmatrix} 1 & a & a^2 \\ 1 & b & b^2 \\ 1 & c & c^2 \end{vmatrix}$

d. $\int g(x)dx = \begin{vmatrix} 1 & a & f(a)\log|x-a| \\ 1 & b & f(b)\log|x-b| \\ 1 & c & f(c)\log|x-c| \end{vmatrix} \div \begin{vmatrix} a^2 & a & 1 \\ b^2 & b & 1 \\ c^2 & c & 1 \end{vmatrix} + k$

10. If $\Delta(x) = \begin{vmatrix} x^2+4x-3 & 2x+4 & 13 \\ 2x^2+5x-9 & 4x+5 & 26 \\ 8x^2-6x+1 & 16x-6 & 104 \end{vmatrix} = ax^3 + bx^2 + cx + d$, then

a. $a = 3$ **b.** $b = 0$

c. $c = 0$ **d.** None of these

11. If $f(\theta) = \begin{vmatrix} \sin\theta & \cos\theta & \sin\theta \\ \cos\theta & \sin\theta & \cos\theta \\ \cos\theta & \sin\theta & \sin\theta \end{vmatrix}$, then

a. $f(\theta) = 0$ has exactly 2 real solutions in $[0, \pi]$

b. $f(\theta) = 0$ has exactly 3 real solutions in $[0, \pi]$

c. range of function $\dfrac{f(\theta)}{1-\sin 2\theta}$ is $\left[-\sqrt{2}, \sqrt{2}\right]$

d. range of function $\dfrac{f(\theta)}{\sin 2\theta - 1}$ is $[-3, 3]$

12. If $f(\theta) = \begin{vmatrix} \sin^2 A & \cot A & 1 \\ \sin^2 B & \cos B & 1 \\ \sin^2 C & \cos C & 1 \end{vmatrix}$, then

a. $\tan A + \tan B + C$ **b.** $\cot A \cot B \cot C$

c. $\sin^2 A + \sin^2 B + \sin^2 C$ **d.** 0

13. The roots of the equation $\begin{vmatrix} {}^xC_r & {}^{n-1}C_r & {}^{n-1}C_{r-1} \\ {}^{x+1}C_r & {}^nC_r & {}^nC_{r-1} \\ {}^{x+2}C_r & {}^{n+1}C_r & {}^{n+1}C_{r-1} \end{vmatrix} = 0$ are

a. $x = n$ **b.** $x = n + 1$

c. $x = n - 1$ **d.** $x = n - 2$

14.

If $f(x) = \begin{vmatrix} 3 & 3x & 3x^2+2a^2 \\ 3x & 3x^2+2a^2 & 3x^3+6a^2x \\ 3x^2+2a^2 & 3x^3+6a^2x & 3x^4+12a^2x^2+2a^4 \end{vmatrix}$, then

a. $f'(x) = 0$

b. $y = f(x)$ is a straight line parallel to x-axis

c. $\int_0^2 f(x)dx = 32a^4$

d. none of these

15. If $\begin{vmatrix} yz-x^2 & zx-y^2 & xy-z^2 \\ xz-y^2 & xy-z^2 & yz-x^2 \\ xy-z^2 & yz-x^2 & zx-y^2 \end{vmatrix} = \begin{vmatrix} r^2 & u^2 & u^2 \\ u^2 & r^2 & u^2 \\ u^2 & u^2 & r^2 \end{vmatrix}$, then

a. $r^2 = x + y + z$ **b.** $r^2 = x^2 + y^2 + z^2$

c. $u^2 = yz + zx + xy$ **d.** $u^2 = xyz$

16. Let $f(n) = \begin{vmatrix} n & n+1 & n+2 \\ {}^nP_n & {}^{n+1}P_{n+1} & {}^{n+2}P_{n+2} \\ {}^nC_n & {}^{n+1}C_{n+1} & {}^{n+2}C_{n+2} \end{vmatrix}$ where the symbols have their usual meanings. Then $f(n)$ is divisible by

a. $n^2 + n + 1$ **b.** $(n + 1)!$

c. $n!$ **d.** none of these

17. If a, b, c are non-zero real numbers such that

$\begin{vmatrix} bc & ca & ab \\ ca & ab & bc \\ ab & bc & ca \end{vmatrix} = 0$, then

a. $\frac{1}{a}+\frac{1}{b\omega}+\frac{1}{c\omega^2}=0$ b. $\frac{1}{a}+\frac{1}{b\omega^2}+\frac{1}{c\omega}=0$

c. $\frac{1}{a\omega}+\frac{1}{b\omega^2}+\frac{1}{c}=0$ d. none of these

18. The values of $k \in R$ for which the system of equations $x + ky + 3z = 0$, $kx + 2y + 2z = 0$, $2x + 3y + 4z = 0$ admits of non-trivial solution is

a. 2 b. 5/2

c. 3 d. 5/4

19. If determinent $\begin{vmatrix} \cos(\theta+\phi) & -\sin(\theta+\phi) & \cos 2\phi \\ \sin\theta & \cos\theta & \sin\phi \\ -\cos\theta & \sin\theta & \cos\phi \end{vmatrix}$ is

a. positive b. independent of θ

c. independent of ϕ d. none of these

Reasoning Type

Each question has four choices a, b, c and d, out of which *only one* is correct. Each question contains Statement 1 and Statement 2.

a. Both the statements are True and Statement 2 is the correct explanation of Statement 1.

b. Both the statements are true but Statement 2 is Not the correct explanation of Statement 1.

c. Statement 1 is True and Statement 2 is False.

d. Statement 1 is False and Statement 2 is True.

1. **Statement 1:** If A, B and C are the angles of a triangle and $\begin{vmatrix} 1 & 1 & 1 \\ 1+\sin A & 1+\sin B & 1+\sin C \\ \sin A+\sin^2 A & \sin B+\sin^2 B & \sin C+\sin^2 C \end{vmatrix} = 0$, then triangle may not be equilateral.

   Statement 2: If any two rows of a determinant are the same, then the value of that determinant is zero.

2. Consider the system of the equations $kx + y + z = 1$, $x + ky + z = k$ and $x + y + kz = k^2$.

   Statement 1: System of equations has infinite solutions when $k=1$.

   Statement 2: If the determinant $\begin{vmatrix} 1 & 1 & 1 \\ k & k & 1 \\ k^2 & 1 & k \end{vmatrix} = 0$, then $k=-1$.

3. Consider the determinant $f(x) = \begin{vmatrix} 0 & x^2-a & x^3-b \\ x^2+a & 0 & x^2+c \\ x^4+b & x-c & 0 \end{vmatrix}$.

   Statement 1: $f(x) = 0$ has one root $x = 0$.

   Statement 2: The value of skew-symmetric determinant of odd-order is always zero.

4. **Statement 1:** If the system of equations $\lambda x + (b-a)y + (c-a)z = 0$, $(a-b)x + \lambda y + (c-b)z = 0$ and $(a-c)x + (b-c)y + \lambda z = 0$ has a non-trivial solution, then the value of λ is 0.

   Statement 2: The value of skew-symmetric matrix of order 3 is zero.

5. **Statement 1:** $\Delta = \begin{vmatrix} my+nz & mq+nr & mb+nc \\ kz-mx & kr-mp & kc-ma \\ -nx-ky & -np-kq & -na-kb \end{vmatrix}$ is equal to 0.

   Statement 2: The value of skew-symmetric matrix of order 3 is zero.

6. **Statement 1:** If $bc + qr = ca + rp = ab + pq = -1$, then $\begin{vmatrix} ap & a & p \\ bq & b & q \\ cr & c & r \end{vmatrix} = 0$ $(abc, pqr \neq 0)$.

   Statement 2: If system of equations $a_1x + b_1y + c_1 = 0$, $a_2x+b_2y+c_2=0$, $a_3x+b_3y+c_3=0$ has non-trivial solutions, $\begin{vmatrix} a_1 & b_1 & c_1 \\ a_2 & b_2 & c_2 \\ a_3 & b_3 & c_3 \end{vmatrix} = 0$

7. Consider the system of equation $x+y+z=6$, $x+2y+3z=10$, and $x + 2y + \lambda z = \mu$.

   Statement 1: If the system has infinite number of solutions, then $\mu = 10$.

   Statement 2: The determinant $\begin{vmatrix} 1 & 1 & 6 \\ 1 & 2 & 10 \\ 1 & 2 & \mu \end{vmatrix} = 0$ for $\mu = 10$.

8. Consider the determinant $\Delta = \begin{vmatrix} a_1+b_1x^2 & a_1x^2+b_1 & c_1 \\ a_2+b_2x^2 & a_2x^2+b_2 & c_2 \\ a_3+b_3x^2 & a_3x^2+b_3 & c_3 \end{vmatrix} = 0$ where $a_i, b_i, c_i \in R$ $(i = 1, 2, 3)$ and $x \in R$.

   Statement 1: The values of x satisfying $\Delta = 0$ are $x = 1, -1$.

   Statement 2: If $\begin{vmatrix} a_1 & b_1 & c_1 \\ a_2 & b_2 & c_2 \\ a_3 & b_3 & c_3 \end{vmatrix} = 0$, then $\Delta = 0$.

Linked Comprehension Type

Based upon each paragraph, three multiple choice questions have to be answered. Each question has four choices a, b, c and d, out of which *only one* is correct.

For Problems 1–3

$f(x) = \begin{vmatrix} x+c_1 & x+a & x+a \\ x+b & x+c_2 & x+a \\ x+b & x+b & x+c_3 \end{vmatrix}$ and $g(x) = (c_1-x)(c_2-x)(c_3-x)$

1. Coefficient of x in $f(x)$ is

a. $\frac{g(a)-f(b)}{b-a}$

b. $\frac{g(-a)-g(-b)}{b-a}$

c. $\frac{g(a)-g(b)}{b-a}$

d. none of these

2. Which of the following is not a constant term in $f(x)$?

a. $\dfrac{bg(a)-ag(b)}{(b-a)}$ b. $\dfrac{bg(a)-af(-b)}{(b-a)}$

c. $\dfrac{bf(-a)-ag(b)}{(b-a)}$ d. none of these

3. Which of the following is not true?

a. $f(-a)=g(a)$ b. $f(-a)=g(-a)$

c. $f(-b)=g(b)$ d. none of these

For Problems 4–6

Consider the function $f(x)=\begin{vmatrix} a^2+x & ab & ac \\ ab & b^2+x & bc \\ ac & bc & c^2+x \end{vmatrix}$

4. Which of the following is true?

a. $f(x)=0$ and $f'(x)=0$ have one positive common root.

b. $f(x)=0$ and $f'(x)=0$ have one negative common root.

c. $f(x)=0$ and $f'(x)=0$ have no common root.

d. None of these.

5. Which of the following is true?

a. $f(x)$ has one positive point of maxima.

b. $f(x)$ has one negative point of minima.

c. $f(x)=0$ has three distinct roots.

d. Local minimum value of $f(x)$ is zero.

6. In which of the following interval $f(x)$ is strictly increasing?

a. $(-\infty,\infty)$ b. $(-\infty,0)$

c. $(0,\infty)$ d. None of these

For Problems 7–9

Given that the system of equations $x=cy+bz$, $y=az+cx$, $z=bx+ay$ has non-zero solutions and at least one of the a, b, c is a proper fraction.

7. $a^2+b^2+c^2$ is

a. >2 c. >3

d. <3 d. <2

8. abc is

a. >-1 b. >1

c. <2 d. <3

9. System has solution such that

a. $x{:}y{:}z\equiv(1-2a^2){:}(1-2b^2){:}(1-2c^2)$

b. $x{:}y{:}z\equiv\dfrac{1}{1-2a^2}{:}\dfrac{1}{1-2b^2}{:}\dfrac{1}{1-2c^2}$

c. $x{:}y{:}z\equiv\dfrac{a}{1-a^2}{:}\dfrac{b}{1-b^2}{:}\dfrac{c}{1-c^2}$

d. $x{:}y{:}z\equiv\sqrt{1-a^2}{:}\sqrt{1-b^2}{:}\sqrt{1-c^2}$

For Problems 10–12

Consider the system of equations

$$x+y+z=6$$
$$x+2y+3z=10$$
$$x+2y+\lambda z=\mu$$

10. The system has unique solution if

a. $\lambda\neq 3$ b. $\lambda=3, \mu=10$

c. $\lambda=3, \mu\neq 10$ d. none of these

11. The system has infinite solutions if

a. $\lambda\neq 3$ b. $\lambda=3, \mu=10$

c. $\lambda=3, \mu\neq 10$ d. none of these

12. The system has no solution if

a. $\lambda\neq 3$ b. $\lambda=3, \mu=10$

c. $\lambda=3, \mu\neq 10$ d. none of these

For Problems 13–15

Let α, β be the roots of the equation $ax^2+bx+c=0$. Let $S_n=\alpha^n+\beta^n$ for $n\geq 1$ and $\Delta=\begin{vmatrix} 3 & 1+S_1 & 1+S_2 \\ 1+S_1 & 1+S_2 & 1+S_3 \\ 1+S_2 & 1+S_3 & 1+S_4 \end{vmatrix}$.

13. If $\Delta<0$, then the equation $ax^2+bx+c=0$ has

a. positive real roots b. negative real roots

c. equal roots d. imaginary roots

14. If a, b, c are rational and one of the roots of the equation is $1+\sqrt{2}$, then the value of Δ is

a. 8 b. 12

c. 30 d. 32

15. If $\Delta>0$, then

a. $f(1)>0$

b. $f(1)<0$

c. $f(1)=0$

d. cannot say anything about $f(1)$

For Problems 16–18

Let $\Delta=\begin{vmatrix} -bc & b^2+bc & c^2+bc \\ a^2+ac & -ac & c^2+ac \\ a^2+ab & b^2+ab & -ab \end{vmatrix}$ and the equation $px^3+qx^2+rx+s=0$ has roots a, b, c, where $a, b, c\in R^+$.

16. The value of Δ is

a. r^2/p^2 b. r^3/p^3

c. $-s/p$ d. none of these

17. The value of Δ is

a. $\leq 9r^2/p^2$ b. $\geq 27s^2/p^2$

c. $\leq 27s^3/p^3$ d. none of these

18. If $\Delta=27$ and $a^2+b^2+c^2=2$, then

a. $3p+2q=0$ b. $4p+3q=0$

c. $3p+q=0$ d. none of these

For Problems 19–21

Consider the polynomial function

$f(x)=\begin{vmatrix} (1+x)^a & (1+2x)^b & 1 \\ 1 & (1+x)^a & (1+2x)^b \\ (1+2x)^b & 1 & (1+x)^a \end{vmatrix}$, a, b being positive integers.

19. The constant term in $f(x)$ is

a. 2 **b.** 1

c. -1 **d.** 0

20. The coefficient of x in $f(x)$ is

a. 2^a **b.** $2^a - 3 \times 2^b + 1$

c. 0 **d.** none of these

21. Which of the following is true?

a. All the roots of the equation $f(x) = 0$ are positive.

b. All the roots of the equation $f(x) = 0$ are negative.

c. At least one of the equation $f(x) = 0$ is repeating one.

d. None of these.

For Problems 22–24

If $x > m, y > n, z > r$ $(x, y, z > 0)$ such that $\begin{vmatrix} x & n & r \\ m & y & r \\ m & n & z \end{vmatrix} = 0$.

22. The value of $\frac{x}{x-m} + \frac{y}{y-n} + \frac{z}{z-r}$ is

a. 1 **b.** -1

c. 2 **d.** -2

23. The value of $\frac{m}{x-m} - 1 + \frac{y}{y-n} - 1 + \frac{r}{z-r}$ is

a. -2 **b.** -4

c. 0 **d.** -1

24. The greatest value of $\frac{xyz}{(x-m)(y-n)(z-r)}$ is

a. 27 **b.** $\frac{8}{27}$

c. $\frac{64}{27}$ **d.** none of these

For Problems 25–27

Suppose $f(x)$ is a function satisfying the following conditions:

i. $f(0) = 2, f(1) = 1,$

ii. f has a minimum value at $x = 5/2$

iii. For all x, $f'(x) = \begin{vmatrix} 2ax & 2ax-1 & 2ax+b+1 \\ b & b+1 & -1 \\ 2(ax+b) & 2ax+2b+1 & 2ax+b \end{vmatrix}$

25. The value of $f(2)$ is

a. 1/4 **b.** 1/2

c. -1 **d.** 3

26. $f(x) = 0$ has

a. both roots positive

b. both roots negative

c. roots of opposite sign

d. imaginary roots

27. Range of $f(x)$ is

a. $[7/16, \infty)$ **b.** $(-\infty, 15/16]$

c. $[3/4, \infty)$ **d.** none of these

Matrix-Match Type

Each question contains statements given in two columns which have to be matched. Statements a, b, c, d in column I have to be matched with statements p, q, r, s in column II. If the correct matches are a → p, s, b → r, c → p, q and d → s, then the correctly bubbled 4 × 4 matrix should be as follows:

	p	q	r	s
a	(p) ●	(q)	(r)	(s) ●
b	(p)	(q) ●	(r) ●	(s)
c	(p) ●	(q) ●	(r)	(s)
d	(p)	(q)	(r)	(s) ●

1.

Column I	Column II
a. Coefficient of x in $f(x) = \begin{vmatrix} x & (1+\sin x)^3 & \cos x \\ 1 & \log(1+x) & 2 \\ x^2 & 1+x^2 & 0 \end{vmatrix}$.	**p.** 10
b. Value of $\begin{vmatrix} 1 & 3\cos\theta & 1 \\ \sin\theta & 1 & 3\cos\theta \\ 1 & \sin\theta & 1 \end{vmatrix}$ is	**q.** 0
c. If a, b, c are in A.P. and $f(x) = \begin{vmatrix} x+a & x^2+1 & 1 \\ x+b & 2x^2-1 & 1 \\ x+c & 3x^2-2 & 1 \end{vmatrix}$, then $f'(0)$ is	**r.** -12
d. If $\begin{vmatrix} x & 2 & x \\ 1 & x & 6 \\ x & x & x+1 \end{vmatrix} = a_4x^4 + a_3x^3 + a_2x^2 + a_1x + a_0$, then a_0 is	**s.** -2

2.

Column I	Column II
a. The value of the determinant $\begin{vmatrix} x+2 & x+3 & x+5 \\ x+4 & x+6 & x+9 \\ x+8 & x+11 & x+15 \end{vmatrix}$ is	**p.** 1
b. If one of the roots of the equation $\begin{vmatrix} 7 & 6 & x^2-13 \\ 2 & x^2-13 & 2 \\ x^2-13 & 3 & 7 \end{vmatrix} = 0$ is $x + 2$, then sum of all other five roots is	**q.** -6

c. The value of $\begin{vmatrix} \sqrt{6} & 2i & 3+\sqrt{6} \\ \sqrt{12} & \sqrt{3}+\sqrt{8}i & 3\sqrt{2}+\sqrt{6}i \\ \sqrt{18} & \sqrt{2}+\sqrt{12}i & \sqrt{27}+2i \end{vmatrix}$ is	**r.** 2
d. If $f(\theta) = \begin{vmatrix} \cos^2\theta & \cos\theta\sin\theta & -\sin\theta \\ \cos\theta\sin\theta & \sin^2\theta & \cos\theta \\ \sin\theta & -\cos\theta & 0 \end{vmatrix}$, then $f(\pi/3)$	**s.** −2

3.

Column I	Column II
a. $\begin{vmatrix} 1/c & 1/c & -(a+b)/c^2 \\ -(b+c)/a^2 & 1/a & 1/a \\ -b(b+c)/a^2c & (a+2b+c)/ac & -b(a+b)/ac^2 \end{vmatrix}$ is	**p.** independent of a
b. $\begin{vmatrix} \sin a\cos b & \sin a\sin b & \cos a \\ \cos a\cos b & \cos a\sin b & -\sin a \\ -\sin a\sin b & \sin a\cos b & 0 \end{vmatrix}$ is	**q.** independent of b
c. $\begin{vmatrix} \frac{1}{\sin a\cos b} & \frac{1}{\sin a\sin b} & \frac{1}{\cos a} \\ \frac{-\cos a}{\sin^2 a\cos b} & \frac{-\cos a}{\sin^2 a\sin b} & \frac{\sin a}{\cos^2 a} \\ \frac{\sin b}{\sin a\cos^2 b} & \frac{-\cos b}{\sin a\sin^2 b} & 0 \end{vmatrix}$ is	**r.** independent of c
d. If a, b, and c are the sides of a triangle and A, B and C are the angles opposite to a, b, and c, respectively, then $\Delta = \begin{vmatrix} a^2 & b\sin A & c\sin A \\ b\sin A & 1 & \cos A \\ c\sin A & \cos A & 1 \end{vmatrix}$	**s.** dependent on a, b

Integer Type

1. The largest value of a third-order determinant whose elements are 0 or 1 is_____.
2. If A, B, and C are the angles of non-right angled triangle ABC, then the value of $\begin{vmatrix} \tan A & 1 & 1 \\ 1 & \tan B & 1 \\ 1 & 1 & \tan C \end{vmatrix}$ is_____.
3. If $e^{i\theta} = \cos\theta + i\sin\theta$, and $D = \begin{vmatrix} 1 & e^{i\pi/3} & e^{i\pi/4} \\ e^{-i\pi/3} & 1 & e^{i2\pi/3} \\ e^{-i\pi/4} & e^{-i2\pi/3} & 1 \end{vmatrix}$, then the value of $[|D|]$ (where $[\cdot]$ represents greatest integer function) is_____.
4. If $a_r = (\cos 2r\pi + i\sin 2r\pi)^{\frac{1}{9}}$, then the value of $\begin{vmatrix} a_1 & a_2 & a_3 \\ a_4 & a_5 & a_6 \\ a_7 & a_8 & a_9 \end{vmatrix}$ is_____.
5. The value of the determinant $\begin{vmatrix} 1 & 1 & 1 & 1 \\ 1 & 2 & 3 & 4 \\ 1 & 3 & 6 & 10 \\ 1 & 4 & 10 & 20 \end{vmatrix}$ is_____.
6. If $A_1B_1C_1$, $A_2B_2C_2$, and $A_3B_3C_3$ are three three–digit numbers, each of which is divisible by 49, then $\Delta = \begin{vmatrix} A_1 & B_1 & C_1 \\ A_2 & B_2 & C_2 \\ A_3 & B_3 & C_3 \end{vmatrix}$ is divisible by_____.
7. If $\begin{vmatrix} 3a & -a+b & -a+c \\ -b+a & 3b & -b+c \\ -c+a & -c+b & 3c \end{vmatrix} = k(a+b+c)(ab+bc+ca)$, then the value of k is_____.
8. If $y = \begin{vmatrix} \sin x & \cos x & \sin x \\ \cos x & -\sin x & \cos x \\ x & 1 & 1 \end{vmatrix}$, then $(dy/dx) =$_____.
9. If $f(x) = \begin{vmatrix} x^n & n! & 2 \\ \cos x & \cos\frac{n\pi}{2} & 4 \\ \sin x & \sin\frac{n\pi}{2} & 8 \end{vmatrix}$, then the value of $\frac{d^n}{dx^n}(f(x))_{x=0}$ is_____.
10. The value of $[k/2]$, (where [.] represents the greatest integer function) for which the system of equations $x + ky + 3z = 0$; $3x + ky - 2z = 0$; $2x + 3y - 4z = 0$ possesses a non-trivial solution is_____.
11. If $2ax - 2y + 3z = 0$, $x + ay + 2z = 0$; and $2x + az = 0$ have a non-trivial solution, then the value of a is_____.
12. If $a \neq p$, $b \neq q$, $c \neq r$ and $\begin{vmatrix} p & b & c \\ a & q & c \\ a & b & r \end{vmatrix} = 0$, then the value of $\frac{p}{p-a} + \frac{q}{q-b} + \frac{r}{r-c}$ is_____.
13. Let λ and α be real. The number of integral values of λ for which the system of linear equations $\lambda x + (\sin\alpha)y + (\cos\alpha)z = 0$; $x + (\cos\alpha)y + (\sin\alpha)z = 0$; $-x + (\sin\alpha)y - (\cos\alpha)z = 0$ has non-trivial solution is_____.
14. Let $p\lambda^4 + q\lambda^3 + r\lambda^2 + s\lambda + t = \begin{vmatrix} \lambda^2+3\lambda & \lambda-1 & \lambda+3 \\ \lambda+1 & -2\lambda & \lambda-4 \\ \lambda-3 & \lambda+4 & 3\lambda \end{vmatrix}$ be an identity in λ, where p, q, r, s, and t are constants. Then the value of t is_____.
15. If $\begin{vmatrix} 1 & x & x^2 \\ x & x^2 & 1 \\ x^2 & 1 & x \end{vmatrix} = 3$, then the value of $\begin{vmatrix} x^3-1 & 0 & x-x^4 \\ 0 & x-x^4 & x^3-1 \\ x-x^4 & x^3-1 & 0 \end{vmatrix}$ is_____.
16. If $\begin{vmatrix} -2a & a+b & a+c \\ b+a & -2b & b+c \\ c+a & c+b & -2c \end{vmatrix} = k(b+c)(c+a)(a+b)$, then the value of k is_____.

ANSWERS AND SOLUTIONS

Objective Type

1. a. $\begin{vmatrix} pa & qb & rc \\ qc & ra & pb \\ rb & pc & qa \end{vmatrix}$

$= pqr(a^3+b^3+c^3-3abc) - abc(p^3+q^3+r^3-3pqr)$

$= pqr(a+b+c)(a^2+b^2+c^2-ab-bc-ca)$

$\quad - abc(p+q+r)(p^2+q^2+r^2-pq-qr-pr)$

$= 0$

2. b. Let $D = \begin{vmatrix} a_{11} & a_{12} & a_{13} \\ a_{21} & a_{22} & a_{23} \\ a_{31} & a_{32} & a_{33} \end{vmatrix}$

Applying $C_2 \to C_2 \frac{a_{12}}{a_{11}} C_1$, $C_3 \to C_3 - \frac{a_{13}}{a_{11}} C_1$, we get

$$D = \begin{vmatrix} a_{11} & 0 & 0 \\ a_{21} & \left(a_{22} - \frac{a_{12}}{a_{11}} \times a_{21}\right) & \left(a_{23} - \frac{a_{13}}{a_{11}} a_{21}\right) \\ a_{31} & \left(a_{32} - \frac{a_{12}}{a_{11}} \times a_{31}\right) & \left(a_{32} - \frac{a_{13}}{a_{11}} \times a_{31}\right) \end{vmatrix},$$

which has minimum value of –4.

3. b. $z = \begin{vmatrix} -5 & 3+4i & 5-7i \\ 3-4i & 6 & 8+7i \\ 5+7i & 8-7i & 9 \end{vmatrix}$

$\Rightarrow \bar{z} = \begin{vmatrix} -5 & 3-4i & 5+7i \\ 3+4i & 6 & 8-7i \\ 5-7i & 8+7i & 9 \end{vmatrix} = \begin{vmatrix} -5 & 3+4i & 5-7i \\ 3-4i & 6 & 8+7i \\ 5+7i & 8-7i & 9 \end{vmatrix} = z$

(Taking transpose)

$\Rightarrow z$ is purely real

4. c. Operation $C_1 \to C_1 + C_2 + C_3$ gives $(\alpha\beta + \beta\gamma + \gamma\alpha)$
$\begin{vmatrix} 1 & \beta\gamma & \gamma\alpha \\ 1 & \gamma\alpha & \alpha\beta \\ 1 & \alpha\beta & \beta\gamma \end{vmatrix}$

From the given equation, $\alpha\beta + \beta\gamma + \gamma\alpha = 0$. So, the value of determinant is 0.

5. c.

$$f(x) = \begin{vmatrix} 1-2\sin^2 x & \sin^2 x & 1-8\sin^2 x(1-\sin^2 x) \\ \sin^2 x & 1-2\sin^2 x & 1-\sin^2 x \\ 1-8\sin^2 x(1-\sin^2 x) & 1-\sin^2 x & 1-2\sin^2 x \end{vmatrix}$$

The required constant term is

$$f(0) = \begin{vmatrix} 1 & 0 & 1 \\ 0 & 1 & 1 \\ 1 & 1 & 1 \end{vmatrix} = \begin{vmatrix} 1 & 0 & 0 \\ 0 & 1 & 1 \\ 1 & 1 & 0 \end{vmatrix} = 1(0-1) = -1$$

6. a. $\Delta = \begin{vmatrix} a & b & c \\ b & c & a \\ c & a & b \end{vmatrix}$

$= (a+b+c)(a^2+b^2+c^2-ab-bc-ca)$

$= \frac{1}{2}(a+b+c)[(a-b)^2+(b-c)^2+(c-a)^2] = 0$

$\Rightarrow a+b+c = 0$ or $a = b = c$

If $a+b+c = 0$, we have

$\cos\theta + \cos 2\theta + \cos 3\theta = 0$ and $\sin\theta - \sin 2\theta + \sin 3\theta = 0$

$\Rightarrow \cos 2\theta(2\cos\theta + 1) = 0$ and $\sin 2\theta(1 - 2\cos\theta) = 0$ (1)

which is not possible as $\cos 2\theta = 0$ gives $\sin 2\theta \neq 0$, $\cos\theta \neq 1/2$. And $\cos\theta = -1/2$ gives $\sin 2\theta \neq 0$, $\cos\theta \neq 1/2$. Therefore, Eq. (1) does not hold simultaneously.

$\therefore a+b+c \neq 0$

$\therefore a = b = c$

or $e^{i\theta} = e^{-2i\theta} = e^{3i\theta}$,

which is satisfied only by $e^{i\theta} = 1$, i.e., $\cos\theta = 1$, $\sin\theta = 0$ so $\theta = 2k\pi$, $k \in Z$.

7. b. Applying $C_1 \to aC_1$ and then $C_1 \to C_1 + bC_2 + cC_3$, and taking $(a^2+b^2+c^2)$ common from C_1, we get

$$\Delta = \frac{(a^2+b^2+c^2)}{a}\begin{vmatrix} 1 & b-c & c+b \\ 1 & b & c-a \\ 1 & b+a & c \end{vmatrix}$$

$$= \frac{(a^2+b^2+c^2)}{a}\begin{vmatrix} 1 & b-c & c+b \\ 0 & c & -a-b \\ 0 & a+c & -b \end{vmatrix}$$

$(R_2 \to R_2 - R_1, R_3 \to R_3 - R_1)$

$$= \frac{(a^2+b^2+c^2)}{a}(-bc + a^2 + ab + ac + bc)$$

(expanding along C_1)

$= (a^2+b^2+c^2)(a+b+c)$

Hence, $\Delta = 0 \Rightarrow a+b+c = 0$

Therefore, line $ax + by + c = 0$ passes through the fixed point (1, 1).

8. b. The degree of the determinant is $n + (n+2) + (n+3) = 3n + 5$ and the degree of the expression on R.H.S. is 2.

$\therefore 3n + 5 = 2 \Rightarrow n = -1$

9. d. $\begin{vmatrix} a & b & c \\ b & c & a \\ c & a & b \end{vmatrix} = -(a^3+b^3+c^3-3abc)$

$= -(a+b+c)(a+b\omega^2+c\omega)(a+b\omega+c\omega^2)$

(where ω is cube roots of unity)

$= -f(\alpha)f(\beta)f(\gamma) \qquad [\because \alpha=1, \beta=\omega, \gamma=\omega^2]$

10. c. $\because -1 \le x < 0 \quad \therefore [x] = -1$

$0 \le y < 1 \quad \therefore [y] = 0$

$1 \le z < 2 \quad \therefore [z] = 1$

Hence, the given determinant is

$\begin{vmatrix} 0 & 0 & 1 \\ -1 & 1 & 1 \\ -1 & 0 & 2 \end{vmatrix} = 1 = [z]$

11. a. Given determinant,

$2a(bc-4a^2)+b(2ac-b^2)+c(2ab-c^2)=0$

$\Rightarrow 6abc-8a^3-b^3-c^3=0$

$\Rightarrow (2a+b+c)[(2a-b)^2+(b-c)^2+(c-2a)^2]=0$

$\Rightarrow 2a+b+c=0 \qquad (\because b \ne c)$

Let $f(x) = 8ax^3+2bx^2+cx$

$f(0)=0$

$f\left(\frac{1}{2}\right) = a+\frac{b}{2}+\frac{c}{2} = \frac{2a+b+c}{2} = 0$

So, $f(x)$ satisfies the Rolle's theorem and hence,

$f'(x)=0$ has at least one root in $\left[0, \frac{1}{2}\right]$.

12. b. The given determinant is

$\begin{bmatrix} 2^{n+1}-2^n+p & 2^{n+2}-2^{n+1}+q & p+r \\ 2^n+p & 2^{n+1} & p+r \\ a^2+2^n+p & b^2+2^n+2q & c^2-r \end{bmatrix}$

(Using $R_1 \to R_1 - R_3$ and $2q = p+r$)

$\begin{bmatrix} 2^n(2-1)+p & 2^{n+1}(2-1)+q & p+r \\ 2^n+p & 2^{n+1}+q & p+r \\ a^2+2^n+p & b^2+2^{n+1}+2q & c^2-r \end{bmatrix}$

$= \begin{bmatrix} 2^n+p & 2^{n+1}+q & p+r \\ 2^n+p & 2^{n+1}+q & p+r \\ a^2+2^n+p & b^2+2^{n+1}+2q & c^2-r \end{bmatrix} = 0 \quad (\because R_1 \equiv R_2)$

13. c. Consider the triangle with vertices $B(x_1, y_1)$, $C(x_2, y_2)$ and $A(x_3, y_3)$, and $AB = c$, $BC = a$ and $AC = b$. Then area of triangle is

$\frac{1}{2}\begin{vmatrix} x_1 & y_1 & 1 \\ x_2 & y_2 & 1 \\ x_3 & y_3 & 1 \end{vmatrix} = \sqrt{s(s-a)(s-b)(s-c)}$ where $2s = a+b+c$

Squaring and simplifying, we get

$4\begin{vmatrix} x_1 & y_1 & 1 \\ x_2 & y_2 & 1 \\ x_3 & y_3 & 1 \end{vmatrix} = (a+b+c)(b+c-a)(c+a-b)(a+b-c)$

Hence, $k = 4$.

14. c. We have

$\begin{vmatrix} ka & k^2+a^2 & 1 \\ kb & k^2+b^2 & 1 \\ kc & k^2+c^2 & 1 \end{vmatrix} = \begin{vmatrix} ka & k^2 & 1 \\ kb & k^2 & 1 \\ kc & k^2 & 1 \end{vmatrix} + \begin{vmatrix} ka & a^2 & 1 \\ kb & b^2 & 1 \\ kc & c^2 & 1 \end{vmatrix}$

$= 0 + k\begin{vmatrix} a & a^2 & 1 \\ b & b^2 & 1 \\ c & c^2 & 1 \end{vmatrix}$

$= k(a-b)(b-c)(c-a)$

15. b. We have,

$\begin{vmatrix} 1 & 1 & 1 \\ a & b & c \\ a^3 & b^3 & c^3 \end{vmatrix} = (a-b)(b-c)(c-a)(a+b+c) \qquad (1)$

Also,

$\begin{vmatrix} 1 & 1 & 1 \\ a & b & c \\ a^3 & b^3 & c^3 \end{vmatrix}$

$= abc\begin{vmatrix} \frac{1}{a} & \frac{1}{b} & \frac{1}{c} \\ 1 & 1 & 1 \\ a^2 & b^2 & c^2 \end{vmatrix}$

[Taking a,b,c common from R_1, R_2, R_3]

$= \begin{vmatrix} bc & ac & ab \\ 1 & 1 & 1 \\ a^2 & b^2 & c^2 \end{vmatrix}$ (Multiplying R_1 by abc)

$= \begin{vmatrix} 1 & 1 & 1 \\ a^2 & b^2 & c^2 \\ bc & ac & ab \end{vmatrix}$

Then,

$D = \begin{vmatrix} 1 & 1 & 1 \\ (x-a)^2 & (x-b)^2 & (x-c)^2 \\ (x-b)(x-c) & (x-c)(x-a) & (x-a)(x-b) \end{vmatrix}$

$= (a-b)(b-c)(c-a)(3x-a-b-c)$

Now given that a, b, c are all different, then $D = 0$.

$\therefore x = \frac{1}{3}(a+b+c)$

16. d. Let, $\Delta = \begin{vmatrix} y^2 & -xy & x^2 \\ a & b & c \\ a' & b' & c' \end{vmatrix}$

Then,

$$\Delta = \frac{1}{xy}\begin{vmatrix} xy^2 & -xy & x^2y \\ ax & b & cy \\ a'x & b' & c'y \end{vmatrix} \quad [\text{Applying } C_1 \to xC_1, C_3 \to yC_3]$$

$$= \frac{1}{xy}\begin{vmatrix} 0 & -xy & 0 \\ ax+by & b & bx+cy \\ a'x+b'y & b' & b'x+c'y \end{vmatrix}$$

[Applying $C_1 \to C_1 + yC_2$, $C_3 \to C_3 + xC_2$]

$$= \frac{1}{xy}\,xy\begin{vmatrix} ax+by & bx+cy \\ a'x+b'y & b'x+c'y \end{vmatrix} \quad [\text{Expanding along } R_1]$$

$$= \begin{vmatrix} ax+by & bx+cy \\ a'x+b'y & b'x+c'y \end{vmatrix}$$

17. b. $\Delta = \begin{vmatrix} b^2+c^2 & ab & ac \\ ab & c^2+a^2 & bc \\ ca & cb & a^2+b^2 \end{vmatrix}$

Applying $R_1 \to aR_1$, $R_2 \to bR_2$, $R_3 \to cR_3$, we get

$$\Delta = \frac{1}{abc}\begin{vmatrix} b^2+c^2 & a^2b & a^2c \\ ab^2 & b(c^2+a^2) & cb^2 \\ ac^2 & bc^2 & c(a^2+b^2) \end{vmatrix}$$

Now, applying $C_1 \to \frac{1}{a}C_1$, $C_2 \to \frac{1}{b}C_2$, $C_3 \to \frac{1}{c}C_3$, we get

$$\Delta = \frac{abc}{abc}\begin{vmatrix} b^2+c^2 & a^2 & a^2 \\ b^2 & c^2+a^2 & b^2 \\ c^2 & c^2 & a^2+b^2 \end{vmatrix}$$

$$= \begin{vmatrix} 0 & -2c^2 & -2b^2 \\ b^2 & c^2+a^2 & b^2 \\ c^2 & c^2 & a^2+b^2 \end{vmatrix} \quad [R_1 \to R_1 - R_2 - R_3]$$

$$= 2\begin{vmatrix} 0 & -c^2 & -b^2 \\ b^2 & a^2 & 0 \\ c^2 & 0 & a^2 \end{vmatrix}$$

[Taking 2 common from R_1 and applying $R_2 \to R_2 + R_1$ and $R_3 \to R_3 + R_1$]

Evaluating along R_1, we get

$\Delta = 2[c^2(a^2b^2) - b^2(-a^2c^2)] = 4a^2b^2c^2$

Hence, $k = 4$.

18. d. Applying $R_1 \to aR_1$, $R_2 \to bR_2$, and $R_3 \to cR_3$, we get

$$\Delta = \frac{1}{abc}\begin{vmatrix} ab^2c^2 & abc & ab+ac \\ a^2bc^2 & abc & bc+ab \\ a^2b^2c & abc & ac+bc \end{vmatrix}$$

$$= \frac{a^2b^2c^2}{abc}\begin{vmatrix} bc & 1 & ab+ac \\ ac & 1 & bc+ab \\ ab & 1 & ac+bc \end{vmatrix}$$

Applying $C_3 \to C_3 + C_1$ and taking $(bc + ca + ab)$ common, we get

$$\Delta = abc(bc+ca+ab)\begin{vmatrix} bc & 1 & 1 \\ ac & 1 & 1 \\ ab & 1 & 1 \end{vmatrix} = 0$$

[$\because C_2$ and C_3 are identical]

19. b. Applying $R_1 \to R_1 - R_2$, $R_2 \to R_2 - R_3$, we get

$$\Delta = \begin{vmatrix} -4-2\sqrt{2} & -2\sqrt{2} & 0 \\ 4\sqrt{2} & 4\sqrt{2} & 0 \\ 3-2\sqrt{2} & 2-2\sqrt{2} & 1 \end{vmatrix}$$

$$= 1(-(4+2\sqrt{2}))\,4\sqrt{2} + 2\sqrt{2} \times 4\sqrt{2}$$

$$= -16\sqrt{2}$$

20. b. The total number of third-order determinants is 9! Since the number of determinants is even and in which there are 9!/2 pairs of determinants which are obtained by changing two consecutive rows, so $\sum_{i=1}^{n} D_i = 0$.

21. a. $\Delta = \begin{vmatrix} a_1+b_1w & a_1w^2+b_1 & c_1+b_1\bar{w} \\ a_2+b_2w & a_2w^2+b_2 & c_2+b_2\bar{w} \\ a_3+b_3w & a_3w^2+b_3 & c_3+b_3\bar{w} \end{vmatrix}$

Operating $C_2 \to wC_2$, we have

$$\Delta = \frac{1}{w}\begin{vmatrix} a_1+b_1w & a_1w^3+b_1w & c_1+b_1\bar{w} \\ a_2+b_2w & a_2w^3+b_2w & c_2+b_2\bar{w} \\ a_3+b_3w & a_3w^3+b_3w & c_3+b_3\bar{w} \end{vmatrix}$$

$$= \frac{1}{w}\begin{vmatrix} a_1+b_1w & a_1+b_1w & c_1+b_1\bar{w} \\ a_2+b_2w & a_2+b_2w & c_2+b_2\bar{w} \\ a_3+b_3w & a_3+b_3w & c_3+b_3\bar{w} \end{vmatrix} \quad (\because \omega^3 = 1)$$

$$= 0$$

22. b. Since x, y, z are in A.P., therefore, $x + z - 2y = 0$. Now,

$$\begin{vmatrix} a+2 & a+3 & a+2x \\ a+3 & a+4 & a+2y \\ a+4 & a+5 & a+2z \end{vmatrix} = \begin{vmatrix} 0 & 0 & 2(x+z-2y) \\ a+3 & a+4 & a+2y \\ a+4 & a+5 & a+2z \end{vmatrix}$$

[Applying $R_1 \to R_1 + R_3 - 2R_2$]

$$= \begin{vmatrix} 0 & 0 & 0 \\ a+3 & a+4 & a+2y \\ a+4 & a+5 & a+2z \end{vmatrix}$$

[$\because x + z - 2y = 0$]

$$= 0$$

23. d. Operating $C_1 \to C_1 + C_2 + C_3$, we get

$$(a+b+c-x)\begin{vmatrix} 1 & c & b \\ 1 & b-x & a \\ 1 & a & c-x \end{vmatrix} = 0$$

$\therefore x = a + b + c = 0$

24. b. Applying $C_1 \to C_1 - C_2$, $C_2 \to C_2 - C_3$, we get

$$\Delta = \begin{vmatrix} 0 & 0 & 1 \\ \cot\frac{A}{2} - \cot\frac{B}{2} & \cot\frac{B}{2} - \cot\frac{C}{2} & \cot\frac{C}{2} \\ \tan\frac{B}{2} - \tan\frac{A}{2} & \tan\frac{C}{2} - \tan\frac{B}{2} & \tan\frac{A}{2} + \tan\frac{B}{2} \end{vmatrix}$$

$$= \begin{vmatrix} 0 & 0 & 1 \\ \cot\frac{A}{2} - \cot\frac{B}{2} & \cot\frac{B}{2} - \cot\frac{C}{2} & \cot\frac{C}{2} \\ \dfrac{\cot\frac{A}{2} - \cot\frac{B}{2}}{\cot\frac{A}{2}\cot\frac{B}{2}} & \dfrac{\cot\frac{B}{2} - \cot\frac{C}{2}}{\cot\frac{B}{2}\cot\frac{C}{2}} & \tan\frac{A}{2} + \tan\frac{B}{2} \end{vmatrix}$$

$$= \left(\cot\frac{A}{2} - \cot\frac{B}{2}\right)\left(\cot\frac{B}{2} - \cot\frac{C}{2}\right)$$

$$\times \begin{vmatrix} 0 & 0 & 1 \\ 1 & 1 & \cot\frac{C}{2} \\ \tan\frac{A}{2}\tan\frac{B}{2} & \tan\frac{B}{2}\tan\frac{C}{2} & \tan\frac{A}{2}\tan\frac{B}{2} \end{vmatrix}$$

$$= \left(\cot\frac{A}{2} - \cot\frac{B}{2}\right)\left(\cot\frac{B}{2} - \cot\frac{C}{2}\right)\left(\tan\frac{C}{2} - \tan\frac{A}{2}\right)\tan\frac{B}{2}$$

Since $\Delta = 0$,

$\cot\frac{A}{2} = \cot\frac{B}{2}$ or $\cot\frac{B}{2} = \cot\frac{C}{2}$ or $\tan\frac{A}{2} = \tan\frac{C}{2}$

Hence, the triangle is definitely isosceles.

25. d. Since a, b, c, d, e, f are in G. P. and if r is the common ratio of the G.P., then

$b = ar$

$c = ar^2$

$d = ar^3$

$e = ar^4$

$f = ar^5$

Therefore, given determinant is

$$\begin{vmatrix} a^2 & a^2r^6 & x \\ a^2r^2 & a^2r^8 & y \\ a^2r^4 & a^2r^{10} & z \end{vmatrix} = a^2a^2r^6 = \begin{vmatrix} 1 & 1 & x \\ r^2 & r^2 & y \\ r^4 & r^4 & z \end{vmatrix}$$

$= a^4r^6\,(0) = 0$ [$\because C_1, C_2$ are identical]

26. c. Operating $C_1 \to C_1 + C_2 + C_3$, we get

$$f(x) = \begin{vmatrix} 1+2x+(a^2+b^2+c^2)\,x & (1+b^2)\,x & (1+c^2)\,x \\ 1+2x+(a^2+b^2+c^2)\,x & 1+b^2\,x & (1+c^2)\,x \\ 1+2x+(a^2+b^2+c^2)\,x & (1+b^2)\,x & 1+x^2\,x \end{vmatrix}$$

$$= \begin{vmatrix} 1 & (1+b^2)\,x & (1+c^2)\,x \\ 1 & 1+b^2\,x & (1+c^2)\,x \\ 1 & (1+b^2)\,x & 1+c^2\,x \end{vmatrix} \quad [\because a^2+b^2+c^2 = -2]$$

$$= \begin{vmatrix} 1 & (1+b^2)\,x & (1+c^2)\,x \\ 0 & 1-x & 0 \\ 0 & 0 & 1-x \end{vmatrix}$$

[Operating $R_2 \to R_2 - R_1$ and $R_3 \to R_3 - R_1$]

$= (1)\,[(1-x)^2 - 0]$

$= (1-x)^2$,

which is a polynomial of degree 2.

27. a. $$\begin{vmatrix} 1 & 1 & 1 \\ {}^mC_1 & {}^{m+1}C_1 & {}^{m+2}C_1 \\ {}^mC_2 & {}^{m+1}C_2 & {}^{m+2}C_2 \end{vmatrix}$$

$$= \begin{vmatrix} 1 & 1 & 1 \\ {}^mC_1 & {}^{m+1}C_1 & {}^{m+1}C_0 + {}^{m+1}C_1 \\ {}^mC_2 & {}^{m+1}C_2 & {}^{m+1}C_1 + {}^{m+1}C_2 \end{vmatrix}$$

$$= \begin{vmatrix} 1 & 1 & 0 \\ {}^mC_1 & {}^{m+1}C_1 & {}^{m+1}C_0 \\ {}^mC_2 & {}^{m+1}C_2 & {}^{m+1}C_1 \end{vmatrix} \quad [\text{Applying } C_3 \to C_3 - C_2]$$

$$= \begin{vmatrix} 1 & 1 & 0 \\ {}^mC_1 & {}^mC_0 + {}^mC_1 & {}^{m+1}C_0 \\ {}^mC_2 & {}^mC_1 + {}^mC_2 & {}^{m+1}C_1 \end{vmatrix}$$

$$= \begin{vmatrix} 1 & 0 & 0 \\ {}^mC_1 & {}^mC_0 & {}^{m+1}C_0 \\ {}^mC_2 & {}^mC_1 & {}^{m+1}C_1 \end{vmatrix} \quad [\text{Applying } C_2 \to C_2 - C_1]$$

$= {}^mC_0\,{}^{m+1}C_1 - {}^{m+1}C_0\,{}^mC_1$

$= m + 1 - m$

$= 1$

28. c. Since each element of C_1 is the sum of two elements, putting the determinant as sum of two determinants, we get

$$\Delta = \begin{vmatrix} x^3 & x^2 & x \\ y^3 & y^2 & y \\ z^3 & z^2 & z \end{vmatrix} + \begin{vmatrix} 1 & x^2 & x \\ 1 & y^2 & y \\ 1 & z^2 & z \end{vmatrix}$$

$$= xyz\begin{vmatrix} x^2 & x & 1 \\ y^2 & y & 1 \\ z^2 & z & 1 \end{vmatrix} + \begin{vmatrix} 1 & x^2 & x \\ 1 & y^2 & y \\ 1 & z^2 & z \end{vmatrix}$$

$$= -(xyz+1)\begin{vmatrix} 1 & x & x^2 \\ 1 & y & y^2 \\ 1 & z & z^2 \end{vmatrix}$$

$$= -(xyz+1)(x-y)(y-z)(z-x)(x+y+z)$$

Since $\Delta = 0$, x, y, z all are distinct, we have $xyz + 1 = 0$ or $xyz = -1$.

29. c. $\begin{vmatrix} 1+x & 1 & 1 \\ 1+y & 1+2y & 1 \\ 1+z & 1+z & 3+3z \end{vmatrix}$

$$= xyz\begin{vmatrix} 1+\frac{1}{x} & \frac{1}{x} & \frac{1}{x} \\ 1+\frac{1}{y} & 2+\frac{1}{y} & \frac{1}{y} \\ 1+\frac{1}{z} & 1+\frac{1}{z} & 3+\frac{1}{z} \end{vmatrix}$$

$$= xyz\left(3+\frac{1}{x}+\frac{1}{y}+\frac{1}{z}\right)\begin{vmatrix} 1 & 1 & 1 \\ 1+\frac{1}{y} & 2+\frac{1}{y} & \frac{1}{y} \\ 1+\frac{1}{z} & 1+\frac{1}{z} & 3+\frac{1}{z} \end{vmatrix}$$

$$= xyz\left(3+\frac{1}{x}+\frac{1}{y}+\frac{1}{z}\right)\begin{vmatrix} 1 & 0 & 0 \\ 1+\frac{1}{y} & 1 & -1 \\ 1+\frac{1}{z} & 0 & 2 \end{vmatrix}$$

$$= 2xyz\left(3+\frac{1}{x}+\frac{1}{y}+\frac{1}{z}\right)$$

Hence, the given equation gives $x^{-1} + y^{-1} + z^{-1} = -3$.

30. b. Let a be the first term and d be the common difference of corresponding A.P. Then

$$\Delta = xyz\begin{vmatrix} 1/x & 1/y & 1/z \\ p & 2q & 3r \\ 1 & 1 & 1 \end{vmatrix}$$

$$= xyz\begin{vmatrix} a+(p-1)d & a+(2q-1)d & a+(3r-1)d \\ p & 2q & 3r \\ 1 & 1 & 1 \end{vmatrix}$$

Applying $R_1 \to R_1 - aR_3$, $R_2 \to R_2 - R_3$ and then taking d common from R_1, we get

$$\Delta = xyzd\begin{vmatrix} (p-1) & (2q-1) & (3r-1) \\ (p-1) & (2q-1) & (3r-1) \\ 1 & 1 & 1 \end{vmatrix} = 0$$

31. a. As $a_1b_1c_1$, $a_2b_2c_2$ and $a_3b_3c_3$ are even natural numbers, each of c_1, c_2, c_3 is divisible by 2. Let $c_i = 2k_i$ for $i = 1, 2, 3$. Thus,

$$\Delta = 2\begin{vmatrix} k_1 & a_1 & b_1 \\ k_2 & a_2 & b_2 \\ k_3 & a_3 & b_3 \end{vmatrix} = 2m,$$

where m is some natural number. Thus, Δ is divisible by 2.That Δ may not be divisible by 4 can be seen by taking the three numbers as 112, 122 and 134. Note that

$$\Delta = \begin{vmatrix} 2 & 1 & 1 \\ 2 & 1 & 2 \\ 4 & 1 & 3 \end{vmatrix} = 2,$$

which is divisible by 2 but not by 4.

32. a. We have,

$$\begin{vmatrix} x & 1 & 1 & \cdots \\ 1 & x & 1 & \cdots \\ 1 & 1 & x & \cdots \\ \cdots & \cdots & \cdots & \cdots \end{vmatrix} = \begin{vmatrix} x & 1 & 1 & \cdots \\ (1-x) & (x-1) & 0 & \cdots \\ (1-x) & 0 & (x-1) & \cdots \\ \cdots & \cdots & \cdots & \cdots \end{vmatrix}$$

[Applying $R_2 \to R_2 - R_1$, $R_3 \to R_3 - R_1$, ..., $R_n \to R_n - R_1$]

$= x(x-1)^{n-1} + |(x-1)^{n-1} + (x-1)^{n-1} + \cdots + (x-1)^{n-1}$ $(n-1)$ times | [Expanding along R_1]

$= x(x-1)^{n-1} + (n-1)(x-1)^{n-1}$

$= (x-1)^{n-1}(x+n-1)$

33. a. Let first term of G.P. is A and common ratio is R. Then, $a = AR^{p-1} \Rightarrow \log a = \log A + (p-1)\log R$, etc.

$$\Rightarrow \begin{vmatrix} \log a & p & 1 \\ \log b & q & 1 \\ \log c & r & 1 \end{vmatrix} = \begin{vmatrix} \log A+(p-1)\log R & p & 1 \\ \log A+(q-1)\log R & q & 1 \\ \log A+(r-1)\log R & r & 1 \end{vmatrix}$$

$$= \begin{vmatrix} (p-1)\log R & p & 1 \\ (q-1)\log R & q & 1 \\ (r-1)\log R & r & 1 \end{vmatrix}$$

$[C_1 \to C_1 - (\log A)C_3]$

$$= \log R\begin{vmatrix} (p-1) & p & 1 \\ (q-1) & q & 1 \\ (r-1) & r & 1 \end{vmatrix}$$

$$= \log R\begin{vmatrix} p & p & 1 \\ q & q & 1 \\ r & r & 1 \end{vmatrix} \quad (C_1 \to C_1 + C_3)$$

$= 0$

34. b. $R_3 \to R_3 - 2R_2$, hence two identical rows $\Rightarrow f(x)$ = constant.

35. b. In each determinant applying $R_1 \to R_1 + R_2 + R_3$ and then taking out $(x+9)$ common, we get

$x + 9 = 0 \Rightarrow x = -9$

36. c. Taking x^5 common from last row, we get

$$x^5\begin{vmatrix} x^n & x^{n+2} & x^{2n} \\ 1 & x^a & a \\ x^n & x^{a+1} & x^{2n} \end{vmatrix} = 0, \forall x \in R$$

$\Rightarrow a+1=n+2 \Rightarrow a=n+1$

(as it will make first and third row is identical)

37. a. Applying $R_1 \to R_1 + R_3 - 2R_2$, we get

$$\Delta = \begin{vmatrix} 0 & 0 & 0 & x+z-2y \\ 4 & 5 & 6 & y \\ 5 & 6 & 7 & z \\ x & y & z & 0 \end{vmatrix}$$

$$= -(x+z-2y)\begin{vmatrix} 4 & 5 & 6 \\ 5 & 6 & 7 \\ x & y & z \end{vmatrix} \quad \text{[Expanding along } R_1]$$

$$= -(x+z-2y)\begin{vmatrix} 0 & -1 & 6 \\ 0 & -1 & 7 \\ x-2y+z & y-z & z \end{vmatrix}$$

[Applying $C_1 \to C_1 + C_3 - 2C_2$ and $C_2 \to C_2 - C_3$]

$$= -(x+z-2y)^2 \begin{vmatrix} -1 & 6 \\ -1 & 7 \end{vmatrix}$$

$$= (x-2y+z)^2$$

Hence, $\Delta = 0 \Rightarrow x, y, z$ are in A.P.

38. c. Applying $R_1 \to R_1 - R_2$ and $R_2 \to R_2 - R_3$ reduce the determinant to

$$\begin{vmatrix} x^2-2x+1 & x-1 & 0 \\ 2x-2 & x-1 & 0 \\ 3 & 3 & 1 \end{vmatrix} = (x-1)^3 - 2(x-1)^2$$

$$= (x-1)^2(x-1-2)$$
$$= (x-1)^2(x-3),$$

which is clearly negative for $x < 1$.

39. d. Applying $R_1 \to R_1 - (R_2 + R_3)$, we get

$$D = \begin{vmatrix} 0 & -2y & -2x \\ x & y+z & x \\ y & y & z+x \end{vmatrix}$$

$$= 2\begin{vmatrix} 0 & -y & -x \\ x & y+z & x \\ y & y & z+x \end{vmatrix}$$

$$= 2\begin{vmatrix} 0 & -y & -x \\ x & z & 0 \\ y & 0 & z \end{vmatrix} \quad (R_2 \to R_2 + R_1 \text{ and } R_3 \to R_3 + R_1)$$

$$= 4xyz$$

40. b. Operation $R_1 \to R_1 - R_2$, gives

$$\Delta = \begin{vmatrix} x-2 & 3(x-2) & -(x-2) \\ 2 & -3x & x-3 \\ -3 & 2x & x+2 \end{vmatrix}$$

$$= (x-2)\begin{vmatrix} 1 & 3 & -1 \\ 2 & -3x & x-3 \\ -3 & 2x & x+2 \end{vmatrix}$$

$$= (x-2)\begin{vmatrix} 1 & 3 & -1 \\ 0 & -3(x+2) & x-1 \\ 0 & 2x+9 & x-1 \end{vmatrix}$$

$[R_2 \to R_2 - 2R_1, R_3 \to R_3 + 3R_1]$

$$= (x-2)\{-(3x+6)(x-1) - (x-1)(2x+9)\}$$

$$= -(x-2)(x-1)(5x+15)$$

Therefore, $\Delta = 0$ gives $x = 2, 1, -3$.

41. b. Applying $C_1 \to C_1 + C_2 + C_3$, we get

$$\Delta = \begin{vmatrix} x+2a & a & a \\ x+2a & x & a \\ x+2a & a & x \end{vmatrix} = (x+2a)\begin{vmatrix} 1 & a & a \\ 1 & x & a \\ 1 & a & x \end{vmatrix}$$

Applying $R_1 \to R_1 - R_2$ and $R_2 \to R_2 - R_3$, we get

$$\Delta = (x+2a)\begin{vmatrix} 0 & a-x & 0 \\ 0 & x-a & a-x \\ 1 & a & x \end{vmatrix} = (x-a)^2(x+2a)$$

42. a.
$$\begin{vmatrix} x & m & n & 1 \\ a & x & n & 1 \\ a & b & x & 1 \\ a & b & c & 1 \end{vmatrix} = 0$$

$[R_1 \to R_1 - R_2, R_2 \to R_2 - R_3, R_3 \to R_3 - R_4]$

$$\Rightarrow \begin{vmatrix} x-a & m-x & 0 & 0 \\ 0 & x-b & n-x & 0 \\ 0 & 0 & x-c & a \\ a & b & c & 1 \end{vmatrix} = 0$$

$$\Rightarrow \begin{vmatrix} x-a & m-x & 0 \\ 0 & x-b & n-x \\ 0 & 0 & x-c \end{vmatrix} = 0$$

$$\Rightarrow (x-a)\begin{vmatrix} (x-b) & n-x \\ 0 & (x-c) \end{vmatrix} = 0$$

$$\Rightarrow (x-a)(x-b)(x-c) = 0$$

$\Rightarrow$ roots are independent of m, n

43. d. Since for $x = 0$, the determinant reduces to the determinant of a skew-symmetric matrix of odd order which is always zero, hence $x = 0$ is the solution of the given equation.

44. b. We have,

$$\begin{vmatrix} b+c & c+a & a+b \\ a+b & b+c & c+a \\ c+a & a+b & b+c \end{vmatrix} = k\begin{vmatrix} a & b & c \\ c & a & b \\ b & c & a \end{vmatrix}$$

$$\Rightarrow \begin{vmatrix} 2(a+b+c) & c+a & a+b \\ 2(a+b+c) & b+c & c+a \\ 2(a+b+c) & a+b & b+c \end{vmatrix} = k \begin{vmatrix} a & b & c \\ c & a & b \\ b & c & a \end{vmatrix}$$

[Applying $C_1 \to C_1 + (C_2 + C_3)$ on L.H.S.]

$$\Rightarrow 2 \begin{vmatrix} a+b+c & -b & -c \\ a+b+c & -a & -b \\ a+b+c & -c & -a \end{vmatrix} = k \begin{vmatrix} a & b & c \\ c & a & b \\ b & c & a \end{vmatrix}$$

[Applying $C_2 \to C_2 - C_1$, $C_3 \to C_3 - C_1$ on L.H.S.]

$$\Rightarrow \begin{vmatrix} a & -b & -c \\ c & -a & -b \\ b & -c & -a \end{vmatrix} = k \begin{vmatrix} a & b & c \\ c & a & b \\ b & c & a \end{vmatrix}$$

[Applying $C_1 \to C_1 + C_2 + C_3$ on L.H.S.]

$$\Rightarrow 2 \begin{vmatrix} a & b & c \\ c & a & b \\ b & c & a \end{vmatrix} = k \begin{vmatrix} a & b & c \\ c & a & b \\ b & c & a \end{vmatrix}$$

$\therefore k = 2$

45. d. $$D' = \begin{vmatrix} a_1 + pb_1 & b_1 + qc_1 & c_1 + ra_1 \\ a_2 + pb_2 & b_2 + qc_2 & c_2 + ra_2 \\ a_3 + pb_3 & b_3 + qc_3 & c_3 + ra_3 \end{vmatrix}$$

$$= \begin{vmatrix} a_1 & b_1 + qc_1 & c_1 + ra_1 \\ a_2 & b_2 + qc_2 & c_2 + ra_2 \\ a_3 & b_3 + qc_3 & c_3 + ra_3 \end{vmatrix} + \begin{vmatrix} pb_1 & b_1 + qc_1 & c_1 + ra_1 \\ pb_2 & b_2 + qc_2 & c_2 + ra_2 \\ pb_3 & b_3 + qc_3 & c_3 + ra_3 \end{vmatrix}$$

In the first determinant, apply $C_3 \to C_3 - rC_1$ and then $C_2 \to C_2 - qC_3$.

In second determinant take p common from C_1 and then apply $C_2 \to C_2 - C_1$. Then take q common from C_2 and apply $C_3 \to C_3 - C_2$.

Finally taking r common from C_3, we have ultimately $D' = (1 + pqr)D$.

46. b. $$\Delta = \begin{vmatrix} 1 & 4 & 9 & 16 \\ 4 & 9 & 16 & 25 \\ 9 & 16 & 25 & 36 \\ 16 & 25 & 36 & 49 \end{vmatrix} \quad [R_3 \to R_3 - R_2, R_4 \to R_4 - R_3]$$

$$= \begin{vmatrix} 1 & 4 & 9 & 16 \\ 4 & 9 & 16 & 25 \\ 5 & 7 & 9 & 11 \\ 15 & 21 & 27 & 33 \end{vmatrix}$$

$$= 3 \begin{vmatrix} 1 & 4 & 9 & 16 \\ 4 & 9 & 16 & 25 \\ 5 & 7 & 9 & 11 \\ 5 & 7 & 9 & 11 \end{vmatrix} = 0 \quad [R_4 \to R_4 - R_3]$$

47. c. Here $a > 0$ and $4b^2 - 4ac < 0$, i.e., $ac - b^2 > 0$.

$\therefore ax^2 + 2bx + c > 0, \forall x \in R$

Now,

$$\Delta = \begin{vmatrix} a & b & ax+b \\ b & c & bx+c \\ 0 & 0 & -(ax^2+2bx+c) \end{vmatrix}$$

[Operating $R_3 \to R_3 - xR_1 - R_2$]

$= -(ax^2 + 2bx + c)(ac - b^2)$

$= -$(positive) (positive) = negative

48. a. We have,

$a_{n+1}^2 = a_n a_{n+2}$

$\Rightarrow 2 \log a_{n+1} = \log a_n + \log a_{n+2}$

Similarly,

$2 \log a_{n+4} = \log a_{n+3} + \log a_{n+5}$

$2 \log a_{n+7} = \log a_{n+6} + \log a_{n+8}$

Substituting these values in second column of determinant, we get

$$\Delta = \frac{1}{2} \begin{vmatrix} \log a_n & \log a_n + \log a_{n+2} & \log a_{n+2} \\ \log a_{n+3} & \log a_{n+3} + \log a_{n+5} & \log a_{n+5} \\ \log a_{n+6} & \log_{n+6} + \log a_{n+8} & \log a_{n+8} \end{vmatrix}$$

$= \frac{1}{2}(0) = 0$ [Using $C_2 \to C_2 - C_1 - C_3$]

49. b.

$$\Delta = \begin{vmatrix} x_1 & x_2 & x_3 \\ y_1 & y_2 & y_3 \\ z_1 & z_2 & z_3 \end{vmatrix} = \begin{vmatrix} x_1 & y_1 & z_1 \\ x_2 & y_2 & z_2 \\ x_3 & y_3 & z_3 \end{vmatrix}$$

$$\Delta^2 = \begin{vmatrix} x_1 & y_1 & z_1 \\ x_2 & y_2 & z_2 \\ x_3 & y_3 & z_3 \end{vmatrix} \begin{vmatrix} x_1 & y_1 & z_1 \\ x_2 & y_2 & z_2 \\ x_3 & y_3 & z_3 \end{vmatrix}$$

$$= \begin{vmatrix} x_1^2 + y_1^2 + z_1^2 & x_1x_2 + y_1y_2 + z_1z_2 & x_1x_3 + y_1y_3 + z_1z_3 \\ x_1x_2 + y_2y_1 + z_2z_1 & x_2^2 + y_2^2 + z_2^2 & x_2x_3 + y_2y_3 + z_2z_3 \\ x_3x_1 + y_3y_1 + z_3z_1 & x_2x_3 + y_2y_3 + z_2z_3 & x_3^2 + y_3^2 + z_3^2 \end{vmatrix}$$

$$= \begin{vmatrix} 1 & 0 & 0 \\ 0 & 1 & 0 \\ 0 & 0 & 1 \end{vmatrix} = 1 \Rightarrow \Delta = \pm 1$$

50. c. Using $C_1 \to C_1 + C_2 + C_3$,

$$\Delta = \begin{vmatrix} \sin x + 2\cos x & \cos x & \cos x \\ \sin x + 2\cos x & \sin x & \cos x \\ \sin x + 2\cos x & \cos x & \sin x \end{vmatrix}$$

$$= (\sin x + 2\cos x) \begin{vmatrix} 1 & \cos x & \cos x \\ 1 & \sin x & \cos x \\ 1 & \cos x & \sin x \end{vmatrix}$$

Applying $R_2 \to R_2 - R_1$ and $R_3 \to R_3 - R_1$, we get

$$\Delta = (\sin x + 2\cos x) \begin{vmatrix} 1 & \cos x & \cos x \\ 0 & \sin x - \cos x & 0 \\ 0 & 0 & \sin x - \cos x \end{vmatrix}$$

$= (\sin x + 2\cos x)(\sin x - \cos x)^2$

Thus, $\Delta = 0 \Rightarrow \tan x = -2$ or $\tan x = 1$

As $-\pi/4 \le x \le \pi/4$, we get $-1 \le \tan x \le 1$

$\therefore \tan x = 1 \Rightarrow x = \pi/4$

51. b. We divide L.H.S. by λ^4 and C_1 by λ^2, C_2 by λ and C_3 by λ on the R.H.S. to obtain

$$p+q\left(\frac{1}{\lambda}\right)+r\left(\frac{1}{\lambda}\right)^2+s\left(\frac{1}{\lambda}\right)^3+t\left(\frac{1}{\lambda}\right)^4$$

$$=\begin{vmatrix} 1+3/\lambda & 1-1/\lambda & 1+3/\lambda \\ 1+1/\lambda^2 & 2/\lambda-1 & 1-3/\lambda \\ 1-3/\lambda^2 & 1+4/\lambda & 3 \end{vmatrix}$$

Taking limit as $\lambda \to \infty$, we get

$$p=\begin{vmatrix} 1 & 1 & 1 \\ 1 & -1 & 1 \\ 1 & 1 & 3 \end{vmatrix}=\begin{vmatrix} 1 & 1 & 1 \\ 0 & -2 & 0 \\ 0 & 0 & 2 \end{vmatrix}=-4$$

[Applying $R_2 \to R_2 - R_1, R_3 \to R_3 - R_1$]

52. d. Applying $R_2 \to R_2 - R_1, R_3 \to R_3 - R_1$, we get

$$\Delta=\begin{vmatrix} a & b-y & c-z \\ -x & y & 0 \\ -x & 0 & z \end{vmatrix}=0$$

Expanding along C_3, we get

$$(c-z)\begin{vmatrix} -x & y \\ -x & 0 \end{vmatrix}+z\begin{vmatrix} a & b-y \\ -x & y \end{vmatrix}=0$$

$\Rightarrow (c-z)(xy)+z(ay+bx-xy)=0$

$\Rightarrow cxy - xyz + ayz + bxz - xyz = 0$

$\Rightarrow ayz + bzx + cxy = 2xyz$

$\Rightarrow \frac{a}{x}+\frac{b}{y}+\frac{c}{z}=2$

53. d. Since $A+B+C=\pi$ and $e^{i\pi}=\cos\pi + i\sin\pi = -1$,

$e^{i(B+c)} = e^{i(\pi-A)} = -e^{iA}$ and $e^{-i(B+C)} = -e^{iA}$

By taking e^{iA}, e^{iB}, e^{iC} common from R_1, R_2 and R_3, respectively, we have

$$\Delta=-\begin{vmatrix} e^{iA} & e^{-i(A+C)} & e^{-i(A+B)} \\ e^{-i(B+C)} & e^{iB} & e^{-i(A+B)} \\ e^{-i(B+C)} & e^{-i(A+C)} & e^{iC} \end{vmatrix}$$

$$=-\begin{vmatrix} e^{iA} & -e^{iB} & -e^{iC} \\ -e^{iA} & e^{iB} & -e^{iC} \\ -e^{iA} & -e^{iB} & e^{iC} \end{vmatrix}$$

By taking e^{iA}, e^{iB}, e^{iC} common from C_1, C_2 and C_3, respectively, we have

$$\Delta=\begin{vmatrix} 1 & -1 & -1 \\ -1 & 1 & -1 \\ -1 & -1 & 1 \end{vmatrix}=-4$$

54. c. $\Delta=(1+x+x^2)\begin{vmatrix} 1 & 1 & 1 \\ x^2 & 1 & x \\ x & x^2 & 1 \end{vmatrix}=(1+x+x^2)(x-1)^2$

Therefore, $\Delta = 0$ has roots 1, 1, ω, ω, ω^2, ω^2.

55. a. Using the sum property, we get

$$\sum_{r=0}^{m}\Delta_r=\begin{vmatrix} \sum_{r=0}^{m}(2r-1) & \sum_{r=0}^{m}{}^mC_r & \sum_{r=0}^{m}1 \\ m^2-1 & 2^m & m+1 \\ \sin^2(m^2) & \sin^2(m) & \sin^2(m+1) \end{vmatrix}$$

But $\sum_{r=0}^{m}(2r-1)=\frac{1}{2}(m+1)(2m-1-1)=m^2-1$,

$\sum_{r=0}^{m}{}^mC_r = 2^m$ and $\sum_{r=0}^{m}1 = m+1$. Therefore,

$$\sum_{r=0}^{m}\Delta_r=\begin{vmatrix} m^2-1 & 2^m & m+1 \\ m^2-1 & 2^m & m+1 \\ \sin^2(m^2) & \sin^2(m) & \sin^2(m+1) \end{vmatrix}=0$$

56. d. $\sum_{k=1}^{n}D_k=56 \Rightarrow \begin{vmatrix} \sum_{k=1}^{n}1 & n & n \\ \sum_{k=1}^{n}2k & n^2+n+1 & n^2+n \\ \sum_{k=1}^{n}(2k-1) & n^2 & n^2+n+1 \end{vmatrix}=56$

$$\Rightarrow \begin{vmatrix} n & n & n \\ n(n+1) & n^2+n+1 & n^2+n \\ n^2 & n^2 & n^2+n+1 \end{vmatrix}=56$$

Applying $C_3 \to C_3 - C_1$ and $C_2 \to C_2 - C_1$, we get

$$\begin{vmatrix} n & 0 & 0 \\ n(n+1) & 1 & 0 \\ n^2 & 0 & n+1 \end{vmatrix}=56 \Rightarrow n(n+1)=56 \Rightarrow n=7$$

57. a. Applying $C_1 \to C_1 + 2C_2 + C_3$, we get

$$S=\sum_{r=2}^{n}(-2)^r\begin{vmatrix} {}^nC_r & {}^{n-2}C_{r-1} & {}^{n-2}C_r \\ 0 & 1 & 1 \\ 0 & -1 & 0 \end{vmatrix}$$

$$=\sum_{r=2}^{n}(-2)^r\,{}^nC_r$$

$$=\sum_{r=0}^{n}(-2)^r\,{}^nC_r-({}^nC_0-2\,{}^nC_1)$$

$$=(1-2)^n-(1-2n)=2n-1+(-1)^n$$

58. b. $\Delta = \begin{vmatrix} {}^nC_{r-1} & {}^nC_r & (r+1)\,{}^{n+2}C_{r+1} \\ {}^nC_r & {}^nC_{r+1} & (r+2)\,{}^{n+2}C_{r+2} \\ {}^nC_{r+1} & {}^nC_{r+2} & (r+3)\,{}^{n+2}C_{r+3} \end{vmatrix}$

Applying $C_1 \to C_1 + C_2$ and using ${}^nC_r = \frac{n}{r}\,{}^{n-1}C_{r-1}$ in C_3, we get

$$\Delta = \begin{vmatrix} {}^{n+1}C_r & {}^nC_r & (n+2)\,{}^{n+1}C_r \\ {}^{n+1}C_{r+1} & {}^nC_{r+1} & (n+2)\,{}^{n+1}C_{r+1} \\ {}^{n+1}C_{r+2} & {}^nC_{r+2} & (n+2)\,{}^{n+1}C_{r+2} \end{vmatrix}$$

$$= (n+2)\begin{vmatrix} {}^{n+1}C_r & {}^nC_r & {}^{n+1}C_r \\ {}^{n+1}C_{r+1} & {}^nC_{r+1} & {}^{n+1}C_{r+1} \\ {}^{n+1}C_{r+2} & {}^nC_{r+2} & {}^{n+1}C_{r+2} \end{vmatrix}$$

$= 0$ (as C_1 and C_3 are identical)

59. a. The given determenant Δ_1 is obtained by corresponding co-factors of determinant Δ_2; hence $\Delta_1 = \Delta_2^2$. Now $\Delta_1\Delta_2 = \Delta_2^2\,\Delta_2 = \Delta_2^3$.

60. c. We have,

$$\Delta^2 = \Delta\,\Delta = \begin{vmatrix} l_1 & m_1 & n_1 \\ l_2 & m_2 & n_2 \\ l_3 & m_3 & n_3 \end{vmatrix}\begin{vmatrix} l_1 & m_1 & n_1 \\ l_2 & m_2 & n_2 \\ l_3 & m_3 & n_3 \end{vmatrix}$$

$$= \begin{vmatrix} l_1^2+m_1^2+n_1^2 & l_1l_2+m_1m_2+n_1n_2 & l_1l_3+m_1m_3+n_1n_3 \\ l_1l_2+m_1m_2+n_1n_2 & l_2^2+m_2^2+n_2^2 & l_2l_3+m_2m_3+n_2n_3 \\ l_1l_3+m_1m_3+n_1n_3 & l_2l_3+m_2m_3+n_2n_3 & l_3^2+m_3^2+n_3^2 \end{vmatrix}$$

$$= \begin{vmatrix} 1 & 0 & 0 \\ 0 & 1 & 0 \\ 0 & 0 & 1 \end{vmatrix} 1 \Rightarrow \Delta = \pm 1 \Rightarrow |\Delta| = 1$$

61. d. The given determinant, on simplification, gives

$$\Delta_1 = \begin{vmatrix} a_1^2 & -2a_1 & 1 & 0 \\ a_2^2 & -2a_2 & 1 & 0 \\ a_3^2 & -2a_3 & 1 & 0 \\ a_4^2 & -2a_4 & 1 & 0 \end{vmatrix} \times \begin{vmatrix} 1 & b_1 & b_1^2 & 0 \\ 1 & b_2 & b_2^2 & 0 \\ 1 & b_3 & b_3^2 & 0 \\ 1 & b_4 & b_4^2 & 0 \end{vmatrix}$$

$= 0 \times 0 = 0$

62. a. Determinant formed by the cofactors of $\begin{vmatrix} a & b & c \\ b & c & a \\ c & a & b \end{vmatrix}$ is

$$\begin{vmatrix} bc-a^2 & ac-b^2 & ab-c^2 \\ ac-b^2 & ab-c^2 & bc-a^2 \\ ab-c^2 & bc-a^2 & ac-b^2 \end{vmatrix}$$

$$\begin{vmatrix} a & b & c \\ b & c & a \\ c & a & b \end{vmatrix}^2$$

63. d. $\begin{vmatrix} 1+x_1 & 1+x_1x & 1+x_1x^2 \\ 1+x_2 & 1+x_2x & 1+x_2x^2 \\ 1+x_3 & 1+x_3x & 1+x_3x^2 \end{vmatrix} = \begin{vmatrix} 1 & x_1 & 0 \\ 1 & x_2 & 0 \\ 1 & x_3 & 0 \end{vmatrix}\begin{vmatrix} 1 & 1 & 0 \\ 1 & x & 0 \\ 1 & x^2 & 0 \end{vmatrix}$

$= 0$

64. c. We observe that the elements in the pre-factor are the cofactors of the corresponding elements of the post-factor. Hence,

$$\begin{vmatrix} \lambda & c & -b \\ -c & \lambda & a \\ b & -a & \lambda \end{vmatrix}^3 = [\lambda(\lambda^2+a^2+b^2+c^2)]^3 = (1+a^2+b^2+c^2)^3$$

$\Rightarrow \lambda = 1$

Alternative solution:

Writing $a = 0$, $b = 0$, $c = 0$ on both sides, we get

$\lambda^6\lambda^3 = 1 \Rightarrow \lambda = 1$

65. c. $f'(x) = \begin{vmatrix} -\sin x & 1 & 0 \\ 2\sin x & x^2 & 2x \\ \tan x & x & 1 \end{vmatrix} + \begin{vmatrix} \cos x & x & 1 \\ 2\cos x & 2x & 2 \\ \tan x & x & 1 \end{vmatrix} + \begin{vmatrix} \cos x & x & 1 \\ 2\sin x & x^2 & 2x \\ \sec^2 x & 1 & 0 \end{vmatrix}$

$$\Rightarrow f'(0) = \begin{vmatrix} 0 & 1 & 0 \\ 0 & 0 & 0 \\ 0 & 0 & 1 \end{vmatrix} + \begin{vmatrix} 1 & 0 & 1 \\ 2 & 0 & 2 \\ 0 & 0 & 1 \end{vmatrix} + \begin{vmatrix} 1 & 0 & 1 \\ 0 & 0 & 0 \\ 1 & 1 & 0 \end{vmatrix} = 0$$

Now, $\lim_{x\to 0}\frac{f(x)}{x} = \lim_{x\to 0} f'(x)$ [as $f(0) = 0$]

$= f'(0) = 0$

66. a. Operating $C_1 \to C_1 + C_2 + C_3$ on the L.H.S. we get

$$\Delta = \begin{vmatrix} 0 & c-a & a-b \\ 0 & c'-a' & a'-b' \\ 0 & c''-a'' & a''-b'' \end{vmatrix} = m\begin{vmatrix} a & b & c \\ a' & b' & c' \\ a'' & b'' & c'' \end{vmatrix}$$

$\Rightarrow m = 0$

67. d. Let the given determinant be equal to $\Delta(x)$. Then,

$5A + 4B + 3C + 2D + E = \Delta(1) + \Delta'(1)$

Now, $\Delta(1) = 0$ as R_2 and R_3 are identical.

$$\Delta'(x) = \begin{vmatrix} 1 & 0 & 1 \\ x^2 & x & 6 \\ x & x & 6 \end{vmatrix} + \begin{vmatrix} x & 2 & x \\ 2x & 1 & 0 \\ x & x & 6 \end{vmatrix} + \begin{vmatrix} x & 2 & x \\ x^2 & x & 6 \\ 1 & 1 & 0 \end{vmatrix}$$

$$\Delta'(1) = \begin{vmatrix} 1 & 2 & 1 \\ 2 & 1 & 0 \\ 1 & 1 & 6 \end{vmatrix} + \begin{vmatrix} 1 & 2 & 1 \\ 1 & 1 & 6 \\ 1 & 1 & 0 \end{vmatrix}$$

$= -17 + (12 + 1 - 1 - 6) = -11$

68. b. $\Delta_1 = x(x^2 - ab) - b(ax - ab) + b(a^2 - ax)$

$= x^3 - 3abx + ab^2 + a^2 b$

$\frac{d}{dx}(\Delta_1) = 3x^2 - 3ab = 3(x^2 - ab) = 3\Delta_2$

69. d. We have $y = \sin mx$, therefore $y_1 = m\cos mx$, $y_2 = -m^2 \sin mx$, etc.

$$\therefore \Delta = \begin{vmatrix} y & y_1 & y_2 \\ y_3 & y_4 & y_5 \\ y_6 & y_7 & y_8 \end{vmatrix}$$

$$= \begin{vmatrix} \sin mx & m\cos mx & -m^2 \sin mx \\ -m^3 \cos mx & m^4 \sin mx & m^5 \cos mx \\ -m^6 \sin mx & -m^7 \cos mx & m^8 \sin mx \end{vmatrix}$$

$$= m^{12} \begin{vmatrix} \sin mx & \cos mx & -\sin mx \\ -\cos mx & \sin mx & \cos mx \\ -\sin mx & -\cos mx & \sin mx \end{vmatrix} = 0$$

70. b. We have,

$$\Delta = \begin{vmatrix} a & 1 & 1 \\ 1 & b & 1 \\ 1 & 1 & c \end{vmatrix} = abc - (a+b+c) + 2$$

$\therefore \Delta > 0 \Rightarrow abc + 2 > a + b + c$

$\Rightarrow abc + 2 > 3(abc)^{1/3}$

$$\left[\because \text{A.M.} > \text{G.M.} \Rightarrow \frac{a+b+c}{3} > (abc)^{1/3}\right]$$

$\Rightarrow x^3 + 2 > 3x$, where $x = (abc)^{1/3}$

$\Rightarrow x^3 - 3x + 2 > 0 \Rightarrow (x-1)^2 (x+2) > 0$

$\Rightarrow x + 2 > 0 \Rightarrow x > -2 \Rightarrow (abc)^{1/3} > -2 \Rightarrow abc > -8$

71. b. $B_2 = a_1c_3 - a_3c_1$, $C_2 = -(a_1b_3 - a_3b_1)$

$B_3 = -(a_1c_2 - a_2c_1)$, $C_3 = a_1b_2 - a_2b_1$

$$\therefore \begin{vmatrix} B_2 & C_2 \\ B_3 & C_2 \end{vmatrix} = \begin{vmatrix} a_1c_3 - a_3c_1 & -a_1b_3 + a_3b_1 \\ -a_1c_2 + a_2c_1 & a_1b_2 - a_2b_1 \end{vmatrix}$$

$$= \begin{vmatrix} a_1c_3 & -a_1b_3 \\ -a_1c_2 & a_1b_2 \end{vmatrix} + \begin{vmatrix} a_1c_3 & a_3b_1 \\ -a_1c_2 & -a_2b_1 \end{vmatrix}$$

$$+ \begin{vmatrix} -a_3c_1 & -a_1b_3 \\ a_2c_1 & a_1b_2 \end{vmatrix} + \begin{vmatrix} -a_3c_1 & a_3b_1 \\ a_2c_1 & -a_2b_1 \end{vmatrix}$$

$$= a_1^2 \begin{vmatrix} c_3 & -b_3 \\ -c_2 & b_2 \end{vmatrix} + a_1b_1 \begin{vmatrix} c_3 & a_3 \\ -c_2 & -a_2 \end{vmatrix}$$

$$+ a_1c_1 \begin{vmatrix} -a_3 & -b_3 \\ a_2 & b_2 \end{vmatrix} + b_1c_1 \begin{vmatrix} -a_3 & a_3 \\ a_2 & -a_2 \end{vmatrix}$$

$= a_1\{a_1(b_2c_3 - b_3c_2) - b_1(a_2c_3 - a_3c_2) + c_1(a_2b_3 - a_3b_2)\}$

$$= a_1 \begin{vmatrix} a_1 & b_1 & c_1 \\ a_2 & b_2 & c_2 \\ a_3 & b_3 & c_3 \end{vmatrix} = a_1\Delta$$

72. a. Applying $C_1 \to C_1 - 2\sin x C_3$ and $C_2 \to C_2 + 2\cos x\, C_3$, we get

$$f(x) = \begin{vmatrix} 2 & 0 & -\sin x \\ 0 & 2 & \cos x \\ \sin x & -\cos x & 0 \end{vmatrix}$$

$= 2\cos^2 x + 2\sin^2 x = 2$

$\therefore f'(x) = 0$

$\therefore \int_0^{\pi/2} [f(x) + f'(x)] = dx = \int_0^{\pi/2} 2dx = \pi$

73. b. $$\begin{vmatrix} x^3+1 & x^2y & x^2z \\ xy^2 & y^3+1 & y^2z \\ xz^2 & yz^2 & z^3+1 \end{vmatrix} = 11$$

Multiplying R_1 by x, R_2 by y and R_3 by z, we get

$$\frac{1}{xyz} \begin{vmatrix} x^4+x & x^3y & x^3z \\ xy^3 & y^4+y & y^3z \\ xz^3 & yz^3 & z^4+z \end{vmatrix} = 11$$

Taking x, y, z common from C_1, C_2, C_3, respectively, we get

$$\begin{vmatrix} x^3+1 & x^3 & x^3 \\ y^3 & y^3+1 & y^3 \\ z^3 & z^3 & z^3+1 \end{vmatrix} = 11$$

Using $R_1 \to R_1 + R_2 + R_3$, we have

$$(x^3 + y^3 + z^3 + 1) \begin{vmatrix} 1 & 1 & 1 \\ y^3 & y^3+1 & y^3 \\ z^3 & z^3 & z^3+1 \end{vmatrix} = 11$$

Using $C_2 \to C_2 - C_1$ and $C_3 \to C_3 - C_1$, we get

$$(x^3 + y^3 + z^3 + 1) \begin{vmatrix} 1 & 0 & 0 \\ y^3 & 1 & 0 \\ z^3 & 0 & 1 \end{vmatrix} = 11$$

Hence,

$x^3 + y^3 + z^3 = 10$

Therefore, the ordered triplets are (2, 1, 1), (1, 2, 1), (1, 1, 2).

74. b. Since the system has non-trivial solution,

$$\therefore \begin{vmatrix} a & 1 & 1 \\ 1 & b & 1 \\ 1 & 1 & c \end{vmatrix} = 0$$

Applying $R_1 \to R_1 - R_2$, $R_2 \to R_2 - R_3$, we get

$$\Delta = \begin{vmatrix} a-1 & 1-b & 0 \\ 0 & b-1 & 1-c \\ 1 & 1 & c \end{vmatrix} = 0$$

$\Rightarrow c(1-a)(1-b)+(1-b)(1-c)-(1-c)(a-1)=0$

Dividing throughout by $(1-a)(1-b)(1-c)$, we get

$$\frac{c}{1-c}+\frac{1}{1-c}+\frac{1}{1-b}=0$$

$$\Rightarrow -1+\frac{1}{1-c}+\frac{1}{1-b}+\frac{1}{1-a}=0$$

$$\Rightarrow \frac{1}{1-c}+\frac{1}{1-a}+\frac{1}{1-b}=1$$

75. a. The given system of linear equations has a unique solution if

$$\begin{vmatrix}1 & 1 & 1\\ 1 & 2 & 3\\ 2 & 5 & \lambda\end{vmatrix}\neq 0$$

i.e., if $\lambda-8\neq 0$ or $\lambda\neq 8$.

76. b. Let, $\Delta=\begin{vmatrix}\cos(\alpha-\beta) & \cos(\beta-\gamma) & \cos(\gamma-\alpha)\\ \cos(\alpha+\beta) & \cos(\beta+\gamma) & \cos(\gamma+\alpha)\\ \sin(\alpha+\beta) & \sin(\beta+\gamma) & \sin(\gamma+\alpha)\end{vmatrix}$

It is clear that either $\alpha=\beta$ or $\beta=\gamma$ or $\gamma=\alpha$ is sufficient to make $\Delta=0$. It is not necessary that triangle is equilateral.

Also, isosceles triangle can be obtuse one.

77. c. $a=x/(y-z)\Rightarrow x-ay+az=0$ (1)

$b=y/(z-x)\Rightarrow bx+y-bz=0$ (2)

$c=z/(x-y)\Rightarrow -cx+cy+z=0$ (3)

Since x, y, z are not all zero, the above system has a non-trivial solution. So,

$$\Delta=\begin{vmatrix}1 & -a & a\\ b & 1 & -b\\ -c & c & 1\end{vmatrix}=0$$

$\therefore\ 1+ab+bc+ca=0$

78. a. $\Delta=\begin{vmatrix}p+a & b & c\\ a & q+b & c\\ a & b & r+c\end{vmatrix}=0$

Applying $R_1\to R_2-R_1$ and $R_3\to R_3-R_1$, we get

$$\begin{vmatrix}p+a & b & c\\ -p & q & 0\\ -p & 0 & r\end{vmatrix}=0$$

$\Rightarrow pqc+[q(p+a)+bp]r=0$

Dividing by pqr, we obtain

$$\frac{a}{p}+\frac{b}{q}+\frac{c}{r}=-1$$

79. b. For no solution or infinitely many solutions

$$\begin{vmatrix}\alpha & -1 & -1\\ 1 & -\alpha & -1\\ 1 & -1 & -\alpha\end{vmatrix}=0$$

$\Rightarrow \alpha(\alpha^2-1)-1(\alpha-1)+1(1-\alpha)=0$

$\Rightarrow \alpha(\alpha^2-1)-2\alpha+2=0$

$\Rightarrow \alpha(\alpha-1)(\alpha+1)-2(\alpha-1)=0$

$\Rightarrow (\alpha-1)(\alpha^2+\alpha-2)=0$

$\Rightarrow (\alpha-1)(\alpha+2)(\alpha-1)=0$

$\Rightarrow (\alpha-1)^2(\alpha+2)=0$

$\Rightarrow \alpha=1, 1, -2$

But for $\alpha=1$, there are infinite solutions. When $\alpha=-2$, we have

$-2x-y-z=-3$

$x+2y-z=-3$

$x-y+2z=-3$

Adding, we get $0=-9$, which is not true. Hence there is no solution.

80. a. $D=\cos\theta-\cos^2.+6>0$. Since $D>0$ only trivial solution is possible.

81. a The given system of equations will have a non-trivial solution if

$$\begin{vmatrix}\alpha+a & \alpha & \alpha\\ \alpha & \alpha+b & \alpha\\ \alpha & \alpha & \alpha+c\end{vmatrix}=0$$

Operating $R_2\to R_2-R_1$ and $R_3\to R_3-R_1$, we get

$$\begin{vmatrix}\alpha+a & \alpha & \alpha\\ -a & b & 0\\ -a & 0 & c\end{vmatrix}=0$$

$\Rightarrow \alpha ab+c(ab+ab+a\alpha)=0$

$\Rightarrow \alpha(bc+ca+ab)+abc=0$

$$\Rightarrow \frac{1}{\alpha}=-\left(\frac{1}{a}+\frac{1}{b}+\frac{1}{c}\right)\quad (\because a, b, c\neq 0)$$

82. c. The given system is consistent.

$$\therefore \Delta=\begin{vmatrix}1 & 1 & -1\\ 2 & -1 & -c\\ -b & 3b & -c\end{vmatrix}=0$$

$\Rightarrow c+bc-6b+b+2c+3bc=0$

$\Rightarrow 3c+4bc-5b=0$

$$\Rightarrow c=\frac{5b}{4b+3}$$

Now,

$c<1$

$$\Rightarrow \frac{5b}{4b+3}<1$$

$$\Rightarrow \frac{5b}{4b+3}-1<0$$

$$\Rightarrow \frac{b-3}{4b+3}<0$$

$$\Rightarrow b\in\left(-\frac{3}{4}, 3\right)$$

83. c. As a, b, c are in G.P. with common ratio r_1 and α, β, γ are in G.P. having common ratio r_2, $a \neq 0$, $\alpha \neq 0$, $b = ar_1$, $c = ar_1^2$, $\beta = ar_2$, $\gamma = ar_2^2$.

Also the system of equations has only zero (trivial) solution.

$$\Delta = \begin{vmatrix} a & \alpha & 1 \\ b & \beta & 1 \\ c & \gamma & 1 \end{vmatrix} \neq 0$$

$$\Rightarrow a\alpha \begin{vmatrix} 1 & 1 & 1 \\ r_1 & r_2 & 1 \\ r_1^2 & r_2^2 & 1 \end{vmatrix} \neq 0$$

$$\Rightarrow a\alpha(r_1 - 1)(r_2 - 1)(r_1 - r_2) \neq 0$$

$$\Rightarrow r_1 \neq 1, r_2 \neq 1 \text{ and } r_1 \neq r_2$$

84. b. For non-trivial solution

$$\begin{vmatrix} a-1 & -1 & -1 \\ 1 & -(b-1) & 1 \\ 1 & 1 & -(c-1) \end{vmatrix} = 0$$

$$\Rightarrow \begin{vmatrix} a-1 & -1 & 0 \\ 1 & -(b-1) & b \\ 1 & 1 & -c \end{vmatrix} = 0$$

$$\Rightarrow (a-1)(bc - c - b) + 1(-c - b) = 0$$

$$\Rightarrow abc - ac - ab - bc + b + c - c - b = 0$$

$$\Rightarrow ab + bc + ac = abc$$

Multiple Correct Answers Type

1. a., b., c.

$$\begin{vmatrix} 8 & 2 & 7 \\ 12 & 3 & 5 \\ 16 & 4 & 3 \end{vmatrix} = \begin{vmatrix} 8 & 2 & 7 \\ 4 & 1 & -2 \\ 4 & 1 & -2 \end{vmatrix} \quad [R_3 \to R_3 - R_2 \text{ and } R_2 \to R_2 - R_1]$$

$$= 0$$

$$\begin{vmatrix} 1/a & a^2 & bc \\ 1/b & b^2 & ac \\ 1/c & c^2 & ab \end{vmatrix} = \frac{1}{abc}\begin{vmatrix} 1 & a^3 & abc \\ 1 & b^3 & abc \\ 1 & c^3 & abc \end{vmatrix}$$

$[R_1 \to aR_1, R_2 \to bR_2, R_3 \to cR_3]$

$$= \frac{abc}{abc}\begin{vmatrix} 1 & a^3 & 1 \\ 1 & b^3 & 1 \\ 1 & c^3 & 1 \end{vmatrix}$$

[taking abc common from C_3]

$$= 0$$

$$\begin{vmatrix} a+b & 2a+b & 3a+b \\ 2a+b & 3a+b & 4a+b \\ 4a+b & 5a+b & 6a+b \end{vmatrix} = \begin{vmatrix} a+b & 2a+b & 3a+b \\ a & a & a \\ 2a & 2a & 2a \end{vmatrix}$$

$[R_3 \to R_3 - R_2, R_2 \to R_2 - R_1]$

$$= 0$$

$$\begin{vmatrix} 2 & 43 & 6 \\ 7 & 35 & 4 \\ 3 & 17 & 2 \end{vmatrix} = \begin{vmatrix} 2 & 1 & 6 \\ 7 & 7 & 4 \\ 3 & 3 & 2 \end{vmatrix} \quad [C_2 \to C_2 - 7C_3]$$

$$= \begin{vmatrix} 1 & 1 & 6 \\ 0 & 7 & 4 \\ 0 & 3 & 2 \end{vmatrix} \quad [C_1 \to C_1 - C_2]$$

$$= 2$$

2. a., c.

$$g(x) = \begin{vmatrix} a^{-x} & e^{\log_e a^x} & x^2 \\ a^{-3x} & e^{\log_e a^{3x}} & x^4 \\ a^{-5x} & e^{\log_e a^{5x}} & 1 \end{vmatrix} = \begin{vmatrix} a^{-x} & a^x & x^2 \\ a^{-3x} & a^{3x} & x^4 \\ a^{-5x} & a^{5x} & 1 \end{vmatrix} \left(e^{\log a^x} = a^x\right)$$

$$\Rightarrow g(-x) = \begin{vmatrix} a^x & a^{-x} & x^2 \\ a^{3x} & a^{-3x} & x^4 \\ a^{5x} & a^{-5x} & 1 \end{vmatrix} = -\begin{vmatrix} a^{-x} & a^x & x^2 \\ a^{-3x} & a^{3x} & x^4 \\ a^{-5x} & a^{5x} & 1 \end{vmatrix}$$

[Interchanging 1st and 2nd columns]

$$= -g(x)$$

$$\Rightarrow g(x) + g(-x) = 0$$

$\Rightarrow g(x)$ is an odd function

Hence, the graph is symmetrical about origin. Also, $g_4(x)$ is an odd function [where $g_4(x)$ is fourth derivative of $g(x)$]. Hence,

$$g_4(x) = -g_4(-x)$$

$$\Rightarrow g_4(0) = -g_4(0)$$

$$\Rightarrow g_4(0) = 0$$

3. c., d.

Applying $C_1 \to C_1 + C_2 + C_3$, we get

$$\Delta = \begin{vmatrix} a+b-x & a & b \\ a+b-x & -x & a \\ a+b-x & b & -x \end{vmatrix} = (a+b-x)\begin{vmatrix} 1 & a & b \\ 1 & -x & a \\ 1 & b & -x \end{vmatrix}$$

$$= (a+b-x)\begin{vmatrix} 1 & a & b \\ 0 & -x-a & a-b \\ 0 & b-a & -x-b \end{vmatrix}$$

[Applying $R_2 \to R_2 - R_1$ and $R_3 \to R_3 - R_1$]

$$= (a+b-x)\,[(x+a)(x+b) + (a-b)^2]$$

[expanding along C_1]

$$= (a+b-x)[x^2 + (a+b)x + a^2 + b^2 - ab]$$

4. b., d.

Applying $C_1 \to C_1 - (\cot \phi)\, C_2$, we get

$$\Delta = \begin{vmatrix} 0 & \sin\theta \sin\phi & \cos\theta \\ 0 & \cos\theta \sin\phi & -\sin\theta \\ -\sin\theta/\sin\phi & \sin\theta \cos\phi & 0 \end{vmatrix}$$

$$= -\frac{\sin\theta}{\sin\phi}\,[-\sin\phi\sin^2\theta - \cos^2\theta\sin\phi]$$

[expanding along C_1]

$= \sin\theta$,

which is independent of ϕ. Also,

$$\frac{d\Delta}{d\theta} = \cos\theta \Rightarrow \left.\frac{d\Delta}{d\theta}\right]_{\theta=\pi/2} = \cos(\pi/2) = 0$$

5. a., c.

$$\begin{vmatrix} \cos\alpha & -\sin\alpha & 1 \\ \sin\alpha & \cos\alpha & 1 \\ \cos(\alpha+\beta) & -\sin(\alpha+\beta) & 1 \end{vmatrix}$$

$$= \begin{vmatrix} \cos\alpha & -\sin\alpha & 1 \\ \sin\alpha & \cos\alpha & 1 \\ 0 & 0 & 1+\sin\beta-\cos\beta \end{vmatrix}$$

[Applying $R_3 \to R_3 - R_1(\cos\beta) + R_2(\sin\beta)$]

$= (1 + \sin\beta - \cos\beta)(\cos^2\alpha + \sin^2\alpha)$

$= 1 + \sin\beta - \cos\beta$, which is independent of α.

6. a., b,

$$\Delta = \frac{1}{a}\begin{vmatrix} a^3+ax & ab & ac \\ a^2b & b^2+x & bc \\ a^2c & bc & c^2+x \end{vmatrix}$$

Applying $C_1 \to C_1 + bC_2 + cC_3$ and taking $a^2 + b^2 + c^2 + x$ common, we get

$$\Delta = \frac{1}{a}(a^2+b^2+c^2+x)\begin{vmatrix} a & ab & ac \\ b & b^2+x & bc \\ c & bc & c^2+x \end{vmatrix}$$

Applying $C_2 \to C_2 - bC_1$ and $C_3 \to C_3 - cC_1$, we get

$$\Delta = \frac{1}{a}(a^2+b^2+c^2+x)\begin{vmatrix} a & 0 & 0 \\ b & x & 0 \\ c & 0 & x \end{vmatrix}$$

$$= \frac{1}{a}(a^2+b^2+c^2+x)(ax^2) = x^2(a^2+b^2+c^2+x)$$

Thus Δ is divisible by x and x^2.

7. a., b., c.

Applying $R_3 \to R_3 - xR_2$ and $R_2 \to R_2 - xR_1$, we get

$$f(x) = \begin{vmatrix} a & -1 & 0 \\ 0 & a+x & -1 \\ 0 & 0 & a+x \end{vmatrix} = a(a+x)^2$$

Hence,

$f(2x) - f(x) = a[(a + 2x)^2 - (a + x)^2]$

$= a(a + 2x - a - x)(a + 2x + a + x)$

$= ax(2a + 3x)$

8. a., b., c,

Operating $C_1 \to C_2 - C_1$, $C_3 \to C_3 - C_1$, we get

$$\Delta = \begin{vmatrix} 1 & ac & bc \\ 1 & ad & bd \\ 1 & ae & be \end{vmatrix} = ab\begin{vmatrix} 1 & c & c \\ 1 & d & d \\ 1 & e & e \end{vmatrix} ab(0) = 0$$

9. a., b.

By partial fractions, we have

$$g(x) = \frac{f(a)}{(x-a)(a-b)(a-c)} + \frac{f(b)}{(b-a)(x-b)(b-c)} + \frac{f(c)}{(c-a)(c-b)(x-c)}$$

$$\Rightarrow g(x) = \frac{1}{(a-b)(b-c)(c-a)} \times \left[\frac{f(a)(c-b)}{(x-a)} + \frac{f(b)(a-c)}{(x-b)} + \frac{f(c)(b-a)}{(x-c)}\right]$$

$$\Rightarrow g(x) = \begin{vmatrix} 1 & a & f(a)/(x-a) \\ 1 & b & f(b)/(x-b) \\ 1 & c & f(c)/(x-c) \end{vmatrix} \div \begin{vmatrix} 1 & a & a^2 \\ 1 & b & b^2 \\ 1 & c & c^2 \end{vmatrix}$$

$$\Rightarrow \int g(x)dx = \begin{vmatrix} 1 & a & f(a)\log|x-a| \\ 1 & b & f(b)\log|x-b| \\ 1 & c & f(c)\log|x-c| \end{vmatrix} \div \begin{vmatrix} 1 & a & a^2 \\ 1 & b & b^2 \\ 1 & c & c^2 \end{vmatrix} + k$$

and

$$\frac{dg(x)}{dx} = \begin{vmatrix} 1 & a & -f(a)(x-a)^{-2} \\ 1 & b & -f(b)(x-b)^{-2} \\ 1 & c & -f(c)(x-c)^{-2} \end{vmatrix} \div \begin{vmatrix} 1 & a & a^2 \\ 1 & b & b^2 \\ 1 & c & c^2 \end{vmatrix}$$

$$= \begin{vmatrix} 1 & a & f(a)(x-a)^{-2} \\ 1 & b & f(b)(x-b)^{-2} \\ 1 & c & f(c)(x-c)^{-2} \end{vmatrix} \div \begin{vmatrix} a^2 & a & 1 \\ b^2 & b & 1 \\ c^2 & c & 1 \end{vmatrix}$$

10. b., c.

$$\Delta'(x) = \begin{vmatrix} 2x+4 & 2x+4 & 13 \\ 4x+5 & 4x+5 & 26 \\ 16x-6 & 16x-6 & 104 \end{vmatrix} + \begin{vmatrix} x^2+4x-3 & 2 & 13 \\ 2x^2+5x-9 & 4 & 26 \\ 8x^2-6x+1 & 16 & 104 \end{vmatrix}$$

$= 0 + 2 \times 13 \times (0) = 0$

$\Rightarrow \Delta(x) = \text{constant} \Rightarrow a = 0, b = 0, c = 0$

11. a., c.

$f(\theta) = \sin^3\theta + \cos^3\theta - \cos\theta\sin\theta(\sin\theta + \cos\theta)$

$= (\sin\theta + \cos\theta)^3 - 4\sin\theta\cos\theta(\sin\theta + \cos\theta)$

$= (\sin\theta + \cos\theta)[1 - \sin 2\theta]$

Now,

$f(\theta) = 0$

$\Rightarrow \tan\theta = -1$ or $\sin 2\theta = 1$

$\Rightarrow f(\theta) = 0$ has 2 real solutions in $[0, \pi]$

Also, $\dfrac{f(\theta)}{1-\sin 2\theta} = \sin\theta + \cos\theta \in [-\sqrt{2}, \sqrt{2}]$

12. d.

Applying $R_2 \to R_2 - R_1$ and $R_3 \to R_3 - R_1$, we get

$$\Delta = \begin{vmatrix} \sin^2 A & \cot A & 1 \\ \sin(B+A)\sin(B-A) & \dfrac{\sin(A-B)}{\sin A \sin B} & 0 \\ \sin(C+A)\sin(C-A) & \dfrac{\sin(A-C)}{\sin A \sin C} & 0 \end{vmatrix}$$

$$\left[\because \cot\alpha - \cot\beta = \frac{\sin(\beta-\alpha)}{\sin\alpha\sin\beta}\right]$$

Expanding along C_3, we get

$$\Delta = \frac{\sin(A-B)\sin(A-C)}{\sin A}\left[-\frac{\sin(B+A)}{\sin C} + \frac{\sin(C+A)}{\sin B}\right]$$

$$= \frac{\sin(A-B)\sin(A-C)}{\sin A}\left[-\frac{\sin(\pi-C)}{\sin C} + \frac{\sin(\pi-B)}{\sin B}\right]$$

$$= \frac{\sin(A-B)\sin(A-C)}{\sin A}\left[-\frac{\sin C}{\sin C} + \frac{\sin B}{\sin B}\right] = 0$$

13. a., c.

$$\begin{vmatrix} {}^xC_r & {}^{n-1}C_r & {}^nC_r \\ {}^{x+1}C_r & {}^nC_r & {}^{n+1}C_r \\ {}^{x+2}C_r & {}^{n+1}C_r & {}^{n+2}C_r \end{vmatrix} = 0 \qquad (1)$$

$$\Rightarrow \begin{vmatrix} \dfrac{x!}{r!(x-r)!} & \dfrac{(n-1)!}{r!(n-r-1)!} & \dfrac{n!}{r!(n-r)!} \\ \dfrac{(x+1)!}{r!(x+1-r)!} & \dfrac{n!}{r!(n-r)!} & \dfrac{(n+1)!}{r!(n-r+1)!} \\ \dfrac{(x+2)!}{r!(x+2-r)!} & \dfrac{(n+1)!}{r!(n+1-r)!} & \dfrac{(n+2)!}{r!(n-r+2)!} \end{vmatrix} = 0$$

Taking $\dfrac{x!}{r!(x+2-r)!}$ common from C_1, we have quadratic equation in x.

Now in (1), if we put $x = n - 1$, C_1 and C_2 are the same, hence $x = n - 1$ is one root of the equation.

If we put $x = n$, then C_1 and C_3 are same. Hence, $x = n$ is the other root.

14. a., b,

Applying $C_3 \to C_3 - xC_2$, $C_2 \to C_2 - xC_1$, we obtain

$$\Delta(x) = \begin{vmatrix} 3 & 0 & 2a^2 \\ 3x & 2a^2 & 4a^2x \\ 3x^2+2a^2 & 4a^2x & 6a^2x^2+2a^2 \end{vmatrix}$$

Applying $C_3 \to C_3 - xC_2$, we get

$$\Delta(x) = 4a^4\begin{vmatrix} 3 & 0 & 1 \\ 3x & 1 & x \\ 3x^2+2a^2 & 2x & x^2+2a^2 \end{vmatrix}$$

Applying $C_1 \to C_1 - 3C_3$, we get

$$\Delta(x) = 4a^4\begin{vmatrix} 0 & 0 & 1 \\ 0 & 1 & x \\ -4a^2 & 2x & x^2+2a^2 \end{vmatrix} = 16a^6$$

15. b., c.

In the left-hand determinant, each element is the cofactor of the elements of the determinant

$$\begin{vmatrix} x & y & z \\ y & z & x \\ z & x & y \end{vmatrix} = \Delta^* \text{ (say)}$$

Hence,

$$\Delta^{*2} = \begin{vmatrix} x & y & z \\ y & z & x \\ z & x & y \end{vmatrix}\begin{vmatrix} x & y & z \\ y & z & x \\ z & x & y \end{vmatrix}$$

$$= \begin{vmatrix} x^2+y^2+z^2 & xy+yz+zx & xz+yx+zy \\ \Sigma xy & \Sigma x^2 & \Sigma xy \\ \Sigma xy & \Sigma xy & \Sigma x^2 \end{vmatrix}$$

$$= \begin{vmatrix} r^2 & u^2 & u^2 \\ u^2 & r^2 & u^2 \\ u^2 & u^2 & r^2 \end{vmatrix} \text{ [Since } x^2+y^2+z^2 = r^2, xy+yz+zx = u^2]$$

16. a., c.

$$f(n) = \begin{vmatrix} n & n+1 & n+2 \\ n! & (n+1)! & (n+2)! \\ 1 & 1 & 1 \end{vmatrix} = \begin{vmatrix} n & 1 & 1 \\ n! & nn! & (n+1)(n+1)! \\ 1 & 0 & 0 \end{vmatrix}$$

[Applying $C_3 \to C_3 - C_2$ and $C_2 \to C_2 - C_1$]

$= (n + 1)(n + 1)! - nn! = n![(n + 1)^2 - n]$

$= n!(n^2 + n + 1)$

Thus, $f(n)$ is divisible by $n!$ and $n^2 + n + 1$.

17. a., b., c.

We have,

$$\begin{vmatrix} bc & ca & ab \\ ca & ab & bc \\ ab & bc & ca \end{vmatrix} = 0$$

$\Rightarrow (ab)^3 + (bc)^3 + (ca)^3 - 3(ab)(bc)(ca) = 0$

$\Rightarrow (ab + bc\omega^2 + ca\omega)(ab\omega + bc\omega^2 + ca)(ab\omega^2 + bc\omega + ca) = 0$

$\Rightarrow ab + bc\omega^2 + ca\omega = 0$, $ab\omega + bc\omega^2 + ca = 0$, $ab\omega^2 + bc\omega + ca = 0$

$\Rightarrow \dfrac{1}{c\omega^2} + \dfrac{1}{a} + \dfrac{1}{b\omega} = 0$, $\dfrac{1}{c\omega} + \dfrac{1}{a} + \dfrac{1}{b\omega^2} = 0$,

$$\frac{1}{c}+\frac{1}{a\omega}+\frac{1}{b\omega^2}=0$$

$$\Rightarrow \frac{1}{a}+\frac{1}{b\omega}+\frac{1}{c\omega^2}=0, \frac{1}{a}+\frac{1}{b\omega^2}+\frac{1}{c\omega}=0,$$

$$\frac{1}{a\omega}+\frac{1}{b\omega^2}+\frac{1}{c}=0$$

18. a., b.

$$\begin{vmatrix} 1 & k & 3 \\ k & 2 & 2 \\ 2 & 3 & 4 \end{vmatrix}=0$$

$\Rightarrow 8 + 4k + 9k - 12 - 4k^2 - 6 = 0$

$\Rightarrow 4k^2 - 13k + 10 = 0$

$\Rightarrow 4k^2 - 8k - 5k + 10 = 0$

$\Rightarrow (2k - 5)(k - 2) = 0$

$\Rightarrow k = 5/2, 2$

19. a., b.

Applying $R_1 \to R_1 + \sin\phi\,(R_2) + \cos\phi\,(R_3)$,

$$f(x) = \Delta = \begin{vmatrix} 0 & 0 & \cos 2\phi + 1 \\ \sin\theta & \cos\theta & \sin\phi \\ -\cos\phi & \sin\theta & \cos\phi \end{vmatrix}$$

$= (\cos 2\phi + 1)(\sin^2\theta + \cos^2\theta)$

$= (1 + \cos 2\phi)$

Hence, Δ is independent of θ.

Reasoning Type

1. a.
$$\begin{vmatrix} 1 & 1 & 1 \\ 1+\sin A & 1+\sin B & 1+\sin C \\ \sin A+\sin^2 A & \sin B+\sin^2 B & \sin C+\sin^2 C \end{vmatrix}=0 \quad (1)$$

Then $A = B$ or $B = C$ or $C = A$, for which any two rows are same. For (1) to hold it is not necessary that all the three rows are same or $A = B = C$.

2. b. The system of equations $kx + y + z = 1$, $x + ky + z = k$, $x + y + kz = k^2$ is inconsistent if $\Delta = \begin{vmatrix} k & 1 & 1 \\ 1 & k & 1 \\ 1 & 1 & k \end{vmatrix} = 0$ and one of $\Delta_1, \Delta_2, \Delta_3$ is non-zero where

$$\Delta_1 = \begin{vmatrix} 1 & 1 & 1 \\ k & k & 1 \\ k^2 & 1 & k \end{vmatrix}, \Delta_2 = \begin{vmatrix} k & 1 & 1 \\ 1 & k & 1 \\ 1 & k^2 & k \end{vmatrix}, \Delta_3 = \begin{vmatrix} k & 1 & 1 \\ 1 & k & k \\ 1 & 1 & k^2 \end{vmatrix}$$

We have,

$\Delta = (k + 2)(k - 1)^2$, $\Delta_1 = -(k + 1)(k - 1)^2$,

$\Delta_2 = -k(k - 1)^2$, $\Delta_3 = (k + 1)^2(k - 1)^2$

The determinant given in statement 2 is $\Delta_1 = 0$, for which $k = 1$ or $k = -1$.

$k = 1$ makes all the determinants zero. But for $k = -1$, all the determinants are not zero.

Hence, both statements are true but statement 2 is not correct explanation of statement 1.

3. a. For $x = 0$, the determinant reduces to the determinant of a skew-symmetric matrix of od order which is always zero. Hence, $x = 0$ is the solution of the given equation.

4. a. As the given system of equations has non-trivial solutions, hence

$$\begin{vmatrix} \lambda & b-a & c-a \\ a-b & \lambda & c-b \\ a-c & b-c & \lambda \end{vmatrix}=0$$

When $\lambda = 0$, then the determinant becomes skew-symmetric of odd order, which is equal to zero. Thus, $\lambda = 0$.

5. a. $\Delta = \begin{vmatrix} x & y & z \\ p & q & r \\ a & b & c \end{vmatrix}\begin{vmatrix} 0 & m & n \\ -m & 0 & k \\ -n & -k & 0 \end{vmatrix}$ where $\begin{vmatrix} 0 & m & n \\ -m & 0 & k \\ -n & -k & 0 \end{vmatrix}$ is skew symmetric.

$\therefore \Delta = 0$

6. a. We are given that

$1 + bc + qr = 0 \quad (1)$

$1 + ca + pr = 0 \quad (2)$

$1 + ab + pq = 0 \quad (3)$

The determinant in the question involves a column consisting the elements ap, bq and cr. So multiplying (1), (2), and (3) by ap, bq and cr, respectively, we get

$ap + abcp + apqr = 0 \quad (4)$

$bq + abcq + bpqr = 0 \quad (5)$

$cq + abcr + cpqr = 0 \quad (6)$

Since abc and pqr occur in all the three equations, putting $abc = x$, $pqr = y$, we get the system

$$\begin{aligned} ap + px + ay &= 0 \\ bq + qx + by &= 0 \\ cr + rx + cy &= 0 \end{aligned} \quad (7)$$

System (7) must have a common solution (i.e., system is consistent). So,

$$\begin{vmatrix} ap & p & a \\ bq & q & b \\ cr & r & c \end{vmatrix}=0$$

$$\Rightarrow \begin{vmatrix} ap & a & p \\ bq & b & q \\ cr & c & r \end{vmatrix}=0$$

7. b. Let $\Delta = \begin{vmatrix} 1 & 1 & 1 \\ 1 & 2 & 3 \\ 1 & 2 & \lambda \end{vmatrix} = 0 \Rightarrow \lambda = 3$. Now,

$$\Delta_1 = \begin{vmatrix} 6 & 1 & 1 \\ 10 & 2 & 3 \\ \mu & 2 & \lambda \end{vmatrix}$$

$$= \begin{vmatrix} 6 & 1 & 1 \\ 10 & 2 & 3 \\ \mu & 2 & 3 \end{vmatrix} = \mu - 10$$

$$\Delta_2 = \begin{vmatrix} 1 & 6 & 1 \\ 1 & 10 & 3 \\ 1 & \mu & \lambda \end{vmatrix} = \begin{vmatrix} 1 & 6 & 1 \\ 1 & 10 & 3 \\ 1 & \mu & 3 \end{vmatrix} = 20 - 2\mu$$

$$\Delta_3 = \begin{vmatrix} 1 & 1 & 6 \\ 1 & 2 & 10 \\ 1 & 2 & \mu \end{vmatrix} = \mu - 10$$

Clearly, for $\mu = 10$, all of Δ_1, Δ_2, Δ_3 are zero.

8. b. $\Delta = \Delta_1 \Delta_2$ where $\Delta_1 = \begin{vmatrix} a_1 & b_1 & c_1 \\ a_2 & b_2 & c_2 \\ a_3 & b_3 & c_3 \end{vmatrix}$ and $\Delta_2 = \begin{vmatrix} 1 & x^2 & 0 \\ x^2 & 1 & 0 \\ 0 & 0 & 1 \end{vmatrix}$

Hence, both the statements are true but statement 2 is not correct explanation of statement 1.

Linked Comprehension Type

For Problems 1–3

1. c., 2. d., 3. b.

Sol. In given determinant applying $C_2 \to C_2 - C_1$ and $C_3 \to C_3 - C_2$, we get

$$f(x) = \begin{vmatrix} x+c_1 & a-c_1 & 0 \\ x+b & c_2-b & a-c_2 \\ x+b & 0 & c_3-b \end{vmatrix}$$

$$= x\begin{vmatrix} 1 & a-c_1 & 0 \\ 1 & c_2-b & a-c_2 \\ 1 & 0 & c_3-b \end{vmatrix} + \begin{vmatrix} c_1 & a-c_1 & 0 \\ b & c_2-b & a-c_2 \\ b & 0 & c_3-b \end{vmatrix}$$

So, $f(x)$ is linear. Let $f(x) = Px + Q$. Then

$f(-a) = -aP + Q, f(-b) = -bP + Q$

Then,

$$f(0) = 0 \times P + Q \Rightarrow Q = \frac{bf(-a) - af(-b)}{(b-a)} \quad (1)$$

Also,

$$f(-a) = \begin{vmatrix} c_1-a & 0 & 0 \\ b-a & c_2-a & 0 \\ b-a & b-a & c_3-a \end{vmatrix}$$

$= (c_1 - a)(c_1 - a)(c_3 - a)$

Similarly,

$f(-b) = (c_1 - b)(c_2 - b)(c_3 - b)$

$g(x) = (c_1 - x)(c_2 - x)(c_3 - x)$

$\Rightarrow g(a) = f(-a)$ and $g(b) = f(-b)$

Now from (1), we get

$$f(0) = \frac{bg(a) - ag(b)}{(b-a)}$$

For Problems 4–6

4. d., 5. d., 6. c.

Sol. $$\Delta = \frac{1}{a}\begin{vmatrix} a^3+ax & ab & ac \\ a^2b & b^2+x & bc \\ a^2c & bc & c^2+x \end{vmatrix}$$

Applying $C_1 \to C_1 + bC_2 + cC_3$ and taking $a^2 + b^2 + c^2 + x$ common, we get

$$\Delta = \frac{1}{a}(a^2+b^2+c^2+x)\begin{vmatrix} a & ab & ac \\ b & b^2+x & bc \\ c & bc & c^2+x \end{vmatrix}$$

Applying $C_2 \to C_2 - bC_1$ and $C_3 \to C_3 - cC_1$, we get

$$\Delta = \frac{1}{a}(a^2+b^2+c^2+x)\begin{vmatrix} a & 0 & 0 \\ b & x & 0 \\ c & 0 & x \end{vmatrix}$$

$= \frac{1}{a}(a^2 + b^2 + c^2 + x)(ax^2)$

$= x^2(a^2 + b^2 + c^2 + x)$

Thus Δ is divisible by x and x^2. Also, graph of $f(x)$ is

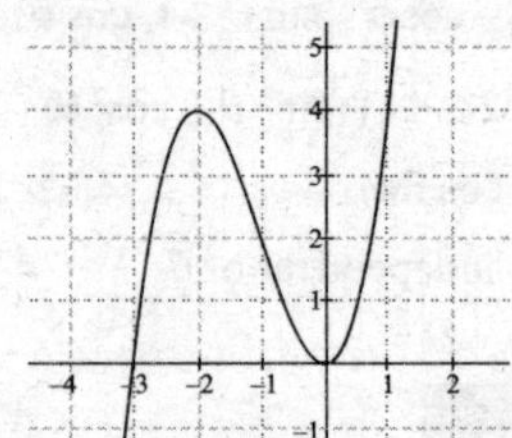

Fig. 7.1

For Problems 7–9

7. c., 8. a., 9. d.

Sol. The system of equations

$-x + cy + bz = 0$ (1)

$cx - y + az = 0$ (2)

$bx + ay - z = 0$ (3)

has a non-zero solution if

$$\Delta = \begin{vmatrix} -1 & c & b \\ c & -1 & a \\ b & a & -1 \end{vmatrix} = 0$$

$\Rightarrow a^2 + b^2 + c^2 + 2abc - 1 = 0$

$\Rightarrow a^2 + b^2 + c^2 + 2abc = 1$ (4)

Then clearly the system has infinitely many solutions. From (1) and (2), we have

$$\frac{x}{ac+b} = \frac{y}{bc+a} = \frac{z}{1-c^2}$$

$$\therefore \frac{x^2}{(ac+b)^2} = \frac{y^2}{(bc+a)^2} = \frac{z^2}{(1-c^2)^2}$$

$$\text{or } \frac{x^2}{(1-a^2)(1-c^2)} = \frac{y^2}{(1-b^2)(1-c^2)} = \frac{z^2}{(1-c^2)^2}$$

[From (4)]

or $\dfrac{x^2}{1-a^2}=\dfrac{y^2}{1-b^2}=\dfrac{z^2}{1-c^2}$ (5)

From (5), we see that $1-a^2$, $1-b^2$, $1-c^2$ are all positive or all negative. Given that one of a, b, c is proper fraction, so

$1-a^2>0$, $1-b^2>0$, $1-c^2>0$, which gives

$a^2+b^2+c^2<3$ (6)

Using (4) and (6), we get

$1<3+2abc$

or $abc>-1$ (7)

For Problems 10–12

10. a., 11. b., 12. c.

Sol. $\Delta=\begin{vmatrix}1&1&1\\1&2&3\\1&2&\lambda\end{vmatrix}=2\lambda+3+2-2-\lambda-6=\lambda-3$

$\Delta_1=\begin{vmatrix}6&1&1\\10&2&3\\\mu&2&\lambda\end{vmatrix}=12\lambda+3\mu+20-2\mu-10\lambda-36$

$=2\lambda+\mu-16$

$\Delta_2=\begin{vmatrix}1&6&1\\1&10&3\\1&\mu&\lambda\end{vmatrix}=10\lambda+18+\mu-10-3\mu-6\lambda$

$=4\lambda-2\mu+8$

$\Delta_3=\begin{vmatrix}1&1&6\\1&2&10\\1&2&\mu\end{vmatrix}=2\mu+10+12-12-\mu-20$

$=\mu-10$

Thus the system has unique solutions if $\Delta\neq0$ or $\lambda\neq3$ and the system has infinite solutions if $\Delta=\Delta_1=\Delta_2=\Delta_3=0$ or $\lambda=3$ and $\mu=10$. System has no solution if $\Delta=0$ and at least one of Δ_1, Δ_2, Δ_3 is non-zero or $\lambda=3$ and $\mu\neq10$.

For Problems 13–15

13. d., 14. d., 15. d.

Sol. $\Delta=\begin{vmatrix}1+1+1&1+\alpha+\beta&1+\alpha^2+\beta^2\\1+\alpha+\beta&1+\alpha^2+\beta^2&1+\alpha^3+\beta^3\\1+\alpha^2+\beta^2&1+\alpha^3+\beta^3&1+\alpha^4+\beta^4\end{vmatrix}$

$=\begin{vmatrix}1&1&1\\1&\alpha&\beta\\1&\alpha^2&\beta^2\end{vmatrix}\times\begin{vmatrix}1&1&1\\1&\alpha&\beta\\1&\alpha^2&\beta^2\end{vmatrix}$ [multiplying row by row]

$=D^2$ (say)

Now,

$D=\begin{vmatrix}1&1&1\\1&\alpha&\beta\\1&\alpha^2&\beta^2\end{vmatrix}$

$=(1-\alpha)(\alpha-\beta)(\beta-1)$

$=(\beta-\alpha)\,[\alpha\beta-\alpha-\beta+1]$

$=(\beta-\alpha)\left(\dfrac{c}{a}+\dfrac{b}{a}+1\right)=\dfrac{(\beta-\alpha)}{a}(a+b+c)$

$\therefore\ \Delta=D^2=\dfrac{(\beta-\alpha)^2}{a^2}(a+b+c)^2$

$=\dfrac{1}{a^2}(a+b+c)^2\left[\dfrac{b^2}{a^2}-4\dfrac{c}{a}\right]$

$=\dfrac{1}{a^4}(a+b+c)^2(b^2-4ac)$

If $\Delta<0$, i.e., $b^2-4ac<0$, then roots are imaginary.

If one root is $1+\sqrt{2}$ and since coefficients are real, the other root is $1-\sqrt{2}$. Hence the equation is $x^2-2x-1=0$. Then the value of Δ is $(1-2-1)^2(4-4(1)(-1))=32$.

If $\Delta>0$, i.e., $b^2-4ac>0$, then roots are real and distinct but nothing can be said about $f(1)$.

For Problems 16–18

16. a., 17. b., 18. c.

Sol. Multiplying R_1, R_2, R_3 by a, b, c, respectively, and then taking a, b, c common from C_1, C_2 and C_3, we get

$\Delta=\begin{vmatrix}-bc&ab+ac&ac+ab\\ab+bc&-ac&bc+ab\\ac+bc&bc+ac&-ab\end{vmatrix}$

Now, using $C_2\to C_2-C_1$ and $C_3\to C_3-C_1$, and then taking $(ab+bc+ca)$ common from C_2 and C_3, we get

$\Delta=\begin{vmatrix}-bc&1&1\\ab+bc&-1&0\\ac+bc&0&-1\end{vmatrix}\times(ab+bc+ca)^2$

Now, applying $R_2\to R_2+R_1$, we get

$\Delta=\begin{vmatrix}-bc&1&1\\ab&0&1\\ac+bc&0&-1\end{vmatrix}(ab+bc+ca)^2$

Expanding along C_2, we get

$\Delta=(ab+bc+ca)^2\,[ac+bc+ab]$

$=(ab+bc+ca)^3$

$=(r/p)^3=r^3/p^3$

Now given a, b, c are all positive, then

A.M. $\geq$ G.M.

$\Rightarrow\dfrac{ab+bc+ac}{3}\geq(ab\times bc\times ac)^{1/3}$

$\Rightarrow(ab+bc+ac)^3\geq27a^2b^2c^2$

$\Rightarrow(ab+bc+ac)^3\geq27(s^2/p^2)$

If $\Delta=27$, then $ab+bc+ca=3$, and given that $a^2+b^2+c^2=3$, from $(a+b+c)^2=a^2+b^2+c^2+2(ab+bc+ca)$, we have $a+b+c=\pm3$

$\Rightarrow a+b+c=3$ (since all the roots are positive)

$\Rightarrow 3p+q=0$

For Problems 19–21

19. d., 20. c., 21. c.

Let,

$$\begin{vmatrix}(1+x)^a & (1+2x)^b & 1\\ 1 & (1+x)^a & (1+2x)^b\\ (1+2x)^b & 1 & (1+x)^a\end{vmatrix} = A + Bx + Cx^2 + \cdots$$

Putting $x = 0$, we get

$$A = \begin{vmatrix}1 & 1 & 1\\ 1 & 1 & 1\\ 1 & 1 & 1\end{vmatrix} = 0$$

Now differentiating both sides with respect to x and putting $x = 0$, we get

$$B = \begin{vmatrix}a & 2b & 0\\ 1 & 1 & 1\\ 1 & 1 & 1\end{vmatrix} + \begin{vmatrix}1 & 1 & 1\\ 0 & a & 2b\\ 1 & 1 & 1\end{vmatrix} + \begin{vmatrix}1 & 1 & 1\\ 1 & 1 & 1\\ 2b & 0 & a\end{vmatrix} = 0$$

Hence coefficient of x is 0. Since $f(x) = 0$ and $f'(0) = 0$, $x = 0$ is a repeating root of the equation $f(x) = 0$.

For Problems 22–24

22. c., 23. d., 24. b.

$$\begin{vmatrix}x & n & r\\ m & y & r\\ m & n & z\end{vmatrix} = 0$$

Applying $R_1 \to R_1 - R_2$ and $R_2 \to R_2 - R_3$, we get

$$\begin{vmatrix}x-m & n-y & 0\\ 0 & y-n & r-z\\ m & n & z\end{vmatrix} = 0$$

$\Rightarrow (x-m)(y-n)z + (n-y)(r-z)m - n(r-z)(x-m) = 0$

Dividing by $(x-m)(y-n)(z-r)$, we have

$$\frac{z}{z-r} + \frac{m}{x-m} + \frac{n}{y-n} = 0$$

$$\Rightarrow \frac{z}{z-r} + \frac{m}{x-m} + \frac{n}{y-n} = 0$$

$$\Rightarrow \frac{z}{z-r} + \frac{m}{x-m} + 1 + \frac{n}{y-n} + 1 = 2$$

$$\Rightarrow \frac{z}{z-r} + \frac{x}{x-m} + \frac{y}{y-n} = 2 = 2$$

$$\Rightarrow \frac{z}{z-r} - 1 + \frac{x}{x-m} - 1 + \frac{y}{y-n} - 1 = -1$$

$$\Rightarrow \frac{m}{x-m} + \frac{n}{y-n} + \frac{r}{z-r} = -1$$

Now,

A.M. $\geq$ G.M.

$$\Rightarrow \frac{\frac{z}{z-r} + \frac{x}{x-m} + \frac{y}{y-n}}{3} \geq \left(\frac{z}{(z-r)}\frac{x}{(x-m)}\frac{y}{(y-n)}\right)^{1/3}$$

$$\Rightarrow \frac{z}{z-r}\frac{x}{x-m}\frac{y}{y-n} \leq \frac{8}{27}$$

For Problems 25–27

25. b., 26. d., 27. a.

Sol. $$f'(x) = \begin{vmatrix}2ax & 2ax-1 & 2ax+b+1\\ b & b+1 & -1\\ 2(ax+b) & 2ax+2b+1 & 2ax+b\end{vmatrix}$$

Applying $C_1 \to C_1 - C_3$, $C_2 \to C_2 - C_3$

$$f'(x) = \begin{vmatrix}-(b+1) & -(b+2) & 2ax+b+1\\ (b+1) & (b+2) & -1\\ b & b+1 & 2ax+b\end{vmatrix}$$

Applying $R_1 \to R_1 + R_2$ and $R_3 \to R_3 - R_2$, we get

$$f'(x) = \begin{vmatrix}0 & 0 & 2ax+b\\ b+1 & b+2 & -1\\ -1 & -1 & 2ax+b+1\end{vmatrix}$$

$= (2ax + b)[-b - 1 + b + 2]$

$\therefore f'(x) = 2ax + b$

$\therefore f(x) = ax^2 + bx + c$

$f(0) = 2 \Rightarrow c = 2$

$f(1) = 1 \Rightarrow a + b + 2 = 1 \Rightarrow a + b = -1$

$f'(5/2) = 0 \Rightarrow 5a + b = 0$

$\Rightarrow a = 1/4, b = -5/4$

Hence, $f(x) = \frac{1}{4}x^2 - \frac{5}{4}x + 2.$

Clearly, discriminant (D) of the equation $f(x) = 0$ is less than 0.

Hence, $f(x) = 0$ has imaginary roots. Also, $f(2) = 1/2$. And minimum value of $f(x)$ is

$$\frac{\frac{25}{16} - 4\frac{1}{4}(2)}{4\frac{1}{4}} = \frac{7}{16}$$

Hence, range of the $f(x)$ is $\left[\frac{7}{16}, \infty\right)$.

Matrix-Match Type

1. a→s; b→p; c→s; d→s.

a. Coefficient of x in $f(x)$ is coefficient of x in $\begin{vmatrix}x & 1 & 1\\ 1 & x & 2\\ x^2 & 1 & 0\end{vmatrix}$

Therefore, coefficient of x is -2.

b. Let $D = \begin{vmatrix} 1 & 3\cos\theta & 1 \\ \sin\theta & 1 & 3\cos\theta \\ 1 & \sin\theta & 1 \end{vmatrix}$

$= (3\cos\theta - \sin\theta)^2$

$\Delta_{max} = 10$

c. $f'(x) = 0$

$\Rightarrow f'(0) = 0$

d. $a_0 = \begin{vmatrix} 0 & 2 & 0 \\ 1 & 0 & 6 \\ 0 & 0 & 1 \end{vmatrix} = -2(1) = -2$

2. a→s; b→r; c→q, r; d→p.

a. The given determinant is $\Delta = \begin{vmatrix} x+2 & x+3 & x+5 \\ x+4 & x+6 & x+9 \\ x+8 & x+11 & x+15 \end{vmatrix}$.

Applying $R_2 \to R_2 - R_1$ and $R_3 \to R_3 - R_2$, we have

$$\Delta = \begin{vmatrix} x+2 & x+3 & x+5 \\ 2 & 3 & 4 \\ 4 & 5 & 6 \end{vmatrix}$$

$$= 2\begin{vmatrix} x & x & x+1 \\ 2 & 3 & 4 \\ 1 & 1 & 1 \end{vmatrix}$$

[Applying $R_1 \to R_1 - R_2$ and $R_3 \to R_3 - R_2$]

$$= 2\begin{vmatrix} x & 0 & 1 \\ 2 & 1 & 1 \\ 1 & 0 & 0 \end{vmatrix}$$

[Applying $C_2 \to C_2 - C_1$ and $C_3 \to C_3 - C_2$]

$= -2$ [Expanding along R_3]

b. $\begin{vmatrix} 7 & 6 & x^2-13 \\ 2 & x^2-13 & 2 \\ x^2-13 & 3 & 7 \end{vmatrix}$

Let $x^2 - 13 = t$. Then

$t^3 - 67t + 126 = 0$

$\Rightarrow t = -9, 2, 7 \Rightarrow x = \pm 2, \pm\sqrt{20}, \pm\sqrt{15}$

Hence sum of other five roots is 2.

c. $\Delta = \begin{vmatrix} \sqrt{6} & 2i & 3+\sqrt{6} \\ \sqrt{12} & \sqrt{3} + \sqrt{8}i & 3\sqrt{2} + \sqrt{6}i \\ \sqrt{18} & \sqrt{2} + \sqrt{12}\,i & \sqrt{27} + 2i \end{vmatrix}$

Taking $\sqrt{6}$ common from C_1, we get

$$\Delta = \sqrt{6}\begin{vmatrix} 1 & 2i & 3+\sqrt{6} \\ \sqrt{2} & \sqrt{3}+2\sqrt{2}i & 3\sqrt{2}+\sqrt{6}i \\ \sqrt{3} & \sqrt{2}+2\sqrt{3}i & 3\sqrt{3}+2i \end{vmatrix}$$

Applying $R_2 \to R_2 - \sqrt{2}R_1$ and $R_3 \to R_3 - \sqrt{3}R_1$, we get

$$\Delta = \sqrt{6}\begin{vmatrix} 1 & 2i & 3+\sqrt{6} \\ 0 & \sqrt{3} & \sqrt{6}i - 2\sqrt{3} \\ 0 & \sqrt{2} & 2i - 3\sqrt{2} \end{vmatrix}$$

$$= \sqrt{6}\begin{vmatrix} \sqrt{3} & \sqrt{6}i - 2\sqrt{3} \\ \sqrt{2} & 2i - 3\sqrt{2} \end{vmatrix}$$

$$= \sqrt{6}\begin{vmatrix} \sqrt{3} & -2\sqrt{3} \\ \sqrt{2} & -3\sqrt{2} \end{vmatrix}$$ [Applying $C_2 \to C_2 - \sqrt{2}i\,C_1$]

$= \sqrt{6}\,(-3\sqrt{6} + 2\sqrt{6})$

$= -6$, which is an integer

d. $f(\theta) = \begin{vmatrix} \cos^2\theta & \cos\theta\sin\theta & -\sin\theta \\ \cos\theta\sin\theta & \sin^2\theta & \cos\theta \\ \sin\theta & -\cos\theta & 0 \end{vmatrix}$

Applying $R_1 \to R_1 + (\sin\theta)R_3$ and $R_2 \to R_2 - (\cos\theta)R_3$, we get

$$f(\theta) = \begin{vmatrix} 1 & 0 & -\sin\theta \\ 0 & 1 & \cos\theta \\ \sin\theta & -\cos\theta & 0 \end{vmatrix}$$

$= \sin^2\theta + \cos^2\theta = 1$

$\Rightarrow f(\pi/3) = 1$

3. a→p, q, r; b→q; c→s; d→p, q, r.

a. Multiplying C_1 by a, C_2 by b and C_3 by c, we obtain

$$\Delta = \frac{1}{abc}\begin{vmatrix} \frac{a}{c} & \frac{b}{c} & -\frac{a+b}{c} \\ -\frac{b+c}{a} & \frac{b}{a} & \frac{c}{a} \\ -\frac{b(b+c)}{ac} & \frac{b(a+2b+c)}{ac} & -\frac{b(a+b)}{ac} \end{vmatrix}$$

Applying $C_1 \to C_1 + C_2 + C_3$, we get

$$\Delta = \frac{1}{abc}\begin{vmatrix} 0 & \frac{b}{c} & -\frac{a+b}{c} \\ 0 & \frac{b}{a} & \frac{c}{a} \\ 0 & \frac{b(a+2b+c)}{ac} & -\frac{b(a+b)}{ac} \end{vmatrix}$$

This shows that Δ is independent of a, b and c.

b. Applying $C_1 \to C_1 - (\cot b)C_2$, we get

$$\Delta = \begin{vmatrix} 0 & \sin a \sin b & \cos a \\ 0 & \cos a \sin b & -\sin a \\ -\sin a/\sin b & \sin a \cos b & 0 \end{vmatrix}$$

$$= -\frac{\sin a}{\sin b}[-\sin b \sin^2 a - \cos^2 a \sin b] \text{ [expanding along } C_1]$$

$$= \sin a$$

c. Taking $1/(\sin a \cos b)$, $1/(\sin a \sin b)$, $1/\cos a$ common from C_1, C_2, C_3, respectively, we get

$$\Delta = \frac{1}{\sin^2 a \ \cos a \ \sin b \cos b}\Delta_1$$

where $\Delta_2 = \begin{vmatrix} 1 & 1 & 1 \\ -\cot a & -\cot a & \tan a \\ \tan b & -\cot b & 0 \end{vmatrix}$

$$= \begin{vmatrix} 0 & 1 & 1 \\ 0 & -\cot a & \tan a \\ 1/\sin b \cos b & -\cot b & 0 \end{vmatrix}$$

Applying $C_1 \to C_1 - C_2$, we get

$$\Delta = \frac{1}{\sin b \cos b}[\tan a + \cot a]$$

$$= \frac{1}{\sin a \cos a \sin b \cos b}$$

d. $\begin{vmatrix} a^2 & b\sin A & c\sin A \\ b\sin A & 1 & \cos A \\ c\sin A & \cos A & 1 \end{vmatrix}$

$$= \begin{vmatrix} a^2 & a\sin B & a\sin C \\ a\sin B & 1 & \cos A \\ a\sin C & \cos A & 1 \end{vmatrix}$$

$$= a^2 \begin{vmatrix} 1 & \sin B & \sin C \\ \sin B & 1 & \cos A \\ \sin C & \cos A & 1 \end{vmatrix}$$

$$= a^2 \begin{vmatrix} 1 & 0 & 0 \\ \sin B & 1 - \sin^2 B & \cos A - \sin B \sin C \\ \sin C & \cos A - \sin B \sin C & 1 - \sin^2 C \end{vmatrix}$$

[Applying $C_2 \to C_2 - (\sin B)C_1$ and $C_3 \to C_3 - (\sin C)C_1$]

$= a^2[\cos^2 B \cos^2 C - (\cos A - \sin B \sin C)^2]$

$= a^2[\cos^2 B \cos^2 C - (\cos(B + C) + \sin B \sin C)^2]$

$= a^2[\cos^2 B \cos^2 C - \cos^2 B \cos^2 C]$

$= 0$

Integer Type

1. (2) Let $\Delta = \begin{vmatrix} a_1 & b_1 & c_1 \\ a_2 & b_2 & c_2 \\ a_3 & b_3 & c_3 \end{vmatrix}$ be a determinant of order 3.

Then, $\Delta = a_1b_2c_3 + a_3b_1c_2 + a_2b_3c_1 - a_1b_3c_2 - a_2b_1c_3 - a_3b_2c_1$

$= (a_1b_2c_3 + a_3b_1c_2 + a_2b_3c_1) - (a_1b_3c_2 + a_2b_1c_3 - a_3b_2c_1)$

Since each element of Δ is either 1 or 0, the value of the determinant cannot exceed 3.

Clearly, the value of Δ is maximum when the value of each term in the first bracket is 1 and the value of each term in the second bracket is zero. But $a_1b_2c_3 = a_3b_1c_1 = 1$ implies that every element of the determinant Δ is 1 and in that case $\Delta = 0$.

Thus, we may have $\Delta = \begin{vmatrix} 0 & 1 & 1 \\ 1 & 0 & 1 \\ 1 & 1 & 0 \end{vmatrix} = 2.$

2. (2) $\begin{vmatrix} \tan A & 1 & 1 \\ 1 & \tan B & 1 \\ 1 & 1 & \tan C \end{vmatrix}$

$= \tan A \tan B \tan C + 1 + 1 - \tan A - \tan B - \tan C$

$= 2$ (as in a triangle, $\tan A + \tan B + \tan C = \tan A \tan B \tan C$)

3. (3) Expanding by Sarrus rule,

$\begin{vmatrix} 1 & e^{i\pi/3} & e^{i\pi/4} \\ e^{-i\pi/3} & 1 & e^{-i2\pi/3} \\ e^{-i\pi/4} & e^{-i2\pi/3} & 1 \end{vmatrix} = 1 + e^{i\pi/3} \times e^{i2\pi/3} \times e^{i\pi/4} + e^{-i\pi/3} \times e^{-i2\pi/3} \times e^{i\pi/4} - (e^{-i\pi/4} \times e^{i\pi/4} + e^{-i\pi/3} \times e^{i\pi/3} + e^{-i2\pi/3} \times e^{i2\pi/3})$

$= 1 + e^{i3\pi/4} + e^{-i3\pi/4} - (1 + 1 + 1)$

$= -2 + 2\cos(3\pi/4)$

$= -2 - \sqrt{2}$

4. (0) $a_r = (\cos 2r\pi + i\sin 2r\pi)^{\frac{1}{9}} = e^{i\frac{2r\pi}{9}}$

$\Rightarrow \begin{vmatrix} a_1 & a_2 & a_3 \\ a_4 & a_5 & a_6 \\ a_7 & a_8 & a_9 \end{vmatrix}$

$$= \begin{vmatrix} e^{i\frac{2\pi}{9}} & e^{i\frac{4\pi}{9}} & e^{i\frac{6\pi}{9}} \\ e^{i\frac{8\pi}{9}} & e^{i\frac{10\pi}{9}} & e^{i\frac{12\pi}{9}} \\ e^{i\frac{14\pi}{9}} & e^{i\frac{16\pi}{9}} & e^{i\frac{18\pi}{9}} \end{vmatrix}$$

$$= e^{i\frac{6\pi}{9}} \begin{vmatrix} e^{i\frac{2\pi}{9}} & e^{i\frac{4\pi}{9}} & e^{i\frac{6\pi}{9}} \\ e^{i\frac{2\pi}{9}} & e^{i\frac{4\pi}{9}} & e^{i\frac{6\pi}{9}} \\ e^{i\frac{14\pi}{9}} & e^{i\frac{16\pi}{9}} & e^{i\frac{18\pi}{9}} \end{vmatrix} \text{ (taking } e^{i\frac{6\pi}{9}} \text{ common from } R_2)$$

$= 0$ (R_1 and R_2 are identical)

5. (1) We have $\begin{vmatrix} 1 & 1 & 1 & 1 \\ 1 & 2 & 3 & 4 \\ 1 & 3 & 6 & 10 \\ 1 & 4 & 10 & 20 \end{vmatrix} = \begin{vmatrix} 1 & 1 & 1 & 1 \\ 0 & 1 & 2 & 3 \\ 0 & 2 & 5 & 9 \\ 0 & 3 & 9 & 19 \end{vmatrix}$ [Applying: $R_2 \to R_2 - R_1, R_3 \to R_3 - R_1, R_4 \to R_4 - R_1$]

$$= \begin{vmatrix} 1 & 2 & 3 \\ 2 & 5 & 9 \\ 3 & 9 & 19 \end{vmatrix} = \begin{vmatrix} 1 & 2 & 3 \\ 0 & 1 & 3 \\ 0 & 3 & 10 \end{vmatrix}$$

[Applying $R_2 \to R_2 - 2R_1$, $R_3 \to R_3 - 3R_1$]

$= (10 - 9) = 1.$

6. (7) Since $A_1B_1C_1$, $A_2B_2C_2$, $A_3B_3C_3$ are divisible by 49, we have $m_1, m_2, m_3 \in Z$ such that

$$100A_1 + 10B_1 + C_1 = 49m_1$$
$$100A_2 + 10B_2 + C_2 = 49m_2$$

and $100A_3 + 10B_3 + C_3 = 49m_3$

Now, $\Delta = \begin{vmatrix} A_1 & B_1 & C_1 \\ A_2 & B_2 & C_2 \\ A_3 & B_3 & C_3 \end{vmatrix}$

$$= \begin{vmatrix} A_1 & B_1 & 100A_1 + 10B_1 + C_1 \\ A_2 & B_2 & 100A_2 + 10B_2 + C_2 \\ A_3 & B_3 & 100A_3 + 10B_3 + C_3 \end{vmatrix} (C_3 \to$$

$(C_3 \to C_3 + 10C_2 + 100C_1)$

$$= \begin{vmatrix} A_1 & B_1 & m_1 \\ A_2 & B_2 & m_2 \\ A_3 & B_3 & m_3 \end{vmatrix} = 49\Delta^*$$

As the elements of Δ^* are integers, Δ^* is an integer. Let $\Delta^* = p$. Then we get $\Delta = 49\,p$, i.e., Δ is divisible by 49.

7. (3) Let $\Delta = \begin{vmatrix} 3a & -a+b & -a+c \\ -b+a & 3b & -b+c \\ -c+a & -c+b & 3c \end{vmatrix}$

Applying $C_1 \to C_1 + C_2 + C_3$, we get

$$\Delta = \begin{vmatrix} a+b+c & -a+b & -a+c \\ a+b+c & 3b & -b+c \\ a+b+c & -c+b & 3c \end{vmatrix}$$

$$= (a+b+c)\begin{vmatrix} 1 & -a+b & -a+c \\ 1 & 3b & -b+c \\ 1 & -c+b & 3c \end{vmatrix}$$

[Applying $C_1 \to C_1 + C_2 + C_3$]

$$= (a+b+c)\begin{vmatrix} 1 & -a+b & -a+c \\ 0 & 2b+a & -b+a \\ 0 & -c+a & 2c+a \end{vmatrix}$$ [Applying $R_2 \to R_2 - R_1$, $R_3 \to R_3 - R_1$]

$= (a + b + c)$ [Expanding along C_1]

$= (a + b + c)\ \{(2b + a)(2c + a) - (-b + a)(-c + a)\}$

$= (a + b + c)\ \{(4bc + 2ab + 2ca + a^2) - (bc - ab - ac + a^2)\}$

$= (a + b + c)\ (3bc + 3ab + 3ca)$

$= 3(a + b + c)\ (ab + bc + ca).$

8. (1) $\dfrac{dy}{dx} = \begin{vmatrix} \cos x & -\sin x & \cos x \\ \sin x & -\sin x & \cos x \\ x & 1 & 1 \end{vmatrix} + \begin{vmatrix} \sin x & \cos x & \sin x \\ -\sin x & -\cos x & -\sin x \\ x & 1 & 1 \end{vmatrix} + \begin{vmatrix} \sin x & \cos x & \sin x \\ \cos x & -\sin x & \cos x \\ 1 & 0 & 0 \end{vmatrix}$

$$= 0 - \begin{vmatrix} \sin x & \cos x & \sin x \\ \sin x & \cos x & \sin x \\ x & 1 & 1 \end{vmatrix} + 1\begin{vmatrix} \cos x & \sin x \\ -\sin x & \cos x \end{vmatrix}$$

$= 0 + (\cos^2 x + \sin^2 x) = 1.$

9. (0) $\dfrac{d^n}{dx^n}[f(x)] = \begin{vmatrix} \dfrac{d^n}{dx^n}(x^n) & n! & 2 \\ \dfrac{d^n}{dx^n}(\cos x) & \cos\dfrac{n\pi}{2} & 4 \\ \dfrac{d^n}{dx^n}(\sin x) & \sin\dfrac{n\pi}{2} & 8 \end{vmatrix}$

$$= \begin{vmatrix} n! & n! & 2 \\ \cos\left(x + \frac{n\pi}{2}\right) & \cos\frac{n\pi}{2} & 4 \\ \sin\left(x + \frac{n\pi}{2}\right) & \sin\frac{n\pi}{2} & 8 \end{vmatrix}$$

$\Rightarrow \dfrac{d^n}{dx^n}[f(x)]_{x=0} = 0.$

10. (8) The system

$x + ky + 3z = 0$

$3x + ky - 2z = 0$

$2x + 3y - 4z = 0$

has non-trivial solution (i.e., non-zero solution) iff the determinant of the coefficients of x, y, and z is zero.

Here $\Delta = \begin{vmatrix} 1 & k & 3 \\ 3 & k & -2 \\ 2 & 3 & -4 \end{vmatrix} = 0$

$\Rightarrow 2k - 33 = 0$ or $k = 33/2$

11. (2) For non-trivial solution, we must have

$$\begin{vmatrix} 2a & -2 & 3 \\ 1 & a & 2 \\ 2 & 0 & a \end{vmatrix} = 0$$

$\Rightarrow 2a(a^2 - 0) + 2(a - 4) + 3(0 - 2a) = 0$

$\Rightarrow 2a^3 + 2a - 8 + 0 - 6a = 0$

$\Rightarrow 2a^3 - 4a - 8 = 0$

$\Rightarrow a^3 - 2a - 4 = 0$

$\Rightarrow a^3 - 2a^2 + 2a^2 - 4a + 2a - 4 = 0$

$\Rightarrow a^2(a - 2) + 2a(a - 2) + 2(a - 2) = 0$

$\Rightarrow (a - 2)\ (A^2 + 2a + 2) = 0$

$\Rightarrow a = 2$

12. (2) Given $= 0 \begin{vmatrix} p & b & c \\ a & q & c \\ a & b & r \end{vmatrix} = 0$

$R_1 \to R_1 - R_2$ and $R_2 \to R_2 - R_3$ reduces the determinant to

$$\begin{vmatrix} p-a & b-q & 0 \\ 0 & q-b & c-r \\ a & b & r \end{vmatrix} = 0$$

$\Rightarrow (p - a)(q - b)r + a(b - q)(c - r) - b(p - a)(c - r) = 0$

Dividing throughout by $(p - a)\ (q - b)\ (r - c)$ we get

$$\frac{r}{r-c} + \frac{a}{p-a} + \frac{b}{q-b} = 0$$

$$\Rightarrow \frac{r}{r-c} + 1 + \frac{a}{p-a} + 1 + \frac{b}{q-b} = 2$$

$$\Rightarrow \frac{r}{r-c} + \frac{p}{p-a} + \frac{q}{q-b} = 2$$

13. (3) The given system has a non-trivial solution if

$$\begin{vmatrix} \lambda & \sin\alpha & \cos\alpha \\ 1 & \cos\alpha & \sin\alpha \\ -1 & \sin\alpha & -\cos\alpha \end{vmatrix} = 0.$$

By expanding the determinant along the first column, we get

$\lambda = \sin 2\alpha + \cos 2\alpha$

Since $-\sqrt{2} \le \sin 2\alpha + \cos 2\alpha \le \sqrt{2}$,

$-\sqrt{2} \le \lambda \le \sqrt{2}$

14. (0) Putting $\lambda = 0$, we have $t = \begin{vmatrix} 0 & -1 & 3 \\ 1 & 0 & -4 \\ -3 & 4 & 0 \end{vmatrix} = 0$ (skew–symmetric determinant)

15. (9) $D_c = \begin{vmatrix} x^3-1 & 0 & x-x^4 \\ 0 & x-x^6 & x^3-1 \\ x-x^4 & x^3-1 & 0 \end{vmatrix}$ is the determinant formed by the cofactors of the determinant

$$D = \begin{vmatrix} 1 & x & x^2 \\ x & x^2 & 1 \\ x^2 & 1 & x \end{vmatrix}$$

Hence $D_c = D^2 = 3^2 = 9$

16. (4) Let $\Delta = \begin{vmatrix} -2a & a+b & a+c \\ b+a & -2b & b+c \\ c+a & c+b & -2c \end{vmatrix}$

Putting $a + b = 0$ or $b = -a$, we get

$$\Delta = \begin{vmatrix} -2a & 0 & a+c \\ 0 & 2a & c-a \\ c+a & c-a & -2c \end{vmatrix}$$

Expanding along R_1, we get

$$-2a\{-4ac-(c-a)^2\} - 0 + (a+c)\{0-2a(c+a)\}$$

$$= 2a(c+a)^2 - 2a(c+a)^2$$

$$= 0$$

Hence, $a + b$ is a factor of Δ. Similarly, $b + c$ and $c + a$ are factors of Δ.

On expanding the determinant, we can see that each term of the determinant is a homogeneous expression in a, b, c of degree 3 and also R.H.S. is a homogeneous expression of degree 3. Therefore,

$$\Delta = k(a+b)(b+c)(c+a)$$

$$= \begin{vmatrix} -2a & a+b & a+c \\ b+a & -2b & b+c \\ c+a & c+b & -2c \end{vmatrix}$$

$$= k(a+b)(b+c)(c+a)$$

Putting $a = 0$, $b = 1$, $c = 2$, we get

$$\begin{vmatrix} 0 & 1 & 2 \\ 1 & -2 & 3 \\ 2 & 3 & -4 \end{vmatrix} = k(0+1)(1+2)(2+0)$$

$$\Rightarrow 0 - 1(-4-6) + 2(3+4) = 6k$$

$$\Rightarrow 24 = 6k$$

$$\Rightarrow k = 4$$

CHAPTER 8

Matrices

Objective Type

Each question has four choices a, b, c, and d, out of which *only one* is correct. Find the correct answer.

1. The inverse of a skew-symmetric matrix of odd order is

   a. a symmetric matrix **b.** a skew symmetric
   c. diagonal matrix **d.** does not exist

2. Let A and B be two 2×2 matrices. Consider the statements

   i. $AB = O \Rightarrow A = O$ or $B = O$
   ii. $AB = I_2 \Rightarrow A = B^{-1}$
   iii. $(A + B)^2 = A^2 + 2AB + B^2$

   Then

   a. (i) and (ii) are false, (iii) is true
   b. (ii) and (iii) are false, (i) is true
   c. (i) is false, (ii) and (iii) are true
   d. (i) and (iii) are false, (ii) is true

3. The equation $\begin{bmatrix} 1 & x & y \end{bmatrix} \begin{bmatrix} 1 & 3 & 1 \\ 0 & 2 & -1 \\ 0 & 0 & 1 \end{bmatrix} \begin{bmatrix} 1 \\ x \\ y \end{bmatrix} = [0]$ has

   i. for $y = 0$ **(p)** rational roots
   ii. for $y = -1$ **(q)** irrational roots
   (r) integral roots

   Then

	(i)	(ii)
a.	(p)	(r)
b.	(q)	(p)
c.	(p)	(q)
d.	(r)	(p)

4. The number of diagonal matrix A of order n for which $A^3 = A$ is

   a. 1 **b.** 0
   c. 2^n **d.** 3^n

5. If $A = \begin{bmatrix} a & b \\ 0 & a \end{bmatrix}$ is n^{th} root of I_2, then choose the correct statements:

   i. if n is odd, $a = 1, b = 0$
   ii. if n is odd, $a = -1, b = 0$
   iii. if n is even, $a = 1, b = 0$
   iv. if n is even, $a = -1, b = 0$

   a. i, ii, iii **b.** ii, iii, iv
   c. i, ii, iii, iv **d.** i, iii, iv

6. A is a 2×2 matrix such that $A\begin{bmatrix} 1 \\ -1 \end{bmatrix} = \begin{bmatrix} -1 \\ 2 \end{bmatrix}$ and $A^2 \begin{bmatrix} 1 \\ -1 \end{bmatrix} = \begin{bmatrix} 1 \\ 0 \end{bmatrix}$.

   The sum of the elements of A is

   a. -1 **b.** 0
   c. 2 **d.** 5

7. The product of matrices $A = \begin{bmatrix} \cos^2\theta & \cos\theta\sin\theta \\ \cos\theta\sin\theta & \sin^2\theta \end{bmatrix}$ and $B = \begin{bmatrix} \cos^2\phi & \cos\phi\sin\phi \\ \cos\phi\sin\phi & \sin^2\phi \end{bmatrix}$ is a null matrix if $\theta - \phi =$

   a. $2n\pi, n \in Z$ **b.** $n\dfrac{\pi}{2}, n \in Z$
   c. $(2n+1)\dfrac{\pi}{2}, n \in Z$ **d.** $n\pi, n \in Z$

8. Let A, B be two matrices such that they commute, then for any positive integer n,

   i. $AB^n = B^n A$ **ii.** $(AB)^n = A^n B^n$

a. Only (i) is correct
b. Both (i) and (ii) are correct
c. Only (ii) is correct
d. None of (i) and (ii) is correct

9. If $A = [a_{ij}]_{4\times4}$, such that $a_{ij} = \begin{cases} 2, & \text{when } i = j \\ 0, & \text{when } i \neq j \end{cases}$, then $\left\{\dfrac{\det(\text{adj}(\text{adj } A))}{7}\right\}$ is (where $\{\cdot\}$ represents fractional part function)

a. 1/7 b. 2/7
c. 3/7 d. none of these

10. If $\begin{bmatrix} \alpha & \beta \\ \gamma & -\alpha \end{bmatrix}$ is to be the square root of two-rowed unit matrix, then α, β and γ should satisfy the relation

a. $1 - \alpha^2 + \beta\gamma = 0$ b. $\alpha^2 + \beta\gamma - 1 = 0$
c. $1 + \alpha^2 + \beta\gamma = 0$ d. $1 - \alpha^2 - \beta\gamma = 0$

11. If A is a square matrix such that $A^2 = A$, then $(I + A)^3 - 7A$ is

a. $3I$ b. O
c. I d. $2I$

12. If A and B are square matrices of order n, then $A - \lambda I$ and $B - \lambda I$ commute for every scalar λ, only if

a. $AB = BA$ b. $AB + BA = O$
c. $A = -B$ d. none of these

13. Matrix A such that $A^2 = 2A - I$, where I is the identity matrix, then for $n \geq 2$, A^n is equal to

a. $2^{n-1}A - (n-1)I$ b. $2^{n-1}A - I$
c. $nA - (n-1)I$ d. $nA - I$

14. Let $A = \begin{bmatrix} 0 & \alpha \\ 0 & 0 \end{bmatrix}$ and $(A+1)^{50} - 50A = \begin{bmatrix} a & b \\ c & d \end{bmatrix}$. Then the value of $a + b + c + d$ is

a. 2 b. 1
c. 4 d. none of these

15. If $A = \begin{bmatrix} i & -i \\ -i & i \end{bmatrix}$ and $B = \begin{bmatrix} 1 & -1 \\ -1 & 1 \end{bmatrix}$, then A^8 equals

a. $4B$ b. $128B$
c. $-128B$ d. $-64B$

16. If $\begin{bmatrix} 2 & -1 \\ 1 & 0 \\ -3 & 4 \end{bmatrix} A = \begin{bmatrix} -1 & -8 & -10 \\ 1 & -2 & -5 \\ 9 & 22 & 15 \end{bmatrix}$, then sum of all the elements of matrix A is

a. 0 b. 1
c. 2 d. -3

17. If $A = \dfrac{1}{\sqrt{3}}\begin{bmatrix} 1 & 1+i \\ 1-i & 1 \end{bmatrix}$, then $A(\bar{A}^T)$ equals

a. O b. I
c. $-I$ d. $2I$

18. Identify the incorrect statement in respect of two square matrices A and B conformable for sum and product:

a. $t_r(A + B) = t_r(A) + t_r(B)$ b. $t_r(\alpha A) = \alpha t_r(A)$, $\alpha \in R$
c. $t_r(A^T) = t_r(A)$ d. none of these

19. A is an involuntary matrix given by $A = \begin{bmatrix} 0 & 1 & -1 \\ 4 & -3 & 4 \\ 3 & -3 & 4 \end{bmatrix}$, then the inverse of $A/2$will be

a. $2A$ b. $\dfrac{A^{-1}}{2}$
c. $\dfrac{A}{2}$ d. A^2

20. If A is a non-singular matrix such that $AA^T = A^TA$ and $B = A^{-1}A^T$, then matrix B is

a. involuntary b. orthogonal
c. idempotent d. none of these

21. If P is an orthogonal matrix and $Q = PAP^T$ and $x = P^TQ^{1000}P$, then x^{-1} is, where A is involutary matrix

a. A b. I
c. A^{1000} d. one of these

22. If n^{th} order square matrix A is a orthogonal, then, $|\text{adj}(\text{adj } A)|$ is

a. always -1 if n is even b. always 1 if n is odd
c. always 1 d. none of these

23. If both $A - \dfrac{1}{2}I$ and $A + \dfrac{1}{2}I$ are orthogonal matrices, then

a. A is orthogonal
b. A is skew-symmetric matrix of even order
c. $A^2 = \dfrac{3}{4}I$
d. none of these

24. In which of the following type of matrix inverse does not exist always

a. idempotent b. orthogonal
c. involuntary d. none of these

25. If A is an orthogonal matrix, then A^{-1} equals

a. A^T b. A
c. A^2 d. none of these

26. If Z is an idempotent matrix, then $(I + Z)^n$

a. $I + 2^nZ$ b. $I + (2^n - 1)Z$
c. $I - (2^n - 1)Z$ d. none of these

27. If A and B are two matrices such that $AB = B$ and $BA = A$, then

a. $(A^5 - B^5)^3 = A - B$ b. $(A^5 - B^5)^3 = A^3 - B^3$
c. $A - B$ is idempotent d. $A - B$ is nilpotent

28. If A is a nilpotent matrix of index 2, then for any positive integer n, $A(I+A)^n$ is equal to

a. A^{-1} **b.** A

c. A^n **d.** I_n

29. Let A be an nth order square matrix and B be its adjoint, then $|AB + KI_n|$ is (where K is a scalar quantity)

a. $(|A| + K)^{n-2}$ **b.** $(|A| + K)^n$

c. $(|A| + K)^{n-1}$ **d.** none of these

30. If $A^2 = I$, then the value of $\det(A - I)$ is (where A has order 3)

a. 1 **b.** −1

c. 0 **d.** cannot say anything

31. If $A = \begin{bmatrix} a & b & c \\ x & y & z \\ p & q & r \end{bmatrix}$, $B = \begin{bmatrix} q & -b & y \\ -p & a & -x \\ r & -c & z \end{bmatrix}$ and if A is invertible, then which of the following is not true?

a. $|A| = |B|$

b. $|A| = -|B|$

c. $|\text{adj } A| = |\text{adj } B|$

d. A is invertible if and only if B is invertible

32. If A and B are two non-singular matrices such that $AB = C$, then $|B|$ is equal to

a. $\dfrac{|C|}{|A|}$ **b.** $\dfrac{|A|}{|C|}$

c. $|C|$ **d.** none of these

33. If A and B are squares matrices such that $A^{2006} = O$ and $AB = A + B$, then $\det(B)$ equals

a. 0 **b.** 1

c. −1 **d.** none of these

34. If matrix A is given by $A = \begin{bmatrix} 6 & 11 \\ 2 & 4 \end{bmatrix}$, then the determinant of $A^{2005} - 6A^{2004}$ is

a. 2^{2006} **b.** $(-11)\,2^{2005}$

c. $-2^{2005} \cdot 7$ **d.** $(-9)\,2^{2004}$

35. If A is a non-diagonal involutory matrix, then

a. $A - I = O$ **b.** $A + I = O$

c. $A - I$ is non-zero singular **d.** none of these

36. If $A = \begin{bmatrix} 0 & 1 & 2 \\ 1 & 2 & 3 \\ 3 & a & 1 \end{bmatrix}$ and $A^{-1} = \begin{bmatrix} 1/2 & 1/2 & 1/2 \\ -4 & 3 & c \\ 5/2 & -3/2 & 1/2 \end{bmatrix}$, then the values of a and c are equal to

a. 1, 1 **b.** 1, −1

c. 1, 2 **d.** −1, 1

37. If A and B are two non-singular matrices of the same order such that $B^r = I$, for some positive integer $r > 1$. Then $A^{-1}B^{r-1}A - A^{-1}B^{-1}A =$

a. I **b.** $2I$

c. O **d.** $-I$

38. For two unimodular complex numbers z_1 and z_2, $\begin{bmatrix} \bar{z}_1 & -z_2 \\ \bar{z}_2 & z_1 \end{bmatrix}^{-1} \begin{bmatrix} z_1 & z_2 \\ -\bar{z}_2 & \bar{z}_1 \end{bmatrix}^{-1}$ is equal to

a. $\begin{bmatrix} z_1 & z_2 \\ \bar{z}_1 & \bar{z}_2 \end{bmatrix}$ **b.** $\begin{bmatrix} 1 & 0 \\ 0 & 1 \end{bmatrix}$

c. $\begin{bmatrix} 1/2 & 0 \\ 0 & 1/2 \end{bmatrix}$ **d.** none of these

39. If $A(\alpha, \beta) = \begin{bmatrix} \cos\alpha & \sin\alpha & 0 \\ -\sin\alpha & \cos\alpha & 0 \\ 0 & 0 & e^{\beta} \end{bmatrix}$, then $A(\alpha, \beta)^{-1}$ is equal to

a. $A(-\alpha, -\beta)$ **b.** $A(-\alpha, \beta)$

c. $A(\alpha, -\beta)$ **d.** $A(\alpha, \beta)$

40. If $A = \begin{bmatrix} a+ib & c+id \\ -c+id & a-ib \end{bmatrix}$ and $a^2 + b^2 + c^2 + d^2 = 1$, then A^{-1} is equal to

a. $\begin{bmatrix} a+ib & -c-id \\ -c+id & a-ib \end{bmatrix}$ **b.** $\begin{bmatrix} a+ib & -c+id \\ -c+id & a-ib \end{bmatrix}$

c. $\begin{bmatrix} a-ib & -c-id \\ -c-id & a+ib \end{bmatrix}$ **d.** none of these

41. If $A^3 = O$, then $I + A + A^2$ equals

a. $I - A$ **b.** $(I + A^1)^{-1}$

c. $(I - A)^{-1}$ **d.** none of these

42. If A is order 3 square matrix such that $|A| = 2$, then $|\text{adj}(\text{adj}(\text{adj } A))|$ is

a. 512 **b.** 256

c. 64 **d.** none of these

43. $(-A)^{-1}$ is always equal to (where A is n^{th} order square matrix)

a. $(-1)^n A^{-1}$ **b.** $-A^{-1}$

c. $(-1)^{n-1} A^{-1}$ **d.** none of these

44. For each real x, $-1 < x < 1$. Let $A(x)$ be the matrix $(1-x)^{-1}\begin{bmatrix} 1 & -x \\ -x & 1 \end{bmatrix}$ and $z = \dfrac{x+y}{1+xy}$. Then

a. $A(z) = A(x)\,A(y)$ **b.** $A(z) = A(x) - A(y)$

c. $(z) = A(x) + A(y)$ **d.** $A(z) = A(x)\,[A(y)]^{-1}$

45. If $\begin{bmatrix} 1/25 & 0 \\ x & 1/25 \end{bmatrix} = \begin{bmatrix} 5 & 0 \\ -a & 5 \end{bmatrix}^{-2}$, then the value of x is

a. $a/125$ **b.** $2a/125$

c. $2a/25$ **d.** none of these

46. If $A=\begin{bmatrix}1 & 2\\ 2 & 1\end{bmatrix}$ and $f(x)=\dfrac{1+x}{1-x}$, then $f(A)$ is

a. $\begin{bmatrix}1 & 1\\ 1 & 1\end{bmatrix}$ b. $\begin{bmatrix}2 & 2\\ 2 & 2\end{bmatrix}$

c. $\begin{bmatrix}-1 & -1\\ -1 & -1\end{bmatrix}$ d. none of these

47. If A is a square matrix of order n such that $|\text{adj}(\text{adj } A)| = |A|^9$, then the value of n can be

a. 4 b. 2

c. either 4 or 2 d. none of these

48. If $A=\begin{bmatrix}1 & \tan x\\ -\tan x & 1\end{bmatrix}$, then A^TA^{-1} is

a. $\begin{bmatrix}-\cos 2x & \sin 2x\\ -\sin 2x & \cos 2x\end{bmatrix}$ b. $\begin{bmatrix}\cos 2x & -\sin 2x\\ \sin 2x & \cos 2x\end{bmatrix}$

c. $\begin{bmatrix}\cos 2x & \cos 2x\\ \cos 2x & \sin 2x\end{bmatrix}$ d. none of these

49. If $A=\begin{bmatrix}0 & -\tan\alpha/2\\ \tan\alpha/2 & 0\end{bmatrix}$ and I is a 2 × 2 unit matrix,

then $(I-A)\begin{bmatrix}\cos\alpha & -\sin\alpha\\ \sin\alpha & \sin\alpha\end{bmatrix}$ is

a. $-I+A$ b. $I-A$

c. $-I-A$ d. none of these

50. The matrix X for which $\begin{bmatrix}1 & -4\\ 3 & -2\end{bmatrix}X=\begin{bmatrix}-16 & -6\\ 7 & 2\end{bmatrix}$ is

a. $\begin{bmatrix}-2 & 4\\ -3 & 1\end{bmatrix}$ b. $\begin{bmatrix}\frac{1}{5} & \frac{2}{5}\\ -\frac{3}{10} & \frac{1}{5}\end{bmatrix}$

c. $\begin{bmatrix}-16 & -6\\ 7 & 2\end{bmatrix}$ d. $\begin{bmatrix}6 & 2\\ \frac{11}{2} & 2\end{bmatrix}$

51. If A and B are square matrices of the same order and A is non-singular, then for a positive integer n, $(A^{-1}BA)^n$ is equal to

a. $A^{-n}B^nA^n$ b. $A^nB^nA^{-n}$

c. $A^{-1}B^nA$ d. $n(A^{-1}BA)$

52. If A is singular matrix, then adj A is

a. singular b. non-singular

c. symmetric d. not defined

53. The inverse of a diagonal matrix is

a. a diagonal matrix

b. askew-symmetric matrix

c. a symmetric matrix

d. none of these

54. If P is non-singular matrix, then value of adj(P^{-1}) in terms of P is

a. $P/|P|$ b. $P|P|$

c. P d. none of these

55. If adj $B=A$, $|P|=|Q|=1$, then adj$(Q^{-1}BP^{-1})$ is

a. PQ b. QAP

c. PAQ d $PA^{-1}Q$

56. If A is non-singular and $(A-2I)(A-4I)=O$, then $\frac{1}{6}A+\frac{4}{3}A^{-1}$ is equal to

a. O b. I

c. $2I$ d. $6I$

57. If $A(\alpha,\beta)=\begin{bmatrix}\cos\alpha & \sin\alpha & 0\\ -\sin\alpha & \cos\alpha & 0\\ 0 & 0 & e^\beta\end{bmatrix}$, then $A(\alpha,\beta)^{-1}$ in terms of function of A is

a. $A(\alpha,-\beta)$ b. $(-\alpha,-\beta)$

c. $A(-\alpha,\beta)$ d. none of these

58. If A and B are two square matrices such that $B=-A^{-1}BA$, then $(A+B)^2$ is equal to

a. A^2+B^2 b. O

c. $A^2+2AB+B^2$ d. $A+B$

59. Let a and b be two real numbers such that $a>1$, $b>1$. If $A=\begin{pmatrix}a & 0\\ 0 & b\end{pmatrix}$, then $\lim_{n\to\infty}A^{-n}$ is

a. unit matrix b. null matrix

c. $2I$ d. none of these

60. Let $f(x)=\dfrac{1+x}{1-x}$. If A is matrix for which $A^3=O$, then $f(A)$ is

a. $I+A+A^2$ b. $I+2A+2A^2$

c. $I-A-A^2$ d. none of these

61. If A and B are two non-zero square matrices of the same order such that the product $AB=O$, then

a. both A and B must be singular

b. exactly one of them must be singular

c. both of them are non-singular

d. none of these

62. If $\begin{bmatrix}2 & 1\\ 3 & 2\end{bmatrix}A\begin{bmatrix}-3 & 2\\ 5 & -3\end{bmatrix}=\begin{bmatrix}1 & 0\\ 0 & 1\end{bmatrix}$, then $A=$

a. $\begin{bmatrix}1 & 1\\ 1 & 0\end{bmatrix}$ b. $\begin{bmatrix}1 & 1\\ 0 & 1\end{bmatrix}$

c. $\begin{bmatrix}1 & 0\\ 1 & 1\end{bmatrix}$ d. $-\begin{bmatrix}1 & 1\\ 1 & 0\end{bmatrix}$

63. If $A^2-A+I=0$, then the inverse of A is

a. A^{-2} b. $A+I$

c. $I-A$ d. $A-I$

64. The number of solutions of the matrix equation $X^2 = \begin{bmatrix} 1 & 1 \\ 2 & 3 \end{bmatrix}$ is

a. more than 2 **b.** 2
c. 0 **d.** 1

65. If A and B are symmetric matrices of the same order and $X = AB + BA$ and $Y = AB - BA$, then $(XY)^T$ is equal to

a. XY **b.** YX
c. $-YX$ **d.** none of these

66. If A is a 3×3 skew-symmetric matrix, then trace of A is equal to

a. -1 **b.** 1
c. $|A|$ **d.** none of these

67. Elements of a matrix A of order 10×10 are defined as $a_{ij} = w^{i+j}$ (where w is cube root of unity), then trace (A) of the matrix is

a. 0 **b.** 1
c. 3 **d.** none of these

68. Let $F(\alpha) = \begin{bmatrix} \cos\alpha & -\sin\alpha & 0 \\ \sin\alpha & \cos\alpha & 0 \\ 0 & 0 & 1 \end{bmatrix}$, where $\alpha \in R$. Then $(F(\alpha))^{-1}$ is equal to

a. $F(\alpha^{-1})$ **b.** $F(-\alpha)$
c. $F(2\alpha)$ **d.** none of these

69. If $F(x) = \begin{bmatrix} \cos x & -\sin x & 0 \\ \sin x & \cos x & 0 \\ 0 & 0 & 1 \end{bmatrix}$ and $G(y) = \begin{bmatrix} \cos y & 0 & \sin y \\ 0 & 1 & 0 \\ -\sin y & 0 & \cos y \end{bmatrix}$, then $[F(x)\, G(y)]^{-1}$ is equal to

a. $F(-x)\, G(-y)$ **b.** $G(-y)\, F(-x)$
c. $F(x^{-1})\, G(y^{-1})$ **d.** $G(y^{-1})\, F(x^{-1})$

70. If A is a skew-symmetric matrix and n is odd positive integer, then A^n is

a. a skew-symmetric matrix **b.** a symmetric matrix
c. a diagonal matrix **d.** none of these

71. If $A = \begin{bmatrix} a & b \\ c & d \end{bmatrix}$ (where $bc \neq 0$) satisfies the equations $x^2 + k = 0$, then

a. $a + d = 0$ **b.** $k = -|A|$
c. $k = |A|$ **d.** none of these

72. If $A, B, A + I, A + B$ are idempotent matrices, then AB is equal to

a. BA **b.** $-BA$
c. I **d.** O

73 The matrix $A = \begin{bmatrix} -5 & -8 & 0 \\ 3 & 5 & 0 \\ 1 & 2 & -1 \end{bmatrix}$ is

a. idempotent matrix **b.** involutory matrix
c. nilpotent matrix **d.** none of these

74. If A is symmetric as well as skew-symmetric matrix, then A is

a. diagonal matrix **b.** null matrix
c. triangular matrix **d.** none of these

75. If A and B are square matrices of the same order and A is non-singular, then for a positive integer n, $(A^{-1}BA)^n$ is equal to

a. $A^{-n}B^nA^n$ **b.** $A^nB^nA^{-n}$
c. $A^{-1}B^nA$ **d.** $n(A^{-1}BA)$

76. Which of the following is an orthogonal matrix?

a. $\begin{bmatrix} 6/7 & 2/7 & -3/7 \\ 2/7 & 3/7 & 6/7 \\ 3/7 & -6/7 & 2/7 \end{bmatrix}$ **b.** $\begin{bmatrix} 6/7 & 2/7 & 3/7 \\ 2/7 & -3/7 & 6/7 \\ 3/7 & 6/7 & -2/7 \end{bmatrix}$

c. $\begin{bmatrix} -6/7 & -2/7 & -3/7 \\ 2/7 & 3/7 & 6/7 \\ -3/7 & 6/7 & 2/7 \end{bmatrix}$ **d.** $\begin{bmatrix} 6/7 & -2/7 & 3/7 \\ 2/7 & 2/7 & -3/7 \\ -6/7 & 2/7 & 3/7 \end{bmatrix}$

77. If $k \in R_0$, then $\det\{\text{adj}(kI_n)\}$ is equal to

a. k^{n-1} **b.** $k^{n(n-1)}$
c. k^n **d.** k

78. Given that matrix $A = \begin{bmatrix} x & 3 & 2 \\ 1 & y & 4 \\ 2 & 2 & z \end{bmatrix}$. If $xyz = 60$ and $8x + 4y + 3z = 20$, then $A(\text{adj } A)$ is equal to

a. $\begin{bmatrix} 64 & 0 & 0 \\ 0 & 64 & 0 \\ 0 & 0 & 64 \end{bmatrix}$ **b.** $\begin{bmatrix} 88 & 0 & 0 \\ 0 & 88 & 0 \\ 0 & 0 & 88 \end{bmatrix}$

c. $\begin{bmatrix} 68 & 0 & 0 \\ 0 & 68 & 0 \\ 0 & 0 & 68 \end{bmatrix}$ **d.** $\begin{bmatrix} 34 & 0 & 0 \\ 0 & 34 & 0 \\ 0 & 0 & 34 \end{bmatrix}$

79. Let $A + 2B = \begin{bmatrix} 1 & 2 & 0 \\ 6 & -3 & 3 \\ -5 & 3 & 1 \end{bmatrix}$ and $2A - B = \begin{bmatrix} 2 & -1 & 5 \\ 2 & -1 & 6 \\ 0 & 1 & 2 \end{bmatrix}$.

Then $\text{tr}(A) - \text{tr}(B)$ has the value equal to

a. 0 **b.** 1
c. 2 **d.** none

80. If $A_1, A_3, \ldots, A_{2n-1}$ are n skew symmetric matrices of same order, then $B = \sum_{r=1}^{n} (2r-1)(A_{2r-1})^{2r-1}$ will be

a. symmetric
b. skew-symmetric
c. neither symmetric nor skew-symmetric
d. data not adequate

81. Let $A=\begin{bmatrix}1 & 2 & 3\\ 2 & 0 & 5\\ 0 & 2 & 1\end{bmatrix}$ and $B=\begin{bmatrix}0\\ -3\\ 1\end{bmatrix}$. Which of the following is true?

a. $AX=B$ has a unique solution
b. $AX=B$ has exactly three solutions
c. $AX=B$ has infinitely many solutions
d. $AX=B$ is inconsistent

82. Consider three matrices $A=\begin{bmatrix}2 & 1\\ 4 & 1\end{bmatrix}$, $B=\begin{bmatrix}3 & 4\\ 2 & 3\end{bmatrix}$, and $C=\begin{bmatrix}3 & -4\\ -2 & 3\end{bmatrix}$. Then the value of the sum $\operatorname{tr}(A)+\operatorname{tr}\left(\frac{ABC}{2}\right)+\operatorname{tr}\left(\frac{A(BC)^2}{4}\right)+\operatorname{tr}\left(\frac{A(BC)^3}{8}\right)+\cdots+\infty$ is

a. 6 **b.** 9
c. 12 **d.** none

Multiple Correct Answers Type

Each question has four choices a, b, c and d, out of which *one or more* answers are correct.

1. If A is unimodular, then which of the following is unimodular?

a. $-A$
b. A^{-1}
c. adj A
d. ωA, where ω is cube root of unity

2. If $A=\begin{bmatrix}1 & -1\\ 2 & 1\end{bmatrix}$, $B=\begin{bmatrix}a & 1\\ b & -1\end{bmatrix}$, and $(A+B)^2=A^2+B^2+2AB$, then

a. $a=-1$ **b.** $a=1$
c. $b=2$ **d.** $b=-2$

3. If $AB=A$ and $BA=B$, then which of the following is/are true?

a. A is idempotent **b.** B is idempotent
c. A^T is idempotent **d.** none of these

4. If $A=\frac{1}{3}\begin{bmatrix}1 & 2 & 2\\ 2 & 1 & -2\\ a & 2 & b\end{bmatrix}$ is an orthogonal matrix, then

a. $a=-2$ **b.** $a=2, b=1$
c. $b=-1$ **d.** $b=1$

5. Let A and B are two non-singular square matrices, A^T and B^T are the transpose matrices of A and B, respectively, then which of the following are correct?

a. B^TAB is symmetric matrix if A is symmetric
b. B^TAB is symmetric matrix if B is symmetric
c. B^TAB is skew-symmetric matrix for every matrix A
d. B^TAB is skew-symmetric matrix if A is skew-symmetric

6. If $A(\theta)=\begin{bmatrix}\sin\theta & i\cos\theta\\ i\cos\theta & \sin\theta\end{bmatrix}$, then which of the following is not true?

a. $A(\theta)^{-1}=A(\pi-\theta)$
b. $A(\theta)+A(\pi+\theta)$ is a null matrix
c. $A(\theta)$ is invertible for all $\theta\in R$
d. $A(\theta)^{-1}=A(-\theta)$

7. If A is a matrix such that $A^2+A+2I=O$, then which of the following is/are true?

a. A is non-singular
b. A is symmetric
c. A cannot be skew-symmetric
d. $A^{-1}=-\frac{1}{2}(A+I)$

8. If $A=\begin{bmatrix}3 & -3 & 4\\ 2 & -3 & 4\\ 0 & -1 & 1\end{bmatrix}$, then

a. adj(adj A) = A **b.** $|$adj(adj A)$|=1$
c. $|$adj $A|=1$ **d.** one of these

9. If $\begin{pmatrix}1 & -\tan\theta\\ \tan\theta & 1\end{pmatrix}\begin{pmatrix}1 & \tan\theta\\ -\tan\theta & 1\end{pmatrix}^{-1}=\begin{bmatrix}a & -b\\ b & a\end{bmatrix}$, then

a. $a=\cos 2\theta$ **b.** $a=1$
c. $b=\sin 2\theta$ **d.** $b=-1$

10. If $A^{-1}=\begin{bmatrix}1 & -1 & 2\\ 0 & 3 & 1\\ 0 & 0 & -1/3\end{bmatrix}$, then

a. $|A|=-1$ **b.** adj $A=\begin{bmatrix}-1 & 1 & -2\\ 0 & -3 & -1\\ 0 & 0 & 1/3\end{bmatrix}$

c. $A=\begin{bmatrix}1 & 1/3 & 7\\ 0 & 1/3 & 1\\ 0 & 0 & -3\end{bmatrix}$ **d.** $A=\begin{bmatrix}1 & -1/3 & -7\\ 0 & -3 & 0\\ 0 & 0 & 1\end{bmatrix}$

11. If B is an idempotent matrix, and $A=I-B$, then

a. $A^2=A$ **b.** $A^2=I$
c. $AB=O$ **d.** $BA=O$

12. Which of the following statements is/are true about square matrix A of order n?

a. $(-A)^{-1}$ is equal to $-A^{-1}$ when n is odd only.
b. If $A^n=O$, then $I+A+A^2+\cdots+A^{n-1}=(I-A)^{-1}$.
c. If A is skew-symmetric matrix of odd order, then its inverse does not exist.
d. $(A^T)^{-1}=(A^{-1})^T$ holds always.

13. If $A_1 = \begin{bmatrix} 0 & 0 & 0 & 1 \\ 0 & 0 & 1 & 0 \\ 0 & 1 & 0 & 0 \\ 1 & 0 & 0 & 0 \end{bmatrix}$, $A_2 = \begin{bmatrix} 0 & 0 & 0 & i \\ 0 & 0 & -i & 0 \\ 0 & i & 0 & 0 \\ -i & 0 & 0 & 0 \end{bmatrix}$, then $A_iA_k + A_kA_i$ is equal to

a. $2I$ if $i = k$ b. O if $i \neq k$
c. $2I$ if $i \neq k$ d. O always

14. If $A = (a_{ij})_{n \times n}$ and f is a function, we define $f(A) = (f(a_{ij}))_{n \times n}$.
Let $A = \begin{pmatrix} \pi/2-\theta & \theta \\ -\theta & \pi/2-\theta \end{pmatrix}$. Then

a. $\sin A$ is invertible b. $\sin A = \cos A$
c. $\sin A$ is orthogonal d. $\sin(2A) = 2\sin A \cos A$

15. If $\begin{bmatrix} 0 & 2b & c \\ a & b & -c \\ a & -b & c \end{bmatrix}$ is orthogonal, then

a. $a = \pm\frac{1}{\sqrt{2}}$ b. $b = \pm\frac{1}{\sqrt{12}}$
c. $c = \pm\frac{1}{\sqrt{3}}$ d. none of these

16. If A and B are two invertible matrices of the same order, then adj(AB) is equal to

a. adj(B) adj(A) b. $|B|\,|A|\,B^{-1}A^{-1}$
c. $|B|\,|A|\,A^{-1}B^{-1}$ d. $|A|\,|B|\,(AB)^{-1}$

17. If A, B, and C are three square matrices of the same order, then $AB = AC \Rightarrow B = C$. Then

a. $|A| \neq 0$ b. A is invertible
c. A may be orthogonal d. A is symmetric

18. Suppose $a_1, a_2, \ldots$ are real numbers, with $a_1 \neq 0$. If $a_1, a_2, a_3, \ldots$ are in A.P., then

a. $A = \begin{bmatrix} a_1 & a_2 & a_3 \\ a_4 & a_5 & a_6 \\ a_5 & a_6 & a_7 \end{bmatrix}$ is singular (where $i = \sqrt{-1}$)

b. the system of equations $a_1x + a_2y + a_3z = 0$, $a_4x + a_5y + a_6z = 0$, $a_7x + a_8y + a_9z = 0$ has an infinite number of solutions

c. $B = \begin{bmatrix} a_1 & ia_2 \\ ia_2 & a_1 \end{bmatrix}$ is non-singular

d. none of these

19. If α, β, γ are three real numbers and

$$A = \begin{bmatrix} 1 & \cos(\alpha-\beta) & \cos(\alpha-\gamma) \\ \cos(\beta-\alpha) & 1 & \cos(\beta-\gamma) \\ \cos(\gamma-\alpha) & \cos(\gamma-\beta) & 1 \end{bmatrix},$$ then which of following is/are true?

a. A is singular b. A is symmetric
c. A is orthogonal d. A is not invertible

20. Let $A = \begin{bmatrix} 1 & 0 \\ 1 & 1 \end{bmatrix}$. Then which of following is not true?

a. $\lim_{n\to\infty} \frac{1}{n^2} A^{-n} = \begin{bmatrix} 0 & 0 \\ -1 & 0 \end{bmatrix}$ b. $\lim_{n\to\infty} \frac{1}{n} A^{-n} = \begin{bmatrix} 0 & 0 \\ -1 & 0 \end{bmatrix}$

c. $A^{-n} = \begin{bmatrix} 1 & 0 \\ -n & 1 \end{bmatrix} \forall n \neq N$ d. none of these

21. If C is skew-symmetric matrix of order n and X is $n \times 1$ column matrix, then X^TCX is

a. singular b. non-singular
c. invertible d. non-invertible

22. If $S = \begin{bmatrix} 0 & 1 & 1 \\ 1 & 0 & 1 \\ 1 & 1 & 0 \end{bmatrix}$ and $A = \begin{bmatrix} b+c & c+a & b-c \\ c-b & c+b & a-b \\ b-c & a-c & a+b \end{bmatrix}$ $(a, b, c \neq 0)$, then SAS^{-1} is

a. symmetric matrix b. diagonal matrix
c. invertible matrix d. singular matrix

23. If D_1 and D_2 are two 3×3 diagonal matrices, then which of the following is/are true?

a. D_1D_2 is diagonal matrix
b. $D_1D_2 = D_2D_1$
c. $D_1^2 + D_2^2$ is a diagonal matrix
d. none of these

24. If A and B are symmetric and commute, then which of the following is/are symmetric?

a. $A^{-1}B$ b. AB^{-1}
c. $A^{-1}B^{-1}$ d. None of these

25. A skew-symmetric matrix A satisfies the relation $A^2 + I = O$, where I is a unit matrix then A is

a. idempotent b. orthogonal
c. of even order d. odd order

26. If $AB = A$ and $BA = B$, then

a. $A^2B = A^2$ b. $B^2A = B^2$
c. $ABA = A$ d. $BAB = B$

27. Let $A = \begin{bmatrix} 1 & 2 & 2 \\ 2 & 1 & 2 \\ 2 & 2 & 1 \end{bmatrix}$. Then

a. $A^2 - 4A - 5I_3 = O$ b. $A^{-1} = \frac{1}{5}(A - 4I_3)$
c. A^3 is not invertible d. A^2 is invertible

Reasoning Type

Each question has four choices a, b, c and d, out of which *only one* is correct. Each question contains Statement 1 and Statement 2.

a. Both the statements are True and Statement 2 is the correct explanation of Statement 1.

b. Both the statements are True but Statement 2 is Not the correct explanation of Statement 1.

c. Statement 1 is True and Statement 2 is False.

d. Statement 1 is False and Statement 2 is True.

1. **Statement 1:** $|\text{adj}(\text{adj}(\text{adj } A))| = |A|^{(n-1)^3}$, where n is order of matrix A.

   Statement 2: $|\text{adj } A| = |A|^n$.

2. **Statement 1:** If $D = \text{diag } [d_1, d_2, \ldots, d_n]$, then $D^{-1} = \text{diag } [d_1^{-1}, d_2^{-1} \ldots, d_n^{-1}]$.

   Statement 2: If $D = \text{diag } [d_1, d_2, \ldots, d_n]$, then $D^n = \text{diag } [d_1^n, d_2^n \ldots, d_n^n]$.

3. **Statement 1:** Matrix 3×3, $a_{ij} = \dfrac{i-j}{i+2j}$ cannot be expressed as a sum of symmetric and skew-symmetric matrix.

   Statement 2: Matrix 3×3, $a_{ij} = \dfrac{i-j}{i+2j}$ is neither symmetric nor skew-symmetric.

4. **Statement 1:** If a, b, c, d are real numbers and $A = \begin{bmatrix} a & b \\ c & d \end{bmatrix}$ and $A^3 = O$, then $A^2 = O$.

   Statement 2: For matrix $A = \begin{bmatrix} a & b \\ c & d \end{bmatrix}$, we have $A^2 - (a+d)A + (ad - bc)I = O$.

5. **Statement 1:** If $F(\alpha) = \begin{bmatrix} \cos\alpha & -\sin\alpha & 0 \\ \sin\alpha & \cos\alpha & 0 \\ 0 & 0 & 1 \end{bmatrix}$, then $[F(\alpha)]^{-1} = F(-\alpha)$.

   Statement 2: For matrix $G(\beta) = \begin{bmatrix} \cos\beta & 0 & \sin\beta \\ 0 & 1 & 0 \\ -\sin\beta & 0 & \cos\beta \end{bmatrix}$, we have $[G(\beta)]^{-1} = G(-\beta)$.

6. **Statement 1:** $A = \begin{bmatrix} 4 & 0 & 4 \\ 2 & 3 & 2 \\ 1 & 2 & 1 \end{bmatrix}$, $B^{-1} = \begin{bmatrix} 1 & 3 & 3 \\ 1 & 4 & 3 \\ 1 & 3 & 4 \end{bmatrix}$. Then $(AB)^{-1}$ does not exist.

   Statement 2: Since $|A| = 0$, $(AB)^{-1} = B^{-1}A^{-1}$ is meaningless.

7. **Statement 1:** The determinant of a matrix $A = [a_{ij}]_{5 \times 5}$ where $a_{ij} + a_{ji} = 0$ for all i and j is zero.

   Statement 2: The determinant of a skew-symmetric matrix of odd order is zero.

8. **Statement 1:** The inverse of the matrix $A = [a_{ij}]_{n \times n}$ where $a_{ij} = 0$, $i \ge j$ is $B = [a_{ij}^{-1}]_{n \times n}$.

   Statement 2: The inverse of singular matrix does not exist.

9. **Statement 1:** For a singular square matrix A, $AB = AC \Rightarrow B = C$.

   Statement 2: If $|A| = 0$, then A^{-1} does not exist.

10. **Statement 1:** If A, B, C are matrices such that $|A_{3\times3}| = 3$, $|B_{3\times3}| = -1$ and $|C_{2\times2}| = +2$, then $|2ABC| = -12$.

    Statement 2: For matrices A, B, C of the same order, $|ABC| = |A|\,|B|\,|C|$.

11. **Statement 1:** If the matrices A, B, $(A + B)$ are non-singular, then $[A(A + B)^{-1} B]^{-1} = B^{-1} + A^{-1}$.

    Statement 2:
    $$\begin{aligned}[A(A + B)^{-1} B]^{-1} &= [A(A^{-1} + B^{-1})B]^{-1} \\ &= [(I + AB^{-1})B]^{-1} \\ &= [(B + AB^{-1}B)]^{-1} \\ &= [(B + AI)]^{-1} \\ &= [(B + A)]^{-1} \\ &= B^{-1} + A^{-1}.\end{aligned}$$

12. **Statement 1:** Let A, B be two square matrices of the same order such that $AB = BA$, $A^m = O$ and $B^n = O$ for some positive integers m, n, then there exists a positive integer r such that $(A + B)^r = O$.

    Statement 2: If $AB = BA$ then $(A + B)^r$ can be expanded as binomial expansion.

13. **Statement 1:** If $A = [a_{ij}]_{n \times n}$ is such that $a_{ij} = \bar{a}_{ji}, \forall\, i, j$ and $A^2 = O$, then matrix A null matrix.

    Statement 2: $|A| = 0$.

14. **Statement 1:** If A is orthogonal matrix of order 2, then $|A| = \pm 1$.

    Statement 2: Every two-rowed real orthogonal matrix is of any one of the forms $\begin{pmatrix} \cos\theta & -\sin\theta \\ \sin\theta & \cos\theta \end{pmatrix}$ or $\begin{pmatrix} \cos\theta & \sin\theta \\ \sin\theta & -\cos\theta \end{pmatrix}$.

Linked Comprehension Type

Based upon each paragraph, some multiple choice questions have to be answered. Each question has four choices a, b, c and d, out of which *only one* is correct.

For Problems 1–3

Let A be matrix of order 2×2 such that $A^2 = O$.

1. $A^2 - (a + d)A + (ad - bc)I$ is equal to

   a. I b. O

   c. $-I$ d. none of these

2. tr(A) is equal to

   a. 1 b. 0

   c. -1 d none of these

3. $(I + A)^{100} =$

   a. $100A$ b. $100(I + A)$

   c. $100I + A$ d. $I + 100A$

For Problems 4–6

If A and B are two square matrices of order 3×3 which satisfy $AB = A$ and $BA = B$, then

4. Which of the following is true?

   a. If matrix A is singular then matrix B is non-singular.

   b. If matrix A is non-singular then matrix B is singular.

   c. If matrix A is singular then matrix B is also singular.

   d Cannot say anything.

5. $(A + B)^7$ is equal to

a. $7(A + B)$ b. $7 \times I_{3\times 3}$

c. $64\,(A + B)$ d. $8\,I$

6. $(A + I)^5$ is equal to (where I is identity matrix)

a. $I + 60I$ b. $I + 16A$

c. $I + 31A$ d. none of these

For Problems 7–8

Consider an arbitrary 3×3 matrix $A = [a_{ij}]$, a matrix $B = [b_{ij}]$ is formed such that b_{ij} is the sum of all the elements except a_{ij} in the ith row of A. Answer the following questions.

7. If there exists a matrix X with constant elements such that $AX = B$, then X is

a. skew-symmetric b. null matrix

c. diagonal matrix d. none of these

8. The value of $|B|$ is equal to

a. $|A|$ b. $|A|/2$

c. $2|A|$ d. none of these

For Problems 9–11

Let $A = \begin{bmatrix} 1 & 0 & 0 \\ 1 & 0 & 1 \\ 0 & 1 & 0 \end{bmatrix}$ satisfies $A^n = A^{n-2} + A^2 - I$ for $n \geq 3$. And trace of a square matrix X is equal to the Sum of elements in its principal diagonal.

Further consider a matrix $\cup_{3\times 3}$ with its column as $\cup_1, \cup_2, \cup_3$ such that

$$A^{50}\cup_1 = \begin{bmatrix} 1 \\ 25 \\ 25 \end{bmatrix},\ A^{50}\cup_2 = \begin{bmatrix} 0 \\ 1 \\ 0 \end{bmatrix},\ A^{50}\cup_3 = \begin{bmatrix} 0 \\ 0 \\ 1 \end{bmatrix}$$

Then answer the following questions.

9. The values of $|A^{50}|$ equals

a. 0 b. 1

c. –1 d. 25

10. Trace of A^{50} equals

a. 0 b. 1

c. 2 d. 3

11. The value of $|\cup|$ equals

a. 0 b. 1

c. 2 d. –1

For Problems 12–14

Let A be a square matrix of order 2 or 3 and I be the identity matrix of the same order. Then the matrix $A - \lambda I$ is called characteristic matrix of the matrix A, where λ is some complex number. The determinant of the characteristic matrix is called characteristic determinant of the matrix A which will of course be a polynomial of degree 3 in λ. The equation $\det(A - \lambda I) = 0$ is called characteristic equation of the matrix A and its roots (the values of λ) are called characteristic roots or eigenvalues. It is also known that every square matrix has its characteristic equation.

12. The eigenvalues of the matrix $A = \begin{bmatrix} 2 & 1 & 1 \\ 2 & 3 & 4 \\ -1 & -1 & -2 \end{bmatrix}$ are

a. 2, 1, 1 b. 2, 3, –2

c. –1, 1, 3 d. none of these

13. Which of the following matrices do not have eigenvalues as 1 and –1?

a. $\begin{bmatrix} 0 & 1 \\ 1 & 0 \end{bmatrix}$ b. $\begin{bmatrix} 0 & -i \\ i & 0 \end{bmatrix}$ (where $i = \sqrt{-1}$)

c. $\begin{bmatrix} 1 & 0 \\ 0 & -1 \end{bmatrix}$ d. $\begin{bmatrix} 1 & 0 \\ 0 & 1 \end{bmatrix}$

14. If one of the eigenvalues of a square matrix A of order 3×3 is zero, then

a. det A must be non-zero b. det A must be zero

c. adj A must be a zero matrix d. none of these

For Problems 15–17

Let A be a $m \times n$ matrix. If there exists a matrix L of type $n \times m$ such that $LA = I_n$, then L is called left inverse of A. Similarly, if there exists a matrix R of type $n \times m$ such that $AR = I_m$, then R is called right inverse of A.

For example, to find right inverse of matrix

$$A = \begin{bmatrix} 1 & -1 \\ 1 & 1 \\ 2 & 3 \end{bmatrix}, \text{ we take } R = \begin{bmatrix} x & y & z \\ u & v & w \end{bmatrix}$$

and solve $AR = I_3$, i.e.,

$$\begin{bmatrix} 1 & -1 \\ 1 & 1 \\ 2 & 3 \end{bmatrix}\begin{bmatrix} x & y & z \\ u & v & w \end{bmatrix} = \begin{bmatrix} 1 & 0 & 0 \\ 0 & 1 & 0 \\ 0 & 0 & 1 \end{bmatrix}$$

$\Rightarrow x - u = 1$ $\quad y - v = 0$ $\quad z - w = 0$

$x + u = 0$ $\quad y + v = 1$ $\quad z + w = 0$

$2x + 3u = 0$ $\quad 2y + 3v = 0$ $\quad 2z + 3w = 1$

As this system of equations is inconsistent, we say there is no right inverse for matrix A.

15. Which of the following matrices is NOT left inverse of matrix $\begin{bmatrix} 1 & -1 \\ 1 & 1 \\ 2 & 3 \end{bmatrix}$?

a. $\begin{bmatrix} \frac{1}{2} & \frac{1}{2} & 0 \\ -\frac{1}{2} & \frac{1}{2} & 0 \end{bmatrix}$ b. $\begin{bmatrix} 2 & -7 & 3 \\ -\frac{1}{2} & \frac{1}{2} & 0 \end{bmatrix}$

c. $\begin{bmatrix} -\frac{1}{2} & \frac{1}{2} & 0 \\ -\frac{1}{2} & \frac{1}{2} & 0 \end{bmatrix}$ d. $\begin{bmatrix} 0 & 3 & -1 \\ -\frac{1}{2} & \frac{1}{2} & 0 \end{bmatrix}$

16. The number of right inverses for the matrix $\begin{bmatrix} 1 & -1 & 2 \\ 2 & -1 & 1 \end{bmatrix}$ is

a. 0 b. 1

c. 2 d. infinite

17. For which of the following matrices, the number of left inverses is greater than the number of right inverses?

a. $\begin{bmatrix} 1 & 2 & 4 \\ -3 & 2 & 1 \end{bmatrix}$ b. $\begin{bmatrix} 3 & 2 & 1 \\ 3 & 2 & 1 \end{bmatrix}$

c. $\begin{bmatrix} 1 & 4 \\ 2 & -3 \\ 5 & 4 \end{bmatrix}$ d. $\begin{bmatrix} 3 & 3 \\ 1 & 1 \\ 4 & 4 \end{bmatrix}$

For Problems 18–20

If e^A is defined as $e^A = I + A + \frac{A^2}{2!} + \frac{A^3}{3!} + \cdots = \frac{1}{2}\begin{bmatrix} f(x) & g(x) \\ g(x) & f(x) \end{bmatrix}$ where $A = \begin{bmatrix} x & x \\ x & x \end{bmatrix}$ and $0 < x < 1$, then I is an identity matrix.

18. $\int \frac{g(x)}{f(x)} dx$ is equal to

a. $\log(e^x + e^{-x}) + c$ b. $\log(e^x - e^{-x}) + c$

c. $\log(e^{2x} - 1) + c$ d. none of these

19. $\int (g(x) + 1)\sin x dx$ is equal to

a. $\frac{e^x}{2}(\sin x - \cos x) + c$ b. $\frac{e^{2x}}{5}(2\sin x - \cos x) + c$

c. $\frac{e^x}{5}(\sin 2x - \cos 2x) + c$ d. none of these

20. $\int \frac{f(x)}{\sqrt{g(x)}} dx$ is equal to

a. $\frac{1}{2\sqrt{e^x - 1}} - \operatorname{cosec}^{-1}(e^x) + c$

b. $\frac{2}{\sqrt{e^x - e^{-x}}} - \sec^{-1}(e^x) + c$

c. $\frac{1}{2\sqrt{e^{2x} - 1}} + \sec^{-1}(e^x) + c$

d. none of these

Matrix-Match Type

Each question contains statements given in two columns which have to be matched. Statements a, b, c, d in column I have to be matched with statements p, q, r, s in column II. If the correct matches are a → p, a → s, b → q, b → r, c → p, c → q and d → s, then the correctly bubbled 4 × 4 matrix should be as follows:

	p	q	r	s
a	(p) filled	(q)	(r)	(s) filled
b	(p)	(q) filled	(r) filled	(s)
c	(p) filled	(q) filled	(r)	(s)
d	(p)	(q)	(r)	(s) filled

1.

Column I	Column II
a. $(I - A)^n$ is if A is idempotent	p. $2^{n-1}(I - A)$
b. $(I - A)^n$ is if A is involuntary	q. $I - nA$
c. $(I - A)^n$ is if A is nilpotent of index 2	r. A
d. If A is orthogonal, then $(A^T)^{-1}$	s. $I - A$

2.

Column I	Column II
a. If A is an idempotent matrix and I is an identity matrix of the same order, then the value of n, such that $(A + I)^n = I + 127$ is	p. 9
b. If $(I - A)^{-1} = I + A + A^2 + \cdots + A^7$, then $A^n = O$ where n is	q. 10
c. If A is matrix such that $a_{ij} = (i + j)(i - j)$, then A is singular if order of matrix is	r. 7
d. If a non-singular matrix A is symmetric, show that A^{-1} is also symmetric, then order of A can be	s. 8

3.

Column I (A, B, C are matrices)	Column II
a. If $\lvert A\rvert = 2$, then $\lvert 2A^{-1}\rvert =$ (where A is of order 3)	p. 1
b. If $\lvert A\rvert = 1/8$, then $\lvert \operatorname{adj}(\operatorname{adj}(2A))\rvert =$ (where A is of order 3)	q. 4
c. If $(A + B)^2 = A^2 + B^2$, and $\lvert A\rvert = 2$, then $\lvert B\rvert =$ (where A and B are of odd order)	r. 24
d. $\lvert A_{2\times 2}\rvert = 2$, $\lvert B_{3\times 3}\rvert = 3$ and $\lvert C_{4\times 4}\rvert = 4$, then $\lvert ABC\rvert$ is equal to	s. 0
	t. does not exist

Integer Type

1. If $A^2 = 3A - 2I$, and $A^8 = pA + qI$, then $p + q$ is____.

2. The value of $1/x$ for which the matrix product $\begin{bmatrix} 2 & 0 & 7 \\ 0 & 1 & 0 \\ 1 & -2 & 1 \end{bmatrix}\begin{bmatrix} -x & 14x & 7x \\ 0 & 1 & 0 \\ x & -4x & -2x \end{bmatrix}$ equals an identity matrix is____.

3. Let $A = \begin{bmatrix} 1 & 2 \\ 3 & 4 \end{bmatrix}$ and $B = \begin{bmatrix} a & b \\ c & d \end{bmatrix}$ be two matrices such that they are commutative and $c \neq 3b$, then the value of $\left|\frac{a-d}{3b-c}\right|$ is____.

4. If $A = \begin{bmatrix} 1 & -1 \\ 2 & -1 \end{bmatrix}$, $B = \begin{bmatrix} a & 1 \\ b & -1 \end{bmatrix}$, and $(A + B)^2 = A^2 + B^2$, then the value of $\lvert a + b\rvert$ is____.

5. If $A = \begin{bmatrix} 1 & 2 & 2 \\ 2 & 1 & -2 \\ a & 2 & b \end{bmatrix}$ is a matrix satisfying $AA^T = 9I_3$, then the value of $|a + b|$ is_____.

6. Let $A = \begin{bmatrix} 1 & -1 & 1 \\ 2 & 1 & -3 \\ 1 & 1 & 1 \end{bmatrix}$ and $(10)\,B = \begin{bmatrix} 4 & 2 & 2 \\ -5 & 0 & \alpha \\ 1 & -2 & 3 \end{bmatrix}$. If B is the inverse of A, then the value of α is _____.

7. If A, B, and C are $n \times n$ matrix and $\det(A) = 2$, $\det(B) = 3$, and $\det(C) = 5$, then the value of $[\det(A^2BC^{-1})]$ (where [.] represents greatest integer function) is _____.

8. Let $A = \begin{bmatrix} 1 & 0 & 0 \\ 2 & 1 & 0 \\ 3 & 2 & 1 \end{bmatrix}$, and U_1, U_2, and U_3 are columns of a 3×3 matrix U. If column matrices U_1, U_2, and U_3 satisfy $AU_1 = \begin{bmatrix} 1 \\ 0 \\ 0 \end{bmatrix}$, $AU_2 = \begin{bmatrix} 2 \\ 3 \\ 0 \end{bmatrix}$, $AU_3 = \begin{bmatrix} 2 \\ 3 \\ 1 \end{bmatrix}$. The value of $|U|$ is_____.

9. Let A be the set of all 3×3 symmetric matrices all of whose entries are either 0 or 1. Five of these entries are 1 and four of them are 0. If n is the number of such matrices, then $n/2$ is_____.

10. If matrix $A = \begin{bmatrix} a & b & c \\ b & c & a \\ c & a & b \end{bmatrix}$, where a, b, c are real positive numbers, $abc = 1$, and $A^TA = I$, then the greatest value of $a^3 + b^3 + c^3$ is_____.

11. If $\begin{bmatrix} 1 & 1 & 1 \\ 2 & 5 & 7 \\ 2 & 1 & -1 \end{bmatrix} \begin{bmatrix} x_1 & y_1 \\ x_2 & y_2 \\ x_3 & y_3 \end{bmatrix} = \begin{bmatrix} 9 & 2 \\ 52 & 15 \\ 0 & 1 \end{bmatrix}$, then the value of $x_1 + x_2 + x_3 - y_1 - y_2 - y_3$ is_____.

ANSWERS AND SOLUTIONS

Objective Type

1. d. Let A be a skew-symmetric matrix of order n. By definition, $A' = -A$

$\Rightarrow |A'| = |-A|$

$\Rightarrow |A| = (-1)^n |A|$

$\Rightarrow |A| = -|A|$ [$\because n$ is odd]

$\Rightarrow 2|A| = 0$

$\Rightarrow |A| = 0$

Hence, A^{-1} does not exist.

2. d. i. is false.

If $A = \begin{bmatrix} 0 & 1 \\ 0 & -1 \end{bmatrix}$ and $B = \begin{bmatrix} 1 & 1 \\ 0 & 0 \end{bmatrix}$, then $AB = \begin{bmatrix} 0 & 0 \\ 0 & 0 \end{bmatrix} = O$

ii. is true as the product AB is an identity matrix, if and only if B is inverse of the matrix A.

iii. is false since matrix multiplication in not commutative.

3. c. $[1\ x\ y]\begin{bmatrix} 1 & 3 & 1 \\ 0 & 2 & -1 \\ 0 & 0 & 1 \end{bmatrix}\begin{bmatrix} 1 \\ x \\ y \end{bmatrix} = 10$

$$\Rightarrow [1\quad 3+2x\quad 1-x+y]\begin{bmatrix} 1 \\ x \\ y \end{bmatrix} = [0]$$

$\Rightarrow [1 + 3x + 2x^2 + y - xy + y^2] = [0]$

$\Rightarrow 2x^2 + y^2 + y + 3x - xy + 1 = 0$

If $y = 0$, $2x^2 + 3x + 1 = 0$

$\Rightarrow (2x - 1)(x + 1) = 0$

$\Rightarrow x = -1/2, -1$ [rational roots]

If $y = -1$, $2x^2 + 4x + 1 = 0$

$\Rightarrow x = \dfrac{-4 \pm \sqrt{12}}{4} = \dfrac{-2 \pm \sqrt{3}}{2}$ [irrational roots]

4. d. $A = \text{diag}(d_1, d_2, ..., d_n)$

Given, $A^3 = A$

$\Rightarrow \text{diag}(d_1^3, d_2^3, ..., d_n^3) = \text{diag}(d_1, d_2, ..., d_n)$

$\Rightarrow d_1^3 = d_1,\ d_2^3 = d_2, ...,\ d_n^3 = d_n$

Hence, all $d_1, d_2, d_3, ..., d_n$ have three possible values ± 1, 0. Each diagonal element can be selected in three ways. Hence, the number of different matrices is 3^n.

5. d. If A is n^{th} root of I_2, then $A^n = I_2$. Now,

$$A^2 = \begin{bmatrix} a & b \\ 0 & a \end{bmatrix}\begin{bmatrix} a & b \\ 0 & a \end{bmatrix} = \begin{bmatrix} a^2 & 2ab \\ 0 & a^2 \end{bmatrix}$$

$$A^3 = A^2A = \begin{bmatrix} a^2 & 2ab \\ 0 & a^2 \end{bmatrix}\begin{bmatrix} a & b \\ 0 & a \end{bmatrix} = \begin{bmatrix} a^3 & 3a^2b \\ 0 & a^3 \end{bmatrix}$$

Thus,

$$A^n = \begin{bmatrix} a^n & na^{n-1}b \\ 0 & a^n \end{bmatrix}$$

Now,

$$A^n = I \Rightarrow \begin{bmatrix} a^n & na^{n-1}b \\ 0 & a^n \end{bmatrix} = \begin{bmatrix} 1 & 0 \\ 0 & 1 \end{bmatrix}$$

$\Rightarrow a^n = 1, b = 0$

6. d. $A\begin{bmatrix} 1 \\ -1 \end{bmatrix} = \begin{bmatrix} -1 \\ 2 \end{bmatrix}$ (1)

$A^2\begin{bmatrix} 1 \\ -1 \end{bmatrix} = \begin{bmatrix} 1 \\ 0 \end{bmatrix}$ (2)

Let A be given by $A = \begin{bmatrix} a & b \\ c & d \end{bmatrix}$. The first equation gives

$a - b = -1$ (3)

$c - d = 2$ (4)

For second equation gives

$$A^2\begin{bmatrix} 1 \\ -1 \end{bmatrix} = A\left(A\begin{bmatrix} 1 \\ -1 \end{bmatrix}\right) = A\left(\begin{bmatrix} -1 \\ 2 \end{bmatrix}\right) = \begin{bmatrix} 1 \\ 0 \end{bmatrix}$$

This gives

$-a + 2b = 1$ (5)

$-c + 2d = 0$ (6)

Eqs. (3) + (5) $\Rightarrow b = 0$ and $a = -1$

Eqs. (4) + (6) $\Rightarrow d = 2$ and $c = 4$

So the sum $a + b + c + d = 5$.

7. c. $AB = \begin{bmatrix} \cos^2\theta & \cos\theta\sin\theta \\ \cos\theta\sin\theta & \sin^2\theta \end{bmatrix}\begin{bmatrix} \cos^2\phi & \cos\phi\sin\phi \\ \cos\phi\sin\phi & \sin^2\phi \end{bmatrix}$

$$= \begin{bmatrix} \cos^2\theta\cos^2\phi + \cos\theta\cos\phi\sin\theta\sin\phi & \cos^2\theta\cos\phi\sin\phi + \cos\theta\sin\theta\sin^2\phi \\ \cos\theta\sin\theta\cos^2\phi + \cos\theta\sin\theta\cos\phi\sin\phi & \cos\theta\cos\phi\sin\theta\sin\phi + \sin^2\theta\sin^2\phi \end{bmatrix}$$

$$= \begin{bmatrix} \cos\theta\cos\phi(\cos(\theta-\phi)) & \cos\theta\sin\phi(\cos(\theta-\phi)) \\ \sin\theta\cos\phi(\cos(\theta-\phi)) & \sin\theta\sin\phi(\cos(\theta-\phi)) \end{bmatrix}$$

$$= (\cos(\theta-\phi))\begin{bmatrix} \cos\theta\cos\phi & \cos\theta\sin\phi \\ \sin\theta\cos\phi & \sin\theta\sin\phi \end{bmatrix}$$

Now, $AB = O \Rightarrow \cos(\theta - \phi) = 0 \Rightarrow \theta - \phi = (2n + 1)\pi/2$, $n \in Z$.

8. b. $AB^n = ABBBBB\cdots B$

$= (AB)BBB\cdots B$

$= B(AB)BBB\cdots B$

$= BB(AB)BB\cdots B$

$\vdots$

$= B^nA$

$$(AB)^n = (AB)(AB)(AB)\cdots(AB)$$
$$= A(BA)(BA)(BA)\cdots(BA)B$$
$$= A(AB)(AB)(AB)\cdots(AB)B$$
$$= A^2(BA)(BA)(BA)\cdots(BA)B^2$$
$$= A^2(AB)(AB)(AB)\cdots(AB)B^2$$
$$= A^3(BA)(BA)(BA)\cdots(BA)B^3$$
$$\vdots$$
$$= A^nB^n$$

9. a. From given data $|A| = 2^4$

$\Rightarrow |\text{adj (adj } A)| = (2^4)^9 = 2^{36}$

$$\Rightarrow \left\{\frac{\det(\text{adj}(\text{adj } A))}{7}\right\} = \left\{\frac{2^{36}}{7}\right\} = \left\{\frac{(7+1)^{12}}{7}\right\} = \frac{1}{7}$$

10. b. We have,

$$\begin{bmatrix} \alpha & \beta \\ \gamma & -\alpha \end{bmatrix}\begin{bmatrix} \alpha & \beta \\ \gamma & -\alpha \end{bmatrix} = \begin{bmatrix} 1 & 0 \\ 0 & 1 \end{bmatrix}$$

$$\Rightarrow \begin{bmatrix} \alpha^2+\beta\gamma & 0 \\ 0 & \alpha^2+\beta\gamma \end{bmatrix} = \begin{bmatrix} 1 & 0 \\ 0 & 1 \end{bmatrix}$$

$$\Rightarrow \alpha^2 + \beta\gamma - 1 = 0$$

11. c. Given $A^2 = A$. Now,

$$(I + A)^3 - 7A = I^3 + 3I^2A + 3IA^2 + A^3 - 7A$$
$$= I + 3A + 3A + A - 7A$$
$$= I + O$$
$$= I$$

12. a. $(A - \lambda I)(B - \lambda I) = (B - \lambda I)(A - \lambda I)$

$\Rightarrow AB - \lambda(A + B)I + \lambda^2I^2 = BA - \lambda(B + A)I + \lambda^2I^2$

$\Rightarrow AB = BA$

13. c. Given, $A^2 = 2A - I$

Now, $A^3 = A(A^2)$

$$= A(2A - I)$$
$$= 2A^2 - A$$
$$= 2(2A - I) - A$$
$$= 3A - 2I$$

$$A^4 = A(A^3)$$
$$= A(3A - 2I)$$
$$= 3A^2 - 2A$$
$$= 3(2A - I) - 2A$$
$$= 4A - 3I$$

Following this, we can say $A^n = nA - (n - 1)I$.

14. a. We have, $A^2 = O, A^k = O, \forall\, k \geq 2$

Thus,

$$(A + I)^{50} = I + 50A$$
$$\Rightarrow (A + I)^{50} = I + 50A$$
$$\Rightarrow a = 1, b = 0, c = 0, d = 1$$

15. b. We have,

$$A = iB$$

$$\Rightarrow A^2 = (iB)^2 = i^2B^2 = -B^2 = \begin{bmatrix} 2 & -2 \\ -2 & 2 \end{bmatrix} = -2B$$

$$\Rightarrow A^4 = (-2B)^2 = 4B^2 = 4(2B) = 8B$$
$$\Rightarrow (A^4)^2 = (8B)^2$$
$$\Rightarrow A^8 = 64B^2 = 128B$$

16. b. Since the product matrix is 3×3 matrix and the pre-multiplier of A is a 3×2 matrix, therefore A is 2×3 matrix. Let,

$A = \begin{bmatrix} l & m & n \\ x & y & z \end{bmatrix}$. Then the given equation becomes

$$\begin{bmatrix} 2 & -1 \\ 1 & 0 \\ -3 & 4 \end{bmatrix}\begin{bmatrix} l & m & n \\ x & y & z \end{bmatrix} = \begin{bmatrix} -1 & -8 & -10 \\ 1 & -2 & -5 \\ 9 & 22 & 15 \end{bmatrix}$$

$$\Rightarrow \begin{bmatrix} 2l-x & 2m-y & 2n-z \\ l & m & x \\ -3l+4x & -3m+4y & -3n+4z \end{bmatrix} = \begin{bmatrix} -1 & -8 & -10 \\ 1 & -2 & -5 \\ 9 & 22 & 15 \end{bmatrix}$$

$\Rightarrow 2l - x = -1,\ 2m - y = -8,\ 2n - z = -10,\ l = 1,\ m = -2,\ n = -5$

$\Rightarrow x = 3, y = 4, z = 0, l = 1, m = -2, n = -5$

$$\Rightarrow A = \begin{bmatrix} l & m & n \\ x & y & z \end{bmatrix} = \begin{bmatrix} 1 & -2 & -5 \\ 3 & 4 & 0 \end{bmatrix}$$

17. b. Let, $A = \dfrac{1}{\sqrt{3}}\begin{bmatrix} 1 & 1+i \\ 1-i & -1 \end{bmatrix}$

$$\therefore A^T = \frac{1}{\sqrt{3}}\begin{bmatrix} 1 & 1-i \\ 1-i & -1 \end{bmatrix}$$

$$\Rightarrow (\bar{A}^T) = \frac{1}{\sqrt{3}}\begin{bmatrix} 1 & 1+i \\ 1-i & -1 \end{bmatrix}$$

$$\therefore A(\bar{A}^T) = \frac{1}{\sqrt{3}}\begin{bmatrix} 1 & 1+i \\ 1-i & -1 \end{bmatrix} \times \frac{1}{\sqrt{3}}\begin{bmatrix} 1 & 1+i \\ 1-i & -1 \end{bmatrix}$$

$$= \frac{1}{3}\begin{bmatrix} 3 & 0 \\ 0 & 3 \end{bmatrix} = \begin{bmatrix} 1 & 0 \\ 0 & 1 \end{bmatrix} = I$$

18. d.

19. a. A is involuntary. Hence,

$A^2 = I \Rightarrow A = A^{-1}$

Also,

$$(kA)^{-1} = \frac{1}{k}(A)^{-1}$$

$$\Rightarrow \left(\frac{1}{2}A\right)^{-1} = 2(A)^{-1} \Rightarrow 2A$$

20. b. Given,

$$B = A^{-1}A^T$$
$$\Rightarrow B^T = (A^{-1}A^T)^T = A \times (A^{-1})^T$$

$\Rightarrow B \times B^T = A^{-1}A^T \times A \times (A^{-1})^T = A^{-1} \times (A^T \times A)(A^{-1})^T$
$= A^{-1}(A \times A^T)(A^{-1})^T$
$= (A^{-1}A) \times (A^{-1}A)^T = I \times I^T = I$

21. b. $P^TP = I$
$Q = PAP^T$
$\therefore x = P^TQ^{1000}P = P^T(PAP^T)^{1000}P$
$= P^TPAP^T(PAP^T)^{999}P$
$= IAP^T\,PAP^T(PAP^T)^{998}P$
$= AIAP^T(PAP^T)^{998}P$
$= A^2P^TPAP^T(PAP^T)^{997}P$
$= A^3P^T((PAP^T)^{997}P$
$\vdots$
$= A^{1000} = I \quad (\because A \text{ is involuntary})$
Hence, $x^{-1} = I$.

22. b. Since A is orthogonal, hence
$AA^T = I$
$\Rightarrow |AA^T| = 1$
$\Rightarrow |A^2| = 1$
$\Rightarrow |A| = \pm 1$
Now, $|\text{adj}(\text{adj } A)| = |A|^{(n-1)^2}$

23. b. $\left(A' - \frac{1}{2}I\right)\left(A - \frac{1}{2}I\right) = I$ and $\left(A' + \frac{1}{2}I\right)\left(A + \frac{1}{2}I\right) = I$
$\Rightarrow A + A' = 0$ (subtracting the two results)
$\Rightarrow A' = -A$
$\Rightarrow A^2 = -\frac{3}{4}I$
$\Rightarrow \left(\frac{-3}{4}\right)^n = (\det(A))^2$
$\Rightarrow n$ is even

24. a. For involuntary matrix,
$A^2 = I$
$\Rightarrow |A^2| = |I| \Rightarrow |A|^2 = 1 \Rightarrow |A| = \pm 1$
For idempotent matrix,
$A^2 = A$
$\Rightarrow |A^2| = |A| \Rightarrow |A|^2 = |A| \Rightarrow |A| = 0$ or 1
For orthogonal matrix,
$AA^T = I$
$\Rightarrow |AA^T| = |I| \Rightarrow |A|\,|A^T| = 1 \Rightarrow |A|^2 = 1 \Rightarrow |A| = \pm 1$
Thus if matrix A is idempotent it may not be invertible.

25. a. $A \times A^T = I$
$\Rightarrow |A \times A^T| = |I|$
$\Rightarrow |A|^2 = 1$
$\Rightarrow |A| = \pm 1$
$\Rightarrow A^{-1}$ exists
$\Rightarrow A^{-1} \times A \times A^T = A^{-1} \times I$
$\Rightarrow A^{-1} = A^T$

26. b. Z is idempotent, then
$Z^2 = Z \Rightarrow Z^3, Z^4, \ldots, Z^n = Z$
$\therefore (I + Z)^n = {}^nC_0I^n + {}^nC_1I^{n-1}Z + {}^nC_2I^{n-2}Z^2 + \cdots + {}^nC_nZ^n$
$= {}^nC_0I + {}^nC_1Z + {}^nC_2Z + {}^nC_3Z + \cdots + {}^nC_nZ$
$= I + ({}^nC_1 + {}^nC_2 + {}^nC_3 + \cdots + {}^nC_n)Z$
$= I + (2^n - 1)Z$

27. d. Since $AB = B$ and $BA = A$, so
$BAB = B^2$
$\Rightarrow (BA)B = B^2$
$\Rightarrow AB = B^2$
$\Rightarrow B = B^2$
Hence, B is idempotent and similarly A.
$(A - B)^2 = A^2 - AB - BA + B^2 = A - B - A + B = O$
Therefore, $A - B$ is nilpotent.

28. b. $A^2 = O, A^3 = A^4 = \cdots = A^n = O$
Then, $A(I + A)^n = A(I + nA) = A + nA^2 = A$

29. b. We have, $AB = A(\text{adj } A) = |A|\, I_n$
$\therefore AB + KI_n = |A|I_n + KI_n$
$\Rightarrow AB + KI_n = (|A| + k)I_n$
$\Rightarrow |AB + KI_n| = |(|A| + k)I_n| \quad (\because |\alpha I_n| = \alpha^n)$
$= (|A| + K)^n$

30. d. $\det(A - I) = \det(A - A^2)$
$= \det A(I - A)$
$= \det A \times \det(I - A)$
$= -\det A \times \det(A - I)$
Now,
$A^2 = I$
$\Rightarrow \det(A^2) = \det(I)$
$\Rightarrow (\det A)^2 = 1$
$\Rightarrow \det(A) = \pm 1$
Thus, det (A) can be 1 or –1, from which we cannot say anything about det($A - I$).

31. a. $|B| = \begin{vmatrix} q & -b & y \\ -p & a & -x \\ r & -c & z \end{vmatrix}$ (Multiplying R_2 by – 1)
$= -\begin{vmatrix} q & -b & y \\ p & -a & x \\ r & -c & z \end{vmatrix}$ (Multiplying C_2 by – 1)
$= \begin{vmatrix} q & b & y \\ p & a & x \\ r & c & z \end{vmatrix}$ (Changing R_1 with R_2)
$= -\begin{vmatrix} p & a & x \\ q & b & y \\ r & c & z \end{vmatrix}$
$= -\begin{vmatrix} a & x & p \\ b & y & q \\ c & z & r \end{vmatrix}$

Hence, $|A| = -|B|$, obviously when $|A| \neq 0$, $|B| \neq 0$. Also, $|\text{adj } B| = |B|^2 = (-|A|)^2 = |A|^2$.

32 a. $AB = C$

$\Rightarrow |AB| = |C|$

$\Rightarrow |A||B| = |C|$

$\Rightarrow |B| = \dfrac{|C|}{|A|}$

33 a. $AB = A + B$

$\Rightarrow B = AB - A = A(B - I)$

$\Rightarrow \det(B) = \det(A)\det(B - I) = 0$

$\Rightarrow \det(B) = 0$

34. b. $|A^{2005} - 6A^{2004}| = |A|^{2004}|A - 6I|$

$$= 2^{2004}\begin{vmatrix} 0 & 11 \\ 2 & -2 \end{vmatrix} = (-22)\,2^{2004} = (-11)\,(2)^{2005}$$

35. c. $A^2 = I$

$\Rightarrow A^2 - I = O$

$\Rightarrow (A + I)(A - I) = 0$

Therefore, either $|A + I| = 0$ or $|A - I| = 0$. If $|A - I| \neq 0$, then $(A + I)(A - I) = O \Rightarrow A - I = O$ which is not so.

$\therefore |A - I| = 0$ and $A - I \neq O$

36. b. We have,

$I = AA^{-1}$

$$= \frac{1}{2}\begin{bmatrix} 0 & 1 & 2 \\ 1 & 2 & 3 \\ 3 & a & 1 \end{bmatrix}\begin{bmatrix} 1 & -1 & 1 \\ -08 & 6 & 2c \\ 5 & -3 & 1 \end{bmatrix}$$

$$= \begin{bmatrix} 1 & 0 & c+1 \\ 0 & 1 & 2\ (c+1) \\ 4(1-a) & 3\ (a-1) & 2+ac \end{bmatrix}$$

Comparing the elements of AA^{-1} with those of I, we have

$c + 1 = 0 \Rightarrow c = -1$

$\Rightarrow c = -1$ and $a - 1 = 0 \Rightarrow a = 1$

37. c. Given $B^r = I \Rightarrow B^rB^{-1} = IB^{-1} \Rightarrow B^{r-1} = B^{-1}$

$\Rightarrow A^{-1}B^{r-1}A - A^{-1}B^{-1}A = A^{-1}B^{-1}A - A^{-1}B^{-1}A = O$

38. c. $\begin{vmatrix} \overline{z}_1 & -z_2 \\ \overline{z}_1 & z_1 \end{vmatrix}^{-1}\begin{bmatrix} z_1 & z_2 \\ -\overline{z}_2 & \overline{z}_1 \end{bmatrix}^{-1}$

$$= \left(\begin{bmatrix} z_1 & z_2 \\ -\overline{z}_2 & \overline{z}_1 \end{bmatrix}\begin{bmatrix} \overline{z}_1 & -z_2 \\ \overline{z}_2 & z_1 \end{bmatrix}\right)^{-1}$$

$$= \begin{bmatrix} z_1\overline{z}_1 + z_2\overline{z}_2 & 0 \\ 0 & z_2\overline{z}_2 + z_1\overline{z}_1 \end{bmatrix}^{-1}$$

$$= \begin{bmatrix} |z_1|^2 + |z_2|^2 & 0 \\ 0 & |z_1|^2 + |z_2|^2 \end{bmatrix}^{-1}$$

$$= \begin{bmatrix} 2 & 0 \\ 0 & 2 \end{bmatrix}^{-1} = \begin{bmatrix} 1/2 & 0 \\ 0 & 1/2 \end{bmatrix}$$

39. a. We have,

$$A(\alpha, \beta)^{-1} = \frac{1}{e^\beta}\begin{bmatrix} e^\beta\cos\alpha & -e^\beta\sin\alpha & 0 \\ e^\beta\sin\ \alpha & e^\beta\cos\alpha & 0 \\ 0 & 0 & 1 \end{bmatrix}$$

$$= A(-\alpha, -\beta)$$

40. a. We have,

$|A| = (a + ib)(a - ib) - (-c + id)(c + id)$

$= a^2 + b^2 + c^2 + d^2 = 1$

and $\text{adj}(A) = \begin{bmatrix} a-ib & -c-id \\ c-id & a+ib \end{bmatrix}$

Then $A^{-1} = \begin{bmatrix} a-ib & -c-id \\ -c+id & a-ib \end{bmatrix}$

41. c. Given

$A^3 = O$

Now,

$(I - A)(I + A + A^2) = I^2 + IA + IA^2 - AI - A^2 - A^3$

$= I - A^3$

$= I$

$\Rightarrow (I - A)^{-1} = I + A + A^2$

42. b. We know that $|\text{adj}(\text{adj } A)| = |A|^{(n-1)^2}$

$\Rightarrow$ $|\text{adj}(\text{adj}(\text{adj } A))| = |\text{adj } A|^{(n-1)^2}$

$= |A|^{(n-1)^3}$

$= 2^8 = 256$

43. b. $(-A)^{-1} = \dfrac{\text{adj}(-A)}{|-A|} = \dfrac{(-1)^{n-1}\text{adj}(A)}{(-1)^n\,|A|} = \dfrac{\text{adj}(A)}{-|A|} = -A^{-1}$

44. a. $A(x)A(y) = (1-x)^{-1}(1-y)^{-1}\begin{bmatrix} 1 & -x \\ -x & 1 \end{bmatrix}\begin{bmatrix} 1 & -y \\ -y & 1 \end{bmatrix}$

$$= (1 + xy - (x + y))^{-1}\begin{bmatrix} 1+xy & -(x+y) \\ -(x+y) & 1+xy \end{bmatrix}$$

$$= \left(1 - \frac{x+y}{1+xy}\right)^{-1}\begin{bmatrix} 1 & -\dfrac{x+y}{1+xy} \\ -\dfrac{x+y}{1+xy} & 1 \end{bmatrix} = A(z)$$

45. b. Let, $A = \begin{bmatrix} 5 & 0 \\ -a & 5 \end{bmatrix}$

$\Rightarrow \text{adj}(A) = \begin{bmatrix} 5 & 0 \\ a & 5 \end{bmatrix}$

$\Rightarrow A^{-1} = \dfrac{1}{|A|}\begin{bmatrix} 5 & 0 \\ a & 5 \end{bmatrix} = \dfrac{1}{25}\begin{bmatrix} 5 & 0 \\ a & 5 \end{bmatrix}$

$\Rightarrow A^{-2} = (A^{-1})^2 = \dfrac{1}{25}\begin{bmatrix} 5 & 0 \\ a & 5 \end{bmatrix}\dfrac{1}{25}\begin{bmatrix} 5 & 0 \\ a & 5 \end{bmatrix}$

$$= \frac{1}{625}\begin{bmatrix} 25 & 0 \\ 10a & 25 \end{bmatrix}$$

$$= \begin{bmatrix} \frac{1}{25} & 0 \\ \frac{2a}{125} & \frac{1}{25} \end{bmatrix}$$

Now,

$$\begin{bmatrix} 1/25 & 0 \\ x & 1/25 \end{bmatrix} = \begin{bmatrix} \frac{1}{25} & 0 \\ \frac{2a}{125} & \frac{1}{25} \end{bmatrix}$$

$\Rightarrow x = 2a/125$

46. c. $f(x) = \dfrac{1+x}{1-x}$

$\Rightarrow (1-x)f(x) = 1 + x$

$\Rightarrow (I - A)f(A) = (I + A)$

$\Rightarrow f(A) = (I - A)^{-1}(I + A)$

$$= \left(\begin{bmatrix} 1 & 0 \\ 0 & 1 \end{bmatrix} - \begin{bmatrix} 1 & 2 \\ 2 & 1 \end{bmatrix}\right)^{-1}\left(\begin{bmatrix} 1 & 0 \\ 0 & 1 \end{bmatrix} + \begin{bmatrix} 1 & 2 \\ 2 & 1 \end{bmatrix}\right)$$

$$\Rightarrow f(A) = \left(\begin{bmatrix} 0 & -2 \\ -2 & 0 \end{bmatrix}^{-1}\begin{bmatrix} 2 & 2 \\ 2 & 2 \end{bmatrix}\right)$$

$$= \frac{\begin{bmatrix} 0 & 2 \\ 2 & 0 \end{bmatrix}\begin{bmatrix} 2 & 2 \\ 2 & 2 \end{bmatrix}}{-4}$$

$$= \frac{\begin{bmatrix} 4 & 4 \\ 4 & 4 \end{bmatrix}}{-4}$$

$$= \begin{bmatrix} -1 & -1 \\ -1 & -1 \end{bmatrix}$$

47. a. We know that in a square matrix of order n,

$|\text{adj } A| = |A|^{n-1}$

$\Rightarrow |\text{adj (adj } A)| = |\text{adj } A|^{n-1} = |A|^{(n-1)^2}$

$\Rightarrow n^2 - 2n - 8 = 0$

$\Rightarrow n = 4$ as $n = -2$ is not possible

48. b. $|A| = \begin{vmatrix} 1 & \tan x \\ -\tan x & 1 \end{vmatrix} = 1 + \tan^2 x \neq 0$

So A is invertible. Also,

$$\text{adj } A = \begin{bmatrix} 1 & \tan x \\ -\tan x & 1 \end{bmatrix}^T = \begin{bmatrix} 1 & -\tan x \\ \tan x & 1 \end{bmatrix}$$

Now,

$$A^{-1} = \frac{1}{|A|}\text{adj } A$$

$$\Rightarrow A^{-1} = \frac{1}{(1+\tan^2 x)}\begin{bmatrix} 1 & -\tan x \\ \tan x & 1 \end{bmatrix}$$

$$= \begin{bmatrix} \frac{1}{1+\tan^2 x} & \frac{-\tan x}{1+\tan^2 x} \\ \frac{\tan x}{1+\tan^2} & \frac{1}{1+\tan^2 x} \end{bmatrix}$$

$$\therefore A^T A^{-1} = \begin{bmatrix} 1 & -\tan x \\ \tan x & 1 \end{bmatrix}\begin{bmatrix} \frac{1}{1+\tan^2 x} & \frac{-\tan x}{1+\tan^2 x} \\ \frac{\tan x}{1+\tan^2} & \frac{1}{1+\tan^2 x} \end{bmatrix}$$

$$= \begin{bmatrix} \frac{1-\tan^2 x}{1+\tan^2 x} & \frac{-2\tan x}{1+\tan^2 x} \\ \frac{2\tan x}{1+\tan^2 x} & \frac{1-\tan^2 x}{1+\tan^2 x} \end{bmatrix}$$

$$= \begin{bmatrix} \cos 2x & -\sin 2x \\ \sin 2x & \cos 2x \end{bmatrix}$$

49. b. Since $I = \begin{bmatrix} 1 & 0 \\ 0 & 1 \end{bmatrix}$ and given $A = \begin{bmatrix} 0 & \tan\alpha/2 \\ -\tan\alpha/2 & 0 \end{bmatrix}$

$$\therefore I - A = \begin{bmatrix} 1 & -\tan\alpha/2 \\ \tan\alpha/2 & 1 \end{bmatrix} \quad (1)$$

Now, $(I - A)\begin{bmatrix} \cos\alpha & -\sin\alpha \\ \sin\alpha & \cos\alpha \end{bmatrix}$

$$= \begin{bmatrix} 1 & \tan\alpha/2 \\ -\tan\alpha/2 & 1 \end{bmatrix}\begin{bmatrix} \cos\alpha & -\sin\alpha \\ \sin\alpha & \cos\alpha \end{bmatrix}$$

$$= \begin{bmatrix} 1 & \tan\alpha/2 \\ -\tan\alpha/2 & 1 \end{bmatrix}\begin{bmatrix} \frac{1-\tan^2\alpha/2}{1+\tan^2\alpha/2} & -\frac{2\tan\alpha/2}{1+\tan^2\alpha/2} \\ \frac{2\tan\alpha/2}{1+\tan^2\alpha/2} & \frac{1-\tan^2\alpha/2}{1+\tan^2\alpha/2} \end{bmatrix}$$

$$= \begin{bmatrix} \frac{1-\tan^2\alpha/2}{1+\tan^2\alpha/2} + \frac{2\tan^2\alpha/2}{1+\tan^2\alpha/2} & -\frac{2\tan\alpha/2}{1+\tan^2\alpha/2} + \frac{\tan\alpha/2\,(1-\tan^2\alpha/2)}{1+\tan^2\alpha/2} \\ \frac{-\tan\alpha/2\,(1-\tan^2\alpha/2)}{1+\tan^2\alpha/2} + \frac{2\tan\alpha/2}{1+\tan^2\alpha/2} & \frac{2\tan^2\alpha/2}{1+\tan^2\alpha/2} + \frac{1-\tan^2\alpha/2}{1+\tan^2\alpha/2} \end{bmatrix}$$

$$= \begin{bmatrix} \frac{(1+\tan^2\alpha/2)}{(1+\tan^2\alpha/2)} & -\frac{\tan\alpha/2\,(1+\tan^2\alpha/2)}{(1+\tan^2\alpha/2)} \\ \frac{\tan\alpha/2(1+\tan^2\alpha/2)}{(1+\tan^2\alpha/2)} & \frac{(1+\tan^2\alpha/2)}{(1+\tan^2\alpha/2)} \end{bmatrix}$$

$$= \begin{bmatrix} 1 & -\tan\alpha/2 \\ \tan\alpha/2 & 1 \end{bmatrix}$$

$= I - A$ [Using (1)]

50. d. Let, $A = \begin{bmatrix} 1 & -4 \\ 3 & -2 \end{bmatrix}$ and $B = \begin{bmatrix} -16 & -6 \\ 7 & 2 \end{bmatrix}$

Then the matrix equation is $AX = B$.

$$\therefore |A| = \begin{vmatrix} 1 & -4 \\ 3 & -2 \end{vmatrix} = -2 + 12 \neq 0$$

So A is an invertible matrix. Also,

$$\text{adj } A = \begin{bmatrix} -2 & -3 \\ 4 & 1 \end{bmatrix}^T = \begin{bmatrix} -2 & 4 \\ -3 & 1 \end{bmatrix}$$

So,

$$A^{-1} = \frac{1}{|A|}\text{adj}A = \frac{1}{10}\begin{bmatrix} -2 & 4 \\ -3 & 1 \end{bmatrix}$$

Now,

$AX = B$

$\Rightarrow A^{-1}(AX) = A^{-1}B$

$\Rightarrow (A^{-1}A)X = A^{-1}B$

$\Rightarrow IX = A^{-1}B$

$\Rightarrow X = A^{1}B$

$$\Rightarrow X = \frac{1}{10}\begin{bmatrix} -2 & 4 \\ -3 & 1 \end{bmatrix}\begin{bmatrix} -16 & -6 \\ 7 & 2 \end{bmatrix} = \begin{bmatrix} 6 & 2 \\ 11/2 & 2 \end{bmatrix}$$

51. c. $(A^{-1}BA)^2 = (A^{-1}BA)(A^{-1}BA)$

$= A^{-1}B(AA^{-1})BA$

$= A^{-1}BIBA = A^{-1}B^2 A$

$\Rightarrow (A^{-1}BA)^3 = (A^{-1}B^2A)(A^{-1}BA)$

$= A^{-1}B^2(AA^{-1})BA$

$= A^{-1}B^2IBA$

$= A^{-1}B^3A$ and so on

$\Rightarrow (A^{-1}BA)^n = A^{-1}B^nA$

52. a. $A \text{ adj } A = |A|\, I$

$\Rightarrow |A \text{ adj } A| = |A|^n$ [If A is of order $n \times n$]

$\Rightarrow |A|\,|\text{adj } A| = |A|^n$

$\Rightarrow |\text{adj } A| = |A|^{n-1}$

Now, A is singular,

$\therefore |A| = 0$

$\Rightarrow |\text{adj } A| = 0$

Hence, adj A is singular.

53. a. $A = \text{diag}(d_1, d_2, d_3, ..., d_n)$

$\Rightarrow |A| = (d_1 \times d_2 \times d_3 \times d_4 \cdots d_n)$

Now,

Cofactor of d_1 is $d_2d_3 \cdots d_n$

Cofactor of d_2 is $d_1 \times d_3 \times d_4 \cdots d_n$

Cofactor of d_3 is $d_1 \times d_2 \times d_4 \cdots d_n$

⋮

Cofactor of d_n is $d_1 \times d_2 \times d_3 \cdots d_{n-1}$

$$\Rightarrow A^{-1} = \frac{1}{|A|}\text{adj}(A) = \text{diag}(d_1^{-1}, d_2^{-1}, d_3^{-1}, ..., d_n^{-1})$$

Hence, A^{-1} is also a diagonal matrix.

54. a. We know that for any non-singular matrix A,

$$A^{-1} = \frac{1}{|A|}\text{adj}(A)$$

Now put $A = P^{-1}$. Then we have

$$(P^{-1})^{-1} = \frac{1}{|P^{-1}|}\text{adj}(P^{-1})$$

$\Rightarrow P = |P|\text{adj}(P^{-1})$

$$\Rightarrow \text{adj}(P^{-1}) = \frac{P}{|P|}$$

55. c. $\text{adj}(Q^{-1}BP^{-1}) = \text{adj}(P^{-1})\text{adj}(B)\text{adj}(Q^{-1})$

$$= \frac{P}{|P|} A \frac{Q}{|Q|}$$

$= PAQ$

56. b. We have,

$(A - 2I)(A - 4I) = O$

$\Rightarrow A^2 - 2A - 4A + 8I = O$

$\Rightarrow A^2 - 6A + 8I = O$

$\Rightarrow A^{-1}(A^2 - 6A + 8I) = A^{-1}O$

$\Rightarrow A - 6I + 8A^{-1} = O$

$\Rightarrow A + 8A^{-1} = 6I$

$$\Rightarrow \frac{1}{6}A + \frac{4}{3}A^{-1} = 1$$

57. b. $|A(\alpha, \beta)| = \cos^2\alpha e^\beta + \sin^2\alpha e^\beta = e^\beta$

Now,

$$A(\alpha, \beta)^{-1} = \frac{1}{e^\beta}\text{adj}(A(\alpha, \beta))$$

$$= \frac{1}{e^\beta}\begin{bmatrix} e^\beta\cos\alpha & -\sin\alpha e^\beta & 0 \\ e^\beta\sin\alpha & \cos\alpha e^\beta & 0 \\ 0 & 0 & 1 \end{bmatrix}$$

$$= \begin{bmatrix} \cos\alpha & -\sin\alpha & 0 \\ \sin\alpha & \cos\alpha & 0 \\ 0 & 0 & e^{-\beta} \end{bmatrix}$$

$= A(-\alpha, -\beta)$

58. a. As $B = -A^{1}BA$, we get

$AB = -BA$ or $AB + BA = O$

Now,

$(A + B)^2 = (A + B)(A + B)$

$= A^2 + BA + BA + B^2$

$= A^2 + O + B^2$

$= A^2 + B^2$

59. b. $A = \begin{pmatrix} a & 0 \\ 0 & b \end{pmatrix}$

$$\Rightarrow A^2 = \begin{pmatrix} a & 0 \\ 0 & b \end{pmatrix}\begin{pmatrix} a & 0 \\ 0 & b \end{pmatrix} = \begin{pmatrix} a^2 & 0 \\ 0 & b^2 \end{pmatrix}$$

$$\Rightarrow A^3 = \begin{pmatrix} a^2 & 0 \\ 0 & b^2 \end{pmatrix}\begin{pmatrix} a & 0 \\ 0 & b \end{pmatrix} = \begin{pmatrix} a^3 & 0 \\ 0 & b^3 \end{pmatrix}$$

$$\Rightarrow A^n = \begin{pmatrix} a^n & 0 \\ 0 & b^n \end{pmatrix}$$

$$\Rightarrow (A^n)^{-1} = \frac{1}{a^n b^n}\begin{pmatrix} b^n & 0 \\ 0 & a^n \end{pmatrix} = \begin{pmatrix} a^{-n} & 0 \\ 0 & b^{-n} \end{pmatrix}$$

$$\Rightarrow \lim_{n\to\infty}(A^n)^{-1} = \begin{pmatrix} 0 & 0 \\ 0 & 0 \end{pmatrix} \text{ as } a > 1 \text{ and } b > 1$$

60. b. $(I - A)f(A) = I + A$

$\Rightarrow f(A) = (I + A)(I - A)^{-1}$

$= (I + A)(I + A + A^2)$

$= I + A + A^2 + A + A^2 + A^3$

$= I + 2A + 2A^2$

61. d. If possible assume that A is non-singular, then A^{-1} exists. Thus,

$AB = O \Rightarrow A^{-1}(AB) = (A^{-1}A)B = O$

$\Rightarrow IB = O$ or $B = O$ × (a contradiction)

Hence, both A and B must be singular.

62. a. $\begin{bmatrix} 2 & 1 \\ 3 & 2 \end{bmatrix} A \begin{bmatrix} -3 & 2 \\ 5 & -3 \end{bmatrix} = \begin{bmatrix} 1 & 0 \\ 0 & 1 \end{bmatrix}$

$$A = \begin{bmatrix} 2 & 1 \\ 3 & 2 \end{bmatrix}^{-1}\begin{bmatrix} 1 & 0 \\ 0 & 1 \end{bmatrix}\begin{bmatrix} -3 & 2 \\ 5 & -3 \end{bmatrix}^{-1}$$

$$\Rightarrow A = \begin{bmatrix} 2 & -1 \\ -3 & 2 \end{bmatrix}\begin{bmatrix} 1 & 0 \\ 0 & 1 \end{bmatrix}\begin{bmatrix} -3 & -2 \\ -5 & -3 \end{bmatrix}$$

$$= \begin{bmatrix} 2 & -1 \\ -3 & 2 \end{bmatrix}\begin{bmatrix} 3 & 2 \\ 5 & 3 \end{bmatrix} = \begin{bmatrix} 1 & 1 \\ 1 & 0 \end{bmatrix}$$

63. c. $A^2 - A + I = 0$

$\Rightarrow I = A - A^2$

$IA^{-1} = AA^{-1} - A^2A^{-1}$

$\Rightarrow A^{-1} = I - A$

64. a. Let, $X = \begin{pmatrix} a & b \\ c & d \end{pmatrix}$

$$\Rightarrow X^2 = \begin{pmatrix} a^2 + bc & ab + bd \\ ac + cd & bc + d^2 \end{pmatrix}$$

$\Rightarrow a^2 + bc = 1$ and $ab + bd = 1 \Rightarrow b(a + d) = 1$

$ac + cd = 2 \Rightarrow c(a + d) = 2 \Rightarrow 2b = c$

Also,

$bc + d^2 = 3 \Rightarrow d^2 - a^2 = 2$

$\Rightarrow (d - a)(a + d) = 2 \Rightarrow d - a = 2b$ (using $bc = 1 - a^2$)

$a + d = 1/b$

$\Rightarrow 2d = 2b + 1/b, \quad 2a = 1/b - 2b$

$d = b + 1/2b, \quad a = 1/(2b) - b$

$c = 2b$

$$\Rightarrow \left(b^2 + \frac{1}{4b^2} + 1\right) + 2b^2 = 3$$

$$\Rightarrow 3b^2 + \frac{1}{4b^2} = 2$$

$$\Rightarrow 3x + \frac{1}{4x} = 2$$

$$\Rightarrow b = \pm\frac{1}{\sqrt{6}} \text{ or } b = \pm\frac{1}{\sqrt{2}}$$

Therefore, matrices are

$$\begin{pmatrix} 0 & 1/\sqrt{2} \\ \sqrt{2} & \sqrt{2} \end{pmatrix}, \begin{pmatrix} 0 & -1/\sqrt{2} \\ -\sqrt{2} & -\sqrt{2} \end{pmatrix}, \begin{pmatrix} 2/\sqrt{6} & -1/\sqrt{6} \\ 2/\sqrt{6} & 4\sqrt{6} \end{pmatrix}$$

65. c. Given that

$X = AB + BA \Rightarrow X = X^T$

and

$Y = AB - BA$

$\Rightarrow Y = -Y^T$

Now, $(XY)^T = Y^TX^T = -YX$.

66. c. As A is a skew-symmetric matrix,

$A^T = -A$

$\Rightarrow a_{ii} = 0, \forall i$

$\Rightarrow \text{tr}(A) = 0$

Also,

$|A| = |A^T| = |-A| = (-1)^3|A|$

$\Rightarrow 2|A| = 0$

$\Rightarrow |A| = 0$

67. d. $\text{tr}(A) = \sum_{i=j} a_{ij}$

$= (a_{11} + a_{22} + a_{33} + \cdots + a_{10\infty10})$

$= (w^2 + w^4 + w^6 + \cdots + w^{20})$

$= w^2(1 + w^2 + w^4 + \cdots + w^{18})$

$= w^2[(1 + w + w^2) + \cdots + (1 + w + w^2) + 1]$

$= w^2 \times 1$

$\Rightarrow \text{tr}(A) = w^2$

68. b. We have,

$$F(\alpha) = \begin{bmatrix} \cos\alpha & -\sin\alpha & 0 \\ \sin\alpha & \cos\alpha & 0 \\ 0 & 0 & 1 \end{bmatrix} \Rightarrow \text{adj}(F(\alpha)) = \begin{bmatrix} \cos\alpha & \sin\alpha & 0 \\ -\sin\alpha & \cos\alpha & 0 \\ 0 & 0 & 1 \end{bmatrix}$$

Also,

$\det(F(\alpha)) = 1$

$$\Rightarrow [F(\alpha)]^{-1} = \begin{bmatrix} \cos(-\alpha) & -\sin(-\alpha) & 0 \\ \sin(-\alpha) & \cos(-\alpha) & 0 \\ 0 & 0 & 1 \end{bmatrix} = F(-\alpha)$$

69. b. We have,

$[F(x)G(y)]^{-1} = [G(y)]^{-1}[F(x)]^{-1}$

$= G(-y)F(-x)$

70. a. Given A is skew-symmetric. Hence,

$A^T = -A$

$\Rightarrow A^n = (-A^T)^n = -(A^T)^n = -(A^n)^T$ (given n is odd)

Hence, A^n is skew-symmetric.

71. c. We have,

$$A^2 = \begin{bmatrix} a & b \\ c & d \end{bmatrix}\begin{bmatrix} a & b \\ c & d \end{bmatrix} = \begin{bmatrix} a^2 + bc & ab + db \\ ac + cd & bc + d^2 \end{bmatrix}$$

As A satisfies $x^2 + k = 0$, therefore

$A^2 + kI = O$

$$\Rightarrow \begin{bmatrix} a^2 + bc + k & (a+d)b \\ (a+d)c & bc + d^2 + k \end{bmatrix} = \begin{bmatrix} 0 & 0 \\ 0 & 0 \end{bmatrix}$$

$\Rightarrow a^2 + bc + k = 0,\ bc + d^2 + k = 0$

and $(a + d)b = (a + d)c = 0$

As $bc \neq 0,\ b \neq 0,\ c \neq 0$, so

$a + d = 0$

$\Rightarrow a = -d$

Also,

$k = -(a^2 + bc)$

$= -(d^2 + bc)$

$= -((-ad) + bc)$

$= |A|$

72. b. Given $A, B, A + I, A + B$ are idempotent. Hence,

$A^2 = A,\ B^2 = B,\ (A + I)^2 = A + I$ and $(A + B)^2 = A + B$

$\Rightarrow A^2 + B^2 + AB + BA = A + B$

$\Rightarrow A + B + AB + BA = A + B$

$\Rightarrow AB + BA = O$

73. b.
$$A^2 = \begin{bmatrix} -5 & -8 & 0 \\ 3 & 5 & 0 \\ 1 & 2 & -1 \end{bmatrix}\begin{bmatrix} -5 & -8 & 0 \\ 3 & 5 & 0 \\ 1 & 2 & -1 \end{bmatrix}$$

$$= \begin{bmatrix} 25 - 24 + 0 & 40 - 40 + 0 & 0 + 0 + 0 \\ -15 + 15 + 0 & -24 + 25 + 0 & 0 + 0 + 0 \\ -5 + 6 - 1 & -8 + 10 - 2 & 0 + 0 + 1 \end{bmatrix}$$

$$= \begin{bmatrix} 1 & 0 & 0 \\ 0 & 1 & 0 \\ 0 & 0 & 1 \end{bmatrix} = I$$

Hence, the matrix A is involutory.

74. b. Let $A = [a_{ij}]$. Since A is skew-symmetric, therefore

$a_{ii} = 0$ and $a_{ij} = -a_{ij}\ (i \neq j)$

A is symmetric as well, so $a_{ij} = a_{ji}$ for all i and j.

$\therefore\ a_{ij} = 0$ for all $i \neq j$

Hence, $a_{ij} = 0$ for all i and j, i.e., A is a null matrix.

75. c. $(A^{-1}BA)^2 = (A^{-1}BA)(A^{-1}BA)$

$= A^{-1}B(AA^{-1})BA$

$= A^{-1}BIBA = A^{-1}B^2A$

$(A^{-1}BA)^3 = (A^{-1}B^2A)(A^{-1}BA)$

$= A^{-1}B^2(AA^{-1})BA$

$= A^{-1}B^3A$ and so on

$\therefore\ (A^{-1}BA)^n = A^{-1}B^nA$

76. a. Matrix $\begin{bmatrix} a_1 & a_2 & a_3 \\ b_1 & b_2 & b_3 \\ c_1 & c_2 & c_3 \end{bmatrix}$ is orthogonal if

$$\sum a_i^2 = \sum b_i^2 = \sum c_i^2 = 1;\ \sum a_ib_i = \sum b_ic_i = \sum c_ia_i = 0$$

77. b. $(kI_n)\ \text{adj}\ (kI_n) = |kI_n|I_n$ [Using $A(\text{adj}\ A) = |A|I$]

$\text{adj}\ (kI_n) = k^{n-1}I_n$

$|\text{adj}\ (kI_n)| = k^{n(n-1)}$

78. c. $A\ \text{adj}\ A = |A|\ I$

$|A| = xyz - 8x - 3(z - 8) + 2(2 - 2y)$

$|A| = xyz - (8x + 3z + 4y) + 28$

$= 60 - 20 + 28$

$= 68$

$$\Rightarrow A\ (\text{adj}\ A) = 68\begin{bmatrix} 1 & 0 & 0 \\ 0 & 1 & 0 \\ 0 & 0 & 1 \end{bmatrix} = \begin{bmatrix} 68 & 0 & 0 \\ 0 & 68 & 0 \\ 0 & 0 & 68 \end{bmatrix}$$

79. c.
$$A + 2B = \begin{bmatrix} 1 & 2 & 0 \\ 6 & -3 & 3 \\ -5 & 3 & 1 \end{bmatrix} \text{ and } 2A - B = \begin{bmatrix} 2 & -1 & 5 \\ 2 & -1 & 6 \\ 0 & 1 & 2 \end{bmatrix}$$

$\Rightarrow \text{tr}(A) + 2\text{tr}(B) = -1$

and $2\ \text{tr}(A) - \text{tr}(B) = 3$

Let $\text{tr}(A) = x$ and $\text{tr}(B) = y$. Then,

$x + 2y = -1$ and $2x - y = 3$

Solving, $x = 1$ and $y = -1$. Hence,

$\text{tr}(A) - \text{tr}(B) = x - y = 2$

80. b. $B = A_1 + 3A_3^3 + \cdots + (2n - 1)\ (A_{2n-1})^{2n-1}$

$B^T = -[A_1 + 3A_3^3 + \cdots + (2n - 1)\ (A_{2r-1})^{2r-1}]$

$= -B$

Hence, B is skew-symmetric.

81. a. $|A| = 1(0 - 10) - 2(2 - 6) + 3(4 - 0)$

$= -10 + 8 + 12 = 10$

$\Rightarrow |A| \neq 0$

$\Rightarrow$ Unique solution

82. a.
$$BC = \begin{bmatrix} 3 & 4 \\ 2 & 3 \end{bmatrix}\begin{bmatrix} 3 & -4 \\ -2 & 3 \end{bmatrix} \Rightarrow B = \begin{bmatrix} 1 & 0 \\ 0 & 1 \end{bmatrix}$$

$$\Rightarrow \text{tr}(A) + \text{tr}\left(\frac{ABC}{2}\right) + \text{tr}\left(\frac{A(BC)^2}{4}\right) + \text{tr}\left(\frac{A(BC)^3}{8}\right) + \cdots + \infty$$

$$= \text{tr}(A) + \text{tr}\left(\frac{A}{2}\right) + \text{tr}\left(\frac{A}{2^2}\right) + \cdots$$

$$= \text{tr}(A) + \frac{1}{2}\text{tr}(A) + \frac{1}{2^2}\text{tr}(A)\cdots$$

$$= \frac{\text{tr}(A)}{1-(1/2)}$$

$$= 2\text{tr}(A) = 2(2+1) = 6$$

Multiple Correct Answers Type

1. b., c.

$\det(-A) = (-1)^n \det(A)$

$\det(A^{-1}) = \dfrac{1}{\det(A)} = 1$

$\det(\text{adj } A) = |A|^{n-1} = 1$

Hence, $|\omega A| = \omega^n |A| = 1$ only when $n = 3k, k \in Z$.

2. a., d.

Given, $(A + B)^2 = A^2 + B^2 + 2AB$

$\Rightarrow (A + B)(A + B) = A^2 + B^2 + 2AB$

$\Rightarrow A^2 + AB + BA + B^2 = A^2 + B^2 + 2AB \Rightarrow BA = AB$

$$\Rightarrow \begin{bmatrix} a & 1 \\ b & -1 \end{bmatrix}\begin{bmatrix} 1 & -1 \\ 2 & 1 \end{bmatrix} = \begin{bmatrix} 1 & -1 \\ 2 & 1 \end{bmatrix}\begin{bmatrix} a & 1 \\ b & -1 \end{bmatrix}$$

$$\Rightarrow \begin{bmatrix} a+2 & -a+1 \\ b-2 & -b-1 \end{bmatrix} = \begin{bmatrix} a-b & 1+1 \\ 2a+b & 2-1 \end{bmatrix}$$

The corresponding elements of equal matrices are equal.

$a + 2 = a - b, -a + 1 = 2 \Rightarrow a = -1$

$b - 2 = 2a + b, -b - 1 = 1 \Rightarrow b = -2$

$\Rightarrow a = -1, b = -2$

3. a., b., c.

Given, $AB = A, BA = B$

$\Rightarrow B \times AB = B \times A$

$\Rightarrow (BA)B = B$

$\Rightarrow B^2 = B$

Also,

$A \times B \times A = AB$

$\Rightarrow (AB)A = A$

$\Rightarrow A^2 = A$

Now $(A^T)^2 = (A^T \times A^T) = (A \times A)^T = (A^2)^T = A^T$

Similarly, $(B^T)^2 = B^T$

$\Rightarrow A^T$ and B^T are idempotent

4. a., c.

A is an orthogonal matrix.

$\therefore AA^T = I$

$$\Rightarrow \frac{1}{3}\begin{bmatrix} 1 & 2 & 2 \\ 2 & 1 & -2 \\ a & 2 & b \end{bmatrix}\frac{1}{3}\begin{bmatrix} 1 & 2 & a \\ 2 & 1 & 2 \\ 2 & -2 & b \end{bmatrix} = \begin{bmatrix} 1 & 0 & 0 \\ 0 & 1 & 0 \\ 0 & 0 & 1 \end{bmatrix}$$

$$\Rightarrow \frac{1}{9}\begin{bmatrix} 1 & 2 & 2 \\ 2 & 1 & -2 \\ a & 2 & b \end{bmatrix}\begin{bmatrix} 1 & 2 & a \\ 2 & 1 & 2 \\ 2 & -2 & b \end{bmatrix} = \begin{bmatrix} 1 & 0 & 0 \\ 0 & 1 & 0 \\ 0 & 0 & 1 \end{bmatrix}$$

$$\Rightarrow \begin{bmatrix} 9 & 0 & a+4+2b \\ 0 & 9 & 2a+2-2b \\ a+4+2b & 2a+2-2b & a^2+4+b^2 \end{bmatrix} = \begin{bmatrix} 9 & 0 & 0 \\ 0 & 9 & 0 \\ 0 & 0 & 9 \end{bmatrix}$$

$\Rightarrow a + 4 + 2b = 0, 2a + 2 - 2b = 0$, and $a^2 + 4 + b^2 = 9$

$\Rightarrow a + 2b + 4 = 0, a - b + 1 = 0$, and $a^2 + b^2 = 5$

$\Rightarrow a = -2, b = -1$

5. a., d.

$(B^TAB)^T = B^TA^T(B^T)^T = B^TA^TB = B^TAB$ if A is symmetric.

Therefore, B^TAB is symmetric if A is symmetric.

Also, $(B^TAB)^T = B^TA^TB = B^T(-A)B = -(B^TA^TB)$

Therefore, B^TAB if A is skew-symmetric if A is skew-symmetric.

6. a., b., c.

We have, $|A(\theta)| = 1$

Hence, A is invertible.

$$A(\pi + \theta) = \begin{bmatrix} \sin(\pi+\theta) & i\cos(\pi+\theta) \\ i\cos(\pi+\theta) & \sin(\pi+\theta) \end{bmatrix}$$

$$= \begin{bmatrix} -\sin\theta & -i\cos\theta \\ -i\cos\theta & -\sin\theta \end{bmatrix} = -A(\theta)$$

$$\text{adj}(A(\theta)) = \begin{bmatrix} \sin\theta & -i\cos\theta \\ -i\cos\theta & \sin\theta \end{bmatrix}$$

$$\Rightarrow A(\theta)^{-1} = \begin{bmatrix} \sin\theta & -i\cos\theta \\ -i\cos\theta & \sin\theta \end{bmatrix} = A(\pi - \theta)$$

7. a., c., d.

Given, $A^2 + A + 2I = O$

$\Rightarrow A^2 + A = -2I$

$\Rightarrow |A^2 + A| = |-2I|$

$\Rightarrow |A||A + I| = (-2)^n$

$\Rightarrow |A| \neq 0$

Therefore, A is non-singular, hence its inverse exists. Also, multiplying the given equation both sides with A^{-1}, we get

$$A^{-1} = -\frac{1}{2}(A + I)$$

8. a., b., c.

$$\because |A| = \begin{vmatrix} 3 & -3 & 4 \\ 2 & -3 & 4 \\ 0 & -1 & 1 \end{vmatrix} = 3(-3+4) + 3(2-0) + 4(-2-0) = 1$$

$\therefore$ adj(adj A) = $|A|^{3-2} A = A$ and $|$adj(adj A)$| = |A| = 1$

Also,

$|\text{adj } A| = |A|^{3-1} = |A|^2 = 1^2 = 1$

9. a., c.

We have,

$$\begin{bmatrix} 1 & \tan\theta \\ -\tan\theta & 1 \end{bmatrix}^{-1} = \frac{1}{1+\tan^2\theta}\begin{bmatrix} 1 & -\tan\theta \\ \tan\theta & 1 \end{bmatrix}$$

$$\Rightarrow \begin{bmatrix} a & -b \\ b & a \end{bmatrix} = \cos^2\theta \begin{bmatrix} 1 & -\tan\theta \\ \tan\theta & 1 \end{bmatrix}\begin{bmatrix} 1 & -\tan\theta \\ \tan\theta & 1 \end{bmatrix}$$

$$= \cos^2\theta \begin{bmatrix} 1-\tan^2\theta & -2\tan\theta \\ 2\tan\theta & 1-\tan^2\theta \end{bmatrix}$$

$$= \begin{bmatrix} \cos 2\theta & -\sin 2\theta \\ \sin 2\theta & \cos 2\theta \end{bmatrix}$$

$\therefore a = \cos 2\theta, b = \sin 2\theta$

10. a., b., c.

$|A^{-1}| = -1 \Rightarrow |A| = -1$

Now, use adj $A = |A|A^{-1}$ and $A = (A^{-1})^{-1}$

11. a., c., d.

B is an idempotent matrix

$\therefore B^2 = B$

Now,

$$\begin{aligned} A^2 &= (I-B)^2 \\ &= (I-B)(I-B) \\ &= I - IB - IB + B^2 \\ &= I - B - B + B^2 \\ &= I - 2B + B^2 \\ &= I - 2B + B \\ &= I - B \\ &= A \end{aligned}$$

Therefore, A is idempotent. Again,

$AB = (I-B)B = IB - B^2 = B - B^2 = B^2 - B^2 = O$

Similarly, $BA = B(I-B) = BI - B^2 = B - B = O$.

12. b., c.

$$(-A)^{-1} = \frac{\text{adj}(-A)}{|-A|} = \frac{(-1)^{n-1}\text{adj}(A)}{(-1)^n |A|} = \frac{\text{adj}(A)}{-|A|} = -A^{-1}$$

[for any value of n]

Given, $A^n = O$

Now,

$(I-A)(I + A + A^2 + \cdots + A^{n-1}) = I - A^n = I$

$\Rightarrow (I-A)^{-1} = I + A + A^2 + \cdots + A^{n-1}$

13. a., b.

Let $I = k = 1$ (say). Then,

$A_iA_k = A_kA_i = A_1A_1$

$$A_iA_k = A_1A_1 = \begin{bmatrix} 0 & 0 & 0 & 1 \\ 0 & 0 & 1 & 0 \\ 0 & 1 & 0 & 0 \\ 1 & 0 & 0 & 0 \end{bmatrix} \times \begin{bmatrix} 0 & 0 & 0 & 1 \\ 0 & 0 & 1 & 0 \\ 0 & 1 & 0 & 0 \\ 1 & 0 & 0 & 0 \end{bmatrix}$$

$$= \begin{bmatrix} 1 & 0 & 0 & 0 \\ 0 & 1 & 0 & 0 \\ 0 & 0 & 1 & 0 \\ 0 & 0 & 0 & 1 \end{bmatrix} = I$$

$$A_2A_2 = \begin{bmatrix} 0 & 0 & 0 & i \\ 0 & 0 & -i & 0 \\ 0 & i & 0 & 0 \\ -i & 0 & 0 & 0 \end{bmatrix} \times \begin{bmatrix} 0 & 0 & 0 & i \\ 0 & 0 & -i & 0 \\ 0 & i & 0 & 0 \\ -i & 0 & 0 & 0 \end{bmatrix} = \begin{bmatrix} 1 & 0 & 0 & 0 \\ 0 & 1 & 0 & 0 \\ 0 & 0 & 1 & 0 \\ 0 & 0 & 0 & 1 \end{bmatrix} = I$$

$\therefore A_iA_k + A_kA_i = I + I = 2I$

If $i \neq k$ let $i = 3$ and $k = 2$, then

$$A_iA = A_1A_2 = \begin{bmatrix} 0 & 0 & 0 & 1 \\ 0 & 0 & 1 & 0 \\ 0 & 1 & 0 & 0 \\ 1 & 0 & 0 & 0 \end{bmatrix} \times \begin{bmatrix} 0 & 0 & 0 & i \\ 0 & 0 & -i & 0 \\ 0 & i & 0 & 0 \\ -i & 0 & 0 & 0 \end{bmatrix} = \begin{bmatrix} -i & 0 & 0 & 0 \\ 0 & i & 0 & 0 \\ 0 & 0 & -i & 0 \\ 0 & 0 & 0 & i \end{bmatrix}$$

$$A_2A_1 = \begin{bmatrix} 0 & 0 & 0 & i \\ 0 & 0 & -i & 0 \\ 0 & i & 0 & 0 \\ -i & 0 & 0 & 0 \end{bmatrix} \times \begin{bmatrix} 0 & 0 & 0 & 1 \\ 0 & 0 & 1 & 0 \\ 0 & 1 & 0 & 0 \\ 1 & 0 & 0 & 0 \end{bmatrix} = \begin{bmatrix} -i & 0 & 0 & 0 \\ 0 & i & 0 & 0 \\ 0 & 0 & -i & 0 \\ 0 & 0 & 0 & -i \end{bmatrix}$$

$\Rightarrow A_1A_2 + A_2A_1 = O$

14. a., c.

$$\sin A = \begin{bmatrix} \cos\theta & \sin\theta \\ -\sin\theta & \cos\theta \end{bmatrix} \text{ and } \cos A = \begin{bmatrix} \sin\theta & \cos\theta \\ \cos\theta & \sin\theta \end{bmatrix}$$

$\therefore |\sin A| = \cos^2\theta + \sin^2\theta = 1$.

Hence, $\sin A$ is invertible.

$$\text{Also, } (\sin A)\times(\sin A)^T = \begin{bmatrix} \cos\theta & \sin\theta \\ -\sin\theta & \cos\theta \end{bmatrix}\begin{bmatrix} \cos\theta & -\sin\theta \\ \sin\theta & \cos\theta \end{bmatrix}$$

$$= \begin{bmatrix} \cos^2\theta + \sin^2\theta & 0 \\ 0 & \cos^2\theta + \sin^2\theta \end{bmatrix}$$

$$= \begin{bmatrix} 1 & 0 \\ 0 & 1 \end{bmatrix}$$

$$= I$$

Hence, $\sin A$ is orthogonal. Also,

$$2\sin A\cos A = 2\begin{bmatrix} \cos\theta & \sin\theta \\ -\sin\theta & \cos\theta \end{bmatrix}\begin{bmatrix} \sin\theta & \cos\theta \\ \cos\theta & \sin\theta \end{bmatrix}$$

$$= 2\begin{bmatrix} 2\sin\theta\cos\theta & \cos^2\theta + \sin^2\theta \\ \cos^2\theta - \sin^2\theta & 0 \end{bmatrix}$$

$$= 2\begin{bmatrix} \sin 2\theta & 1 \\ \cos 2\theta & 0 \end{bmatrix}$$

$$\neq \sin 2A$$

15. a., c. Let,

$$A = \begin{bmatrix} 0 & 2b & c \\ a & b & -c \\ a & -b & c \end{bmatrix}$$

Now,

$$A^T = \begin{bmatrix} 0 & a & a \\ 2b & b & -b \\ c & -c & c \end{bmatrix}$$

Hence, A is orthogonal. Therefore,

$$AA^{\mathrm{T}} = I \Rightarrow \begin{bmatrix} 0 & 2b & c \\ a & b & -c \\ a & -b & c \end{bmatrix}\begin{bmatrix} 0 & a & a \\ 2b & b & -b \\ c & -c & c \end{bmatrix} = \begin{bmatrix} 1 & 0 & 0 \\ 0 & 1 & 0 \\ 0 & 0 & 1 \end{bmatrix}$$

Equating the corresponding elements, we get

$4b^2 + c^2 = 1$ (1)

$2b^2 - c^2 = 0$ (2)

$a^2 + b^2 + c^2 = 1$ (3)

Solving Eqs. (1), (2), and (3), we get

$$a = \pm\frac{1}{\sqrt{2}}, b = \pm\frac{1}{\sqrt{6}}, c = \pm\frac{1}{\sqrt{3}}$$

16. a., b., d.

17. a., b., c.

If $|A| \neq 0$, then

$AB = AC$

$\Rightarrow A^{-1}AB = A^{-1}AC$

$\Rightarrow B = C$

Also if A is an orthogonal matrix, then, $AA^T = I$

$\Rightarrow |AA^T| = 1 \Rightarrow |A|^2 = 1 \Rightarrow A$ is invertible

18. a., b., c.

Applying $R_3 \to R_3 - R_2$, $R_2 \to R_2 - R_1$, we get

$$|a| = 3\begin{vmatrix} a_1 & a_2 & a_3 \\ d & d & d \\ d & d & d \end{vmatrix} = 0,$$

where d is the common difference of the A.P.

Therefore, the given system of equations has an infinite number of solutions. Also,

$|B| = a_1^{\,2} + a_2^{\,2} \neq 0$

19. a., b., d.

$$A = \begin{bmatrix} \cos\alpha & \sin\alpha & 0 \\ \cos\beta & \sin\beta & 0 \\ \cos\gamma & \sin\gamma & 0 \end{bmatrix}\begin{bmatrix} \cos\alpha & \cos\beta & \cos\gamma \\ \sin\alpha & \sin\beta & \sin\gamma \\ 0 & 0 & 0 \end{bmatrix}$$

Thus, A is symmetric and $|A| = 0$, hence singular and not invertible. Also,

$AA^T \neq I$

20. b., c.

$$A^{-1} = \begin{bmatrix} 1 & 0 \\ -1 & 1 \end{bmatrix}$$

$$\Rightarrow (A^{-1})^2 = \begin{bmatrix} 1 & 0 \\ -1 & 1 \end{bmatrix}\begin{bmatrix} 1 & 0 \\ -1 & 1 \end{bmatrix} = \begin{bmatrix} 1 & 0 \\ -2 & 1 \end{bmatrix}$$

Similarly,

$$(A^{-1})^3 = \begin{bmatrix} 1 & 0 \\ -2 & 1 \end{bmatrix}\begin{bmatrix} 1 & 0 \\ -1 & 1 \end{bmatrix} = \begin{bmatrix} 1 & 0 \\ -3 & 1 \end{bmatrix} \text{ and}$$

$$(A^{-1})^n = \begin{bmatrix} 1 & 0 \\ -n & 1 \end{bmatrix} \Rightarrow \lim_{n\to\infty}\frac{1}{n}A^{-n} = \lim_{n\to\infty}\begin{bmatrix} 1/n & 0 \\ -1 & 1/n \end{bmatrix} = \begin{bmatrix} 0 & 0 \\ -1 & 0 \end{bmatrix}$$

$$\lim_{n\to\infty}\frac{1}{n^2}A^{-n} = \lim_{n\to\infty}\begin{bmatrix} 1/n^2 & 0 \\ -1/n & 1/n^2 \end{bmatrix} = \begin{bmatrix} 0 & 0 \\ 0 & 0 \end{bmatrix}$$

21. a., d.

Here X is a $n \times 1$ matrix, C is a $n \times n$ matrix and X^T is a $1 \times n$ matrix. Hence, X^TCX is an 1×1 matrix. Let $X^TCX = k$. Then,

$(X^TCX)^T = X^TC^T(X^T)^T = X^T(-C)X = -X^TCX = -k$

$\Rightarrow k = -k$

$\Rightarrow k = 0$

$\Rightarrow X^TCX$ is null matrix

22. a., b., c.

$$S = \begin{bmatrix} 0 & 1 & 1 \\ 1 & 0 & 1 \\ 1 & 1 & 0 \end{bmatrix} \Rightarrow S^{-1} = \frac{1}{2}\begin{bmatrix} -1 & 1 & 1 \\ 1 & -1 & 1 \\ 1 & 1 & -1 \end{bmatrix}$$

We have,

$$SA = \begin{bmatrix} 0 & 1 & 1 \\ 1 & 0 & 1 \\ 1 & 1 & 0 \end{bmatrix}\begin{bmatrix} b+c & c+a & b-c \\ c-b & c+a & a-b \\ b-c & a-c & a+b \end{bmatrix}$$

$$= \begin{bmatrix} 0 & 2a & 2a \\ 2b & 0 & 2b \\ 2c & 2c & 0 \end{bmatrix}$$

$$\therefore SAS^{-1} = \begin{bmatrix} 0 & 2a & 2a \\ 2b & 0 & 2b \\ 2c & 2c & 0 \end{bmatrix}\frac{1}{2}\begin{bmatrix} -1 & 1 & 1 \\ 1 & -1 & 1 \\ 1 & 1 & -1 \end{bmatrix}$$

$$= \begin{bmatrix} 0 & a & a \\ b & 0 & b \\ c & c & 0 \end{bmatrix}\begin{bmatrix} -1 & 1 & 1 \\ 1 & -1 & 1 \\ 1 & 1 & -1 \end{bmatrix}$$

$$= \begin{bmatrix} 2a & 0 & 0 \\ 0 & 2b & 0 \\ 0 & 0 & 2c \end{bmatrix}$$

$= \text{diag}(2a, 2b, 2c)$

23. a., b., c.

All are properties of diagonal matrix.

24. a., b., c.

Given that A and B commute, we have

$AB = BA$ [$\because$ A and B are symmetric] (1)

Also,

$A^T = A, B^T = B$ (2)

$(A^{-1}B)^T = B^T(A^{-1})^T = BA^{-1}$

[$\because$ if A is symmetric, A^{-1} is also symmetric]

Also from Eq. (1),

$ABA^{-1} = B$ (2)

$\Rightarrow A^{-1}ABA^{-1} = A^{-1}B$

$\Rightarrow IBA^{-1} = A^{-1}B$

$\Rightarrow BA^{-1} = A^{-1}B$

Hence, from Eq. (2),

$(A^{-1}B)^T = A^{-1}B$

Thus, $A^{-1}B$ is symmetric. Similarly, AB^{-1} is also symmetric.

Also,

$BA = AB$

$\Rightarrow (BA)^{-1} = (AB)^{-1}$

$\Rightarrow A^{-1}B^{-1} = B^{-1}A^{-1}$

$\Rightarrow (A^{-1}B^{-1})^T = (B^{-1}A^{-1})^T$

$= (A^{-1})^T(B^{-1})^T$

$= A^{-1}B^{-1}$

Hence, $A^{-1}B^{-1}$ is symmetric.

25. b., c.

Since A is skew-symmetric, $A^T = -A$. We have,

$A^2 + I = O$

$\Rightarrow A^2 = -I$ or $AA = -I$

$\Rightarrow A(-A) = I$

$\Rightarrow AA^T = I$

Again, we know that

$|A| = |A^T|$ and $|kA| = k^n|A|$

where n is the order of A. Now,

$A^T = (-1)^n \times A$

$\Rightarrow |A^T| = (1)^n |A|$

$\Rightarrow [1 - (-1)^n] |A| = 0$

Hence, either $|A| = 0$ or $1 - (-1)^n = 0$, i.e., n is even. But

$A^2 = O - I = -I$

$\Rightarrow |A|^2 = (-1)^n |I| = (-1)^n \neq 0$

Hence, the only possibility is that A is of even order.

26. a., b., c., d.

We have, $A^2B = A(AB) = AA = A^2$, $B^2A = B(BA) = BB = B^2$,

$ABA = A(BA) = AB = A$ and $BAB = B(AB) = BA = B$.

27. a., b., d.

$$A^2 = \begin{bmatrix} 1 & 2 & 2 \\ 2 & 1 & 2 \\ 2 & 2 & 1 \end{bmatrix}\begin{bmatrix} 1 & 2 & 2 \\ 2 & 1 & 2 \\ 2 & 2 & 1 \end{bmatrix} = \begin{bmatrix} 9 & 8 & 8 \\ 8 & 9 & 8 \\ 8 & 8 & 9 \end{bmatrix}$$

We have, $A^2 - 4A - 5I$

$$A^2 - 4A - 5I_3 = \begin{bmatrix} 9 & 8 & 8 \\ 8 & 9 & 8 \\ 8 & 8 & 9 \end{bmatrix} - 4\begin{bmatrix} 1 & 2 & 2 \\ 2 & 1 & 2 \\ 2 & 2 & 1 \end{bmatrix} - 5\begin{bmatrix} 1 & 0 & 0 \\ 0 & 1 & 0 \\ 0 & 0 & 1 \end{bmatrix} = O$$

$\Rightarrow 5I_3 = A^2 - 4A = A(A - 4I_3)$

$\Rightarrow I_3 = \frac{1}{5}(A - 4I_3) \Rightarrow A^{-1} = \frac{1}{5}(A - 4I_3)$

Note that $|A| = 5$. Since $|A^3| = |A|^3 = 5^3 \neq 0$, A^3 is invertible. Similarly, A^2 is invertible.

Reasoning Type

1. c. We know that $|\text{adj } A| = |A|^{n-1}$. Hence, statement 2 is false.

Now,

$|\text{adj}(\text{adj } A)| = |\text{adj } A|^{n-1} = ||A|^{n-1}|^{n-1} = |A|^{(n-1)^2}$

Then,

$|\text{adj}(\text{adj}(\text{adj } A))| = |\text{adj}(\text{adj } A))|^{n-1}$

$= ||A|^{(n-1)^2}|^{n-1}$

$= |A|^{(n-1)^3}$

Hence, statement 1 is true.

2. b. Both the statements are true as both are standard properties of diagonal matrix. But statement 2 does not explain statement 1.

3. d. Matrix $a_{ij} = \dfrac{i-j}{i+2j}$ is $A = \begin{bmatrix} 0 & -\frac{1}{5} & -\frac{2}{7} \\ \frac{1}{4} & 0 & -\frac{1}{8} \\ \frac{2}{5} & \frac{1}{7} & 0 \end{bmatrix}$ which is neither symmetric nor skew-symmetric. But this is not the reason for which A cannot be expressed as sum of symmetric and skew-symmetric matrix. In fact any matrix can be expressed as a sum of symmetric and skew-symmetric matrix. Hence, statement 1 is false but statement 2 is true.

4. a. Given,

$$A = \begin{bmatrix} a & b \\ c & d \end{bmatrix}$$

$$\Rightarrow A^2 = \begin{bmatrix} a & b \\ c & d \end{bmatrix}\begin{bmatrix} a & b \\ c & d \end{bmatrix} = \begin{bmatrix} a^2 + bc & ab + bd \\ ac + cd & bc + d^2 \end{bmatrix}$$

Hence,

$A^2-(a+d)A+(ad-bc)I$

$$=\begin{bmatrix} a^2+bc & ab+bd \\ ac+cd & bc+d^2 \end{bmatrix}-(a+d)\begin{bmatrix} a & b \\ c & d \end{bmatrix}+(ad-bc)\begin{bmatrix} 1 & 0 \\ 0 & 1 \end{bmatrix}$$

$$=\begin{bmatrix} a^2+bc-(a^2+ad)+(ad-bc) & ab+bd-(ab+bd) \\ ac+cd-(ac+cd) & bc+d^2-(ad+d^2)+(ad-bc) \end{bmatrix}$$

$$=\begin{bmatrix} 0 & 0 \\ 0 & 0 \end{bmatrix}$$

$= O$

Given,

$A^3 = O$

$\Rightarrow |A| = 0$ or $ad-bc=0$

$\Rightarrow A^2-(a+d)A = O$ or $A^2=(a+d)A$ (1)

Case (1)

$a+d=0$

From Eq. (1),

$A^2 = O$

Case (2)

$a+d \neq 0$

Given,

$A^3 = O$

$\Rightarrow A^2A = O$

$\Rightarrow (A+d)A \times A = O$

$\Rightarrow A^2 = O$

5. b. $\text{adj}(F(\alpha)) = \begin{bmatrix} \cos\alpha & \sin\alpha & 0 \\ -\sin\alpha & \cos\alpha & 0 \\ 0 & 0 & 1 \end{bmatrix}$

Also,

$|F(\alpha)| = 1$

Then,

$$[F(\alpha)]^{-1} = \begin{bmatrix} \cos\alpha & -\sin\alpha & 0 \\ -\sin\alpha & \cos\alpha & 0 \\ 0 & 0 & 1 \end{bmatrix}$$

$$= \begin{bmatrix} \cos(-\alpha) & \sin(-\alpha) & 0 \\ \sin(-\alpha) & \cos(-\alpha) & 0 \\ 0 & 0 & 1 \end{bmatrix}$$

$= F(-\alpha)$

Similarly, we can prove that $[G(\beta)]^{-1} = G(-\beta)$.

But again given matrices $F(\alpha)$ and $G(\beta)$ are special matrices for which this type of result holds.

In general, such result is not true. You can verify with any other matrix. Hence, both statements are true but statement 2 is not correct explanation of statement 1.

6. a. Statement 1 is true as $|A| = 0$. Since $|B| \neq 0$, statement 2 is also true and correct explanation of statement 1.

7. a. $A = -A^T \Rightarrow |A| = -|A^T| = -|A|$

$\Rightarrow 2|A| = 0$

$\Rightarrow |A| = 0$

8. d. Statement 1 is false.

$\because A = [A_{ij}]_{n\times n}$ where $a_{ij} = 0, i \geq j$

Therefore, $|A| = 0$ and hence A is singular. So, inverse of A is not defined.

In statement 2, $|A| = 0$. Therefore, inverse of A is not defined.

9. d. A^{-1} exists only for non-singular matrix.

$\therefore AB = AC \Rightarrow B = C$ if A^{-1} exists

10. d. ABC is not defined, as order of A, B and C are such that they are not conformable for multiplication.

11. c. $[A(A+B)^{-1}B]^{-1} = B^{-1}((A+B)^{-1})^{-1}A^{-1}$

$= B^{-1}(A+B)A^{-1} = (B^{-1}A+I)A^{-1} = B^{-1}I + IA^{-1} = B^{-1}+A^{-1}$

Hence, statement 1 is true. Statement 2 is false as $(A+B)^{-1} = A^{-1}+B^{-1}$ is not true.

12. b. Since $AB = BA$, we have

$(A+B)^r = {}^rC_0A^r + {}^rC_1A^{r-1}B + {}^rC_2A^{r-2}B^2 + \cdots + {}^rC_sA^{r-s}B^s + \cdots + {}^rC_{r-1}AB^{r-1} + {}^rC_rB^r$

If $r = m+n$, then

$A^{r-p}B^p = A^mB^{r-p-m}B^p = O$ if $p \leq n$

and $A^{r-p}B^p = A^{r-p}B^nB^{p-n} = O$ if $p > n$

Then, $(A+B)^r = O$, for $r = m+n$

Thus, both the statements are correct but statement 2 is not correctly explaining statement 1.

13. b. Let, $A = \begin{bmatrix} d_1 & z_1 & z_2 \\ \overline{z}_1 & d_2 & z_3 \\ \overline{z}_2 & \overline{z}_3 & d_3 \end{bmatrix}$.

$A^2 = O$

$$\Rightarrow \begin{bmatrix} d_1 & z_1 & z_2 \\ \overline{z}_1 & d_2 & z_3 \\ \overline{z}_2 & \overline{z}_3 & d_3 \end{bmatrix}\begin{bmatrix} d_1 & z_1 & z_2 \\ \overline{z}_1 & d_2 & z_3 \\ \overline{z}_2 & \overline{z}_3 & d_3 \end{bmatrix}$$

$$= \begin{bmatrix} d_1^2+|z_1|^2+|z_2|^2 & d_1z_1+d_2z_1+z_2\overline{z}_3 & d_1z_2+z_1z_3+z_2d_3 \\ d_1\overline{z}_1+d_2\overline{z}_1+z_3\overline{z}_2 & d_2^2+|z_1|^2+|z_3|^2 & \overline{z}_1z_2+d_2z_3+z_3d_3 \\ d_1\overline{z}_2+\overline{z}_3\overline{z}_1+d_3\overline{z}_2 & z_1\overline{z}_2+d_2\overline{z}_3+d_3\overline{z}_3 & d_3^2+|z_1|^2+|z_2|^2 \end{bmatrix} = O$$

$\Rightarrow$ Diagonal elements $d_1 = d_2 = d_3 = 0$ and $|z_1| = |z_2| = |z_3| = 0$

$\Rightarrow z_1 = z_2 = z_3 = 0$

$\Rightarrow A =$ Null matrix

Thus, statement 1 is true. Also,

$A^2 = O \Rightarrow |A|^2 = 0$ or $|A| = 0$

Thus, statement 2 is true but it does not explain statement 1.

14. a. $\begin{bmatrix} a & b \\ c & d \end{bmatrix}\begin{bmatrix} a & c \\ b & d \end{bmatrix} = \begin{bmatrix} 1 & 0 \\ 0 & 1 \end{bmatrix}$

$$\Rightarrow \begin{bmatrix} a^2+b^2 & ac+bd \\ ac+bd & c^2+d^2 \end{bmatrix} = \begin{bmatrix} 1 & 0 \\ 0 & 1 \end{bmatrix}$$

$\Rightarrow a^2 + b^2 = 1$ (1)

$c^2 + d^2 = 1$ (2)

$ac + bd = 0$ (3)

$\Rightarrow \dfrac{a}{d} = \dfrac{-b}{c} = k$ (let)

$\Rightarrow c^2 + d^2 = 1/k^2$ or $k^2 = 1$ or $k = \pm 1$

$\Rightarrow \dfrac{a}{d} = \dfrac{-b}{c} = \pm 1$

Also, we must have $a, b, c, d \in [-1, 1]$ for Eqs. (1) and (2) to get defined. Hence, without loss of generality, we can assume $a = \cos\theta$ and $b = \sin\theta$.

So for

$\dfrac{a}{d} = \dfrac{-b}{c} = 1$, we have $\begin{bmatrix} a & b \\ c & d \end{bmatrix} \equiv \begin{bmatrix} \cos\theta & -\sin\theta \\ \sin\theta & \cos\theta \end{bmatrix}$ and

for $\dfrac{a}{d} = \dfrac{-b}{c} = -1$, we have $\begin{bmatrix} a & b \\ c & d \end{bmatrix} \equiv \begin{bmatrix} \cos\theta & \sin\theta \\ \sin\theta & -\cos\theta \end{bmatrix}$

Linked Comprehension Type

For Problems 1–3

1. b., 2. b., 3. d.

Sol.

1. Let,

$$a = \begin{bmatrix} a & b \\ c & d \end{bmatrix}$$

$\Rightarrow A^2 - (a+d)A + (ad-bc)I$

$$= \begin{bmatrix} a & b \\ c & d \end{bmatrix}\begin{bmatrix} a & b \\ c & d \end{bmatrix} - (a+d)\begin{bmatrix} a & b \\ c & d \end{bmatrix} + (ad-bc)\begin{bmatrix} 1 & 0 \\ 0 & 1 \end{bmatrix}$$

$$= \begin{bmatrix} a^2+bc & ab+bd \\ ac+cd & bc+d^2 \end{bmatrix} - \begin{bmatrix} a^2+ad & ab+bd \\ ac+cd & ad+d^2 \end{bmatrix} + \begin{bmatrix} ad-bc & 0 \\ 0 & ad-bc \end{bmatrix}$$

$= O$

2. If $A = O$, tr$(A) = 0$. Suppose $A \neq O$ and $A = \begin{bmatrix} a & b \\ c & d \end{bmatrix}$. Then

$|A| = 0$ and $A^2 - (A+d)A + (ad-bc)I = 0$

$\Rightarrow a + d = 0$

3. $(I+A)^{100} = {}^{100}C_0 I^{100} + {}^{100}C_1 I^{99} A + {}^{100}C_2 I^{98} A^2 + \cdots + {}^{100}C_{100} A^{100}$

$= I + 100A + O + O + \cdots + O$

$= I + 100A$

For Problems 4–6

4. c., 5. c., 6. c.

Sol.

4. $AB = A \Rightarrow |AB| = |A|$ (1)

$\Rightarrow |A| = 0$ or $|B| = 1$

$BA = B \Rightarrow |BA| = |B|$ (2)

$\Rightarrow |A| = 1$ or $|B| = 0$

If $|A| = 0$, then from Eq. (2), $|B| = 0$

If $|B| = 0$, then from Eq. (1), $|A| = 0$

5. $AB = A$, $BA = B$

$ABA = A^2 \Rightarrow A(BA) = A^2 \Rightarrow AB = A^2 \Rightarrow A = A^2$

Similarly, $B^2 = B$

$(A+B)^2 = A^2 + B^2 + AB + BA$

$\ldots = A + B + A + B = 2(A+B)$

$(A+B)^3 = (A+B)^2(A+B) = 2(A+B)^2 = 2^2(A+B)$

$\Rightarrow (A+B)^7 = 2^6(A+B) = 64\,(A+B)$

6. $(A+I)^5 = I + 5A + 10A^2 + 10A^3 + 5A^4 + A^5$

$= I + 5A + 10A + 10A + 5A + A$

$[\because A^2 = A \Rightarrow A^3 = A^4 = A^5 = \cdots = A]$

$= I + 31A$

For Problems 7–8

7. d., 8. c.

Sol.

$$A = \begin{bmatrix} a_{11} & a_{12} & a_{13} \\ a_{21} & a_{22} & a_{23} \\ a_{31} & a_{32} & a_{33} \end{bmatrix}$$

$$\Rightarrow B = \begin{bmatrix} a_{12}+a_{13} & a_{11}+a_{13} & a_{11}+a_{12} \\ a_{22}+a_{23} & a_{21}+a_{23} & a_{21}+a_{22} \\ a_{32}+a_{33} & a_{31}+a_{33} & a_{31}+a_{32} \end{bmatrix}$$

$\Rightarrow X = A^{-1}B$

$$= \frac{1}{|A|}\begin{bmatrix} C_{11} & C_{21} & C_{31} \\ C_{12} & C_{22} & C_{32} \\ C_{13} & C_{23} & C_{33} \end{bmatrix}\begin{bmatrix} a_{12}+a_{13} & a_{11}+a_{13} & a_{11}+a_{12} \\ a_{22}+a_{23} & a_{21}+a_{23} & a_{21}+a_{22} \\ a_{32}+a_{33} & a_{31}+a_{33} & a_{31}+a_{32} \end{bmatrix}$$

$$= \frac{1}{|A|}\begin{bmatrix} 0 & |A| & |A| \\ |A| & 0 & |A| \\ |A| & |A| & 0 \end{bmatrix} = \begin{bmatrix} 0 & 1 & 1 \\ 1 & 0 & 1 \\ 1 & 1 & 0 \end{bmatrix}$$

$\Rightarrow |A^{-1}B| = 2$

$\Rightarrow |A^{1}||B| = 2$

$\Rightarrow |B| = 2|A|$

For Problems 9–11

9. b., 10. d., 11. b.

Sol.

$A^n - A^{n-2} = A^2 - I \Rightarrow A^{50} = A^{48} + A^2 - I$

Further,

$A^{48} = A^{46} + A^2 - I$

$A^{46} = A^{44} + A^2 - I$

$\vdots \quad \vdots \quad \vdots$

$A^4 = A^2 + A^2 - I$

$A^{50} = 25A^2 - 24I$

Here,

$$A^2 = \begin{bmatrix} 1 & 0 & 0 \\ 1 & 0 & 1 \\ 0 & 1 & 0 \end{bmatrix}\begin{bmatrix} 1 & 0 & 0 \\ 1 & 0 & 1 \\ 0 & 1 & 0 \end{bmatrix} = \begin{bmatrix} 1 & 0 & 0 \\ 1 & 1 & 0 \\ 1 & 0 & 1 \end{bmatrix}$$

$$\Rightarrow A^{50} = \begin{bmatrix} 25 & 0 & 0 \\ 25 & 25 & 0 \\ 25 & 0 & 25 \end{bmatrix} - 24\begin{bmatrix} 1 & 0 & 0 \\ 0 & 1 & 0 \\ 0 & 0 & 1 \end{bmatrix} = \begin{bmatrix} 1 & 0 & 0 \\ 25 & 1 & 0 \\ 25 & 0 & 1 \end{bmatrix}$$

$\therefore |A^{50}| = 1$

Also, $\text{tr}(A^{50}) = 1 + 1 + 1 = 3$. Further,

$$\begin{bmatrix} 1 & 0 & 0 \\ 25 & 1 & 0 \\ 25 & 0 & 1 \end{bmatrix}\begin{bmatrix} x \\ y \\ z \end{bmatrix} = \begin{bmatrix} 1 \\ 25 \\ 25 \end{bmatrix} \Rightarrow \begin{bmatrix} x \\ y \\ z \end{bmatrix} \cup_1 = \begin{bmatrix} 1 \\ 0 \\ 0 \end{bmatrix}$$

Similarly,

$$\cup_2 = \begin{bmatrix} 0 \\ 1 \\ 0 \end{bmatrix} \text{ and } \cup_3 = \begin{bmatrix} 0 \\ 0 \\ 1 \end{bmatrix} \Rightarrow \cup \begin{bmatrix} 1 & 0 & 0 \\ 0 & 1 & 0 \\ 0 & 0 & 1 \end{bmatrix}, \text{ i.e., } |\cup| = 1$$

For Problems 12–14

12. c., 13. d., 14. b.

Sol.

12. $$A = \begin{bmatrix} 2 & 1 & 1 \\ 2 & 3 & 4 \\ -1 & -1 & -2 \end{bmatrix}$$

$$\Rightarrow \lambda I = \begin{bmatrix} \lambda & 0 & 0 \\ 0 & \lambda & 0 \\ 0 & 0 & \lambda \end{bmatrix}$$

$$\Rightarrow A - \lambda I = \begin{bmatrix} 2-\lambda & 1 & 1 \\ 2 & 3-\lambda & 4 \\ -1 & -1 & -2-\lambda \end{bmatrix}$$

$\Rightarrow \det(A - \lambda I) = -(\lambda - 1)(\lambda + 1)(\lambda - 3)$

Thus, the characteristic roots are –1, 1 and 3.

13. Option (a) is not correct since its characteristic determinant is $\begin{vmatrix} -\lambda & 1 \\ 1 & -\lambda \end{vmatrix}$

The characteristic equation is $\lambda^2 - 1 = 0$. Therefore, $\lambda = 1, -1$

Hence, eigenvalues are 1 and –1.

We similarly note that matrices given in options (b) and (c) have eigenvalues 1 and –1. Hence, they are not correct.

Option (d) has characteristic equation $(1 - \lambda)^2 = 0$. Hence, eigenvalues are not 1 and –1.

14. b. Let, $A = \begin{bmatrix} a_1 & b_1 & c_1 \\ a_2 & b_2 & c_2 \\ a_3 & b_3 & c_3 \end{bmatrix}$

$$\Rightarrow A - \lambda I = \begin{bmatrix} a_1-\lambda & b_1 & c_1 \\ a_2 & b_2-\lambda & c_2 \\ a_3 & b_3 & c_3-\lambda \end{bmatrix}$$

$\Rightarrow \det(A - \lambda I) = (a_1 - \lambda)[(b_2 - \lambda)(c_3 - \lambda) - b_3c_2] - b_1[a_2(c_3 - \lambda) - a_3c_2] + c_1[a_2b_3 - a_3(b_2 - \lambda)]$

Now one of the eigenvalues is zero, so one root of equation should be zero. Therefore, constant term in the above polynomial is zero.

$\therefore a_1b_2c_3 - a_1b_3c_2 - b_1a_2c_3 + b_1c_2a_3 + c_1a_2b_3 - c_1a_3b_2 = 0$

[Collecting constant terms]

But this value is value of determinant of A.

$\therefore \det A = 0$

For Problems 15–17

15. c., 16. d., 17. c.

Sol.

15. As second row of all the options is same, we are to look at the elements of the first row. Let the left inverse be $\begin{bmatrix} a & b & c \\ d & e & f \end{bmatrix}$. Then,

$$\begin{bmatrix} a & b & c \\ d & e & f \end{bmatrix}\begin{bmatrix} 1 & -1 \\ 1 & 1 \\ 2 & 3 \end{bmatrix} = \begin{bmatrix} 1 & 0 \\ 0 & 1 \end{bmatrix}$$

$\therefore a + b + 2c = 1$

$-a + b + 3c = 1$, i.e., $b = \dfrac{1-5c}{2}, a = \dfrac{1+c}{2}$

Thus, matrices in the options (a), (b), and (d) are the inverses and matrix in option (c) is not the left inverse.

16. Let right inverse be $\begin{bmatrix} a & b \\ c & d \\ e & f \end{bmatrix}$. Then,

$$\begin{bmatrix} 1 & -1 & 2 \\ 2 & -1 & 1 \end{bmatrix}\begin{bmatrix} a & b \\ c & d \\ e & f \end{bmatrix} = \begin{bmatrix} 1 & 0 \\ 0 & 1 \end{bmatrix}$$

Now, $a - c + 2e = 1$

$b - d + 2f = 0$

$2a - c + e = 0$

$2b - d + f = 1$

This system of equations has infinite solutions.

17. By observation there cannot be any left inverse for options (b) and (d). So we will check for (a) and (c) only.

For option (a), let the left inverse be $\begin{bmatrix} a & b \\ c & d \\ e & f \end{bmatrix}$. Then,

$$\begin{bmatrix} a & b \\ c & d \\ e & f \end{bmatrix}\begin{bmatrix} 1 & 2 & 4 \\ -3 & 2 & 1 \end{bmatrix} = \begin{bmatrix} 1 & 0 & 0 \\ 0 & 1 & 0 \\ 0 & 0 & 1 \end{bmatrix}$$

Now, $a - 3b = 1$, $2a + 2b = 0$, and $4a + b = 0$ which is not possible. For option (c),

$$\begin{bmatrix} a & b & c \\ d & e & f \end{bmatrix}\begin{bmatrix} 1 & 4 \\ 2 & -3 \\ 5 & 4 \end{bmatrix} = \begin{bmatrix} 1 & 0 \\ 0 & 1 \end{bmatrix}$$

$\Rightarrow a + 2b + 5c = 1$, $4a - 3b + 4c = 0$, $d + 2e + 5f = 0$, $4d - 3e + 4f = 1$

Therefore, there are infinite number of left inverses.

$$\begin{bmatrix} 1 & 4 \\ 2 & -3 \\ 5 & 4 \end{bmatrix}\begin{bmatrix} a & b & c \\ d & e & f \end{bmatrix} = \begin{bmatrix} 1 & 0 & 0 \\ 0 & 1 & 0 \\ 0 & 0 & 1 \end{bmatrix}$$

$\Rightarrow a + 4d = 1$, $2a - 3d = 0$, and $5a + 4d = 0$

which is not possible. Therefore, there is no right inverse.

For Problems 18–20

18. a., 19. b., 20. c.

Sol.

$$A = \begin{bmatrix} x & x \\ x & x \end{bmatrix} \Rightarrow A^2 = \begin{bmatrix} 2x^2 & 2x^2 \\ 2x^2 & 2x^2 \end{bmatrix}, A^3 = \begin{bmatrix} 2^2x^2 & 2^2x^2 \\ 2^2x^2 & 2^2x^2 \end{bmatrix}$$

and so on. Then

$$e^A = I + A + \frac{A^2}{2!} + \frac{A^3}{3!} + \cdots$$

$$= \begin{bmatrix} 1+x+\frac{2x^2}{2!}+\frac{2^2x^3}{3!}+\cdots & x+\frac{2x^2}{2!}+\frac{2^2x^3}{3!}+\cdots \\ x+\frac{2x^2}{2!}+\frac{2^2x^3}{3!}+\cdots & 1+x+\frac{2x^2}{2!}+\frac{2^2x^3}{3!}+\cdots \end{bmatrix}$$

$$= \begin{bmatrix} \frac{1}{2}\left(1+2x+\frac{2^2x^2}{2!}+\frac{2^3x^3}{3!}+\cdots\right)+\frac{1}{2} & \frac{1}{2}\left(1+2x+\frac{2^2x^2}{2!}+\cdots\right)-\frac{1}{2} \\ \frac{1}{2}\left(1+2x+\frac{2^2x^2}{2!}+\frac{2^3x^3}{3!}+\cdots\right)-\frac{1}{2} & \frac{1}{2}\left(1+2x+\frac{2^2x^2}{2!}+\cdots\right)+\frac{1}{2} \end{bmatrix}$$

$$= \frac{1}{2}\begin{bmatrix} e^{2x}+1 & e^{2x}-1 \\ e^{2x}-1 & e^{2x}+1 \end{bmatrix}$$

$\Rightarrow f(x) = e^{2x} + 1$ and $g(x) = e^{2x} - 1$

18. $\int \frac{e^{2x}-1}{e^{2x}+1}dx = \int \frac{e^x - e^{-x}}{e^x + e^{-x}}dx = \log_e(e^x + e^{-x}) + c$

19.

$$\int (g(x)+1)\sin x\,dx = \int e^{2x}\sin x\,dx = \frac{e^{2x}}{5}(2\sin x - \cos x) + c$$

20. $\int \frac{e^{2x}+1}{\sqrt{e^{2x}-1}}dx = \int \frac{e^{2x}}{\sqrt{e^{2x}-1}}dx + \int \frac{1}{\sqrt{e^{2x}-1}}dx$

$$= \int \frac{e^{2x}}{\sqrt{e^{2x}-1}}dx + \int \frac{e^x}{e^x\sqrt{e^{2x}-1}}dx$$

$$= \frac{1}{2\sqrt{e^{2x}-1}} + \sec^{-1}(e^x) + c$$

Matrix-Match Type

1. **a → s; b → p; c → q; d → r.**

a. Since A is idempotent, hence,

$A^2 = A$

$\Rightarrow A^3 = AA^2 = AA = A^2 = A,\ A^4 = AA^3 = AA = A^2 = A$

$\Rightarrow A^n = A$

$\Rightarrow (I - A)^n = {}^nC_0I - {}^nC_1A + {}^nC_2A^2 - {}^nC_3A^3 + \cdots$

$= I + (-{}^nC_1 + {}^nC_2 - {}^nC_3 + \cdots)A$

$= I + [({}^nC_0 - {}^nC_1 + {}^nC_2 - {}^nC_3 + \cdots) - {}^nC_0]A = I - A$

b. A is involuntary. Hence,

$A^2 = I$

$\Rightarrow A^3 = A^5 = \cdots = A$ and $A^2 = A^4 = A^6 = \cdots = I$

$\Rightarrow (I - A)^n = {}^nC_0I - {}^nC_1A + {}^nC_2A^2 - {}^nC_3A^3 + \cdots$

$= {}^nC_0I - {}^nC_1A + {}^nC_2I - {}^nC_3A + {}^nC_4I - \cdots$

$= ({}^nC_0 + {}^nC_2 + {}^nC_4 + \cdots)I - ({}^nC_1A + {}^nC_3 + {}^nC_5 + \cdots)$

$A = 2^{n-1}(I - A)$

$\Rightarrow [(I - A)^n]A^{-1} = 2^{n-1}(I - a)A^{-1} = 2^{n-1}(A^{-1} - I)$

c. If A is nilpotent of index 2, then

$A^2 = A^3 = A^4 \cdots = A^n = O$

$\Rightarrow (I - A)^n = {}^nC_0I - {}^nC_1A + {}^nC_2A^2 - {}^nC_3A^3 + \cdots$

$= I - nA + O + O + \cdots$

$= I - nA$

d. A is orthogonal. Hence,

$AA^T = I$

$\Rightarrow (A^T)^{-1} = A$

2. **a → r; b → s; c → p, r; d → p, q, r, s.**

a. Since A is idempotent, $A^2 = A^3 = A^4 = \cdots = A$. Now,

$(A + I)^n = I + {}^nC_1A + {}^nC_2A^2 + \cdots + {}^nC_nA^n$

$= I + {}^nC_1A + {}^nC_2A + \cdots + {}^nC_nA$

$= I + {}^nC_1A + {}^nC_2A + \cdots + {}^nC_nA$

$= I + ({}^nC_1 + {}^nC_2 + \cdots + {}^nC_n)A$

$= I + (2^n - 1)A$

$\Rightarrow 2^n - 1 = 127$

$\Rightarrow n = 7$

b. We have,

$(I - A)(I + A + A^2 + \cdots + A^7)$

$= I + A + A^2 + \cdots + A^7 + (-A - A^2 - A^3 - A^4 \cdots - A^8)$

$= I - A^8$

$= I$ (if $A^8 = O$)

c. Here matrix A is skew-symmetric and since $|A| = |A^T| = (-1)^n\,|A|$, so $|A|(1 - (-1)^n) = 0$. As n is odd, hence $|A| = 0$. Hence A is singular.

d. If A is symmetric, A^{-1} is also symmetric for matrix of any order.

3. **a → q; b → p; c → s; d → r.**

a. $|A| = 2 \Rightarrow |2A^{-1}| = 2^3/|A| = 4$

b. $|\text{adj}(\text{adj}(2A))| = |2A|^4 = 2^{12}|A|^4 = 2^{12}/\,2^{12} = 1$

c. $(A + B)^2 = A^2 + B^2$

$\Rightarrow AB + BA = O$

$\Rightarrow |AB| = |-BA| = -|BA| = -|AB|$

$\Rightarrow |AB| = 0$

$\Rightarrow |B| = 0$

d. Product ABC is not defined.

Integer Type

1. **(1)** $A^2 = 3A - 2I$

Now $A^4 = (A^2)^2 = (3A - 2I)^2$

$= 9A^2 - 12A \times I + 4I^2$

$= 9A^2 - 12A + 4I$

$= 9(3A - 2I) - 12A + 4I$

$= 15A - 14I$

$A^8 = (A^4)^2 = (15A - 14I)^2$

$= 225A^2 - 420A \times I + 196I^2$

$= 225(3A - 2I) - 420A + 196I$

$= 255A - 254I$

2. **(5)** $\begin{bmatrix} 2 & 0 & 7 \\ 0 & 1 & 0 \\ 1 & -2 & 1 \end{bmatrix}\begin{bmatrix} -x & 14x & 7x \\ 0 & 1 & 0 \\ x & -4x & -2x \end{bmatrix}$

$= \begin{bmatrix} 5x & 0 & 0 \\ 0 & 1 & 0 \\ 0 & 10x - 2 & 5x \end{bmatrix} = \begin{bmatrix} 1 & 0 & 0 \\ 0 & 1 & 0 \\ 0 & 0 & 1 \end{bmatrix}$ (given)

$\Rightarrow 5x = 1$ and $10x - 2 = 0$

Therefore, $x = \frac{1}{5}$

3. **(1)** $AB = \begin{bmatrix} 1 & 2 \\ 3 & 4 \end{bmatrix}\begin{bmatrix} a & b \\ c & d \end{bmatrix}$

$= \begin{bmatrix} a + 2c & b + 2d \\ 3a + 4c & 3b + 4d \end{bmatrix},$

$BA = \begin{bmatrix} a & b \\ c & d \end{bmatrix}\begin{bmatrix} 1 & 2 \\ 3 & 4 \end{bmatrix} = \begin{bmatrix} a + 3b & 2a + 4b \\ c + 3d & 2c + 4d \end{bmatrix}$

If $AB = BA$, then $a + 2c = a + 3b \Rightarrow 2c = 3b \Rightarrow b \neq 0$

$b + 2d = 2a + 4b$

$\Rightarrow 2a - 2d = -3b$

$\Rightarrow 2a - 2d = -3b$

$\Rightarrow \dfrac{a - d}{3b - c} = \dfrac{-\frac{3b}{2}}{3b - \frac{3}{2}b} = -1$

4. **(5)** $(A + B)^2 = A^2 + B^2$

$\Rightarrow AB = BA$

$\Rightarrow \begin{bmatrix} 1 & -1 \\ 2 & -1 \end{bmatrix}\begin{bmatrix} a & 1 \\ b & -1 \end{bmatrix} = \begin{bmatrix} a & 1 \\ b & -1 \end{bmatrix}\begin{bmatrix} 1 & -1 \\ 2 & -1 \end{bmatrix}$

$\Rightarrow \begin{bmatrix} a - b & 2 \\ 2a - b & 3 \end{bmatrix} = \begin{bmatrix} a + 2 & -a - 1 \\ b - 2 & -b + 1 \end{bmatrix}$

$\Rightarrow a - b = a + 2$

$2 = -a - 1$

$2a - b = b - 2$

$3 = -b + 1$

$\Rightarrow b = -2$ and $a = -3$

5. **(3)** We have

$A = \begin{bmatrix} 1 & 2 & 2 \\ 2 & 1 & -2 \\ a & 2 & b \end{bmatrix} \Rightarrow A^T = \begin{bmatrix} 1 & 2 & a \\ 2 & 1 & 2 \\ 2 & -2 & b \end{bmatrix}$

$AA^T = 9I_3$

$\Rightarrow \begin{bmatrix} 1 & 2 & 2 \\ 2 & 1 & -2 \\ a & 2 & b \end{bmatrix}\begin{bmatrix} 1 & 2 & a \\ 2 & 1 & 2 \\ 2 & -2 & b \end{bmatrix} = 9\begin{bmatrix} 1 & 0 & 0 \\ 0 & 1 & 0 \\ 0 & 0 & 1 \end{bmatrix}$

$\Rightarrow \begin{bmatrix} 9 & 0 & a + 2b + 4 \\ 0 & 9 & 2a + 2 - 2b \\ a + 2b + 4 & 2a + 2 - 2b & a^2 + 4 + b^2 \end{bmatrix} = \begin{bmatrix} 9 & 0 & 0 \\ 0 & 9 & 0 \\ 0 & 0 & 9 \end{bmatrix}$

$\Rightarrow a + 2b + 4 = 0,\ 2a + 2 - 2b = 0$, and $a^2 + 4 + b^2 = 9$

$\Rightarrow a + 2b + 4 = 0,\ a - b + 1 = 0$, and $a^2 + b^2 = 5$

Solving $a + 2b + 4 = 0$ and $a - b + 1 = 0$, we get

$a = -2,\ b = -1$

Clearly, these values satisfy $a^2 + b^2 = 5$

Hence, $a = -2$ and $b = -1$

6. **(5)** Here $A = \begin{bmatrix} 1 & -1 & 1 \\ 2 & 1 & -3 \\ 1 & 1 & 1 \end{bmatrix}$.

Therefore, $|A| = \begin{vmatrix} 1 & -1 & 1 \\ 2 & 1 & -3 \\ 1 & 1 & 1 \end{vmatrix}$

$= 1(1 + 3) + 1\,(2 + 3) + 1\,(2 - 1)$

$= 4 + 5 + 1 = 10.$

$\Rightarrow \text{adj } A = \begin{bmatrix} 4 & -5 & 1 \\ 2 & 0 & -2 \\ 2 & 5 & 3 \end{bmatrix}^T = \begin{bmatrix} 4 & 2 & 2 \\ -5 & 0 & 5 \\ 1 & -2 & 3 \end{bmatrix}$

$\Rightarrow B = A^{-1} = \frac{1}{10}\begin{bmatrix} 4 & 2 & 2 \\ -5 & 0 & 5 \\ 1 & -2 & 3 \end{bmatrix}$

$\Rightarrow (10)\,B = \begin{bmatrix} 4 & 2 & 2 \\ -5 & 0 & 5 \\ 1 & -2 & 3 \end{bmatrix}$

Hence, $\alpha = 5$.

7. **(2)** $|A| = 2$; $|B| = 3$; $|C| = 5$

$\det(A^2BC^{-1}) = |A^2BC^{-1}| = \frac{|A|^2|B|}{|C|} = \frac{4 \times 3}{5} = \frac{12}{5}$

8. **(3)** Let $U_1 = \begin{bmatrix} a \\ b \\ c \end{bmatrix}$, then $\begin{bmatrix} 1 & 0 & 0 \\ 2 & 1 & 0 \\ 3 & 2 & 1 \end{bmatrix}\begin{bmatrix} a \\ b \\ c \end{bmatrix} = \begin{bmatrix} 1 \\ 0 \\ 0 \end{bmatrix}$

$\Rightarrow \begin{bmatrix} a \\ 2a + b \\ 3a + 2b + c \end{bmatrix} = \begin{bmatrix} 1 \\ 0 \\ 0 \end{bmatrix}$

$\Rightarrow a = 1, b = -2, c = 1$

Therefore, $U_1 = \begin{bmatrix} 1 \\ -2 \\ 1 \end{bmatrix}$. Smililarly, $U_2 = \begin{bmatrix} 2 \\ -1 \\ -4 \end{bmatrix}$ and $U_3 \begin{bmatrix} 2 \\ -1 \\ -3 \end{bmatrix}$

$U = \begin{bmatrix} 1 & 2 & 2 \\ -2 & -1 & -1 \\ 1 & -4 & -3 \end{bmatrix} \Rightarrow |U| = 3$

9. **(6)** Let the matrix be $\begin{bmatrix} a & b & c \\ p & q & r \\ x & y & z \end{bmatrix}$

We have five entries 1 and the remaining four entries are 0.

Since the matrix is symmetric, we must have an even number of zeros for $i \neq j$.

Hence, we have two cases:

(i) Two entries in the diagonal are zero.

We can select two places from three (in diagonal) in ${}^3C_2$ ways.

Now we have to select elements for the upper triangle.

For the upper triangle, we have three places of which one entry is 0 and two entries are 1.

One place from three can be selected in ${}^3C_1$ ways.

Hence, the number of matrices is ${}^3C_2 \times {}^3C_1 = 9$.

(ii) If all the entries in the principle diagonal is 1,

We have two 0 and one 1 in the upper triangle.

Hence, the number of matrices is 3.

Thus, total matrices = 12

10. **(4)** $A^TA = I$

$\Rightarrow \begin{bmatrix} a & b & c \\ b & c & a \\ c & a & b \end{bmatrix}\begin{bmatrix} a & b & c \\ b & c & a \\ c & a & b \end{bmatrix} = \begin{bmatrix} 1 & 0 & 0 \\ 0 & 1 & 0 \\ 0 & 0 & 1 \end{bmatrix}$

$\Rightarrow \begin{bmatrix} a^2 + b^2 + c^2 & ab + bc + ca & ab + bc + ca \\ ab + bc + ca & a^2 + b^2 + c^2 & ab + bc + ca \\ ab + bc + ca & ab + bc + ca & a^2 + b^2 + c^2 \end{bmatrix} = \begin{bmatrix} 1 & 0 & 0 \\ 0 & 1 & 0 \\ 0 & 0 & 1 \end{bmatrix}$

$\Rightarrow a^2 + b^2 + c^2 = 1$ (1)

and $ab + bc + ca = 0$ (2)

Now $a^3 + b^3 + c^3 = (a + b + c)\,(a^2 + b^2 + c^2 - ab - bc - ca)$

$= 3abc = (a + b + c) + 3$ (3)

Now $(a + b + c)^2 = a^2 + b^2 = c^2 + 2\,(ab + bc + ca)$

$= 1 + 2 \times 0 = 1$

$\Rightarrow a + b + c = 1$ ($\because$ *a*, *b*, *c* are real positive numbers)

Now from (3), $a^3 + b^3 + c^3 = 1 + 3 = 4$

11. **(7)** We have $\begin{bmatrix} 1 & 1 & 1 \\ 2 & 5 & 7 \\ 2 & 1 & -1 \end{bmatrix}\begin{bmatrix} x_1 & y_1 \\ x_2 & y_2 \\ x_3 & y_3 \end{bmatrix} = \begin{bmatrix} 9 & 2 \\ 52 & 15 \\ 0 & 1 \end{bmatrix}$

$\Rightarrow AX = B$ (1)

Clearly, $|A| = -4 \neq 0$

$\text{adj } A = \begin{bmatrix} -12 & 16 & -8 \\ 2 & -3 & 1 \\ 2 & 1 & 3 \end{bmatrix}^T = \begin{bmatrix} -12 & 2 & 2 \\ 16 & -3 & 1 \\ -8 & 1 & 3 \end{bmatrix}$

$A^{-1} = \frac{\text{adj } A}{|A|} = \frac{-1}{4}\begin{bmatrix} -12 & 2 & 2 \\ 16 & -3 & 1 \\ -8 & 1 & 3 \end{bmatrix}$

Now, $A^{-1}B = \frac{-1}{4}\begin{bmatrix} -12 & 2 & 2 \\ 16 & -3 & 1 \\ -8 & 1 & 3 \end{bmatrix}\begin{bmatrix} 9 & 2 \\ 52 & 15 \\ 0 & -1 \end{bmatrix} = \frac{-1}{4}\begin{bmatrix} -4 & 4 \\ -12 & -8 \\ -20 & -4 \end{bmatrix}$

$= \begin{bmatrix} 1 & -1 \\ 3 & 2 \\ 5 & 1 \end{bmatrix}$

From equation (1), $X = A^{-1}B \Rightarrow \begin{bmatrix} x_1 & y_1 \\ x_2 & y_2 \\ x_3 & y_3 \end{bmatrix} = \begin{bmatrix} 1 & -1 \\ 3 & 2 \\ 5 & 1 \end{bmatrix}$

$\Rightarrow x_1 = 1, x_2 = 3, x_3 = 5$ or $y_1 = -1, y_2 = 2, y_3 = 1$

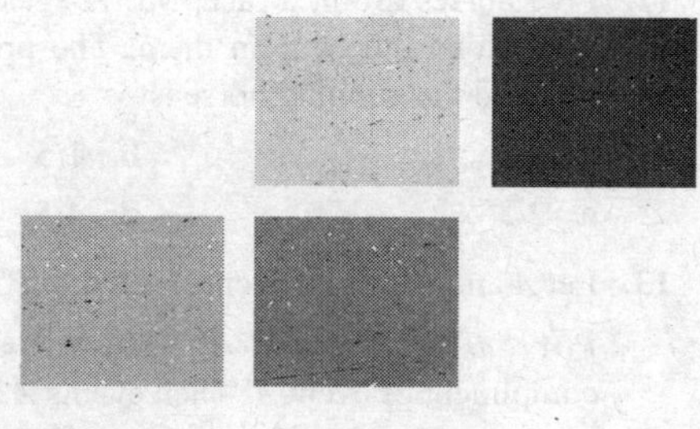

CHAPTER 9

Probability

Objective Type

Each question has four choices a, b, c and d, out of which *only one* is correct. Find the correct answer.

1. Given two events A and B. If odds against A are as 2:1 and those in favour of $A \cup B$ are as 3:1, then

   a. $1/2 \le P(B) \le 3/4$ b. $5/12 \le P(B) \le 3/4$
   c. $1/4 \le P(B) \le 3/5$ d. none of these

2. The probability that a marksman will hit a target is given as $1/5$. Then the probability that at least once hit in 10 shots is

   a. $1-(4/5)^{10}$ b. $1/5^{10}$
   c. $1-(1/5^{10})$ d. $(4/5)^{10}$

3. A six-faced dice is so biased that it is twice as likely to show an even number as an odd number when thrown. It is thrown twice, the probability that the sum of two numbers thrown is even is

   a. 1/12 b. 1/6
   c. 1/3 d. 5/9

4. A draws a card from a pack of n cards marked 1, 2, ..., n. The card is replaced in the pack and B draws a card. Then the probability that A draws a higher card than B is

   a. $(n+1)2n$ b. $1/2$
   c. $(n-1)\,2n$ d. none of these

5. A student appears for tests I, II and III. The student is successful if he passes either in tests I and II or tests I and III. The probabilities of the student passing in tests I, II and III are, respectively, p, q and 1/2. If the probability that the student is successful is 1/2, then $p(1+q) =$

   a. 1/2 b. 1
   c. 3/2 d. 3/4

6. A problem in mathematics is given to three students A, B, C and their respective probability of solving the problem is 1/2, 1/3 and 1/4. Probability that the problem is solved is

   a. 3/4 b. 1/2
   c. 2/3 d. 1/3

7. The probability that in a family of 5 members, exactly two members have birthday on Sunday is

   a. $(12 \times 5^3)\, 7^5$ b. $(10 \times 6^2)\, 7^5$
   c. 2/5 d. $(10 \times 6^3)\, 7^5$

8. Three houses are available in a locality. Three persons apply for the houses. Each applies for one houses without consulting others. The probability that all three apply for the same houses is

   a. 1/9 b. 2/9
   c. 7/9 d. 8/9

9. The numbers 1, 2, 3, ..., n are arrange in random order. The probability that the digits 1, 2, 3, ..., k $(k < n)$ appear as neighbours in that order is

   a. $1/n!$ b. $k!/n!$
   c. $(n-k)!\, n!$ d. $(n-k+1)!\, n!$

10. A die is thrown a fixed number of times. If probability of getting even number 3 times is same as the probability of getting even number 4 times, then probability of getting even number exactly once is

    a. 1/6 b. 1/9
    c. 5/36 d. 7/128

11. A pair of four dice is thrown independently three times. The probability of getting a score of exactly 9 twice is

    a. 8/9 b. 8/729
    c. 8/243 d. 1/729

12. Five horses are in a race. Mr. A selects two of the horses at random and bets on them. The probability that Mr. A selected the winning horse is

a. 3/5 **b.** 1/5
c. 2/5 **d.** 4/5

13. Let A and B be two events such that $P(\overline{A \cup B}) = 1/6$, $P(A \cap B) = 1/4$, and $P(\bar{A}) = 1/4$ where $\bar{A}$ stands for complment of event A. Then events A and B are

a. equally likely but not independent
b. equally likely and mutually exclusive
c. mutually exclusive and independent
d. independent but not equally likely

14. Words from the letters of the word PROBABILITY are formed by taking all letters at a time. The probability that both B's are not together and both I's are not together is

a. 52/55 **b.** 53/55
c. 54/55 **d.** none of these

15. A class consists of 80 students, 25 of them are girls and 55 are boys. If 10 of them are rich and the remaining are poor and also 20 of them are intelligent, then the probability of selecting an intelligent rich girl is

a. 5/128 **b.** 25/128
c. 5/512 **d.** none of these

16. Let A, B, C, D be independent events such that $P(A)=1/2$, $P(B) = 1/3$, $P(C) = 1/5$ and $P(D) = 1/6$. Then the probability that none of A, B, C and D occurs

a. 1/180 **b.** 1/45
c. 1/18 **d.** none of these

17. A sample space consists of 3 sample points with associated probabilities given as $2p, p^2, 4p - 1$. Then the value of p is

a. $p = \sqrt{11} - 3$ **b.** $\sqrt{10} - 3$
c. $\frac{1}{4} < p < \frac{1}{2}$ **d.** none

18. South African cricket captain lost the toss of a coin 13 times out of 14. The chance of this happening was

a. $7/2^{13}$ **b.** $1/2^{13}$
c. $13/2^{14}$ **d.** $13/2^{13}$

19. Events A and C are independent. If the probabilities relating A, B and C are $P(A) = 1/5$, $P(B) = 1/6$; $P(A \cap C) = 1/20$; $P(B \cup C) = 3/8$. Then

a. events B and C are independent
b. events B and C are mutually exclusive
c. events B and C are neither independent nor mutually exclusive
d. events B and C are equiprobable

20. There are only two women among 20 persons taking part in a pleasure trip. The 20 persons are divided into two groups, each group consisting of 10 persons. Then the probability that the two women will be in the same group is

a. 9/19 **b.** 9/38
c. 9/35 **d.** none

21. Let A and B be two events. Suppose $P(A) = 0.4$, $P(B) = p$, and $P(A \cup B) = 0.7$. The value of p for which A and B are independent is

a. 1/3 **b.** 1/4
c. 1/2 **d.** 1/5

22. A man has 3 pairs of black socks and 2 pairs of brown socks kept together in a box. If he dressed hurriedly in the dark, the probability that after he has put on a black sock, he will then put on another black sock is

a. 1/3 **b.** 2/3
c. 3/5 **d.** 2/15

23. Five different games are to be distributed among 4 children randomly. The probability that each child get atleast one game is

a. 1/4 **b.** 15/64
c. 21/64 **d.** none of these

24. A drawer contains 5 brown socks and 4 blue socks well mixed. A man reaches the drawer and pulls out socks at random. What is the probability that they match?

a. 4/9 **b.** 5/8
c. 5/9 **d.** 7/12

25. Two dices are rolled one after the other. The probability that the number on the first is smaller than the number on the second is

a. 1/2 **b.** 7/18
c. 3/4 **d.** 5/12

26. A natural number is chosen at random from the first 100 natural numbers. The probability that $x + \frac{100}{x} > 50$ is

a. 1/10 **b.** 11/50
c. 11/20 **d.** none of these

27. A four figure number is formed of the figures 1, 2, 3, 5 with no repetitions. The probability that the number is divisible by 5 is

a. 3/4 **b.** 1/4
c. 1/8 **d.** none of these

28. Twelve balls are distributed among three boxes. The probability that the first box contains three balls is

a. $\frac{110}{9}\left(\frac{2}{3}\right)^{10}$ **b.** $\frac{9}{110}\left(\frac{2}{3}\right)^{10}$
c. $\frac{{}^{12}C_3}{12^3} \times 2^9$ **d.** $\frac{{}^{12}C_3}{3^{12}}$

29. A box contains two black, four white and three red balls. One ball is drawn at random from the box and kept aside. From the remaining balls in the box, another ball is drawn at random and kept aside the first. This process is repeated till all the balls are drawn from the box. The probability that the balls drawn are in the sequence of two black, four white and three red is

a. 1/1260 **b.** 1/7560
c. 1/126 **d.** none of these

30. A cricket club has 15 members, of whom only five can bowl. If the names of 15 members are put into a box and 11 are drawn at random, then the probability of getting an eleven containing at least three bowlers is

a. 7/13 **b.** 6/13
c. 11/15 **d.** 12/13

31. A speaks truth in 60% cases and B speaks truth in 70% cases. The probability that they will say the same thing while describing a single event is

a. 0.56 **b.** 0.54
c. 0.38 **d.** 0.94

32. Three integers are chosen at random from the first 20 integers. The probability that their product is even is

a. 2/19 **b.** 3/29
c. 17/19 **d.** 4/29

33. There are 20 cards. Ten of these cards have the letter I printed on them and the other 10 have the letter T printed on them. If three cards are picked up at random and kept in the same order, the probability of making word IIT is

a. 4/27 **b.** 5/38
c. 1/8 **d.** 9/80

34. The chance of an event happening is the square of the chance of a second event but the odds against the first are the cube of the odds against the second. The chances of the events are

a. 1/9, 1/3 **b.** 1/16, 1/4
c. 1/4, 1/2 **d.** none of these

35. If a party of n persons sit at a round table, then the odds against two specified individuals sitting next to each other are

a. $2:(n-3)$ **b.** $(n-3):2$
c. $(n-2):2$ **d.** $2:(n-2)$

36. The sum of two positive quantities is equal to $2n$. The probability that their product is not less than 3/4 times their greatest product is

a. 3/4 **b.** 1/2
c. 1/4 **d.** none of these

37. A bag contains an assortment of blue and red balls. If two balls are drawn at random, the probability of drawing two red balls is five times the probability of drawing two blue balls. Furthermore, the probability of drawing one ball of each colour is six times the probability of drawing two blue balls. The number of red and blue balls in the bag is

a. 6, 3 **b.** 3, 6
c. 2, 7 **d.** none of these

38. Dialing a telephone number an old man forgets the last two digits remembering only that these are different dialed at random. The probability that the number is dialed correctly is

a. 1/45 **b.** 1/90
c. 1/100 **d.** none of these

39. The probability that a teacher will give an unannounced test during any class meeting is 1/5. If a student is absent twice, then the probability that the student will miss at least one test is

a. 4/5 **b.** 2/5
c. 7/25 **d.** 9/25

40. A box contains tickets numbered from 1 to 20. Three tickets are drawn from the box with replacement. The probability that the largest number on the tickets is 7 is

a. 2/19 **b.** 7/20
c. $1-(7/20)^3$ **d.** none of these

41. An unbiased coin is tossed n times. If the probability that head occurs six times is equal to the probability that head occurs eight times, then $n =$

a. 7 **b.** 14
c. 16 **d.** 19

42. One ticket is selected at random from 100 tickets numbered 00, 01, 02, …, 98, 99. If x_1 and x_2 denotes the sum and product of the digits on the tickets, then $P(x_1 = 9/x_2 = 0)$ is equal to

a. 2/19 **b.** 19/100
c. 1/50 **d.** none of these

43. Let A and B be two events such that $P(A \cap B') = 0.20$, $P(A' \cap B) = 0.15$, $P(A' \cap B') = 0.1$, then $P(A/B)$ is equal to

a. 11/14 **b.** 2/11
c. 2/7 **d.** 1/7

44. A and B toss a fair coin each simultaneously 50 times. The probability that both of them will not get tail at the same toss is

a. $(3/4)^{50}$ **b.** $(2/7)^{50}$
c. $(1/8)^{50}$ **d.** $(7/8)^{50}$

45. Cards are drawn one-by-one at random from a well-shuffled pack of 52 playing cards until 2 aces are obtained from the first time. The probability that 18 draws are required for this is

a. 3/34 **b.** 17/455
c. 561/15925 **d.** none of these

46. A father has 3 children with at least one boy. The probability that he has 2 boys and 1 girl is

a. 1/4 **b.** 1/3
c. 2/3 **d.** None of these

47. Two players toss 4 coins each. The probability that they both obtain the same number of heads is

a. 5/256 **b.** 1/16
c. 35/128 **d.** none of these

48. In a game called odd man out m $(m > 2)$ persons toss a coin to determine who will buy refreshments for the entire group. A person who gets an outcome different from that of the rest of the members of the group is called the odd man out. The probability that there is a loser in any game is

a. $1/2m$ **b.** $m/2^{m-1}$
c. $2/m$ **d.** none of these

49. If a is an integer lying in $[-5, 30]$, then the probability that the graph of $y = x^2 + 2(a+4)x - 5a + 64$ is strictly above the x-axis is

a. 1/6 **b.** 7/36

c. 2/9 **d.** 3/5

50. $2n$ boys are randomly divided into two subgroups containing n boys each. The probability that the two tallest boys are in different groups is

a. $n/(2n-1)$ **b.** $(n-1)/(2n-1)$

c. $(n-1)/4n^2$ **d.** none of these

51. A bag contains some white and some black balls, all combinations of balls being equally likely. The total number of balls in the bag is 10. If three balls are drawn at random without replacement and all of them are found to be black, the probability that the bag contains 1 white and 9 black balls is

a. 14/55 **b.** 12/55

c. 2/11 **d.** 8/55

52. Three ships A, B and C sail from England to India. If the ratio of their arriving safely are 2:5, 3:7 and 6:11, respectively, then the probability of all the ships for arriving safely is

a. 18/595 **b.** 6/17

c. 3/10 **d.** 2/7

53. The probability of solving a question by three students are 1/2, 1/4, 1/6, respectively. Probability of question being solved will be

a. 33/48 **b.** 35/48

c. 31/48 **d.** 37/48

54. Let A, B, C be three mutually independent events. Consider the two statements S_1 and S_2.

S_1: A and $B \cup C$ are independent.

S_2: A and $B \cap C$ are independent.

Then

a. both S_1 and S_2 are true **b.** only S_1 is true

c. only S_2 is true **d.** neither S_1 nor S_2 is true

55. If the papers of four students can be checked by any one of the seven teachers, then the probability that all the four papers are checked by exactly two teachers is

a. 2/7 **b.** 12/49

c. 32/343 **d.** none of these

56. A and B play a game of tennis. The situation of the game is as follows: if one scores two consecutive points after a deuce, he wins, if loss of a point is followed by win of a point, it is deuce. The chance of a server to win a point is 2/3. The game is at deuce and A is serving. Probability that A will win the match is (serves are changed after each game)

a. 3/5 **b.** 2/5

c. 1/2 **d.** 4/5

57. A coin is tossed $2n$ times. The chance that the number of times one gets head is not equal to the number of times one gets tails is

a. $\frac{(2n!)}{(n!)^2}\left(\frac{1}{2}\right)^{2n}$ **b.** $1-\frac{(2n!)}{(n!)^2}$

c. $1-\frac{(2n!)}{(n!)^2}\frac{1}{4^n}$ **d.** none of these

58. The probability that a bulb produced by a factory will fuse after 150 days if used is 0.50. What is the probability that out of 5 such bulbs none will fuse after 150 days of use

a. $1-(19/20)^5$ **b.** $(19/20)^5$

c. $(3/4)^5$ **d.** $90\,(1/4)^5$

59. If $\bar{E}$ and $\bar{F}$ are the complementary events of events E and F, respectively, and if $0 < P(F) < 1$, then

a. $P(E/F) + P(\bar{E}/F) = 1/2$ **b.** $P(E/F) + P(E/\bar{F}) = 1$

c. $P(\bar{E}/F) + P(E/\bar{F}) = 1$ **d.** $P(E/\bar{F}) + P(\bar{E}/\bar{F}) = 1$

60. In a certain town, 40% of the people have brown hair, 25% have brown eyes and 15% have both brown hair and brown eyes. If a person selected at random from the town, has brown hair, the probability that he also has brown eyes is

a. 1/5 **b.** 3/8

c. 1/3 **d.** 2/3

61. A letter is known to have come either from LONDON or CLIFTON; on the postmark only the two consecutive letters ON are legible. The probability that it came from LONDON is

a. 1/17 **b.** 12/17

c. 17/30 **d.** 3/5

62. In a game a coin is tossed $2n + m$ times and a player wins if he does not get any two consecutive outcomes same for atleast $2n$ times in a row. The probability that player wins the game is

a. $\frac{m+2}{2^{2n}+1}$ **b.** $\frac{2n+2}{2^{2n}}$

c. $\frac{2n+2}{2^{2n+1}}$ **d.** $\frac{m+2}{2^{2n}}$

63. Words from the letters of the word PROBABILITY are formed by taking all letters at a time. The probability that both B's are not together and both I's are not together is

a. 52/55 **b.** 53/55

c. 54/55 **d.** none of these

64. The probabilities of winning a race by three persons A, B, and C are 1/2, 1/4, and 1/4, respectively. They run two races. The probability of A winning the second race when B wins the first race is

a. 1/3 **b.** 1/2

c. 1/4 **d.** 2/3

65. A die is rolled four times. The probability of getting a larger number than the previous number each time is

a. 17/216 **b.** 5/432

c. 15/432 **d.** none of these

66. Four die are thrown simultaneously. The probability that 4 and 3 appear on two of the die given that 5 and 6 have appeared on other two die is

a. 1/6 **b.** 1/36

c. 12/151 **d.** none of these

67. A fair coin is tossed five times, then the probability that no two consecutive heads occur is

a. 11/32 **b.** 15/32
c. 13/32 **d.** none of these

68. A $2n$ digit number starts with 2 and all its digits are prime, then the probability that the sum of all 2 consecutive digits of the number is prime is

a. 4×2^{3n} **b.** 4×2^{-3n}
c. 2^{3n} **d.** none of these

69. The numbers (a, b, c) are selected by throwing a dice thrice, then the probability that (a, b, c) are in A.P. is

a. 1/12 **b.** 1/6
c. 1/4 **d.** none of these

70. In a n-sided regular polygon, the probability that the two diagonal chosen at random will intersect inside the polygon is

a. $\dfrac{2\,{}^nC_2}{{}^{({}^nC_2-n)}C_2}$ **b.** $\dfrac{{}^{n(n-1)}C_2}{{}^{({}^nC_2-n)}C_2}$
c. $\dfrac{{}^nC_4}{{}^{({}^nC_2-n)}C_2}$ **d.** none of these

71. A three-digit number is selected at random from the set of all three-digit numbers. The probability that the number selected has all the three digits same is

a. 1/9 **b.** 1/10
c. 1/50 **d.** 1/100

72. Two numbers a, b are chosen from the set of integers 1, 2, 3, ..., 39. Then probability that the equation $7a - 9b = 0$ is satisfied is

a. 1/247 **b.** 2/247
c. 4/741 **d.** 5/741

73. Two numbers x and y are chosen at random (without replacement) from amongst the numbers 1, 2, 3, ..., 2004. The probability that $x^3 + y^3$ is divisible by 3 is

a. 1/3 **b.** 2/3
c. 1/6 **d.** 1/4

74. One mapping is selected at random from all mappings of the set $S = \{1, 2, 3, ..., n\}$ into itself. If the probability that the mapping is one-one is 3/32, then the value of n is

a. 2 **b.** 3
c. 4 **d.** none of these

75. A bag contains 20 coins. If the probability that thebag contains exactly 4 biased coin is 1/3 and that of exactly 5 biased coin is 2/3, then the probability that all the biased coin are sorted out from the bag in exactly 10 draws is

a. $\dfrac{5}{10}\dfrac{{}^{16}C_6}{{}^{20}C_9}+\dfrac{1}{11}\dfrac{{}^{15}C_5}{{}^{20}C_9}$ **b.** $\dfrac{2}{33}\left[\dfrac{{}^{16}C_6+5\,{}^{15}C_5}{{}^{20}C_9}\right]$
c. $\dfrac{5}{33}\dfrac{{}^{16}C_7}{{}^{20}C_9}+\dfrac{1}{11}\dfrac{{}^{15}C_6}{{}^{20}C_9}$ **d.** none of these

76. Cards are drawn one by one without replacement from a pack of 52 cards. The probability that 10 cards will precede the first ace is

a. 241/1456 **b.** 164/4165
c. 451/884 **d.** none of these

77. If any four numbers are selected and they are multiplied, then the probability that the last digit will be 1, 3, 5 or 7 is

a. 4/625 **b.** 18/625
c. 16/625 **d.** none of these

78. Four numbers are multiplied together. Then the probability that the product will be divisible by 5 or 10 is

a. 369/625 **b.** 399/625
c. 123/625 **d.** 133/625

79. A coin is tossed seven times. Then the probability that at least four consecutive heads appear is

a. 3/16 **b.** 5/32
c. 3/16 **d.** 1/8

80. If odds against solving a question by three students are 2:1, 5:2 and 5:3, respectively, then probability that the question is solved only by one student is

a. 31/56 **b.** 24/56
c. 25/56 **d.** none of these

81. An unbiased coin is tossed six times. The probability that third head appears on the sixth trial is

a. 5/16 **b.** 5/32
c. 5/8 **d.** 5/64

82. There are two urns A and B. Urn A contains five red, three blue and two white balls, urn B contains four red, three blue and 3 white balls. An urn is choosen at random and a ball is drawn. Probability that the ball drawn is red is

a. 9/10 **b.** 1/2
c. 11/20 **d.** 9/20

83. Three critics review a book. Odds in favour of the book are 5:2, 4:3 and 3:4, respectively, for the three critics. The probability that majority are in favour of the book is

a. 35/49 **b.** 125/343
c. 164/343 **d.** 209/343

84. Let A and B are events of an experiment and $P(A) = 1/4$, $P(A \cup B) = 1/2$, then value of $P(B/A^c)$ is

a. 2/3 **b.** 1/3
c. 5/6 **d.** 1/2

85. The probability that an automobile will be stolen and found within one week is 0.0006. The probability that an automobile will be stolen is 0.0015. The probability that a stolen automobile will be found in one week is

a. 0.3 **b.** 0.4
c. 0.5 **d.** 0.6

86. A pair of numbers is picked up randomly (without replacement) from the set {1, 2, 3, 5, 7, 11, 12, 13, 17, 19}. The probability that the number 11 was picked given that the sum of the numbers was even is nearly

a. 0.1 **b.** 0.125
c. 0.24 **d.** 0.18

87. An unbiased cubic die marked with 1, 2, 2, 3, 3, 3 is rolled three times. The probability of getting a total score of 4 or 6 is

a. 16/216 b. 50/216
c. 60/216 d. none of these

88. A bag contains three red and three green balls and a person draws out three at random. He then drops three blue balls into the bag and again draws out thee at random. The chance that the three later balls being all of different colours is

a. 15% b. 20%
c. 27% d. 40%

89. A bag contains 20 coins. If the probability that bag contains exactly four biased coin is 1/3 and that of exactly five biased coin is 2/3, then the probability that all the biased coin are sorted out from the bag in exactly 10 draws is

a. $\frac{5}{33}\frac{{}^{16}C_6}{{}^{20}C_9}+\frac{1}{11}\frac{{}^{15}C_5}{{}^{20}C_9}$ b. $\frac{2}{33}\left[\frac{{}^{16}C_6+5\,{}^{15}C_5}{{}^{20}C_9}\right]$

c. $\frac{5}{33}\frac{{}^{16}C_7}{{}^{20}C_9}+\frac{1}{11}\frac{{}^{15}C_6}{{}^{20}C_9}$ d. none of these

90. A doctor is called to see a sick child. The doctor knows (prior to the visit) that 90% of the sick children in that neighbourhood are sick with the flu, denoted by F, while 10% are sick with the measles, denoted by M. A well-known symptom of measles is a rash, denoted by R. The probability of having a rash for a child sick with the measles is 0.95. However, occasionally children with the flu also develop a rash, with conditional probability 0.08. Upon examination the child, the doctor finds a rash. Then what is the probability that the child has the measles?

a. 91/165 b. 90/163
c. 82/161 d. 95/167

91. A dice is thrown six times, it being known that each time a different digit is shown. The probability that a sum of 12 will be obtained in the first three throws is

a. 5/24 b. 25/216
c. 3/20 d. 1/12

92. A composite number is selected at random from the first 30 natural numbers and it is divided by 5. The probability that there will be a remainder is

a. 14/19 b. 5/19
c. 5/6 d. 7/15

93. Let E be an event which is neither a certainty nor an impossibility. If probability is such that $P(E) = 1 + \lambda + \lambda^2$ and $P(E') = (1 + \lambda)^2$ in terms of an unknown λ. Then $P(E)$ is equal to

a. 1 b. 3/4
c. 1/4 d. none of these

94. A student can solve two out of four problems of mathematics, three out of five problem of physics and four out of five problems of chemistry. There are equal number of books of math, physics and chemistry in his shelf. He selects one book randomly and attempts 10 problems from it. If he solves the first problem, then the probability that he will be able to solve the second problem is

a. 2/3 b. 25/38
c. 13/21 d. 14/23

95. A fair die is tossed repeatedly. A wins if it is 1 or 2 on two consecutive tosses and B wins if it is 3, 4, 5 or 6 on two consecutive tosses. The probability that A wins if the die is tossed indefinitely is

a. 1/3 b. 5/21
c. 1/4 d. 2/5

96. Whenever horses a, b, c race together, their respective probabilities of winning the race are 0.3, 0.5 and 0.2, respectively. If they race three times the probability that 'the same horse wins all the three races, and the probability that a, b, c each wins one race are, respectively

a. 8/50, 9/50 b. 16/100, 3/100
c. 12/50, 15/50 d. 10/50, 8/50

97. Five different games are to be distributed among four children randomly. The probability that each child get atleast one game is

a. 1/4 b. 15/64
c. 21/64 d. none of these

98. Forty teams play a tournament. Each team plays every other team just once. Each game results in a win for one team. If each team has a 50% chance of winning each game, the probability that at the end of the tournament, every team has won a different number of games is

a. 1/780 b. $40!/2^{780}$
c. $40!/3^{780}$ d. none of these

99. If three squares are selected at random from chessboard, then the probability that they form the letter L is

a. $196/{}^{64}C_3$ b. $49/{}^{64}C_3$
c. $36/{}^{64}C_3$ d. $98/{}^{64}C_3$

100. A bag has 10 balls. Six balls are drawn in an attempt and replaced. Then another draw of five balls is made from the bag. The probability that exactly two balls are common to both the draw is

a. 5/21 b. 2/21
c. 7/21 d. 3/21

101. There are three bags. Bag 1 contains two red and $a^2 - 4a + 8$ black balls, bag two contains one red and $a^2 - 4a + 9$ black balls and bag three containa three red and $a^2 - 4a + 7$ black balls. A ball is drawn at random from at random chosen bag. Then the maximum value of probability that it is a red ball is

a. 1/3 b. 1/2
c. 2/9 d. 4/9

102. The probability that a random chosen three-digit number has exactly 3 factors is

a. 2/225 b. 7/900
c. 1/800 d. none of these

103. Let p, q be chosen one by one from the set $\{1, \sqrt{2}, \sqrt{3}, 2, e, \pi\}$ with replacement. Now a circle is drawn taking (p, q) as its centre. Then the probability that at the most two rational points exist on the circle is (rational points are those points whose both the coordinates are rational)

a. 2/3 **b.** 7/8
c. 8/9 **d.** none of these

104. An event X can take place in conjuction with any one of the mutually exclusive and exhaustive events A, B and C. If A, B, C are equiprobable and the probability of X is 5/12, and the probability of X taking place when A has happened is 3/8, while it is 1/4 when B has taken place, then the probability of X taking place in conjuction with C is

a. 5/8 **b.** 3/8
c. 5/24 **d.** none of these

105. On a Saturday night, 20% of all drivers in U.S.A. are under the influence of alcohol. The probability that a driver under the influence of alcohol will have an accident is 0.001. The probability that a sober driver will have an accident is 0.0001. If a car on a Saturday night smashed into a tree, the probability that the driver was under the influence of alcohol is

a. 3/7 **b.** 4/7
c. 5/7 **d.** 6/7

106. A purse contains two six-sided dice. One is a normal fair die, while the other has two 1's, two 3's, and two 5's. A die is picked up and rolled. Because of some secret magnetic attraction of the unfair die, there is 75% chance of picking the unfair die and a 25% chance of picking a fair die. The die is rolled and shows up the face 3. The probability that a fair die was picked up is

a. 1/7 **b.** 1/4
c. 1/6 **d.** 1/24

107. Five different marbles are placed in five different boxes randomly. Then the probability that exactly two boxes remain empty is (each box can hold any number of marbles)

a. 2/5 **b.** 12/25
c. 3/5 **d.** none of these

108. There are 10 prizes, five A's, three B's and two C's, placed in identical sealed envelopes for the top 10 contestants in a mathematics contest. The prizes are awarded by allowing winners to select an envelope at random from those remaining. When the 8th contestant goes to select the prize, the probability that the remaining three prizes are one A, one B and one C is

a. 1/4 **b.** 1/3
c. 1/12 **d.** 1/10

109. Thirty-two players ranked 1 to 32 are playing in a knockout tournament. Assume that in every match between any two players, the better-ranked player wins, the probability that ranked 1 and ranked 2 players are winner and runner up, respectively, is

a. 16/31 **b.** 1/2
c. 17/31 **d.** none of these

110. Three integers are chosen at random from the set of first 20 natural numbers. The chance that their product is a multiple of 3 is

a. 194/285 **b.** 1/57
c. 13/19 **d.** 3/4

111. A car is parked among N cars standing in a row, but not at either end. On his return, the owner finds that exactly r of the N places are still occupied. The probability that the places neighbouring his car are empty is

a. $\dfrac{(r-1)!}{(N-1)!}$ **b.** $\dfrac{(r-1)!\,(N-r)!}{(N-1)!}$

c. $\dfrac{(N-r)(N-r-1)}{(N+1)(N+2)}$ **d.** $\dfrac{{}^{N-r}C_2}{{}^{N-1}C_2}$

112. One ticket is selected at random from 100 tickets numbered 00, 01, 02, …, 99. Suppose A and B are the sum and product of the digit found on the ticket. Then $P((A = 7)/(B = 0))$ is given by

a. 2/13 **b.** 2/19
c. 1/50 **d.** none of these

113. A fair coin is tossed 100 times. The probability of getting tails 1, 3, …, 49 times is

a. 1/2 **b.** 1/4
c. 1/8 **d.** 1/16

114. A pair of unbiased dice is rolled together till a sum of either 5 or 7 is obtained. The probability that 5 comes before 7 is

a. 2/5 **b.** 3/5
c. 4/5 **d.** none of these

115. If n integers taken at random are multiplied together, then the probability that the last digit of the product is 1, 3, 7 or 9 is

a. $2^n/5^n$ **b.** $4^n - 2^n/5^n$
c. $4^n/5^n$ **d.** none of these

116. If A and B each toss three coins. The probability that both get the same number of heads is

a. 1/9 **b.** 3/16
c. 5/16 **d.** 3/8

117. Let A be a set containing n elements. A subset P of the set A is chosen at random. The set A is reconstructed by replacing the elements of P, and another subset Q of A is chosen at random. The probability that $P \cap Q$ contains exactly $m(m < n)$ elements is

a. $3^{n-m}/4^n$ **b.** ${}^nC_m \times 3^m/4^n$
c. ${}^nC_m \times 3^{n-m}/4^n$ **d.** none of these

118. A fair die is thrown 20 times. The probability that on the 10th throw, the fourth six appears is

a. ${}^{20}C_{10} \times 5^6/6^{20}$ **b.** $120 \times 5^7/6^{10}$
c. $84 \times 5^6/6^{10}$ **d.** none of these

119. If p is the probability that a man aged x will die in a year, then the probability that out of n men $A_1, A_2, \ldots, A_n$ each aged x, A_1 will die in an year and be the first to die is

a. $1-(1-p)^n$ **b.** $(1-p)^n$
c. $1/n\,[1-(1-p)^n]$ **d.** $1/n\,(1-p)^n$

120. A bag contains n white and n black balls. Pairs of balls are drawn without replacement until the bag is empty. The probability that each pair consists of one white and one black ball is

a. $1/{}^{2n}C_n$ b. $2n/{}^{2n}C_n$
c. $2n/n!$ d. $2n/(2n!)$

121. A man alternately tosses a coin and throws a die beginning with the coin. The probability that he gets a head in the coin before he gets a 5 or 6 in the dice is

a. 3/4 b. 1/2
c. 1/3 d. none of these

122. There are three bags which are known to contain two white and three black, four white and one black, and three white and seven black balls, respectively. A ball is drawn at random from one of the bags and found to be a black ball. Then the probability that it was drawn from the bag containing the most black ball is

a. 7/15 b. 5/19
c. 3/4 d. None of these

123. Consider $f(x) = x^3 + ax^2 + bx + c$. Parameters a, b, c are chosen, respectively, by throwing a die three times. Then the probability that $f(x)$ is an increasing function is

a. 5/36 b. 8/36
c. 4/9 d. 1/3

124. A fair coin is tossed 10 times. Then the probability that two heads do not occur consecutively is

a. 7/64 b. 1/8
c. 9/16 d. 9/64

125. If a and b are chosen randomly from the set consisting of numbers 1, 2, 3, 4, 5, 6 with replacement. Then the probability that $\lim_{x\to 0} [(a^x + b^x)/2]^{2/x} = 6$ is

a. 1/3 c. 1/4
c. 1/9 d. 2/9

126. An artillery target may be either at point I with probability 8/9 or at point II with probability 1/9. We have 55 shells, each of which can be fired either rat point I or II. Each shell may hit the target, independent of the other shells, with probability 1/2. Maximum number of shells must be fired at point I to have maximum probability is

a. 20 b. 25
c. 29 d. 35

127. An urn contains three red balls and n white balls. Mr. A draws two balls together from the urn. The probability that they have the same colour is 1/2. Mr. B draws one balls form the urn, notes its colour and replaces it. He then draws a second ball from the urn and finds that both balls have the same colour is 5/8. The possible value of n is

a. 9 b. 6
c. 5 d. 1

128. A hat contains a number of cards with 30% white on both sides, 50% black on one side and white on the other side, 20% black on both sides. The cards are mixed up, and a single card is drawn at random and placed on the table. Its upper side shows up black. The probability that its other side is also black is

a. 2/9 b. 4/9
c. 2/3 d. 2/7

129. All the jacks, queens, kings and aces of a regular 52 cards deck are taken out. The 16 cards are thoroughly shuffled and my opponent, a person who always tells the truth, simultaneously draws two cards at random and says, 'I hold at least one ace'. The probability that he holds two aces is

a. 2/8 b. 4/9
c. 2/3 d. 1/9

130. Mr. A lives at origin on the Cartesian plane and has his office at (4, 5). His friend lives at (2, 3) on the same plane. Mr. A can go to his office travelling one block at a time either in the $+y$ or $+x$ direction. If all possible paths are equally likely then the probability that Mr. A passed his friends house is (shortest path for any event must be considered)

a. 1/2 b. 10/21
c. 1/4 d. 11/21

Multiple Correct Answers Type

Each question has four choices a, b, c and d, out of which *one or more* answers are correct.

1. If A and B are two independent events such that $P(A) = 1/2$ and $P(B) = 1/5$, then

a. $P(A \cup B) = 3/5$ b. $P(A/B) = 1/4$
c. $P(A/A \cup B) = 5/6$ d. $P(A \cap B/\bar{A} \cup \bar{B}) = 0$

2. Let A and B be two events such that $P(A \cap B') = 0.20$, $P(A' \cap B) = 0.15$ and $P(A$ and B both fail$) = 0.10$. Then

a. $P(A/B) = 2/7$ b. $P(A) = 0.3$
c. $P(A \cup B) = 0.55$ d. $P(B/A) = 1/2$

3. If A and B are two events such that $P(A) = 3/4$ and $P(B) = 5/8$, then

a. $P(A \cup B) \geq 3/4$ b. $P(A' \cap B) \leq 1/4$
c. $3/8 \leq P(A \cap B) \leq 5/8$ d. $3/8 \leq P(A \cap B) \leq 5/8$

4. If A and B are two mutually exclusive events, then

a. $P(A) \leq P(\bar{B})$ b. $P(A) > P(B)$
c. $P(B) \leq P(\bar{A})$ d. $P(A) > P(B)$

5. The probability that a married man watches a certain TV show is 0.4 and the probability that a married woman watches the show is 0.5. The probability that a man watches the show, given that his wife does, is 0.7. Then

a. the probability that married couple watchs the show is 0.35

b. the probability that a wife watches the show given that her husband does is 7/8

c. the probability that atleat one person of a married couple will watch the show is 0.55

d. none of these

6. A and B are two events defined as follows:

A: It rains today with $P(A) = 40\%$

B: It rains tomorrow with $P(B) = 50\%$

Also, P(it rains today and tomorrow) = 30%

Also, E_1: $P((A \cap B)/(A \cup B))$ and E_2: $P(\{(A \cap \bar{B})$or $(B \cup \bar{A})\}/(A \cup B))$. Then which of the following is/are true?

a. A and B are independent

b. $P(A/B) < P(B/A)$

c. E_1 and E_2 are equiprobable

d. $P(A/(A \cup B)) = P(B/(A \cup B))$

7. Two numbers are randomly selected and multiplied. Consider two events E_1 and E_2 defined as

E_1: Their product is divisible by 5

E_2: Unit's places in their product is 5

Which of the following statement is/are correct?

a. E_1 is twice as likely to occur as E_2

b. E_1 and E_2 are disjoint

c. $P(E_2/E_1) = 1/4$

d. $P(E_1/E_2) = 1$

8. Probability if n heads in $2n$ tosses of a fair coin can be given by

a. $\prod_{r=1}^{n}\left(\frac{2r-1}{2r}\right)$ **b.** $\prod_{r=1}^{n}\left(\frac{n+r}{2r}\right)$

c. $\sum_{r=0}^{n}\left(\frac{{}^nC_r}{2^n}\right)^2$ **d.** $\frac{\sum_{r=0}^{n}({}^nC_r)^2}{\left(\sum_{r=0}^{2n}{}^{2n}C_r\right)}$

9. The probability that a 50-year-old man will be alive at 60 is 0.83 and the probability that a 45-year-old woman will be alive at 55 is 0.87. Then

a. the probability that both will be alive is 0.7221

b. at least one of them will alive is 0.9779

c. at least one of them will alive is 0.8230

d. the probability that both will be alive is 0.6320

10. Which of the following statement is/are correct?

a. Three coins are tossed once. At least two of them must land the same way. No mater whether they land heads or tails, the third coin is equally likely to land either the same way or oppositely. So, the chance that all the three coins land the same way is 1/2.

b. Let $0 < P(B) < 1$ and $P(A/B) = P(A/B^C)$. Then A and B are independent.

c. Suppose an urn contains w white and b black balls and a ball is drawn from it and is replaced along with d additional balls of the same colour. Now a second ball is drawn from it. The probability that the second drawn ball is white is independent of the value of d.

d. A, B, C simultaneously satisfy

$P(ABC) = P(A)\,P(B)\,P(C)$

$P(AB\bar{C}) = P(A)\,P(B)\,P(\bar{C})$

$P(A\bar{B}C) = P(A)\,P(\bar{B})\,P(C)$

$P(A\text{–}BC) = P(\bar{A})\,P(B)\,P(C)$

Then A, B, C are independent.

11. A bag initially contains one red and two blue balls. An experiment consisting of selecting a ball at random, noting its colour and replacing it together with an additional ball of the same colour. If three such trials are made, then

a. probability that at least one blue ball is drawn is 0.9

b. probability that exactly one blue ball is drawn is 0.2

c. probability that all the drawn balls are red given that all the drawn balls are of same colour is 0.2

d. probability that atleast one red ball is drawn is 0.6

12. $P(A) = 3/8$; $P(B) = 1/2$; $P(A \cup B) = 5/8$, which of the following do/does hold good?

a. $P(A^C/B) = 2P(A/B^C)$ **b.** $P(B) = P(A/B)$

c. $15\,P(A^C/B^C) = 8P(B/A^C)$ **d.** $P(A/B^C) = (A \cap B)$

13. The chance of an event happening is the square of the chance of a second event but the odds against the first are the cube of the odds against the second. The chances of the events are

a. $p_1 = 1/9$ $p_1 = 1/16$

c. $p_2 = 1/3$ **d.** $p_2 = 1/4$

14. A bag contains b blue balls and r red balls. If two balls are drawn at random, the probability of drawing two red balls is five times the probability of drawing two blue balls. Furthermore, the probability of drawing one ball of each colour is six times the probability of drawing two blue balls. Then

a. $b + r = 9$ **b.** $br = 18$

c. $|b - r| = 4$ **d.** $b/r = 2$

15. In a precision bombing attack, there is a 50% chance that any one bomb will strike the target. Two direct hits are required to destroy the target completely. The number of bombs which should be dropped to give a 99% chance or better of completely destroying the target can be

a. 12 **b.** 11

c. 10 **d.** 13

16. If A and B are two events, the probability that exactly one of them occurs is given by

a. $P(A) + P(B) - 2P(A \cap B)$

b. $P(A \cap \bar{B}) + P(\bar{A} \cap B)$

c. $P(A \cup B) - P(A \cap B)$

d. $P(\bar{A}) + P(\bar{B}) - 2P(\bar{A} \cap \bar{B})$

17. If A and B are two events, then which one of the following is/are always true?

a. $P(A \cap B) \geq P(A) + P(B) - 1$

b. $P(A \cap B) \leq P(A)$

c. $P(A' \cap B') \geq P(A') + P(B') - 1$

d. $P(A \cap B) = P(A)\,P(B)$

18. If A and B are two independent events such that $P(A) = 1/2$, $P(B) = 1/5$, then

a. $P(A/B) = 1/2$ **b.** $P\left(\frac{A}{A \cup B}\right) = \frac{5}{6}$

c. $P\left(\frac{A \cap B}{A' \cup B'}\right) = 0$ **d.** none of these

19. If A and B are two independent events such that $P(\bar{A} \cap B) = 2/15$ and $P(A \cap \bar{B}) = 1/6$, then $P(B)$ is

a. 1/5 **b.** 1/6

c. 4/5 **d.** 5/6

20. Two buses A and B are scheduled to arrive at a town central bus station at noon. The probability that bus A will be late is 1/5. The probability that bus B will be late is 7/25. The probability that the bus B is late given that bus A is late is 9/10. Then,

a. probability that neither bus will be late on a particular day is 7/10

b. probability that bus A is late given that bus B is late is 18/28

c. probability that at least one bus is late is 3/10

d. probability that at least one bus is in time is 4/5

21. If p and q are chosen randomly from the set {1, 2, 3, 4, 5, 6, 7, 8, 9, 10} with replacement, then the probability that the roots of the equation $x^2 + px + q = 0$

a. are real is 33/50 **b.** are imaginary is 19/50

c. are real and equal is 3/50 **d.** are real and distinct is 3/5

22. Two numbers are chosen from {1, 2, 3, 4, 5, 6, 7, 8} one after another without replacement. Then the probability that

a. the smaller value of two is less than 3 is 13/28

b. the bigger value of two is more than 5 is 9/14

c. product of two number is even is 11/14

d. none of these

23. A fair coin is tossed 99 times. Let X be the number of times heads occurs. Then $P(X = r)$ is maximum when r is

a. 49 **b.** 52

c. 51 **d.** 50

24. If the probability of choosing an integer k out of $2m$ integers 1, 2, 3, …, $2m$ is inversely proportional to k^4 $(1 \le k \le m)$. If x_1 is the probability that chosen number is odd and x_2 is the probability that chosen number is even, then

a. $x_1 > 1/2$ **b.** $x_1 > 2/3$

c. $x_2 < 1/2$ **d.** $x_2 < 2/3$

Reasoning Type

Each question has four choices a, b, c, and d, out of which *only one* is correct. Each question contains Statement 1 and Statement 2.

a. Both the statements are True and Statement 2 is the correct explanation of Statement 1.

b. Both the statements are True but Statement 2 is Not the correct explanation of Statement 1.

c. Statement 1 is True and Statement 2 is False.

d. Statement 1 is False and Statement 2 is True.

1. Statement 1: For events A and B of sample space if $P\left(\frac{A}{B}\right) \ge P(A)$, then $P\left(\frac{B}{A}\right) \ge P(B)$.

Statement 2: $P\left(\frac{A}{B}\right) = \frac{P(A \cap B)}{P(B)}$ $(P(B) \ne 0)$.

2. A fair die is rolled once.

Statement 1: The probability of getting a composite number is 1/3.

Statement 2: There are three possibilities for the obtained number: (i) the number is a prime number, (ii) the number is a composite number and (iii) the number is 1. Hence probability of getting a prime number is 1/3.

3. Statement 1: If $P(A) = 0.25$, $P(B) = 0.50$ and $P(A \cap B) = 0.14$, then the probability that neither A nor B occurs is 0.39.

Statement 2: $\overline{(A \cup B)} = \bar{A} \cup \bar{B}$.

4. Let A and B be two events such that $P(A) = 3/5$ and $P(B) = 2/3$. Then

Statement 1: $4/15 \le P(A \cap B) \le 3/5$.

Statement 2: $2/5 \le P(A/B)) \le 9/10$.

5. Statement 1: If A, B, C be three mutually independent events, then A and $B \cup C$ are also independent events.

Statement 2: Two events A and B are independent if and only if $P(A \cap B) = P(A)\, P(B)$.

6. Statement 1: Out of five tickets consecutively numbered, three are drawn at random. The chance that the numbers on them are in A.P. is 2/15.

Statement 2: Out of $2n + 1$ tickets consecutively numbered, three are drawn at random, the chance that the numbers on them are in A.P. is $3n/(4n^2 - 1)$.

7. Let A and B be two event such that $P(A \cup B) \ge 3/4$ and $1/8 \le P(A \cap B) \le 3/8$.

Statement 1: $P(A) + P(B) \ge 7/8$.

Statement 2: $P(A) + P(B) \le 11/8$.

8. Statement 1: The probability of drawing either an ace or a king from a pack of card in a single draw is 2/13.

Statement 2: For two events A and B which are not mutually exclusive, $P(A \cup B) = P(A) + P(B) - P(A \cap B)$.

9. Let A and B be two independent events.

Statement 1: If $P(A) = 0.4$ and then $P(A \cup \bar{B}) = 0.9$, then $P(B)$ is 1/6.

Statement 2: If A and B are independent, then $P(A \cap B) = P(A)P(B)$.

10. Consider an event for which probability of success is 1/2.

Statement 1: Probability that in n trials, there are r successes where $r = 4k$ and k is an integer is $\frac{1}{4} + \frac{1}{2^{n/2+1}} \cos\left(\frac{n\pi}{4}\right)$

Statement 2: ${}^nC_0 + {}^nC_4 + {}^nC_8 + \cdots = 2^{n/2} \sin\left(\frac{n\pi}{4}\right)$.

11. Statement 1: If A and B are two events such that $0 < P(A)$, $P(B) < 1$, then $P(A/\bar{B}) + P(\bar{A}/\bar{B}) = 3/2$.

Statement 2: If A and B are two events such that $0 < P(A)$, $P(B) < 1$, then $P(A/B)$

$= P(A \cap B)/P(B)$ and $P(\bar{B}) = P(A \cap \bar{B}) + P(\bar{A} \cap \bar{B})$

12. **Statement 1:** If a fair coin is tossed 15 times, then the probability of getting head as many times in the first 10 throws as in the last five is 3003/32768.

Statement 2: Sum of the series ${}^{m}C_r{}^{n}C_0 + {}^{m}C_{r-1}{}^{n}C_1 + \cdots + {}^{m}C_0{}^{n}C_r = {}^{m+n}C_r$.

13. **Statement 1:** If $A = \{2, 4, 6\}$, $B = \{1, 2, 3\}$ where A and B are the events of numbers occurring on a dice, then $P(A) + P(B) = 1$.

Statement 2: If $A_1, A_2, A_3, \ldots, A_n$ are all mutually exclusive events, then $P(A_1) + P(A_2) + \cdots + P(A_n) = 1$.

14. **Statement 1:** There are 4 addressed envelopes and 4 letters for each one of them. The probability that no letter is mailed in its correct envelops is 3/8.

Statement 2: The probability that all letters are not mailed in their correct envelope is 23/24.

15. Let A and B be two independent events.

Statement 1: If $P(A) = 0.3$ and $P(A \cup \bar{B}) = 0.8$, then $P(B)$ is 2/7.

Statement 2: $P(\bar{E}) = 1 - P(E)$, where E is any event.

Linked Comprehension Type

Based upon each paragraph, three multiple choice questions have to be answered. Each question has four choices a, b, c and d, out of which *only one* is correct.

For Problems 1–3

In a class of 10 students, probability of exactly i students passing an examination is directly proportional to i^2. Then answer the following questions:

1. The probability that exactly five students passing an examination is

   a. 1/11 b. 5/77
   c. 25/77 d. 10/77

2. If a student is selected at random, then the probability that he has passed the examination is

   a. 1/7 b. 11/35
   c. 11/14 d. none of these

3. If a students selected at random is found to have passed the examination, then the probability that he was the only student who has passed the examination is

   a. 1/3025 b. 1/605
   c. 1/275 d. 1/121

For Problems 4–6

A shoping mall is running a scheme: "Each packet of detergent 'SURF' contains a coupon which bears letter of the word 'SURF', if a person buys at least four packets of detergent 'SURF', and produce all the letters of the word 'SURF', then he gets one free packet of detergent.

4. If a person buys eight such packets at a time, then number of different combinations of coupon he has

   a. 4^8 b. 8^4
   c. ${}^{11}C_3$ d. ${}^{12}C_4$

5. If person buys eight such packets, then the probability that he gets exactly one free packets is

   a. 7/33 b. 102/495
   c. 13/55 d. none of these

6. If a person buys eight such packets, then the probability that he gets two free packets is

   a. 1/7 b. 1/5
   c. 1/42 d. 1/165

For Problems 7–9

In an objective paper, there are two sections of 10 questions each. For 'section 1', each question has five options and only one option is correct and 'section 2' has four options with multiple answers and marks for a question in this section is awarded only if he ticks all correct answers. Marks for each question in 'section 1' is 1 and in 'section 2' is 3. (There is no negative marking.)

7. If a candidate attempts only two questions by guessing, one from 'section 1' and one from 'section 2', the probability that he scores in both questions is

   a. 74/75 b. 1/25
   c. 1/15 d. 1/75

8. If a candidate in total attempts four questions all by guessing, then the probability of scoring 10 marks is

   a. $1/15(1/15)^3$ b. $4/5(1/15)^3$
   c. $1/5(14/15)^3$ d. none of these

9. The probability of getting a score less than 40 by answering all the questions by guessing in this paper is

   a. $(1/75)^{10}$ b. $1-(1/75)^{10}$
   c. $(74/75)^{10}$ d. none of these

For Problems 10–12

There are two die A and B both having six faces. Die A has three faces marked with 1, two faces marked with 2 and one face marked with 3. Die B has one face marked with 1, two faces marked with 2 and three faces marked with 3. Both dices are thrown randomly once. If E be the event of getting sum of the numbers appearing on top faces equal to x and let $P(E)$ be the probability of even E, then

10. $P(E)$ is maximum when x equal to

    a. 5 b. 3
    c. 4 d. 6

11. $P(E)$ is minimum when x equals to

    a. 3 b. 4
    c. 5 d. 6

12. When $x = 4$, then $P(E)$ is equal to

    a. 5/9 b. 6/7
    c. 7/18 d. 8/19

For Problems 13–15

A JEE aspirant estimates that she will be successful with an 80% chance if she studies 10 hours per day, with a 60% chance if she studies seven hours per day and with a 40% chance if she studies four hours per day. She further believes that she will study 10 hours, seven hours and four hours per day with probabilities 0.1, 0.2 and 0.7, respectively.

13. The chance she will be successful is

a. 0.28 **b.** 0.38
c. 0.48 **d.** 0.58

14. Given that she is successful, the chance that she studied for four hours is

a. 6/12 **b.** 7/12
c. 8/12 **d.** 9/12

15. Given that she does not achieve success, the chance she studied for four hours is

a. 18/26 **b.** 19/26
c. 20/26 **d.** 21/26

For Problems 16–18

Let S and T are two events defined on a sample space with probabilities
$P(S) = 0.5$, $P(T) = 0.69$, $P(S/T) = 0.5$

16. Events S and T are

a. mutually exclusive
b. independent
c. mutually exclusive and independent
d. neither mutually exclusive nor independent

17. The value of $P(S \text{ and } T)$ is

a. 0.3450 **b.** 0.2500
c. 0.6900 **d.** 0.350

18. The value of $P(S \text{ or } T)$ is

a. 0.6900 **b.** 1.19
c. 0.8450 **d.** 0

For Problems 19–21

An amoeba either splits into two or remains the same or eventually dies out immediately after completion of every second with probabilities, respectively, 1/2, 1/4 and 1/4. Let the initial amoeba be called as mother amoeba and after every second, the amoeba, if it is distinct from the previous one, be called as 2nd, 3rd, ... generations.

19. The probability that immediately after completion of 2 s all the amoeba population dies out is

a. 9/32 **b.** 11/32
c. 1/2 **d.** 3/32

20. The probability that after 2 s exactly four amoeba are alive is

a. 1/16 **b.** 1/8
c. 1/4 **d.** 1/2

21. The probability that amoeba population will be maximum after completion of 3 s is

a. $1/2^7$ **b.** $1/2^6$
c. $1/2^8$ **d.** none of these

For Problems 22–24

A cube having all of its sides painted is cut by two horizontal, two vertical and other two planes so as to form 27 cubes all having the same dimensions. Of these cubes, a cube is selected at random.

22. The probability that the cube selected has none of its sides painted is

a. 1/9 **b.** 1/27
c. 1/18 **d.** 5/54

23. The probability that the cube selected has two sides painted is

a. 1/9 **b.** 4/9
c. 8/27 **d.** none of these

24. The total number of cubes having at least one of its sides painted is

a. 8 **b.** 53
c. 49 **d.** 26

For Problems 25–27

There are some experiments in which the outcomes cannot be identified discretely. For example, an ellipse of eccentricity $2\sqrt{2}/3$ is inscribed in a circle and a point within the circle is chosen at random. Now, we want to find the probability that this point lies outside the ellipse. Then, the point must lie in the shaded region shown in Fig. 9.1. Let the radius of the circle be a and length of minor axis of the ellipse be $2b$. Given that

$$1-\frac{b^2}{a^2}=\frac{8}{9} \Rightarrow \frac{b^2}{a^2}=\frac{1}{9}$$

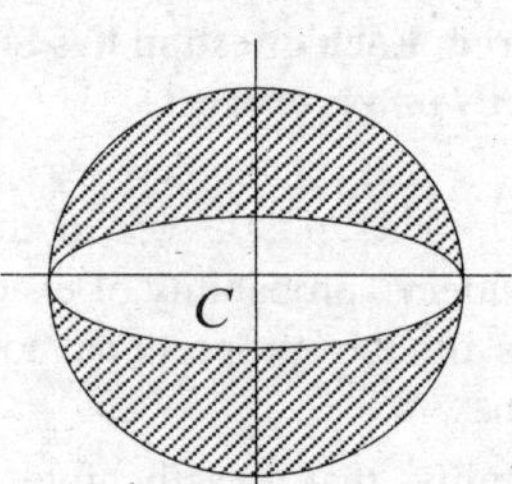

Fig. 9.1

Then, the area of circle serves as sample space and area of the shaded region represents the area for favourable cases. Then, required probability is

$$p=\frac{\text{area of shaded region}}{\text{area of circle}}$$

$$=\frac{\pi a^2-\pi ab}{\pi a^2}$$

$$=1-\frac{b}{a}$$

$$=1-\frac{b}{a}$$

$$=1-\frac{1}{3}$$

$$=\frac{2}{3}$$

Now answer the following questions.

25. A point is selected at random inside a circle. The probability that the point is closer to the centre of the circle than to its circumference is

a. 1/4 b. 1/2

c. 1/3 d. $1/\sqrt{2}$

26. Two persons A and B agree to meet at a place between 5 and 6 pm. The first one to arrive waits for 20 min and then leave. If the time of their arrival be independent and at random, then the probability that A and B meet is

a. 1/3 b. 1/3

c. 2/3 d. 5/9

27. If points x, y are chosen randomly from the intervals [0, 2] and [0, 1], respectively, then the probability that $y \le x^2$ is

a. 1/2 b. 2/3

c. 3/4 d. 1/4

For Problems 28–30

If the squares of a 8 × 8 chessboard are painted either red or black at random.

28. The probability that not all the squares in any column are alternating in colour is

a. $(1 - 1/2^7)^8$ b. $1/2^{56}$

c. $1 - 1/2^7$ d. none of these

29. The probability that the chessboard contains equal number of red and black squares is

a. $\frac{{}^{64}C_{32}}{2^{64}}$ b. $\frac{64!}{322^{64}!}$

c. $\frac{2^{32}-1}{2^{64}}$ d. none of these

30. The probability that all the squares in any column are of same colour and that of a row are of alternating colour is

a. $1/2^{64}$ b. $1/2^{63}$

c. 1/2 d. none of these

For Problems 31–33

Two fair dice are rolled. Let $P(A_i) > 0$ denote the event that the sum of the faces of the dice is divisible by i.

31. Which one of the following events is most probable?

a. A_3 b. A_4

c. A_5 d. A_6

32. For which one of the following pairs (i, j) are the events A_i and A_j independent?

a. (3, 4) b. (4, 6)

c. (2, 3) d. (4, 2)

33. The number of all possible ordered pairs (i, j) for which the events A_i and A_j are independent is

a. 6 b. 12

c. 13 d. 25

For Problems 34–36

A player tosses a coin and scores one point for every head and two points for every tail that turns up. He plays on until his score reaches or passes n. P_n denotes the probability of getting a score of exactly n.

34. The value of P_n is equal to

a. $(1/2)[P_{n-1} + P_{n-2}]$ b. $(1/2)[2P_{n-1} + P_{n-2}]$

c. $(1/2)[P_{n-1} + 2P_{n-2}]$ d. none of these

35. The value of $P_n + (1/2)P_{n-1}$ is equal to

a. 1/2 b. 2/3

c. 1 d. none of these

36. Which of the following is not true?

a. $P_{100} > 2/3$ b. $P_{101} < 2/3$

c. $P_{100}, P_{101} > 2/3$ d. none of these

For Problems 37–39

The probability that a family has exactly n children is αp^n, $n \ge 1$. All sex distributions of n children in a family have the same probability.

37. The probability that a family contains exactly k boys is (where $k \ge 1$)

a. $\alpha p^k(1 - p)^{-k-1}$ b. $2\alpha p^k(2 - p)^{-k-1}$

c. $2\alpha p^k(2 - p)^{-k}$ d. $2\alpha p^{k-1}(2 - p)^{-k-1}$

38. The probability that a family includes at least one boy is

a. $\frac{\alpha^2 p}{(2-p)(1-p)}$ b. $\frac{\alpha p^2}{(2-p)(1-p)}$

c. $\frac{\alpha p}{(2-p)(1-p)}$ d. $\frac{2\alpha p}{(2-p)(1-p)}$

39. Given that a family includes at least one boy, show that the probability that there are two or more boys is

a. $p/(2 - p)$ b. $p/(1 - p)$

c. $p/(2 - p)^2$ d. $p/(1 - p)^2$

Matrix-Match Type

Each question contains statements given in two columns which have to be matched. Statements a, b, c, d in column I have to be matched with statements p, q, r, s in column II. If the correct matches are a→p, a→s, b→q, b→r, c→p, c→q and d→s, then the correctly bubbled 4 × 4 matrix should be as follows:

	p	q	r	s
a	**(p)**	(q)	(r)	**(s)**
b	(p)	**(q)**	**(r)**	(s)
c	**(p)**	**(q)**	(r)	(s)
d	(p)	(q)	(r)	**(s)**

1. An urn contains four black and eight white balls. Three balls are drawn from the urn without replacement. Three events are defined on this experiment.

A: Exactly one black ball is drawn.

B: All balls are drawn are of the same colour.

C: Third drawn ball is black.

Match the entries of column I with none, one or more entries of column II.

Column I	Column II
a. The events A and B are	p. mutually exclusive
b. The events B and C are	q. independent
c. The events C and A are	r. neither independent nor mutually exclusive
d. The events A, B and C are	s. exhaustive

2.

Column I	Column II
a. If the probability of getting at least one head is at least 0.8 is n trials then value of n can be	p. 2
b. One mapping is selected at random from all mappings of the set $s = \{1, 2, 3, \ldots, n\}$ into itself. If the probability that the mapping being one-one is 3/32, then find the value of n is	q. 3
c. If m is selected at random from set $\{1, 2, \ldots, 10\}$ and the probability that the quadratic equation $2x^2 + 2mx + m + 1 = 0$ has real roots is k, then value of $5k$ is more than	r. 4
d. A man firing at a distant target as 20% chance of hitting the target in one shoot. If P be the probability of hitting the target in 'n' attempts, where $20P^2 - 13P + 2 \le 0$, then the ratio of maximum and minimum value of n is less than	s. 5

3.

Column I	Column II
a. The probability of a bomb hitting a bridge is 1/2. Two direct hits are needed to destroy it. The number of bombs required so that the probability of the bridge being destroyed is greater than 0.9 can be	p. 4
b. A bag contains two red, three white and five black balls, a ball is drawn its colour is noted and replaced. The number of times, a ball can be drawn so that the probability of getting a red ball for the first time is at least 1/2	q. 6
c. A drawer contains a mixture of red socks and blue socks, at most 17 in all. It so happens that when two socks are selected randomly without replacement, there is a probability of exactly 1/2 that both are red or both are blue. Then number of red socks in the drawer can be	r. 7
d. There are two red, two blue, two white and certain number (greater than 0) of green socks in a drawer. If two socks are taken at random from the drawer without replacement, the probability that they are of the same colour is 1/5, then the number of green socks are	s. 10

4. Let A and B are two independent events such that $P(A) = 1/3$ and $P(B) = 1/4$.

Column I	Column II
a. $P(A \cup B)$ is equal to	p. 1/12
b. $P(A/A \cup B)$ is equal to	q. 1/2
c. $P(B/A' \cap B')$ is equal to	r. 2/3
d. $P(A'/B)$ is equal to	s. 0

5. A bag contains some white and some black balls, all combinations being equally likely. The total number of balls in the bag is 12. Four balls are drawn at random from the bag at random without replacement.

Column I	Column II
a. Probability that all the four balls are black is equal to	p. 14/33
b. If the bag contains 10 black and two white balls then the probability that all four balls are black is equal to	q. 1/3
c. If all the four balls are black, then the probability that the bag contains 10 black balls is equal to	r. 70/429
d. Probability that two balls are black and two are white is	s. 13/165

6.

Column I	Column II
a. Six different balls are put in three different boxes, none being empty. The probability of putting the balls equal number is	p. 20/27
b. Six letters are posted in three letter boxes. The probability that no letter box remains empty is	q. 1/6
c. Two persons A and B throw two dice each. If A throw a sum of 9, then the probability of B throwing a sum greater than A is	r. 1/3
d. If A and B are independent and $P(A) = 0.3$ and $P(A \cup \bar{B}) = 0.8$, then $P(B)$ is equal to	s. 2/7

7. An urn contains r red balls and b black balls.

Column I	Column II
a. If the probability of getting two red balls in first two draws (without replacement) is 1/2, then value of r can be	p. 10
b. If the probability of getting two red balls in first two draws (without replacement) is 1/2 and b is an even number, then r can be	q. 3

c. If the probability of getting exactly two red balls in four draws (with replacement) from the urn is 3/8 and $b = 10$, then r can be	**r.** 8
d. If the probability of getting exactly n red balls in $2n$ draw (with replacement) is equal to probability of getting exactly n black balls in $2n$ draws (with replacement), then the ratio r/b can be	**s.** 2

8. 'n' whole numbers are randomly chosen and multiplied,

Column I	Column II
a. The probability that the last digit is 1, 3, 7 or 9 is	**p.** $\frac{8^n - 4^n}{10^n}$
b. The probability that the last digit is 2, 4, 6, 8 is	**q.** $\frac{5^n - 4^n}{10^n}$
c. The probability that the last digit is 5 is	**r.** $\frac{4^n}{10^n}$
d. The probability that the last digit is zero is	**s.** $\frac{10^n - 8^n - 5^n + 4^n}{10^n}$

Integer Type

1. A rifleman is firing at a distance target and hence there is only 10% chance of hitting it. The number of rounds he must fire in order to have more than 50% chance of hitting it at least once is __________.

2. If A and B are two independent events, the probability that both A and B occur is 1/8 and the probability that neither of them occurs is 3/8. If the probability of the occurrence of A is p, then the greatest value of $1/p$ is __________.

3. A bag contains $n + 1$ coins. It is known that one of these coins shows heads on both sides, whereas the other coins are fair. One coin is selected at random and tossed. If the probability that toss results in heads is 7/12, then the value of n is __________.

4. Each of the n urns contains four white and six black balls. The $(n+1)^{\text{th}}$ urn contains five white and five black balls. one of the $(n + 1)$ urns is chosen at random and two balls are drawn from it without replacement. Both the balls turn out to be black. If the probability that the $(n + 1)^{\text{th}}$ urn was chosen to draw the balls is 1/16, then the value of $n/5$ is ______.

5. Die A has four red and two white faces whereas die B has two red and four white faces. A coin is flipped once. If it falls a head, the game continues by throwing die A, if it falls tail then die B is to be used. If the probability that die A used is 32/33 when it is given that red turns up every time in first n throws, then the value of $n/2$ is __________.

6. A fair coin is tossed n times. If the probability that head occurs six times is equal to the probability that head occurs eight times, then the value of $n/2$ is __________.

7. A bag contains a white and b black balls. Two players, A and B alternately draw a ball from the bag replacing the ball each time after the draw till one of them draws a white ball and wins the game. A begins the game. If the probability of A winning the game is three times that of B, the value of $a : b$ is __________.

8. If $(1 + 3p)/3$, $(1 - p)/4$, and $(1 - 2p)/2$ are the probabilities of three mutually exclusive events, then number of integral values of $1/p$ is __________.

9. Let $A = \{0, 5, 10, 15,\ldots, 195\}$. Let B e any subset of A with atleast 15 elements. If p is the probability that B has at least one pair of elements whose sum is divisible by 15, then the value of $3p$ is __________.

10. Suppose A and B shoot independently until each hits his target. They have probabilities 3/5 and 5/7 of hitting the targets at each shot. If p is the probability that B will require more shots than A, then the value of $31p$ is __________.

11. Two natural numbers x and y are chosen at random. If p is the probability that $x^2 + y^2$ is divisible by 5 and 7, then the value of $25p$ is __________.

12. $8n$ players $P_1, P_2, P_3, \ldots P_{8n}$ play a knock out tournament. It is known that all the players are of equal strength. The tournament is held in three rounds where the players are paired at random in each round. If it is given that P_1 wins in the third round. If the probability that P_2 looses in the second round is 8/31, then the value of n is __________.

13. A tennis match of best of five sets is played by two players A and B. The probability that first set is won by A is 1/2 and if he looses the first then the probability of his winning of next set is 1/4 otherwise it remains same. If the probability that A wins the match is p, then the value of $16p$ is __________.

14. An urn contains two white and two black balls. A ball is drawn at random. If it is white, it is not replaced into urn otherwise it is replaced along with another ball of the same colour. The process is repeated. If p is the probability that the third ball drawn is black, then the value of $23p/5$ is __.

15. In a test an examinee either guesses or copies or knows the answer to a multiple choice question with four choices. The probability that he makes a guess is 1/3 and the probability that he copies the answer is 1/6. The probability that his answer is correct given that he copied it is 1/8. If p is the probability that he knew the answer to the question, given that he correctly answered it, then the value of $29p/4$ is __.

16. A person goes to office by car, scooter, bus or train, the probability of which being 1/7, 3/7, 2/7 and 1/7, respectively. Probability that he reaches office late, if he takes car, scooter, bus or train is 2/9, 1/9, 4/9 and 1/9, respectively. Given that he reached office in time, then if the probability that he travelled by a car is p, then the value of $7p$ is __________.

17. If the probability that the 3 N's come consecutively in the arrangement of the letters of the word CONSTANTINOPLE is p, then the value of $3/13p$ is ________.

18. A die is loaded so that the probability of a face i is proportional to i, $i = 1, 2, \ldots, 6$. If the probability of an even number occurring when die is rolled is p, then the value of $7p$ is ___.

19. A bag contains three white, three black, and two red balls. One by one three balls are drawn without replacing them. If the probability that the third ball is red is p, then the value of $4p$ is ________.

20. A box contains 10 mangoes out of which four are rotten. two mangoes are taken out together. If one of them is found to be good. If p is the probability that the other is also good, then the value of $13p$ is ________.

21. An urn contains six white and four black balls. A fair die is rolled and that number of balls are chosen from the urn. If the probability that the balls selected are white is p, then the value of $9p$ is ________.

22. An urn contains 3 red balls and n white balls. Mr. A draws two balls together from the urn. The probability that they have the same colour is 1/2. Mr. B draws one ball from the urn, notes its colour and replaces it. He then draws a second ball from the urn and finds that both balls have the same colour is, 5/8. The possible value of n is ________.

ANSWERS AND SOLUTIONS

Objective Type

1. b. $P(A)=\frac{1}{3}, P(A\cup B)=\frac{3}{4}$

Now,

$P(A\cup B)=P(A)+P(B)-P(A\cap B)\le P(A)+P(B)$

$\Rightarrow \frac{3}{4}\le\frac{1}{3}+P(B)$

$\Rightarrow \frac{5}{12}\le P(B)$

Again we have $B\subseteq A\cup B$.

$\therefore\ P(B)\le P(A\cup B)=\frac{3}{4}$

Hence, $5/12\le P(B)\le 3/4$.

2. a. The probability of hitting a target is $p = 1/5$. Therefore, the probability of not hitting a target is $q = 1 - 1/5 = 4/5$. Hence, the required probability is $1-(4/5)^{10}$.

3. d. Let the probability for getting an odd number be p. Therefore, the probability for getting an even number is $2p$.

$\therefore\ p+2p=1\Rightarrow 3p=1\Rightarrow p=\frac{1}{3}$

Sum of two nunbers is even means either both are odd or both are even. Therefore, the required probability is

$\frac{1}{3}\times\frac{1}{3}+\frac{2}{3}\times\frac{2}{3}=\frac{1}{9}+\frac{4}{9}=\frac{5}{9}$

4. c. If A draws card higher than B, then number of favourable cases is $(n-1)+(n-2)+\cdots+3+2+1$ (as when B draws card number 1, then A can draw any card from 2 to n and so on). Therefore, the required probability is

$$\frac{\frac{n(n-1)}{2}}{n^2}=\frac{n-1}{2n}$$

5. b. If A, B, C represent events that the student is successful in tests I, II, III, respectively, Then the probability tha the student is successful is

$P[(A\cap B\cap C')\cup(A\cap B'\cap C)\cup(A\cap B\cap C)]$

$=P(A\cap B\cap C')+P(A\cap B'\cap C)+P(A\cap B\cap C)$

$=P(A)\,P(B)\,P(C')+P(A)\,P(B')\,P(C)+P(A)\,P(B)\,P(C)$

[$\therefore$ A, B, C are independent events]

$=pq\left(1-\frac{1}{2}\right)+p(1-q)\frac{1}{2}+pq\frac{1}{2}$

$=pq+\frac{1}{2}p-\frac{1}{2}pq$

$=\frac{1}{2}(pq+p)$

$\therefore\ \frac{1}{2}p(1+q)=\frac{1}{2}$

$\Rightarrow p(1+q)=1$

6. a. $P(A)=\frac{1}{2}, P(B)=\frac{1}{3}, P(C)=\frac{1}{4}$

$\therefore\ P(\bar{A})=1-\frac{1}{2}=\frac{1}{2}, P(\bar{B})=1-\frac{1}{3}=\frac{2}{3}$

$P(\bar{C})=1-\frac{1}{4}=\frac{3}{4}$

Therefore, the required probability is

$1-P(\bar{A})\,P(\bar{B})\,P(\bar{C})=1-\frac{1}{2}\times\frac{2}{3}\times\frac{3}{4}=1-\frac{1}{4}=\frac{3}{4}$

7. d. A person can have his/her birthday on any one of the seven days of the week. So five persons can have their birthdays in 7^5 ways. Out of five, three persons can have their birthdays on days other than Sundays in 6^3 ways and other two on Sundays. Hence, the required probability is

$$\frac{{}^5C_2\times 6^3}{7^5}=\frac{10\times 6^3}{7^5}$$

(Note that two persons can be selected out of five in ${}^5C_2$ ways.)

8. a. Required probability $=\frac{\text{No. of favourable cases}}{\text{Total no. of exhaustive cases}}$

$=\frac{3}{3\times 3\times 3}=\frac{1}{9}$

9. d. The number of ways of arranging n numbers is $n!$ In each order obtained, we must now arrange the digits 1, 2, …, k as group and the $n-k$ remaining digits. This can be done in $(n-k+1)!$ ways. Therefore, the probability for the required event is $(n-k+1)!/n!$

10. d. According to the given çondition

$${}^nC_3\left(\frac{1}{2}\right)^n={}^nC_4\left(\frac{1}{2}\right)^n$$

where n is the number of times die is thrown.

$\therefore\ {}^nC_3={}^nC_4\Rightarrow n=7$

Thus, the required probability is

$${}^7C_1\left(\frac{1}{2}\right)^7=\frac{7}{2^7}=\frac{7}{128}$$

11. c. Possibilities of getting 9 are (5, 4), (4, 5), (6, 3), (3, 6).

$\therefore\ p=\frac{4}{36}=\frac{1}{9}$ and $q=1-\frac{1}{9}=\frac{8}{9}$

Therefore, the required probability is

$${}^3C_2q^1p^2=(3)\left(\frac{8}{9}\right)\left(\frac{1}{9}\right)^2=\frac{8}{243}$$

12. c. Out of five horses, only one is the wining horse. The probability that Mr. A selected that losing horse is $4/5\times 3/4$. Therefore, the required probability is

$1-\frac{4}{5}\times\frac{3}{4}=1-\frac{3}{5}=\frac{2}{5}$

13. d. We have,

$P(\overline{A\cup B})=\frac{1}{4}, P(A\cap B)=\frac{1}{4}$

$P(\bar{A})=\frac{1}{4}\Rightarrow P(A)=1-\frac{1}{4}=\frac{3}{4}$

$\therefore P(\overline{A \cup B}) = 1 - P(A \cup B)$

$= 1 - [P(A) + P(B) - P(A \cap B)]$

$\Rightarrow \frac{1}{6} = 1 - \frac{3}{4} - P(B) + \frac{1}{4}$

$\Rightarrow P(B) = 1 - \frac{1}{2} - \frac{1}{6} = \frac{6-3-1}{6} = \frac{2}{6} = \frac{1}{3}$

Since $P(A \cap B) = P(A)P(B)$ and $P(A) \neq P(B)$, A and B are independent but not equally likely.

14. b. The total number of cases is $11!/2! \times 2!$. The number of favourable cases is $[11!/(2! \times 2!)] - 9!$ Therefore, the required probability is

$1 - \frac{9! \times 4}{11!} = \frac{53}{55}$

15. c. Total number of the students is 80. Total number of girls is 25. Total number of boys is 55. There are 10 rich, 70 poor, 20 intelligent students in the class. Therefore, required probability is

$\underset{(I)}{\frac{1}{4}} \times \underset{(R)}{\frac{1}{8}} \times \underset{(G)}{\frac{25}{80}} = \frac{5}{512}$

16. b. $P(A' \cap B \cap C' \cap D) = P(A')\,P(B)\,P(C')\,P(D)$

$= \left(1 - \frac{1}{2}\right)\frac{1}{3}\left(1 - \frac{1}{5}\right)\left(\frac{1}{6}\right)$

$= \frac{1}{2} \times \frac{1}{3} \times \frac{4}{5} \times \frac{1}{6} = \frac{1}{45}$

17. a. $p^2 + 2p + 4p - 1 = 1$ (Exhaustive)

$p^2 + 6p - 2 = 0$

$\Rightarrow p = -3 \pm \sqrt{11}$

$\Rightarrow p = \sqrt{11} - 3$

18. a. L and W can be filled at 14 places in 2^{14} ways.

$\therefore n(S) = 2^{14}$

Now 13 L's and 1 W can be arranged at 14 places in 14 ways.

Hence, $n(A) = 14$.

$\therefore p = \frac{14}{2^{14}} = \frac{7}{2^{13}}$

19. a. $P(A \cap C) = P(A)\,P(C)$

$\Rightarrow \frac{1}{20} = \frac{1}{5}P(C)$

$\Rightarrow P(C) = \frac{1}{4}$

Now,

$P(B \cup C) = \frac{1}{6} + \frac{1}{4} - P(B \cap C)$

$\Rightarrow P(B \cup) = \frac{3}{8} - \frac{1}{3} = \frac{1}{24} = \frac{1}{24} = P(B)P(C)$

Therefore, B and C are independent.

20. a. The number of ways in which 20 people can be divided into two equal groups is

$n(s) = \frac{20!}{10!\ 10!\ 2!}$

The number of ways in which 18 people can be divided into groups of 10 and 8 is

$n(A) = \frac{18!}{10!\ 8!}$

$\therefore P(E) = \frac{18!}{10!\,8!}\frac{10!\,10!\,2}{20!} = \frac{10 \times 9 \times 2}{20 \times 19} = \frac{9}{19}$

21. c. $P(A \cup B) = P(A) + P(B) - P(A) \times P(B)$

$\Rightarrow 0.7 = 0.4 + p - 0.4p$

$\therefore 0.6p = 0.3 \Rightarrow p = \frac{1}{2}$

22. a. $P(B_1) = \frac{{}^6C_1}{{}^{10}C_1} = \frac{6}{10} = \frac{3}{5}$

$P(B_2 / B_1) = \frac{5}{9}$ (B_2 = black)

$\therefore P(B_1 \cap B_2) = P(B_1)\,P(B_2 / B_1) = \frac{3}{5} \times \frac{5}{9} = \frac{1}{3}$

23. b. Total number of ways of distribution is 4^5.

$\therefore n(S) = 4^5$

Total number of ways of distribution so that each child gets at least one game is

$4^5 - {}^4C_1\,3^5 + {}^4C_2\,2^5 - {}^4C_3 = 1024 - 4 \times 243 + 6 \times 32 - 4 = 240$

$\therefore n(E) = 240$

Therefore, the required probability is

$\frac{n(E)}{n(S)} = \frac{240}{4^5} = \frac{15}{64}$

24. a. Out of nine socks, two can be drawn in ${}^9C_2$ ways. Therefore, the total number of cases is ${}^9C_2$. Two socks drawn from the drawer will match if either both are brown or both are blue. Therefore, favourable number of cases is ${}^5C_2 + {}^4C_2$. Hence, the required probability is

$\frac{{}^5C_2 + {}^4C_2}{{}^9C_2} = \frac{4}{9}$

25. d. Consider two events as follows:

A_i: getting number i on first die

B_i: getting a number more than i on second die

The required probability is

$P(A_1 \cap B_1) + P(A_2 \cap B_2) + P(A_3 \cap B_3) + P(A_4 \cap B_4) + P(A_5 \cap B_5) = \sum_{i=1}^{5} P(A_i \cap B_i) = \sum_{i=1}^{5} P(A_i)\,P(B_i)$

[$\because A_i, B_i$ are independent]

$= \frac{1}{6}[P(B_1) + P(B_2) + \cdots + P(B_5)]$

$= \frac{1}{6}\left(\frac{5}{6} + \frac{4}{6} + \frac{3}{6} + \frac{2}{6} + \frac{1}{6}\right) = \frac{5}{12}$

26. d. We have, $x + \frac{100}{x} > 50$

$\Rightarrow x^2 + 100 > 50x$

$\Rightarrow (x - 25)^2 > 525$

$\Rightarrow x - 25 < \sqrt{525}$ or $x - 25 > \sqrt{525}$

$\Rightarrow x < 25 - \sqrt{525}$ or $25 + \sqrt{525}$

As x is a positive integer and $\sqrt{525} = 22.91$, we must have $x \le 2$ or $x \ge 48$. Thus, the favourable number of cases is 2 + 53 = 55. Hence, the required probability is 55/100 = 11/20.

27. b. The total number of ways in which four-figure numbers can be formed is 4! = 24. A number is divisible by 5 if at unit's place we have 5. Therefore, unit's place can be filled in one way and the remaining three places can be filled with the other digits in 3! ways. Hence, total number of numbers divisible by 5 is 3! = 6. So, the required probability is 6/24 = 1/4.

28. a. Since each ball can be placed in any one of the three boxes, therefore there are three ways in which a ball can be placed in any one of the three boxes. Thus, there are 3^{12} ways in which 12 balls can be placed in three boxes. The number of ways in which three balls out of 12 can be put in the box is ${}^{12}C_3$. The remaining nine balls can be placed in two boxes in 2^9 ways. So, required probability is

$$\frac{{}^{12}C_3\, 2^9}{3^{12}} = \frac{110}{9}\left(\frac{2}{3}\right)^{10}$$

29. a. The required probability is

$$\frac{2}{9}\times\frac{1}{8}\times\frac{4}{7}\times\frac{3}{6}\times\frac{2}{5}\times\frac{1}{4}\times 1\times 1\times 1 = \frac{1}{1260}$$

30. d. The total number of ways of choosing 11 players out of 15 is ${}^{15}C_{11}$. A team of 11 players containing at least three bowlers can be chosen in the following mutually exclusive ways:

(I) Three bowlers out of five bowlers and eight other players out of the remaining 10 players.

(II) Four bowlers out of five bowlers and seven other players out of the remaining 10 players.

(III) Five bowlers out of five bowlers and six other players out of the remaining 10 players.

So, required probability is

$$P(\text{I}) + P(\text{II}) + P(\text{III}) = \frac{{}^5C_3 \times {}^{10}C_8}{{}^{15}C_{11}} + \frac{{}^5C_4 \times {}^{10}C_7}{{}^{15}C_{11}} + \frac{{}^5C_5 \times {}^{10}C_6}{{}^{15}C_{11}}$$

$$= \frac{1260}{1365} = \frac{12}{13}$$

31. b. Consider the following events:

A_1: A speaks truth

A_2: B speaks truth

Then, $P(A_1) = 60/100 = 3/5$, $P(A_2) = 70/100 = 7/10$.

For the required event, either both of them should speak the truth or both of them should tell a lie. Thus, the required probability is

$$P((A_1 \cap A_2) \cup (\bar{A}_1 \cap \bar{A}_2)) = P(A_1 \cap A_2) + P(\bar{A}_1 \cap \bar{A}_2)$$

$$= P(A_1)\, P(A_2) + P(\bar{A}_1)\,(\bar{A}_2)$$

$$= \frac{3}{5}\times\frac{7}{10} + \left(1 - \frac{3}{5}\right)\left(1 - \frac{7}{10}\right)$$

$$= 0.54$$

32. c. The total number of ways in which three integers can be chosen from first 20 integers is ${}^{20}C_3$. The product of three integers will be even if at least one of the integers is even. Therefore, the required probability is

1 – Probability that none of the three integers is even

$$= 1 - \frac{{}^{10}C_3}{{}^{20}C_3} = 1 - \frac{2}{19} = \frac{17}{19}$$

33. b. Consider the following events:

A: getting a card with mark I in first draw

B: getting a card with mark I in second draw

C: getting a card with mark T in this draw

Then, the required probability is

$$P(A \cap B \cap C) = P(A)\, P(B/A)\, P(C/A \cap B)$$

$$= \frac{10}{20}\times\frac{9}{19}\times\frac{10}{18} = \frac{5}{38}$$

34. a. Let p_1 and p_2 be the chances of happening of the first and second events, respectively, then according to the given conditions, we have

$$p_1 = p_2^2 \text{ and } \frac{1 - p_1}{p_1} = \left(\frac{1 - p_2}{p_2}\right)^3$$

$$\Rightarrow \frac{1 - p_2^2}{p_2^2} = \left(\frac{1 - p_2}{p_2}\right)^3$$

$$\Rightarrow p_2(1 + p_2) = (1 - p_2)^2$$

$$\Rightarrow p_2 = \frac{1}{3}$$

and so

$$p_1 = \frac{1}{9}$$

35. b. The total number of ways in which n persons can sit at a round table is $(n - 1)!$ So, total number of cases is $(n - 1)!$.

Let A and B be two specified persons. Considering these two as one person, the total number of ways in which $n - 1$ persons, $n - 2$ other persons and one AB can sit at a round table is $(n - 2)!$ So, favourable number of cases is $2!(n - 2)!$ Thus, the required probability is

$$p = \frac{2!\,(n-2)!}{(n-1)!} = \frac{2}{n-1}$$

Hence, the required odds are $(1 - p):p$ or $(n - 3):2$.

36. b. Let one of the quantities be x. Then the other is $2n - x$. Their product will be greatest when they are equal, i.e., each is n in which case the product is n^2. According to the proposition,

$$x(2n - x) \ge \frac{3}{4}n^2$$

$$\Rightarrow 4x^2 - 8nx + 3n^2 \le 0$$

$$\Rightarrow (2x - 3n)(2x - n) \le 0$$

$$\Rightarrow \frac{n}{2} \le x \le \frac{3}{2}n$$

So, favourable number of cases is $3/2n - n - 2 = n$. Hence, the required probability is $n/2n = 1/2$.

37. a. Let the number of red and blue balls be r and b, respectively.

Then, the probability of drawing two red balls is

$$p_1 = \frac{{}^rC_2}{{}^{r+b}C_2} = \frac{r(r-1)}{(r+b)(r+b-1)}$$

The probability of drawing two blue balls is

$$p_2 = \frac{{}^bC_2}{{}^{r+b}C_2} = \frac{b(b-1)}{(r+b)(r+b-1)}$$

The probability of drawing one red and one blue ball is

$$p_3 = \frac{{}^rC_1\,{}^bC_1}{{}^{r+b}C_2} = \frac{2br}{(r+b)(r+b-1)}$$

By hypothesis, $p_1 = 5p_2$ and $p_3 = 6p_2$.

$\therefore r(r-1) = 5b(b-1)$ and $2br = 6b(b-1)$

$\Rightarrow r = 6, b = 3$

38. b. There are 10 digits 0, 1, 2, 3, 4, 5, 6, 7, 8, 9. The last two digits can be dialled in ${}^{10}P_2 = 90$ ways out of which only one way is favourable, thus, the required probability is 1/90.

39. d. The probability that one test is held is $2 \times (1 \times 5) \times (4 \times 5) = 8/25$. The probability that test is held on both days is $(1 \times 5) \times (1 \times 5) = 1/25$.

Therefore, the probability that the student misses at least one test is $8/25 + 1/25 = 9/25$.

40. d. Let X denote the largest number on the 3 tickets drawn.

Then, $P(X \le 7) = (7/20)^3$ and $P(X \le 6) = (6/20)^3$. Then, the required probability is

$$P(X=7) = \left(\frac{7}{10}\right)^3 - \left(\frac{6}{20}\right)^3$$

41. b. Let X denote the number of heads in n trials. Then X is a binomial variant with $p = q = 1/2$. Therefore,

$$P(X=r) = {}^nC_r\left(\frac{1}{2}\right)^n$$

Now,

$P(X=6) = P(X=8)$

$$\Rightarrow {}^nC_6\left(\frac{1}{2}\right)^n = {}^nC_8\left(\frac{1}{2}\right)^n$$

$\Rightarrow {}^nC_6 = {}^nC_8 \Rightarrow n = 14$

42. a. Let the number selected be xy. Then

$x + y = 9, 0 \le x, y \le 9$

and

$xy = 0 \Rightarrow x = 0, y = 9$

or

$y = 0, x = 9$

$$P(x_1 = 9/x_2 = 0) = \frac{P(x_1 = 9 \cap x_2 = 0)}{P(x_2 = 0)}$$

Now,

$$P(x_2 = 0) = \frac{19}{100}$$

and

$$P(x_1 = 9 \cap x_2 = 0) = \frac{2}{100}$$

$$\Rightarrow P(x_1 = 9/x_2 = 0) = \frac{2/100}{19/100} = \frac{2}{19}$$

43. a. $P(A \cap B') = P(A) - P(A \cap B) = 0.20$

Also,

$P(A' \cap B) = P(B) - P(A \cap B) = 0.15$

$\Rightarrow P(A) + P(B) - 2P(A - B) = 0.35$

Now,

$P(A' \cap B') = 1 - P(A \cup B)$

$\Rightarrow 0.1 = 1 - P(A) - P(B) + P(A \cap B)$

$\Rightarrow P(A) + P(B) - P(A \cap B) = 0.9$

$\Rightarrow P(A \cap B) = 0.9 - 0.35 = 0.55$

and

$P(A) = 0.75, P(B) = 0.70$

Now,

$$P(A/B) = \frac{P(A \cap B)}{P(B)} = \frac{0.55}{0.70}$$

44. a. For each toss, there are four choices:

(i) A gets head, B gets head

(ii) A gets tail, B gets head

(iii) A gets head, B gets tail

(iv) A gets tail, B gets tail

Thus, exhaustive number of ways is 4^{50}. Out of the four choices listed above, (iv) is not favourable to the required event in a toss. Therefore, favourable number of cases is 3^{50}. Hence, the required probability is $(3/4)^{50}$.

45. c. 18 draws are required for two aces means in the first 17 draws, there is one ace and 16 other cards and 18th draw produces an ace. So, the required probability is

$$\frac{{}^{48}C_{16} \times {}^4C_1}{{}^{52}C_{17}} \times \frac{3}{35} = \frac{561}{15925}$$

46. b. Consider the following events:

A: Father has at least one boy

B: Father has two boys and one girl

Then,

A = one boy and two girls, two boys and one girl, three boys and no girl

$A \cap B$ = 2 boys and one girl

Now, the required probability is

$$P(A/B) = \frac{P(A \cap B)}{P(A)} = \frac{1}{3}$$

47. c. The required probability is

$$\left[{}^4C_0\left(\frac{1}{2}\right)^0\left(\frac{1}{2}\right)^4\right]^2 + \left[{}^4C_1\left(\frac{1}{2}\right)^1\left(\frac{1}{2}\right)^3\right]^2$$

$$+\left[{}^4C_2\left(\frac{1}{2}\right)^2\left(\frac{1}{2}\right)^2\right]^2+\left[{}^4C_3\left(\frac{1}{2}\right)^3\left(\frac{1}{2}\right)^1\right]^2$$

$$+\left[{}^4C_4\left(\frac{1}{2}\right)^4\left(\frac{1}{2}\right)^0\right]^2=\frac{35}{128}$$

48. b. Let A denote the event that there is an odd man out in a game. The total number of possible cases is 2^m. A person is odd man out if he is alone in getting a head or a tail.

The number of ways in which there is exactly one tail (head) and the rest are heads (tails) is ${}^mC_1 = m$. Thus, the number of favourable ways is $m + m = 2m$. Therefore,

$$P(A)=\frac{2m}{2^m}=\frac{m}{2^{m-1}}$$

49. c. $x^2 + 2(a+4)x - 5a + 64 \geq 0$

If $D \leq 0$, then

$(a+4)^2 - (-5a + 64) < 0$

$\Rightarrow a^2 + 13a - 48 < 0$

$\Rightarrow (a+16)(a-3) < 0$

$\Rightarrow -16 < a < 3 \Leftrightarrow -5 \leq a \leq 2$

Then, the favorable cases is equal to the number of integers in the interval $[-5, 2]$, i.e., 8.

Total number of cases is equal to the number of integers in the interval $[-5, 30]$, i.e., 36.

Hence, the required probability is $8/36 = 2/9$.

50. a. The total number of ways in which $2n$ boys can be divided into two equal groups is

$$\frac{(2n)!}{(n!)^2\,2!}$$

Now, the number of ways in which $2n-2$ boys other than the two tallest boys can be divided into two equal groups is

$$\frac{(2n-2)!}{((n-1)!)^2\,2!}$$

Two tallest boys can be put in different groups in ${}^2C_1$ ways. Hence, the required probability is

$$\frac{2\dfrac{(2n-2)!}{((n-1)!)^2\,2!}}{\dfrac{(2n)!}{(n!)^2\,2!}}=\frac{n}{2n-1}$$

51. a. Let E_i denote the event that the bag contains i black and $(10-i)$ white balls $(i = 0, 1, 2, \ldots, 10)$. Let A denote the event that the three balls drawn at random from the bag are black. We have,

$$P(E_i)=\frac{1}{11}\ (i=0,1,2,\ldots,10)$$

$P(A/E_i) = 0$ for $i = 0, 1, 2$ and $P(A/E_i) = {}^iC_3/{}^{10}C_3$ for $i \geq 3$

$$\Rightarrow P(A)=\frac{1}{11}\times\frac{1}{{}^{10}C_3}\left[{}^3C_3+{}^4C_3+\cdots+{}^{10}C_3\right]$$

But

${}^3C_3 + {}^4C_3 + {}^5C_3 + \cdots + {}^{10}C_3 = {}^4C_4 + {}^4C_3 + {}^5C_3 + \cdots + {}^{10}C_3$
$= {}^5C_4 + {}^5C_3 + {}^6C_3 + \cdots + {}^{10}C_3$

$\vdots$

$= {}^{11}C_4$

$$\Rightarrow p(A)=\frac{1}{11}\times\frac{1}{{}^{10}C_3}\times{}^{11}C_4$$

$$=\frac{\dfrac{11\times10\times9\times8}{4!}}{11\times\dfrac{10\times9\times8}{3!}}=\frac{1}{4}$$

$$\therefore\ P(E_9/A)=\frac{P(E_9)\,P(A/E_9)}{P(A)}$$

$$=\frac{\dfrac{1}{11}\times\dfrac{{}^9C_3}{{}^{10}C_3}}{\dfrac{1}{4}}$$

$$=\frac{14}{55}$$

52. a. We have ratio of the ships A, B and C for arriving safely are 2:5, 3:7 and 6:11, respectively. Therefore, the probability of ship A for arriving safely is $2/(2+5) = 2/7$.

Similarly, for B the probability is $3/(3+7) = 3/10$ and for C the probability is $C = 6/(6+11) = 6/17$.

Therefore, the probability of all the ships for arriving safely is $(2/7) \times (3/10) \times (6/17)$ 18/595.

53. a. i. This question can also be solved by one student.

ii. This question can be solved by two students simultaneously.

iii. This question can be solved by three students all together.

$$P(A)=\frac{1}{2},\ P(B)=\frac{1}{4},\ P(C)=\frac{1}{6}$$

$$P(A\cup B\cup C)=P(A)+P(B)+P(C)-[P(A)\,P(B)+P(B)\,P(C)+P(C)\,P(A)]+[P(A)\,P(B)\,P(C)]$$

$$=\frac{1}{2}+\frac{1}{4}+\frac{1}{6}-\left[\frac{1}{2}\times\frac{1}{4}+\frac{1}{4}\times\frac{1}{6}+\frac{1}{6}\times\frac{1}{2}\right]+\left[\frac{1}{2}\times\frac{1}{4}\times\frac{1}{6}\right]$$

$$=\frac{33}{48}$$

Alternative solution:

We have,

$$P(\bar{A})=\frac{1}{2},\ P(\bar{B})=\frac{3}{4},\ P(\bar{C})=\frac{5}{6}$$

Then the probability that the problem is not solved is

$$P(\bar{A})\,P(\bar{B})\,P(\bar{C})=\left(\frac{1}{2}\right)\left(\frac{3}{4}\right)\left(\frac{5}{6}\right)=\frac{5}{16}$$

Hence probability that problem is solved is $1 - 5/16 = 11/16$.

54. a. We are given that

$P(A \cap B) = P(A)\,P(B)$

$P(B \cap C) = P(B)P(C)$

$P(C \cap A) = P(C)P(A)$

$P(A \cap B \cap C) = P(A)(B)P(C)$

We have,

$P(A \cap (B \cap C)) = P(A \cap B \cap C) = P(A)\ P(B)\ P(C)$
$= P(A)P(B \cap C)$

$\Rightarrow A$ and $B \cap C$ are indepedent

Therefore, S_2 is true. Also,

$P[A \cap (B \cup C)] = P[(A \cap B) \cup (A \cap C)]$
$= P(A \cap B) + P(A \cap C) - P[(A \cap B) \cap (A \cap C)]$
$= P(A \cap B) + P(A \cap C) - P(A \cap B \cap C)$
$= P(A)P(B) + P(A)\ P(C) - P(A)\ P(B)\ P(C)$
$= P(A)[P(B) + P(C) - P(B)P(C)]$
$= P(A)[P(B) + P(C) - P(B \cap C)]$
$= P(A)P(B \cup C)$

Therefore, A and $B \cup C$ are independent.

55. d. The total number of ways in which papers of four students can be checked by seven teachers is 7^4. The number of ways of choosing two teachers out of seven is ${}^7C_2$. The number of ways in which they can check four papers is 2^4. But this includes two ways in which all the papers will be checked by a single teacher. Therefore, the number of ways in which four papers can be checked by exactly two teachers is $2^4 - 2 = 14$. Therefore, the number of favourable ways is $({}^7C_2)$ (14) = (21) (14). Thus, the required probability is $(21)(14)/7^4 = 6/49$.

56. c. Let us assume that A wins after n deuces, $n = 0, 1, 2, 3, \ldots$. The probability of a deuce is (2/3) × (2/3) + (1/3) × (1/3) = (5/9). [A wins his serve, then B wins his serve or A loses his serve.] So, the probability that 'A' wins game after n deuces is $(5/9)^n \times (2/3) \times (1/3)$. [After n^{th} deuce, A serves and wins, then B serves and loses.] Therefore, the required probability of 'A' winning the game is

$$\sum_{n=0}^{\infty}\left(\frac{5}{9}\right)^n \times \frac{2}{3} \times \frac{1}{3} = \frac{1}{1-\frac{5}{9}} \times \frac{2}{9} = \frac{1}{2}$$

57. c. The required probability is

1 – Probability of getting equal number of heads and tails

$$= 1 - {}^{2n}C_n\left(\frac{1}{2}\right)^n\left(\frac{1}{2}\right)^{2n-n}$$

$$= 1 - \frac{(2n)!}{n!n!}\left(\frac{1}{4}\right)^n$$

$$= 1 - \frac{(2n)!}{(n!)^2} \times \frac{1}{4^n}$$

58. b. Here $p = 19/20$, $q = 1/20$, $n = 5$, $r = 5$. The required probability is

$${}^5C_5\left(\frac{19}{20}\right)^5\left(\frac{1}{20}\right)^0 = \left(\frac{19}{20}\right)^5$$

59. d. $$P(E/F) + P(\bar{E}/F) = \frac{P(E \cap F) + P(\bar{E} \cap F)}{P(F)}$$

$$= \frac{P\{(E \cap F) \cup (\bar{E} \cap F)\}}{P(F)}$$

[$\because E \cap F$ **and** $\bar{E} \cap F$ **are disjoint**]

$$= \frac{P\{(E \cup \bar{E}) \cap F\}}{P(F)} = \frac{P(F)}{P(F)} = 1$$

Similarly, we can show that (b) and (c) are not true while (d) is true.

$$P\left(\frac{E}{\bar{F}}\right) + P\left(\frac{\bar{E}}{\bar{F}}\right) = \frac{P(E \cap \bar{F})}{P(\bar{F})} + \frac{P(\bar{E} \cap \bar{F})}{P(\bar{F})} = \frac{P(\bar{F})}{P(\bar{F})} = 1$$

60. b. We have,

$$P(A) = \frac{40}{100}, P(B) = \frac{25}{100}, \text{ and } P(A \cap B) = \frac{15}{100}$$

So,

$$P\left(\frac{B}{A}\right) = \frac{P(A \cap B)}{P(A)} = \frac{15/100}{40/100} = \frac{3}{8}$$

61. b. We define the following events:

A_1: Selecting a pair of consecutive letters from the word LONDON

A_2: Selecting a pair of consecutive letters from the word CLIFTON

E: Selecting a pair of letters ON

Then, $P(A_1 \cap E) = 2/5$ as there are five pairs of consecutive letters out of which two are ON and $P(A_2 \cap E) = 1/6$ as there are six pairs of consecutive letters of which 1 is ON. Therefore, the required probability is

$$P\left(\frac{A_1}{E}\right) = \frac{P(A_1 \cap E)}{P(A_1 \cap E) + P(A_2 \cap E)} = \frac{\frac{2}{5}}{\frac{2}{5} + \frac{1}{6}} = \frac{12}{17}$$

62. d. Player should get ($HT, HT, HT, \ldots$) or ($TH, TH, \ldots$) at least $2n$ times. If the sequence starts from first place, then the probability is $1/2^{2n}$ and if starts from any other place, then the probability is $1/2^{2n+1}$. Hence, required probability is

$$2\left(\frac{1}{2^{2n}} + \frac{m}{2^{2n+1}}\right) = \frac{m+2}{2^{2n}}$$

63. b. The total number of cases is 11!/2! × 2! The number of favourable cases is 11!/(2! × 2!) – 9! Therefore, required probability is

$$1 - \frac{9! \times 4}{11!} = \frac{53}{55}$$

64. b. The probability of winning of A the second race is 1/2 (since both events are independent).

65. b. Given that $n(S) = 6 \times 6 \times 6 \times 6 = 6^4$. The number of favourable ways is ${}^6C_4 = 6 \times 5/2 = 15$. Therefore, the required probability is

$$\frac{15}{6 \times 216} = \frac{5}{2 \times 216} = \frac{5}{432}$$

66. c. Given that 5 and 6 have appeared on two of the dice, the sample space reduces to $6^4 - 2 \times 5^4 + 4^4$ (inclusion–exclusion principle). Also, the number of favourable cases are 4! = 24. So, the required probability is 24/302 = 12/151.

67. c. Let a_n be the number of strings of H and T of length n with no two adjacent H's. Then $a_1 = 2$, $a_2 = 3$. Also,

$a_{n+2} = a_{n+1} + a_n$ (since the string must begin with T or HT)

So,

$a_3 = 5, a_4 = 8, a_5 = 8 + 5 = 13$

Therefore, the required probability is $13/2^5 = 13/32$.

68. b.

2							

The prime digits are 2, 3, 5, 7. If we fix 2 at first place, then other $2n - 1$ places are filled by all four digits. So the total number of cases is 4^{2n-1}.

Now, sum of two consecutive digits is prime when consecutive digits are (2, 3) or (2, 5). Then 2 will be fixed at all alternative places.

2		2		2		2	

So favourable number of cases is 2^n. Therefore, probability is

$$\frac{2^n}{4^{2n-1}} = 2^n\, 2^{-4n+2} = 2^2\, 2^{-3n} = \frac{4}{2^{3n}}$$

69. d. Since a, b, c are in A.P., $2b = a + c$. The possible cases are tabulated as follows.

b	a	c	Number of ways
1	1	1	1
2	2	2	1
2	1	3	6
3	3	3	1
3	1	5	6
3	2	4	6

Total number of ways is 21. So, required probability is 21/216 = 7/72.

70. c. When four points are selected, we get one intersecting point. So, probability is

$$\frac{{}^nC_4}{{}^{({}^nC_2 - n)}C_2}$$

71. d. Three-digit numbers are 100, 101, …, 999. Total number of such numbers is 900. The three-digit numbers (which have all same digits) are 111, 222, 333, …, 999. Favorable number of cases is 9. Therefore, the required probability is 9/900 = 1/100.

72. c. Given,

$$7a - 9b = 0 \Rightarrow b = \frac{7}{9}a$$

Hence, number of pairs (a, b) can be (9, 7); (18, 14); (27, 21); (36, 28). Hence, the required probability is $4/{}^{39}C_2 = 4/741$.

73. a. Let $E_1 = 1, 4, 7, \ldots$ (n each)

$E_2 = 2, 5, 8, \ldots$ (n each)

$E_3 = 3, 6, 9, \ldots$ (n each)

x and y belong to (E_1, E_2), (E_2, E_1) or (E_3, E_3). So, the required probability is

$$\frac{n^2 + {}^nC_2}{{}^{3n}C_2} = \frac{1}{3}$$

74. c. The total number of mapping is n^n. The number of one-one mapping is ${}^nC_1\, {}^{n-1}C_1 \cdots {}^1C_1 = n!$ Hence, the probability is

$$\frac{n!}{n^n} = \frac{3}{32} = \frac{4!}{4^4}$$

Comparing, we get $n = 4$.

75. b. $P(\text{4 biased coin}) = \frac{1}{3}$

$P(\text{5 biased coin}) = \frac{1}{4}$

Hence, the required probability is

$$\frac{1}{3}\frac{{}^4C_3\, {}^{16}C_6}{{}^{20}C_9} + \frac{2}{3}\frac{{}^5C_4\, {}^{15}C_5}{{}^{20}C_9}\frac{1}{{}^{11}C_1}$$

$$= \frac{2}{33}\left[\frac{{}^{16}C_6 + 5\, {}^{15}C_5}{{}^{20}C_9}\right]$$

76. b. Let event A be drawing nine cards which are not ace and B be drawing an ace card. Therefore, the required probability is $P(A \cap B) = P(A) \times P(B)$

Now, there are four aces and 48 other cards. Hence,

$$P(A) = \frac{{}^{48}C_9}{{}^{52}C_9}$$

After having drawn nine non-ace cards, the 10th card must be ace. Hence,

$$P(B) = \frac{{}^4C_1}{{}^{42}C_1} = \frac{4}{42}$$

Hence,

$$P(A \cap B) = \frac{{}^{48}C_9}{{}^{52}C_9}\frac{4}{42}$$

77. c. The total number of digits in any number at the unit's place is 10.

$\therefore n(S) = 10$

If the last digit in product is 1, 3, 5 or 7, then it is necessary that the last digit in each number must be 1, 3, 5 or 7.

$\therefore n(A) = 4$

$\therefore P(A) = \frac{4}{10} = \frac{2}{5}$

Hence, the required probability is $(2/5)^4 = 16/625$.

78. a. The divisibility of the product of four numbers depends upon the value of the last digit of each number. The last digit of a number can be any one of the 10 digits 0, 1, 2, …, 9. So, the total number of ways of selecting last digits of four numbers is $10 \times 10 \times 10 \times 10 = 10^4$. If the product of the 4 numbers is not divisible by 5 or 10, then the number of choices for the last digit of each number is 8 (excluding 0 or 5). So, favourable number of ways is 8^4. Therefore, the probability that the product is not divisible by 5 or 10 is $(8/10)^4$. Hence, the required probability is $1 - (8/10)^4 = 369/625$.

79. b. Let H denote the head, T the tail and * any of the head or tail. Then, $P(H) = 1/2$, $P(T) = 1/2$ and $P(*) = 1$. For at least four consecutive heads, we should have any of the following patterns:

	Probability
(i) $HHHH***$	$(1/2)^4 \times 1 = 1/16$
(ii) $THHHH**$	$(1/2)^5 = 1/32$
(iii) $*THHHH*$	$(1/2)^5 = 1/32$
(iv) $**THHHH$	$(1/2)^5 = 1/32$

Since all the above cases are mutually exclusive, the probability of getting at least four consecutive heads (on adding) is $1/16 + 3/32 = 5/32$.

80. c. The probabilities of solving the question by these three students are 1/3, 2/7 and 3/8, respectively.

$$\therefore P(A) = \frac{1}{3}; P(B) = \frac{2}{7}; P(C) = \frac{3}{8}$$

Then probability of question solved by only one student is

$$P((A\bar{B}\bar{C} \text{ or } \bar{A}B\bar{C} \text{ or } \bar{A}\bar{B}C)) = P(A)P(\bar{B})P(\bar{C}) + P(\bar{A})P(B)P(\bar{C}) + P(\bar{A})P(\bar{B})P(C)$$

$$= \frac{1}{3}\times\frac{5}{7}\times\frac{5}{8}+\frac{2}{3}\times\frac{2}{7}\times\frac{5}{8}+\frac{2}{3}\times\frac{5}{7}\times\frac{3}{8}$$

$$= \frac{25+20+30}{168} = \frac{25}{56}$$

81. b. Probability of getting two heads in the first five trials is

$$^5C_2\left(\frac{1}{2}\right)^2\left(\frac{1}{2}\right)^3 = \frac{10}{32} = \frac{5}{16}$$

Therefore, the probability that third head appears on the sixth trial is $5/16 \times 1/2 = 5/32$.

82. d. Let A and B, respectively, be the events that urn A and urn B are selected. Let R be the event that the selected ball is red. Since the urn is chosen at random, Therefore,

$$P(A) = P(B) = \frac{1}{2}$$

and

$$P(R) = P(A)P(R/A) + P(B)P(R/B)$$

$$= \frac{1}{2}\times\frac{5}{10}+\frac{1}{2}\times\frac{4}{10} = \frac{9}{20}$$

83. d. The probability that the first critic favours the book is

$$P(E_1) = \frac{5}{5+2} = \frac{5}{7}$$

The probability that the second critic favours the book is

$$P(E_2) = \frac{4}{4+3} = \frac{4}{7}$$

The probability that the third critic favours the book is

$$P(E_3) = \frac{3}{3+4} = \frac{3}{7}$$

Majority will be in favour of the book if at least two critics favour the book. Hence, the probability is

$$P(E_1 \cap E_2 \cap \bar{E}_3) + P(E_1 \cap \bar{E}_2 \cap E_3) + P(\bar{E}_1 \cap E_2 \cap E_3) + P(E_1 \cap E_2 \cap E_3)$$

$$= P(E_1)P(E_2)P(\bar{E}_3) + P(E_1)P(\bar{E}_2)P(E_3) + P(\bar{E}_1 \cap E_2 \cap E_3) + P(E_1)P(E_2)P(E_3)$$

$$= \frac{5}{7}\times\frac{4}{7}\times\left(1-\frac{3}{7}\right)+\frac{5}{7}\times\left(1-\frac{4}{7}\right)\times\frac{3}{7} + \left(1-\frac{5}{7}\right)\times\frac{4}{7}\times\frac{3}{7}+\frac{5}{7}\times\frac{4}{7}\times\frac{3}{7} = \frac{209}{343}$$

84. b.

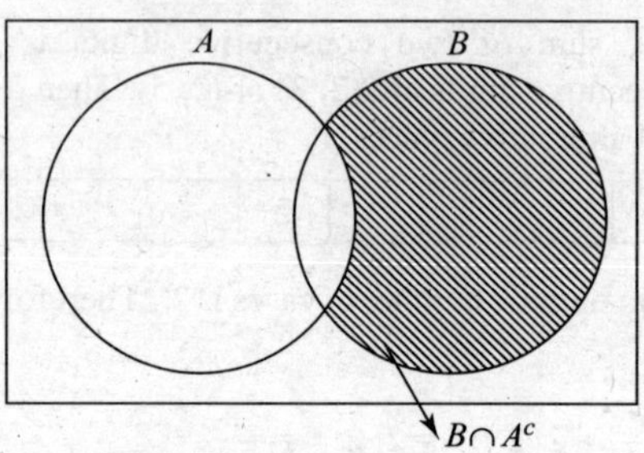

Fig. 9.2

$$P(A) = \frac{1}{4}, P(A \cup B) = \frac{1}{2}$$

$$P\left(\frac{B}{A^C}\right) = \frac{P(B \cap A^C)}{P(A^C)}$$

$$= \frac{P(A \cup B) - P(A)}{1 - P(A)}$$

$$[\because P(A \cup B) + P(B) - P(A \cap B)]$$

$$= \frac{\frac{1}{2}-\frac{1}{4}}{1-\frac{1}{4}} = \frac{1}{4}\times\frac{4}{3} = \frac{1}{3}$$

85. b. $P(S \cap F) = 0.0006$, where S is the event that the motor cycle is stolen and F is the event that the moter cycle is found. Therefore,

$$P(S) = 0.0015$$

$$P(F/S) = \frac{P(F \cap S)}{P(S)} = \frac{6\times10^{-4}}{15\times10^{-4}} = \frac{2}{5}$$

86. c. Let A be the event that 11 is picked and B be the event that sum is even. The number of ways of selecting 11 along with one more odd number is $n(A \cap B) = {}^7C_1$.

The number of ways of selecting either two even numbers or selecting two odd numbers is $n(B) = 1 + {}^8C_2$.

$$\therefore P(A/B) = \frac{P(A \cap B)}{P(B)}$$

$$= \frac{7}{29} = 0.24$$

87. b. Die marked with 1, 2, 2, 3, 3, 3 is thrown three times.

$$P(1) = \frac{1}{6}, P(2) = \frac{2}{6}, P(3) = \frac{3}{6}$$

$$P(S) = P(4 \text{ or } 6)$$

$$= P(112(3 \text{ cases}) \text{ or } 123 \text{ (6 cases) or } 222)$$

$$= 3\times\frac{1}{6}\times\frac{1}{6}\times\frac{2}{6}+6\frac{1}{6}\times\frac{2}{6}\times\frac{3}{6}+\frac{2}{6}\times\frac{2}{6}\times\frac{2}{6}$$

$$=\frac{6+36+8}{216}=\frac{50}{216}=\frac{25}{108}$$

88. c.

3R 3G:
- $1R+2G \to$ [$2R+1G+3B$]
- $2R+1G \to$ [$1R+2G+3B$]
- $3R \to$ [$3G+3B$] zero $\to$
- $3G \to$ [$3R+3B$] zero $\to$

Fig. 9.3

The required probability is

$$\frac{{}^3C_1\,{}^3C_2}{{}^6C_3}\frac{{}^2C_1\,{}^1C_1\,{}^3C_1}{{}^6C_3}+\frac{{}^3C_2\,{}^3C_1}{{}^6C_3}\frac{{}^1C_1\,{}^2C_1\,{}^3C_1}{{}^6C_3}$$

$$=2\frac{9}{20}\times\frac{6}{20}$$

$$=\frac{27}{100}$$

89. c. $P(\text{4 biased coins})=\frac{1}{3}$

$P(\text{5 biased coins})=\frac{1}{4}$

The required probability is

$$\frac{1}{3}\frac{{}^4C_3\,{}^{16}C_6}{{}^{20}C_9}\frac{1}{{}^{11}C_1}+\frac{2}{3}\frac{{}^5C_4\,{}^{15}C_5}{{}^{20}C_9}\frac{1}{{}^{11}C_1}$$

$$=\frac{2}{33}\left[\frac{{}^{16}C_6+5\,{}^{15}C_5}{{}^{20}C_9}\right]$$

90. d. A: Doctor finds a rash

B_1: Child has measles

S: Sick children

$P(S/F)=0.9$

B_2: Child has flu $\Rightarrow P(B_2)=9/10$

$P(S/M)=0.10$

$P(A/B_1)=0.95$

$P(R/M)=0.95$

$P(A/B_2)=0.08$

$P(R/F)=0.08$

$$P(B_1/A)=\frac{0.1\times0.95}{0.1\times0.95+0.9\times0.08}$$

$$=\frac{0.095}{0.095+0.072}$$

$$=\frac{0.095}{0.167}=\frac{95}{167}$$

91. c. The sum is 12 in the first three throws if they are (1, 5, 6) in any order or (2, 4, 6) in any order or (3, 4, 5) in any order. Therefore, the required probability is

$$\frac{1}{6}\times\frac{1}{5}\times\frac{1}{4}\times3!+\frac{1}{6}\times\frac{1}{5}\times\frac{1}{4}\times3!+\frac{1}{6}\times\frac{1}{5}\times\frac{1}{4}\times3!=\frac{3}{20}$$

(because after throwing 1, in the next throw 1 cannot come, etc.)

92. a. The number of composite numbers in 1 to 30 is $n(S)=19$.

The number of composite number when divided by 5 leaves a remainder is $n(E)=14$. Therefore, the required probability is 14/19.

93. b. $P(E)+P(E')=1=1+\lambda+\lambda^2+(1+\lambda)^2$

$\Rightarrow 2\lambda^2+3\lambda+1=0$

$\Rightarrow (2\lambda+1)(\lambda+1)=0$

$\Rightarrow \lambda=-1,\ -\frac{1}{2}$

Then,

$P(E)=1+(-1)+(-1)^2=1$ (not possible)

$$\Rightarrow P(E)=1+\left(-\frac{1}{2}\right)+\left(-\frac{1}{2}\right)^2=\frac{1}{2}+\frac{1}{4}=\frac{3}{4}$$

94. b. Let $P(m)$, $P(p)$, $P(c)$ be the probability of selecting a book of maths, physics and chemistry, respectively. Clearly,

$$P(m)=P(p)=P(c)=\frac{1}{3}$$

Again let $P(s_1)$ and $P(s_2)$ be the probability that he solves the first as well as second problem, respectively. Then,

$$P(s_1)=P(m)\times P\left(\frac{s_1}{m}\right)+P(p)\times P\left(\frac{s_1}{p}\right)+P(c)\times P\left(\frac{s_1}{c}\right)$$

$$\Rightarrow P(s_1)=\frac{1}{3}\times\frac{1}{2}+\frac{1}{3}\times\frac{3}{5}+\frac{1}{3}\times\frac{4}{5}=\frac{19}{30}$$

Similarly,

$$P(S_2)=\frac{1}{3}\times\left(\frac{1}{2}\right)^2+\frac{1}{3}\times\left(\frac{3}{5}\right)^2+\frac{1}{3}\times\left(\frac{4}{5}\right)^2=\frac{125}{300}$$

$$\Rightarrow P\left(\frac{S_2}{S_1}\right)=\frac{\frac{125}{300}}{\frac{19}{30}}=\frac{25}{38}$$

95. b. Let,

$P(S)=P(1\text{ or }2)=1/3$

$P(F)=P(3\text{ or }4\text{ or }5\text{ or }6)=2/3$

$P(A\text{ wins})=P[(S\,S\text{ or }S\,F\,S\,S\text{ or }S\,F\,S\,F\,S\,S\text{ or }\ldots)$

or $(F\,S\,S\text{ or }F\,S\,F\,S\,S\text{ or }\ldots)]$

$$=\frac{\frac{1}{9}}{1-\frac{2}{9}}+\frac{\frac{2}{27}}{1-\frac{2}{9}}$$

$$=\frac{1}{9}\times\frac{9}{7}+\frac{2}{27}\times\frac{9}{7}$$

$$=\frac{1}{7}+\frac{2}{21}=\frac{3+2}{21}=\frac{5}{21}$$

$$P(A\text{ winning})=\frac{5}{21},\ P(B\text{ winning})=\frac{16}{21}$$

96. a. $P(a)=0.3$, $P(b)=0.5$, $P(c)=0.2$. Hence, a, b, c are exhaustive.

$P(\text{same horse wins all the three races})=P(aaa\text{ or }bbb\text{ or }ccc)$

$= (0.3)^3 + (0.5)^3 + (0.2)^3$

$= \frac{27+125+8}{1000} = \frac{160}{1000}$

$= \frac{4}{25}$

P(each horse wins exactly one race)

$= P(abc \text{ or } acb \text{ or } bca \text{ or } bac \text{ or } cab \text{ or } cba)$

$= 0.3 \times 0.5 \times 0.2 \times 6 = 0.18 = \frac{9}{50}$

97. b. The total number of ways of distribution is 4^5.

$\therefore n(S) = 4^5$

The total number of ways of distribution so that each child gets at least one game is $4^5 - {}^4C_1\, 3^5 + {}^4C_2\, 2^5 - {}^4C_3$.

$\therefore n(E) = 240$

Hence, the required probability is

$\frac{n(E)}{n(S)} = \frac{240}{4^5} = \frac{15}{64}$

98. b. Team totals must be 0, 1, 2, …, 39. Let the teams be T_1T_2, …, T_{40}, so that T_1 loses to T_1 for $i < j$. In other words, this order uniquely determines the result of every game. There are 40! such orders and 780 games, so 2^{780} possible outcomes for the games. Hence, the probability is $40!/2^{780}$.

99. a. We have,

$n(S) = {}^{64}C_3$

Let E be the event of selecting three squares which form the letter L.

The number of ways of selecting squares consisting of four unit squares is $7 \times 7 = 49$.

Each square with four unit squares form four L-shapes consisting of three unit squares.

$\therefore n(E) = 7 \times 7 \times 4 = 196$

$\therefore P(E) = \frac{196}{{}^{64}C_3}$

100.a. The total number of ways of making the second draw is ${}^{10}C_5$.

The number of draws of five balls containing two balls common with first draw of six balls is ${}^6C_2\, {}^4C_3$. Therefore, the probability is

$\frac{{}^6C_2\, {}^4C_3}{{}^{10}C_5} = \frac{5}{21}$

101. a. The required probability is

$P(A) = \frac{1}{3}\frac{6}{a^2 - 4a + 10}$

$(P(A))_{\max} = \frac{1}{3} \times \frac{6}{6} = \frac{1}{3}$

102. b. A number has exactly three factors if the number is squares of a prime number. Squares of 11, 13, 17, 19, 23, 29, 31 are 3-digit numbers. Hence, the required probability is 7/900.

103. c. Suppose, there exist three rational points or more on the circle $x^2 + y^2 + 2gx + 2fy + c = 0$. Therefore, if (x_1, y_1), (x_2, y_2) and (x_3, y_3) be those three points, then

$$x_1^2 + y_1^2 + 2gx_1 + 2fy_1 + c = 0 \quad (1)$$

$$x_2^2 + y_2^2 + 2gx_2 + 2fy_1 + c = 0 \quad (2)$$

$$x_3^2 + y_3^2 + 2gx_3 + 2fy_3 + c = 0 \quad (3)$$

Solving Eqs. (1), (2) and (3), we will get g, f, c as rational. Thus, centre of the circle $(-g, -f)$ is a rational point. Therefore, both the coordinates of the centre are rational numbers. Obviously, the possible values of p are 1, 2. Similarly, the possible values of q are 1, 2. Thus for this case, (p, q) may be chosen in 2×2, i.e., 4 ways. Now, (p, q) can be, without restriction, chosen in 6×6, i.e., 36 ways.

Hence, the probability that at the most two rational points exist on the circle is $(36 - 4)/36 = 32/36 = 8/9$.

104. a. $P(A) = P(B) = P(C)$ and $P(A) + P(B) + P(C) = 1$

$\therefore P(A) = P(B) = P(C) = \frac{1}{3}$

Also,

$P(X) = \frac{5}{12}, P(X/A) = \frac{3}{8}, P(X/B) = \frac{1}{4}$

We have,

$P(X) = P(A)\,P(X/A) + P(B)\,P(X/B) + P(C)\,P(X/C)$

$\therefore \frac{5}{12} = \frac{1}{3}\left\{\frac{3}{8} + \frac{1}{4} + P(X/C)\right\}$

$\Rightarrow P(X/C) = \frac{5}{8}$

105. c. A: car met with an accident

B_1: driver was alcoholic, $P(B_1) = 1/5$

B_2: driver was sober, $P(B_2) = 4/5$

$P(A/B_1) = 0.001$; $P(A/B_2) = 0.0001$

$P(B_1/A) = \frac{(0.2)\,(0.001)}{(0.2)\,(0.001) + (0.8)(0.0001)} = 5/7$

106. a.

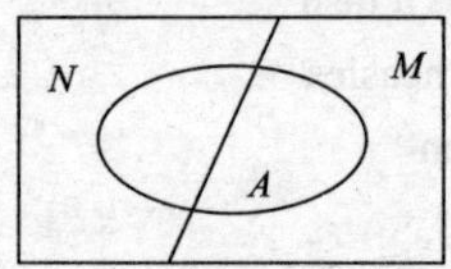

Fig. 9.4

Let N be the event of picking up a normal die; $P(N) = 1/4$. Let M be the event of picking up a magnetic die; $P(M) = 3/4$. Let A be the event that die shows up 3.

$\therefore P(A) = P(A \cap N) + P(A \cap M)$

$= P(N)P(A/N) + P(M)\,P(A/M)$

$= \frac{1}{4} \times \frac{1}{6} + \frac{3}{4} \times \frac{2}{3} = \frac{7}{24}$

$P(N/A) = \frac{P(N \cap A)}{P(A)} = \frac{(1/4)(1/6)}{7/24} = \frac{1}{7}$

107. c. We have,

$n(S) = 5^5$

For computing favourable outcomes, two boxes which are to remain empty, can be selected in ${}^5C_2$ ways and five marbles can be placed in the remaining three boxes in groups of 221 or 311 in

$$3!\left[\frac{5!}{2!2!2!}+\frac{5!}{3!2!}\right]=150 \text{ ways} \Rightarrow n(A)={}^5C_2\times 150$$

Hence, $P(E)={}^5C_2\times\dfrac{150}{5^5}=\dfrac{60}{125}=\dfrac{12}{25}$

108. a. $n(S)={}^{10}C_7=120$

$$n(A)={}^5C_4\times{}^3C_2\times{}^2C_1$$

$$P(E)=\frac{5\times3\times2}{120}=\frac{1}{4}$$

109. a. For ranked 1 and 2 players to be winners and runners up, respectively, they should not be paired with each other in any round. Therefore, the required probability is $30/31\times14/15\times6/7\times2/3=16/31$.

110. a. The total number of ways of selecting three integers from 20 natural numbers is ${}^{20}C_3=1140$. Their product is a multiple of 3 means at least one number is divisible by 3. The numbers which are divisible by 3 are 3, 6, 9, 12, 15, 18 and the number of ways of selecting at least one of them is ${}^6C_1\times{}^{14}C_2+{}^6C_2\times{}^{14}C_1+{}^6C_3=776$. Hence, the required probability is $776/1140=194/285$.

111. d. Since there are r cars in N places, total number of selection of places out of $N-1$ places for $r-1$ cars (excepting the owner's car) is

$${}^{N-1}C_{r-1}=\frac{(N-1)!}{(r-1)!(N-r)!}$$

If neighbouring places are empty, then $r-1$ cars must be parked in $N-3$ places. So, the favourable number of cases is

$${}^{N-3}C_{r-1}=\frac{(n-3)!}{(r-1)!(N-r-2)!}$$

Therefore, the required probability is

$$\frac{(N-3)!}{(r-1)!(N-r-2)!}\times\frac{(r-1)!(N-r)!}{(N-1)!}$$

$$=\frac{(N-r)(N-r-1)}{(N-1)(N-2)}=\frac{{}^{N-r}C_2}{{}^{N-1}C_2}$$

112. b. The sum of the digits can be 7 in the following ways: 07, 16, 25, 34, 43, 52, 61, 70.

$\therefore (A=7)=\{07, 16, 25, 34, 43, 52, 61, 70\}$

Similarly,

$(B=0)=\{00, 01, 02, \ldots, 10, 20, 30, \ldots, 90\}$

Thus,

$(A=7)\cap(B=0)=\{09, 70\}$

$\therefore P((A=7)\cap(B=0))=\dfrac{2}{100}, P((B=0))=\dfrac{19}{100}$

Hence,

$$P(A=7\mid B=0)=\frac{P((A=7)\cap(B=0))}{P(B=0)}$$

$$=\frac{\frac{2}{100}}{\frac{19}{100}}=\frac{2}{19}$$

113. b. Let the probability of getting a tail in a single trial be $p=1/2$. The number of trials be $n=100$ and the number of trials in 100 trials be X. We have,

$$P(X=r)={}^{100}C_r p^r q^{n-r}$$

$$={}^{100}C_r\left(\frac{1}{2}\right)^r\left(\frac{1}{2}\right)^{100-r}$$

$$={}^{100}C_r\left(\frac{1}{2}\right)^{100}$$

Now,

$$P(X=1)+P(X=3)+\cdots+P(X=49)$$

$$={}^{100}C_1\left(\frac{1}{2}\right)^{100}+{}^{100}C_3\left(\frac{1}{2}\right)^{100}+\cdots+{}^{100}C_{49}\left(\frac{1}{2}\right)^{100}$$

$$=({}^{100}C_1+{}^{100}C_3+\cdots+{}^{100}C_{49})\left(\frac{1}{2}\right)^{100}$$

But

$${}^{100}C_1+{}^{100}C_3+\cdots+{}^{100}C_{99}=2^{99}$$

Also,

$${}^{100}C_{99}={}^{100}C_1$$

$${}^{100}C_{97}={}^{100}C_3, \ldots {}^{100}C_{51}={}^{100}C_{49}$$

Thus,

$$2({}^{100}C_1+{}^{100}C_3+\cdots+{}^{100}C_{49})=2^{99}$$

$$\Rightarrow {}^{100}C_1+{}^{100}C_3+\cdots+{}^{100}C_{49}=2^{98}$$

Therefore, probability of required event is

$$\frac{2^{98}}{2^{100}}=\frac{1}{4}\quad 2^{98}/2^{100}=1/4$$

114. a. Let A denote the event that a sum of 5 occurs, B the event that a sum of 7 occurs and C the event that neither a sum of 5 nor a sum of 7 occurs. We have,

$$P(A)=\frac{4}{36}=\frac{1}{9}, P(B)=\frac{6}{36}=\frac{1}{6}$$

$$P(C)=\frac{26}{36}=\frac{13}{18}$$

Thus, probability that A occurs before B is

$$P[A \text{ or } (C\cap A) \text{ or } (C\cap C\cap A) \text{ or } \ldots]$$

$$=P(A)+P(C\cap A)+P(C\cap C\cap A)+\cdots$$

$$=P(A)+P(C)P(A)+P(C)^2P(A)+\cdots$$

$$=\frac{1}{9}+\left(\frac{13}{18}\right)\times\frac{1}{9}+\left(\frac{13}{18}\right)^2\frac{1}{9}+\cdots$$

$$=\frac{1/9}{1-13/18}=\frac{2}{5}$$

115. a. In any number the last digits can be 0, 1, 2, 3, 4, 5, 6, 7, 8, 9. Therefore, last digit of each number can be chosen in 10 ways. Thus, exhaustive number of ways is 10^n. If the last digit be 1, 3, 7 or 9, then none of the numbers can be even or end in 0 or 5. Thus, we have a choice of four digits, viz., 1, 3, 7 or 9 with which each of n numbers should end.

So favourable number of ways is 4^n. Hence, the required probability is

$$\frac{4^n}{10^n}=\left(\frac{2}{5}\right)^n$$

116. c. The probability that A gets r heads in three tosses of a coin is

$$P(X=r)={}^3C_r\left(\frac{1}{2}\right)^r\left(\frac{1}{2}\right)^{3-r}={}^3C_r\left(\frac{1}{2}\right)^3$$

The probability that A and B both get r heads in three tosses of a coin is

$${}^3C_r\left(\frac{1}{2}\right)^3{}^3C_r\left(\frac{1}{2}\right)^3=({}^3C_r)^2\left(\frac{1}{2}\right)^6$$

Hence, the required probability is

$$\sum_{r=0}^{3}({}^3C_r)^2\left(\frac{1}{2}\right)^6=\left(\frac{1}{2}\right)^6\{1+9+9+1\}=\frac{20}{64}=\frac{5}{16}$$

117. c. We know that the number of subsets of a set containing n elements is 2^n. Therefore, the number of ways of choosing P and Q is

$${}^{2^n}C_1\times{}^{2^n}C_1=2^n\times 2^n=4^n$$

Out of n elements, m elements are chosen and then from the remaining $n-m$ elements either an element belongs to P or Q. But not both P and Q. Suppose P contains r elements from the remaining $n-m$ elements. Then, Q may contain any number of elements from the remaining $(n-m)-r$ elements Therefore, P and Q can be chosen in ${}^{n-m}C_r2^{(n-m)-r}$ ways.

But r can vary from 0 to $n-m$. So, in general the number of ways in which P and Q can be chosen is

$$\left(\sum_{r=0}^{n-m}{}^{n-m}C_r2^{(n-m)-r}\right){}^nC_m=(1+2)^{n-m}\,{}^nC_m={}^nC_m3^{n-m}$$

Hence, the required probability is ${}^nC_m3^{n-m}/4^n$.

118. c. In the first nine throws, we should have three sixes and six non-sixes; and a six in the 10th throw, and thereafter it does not matter whatever face appears. So, the required probability is

$${}^9C_3\left(\frac{1}{6}\right)^3\times\left(\frac{5}{6}\right)^6\times\frac{1}{6}\times\underbrace{1\times1\times1\times\cdots\times1}_{10\text{ times}}$$

$$=\frac{84\times5^6}{6^{10}}$$

119. c. Let the probability that a man aged x dies in a year p. Thus the probability that a man aged x does not die in a year $=1-p$. The probability that all n men aged x do not die in a year is $(1-p)^n$. Therefore, the probability that at least one man dies in a year is $1-(1-p)^n$. The probability that out of n men, A_1 dies first is $1/n$. Since this event is independent of the event that at least one man dies in a year, hence, the probability that A_1 dies in the year and he is first one to die is $1/n.[1-(1-p)^n]$.

120. b. The required probability is

$$\frac{n^2}{{}^{2n}C_2}\cdot\frac{(n-1)^2}{{}^{2n-2}C_2}\cdot\frac{(n-2)^2}{{}^{2n-4}C_2}\cdots\frac{2^2}{{}^4C_2}\cdot\frac{1^2}{{}^2C_2}$$

$$=\frac{(1\times2\times3\times4\times\cdots\times(n-1)n)^2}{\frac{(2n)!}{2^n}}=\frac{2^n(n!)^2}{(2n)!}=\frac{2^n}{{}^{2n}C_n}$$

121. a. The probability of getting a head in a single toss of a coin is $p=1/2$ (say). The probability of getting 5 or 6 in a single throw of a die is $q=2/6=1/3$ (say). Therefore, the required probability is

$$p+(1-p)(1-q)p+(1-p)(1-q)(1-p)(1-q)p+\cdots$$

$$=p+(1-p)(1-q)p+(1-p)^2(1-q)^2p+\cdots$$

$$=\frac{p}{1-(1-p)(1-q)}$$

$$=\frac{1/2}{1-1/2\times2/3}=\frac{3}{4}$$

122. a. Consider the following events.

A: ball drawn is black

E_1: bag I is chosen

E_2: bag II is chosen

E_3: bag III is chosen

Then,

$$P(E_1)=(E_2)=P(E_3)=\frac{1}{3}$$

$$P(A/E_1)=\frac{3}{5},\ P(A/E_2)=\frac{1}{5},\ P(A/E_3)=\frac{7}{10}$$

Therefore, the required probability is

$P(E_3/A)$

$$=\frac{P(E_3)P(A/E_3)}{P(E_1)P(A/E_1)+P(E_2)P(A/E_2)+P(E_3)P(A/E_3)}=\frac{7}{15}$$

123. c. $f'(x)=3x^2+2ax+9$

$y=f(x)$ is increasing

$\Rightarrow f'(x)\ge 0,\ \forall\ x$ and for $f'(x)=0$ should not form an interval

$\Rightarrow (2a)^2-4\times3\times b\le0\Rightarrow a^2-3b\le0$

This is true for exactly 16 ordered pairs (a, b), $1\le a, b\le 6$, namely (1, 1), (1, 2), (1, 3), (1, 4), (1, 5), (1, 6), (2, 2), (2, 3), (2, 4), (2, 5), (2, 6), (3, 3), (3, 4), (3, 6) and (4, 6). Thus, the required probability is $16/36=4/9$

124. d. Let p_i denote the probability that out of 10 tosses, head occurs i times and no two heads occur consecutively. It is clear that $i>5$.

For $i=0$, i.e., no head, $p_0=1/2^{10}$.

For $i=1$, i.e., one head, $p_1={}^{10}C_1\ (1/2)^1\ (1/2)^9=10/2^{10}$.

Now for $i=2$, we have 2 heads and 8 tails. Then, we have 9 possible places for heads. For example, see the construction:

$x\,T\,x\,T\,x\,T\,x\,T\,x\,T\,x\,T\,x\,T\,x\,T\,x$

Here x represents possible places for heads.

$$\therefore\ p_2 = {}^9C_2\left(\frac{1}{2}\right)^2 (1/2)^8 = 36/2^{10}$$

Similarly,

$p_3 = {}^8C_3/2^{10} = 56/2^{10}$

$p_4 = {}^7C_4/2^{10} = 35/2^{10}$

$p_5 = {}^6C_5/2^{10} = 6/2^{10}$

$\therefore\ p = p_0 + p_1 + p_2 + p_3 + p_4 + p_5$

$$= \frac{1+10+36+56+35+6}{2^{10}} = \frac{144}{2^{10}} = \frac{9}{64}$$

125. c. Given limit,

$$\lim_{x\to 0}\left(\frac{a^x+b^x}{2}\right)^{\frac{2}{x}}$$

$$= \lim_{x\to 0}\left(1+\frac{a^x+b^x-2}{2}\right)^{\frac{2}{a^x+b^x-2}\ \lim\limits_{x\to 0}\left(\frac{a^x-1+b^x-1}{x}\right)}$$

$$= e^{\log ab} = ab = 6$$

Total number of possible ways in which a, b can take values is $6 \times 6 = 36$. Total possible ways are (1, 6), (6, 1), (2, 3), (3, 2). The total number of possible ways is 4. Hence, the required probability is $4/36 = 1/9$.

126. c. Let A denote the event that target is hit when x shells are fired at point I. Let P_1 and P_2 denote the event that the target is at point I and II, respectively. We have $P(P_1) = 8/9$, $P(P_2) = 1/9$, $P(A/P_1) = 1 - (1/2)^x$, $P(A/P_2) = 1 - (1/2)^{55-x}$.

Now from total probability theorem,

$P(A) = P(P_1)\,P(A/P) + P(P_2)\,P(A/P_2)$

$$= \frac{1}{9}\left(8 - 8\left(\frac{1}{2}\right)^x + 1 - \left(\frac{1}{2}\right)^{55-x}\right)$$

$$= \frac{1}{9}\left(9 - 8\left(\frac{1}{2}\right)^x - \left(\frac{1}{2}\right)^{55-x}\right)$$

Now,

$$\frac{dP(A)}{dx} = \frac{1}{9}\left(-8\left(\frac{1}{2}\right)^x \ln\left(\frac{1}{2}\right) + \left(\frac{1}{2}\right)^{55-x}\ln\left(\frac{1}{2}\right)\right)$$

(Note that in this step, it is being assumed that $x \in R^+$)

$$= \frac{1}{9}\ln\left(\frac{1}{2}\right)\left(\frac{1}{2}\right)^{55-x}\left(1 - \left(\frac{1}{2}\right)^{2x-58}\right)$$

> 0 if $x < 29$

< 0 if $x > 29$

Therefore, $P(A)$ is maximum at $x = 29$. Thus, '29' shells must be fired at point I.

127. d. In the first case, the urn contains three red and n white balls. The probability that colour of both the balls matches is

$$\frac{{}^3C_2 + {}^nC_2}{{}^{n+3}C_2} = \frac{1}{2}$$

$$\Rightarrow \frac{6 + n(n-1)}{(n+3)(n+2)} = \frac{1}{2}$$

$\Rightarrow 2(n^2 - n + 6) = n^2 + 5n + 6$

$\Rightarrow n^2 - 7n + 6 = 0$

$\Rightarrow n = 1$ or 6 (1)

In the second case,

$$\frac{3}{n+3}\frac{3}{n+3} + \frac{n}{n+3}\frac{n}{n+3} = \frac{5}{8}$$

Solving, we get

$n^2 - 10n + 9 = 0$

$\Rightarrow n = 9$ or 1. (2)

From Eqs. (1) and (2), we have $n = 1$.

128. b. Let us consider the following events.

A: card shows up black

B_1: card with both sides black

B_2: card with both sides white

B_3: card with one side white and one black

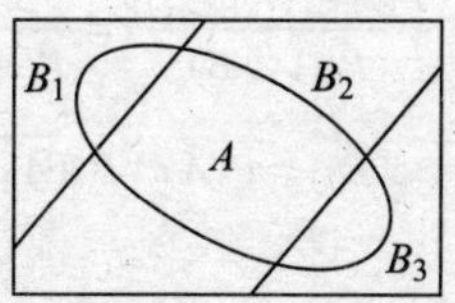

Fig. 9.5

$$P(B_1) = \frac{2}{10},\ P(B_2) = \frac{3}{10},\ P(B_3) = \frac{5}{10}$$

$$P(A/B_1) = 1,\ P(A/B_2) = 0,\ P(A/B_3) = \frac{1}{2}$$

$$P(B_1/A) = \frac{\frac{2}{10}\times 1}{\frac{2}{10}\times 1 + \frac{3}{10}\times(0) + \frac{5}{10}\times\frac{1}{2}} = \frac{4}{4+5} = \frac{4}{9}$$

129. d. A: exactly one ace

B: both aces

E: $A \cup B$

$$P(B/A \cup B)) = \frac{{}^4C_2}{{}^4C_1\,{}^{12}C_1 + {}^4C_2} = \frac{6}{54} = \frac{1}{9}$$

130. b. $n(S) = \dfrac{9!}{4!.5!} = 126$

$n(A)$ = 0 to F and F to P

$$= \frac{5!}{2!\times 3!} \times \frac{4}{2!\times 2!} = 60$$

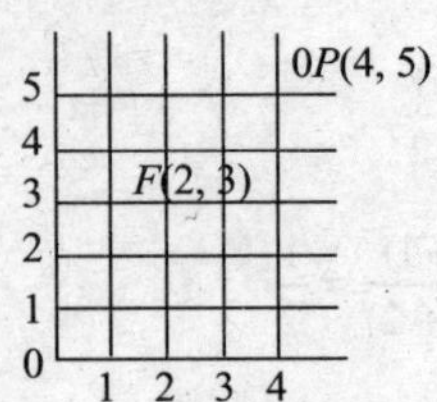

Fig. 9.6

$$\Rightarrow P(A) = \frac{60}{126} = \frac{10}{21}$$

Multiple Correct Answers Type

1. a., c., d.

Since A and B are independent events,

$$P(A \cap B) = P(A)\,P(B) = \frac{1}{2} \times \frac{1}{5} = \frac{1}{10}$$

$$P(A/B) = P(A) = \frac{1}{2}$$

Now,

$$P(A \cup B) = P(A) + P(B) - P(A \cap B)$$

$$= \frac{1}{2} + \frac{1}{5} - \frac{1}{10} = \frac{3}{5}$$

Now,

$$P(A/A \cup B) = \frac{P[A \cap (A \cup B)]}{P(A \cup B)} = \frac{P(A)}{P(A \cup B)} = \frac{1/2}{3/5} = \frac{5}{6}$$

$$P[(A \cap B)/(\bar{A} \cup \bar{B})] = P(A \cap B/(\overline{A \cap B})) = 0$$

2. a., b., c.

We are given that

$$P(A \cap B') = 0.20,\ P(A' \cap B) = 0.15,\ P(A \cap B) = 0.10$$

Now,

$$P(B) = P(A' \cap B) + P(A \cap B) = 0.15 + 0.20 = 0.35$$

$$\therefore P(A/B) = \frac{P(A \cap B)}{P(B)} = \frac{0.10}{0.35} = \frac{2}{7}$$

$$P(A) = P(A \cap B') + P(A \cap B) = 0.3$$

$$P(A \cup B) = P(A) + P(B) - P(A \cap B) = 0.55$$

$$P(B/A) = \frac{P(B \cap A)}{P(A)} = \frac{1}{3}$$

3. a., b., c., d.

$$A \subseteq A \cup B$$

$$\Rightarrow P(A) \le P(A \cup B) \Rightarrow P(A \cup B) \ge \frac{3}{4}$$

Also,

$$P(A \cap B) = P(A) + P(B) - P(A \cup B)$$

$$\ge P(A) + P(B) - 1$$

$$= \frac{3}{4} + \frac{5}{8} - 1 = \frac{3}{8}$$

Now,

$$A \cap B \subseteq B$$

$$\Rightarrow P(A \cap B) \le P(B) = \frac{5}{8}$$

$$\therefore \frac{3}{8} \le P(A \cap B) \le \frac{5}{8} \qquad (1)$$

and

$$P(A \cap B') = P(A) - P(A \cap B)$$

$$\Rightarrow \frac{3}{4} - \frac{5}{8} \le P(A \cap B') \le \frac{3}{4} - \frac{3}{8}$$

$$\Rightarrow \frac{1}{8} \le P(A \cap B') \le \frac{3}{8}$$

$$\therefore P(A \cap B) = P(B) - P(A' \cap B) \qquad \text{[Using Eq. (1)]}$$

$$\Rightarrow \frac{3}{8} \le P(B) - P(A' \cap B) \le \frac{5}{8}$$

$$\Rightarrow 0 \le P(A' \cap B) \le \frac{1}{4}$$

4. a., c.

Given that A and B are mutually exclusive events.

$$\therefore A \cap B = \phi$$

$$\Rightarrow A \subseteq \bar{B} \text{ and } B \subseteq \bar{A}$$

$$\Rightarrow P(A) \le P(\bar{B}) \text{ and } P(B) \le P(\bar{A})$$

5. a., b., c.

Let H be the event that married man watches the show and W be the probability that married woman watches the show.

$$\therefore P(H) = 0.4,\ P(W) = 0.5,\ P(H/W) = 0.7$$

a. $P(H \cap W) = P(W)\,P(H/W) = 0.5 \times 0.7 = 0.35$

b. $P(W/H) = \dfrac{P(H \cap W)}{P(H)} = \dfrac{0.35}{0.4} = \dfrac{7}{8}$

c. $P(H \cup W) = P(H) + P(W) - P(H \cap W)$

$$= 0.4 + 0.5 - 0.35 = 0.55$$

6. b., c.

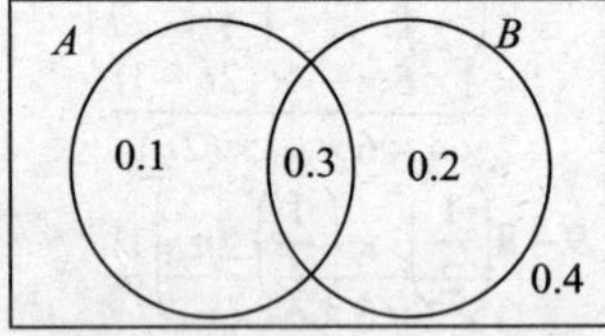

Fig. 9.7

$$P(A) = 0.4,\ P(B) = 0.5,\ P(A \cap B) = 0.3$$

$$\Rightarrow P(A \cup B) = P(A) + P(B) - P(A \cap B) = 0.6$$

Now,

$$P(E_1) = \frac{P(A \cap B)}{P(A \cup B)} = \frac{0.3}{0.6} = \frac{1}{2}$$

$$P(E_2) = \frac{P(A \cap \bar{B}) + P(\bar{A} \cap B)}{P(A \cup B)} = \frac{0.1 + 0.2}{0.6} = \frac{1}{2}$$

$$P(A/B) = \frac{P(A \cap B)}{P(B)} = \frac{0.3}{0.5} = \frac{3}{5} = 0.60;$$

$$P(B/A) = \frac{0.3}{0.4} = \frac{3}{4} = 0.75 = 0.75$$

$$P(A/(A \cup B)) = \frac{P(A)}{P(A \cup B)} = \frac{0.4}{0.6} = \frac{2}{3}$$

$$\frac{P(B)}{P(A \cup B)} = \frac{0.5}{0.6} = \frac{5}{6}$$

7. c., d.

$P(E_1) = 1 - P$ (unit's place in both is 1, 2, 3, 4, 6, 7, 8, 9)

$$P(E_1 = 0 \text{ or } 5) = 1 - \left(\frac{4}{5}\right)^2 = \frac{9}{25}$$

$$P(E_2 : 5) = P(1\ 3\ 5\ 7\ 9) - P(1\ 3\ 7\ 9)$$

$$= \frac{1}{4} - \frac{4}{25}$$

$$= \frac{25 - 16}{100} = \frac{9}{100}$$

$$\frac{P(E_2)}{P(E_1)} = \frac{9}{100} \times \frac{25}{9} = \frac{1}{4}$$

$$P(E_1) = 4P(E_2)$$

$$P(E_2/E_1) = \frac{P(E_2 \cap E_1)}{P(E_1)} = \frac{P(E_2)}{P(E_1)} = \frac{1}{4}$$

$$P(E_1/E_2) = \frac{P(E_1 \cap E_2)}{P(E_2)} = \frac{P(E_2)}{P(E_2)} = 1$$

8. a., c., d.

$$P(E) = \frac{{}^{2n}C_n}{2^{2n}} = \frac{(2n)!}{n!\,n!\,2^n\,2^n}$$

$$= \frac{1 \times 2 \times 3 \times \cdots \times (2n)}{n!\,n!\,2^n\,2^n}$$

$$= \frac{1 \times 3 \times 5 \times \cdots \times (2n-1)}{n!\ 2^n}$$

Now,

$$\prod_{r=1}^{n}\left(\frac{2r-1}{2r}\right) = \frac{1 \times 3 \times 5 \times \cdots \times (2n-1)}{2 \times 4 \times 6 \times \cdots \times (2n)}$$

$$= \frac{1 \times 3 \times 5 \times \cdots \times (2n-1)}{(1 \times 2 \times 3 \times \cdots \times n)2^n}$$

$$\sum_{r=0}^{n}\left(\frac{{}^nC_r}{2^n}\right)^2 = \frac{1}{2^n 2^n}\sum_{r=0}^{n}({}^nC_r)^2$$

$$= \frac{1}{2^n 2^n}\,{}^{2n}C_n$$

Also,

$$\frac{\sum_{r=0}^{n}({}^nC_r)^2}{\left(\sum_{r=0}^{2n}{}^{2n}C_r\right)} = \frac{{}^{2n}C_n}{2^{2n}}$$

9. a., b. The probability that both will be alive for 10 years, hence, i.e., the probability that the man and his wife both will be alive 10 years hence is 0.83 × 0.87 = 0.7221. The probability that at least one of them will be alive is

$1 - P$ {That none of them remains alive 10 years hence}

$$= 1 - (1 - 0.83)(1 - 0.87) = 1 - 0.17 \times 0.13$$

$$= 0.9779$$

10. b., c., d.

a. False.

$$P(TTT \text{ or } HHH) = \frac{1}{8} + \frac{1}{8} = \frac{1}{4}$$

b. $\dfrac{P(A \cap B)}{P(B)} = \dfrac{P(A \cap B^c)}{1 - P(B)} = \dfrac{P(A) - P(A \cap B)}{1 - P(B)}$

$$P(A \cap B)[1 - P(B)] = P(B)\,P(A) - P(B)\,P(A \cap B)$$

$$P(A \cap B) = P(A)\,P(B)$$

Hence, the given statement is true.

c. Let E_1 be the white ball is drawn in first draw; E_2 be the event that black ball is drawn in second draw; E be the event that white ball is drawn in second draw.

$$\therefore P(E) = P(E/E_1)\,P(E_1) + P(E/E_2)\,P(E_2)$$

$$= \frac{d+w}{w+b+d}\left(\frac{w}{w+b}\right) + \frac{w}{w+b+d}\left(\frac{b}{w+b}\right)$$

$$= \left(\frac{w}{w+b}\right)\left(\frac{d+w}{w+b+d} + \frac{b}{w+b+d}\right)$$

$$= \left(\frac{w}{w+b}\right),$$

which is independent of d.

d. To prove that A, B, C are pairwise independent only. Now,

$$P(A \cap B) = P((A \cap B \cap C) \cup (A \cap B \cap \bar{C}))$$

$$= P(A \cap B \cap C) + P(A \cap B \cap \bar{C})$$

$$= P(A)\,P(B)\,P(\bar{C}) + P(A)\,P(B)\,P(C) \quad \text{(given)}$$

$$= P(A) \times P(B)\,[P(C') + P(\bar{C})]$$

$$= P(A) \times P(B)$$

Similarly, for the other two. Hence, this statement is correct.

11. a., b., c., d.

a. $P(E_1) = 1 - P(RRR)$

$$= 1 - \left[\frac{1}{3} \times \frac{2}{4} \times \frac{3}{5}\right] = 0.9$$

b. $P(E_2) = 3P(BRR) = 3 \times \dfrac{2}{3} \times \dfrac{1}{4} \times \dfrac{2}{5} = 0.2$

c. $P(E_3) = P(RRR/(RRR \cup BBB))$

$$= \frac{P(RRR)}{P(RRR) + P(BBB)}$$

$$= \frac{0.1}{0.1 + \frac{2}{3} \times \frac{3}{4} \times \frac{4}{5}}$$

$$= \frac{0.1}{0.1 + 0.4} = 0.2$$

d. $P(E_4) = 1 - P(BBB) = 1 - \dfrac{2}{5} = 0.6$

12. a., b., c., d.

$$P(A \cup B) = P(A) + P(B) - P(A \cap B)$$

$$\Rightarrow \frac{5}{8} = \frac{3}{8} + \frac{4}{8} - P(A \cap B)$$

$$\Rightarrow P(A \cap B) = \frac{2}{8} = \frac{1}{4}$$

Now,

$$P(A^c/B) = \frac{P(A^c \cap B)}{P(B)}$$

$$= \frac{P(B) - P(A \cap B)}{P(B)}$$

$$= 1 - 2\left(\frac{1}{4}\right)$$

$$= \frac{1}{2}$$

$$2P(A/B^c) = \frac{2P(A \cap B^C)}{P(B^C)}$$

$$= \frac{2(P(A) - P(A \cap B))}{1 - P(B)}$$

$$= 4\left(\frac{3}{8} - \frac{2}{8}\right) = \frac{1}{2}$$

Hence option (a) is correct.

$$P(A/B) = \frac{P(A \cap B)}{P(B)} = \frac{1}{4} \times \frac{2}{1} = \frac{1}{2} = P(B)$$

Hence (b) is correct. Again,

$$P(A^c/B^c) = \frac{P(A^C \cap B^C)}{P(B^C)}$$

$$= \frac{1 - P(A \cup B)}{1 - P(B)}$$

$$= 2\left(1 - \frac{5}{8}\right) = \frac{3}{4}$$

$$P(B/A^c) = \frac{P(B \cap A^C)}{1 - P(A)}$$

$$= \frac{P(B) - P(A \cap B)}{5/8}$$

$$= \frac{1/2 - 1/4}{5/8}$$

$$= \frac{1}{4} \times \frac{8}{5}$$

$$= \frac{2}{5}$$

Hence,

$8P(A^c/B^c) = 15P(B/A^c)$

Hence, (c) is not correct. Again,

$$2P(A/B^c) = \frac{1}{2}$$

$$\Rightarrow P(A/B^c) = \frac{1}{4} = P(A \cap B)$$

Hence (d) is correct.

13. a., c.

Let p_1, p_2 be the chances of happening of the first and second events, respectively. Then according to the given conditions, we have

$$p_1 = p_2^2$$

and

$$\frac{1-p_1}{p_1} = \left(\frac{1-p_2}{p_2}\right)^3$$

Hence,

$$\frac{1-p_2^2}{p_2^2} = \left(\frac{1-p_2}{p_2}\right)^3 \Rightarrow p_2(1+p_2) = (1-p_2)^2$$

$$\Rightarrow 3p_2 = 1 \Rightarrow p_2 = \frac{1}{3}$$

and so

$$p_1 = \frac{1}{9}$$

14. a., b.

Let the number of red and blue balls be r and b, respectively. Then, the probability of drawing two red balls is

$$p_1 = \frac{{}^rC_2}{{}^{r+b}C_2} = \frac{r(r-1)}{(r+b)(r+b-1)}$$

The probability of drawing two blue balls is

$$p_2 = \frac{{}^bC_2}{{}^{r+b}C_2} = \frac{b(b-1)}{(r+b)(r+b-1)}$$

The probability of drawing one red and one blue ball is

$$p_3 = \frac{{}^rC_1 \times {}^bC_1}{{}^{r+b}C_2} = \frac{2br}{(r+b)(r+b-1)}$$

By hypothesis, $p_1 = 5p_2$ and $p_3 = 6p_2$.

$\therefore r(r-1) = 5b(b-1)$ and $2br = 6b(b-1)$

$\Rightarrow r = 6, b = 3$

15. a., b., d.

We have, the probability that the bomb strikes the target is $p = 1/2$. Let n be the number of bombs which should be dropped to ensure 99% chance or better of completely destroying the target. Then, the probability that out of n bombs at least two bombs strike the target is greater than 0.99. Let X denote the number of bombs striking the target. Then

$$P(X = r) = {}^nC_r\left(\frac{1}{2}\right)^r\left(\frac{1}{2}\right)^{n-r} = {}^nC_r\left(\frac{1}{2}\right)^n, r = 0, 1, 2, \ldots, n$$

We should have

$$P(X \geq 2) \geq 0.99$$

$$\Rightarrow \{1 - P(X < 2)\} \geq 0.99$$

$$\Rightarrow 1 - \{P(X = 0) + P(X = 1)\} \geq 0.99$$

$$\Rightarrow 1 - \left\{(1+n)\frac{1}{2^n}\right\} \geq 0.99$$

$$\Rightarrow 0.001 \geq \frac{1+n}{2^n}$$

$\Rightarrow 2^n > 100 + 100n \Rightarrow n \geq 11$

Thus, the minimum number of bombs is 11.

16. a., b., c., d.

We have,

P(exactly one of A, B occurs)

$= P[(A \cap \bar{B}) \cup \bar{A} \cap B)]$

$= P(A \cap \bar{B}) + P(\bar{A} \cap B)$

$= P(A) - P(A \cap B) + P(B) - P(A \cap B)$

$= P(A) + P(B) - 2P(A \cap B)$

$= P(A \cup B) - P(A \cap B)$

Also,

P(exactly one of A, B occurs)

$= [1 - P(\bar{A} \cap \bar{B})] - [1 - P(\bar{A} \cup \bar{B})]$

$= P(\bar{A} \cup \bar{B}) - P(\bar{A} \cap \bar{B})] = P(\bar{A}) + P(\bar{B}) - 2P(\bar{A} \cap \bar{B})$

17. a., b., c.

Option (d) is true if and only if A and B are independent.

18. a., b., c.

$$P(A/B) = P(A) = \frac{1}{2}$$

$$P\{A/(A \cup B)\} = \frac{P[A \cap (A \cup B)]}{P(A \cup B)}$$

$$[\because A \cap (A \cup B) = A \cap (A - B - A \cap B)]$$
$$= A - A \cap B - A \cap B = a\}$$

$$\Rightarrow P\left(\frac{A}{A \cup B}\right) = \frac{P(A)}{P(A \cup B)}$$

$$= \frac{\frac{1}{2}}{\frac{1}{2} - \frac{1}{5} - \frac{1}{10}} = \frac{\frac{1}{2}}{\frac{6}{10}} = \frac{5}{6}$$

and similarly,

$$P\left(\frac{A \cap B}{A' \cup B'}\right) = 0$$

19. b., c.

Let $P(A) = x$ and $P(B) = y$. Since A and B are independent events, therefore,

$P(\bar{A} \cap B) = 2/15 \Rightarrow P(\bar{A})P(B) = 2/15$

$\Rightarrow (1 - P(A))\, P(B) = 2/15$

$\Rightarrow (1 - x)y = 2/15$ (1)

and

$$P(A \cap \bar{B}) = \frac{1}{6} \Rightarrow P(A)\, P(\bar{B}) = \frac{1}{6}$$

$$\Rightarrow x(1 - y) = \frac{1}{6}$$

$$\Rightarrow x - xy = \frac{1}{6} \quad (2)$$

Subtracting Eq. (1) from Eq. (2), we get

$$x - y = \frac{1}{30} \Rightarrow x = \frac{1}{30} + y$$

Putting this value of x in Eq. (1), we get

$$y - y\left(\frac{1}{30} + y\right) = \frac{2}{15}$$

$\Rightarrow 30y - y - 30y^2 = 2/5$

$\Rightarrow 30y^2 - 29y + 4 = 0$

$\Rightarrow (6y - 1)(5y - 4) = 0$

$\Rightarrow y = 1/6$ or $y = 4/5$

$\Rightarrow P(B) = 1/6$ or $P(B) = 4/5$

20. a., b., c.

a. $P(A) = \frac{1}{5}, P(B) = \frac{7}{25}, P(B/A) = \frac{9}{10}$

$$P(\bar{A} \cap \bar{B}) = 1 - P(A \cup B)$$
$$= 1 - [P(A) + P(B) - P(A \cap B)]$$
$$= 1 - \left[\frac{1}{5} + \frac{7}{25} - P(A)\,P(B/A)\right]$$
$$= 1 - \left[\frac{1}{5} + \frac{7}{25} - \frac{1}{5} \times \frac{7}{25}\right] = \frac{7}{10}$$

b. $P(A/B) = \dfrac{P(A \cap B)}{P(B)}$

$$= \frac{P(A)\,P(B/A)}{P(B)}$$
$$= \frac{\frac{1}{5} \times \frac{9}{10}}{\frac{7}{25}}$$
$$= \frac{9}{50} \times \frac{25}{7} = \frac{9}{14} = \frac{18}{28}$$

c. $P(A \cup B) = P(A) + P(B) - P(A \cap B)$

$$= P(A) + P(B) - P(A)P(B/A)$$
$$= \frac{1}{5} + \frac{7}{25} - \frac{1}{5} \times \frac{9}{10}$$
$$= \frac{10 + 14 - 9}{50}$$
$$= \frac{3}{10}$$

d. $P(A' \cup B') = 1 - P(A \cap B)$

$$= 1 - P(A)P(B/A)$$
$$= 1 - \frac{1}{5} \times \frac{9}{10}$$
$$= \frac{41}{50}$$

21. b., c., d.

Roots of $x^2 + px + q = 0$ will be real if $p^2 \geq 4q$.

The possible selections are as follows:

p	q
1	–
2	1
3	1, 2

4	1, 2, 3, 4
5	1, 2, 3, 4, 5, 6
6	1, 2, ..., 9
7	1, 2, ..., 10
8	1, 2, ..., 10
9	1, 2, ..., 10
10	1, 2, ..., 10
Total	62

Therefore, number of favourable ways is 62 and total number of ways is $10^2 = 100$. Hence, the required probability is 62/100 = 31/50.

The probability that the roots are imaginary is 1 – 31/50 = 19/50.

Roots are equal when $(p, q) \equiv (2, 1), (4, 4), (9, 6)$. The probability that the roots are real and equal is 3/50. Hence, probability that the roots are real and distinct is 3/5.

22. a., b., c.

Here total number of cases is $^8C_2 = 28$.

a. Favourable number of case is 13.

For $2 \to 6$ choices

For $1 \to 7$ choices

b. For $6 \to 5$ choices

For $7 \to 6$ choices

For $8 \to 7$ choices

c. For $1 \to 4$ choices (2, 4, 6, 8)

For $2 \to 6$ choices (3, 4, 5, 6, 7, 8)

For $3 \to 3$ choices (4, 6, 8)

For $4 \to 4$ choices (5, 6, 7, 8)

For $5 \to 2$ choices (6, 8)

For $6 \to 2$ choices (7, 8)

For $7 \to 1$ choice (8)

Alternative solution:

a. $$\frac{^8C_2 - {^6C_2}}{^8C_2} = \frac{13}{28}$$

b. $$\frac{^8C_2 - {^5C_2}}{^8C_2} = \frac{9}{14}$$

c. $$\frac{^8C_2 - {^4C_2}}{^8C_2} = \frac{11}{14}$$

23. a., d.

The probability that head appears r times is

$$^{99}C_r\left(\frac{1}{2}\right)^r\left(\frac{1}{2}\right)^{99-r}$$

which is maximum when $r = 49$ or 50.

24. a., c.

Let one probability of choosing one integer k be $P(k) = \lambda/k^4$. (λ is one constant of proportionality). Then,

$$\sum_{k=1}^{2m}\frac{\lambda}{k^4} = 1$$

$$\Rightarrow \lambda\sum_{k=1}^{2m}\frac{1}{k^4} = 1$$

Let x_1 be the probability of choosing the odd number. Then,

$$x_1 = \sum_{k=1}^{m} P(2k-1) = \lambda\sum_{k=1}^{m}\frac{1}{(2k-1)^4}$$

Also,

$$1 - x_1 = \sum_{k=1}^{m} P(2k)$$

$$= \lambda\sum_{k=1}^{m}\frac{1}{(2k)^4}$$

$$< \lambda\sum_{k=1}^{m}\frac{1}{(2k-1)^4}$$

$$\Rightarrow 1 - x_1 < x_1$$

$$\Rightarrow x_1 > 1/2$$

$$\Rightarrow x_2 < 1/2$$

Reasoning Type

1. a. $P(A/B) \geq P(A)$

$$\Rightarrow \frac{P(A\cap B)}{P(B)} \geq P(A)$$

$$\Rightarrow \frac{P(A\cap B)}{P(A)} \geq P(B)$$

$$\Rightarrow P(B/A) \geq P(B)$$

2. c. Statement 1 is true as there are six equally likely possibilities of which only two are favourable (4 and 6). Hence, probability that the obtained number is composite) is 2/6 = 1/3.

Statement 2 is not true, as the three possibilities are not equally likely.

3. c. The required probability is

$$P(\bar{A}\cap\bar{B}) = 1 - P(A\cup B)$$
$$= 1 - [P(A) + P(B) - P(A\cap B)]$$
$$= 0.39$$

4. a. $\because P(A\cap B) = P(A) + P(B) - P(A\cup B)$

$$\geq P(A) + P(B) - 1$$

$$\therefore P(A\cap B) \geq \frac{3}{5} + \frac{2}{3} - 1$$

$$\Rightarrow P(A\cap B) \geq \frac{4}{15} \quad (1)$$

$$P(A\cap B) \leq P(A)$$

$$\Rightarrow P(A\cap B) \leq \frac{3}{5} \quad (2)$$

From Eqs. (1) and (2),

$$\frac{4}{15} \leq P A\cap B) \leq \frac{3}{5} \quad (3)$$

From (3),

$$\frac{4}{15P(B)} \leq \frac{P(A\cap B)}{P(B)} \leq \frac{3}{5P(B)}$$

$$\Rightarrow \frac{2}{5} \leq P\left(\frac{A}{B}\right) \leq \frac{9}{10}$$

5. a. $P\{A\cap(B\cap C)\} = P(A\cap B\cap C) = P(A)P(B)P(C)$

$\therefore P[A \cap (B \cup C)]$
$= P[(A \cap B) \cup (A \cap C)]$
$= P[(A \cap B) + (A \cap C) - P[(A \cap B) \cap (A \cap C)]$
$= p(A \cap B) + P(A \cap C) - P(A \cap B \cap C)$
$= P(A)P(B) + P(A)P(C) - P(A)P(B)P(C)$
$= P(A)[P(B) + P(C) - P(B)P(C)]$
$= P(A)\, P(B \cup C)$

Therefore, A and $B \cup C$ are independent events.

6. a. $2n + 1 = 5, n = 2$

$$P(E) = \frac{3n}{4n^2 - 1} = \frac{6}{15} = \frac{2}{5}$$

As a, b, c are in A.P., so

$a + c = 2b$

$\Rightarrow a + c$ is even

Therefore, a and c are both even or both odd. So, the number of ways of choosing a and c is ${}^nC_2 + {}^{n+1}C_2 = n^2$.

$$\therefore\ P(E) = \frac{n^2}{{}^{2n+1}C_3} = \frac{3n}{4n^2 - 1}$$

7. d. $P(A \cup B) = P(A) + P(B) - P(A \cap B)$

$\therefore 1 \ge P(A) + P(B) - P(A \cap B) \ge 3/4$

$\Rightarrow P(A) + P(B) - 1/8 \ge 3/4$

[since minimum value of $P(A \cap B)$ is 1/8)

$\Rightarrow P(A) + P(B) \le 1/8 + 3/4 = 7/8$

As the maximum value of $P(A \cap B)$ is 3/8, we get

$1 \ge P(A) + P(B) - 3/8$

$\Rightarrow P(A) + P(B) \le 1 + 3/8 = 11/8$

8. b. Clearly both are correct but statement 2 is not the correct explanation for statement 1.

9. b. $P(A \cup \bar{B}) = 1 - \overline{(A \cup \bar{B})} = 1 - (\bar{A} \cap B) = 1 - P(\bar{A})P(B)$

$\Rightarrow 0.9 = 1 - 0.6 \times P(B)$

$$\Rightarrow P(B) = \frac{1}{6}$$

Clearly, statement 2 is not correct explanation of statement 1.

10. c. According to statement 1, the required probability is

$${}^nC_0\left(\frac{1}{2}\right)^n + {}^nC_4\left(\frac{1}{2}\right)^4\left(\frac{1}{2}\right)^{n-4} + {}^nC_8\left(\frac{1}{2}\right)^8\left(\frac{1}{2}\right)^{n-8} + \cdots$$

$$= ({}^nC_0 + {}^nC_4 + {}^nC_8 + \cdots)\left(\frac{1}{2}\right)^n$$

Now consider the binomial expansion,

$(1 + x)^n = {}^nC_0 + {}^nC_1x + {}^nC_2x^2 + \cdots$

Putting $x + i$, where $i = \sqrt{-1}$, we get

$(1 + i)^n = ({}^nC_0 - {}^nC_2 + {}^nC_4 - \cdots) + i({}^nC_1 - {}^nC_3 + {}^nC_5 - \cdots)$

$$\Rightarrow \left[\sqrt{2}\left(\cos\frac{\pi}{4} + i\sin\frac{\pi}{4}\right)\right]^n = ({}^nC_0 - {}^nC_2 + {}^nC_4 - \cdots) + i({}^nC_1 - {}^nC_3 + {}^nC_5 - \cdots)$$

$$\Rightarrow {}^nC_0 - {}^nC_2 + {}^nC_4 - \cdots = 2^{n/2}\cos\frac{n\pi}{4}$$

Also we know that

${}^nC_0 + {}^nC_2 + {}^nC_4 + \cdots = 2^{n-1}$

$$\Rightarrow 2({}^nC_0 + {}^nC_4 + {}^nC_8 + \cdots) = 2^{n-1} + 2^{n/2}\cos\frac{n\pi}{4}$$

Hence, the required probability is

$$\frac{1}{4} + \frac{1}{2^{n/2+1}}\cos\left(\frac{n\pi}{4}\right).$$

11. d. $P\left(\frac{A}{B}\right) = \frac{P(A \cap B)}{P(B)}$ [by definition]

$\Rightarrow P(\bar{B}) = P((A \cup \bar{A}) \cap \bar{B}) = P((A \cap \bar{B}) \cup (\bar{A} \cap \bar{B}))$

Hence, statement 2 is true. Now,

$$P(A/\bar{B}) + P(\bar{A}/\bar{B}) = \frac{P(A \cap \bar{B})}{P(\bar{B})} + \frac{P(\bar{A} \cap \bar{B})}{P(\bar{B})} = \frac{P(A \cap \bar{B}) + P(\bar{A} \cap \bar{B})}{P(\bar{B})} = \frac{P(\bar{B})}{P(\bar{B})} = 1$$

Therefore, statement 1 is false.

12. a. In binomial theorem, we have proved statement 2. Now, there may be 0, 1, 2, 3, 4 or 5 heads in the last five throws and the same can be for the first 10 throws. The number of cases thus may be given by

$m = {}^5C_0{}^{10}C_0 + {}^5C_1{}^{10}C_1 + {}^5C_2{}^{10}C_2 + {}^5C_3{}^{10}C_3 + {}^5C_4{}^{10}C_4 + {}^5C_5{}^{10}C_5$
$= {}^5C_0{}^{10}C_{10} + {}^5C_1{}^{10}C_9 + {}^5C_2{}^{10}C_8 + {}^5C_3{}^{10}C_7 + {}^5C_4{}^{10}C_6 + {}^5C_5{}^{10}C_5$
$= {}^{10+5}C_{10} = {}^{15}C_{10} = 3003$

The total number of ways (N) is $2^{15} = 32768$. Hence, the required probability is $m/N = 3003/32768$.

13. c. $P(A) + P(B) = 1$ is true as A and B are mutually exclusive and exhaustive events, but statement 2 is false as it is not given that the events are exhaustive.

14. b. The total number of cases, $n(S) = 4!$. Let E be the event that no letter is mailed in its correct envelope. Then the favourable number of cases is

$$4!\left(1 - \frac{1}{1!} + \frac{1}{2!} - \frac{1}{3!} + \frac{1}{4!}\right) = 9$$

Hence, the required probability is

$$P(E) = \frac{9}{24} = \frac{3}{8}$$

Also, the probability that all the letters are placed in the correct envelope is 1/24.

Hence, the probability that all the letters are not placed in the correct envelope is 23/24.

Hence, statement 2 is correct but does not explain statement 1.

15. a. We have,

$$P(A \cup \bar{B}) = 1 - \overline{(A \cup \bar{B})}$$
$$= 1 - (\bar{A} \cap B)$$
$$= 1 - P(\bar{A})\,P(B)\ 0.8 = 1 - 0.7 \times P(B)$$
$$\Rightarrow P(B) = \frac{2}{7}$$

Linked Comprehension Type

For Problems 1–3

1. b., 2. c., 3. a.

Sol. Let $P(i)$ be the probability that exactly i students are passing an examination. Now given that

$P(A_i) = \lambda i^2$ (where λ is constant)

$$\Rightarrow \sum_{i=1}^{10} P(A_i) = \sum_{i=1}^{10} \lambda i^2 = \lambda \frac{10 \times 11 \times 21}{6} = \lambda 385 = 1$$

$\Rightarrow \lambda = 1/385$

Now, $P(5) = 25/385 = 5/77$.

Let A represent the event that selected students have passed the examination.

$$\therefore P(A) = \sum_{i=1}^{10} P(A/A_i)P(A_i)$$
$$= \sum_{i=1}^{10} \frac{i}{10}\frac{i^2}{385}$$
$$= \frac{1}{3850}\sum_{i=1}^{10} i^3$$
$$= \frac{10^2 11^2}{4 \times 3850} = \frac{11}{14}$$

Now,

$$P(A_1/A) = \frac{P(A/A_1)P(A_1)}{P(A)}$$
$$= \frac{\frac{1}{385}\frac{1}{10}}{\frac{11}{14}}$$
$$= \frac{1}{11 \times 55}\frac{1}{5}$$
$$= \frac{1}{3025}$$

For Problems 4–6

4. c., 5. b., 6. d.

Sol. Let in eight coupons S, U, R, F appears x_1, x_2, x_3, x_4 times. Then $x_1 + x_2 + x_3 + x_4 = 8$, where $x_1, x_2, x_3, x_4 \geq 0$.

We have to find non-negative integral solutions of the equation. The total number of such solutions is ${}^{8+4-1}C_{4-1} = {}^{11}C_3 = 165$.

If a person gets at least one free packet, then he must get each coupon at least once, which is equal to number of positive integral solutions of the equation. The number of such solutions is ${}^{8-1}C_{4-1} = {}^{7}C_3 = 35$. Then, the probability that he gets exactly one free packet is

$(35 - 1)/165 = 102/495$.

The probability that he gets two free packets is $1/{}^{11}C_3 = 1/165$.

For Problems 7–9

7. d., 8. d., 9. b.

Sol. 7. Let p_1 be the probability of being an answer correct from section 1. Then $p_1 = 1/5$. Let p_2 be the probability of being an answer correct from section 2. Then $p_2 = 1/15$. Hence, the required probability is $1/5 \times 1/15 = 1/75$.

8. Scoring 10 marks from four questions can be done in $3 + 3 + 3 + 1 = 10$ ways so as to answer three questions from section 2 and 1 question from section 1 correctly. Hence the required probability is

$$\frac{{}^{10}C_3\,{}^{10}C_1}{{}^{20}C_4}\frac{1}{5}\left(\frac{1}{15}\right)^3$$

9. To get 40 marks, he has to answer all questions correctly and its probability is $(1/5)^{10}\,(1/15)^{10}$. Hence, probability of getting a score less than 40 is

$$1 - \left(\frac{1}{5}\right)^{10}\left(\frac{1}{15}\right)^{10} = 1 - \left(\frac{1}{75}\right)^{10}$$

For Problems 10–12

10. c., 11. d., 12.c.

Sol. 10. x can be 2, 3, 4, 5, 6. The number of ways in which sum of 2, 3, 4, 5, 6 can occur is given by the coefficients of x^2, x^3, x^4, x^5, x^6 in

$$(3x + 2x^2 + x^3)\,(x + 2x^2 + 3x^3) = 3x^2 + 8x^3 + 14x^4 + 8x^5 + 3x^6$$

This shows that sum that occurs most often is 4.

11. Sum that occurs for minimum times is 2 or 6.

12. The number of ways in which different sums can occur is $(3 + 2 + 1)\,(1 + 2 + 3) = 36$. The probability of 4 is $14/36 = 7/18$.

For Problems 13–15

13. c., 14. b., 15. d.

Sol. A: She gets a success

T: She studies 10 h: $P(T) = 0.1$

S: She studies 7 h: $P(S) = 0.2$

F: She studies 4 h: $P(F) = 0.7$

$P(A/T) = 0.8, P(A/S) = 0.6, P(A/F) = 0.4$

$$P(A) = P(A \cap T) + P(A \cap S) + P(A \cap F)$$
$$= P(T)\,P(A/T) + P(S)\,P(A/S) + P(F)\,P(A/F)$$
$$= (0.1)(0.8) + (0.2)(0.6) + (0.7)(0.4)$$
$$= 0.08 + 0.12 + 0.28 = 0.48$$
$$= P(F/A) = \frac{P(F \cap A)}{P(A)}$$
$$= \frac{(0.7)\,(0.4)}{0.48}$$
$$= \frac{0.28}{0.48} = \frac{7}{12}$$

$$P(F/\bar{A})=\frac{P(F\cap\bar{A})}{P(\bar{A})}$$

$$=\frac{P(F)-P(F\cap A)}{0.52}$$

$$=\frac{(0.7)-0.28}{0.52}$$

$$=\frac{0.42}{0.52}=\frac{21}{26}$$

For Problems 16–18

16. b., 17. a., 18. c.

Sol. $P(S/T)=\frac{P(S\cap T)}{P(T)}$

$$\Rightarrow 0.5=\frac{P(S\cap T)}{0.69}$$

$$\Rightarrow P(S\cap T)=0.5\times 0.69=P(S)\,P(T)$$

Therefore, S and T are independent.

$\therefore P(S \text{ and } T)=P(S)P(T)$

$=0.69\times 0.5=0.345$

$P(S \text{ or } T)=P(S)+P(T)-P(S\cap T)$

$=0.5+0.69-0.345$

$=0.8450$

For Problems 19–21

19. d., 20. b., 21. a.

Sol. 19. Let E be the event that all the amoeba population dies out;

E_1 be the event that after first second amoeba splits into two; E_2 be the event that after first second amoeba remains the same Then,

$$P(E)=P(E_1)P(E/E_1)+P(E_2)P(E/E_2)$$

$$=\frac{1}{2}\frac{1}{4}\frac{1}{4}+\frac{1}{4}\frac{1}{4}$$

$$=\frac{3}{32}$$

20. b. After 2s, exactly four amoeba are alive, i.e. initially amoeba must split into two and in 2nd second again amoeba must split into two. Hence, the required probability is $(1/2)\times(1/2)\times(1/2)=1/8$.

21. a. In first second, mother amoeba splits into two with the probability 1/2.

In 2nd second, two amoeba split into two with probability $(1/2)\times(1/2)=1/4$.

In 3rd second, four amoeba split into two with probability $(1/2)\times(1/2)\times(1/2)\times(1/2)=1/16$.

Hence, the probability that the population is maximum after 3s is $(1/2)\times(1/4)\times(1/16)=1/2^7$.

For Problems 22–24

22. b., 23. b., 24. d.

Sol. The number of cubes having at least one side painted is $9+9+3+3+1+1=26$. The number of cubes having two sides painted is $4+4+1+1+1+1=12$.

For Problems 25–27

25. a., 26. d., 27. b.

Sol.

25. a. For the favourable cases, the points should lie inside the concentric circle of radius $r/2$. So the desired probability is given by

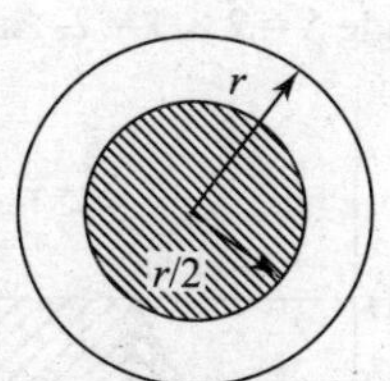

Fig. 9.8

$$\frac{\text{Area of smaller circle}}{\text{Area of larger circle}}=\frac{\pi\left(\frac{r}{2}\right)^2}{\pi r^2}=\frac{1}{4}$$

26. d.

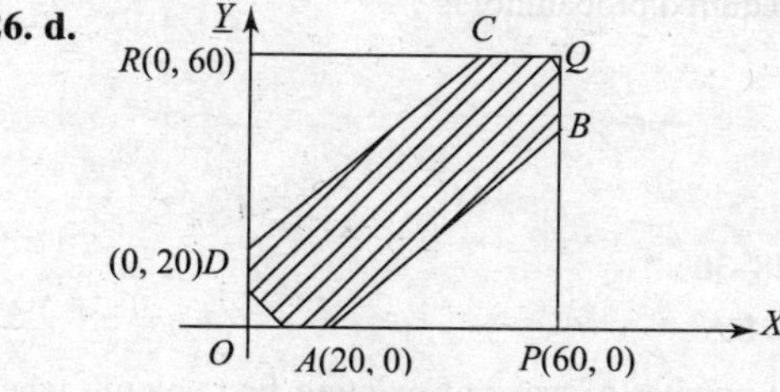

Fig. 9.9

Let A and B arrive at the place of their meeting a minutes and 'b' minutes after 5 pm. Their meeting is possible only if $|a-b|\le 20$ (1)

Clearly, $0\le a\le 60$ and $0\le b\le 60$. Therefore, a and b can be selected as an ordered pair (a, b) from the set $[0, 60]\times[0, 60]$.

Alternatively, it is equivalent to select a point (a, b) from the square $OPQR$, where P is $(60, 0)$ and R is $(0, 60)$ in the Cartesian plane. Now,

$|a-b|\le 20 \Rightarrow -20\le a-b\le 20$

Therefore, points (a, b) satisfy the equation $-20\le x-y\le 20$.

Hence, favourable condition is equivalent to selecting a point from the region bounded by $y\le x+20$ and $y\ge x-20$. Therefore, the required probability is

$$\frac{\text{Area of } OABQCDO}{\text{Area of square } OPQR}=\frac{[\text{Ar}(OPQR)-2\text{Ar}(\Delta\, APB)]}{\text{Ar}(OPQR)}$$

$$=\frac{60\times 60-\frac{2}{2}\times 40\times 40}{60\times 60}=\frac{5}{9}$$

27. b. Picking two points x and y randomly from the intervals [0, 2] and [0, 1] is equivalent to picking a single point (x, y) randomly from the rectangle S shown in the following figure, which has vertices at (0, 0) (2, 0), (2, 1) and (0, 1). So, we take S as our sample space. Now, the condition $y \le x^2$ is satisfied if and only if the point (x, y) lies in the shaded region. It is the portion of the rectangle lying below the parabola $y = x^2$. Therefore, the required probability is given by

$$\frac{\text{Area of the shaded region}}{\text{Area of the rectangle } S}$$

Area of rectangle $S = 2 \times 1 = 2$. Area of shaded region is

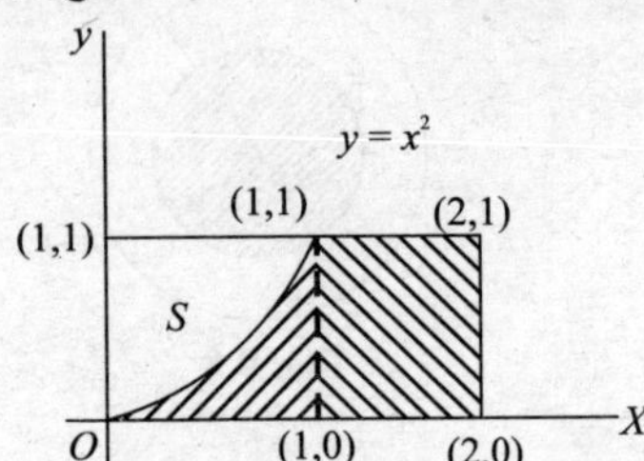

Fig. 9.10

$$\int_0^1 x^2\,dx + 1\times 1 = \frac{1}{3} + \frac{4}{3}$$

Hence, required probability is

$$\frac{\frac{4}{3}}{2} = \frac{2}{3}$$

For Problems 28–30

28. a., 29. a., 30. b.

Sol. 28. The total number of ways of painting first column when colours are not alternating is $2^8 - 2$. The total number of ways when no column has alternating colours is $(2^8 - 2)^8/2^{64}$.

29. The number of ways the square has equal number of red and black squares is ${}^{64}C_{32}$. Hence, the required probability is ${}^{64}C_{32}/2^{64}$.

30. This is possible only when the column are alternating red and black. Hence, the required probability is $2/2^{64} = 1/2^{63}$

For Problems 31–33

31. a., 32. c., 33. d.

31. a. $P(A_2) = \dfrac{18}{36}$

$$P(A_3) = \frac{12}{36} = \frac{1}{3}$$

$$P(A_4) = \frac{9}{36} = \frac{1}{4}$$

$$P(A_5) = \frac{7}{36}$$

$$P(A_6) = \frac{6}{36} = \frac{1}{6}$$

Hence, A_3 is most probable.

32. c. $P(A_2) = \dfrac{1}{2}, P(A_3) = \dfrac{1}{3}, P(A_6) = \dfrac{1}{6}$

$$\therefore P(A_2 \cap A_3) = P(A_2)\,P(A_3)$$

$$\Rightarrow P(A_6) = P(A_2)\,P(A_3)$$

$$\frac{6}{36} = \frac{1}{2} \times \frac{1}{3}$$

Hence, A_2 and A_3 are independent.

33. d. Note that A_1 is independent with all events $A_1, A_2, A_3, A_4, \ldots, A_{12}$. Now, total number of ordered pairs is 23.

$$\underbrace{(1, 1), (1, 2), (1, 3), \ldots, (1, 11) + (1, 12)}_{22}$$

Also, A_2, A_3 and A_3, A_2 are independent. Hence, there are 25 ordered pairs.

For Problems 34–36

34. a., 35. c., 36. c.

Sol. The scores of n can be reached in the following two mutually exlcusive events:

i. by throwing a head when the score is $(n - 1)$;

ii. by throwing a tail when the score is $(n - 2)$

Hence,

$$P_n = P_{n-1} \times \frac{1}{2} + P_{n-2} \times \frac{1}{2} \quad [\because P(\text{head}) = P(\text{tail}) = 1/2]$$

$$\Rightarrow P_n = \frac{1}{2}[P_{n-1} + P_{n-2}] \quad (1)$$

$$\Rightarrow P_0 + \frac{1}{2}P_{n-1} = P_{n-1} + \frac{1}{2}P_{n-2}$$

[Adding (1/2) P_{n-1} on both sides]

$$= P_{n-2} + \frac{1}{2}P_{n-3}$$

$$\vdots$$

$$= P_2 + \frac{1}{2}P_1 \quad (2)$$

Now, a score of 1 can be obtained by throwing a head at a single toss.

$$\therefore P_1 = \frac{1}{2}$$

And a score of 2 can be obtained by throwing either a tail at a single toss or a head at the first toss as well as second toss.

$$\therefore P_2 = \frac{1}{2} + \left(\frac{1}{2} \times \frac{1}{2}\right) = \frac{3}{4}$$

From Eq. (2), we have

$$P_n + \frac{1}{2}P_{n-1} = \frac{3}{4} + \frac{1}{2}\left(\frac{1}{2}\right) = 1$$

$$\Rightarrow P_n = 1 - \frac{1}{2}P_{n-1}$$

$$\Rightarrow P_n - \frac{2}{3} = 1 - \frac{1}{2}P_{n-1} - \frac{2}{3}$$

$$\Rightarrow P_n - \frac{2}{3} = -\frac{1}{2}\left(P_{n-1} - \frac{2}{3}\right)$$

$$= \left(-\frac{1}{2}\right)^2 \left(P_{n-2} - \frac{2}{3}\right)$$

$$= \left(-\frac{1}{2}\right)^3 \left(P_{n-3} - \frac{2}{3}\right)$$

$$= \left(-\frac{1}{2}\right)^{n-1} \left(P_1 - \frac{2}{3}\right)$$

$$= \left(-\frac{1}{2}\right)^{n-1} \left(\frac{1}{2} - \frac{2}{3}\right)$$

$$= \left(-\frac{1}{2}\right)^{n-1} \left(-\frac{1}{6}\right)$$

$$= \left(-\frac{1}{2}\right)^{n} \frac{1}{3}$$

$$\Rightarrow P_n = \frac{2}{3} + \frac{(-1)^n}{2^n}\frac{1}{3} = \frac{1}{3}\left\{2 + \frac{(-1)^n}{2^n}\right\}$$

Now,

$$P_{100} = \frac{2}{3} + \frac{1}{3\times 2^{100}} > \frac{2}{3} \text{ and } P_{101} = \frac{2}{3} - \frac{1}{3\times 2^{101}} < \frac{2}{3}$$

$$\Rightarrow P_{101} < \frac{2}{3} < P_{100}$$

For Problems 37–39

37. b., 38. c., 39. a.

Sol. 37. If a family of n children contains exactly k boys, then, by binomial distribution, its probability is

$${}^nC_k \left(\frac{1}{2}\right)^k \left(\frac{1}{2}\right)^{n-k}$$

Hence, by total probability law, the probability of a family of n children having exactly k boys is given by

$$\alpha p^n \, {}^nC_k \left(\frac{1}{2}\right)^k \left(\frac{1}{2}\right)^{n-k} \quad \text{(where } n \ge k\text{)}$$

Therefore, the required probability is

$$= \sum_{n=k}^{\infty} \alpha p^n \, {}^nC_k \left(\frac{1}{2}\right)^k \left(\frac{1}{2}\right)^{n-k}$$

$$= \alpha \left(\frac{1}{2}\right)^k p^k \sum_{n=k}^{\infty} {}^nC_k \left(\frac{1}{2}\right)^{n-k} (p^{n-k})$$

$$= \alpha \left(\frac{1}{2}\right)^k p^k \left[1 + {}^{(k+1)}C_1 \left(\frac{p}{2}\right) + {}^{(k+2)}C_2 \left(\frac{p}{2}\right)^2 + \cdots\right]$$

$$= \alpha \left(\frac{1}{2}\right)^k p^k \left(1 - \frac{p}{2}\right)^{-(k+1)}$$

$$= \alpha p^k (2-p)^{-(k+1)}$$

$$= \frac{2\alpha}{2-p}\left(\frac{p}{2-p}\right)^k, \; k \ge 1$$

38. c. Let A denote the event of a family including at least one boy. Then,

$$P(A) = \frac{2\alpha}{2-p} \sum_{k=1}^{\infty} \left(\frac{p}{2-p}\right)^k$$

$$= \frac{2\alpha}{2-p} \frac{\left(\frac{p}{2-p}\right)}{1 - \left(\frac{p}{2-p}\right)} \quad \text{(sum of infinite terms of G.P.)}$$

$$= \frac{\alpha p}{(2-p)(1-p)} \qquad (1)$$

39. a. Let B denote the event of a family including at least two or more boys. Then,

$$P(B) = \frac{2\alpha}{2-p} \sum_{k=2}^{\infty} \left(\frac{p}{2-p}\right)^k$$

$$= \frac{\alpha p^2}{(2-p)^2(1-p)} \qquad (2)$$

$$\therefore P\left(\frac{B}{A}\right) = \frac{P(A\cap B)}{P(A)} = \frac{P(B)}{P(A)} \qquad \text{(Since } B \subset A\text{)}$$

$$= \frac{p}{2-p} \qquad \text{[From Eqs. (1) and (2)]}$$

Matrix-Match Type

1. a → p; b → r; c → q; d → p.

$$P(A) = \frac{{}^4C_1 \, {}^8C_2}{{}^{12}C_3} = \frac{4\,28}{220} = \frac{112}{220} = \frac{28}{55}$$

$$P(B) = \frac{{}^4C_3 + {}^8C_3}{{}^{12}C_3} = \frac{4+56}{220} = \frac{60}{220} = \frac{3}{11}$$

$$P(C) = P(WBB \text{ or } BWB \text{ or } WWB \text{ or } BBB)$$

$$= \frac{8}{12}\times\frac{4}{11}\times\frac{3}{10} + \frac{4}{12}\times\frac{8}{11}\times\frac{3}{10} + \frac{8}{12}\times\frac{7}{11}\times\frac{4}{10} + \frac{4}{12}\times\frac{3}{11}\times\frac{2}{10}$$

$$= \frac{96+96+224+24}{12\times 110} = \frac{440}{12\times 110} = \frac{1}{3}$$

$A \cap B = \phi \Rightarrow A$ and B are mutually exclusive

$$P(B\cap C) = P(BBB) = \frac{4\times 3\times 2}{12\times 110} = \frac{1}{55}$$

Also,

$$P(B)\,P(C) = \frac{3}{11}\times\frac{1}{3} = \frac{1}{11}$$

Hence, B and C are neither independent nor mutually exclusive.

$$P(C\cap A) = P(WWB)\frac{8\times 7\times 4}{12\times 11\times 10} = \frac{28}{3\times 55}$$

$$P(C)\,P(A) = \frac{1}{3}\times\frac{112}{220} = \frac{28}{3\times 55}$$

$\Rightarrow C$ and A are independent

Also, A, B, C are mutually exclusive as A and B are mutually exclusive.

2. a → q, r, s; b → r; c → p, q; d → p, q, r, s.

a. Suppose the coin is tossed n times. The probability of getting head or tail is 1/2. The probability of not getting any head in n tosses is $(1/2)^n$. The probability of getting at least one head is $1-(1/2)^n$. Now given that

$$1-(1/2)^n \geq 0.8$$

$$\Rightarrow \left(\frac{1}{2}\right)^n \leq 0.2$$

$$\Rightarrow 2^n \geq 5$$

Therefore, the least value of n is 3.

b. The total number of mappings is n^n. The number of one-one mappings is $n!$ Hence the probability is

$$\frac{n!}{n^n} = \frac{3}{32} = \frac{6}{64} = \frac{3!}{4^3} = \frac{4!}{4^4}$$

Comparing, we get $n = 4$.

c. Given equation is

$$2x^2 + 2mx + m + 1 = 0$$

$$D = 4m^2 - 8(m+1) \geq 0$$

$$m^2 - 2m - 2 \geq 0$$

$$(m-1)^2 - 3 \geq 0$$

$$\Rightarrow m = 3, 4, 5, 6, 7, 8, 9, 10$$

Also, the number of ways of choosing m is 10. Therefore, the required probability is 4/5.

$$\therefore 5k = 4$$

d. $20P^2 - 13P + 2 \leq 0$

$$\Rightarrow (4P-1)(5P-2) \leq 0$$

$$\Rightarrow \frac{1}{4} \leq P \leq \frac{2}{5}$$

$$\Rightarrow \frac{1}{4} \leq \frac{1}{5} + \frac{1}{5}\left(\frac{4}{5}\right) + \frac{1}{5}\left(\frac{4}{5}\right)^2 + \cdots + \frac{1}{5}\left(\frac{4}{5}\right)^{n-1} \leq \frac{2}{5}$$

$$\Rightarrow n = 2$$

Hence, maximum as well as minimum value of n is 2.

3. a → r, s; b → p, q, r, s; c → p, q, r, s; d → p.

a. $P(\text{success}) = 1/2$; $P(\text{failure}) = 1/2$

Suppose n bombs are to dropped. Let E be the event that the bridge is destroyed. Then,

$$P(E) = 1 - P(0 \text{ or } 1 \text{ success})$$

$$= 1 - \left(\left(\frac{1}{2}\right)^n + {}^nC_1 \frac{1}{2}\left(\frac{1}{2}\right)^{n-1}\right) = 1 - \left(\frac{1}{2^n} + \frac{n}{2^n}\right) \geq 0.9$$

$$\Rightarrow \frac{1}{10} \geq \frac{n+1}{2^n} \text{ or } \frac{2^n}{10(n+1)} \geq 1$$

b. The bag contains two red, three white and five black balls. Hence, $P(S) = 1/5$; $P(F) = 4/5$; Let E be the even of getting a red ball.

$$P(E) = P(S \text{ or } FS \text{ or } FFS \text{ or } \ldots] \geq \frac{1}{2}$$

$$\therefore P(F)^n \leq \frac{1}{2}; \left(\frac{4}{5}\right)^n \leq \frac{1}{2}$$

The value of n consistent is 4.

c. Let there be x red socks and y blue socks and $x > y$. Then

$$\frac{{}^xC_2 + {}^yC_2}{{}^{x+y}C_2} = \frac{1}{2}$$

or

$$\frac{x(x-1) + y(y-1)}{(x+y)(x+y-1)} = \frac{1}{2}$$

Multiplying both sides by $2(x+y)(x+y-1)$ and expanding, we get

$$2x^2 - 2x + 2y^2 - 2y = x^2 + 2xy + y^2 - x - y$$

Rearranging, we have

$$x^2 - 2xy + y^2 = x + y$$

$$\Rightarrow (x-y)^2 = x + y$$

$$\Rightarrow |x-y| = x + y$$

Now,

$$x + y \leq 17$$

$$x - y \leq \sqrt{17}$$

As $x - y$ must be an integer, so

$$x - y = 4$$

$$\therefore x + y = 16$$

Adding both together and dividing by 2 yields $x \leq 10$.

d. Let the number of green socks be $x > 0$. Let E: be the event that two socks drawn are of the same colour.

$$P(E) = P(RR \text{ or } BB \text{ or } WW \text{ or } GG)$$

$$= \frac{3}{{}^{6+x}C_2} + \frac{{}^xC_2}{{}^{6+x}C_2}$$

$$= \frac{6}{(x+6)(x+5)} + \frac{x(x-1)}{(x+6)(x+5)}$$

$$= \frac{1}{5}$$

$$\Rightarrow 5(x^2 - x + 6) = x^2 + 11x + 30$$

$$\Rightarrow 4x^2 - 16x = 0$$

$$\Rightarrow x = 4$$

4. a → q; b → r; c → s; d → r.

We have,

$$P(A \cap B) = P(A)\,P(B) = \frac{1}{12}$$

a. $P(A \cup B) = P(A) + P(B) - P(A \cap B) = \frac{1}{3} + \frac{1}{4} - \frac{1}{12} = \frac{1}{2}$

b. $P\left(\frac{A}{A \cup B}\right) = \frac{P(A)}{P(A \cup B)} = \frac{2}{3}$

c. (C) $P\left(\dfrac{B}{A'\cap B'}\right)=\dfrac{P(B\cap(A'\cap B'))}{P(A'\cap B')}=\dfrac{P(\phi)}{P(A'\cap B')}=0$

d. $P\left(\dfrac{A'}{B}\right)=\dfrac{P(A'\cap B)}{P(B)}=P(A')=\dfrac{2}{3}$

5. a → q; b → p; c → r; d → q.

Let E_i denote the event that the bag contains i black and $(12-i)$ white balls ($i = 0, 1, 2, \ldots, 12$) and A denote the event that the four balls drawn are all black. Then

$$P(E_i)=\frac{1}{13}\ (i=0,1,2,\ldots,12)$$

$$P\left(\frac{A}{E_i}\right)=0 \text{ for } i=0,1,2,3$$

$$P\left(\frac{A}{E_i}\right)=\frac{{}^iC_4}{{}^{12}C_4} \text{ for } i\geq 4$$

a. $P(A)=\sum_{i=0}^{12}P(E_i)\,P\left(\dfrac{A}{E_i}\right)$

$$=\frac{1}{13}\times\frac{1}{{}^{12}C_4}\left[{}^4C_4+{}^5C_4+\cdots+{}^{12}C_4\right]$$

$$=\frac{{}^{13}C_5}{13\times{}^{12}C_4}=\frac{1}{5}$$

b. Clearly,

$$P\left(\frac{A}{E_{10}}\right)=\frac{{}^{10}C_4}{{}^{12}C_4}=\frac{14}{33}$$

c. By Bayer's theorem,

$$P\left(\frac{E_{10}}{A}\right)=\frac{P(E_{10})\,P\left(\dfrac{A}{E_{10}}\right)}{P(A)}$$

$$=\frac{\dfrac{1}{13}\times\dfrac{14}{33}}{\dfrac{1}{5}}=\frac{70}{429}$$

d. Let B denote the probability of drawing two white and two black balls. Then

$$P\left(\frac{B}{E_i}\right)=0 \text{ if } i=0,1 \text{ or } 11,12$$

$$P\left(\frac{B}{E_i}\right)=\frac{{}^iC_2\times{}^{12-i}C_2}{{}^{12}C_4} \text{ for } i=2,3,\ldots,10$$

$$\therefore P(B)=\sum_{i=0}^{12}P(E_i)\,P\left(\frac{B}{E_i}\right)$$

$$=\frac{1}{13}\times\frac{1}{{}^{12}C_4}[2\{{}^2C_2\times{}^{10}C_2+{}^3C_2+\cdots+{}^{10}C_2\times{}^2C_2]$$

$$=\frac{1}{13}\times\frac{1}{{}^{12}C_4}[2\{{}^2C_2\,{}^{10}C_2+{}^3C_2\,{}^9C_2+\cdots+{}^5C_2\times{}^7C_2\}+{}^6C_2\times{}^6C_2]$$

$$=\frac{1}{13}\times\frac{1}{495}(1287)$$

$$=\frac{1}{5}$$

6. a → q; b → p; c → q; d → s.

a. When no box remain empty, then

$$n(S)=3^6-{}^3C_12^6+{}^3C_2=3(243-64+1)$$
$$=540$$

When each box contains equal number of balls, then

$$n(E)=\frac{6!}{2!^3\,3!}3!=90$$

Therefore, the required probability is 90/540 = 1/6.

b. The required probability is

$$\frac{3^6-{}^3C_12^6+{}^3C_2}{3^6}=\frac{20}{27}$$

c. Let A be the event that A is throwing sum of 9 and B be the event that B throws a number greater than that thrown by A. We have to find $P(B/A)=P(A\cap B)/P(A)=P(B)$ (as A and B are independent). The probability that is throwing dice so that sum is higher than 9 is

$P(B)$ = P((4, 6) or (6, 4) or (5, 5) or (6, 5) or (5, 6) or (6, 6))

$$=\frac{6}{36}=\frac{1}{6}$$

d. $P(A\cup\bar{B})=P(A)+P(\bar{B})-P(A\cap\bar{B})$

$$=P(A)+P(\bar{B})-P(A)P(\bar{B})$$
$$\Rightarrow 0.8=0.3+P(\bar{B})-0.3\,P(\bar{B})$$
$$\Rightarrow 0.5=0.7\,P(\bar{B})$$
$$\Rightarrow P(\bar{B})=\frac{5}{7}$$
$$\Rightarrow P(B)=1-\frac{5}{7}=\frac{2}{7}$$

7. a → q, r; b → r; c → p; d → a, b, c, d.

a. $\dfrac{{}^rC_2}{{}^{r+b}C_2}=\dfrac{1}{2}\;2r(r-1)=(r-b)(r+b-1)$

$$=2r(r-1)=(r+b)(r+b-1)$$
$$\Rightarrow 2r^2-2r=r^2+(2b-1)r+b^2-1$$
$$\Rightarrow r^2-(1+2b)r+1-b^2=0$$
$$\Rightarrow b^2+2br+r-r^2-1=0$$
$$\Rightarrow b=\frac{-2r\pm\sqrt{4r^2-4(r-r^2-1)}}{2}$$
$$=-r\pm\sqrt{2r^2-r+1}$$

Since b is integer, possible values of r are 3 and 8.

b. ${}^4C_2\left(\dfrac{r}{r+10}\right)^2\left(\dfrac{10}{r+10}\right)^2=\dfrac{3}{8}$

c. $\left(\frac{r}{r+10}\right)^2\left(\frac{10}{r+10}\right)^2=\frac{1}{16}$

$\Rightarrow r = 10$

d. Probability of getting exactly n red balls in $2n$ draws is always equal to probability of getting exactly n black balls in $2n$ draws for any value of r and b, hence the ratio r/b can be 10, 3, 8, 2.

8. a → r; b → p; c. → q; d → s.

a. The required event will occur if last digit in all the chosen numbers is 1, 3, 7 or 9. Therefore, the required probability is $(4/10)^n$.

b. The required probability is equal to the probability that the last digit is 2, 4, 6, 8 and is given by P (last digit is 1, 2, 3, 4, 6, 7, 8, 9) $-P$ (last digit is 1, 3, 7, 9) $=\frac{8^n-4^n}{10^n}$

c. $P(1, 3, 5, 7, 9) - P(1, 3, 7, 9) = \frac{5^n - 4^n}{10^n}$

d. The required probability is

$$P(0, 5) - P(5) = \frac{(10^n - 8^n) - (5^n - 4^n)}{10^n}$$

$$= \frac{10^n - 8^n - 5^n + 4^n}{10^n}$$

Integer Type

1. (7) We have $p=\frac{10}{100}=\frac{1}{10}\ \therefore\ q=\frac{9}{10}$

By the given condition

$1 - q^n > \frac{1}{2}$

$\Rightarrow q^n < \frac{1}{2}$

$\Rightarrow \left(\frac{9}{10}\right)^n < \frac{1}{2},$

which is possible if n is at least 7.

$\therefore\ n = 7.$

2. (4) We have $P(A \cap B) = \frac{1}{8}$ and $P(\bar{A} \cap \bar{B}) = \frac{3}{8}$

$\therefore\ P(A)\,P(B) = \frac{1}{8}$ and $P(\bar{A})\,P(\bar{B})\ \frac{3}{8}$

[$\because$ A and B are independent]

Now, $P(\bar{A} \cap \bar{B}) = \frac{3}{8} \Rightarrow 1 - P(A \cup B) = \frac{3}{8}$

$\Rightarrow 1 - (P(A) + P(B) - P(A \cap B)) = \frac{3}{8}$

$\Rightarrow 1 - (P(A) + P(B)) + \frac{1}{8} = \frac{3}{8}$

$\Rightarrow P(A) + P(B) = \frac{3}{4}$

The quadratic equation whose roots are $P(A)$ and $P(B)$ is

$x^2 - x\{P(A) + P(B)\} + P(A)\,P(B) = 0$

$\Rightarrow x^2 - \frac{3}{4}x + \frac{1}{8} = 0$

$\Rightarrow 8x^2 - 6x + 1 = 0$

$\Rightarrow x = \frac{1}{2}, \frac{1}{4}$

3. (5) Let E_1 denote the event a coin with two heads is selected and E_2 denote the event a fair coin is selected. Let A be the event the toss results in heads. Then,

$P(E_1) = \frac{1}{n+1},\ P(E_2) = \frac{n}{n+1},$

$P(A/E_1) = 1$ and $P(A/E_2) = 1/2$

$\therefore\ P(A) = P(E_1)\,P(A/E_1) + P(E_2)\,P(A/E_2)$

$\Rightarrow \frac{7}{12} = \frac{1}{n \times 1} \times 1 + \frac{n}{n+1} \times \frac{1}{2}$

$\Rightarrow 12 + 6n = 7n + 7$

$\Rightarrow n = 5$

4. (2) Let E_1 denote the event that one of the first n urns is chosen and E_2 denote the event that $(n + 1)^{\text{th}}$ urn is selected

A denotes the event that two balls drawn are black, then

$p(E_1) = \frac{n}{n+1},\ p(E_2) = \frac{1}{n+1}$

$P(A/E_1) = \frac{{}^6C_2}{{}^{10}C_2} = \frac{1}{3}$ and $P(A/E_2) = \frac{{}^5C_2}{{}^{10}C_2} = \frac{2}{9}$

Using Bayes, theorem, required probability is

$$P(E_2/A) = \frac{P(E_2)\,P(A/E_2)}{p(E_1)P(A/E_1) + P(E_2)P(A/E_2)}$$

$$\Rightarrow \frac{1}{16} = \frac{\left(\frac{1}{n+1}\right)\frac{2}{9}}{\left(\frac{n}{n+1}\right)\left(\frac{1}{3}\right) + \left(\frac{1}{n+1}\right)\left(\frac{2}{9}\right)}$$

$\Rightarrow n = 10$

5. (7) Let R be the event that a red face appears in each of the first n throws.

E_1 : Die A is used when head has already fallen

E_2 : Die B is used when tail has already fallen.

$\therefore\ P(R/E_1) = \left(\frac{2}{3}\right)^n$ and $P\left(\frac{R}{E_2}\right) = \left(\frac{1}{3}\right)^n$

As per the given condition

$$\frac{P(E_1) \times P(R/E_1)}{P(E_1) \times P(R/E_1) + P(E_2) \times P(R/E_2)} = \frac{32}{33}$$

$$\Rightarrow \frac{1/2 \times (2/3)^n}{(1/2) \times (2/3)^n + (1/3)^n} = \frac{32}{33}$$

$$\Rightarrow \frac{2^n}{2^n + 1} = \frac{32}{33}$$

$\Rightarrow n = 5.$

6. **(7)** According to the question

$${}^nC_6\left(\frac{1}{2}\right)^6\left(\frac{1}{2}\right)^{n-6} = {}^nC_8\left(\frac{1}{2}\right)^8\left(\frac{1}{2}\right)^{n-8}$$

$$\Rightarrow {}^nC_6\left(\frac{1}{2}\right)^n = {}^nC_8\left(\frac{1}{2}\right)^n$$

$$\Rightarrow {}^nC_6 = {}^nC_8 = {}^nC_{n-8}$$

$\Rightarrow 6 = n - 8$ or $n = 14$

7. **(2)** Let W denote the event of drawing a white ball at any draw and B that for a black ball. Then

$$P(W) = \frac{a}{a+b} \text{ and } P(B) = \frac{b}{a+b}$$

$P(A$ wins the game$) = P(W$ or BBW or $BBBBW$ or ...)

$= P(W) + P(BBW) + P(BBBBW) + \ldots$

$= P(W) + P(B)\,P(B)\,P(W) + P(B)\,P(B)$

$P(B)\,P(B)\,P(W) + \ldots$

$= P(W) + P(W)\,P(B)^2 + P(W)\,P(B)^4 + \cdots$

$$= \frac{P(W)}{1 - P(B)^2} = \frac{a(a+b)}{a^2 + 2ab} = \frac{a+b}{a+2b}$$

Also $P(B$ wins the game$) = 1 - \dfrac{a+b}{a+2b} = \dfrac{a}{a+2b}$

According to the given condition,

$$\Rightarrow \frac{a+b}{a+2b} = 3 \times \frac{b}{a+2b} \Rightarrow a = 2b \Rightarrow a : b = 2 : 1$$

8. **(2)** Let $P(A) = \dfrac{1+3p}{3}$, $P(B) = \dfrac{1-p}{4}$, $P(C) = \dfrac{1-2p}{2}$

As A, B, and C are three mutually exclusive events. Therefore,

$P(A) + P(B) + P(C) \le 1$

$$\Rightarrow \frac{1+3p}{3} + \frac{1-p}{4} + \frac{1-2p}{2} \le 1$$

$\Rightarrow 4 + 12P + 3 - 3p + 6 - 12p \le 12$

$\Rightarrow p \le 1/3$ (1)

Also $0 \le P(A) \le 1 \Rightarrow 0 \le \dfrac{1+3p}{3} \le 1$

$0 \le 1 + 3p \le 3$

$\Rightarrow -\frac{1}{3} \le p \le \frac{2}{3}$ (2)

$0 \le P(B) \le 1 \Rightarrow 0 \le \dfrac{1-p}{4} \le 1$

$\Rightarrow 0 \le 1 - p \le 4$

$\Rightarrow -3 \le p \le 1$ (3)

$0 \le P(C) \le 1 \Rightarrow 0 \le \dfrac{1-p}{4} \le 1$

$\Rightarrow -\frac{1}{2} \le p \le \frac{1}{2}$ (4)

Combining (1), (2), (3), and (4), we get

$\frac{1}{3} \le p \le \frac{1}{2}$

9. **(3)** The elements of A are all multiples of 5. Sum of every pair of elements of A is divisible by 5. Therefore, we have to find the probability that B has two distinct elements whose sum is divisible by 3.

Let A_0 = set of elements of A of the of the form $3k = \{0, 15, 30, \ldots, 195\}$

A_1 = Set of elements of A of the form $3k + 1 = \{10, 25, \ldots, 190\}$

A_2 = Set of elements of A of the form $3k + 2 = \{5, 20, 35, 185\}$

$n(A_0) = 14$, $n(A_1) = n(A_2) = 13$

If B has at least two elements from A_0, then we are done.

If B contains at most one element of A_0, then it must have at least one element from each of A_1 and A_2 for which the sum of these two elements will be divisible by 3.

So, required probability is 1.

10. **(6)** Let X be the number of times A shoots at the target to hit if for the first time

Y be the number of times B shoots at the target to hit for the first time

Then $P(X = m) = \left(\frac{2}{5}\right)^{m-1}\left(\frac{3}{5}\right)$ and $P(Y = n)\left(\frac{2}{7}\right)^{n-1}\left(\frac{5}{7}\right)$

We have

$$P(Y > X) = \sum_{m=1}^{\infty}\sum_{n=m+1}^{\infty} P(X = m)\,P(Y = n)$$

[$\because X$ and Y are independent]

$$= \sum_{m=1}^{\infty}\left[\left\{\left(\frac{2}{5}\right)^{m-1}\left(\frac{3}{5}\right)\right\}\sum_{n=m+1}^{\infty}\left\{\left(\frac{2}{7}\right)^{n-1}\left(\frac{5}{7}\right)\right\}\right]$$

$$= \sum_{m=1}^{\infty}\left(\frac{2}{5}\right)^{m-1}\left(\frac{3}{5}\right)\left\{\frac{5}{7}\,\frac{\left(\frac{2}{7}\right)^m}{1 - \frac{2}{7}}\right\}$$

$$= \sum_{m=1}^{\infty}\left(\frac{2}{5}\right)^{m-1}\left(\frac{3}{5}\right)\left(\frac{2}{7}\right)^m$$

$$= \frac{6}{35}\sum_{m=1}^{\infty}\left(\frac{4}{35}\right)^{m-1}$$

$$= \frac{6}{35}\,\frac{1}{1 - \frac{4}{35}} = \frac{6}{31}$$

11. **(9)** (i) Let r and s be the remainders when x and y are divided by 5 so $x = 5p + r$, $y = 5q + s$, $p, q \in N$,

$0 \le r \le 4$, $0 \le s \le 4 \therefore x^2 + y^2 = 25(p^2 + q^2) + 10(pr + qs) + r^2 + s^2 = 5k + (r^2 + s^2)$, $k \in N$

Thus $x^2 + y^2$ will be divisible by 5 iff $r^2 + s^2$ is divisible by 5. Therefore, total number of ways is number of ways of selecting r and s = $5^2 = 25$.

For favourable ways, we should have $r^2 + s^2$ divisible by 5 and $0 \le r^2 + s^2 \le 32$ so $r^2 + s^2 = 0, 5, 10, 20, 25$ or 30. Thus

r	0	1	2	3	4
s	0	2, 3	1, 4	1, 4	2, 3

So the number of favourable ways is 9.
Hence, the probability is 9/25.

12. (4) Let A be the event of P_1 winning in 3rd round and B be the event of P_2 winning in 1st but loosing in 2nd.

To find $P\left(\frac{B}{A}\right) = \frac{P(B \cap A)}{P(A)}$

$$P(A) = \frac{{}^{8n-1}C_{n-1}}{{}^{8}{}_{n}C_{n}} = \frac{1}{8} \text{ and } P(B \cap A)$$

= Probability of both P_1 and P_2 winning in 1st round × probability of P_1 winning and P_2 loosing in 2nd round × probability of P_1 winning in 3rd round.

$$= \frac{{}^{8n-2}C_{4n-2}}{{}^{8n}C_{4n}} \times \frac{{}^{4n-2}C_{2n-1}}{{}^{4n}C_{2n}} \times \frac{{}^{2n-1}C_{n-1}}{{}^{2n}C_{n}}$$

$$= \frac{n}{4(8n-1)}$$

Hence, $P\left(\frac{B}{A}\right) = \frac{2n}{8n-1}$.

Alternative Solution:

Probability that P_2 wins in first round is given by

$$P_1 \text{ wins} = \frac{{}^{8n-2}C_{4n-2}}{{}^{8n-1}C_{4n-1}} = \frac{4n-1}{8n-1}$$

In second round probability that P_2 looses in second round is given by

$$P_1 \text{ wins} = 1 - \frac{2n-1}{4n-1} = \frac{2n}{4n-1}$$

Hence, probability that P_2 looses in second round. Given P_1 wins in 3rd round is $\frac{4n-1}{8n-1} \times \frac{2n}{4n-1} = \frac{2n}{8n-1}$.

13. (5) In the tennis match of best of five sets, A can win the match if the score of A against the score of B is (3, 0), (3, 1) or (3, 2).

The probability of A's doing the score of (3, 0) is $\frac{1}{2} \times \frac{1}{2} \times \frac{1}{2}$ $= \frac{1}{8}$.

The probability of A's winning by the score of (3, 1) is

$P(A$ wins the first III sets)

+ $P(A$ wins I, loses II and wins III, IV sets)

+ $P(A$ wins set I, II, loses III and wins IV set) 2

$$= \frac{1}{4}\left(\frac{1}{2}\right)^3 + \frac{1}{2} \times \left(\frac{1}{2}\right) \times \left(\frac{1}{4}\right)\left(\frac{1}{2}\right) + \frac{1}{2} \times \frac{1}{2} \times \frac{1}{2} \times \frac{1}{4} = \frac{3}{32}$$

The probability of A's winning by the score of (3, 2) is

$P(A$ looses 1st and 2nd sets)

+ $P(A$ looses 1st and 3rd sets)

+ $P(A$ looses 1st and 4th sets)

+ $P(A$ looses 2nd and 3rd sets)

+ $P(A$ looses 2nd and 4th sets)

+ $P(A$ looses 3rd and 4th sets)

$$= \frac{1}{2}\left(\frac{3}{4}\right)\left(\frac{1}{4}\right)\left(\frac{1}{2}\right)^2 + \frac{1}{2}\left(\frac{1}{4}\right)\left(\frac{1}{2}\right)\left(\frac{1}{4}\right)\left(\frac{1}{2}\right)$$

$$+ \frac{1}{2}\left(\frac{1}{4}\right)\left(\frac{1}{2}\right)\left(\frac{1}{2}\right)\left(\frac{1}{4}\right) + \frac{1}{2}\left(\frac{1}{2}\right)\left(\frac{3}{4}\right)\left(\frac{1}{4}\right)\left(\frac{1}{2}\right)$$

$$+ \frac{1}{2}\left(\frac{1}{2}\right)\left(\frac{1}{4}\right)\left(\frac{1}{2}\right)\left(\frac{1}{4}\right) + \frac{1}{2}\left(\frac{1}{2}\right)\left(\frac{1}{2}\right)\left(\frac{3}{4}\right)\frac{1}{4}$$

$$= \frac{3}{128} + \frac{1}{128} + \frac{1}{128} + \frac{3}{128} + \frac{1}{128} + \frac{3}{128}$$

$$= \frac{12}{128}$$

The probability that A wins the match is

$$\frac{1}{8} + \frac{3}{32} + \frac{12}{128} = \frac{16 + 12 + 12}{128} = \frac{40}{128} = \frac{5}{16}.$$

14. (6) For the first two draws, following events may occur

E_1 = Both balls are white

E_2 = First is white and second is black

E_3 = First is black and second is white

E_4 = Both balls are black

Let E represent the event that the 3rd ball is black.

Then

$$P(E_1) = \frac{2}{4} \times \frac{1}{2} = \frac{1}{6};$$

$$P(E_2) = \frac{2}{4} \times \frac{2}{3} = \frac{1}{3};$$

$$P(E_3) = \frac{2}{4} \times \frac{2}{5} = \frac{1}{5};$$

$$P(E_4) = \frac{2}{4} \times \frac{3}{5} = \frac{3}{10}$$

The four events E_1, E_2, E_3, and E_4 are mutually exclusive and exhaustive. Using the theorem of total probability,

$$P(E) = P(E_1)\, P(E/E_1) + P(E_2)\, P(E/E_2) + P(E_3)\, P(E/E_3) + P(E_4)\, P(E/E_4)$$

$$= \frac{1}{6} \times \frac{3}{2} \cdot \frac{1}{3} + \frac{1}{3} \times \frac{3}{4} + \frac{1}{5} \times \frac{3}{4} + \frac{3}{10} \times \frac{4}{6} = \frac{23}{30}$$

15. (6) Let G denote the event the examinee guesses, C denote the event the examinee copies, and K the event the examinee knows, then given that $P(G) = 1/3$, $P(C) = 1/6$, and $P(K) = 1 - 1/3 - 1/6 = 1/2$.

Here it has been assumed that the events G, C, and K are mutually exclusive and totally exhaustive.

If R denotes the event that the answer is right, then $P(R/G) = 1/4$. Out of the four choices only one is correct. So $P(R/C) = 1/8$ (given). Also $P(R/K) = 1$

Since the probability of answering correctly when one knows the answer is equal to 1.

Now by Bayes' theorem, the probability that he know the answer, given that he answered correctly is given by

$$P(K/R) = \frac{P(K)\, P(R/K)}{p(G)\, P(R/G) + P(C)P(R/C) + P(K)P(R/K)}$$

$$= \frac{(1/2).1}{(1/3)\,(1/4) + (1/6)\,(1/8) + (1/2).\,1} = \frac{24}{29}$$

16. (1) Let us define the following events

$C \equiv$ person goes by car,

$S \equiv$ person goes by scooter

$B \equiv$ person goes by bus,

$T \equiv$ person goes by train,

$L \equiv$ person reaches late

Then we are given in the question

$P(C) = \frac{1}{7}; P(S) = \frac{3}{7}; P(B) = \frac{2}{7}; P(T) = \frac{1}{7}$

$P(L/C) = \frac{2}{9}; P(L/S) = \frac{1}{9}; P(L/B) = \frac{4}{9}; P(L/T) = \frac{1}{9}$

To find the probability $P(C/\bar{L})$

[$\because$ reaches in time $\equiv$ not late]

Using Bayes' theorem, we get

$$P(C/\bar{L}) = \frac{P(\bar{L}/C)P(C)}{P(\bar{L}/C)P(C) + P(\bar{L}/S)P(S) + P(\bar{L}/B) + A(\bar{L}/T)P(T)} \quad (1)$$

Now, $P(\bar{L}/C) = 1 - \frac{2}{9} = \frac{7}{9}; P(\bar{L}/S) = 1 - \frac{1}{9} = \frac{8}{9}$

$P(\bar{L}/B) = 1 - \frac{4}{9} = \frac{5}{9}; P(\bar{L}/T) = 1 - \frac{1}{9} = \frac{8}{9}$

Substituting these values in Eq. (1), we get

$$P(C/\bar{L}) = \frac{\frac{7}{9} \times \frac{1}{7}}{\frac{7}{9} \times \frac{1}{7} \times \frac{8}{9} \times \frac{3}{7} + \frac{5}{9} \times \frac{2}{7} + \frac{8}{9} \times \frac{1}{7}}$$

$$= \frac{7}{7 + 24 + 10 + 8} = \frac{7}{49} = \frac{1}{7}$$

17. (7) Total number of arrangement of the letters of the word CONSTANTINOPLE is

$$\frac{(14)!}{3! \cdot 2! \cdot 2!} \quad (1)$$

3N's are consecutive then considering all the 3N's as single letter, total arrangements is

$$\frac{(12)!}{2! \cdot 2!}. \quad (2)$$

Therefore, required probability is

$$\frac{(12)!/(2! \cdot 2!)}{(14)!/(3!\ 2!\ 2!)} = \frac{3!}{14 : 13} = \frac{3}{91}.$$

18. (4) Since the probability of the faces are proportional to the numbers on them, we can take the probabilities of faces 1, 2, …, 6 as $k, 2k, \ldots, 6k$, respectively.

Since one of the faces must occur,

we have $k + 2k + 3k + 4k + 5k + 6k = 1$ or $k = 1/21$

Therefore, the probability of an even number is

$2k + 4k + 6k = 12k = 12/21 = 4/7$.

19. (1) Let R stand for drawing a red ball, B for drawing a black ball, and W for drawing a white ball.

Then required probability is

$$P(WWR) + P(BBR) + P(WBR) + P(BWR) + P(WRR) + P(BRR) + P(RWR) + P(RBR)$$

$$= \frac{3 \times 2 \times 2}{8 \times 7 \times 6} + \frac{3 \times 2 \times 2}{8 \times 7 \times 6} + \frac{3 \times 3 \times 2}{8 \times 7 \times 6} + \frac{3 \times 3 \times 3}{8 \times 7 \times 6}$$

$$+ \frac{3 \times 2 \times 1}{8 \times 7 \times 6} + \frac{3 \times 2 \times 1}{8 \times 7 \times 6} + \frac{2 \times 3 \times 1}{8 \times 7 \times 6} + \frac{2 \times 3 \times 1}{8 \times 7 \times 6}$$

$$= \frac{2}{56} + \frac{2}{56} + \frac{3}{56} + \frac{3}{56} + \frac{1}{56} + \frac{1}{56} + \frac{1}{56} + \frac{1}{56}$$

$$= \frac{1}{4}$$

20. (5) Let A be the event that the first mango is good and B denotes the event that the second is good. Then

$$P(B/A) = \frac{P(A \cap B)}{P(A)}$$

Now, $P(A \cap B)$ = Probability that both mangoes are good

$$= \frac{{}^6C_2}{10C_2}$$

$P(A)$ = Probability that first mango is good

$$= \frac{{}^6C_2}{{}^{10}C_2} + \frac{{}^6C_1 \times {}^4C_1}{{}^{10}C_2}.$$

Hence, $P(B/A) = \frac{{}^6C_2}{{}^6C_2 + {}^6C_1 \times {}^4C_1}$

$$= \frac{15}{15 + 24}$$

$$= \frac{5}{13}.$$

21. (3) Let A_i denote the event that the number i appears on the die and let E denote the event that only white balls are drawn.

Then $P(A_i) = \frac{1}{6}$ for $i = 1, 2, \ldots, 6$

and $P(E/A_i) = \frac{{}^6C_i}{{}^{10}C_i}, i = 1, 2, \ldots, 6$

Then required probability is

$$P(E) = P\left(\bigcup_{i=1}^{6} (E \cap A_i)\right)$$

$$= \sum_{i=1}^{6} P(E \cap A_i)$$

$$= \sum_{i=1}^{6} P(A_i)P(E/A_i)$$

$$= \frac{1}{6}\left[\frac{6}{10} + \frac{15}{45} + \frac{20}{120} + \frac{15}{210} + \frac{6}{252} + \frac{1}{210}\right] = \frac{1}{3}$$

22. (1) In the 1st case

$$P(\text{they match}) = \frac{{}^3C_2 + {}^nC_2}{{}^{n+3}C_2} = \frac{1}{2}; \frac{6 + n(n-1)}{(n+3)(n+2)} = \frac{1}{2}$$

$\Rightarrow 2(n^2 - n + 6) = n^2 + 5n + 6$

$\Rightarrow n^2 - 7n + 6 = 0$

$\Rightarrow n = 1$ or 6

In the 2nd case (1)

$$\frac{3}{n+3} \times \frac{3}{n+3} + \frac{n}{n+3} \times \frac{n}{n+3} = \frac{5}{8}$$

Solving, $n^2 - 10n + 9 = 0$

n = 9 or 1 (2)

form (1) and (2), $n = 1$

Section C

Coordinate Geometry

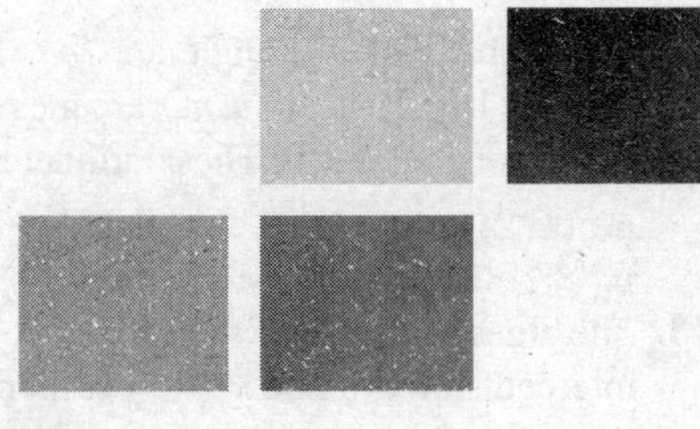

CHAPTER 1

Straight Lines

Objective Type

Each question has four choices a, b, c and d, out of which *only one* is correct. Find the correct answer.

1. Equations of diagonals of square formed by lines $x = 0$, $y = 0$, $x = 1$ and $y = 1$ are

   a. $y = x, y + x = 1$ **b.** $y = x, x + y = 2$
   c. $2y = x, y + x = 1/3$ **d.** $y = 2x, y + 2x = 1$

2. If each of the points $(x_1, 4)$, $(-2, y_1)$ lies on the line joining the points $(2, -1)$, $(5, -3)$, then the point $P(x_1, y_1)$ lies on the line

   a. $6(x + y) - 25 = 0$ **b.** $2x + 6y + 1 = 0$
   c. $2x + 3y - 6 = 0$ **d.** $6(x + y) + 25 = 0$

3. If
$$\begin{vmatrix} x_1 & y_1 & 1 \\ x_2 & y_2 & 1 \\ x_3 & y_3 & 1 \end{vmatrix} = \begin{vmatrix} a_1 & b_1 & 1 \\ a_2 & b_2 & 1 \\ a_3 & b_3 & 1 \end{vmatrix}$$
   then the two triangles with vertices (x_1, y_1), (x_2, y_2), (x_3, y_3), and (a_1, b_1), (a_2, b_2), (a, b_3) are

   a. equal in area **b.** similar
   c. congruent **d.** none of these

4. The area of a parallelogram formed by the lines $ax \pm bx \pm c = 0$ is

   a. $c^2/(ab)$ **b.** $2c^2/(ab)$
   c. $c^2/2ab$ **d.** none of these

5. The straight line $ax + by + c = 0$ where $abc \neq 0$ will pass through the first quadrant if

   a. $ac > 0, bc > 0$ **b.** $c > 0$ and $bc < 0$
   c. $bc > 0$ and/or $ac > 0$ **d.** $ac < 0$ and/or $bc < 0$

6. If the point $(x_1 + t(x_2 - x_1), y_1 + t(y_2 - y_1))$ divides the join of (x_1, y_1) and (x_2, y_2) internally, then

   a. $t < 0$ **b.** $0 < t < 1$
   c. $t > 1$ **d.** $t = 1$

7. A square of side a lies above the x-axis and has one vertex at the origin. The side passing trough the origin makes an angle $\alpha (0 < \alpha < \pi/4)$ with the positive direction of x-axis. The equation of its diagonal not passing through the origin is

   a. $y(\cos\alpha + \sin\alpha) + x(\sin\alpha - \cos\alpha) = a$
   b. $y(\cos\alpha + \sin\alpha) + x(\sin\alpha + \cos\alpha) = a$
   c. $y(\cos\alpha + \sin\alpha) + x(\cos\alpha - \sin\alpha) = a$
   d. $y(\cos\alpha - \sin\alpha) - x(\sin\alpha - \cos\alpha) = a$

8. If sum of the distances of a point from two perpendicular lines in a plane is 1, then its locus is

   a. a square **b.** a circle
   c. a straight line **d.** two intersecting lines

9. The area enclosed by $2|x| + 3|y| \leq 6$ is

   a. 3 sq. units **b.** 4 sq. units
   c. 12 sq. units **d.** 24 sq. units

10. If two vertices of a triangle are $(-2, 3)$ and $(5, -1)$, orthocentre lies at the origin and centroid on the line $x + y = 7$, then the third vertex lies at

    a. (7, 4) **b.** (8, 14)
    c. (12, 21) **d.** none of these

11. $OPQR$ is a square and M, N are the middle points of the sides PQ and QR, respectively, then the ratio of the areas of the square and the triangle OMN is

    a. 4:1 **b.** 2:1
    c. 8:3 **d.** 7:3

12. The vertices of a triangle are $(pq, 1/(pq))$, $(qr, 1/(qr))$ and $(rq, 1/(rp))$ where p, q, r are the roots of the equation $y^3 - 3y^2 + 6y + 1 = 0$. The coordinates of its centroid are

a. (1, 2) b. (2, −1)
c. (1, − 1) d. (2, 3)

13. The lines $y = m_1x$, $y = m_2x$ and $y = m_3x$ make equal intercepts on the line $x + y = 1$, then

a. $2(1 + m_1)(1 + m_3) = (1 + m_2)(2 + m_1 + m_3)$
b. $(1 + m_1)(1 + m_3) = (1 + m_2)(1 + m_1 + m_3)$
c. $(1 + m_1)(1 + m_2) = (1 + m_3)(2 + m_1 + m_3)$
d. $2(1 + m_1)(1 + m_3) = (1 + m_2)(1 + m_1 + m_3)$

14. The condition on a and b, such that the portion of the line $ax + by - 1 = 0$, intercepted between the lines $ax + y = 0$ and $x + by = 0$, subtends a right angle at the origin is

a. $a = b$ b. $a + b = 0$
c. $a = 2b$ d. $2a = b$

15. One diagonal of a square is along the line $8x - 15y = 0$ and one of its vertex is (1, 2). Then the equations of the sides of the square passing through this vertex are

a. $23x + 7y = 9$, $7x + 23y = 53$
b. $23x - 7y + 9 = 0$, $7x + 23y + 53 = 0$
c. $23x - 7y - 9 = 0$, $7x + 23y - 53 = 0$
d. none of these

16. If $u = a_1x + b_1y + c_1 = 0$, $v = a_2x + b_2y + c_2 = 0$ and $a_1/a_2 = b_1/b_2 = c_1/c_2$, then the curve $u + kv = 0$ is

a. the same straight line u
b. different straight line
c. not a straight line
d. none of these

17. The point $A(2, 1)$ is translated parallel to the line $x - y = 3$ by a distance of 4 units. If the new position A' is in third quadrant, then the coordinates of A' are

a. $(2 + 2\sqrt{2}, 1 + 2\sqrt{2})$ b. $(-2 + \sqrt{2}, -1 - 2\sqrt{2})$
c. $(2 - 2\sqrt{2}, 1 - 2\sqrt{2})$ d. none of these

18. The area of the triangle formed by the lines $y = ax$, $x + y - a = 0$, and the y-axis is equal to

a. $\frac{1}{2|1 + a|}$ b. $\frac{a^2}{|1 + a|}$
c. $\frac{1}{2}\left|\frac{a}{1 + a}\right|$ d. $\frac{a^2}{2|1 + a|}$

19. An equation of a line through the point (1, 2) whose distance from the point (3, 1) has the greatest value is

a. $y = 2x$ b. $y = x + 1$
c. $x + 2y = 5$ d. $y = 3x - 1$

20. A rectangle $ABCD$ has its side AB parallel to line $y = x$ and vertices A, B and D lie on $y = 1$, $x = 2$ and $x = -2$, respectively. Locus of vertex C is

a. $x = 5$ b. $x - y = 5$
c. $y = 5$ d. $x + y = 5$

21. The centroid of an equilateral triangle is (0, 0). If two vertices of the triangle lie on $x + y = 2\sqrt{2}$ then one of them will have its coordinates

a. $(\sqrt{2} + \sqrt{6}, \sqrt{2} - \sqrt{6})$ b. $(\sqrt{2} + \sqrt{3}, \sqrt{2} - \sqrt{3})$
c. $(\sqrt{2} + \sqrt{5}, \sqrt{2} - \sqrt{5})$ d. none of these

22. Vertices of a variable triangle are (3, 4), $(5 \cos \theta, 5 \sin \theta)$ and $(5 \sin \theta, -5 \cos \theta)$, where $\theta \in R$. Locus of its orthocentre is

a. $(x + y - 1)^2 + (x - y - 7)^2 = 100$
b. $(x + y - 7)^2 + (x - y - 1)^2 = 100$
c. $(x + y - 7)^2 + (x + y - 1)^2 = 100$
d. $(x + y - 7)^2 + (x - y + 1)^2 = 100$

23. The line $x/3 + y/4 = 1$ meets the y-axis and x-axis at A and B, respectively. A square $ABCD$ is constructed on the line segment AB away from the origin. The coordinates of the vertex of the square farthest from the origin are

a. (7, 3) b. (4, 7)
c. (6, 4) d. (3, 8)

24. A line of fixed length 2 units moves so that its ends are on the positive x-axis and that part of the line $x + y = 0$ which lies in the second quadrant. Then the locus of the midpoint of the line has the equation

a. $x^2 + 5y^2 + 4xy - 1 = 0$ b. $x^2 + 5y^2 + 4xy + 1 = 0$
c. $x^2 + 5y^2 - 4xy - 1 = 0$ d. $4x^2 + 5y^2 + 4xy + 1 = 0$

25. Consider the points $A(0, 1)$ and $B(2, 0)$, and P be a point on the line $4x + 3y + 9 = 0$. Coordinates of P such that $|PA - PB|$ is maximum are

a. (− 12/5, 17/5) b. (− 84/5, 13/5)
c. (− 6/5, 17/5) d. (0, − 3)

26. Consider points $A(3, 4)$ and $B(7, 13)$. If P be a point on the line $y = x$ such that $PA + PB$ is minimum, then coordinates of P are

a. (12/7, 12/7) b. (13/7, 13/7)
c. (31/7, 31/7) d. (0, 0)

27. A line L is drawn from $P(4, 3)$ to meet the lines L_1 and L_2 given by $3x + 4y + 5 = 0$ and $3x + 4y + 15 = 0$ at points A and B, respectively. From A, a line perpendicular to L is drawn meeting the line L_2 at A_1. Similarly from point B, a line perpendicular to L is drawn meeting the line L_1 at B_1. Thus a parallelogram AA_1BB_1 is formed. Then the the equation of L so that the area of the parallelogram AA_1BB_1 is least is

a. $x - 7y + 17 = 0$ b. $7x + y + 31 = 0$
c. $x - 7y - 17 = 0$ d. $x + 7y - 31 = 0$

28. Line L has intercepts a and b on the coordinate axes. When the axes are rotated through a given angle keeping the origin fixed, the same line L has intercepts p and q. Then,

a. $a^2 + b^2 = P^2 + q^2$ **b.** $\frac{1}{a^2} + \frac{1}{b^2} = \frac{1}{p^2} + \frac{1}{q^2}$

c. $a^2 + p^2 = b^2 + q^2$ **d.** $\frac{1}{a^2} + \frac{1}{p^2} = \frac{1}{b^2} + \frac{1}{q^2}$

29. If the straight lines $2x + 3y - 1 = 0$, $x + 2y - 1 = 0$ and $ax + by - 1 = 0$ form a triangle with origin as orthocentre, then (a, b) is given by

a. (6, 4) **b.** (−3, 3)

c. (−8, 8) **d.** (0, 7)

30. A light ray emerging from the point source placed at $P(2, 3)$ is reflected at a point Q on the y-axis and then passes through the point $R(5, 10)$. Coordinates of Q are

a. (0, 3) **b.** (0, 2)

c. (0, 5) **d.** none of these

31. A straight line passing through $P(3, 1)$ meets the coordinate axes at A and B. It is given that distance of this straight line from the origin O is maximum. Area of triangle OAB is equal to

a. 50/3 sq. units **b.** 25/3 sq. units

c. 20/3 sq. units **d.** 100/3 sq. units

32. Let O be the origin. If $A(1, 0)$ and $B(0, 1)$ and $P(x, y)$ are points such that $xy > 0$ and $x + y < 1$, then

a. P lies either inside the triangle OAB or in the third quadrant

b. P cannot lie inside the triangle OAB

c. P lies inside the triangle OAB

d. P lies in the first quadrant only

33. The graph $y^2 + 2xy + 40|x| = 400$ divides the plane into regions. Then the area of bounded region is

a. 200 sq. units **b.** 400 sq. units

c. 800 sq. units **d.** 500 sq. units

34. In a triangle ABC, $A \equiv (\alpha, \beta)$, $B \equiv (2, 3)$ and $C \equiv (1, 3)$ and point A lies on line $y = 2x + 3$ where $\alpha \in I$. Area of ΔABC, Δ, is such that $[\Delta] = 5$. Possible coordinates of A are (where $[\cdot]$ represents greatest integer function)

a. (2, 3) **b.** (5, 13)

c. (−5, −7) **d.** (−3, −5)

35. If $a/bc - 2 = \sqrt{b/c} + \sqrt{c/b}$, where $a, b, c > 0$, then family of lines $\sqrt{a}x + \sqrt{b}y + \sqrt{c} = 0$ passes though the fixed point given by

a. (1, 1) **b.** (1, − 2)

c. (−1, 2) **d.** (−1, 1)

36. $P(m, n)$ (where m, n are natural numbers) is any point in the interior of the quadrilateral formed by the pair of lines $xy = 0$ and the lines $2x + y - 2 = 0$ and $4x + 5y = 20$. The possible number of positions of the point P is

a. 7 **b.** 5

c. 4 **d.** 6

37. If the area of the rhombus enclosed by the lines $lx \pm my \pm n = 0$ be 2 sq. units, then

a. l, m, n are in G.P. **b.** l, n, m are in G.P.

c. $lm = n$ **d.** $ln = m$

38. P, Q, R, and S are the points of intersection with the coordinate axes of the lines $px + qy = pq$ and $qx + py = pq$, then $(P, Q > 0)$

a. P, Q, R, S from a parallelogram

b. P, Q, R, S from a rhombus

c. P, Q, R, S are concyclic

d. none of these

39. In a triangle ABC if $A \equiv (1, 2)$ and internal angle bisectors through B and C are $y = x$ and $y = -2x$, then the inradius r of the ΔABC is

a. $1/\sqrt{3}$ **b.** $1/\sqrt{2}$

c. 2/3 **d.** none of these

40. In a triangle ABC, the bisectors of angles B and C lie along the lines $x = y$ and $y = 0$. If A is (1, 2), then the equation of line BC is

a. $2x + y = 1$ **b.** $3x - y = 5$

c. $x - 2y = 3$ **d.** $x + 3y = 1$

41. If the vertices of a triangle are $(\sqrt{5}, 0)$, $(\sqrt{3}, \sqrt{2})$, and $(2, 1)$, then the orthocentre of the triangle is

a. $(\sqrt{5}, 0)$ **b.** (0, 0)

c. $(\sqrt{5} + \sqrt{3} + 2, \sqrt{2} + 1)$ **d.** none of these

42. One of the diagonals of a square is the portion of the line $x/2 + y/3 = 2$ intercepted between the axes. Then the extremities of the other diagonal are

a. (5, 5), (−1, 1) **b.** (0, 0), (4, 6)

c. (0, 0), (−1, 1) **d.** (5, 5), (4, 6)

43. Two sides of a triangle are along the coordinate axes and the medians through the vertices (other than the origin) are mutually perpendicular. The number of such triangles is/are

a. zero **b.** two

c. four **d.** infinite

44. A rectangular billiard table has vertices at P (0, 0), Q (0, 7), R (10, 7), and S (10, 0). A small billiard ball starts at M (3, 4) and moves in a straight line to the top of the table, bounces to the right side of the table, then comes to rest at N (7, 1). The y-coordinate of the point where it hits the right side, is

a. 3.7 **b.** 3.8 **c.** 3.9 **d.** 4

45. In ΔABC the coordinates of the vertex A are (4, −1) and lines $x - y - 1 = 0$ and $2x - y = 3$ are internal bisectors of angles B and C. Then, radius of incircle of triangle ABC is

a. $4/\sqrt{5}$ b. $3/\sqrt{5}$
c. $6/\sqrt{5}$ d. $7/\sqrt{5}$

46. Distance of origin from line $(1+\sqrt{3})y+(1-\sqrt{3})x = 10$ along the line $y = \sqrt{3}x + k$ is
a. $5/\sqrt{2}$ b. $5\sqrt{2} + k$
c. 10 d. 0

47. If it is possible to draw a line which belongs to all the given family of lines $y - 2x + 1 + \lambda_1(2y - x - 1) = 0$, $3y - x - 6 + \lambda_2(y - 3x + 6) = 0$, $ax + y - 2 + \lambda_3(6x + ay - a) = 0$, then
a. $a = 4$ b. $a = 3$
c. $a = -2$ d. $a = 2$

48. If the equation of any two diagonals of a regular pentagon belongs to family of lines $(1 + 2\lambda)y - (2 + \lambda)x + 1 - \lambda = 0$ and their lengths are sin 36°, then locus of centre of circle circumscribing the given pentagon (the triangles formed by these diagonals with sides of pentagon have no side common) is
a. $x^2 + y^2 - 2x - 2y + 1 + \sin^2 72° = 0$
b. $x^2 + y^2 - 2x - 2y + \cos^2 72° = 0$
c. $x^2 + y^2 - 2x - 2y + 1 + \cos^2 72° = 0$
d. $x^2 + y^2 - 2x - 2y + \sin^2 72° = 0$

49. ABC is a variable triangle such that A is (1, 2), B and C lie on line $y = x + \lambda$ (where λ is a variable), then locus of the orthocentre of triangle ABC is
a. $(x - 1)^2 + y^2 = 4$ b. $x + y = 3$
c. $2x - y = 0$ d. none of these

50. Locus of the image of the point (2, 3) in the line $(x - 2y + 3) + \lambda(2x - 3y + 4) = 0$ is ($\lambda \in R$)
a. $x^2 + y^2 - 3x - 4y - 4 = 0$
b. $2x^2 + 3y^2 + 2x + 4y - 7 = 0$
c. $x^2 + y^2 - 2x - 4y + 4 = 0$
d. none of these

51. If one side of a rhombus has end points (4, 5) and (1, 1) then the maximum area of the rhombus is
a. 50 sq. units b. 25 sq. units
c. 30 sq. units d. 20 sq. units

52. The equations of the sides of a triangle are $x + y - 5 = 0$, $x - y + 1 = 0$, and $y - 1 = 0$. Then the coordinates of the circumcentre are
a. (2, 1) b. (1, 2)
c. (2, −2) d. (1, −2)

53. Two vertices of a triangle are (4, −3) and (−2, 5). If the orthocentre of the triangle is at (1, 2), then the third vertex is
a. (−33, −26) b. (33, 26)
c. (26, 33) d. none of these

54. If $\Sigma_{i=1}^{4}(x_i^2 + y_i^2) \le 2x_1x_3 + 2x_2x_4 + 2y_2y_3 + 2y_1y_4$, the points $(x_1, y_1), (x_2, y_2), (x_3, y_3), (x_4, y_4)$ are
a. the vertices of a rectangle
b. collinear
c. trapezium
d. none of these

55. Locus of a point that is equidistant from the lines $x + y - 2\sqrt{2} = 0$ and $x + y - \sqrt{2} = 0$ is
a. $x + y - 5\sqrt{2} = 0$ b. $x + y - 3\sqrt{2} = 0$
c. $2x + 2y - 3\sqrt{2} = 0$ d. $2x + 2y - 5\sqrt{2} = 0$

56. If the intercept made on the line $y = mx$ by lines $y = 2$ and $y = 6$ is less than 5, then the range of values of m is
a. $(-\infty, -4/3) \cup (4/3, +\infty)$
b. $(-4/3, 4/3)$
c. $(-3/4, 4/3)$
d. none of these

57. If $P\,(1 + t/\sqrt{2}, 2 + t/\sqrt{2}\,)$ be any point on a line, then the range of the values of t for which the point P lies between the parallel lines $x + 2y = 1$ and $2x + 4y = 15$ is
a. $-4\sqrt{2}/3 < t < 5\sqrt{2}/6$ b. $0 < t < 5\sqrt{2}/6$
c. $4\sqrt{2}/ < t < 0$ d. none of these

58. P is a point on the line $y + 2x = 1$ and, Q and R are two points on the line $3y + 6x = 6$ such that triangle PQR is an equilateral triangle. The length of the side of the triangle is
a. $2/\sqrt{5}$ b. $3/\sqrt{5}$
c. $4/\sqrt{5}$ d. none of these

59. The coordinates of two consecutive vertices A and B of a regular hexagon $ABCDEF$ are (1, 0) and (2, 0), respectively. The equation of the diagonal CE is
a. $\sqrt{3}x + y = 4$ b. $x + \sqrt{3}y + 4 = 0$
c. $x + \sqrt{3}y = 4$ d. none of these

60. The foot of the perpendicular on the line $3x + y = \lambda$ drawn from the origin is C. If the line cuts the x-axis and y-axis at A and B, respectively, then $BC:CA$ is
a. 1:3 b. 3:1
c. 1:9 d. 9:1

61. If x_1, x_2, x_3 as well as y_1, y_2, y_3 are in G.P. with same common ratio, then the points $P\,(x_1, y_1)$, $Q\,(x_2, y_2)$ and $R\,(x_3, y_3)$

a. lie on a straight line b. lie on an ellipse
c. lie on a circle d. are vertices of a triangle

62. If the quadrilateral formed by the lines $ax + by + c = 0$, $a'x + b'y + c = 0$, $ax + by + c' = 0$, $a'x + b'y + c' = 0$ have perpendicular diagonals, then

a. $b^2 + c^2 = b'^2 + c'^2$ b. $c^2 + a^2 = c'^2 + a'^2$
c. $a^2 + b^2 = a'^2 + b'^2$ d. none of these

63. The straight lines $7x - 2y + 10 = 0$ and $7x + 2y - 10 = 0$ form an isosceles triangle with the line $y = 2$. Area of this triangle is equal to

a. 15/7 sq. units b. 10/7 sq. units
c. 18/7 sq. units d none of these

64. A rectangle $ABCD$, where $A \equiv (0, 0)$, $B \equiv (4, 0)$, $C \equiv (4, 2)$, $D \equiv (0, 2)$, undergoes the following transformations successively: (i) $f_1(x, y) \to (y, x)$, (ii) $f_2(x, y) \to (x + 3y, y)$, (iii) $f_3(x, y) \to ((x - y)/2, (x + y)/2)$. The final figure will be

a. a square b. a rhombus
c. a rectangle d. a parallelogram

65. θ_1 and θ_2 are the inclination of lines L_1 and L_2 with x-axis. If L_1 and L_2 pass through $P(x_1, y_1)$, then equation of one of the angle bisector of these lines is

a. $\dfrac{x - x_1}{\cos\left(\dfrac{\theta_1 - \theta_2}{2}\right)} = \dfrac{y - y_1}{\sin\left(\dfrac{\theta_1 - \theta_2}{2}\right)}$

b. $\dfrac{x - x_1}{-\sin\left(\dfrac{\theta_1 - \theta_2}{2}\right)} = \dfrac{y - y_1}{\cos\left(\dfrac{\theta_1 - \theta_2}{2}\right)}$

c. $\dfrac{x - x_1}{\sin\left(\dfrac{\theta_1 - \theta_2}{2}\right)} = \dfrac{y - y_1}{\cos\left(\dfrac{\theta_1 - \theta_2}{2}\right)}$

d. $\dfrac{x - x_1}{-\sin\left(\dfrac{\theta_1 - \theta_2}{2}\right)} = \dfrac{y - y_1}{\cos\left(\dfrac{\theta_1 + \theta_2}{2}\right)}$

66. A light ray coming along the line $3x + 4y = 5$ gets reflected from the line $ax + by = 1$ and goes along the line $5x - 12y = 10$. Then,

a. $a = 64/115$, $b = 112/15$
b. $a = 14/15$, $b = -8/115$
c. $a = 64/115$, $b = -8/115$
d. $a = 64/15$, $b = 14/15$

67. Line $ax + by + p = 0$ makes angle $\pi/4$ with $x \cos\alpha + y \sin\alpha = p$, $p \in R^+$. If these lines and the line $x \sin\alpha - y \cos\alpha = 0$ are concurrent, then

a. $a^2 + b^2 = 1$ b. $a^2 + b^2 = 2$
c. $2(a^2 + b^2) = 1$ d. none of these

68. A line is drawn perpendicular to line $y = 5x$, meeting the coordinate axes at A and B. If the area of triangle OAB is 10 sq. units where O is the origin, then the equation of drawn line is

a. $3x - y - 9$ b. $x - 5y = 10$
c. $x + 4y = 10$ d. $x - 4y = 10$

69. If $x - 2y + 4 = 0$ and $2x + y - 5 = 0$ are the sides of a isosceles triangle having area 10 sq. units the equation of third side is

a. $3x - y = -9$ b. $3x - y + 11 = 0$
c. $x - 3y = 19$ d. $3x - y + 15 = 0$

70. The number of values of a for which the lines

$2x + y - 1 = 0$,
$ax + 3y - 3 = 0$, and
$3x + 2y - 2 = 0$
are concurrent is

a. 0 b. 1 c. 2 d. infinite

71. The extremities of the base of an isosceles triangle are (2, 0) and (0, 2). If the equation of one of the equal side is $x = 2$, then equation of other equal side is

a. $x + y = 2$ b. $x - y + 2 = 0$
c. $y = 2$ d. $2x + y = 2$

72. If the equation of the locus of a point equidistant from the points (a_1, b_1) and (a_2, b_2) is $(a_1 - a_2)x + (b_1 - b_2)y + c = 0$, then the value of c is

a. $a_1^2 - a_2^2 + b_1^2 - b_2^2$ b. $\sqrt{a_1^2 + b_1^2 - a_2^2 - b_2^2}$
c. $\frac{1}{2}(a_1^2 + a_2^2 + b_1^2 + b_2^2)$ d. $\frac{1}{2}(a_2^2 + b_2^2 - a_1^2 - b_1^2)$

73. The line PQ whose equation is $x - y = 2$ cuts the x-axis at P and Q is (4, 2). The line PQ is rotated about P through 45° in the anticlockwise direction. The equation of the line PQ in the new position is

a. $y = -\sqrt{2}$ b. $y = 2$ c. $x = 2$ d. $x = -2$

74. The equation of straight line passing through $(-a, 0)$ and making the triangle with axes of area T is

a. $2Tx + a^2y + 2aT = 0$ b. $2Tx - a^2y + 2aT = 0$
c. $2Tx - a^2y - 2aT = 0$ d. none of these

75. If the equation of base of an equilateral triangle is $2x - y = 1$ and the vertex is $(-1, 2)$, then the length of the sides of the triangle is

a. $\sqrt{\frac{20}{3}}$ b. $\frac{2}{\sqrt{15}}$ c. $\sqrt{\frac{8}{15}}$ d. $\sqrt{\frac{15}{2}}$

76. The number of integral values of m for which the x-coordinate of the point of intersection of the lines $3x + 4y = 9$ and $y = mx + 1$ is also an integer is

a. 2 b. 0 c. 4 d. 1

77. Equation of a straight line on which length of perpendicular from the origin is four units and the line makes an angle of 120° with the x-axis is

a. $x\sqrt{3}+y+8=0$ **b.** $x\sqrt{3}-y=8$

c. $x\sqrt{3}-y=8$ **d.** $x-\sqrt{3}y+8=0$

78. The equation to the straight line passing through the point $(a\cos^3\theta, a\sin^3\theta)$ and perpendicular to the line $x\sec\theta+y\operatorname{cosec}\theta=a$ is

a. $x\cos\theta-y\sin\theta=a\cos 2\theta$

b. $x\cos\theta+y\sin\theta=a\cos 2\theta$

c. $x\sin\theta+y\cos\theta=a\cos 2\theta$

d. none of these

79. The line parallel to the x-axis and passing through the intersection of the lines $ax+2by+3b=0$ and $bx-2ay-3a=0$, where $(a, b)\neq(0, 0)$, is

a. above the x-axis at a distance of 3/2 units from it

b. above the x-axis at a distance of 2/3 units from it

c. below the x-axis at a distance of 3/2 units from it

d. below the x-axis at a distance of 2/3 units from it

80. If a straight line through origin bisects the line passing through the given points $(a\cos\alpha, a\sin\alpha)$ and $(a\cos\beta, a\sin\beta)$, then the lines

a. are perpendicular

b. are parallel

c. have an angle between them of $\pi/4$

d. none of these

81. The straight lines $4ax+3by+c=0$, where $a+b+c=0$, are concurrent at the point

a. (4, 3) **b.** (1/4, 1/3)

c. (1/2, 1/3) **d.** none of these

82. The straight lines $x+2y-9=0$, $3x+5y-5=0$ and $ax+by-1=0$ are concurrent, if the straight line $35x-22y+1=0$ passes through the point

a. (a, b) **b.** (b, a)

c. $(-a, -b)$ **d.** none of these

83. The coordinates of the foot of the perpendicular from the point (2, 3) on the line $-y+3x+4=0$ are given by

a. (37/10 , – 1/10) **b.** (– 1/10 , 37/10)

c. (10/37, – 10) **d.** (2/3, – 1/3)

84. If the extremities of the base of an isosceles triangle are the points $(2a, 0)$ and $(0, a)$ and the equation of one of the sides is $x=2a$, then the area of the triangle is

a. $5a^2$ sq. units **b.** $5a^2/2$ sq. units

c. $25a^2/2$ sq. units **d.** none of these

85. The combined equation of straight lines that can be obtained by reflecting the lines $y=|x-2|$ in the y-axis is

a. $y^2+x^2+4x+4=0$ **b.** $y^2+x^2-4x+4=0$

c. $y^2-x^2+4x-4=0$ **d.** $y^2-x^2-4x-4=0$

86. If the slope of one line represented by $a^3x^2-2hxy+b^3y^2=0$ is square of the slope of another line, then

a. $h=2ab(a+b)$ **b.** $h=ab(a+b)$

c. $3h=2ab(a+b)$ **d.** $2h=ab(a+b)$

87. $A\equiv(-4, 0)$, $B\equiv(4, 0)$. M and N are the variable points of y-axis such that M lies below N and $MN=4$. Line joining AM and BN intersect at P. Locus of P is

a. $2xy-16-x^2=0$ **b.** $2xy+16-x^2=0$

c. $2xy+16+x^2=0$ **d.** $2xy-16+x^2=0$

88. Let A_r, $r=1, 2, 3, \ldots$ be points on the number line such that $OA_1, OA_2, OA_3, \ldots$ are in G.P. where O is the origin and the common ratio of the G.P. be a positive proper fraction. Let M_r be the middle point of the line segment A_rA_{r+1}. Then the value of $\Sigma_{r=1}^{\infty} OM_r$ is equal to

a. $\dfrac{OA_1(OSA_1-OA_2)}{2(OA_1+OA_2)}$ **b.** $\dfrac{OA_1(OA_1-OA_2)}{2(OA_1+OA_2)}$

c. $\dfrac{OA_1}{2(OA_1-OA_2)}$ **d.** ∞

89. The number of triangles that the four lines $y=x+3$, $y=2x+3$, $y=3x+2$, and $y+x=3$ form is

a. 4 **b.** 2 **c.** 3 **d.** 1

90. Let $A=(3, -4)$, $B=(1, 2)$, let $P=(2k-1, 2k+1)$ be a variable point such that $PA+PB$ is the minimum. Then k is

a. 7/9 **b.** 0

c. 7/8 **d.** none of these

91. L_1 and L_2 are two lines. If the reflection of L_1 in L_2 and the reflection of L_2 in L_1 coincide, then the angle between the lines is

a. 30° **b.** 60° **c.** 45° **d.** 90°

92. If the straight lines $x+y-2=0$, $2x-y+1=0$ and $ax+by-c=0$ are concurrent, then the family of lines $2ax+3by+c=0$ (a, b, c are nonzero) is concurrent at

a. (2, 3) **b.** (1/2 , 1/3)

c. (–1/6 , –5/9) **d.** (2/3 , –7/5)

93. Two medians drawn from acute angles of a right angled triangle intersect at an angle $\pi/6$. If the length of the hypotenuse of the triangle is 3 units, then area of the triangle (in sq. units) is

a. $\sqrt{3}$ **b.** 3 **c.** $\sqrt{2}$ **d.** 9

94. A variable line $x/a + y/b = 1$ moves in such a way that harmonic mean of a and b is 8. Then the least area of triangle made by the line with the coordinate axes is

a. 8 sq. unit **b.** 16 sq. unit
c. 32 sq. unit **d.** 64 sq. unit

95. The number of integral points (x, y) (i.e., x and y both are integers) which lie in the first quadrant but not on the coordinate axes and also on the straight line $3x + 5y = 2007$ is equal to

a. 133 **b.** 135 **c.** 138 **d.** 140

96. The number of straight lines equidistant from three non-collinear points in the plane of the points is equal to

a. 0 **b.** 1 **c.** 2 **d.** 3

97. The locus of the point which moves such that its distance from the point (4, 5) is equal to its distance from the line $7x - 3y - 13 = 0$ is

a. a straight line **b.** a circle
c. a parabola **d.** an ellipse

98. In $\Delta\, ABC$ if orthocentre be (1, 2) and circumcentre be (0, 0), then centroid of $\Delta\, ABC$ is

a. (1/2, 2/3) **b.** (1/3 , 2/3)
c. (2/3 , 1) **d.** none of these

99. The line $x + 3y - 2 = 0$ bisects the angle between a pair of straight lines of which one has equation $x - 7y + 5 = 0$. The equation of the other line is

a. $3x + 3y - 1 = 0$ **b.** $x - 3y + 2 = 0$
c. $5x + 5y - 3 = 0$ **d.** none of these

100. Through a point A on the x-axis a straight line is drawn parallel to the y-axis so as to meet the pair of straight lines $ax^2 + 2hxy + by^2 = 0$ in B and C. If $AB = BC$, then

a. $h^2 = 4ab$ **b.** $8h^2 = 9ab$ **c.** $9h^2 = 8ab$ **d.** $4h^2 = ab$

101. If $A(1, p^2)$, $B(0, 1)$, and $C(p, 0)$ are the coordinates of three points, then the value of p for which the area of the triangle ABC is minimum is

a. $1/\sqrt{3}$ **b.** $-1/\sqrt{3}$
c. $1/\sqrt{2}$ **d.** none of these

102. m, n are integers with $0 < n < m$. A is the point (m, n) on the Cartesian plane. B is the reflection of A in the line $y = x$. C is the reflection of B in the y-axis, D is the reflection of C in the x-axis and E is the reflection of D in the y-axis. The area of the pentagon $ABCDE$ is

a. $2m(m + n)$ **b.** $m(m + 3n)$
c. $m(2m + 3n)$ **d.** $2m(m + 3n)$

103. If the ends of the base of an isosceles triangle are at (2, 0) and (0, 1) and the equation of one side is $x = 2$, then the orthocentre of the triangle is

a. (3/2, 3/2) **b.** (5/4, 1)
c. (3/4, 1) **d.** (4/3, 7/12)

104. Vertices of a parallelogram $ABCD$ are A (3, 1), B (13, 6), C (13, 21) and D (3, 16). If a line passing through the origin divides the parallelogram into two congruent parts, then the slope of the line is

a. 11/12 **b.** 11/8 **c.** 25/8 **d.** 13/8

105. The distance between the two parallel lines is 1 unit. A point A is chosen to lie between the lines at a distance d from one of them. Triangle ABC is equilateral with B on one line and C on the other parallel line. The length of the side of the equilateral triangle is

a. $(2/3)\sqrt{d^2 + d + 1}$ **b.** $2\sqrt{(d^2 - d + 1)/3}$
c. $2\sqrt{d^2 - d + 1}$ **d.** $\sqrt{d^2 - d + 1}$

106. Given A (0, 0) and B (x, y) with $x \in (0, 1)$ and $y > 0$. Let the slope of the line AB equal to m_1. Point C lies on the line $x = 1$ such that the slope of BC equal to m_2 where $0 < m_2 < m_1$. If the area of the triangle ABC can be expressed as $(m_1 - m_2) f(x)$, then the largest possible value of x is

a. 1 **b.** 1/2 **c.** 1/4 **d.** 1/8

107. If the vertices P and Q of a triangle PQR are given by (2, 5) and (4, –11), respectively, and the point R moves along the line N given by $9x + 7y + 4 = 0$, then the locus of the centroid of the triangle PQR is a straight line parallel to

a. PQ **b.** QR **c.** RP **d.** N

108. Given $A \equiv (1, 1)$ and AB is any line through it cutting the x-axis in B. If AC is perpendicular to AB and meets the y-axis in C, then the equation of locus of midpoint P of BC is

a. $x + y = 1$ **b.** $x + y = 2$
c. $x + y = 2xy$ **d.** $2x + 2y = 1$

109. The number of possible straight lines, passing through (2, 3) and forming a triangle with coordinate axes, whose area is 12 sq. units, is

a. one **b.** two **c.** three **d.** four

110. In a triangle ABC, if A is (2, –1), and $7x - 10y + 1 = 0$ and $3x - 2y + 5 = 0$ are equations of an altitude and an angle bisector, respectively, drawn from B, then equation of BC is

a. $x + y + 1 = 0$ **b.** $5x + y + 17 = 0$
c. $4x + 9y + 30 = 0$ **d.** $x - 5y - 7 = 0$

111. A triangle is formed by the lines $x + y = 0$, $x - y = 0$ and $lx + my = 1$. If l and m vary subject to the condition $l^2 + m^2 = 1$, then the locus of its circumcentre is

a. $(x^2 - y^2)^2 = x^2 + y^2$ **b.** $(x^2 + y^2)^2 = (x^2 - y^2)$
c. $(x^2 + y^2) = 4x^2 y^2$ **d.** $(x^2 - y^2)^2 = (x^2 + y^2)^2$

112. The line $x+y=p$ meets the x- and y-axes at A and B, respectively. A triangle APQ is inscribed in the triangle OAB, O being the origin, with right angle at Q. P and Q lie, respectively, on OB and AB. If the area of the triangle APQ is $3/8^{th}$ of the area of the triangle OAB, then AQ/BQ is equal to

a. 2 **b.** 2/3 **c.** 1/3 **d.** 3

113. A is a point on either of two lines $y+\sqrt{3}|x|=2$ at a distance of $4\sqrt{3}$ units from their point of intersection. The coordinates of the foot of perpendicular from A on the bisector of the angle between them are

a. $(2/\sqrt{3}, 2)$ **b.** $(0, 0)$
c. $(2/\sqrt{3}, 2)$ **d.** $(0, 4)$

114. A pair of perpendicular straight lines is drawn through the origin forming with the line $2x+3y=6$ an isosceles triangle right angled at the origin. The equation to the line pair is

a. $5x^2-24xy-5y^2=0$ **b.** $5x^2-26xy-5y^2=0$
c. $5x^2+24xy-5y^2=0$ **d.** $5x^2+26xy-5y^2=0$

115. Points A and B are in the first quadrant. Point O is the origin. If the slope of OA is 1, slope of OB is 7 and $OA=OB$, then the slope of AB is

a. $-1/5$ **b.** $-1/4$
c. $-1/3$ **d.** $-1/2$

116. $OPQR$ is a square and M, N are the midpoints of the sides PQ and QR, respectively. If the ratio of the areas of the square and the triangle OMN is $\lambda : 6$, then $\lambda/4$ is equal to

a. 2 **b.** 4
c. 2 **d.** 16

117. A triangle ABC with vertices $A(-1, 0)$, $B(-2, 3/4)$, and $C(-3, -7/6)$ has its orthocentre H. Then the orthocentre of triangle BCH will be

a. $(-3, -2)$ **b.** $(1, 3)$
c. $(-1, 2)$ **d.** none of these

118. If a pair of perpendicular straight lines drawn through the origin forms an isosceles triangle with the line $2x+3y=6$, then area of the triangle so formed is

a. 36/13 **b.** 12/17
c. 13/5 **d.** 17/13

119. The image of $P(a, b)$ in the line $y=-x$ is Q and the image of Q in the $y=x$ is R. Then the midpoint of PR is

a. $(a+b, b+a)$ **b.** $((a+b)/2, (b+2)/2)$
c. $(a-b, b-a)$ **d.** $(0, 0)$

120. Let ABC be a triangle. Let A be the point $(1, 2)$, $y=x$ be the perpendicular bisector of AB and $x-2y+1=0$ be the angle bisector of $\angle C$. If equation of BC is given by $ax+by-5=0$, then the value of $a+b$ is

a. 1 **b.** 2 **c.** 3 **d.** 4

121. If in triangle ABC, $A\equiv(1, 10)$, circumcentre $\equiv(-1/3, 2/3)$ and orthocentre $\equiv(11/4, 4/3)$, then the coordinates of mid-point of side opposite to A is

a. $(1, -11/3)$ **b.** $(1, 5)$
c. $(1, -3)$ **d.** $(1, 6)$

122. In the ΔABC, the coordinates of B are $(0, 0)$, $AB=2$, $\angle ABC=\pi/3$ and the middle point of BC has the coordinates $(2, 0)$. The centroid of the triangle is

a. $(1/2, \sqrt{3}/2)$ **b.** $(5/3, 1/\sqrt{3})$
c. $(4+\sqrt{3}/3, 1/3)$ **d.** none of these

123. A beam of light is sent along the line $x-y=1$, which after refracting from the x-axis enters the opposite side by turning through 30° towards the normal at the point of incidence on the x-axis. Then the equation of the refracted ray is

a. $(2-\sqrt{3})x-y=2+\sqrt{3}$
b. $(2+\sqrt{3})x-y=2+\sqrt{3}$
c. $(2-\sqrt{3})x+y=(2+\sqrt{3})$
d. $y=(2-\sqrt{3})(x-1)$

124. The equation of the straight line which passes through the point $(-4, 3)$ such that the portion of the line between the axes is divided internally by the point in the ratio $5:3$ is

a. $9x-20y+96=0$ **b.** $9x+20y=24$
c. $20x+9y+53=0$ **d.** none of these

125. $ABCD$ is a square $A\equiv(1, 2)$, $B\equiv(3, -4)$. If line CD passes through $(3, 8)$ then midpoint of CD is

a. $(2, 6)$ **b.** $(6, 2)$ **c.** $(2, 5)$ **d.** $(24/5, 1/5)$

126. If the equations $y=mx+c$ and $x\cos\alpha+y\sin\alpha=p$ represent the same straight line, then

a. $p=c\sqrt{1+m^2}$ **b.** $c=p\sqrt{1+m^2}$
c. $cp=\sqrt{1+m^2}$ **d.** $p^2+c^2+m^2=1$

127. The equation of the bisector of the acute angle between the lines $2x-y+4=0$ and $x-2y=1$ is

a. $x+y+5=0$ **b.** $x-y+1=0$
c. $x-y=5$ **d.** none of these

128. The equation of the line segment AB is $y=x$. If A and B lie on the same side of the line mirror $2x-y=1$, then the image of AB has the equation

a. $x+y=2$ **b.** $8x+y=9$
c. $7x-y=6$ **d.** none of these

129. The line $L_1\equiv 4x+3y-12=0$ intersects the x- and the y-axes at A and B, respectively. A variable line perpendicular to L_1 intersects the x- and the y-axes at P and Q, respectively. Then the locus of the circumcentre of triangle ABQ is

a. $3x-4y+2=0$ **b.** $4x+3yy+7=0$
c. $6x-8y+7=0$ **d.** none of these

130. Consider three lines as follows.
$L_1: 5x - y + 4 = 0$
$L_2: 3x - y + 5 = 0$
$L_3: x + y + 8 = 0$
If these lines enclose a triangle ABC and sum of the squares of the tangent to the interior angles can be expressed in the form p/q where p and q are relatively prime numbers, then the value of $p + q$ is

a. 500 **b.** 450
c. 230 **d.** 465

131. If the lines $ax + y + 1 = 0$, $x + by + 1 = 0$ and $x + y + c = 0$ (a, b, c being distinct and different from 1) are concurrent, then $\left(\frac{1}{1-a}\right)+\left(\frac{1}{1-b}\right)+\left(\frac{1}{1-c}\right)=$

a. 0 **b.** 1
c. $1/(a+b+c)$ **d.** none of these

132. If the pairs of lines $x^2 + 2xy + ay^2 = 0$ and $ax^2 + 2xy + y^2 = 0$ have exactly one line in common, then the joint equation of the other two lines is given by

a. $3x^2 + 8xy - 3y^2 = 0$ **b.** $3x^2 + 10xy + 3y^2 = 0$
c. $y^2 + 2xy - 3x^2 = 0$ **d.** $x^2 + 2xy - 3y^2 = 0$

133. If the origin is shifted to the point $(ab/(a - b), 0)$ without rotation, then the equation $(a - b)(x^2 + y^2) - 2abx = 0$ becomes

a. $(a - b)(x^2 + y^2) - (a + b)xy + abx = a^2$
b. $(a + b)(x^2 + y^2) = 2ab$
c. $(x^2 + y^2) = (a^2 + b^2)$
d. $(a - b)^2 (x^2 + y^2) = a^2b^2$

134. The straight lines represented by $(y - mx)^2 = a^2(1 + m^2)$ and $(y - nx)^2 = a^2(1 + n^2)$ form a

a. rectangle **b.** rhombus
c. trapezium **d.** none of these

135. The condition that one of the straight lines given by the equation $ax^2 + 2hxy + by^2 = 0$ may coincide with one of those given by the equation $a'x^2 + 2h'xy + b'y^2 = 0$ is

a. $(aB' - a'b)^2 = 4(ha' - h'a)(bh' - b'h)$
b. $(aB' - a'b)^2 = (ha' - h'a)(bh' - b'h)$
c. $(ha' - h'a)^2 = 4(aB' - a'b)(bh' - b'h)$
d. $(bh' - b'h)^2 = 4(aB' - a'b)(ha' - h'a)$

136. The angle between the pair of lines whose equation is $4x^2 + 10xy + my^2 + 5x + 10y = 0$ is

a. $\tan^{-1}(3/8)$
b. $\tan^{-1}(3/4)$
c. $\tan^{-1}(2\sqrt{25 - 4m}/m + 4)$, $m \in R$
d. none of these

137. If one of the lines of $my^2 + (1 - m^2)xy - mx^2 = 0$ is a bisector of the angle between the lines $xy = 0$, then m is

a. 3 **b.** 2 **c.** $-1/2$ **d.** -1

138. The pair of lines represented by $3ax^2 + 5xy + (a^2 - 2)y^2 = 0$ are perpendicular to each other for

a. two values of a **b.** a
c. for one value of a **d.** for no value of a

139. If two of the lines represented by $x^4 + x^3y + cx^2y^2 - xy^3 + y^4 = 0$ bisect the angle between the other two, then the value of c is

a. 0 **b.** -1 **c.** 1 **d.** -6

140. If the lines represented by the equation $3y^2 - x^2 + 2\sqrt{3}x - 3 = 0$ are rotated about the point $(\sqrt{3}, 0)$ through an angle 15°, one clockwise direction and other in anti clockwise direction, so that they become perpendicular, then the equation of the pair of lines in the new position is

a. $y^2 - x^2 + 2\sqrt{3}x + 3 = 0$ **b.** $y^2 - x^2 + 2\sqrt{3}x - 3 = 0$
c. $y^2 - x^2 - 2\sqrt{3}x + 3 = 0$ **d.** $y^2 - x^2 + 3 = 0$

141. If the equation of the pair of straight lines passing through the point (1, 1), one making an angle θ with the positive direction of the x-axis and the other making the same angle with the positive direction of the y-axis, is $x^2 - (a + 2)xy + y^2 + a(x + y - 1) = 0$, $a \neq -2$, then the value of $\sin 2\theta$ is

a. $a - 2$ **b.** $a + 2$
c. $2/(a + 2)$ **d.** $2/a$

142. Equation of a line which is parallel to the line common to the pair of lines given by $6x^2 - xy - 12y^2 = 0$ and $15x^2 + 14xy - 8y^2 = 0$ and at a distance 7 from it is

a. $3x - 4y = -35$ **b.** $5x - 2y = 7$
c. $3x + 4y = 35$ **d.** $2x - 3y = 7$

143. The equation $x^2y^2 - 9y^2 + 6x^2y + 54y = 0$ represents

a. a pair of straight lines and a circle
b. a pair of straight lines and a parabola
c. a set of four straight lines forming a square
d. none of these

144. The combined equation of the lines l_1, l_2 is $2x^2 + 6xy + y^2 = 0$ and that of the lines m_1, m_2 is $4x^2 + 18xy + y^2 = 0$. If the angle between l_1 and m_2 be α, then the angle between l_2 and m_1 will be

a. $\pi/2 - \alpha$ **b.** 2α
c. $\pi/4 + \alpha$ **d.** α

145. The equations $x - y = 4$ and $x^2 + 4xy + y^2 = 0$ represent the sides of

a. an equilateral triangle b. a right-angled triangle
c. an isosceles triangle d. none of these

146. The equations $a^2x^2 + 2h(a + b)xy + b^2y^2 = 0$ and $ax^2 + 2hxy + by^2 = 0$ represent

a. two pairs of perpendicular straight lines
b. two pairs of parallel straight lines
c. two pairs of straight lines which are equally inclined to each other
d. none of these

147. The distance between the two lines represented by the equation $9x^2 - 24xy + 16y^2 - 12x + 16y - 12 = 0$ is

a. 8/5 b. 6/5
c. 11/5 d. none of these

Multiple Correct Answers Type

Each question has four choices a, b, c and d, out of which *one or more* answers are correct.

1. If P is a point (x, y) on the line $y = -3x$ such that P and the point (3, 4) are on the opposite sides of the line $3x - 4y = 8$, then

a. $x > 8/15$ b. $x > 8/5$
c. $y < -8/5$ d. $y < -8/15$

2. If (–6, –4), (3, 5), (–2, 1) are the vertices of a parallelogram, then remaining vertex can be

a. (0, –1) b. (7, 9)
c. (–1, 0) d. (–11, –8)

3. The lines $x + 2y + 3 = 0$, $x + 2y - 7 = 0$ and $2x - y - 4 = 0$ are the sides of a square. Equation of the remaining side of the square can be

a. $2x - y + 6 = 0$ b. $2x - y + 8 = 0$
c. $2x - y - 10 = 0$ d. $2x - y - 14 = 0$

4. Let $0 \equiv (0, 0)$, $A \equiv (0, 4)$, $B \equiv (6, 0)$. P be a moving point such that the area of triangle POA is two times the area of triangle POB. Locus of P will be straight line whose equation can be

a. $x + 3y = 0$ b. $x + 2y = 0$
c. $2x - 3y = 0$ d. $3y - x = 0$

5. If (α, α^2) lies inside the triangle formed by the lines $2x + 3y - 1 = 0$, $x + 2y - 3 = 0$, $5x - 6y - 1 = 0$, then

a. $2\alpha + 3\alpha^2 - 1 > 0$ b. $\alpha + 2\alpha^2 - 3 < 0$
c. $\alpha + 2\alpha^2 - 3 < 0$ d. $6\alpha^2 - 5\alpha + 1 > 0$

6. Angles made with the x-axis by a straight line drawn through (1, 2) so that it intersects $x + y = 4$ at a distance $\sqrt{6}/3$ from (1, 2) are

a. 105° b. 75° c. 60° d. 15°

7. Given three straight lines $2x + 11y - 5 = 0$, $24x + 7y - 20 = 0$ and $4x - 3y - 2 = 0$. Then,

a. they form a triangle
b. they are concurrent
c. one line bisects the angle between the other two
d. two of them are parallel

8. If (–4, 0) and (1, –1) are two vertices of a triangle of area 4 sq. units, then its third vertex lies on

a. $y = x$ b. $5x + y + 12 = 0$
c. $x + 5y - 4 = 0$ d. $x + 5y + 12 = 0$

9. The equation $x^3 + x^2y - xy^2 = y^3$ represents

a. three real straight lines
b. lines in which two of them are perpendicular to each other
c. lines in which two of them are coincident
d. none of these

10. The straight line represented by $x^2 + mxy - 2y^2 + 3y - 1 = 0$ meets at

a. (– 1/3, 2/3) b. (– 1/3, – 2/3)
c. (1/3, 2/3) d. none of these

11. If m_1 and m_2 are the roots of the equation $x^2 - ax - a - 1 = 0$, then the area of the triangle formed by the three straight lines $y = m_1x$, $y = m_2x$ and $y = a$ $(a \neq -1)$ is

a. $\dfrac{a^2(a+2)}{2(a+1)}$ if $a > -1$

b. $\dfrac{-a^2(a+2)}{2(a+1)}$ if $a < -1$

c. $\dfrac{-a^2(a+2)}{2(a+1)}$ if $-2 < a - 1$

d. $\dfrac{a^2(a+2)}{2(a+1)}$ if $a < -2$

12. If the points $\left(a^3/(a-1), (a^2-3)/(a-1)\right)$, $\left(b^3/(b-1), (b^2-3)/(b-1)\right)$, and $\left(c^3/(c-1), (c^2-3)/(c-1)\right)$, where a, b, c are different from 1, lie on the line $lx + my + n = 0$, then

a. $a + b + c = -\dfrac{m}{l}$

b. $ab + bc + ca = \dfrac{n}{l}$

c. $abc = \dfrac{(m+n)}{l}$

d. $abc - (bc + ca + ab) + 3(a + b + c) = 0$

13. If one of the lines of $my^2 + (1 - m^2)xy - mx^2 = 0$ is a bisector of the angle between the lines $xy = 0$, then m is

a. 1 b. 2 c. – 1/2 d. – 1

14. If the chord $y = mx + 1$ of the circle $x^2 + y^2 = 1$ subtends an angle of measure 45° at the major segment of the circle, then the value of m is
a. 2 **b.** 1
c. –1 **d.** none of these

15. Two sides of a rhombus *OABC* (lying entirely in first quadrant or third quadrant) of area equal to 2 sq. units are $y = x/\sqrt{3}$, $y = \sqrt{3}\,x$. Then possible coordinates of *B* is/are (*O* being origin)
a. $(1+\sqrt{3},\ 1+\sqrt{3})$ **b.** $(-1-\sqrt{3}, -1-\sqrt{3})$
c. $(3+\sqrt{3}, 3+\sqrt{3})$ **d.** $(\sqrt{3}-1, \sqrt{3}-1)$

16. If $(x/a) + (y/b) = 1$ and $(x/c) + (y/d) = 1$ intersect the axes at four concyclic points and $a^2 + c^2 = b^2 + d^2$, then these lines can intersect at $(a, b, c, d > 0)$
a. (1, 1) **b.** (1, –1) **c.** (2, –2) **d.** (3, 3)

17. The straight line $3x + 4y - 12 = 0$ meets the coordinate axes at *A* and *B*. An equilateral triangle *ABC* is constructed. The possible coordinates of vertex '*C*' are
a. $\left(2\left(1-\frac{3\sqrt{3}}{4}\right), \frac{3}{2}\left(1-\frac{4}{\sqrt{3}}\right)\right)$
b. $(-2(1+\sqrt{3}), 3/2\,(1-\sqrt{3}))$
c. $(2(1+\sqrt{3}), 3/2\,(1+\sqrt{3}))$
d. $\left(2\left(1+\frac{3\sqrt{3}}{4}\right), \frac{3}{2}\left(1+\frac{4}{\sqrt{3}}\right)\right)$

18. The equation of the lines passing through the point (1, 0) and at a distance $\sqrt{3}/2$ from the origin are
a. $\sqrt{3}x + y - \sqrt{3} = 0$ **b.** $x + \sqrt{3}y - \sqrt{3} = 0$
c. $\sqrt{3}x - y - \sqrt{3} = 0$ **d.** $x - \sqrt{3}y - \sqrt{3} = 0$

19. The sides of a triangle are the straight lines $x + y = 1$, $7y = x$, and $\sqrt{3}y + x = 0$. Then which of the following is an interior point of the triangle?
a. circumcentre **b.** centroid
c. incentre **d.** orthocentre

In an obtused-angled triangle orthocentre and cicumcentre are exterior to the triangle.

20. If (x, y) be a variable point on the line $y = 2x$ lying between the lines $2(x + 1) + y = 0$ and $x + 3(y - 1) = 0$, then
a. $x \in (-1/2, 6/7)$ **b.** $x \in (-1/2, 3/7)$
c. $y \in (-1, 3/7)$ **d.** $y \in (-1, 6/7)$

21. The *x*-coordinates of the vertices of a square of unit area are the roots of the equation $x^2 - 3|x| + 2 = 0$ and the *y*-coordinates of the vertices are the roots of the equation $y^2 - 3y + 2 = 0$. Then the possible vertices of the square is/are
a. (1, 1), (2, 1), (2, 2), (1, 2)
b. (– 1, 1), (– 2, 1), (– 2, 2), (– 1, 2)
c. (2, 1), (1, – 1), (1, 2), (2, 2)
d. (– 2, 1), (– 1, – 1), (– 1, 2), (– 2, 2)

22. The area of triangle *ABC* is 20 cm². The coordinates of vertex *A* are (– 5, 0) and those of *B* are (3, 0). The vertex *C* lies on the line $x - y = 2$. The coordinates of *C* are
a. (5, 3) **b.** (– 3, – 5)
c. (– 5, – 7) **d.** (7, 5)

23. If the straight line $ax + cy = 2b$ where $a, b, c > 0$ makes a triangle of area 2 sq. units with coordinate axes, then
a. a, b, c are in G.P. **b.** $a, -b, c$ are in G.P.
c. $a, 2b, c$ are in G.P. **d.** $a, -2b, c$ are in G.P.

24. If $(a\cos\theta_1, a\sin\theta_1)$, $(a\cos\theta_2, a\sin\theta_2)$, and $(a\cos\theta_3, a\sin\theta_3)$ represents the vertices of an equilateral triangle inscribed in a circle, then
a. $\cos\theta_1 + \cos\theta_2 + \cos\theta_3 = 0$
b. $\sin\theta_1 + \sin\theta_2 + \sin\theta_3 = 0$
c. $\tan\theta_1 + \tan\theta_2 + \tan\theta_3 = 0$
d. $\cot\theta_1 + \cot\theta_2 + \cot\theta_3 = 0$

25. Consider the equation $y - y_1 = m(x - x_1)$. If m and x_1 are fixed and different lines are drawn for different values of y_1, then
a. the lines will pass through a fixed point
b. there will be a set of parallel lines
c. all the lines intersect the line $x = x_1$
d. all the lines will be parallel to the line $y = x_1$

26. The points *A*(0, 0), *B*(cos *α*, sin *α*), and *C*(cos *β*, sin *β*) are the vertices of a right-angled triangle if
a. $\sin\frac{\alpha-\beta}{2} = \frac{1}{\sqrt{2}}$ **b.** $\cos\frac{\alpha-\beta}{2} = -\frac{1}{\sqrt{2}}$
c. $\cos\frac{\alpha\ \beta}{2} = \frac{1}{\sqrt{2}}$ **d.** $\sin\frac{\alpha-\beta}{2} = -\frac{1}{\sqrt{2}}$

27. The ends of a diagonal of a square are (2, –3) and (–1, 1). Another vertex of the square can be
a. (–3/2, –5/2) **b.** (5/2, 1/2)
c. (1/2, 5/2) **d.** none of these

28. If each of the vertices of a triangle has integral coordinates, then the triangle may be
a. right angled **b.** equilateral
c. isosceles **d.** none of these

29. If $A(x_1, y_1)$, $B(x_2, y_2)$, $C(x_3, y_3)$ are the vertices of a triangle, then the equation

$$\begin{vmatrix} x & y & 1 \\ x_1 & y_1 & 1 \\ x_2 & y_2 & 1 \end{vmatrix} + \begin{vmatrix} x & y & 1 \\ x_1 & y_1 & 1 \\ x_3 & y_3 & 1 \end{vmatrix} = 0$$

represents

a. the median through A

b. the altitude through A

c. the perpendicular bisector of BC

d. the line joining the centroid with a vertex

30. In a $\Delta\, ABC$, $A \equiv (\alpha, \beta)$, $B \equiv (1, 2)$, $C \equiv (2, 3)$ and point A lies on the line $y = 2x + 3$ where $\alpha, \beta \in$ integer and area of the triangle is S such that $[S] = 2$, where $[\cdot]$ denotes the greatest integer function. Then all possible coordinates of A.

a. $(-7, -11)$ b. $(-6, -9)$

c. $(2, 7)$ d. $(3, 9)$

31. Let $P(\sin\theta, \cos\theta)$ $(0 \le \theta \le 2\pi)$ be a point in triangle with vertices $(0, 0)$, $(\sqrt{3/2}, 0)$ and $(0, \sqrt{3/2})$. Then,

a. $0 < \theta < \pi/12$ b. $5\pi/2 < \theta < \pi/2$

c. $0 < \theta < 5\pi/2$ d. $5\pi/2 < \theta < \pi$

32. Equation(s) of the straight line(s), inclined at 30° to the x-axis such that the length of its (each of their) line segment(s) between the coordinate axes is 10 units is are

a. $x + \sqrt{3}\,y + 5\sqrt{3} = 0$

b. $x - \sqrt{3}y + 5\sqrt{3} = 0$

c. $x + \sqrt{3}y - 5\sqrt{3} = 0$

d. $x - \sqrt{3}y - 5\sqrt{3} = 0$

33. The lines $x + y - 1 = 0$, $(m - 1)x + (m^2 - 7)y - 5 = 0$, and $(m - 2)x + (2m - 5)y = 0$ are

a. concurrent for three values of m

b. concurrent for one value of m

c. concurrent for no value of m

d. are parallel for $m = 3$

34. Equation of a straight line passing through the point $(2, 3)$ and inclined at an angle of $\tan^{-1}(1/2)$ with the line $y + 2x = 5$ is

a. $y = 3$ b. $x = 2$

c. $3x + 4y - 18 = 0$ d. $4x + 3y - 17 = 0$

35. The equation of the lines on which the perpendiculars from the origin make 30° angle with the x-axis and which form a triangle of area $50/\sqrt{3}$ with axes are

a. $\sqrt{3}x + y - 10 = 0$ b. $\sqrt{3}x + y + 10 = 0$

c. $x + \sqrt{3}y - 10 = 0$ d. $x - \sqrt{3}y - 10 = 0$

36. Sides of a rhombus are parallel to the lines $x + y - 1 = 0$ and $7x - y - 5 = 0$. It is given that diagonals of the rhombus intersect at $(1, 3)$ and one vertex A of the rhombus lies on the line $y = 2x$. Then the coordinates of the vertex A are

a. $(8/5, 16/5)$ b. $(7/15, 14/15)$

c. $(6/5, 12/5)$ d. $(4/15, 8/15)$

37. The equations of two equal sides AB and AC of an isosceles triangle ABC are $x + y = 5$ and $7x - y = 3$, respectively. Then the equations of the sides BC if $ar(\Delta ABC) = 5$ sq. units.

a. $x - 3y + 1 = 0$ b. $x - 3y - 21 = 0$

c. $3x + y + 2 = 0$ d. $3x + y - 12 = 0$

38. Two straight lines $u = 0$ and $v = 0$ pass through the origin and angle between them is $\tan^{-1}(7/9)$. If the ratio of the slope of $v = 0$ and $u = 0$ is $9/2$ then their equations are

a. $y + 3x = 0$ and $3y + 2x = 0$

b. $2y + 3x = 0$ and $3y + x = 0$

c. $2y = 3x$ and $3y = x$

d. $y = 3x$ and $3y = 2x$

39. Let $u \equiv ax + by + a\sqrt[3]{b} = 0$, $v \equiv bx - ay + b\sqrt[3]{a} = 0$, $a, b \in R$ be two straight lines. The equations of the bisectors of the angle formed by $k_1u - k_2v = 0$ and $k_1u + k_2v = 0$ for nonzero real k_1 and k_2 are

a. $u = 0$ b. $k_2u + k_1v = 0$

c. $k_2u - k_1v = 0$ d. $v = 0$

40. If $x^2 + 2hxy + y^2 = 0$ represents the equation of the straight lines through the origin which make an angle α with the straight line $y + x = 0$, then

a. $\sec 2\alpha = h$ b. $\cos\alpha = \sqrt{(1 + h)/(2h)}$

c. $2\sin\alpha = \sqrt{(1 + h)/h}$ d. $\cot\alpha = \sqrt{(h + 1)/(h - 1)}$

41. The combined equation of three sides of a triangle is $(x^2 - y^2)(2x + 3y - 6) = 0$. If $(-2, a)$ is an interior point and $(b, 1)$ is an exterior point of the triangle, then

a. $2 < a < 10/3$ b. $-2 < a < 10/3$

c. $-1 < b < 9/2$ d. $-1 < b < 1$

42. If one of the lines given by the equation $2x^2 + pxy + 3y^2 = 0$ coincide with one of those given by $2x^2 + qxy - 3y^2 = 0$ and the other lines represented by them be perpendicular, then

a. $p = 5$ b. $p = -5$ c. $q = -1$ d. $q = 1$

43. The lines joining the origin to the point of intersection of $3x^2 + mxy - 4x + 1 = 0$ and $2x + y - 1 = 0$ are at right angles. Then which of the following is not possible value of m?

a. -4 b. 4 c. 7 d. 3

Reasoning Type

Each question has four choices, a, b, c and d, out of which *only one* is correct. Each question contains Statement 1 and Statement 2.

a. Both the statements are True but Statement 2 is the correct explanation of Statement 1.

b. Both the statements are True but Statement 2 is not the correct explanation of Statement 1.

c. Statement 1 is True and Statement 2 is False.

d. Statement 1 is False and Statement 2 is True.

1. **Statement 1:** The lines $(a + b)x + (a - 2b)y = a$ are concurrent at the point (2/3, 1/3).
   Statement 2: The lines $x + y - 1 = 0$ and $x - 2y = 0$ intersect at the point (2/3, 1/3).
2. **Statement 1:** If the vertices of a triangle are having rational coordinates then its centroid, circumcentre, and orthocentre are rational.
   Statement 2: In any triangle, orthocentre, centroid, and circumcentre are collinear and centroid divides the line joining orthocentre and circumcentre in the ratio 2 : 1.
3. **Statement 1:** If sum of algebraic distances from points A (1, 1), B (2, 3), C (0, 2) is zero on the line $ax + by + c = 0$, then $a + 3b + c = 0$.
   Statement 2: The centroid of triangle is (1, 2).
4. **Statement 1:** Each point on the line $y - x + 12 = 0$ is equidistant from the lines $4y + 3x - 12 = 0$, $3y + 4x - 24 = 0$.
   Statement 2: The locus of a point which is equidistant from two given lines is the angular bisector of the two lines.
5. **Statement 1:** If the point $(2a - 5, a^2)$ is on the same side of the line $x + y - 3 = 0$ as that of the origin, then $a \in (2, 4)$.
   Statement 2: The points (x_1, y_1) and (x_2, y_2) lie on the same or opposite sides of the line $ax + by + c = 0$, as $ax_1 + by_1 + c$ and $ax_2 + by_2 + c$ have the same or opposite signs.
6. **Statement 1:** Lines passing through the given point and is equally inclined to the given two lines are always perpendicular.
   Statement 2: Angle bisectors of the given two lines are always perpendicular.
7. **Statement 1:** The area of the triangle formed by the points A (1000, 1002), B (1001, 1004) C (1002, 1003) is same as the area formed by A' (0, 0), B' (1, 2), C' (2, 1).
   Statement 2: The area of the triangle is constant with respect to translation of axes.
8. **Statement 1:** If the diagonals of the quadrilateral formed by the lines $px + qy + r = 0$, $p'x + q'y + r = 0$ are at right angles, then $p^2 + q^2 = p'^2 + q'^2$.
   Statement 2: Diagonals of a rhombus are bisected and perpendicular to each other.
9. **Statement 1:** The joint equation of lines $y = x$ and $y = -x$ is $y^2 = -x^2$, i.e., $x^2 + y^2 = 0$.
   Statement 2: The joint equation of lines $ax + by = 0$ and $cx + dy = 0$ is $(ax + by)(cx + dy) = 0$, where a, b, c, d are constant.
10. **Statement 1:** The internal angle bisector of angle C of a triangle ABC with sides AB, AC and BC as $y = 0$, $3x + 2y = 0$ and $2x + 3y + 6 = 0$, respectively, is $5x + 5y + 6 = 0$.
    Statement – 2: Image of point A with respect to $5x + 5y + 6 = 0$ lies on side BC of the triangle.
11. **Statement 1:** If the lines $2x + 3y + 19 = 0$ and $9x + 6y - 17 = 0$ cut the axis in A, B and y-axis at C, D, then points A, B, C, D are concyclic.
    Statement 2: Since $OA \times OB = OC \times OD$, where O is origin, therefore A, B, C, D are concyclic.
12. **Statement 1:** Let the vertices of a ΔABC are $A(-5, -2)$, $B(7, 6)$ and $C(5, -4)$. Then coordinates of circumcentre are (1, 2).
    Statement 2: In a right angle triangle, midpoint of hypotenuse is the circumcenter of the triangle.
13. **Statement 1:** The incentre of a triangle formed by the lines $x \cos(\pi/9) + y \sin(\pi/9) = \pi$, $x \cos(8\pi/9) + y \sin(8\pi/9) = \pi$, $x \cos(13\pi/9) + y \sin(13\pi/9) = \pi$ is (0, 0).
    Statement 2: Any point equidistant from the given three non-concurrent straight lines in the plane is incentre of the triangle.
14. **Statement 1:** If $(a_1x + b_1y + c_1) + (a_2x + b_2y + c_2) + (a_3x + b_3y + c_3) = 0$, then lines $a_1x + b_1y + c_1 = 0$, $a_2x + b_2y + c_2 = 0$ and $a_3x + b_3y + c_3 = 0$ cannot be parallel.
    Statement 2: If sum of the equations for three straight lines is identically zero, then they are either concurrent or parallel.
15. **Statement 1:** The lines $(a + b)x + (a - b)y - 2ab = 0$, $(a - b)x + (a + b)y - 2ab = 0$, and $x + y = 0$ form an isosceles triangle.
    Statement 2: If internal bisector of any angle of triangle is perpendicular to the opposite side, then the given triangle is isosceles.
16. **Statement 1:** If $-2h = a + b$, then one line of the pair of lines $ax^2 + 2hxy + by^2 = 0$ bisects the angle between coordinate axes in positive quadrant.
    Statement 2: If $ax + y(2h + a) = 0$ is a factor of $ax^2 + 2hxy + by^2 = 0$, then $b + 2h + a = 0$.

Linked Comprehension Type

Based upon each paragraph, some multiple choice questions have to be answered. Each question has four choices a, b, c and d, out of which *only one* is correct.

For Problems 1–3

Let L be the line belonging to the family of the straight lines $(a+2b)x+(a-3b)y+a-8b=0$, $a, b \in R$, which is farthest from the point (2, 2).

1. The equation of line L is

   a. $x+4y+7=0$ b. $2x+3y+4=0$
   c. $4x-y-6=0$ d. none of these

2. Area formed by the line L with coordinate axis is

   a. 4/3 sq. units b. 9/2 sq. units
   c. 49/8 sq. units d. none of these

3. If L is concurrent with the lines $x-2y+1=0$ and $3x-4y+\lambda=0$, then the value of λ is

   a. 2 b. 1 c. -4 d. 5

For Problems 4–6

The equation of an altitude of an equilateral triangle is $\sqrt{3}x+y=2\sqrt{3}$ and one of the vertices is $(3,\sqrt{3})$.

4. The possible number of triangles is

   a. 1 b. 2 c. 3 d. 4

5. Which of the following is not one of the possible vertices the triangle?

   a. (0, 0) b. $(0, 2\sqrt{3})$
   c. $(3, -\sqrt{3})$ d. none of these

6. Which of the following is one of the orthocentres of the triangle?

   a. $(1, \sqrt{3})$ b. $(0, \sqrt{3})$
   c. (0, 2) d. none of these

For Problems 7–9

For points $P=(x_1, y_1)$ and $Q=(x_2, y_2)$ of the coordinate plane, a new distance $d(P, Q)$ is defined by $d(P, Q)=|x_1-x_2|+|y_1-y_2|$. Let $O=(0, 0)$ and $A=(3, 2)$. Consider the set of points P in the first quadrant which are equidistant (with respect to the new distance) from O and A.

7. The set of points P consists of

   a. one straight line only
   b. union of two line segments
   c. union of two infinite rays
   d. union of a line segment of finite length and an infinite ray

8. The area of the region bounded by locus of P and line $y=4$ in first quadrant is

   a. 2 sq. units b. 4 sq. units
   c. 6 sq. units d. none of these

9. The locus of points P is

   a. one-one and onto function
   b. many-one and onto function
   c. one-one and into function
   d. relation but not function

For Problems 10–12

A variable line L is drawn through $O(0, 0)$ to meet the lines L_1 and L_2 given by $y-x-10=0$ and $y-x-20=0$ at the points A and B, respectively.

10. A point P is taken on L such that $2/OP=(1/OA)+(1/OB)$. Then the locus of P is

    a. $3x+3y=40$ b. $3x+3y+40=0$
    c. $3x-3y=40$ d. $3y-3x=40$

11. Locus of P, if $OP^2=OA \times OB$, is

    a. $(y-x)^2=100$ b. $(y+x)^2=50$
    c. $(y-x)^2=200$ d. none of these

12. Locus of P, if $(1/OP^2)=(1/OB^2)+(1/OA^2)$, is

    a. $(y-x)^2=80$ b. $(y-x)^2=100$
    c. $(y-x)^2=64$ d. none of these

For Problems 13–15

The line $6x+8y=48$ intersects the coordinate axes at A and B, respectively. A line L bisects the area and the perimeter of the triangle OAB where O is the origin.

13. The number of such lines possible is

    a. 1 b. 2
    c. 3 d. more than 3

14. The slope of the line L can be

    a. $(10+5\sqrt{6})/10$ b. $(10-5\sqrt{6})/10$
    c. $(8+3\sqrt{6})/10$ d. none of these

15. The line L

    a. does not intersect AB
    b. does not intersect OB
    c. does not intersect OA
    d. can intersect all the sides

For Problems 16–18

$A(1, 3)$ and $C(-2/5, -2/5)$ are the vertices of a triangle ABC and the equation of the internal angle bisector of $\angle ABC$ is $x+y=2$.

16. Equation of side BC is

    a. $7x+3y-4=0$ b. $7x+3y+4=0$
    c. $7x-3y+4=0$ d. $7x-3y-4=0$

17. Coordinates of vertex B are

a. (3/10, 17/10) **b.** (17/10, 3/10)
c. (–5/2, 9/2) **d.** (1, 1)

18. Equation of side AB is

a. $3x + 7y = 24$ **b.** $3x + 7y + 24 = 0$
c. $13x + 7y + 8 = 0$ **d.** $13x - 7y + 8 = 0$

For Problems 19–21

Let $ABCD$ be a parallelogram whose equations for the diagonals AC and BD are $x + 2y = 3$ and $2x + y = 3$, respectively. If length of diagonal $AC = 4$ units and area of parallelogram $ABCD = 8$ sq. units, then

19. The length of other diagonal BD is

a. 10/3 **b.** 2
c. 20/3 **d.** none of these

20. The length of side AB is equal to

a. $2\sqrt{58}/3$ **b.** $4\sqrt{58}/9$
c. $3\sqrt{58}/9$ **d.** $4\sqrt{58}/9$

21. The length of BC is equal to

a. $2\sqrt{10}/3$ **b.** $4\sqrt{10}/3$
c. $8\sqrt{10}/3$ **d.** none of these

For Problems 22–24

Consider a triangle PQR with coordinates of its vertices as $P(-8, 5)$; $Q(-15, -19)$ and $R(1, -7)$. The bisector of the interior angle of P has the equation which can be written in the form $ax + 2y + c = 0$.

22. The distance between the orthocentre and the circumcentre of the triangle PQR is

a. 25/2 **b.** 29/2
c. 37/2 **d.** 51/2

23. Radius of the incircle of the triangle PQR is

a. 4 **b.** 5
c. 6 **d.** 8

24. The sum of $a + c$ is

a. 129 **b.** 78
c. 89 **d.** none of these

For Problems 25–27

Let us consider the situation when axes are inclined at an angle ω. If coordinates of a point P are (x_1, y_1), then $PN = x_1$, $PM = y_1$, where PM is parallel to y-axis and PN is parallel x-axis. Straight line through P that makes an angle θ with x-axis is

$$RQ = y - y_1,\ PQ = x - x_1$$

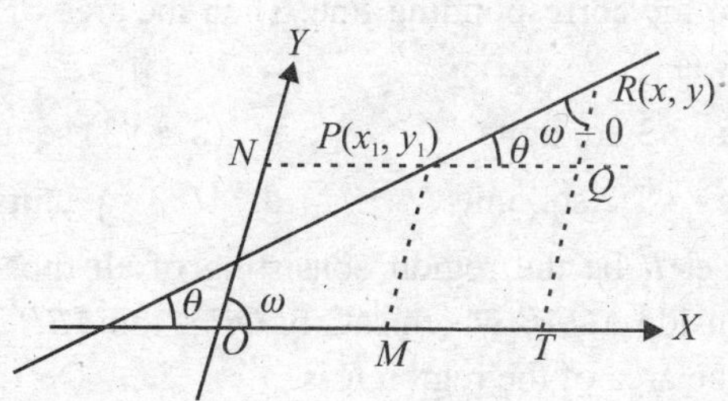

Fig. 1.1

From ΔPQR, we have

$$\frac{PQ}{\sin(\omega - \theta)} = \frac{RQ}{\sin\theta}$$

$$\Rightarrow y - y_1 = \frac{\sin\theta}{\sin(\omega - \theta)}(x - x_1)$$

written in the form

$y - y_1 = m(x - x_1)$, where

$$m = \frac{\sin\theta}{\sin(\omega - \theta)}$$

Therefore, if slope of the line is m, then angle of inclination of the line with the x-axis is given by

$$\tan\theta = \left(\frac{m\sin\omega}{1 + m\cos\omega}\right)$$

25. The axes being inclined at an angle of 60°, the inclination of the straight line $y = 2x + 5$ with the x-axis is

a. 30° **b.** $\tan^{-1}(\sqrt{3}/2)$
c. $\tan^{-1}2$ **d.** 60°

26. The axes being inclined at an angle of 60°, the angle between the two straight lines $y = 2x + 5$ and $2y + x + 7 = 0$ is

a. 90° **b.** $\tan^{-1}(5/3)$
c. $\tan^{-1}(\sqrt{3}/2)$ **d.** $\tan^{-1}(5/\sqrt{3})$

27. The axes being inclined at an angle of 30°, the equation of straight line which makes an angle of 60° with the positive direction of x-axis and x-intercept 2 is

a. $y - \sqrt{3}x = 0$ **b.** $\sqrt{3}y = x$
c. $y + \sqrt{3}x = 2\sqrt{3}$ **d.** $y + 2x = 0$

For Problems 28–29

Consider the triangle having vertices O (0, 0), A (2, 0) and $B(1, \sqrt{3})$. Also $b \le \min\{a_1, a_2, a_3, \ldots, a_n\}$ means $b \le a_1$ when a_1 is least; $b \le a_2$ when a_2 is least and so on. From this we can say $b \le a_1, b \le a_2, \ldots, b \le a_n$.

28. Let R be the region consisting of all those points P inside ΔOAB which satisfy $d(P, OA) \le \min[d(P, OB), d(P, AB)]$, where d denotes the distance from the point to the corresponding line. Then the area of the region R is

a. $\sqrt{3}$ sq. units b. $(2+\sqrt{3})$ sq. units

c. $\sqrt{3}/2$ sq. units d. $1/\sqrt{3}$ sq. units

29. Let R be the region consisting of all those points P inside ΔOAB which satisfy $OP \le \min[BP, AP]$. Then the area of the region R is

a. $\sqrt{3}$ sq. units b. $1/(2\sqrt{3})$ sq. units

c. $\sqrt{3}/2$ sq. units d. none of these

For Problems 30–32

Let $ABCD$ is a square with sides of unit length. Points E and F are taken on sides AB and AD respectively so that $AE = AF$. Let P be a point inside the square $ABCD$.

30. The maximum possible area of quadrilateral $CDFE$ is

a. 1/8 b. 1/4

c. 5/8 d. 3/8

31. The value of $(PA)^2 - (PB)^2 + (PC)^2 - (PD)^2$ is

a. 3 b. 2

c. 1 d. 0

32. Let a line passing through point A divides the square $ABCD$ into two parts so that area of one portion is double the other, then the length of portion of line inside the square is

a. $\sqrt{10}/3$ b. $\sqrt{13}/3$

c. $\sqrt{11}/3$ d. $2/\sqrt{3}$

Matrix-Match Type

Each question contains statements given in two columns which have to be matched.

Statements (a, b, c, d) in column I have to be matched with statements (p, q, r, s) in column II. If the correct matches are a → p, a → s; b → q, b → r; c → p, c → q; and d → s; then the correctly bubbled 4 × 4 matrix should be as follows:

	p	q	r	s
a	(p) filled	(q)	(r)	(s) filled
b	(p)	(q) filled	(r) filled	(s)
c	(p) filled	(q) filled	(r)	(s)
d	(p)	(q)	(r)	(s) filled

1. Consider the lines represented by equation $(x^2 + xy - x)(x - y) = 0$, forming a triangle. Then match the following.

Column I	Column II
a. Orthocentre of triangle	**p.** (1/6, 1/2)
b. Circumcentre	**q.** $(1/(2+2\sqrt{2}), 1/2)$
c. Centroid	**r.** (0, 1/2)
d. Incentre	**s.** (1/2, 1/2)

2. Consider the triangle formed by the lines

$y + 3x + 2 = 0$, $3y - 2x - 5 = 0$, $4y + x - 14 = 0$

Column I	Column II
a. Values of α if $(0, \alpha)$ lies inside triangle	**p.** $(-\infty, 7/3) \cup (13/4, \infty)$
b. Values of α if $(\alpha, 0)$ lies inside triangle	**q.** $-4/3 < \alpha < 1/2$
c. Values of α if $(\alpha, 2)$ lies inside triangle	**r.** No value of α
d. Value of α if $(1, \alpha)$ lies outside triangle	**s.** $5/3 < \alpha < 7/2$

3.

Column I	Column II
a. A straight line with negative slope passing through (1, 4) meets the coordinate axes at A and B. The minimum length of $OA + OB$, O being the origin, is	**p.** $5\sqrt{2}$
b. If the point P is symmetric to the point $Q(4, -1)$ with respect to the bisector of the first quadrant, then the length of PQ is	**q.** $3\sqrt{2}$
c. On the portion of the straight line $x + y = 2$ between the axis a square is constructed away from the origin, with this portion as one of its sides. If 'd' denotes the perpendicular distance of a side of this square from the origin then the maximum value of 'd' is	**r.** 9/2
d. If the parametric equation of a line is given by $x = 4 + (\lambda/\sqrt{2})$ and $y = -1 + \sqrt{2}\lambda$ where λ is a parameter, then the intercept made by the line on the x-axis is	**s.** 9

4.

Column I	Column II
a. If lines $3x + y - 4 = 0$, $x - 2y - 6 = 0$ and $\lambda x + 4y + \lambda^2 = 0$ are concurrent, then value of λ is	**p.** -4
b. If the points $(\lambda + 1, 1)$, $(2\lambda + 1, 3)$, and $(2\lambda + 2, 2\lambda)$ are collinear, then the value of λ is	**q.** $-1/2$

c. If line $x + y - 1 - \|\lambda/2\| = 0$, passing through the intersection of $x - y + 1 = 0$ and $3x + y - 5 = 0$, is perpendicular to one of them, then the value of λ is	**r.** 4
d. If line $y - x - 1 + \lambda = 0$ is equidistant from the points (1, −2) and (3, 4), then λ is	**s.** 2

5.

Column I	Column II
a. Two vertices of a triangle are (5, −1) and (−2, 3). If orthocentre is the origin, then coordinates of the third vertex are	**p.** (−4, −7)
b. A point on the line $x + y = 4$ which lies at a unit distance from the line $4x + 3y = 10$ is	**q.** (−7, 11)
c. Orthocentre of the triangle formed by the lines $x + y - 1 = 0$, $x - y + 3 = 0$, $2x + y = 7$ is	**r.** (2, −2)
d. If $2a$, b, c are in A.P., then lines $ax + by = c$ are concurrent at	**s.** (−1, 2)

6.

Column I	Column II
a. Four lines $x + 3y - 10 = 0$, $x + 3y - 20 = 0$, $3x - y + 5 = 0$ and $3x - y - 5 = 0$ form a figure which is	**p.** a quadrilateral which is neither a parallelogram nor a trapezium
b. The points $A(1, 2)$, $B(2, -3)$, $C(-1, -5)$ and $D(-2, 4)$ in order are the vertices of	**q.** a parallelogram
c. The lines $7x + 3y - 33 = 0$, $3x - 7y + 19 = 0$, $3x - 7y - 10$ and $7x + 3y - 4 = 0$ form a figure which is	**r.** a rectangle of area 10 sq. units
d. Four lines $4y - 3x - 7 = 0$, $3y - 4x + 7 = 0$, $4y - 3x - 21 = 0$, $3y - 4x + 14 = 0$ form a figure which is	**s.** a square

7.

Column I	Column II
a. The distance between the lines $(x + 7y)^2 + 4\sqrt{2}\,(x + 7y) - 42 = 0$ is	**p.** 2
b. If the sum of the distances of a point from two perpendicular lines in a plane is 1, then its locus is $\|x\| + \|y\| = k$, where k is equal to	**q.** 7
c. If $6x + 6y + m = 0$ is acute angle bisector of line $x + 2y + 4 = 0$ and $4x + 2y - 1 = 0$, then m is equal to	**r.** 3
d. Area of the triangle formed by the lines $y^2 - 9xy + 18x^2 = 0$ and $y = 6$ is	**s.** 1

8.

Column I	Column II
a. The value k for which $4x^2 + 8xy + ky^2 = 9$ is the equation of a pair of straight lines is	**p.** 3
b. If the sum of the slopes of the lines given by $x^2 - 2cxy - 7y^2 = 0$ is four times their product, then the value of c is	**q.** −3
c. If the gradient of one of the lines $x^2 + hxy + 2y^2 = 0$ is twice that of the other, then $h =$	**r.** 2
d. If the lines $ax^2 + 2hxy + by^2 = 0$ are equally inclined to the lines $ax^2 + 2hxy + by^2 + \lambda(x^2 + y^2) = 0$, then the value of λ can be	**s.** 4

9. O is origin and B is a point on the x-axis at a distance of 2 units from the origin.

Column I	Column II
a. If ΔAOB is equilateral triangle, then the coordinates of A can be	**p.** $(-1, \sqrt{3})$
b. If ΔAOB is isosceles such that $\angle OAB$ is 30°, then coordinate of A can be	**q.** $(-1, 2 - \sqrt{3})$
c. If OB is one side of rhombus of area $\sqrt{3}$ units, then other vertices of rhombus can be	**r.** $(-3, -\sqrt{3})$
d. If OB is a chord of circle with radius equal to OB, then coordinates of point A on the circumference of the circle such that ΔOAB is isosceles can be	**s.** $(1, 2 + \sqrt{3})$

10.

Column I	Column II
a. The lines $y = 0$, $y = 1$, $x - 6y + 4 = 0$, and $x + 6y - 9 = 0$ constitute a figure which is	**p.** a cyclic quadratieral
b. The points $A(a, 0)$, $B(0, b)$, $C(c, 0)$, and $D(0, d)$ are such that $ac = bd$ and a, b, c, d are all non-zero. The points A, B, C and D always constitute	**q.** a rhombus

c.	The figure formed by the four lines $ax \pm by \pm c = 0$ $(a \neq b)$ is	r. a square
d.	The line pairs $x^2 - 8x + 12 = 0$ and $y^2 - 14y + 45 = 0$ constitue a figure which is	s. a trapezium

Integer Type

1. $P(3, 1)$, $Q(6, 5)$, and $R(x, y)$ are three points such that the angle PRQ is a right angle and the area of ΔRQP is 7. Then the number of such points R is ________.
2. If $\lambda : 1$ is the ratio in which the line joining the points (2, 3) and (4, 1) divides the segment joining the points (1, 2) and (4, 3), then the value of λ is ________.
3. The number of integral values of a for which the point $P(a^2, a)$ lies in the region corresponding to the acute angle between the lines $2y = x$ and $4y = x$ is ________.
4. The value of λ if $2x^2 + 7xy + 3y^2 + 8x + 14y + \lambda = 0$ represents a pair of straight lines is ________.
5. Let D, E, F be three points on the sides BC, AC, and AB of a triangle ABC such that AD, BE, and CF are concurrent. If $BD \times CE \times AF = k\, DC \times EA \times FB$, then the value of k is ________.
6. Consider two lines $L_1: x - y = 0$ and $L_2: x + y = 0$ and a moving point $P(x, y)$. Let $d(P, L_i)$, $i = 1, 2$, represents the distance of point P from the line L_i. The point P moves in a certain region R in such a way that $2 \le d(P, L_1) + d(P, L_2) \le 4$. If the area is A sq. units, then the value of $A/6$ is ________.
7. The number of integral values of m for which the x-coordinate of the point of intersection of the lines $3x + 4y = 9$ and $y = mx + 1$ is also an integer is ________.
8. One of the diameters of the circle circumscribing the rectangle $ABCD$ is $4y = x + 7$. If A and B are the points (–3, 4) and (5, 4), respectively, and the area of the rectangle is K sq. units, then the value of $K/4$ is ________.
9. All chords of the curve $3x^2 - y^2 - 2x + 4y = 0$, which subtend a right angle at the origin, pass through a fixed point. The square of its distance from the origin is ________.
10. A rectangular billiard table has vertices at $P(0, 0)$, $Q(0, 7)$, $R(10, 7)$, and $S(10, 0)$. A small billiard ball starts at $M(3, 4)$ and moves in a straight line to the top of the table, bounces to the right side of the table, then comes to rest at $N(7, 1)$. If the y-coordinate of the point where it hits the right side is p, then the value of $[p]$, where [.] denotes the greatest integer function, is ________.
11. A straight line L with negative slope passes through the point (8, 2) and cuts the positive coordinate axes at the points P and Q. As L varies, the absolute minimum value of $(OP + OQ)/2$ (O is origin) is ________.
12. If the lines $x = a + m$, $y = -2$, and $y = mx$ are concurrent, then the least value of a^2 is ________.
13. The number or integral values of b for which the origin and the point (1, 1) lie on the same side of the straight line $a^2x\ aby + 1 = 0$, for all $a \in R - \{0\}$, is ________.
14. The line joining $A(b \cos \alpha, b \sin \alpha)$ and $B(a \cos \beta, a \sin \beta)$ is produced to the point $M(x, y)$ so that $AM : MB = b : a$, then the value of $x \cos \dfrac{\alpha+\beta}{2} + y \sin \dfrac{\alpha+\beta}{2}$ is ________.
15. The line L has intercepts a and b on the coordinate axes. The coordinate axes are rotated through a fixed angle, keeping the origin fixed. If p and q are the intercepts of the line L on the new axes, then the value of $\dfrac{1/a^2 + 1/b^2}{1/p^2 + 1/q^2}$ is ________.

ANSWERS AND SOLUTIONS

Objective Type

1. a. Coordinates of the vertices of the square are $A(0, 0)$, $B(0, 1)$, $C(1, 1)$, and $D(1, 0)$.

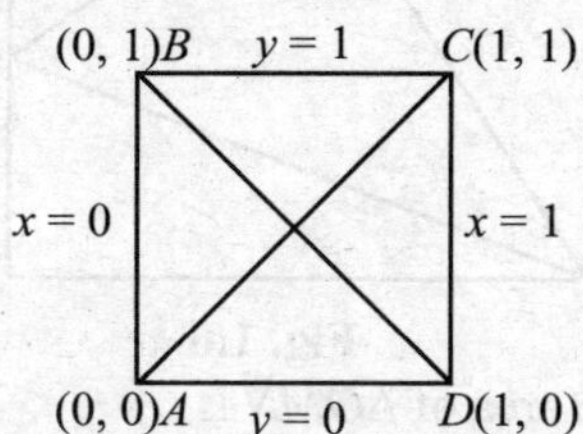

Fig. 1.2

Now the equation of AC is

$$y = x$$

and that of BD is

$$y - 1 = -\frac{1}{1}(x - 0)$$

$$\Rightarrow \quad x + y = 1$$

2. b. The equation of the line joining the points (2, –1) and (5, –3) is given by

$$y + 1 = \frac{-1+3}{2-5}(x - 2) \text{ or}$$

$$2x + 3y - 1 = 0 \qquad (1)$$

Since $(x_1, 4)$ and $(-2, y_1)$ lie on $2x + 3y - 1 = 0$, therefore

$$2x_1 + 12 - 1 = 0 \Rightarrow x_1 = -\frac{11}{2}$$

and

$$-4 + 3y_1 - 1 = 0 \Rightarrow y_1 = \frac{5}{3}$$

Thus, (x_1, y_1) satisfies $2x + 6y + 1 = 0$.

3. a. We have,

$$\begin{vmatrix} x_1 & y_1 & 1 \\ x_2 & y_2 & 1 \\ x_3 & y_3 & 1 \end{vmatrix} = \begin{vmatrix} a_1 & b_1 & 1 \\ a_2 & b_2 & 1 \\ a_3 & b_3 & 1 \end{vmatrix}$$

$$\Rightarrow \frac{1}{2}\begin{vmatrix} x_1 & y_1 & 1 \\ x_2 & y_2 & 1 \\ x_3 & y_3 & 1 \end{vmatrix} = \frac{1}{2}\begin{vmatrix} a_1 & b_1 & 1 \\ a_2 & b_2 & 1 \\ a_3 & b_3 & 1 \end{vmatrix}$$

Hence, the area of triangle with vertices (x_1, y_1), (x_2, y_2), (x_3, y_3) is same as the area of triangle with vertices (a_1, b_1), (a_2, b_2), (a_3, b_3). Hence, the two triangles are equal in area.

4. b. The given lines

$$ax \pm by \pm c = 0$$

$$\Rightarrow \frac{x}{\pm c/a} + \frac{y}{\pm c/b} = 1$$

the vertex at A $(c/a, 0)$, C $(-c/a, 0)$, B $(0, c/b)$, D $(0, -c/b)$. Therefore, the diagonals AC and BD of quadrilateral $ABCD$ are perpendicular. Hence, it is a rhombus whose area is given by

$$\frac{1}{2} \times AC \times BD = \frac{1}{2} \times \frac{2c}{a} \times \frac{2c}{b} = \frac{2c^2}{ab}$$

5. d. If the line meets the x-and y-axes at A and B, then $A \equiv (-c/a, 0)$, $B \equiv (0, -c/b)$. The line will pass through the first quadrant if

$$-c/a > 0 \text{ and/or } -c/b > 0$$

$$\Rightarrow ac < 0 \text{ and/or } bc < 0$$

6. b. $(x_1 + t(x_2 - x_1), y_1 + t(y_2 - y_1)) = (x_1(1 - t) + tx_2, y_1(1 - t) + ty_2)$ is the point which divides the join of (x_1, y_1) and (x_2, y_2) in the ratio $t : (1 - t)$ which is positive if $0 < t < 1$.

7. c. The coordinates of A and B are as shown in the figure.

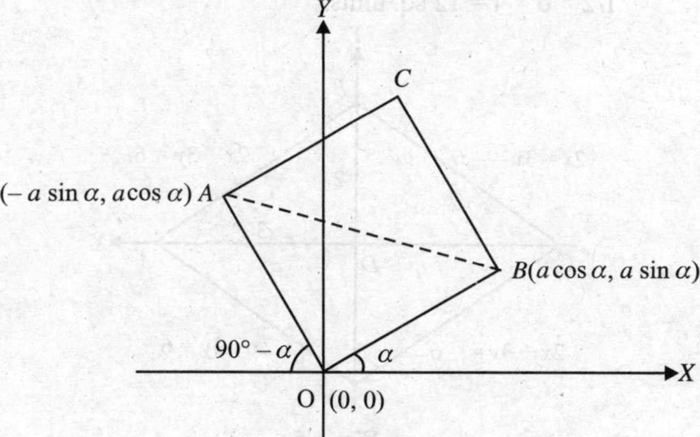

Fig. 1.3

The equation of the diagonal AB is

$$y - a\sin\alpha = \frac{a\cos\alpha - a\sin\alpha}{-a\sin\alpha - a\cos\alpha}(x - a\cos\alpha)$$

$$\Rightarrow y(\cos\alpha + \sin\alpha) - a(\sin\alpha\cos\alpha + \sin^2\alpha)$$

$$= -(\cos\alpha - \sin\alpha)x + a\cos\alpha(\cos\alpha - \sin\alpha)$$

$$\Rightarrow \quad y(\cos\alpha + \sin\alpha) + x(\cos\alpha - \sin\alpha) = a.$$

8. a. Let the two perpendicular lines be taken as the coordinate axes. If (h, k) be any point on the locus, then according to the given condition $|h| + |k| = 1$. Hence, the locus of (h, k) is $|x| + |y| = 1$. This consists of four line segments enclosing a square as shown in the figure below.

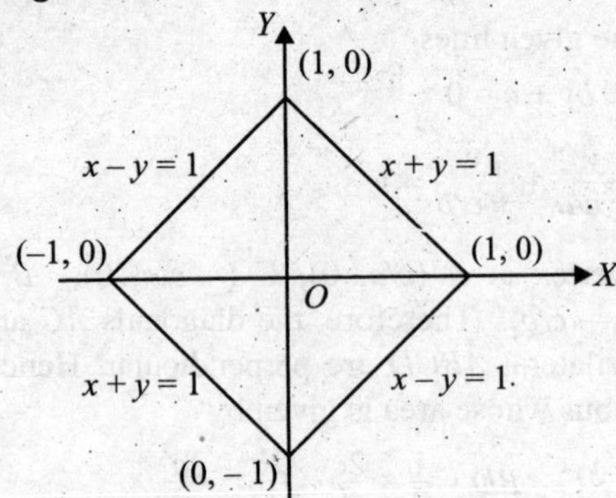

Fig. 1.4

9. c. The given inequality is equivalent to the following system of inequalities.

$2x + 3y \le 6$, when $x \ge 0, y \ge 0$

$2x - 3y \le 6$, when $x \ge 0, y \le 0$

$-2x + 3y \le 6$, when $x \le 0, y \ge 0$

$-2x - 3y \le 6$, when $x \le 0, y \le 0$

which represents a rhombus with sides

$2x \pm 3y = 6$ and $2x \pm 3y = -6$

Length of the diagonals is 6 and 4 units along x- and y-axes. Therefore, the required area is

$1/2 \times 6 \times 4 = 12$ sq. units.

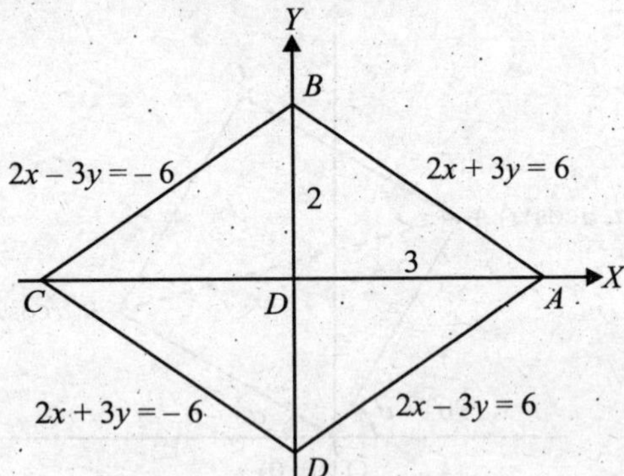

Fig. 1.5

10. d. Given $O(0, 0)$ is the orthocentre. Let $A(h, k)$ be the third vertex, $B(-2, 3)$, and $C(5, -1)$ the other two vertices. Then the slope of the line through A and O is k/h, while the lines through B and C has the slope $-4/7$. By the property of the orthocentre, these two lines must be perpendicular, so we have

$$(k/h)\left(-\frac{4}{7}\right) = -1 \Rightarrow k/h = \frac{7}{4} \quad (1)$$

Also,

$$\frac{5 - 2 + h}{3} + \frac{-1 + 3 + k}{3} = 7$$

$$\Rightarrow h + k = 16 \quad (2)$$

which is not satisfied by the points given in (*a*), (*b*), or (*c*).

11. c. Let the coordinates of vertices O, P, Q, R be $(0, 0)$, $(a, 0)$, (a, a), $(0, a)$, respectively. Then, we get the coordinates of M as $(a, a/2)$ and those of N as $(a/2, a)$.

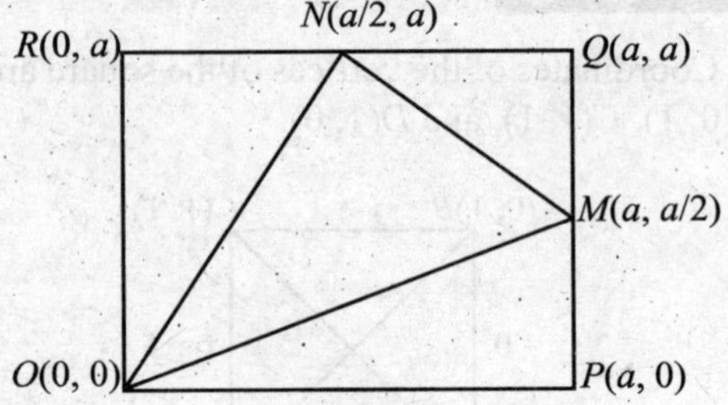

Fig. 1.6

Therefore, area of ΔOMN is

$$\frac{1}{2}\begin{vmatrix} 0 & 0 & 1 \\ a & a/2 & 1 \\ a/2 & a & 1 \end{vmatrix} = \frac{3a^2}{8}$$

Area of the square is a^2. Hence, the required ratio is 8 : 3.

12. b. p, q, r are the roots of equation $y^3 - 3y^2 + 6y + 1 = 0$. So, $p + q + r = 3$, $pq + qr + rp = 6$ and $pqr = -1$. Now, the centroid of the triangle is

$$\left(\frac{pq + qr + rp}{3}, \frac{\frac{1}{pq} + \frac{1}{qr} + \frac{1}{rp}}{3}\right) \text{ i.e.,}$$

$$\left(\frac{pq + qr + rp}{3}, \frac{p + q + r}{3pqr}\right) = \left(\frac{6}{3}, \frac{3}{-3}\right) \text{ or } (2, -1)$$

13. a.

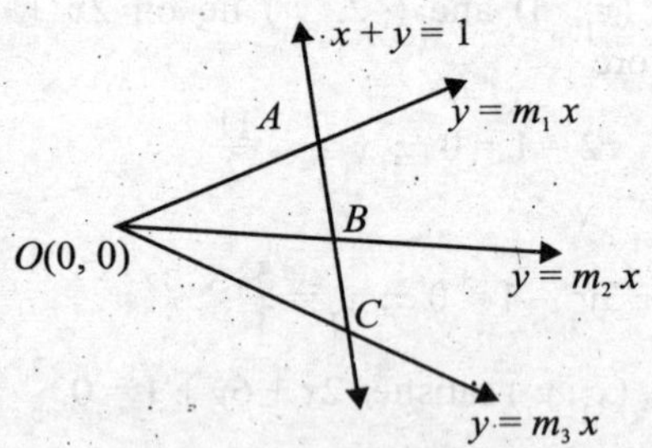

Fig. 1.7

Solving the equations of the lines, we get

$$A = \left(\frac{1}{1 + m_1}, \frac{m_1}{1 + m_1}\right), C = \left(\frac{1}{1 + m_3}, \frac{m_3}{1 + m_3}\right)$$

If $AB = BC$, then midpoint of AC lies on $y = m_2 x$

$$\Rightarrow \frac{\frac{m_1}{1 + m_1} + \frac{m_3}{1 + m_3}}{2} = m_2\left[\frac{\frac{1}{1 + m_1} + \frac{1}{1 + m_3}}{2}\right]$$

14. b. Lines $ax + y = 0$ and $x + by = 0$ intersect at $O(0, 0)$.

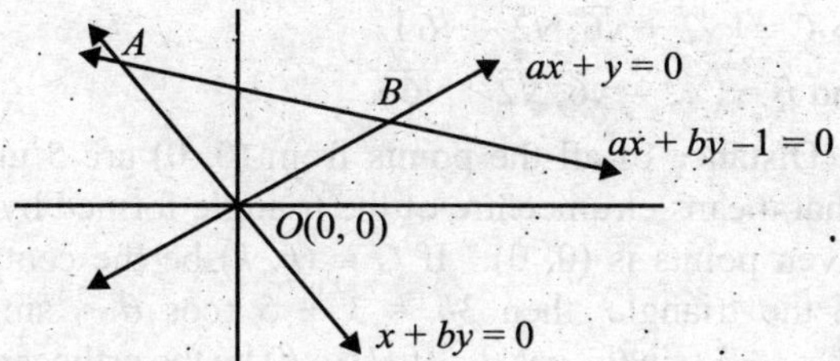

Fig. 1.8

Hence, if AB subtends right angle at $O(0, 0)$, then $ax + y = 0$ and $x + by = 0$ are perpendicular to each other. So,

$$(-a)\left(-\frac{1}{b}\right) = -1$$

$$\Rightarrow a + b = 0$$

15. c.

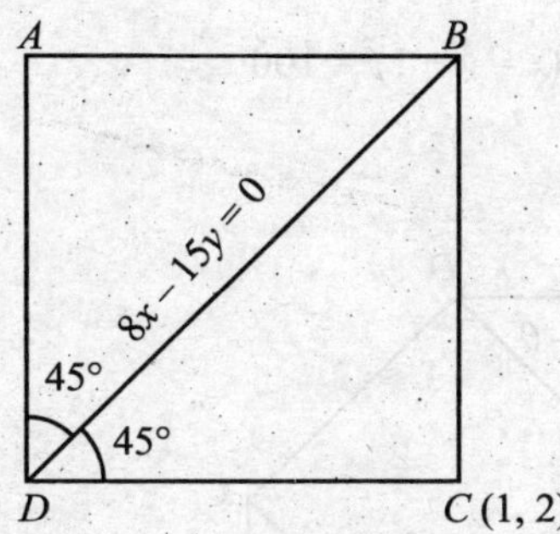

Fig. 1.9

Slope of BD is 8/15 and angle made by BD with DC and BC is 45°. So let slope of DC be m. Then,

$$\tan 45° = \pm \frac{m - \frac{8}{15}}{1 + \frac{8}{15}m}$$

$$\Rightarrow (15 + 8m) = \pm (15m - 8)$$

$$\Rightarrow m = \frac{23}{7} \text{ and } -\frac{7}{23}$$

Hence, the equations of DC and BC are

$$y - 2 = \frac{23}{7}(x - 1)$$

$$\Rightarrow 23x - 7y - 9 = 0$$

and

$$y - 2 = -\frac{7}{23}(x - 1)$$

$$\Rightarrow 7x + 23y - 53 = 0$$

16. a. Since $a_1/a_2 = b_1/b_2 = c_1/c_2$, then $u = 0$ and $v = 0$ are the same straight line. Hence, $u + kv = 0$ is also the same straight line.

17. c. Since the point $A(2, 1)$ is translated parallel to $x - y = 3$, therefore AA' has the same slope as that of $x - y = 3$. Therefore, AA' passes through (2, 1) and has the slope of 1. Here,

$\tan\theta = 1 \Rightarrow \cos\theta = 1/\sqrt{2}, \sin\theta = 1/\sqrt{2}$

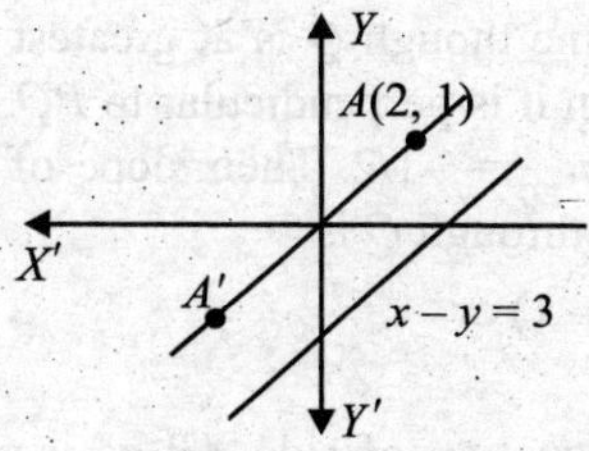

Fig. 1.10

Thus, the equation of AA' is

$$\frac{x - 2}{\cos(\pi/4)} = \frac{y - 1}{\sin(\pi/4)}$$

Since $AA' = 4$, therefore the coordinates of A' are given by

$$\frac{x - 2}{\cos(\pi/4)} = \frac{y - 1}{\sin(\pi/4)} = -4$$

$$\Rightarrow x = 2 - 4\cos\frac{\pi}{4}, y = 1 - 4\sin\frac{\pi}{4}$$

$$\Rightarrow x = 2 - 2\sqrt{2}, y = 1 - 2\sqrt{2}$$

Hence, the coordinates of A' are $(2 - 2\sqrt{2}, 1 - 2\sqrt{2})$.

18. d.

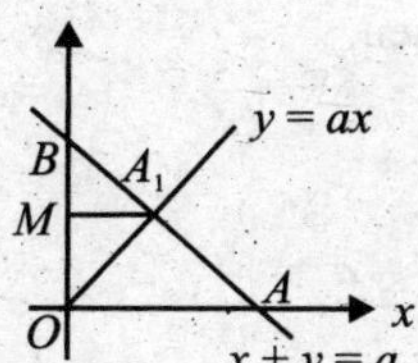

Fig. 1.11

$$B \equiv (0, a), A_1 \equiv \left(\frac{a}{1+a}, \frac{a^2}{1+a}\right)$$

$$\Delta_{OA_1B} = \frac{1}{2}(OB)\,A_1M$$

$$= \frac{1}{2}|a|\left|\frac{a}{1+a}\right| = \frac{1}{2}\frac{a^2}{|1+a|}$$

19. a.

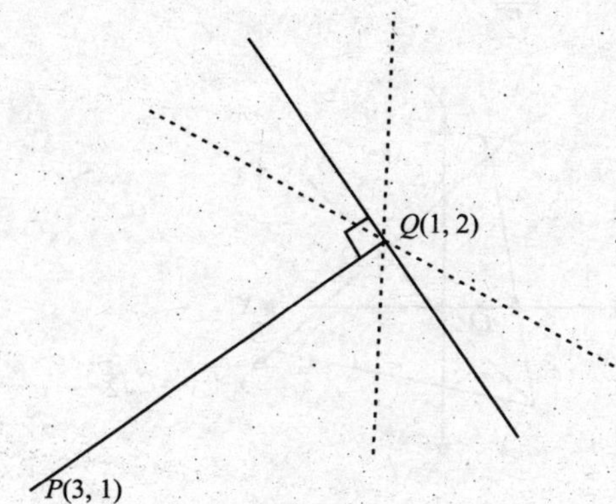

Fig. 1.12

Obviously line though Q is at greatest distance from point P when it is perpendicular to PQ. Now slope of line PQ is $m_{PQ} = -1/2$. Then slope of perpendicular line passing through Q is

$y - 2 = 2(x - 1)$ or

$2x - y = 0$

20. c. Let the equation of side AB be $y = x + a$. Then, $A \equiv (1 - a, 1)$, $B \equiv (2, 2 + a)$. Equation of side AD is $y - 1 = -(x - (1 - a))$. Hence, $D \equiv (-2, 4 - a)$.

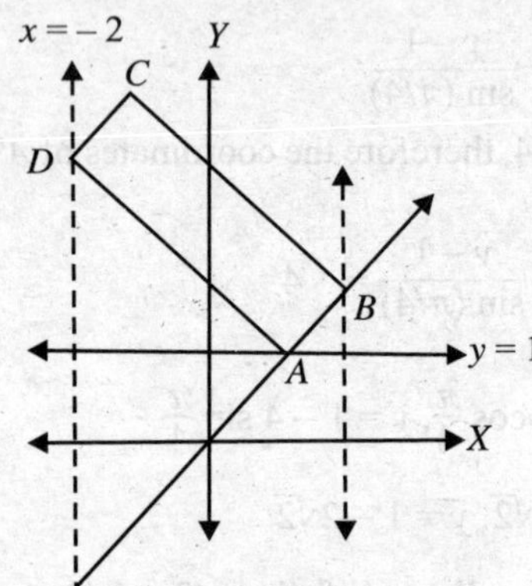

Fig. 1.13

Let $C \equiv (h, k)$. Then,

$h + 1 - a = 2 - 2$

$\Rightarrow h = a - 1$ and

$k + 1 = 2 + a + 4 - a$

$\Rightarrow k = 5$

Thus locus of C is $y = 5$.

21. a. Let the vertices B and C lie on the given line. Then,

$OD = \dfrac{2\sqrt{2}}{\sqrt{2}} = 2$. Equation of OD is

$y = x \Rightarrow x = y = \sqrt{2}$ (for point D)

Also, $BD = OD \times \tan 60° = 2\sqrt{3}$ for the coordinates of B and C. Using parametric equation of line, we get

$$\frac{x - \sqrt{2}}{-\frac{1}{\sqrt{2}}} = \frac{y - \sqrt{2}}{\frac{1}{\sqrt{2}}} = \pm 2\sqrt{3}$$

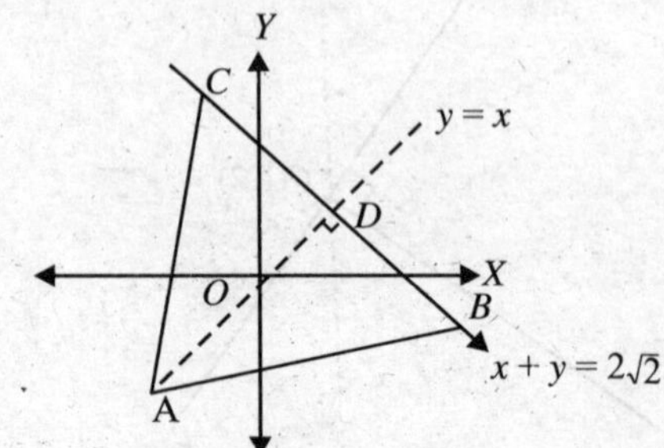

Fig. 1.14

$\Rightarrow C \equiv (\sqrt{2} + \sqrt{6}, \sqrt{2} - \sqrt{6})$

and $B \equiv (\sqrt{2} - \sqrt{6}, \sqrt{2} + \sqrt{6})$

22. d. Distance of all the points from (0, 0) are 5 units. That means cirumcentre of the triangle formed by the given points is (0, 0). If $G \equiv (h, k)$ be the centroid of the triangle, then $3h = 3 + 5(\cos\theta + \sin\theta)$, $3k = 4 + 5(\sin\theta - \cos\theta)$. If $H(\alpha, \beta)$ be the orthocentre, then

$OG{:}GH = 1{:}2 \Rightarrow \alpha = 3h, \beta = 3k$

$\cos\theta + \sin\theta = \dfrac{\alpha - 3}{5}$, $\sin\theta - \cos\theta = \dfrac{\beta - 4}{5}$

$\Rightarrow \sin\theta = \dfrac{\alpha + \beta - 7}{10}$, $\cos\theta = \dfrac{\alpha - \beta + 1}{10}$

Thus the locus of (α, β) is

$(x + y - 7)^2 + (x - y + 1)^2 = 100$

23. b.

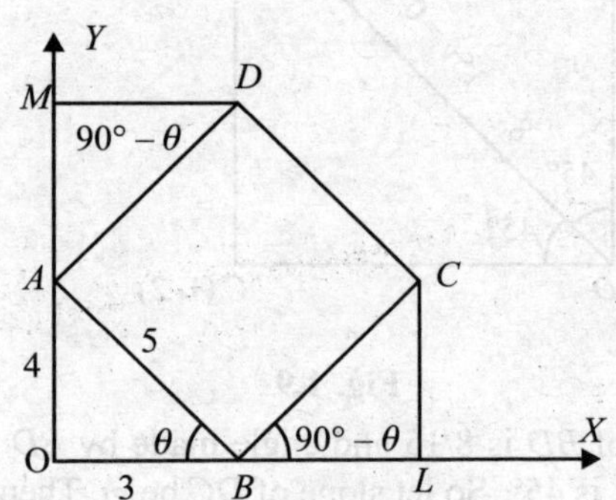

Fig. 1.15

The coordinates of A are (0, 4) and those of B are (3, 0).

$BC = AB = \sqrt{3^2 + 4^2} = 5$

$\Rightarrow BL = BC \sin\theta$ and $CL = BC \cos\theta$

$\Rightarrow BL = 5 \times \dfrac{4}{5} = 4$

and $CL = 5 \times \dfrac{3}{5} = 3$

Similarly, $MD = 4$ and $AM = 3$. So the coordinates of C are $(OB + BL, CL) = (7, 3)$ and those of D are $(MD, OA + AM) = (4, 7)$.

The coordinates of the vertex farthest from the origin is (4, 7).

24. a.

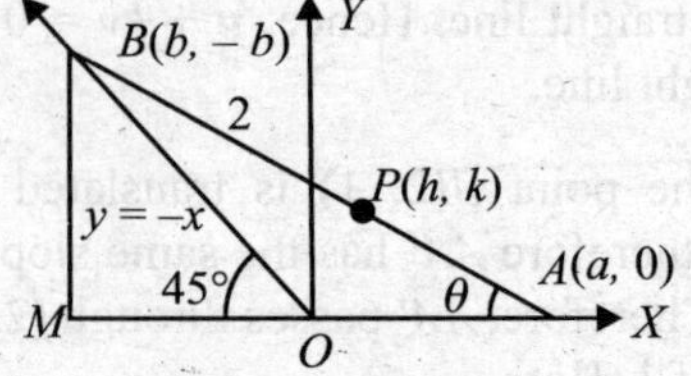

Fig. 1.16

If $\angle BAO = \theta$, then $BM = 2\sin\theta$ and $MO = BM = 2\sin\theta$, $MA = 2\cos\theta$. Hence, $A = (2\cos\theta - 2\sin\theta, 0)$ and $B = (-2\sin\theta, 2\sin\theta)$. Since $P(x, y)$ is the midpoint of AB, so

$2x = (2\cos\theta) + (-4\sin\theta)$ or

$\cos\theta - 2\sin\theta = x$

$2y = (2\sin\theta)$ or $\sin\theta = y$

Eliminating θ, we have

$(x + 2y)^2 + y^2 = 1$ or $x^2 + 5y^2 + 4xy - 1 = 0$

25. b.

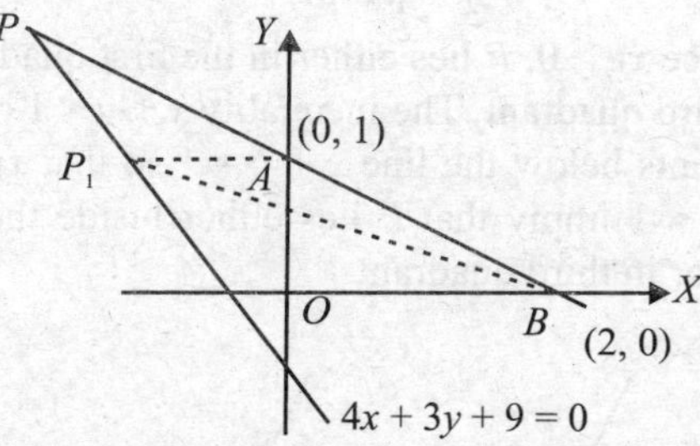

Fig. 1.17

Equation of AB is

$$y - 1 = \frac{0 - 1}{2 - 1}x \Rightarrow x + 2y - 2 = 0$$

$|PA - PB| \le AB$

Thus, $|PA - PB|$ is maximum if points A, B and P are collinear.

Hence, solving $x + 2y - 2 = 0$ and $4x + 3y + 9 = 0$, we get point $P = (-84/5, 13/5)$.

26. c.

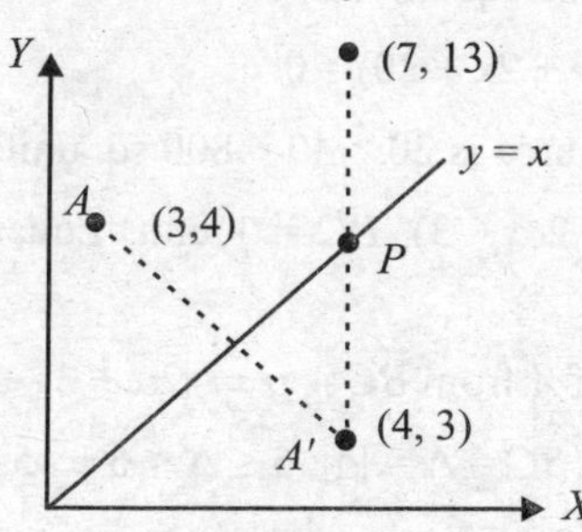

Fig. 1.18

Consider a point A', the image of A in $y = x$. Therefore, the coordinates of A' are $(4, 3)$.

or

(Notice that A and B lie are the same side with respect to $y = x$). Then $PA = PA'$. Thus, $PA + PB$ is minimum, if $PA' + PB$ is minimum, i.e., if P, A', B are collinear. Now, the equation of AB is

$$y - 3 = \frac{13 - 3}{7 - 4}(x - 4)$$

$\Rightarrow 3y - 10x + 31 = 0$

It intersects $y = x$ at $(31/7, 31/7)$, which is the required point P.

27. a.

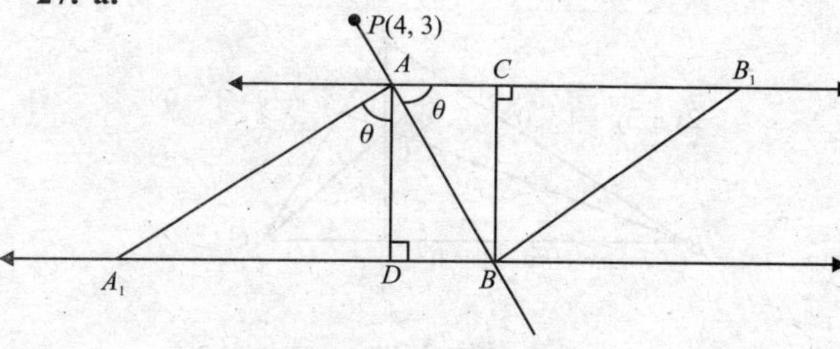

Fig. 1.19

The given lines (L_1 and L_2) are parallel and distance between them (BC or AD) is $(15 - 5)/5 = 2$ units. Let $\angle BCA = \theta \Rightarrow AB = BC \operatorname{cosec}\theta = 2\operatorname{cosec}\theta$ and $AA_1 = AD\sec\theta = 2\sec\theta$. Now area of parallelogram AA_1BB_1 is

$$\Delta = AB \times AA_1 = 4\sec\theta\operatorname{cosec}\theta$$

$$= \frac{8}{\sin 2\theta}$$

Clearly, Δ is least for $\theta = \pi/4$. Let slope of AB be m.

$$\text{Then, } 1 = \left|\frac{m + 3/4}{1 - \frac{3m}{4}}\right|$$

$\Rightarrow 4m + 3 = \pm(4 - 3m) \Rightarrow m = 1/7$ or -7

Hence, the equation of L is

$x - 7y + 17 = 0$ or

$7x + y - 31 = 0$

28. b. Suppose we rotate the coordinate axes in the anticlockwise direction through an angle α. The equation of the line L with respect to old axes is

$$\frac{x}{a} + \frac{y}{b} = 1$$

In this equation replacing x by $x\cos\alpha - y\sin\alpha$ and y by $x\sin\alpha + y\cos\alpha$, the equation of the line with respect to new axes is

$$\frac{x\cos\alpha - y\sin\alpha}{a} + \frac{x\sin\alpha + y\cos\alpha}{b} = 1$$

$$\Rightarrow x\left(\frac{\cos\alpha}{a} + \frac{\sin\alpha}{b}\right) + y\left(\frac{\cos\alpha}{b} - \frac{\sin\alpha}{a}\right) = 1 \quad (1)$$

The intercepts made by (1) on the coordinate axes are given as p and q. Therefore,

$$\frac{1}{p} = \frac{\cos\alpha}{a} + \frac{\sin\alpha}{b} \text{ and}$$

$$\frac{1}{q} = \frac{\cos\alpha}{b} - \frac{\sin\alpha}{a}$$

Squaring and adding, we get

$$\frac{1}{p^2} + \frac{1}{q^2} = \frac{1}{a^2} + \frac{1}{b^2}$$

29. c.

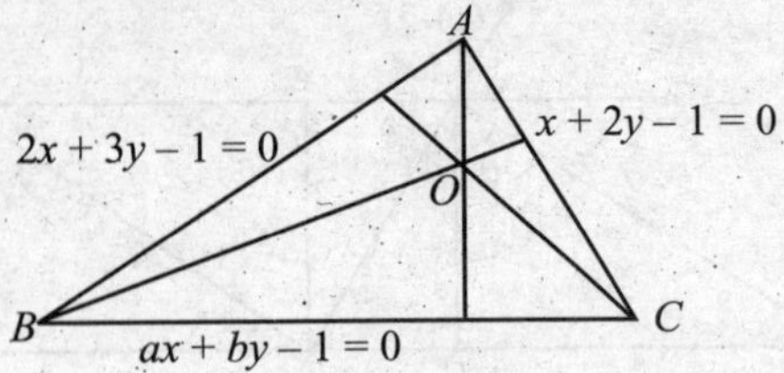

Fig. 1.20

Equation of AO is $2x + 3y - 1 + \lambda (x + 2y - 1) = 0$, where $\lambda = -1$ since the line passes through the origin. So, $x + y = 0$. Since AO is perpendicular to BC, so

$$(-1)\left(-\frac{a}{b}\right) = -1$$

$\Rightarrow a = -b$

Similarly, $(2x + 3y - 1) + \mu(ax - ay - 1) = 0$ will be the equation of BO for $\mu = -1$. Now, BO is perpendicular to AC. Hence,

$$\left\{-\frac{(2-a)}{3+a}\right\}\left(-\frac{1}{2}\right) = -1$$

Hence, $a = -8$, $b = 8$

30. c. If P_1 be the reflection of P in y-axis, then $P_1 \equiv (-2, 3)$.

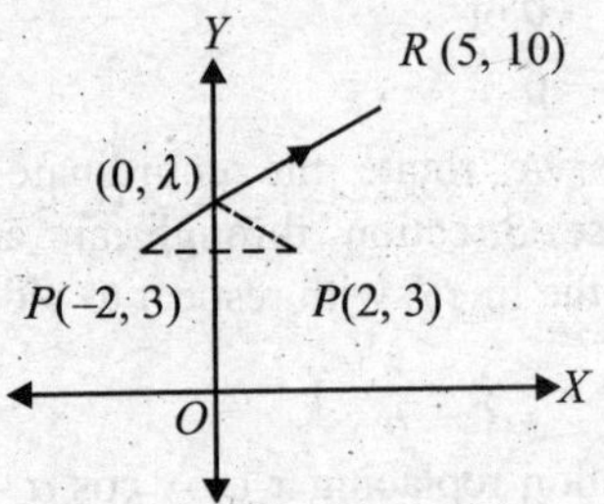

Fig. 1.21

Equation of line P_1R is

$$(y - 3) = \frac{10-3}{5+2}(x + 2)$$

$\Rightarrow y = x + 5$

It meets the y-axis at $(0, 5) \Rightarrow Q \equiv (0, 5)$.

31. a.

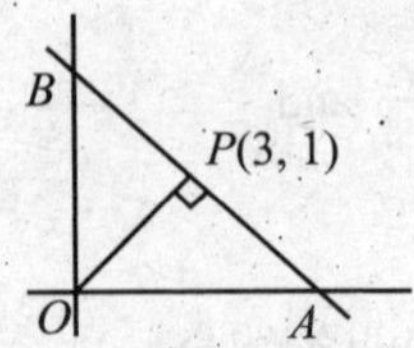

Fig. 1.22

Line AB will be farthest from the origin if OP is right angled to the line drawn. Hence,

$m_{OP} = \frac{1}{3} \Rightarrow m_{AB} = -3$

Thus equation of AB is

$(y - 1) = -3(x - 3)$

$\Rightarrow A \equiv \left(\frac{10}{3}, 0\right), B \equiv (0, 10)$

$\Rightarrow \Delta_{OAB} = \frac{1}{2}(OA)(OB) = \frac{1}{2} \times \frac{10}{3} \times 10$

$= \frac{100}{6} = \frac{50}{3}$ sq. units

32. a. Since $xy > 0$, P lies either in the first quadrant or in the third quadrant. The inequality $x + y < 1$ represents all points below the line $x + y = 1$ so that $xy > 0$ and $x + y < 1$ imply that P lies either inside the triangle OAB or in third quadrant.

33. c.

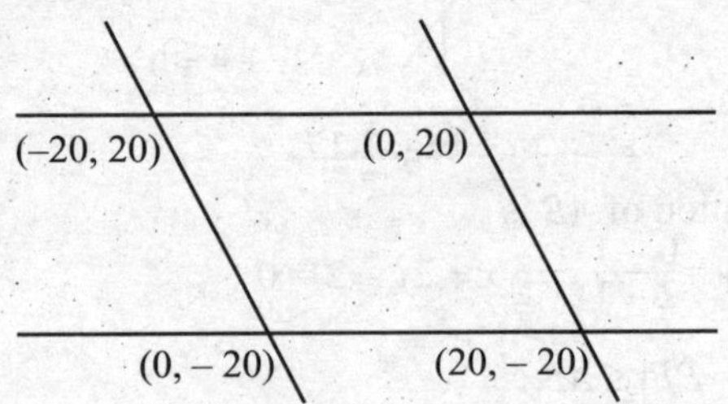

Fig. 1.23

For $x \geq 0$, the equation is

$(y + 20)(y + 2x - 20) = 0$

For $x \leq 0$, the equation is

$(y - 20)(y + 2x + 20) = 0$

Hence, the area is $20 \times 40 = 800$ sq. units.

34. b. $A \equiv (\alpha, 2\alpha + 3)$, $BC = 1$ unit. Equation of BC is $y - 3 = 0$.

Distance of A from BC is $p \Rightarrow |2\alpha + 3 - 3|$.

Area of $\Delta ABC = \Delta = |\alpha|$; $5 \leq \Delta < 6 \Rightarrow 5 \leq |\alpha| < 6$.

35. d. $\frac{a}{\sqrt{bc}} - 2 = \sqrt{\frac{b}{c}} + \sqrt{\frac{c}{b}}$

$\Rightarrow a = b + c + 2\sqrt{bc}$

$\Rightarrow a = (\sqrt{b} + \sqrt{c})^2$

$\Rightarrow (\sqrt{a} - \sqrt{b} - \sqrt{c})(\sqrt{a} + \sqrt{b} + \sqrt{c}) = 0$

$\Rightarrow \sqrt{a} - \sqrt{b} - \sqrt{c} = 0$

Since $\sqrt{a} + \sqrt{b} + \sqrt{c} \neq 0$, $(\because a, b, c > 0)$.
comparing with $\sqrt{a}x + \sqrt{b}y = \sqrt{c} = 0$, we have $x = -1$, $y = 1$.

36. b. There are clearly five points.

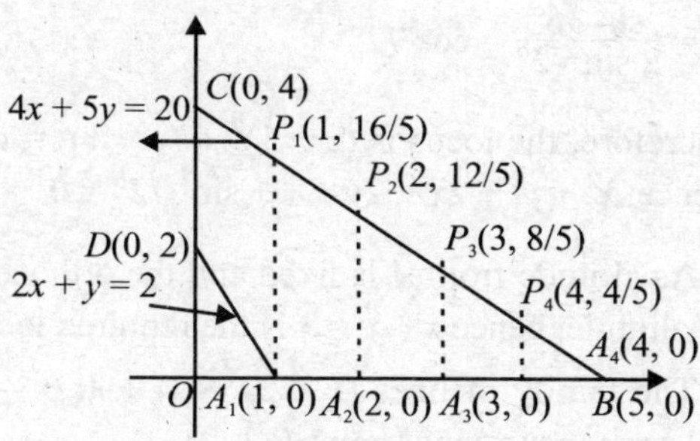

Fig. 1.24

37. b. By solving the sides of the rhombus, its vertices are $(0, -n/m)$, $(-n/l, 0)$, $(0, n/m)$, and $(n/l, 0)$. Hence, the

area is $\frac{1}{2} \times \frac{2n}{m} \times \frac{2n}{l} = 2$

$\Rightarrow n^2 = lm.$

Therefore, l, n, m are in G.P.

38. c. If the point of intersection of two lines with coordinate axes be concyclic, then product of intercepts on the x-axis is equal to product of intercepts on y-axis by these lines. This is a geometric property. The intercepts on x-axis are p and q and whose product is pq. Also, the intercepts on y-axis are p, and q, whose product is also pq. Hence, the four points are concyclic

39. b.

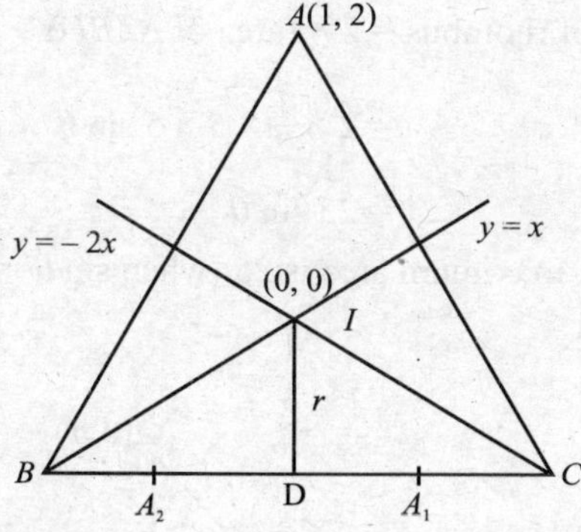

Fig. 1.25

Images of A about $y = x$, $y = -2x$ are A_1 and A_2 which lie on BC. Now, $A_1 \equiv (2, 1)$ and $A_2 \equiv (-11/5, 2/5)$. Equation of BC is $x - 7y + 5 = 0$. Hence,

$$r = ID = \left|\frac{5}{\sqrt{1+49}}\right| = \frac{5}{2\sqrt{5}} = \frac{1}{\sqrt{2}}$$

40. b. Reflections of A in the two angle bisectors will lie on the line BC, so (2, 1) and (1, −2) will lie on BC. Equation of BC will be

$$y + 2 = \left(\frac{1+2}{2-1}\right)(x - 1), \text{ i.e., } 3x - y = 5.$$

41. c. The circumcentre of the triangle is (0, 0) as all the vertices lie on the circle $x^2 + y^2 = 5$. So the orthocentre will be (sum of x coordinates, sum of y coordinates).

42. a. Extremities of the given diagonal are (4, 0) and (0, 6). Hence, slope of this diagonal is $-3/2$ and slope of other diagonal is 2/3. The equation of the other diagonal is

$$\frac{x-2}{\frac{3}{\sqrt{13}}} = \frac{y-3}{\frac{2}{\sqrt{13}}} = r$$

For the extremities of the diagonal, $r = \pm\sqrt{13}$. Hence,

$x - 2 = \pm 3, y - 3 = \pm 2$

$x = 5, -1$ and $y = 5, 1$

Therefore, the extremities of the diagonal are (5, 5) and (−1, 1).

43. a. No such triangle is possible as the medians through the vertices of a right-angled triangle (other than right angle) cannot be perpendicular to each other.

44. a. $\tan\theta = \frac{a-1}{3} = \frac{7-a}{10-b}$

also $\tan\theta = \frac{7-4}{b-3} = \frac{3}{b-3}$

Hence, $\frac{3}{b-3} = \frac{a-1}{3} = \frac{7-a}{10-b}$

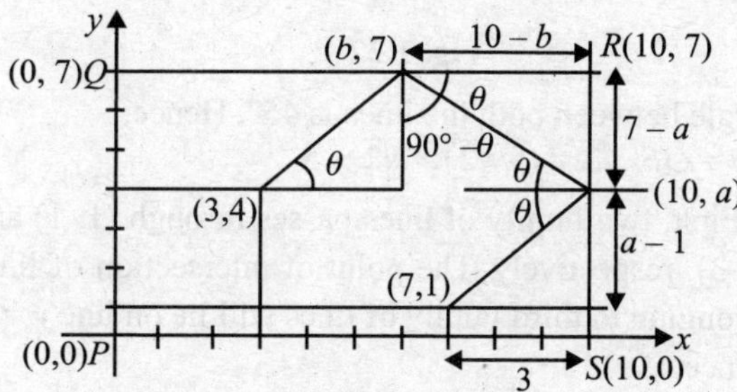

Fig. 1.26

From 1st two relations, we get

$9 = ab - b - 3a = 3$

$3a + 6 = ab - b$ (1)

From last two

$10a - ab - 10 + b = 21 - 3a$

$13a - ab + b = 31$

or $ab - b = 13a - 31$ (2)

Hence, from (1) and (2)

$3a + 6 = 13a - 31 \Rightarrow 10a = 37$

$\Rightarrow a = 3.7$

45. c. Let the incentre be (2, 1), then

$IA = \sqrt{2^2 + 2^2} = 2\sqrt{2}$

Also, $AI = r \operatorname{cosec} \frac{1}{2}$

$$\angle BIC = \frac{\pi}{2} + \frac{A}{2}$$

$$\Rightarrow \tan\left(\frac{\pi}{2} + \frac{A}{2}\right) = \frac{1-2}{1+2}$$

$$= \frac{1}{3} \cot \frac{1}{2} = 1$$

$$\text{So } r = \frac{AI}{\operatorname{coses} \frac{A}{2}}$$

$$= \frac{2\sqrt{2}}{\sqrt{1+\frac{1}{9}}}$$

$$\frac{2\sqrt{2} \times 3}{\sqrt{10}} = \frac{6}{\sqrt{5}}$$

46. d.

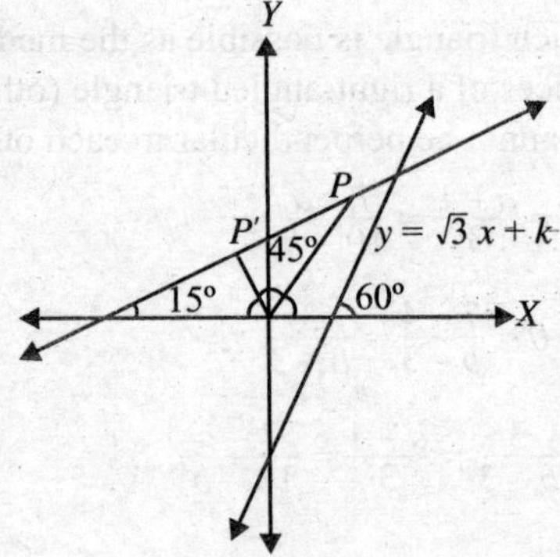

Fig. 1.27

Angle between both the lines is 45°. Hence, $OP = OP' \sqrt{2} = (5/\sqrt{2}) \times \sqrt{2} = 5$.

47. a. First, two family of lines passes through (1, 1) and (3, 3), respectively. The point of intersection of lines belonging to third family of lines will lie on line $y = x$. Hence,

$\Rightarrow ax + x - 2 = 0$ and $6x + ax - a = 0$

$$\text{or } \frac{2}{a+1} = \frac{a}{6+a}$$

$$\Rightarrow a^2 - a - 12 = 0 \Rightarrow (a-4)(a+3) = 0$$

48. a.

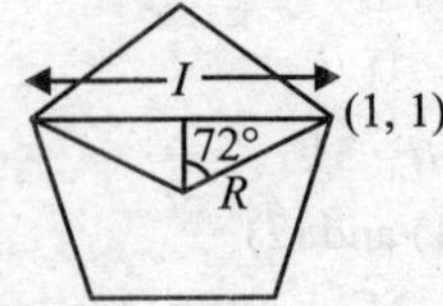

Fig. 1.28

The point of intersection of diagonals, i.e., (1, 1), lies on the circumcircle. Hence,

$$\Rightarrow I = 2R \sin 72°$$

$$R = \frac{\sin 36°}{2 \sin 72°} = \cos 72°$$

Therefore, the locus is $(x-1)^2 + (y-1)^2 = \cos^2 72°$.
Hence, $x^2 + y^2 - 2x - 2y + 1 + \sin^2 72° = 0$

49. b. As altitude from A is fixed and the orthocentre lies on altitude, hence $x + y = 3$ is the required locus.

50. c. The family of lines $(x - 2y + 3) + \lambda(2x - 3y + 4) = 0$ are concurrent at point $P(1, 2)$

If image of point $A(2, 3)$ in the above variable line is $B(h, k)$ then $AP = BP$

$\Rightarrow (h-1)^2 + (k-2)^2 = (2-1)^2 + (3-2)^2$

Hence, locus of point P is $x^2 + y^2 - 2x - 4y + 4 = 0$.

51. b. From the figure

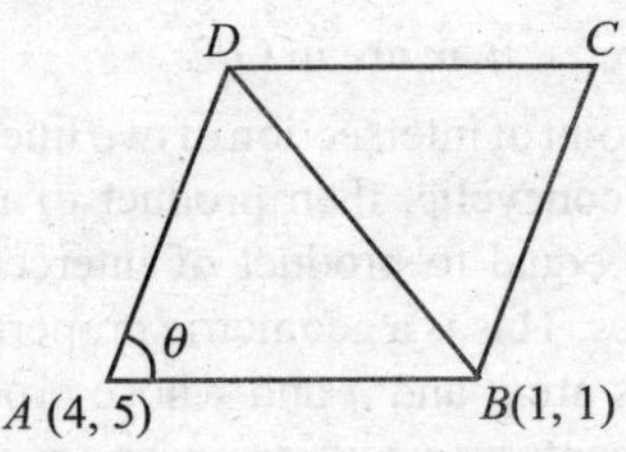

Fig. 1.29

Area of rhombus = 2 × (area of ΔABD)

$$= 2 \times \frac{1}{2} \times 5 \times 5 \sin\theta$$

$$= 25 \sin\theta$$

Hence, maximum area is 25 (when $\sin\theta = 1$).

52. a.

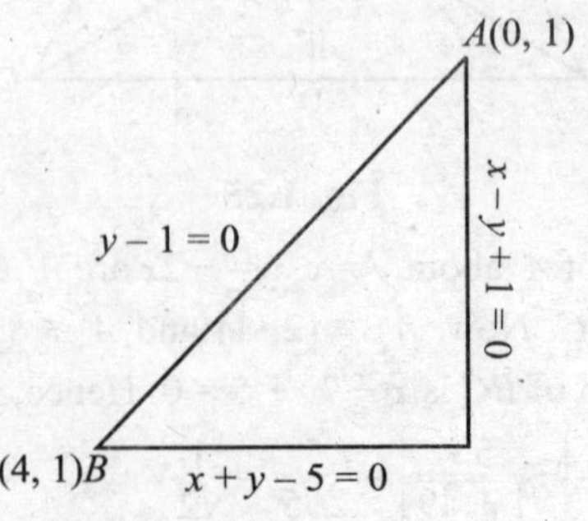

Fig. 1.30

Since the triangle is right angled, so the circumcentre will be the middle point of hypotenuse, i.e., (2, 1).

53. b. Let the third vertex be (h, k).

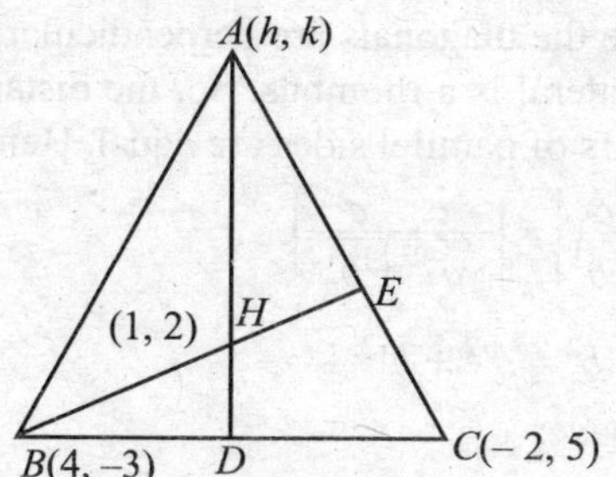

Fig. 1.31

Now slope of AD is $(k - 2)/(h - 1)$ slope of BC is $(5 + 3)/(-2 - 4) = -4/3$, slope of BE is $(-3 - 2)/(4 - 1) = 5/3$ and slope of AC is $(k - 5)/(h + 2)$. Since $AD \perp BC$, so

$$\frac{k-2}{h-1} \times \frac{-4}{3} = -1$$

$$\Rightarrow 3h - 4k + 5 = 0 \quad (1)$$

Again since $BE \perp AC$, so

$$-\frac{5}{3} \times \frac{k-5}{h+2} = -1$$

$$\Rightarrow 3h - 5k + 31 = 0 \quad (2)$$

On solving (1) and (2) we get $h = 33$, $k = 26$. Hence, the third vertex is (33, 26).

54. a. Let $A \equiv (x_1, y_1)$, $B \equiv (x_2, y_2)$, $C \equiv (x_3, y_3)$, $D \equiv (x_4, y_4)$

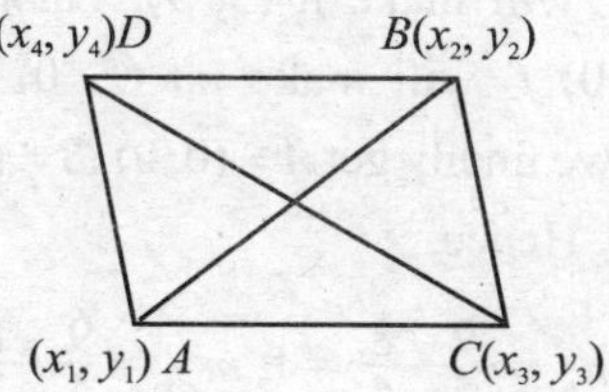

Fig. 1.32

Given,

$x_1^2 + x_2^2 + x_3^2 + x_4^2 + y_1^2 + y_2^2 + y_3^2 + y_4^2 - 2x_1x_3 - 2x_2x_4 - 2y_2y_3 - 2y_1y_4 \le 0$

$$\Rightarrow (x_1 - x_3)^2 + (x_2 - x_4)^2 + (y_2 - y_3)^2 + (y_1 - y_4)^2 \le 0$$

$$\Rightarrow x_1 = x_3, x_2 = x_4, y_2 = y_3, y_1 = y_4$$

$$\Rightarrow \frac{x_1 + x_2}{2} = \frac{x_3 + x_4}{2} \text{ and } \frac{y_1 + y_2}{2} = \frac{y_4 + y_3}{2}$$

Hence, AB and CD bisect each other.

Therefore, $ACBD$ is a parallelogram

Also

$$AB^2 = (x_1 - x_2)^2 + (y_1 - y_2)^2$$
$$= (x_3 - x_4)^2 + (y_4 - y_3)^2$$
$$= CD^2$$

Thus $ACBD$ is a parallelogram and $AB = CD$, hence is a rectangle

55. c. For any point $P(x, y)$ that is equidistant from given line, we have

$$x + y - \sqrt{2} = -\left(x + y - 2\sqrt{2}\right)$$

$$\Rightarrow 2x + 2y - 3\sqrt{2} = 0$$

56. a. The distance between $(2/m, 2)$ and $(6/m, 6)$ is less than 5. Hence,

$$\left(\frac{2}{m} - \frac{6}{m}\right)^2 + (2 - 6)^2 < 25$$

$$\Rightarrow \frac{16}{m^2} < 9$$

$$\Rightarrow m^2 > \frac{16}{9}$$

$$\Rightarrow m > \frac{4}{3} \text{ or } m < \frac{-4}{3}$$

57. a.

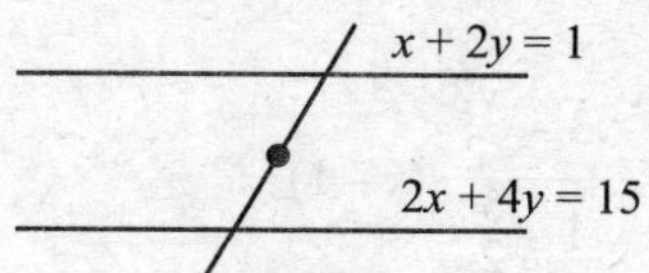

Fig. 1.33

Let P be on $x + 2y = 1$. Then,

$$1 + \frac{t}{\sqrt{2}} + 2\left(2 + \frac{t}{\sqrt{2}}\right) = 1 \text{ or}$$

$$t = \frac{-4\sqrt{2}}{3}$$

Let P be on $2x + 4y = 15$. Then,

$$2\left(1 + \frac{t}{\sqrt{2}}\right) + 4\left(2 + \frac{t}{\sqrt{2}}\right) = 15 \text{ or}$$

$$t = \frac{5\sqrt{2}}{6}$$

Since point lies between the lines and $x = t$, then

$$t \in \left(\frac{-4\sqrt{2}}{3}, \frac{5\sqrt{2}}{3}\right).$$

58. a.

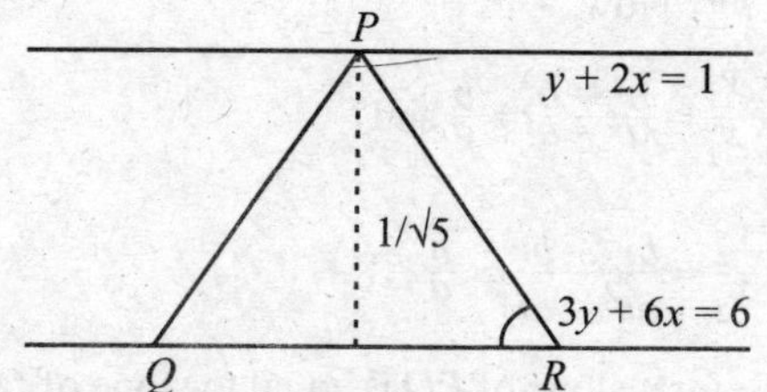

Fig. 1.34

The given lines are $y + 2x = 1$ and $y + 2x = 2$. The distance between the lines is $(2-1)/\sqrt{5} = 1/\sqrt{5}$.

The side length of the triangle is $(1/\sqrt{5})$ cosec $60° = 2/\sqrt{5}$.

59. c.

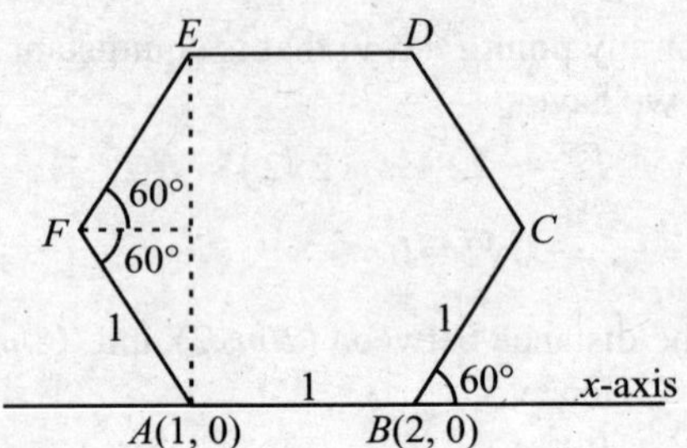

Fig. 1.35

$AB = 1$

$$\Rightarrow C \equiv (2 + 1 \times \cos 60°, 1 \times \sin 60°) = \left(\frac{5}{2}, \frac{\sqrt{3}}{2}\right)$$

$$E \equiv (1, 1 \times \sin 60° + 1 \times \sin 60°) = (1, \sqrt{3})$$

Therefore, the equation of CE is

$$y - \sqrt{3} = \frac{\sqrt{3} - \frac{\sqrt{3}}{2}}{1 - \frac{5}{2}}(x - 1)$$

60. d.

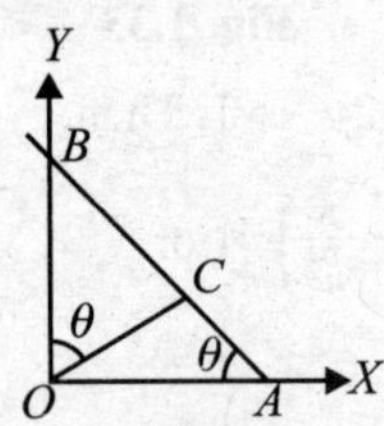

Fig. 1.36

$\tan(180° - \theta) = \text{slope of } AB = -3$

$\therefore \tan\theta = 3$

$\therefore \frac{OC}{AC} = \tan\theta, \frac{OC}{BC} = \cot\theta$

$$\Rightarrow \frac{BC}{AC} = \frac{\tan\theta}{\cot\theta} = \tan^2\theta = 9$$

61. a. Let $x_1 = a$, $x_2 = ar$, and $x_3 = ar^2$; $y_1 = b$, $y_2 = br$, and $y_3 = br^2$. Now,

$$\frac{y_2 - y_1}{x_2 - x_1} = \frac{br - b}{ar - a} = \frac{b}{a} \text{ and}$$

$$\frac{y_3 - y_2}{x_3 - x_2} = \frac{br^2 - br}{ar^2 - ar} = \frac{b}{a}$$

Therefore, slope of PQ is equal to slope of QR. Hence, points P, Q, R are collinear.

62. c. Since the diagonals are perpendicular, so the given quadrilateral is a rhombus. So, the distances between two pairs of parallel sides are equal. Hence,

$$\left|\frac{c' - c}{\sqrt{a^2 + b^2}}\right| = \left|\frac{c' - c}{\sqrt{a'^2 + b'^2}}\right|$$

$$\Rightarrow a^2 + b^2 = a'^2 + b'^2$$

63. c. We have,

$$B \equiv \left(\frac{6}{7}, 2\right), C \equiv \left(-\frac{6}{7}, 2\right)$$

$$\Rightarrow BC = \frac{12}{7}, AD = 3$$

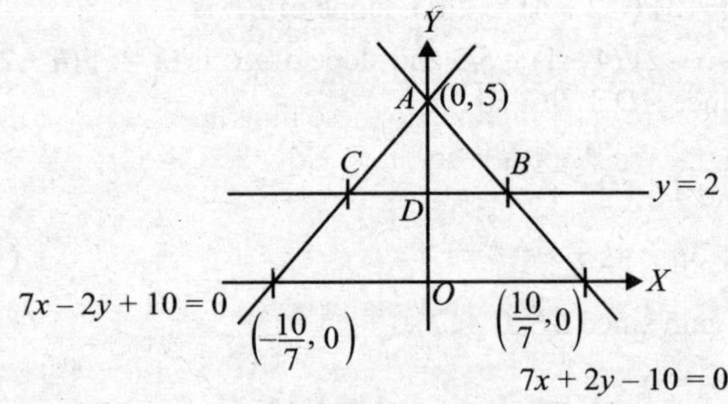

Fig. 1.37

$$\therefore \Delta_{ABC} = \frac{1}{2} \times \frac{12}{7} \times 3 = \frac{18}{7} \text{ sq. units}$$

64. d. Clearly, A will remain as (0, 0); f_1 will make B as (0, 4); f_2 will make it (12, 4) and f_3 will make it (4, 8); f_1 will make 'C' as (2, 4); f_2 will make it (14, 4); f_3 will make it (5, 9). Finally f_1 will make D as (2, 0) f_2 will make ixt (2, 0) f_3 will make it (1, 1). So we finally get $A \equiv (0, 0)$, $B \equiv (4, 8)$, $C \equiv (5, 9)$, $D \equiv (1, 1)$. Hence,

$$m_{AB} = \frac{8}{4}, m_{BC} = \frac{9-8}{5-4} = 1, m_{CD} = \frac{9-1}{5-1} = \frac{8}{4}, m_{AD} = 1,$$

$$m_{AC} = \frac{9}{5}, m_{BD} = \frac{8-1}{4-1} = 7/3$$

Hence, the final figure will be a parallelogram.

65. d.

Angle bisector will make the angles $(\theta_1 + \theta_2)/2$ and $(\pi/2 + (\theta_1 + \theta_2)/2)$ with the x-axis. Hence, their equations are

$$\frac{x - x_1}{\cos\left(\frac{\theta_1 + \theta_2}{2}\right)} = \frac{y - y_1}{\sin\left(\frac{\theta_1 + \theta_2}{2}\right)} \text{ or}$$

$$\frac{x - x_1}{-\sin\left(\frac{\theta_1 + \theta_2}{2}\right)} = \frac{y - y_1}{\cos\left(\frac{\theta_1 + \theta_2}{2}\right)}$$

66. c. $ax + by = 1$ will be one of the bisectors of the given line. Equation of bisectors of the given lines are

$$\frac{3x+4y-5}{5} = \pm\left(\frac{5x-12y-10}{13}\right)$$

$\Rightarrow 64x - 8y = 115$

or $14x + 112y = 15$

$\Rightarrow a = \frac{64}{115}, b = -\frac{8}{115}$

or $a = \frac{14}{15}, b = \frac{12}{115}$

67. b. Lines $x \cos \alpha + y \sin \alpha = p$ and $x \sin \alpha - y \cos \alpha = 0$ are mutually perpendicular. Thus $ax + by + p = 0$ will be equally inclined to these lines and would be the angle bisector of these lines. Now equations of angle bisectors is

$x \sin \alpha - y \cos \alpha = \pm (x \cos \alpha + y \sin \alpha - p)$

$\Rightarrow x (\cos \alpha - \sin \alpha) + y (\sin \alpha + \cos \alpha) = p$

or $x (\sin \alpha + \cos \alpha) - y (\cos \alpha - \sin \alpha) = p$

Comparing these lines with $ax + by = p = 0$, we get

$$\frac{a}{\cos \alpha - \sin \alpha} = \frac{b}{\sin \alpha + \cos \alpha} = 1$$

$\Rightarrow a^2 + b^2 = 2$ or

$$\frac{a}{\sin \alpha + \cos \alpha} = \frac{b}{\cos \alpha - \sin \alpha} = 1$$

$\Rightarrow a^2 + b^2 = 2$

68. a. Let the equation of line be $x/a + y/b = 1$.

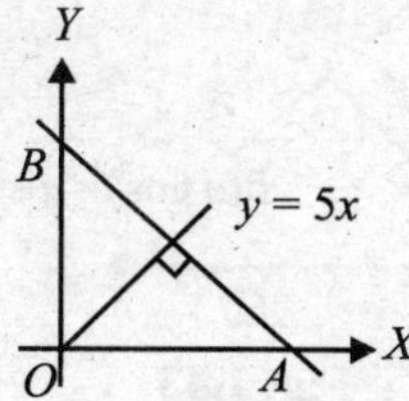

Fig. 1.38

AB is perpendicular to $y = 5x$. Hence,

$-b/a \times 5 = -1 \Rightarrow 5b = a$

Area of $\Delta_{OAB} = \frac{1}{2}|ab|$

$\Rightarrow 10 = \frac{1}{2}|5b^2|$

$\Rightarrow b^2 = 4 \Rightarrow b = \pm 2, a = \pm 10$

The line can be $x/10 + y/2 = 1$ or $x/10 + y/2 = -1$.

69. a. Given lines are mutually perpendicular and intersect at (6/5, 13/5).

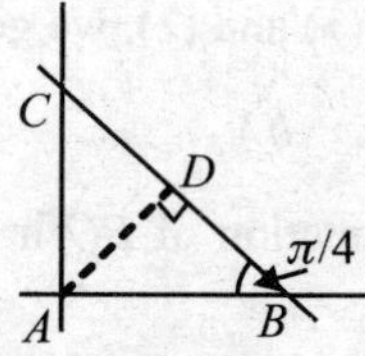

Fig. 1.39

Equations of angle bisectors of the given lines are $x - 2y + 4 = \pm (2x + y - 5)$,

i.e., $x + 3y = 9$ and $3x - y = 1$.

Side BC will be parallel to these bisectors. Let $AD = a$

$\Rightarrow AB = a\sqrt{2}$ and area of $\Delta\, ABC$ is

$\Delta_{ABC} = \frac{1}{2} \times (a\sqrt{2})^2 = a^2 = 10$

$\Rightarrow a = \sqrt{10}$

Let equation of BC be $x + 3y = \lambda$. Then,

$$\sqrt{10} = \frac{\frac{6}{5} + \frac{39}{5} - \lambda}{\sqrt{10}} \Rightarrow \lambda = -1, 19$$

Therefore, equation of BC is $x + 3y = -1$ or $x + 3y = 19$. If equation of BC is $3x - y = \lambda$, then

$$\sqrt{10} = \frac{\frac{18}{5} - \frac{13}{5} - \lambda}{\sqrt{10}} \Rightarrow \lambda = -9, 11$$

Hence, equation of BC is $3x - y = -9$ or $3x - y = 11$.

70. d. All values of a.

71. c.

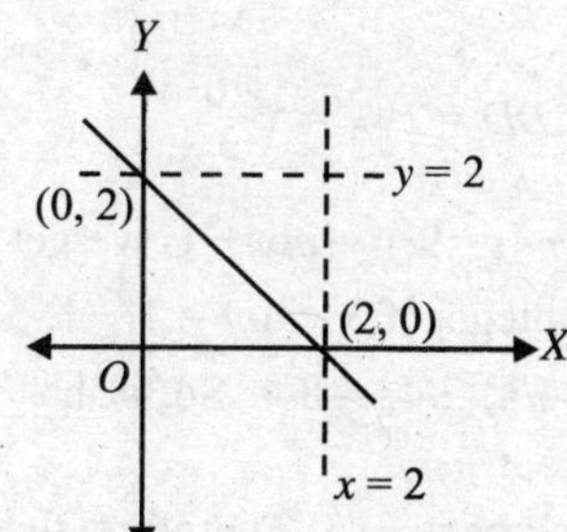

Fig. 1.40

Above figure represents the given isosceles triangle. Clearly the equation of other equal side is $y = 2$.

72. d. Let (h, k) be the point on the locus. Then by the given conditions,

$(h - a_1)^2 + (k - b_1)^2 = (h - a_2)^2 + (k - b_2)^2$

$\Rightarrow 2h(a_1 - a_2) + 2k(b_1 - b_2) + a_2^2 + b_2^2 - a_1^2 - b_1^2 = 0$

$$\Rightarrow h(a_1 - a_2) + k(b_1 - b_2) + \frac{1}{2}(a_2^2 + b_2^2 - a_1^2 - b_1^2) = 0 \quad (1)$$

Also since (h, k) lies on the given locus,

$$(a_1 - a_2)h + (b_1 - b_2)k + c = 0 \quad (2)$$

Comparing Eqs. (1) and (2), we get

$$c=\frac{1}{2}(a_2^2+b_2^2-a_1^2-b_1^2)$$

73. c. Clearly, the equation of PQ in the new position is $x = 2$.

74. b. If the line cuts off the axes at A and B, then area of triangle is $\frac{1}{2}\times OA\times OB = T$.

$$\Rightarrow \frac{1}{2}\times a\times OB = T \Rightarrow OB=\frac{2T}{a}$$

Hence, the equation of line is

$$\frac{x}{-a}+\frac{y}{2T/a}=1$$

$$\Rightarrow \quad 2Tx - a^2y + 2aT = 0$$

75. a.

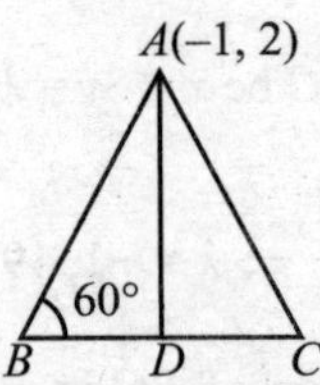

Fig. 1.41

$$AD=\left|\frac{-2-2-1}{\sqrt{(2)^2+(-1)^2}}\right|=\left|\frac{-5}{\sqrt{5}}\right|=\sqrt{5}$$

$$\tan 60° = \frac{AD}{BD} \Rightarrow \sqrt{3}=\frac{\sqrt{5}}{BD}$$

$$\Rightarrow BD=\sqrt{\frac{5}{3}}$$

Hence, $BC = 2BD = 2\sqrt{\frac{5}{3}}=\sqrt{\frac{20}{3}}$

76. a. Solving $3x+4y=9, y = mx + 1$, we get $x=\frac{5}{3+4m}$. Here, x is an integer if $3 + 4m = 1, -1, 5, -5$. Hence, $m = -2/4, -4/4, 2/4, -8/4$. So, m has two integral values.

77. a. Given that slope is $-\sqrt{3}$. Therefore, the line is

$$y=-\sqrt{3}\,x+c$$

$$\Rightarrow \sqrt{3}\,x+y=c$$

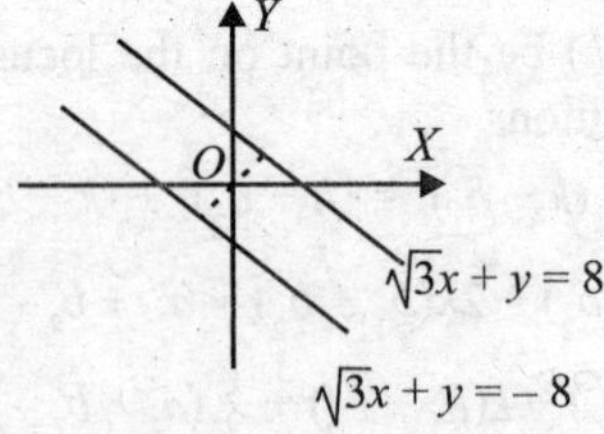

Fig. 1.42

Now, $\left|\frac{c}{2}\right|=4 \Rightarrow c=\pm 8 \Rightarrow x\sqrt{3}+y=\pm 8$

78. a. Line perpendicular to $x \sec\theta + y \operatorname{cosec}\theta = a$ is

$$x \operatorname{cosec}\theta - y\sec\theta = \lambda.$$

This line passes through the point $(a\cos^3\theta, a\sin^3\theta)$. Then, $(a\cos^3\theta)\operatorname{cosec}\theta - (a\sin^3\theta)\sec\theta = \lambda$

$$\Rightarrow \lambda = a\left(\frac{\cos^3\theta}{\sin\theta}-\frac{\sin^3\theta}{\cos\theta}\right)$$

$$= a\frac{\cos 2\theta}{\cos\theta\sin\theta}$$

Hence, the equation of line is $x\cos\theta - y\sin\theta = a\cos 2\theta$.

79. c. The line passing through the intersection of the lines $ax + 2by + 3b = 0$ and $bx - 2ay - 3a = 0$ is

$$ax + 2by + 3b + \lambda(bx - 2ay - 3a) = 0$$

$$\Rightarrow (a+b\lambda)x + (2b-2a\lambda)y + 3b - 3\lambda a = 0 \qquad (1)$$

Line Eq. (1) is parallel to x-axis. Therefore,

$$a+b\lambda = 0 \Rightarrow \lambda = -\frac{a}{b}=0$$

Putting the value of λ in Eq. (1), we get

$$y\left(2b+\frac{2a^2}{b}\right)+3b+\frac{3a^2}{b}=0$$

$$\Rightarrow y=-\frac{3}{2}$$

So, it is 3/2 units below x-axis.

80. a. Midpoint of $(a\cos\alpha, a\sin\alpha)$ and $(a\cos\beta, a\sin\beta)$ is

$$P\left(\frac{a(\cos\alpha+\cos\beta)}{2},\frac{a(\sin\alpha+\sin\beta)}{2}\right).$$

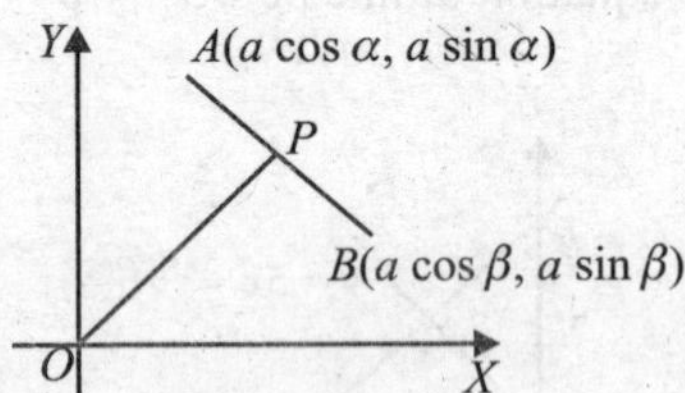

Fig. 1.43

Slope of line AB is

$$\frac{a\sin\beta - a\sin\alpha}{a\cos\beta - a\cos\alpha}=\frac{\sin\beta-\sin\alpha}{\cos\beta-\cos\alpha}=m_1$$

and slope of OP is

$$\frac{\sin\alpha+\sin\beta}{\cos\alpha+\cos\beta}=m_2$$

$$\text{Now, } m_1\times m_2=\frac{\sin^2\beta-\sin^2\alpha}{\cos^2\beta-\cos^2\alpha}=-1$$

Hence, the lines are perpendicular.

81. b. The set of lines is $4ax + 3by + c = 0$, where $a + b + c = 0$. Eliminating c, we get

$4ax + 3by - (a+b) = 0$

$\Rightarrow a(4x-1)+b(3y-1)=0$

This passes through the intersection of the lines $4x - 1 = 0$ and $3y - 1 = 0$, i.e., $x = 1/4$, $y = 1/3$, i.e., $(1/4, 1/3)$.

82. a. The three lines are concurrent, if

$$\begin{vmatrix} 1 & 2 & -9 \\ 3 & 5 & -5 \\ a & b & -1 \end{vmatrix} = 0$$

$\Rightarrow 35a - 22b + 1 = 0,$

which is true if the line $35x - 22y + 1 = 0$ passes through (a, b).

83. b. The line passing through (2, 3) and perpendicular to $-y + 3x + 4 = 0$ is

$$\frac{y-3}{x-2} = -\frac{1}{3}$$

or $3y + x - 11 = 0$

Therefore, foot is $x = -1/10$, $y = 37/10$.

84. b. Let the coordinates of the third vertex be $(2a, t)$. Now, $AC = BC$. Hence,

$$t = \sqrt{4a^2 + (a-t)^2} \Rightarrow t = \frac{5a}{2}$$

So the coordinates of third vertex C are $(2a, 5a/2)$. Therefore, area of the triangle is

$$\pm\frac{1}{2}\begin{vmatrix} 2a & \frac{5a}{2} & 1 \\ 2a & 0 & 1 \\ 0 & a & 1 \end{vmatrix} = \begin{vmatrix} a & \frac{5a}{2} & 1 \\ 0 & -\frac{5a}{2} & 0 \\ 0 & a & 1 \end{vmatrix} = \frac{5a^2}{2} \text{ sq. units}$$

85. d. If we reflect $y = |x - 2|$ in y-axis, it will become $y = |-x-2| = |x+2|$. The reflected lines are $y = x + 2$, $y = -x - 2$. Their combined equation is

$(y - x - 2)(y + x + 2) = 0$

$\Rightarrow y^2 - (x+2)^2 = 0$

$\Rightarrow y^2 - x^2 - 4x - 4 = 0$

86. d. $a^3x^2 - 2hxy + b^3y^2 = 0$

Let the slope of lines be m_1 and m_2. Then,

$$m_1 + m_2 = \frac{2h}{b^3},\ m_1m_2 = \frac{a^3}{b^3}$$

Given $m_2^2 = m_1 \Rightarrow m_2^3 = \frac{a^3}{b^3}$

$$\Rightarrow m_2 = \frac{a}{b}$$

Also $m_2^2 + m_2 = \frac{2h}{b^3}$

$$\Rightarrow \frac{2h}{b^3} = \frac{a}{b} + \frac{a^2}{b^2}$$

$$\Rightarrow ab + a^2 = \frac{2h}{b}$$

$\Rightarrow 2h = a^2b + ab^2 = ab(a+b)$

87. d.

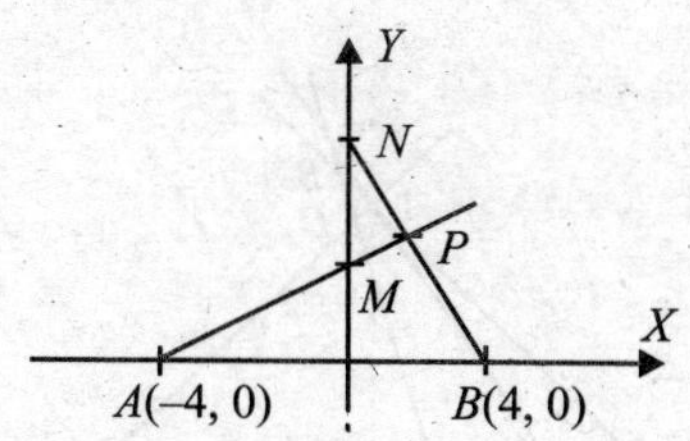

Fig. 1.44

Let $M = (0, h)$

$\Rightarrow N = (0, h+4)$.

Equation of AM is

$$\frac{x}{-4} + \frac{y}{h} = 1$$

$$\Rightarrow \frac{y}{h} = \frac{4+x}{4} \Rightarrow h = \frac{4y}{4+x}$$

Equation of BN is

$$\frac{x}{4} + \frac{y}{h+4} = 1$$

$$\Rightarrow \frac{y}{h+4} = \frac{4-x}{4}$$

$$\Rightarrow h + 4 = \frac{4y}{4-x}$$

$$\Rightarrow h = \frac{4y - 16 + 4x}{4-x}$$

$$\Rightarrow \frac{4(y-4+x)}{4-x} = \frac{4y}{4+x} \text{ (eliminating } h\text{)}$$

$\Rightarrow 2xy - 16 + x^2 = 0$, which is a required locus.

88. b.

$$OM_r = OA_r + \frac{OA_{r+1} - OA_r}{2}$$

$$= \frac{OA_r + OA_{r+1}}{2}$$

$$= \frac{1}{2}\{OA_1 \times k^{r-1} + OA_1 \times k^r\}$$

$$= \frac{OA_1}{2}(1+k)k^{r-1}$$

$$\therefore \sum_{r=1}^{\infty} OM_r = \frac{OA_1}{2}(1+k)\sum_{r=1}^{\infty} k^{r-1}$$

$$= \frac{OA_1}{2}(1+k) \times \frac{1}{1-k}$$

$$= \frac{OA_1}{2} \times \frac{1 + \frac{OA_2}{OA_1}}{1 - \frac{OA_2}{OA_1}}$$

$$= \frac{OA_1(OA_1 + OA_2)}{2(OA_1 - OA_2)}$$

89. c. A rough sketch of the lines is given below. There are three triangles namely ABC, ABD, ACD.

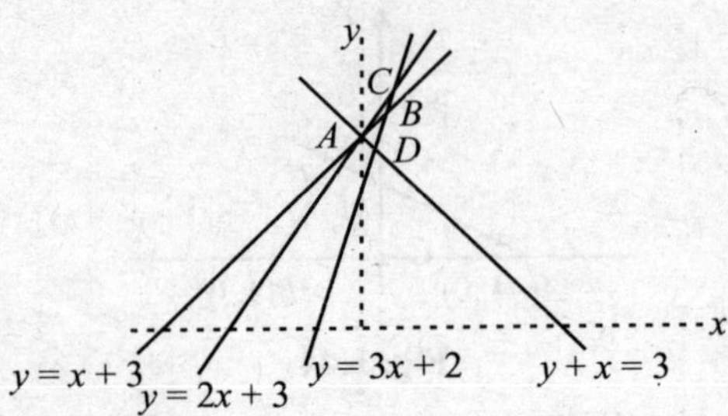

Fig. 1.45

90. c. We know that $PA + PB \geq AB$ (by triangle inequality). So, $PA + PB$ is the minimum if $PA + PB = AB$, i.e., A, P, B are collinear.

$$\therefore \begin{vmatrix} 3 & -4 & 1 \\ 1 & 2 & 1 \\ 2k-1 & 2k+1 & 1 \end{vmatrix} = 0$$

or $3(2 - 2k - 1) + 4(1 - 2k + 1) + 1(2k + 1 - 4k + 2) = 0$

or $3 - 6k + 8 - 8k + 3 - 2k = 0$

or $14 - 16k = 0$

$$\therefore k = \frac{7}{8}$$

91. b.

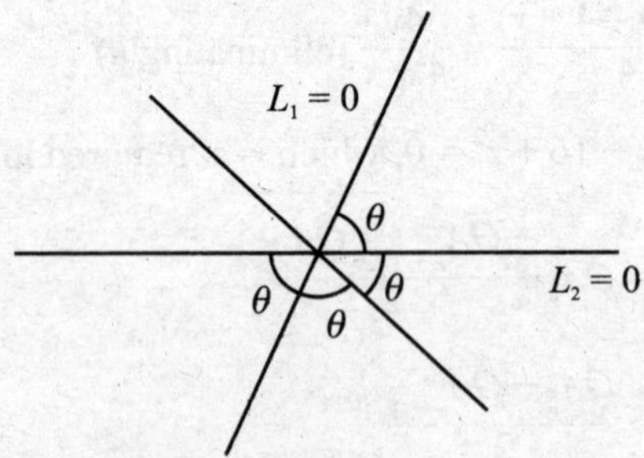

Fig. 1.46

From the figure, we have

$3\theta = 180 \Rightarrow \theta = 60°$

92. c.

$$\begin{vmatrix} 1 & 1 & -2 \\ 2 & -1 & 1 \\ a & b & -c \end{vmatrix} = 0$$

$$\Rightarrow a + 5b - 3c = 0$$

$$\Rightarrow -\frac{a}{3} - \frac{5}{3}b + c = 0$$

Hence, $2ax + 3by + c = 0$ is concurrent at $2x = -1/3$ and $3y = -5/3$. So, $x = -1/6$, $y = -5/9$.

93. a. Slope of is $AG = -b/(2a)$. Now,

$$\tan 30° = \frac{\frac{3b}{2a}}{1 + \frac{b^2}{a^2}}$$

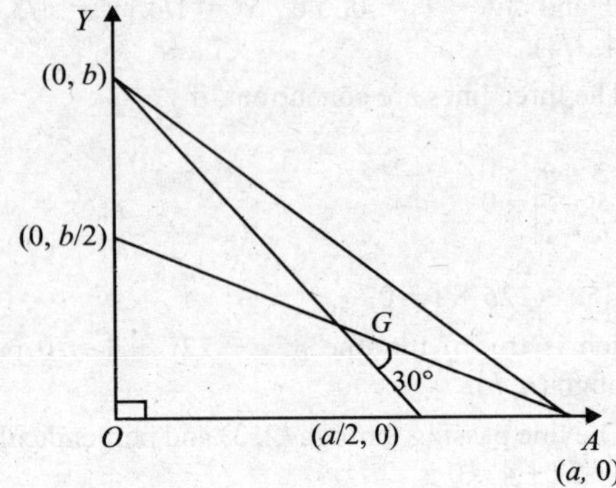

Fig. 1.47

$$\Rightarrow \frac{1}{\sqrt{3}} = \frac{3ba}{2(a^2 + b^2)}$$

$$\Rightarrow \frac{1}{2}ab = \left(\frac{a^2 + b^2}{3\sqrt{3}}\right)$$

$$= 9/3\sqrt{3} = \sqrt{3} \text{ (Putting } a^2 + b^2 = 9)$$

94. c. a, 8, b are in $H.P$

$$\Rightarrow \frac{1}{a} + \frac{1}{b} = 1/4$$

$$\Rightarrow b = \frac{4a}{a-4}$$

$$\Rightarrow \text{area, } A = \frac{4a^2}{2(a-4)}$$

A is minimum at $a = 8$. Hence, minimum value of A is 32 sq. units.

95. a. We have,

$$3x + 5y = 2007 \Rightarrow x + \frac{5y}{3} = 669$$

Clearly, 3 must divide $5y$ and so $y = 3k$, for some $k \in N$. Thus,

$x + 5k = 669$

$\Rightarrow 5k \leq 668$

$$\Rightarrow k \leq \frac{668}{5} \Rightarrow k \leq 133$$

96. d. Three non-collinear points form a triangle and the line joining the midpoints of any two sides is equidistant from all the three vertices.

97. a. The point (4, 5) lies on the given line $7x - 3y - 13 = 0$. The locus of the point equidistant from the given point and the line is a line perpendicular to $7x - 3y - 13 = 0$ at (4, 5).

98. b.

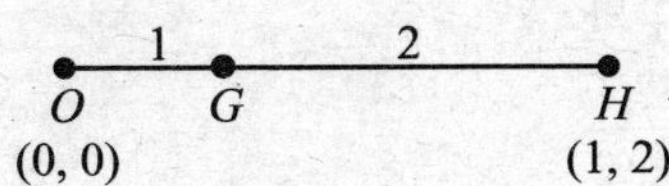

Fig. 1.48

$$G \equiv \left(\frac{1 \times 1 + 2 \times 0}{3}, \frac{2 \times 1 + 2 \times 0}{3}\right)$$

$$= \left(\frac{1}{3}, \frac{2}{3}\right)$$

99. c. Family of line through the given lines is

$L \equiv x - 7y + 5 + \lambda(x + 3y - 2) = 0$ (1)

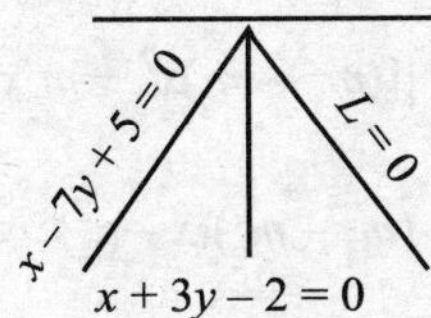

Fig. 1.49

For line $L = 0$ in the diagram, distance of any point say (2, 0) on the line $x + 3y - 2 = 0$ from the line $x - 7y + 5 = 0$ and the line $L = 0$ must be same.

$$\Rightarrow \left|\frac{2+5}{\sqrt{50}}\right| = \left|\frac{2 + 2\lambda + 5 - 2\lambda}{\sqrt{(1+\lambda)^2 + (3\lambda - 7)^2}}\right|$$

$\Rightarrow 10\lambda^2 - 40\lambda = 0$

$\Rightarrow \lambda = 4$ or 0

Hence, $L = 0, \lambda = 4$

$\Rightarrow$ Required line is $5x + 5y - 3 = 0$

100. b. Given $AB = BC$.

$$\tan\theta = \frac{AB}{OA} = m_1$$

$$\tan\alpha = \frac{2\,AB}{OA} = m_2$$

$$\frac{m_2}{m_1} = 2 \Rightarrow \frac{m_2 + m_1}{m_2 - m_1} = \frac{2+1}{2-1} = 3$$

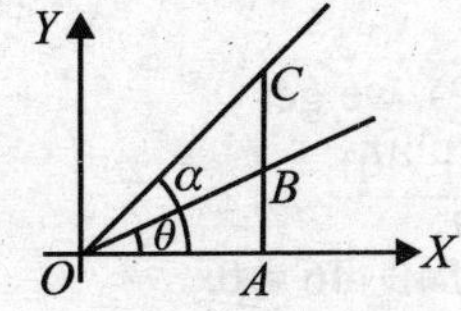

Fig. 1.50

$$\Rightarrow \frac{m_1 + m_2}{\sqrt{(m_1 + m_2)^2 - 4m_1m_2}} = 3$$

$$\Rightarrow -\frac{\frac{2h}{b}}{\sqrt{\frac{4h}{b^2} - \frac{4a}{b}}} = 3$$

$$\Rightarrow \frac{4h^2}{b^2} - \frac{4a}{b} = \frac{4h^2}{9b^2}$$

$$\Rightarrow \frac{4h^2}{b^2} \times \frac{8}{9} = \frac{4a}{b}$$

$\Rightarrow 8h^2 = 9ab$

101. d.

$$A = \frac{1}{2}\begin{vmatrix} 1 & p^2 & 1 \\ 0 & 1 & 1 \\ p & 0 & 1 \end{vmatrix}$$

$$= \frac{1}{2}[1(1-0) + p(p^2 - 1)]$$

$$= \frac{1}{2}(p^3 - p + 1)$$

Hence, $A = \frac{1}{2}|p^3 - p + 1|$. Now, minimum value of modulus is zero. Since $A(p)$ is cubic, it must vanish for some p other than given in (a), (b), (c).

102. b.

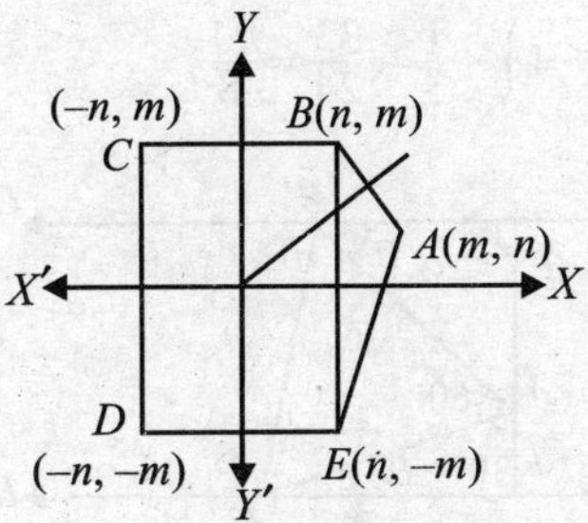

Fig. 1.51

Area of rectangle $BCDE$ is $4mn$. Area of ΔABE is $\frac{2m(m-n)}{2} = m^2 - mn$.

Therefore, the area of pentagon is $4mn + m^2 - mn = m^2 + 3mn$.

103. b.

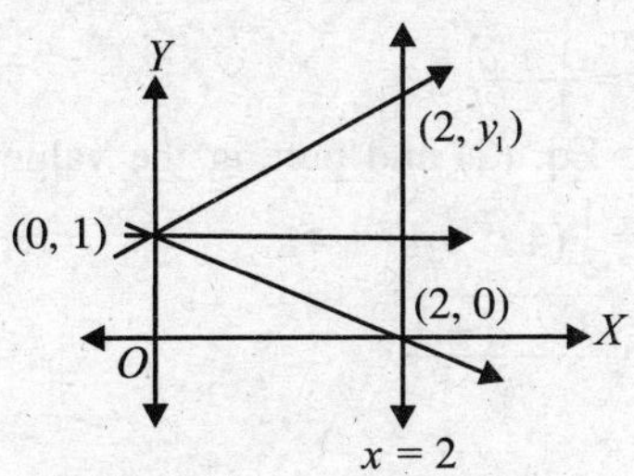

Fig. 1.52

From the figure,

$2^2 + (y_1 - 1)^2 = y_1^2$

$4 + y_1^2 + 1 - 2y_1 = y_1^2$

$5 = 2y_1$ or $y_1 = 5/2$

Equation of the line from (2, 5/2) to the given base is
$y - 5/2 = 2(x - 2)$ or
$2y - 5 = 4(x - 2)$
at $y = 1$, $-3/4 = x - 2$
or $x = 5/4$.

104. b.

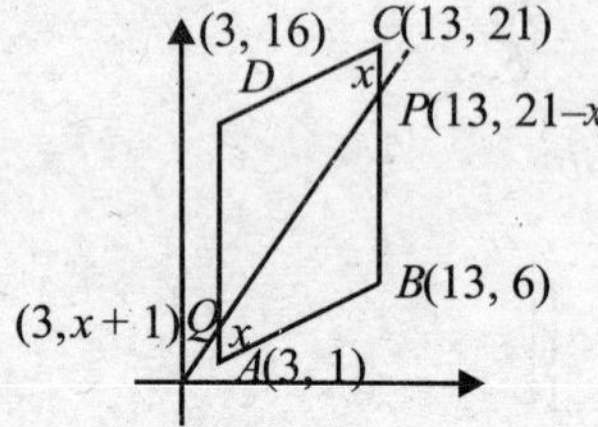

Fig. 1.53

$$m = \frac{21 - x}{13} = \frac{x + 1}{3}$$

$\Rightarrow 63 - 3x = 13x + 13$
$\Rightarrow 16x = 50$
$\Rightarrow x = \frac{25}{8}$

Hence,

$$m = \left(\frac{25}{8} + 1\right) \times \frac{1}{3} = \frac{33}{24} = \frac{11}{8}$$

105. b.

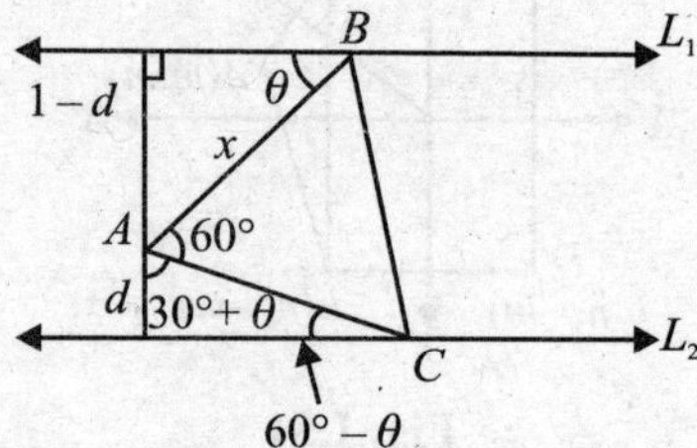

Fig. 1.54

From the figure,

$x \cos(\theta + 30°) = d$ (1)

$x \sin \theta = 1 - d$ (2)

Dividing (1) by (2), we have

$$\sqrt{3} \cot \theta = \frac{1 + d}{1 - d}$$

Squaring Eq. (2) and putting the value of cot θ, we have $x^2 = \frac{1}{3}(4d^2 - 4d + 4)$

$$\Rightarrow x = 2\sqrt{\frac{d^2 - d + 1}{3}}$$

106. d.

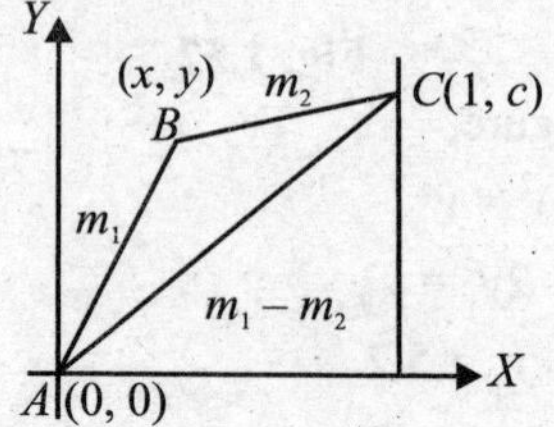

Fig. 1.55

Let the coordinates of C be $(1, c)$. Then,

$$m_2 = \frac{c - y}{1 - x} \text{ or}$$

$$m_2 = \frac{c - m_1 x}{1 - x}$$

$\Rightarrow m_2 - m_2 x = c - m_1 x$

$\Rightarrow (m_1 - m_2)x = c - m_2$

$\Rightarrow c = (m_1 - m_2)x + m_2$ (1)

Now area of ΔABC is

$$\frac{1}{2}\begin{vmatrix} 0 & 0 & 1 \\ x & m_1 x & 1 \\ 1 & c & 1 \end{vmatrix} = \frac{1}{2}[cx - m_1 x]$$

$$= \frac{1}{2}\left|[((m_1 - m_2)x + m_2)x - m_1 x]\right|$$

$$= \frac{1}{2}\left|[(m_1 - m_2)x^2 + m_2 x - m_1 x]\right|$$

$$= \frac{1}{2}(m_1 - m_2)(x - x^2)$$

$[\because x > x^2 \text{ in } (0, 1)]$

Hence, $f(x) = \frac{1}{2}(x - x^2)$

$f(x)_{\max} = \frac{1}{8}$ when $x = 1/2$.

107. d. $R(x, y)$ lies on $9x + 7y + 4 = 0$

Hence, $R\left(a \frac{-(4 + 9a)}{7}\right), a \in R$

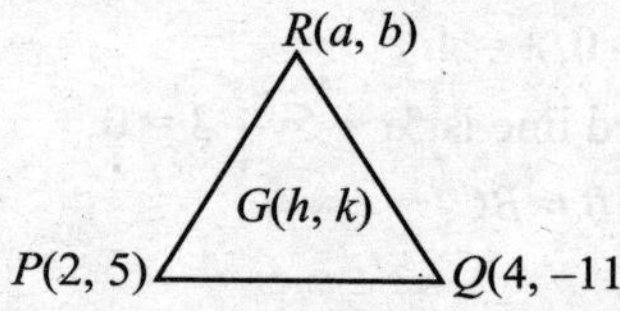

Fig. 1.56

$$h = \left(\frac{2 + 4 + a}{3}\right) = \frac{6 + a}{3} \quad (1)$$

$$k = \frac{5 - 11 - \frac{(4 + 9a)}{7}}{3}$$

$$= \frac{-46 - 9a}{7 \times 3} \quad (2)$$

From (1) and (2), we get

$$3h - 6 = \frac{-(21k - 46)}{9}$$

$\Rightarrow 27h + 21k - 54 + 46 = 0$

Hence, the locus is $9x + 7y - 8/3 = 0$. This line is parallel to N.

108. a. Equation of line AB is $y - 1 = m(x - 1)$

$\Rightarrow$ Equation of line AC is $y - 1 = -\frac{1}{m}(x - 1)$

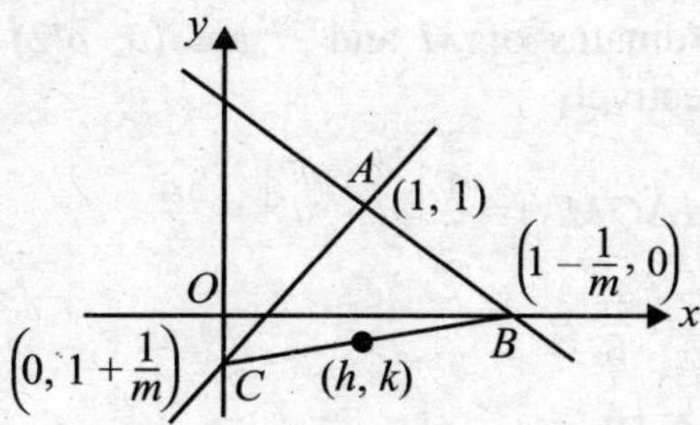

Fig. 1.57

$2h = 1 - \frac{1}{m}$

$2k = 1 + \frac{1}{m}$

Eliminating m we have locus $x + y = 1$.

109. c. Equation of any line through (2, 3) is

$y - 3 = m(x - 2)$

$\Rightarrow y = mx - 2m + 3$

From the figure, area of ΔOAB is $\pm$ 12. That is,

$\frac{1}{2}\left(\frac{2m-3}{m}\right)(3-2m) = \pm 12$

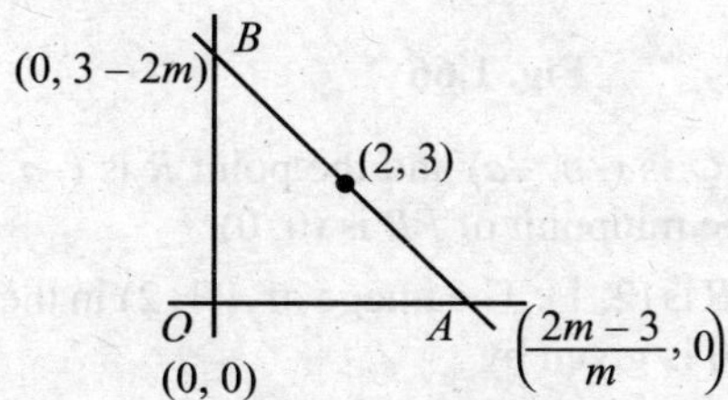

Fig. 1.58

Taking positive sign, we get $(2m + 3)^2 = 0$. This gives one value of m as $-3/2$. Taking negative sign, we get $4m^2 - 36m + 9 = 0$ $(D > 0)$

This is a quadratic in m which gives two values of m. Hence, three straight lines are possible.

110. b. BD and BE intersect at B. Coordinates of B are (–3, –2).

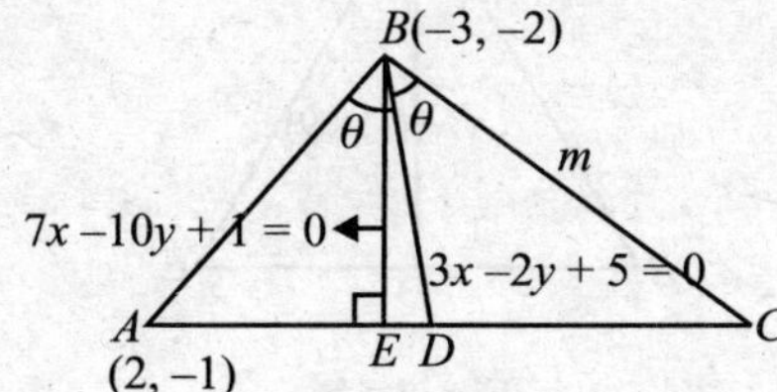

Fig. 1.59

$m_{AB} = \frac{1}{5}$

$\Rightarrow \tan\theta = \left|\frac{\frac{3}{2}-\frac{1}{5}}{1+\frac{3}{10}}\right| = \left|\frac{\frac{3}{2}-m}{1+\frac{3m}{2}}\right|$

$\Rightarrow 1 = \left|\frac{3-2m}{2+3m}\right|$

$\Rightarrow \pm 1 = \frac{3-2m}{2+3m}$

$\Rightarrow m = 1/5$ (rejected) or -5

Equation of BC is

$y + 2 = -5(x + 3)$

$\Rightarrow 5x + y + 17 = 0$

Alternative solution:

Take image of (2, –1) in the line BD to get a point on BC.

111. a. Coordinates of circumcentre are $l/(l^2 - m^2)$, $m/(m^2 - l^2)$. Hence,

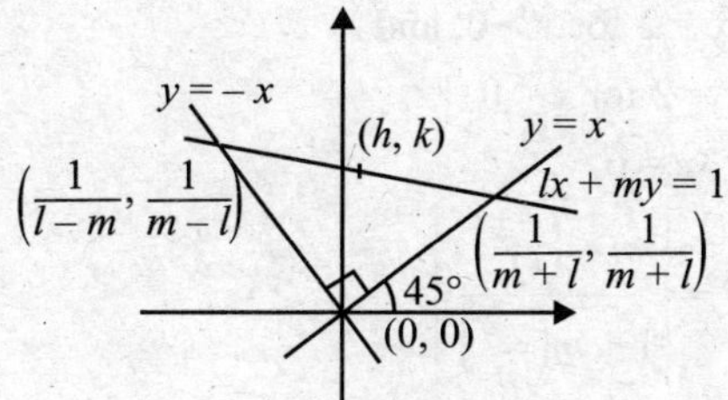

Fig. 1.60

$$h = \frac{l}{l^2 - m^2} \quad (1)$$

$$k = -\frac{m}{l^2 - m^2} \quad (2)$$

Squaring and adding (1) and (2), we get

$$h^2 + k^2 = \frac{l^2 + m^2}{(l^2 - m^2)^2} = \frac{1}{(l^2 - m^2)^2}$$

(putting $l^2 + m^2 = 1$)

Hence, $\frac{1}{(l^2 - m^2)^2} = (h^2 - k^2)^2$

Therefore, the locus is $x^2 + y^2 = (x^2 - y^2)^2$.

112. d. $\frac{\Delta AQP}{\Delta AOB} = \frac{3}{8}$ or

$$\frac{\frac{p^2\lambda}{(\lambda+1)^2}}{\frac{1}{2}p^2} = \frac{3}{8}$$

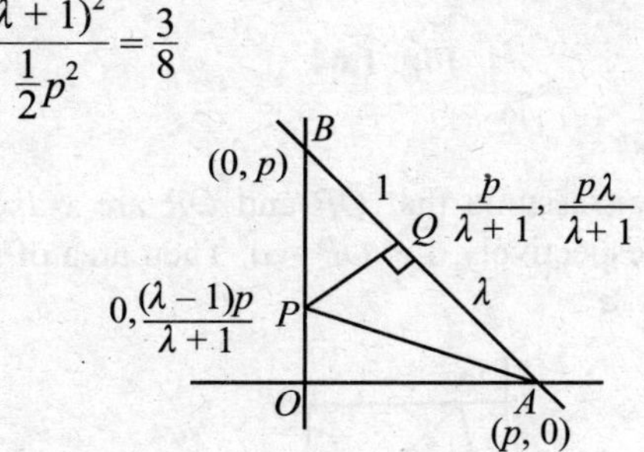

Fig. 1.61

$\Rightarrow \lambda = 3, \frac{1}{3}$

$\Rightarrow \frac{AQ}{BQ} = 3$ or $\frac{1}{3}$

The value 1/3 is rejected because this gives negative coordinates of P and it is given that P lies on OB.

113. b.

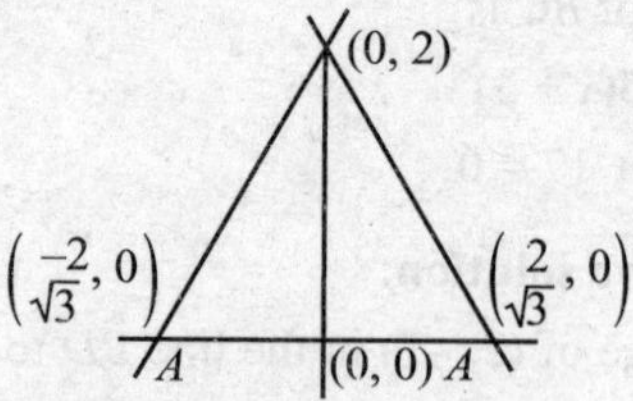

Fig. 1.62

From the figure, we have

$y + \sqrt{3}x = 2$ for $x > 0$, and

$y - \sqrt{3}x = 2$ for $x < 0$

114. a. $2x + 3y = 6$

$$\tan 45° = \left|\frac{m - \left(\frac{-2}{3}\right)}{1 + m\left(\frac{-2}{3}\right)}\right|$$

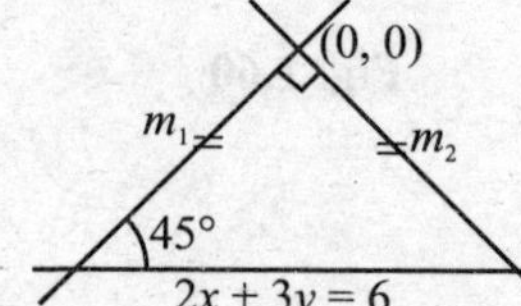

Fig. 1.63

Hence, $m_1 = -5$, $m_2 = 1/5$.

115. d. $\tan\theta = 7$

$OA = OB = r$

$\sin\theta = \frac{7}{5\sqrt{2}}$, $\cos\theta = \frac{1}{5\sqrt{2}}$

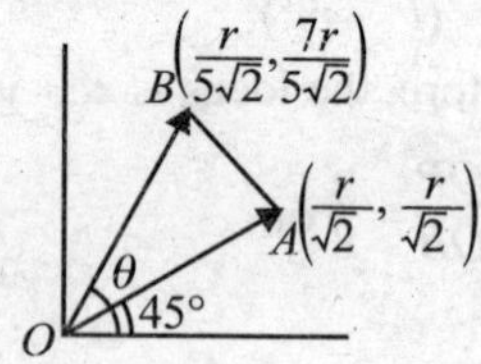

Fig. 1.64

Now, $m_{AB} = -1/2$.

116. b. We can assume that OP and OR are x-axes and y-axes, respectively. Let $OP = a$. Then area of square $OPQR$ is a^2.

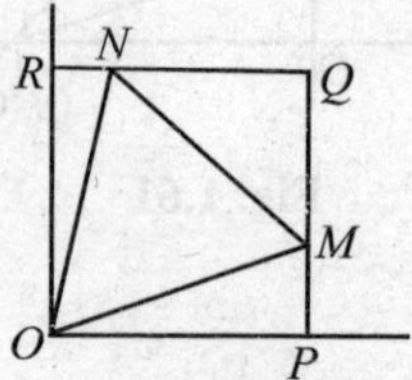

Fig. 1.65

Coordinates of M and N are $(a, a/2)$ and $(a/2, a)$, respectively.

$$\therefore \text{ar}(\Delta OMN) = \frac{1}{2}\begin{vmatrix} a & a/2 \\ a/2 & a \end{vmatrix} = \frac{3a^2}{8}$$

$$\therefore \frac{8}{3} = \frac{\lambda}{6}$$

$\Rightarrow \lambda = 16$

117. d. Orthocentre of triangle BCH is the vertex $A(-1, 0)$.

118. a. Distance of (0, 0) from the line $2x + 3y - 6 = 0$ is $6/\sqrt{4+9} = 6/\sqrt{13}$. The area of the triangle is $(6/\sqrt{13})^2 = 36/13$.

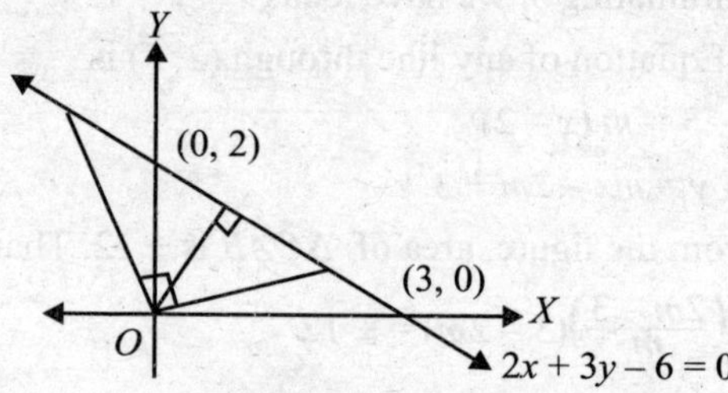

Fig. 1.66

119. d. The point Q is $(-b, -a)$ and the point R is $(-a, -b)$. Therefore, the midpoint of PR is (0, 0).

120. b. The point B is (2, 1). The image of $A(1, 2)$ in the line $x - 2y + 1 = 0$ is given by

$$\frac{x-1}{-1} = \frac{y-2}{-2} = \frac{4}{5}$$

Hence, coordinates of the point are (9/5, 2/5).

Since this point lies on BC, therefore the equation of BC is $3x - y - 5 = 0$. Hence, $a + b = 2$.

121. a.

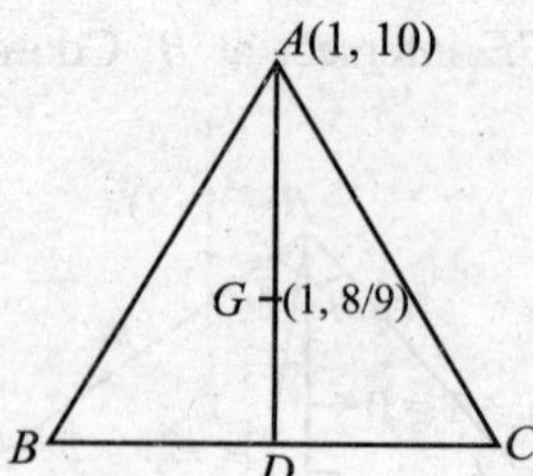

Fig. 1.67

Circumcentre $O \equiv (-1/3, 2/3)$ and orthocentre $H \equiv (11/3, 4/3)$.

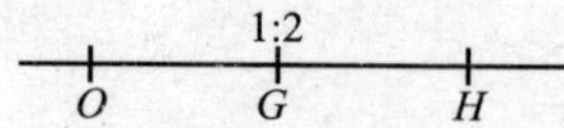

Fig. 1.68

Therefore, the coordinates of G are (1, 8/9). Now, the point A is (1, 10) as G is (1, 8/9). Hence,

$AD:DG = 3:1$

$\therefore D_x = \frac{3-1}{2} = 1, D_y = \frac{\frac{8}{3}-10}{2} = -\frac{11}{3}$

Hence, the coordinates of the midpoint of BC are $(1, -11/3)$.

122. b. (2, 0) is midpoint of $B(0, 0)$ and C, then C has coordinates (4, 0). Also, A has coordinates $(0 + 2\cos\pi/3, 0 + 2\sin\pi/3) \equiv (1, \sqrt{3})$. Then centroid is $(5/3, 1/\sqrt{3})$.

123. d.

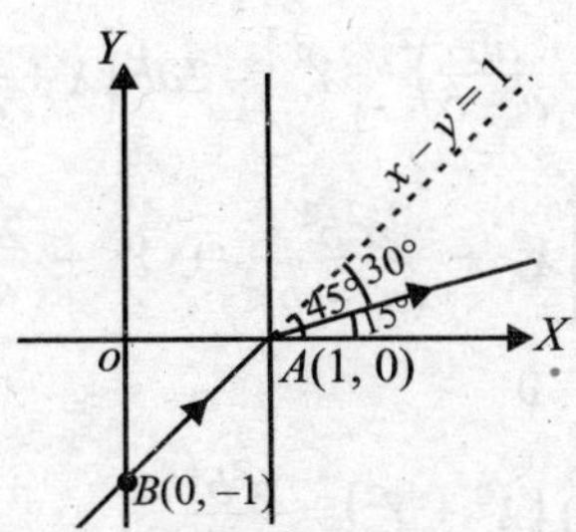

Fig. 1.69

From figure refracted ray makes an angle of 15° with positive direction of x-axis and passes through the point (1, 0). Its equation is

$(y - 0) = \tan(45° - 30°)(x - 1)$ or

$y = (2 - \sqrt{3})(x - 1)$

124. a. Let the line cut the axis at points $A(a, 0)$ and $B(0, b)$. Now given that (−4, 3) divides AB in ratio 5:3. Then, $-4 = 3a/8$ and $3 = 5b/8$. Therefore, $a = -32/3$ and $b = 24/5$. Then using the intercept form $x/a + y/b = 1$, the equation of line is

$$-\frac{3x}{32} + \frac{5y}{24} = 1 \text{ or}$$

$$9x - 20y + 96 = 0$$

125. d.

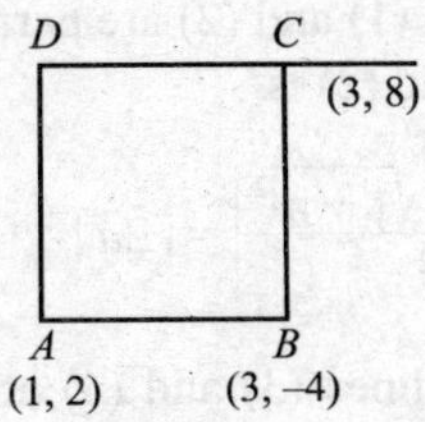

Fig. 1.70

$$m_{AB} = \frac{-4-2}{3-1} = -3$$

Thus equation of CD is $y - 8 = -3(x - 3)$, i.e., $y + 3x = 17$. Equation of right bisector of AB is

$$y + 1 = \frac{1}{3}(x - 2)$$

$$\Rightarrow 3y = x - 5$$

Solving it with line CD, we get

$x = 24/5, y = 1/5$. Thus midpoint of CD is (24/5, 1/5).

126. b. If the given lines represent the same line, then the lengths of the perpendiculars from the origin to the lines are equal, so that

$$\frac{c}{\sqrt{1+m^2}} = \frac{p}{\sqrt{\cos^2\alpha + \sin^2\alpha}}$$

$$\Rightarrow c = p\sqrt{1+m^2}$$

127. b.

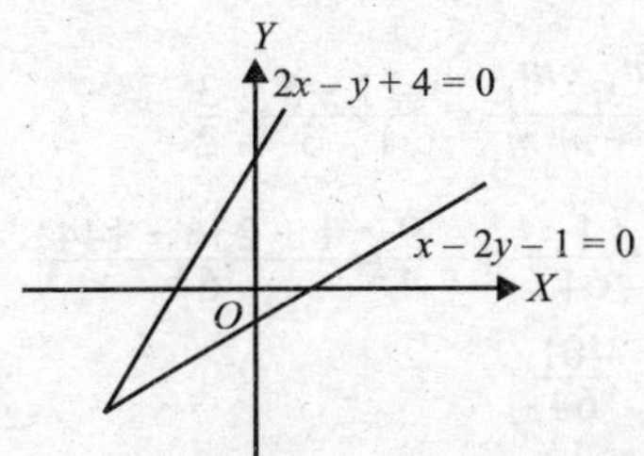

Fig. 1.71

Clearly, from the figure, the origin is contained in the acute angle. Writing the equations of the lines as $2x - y + 4 = 0$ and $-x + 2y + 1 = 0$, the required bisector is

$$\frac{2x - y + 4}{\sqrt{5}} = \frac{-x + 2y + 1}{\sqrt{5}}$$

128. c. We have to find locus of the point (h, k) whose image in the line $2x - y - 1 = 0$ lies on the line $y = x$. Now, image of (h, k) in the line $2x - y - 1 = 0$ is given by

$$\frac{x_2 - h}{2} = \frac{y_2 - k}{-1} = -\frac{2(2h - k - 1)}{5}$$

$$\Rightarrow x_2 = \frac{-3h + 4k + 4}{5}$$

$$\text{and } y_2 = \frac{4h + 3k - 2}{5}$$

This point lies on $y = x$. Then,

$$\frac{-3h + 4k + 4}{5} = \frac{4h + 3k - 2}{5} \Rightarrow 7h - k = 6$$

129. c.

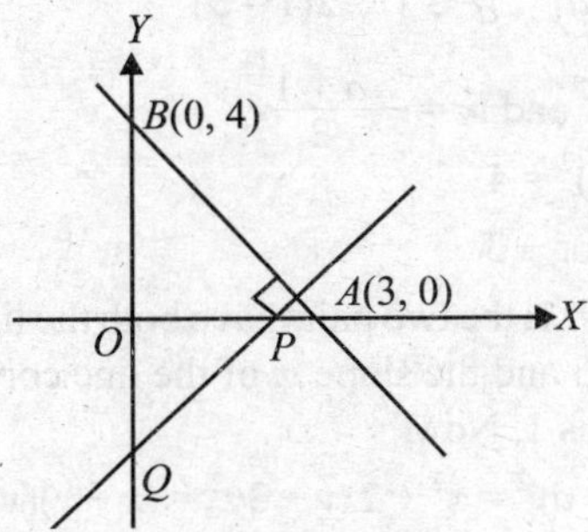

Fig. 1.72

Clearly, circumcentre of triangle ABQ will lie on the perpendicular bisector of line AB. Now equation of perpendicular bisector of line AB is $3x - 4y + 7/2 = 0$.

Hence, locus of circumcentre is $6x - 8y + 7 = 0$.

130. d. Arranging the lines in descending order of slope, we have $m_1 = 5$, $m_2 = 3$ and $m_3 = -1$

$$\therefore \tan A = \frac{m_1 - m_2}{1 + m_1 m_2} = \frac{2}{1+15} = \frac{1}{8}$$

$$\tan B = \frac{m_2 - m_3}{1 + m_2 m_3} = \frac{3+1}{1-3} = -2$$

$$\tan C = \frac{m_3 - m_1}{1 + m_3 m_1} = \frac{-1-5}{1-5} = \frac{3}{2}$$

$$\sum \tan^2 A = \frac{1}{64} + 4 + \frac{9}{4} = \frac{1 + 256 + 144}{64}$$

$$= \frac{401}{64}$$

$$\Rightarrow p + q = 465$$

131. b. If the given lines are concurrent, then

$$\begin{vmatrix} a & 1 & 1 \\ 1 & b & 1 \\ 1 & 1 & c \end{vmatrix} = 0$$

$$\Rightarrow \begin{vmatrix} a & 1-a & 1-a \\ 1 & b-1 & 0 \\ 1 & 0 & c-1 \end{vmatrix} = 0 \text{ (Applying } C_2 \to C_2 - C_1 \text{ and } C_3 \to C_3 - C_1)$$

$$\Rightarrow a(b-1)(c-1) - (c-1)(1-a) - (b-1)(1-a) = 0$$

$$\Rightarrow \frac{a}{1-a} + \frac{1}{1-b} + \frac{1}{1-c} = 0$$

[Dividing by $(1-a)(1-b)(1-c)$]

Adding 1 on both sides, we get

$$\Rightarrow \frac{1}{1-a} + \frac{1}{1-b} + \frac{1}{1-c} = 1$$

132. b. Let $y = mx$ be a line common to the given pairs of lines. Then

$$am^2 + 2m + 1 = 0 \text{ and } m^2 + 2m + a = 0$$

$$\Rightarrow \frac{m^2}{2(1-a)} = \frac{m}{a^2 - 1} = \frac{1}{2(1-a)}$$

$$\Rightarrow m^2 = 1 \text{ and } m = -\frac{a+1}{2}$$

$$\Rightarrow (a+1)^2 = 4$$

$$\Rightarrow a = 1 \text{ or } -3$$

But for $a = 1$, the two pairs have both the lines common. So $a = -3$ and the slope m of the line common to both the pairs is 1. Now,

$$x^2 + 2xy + ay^2 = x^2 + 2xy - 3y^2 = (x - y)(x + 3y) \text{ and}$$

$$ax^2 + 2xy + y^2 = -3x^2 + 2xy + y^2 = -(x - y)(3x + y)$$

So the equation of the required lines is

$$(x + 3y)(3x + y) = 0$$

$$\Rightarrow \quad 3x^2 + 10xy + 3y^2 = 0$$

133. d. The given equation is

$$(a - b)(x^2 + y^2) - 2abx = 0 \qquad (1)$$

The origin is shifted to $(ab/(a - b), 0)$. Any point (x, y) on the curve (1) must be replaced with new point (X, Y) with reference to new axes, such that

$$x = X + \frac{ab}{a-b} \text{ and } y = Y + 0$$

Substituting these in (1), we get

$$(a-b)\left[\left(X + \frac{ab}{a-b}\right)^2 + Y^2\right] - 2ab\left[X + \frac{ab}{a-b}\right] = 0$$

$$\Rightarrow (a-b)\left[X^2 + \frac{a^2b^2}{(a-b)^2} + Y^2 + \frac{2abX}{a-b}\right] - 2abX - \frac{2a^2b^2}{a-b} = 0$$

$$\Rightarrow \quad (a-b)(X^2 + Y^2) = \frac{a^2b^2}{a-b}$$

$$\Rightarrow \quad (a-b)^2 (X^2 + Y^2) = a^2b^2$$

134. b. The straight lines represented by $(y - mx)^2 = a^2(1 + m^2)$ are

$$y - mx = \pm a\sqrt{1 + m^2} \text{ i.e.,}$$

$$y - mx = a\sqrt{1 + m^2} \qquad (1)$$

and

$$y - mx = -a\sqrt{1 + m^2} \qquad (2)$$

Similarly, the straight lines represented by $(y - nx)^2 = a^2(1 + n^2)$ are

$$y - nx = a\sqrt{1 + n^2} \qquad (3)$$

$$\text{and } y - nx = -a\sqrt{1 + n^2} \qquad (4)$$

Since the lines (1) and (2) are parallel, so the distance between (1) and (2) is

$$\left|\frac{a\sqrt{1+m^2} + a\sqrt{1+m^2}}{\sqrt{1+m^2}}\right| = |2a|$$

Similarly, the lines (3) and (4) are parallel lines and the distance between them is $|2a|$. Since the distances between parallel lines are same, hence the four lines form a rhombus.

135. a. Let the common line be $y = mx$. Then it must satisfy both the equations. Therefore, we have

$$bm^2 + 2hm + a = 0 \qquad (1)$$

$$b'm^2 + 2h'm + a' = 0 \qquad (2)$$

Solving Eqs. (1) and (2), we get

$$\frac{m^2}{2(ha' - h'a)} = \frac{m}{ab' - a'b} = \frac{a}{2(bh' - b'h)}$$

Eliminating m, we get

$$\left[\frac{ab' - a'b}{2(bh' - b'h)}\right]^2 = \frac{ha' - h'a}{bh' - b'h}$$

$\Rightarrow (ab' - a'b)^2 = 4(ha' - h'a)(bh' - b'h)$

136. b. Being a pair of lines, $abc + 2fgh - af^2 - bg^2 - ch^2 = 0$. This gives $m = 4$. Now find angle between lines.

137. d. Here $my(y - mx) + x(y - mx) = 0$, i.e., $(y - mx)(my + x) = 0$. So the lines are $y = mx$ and $y = (-1/m)x$. Bisectors between the lines $xy = 0$ are $y = x$ and $y = -x$. Therefore, $m = 1$ or -1.

138. a. $3a + a^2 - 2 = 0$

$\Rightarrow a^2 + 3a - 2 = 0;$

$\Rightarrow a = \dfrac{-3 \pm \sqrt{9+8}}{2} = \dfrac{-3 \pm \sqrt{17}}{2}$

$\Rightarrow$ two values of a

139. d. Since the product of the slope of the four lines represented by the given equation is 1 and a pair of lines represents the bisectors of the angles between the other two, the product of the slopes of each pair is -1. So let the equation of one pair be $ax^2 + 2hxy - ay^2 = 0$. Then the equation of its bisectors is

$$\frac{x^2 - y^2}{2a} = \frac{xy}{h}$$

By hypothesis,

$$x^4 + x^3y + cx^2y^2 - xy^3 + y^4 = (ax^2 + 2hxy + ay^2)(hx^2 - 2axy - hy^2)$$

$$= ah(x^4 + y^4) + 2(h^2 - a^2)(x^3y - xy^3) - 6ahx^2y^2$$

140. b. The given equation of pair of straight lines can be rewritten as $(\sqrt{3}\,y - x + \sqrt{3})(\sqrt{3}\,y + x - \sqrt{3}) = 0$

Their separate equations are $\sqrt{3}y - x + \sqrt{3} = 0$ and $\sqrt{3}\,y + x - \sqrt{3} = 0$

or $y = \dfrac{1}{\sqrt{3}}x - 1$ and $y = -\dfrac{1}{\sqrt{3}}x + 1$

or $y = (\tan 30°)\,x - 1$ and $y = (\tan 150°)\,x + 1$

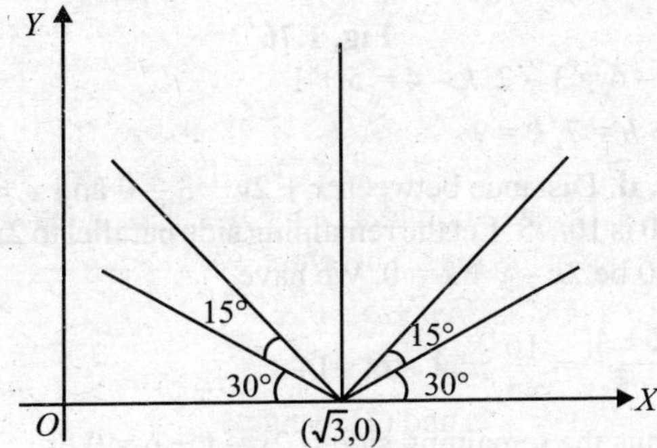

Fig. 1.73

After rotation through an angle of 15°, the lines are $(y - 0) = \tan 45° (x - \sqrt{3})$ and $(y - 0) = \tan 135° (x - \sqrt{3})$ or $y = x - \sqrt{3}$ and $y = -x + \sqrt{3}$

Their combined equation is

$(y - x + \sqrt{3})(y + x - \sqrt{3}) = 0$

or $y^2 - x^2 + 2\sqrt{3}x - 3 = 0$

141. c. Equations of the given lines are

$y - 1 = \tan\theta\,(x - 1)$ and

$y - 1 = \cot\theta\,(x - 1)$

So their joint equation is

$[(y - 1) - \tan\theta\,(x - 1)]\,[(y - 1) - \cot\theta\,(x - 1)] = 0$

$\Rightarrow (y - 1)^2 - (\tan\theta + \cot\theta)(x - 1)(y - 1) + (x - 1)^2 = 0$

$\Rightarrow x^2 - (\tan\theta + \cot\theta)\,xy + y^2 + (\tan\theta + \cot\theta - 2)(x + y - 1) = 0$

Comparing with the given equation, we get

$\tan\theta + \cot\theta = a + 2$

$\Rightarrow \dfrac{1}{\sin\theta\cos\theta} = a + 2$

$\Rightarrow \sin 2\theta = \dfrac{2}{a+2}$

142. c. We have, $6x^2 - xy - 12y^2 = 0$

$\Rightarrow (2x - 3y)(3x + 4y) = 0$ (1)

and $15x^2 + 14xy - 8y^2 = 0$

$\Rightarrow (5x - 2y)(3x + 4y) = 0$ (2)

Equation of the line common to (1) and (2) is

$3x + 4y = 0$ (3)

Equation of any line parallel to (3) is

$3x + 4y = k$

Since its distance from (3) is 7, so

$$\left|\frac{k}{\sqrt{3^2 + 4^2}}\right| = 7 \Rightarrow k = \pm 35$$

143. c. We have,

$x^2y^2 - 9y^2 - 6x^2y + 54y = 0$

$\Rightarrow y^2(x^2 - 9) - 6y(x^2 - 9) = 0$

$\Rightarrow y(y - 6)(x - 3)(x + 3) = 0$

$\Rightarrow y = 0, y = 6, x = 3, x = -3$

So, the given equation represents four straight lines which form a square.

144. d. The combined equation of bisectors of angles between the lines of the first pair is

$$\frac{x^2 - y^2}{2 - 1} = \frac{xy}{9}$$

As these equations are the same, the two pairs are equally inclined to each other.

145. a. Acute angle between the lines $x^2 + 4xy + y^2 = 0$ is $\tan^{-1}(2\sqrt{4-1})/(1+1) = \tan^{-1}\pi/3$. Angle bisectors of $x^2 + 4xy + y^2 = 0$ are given by

$$\frac{x^2 - y^2}{1-1} = \frac{xy}{2}$$

$$\Rightarrow x^2 - y^2 = 0 \Rightarrow x = \pm y$$

As $x + y = 0$ is perpendicular to $x - y = 4$, the given triangle is isosceles with vertical angle equal to $\pi/3$ and hence it is equilateral.

146. c. The given equations are

$a^2x^2 + 2h(a+b)xy + b^2y^2 = 0$ (1)

and $ax^2 + 2hxy + by^2 = 0$ (2)

The equation of the bisectors of the angles between the lines represented by (1) is

$$\frac{x^2 - y^2}{a^2 - b^2} = \frac{xy}{h(a+b)} \text{ or}$$

$$\frac{x^2 - y^2}{a-b} = \frac{xy}{h}$$

which is same as equation of the bisectors of angles between the line pair (2). Thus, two line pairs are equally inclined to each other.

147. a. $9x^2 - 24xy + 16y^2 - 12x + 16y - 12 = 0$

$$\Rightarrow (3x - 4y + 2)(3x - 4y - 6) = 0$$

Hence, distance between lines is $\dfrac{|6-(-2)|}{5} = 8/5$.

Multiple Correct Answers Type

1. a., c.

Let $L = 3x - 4y - 8$. Then the value of L at (3, 4) is $3 \times 3 - 4 \times 4 - 8 = -15 < 0$. Hence, for the point $P(x, y)$, we should have $L > 0$

$\Rightarrow 3x - 4y - 8 > 0$

$\Rightarrow 3x - 4(-3x) - 8 > 0$

[$\because P(x, y)$ lies on $y = -3x$]

$\Rightarrow x > 8/15$

and $-y - 4y - 8 > 0$

$\Rightarrow y < -8/5$

2. b., c., d. If the remaining vertex is (h, k), then

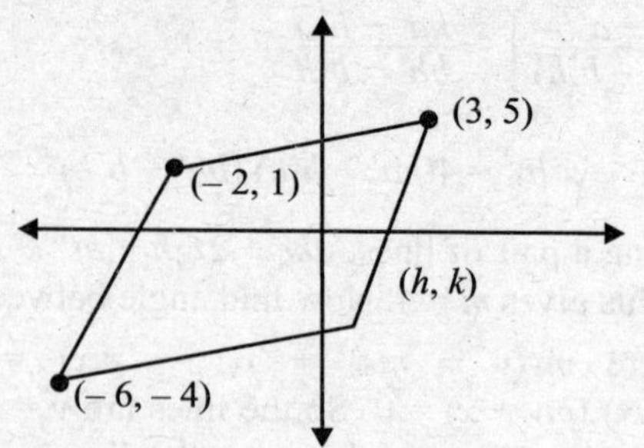

Fig. 1.74

$h - 2 = -6 + 4, k + 1 = 5 - 4$

$\Rightarrow h = -1, k = 0$

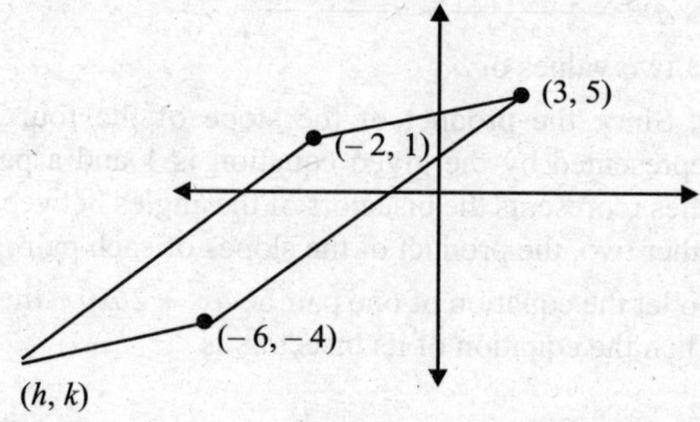

Fig. 1.75

$h + 3 = -6 - 2, k + 5 = -4 + 1$

$\Rightarrow h = -11, k = -8$

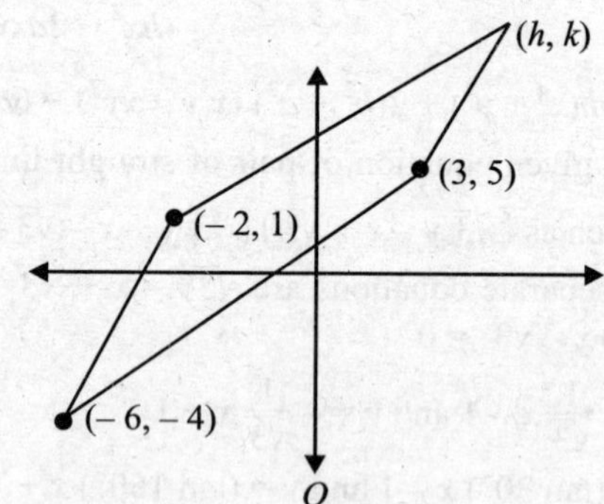

Fig. 1.76

$h - 6 = 3 - 2, k - 4 = 5 + 1$

$\Rightarrow h = 7, k = 9$

3. a., d. Distance between $x + 2y + 3 = 0$ and $x + 2y - 7 = 0$ is $10/\sqrt{5}$. Let the remaining side parallel to $2x - y - 4 = 0$ be $2x - y + \lambda = 0$. We have,

$$\frac{|\lambda + 4|}{\sqrt{5}} = \frac{10}{\sqrt{5}} \Rightarrow \lambda = 6, -14$$

Thus the remaining side is $2x - y + 6 = 0$ or

$2x - y - 14 = 0$.

4. a., d. Let point P be (x, y)

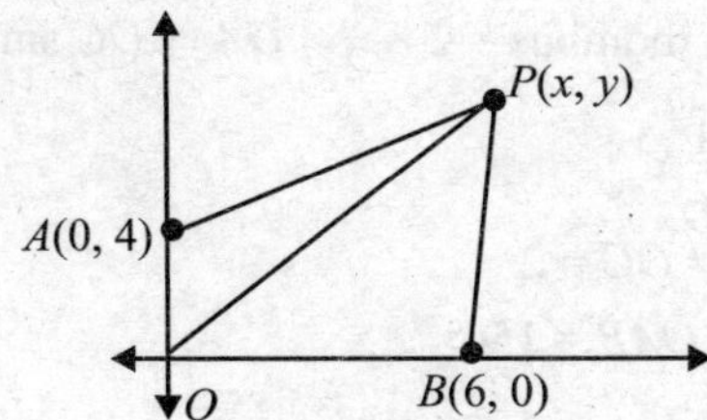

Fig. 1.77

$\Delta_{POA} = \frac{1}{2}(OA)|x| = 2|x|$

$\Delta_{POB} = \frac{1}{2}(OB)|y| = 2|y|,$

where $P \equiv (x, y) \Rightarrow 2|x| = 6|y|$

$\Rightarrow |x| = 3|y| \Rightarrow 3y - x = 0$

or $3y + x = 0$

5. a., c., d.

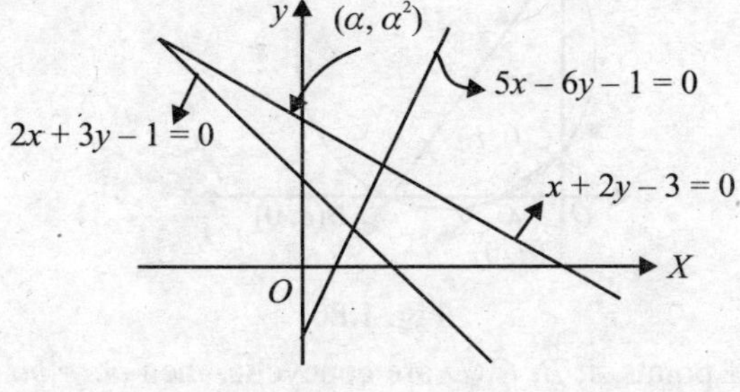

Fig. 1.78

O and the point (α, α^2) lie on the opposite sides w.r.t. $2x + 3y - 1 = 0$. Hence,

$\Rightarrow 2\alpha + 3\alpha^2 - 1 > 0$ (1)

O and the point (α, α^2) lie to the same side w.r.t. $x + 2y - 3 = 0$. Hence,

$\Rightarrow \alpha + 2\alpha^2 - 3 < 0$ (2)

Again O and the point (α, α^2) lie to the same side w.r.t. $5x - 6y - 1 = 0$. Hence,

$5\alpha - 6\alpha^2 - 1 < 0$

$\Rightarrow 6\alpha^2 - 5\alpha + 1 > 0$

6. b., d. Let the angle be θ. Then, equation of the given line is

$$\frac{x-1}{\cos\theta} = \frac{y-2}{\sin\theta} \quad (1)$$

The coordinates of a point on (1) at a distance $\sqrt{6}/3$ from (1, 2) are $(1 + \sqrt{6}/3 \cos\theta, 2 + \sqrt{6}/3 \sin\theta)$. This point lies on $x + y = 4$. Therefore,

$$1 + \frac{\sqrt{6}}{3}\cos\theta + 2 + \frac{\sqrt{6}}{3}\sin\theta = 4$$

$$\Rightarrow \cos\theta + \sin\theta = \sqrt{\frac{3}{2}}$$

$$\Rightarrow \frac{1}{\sqrt{2}}\cos\theta + \frac{1}{\sqrt{2}}\sin\theta = \frac{\sqrt{3}}{2}$$

$\Rightarrow \cos(\theta - \pi/4) = \cos(\pm\pi/6)$

$\Rightarrow \theta - \pi/4 = \pm\pi/6$

$\Rightarrow \theta = 75°$ or $\theta = 15°$

7. b., c. For the two lines $24x + 7y - 20 = 0$ and $4x - 3y - 2 = 0$, the angle bisectors are given by

$$\frac{24x + 7y - 20}{25} = \pm\frac{4x - 3y - 2}{5}$$

Taking positive sign, we get

$$2x + 11y - 5 = 0$$

8. c., d. Let the third vertex be (x, y). Then,

$$\frac{1}{2}\begin{vmatrix} x & y & 1 \\ -4 & 0 & 1 \\ 1 & -1 & 1 \end{vmatrix} = \pm 4$$

$\Rightarrow x + 5y + 4 = \pm 8$

$\Rightarrow x + 5y + 12 = 0$ or $x + 5y - 4 = 0$

Hence, the third vertex lies on $x + 5y + 12 = 0$ or $x + 5y - 4 = 0$

9. a., b., c. The equation is

$$x^2(x + y) - y^2(x + y) = 0$$

or $(x + y)^2 (x - y) = 0$

It represents the lines $x + y = 0$, $x + y = 0$, $x - y = 0$.

10. a., c. The equation represents a pair of straight lines. Hence,

$$1 \times (-2)(-1) + 2\left(\frac{3}{2}\right) \times 0 \times \frac{m}{2} - 1 \times \left(\frac{3}{2}\right)^2 - (-2) \times 0^2 - (-1) \times \left(\frac{m}{2}\right)^2 = 0$$

$\Rightarrow m = 1, -1$

The points of intersection of the pair of lines are obtained by solving

$$\frac{\partial S}{\partial x} \equiv 2x + my = 0$$

and $\frac{\partial S}{\partial y} \equiv mx - 4y + 3 = 0$

When $m = 1$, the required point is the intersection of $2x + y = 0$, $x - 4y + 3 = 0$. When $m = -1$, the required point is the intersection of $2x - y = 0$, $-x - 4y + 3 = 0$.

11. a, c., d. Vertices of the given triangle are $(0, 0)$, $(a/m_1, a)$, and $(a/m_2, a)$. So the area of the triangle is equal to $a^2(m_2 - m_1)/(2m_1m_2)$. Since m_1, m_2 are the roots of $x^2 - ax - a - 1 = 0$, so

$m_1 + m_2 = a$; $m_1 m_2 = -(a + 1)$

$$\Rightarrow (m_1 - m_2)^2 = (m_1 + m_2)^2 - 4m_1m_2$$
$$= a^2 + 4(a + 1)$$
$$= (a + 2)^2$$

$\Rightarrow m_1 - m_2 = \pm(a + 2)$

So the required area is

$$\Delta = \pm \frac{a^2(a+2)}{-2(a+1)} = \pm \frac{a^2(a+2)}{2(a+1)}$$

Since area is a positive quantity, so

$$\Delta = \frac{a^2(a+2)}{2(a+1)} \text{ if } a > -1 \text{ or } a < -2$$

$$\text{and } \Delta = -\frac{a^2(a+2)}{2(a+1)} \text{ if } -2 < a < -1$$

12. a., b., d. Since the given point lies on the line $lx + my + n = 0$, so a, b, c are the roots of the equation

$$l\left(\frac{t^3}{t-1}\right) + m\left(\frac{t^2-3}{t-1}\right) + n = 0 \text{ or}$$

$$lt^3 + mt^2 + nt - (3m+n) = 0 \quad (1)$$

$$\text{Hence, } a+b+c = -\frac{m}{l}$$

$$ab + bc + ca = \frac{n}{l} \quad (2)$$

$$abc = \frac{3m+n}{l} \quad (3)$$

So from Eqs. (1), (2) and (3), we get

$$abc - (bc + ca + ab) + 3(a+b+c) = 0$$

13. a.,d. Here,

$$my(y - mx) + x(y - mx) = 0$$

$$\Rightarrow (y - mx)(my + x) = 0$$

So, the lines are $y = mx$ or $y = (-1/m)x$. Bisectors between the lines $xy = 0$ are $y = x$ and $y = -x$. Therefore, $m = 1, -1$.

14. b., c. The chord subtends 90° at the centre (0, 0). Making $x^2 + y^2 = 1$ homogeneous in the second degree with the help of $y = mx + 1$, we get

$$x^2 + y^2 = (y - mx)^2$$

$$\text{or } (1 - m^2)x^2 + 2mxy = 0$$

The angle between these lines is 90° if $1 - m^2 + 0 = 0$, i.e., $m = \pm 1$.

15. a., b.

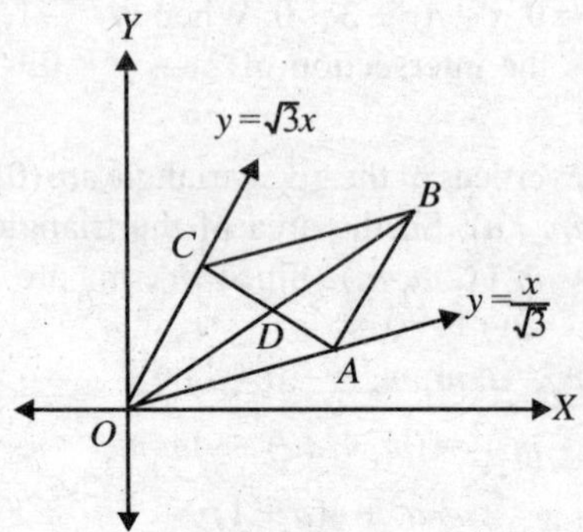

Fig. 1.79

Here, $\angle COA = 30°$.

Area of rhombus $= 2 \times \frac{1}{2} \times OA \times OC \sin 30°$

$$\Rightarrow 2 = \frac{1}{2}x^2$$

$$\Rightarrow OA = OC = 2$$

Also, $\angle OAB = 150°$

$$\Rightarrow \cos 150° = \frac{OA^2 + AB^2 - OB^2}{2\, OA \times AB}$$

$$OB^2 = 8 + 4\sqrt{3}$$

$$\Rightarrow OB = \sqrt{2}\,(\sqrt{3} + 1)$$

Hence, the coordinates of B are $(\pm\sqrt{2}\,(\sqrt{3}+1)\cos 45°, \pm\sqrt{2}(\sqrt{3}+1)\sin 45°)$.

16. a., b., c., d.

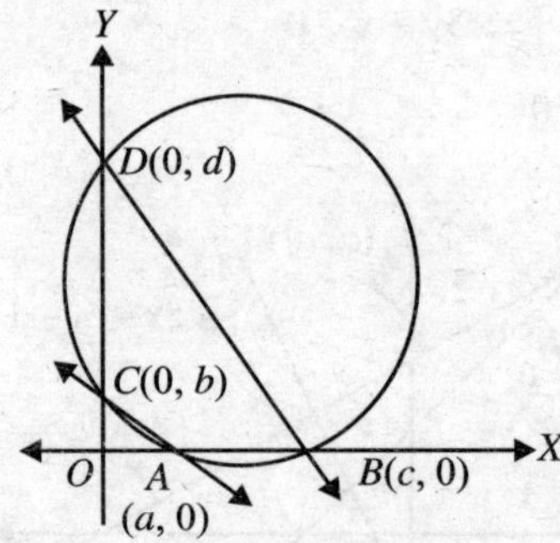

Fig. 1.80

If points A, B, C, D are concyclic, then $ac = bd$. The coordinates of the points of intersection of lines are

$$\left(\frac{ac(b-d)}{bc-ad}, \frac{bd(c-a)}{bc-ad}\right)$$

Let coordinates of the point of intersection be (h, k).

$$\text{Then } h = \frac{ac(b-d)}{bc-ad}, k = \frac{bd(c-a)}{bc-ad}$$

Given $c^2 + a^2 = b^2 + d^2$. Since $ac = bd$, so

$$(c-a)^2 = (b-d)^2$$

$$\text{or } (c-a) = \pm(b-d)$$

Then the locus of the points of intersection is $y = \pm x$

17. a., d.

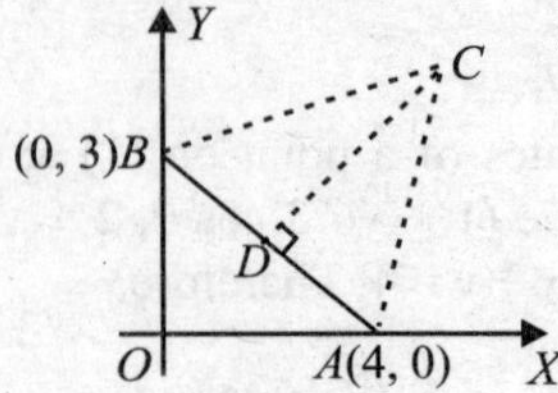

Fig. 1.81

$$AB = 5, D \equiv \left(2, \frac{3}{2}\right)$$

$CD = 5\times\frac{\sqrt{3}}{2} = \frac{5\sqrt{3}}{2}$, slope of AB is $-3/4$, slope of CD is $4/3$.

If $C \equiv (h, k)$, then

$$\frac{h-2}{3/5} = \frac{k-3/2}{4/5} = \pm\frac{5\sqrt{3}}{2}$$

$$\Rightarrow h = 2\left(1-\frac{3\sqrt{3}}{4}\right),\ k = \frac{3}{2}\left(1-\frac{4}{\sqrt{3}}\right)$$

$$\text{or } h = 2\left(1+\frac{3\sqrt{3}}{4}\right),\ k = \frac{3}{2}\left(1+\frac{4}{\sqrt{3}}\right)$$

18. a., c. The equations of lines passing through (1, 0) are given by $y = m(x-1)$. Its distance from origin is $\sqrt{3}/2$. Hence,

$$\left|\frac{-m}{\sqrt{1+m^2}}\right| = \sqrt{3}/2 \Rightarrow m = \pm\sqrt{3}$$

Hence, the lines are $\sqrt{3}\,x + y - \sqrt{3} = 0$ and $\sqrt{3}\,x - y - \sqrt{3} = 0$.

19. b., c.

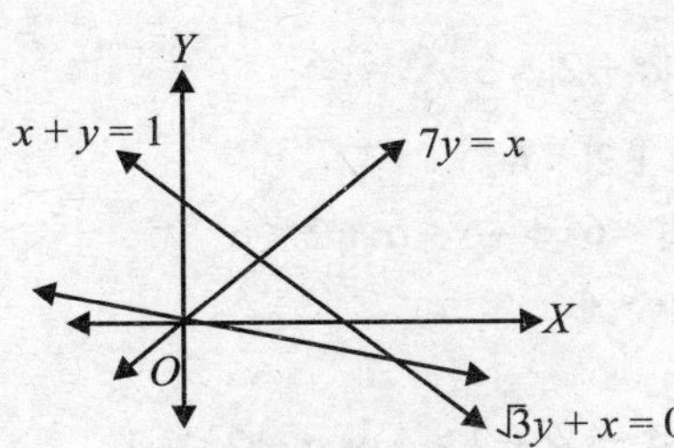

Fig. 1.82

Clearly, from the figure above, the triangle is obtuse angled. Hence, centroid and incentre lie inside the triangle. Orthocentre and circumcentre lie outside the triangle.
Therefore, it is an obtuse angled triangle.

20. b., d.

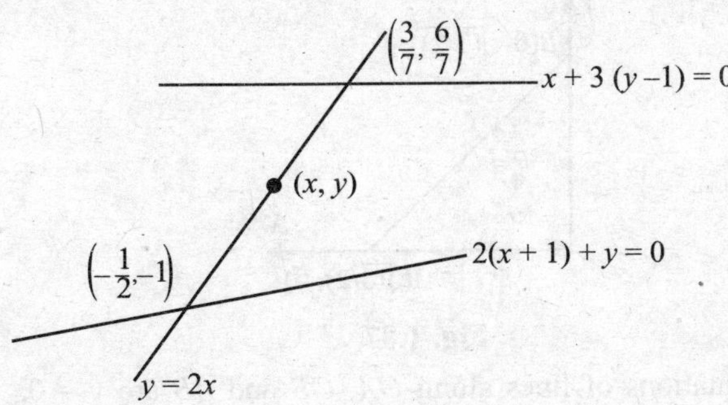

Fig. 1.83

Solving $y = 2x$, $2(x+1) + y = 0$, we get $x = -1/2$, $y = -1$. Solving $y = 2x$, $x + 3(y-1) = 0$, we get $x = 3/7$, $y = 6/7$.

21. a., b.

$$x^2 - 3|x| + 2 = 0$$

$$\Rightarrow (|x|-1)(|x|-2) = 0 \Rightarrow x = \pm 1, \pm 2$$

$$y^2 - 3x + 2 = 0$$

$$\Rightarrow (y-1)(y-2) = 0 \Rightarrow y = 1, 2$$

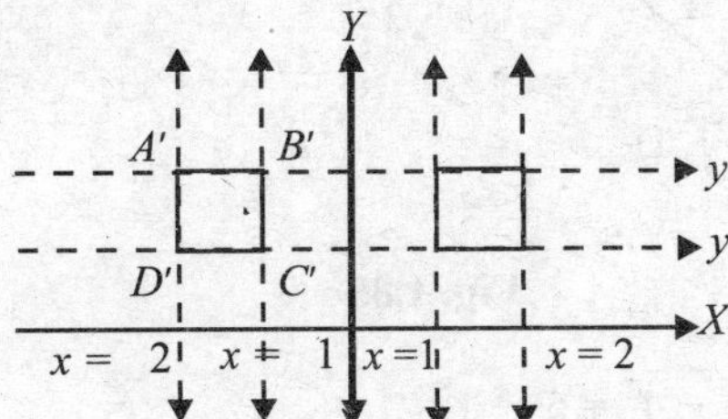

Fig. 1.84

From the figure two such squares are possible whose coordinates are $A(1, 2)$, $B(2, 2)$, $C(2, 1)$, $D(1, 1)$ and $A'(-2, 2)$, $B'(-1, 2)$, $C'(-1, 1)$, $D'(-2, 1)$.

22. b., d. Let any point on the line $x - y = 2$ be $C(h, h-2)$. Given area of ΔABC is

$$\left|\frac{1}{2}\begin{vmatrix} h & h-2 & 1 \\ -5 & 0 & 1 \\ 3 & 0 & 1\end{vmatrix}\right| = 20$$

$$\Rightarrow |8(h-2)| = 40$$

$$\Rightarrow h - 2 = \pm 5$$

$$\Rightarrow h = 7, -3$$

Hence, the points are (7, 5) and (–3, –5).

23. a., b. The area of the triangle is

$$\frac{1}{2}\times\frac{2b}{a}\times\frac{2b}{c} = \frac{2b^2}{ac} = 2$$

$$\Rightarrow b^2 = ac$$

$\Rightarrow$ a, b, c are in G.P. So, $a, -b, c$ are in G.P.

24. a., b. Vertices $(a\cos\theta_1, a\sin\theta_1)$, $(a\cos\theta_2, a\sin\theta_2)$ and $(a\cos\theta_3, a\sin\theta_3)$ are equidistant from origin (0, 0). Hence, the origin is circumcentre (centroid) of circumcircle. Therefore, the coordinates of centroid are

$$\left(\frac{a(\cos\theta_1+\cos\theta_2+\cos\theta_3)}{3}, \frac{a(\sin\theta_1+\sin\theta_2+\sin\theta_3)}{3}\right)$$

But as the centroid is the origin (0, 0), $\cos\theta_1 + \cos\theta_2 + \cos\theta_3 = 0$ and $\sin\theta_1 + \sin\theta_2 + \sin\theta_3 = 0$.

25. b., c. Draw the diagram and verify.

26. a., c. Since $AB = AC = 1$, the triangle is right angled at point A. We have,

$$\tan\alpha\tan\beta = -1$$

$$\Rightarrow \cos(\alpha-\beta) = 0 \Rightarrow \alpha - \beta = \frac{\pi}{2}$$

27. a., b.

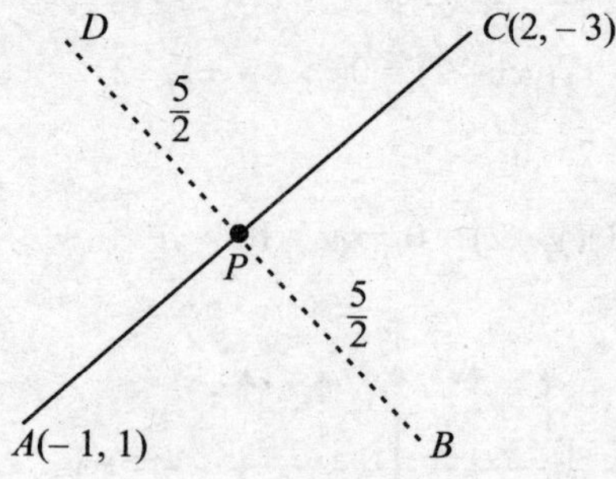

Fig. 1.85

$AC = \sqrt{3^2 + 4^2} = 5$

The midpoint P of AC is $\left(\frac{1}{2}, -1\right)$.

Slope of AC is $\frac{-4}{3}$.

Therefore, slope of BD is $\frac{3}{4} = \tan\theta$

Therefore, coordinates of B and D are

$$\left(1/2 \pm \frac{5}{2}\cos\theta, -1 \pm \frac{5}{2}\sin\theta\right)$$

28. a., c.

29. a., d. Equation of line passing though two given points (x_1, y_1) and (x_2, y_2) is

$$\begin{vmatrix} x & y & 1 \\ x_1 & y_1 & 1 \\ x_2 & y_2 & 1 \end{vmatrix} = 0$$

Now given expression is

$$\begin{vmatrix} x & y & 1 \\ x_1 & y_1 & 1 \\ x_2 & y_2 & 1 \end{vmatrix} + \begin{vmatrix} x & y & 1 \\ x_1 & y_1 & 1 \\ x_3 & y_3 & 1 \end{vmatrix} = 0$$

$$\begin{vmatrix} x & y & 1 \\ x_1 & y_1 & 1 \\ x_2 + x_3 & y_2 + y_3 & 1 \end{vmatrix} = 0$$

$$\text{or} \begin{vmatrix} x & y & 1 \\ x_1 & y_1 & 1 \\ \frac{x_2 + x_3}{2} & \frac{y_2 + y_3}{2} & 1 \end{vmatrix} = 0$$

This is the equation of the line passing through the points (x_1, y_1) and $\left((x_2 + x_3)/2, (y_2 + y_3)/2\right)$. This is an equation of median through vertex A.

30. a., b., c., d.

The point $A(\alpha, \beta)$ lies on $y = 2x + 3$. Hence,

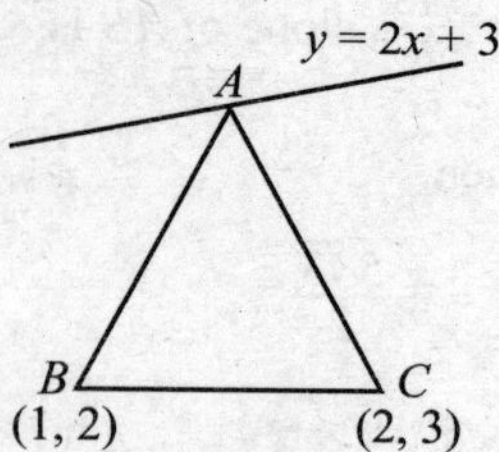

Fig. 1.86

$\beta = 2\alpha + 3$

$A \equiv (\alpha, 2\alpha + 3)$

Area of ΔABC is

$$\left|\frac{1}{2}\begin{vmatrix} \alpha & 2\alpha + 3 & 1 \\ 1 & 2 & 1 \\ 2 & 3 & 1 \end{vmatrix}\right|$$

$$= \left|\frac{1}{2}[\alpha(2-3) + (2\alpha+3)(2-1) + 1(3-4)]\right|$$

$$= \frac{1}{2}|-\alpha + 2\alpha + 3 - 1| = \frac{1}{2}|\alpha + 2| = S$$

$[S] = 2 \Rightarrow 2 \le S < 3$

$\therefore 2 \le \frac{1}{2}|\alpha + 2| < 3$

$\Rightarrow 4 \le |\alpha + 2| < 6$

$\Rightarrow |\alpha + 2| < 6 \Rightarrow -6 < \alpha + 2 < 6$

$\Rightarrow -8 < \alpha < 4$ (1)

and

$|\alpha + 2| \ge 4 \Rightarrow \alpha + 2 \ge 4$ or $\alpha + 2 \le -4$

$\Rightarrow \alpha \ge 2$ or $\alpha \le -6$ (2)

From Eqs. (1) and (2),

$-8 < \alpha \le -6$ or $2 \le \alpha < 4$

$\Rightarrow \alpha = -7, -6, 2, 3$

Possible coordinates of A are $(-7, -11)$, $(-6, -9)$, $(2, 7)$, $(3, 9)$.

31. a., c.

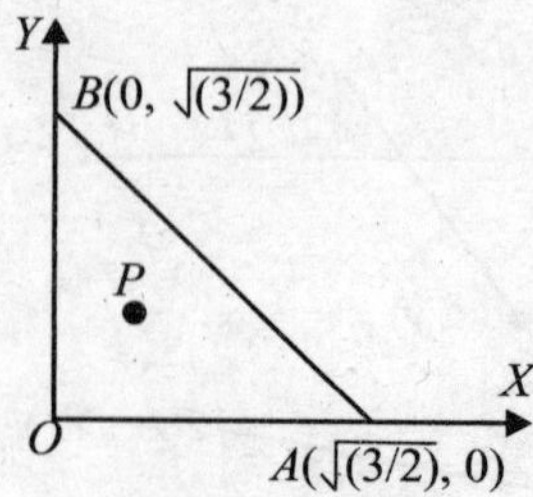

Fig. 1.87

Equations of lines along OA, OB and AB are $y = 0$, $x = 0$, and $x + y = \sqrt{3/2}$, respectively.

Now P and B will lie on the same side of $y = 0$ if $\cos\theta > 0$. Similarly, P and A will lie on the same side of $x = 0$ if $\sin\theta > 0$ and P and O will lie on the same side of $x + y = \sqrt{3/2}$ if $\sin\theta + \cos\theta < \sqrt{3/2}$. Hence, P

will lie inside ΔABC if $\sin\theta > 0$, $\cos\theta > 0$ and $\sin\theta + \cos\theta < \sqrt{3/2}$. Now,

$$\sin\theta + \cos\theta < \sqrt{\frac{3}{2}}$$

$$\Rightarrow \sin(\theta + \pi/4) < \sqrt{\frac{3}{4}}$$

Since $\sin\theta > 0$ and $\cos\theta > 0$, so $0 < \theta < \pi/12$ or $5\pi/12 < \theta < \pi/2$.

32. b., d.

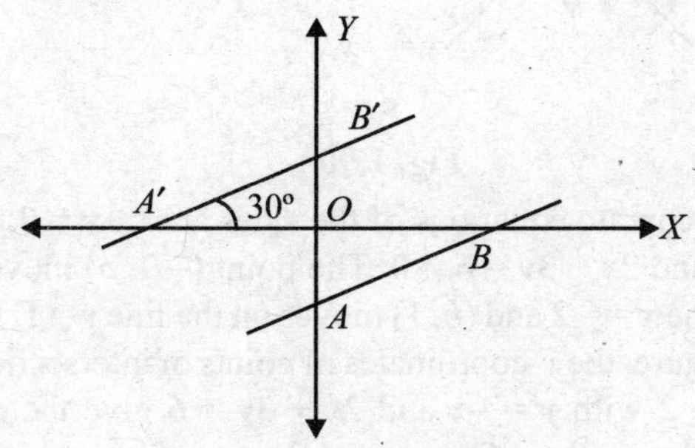

Fig. 1.88

According to the question $AB = 10$.

So, $OA = 10 \sin 30° = 5$.

Then equation of line is

$$y = \frac{1}{\sqrt{3}}x \pm 5$$

$$\Rightarrow x - \sqrt{3}\,y \pm 5\sqrt{3} = 0$$

33. c., d. If lines $x + y - 1 = 0$, $(m - 1)x + (m^2 - 7)y - 5 = 0$, and $(m - 2)x + (2m - 5)y = 0$ are concurrent, then

$$\Delta = 0$$

$$\Rightarrow \begin{vmatrix} 1 & 1 & -1 \\ m-1 & m^2-7 & -5 \\ m-2 & 2m-5 & 0 \end{vmatrix} = 0$$

$$\Rightarrow (m - 2)(-5 + m^2 - 7) - (2m - 5)(-5 + m - 1) + 0 = 0$$

$$\Rightarrow (m - 2)(m^2 - 12) - (2m - 5)(m - 6) = 0$$

$$\Rightarrow m^3 - 4m^2 + 5m - 6 = 0$$

$$\Rightarrow (m - 3)(m^2 - m + 2) = 0$$

$\Rightarrow m = 3$ but $m^2 - m + 2 = 0$ has no real roots. If $m = 3$, then the two lines are parallel.

34. b., c. Let slope of line is m. Then

$$\frac{1}{2} = \left|\frac{m - (-2)}{1 + (-2)m}\right|$$

$$\Rightarrow m = -3/4 \text{ and } \infty$$

Hence, equation of line is $y - 3 = -3/4\ (x - 2)$ and $x = 2$.

35. a., b. Let p be the length of the perpendicular from the origin on the given line. Then its equation in normal form is

$$x \cos 30° + y \sin 30° = p$$

or $\sqrt{3}\,x + y = 2p$

This meets the coordinate axes at $A\left(\frac{2p}{\sqrt{3}}, 0\right)$ and $B(0, 2p)$. Therefore, area of ΔAOB is

$$\frac{1}{2}\left(\frac{2p}{\sqrt{3}}\right)2p = \frac{2p^2}{\sqrt{3}}$$

By hypothesis,

$$\frac{2p^2}{\sqrt{3}} = \frac{50}{\sqrt{3}} \Rightarrow p = \pm 5$$

Hence, the lines are $\sqrt{3}x + y \pm 10 = 0$.

36. a, c. It is clear that diagonals of the rhombus will be parallel to the bisectors of the given lines and will pass through (1, 3). Equations of bisectors of the given lines are

$$\frac{x + y - 1}{\sqrt{2}} = \pm\left(\frac{7x - y - 5}{5\sqrt{2}}\right)$$

or $2x - 6y = 0$, $6x + 2y = 5$

Therefore, the equations of diagonals are $x - 3y + 8 = 0$ and $3x + y - 6 = 0$. Thus the required vertex will be the point where these lines meet the line $y = 2x$. Solving these lines we get possible coordinates as (8/5, 16/5) and (6/5, 12/5).

37. a., b., c., d.

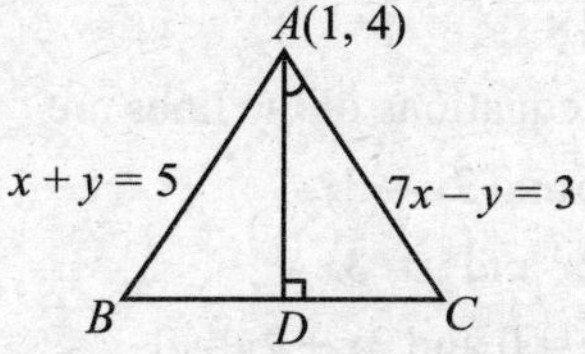

Fig. 1.89

Side BC will be perpendicular to the bisector of the angle BAC. Now equations of the bisectors of lines AB and AC are

$$\frac{(x + y - 5)}{\sqrt{2}} = \pm\frac{(7x - y - 3)}{5\sqrt{2}}$$

$$\Rightarrow x - 3y + 11 = 0 \text{ or } 3x + y - 7 = 0$$

Let equation of side BC be $x - 3y + \lambda = 0$ and altitude through vertex A be AD. Then equation of AD is $3x + y - 7 = 0$. If $AD = \lambda$, then

$$\Delta ABC = \frac{1}{2} \times |BC| = \frac{1}{2} \times \lambda \times 2\lambda\ |\tan\theta| = I^2|\tan\theta|.$$

Hence,

$$\lambda^2 \times \frac{1}{2} = 5 \Rightarrow \lambda^2 = 10$$

$$\Rightarrow (11 - \lambda)^2 = 100$$

$$\Rightarrow 11 - \lambda = \pm 10 \Rightarrow \lambda = 1, -21$$

Hence, equation of BC is $x - 3y + 1 = 0$ or $x - 3y - 21 = 0$. Similarly, if equation of BC is $3x + y + \lambda = 0$, then equation of AD will be $x - 3y + 11 = 0$. Therefore,

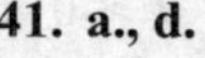

$$|\tan\theta| = \left|\frac{7-\frac{1}{3}}{1+\frac{7}{3}}\right| = 2$$

$\Rightarrow \lambda^2\,|\tan\theta| = 2\lambda^2 = 5 \Rightarrow \lambda^2 = \frac{5}{2}$

Also,

$$\frac{5}{2} = \frac{(3+4+\lambda)^2}{10}$$

$\Rightarrow 7+\lambda = \pm 5 \Rightarrow \lambda = 2, -12$

Hence, equation of BC is $3x + y + 2 = 0$ or $3x + y - 12 = 0$. Finally, there are four possible equations of side BC, viz., $x - 3y + 1 = 0$, $x - 3y - 21 = 0$, $3x + y + 2 = 0$ or $3x + y - 12 = 0$.

38. a., b., c., d.

Let the slope of $u = 0$ be m. Then slope of $v = 0$ is $\frac{9m}{2}$.

$$\text{Therefore, } \frac{7}{9} = \left|\frac{m - \frac{9m}{2}}{1 + m\times\frac{9m}{2}}\right| = \left|\frac{-7m}{2+9m^2}\right|$$

$\Rightarrow 9m^2 - 9m + 2 = 0$ or $9m^2 + 9m + 2 = 0$

$$\Rightarrow m = \frac{9\pm\sqrt{81-72}}{18} = \frac{9\pm 3}{18} = \frac{2}{3}, \frac{1}{3}$$

$$\text{or } m = \frac{-9\pm 3}{18} = -\frac{2}{3}, -\frac{1}{3}$$

Therefore, equations of the lines are

i. $3y = x$ and $2y = 3x$

ii. $3y = 2x$ and $y = 3x$

iii. $x + 3y = 0$ and $3x + 2y = 0$

iv. $2x + 3y = 0$ and $3x + y = 0$

39. a, d. Note that the lines are perpendicular. Assume the coordinate axes to be directed along $u = 0$ and $v = 0$. Now the lines $k_1 u - k_2 v = 0$ and $k_1 u + k_2 v = 0$ are equally inclined with $u\,v$ axes. Hence, the bisectors are $u = 0$ and $v = 0$.

40. a., b., d. Equation of the lines given by $x^2 + 2hxy + y^2 = 0$ be $y = m_1 x$ and $y = m_2 x$. Since these make an angle α with $y + x = 0$ whose slope is -1, so

$$\frac{m_1+1}{1-m_1} = \tan\alpha = \frac{-1-m_2}{1-m_2}$$

$$\Rightarrow m_1 + m_2 = \frac{(\tan\alpha - 1)^2 + (\tan\alpha + 1)^2}{\tan^2\alpha - 1}$$

$$= \frac{-2\sec^2\alpha\times\cos^2\alpha}{\cos 2\alpha}$$

$\therefore -2\sec 2\alpha = -2h$

$\Rightarrow \sec 2\alpha = h$

$\Rightarrow \cos 2\alpha = \frac{1}{h} \Rightarrow 2\cos^2\alpha - 1 = \frac{1}{h}$

$$\Rightarrow \cos\alpha = \sqrt{\frac{1+h}{2h}} \text{ and } \cot\alpha = \sqrt{\frac{h+1}{h-1}}$$

41. a., d.

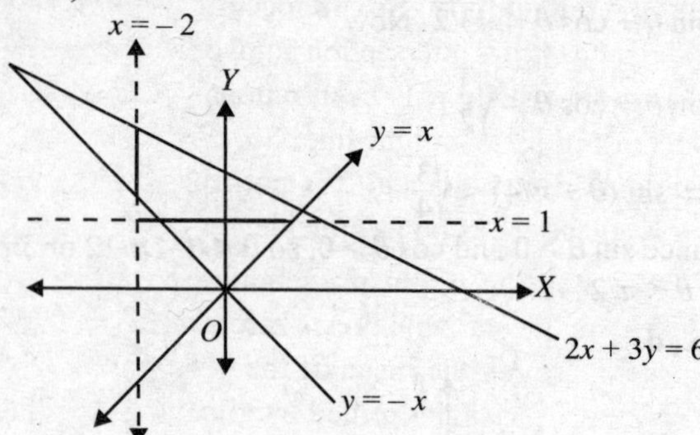

Fig. 1.90

The separate equations of the sides are $x - y = 0$, $x + y = 0$, and $2x + 3y - 6 = 0$. The point $(-2, a)$ moves on the line $x = -2$ and $(b, 1)$ moves on the line $y = 1$. From the figure, the y-coordinates of points of intersection of $x = -2$ with $y = -x$ and $2x + 3y = 6$ give the range of values of a. The x-coordinates of the points of intersection of $y = 1$ with $y = -x$ and $y = x$ give the range of values of b.

42. a., b., c., d. Let,

$$\frac{2}{3}x^2 + \frac{p}{3}xy + y^2 = (y - mx)(y - m'x)$$

and

$$\frac{2}{-3}x^2 + \frac{q}{-3}xy + y^2 = \left(y + \frac{1}{m}x\right)(y - m'x)$$

Then,

$$m + m' = -\frac{p}{3},\; mm' = \frac{2}{3} \quad (1)$$

$$\frac{1}{m} - m' = \frac{-q}{3},\; -\frac{m'}{m} = -\frac{2}{3} \quad (2)$$

$\Rightarrow m^2 = 1 \Rightarrow m = \pm 1$

If $m = 1$, $m' = 2/3$ and so $p = -5$, $q = -1$. If $m = -1$, $m' = -2/3$ and so $p = 5$, $q = 1$.

43. a., b., c., d. Equation of the lines joining the origin to the points of intersection of the given lines is

$3x^2 + mxy - 4x(2x + y) + 1(2x + y)^2 = 0$

(by homogenization)

$\Rightarrow x^2 - mxy - y^2 = 0,$

which are perpendiculars for all values of m.

Reasoning Type

1. a. Statement 1 is true and follows from statement 2 as the family of lines can be written as $a(x + y - 1) + b(x - 2y) = 0$.

2. c. Centroid $\left(\frac{x_1 + x_2 + x_3}{3}, \frac{y_1 + y_2 + y_3}{3}\right)$ is a rational point. Orthocentre is intersection point of two altitudes which will bear rational coefficients when expressed as *a* straight line. So orthocentre is also rational. Circumcentre is intersection point of two perpendicular bisectors which will bear rational coefficient when expressed as a straight line. So, circumcentre is also rational. But statement 2 is not true as in equilateral triangle all the centres coincide.

3. d. We know that if sum of algebraic distances from three points on the variable line is zero, then the line always passes through the mean of the given point, which is centroid of triangle formed by given three points. But centroid of triangle is (1, 2). Hence, the line must pass through it, for which $a + 2b + c = 0$. Therefore, statement 1 is false and statement 2 is true.

4. a. Bisectors of the given lines are $(3x + 4y - 12)/5 = \pm (4x + 3y - 24)/5$, of which one the bisectors is $y - x + 12 = 0$. Also any point on the bisector is always equidistant from the given lines.

5. d. According to given data $2a - 5 + a^2 - 3 < 0$ or $a^2 + 2a - 8 < 0$ or $(a - 2)(a + 4) < 0$ or $a \in (-4, 2)$.

6. a. Any line equally inclined to given lines is always parallel to angle bisectors.

7. a. Area of triangles is unaltered by shifting origin to any point. If origin is shifted to (1000, 1002), *A*, *B*, *C* become *P* (0, 0) *Q* (1, 2), *R* (2, 1). Both are true.

8. a. The quadrilateral is obviously a parallelogram and if the diagonals are at right angles, it must be a rhombus. Hence, the distance between the pairs of opposite sides must be the same, i.e.,

$$\frac{|r - r'|}{\sqrt{p^2 + q^2}} = \frac{|r - r'|}{\sqrt{p'^2 + q'^2}}$$

$$\Rightarrow p^2 + q^2 = p'^2 + q'^2$$

9. d. The joint equation of $y = x$ and $y = -x$ is $(x - y)(x + y) = 0$, i.e., $x^2 - y^2 = 0$.

10. b. Bisectors of angle *C* are

$$\frac{3x + 2y}{\sqrt{13}} = \pm \frac{2x + 3y + 6}{\sqrt{13}}$$

$$\Rightarrow x - y - 6 = 0 \text{ and } 5x + 5y + 6 = 0$$

According to given equations of sides, internal angle bisector at *C* will have negative slope. Also, image of *A* will lie on *BC* respect to both bisectors, from which we can conclude that $5x + 5y + 6 = 0$ is internal angle bisector. Hence, statement 2 is not correct explanation of statement 1.

11. a. From the figure, both the statements are true and statement 2 correctly explains statement 1.

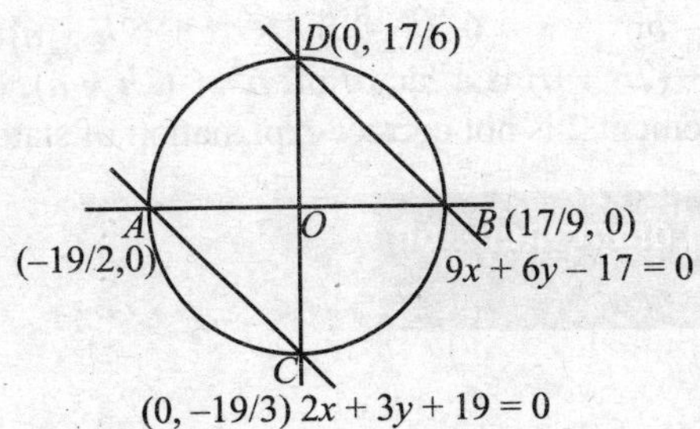

Fig. 1.91

12. a. We have,

$$(m_{AC})(m_{BC}) = \left(\frac{-4 + 2}{5 + 5}\right)\left(\frac{-4 - 6}{5 - 7}\right) = -1$$

Therefore, *ABC* is a right-angled triangle with *C* as the right angle. Hence, circumcentre is midpoint of *AB*, i.e., (1, 2).

13. c. Statement 2 is false as point satisfying such property can be ex-centre of the triangle. However, statement 1 is true as (0, 0) is at distance π from all the lines and it lies inside the triangle.

14. d. Statement 1 is false since $(x - 2) + (2x - 4) + (6 - 3x) = 0$ but the lines $x - 2 = 0$, $2x - 4 = 0$, and $6 - 3x = 0$ are parallel. Statement 2 is a standard result whose more general form as follows. Let $L_1 = 0, L_2 = 0, L_3 = 0$ be three lines. Now, if we can find λ, μ, ν (not all zero) such that $\lambda L_1 + \mu L_2 + \nu L_3 = 0$, then the three lines $L_1 = 0$, $L_2 = 0$, $L_3 = 0$ are either concurrent or parallel.

15. a. The given lines are

$$(a + b)x + (a - b)y - 2ab = 0 \quad (1)$$

$$(a - b)x + (a + b)y - 2ab = 0 \quad (2)$$

$$x + y = 0 \quad (3)$$

The triangle formed by the lines (1), (2) and (3) is an isosceles triangle if the internal bisector of the vertical angle is perpendicular to the third side. Now equations of bisectors of the angle between lines (1) and (2) are

$$\frac{(a + b)x + (a - b)y - 2ab}{\sqrt{[(a + b)^2 + (a - b)^2]}} = \pm \frac{(a - b)x + (a + b)y - 2ab}{\sqrt{[(a - b)^2 + (a + b)^2]}}$$

$$\Rightarrow x - y = 0 \quad (4)$$

$$\text{and } x + y = 2b \quad (5)$$

Obviously the bisector (4) is perpendicular to the third side of the triangle. Hence, the given lines form an isosceles triangle.

16. b. Put $2h = -(a + b)$ in $ax^2 + 2hxy + by^2 = 0$. Then,

$$ax^2 - (a + b)xy + by^2 = 0$$

$$\Rightarrow (x - y)(ax - by) = 0$$

Therefore, one of the lines bisects the angle between coordinates axes in positive quadrant. Also putting $b = -2h - 1$ in $ax - by$, we have

$ax - by = ax - (-2h - a)y = ax + (2h + a)y$. Hence, $ax + (2h + a)$ is a factor of $ax^2 + (2h + a)$. However, statement 2 is not correct explanation of statement 1.

Linked Comprehension Type

For Problems 1–3

1. a., 2. c., 3. d.

Sol. 1. a.

Given lines $(x+y+1)+b(2x-3y-8)=0$ are concurrent at point of intersection of the lines $x + y + 1 = 0$ and $2x - 3y - 8 = 0$, which is $(1, -2)$. Now line through $A(1, -2)$ which is farthest from the point $B(2, 2)$, is perpendicular to AB. Now, slope of AB is 4. Then the required line is $y + 2 = -(1/4)(x - 1)$ or $x + 4y + 7 = 0$.

2. c.

Also this line $x + 4y + 7 = 0$ meets axis at $C(-7, 0)$ and $D(0, -7/4)$. Then area of triangle OCD is $1/2\,|(-7)(-7/4)|$ or $49/8$.

3. d. Lines $x + 4y + 7 = 0$ and $x - 2y + 1 = 0$ intersect at $(-3, -1)$ which must satisfy the line $3x - 4y + \lambda =$. Then $-9 + 4 + \lambda = 0$ or $\lambda = 5$.

For Problems 4–6

4. b., 5. d., 6. a.

Sol. 4. b.

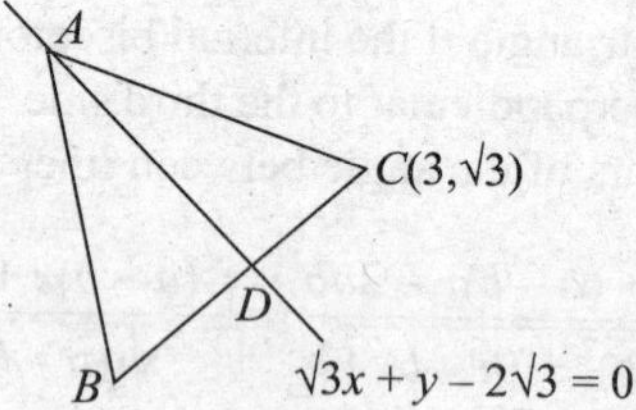

Fig. 1.92

Let the triangle be ABC with $C \equiv (3, \sqrt{3})$ and altitude drawn through vertex (meeting BC at D) be $\sqrt{3}x + y - 2\sqrt{3} = 0$. If B is (x_b, y_b), then we have

$$\frac{2(x_b - 3)}{\sqrt{3}} = \frac{y_b - \sqrt{3}}{2} = -\frac{2(3\sqrt{3} + \sqrt{3} - 2\sqrt{3})}{2} = -2\sqrt{3}$$

$\Rightarrow x_b = 0, y_b = 0$

and coordinates of D is $(3/2, \sqrt{3}/2)$. Let coordinates of vertex A be (x_a, y_a). Then,

$$\frac{x_a - \frac{3}{2}}{-1/2} = \frac{y_a - \sqrt{3}/2}{\sqrt{3}/2} = \pm 3$$

$\Rightarrow (x_a, y_a) \equiv (0, 2\sqrt{3})$ or $(3, -\sqrt{3})$

Hence, the remaining vertices are $(0, 0)$ and $(0, 2\sqrt{3})$ or $(0, 0)$ and $(3, -\sqrt{3})$. Also, the orthocentre is $(1, \sqrt{3})$ or $(2, 0)$.

For Problems 7–9

7. d., 8. b., 9. d.

Sol.

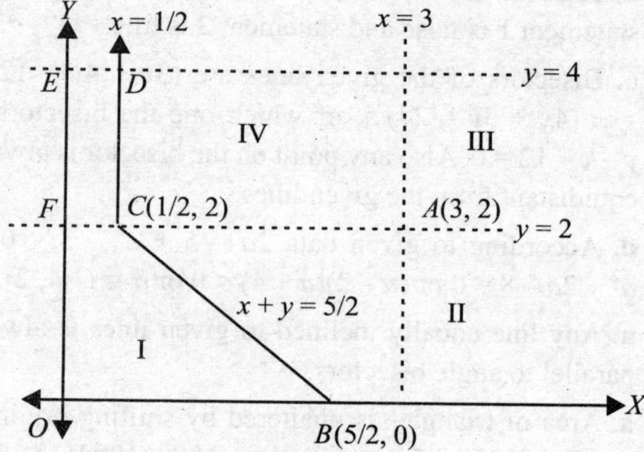

Fig. 1.93

Let $P = (h, k)$ be a general point in the first quadrant such that

$d(P, A) = d(P, O)$

$$|h - 3| + |k - 2| = |h| + |k| = h + k \quad (1)$$

[h and k are positive, point $P(h, k)$ being in first quadrant]

If $h < 3, k < 2$, then (h, k) lies in region *I*. Then,

$3 - h + 2 - k = h + k \Rightarrow h + k = 5/2$

If $h > 3, k < 2$, then (h, k) lies in region II. Then,

$h - 3 + 2 - k = h + k$

$\Rightarrow k = -1/2$ (not possible)

If $h > 3, k > 2$ then (h, k) lies in region III. Then,

$h - 3 + k - 2 = h + k$

$\Rightarrow -5 = 0$ (not possible)

If $h < 3, k > 2$, then (h, k) lies in region IV. Then,

$3 - h + k - 2 = h + k \Rightarrow h = 1/2$

Hence, the required set consists of line segment $x + y = 5/2$ of finite length as shown in the first region and the ray $x = 1/2$ in the fourth region.

7. d. Obviously, locus of P is union of line segment and one infinite ray.

8. b. Area of region $OBCDEFO$

A = area of trapezium $OBCF$ + area of rectangle $FCDE$

$$= \frac{1}{2} \times \left(\frac{5}{2} + \frac{1}{2}\right) \times 2 + \frac{1}{2} \times 2$$

$$= 4$$

9. d. Obviously locus of P is a relation but not a function.

For Problems 10–12

10. d., 11. c., 12. a.

Sol. 10. d.

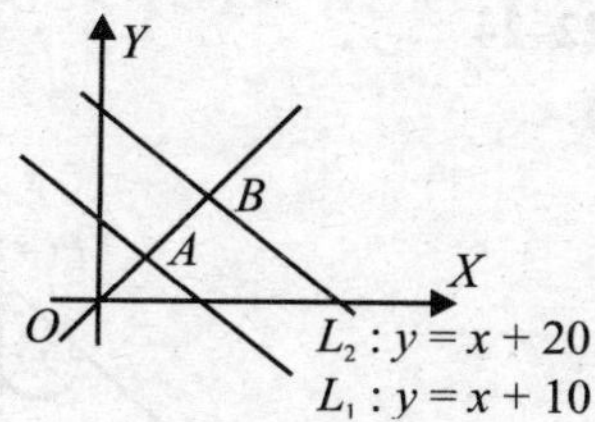

Fig. 1.94

Let the parametric equation of drawn line be

$$\frac{x}{\cos\theta} = \frac{y}{\sin\theta} = r$$

$\Rightarrow x = r\cos\theta,\ y = r\sin\theta$

Putting it in L_1, we get

$$r\sin\theta = r\cos\theta + 10$$

$$\Rightarrow \frac{1}{OA} = \frac{\sin\theta - \cos\theta}{10}$$

Similarly, putting the general point of drawn line in the equation of L_2, we get

$$\frac{1}{OB} = \frac{\sin\theta - \cos\theta}{20}$$

Let $P = (h, k)$ and $OP = r \Rightarrow r\cos\theta = h,\ r\sin\theta = k$, we have

$$\frac{2}{r} = \frac{\sin\theta - \cos\theta}{10} + \frac{\sin\theta - \cos\theta}{20}$$

$$\Rightarrow 40 = 3r\sin\theta - 3r\cos\theta$$

$$\Rightarrow 3y - 3x = 40$$

11. c.

$$r^2 = \frac{10 \times 20}{(\sin\theta - \cos\theta)^2}$$

$$\Rightarrow (r\sin\theta - r\cos\theta)^2 = 200$$

Hence, locus is $(y - x)^2 = 200$.

12. a. $\dfrac{1}{r^2} = \dfrac{(\sin\theta - \cos\theta)^2}{100} + \dfrac{(\sin\theta - \cos\theta)^2}{400}$

$$\Rightarrow 400 = 5(r\sin\theta - r\cos\theta)^2$$

Hence, the locus is $400 = 5(x - y)^2$, i.e., $(x - y)^2 = 80$.

For Problems 13–15

13. a., 14. b., 15. c.

Sol.

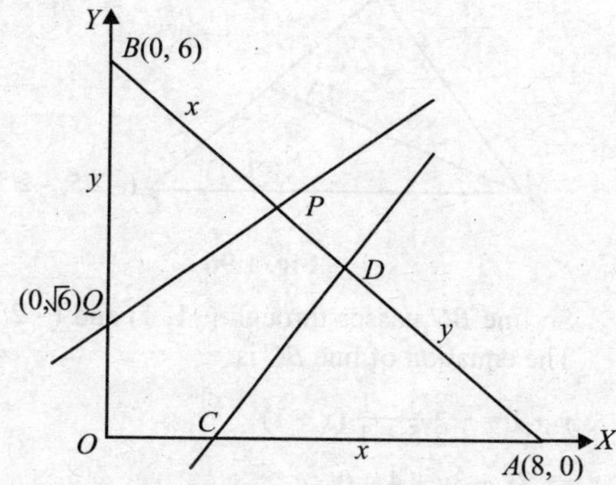

Fig. 1.95

Case I: Let the line L cut AO and AB at distances x and y from A. Then, the area of the triangle with sides x and y is

$$\frac{1}{2}xy\sin(\angle CAD) = \frac{1}{2}xy\frac{3}{5} = \frac{3xy}{10} = 12$$

$$\Rightarrow xy = 40$$

Also, $x + y = 12$ (from perimeter bisection). Then x and y are roots of $r^2 - 12x + 40 = 0$ which has imaginary roots.

Case II: If the line L cuts OB and BA at distances y and x from B, then we have $xy = 30$ and $x + y = 12$.

Solving, we get $x = 6 + \sqrt{6}$ and $y = 6 - \sqrt{6}$.

Case III: If the line L cuts the sides OA and OB at distances x and y from O, then

$x + y = 12$ and $xy = 24$

$\therefore x, y = 6 \pm 2\sqrt{3}$ (not possible)

So there is a unique line possible. Let point P be (α, β).

Using parametric equation of AB, we have

$\beta = 6 - \frac{3}{5}(6 + \sqrt{6})$

and $\alpha = \frac{4}{5}(6 + \sqrt{6})$

Hence, slope of PQ is

$$\frac{\beta - \sqrt{6}}{\alpha - 0} = \frac{10 - 5\sqrt{6}}{10}$$

For Problems 16–18

16. b., 17. c., 18. a.

Sol. 16. b.

Image of $A(1, 3)$ in line $x + y = 2$ is $(1 - 2(2)/2, 3 - 2(2)/2) \equiv (-1, 1)$.

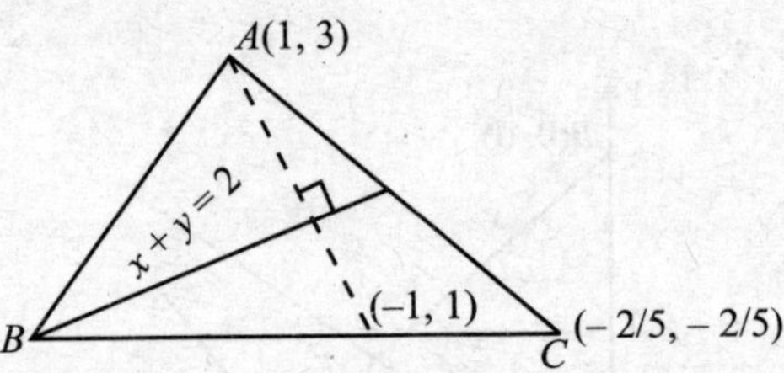

Fig. 1.96

So line BC passes through $(-1, 1)$ and $(-2/5, -2/5)$. The equation of line BC is

$$y - 1 = \frac{-2/5 - 1}{-2/5 + 1}(x + 1)$$

$\Rightarrow 7x + 3y + 4 = 0$

17. c. Vertex B is point of intersection of $7x + 3y + 4 = 0$ and $x + y = 2$, i.e., $B \equiv (-5/2, 9/2)$.

18. a. Line AB is $y - 3 = \frac{3 - 9/2}{1 + 5/2}(x - 1)$

$\Rightarrow 3x + 7y = 24$

For Problems 19–21

19. c., 20. a., 21. a.

Sol. 19. c.

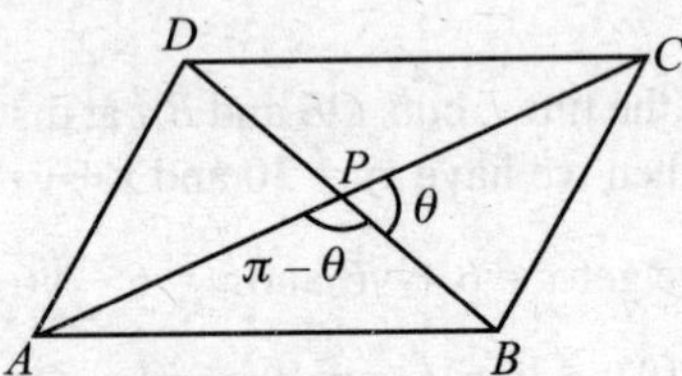

Fig. 1.97

Angle between the diagonals is given by

$$\tan\theta = \left|\frac{-\frac{1}{2} + 2}{1 + 1}\right| = \frac{3}{4}$$

$\Rightarrow \sin\theta = \frac{3}{5}$

Area of ΔCPB is

$$\frac{1}{2} \times PC \times PB \sin\theta = 2 \Rightarrow PB = \frac{10}{3}$$

$\Rightarrow BD = \frac{20}{3}$

20. a. $\cos(\pi - \theta) = \frac{AP^2 + PB^2 - AB^2}{2AP \times PB}$

$$\Rightarrow -\frac{4}{5} = \frac{4 + \frac{100}{9} - AB^2}{2 \times 2 \times \frac{10}{3}}$$

$\Rightarrow AB = \frac{2\sqrt{58}}{3}$

21. a. In ΔCPB,

$$\cos\theta = \frac{PC^2 + PB^2 - BC^2}{2PC \times PB}$$

$\Rightarrow BC = \frac{2\sqrt{10}}{3}$

For Problems 22–24

22. a., 23. b., 24. c.

Sol. 22. a.

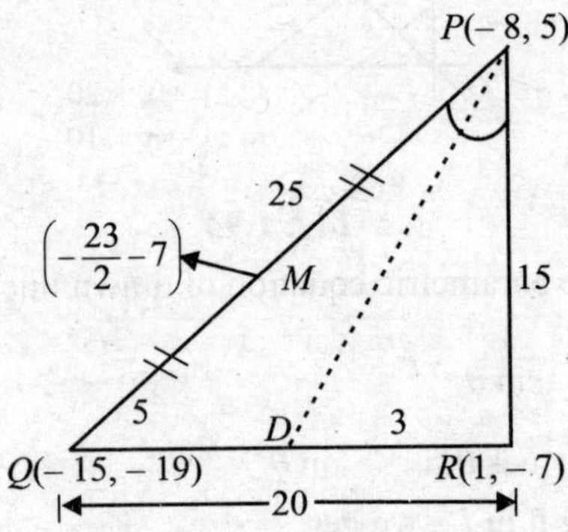

Fig. 1.98

Since triangle is right angled, circumcentre is the mid point of PQ and orthocentre is $R(1, -7)$. Hence,

$$RM = \left|\sqrt{\left(\frac{23}{2} + 1\right)^2}\right| = 12\frac{1}{2} = \frac{25}{2}$$

23. b. The coordinates of the incentre are given as follows:

$$\bar{x} = \frac{20(-8) + 15(-15) + 25(1)}{20 + 15 + 25}$$

$= -6$

$$\bar{y} = \frac{20(5) + 15(-19) + 25(-7)}{60}$$

$= -6$

Hence, incentre I is $(-6, -6)$.

Now, equation of side PR is

$$y + 7 = \frac{12}{-9}(x - 1)$$

or $4x - 4 = -3y - 21$

or $4x + 3y + 17 = 0$

Inradius is given as the distance of I from side PR, i.e.,

$$\frac{|-24-18+17|}{5}=5.$$

24. c. Coordinates of D using section formula are $(-5,-23/2)$ and $m_{PD}=-11/2$. Equation PD is

$$11x+2y+78=0$$

$$\Rightarrow a+c=89$$

For Problems 25–27

25. b., 26. d., 27.

Sol. 25. b.

$\theta=60°, m=2$

$$\tan\theta=\frac{m\sin\omega}{1+m\cos\omega}$$

$$=\frac{2\sin 60°}{1+2\cos 60°}$$

$$=\frac{2\times\sqrt{3}/2}{1+2\times 1/2}=\frac{\sqrt{3}}{2}$$

$$\Rightarrow\theta=\tan^{-1}\left(\frac{\sqrt{3}}{2}\right)$$

26. d. $\omega=60°, m_1=2, m_2=-\frac{1}{2}$

$$\tan\theta_1=\frac{m_1\sin\omega}{1+m_1\cos\omega}=\frac{2\times\sqrt{3}/2}{1+2\times 1/2}$$

$$=\frac{\sqrt{3}}{2}$$

$$\tan\theta_2=\frac{-1/2\times\sqrt{3}/2}{1-1/2\times 1/2}$$

$$=\frac{-\sqrt{3}}{4}\times\frac{4}{3}=\frac{-1}{\sqrt{3}}$$

Let angle between the lines be ϕ. Then,

$$\tan\phi=\left|\frac{\tan\theta_1-\tan\theta_2}{1+\tan\theta_1\tan\theta_2}\right|$$

$$=\left|\frac{\frac{\sqrt{3}}{2}+\frac{1}{\sqrt{3}}}{1-\frac{\sqrt{3}}{2}\times\frac{1}{\sqrt{3}}}\right|$$

$$\Rightarrow\phi=\tan^{-1}\left(\frac{5}{\sqrt{3}}\right)$$

27. c. $m=\dfrac{\sin 60°}{\sin(30°-60°)}=-\sqrt{3}$

Therefore, the equation of the line is

$y-0=-\sqrt{3}(x-2)$, i.e. $\sqrt{3}x+y=2\sqrt{3}$.

For Problems 28–29

28. d., 29. b.

Sol. 28. d.

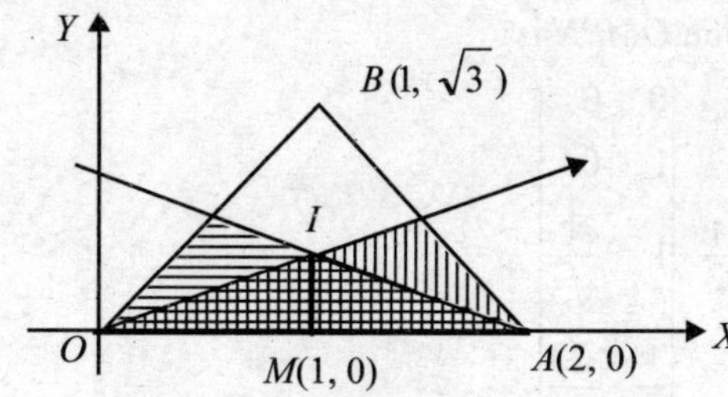

Fig. 1.99

$d(P, OA)\le\min[d(P, OB), d(P, AB)]$

$\Rightarrow d(P, OA)\le d(P, OB)$

and $d(P, OA)\le d(P, AB)$

When $d(P, OA)=d(P, OB)$, P is equidistant from OA and OB, or P lies on angle bisector of lines OA and OB.

Hence, when $d(P, OA)\le d(P, OB)$, point P is nearer to OA than OB or lies below bisector of OA and OB. Similarly, when $d(P, OA)\le d(P, AB)$, P is nearer to OA than AB, or lies below bisector of OA and AB. Therefore, the required area is equal to the area of ΔOIA.

Now,

$$\tan\angle BOA=\frac{\sqrt{3}}{1}=\sqrt{3}$$

$$\Rightarrow\angle BOA=60°$$

Hence, triangle is equilateral. Then I coincides with centroid, which is $(1, 1/\sqrt{3})$.

Therefore, area of ΔOIA is

$$\frac{1}{2}OA\times IM=(1/2)\times 2\times(1/\sqrt{3})$$
$$=1/\sqrt{3}\text{ sq. units.}$$

29. b.

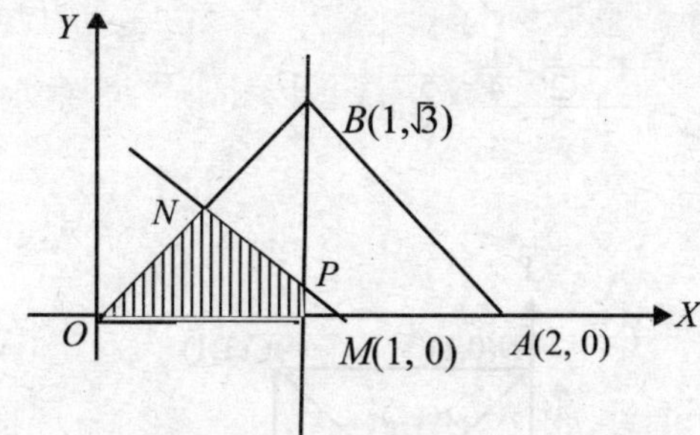

Fig. 1.100

$OP\le\min[BP, AP]$

$\Rightarrow OP\le BP$ (when $BP<AP$)

$OP\le AP$ (when $AP<BP$)

Let $OP = BP$. Then P lies on the perpendicular bisector of OB. For $OP = AP$, P lies on the perpendicular bisector of OA. Then for required condition, P lies in the region as shown in the diagram. The area of the region $OMPN$ is

$$= \frac{1}{2}\begin{vmatrix} 0 & 0 \\ 1 & 0 \\ 1 & \frac{1}{\sqrt{3}} \\ \frac{1}{2} & \frac{\sqrt{3}}{2} \\ 0 & 0 \end{vmatrix}$$

$$= \frac{1}{2}\left[\frac{1}{\sqrt{3}} + \frac{\sqrt{3}}{2} - \frac{1}{2\sqrt{3}}\right]$$

$$= \frac{1}{2}\left[\frac{\sqrt{3}}{2} - \frac{1}{\sqrt{3}}\right]$$

$$= \frac{1}{2\sqrt{3}}$$

For Problems 30–32

30. c., 31. d., 32. b.

Sol. 30. c.

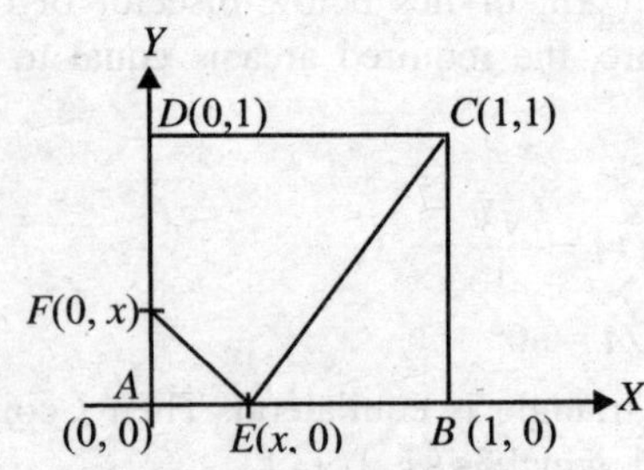

Fig. 1.101

Area of $CDFE$,

$$A = 1 - \frac{1}{2}x^2 - \frac{1}{2}(1 - x)$$

$$= \frac{2 - x^2 - 1 + x}{2} = \frac{1 + x - x^2}{2}$$

$$A_{max} = \frac{1 + \frac{1}{2} - \frac{1}{4}}{2} = \frac{5}{8} \text{ at } x = \frac{1}{2}$$

31. d.

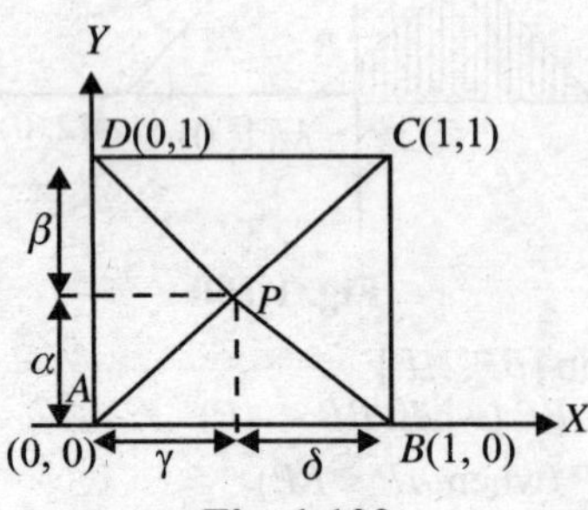

Fig. 1.102

$$(PA)^2 - (PB)^2 + (PC)^2 - (PD)^2$$

$$= (\alpha^2 + \gamma^2) - (\alpha^2 + \delta^2) + (\delta^2 + \beta^2) - (\gamma^2 + \beta^2)$$

32. b.

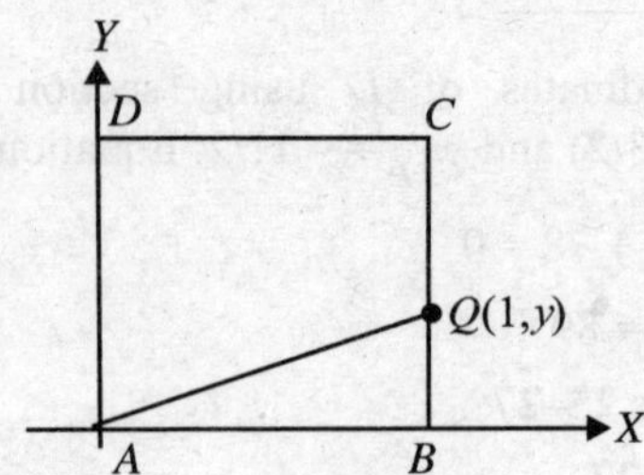

Fig. 1.103

$$\frac{1}{2}\,y\,(1) = \frac{1}{3}(1)$$

$$\Rightarrow y = \frac{2}{3}$$

$$LAQ = \sqrt{(1)^2 + \left(\frac{2}{3}\right)^2} = \frac{\sqrt{13}}{3}$$

Matrix-Match Type

1. a → s; b → r; c → p; d → q.

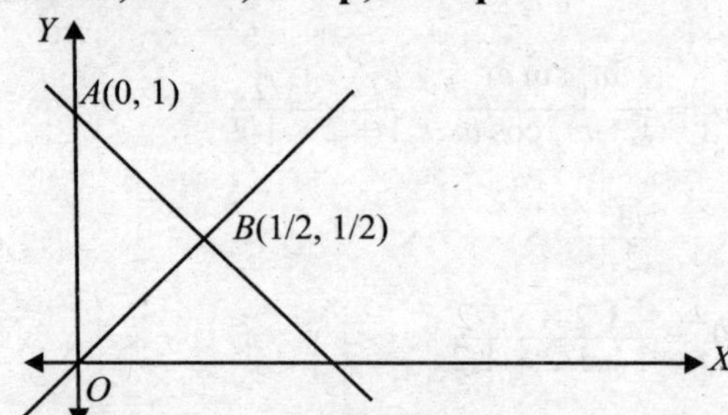

Fig. 1.104

The given lines are $x(x + y - 1)(x - y) = 0$. In other words, lines $x = 0$, $x + y - 1 = 0$ and $x - y = 0$ form triangle OAB as shown in the above diagram.

The triangle is right angled at point B, hence orthocentre is (1/2, 1/2). Also, circumcentre is midpoint of OA which is (0, 1/2). The centroid is

$$\left(\frac{0 + \frac{1}{2} + 0}{3}, \frac{0 + \frac{1}{2} + 1}{3}\right) \text{ or } \left(\frac{1}{6}, \frac{1}{2}\right)$$

Also, $OA = 1$, $OB = OC = 1/\sqrt{2}$. Hence, the incentre is

$$\left(\frac{0\left(\frac{1}{\sqrt{2}}\right) + \frac{1}{2}(1) + 0\left(\frac{1}{\sqrt{2}}\right)}{\frac{1}{\sqrt{2}} + 1 + \frac{1}{\sqrt{2}}}, \frac{0\left(\frac{1}{\sqrt{2}}\right) + \frac{1}{2}(1) + 1\left(\frac{1}{\sqrt{2}}\right)}{\frac{1}{\sqrt{2}} + 1 + \frac{1}{\sqrt{2}}}\right)$$

$$\equiv \left(\frac{1}{2 + 2\sqrt{2}}, \frac{1}{2}\right)$$

2. a → s; b → r; c → q; d → p.

a.

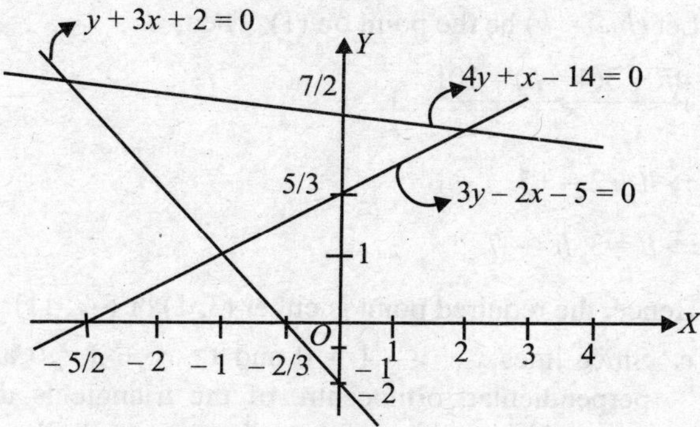

Fig. 1.105

b. Clearly, point $(\alpha, 0)$ lies on the x-axis, which is not intersecting any side of triangle, hence no such α exists.

c.

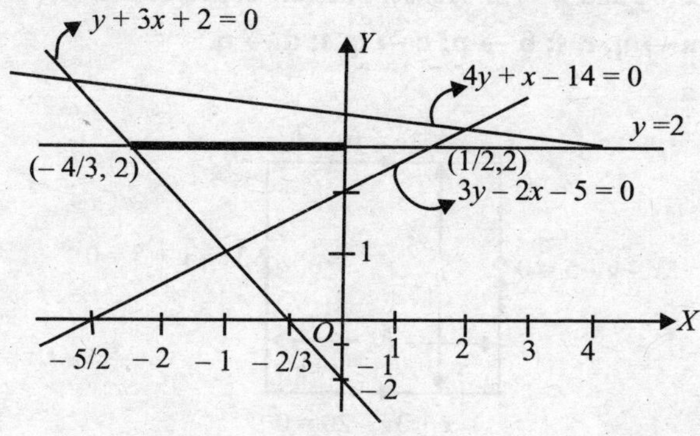

Fig. 1.106

d.

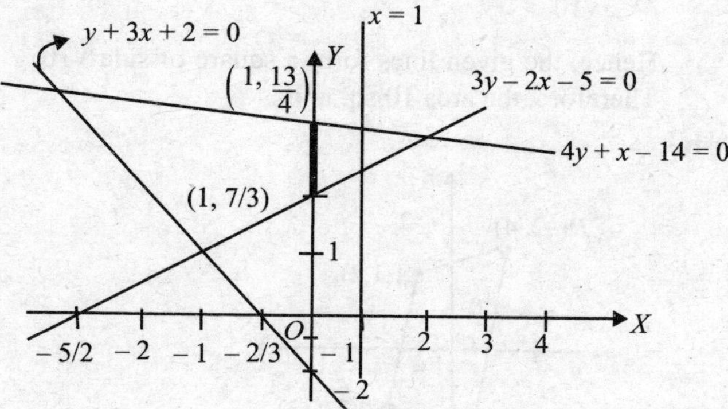

Fig. 1.107

3. a → s; b → p; c → q; d → r.

a.

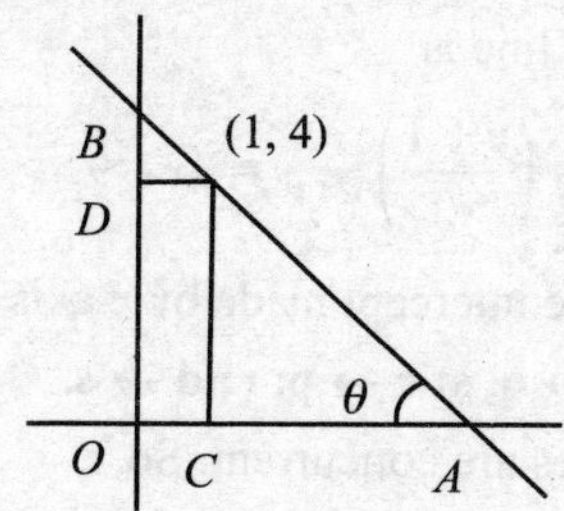

Fig. 1.108

$OA = 1 + 4 \cot \theta$

$OB = 4 + \tan \theta$

$OA + OB = 5 + 4 \cot \theta + \tan \theta$

$\geq 5 + 2\sqrt{4 \cot \theta \tan \theta}$

$= 5 + (2 \times 2) = 9$

b.

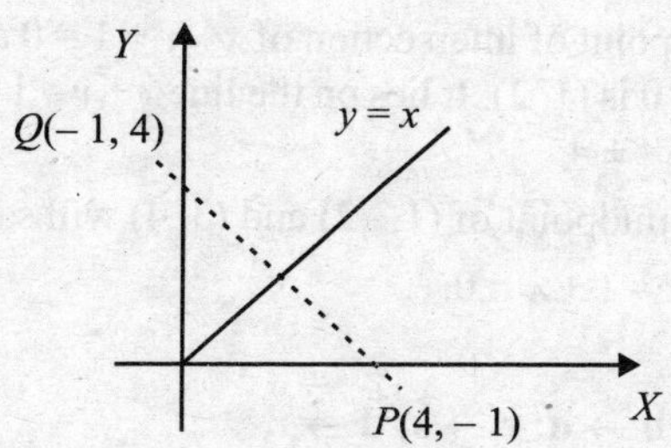

Fig. 1.109

Reflection of $P(4, -1)$ in $y = x$ is $Q(-1, 4)$. Hence,

$PQ = \sqrt{(4+1)^2 + (-1-4)^2}$

$= \sqrt{50} = 5\sqrt{2}$

c.

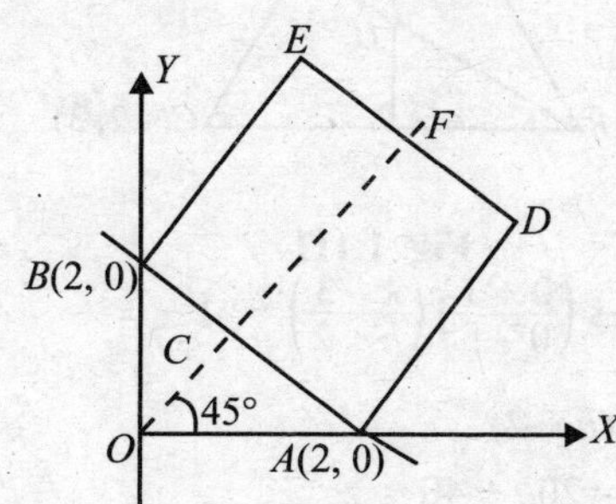

Fig. 1.110

$AB = 2\sqrt{2}$

$OC = \sqrt{2}$

Maximum value of d is

$OF = \sqrt{2} + 2\sqrt{2}$

$= 3\sqrt{2}$

d. The given line is

$$x = 4 + \frac{1}{\sqrt{2}}\left(\frac{y+1}{\sqrt{2}}\right) \Rightarrow y = 2x - 9$$

Hence, the intercept made by x-axis is 9/2.

4. a → p, s; b → q, s; c → p, r; d → s.

a. Given lines are concurrent. So,

$$\begin{vmatrix} 3 & 1 & -4 \\ 1 & -2 & -6 \\ \lambda & 4 & \lambda^2 \end{vmatrix} = 0$$

$\Rightarrow \lambda^2 + 2\lambda - 8 = 0$

$\Rightarrow \lambda = 2, -4$

b. Points are collinear. Hence,

$$\begin{vmatrix} \lambda+1 & 1 & 1 \\ 2\lambda+1 & 3 & 1 \\ 2\lambda+2 & 2\lambda & 1 \end{vmatrix} = 0$$

$\Rightarrow 2\lambda^2 - 3\lambda - 2 = 0 \Rightarrow \lambda = 2, -1/2$

c. The point of intersection of $x - y + 1 = 0$ and $3x + y - 5 = 0$ is (1, 2). It lies on the line $x + y - 1 - |\lambda/2| = 0$

$\Rightarrow \lambda = \pm 4$.

d. The midpoint of (1, –2) and (3, 4) will satisfy

$y - x - 1 + \lambda = 0$.

$\Rightarrow \lambda = 2$

5. a → p; b → q; c → s; d → r.

a. $AH \perp BC \Rightarrow \left(\frac{k}{h}\right)\left(\frac{3+1}{-2-5}\right) = -1$

$\therefore 4k = 7h$ (1)

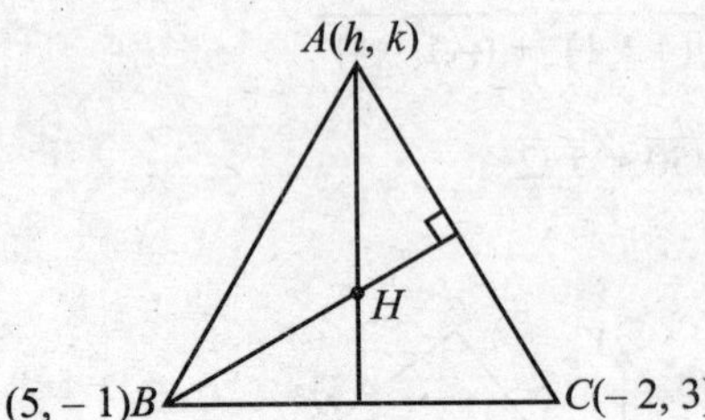

Fig. 1.111

$BH \perp AC \Rightarrow \left(\frac{0+1}{0-5}\right)\left(\frac{k-3}{h+2}\right) = -1$

$\therefore k - 3 = 5(h + 2)$ (2)

$\Rightarrow 7h - 12 = 20h + 40$

$\Rightarrow 13h = -52$

$\Rightarrow h = -4$

$\therefore k = -7$

Hence, point, A is (–4, –7)

b. $x + y - 4 = 0$ (1)

$4x + 3y - 10 = 0$ (2)

Let $(h, 4 - h)$ be the point on (1). Then,

$$\left|\frac{4h + 3(4-h) - 10}{5}\right| = 1$$

$\Rightarrow h + 2 = \pm 5$.

$\Rightarrow h = 3, h = -7$

Hence, the required point is either (3, 1) or (–7, 11).

c. Since lines $x + y - 1 = 0$ and $x - y + 3 = 0$ are perpendicular, orthocentre of the triangle is the point of intersection of these lines, i.e., (–1, 2).

d. Since $2a, b, c$ are in A.P., so

$$b = \frac{2a + c}{2}$$

$\Rightarrow 2a - 2b + c = 0$

Comparing with the line $ax + by + c = 0$, we have $x = 2$ and $y = -2$. Hence, lines are concurrent at (2, – 2).

6. a → q, r, s; b → p; c → q, s; d → q.

a.

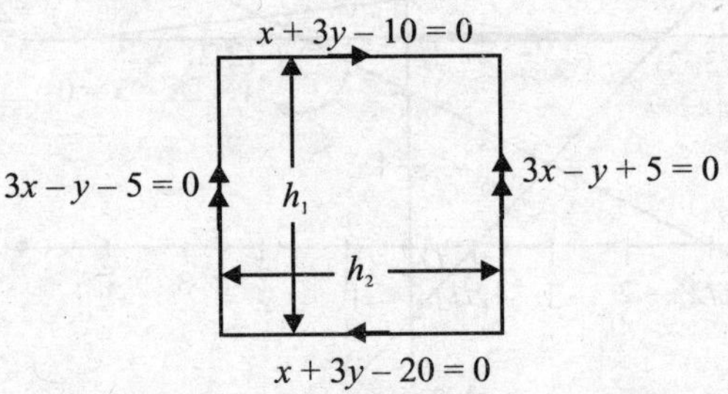

Fig. 1.112

$$h_1 = \left|\frac{10}{\sqrt{10}}\right| = \sqrt{10}$$

$$h_2 = \frac{10}{\sqrt{10}} = \sqrt{10}$$

Hence, the given lines form a square of side $\sqrt{10}$. Therefore, the area 10 sq. units.

b.

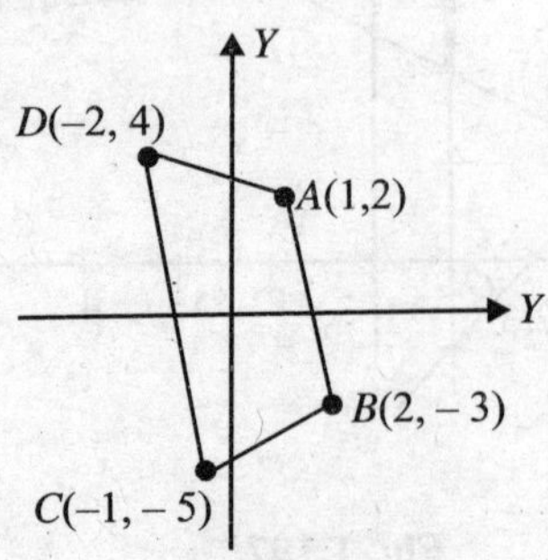

Fig. 1.113

$$m_{AB}=\frac{5}{-1}=-5$$

$$m_{DC}=\frac{9}{-1}=-9$$

Hence, the figure is not a parallelogram.

c. Lines $7x+3y-33=0$, $7x+3y-4=0$ are parallel and distance between them is $|29/\sqrt{58}|$. Lines $3x-7y+19=0$, $3x-7y-10$ are parallel and distance between them $|29/\sqrt{58}|$. Also, lines $7x+3y-33=0$ and $3x-7y+19=0$ are perpendicular. Hence, given lines form a square.

d. Lines $4y-3x-7=0$ and $4y-3x-21=0$ are parallel. Lines $3y-4x+7=0$, $3y-4x+14=0$ are parallel. Also, lines $4y-3x-7=0$ and $3y-4x+14=0$ are not perpendicular. Hence, given lines form parallelogram.

7. a → p; b → s; c → q; d → r.

a. $(x+7y)^2+4\sqrt{2}(x+7y)-42=0$

$$\Rightarrow (x+7y)^2+7\sqrt{2}(x+y)-3\sqrt{2}(x+y)-42=0$$

$$\Rightarrow (x+y)[x+7y+7\sqrt{2}]-3\sqrt{2}(x-7y+7\sqrt{2})=0$$

$$\Rightarrow (x+7y+7\sqrt{2})(x+7y-3\sqrt{2})=0$$

$$\Rightarrow x+7y+7\sqrt{2}=0 \text{ and } x+7y-3\sqrt{2}=0$$

$$\Rightarrow d=\left|\frac{7\sqrt{2}+3\sqrt{2}}{\sqrt{1+49}}\right|=\frac{10\sqrt{2}}{\sqrt{50}}=2$$

b.

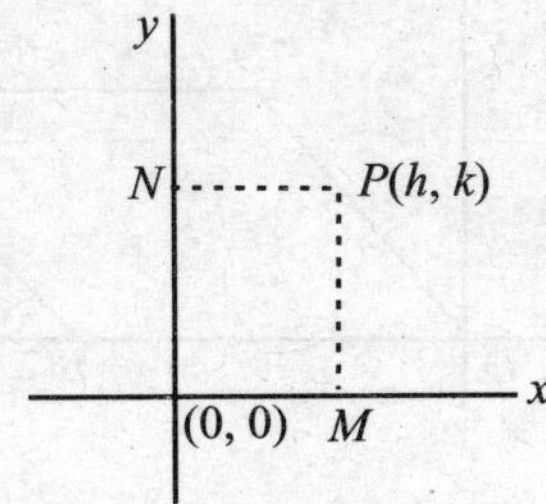

Fig. 1.114

Let two perpendicular lines are coordinate axes. Then,

$PM+PN=1$

$\Rightarrow h+k=1$

Hence, the locus is $x+y=1$.

But if the point lies in other quadrants also, then $|x|+|y|=1$. Hence, value of k is 1.

c. Angle bisector between the lines $x+2y+4=0$ and $4x+2y-1=0$ is

$$\frac{x+2y+4}{\sqrt{1+4}}=\pm\frac{(-4x+2y+1)}{\sqrt{16+4}}$$

$$\Rightarrow x+2y+4=\pm\frac{(-4x-2y+1)}{2}$$

$$\Rightarrow 2(x+2y+4)=\pm(-4x-2y+1)$$

Since $AA'+BB'<0$, so positive sign gives acute angle bisector. Hence,

$2x+4y+8=-4x-2y+1$

$\Rightarrow 6x+6y+7=0$

$\Rightarrow m=7$

d. We have,

$y^2-9xy+18x^2=0$

or $y^2-6xy-3xy+18x^2=0$

$\Rightarrow y(y-6x)-3x(y-6x)=0$

$\Rightarrow (y-3x)=0$ and $y-6x=0$

The third line is $y=6$. Therefore, area of the triangle formed by these lines is

$$=\frac{1}{2}\begin{vmatrix}0 & 0 & 1\\ 1 & 6 & 1\\ 2 & 6 & 1\end{vmatrix}=\frac{1}{2}|6-12|$$

$=3$ units^2

8. a → s; b → r; c → p, q; d → p, q, r, s.

a. The equation $4x^2+8xy+ky^2-9=0$ represents a pair of straight lines if $(4)(k)(-a)-(-a)(4)^2=0$

$\Rightarrow k=4$.

b. $m_1+m_2=4m_1m_2$

$$\Rightarrow -\frac{2h}{b}=\frac{4a}{b}$$

$$\Rightarrow -\frac{2(-c)}{-7}=\frac{4\times 1}{-7}$$

$$\Rightarrow 2c=4\Rightarrow c=2$$

c. Let m_1, m_2 be the slopes of the lines $x^2+hxy+2y^2=0$. Then,

$$m_1+m_2=-\frac{h}{2},\ m_1m_2=\frac{1}{2}$$

But $m_1=2m_2$ (given). Therefore,

$3m_2=-h/2$ and $2m_2^2=1/2$,

i.e., $m_2^2=\frac{1}{4}$. Also, $m_2=-h/6$. Therefore,

$$\frac{h^2}{36}=\frac{1}{4}\Rightarrow h^2=9\Rightarrow h=\pm 3$$

d. Equation of the bisectors of the angle between the lines $ax^2+2hxy+by^2+\lambda(x^2+y^2)=0$ is

$$\frac{x^2 - y^2}{(a+\lambda)-(b+\lambda)} = \frac{xy}{h}$$

or $\dfrac{x^2 - y^2}{a-b} = \dfrac{xy}{h}$,

which is same as the equation of the bisectors of angles between the lines $ax^2 + 2hxy + by^2 = 0$.

Thus, the two line pairs are equally inclined to each other for any value of λ.

9. a → p; b → p, s; c → p, r; d → q, s.

a. Given that points are $O(0, 0)$ and $B(2, 0)$.

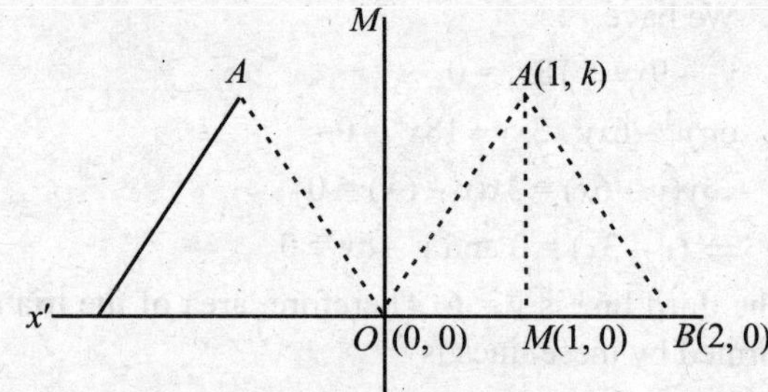

Fig. 1.115

From figure, ΔABC is equilateral.

Hence, $\tan 60° = k$

or $k = \sqrt{3}$ (for first quadrant) or $k = -\sqrt{3}$ (for fourth quadrant). Then possible coordinates are $(1, \pm\sqrt{3})$. Similarly, for second quadrant, the point is $(-1, \sqrt{3})$ and for third quadrant, the point is $(-1, -\sqrt{3})$.

b. Case (i)

If $OA = AB$, then $\angle A = 30°$.

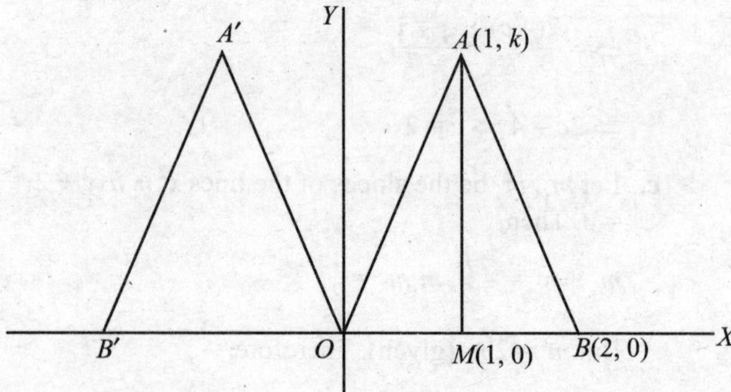

Fig. 1.116

$\therefore \angle AOB = 75°$

$\therefore \dfrac{AM}{OM} = \tan 75°$

$AM = OM \tan 75°$

$k = 1 \times (2 + \sqrt{3})$

$\therefore k = 2 + \sqrt{3}$

Hence, point A is $(1, 2 + \sqrt{3})$. By symmetry, all possible points are $(\pm 1, \pm(2 + \sqrt{3}))$.

Case (ii)

$AO = OB$

$\therefore \angle AOB = 120°$

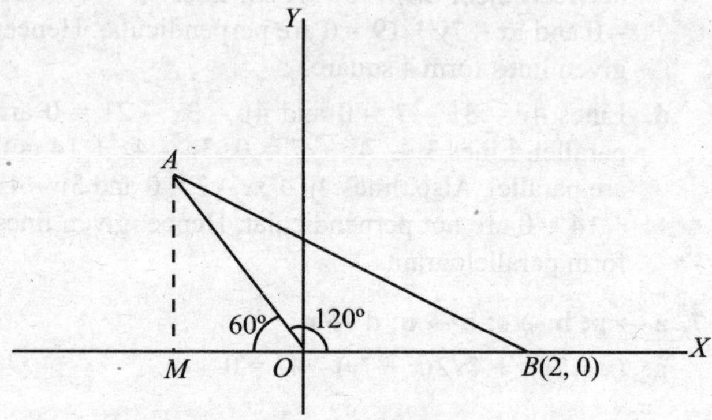

Fig. 1.117

$AM = 2 \sin 60° = \sqrt{3}$

and $OM = 2 \cos 60° = 1$

Hence, point A is $(1, -\sqrt{3})$. By symmetry, all possible points are $(\pm 1, \pm\sqrt{3})$.

c.

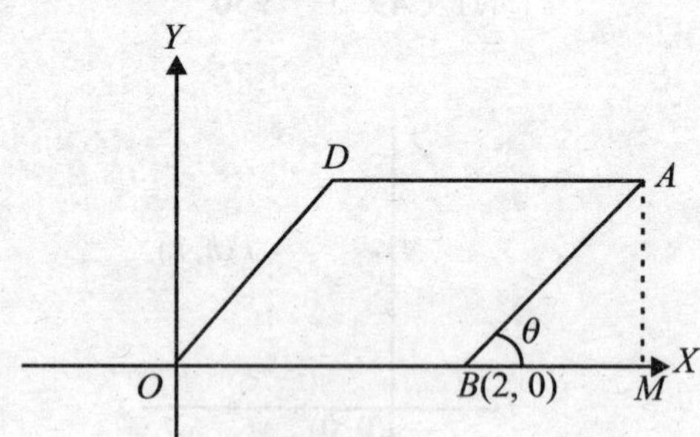

Fig. 1.118

Let $\angle DOB = \angle ABM = \theta$.

Area of ΔOAB is $\dfrac{1}{2} \times OB \times AM = \dfrac{1}{2} \times \sqrt{3}$

$\Rightarrow 2 \times 2 \sin\theta = \sqrt{3}$

$\Rightarrow \sin\theta = \dfrac{\sqrt{3}}{4} \Rightarrow AM = \sqrt{3}$ and $BM = 1$

Hence, A has coordinates $(3, \sqrt{3})$. By symmetry, all possible coordinates are $(\pm 3, \pm\sqrt{3})$.

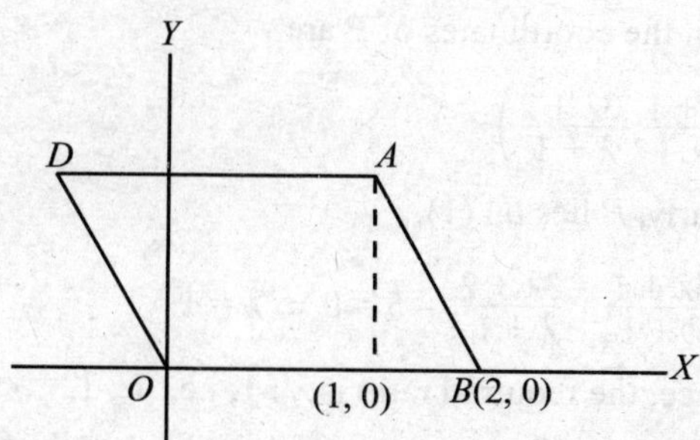

Fig. 1.119

From the above figure, A has coordinates $(1, \sqrt{3})$.

By symmetry, all possible coordinates are $(\pm 1, \pm\sqrt{3})$.

d.

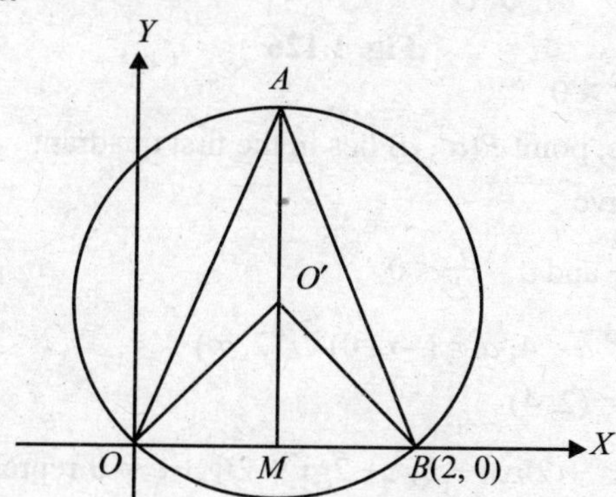

Fig. 1.120

$OB = 2$ units $= OO' =$ radius

$\Rightarrow OM = \frac{2}{2} = 1$ unit

In $\Delta OO'M$,

$O'M = \sqrt{4-1} = \sqrt{3}$

Since ΔOAB is isosceles hence point A lies on perpendicular bisector of OB.

$\therefore AM = \sqrt{3} + 2 = O'M + OA$

Hence, the coordinate A will be $(1, 2+\sqrt{3})$ in first quadrant. By symmetry, all possible coordinates of A are $(\pm 1, \pm(2+\sqrt{3}))$.

10. p → a, d; q → a; r → b; s → a, b, c

p.

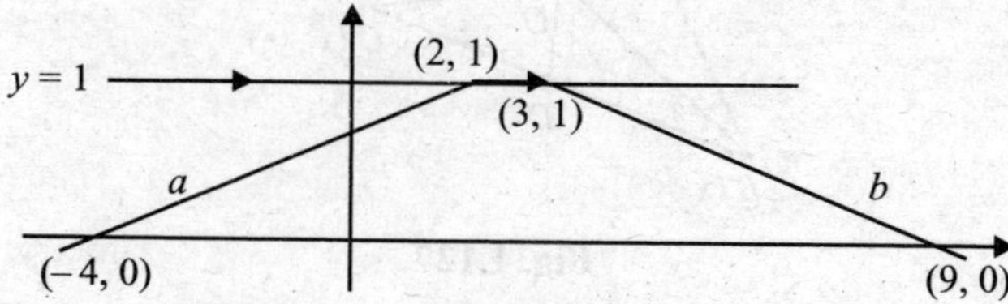

Fig. 1.121

Obviously, trapezium

$$\left.\begin{aligned} a &= \sqrt{37} \\ b &= \sqrt{37} \end{aligned}\right\} \Rightarrow a = b$$

Hence, isosceles trapezium

$\Rightarrow$ a cyclic quadrilateral

q.

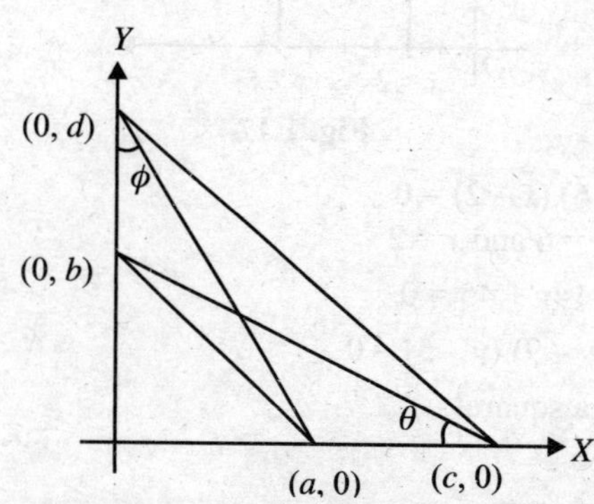

Fig. 1.122

$ac = bd$

$\Rightarrow \frac{b}{c} = \frac{a}{d}$

$$\left.\begin{aligned} \tan\theta &= \frac{b}{c} \\ \tan\phi &= \frac{a}{d} \end{aligned}\right\} \Rightarrow \theta = \phi$$

Hence, cyclic quadrilateral.

c.

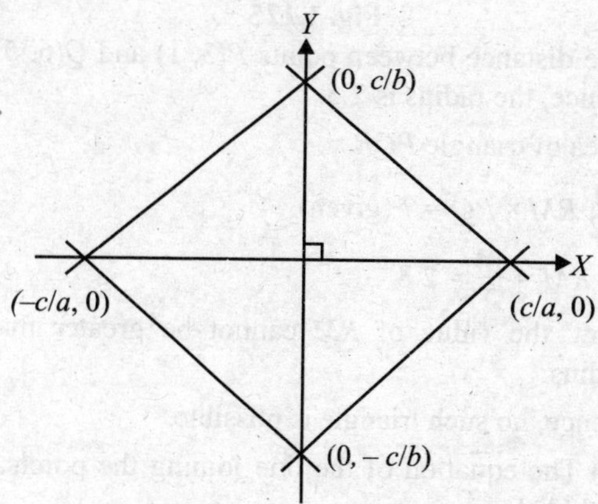

Fig. 1.123

$ax \pm by \pm c = 0$

If $y = 0$, $y = \pm\frac{c}{a}$

If $x = 0$, $y = \pm\frac{c}{d}$

$\Rightarrow$ rhombus

d.

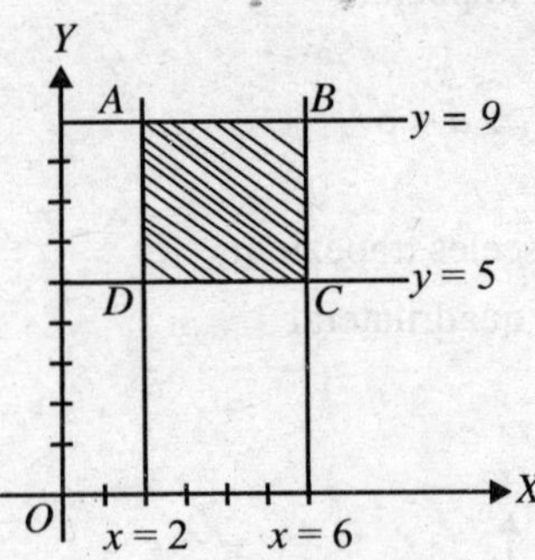

Fig. 1.124

$(x-6)(x-2)=0$
$\Rightarrow x=6$ and $x=2$
$y^2-14y+45=0$
$\Rightarrow (y-9)(y-5)=0$
$\Rightarrow$ a square.

Integer Type

1. **(0)** Obviously R lies on the circle with P and Q as end points of a diameter.

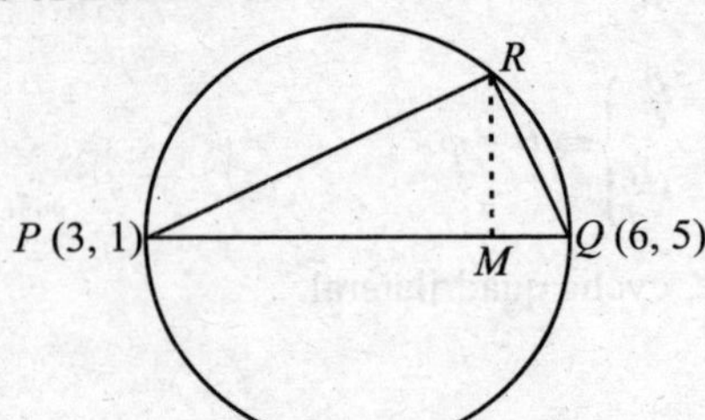

Fig. 1.125

The distance between points $P(3, 1)$ and $Q(6, 5)$ is 5.
Hence, the radius is 2.5.
Area of triangle PQR

$=\frac{1}{2} RM \times PQ = 7$ (given)

$\Rightarrow RM = \frac{14}{5} = 2.8$

Now the value of RM cannot be greater than the radius.

Hence, no such triangle is possible.

2. **(1)** The equation of the line joining the points (2, 3) and (4, 1) is

$$y-3=\frac{1-3}{4-2}(x-2)$$

$\Rightarrow y-3=-x+2$

$\Rightarrow x+y-5=0 \qquad (1)$

Suppose the line joining (2, 3) and (4, 1) divides the segment joining (1, 2) and (4, 3) at the point P in the ratio of $\lambda : 1$.

Then the coordinates of P are

$$\left(\frac{4\lambda+1}{\lambda+1}, \frac{3\lambda+2}{\lambda+1}\right).$$

Clearly, P lies on (1),

$$\Rightarrow \frac{4\lambda+1}{\lambda+1}+\frac{3\lambda+2}{\lambda+1}-5=0 \Rightarrow \lambda=1$$

Hence, the required ratio is $\lambda : 1$, i.e., $1 : 1$.

3. **(1)** Acute angles are formed by the lines in the first and third quadrants.

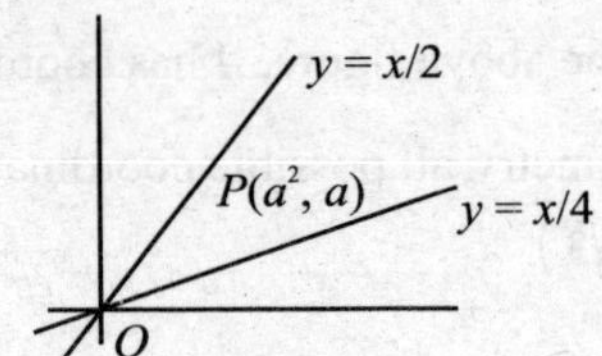

Fig. 1.126

But $a^2 > 0$

Hence, point $P(a^2, a)$ lies in the first quadrant.

We have

$$a-\frac{a^2}{4} \text{ and } a-\frac{a^2}{2}<0$$

$\Rightarrow 0<a<4, a \in (-\infty, 0) \cup (2, \infty)$

$\Rightarrow a \in (2, 4)$

4. **(8)** $ax^2+2hxy+by^2+2gx+2fy+c=0$ represents a pair of lines if

$abc+2fgh-af^2-bg^2-bc^2=0$, i.e., if

$$6\lambda+2(7)(4)\left(\frac{7}{2}\right)-2(7)^2-3(4)^2-\lambda\left(\frac{7}{2}\right)^2=0$$

$$\Rightarrow 6\lambda+196-98-48-\frac{49\lambda}{4}=0$$

$$\Rightarrow \frac{49\lambda}{4}-6\lambda=196-146=50$$

$$\Rightarrow \frac{25\lambda}{4}=50.$$

Hence, $\lambda=\frac{200}{25}=8$.

5. **(1)**

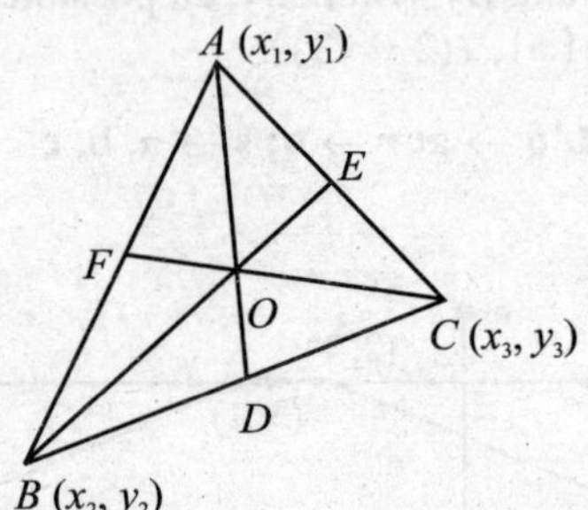

Fig. 1.127

Let AD, BE, and CF meet at O. We take O as the origin.

Let the coordinates of points A, B, and C be (x_1, y_1), (x_2, y_2), and (x_3, y_3), respectively.

Let D divide BC in the ratio of $k : 1$, i.e., $BD/DC = k/1$.

Then $D \equiv \left(\dfrac{kx_3 + x_2}{k+1}, \dfrac{ky_3 + y_2}{k+1}\right)$

Also the equation of line AD is

$$y - 0 = \frac{y_1 - 0}{x_1 - 0}(x - 0) \text{ or } y = \frac{y_1}{x_1}x \quad (1)$$

Since D lies on AD,

$$\frac{ky_3 + y_2}{k+1} = \frac{y_1}{x_1}\left(\frac{kx_3 + x_2}{k+1}\right)$$

or $k(x_1y_3 - x_3y_1) = x_2y_1 - x_1y_2$

$$\Rightarrow k = \frac{BD}{DC} = \frac{x_2y_1 - x_1y_2}{x_1y_3 - x_3y_1} \quad (2)$$

Similarly, $\dfrac{CE}{EA} = \dfrac{x_3y_2 - x_2y_3}{x_2y_1 - x_1y_2}$ (3)

and $\dfrac{AF}{FB} = \dfrac{x_1y_3 - x_3y_1}{x_3y_2 - x_2y_3}$ (4)

From (1), (2), and (3), we get

$$\frac{BD}{DC} \times \frac{CE}{EA} \times \frac{AF}{FB} = 1$$

or $BD \times CE \times AF = DC \times EA \times FB$

6. (4)

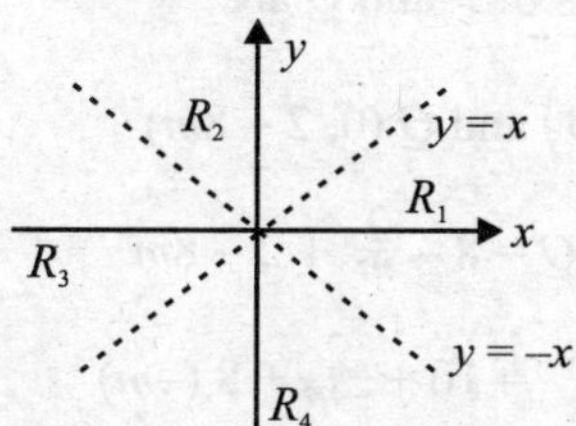

Fig. 1.128

$$d(P, L_1) = \frac{|x - y|}{\sqrt{2}} \text{ and } d(P, L_2) = \frac{|x + y|}{\sqrt{2}}$$

We have $2 \le d(P, L_1) + d(P, L_2) \le 4$

$$\Rightarrow 2\sqrt{2} \le |x - y| + |x + y| \le 4\sqrt{2} \quad (1)$$

Let us consider the four regions, namely, R_1, R_2, R_3, and R_4 in the lines L_1 and L_2 dividing the coordinate plane.

In R_1, we have $y < x$, $y > -x$. In R_2, we have $y > x$, $y > -x$. Similarly in R_3, we have $y > x$, $y < -x$.

Finally, in R_4 we have $y < x$, $y < -x$.

Thus for R_1, equation (1) becomes

$2\sqrt{2} \le x - y + x + y \le 4\sqrt{2} \Rightarrow \sqrt{2} \le x \le 2\sqrt{2}$

Similarly for R_2, equation (1) becomes

$2\sqrt{2} \le y - x + x + y \le 4\sqrt{2} \Rightarrow \sqrt{2} \le y \le 2\sqrt{2}$

For R_3, equation (1) becomes

$2\sqrt{2} \le y - x - x - y \le 4\sqrt{2} \Rightarrow -\sqrt{2} \le y \le -2\sqrt{2}$

For R_4, equation (1) becomes

$2\sqrt{2} \le x - y - x - y \le 4\sqrt{2} \Rightarrow -\sqrt{2} \le y \le -2\sqrt{2}$

Thus region R will be between concentric squares formed by the lines

$x = \pm 2\sqrt{2}$, $y = \pm 2\sqrt{2}$ and $x = \pm 2\sqrt{2}$, $y = \pm\sqrt{2}$.

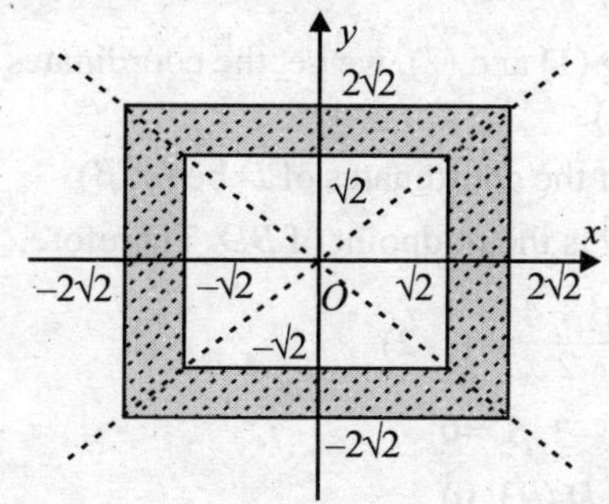

Fig. 1.129

Thus the required area is $(4\sqrt{2})^2 - (2\sqrt{2})^2 = 24$ sq. units.

7. (2) Intersection of $3x + 4y = 9$ and $y = mx + 1$.

For x-coordinate,

$3x + 4(mx + 1) = 9$

$\Rightarrow (3 + 4m)x = 5$

$\Rightarrow x = \dfrac{5}{3 + 4m}$

For x to be an integer, $3 + 4m$ should be a divisor of 5 i.e.,

1, – 1, 5, or – 5.

$3 + 4m = 1 \Rightarrow m = 1/2$ (not an integer)

$3 + 4m = -1 \Rightarrow m = -1$ (integer)

$3 + 4m = 5 \Rightarrow m = 1/2$ (not an integer)

$3 + 4m = -5 \Rightarrow m = -2$ (integer)

Hence, there are two integral values of m.

8. (8) Let O be the centre of the circle. M is the midpoint of AB. Then $OM \perp AB$.

Let OM when produced meet the circle at P and Q.

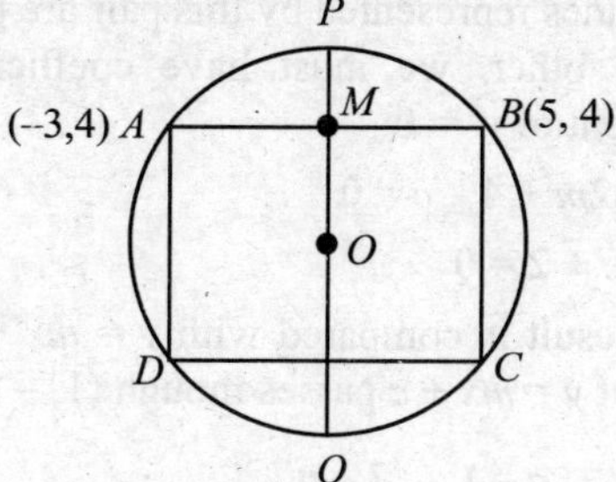

Fig. 1.130

PQ is a diameter perpendicular to AB and passing through M.

$M = \left(\frac{-3+5}{2}, \frac{4+4}{2}\right) = (1, 4)$

Slope of $AB = \frac{4-4}{5+3} = 0$

So PQ is a line parallel to y-axis passing through (1, 4).

Its equation is $x = 1$ (1)

Also equation of the given diameter is

$4y = x + 7$ (2)

Solving (1) and (2), we get the coordinates of centre O as (1, 2).

Also let the coordinates of D be (α, β).

Then O is the midpoint of BD. Therefore,

$\frac{\alpha+5}{2}, \frac{\beta+4}{2} = (1, 2)$

$\Rightarrow \alpha = -3, \beta = 0$

Hence, $D(-3, 0)$

Using the distance formula, we get

$AD = \sqrt{(-3+3)^2 + (4-0)^2} = 4$

$AB = \sqrt{(5+3)^2 + (4-4)^2} = 8$

Area of rectangle $ABCD = AB \times AD$

$= 8 \times 4$

$= 32$ sq. units

9. (5) The given curve is $3x^2 - y^2 - 2x + 4y = 0$ (1)

Let $y = mx + c$ be the chord of curve (1) which subtends an angle of 90° at the origin.

Then the combined equations of lines joining the points of intersection of the curve and the chord $y = mx + c$ to the origin can be obtained by making the equation of the curve homogeneous with the help of the equation of the chord as follows.

$3x^2 - y^2 - 2x\left(\frac{y-mx}{c}\right) + 4y\left(\frac{y-mx}{c}\right) = 0$

$\Rightarrow 3cx^2 - cy^2 - 2xy + 2mx^2 + 4y^2 - 4mxy = 0$

$\Rightarrow (3c + 2m)x^2 - 2(1 + 2m)xy + (4 - c)y^2 = 0.$

As the lines represented by this pair are perpendicular to each other, we must have coefficient of x^2 + coefficient of $y^2 = 0$

$\Rightarrow 3c + 2m + 4 - c = 0$

$\Rightarrow m + c + 2 = 0$

If this result is compared with $y = mx + c$, it can be seen that $y = mx + c$ passes though (1, – 2).

10. (3) $\tan\theta = \frac{a-1}{3} = \frac{7-a}{10-b}$

Also $\tan\theta = \frac{7-4}{b-3} = \frac{3}{b-3}$

Hence, $\frac{3}{b-3} = \frac{a-1}{3} = \frac{7-a}{10-b}$.

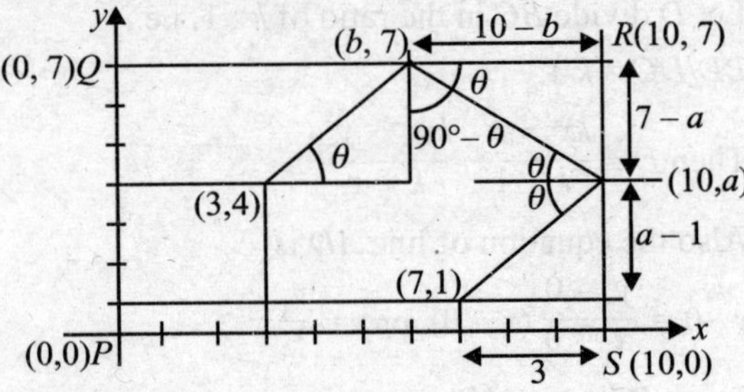

Fig. 1.131

From the first two relations, we get

$9 = AB - b - 3a = 3$

$\Rightarrow 3a + 6 = ab - b$ (1)

From the last two, we get

$10a - ab - 10 + b = 21 - 3a$

$\Rightarrow 13a - ab + b = 31$

or $ab - b = 13a - 31$ (2)

Hence, from (1) and (2) we get

$3a + 6 = 13a - 31$

$\Rightarrow 10a = 37 \Rightarrow a = 3.7.$

11. (9) Let the equation of the line L be

$y - 2 = m(x - 8), m < 0$

Coordinates of P and Q are

$P\left(8 - \frac{2}{m}, 0\right)$ and $Q(0, 2 - 8m)$

$\Rightarrow OP + OQ = 8 - \frac{2}{m} + 2 - 8m$

$= 10 + \frac{2}{-m} + 8(-m)$

$\geq 10 + 2\sqrt{\frac{2}{-m} \times 8(-m)}$

≥ 18

$\Rightarrow$ Absolute minimum value of $(OP + OQ)/2$ is 9.

12. (8) Eliminating x and y from the three equations, we get

$-2 = m(a + m)$

$\Rightarrow m^2 + am + 2 = 0.$

Since $m \in R$,

discriminant ≥ 0

$\Rightarrow a^2 - 8 \geq 0$

$\Rightarrow a^2 \geq 8$

13. (3) Let $L = a^2x\ aby + 1$

$L(0, 0) = 1$

If (0, 0) and (1, 1) lie on the same side, then

$L(1, 1) = a^2 + ab + 1 > 0$

Since $a \in R$,

$D < 0 \Rightarrow b^2 - 4 < 0$

$\Rightarrow -2 < b < 2$

$\Rightarrow b = -1, 0, 1$

14. (0) As M divides AB externally in the ratio of b: a,

$$x = \frac{b(a \cos \beta) - a(b \cos \alpha)}{b - a} \text{ and}$$

$$y = \frac{b(a \sin \beta) - a(b \sin \alpha)}{b - a}$$

$$\Rightarrow \frac{x}{y} = \frac{\cos \beta - \cos \alpha}{\sin \beta - \sin \alpha} = \frac{2 \sin \frac{\alpha + \beta}{2} \sin \frac{\alpha - \beta}{2}}{2 \cos \frac{\alpha + \beta}{2} \sin \frac{\beta - \alpha}{2}}$$

$$\Rightarrow x \cos \frac{\alpha + \beta}{2} + y \sin \frac{\alpha + \beta}{2} = 0$$

15. (1) The equations of the line L in the two coordinate systems are

$$\frac{x}{a} + \frac{y}{b} = 1 \text{ and } \frac{X}{p} + \frac{Y}{q} = 1$$

where (X, Y) are the new coordinates of a point (x, y) when the axes are rotated through a fixed angle, keeping the origin fixed.

As the length of the perpendicular from the origin has not changed,

$$\frac{1}{\sqrt{\frac{1}{a^2} + \frac{1}{b^2}}} = \frac{1}{\sqrt{\frac{1}{p^2} + \frac{1}{q^2}}} \Rightarrow \frac{1}{a^2} + \frac{1}{b^2} = \frac{1}{p^2} + \frac{1}{q^2}$$

$$\Rightarrow \frac{\frac{1}{a^2} + \frac{1}{b^2}}{\frac{1}{p^2} + \frac{1}{q^2}} = 1$$

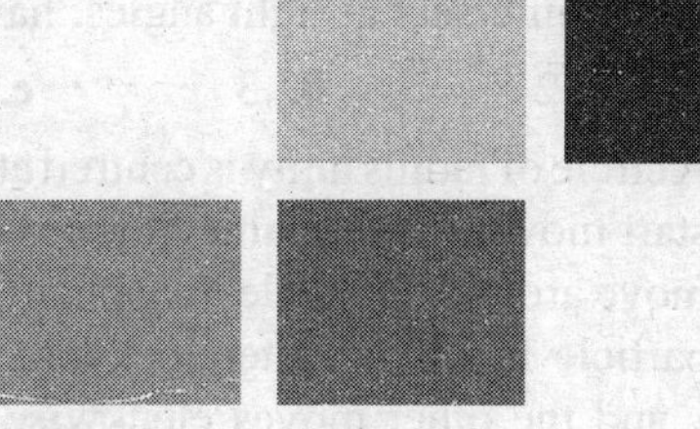

CHAPTER 2

Circle

Objective Type

Each question has four choices a, b, c and d, out of which *only one* answer is correct.

1. The number of rational point(s) (a point (a, b) is called rational, if a and b both are rational numbers) on the circumference of a circle having centre (π, e) is

   a. at most one **b.** at least two
   c. exactly two **d.** infinite

2. If the equation of any two diagonals of a regular pentagon belongs to family of lines $(1 + 2\lambda)y - (2 + \lambda)x + 1 - \lambda = 0$ and their lengths are $\sin 36°$, then locus of centre of circle circumscribing the given pentagon (the triangles formed by these diagonals with sides of pentagon have no side common) is

   a. $x^2 + y^2 - 2x - 2y + 1 + \sin^2 72° = 0$
   b. $x^2 + y^2 - 2x - 2y + \cos^2 72° = 0$
   c. $x^2 + y^2 - 2x - 2y + 1 + \cos^2 72° = 0$
   d. $x^2 + y^2 - 2x - 2y + \sin^2 72° = 0$

3. If OA and OB are equal perpendicular chord of the circles $x^2 + y^2 - 2x + 4y = 0$, then equations of OA and OB are where O is origin.

   a. $3x + y = 0$ and $3x - y = 0$
   b. $3x + y = 0$ or $3y - x = 0$
   c. $x + 3y = 0$ and $y - 3x = 0$
   d. $x + y = 0$ or $x - y = 0$

4. Equation of chord of the circle $x^2 + y^2 - 3x - 4y - 4 = 0$, which passes through the origin such that the origin divides it in the ratio 4 : 1, is

   a. $x = 0$ **b.** $24x + 7y = 0$
   c. $7x + 24y = 0$ **d.** $7x - 24y = 0$

5. The line $2x - y + 1 = 0$ is tangent to the circle at the point (2, 5) and the centre of the circles lies on $x - 2y = 4$. The radius of the circle is

   a. $3\sqrt{5}$ **b.** $5\sqrt{3}$ **c.** $2\sqrt{5}$ **d.** $5\sqrt{2}$

6. In a triangle ABC, right angled at A, on the leg AC as diameter, a semicircle is described. If a chord joins A with the point of intersection D of the hypotenuse and the semicircle, then the length of AC equals to

   a. $\dfrac{AB \times AD}{\sqrt{AB^2 + AD^2}}$ **b.** $\dfrac{AB \times AD}{AB + AD}$
   c. $\sqrt{AB \times AD}$ **d.** $\dfrac{AB \times AD}{\sqrt{AB^2 - AD^2}}$

7. A rhombus is inscribed in the region common to the two circles $x^2 + y^2 - 4x - 12 = 0$ and $x^2 + y^2 + 4x - 12 = 0$ with two of its vertices on the line joining the centres of the circles. The area of the rhombus is

   a. $8\sqrt{3}$ sq. units **b.** $4\sqrt{3}$ sq. units
   c. $6\sqrt{3}$ sq. units **d.** None

8. The locus of the centre of the circles such that the point (2, 3) is the midpoint of the chord $5x + 2y = 16$ is

   a. $2x - 5y + 11 = 0$ **b.** $2x + 5y - 11 = 0$
   c. $2x + 5y + 11 = 0$ **d.** None

9. Two congruent circles with centres at (2, 3) and (5, 6), which intersect at right angles, have radius equal to

a. $2\sqrt{2}$ **b.** 3 **c.** 4 **d.** None

10. A circle of radius unity is centred at origin. Two particles start moving at the same time from the point (1, 0) and move around the circle in opposite direction. One of the particle moves counterclockwise with constant speed v and the other moves clockwise with constant speed $3v$. After leaving (1, 0), the two particles meet first at a point P, and continue until they meet next at point Q. The coordinates of the point Q are

a. (1, 0) **b.** (0, 1) **c.** (0, –1) **d.** (–1, 0)

11. The value of 'c' for which the set $\{(x, y) \mid x^2 + y^2 + 2x \le 1\} \cap \{(x, y) \mid x - y + c \ge 0\}$ contains only one point in common is

a. $(-\infty, -1] \cup [3, \infty)$ **b.** $\{-1, 3\}$

c. $\{-3\}$ **d.** $\{-1\}$

12. A circle is inscribed into a rhombous $ABCD$ with one angle 60°. The distance from the centre of the circle to the nearest vertex is equal to 1. If P is any point of the circle, then $|PA|^2 + |PB|^2 + |PC|^2 + |PD|^2$ is equal to

a. 12 **b.** 11

c. 9 **d.** None of these

13. Two circles with radii a and b touch each other externally such that θ is the angle between the direct common tangents $(a > b \ge 2)$, then

a. $\theta = 2\cos^{-1}\left(\dfrac{a-b}{a+b}\right)$

b. $\theta = 2\tan^{-1}\left(\dfrac{a+b}{a-b}\right)$

c. $\theta = 2\sin^{-1}\left(\dfrac{a+b}{a-b}\right)$

d. $\theta = 2\sin^{-1}\left(\dfrac{a-b}{a+b}\right)$

14. B and C are fixed points having coordinates (3, 0) and (– 3, 0), respectively. If the vertical angle BAC is 90°, then the locus of the centroid of the ΔABC has the equation

a. $x^2 + y^2 = 1$ **b.** $x^2 + y^2 = 2$

c. $9(x^2 + y^2) = 1$ **d.** $9(x^2 + y^2) = 4$

15. $ABCD$ is a square of unit area. A circle is tangent to two sides of $ABCD$ and passes through exactly one of its vertices. The radius of the circle is

a. $2 - \sqrt{2}$ **b.** $\sqrt{2} - 1$

c. $\dfrac{1}{2}$ **d.** $\dfrac{1}{\sqrt{2}}$

16. A pair of tangents are drawn to a unit circle with centre at the origin and these tangents intersect at A enclosing an angle of 60°. The area enclosed by these tangents and the arc of the circle is

a. $\dfrac{2}{\sqrt{3}} - \dfrac{\pi}{6}$ **b.** $\sqrt{3} - \dfrac{\pi}{3}$

c. $\dfrac{\pi}{3} - \dfrac{\sqrt{3}}{6}$ **d.** $\sqrt{3}\left(1 - \dfrac{\pi}{6}\right)$

17. A straight line with slope 2 and y-intercept 5 touches the circle, $x^2 + y^2 + 16x + 12y + c = 0$ at a point Q. Then the co-ordinates of Q are

a. (–6, 11) **b.** (–9, –13)

c. (–10, –15) **d.** (–6, –7)

18. A circle of constant radius a passes through origin O and cuts the axes of coordinates in points P and Q, then the equation of the locus of the foot of perpendicular from O to PQ is

a. $(x^2 + y^2)\left(\dfrac{1}{x^2} + \dfrac{1}{y^2}\right) = 4a^2$

b. $(x^2 + y^2)^2\left(\dfrac{1}{x^2} + \dfrac{1}{y^2}\right) = a^2$

c. $(x^2 + y^2)^2\left(\dfrac{1}{x^2} + \dfrac{1}{y^2}\right) = 4a^2$

d. $(x^2 + y^2)\left(\dfrac{1}{x^2} + \dfrac{1}{y^2}\right) = a^2$

19. A line meets the coordinate axes in A and B. A circle is circumscribed about the triangle OAB. If d_1 and d_2 are the distances of the tangent to the circle at the origin O from the points A and B, respectively, then the diameter of the circle is

a. $\dfrac{2d_1 + d_2}{2}$ **b.** $\dfrac{d_1 + 2d_2}{2}$

c. $d_1 + d_2$ **d.** $\dfrac{d_1 d_2}{d_1 + d_2}$

20. If a circle of constant radius $3k$ passes through the origin O and meets coordinate axes at A and B, then the locus of the centroid of the triangle OAB is

a. $x^2 + y^2 = (2k)^2$ **b.** $x^2 + y^2 = (3k)^2$

c. $x^2 + y^2 = (4k)^2$ **d.** $x^2 + y^2 = (6k)^2$

21. The equation of a line inclined at an angle $\pi/4$ to the x-axis, such that the two circles $x^2 + y^2 = 4$, $x^2 + y^2 - 10x - 14y + 65 = 0$ intercept equal lengths on it, is

a. $2x - 2y - 3 = 0$ **b.** $2x - 2y + 3 = 0$

c. $x - y + 6 = 0$ **d.** $x - y - 6 = 0$

22. Let C be a circle with two diameters intersecting at an angle of 30°. A circle S is tangent to both the diameters and to C, and has radius unity. The largest radius of C is

a. $1 + \sqrt{6} + \sqrt{2}$ **b.** $1 + \sqrt{6} - \sqrt{2}$

c. $\sqrt{6} + \sqrt{2} - 11$ **d.** None of these

23. A straight line l_1 with equation $x - 2y + 10 = 0$ meets the circle with equation $x^2 + y^2 = 100$ at B in the first quadrant. A line through B, perpendicular to l_1 cuts y-axis at $P(0, t)$. The value of 't' is

a. 12 **b.** 15 **c.** 20 **d.** 25

24. Let a and b represent the length of a right triangle's legs. If d is the diameter of a circle inscribed into the triangle, and D is the diameter of a circle circumscribed on the triangle, then $d + D$ equals

a. $a + b$ **b.** $2(a + b)$

c. $\frac{1}{2}(a + b)$ **d.** $\sqrt{a^2 + b^2}$

25. If the chord $y = mx + 1$ of the circle $x^2 + y^2 = 1$ subtends an angle 45° at the major segment of the circle, then value of m is

a. 2 **b.** −2

c. −1 **d.** None of these

26. A variable chord of circle $x^2 + y^2 = 4$ is drawn from the point $P(3, 5)$ meeting the circle at the points A and B. A point Q is taken on this chord such that $2PQ = PA + PB$. Locus of Q is

a. $x^2 + y^2 + 3x + 4y = 0$

b. $x^2 + y^2 = 36$

c. $x^2 + y^2 = 16$

d. $x^2 + y^2 - 3x - 5y = 0$

27. In triangle ABC, equation of side BC is $x - y = 0$. Circumcentre and orthocentre of the triangle are (2, 3) and (5, 8), respectively. Equation of circumcircle of the triangle is

a. $x^2 + y^2 - 4x + 6y - 27 = 0$

b. $x^2 + y^2 - 4x - 6y - 27 = 0$

c. $x^2 + y^2 + 4x + 6y - 27 = 0$

d. $x^2 + y^2 + 4x + 6y - 27 = 0$

28. The range of values of r for which the point $\left(-5 + \frac{r}{\sqrt{2}}, -3 + \frac{r}{\sqrt{2}}\right)$ is an interior point of the major segment of the circle $x^2 + y^2 = 16$, cut-off by the line $x + y = 2$, is

a. $(-\infty, 5\sqrt{2})$

b. $(4\sqrt{2} - \sqrt{14}, 5\sqrt{2})$

c. $(4\sqrt{2} - \sqrt{14}, 4\sqrt{2} + \sqrt{14})$

d. None of these

29. A square is inscribed in the circle $x^2 + y^2 - 2x + 4y - 93 = 0$ with its sides parallel to the coordinate axis. The coordinates of its vertices are

a. (−6, −9), (−6, 5), (8, −9), (8, 5)

b. (−6, 9), (−6, −5), (8, −9), (8, 5)

c. (−6, −9), (−6, 5), (8, 9), (8, 5)

d. (−6, −9), (−6, 5), (8, −9), (8, −5)

30. (− 6, 0), (0, 6) and (− 7, 7) are the vertices of a ΔABC. The incircle of the triangle has the equation

a. $x^2 + y^2 - 9x - 9y + 36 = 0$

b. $x^2 + y^2 + 9x - 9y + 36 = 0$

c. $x^2 + y^2 + 9x + 9y - 36 = 0$

d. $x^2 + y^2 + 18x - 18y + 36 = 0$

31. If O is the origin and OP, OQ are the tangents from the origin to the circle $x^2 + y^2 - 6x + 4y + 8 = 0$, then circumcenter of the triangle OPQ is

a. $(3, -2)$ **b.** $\left(\frac{3}{2}, -1\right)$

c. $\left(\frac{3}{4}, -\frac{1}{2}\right)$ **d.** $\left(-\frac{3}{2}, 1\right)$

32. Radii of the smallest and the largest circle passing though a point lying on the sides of a rectangle with vertices $(\pm 2, \pm 1)$ and touching the circle $x^2 + y^2 = 9$, are r_1 and r_2, respectively. Let $d = |r_1 - r_2|$, then minimum value of d is

a. $\frac{1}{2}$ **b.** 1 **c** 2 **d.** 3

33. The difference between the radii of the largest and the smallest circles which have their centre on the circumference of the circle $x^2 + y^2 + 2x + 4y - 4 = 0$ and pass through the point (a, b) lying outside the given circle, is

a. 6 **b.** $\sqrt{(a + 1)^2 + (b + 2)^2}$

c. 3 **d.** $\sqrt{(a + 1)^2 + (b + 2)^2} - 3$

34. An isosceles triangles ABC is inscribed in a circle $x^2 + y^2 = a^2$ with the vertex A at $(a, 0)$ and the base angle B and C each equal to 75°, then coordinates of an end point of the base are

a. $\left(\frac{-\sqrt{3}a}{2}, \frac{a}{2}\right)$ **b.** $\left(-\frac{\sqrt{3}a}{2}, a\right)$

c. $\left(\frac{a}{2}, \frac{\sqrt{3}a}{2}\right)$ **d.** $\left(\frac{\sqrt{3}a}{2}, -\frac{a}{2}\right)$

35. The equations of four circles are $(x \pm a)^2 + (y \pm a)^2 = a^2$. The radius of a circle touching all the four circles is

a. $(\sqrt{2} + 2)a$ **b.** $2\sqrt{2}a$

c. $(\sqrt{2} + 1)a$ **d.** $(2 + \sqrt{2})a$

36. The locus of a point which moves such that the sum of the squares of its distance from three vertices of a triangle is constant is a/an

a. circle **b.** straight line

c. ellipse **d.** None of these

37. A circle passes through the points $A(1, 0)$, $B(5, 0)$ and touches the y-axis at $C(0, h)$. If $\angle ACB$ is maximum then

a. $h = 3\sqrt{5}$ **b.** $h = 2\sqrt{5}$

c. $h = \sqrt{5}$ **d.** $h = 2\sqrt{10}$

38. A circle with centre (a, b) passes through the origin. The equation of the tangent to the circle at the origin is

a $ax - by = 0$ **b.** $ax + by = 0$
c. $bx - ay = 0$ **d.** $bx + ay = 0$

39. The area of the triangle formed by joining the origin to the points of intersection of the line $x\sqrt{5} + 2y = 3\sqrt{5}$ and circle $x^2 + y^2 = 10$ is

a. 3 **b.** 4 **c.** 5 **d.** 6

40. If (α, β) is a point on the circle whose centre is on the x-axis and which touches the line $x + y = 0$ at $(2, -2)$, then the greatest values of α is

a. $4 - \sqrt{2}$ **b.** 6
c. $4 + 2\sqrt{2}$ **d.** $4 + \sqrt{2}$

41. A region in the x–y plane is bounded by the curve $y = \sqrt{25 - x^2}$ and the line $y = 0$. If the point $(a, a + 1)$ lies in the interior of the region, then

a. $a \in (-4, 3)$ **b.** $a \in (-\infty, -1) \in (3, \infty)$
c. $a \in (-1, 3)$ **d.** None of these

42. There are two circles whose equations are $x^2 + y^2 = 9$ and $x^2 + y^2 - 8x - 6y + n^2 = 0, n \in Z$. If the two circles have exactly two common tangents then the number of possible values of n is

a. 2 **b.** 8
c. 9 **d.** None of these

43. C_1 is a circle of radius 1 touching the x-axis and the y-axis. C_2 is another circle of radius > 1 and touching the axes as well as the circle C_1. Then, the radius of C_2 is

a. $3 - 2\sqrt{2}$ **b.** $3 + 2\sqrt{2}$
c. $3 + 2\sqrt{3}$ **d.** None of these

44. Equation of incircle of equilateral triangle ABC where $B \equiv (2, 0)$ $C \equiv (4, 0)$ and A lies in fourth quadrant is

a. $x^2 + y^2 - 6x + \frac{2y}{\sqrt{3}} + 9 = 0$

b. $x^2 + y^2 - 6x - \frac{2y}{\sqrt{3}} + 9 = 0$

c. $x^2 + y^2 + 6x + \frac{2y}{\sqrt{3}} + 9 = 0$

d. None of these

45. $f(x, y) = x^2 + y^2 + 2ax + 2by + c = 0$ represents a circle. If $f(x, 0) = 0$ has equal roots, each being 2 and $f(0, y) = 0$ has 2 and 3 as it's roots, then centre of circle is

a. (2, 5/2) **b.** Data are not sufficient
c. (– 2, – 5/2) **d.** Data are inconsistent

46. The area bounded by the circles $x^2 + y^2 = 1$, $x^2 + y^2 = 4$ and the pair of lines $\sqrt{3}\,(x^2 + y^2) = 4xy$, is equal to

a. $\frac{\pi}{2}$ **b.** $\frac{5\pi}{2}$ **c.** 3π **d.** $\frac{\pi}{4}$

47. The straight line $x \cos\theta + y \sin\theta = 2$ will touch the circle $x^2 + y^2 - 2x = 0$, if

a. $\theta = n\pi, n \in I$ **b.** $A = (2n + 1)\pi, n \in I$
c. $\theta = 2n\pi, n \in I$ **d.** None of these

48. Let $ABCD$ be a quadrilateral with area 18, with side AB parallel to the side CD and $AB = 2CD$. Let AD be perpendicular to AB and CD. If a circle is drawn inside the quadrilateral $ABCD$ touching all the side, then its radius is

a. 3 **b.** 2 **c.** $\frac{3}{2}$ **d.** 1

49. Consider a family of circle which are passing through the point $(-1, 1)$ and are tangent to the x-axis. If (h, k) are the coordinates of the centre of the circles, then the set of values of k is given by the interval

a. $k \ge \frac{1}{2}$ **b.** $-\frac{1}{2} \le k \le \frac{1}{2}$
c. $k \le \frac{1}{2}$ **d.** $0 < k < \frac{1}{2}$

50. If the conics whose equations are $S \equiv \sin^2\theta\, x^2 + 2hxy + \cos^2\theta\, y^2 + 32x + 16y + 19 = 0$, $S' \equiv \cos^2\theta\, x^2 + 2h'xy + \sin^2\theta\, y^2 + 16x + 32y + 19 = 0$ intersects in four concyclic points then, (where $\theta \in R$)

a. $h + h' = 0$ **b.** $h = h'$
c. $h + h' = 1$ **d.** None of these

51. The range of values of $\lambda (\lambda > 0)$ such that the angle θ between the pair of tangents drawn from $(\lambda, 0)$ to the circle $x^2 + y^2 = 4$ lies in $\left(\frac{\pi}{2}, \frac{2\pi}{3}\right)$ is

a. $\left(\frac{4}{\sqrt{3}}, 2\sqrt{2}\right)$ **b.** $(0, \sqrt{2})$
c. (1, 2) **d.** None of these

52. The circles which can be drawn, to pass through (1, 0) and (3, 0) and to touch the y-axis, intersect at an angle θ, then $\cos\theta$ is equal to

a. $\frac{1}{2}$ **b.** $-\frac{1}{2}$ **c.** $\frac{1}{4}$ **d.** $-\frac{1}{4}$

53. Locus of midpoints of the chords of contact of $x^2 + y^2 = 2$ from the points on the line $3x + 4y = 10$ is a circle with centre P. If O be the origin then OP is equal to

a. 2 **b.** 3 **c.** $\frac{1}{2}$ **d.** $\frac{1}{3}$

54. Suppose $ax + by + c = 0$, where a, b, c are in A.P. be normal to a family of circles. The equation of the circle of the family intersects the circle $x^2 + y^2 - 4x - 4y - 1 = 0$ orthogonally is

a. $x^2 + y^2 - 2x + 4y - 3 = 0$
b. $x^2 + y^2 + 2x - 4y - 3 = 0$
c. $x^2 + y^2 - 2x + 4y - 5 = 0$
d. $x^2 + y^2 - 2x - 4y + 3 = 0$

55. If a circle of radius r is touching the lines $x^2 - 4xy + y^2 = 0$ in the first quadrant at points A and B, then area of triangle OAB (O being the origin) is

a. $\frac{3\sqrt{3}\,r^2}{4}$ **b.** $\frac{\sqrt{3}\,r^2}{4}$

c. $\frac{3r^2}{4}$ **d.** r^2

56. Let P be any moving point on the circle $x^2 + y^2 - 2x = 1$. AB be the chord of contact of this point w.r.t. the circle $x^2 + y^2 - 2x = 0$. The locus of the circumcnetre of the triangle CAB (C being centre of the circles) is

a. $2x^2 + 2y^2 - 4x + 1 = 0$

b. $x^2 + y^2 - 4x + 2 = 0$

c. $x^2 + y^2 - 4x + 1 = 0$

d. $2x^2 + 2y^2 - 4x + 3 = 0$

57. Two circles of radii a and b touching each other externally, are inscribed in the area bounded by $y = \sqrt{1 - x^2}$ and the x-axis. If $b = 1/2$, then a is equal to

a. $\frac{1}{4}$ **b.** $\frac{1}{8}$ **c.** $\frac{1}{2}$ **d.** $\frac{1}{\sqrt{2}}$

58. From an arbitrary point P on the circle $x^2 + y^2 = 9$, tangents are drawn to the circle $x^2 + y^2 = 1$, which meet $x^2 + y^2 = 9$ at A and B. Locus of the point of intersection of tangents at A and B to the circle $x^2 + y^2 = 9$ is

a. $x^2 + y^2 = \left(\frac{27}{7}\right)^2$ **b.** $x^2 - y^2 = \left(\frac{27}{7}\right)^2$

c. $y^2 - x^2 = \left(\frac{27}{7}\right)^2$ **d.** None of these

59. If r_1 and r_2 are the radii of smallest and largest circles which passes through (5, 6) and touches the circle $(x - 2)^2 + y^2 = 4$, then $r_1 r_2$ is

a. $\frac{4}{41}$ **b.** $\frac{41}{4}$ **c.** $\frac{5}{41}$ **d.** $\frac{41}{6}$

60. Minimum radius of circle which is orthogonal with both the circles $x^2 + y^2 - 12x + 35 = 0$ and $x^2 + y^2 + 4x + 3 = 0$ is

a. 4 **b.** 3 **c.** $\sqrt{15}$ **d.** 1

61. If $C_1 : x^2 + y^2 = (3 + 2\sqrt{2})^2$ be a circle and PA and PB are pair of tangents on C_1, where P is any point on the director circle of C_1, then the radius of smallest circle which touch C_1 externally and also the two tangents PA and PB is

a. $2\sqrt{3} - 3$ **b.** $2\sqrt{2} - 1$

c. $2\sqrt{2} - 1$ **d.** 1

62. From a point $R(5, 8)$ two tangents RP and RQ are drawn to a given circle $S = 0$ whose radius is 5. If circumcentre of the triangle PQR is (2, 3), then the equation of circle $S = 0$ is

a. $x^2 + y^2 + 2x + 4y - 20 = 0$

b. $x^2 + y^2 + x + 2y - 10 = 0$

c. $x^2 + y^2 - x - 2y - 20 = 0$

d. $x^2 + y^2 - 4x - 6y - 12 = 0$

63. On the line segment joining (1, 0) and (3, 0), an equilateral triangle is drawn having its vertex in the fourth quadrant, then radical centre of the circles described on its sides as diameter is

a. $\left(3, -\frac{1}{\sqrt{3}}\right)$ **b.** $(3, -\sqrt{3})$

c. $\left(2, -\frac{1}{\sqrt{3}}\right)$ **d.** $(2, -\sqrt{3})$

64. If the tangents are drawn from any point on the line $x + y = 3$ to the circle $x^2 + y^2 = 9$, then the chord of contact passes through the point

a. (3, 5) **b.** (3, 3)

c. (5, 3) **d.** None of these

65. If the radius of the circumcircle of the triangle TPQ, where PQ is chord of contact corresponding to point T with respect to circle $x^2 + y^2 - 2x + 4y - 11 = 0$, is units, then minimum distance of T from the director circle of the given circle is

a. 6 **b.** 12

c. $6\sqrt{2}$ **d.** $12 - 4\sqrt{2}$

66. P is a point (a, b) in the first quadrant. If the two circles which pass through P and touch both the coordinates axes cut at right angles, then

a. $a^2 - 6ab + b^2 = 0$ **b.** $a^2 + 2ab - b^2 = 0$

c. $a^2 - 4ab + b^2 = 0$ **d.** $a^2 - 8ab + b^2 = 0$

67. The number of common tangent(s) to the circles $x^2 + y^2 + 2x + 8y - 23 = 0$ and $x^2 + y^2 - 4x - 10y - 19 = 0$ is

a. 1 **b.** 2 **c.** 3 **d.** 4

68. Two circles of radii 4 cm and 1 cm touch each other externally and θ is the angle contained by their direct common tangents. Then $\sin\theta$ is equal to

a. $\frac{24}{25}$ **b.** $\frac{12}{25}$

c. $\frac{3}{4}$ **d.** None of these

69. The locus of the midpoints of the chords of the circle $x^2 + y^2 - ax - by = 0$ which subtend a right angle at $\left(\frac{a}{2}, \frac{b}{2}\right)$ is

a. $ax + by = 0$

b. $ax + by = a^2 + b^2$

c. $x^2 + y^2 - ax - by + \frac{a^2 + b^2}{8} = 0$

d. $x^2 + y^2 - ax - by - \frac{a^2 + b^2}{8} = 0$

70. From points (3 , 4), chords are drawn to the circle $x^2 + y^2 - 4x = 0$. The locus of the midpoints of the chords is

a. $x^2 + y^2 - 5x - 4y + 6 = 0$
b. $x^2 + y^2 + 5x - 4y + 6 = 0$
c. $x^2 + y^2 - 5x + 4y + 6 = 0$
d. $x^2 + y^2 - 5x - 4y - 6 = 0$

71. The locus of the centers of the circles which cut the circles $x^2 + y^2 + 4x - 6y + 9 = 0$ and $x^2 + y^2 - 5x + 4y - 2 = 0$ orthogonally is

a. $9x + 10y - 7 = 0$ **b.** $x - y + 2 = 0$
c. $9x - 10y + 11 = 0$ **d.** $9x + 10y + 7 = 0$

72. The angle at which the circles $(x - 1)^2 + y^2 = 10$ and $x^2 + (y - 2)^2 = 5$ intersect is

a. $\frac{\pi}{6}$ **b.** $\frac{\pi}{4}$ **c.** $\frac{\pi}{3}$ **d.** $\frac{\pi}{2}$

73. A tangent at a point on the circle $x^2 + y^2 = a^2$ intersects a concentric circle C at two points P and Q. The tangents to the circle X at P and Q meet at a point on the circle $x^2 + y^2 = b^2$, then the equation of circle is

a. $x^2 + y^2 = ab$ **b.** $x^2 + y^2 = (a - b)^2$
c. $x^2 + y^2 = (a + b)^2$ **d.** $x^2 + y^2 = a^2 + b^2$

74. Tangents are drawn to the circle $x^2 + y^2 = 1$ at the points where it is met by the circles $x^2 + y^2 - (\lambda + 6)x + (8 - 2\lambda)y - 3 = 0$, λ being the variable. The locus of the point of intersection of these tangents is

a. $2x - y + 10 = 0$ **b.** $x + 2y - 10 = 0$
c. $x - 2y + 10 = 0$ **d.** $2x + y - 10 = 0$

75. If point A and B are (1, 0) and B (0, 1).If point C is on the circle $x^2 + y^2 = 1$, then locus of the orthocentre of the triangle ABC is

a. $x^2 + y^2 = 4$
b. $x^2 + y^2 - x - y = 0$
c. $x^2 + y^2 - 2x - 2y + 1 = 0$
d. $x^2 + y^2 + 2x - 2y + 1 = 0$

76. If the line $x \cos\theta + y \sin\theta = 2$ is the equation of a transverse common tangent to the circles $x^2 + y^2 = 4$ and $x^2 + y^2 - 6\sqrt{3}x - 6y + 20 = 0$, then the value of θ is

a. $5\pi/6$ **b.** $2\pi/3$ **c.** $\pi/3$ **d.** $\pi/6$

77. Any circle through the point of intersection of the lines $x + \sqrt{3}y = 1$ and $\sqrt{3}x - y = 2$ if intersects these lines at points P and Q, then the angle subtended by the arc PQ at its centre is

a. 180°
b. 90°
c. 120°
d. Depends on centre and radius

78. If the angle between tangents drawn to $x^2 + y^2 + 2gx + 2fy + c = 0$ form (0, 0) is $\pi/2$, then

a. $g^2 + f^2 = 3c$ **b.** $g^2 + f^2 = 2c$
c. $g^2 + f^2 = 5c$ **d.** $g^2 + f^2 = 4c$

79. The common chord of the circle $x^2 + y^2 + 6x + 8y - 7 = 0$ and a circle passing through the origin, and touching the line $y = x$, always passes through the point

a. (−1/2, 1/2) **b.** (1, 1)
c. (1/2, 1/2) **d.** None of these

80. The chord of contact of tangents from three points A, B, C to the circle $x^2 + y^2 = a^2$ are concurrent, then A, B, C will

a. be concyclic
b. be collinear
c. form the vertices of a triangle
d. None of these

81. The chord of contact of tangents from a point P to a circle passes through Q. If l_1 and l_2 are the lengths of the tangents from P and Q to the circle, then PQ is equal to

a. $\frac{l_1 + l_2}{2}$ **b.** $\frac{l_1 - l_2}{2}$
c. $\sqrt{l_1^2 + l_2^2}$ **d.** $2\sqrt{l_1^2 + l_2^2}$

82. If $C_1: x^2 + y^2 - 20x + 64 = 0$ and $C_2: x^2 + y^2 + 30x + 144 = 0$. Then the length of the shortest line segment PQ which touches C_1 at P and to C_2 at Q is

a. 20 **b.** 15 **c.** 22 **d.** 27

83. The ends of a quadrant of a circle have the coordinates (1, 3) and (3, 1) . Then the centre of such a circle is

a. (2, 2) **b.** (1,1) **c.** (4, 4) **d.** (2, 6)

84. If the line $ax + by = 2$ is a normal to the circle $x^2 + y^2 - 4x - 4y = 0$ and a tangent to the circle $x^2 + y^2 = 1$, then

a. $a = \frac{1}{2}, b = \frac{1}{2}$
b. $a = \frac{1 + \sqrt{7}}{2}, b = \frac{1 - \sqrt{7}}{2}$
c. $a = \frac{1}{4}, b = \frac{3}{4}$
d. $a = 1, b = \sqrt{3}$

85. Radius of the tangent circle that can be dawn to pass through the point (0, 7), (0, 6) and touching the x-axis is

a. $\frac{5}{2}$ **b.** $\frac{3}{2}$ **c.** $\frac{7}{2}$ **d.** $\frac{9}{2}$

86. The equation of the tangent to the circle $x^2 + y^2 = a^2$, which makes a triangle of area a^2 with the coordinate axes, is

a. $x \pm y = a\sqrt{2}$ **b.** $x \pm y = \pm a\sqrt{2}$
c. $x \pm y = 2a$ **d.** $x + y = \pm 2a$

87. If the circle $x^2 + y^2 + 2gx + 2fy + c = 0$ is touched by $y = x$ at P such that $OP = 6\sqrt{2}$, then the value of c is

a. 36 **b.** 144

c. 72 **d.** None of these

88. The circle $x^2 + y^2 = 4$ cuts the line joining the points $A(1, 0)$ and $B(3, 4)$ in two points P and Q. Let $\frac{BP}{PA} = \alpha$ and $\frac{BQ}{QA} = \beta$. Then α and β are roots of the quadratic equation

a. $3x^2 + 2x - 21 = 0$ **b.** $3x^2 + 2x + 21 = 0$

c. $2x^2 + 3x - 21 = 0$ **d.** None of these

89. The area of the triangle formed by the positive x-axis and the normal and tangent to the circle $x^2 + y^2 = 4$ at $(1, \sqrt{3})$ is

a. $2\sqrt{3}$ sq. units **b.** $3\sqrt{2}$ sq. units

c. $\sqrt{6}$ sq. units **d.** None of these

90. Let P be point on the circle $x^2 + y^2 = 9$, Q a point on the line $7x + y + 3 = 0$, and the perpendicular bisector of PQ be the line $x - y + 1 = 0$. Then the coordinate of P are

a. $(0, -3)$ **b.** $(0, 3)$

c. $\left(\frac{72}{25}, -\frac{21}{25}\right)$ **d.** $\left(-\frac{72}{25}, \frac{21}{25}\right)$

91. A straight line moves such that the algebraic sum of the perpendicular drawn to it from two fixed points is equal to $2k$. Then, the straight line always touches a fixed circle of radius

a. $2k$ **b.** $k/2$

c. k **d.** None of these

92. The coordinates of the middle point of the chord cut-off by $2x - 5y + 18 = 0$ by the circle

$x^2 + y^2 - 6x + 2y - 54 = 0$ are

a. (1, 4) **b.** (2, 4) **c.** (4, 1) **d.** (1, 1)

93. The locus of a point from which the lengths of the tangents to the circles $x^2 + y^2 = 4$ and $2(x^2 + y^2) - 10x + 3y - 2 = 0$ are equal to

a. a straight line inclined at $\pi/4$ with the line joining the centres of the circles

b. a circle

c. an ellipse

d. a straight line perpendicular to the line joining the centres of the circles

94. The equation of circumcircle of an equilateral triangle is $x^2 + y^2 + 2gx + 2fy + c = 0$ and one vertex of the triangle is (1, 1). The equation of incircle of the triangle is

a. $4(x^2 + y^2) = g^2 + f^2$

b. $4(x^2 + y^2) + 8gx + 8fy = (1 - g)(1 + 3g) + (1 - f)(1 + 3f)$

c. $4(x^2 + y^2) + 8gx + 8fy = g^2 + f^2$

d. None of these

95. A light ray gets reflected from the $x = -2$. If the reflected ray touches the circle $x^2 + y^2 = 4$ and point of incident is $(-2, -4)$, then equation of incident ray is

a. $4y + 3x + 22 = 0$

b. $3y + 4x + 20 = 0$

c. $4y + 2x + 20 = 0$

d. $y + x + 6 = 0$

96. Tangents PA and PB drawn to $x^2 + y^2 = 9$ from any arbitrary point P on the line $x + y = 25$. Locus of midpoint of chord AB is

a. $25(x^2 + y^2) = 9(x + y)$

b. $25(x^2 + y^2) = 3(x + y)$

c. $5(x^2 + y^2) = 3(x + y)$

d. None of these

97. The circles having radii r_1 and r_2 intersect orthogonally. Length of their common chord is

a. $\frac{2r_1r_2}{\sqrt{r_1^2 + r_2^2}}$ **b.** $\frac{\sqrt{r_1^2 + r_2^2}}{2r_1r_2}$

c. $\frac{r_1r_2}{\sqrt{r_1^2 + r_2^2}}$ **d.** $\frac{\sqrt{r_1^2 + r_2^2}}{r_1r_2}$

98. If the pair of straight line $x\,y\sqrt{3} - x^2 = 0$ is tangent to the circle at P and Q from origin O such that area of smaller sector formed by CP and CQ is 3π sq. unit, where C is the centre of circle, then OP equals to

a. $(3\sqrt{3})/2$ **b.** $3\sqrt{3}$ **c.** 3 **d.** $\sqrt{3}$

99. The two circles which passes through $(0, a)$ and $(0, -a)$ and touch the line $y = mx + c$ will intersect each other at right angle, if

a. $a^2 = c^2(2m + 1)$ **b.** $a^2 = c^2(2 + m^2)$

c. $c^2 = a^2(2 + m^2)$ **d.** $c^2 = a^2(2m + 1)$

100. The condition that the chord $x \cos \alpha + y \sin \alpha - p = 0$ of $x^2 + y^2 - a^2 = 0$ may subtend a right angle at the centre of the circle is

a. $a^2 = 2p^2$ **b.** $p^2 = 2a^2$ **c.** $a = 2p$ **d.** $p = 2a$

101. The locus of the midpoints of a chord of the circle $x^2 + y^2 = 4$, which subtends a right angle at the origin, is

a. $x + y = 2$ **b.** $x^2 + y^2 = 1$

c. $x^2 + y^2 = 2$ **d.** $x + y = 1$

102. If the chord of contact of tangents from a point P to a given circle passes through Q, then the circle on PQ as diameter

a. cuts the given circle orthogonally

b. touches the given circle externally

c. touches the given circle internally

d. None of these

103. Let these base AB of a triangle ABC be fixed and the vertex C lie on a fixed circle of radius r. Lines through A and B are drawn to intersect CB and CA, respectively, at E and F such that $CE : EB = 1 : 2$ and $CF : FA = 1 : 2$. If the point of intersection P of these lines lies on the median through AB for all positions of AB then the locus of P is

a. a circle of radius $\frac{r}{2}$

b. a circle of radius $2r$

c. a parabola of latus rectum $4r$

d. a rectangular hyperbola

104. The number of integral values of y for which the chord of the circle $x^2 + y^2 = 125$ passing through the point $P(8, y)$ gets bisected at the point $P(8, y)$ and has integral slope is

a. 8 **b.** 6 **c.** 4 **d.** 2

105. Consider a square $ABCD$ of side length 1. Let P be the set of all segments of length 1 with end points on adjacent sides of square $ABCD$. The midpoints of segments in P enclose a region with area A, the value of A is

a. $\frac{\pi}{4}$ **b.** $1-\frac{\pi}{4}$

c. $4-\frac{\pi}{4}$ **d.** None of these

106. The range of values of α for which the line $2y = gx + \alpha$ is a normal to the circle $x^2 + y^2 + 2gx + 2gy - 2 = 0$ for all values of g is

a. $[1, \infty)$ **b.** $[-1, \infty)$ **c.** $(0, 1)$ **d.** $(-\infty, 1]$

107. Six points (x_i, y_i), $i = 1, 2, \ldots, 6$ are taken on the circle $x^2 + y^2 = 4$ such that $\sum_{i=1}^{6} x_i = 8$ and $\sum_{i=1}^{6} y_i = 4$. The line segment joining orthocentre of a triangle made by any three points and the centroid of the triangle made by other three points passes through a fixed points (h, k), then $h + k$ is

a. 1 **b.** 2 **c.** 3 **d.** 4

108. A circle with radius $|a|$ and centre on the y-axis slides along it and a variable lines through $(a, 0)$ cuts the circle at points P and Q. The region in which the point of intersection of tangents to the circle at points P and Q lies is represented by

a. $y^2 \geq 4(ax - a^2)$ **b.** $y^2 \leq 4(ax - a^2)$

c. $y \geq 4(ax - a^2)$ **d.** $y \leq 4(ax - a^2)$

109. If the angle of intersection of the circles $x^2 + y^2 + x + y = 0$ and $x^2 + y^2 + x - y = 0$ is θ, then equation of the line passing through (1, 2) and making an angle θ with the y-axis is

a. $x = 1$ **b.** $y = 2$

c. $x + y = 3$ **d.** $x - y = 3$

110. The centres of a set of circle, each of radius 3, lies on the circle $x^2 + y^2 = 25$. The locus of any point in the set is

a. $4 \leq x^2 + y^2 \leq 64$ **b.** $x^2 + y^2 \leq 25$

c. $x^2 + y^2 \geq 25$ **d.** $3 \leq x^2 + y^2 \leq 9$

111. The coordinates of two points P and Q are (x_1, y_1) and (x_2, y_2) and O is the origin. If circles be described on OP and OQ as diameters then length of their common chord is

a. $\frac{|x_1y_2 + x_2y_1|}{PQ}$ **b.** $\frac{|x_1y_2 - x_2y_1|}{PQ}$

c. $\frac{|x_1x_2 + y_1y_2|}{PQ}$ **d.** $\frac{|x_1x_2 + y_1y_2|}{PQ}$

112. Consider a circle $x^2+y^2+ax+by+c=0$ lying completely in first quadrant. If m_1 and m_2 are the maximum and minimum values of y/x for all ordered pairs (x, y) on the circumference of the circle, then the value of $(m_1 + m_2)$ is

a. $\frac{a^2-4c}{b^2-4c}$ **b.** $\frac{2ab}{b^2-4c}$

c. $\frac{2ab}{4c-b^2}$ **d.** $\frac{2ab}{b^2-4ac}$

113. If the circumference of the circle $x^2 + y^2 + 8x + 8y - b = 0$ is bisected by the circle $x^2 + y^2 - 2x + 4y + a = 0$, then $a + b$ equals to

a. 50 **b.** 56 **c.** −56 **d.** −34

114. The equation of the circle passing through the point of intersection of the circle $x^2 + y^2 = 4$ and the line $2x + y = 1$ and having minimum possible radius is

a. $5x^2 + 5y^2 + 18x + 6y - 5 = 0$

b. $5x^2 + 5y^2 + 9x + 8y - 15 = 0$

c. $5x^2 + 5y^2 + 4x + 9y - 5 = 0$

d. $5x^2 + 5y^2 - 4x - 2y - 18 = 0$

115. The locus of the centre of the circle touching the line $2x - y = 1$ at (1, 1) is

a. $x + 3y = 2$ **b.** $x + 2y = 0$

c. $x + y = 2$ **d.** None of these

116. The distance from the centre of the circle $x^2 + y^2 = 2x$ to the common chord of the circles $x^2 + y^2 + 5x - 8y + 1 = 0$ and $x^2 + y^2 - 3x + 7y - 25 = 0$ is

a. 2 **b.** 4 **c.** $\frac{34}{13}$ **d.** $\frac{26}{17}$

117. The equation of the circle passing through the point of intersection of the circles $x^2 + y^2 - 4x - 2y = 8$ and $x^2 + y^2 - 2x - 4y = 8$ and the point (−1, 4) is

a. $x^2 + y^2 + 4x + 4y - 8 = 0$

b. $x^2 + y^2 - 3x + 4y + 8 = 0$

c. $x^2 + y^2 + x + y - 8 = 0$

d. $x^2 + y^2 - 3x - 3y - 8 = 0$

118. If the radius of the circle $(x-1)^2+(y-2)^2=1$ and $(x-7)^2+(y-10)^2=4$ are increasing uniformly w.r.t. time as 0.3 and 0.4 unit/sec, then they will touch each other at t equal to

a. 45 sec b. 90 sec c. 11 sec d. 135 sec

119. The equation of a circle which has normals $(x-1)(y-2)=0$ and a tangent $3x+4y=6$ is

a. $x^2+y^2-2x-4y+4=0$
b. $x^2+y^2-2x-4y+5=0$
c. $x^2+y^2=5$
d. $(x-3)^2+(y-4)^2=5$

Multiple Correct Answers Type

Each question has four choices a, b, c, and d, out of which *one or more* answers are correct.

1. The circle $x^2+y^2+2a_1x+c=0$ lies completely inside the circle $x^2+y^2+2a_2x+c=0$, then

a. $a_1a_2>0$ b. $a_1a_2<0$ c. $c>0$ d. $c<0$

2. Consider the circle $x^2+y^2-10x-6y+30=0$. Let O be the centre of the circle and tangent at $A(7, 3)$ and $B(5, 1)$ meet at C. Let $S=0$ represents family of circles passing through A and B, then

a. area of quadrilateral $OACB=4$
b. the radical axis for the family of circles $S=0$ is $x+y=10$
c. the smallest possible circle of the family $S=0$ is $x^2+y^2-12x-4y+38=0$
d. the coordinates of point C are (7, 1)

3. If the circle $x^2+y^2+2x+2ky+6=0$ and $x^2+y^2+2ky+k=0$ intersect orthogonally, then k is

a. 2 b. -2 c. $-\frac{3}{2}$ d. $\frac{3}{2}$

4. $A\left(\frac{1}{\sqrt{2}}, \frac{1}{\sqrt{2}}\right)$ is a point on the circle $x^2+y^2=1$ and B is another point on the circle such that arc length $AB=\frac{\pi}{2}$ units. Then, coordinates of B can be

a. $\left(\frac{1}{\sqrt{2}}, -\frac{1}{\sqrt{2}}\right)$ b. $\left(-\frac{1}{\sqrt{2}}, \frac{1}{\sqrt{2}}\right)$
c. $\left(-\frac{1}{\sqrt{2}}, \frac{-1}{\sqrt{2}}\right)$ d. None of these

5. Let x, y be real variable satisfying the $x^2+y^2+8x-10y-40=0$. Let $a=\max\{(x+2)^2+(y-3)^2\}$ and $b=\min\{(x+2)^2+(y-3)^2\}$, then

a. $a+b=18$ b. $a+b=\sqrt{2}$
c. $a-b=4\sqrt{2}$ d. $a\times b=73$

6. Three sides of a triangle have the equations $L_i \equiv y-m_ix=0$; $i=1, 2, 3$. Then $L_1L_2+\lambda L_2L_3+\mu L_3L_1=0$, where $\lambda\neq 0$, $\mu\neq 0$, is the equation of the circumcircle of the triangle if

a. $1+\lambda+\mu=m_1m_2+\lambda m_2m_3+\lambda m_3m_1$
b. $m_1(1+\mu)+m_2(1+\lambda)+m_3(\mu+\lambda)=0$
c. $\frac{1}{m_3}+\frac{1}{m_1}+\frac{1}{m_2}=1+\lambda+\mu$
d. None of these

7. If equation $x^2+y^2+2hxy+2gx+2fy+c=0$ represents a circle, then the condition for that circle to pass through three quadrants only but not passing through the origin is

a. $f^2>c$ b. $g^2>c$
c. $c>0$ d. $h=0$

8. The points on the line $x=2$ from which the tangents drawn to the circle $x^2+y^2=16$ are at right angles is (are)

a. $(2, 2\sqrt{7})$ b. $(2, 2\sqrt{5})$
c. $(2, -2\sqrt{7})$ d. $(2, -2\sqrt{5})$

9. Coordinates of the centre of a circle, whose radius is 2 unit and which touches the line pair $x^2-y^2-2x+1=0$, are

a. (4, 0) b. $(1+2\sqrt{2}, 0)$
c. (4, 1) d. $(1, 2\sqrt{2})$

10. If the circles $x^2+y^2-9=0$ and $x^2+y^2+2\alpha x+2y+1=0$ touch each other, then α is

a. $-\frac{4}{3}$ b. 0 c. 1 d. $\frac{4}{3}$

11. Point M moved on the circle $(x-4)^2+(y-8)^2=20$. Then it broke away from it and moving along a tangent to the circle, cuts the x-axis at the point $(-2, 0)$. The co-ordinates of a point on the circle at which the moving point broke away is

a. $\left(\frac{42}{5}, \frac{36}{5}\right)$ b. $\left(-\frac{2}{5}, \frac{44}{5}\right)$
c. (6, 4) d. (2, 4)

12. The equation of a tangent to the circle $x^2+y^2=25$ passing through $(-2, 11)$ is

a. $4x+3y=25$ b. $3x+4y=38$
c. $24x-7y+125=0$ d. $7x+24y=250$

13. If the area of the quadrilateral formed by the tangents from the origin to the circle $x^2+y^2+6x-10y+c=0$ and the radii corresponding to the points of contact is 15, then a value of c is

a. 9 b. 4 c. 5 d. 25

14. The equation of the circle which touches the coordinate axes and the line $\frac{x}{3}+\frac{y}{4}=1$ and whose centre lies in the first quadrant is $x^2+y^2-2cx-2cy+c^2=0$, where c is

a. 1 b. 2 c. 3 d. 6

15. The equations of tangents to the circle $x^2+y^2-6x-6y+9=0$ drawn from the origin are

a. $x=0$ b. $x=y$

c. $y=0$ d. $x+y=0$

16. If a circle passes through the point of intersection of the lines $x+y+1=0$ and $x+\lambda y-3=0$ with the coordinate axes, then

a. $\lambda=-1$

b. $\lambda=1$

c. $\lambda=2$

d. λ can have any real value

17. Which of the following lines have intercepts of equal lengths on the circle, $x^2+y^2-2x+4y=0$?

a. $3x-y=0$

b. $x+3y=0$

c. $x+3y+10=0$

d. $3x-y-10=0$

18. The equation of the lines parallel to $x-2y=1$ which touches (touch) the circle $x^2+y^2-4x-2y-15=0$ is (are)

a. $x-2y+2=0$

b. $x-2y-10=0$

c. $x-2y-5=0$

d. $x-2y+10=0$

19. The circles $x^2+y^2-2x-4y+1=0$ and $x^2+y^2+4x+4y-1=0$

a. touch internally

b. touch externally

c. have $3x+4y-1=0$ as the common tangent at the point of contact

d. have $3x+4y+1=0$ as the common tangent at the point of contact

20. The circles $x^2+y^2+2x+4y-20=0$ and $x^2+y^2+6x-8y+10=0$

a. are such that the number of common tangents on them is 2

b. are orthogonal

c. are such that the length of their common tangent is $5(12/5)^{1/4}$

d. are such that the length of their common chord is $5\sqrt{\frac{3}{2}}$

21. A point $P(\sqrt{3},1)$ moves on the circle $x^2+y^2=4$ and after covering a quarter of the circle leaves it tangentially. The equation of a line along which the point moves after leaving the circle is

a. $y=\sqrt{3}\,x+4$ b. $\sqrt{3}\,y=x+4$

c. $y=\sqrt{3}x-4$ d. $\sqrt{3}y=x-4$

22. The equation of a circle of radius 1 touching the circles $x^2+y^2-2|x|=0$ is

a. $x^2+y^2+2\sqrt{2}\,x+1=0$

b. $x^2+y^2-2\sqrt{3}\,y+2=0$

c. $x^2+y^2+2\sqrt{3}\,y+2=0$

d. $x^2+y^2-2\sqrt{2}+1=0$

23. If the circles $x^2+y^2-9=0$ and $x^2+y^2+2ax+2y+1=0$ touch each other, then $a=$

a. $-4/3$ b. 0 c. 1 d. 4/3

24. The range of values of a such that angle θ between the pair of tangent drawn from $(a, 0)$ to the circle $x^2+y^2=1$ satisfies $\frac{\pi}{2}<\theta<\pi$, lies in

a. (1, 2) b. $(1, \sqrt{2})$

c. $(-\sqrt{2},-1)$ d. $(-2,-1)$

25. Let L_1 be a straight line passing through the origin and L_2 be the straight line $x+y=1$. If the intercepts made by the circle $x^2+y^2-x+3y=0$ on L_1 and L_2 are equal, then which of the following equations can represent L_1

a. $x+y=0$ b. $x-y=0$

c. $x+7y=0$ d. $x-7y=0$

26. The centre(s) of the circle(s) passing through the points (0, 0), (1, 0) and touching the circle $x^2+y^2=9$ is/are

a. $\left(\frac{3}{2},\frac{1}{2}\right)$ b. $\left(\frac{1}{2},\frac{3}{2}\right)$

c. $\left(\frac{1}{2},2^{1/2}\right)$ d. $\left(\frac{1}{2},-2^{1/2}\right)$

Reasoning Type

Each question has four choices a, b, c, and d, out of which *only one* is correct. Each question contains Statement 1 and Statement 2.

a. Both the statements are True and Statement 2 is the correct explanation of Statement 1.

b. Both the statements are True but Statement 2 is not the correct explanation of Statement 1.

c. Statement 1 is True and Statement 2 is False.

d. Statement 1 is False and Statement 2 is True.

1. **Statement 1 :** The number of circles that pass through the points $(1,-7)$ and $(-5, 1)$ and of radius 4 is two.

Statement 2: The centre of any circle that pass through the points A and B lies on the perpendicular bisector of AB.

2. **Statement 1:** The chord of contact of tangent from three points A, B, C to the circle $x^2 + y^2 = a^2$ are concurrent, then A, B, C will be collinear.

   Statement 2: Lines $(a_1x + b_1y + c_1) + k(a_2x + b_2y + c_2) = 0$ always pass through a fixed point for $k \in R$.

3. **Statement 1:** Circles $x^2 + y^2 = 144$ and $x^2 + y^2 - 6x - 8y = 0$ do not have any common tangent.

   Statement 2: If two circles are concentric, then they do not have common tangents.

4. **Statement 1:** The least and greatest distances of the point $P(10, 7)$ from the circle $x^2 + y^2 - 4x - 2y - 20 = 0$ are 5 and 15 units, respectively.

   Statement 2: A point (x_1, y_1) lies outside a circle $S = x^2 + y^2 + 2gx + 2fy + c = 0$, if $S_1 > 0$, where $S_1 = x_1^2 + y_1^2 + 2gx_1 + 2fy_1 + c$.

5. **Statement 1:** Number of circles passing through (1, 2), (4, 8), and (0, 0) is one.

   Statement 2: Every triangle has one circumcircle.

6. Let C_1 be the circle with centre O_1 (0, 0) and radius 1 and C_2 be the circle with centre O_2 $(t, t^2 + 1)$ $(t \in R)$ and radius 2.

   Statement 1: Circles C_1 and C_2 always have at least one common tangent for any value of t.

   Statement 2: For the two circles, $O_1O_2 \geq |r_1 - r_2|$, where r_1 and r_2 are their radii for any value of t.

7. Tangents are drawn from the origin to the circle $x^2 + y^2 - 2hx - 2hy + h^2 = 0$ $(h \geq 0)$.

   Statement 1: Angle between the tangents is $\pi/2$.

   Statement 2: The given circle is touching the coordinate axes.

8. Consider two circles $x^2 + y^2 - 4x - 6y - 8 = 0$ and $x^2 + y^2 - 2x - 3 = 0$.

   Statement 1: Both circles intersect each other at two distinct points.

   Statement 2: Sum of radii of two circles in greater than distance between the centres of two circles.

9. From the point P $(\sqrt{2}, \sqrt{6})$, tangents PA and PB are drawn to the circle $x^2 + y^2 = 4$.

   Statement 1: Area of the quadrilateral $OAPB$ (being origin) is 4.

   Statement 2: Area of square is a^2 where a is length of side.

10. **Statement 1:** Centre of the circle having $x + y = 3$ and $x - y = 1$ as its normal is (1, 2).

    Statement 2: Normals to the circle always passes through its centre.

11. **Statement 1:** The circle having equation $x^2 + y^2 - 2x + 6y + 5 = 0$ intersects both the coordinate axes.

    Statement 2: The lengths of x and y intercepts made by the circle having equation $x^2 + y^2 + 2gx + 2fy + c = 0$ are $2\sqrt{g^2 - c}$ and $2\sqrt{f^2 - c}$, respectively.

12. **Statement 1:** Number of circles touching lines $x + y = 1$, $2x - y = 5$, and $3x + 5y - 1 = 0$ is four.

    Statement 2: In any triangle, four circles can be drawn touching all the three sides of triangle.

13. **Statement 1:** Chord of contact of the circle $x^2 + y^2 = 1$ w.r.t. points (2, 3). (3, 5) and (1, 1) are concurrent.

    Statement 2: Points (1, 1), (2, 3), and (3, 5) are collinear.

14. **Statement 1:** The equation $x^2 + y^2 - 2x - 2ay - 8 = 0$ represents, for different values of a, a system of circles passing through two fixed points lying on the x-axis.

    Statement 2: $S = 0$ is a circle and $L = 0$ is a straight line, then $S + \lambda L = 0$ represents the family of circles passing through the points of intersection of circle and straight line (where λ is arbitrary parameter).

15. **Statement 1:** The chord of contact of tangent from three points A, B, C to the circle $x^2 + y^2 = a^2$ are concurrent, then A, B, C will be collinear.

    Statement 2: A, B, C always lies on the normal to the circle $x^2 + y^2 = a^2$

16. **Statement 1:** Points $A(1, 0)$, $B(2, 3)$, $C(5, 3)$and $D(6, 0)$ are concyclic.

    Statement 2: Points A, B, C, D forms isosceles trapezium or AB and CD meet in E, then $EA \times EB = EC \times ED$.

17. **Statement 1:** The equations of the straight lines joining origin to the points of intersection of $x^2 + y^2 - 4x - 2y = 4$ and $x^2 + y^2 - 2x - 4y - 4 = 0$ is $x - y = 0$.

    Statement 2: $y + x = 0$ is common chord of $x^2 + y^2 - 4x - 2y = 4$ and $x^2 + y^2 - 2x - 4y - 4 = 0$.

18. **Statement 1:** Two orthogonal circles intersect to generate a common chord which subtends complimentary angles at their circumferences.

    Statement2: Two orthogonal circles intersect to generate a common chord which subtends supplementary angles at their centres.

19. **Statement 1:** The point $(a, -a)$ lies inside the circle $x^2 + y^2 - 4x + 2y - 8 = 0$ whenever $a \in (-1, 4)$

    Statement 2: Point (x_1, y_1) lies inside the circle $x^2 + y^2 + 2gx + 2fy + c = 0$, if $x_1^2 + y_1^2 + 2gx_1 + 2fy_1 + c < 0$.

20. **Statement 1:** If circle with centre $P(t, 4 - 2t)$, $t \in R$ cuts the circles $x^2 + y^2 = 16$ and $x^2 + y^2 - 2x - y - 12 = 0$; then both the intersections are orthogonal.

Statement 2: Length of tangent from P for $t \in R$ is same for both the given circles.

21. Statement 1: Let $S_1 : x^2 + y^2 - 10x - 12y - 39 = 0$
$S_2 : x^2 + y^2 - 2x - 4y + 1 = 0$
and $S_3 : 2x^2 + 2y^2 - 20x - 24y + 78 = 0$
The radical centre of these circles taken pairwise as $(-2, -3)$.

Statement 2: Point of intersection of three radical axis of three circles taken in pairs is known as radical centre.

22. Statement 1: The equation of chord through the point $(-2, 4)$ which is farthest from the centre of the circle $x^2 + y^2 - 6x + 10y - 9 = 0$ is $x + y - 2 = 0$.

Statement 2: In notations, the equation of such chord of the circle $S = 0$ bisected at (x_1, y_1) must be $T = S_1$.

23. Statement 1: If two circles $x^2 + y^2 + 2gx + 2fy = 0$ and $x^2 + y^2 + 2g'x + 2f'y = 0$ touch each other, then $f'g = fg'$.

Statement 2: Two circles touch other, if line joining their centres is perpendicular to all possible common tangents.

24. Statement 1: The circles $x^2 + y^2 + 2px + r = 0$, and $x^2 + y^2 + 2qy + r = 0$ touch if $\frac{1}{p^2} + \frac{1}{q^2} = \frac{1}{r}$.

Statement 2: Two circles with centre C_1, C_2 and radii r_1, r_2 touch each other if $|r_1 \pm r_2| = c_1c_2$.

Linked Comprehension Type

Based upon each paragraph, three multiple choice questions have to be answered. Each face equation has four choices a, b, c, and d, out of which *only one* is correct.

For Problems 1–3

Each side of a square has length 4 units and its centre is at (3, 4). If one of the diagonals is parallel to the line $y = x$, then answer the following questions.

1. Which of the following is not the vertex of the square?

a. (1, 6) **b.** (5, 2) **c.** (1, 2) **d.** (4, 6)

2. The radius of the circle inscribed in the triangle formed by any three vertices is

a. $2\sqrt{2}(\sqrt{2} + 1)$ **b.** $2\sqrt{2}(\sqrt{2} - 1)$
c. $2(\sqrt{2} + 1)$ **d.** None of these

3. The radius of the circle inscribed in the triangle formed by any two vertices of square and the centre is

a. $2(\sqrt{2} - 1)$ **b.** $2(\sqrt{2} + 1)$
c. $\sqrt{2}(\sqrt{2} - 1)$ **d.** None of these

For Problems 4–6

Tangents PA and PB are drawn to the circle $(x - 4)^2 + (y - 5)^2 = 4$ from the point P on the curve $y = \sin x$, where A and B lie on the circle. Consider the function $y = f(x)$ represented by the locus of the center of the circumcircle of triangle PAB, then answer the following questions.

4. Range of $y = f(x)$ is

a. $[-2, 1]$ **b.** $[-1, 4]$ **c.** $[0, 2]$ **d.** $[2, 3]$

5. Period of $y = f(x)$ is

a. 2π **b.** 3π
c. π **d.** Not defined

6. Which of the following is true?

a. $f(x) = 4$ has real roots
b. $f(x) = 1$ has real roots
c. Range of $y = f^{-1}(x)$ is $\left[-\frac{\pi}{4} + 2, \frac{\pi}{4} + 2\right]$
d. None of these

For Problems 7–9

Consider a family of circles passing through the points (3, 7) and (6, 5). Answer the following questions.

7. Number of circles which belong to the family and also touching x-axis are

a. 0 **b.** 1 **c.** 2 **d.** Infinite

8. If each circle in the family cuts the circle $x^2 + y^2 - 4x - 6y - 3 = 0$, then all the common chords pass through the fixed point which is

a. (1, 23) **b.** (2, 23/2)
c. (– 3, 3/2) **d.** None of these.

9. If circle which belong the given family cuts the circle $x^2 + y^2 = 29$ orthogonally then centre of that circle is

a. (1/2, 3/2) **b.** (9/2, 7/2)
c. (7/2, 9/2) **d.** (3, –7/9)

For Problems 10–12

Consider the relation $4l^2 - 5m^2 + 6l + 1 = 0$, where $l, m \in R$, then the line $lx + my + 1 = 0$ touches a fixed circle whose

10. Centre and radius of circle one

a. (2, 0), 3 **b.** (–3, 0), $\sqrt{3}$
c. (3, 0), $\sqrt{5}$ **d.** None of these

11. Tangents PA and PB are drawn to the above fixed circle from the points P on the line $x + y - 1 = 0$. Then chord of contact AB passes though the fixed point

a. $\left(\frac{1}{2}, -\frac{5}{2}\right)$ **b.** $\left(\frac{1}{3}, \frac{4}{3}\right)$
c. $\left(-\frac{1}{2}, \frac{3}{2}\right)$ **d.** None of these

12. Number of tangents which can be drawn from the point (2, – 3) are

a. 0 **b.** 1 **c.** 2 **d.** 1 or 2

For Problems 13–15

A circle C whose radius is 1 unit, touches the x-axis at point A. The centre Q of C lies in first quadrant. The tangent from origin O to the circle touches it at T and a point P lies on it such that ΔOAP is a right-angled triangle at A and its perimeter is 8 units.

13. The length of PQ is

a. $\frac{1}{2}$ **b.** $\frac{4}{3}$

c. $\frac{5}{3}$ **d.** None of these

14. Equation of circle C is

a. $\{x-(2+\sqrt{3})\}^2+(y-1)^2=1$

b. $\{x-(\sqrt{3}-\sqrt{2})\}^2+(y-1)^2=1$

c. $(x-\sqrt{3})^2+(y-1)^2=1$

d. None of these

15. Equation of tangent OT is

a. $x-\sqrt{3}y=0$ **b.** $x-\sqrt{2}y=0$

c. $y-\sqrt{3}x=0$ **d.** None of these

For Problems 16–18

P is a variable point on the line $L=0$. Tangents are drawn to the circles $x^2+y^2=4$ from P to touch it at Q and R. The parallelogram $PQSR$ is completed.

16. If $L\equiv 2x+y-6=0$, then the locus of circumcentre of ΔPQR is

a. $2x-y=4$ **b.** $2x+y=3$

c. $x-2y=4$ **d.** $x+2y=3$

17. If $P\equiv(6, 8)$, then the area of ΔQRS is

a. $\frac{(6)^{3/2}}{25}$ sq. units **b.** $\frac{(24)^{3/2}}{25}$ sq. units

c. $\frac{48\sqrt{6}}{25}$ sq. units **d.** $\frac{192\sqrt{6}}{25}$ sq. units

18. If $P\equiv(3, 4)$, then coordinates of S are

a. $\left(-\frac{46}{25}, \frac{63}{25}\right)$ **b.** $\left(\frac{-51}{25}, \frac{-68}{25}\right)$

c. $\left(-\frac{46}{25}, \frac{68}{25}\right)$ **d.** $\left(-\frac{68}{25}, \frac{51}{25}\right)$

For Problems 19–21

To the circle $x^2+y^2=4$, two tangents are drawn from $P(-4, 0)$, which touches the circle at T_1 and T_2, a rhombus $PT_1P'T_2$ is completed.

19. Circumcentre of the triangle PT_1T_2 is at

a. $(-2, 0)$ **b.** $(2, 0)$

c. $\left(\frac{\sqrt{3}}{2}, 0\right)$ **d.** None of these

20. Ratio of the area of triangle PT_1P' to that the $P'T_1T_2$ is

a. $2:1$ **b.** $1:2$

c. $\sqrt{3}:2$ **d.** None of these

21. If P is taken to be at $(h, 0)$ such that P' lies on the circle, the area of the rhombus is

a. $6\sqrt{3}$ **b.** $2\sqrt{3}$

c. $3\sqrt{3}$ **d.** None of these

For Problems 22–24

Let α chord of a circle be that chord of the circle which subtends an angle α at the centre.

22. If $x+y=1$ is a chord of $x^2+y^2=1$, then α is equal to

a. $\frac{\pi}{4}$ **b.** $\frac{\pi}{2}$

c. $\frac{\pi}{6}$ **d.** $x+y=1$ is not a chord

23. If slope of a $\frac{\pi}{3}$ chord of $x^2+y^2=4$ is 1, then its equation is

a. $x-y+\sqrt{6}=0$ **b.** $x-y=2\sqrt{3}$

c. $x-y=\sqrt{3}$ **d.** $x-y+\sqrt{3}=0$

24. Distance of $\frac{2\pi}{3}$ chord of $x^2+y^2+2x+4y+1=0$ from the centre is

a. 1 **b.** 2 **c.** $\sqrt{2}$ **d.** $\frac{1}{\sqrt{2}}$

For Problems 25–27

Two variable chords AB and BC of a circles $x^2+y^2=a^2$ are such that $AB=BC=a$, and M and N are the midpoints of AB and BC, respectively, such that line joining MN intersect the circles at P and Q, where P is closer to AB and Q is the centre of the circle.

25. $\angle OAB$ is

a. 30° **b.** 60° **c.** 45° **d.** 15°

26. Angles between tangents at A and C is

a. 90° **b.** 120° **c.** 60° **d.** 150°

27. Locus of point of intersection of tangents at A and C is

a. $x^2+y^2=a^2$ **b.** $x^2+y^2=2a^2$

c. $x^2+y^2=4a^2$ **d.** $x^2+y^2=8a^2$

For Problems 28–30

Given two circles intersecting orthogonally having length of common chord 24/5 units. Radius of one of the circles is 3 units.

28. Radius of other circle is

a. 6 units **b.** 5 units **c.** 2 units **d.** 4 units

29. Angle between direct common tangent is

a. $\sin^{-1}\frac{24}{25}$ **b.** $\sin^{-1}\frac{4\sqrt{6}}{25}$

c. $\sin^{-1}\frac{4}{5}$ **d.** $\sin^{-1}\frac{12}{25}$

30. Length of direct common tangent is

a. $\sqrt{12}$ **b.** $4\sqrt{3}$ **c.** $2\sqrt{6}$ **d.** $3\sqrt{6}$

Matrix-Match Type

Each question contains statements given in two columns which have to be matched. Statements (a, b, c, d) in column I have to be matched with statements (p, q, r, s) in column II. If the correct matches are a → p, a → s; b → q, b → r; c → p; c → q and d → s, then the correctly bubbled 4 × 4 matrix should be as follows:

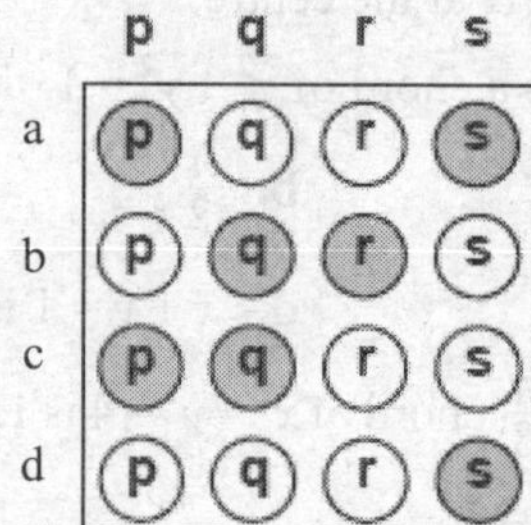

1.

Column I	Column II
a. Number of circles touching given three non-concurrent lines	**p.** 1
b. Number of circles touching $y = x$ at (2, 2) and also touching line $x + 2y - 4 = 0$	**q.** 2
c. Number of circles touching lines $x \pm y = 2$ and passing thorough the point (4, 3)	**r.** 4
d. Number of circle intersecting given three circles orthogonally	**s.** Infinite

2. Let $x^2 + y^2 + 2gx + 2fy + c = 0$ be an equation of circle.

Column I	Column II
a. If circle lie in first quadrant, then	**p.** $g < 0$
b. If circle lie above the x-axis, then	**q.** $g > 0$
c. If circle lie on the left of the y-axis, then	**r.** $g^2 - c < 0$
d. If circle touches the positive x-axis and does not intersect the y-axis, then	**s.** $c > 0$

3.

Column I	Column II
a. If $ax + by - 5 = 0$ is the equation of the chord of the circle $(x - 3)^2 + (y - 4)^2 = 4$, which passes through (2, 3) and at the greatest distance from the centre of the circle, then $\|a + b\|$ is equal to	**p.** 6
b. Let O be the origin and P be a variable point on the circle $x^2 + y^2 + 2x + 2y = 0$. If the locus of midpoint of OP is $x^2 + y^2 + 2gx + 2fy + c = 0$, then $(g + f)$ is equal to	**q.** 3
c. The x-coordinate of the centre of the smallest circle which cuts the circles $x^2 + y^2 - 2x - 4y - 4 = 0$ and $x^2 + y^2 - 10x + 12y + 52 = 0$ orthogonally is	**r.** 2
d. If θ be the angle between two tangents which are drawn to the circles $x^2 + y^2 - 6\sqrt{3}x - 6y + 27 = 0$ from the origin, then $2\sqrt{3}\tan\theta$ equals to	**s.** 1

4.

Column I	Column II
a. The length of the common chord of two circles of radii 3 and 4 units which intersect orthogonally is $\frac{k}{5}$, then k equals to	**p.** 1
b. The circumference of the circle $x^2 + y^2 + 4x + 12y + p = 0$ is bisected by the circle $x^2 + y^2 - 2x + 8y - q = 0$, then $p + q$ is equal to	**q.** 24
c. Number of distinct chords of the circle $2x(x - \sqrt{2}) + y(2y - 1) = 0$; chords are passing through the point $\left(\sqrt{2}, \frac{1}{2}\right)$ and are bisected on x-axis is	**r.** 32
d. One of the diameters of the circle circumscribing the rectangle $ABCD$ is $4y = x + 7$. If A and B are the points (− 3, 4) and (5, 4), respectively, then the area of the rectangle is	**s.** 36

5. Let C_1 and C_2 be two circles whose equations are $x^2 + y^2 - 2x = 0$ and $x^2 + y^2 + 2x = 0$. $P(\lambda_1, \lambda)$ is a variable point. Then match the following.

Column I	Column II
a. P lies inside C_1 but outside C_2	**p.** $\lambda \in (-\infty, -1) \cup (0, \infty)$
b. P lies inside C_2 but outside C_1	**q.** $\lambda \in (-\infty, -1) \cup (1, \infty)$
c. P lies outside C_1 and outside C_2	**r.** $\lambda \in (-1, 0)$
d. P does not lie inside C_2	**s.** $\lambda \in (0, 1)$

6.

Column I	Column II
a. If two circles $x^2 + y^2 + 2a_1x + b = 0$ and $x^2 + y^2 + 2a_2x + b = 0$ touch each other then triplet (a_1, a_2, b) can be	**p.** (2, 2, 2)

b. If two circles $x^2 + y^2 + 2a_1x + b = 0$ and $x^2 + y^2 + 2a_2y + b = 0$ touch each other then triplet (a_1, a_2, b) can be	**q.** $\left(1, 1, \frac{1}{2}\right)$
c. If the straight line $a_1x - by + b^2 = 0$ touches the circle $x^2 + y^2 = a_2x + by$, then triplet (a_1, a_2, b) can be	**r.** (2, 1, 0)
d. If the line $3x + 4y - 4 = 0$ touches the circle $(x - a_1)^2 + (y - a_2)^2 = b^2$, then triplet (a_1, a_2, b) can be	**s.** $\left(1, 1, \frac{3}{5}\right)$

Integer Type

1. The circle which is touched by $y = x$ has its centre on the positive direction of the x-axis and cuts off a chord of length 2 units along the line $\sqrt{3}y - x = 0$. The distance of the centre from the origin is________.

2. The number of integral values of α for which the point $(\alpha - 1, \alpha + 1)$ lies in the larger segment of the circles $x^2 + y^2 - x - y - 6 = 0$ made by the chord whose equation is $x + y - 2 = 0$ is________.

3. The number of integral values of k for which the circle $x^2 + y^2 - 6x - 10y + k = 0$ does not touch or intersect the coordinate axes and the point (1, 4) inside the circle is________.

4. If the length of the tangents from any point on the circle $15x^2 + 15y^2 - 48x + 64y = 0$ to the two circles $5x^2 + 5y^2 - 24x + 32y + 75 = 0$ and $5x^2 + 5y^2 - 48x + 64y + 300 = 0$ is PT_1 and PT_2, then the value of PT_2/PT_1 is________.

5. Tangents are drawn to $x^2 + y^2 = 1$ from any arbitrary point P on the line $2x + y - 4 = 0$. The corresponding chord of contact passes through a fixed point (a, b), then the value of a/b is________.

6. The area of the triangle formed by the tangents from the point (4, 3) to the circle $x^2 + y^2 = 9$ and the line joining their points of contact is S sq. units. The value of $[S]$, where $[\cdot]$ represents the greatest integer function, is________.

7. Let the radii of the circles $(x - 1)^2 + (y - 2)^2 = 1$ and $(x - 7)^2 + (y - 10)^2 = 4$ increase uniformly by 0.3 and 0.4 unit/sec with respect to time. If t_1 and t_2 are the time when they touch externally and internally, then the value of t_2/t_1 is________.

8. The length of the common chord of the circles $x^2 + y^2 + 5x + 7y + 9 = 0$ and $x^2 + y^2 + 7x + 5y + 9 = 0$ is________.

9. A circle passes through the vertex C of a rectangle $ABCD$ and touches its sides AB and AD at M and N, respectively. If the distance from C of the line segment MN is equal to 5 units and the area of the rectangle $ABCD$ is S sq. units, then the value of $\sqrt{S}$ is________.

10. Let $A = (-1, 0)$, $B = (3, 0)$, and PQ be any line passing through (4, 1) and having slope m. Then the number of integral values of m for which there exist two points on PQ at which AB subtends a right angle is________.

11. From a point P on the normal $y = x + c$ of the circle $x^2 + y^2 - 2x - 4y + 5 - \lambda^2 = 0$, two tangents are drawn to the same circle touching it at points B and C. If the area of the quadrilateral $OBPC$ (where O is the centre of the circle) is 36 sq. units, the possible positive value of λ, if it is given that the point P is at a distance of $|\lambda|(\sqrt{2} - 1)$ from the circle, is________.

12. The distance of the centre of the smallest circle which cuts circles $x^2 + y^2 = 1$ and $x^2 + y^2 + 8x + 8y - 33 = 0$ orthogonally from the point (7, 2) is________.

13. Let $x^2 + y^2 - 4x - 2y - 11 = 0$ be a circle. A pair of tangents from the point (4, 5) with a pair of radii form a quadrilateral of area________.

14. If a circle passes through the points of intersection of the coordinate axes with the lines $\lambda x - y + 1 = 0$ and $x - 2y + 3 = 0$, then the value of λ is________.

15. The number of integral values of r for which the circles $(x - 1)^2 + (y - 3)^2 = r^2$ and $x^2 + y^2 - 8x + 2y + 8 = 0$ intersect at two distinct points is________.

16. If one of the diameters of the circle $x^2 + y^2 - 2x - 6y + 6 = 0$ is a chord to the circle with centre (2, 1), then the radius of the circle is________.

17. The lines $5x + 12y - 10 = 0$ and $5x - 12y - 40 = 0$ touch a circle C_1 of diameter 6. If the centre of C_1 lies in the first quadrant, then the radius of the circle C_2 which is concentric with C_1 and cuts intercepts of length 8 on the given lines is________.

18. If $(m_i, 1/m_i)$, $m_i > 0$, $i = 1, 2, 3, 4$, are four distinct points on a circle, then the value of $m_1m_2m_3m_4$ is________.

19. For the circle $x^2 + y^2 = r^2$, the value of r for which the area enclosed by the tangents drawn from the point P (6, 8) to the circle and the chord of contact is maximum is________.

20. The centres of two circles C_1 and C_2 of unit radius are at a distance of 6 units. Let P be the midpoint of the line segment joining the centres and C be a circle touching C_1 and C_2 externally. If a common tangent to C_1 and C passing through P is also a common tangent to C_2 and C, then the radius of the circle C is________.

ANSWERS AND SOULTIONS

Objective Type

1. a. If there are more than one rational points on the circumference of the circle $x^2 + y^2 - 2\pi x - 2ey + c = 0$ (as (π, e) is the centre), then e will be a rational multiple of π, which is not possible. Thus, the number of rational points on the circumference of the circle is at most one.

2. a. Point of intersection of diagonals lie on circumcircle i.e., (1, 1), since $(y - 2x + 1) + \lambda(2y - x - 1) = 0$

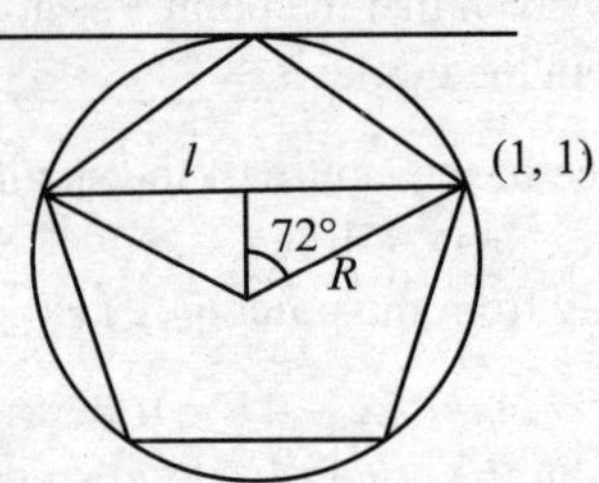

Fig. 2.1

$$l = 2R \sin 72°$$

$$R = \frac{\sin 36°}{2 \sin 72°} = \cos 72°$$

$$\Rightarrow \text{Locus is } (x - 1)^2 + (y - 1)^2 = \cos^2 72°$$

$$\Rightarrow x^2 + y^2 - 2x - 2y + 1 + \sin^2 72° = 0$$

3. c. Let the equation of the chord OA of the circle

$$x^2 + y^2 - 2x + 4y = 0 \qquad (1)$$

be $y = mx$ (2)

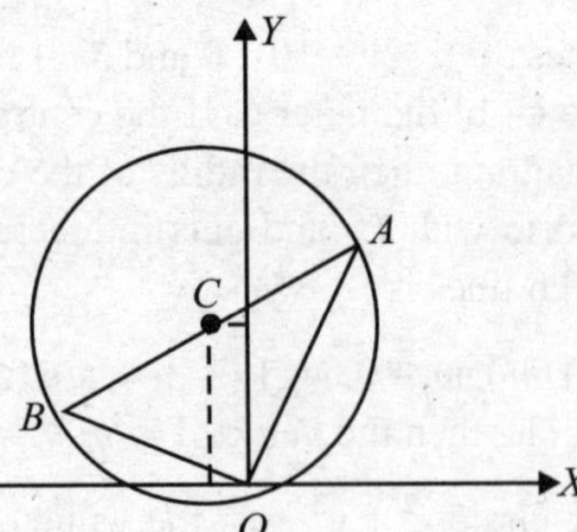

Fig. 2.2

Solving (1) and (2), we get

$$\Rightarrow x^2 + m^2x^2 - 2x + 4mx = 0$$

$$\Rightarrow (1 + m^2)\, x^2 - (2 - 4m)\, x = 0$$

$$\Rightarrow x = 0 \text{ and } x = \frac{2 - 4m}{1 + m^2}$$

Hence, the points of intersection are

$$(0, 0) \text{ and } A\left(\frac{2 - 4m}{1 + m^2}, \frac{m(2 - 4m)}{1 + m^2}\right).$$

$$\Rightarrow OA^2 = \left(\frac{2 - 4m}{1 + m^2}\right)^2 (1 + m^2)$$

$$= \frac{(2 - 4m)^2}{1 + m^2}$$

Since OAB is an isosceles right-angled triangle $OA^2 = \frac{1}{2} AB^2$

where AB is a diameter of the given circle $OA^2 = 10$

$$\Rightarrow \frac{(2 - 4m)^2}{1 + m^2} = 10$$

$$\Rightarrow 4 - 16m + 16m^2 = 10\,(1 + m)^2$$

$$\Rightarrow 3m^2 - 8m - 3 = 0$$

$$\Rightarrow m = 3 \text{ or } -\frac{1}{3}$$

Hence, the required equations are $y = 3x$ or $x + 3y = 0$.

4. b. $y = mx$ be the chord.

Then point of intersection are given by

$x^2(1 + m^2) - x(3 + 4m) - 4 = 0$. Therefore,

$$x_1 + x_2 = \frac{3 + 4m}{1 + m^2}$$

$$\text{and } x_1 x_2 = \frac{-4}{1 + m^2}$$

Since (0, 0) divides the chord in the ratio 1 : 4. Therefore,

$$x_2 = -4x_1$$

$$\therefore -3x_1 = \frac{3 + 4m}{1 + m^2}$$

$$\text{and } 4x_1^2 = -\frac{-4}{1 + m^2}$$

$$\therefore 9 + 9m^2 = 9 + 16m^2 + 24m$$

$$\text{i.e., } m = 0, -\frac{24}{7}$$

Therefore, the lines are $y = 0$ and $y + 24x = 0$.

5. a.

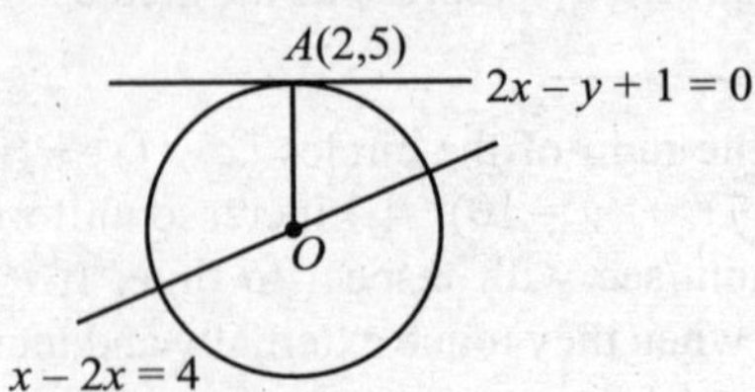

Fig. 2.3

$2x - y + 1 = 0$ is tangent

$$\text{Slope of line } OA = -\frac{1}{2}$$

$$\text{Equation of } OA,\ (y - 5) = -\frac{1}{2}(x - 2)$$

or $x + 2y = 12$

Intersection with $x - 2y = 4$ will give coordinates of centre which are (8, 2). Hence,

$$r = OA = \sqrt{(8-2)^2 + (2-5)^2}$$

$$= 3\sqrt{5}$$

6. d.

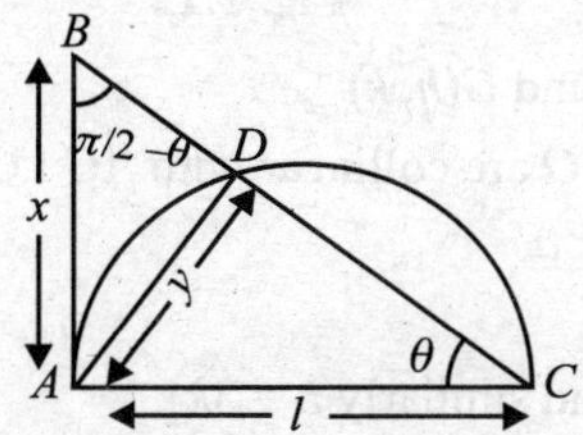

Fig. 2.4

ΔABC and ΔDBA are similar

$\Rightarrow lx = y\sqrt{l^2 + x^2}$

$\Rightarrow l^2x^2 = y^2(l^2 + x^2)$

$\Rightarrow l^2(x^2 - y^2) = x^2y^2$

$\Rightarrow l = \dfrac{xy}{\sqrt{x^2 - y^2}} = \dfrac{AB \times AD}{\sqrt{AB^2 - AD^2}}$

7. a.

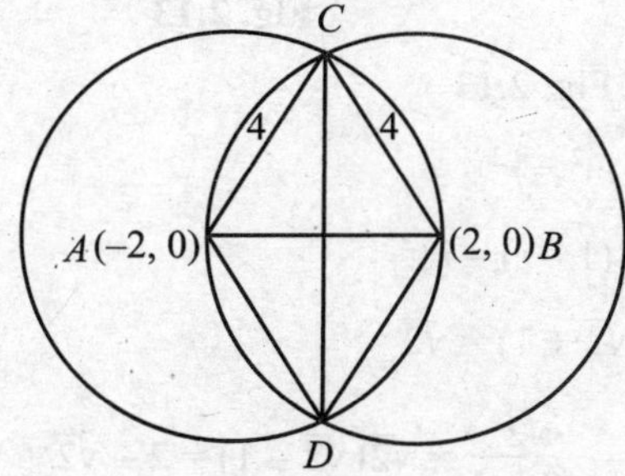

Fig. 2.5

Circles with centres (2, 0) and (– 2, 0) each with radius 4.

$\Rightarrow$ y-axis is their common chord.

ΔABC is equilateral.

Hence, area of $ADBC$ is $\dfrac{2\sqrt{3}}{4}(4)^2 = 8\sqrt{3}$.

8. a.

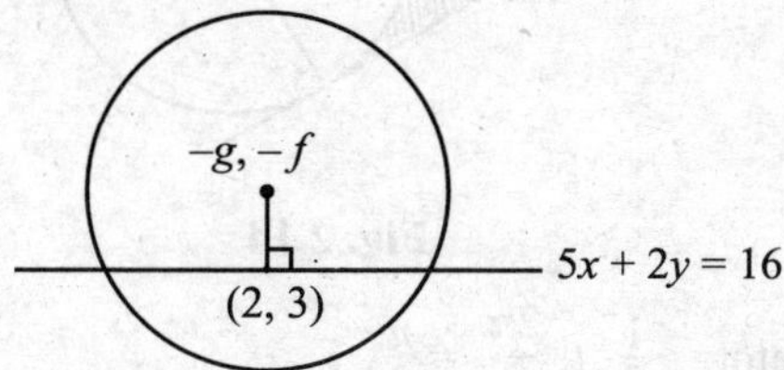

Fig. 2.6

Slope of the given line $= -\dfrac{5}{2}$

$\Rightarrow \left(\dfrac{5}{2}\right)\left(\dfrac{3+f}{2+g}\right) = -1$

$\Rightarrow 15 + 5f = 4 + 2g$

$\Rightarrow$ Locus is $2x - 5y + 11 = 0$

9. b.

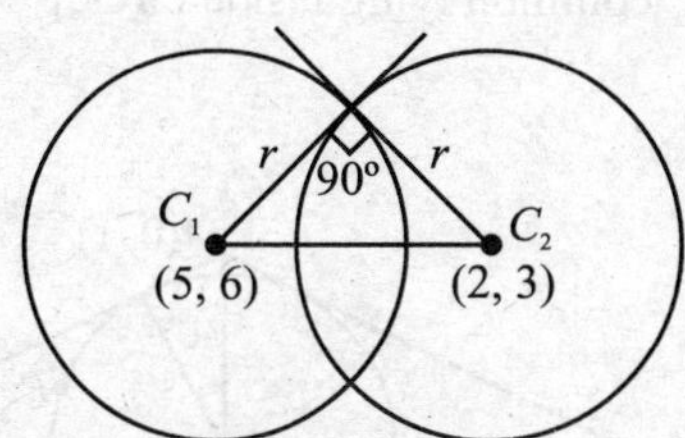

Fig. 2.7

$2r^2 = 3^2 + 3^2 = 18 \Rightarrow r^2 = 9$

$\Rightarrow r = 3$

10. d.

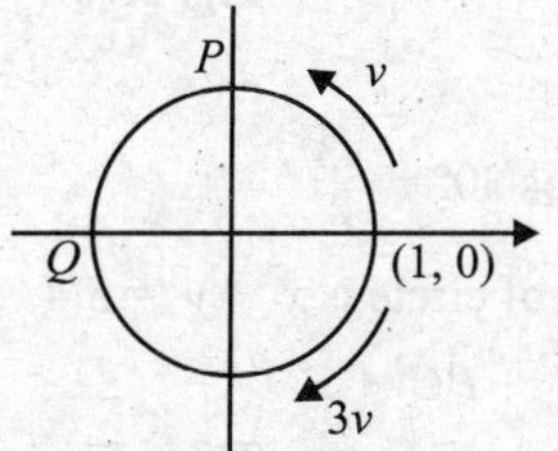

Fig. 2.8

The particle which moves clockwise is moving three times as fast as the particle moving anticlockwise. This means the clockwise particle travels (3/4)th of the way around the circle, the anticlockwise particle will travel (1/4)th of the way around the circle and so the second particle will meet at $P(0, 1)$.

Using the same logic they will meet at $Q(-1, 0)$ when they meet the second time.

11. d.

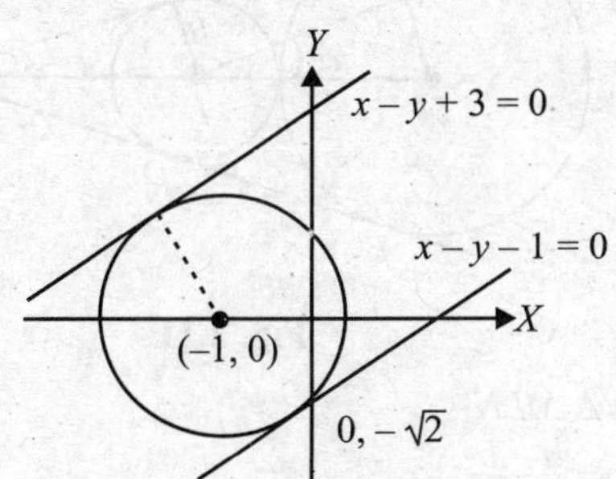

Fig. 2.9

$x^2 + y^2 + 2x - 1 = 0$

Centre (–1, 0) and radius $= \sqrt{2}$

Line $x - y + c = 0$ must be tangent to the circle.

$\Rightarrow \left|\frac{-1+c}{\sqrt{2}}\right| = \sqrt{2}$

$\Rightarrow |c-1| = 2$

$\Rightarrow c-1 = \pm 2$

$\Rightarrow c = 3$ or -1

$\Rightarrow c = 1$ [$\because$ For $c = 3$ there will be infinite points common lying inside circle]

12. b.

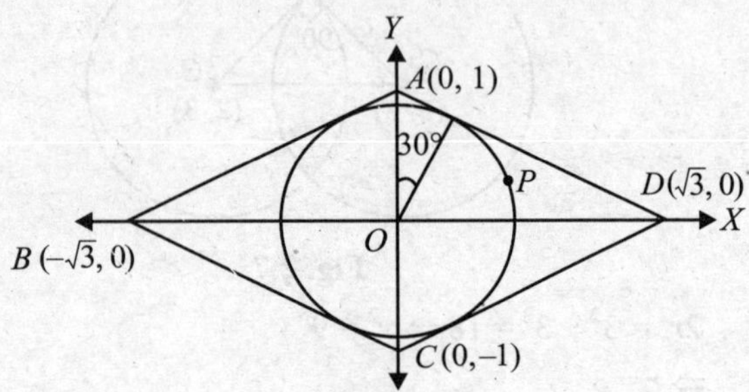

Fig. 2.10

$OA = 1$

$r = OA \cos 30° = \frac{\sqrt{3}}{2}$

Equation of circle is $x^2 + y^2 = 3/4$

$PA^2 + PB^2 + PC^2 + PD^2$

$= x_1^2 + (y_1-1)^2 + (x_1+\sqrt{3})^2 + y_1^2 + x_1^2 + (y_1+1)^2 + (x_1-\sqrt{3})^2 + y_1^2$

$= 4x_1^2 + 4y_1^2 + 8 = 4(x_1^2 + y_1^2) + 8$

$= 4 \times \frac{3}{4} + 8$

$= 11$

13. d.

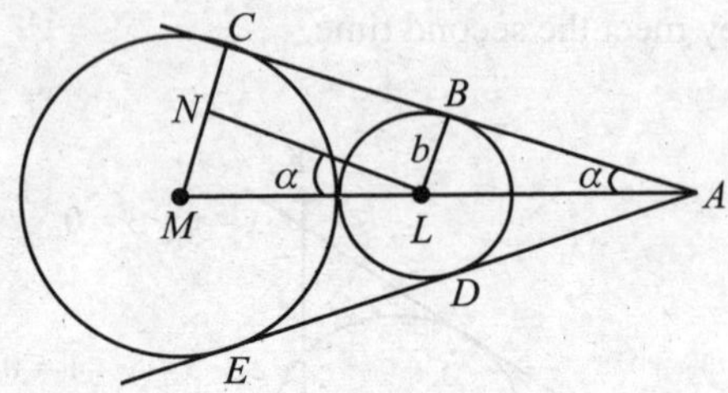

Fig. 2.11

From Δ *MLN*

$\sin\alpha = \frac{a-b}{a+b}$

$\therefore a = \sin^{-1}\left(\frac{a-b}{a+b}\right)$

$\therefore$ Angle between *AB* and *AD*

$= 2\alpha = \sin^{-1}\left(\frac{a-b}{a+b}\right)$

14. a.

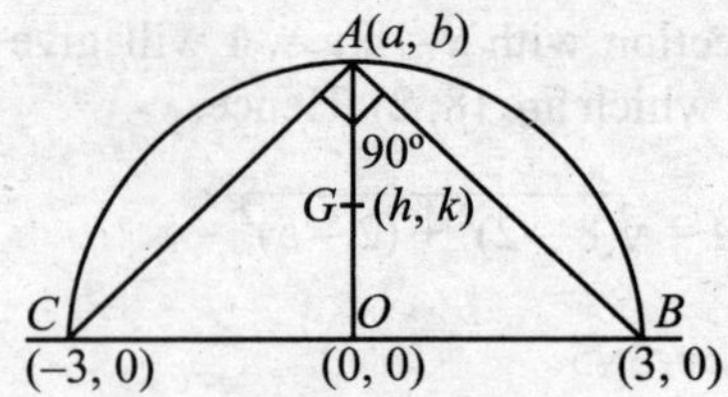

Fig. 2.12

Let $A(a, b)$ and $G(h, k)$

Now A, G, O are collinear with $AG : GO \equiv 2 : 1$

$\Rightarrow h = \frac{2.0 + a}{3}$

$\Rightarrow a = 3h$ and similarly $b = 3k$

Now (a, b) lies on the circle $x^2 + y^2 = 9$

Therefore, locus of (h, k) is $x^2 + y^2 = 1$.

15. a.

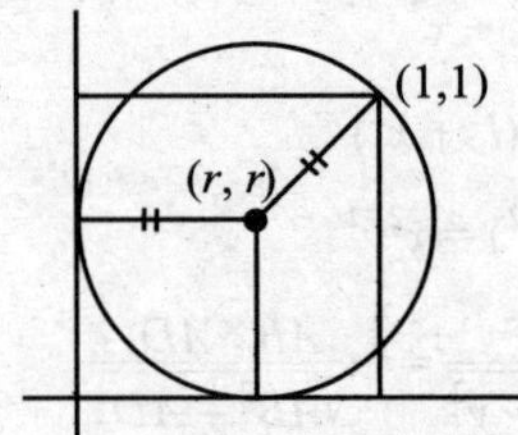

Fig. 2.13

From Fig. 2.13

$2(1-r)^2 = r^2$

$\Rightarrow \sqrt{2}(1-r) = r$

$\Rightarrow r(\sqrt{2}+1) = \sqrt{2}$

$\Rightarrow r = \frac{\sqrt{2}}{\sqrt{2}+1} = \sqrt{2}(\sqrt{2}-1) = 2-\sqrt{2}$

16. b. $r = 1$; $L = \sqrt{3}$

Area of quadrilateral $= \sqrt{3}$

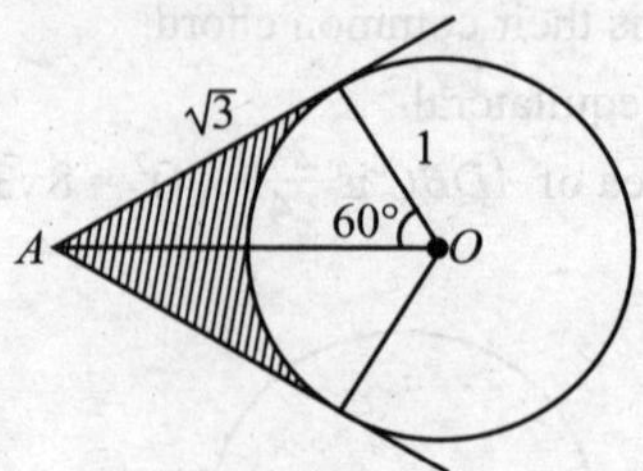

Fig. 2.14

Sector $= \frac{1}{2}.1.\frac{2\pi}{3} = \frac{\pi}{3}$

Shaded region $= \sqrt{3} - \frac{\pi}{3}$

17. d.

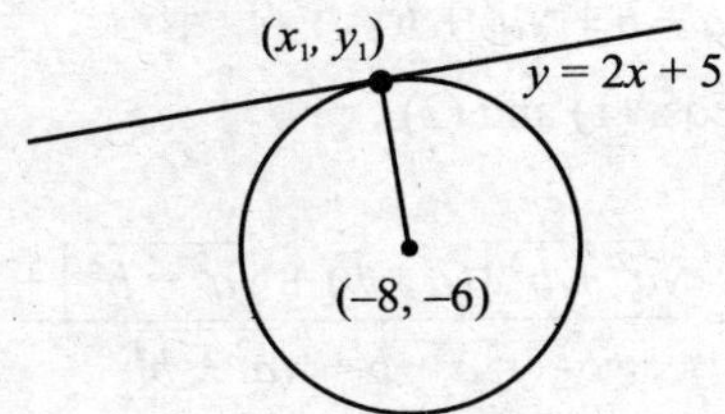

Fig. 2.15

From the figure, we have

$y_1 = 2x_1 + 5$ and $\frac{y_1 + 6}{x_1 + 8} \times 2 = -1$

$\Rightarrow x_1 = -6$ and $y_1 = -7$

18. c.

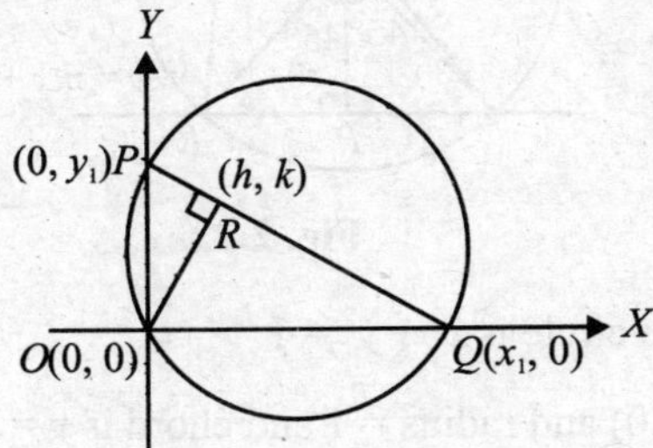

Fig. 2.16

Equation of line PQ is $y - k = -\frac{h}{k}(x - h)$

or $hx + ky = h^2 + k^2$

$\Rightarrow$ Points $Q\left(\frac{h^2 + k^2}{h}, 0\right)$ and $P\left(0, \frac{h^2 + k^2}{k}\right)$

Also $2a = \sqrt{x_1^2 + y_1^2}$

$\Rightarrow x_1^2 + y_1^2 = 4a^2$

Eliminating x_1 and y_1 we have

$(x^2 + y^2)^2\left(\frac{1}{x^2} + \frac{1}{y^2}\right) = 4a^2$

19. c.

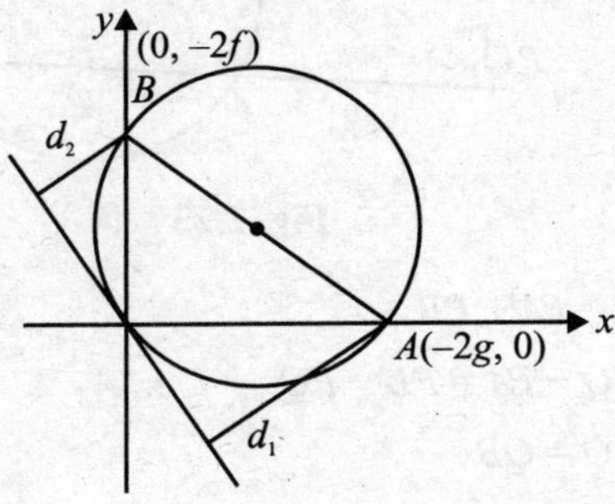

Fig. 2.17

Let the circle be $x^2 + y^2 + 2gx + 2fy = 0$

Tangent at the origin is

$gx + fy = 0$

$d_1 = \frac{2g^2}{\sqrt{g^2 + f^2}}$

and $d^2 = \frac{2f^2}{\sqrt{g^2 + f^2}}$

$\Rightarrow d_1 + d_2 = 2\sqrt{g^2 + f^2}$

= diameter of the circle

20. a.

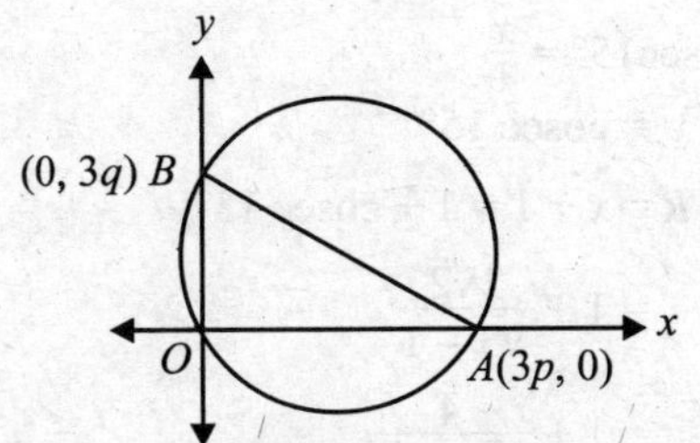

Fig. 2.18

Let the centroid of triangle OAB is (p, q).

Hence, points A and B are $(3p, 3q)$.

But diameter of triangle $AB = 6k$

Hence, $\sqrt{9p^2 + 9q^2} = 6k$

Therefore, locus of (p, q) is $x^2 + y^2 = 4k^2$.

21. a.

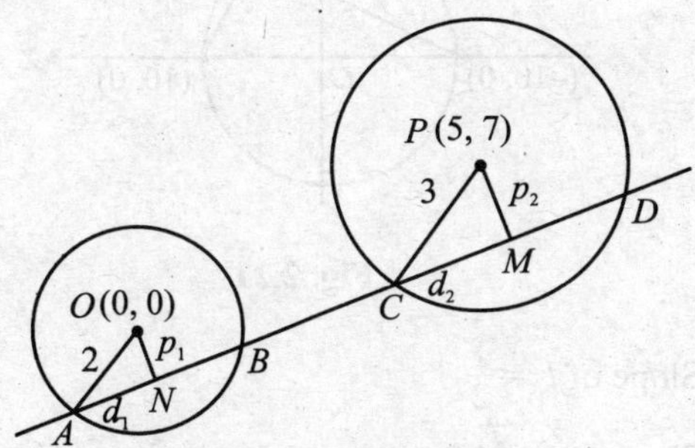

Fig. 2.19

Let equation of line be $y = x + c$ or $y - x = c$ (1)

Perpendicular from (0, 0) on line (1) is $\left|\frac{-c}{\sqrt{2}}\right| = \frac{c}{\sqrt{2}}$

In ΔAON, $\sqrt{2^2 - \left(\frac{c}{\sqrt{2}}\right)^2} = AN$

and in ΔCPM, $\sqrt{3^2 - \left(\frac{2 - c}{\sqrt{2}}\right)^2} = CM$.

Given $AN = CM$

$\Rightarrow 4 - \frac{c^2}{2} = 9 - \frac{(2 - c)^2}{2}$

$\Rightarrow c = -\frac{3}{2}$

Therefore, equation of the line is $y = x - \frac{3}{2}$ or $2x - 2y - 3 = 0$

22. a.

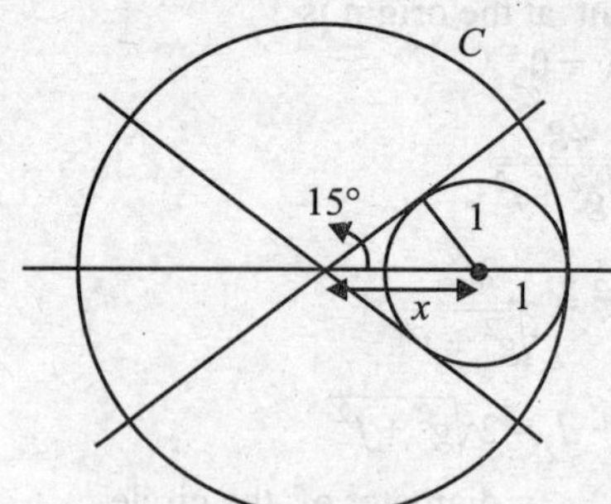

Fig. 2.20

$\text{cosec}15° = \frac{x}{1}$

$\Rightarrow x = \text{cosec } 15°$

$\Rightarrow R = x + 1 = 1 + \text{cosec } 15°$

$= 1 + \frac{2\sqrt{2}}{\sqrt{3} - 1}$

$= 1 + \frac{4}{\sqrt{6} - \sqrt{2}}$

$= 1 + \sqrt{6} + \sqrt{2}$

23. c.

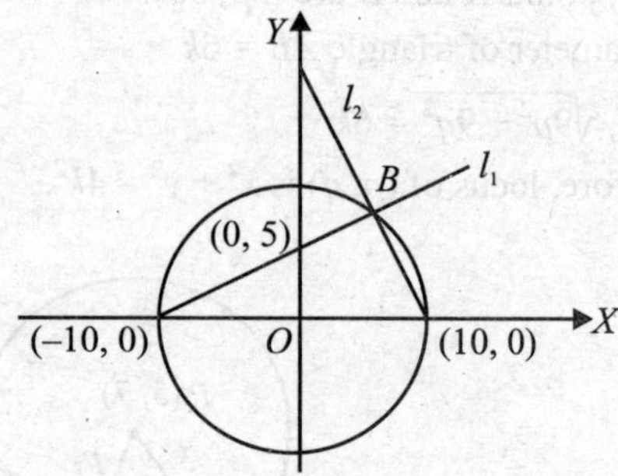

Fig. 2.21

Slope of $l_1 = \frac{1}{2}$

Slope of $l_2 = -2$

Equation of l_2, $y = -2(x - 10)$

$\Rightarrow y + 2x = 20$

Hence, $t = 20$

24. a. $AB = \sqrt{a^2 + b^2}$

Hence, $D = \sqrt{b^2 + a^2}$ (1)

Now, $\frac{d}{2} = \frac{\Delta}{s} = \frac{ab}{2s}$

[Where s is semi-perimeter]

$\therefore \frac{d}{2} = \frac{ab}{a + b + \sqrt{a^2 + b^2}}$

or $d = \frac{2ab}{a + b + \sqrt{a^2 + b^2}}$ (2)

From Eqs. (1) and (2)

$$d + D = \frac{\sqrt{a^2 + b^2}\left[(a + b) + \sqrt{a^2 + b^2}\right] + 2ab}{a + b + \sqrt{a^2 + b^2}}$$

$$= \frac{(a + b)^2 + (a + b)\sqrt{a^2 + b^2}}{a + b + \sqrt{a^2 + b^2}} = a + b$$

25. c.

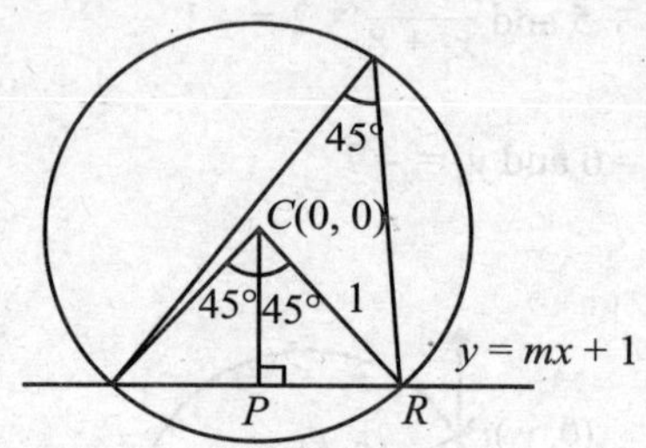

Fig. 2.22

Given circle is $x^2 + y^2 = 1$.

C (0, 0) and radius = 1 and chord is $y = mx + 1$

$\cos 45° = \frac{CP}{CR}$

CP = perpendicular distance from (0, 0) to chord $y = mx + 1$

$CP = \frac{1}{\sqrt{m^2 + 1}}$ (CR = radius = 1)

$\Rightarrow \cos 45° = \frac{1/\sqrt{m^2 + 1}}{1}$

$\Rightarrow \frac{1}{\sqrt{2}} = \frac{1}{\sqrt{m^2 + 1}}$

$\Rightarrow m^2 + 1 = 2$

$\Rightarrow m = \pm 1$

26. d.

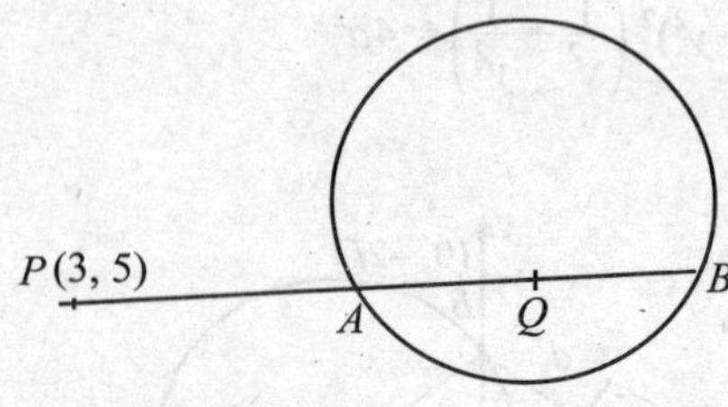

Fig. 2.23

$2PQ = PA + PB$

$\Rightarrow PQ - PA = PB - PQ$

$\Rightarrow AQ = QB$

$\Rightarrow Q$ is midpoint of AB.

Let Q has coordinates (h, k).

Then equation of chord AB is given by $T = S_1$

or $hx + ky - 4 = h^2 + k^2 - 4$

This variable chord passes through the point $P(3, 5)$.

$\Rightarrow 3h + 5k = h^2 + k^2$

$\Rightarrow x^2 + y^2 - 3x - 5y = 0$

which is required locus.

27. b.

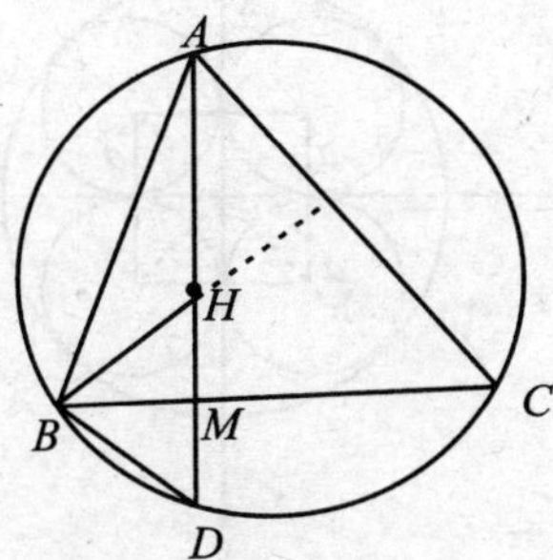

Fig. 2.24

Let the orthocenter be $H(5, 8)$.

Now, $\angle HBM = \pi/2 - C$

Also, $\angle DBC = \angle DAC = \pi/2 - C$

Hence, ΔBMH and ΔBMD are congruent.

$\Rightarrow HM = MD$

$\Rightarrow D$ is the image of H in the line $x - y = 0$ which is $D(8, 5)$.

Thus, equation of circumcircle is

$(x - 2)^2 + (y - 3)^2 = (8 - 2)^2 + (5 - 3)^2$

i.e., $x^2 + y^2 - 4x - 6y - 27 = 0$

28. b. The given point is an interior point

$\Rightarrow \left(-5 + \frac{r}{\sqrt{2}}\right)^2 + \left(-3 + \frac{r}{\sqrt{2}}\right)^2 - 16 < 0$

$\Rightarrow r^2 - 8\sqrt{2}\, r + 18 < 0$

$\Rightarrow 4\sqrt{2} - \sqrt{14} < r < 4\sqrt{2} + \sqrt{14}$ (1)

The point is on the major segment

$\Rightarrow$ The centre and the point are on the same side of the line $x + y = 2$

$\Rightarrow -5 + \frac{r}{\sqrt{2}} - 3 + \frac{r}{\sqrt{2}} - 2 < 0$

$\Rightarrow r < 5\sqrt{2}$ (2)

From Eqs. (1) and (2), $4\sqrt{2} - \sqrt{14} < r < 5\sqrt{2}$

29. a. Let $x = a$, $x = b$, $y = c$, $y = d$ be the sides of the square.

The length of each diagonal of the square is equal to the diameter of the circle, i.e., $2\sqrt{98}$.

Let l be the length of each side of the square. Then, $2l^2 = (\text{Diagonal})^2$

$\Rightarrow l = 14$

Therefore, each side of the square is at a distance 7 from the centre $(1, -2)$ of the given circle. This implies that $a = -6$, $b = 8$, $c = -9$, $d = 5$

Hence, the vertices of the square are $(-6, -9)$, $(-6, 5)$, $(8, -9)$, $(8, 5)$.

30. b.

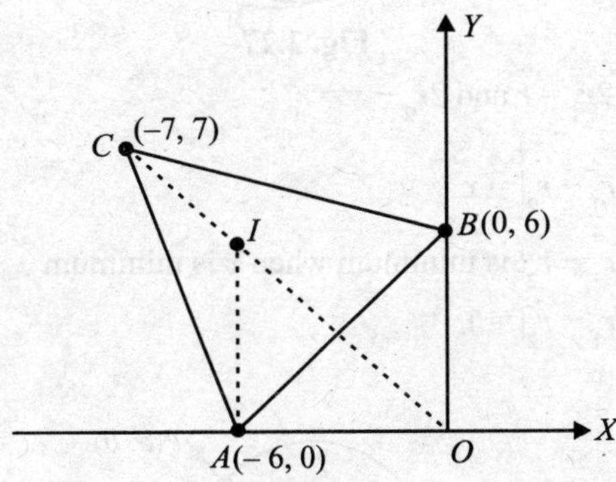

Fig. 2.25

The triangle is evidently isosceles and, therefore, the median through C is the angle bisector of $\angle C$.

The equation of the angle bisector is $y = -x$ and in centre $I = (-a, a)$ where a is positive.

Equation of AC is $y - 0 = -7(x + 6)$ or $7x + y + 42 = 0$ and equation of AB is $x - y + 6 = 0$.

The length of the perpendicular from I to AB and AC are equal. Therefore,

$\left|\frac{-7a + a + 42}{\sqrt{50}}\right| = \left|\frac{-a - a + 6}{\sqrt{2}}\right|$

$\Rightarrow a = \frac{9}{2}$ ($\because a > 0$)

Centre is $\left(-\frac{9}{2}, \frac{9}{2}\right)$ and radius $= \frac{3}{\sqrt{2}}$.

The equation of the circle is $\left(x + \frac{9}{2}\right)^2 + \left(y - \frac{9}{2}\right)^2 = \frac{9}{2}$

$\Rightarrow x^2 + y^2 + 9x - 9y + 36 = 0$

31. b.

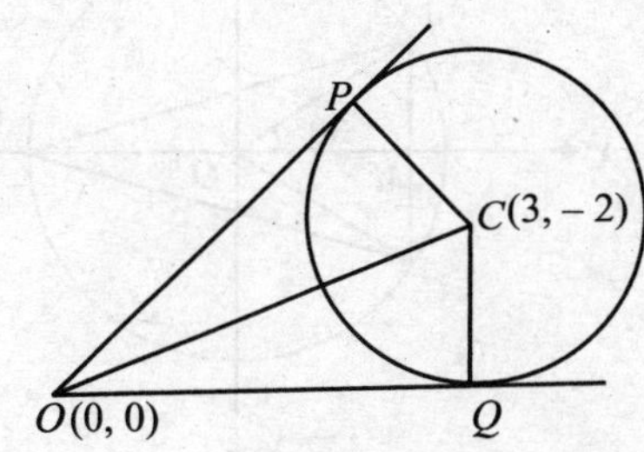

Fig. 2.26

Clearly, $OPCQ$ is cyclic quadrilateral, then the circumcircle of ΔOPQ passes through the point C.

For this circle, OC is diameter, then centre is midpoint of OC which is $\left(\frac{3}{2}, -1\right)$.

32. b.

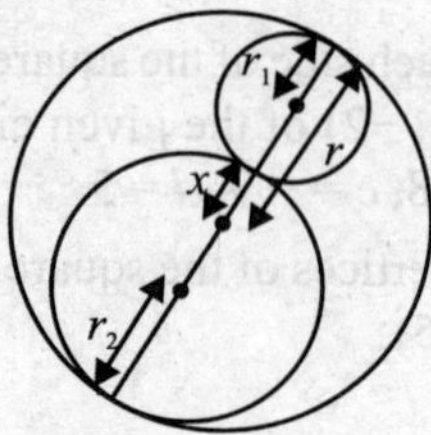

Fig. 2.27

$x + 2r_1 = r$ and $2r_2 - x = r$

$\Rightarrow |r_1 - r_2| = x$

$\Rightarrow |r_1 - r_2|$ is minimum when x is minimum

$\Rightarrow |r_1 - r_2| = 1$

33. a.

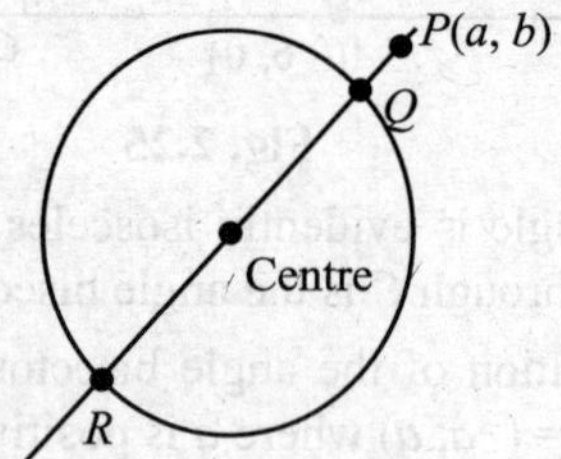

Fig. 2.28

The given circle is $(x + 1)^2 = (y + 2)^2 = 9$, which has radius = 3.

The points on the circle which are nearest and farthest to the point $P(a, b)$ are Q and R, respectively.

Thus, the circle centred at Q having radius PQ will be the smallest circle while the circle centred at R having radius PR will be the largest required circle.

Hence, difference between their radii is

$PR - PQ = QR = 6$

34. a.

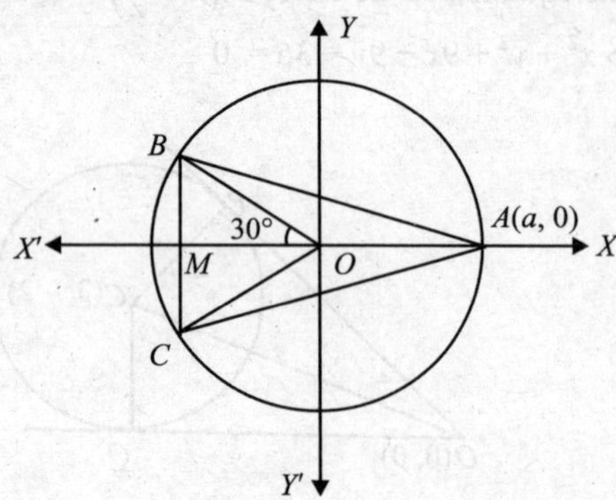

Fig. 2.29

Since $\angle B = \angle C = 75°$,

$\angle BAC = 30°$

$\Rightarrow \angle BOC = 60°$

$\Rightarrow B$ has coordinates $(-a\cos 30°, a\sin 30°)$

or $\left(\frac{-\sqrt{3}a}{2}, \frac{a}{2}\right)$ and those of C are $\left(-\frac{\sqrt{3}a}{2}, -\frac{a}{2}\right)$.

35. c.

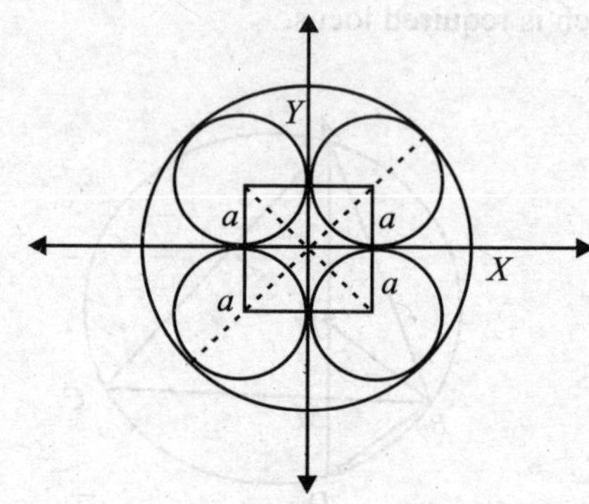

Fig. 2.30

The four circles are as shown in Fig. 2.30.

The smallest circle touching all of them has the radius $= \sqrt{2}\,a - a$ and the greatest circle touching all of them has the radius $= \sqrt{2}\,a + a$.

36. a. Let $A(x_1, y_1)$, $B(x_2, y_2)$, and $C(x_3, y_3)$ be the vertices of the triangle ABC, and let $P(h, k)$ be any point on the locus.

Then, $\quad PA^2 + PB^2 + PC^2 = c$ (constant)

$$\Rightarrow \sum_{i=1}^{3} (h - x_i)^2 + (k - y_i)^2 = c$$

$$\Rightarrow h^2 + k^2 - \frac{2h}{3}(x_1 + x_2 + x_3) - \frac{2k}{3}(y_1 + y_2 + y_3) + \sum_{i=1}^{3}(x_i^2 + y_i^2) - c = 0$$

So locus of (h, k) is

$$x^2 + y^2 - \frac{2x}{3}(x_1 + x_2 + x_3) - \frac{2y}{3}(y_1 + y_2 + y_3) + \lambda = 0,$$

where $\lambda = \sum_{i=1}^{3}(x_i^2 + y_i^2) - c = 0$ constant.

37. c. Equation of any circle passing through (1, 0) and (5, 0) is

$$y^2 + (x - 1)(x - 5) + \lambda y = 0$$

i.e. $\quad x^2 + y^2 + \lambda y - 6x + 5 = 0$

If $\angle ACB$ is maximum, then this circle must touch the y-axis at $(0, h)$. Putting $x = 0$ in the equation of circle, we get $y^2 + \lambda y + 5 = 0$. It should have $y = h$ as it is a repeated root.

$\Rightarrow h^2 = 5$ and $\lambda = -2h$

$\Rightarrow |h| = \sqrt{5}$

38. b.

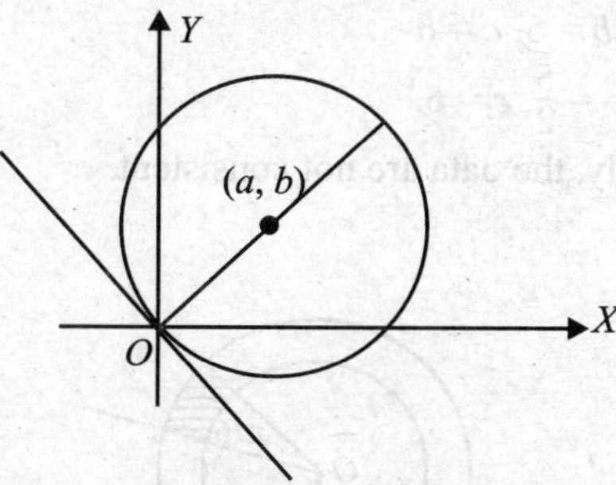

Fig. 2.31

Obviously, the slope of the tangent will be

$-\left(\frac{1}{b/a}\right)$, i.e., $-\frac{a}{b}$.

Hence, the equation of the tangent is $y = -\frac{a}{b}x$, i.e., $by + ax = 0$.

39. c.

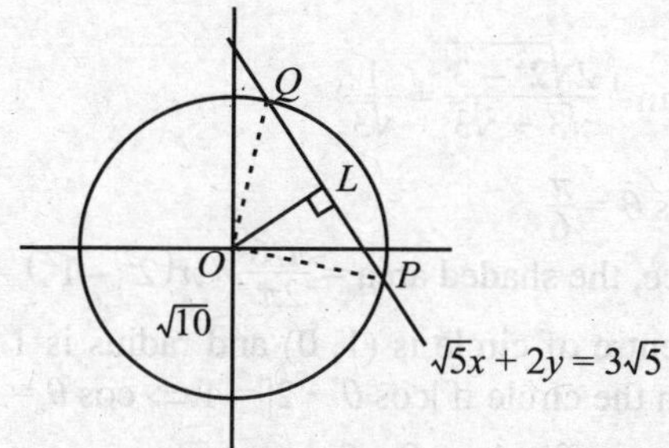

Fig. 2.32

Length of perpendicular from origin to the line $x\sqrt{5} + 2y = 3\sqrt{5}$ is

$$OL = \frac{3\sqrt{5}}{\sqrt{(\sqrt{5})^2 + 2^2}} = \frac{3\sqrt{5}}{\sqrt{9}}$$

$$= \sqrt{5}$$

Radius of the given circle $= \sqrt{10} = OQ = OP$

$$\Rightarrow PQ = 2QL = 2\sqrt{OQ^2 - OL^2}$$

$$= 2\sqrt{10 - 5} = 2\sqrt{5}$$

Thus, area of $\Delta OPQ = \frac{1}{2} \times PQ \times OL$

$$= \frac{1}{2} \times 2\sqrt{5} \times \sqrt{5} = 5$$

40. c.

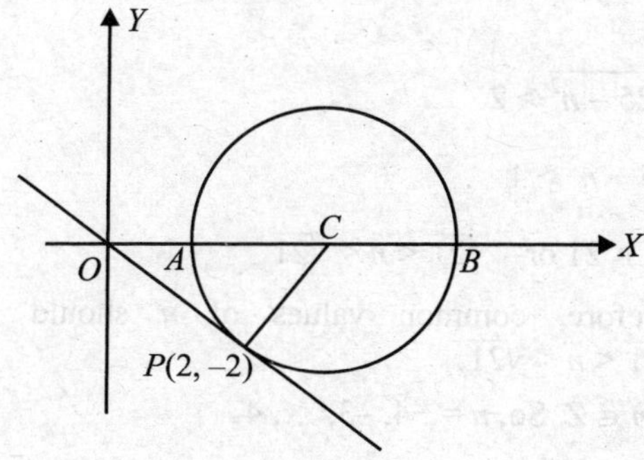

Fig. 2.33

If $(a, 0)$ is the centre C and P is $(2, -2)$, then $\angle COP = 45°$

Since the equation of OP is $x + y = 0$

$\therefore OP = 2\sqrt{2} = CP$. Hence, $OC = 4$

The point on the circle with the greatest x-coordinate is B.

$\alpha = OB = OC + CB = 4 + 2\sqrt{2}$

41. c.

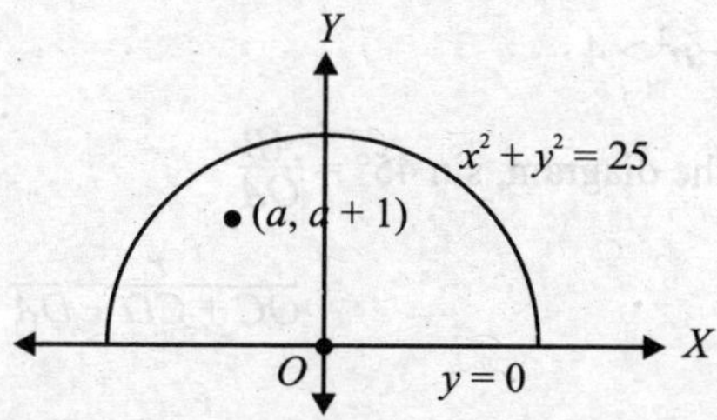

Fig. 2.34

$y = \sqrt{25 - x^2}$, $y = 0$ bound the semi-circle above the x-axis. Therefore,

$a + 1 > 0$ (1)

and $a^2 + (a + 1)^2 - 25 < 0 \Rightarrow 2a^2 + 2a - 24 < 0$

$\Rightarrow a^2 + a - 12 < 0$

$\Rightarrow -4 < a < 3$ (2)

From Eqs.(1) and (2), we get

$-1 < a < 3$

42. c. For $x^2 + y^2 = 9$, the centre $= (0, 0)$ and the radius $= 3$

For $x^2 + y^2 - 8x - 6y + n^2 = 0$,

the centre $= (4, 3)$ and the radius $= \sqrt{4^2 + 3^2 - n^2}$

$\therefore 4^2 + 3^2 - n^2 > 0$ or $n^2 < 5^2$ or $-5 < n < 5$

Circles should cut to have exactly two common tangents.

So, $r_1 + r_2 > d$

[distance between centres]

$\therefore 3 + \sqrt{25 - n^2} > \sqrt{4^2 + 3^2}$

or $\sqrt{25 - n^2} > 2$

or $25 - n^2 > 4$

$\therefore n^2 < 21$ or $-\sqrt{21} < n < \sqrt{21}$

Therefore, common values of n should satisfy $-\sqrt{21} < n < \sqrt{21}$.

But $n \in Z$. So, $n = -4, -3, \ldots, 4$

43. b.

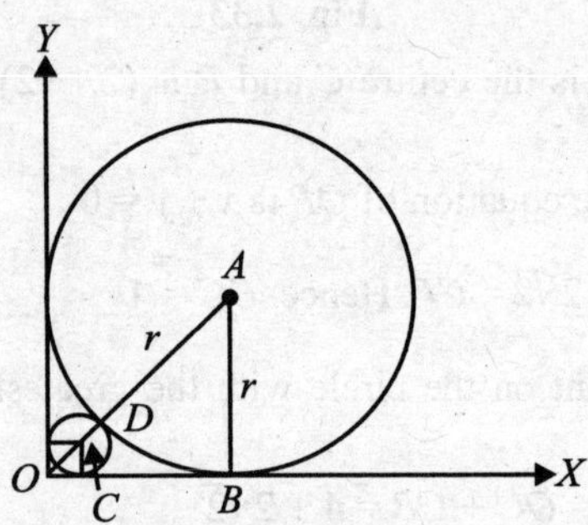

Fig. 2.35

or $25 - n^2 > 4$

From the diagram, $\sin 45° = \frac{AB}{OA}$

$= \frac{r}{OC + CD + DA}$

$= \frac{r}{\sqrt{2} + 1 + r}$

$\Rightarrow \sqrt{2} + 1 + r = \sqrt{2}\, r$

$\Rightarrow r = \frac{\sqrt{2} + 1}{\sqrt{2} - 1} = 3 + 2\sqrt{2}$

44. a.

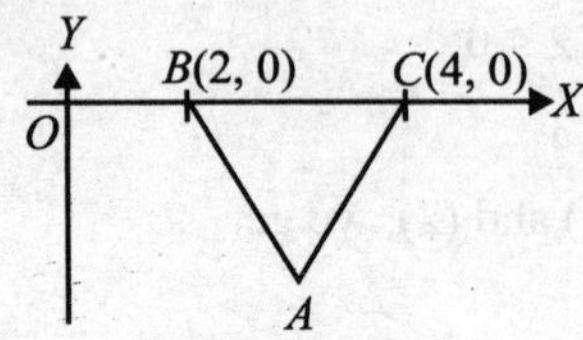

Fig. 2.36

Clearly, $A \equiv (3, -\sqrt{3})$.

Centroid of triangle ABC is $\left(3, -\frac{1}{\sqrt{3}}\right)$, thus equation of incircle is

$(x - 3)^2 + \left(y + \frac{1}{\sqrt{3}}\right)^2 = \frac{1}{3}$

$\Rightarrow x^2 + y^2 - 6x + \frac{2y}{\sqrt{3}} + 9 = 0$

45. d. $x^2 + 2ax + c = (x - 2)^2$

$\Rightarrow -2a = 4, c = 4$

$\Rightarrow a = -2, c = 4,$

$y^2 + 2by + c = (y - 2)(y - 3)$

$\Rightarrow -2b = 5, c = 6$

$\Rightarrow b = -\frac{5}{2}, c = 6.$

Clearly, the data are not consistent.

46. d.

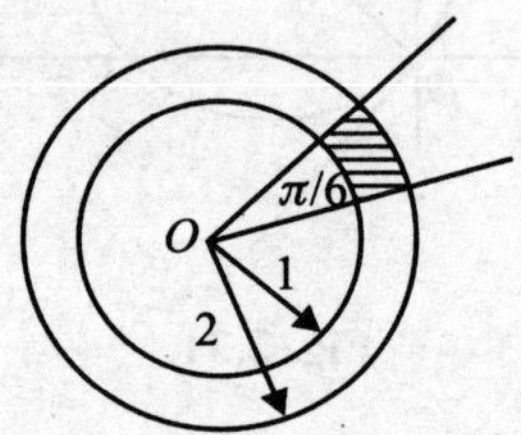

Fig. 2.37

The angle θ between the lines represented by $\sqrt{3}x^2 - 4xy + \sqrt{3}y^2 = 0$ is given by

$\theta = \tan^{-1} \frac{\sqrt{h^2 - ab}}{|a + b|}$

$= \tan^{-1} \frac{2\sqrt{2^2 - 3}}{\sqrt{3} + \sqrt{3}} = \frac{1}{\sqrt{3}}$

Gives $\theta = \frac{\pi}{6}$

Hence, the shaded area $= \frac{\pi/6}{2\pi} \times \pi(2^2 - 1^2) = \frac{\pi}{4}$.

47. c. Centre of circle is (1, 0) and radius is 1. Line will touch the circle if $|\cos\theta - 2| = 1 \Rightarrow \cos\theta = 1, 3$.

Thus, $\cos\theta = 1 \Rightarrow \theta = 2n\pi, n \in I$

48. b.

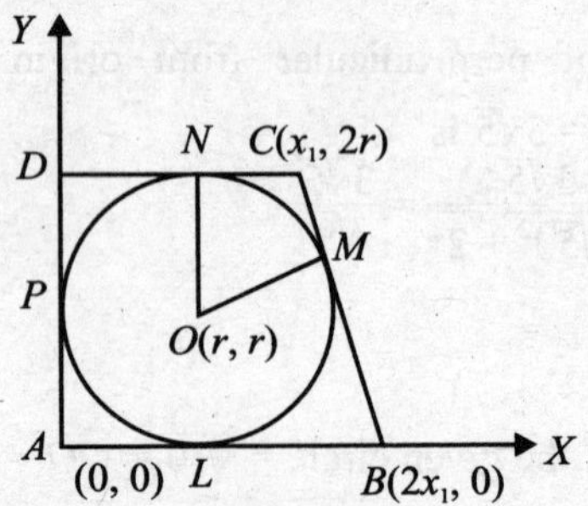

Fig. 2.38

Here $\frac{1}{2} \times 3x_1 \times 2r = 18$

$\Rightarrow x_1 \times r = 6$ (1)

Equation of BC is $y = -\frac{2r}{x_1}(x - 2x_1)$

BC is tangent to the circle $(x - r)^2 + (y - r)^2 = r^2$

Perpendicular distance of BC from centre = radius

$$\Rightarrow \frac{\left|r + \frac{2r}{x_1}(r - 2x_1)\right|}{\sqrt{1 + \frac{4r^2}{x_1^2}}} = r$$

$$\Rightarrow \frac{2r^2}{x_1} - 3r = r\sqrt{1 + 4\frac{r^2}{x_1^2}}$$

$$\Rightarrow (2r - 3x_1)^2 = x_1^2 + 4r^2$$

$$\Rightarrow r x_1 = \frac{2}{3} x_1^2 \Rightarrow 3r = 2x_1 \quad (2)$$

From Eqs. (1) and (2), $r = 2$ units

49. a.

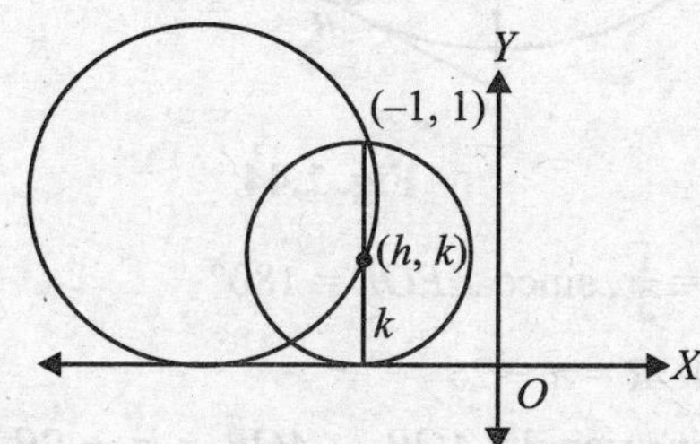

Fig. 2.39

From the figure, $k \geq \frac{1}{2}$

50. a. Curve passing through point of intersection of S and S' is

$$\Rightarrow S + \lambda S' = 0$$

$$\Rightarrow x^2(\sin^2\theta + \lambda\cos^2\theta) + y^2(\cos^2\theta + \lambda\sin^2\theta) + 2xy(h + \lambda h') + x(32 + 16\lambda) + y(16 + 32\lambda) + 19(1 + \lambda) = 0$$

For this equation to be a circle,

$$\sin^2\theta + \lambda\cos^2\theta = \cos^2\theta + \lambda\sin^2\theta$$

$$\Rightarrow \lambda = 1$$

and $h + \lambda h' = 0 \Rightarrow h + h' = 0$

51. a.

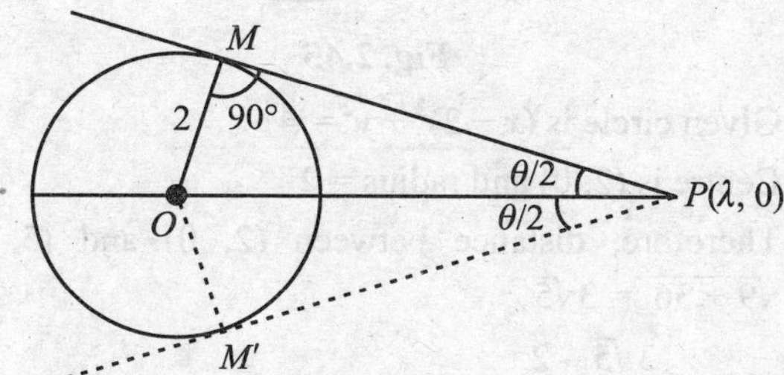

Fig. 2.40

We have $\frac{\pi}{2} < \theta < \frac{2\pi}{3}$, i.e., $\frac{\pi}{4} < \frac{\theta}{2} < \frac{\pi}{3}$

$$\Rightarrow \frac{1}{\sqrt{2}} < \sin\frac{\theta}{2} < \frac{\sqrt{3}}{2}$$

But, $\sin\frac{\theta}{2} = \frac{2}{\lambda} \Rightarrow \frac{1}{\sqrt{2}} < \frac{2}{\lambda} < \frac{\sqrt{3}}{2}$

$$\Rightarrow \frac{4}{\sqrt{3}} < \lambda < 2\sqrt{2}$$

52. a.

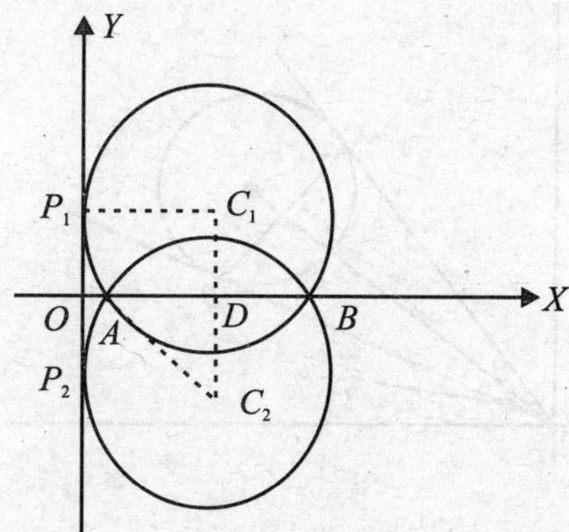

Fig. 2.41

Let $A \equiv (1, 0)$, $B \equiv (3, 0)$ and C_1, C_2 be the centre of circles passing through A, B and touching the y-axis at P_1 and P_2. If r be the radius of circle (here radius of both circles will be same), $C_1A = C_2A = r = OD = 2$ and $C_1 \equiv (2, h)$

where $h^2 = AC_1^2 - AD^2 = 4 - 1 = 3$

$$\Rightarrow C_1 \equiv (2, \sqrt{3}),\ C_2 \equiv (2, -\sqrt{3})$$

If $\angle C_1AC_2 = \theta$

$$\Rightarrow \cos\theta = \frac{AC_1^2 + AC_2^2 - C_1C_2^2}{2AC_1 \cdot AC_2} = \frac{1}{2}$$

53. c. Chord with midpoint (h, k) is

$$hx + ky = h^2 + k^2 \quad (1)$$

Chord of contact of (x_1, y_1) is

$$xx_1 + yy_1 = 2 \quad (2)$$

Comparing, we get

$$x_1 = \frac{2h}{h^2 + k^2} \text{ and } y_1 = \frac{2k}{h^2 + k^2}$$

(x_1, y_1) lies on $3x + 4y = 10$

$$\Rightarrow 6h + 8k = 10(h^2 + k^2)$$

$\therefore$ Locus of (h, k) is $x^2 + y^2 - \frac{3}{5}x - \frac{4}{5}y = 0$,

which is circle with centre $P\left(\frac{3}{10}, \frac{4}{10}\right)$.

Hence, $OP = \frac{1}{2}$

54. a. a, b, c are in A.P. So $ax + by + c = 0$ represents a family of lines passing through the point $(1, -2)$. So, the family of circles (concentric) will be given by $x^2 + y^2 - 2x + 4y + c = 0$. It intersects given circle orthogonally.

$\Rightarrow 2(-1 \times -2) + (2 \times -2) = -1 + c \Rightarrow c = -3$

55. d.

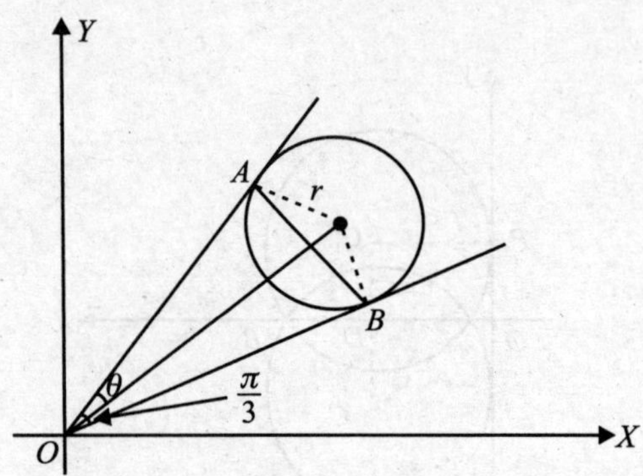

Fig. 2.42

Here $\tan 2\theta = \dfrac{2\sqrt{4-1}}{2} = \sqrt{3}$

$\Rightarrow \theta = \pi/6$

Area of $\Delta OAB = \frac{1}{2}(r \cot \theta)^2 (\sin 2\theta)$

$= \frac{1}{2}(r\sqrt{3})^2 \frac{\sqrt{3}}{2}$

56. a. Let P be $(1 + \sqrt{2}\cos\theta, \sqrt{2}\sin\theta)$ and C is $(1, 0)$. Circumcentre of triangle ABC is midpoint of PC.

$\Rightarrow 2h = 1 + \sqrt{2}\cos\theta + 1$

and $2k = \sqrt{2}\sin\theta$

$\Rightarrow [2(h-1)]^2 + (2k)^2 = 2$

$\Rightarrow 2(h-1)^2 + k^2 - 1 = 0$

$\Rightarrow 2x^2 + 2y^2 - 4x + 1 = 0$

57. a.

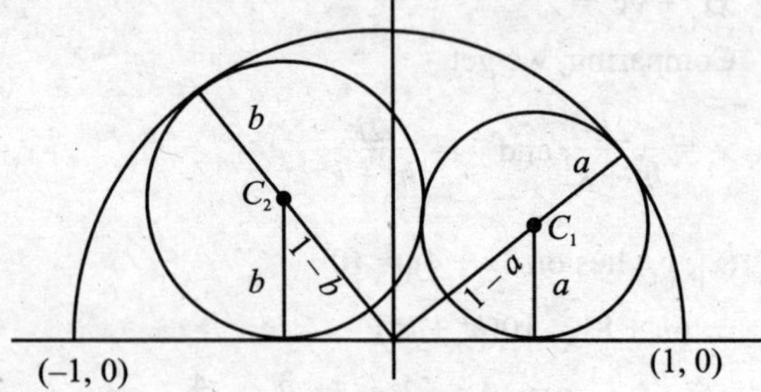

Fig. 2.43

Let the centres of the circles be C_1 and C_2.

$\Rightarrow C_1$ is $(\sqrt{1-2a}, a)$ and C_2 is $(\sqrt{1-2b}, b)$

Now $C_1C_2 = a + b = a + \frac{1}{2}$

$\Rightarrow 1 - 2a + \left(a - \frac{1}{2}\right)^2 = \left(a + \frac{1}{2}\right)^2$

$\Rightarrow a = \frac{1}{4}$

58. a. Since ΔPOQ and ΔAOQ are congruent

Hence, $\angle POQ = \angle QOA = \theta$

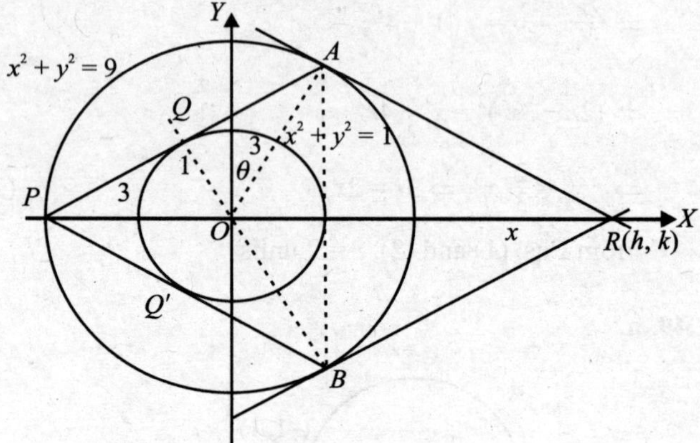

Fig. 2.44

$\cos\theta = \frac{1}{3}$, since $\angle POR = 180°$

$\Rightarrow \angle AOR = \pi - 2\theta$

Now in triangle AOR, $\angle AOR = \pi - 2\theta$ and $AO = 3$ unit

$\Rightarrow \cos(\pi - 2\theta) = \dfrac{OA}{OR} = \dfrac{3}{\sqrt{h^2 + k^2}}$

$\Rightarrow \sqrt{h^2 + k^2} = \dfrac{27}{7}$

$\Rightarrow x^2 + y^2 = \left(\dfrac{27}{7}\right)^2$

59. b.

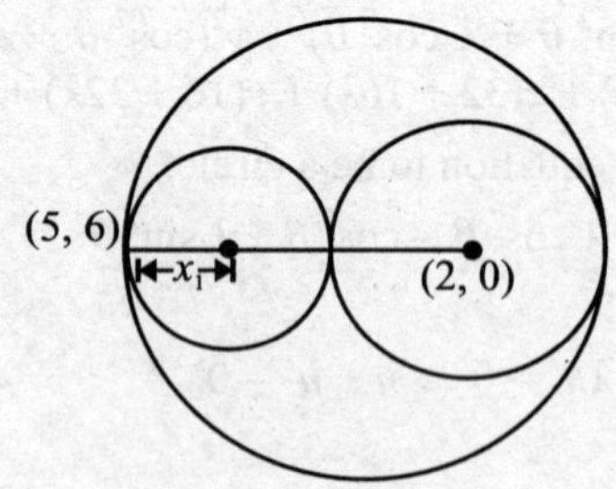

Fig. 2.45

Given circle is $(x-2)^2 + y^2 = 4$

Centre is $(2, 0)$ and radius $= 2$

Therefore, distance between $(2, 0)$ and $(5, 6)$ is $\sqrt{9 + 36} = 3\sqrt{5}$

$\Rightarrow r_1 = \dfrac{3\sqrt{5} - 2}{2}$

and $r_2 = \dfrac{3\sqrt{5} + 2}{2}$

$= r_1 r_2 = \dfrac{41}{4}$

60. c. $x^2 + y^2 - 12x + 35 = 0$ (1)

$x^2 + y^2 + 4x + 3 = 0$ (2)

Equation of radical axis of circles (1) and (2) is

$-16x + 32 = 0 \Rightarrow x = 2$

It intersects the line joining the centres, i.e. $y = 0$ at the point (2, 0)

$\therefore$ Required radius $= \sqrt{4 - 24 + 35} = \sqrt{15}$ [length of tangent from (2, 0)].

61. d.

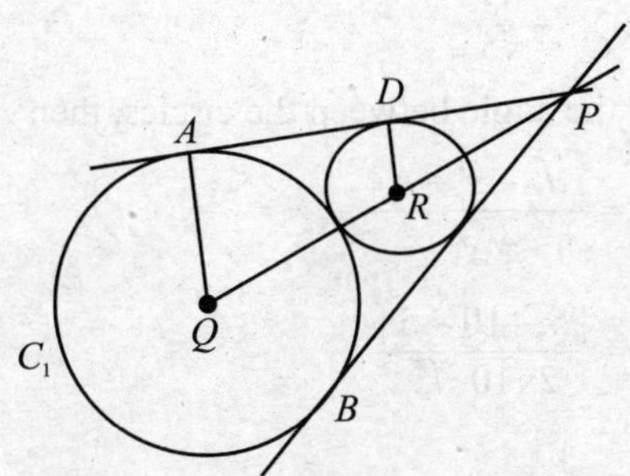

Fig. 2.46

$AQ = 3 + 2\sqrt{2}$

$PQ = 3\sqrt{2} + 4$

Let r be required radius. Therefore,

$3\sqrt{2} + 4 = 3 + 2\sqrt{2} + r + r\sqrt{2}$

$\left[\because \angle RPD = \frac{\pi}{4}\right]$

$\sqrt{2} + 1 = r(1 + \sqrt{2}) \Rightarrow r = 1$

62. a.

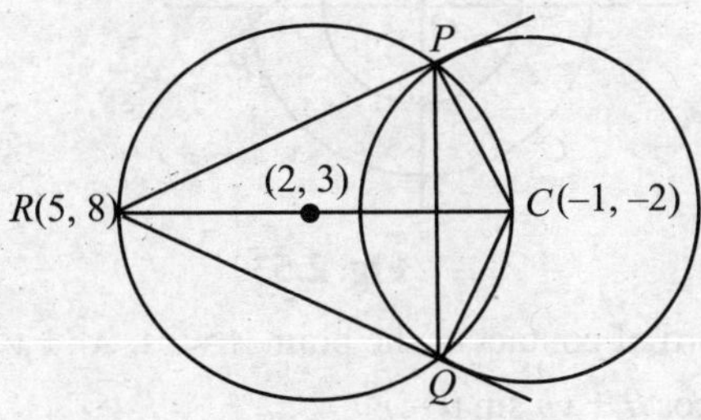

Fig. 2.47

Let C be the centre of the given circle.

Then, circumcircle of the ΔRPQ passes through C. (2, 3) is the midpoint of RC. Coordinates of C are (−1, −2). Hence, equation of the circle is $x^2 + y^2 + 2x + 4y - 20 = 0$.

63. c.

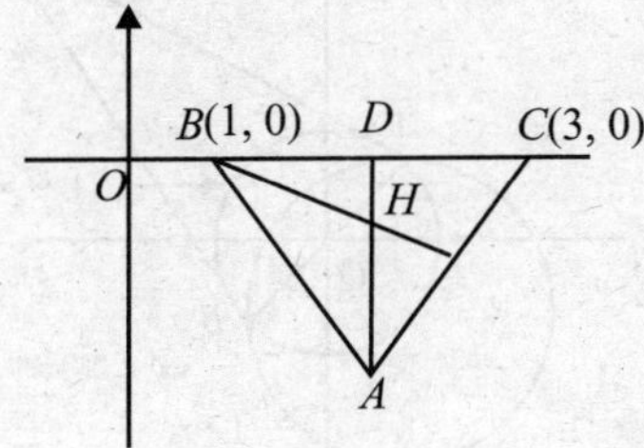

Fig. 2.48

Radical centre of the circles described on the sides of a triangle as diameters is the orthocentre of the triangle.

$D = (2, 0)$

$DH = -BD \tan\frac{\pi}{6} = -\frac{1}{\sqrt{3}}$

Coordinates of H are $\left(2, -\frac{1}{\sqrt{3}}\right)$.

64. b. Let $(\alpha, 3 - \alpha)$ be any point on $x + y = 3$.

Equation of chord of contact is $\alpha x + (3 - \alpha)y = 9$

i.e., $\alpha(x - y) + 3y - 9 = 0$

The chord passes though the point (3, 3) for all values of α.

65. d. Given circle $(x - 1)^2 + (y + 2)^2 = 16$

Its director circle is $(x - 1)^2 + (y + 2)^2 = 32$

$\Rightarrow OS = 4\sqrt{2}$

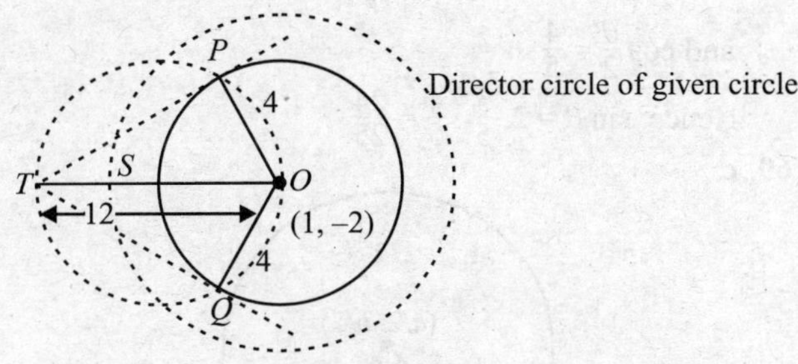

Fig. 2.49

Therefore, required distance.

$TS = OT - SO = 12 - 4\sqrt{2}$.

66. c. Equation of the two circles be $(x - r)^2 + (y - r)^2 = r^2$

i.e., $x^2 + y^2 - 2rx - 2ry + r^2 = 0$, where $r = r_1$ and r_2. Condition of orthogonality gives

$2r_1r_2 + 2r_1r_2 = r_1^2 + r_2^2 \Rightarrow 4r_1r_2 = r_1^2 + r_2^2$

Circle passes through (a, b)

$\Rightarrow a^2 + b^2 - 2ra - 2rb + r^2 = 0$

i.e., $r^2 - 2r(a + b) + a^2 + b^2 = 0$

$r_1 + r_2 = 2(a + b)$ and $r_1r_2 = a^2 + b^2$

$4(a^2 + b^2) = 4(a + b)^2 - 2(a^2 + b^2)$

i.e., $a^2 - 4ab + b^2 = 0$

67. c. $C_1 = (-1, -4)$; $C_2 = (2, 5)$;

$r_1 = \sqrt{1 + 16 + 23} = 2\sqrt{10}$;

$r_2 = \sqrt{4 + 25 + 19} = \sqrt{10}$;

$C_1C_2 = \sqrt{9 + 18} = 3\sqrt{10}$

$\Rightarrow C_1C_2 = r_1 + r_2$.

Hence, circles touch externally.

68. a.

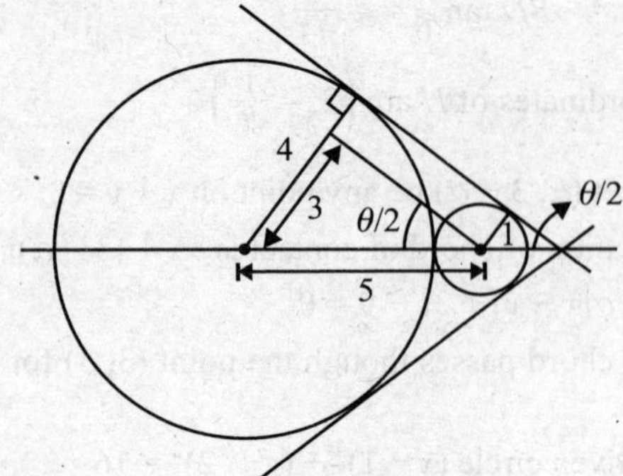

Fig. 2.50

$\sin\frac{\theta}{2} = \frac{3}{5}$

and $\cos\frac{\theta}{2} = \frac{4}{5}$

Hence, $\sin\theta = 2.\frac{3}{5}\cdot\frac{4}{5} = \frac{24}{25}$

69. c.

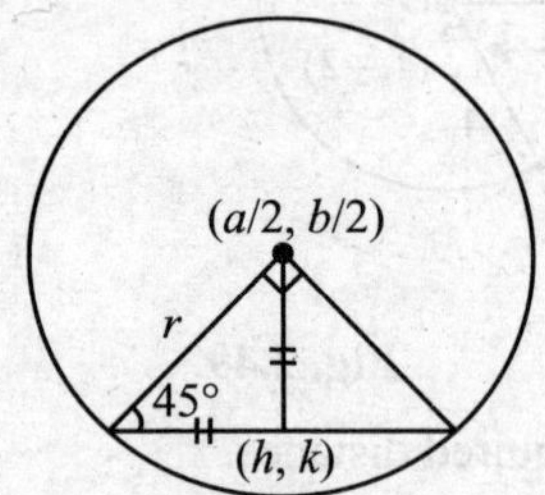

Fig. 2.51

$$r = \sqrt{\frac{a^2}{4} + \frac{b^2}{4}} = \frac{\sqrt{a^2 + b^2}}{2}$$

$$\sin 45° = \frac{\sqrt{\left(h - \frac{a}{2}\right)^2 + \left(h - \frac{b}{2}\right)^2}}{\frac{\sqrt{a^2 + b^2}}{2}}$$

$$\Rightarrow \frac{1}{2} = 4\left[\frac{\frac{(2h - a)^2}{4} + \frac{(2k - b)^2}{4}}{a^2 + b^2}\right]$$

Simplify to get locus $x^2 + y^2 - ax - by - \frac{a^2 + b^2}{8} = 0$.

70. a.

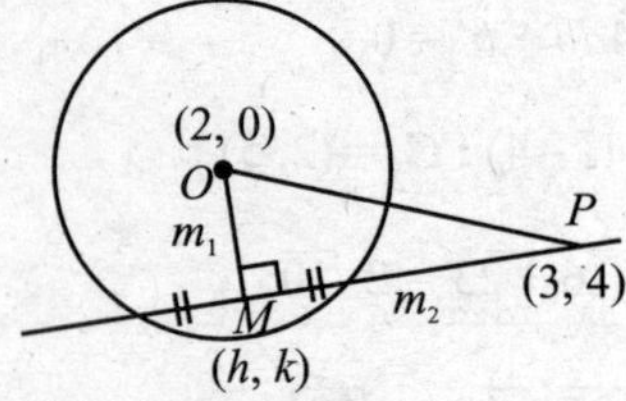

Fig. 2.52

$\Rightarrow m_1m_2 = -1$

$$\Rightarrow \left(\frac{4 - k}{3 - h}\right)\left(\frac{k - 0}{h - 2}\right) = -1$$

Hence, locus is $x^2 + y^2 - 5x - 4y + 6 = 0$.

71. c. Locus of the centre of the circle cutting $S_1 = 0$ and $S_2 = 0$ orthogonally is the radical axis between $S_1 = 0$ and $S_2 = 0$, i.e., $S_1 - S_2 = 0$ or $9x - 10y + 11 = 0$.

72. b. For given $r_1 = \sqrt{10}, C_1(1, 0)$

and $r_2 = \sqrt{5}, C_2(0, 2)$

$d = C_1C_2 = \sqrt{5}$

If θ is the angle between the circles, then

$$\cos\theta = \frac{|d^2 - r_1^2 - r_2^2|}{2r_1r_2}$$

$$= \frac{|5 - 10 - 5|}{2\sqrt{10}\sqrt{5}}$$

$$= \frac{1}{\sqrt{2}}$$

Hence, $\theta = \frac{\pi}{4}$

73. a.

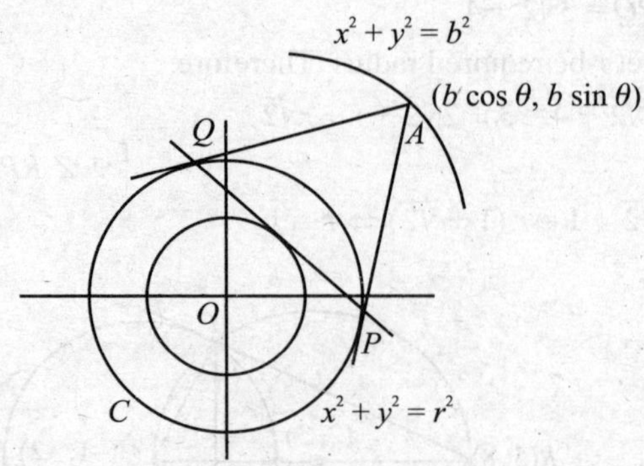

Fig. 2.53

Chord of contact of the point A w.r.t. $x^2 + y^2 = r^2$ is

$xb\cos\theta + yb\sin\theta = r^2$ (1)

This must be a tangent to the circle $x^2 + y^2 = a^2$

$$\Rightarrow \left[\frac{r^2}{\sqrt{b^2\cos^2\theta + b^2\sin^2\theta}}\right] = a \Rightarrow r^2 = ab$$

Hence, equation of circle is $x^2 + y^2 = ab$.

74. a.

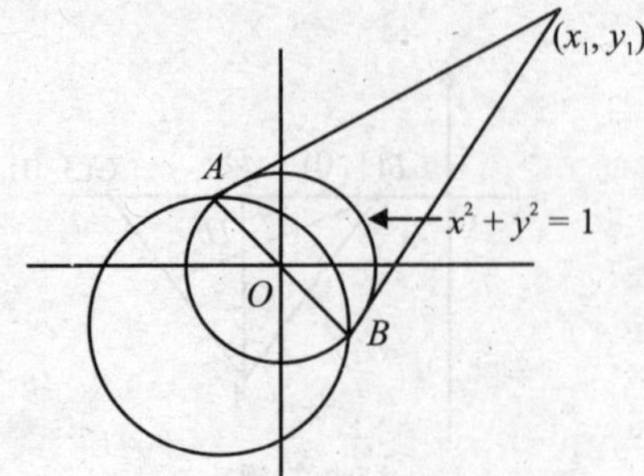

Fig. 2.54

Locus of point of intersection of tangents chord of contact of (x_1, y_1) w.r.t.

$x^2 + y^2 = 1$ is $xx_1 + yy_1 = 1$ (AB) (1)

AB is also common chord between two circles. So,

$-1 + (\lambda + 6)x - (8 - 2\lambda)y + 3 = 0$

$\Rightarrow (\lambda + 6)x - (8 - 2\lambda)y + 2 = 0$ (2)

Comparing Eqs. (1) and (2), we get

$$\frac{x_1}{\lambda + 6} = \frac{y_1}{2\lambda - 8} = \frac{-1}{2}$$

Eliminate $\lambda \Rightarrow 2x - y + 10 = 0$, which is the required locus.

75. c.

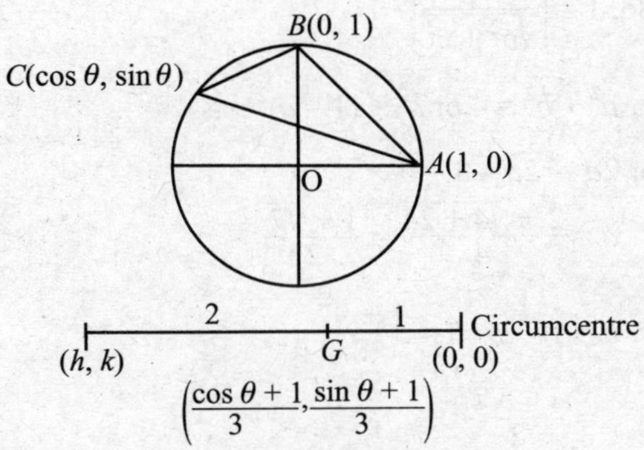

Fig. 2.55

Let C $(\cos\theta, \sin\theta)$. H (h, k) is the orthocentre of the ΔABC.

Since circumcentre of the triangle is (0, 0), for orthocentre $h = 1 + \cos\theta$ and $k = 1 + \sin\theta$.

Eliminating θ, $(x - 1)^2 + (y - 1)^2 = 1$. Hence,

$x^2 + y^2 - 2x - 2y + 1 = 0$

76. d.

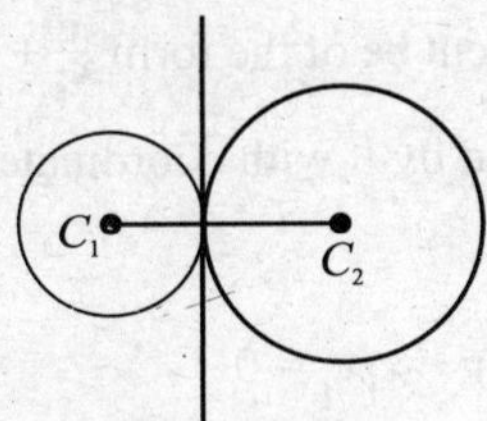

Fig. 2.56

$C_1C_2 = r_1 + r_2$

$C_1 = (0, 0)$; $C_2 = (3\sqrt{3}, 3)$

and $r_1 = 2, r_2 = 4$

$\Rightarrow$ Circles touch each other externally.

Equation of common tangent is $\sqrt{3}x + y - 4 = 0$ (1)

Comparing it with $x\cos\theta + y\sin\theta = 2$, we get

$\theta = \frac{\pi}{6}$

77. a.

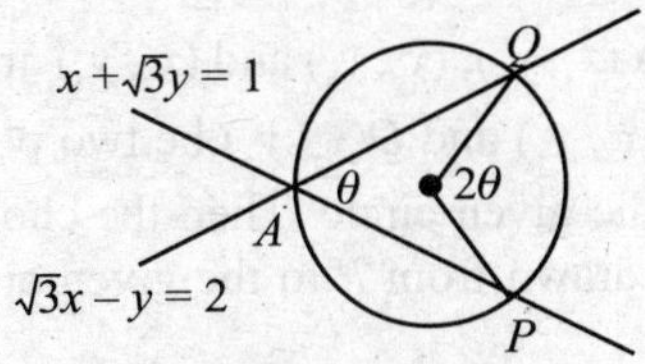

Fig. 2.57

Let the point of intersection of two lines is A.

$\therefore$ The angle subtended by PQ on centre C

$= 2 \times$ the angle subtended by PQ on point A.

For $x + \sqrt{3}y = 1$, $m_1 = \frac{-1}{\sqrt{3}}$ and for $\sqrt{3}x - y = 2$, $m_2 = \sqrt{3}$

$\because m_1 \times m_2 = \frac{-1}{\sqrt{3}} \times \sqrt{3} = -1$, So

$\angle A = 90°$

$\therefore$ The angle subtended by arc PQ at its centre

$= 2 \times 90° = 180°$

78. b. Clearly, (0, 0) lies on the director circle of the given circle.

Now, equation of director circle is

$(x + g)^2 + (y + f)^2 = 2(g^2 + f^2 - c)$

If (0, 0) lies on it, then

$g^2 + f^2 = 2(g^2 + f^2 - c)$

$\Rightarrow g^2 + f^2 = 2c$

79. c. Let the second circle be $x^2 + y^2 + 2gx + 2fy = 0$.

But $y = x$ touches the circle.

Hence, $x^2 + x^2 + 2gx + 2fx = 0$ has equal roots, i.e., $f + g = 0$

Therefore, the equation of the common chord is $2(g - 3)x + 2(-g - 4)y + 7 = 0$

or $(-6x - 8y + 7) + g(2x - 2y) = 0$, which passes through the point of intersection of

$-6x - 8y + 7 = 0$ and $2x - 2y = 0$, which is (1/2, 1/2).

80. b. Let the coordinates of A, B, and C be (x_1, y_1), (x_2, y_2) and (x_3, y_3), respectively. Then, the chords of contact of tangents drawn from A, B and C are $xx_1 + yy_1 = a^2$, $xx_2 + yy_2 = a^2$, and $xx_3 + yy_3 = a^2$, respectively. These three lines will be concurrent, if

$$\begin{vmatrix} x_1 & y_1 & -a^2 \\ x_2 & y_2 & -a^2 \\ x_3 & y_3 & -a^2 \end{vmatrix} = 0$$

$$\Rightarrow -a^2\begin{vmatrix} x_1 & y_1 & 1 \\ x_2 & y_2 & 1 \\ x_3 & y_3 & 1 \end{vmatrix} = 0$$

$$\Rightarrow \begin{vmatrix} x_1 & y_1 & 1 \\ x_2 & y_2 & 1 \\ x_3 & y_3 & 1 \end{vmatrix} = 0$$

$\Rightarrow$ Points (x_1, y_1), (x_2, y_2) and (x_3, y_3) are collinear.

81. c. Let $P(x_1, y_1)$ and $Q(x_2, y_2)$ be two points and $x^2 + y^2 = a^2$ be the given circle. Then the chord of contact of tangents drawn from P to the given circle is $xx_1 + yy_1 = a^2$.

It will pass through $Q(x_2, y_2)$, if

$$x_1x_2 + y_1y_2 = a^2 \quad (1)$$

Now, $l_1 = \sqrt{x_1^2 + y_1^2 - a^2}$,

$l_2 = \sqrt{x_2^2 + y_2^2 - a^2}$

and $PQ = \sqrt{(x_2 - x_1)^2 + (y_2 - y_1)^2}$

$= \sqrt{(x_2^2 + y_2^2) + (x_1^2 + y_1^2) - 2(x_1x_2 + y_1y_2)}$

$\therefore \quad PQ = \sqrt{[(x_2^2 + y_2^2) + (x_1^2 + y_1^2) - 2a^2]}$

[Using Eq. (1)]

$\Rightarrow \quad PQ = \sqrt{(x_1^2 + y_1^2 - a^2) + (x_2^2 + y_2^2 - a^2)}$

$\Rightarrow \quad PQ = \sqrt{l_1^2 + l_2^2}$

82. a. Centres are (10, 0) and (– 15, 0)

and radii are $r_1 = 6$ and $r_2 = 9$

Also $d = 25$

$r_1 + r_2 < d$

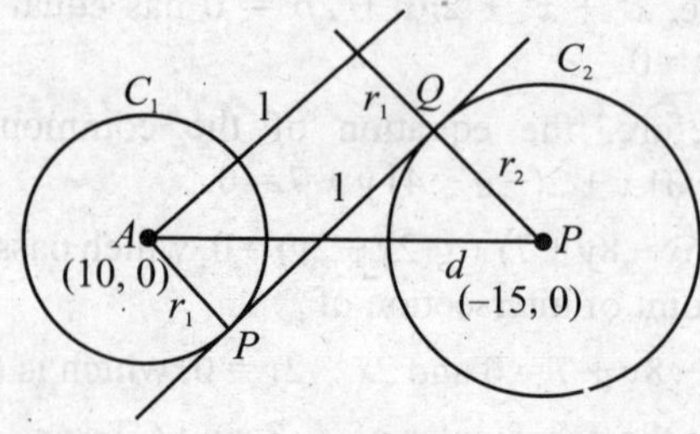

Fig. 2.58

$\Rightarrow$ circles are neither intersecting nor touching

$PQ = \sqrt{d^2 - (r_1 + r_2)^2}$

$= \sqrt{625 - 225}$

$= 20$

83. b. If (α, β) is the centre

then $(\alpha - 1)^2 + (\beta - 3)^2 = (\alpha - 3)^2 + (\beta - 1)^2 \quad (1)$

and $\dfrac{\beta - 3}{\alpha - 1} \cdot \dfrac{\beta - 1}{\alpha - 3} = -1$

or $(\alpha - 1)(\alpha - 3) + (\beta - 1)(\beta - 3) = 0 \quad (2)$

$(1) \Rightarrow 4\alpha - 4\beta = 0 \qquad \therefore \alpha = \beta$

$(2) \Rightarrow 2(\alpha - 1)(\alpha - 3) = 0 \qquad \therefore \alpha = 1, 3$

Hence, $(\alpha, \beta) = (1, 1), (3, 3)$.

84. b. The centre of $x^2 + y^2 - 4x - 4y = 0$ is (2, 2).

It is $ax + by = 2$

$\therefore 2a + 2b = 2$ or $a + b = 1$

$ax + by = 2$ touches $x^2 + y^2 = 1$.

So, $1 = \left|\dfrac{-2}{\sqrt{a^2 + b^2}}\right|$

$\therefore a^2 + b^2 = 4$ or $a^2 + (1 - a)^2 = 4$

or $2a^2 - 2a - 3 = 0$

$\therefore a = \dfrac{2 \pm \sqrt{4 + 24}}{4} = \dfrac{1 \pm \sqrt{7}}{2}$

$\therefore b = 1 - a = 1 - \dfrac{1 \pm \sqrt{7}}{2}$

$= \dfrac{1 \mp \sqrt{7}}{2}$

85. c. Equation of any circles through (0, 1) and (0, 6) is

$x^2 + (y - 1)(y - 6) + \lambda x = 0$

$\Rightarrow x^2 + y^2 + \lambda x - 7y + 6 = 0$

If it touches the x-axis, then $x^2 + \lambda x + 6 = 0$ should have equal roots

$\Rightarrow \lambda^2 = 24 \Rightarrow \lambda = \pm\sqrt{24}$

Radius of these circles $= \sqrt{6 + \dfrac{49}{4} - 6} = \dfrac{7}{2}$ units.

That means we can draw two circles but radius of both circles is $\dfrac{7}{2}$.

86. b. Let the tangent be of the form $\dfrac{x}{x_1} + \dfrac{y}{y_1} = 1$ and area of Triangle formed by it with coordinate axes is

$$\frac{1}{2}x_1y_1 = a^2 \quad (1)$$

Again, $y_1x + x_1y - x_1y_1 = 0$

Applying conditions of tangency

$$\frac{|-x_1y_1|}{\sqrt{x_1^2 + y_1^2}} = a \text{ or } (x_1^2 + y_1^2) = \frac{x_1^2\, y_1^2}{a^2} \quad (2)$$

From Eqs. (1) and (2), we get x_1, y_1, which gives equation of tangent as $x \pm y = \pm a\sqrt{2}$.

87. c. The equation of the line $y = x$ in distance form is $\frac{x}{\cos\theta} = \frac{y}{\sin\theta} = r$, where $\theta = \frac{\pi}{4}$.

For point P, $r = 6\sqrt{2}$. Therefore, coordinates of P are given by $\frac{x}{\cos\frac{\pi}{4}} = \frac{y}{\sin\frac{\pi}{4}} = 6\sqrt{2} \Rightarrow x = 6, y = 6$.

Since $P(6, 6)$ lies on $x^2 + y^2 + 2gx + 2fy + c = 0$,

$72 + 12(g + f) + c = 0$ (1)

Since $y = x$ touches the circle, the equation $2x^2 + 2x(g + f) + c = 0$ has equal roots.

$\Rightarrow 4(g + f)^2 = 8c$

$\Rightarrow (g + f)^2 = 2c$ (2)

From (1), we get

$[12(g + f)]^2 = [-(c + 72)]^2$

$\Rightarrow 144(g + f)^2 = (c + 72)^2$

$\Rightarrow 144(2c) = (c + 72)^2$

$\Rightarrow (c - 72)^2 = 0 \Rightarrow c = 72$

88. a.

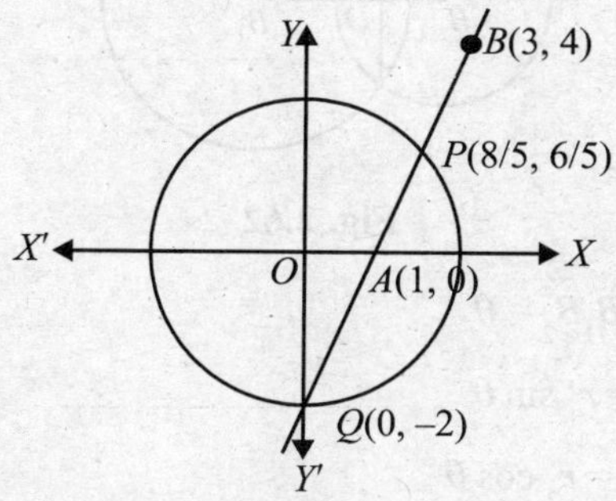

Fig. 2.59

The equation of the line joining $A(1, 0)$ and $B(3, 4)$ is $y = 2x - 2$.

This cuts the circle $x^2 + y^2 = 4$ at $Q(0, -2)$ and $P\left(\frac{8}{5}, \frac{6}{5}\right)$.

We have $BQ = 3\sqrt{5}$, $QA = \sqrt{5}$, $BP = \frac{7}{\sqrt{5}}$ and $PA = \frac{3}{\sqrt{5}}$

$\therefore \alpha = \frac{BP}{PA} = \frac{7/\sqrt{5}}{3/\sqrt{5}} = \frac{7}{3}$

and $\beta = \frac{BQ}{QA} = \frac{3\sqrt{5}}{-\sqrt{5}} = -3$

Hence, α, β are roots of the equation $x^2 - x(\alpha + \beta) + \alpha\beta = 0$

i.e., $x^2 - x\left(\frac{7}{3} - 3\right) + \frac{7}{3}(-3) = 0$

or $3x^2 + 2x - 21 = 0$

89. a.

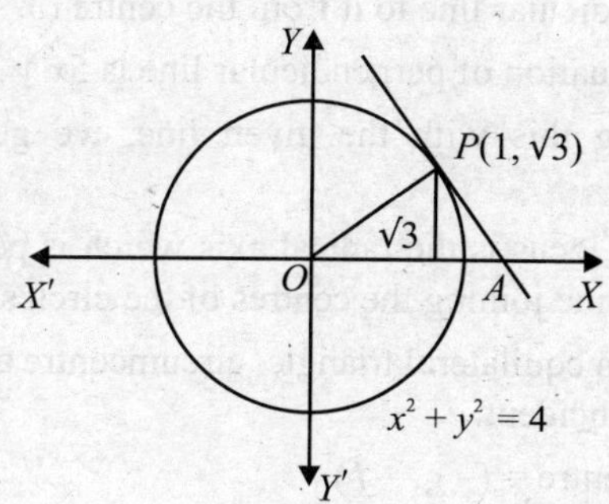

Fig. 2.60

The equations of the tangent and normal to $x^2 + y^2 = 4$ at $(1, \sqrt{3})$ are $x + \sqrt{3}y = 4$ and $y = \sqrt{3}x$

The tangent meets the x-axis at $(4, 0)$.

Therefore, area of $\Delta OAP = \frac{1}{2}(4)\sqrt{3} = 2\sqrt{3}$ sq. units

90. d. Any point on the line $7x + y + 3 = 0$ is $Q(t, -3 - 7t)$, $t \in R$

Now $P(h, k)$ is image of point Q in the line $x - y + 1 = 0$

Then, $\frac{h - t}{1} = \frac{k - (-3 - 7t)}{-1}$

$= -\frac{2(t - (-3 - 7t) + 1)}{1 + 1}$

$= -8t - 4$

$\Rightarrow (h, k) \equiv (-7t - 4, t + 1)$

This point lies on the circle $x^2 + y^2 = 9$. So

$(-7t - 4)^2 + (t + 1)^2 = 9$

$\Rightarrow 50t^2 + 58t + 8 = 0$

$\Rightarrow 25t^2 + 29t + 4 = 0$

$\Rightarrow (25t + 4)(t + 1) = 0$

$\Rightarrow t = -4/25, t = -1$

$\Rightarrow (h, k) \equiv \left(-\frac{72}{25}, \frac{21}{25}\right)$ or $(3, 0)$

91. c. Let the coordinates be $A(a, 0)$ and $B(-a, 0)$ and let the straight line be $y = mx + c$. Then,

$\frac{mx + c}{\sqrt{1 + m^2}} + \frac{-mx + c}{\sqrt{1 + m^2}} = 2k$

$\Rightarrow c = k\sqrt{1 + m^2}$

So the straight line is $y = mx + k\sqrt{1 + m^2}$.

Clearly, it touches the circle $x^2 + y^2 = k^2$ of radius k.

92. a. The midpoint is the intersection of the chord and perpendicular line to it from the centre (3, –1).

The equation of perpendicular line is $5x + 2y - 13 = 0$.

Solving this with the given line, we get the point (1, 4).

93. d. The locus is the radical axis which is perpendicular to the line joining the centres of the circles.

94. b. In an equilateral triangle, circumcentre and in centre are coincident.

$\therefore$ Incentre $= (-g, -f)$

(1, 1) lies on the circle

$\Rightarrow 1^2 + 1^2 + 2g + 2f + c = 0$

$\Rightarrow c = -2(g + f + 1)$

Also, in an equilateral triangle,

Circumradius = 2 × inradius

Therefore, inradius $= \frac{1}{2} \times \sqrt{g^2 + f^2 - c}$

Therefore, the equation of the incircle is

$$(x+g)^2 + (y+f)^2 = \frac{1}{4}(g^2 + f^2 - c)$$

$$= \frac{1}{4}(g^2 + f^2 + 2(g + f + 1))$$

95. a.

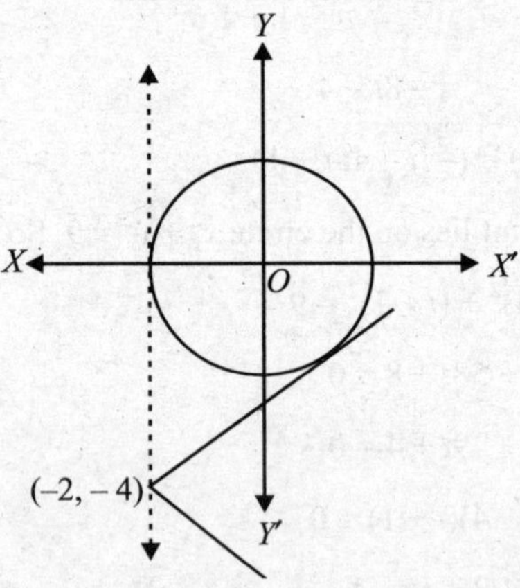

Fig. 2.61

Any tangent of $x^2 + y^2 = 4$ is $y = mx \pm 2\sqrt{1 + m^2}$ if it passes through (–2, –4) then $(2m - 4)^2 = 4(1 + m^2)$

$\Rightarrow 4m^2 + 16 - 16m = 4 + 4m^2$

$\Rightarrow m = \infty, m = \frac{3}{4}$.

Hence, slope of reflected ray is $\frac{3}{4}$.

Thus, equation of incident ray is $(y + 4) = -\frac{3}{4}(x + 2)$, i.e., $4y + 3x + 22 = 0$.

96. a. Any point on the line $x + y = 25$ is $P \equiv (a, 25 - a)$, $a \in R$

Equation of chord AB is $T = 0$,

i.e., $xa + y(25 - a) = 9$ (1)

If midpoint of chord AB is $C(h, k)$, then equation of chord AB is

$T = S_1$, i.e., $xh + yk = h^2 + k^2$ (2)

Comparing the ratio of coefficients of Eqs. (1) and (2), we get

$$\frac{a}{h} = \frac{25 - a}{k} = \frac{9}{h^2 + k^2}$$

$$\Rightarrow \frac{a + 25 - a}{h + k} = \frac{9}{h^2 + k^2}$$

Thus, locus of C is $25(x^2 + y^2) = 9(x + y)$.

97. a.

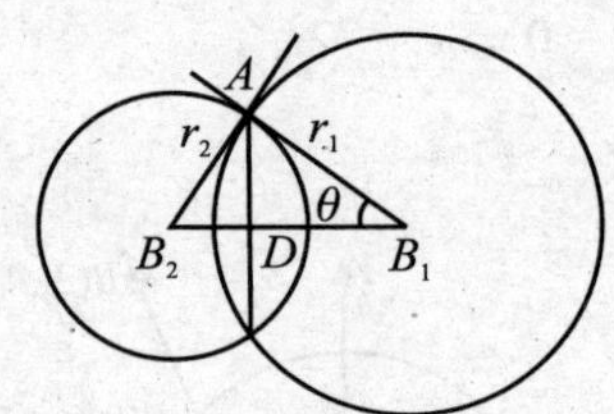

Fig. 2.62

Let $\angle AB_1B_2 = \theta$

$\Rightarrow AD = r_1 \sin\theta$

and $AD = r_2 \cos\theta$

$$\Rightarrow AD^2\left(\frac{1}{r_1^2} + \frac{1}{r_2^2}\right) = 1$$

$$\Rightarrow AD = \frac{r_1 r_2}{\sqrt{r_1^2 + r_2^2}}$$

Thus, length of common chord is $\frac{2r_1 r_2}{\sqrt{r_1^2 + r_2^2}}$.

98. b.

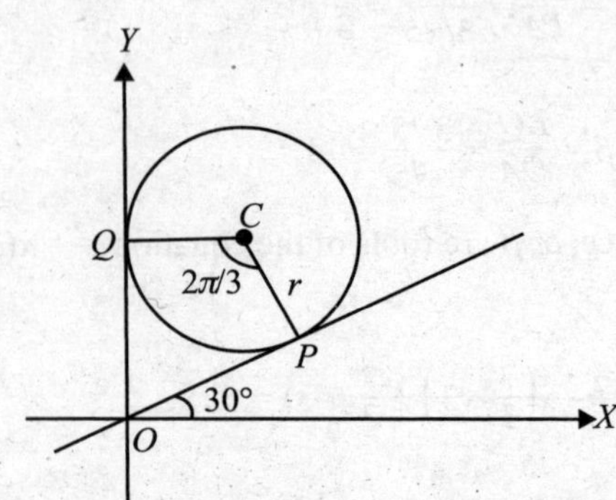

Fig. 2.63

$\tan 30^\circ = \frac{r}{OP}$

$\Rightarrow OP = r\sqrt{3}$

Also $r^2 \frac{1}{2}\frac{2\pi}{3} = 3\pi$

$\therefore r = 3$

$\therefore OP = 3\sqrt{3}$

99. c.

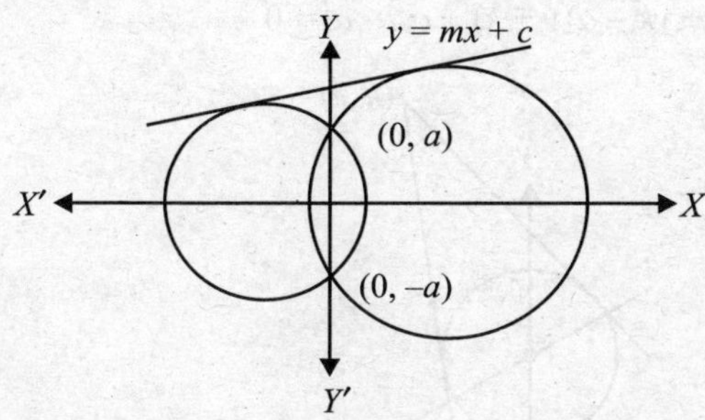

Fig. 2.64

Equation of family circles through (0, a) and (0, $-a$) is

$[x^2 + (y - a)(y + a)] + \lambda x = 0, \lambda \in R$

$\Rightarrow x^2 + y^2 + \lambda x - a^2 = 0$

and $\sqrt{\left(\frac{\lambda}{2}\right)^2 + a^2} = \frac{-\frac{m\lambda}{2} + c}{\sqrt{1 + m^2}}$

$\Rightarrow (1 + m^2)\left[\frac{\lambda^2}{4} + a^2\right] = \left(\frac{m\lambda}{2} - c\right)^2$

$\Rightarrow (1 + m^2)\left[\frac{\lambda^2}{4} + a^2\right] = \frac{m^2\lambda^2}{4} - mc\lambda + c^2$

$\Rightarrow \lambda^2 + 4mc\lambda + 4a^2(1 + m^2) - 4c^2 = 0$

$\therefore \lambda_1\lambda_2 = 4[a^2(1 + m^2) - c^2]$

$\Rightarrow g_1g_2 = [a^2(1 + m^2) - c^2]$

and $g_1g_2 + f_1f_2 = \frac{c_1 + c_2}{2}$

$\Rightarrow a^2(1 + m^2) - c^2 = -a^2$

Hence, $c^2 = a^2(2 + m^2)$

100. a. Distance of given line from the centre of the circle is $|p|$. Now the line subtends a right angle at the centre.

Hence, radius $= \sqrt{2}\,|p|$

$\Rightarrow a = \sqrt{2}\,|p|$

$\Rightarrow a^2 = 2p^2$

101. c. Let the midpoint of the chord be $P(h, k)$.

Then $CP = \sqrt{h^2 + k^2}$, where C is the centre of the circle.

Since chord subtends right angle at the centre.

Radius $= \sqrt{2}\sqrt{h^2 + k^2}$

$\Rightarrow 2 = \sqrt{2}\sqrt{h^2 + k^2}$

$\Rightarrow$ locus of P is $x^2 + y^2 = 2$

102. a. Let $P(x_1, y_1)$ and $Q(x_2, y_2)$ be the given points and $x^2 + y^2 = a^2$ be the circle.

The chord of contact of tangents drawn from $P(x_1, y_1)$ to $x^2 + y^2 = a^2$ is $xx_1 + yy_1 = a^2$

If it passes through $Q(x_2, y_2)$, then

$$x_1x_2 + y_1y_2 = a^2 \quad (1)$$

The equation of the circle on PQ as diameter is

$(x - x_1)(x - x_2) + (y - y_1)(y - y_2) = 0$

$\Rightarrow x^2 + y^2 - x(x_1 + x_2) - y(y_1 + y_2) + x_1x_2 + y_1y_2 = 0$

This circle will cut the given circle orthogonally, if

$0(x_1 + x_2) + 0(y_1 + y_2) = -a^2 + x_1x_2 + y_1y_2$

$\Rightarrow x_1x_2 + y_1y_2 - a^2 = 0$, which is true by Eq. (1).

103. a.

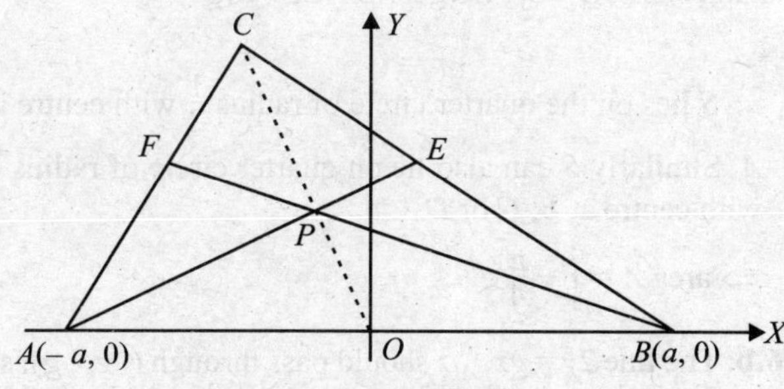

Fig. 2.65

Let A and B be $(-a, 0)$ and $(a, 0)$. Also let P be (h, k).

Then by geometry, we know $\frac{CP}{PO} = \frac{CF}{FA} + \frac{CE}{EB}$

$\therefore \frac{CP}{PO} = 1$

If $C(\alpha, \beta)$ lies on $x^2 + y^2 + 2gx + 2fy + c = 0$, then $\alpha = 2h$ and $\beta = 2k$

$\Rightarrow 4(h^2 + k^2 + gh + fk) + c = 0$

Therefore, Locus of $P(h, k)$ is $x^2 + y^2 + gx + fy + \frac{c}{4} = 0$,

which is a circle of radius $= \sqrt{\left(\frac{g}{2}\right)^2 + \left(\frac{f}{2}\right)^2 - \frac{c}{4}}$

$= \frac{1}{2}\sqrt{g^2 + f^2 - c}$

$= \frac{r}{2}$

104. b. The slope of the chord is $m = -\frac{8}{y}$

$\Rightarrow y = \pm 1, \pm 2, \pm 4, \pm 8$

But $(8, y)$ must also lie inside the circle $x^2 + y^2 = 125$

$\Rightarrow y$ can be equal to $\pm 1, \pm 2, \pm 4 \Rightarrow 6$ values.

105. b.

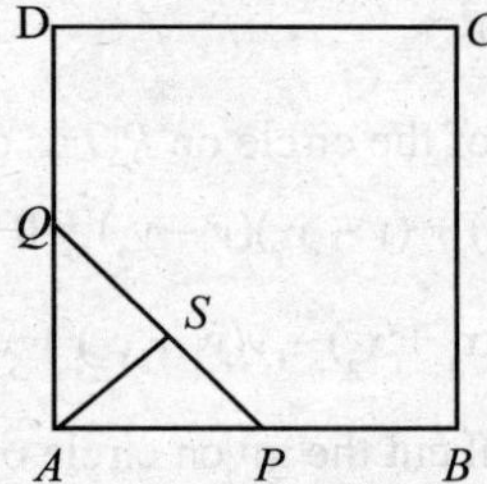

Fig. 2.66

Let S be the midpoint of PQ.

Since $\angle PAQ = \frac{\pi}{2}$, we get $AS = SP = SQ = \frac{1}{2}$

$\Rightarrow S$ lies on the quarter circle of radius $\frac{1}{2}$ with centre at A. Similarly S can also lie on quarter circle of radius $\frac{1}{2}$ with centre at B, C or D.

$\Rightarrow$ area $A = 1 - \frac{\pi}{4}$

106. b. The line $2y = gx + \alpha$ should pass through $(-g, -g)$, so $-2g = -g^2 + \alpha \Rightarrow \alpha = g^2 - 2g = (g-1)^2 - 1 \geq -1$.

107. c. Let $\sum_{i=1}^{6} x_i = \alpha$ and $\sum_{i=1}^{6} y_i = \beta$

Let O be the orthocentre of the triangle made by (x_1, y_1), (x_2, y_2) and (x_3, y_3)

$\Rightarrow O$ is $(x_1 + x_2 + x_3, y_1 + y_2 + y_3) \equiv (\alpha_1, \beta_1)$

Similarly let G be the centroid of the triangle made by other three points

$\Rightarrow \quad G$ is $\left(\frac{x_4 + x_5 + x_6}{3}, \frac{y_4 + y_5 + y_6}{3}\right)$

$\Rightarrow \quad G$ is $\left(\frac{\alpha - \alpha_1}{3}, \frac{\beta - \beta_1}{3}\right)$

The point dividing OG in the ratio 3 : 1 is $\left(\frac{\alpha}{4}, \frac{\beta}{4}\right)$ $\equiv (2, 1)$

$\Rightarrow h + k = 3$

108. a. Let the centre be $(0, \alpha)$. Equation of the circle is $x^2 + (y - \alpha)^2 = |a|^2$

Equation of chord of contact for $P(h, k)$ is $xh + yk - \alpha(y + k) + \alpha^2 - a^2 = 0$

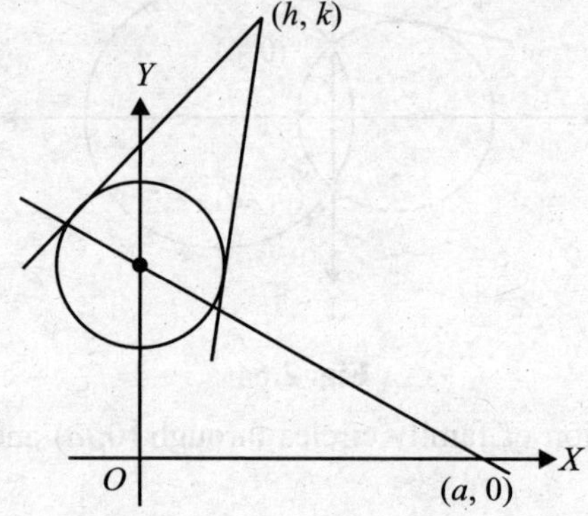

Fig. 2.67

It passes through $(a, 0)$

$\Rightarrow \alpha^2 - \alpha k + ah - a^2 = 0$

As α is real

$\Rightarrow k^2 - 4(ah - a^2) \geq 0$

109. b. Let A and B be the centres and r_1 and r_2 the radii of the two circles, then

$A \equiv \left(-\frac{1}{2}, -\frac{1}{2}\right), B \equiv \left(-\frac{1}{2}, \frac{1}{2}\right)$,

$r_1 = \frac{1}{\sqrt{2}}, r_2 = \frac{1}{\sqrt{2}}$

$\cos\theta = \frac{r_1^2 + r_2^2 - AB^2}{2r_1 r_2}$

$= \frac{\frac{1}{2} + \frac{1}{2} - 1}{2 \times \frac{1}{\sqrt{2}} \times \frac{1}{\sqrt{2}}} = 0$

$\therefore \theta = \frac{\pi}{2}$

Required line is parallel to the x-axis and since it passes through $(1, 2)$, its equation will be $y = 2$.

110. a.

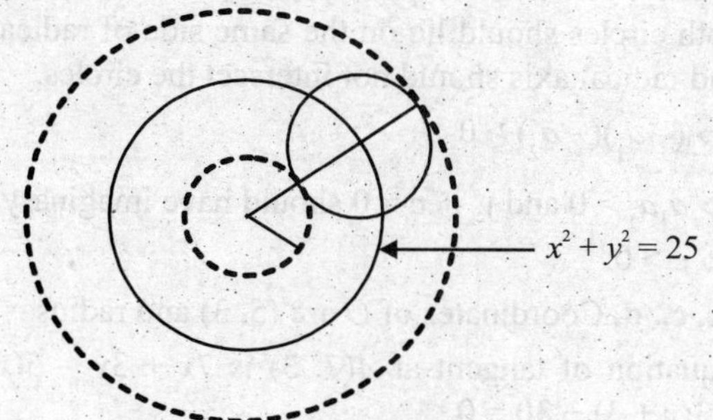

Fig. 2.68

Let (h, k) be any point in the set, then equation of circle is

$(x - h)^2 + (y - k)^2 = 9$

But (h, k) lies on $x^2 + y^2 = 25$,

then $h^2 + k^2 = 25$

$\therefore$ $2 \le$ Distance between the centres of two circles ≤ 8

$\Rightarrow 4 \le h^2 + k^2 \le 64$

Therefore, locus of (h, k) is $4 \le x^2 + y^2 \le 64$.

111. b. $OR = \dfrac{2 \text{ area of } \Delta OPQ}{PQ}$

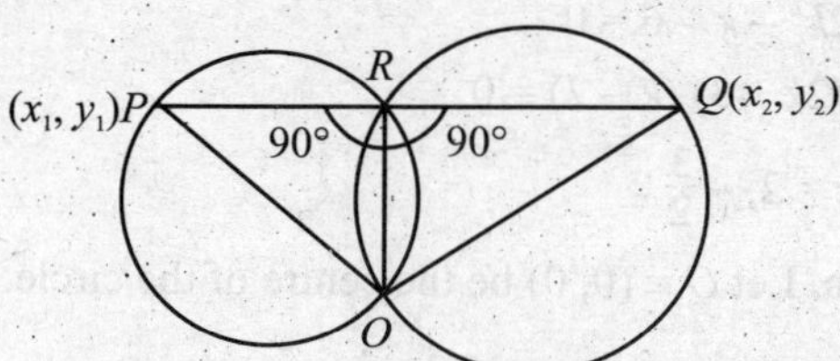

Fig. 2.69

$$= \frac{2 \times \left|\frac{1}{2}(x_1 y_2 - x_2 y_1)\right|}{PQ}$$

$$= \frac{|x_1 y_2 - x_2 y_1|}{PQ}$$

112. c. Substituting $y = mx$ in the equation of circle, we get $x^2 + m^2x^2 + ax + bmx + c = 0$ (y/x denotes the slope of the tangent from the origin on the circle)

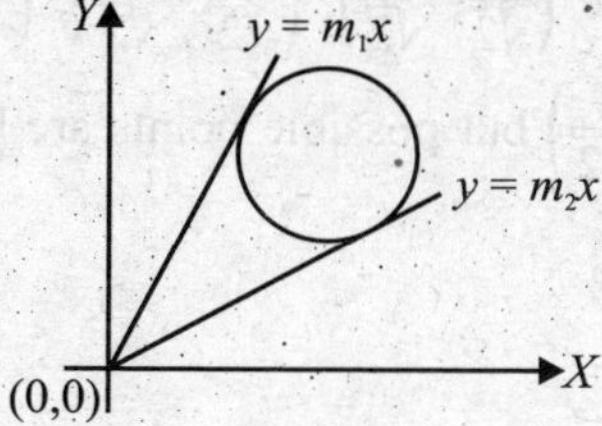

Fig. 2.70

Since the line touches the circle, we must have discriminant

$\Rightarrow \quad (a + bm)^2 - 4c(1 + m^2) = 0$

$\Rightarrow a^2 + b^2m^2 + 2abm - 4c - 4cm^2 = 0$

$\Rightarrow \quad m^2 (b^2 - 4c) + 2abm + a^2 - 4c = 0$

This equation has two roots m_1 and m_2,

$\Rightarrow m_1 + m_2 = -\dfrac{2ab}{b^2 - 4c} = \dfrac{2ab}{4c - b^2}$

113. c. Equation of radical axis (i.e. common chord) of the two circles is

$10x + 4y - a - b = 0$ (1)

Centre of first circle is $H(-4, -4)$.

Since second circle bisects the circumference of the first circle, centre $H(-4, -4)$ of the first circle must lie on the common chord Eq. (1).

$\therefore -40 - 16 - a - b = 0$

$\Rightarrow a + b = -56$

114. d. Let the equation of circle be

$x^2 + y^2 - 4 + k(2x + y - 1) = 0,$

where k is a real number.

Radius $= \sqrt{\dfrac{5k^2}{4} + 4 + k}$

Radius is minimum when $k = -\dfrac{2}{5}$

The required equation is

$5x^2 + 5y^2 - 4x - 2y - 18 = 0$

115. b.

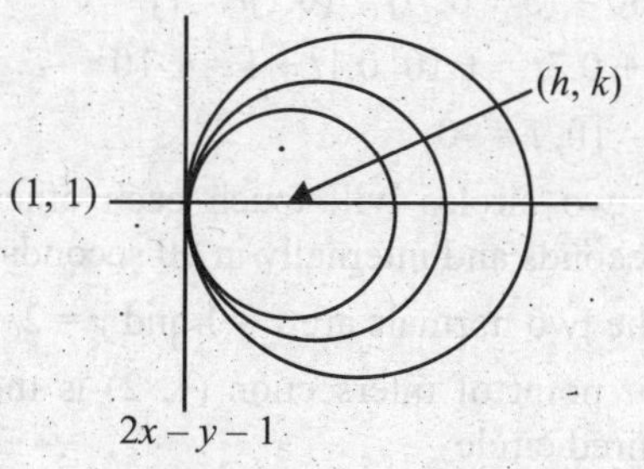

Fig. 2.71

Obviously, locus of centre is line perpendicular to the given line.

Hence, locus is $\dfrac{k-1}{h-1} = -\dfrac{1}{2}$ or $x + 2y = 0$.

116. a. Centre of the circle $x^2 + y^2 = 2x$ is $(1, 0)$.

Common chord of the other two circles is

$8x - 15y + 26 = 0.$

Distance of $(1, 0)$ from $8x - 15y + 26 = 0$ is

$$\frac{|8 + 26|}{\sqrt{15^2 + 8^2}} = 2$$

117. d. Equation of any circle though the points of intersection of given circles is

$$x^2 + y^2 - 4x - 2y - 8 + k(x^2 + y^2 - 2x - 4y - 8) = 0 \quad (1)$$

Since circle Eq. (1) passes through (–1, 4)

$\therefore k = 1$

Required circle is

$x^2 + y^2 - 3x - 3y - 8 = 0$

118. b. Given circles are

$(x-1)^2 + (y-2)^2 = 1 \quad (1)$

and $(x-7)^2 + (y-10)^2 = 4 \quad (2)$

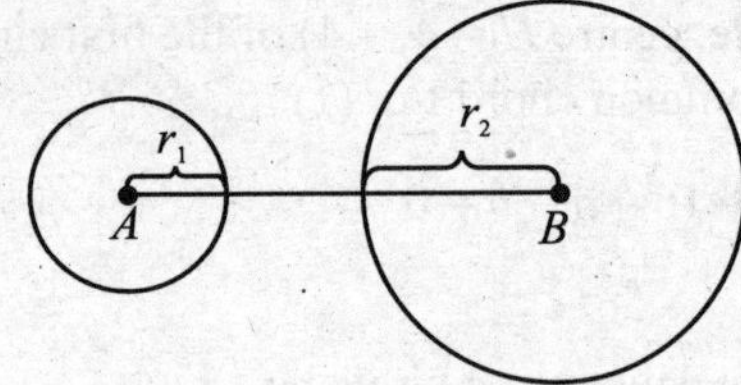

Fig. 2.72

Let $A \equiv (1, 2)$, $B \equiv (7, 10)$, $r_1 = 1$, $r_2 = 2$

$AB \equiv 10$, $r_1 + r_2 = 3$

$AB > r_1 + r_2$, hence the two circles are separated.

Radius of the two circles at time t are $(1 + 0.3t)$ and $(2 + 0.4t)$

For the two circle to touch each other

$AB^2 = [(r_1 + 0.3t) \pm (r_2 + 0.4t)]^2$

or $100 = [(1 + 0.3t) \pm (2 + 0.4t)]^2$

or $100 = (3 + 0.7t)^2, [(0.1)t + 1]^2$

or $3 + 0.7t = \pm 10$, $0.1t + 1 = \pm 10$

$\therefore t = 10, t = 90 \quad [\because t > 0]$

The two circles will touch each other externally in 10 seconds and internally in 90 seconds.

119. a. The two normals are $x = 1$ and $y = 2$

Their point of intersection (1, 2) is the centre of the required circle

Radius $= \dfrac{|3 + 8 - 6|}{5} = 1$

Required circle is

$(x-1)^2 + (y-2)^2 = 1$

i.e., $x^2 + y^2 - 2x - 4y + 4 = 0$

Multiple Correct Answers Type

1. a., c. Equation of radical axis of the given circle is $x = 0$.

If one circle lies completely inside the other, centre of both circles should lie on the same side of radical axis and radical axis should not intersect the circles.

$\Rightarrow (-a_1)(-a_2) > 0$

$\Rightarrow a_1 a_2 > 0$ and $y^2 + c = 0$ should have imaginary roots

$\Rightarrow c > 0$.

2. a., c., d. Coordinates of O are (5, 3) and radius = 2

Equation of tangent at $A(7, 3)$ is $7x + 3y - 5(x + 7) - 3(y + 3) + 30 = 0$

i.e., $2x - 14 = 0$, i.e., $x = 7$

Equation of tangent at $B(5, 1)$ is $5x + y - 5(x + 5) - 3(y + 1) + 30 = 0$, i.e., $-2y + 2 = 0$, i.e. $y = 1$

Therefore, Coordinates of C are (7, 1)

Therefore, Area of $OACB$ is 4.

Equation of AB is $x - y = 4$ (radical axis).

Equation of the smallest circles is

$(x-7)(x-5) + (y-3)(y-1) = 0$

i.e., $x^2 + y^2 - 12x - 4y + 38 = 0$

3. a., c. $2gg' + 2ff' = c + c'$

$\Rightarrow 2 \times 1 \times 0 + 2 \times k \times k = 6 + k$

$\Rightarrow 2k^2 - k - 6 = 0$

$\Rightarrow (2k + 3)(k - 2) = 0$

$\therefore k = 2, -\dfrac{3}{2}$

4. a., b. Let $O \equiv (0, 0)$ be the centre of the circle.

$\because$ Arc length $AB = \dfrac{\pi}{2} = \dfrac{1}{4}$ (circumference of the circle)

$\therefore \angle AOB = \dfrac{\pi}{2}$

Slope of $OB = -\dfrac{1}{\text{slope of } OA}$

$\Rightarrow$ slope of $OB = -\dfrac{1}{1} = -1 \quad (1)$

Let $B \equiv \left(\alpha, \pm\sqrt{1-\alpha^2}\right)$

$\therefore \pm\dfrac{\sqrt{1-\alpha^2}}{\alpha} = -1 \quad$ [From Eq. (1)]

$\therefore B$ can be $\left(\dfrac{1}{\sqrt{2}}, \dfrac{1}{\sqrt{2}}\right), \left(\dfrac{1}{\sqrt{2}}, \dfrac{-1}{\sqrt{2}}\right), \left(-\dfrac{1}{\sqrt{2}}, \dfrac{1}{\sqrt{2}}\right)$, and $\left(-\dfrac{1}{\sqrt{2}}, -\dfrac{1}{\sqrt{2}}\right)$ but possible points are $\left(\dfrac{1}{\sqrt{2}}, -\dfrac{1}{\sqrt{2}}\right)$ and $\left(-\dfrac{1}{\sqrt{2}}, \dfrac{1}{\sqrt{2}}\right)$.

5. a., c., d.

$$x^2 + y^2 + 8x - 10y - 40 = 0$$

Centre of the circle is (– 4, 5). Its radius is 9.

Distance of the centre (– 4, 5) from the point (– 2, 3) is $\sqrt{4+4} = 2\sqrt{2}$

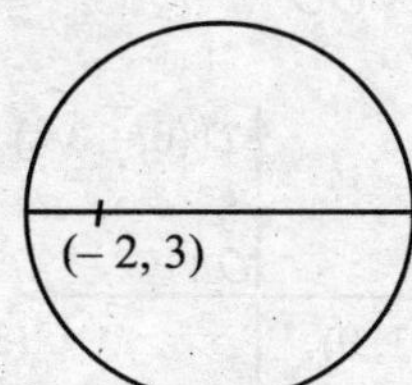

Fig. 2.73

$\therefore a = 2\sqrt{2} + 9$ and $b = -2\sqrt{2} + 9$

$\therefore a + b = 18$

$a - b = 4\sqrt{2}$

$ab = 81 - 8 = 73$

6. a., b. $L_1L_2 + \lambda L_2L_3 + \mu L_3L_1 = 0$

$\Rightarrow (y - m_1x)(y - m_2x) + \lambda(y - m_2x)(y - m_3x)$
$+ \mu(y - m_3x)(y - m_1x) = 0$ (1)

Clearly, Eq. (1) represents a curve passing through points of intersection of lines L_1, L_2 and L_3.

Equation (1) will represent a circle if coefficient of x^2 = coefficient of y^2 and coefficient of $xy = 0$

$\therefore 1 + \lambda + \mu = m_1m_2 + \lambda m_2m_3 + \mu m_1m_3$

and $m_1(1 + \mu) + m_2(1 + \lambda) + m_3(\mu + \lambda) = 0$

7. a., b., c., d. Given circle is

$x^2 + y^2 + 2hxy + 2gx + 2fy + c = 0$ (1)

For Eq. (1) to represent a circle, $h = 0$

Therefore, given circle is

$x^2 + y^2 + 2gx + 2fy + c = 0$ (2)

Fig. 2.74

For circle Eq. (2) to pass through three quadrants only

i. $AB > 0 \therefore g^2 - c > 0$

ii. $CD > 0 \therefore g^2 - c > 0$

iii. Origin should lie outside circle Eq. (2)

$\therefore c > 0$

Therefore, required conditions are $g^2 > c, f^2 > c, c > 0$, $h = 0$

8. a., c. The point from which the tangents drawn are at right angle lie on the director circle.

Equation of director circle is $x^2 + y^2 = 2 \times 16 = 32$

Putting $x = 2$, we get

$y^2 = 28$

$\Rightarrow y = \pm 2\sqrt{7}$

The points can be $(2, 2\sqrt{7})$ or $(2, -2\sqrt{7})$.

9. b., d. Line pair is $(x - 1)^2 - y^2 = 0$, i.e., $x + y - 1 = 0$, $x - y - 1 = 0$.

Let the centre be $(\alpha, 0)$, then its distance from $x + y - 1 = 0$ is

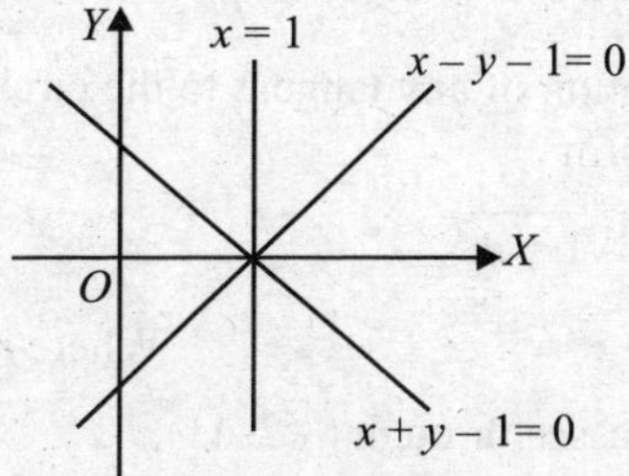

Fig. 2.75

$$\left|\frac{\alpha - 1}{\sqrt{2}}\right| = 2\text{(radius)}$$

i.e., $\alpha = 1 \pm 2\sqrt{2}$

Centre may be $(1 + 2\sqrt{2}, 0)$, $(1 - 2\sqrt{2}, 0)$.

Now let the centre be $(1, \beta)$, then

$$\left|\frac{1 + \beta - 1}{\sqrt{2}}\right| = 2$$

$\Rightarrow \beta = \pm 2\sqrt{2}$

Centre may be $(1, 2\sqrt{2})$, $(1, -2\sqrt{2})$.

10. a., d. Equation of the radical axis is

$2\alpha x + 2y + 10 = 0$

i.e., $\alpha x + y + 5 = 0$ (1)

Putting the value of y from Eq. (1) in the circle $x^2 + y^2 = 9$, we get

$(1 + \alpha^2)x^2 + 10\alpha x + 16 = 0$

Since, radical axis is a tangent,

$D = 0$

$\Rightarrow 36\alpha^2 - 64 = 0$

$\Rightarrow \alpha = \pm \frac{4}{3}$

11. b., c.

$x^2 + y^2 - 8x - 16y + 60 = 0$ (1)

Equation of chord of contact from (– 2, 0) is $-2x - 4(x - 2) - 8y + 60 = 0$

$\Rightarrow 3x + 4y - 34 = 0$ (2)

Solving Eqs. (1) and (2),

$$x^2 + \left(\frac{34 - 3x}{4}\right)^2 - 8x - 16\left(\frac{34 - 3x}{4}\right) + 60 = 0$$

$\Rightarrow 16x^2 + 1156 - 204x + 9x^2 - 128x - 2176 + 192x + 960 = 0$

$\Rightarrow 5x^2 - 28x - 12 = 0$

$\Rightarrow (x-6)(5x+2) = 0$

$\Rightarrow x = 6, -\frac{2}{5}$

$\Rightarrow$ Points are $(6, 4), \left(-\frac{2}{5}, \frac{44}{5}\right)$.

12. a., c. Equation of any tangent to the circle $x^2 + y^2 = 25$ is of the form

$y = mx + 5\sqrt{1+m^2}$,

[Where m is the slope]

Since, it passes through $(-2, 11)$,

$11 = -2m + 5\sqrt{1+m^2}$

$\Rightarrow (11+2m)^2 = 25(1+m^2)$

$\Rightarrow m = \frac{24}{7}, -\frac{4}{3}$

Therefore, equation of the tangents are

$24x - 7y + 125 = 0$

or $4x + 3y = 25$

13. a., d. Area of the quadrilateral is $\sqrt{c} \times \sqrt{9+25-c} = 15$

$\therefore c = 9, 25$

14. a., d.

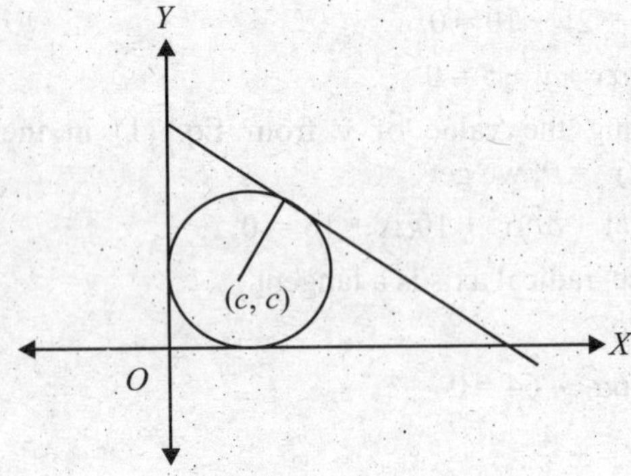

Fig. 2.76

We must have $\left|\frac{\frac{c}{3}+\frac{c}{4}-1}{\sqrt{\frac{1}{3^2}+\frac{1}{4^2}}}\right| = c$

$\Rightarrow c = 6, 1$

15. a., c. Since the given circle $(x-3)^2 + (y-3)^2 = 9$ is touches both axis, tangents from the origin are x-axis and y-axis or $y = 0$ and $x = 0$.

16. b. Since A, B, C, D are concyclic,

$OA \times OC = OB \times OD$

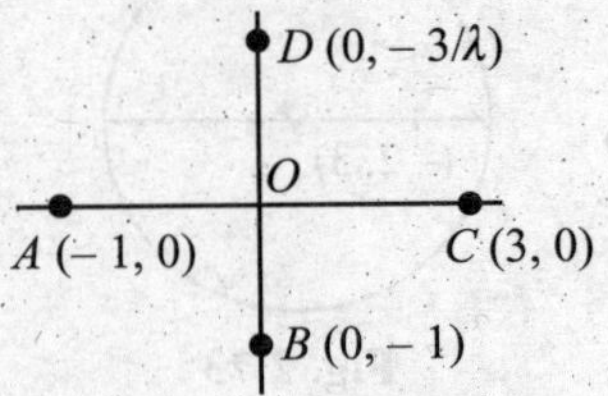

Fig. 2.77

$\Rightarrow 1 \times 3 = 1 \times \left|\frac{3}{\lambda}\right|$

$\Rightarrow \lambda = \pm 1$

But when $\lambda = -1$, B and D will not lie on the circle simultaneously.

Hence, $\lambda = 1$

17. a., b., c., d. Chords equidistance from the centre are equal.

18. b., d. Let the equation of the tangent be

$x - 2y = k$ (1)

Since, line Eq. (1) touches the circle,

distance from centre to line (1) is

radius of the circle

$\therefore \frac{|2-2-k|}{\sqrt{5}} = \sqrt{4+1+15}$

$|k| = 10 \Rightarrow k = \pm 10$

The tangents can be $x - 2y \pm 10 = 0$.

19. b., c. For given circle $S_1 : x^2 + y^2 - 2x - 4y + 1 = 0$ and $S_2 : x^2 + y^2 + 4x + 4y - 1 = 0$

$C_1(1, 2), r_1 = 2$ and $C_2(-2, -2), r_2 = 3$

Now $r_1 + r_2 = 5$ and $C_1C_2 = 5$.

Hence, circles touch externally. Also common tangent at point of contact is $S_1 - S_2 = 0$ or $3x + 4y - 1 = 0$.

20. a., b., c., d.

$r_1 = 5, r_2 = \sqrt{15}, C_1C_2 = \sqrt{40}$

$\Rightarrow r_1 + r_2 > C_1C_2 > r_1 - r_2$

Hence, circles intersect in two distinct points.

There are two common tangents.

Also $2g_1g_2 + 2f_1f_2 = 2(1)(3) + 2(2)(-4) = -10$

and $c_1 + c_2 = -20 + 10 = -10$

Thus, two circle are orthogonal.

Length of common chord is $\frac{2r_1r_2}{\sqrt{r_1^2+r_2^2}} = 5\sqrt{\frac{3}{2}}$

Length of common tangent is

$\sqrt{C_1C_2^2-(r_1-r_2)^2} = 5\left(\frac{12}{5}\right)^{\frac{1}{4}}$.

21. b., d.

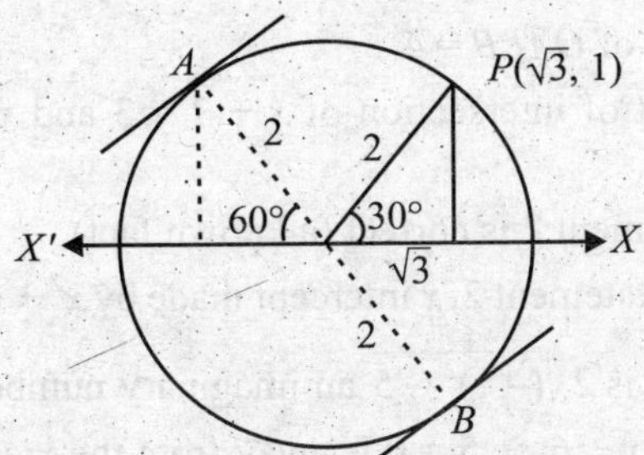

Fig. 2.78

Clearly, $A = (-2\cos 60°, 2\sin 60°)$ and $B = (2\cos 60°, -2\sin 60°)$

The tangent at A is $x(-2\cos 60°) + y(2\sin 60°) = 4$ and the tangent at B is $x(2\cos 60°) + y(-2\sin 60°) = 4$.

22. b., c.

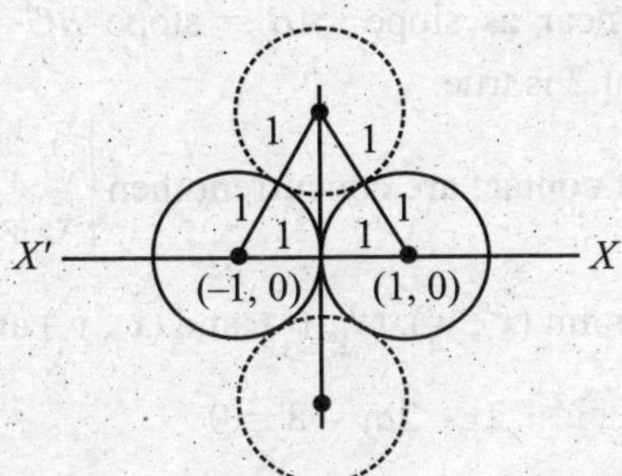

Fig. 2.79

The given circles are $x^2 + y^2 - 2x = 0$, $x > 0$, and $x^2 + y^2 + 2x = 0$, $x < 0$

From the above figure, the centres of the required circles will be $(0, \sqrt{3})$ and $(0, -\sqrt{3})$

The equations of the circles are

$(x-0)^2 + (y \mp \sqrt{3})^2 = 1^2$

23. a., d. When two circle touch each other externally, then

$r_1 + r_2 = \sqrt{\{0-(-a)\}^2 + \{0-(-1)\}^2}$

$\Rightarrow 3 + a = \sqrt{a^2+1}$

$\Rightarrow a = -\frac{4}{3}$

When two circle touch each other internally,

$|r_1 - r_2| = \sqrt{\{0-(-a)\}^2 + \{0-(-1)\}^2}$

$\Rightarrow |3-a| = \sqrt{a^2+1}$

$\Rightarrow a = \frac{4}{3}$

24. b., c. Equation of pair of tangent by $SS' = T^2$ is

$(ax+0-1)^2 = (x^2+y^2-1)(a^2+0-1)$

or $(a^2-1)y^2 - x^2 + 2ax - a^2 = 0$

If θ be the angle between the tangents, then

$\tan\theta = \frac{2\sqrt{H^2-AB}}{A+B}$

$= \frac{2\sqrt{-(a^2-1)(-1)}}{a^2-2}$

$= \frac{2\sqrt{a^2-1}}{a^2-2}$

If θ lies in second quadrant, then $\tan\theta < 0$

$\Rightarrow \frac{2\sqrt{a^2-1}}{a^2-2} < 0$

$\Rightarrow a^2 - 1 > 0$

and $a^2 - 2 < 0$

$\Rightarrow |a| > 1$ and $|a| < 2$

$\Rightarrow a \in (-\sqrt{2}, -1) \cup (1, \sqrt{2})$

25. b,. c.

Distance of line $x + y - 1 = 0$ from the centre $\left(\frac{1}{2}, -\frac{3}{2}\right)$

is $\frac{\left|\frac{1}{2}-\frac{3}{2}-1\right|}{\sqrt{2}} = \sqrt{2}$.

Now distance of line in options (b) and (c) from the centre is also $\sqrt{2}$.

Hence, given lines are $x - y = 0$ and $x + 7y = 0$.

26. c., d.

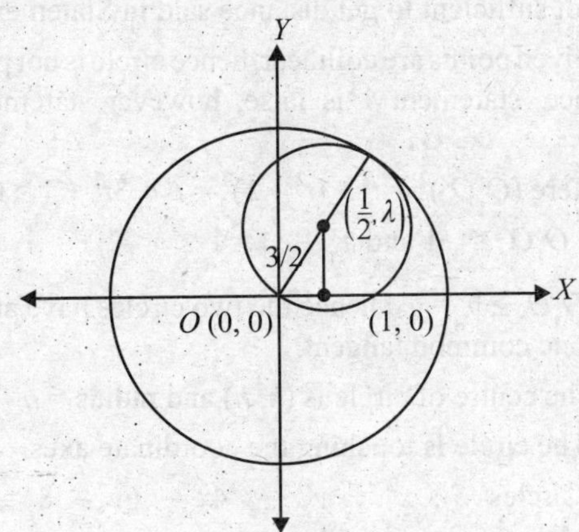

Fig. 2.80

From the diagram,

$$\sqrt{\left(\frac{1}{2}\right)^2 + \lambda^2} = \frac{3}{2} \Rightarrow 1 = \pm\sqrt{2}$$

Hence, centres of the circle are $\left(\frac{1}{2}, \pm\sqrt{2}\right)$.

Reasoning Type

1. **d.** Statement 2 is true as the centre is equidistant from A and B, hence lies on the perpendicular bisector of AB.

   Statement 1 is false as the distance between the given points is 10 and hence any circle through A and B has radius more than or equal to 5, and hence there is no circle of radius 4 through A and B is possible.

2. **a.** We know that chords of contact of given circle generated by any point on given line passes through the fixed point, as they form family of straight lines, hence both the statements are true and statement 2 is the correct explanation of statement 1.

3. **b.** For circle $x^2 + y^2 = 144$, centre $C_1(0, 0)$ and radius $r_1 = 12$.

   For circle $x^2 + y^2 - 6x - 8y = 0$, centre $C_2 = (3, 4)$ and radius $r_2 = 5$.

   Now $C_1C_2 = 5$ and $r_1 - r_2 = 7$, thus $C_1C_2 < r_1 - r_2$, hence one circle is completely lying inside other without touching it, hence there is no common tangent. Therefore, statement 1 is true. Therefore, both the statements are true but statement 2 is not correct explanation of statement 1.

4. **b.** Centre of the circle $C(2, 1)$ and radius $r = 5$.

   Distance of $P(10, 7)$ from $C(2, 1)$ is 10 units, hence required distances are 5, 15, respectively. Therefore, Statement 1 is true. Statement 2 is true but not the correct explanation of statement 1, as the information is not sufficient to get distance said in Statement 1.

5. **d.** Given points are collinear, hence circle is not possible. Hence, statement 1 is false, however statement 2 is true.

6. **a.** Here $(O_1O_2)^2 = t^2 + (t^2 + 1)^2 = t^4 + 3t^2 + 1 \geq 0$

   $\Rightarrow O_1O_2 \geq 1$ and $|r_1 - r_2| = 1$

   $\Rightarrow O_1O_2 \geq |r_1 - r_2|$ hence the two circles have at least one common tangent.

7. **a.** The centre of circle is (h, h) and radius $= h$

   $\Rightarrow$ The circle is touching the coordinate axes.

8. **b.** Circles S_1: $x^2 + y^2 - 4x - 6y - 8 = 0$ and S_2: $x^2 + y^2 - 2x - 3 = 0$

   $C_1(2, 3), r_1 = \sqrt{21}, C_2(1, 0), r_2 = 2$

   $\therefore C_1C_2 = \sqrt{10}, r_1 + r_2 = 2 + \sqrt{21}, r_2 - r_1 = \sqrt{21} - 2$

   Here $r_2 - r_1 < C_1C_2 < r_1 + r_2$. Hence, two circles intersect at two distinct points. Statement 2 is true, but does not explain statement 1.

9. **a.** Clearly $(\sqrt{2}, \sqrt{6})$ lies on $x^2 + y^2 = 8$, which is the director circle of $x^2 + y^2 = 4$.

   $\Rightarrow$ Tangents PA and PB are perpendicular to each other.

   $\therefore$ $(OAPB)$ is a square.

   $\therefore$ Area of $OAPB = 4$.

10. **d.** Point of intersection of $x + 7 = 3$ and $x - y = 1$ is $(2, 1)$.

11. **d.** Statement 2 is correct (a known fact).

    Using statement 2, x intercept made by $x^2 + y^2 - 2x + 6y + 5 = 0$ is $2\sqrt{(-1)^2 - 5}$ an imaginary number. Thus, $x^2 + y^2 - 2x + 6y + 5 = 0$ is away from the x-axis. Hence, statement 1 is false.

12. **d.** Statement 2 is true as in any triangle in circle and three excircles touches the three sides of the triangle. But Statement 1 is false as given lines are concurrent, hence triangle is not formed.

13. **a.** Given points are $A(1, 1)$, $B(2, 3)$ and $C(3, 5)$ which are collinear as slope AB = slope BC = 2. Hence, statement 2 is true.

    Chord of contact are concurrent then $\begin{vmatrix} x_1 & y_1 & 1 \\ x_2 & y_2 & 1 \\ x_3 & y_3 & 1 \end{vmatrix} = 0$

    Hence, point (x_1, y_1), (x_1, y_1), and (x_3, y_3) are collinear.

14. **a.** $x^2 + y^2 - 2x - 2ay - 8 = 0$

    $\Rightarrow (x^2 + y^2 - 2x - 8) - 2a(y) = 0$

    $S + \lambda L = 0$

    Solving circle $x^2 + y^2 - 2x - 8 = 0$ and line $y = 0$, we get

    $x^2 - 4x + 2x - 8 = 0$

    $\therefore x = 4, x = -2$

    So $(4, 0)$, $(-2, 0)$ are the points of intersection which lie on the x-axis.

15. **c.** Equation of chord of contact from $A(x_1, y_1)$ is

    $xx_1 + yy_1 - a^2 = 0$

    $xx_2 + yy_2 - a^2 = 0$

    $xx_3 + yy_3 - a^2 = 0$

    i.e., $\begin{vmatrix} x_1 & y_1 & 1 \\ x_2 & y_2 & 1 \\ x_3 & y_3 & 1 \end{vmatrix} = 0$

    $\Rightarrow A, B, C$ are collinear.

16. a.

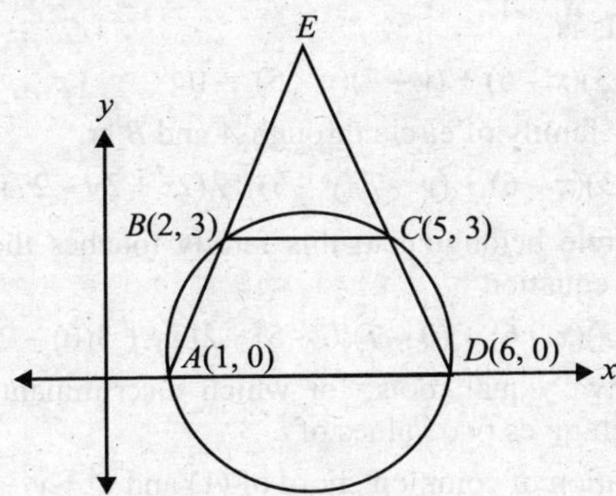

Fig. 2.81

From the figure, it is clear that $ABCD$ is isosceles trapezium as $AB = CD$. Also ΔEAD is isosceles $\Rightarrow EA \times EB = EC \times ED$

Hence, both the statements are correct and statement 1 is correct explanation of statement 1.

17. c. Statement 1 is true because common chord itself passes through origin.

Statement 2 is false (common chord is $x - y = 0$).

18. a. Common chord of two orthogonal circles subtends supplementary angles at the centre and so complemtary angles on the circumferences of the two circle.

So both statements are correct and statement 2 is the correct explanation of statement 1.

19. a. Since point lies inside the circle

$\Rightarrow a^2 + a^2 - 4a - 2a - 8 < 0$

$\Rightarrow a^2 - 3a - 4 < 0$

$\Rightarrow -1 < a < 4$

20. a. We know that the radical axis of the circle is the locus of point from which length of tangents to given two circles is same, also it the locus of the centre of the circle which intersect the given two circles orthogonally.

Now radical axis of the given two circles is $2x + y - 4 = 0$. Any point on this line is $(t, 4 - 2t)$, $t \in R$.

Hence, both the statements are true and statement 2 is correct explanation of statement.

21. d. Since $S_1 = 0$ and $S_3 = 0$ has no radical axis. Therefore, radical centre does not exist.

22. d. The statement 2 is well-known result, but if applied to the data given in statement 1 will yield $5x - 9y + 46 = 0$.

$\Rightarrow$ Statement 1 is false, statement 2 is true.

23. c. Statement 2 is false because line joining centres may not be parallel to common tangents.

Statement 1 can be proved easily by using distance between centres = sum of radii.

24. a. Two circles touch each other $C_1C_2 = |r_1 \pm r_2|$

$\Rightarrow \sqrt{p^2 + q^2} = \sqrt{p^2 - r} = \sqrt{q^2 - r}$

$\Rightarrow p^2 + q^2 = p^2 - r + q^2 - r + 2\sqrt{(p^2 - r)(q^2 - r)}$

$\Rightarrow \frac{1}{r} = \frac{1}{p^2} + \frac{1}{q^2}$

Linked Comprehension Type

For Problems 1–3

1. a, 2. b, 3. a.

Sol. It is given that one of the diagonals of the square is parallel to the line $y = x$.

Also the length of the diagonal of the square is $4\sqrt{2}$.

Hence, the equation of the one of diagonals is

$$\frac{x-3}{\frac{1}{\sqrt{2}}} = \frac{y-4}{\frac{1}{\sqrt{2}}} = r = \pm 2\sqrt{2}$$

Hence, $x - 3 = y - 4 = \pm 2$

$\Rightarrow x = 5, 1$ and $y = 6, 2$

Hence, two of the vertices are (1, 2) and (5, 6).

The other diagonal is parallel to the line $y = -x$, so that its equation is

$$\frac{x-3}{\frac{1}{\sqrt{2}}} = \frac{y-4}{-\frac{1}{\sqrt{2}}} = r = \pm 2\sqrt{2}$$

Hence, the two vertices on this diagonal are (1, 6) and (5, 2).

(a)

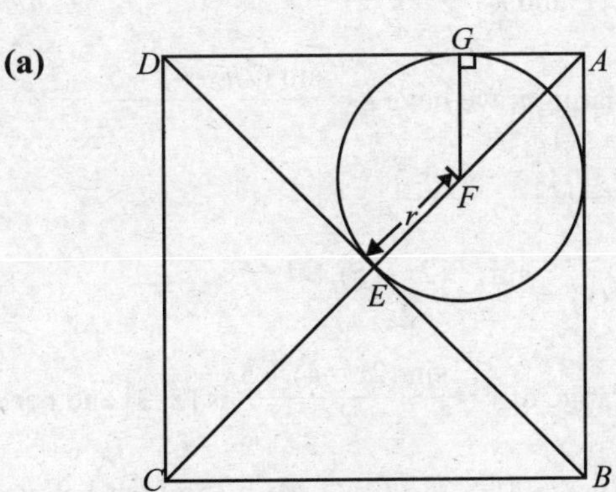

(b)

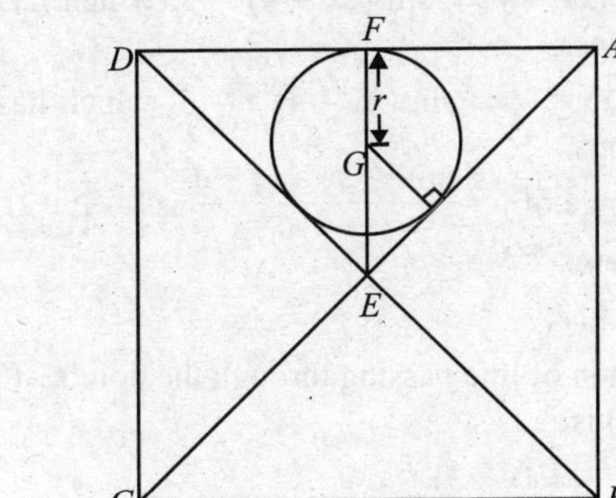

Fig. 2.82

$AB = 4, AC = 4\sqrt{2}$

$\Rightarrow AE = 2\sqrt{2}$

In Fig. (a), $EF + FA = AE$

$\Rightarrow r + \sqrt{2}r = 2\sqrt{2}$

$\Rightarrow r = \frac{2\sqrt{2}}{\sqrt{2}+1} = 2\sqrt{2}(\sqrt{2}-1)$

In Fig. (b), $EG + GF = EF$

$\Rightarrow \sqrt{2}\, r + r = 2$

$\Rightarrow r = \frac{2}{\sqrt{2}+1} = 2(\sqrt{2}-1)$

For Problems 4–6

4. d., 5.c., 6. c.,

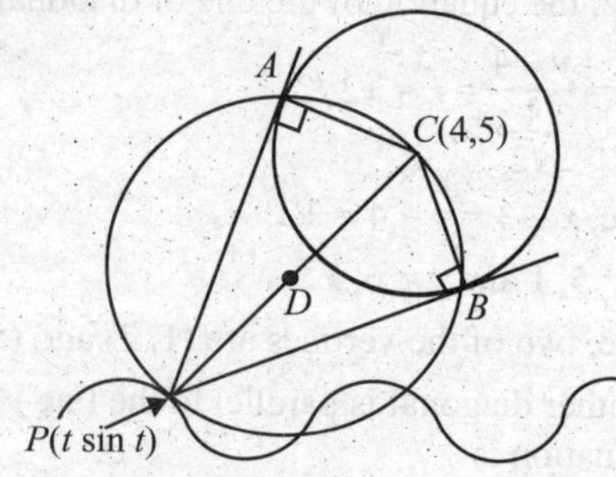

Fig. 2.83

Sol. Centre of the given circle is $C(4, 5)$. Points P, A, C, B are concyclic such that PC is diameter of the circle. Hence, centre D of the circumcircle of ΔABC is midpoint of PC, then we have

$h = \frac{t+4}{2}$ and $k = \frac{\sin t - 5}{2}$.

Eliminating t, we have $k = \frac{\sin(2h-4)+5}{2}$

or $y = \frac{\sin(2x-4)+5}{2}$

$\Rightarrow f^{-1}(x) = \frac{\sin^{-1}(2x-5)+4}{2}$

Thus range of $y = \frac{\sin(2x-4)+5}{2}$ is $[2, 3]$ and period is π.

Also $f(x) = 4 \Rightarrow \sin(2x-4) = 3$, which has no real solutions.

For $f(x) = 1 \Rightarrow \sin(2x-4) = -3$, which has no real solutions.

But range of $y = \frac{\sin^{-1}(2x-5)+4}{2}$ is $\left[-\frac{\pi}{4}+2, \frac{\pi}{4}+2\right]$.

For Problems 7–9

7. c., 8. d., 9. c.

Sol. Equation of line passing through the points $A(3, 7)$ and $B(6, 5)$ is

$y - 7 = -\frac{2}{3}(x-3)$

or $2x + 3y - 27 = 0$

Also equation of circle with A and B as diameter end points is

$(x-3)(x-6) + (y-7)(y-5) = 0$

Now family of circle through A and B is

$(x-3)(x-6) + (y-7)(y-5) + \lambda(2x+3y-27) = 0$ (1)

If circle belonging to this family touches the x-axis, then equation

$(x-3)(x-6) + (0-7)(0-5) + \lambda(2x+3(0)-27) = 0$

has two equal roots, for which discriminant $D = 0$, which gives two values of λ.

Equation of common chord of (1) and $x^2 + y^2 - 4x - 6y - 3 = 0$ is radical axis, which is

$[(x-3)(x-6) + (y-7)(y-5) + \lambda(2x+3y-27)] - [x^2+y^2-4x-6y-3] = 0$

or $(2\lambda - 5)x + (3\lambda - 6)y + (-27\lambda + 56) = 0$

or $(-5x - 6y + 56) + \lambda(2x + 3y - 27) = 0$

This is family of lines which passes through the point of intersection of $-5x - 6y + 56 = 0$ and $2x + 3y - 27 = 0$ which is $(2, 23/3)$.

If circle (1) cuts $x^2 + y^2 = 29$ orthogonally, then $0 + 0 = -29 + 56 - 27\lambda = 0 \Rightarrow \lambda = 1$

$\Rightarrow$ Required circle is $x^2 + y^2 - 7x - 9y + 26 = 0$, centre is $(7/2, 9/2)$.

For Problems 10–12

10. c., 11. a., 12.c.

Sol. Let the equation of the circle be

$$x^2 + y^2 + 2gx + 2fy + c = 0 \quad (1)$$

The line $lx + my + 1 = 0$, will touch circle (1), if the length of the perpendicular from the centre $(-g, -f)$ of the circle on the line is equal to its radius,

i.e., $\frac{|-gl - mf + 1|}{\sqrt{(l^2+m^2)}} = \sqrt{(g^2+f^2-c)}$

$\Rightarrow (gl + mf - 1)^2 = (l^2 + m^2)(g^2 + f^2 - c)$

$\Rightarrow (c - f^2)\, l^2 + (c - g^2)m^2 - 2gl - 2fm + 2gf\, lm + 1 = 0$ (2)

But the given condition of tangency is

$4l^2 - 5m^2 + 6l + 1 = 0$ (3)

Comparing Eqs. (2) and (3), we get

$c - f^2 = 4, c - g^2 = -5, -2g = 6, -2f = 0, 2gf = 0$.

Solving, we get $f = 0, g = -3, c = 4$.

Substituting these values in Eq. (1), the equation of the circle is $x^2 + y^2 - 6x + 4 = 0$. Any point on the line $x + y - 1 = 0$ is $(t, 1-t), t \in R$.

Chord of contact generated by this point for the circle is $tx + y(1-t) - 3(t+x) + 4 = 0$ or $t(x - y - 3) + (-3x + y + 4) = 0$, which are concurrent at point of intersection of the lines $x - y - 3 = 0$ and $-3x + y + 4 = 0$ for all values of t. Hence, lines are concurrent at $\left(\frac{1}{2}, -\frac{5}{2}\right)$.

Also point (2, –3) lies outside the circle from which two tangents can be drawn.

For Problems 13–15

13. c., 14. a., 15. a.

Sol. 13. c. Given $QT = QA = 1$

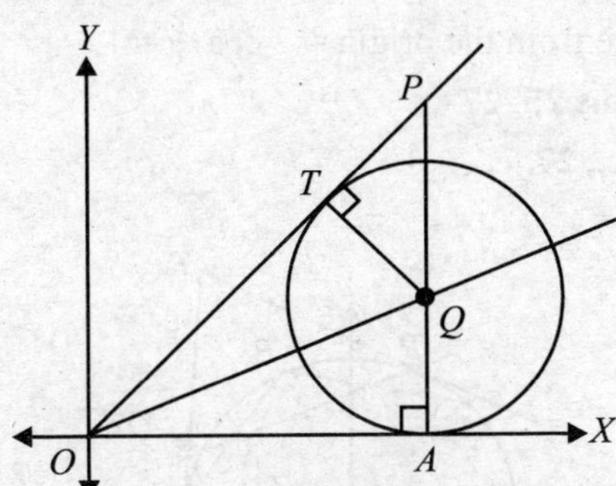

Fig. 2.84

Let $PQ = x$, then $PT = \sqrt{x^2 - 1}$

ΔTQP and ΔAPO are similar triangles

Then, $OT = OA = \dfrac{x+1}{\sqrt{x^2-1}}$.

$\Rightarrow 1 + x + \dfrac{2(x+1)}{\sqrt{x^2-1}} + \sqrt{x^2-1} = 8$

$\Rightarrow x = \dfrac{5}{3}$

14. a. $AP = \dfrac{8}{3}$, $OP = \dfrac{16}{3}$

Let $\angle AOP = 2\theta$, then $\sin 2\theta = \dfrac{1}{2}$

From ΔOAQ, $\tan\theta = \dfrac{1}{OA}$

$\Rightarrow OA = \dfrac{1}{\tan\theta}$

From $\sin 2\theta = \dfrac{2\tan\theta}{1+\tan^2\theta} = \dfrac{1}{2}$

$\Rightarrow \tan\theta = 2 - \sqrt{3}$

Hence, $OA = 2 + \sqrt{3}$

Hence, equation of circle is

$(x - (2+\sqrt{3}))^2 + (y-1)^2 = 1$

15. a. Equation of tangent OT is

$\dfrac{x-0}{\cos 2\theta} = \dfrac{y-0}{\sin 2\theta}$

$\Rightarrow x - \sqrt{3}y = 0.$

For Problems 16–18

16. b., 17.d., 18.b.

Sol. 16. b. $\because PQ = PR$, i.e., parallelogram $PQRS$ is a rhombus. Therefore, midpoint of QR = midpoint of PS and $QR \perp PS$

S is the mirror image of P w.r.t. QR.

$\because L \equiv 2x + y = 6$,

let $P \equiv (k, 6 - 2k)$.

$\because \angle PQO = \angle PRO = \dfrac{\pi}{2}$

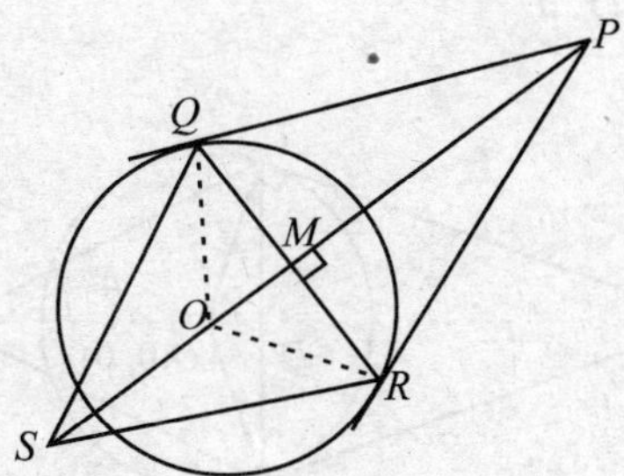

Fig. 2.85

Therefore, OP is diameter of circumcircle PQR, then centre is $\left(\dfrac{k}{2}, 3-k\right)$

$\therefore x = \dfrac{k}{2} \Rightarrow k = 2x$ and $y = 3 - k$

Hence, required locus is $2x + y = 3$.

17. d. $P(6, 8)$

Equation of QR (chord of contact) is $6x + 8y = 4$

$\Rightarrow 3x + 4y - 2 = 0$

$\therefore PM = \dfrac{48}{5}$ and $PQ = \sqrt{96}$

$QM = \sqrt{96 - \dfrac{48^2}{25}} = \sqrt{\dfrac{96}{25}}$

$\therefore QR = 2\sqrt{\dfrac{96}{25}}$

So area of $\Delta PQR = \dfrac{1}{2} \times PM \times QR = \dfrac{192\sqrt{6}}{25}$

Since $PQRS$ is a rhombus,

area of ΔQRS = Area of ΔPQR

$= \dfrac{192\sqrt{6}}{25}$ sq. units

18. b. As $P \equiv (3, 4)$,

Equation of QR is $3x + 4y = 4$ (1)

Let $S \equiv (x_1, y_1)$

Since S is the mirror image of P w.r.t. Eq. (1),

$\dfrac{x_1 - 3}{3} = \dfrac{y_1 - 4}{4}$

$= \dfrac{-2(3 \times 3 + 4 \times 4 - 4)}{3^2 + 4^2}$

$= -\dfrac{42}{25}$

$\therefore x_1 = -\dfrac{51}{25}, y_1 = -\dfrac{68}{25}$

$S\left(-\dfrac{51}{25}, -\dfrac{68}{25}\right)$

For Problems 19–21

19. a., 20. d., 21. a.

Sol. 19. a.

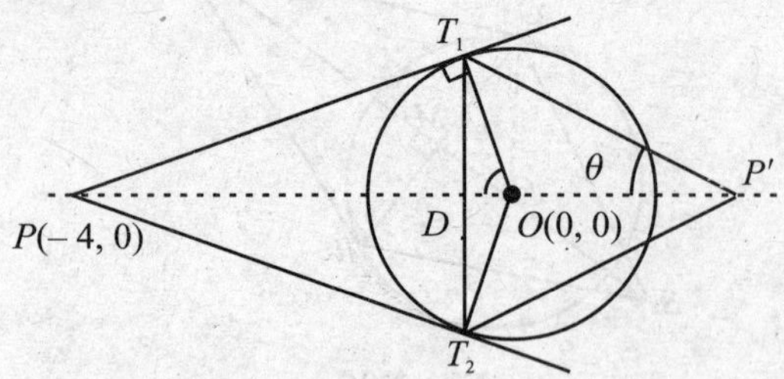

Fig. 2.86

$$PT_2 = PT_1 = \sqrt{(-4)^2 + 0^2 - 4} = 2\sqrt{3}$$

Circumcentre of triangle PT_1T_2 is midpoint of PO as

$\angle PT_1O = \angle PT_2O = 90°$

So, $\left(\dfrac{-4+0}{2}, \dfrac{0+0}{2}\right) = (-2, 0)$.

20. d.

21. a. Let P' be a point on the circle.

$3\theta = \dfrac{\pi}{2}$

$\Rightarrow \theta = \dfrac{\pi}{6}$

Area of the rhombus $= (2\sqrt{3})(2\sqrt{3})\sin\dfrac{\pi}{3} = 6\sqrt{3}$

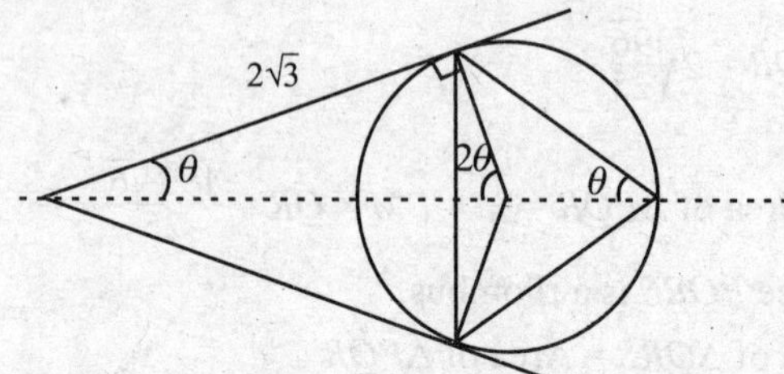

Fig. 2.87

For Problems 22–24

22. b., 23.a., 24. a.

Sol. 22. b.

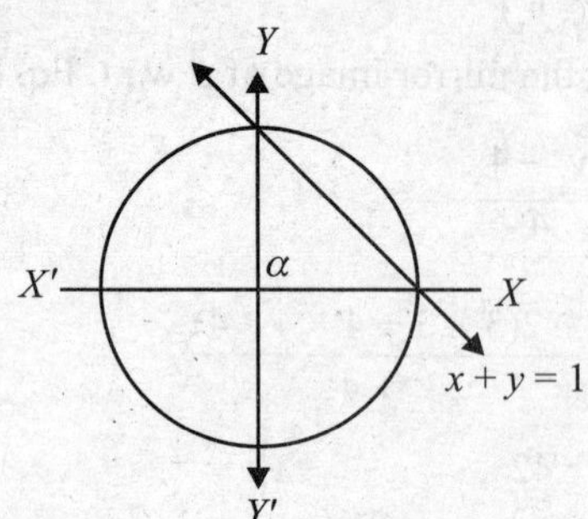

Fig. 2.88

23. a. Slope of chord = 1

Since the chord is $\dfrac{\pi}{3}$ chord,

distance from the origin is $\sqrt{3}$.

Let the equation of the chord be $x - y + k = 0$

$\therefore \left|\dfrac{k}{\sqrt{3}}\right| = \sqrt{3}$ i.e., $k = \pm\sqrt{6}$

24. a. Radius of the circle = 2

Distance from the origin $= 2\cos\dfrac{\pi}{3} = 1$.

For Problems 25–27

25. b., 26. c., 27. c.

Sol.

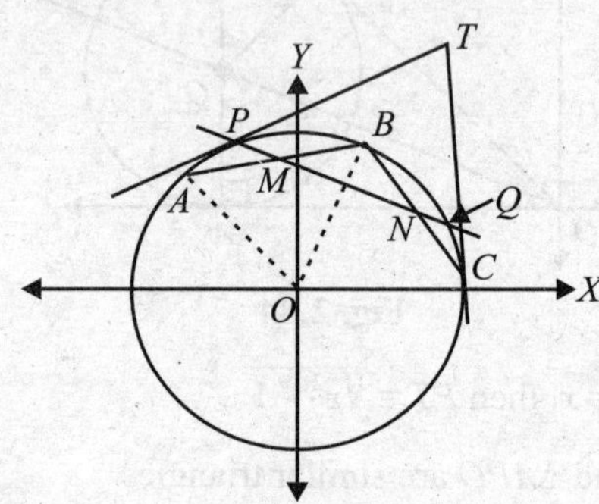

Fig. 2.89

25. b. From the figure, since ΔOAB is an equilateral triangle, $\angle OAB = 60°$.

26. c. Let T be the point of intersection of tangents.

Since $\angle AOC = 120°$,

angle between tangents is 60°.

27. c. Locus of point of intersection of tangents at A and C is a circle whose centre is $O(0, 0)$ and radius is

$OT = a \operatorname{cosec}\dfrac{\pi}{6} = 2a$

$\Rightarrow$ locus is $x^2 + y^2 = 4a^2$.

For Problems 28–30

28. b., 29. b., 30. c.

Sol.

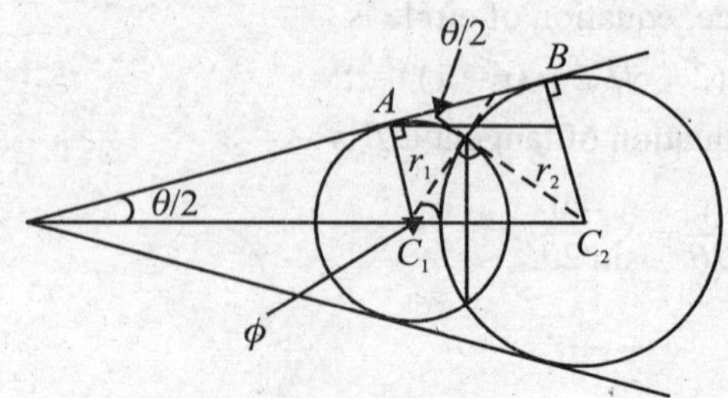

Fig. 2.90

We have $\sin\phi = \dfrac{d}{r_1}$, $\cos\phi = \dfrac{d}{r_2}$, (where $2d$ = length of common chord)

$\Rightarrow 1 = \dfrac{d^2}{r_1^2} + \dfrac{d^2}{r_2^2}$

$$\Rightarrow d = \frac{r_1 r_2}{\sqrt{r_1^2 + r_2^2}}$$

$$\Rightarrow 2d = \frac{2r_1 r_2}{\sqrt{r_1^2 + r_2^2}} = \frac{24}{5},$$

where $r_1 = 3$.

$$\Rightarrow d = \frac{6r_2}{\sqrt{9 + r_2^2}} = \frac{24}{5}$$

$$\Rightarrow r_2 = 4$$

From the figure, $\sin\frac{\theta}{2} = \frac{r_2 - r_1}{C_1C_2}$

where $C_1^2C_2^2 = r_1^2 + r_1^2$

$$\Rightarrow C_1C_2 = 5$$

$$\therefore \sin\frac{\theta}{2} = \frac{1}{5}$$

$$\Rightarrow \cos\frac{\theta}{2} = \frac{\sqrt{24}}{5}$$

$$\Rightarrow \sin\theta = 2 \times \frac{1}{5} \times \frac{\sqrt{24}}{5}$$

$$= \frac{4\sqrt{6}}{25}$$

$$\Rightarrow \theta = \sin^{-1}\frac{4\sqrt{6}}{25}$$

$$\Rightarrow \text{Also } AB = \sqrt{C_1C_2^2 - (r_1 - r_2)^2}$$

$$= \sqrt{25 - 1} = \sqrt{24}$$

Matrix-Match Type

1. a → r; b → q; c → q; d → p.

a.

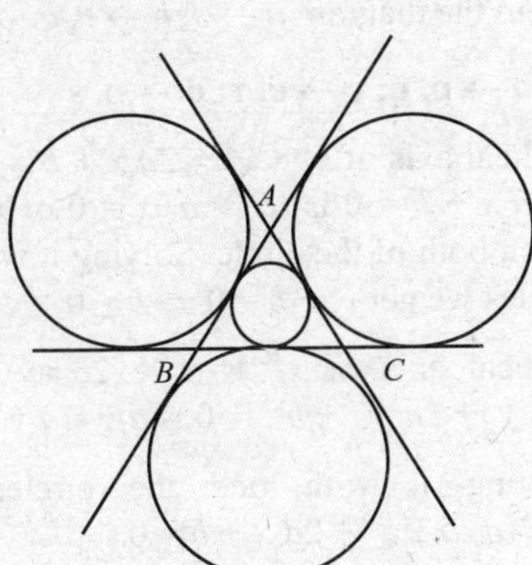

Fig. 2.91

b.

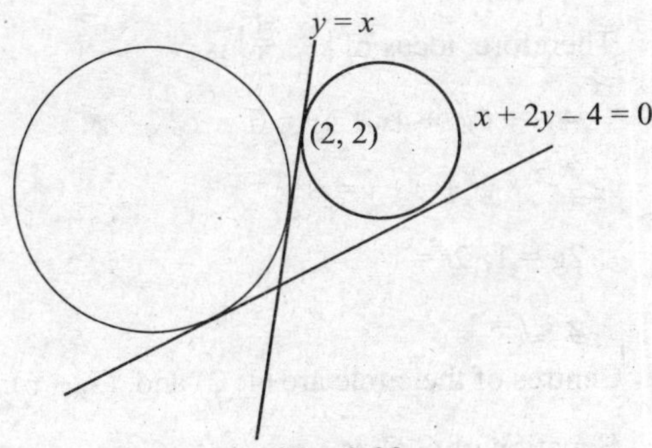

Fig. 2.92

c.

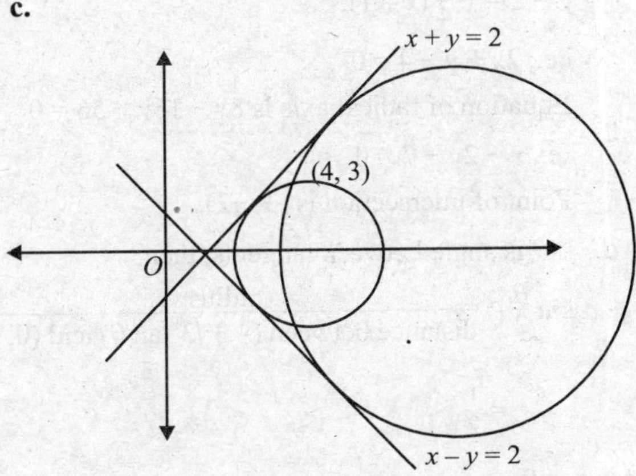

Fig. 2.93

d. Obviously one

2. a → p, r, s; b → r, s; c → q, s; d → p, s.

a. $(-g, -f)$ lies in first quadrant, then $g < 0$ and $f < 0$; also x-axis and y-axis must not cut the circle.

Solving circle and x-axis, we have $x^2 + 2gx + c = 0$, which must have imaginary roots; then $g^2 - c < 0$, then c must be positive. Also $f^2 - c < 0$.

b. If circle lies above x-axis then $x^2 + 2gx + c = 0$ must have imaginary roots, then $g^2 - c < 0$ and $c > 0$.

c. $(-g, -f)$ lies in third or fourth quadrant, then $g > 0$, Also y-axis must not cut the circle.

Solving circle and y-axis, we have $y^2 + 2gy + c = 0$, which must have imaginary roots, then $f^2 - c < 0$, then c must be positive.

d. $x^2 + 2gx + c = 0$ must have equal roots, then $g^2 = c$, hence $c > 0$.

Also $-g > 0 \Rightarrow g < 0$

3. a → r; b → s; c → q; d → p.

a. Since (2, 3) lies inside circle, such chord is bisected at (2, 3), which has equation $y - 3 = -(x - 2)$

or $x + y - 5 = 0 \Rightarrow a = b = 1$

b. Let P be the point (α, β), then $\alpha^2 + \beta^2 + 2\alpha + 2\beta = 0$

Midpoint of OP is $\left(\frac{\alpha}{2}, \frac{\beta}{2}\right)$.

Therefore, locus of $\left(\frac{\alpha}{2}, \frac{\beta}{2}\right)$ is

$$4x^2 + 4y^2 + 4x + 4y = 0$$

i.e., $x^2 + y^2 + x + y = 0$

$\therefore 2g = 1, 2f = 1$

$\therefore g + f = 1$

c. Centres of the circle are (1, 2) and, (5, – 6).

Equation of C_1C_2 is

$y - 2 = -\frac{8}{4}(x - 1)$,

i.e., $2x + y - 4 = 0$

Equation of radical axis is $8x - 16y - 56 = 0$

i.e., $x - 2y - 7 = 0$

Point of intersection is (3, –2).

d. If θ is angle between tangents, then

$$\sin\frac{\theta}{2} = \frac{\text{radius}}{\text{distance between } (-3\sqrt{3}\tan\theta) \text{ and } (0, 0)}$$

$$= \frac{1}{2}$$

$$\Rightarrow \frac{\theta}{2} = \frac{\pi}{6} \Rightarrow \theta = \frac{\pi}{3} = 2\sqrt{3}\tan\theta = 6$$

4. a → q; b → s; c → p; d → r.

a. Let length of common chord be $2a$, then

$\sqrt{9 - a^2} + \sqrt{16 - a^2} = 5$

$\Rightarrow \sqrt{16 - a^2} = 5 - \sqrt{9 - a^2}$

$\Rightarrow 16 - a^2 = 25 + 9 - a^2 - 10\sqrt{9 - a^2}$

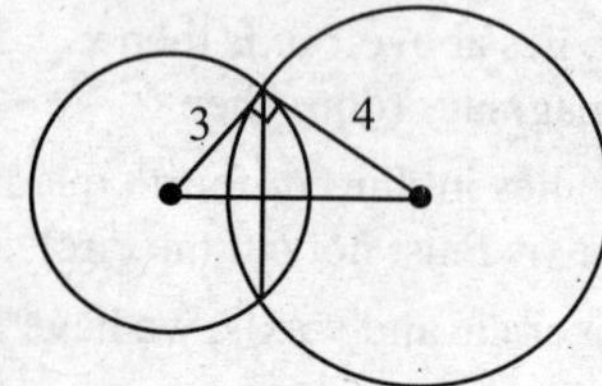

Fig. 2.94

$\Rightarrow 10\sqrt{9 - a^2} = 18$

$\Rightarrow 100(9 - a^2) = 324$, i.e., $100a^2 = 576$

$\therefore a = \sqrt{\frac{576}{100}} = \frac{24}{10}$

$\therefore 2a = \frac{24}{5} = \frac{k}{5} \Rightarrow k = 24$

b. Equation of common chord is $6x + 4y + p + q = 0$.

Common chord passes through centre (–2, –6) of circle $x^2 + y^2 + 4x + 12y + p = 0$

$\therefore p + q = 36$

c. Equation of the circle is $2x^2 + 2y^2 - 2\sqrt{2}\,x - y = 0$.

Let $(\alpha, 0)$ be midpoint of a chord. Then, equation of the chord is

$$2\alpha x - \sqrt{2}(x + \alpha) - \frac{1}{2}(y + 0) = 2\alpha^2 - 2\sqrt{2}\alpha$$

Since it passes through the point $\left(\sqrt{2}, \frac{1}{2}\right)$,

$2\sqrt{2}\alpha - \sqrt{2}(\sqrt{2} + \alpha) - \frac{1}{4} = 2\alpha^2 - 2\sqrt{2}\alpha$

i.e., $8\alpha^2 - 12\sqrt{2}\alpha + 9 = 0$,

i.e., $(2\sqrt{2}\alpha - 3)^2 = 0$

i.e., $\alpha = \frac{3}{2\sqrt{2}}, \frac{3}{2\sqrt{2}}$.

Hence number of chords is 1.

d. Midpoint of AB = (1, 4).

Equation perpendicular bisector ofAB is $x = 1$.

A diameter is $4y = x + 7$. Therefore, centre of the circle is (1, 2). Sides of the rectangle are 8 and 4. Hence, area = 32.

5. a → s; b → r; c → q; d → p.

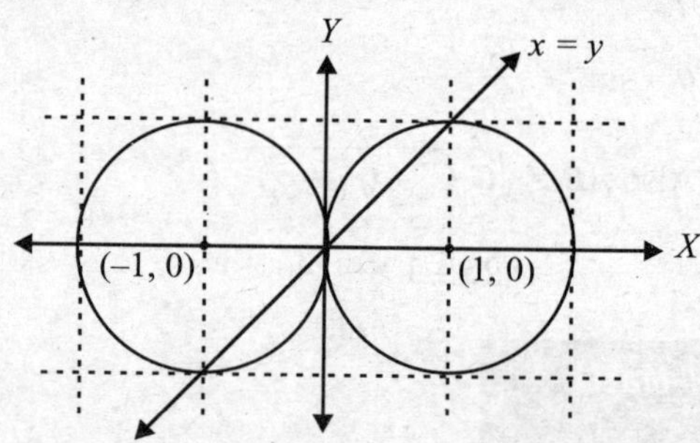

Fig. 2.95

(λ, λ) lies on the line $y = x$.

From the diagram $a \to s, b \to r, c \to q, d \to p$.

6. a → r; b → p, q; c → q, r; d → p, s

a. Radical axis of $x^2 + y^2 + 2a_1x + b = 0$ and $x^2 + y^2 + 2a_2x + b = 0$ is $(a_1 - a_2)x = 0$ or $x = 0$, it must touch both of the circle. Solving it with one of the circles we get $y^2 + b = 0 \Rightarrow b \le 0$

b. Radical axis of $x^2 + y^2 + 2a_1x + b = 0$ and $x^2 + y^2 + 2a_2y + b = 0$ is $a_1x - a_2y = 0$

Solving it with one the circles, we have $x^2 + (a_1/a_2)^2x^2 + 2a_1x + b = 0$.

This equation must have equal roots.

Hence, $4a_1^2 - 4b\left[1 + (a_1^2/a_2^2)\right] = 0$

$\Rightarrow a_1^2 - b\left[1 + (a_1^2/a_2^2)\right] = 0$

Options p and q satisfy this condition.

c. If the straight line $a_1x - by + b^2 = 0$ touches the circle $x^2 + y^2 = a_2x + by$, then

$$\frac{\left|a_1\frac{a_2}{2} - b\frac{b}{2} + b^2\right|}{\sqrt{a_1^2 + b^2}} = \sqrt{\frac{a_2^2}{4} + \frac{b^2}{4}}$$

$$\Rightarrow \frac{|a_1a_2 + b^2|}{\sqrt{a_1^2 + b^2}} = \sqrt{a_2^2 + b^2}$$

$$\Rightarrow a_1^2a_2^2 + 2b^2a_1a_2 + b^4 = a_1^2a_2^2 + a_1^2b^2 + a_2^2b^2 + b^4$$

$$\Rightarrow b^2 = 0 \text{ or } 2a_1a_2 = a_1^2 + a_2^2$$

$\Rightarrow$ Options (q) and (r)

d. Line $3x + 4y - 4 = 0$ touches the circle $(x - a_1)^2 + (y - a_2)^2 = b^2$. So

$$\frac{|3a_1 + 4a_2 - 4|}{5} = b$$

Integer Type

1. (2) Since the required circle has its centre on the x-axis. Let the coordinates of the centre be $(a, 0)$. The circle touches $y = x$. Therefore, radius = length of the perpendicular from $(a, 0)$ on $x - y = 0 = \frac{a}{\sqrt{2}}$

The circle cuts off a chord of length 2 units along $x - \sqrt{3}y = 0$.

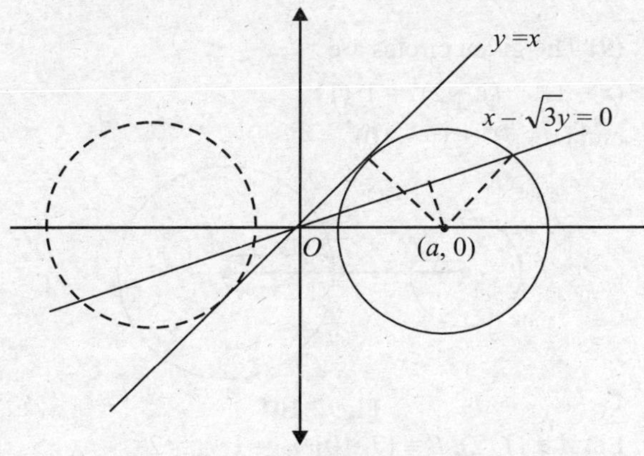

Fig. 2.96

$$\left(\frac{a}{\sqrt{2}}\right)^2 = 1^2 + \left(\frac{a - \sqrt{3} \times 0}{\sqrt{1^2 + (\sqrt{3})^2}}\right)^2$$

$$\Rightarrow \frac{a^2}{2} = 1 + \frac{a^2}{4} \Rightarrow a = 2$$

Thus the centre of the circle is at (2, 0) and radius is $a/\sqrt{2} = \sqrt{2}$.

2. (1) The given circle

$S(x, y) \equiv x^2 + y^2 - x - y - 6 = 0$ (1)

has centre at $C \equiv \left(\frac{1}{2}, \frac{1}{2}\right)$.

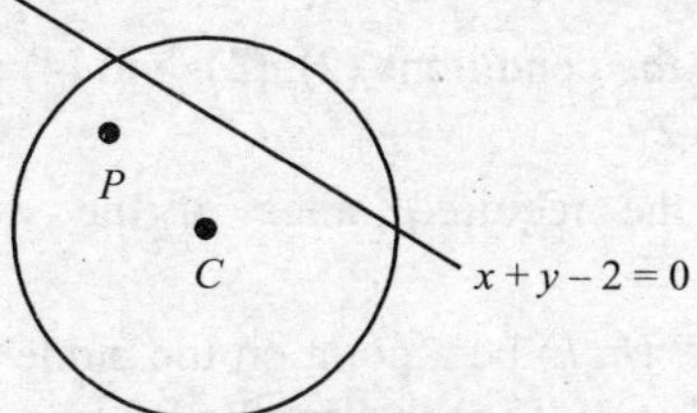

Fig. 2.97

According to the required conditions, the given point $P(\alpha - 1, \alpha + 1)$ must lie inside the given circle,

i.e., $S(\alpha - 1, \alpha + 1) < 0$

$\Rightarrow (\alpha - 1)^2 + (\alpha + 1)^2 - (\alpha - 1) - (\alpha + 1) - 6 < 0$

$\Rightarrow \alpha^2 - \alpha - 2 < 0$,

i.e., $(\alpha - 2)(\alpha + 1) < 0$

$\Rightarrow -1 < \alpha < 2$ (Using sign scheme from algebra) (2)

Also P and C must lie on the same side of the line (see figure)

$L(x, y) \equiv x + y - 2 = 0$ (3)

i.e., $L(1/2, 1/2)$ and $L(\alpha - 1, \alpha + 1)$ must have the same sign.

Since $L\left(\frac{1}{2}, \frac{1}{2}\right) = \frac{1}{2} + \frac{1}{2} - 2 < 0$,

$L(\alpha - 1, \alpha + 1) = (\alpha - 1) + (\alpha + 1) - 2 < 0$, i.e., $\alpha < 1$ (4)

Inequalities (2) and (4) together give the permissible values of α as $-1 < \alpha < 1$.

3. (3) The equation of the circle is

$x^2 + y^2 - 6x - 10y + k = 0$ (1)

Its centre is $C(3, 5)$ and radius r is $\sqrt{34 - k}$.

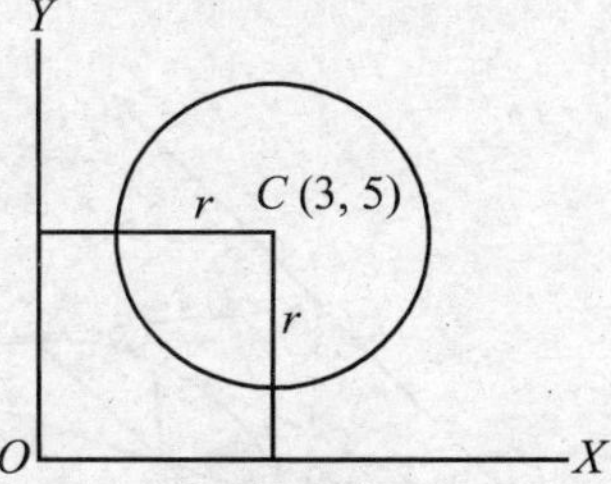

Fig. 2.98

If the circle does not touch or intersect the x-axis, then radius $r <$ y-coordinate of centre C

or $\sqrt{(34 - k)} < 5$

$\Rightarrow 34 - k < 25 \Rightarrow k > 9$ (2)

Also if the circle does not touch or intersect the y-axis, then radius $r <$ x-coordinate of centre C

or $\sqrt{(34 - k)} < 3$

$\Rightarrow 34 - k < 9 \Rightarrow k > 25$ (3)

If the point (1, 4) is inside the circle, then its distance from centre $C < r$ (radius),

or $\sqrt{[(3-1)^2 + (5-4)^2]} < \sqrt{(34-k)}$

$\Rightarrow 5 < 34 - k \Rightarrow k < 29$ (4)

Now all the conditions (2), (3), and (4) are satisfied if $25 < k < 29$.

Hence, the required range of the values of k is $25 < k < 29$.

4. (2) Let $P(h, k)$ be a point on the circle $15x^2 + 15y^2 - 48x + 64y = 0$. Then the lengths PT_1 and PT_2 of the tangent from $P(h, k)$ to $5x^2 + 5y^2 - 24x + 32y + 75 = 0$ and $5x^2 + 5y^2 - 48x + 64y + 300 = 0$, respectively, are

$$PT_1 = \sqrt{h^2 + k^2 - \frac{24}{5}h + \frac{32}{5}k + 15} \text{ and}$$

$$PT_2 = \sqrt{h^2 + k^2 - \frac{48}{5}h + \frac{64}{5}k + 60}.$$

Since (h, k) lies on $15x^2 + 5y^2 - 48x + 64y = 0$,

$h^2 + k^2 - \frac{48}{15} + \frac{64}{15}k = 0$. Therefore,

$$PT_1 = \sqrt{\frac{48}{15}h - \frac{64}{15}k - \frac{24}{5}h + \frac{32}{5}k + 15}$$

$$= \sqrt{\frac{32k}{15} - \frac{24}{15}h + 15}$$

$$\text{and } PT_2 = \sqrt{\frac{48}{15}h - \frac{64}{15}k - \frac{48}{15}h + \frac{64}{5}k + 60}$$

$$= \sqrt{-\frac{96}{15}h + \frac{128}{15}k + 60}$$

$$= 2\sqrt{-\frac{24}{15}h + \frac{32}{15}k + 15} = 2(PT_1)$$

Hence, $PT_1 : PT_2 = 1 : 2$.

5. (2)

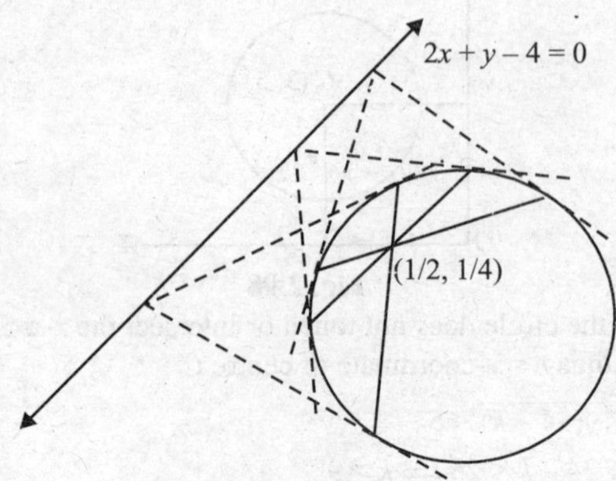

Fig. 2.99

Let any point on the line $2x + y - 4 = 0$ be $P \equiv (a, 4 - 2a)$.

The equation of the chord of contact of the circle $x^2 + y^2 = 1$ with respect to point P is $xa + y(4 - 2a) = 1$

$\Rightarrow (4y - 1) + a(x - 2y) = 0$.

It will always pass through the point of intersection of the lines $4y - 1 = 0$ and $x - 2y = 0$, which is a fixed point whose coordinates are $y = \frac{1}{4}$ and $x = 2y = \frac{1}{2}$.

6. (7)

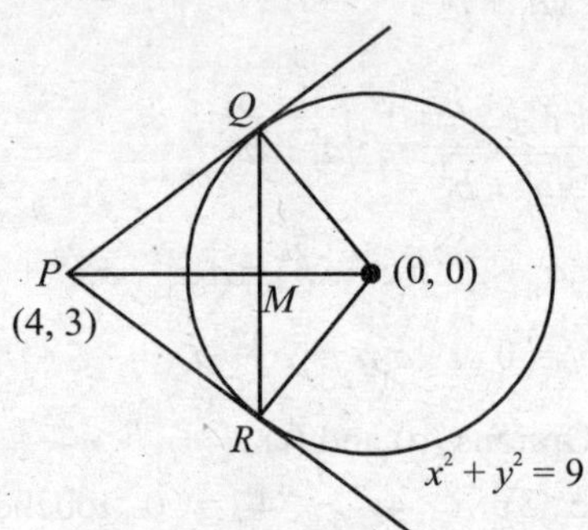

Fig. 2.100

The equation of the chord of contact of tangents drawn from $P(4, 3)$ to $x^2 + y^2 = 9$ is $4x + 3y = 9$.

The equation of PQ is $y = (3/4)x$.

Now OM = length of the perpendicular from (0, 0) on $4x + 3y - 9 = 0 = \frac{9}{5}$

$QR = 2\,QM = 2\sqrt{OQ2 - OM^2} = 2\sqrt{9 - \frac{81}{25}} = \frac{24}{5}$

Now $PM = OP - OM = 5 - \frac{9}{5} = \frac{16}{5}$.

So area of $\Delta PQR = \frac{1}{2}\left(\frac{24}{5}\right)\left(\frac{16}{5}\right) = \frac{192}{25}$ sq. units

7. (9) The given circles are

$(x - 1)^2 + (y - 2)^2 = 1$ (1)

and $(x - 7)^2 + (y - 10)^2 = 4$ (2)

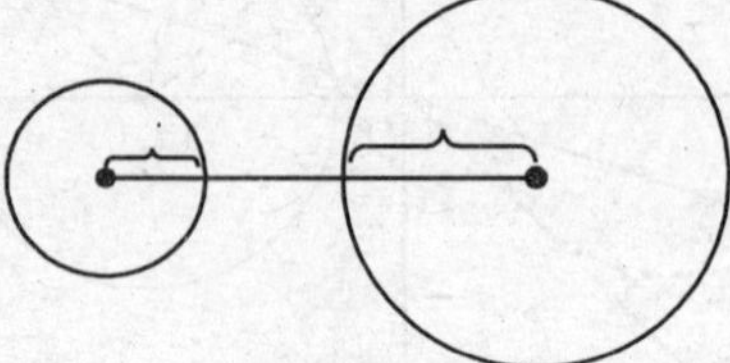

Fig. 2.101

Let $A \equiv (1, 2)$, $B \equiv (7, 10)$, $r_1 = 1$, $r_2 = 2$

$AB \equiv 10$, $r_1 + r_2 = 3$

Since $AB > r_1 + r_2$, the two circles are non-intersecting.

Radii of the two circles at time t are $1 + 0.3t$ and $2 + 0.4t$.

For the two circles to touch each other,

$AB^2 = [(r_1 + 0.3t) \pm (r_2 + 0.4t)]^2$

$\Rightarrow 100 = [(1 + 0.3t) \pm (2 + 0.4t)]^2$

$\Rightarrow 100 = (3 + 0.7t)^2, [(0.1)\, t + 1]^2$

$\Rightarrow 3 + 0.7t = \pm 10,\ 0.1t + 1 = \pm 10$

$\Rightarrow t = 10, t = 90 \qquad (\because t > 0)$

The two circles will touch each other externally in 10 seconds and internally in 90 seconds.

8. (6) The equation of the common chord is $S_1 - S_2 = 0$

$\Rightarrow 2x - 2y = 0$, i.e., $x - y = 0$

Since the length of the perpendicular drawn from C_1 to $x - y = 0$ is $1/\sqrt{2}$, length of common chord $= 2\sqrt{\frac{19}{2} - \frac{1}{2}} = 6$.

9. (5)

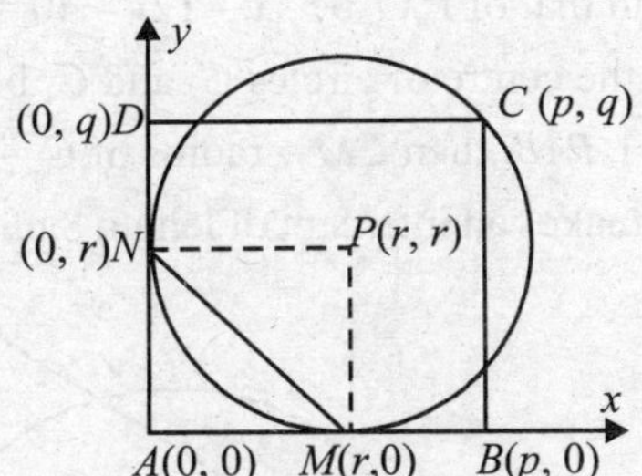

Fig. 2.102

Let us take AB and AD as coordinate axes.

If r is the radius of the circle, then its centre is $P(r, r)$.

The equation of the circle is $(x - r)^2 + (y - r)^2 = r^2$

$\Rightarrow x^2 + y^2 - 2rx - 2ry + r^2 = 0$

Let the coordinates of $C \equiv (p, q)$.

The equation of MN is $x + y = r$.

Its distance from C is 5 units

$\Rightarrow \frac{|p + q - r|}{\sqrt{2}} = 5$

$\Rightarrow (p + q - r)^2 = 50$

Since (p, q) lies on the circle,

$p^2 + q^2 - 2rp - 2rq + r^2 = 0$

$\Rightarrow (p + q - r)^2 - 2pq = 0$

$\Rightarrow 50 - 2pq = 0 \Rightarrow pq = 25$

Hence, area of rectangle = 25 sq. units.

10. (2) The equation of line PQ is $(y - 1) = m(x - 4)$ or $y - mx + 4m - 1 = 0$.

For the required m, we have to make sure that the line PQ meets the circle having diameter AB at real and distinct points.

The equation of the circle having AB as diameter is $x^2 + y^2 - 2x - 3 = 0$.

Thus we have to make sure that

$\frac{|0 - m + 4m - 1|}{\sqrt{1+m^2}} < 2$

$\Rightarrow 5m^2 - 6m - 3 > 0$

$\Rightarrow m \in \left(\frac{3 - 2\sqrt{6}}{5}, \frac{3 + 2\sqrt{6}}{5}\right)$

11. (6)

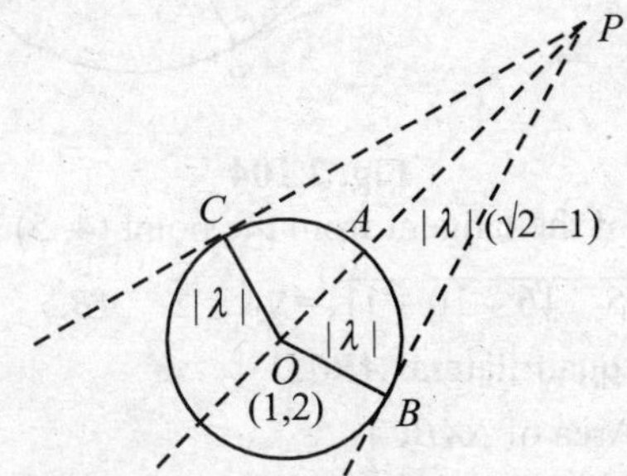

Fig. 2.103

Since the line $y = x + c$ is normal to the given circle, $c = 1$.

So the equation of line is $y = x + 1$ (1)

Also the radius of the circle is $|\lambda|$.

Given $AP = |\lambda|(\sqrt{2} - 1)$

$\Rightarrow OP = \sqrt{2}|\lambda|$

$\Rightarrow PC = |\lambda|$

$\Rightarrow$ Area of quadrilateral $OBPC$

$= 2 \times \frac{1}{2}|\lambda|^2 = 36 \Rightarrow \lambda = \pm 6$

12. (5) Let the equation of the required circle be $x^2 + y^2 + 2gx + 2fy + c = 0$.

The condition for orthogonal intersection is

$2g \times 0 + 2f \times 0 = -1 + c \Rightarrow c = 1$

and $2 \times 4 \times g + 2 \times 4 \times f$

$= -33 + c$

$= -33 + 1$

$= -32$

$\Rightarrow g + f = -4$.

Hence, radius $= \sqrt{g^2 + f^2 - c}$

$= \sqrt{g^2 + (g + 4)^2 - 1}$

i.e., $r = \sqrt{2g^2 + 8g + 15}$

$= \sqrt{2(g + 2)^2 + 7}$

For minimum r,

$g + 2 = 0 \Rightarrow g = -2$.

Hence, the centre is (2, 2).

13. (8) The equation of the circle is $x^2 + y^2 - 4x - 2y - 11 = 0$.

Its centre is (2, 1) and radius $= \sqrt{4 + 1 + 1} = 4 = BC$.

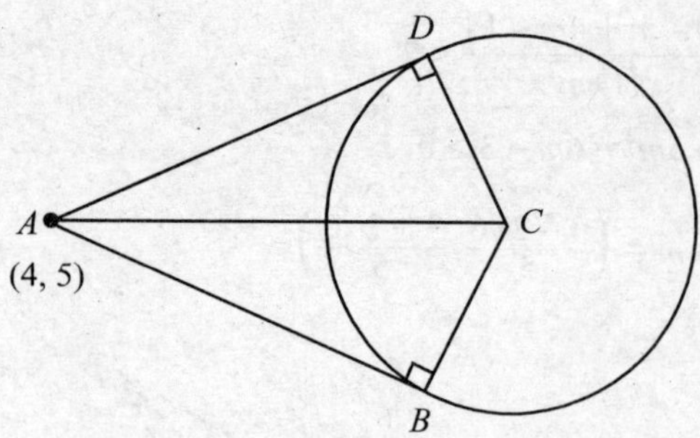

Fig. 2.104

Length of the tangent from the point (4, 5) is

$\sqrt{16+25-16-10-11} = \sqrt{4} = 2 = AB.$

Area of quadrilateral $ABCD$

$= 2$ (Area of ΔABC)

$= 2 \times \frac{1}{2} \times AB \times BC$

$= 2 \times \frac{1}{2} \times 2 \times 4$

$= 8$ sq. units

14. (2) The given lines are $\lambda x - y + 1 = 0$ and $x - 2y + 3 = 0$, which meet x-axis at

$A(-1/\lambda, 0)$ and $B(-3, 0)$ and y-axis at $C(0, 1)$ and $D(0, 3/2)$, respectively.

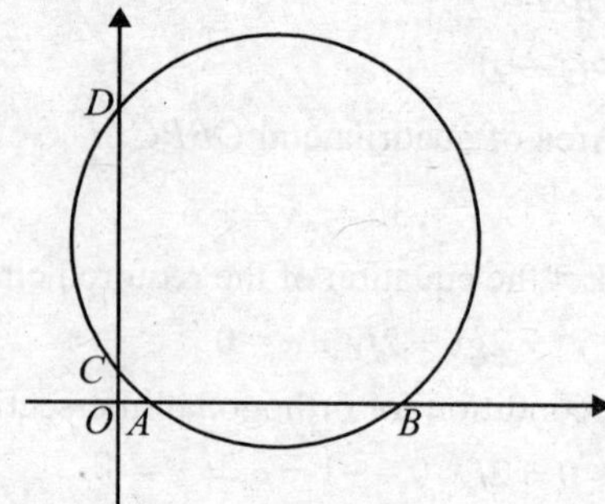

Fig. 2.105

Then we must have,

$OA \times OB = OC \times OD$

$\Rightarrow \left(-\frac{1}{\lambda}\right)(-3) = 1 \times \frac{3}{2}$

$\Rightarrow \lambda = 2$

15. (5) If d is the distance between the centres of the two circles of radii r_1 and r_2, then they intersect at two distinct points iff $|r_1 - r_2| < d < r_1 + r_2$.

Here radii of the two circles are r and 3 and distance between the centres is 5.

Thus $|r - 3| < 5 < r + 3$

$\Rightarrow -2 < r < 8$ and $r > 2$

$\Rightarrow 2 < r < 8.$

16. (3) The given circle is $x^2 + y^2 - 2x - 6y + 6 = 0.$

Its centre is $H(1, 3)$ and radius is 2.

So $AH = 2$

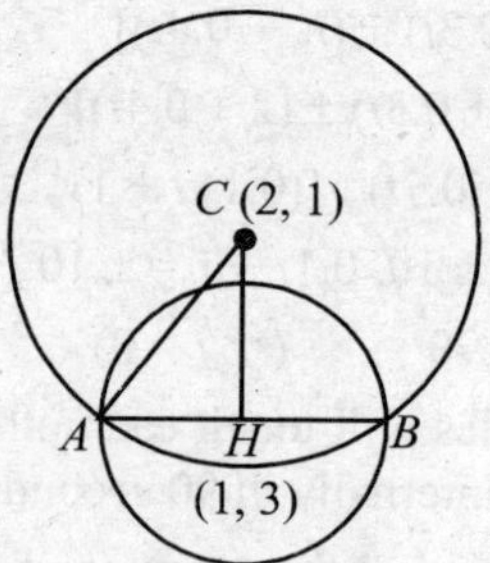

Fig. 2.106

Radius of the required circle

$= AC = \sqrt{AH^2 + CH^2}$

$= \sqrt{2^2 + 5} = 3$

17. (5) Let the equation of the tangent PAB be $5x + 12y - 10 = 0$ and that of PXY be $5x - 12y - 40 = 0.$

Now let the centre of circles C_1 and C_2 be $C(h, k)$.

Let $CM \perp PAB$, then $CM =$ radius of $C_1 = 3.$

Also C_2 makes an intercept of length 8 units on PAB. So $AM = 4.$

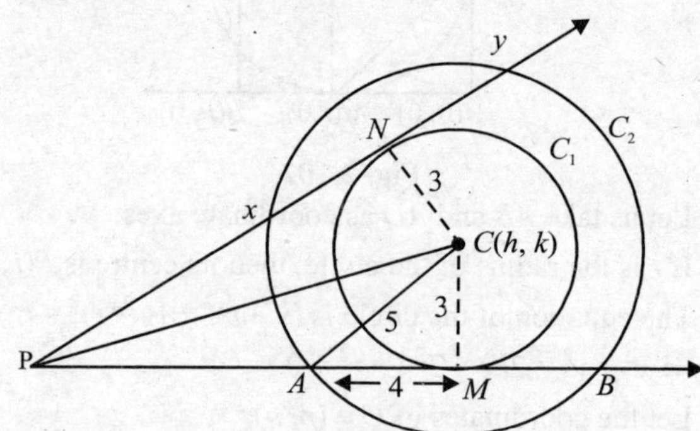

Fig. 2.107

Then in ΔAMC, we get $AC = \sqrt{4^2 + 3^2} = 5$

Hence, radius of C_2 is 5 units.

Also as $5x + 12y - 10 = 0$ and $5x - 12y - 40 = 0$ are tangents to C_1, length of the perpendicular form C to AB is 3 units. Therefore,

$\frac{5h + 12k - 10}{13} = \pm 3$

$\Rightarrow 5h + 12k - 49 = 0$ (1)

or $5h + 12k + 29 = 0$ (2)

Similarly, $\frac{5h - 12k - 40}{13} = \pm 3$

$\Rightarrow 5h - 12k - 79 = 0$ (3)

or $5h - 12k - 1 = 0$ (4)

As C lies in the first quadrant, h, k are positive.

So equation (2) is not possible.

Solving (1) and (3), we get $h = 64/5$, $k = -5/4$. This is also not possible.

Now solving (1) and (4), we get $h = 5, k = 2.$

Thus the centre of C_2 is (5, 2) and radius is 5.

Hence, the equation of C_2 is $(x-5)^2+(y-2)^2=5^2$.

18. (1) Given that $(m_i, 1/m_i)$, $m_i > 0$, $i = 1, 2, 3, 4$, are four distinct points on a circle.

Let the equation of the circle be $x^2+y^2+2gx+2fy+c=0$.

As the point $(m, 1/m)$ lies on it, we have

$$m^2+\frac{1}{m^2}+2\,gm+\frac{2f}{m}+c=0$$

$$\Rightarrow m^4+2gm^3+cm^2+2fm+1=0$$

Since m_1, m_2, m_3, m_4 are roots of this equation, the product of roots = 1

$$\Rightarrow m_1\,m_2\,m_3\,m_4=1$$

19. (5) The given circle is $x^2+y^2=r^2$.

From the point (6, 8), tangents are drawn to this circle.

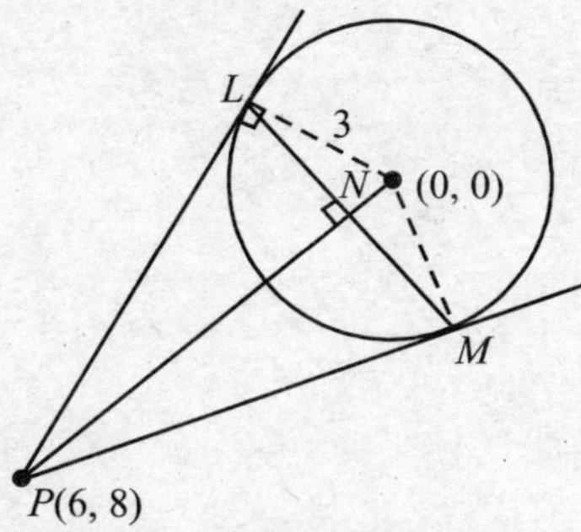

Fig. 2.108

Then the length of tangent PL is

$\sqrt{6^2+8^2-r^2}=\sqrt{100-r^2}$.

Also the equation of the chord of contact LM is $6x+8y-r^2=0$.

PN = length of perpendicular from P to LM

$$=\frac{36+64-r^2}{\sqrt{36+64}}$$

$$=\frac{100-r^2}{10}$$

Now in the right-angled ΔPLN,

$$LN^2=PL^2-PN^2$$

$$=(100-r^2)-\frac{(100-r^2)^2}{100}$$

$$=\frac{(100-r^2)\,r^2}{100}$$

$$\Rightarrow LN=\frac{r\sqrt{100-r^2}}{10}$$

So $LM=\dfrac{r\sqrt{100-r^2}}{5}$. ($\because LM=2LN$)

$$\text{Area of } \Delta PLM = A \times \frac{1}{2}\times LM\times PN$$

$$=\frac{1}{2}\times\frac{r\sqrt{100-r^2}}{5}\times\frac{100-r^2}{10}$$

$$=\frac{1}{100}\left[r\,(100-r^2)^{3/2}\right]$$

For the maximum value of A, we should have $\dfrac{dA}{dr}=0$

$$\Rightarrow\frac{1}{100}\left[(100-r^2)^{3/2}+r\,\frac{3}{2}\,(100-r^2)^{1/2}\,(-2r)\right]=0$$

$$\Rightarrow(100-r^2)^{1/2}\,[100-r^2-3r^2]=0$$

$$\Rightarrow r=10 \text{ or } r=5$$

But $r = 10$ gives the length of tangent $PL = 0$. Therefore, $r \neq 10$.

Hence, $r = 5$.

20. (8)

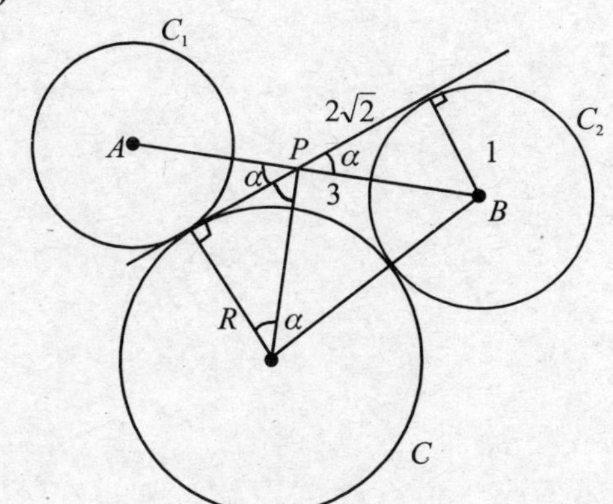

Fig. 2.109

$$\cos\alpha=\frac{2\sqrt{2}}{3}$$

$$\sin\alpha=\frac{1}{3}$$

$$\tan\alpha=\frac{2\sqrt{2}}{R}$$

$$\Rightarrow R=\frac{2\sqrt{2}}{\tan\alpha}=8 \text{ units.}$$

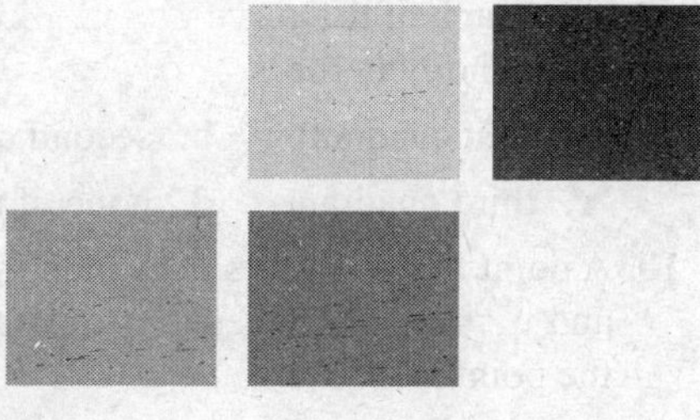

CHAPTER 3

Parabola

Objective Type

Each question has four choices a, b, c and d, out of which *only one* answer is correct. Find the correct answer.

1. The equation of the parabola whose vertex and focus lie on the axis of x at distances a and a_1 from the origin respectively is

   a. $y^2 = 4(a_1 - a)x$
   b. $y^2 = 4(a_1 - a)(x - a)$
   c. $y^2 = 4(a_1 - a)(x - a_1)$
   d. none of these

2. The vertex of a parabola is the point (a, b) and latus rectum is of length l. If the axis of the parabola is along the positive direction of y-axis, then its equation is

   a. $(x + a)^2 = \frac{l}{2}(2y - 2b)$
   b. $(x - a)^2 = \frac{l}{2}(2y - 2b)$
   c. $(x + a)^2 = \frac{l}{4}(2y - 2b)$
   d. $(x - a)^2 = \frac{l}{8}(2y - 2b)$

3. Which one of the following equations represents parametric equation of a parabolic curve?

   a. $x = 3\cos t, y = 4\sin t$
   b. $x^2 - 2 = 2\cos t, y = 4\cos^2\frac{t}{2}$
   c. $\sqrt{x} = \tan t, \sqrt{y} = \sec t$
   d. $x = \sqrt{1 - \sin t}, y = \sin\frac{t}{2} + \cos\frac{t}{2}$

4. The equation of the parabola whose focus is the point $(0, 0)$ and the tangent at the vertex is $x - y + 1 = 0$ is

   a. $x^2 + y^2 - 2xy - 4x - 4y - 4 = 0$
   b. $x^2 + y^2 - 2xy + 4x - 4y - 4 = 0$
   c. $x^2 + y^2 + 2xy - 4x + 4y - 4 = 0$
   d. $x^2 + y^2 + 2xy - 4x - 4y + 4 = 0$

5. The curve represented by the equation $\sqrt{px} + \sqrt{qy} = 1$, where $p, q \in R, p, q > 0$ is

   a. a circle **b.** a parabola
   c. an ellipse **d.** a hyperbola

6. If parabola $y^2 = \lambda x$ and $25[(x - 3)^2 + (y + 2)^2] = (3x - 4y - 2)^2$ are equal, then value of λ is

   a. 9 **b.** 3 **c.** 7 **d.** 6

7. The length of the latus rectum of the parabola whose focus is $\left(\frac{u^2}{2g}\sin 2\alpha, -\frac{u^2}{2g}\cos 2\alpha\right)$ and directrix is $y = \frac{u^2}{2g}$ is

   a. $\frac{u^2}{g}\cos^2\alpha$ **b.** $\frac{u^2}{g}\cos 2\alpha$
   c. $\frac{2u^2}{g}\cos 2\alpha$ **d.** $\frac{2u^2}{g}\cos^2\alpha$

8. If the segment intercepted by the parabola $y = 4ax$ with the line $lx + my + n = 0$ subtends a right angle at the vertex, then

   a. $4al + n = 0$ **b.** $4al + 4am + n = 0$
   c. $4am + n = 0$ **d.** $al + n = 0$

9. The graph of the curve $x^2 + y^2 - 2xy - 8x - 8y + 32 = 0$ falls wholly in the

a. first quadrant b. second quadrant

c. third quadrant d. none of these

10. A point $P(x, y)$ moves in xy plane such that $x = a\cos^2\theta$ and $y = 2a\sin\theta$, where θ is a parameter. The locus of the point P is

a. circle

b. ellipse

c. unbounded parabola

d. part of the parabola

11. Locus of the point $\sqrt{3h}$, $\sqrt{3k+2}$ if it lies on the line $x - y - 1 = 0$ is a

a. straight line b. circle

c. parabola d. none of these

12. A water jet from a fountain reaches its maximum height of 4 m at a distance 0.5 m from the vertical passing through the point O of water outlet. The height of the jet above the horizontal OX at a distance of 0.75 m from the point O is

a. 5 m b. 6 m c. 3 m d. 7 m

13. Vertex of the parabola whose parametric equation is $x = t^2 - t + 1, y = t^2 + t + 1, t \in R$, is

a. (1, 1) b. (2, 2)

c. $\left(\frac{1}{2}, \frac{1}{2}\right)$ d. (3, 3)

14. The ratio in which the line segment joining the points (4, –6) and (3, 1) is divided by the parabola $y^2 = 4x$ is

a. $\frac{-20 \pm \sqrt{155}}{11} : 1$ b. $\frac{-2 \pm 2\sqrt{155}}{11} : 2$

c. $-20 \pm 2\sqrt{155} : 11$ d. $-20 \pm \sqrt{155} : 11$

15. If (a, b) is the midpoint of a chord passing through the vertex of the parabola $y^2 = 4x$, then

a. $a = 2b$ b. $2a = b$ c. $a^2 = 2b$ d. $2a = b^2$

16. A set of parallel chords of the parabola $y^2 = 4ax$ have their midpoints on

a. any straight line through the vertex

b. any straight line through the focus

c. a straight line parallel to the axis

d. another parabola

17. A line L passing through the focus of the parabola $y^2 = 4(x - 1)$ intersects the parabola in two distinct points. If m is the slope of the line L then

a. $-1 < m < 1$ b. $m < -1$ or $m > 1$

c. $m \in R$ d. none of these

18. If PSQ is the focal chord of the parabola $y^2 = 8x$ such that $SP = 6$. Then the length of SQ is

a. 6 b. 4

c. 3 d. none of these

19. The circle $x^2 + y^2 + 2\lambda x = 0, \lambda \in R$, touches the parabola $y^2 = 4x$ externally. Then

a. $\lambda > 0$ b. $\lambda < 0$

c. $\lambda > 1$ d. none of these

20. If y_1, y_2, and y_3, are the ordinates of the vertices of a triangle inscribed in the parabola $y^2 = 4ax$, then its area is

a. $\frac{1}{2a}(y_1 - y_2)(y_2 - y_3)(y_3 - y_1)$

b. $\frac{1}{4a}(y_1 - y_2)(y_2 - y_3)(y_3 - y_1)$

c. $\frac{1}{8a}(y_1 - y_2)(y_2 - y_3)(y_3 - y_1)$

d. none of these

21. If $a \neq 0$ and the line $2bx + 3cy + 4d = 0$ passes through the points of intersection of the parabolas $y^2 = 4ax$ and $x^2 = 4ay$, then

a. $d^2 + (2b + 3c)^2 = 0$

b. $d^2 + (3b + 2c)^2 = 0$

c. $d^2 + (2b - 3c)^2 = 0$

d. none of these

22. Let P be the point (1, 0) and Q a point on the locus $y^2 = 8x$. The locus of the midpoint of PQ is

a. $y^2 + 4x + 2 = 0$ b. $y^2 - 4x + 2 = 0$

c. $x^2 - 4y + 2 = 0$ d. $x^2 + 4y + 2 = 0$

23. The locus of the vertex of the family of parabolas $y = \frac{a^3x^2}{3} + \frac{a^2x}{2} - 2a$ is

a. $xy = \frac{105}{64}$ b. $xy = \frac{3}{4}$

c. $xy = \frac{35}{16}$ d. $xy = \frac{64}{105}$

24. A circle touches the x-axis and also touches the circle with centre (0, 3) and radius 2. The locus of the centre of the circle is

a. a circle b. an ellipse

c. a parabola d. a hyperbola

25. Parabolas $y^2 = 4a(x - c_1)$ and $x^2 = 4a(y - c_2)$, where c_1 and c_2 are variable, are such that they touch each other. Locus of their point of contact is

a. $xy = 2a^2$ b. $xy = 4a^2$

c. $xy = a^2$ d. none of these

26. The locus of a point on the variable parabola $y^2 = 4ax$, whose distance from focus is always equal to k, is equal to (a is parameter)

a. $4x^2 + y^2 - 4kx = 0$

b. $x^2 + y^2 - 4kx = 0$

c. $2x^2 + 4y^2 - 8kx = 0$

d. $4x^2 - y^2 + 4kx = 0$

27. If the line $y - \sqrt{3}x + 3 = 0$ cuts the parabola $y^2 = x + 2$ at P and Q, then $AP \times AQ$ is equal to [where $A \equiv (\sqrt{3}, 0)$]

a. $\dfrac{2(\sqrt{3}+2)}{3}$ **b.** $\dfrac{4\sqrt{3}}{2}$

c. $\dfrac{4(2-\sqrt{2})}{3}$ **d.** $\dfrac{4(\sqrt{3}+2)}{3}$

28. A line is drawn form A $(-2, 0)$ to intersect the curve $y^2 = 4x$ in P and Q in the first quadrant such that $\dfrac{1}{AP} + \dfrac{1}{AQ} < \dfrac{1}{4}$, then slope of the line is always

a. $> \sqrt{3}$ **b.** $< \dfrac{1}{\sqrt{3}}$ **c.** $> \sqrt{2}$ **d.** $> \dfrac{1}{\sqrt{3}}$

29. Let $y = f(x)$ be a parabola, having its axis parallel to y-axis, which is touched by the line $y = x$ at $x = 1$, then

a. $2f(0) = 1 - f'(0)$ **b.** $f(0) + f'(0) + f''(0) = 1$

c. $f'(1) = 1$ **d.** $f'(0) = f'(1)$

30. An equilateral triangle SAB is inscribed in the parabola $y^2 = 4ax$ having its focus at S. If chord AB lies towards the left of S, then side length of this triangle is

a. $2a(2 - \sqrt{3})$ **b.** $4a(2 - \sqrt{3})$

c. $a(2 - \sqrt{3})$ **d.** $8a(2 - \sqrt{3})$

31. Let S be the focus of $y^2 = 4x$ and a point P is moving on the curve such that its abscissa is increasing at the rate of 4 units/sec, then the rate of increase of projection of SP on $x + y = 1$ when P is at (4, 4) is

a. $\sqrt{2}$ **b.** -1

c. $-\sqrt{2}$ **d.** $-\dfrac{3}{\sqrt{2}}$

32. Two parabolas have the same focus. If their directrices are the x-axis and the y-axis, respectively, then the slope of their common chord is

a. ± 1 **b.** $\dfrac{4}{3}$

c. $\dfrac{3}{4}$ **d.** none of these

33. C is the centre of the circle with centre (0, 1) and radius unity. P is the parabola $y = ax^2$. The set of values of a for which they meet at a point other than the origin, is

a. $a > 0$ **b.** $a \in \left(0, \dfrac{1}{2}\right)$

c. $\left(\dfrac{1}{4}, \dfrac{1}{2}\right)$ **d.** $\left(\dfrac{1}{2}, \infty\right)$

34. The length of the chord of the parabola $y^2 = x$ which is bisected at the point (2, 1) is

a. $2\sqrt{3}$ **b.** $4\sqrt{3}$ **c.** $3\sqrt{2}$ **d.** $2\sqrt{5}$

35. The circle $x^2 + y^2 = 5$ meets the parabola $y^2 = 4x$ at P and Q. Then the length PQ is equal to

a. 2 **b.** $2\sqrt{2}$

c. 4 **d.** none of these

36. The triangle PQR of area A is inscribed in the parabola $y^2 = 4ax$ such that the vertex P lies at the vertex of the parabola and the base QR is a focal chord. The modules of the difference of the ordinates of the points Q and R is

a. $\dfrac{A}{2a}$ **b.** $\dfrac{A}{a}$ **c.** $\dfrac{2A}{a}$ **d.** $\dfrac{4A}{a}$

37. If A_1B_1 and A_2B_2 are two focal chords of the parabola $y^2 = 4ax$, then the chords A_1A_2 and B_1B_2 intersect on

a. directrix **b.** axis

c. tangent at vertex **d.** none of these

38. If a line $y = 3x + 1$ cuts the parabola $x^2 - 4x - 4y + 20 = 0$ at A and B, then the tangent of the angle subtended by line segment AB at origin is

a. $\dfrac{8\sqrt{3}}{205}$ **b.** $\dfrac{8\sqrt{3}}{209}$

c. $\dfrac{8\sqrt{3}}{215}$ **d.** none of these

39. $P(x, y)$ is a variable point on the parabola $y^2 = 4ax$ and $Q(x + c, y + c)$ is another variable point, where c is a constant. The locus of the midpoint of PQ is

a. parabola **b.** ellipse

c. hyperbola **d.** circle

40. If a and c are the lengths of segments of any focal chord of the parabola $y^2 = 2bx$ $(b > 0)$, then the roots of the equation $ax^2 + bx + c = 0$ are

a. real and distinct **b.** real and equal

c. imaginary **d.** none of these

41. AB is a chord of the parabola $y^2 = 4ax$ with vertex A. BC is drawn is drawn perpendicular to AB meeting the axis at C. The projection of BC on the axis of the parabola is

a. a **b.** $2a$ **c.** $4a$ **d.** $8a$

42. Set of values of α for which the point $(\alpha, 1)$ lies inside the curves $c_1: x^2 + y^2 - 4 = 0$ and $c_2: y^2 = 4x$ is

a. $|\alpha| < \sqrt{3}$ **b.** $|\alpha| < 2$

c. $\dfrac{1}{4} < \alpha < \sqrt{3}$ **d.** none of these

43. If P is a point on the parabola $y^2 = 3(2x - 3)$ and M is the foot perpendicular drawn from P on the directrix of the parabola, then length of each side of an equilateral triangle SMP, where S is focus of the parabola is

a. 2 **b.** 4 **c.** 6 **d.** 8

44. If $y = mx + c$ touches the parabola $y^2 = 4a(x + a)$, then

a. $c = \frac{a}{m}$ **b.** $c = am + \frac{a}{m}$

c. $c = a + \frac{a}{m}$ **d.** none of these

45. The angle between the tangents to the parabola $y^2 = 4ax$ at the points where it intersects with the line $x - y - a = 0$ is

a. $\frac{\pi}{3}$ **b.** $\frac{\pi}{4}$ **c.** $\frac{\pi}{6}$ **d.** $\frac{\pi}{2}$

46. The area of the triangle formed by the tangent and the normal to the parabola $y^2 = 4ax$, both drawn at the same end of the latus rectum, and the axis of the parabola is

a. $2\sqrt{2}a^2$ **b.** $2a^2$

c. $4a^2$ **d.** none of these

47. Double ordinate AB of the parabola $y^2 = 4ax$ subtends an angle $\pi/2$ at the focus of the parabola, then tangents drawn to parabola at A and B will intersect at

a. $(-4a, 0)$ **b.** $(-2a, 0)$

c. $(-3a, 0)$ **d.** none of these

48. $y = x + 2$ is any tangent to the parabola $y^2 = 8x$. The point P on this tangent is such that the other tangent from it which is perpendicular to it is

a. (2, 4) **b.** (−2, 0)

c. (−1, 1) **d.** (2, 0)

49. The tangent and normal at the point $P(at^2, 2at)$ to the parabola $y^2 = 4ax$ meet the x-axis at T and G, respectively. Then the angle at which the tangent at P to the parabola is inclined to the tangent at P to the circle through P, T, G is

a. $\tan^{-1}(t^2)$ **b.** $\cot^{-1}(t^2)$ **c.** $\tan^{-1}(t)$ **d.** $\cot^{-1}(t)$

50. Angle between the tangents to the curve $y = x^2 - 5x + 6$ at the points (2, 0) and (3, 0) is

a. $\frac{\pi}{2}$ **b.** $\frac{\pi}{3}$ **c.** $\frac{\pi}{6}$ **d.** $\frac{\pi}{4}$

51. Radius of the circle that passes through origin and touches the parabola $y^2 = 4ax$ at the point $(a, 2a)$ is

a. $\frac{5}{\sqrt{2}}a$ **b.** $2\sqrt{2}\,a$ **c.** $\sqrt{\frac{5}{2}}\,a$ **d.** $\frac{3}{\sqrt{2}}a$

52. If the line $x + y = 1$ touches the parabola $y^2 - y + x = 0$, then the coordinates of the point of contact are

a. (1, 1) **b.** $\left(\frac{1}{2}, \frac{1}{2}\right)$ **c.** (0, 1) **d.** (1, 0)

53. Two straight lines are perpendicular to each other. One of them touches the parabola $y^2 = 4a(x + a)$ and the other touches $y^2 = 4b(x + b)$. Their point of intersection lies on the line

a. $x - a + b = 0$ **b.** $x + a - b = 0$

c. $x + a + b = 0$ **d.** $x - a - b = 0$

54. The mirror image of the parabola $y^2 = 4x$ in the tangent to the parabola at the point (1, 2) is

a. $(x - 1)^2 = 4(y + 1)$

b. $(x + 1)^2 = 4(y + 1)$

c. $(x + 1)^2 = 4(y - 1)$

d. $(x - 1)^2 = 4(y - 1)$

55. Consider the parabola $y^2 = 4x$. $A \equiv (4, -4)$ and $B \equiv (9, 6)$ be two fixed points on the parabola. Let C be a moving point on the parabola between A and B such that the area of the triangle ABC is maximum, then coordinate of C is

a. $\left(\frac{1}{4}, 1\right)$ **b.** (4, 4)

c. $(3, 2\sqrt{3})$ **d.** $(3, -2\sqrt{3})$

56. $\min\left[(x_1 - x_2)^2 + \left(5 + \sqrt{1 - x_1^2} - \sqrt{4x_2}\right)^2\right], \forall x_1, x_2 \in R$ is

a. $4\sqrt{5} + 1$ **b.** $4\sqrt{5} - 1$

c. $\sqrt{5} + 1$ **d.** $\sqrt{5} - 1$

57. Two mutually perpendicular tangents of the parabola $y^2 = 4ax$ meet the axes at P_1 and P_2. If S is the focus of the parabola, then $\frac{1}{(SP_1)} + \frac{1}{(SP_2)}$ is equal to

a. $\frac{4}{a}$ **b.** $\frac{2}{a}$ **c.** $\frac{1}{a}$ **d.** $\frac{1}{4a}$

58. A tangent is drawn to the parabola $y^2 = 4ax$ at the point P whose abscissa lies in the interval (1, 4). The maximum possible area of the triangle formed by the tangent at P ordinates of the point P and the x-axis is equal to

a. 8 **b.** 16 **c.** 24 **d.** 32

59. A parabola $y = ax^2 + bx + c$ crosses the x-axis at $(\alpha, 0)$ $(\beta, 0)$ both to the right of the origin. A circle also passes through these two points. The length of a tangent from the origin to the circle is

a. $\sqrt{\frac{bc}{a}}$ **b.** ac^2 **c.** $\frac{b}{a}$ **d.** $\sqrt{\frac{c}{a}}$

60. The straight line joining any point P on the parabola $y^2 = 4ax$ to the vertex and perpendicular from the focus to the tangent at P, intersect at R, then the equation of the locus of R is

a. $x^2 + 2y^2 - ax = 0$

b. $2x^2 + y^2 - 2ax = 0$

c. $2x^2 + 2y^2 - ay = 0$

d. $2x^2 + y^2 - 2ay = 0$

61. Through the vertex O of the parabola $y^2 = 4ax$, two chords OP and OQ are drawn and the circles on OP and OQ as diameters intersect in R. If θ_1, θ_2 and ϕ are the angles made with axis by the tangents at P and Q on the parabola and by OR, then the value of $\cot\theta_1 + \cot\theta_2$ is

a. $-2\tan\phi$ b. $-2\tan(\pi - \phi)$

c. 0 d. $2\cot\phi$

62. A line of slope $\lambda(0 < \lambda < 1)$ touches parabola $y + 3x^2 = 0$ at P. If S is the focus and M is the foot of the perpendicular of directrix from P, then $\tan\angle MPS$ equals

a. 2λ b. $\frac{2\lambda}{-1+\lambda^2}$

c. $\frac{1-\lambda^2}{1+\lambda^2}$ d. none of these

63. If $y = 2x - 3$ is a tangent to the parabola $y^2 = 4a\left(x - \frac{1}{3}\right)$, then a is equal to

a. $\frac{22}{3}$ b. -1 c. $\frac{14}{3}$ d. $\frac{-14}{3}$

64. AB is a double ordinate of the parabola $y^2 = 4ax$. Tangents drawn to the parabola at A and B meet y-axis at A_1 and B_1, respectively. If the area of trapezium AA_1B_1B is equal to $12a^2$, then angle subtended by A_1B_1 at the focus of the parabola is equal to

a. $2\tan^{-1}(3)$ b. $\tan^{-1}(3)$

c. $2\tan^{-1}(2)$ d. $\tan^{-1}(2)$

65. If $y + 3 = m_1(x + 2)$ and $y + 3 = m_2(x + 2)$ are two tangents to the parabola $y^2 = 8x$, then

a. $m_1 + m_2 = 0$ b. $m_1m_2 = -1$

c. $m_1m_2 = 1$ d. none

66. The tangent at any point P on the parabola $y^2 = 4ax$ intersects the y-axis at Q. The tangent to the circumcircle of triangle PQS (S is the focus) at Q is

a. a line parallel to the x-axis

b. y-axis

c. a line parallel to the y-axis

d. none of these

67. If $y = m_1x + c$ and $y = m_2x + c$ are two tangents to the parabola $y^2 + 4a(x + a) = 0$, then

a. $m_1 + m_2 = 0$ b. $1 + m_1 + m_2 = 0$

c. $m_1m_2 - 1 = 0$ d. $1 + m_1m_2 = 0$

68. If the parabola $y = ax^2 - 6x + b$ passes through (0, 2) and has its tangent at $x = \frac{3}{2}$ parallel to the$x$-axis then

a. $a = 2, b = -2$ b. $a = 2, b = 2$

c. $a = -2, b = 2$ d. $a = -2, b = -2$

69. If the angle between the tangents from the point $(\lambda, 1)$ to the parabola $y^2 = 16x$ be $\frac{\pi}{2}$ then λ is

a. 4 b. -4 c. -1 d. 2

70. Minimum area of circle which touches the parabolas $y = x^2 + 1$ and $y^2 = x - 1$ is

a. $\frac{9\pi}{16}$ sq. unit b. $\frac{9\pi}{32}$ sq. unit

c. $\frac{9\pi}{8}$ sq. unit d. $\frac{9\pi}{4}$ sq. unit

71. If the locus of middle of point of contact of tangent drawn to the parabola $y^2 = 8x$ and foot of perpendicular drawn from its focus to the tangents is a conic then length of latus rectum of this conic is

a. $\frac{9}{4}$ b. 9 c. 18 d. $\frac{9}{2}$

72. If d is the distance between parallel tangents with positive slope to $y^2 = 4x$ and $x^2 + y^2 - 2x + 4y - 11 = 0$, then

a. $10 < d < 2$ b. $4 < d < 6$

c. $d < 4$ d. none of these

73. If bisector of the angle APB, where PA and PB are the tangents to the parabola $y^2 = 4ax$, is equally inclined to the coordinate axes, then the point P lies on

a. tangent at vertex of the parabola

b. directrix of the parabola

c. circle with centre at the origin and radius a

d. the line of latus rectum

74. The locus of the centre of a circle which cuts orthogonally the parabola $y^2 = 4x$ at (1, 2) will pass though points

a. (3, 4) b. (4, 3) c. (5, 3) d. (2, 4)

75. From a point $A(t)$ on the parabola $y^2 = 4ax$, a focal chord and a tangent is drawn. Two circles are drawn in which one circle is drawn taking focal chord AB as diameter and other is drawn by taking intercept of tangent between point A and point P on the directrix, as diameter. Then the common chord of the circles is

a. line joining focus and P

b. line joining focus and A

c. tangent to the parabola at point A

d. none of these

76. The normal at the point $P(ap^2, 2ap)$ meets the parabola $y^2 = 4ax$ again at $Q(aq^2, 2aq)$ such that the lines joining the origin to P and Q are at right angle. Then

a. $p^2 = 2$ b. $q^2 = 2$ c. $p = 2q$ d. $q = 2p$

77. The set of points on the axis of the parabola $y^2 = 4x + 8$ from which the three normals to the parabola are all real and different is

a. $\{(k, 0) \mid k \le -2\}$ **b.** $\{(k, 0) \mid k > -2\}$

c. $\{(0, k) \mid k > -2\}$ **d.** none of these

78. Tangents and normal drawn to parabola $y^2 = 4ax$ at point $P(at^2, 2at)$, $t \neq 0$, meet the x-axis at points T and N, respectively. If S is the focus of the parabola, then

a. $SP = ST \neq SN$ **b.** $SP \neq ST = SN$

c. $SP = ST = SN$ **d.** $SP \neq ST \neq SN$

79. Locus of the midpoint of any normal chords of $y^2 = 4ax$ is

a. $x = a\left(\frac{4a^2}{y^2} - 2 + \frac{y^2}{2a^2}\right)$

b. $x = a\left(\frac{4a^2}{y^2} + 2 + \frac{y^2}{2a^2}\right)$

c. $x = a\left(\frac{4a^2}{y^2} - 2 - \frac{y^2}{2a^2}\right)$

d. $x = a\left(\frac{4a^2}{y^2} + 2 - \frac{y^2}{2a^2}\right)$

80. Normals AO, AA_1, AA_2 are drawn to the parabola $y^2 = 8x$ from the point $A(h, 0)$. If triangle OA_1A_2 is equilateral, then possible values of h is

a. 26 **b.** 24

c. 28 **d.** none of these

81. If $2x + y + \lambda = 0$ is a normal to the parabola $y^2 = -8x$, then λ is

a. 12 **b.** -12 **c.** 24 **d.** -24

82. At what point on the parabola $y^2 = 4x$ the normal makes equal angle with axes?

a. (4, 4) **b.** (9, 6) **c.** (4, -4) **d.** (1, $\pm$ 2)

83. If the normals to the parabola $y^2 = 4ax$ at three points P, Q, and R meet at A and S be the focus, then $SP \times SQ \times SR$ is equal to

a. a^2SA **b.** SA^3

c. aSA^2 **d.** none of these

84. If the normals to the parabola $y^2 = 4ax$ at the ends of the latus rectum meet the parabola at Q, Q', then QQ' is

a. $10a$ **b.** $4a$ **c.** $20a$ **d.** $12a$

85. From a point $(\sin\theta, \cos\theta)$ if three normals can be drawn to the parabola $y^2 = 4ax$, then the value of a is

a. $\left(\frac{1}{2}, 1\right)$ **b.** $\left[-\frac{1}{2}, 0\right)$

c. $\left[\frac{1}{2}, 1\right]$ **d.** $\left(-\frac{1}{2}, 0\right) \cup \left(0, \frac{1}{2}\right)$

86. If the normals at points t_1 and t_2 meet on the parabola, then

a. $t_1t_2 = -1$ **b.** $t_2 = -t_1 - \frac{2}{t_1}$

c. $t_1t_2 = 2$ **d.** none of these

87. Length of the normal chord of the parabola $y^2 = 4x$ which makes an angle of $\frac{\pi}{4}$ with the axis of x is

a. 8 **b.** $8\sqrt{2}$ **c.** 4 **d.** $4\sqrt{2}$

88. If the normal to a parabola $y^2 = 4ax$ at P meets the curve again in Q and if PQ and the normal at Q makes angle α and β, respectively, with the x-axis then $\tan\alpha$ ($\tan\alpha + \tan\beta$) has the value equal to

a. 0 **b.** -2 **c.** $-\frac{1}{2}$ **d.** -1

89. If two normals to a parabola $y^2 = 4ax$ intersect at right angles, then the chord joining their feet passes through a fixed point whose coordinates are

a. $(-2a, 0)$ **b.** $(a, 0)$ **c.** $(2a, 0)$ **d.** none

90. PQ is a normal chord of the parabola $y^2 = 4ax$ at P, A being the vertex of the parabola. Through P a line is drawn parallel to AQ meeting the x-axis in R. Then line length of AR is

a. equal to the length of the latus rectum

b. equal to the focal distance of the point P

c. equal to twice the focal distance of the point P

d. equal to the distance of the point P from the directrix

91. Tangent and normal drawn to parabola at $A(at^2, 2at)$, $t \neq 0$ meet the x-axis at point B and D, respectively. If the rectangle $ABCD$ is completed, then locus of C is

a. $y = 2a$ **b.** $y + 2a = c$

b. $x = 2a$ **d.** $x + 2a = 0$

92. An equation for the line that passes through (10, $-$ 1) and is perpendicular to $y = \frac{x^2}{4} - 2$ is

a. $4x + y = 39$ **b.** $2x + y = 19$

c. $x + y = 9$ **d.** $x + 2y = 8$

93. P, Q, R are the feet of the normals drawn to a parabola $(y - 3)^2 = 8(x - 2)$. A circle cuts the above parabola in points P, Q, R, and S. Then this circle always passes through the point

a. (2, 3) **b.** (3, 2) **c.** (0, 3) **d.** (2, 0)

94. Normals at two point (x_1, y_1) and (x_2, y_2) of parabola $y^2 = 4x$ meet again on the parabola where $x_1 + x_2 = 4$, then $|y_1 + y_2|$ equals to

a. $\sqrt{2}$ **b.** $2\sqrt{2}$

c. $4\sqrt{2}$ **d.** none of these

95. The end points of two normal chords of a parabola are concyclic, then the tangents at the feet of the normals will intersect at

a. tangent at vertex of the parabola

b. axis of the parabola

c. directrix of the parabola

d. none of these

96. The set of points on the axis of the parabola $(x-1)^2 = 8(y+2)$, from where three distinct normals can be drawn to the parabola is the set (h, k) of points satisfying

a. $h>2$ **b.** $h>1$

c. $k>2$ **d.** none of these

97. The shortest distance between the parabolas $2y^2 = 2x - 1$, $2x^2 = 2y - 1$ is

a. $2\sqrt{2}$ **b.** $\frac{1}{2\sqrt{2}}$ **c.** 4 **d.** $\sqrt{\frac{36}{5}}$

98. A tangent and normal is drawn at the point $P \equiv (16, 16)$ of the parabola $y^2 = 16x$ which cut the axis of the parabola at the points A and B, respectively. If the centre of the circle through P, A and B is C, then the angle between PC and the axis of x is

a. $\tan^{-1}\frac{1}{2}$ **b.** $\tan^{-1}2$ **c.** $\tan^{-1}\frac{3}{4}$ **d.** $\tan^{-1}\frac{4}{3}$

99. Length of the shortest normal chord of the parabola $y^2 = 4ax$ is

a. $2a\sqrt{27}$ **b.** $9a$

c. $a\sqrt{54}$ **d.** None of these

100. From the point (15, 12), three normals are drawn to the parabola $y^2 = 4x$, then centroid of triangle formed by three co-normal points is

a. $\left(\frac{16}{3}, 0\right)$ **b.** (4, 0) **c.** $\left(\frac{26}{3}, 0\right)$ **d.** (6, 0)

101. t_1 and t_2 are two points on the parabola $y^2 = 4ax$. If the focal chord joining them coincides with the normal chord, then

a. $t_1(t_1 + t_2) + 2 = 0$ **b.** $t_1 + t_2 = 0$

c. $t_1 t_2 = -1$ **d.** none of these

102. The line $x - y = 1$ intersects the parabola $y^2 = 4x$ at A and B. Normals at A and B intersect at C. If D is the point at which line CD is normal to the parabola, then coordinates of D are

a. $(4, -4)$ **b.** $(4, 4)$

c. $(-4, -4)$ **d.** none of these

103. If normals are drawn from a point $P(h, k)$ to the parabola $y^2 = 4ax$, then the sum of the intercepts which the normals cutoff from the axis of the parabola is

a. $(h + a)$ **b.** $3(h + a)$

c. $2(h + a)$ **d.** none of these

104. The radius of circle touching parabola $y^2 = x$ at (1, 1) and having directrix of $y^2 = x$ as its normal is

a. $\frac{5\sqrt{5}}{8}$ **b.** $\frac{10\sqrt{5}}{3}$

c. $\frac{5\sqrt{5}}{4}$ **d.** none of these

105. Normals drawn to $y^2 = 4ax$ at the points where it is intersected by the line $y = mx + c$ intersect at P. Foot of the another normal drawn to the parabola from the point P is

a. $\left(\frac{a}{m^2}, -\frac{2a}{m}\right)$ **b.** $\left(\frac{9a}{m}, \frac{-6a}{m}\right)$

c. $(am^2, -2am)$ **d.** $\left(\frac{4a}{m^2}, -\frac{4a}{m}\right)$

106. If two different tangents of $y^2 = 4x$ are the normals to $x^2 = 4by$ then

a. $|b| > \frac{1}{2\sqrt{2}}$ **b.** $|b| < \frac{1}{2\sqrt{2}}$

c. $|b| > \frac{1}{\sqrt{2}}$ **d.** $|b| < \frac{1}{\sqrt{2}}$

107. Maximum number of common normals of $y^2 = 4ax$ and $x^2 = 4by$ is equal to

a. 3 **b.** 4 **c.** 6 **d.** 5

108. A ray of light travels along a line $y = 4$ and strikes the surface of a curves $y^2 = 4(x + y)$, then equations of the line along which reflected ray travel is

a. $x = 0$ **b.** $x = 2$

c. $x + y = 4$ **d.** $2x + y = 4$

109. The largest value of a for which the circle $x^2 + y^2 = a^2$ falls totally in the interior of the parabola $y^2 = 4(x + 4)$ is

a. $4\sqrt{3}$ **b.** 4 **c.** $4\frac{\sqrt{6}}{7}$ **d.** $2\sqrt{3}$

110. If two chords drawn from the point $A(4, 4)$ to the parabola $x^2 = 4y$ are bisected by line $y = mx$, the interval in which m lies is

a. $(-2\sqrt{2}, 2\sqrt{2})$

b. $(-\infty, -\sqrt{2}) \cup (\sqrt{2}, \infty)$

c. $(-\infty, -2\sqrt{2}-2) \cup (2\sqrt{2}-2, \infty)$

d. none of these

111. The point of intersection of the tangents of the parabola $y^2 = 4x$, drawn at end points of the chord $x + y = 2$ lies on

a. $x - 2y = 0$ b. $x + 2y = 0$
c. $y - x = 0$ d. $x + y = 0$

112. The number of common chords of the parabolas $x = y^2 - 6y + 11$ and $y = x^2 - 6x + 1$ are

a. 1 b. 2 c. 4 d. 6

113. Two parabola have the focus (3, −2). Their directrices are the x-axis, and the y-axis, respectively. Then the slope of their common chord is

a. − 1 b. $-\frac{1}{2}$
c. $-\frac{\sqrt{3}}{2}$ d. none of these

114. If the tangent at the point $P(2, 4)$ to the parabola $y^2 = 8x$ meets the parabola $y^2 = 8x + 5$ at Q and R, then the mid-point of QR is

a. (4, 2) b. (2, 4)
c. (7, 9) d. None of these

Multiple Correct Answers Type

Each question has four choices a, b, c and d, out of which *one or more* answers are correct.

1. The equation of the directrix of the parabola with vertex at the origin and having the axis along the x-axis and a common tangent of slope 2 with the circle $x^2 + y^2 = 5$ is/are

a. $x = 10$ b. $x = 20$ c. $x = -10$ d. $x = -20$

2. A tangent is drawn at any point (x_1, y_1) other than vertex on the parabola $y^2 = 4ax$. If tangents are drawn from any point on this tangent to the circle $x^2 + y^2 = a^2$ such that all the chords of contact pass through a fixed point (x_2, y_2), then

a. x_1, a, x_2 are in G.P.

b. $\frac{y_1}{2}, a, y_2$ are in G.P.

c. $-4, \frac{y_1}{y_2}, \frac{x_1}{x_2}$ are in G.P.

d. $x_1x_2 + y_1y_2 = a^2$

3. If the focus of the parabola $x^2 - ky + 3 = 0$ is (0, 2), then a value of k is/are

a. 4 b. 6 c. 3 d. 2

4. Let P be a point whose coordinates differ by unity and the point does not lie on any of the axes of reference. If the parabola $y^2 = 4x + 1$ passes through P, then the ordinate of P may be

a. 3 b. − 1 c. 5 d. 1

5. If $y = 2$ be the directrix and (0, 1) be the vertex of the parabola $x^2 + \lambda y + \mu = 0$ then

a. $\lambda = 4$ b. $\mu = 8$ c. $\lambda = -8$ d. $\mu = 4$

6. The extremities of latus rectum of a parabola are (1, 1) and (1, −1), then the equation of the parabola can be

a. $y^2 = 2x - 1$ b. $y^2 = 1 - 2x$
c. $y^2 = 2x - 3$ d. $y^2 = 2x - 4$

7. Parabola $y^2 = 4x$ and the circle having it's centre at (6, 5) intersect at right angle. Possible point of intersection of these curves can be

a. (9, 6) b. $(2, \sqrt{8})$ c. (4, 4) d. $(3, 2\sqrt{3})$

8. A normal drawn to the parabola $y^2 = 4ax$ meets the curve again at Q such that angle subtended by PQ at vertex is 90°, then coordinates of P can be

a. $(8a, 4\sqrt{2}a)$ b. $(8a, 4a)$
c. $(2a, -2\sqrt{2}a)$ d. $(2a, 2\sqrt{2}a)$

9. A quadrilateral is inscribed in parabola, then

a. quadrilateral may be cyclic.
b. diagonal of the quadrilateral may be equal.
c. all possible pairs of adjacent sides may be perpendicular.
d. none of these

10. The locus of the midpoint of the focal distance of a variable point moving on the parabola $y^2 = 4ax$ is a parabola whose

a. latus rectum is half the latus rectum of the original parabola.
b. vertex is $(\frac{a}{2}, 0)$.
c. directrix is y-axis.
d. focus has the coordinates $(a, 0)$.

11. Which of the following line can be tangent to parabola $y^2 = 8x$?

a. $x - y + 2 = 0$ b. $9x - 3y + 2 = 0$
c. $x + 2y + 8 = 0$ d. $x + 3y + 12 = 0$

12. Which of the following line can be normal to parabola $y^2 = 12x$?

a. $x + y - 9 = 0$ b. $2x - y - 32 = 0$
c. $2x + y - 36 = 0$ d. $3x - y - 72 = 0$

13. A square has one vertex at the vertex of the parabola $y^2 = 4ax$ and the diagonal through the vertex lies along the axis of the parabola. If the ends of the other diagonal lie on the parabola, the coordinates of the vertices of the square are

a. $(4a, 4a)$ b. $(4a, -4a)$
c. (0, 0) d. $(8a, 0)$

Reasoning Type

Each question has four choices a, b, c and d, out of which *only one* is correct. Each question contains Statement 1 and Statement 2 .

a. Both the statements are True and Statement 1 is the correct explanation of Statement 2

b. Both the statements are True but Statement 1 is not the correct explanation of Statement 2

c. Statement 1 is True and Statement 2 is False

d. Statement 1 is False and Statement 2 is True

1. Statement 1: Slope of tangents drawn from (4, 10) to parabola $y^2 = 9x$ are $\frac{1}{4}, \frac{9}{4}$.

Statement 2: Two tangents can be drawn to parabola from any point lying outside parabola.

2. Statement 1: Through $(\lambda, \lambda + 1)$ there can't be more than one normal to the parabola $y^2 = 4x$, if $\lambda < 2$.

Statement 2: The point $(\lambda, \lambda + 1)$ lies outside the parabola for all $\lambda \neq 1$.

3. Statement 1: In parabola $y^2 = 4ax$, the circle drawn taking focal radii as diameter touches y-axis.

Statement 2: The portion of the tangent intercepted between point of contact and directix subtends 90° angle at focus.

4. Statement 1: The line joining points (8, – 8) and $(\frac{1}{2}, 2)$, which are on parabola $y^2 = 8x$, passes through focus of parabola.

Statement 2: Tangents drawn at (8, – 8) and $(\frac{1}{2}, 2)$ on the parabola $y^2 = 4ax$ are perpendicular.

5. Statement 1: The normals at the point (4, 4) and $\left(\frac{1}{4}, -1\right)$ of the parabola $y^2 = 4x$ are perpendicular.

Statement 2: The tangents to the parabola at the end of a focal chord are perpendicular.

6. Statement 1: The line $y = x + 2a$ touches the parabola $y^2 = 4a(x + a)$.

Statement 2: The line $y = mx + am + a/m$ touches $y^2 = 4a(x + a)$ for all real values of m.

7. Statement 1: Equation $(5x - 5)^2 + (5y + 10)^2 = (3x + 4y + 5)^2$ is parabola.

Statement 2: If distance of the point from the given line and from the given point (not lying on the given line) is equal, then locus of variable point is parabola.

8. Statement 1: Length of focal chord of a parabola $y^2 = 8x$ making on angle of 60° with x-axis is 32.

Statement 2: Length of focal chord of a parabola $y^2 = 4ax$ making an angle a with x-axis is $4a \operatorname{cosec}^2 \alpha$.

9. Statement 1: Circumcircle of a triangle formed by the lines $x = 0$, $x + y + 1 = 0$ and $x - y + 1 = 0$ also passes through the point (1, 0).

Statement 2: Circumcircle of a triangle formed by three tangents of a parabola passes through its focus.

10. Statement 1 : The point of intersection of the tangents at three distinct points A, B, C on the parabola $y^2 = 4x$ can be collinear.

Statement 2: If a line L does not intersect the parabola $y^2 = 4x$, then from every point of the line two tangents can be drawn to the parabola.

11. Statement 1: There are no common tangents between circle $x^2 + y^2 - 4x + 3 = 0$ and parabola $y^2 = 2x$.

Statement 2: Given circle and parabola do not intersect.

12. Statement 1: The line $ax + by + c = 0$ is a normal to the parabola $y^2 = 4ax$, then the equation of tangent at the foot of this normal is $y = (b/a)x + (a^2/b)$.

Statement 2: Equation of normal at any point $P(at^2, 2at)$ to the parabola $y^2 = 4ax$ is $y = -tx + 2at + at^3$.

13. Statement 1: The values of α for which the point (α, α^2) lies inside the triangle formed by the lines $x = 0$, $x + y = 2$ and $3y = x$ is (0, 1).

Statement 2: Parabola $y = x^2$ meets the line $x + y = 2$ at (1, 1).

14. Statement 1: If there exists points on the circle $x^2 + y^2 = a^2$ from which two perpendicular tangents can be drawn to parabola $y^2 = 2x$, then $a \geq 1/2$.

Statement 2: Perpendicular tangents to parabola meet on the directrix.

15. Statement 1: If the straight line $x = 8$ meets the parabola $y^2 = 8x$ at P and Q, then PQ substends a right angle at the origin.

Statement 2: Double ordinate equal to twice of latus rectum of a parabola subtends a right angle at the vertex.

16. Statement 1: Normal chord drawn at the point (8, 8) of the parabola $y^2 = 8x$ subtends a right angle at the vertex of the parabola.

Statement 2: Every chord of the parabola $y^2 = 4ax$ passing through the point $(4a, 0)$ subtends a right angle at the vertex of the parabola.

17. Statement 1: If end points of two normal chords AB and CD (normal at A and C) of a parabola $y^2 = 4ax$ are concyclic, then the tangents at A and C will intersect on the axis of the parabola.

Statement 2: If four points on the parabola $y^2 = 4ax$ are concyclic, then sum of their ordinates is zero.

18. Statement 1: If parabola $y^2 = 4ax$ and circle $x^2 + y^2 + 2bx = 0$ touch each other externally, then roots of the equation $f(x) = x^2 - (b + a + 1)x + a = 0$ has real roots.

Statement 2: For parabola and circle externally touching a and b must have the same sign.

19. Statement 1: Line $x - y - 5 = 0$ cannot be normal to parabola $(5x - 15)^2 + (5y + 10)^2 = (3x - 4y + 2)^2$.

Statement 2: Normal to parabola never passes through its focus.

20. Statement 1: AA' and BB' are double ordinates of the parabola. Then points A, A', B, B' are concyclic.

Statement 2: Circle can cut parabola in maximum four points.

Linked Comprehension Type

Based upon each paragraph, three multiple choice questions have to be answered. Each question has four choices a, b, c and d, out of which *only one* is correct.

For Problems 1–3

A tangent is drawn at any point $P(t)$ on the parabola $y^2 = 8x$ and on it is taken a point $Q(\alpha, \beta)$ from which pair of tangents QA and QB are drawn to the circle $x^2 + y^2 = 4$. Using this information answer the following questions

1. The locus of the point of concurrency of the chord of contact AB of the circle $x^2 + y^2 = 4$ is

a. $y^2 - 2x = 0$ **b.** $y^2 - x^2 = 4$
c. $y^2 + 4x = 0$ **d.** $y^2 - 2x^2 = 4$

2. The point from which perpendicular tangents can be drawn both to the given circle and the parabola is

a. $(4, \pm\sqrt{3})$ **b.** $(-1, \sqrt{2})$
c. $(-\sqrt{2}, -\sqrt{2})$ **d.** $(+2, \pm 2\sqrt{3})$

3. The locus of circumcentre of ΔAQB if $t = 2$ is

a. $x - 2y + 2 = 0$ **b.** $x + 2y - 4 = 0$
c. $x - 2y - 4 = 0$ **d.** $x + 2y + 4 = 0$

For Problems 4–6

Tangent to the parabola $y = x^2 + ax + 1$, at the point of intersection of y-axis also touches the circle $x^2 + y^2 = r^2$. Also no point of the parabola is below the x-axis.

4. The radius of circle when a attains its maximum value

a. $\frac{1}{\sqrt{10}}$ **b.** $\frac{1}{\sqrt{5}}$ **c.** 1 **d.** $\sqrt{5}$

5. The slope of the tangent when radius of the circle is maximum is

a. -1 **b.** 1 **c.** 0 **d.** 2

6. The minimum area bounded by the tangent and the coordinate axes is

a. 1 **b.** $\frac{1}{3}$ **c.** $\frac{1}{2}$ **d.** $\frac{1}{4}$

For Problems 7–9

If the locus of the circumcentre of a variable triangle having sides y-axis, $y = 2$ and $lx + my = 1$, where (l, m) lies on the parabola $y^2 = 4x$ is a curve C, then

7. Coordinates of the vertex of this curve C is

a. $\left(-2, \frac{3}{2}\right)$ **b.** $\left(-2, -\frac{3}{2}\right)$
c. $\left(2, \frac{3}{2}\right)$ **d.** $\left(-2, -\frac{3}{2}\right)$

8. The length of smallest focal chord of this curve C is

a. $\frac{1}{4}$ **b.** $\frac{1}{12}$ **c.** $\frac{1}{8}$ **d.** $\frac{1}{16}$

9. The curve C is symmetric about the line

a. $x = \frac{3}{2}$ **b.** $y = -\frac{3}{2}$ **c.** $x = -\frac{3}{2}$ **d.** $y = \frac{3}{2}$

For Problems 10–12

$y = x$ is tangent to the parabola $y = ax^2 + c$

10. If $a = 2$, then the value of c is

a. 1 **b.** $-\frac{1}{2}$ **c.** $\frac{1}{2}$ **d.** $\frac{1}{8}$

11. If (1, 1) is point of contact, then a is

a. $\frac{1}{4}$ **b.** $\frac{1}{3}$ **c.** $\frac{1}{2}$ **d.** $\frac{1}{6}$

12. If $c = 2$, then point of contact is

a. (3, 3) **b.** (2, 2) **c.** (6, 6) **d.** (4, 4)

For Problems 13–15

If l, m are variable real numbers such that $5l^2 + 6m^2 - 4lm + 3l = 0$, then variable line $lx + my = 1$ always touches a fixed parabola, whose axes is parallel to x-axis.

13. Vertex of the parabola is

a. $\left(-\frac{5}{3}, \frac{4}{3}\right)$ **b.** $\left(-\frac{7}{4}, \frac{3}{4}\right)$
c. $\left(\frac{5}{6}, -\frac{7}{6}\right)$ **d.** $\left(\frac{1}{2}, -\frac{3}{4}\right)$

14. Focus of the parabola is

a. $\left(\frac{1}{6}, -\frac{7}{6}\right)$ **b.** $\left(\frac{1}{3}, \frac{4}{3}\right)$
c. $\left(\frac{3}{2}, -\frac{3}{2}\right)$ **d.** $\left(-\frac{3}{4}, \frac{3}{4}\right)$

15. Directrix of the parabola is

a. $6x + 7 = 0$ **b.** $4x + 11 = 0$
c. $3x + 11 = 0$ **d.** none of these

For Problems 16–18

Consider the parabola whose focus is at (0, 0) and tangent at vertex is $x - y + 1 = 0$.

16. The length of latus rectum is

a. $4\sqrt{2}$ **b.** $2\sqrt{2}$ **c.** $8\sqrt{2}$ **d.** $3\sqrt{2}$

17. The length of the chord of parabola on the x-axis is

a. $4\sqrt{2}$ **b.** $2\sqrt{2}$ **c.** $8\sqrt{2}$ **d.** $3\sqrt{2}$

18. Tangents drawn to the parabola at the extremities of the chord $3x + 2y = 0$ intersect at an angle

a. $\frac{\pi}{6}$ **b.** $\frac{\pi}{3}$

c. $\frac{\pi}{2}$ **d.** none of these

For Problems 19–21

Two tangents on a parabola are $x - y = 0$ and $x + y = 0$. If (2, 3) is focus of the parabola, then

19. The equation of tangent at vertex is

a. $4x - 6y + 5 = 0$ **b.** $4x - 6y + 3 = 0$

c. $4x - 6y + 1 = 0$ **d.** $4x - 6y + 3/2 = 0$

20. Length of latus rectum of the parabola is

a. $\frac{6}{\sqrt{3}}$ **b.** $\frac{10}{\sqrt{13}}$

c. $\frac{2}{\sqrt{13}}$ **d.** none of these

21. If P, Q are ends of focal chord of the parabola, then $\frac{1}{SP} + \frac{1}{SQ} =$

a. $\frac{2\sqrt{13}}{3}$ **b.** $2\sqrt{13}$

c. $\frac{2\sqrt{13}}{5}$ **d.** none of these

For Problems 22–24

$y^2 = 4x$ and $y^2 = -8(x - a)$ intersect at points A and C. Points $O(0, 0)$, A, $B(a, 0)$, C are concyclic.

22. The length of common chord of parabolas is

a. $2\sqrt{6}$ **b.** $4\sqrt{3}$ **c.** $6\sqrt{5}$ **d.** $8\sqrt{2}$

23. The area of cyclic quadrilateral $OABC$ is

a. $24\sqrt{3}$ **b.** $48\sqrt{2}$ **c.** $12\sqrt{6}$ **d.** $18\sqrt{5}$

24. Tangents to parabola $y^2 = 4x$ at A and C intersect at point D and tangents to parabola $y^2 = -8(x - a)$ intersect at point E, then the area of quadrilateral $DAEC$ is

a. $96\sqrt{2}$ **b.** $48\sqrt{3}$ **c.** $54\sqrt{5}$ **d.** $36\sqrt{6}$

For Problems 25–27

PQ is double ordinate of the parabola $y^2 = 4x$ which passes through the focus S. ΔPQA is an isosceles right angled triangle, where A is on the axis of the parabola. Line PA meets the parabola at C and QA meets the parabola at B.

25. Area of the trapezium $PBCQ$ is

a. 96 sq. units **b.** 64 sq. units

c. 72 sq. units **d.** 48 sq. units

26. Circumradius of trapezium $PBCQ$ is

a. $6\sqrt{5}$ **b.** $3\sqrt{6}$ **c.** $2\sqrt{10}$ **d.** $5\sqrt{3}$

27. Ratio of inradius of ΔABC and that of ΔPAQ is

a. 2 : 1 **b.** 3 : 2

c. 4 : 3 **d.** 3 : 1

For Problems 28–30

Consider the inequality $9^x - a\,3^x - a + 3 \le 0$, where a is a real parameter.

28. The given inequality has at least one negative solution for $a \in$

a. $(-\infty, 2)$ **b.** $(3, \infty)$ **c.** $(-2, \infty)$ **d.** $(2, 3)$

29. The given inequality has at least one positive solution for $a \in$

a. $(-\infty, -2)$ **b.** $[3, \infty)$ **c.** $(2, \infty)$ **d.** $[-2, \infty)$

30. The given inequality has at least one real solution for $a \in$

a. $(-\infty, 3)$ **b.** $[2, \infty)$ **c.** $(3, \infty)$ **d.** $[-2, \infty)$

Matrix-Match Type

Each question contains statements given in two columns which have to be matched. Statements (a, b, c, d) in column I have to be matched with statements (p, q, r, s) in column II. If the correct matches are a → p, a → s; b → q, b → r; c → p, c → q and d → s; then the correctly bubbled 4 × 4 matrix should be as follows:

	p	q	r	s
a	(p) filled	(q)	(r)	(s) filled
b	(p)	(q) filled	(r) filled	(s)
c	(p) filled	(q) filled	(r)	(s)
d	(p)	(q)	(r)	(s) filled

1. Consider the parabola $(x - 1)^2 + (y - 2)^2 = \frac{(12x - 5y + 3)^2}{169}$

Column I	Column II
a. Locus of point of intersection of perpendicular tangent	**p.** $12x - 5y - 2 = 0$
b. Locus of foot of perpendicular from focus upon any tangent	**q.** $5x + 12y - 29 = 0$
c. Line along which minimum length of focal chord occurs	**r.** $12x - 5y + 3 = 0$
d. Line about which parabola is symmetrical	**s.** $24x - 10y + 1 = 0$

2. Consider the parabola $y^2 = 12x$.

Column I	Column II
a. Equation of tangent can be	**p.** $2x+y-6=0$
b. Equation of normal can be	**q.** $3x-y+1=0$
c. Equation of chord of contact w.r.t. any point on the directrix can be	**r.** $x-2y-12=0$
d. Equation of chord which subtends right angle at the vertex can be -	**s.** $2x-y-36=0$

3.

Column I	Column II
a. Tangents are drawn from point (2, 3) to the parabola $y^2 = 4x$, then points of contact are	**p.** (9, – 6)
b. From a point P on the circle $x^2+y^2 = 5$, the equation of chord of contact to the parabola $y^2 = 4x$ is $y = 2(x - 2)$, then the coordinate of point P will be	**q.** (1, 2)
c. $P(4, -4)$, Q are points on parabola $y^2 = 4x$ such that area of ΔPOQ is 6 sq. units where O is the vertex, then coordinates of Q may be	**r.** (–2, 1)
d. The common chord of circle $x^2+y^2=5$ and parabola $6y = 5x^2 + 7x$ will pass through point(s)	**s.** (4, 4)

4.

Column I	Column II
a. Points from which perpendicular tangents can be drawn to parabola $y^2 = 4x$	**p.** (–1, 2)
b. Points from which only one normal can be drawn to parabola $y^2 = 4x$	**q.** (3, 2)
c. Point at which chord $x - y + 1 = 0$ of parabola $y^2 = 4x$ is bisected	**r.** (–1, –5)
d. Points from which tangents cannot be drawn to parabola $y^2 = 4x$	**s.** (5, –2)

Integer Type

1. A beam is supported at its ends by supports 12m apart. Since the load is concentrated at its centre, there is a deflection of 3cm at the centre and the deflected beam is in the shape of a parabola. If L is the distance from the centre where the deflection is 1cm, then the value of $[L]$, where $[\cdot]$ represents the greatest integer function, is________.

2. The number of parabolas that can be drawn passing through the two ends of a latus rectum are________.

3. If M is the foot of the perpendicular from a point P on the parabola $y^2 = 4x$ to its directrix, and SPM is an equilateral triangle, where S is the focus, then SP is equal to________.

4. An equilateral triangle is inscribed in the parabola $y^2 = x$ in such a way that one of its vertices coincides with the vertex of the parabola. If L is the length of the sides of the triangle, then the value of $[L]$, where $[\cdot]$ represents the greatest integer function, is________.

5. The point $(a, 2a)$ is an interior point of the region bounded by the parabola $y^2 = 16x$ and the double ordinate through the focus. Then the number of integral values of a is________.

6. Latus rectum of the parabola which has axis parallel to y-axis and which passes through points (0, 2), (– 1, 0), and (1, 6) is________.

7. If the equation $(x-1)^2+(y-2)^2=\lambda(x+y+3)^2$ represents a parabola, then the value of $1/\lambda$ is________.

8. The line passing through the focus S of the parabola $y = ax^2 + bx + c$ meets the parabola at P and Q such that $SP = 4$ and $SQ = 6$. If L is the latus rectum, then the value of $[L]$, where $[\cdot]$ represents the greatest integer function, is________.

9. If a focal chord of $y^2 = 2x$ makes an angle $\alpha \in (0, \pi/4]$ with the positive direction of the x-axis, then the minimum length of the focal chord is________.

10. Two tangents are drawn from the point (–2, –1) to the parabola $y^2 = 4x$. If α is the angle between the tangents, then the value of $\tan\alpha$ is________.

11. Two tangents are drawn from the point (α, β) to the parabola $y^2 = 4x$. If the slope of one tangent is double that of the other, then the value of $2\beta^2/\alpha$ is________.

12. Tangents are drawn from the point (–1, 2) on the parabola $y^2 = 4x$. If the length of the intercept of these tangents on the line $x = 2$ is L, then the value of $[L]$, where $[\cdot]$ represents the greatest integer function, is________.

13. Tangents PA and PB are drawn from a point P on the directrix of the parabola $(x-2)^2 + (y-3)^2 = (5x - 12y + 3)^2/169$. If L is the least radius of the circumcircle of the triangle PAB, then the value of $[L]$, where $[\cdot]$ represents the greatest integer function, is________.

14. The maximum slope of the normal to $y^2 = 4ax$ passing through the point (15, 12) is________.

15. If L is the minimum distance between the curves $y^2 = 4x$ and $x^2 + y^2 - 12x + 31 = 0$, then the value of $[L]$, where $[\cdot]$ represents the greatest integer function, is__________.

16. If L is the length of the normal chord which subtends an angle of 90° at the vertex of the parabola $y^2 = 4x$, then the value of $[L/2]$, where $[\cdot]$ represents the greatest integer function, is__________.

17. If three distinct normals can be drawn to the parabola $y^2 - 2y = 4x - 9$ from the point $(2a, b)$, then the least integral value of a is__________.

18. If (h, k) is a point on the axis of the parabola $2(x - 1)^2 + 2(y - 1)^2 = (x + y + 2)^2$ from where three distinct normals can be drawn, then the least integral value of h is__________.

19. If the parabolas $y^2 = 4ax$ and $y^2 = 4c\ (x - b)$ have a common normal other than the x-axis (a, b, and c being distinct positive real numbers), then the least integral value of $b/(a - c)$ is__________.

20. If the tangent at the point P (2, 4) to the parabola $y^2 = 8x$ meets the parabola $y^2 = 8x + 5$ at Q and R, then the sum of the abscissa and ordinate of the midpoint of QR is__________.

21. PN is the ordinate of the parabola $y^2 = 4ax$. A straight line is drawn parallel to its axis which bisects PN and intersects the curve at Q. If NO meets the tangent at the vertex at the point T, then the value of $(2PN)/AT$ is__________.

22. If A is the area of the trapezium whose vertices lie on the parabola $y^2 = 4x$ and its diagonals pass through (1, 0) and having length 25/4 unit each, then the value of $4A/15$ is__________.

23. The radius of the largest circle which passes through the focus of the parabola $y^2 = 4(x + y)$ and contained in it is__________.

24. Three normals are drawn from the point $(c, 0)$ to the curve $y^2 = x$. If two normals other than the x-axis are perpendicular, then the value of $4c$ is__________.

25. Through the vertex O of the parabola $y^2 = 4x$, chords OP and OQ are drawn at right angles to one another. If the locus of the middle point of PQ is a parabola, then the latus rectum of the parabola is__________.

26. If $x + y = k$ is a normal to $y^2 = 12x$, then the value of k is__________.

ANSWERS AND SOLUTIONS

Objective Type

1. b. The coordinates of the focus and vertex of the required parabola are $S(a_1, 0)$ and $A(a, 0)$, respectively. Therefore, the distance between the vertex and focus is $AS = a_1 - a$ and so the length of the latus rectum is $4(a_1 - a)$.

Thus, the equation of the parabola is

$y^2 = 4(a_1 - a)(x - a)$.

2. d. Parabola having axis parallel to y-axis is

$(x - a)^2 = 4A(y - b)$.

According to question, length of latus rectum $4A = l$.

Hence, equation of parabola is

$(x - a)^2 = \frac{l}{4}(y - b)$

or $(x - a)^2 = \frac{l}{8}(2y - 2b)$

3. b. $x = 3\cos t, y = 4\sin t$

Eliminating t, we have

$\frac{x^2}{9} + \frac{y^2}{16} = 1$, which is an ellipse

$\therefore x^2 - 2 = 2\cos t;\ y = 4\cos^2\frac{t}{2}$

$\Rightarrow y = 2(1 + 2\cos t)$

$\Rightarrow y = 2\left(1 + \frac{x^2 - 2}{2}\right)$,

which is a parabola.

$\sqrt{x} = \tan t, \sqrt{y} = \sec t$

Eliminating t, we have

$y - x = 1$, which is a straight line

$x = \sqrt{1 - \sin t}, y$

$y = \sin\frac{t}{2} + \cos\frac{t}{2}$

Eliminating t, we have

$x^2 + y^2 = 1 - \sin t + 1 + \sin t = 2$, which is circle.

4. c. Tangent at the vertex is

$x - y + 1 = 0$ (1)

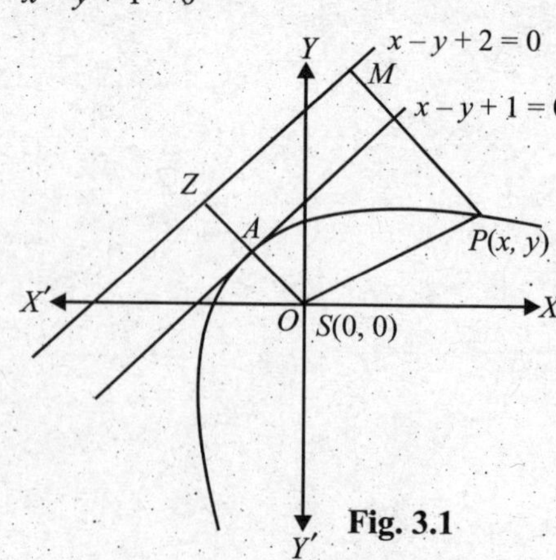

Fig. 3.1

Therefore, equation of axis of the parabola is

$x + y = 0$ (2)

Now solving Eqs. (1) and (2), we get $A\left(-\frac{1}{2}, \frac{1}{2}\right)$

$\therefore Z$ is $(-1, 1)$.

Now directrix is

$x - y + k = 0$

But this passes through $Z(-1, 1)$, so

$k = 2$

$\Rightarrow$ Directrix is $x - y + 2 = 0$.

Therefore, by definition equation of parabola is given by

$OP = PM$

$\Rightarrow OP^2 = PM^2$

$\Rightarrow \left(\frac{x - y + 2}{\sqrt{2}}\right)^2 = x^2 + y^2$

$\Rightarrow (x - y + 2)^2 = 2x^2 + 2y^2$

$\Rightarrow x^2 + y^2 + 4 - 2xy + 4x - 4y = 2x^2 + 2y^2$

$\Rightarrow x^2 + y^2 + 2xy - 4x + 4y - 4 = 0$

5. b. We have, $\sqrt{px} + \sqrt{qy} = 1$

$\Rightarrow (\sqrt{px} + \sqrt{qy})^2 = 1$

$\Rightarrow px + qy + 2\sqrt{(pq)(xy)} = 1$

$\Rightarrow (px + qy - 1)^2 = 4(pq)(xy)$

$\Rightarrow p^2x^2 - 2(pq)(xy) + q^2y^2 - 2px - 2qy + 1 = 0$

On comparing this equation with the equation

$ax^2 + 2hxy + by^2 + 2gx + 2fy + c = 0$, we get

$a = p^2, b = q^2, c = 1, g = -p, f = -q$ and $h = -pq$.

$\therefore \Delta = abc + 2fgh - af^2 - bg^2 - ch^2$

$= p^2q^2 - 2p^2q^2 - p^2q^2 - p^2q^2 - p^2q^2$

$= -4p^2q^2 \neq 0$

and, $h^2 - ab = p^2q^2 - p^2q^2 = 0$

Thus, we have $\Delta \neq 0$ and $h^2 = ab$

Hence, the given curve is a parabola.

6. d. Two parabolas are equal if the length of their latus rectum are equal.

Length of the latus rectum of $y^2 = \lambda x$ is λ.

The equation of the second parabola is

$25\{(x - 3)^2 + (y + 2)^2\} = (3x - 4y - 2)^2$

$\Rightarrow \sqrt{(x-3)^2+(y+2)^2} = \frac{|3x-4y-2|}{\sqrt{3^2+4^2}}$

Here focus is (3, – 2) and equation of the directrix is $3x - 4y - 2 = 0$.

Therefore, length of the latus rectum is 2 × distance between focus and directrix $= 2\left|\frac{3\times 3 - 4\times(-2) - 2}{\sqrt{3^2+(-4)^2}}\right| = 6$

Thus, the two parabolas are equal if $\lambda = 6$.

7. d. Length of latus rectum

= 2 × distance of focus form directrix

$= 2 \times \left|\frac{-\frac{u^2}{2g}\cos 2\alpha - \frac{u^2}{2g}}{\sqrt{1}}\right|$

$= \frac{2u^2}{g}\cos^2\alpha$

8. a. Make homogeneous and put $A + B = 0$.

$y^2 = 4ax\left(\frac{lx+my}{-n}\right)$

$\therefore 4al + n = 0$

9. a. $x^2 + y^2 - 2xy - 8x - 8y + 32 = 0$

$\Rightarrow (x-y)^2 = 8(x+y-4)$

is a parabola whose axis is $x - y = 0$ and the tangent at the vertex is $x + y - 4 = 0$.

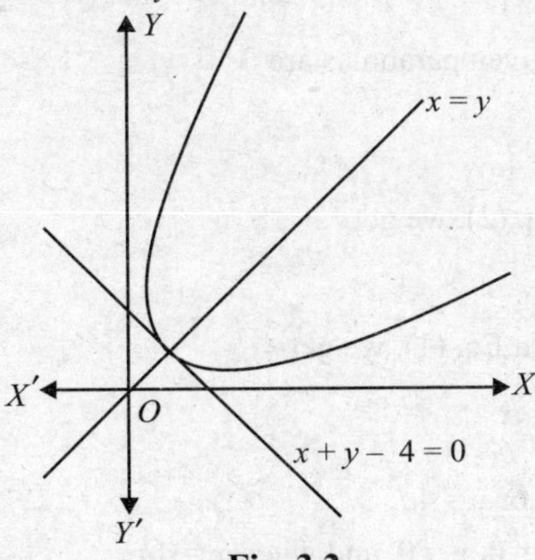

Fig. 3.2

Also, when $y = 0$, we have

$x^2 - 8x + 32 = 0$, which gives no real values of x.

When $x = 0$, we have $y^2 - 8y + 32 = 0$ which gives no real values of y.

So the parabola does not intersect the axes. Hence, the graph falls in the first quadrant.

10. d. Eliminating θ from the given equations, we get

$y^2 = -4a(x - a)$;

which is a parabola. But $0 \le \cos^2\theta \le 1$,

$0 \le x \le a$

and $-1 \le \sin\theta \le 1$

$\Rightarrow -2a \le y \le 2a$

Hence, the locus of the point P is not exactly the parabola, rather it is a part of the parabola.

11. c. $(\sqrt{3h}, \sqrt{3k+2})$ lies on the line $x - y - 1 = 0$

$\Rightarrow (\sqrt{3h})^2 = (\sqrt{3k+2}+1)^2$

$\Rightarrow 3h = 3k + 2 + 1 + 2\sqrt{3k+2}$

$\Rightarrow 3^2(h-k-1)^2 = 2^2(\sqrt{3k+2})^2$

$\Rightarrow 9(h^2 + k^2 + 1 - 2hk - 2h + 2k) = 4(3k+2)$

$\Rightarrow 9(x^2+y^2) - 18xy - 18x + 6y + 1 = 0$

Now $h^2 = ab$ and $\Delta \ne 0$

Therefore, locus is a parabola.

12. c.

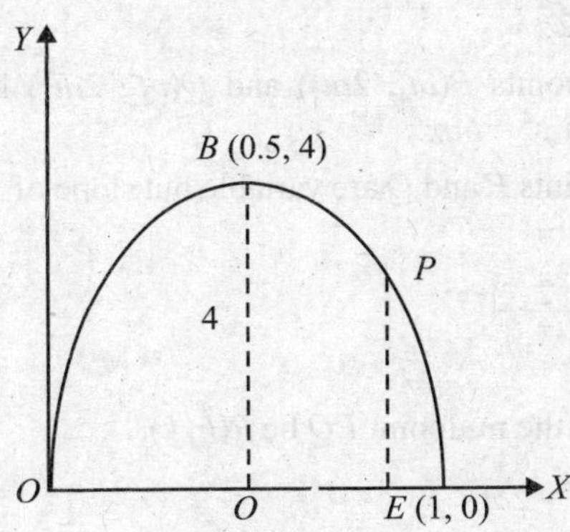

Fig. 3.3

The path of the water jet is a parabola.

Let its equation be $y = ax^2 + bx + c$

It should pass through (0, 0), (0.5, 4), (1, 0)

$\Rightarrow c = 0, a = -16, b = 16$

$\Rightarrow y = -16x^2 + 16x$

If $x = 0.75$, we get $y = 3$.

13. a. $x = t^2 - t + 1, y = t^2 + t + 1$

$\Rightarrow x + y = 2(t^2 + 1)$ and $y - x = 2t$

$\Rightarrow \frac{x+y}{2} = 1 + \left(\frac{y-x}{2}\right)^2$

$\Rightarrow (y-x)^2 = 2(x+y) - 4$

$\Rightarrow (y-x)^2 = 2(x+y-2)$

Vertex will be the point where lines $y - x = 0$ and $x + y - 2 = 0$ meet, i.e., the point (1, 1).

14. c. Let the point $P(h, k)$ on the parabola divide the line joining $A(4, -6)$ and $B(3, 1)$ in ratio λ.

Then, we have $(h, k) \equiv \left(\frac{3\lambda+4}{\lambda+1}, \frac{\lambda-6}{\lambda+1}\right)$

This point lies on the parabola, so

$$\left(\frac{\lambda-6}{\lambda+1}\right)^2 = 4\left(\frac{3\lambda+4}{\lambda+1}\right)$$

$$\Rightarrow (\lambda-6)^2 = 4(3\lambda+4)(\lambda+1)$$

$$\Rightarrow 11\lambda^2 + 40\lambda - 20 = 0$$

$$\Rightarrow \lambda = \frac{-20 \pm 2\sqrt{155}}{11} : 1$$

15. d. Let $P(x, y)$ be the coordinates of the other end of the chord. OP.

Then $\frac{x+0}{2} = a, \ \frac{y+0}{2} = b$

But (x, y) lies on the parabola. So

$$y^2 = 4x$$

$$\Rightarrow (2b)^2 = 4(2a)$$

$$\Rightarrow b^2 = 2a$$

16. c. Let points $P(at_1^2, 2at_1)$ and $Q(at_2^2, 2at_2)$ lie on the parabola $y^2 = 4ax$.

Here points P and Q are variable but slope of the chord PQ,

$$m_{PQ} = \frac{2}{t_1 + t_2}.$$

Now let the midpoint PQ be $R(h, k)$,

$$k = \frac{2at_1 + 2at_2}{2}$$

or $k = a(t_1 + t_2) = \frac{2}{m}$

$$\Rightarrow y = \frac{2}{m},$$

which is a line parallel to the axis of the parabola.

17. d. Any line passing through focus other than axis always meets parabola in two distinct points.

Hence, $m \in R - \{0\}$.

18. c. Since the semi-latus rectum of a parabola is the harmonic mean between the segments of any focal chord of a parabola, therefore SP, 4, SQ are in H.P.

$$\Rightarrow 4 = \frac{2SP \times SQ}{SP + SQ}$$

$$\Rightarrow 4 = \frac{2(6)(SQ)}{6 + SQ}$$

$$\Rightarrow 24 + 4(SQ) = 12(SQ)$$

$$\Rightarrow SQ = 3$$

19. a. The graph shows $\lambda > 0$.

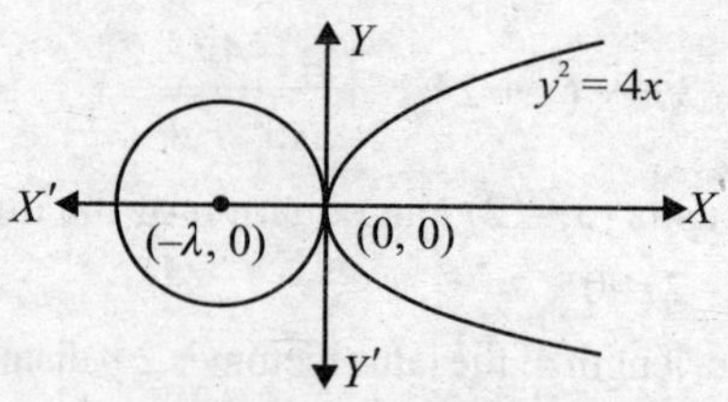

Fig. 3.4

20. c. Let x_1, x_2, and x_3 be the abscissae of the points on the parabola whose ordinates are y_1, y_2 and y_3, respectively. Then $y_1^2 = 4ax_1$, $y_2^2 = 4ax_2$ and $y_3^2 = 4ax_3$. Therefore, the area of the triangle whose vertices are (x_1, y_1) (x_2, y_2) and (x_3, y_3) is

$$\Delta = \frac{1}{2}\begin{vmatrix} x_1 & y_1 & 1 \\ x_2 & y_2 & 1 \\ x_3 & y_3 & 1 \end{vmatrix}$$

$$= \frac{1}{2}\begin{vmatrix} \frac{y_1^2}{4a} & y_1 & 1 \\ \frac{y_2^2}{4a} & y_2 & 1 \\ \frac{y_3^2}{4a} & y_3 & 1 \end{vmatrix}$$

$$= \frac{1}{8a}\begin{vmatrix} y_1^2 & y_1 & 1 \\ y_2^2 & y_2 & 1 \\ y_3^2 & y_3 & 1 \end{vmatrix}$$

$$= \frac{1}{8a}(y_1 - y_2)(y_2 - y_3)(y_3 - y_1)$$

21. d. The given parabolas are

$$y^2 = 4ax \quad (1)$$

and $x^2 = 4ay$ (2)

From Eq. (2), we get

$$y = \frac{x^2}{4a}$$

Putting in Eq. (1), we get

$$\frac{x^4}{16a^2} = 4ax$$

$$\Rightarrow x = 0 \text{ or } x = 4a$$

When $x = 0$, $y = 0$, and when $x = 4a$,

$$y = \frac{16a^2}{4a} = 4a$$

Thus, Eqs. (1) and (2) meet at $(0, 0)$ and $(4a, 4a)$.

Now $2bx + 3cy + 4d = 0$ passes through $(4a, 4a)$ and $(0, 0)$. So $d = 0$ and $2b(4a) + 3c(4a) = 0$. Therefore,

$$2b + 3c = 0$$

$$\Rightarrow d^2 + (2b + 3c)^2 = a^2$$

22. b. Let $R(h, k)$ be the midpoint of PQ. Therefore, Q is $(2h - 1, 2k)$

Since Q lies on $y^2 = 8x$, $(2k)^2 = 8(2h - 1)$

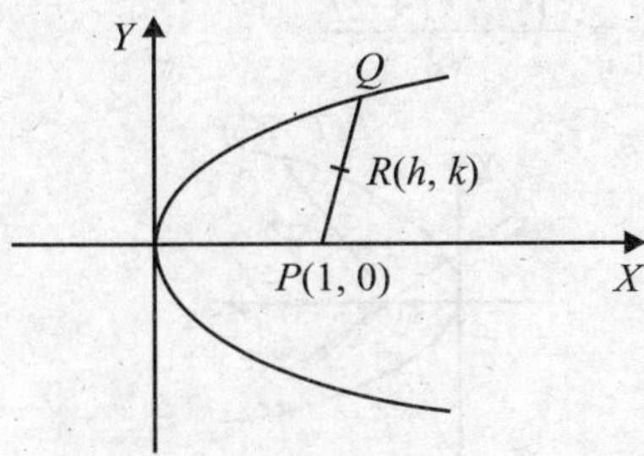

Fig. 3.5

Hence, locus of $Q(h, k)$ is

$y^2 = 2(2x - 1)$

or $y^2 = 4x - 2$

$\Rightarrow y^2 - 4x + 2 = 0$

23. a. The family of parabolas is $y = \dfrac{a^3x^2}{3} + \dfrac{a^2x}{2} - 2a$ and the vertex is $A\left(\dfrac{-B}{2A}, \dfrac{-D}{4A}\right) = (h, k)$

$$\Rightarrow h = -\frac{\frac{a^2}{2}}{2\frac{a^3}{3}} = -\frac{3}{4a}$$

$$\text{and } k = \frac{\left(\frac{a^2}{2}\right)^2 - \frac{4a^3(-2a)}{3}}{4\frac{a^3}{3}}$$

$$\Rightarrow h = -\frac{3}{4a} \text{ and } k = -\frac{35a}{16}$$

Eliminating a, we have $hk = 105/64$.

Hence, the required locus is $xy = 105/64$.

24. c. Let $C_1(h, k)$ be the centre of the circle.

Circle touches the x-axis then its radius is $r_1 = k$.

Also circle touches the circle with centre $C_2(0, 3)$ and radius $r_2 = 2$.

$\therefore |C_1 C_2| = r_1 + r_2$

$$\Rightarrow \sqrt{(h-0)^2 + (k-3)^2} = |k+2|$$

Squaring, we get

$h^2 - 10k + 5 = 0$

$\Rightarrow$ Locus is $x^2 - 10y + 5 = 0$, which is a parabola.

25. b. Let $P(x, y)$ be the point of contact. At P both of them must have same slope. We have,

$$2y\frac{dy}{dx} = 4a,\ 2x = 4a\frac{dy}{dx}$$

Eliminating $\dfrac{dy}{dx}$, we get $xy = 4a^2$.

26. a. Let the point be $P(at^2, 2at)$.

Then according to question, $SP = at^2 + a = k$ (1)

Let (α, β) be the moving point, then $\alpha = at^2, \beta = 2at$

$$\Rightarrow \frac{\alpha}{\beta} = \frac{t}{2}$$

and $a = \dfrac{\beta^2}{4\alpha}$ ($\because$ point (α, β) lies on $y^2 = 4ax$)

On substituting these values in Eq. (1), we get

$$\frac{\beta^2}{4\alpha}\left(1 + \frac{4\alpha^2}{\beta^2}\right) = k$$

$\Rightarrow \beta^2 + 4\alpha^2 = 4k\alpha$

$\Rightarrow 4x^2 + y^2 - 4kx = 0$ is the required locus.

27. d. $y - \sqrt{3}\,x + 3 = 0$ can be rewritten as

$$\frac{y - 0}{\sqrt{3}/2} = \frac{x - \sqrt{3}}{1/2} = r \qquad (1)$$

On solving Eq. (1) with the parabola $y^2 = x + 2$, we get

$$\frac{3r^2}{4} = \frac{r}{2} + \sqrt{3} + 2$$

$\Rightarrow 3r^2 - 2r - (4\sqrt{3} + 8) = 0$

$\Rightarrow AP \times AQ = |r_1 r_2|$

$$= \frac{4(\sqrt{3} + 2)}{3} \quad \text{(product of roots)}$$

28. a. Let $P(-2 + r\cos\theta, r\sin\theta)$ and Q lie on the parabola.

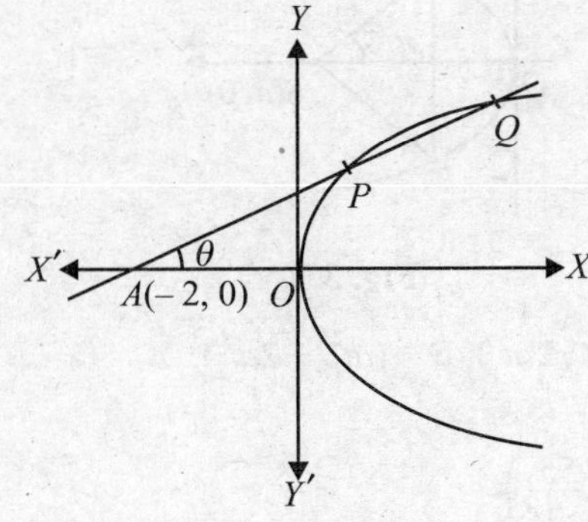

Fig. 3.6

$\Rightarrow r^2\sin^2\theta - 4(-2 + r\cos\theta) = 0$

$$\Rightarrow r_1 + r_2 = \frac{4\cos\theta}{\sin^2\theta},\ r_1r_2 = \frac{8}{\sin^2\theta}$$

$$\text{Now } \frac{1}{AP} + \frac{1}{AQ} = \frac{r_1 + r_2}{r_1r_2} = \frac{\cos\theta}{2}$$

Given that $\dfrac{1}{AP} + \dfrac{1}{AQ} < \dfrac{1}{4}$. So,

$\cos\theta < \frac{1}{2}$

$\Rightarrow \tan\theta > \sqrt{3}$

[$\because$ $\cos\theta$ is decreasing and $\tan\theta$ is increasing in $(0, \pi/2)$]

$\Rightarrow m > \sqrt{3}$

29. a. The general equation of a parabola having its axis parallel to y-axis is

$y = ax^2 + bx + c$ (1)

This is touched by the line $y = x$ at $x = 1$.

Therefore, slope of the tangent at (1, 1) is 1 and $x = ax^2 + bx + c$ must have equal roots.

$\Rightarrow \left(\frac{dy}{dx}\right)_{(1,1)} = 1$ and $(b-1)^2 = 4ac$

$\Rightarrow 2a + b = 1$ and $(b-1)^2 = 4ac$

Also, (1, 1) lies on Eq. (1)

$\Rightarrow a + b + c = 1$

From $2a + b = 1$ and $a + b + c = 1$,

$a - c = 0$.

$\Rightarrow a = c$

Then from $a + b + c = 1$, and $2c + b = 1$, we get

$2f(0) + f'(0) = 1$. [$\because f(0) = c$ and $f'(0) = b$]

30. b.

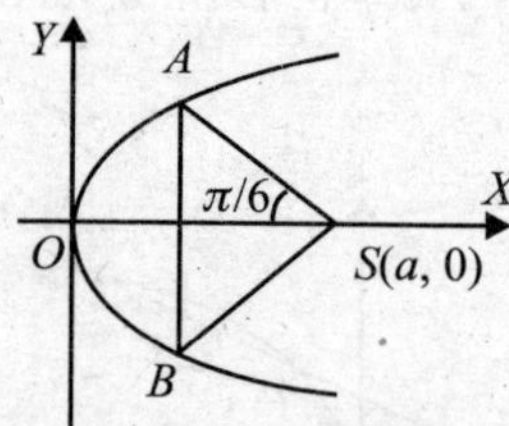

Fig. 3.7

Let $A = (at_1^2, 2at_1)$, $B \equiv (at_1^2, -2at_1)$.

We have

$m_{AS} = \tan\left(\frac{5\pi}{6}\right)$

$\Rightarrow \frac{2at_1}{at_1^2 - a} = -\frac{1}{\sqrt{3}}$

$\Rightarrow t_1^2 + 2\sqrt{3}t_1 - 1 = 0$

$\Rightarrow t_1 = -\sqrt{3} \pm 2$

Clearly, $t_1 = -\sqrt{3} - 2$ is rejected.

Thus $t_1 = (2 - \sqrt{3})$

Hence, $AB = 4at_1 = 4a(2 - \sqrt{3})$

31. c. $\vec{V} = (T^2 - 1)\hat{i} + 2T\hat{j}$

$\vec{n} = \hat{j} - \hat{i}$

Projection of $\vec{V}$ on $\vec{n}$,

$y = \frac{\vec{V}\cdot\vec{n}}{|\vec{n}|} = \frac{(1 - T^2) + 2T}{\sqrt{2}}$

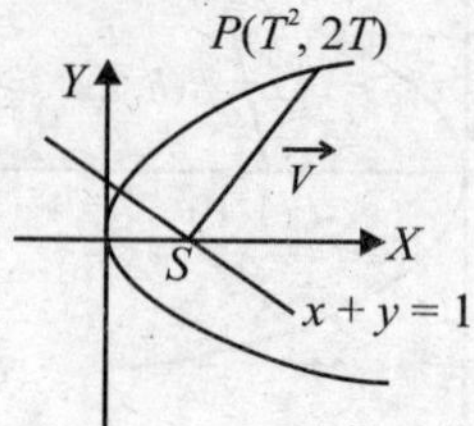

Fig. 3.8

Given $\frac{dx}{dt} = 4$. But $x = T^2$, so

$\frac{dx}{dt} = 2T\frac{dT}{dt}$

When $P(4, 4)$, $T = 2$

$\Rightarrow 4 = 2(2)\frac{dT}{dt}$

$\Rightarrow \frac{dT}{dt} = 1$

Now $\frac{dy}{dt} = \left(\frac{-2T + 2}{\sqrt{2}}\right)\frac{dT}{dt}$

Therefore, at $T = 2$,

$\sqrt{2}\frac{dy}{dt} = -4 + 2 = -2$

$\Rightarrow \frac{dy}{dt} = -\sqrt{2}$

32. a. Let focus be (a, b).

Equations are

$S_1 : (x-a)^2 + (y-b)^2 = x^2$

and $S_2 : (x-a)^2 + (y-b)^2 = y^2$

Common chord $S_1 - S_2 = 0$ gives $x^2 - y^2 = 0$

$\Rightarrow y = \pm x$.

33. d. Put $x^2 = \frac{y}{a}$ in circle, $x^2 + (y-1)^2 = 1$, we get (Note that for $a < 0$ they cannot intersect other than origin) $\frac{y}{a} + y^2 - 2y = 0$. Hence, we get $y = 0$ or $y = 2 - \frac{1}{a}$

Substituting $y = 2 - \frac{1}{a}$ in $y = ax^2$, we get

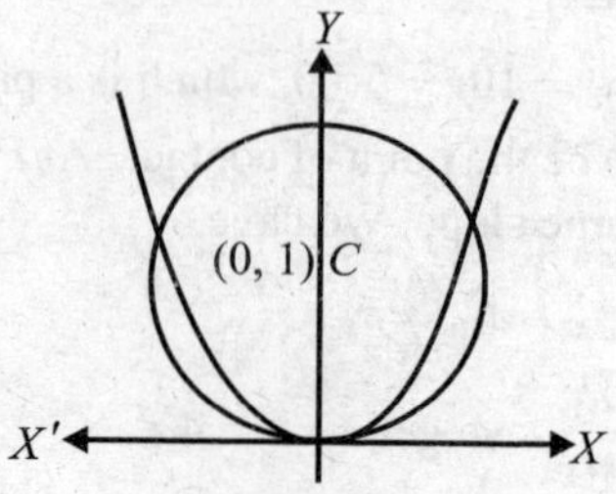

Fig. 3.9

$ax^2 = 2 - \frac{1}{a}$

$\Rightarrow x^2 = \frac{2a-1}{a^2} > 0$

$\Rightarrow a > \frac{1}{2}$

34. d. Chord through (2, 1) is $\frac{x-2}{\cos\theta} = \frac{y-1}{\sin\theta} = r$ (1)

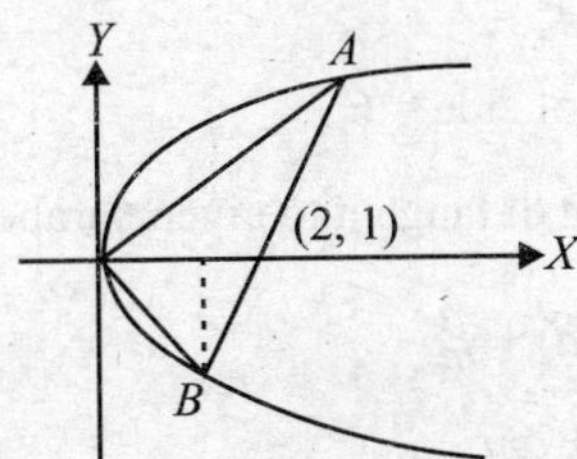

Fig. 3.10

Solving Eq. (1) with parabola $y^2 = x$, we have

$(1 + r\sin\theta)^2 = 2 + r\cos\theta$

$\Rightarrow \sin^2\theta\, r^2 + (2\sin\theta - \cos\theta)r - 1 = 0$

This equation has two roots $r_1 = AC$ and $r_2 = -BC$.

Then, sum of roots $r_1 + r = 0$

$\Rightarrow 2\sin\theta - \cos\theta = 0 \Rightarrow \tan\theta = \frac{1}{2}$.

$AB = |r_1 - r_2|$

$= \sqrt{(r_1 + r_2)^2 - 4r_1 r_2}$

$= \sqrt{4\frac{1}{\sin^2\theta}} = 2\sqrt{5}$

35. c. Solving circle $x^2 + y^2 = 5$ and parabola $y^2 = 4x$, we have

$x^2 + 4x - 5 = 0$

$\Rightarrow x = 1$

or $x = -5$ (not possible)

$\Rightarrow y^2 = 4$ or $y = \pm 2$

$\Rightarrow$ Points of intersection are $P(1, 2)$; $Q(1, -2)$

Hence, $PQ = 4$.

36. c. Difference of the ordinate

$d = \left|2at + \frac{2a}{t}\right| = 2a\left|t + \frac{1}{t}\right|$

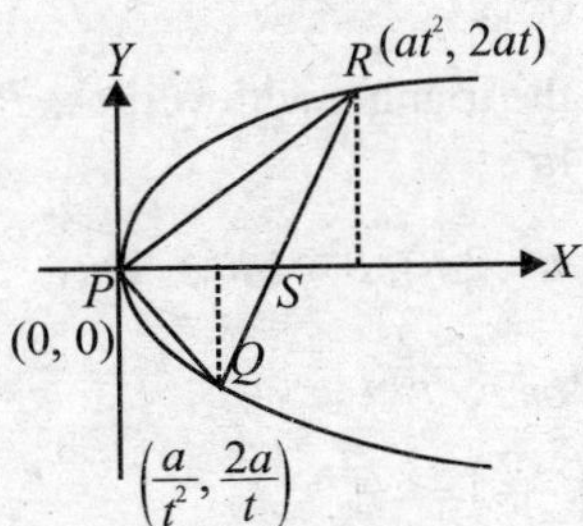

Fig. 3.11

Now area $A = \frac{1}{2}\begin{vmatrix} at^2 & 2at & 1 \\ \frac{a}{t^2} & -\frac{2a}{t} & 1 \\ 0 & 0 & 1 \end{vmatrix}$

$= a^2\left(t + \frac{1}{t}\right)$

$\Rightarrow 2a\left(t + \frac{1}{t}\right) = \frac{2A}{a}$

37. a. A_1B_1 is a focal chord, then $A_1(at_1^2, 2at_1)$ and $B_1\left(\frac{a}{t_1^2}, \frac{-2a}{t_1}\right)$.

A_2B_2 is a focal chord, then $A_2(at_2^2, 2at_2)$ and $B_2\left(\frac{a}{t_2^2}, \frac{-2a}{t_2}\right)$.

Now equation of chord A_1A_2 is

$y(t_1 + t_2) - 2x - 2at_1t_2 = 0$ (1)

Chord B_1B_2 is

$y\left(-\frac{1}{t_1} - \frac{1}{t_2}\right) - 2x - 2a\left(-\frac{1}{t_1}\right)\left(-\frac{1}{t_2}\right) = 0$

or $y(t_1 + t_2) + 2x\,t_1t_2 + 2a = 0$ (2)

For their intersection, we subtract them and get

$2x(t_1t_2 + 1) + 2a(t_1t_2 + 1) = 0$

or $(x + a)(1 + t_1t_2) = 0$

$\Rightarrow x + a = 0$

Hence, they intersect on the directrix.

38. b. Joint equation of OA and OB is

$x^2 - 4x(y - 3x) - 4y(y - 3x) + 20(y - 3x)^2 = 0$

Making equation of parabola homogeneous using straight line.

$\Rightarrow x^2(1 + 12 + 180) - y^2(4 - 20) - xy(4 - 12 + 120) = 0$

$\Rightarrow 193x^2 + 16y^2 - 112xy = 0$

$\tan\theta = \frac{2\sqrt{h^2 - ab}}{a + b}$

$= \frac{2\sqrt{56^2 - 193 \times 16}}{193 + 16}$

$= \frac{8\sqrt{3}}{209}$

39. a. Let the midpoint of PQ be (α, β)

$\Rightarrow \alpha = x + \frac{c}{2}$ and $\beta = y + \frac{c}{2}$

$\Rightarrow \left(\beta - \frac{c}{2}\right)^2 = 4a\left(\alpha - \frac{c}{2}\right)$

$\Rightarrow \left(y - \frac{c}{2}\right)^2 = 4a\left(x - \frac{c}{2}\right)$,

which is the required locus.

40. c. Latus rectum of $y^2 = 2bx$ is $2b$. Semi-latus rectum is b. We know that semi-latus rectum is H.M. of segments of focal chord. Therefore,

$\frac{1}{a} + \frac{1}{c} = \frac{2}{b}$

$\Rightarrow b = \frac{2ac}{a+c}$

Now for $ax^2 + bx + c = 0$,

$D = b^2 - 4ac$

$= \left(\frac{2ac}{a+c}\right)^2 - 4ac$

$= -4ac\left(\frac{a^2 + c^2 - ac}{(a+c)^2}\right)$

< 0

Hence, roots are imaginary.

41. c. $\tan\theta = \frac{y}{x}$

Projection of BC on the x-axis

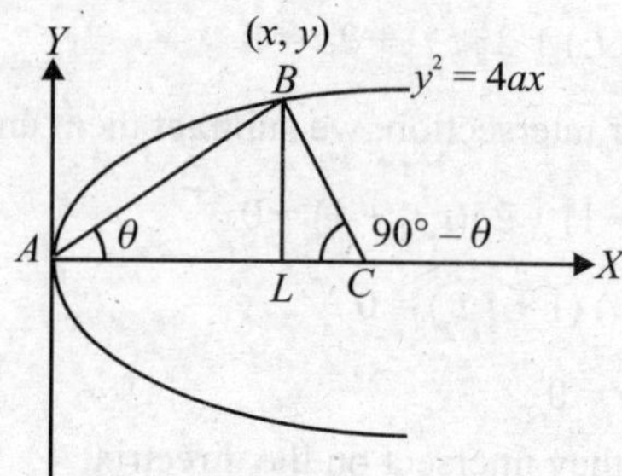

Fig. 3.12

$LC = \frac{y}{\tan(90° - \theta)} = y\tan\theta$

$\frac{y^2}{x} = 4a$

42. c. $\alpha^2 + 1 - 4 < 0$

$\Rightarrow \alpha^2 < 3, |\alpha| < \sqrt{3}$

$\Rightarrow 1 - 4\alpha < 0$

$\Rightarrow \alpha > \frac{1}{4}$

43. c. $y^2 = 6\left(x - \frac{3}{2}\right)$

Equation of directrix is

$x - \frac{3}{2} = -\frac{3}{2}$, i.e, $x = 0$

Let coordinates of P be $\left(\frac{3}{2} + \frac{3}{2}t^2, 3t\right)$.

Therefore, coordinate of M are $(0, 3t)$

$\Rightarrow MS = \sqrt{9 + 9t^2}$ and

$MP = \frac{3}{2} + \frac{3}{2}t^2$

$\therefore 9 + 9t^2 = \left(\frac{3}{2} + \frac{3}{2}t^2\right)^2$

$= \frac{9}{4}(1 + t^2)^2$

$\therefore 4 = 1 + t^2$

$\therefore$ Length of side = 6

44. d. Equation of tangent to given parabola having slope m is

$y = m(x + a) + \frac{a}{m}$

or $y = mx + am + \frac{a}{m}$

Comparing Eq. (1) with $y = mx + c$, we have

$c = am + \frac{a}{m}$.

45. d. The coordinates of the focus of the parabola $y^2 = 4ax$ are $(a, 0)$. The line $y - x - a = 0$ passes through this point. Therefore, it is a focal chord of the parabola. Hence, the tangent intersect at right angle.

46. c.

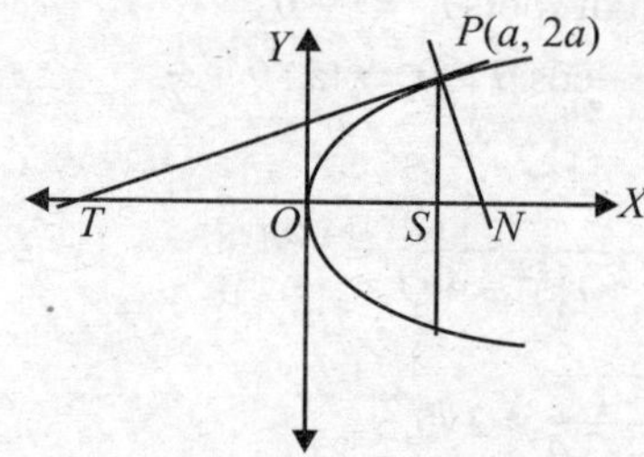

Fig. 3.13

One end of the latus rectum is $P(a, 2a)$.

The equation of the tangent PT at $P(a, 2a)$ is

$2ya = 2a(x + a)$, i.e., $y = x + a$

The equation of normal PN at $(a, 2a)$ is

$y + x = 2a + a$, i.e., $y + x = 3a$

Solving $y = 0$ and $y = x + a$, we get

$x = -a, y = 0$.

Solving $y = 0, y + x = 3a$, we get

$x = 3a, y = 0$.

The area of the triangle with vertices $P(a, 2a)$, $T(-a, 0)$, $N(3a, 0)$ is $4a^2$.

47. a. Let $A \equiv (at^2, 2at)$, $B \equiv (at^2, -2at)$.

$m_{OA} = \frac{2}{t}, m_{OB} = \frac{-2}{t}$

Thus, $\left(\frac{2}{t}\right)\left(\frac{-2}{t}\right) = -1$

$\Rightarrow t^2 = 4.$

Thus, tangents will intersect at $(-4a, 0)$.

48. b. Clearly P is the point of intersection of two perpendicular tangents to the parabola $y^2 = 8x$.

Hence, P must lie on the directrix $x + a = 0$ or $x + 2 = 0$

$\therefore x = -2.$

Hence, the point is $(-2, 0)$.

49. c.

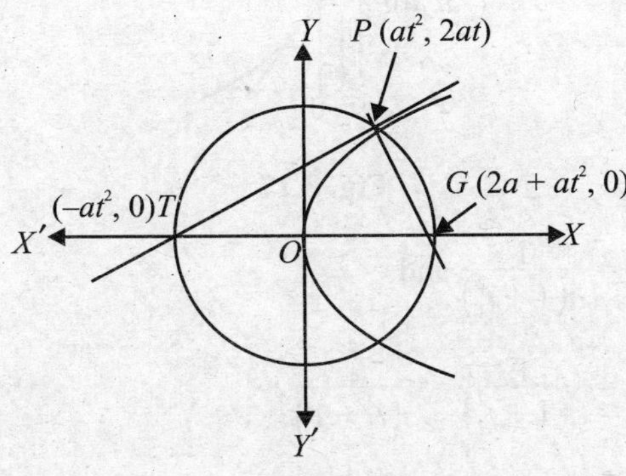

Fig. 3.14

Tangent and normal at $P(at^2, 2at)$ to the parabola $y^2 = 4ax$ is

$ty = x + at^2$ (1)

and $y = -tx + 2at + at^3$ (2)

Equations (1) and (2) meet the x-axis where $y = 0$.

From Eq. (1), we get $x = -at^2$. So T is $(-at^2, 0)$.

From Eq. (2), we get $tx = 2at + at^3$. So G is $(2a + at^2, 0)$.

Midpoint of TG is $\left(\dfrac{2a + at^2 - at^2}{2}, 0\right) = O\,(a, 0)$

Since $\angle TPG = 90°$, centre of the circle through PTG is $(a, 0)$.

If θ is the angle between tangents at P to the parabola and circle through P, T, G, then $(90° - \theta)$ is the angle between PT and OP.

Slope of $PT = \dfrac{2at}{2at^2} = \dfrac{1}{t}.$

Slope of $OP = \dfrac{2at}{a(t^2 - 1)} = \dfrac{2t}{t^2 - 1}.$

$$\therefore \tan(90° - \theta) = \left|\frac{\frac{1}{t} - \frac{2t}{t^2 - 1}}{1 + \frac{1}{t}\left(\frac{2t}{t^2 - 1}\right)}\right| = \frac{1}{t}$$

$\therefore \cot\theta = \dfrac{1}{t} \Rightarrow \tan\theta = t$

$\Rightarrow \theta = \tan^{-1}(t)$

50. a. $\dfrac{dy}{dx} = 2x - 5$

$\therefore m_1 = \left(\dfrac{dy}{dx}\right)_{(2,\,0)} = 4 - 5 = -1$

and $m_2 = \left(\dfrac{dy}{dx}\right)_{(3,\,0)} = 6 - 5 = 1$

$\Rightarrow m_1 m_2 = -1 \Rightarrow$ angle between tangents $= \dfrac{\pi}{2}.$

51. a.

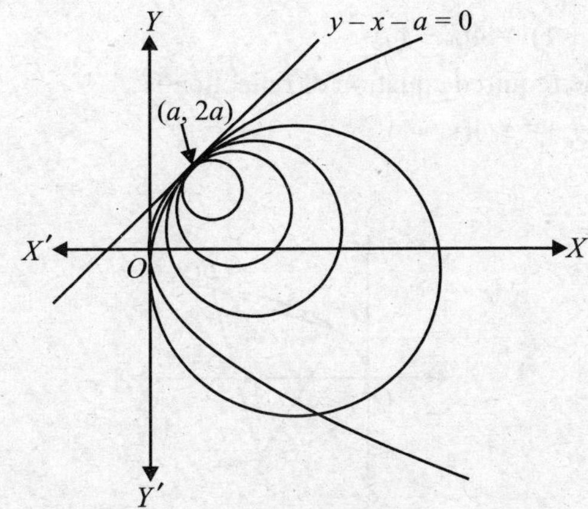

Fig. 3.15

Equation of tangent of parabola at $(a, 2a)$ is $2ya = 2a\,(x + a)$, i.e., $y - x - a = 0$.

Equation of circle touching the parabola at $(a, 2a)$ is $(x - a)^2 + (y - 2a)^2 + \lambda(y - x - a) = 0$.

It passes through $(0, 0)$. So,

$a^2 + 4a^2 + \lambda(-a) = 0 \Rightarrow \lambda = 5a$

Thus the required circle is $x^2 + y^2 - 7ax - ay = 0$.

Its radius is $\sqrt{\dfrac{49}{4}a^2 + \dfrac{a^2}{4}} = \dfrac{5}{\sqrt{2}}\,a.$

52. c. The required point is obtained by solving $x + y = 1$ and $y^2 - y + x = 0$.

53. c. Any tangent to $y^2 = 4a(x + a)$ is

$y = m\,(x + a) + \dfrac{a}{m}$ (1)

Any tangent to $y^2 = 4b(x + b)$ which is perpendicular to Eq. (1) is

$y = -\dfrac{1}{m}x + (x + b) - bm$ (2)

Subtracting, we get

$$\left(m + \frac{1}{m}\right)x + (a + b)\left(m + \frac{1}{m}\right) = 0$$

or $x + a + b = 0$, which is a locus of their point of intersection.

54. c. Any point on the given parabola is $(t^2, 2t)$.

The equation of the tangent at (1, 2) is $x - y + 1 = 0$

The image (h, k) of the point $(t^2, 2t)$ in $x - y + 1 = 0$ is given by

$$\frac{h-t^2}{1} = \frac{k-2t}{-1} = -\frac{2(t^2-2t+1)}{1+1}$$

$\therefore h = t^2 - t^2 + 2t - 1 = 2t - 1$

and $k = 2t + t^2 - 2t + 1 = t^2 + 1$

Eliminating t from $h = 2t - 1$ and $k = t^2 + 1$, we get $(h + 1)^2 = 4(k - 1)$.

The required equation of reflection is $(x + 1)^2 = 4(y - 1)$.

55. a.

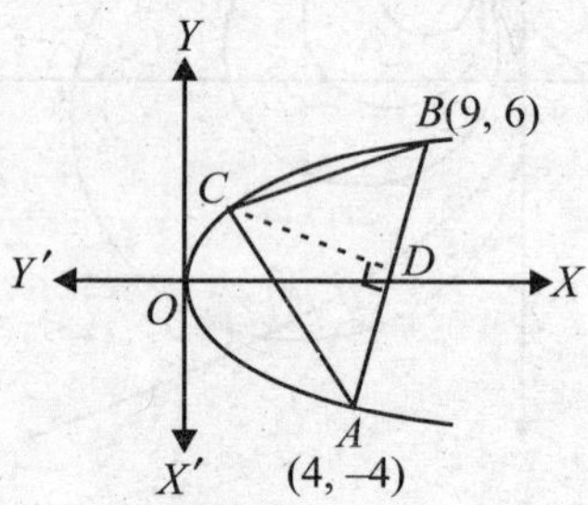

Fig. 3.16

Area of triangle ABC is maximum if CD is maximum, because AB is fixed.

That means tangent drawn to parabola at C should be parallel to AB.

Slope of $AB = \frac{6+4}{9-4} = 2$.

For $y^2 = 4x$, $\frac{dy}{dx} = \frac{2}{y} = 2$

$\Rightarrow y = 1$

$\Rightarrow x = \frac{1}{4}$

56. b. Let $y_1 = 5 + \sqrt{1 - x_1^2}$ and $y_2 = \sqrt{4x_2}$ or

$x_1^2 + (y_1 - 5)^2 = 1$ and $y_2^2 = 4x_2$

Thus (x_1, y_1) lies on the circle $x^2 + (y - 5)^2 = 1$, and (x_2, y_2) lies on the parabola $y^2 = 4x$.

Thus, given expression is the shortest distance between the curves $x^2 + (y - 5)^2 = 1$ and $y^2 = 4x$.

Now the shortest distance always occur along common normal to the curves and normal to circle passes through the centre of the circle.

Normal to the parabola $y^2 = 4x$ is $y = mx - 2m - m^3$ passes through (0, 5) gives $m^3 + 2m + 5 = 0$, which has only one root $m = -2$.

Hence, corresponding point on the parabola is (4, 4).

Thus required minimum distance

$= \sqrt{4^2 + 8^2} - 1 = 4\sqrt{5} - 1.$

57. c. $SP_1 = a(1 + t_1^2)$, $SP_2 = a(1 + t_2^2)$

$\Rightarrow t_1 t_2 = -1$

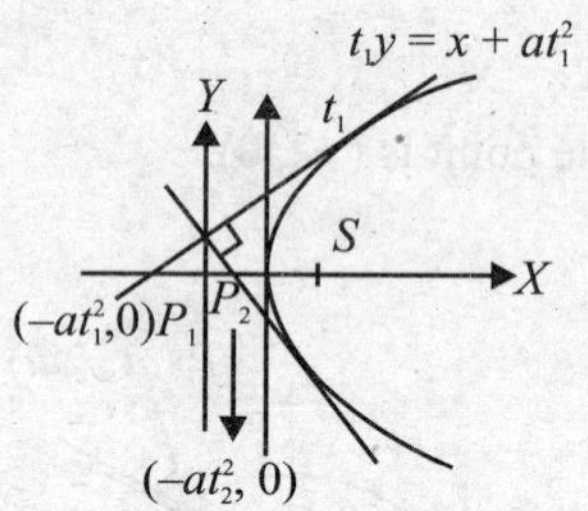

Fig. 3.17

$\frac{1}{SP_1} = \frac{1}{a(1 + t_1^2)}$ and

$\frac{1}{SP_2} = \frac{1}{a(1 + t_2^2)} = \frac{t_1^2}{a(t_1^2 + 1)}$

$\therefore \frac{1}{SP_1} + \frac{1}{SP_2} = \frac{1}{a}$

58. b. Tangent at point P is $ty = x + t^2$, where slope of tangent is $\tan\theta = \frac{1}{t}$.

Now required area is $A = \frac{1}{2}(AN)(PN) = \frac{1}{2}(2t^2)(2t)$

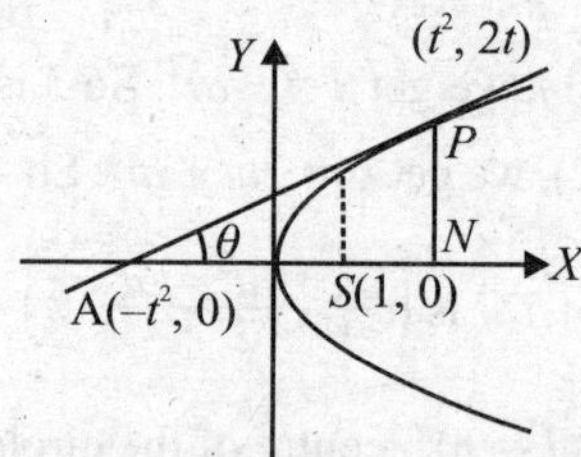

Fig. 3.18

$A = 2t^3 = 2(t^2)^{3/2}$

Now $t^2 \in [1, 4]$, then A_{max} occurs when $t^2 = 4$. So $A_{max} = 16$

59. d. $OT^2 = OA \times OB = \alpha\beta = \frac{c}{a} \Rightarrow OT = \sqrt{\frac{c}{a}}$

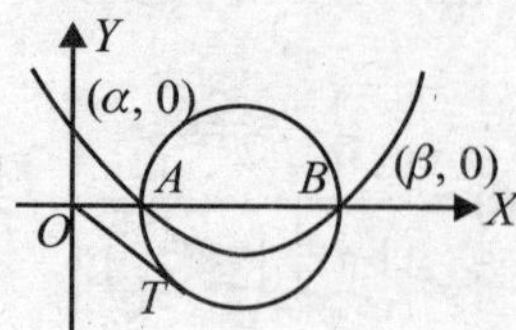

Fig. 3.19

60. b. Tangent at point P is $ty = x + at^2$. (1)

Line perpendicular to Eq. (1) passes through $(a, 0)$

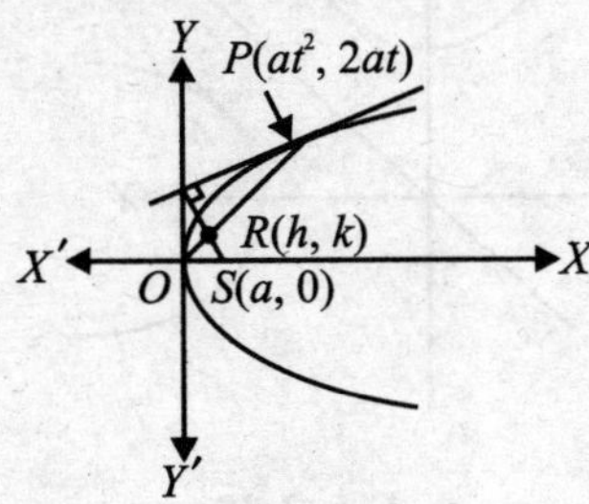

Fig. 3.20

$\therefore y - 0 = -t(x - a)$ or $tx + y = ta$ or $y = t(a - x)$ (2)

Equation of OP is

$y - \frac{2}{t}x = 0$ or $y = \frac{2}{t}x$ (3)

From Eqs. (2) and (3), eliminating t, we get

$y^2 = 2x(a - x)$

or $2x^2 + y^2 - 2ax = 0$

61. a. Slope of tangent at P is $\frac{1}{t_1}$ and at Q is $\frac{1}{t_2}$.

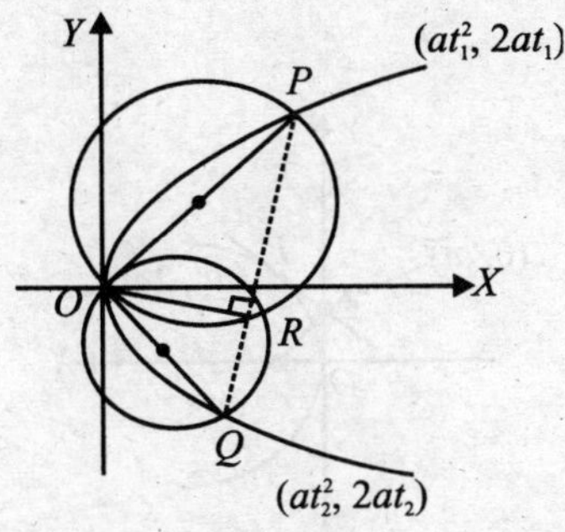

Fig. 3.21

$\Rightarrow \cot\theta_1 = t_1$ and $\cot\theta_2 = t_2$

Slope of $PQ = \frac{2}{t_1 + t_2}$

$\Rightarrow$ Slope of OR is $-\frac{t_1 + t_2}{2} = \tan\phi$

[Note that angle in a semicircle is 90°]

$\Rightarrow \tan\phi = -\frac{1}{2}(\cot\theta_1 + \cot\theta_2)$

$\Rightarrow \cot\theta_1 + \cot\theta_2 = -2\tan\phi$

62. b.

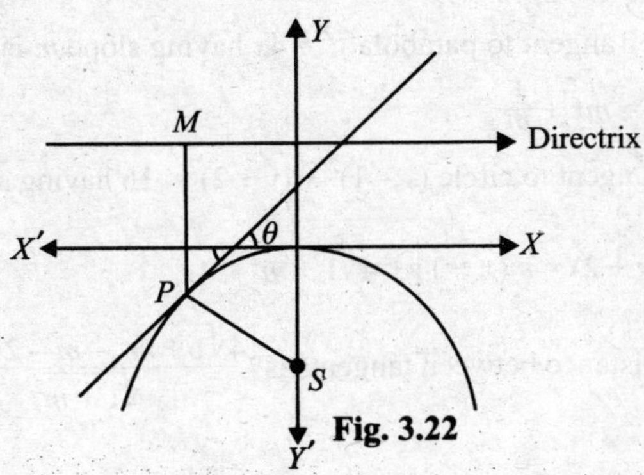

Fig. 3.22

Slope of line is $\lambda = \tan\theta$.

$\Rightarrow \tan(\angle MPS) = \tan 2\left(\frac{\pi}{2} - \theta\right) = \tan(\pi - 2\theta) = -\tan 2\theta$

$= \frac{2\lambda}{\lambda^2 - 1}$

63. c. Solving $y = 2x - 3$ and $y^2 = 4a\left(x - \frac{1}{3}\right)$, we have

$(2x - 3)^2 = 4a\left(x - \frac{1}{3}\right)$

$\Rightarrow 4x^2 + 9 - 12x = 4ax - \frac{4a}{3}$

$\Rightarrow 4x^2 - 4(3 + a)x + 9 + \frac{4a}{3} = 0$

This equation must have equal roots $\Rightarrow D = 0$

$\Rightarrow 16(3 + a)^2 - 16\left(9 + \frac{4a}{3}\right) = 0$

$\Rightarrow 9 + a^2 + 6a = 9 + \frac{4a}{3}$

$\Rightarrow a^2 + \frac{14a}{3} = 0$

$\Rightarrow a = 0$ or $a = \frac{14}{3}$

64. c.

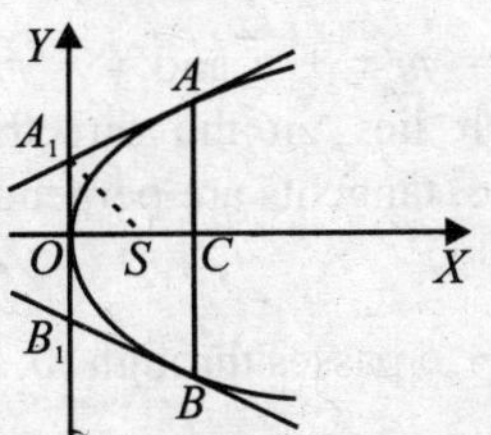

Fig. 3.23

Let $A \equiv (at_1^2, 2at_1)$ and $B \equiv (at_1^2, -2at_1)$.

Equation of tangents at A and B are,

$t_1 y = x + at_1^2$

and $-t_1 y = x + at_1^2$, respectively.

These tangents meet y-axis at

$A_1 \equiv (0, at_1)$

and $B_1 \equiv (0, -at_1)$.

Area of trapezium $AA_1B_1B = \frac{1}{2}(AB + A_1B_1) \times OC$

$\Rightarrow 24a^2 = \frac{1}{2}(4at_1 + 2at_1)(at_1^2)$

$\Rightarrow t_1^3 = 8 \Rightarrow t_1 = 2$

$\Rightarrow A_1 \equiv (0, 2a)$

If $\angle OSA_1 = \theta$, then $\tan\theta = \frac{2a}{a} = 2$

$\Rightarrow \theta = \tan^{-1}(2)$

Thus, required angle is $2\tan^{-1}(2)$.

65. b. Here $a = 2$ for parabola and the two tangents pass through the points $(-2, -3)$, which lie on the directrix, then tangents are perpendicular or $m_1 m_2 = -1$.

66. b.

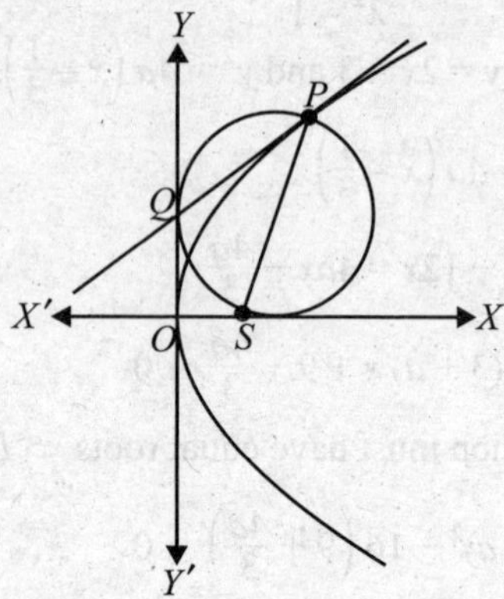

Fig. 3.24

Tangent at P intersects the y-axis at $Q \equiv (0, at)$.

Also circle with PS as diameter touches the y-axis at $(0, at)$. So y-axis is the tangent to circumcircle of ΔPQS at Q.

67. d. Tangents $y = m_1 x + c$ and $y = m_2 x + c$ intersect at $(0, c)$ which lies on the directrix of the given parabola. Hence, tangents are perpendicular for which $m_1 m_2 = -1$.

68. b. $y = ax^2 - 6x + b$ passes through $(0, 2)$.

Here, $2 = a(0^2) - 6(0) + b$

$\therefore b = 2$.

Also, $\frac{dy}{dx} = 2ax - 6$

$$\therefore \left(\frac{dy}{dx}\right)_{x=\frac{3}{2}} = 2a\left(\frac{3}{2}\right) - 6$$

$$= 3a - 6 = 0$$

$\therefore a = 2$

69. b. Since tangents are perpendicular, they intersect on the directrix.

$\Rightarrow (\lambda, 1)$ lies on the line $x = -4$

$\Rightarrow \lambda = -4$

70. b. Parabolas $y = x^2 + 1$ and $x = y^2 + 1$ are symmetrical about $y = x$.

Therefore, tangent at point A is parallel to $y = x$.

$$\Rightarrow \frac{dy}{dx} = 2x \Rightarrow 2x = 1$$

$$\Rightarrow x = \frac{1}{2} \text{ and } y = \frac{5}{4}$$

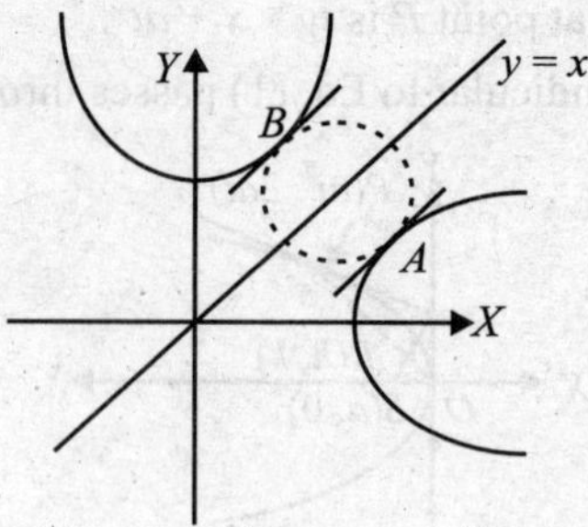

Fig. 3.25

$A\left(\frac{1}{2}, \frac{5}{4}\right)$ and $B\left(\frac{5}{4}, \frac{1}{2}\right)$

Therefore, radius $= \frac{1}{2}\sqrt{\left(\frac{1}{2} - \frac{5}{4}\right)^2 + \left(\frac{5}{4} - \frac{1}{2}\right)^2}$

$$= \frac{1}{2}\sqrt{\frac{9}{16} + \frac{9}{16}} = \frac{3}{8}\sqrt{2}$$

Therefore, area $= \frac{9\pi}{32}$.

71. b.

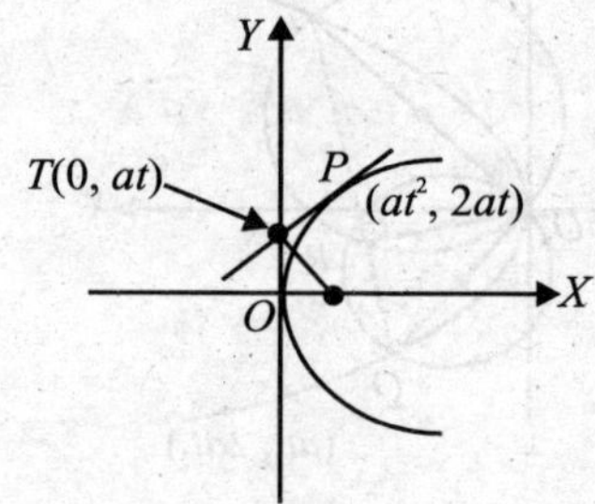

Fig. 3.26

Let the middle point of P and T be (h, k)

$\therefore 2h = at^2$

and $2k = 3at$

$\therefore 2h = a \times \frac{4k^2}{9a^2}$

Locus of (h, k) is $2y^2 = 9ax$.

As $a = 2$, $y^2 = 9x$.

72. c. Tangent to parabola $y^2 = 4x$ having slope m is

$y = mx + \frac{1}{m}$.

Tangent to circle $(x - 1)^2 + (y + 2)^2 = 16$ having slope m is

$(y + 2) = m(x - 1) + 4\sqrt{1 + m^2}$

Distance between tangents is $\left|\frac{4\sqrt{1 + m^2} - m - 2 - 1/m}{\sqrt{1 + m^2}}\right|$

$$= \left| 4 - \frac{2}{\sqrt{1+m^2}} - \frac{\sqrt{m^2+1}}{m} \right|$$

As $m > 0 \Rightarrow d < 4$

73. d.

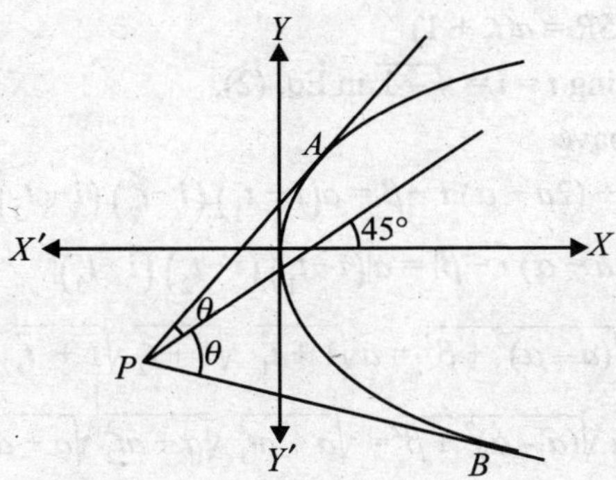

Fig. 3.27

Here $\frac{1}{t_1} = \tan\left(\frac{\pi}{4} + \theta\right)$

and $\frac{1}{t_2} = \tan\left(\frac{\pi}{4} - \theta\right)$

So $t_1 t_2 = 1$

$\Rightarrow$ the x-coordinate of $P = at_1t_2 = a$.

74. a.

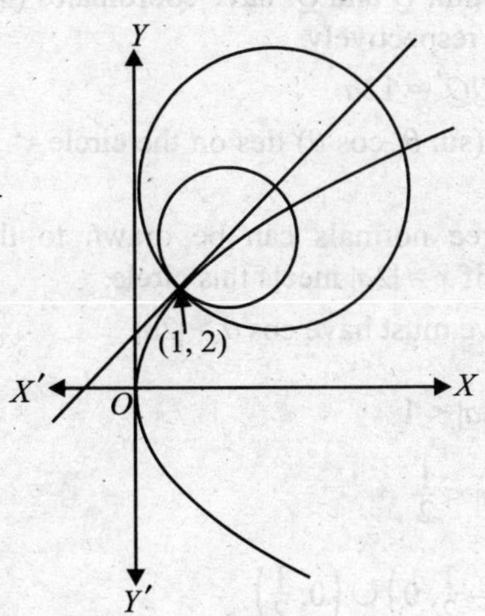

Fig. 3.28

Tangent to parabola $y^2 = 4x$ at (1, 2) will be the locus

i.e., $2y = 2(x + 1)$

$\Rightarrow y = x + 1$

75. c.

Circle S_2, taking focal chord AB as diameter, will touch directrix at point P and circle S_1, taking AP as diameter, will pass through focus S (since AP subtends angle 90° at focus of parabola).

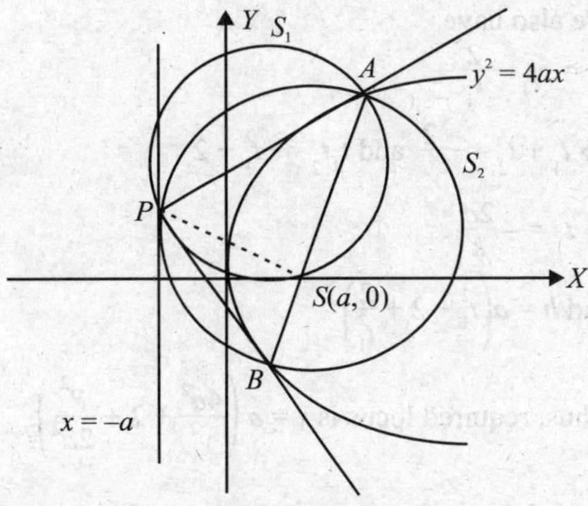

Fig. 3.29

Hence, common chord of given circles is line AP (which is intercept of tangent at point A between point A and directrix).

76. a. Since the normal at $(ap^2, 2ap)$ to $y^2 = 4ax$ meets the parabola at $(aq^2, 2aq)$,

$$q = -p - \frac{2}{p} \qquad (1)$$

Since $OP \perp OQ$,

$$\frac{2ap - 0}{ap^2 - 0} \times \frac{2aq - 0}{aq^2 - 0} = -1 \Rightarrow pq = -4$$

$$\Rightarrow p\left(-p - \frac{2}{p}\right) = -4 \qquad \text{[Using (1)]}$$

$$\Rightarrow p^2 = 2$$

77. d. Given parabola is $y^2 = 4x + 8$ or $y^2 = 4(x + 2)$.

Equation of normal to parabola at any point $P(t)$ is $y = -t(x + 2) + 2t + t^3$.

It passes through $(k, 0)$ if $tk = t^3 \Rightarrow t(t^2 - k) = 0$.

Hence, it has three real values of t if $k > 0$.

78. c. Equations of tangent and normal at $P(at^2, 2at)$ are $ty = x + at^2$ and $y = -tx + 2at + at^3$, respectively.

Thus, $T \equiv (-at^2, 0)$ and $N \equiv (2a + at^2, 0)$. Also, $S \equiv (a, 0)$.

Hence, $SP = a + at^2$, $ST = a + at^2$

and $SN = a + at^2$

Thus, $SP = ST = SN$.

79. b. Let AB be a normal chord where $A \equiv (at_1^2, 2at_1)$, $B \equiv (at_2^2, 2at_2)$. If its midpoint is $P(h, k)$, then

$$2h = a\left(t_1^2 + t_2^2\right)$$

$$= a\left[(t_1 + t_2)^2 - 2t_1t_2\right]$$

and $2k = 2a(t_1 + t_2)$.

We also have,

$t_2 = -t_1 - \frac{2}{t_1}$

$\Rightarrow t_1 + t_2 = \frac{-2}{t_1}$ and $t_1 t_2 = -t_1^2 - 2$

$\Rightarrow t_1 = -\frac{2a}{k}$

and $h = a\left(t_1^2 + 2 + \frac{2}{t_1^2}\right)$.

Thus, required locus is $x = a\left(\frac{4a^2}{y^2} + 2 + \frac{y^2}{2a^2}\right)$.

80. c.

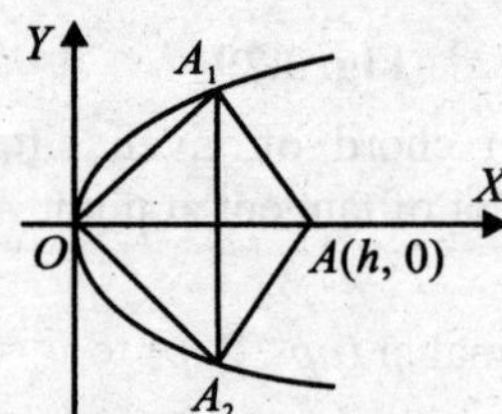

Fig. 3.30

Let $A_1 \equiv (2at_1^2, 4t_1)$ and $A_2 \equiv (2t_1^2, -4t_1)$.

Clearly, $\angle A_1OA = \frac{\pi}{6}$

$\Rightarrow \frac{2}{t_1} = \frac{1}{\sqrt{3}}$

$\Rightarrow t_1 = 2\sqrt{3}$

Equation of normal at A_1 is $y = -t_1x + 4t_1 + 2t_1^3$

$\Rightarrow h = 4 + 2t_1^2 = 4 + 2\,(12)$

$= 28$

81. c. $y = mx + c$ is a normal to $y^2 = 4ax$ if $c = -2am - am^3$, $y = -2x - \lambda$.

$\Rightarrow m = -2, a = -2$

$\Rightarrow -\lambda = -2am - am^3 = -2(-2)(-2) - (-2)(-2)^3 = -24$

$\Rightarrow \lambda = 24$

82. d. We have $a = 1$.

Normal at $(m^2, -2m)$ is $y = mx - 2m - m^3$.

Given that normal makes equal angle with axes, then its slope $m = \pm 1$.

Therefore, point P is $(m^2, -2m) = (1, \pm 2)$.

83. c. Let $A = (\alpha, \beta)$

The normal at $(at^2, 2at)$ is $y = -tx + 2at + at^3$

$\therefore at^3 + (2a - \alpha)t - \beta = 0$ (1)

Let t_1, t_2, t_3 be roots of Eq. (1), then

$at^3 + (2a - \alpha)t - \beta = a(t - t_1)(t - t_2)(t - t_3)$ (2)

Let $P = (at_1^2, 2at_1)$, $Q = (at_2^2, 2at_2)$, and $R = (at_3^2, 2at_3)$.

Since the focus S is $(a, 0)$,

$SP = a(t_1^2 + 1)$

Similarly, $SQ = a(t_2^2 + 1)$,

and $SR = a(t_3^2 + 1)$

Putting $t = i = \sqrt{-1}$ in Eq. (2),

we have

$-ai + (2a - \alpha)\,i - \beta = a(i - t_1)(i - t_2)(i - t_3)$

$\Rightarrow |(a - \alpha)\,i - \beta| = a|(i - t_1)(i - t_2)(i - t_3)|$

$\Rightarrow \sqrt{(a - \alpha)^2 + \beta^2} = a\sqrt{1 + t_1^2}\,\sqrt{1 + t_2^2}\,\sqrt{1 + t_3^2}$

$\Rightarrow a\sqrt{(a - \alpha)^2 + \beta^2} = \sqrt{a + at_1^2}\,\sqrt{a + at_2^2}\,\sqrt{a + at_3^2}$

$\Rightarrow aSA^2 = SP \times SQ \times SR$

84. d. Ends of latus rectum are $P(a, 2a)$ and $P'(a, -2a)$.

Point P has parameter $t_1 = 1$ and point P' has parameter $t_2 = -1$.

Normal at point P meets the curve again at point Q whose parameter $t_1' = -t_1 - \frac{2}{t_1} = -3$.

Normal at point P' meets the curve again at point Q' whose parameter $t_2' = -t_2 - \frac{2}{t_2} = 3$.

Hence, point Q and Q' have coordinates $(9a, -6a)$ and $(9a, 6a)$, respectively.

Hence, $QQ' = 12a$

85. d. Point $(\sin\theta, \cos\theta)$ lies on the circle $x^2 + y^2 = 1$ for $\forall\,\theta \in R$.

Now three normals can be drawn to the parabola $y^2 = 4ax$ if $x = |2a|$ meets this circle.

Hence, we must have $\cos\theta > |2a|$.

$\Rightarrow 0 < |2a| < 1$

$\Rightarrow 0 < |a| < \frac{1}{2}$

$\Rightarrow a \in \left(-\frac{1}{2}, 0\right) \cup \left(0, \frac{1}{2}\right)$

86. c. Normal at point $P(t_1)$ meets the parabola again at point $R(t_3)$, then

$t_3 = -t_1 - \frac{2}{t_1}$

Also normal at point $Q(t_2)$ meets the parabola at the same point $R(t_3)$, then

$t_3 = -t_2 - \frac{2}{t_2}$

Comparing these values of t_3, we have

$-t_1 - \frac{2}{t_1} = -t_2 - \frac{2}{t_2}$ or $t_1 t_2 = 2$.

87. b. Normal at point $(t^2, 2t)$ is $y = -tx + 2t + t^3$.

Slope of the tangent is 1.

Hence, $-t = 1 \Rightarrow t = -1$.

$\Rightarrow$ Coordinates of P are $(1, -2)$.

Hence, parameter at Q is $t_2 = -t_1 - 2/t_1 = 1 + 2 = 3$

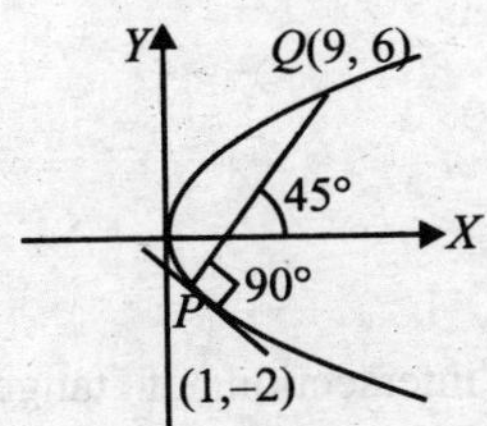

Fig. 3.31

Therefore, coordinates at Q are $(9, 6)$.

$\therefore l(PQ) = \sqrt{64 + 64} = 8\sqrt{2}$

88. b. $\tan \alpha = -t_1$ and $\tan \beta = -t_2$

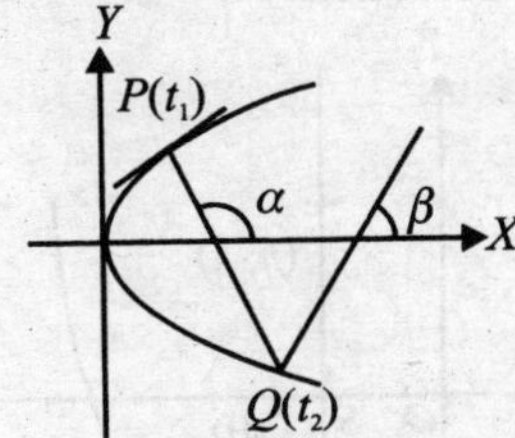

Fig. 3.32

also $t_2 = -t_1 - \frac{2}{t_1}$

$\Rightarrow t_1 t_2 + t_1^2 = -2$

$\Rightarrow \tan \alpha \tan \beta + \tan^2 \alpha = -2$

89. b. Slope of normal at point $P(t_1)$ and $Q(t_2)$ is $-t_1$ and $-t_2$, respectively.

Equation of chord joining $P(t_1)$ and $Q(t_2)$ is

$y - 2at_1 = \frac{2}{t_1 + t_2}(x - at_1^2)$

or $2x - (t_1 + t_2)y + 2at_1 t_2 = 0$

But $t_1 t_2 = -1$.

Chord PQ is $2x - (t_1 + t_2)y - 2a = 0$

or $(2x - 2a) - (t_1 + t_2)y = 0$,

which passes through the fixed point $(a, 0)$.

90. c. $t_2 = -t_1 - \frac{2}{t_1} \Rightarrow t_1 t_2 = -t_1^2 - 2$

Equation of the line through P parallel to AQ.

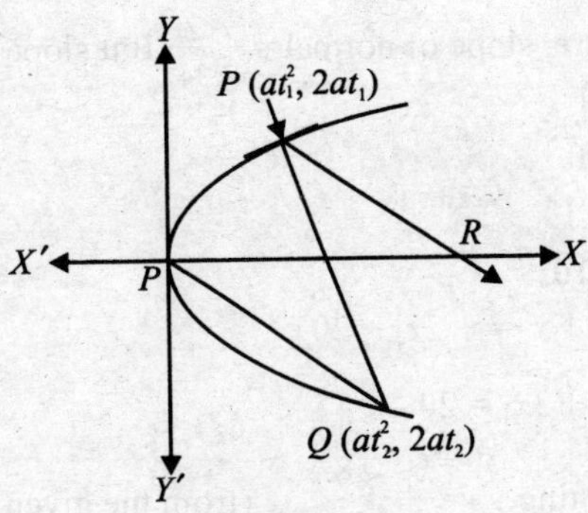

Fig. 3.33

$y - 2at_1 = \frac{2}{t_2}(x - at_1^2)$

Putting $y = 0$,

$x = at_1^2 - at_1 t_2$

$= at_1^2 - a(-2 - t_1^2)$

$= 2a + 2at_1^2$

$= 2(a + at_1^2)$

= twice the focal distance of P.

91. b. Equations of tangent and normal at A are $yt = x + at^2$ and $y = -tx + 2at + at^3$

$\Rightarrow B \equiv (-at^2, 0), D \equiv (2a + at^2, 0)$. If $ABCD$ is a rectangle, then midpoints of BD and AC will be coincident.

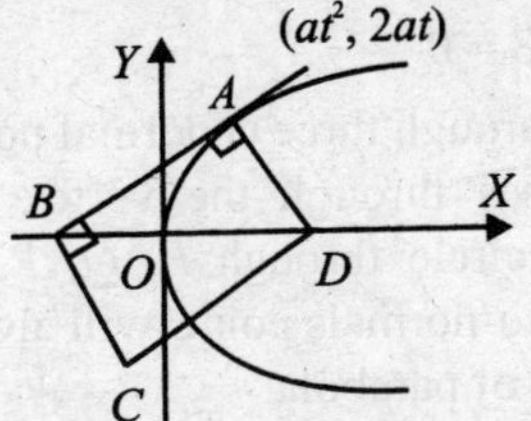

Fig. 3.34

$\Rightarrow h + at^2 = 2a + at^2 - at^2$,

$k + 2at = 0$

$\Rightarrow h = 2a, t = -\frac{k}{2a}$

92. d. $4y = x^2 - 8$

$4\frac{dy}{dx} = 2x$

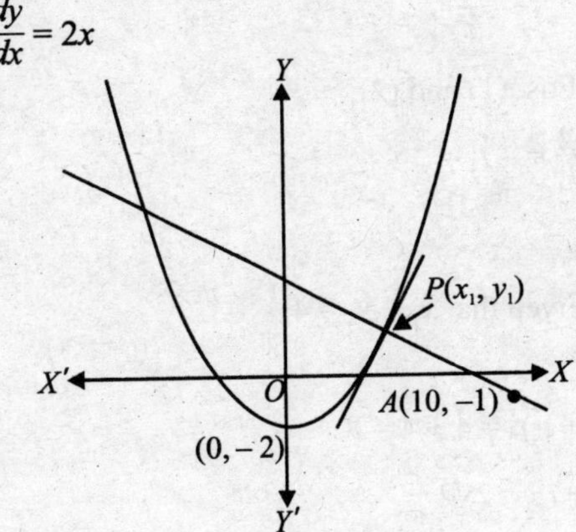

Fig. 3.35

Therefore, slope of normal $= -\frac{2}{x_1}$. But slope of normal $= \frac{y_1+1}{x_1-10}$.

$\therefore \frac{y_1+1}{x_1-10} = -\frac{2}{x_1}$

$\Rightarrow x_1y_1 + x_1 = -2x_1 + 20$

$\Rightarrow x_1y_1 + 3x_1 = 20$

Substituting $y_1 = \frac{x_1^2-8}{4}$ (from the given equation)

$x_1\left(\frac{x_1^2-8}{4}+3\right) = 20$

$\Rightarrow x_1(x_1^2+4) = 80$

$\Rightarrow x_1^3 + 4x_1 - 80 = 0$,

which has one root $x_1 = 4$.

Hence, $x_1 = 4; y_1 = 2$

$\therefore P = (4, 2)$.

Therefore, equation of PA is

$y+1 = -\frac{1}{2}(x-10)$

$\Rightarrow 2y + 2 = -x + 10$

$\Rightarrow x + 2y - 8 = 0$

93. a. A circle through three conormal points of a parabola always passes through the vertex of the parabola. Hence, the circle through P, Q, R, S out of which P, Q, R are co-normals points will always pass through vertex (2, 3) of parabola.

94. c. Normal at point $P(x_1, y_1) \equiv (at_1^2, 2at_1)$ meets the parabola at $R(at^2, 2at)$

$\Rightarrow t = -t_1 - \frac{2}{t_1}$ (1)

Normal at point $Q(x_2, y_2) \equiv (at_2^2, 2at_2)$ meets the parabola at $R(at^2, 2at)$

$\Rightarrow t = -t_2 - \frac{2}{t_2}$ (2)

From Eqs. (1) and (2),

$-t_1 - \frac{2}{t_1} = -t_2 - \frac{2}{t_2}$

$\Rightarrow t_1t_2 = 2$

Now given that $x_1 + x_2 = 4$

$\Rightarrow t_1^2 + t_2^2 = 4$

$\Rightarrow (t_1+t_2)^2 = 4 + 4 = 8$

$\Rightarrow |t_1 + t_2| = 2\sqrt{2}$

$\Rightarrow |y_1 + y_2| = 4\sqrt{2}$

95. b. Let the concyclic points be t_1, t_2, t_3 and t_4

$\Rightarrow t_1 + t_2 + t_3 + t_4 = 0$

Here, t_1 and t_3 are feet of the normals. So

$t_2 = -t_1 - \frac{2}{t_1}$

and $t_4 = -t_3 - \frac{2}{t_3}$

$\Rightarrow t_1 + t_2 = -\frac{2}{t_1}$ and $t_4 + t_3 = -\frac{2}{t_3}$

Adding, we get

$-2\left(\frac{1}{t_1}+\frac{1}{t_3}\right) = 0$

$\Rightarrow t_1 + t_3 = 0$

$\Rightarrow$ Point of intersection of tangents at t_1 and t_3 $(at_1t_3, a(t_1+t_3)) \equiv (at_1t_3, 0)$.

$\Rightarrow$ This point lies on the axis of the parabola.

96. c. Axis of the parabola is $x = 1$. Any point on it is $(1, k)$. Now distance of $(1, k)$ from $(1, -2)$ should be more than the semi-latus rectum and $(1, k)$ should be inside the parabola, hence $k > 2$.

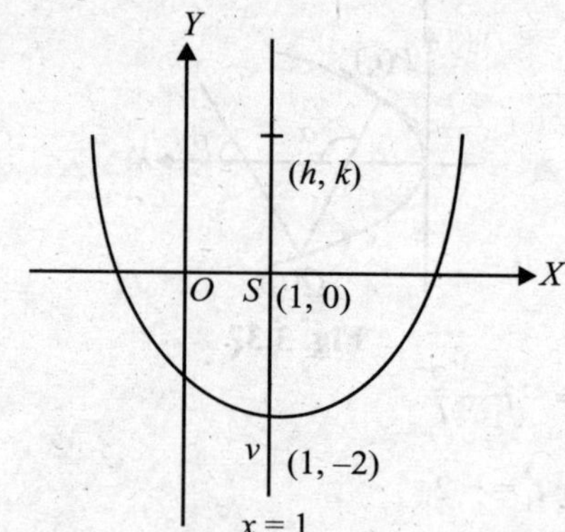

Fig. 3.36

97. b.

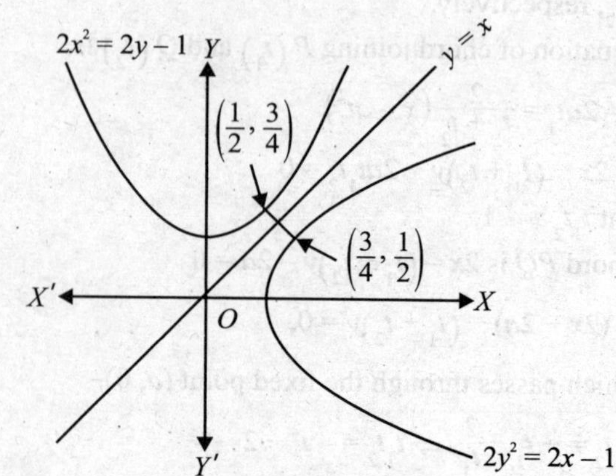

Fig. 3.37

Given parabolas $2y^2 = 2x - 1$ and $2x^2 = 2y - 1$ are symmetrical about the line $y = x$.

Also shortest distance occurs along the common normal which perpendicular to the line $y = x$.

Differentiating $2y^2 = 2x - 1$ w.r.t. x, we get

$$2y\frac{dy}{dx} = 1$$

$$\Rightarrow \frac{dy}{dx} = \frac{1}{2y} = 1 \Rightarrow y = \frac{1}{2}$$

Hence, points are as shown in the figure.

Then, the shortest distance

$$d = \sqrt{\frac{1}{16} + \frac{1}{16}} = \frac{1}{2\sqrt{2}}.$$

98. d.

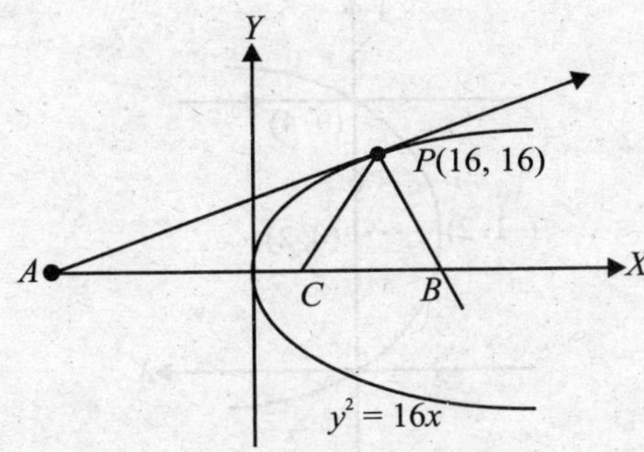

Fig. 3.38

By property, centre of circle coincides with focus of parabola

$\Rightarrow C \equiv (4, 0)$

$\tan \alpha =$ slope of $PC = \frac{16}{12}$

$$\Rightarrow \alpha = \tan^{-1}\left(\frac{4}{3}\right)$$

99. a. Let AB be a normal chord where $A \equiv (at^2, 2at)$, $B \equiv (at_1^2, 2at_1)$.

We have $t_1 = -t - \frac{2}{t}$.

Now, $AB^2 = [a^2(t^2 - t_1^2)]^2 + 4a^2(t - t_1)^2$

$$= a^2(t - t_1)^2 [(t + t_1)^2 + 4]$$

$$= a^2\left(t + t + \frac{2}{t}\right)^2 \left(\frac{4}{t^2} + 4\right)$$

$$= \frac{16a^2 (1 + t^2)^3}{t^4}$$

$$\Rightarrow \frac{d(AB)^2}{dt} = 16a^2\left(\frac{t^4[3(1 + t^2)^2 2t] - (1 + t^2)^3 4t^3}{t^8}\right)$$

$$= 32a^2 (1 + t^2)^2 \left(\frac{3t^2 - 2 - 2t^2}{t^5}\right)$$

$$= \frac{a^2 \times 32(1 + t^2)^2}{t^5}(t^2 - 2)$$

For $\frac{d(AB^2)}{dt} = 0 \Rightarrow t = \sqrt{2}$ for which AB^2 is minimum.

Thus, $AB_{\min} = \frac{4a}{2}(1 + 2)^{3/2} = 2a\sqrt{27}$ units

100. c. The equation of any normal be $y = -tx + 2t + t^3$.

Since it passes through the points (15, 12),

$12 = -15t + 2t + t^3$

$\Rightarrow t^3 - 13t - 12 = 0$

One root is -1, then

$(t + 1)(t^2 + t - 12) = 0$

$\Rightarrow t = 1, 3, 4$

Therefore, the conormal points are $(1, -2)$, $(9, -6)$, and $(16, 8)$.

Therefore, centroid is $\left(\frac{26}{3}, 0\right)$.

101.d. For a focal chord $t_1 t_2 = -1$ and for the normal $t_1(t_1 + t_2) + 2 = 0$.

$\therefore t_1^2 + t_1t_2 + 2 = 0 \Rightarrow t_1^2 = -1$.

Therefore, t_1 is imaginary.

102.b. Solving the line $y = x - 1$ and parabola $y^2 = 4x$, we have

$(x - 1)^2 = x$

$\Rightarrow x^2 - 6x + 1 = 0$

$\Rightarrow x = 3 \pm \sqrt{8}$

$\therefore y = 2 \pm \sqrt{8}$

Suppose point D is (x_3, y_3), then

$y_1 + y_2 + y_3 = 0$

$\Rightarrow 2 + \sqrt{8} + 2 - \sqrt{8} + y_3 = 0$

$\Rightarrow y_3 - 4$, then $x_3 = 4$.

Therefore, the point is (4, 4).

103.c. Equation of the normal is $y = mx - 2am - am^3$

Putting $y = 0$, we get

$x_1 = 2a + am_1^2$

$x_2 = 2a + am_2^2$

$x_3 = 2a + am_3^2$,

where x_1, x_2, x_3 are the intercepts on the axis of the parabola.

The normal passes through (h, k).

$\Rightarrow am^3 + (2a - h)m + k = 0$,

which has roots m_1, m_2, m_3 which are slopes of the normals.

$\Rightarrow m_1 + m_2 + m_3 = 0$

and $m_1m_2 + m_2m_3 + m_3m_1 = \frac{2a - h}{a}$

$\Rightarrow m_1^2 + m_2^2 + m_3^2 = (m_1 + m_2 + m_3)^2 - 2(m_1m_2 + m_2m_3 + m_3m_1)$

$= -\dfrac{2(2a-h)}{a}$

$\Rightarrow x_1 + x_2 + x_3 = 6a - 2(2a - h) = 2(h + a)$

104c.

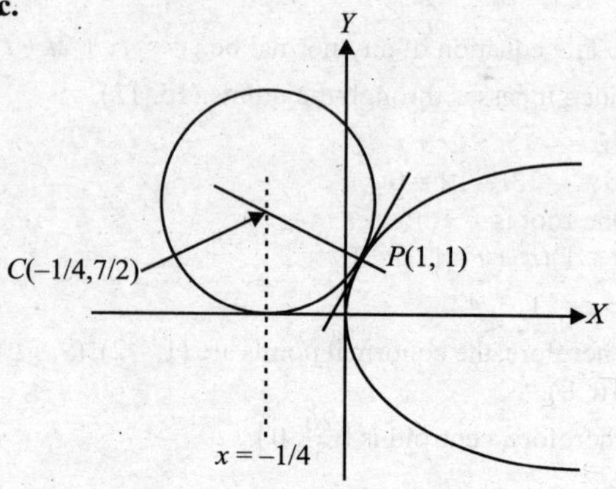

Fig. 3.39

Equation of normal at $P(1, 1)$ is

$y - 1 = -2(x - 1)$

or $y + 2x = 3$ (1)

Directrix of the parabola $y^2 = x$ is

$x = -\dfrac{1}{4}$. (2)

Centre of the circle is intersection of two normals to the circle, i.e., Eqs. (1) and (2) which is $\left(-\dfrac{1}{4}, \dfrac{7}{2}\right)$.

Hence, radius of the circle is

$$\sqrt{\left(1+\frac{1}{4}\right)^2 + \left(1-\frac{7}{2}\right)^2} = \sqrt{\frac{25}{16} + \frac{25}{4}} = \frac{5\sqrt{5}}{4}$$

105.d. Let $y = mx + c$ intersect $y^2 = 4ax$ at $A(at_1^2, 2at_1)$ and $B(at_2^2, 2at_2)$.

Then, $\dfrac{2}{t_1 + t_2} = m$

$\Rightarrow t_1 + t_2 = \dfrac{2}{m}$

Let the foot of another normal be $C(at_3^2, 2at_3)$.

Then,

$t_1 + t_2 + t_3 = 0$

$\Rightarrow t_3 = -(t_1 + t_2) = -\dfrac{2}{m}$

Thus, other foot is $\left(\dfrac{4a}{m^2}, \dfrac{-4a}{m}\right)$.

106.b. Tangent to $y^2 = 4x$ in terms of m is

$y = mx + \dfrac{1}{m}$.

Normal to $x^2 = 4by$ in terms of m is

$y = mx + 2b + \dfrac{b}{m^2}$.

If these are same lines, then

$\dfrac{1}{m} = 2b + \dfrac{b}{m^2}$

$\Rightarrow 2bm^2 - m + b = 0$

For two different tangents

$1 - 8b^2 > 0$

$\Rightarrow |b| < \dfrac{1}{\sqrt{8}}$

107.d. Normals to $y^2 = 4ax$ and $x^2 = 4by$ in terms of m are

$y = mx - 2am - am^3$

and $y = mx + 2b + \dfrac{b}{m^2}$.

For a common normal,

$2b + b/m^2 + 2am + am^3 = 0$

$\Rightarrow am^5 + 2am^3 + 2bm^2 + b = 0$

This means there can be most five common normals.

108. a.

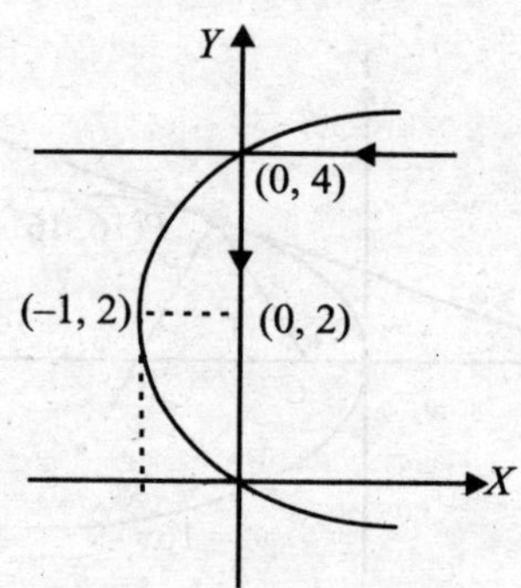

Fig. 3.40

Given curve is $(y - 2)^2 = 4(x + 1)$

Focus is $(0, 2)$.

Point of intersection of the curve and $y = 4$ is $(0, 4)$.

From the reflection property of parabola, reflected ray passes through the focus.

$\therefore x = 0$ is required line.

109.d. Solving the equations, we get

$x^2 + 4(x + 4) = a^2$

If circle and parabola touch each other, then

$D = 0$

$\Rightarrow 16 - 4(16 - a^2) = 0$

$\Rightarrow a = 2\sqrt{3}$

110.c.

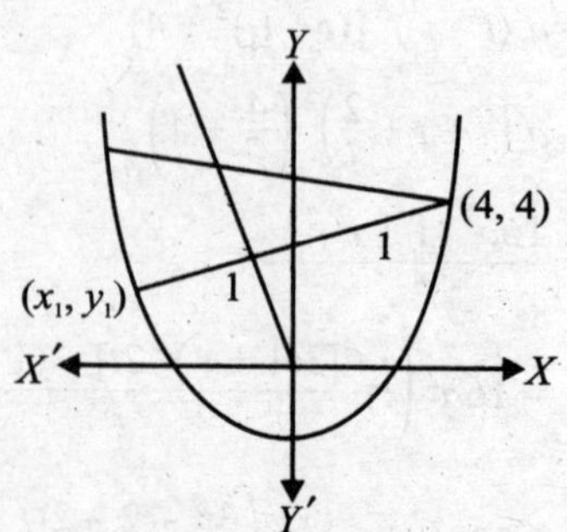

Fig. 3.41

Point (4, 4) lies on the parabola.

Let the point of intersection of the line $y = mx$ with the chords be $(\alpha, m\alpha)$, then

$$\alpha = \frac{4 + x_1}{2}$$

$\Rightarrow x_1 = 2\alpha - 4$

and

$$m\alpha = \frac{4 + y_1}{2}$$

$\Rightarrow y_1 = 2m\alpha - 4$

As (x_1, y_1) lies on the curve,

$(2\alpha - 4)^2 = 4(2m\alpha - 4)$

$\Rightarrow 4\alpha^2 + 16 - 16\alpha = 8(m\alpha - 2)$

$\Rightarrow 4\alpha^2 - 8\alpha(2 + m) + 32 = 0$

For two distinct chords,

$D > 0$

$\Rightarrow (8(2 + m))^2 - 4(4)(32) > 0$

$\Rightarrow (2 + m)^2 - 8 > 0$

$\Rightarrow 2 + m > 2\sqrt{2}$

or $2 + m < -2\sqrt{2}$

$\Rightarrow m > 2\sqrt{2} - 2$

or $m < -2\sqrt{2} - 2$

111. c. Let point of intersection be (α, β).

Therefore, chord of contact w.r.t. this point is

$\beta y = 2x + 2\alpha$ which is same as $x + y = 2$. Therefore, $\alpha = \beta = -2$, which satisfy $y - x = 0$.

112. d. The given parabolas are symmetrical about the line $y = x$ as shown in the figure.

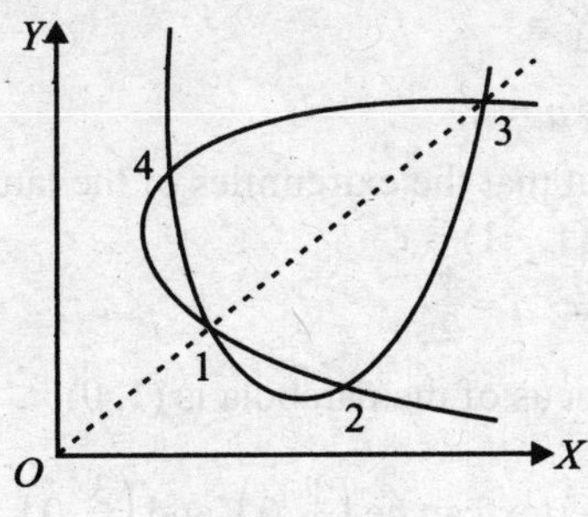

Fig. 3.42

They intersect with each other at four distinct points.

Hence, the number of common chords is

$$4C_2 = \frac{4 \times 3}{2} = 6$$

113. a. Let the focus be F. The parabolas are open down and open right, respectively. Let the parabolas intersect at points P and Q. From P perpendiculars are drawn on the x-axis and y-axis at A and B, respectively, then

$PA = PF = PB$

$\Rightarrow P$ lies on the line $y = -x$.

Similarly, Q lies on the line $y = -x$

$\Rightarrow$ slope of $PQ = -1$

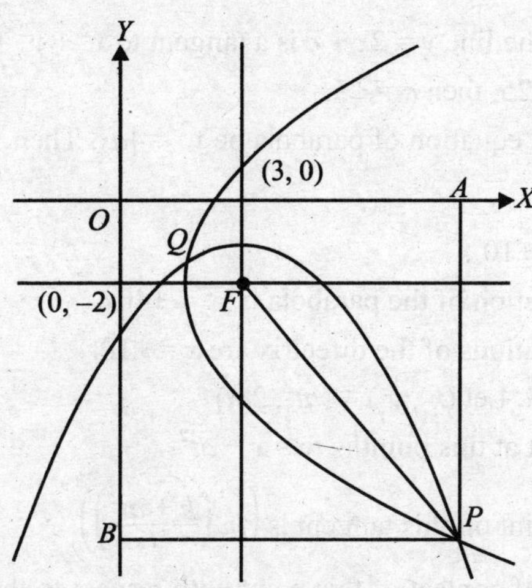

Fig. 3.43

114. b.

Equation of tangent to the parabola $y^2 = 8x$ at $P(2, 4)$ is

$4y = 4(x + 2)$

or $x - y + 2 = 0$ (1)

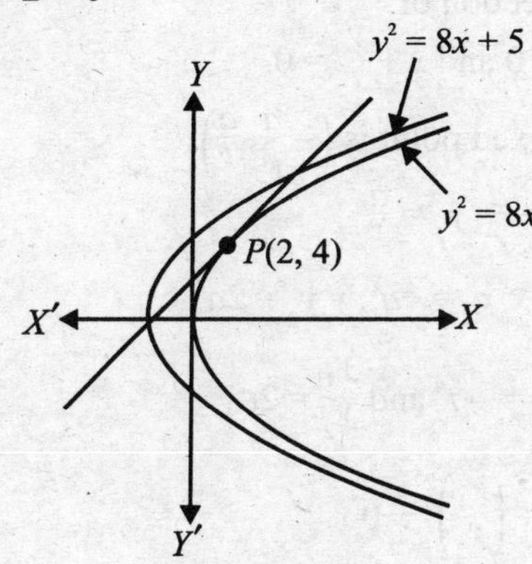

Fig. 3.44

Equation of chord of parabola $y^2 = 8x + 5$ whose middle point is (h, k) is $T = S_1$.

i.e., $ky - 4(x + h) - 5 = k^2 - 8h - 5$

or $4x - ky + k^2 - 4h = 0$ (2)

Equations (1) and (2) must be identical. Therefore,

$$\frac{4}{1} = \frac{k}{1} = \frac{k^2 - 4h}{2}$$

By comparing Eqs. (1), (2), we get

$k = 4$

and $8 = k^2 - 4h$.

Hence, the required point is (2, 4).

Multiple Correct Answers Type

1. a., c. The line $y = 2x + c$ is a tangent to $x^2 + y^2 = 5$.

If $c^2 = 25$, then $c = \pm 5$.

Let the equation of parabola be $y^2 = 4ax$. Then

$\frac{a}{2} = \pm 5$

$\Rightarrow a = \pm 10$

$\Rightarrow$ Equation of the parabola is $y^2 = \pm 40x$.

$\Rightarrow$ Equations of the directrix are $x = \pm 10$.

2. b., c., d. Let $(x_1, y_1) \equiv (at^2, 2at)$

Tangent at this point is $ty = x + at^2$.

Any point on this tangent is $\left(h, \left(\frac{h + at^2}{t}\right)\right)$.

Chord of contact of this point with respect to the circle $x^2 + y^2 = a^2$ is

$hx + \left(\frac{h + at^2}{t}\right)y = a^2$

or $(aty - a^2) + h\left(x + \frac{y}{t}\right) = 0$,

which is a family of straight lines passing through point of intersection of

$ty - a = 0$ and $x + \frac{y}{t} = 0$.

So the fixed point is $\left(-\frac{a}{t^2}, \frac{a}{t}\right)$.

$\therefore x_2 = -\frac{a}{t^2}, y_2 = \frac{a}{t}$

Clearly, $x_1x_2 = -a^2, y_1y_2 = 2a^2$

Also, $\frac{x_1}{x_2} = -t^4$ and $\frac{y_1}{y_2} = 2t^2$.

$\Rightarrow 4\frac{x_1}{x_2} + \left(\frac{y_1}{y_2}\right)^2 = 0$

3. b., d. Given parabola is

$x^2 - ky + 3 = 0$

or $x^2 = k\left(y - \frac{3}{k}\right)$

Let $x = Y, y - \frac{3}{k} = X$.

Then the parabola is $Y^2 = kX$,

whose focus is $\left(0, \frac{k}{4}\right)$.

Therefore, the focus of $x^2 = k\left(y - \frac{3}{k}\right)$ is

$\left(0, \frac{3}{k} + \frac{k}{4}\right) \equiv (0, 2)$

$\therefore \frac{3}{k} + \frac{k}{4} = 2$

$\Rightarrow 12 + k^2 = 8k$

$\Rightarrow k^2 - 8k + 12 = 0$

$\Rightarrow (k - 6)(k - 2) = 0$

$\Rightarrow k = 2, 6$

4. a., c. $P = (\alpha, \alpha + 1)$ where $\alpha \neq 0, -1$

or $P = (\alpha, \alpha - 1)$

where $\alpha \neq 0, 1$

$(\alpha, \alpha + 1)$ is on $y^2 = 4x + 1$

$(\alpha + 1)^2 = 4\alpha + 1$

$\Rightarrow \alpha^2 - 2\alpha = 0$

$\Rightarrow \alpha = 2$ $[\because \alpha \neq 0]$

Therefore, ordinate of P is 3.

$(\alpha, \alpha - 1)$ is on $y^2 = 4x + 1$. So

$(\alpha - 1)^2 = 4\alpha + 1$

$\Rightarrow \alpha^2 - 6\alpha = 0$

$\Rightarrow \alpha = 6$ $[\because \alpha \neq 0]$

Therefore, ordinate of P is 5

5. a., d. Here $x^2 = -\lambda\left(y + \frac{\mu}{\lambda}\right)$

Therefore, vertex $= \left(0, -\frac{\mu}{\lambda}\right)$

And the directrix is

$\left(y + \frac{\mu}{\lambda}\right) + \frac{-\lambda}{4} = 0$.

Comparing with the given data,

$\frac{\mu}{\lambda} = 1$ and $\frac{\mu}{\lambda} - \frac{\lambda}{4} = -2$

$\therefore -1 - \frac{\lambda}{4} = -2$

or $\lambda = 4 \Rightarrow \mu = 4$.

6. a., c. Given that the extremities of the latus rectum are $(1, 1)$ and $(1, -1)$

$\Rightarrow 4a = 2 \Rightarrow a = \frac{1}{2}$

$\Rightarrow$ The focus of the parabola is $(1, 0)$.

$\Rightarrow$ The vertex can be $\left(\frac{1}{2}, 0\right)$ and $\left(\frac{3}{2}, 0\right)$.

$\Rightarrow$ The equations of the parabola can be

$y^2 = 2\left(x - \frac{1}{2}\right)$

or $y^2 = 2\left(x - \frac{3}{2}\right)$

$\Rightarrow y^2 = 2x - 1$

or $y^2 = 2x - 3$

7. a., c. Let the possible point be $(t^2, 2t)$. Equation of tangent at this point is $yt = x + t^2$.

It must pass through (6, 5). (Since normal to circle always passes through its centre)

$\Rightarrow t^2 - 5t + 6 = 0$

$\Rightarrow t = 2, 3$

$\Rightarrow$ Possible points are (4, 4), (9, 6).

8. c., d.

$$t_2 = -t_1 - \frac{2}{t_1}$$

Also, $\frac{2at_1}{at_1^2} \times \frac{2at_2}{at_2^2} = -1$

$\Rightarrow t_1 t_2 = -4$

$\therefore \frac{-4}{t_1} = -t_1 - \frac{2}{t_1}$

$\Rightarrow t_1^2 + 2 = 4$ and $t_1 = \pm\sqrt{2}$.

So point can be $(2a, \pm 2\sqrt{2}a)$.

9. a., b. As a circle can intersect a parabola in four points, so quadrilateral may be cyclic.

The diagonals of the quadrilateral may be equal as the quadrilateral may be an isosceles trapezium.

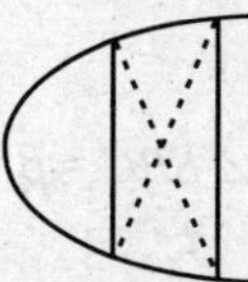

Fig. 3.45

A rectangle cannot be inscribed in a parabola. So (*c*) is wrong.

10. a., b., c., d. Any point on the parabola is $P(at^2, 2at)$.

Therefore, midpoint of $S(a, 0)$ and $P(at^2, 2at)$ is

$$R\left(\frac{a + at^2}{2}, at\right) \equiv (h, k).$$

$$\therefore h = \frac{a + at^2}{2}, k = at.$$

Eliminating t, we get

$$2x = a\left(1 + \frac{y^2}{a^2}\right) = a + \frac{y^2}{a}$$

$\Rightarrow 2ax = a^2 + y^2$

$\Rightarrow y^2 = 2a\left(x - \frac{a}{2}\right)$

It's a parabola with vertex at $\left(\frac{a}{2}, 0\right)$, latus rectum = $2a$.

Directrix is

$$x - \frac{a}{2} = -\frac{a}{2},$$

i.e., $x = 0$

Focus is

$$x - \frac{a}{2} = \frac{a}{2},$$

i.e., $x = a$

i.e., $(a, 0)$.

11. a., b., c. Equation of tangent to parabola $y^2 = 8x$ having slope m is $y = mx + \frac{2}{m}$.

Options (*a*), (*b*), (*c*) are tangents for $m = 1, 3, -\frac{1}{2}$, respectively.

12. a., c., d. Equation of normal to parabola $y^2 = 12x$ having slope m is $y = mx - 6m - 3m^3$. Options (*a*), (*c*), (*d*) are normal for $m - 1, -2$ and 3, respectively.

13. a., b., c., d.

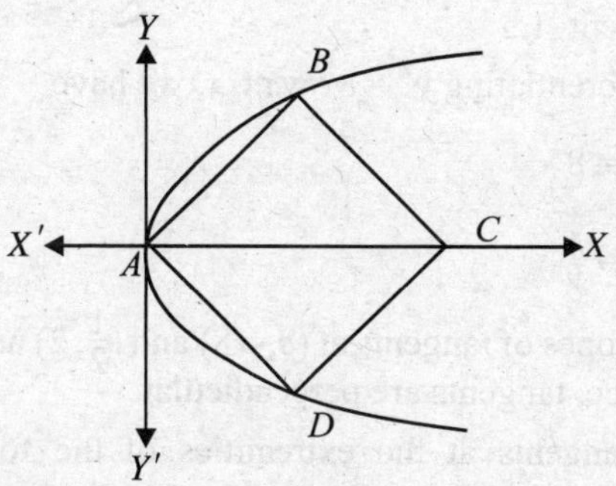

Fig. 3.46

AC is one diagonal along the x-axis, then the other diagonal is BD where both B and D lie on parabola. Also slope of AB is $\tan\frac{\pi}{4} = 1$. If B is $(at^2, 2at)$ then the slope of AB is

$$\frac{2at}{at^2} = \frac{2}{t} = 1$$

$\therefore t = 2$

Therefore, B is $(4a, 4a)$ and hence D is $(4a, -4a)$.

Clearly, C is $(8a, 0)$.

Reasoning Type

1. b. Any tangent having slope m is

$$y = mx + \frac{a}{m}$$

or $y = mx + \frac{9/4}{m}$

It passes through the point (4, 10), then

$$10 = 4m + \frac{9/4}{m}$$

$\Rightarrow 16m^2 - 40m + 9 = 0$

$\Rightarrow m_1 = \frac{1}{4}, m_2 = \frac{9}{4}$

$\Rightarrow$ Statement 1 is correct.

Also statement 2 is correct but it does not say anything about slope of the tangents, hence it is not correct explanation of statement 1.

2. b. Any normal to $y^2 = 4x$ is

$y = - tx + 2t + t^3$

If only one normal can be drawn to parabola from $(\lambda, \lambda + 1)$, then $\lambda < 2$.

Hence, statement 1 is true.

Statement 2 is also true as $(\lambda + 1)^2 > 4\lambda$ is true, $\forall \lambda \in R - \{1\}$, but does not explain statement 1, as it is not necessary that from every outside points only one normal can be drawn.

3. b. Both the statements are true (see properties of focal chord), but statement 2 is not correct explanation of statement 1.

4. a. Differentiating $y^2 = 8x$ w.r.t. x, we have

$2y\dfrac{dy}{dx} = 8$

$\Rightarrow \dfrac{dy}{dx} = \dfrac{4}{y}$

Now slopes of tangents at $(8, -8)$ and $(\frac{1}{2}, 2)$ are $-\frac{1}{2}$ and 2. Hence, tangents are perpendicular.

Also tangents at the extremities of the focal chord are perpendicular and meet on the directrix. Hence, both the statements are true and statement 2 is correct explanation of statement 1.

5. a. For parabola $y^2 = 4x$, $(4, 4)$ and $\left(\frac{1}{4}, -1\right)$ are extremities of the focal chord. Hence, tangents are perpendicular.

Then obviously normals at these points are also perpendicular.

6. c. Any tangent having slope m is

$y = m(x + a) + \dfrac{a}{m}$

or $y = mx + am + \dfrac{a}{m}$

is tangent to the given parabola for all $m \in R - \{0\}$.

Hence, statement 2 is false. However, statement 1 is true as when $m = 1$, tangent is $y = x + 2a$.

7. d. Statement 2 is true as it is the definition of parabola.

From statement 1, we have

$\sqrt{(x-1)^2 + (y+2)^2} = \dfrac{|3x + 4y + 5|}{5}$,

which is not a parabola as point $(1, -2)$ lies on the line $3x + 4y + 5 = 0$. Hence, statement 1 is false.

8. d. Statement 2 is correct (see properties of the focal chord). Then length of the focal chord according to the statement 1 is

$4(2)\left(\dfrac{4}{3}\right) = \dfrac{32}{3}$.

9. a Let parabola be $y^2 = 4x$

Clearly $x = 0$ is tangent to the parabola at $(0, \theta)$.

And lines $y = -x - 1$ and $y = x + 1$ are tangents to the parabola at $(1, 2)$ and $(1, -2)$ which are extremities of the latus rectum. These tangents meet on the directrix at right angle at $(-a, 0)$. Hence, circle passing through the point A, B, C also passes through its focus, as shown in the figure.

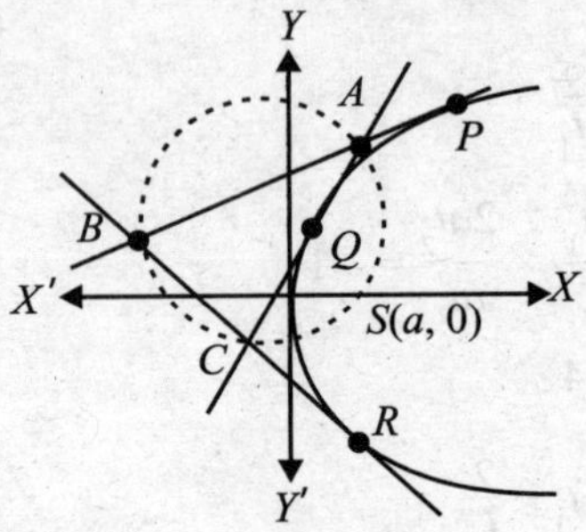

Fig. 3.47

Now consider a parabola $y^2 = 4ax$.

Let $P(t_1)$, $Q(t_2)$ and $R(t_3)$ be three points on it.

Tangents are drawn at these points which intersect at

$A \equiv (at_1t_2, a(t_1 + t_2))$

$B \equiv (at_1t_3, a(t_1 + t_3))$

$C \equiv (at_2t_3, a(t_2 + t_3))$

Let $\angle SAC = \alpha$ and $\angle SBC = \beta$

$$\Rightarrow \tan\alpha = \left|\frac{\frac{1}{t_2} - \frac{t_1 + t_2}{t_1t_2 - 1}}{1 + \frac{1}{t_2}\left(\frac{t_1 + t_2}{t_1t_2 - 1}\right)}\right|$$

$$= \left|\frac{1}{t_1}\right|$$

Similarly $\tan\beta = \left|\dfrac{1}{t_1}\right|$

$\Rightarrow \alpha = \beta$ or $\alpha + \beta = \pi$

$\Rightarrow A, B, C$ and S are concyclic.

10. d. Area of the triangle formed by the intersection points of tangents at point $A(t_1)$, $B(t_2)$ and $C(t_3)$ is

$\frac{1}{2}|t_1 - t_2|\,|t_2 - t_3|\,|t_3 - t_1| \neq 0$.

Hence, statement 1 is wrong. However, statement 2 is correct.

11. b. Statement 2 is true as circle is lying inside parabola without intersecting it.

But this cannot be considered the explanation of the statement 1, as even if they are not intersecting we can have common tangents as shown in the figure.

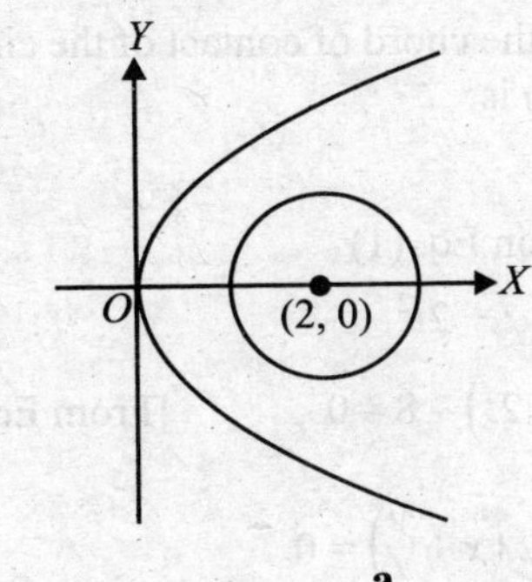

a.

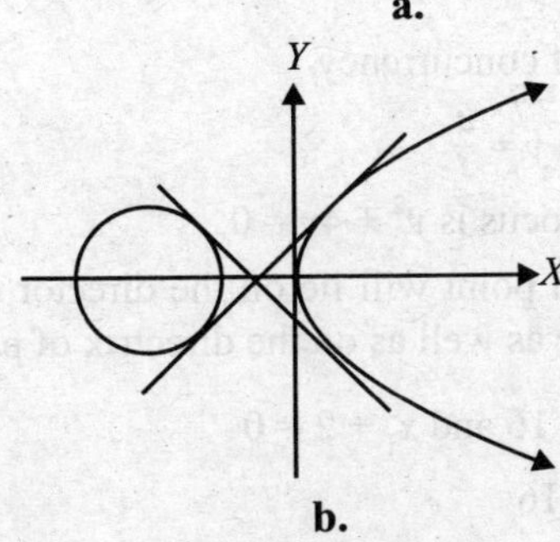

b.

Fig. 3.48

12. a. Let the foot of normal be $P(at^2, 2at)$, then

$ax + by + c = 0$ and $y = -\ tx + 2at + at^3$ are identical.

So

$$\frac{1}{b} = \frac{t}{a} = \frac{2at + at^3}{-c}$$

$$\Rightarrow t = \frac{a}{b}.$$

Thus P is $\left(\frac{a^3}{b^2}, \frac{2a^2}{b}\right)$.

Hence, equation of required tangent is $ty = x + at^2$

or $\frac{a}{b} y = x + a\left(\frac{a}{b}\right)^2$

or $y = \frac{b}{a} x + \frac{a^2}{b}$

13. a.

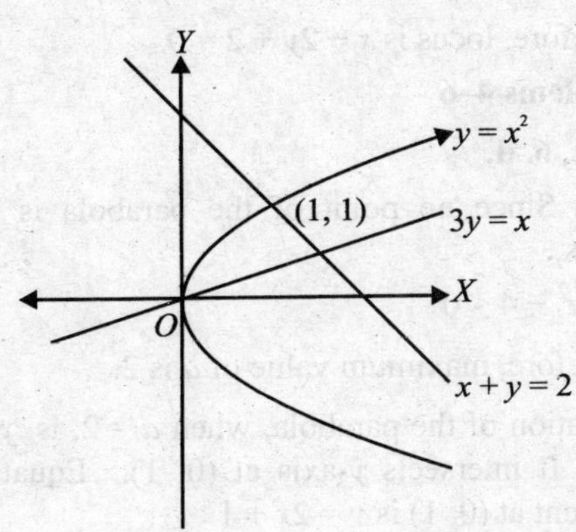

Fig. 3.49

Point (α, α^2) lies on the parabola $y^2 = x$

As shown in the figure, we have to find the value of α for which the part of the parabola lies inside the triangle formed by three lines.

Now line $x + y = 2$ meets the parabola at point (0, 0) and (1, 1).

Hence, $\alpha \in (0, 1)$

14. a.

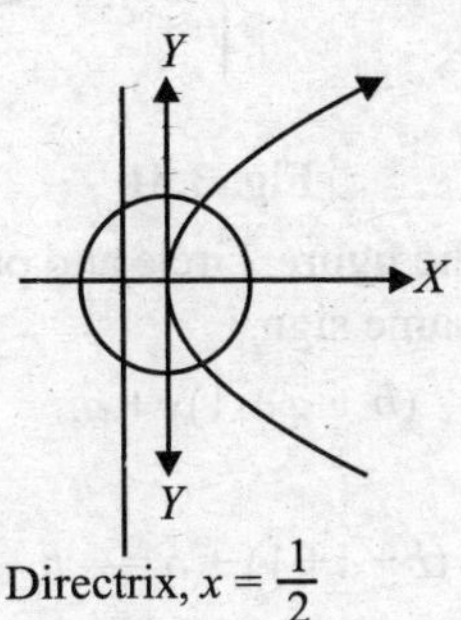

Fig. 3.50

Statement 2 is true as it is the property of the parabola. Now such points exists on the circle $x^2 + y^2 = a^2$ if it meets the directrix at least one point, for which radius of the circle $a \geq 1/2$.

15. a. Let $y^2 = 4ax$ be a parabola. Consider a line $x = 4a$ (this is a double ordinate which is twice of latus rectum), which cuts the parabola at $A(4a, 4a)$ and $B(4a, -4a)$.

Slope of OA is 1 and slope of OB is $-$ 1, where O is given. Therefore, AB subtends 90° at the origin. Statement 2 is correct and it clearly explains Statement 1.

16. d. Statement 1 is false. Since here $t^2 = 4$, the normal chord subtends a right angle at the focus (not at the vertex). However, statement 2 true (a standard result).

17. a. Let normals at points $A(at_1^2, 2at_1)$ and $C(at_3^2, 2at_3)$ meet the parabola again at points $B(at_2^2, 2at_2)$ and $D(at_4^2, 2at_4)$, then $t_2 = -t_1 - \frac{2}{t_1}$ and $t_4 = -t_3 - \frac{2}{t_3}$

Adding, we get

$$t_2 + t_4 = -t_1 - t_3 - \frac{2}{t_1} - \frac{2}{t_3}$$

$$\Rightarrow t_1 + t_2 + t_3 + t_4 = -\frac{2}{t_1} - \frac{2}{t_3}$$

$$\Rightarrow \frac{1}{t_1} + \frac{1}{t_3} = 0$$

$$\Rightarrow t_1 + t_3 = 0$$

Now, point of intersection of tangent at A and C will be $(at_1 t_3, a(t_1 + t_3))$. Since $t_1 + t_3 = 0$, this point will lie on x-axis, which is axis of parabola.

18. a.

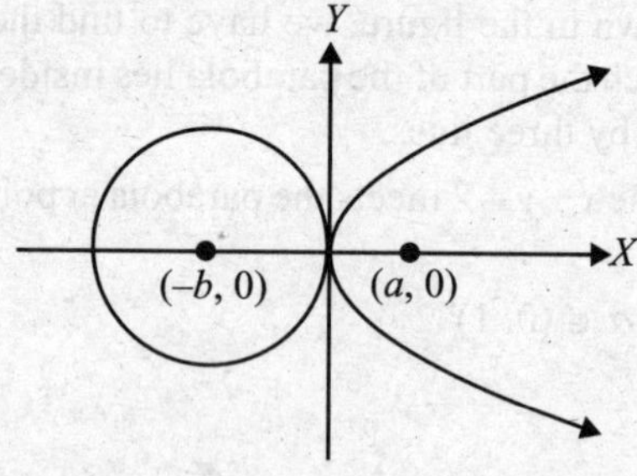

Fig. 3.51

As shown in the figure, circle and parabola touch when a and b have same sign.

Now $f(x) = x^2 - (b + a + 1)x + a,$

$\Rightarrow f(0) = a$

and $f(1) = 1 - (b+a+1) + a = -b$

$\Rightarrow f(0) f(1) = -ab < 0$

Hence, one root lies in (0, 1).

$\Rightarrow$ Both statements are true and statement 2 is the correct explanation of statement 1.

19. c. Statement 2 is false, as axis of parabola is normal to parabola which passes through the focus. However, normal other than axis never passes through focus. Statement 1 is correct as $x - y - 5 = 0$ passes through focus (3, –2), hence it cannot be normal.

20. b. Obviously, statement 2 is true, but it is not the correct explanation of statement 1 as A, A', B, B' form an isosceles trapezium, hence points are concyclic.

Linked Comprehension Type

For Problems 1–3

1. c., 2. d., 3. a.

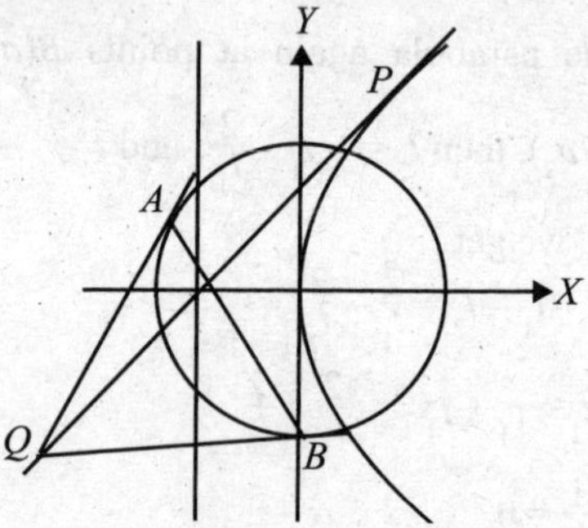

Fig. 3.52

Sol. 1. c. Equation of the tangent at point P of the parabola $y^2 = 8x$ is

$$yt = x + 2t^2 \quad (1)$$

Equation of the chord of contact of the circle $x^2 + y^2 = 8$ w.r.t. $Q(\alpha, \beta)$ is

$$x\alpha + y\beta = 8 \quad (2)$$

$Q(\alpha, \beta)$ lies on Eq. (1)

Hence, $\beta t = \alpha + 2t^2 \quad (3)$

$x\alpha + y\left(\frac{\alpha}{t} + 2t\right) - 8 = 0$ [From Eqs. (2) and (3)]

$2(ty - 4) + \alpha\left(x + \frac{y}{t}\right) = 0$

For point of concurrency,

$x = -\frac{y}{t}$ and $y = \frac{4}{t}$

Therefore, locus is $y^2 + 4x = 0$.

2. d. Required point will lie on the director circle of the given circle as well as on the directrix of parabola.

$\Rightarrow x_1^2 + y_1^2 = 16$ and $x_1 + 2 = 0$

$\Rightarrow 4 + y_1^2 = 16$

$\Rightarrow y = \pm 2\sqrt{3}$

Therefore, point is $(-2, \pm 2\sqrt{3})$.

3. a. Equation of circumcentre of ΔAQB is

$x^2 + y^2 - 4 + \lambda(x\alpha + y\beta - 8) = 0.$

Because it passes through (0, 0), i.e., centre of circle

$\Rightarrow \lambda = -\frac{1}{2}$

Let circumcentre be (h, k).

$\therefore h = \frac{\alpha}{4}, k = \frac{\beta}{4}$

$\Rightarrow \alpha = 4h, \beta = 4k$

Also $\beta t = \alpha + 2t^2$

or $\alpha - 2\beta + 8 = 0$ $[\because t = 2]$

Substituting $\alpha = 4h$ and $\beta = 4k$, we get

$h - 2k + 2 = 0$

Therefore, locus is $x - 2y + 2 = 0$.

For Problems 4–6

4. b., 5. c., 6. d.

Sol. 4. b. Since no point of the parabola is below the x-axis,

$D = a^2 - 4 \le 0$

Therefore, maximum value of a is 2.

Equation of the parabola, when $a = 2$, is $y = x^2 + 2x + 1$. It intersects y-axis at (0, 1). Equation of the tangent at (0, 1) is $y = 2x + 1$.

Since $y = 2x + 1$ touches the circle $x^2 + y^2 = r^2$,

$r = \frac{1}{\sqrt{5}}$

5. c. Equation of the tangent at (0, 1) to the parabola $y = x^2 + ax + 1$ is $y - 1 = a(x - 0)$ or $ax - y + 1 = 0$

As it touches the circle,

$$r = \frac{1}{\sqrt{a^2+1}}$$

Radius is maximum when $a = 0$.
Therefore, equation of the tangent is $y = 1$.
Therefore, slope of the tangent is 0.

6. d. Equation of tangent is $y = ax + 1$. Intercepts are $-1/a$ and 1. Therefore, area of the triangle bounded by tangent and the axes is

$$\frac{1}{2}\left|-\frac{1}{a}\times 1\right| = \frac{1}{2|a|}$$

It is minimum when $a = 2$. Therefore, minimum area is 1/4.

For Problems 7–9

7. a., 8. c., 9. d.

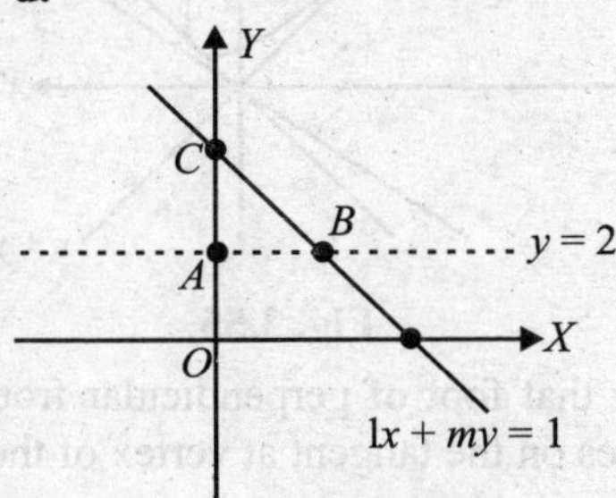

Fig. 3.53

Sol. $C \equiv \left(0, \frac{1}{m}\right)$, $B \equiv \left(\frac{1-2m}{l}, 2\right)$, $A \equiv (0, 2)$

Let (h, k) be the circumcentre of $\Delta\ ABC$ which is midpoint of BC

$$\Rightarrow h = \frac{1-2m}{2l},\ k = \frac{1+2m}{2m}$$

$$\Rightarrow m = \frac{1}{2k-2},\ l = \frac{k-2}{2h(k-1)}$$

Given that (l, m) lies on $y^2 = 4x$. Therefore,

$$m^2 = 4l$$

$$\Rightarrow \left(\frac{1}{2k-2}\right)^2 = 4\left\{\frac{k-2}{2h(k-1)}\right\}$$

$$\Rightarrow h = 8(k^2 - 3k + 2)$$

Therefore, locus of (h, k) is $x = 8(y^2 - 3y + 2)$

or $\left(y - \frac{3}{2}\right)^2 = \frac{1}{8}(x + 2)$

Therefore, vertex is $\left(-2, \frac{3}{2}\right)$.

Length of smallest focal chord = length of latus rectum $= \frac{1}{8}$.

From the equation of curve C, it is clear that it is symmetric about line $y = \frac{3}{2}$.

For Problems 10–12

10. d., 11. c., 12. d.

Sol. 10. d. $y = ax^2 + c$

$$\therefore \frac{dy}{dx} = 2ax = 1$$

Therefore, point of contact of the tangent is $\left(\frac{1}{2a}, \frac{1}{4a} + c\right)$.

Since it lies on $y = x$,

$$c = \frac{1}{4a}.$$

Thus $c = 1/8$

for $a = 2$.

11. c. If (1, 1) is point of contact, then $a = 1/2$

12. d. If $c = 2$, then point of contact is

$$\left(\frac{1}{2a}, \frac{1}{4a} + 2\right).$$

Since it lies on line $y = x$,

$$\frac{1}{2a} = \frac{1}{4a} + 2,$$

i.e., $a = \frac{1}{8}$

Therefore, point of contact is (4, 4).

For Problems 13–15

13. a., 14. b., 15. c.

Sol. Any parabola whose axes is parallel to x-axis will be of the form

$$(y - a)^2 = 4b(x - c) \quad (1)$$

Now, $lx + my = 1$, can be rewritten as

$$y - a = -\frac{l}{m}(x - c) + \frac{1 - am - lc}{m} \quad (2)$$

Eq. (2) will touch Eq. (1) if

$$\frac{1 - am - lc}{m} = \frac{b}{-l/m}$$

$$\Rightarrow -\frac{l}{m} = \frac{bm}{1 - am - lc}$$

$$\Rightarrow cl^2 - bm^2 + alm - l = 0 \quad (3)$$

But given that $5l^2 + 6m^2 - 4lm + 3l = 0$ (4)

Comparing Eqs. (3) and (4), we get

$$\frac{c}{5} = \frac{-b}{6} = \frac{a}{-4} = \frac{-1}{3}$$

$\Rightarrow c = \frac{-5}{3}$, $b = 2$, and $a = \frac{4}{3}$

So parabola is

$$\left(y - \frac{4}{3}\right)^2 = 8\left(x + \frac{5}{3}\right),$$

whose focus is $\left(\frac{1}{3}, \frac{4}{3}\right)$ and directrix is $3x + 11 = 0$.

For Problems 16–18

16. b., 17. a., 18. c.

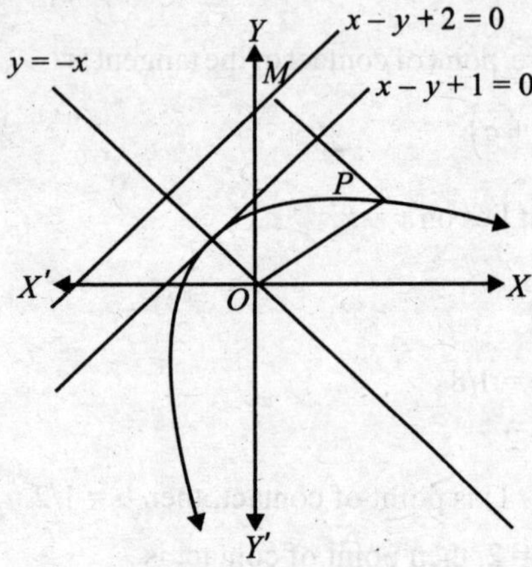

Fig 3.54

Sol. The distance between the focus and the tangent at the vertex $= \frac{|0 - 0 + 1|}{\sqrt{1^2 + 1^2}} = \frac{1}{\sqrt{2}}$.

The directrix is the line parallel to the tangent at vertex and at a distance $2 \times \frac{1}{\sqrt{2}}$ from the focus.

Let equation of directrix is

$x - y + \lambda = 0$,

where $\frac{\lambda}{\sqrt{1^2 + 1^2}} = \frac{2}{\sqrt{2}}$

$\Rightarrow \lambda = 2$

Let $P(x, y)$ be any moving point on the parabola, then $OP = PM$

$$\Rightarrow x^2 + y^2 = \left(\frac{x - y + 2}{\sqrt{1^2 + 1^2}}\right)^2$$

$$\Rightarrow 2x^2 + 2y^2 = (x - y + 2)^2$$

$$\Rightarrow x^2 + y^2 + 2xy - 4x + 4y - 4 = 0.$$

Latus rectum length = 2 × (distance of focus from directrix)

$$= 2\left|\frac{0 - 0 + 2}{\sqrt{1^2 + 1^2}}\right|$$

$$= 2\sqrt{2}$$

Solving parabola with the x-axis, we get

$$x^2 - 4x - 4 = 0$$

$$\Rightarrow x = \frac{4 \pm \sqrt{32}}{2} = 2 \pm 2\sqrt{2}$$

$\Rightarrow$ Length of chord on the x-axis is $4\sqrt{2}$.

Since the chord $3x + 2y = 0$ passes through the focus, it is a focal chord.

Hence, tangents at the extremities of chord are perpendicular.

For Problems 19–21

19. a., 20. b., 21. c.

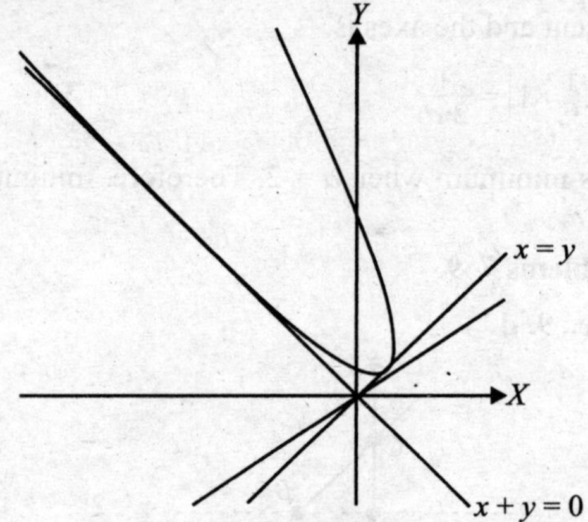

Fig. 3.55

Sol. We know that foot of perpendicular from focus upon tangent lies on the tangent at vertex of the parabola.

Now, if foot of perpendicular of (2, 3) on the line $x - y = 0$ is (x_1, y_1), then

$$\frac{x_1 - 2}{1} = \frac{y_1 - 3}{-1} = \frac{2 - 3}{2}$$

$$\Rightarrow x_1 = \frac{5}{2} \text{ and } y_1 = \frac{5}{2}$$

If foot of perpendicular of (2, 3) on the line $x + y = 0$ is (x_2, y_2), then

$$\frac{x_2 - 2}{1} = \frac{y_2 - 3}{1} = -\frac{2 + 3}{2}$$

$$\Rightarrow x_2 = -\frac{1}{2} \text{ and } y_2 = \frac{1}{2}$$

Now tangent at vertex passes through the points $\left(\frac{5}{2}, \frac{5}{2}\right)$ and $\left(-\frac{1}{2}, \frac{1}{2}\right)$. Then, its equation is

$$y - \frac{1}{2} = \frac{2}{3}\left(x + \frac{1}{2}\right)$$

or $4x - 6y + 5 = 0$

Latus rectum of the parabola = 4× (distance of focus from tangent at vertex)

$$= 4 \times \left|\frac{8 - 18 + 5}{\sqrt{52}}\right| = \frac{10}{\sqrt{13}}$$

Also, distance between the focus and tangent at vertex is $\frac{5}{\sqrt{13}}$

Since tangents $x + y = 0$ and $x - y = 0$ are perpendicular, they meet at (0, 0) which lies on the directrix.

Also, it is parallel to the tangent at vertex, hence its equation is $4x - 6y = 0$.

We know that $\frac{1}{SP} + \frac{1}{SQ} = \frac{1}{a}$, where a is $\frac{1}{4}$ th of latus rectum.

$\Rightarrow \frac{1}{SP} + \frac{1}{SQ} = \frac{2\sqrt{13}}{5}$

For Problems 22–24

22. d., 23. b., 24. a.

Sol. Solving the given parabolas, we have

$-8(x - a) = 4x$

$\Rightarrow x = \frac{2a}{3}$

Points of intersection are $\left(\frac{2a}{3}, \pm\sqrt{\frac{8a}{3}}\right)$.

Now $OABC$ is concyclic.

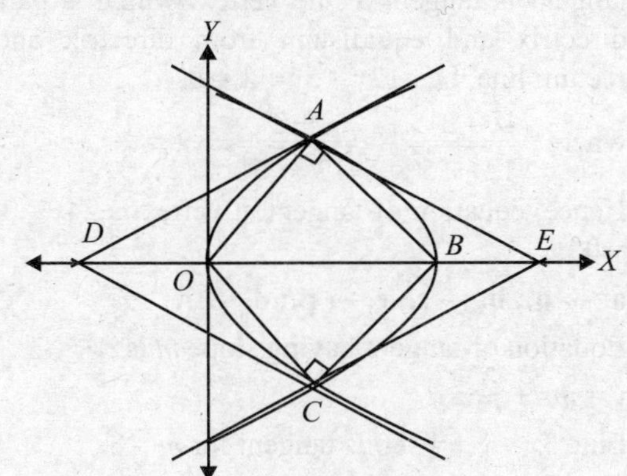

Fig. 3.56

Hence, $\angle OAB$ must be a right angle.

$\Rightarrow$ Slope of OA × Slope of $AB = -1$

$\Rightarrow \frac{\sqrt{\frac{8a}{3}}}{\frac{2a}{3}} \times \frac{\sqrt{\frac{8a}{3}}}{a - \frac{2a}{3}} = -1$

$\Rightarrow a = 12$

$\Rightarrow$ Coordinates of A and B are $(8, 4\sqrt{2})$ and $(8, -4\sqrt{2})$, respectively.

$\Rightarrow$ Length of common chord $= 8\sqrt{2}$.

Area of quadrilateral $= \frac{1}{2} OB \times AC$

$= \frac{1}{2} \times 12 \times 8\sqrt{2}$

$= 48\sqrt{2}$

Tangent to parabola $y^2 = 4x$ at point $(8, 4\sqrt{2})$ is $4\sqrt{2}y = 2(x + 8)$ or $x - 2\sqrt{2}y + 8 = 0$, which meets the x-axis at $D(-8, 0)$.

Tangent to parabola $y^2 = -8(x - 12)$ at point $(8, 4\sqrt{2})$ is $4\sqrt{2}y = -4(x + 8) + 96$ or $x + \sqrt{2}y - 16 = 0$, which meets the x-axis at $E(16, 0)$.

Hence, area of quadrilateral $DAEC = \frac{1}{2} DE \times AC$

$= \frac{1}{2} \times 24 \times 8\sqrt{2}$

$= 96\sqrt{2}$

For Problems 25–27

25.b., 26. c., 27. d.

Sol. For $y^2 = 4x$, coordinates of end of latus rectum are $P(1, 2)$ and $Q(1, -2)$.

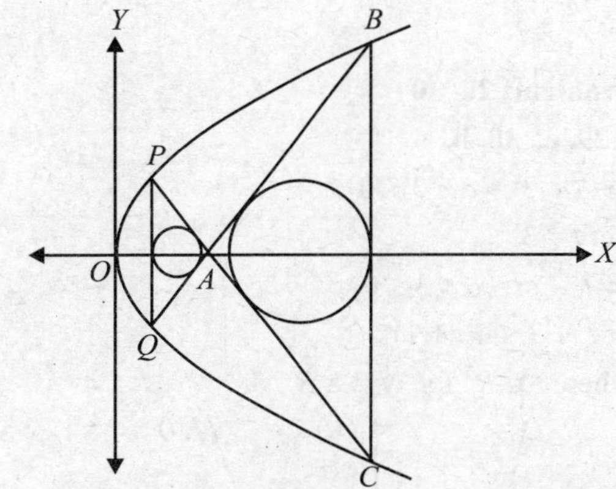

Fig. 3.57

ΔPAQ is isosceles right angled. Therefore, slope of PA is -1 and its equation is $y - 2 = -(x - 1)$ or $x + y - 3 = 0$.

Similarly, equation of line QB is $x - y - 3 = 0$.

Solving $x + y - 3 = 0$ with the parabola $y^2 = 4x$, we have

$(3 - x)^2 = 4x$ or $x^2 - 10x + 9 = 0$

$\therefore x = 1, 9$.

Therefore, coordinates of B and C are $(9, -6)$ and $(9, 6)$, respectively.

Area of trapezium $PBCQ = \frac{1}{2} \times (12 + 4) \times 8$

$= 64$ sq. units

Let the circumcentre of trapezium $PBCQ$ be $T(h, 0)$

Then $PT = PB$

$\Rightarrow \sqrt{(h-1)^2 + 4} = \sqrt{(h-9)^2 + 36}$

$\Rightarrow -2h + 5 = -18h + 81 + 36$

$\Rightarrow 16h = 112$

$\Rightarrow h = 7$

Hence, radius is $\sqrt{40} = 2\sqrt{10}$.

Let inradius of ΔAPQ be r_1, then $r_1 = \frac{\Delta_1}{s_1}$

$$= \frac{\frac{1}{2} \times 4 \times 2}{4 + 2\sqrt{4+4}}$$
$$= \frac{1}{1+\sqrt{2}} = \sqrt{2} - 1$$

Let inradius of ΔABC be r_2, then

$$r_2 = \frac{\Delta_2}{s_2}$$
$$= \frac{\frac{1}{2} \times 12 \times 6}{12 + 2\sqrt{36+36}}$$
$$= \frac{3}{1+\sqrt{2}} = 3(\sqrt{2}-1)$$
$$\Rightarrow \frac{r_2}{r_1} = 3$$

For Problems 28–30

28.d., 29. c., 30. b.

Sol. $9x - a.\, 3^x - a + 3 \le 0$

Let $t = 3^x$

$\Rightarrow t^2 - at - a + 3 \le 0$

or $t^2 + 3 \le a(t+1)$

where $t \in R^+$ for $\forall\, x \in R$

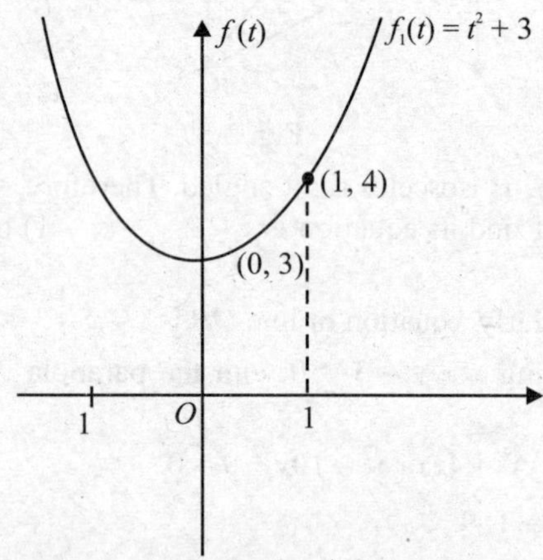

Fig. 3.58

Let $f_1(t)$ be $t^2 + 3$ and $f_2(t)$ be $a(t+1)$.

28. d. From $x < 0$, $t \in (0, 1)$. That means (1) should have at least one solution in $t \in (0, 1)$.

From (1), it is obvious that $a \in R^+$.

Now $f_2(t) = a(t+1)$ represents a straight line. It should meet the curve.

$f_1(t) = t^2 + 3$, at least once in $t \in (0, 1)$.

$f_1(0) = 3, f_1(1) = 4, f_2(0) = a, f_2(1) = 2a$.

If $f_1(0) = f_2(0) \Rightarrow a = 3$; if $f_1(1) = f_2(1) = a = 2$

Hence, required $a \in (2, 3)$.

29. c. For at least one positive solution, $t \in (1, \infty)$. That means graphs of $f_1(t) = t^2 + 3$ and $f_2(t) = a(t+1)$ should meet at least once in $t \in (1, \infty)$.

If $a = 2$, both curve touch each other at (1, 4). Hence, the required $a \in (2, \infty)$.

30 b. In this case, both graphs should meet at least once in $t \in (0, \infty)$. For $a = 2$, both curves touch; hence, the required $a \in [2, \infty)$.

Matrix-Match Type

1. a. → r.; b. → s.; c. → p.; d. → q.

Sol. Locus of point of intersection of perpendicular tangent is directrix which is $12x - 5y + 3 = 0$.

Parabola is symmetrical about its axis, which is a line passing through the focus (1, 2) and perpendicular to the directrix, which has equation $5x + 12y - 29 = 0$.

Minimum length of focal chord occurs along the latus rectum line, which is a line passing through the focus and parallel to directrix, i.e,. $12x - 5y - 2 = 0$.

Locus of foot of perpendicular from foucs upon any tangent is tangent at the vertex, which is parallel to directrix and equidistant from directrix and latus rectum line, i.e., $12x - 5y + \lambda = 0$,

where $\frac{|\lambda - 3|}{\sqrt{12^2 + 5^2}} = \frac{|\lambda + 2|}{\sqrt{12^2 + 5^2}} \Rightarrow \lambda = \frac{1}{2}$.

Hence, equation of tangent at vertex is $24x - 10y + 1 = 0$.

2. a. → q.; b. → s.; c. → p.; d. → r.

Equation of tangent having slope m is

$y = mx + \frac{3}{m}$.

Line $3x - y + 1 = 0$ is tangent for $m = 3$.

Equation of normal having slope m is

$y = mx - 6m - 3m^3$.

Line $2x - y - 36 = 0$ is normal for $m = 2$.

Chord of contact w.r.t. any point on the directrix is the focal chord which passes through the focus (3, 0).

Line $2x - y - 36 = 0$ passes through the focus.

Chords which subtend right angle at the vertex are concurrent at point $(4 \times 3, 0)$ or (12, 0).

Line $x - 2y - 12 = 0$ passes through the point (12, 0).

3. a. → q., s; b. → r.; c. → p., q.; d. → q., r.

a. Tangent to parabola having slope m is $ty = x + t^2$, it passes through point (2, 3) then $3t = 2 + t^2 \Rightarrow t = 1$ or $2 \Rightarrow$ point of contact $(t^2 + 2t) \equiv (1, 2)$ or (4, 4)

b. Let point on the circle be $P(x_1, y_1)$, then chord of contact of parabola w.r.t. P is $yy_1 = 2(x + x_1)$. Comparing with $y = 2(x - 2)$, we have $y_1 = 1$ and $x_1 = 2$, which also satisfy the circle.

c. Point Q on the parabola $Q(t^2, 2t)$

Now area of triangle OPQ is

is $\frac{1}{2}\left\|\begin{matrix}0 & 0\\ 4 & -4\\ t^2 & 2t\\ 0 & 0\end{matrix}\right\| = 6 \Rightarrow 8t + 4t^2 = \pm 12$

For $t^2 + 2t - 3 = 0$, $(t - 1)(t + 3) = 0$, then $t = 1$ or $t = -3$.

Then point Q are (1, 2) or (9, – 6).

d. Points (1, 2) and (– 2, 1) satisfy both the curves.

4. a. → p, r.; b. → p., r.; c. → q; d. → q., s.

Points through which perpendicular tangent can be drawn to the parabola $y^2 = 4x$ lie on the directrix. Points (–1, 2) and (–1, –5) lie on the directirx. Also from these points only one normal can be drawn.

Integer Type

1.(6)

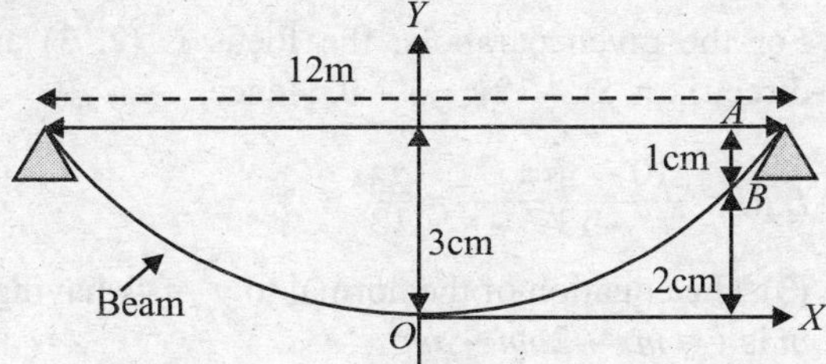

Fig. 3.59

Let the vertex be at the lowest point and the axis vertical. Let the coordinate axes be chosen as shown in the figure. The equation of the parabola takes the form $x^2 = 4ay$. Since it passes through (6, 3/100), we have $6^2 = 4a\,(3/100)$. So $a = 300$m.

Let AB be the deflection of the beam which is (1/100) m.

The coordinates of B are (x, 2/100). Therefore,

$x^2 = 4 \times 300 \times \frac{2}{100} = 24.$

$\Rightarrow x = 2\sqrt{6}$m.

2. (2)

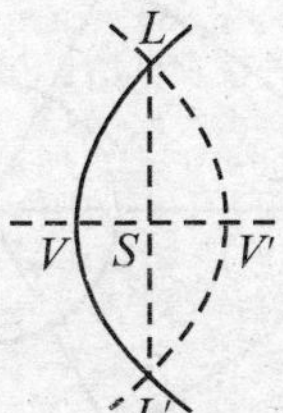

Fig. 3.60

Let L and L' be the ends of the latus rectum. S bisects LL'. VSV' is the perpendicular bisector of LL', where $VS = (1/4)\,LL' = V'S$. Clearly, two parabolas are possible.

3. (4)

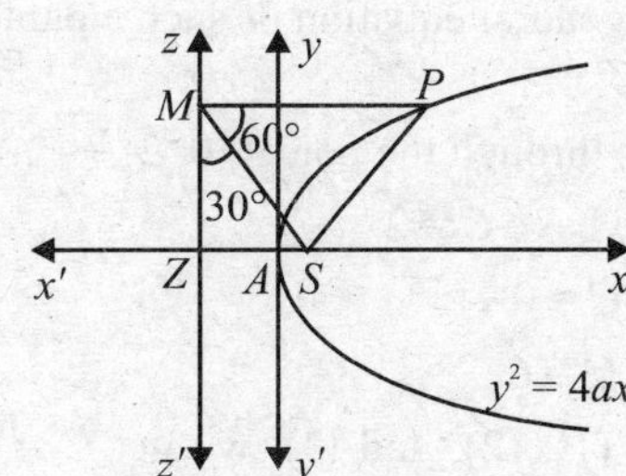

Fig. 3.61

From the definition of a parabola, we have $SP = PM$. SPM is an equilateral triangle. Therefore,

$SP = PM = SM$

$\Rightarrow \angle PMS = 60°$

$\Rightarrow \angle SMZ = 30°$

In ΔSMZ, we have

$\sin 30° = \frac{SZ}{SM}$

$\Rightarrow \frac{1}{2} = \frac{2}{SM} \Rightarrow SM = 4$

Hence, $SP = SM = 4$.

4.

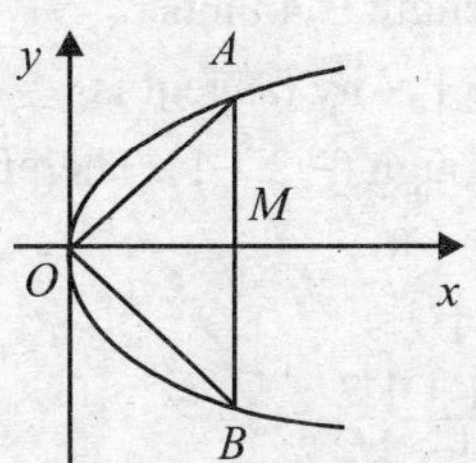

Fig. 3.62

Since triangle OAB is equilateral, $OA = OB = AB = p$. Thus AB will be a double ordinate of the parabola. Thus $\angle AOM = \angle MOB = \frac{\pi}{6}$.

$\Rightarrow OM = p\cos\frac{\pi}{6}$ and $AM = p\sin\frac{\pi}{6}$.

Then the co-ordinates of A are $(\sqrt{3}p/2, p/2)$. A lies on the parabola. So

$\frac{p^2}{4} = \frac{\sqrt{3}p}{2} \Rightarrow p = 2\sqrt{3}$

5. (3)

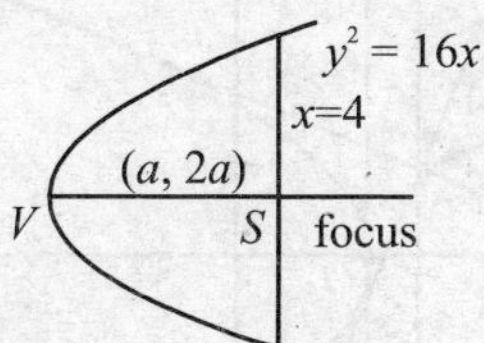

Fig. 3.63

$(a, 2a)$ is an interior point of $y^2 - 16x = 0$ if $(2a)^2 - 16a < 0$, i.e., $a^2 - 4a < 0$.

V (0, 0) and $(a, 2a)$ are on the same side of $x - 4 = 0$. So $a - 4 < 0$, i.e., $a < 4$.

Now $a^2 - 4a < 0 \Rightarrow 0 < a < 4$

6. **(1)** The general equation of such a parabola is $y = Ax^2 + Bx + C$.

It passes through the points (0, 2), (– 1, 0), and (1, 6).

Then we have $C = 2$ (1)

$A - B + C = 0$ (2)

$A + B + C = 6$ (3)

Solving (1), (2), and (3), we get $C = 2$, $A = 1$, and $B = 3$.

Hence, the equation of the parabola is $y = x^2 + 3x + 2$.

7. **(7)** $(x-1)^2+(y-2)^2 = \lambda (x + y + 3)^2 = 2\lambda \left(\frac{x+y+3}{\sqrt{2}}\right)^2$.

Comparing it with $PS^2 = PM^2$, we get $2\lambda = 1 \Rightarrow \lambda = 1/2$.

8. **(9)** The length of the latus rectum of $y = ax^2 + bx + c$ is $1/a$. Now the semi-latus rectum is the *H.M.* of *SP* and *SQ*. Therefore, we have $\frac{1}{SP} + \frac{1}{SQ} = \frac{2}{1/2a}$

$\Rightarrow 4a = \frac{1}{4} + \frac{1}{6} = \frac{5}{12} \Rightarrow a = \frac{5}{48}$. Hence Latus

9. **(4)** The length of the focal chord making an angle α with the *x*-axis is $2 \operatorname{cosec}^2 \alpha$. For $\alpha \in (0, \pi/4]$, its maximum length is 4 units.

10. **(3)** Here $a = 1$. Any tangent is $y = mx + 1/m$.

It passes through (– 2, –1). Therefore,

$2m^2 - m - 1 = 0$

$\Rightarrow m = 1, - 1/2$

$\Rightarrow \tan \alpha = \frac{1 + 1/2}{1 - 1/2} = 3$.

11. **(9)** Any tangent to the parabola having slope m is $y = mx + (1/m)$. It passes through (α, β). Therefore,

$\beta = m\alpha + \frac{1}{m}$ or $\alpha m^2 - \beta m + 1 = 0$.

According to the question, its roots are $m_1, 2m_1$. Now

$m_1 + 2m_1 = \frac{\beta}{\alpha}$ and $m_1 \times 2m_1 = \frac{1}{\alpha}$. So $2\left(\frac{\beta}{3\alpha}\right)^2 = \frac{1}{\alpha}$

$\Rightarrow \alpha = \frac{2}{9}\beta^2$

12. **(8)**

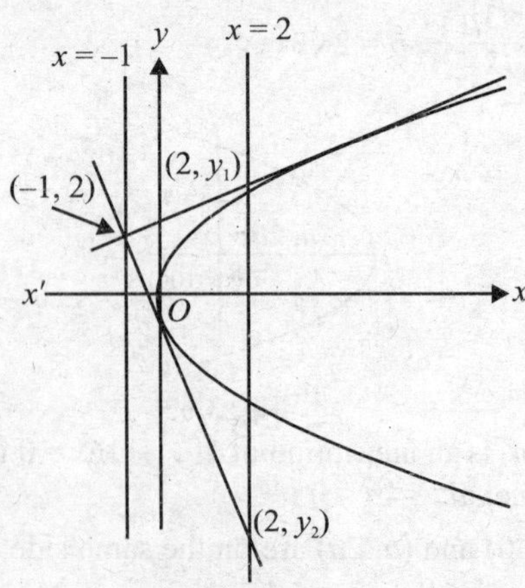

Fig. 3.64

The pair of tangents from the point (–1, 2) is $SS_1 = T^2$ or $(y^2 - 4x)(y^2 - 4x_1) = [yy_1 - 2(x + x_1)]^2$.

Then the pair of tangents from (–1, 2) is $(y^2 - 4x)(4 + 4) = [2y - 2(x - 1)]^2 = 4(y - x + 1)^2$

or $2(y^2 - 4x) = (y - x + 1)^2$

Solving with the line $x = 2$, we get

$2(y^2 - 8) = (y - 1)^2$

or $y^2 + 2y - 17 = 0$,

where $y_1 + y_2 = - 2$ and $y_1y_2 = - 17$

Now $|y_1 - y_2|^2 = (y_1 + y_2)^2 - 4y_1y_2$

$= 4 - 4(-17) = 72$

Hence, $(y_1 - y_2) = \sqrt{72} = 6\sqrt{2}$.

13. **(1)** Tangents from any point on the directrix are perpendicular and the corresponding chord of contact is the focal chord, which is a diameter of the circumcircle of the triangle *PAB*. The least value of the diameter is the latus rectum.

For the given parabola, the focus is (2, 3) and the directrix is $5x - 12y + 3 = 0$. Hence,

$L.R. = \frac{|10 - 36 + 3|}{13} = \frac{23}{13}$.

14. **(3)** The equation of the normal to $y^2 = 4x$ having slope m is $y = mx - 2am - am^3$.

If it passes through the point (15, 12), then

$12 = 15m - 2m - m^3$

$\Rightarrow m^3 - 13m + 12 = 0$

$\Rightarrow (m - 1)(m - 3)(m + 4) = 0$

$\Rightarrow m = 1, 3, -4$

15. **(2)** The centre and radius of the given circle are $P(6, 0)$ and $\sqrt{5}$, respectively.

The shortest distance always occurs along the common normal.

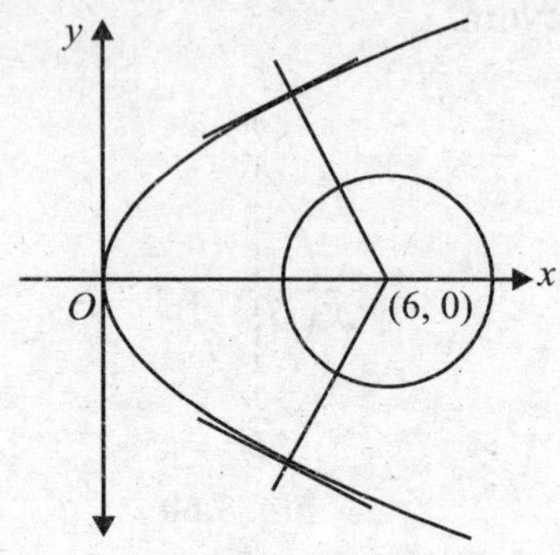

Fig. 3.65

Differentiating $y^2 = 4x$ w.r.t. x, we get $\frac{dy}{dx} = \frac{2}{y}$.

Then the slope of the normal at point $A\ (y_1^2/4, y_1)$ is $-y_1/2$.

Also the slope of AP is

$$\frac{\frac{y_1 - 0}{2}}{\frac{y_1^2}{4} - 6} = -\frac{y_1}{2}$$

$\Rightarrow y_1 = 0$ or $y_1 = \pm 4$.

So the points are $0(0, 0)$, $A(4, 4)$, and $C(4, -4)$. Hence, the shortest distance is

$AP - \sqrt{5} = \sqrt{20} - \sqrt{5} = \sqrt{5}$.

16. (5)

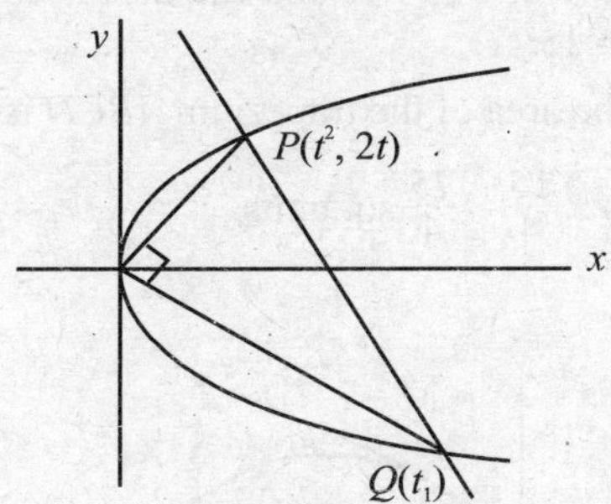

Fig. 3.66

We have $t_1 = -t - (2/t)$ and $tt_1 = -4$. Therefore,

$t_1 = -\frac{4}{t} \Rightarrow -\frac{4}{t} = -2 - \frac{2}{t} \Rightarrow \frac{2}{t} = 2 \Rightarrow t = 1$

$\Rightarrow t_1 = -3 \Rightarrow Q \equiv (9, -6), P \equiv (1, 2)$.

Hence, $PQ = \sqrt{8^2 + 8^2} = 8\sqrt{2}$.

17. (3) The given parabola is $(y - 1)^2 = 4(x - 2)$. Hence, the vertex is (1, 2).

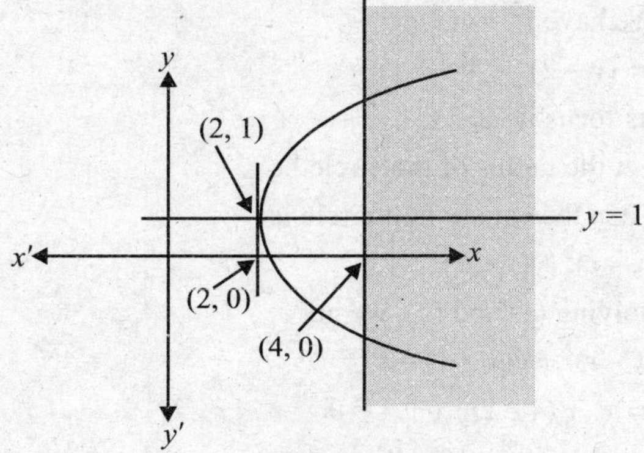

Fig. 3.67

The focus is (2, 2). Hence, all points $(2a, b)$ must lie in the shaded region. Therefore, $2a > 4 \Rightarrow a > 2$.

18. (3)

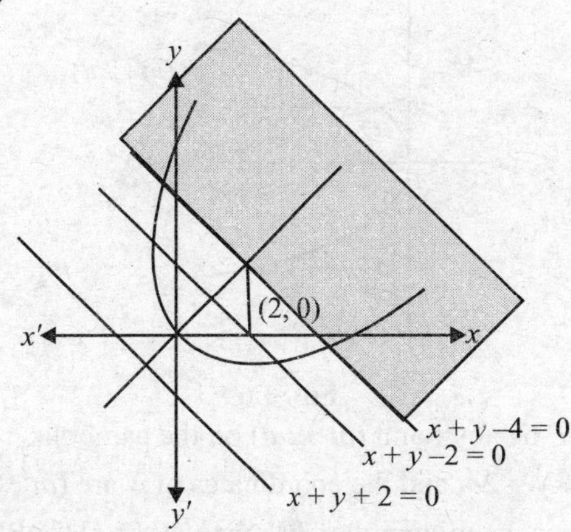

Fig. 3.68

We have $2(x - 1)^2 + 2(y - 1)^2 = (x + y + 2)^2$

$\Rightarrow \sqrt{(x-1)^2 + (y-1)^2} = \left|\frac{x + y + 2}{\sqrt{1+1}}\right|$

The focus is (1, 1) and the directrix is $x + y + 2 = 0$.

The equation of the latus rectum is $x + y - 2 = 0$.

Then the required points lie above the line $x + y - 4 = 0$, which intersects the line $x = y$ at (2, 2).

Hence, $h > 2$.

19. (3) For $y^2 = 4ax$, the normal is $y = mx - 2am - am^3$ (1)

For $y^2 = 4c(x - b)$, the normal is $y = m(x - b) - 2cm - cm^3$ (2)

If two parabolas have common normal, then (1) and (2) must be identical.

After comparing the coefficients, we get

$m = \pm\sqrt{\frac{2(a-c) - b}{(c-a)}}$,

which is real if,

$-2 - \frac{b}{c-a} > 0 \Rightarrow \frac{b}{a-c} > 2$

20. (6) The equation of the tangent to $y^2 = 8x$ at $P(2, 4)$ is

$4y = 4(x + 2)$ or $x - y + 2 = 0$ (1)

Let (x_1, y_1) be the midpoint of the chord QR. Then the equation of QR is

$yy_1 - 4(x + x_1) - 5 = y_1^2 - 8x_1 - 5$

$\Rightarrow 4x - yy_1 - 4x_1 + y_1^2 = 0$ (2)

Clearly, (1) and (2) represent the same line. So,

$\frac{4}{1} = -\frac{y_1}{-1} = \frac{-4x_1 + y_1^2}{2}$

$\Rightarrow y_1 = 4$ and $8 = 4x_1 + y_1^2$

$\Rightarrow y_1 = 4$ and $x_1 = 2$.

21. (3) The tangent at the vertex is the y-axis or $x = 0$ (1)

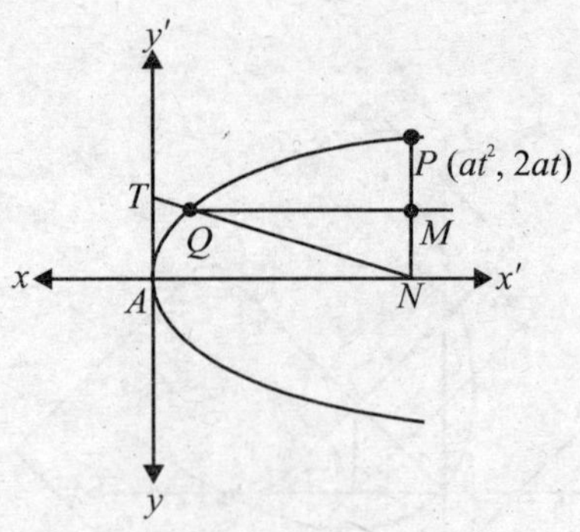

Fig. 3.69

Let P be any point $(at^2, 2at)$ on the parabola.

So $PN = 2at$ and the coordinates of N are $(at^2, 0)$.

If M is the midpoint of PN, then $NM = (1/2)\ PN = at$.

As MQ is parallel to the x-axis, its equation is $y = at$ (2)

Solving (3) with the parabola, we get

$a^2t^2 = 4ax \Rightarrow x = \frac{1}{4}at^2$

So the coordinates of Q are $\left(\frac{1}{4}at^2, at\right)$.

The equation of NQ will be

$$(y - 0) = \frac{at - 0}{\frac{1}{4}at^2 - at^2}(x - at^2)$$

$$\Rightarrow y = \frac{-4}{3t}(x - at^2) \quad (3)$$

If NQ meets the tangent at the vertex, i.e., $x = 0$ at T, then on putting $x = 0$ in equation (3), we get $y = \frac{4at}{3}$.

So the coordinates of T are $\left(0, \frac{4at}{3}\right)$ or $AT = \frac{4at}{3}$.

$\Rightarrow AT = \frac{2}{3} \times PN$

22. (5) $AS = 1 + t^2$ (Focal distance of any point on the parabola $y^2 = 4ax$ is $a + at^2$.)

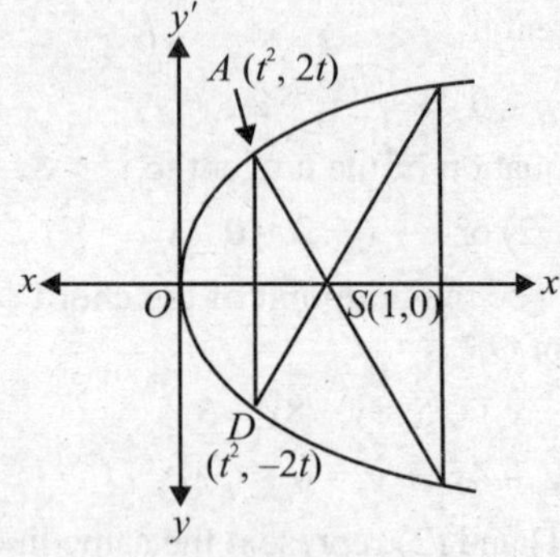

Fig. 3.70

$$CS = AC - AS = \frac{25}{4} - (1 + t^2)$$

If PQ is a focal chord of the parabola $y^2 = 4ax$, then

$$\frac{1}{SP} + \frac{1}{SQ} = \frac{1}{a}$$

$$\Rightarrow \frac{1}{AS} + \frac{1}{CS} = \frac{1}{a}$$

$$\Rightarrow \frac{1}{1 + t^2} + \frac{1}{\frac{25}{4} - (1 + t^2)} = 1$$

$$\Rightarrow \frac{1}{1 + t^2} + \frac{4}{25 - 4(1 + t^2)} = 1$$

$$\Rightarrow 25 - 4(1 + t^2) + 4(1 + t^2) = (1 + t^2)[25 - 4(1 + t^2)]$$

$$\Rightarrow 4(1 + t^2)^2 - 25(1 + t^2) + 25 = 0$$

$$\Rightarrow [(1 + t^2) - 5][4(1 + t^2) - 5] = 0$$

$$\Rightarrow 1 + t^2 = 5, \frac{5}{4}$$

$$\Rightarrow t^2 = 4, \frac{1}{4} \Rightarrow t = \pm 2, \pm\frac{1}{2}$$

Thus the coordinates of A, B, C, and D are $A \equiv (1/4, 1)$, $B \equiv (4, 4)$, $C \equiv (4, -4)$, and $D \equiv (1/4, -1)$.

Thus $AD = 2$, $BC = 8$, and the distance between AD and $BC = 15/4$.

Hence, the area of the trapezium $ABCD$ is

$$\frac{1}{2}(2 + 8) \times \frac{15}{4} = \frac{75}{4} \text{ sq. units.}$$

23. (4)

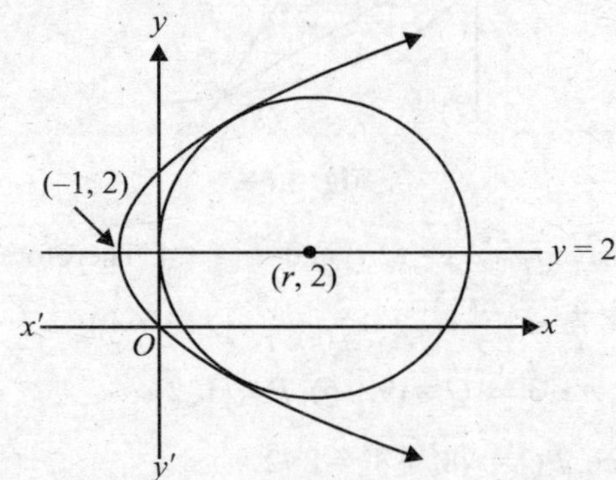

Fig. 3.71

We have $y^2 = 4(x + y)$

$\Rightarrow (y - 2)^2 = 4(x + 1)$ (1)

Its focus is (0, 2).

Let the radius of the circle be r.

Then the equation of circle is

$(x - r)^2 + (y - 2)^2 = r^2$ (2)

Solving (1) and (2), we get

$(x - r)^2 + 4(x + 1) = r^2$

$\Rightarrow x^2 + (4 - 2r)x + 4 = 0$

$\Rightarrow (4 - 2r)^2 - 16 = 0$

$\Rightarrow 4 - 2r = \pm 4$

$\Rightarrow r = 4$

24. (3) We know that from any point, normal to $y^2 = 4ax$ is given by $y = mx - 2am - am^3$. Here $a = 1/4$. Therefore, the normal is $y = mx - \frac{m}{2} - \frac{m^3}{4}$.

This normal passes through $(c, 0)$. Therefore,

$$mc - \frac{m}{2} - \frac{m^3}{4} = 0 \quad (1)$$

Now for two normals to be perpendicular to each other, we must have $m_1m_2 = -1$, i.e., the product of the roots of the equation $(m^2/4) + (1/2) - c = 0$ is -1. Thus

$$\frac{\left(\frac{1}{2} - c\right)}{1/4} = -1$$

$$\Rightarrow \frac{1}{2} - c = -\frac{1}{4}$$

$$\Rightarrow c = 3/4$$

25. (2)

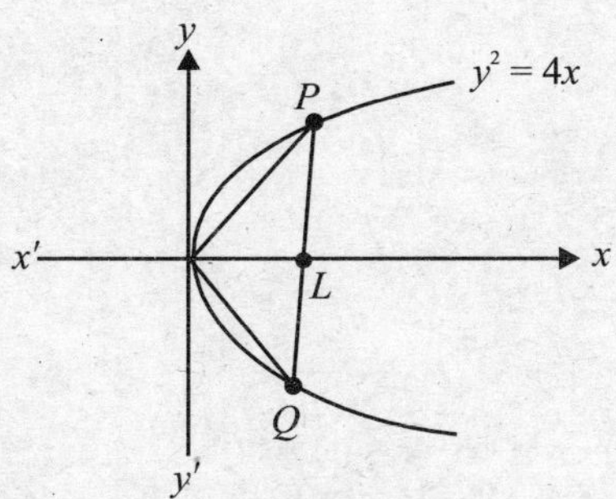

Fig. 3.72

The chord joining $P(t_1^2, 2t_1)$ and $Q(t_2^2, 2t_2)$ subtends a right angle at the origin. Thus $t_1t_2 = -4$.

Also the slope of the chord is $2/(t_1 + t_2)$. So the equation of chord AB is

$$y - 2t_1 = \frac{2}{t_1 + t_2}(x - t_1^2)$$

or $(t_1 + t_2)\,y - 2t_1^2 - 2t_1t_2 = 2(x - t_1^2)$

or $(t_1 + t_2)\,y - 2t_1^2 + 8 = 2(y - t_1^2)$

or $(t_1 + t_2)\,y + 8 = 2x$

or $2\,(x - 4) = (t_1 + t_2)\,y$,

which always passes through the point $(4, 0)$. If (h, k) is the midpoint of PQ, then $h = \frac{t_1^2 + t_2^2}{2}$ and $k = t_1 + t_2$

$$\Rightarrow h = \frac{(t_1 + t_2)^2 - 2\,(t_1t_2)}{2} = \frac{k^2 + 8}{2}$$

or $y^2 = 2(x - 4)$

26. (9) $y = mx + c$ is a normal to the parabola $y^2 = 4ax$ if $c = -2am - am^3$.

Here $m = -1$ and $c = k$ and $a = 3$. Hence,

$c = k = -2(3)\,(-1) - 3\,(-1)^3 = 9$

CHAPTER 4

Ellipse

Objective Type

Each question has four choices a, b, c, d, out of which *only one* answer is correct. Find the correct answer.

1. P and Q are the foci of the ellipse $\frac{x^2}{a^2}+\frac{y^2}{b^2}=1$ and B is an end of the minor axis. If PBQ is an equilateral triangle, then eccentricity of the ellipse is

   a. $\frac{1}{\sqrt{2}}$ **b.** $\frac{1}{3}$ **c.** $\frac{1}{2}$ **d.** $\frac{\sqrt{3}}{2}$

2. An ellipse having foci at (3, 3) and (– 4, 4) and passing through the origin has eccentricity equal to

   a. $\frac{3}{7}$ **b.** $\frac{2}{7}$ **c.** $\frac{5}{7}$ **d.** $\frac{3}{5}$

3. If the eccentricity of the ellipse $\frac{x^2}{a^2+1}+\frac{y^2}{a^2+2}=1$ is $\frac{1}{\sqrt{6}}$, then latus rectum of ellipse is

   a. $\frac{5}{\sqrt{6}}$ **b.** $\frac{10}{\sqrt{6}}$
   c. $\frac{8}{\sqrt{6}}$ **d.** none of these

4. If PQR is an equilateral triangle inscribed in the auxiliary circle of the ellipse $\frac{x^2}{a^2}+\frac{y^2}{b^2}=1$ $(a > b)$ and $P'Q'R'$ is corresponding triangle inscribed within the ellipse then centroid of the triangle $P'Q'R'$ lies at

   a. centre of ellipse
   b. focus of ellipse
   c. between focus and centre on major axis
   d. none of these

5. S_1, S_2 are foci of an ellipse of major axis of length 10 units and P is any point on the ellipse such that perimeter of triangle PS_1S_2 is 15. Then eccentricity of the ellipse is

   a. 0.5 **b.** 0.25 **c.** 0.28 **d.** 0.75

6. If the ellipse $\frac{x^2}{4}+y^2=1$ meets the ellipse $x^2+\frac{y^2}{a^2}=1$ in four distinct points and $a=b^2-5b+7$, then b does not lie in

   a. [4, 5] **b.** $(-\infty, 2)\cup(3, \infty)$
   c. $(-\infty, 0)$ **d.** [2, 3]

7. With a given point and line as focus and directrix, a series of ellipses are described, the locus of the extremities of their minor axis is

   a. ellipse **b.** parabola
   c. hyperbola **d.** none of these

8. A line of fixed length $a+b$ moves so that its ends are always on two fixed perpendicular straight lines, then the locus of a point, which divides this line into portions of lengths a and b is a/an

   a. ellipse **b.** parabola
   c. straight line **d.** none of these

9. The length of the major axis of the ellipse $(5x-10)^2+(5y+15)^2=\frac{(3x-4y+7)^2}{4}$ is

   a. 10 **b.** $\frac{20}{3}$ **c.** $\frac{20}{7}$ **d.** 4

10. Angle subtended by common tangents of two ellipses $4(x-4)^2+25y^2=100$ and $4(x+1)^2+y^2=4$ at origin is

    a. $\frac{\pi}{3}$ **b.** $\frac{\pi}{4}$ **c.** $\frac{\pi}{6}$ **d.** $\frac{\pi}{2}$

11. A circle has the same centre as an ellipse and passes through the focii F_1 and F_2 of the ellipse, such that the two curves intersect in 4 points. Let 'P' be any one of their point of intersection. If the major axis of the ellipse is 17 and the area of the triangle PF_1F_2 is 30, then the distance between the focii is

a. 13 **b.** 10 **c.** 11 **d.** None of these

12. The line $x = t^2$ meets the ellipse $x^2 + \frac{y^2}{9} = 1$ in the real and distinct points if and only if

a. $|t| < 2$ **b.** $|t| < 1$
c. $|t| > 1$ **d.** None of these

13. The eccentric angle of a point on the ellipse $\frac{x^2}{4} + \frac{y^2}{3} = 1$ at a distance of $\frac{5}{4}$ units from the focus on the positive x-axis, is

a. $\cos^{-1}\left(\frac{3}{4}\right)$ **b.** $\pi - \cos^{-1}\left(\frac{3}{4}\right)$
c. $\pi + \cos^{-1}\left(\frac{3}{4}\right)$ **d.** None of these

14. If S and S' are the foci of the ellipse $\frac{x^2}{25} + \frac{y^2}{16} = 1$, and P is any point on it then range of values of $SP \times S'P$ is

a. $9 \le f(\theta) \le 16$ **b.** $9 \le f(\theta) \le 25$
c. $16 \le f(\theta) \le 25$ **d.** $1 \le f(\theta) \le 16$.

15. Let d_1 and d_2 be the lengths of the perpendiculars drawn from foci S and S' of the ellipse $\frac{x^2}{a^2} + \frac{y^2}{b^2} = 1$ to the tangent at any point P on the ellipse. Then $SP:S'P$ =

a. $d_1:d_2$ **b.** $d_2:d_1$ **c.** $d_1^2 : d_2^2$ **d.** $\sqrt{d_1}/\sqrt{d_2}$

16. The auxiliary circle of a family of ellipse passes through origin and makes intercept of 8 and 6 units on the x-axis and the y-axis, respectively. If eccentricity of all such family of ellipse is 1/2, then locus of the focus will be

a. $\frac{x^2}{16} + \frac{y^2}{9} = 25$

b. $4x^2 + 4y^2 - 32x - 24y + 75 = 0$

c. $\frac{x^2}{16} + \frac{y^2}{9} = 25$

d. none of these

17. A man running round a race course notes that the sum of the distances of two flagposts from him is always 10 m and the distance between the flag posts is 8 m. Then the area of the path he encloses in square metres is

a. 15π **b.** 20π **c.** 27π **d.** 30π

18. There are exactly two points on the ellipse $\frac{x^2}{x^2} + \frac{y^2}{b^2} = 1$ whose distance from its centre is the same and is equal to $\frac{\sqrt{a^2 + 2b^2}}{2}$. Then the eccentricity of the ellipse is

a. $\frac{1}{2}$ **b.** $\frac{1}{\sqrt{2}}$ **c.** $\frac{1}{3}$ **d.** $\frac{1}{3\sqrt{2}}$

19. The eccentricity of locus of point $(3h + 2, k)$ where (h, k) lies on the circle $x^2 + y^2 = 1$ is

a. $\frac{1}{3}$ **b.** $\frac{\sqrt{2}}{3}$ **c.** $\frac{2\sqrt{2}}{3}$ **d.** $\frac{1}{\sqrt{3}}$

20. Let S and S' be two foci of the ellipse $\frac{x^2}{a^2} + \frac{y^2}{b^2} = 1$. If a circle described on SS' as diameter intersects the ellipse in real and distinct points, then the eccentricity e of the ellipse satisfies

a. $e = 1\sqrt{2}$ **b.** $e \in (1/\sqrt{2}, 1)$
c. $e \in (0, 1/\sqrt{2})$ **d.** none of these

21. If the curves $\frac{x^2}{4} + y^2 = 1$ and $\frac{x^2}{a^2} + y^2 = 1$ for suitable value of a cut on four concyclic points, the equation of the circle passing through these four points is

a. $x^2 + y^2 = 2$ **b.** $x^2 + y^2 = 1$
c. $x^2 + y^2 = 4$ **d.** none of these

22. From any point P lying in first quadrant on the ellipse $\frac{x^2}{25} + \frac{y^2}{16} = 1$, PN is drawn perpendicular to the major axis and produced at Q so that NQ equals to PS, where S is a focus. Then the locus of Q is

a. $5y - 3x - 25 = 0$
b. $3x + 5y + 25 = 0$
c. $3x - 5y - 25 = 0$
d. none of these

23. Locus of the point which divides double ordinates of the ellipse $\frac{x^2}{a^2} + \frac{y^2}{b^2} = 1$ in the ratio 1:2 internally is

a. $\frac{x^2}{a^2} + \frac{9y^2}{b^2} = 1$ **b.** $\frac{x^2}{a^2} + \frac{9y^2}{b^2} = \frac{1}{9}$
c. $\frac{9x^2}{a^2} + \frac{9y^2}{b^2} = 1$ **d.** none of these

24. The slopes of the common tangents of the ellipse $\frac{x^2}{4} + \frac{y^2}{1} = 1$ and the circle $x^2 + y^2 = 3$ are

a. ± 1 **b.** $\pm\sqrt{2}$
c. $\pm\sqrt{3}$ **d.** none of these

25. Tangents are drawn to the ellipse $\frac{x^2}{a^2}+\frac{y^2}{b^2}=1$ $(a>b)$ and the circle $x^2+y^2=a^2$ at the points where a common ordinate cuts them (on the same side of the x-axis). Then the greatest acute angle between these tangents is given by

a. $\tan^{-1}\left(\frac{a-b}{2\sqrt{ab}}\right)$ **b.** $\tan^{-1}\left(\frac{a+b}{2\sqrt{ab}}\right)$

c. $\tan^{-1}\left(\frac{2ab}{\sqrt{a-b}}\right)$ **d.** $\tan^{-1}\left(\frac{2ab}{\sqrt{a+b}}\right)$

26. The point of intersection of the tangents at the point P on the ellipse $\frac{x^2}{a^2}+\frac{y^2}{b^2}=1$ and its corresponding point Q on the auxiliary circle meet on the line

a. $x=a/e$ **b.** $x=0$

c. $y=0$ **d.** none of these

27. If the tangents to the ellipse $\frac{x^2}{a^2}+\frac{y^2}{b^2}=1$ make angles α and β with the major axis such that $\tan\alpha+\tan\beta=\lambda$, then the locus of their point of intersection is

a. $x^2+y^2=a^2$ **b.** $x^2+y^2=b^2$

c. $x^2-a^2=2\lambda\, xy$ **d.** $\lambda\,(x^2-a^2)=2xy$

28. If $\alpha-\beta=$ constant, then the locus of the point of intersection of tangents at P $(a\cos\alpha, b\sin\alpha)$ and $Q(a\cos\beta, b\sin\beta)$ to the ellipse $\frac{x^2}{a^2}+\frac{y^2}{b^2}=1$ is

a. a circle **b.** a straight line

c. an ellipse **d.** a parabola

29. The locus of the point of intersection of tangents to an ellipse at two points, sum of whose eccentric angles is constant, is a/an

a. parabola **b.** circle

c. ellipse **d.** straight line

30. The sum of the squares of the perpendiculars on any tangents to the ellipse $\frac{x^2}{a^2}+\frac{y^2}{b^2}=1$ from two points on the minor axis each at a distance ae from the centre is

a. $2a^2$ **b.** $2b^2$ **c.** a^2+b^2 **d.** a^2-b^2

31. A tangent to the ellipse $\frac{x^2}{25}+\frac{y^2}{16}=1$ at any point P meets the line $x=0$ at a point Q. Let R be the image of Q in the line $y=x$, then the circle whose extremities of a diameter are Q and R passes through a fixed point. The fixed point is

a. $(3, 0)$ **b.** $(5, 0)$ **c.** $(0, 0)$ **d.** $(4, 0)$

32. For the ellipse $\frac{x^2}{9}+\frac{y^2}{4}=1$ with vertices A and A', tangent drawn at the point P in the first quadrant meets the y-axis at Q and the chord $A'P$ meets the y-axis at M. If O is the origin, then OQ^2-MQ^2 equals to

a. 9 **b.** 13 **c.** 4 **d.** 5

33. A tangent having slope of $-\frac{4}{3}$ to the ellipse $\frac{x^2}{18}+\frac{y^2}{32}=1$ intersects the major and minor axes at points A and B respectively. If C is the centre of the ellipses, then the area of the triangle ABC is

a. 12 sq. units **b.** 24 sq. units

c. 36 sq. units **d.** 48 sq. units

34. Let P be any point on a directrix of an ellipse of eccentricity e. S be the corresponding focus and C the centre of the ellipse. The line PC meets the ellipse at A. The angle between PS and tangent at A is α, then α is equal to

a. $\tan^{-1} e$ **b.** $\frac{\pi}{2}$

c. $\tan^{-1}(1-e^2)$ **d.** none of these

35. If a tangent of slope 2 of the ellipse $\frac{x^2}{a^2}+\frac{y^2}{b^2}=1$ is normal to the circle $x^2+y^2+4x+1=0$, then the maximum value of ab is

a. 4 **b.** 2

c. 1 **d.** none of these

36. If $(\sqrt{3})\,bx+ay=2ab$ touches the ellipse $\frac{x^2}{a^2}+\frac{y^2}{b^2}=1$, then the eccentric angle of the point of contact is

a. $\frac{\pi}{6}$ **b.** $\frac{\pi}{4}$ **c.** $\frac{\pi}{3}$ **d.** $\frac{\pi}{2}$

37. If the ellipse $\frac{x^2}{a^2-7}+\frac{y^2}{13-5a}=1$ is inscribed in a square of side length $\sqrt{2}a$, then a is equal to

a. $\frac{6}{5}$

b. $(-\infty, -\sqrt{7})\cup(\sqrt{7}, 13/5)$

c. $(-\infty, -\sqrt{7})\cup(13/5, \sqrt{7})$

d. no such a exists

38. Locus of the point of intersection of the tangent at the end points of the focal chord of the ellipse $\frac{x^2}{a^2}+\frac{y^2}{b^2}=1$, $(b<a)$ is a/an

a. circle **b.** ellipse

c. hyperbola **d.** pair of straight lines

39. The normal at a variable point P on the ellipse $\frac{x^2}{a^2}+\frac{y^2}{b^2}=1$ of eccentricity e meets the axes of the ellipse at Q and R, then the locus of the midpoint of QR is a conic with an eccentricity e' such that

a. e' is independent of e

b. $e'=1$

c. $e'=e$

d. $e'=1/e$

40. Any ordinate MP of the ellipse $\frac{x^2}{25}+\frac{y^2}{9}=1$ meets the auxiliary circle at Q, then locus of the point of intersection of normals at P and Q to the respective curves is

a. $x^2+y^2=8$ **b.** $x^2+y^2=34$
c. $x^2+y^2=64$ **d.** $x^2+y^2=15$

41. Number of distinct normal lines that can be drawn to the ellipse $\frac{x^2}{169}+\frac{y^2}{25}=1$ from the point $P(0, 6)$ is

a. one **b.** two **c.** three **d.** four

42. An ellipse has the points $(1, -1)$ and $(2, -1)$ as its foci and $x+y-5=0$ as one of its tangents. Then the point where this line touches the ellipse from origin is

a. $\left(\frac{32}{9}, \frac{22}{9}\right)$ **b.** $\left(\frac{23}{9}, \frac{2}{9}\right)$
c. $\left(\frac{34}{9}, \frac{11}{9}\right)$ **d.** none of these

43. If tangents PQ and PR are drawn from a point on the circle $x^2+y^2=25$ to the ellipse $\frac{x^2}{4}+\frac{y^2}{b^2}=1$, $(b<4)$, so that the fourth vertex S of parallelogram $PQSR$ lies on the circumcircle of triangle PQR, then eccentricity of the ellipse is

a. $\frac{\sqrt{5}}{4}$ **b.** $\frac{\sqrt{7}}{3}$
c. $\frac{\sqrt{7}}{4}$ **d.** $\frac{\sqrt{5}}{3}$

44. If the ellipse $\frac{x^2}{a^2}+\frac{y^2}{b^2}=1$ is inscribed in a rectangle whose length to breadth ratio is 2:1 then the area of the rectangle is

a. $4\frac{a^2+b^2}{7}$ **b.** $4\frac{a^2+b^2}{3}$
c. $12\frac{a^2+b^2}{5}$ **d.** $8\frac{a^2+b^2}{5}$

45. If the normals at $P(\theta)$ and $Q(\pi/2+\theta)$ to the ellipse $\frac{x^2}{a^2}+\frac{y^2}{b^2}=1$ meet the major axis at G and g, respectively, then $PG^2+Qg^2=$

a. $b^2(1-e^2)(2-e^2)$
b. $a^2(e^4-e^2+2)$
c. $a^2(1+e^2)(2+e^2)$
d. $b^2(1+e^2)(2+e^2)$

46. The line $y=mx-\frac{(a^2-b^2)m}{\sqrt{a^2+b^2m^2}}$ is normal to the ellipse $\frac{x^2}{a^2}+\frac{y^2}{b^2}=1$ for all values of m belongs to

a. $(0, 1)$ **b.** $(0, \infty)$
c. R **d.** none of these

47. The length of the sides of a square which can be made by four perpendicular tangents to the ellipse $\frac{x^2}{7}+\frac{2y^2}{11}=1$ is

a. 10 units **b.** 8 units **c.** 6 units **d.** 5 units

48. From point $P(8, 27)$, tangent PQ and PR are drawn to the ellipse $\frac{x^2}{4}+\frac{y^2}{9}=1$. Then the angle subtended by QR at origin is

a. $\tan^{-1}\frac{\sqrt{6}}{65}$ **b.** $\tan^{-1}\frac{4\sqrt{6}}{65}$
c. $\tan^{-1}\frac{8\sqrt{2}}{65}$ **d.** $\tan^{-1}\frac{48\sqrt{6}}{455}$

49. Let P be any point on any directrix of an ellipse. Then chords of contact of point P with respect to the ellipse and its auxiliary circle intersect at

a. some point on the major axis depending upon the position of point P
b. midpoint of the line segment joining the centre to the corresponding focus
c. corresponding focus
d. none of these

50. The equation of the line passing through the centre and bisecting the chord $7x+y-1=0$ of the ellipse $\frac{x^2}{1}+\frac{y^2}{7}=1$ is

a. $x=y$ **b.** $2x=y$ **c.** $x=2y$ **d.** $x+y=0$

51. Equation of the chord of contact of pair of tangents drawn to the ellipse $4x^2+9y^2=36$ from the point (m, n) where $mn=m+n$, m, n being non-zero positive integers is

a. $2x+9y=18$ **b.** $2x+2y=1$
c. $4x+9y=18$ **d.** none of these

52. The locus of the point which is such that the chord of contact of tangents drawn from it to the ellipse $\frac{x^2}{a^2}+\frac{y^2}{b^2}=1$ forms a triangle of constant area with the coordinate axes is

a. a straight line **b.** a hyperbola
c. an ellipse **d.** a circle

53. An ellipse is sliding along the coordinate axes. If the foci of the ellipse are $(1, 1)$ and $(3, 3)$, then area of the director circle of the ellipse (in sq. units) is

a. 2π **b.** 4π **c.** 6π **d.** 8π

54. The equation of the ellipse whose axes are coincident with the coordinates axes and which touches the straight lines $3x-2y-20=0$ and $x+6y-20=0$ is

a. $\frac{x^2}{40}+\frac{y^2}{10}=1$ b. $\frac{x^2}{5}+\frac{y^2}{8}=1$

c. $\frac{x^2}{10}+\frac{y^2}{40}=1$ d. $\frac{x^2}{40}+\frac{y^2}{30}=1$

55. An ellipse with major and minor axes length as $2a$ and $2b$ touches coordinate axes in first quadrant and having foci (x_1, y_1) and (x_2, y_2) then the value of x_1x_2 and y_1y_2 is

a. a^2 b. b^2 c. a^2b^2 d. a^2+b^2

56. Let P_i and P'_i be the feet of the perpendiculars drawn from foci S, S' on a tangent T_i to an ellipse whose length of semi-major axis is 20, if $\sum_{i=1}^{10}(SP_i)(S'P'_i) = 2560$, then the value of eccentricity is

a. $\frac{1}{5}$ b. $\frac{2}{5}$ c. $\frac{3}{5}$ d. $\frac{4}{5}$

57. Number of points on the ellipse $\frac{x^2}{50}+\frac{y^2}{20}=1$ from which pair of perpendicular tangents are drawn to the ellipse $\frac{x^2}{16}+\frac{y^2}{9}=1$ is

a. 0 b. 2 c. 1 d. 4

58. A parabola is drawn with focus is at one of the foci of the ellipse $\frac{x^2}{a^2}+\frac{y^2}{b^2}=1$ (where $a > b$) and directrix passing through the other focus and perpendicular to the major axes of the ellipse. If latus rectum of the ellipse and the parabola are same, then the eccentricity of the ellipse is

a. $1-\frac{1}{\sqrt{2}}$ b. $2\sqrt{2}-2$

c. $\sqrt{2}-1$ d. None of these

59. If maximum distance of any point on the ellipse $x^2+2y^2+2xy=1$ from its centre be r, then r is equal to

a. $3+\sqrt{3}$ b. $2+\sqrt{2}$

c. $\frac{\sqrt{2}}{\sqrt{3-\sqrt{5}}}$ d. $\sqrt{2-\sqrt{2}}$

60. The set of values of m for which it is possible to draw the chord $y = \sqrt{m}x + 1$ to the curve $x^2+2xy+(2+\sin^2\alpha)y^2=1$, which subtends a right angle at the origin for some value of α is

a. [2, 3] b. [0, 1]

c. [1, 3] d. none of these

Multiple Correct Answers Type

Each question has four choices a, b, c, and d, out of which *one or more* answers are correct.

1. $\frac{x^2}{r^2-r-6}+\frac{y^2}{r^2-6r+5}=1$ will represents the ellipse, if r lies in the interval

a. $(-\infty, -2)$ b. $(3, \infty)$ c. $(5, \infty)$ d. $(1, \infty)$

2. On the x-y plane, the eccentricity of an ellipse is fixed (in size and position) by

a. both foci
b. both directrices
c. one focus and corresponding directrix
d. length of major axis

3. The distance of a point on the ellipse $\frac{x^2}{6}+\frac{y^2}{2}=1$ from the centre is 2. Then the eccentric angle of the point is

a. $\frac{\pi}{4}$ b. $\frac{3\pi}{4}$ c. $\frac{5\pi}{6}$ d. $\frac{\pi}{6}$

4. If the equation of the ellipse is $3x^2+2y^2+6x-8y+5=0$, then which of the following is/are true?

a. $e=\frac{1}{\sqrt{3}}$
b. centre is $(-1, 2)$
c. foci are $(-1, 1)$ and $(-1, 3)$
d. directrices are $y = 2 \pm \sqrt{3}$

5. If the tangent at the point $P(\theta)$ to the ellipse $16x^2+11y^2=256$ is also a tangent to the circle $x^2+y^2-2x=15$, then $\theta =$

a. $\frac{2\pi}{3}$ b. $\frac{4\pi}{3}$ c. $\frac{5\pi}{3}$ d. $\frac{\pi}{3}$

6. The coordinates (2, 3) and (1, 5) are the foci of an ellipse which passes through the origin, then the equation of

a. tangent at the origin is
$(3\sqrt{2}-5)x+(1-2\sqrt{2})y=0$
b. tangent at the origin is
$(3\sqrt{2}+5)x+(1+2\sqrt{2}y)=0$
c. normal at the origin is
$(3\sqrt{2}+5)x-(2\sqrt{2}+1)y=0$
d. normal at the origin is
$x(3\sqrt{2}-5)-y(1-2\sqrt{2})=0$

7. If the chord through the points whose eccentric angles are θ and ϕ on the ellipse $\frac{x^2}{25}+\frac{y^2}{9}=1$ passes through a focus, then the value of $\tan(\theta/2)\tan(\phi/2)$ is

a. $\frac{1}{9}$ b. -9 c. $-\frac{1}{9}$ d. 9

8. The equation $3x^2 + 4y^2 - 18x + 16y + 43 = k$

a. represents empty set, if $k < 0$

b. represents an ellipse, if $k > 0$

c. a point, if $k = 0$

d. cannot represent a real pair of straight lines for any value of k

9. If a pair of variable straight lines $x^2 + 4y^2 + \alpha xy = 0$ (where α is a real parameter) cut the ellipse $x^2 + 4y^2 = 4$ at two points A and B, then the locus of the point of intersection of tangents at A and B is

a. $x - 2y = 0$ **b.** $2x - y = 0$

c. $x + 2y = 0$ **d.** $2x + y = 0$

10. Which of the following is/are true?

a. There are infinite positive integral values of a for which $(13x - 1)^2 + (13y - 2)^2 = \left(\frac{5x + 12y - 1}{a}\right)^2$ represents an ellipse

b. The minimum distance of a point (1, 2) from the ellipse $4x^2 + 9y^2 + 8x - 36y + 4 = 0$ is 1

c. If from a point $P(0, \alpha)$ two normals other than axes are drawn to the ellipse $\frac{x^2}{25} + \frac{y^2}{16} = 1$, then $|\alpha| < \frac{9}{4}$

d. If the length of latus rectum of an ellipse is one-third of its major axis, then its eccentricity is equal to $\frac{1}{\sqrt{3}}$

11. Which of the following is/are true about the ellipse $x^2 + 4y^2 - 2x - 16y + 13 = 0$?

a. The latus rectum of the ellipse is 1

b. Distance between foci of the ellipse is $4\sqrt{3}$

c. Sum of the focal distances of a point $P(x, y)$ on the ellipse is 4

d. $y = 3$ meets the tangents drawn at the vertices of the ellipse at points P and Q then PQ subtends a right angle at any of its foci

12. A point on the ellipse $x^2 + 3y^2 = 37$ where the normal is parallel to the line $6x - 5y = 2$ is

a. (5, −2) **b.** (5, 2) **c.** (−5, 2) **d.** (−5, −2)

13. The locus of the image of the focus of the ellipse $\frac{x^2}{25} + \frac{y^2}{9} = 1$ $(a > b)$ with respect to any of the tangents to the ellipse is

a. $(x + 4)^2 + y^2 = 100$ **b.** $(x + 2)^2 + y^2 = 50$

c. $(x - 4)^2 + y^2 = 100$ **d.** $(x - 2)^2 + y^2 = 50$

14. Let E_1 and E_2 be two ellipses $\frac{x^2}{a^2} + y^2 = 1$ and $x^2 + \frac{y^2}{a^2} = 1$ (where a is a parameter). Then the locus of the points of intersection of the ellipses E_1 and E_2 is a set of curves comprising

a. two straight lines

b. one straight line

c. one circle

d. one parabola

15. Consider the ellipse $\frac{x^2}{f(k^2 + 2k + 5)} + \frac{y^2}{f(k + 11)} = 1$ and $f(x)$ is a positive decreasing function, then

a. the set of values of k, for which the major axis is x-axis is $(-3, 2)$

b. the set of values of k, for which the major axis is y-axis is $(-\infty, 2)$

c. the set of values of k, for which the major axis is y-axis is $(-\infty, -3) \cup (2, \infty)$

d. the set of values of k, for which the major axis is y-axis is $(-3, \infty)$

16. If two concentric ellipses are such that the foci of one are on the other and their major axes are equal. Let e and e' be their eccentricities, then

a. the quadrilateral formed by joining the foci of the two ellipses is a parallelogram

b. the angle θ between their axes is given by $\theta = \cos^{-1}\sqrt{\frac{1}{e^2} + \frac{1}{e'^2} - \frac{1}{e^2 e'^2}}$

c. if $e^2 + e'^2 = 1$, then the angle between the axes of the two ellipses is 90°

d. none of these

17. If the tangent drawn at point $(t^2, 2t)$ on the parabola $y^2 = 4x$ is same as the normal drawn at point $(\sqrt{5}\cos\theta, 2\sin\theta)$ on the ellipse $4x^2 + 5y^2 = 20$. Then

a. $\theta = \cos^{-1}\left(-\frac{1}{\sqrt{5}}\right)$ **b.** $\theta = \cos^{-1}\left(\frac{1}{\sqrt{5}}\right)$

c. $t = -\frac{2}{\sqrt{5}}$ **d.** $t = -\frac{1}{\sqrt{5}}$

Reasoning Type

Each question has four choices a, b, c and d, out of which *only one* is correct. Each question contains Statement 1 and Statement 2.

a. Both the statements are True and Statement 2 is the correct explanation of Statement 1.

b. Both the statements are True but Statement 2 is NOT the correct explanation of Statement 1.

c. Statement 1 is True and Statement 2 is False.

d. Statement 1 is False and Statement 2 is True.

1. Statement 1: The locus of a moving point (x, y) satisfying $\sqrt{(x-2)^2 + y^2} + \sqrt{(x-2)^2 + y^2} = 4$ is an ellipse.

Statement 2: Distance between (−2, 0) and (2, 0) is 4.

2. **Statement 1:** In a triangle ABC, if base BC is fixed and perimeter of the triangle is constant, then vertex A moves on an ellipse.

   Statement 2: If the sum of distances of a point P from two fixed points is constant, then locus of P is a real ellipse.

3. **Statement 1:** In an ellipse the sum of the distances between foci is always less than the sum of focal distances of any point on it.

   Statement 2: The eccentricity of any ellipse is less than 1.

4. **Statement 1:** The equation of the tangents drawn at the ends of the major axis of the ellipse $9x^2 + 5y^2 - 30y = 0$ is $y = 0, y = 6$.

   Statement 2: The equation of the tangent drawn at the ends of major axis of the ellipse $\frac{x^2}{a^2} + \frac{y^2}{b^2} = 1$ is always parallel to y-axis.

5. **Statement 1:** There can be maximum two points on the line $px + qy + r = 0$, from which perpendicular tangents can be drawn to the ellipse $\frac{x^2}{a^2} + \frac{y^2}{b^2} = 1$.

   Statement 2: Circle $x^2 + y^2 = a^2 + b^2$ and the given line can intersect in maximum two distinct points.

6. **Statement 1:** Circle $x^2 + y^2 = 9$, and the circle $(x - \sqrt{5})(\sqrt{2}x - 3) + y(\sqrt{2}y - 2) = 0$ touches each other internally.

   Statement 2: Circle described on the focal distance as diameter of the ellipse $4x^2 + 9y^2 = 36$ touches the auxiliary circle $x^2 + y^2 = 9$ internally.

7. **Statement 1:** Locus of the centre of a variable circle touching two circles $(x - 1)^2 + (y - 2)^2 = 25$ and $(x - 2)^2 + (y - 1)^2 = 16$ is an ellipse.

   Statement 2: If a circle $S_2 = 0$ lies completely inside the circle $S_1 = 0$, then locus of the centre of a variable circle $S = 0$ that touches both the circles is an ellipse.

8. **Statement 1:** For the ellipse $\frac{x^2}{5} + \frac{y^2}{3} = 1$ the product of the perpendiculars drawn from foci on any tangent is 3.

   Statement 2: For ellipse $\frac{x^2}{5} + \frac{y^2}{3} = 1$, the foot of the perpendiculars drawn from foci on any tangent lies on the circle $x^2 + y^2 = 5$ which is an auxiliary circle of the ellipse.

9. **Statement 1:** If there is exactly one point on the line $3x + 4y + 5\sqrt{5} = 0$, from which perpendicular tangents can be drawn to the ellipse $\frac{x^2}{a^2} + y^2 = 1$ $(a > 1)$, then the eccentricity of the ellipse is $\frac{1}{3}$.

   Statement 2: For the condition given in statement 1, given line must touch the circle $x^2 + y^2 = a^2 + 1$.

10. **Statement 1:** Any chord of the conic $x^2 + y^2 + xy = 1$ through $(0, 0)$ is bisected at $(0, 0)$.

    Statement 2: The centre of a conic is a point through which every chord is bisected.

11. **Statement 1:** The area of the ellipse $2x^2 + 3y^2 = 6$ is more than the area of the circle $x^2 + y^2 - 2x + 4y + 4 = 0$.

    Statement 2: the length of semi-major axes of an ellipse is more than the radius of the circle.

12. **Statement 1:** If line $x + y = 3$ is a tangent to an ellipse with foci $(4, 3)$ and $(6, y)$ at the point $(1, 2)$, then $y = 17$.

    Statement 2: Tangent and normal to the ellipse at any point bisects the angle subtended by foci at that point.

13. **Statement 1:** Diagonals of any parallelogram inscribed in an ellipse always intersect at the centre of the ellipse

    Statement 2: Centre of the ellipse is the point at which chord passing through the centre of the ellipse gets bisected at the centre.

14. **Statement 1:** A triangle ABC right angled at A moves so that its perpendicular sides touch the curve $\frac{x^2}{a^2} + \frac{y^2}{b^2} = 1$ all the time. Then loci of the points A, B, and C are circle.

    Statement 2: Locus of point of intersection of two perpendicular tangents to the curve is a director circle.

15. **Statement 1:** Tangents are drawn to the ellipse $\frac{x^2}{4} + \frac{y^2}{2} = 1$ at the points, where it is intersected by the line $2x + 3y = 1$. Point of intersection of these tangents is $(8, 6)$.

    Statement 2: Equation of the chord of contact to the ellipse $\frac{x^2}{a^2} + \frac{y^2}{b^2} = 1$ from an external point is given by $\frac{xx_1}{a^2} + \frac{yy_1}{b^2} - 1 = 0$.

16. **Statement 1:** If tangent at point P (in first quadrant) to the ellipse $\frac{x^2}{a^2} + \frac{y^2}{b^2} = 1$, $(a > b)$, meets corresponding directrix $x = a/e$ at point Q, then circle with minimum radius having PQ as chord passes through the corresponding focus.

    Statement 2: PQ subtends right angle at corresponding focus.

17. **Statement 1:** If a, b are real numbers and $c > 0$, then the locus represented by the equation $|ay - bx| = c\sqrt{(x - a)^2 + (y - b)^2}$ is an ellipse.

Statement 2: An ellipse is the locus of a point which moves in a plane such that ratio of its distances from a fixed point (i.e., focus) to the fixed line (i.e., directrix) is constant and less than 1.

Linked Comprehension Type

Based upon each paragraph, three multiple choice questions have to be answered. Each question has 4 choices a, b, c and d, out of which *only one* is correct.

For Problems 1–3

An ellipse (E) $\frac{x^2}{a^2}+\frac{y^2}{b^2}=1$, centred at point O have AB and CD as its major and minor axes, respectively, let the S_1 be one of the foci of the ellipse, radius of incircle of triangle OCS_1 be 1 unit and $OS_1 = 6$ units.

1. Perimeter of ΔOCS_1 is

a. 20 units **b.** 10 units **c.** 15 units **d.** 25 units

2. Equation of director the director circle of ellipse (E) is

a. $x^2+y^2=(48.5)$ **b.** $x^2+y^2=\sqrt{97}$

c. $x^2+y^2=97$ **d.** $x^2+y^2=\sqrt{48.5}$

3. Area of ellipse (E) is

a. $\frac{65\pi}{4}$ **b.** $\frac{64\pi}{5}$ **c.** 64π **d.** 65π

For Problems 4–6

Consider the ellipse whose major and minor axes are x-axis and y-axis, respectively. If ϕ is the angle between the CP and the normal at point P on the ellipse, and the greatest value $\tan\phi$ is $\frac{3}{2}$ (where C is the centre of the ellipse). Also semi-major axis is 10 units.

4. The eccentricity of the ellipse is

a. $\frac{1}{2}$ **b.** $\frac{1}{3}$

c. $\frac{\sqrt{3}}{2}$ **d.** none of these

5. A rectangle is inscribed in the ellipse whose sides are parallel to the coordinates axes, then maximum area of rectangle is

a. 50 units **b.** 100 units

c. 25 units **d.** none of these

6. Locus of the point of intersection of perpendicular tangents to the ellipse is

a. $x^2+y^2=125$ **b.** $x^2+y^2=150$

c. $x^2+y^2=200$ **d.** none of these

For Problems 7–9

A curve is represented by $C = 21x^2 - 6xy + 29y^2 + 6x - 58y - 151 = 0$.

7. Eccentricity of curve is

a. 1/3 **b.** $1/\sqrt{3}$ **c.** 2/3 **d.** $2/\sqrt{5}$

8. The lengths of axes

a. $6, 2\sqrt{6}$ **b.** $5, 2\sqrt{5}$

c. $4, 4\sqrt{5}$ **d.** none of these

9. The centre of the conic C is

a. (1, 0) **b.** (0,0)

c. (0, 1) **d.** none of these

For Problems 10–12

For all real p, the line $2px + my\sqrt{1-p^2} = 1$ touches a fixed ellipse whose axes are coordinate axes.

10. The eccentricity of the ellipse is

a. $\frac{2}{3}$ **b.** $\frac{\sqrt{3}}{2}$ **c.** $\frac{1}{\sqrt{3}}$ **d.** $\frac{1}{2}$

11. The foci of ellipse are

a. $(0, \pm\sqrt{3})$ **b.** $(0, \pm 2/3)$

c. $(\pm\sqrt{3}/2, 0)$ **d.** none of these

12. The locus of point of intersection of perpendicular tangents is

a. $x^2+y^2=\frac{5}{4}$ **b.** $x^2+y^2=\frac{3}{2}$

c. $x^2+y^2=2$ **d.** none of these

For Problems 13–15

Let S, S' be the foci of the ellipse $\frac{x^2}{a^2}+\frac{y^2}{b^2}=1$ whose eccentricity is e. P is a variable point on the ellipse. Consider the locus of the incentre of the $\Delta PSS'$.

13. The locus of incentre is

a. ellipse **b.** hyperbola

c. parabola **d.** circle

14. The eccentricity of locus of P is

a. $\sqrt{\frac{2e}{1-e}}$ **b.** $\sqrt{\frac{2e}{1+e}}$

c. 1 **d.** none of these

15. Maximum area of rectangle inscribed in the locus is

a. $\frac{2abe^2}{1+e}$ **b.** $\frac{2abe}{1-e}$

c. $\frac{abe}{1+e}$ **d.** none of these

For Problems 16–18

$C_1: x^2+y^2=r^2$ and $C_2: \frac{x^2}{16}+\frac{y^2}{9}=1$ intersect at four distinct points A, B, C, and D. Their common tangents form a parallelogram $A'B'C'D'$.

16. If $ABCD$ is a square then r is equal to

a. $\frac{12}{5}\sqrt{2}$ **b.** $\frac{12}{5}$

c. $\frac{12}{5\sqrt{5}}$ **d.** none of these

17. If A', B', C', D' is a square then r is equal to

a. $\sqrt{20}$ **b.** $\sqrt{12}$

c. $\sqrt{15}$ **d.** none of these

18. If $A'B'C'D'$ is a square, then the ratio of area of the circle C_1 to the area of the circumcircle of $\Delta A'B'C'$ is

a. $\frac{9}{16}$ **b.** $\frac{3}{4}$

c. $\frac{1}{2}$ **d.** none of these

For Problems 19–21

A coplanar beam of light emerging from a point source has the equation $\lambda x - y + 2(1 + \lambda) = 0$, $\lambda \in R$, the rays of the beam strike an elliptical surface and get reflected. The reflected rays form another convergent beam having equation $\mu x - y + 2(1 - \mu) = 0, \mu \in R$. Further it is found that the foot of the perpendicular from the point (2, 2) upon any tangent to the ellipse lies on the circle $x^2 + y^2 - 4y - 5 = 0$.

19. The eccentricity of the ellipse is equal to

a. $\frac{1}{3}$ **b.** $\frac{1}{\sqrt{3}}$ **c.** $\frac{2}{3}$ **d.** $\frac{1}{2}$

20. The area of the largest triangle that an incident ray and the corresponding reflected ray can enclose with axis of the ellipse is equal to

a. $4\sqrt{5}$ **b.** $2\sqrt{5}$

c. $\sqrt{5}$ **d.** none of these

21. Total distance travelled by an incident ray and the corresponding reflected ray is the least if the point of incidence coincides with

a. an end of the minor with

b. an end of the major axis

c. an end of this latus rectum

d. none of these

For Problems 22–24

The tangent at any point P of the circle $x^2 + y^2 = 16$ meets the tangent at a fixed point A at T, and T is joined to B, the other end of the diameter through A.

22. The locus of the intersection of AP and BT is conic whose eccentricity is

a. $\frac{1}{2}$ **b.** $\frac{1}{\sqrt{2}}$ **c.** $\frac{1}{3}$ **d.** $\frac{1}{\sqrt{3}}$

23. Sum of focal distances of any point on the curve is

a. 12 **b.** 16 **c.** 20 **d.** 8

24. Which of the following does not change by changing the radius of the circle?

a. coordinates of focii

b. length of major axis

c. eccentricity

d. length of minor axis

For Problems 25–27

The ellipse $\frac{x^2}{a^2} + \frac{y^2}{b^2} = 1$ is such that it has the least area but contains the circle $(x - 1)^2 + y^2 = 1$.

25. The eccentricity of the ellipse is

a. $\sqrt{\frac{2}{3}}$ **b.** $\frac{1}{\sqrt{3}}$

c. $\frac{1}{2}$ **d.** none of these

26. Equation of auxiliary circle of ellipse is

a. $x^2 + y^4 = 6.5$ **b.** $x^2 + y^4 = 5$

c. $x^2 + y^4 = 45$ **d.** none of these

27. Length of latus-rectum of the ellipse is

a. 2 units **b.** 1 unit **c.** 3 unit **d.** 2.5 unit

Matrix-Match Type

Each question contains statements given in two columns which have to be matched.

Statements (a, b, c, d) in column I have to be matched with statements (p, q, r, s) in column II. If the correct matches are a → p, a → s, , b → q, b → r, c → p, c → q and d → s, then the correctly bubbled 4 × 4 matrix should be as follows:

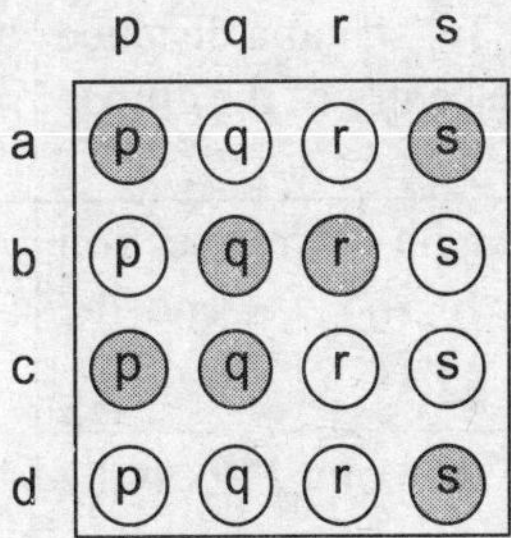

Fig. 4.35

1.

Column I	Column II
a. Distance between the points on the curve $4x^2 + 9y^2 = 1$, where tangent is parallel to the line $8x = 9y$, is less than	**p.** 1
b. Sum of distance between the foci of the curve $25(x + 1)^2 + 9(y + 2)^2 = 225$ from (–1, 0) is more than	**q.** 4

c. Sum of distances from the x-axis of the points on the ellipse $\frac{x^2}{9}+\frac{y^2}{4}=1$, where the normal is parallel to the line $2x+y=1$, is less than	**r.** 7
d. Tangents are drawn from points on the line $x-y+2=0$ to the ellipse $x^2+2y^2=2$, then all the chords of contact pass through the point whose distance from (2, 1/2) is more than	**s.** 5

2. The tangents drawn from a point P to the ellipse $\frac{x^2}{a^2}+\frac{y^2}{b^2}=1$ make angle α and β with the major axis.

Column I	Column II
a. If $\alpha+\beta=\frac{c\pi}{2}$ $(c \in N)$	**p.** circle
b. If $\tan\alpha\tan\beta=c$ {where $c \in R$}, then locus of P can be	**q.** ellipse
c. If $\tan\alpha+\tan\beta=c$ {where $c \in R$}, then locus of P can be	**r.** hyperbola
d. If $\cot\alpha+\cot\beta=c$ {where $c \in R$}, then locus of P can be	**s.** pair of straight lines

3.

Column I	Column II
a. If the tangent to the ellipse $x^2+4y^2=16$ at the point $P(\phi)$ is a normal to the circle $x^2+y^2-8x-4y=0$, then $\frac{\phi}{2}$ may be	**p.** 0
b. The eccentric angle(s) of a point on the ellipse $x^2+3y^2=6$ at a distance 2 units from the centre of the ellipse is/are	**q.** $\cos^{-1}\left(-\frac{2}{3}\right)$
c. The eccentric angle of intersection of the ellipse $x^2+4y^2=4$ and the parabola $x^2+1=y$ is	**r.** $\frac{\pi}{4}$
d. If the normal at the point $P(\theta)$ to the ellipse $\frac{x^2}{14}+\frac{y^2}{5}=1$ intersects it again at the point $Q(2\theta)$, then θ is	**s.** $\frac{5\pi}{4}$

4.

Column I	Column II
a. A stick of length 10 m slides on coordinate axes, then locus of a point dividing this stick from x-axis in the ratio 6 : 4 is a curve whose eccentrictly is e, then $3e$ is equal to	**p.** $\sqrt{6}$
b. AA' is a major axis of an ellipse $3x^2+2y^2+6x-4y-1=0$ and P is a variable point on it, then greatest area of triangle APA' is	**q.** $2\sqrt{7}$
c. Distance between foci of the curve represented by the equation $x=1+4\cos\theta$, $y=2+3\sin\theta$ is	**r.** $\frac{128}{3}$
d. Tangents are drawn to the ellipse $\frac{x^2}{16}+\frac{y^2}{7}=1$ at end points of the latus rectum. The area of equadrilateral so formed is	**s.** $\sqrt{5}$

5.

Column I	Column II
a. An ellipse passing through the origin has its foci (3, 4) and (6, 8), then length of its minor axis is	**p.** 8
b. If PQ is focal chord of ellipse $\frac{x^2}{25}+\frac{y^2}{16}=1$ which passes through $S\equiv(3,0)$ and $PS=2$, then length of chord PQ is	**q.** $10\sqrt{2}$
c. If the line $y=x+K$ touches the ellipse $9x^2+16y^2=144$, then the difference of values of K is	**r.** 10
d. Sum of distances of a point on the ellipse $\frac{x^2}{9}+\frac{y^2}{16}=1$ from the foci	**s.** 12

6.

Column I	Column II
a. If vertices of a rectangle of maximum area inscribed in the ellipse $\frac{x^2}{a^2}+\frac{y^2}{b^2}=1$ are extremities of latus rectum Then eccentricity of ellipse is	**p.** $\frac{2}{\sqrt{5}}$
b. If extremities of diameter of the circle $x^2+y^2=16$ are foci of a ellipse, then eccentricity of the ellipse, if its size is just sufficient to contain the circle, is	**q.** $\frac{1}{\sqrt{2}}$
c. If normal at point (6, 2) to the ellipse passes through its nearest focus (5, 2), having centre at (4, 2) then its eccentricity is	**r.** $\frac{1}{3}$
d. If extremities of latus rectum of the parabola $y^2=24x$ are foci of ellipse and if ellipse passes through the vertex of the parabola, then its eccentricity is	**s.** $\frac{1}{2}$

Integer Type

1. A ladder 12 units long slides in a vertical plane with its ends in contact with a vertical wall and a horizontal floor along the x-axis. If the locus of a point on the ladder 4 units from its foot is an ellipse, then the ratio of the length of the major axis to the minor axis is ________.

2. The coordinates of the vertices B and C of a triangle ABC are (2, 0) and (8, 0), respectively. The vertex A varies in such a way that $4 \tan (B/2) \tan (C/2) = 1$. If the locus of A is a conic, then the difference between the semi-major axis and the semi-minor axis is ________.

3. If $P(x, y)$ be any point on the ellipse $16x^2 + 25y^2 = 400$ and $F_1 = (3, 0)$, $F_2 = (-3, 0)$, then find the value of $(PF_1 + PF_2)/2$ is ________.

4. If (5, 12) and (24, 7) are the foci of an ellipse passing through the origin and if e is the eccentricity, then the value of $[1/e]$, where [.] represents the greatest integer function, is ________.

5. An ellipse is drawn with major and minor axes of lengths 10 and 8, respectively. Using one focus as the centre, a circle is drawn making tangent to the ellipse, with no part of the circle outside the ellipse. Then the radius of the circle is ________.

6. If the equation $(5x - 1)^2 + (5y - 2)^2 = (\lambda^2 - 2\lambda + 1) \times (3x + 4y - 1)^2$ represents an ellipse, then the number of integral values of λ is ________.

7. The area of the greatest rectangle that can be inscribed in the ellipse $(x^2/9) + y^2 = 1$ is ________.

8. If the line $lx + my + n = 0$ intersects the ellipse $(x^2/a^2) + (y^2/25) = 1$ at points whose eccentric angles differ by $\pi/2$, then find the value of $(a^2l^2 + b^2m^2)/n^2$.

9. The area of the greatest isosceles triangle that can be inscribed in an ellipse $(x^2/16) + (y^2/3) = 1$ having its vertex coincident with an extremity of the major axis is ________.

10. If the chord joining the points $P(\alpha)$ and $Q(\beta)$ on the ellipse $(x^2/64) + (y^2/16) = 1$ subtends a right angle at the vertex $A(8, 0)$, then the value of $|\tan (\alpha/2) \tan (\beta/2)|$ is ________.

11. If PSQ is a focal chord of the ellipse $16x^2 + 25y^2 = 400$ such that $SP = 8$, then the length of SQ is ________.

12. If F_1 and F_2 are the feet of the perpendiculars from the foci S_1 and S_2 of an ellipse $(x^2/25) + (y^2/16) = 1$ on the tangent at any point P on the ellipse, then the least value of $S_1F_1 + S_2F_2$ is ________.

13. If S and S' are the foci of the ellipse $16x^2 + 25y^2 = 400$ and PSQ is a focal chord such that $SP = 16$, then the value of $[S'Q]$, where [.] represents the greatest integer function, is ________.

14. If the locus of the middle point of the portion of a tangent on the ellipse $(x^2/a^2) + (y^2/b^2) = 1$ included between the axes is $(a^2/x^2) + (b^2/y^2) = k$, then the value of k is ________.

15. If θ is the angle between the pair of tangents from the point (1, 2) to the ellipse $3x^2 + 2y^2 = 5$, then the value of $[\tan \theta]$, where [.] represents the greatest integer function, is ________.

16. A tangent with slope $-$ 4/3 on the ellipse $(x^2/18) + (y^2/32) = 1$ intersects the major and minor axes at the points A and B, respectively. If C is the centre of the ellipse and the area of the triangle ABC is K, then the value of $K/4$ is ________.

17. A circle is concentric with the ellipse $(x^2/a^2) + (y^2/b^2) = 1$ and passes through the foci F_1 and F_2 of the ellipse. The two curves intersect at four points. Let P be any point of intersection. If the major axis of the ellipse is 15 and the area of the triangle PF_1F_2 is 26, then the value of $2(a^2 - b^2)^{1/2} - 4$ is ________.

18. The number of integral values of α for which three distinct chords drawn from $(\alpha, 0)$ to the ellipse $x^2 + 2y^2 = 1$ are bisected by the parabola $y^2 = 4x$ is ________.

19. A tangent is drawn to the ellipse $(x^2/a^2) + (y^2/b^2) = 1$ to cut the ellipse $(x^2/c^2) + (y^2/d^2) = 1$ at the points P and Q. If the tangents at P and Q to the ellipse $(x^2/c^2) + (y^2/d^2) = 1$ intersect at right angles, then the value of $(a^2/c^2) + (b^2/d^2)$ is ________.

20. From a point, common tangents are drawn to the curves $x^2 + y^2 = 16$ and $(x^2/25) + (y^2/4) = 1$. If the length of the intercept between the coordinate axes is L, then the value of $[L]$, where [.] represents the greatest integer function, is ________.

ANSWERS AND SOLUTIONS

Objective Type

1. c. We have $PQ = BP$

$\Rightarrow 2ae = \sqrt{a^2e^2 + b^2} = \sqrt{a^2} = a$

$\Rightarrow e = \frac{1}{2}$

2. c. Ellipse passing through $O(0, 0)$ and having foci $P(3, 3)$ and $Q(-4, 4)$,

then $e = \dfrac{PQ}{OP + OQ}$

$= \dfrac{\sqrt{50}}{3\sqrt{2} + 4\sqrt{2}}$

$= \dfrac{5}{7}$

3. b. Here $a^2 + 2 > a^2 + 1$

$\Rightarrow a^2 + 1 = (a^2 + 2)(1 - e^2)$

$\Rightarrow a^2 + 1 = (a^2 + 2)\frac{5}{6}$

$\Rightarrow 6a^2 + 6 = 5a^2 + 10$

$\Rightarrow a^2 = 10 - 6 = 4$

$\Rightarrow a = \pm 2$

Latus rectum $= \dfrac{2(a^2 + 1)}{\sqrt{a^2 + 2}} = \dfrac{2 \times 5}{\sqrt{6}} = \dfrac{10}{\sqrt{6}}$

4. a. Let $P(\theta)$, $Q\left(\theta + \frac{2\pi}{3}\right)$, $R\left(\theta + \frac{4\pi}{3}\right)$

then $P' \equiv (a\cos\theta, b\sin\theta)$,

$Q' \equiv \left(a\cos\left(\theta + \frac{2\pi}{3}\right), b\sin\left(\theta + \frac{2\pi}{3}\right)\right)$

$R' \equiv \left(a\cos\left(\theta + \frac{4\pi}{3}\right), b\sin\left(\theta + \frac{4\pi}{3}\right)\right)$

Let centroid of $\Delta P'Q'R' \equiv (x', y')$

$= a\left[\dfrac{\cos\theta + \cos\left(\theta + \frac{2\pi}{3}\right) + \cos\left(\theta + \frac{4\pi}{3}\right)}{3}\right]$

$= \frac{a}{3}\left[\cos\theta + 2\cos(\theta + \pi)\cos\frac{\pi}{3}\right] = 0$

$y' \equiv \frac{a}{3}\left[\sin\theta + \sin\left(\theta + \frac{2\pi}{3}\right) + \sin\left(\theta + \frac{4\pi}{3}\right)\right]$

$= \frac{a}{3}\left[\sin\theta + 2\sin(\theta + \pi)\sin\frac{\pi}{3}\right] = 0$

$= 0$

5. a. $2a(1 + e) = 15$

$1 + e = \frac{3}{2}$

$e = 0.5$

6. d. For the two ellipses to intersect at four distinct points, $a > 1$

$\Rightarrow b^2 - 5b + 7 > 1$

$\Rightarrow b^2 - 5b + 6 > 0$

$\Rightarrow b \in (-\infty, 2) \cup (3, \infty)$

$\Rightarrow b$ does not lie in $[2, 3]$

7. b. Let S be the given focus and ZM be the given line.

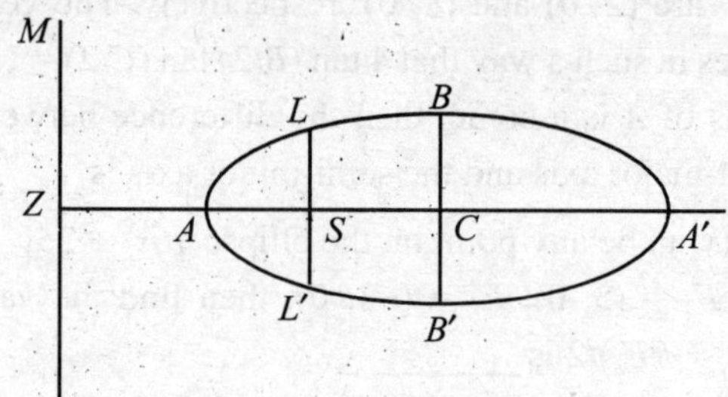

Fig. 4.1

Then $SZ = \frac{a}{e} - ae$

$= \frac{a}{e}(1 - e^2)$

$= \frac{b^2}{ae} = k$ (say)

As $b^2 = a^2(1 - e^2)$

Now take SC as x-axis and LSL' as y-axis. Let (x, y) be the coordinates of B with respect to these axes, then $x = SC = ae$, $y = CB = b$

Hence, $\frac{y^2}{x} = \frac{b^2}{ae} = SZ$, which is constant.

Therefore, $y^2 = kx$ is the required locus, which is a parabola.

8. a. Let AB be the line.

Let $AP = a$, $PB = b$,

so that $AB = a + b$.

If AB makes an angle θ with the x-axis and coordinates of P are (x, y), then in ΔAPL, $x = a\cos\theta$, and in ΔPBQ, $y = b\sin\theta$.

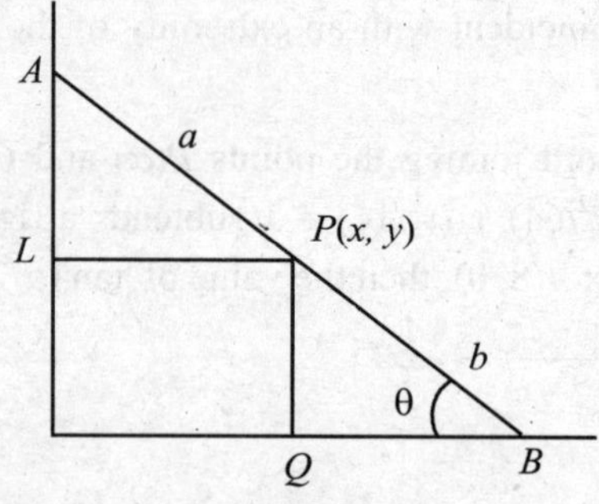

Fig. 4.2

Therefore, locus of $P(x, y)$ is

$$\frac{x^2}{a^2}+\frac{y^2}{b^2}=1,$$

which is an ellipse

9. b. $(5x-10)^2+(5y+15)^2=\frac{(3x-4y+7)^2}{4}$

$$\Rightarrow (x-2)^2+(y+3)^2=\left(\frac{1}{2}\,\frac{3x-4y-7}{5}\right)^2$$

$\Rightarrow \sqrt{(x-2)^2+(y+3)^2}=\frac{1}{2}\,\frac{|3x-4y-7|}{5}$ is an ellipse, whose focus is $(2,-3)$, directrix $3x-4y+7=0$, and eccentricity is $\frac{1}{2}$.

Length of Perpendicular from focus to directrix is

$$\frac{|3\times 2-4(-3)+7|}{5}=5$$

$$\Rightarrow \frac{a}{e}-ae=5$$

$$\Rightarrow 2a-\frac{a}{2}=5$$

$$\Rightarrow a=\frac{10}{3}$$

So length of major axis is $\frac{20}{3}$.

10. d.

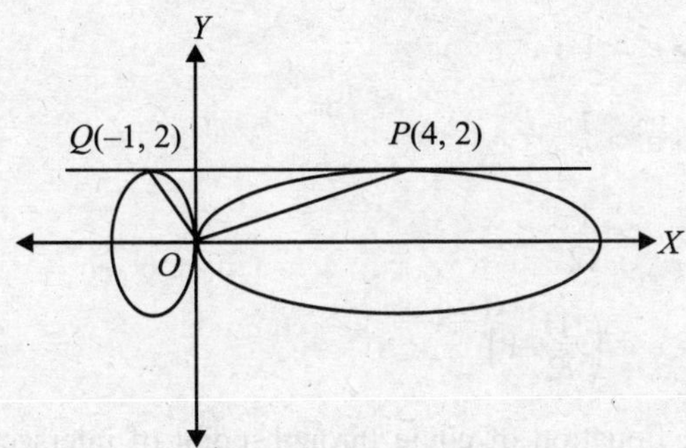

Fig. 4.3

$$\frac{(x-4)^2}{25}+\frac{y^2}{4}=1 \text{ and } (x-1)^2+\frac{y^2}{4}=1$$

Clearly, $m_{OP}\times m_{OQ}=-1$

$\Rightarrow OP$ and OQ are perpendicular to each other.

11. a. Let ellipse $\frac{x^2}{a^2}+\frac{y^2}{b^2}=1$ and circle $x^2+y^2=a^2e^2$

Radius of circle $= ae$.

Point of intersection of circle and ellipse is $\left[\frac{a}{e}\sqrt{2e^2-1}, \frac{a}{e}(1-e^2)\right]$.

Now area of ΔPF_1F_2

$$=\frac{1}{2}\begin{vmatrix}\frac{a}{e}\sqrt{2e^2-1} & \frac{a}{e}(1-e^2) & 1\\ ae & 0 & 1\\ -ae & 0 & 1\end{vmatrix}=\frac{1}{2}\left|\frac{a}{e}(1-e^2)(2ae)\right|=30$$

$$\Rightarrow a^2(1-e^2)=30 \text{ (given)}$$

$$\Rightarrow a^2e^2=a^2-30=\left(\frac{17}{2}\right)^2-30=\frac{169}{4}$$

$$\Rightarrow 2ae=13$$

12. b. Solving the given line and the ellipse, we get

$$t^2+\frac{y^2}{9}=1$$

$$\Rightarrow y^2=9(1-t^2),$$

which gives real and distinct values of y, if $1-t^2>0$

$$\Rightarrow t\in(-1,1).$$

13. a. Any point on the ellipse is $(2\cos\theta, \sqrt{3}\sin\theta)$.

The focus on the positive x-axis is $(1, 0)$.

Given that $(2\cos\theta-1)^2+3\sin^2\theta=\frac{25}{16}$

$$\Rightarrow \cos\theta=\frac{3}{4}.$$

14. c. Let $P(5\cos\theta, 4\sin\theta)$ be any point on the ellipse

Then $SP=5+5e\cos\theta$ and $S'P=5-5e\cos\theta$

$$\therefore SP\times S'P=25-25e^2\cos^2\theta$$
$$=25\sin^2\theta+16\cos^2\theta$$
$$=16+9\sin^2\theta=f(\theta) \text{ (say)}$$

$$\Rightarrow 16\le f(\theta)\le 25.$$

15. a. Tangent at $P(a\cos\alpha, b\sin\alpha)$ is

$$\frac{x}{a}\cos\alpha+\frac{y}{b}\sin\alpha=1 \quad (1)$$

Distance of focus $S(ae, 0)$ from this tangent is

$$d_1=\frac{|e\cos\alpha-1|}{\sqrt{\frac{\cos^2\alpha}{a^2}+\frac{\sin^2\alpha}{b^2}}}$$

$$=\frac{1-e\cos\alpha}{\sqrt{\frac{\cos^2\alpha}{a^2}+\frac{\sin^2\alpha}{b^2}}}$$

Distance of focus $S'(-ae, 0)$ from this line

$$d_2=\frac{1+e\cos\alpha}{\sqrt{\frac{\cos^2\alpha}{a^2}+\frac{\sin^2\alpha}{b^2}}}$$

$$\Rightarrow \frac{d_1}{d_2}=\frac{1-e\cos\alpha}{1+e\cos\alpha}$$

Now $SP=a-ae\cos\alpha$ and $S'P=a+ae\cos\alpha$

$$\Rightarrow \frac{SP}{S'P} = \frac{1 - e\cos\alpha}{1 + e\cos\alpha}$$

$$\Rightarrow \frac{SP}{S'P} = \frac{d_1}{d_2}$$

16. b.

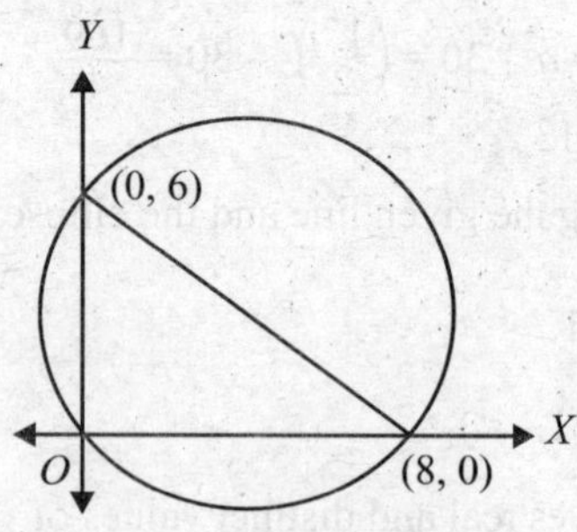

Fig. 4.4

Centre of family of ellipse is (4, 3) and distance of focus from centre $= ae = \frac{5}{2}$.

Hence, locus $(x - 4)^2 + (y - 3)^2 = \frac{25}{4}$.

17. a. Let $P(x, y)$ be the position of the man at any time.

Let S (4, 0) and S' (– 4, 0) be the fixed flag, post, with C as the origin.

Since $SP + S'P = 10$ m i.e., a constant, the locus of P is an ellipse with S and S' as foci

$\Rightarrow ae = 4$, and $2a = 10$

$$\Rightarrow e = \frac{4}{5}$$

Now $b^2 = a^2(1 - e^2)$

$$\Rightarrow b^2 = 25\left(1 - \frac{16}{25}\right) = 9$$

$\Rightarrow b = 3$

Hence, the area of the ellipse $= \pi ab = \pi \times 5 \times 3 = 15\pi$.

18. b. Since there are exactly two points on the ellipse $\frac{x^2}{a^2} + \frac{y^2}{b^2} = 1$, whose distance from centre is same, the points would be either end points of the major axis or of the minor axis.

But $\sqrt{\frac{a^2 + 2b^2}{2}} > b$, so the points are the vertices of major axis.

Hence, $a = \sqrt{\frac{a^2 + 2b^2}{2}}$

$\Rightarrow a^2 = 2b^2$

$$\Rightarrow e = \sqrt{1 - \frac{b^2}{a^2}} = \frac{1}{\sqrt{2}}$$

19. c. Let $p = 3h + 2$ and $q = k$

$\Rightarrow h = \frac{p-2}{3}$ and $k = q$

Since (h, k) lies on $x^2 + y^2 = 1$,

$h^2 + k^2 = 1$

$$\Rightarrow \left(\frac{p-2}{3}\right)^2 + q^2 = 1$$

Locus is $\left(\frac{x-2}{3}\right)^2 + y^2 = 1$,

which has eccentricity $e = \sqrt{1 - \frac{1}{9}} = \frac{2\sqrt{2}}{3}$

20. b.

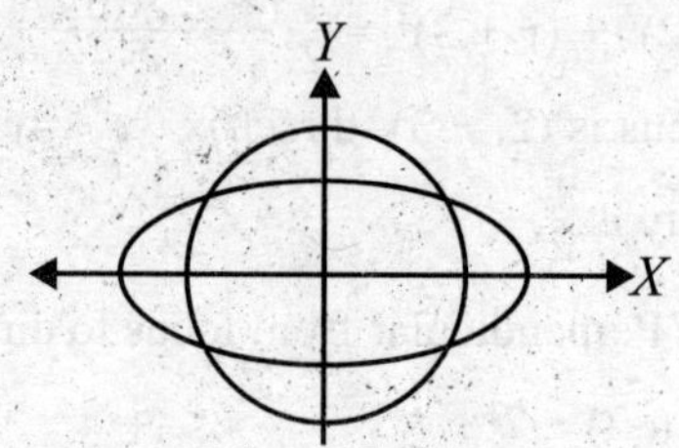

Fig. 4.5

Radius of the circle having SS' as diameter is $r = ae$.

If it cuts an ellipse, then $r > b$

$\Rightarrow ae > b$

$$\Rightarrow e^2 > \frac{b^2}{a^2}$$

$\Rightarrow e^2 > 1 - e^2$

$$\Rightarrow e^2 > \frac{1}{2}$$

$$\Rightarrow e > \frac{1}{\sqrt{2}}$$

$$\Rightarrow e \in \left(\frac{1}{\sqrt{2}}, 1\right)$$

21. b. Equation of conic through point of intersection of given two ellipse is

$$\left(\frac{x^2}{4} + y^2 - 1\right) + \lambda\left(\frac{x^2}{a^2} + y^2 - 1\right) = 0$$

$$\Rightarrow x^2\left(\frac{1}{4} + \frac{\lambda}{a^2}\right) + y^2(1 + \lambda) = 1 + \lambda$$

$$\Rightarrow x^2\left(\frac{a^2 + 4\lambda}{4a^2(1 + \lambda)}\right) + y^2 = 1$$

This equation is a circle if $\frac{a^2 + 4\lambda}{4a^2(1 + \lambda)} = 1$.

Circle is $x^2 + y^2 = 1$.

22. b.

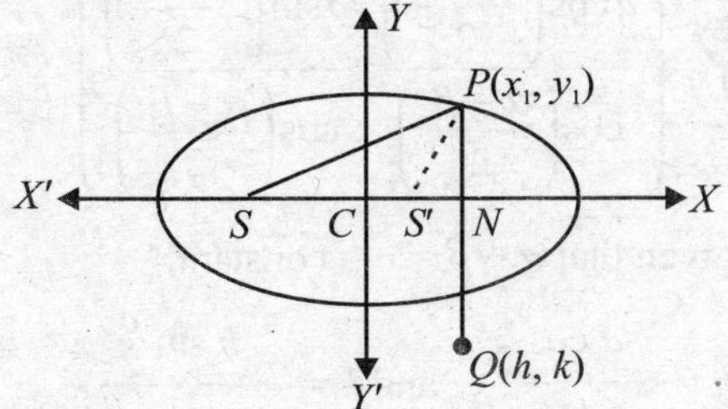

Fig. 4.6

$a^2 = 25$ and $b^2 = 16$

$\Rightarrow e = \sqrt{1 - \frac{16}{25}}$

$= \frac{3}{5}$

Let point Q be (h, k), where $k < 0$.

Given that $k = SP = a + ex_1$, where $P(x_1, y_1)$ lies on the ellipse

$\Rightarrow |k| = a + eh$ (as $x_1 = h$)

$\Rightarrow -y = a + ex$

$\Rightarrow 3x + 5y + 25 = 0$

23. a.

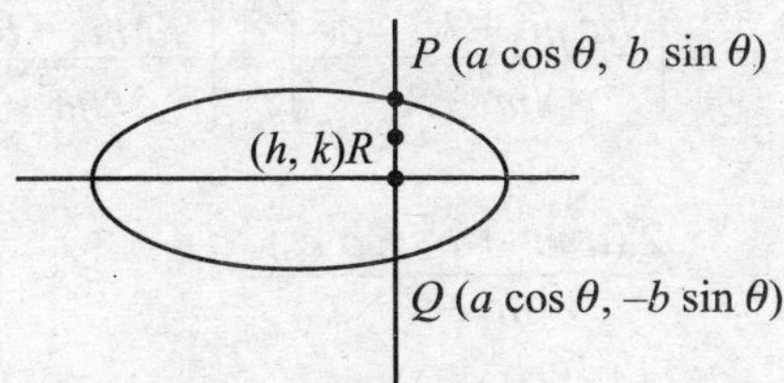

Fig. 4.7

Let $P(a \cos \theta, b \sin \theta)$, $Q(a \cos \theta, -b \sin \theta)$. So

$PR : RQ = 1:2$

Therefore, $h = a \cos \theta$

$\Rightarrow \cos \theta = \frac{h}{a}$ (1)

and $k = \frac{b}{3} \sin \theta$

$\Rightarrow \sin \theta = \frac{3k}{b}$ (2)

On squaring and adding Eqs. (1) and (2), we get

$\frac{x^2}{a^2} + \frac{9y^2}{b^2} = 1.$

24. b. Let m be the slope of the common tangent, then

$\pm\sqrt{3}\sqrt{1 + m^2} = \pm\sqrt{4m^2 + 1}$

$\Rightarrow 3 + 3m^2 = 4m^2 + 1$

$\Rightarrow m^2 = 2$

$\Rightarrow m = \pm\sqrt{2}$

25. a.

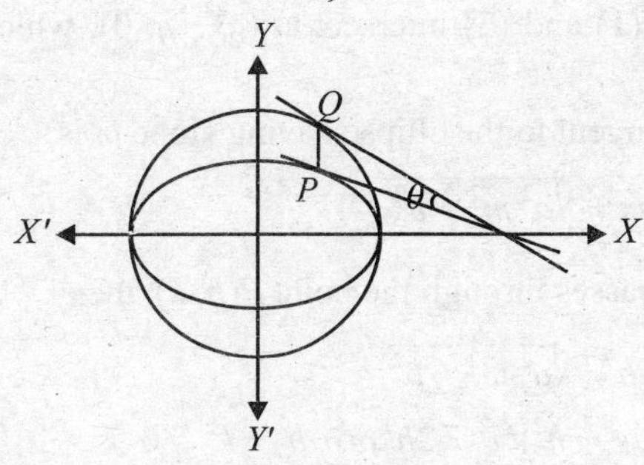

Fig. 4.8

Tangent to the ellipse at $P(a \cos \alpha, b \sin \alpha)$ is

$\frac{x}{a} \cos \alpha + \frac{y}{b} \sin \alpha = 1$ (1)

Tangent to the circle at $Q(a \cos \alpha, a \sin \alpha)$ is

$\cos \alpha x + \sin \alpha y = a$ (2)

Now angle between tangents is θ,

then $\tan \theta = \left| \frac{-\frac{b}{a} \cot \alpha - (-\cot \alpha)}{1 + \left(-\frac{b}{a} \cot \alpha\right)(-\cot \alpha)} \right|$

$= \left| \frac{\cot \alpha \left(1 - \frac{b}{a}\right)}{1 + \frac{b}{a} \cot^2 \alpha} \right|$

$= \left| \frac{a - b}{a \tan \alpha + b \cot \alpha} \right|$

$= \left| \frac{a - b}{(\sqrt{a} \tan \alpha - \sqrt{b} \cot \alpha)^2 + 2\sqrt{ab}} \right|$

Now the greatest value of the above expression is $\left| \frac{a - b}{2\sqrt{ab}} \right|$ when $\sqrt{a} \tan \alpha = \sqrt{b} \cot \alpha$

$\Rightarrow \theta_{\max} = \tan^{-1}\left(\frac{a - b}{2\sqrt{ab}}\right).$

26. c.

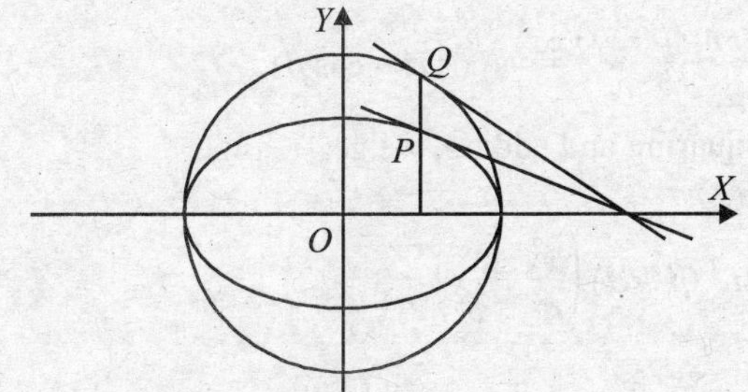

Fig. 4.9

Tangent to the ellipse at point $P(a \cos \theta, b \sin \theta)$ is

$\frac{x}{a} \cos \theta + \frac{y}{b} \sin \theta = 1$ (1)

Tangent to the circle at point $Q(a \cos \theta, a \sin \theta)$ is

$x \cos\theta + y \sin\theta = a$ (2)

Eqs. (1) and (2) intersect at $(\frac{a}{\cos\theta}, 0)$, which lies on $y = 0$.

27. d. Tangent to the ellipse having slope m is

$y = mx + \sqrt{a^2 m^2 + b^2}$.

If it passes through the point $P(h, k)$, then

$k = mh + \sqrt{a^2 m^2 + b^2}$.

or $(a^2 - h^2)m^2 + 2hkm + b^2 - k^2 = 0$

Now given $\tan\alpha + \tan\beta = \lambda$

$\Rightarrow m_1 + m_2 = \lambda$

$\Rightarrow \frac{-2hk}{a^2 - h^2} = \lambda$

$\Rightarrow$ locus is $\lambda(x^2 - a^2) = 2xy$

28. c. Tangent to the ellipse at P and Q are

$\frac{x}{a}\cos\alpha + \frac{y}{b}\sin\alpha = 1$ (1)

and $\frac{x}{a}\cos\beta + \frac{y}{b}\sin\beta = 1$ (2)

Solving Eqs. (1) and (2), we get

$$\frac{x}{\begin{vmatrix} \frac{\sin\alpha}{b} & 1 \\ \frac{\sin\beta}{b} & 1 \end{vmatrix}} = \frac{-y}{\begin{vmatrix} \frac{\cos\alpha}{a} & 1 \\ \frac{\cos\beta}{a} & 1 \end{vmatrix}} = \frac{1}{\begin{vmatrix} \frac{\cos\alpha}{a} & \frac{\sin\alpha}{b} \\ \frac{\cos\beta}{a} & \frac{\sin\beta}{b} \end{vmatrix}}$$

$\Rightarrow x = \frac{a(\sin\alpha - \sin\beta)}{\sin(\beta - \alpha)}$,

$y = \frac{-b(\cos\alpha - \cos\beta)}{\sin(\beta - \alpha)}$

$\Rightarrow \frac{x\sin(\beta - \alpha)}{a} = \sin\alpha - \sin\beta$,

$\frac{y\sin(\beta - \alpha)}{b} = -(\cos\alpha - \cos\beta)$

Squaring and adding, we get

$\sin^2(\beta - \alpha)\left(\frac{x^2}{a^2} + \frac{y^2}{b^2}\right) = 2$

$\Rightarrow \frac{x^2}{a^2} + \frac{y^2}{b^2} = \frac{2}{\sin^2 c}$,

[where $\beta - \alpha = c$ (constant) given]

which is an ellipse.

29. d. As in the above question point of intersection is

$$(h, k) \equiv \left(\frac{a\cos\left(\frac{\alpha+\beta}{2}\right)}{\cos\left(\frac{\alpha-\beta}{2}\right)}, \frac{b\sin\left(\frac{\alpha+\beta}{2}\right)}{\cos\left(\frac{\alpha-\beta}{2}\right)}\right)$$

It is given that $\alpha + \beta = c$ = constant.

$\Rightarrow h = \frac{a\cos\frac{c}{2}}{\cos\left(\frac{\alpha-\beta}{2}\right)}$ and $k = \frac{b\sin\frac{c}{2}}{\cos\left(\frac{\alpha-\beta}{2}\right)}$

$\Rightarrow \frac{h}{k} = \frac{a}{b}\cot\left(\frac{c}{2}\right)$

$\Rightarrow k = \frac{b}{a}\tan\left(\frac{c}{2}\right)h$

$\Rightarrow$ (h, k) lies on the straight line

30. a. Any tangent to the ellipse $\frac{x^2}{a^2} + \frac{y^2}{b^2} = 1$ having slope m is

$y = mx + \sqrt{a^2m^2 + b^2}$

Points on the minor axis are $(0, ae)$, $(0, -ae)$.

Sum of the squares of the perpendicular on the tangent from $(0, ae)$ and $(0, -ae)$

$$= \left[\frac{\sqrt{a^2m^2 + b^2} - ae}{\sqrt{m^2 + 1}}\right]^2 + \left[\frac{\sqrt{a^2m^2 + b^2} + ae}{\sqrt{m^2 + 1}}\right]^2$$

$$= \frac{2(a^2m^2 + b^2 + a^2e^2)}{m^2 + 1}$$

$$= \frac{2(a^2m^2 + a^2 - a^2e^2 + a^2e^2)}{m^2 + 1}$$

$$= \frac{2a^2(m^2 + 1)}{m^2 + 1} = 2a^2$$

31. c. Equation of the tangent to the ellipse at $P(5\cos\theta, 4\sin\theta)$ is

$$\frac{x\cos\theta}{5} + \frac{y\sin\theta}{4} = 1$$

It meets the line $x = 0$ at $Q(0, 4\operatorname{cosec}\theta)$.

Image of Q in the line $y = x$ is $R(4\operatorname{cosec}\theta, 0)$.

Therefore, equation of the circle is

$x(x - 4\operatorname{cosec}\theta) + y(y - \operatorname{cosec}\theta) = 0$

i.e., $x^2 + y^2 - 4(x + y)\operatorname{cosec}\theta = 0$

Therefore, each member of the family passes through the intersection of $x^2 + y^2 = 0$ and $x + y = 0$, i.e., the point $(0, 0)$.

32. c.

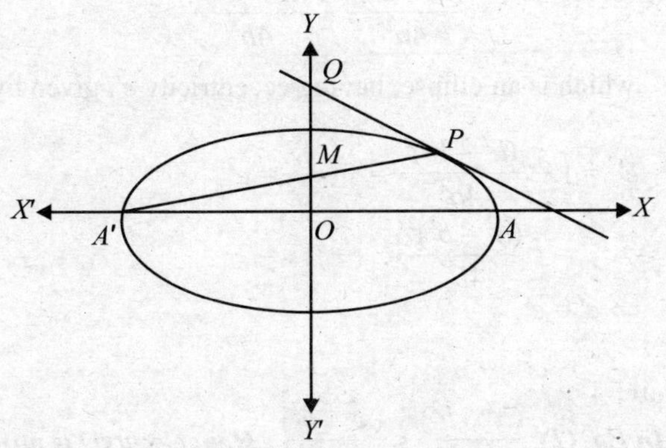

Fig. 4.10

Let point P be $(a\cos\theta, b\sin\theta)$.

Equation of the tangent at point P is

$\frac{x}{a}\cos\theta + \frac{y}{b}\sin\theta = 1.$

Then point Q is $(b\operatorname{cosec}\theta, 0)$.

Equation of chord $A'P$ is

$$y - 0 = \frac{b\sin\theta}{a\cos\theta + a}(x + a)$$

Putting $x = 0$, we have $y = \frac{b\sin\theta}{\cos\theta + 1}$.

Then

$$OQ^2 - MQ^2 = b^2\operatorname{cosec}^2\theta - \left(b\operatorname{cosec}\theta - \frac{b\sin\theta}{\cos\theta + 1}\right)^2$$

$$= \frac{2b^2}{\cos\theta + 1} - \frac{b^2\sin^2\theta}{(\cos\theta + 1)^2}$$

$$= \frac{b^2}{\cos\theta + 1}\left(\frac{2\cos\theta + 2 - \sin^2\theta}{(\cos\theta + 1)}\right)$$

$$= \frac{b^2}{\cos\theta + 1}\left(\frac{2\cos\theta + 1 + \cos^2\theta}{(\cos\theta + 1)}\right)$$

$$= b^2 = 4$$

33. b. One of the tangents of slope m to the given ellipse is

$y = mx + \sqrt{18m^2 + 32}$

For $m = -\frac{4}{3}$, we have $y = -\frac{4}{3}x + 8$.

Then points on the axis where tangents meet are $A(6, 0)$ and $B(0, 8)$.

Then area of triangle ABC is $\frac{1}{2}(6)(8) = 24$ units.

34. b.

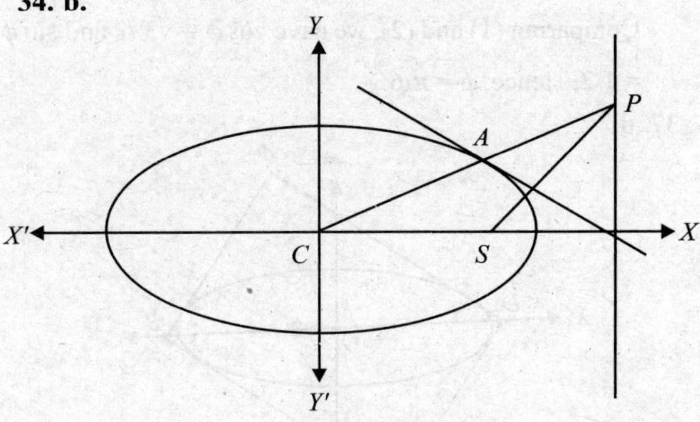

Fig. 4.11

Let the ellipse be $\frac{x^2}{a^2} + \frac{y^2}{b^2} = 1$

and let $A \equiv (a\cos\theta, b\sin\theta)$

Equation of AC will be $y = \frac{b}{a}\tan\theta\, x$.

Solving with $x = \frac{a}{e}$, we get

$P \equiv \left(\frac{a}{e}, \frac{b}{e}\tan\theta\right)$

Slope of tangent at A is $-\frac{b}{a\tan\theta}$.

$$\text{Slope of } PS = \frac{\frac{b}{e}\tan\theta}{\frac{a}{e} - ae} = \frac{b\tan\theta}{a(1 - e^2)}$$

$$= \frac{a}{b}\tan\theta$$

So $\alpha = \frac{\pi}{2}$.

35. a. A tangent of slope 2 to the ellipse $\frac{x^2}{a^2} + \frac{y^2}{b^2} = 1$ is

$y = 2x \pm \sqrt{4a^2 + b^2}$ (1)

This is normal to the circle $x^2 + y^2 + 4x + 1 = 0$

$\Rightarrow$ Eq. (1) passes through $(-2, 0)$

$\Rightarrow 0 = -4 \pm \sqrt{4a^2 + b^2}$

$\Rightarrow 4a^2 + b^2 = 16$

Using A.M. $\geq$ G.M., we get

$\frac{4a^2 + b^2}{2} \geq \sqrt{4a^2 b^2}$

$\Rightarrow ab \leq 4$

36. a. Equation of tangent $\frac{x}{a}\frac{\sqrt{3}}{2} + \frac{y}{b}\frac{1}{2} = 1$ (1)

and equation of tangent at the point $(a\cos\phi, b\sin\phi)$ is

$\frac{x}{a}\cos\phi + \frac{y}{b}\sin\phi = 1$ (2)

Comparing (1) and (2), we have $\cos\phi = \sqrt{3}/2$ and $\sin\phi = 1/2$. Hence, $\phi = \pi/6$.

37. d.

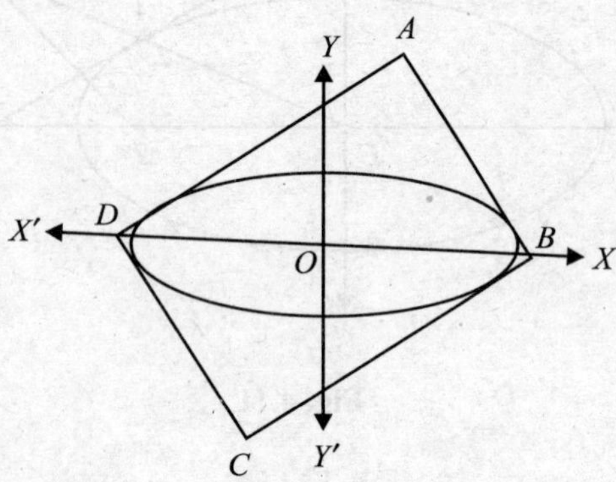

Fig. 4.12

Since sides of the square are tangent and perpendicular to each other, so 2the vertices lie on director circle

$\Rightarrow x^2 + y^2 = (a^2 - 7) + (13 - 5a)$

$= a^2$ ($\sqrt{2}\,a$ is side of the square)

$\Rightarrow (a^2 - 7) + (13 - 5a) = a^2$

$\Rightarrow a = \dfrac{6}{5}$

But for an ellipse to exist $a^2 - 7 > 0$ and $13 - 5a > 0$

$\Rightarrow a \in (-\infty, -\sqrt{7})$

Hence, $a \neq \dfrac{6}{5}$

Hence, no such a exists.

38. d. Since locus of the point of intersection of the tangent at the end points of a focal chord is directrix, required locus is

$x = \pm\dfrac{a}{e}$, which is pair of straight lines.

39. c. Normal at point $P\,(a\cos\theta, b\sin\theta)$ is

$$\frac{ax}{\cos\theta} - \frac{by}{\sin\theta} = a^2 - b^2 \quad (1).$$

It meets axes at $Q\left(\dfrac{(a^2-b^2)\cos\theta}{a}, 0\right)$

and $R\left(0, -\dfrac{(a^2-b^2)\sin\theta}{b}\right)$

Let $T\,(h, k)$ is the midpoint of QR.

Then $2h = \dfrac{(a^2-b^2)\cos\theta}{a}$

and $2k = -\dfrac{(a^2-b^2)\sin\theta}{b}$

$$\Rightarrow \cos^2\theta + \sin^2\theta = \frac{4h^2a^2}{(a^2-b^2)^2} + \frac{4k^2b^2}{(a^2-b^2)^2} = 1$$

$$\Rightarrow \text{Locus is } \frac{x^2}{\frac{(a^2-b^2)^2}{4a^2}} + \frac{y^2}{\frac{(a^2-b^2)^2}{4b^2}} = 1, \quad (2)$$

which is an ellipse, having eccentricity e', given by

$$e'^2 = 1 - \frac{\frac{(a^2-b^2)^2}{4a^2}}{\frac{(a^2-b^2)^2}{4b^2}} = 1 - \frac{b^2}{a^2} = e^2$$

$\Rightarrow e' = e$

> **Note:**
> ***In Eq. (2),*** $\dfrac{(a^2-b^2)}{4a^2} < \dfrac{(a^2-b^2)}{4b^2}$***. Hence, x-axis is minor axis.***

40. c. Equation of normal to the ellipse at P is

$5x\sec\theta - 3y\,\mathrm{cosec}\,\theta = 16$ (1)

Equation of normal to the circle $x^2 + y^2 = 25$ at point Q is

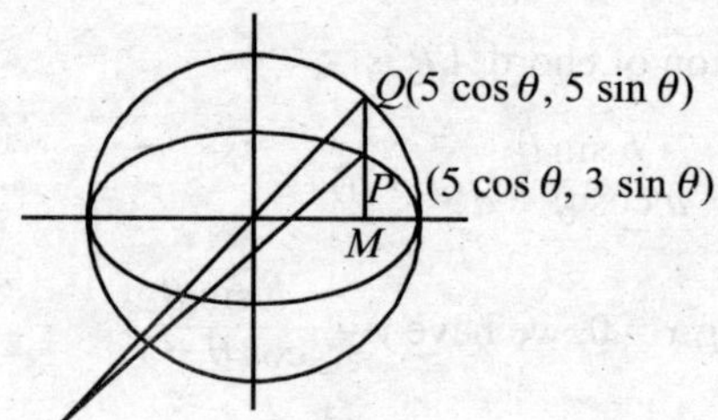

Fig. 4.13

$y = x\tan\theta$ (2)

Eliminating θ from (1) and (2), we get $x^2 + y^2 = 64$.

41. c. $\dfrac{x^2}{169} + \dfrac{y^2}{25} = 1$

Equation of normal at the point $(13\cos\theta, 5\sin\theta)$ is

$\dfrac{13x}{\cos\theta} - \dfrac{5y}{\sin\theta} = 144$. It passes through (0, 6), So

$(15 + 72\sin\theta) = 0$

$\Rightarrow \sin\theta = -\dfrac{5}{24}$

$\Rightarrow \theta = 2\pi - \sin^{-1}\left(\dfrac{5}{24}\right)$,

and $\pi + \sin^{-1}\dfrac{5}{24}$

Also the y-axis is one of the normals.

42. c. Equation has three roots. Hence, three normals can be drawn.

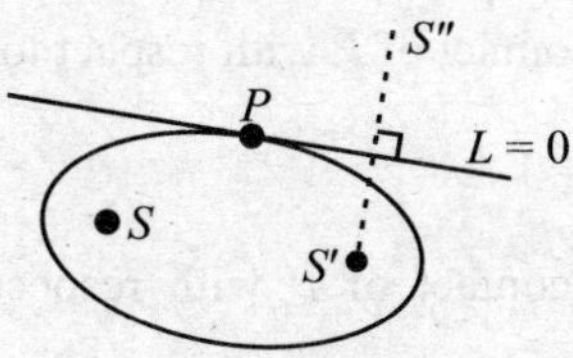

Fig. 4.14

Let image of S' be with respect to $x + y - 5 = 0$. So

$$\frac{h-2}{1} = \frac{k+1}{1} = \frac{-2(-4)}{2}$$

$\Rightarrow S'' = (6, 3)$

Let P be the point of contact.

Because the line $L = 0$ is tangent to the ellipse, there exists a point P uniquely on the line such that $PS + PS' = 2a$.

Since $PS' = PS''$, there exists one and only one point P on $L = 0$ such that $PS + PS'' = 2a$.

Hence, P should be the collinear with SS''.

Hence, P is a point of intersection of SS'' $(4x - 5y = 9)$, and $x + y - 5 = 0$, i.e., $P = \left(\frac{34}{9}, \frac{11}{9}\right)$.

43. c. A cyclic parallelogram will be a rectangle or square

So, $\angle QPR = 90°$

$\Rightarrow$ P lies on director circle of the ellipse $\frac{x^2}{16} + \frac{y^2}{b^2} = 1$

$\Rightarrow$ $x^2 + y^2 = 25$ is the director circle of $\frac{x^2}{16} + \frac{y^2}{b^2} = 1$

$\Rightarrow 16 + b^2 = 25$

$\Rightarrow b^2 = 9$

$\Rightarrow a^2(1 - e^2) = 9$

$\Rightarrow 1 - e^2 = \frac{9}{16}$

$\Rightarrow e^2 = \frac{7}{16}$

$\Rightarrow e = \frac{\sqrt{7}}{4}$

44. d.

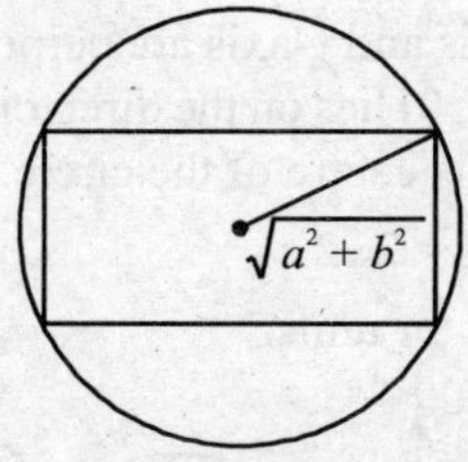

Fig. 4.15

Since mutually perpendicular tangents can be drawn from vertices of rectangle. So all the vertices of rectangle should lie on director circle $x^2 + y^2 = a^2 + b^2$.

Let breadth = $2l$ and length = $4l$, then

$l^2 + (2l)^2 = a^2 + b^2$

$\Rightarrow l^2 = \frac{a^2 + b^2}{5}$

$\Rightarrow \text{Area} = 4l \times 2l = 8\,\frac{a^2 + b^2}{5}$.

45. b. Normal at $P(\theta)$ is $\frac{ax}{\cos\theta} - \frac{by}{\sin\theta} = a^2 - b^2$ (1)

Normal at $P\left(\frac{\pi}{2} + \theta\right)$ is

$$\frac{ax}{\cos\left(\frac{\pi}{2} + \theta\right)} - \frac{by}{\sin\left(\frac{\pi}{2} + \theta\right)} = a^2 - b^2$$

or $-\frac{ax}{\sin\theta} - \frac{by}{\cos\theta} = a^2 - b^2$ (2)

Eqs. (1) and (2) meet the major axis at G $\left(\frac{(a^2 - b^2)\cos\theta}{a}, 0\right)$ and $g\left(-\frac{(a^2 - b^2)\sin\theta}{a}, 0\right)$.

Now $PG^2 + Qg^2 = \left(\frac{(a^2 - b^2)\cos\theta}{a} - a\cos\theta\right)^2 + (0 - b\sin\theta)^2 + \left(-\frac{(a^2 - b^2)\sin\theta}{a} - a\sin\theta\right)^2 + (0 - b\cos\theta)^2$

$$= \frac{(a^2 - b^2)^2}{a^2} + b^2 + a^2$$

$$= a^2\left(\frac{(a^2 - b^2)^2}{a^4} + \frac{b^2}{a^2} + 1\right)$$

$$= a^2\left(\left(1 - \frac{b^2}{a^2}\right)^2 + \frac{b^2}{a^2} + 1\right)$$

$$= a^2(e^4 + 2 - e^2)$$

46. c. The equation of the normal to the given ellipse at the point $P(a\cos\theta, b\sin\theta)$ is $ax\sec\theta - by\,\text{cosec}\,\theta = a^2 - b^2$.

$\Rightarrow y = \left(\frac{a}{b}\tan\theta\right)x - \frac{(a^2 - b^2)}{b}\sin\theta$ (1)

Let $\frac{a}{b}\tan\theta = m$, so that

$$\sin\theta = \frac{bm}{\sqrt{a^2 + b^2m^2}}.$$

Hence, the equation of the normal Eq. (1) becomes

$$y = mx - \frac{(a^2 - b^2)m}{\sqrt{a^2 + b^2m^2}}$$

$\therefore m \in R$, as $m = \frac{a}{b}\tan\theta \in R$.

47. d.

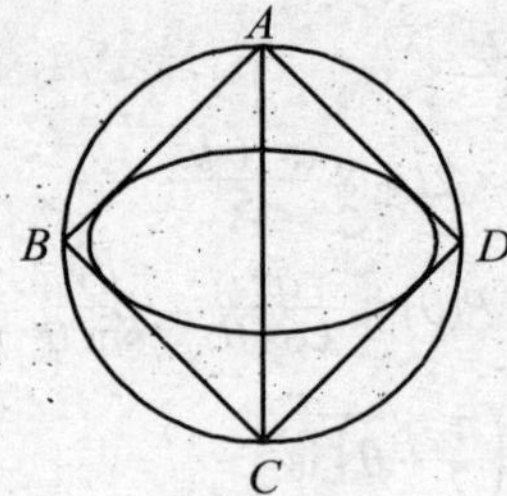

Fig. 4.16

Clearly, vertices of the square lie on the director circle of the ellipse $\frac{x^2}{7}+\frac{2y^2}{11}=1$,

which is $x^2+y^2=7+\frac{11}{2}$ or $x^2+y^2=\frac{25}{2}$

Clearly, $AC=2\sqrt{\frac{25}{2}}$

Now $AB = BC = CD = AD$

and in ΔACD, $AC^2 = CD^2 + AD^2$

$\Rightarrow 2AD^2=\left(2\sqrt{\frac{25}{2}}\right)^2$

$\Rightarrow AD^2 = 25$

$\Rightarrow AD = 5$ units

48. d. Equation of QR is $T = 0$ (chord of contact)

$\frac{8x}{4}+\frac{27y}{9}=1$

$\Rightarrow 2x + 3y = 1$ (1)

Now, equation of the pair of lines passing through origin and points Q, R is given by

$\left(\frac{x^2}{4}+\frac{y^2}{9}\right)=(2x+3y)^2$ (Making equation of ellipse homogeneous using Eq. (1))

$\Rightarrow 9x^2 + 4y^2 = 36\,(4x^2 + 12xy + 9y^2)$

$\Rightarrow 135x^2 + 432xy + 320y^2 = 0$

Hence, required angle is $\tan^{-1}\frac{2\sqrt{216^2-135\times 320}}{455}$

$=\tan^{-1}\frac{8\sqrt{2916-2700}}{455}$

$=\tan^{-1}\frac{8\sqrt{216}}{455}$

$=\tan^{-1}\frac{48\sqrt{6}}{455}$

49. c. Let the ellipse be $\frac{x^2}{a^2}+\frac{y^2}{b^2}=1$. Any point on the directrix is $P\left(\frac{a}{e},k\right)$.

Chord of contact of P with respect to the ellipse is

$\frac{\frac{a}{e}x}{a^2}+\frac{ky}{b^2}=1$ (1)

Chord of contact of P with respect to the auxiliary circle is

$\frac{a}{e}x+ky=a^2$ (2)

Eqs. (1) and (2) intersect at $(ae, 0)$

50. a. Let (h, k) be the midpoint of the chord $7x + y - 1 = 0$

$\Rightarrow \frac{hx}{1}+\frac{ky}{7}=\frac{h^2}{1}+\frac{k^2}{7}$ (1)

and $7x + y = 1$ (2)

represents same straight line.

$\Rightarrow \frac{h}{7}=\frac{k}{7}\Rightarrow h=k$

$\Rightarrow$ Equation of the line joining $(0, 0)$ and (h, k) is $y - x = 0$.

51. c. Given $m(n - 1) = n$.

n is divisible by $n - 1$

$\Rightarrow n = 2 \Rightarrow m = 2$

Hence, chord of contact of tangents drawn from $(2, 2)$ to

$\frac{x^2}{9}+\frac{y^2}{4}=1$ is $\frac{2x}{9}+\frac{2y}{4}=1$.

$\Rightarrow 4x + 9y = 18$

52. b. The chord of contact of tangents from (h, k) is

$\frac{xh}{a^2}+\frac{yk}{b^2}=1$.

It meets the axes at points $\left(\frac{a^2}{h},0\right)$ and $\left(0,\frac{b^2}{k}\right)$.

Area of the triangle $=\frac{1}{2}\times\frac{a^2}{h}\times\frac{b^2}{k}=c$ (constant)

$\Rightarrow hk=\frac{a^2b^2}{2c}$ (c is constant)

$xy = c^2$ is the required locus.

53. d. Since x-axis and y-axis are perpendicular tangents to the ellipse. $(0, 0)$ lies on the director circle and midpoint of foci $(2, 2)$ is centre of the circle.

Hence, radius $= 2\sqrt{2}$

$\Rightarrow$ the area is 8π units.

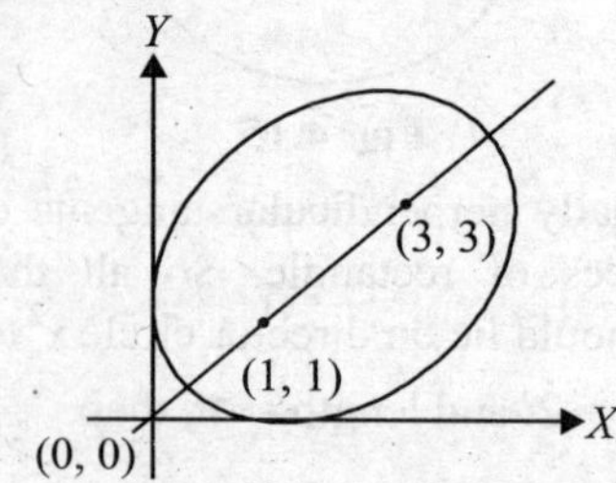

Fig. 4.17

54. a. Let the equation of the ellipse be

$$\frac{x^2}{a^2}+\frac{y^2}{b^2}=1$$

We know that the general equation of the tangent to the ellipse is

$$y = mx \pm \sqrt{a^2m^2 + b^2} \quad (1)$$

Since $3x - 2y - 20 = 0$ or $y = \frac{3}{2}x - 10$ is tangent to the ellipse, comparing with Eq. (1),

$m = \frac{3}{2}$ and $a^2m^2 + b^2 = 100$

$\Rightarrow a^2 \times \frac{9}{4} + b^2 = 100$

$$\Rightarrow 9a^2 + 4b^2 = 400 \quad (2)$$

Similarly, since $x + 6y - 20 = 0$, i.e., $y = -\frac{1}{6}x + \frac{10}{3}$ is tangent to the ellipse, comparing with Eq. (1),

$m = \frac{1}{6}$ and $a^2m^2 + b^2 = \frac{100}{9}$

$\Rightarrow \frac{a^2}{36} + b^2 = \frac{100}{9}$

$$\Rightarrow a^2 + 36b^2 = 400 \quad (3)$$

Solving Eqs. (2) and (3), we get $a^2 = 40$ and $b^2 = 10$.

Therefore, the required equation of the ellipse is

$$\frac{x^2}{40}+\frac{y^2}{10}=1.$$

55. b. We know that product of length of perpendiculars from foci upon any tangents to ellipse is b^2.

Hence, from the diagram, x_1 and x_2 are length of perpendiculars from foci upon tangent y-axis of the given ellipse, hence $x_1x_2 = b^2$.

Similarly $y_1y_2 = b^2$

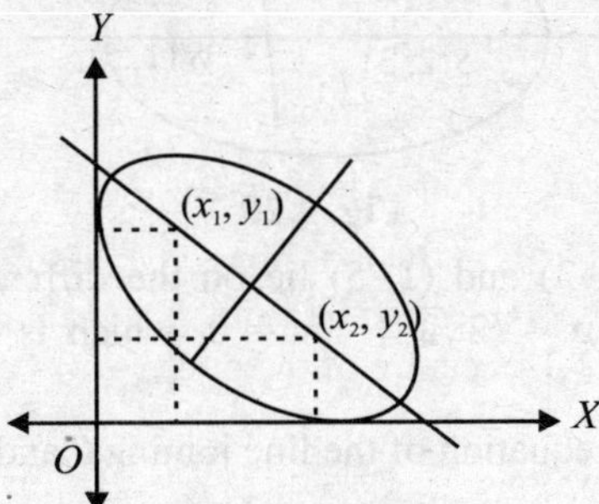

Fig. 4.18

56. c. $\sum_{i=1}^{10}(SP_i)(S'P'_i) = 2560$

$\Rightarrow 10b^2 = 2560$

$\Rightarrow b^2 = 256$

$\Rightarrow b = 16$

$\Rightarrow 256 = 400\,(1 - e^2)$

$\Rightarrow 1 - e^2 = \frac{16}{25}$

$\Rightarrow e = \frac{3}{5}$

57. d. For the ellipse $\frac{x^2}{16}+\frac{y^2}{9}=1$, equation of director circle is $x^2 + y^2 = 25$. The director circle will cut the ellipse $\frac{x^2}{50}+\frac{y^2}{20}=1$ at four points. Hence, number of points is 4.

58. c. Equation of the ellipse is $\frac{x^2}{a^2}+\frac{y^2}{b^2}=1$.

Equation of the parabola with focus $S\,(ae, 0)$ and directrix $x + ae = 0$ is $y^2 = 4aex$.

Now length of latus rectum of the ellipse is $\frac{2b^2}{a}$ and that of the parabola is $4ae$.

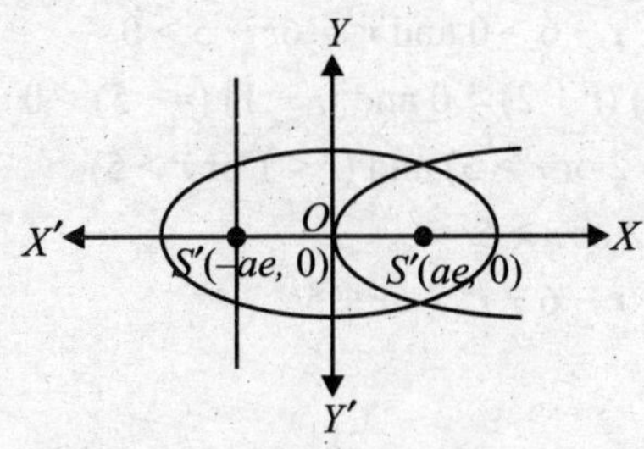

Fig. 4.19

For the two latus recta to be equal, we get

$\frac{2b^2}{a} = 4ae$

$\Rightarrow \frac{2a^2(1-e^2)}{a} = 4ae$

$\Rightarrow 1 - e^2 = 2e$

$\Rightarrow e^2 + 2e - 1 = 0$

Therefore, $e = -\frac{2 \pm \sqrt{8}}{2} = -1 \pm \sqrt{2}$

Hence, $e = \sqrt{2} - 1$.

59. c. Here centre of the ellipse is (0, 0).

Let $P(r\cos\theta, r\sin\theta)$ be any point on the given ellipse then $r^2\cos^2\theta + 2r^2\sin^2\theta + 2r^2\sin\theta\cos\theta = 1$.

$$\Rightarrow r^2 = \frac{1}{\cos^2\theta + 2\sin^2\theta + \sin 2\theta}$$

$$= \frac{1}{\sin^2\theta + 1 + \sin 2\theta}$$

$$= \frac{2}{1 - \cos 2\theta + 2 + 2\sin 2\theta}$$

$$= \frac{2}{3 - \cos 2\theta + 2\sin 2\theta}$$

$$\Rightarrow r_{\max} = \frac{\sqrt{2}}{\sqrt{3-\sqrt{5}}}$$

60. a. Combined equation of a pair of lines through the origin joining the points of intersection of line $y = \sqrt{m}x + 1$ with the given curve is $x^2 + 2xy + (2 + \sin^2\alpha)y^2 - (y - \sqrt{m}x)^2 = 0$.

For the chord to subtend a right angle at the origin, $(1 - m) + (2 + \sin^2\alpha - 1) = 0$ (as sum of the coefficients of $x^2 + y^2 = 10$)

$\Rightarrow \sin^2\alpha = m - 2$

$\Rightarrow 0 \le m - 2 \le 1$

$\Rightarrow 2 \le m \le 3$

Multiple Correct Answers Type

1. a., c. $r^2 - r - 6 > 0$ and $r^2 - 6r + 5 > 0$

$\Rightarrow (r - 3)(r + 2) > 0$ and $(r - 1)(r - 5) > 0$

$\Rightarrow (r < -2$ or $r > 3)$ and $(r < 1$ or $r > 5)$

$\Rightarrow r < -2$ or $r > 5$

Also $r^2 - r - 6 \ne r^2 - 6r + 5$

$\Rightarrow r \ne \frac{11}{5}$

2. a., c.

If both foci are fixed, then the ellipse is fixed, that is, both the directrices can be decided (eccentricity is given). Similar is the case for option (c). Thus (a) and (c) are the correct choices. In the remaining cases, size of the ellipse is fixed, but its position is not fixed.

3. a., b. Any point on this ellipse is $(\sqrt{6}\cos\phi, \sqrt{2}\sin\phi)$.

Here centre is (0, 0), so $6\cos^2\phi + 2\sin^2\phi = 4$

$\Rightarrow 2\cos^2\phi = 1$

$\Rightarrow \cos^2\phi = \left(\frac{1}{\sqrt{2}}\right)^2 = \cos^2\frac{\pi}{4}$

$\Rightarrow \phi = \frac{\pi}{4}$ or $\frac{3\pi}{4}$

4. a., b., c. $3x^2 + 2y^2 + 6x - 8y + 5 = 0$

$$\Rightarrow \frac{(x+1)^2}{2} + \frac{(y-2)^2}{3} = 1$$

Therefore, centre is (−1, 2) and ellipse is vertical ($\because b > a$)

$a^2 = 2, b^2 = 3$

Now $2 = 3(1 - e^2)$

$\Rightarrow e = \frac{1}{\sqrt{3}}$

Foci are $(-1, 2 \pm be)$ and $(-1, 2 \pm 1) \equiv (-1, 3)$ and $(-1, 1)$.

And directrix are $y = 2 \pm \frac{b}{e} \Rightarrow y = 5$ and $y = -1$.

5. c., d. Ellipse is $16x^2 + 11y^2 = 256$.

Equation of tangent at $\left(4\cos\theta, \frac{16}{\sqrt{11}}\sin\theta\right)$ is

$$16x(4\cos\theta) + 11y\left(\frac{16}{\sqrt{11}}\sin\theta\right) = 256$$

It touches $(x - 1)^2 + y^2 = 4^2$ if

$$\left|\frac{4\cos\theta - 16}{\sqrt{16\cos^2\theta + 11\sin^2\theta}}\right| = 4$$

$\Rightarrow (\cos\theta - 4)^2 = 16\cos^2\theta + 11\sin^2\theta$

$\Rightarrow 4\cos^2\theta + 8\cos\theta - 5 = 0$

$\Rightarrow \cos\theta = \frac{1}{2}$

$\therefore \theta = \frac{\pi}{3}, \frac{5\pi}{3}$

6. a., c. Tangent and normal are bisectors of $\angle SPS'$.

Now equation of SP is $y = 3x/2$ and that of $S'P$ is $y = 5x$.

Then their equations are

$$\frac{3x - 2y}{\sqrt{13}} = \pm\frac{5x - y}{\sqrt{26}}$$

or $3x - 2y = \pm\frac{5x - y}{\sqrt{2}}$

$\Rightarrow$ lines are $(3\sqrt{2} - 5)x + (1 - 2\sqrt{2})y = 0$ and $(3\sqrt{2} + 5)x - (2\sqrt{2} + 1)y = 0$

Now (2, 3) and (1, 5) lie on the same side of $(3\sqrt{2} - 5)x + (1 - 2\sqrt{2})y = 0$, which is equation of tangent.

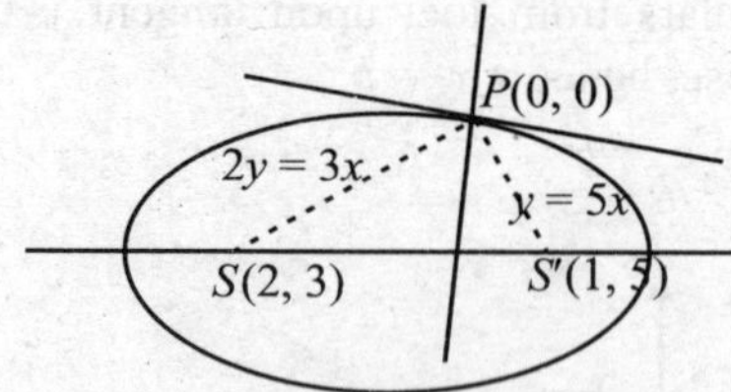

Fig. 4.20

Points (2, 3) and (1, 5) lie on the different sides of $(3\sqrt{2} + 5)x - (2\sqrt{2} + 1)y = 0$, which is equation of normal.

7. c., d. The equation of the line joining θ and ϕ is

$$\frac{x}{5}\cos\left(\frac{\theta+\phi}{2}\right) + \frac{y}{3}\sin\left(\frac{\theta+\phi}{2}\right) = \cos\left(\frac{\theta-\phi}{2}\right)$$

If it passes through the point (4, 0), then

$$\frac{4}{5}\cos\left(\frac{\theta+\phi}{2}\right) = \cos\left(\frac{\theta-\phi}{2}\right)$$

$$\Rightarrow \frac{4}{5} = \frac{\cos\left(\frac{\theta-\phi}{2}\right)}{\cos\left(\frac{\theta+\phi}{2}\right)}$$

$$\Rightarrow \frac{4+5}{4-5} = \frac{\cos\left(\frac{\theta-\phi}{2}\right) + \cos\left(\frac{\theta-\phi}{2}\right)}{\cos\left(\frac{\theta-\phi}{2}\right) - \cos\left(\frac{\theta+\phi}{2}\right)}$$

$$= \frac{2\cos\frac{\theta}{2}\cos\frac{\phi}{2}}{2\sin\frac{\phi}{2}\sin\frac{\theta}{2}}$$

$$\Rightarrow \tan\frac{\theta}{2}\tan\frac{\phi}{2} = -\frac{1}{9}.$$

If it passes through the point $(-5, 0)$,

then $\tan\frac{\phi}{2}\tan\frac{\theta}{2} = 9$.

8. a., b., c., d.

$(\sqrt{3}x - 3\sqrt{3})^2 + (2y+4)^2 = k$

So no locus for $k < 0$

Ellipse for $k > 0$ and point for $k = 0$

9. a., c.

Let the point of equation of intersection of tangents A and B be $P(h, k)$, then equation of AB is

$$\frac{xh}{4} + \frac{yk}{1} = 1 \quad (1)$$

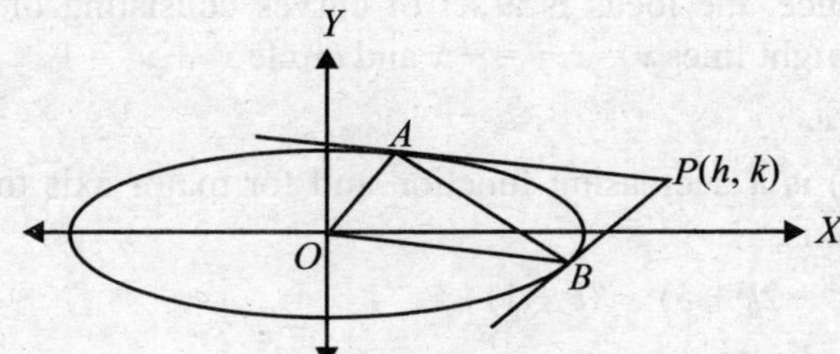

Fig. 4.21

Homogenizing the equation of ellipse using Eq. (1), we get

$$\frac{x^2}{4} + \frac{y^2}{1} = \left(\frac{xh}{4} + \frac{yk}{1}\right)^2$$

$$\Rightarrow x^2\left(\frac{h^2-4}{16}\right) + y^2(k^2-1) + \frac{2hk}{4}xy = 0 \quad (2)$$

Given equation of OA and OB is

$$x^2 + 4y^2 + \alpha xy = 0 \quad (3)$$

Since Eqs. (2) and (3) represent the same line,

$$\frac{h^2-4}{16} = \frac{k^2-1}{4} = \frac{hk}{2a}$$

$$\Rightarrow h^2 - 4 = 4(k^2-1)$$

$$\Rightarrow h^2 - 4k^2 = 0$$

$$\Rightarrow \text{Locus } (x-2y)(x+2y) = 0$$

10. a., b., c.

a. The given equation is $\left(x - \frac{1}{13}\right)^2 + \left(y - \frac{2}{13}\right)^2$

$$= \frac{1}{a^2}\left(\frac{5x+12y-1}{13}\right)^2$$

It represents ellipse if $\frac{1}{a^2} < 1 \Rightarrow a^2 > 1 \Rightarrow a > 1$

b. $4x^2 + 8x + 9y^2 - 36y = -4$

$$\Rightarrow 4(x^2+2x+1) + 9(y^2-4y+4) = 36$$

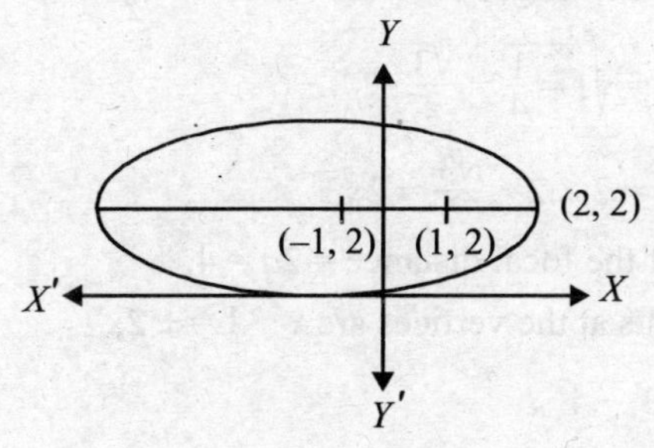

Fig. 4.22

$$\Rightarrow \frac{(x+1)^2}{9} + \frac{(y-2)^2}{4} = 1$$

Hence, $(-1, 2)$ is focus and $(1, 2)$ lies on the major axis. Then required minimum distance is 1.

c. Equation of normal at $P(\theta)$ is $5\sec\theta\, x - 4\operatorname{cosec}\theta\, y = 25 - 16$, and it passes through $P(0, \alpha)$

$$\therefore \alpha = \frac{-9}{4\operatorname{cosec}\theta}$$

$$\Rightarrow \alpha = \frac{-9}{4}\sin\theta$$

$$\Rightarrow |\alpha| < \frac{9}{4}$$

d. $\frac{2b^2}{a} = \frac{2a}{3} \Rightarrow 3b^2 = a^2$

$\Rightarrow$ From $b^2 = a^2(1-e^2)$, $1 = 3(1-e^2) \Rightarrow e = \sqrt{2/3}$

11. a., c., d.

$$x^2 + 4y^2 - 2x - 16y + 13 = 0$$

$$\Rightarrow (x^2 - 2x + 1) + 4(y^2 - 4y + 4) = 4$$

$$\Rightarrow \frac{(x-1)^2}{4} + \frac{(y-2)^2}{1} = 1$$

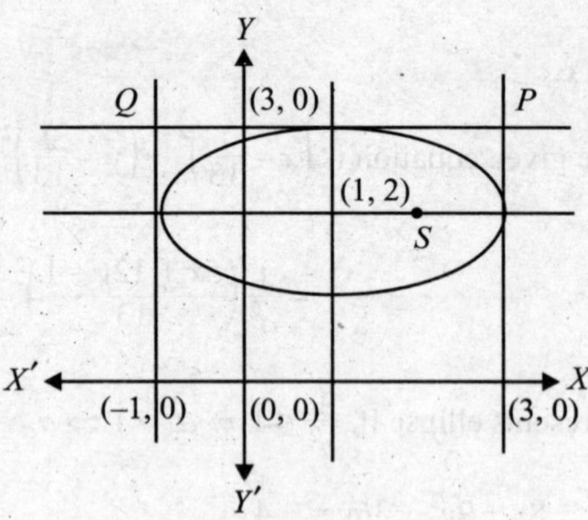

Fig. 4.23

Therefore, length of latus rectum $= \frac{2 \times 1}{2} = 1$

Also $e = \sqrt{1 - \frac{1}{4}} = \frac{\sqrt{3}}{2}$.

$\Rightarrow 2ae = 2 \times 2 \times \frac{\sqrt{3}}{2} = 2\sqrt{3}$

Sum of the focal distance $= 2a = 4$.

Tangents at the vertices are $x - 1 = \pm 2$,

or $x = 3, -1$

Therefore, the line $y = 3$ intersect these at points $P(3, 3)$ and $Q(-1, 3)$.

Coordinate of focus are $S(\sqrt{3} + 1, 2)$

Slope of PS is $\frac{1}{2 - \sqrt{3}}$, slope of QS is $\frac{1}{-2 - \sqrt{3}}$

$\Rightarrow$ product of slopes $= \frac{1}{2 - \sqrt{3}} \times \frac{1}{-2 - \sqrt{3}} = -1$

12. b., d.

Differentiating the equation of ellipse $x^2 + 3y^2 = 37$ w.r.t. x, we get

$$\frac{dy}{dx} = -\frac{x}{3y}$$

Slope of the given line is $\frac{6}{5}$, which is normal to the ellipse.

Hence, $\frac{3y}{5} = \frac{6}{5}$ or $2x = 5y$.

Points in the option (b) and (d) are satisfying the above relation.

13. a., c.

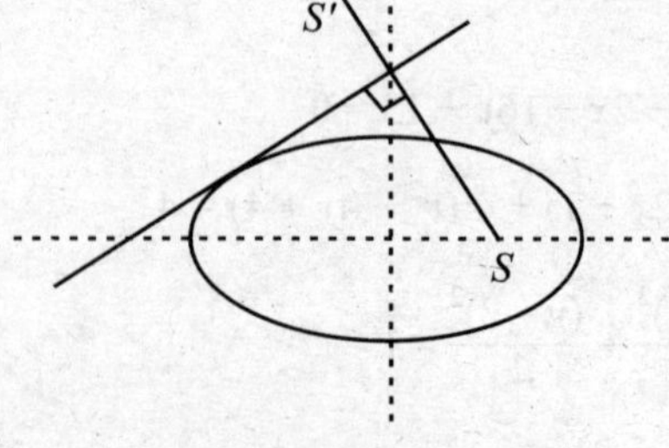

Fig. 4.24

Let $S'(h, k)$ be the image.

SS' cuts a tangent at a point which lies on the auxiliary circle of the ellipse

$$\Rightarrow \left(\frac{h \pm 4}{2}\right)^2 + \frac{k^2}{4} = 25$$

$\Rightarrow$ locus is $(x \pm 4)^2 + y^2 = 100$.

14. a., c.

Let $P(h, k)$ be the point of intersection of E_1 and E_2

$$\Rightarrow \frac{h^2}{a^2} + k^2 = 1$$

$$\Rightarrow h^2 = a^2(1 - k^2) \quad (1)$$

and $\frac{h^2}{1} + \frac{k^2}{a^2} = 1$

$$\Rightarrow k^2 = a^2(1 - h^2) \quad (2)$$

Eliminating a from Eqs. (1) and (2), we get

$$\frac{h^2}{1 - k^2} = \frac{k^2}{1 - h^2}$$

$$\Rightarrow h^2(1 - h^2) = k^2(1 - k^2)$$

$$\Rightarrow (h - k)(h + k)(h^2 + k^2 - 1) = 0$$

Hence, the locus is a set of curves consisting of the straight lines $y = x$, $y = -x$ and circle $x^2 + y^2 = 1$.

15. a., c.

$f(x)$ is a decreasing function and for major axis to be x-axis,

$f(k^2 + 2k + 5) > f(k + 11)$

$\Rightarrow k^2 + 2k + 5 < k + 11$

$\Rightarrow k \in (-3, 2)$

Then for the remaining values of k, i.e., $k \in (-\infty, -3) \cup (2, \infty)$, major axis is y-axis.

16. a., b., c.

Clearly O is the midpoint of SS' and HH'

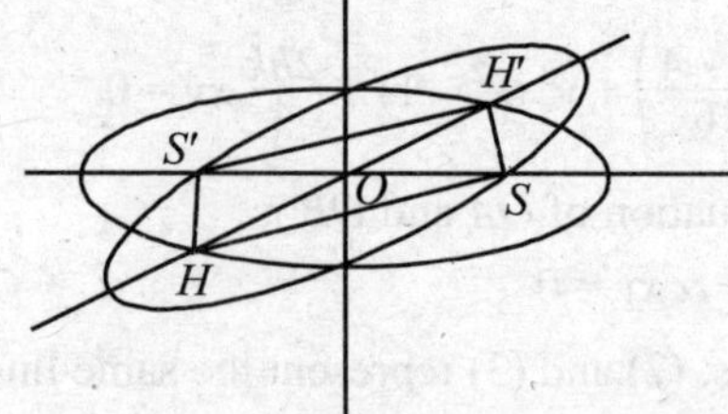

Fig. 4.25

$\Rightarrow$ Diagonals of quadrilateral $HSH'S'$ bisect each other. So it is a parallelogram.

Let $H'OH = 2r \Rightarrow OH = r = ae'$

H lies on $\frac{x^2}{a^2} + \frac{y^2}{b^2} = 1$ (suppose)

$$\therefore \frac{r^2 \cos^2\theta}{a^2} + \frac{r^2 \sin^2\theta}{b^2} = 1$$

$$e'^2 \cos^2\theta + \frac{e'^2 \sin^2\theta}{1-e^2} = 1 \qquad [\because b^2 = a^2(1-e'^2)]$$

$$\Rightarrow e'^2 \cos^2\theta - \frac{e'^2\cos^2\theta}{1-e^2} = 1 - \frac{e'^2}{1-e^2}$$

$$\Rightarrow \cos^2\theta = \frac{1}{e^2} + \frac{1}{e'^2} - \frac{1}{e^2e'^2}$$

$$\Rightarrow \theta = \cos^{-1}\sqrt{\frac{1}{e^2} + \frac{1}{e'^2} - \frac{1}{e^2e'^2}}$$

For $\theta = 90°$, $\frac{e^2+e'^2}{e^2e'^2} = \frac{1}{e^2e'^2}$

$$\Rightarrow e^2 + e'^2 = 1$$

17. a., d. The equation of the tangent at $(t^2, 2t)$ to the parabola $y^2 = 4x$ is

$2ty = 2(x + t^2)$

$\Rightarrow ty = x + t^2$

$\Rightarrow x - ty + t^2 = 0$ (1)

The equation of the normal at the point $(\sqrt{5}\cos\theta, 2\sin\theta)$ on the ellipse $5x^2 + 5y^2 = 20$ is

$(\sqrt{5}\sec\theta)x - (2\,\mathrm{cosec}\,\theta)y = 5 - 4$

$\Rightarrow (\sqrt{5}\sec\theta)x - (2\,\mathrm{cosec}\,\theta)y = 1$ (2)

Given that Eqs. (1) and (2) represent the same line. So

$$\frac{\sqrt{5}\sec\theta}{1} = \frac{-2\text{ coses }\theta}{-t} = \frac{-1}{t^2}$$

$$\Rightarrow t = \frac{2}{\sqrt{5}}\cot\theta \text{ and } t = -\frac{1}{2}\sin\theta$$

$$\Rightarrow \frac{2}{\sqrt{5}}\cot\theta = -\frac{1}{2}\sin\theta$$

$$\Rightarrow 4\cos\theta = -\sqrt{5}\sin^2\theta$$

$$\Rightarrow 4\cos\theta = -\sqrt{5}(1-\cos^2\theta)$$

$$\Rightarrow \sqrt{5}\cos^2\theta - 4\cos\theta - \sqrt{5} = 0$$

$$\Rightarrow (\cos\theta - \sqrt{5})(\sqrt{5}\cos\theta + 1) = 0$$

$$\Rightarrow \cos\theta = -\frac{1}{\sqrt{5}} \qquad [\because \cos\theta \neq -\sqrt{5}]$$

$$\Rightarrow \theta = \cos^{-1}\left(-\frac{1}{\sqrt{5}}\right)$$

Putting $\cos\theta = -\frac{1}{\sqrt{5}}$ in $t = -\frac{1}{2}\sin\theta$, we get

$$t = -\frac{1}{2}\sqrt{1-\frac{1}{5}} = -\frac{1}{\sqrt{5}}$$

Hence, $\theta = \cos^{-1}\left(-\frac{1}{\sqrt{5}}\right)$ and $t = -\frac{1}{\sqrt{5}}$.

Reasoning Type

1. d. Statement 1 is false as locus of (x, y) is a line segment joining points (2, 0) and (–2, 0).

2. c. Statement 2 is false (locus of P may be a line segment also). statement 1 is true.

3. a. It is fundamental property of an ellipse.

4. c. $\frac{x^2}{5} + \frac{(y-3)^2}{9} = 1$

Ends of the major axis are (0, 6) and (0, 0).

Equation of tangent at (0, 6) and (0, 0) is $y = 6$ and $y = 0$.

Hence, statement 1 is true.

But statement 2 is false, as tangents at the ends of major axis may be lines parallel to y-axis when $a < b$.

5. a. Locus of point of intersection of perpendicular tangents is director circle, which is $x^2 + y^2 = a^2 + b^2$.

Now line $px + qy + r = 0$ may intersect this circle maximum at two points.

Thus there can be maximum two points on the line from which perpendicular tangents can be drawn to the ellipse.

6. a. Statement 2 is true as it is one of the properties of ellipse.

Ellipse is $\frac{x^2}{9} + \frac{y^2}{4} = 1$

Focus $\equiv (\sqrt{5}, 0)$, $e = \frac{\sqrt{5}}{3}$

One of the points on the ellipse $\equiv \left(\frac{3}{\sqrt{2}}, \frac{2}{\sqrt{2}}\right)$

Equation of the circle as the diameter joining the points $(3/\sqrt{2}, 2/\sqrt{2})$ and focus $(\sqrt{5}, 0)$ is

$(x - \sqrt{5})(\sqrt{2}x - 3) + y(\sqrt{2}y - 2) = 0$

Hence, statement 1 is true and statement 2 is correct explanation of statement 1.

7. d. Let C_1, C_2 be the centres and r_1, r_2 be the radii of the two circles. Let $S_1 = 0$ lie completely inside the circle. $S_2 = 0$. Let C and r be centre and radius of the variable circle, respectively.

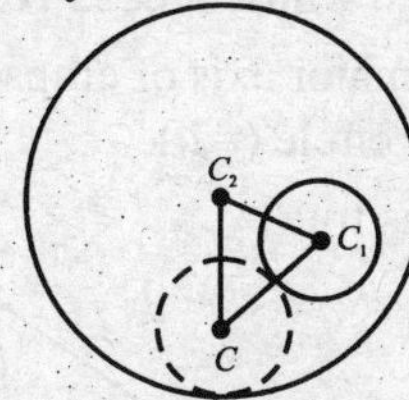

Fig. 4.26

Then $CC_2 = r_2 - r$ and $C_1C = r_1 + r$

$\Rightarrow C_1C + C_2C = r_1 + r_2$ (constant)

$\Rightarrow$ Locus of C is an ellipse

$\Rightarrow S_2$ is true.

Statement 1 is false (two circles are intersecting).

8. b. By formula $p_1p_2 = b^2 = 3$. Also foot of perpendicular lies on auxiliary circle of the ellipse. Thus both the statements are true. But statement 2 is not correct explanation of statement 1.

9. d. Locus of point of intersection of perpendicular tangents is director circle. If there exists exactly one such point on the line $3x + 4y + 5\sqrt{5} = 0$, then it must touch the director circle

$x^2 + y^2 = a^2 + 1$

$\Rightarrow 5 = a^2 + 1$

$\Rightarrow a^2 = 4$

$\Rightarrow a = 2$

Hence, eccentricity $= \sqrt{1 - \frac{1}{4}} = \frac{\sqrt{3}}{2}$.

10. a. Let $y = mx$ be any chord through (0, 0). This will meet conic at points whose x-coordinates are given by

$x^2 + m^2x^2 + mx^2 = 1$

$\Rightarrow (1 + m + m^2)\, x^2 - 1 = 0$

$\Rightarrow x_1 + x_2 = 0$

$\Rightarrow \frac{x_1 + x_2}{2} = 0$

Also $y_1 = mx_1, y_2 = mx_2$

$\Rightarrow y_1 + y_2 = m\,(x_1 + x_2) = 0$

$\Rightarrow \frac{y_1 + y_2}{2} = 0$

$\Rightarrow$ Midpoint of chord is (0, 0); $\forall m$.

Hence, statement 1 is true as (0, 0) is also centre of the ellipse. Statement 2 is fundamental property of the ellipse. Hence statement 2 is correct explanation of statement 1.

11. b. Given ellipse is $\frac{x^2}{3} + \frac{y^2}{2} = 1$, whose area is $= \pi\sqrt{3}\sqrt{2} = \pi\sqrt{6}$.

Circle is $x^2 + y^2 - 2x + 4y + 4 = 0$ or

$(x - 1)^2 + (y - 2)^2 = 1$.

Its area is π. Hence, statement 1 is true.

Also statement 2 is true but it is not the correct explanation of statement 1.

Consider the ellipse $\frac{x^2}{25} + \frac{y^2}{1} = 1$, whose area is 5π and circle $x^2 + y^2 = 16$ whose area is 16π.

Also here semi-major axis of ellipse (= 5) is more than the radius of the circle (= 4).

12. d. Statement 2 is true

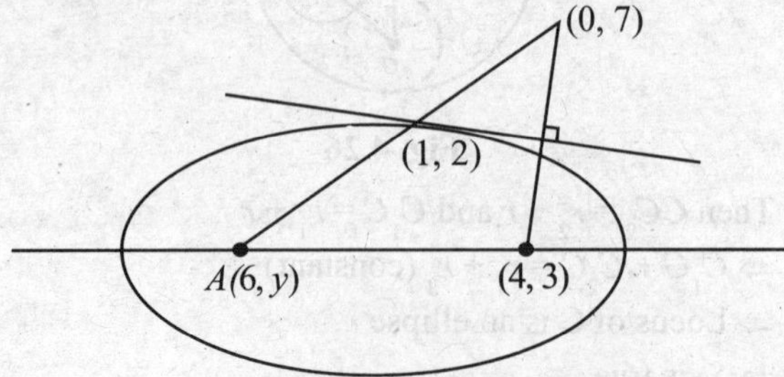

Fig. 4.27

Image of (4, 3) in the line $x + y - 3 = 0$ is

$\frac{x-4}{1} = \frac{y-3}{1} = -2\,\frac{(4+3-3)}{2} = -4$

$\Rightarrow x = 0$ and $y = 7$

Now points (0, 7), (1, 2), and (6, y) are collinear.

$\Rightarrow \frac{7-2}{0-1} = \frac{y-2}{6-1}$

$\Rightarrow y = 27$

13. a. Statement 2 is correct as ellipse is a central conic and it also explains statement 1.

14. d. Statement 1: Locus of point of intersection of only perpendicular lines is a circle, and other vertices B and C do not form a circle

Statement 2 is obvious (standard definition).

15. a. Chord of contact of the ellipse $\frac{x^2}{4} + \frac{y^2}{2} = 1$ with respect to point (8, 6) is

$\frac{8x}{4} + \frac{6x}{2} = 1$

or $2x + 3y = 1$.

Hence, statement 1 is correct, also statement 2 is correct and explains the statement 1.

16. a. Statement 2 is true as it is one of the properties of the ellipse.

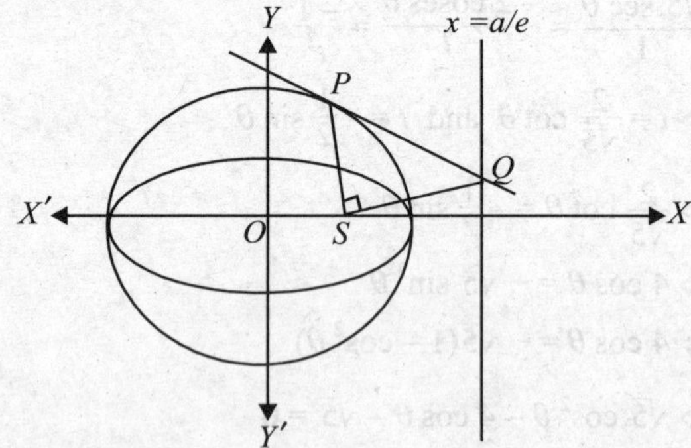

Fig. 4.28

Circle with minimum radius having PQ as chord when PQ is diameter of the circle, hence as shown in the figure it passes through the focus.

17. d. $|\,ay - bx\,| = c\,\sqrt{(x-a)^2 + (y-b)^2}$

$\Rightarrow \frac{|\,ay - bx\,|}{\sqrt{a^2 + b^2}} = \frac{c}{\sqrt{a^2 + b^2}}\sqrt{(x-a)^2 + (y-b)^2}$

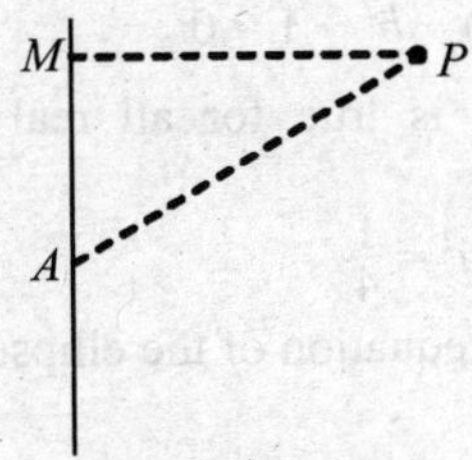

Fig. 4.29

or $PM = kPA$, where m is the length of perpendicular from P on the line $ay - bx = 0$ and PA is the length of line segment joining P to the point A (a, b) and A lies on the line, so the locus of P is a straight line through A inclined at an angle $\sin^{-1}\dfrac{c}{\sqrt{a^2+b^2}}$ to the given line (provided $c < \sqrt{a^2+b^2}$).

Linked Comprehension Type

For Problems 1–3

1. c., 2. a., 3. b.

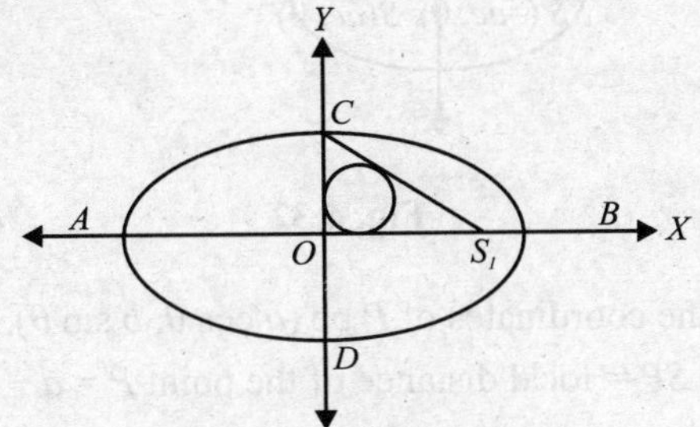

Fig. 4.30

Sol. $\because OS_1 = ae = 6, OC = b$

Also $CS_1 = a$

$\Rightarrow$ Area of $\Delta OCS_1 = \frac{1}{2}(OS_1) \times (OC) = 3b$

$\Rightarrow$ Semi-perimeter of $\Delta OCS_1 = \frac{1}{2}(OS_1 + OC + CS_1)$

$= \frac{1}{2}(6 + a + b)$ (1)

$\Rightarrow$ In radius of $\Delta OCS_1 = 1$

$\Rightarrow \dfrac{3b}{\frac{1}{2}(6+a+b)} = 1$

$\Rightarrow 5b = 6 + a$ (2)

Also $b^2 = a^2 - a^2e^2$

$= a^2 - 36$ (3)

$\Rightarrow$ From (2), we get

$25(a^2 - 36) = 36 + a^2 + 12a$

$\Rightarrow 2a^2 - a - 78 = 0$

$\Rightarrow a = \frac{13}{2}, -6$

$\Rightarrow a = \frac{13}{2}$

and $b = \frac{5}{2}$

Area of ellipse $= \pi ab = \dfrac{65\pi}{4}$ sq. unit

Perimeter of $\Delta\, OCS_1 = 6 + a + b = 6 + \frac{13}{2} + \frac{5}{2} = 15$ units

Equation director circle is $x^2 + y^2 = a^2 + b^2$

or $x^2 + y^2 = \dfrac{97}{2} = r^2$

For Problems 4–6

4. c., 5. b., 6. a.

Sol.

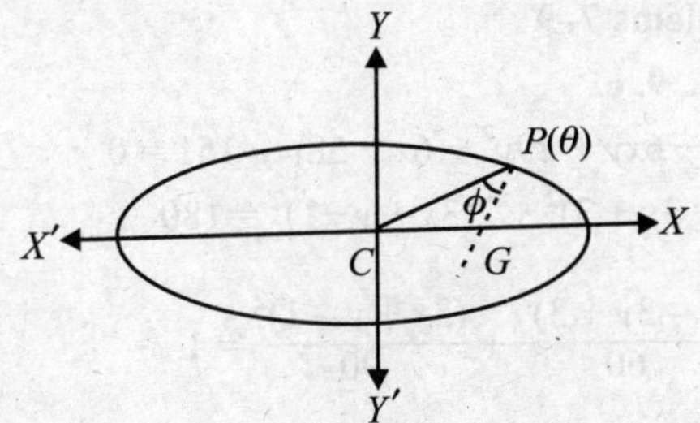

Fig. 4.31

Any point P on the ellipse at $(a\cos\theta, b\sin\theta)$,

Therefore, equation of CP is $y = \left(\frac{b}{a}\tan\theta\right)x$

The normal to the ellipse at P is

$$\frac{ax}{\cos\theta} - \frac{bx}{\sin\theta} = a^2 - b^2$$

Slope of the lines CP and the normal GP are $\frac{b}{a}\tan\theta$ and $\frac{a}{b}\tan\theta$, respectively.

$$\therefore \tan\phi = \frac{\frac{a}{b}\tan\theta - \frac{b}{a}\tan\theta}{1 + \frac{a}{b}\tan\theta\,\frac{b}{a}\tan 0}$$

$$= \frac{a^2 - b^2}{ab}\,\frac{\tan\theta}{\sec^2\theta}$$

$$= \frac{a^2 - b^2}{ab}\sin\theta\cos\theta = \frac{a^2 - b^2}{2ab}\sin 2\theta$$

Therefore, the greatest value of $\tan\phi = \dfrac{a^2 - b^2}{2ab} \times 1$

$$= \frac{a^2 - b^2}{2ab}$$

Given that $\frac{a^2-b^2}{2ab}=\frac{3}{2}$. Let $\frac{a}{b}=t$. Then

$$t-\frac{1}{t}=\frac{3}{2}$$

$$\Rightarrow 2t^2-3t-2=0$$

$$\Rightarrow 2t^2-4t+t-2=0$$

$$\Rightarrow (2t+1)(t-2) \Rightarrow \frac{a}{b}=2$$

$$\Rightarrow e^2=1-\frac{1}{4}$$

$$\Rightarrow e=\frac{\sqrt{3}}{2}$$

Rectangle inscribed in the ellipse whose one vertex is $(a\cos\theta, b\sin\theta)$ is $(2a\cos\theta)(2b\sin\theta)=2ab\sin(2\theta)$, which has maximum value $2ab$. Given that $a=10$, then $b=5 \Rightarrow$ maximum area is 100.

Locus of intersection point of perpendicular tangents is $x^2+y^2=10^2+5^2$ or $x^2+y^2=125$ (director circle).

For Problems 7–9

7. b., 8. a., 9. c.

$$21x^2-6xy+29y^2+6x-58y-151=0$$

$$3(x-3y+3)^2+2(3x+y-1)^2=180$$

$$\Rightarrow \frac{(x-3y+3)^2}{60}+\frac{(3x+y-1)^2}{90}=1$$

$$\Rightarrow \left(\frac{x-3y+3}{\sqrt{1+3^2}\sqrt{6}}\right)^2+\left(\frac{3x+y-1}{\sqrt{1+3^2}3}\right)^2=1$$

Thus C is an ellipse whose lengths of axes are 6 and $2\sqrt{6}$.

The minor and the major axes are $x-3y+3=0$ and $3x+y-1=0$, respectively.

Their point of intersection gives the centre of the conic. Hence, centre $\equiv (0, 1)$.

For Problems 10–12

10. b., 11. d., 12. a.

Let the ellipse be $\frac{x^2}{a^2}+\frac{y^2}{b^2}=1$.

The line $y=mx\pm\sqrt{a^2m^2+b^2}$ touches the ellipse for all m.

Hence, it is identical with

$$y=-\frac{2px}{\sqrt{1-p^2}}+\frac{1}{\sqrt{1-p^2}}$$

Hence, $m=-\frac{2p}{\sqrt{1-p^2}}$

and $a^2m^2+b^2=\frac{1}{1-p^2}$

$$\Rightarrow a^2\frac{4p^2}{1-p^2}+b^2=\frac{1}{1-p^2}$$

$$\Rightarrow p^2(4a^2-b^2)+b^2-1=0$$

This equation is true for all real p if $b^2=1$ and $4a^2=b^2$

$$\Rightarrow b^2=1 \text{ and } a^2=\frac{1}{4}$$

Therefore, the equation of the ellipse is

$$\frac{x^2}{1/4}+\frac{y^2}{1}=1.$$

If e is its eccentricity, then

$$\frac{1}{4}=1-e^2 \Rightarrow e^2=\Rightarrow e=\frac{\sqrt{3}}{2}$$

$be=\frac{\sqrt{3}}{2}$, hence foci are $\left(0, \pm\frac{\sqrt{3}}{2}\right)$.

Equation of director circle is $x^2+y^2=\frac{5}{4}$.

For Problems 13–15

13. a., 14. b., 15. a.

Sol.

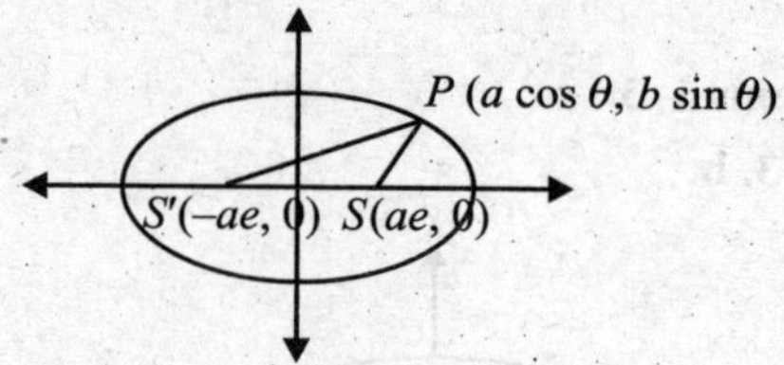

Fig. 4.32

Let the coordinates of P be $(a\cos\theta, b\sin\theta)$.

Here SP = focal distance of the point $P=a-ae\cos\theta$

$S'P=a+ae\cos\theta$ and $SS'=2ae$.

If (h, k) are the coordinates of the incentre of $\Delta PSS'$, then

$$h=\frac{2ae(a\cos\theta)+a(1-e\cos\theta)(-ae)+a(1+e\cos\theta)ae}{2ae+a(1-e\cos\theta)+a(1+e\cos\theta)}$$

$$\Rightarrow h=ae\cos\theta \quad (1)$$

and

$$k=\frac{2ae(b\sin\theta)+a(1-e\cos\theta)\times 0+a(1+e\cos\theta)\times 0}{2ae+a(1-e\cos\theta)+a(1+e\cos\theta)}$$

$$\Rightarrow k=\frac{eb\sin\theta}{(e+1)} \quad (2)$$

Eliminating θ from (1) and (2), we get

$$\frac{x^2}{a^2e^2}+\frac{y^2}{\left(\frac{be}{e+1}\right)^2}=1,$$

which clearly represents an ellipse. Let e_1 be its eccentricity. Then

$$\frac{b^2e^2}{(e+1)^2} = a^2e^2(1-e_1^2)$$

$$\Rightarrow e_1^2 = 1 - \frac{b^2}{a^2(e+1)^2}$$

$$\Rightarrow e_1^2 = 1 - \frac{1-e^2}{(e+1)^2} = 1 - \frac{1-e}{1+e}$$

$$\Rightarrow e_1^2 = \frac{2e}{e+1} \Rightarrow e_1 = \sqrt{\frac{2e}{e+1}}$$

Maximum area of rectangle is $2(ae)\left(\frac{be}{e+1}\right) = \frac{2abe^2}{e+1}$.

For Problems 16–18

16. a., 17. d., 18. c.

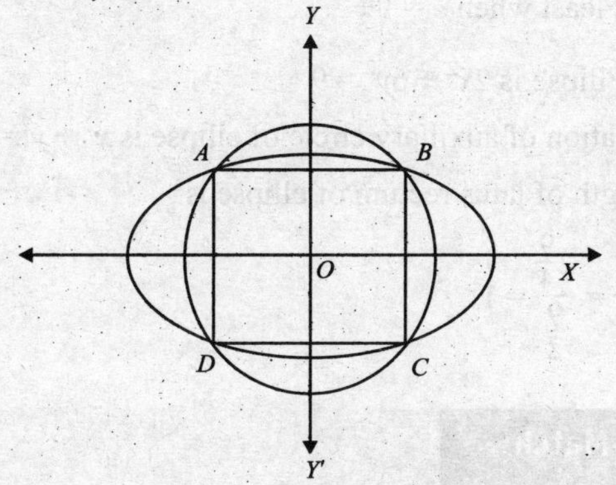

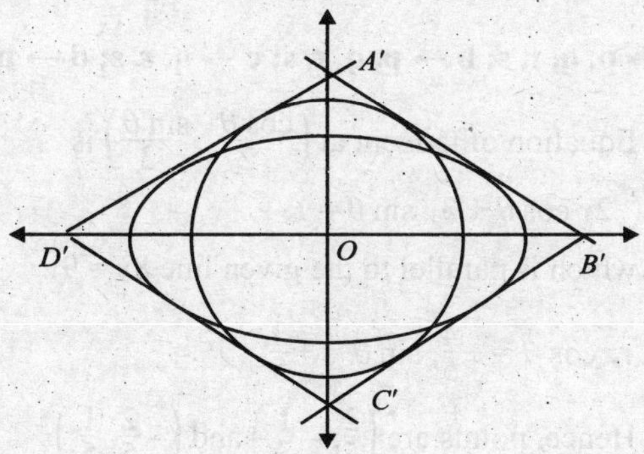

Fig. 4.33

16. a. Solving the curves given (eliminating x^2), we have

$$\frac{r^2-y^2}{16} + \frac{y^2}{9} = 1$$

$$\Rightarrow y^2 = \frac{144-9r^2}{7}$$

Solving the curves given (eliminating y^2), we have

$$\frac{x^2}{16} + \frac{r^2-x^2}{9} = 1$$

$$\Rightarrow x^2 = \frac{16r^2-144}{7}$$

If $ABCD$ is a square, then

$x^2 = y^2$

or $\frac{144-9r^2}{7} = \frac{16r^2-144}{7}$

or $25r^2 = 288$

or $r = \frac{12}{5}\sqrt{2}$

17. d. Tangents of slope m to circle and ellipse is

$y = mx \pm \sqrt{r^2m^2 + r^2}$

and $y = mx \pm \sqrt{16m^2 + 9}$, respectively.

For common tangent,

$r^2m^2 + r^2 = 16m^2 + 9$

Also if $A'\,B'\,C'\,D'$ is a square, then

$m = \pm 1$

$\Rightarrow r^2 + r^2 = 25$

$\Rightarrow r = 5/\sqrt{2}$

18. c. If $A'B'C'D'$ is a square, then tangents are $y = \pm x \pm 5$, for which diagonal length $A'C' = 10$.

Then area of $\Delta A'B'C'$ is 25π.

Also area of circle C_1 is $25\pi/2$. Hence, the required ratio is 1/2.

For Problems 19–21

19. c., 20. b., 21. d.

19. c. $\lambda x - y + 2(1+\lambda) = 0$

$\Rightarrow \lambda(x+2) - (y-2) = 0$

This line passes through $(-2, 2)$. So

$\mu x - y + 2(1-\mu) = 0$

$\Rightarrow \mu(x-2) - (y-2) = 0$

This line passes through $(2, 2)$.

Clearly these represent the foci of the ellipse. So $2ae = 4$.

The circle $x^2 + y^2 - 4y - 5 = 0$ or $x^2 + (y-2)^2 = 9$ represents auxiliary circle. Thus $a^2 = 9 \Rightarrow e = \frac{2}{3}$ and $b^2 = 5$.

20. c. Required area $= \frac{1}{2}(2ae)(b) = 3\frac{2}{3}\sqrt{5} = 2\sqrt{5}$.

For Problems 22–24

22. b., 23. d., 24. c.

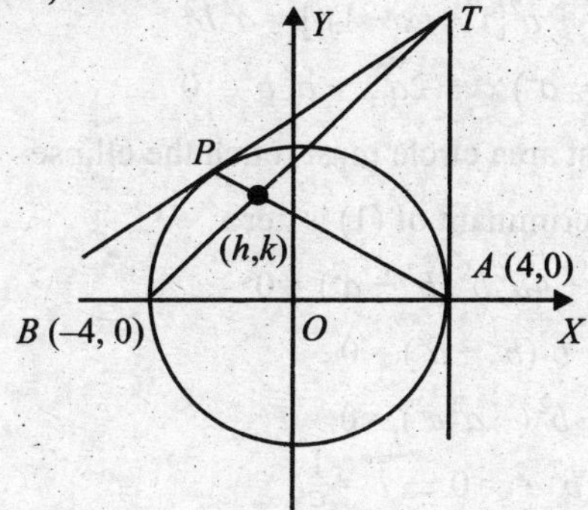

Fig. 4.34

Tangents at P (4 cos θ, 4 sin θ) to $x^2 + y^2 = 16$ is

$$x \cos \theta + y \sin \theta = 4 \quad (1)$$

Equation of AP is

$$y = \frac{\sin \theta}{\cos \theta - 1}(x - 4) \quad (2)$$

From (1), coordinates of the point T are given by

$$\left(4, \frac{4(1 - \cos \theta)}{\sin \theta}\right).$$

Equation of BT is

$$y = \frac{1 - \cos \theta}{2 \sin \theta}(x + 4) \quad (3)$$

Let (h, k) be the point of intersection of the lines (2) and (3). Then we have

$$k^2 = -\frac{1}{2}(h^2 - 16)$$

$$\Rightarrow \frac{h^2}{16} + \frac{k^2}{8} = 1$$

Therefore, locus of (h, k) is

$$\frac{x^2}{16} + \frac{y^2}{8} = 1,$$

which is an ellipse with eccentrically $e = \frac{1}{\sqrt{2}}$.

Sum of focal distances of any point $= 2a = 8$.

Considering circle $x^2 + y^2 = a^2$, we find that the eccentricity of the ellipse is $1/\sqrt{2}$ which is constant and does not change by changing the radius of the circle.

For Problems 25–27

25. a, 26. c., 27. b

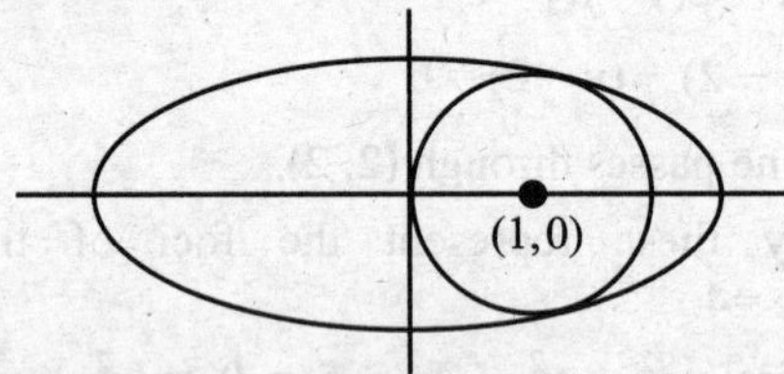

Fig. 4.35

Solving both equations, we have

$$\frac{x^2}{a^2} + \frac{1 - (x - 1)^2}{b^2} = 1$$

$$\Rightarrow b^2x^2 + a^2[1 - (x - 1)^2] = a^2 b^2$$

$$\Rightarrow (b^2 - a^2)\, x^2 + 2a^2x - a^2 b^2 = 0 \quad (1)$$

for least area circle must touch the ellipse

$\Rightarrow$ Discriminant of (1) is zero

$$\Rightarrow 4a^4 + 4a^2 b^2 (b^2 - a^2) = 0$$

$$\Rightarrow a^2 + b^2(b^2 - a^2) = 0$$

$$\Rightarrow a^2 + b^2(- a^2 e^2) = 0$$

$$\Rightarrow 1 - b^2 e^2 = 0 \Rightarrow b = \frac{1}{e}$$

Also $a^2 = \frac{b^2}{1 - e^2} = \frac{1}{e^2(1 - e^2)}$

$$\Rightarrow a = \frac{1}{e\sqrt{1 - e^2}}$$

Let S be the area of the ellipse. Then

$$S = \pi ab = \frac{\pi}{e^2 \sqrt{1 - e^2}}$$

$$= \frac{\pi}{\sqrt{e^4 - e^6}}$$

Area is minimum if $f(e) = e^4 - e^6$ is maximum, when $f'(e) = 4e^3 - 6e^5 = 0$,

or $e = \sqrt{2/3}$ (which is point of maxima for $f(e)$).

S is least when $e = \sqrt{\frac{2}{3}}$

$\Rightarrow$ Ellipse is $2x^2 + 6y^2 = 9$.

Equation of auxiliary circle of ellipse is $x^2 + y^2 = 4.5$.

Length of latus rectum of ellipse is

$$\frac{2b^2}{a} = \frac{2\frac{9}{4}}{\frac{9}{2}} = 1$$

Matrix-Match Type

1. a. → p, q, r, s; b → p, q, r, s; c → q, r, s; d → p.

a. Equation of tangent at $\left(\frac{\cos \theta}{2}, \frac{\sin \theta}{3}\right)$ is

$$2x \cos \theta + 3y \sin \theta = 1,$$

which is parallel to the given line $8x = 9y$.

$$\therefore \cos \theta = \pm\frac{4}{5}, \sin \theta = \mp\frac{3}{5}$$

Hence, points are $\left(\frac{2}{5}, -\frac{1}{5}\right)$ and $\left(-\frac{2}{5}, \frac{1}{5}\right)$.

Distance between the points is

$$\sqrt{\frac{16}{25} + \frac{4}{25}} = \frac{2}{\sqrt{5}},$$

which is less than 1.

b. The given equation is

$$\frac{(x + 1)^2}{9} + \frac{(y + 2)^2}{25} = 1$$

$$\Rightarrow e^2 = 1 - \frac{9}{25} = \frac{16}{25} \Rightarrow e = \frac{4}{5}$$

Hence, the foci are $S, S' \equiv (-1, -2 \pm 4) \equiv S(-1, 2)$ and $S'(-1, -6)$.

The required sum of distances $= 2 + 6 = 8$.

c. Equation of normal at $(3\cos\theta, 2\sin\theta)$ is

$3x\sec\theta - 2y\operatorname{cosec}\theta = 5$,

which is parallel to the given line $2x + y = 1$. Therefore,

$\cos\theta = \mp\frac{3}{5}, \sin\theta = \pm\frac{4}{5}$

Hence, points are $\left(\frac{-9}{5}, \frac{8}{5}\right)$ and $\left(\frac{9}{5}, -\frac{8}{5}\right)$.

The required sum of distances $= \frac{16}{5}$.

d. Consider any point $(t, t+2)$, $t \in R$, on the line $x - y + 2 = 0$.

The chord of contact of ellipse w.r.t. this point is

$xt + 2y(t+2) = 2$

$\Rightarrow (4y - 2) + t(x + 2y) = 0$

This line passes through point of intersection of lines $4y - 2 = 0$ and $x + 2y = 0$. Therefore, $x = -1$.

Hence, the point is $(-1, 1/2)$, whose distance from $(2, 1/2)$ is 3.

2. a → r, s; b → p, q, r, s; c → r, s; d → r, s.

Equation of any tangent to

$$\frac{x^2}{a^2} + \frac{y^2}{b^2} = 1 \quad (1)$$

is given by

$y = mx + \sqrt{a^2m^2 + b^2}$.

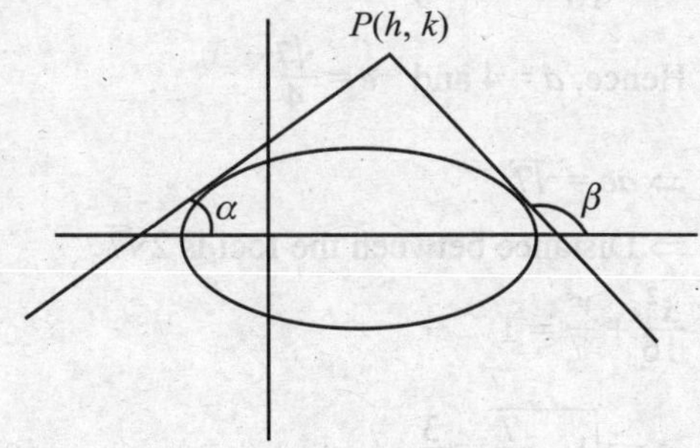

Fig. 4.36

Since it passes through $P(h, k)$,

$k = mh + \sqrt{a^2m^2 + b^2}$

$$\Rightarrow m^2(h^2 - a^2) - 2kmh + (k^2 - b^2) = 0 \quad (2)$$

As (2) is quadratic in m, having two roots m_1 and m_2 (say). Therefore,

$$m_1 + m_2 = \frac{2hk}{h^2 - a^2}, m_1m_2 = \frac{k^2 - b^2}{h^2 - a^2} \quad (3)$$

$$\Rightarrow \tan(\alpha + \beta) = \frac{\tan\alpha + \tan\beta}{1 - \tan\alpha\tan\beta}$$

$$= \frac{m_1 + m_2}{1 - m_1m_2}$$

$$= \frac{\frac{2hk}{h^2 - a^2}}{1 - \frac{k^2 - b^2}{h^2 - a^2}}$$

$$= \frac{2hk}{h^2 - k^2}$$

a. $\alpha + \beta = \frac{c\pi}{2}$

When c is even,

$m_1 + m_2 = 0$

$\Rightarrow \frac{2kh}{h^2 - a^2} = 0 \Rightarrow 2kh = 0$

$\Rightarrow xy = 0$, which is the equation of a pair of straight lines.

When c is odd,

$1 - m_1m_2 = 0$

$\Rightarrow \frac{k^2 - b^2}{h^2 - b^2} = 1$

Therefore, the locus of (h, k) is $y^2 - b^2 = x^2 - a^2$, which is a hyperbola.

b. $m_1m_2 = c$

$\Rightarrow \frac{k^2 - b^2}{h^2 - a^2} = c.$

When $c = 0$, $k = \pm b$, the locus is a pair of straight lines.

When $c = 1$, $h^2 - k^2 = a^2 - b^2$, the locus is a hyperbola.

When $c = -1$, $h^2 + k^2 = a^2 + b^2$, the locus is a circle.

When $c = -2$, $2h^2 + k^2 = 2a^2 + b^2$, the locus is an ellipse.

c. $\tan\alpha + \tan\beta = c$

$\Rightarrow m_1 + m_2 = c$

$\Rightarrow \frac{2kh}{h^2 - a^2} = c$

When $c = 0$, $kh = 0$, the locus is a pair of straight lines.

When $c \neq 0$ $c(h^2 - a^2) - 2kh = 0$

Locus of (h, k) is

$cx^2 - 2xy - ca^2 = 0$

$\Delta = -ca^2 \neq 0$

Also,

$h^2 - ab = 1 > 0$

Therefore, the locus is a hyperbola for $c \neq 0$.

d. $\cot\alpha + \cot\beta = c$

$\Rightarrow \frac{1}{m_1} + \frac{1}{m_2} = c$

$$\Rightarrow \frac{m_1 + m_2}{m_1 m_2} = c$$

$$\Rightarrow \frac{2kh}{k^2 - b^2} = c$$

$$\Rightarrow c(k^2 - b^2) - 2kh = 0$$

When $c = 0$, locus is a pair of straight lines.

When $c \neq 0$, locus is a hyperbola (as in the previous case c).

3. a. → p, r; b → r, s; c → p; d → q.

Tangent to ellipse at $P(\phi)$ is $\frac{x}{4}\cos\phi + \frac{y}{2}\sin\phi = 1$.

It must pass through the centre of the circle. Hence,

$$\frac{4}{4}\cos\phi + \frac{2}{2}\sin\phi = 1$$

$$\Rightarrow \cos\phi + \sin\phi = 1$$

$$\Rightarrow 1 + \sin 2\phi = 1$$

or $\sin 2\phi = 0$

$$\Rightarrow 2\phi = 0 \text{ or } \pi$$

$$\Rightarrow \frac{\phi}{2} = 0 \text{ or } \frac{\pi}{4}$$

b. Consider any point $P(\sqrt{6}\cos\theta, \sqrt{2}\sin\theta)$ on the ellipse $\frac{x^2}{6} + \frac{y^2}{2} = 1$.

Given that $OP = 2$. So

$$6\cos^2\theta + 2\sin^2\theta = 4$$

$$\Rightarrow 4\cos^2\theta = 2$$

$$\Rightarrow \cos\theta = \pm\frac{1}{\sqrt{2}}$$

$$\Rightarrow \theta = \frac{\pi}{4} \text{ or } \frac{5\pi}{4}$$

c. Solving the equation of ellipse and parabola (eliminating x^2), we have

$$y - 1 + 4y^2 = 4$$

$$\Rightarrow 4y^2 + y - 5 = 0$$

$$\Rightarrow (4y + 5)(y - 1) = 0$$

$$\Rightarrow y = 1, x = 0$$

The curves touch at (0, 1). So the angle of intersection is 0.

d. The normal at $P(a\cos\theta, b\sin\theta)$ is

$$\frac{ax}{\cos\theta} - \frac{bx}{\sin\theta} = a^2 - b^2,$$

where $a^2 = 14$, $b^2 = 5$.

It meets the curve again at $Q(2\theta)$, i.e. $(a\cos 2\theta, b\sin 2\theta)$. Hence,

$$\frac{a}{\cos\theta} a\cos 2\theta - \frac{b}{\sin\theta}(b\sin 2\theta) = a^2 - b^2$$

$$\Rightarrow \frac{14}{\cos\theta}\cos 2\theta - \frac{5}{\sin\theta}(\sin 2\theta) = 14 - 5$$

$$\Rightarrow 28\cos^2\theta - 14 - 10\cos^2\theta = 9\cos\theta$$

$$\Rightarrow 18\cos^2\theta - 9\cos\theta - 14 = 0$$

$$\Rightarrow (6\cos\theta - 7)(3\cos\theta - 2) = 0$$

$$\Rightarrow \cos\theta = -\frac{2}{3}$$

4. a → s; b → p; c → q; d → r.

a. The locus is

$$\Rightarrow \frac{x^2}{16} + \frac{y^2}{36} = 1$$

$$\Rightarrow e = \sqrt{1 - \frac{16}{36}} = \sqrt{\frac{20}{36}} = \frac{\sqrt{5}}{3}$$

$$\Rightarrow 3e = \sqrt{5}$$

b. $3(x^2 + 2x + 1) + 2(y^2 - 2y + 1) = 3 + 2 + 1$

$$\Rightarrow \frac{(x+1)^2}{2} + \frac{(y-1)^2}{3} = 1$$

$$\Rightarrow e = \sqrt{1 - \frac{2}{3}} = \frac{1}{\sqrt{3}}$$

$$\therefore a = \sqrt{3}, b = \sqrt{2}$$

$$\therefore \text{Area} = \frac{1}{2} \times 2\sqrt{3} \times \sqrt{2} = \sqrt{6}$$

c. Eliminating θ from $x = 1 + 4\cos\theta$, $y = 2 + 3\sin\theta$, we have

$$\frac{(x-1)^2}{16} + \frac{(y-2)^2}{9} = 1$$

Hence, $a = 4$ and $e = \frac{\sqrt{7}}{4}$

$$\Rightarrow ae = \sqrt{7}$$

⇒ Distance between the foci is $2\sqrt{7}$.

d. $\frac{x^2}{16} + \frac{y^2}{7} = 1$

$$e = \sqrt{1 - \frac{7}{16}} = \frac{3}{4}$$

One end of latus rectum is

$$\left(ae, \frac{b^2}{a}\right) = \left(3, \frac{7}{4}\right).$$

Therefore, equation of tangent is

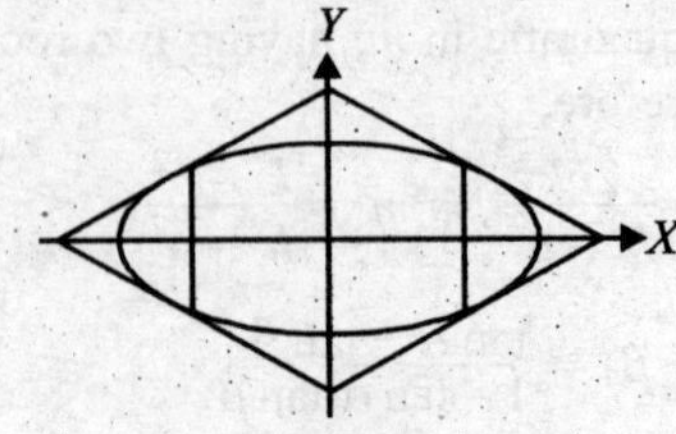

Fig. 4.37

$$\frac{3x}{16}+\frac{7}{4}\frac{y}{7}=1$$

or $\dfrac{3x}{16}+\dfrac{y}{4}=1$

It meets the x-axis at $\left(\frac{16}{3}, 0\right)$ and the y-axis at (0, 4).

Hence, the area of quadrilateral is

$$2\times\frac{16}{3}\times 4=\frac{128}{3}.$$

5. a → q; b → r; c → r; d → p.

a. Points are $O(0, 0)$, $P(3, 4)$, and $Q(6, 8)$

$$2a = OP + OQ$$
$$= 5 + 10 = 15$$
$$\Rightarrow a=\frac{15}{2}$$

Also distance between foci is

$$2ae=\sqrt{(6-3)^2+(8-4)^2}=5$$
$$\Rightarrow e=\frac{1}{3}$$
$$\Rightarrow b^2=\frac{225}{4}\left(1-\frac{1}{9}\right)=50$$
$$\Rightarrow b=5\sqrt{2}$$
$$\Rightarrow 2b=10\sqrt{2}$$

b. We know that $\dfrac{1}{SP}+\dfrac{1}{SQ}=\dfrac{2a}{b^2}$

$$\Rightarrow \frac{1}{2}+\frac{1}{SQ}=\frac{10}{16}$$
$$\Rightarrow SQ=8$$
$$\Rightarrow PQ=10$$

c. If the line $y = x + k$ touches the ellipse $9x^2 + 16y^2 = 144$, then

$$k^2 = 16(1)^2 + 9$$
$$\Rightarrow k = \pm 5.$$

d. Sum of the distances of a point on the ellipse from the foci $= 2a = 8$.

6. a → q; b → q; c → s; d → p.

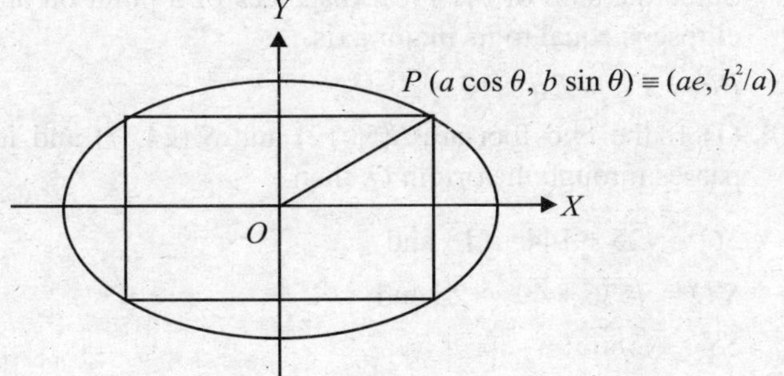

Fig. 4.38

a. Let one of the vertices of the rectangle be $P(a\cos\theta, b\sin\theta)$. Then its area $A = (2a\cos\theta)(2b\sin\theta) = 2ab\sin 2\theta$. Hence, $A_{max} = 2ab$.
Now area of rectangle formed by extremities of LR $= (2ae)(2b^2/a) = 4eb^2$.

Given that

$$2ab = 4eb^2 \Rightarrow \frac{2b}{a}e=1$$
$$\Rightarrow \frac{4b^2}{a^2}e^2=1$$
$$\Rightarrow 4(1-e^2)e^2=1$$
$$\Rightarrow 4e^4-4e^2+1=0$$
$$\Rightarrow (2e^2-1)^2=0$$
$$\Rightarrow e=\frac{1}{\sqrt{2}}$$

b.

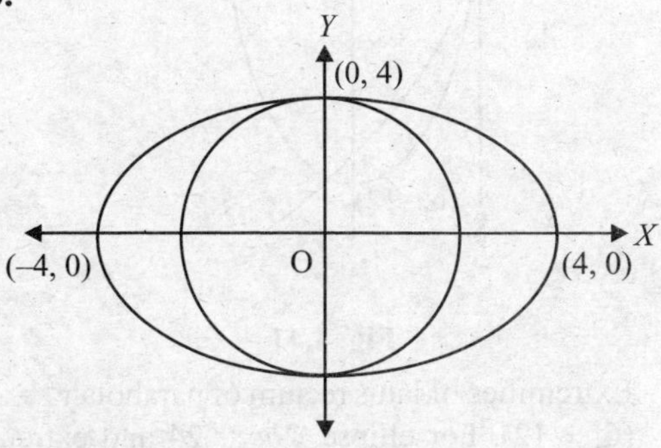

Fig. 4.39

For ellipse, distance between the foci is $2ae = 8$ and length of semi-minor axis, is $b = 4$. Now,

$$b^2 = a^2 - a^2e^2$$
$$\Rightarrow 16 = a^2 - 16$$
$$\Rightarrow a^2 = 32$$
$$\Rightarrow e=\sqrt{1-\frac{16}{32}}=\frac{1}{\sqrt{2}}$$

c. Normal at point $P(6, 2)$ to the ellipse passes through its focus $Q(5, 2)$. Then P must be extremity of the major axis. Now $ae = QR = 1$ (where R is centre) and $a - ae = 1$.

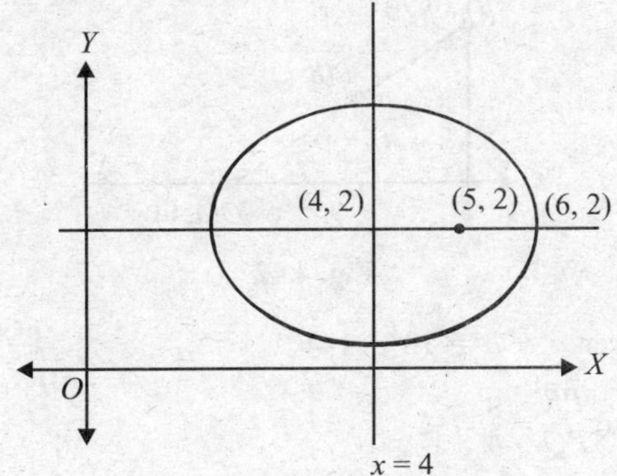

Fig. 4.40

$\therefore a = 2$

$b^2 = a^2 - a^2e^2 = 4 - 1 = 3$

$\Rightarrow e = \sqrt{1 - \frac{3}{4}} = \frac{1}{2}$

d.

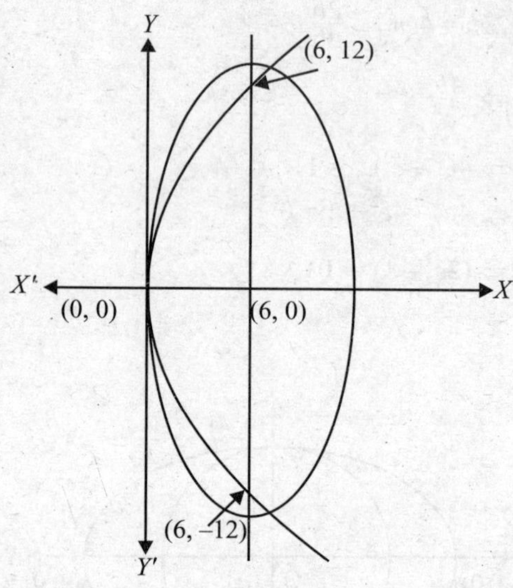

Fig. 4.41

Extremities of latus rectum of parabola $y^2 = 24x$ are $(6, \pm 12)$. For ellipse, $2be = 24$ and extremity of minor axis is (0, 0). Hence, $a = 6$.

Now, $a^2 = b^2 - b^2e^2$

$\Rightarrow b^2 = 36 + 144 = 180$

$\Rightarrow e^2 = \sqrt{1 - \frac{36}{180}} = \sqrt{1 - \frac{1}{5}} = \frac{2}{\sqrt{5}}$.

Integer Type

1. (2)

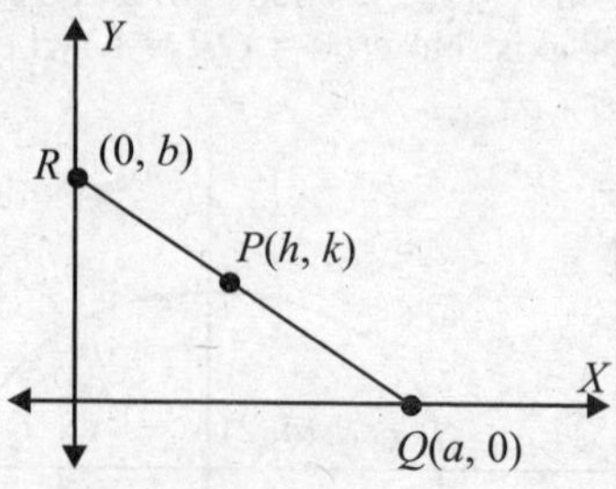

Fig. 4.42

Given $a^2 + b^2 = 12^2 = 144$ (1)

Also $\frac{PR}{PQ} = \frac{8}{4} = \frac{2}{1}$

$\Rightarrow h = \frac{2a}{3}$ and $k = \frac{b}{3}$

$\Rightarrow a = \frac{3h}{2}$ and $b = 3k$

From (1), we get

$\frac{9h^2}{4} + 9k^2 = 144$

$\Rightarrow \frac{x^2}{64} + \frac{y^2}{16} = 1$

2. (1) $4 \tan\frac{B}{2} \tan\frac{C}{2} = 1$

$\Rightarrow \sqrt{\frac{(s-c)(s-a)}{s(s-b)} \times \frac{(s-a)(s-b)}{s(s-c)}} = \frac{1}{4}$

$\Rightarrow \frac{s-a}{s} = \frac{1}{4}$

$\Rightarrow \frac{2s-a}{a} = \frac{5}{3}$

$\Rightarrow b + c = \frac{5}{3} \times 6 = 10$ $(\because a = BC = 6)$

Thus the sum of the distance of the variable point A from the two given fixed points B and C is always 10. Therefore, the equation of the locus of A is an ellipse. Also its centre is the midpoint of BC, which is (5, 0).

So $2ae = BC = 6$. Since the sum of the focal distance from any point on the ellipse is 10,

$e = \frac{6}{10} = \frac{3}{5}$.

Length of semi-major axis = 5.

Length of semi-minor axis = 4.

3. (5) We have $16x^2 + 25y^2 = 400$

$\Rightarrow \frac{x^2}{25} + \frac{y^2}{16} = 1 \Rightarrow a^2 = 25$ and $b^2 = 16$

Then eccentricity is given by

$e^2 = 1 - \frac{b^2}{a^2} = 1 - \frac{16}{25} = \frac{9}{25}$

$\Rightarrow e = 3/5$

So the coordinates of the foci are $(\pm ae, 0)$ or $(\pm 3, 0)$.

Thus F_1 and F_2 are the foci of the ellipse.

Since the sum of the focal distances of a point on an ellipse is equal to its major axis,

$PF_1 + PF_2 = 2a = 10$.

4. (1) If the two foci are $S(5, 12)$ and $S^1(24, 7)$ and it passes through the origin O, then

$SO = \sqrt{25 + 144} = 13$ and

$S'O = \sqrt{576 + 49} = 25$ and

$SS' = \sqrt{386}$

If the conic is an ellipse, then $SO + S'O = 2a$ and $SS' = 2ae$. Therefore,

$$e = \frac{SS'}{S'O + SO} = \frac{\sqrt{386}}{38}$$

5. (2)

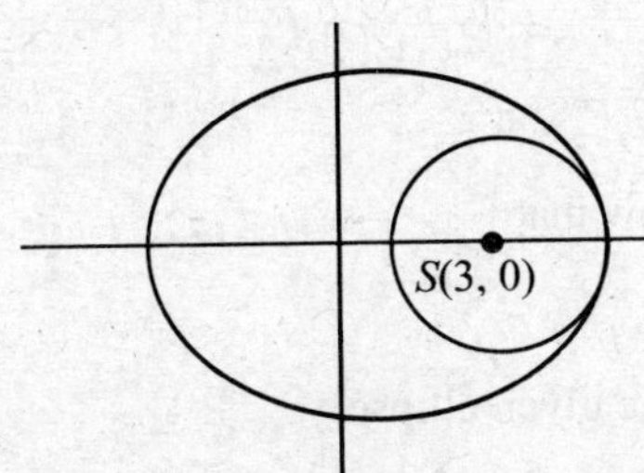

Fig. 4.43

Given ellipse

$$\frac{x^2}{25} + \frac{y^2}{16} = 1 \quad (1)$$

$$\Rightarrow \quad e^2 = 1 - \frac{16}{25} \Rightarrow e = \frac{3}{5}$$

Then foci are $(\pm ae, 0) \equiv (\pm 3, 0)$.

Now circle having centre (6, 0) is $(x - 6)^2 + y^2 = r^2$ (2)

Eliminating y^2 from (1) and (2), we get

$$\frac{x^2}{25} + \frac{r^2 - (x-6)^2}{16} = 1$$

or $16x^2 + 25r^2 - 25(x^2 - 12x + 36) = 400$

or $-9x^2 + 300x + 25r^2 - 1300 = 0$

Since circle touches ellipse, above equation has equal roots. Hence, $D = 0 \Rightarrow 90000 + 36(25r^2 - 1300) = 0$.

r is not possible.

Then circle will touch the ellipse at the end of the major axis (5, 0). Hence, radius is 2.

6. (0) The given equation is

$$\left(x - \frac{1}{5}\right)^2 + \left(y - \frac{2}{5}\right)^2 = (\lambda - 1)\left[\frac{3x + 4y - 1}{\sqrt{3^2 + 4^2}}\right]^2$$

or $$\sqrt{\left(x - \frac{1}{5}\right)^2 + \left(y - \frac{2}{5}\right)^2} = |\lambda - 1| \frac{|3x + 4y - 1|}{\sqrt{3^2 + 4^2}}$$

For the ellipse, $SP = ePM$, where $0 < e < 1$. So

$0 < |\lambda - 1| < 1$

$\Rightarrow \lambda \in (0, 2) - \{1\}$

7. (6) Let $PQRS$ be a rectangle, where P is $(3\cos\theta, \sin\theta)$.

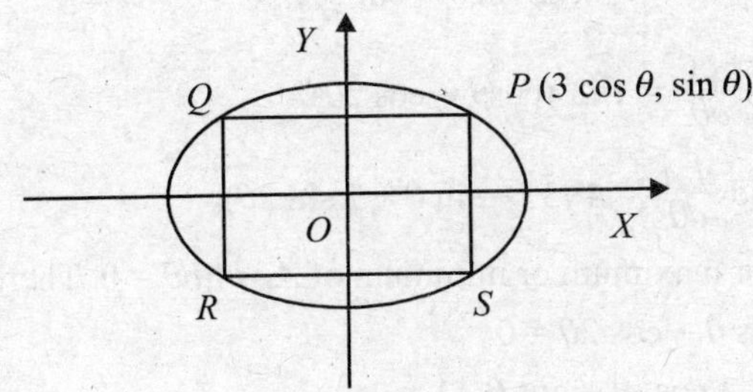

Fig. 4.44

Area of rectangle $= 4(3\cos\theta)(\sin\theta)$

$= 6\sin 2\theta$

This is maximum when $\sin 2\theta = 1$.

Hence, maximum area = 6.

8. (2) Let the points of intersection of the line and the ellipse be $(a\cos\theta, b\sin\theta)$ and $\left(a\cos\left(\frac{\pi}{2} + \theta\right), b\sin\left(\frac{\pi}{2} + \theta\right)\right)$.

Since they lie on the given line $lx + my + n = 0$,

$la\cos\theta + mb\sin\theta + n = 0$

$\Rightarrow la\cos\theta + mb\sin\theta = -n$

and $-la\sin\theta + mb\cos\theta + n = 0$

$\Rightarrow la\sin\theta - mb\cos\theta = n.$

Squaring and adding, we get

$a^2l^2 + b^2m^2 = 2n^2$

$$\Rightarrow \frac{a^2l^2 + b^2m^2}{n^2} = 2.$$

9. (9) Let APQ be an isosceles triangle inscribed in the ellipse $(x^2/16) + (y^2/3) = 1$ with the vertex A at the extremity (4, 0) of the major axis. If the coordinates of P are $(4\cos\theta, \sqrt{3}\sin\theta)$, then those of Q will be $(4\cos\theta, -\sqrt{3}\sin\theta)$.

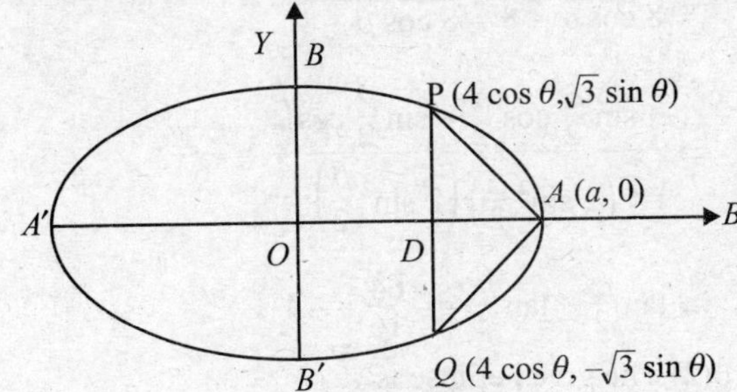

Fig. 4.45

So the area of ΔAPQ is given by

$A = \frac{1}{2} PQ \times AD$

$= PD \times AD$

$= PD \times (OA - OD)$

$= \sqrt{3}\sin\theta\,(4 - 4\cos\theta)$

$= \frac{1}{2} \times 4\sqrt{3}\ (2 \sin\theta - \sin 2\theta),\ 0 < \theta < \pi.$

So $\frac{dA}{d\theta} = 4\sqrt{3}\ (\cos\theta - \cos 2\theta),$

and $\frac{d^2A}{d\theta} = 4\sqrt{3}\ (-\sin\theta + 2\sin 2\theta).$

For maximum or minimum of A, $dA/d\theta = 0$. Therefore,

$\cos\theta - \cos 2\theta = 0$

$\Rightarrow 2\cos^2\theta - \cos\theta - 1 = 0$

$\Rightarrow (2\cos\theta + 1)(\cos\theta - 1) = 0$

$\Rightarrow \cos\theta = -1/2$ $\quad [\because \cos\theta \neq 1 \text{ as } \theta \neq 0]$

$\Rightarrow \theta = 2\pi/3$

When $\theta = 2\pi/3$,

$\frac{d^2A}{d\theta^2} = -\frac{1}{2}\ (3\sqrt{3})ab$ (negative)

Therefore, A is maximum when $\theta = 2\pi/3$.

Hence, maximum area

$A = \frac{1}{2} \times 4\sqrt{3}\left[2\sin\frac{2\pi}{3} - \sin\frac{4\pi}{3}\right]$

$= 9$

10. (4)

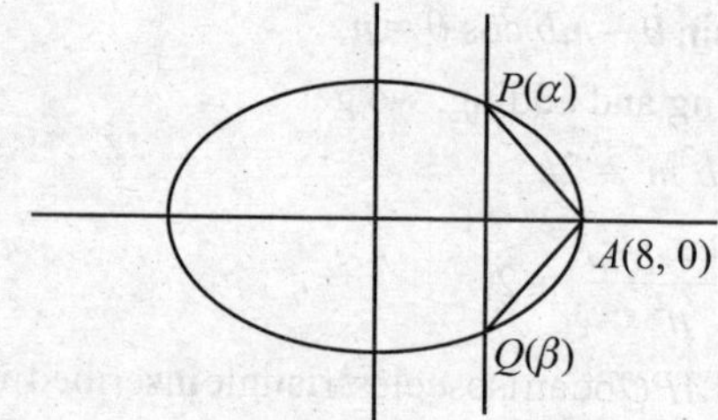

Fig. 4.46

$m_{AP} \times m_{QA} = -1$

$\Rightarrow \frac{4\sin\alpha}{8\cos\alpha - 8} \times \frac{4\sin\beta}{8\cos\beta - 8} = -1$

$\Rightarrow \frac{4\sin\frac{\alpha}{2}\cos\frac{\alpha}{2} \times \sin\frac{\beta}{2}\cos\frac{\beta}{2}}{\left(2\sin^2\frac{\alpha}{2}\right)\left(2\sin^2\frac{\beta}{2}\right)} = \frac{-16}{64}$

$\Rightarrow \tan\frac{\alpha}{2} \times \tan\frac{\beta}{2} = -\frac{64}{16} = -4$

11. (2) The given ellipse is

$\frac{x^2}{25} + \frac{y^2}{16} = 1.$

We know that

$\frac{1}{SP} + \frac{1}{SQ} = \frac{2a}{b^2}.$

Then

$\frac{1}{8} + \frac{1}{SQ} = \frac{2 \times (5)}{16} = \frac{5}{8} \Rightarrow SQ = 2$

12. (8) We known that the product of the perpendiculars from the two foci of an ellipse on any tangent is equal to the square of the semi-minor axis.

Thus $(S_1F_1)(S_2F_2) = 16$

Now A.M. ≥ G.M. So

$\Rightarrow \frac{S_1F_1 + S_2F_2}{2} \geq \sqrt{S_1F_1 \times S_2F_2}$

$\Rightarrow S_1F_1 + S_2F_2 \geq 8$

13. (8) We know that

$\frac{1}{SP} + \frac{1}{SQ} = \frac{2a}{b^2}$

Then for the given ellipse,

$\frac{1}{16} + \frac{1}{SQ} = 2 \times \frac{5}{16} = \frac{5}{8} \Rightarrow \frac{1}{SQ} = \frac{9}{16}$

Now $SQ + SQ' = 2a = 10$

$\Rightarrow SQ' = 10 - \frac{16}{9} = \frac{74}{9}$

14. (4)

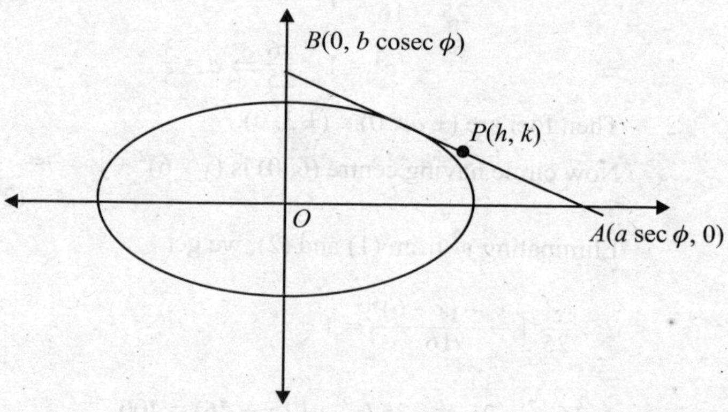

Fig. 4.47

The equation of the tangent at any point ϕ of the ellipse is

$\frac{x}{a}\cos\phi + \frac{y}{b}\sin\phi = 1.$

It meets the axes at the point $A(a \sec\phi, 0)$ and $B(0, b \operatorname{cosec}\phi)$.

Let $P(h, k)$ be the midpoint of AB.

Then

$h = \frac{a\sec\phi}{2}, k = \frac{b\operatorname{cosec}\phi}{2}$. So

$\cos\phi = \frac{a}{2h}$ and $\sin\phi = \frac{b}{2k}$

$\Rightarrow \frac{a^2}{4h^2} + \frac{b^2}{4k^2} = 1$

Thus the locus of (h, k) is

$\frac{a^2}{x^2} + \frac{b^2}{y^2} = 4.$

15. (5) The combined equation of the pair of tangents drawn from (1, 2) on the ellipse $3x^2 + 2y^2 = 5$ is $(3x^2 +$

$2y^2 - 5)(3 + 8 - 5) = (3x + 4y - 5^2)$ [Using $SS' = T^2$]
$\Rightarrow 9x^2 - 24xy - 4y^2 + \cdots = 0$

If the angle between these lines is θ, then

$\tan\theta = \dfrac{2\sqrt{h^2 - ab}}{a + b}$, where $a = 9, h = -12, b = -4$

$\Rightarrow \tan\theta = \dfrac{12}{\sqrt{5}}$

16. (6) For the given ellipse, the equation of a tangent with slope m is

$y = mx + \sqrt{18m^2 + 32}$.

For $m = -4/3$, the tangent is

$y = -\dfrac{4}{3}x + \sqrt{18 \times \dfrac{16}{9} + 3^2} = -\dfrac{4}{3}x + 8$ or $4x + 3y = 24$

This intersects the major and minor axes at $A(6, 0)$ and $B(0, 8)$. Then the area of ΔACB is $\dfrac{1}{2} \times 6 \times 8 = 34$.

17. (7)

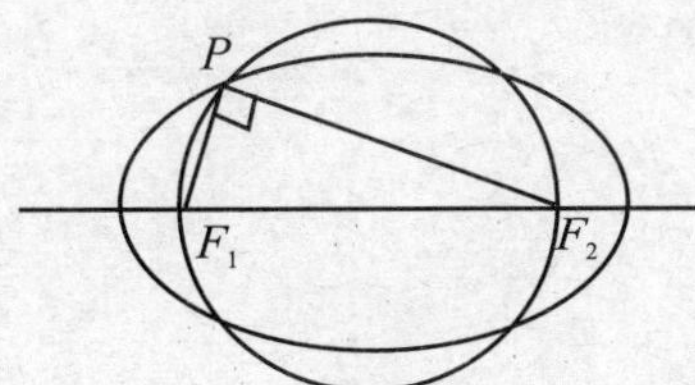

Fig. 4.48

$F_1 + PF_2 = 15$
$PF_1 \times PF_2 = 52$
$(F_1F_2)^2 = (PF_1 + PF_2)^2 - 2PF_1 \times PF_2 = 225 - 104$
$= 121$
$\Rightarrow F_1F_2 = 11$
$\Rightarrow 2ae = 11 \Rightarrow ae = 11/2$

Also $b^2 = a^2(1 - e^2)$
$\Rightarrow b^2 = a^2 - a^2e^2$
$\Rightarrow a^2 - b^2 = (ae)^2$
$\Rightarrow 4(a^2 - b^2) = 4\left(\dfrac{11}{2}\right)^2 = 121$

18. (1) Let the middle point of the chord be $(t^2, 2t)$.

The midpoint of a chord must lie inside the ellipse. So

$t^4 + 8t^2 - 1 < 0$

$\Rightarrow t^2 \in (0, -4 + \sqrt{17})$ (1)

Also the equation of the chord is $T = S_1$. Thus

$t^2x + 4ty = t^4 + 8t^2$

Let this pass through $(\alpha, 0)$. Then

$\alpha t^2 = t^4 + 8t^2$
$\Rightarrow t^4 + (8 - \alpha)t^2 = 0$
$\Rightarrow t^2 = 0$ or $t^2 = \alpha - 8$
$\Rightarrow \alpha = t^2 + 8 \Rightarrow \alpha \in (8, 4 + \sqrt{17})$

19. (1) The equation of any tangent PQ to the ellipse $x^2/a^2 + y^2/b^2 = 1$ is

$\dfrac{x}{a}\cos\theta + \dfrac{y}{b}\sin\theta = 1$ (1)

This tangent cuts the ellipse $x^2/c^2 + y^2/d^2 = 1$ at the points P and Q.

Let the tangents at P and Q intersect at the point $R(h, k)$.

Then PQ becomes the chord of contact w.r.t. to the point R for the ellipse $x^2/c^2 + y^2/d^2 = 1$, i.e., the equation of PQ is

$\dfrac{hx}{c^2} + \dfrac{ky}{d^2} = 1$ (2)

The equations (1) and (2) represent the same straight line. Therefore,

$\dfrac{(\cos\theta)/a}{h/c^2} = \dfrac{(\sin\theta)/b}{k/d^2} = 1$

$\Rightarrow \cos\theta = \dfrac{ah}{c^2}$ and $\sin\theta = \dfrac{bk}{d^2}$

Squaring and adding, we get

$\dfrac{a^2h^2}{c^4} + \dfrac{b^2k^2}{d^4} = 1$

$\Rightarrow \dfrac{a^2x^2}{c^4} + \dfrac{b^2y^2}{d^4} = 1,$ (3)

which is the locus of the point $R(h, k)$.

If $R(h, k)$ is the point of intersection of the two perpendicular tangents, then the locus of R should be the director circle of the ellipse $x^2/c^2 + y^2/d^2 = 1$, i.e., $x^2 + y^2 = c^2 + d^2$.

i.e., $\dfrac{x^2}{c^2 + d^2} + \dfrac{y^2}{c^2 + d^2} = 1$ (4)

The equations (3) and (4) represent the same locus. Therefore,

$\dfrac{a^2}{c^4} = \dfrac{1}{c^2 + d^2}$ and $\dfrac{b^2}{b^4} = \dfrac{1}{c^2 + d^2}$

$\Rightarrow \dfrac{a^2}{c^2} + \dfrac{b^2}{d^2} = 1$

20. (8) Any tangent on the ellipse is $y = mx \pm \sqrt{25m^2 + 4}$.

If it is also a tangent on the circle, then

$\left|\dfrac{0 - m \times 0 \pm \sqrt{25m^2 + 4}}{\sqrt{1 + m^2}}\right|$

$\Rightarrow m = \pm\dfrac{2}{\sqrt{5}}$

Since the common tangent is in the first quadrant, so $m = -(2/\sqrt{3})$.

So the common tangent in the first quadrant is $\sqrt{3}y + 2x = 4\sqrt{7}$ (1)

The points of intersection of this tangent with the x-axis and y-axis are $(2\sqrt{7}, 0)$ and $\left(0, \dfrac{4\sqrt{7}}{\sqrt{3}}\right)$.

Hence, the length of the intercept is

$\sqrt{(2\sqrt{7} - 0)^2 + \left(0 - \dfrac{4\sqrt{7}}{\sqrt{3}}\right)^2} = \dfrac{14}{\sqrt{3}}$.

CHAPTER 5

Hyperbola

Objective Type

Each question has four choices a, b, c, d, out of which *only one* answer is correct. Find the correct answer.

1. If the distance between two parallel tangents drawn to the hyperbola $\frac{x^2}{9}-\frac{y^2}{49}=1$ is 2, then their slope is equal to

   a. $\pm\frac{5}{2}$ **b.** $\pm\frac{4}{5}$
   c. $\pm\frac{7}{2}$ **d.** none of these

2. If the distance between the foci and the distance between the two directries of the hyperbola $\frac{x^2}{a^2}-\frac{y^2}{b^2}=1$ are in the ratio 3:2, then $b{:}a$ is

   a. $1{:}\sqrt{2}$ **b.** $\sqrt{3}{:}\sqrt{2}$ **c.** 1:2 **d.** 2:1

3. A tangent drawn to hyperbola $\frac{x^2}{a^2}-\frac{y^2}{b^2}=1$ at $P\left(\frac{\pi}{6}\right)$ forms a triangle of area $3a^2$ square units, with the coordinate axes, then the square of its eccentricity is

   a. 15 **b.** 24 **c.** 17 **d.** 14

4. The length of the transverse axis of the rectangular hyperbola $xy = 18$ is

   a. 6 **b.** 12 **c.** 18 **d.** 9

5. A straight line has its extremities on two fixed straight lines and cuts off from them a triangle of constant area c^2. Then the locus of the middle point of the line is

   a. $2xy = c^2$ **b.** $xy + c^2 = 0$
   c. $4x^2y^2 = c$ **d.** none of these

6. If a variable line has its intercepts on the coordinate axes e, e', where $\frac{e}{2}$, $\frac{e'}{2}$ are the eccentricities of a hyperbola and its conjugate hyperbola, then the line always touches the circle $x^2 + y^2 = r^2$, where $r =$

   a. 1 **b.** 2
   c. 3 **d.** cannot be decided

7. The equation of the transverse axis of the hyperbola $(x-3)^2 + (y+1)^2 = (4x+3y)^2$ is

   a. $x + 3y = 0$ **b.** $4x + 3y = 9$
   c. $3x - 4y = 13$ **d.** $4x + 3y = 0$

8. The family of curves for which the length of the normal at any point is equal to the radius vector of that point is

   a. hyperbola **b.** straight line
   c. parabola **d.** ellipse

9. The eccentricity of the conic represented by $x^2 - y^2 - 4x + 4y + 16 = 0$ is

   a. 1 **b.** $\sqrt{2}$ **c.** 2 **d.** $\frac{1}{2}$

10. The equation $16x^2 - 3y^2 - 32x + 12y - 44 = 0$ represents a hyperbola

    a. the length of whose transverse axis is $4\sqrt{3}$
    b. the length of whose conjugate axis is 4
    c. whose centre is $(-1, 2)$
    d. whose eccentricity is $\sqrt{\frac{19}{3}}$

11. If the vertex of a hyperbola bisects the distance between its centre and the corresponding focus, then ratio of square of its conjugate axis to the square of its transverse axis is

a. 2 b. 4 c. 6 d. 3

12. The eccentricity of the hyperbola whose latus rectum is 8 and conjugate axis is equal to half the distance between the foci is (axes are coordinate axes)

a. $\frac{4}{3}$ b. $\frac{4}{\sqrt{3}}$

c. $\frac{2}{\sqrt{3}}$ d. none of these

13. Let LL' be the latus rectum through the focus of the hyperbola $\frac{x^2}{a^2} - \frac{y^2}{b^2} = 1$ and A' be the farther vertex. If $\Delta A'LL'$ is equilateral, then the eccentricity of the hyperbola is (axes are coordinate axes)

a. $\sqrt{3}$ b. $\sqrt{3}+1$

c. $\frac{\sqrt{3}+1}{\sqrt{2}}$ d. $\frac{\sqrt{3}+1}{\sqrt{3}}$

14. The latus rectum of the hyperbola $9x^2 - 16y^2 - 18x - 32y - 151 = 0$ is

a. $\frac{9}{4}$ b. 9 c. $\frac{3}{2}$ d. $\frac{9}{2}$

15. The eccentricity of the conjugate hyperbola of the hyperbola $x^2 - 3y^2 = 1$ is

a. 2 b. $\frac{2}{\sqrt{3}}$ c. 4 d. $\frac{4}{5}$

16. The equations of the transverse and conjugate axes of a hyperbola are respectively $x + 2y - 3 = 0$, $2x - y + 4 = 0$, and their respective lengths are $\sqrt{2}$ and $\frac{2}{\sqrt{3}}$. The equation of the hyperbola is

a. $\frac{2}{5}(x + 2y - 3)^2 - \frac{3}{5}(2x - y + 4)^2 = 1$

b. $\frac{2}{5}(2x - y + 4)^2 - \frac{3}{5}(x + 2y - 3)^2 = 1$

c. $2(2x - y + 4)^2 - 3(x + 2y - 3)^2 = 1$

d. $2(x + 2y - 3)^2 - 3(2x - y + 4)^2 = 1$

17. The locus of the point of intersection of the lines $\sqrt{3}x - y - 4\sqrt{3}t = 0$ and $\sqrt{3}tx + ty - 4\sqrt{3} = 0$ (where t is a parameter) is a hyperbola whose eccentricity is

a. $\sqrt{3}$ b. 2 c. $\frac{2}{\sqrt{3}}$ d. $\frac{4}{3}$

18. If the eccentricity of the hyperbola $x^2 - y^2 \sec^2 \alpha = 5$ is $\sqrt{3}$ times the eccentricity of the ellipse $x^2 \sec^2 \alpha + y^2 = 25$, then a value of α is

a. $\frac{\pi}{6}$ b. $\frac{\pi}{4}$ c. $\frac{\pi}{3}$ d. $\frac{\pi}{2}$

19. With one focus of the hyperbola $\frac{x^2}{9} - \frac{y^2}{16} = 1$ as the centre, a circle is drawn which is tangent to the hyperbola with no part of the circle being outside the hyperbola. The radius of the circle is

a. less than 2 b. 2

c. 1/3 d. none of these

20. If $ax + by = 1$ is tangent to the hyperbola $\frac{x^2}{a^2} - \frac{y^2}{b^2} = 1$, then $a^2 - b^2$ equals to

a. $\frac{1}{a^2e^2}$ b. a^2e^2

c. b^2e^2 d. none of these

21. Locus of a point whose chord of contact with respect to the circle $x^2 + y^2 = 4$ is a tangent to the hyperbola $xy = 1$ is a/an

a. ellipse b. circle

c. hyperbola d. parabola

22. The sides AC and AB of the ΔABC touch the conjugate hyperbola of the hyperbola $\frac{x^2}{a^2} - \frac{y^2}{b^2} = 1$. If the vertex A lies on the ellipse $\frac{x^2}{a^2} + \frac{y^2}{b^2} = 1$, then the side BC must touch

a. parabola b. circle

c. hyperbola d. ellipse

23. The tangent at a point P on the hyperbola $\frac{x^2}{a^2} - \frac{y^2}{b^2} = 1$ passes through the point $(0, -b)$ and the normal at P passes through the point $(2a\sqrt{2}, 0)$ then eccentricity of the hyperbola is

a. 2 b. $\sqrt{2}$ c. 3 d. $\sqrt{3}$

24. If values of m for which the line $y = mx + 2\sqrt{5}$ touches the hyperbola $16x^2 - 9y^2 = 144$ are roots of the equation $x^2 - (a + b)x - 4 = 0$, then value of $(a + b)$ is equal to

a. 2 b. 4

c. zero d. none of these

25. Portion of asymptote of hyperbola $\frac{x^2}{a^2} - \frac{y^2}{b^2} = 1$ (between centre and the tangent at vertex) in the first quadrant is cut by the line $y + \lambda(x - a) = 0$ (λ is a parameter) then

a. $\lambda \in R$ b. $\lambda \in (0, \infty)$

c. $\lambda \in (-\infty, 0)$ d. none of these

26. If angle between asymptotes of hyperbola $\frac{x^2}{a^2} - \frac{y^2}{b^2} = 1$ is 120° and product of perpendiculars drawn from foci upon its any tangent is 9, then locus of point of intersection of perpendicular tangents of the hyperbola can be

a. $x^2+y^2=6$ b. $x^2+y^2=9$
c. $x^2+y^2=3$ d. $x^2+y^2=18$

27. The coordinates of a point on the hyperbola, $\frac{x^2}{24}-\frac{y^2}{18}=1$, which is nearest to the line $3x+2y+1=0$ are

a. (6, 3) b. (– 6, – 3) c. (6, – 3) d.(– 6, 3)

28. The number of possible tangents which can be drawn to the curve $4x^2-9y^2=36$, which are perpendicular to the straight line $5x+2y-10=0$ is

a. zero b. 1 c. 2 d. 4

29. Locus of the feet of the perpendiculars drawn from either focus on a variable tangent to the hyperbola $16y^2-9x^2=1$ is

a. $x^2+y^2=9$ b. $x^2+y^2=\frac{1}{9}$
c. $x^2+y^2=\frac{7}{144}$ d. $x^2+y^2=\frac{1}{16}$

30. The locus of the foot of the perpendicular form the centre of the hyperbola $xy=1$ on a variable tangent is

a. $(x^2-y^2)^2=4xy$ b. $(x^2+y^2)^2=2xy$
c. $(x^2+y^2)=4xy$ d. $(x^2+y^2)^2=4xy$

31. P is a point on the hyperbola $\frac{x^2}{a^2}-\frac{y^2}{b^2}=1$, N is the foot of the perpendicular from P on the transverse axis. The tangent to the hyperbola at P meets the transverse axis at T. If O is the centre of the hyperbola, the $OT\times ON$ is equal to

a. e^2 b. a^2 c. b^2 d. $\frac{b^2}{a^2}$

32. The tangent at a point P on the hyperbola $\frac{x^2}{a^2}-\frac{y^2}{b^2}=1$ meets one of the directrix in F. If PF subtends an angle θ at the corresponding focus, then θ equals

a. $\frac{\pi}{4}$ b. $\frac{\pi}{2}$ c. $\frac{3\pi}{4}$ d. π

33. The locus of a point from where tangents to the rectangular hyperbola $x^2-y^2=a^2$ contain an angle of 45°, is

a. $(x^2+y^2)^2+a^2(x^2-y^2)=4a^2$
b. $2(x^2+y^2)^2+4a^2(x^2-y^2)=4a^2$
c. $(x^2+y^2)^2+4a^2(x^2-y^2)=4a^2$
d. $(x^2+y^2)^2+a^2(x^2-y^2)=a^4$

34. If tangents PQ and PR are drawn from variable point P to the hyperbola $\frac{x^2}{a^2}-\frac{y^2}{b^2}=1$ $(a>b)$ so that the fourth vertex S of parallelogram $PQSR$ lies on circumcircle of triangle PQR, then locus of P is

a. $x^2+y^2=b^2$ b. $x^2+y^2=a^2$
c. $x^2+y^2=(a^2-b^2)$ d. none of these

35. Number of points on the hyperbola $\frac{x^2}{a^2}-\frac{y^2}{b^2}=3$ from which mutually perpendicular tangents can be drawn to the circle $x^2+y^2=a^2$, is /are

a. 0 b. 2 c. 3 d. 4

36. A normal to the hyperbola $\frac{x^2}{4}-\frac{y^2}{1}=1$ has equal intercepts on positive x- and y-axes. If this normal touches the ellipse $\frac{x^2}{a^2}+\frac{y^2}{b^2}=1$, then a^2+b^2 is equal to

a. 5 b. 25
c. 16 d. none of these

37. If the normal to the given hyperbola at the point $(ct, \frac{c}{t})$ meets the curve again at $(ct', \frac{c}{t'})$, then

a. $t^3t'=1$ b. $t^3t'=-1$
c. $tt'=1$ d. $tt'=-1$

38. If the sum of the slopes of the normal from a point P to the hyperbola $xy=c^2$ is equal to λ ($\lambda\in R^+$), then locus of point P is

a. $x^2=\lambda c^2$ b. $y^2=\lambda c^2$
c. $xy=\lambda c^2$ d. none of these

39. If a ray of light incident along the line $3x+(5-4\sqrt{2})y=15$ gets reflected from the hyperbola $\frac{x^2}{16}-\frac{y^2}{9}=1$, then its reflected ray goes along the line

a. $x\sqrt{2}-y+5=0$ b. $\sqrt{2}y-x+5=0$
c. $\sqrt{2}\,y-x-5=0$ d. none of these

40. Let any double ordinate PNP' of the hyperbola $\frac{x^2}{25}-\frac{y^2}{16}=1$ be produced on both sides to meet the asymptotes in Q and Q', then $PQ\times P'Q$ is equal to

a. 25 b. 16
c. 41 d. none of these

41. For a hyperbola whose centre is at (1, 2) and asymptotes are parallel to lines $2x+3y=0$ and $x+2y=1$, then equation of hyperbola passing through (2, 4) is

a. $(2x+3y-5)(x+2y-8)=40$
b. $(2x+3y-8)(x+2y-5)=40$
c. $(2x+3y-8)(x+2y-5)=30$
d. none of these

42. The chords of contact of a point P w.r.t. a hyperbola and its auxiliary circle are at right angle, then the point P lies on

a. conjugate hyperbola
b. one of the directirx
c. one of the asymptotes
d. none of these

43. Asymptotes of the hyperbolas $\frac{x^2}{a_1^2} - \frac{y^2}{b_1^2} = 1$ and $\frac{x^2}{a_2^2} - \frac{y^2}{b_2^2} = 1$ are perpendicular to each other, then

a. $\frac{a_1}{a_2} = \frac{b_1}{b_2}$ **b.** $a_1 a_2 = b_1 b_2$

c. $a_1 a_2 + b_1 b_2 = 0$ **d.** $a_1 - a_2 = b_1 - b_2$

44. If $S = 0$ be the equation of the hyperbola $x^2 + 4xy + 3y^2 - 4x + 2y + 1 = 0$, then the value of k for which $S + K = 0$ represents its asymptotes is

a. 20 **b.** -16 **c.** -22 **d.** 18

45. If two distinct tangents can be drawn from the point $(\alpha, 2)$ on different branches of the hyperbola $\frac{x^2}{9} - \frac{y^2}{16} = 1$, then

a. $|\alpha| < \frac{3}{2}$ **b.** $|\alpha| > \frac{2}{3}$

c. $|\alpha| > 3$ **c.** none of these

46. A hyperbola passes through (2, 3) and has asymptotes $3x - 4y + 5 = 0$ and $12x + 5y - 40 = 0$, then the equation of its transverse axis is

a. $77x - 21y - 265 = 0$

b. $21x - 77y + 265 = 0$

c. $21x - 77y - 265 = 0$

d. $21x + 77y - 265 = 0$

47. From any point on the hyperbola $\frac{x^2}{a^2} - \frac{y^2}{b^2} = 1$, tangents are drawn to the hyperbola $\frac{x^2}{a^2} - \frac{y^2}{b^2} = 2$. Then area cut-off by the chord of contact on the asymptotes is equal to

a. $\frac{a}{2}$ **b.** ab **c.** $2ab$ **d.** $4ab$

48. From a point P (1, 2) two tangents are drawn to a hyperbola H in which one tangent is drawn to each arm of the hyperbola. If the equations of asymptotes of hyperbola H are $\sqrt{3}x - y + 5 = 0$ and $\sqrt{3}x + y - 1 = 0$, then eccentricity of H is

a. 2 **b.** $\frac{2}{\sqrt{3}}$ **c.** $\sqrt{2}$ **d.** $\sqrt{3}$

49. The combined equation of the asymptotes of the hyperbola $2x^2 + 5xy + 2y^2 + 4x + 5y = 0$ is

a. $2x^2 + 5xy + 2y^2 + 4x + 5y + 2 = 0$

b. $2x^2 + 5xy + 2y^2 + 4x + 5y - 2 = 0$

c. $2x^2 + 5xy + 2y^2 = 0$

d. none of these

50. The asymptotes of the hyperbola $xy = hx + ky$ are

a. $x - k = 0$ and $y - h = 0$

b. $x + h = 0$ and $y + k = 0$

c. $x - k = 0$ and $y + h = 0$

d. $x + k = 0$ and $y - h = 0$

51. The asymptote of the hyperbola $\frac{x^2}{a^2} - \frac{y^2}{b^2} = 1$ form with any tangent to the hyperbola a triangle whose area is $a^2 \tan \lambda$ in magnitude, then its eccentricity is

a. $\sec \lambda$ **b.** $\operatorname{cosec} \lambda$ **c.** $\sec^2 \lambda$ **d.** $\operatorname{cosec}^2 \lambda$

52. The centre of a rectangular hyperbola lies on the line $y = 2x$. If one of the asymptotes is $x + y + c = 0$, then the other asymptote is

a. $x - y - 3c = 0$ **b.** $2x - y + c = 0$

c. $x - y - c = 0$ **d.** none of these

Hence, equation of other asymptote is $x + y - 3c = 0$.

53. Equation of a rectangular hyperbola whose asymptotes are $x = 3$ and $y = 5$ and passing through (7, 8) is

a. $xy - 3y + 5x + 3 = 0$

b. $xy + 3y + 4x + 3 = 0$

c. $xy - 3y + 5x - 3 = 0$

d. $xy - 3y - 5x + 3 = 0$

54. If foci of hyperbola lie on $y = x$ and one of the asymptote is $y = 2x$, then equation of the hyperbola, given that it passes through (3, 4) is

a. $x^2 - y^2 - \frac{5}{2}xy + 5 = 0$

b. $2x^2 - 2y^2 + 5xy + 5 = 0$

c. $2x^2 + 2y^2 - 5xy + 10 = 0$

d. none of these

55. $(x - 1)(y - 2) = 5$ and $(x - 1)^2 + (y + 2)^2 = r^2$ intersect at four points A, B, C, D and if centroid of ΔABC lies on line $y = 3x - 4$, then locus of D is

a. $y = 3x$ **b.** $x^2 + y^2 + 3x + 1 = 0$

c. $3y = x + 1$ **d.** $y = 3x + 1$

56. If tangents OQ and OR are dawn to variable circles having radius r and the centre lying on the rectangular hyperbola $xy = 1$, then locus of circumcentre of triangle OQR is (O being the origin)

a. $xy = 4$ **b.** $xy = \frac{1}{4}$

c. $xy = 1$ **d.** none of these

57. Four points are such that the line joining any two points is perpendicular to the line joining other two points. If three points out of these lie on a rectangular hyperbola then the fourth point will lie on

a. the same hyperbola

b. conjugate hyperbola

c. one of the directrix

d. one of the asymptotes

58. Equation of conjugate axis of hyperbola $xy - 3y - 4x + 7 = 0$ is

a. $y + x = 3$ **b.** $y + x = 7$
c. $y - x = 3$ **d.** none of these

59. If S_1 and S_2 are the foci of the hyperbola whose transverse axis length is 4 and conjugate axis length is 6, S_3 and S_4 are the foci of the conjugate hyperbola, then the area of the quadrilateral $S_1S_3S_2S_4$ is

a. 24 **b.** 26
c. 22 **d.** None of these

60. Suppose the circle having equation $x^2 + y^2 = 3$ intersects the rectangular hyperbola $xy = 1$ at the points A, B, C, and D. The equation $x^2 + y^2 - 3 + \lambda(xy - 1) = 0$, $\lambda \in R$, represents

a. a pair of lines through origin for $\lambda = -3$
b. an ellipse through A, B, C and D for $\lambda = -3$
c. a parabola through A, B, C and D for $\lambda = -3$
d. a circle for any $\lambda \in R$

61. The family of the curves which intersects the family of rectangular hyperbola $xy = c^2$ orthogonally is

a. family of parabola
b. family of ellipse
c. family of circle
d. family of rectangular hyperbola

62. If two points P and Q on the hyperbola $\frac{x^2}{a^2} - \frac{y^2}{b^2} = 1$, whose centre C be such that CP is perpendicular to CQ, $a < b$, then the value of $\frac{1}{CP^2} + \frac{1}{CQ^2}$ is

a. $\frac{b^2 - a^2}{2ab}$ **b.** $\frac{1}{a^2} + \frac{1}{b^2}$ **c.** $\frac{2ab}{b^2 - a^2}$ **d.** $\frac{1}{a^2} - \frac{1}{b^2}$

63. The equation of the chord joining two points (x_1, y_1) and (x_2, y_2) on the rectangular hyperbola $xy = c^2$ is

a. $\frac{x}{x_1 + x_2} + \frac{y}{y_1 + y_2} = 1$
b. $\frac{x}{x_1 - x_2} + \frac{y}{y_1 - y_2} = 1$
c. $\frac{x}{y_1 + y_2} + \frac{y}{x_1 + x_2} = 1$
d. $\frac{x}{y_1 - y_2} + \frac{y}{x_1 - x_2} = 1$

64. If $P(x_1, y_1)$, $Q(x_2, y_2)$, $R(x_3, y_3)$, and $S(x_4, y_4)$ are four concyclic points on the rectangular hyperbola $xy = c^2$, then coordinates of the orthocentre of the triangle PQR is

a. $(x_4, -y_4)$ **b.** (x_4, y_4)
c. $(-x_4, -y_4)$ **d.** $(-x_4, y_4)$

65. The chord PQ of the rectangular hyperbola $xy = a^2$ meets the axis of x at A; C is the midpoint of PQ and O is the origin. Then the ΔACO is

a. equilateral **b.** isosceles
c. right angled **d.** right isosceles

66. The curve $xy = c$ $(c > 0)$ and the circle $x^2 + y^2 = 1$ touch at two points, then distance between the points of contact is

a. 1 **b.** 2
c. $2\sqrt{2}$ **d.** none of these

67 Let C be a curve which is locus of the point of the intersection of lines $x = 2 + m$ and $my = 4 - m$. A circle $S \equiv (x - 2)^2 + (y + 1)^2 = 25$ intersects the curve C at four points P, Q, R and S. If O is centre of the curve C, then $OP^2 + OQ^2 + OR^2 + OS^2$ is

a. 50 **b.** 100 **c.** 25 **d.** $\frac{25}{2}$

68. The ellipse $4x^2 + 9y^2 = 36$ and the hyperbola $a^2x^2 - y^2 = 4$ intersect at right angles then the equation of the circle through the points of intersection of two conic is

a. $x^2 + y^2 = 5$
b. $\sqrt{5}(x^2 + y^2) - 3x - 4y = 0$
c. $\sqrt{5}(x^2 + y^2) + 3x + 4y = 0$
d. $x^2 + y^2 = 25$

69. The locus of the point which is such that the chord of contact of tangents drawn from it to the ellipse $\frac{x^2}{a^2} + \frac{y^2}{b^2} = 1$ forms a triangle of constant area with the coordinate axe is

a. a straight line **b.** a hyperbola
c. an ellipse **d.** a circle

70. The angle between lines joining the origin to the points of intersection of the line $\sqrt{3}x + y = 2$ and the curve $y^2 - x^2 = 4$ is

a. $\tan^{-1}\left(\frac{2}{\sqrt{3}}\right)$ **b.** $\frac{\pi}{6}$
c. $\tan^{-1}\left(\frac{\sqrt{3}}{2}\right)$ **d.** $\frac{\pi}{2}$

71. A variable chord of the hyperbola $\frac{x^2}{a^2} - \frac{y^2}{b^2} = 1$ $(b > a)$ subtends a right angle at the centre of the hyperbola, if this chord touches

a. a fixed circle concentric with the hyperbola
b. a fixed ellipse concentric with the hyperbola
c. a fixed hyperbola concentric with the hyperbola
d. a fixed parabola having vertex at (0, 0)

72. The exhaustive set of values of α^2 such that there exists a tangent to the ellipse $x^2 + \alpha^2y^2 = \alpha^2$ such that the portion of the tangent intercepted by the hyperbola $\alpha^2x^2 - y^2 = 1$ subtends a right angle at the centre of the curves is

a. $\left[\frac{\sqrt{5}+1}{2}, 2\right]$ b. $(1, 2]$

c. $\left[\frac{\sqrt{5}-1}{2}, 1\right)$ d. $\left[\frac{\sqrt{5}-1}{2}, 1\right) \cup \left(1, \frac{\sqrt{5}+1}{2}\right]$

Multiple Correct Answers Type

Each question has four choices a, b, c, and d, out of which *one or more* answers are correct.

1. The equation $\left|\sqrt{x^2+(y-1)^2} - \sqrt{x^2+(y+1)^2}\right| = K$ will represent a hyperbola for

a. $K \in (0, 2)$ b. $K \in (-2, 1)$

c. $K \in (1, \infty)$ d. $K \in (0, \infty)$

2. If $x, y \in R$ then the equation $3x^4 - 2(19y + 8)x^2 + (361y^2 + 2(100 + y^4) + 64) = 2(190y + 2y^2)$ represents in rectangular cartesian system

a. parabola b. hyperbola

c. circle d. ellipse

3. If (5, 12) and (24, 7) are the foci of a hyperbola passing through the origin, then

a. $e = \frac{\sqrt{386}}{12}$ b. $e = \frac{\sqrt{386}}{13}$

c. $LR = \frac{121}{6}$ d. $LR = \frac{121}{3}$

4. If (5, 12) and (24, 7) are the foci of a conic passing through the origin then the eccentricity of conic is

a. $\frac{\sqrt{386}}{12}$ b. $\frac{\sqrt{386}}{13}$

c. $\frac{\sqrt{386}}{25}$ d. $\frac{\sqrt{386}}{38}$

5. For which of the hyperbolas, we can have more than one pair of perpendicular tangents?

a. $\frac{x^2}{4} - \frac{y^2}{9} = 1$ b. $\frac{x^2}{4} - \frac{y^2}{9} = -1$

c. $x^2 - y^2 = 4$ d. $xy = 44$

6. For the hyperbola $9x^2 - 16y^2 - 18x + 32y - 151 = 0$

a. one of the directrix is $x = \frac{21}{5}$

b. length of latus rectum $= \frac{9}{2}$

c. focii are (6, 1) and (– 4, 1)

d. eccentricity is $\frac{5}{4}$

7. If foci of $\frac{x^2}{a^2} - \frac{y^2}{b^2} = 1$ coincide with the foci of $\frac{x^2}{25} + \frac{y^2}{9} = 1$ and eccentricity of the hyperbola is 2, then

a. $a^2 + b^2 = 16$

b. there is no director circle of the hyperbola

c. centre of the director circle is (0, 0)

d. length of latus rectum of the hyperbola = 12

8. If the circle $x^2 + y^2 = a^2$ intersects the hyperbola $xy = c^2$ in four points $P(x_1, y_1)$, $Q(x_2, y_2)$, $R(x_3, y_3)$, $S(x_4, y_4)$ then

a. $x_1 + x_2 + x_3 + x_4 = 0$

b. $y_1 + y_2 + y_3 + y_4 = 0$

c. $x_1x_2x_3x_4 = c^4$

d. $y_1y_2y_3y_4 = c^4$

9. The differential equation $\frac{dy}{dx} = \frac{3y}{2x}$ represents a family of hyperbolas (except when it represents a pair of lines) with eccentricity

a. $\sqrt{\frac{3}{5}}$ b. $\sqrt{\frac{5}{3}}$

c. $\sqrt{\frac{2}{5}}$ d. $\sqrt{\frac{5}{2}}$

10. Circles are drawn on chords of the rectangular hyperbola $xy = 4$ parallel to the line $y = x$ as diameters. All such circles pass through two fixed points whose coordinates are

a. (2, 2) b. (2, –2) c. (–2, 2) d. (–2, –2)

11. The equation $(x - \alpha)^2 + (y - \beta)^2 = k(lx + my + n)^2$ represents

a. a parabola for $k < (l^2 + m^2)^{-1}$

b. an ellipse for $0 < k < (l^2 + m^2)^{-1}$

c. a hyperbola for $k > (l^2 + m^2)^{-1}$

d. a point circle for $k = 0$

12. If P is a point on a hyperbola, then

a. locus of excentre of the circle described opposite to $\angle P$ for $\Delta PSS'$ (S, S' are foci) is tangents at vertex

b. locus of excentre of the circle described opposite to $\angle S'$ is hyperbola

c. locus of excentre of the circle described opposite to $\angle P$ for $\Delta PSS'$ (S, S' are foci), is hyperbola

d. locus of excentre of the circle described opposite to $\angle S'$, is tangent at vertex

13. The lines parallel to normal to the curve $xy = 1$ is/are

a. $3x + 4y + 5 = 0$ b. $3x - 4y + 5 = 0$

c. $4x + 3y + 5 = 0$ d. $3y - 4x + 5 = 0$

14. From point (2, 2) tangents are drawn to the hyperbola $\frac{x^2}{16} - \frac{y^2}{9} = 1$, then the point of contact lies in

a. First quadrant **b.** Second quadrant
c. Third quadrant **d.** Fourth quadrant

15. If the two intersecting lines intersect the hyperbola and neither of them is a tangent to it, then number of intersecting points are

a. 1 **b.** 2 **c.** 3 **d.** 4

16. For hyperbola $\frac{x^2}{a^2}-\frac{y^2}{b^2}=1$, let n be the number of points on the plane through which perpendicular tangents are drawn.

a. if $n = 1$, then $e = \sqrt{2}$
b. if $n > 1$, then $0 < e < \sqrt{2}$
c. if $n = 0$, then $e > \sqrt{2}$
d. none of these

17. If the normal at P to the rectangular hyperbola $x^2 - y^2 = 4$ meets the axes in G and g and C is the centre of the hyperbola, then

a. $PG = PC$ **b.** $Pg = PC$
c. $PG = Pg$ **d.** $Gg = 2PC$

Reasoning Type

Each question has four choices a, b, c and d, out of which *only one* is correct. Each question contains Statement 1 and Statement 2.

a. Both the statements are True and Statement 2 is the correct explanation of Statement 1.
b. Both the statements are True but Statement 2 is not the correct explanation of Statement 1.
c. Statement 1 is True and Statement 2 is False.
d. Statement 1 is False and Statement 2 is True.

1. Statement 1: Asymptotes of hyperbola $3x + 4y = 2$ and $4x - 3y = 5$ are bisectors of transverse and conjugate axes of hyperbola.

Statement 2: Transverse and conjugate axes of hyperbola are bisectors of the asymptotes.

2. Statement 1: Every line which cuts the hyperbola in two distinct points has slope lies in $(-2, 2)$.

Statement 2: Slope of tangents of hyperbola lies in $(-\infty, -2) \cup (2, \infty)$.

3. Statement 1: A bullet is fired and it hits a target. An observer in the same plane heard two sounds, the crack of the rifle and the thud of the bullet striking the target at the same instant, then locus of the observer is hyperbola where velocity of sound is smaller than velocity of bullet.

Statement 2: If difference of distances of a point P from the two fixed points is constant and less than the distance between the fixed points then locus of P is a hyperbola.

4. Statement 1: Ellipse $\frac{x^2}{25}+\frac{y^2}{16}=1$ and $12x^2 - 4y^2 = 27$ intersect each other at right angle.

Statement 2: Given ellipse and hyperbola have same foci.

5. Statement 1: If a circle $S = 0$ intersects a hyperbola $xy = 4$ at four points. Three of them are (2, 2), (4, 1) and (6, 2/3), then coordinates of the fourth point are (1/4, 16).

Statement 2: If a circle $S = 0$ intersects a hyperbola $xy = c^2$ at t_1, t_2, t_3, t_4, then $t_1 - t_2 - t_3 - t_4 = 1$.

6. Statement 1: If a point (x_1, y_1) lies in the shaded region $\frac{x^2}{a^2}-\frac{y^2}{b^2}=1$, shown in the figure, then $\frac{x_1^2}{a^2}-\frac{y_1^2}{b^2}<0$.

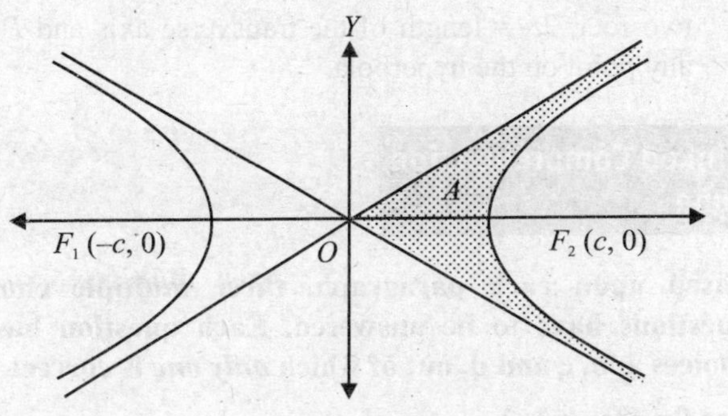

Fig. 5.1

Statement 2: If $P(x_1, y_1)$ lies outside the a hyperbola $\frac{x^2}{a^2}-\frac{y^2}{b^2}=1$, then $\frac{x_1^2}{a^2}-\frac{y_1^2}{b^2}<1$.

7. Statement 1: Equations of tangents to the hyperbola $2x^2 - 3y^2 = 6$ which is parallel to the line $y = 3x + 4$ is $y = 3x - 5$ and $y = 3x + 5$.

Statement 2: For given slope two parallel tangents can be drawn to the hyperbola.

8. Statement 1: There are infinite points from which two mutually perpendicular tangents can be drawn to the hyperbola $\frac{x^2}{9}-\frac{y^2}{16}=1$.

Statement 2: The locus of point of intersection of perpendicular tangents lies on the circle.

9. Statement 1: If from any point $P(x_1, y_1)$ on the hyperbola $\frac{x^2}{a^2}-\frac{y^2}{b^2}=-1$, tangents are draws to the hyperbola $\frac{x^2}{a^2}-\frac{y^2}{b^2}$ $= 1$, then corresponding chord of contact lies on another branch of the hyperbola $\frac{x^2}{a^2}-\frac{y^2}{b^2}=-1$.

Statement 2: From any point outside the hyperbola, two tangents can be drawn to the hyperbola.

10. Statement 1: If (3, 4) is a point on a hyperbola having focus (3, 0) and (λ, 0) and length of the transverse axis being 1 unit then λ can take the value 0 or 3.

Statement 2: $|S'P - SP| = 2a$, where S and S' are the two foci, $2a$ = length of the transverse axis and P be any point on the hyperbola.

11. Statement 1: Given the base BC of the triangle and the ratio radius of the ex-circles opposite to the angles B and C. Then locus of the vertex A is hyperbola.

Statement 2: $|S'P - SP| = 2a$, where S and S' are the two foci, $2a$ = length of the transverse axis and P be any point on the hyperbola.

Linked Comprehension Type

Based upon each paragraph, three multiple choice questions have to be answered. Each question has 4 choices a, b, c and d, out of which *only one* is correct.

For Problems 1–3

A conic passes through the point (2, 4) and is such that the segment of any of its tangents at any point contained between the coordinate axes is bisected at the point of tangency.

1. The eccentricity of the conic is

a. 2 **b.** $\sqrt{2}$ **c.** $\sqrt{3}$ **d.** $\sqrt{\frac{3}{2}}$

2. The foci of the conic are

a. $(2\sqrt{2}, 0)$ and $(-2\sqrt{2}, 0)$

b. $(2\sqrt{2}, 2\sqrt{2})$ and $(-2\sqrt{2}, -2\sqrt{2})$

c. (4, 4) and (– 4, – 4)

d. $(4\sqrt{2}, 4\sqrt{2})$ and $(-4\sqrt{2}, -4\sqrt{2})$

3. The equations of directrix are

a. $x + y = \pm 8$ **b.** $x + y = \pm 4$

c. $x + y = \pm 4\sqrt{2}$ **d.** none of these

For Problems 4–6

The locus of foot of perpendicular from any focus of a hyperbola upon any tangent to the hyperbola is the auxiliary circle of the hyperbola. Consider the foci of a hyperbola as (–3, –2) and (5, 6) and the foot of perpendicular from the focus (5, 6) upon a tangent to the hyperbola as (2, 5).

4. The conjugate axis of the hyperbola is

a. $4\sqrt{11}$ **b.** $2\sqrt{11}$ **c.** $4\sqrt{22}$ **d.** $2\sqrt{22}$

5. The directrix of the hyperbola corresponding to the focus (5, 6) is

a. $2x + 2y - 1 = 0$ **b.** $2x + 2y - 11 = 0$

c. $2x + 2y - 7 = 0$ **d.** $2x + 2y - 9 = 0$

6. The point of contact of the tangent with the hyperbola is

a. $\left(\frac{2}{9}, \frac{31}{3}\right)$ **b.** $\left(\frac{7}{4}, \frac{23}{4}\right)$ **c.** $\left(\frac{2}{3}, 9\right)$ **d.** $\left(\frac{7}{9}, 7\right)$

For Problems 7–9

Let $P(x, y)$ is a variable point such that

$\left|\sqrt{(x-1)^2+(y-2)^2}-\sqrt{(x-5)+(y-5)^2}\right| = 3$, which represents hyperbola.

7. The eccentricity e' of the corresponding conjugate hyperbola is

a. $\frac{5}{3}$ **b.** $\frac{4}{3}$ **c.** $\frac{5}{4}$ **d.** $\frac{3}{\sqrt{7}}$

8. Locus of intersection of two perpendicular tangents to the hyperbola is

a. $(x-3)^2+\left(y-\frac{7}{2}\right)^2=\frac{55}{4}$

b. $(x-3)^2+\left(y-\frac{7}{2}\right)^2=\frac{25}{4}$

c. $(x-3)^2+\left(y-\frac{7}{2}\right)^2=\frac{7}{4}$

d. none of these

9. If origin is shifted to point $\left(3, \frac{7}{2}\right)$ and the axes are rotated through an angle θ is clockwise sense so that equation of given hyperbola changes to the standard form $\frac{x^2}{a^2}-\frac{y^2}{b^2}=1$, then θ is

a. $\tan^{-1}\left(\frac{4}{3}\right)$ **b.** $\tan^{-1}\left(\frac{3}{4}\right)$

c. $\tan^{-1}\left(\frac{5}{3}\right)$ **d.** $\tan^{-1}\left(\frac{3}{5}\right)$

For Problems 10–12

In hyperbola portion of tangent intercepted between asymptotes is bisected at the point of contact.

Consider a hyperbola whose centre is at origin. A line $x + y = 2$ touches this hyperbola at $P(1, 1)$ and interests the asymptotes at A and B such that $AB = 6\sqrt{2}$ units.

10. Equation of asymptotes are

a. $5xy + 2x^2 + 2y^2 = 0$

b. $3x^2 + 4y^2 + 6xy = 0$

c. $2x^2 + 2y^2 - 5xy = 0$

d. none of these

11. Angle subtended by AB at centre of the hyperbola is

a. $\sin^{-1}\frac{4}{5}$ **b.** $\sin^{-1}\frac{2}{5}$

c. $\sin^{-1}\frac{3}{5}$ **d.** none of these

12. Equation of the tangent to the hyperbola at $\left(-1, \frac{7}{2}\right)$ is

a. $5x + 2y = 2$ **b.** $3x + 2y = 4$

c. $3x + 4y = 11$ **d.** none of these

For Problems 13–15

A point P moves such that sum of the slopes of the normals drawn from it to the hyperbola $xy = 16$ is equal to the sum of ordinates of feet of normals. The locus of P is a curve C.

13. The equation of the curve C is

a. $x^2 = 4y$ **b.** $x^2 = 16y$

c. $x^2 = 12y$ **d.** $y^2 = 8x$

14. If the tangent to the curve C cuts the coordinate axes at A and B, then the locus of the middle point of AB is

a. $x^2 = 4y$ **b.** $x^2 = 2y$

c. $x^2 + 2y = 0$ **d.** $x^2 + 4y = 0$

15. Area of the equilateral triangle, inscribed in the curve C, having one vertex as the vertex of curve C is

a. $772\sqrt{3}$ sq. units **b.** $776\sqrt{3}$ sq. units

c. $760\sqrt{3}$ sq. units **d.** $768\sqrt{3}$ sq. units

For Problems 16–18

The vertices of ΔABC lie on a rectangular hyperbola such that the orthocentre of the triangle is (3, 2) and the asymptotes of the rectangular hyperbola are parallel to the coordinate axes. The two perpendicular tangents of the hyperbola intersect at the point (1, 1).

16. The equation of the asymptotes is

a. $xy - 1 = x - y$ **b.** $xy + 1 = x + y$

c. $2xy = x + y$ **d.** none of these

17. Equation of the rectangular hyperbola is

a. $xy = 2x + y - 2$ **b.** $2xy = x + 2y + 5$

c. $xy = x + y + 1$ **d.** none of these

18. Number of real tangents that can be drawn from the point (1, 1) to the rectangular hyperbola is

a. 4 **b.** 0 **c.** 3 **d.** 2

Matrix-Match Type

Each question contains statements given in two columns which have to be matched.

Statements (a, b, c, d) in column I have to be matched with statements (p, q, r, s) in column II. If the correct matches are a → p, a → s, b → q, b → r, c → p, c → q, and d → s, then the correctly bubbled 4 × 4 matrix should be as follows:

	p	q	r	s
a	(p) filled	(q)	(r)	(s) filled
b	(p)	(q) filled	(r) filled	(s)
c	(p) filled	(q) filled	(r)	(s)
d	(p)	(q)	(r)	(s) filled

1. Let the foci of the hyperbola $\frac{x^2}{A^2} - \frac{y^2}{B^2} = 1$ be the vertices of the ellipse $\frac{x^2}{a^2} + \frac{y^2}{b^2} = 1$ and the foci of the ellipse be the vertices of the hyperbola. Let the eccentricities of the ellipse and hyperbola be e_E and e_H, respectively, then match the following.

Column I	Column II
a. $\frac{b}{B}$ is equal to	**p.** 1
b. $e_H + e_E$ is always greater than	**q.** 2
c. if angle between the asymptotes of hyperbola is $\frac{2\pi}{3}$, then $4e_E$ is equal to	**r.** 3
d. If $e_E^2 = \frac{1}{2}$ and (x, y) is point of intersection of ellipse and the hyperbola then $\frac{9x^2}{2y^2}$ is	**s.** 4

2.

Column I	Column II
a. The points common to the hyperbola $x^2 - y^2 = 9$ and circle $x^2 + y^2 = 41$ are	**p.** (−5, −4)
b. Tangents are drawn from point $\left(0, -\frac{9}{4}\right)$ to the hyperbola $x^2 - y^2 = 9$, then the point of tangency may have coordinate(s)	**q.** (5, 4)
c. The point which is diametrically opposite of point (5, 4) with respect to the hyperbola $x^2 - y^2 = 9$ is	**r.** (−5, 4)
d. If P and Q lie on the hyperbola $x^2 - y^2 = 9$ such that area of the isosceles triangle PQR where $PR = QR$ is 10 sq. units, where $R \equiv (0, -6)$, then P can have the coordinate(s)	**s.** (5, −4)

3. $A(-2, 0)$ and $B(2, 0)$ are the two fixed points and P is a point such that $PA - PB = 2$. Let S be the circle $x^2 + y^2 = r^2$, then match the following.

Column I	Column II
a. If $r = 2$, then the number of points P satisfying $PA - PB = 2$ and lying on $x^2 + y^2 = r^2$ is	**p.** 2
b. If $r = 1$, then he number of points satisfying $PA - PB = 2$ and lying on $x^2 + y^2 = r^2$ is	**q.** 4
c. For $r = 2$ the number of common tangents is	**r.** 0
d. For $r = 1/2$, the number of common tangents is	**s.** 1

4.

Column I	Column II
a. If z is a complex number such that Im $(z^2) = 3$, then eccentricity of the locus is	**p.** $\sqrt{3}$
b. If the latus rectum of a hyperbola through one focus subtends 60° angle at the other focus, then its eccentricity is	**q.** 2
c. If $A(3, 0)$ and $B(-3, 0)$ and $PA - PB = 4$, then eccentricity of conjugate hyperbola is	**r.** $\sqrt{2}$
d. If the angle between the asymptotes of hyperbola is $\pi/3$, then the eccentricity of its conjugate hyperbola is	**s.** $\frac{3}{\sqrt{5}}$

5. If e_1 and e_2 are the roots of the equation $x^2 - ax + 2 = 0$, then match the following.

Column I	Column II
a. If e_1 and e_2 are the eccentricities of the ellipse and hyperbola, respectively then the values of a are	**p.** 6
b. If both e_1 and e_2 are the eccentricities of the hyperbolas, then values of a are	**q.** $\frac{5}{2}$
c. If e_1 and e_2 are eccentricities of hyperbola and conjugate hyperbola, then values of a are	**r.** $2\sqrt{2}$
d. If e_1 is the eccentricity of the hyperbola for which there exists infinite points from which perpendicular tangents can be drawn and e_2 is the eccentricity of the hyperbola in which no such points exist then the values of a are	**s.** 5

Integer Type

1. If the foci of the ellipse $(x^2/16) + (y^2/b^2) = 1$and the hyperbola $(x^2/144) - (y^2/81) = 1/25$ coincide, then the value of b^2 is ______.

2. If PQ is a double ordinate of the hyperbola $(x^2/a^2) - (y^2/b^2) = 1$ such that OPQ is an equilateral triangle, O being the centre of the hyperbola, then the least value of eccentricity is ______.

3. The latus rectum subtends a right angle at the centre of the hyperbola $(x^2/a^2) - (y^2/b^2) = 1$. If the eccentricity is e, then the value of $[e]$, where [.] represents the greatest integer function, is ______.

4. An ellipse and a hyperbola are confocal (have the same focus) and the conjugate axis of the hyperbola is equal to the minor axis of the ellipse. If e_1 and e_2 are the eccentricities of the ellipse and hyperbola, then the value of $(1/e_1^2) + (1/e_2^2)$ is ______.

5. If the product of the lengths of perpendiculars drawn from any point on the hyperbola $X^2 - 2y^2 - 2 = 0$ to its asymptotes is p, then the value of $[p]$, where [.] represents the greatest integer function, is ______.

6. PQ and RS are two perpendicular chords of the rectangular hyperbola $xy = c^2$. If C is the centre of the rectangular hyperbola, then the value of the product of the slopes of CP, CQ, CR, and CS is ___.

7. The number of integral values of a for which a unique circle passes through the points of intersection of the rectangular hyperbola $x^2 - y^2 = a^2$ and the parabola $y = x^2$ is______.

8. Two tangents to the hyperbola $(x^2/a^2) - (y^2/b^2) = 1$ having slopes m_1 and m_2 intersect the axes at four concyclic points. Then the value of m_1m_2 is ______.

9. The slope of one of the common tangents to the hyperbolas $(x^2/a^2) - (y^2/b^2) = 1$ and $(y^2/a^2) - (x^2/b^2) = 1$ is______.

ANSWERS AND SOLUTIONS

Objective Type

1. a. According to the question

$$\frac{2\sqrt{9m^2-49}}{\sqrt{1+m^2}}=2$$

$\Rightarrow 9m^2-49=1+m^2$

$\Rightarrow 8m^2=50$

$\Rightarrow m=\pm\frac{5}{2}$

2. a. Given that

$$\frac{\text{Distance between foci}}{\text{Distance between two directrix}}=\frac{3}{2}$$

$\Rightarrow \frac{2ae}{2\frac{a}{e}}=\frac{3}{2}$

$\Rightarrow e^2=\frac{3}{2}$

$\Rightarrow 1+\frac{b^2}{a^2}=\frac{3}{2}$

$\Rightarrow \frac{b}{a}=\frac{1}{\sqrt{2}}$

3. c. $P\left(a\sec\frac{\pi}{2}, b\tan\frac{\pi}{6}\right)\equiv P\left(\frac{2a}{\sqrt{3}},\frac{b}{\sqrt{3}}\right)$

Therefore, equation of tangent at P is

$$\frac{x}{\frac{\sqrt{3}a}{2}}-\frac{y}{\sqrt{3}b}=1$$

Area of the triangle $=\frac{1}{2}\times\frac{\sqrt{3}a}{2}\times\sqrt{3}b=3a^2$

$\Rightarrow \frac{b}{a}=4$

$\therefore e^2=1+\frac{b^2}{a^2}=17$

4. b. Transverse axis is along the line $y=x$.

Solving $y=x$ and $xy=18$, we have $x^2=18$ or $x=\pm3\sqrt{2}$.

Then two vertices of the hyperbola are $(\pm3\sqrt{2},\pm3\sqrt{2})$.

Distance between them $=\sqrt{72+72}=12$.

5. a. Let the given straight line be axis of coordinates and let the equation of the variable line be

$$\frac{x}{a}+\frac{y}{b}=1$$

This line cuts the coordinate axis at the point $A\ (a, 0)$ and $B\ (0, b)$.

Therefore, the area of ΔAOB is

$\frac{1}{2}ab=c^2$

$\Rightarrow ab=2c^2 \quad (1)$

If (h, k) are the coordinates of the middle point of AB, then

$h=\frac{a}{2}$ and $k=\frac{b}{2} \quad (2)$

On eliminating a and b from Eqs. (1) and (2), we get

$2hk=c^2$

Hence, the locus of (h, k) is $2xy=c^2$.

6. b. Since $\frac{e}{2}$ and $\frac{e'}{2}$ are eccentricities of a hyperbola and its conjugate hyperbola,

$$\frac{4}{e^2}+\frac{4}{e'^2}=1$$

$\Rightarrow 4=\frac{e^2e'^2}{e'^2+e'^2}$

The line passing through the points $(e, 0)$ and $(0, e')$ is $e'x+ey-ee'=0$.

It is tangent to the circle $x^2+y^2=r^2$.

Hence, $\frac{ee'}{\sqrt{e^2+e'^2}}=r$

$\Rightarrow r^2=\frac{e^2e'^2}{e^2+e'^2}=4$

$\Rightarrow r=2$

7. c. $(x-3)^2+(y+1)^2=(4x+3y)^2$

$\Rightarrow (x-3)^2+(y+1)^2=25\left(\frac{4x+3y}{5}\right)^2$

$\Rightarrow PS=5PM$

$\Rightarrow$ directrix is $4x+3y=0$ and focus $(3, -1)$

So equation of transverse axis is $y+1=\frac{3}{4}(x-3)$

$\Rightarrow 3x-4y=13$

8. a. Given that

$$y\sqrt{1+\left(\frac{dy}{dx}\right)^2}=\sqrt{x^2+y^2}$$

$\Rightarrow y^2\left(\frac{dy}{dx}\right)=x^2$

$\Rightarrow ydy=\pm xdx$

$\Rightarrow y^2\pm x^2=k^2$

$\Rightarrow$ Family of curves may be a circle or rectangular hyperbola.

9. b. We have

$x^2-y^2-4x+4y+16=0$

$\Rightarrow (x-2)^2-(y-2)^2=-16$

$\Rightarrow \frac{(x-2)^2}{4^2}-\frac{(y-2)^2}{4^2}=-1$

This is a rectangular hyperbola, whose eccentricity is always $\sqrt{2}$.

10. d. We have

$16(x^2 - 2x) - 3(y^2 - 4y) = 44$

$\Rightarrow 16(x-1)^2 - 3(y-2)^2 = 48$

$\Rightarrow \frac{(x-1)^2}{3} - \frac{(y-2)^2}{16} = 1$

This equation represents a hyperbola with eccentricity

$e = \sqrt{1 + \frac{16}{3}} = \sqrt{\frac{19}{3}}.$

11. c. Let the hyperbola be $\frac{x^2}{a^2} - \frac{y^2}{b^2} = 1$, then

$2a = ae$, i.e., $e = 2$

$\therefore \frac{b^2}{a^2} = e^2 - 1 = 3$

$\therefore \frac{(2b)^2}{(2a)^2} = 3$

12. c. We have

$\frac{2b^2}{a} = 8$

and $2b = \frac{1}{2}(2ae)$

$\therefore \frac{2}{a}\left(\frac{ae}{2}\right)^2 = 8$

$\Rightarrow ae^2 = 16$ (1)

Also, $2\frac{b^2}{a} = 8$

$\Rightarrow b^2 = 4a$

$\Rightarrow a^2(e^2 - 1) = 4a$

$\Rightarrow ae^2 - a = 4$ (2)

From (1) and (2), we have

$16 - \frac{16}{e^2} = 4$

$\Rightarrow \frac{16}{e^2} = 12$

$\Rightarrow e = \frac{2}{\sqrt{3}}$

13. d.

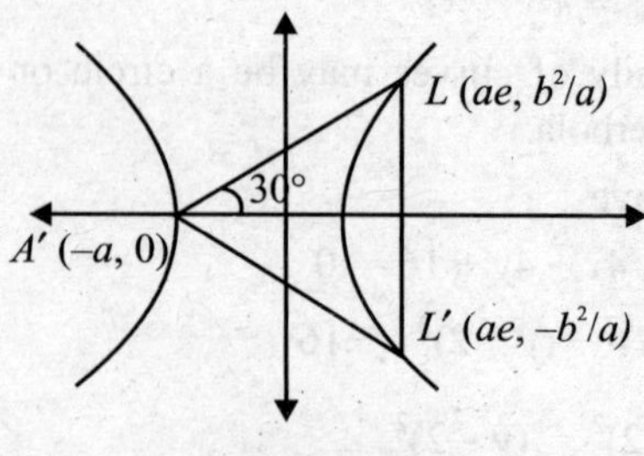

Fig. 5.2

$\tan 30° = \frac{\frac{b^2}{a}}{a + ae}$

$\Rightarrow \frac{1+e}{\sqrt{3}} = e^2 - 1$

$\Rightarrow e - 1 = \frac{1}{\sqrt{3}}$

$\Rightarrow e = \frac{\sqrt{3}+1}{\sqrt{3}}$

14. d. Hyperbola $9x^2 - 16y^2 - 18x - 32y - 151 = 0$ can be written as

$9(x^2 - 2x) - 16(y^2 + 2y) = 151$

$\Rightarrow 9(x-1)^2 - 16(y+1)^2 = 151 + 9 - 16 = 144$

$\Rightarrow \frac{(x-1)^2}{16} - \frac{(y+1)^2}{9} = 1$

or $\frac{X^2}{16} - \frac{Y^2}{9} = 1$ [where $X = x - 1$, $Y = y + 1$]

Here $a^2 = 16$, $b^2 = 9$

Latus rectum $= 2\frac{b^2}{a} = \frac{2(9)}{4} = \frac{9}{2}.$

15. a. Given hyperbola is

$\frac{x^2}{1} - \frac{y^2}{1/3} = 1$

Its eccentricity e is given by

$\frac{1}{3} = 1(e^2 - 1)$

Hence, eccentricity e' of the conjugate hyperbola is given by

$1 = \frac{1}{3}(e'^2 - 1)$

$\Rightarrow e'^2 = 4$

$\Rightarrow e' = 2$

16. b. The equation of the hyperbola is

$$\frac{\left(\frac{2x - y + 4}{\sqrt{5}}\right)^2}{\frac{1}{2}} - \frac{\left(\frac{x + 2y - 3}{\sqrt{5}}\right)^2}{\frac{1}{3}} = 1$$

or $\frac{2}{5}(2x - y + 4)^2 - \frac{3}{5}(x + 2y - 3)^2 = 1$

17. b. Eliminating t from the given two equation, we have

$\frac{x^2}{16} - \frac{y^2}{48} = 1,$

whose eccentricity is

$e = \sqrt{1 + \frac{48}{16}} = 2.$

18. b. For hyperbola

$\frac{x^2}{5} - \frac{y^2}{5\cos^2\alpha} = 1,$

we have

$$e_1^2 = 1 + \frac{b^2}{a^2} = 1 + \frac{5\cos^2\alpha}{5}$$

$$= 1 + \cos^2\alpha$$

For the ellipse

$$\frac{x^2}{25\cos^2\alpha} + \frac{y^2}{25} = 1,$$

we have

$$e_2^2 = 1 - \frac{25\cos^2\alpha}{25} = \sin^2\alpha$$

Given that $e_1 = \sqrt{3}e_2$. So

$$e_1^2 = 3e_2^2$$

$$\Rightarrow 1 + \cos^2\alpha = 3\sin^2\alpha$$

$$\Rightarrow 2 = 4\sin^2\alpha$$

$$\Rightarrow \sin\alpha = \frac{1}{\sqrt{2}}$$

19. b. Given hyperbola is

$$\frac{x^2}{9} - \frac{y^2}{16} = 1 \quad (1)$$

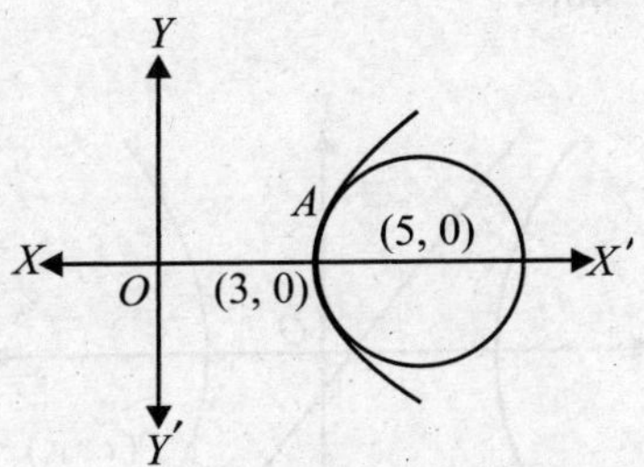

Fig. 5.3

$$\Rightarrow e^2 = 1 + \frac{16}{9} = \frac{25}{9}$$

$$\Rightarrow e = \frac{5}{3}$$

Hence, its foci are $(\pm 5, 0)$.

The equation of the circle with (5, 0) as centre is

$$(x-5)^2 + y^2 = r^2 \quad (2)$$

Solving (1) and (2), we have

$$16x^2 - 9[r^2 - (x-5)^2] = 144$$

or $25x^2 - 90x - 9r^2 + 81 = 0$

Since the circle touches the hyperbola, above equation must have equal roots. Hence,

$$90^2 - 4(25)(81 - 9r^2) = 0$$

$$\Rightarrow 9 - (9 - r^2) = 0$$

$\Rightarrow r = 0$, which is not possible.

Hence, the circle cannot touch at two points.

It can only be tangent at the vertex. Hence,

$r = 5 - 3 = 2.$

20. a. Any tangent to hyperbola is

$$\frac{x}{a}\sec\theta - \frac{y}{b}\tan\theta = 1 \quad (1)$$

Given tangent is

$$ax + by = 1 \quad (2)$$

Comparing Eqs. (1) and (2), we have

$\sec\theta = a^2$ and $\tan\theta = -b^2$

Eliminating θ, we have

$$a^4 - b^4 = 1$$

$$\Rightarrow (a^2 - b^2)(a^2 + b^2) = 1$$

Also $a^2 + b^2 = a^2e^2$

$$\Rightarrow a^2 - b^2 = \frac{1}{a^2e^2}$$

21. c. Let the point be (h, k).

Then equation of the chord of contact is $hx + ky = 4$.

Since $hx + ky = 4$ is tangent to $xy = 1$

$\therefore x\left(\frac{4 - hx}{k}\right) = 1$ has two equal roots

$$\Rightarrow hx^2 - 4x + k = 0$$

$$\Rightarrow hk = 4$$

$\Rightarrow$ locus of (h, k) is $xy = 4$

$$\Rightarrow c^2 = 4$$

22. d. Let the vertex A be $(a\cos\theta, b\sin\theta)$.

Since AC and AB touch the hyperbola, $\frac{x^2}{a^2} - \frac{y^2}{b^2} = -1$,

BC is the chord of contact. Its equation is

$$\frac{x\cos\theta}{a} - \frac{y\sin\theta}{b} = -1$$

or $-\frac{x\cos\theta}{a} + \frac{y\sin\theta}{b} = 1$

or $\frac{x\cos(\pi - \theta)}{a} + \frac{y\sin(\pi - \theta)}{b} = 1$

which is the equation of the tangent to the ellipse $\frac{x^2}{a^2} + \frac{y^2}{b^2} = 1$ at the point $(a\cos(\pi - \theta), b\sin(\pi - \theta))$.

Hence, BC touches the given ellipse.

23. b. Equation of tangent at (x_1, y_1) is

$$\frac{xx_1}{a^2} - \frac{yy_1}{b^2} = 1$$

It passing through $(0, -b)$. So,

$$0 + \frac{y_1}{b} = 1 \Rightarrow y_1 = b$$

Equation of normal is

$$\frac{a^2x}{x_1}+\frac{b^2y}{y_1}=a^2e^2,$$

which passes through $(2a\sqrt{2}, 0)$. Hence,

$$\frac{a^2 2a\sqrt{2}}{x_1}=a^2e^2$$

$$\Rightarrow x_1=\frac{2a\sqrt{2}}{e^2}$$

Now (x_1, y_1) lies on the hyperbola

$$\frac{x_1^2}{a^2}-\frac{y_1^2}{b^2}=1$$

$$\Rightarrow \frac{8}{e^4}-1=1$$

$$\Rightarrow e^2=2$$

24. c. Equation of the hyperbola is

$$\frac{x^2}{9}-\frac{y^2}{16}=1$$

Equation of tangent is $y=mx+\sqrt{9m^2-16}$

$$\Rightarrow \sqrt{9m^2-16}=2\sqrt{5}$$

$$\Rightarrow m=\pm 2$$

$\Rightarrow a+b=$ sum of roots $=0.$

25. b.

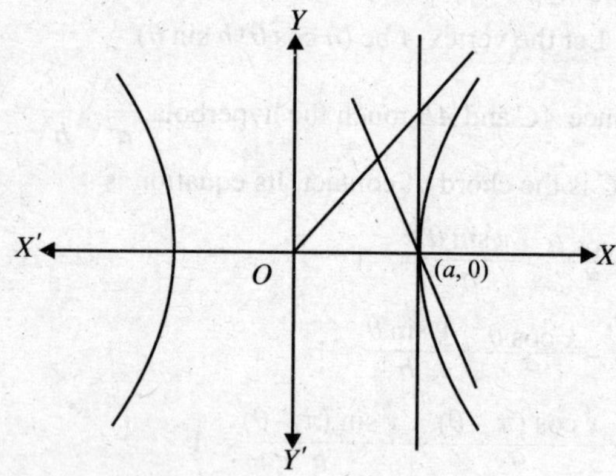

Fig. 5.4

The line $y+\lambda(x-a)=0$ will intersect the portion of the asymptote in the first quadrant only if its slope is negative. Hence, $-\lambda<0$

$$\Rightarrow \lambda>0$$

$$\therefore \lambda\in(0,\infty)$$

26. d. Product of perpendiculars drawn from foci upon any of its tangents $=9$

$$\Rightarrow b^2=9$$

Also, $\dfrac{b}{a}=\tan 30°=\dfrac{1}{\sqrt{3}}$

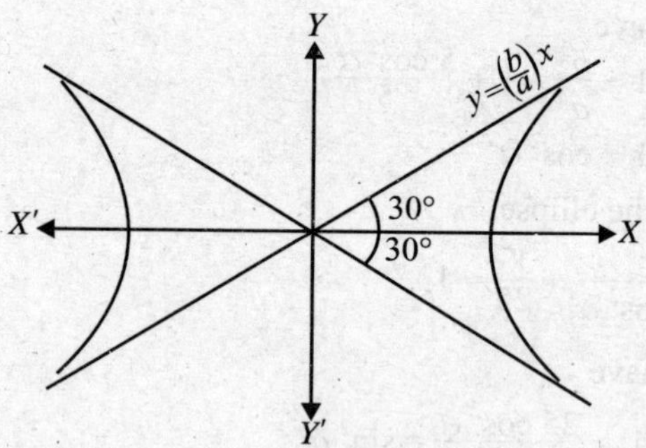

Fig. 5.5

$$\therefore a^2=3b^2=27$$

Therefore, required locus is the director circle of the hyperbola which is given by $x^2+y^2=27-9$ or $x^2+y^2=18$

If $\dfrac{b}{a}=\tan 60°$, then

$$a^2=\frac{b^2}{3}=\frac{9}{3}=3.$$

Hence, the required locus is $x^2+y^2=3-9=-6$, which is not possible.

27. c.

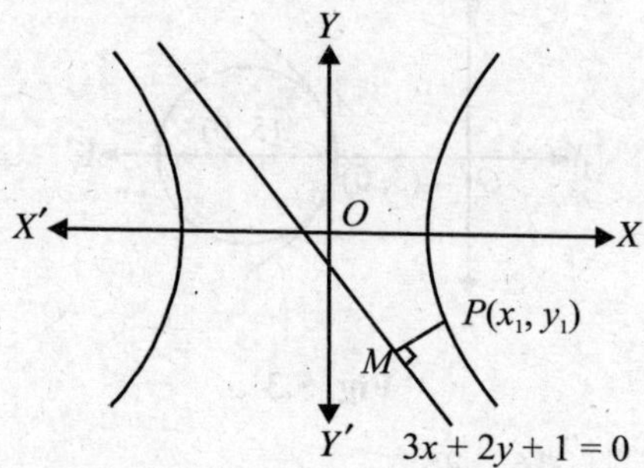

Fig. 5.6

Point P is nearest to the given line if tangent at P is parallel to the given line.

Now slope of tangent at $P(x_1, y_1)$ is

$\left(\dfrac{dy}{dx}\right)_{(x_1,y_1)}=\dfrac{18y_1}{24x_1}=\dfrac{3}{4}\dfrac{y_1}{x_1}$, which must be equal to $-\dfrac{3}{2}$

$$\Rightarrow \frac{3}{4}\frac{y_1}{x_1}=-\frac{3}{2}$$

$$\Rightarrow y_1=-2x_1 \quad (1)$$

Also (x_1, y_1) lies on the curve. Hence,

$$\frac{x_1^2}{24}-\frac{y_1^2}{18}=1 \quad (2)$$

Solving (1) and (2) we get two points $(6, -3)$ and $(-6, 3)$ of which $(6, -3)$ is nearest.

28. a. Tangent to $\dfrac{x^2}{9}-\dfrac{y^2}{4}=1$ at $P(3\sec\theta, 2\tan\theta)$ is

$$\frac{x}{3}\sec\theta-\frac{y}{2}\tan\theta=1$$

This is perpendicular to

$5x + 2y - 10 = 0$

$\Rightarrow \frac{2 \sec \theta}{3 \tan \theta} = \frac{2}{5}$

$\Rightarrow \sin \theta = \frac{5}{3}$ which is not possible.

Hence, there is no such tangent.

29. d. $\frac{y^2}{\frac{1}{16}} - \frac{x^2}{\frac{1}{9}} = 1$

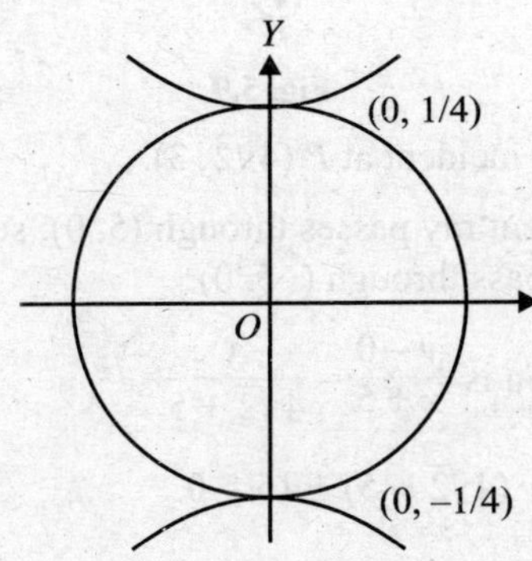

Fig. 5.7

Locus will be the auxiliary circle $x^2 + y^2 = 1/16$.

30. d. Let the foot of perpendicular from $O(0, 0)$ to tangent to hyperbola is $P(h, k)$. Slope of $OP = k/h$.

Then equation of tangent to hyperbola is

$y - k = -\frac{h}{k}(x - h)$

or $hx + ky = h^2 + k^2$

Solving it with $xy = 1$, we have

$hx + \frac{k}{x} = h^2 + k^2$

or $hx^2 - (h^2 + k^2)x + k = 0$

This equation must have real and equal roots. Hence, $D = 0$

$\Rightarrow (h^2 + k^2)^2 - 4hk = 0$

$\Rightarrow (x^2 + y^2)^2 = 4xy$

31. b.

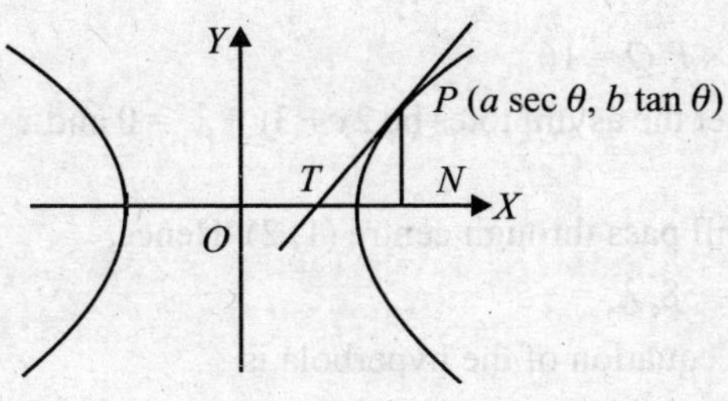

Fig. 5.8

Tangent at point P is

$\frac{x}{a} \sec \theta - \frac{y}{b} \tan \theta = 1.$

It meets the x-axis at point T $(a \cos \theta, 0)$ and foot of perpendicular from P to x-axis is N $(a \sec \theta, 0)$.

From the diagram, we have

$OT = a \cos \theta$ and $ON = a \sec \theta$

$\Rightarrow OT \times ON = a^2$

32. b. Let directrix be $x = a/e$ and focus be S $(ae, 0)$. Let $P(a \sec \theta, b \tan \theta)$ be any point on the curve.

Equation of tangent at P is

$\frac{x \sec \theta}{a} - \frac{y \tan \theta}{b} = 1.$

Let F be the intersection point of tangent of directrix.

Then $F = \left(\frac{a}{e}, \frac{b(\sec \theta - e)}{e \tan \theta}\right)$

$\Rightarrow m_{SF} = \frac{b(\sec \theta - e)}{-e \tan \theta (a^2 - 1)},$

and $m_{PS} = \frac{b \tan \theta}{a(\sec \theta - e)}$

$\Rightarrow m_{SF} \times m_{PS} = -1$

33. c. Let $y = mx \pm \sqrt{m^2 a^2 - a^2}$ be two tangents passing through (h, k). Then

$(k - mh)^2 = m^2a^2 - a^2$

$\Rightarrow m^2(h^2 - a^2) - 2khm + k^2 + a^2 = 0$

$\Rightarrow m_1 + m_2 = \frac{2kh}{h^2 - a^2}$

and $m_1 m_2 = \frac{k^2 + a^2}{h^2 - a^2},$

Now, $\tan 45° = \frac{m_1 - m_2}{1 + m_1 m_2}$

$\Rightarrow 1 = \frac{(m_1 + m_2)^2 - 4m_1 m_2}{(1 + m_1 m_2)^2}$

$\Rightarrow \left(1 + \frac{k^2 + a^2}{h^2 - a^2}\right)^2 = \left(\frac{2kh}{h^2 - a^2}\right)^2 - 4\left(\frac{k^2 + a^2}{h^2 - a^2}\right)$

$\Rightarrow (h^2 + k^2)^2 = 4h^2k^2 - 4(k^2 + a^2)(h^2 - a^2)$

$\Rightarrow (x^2 + y^2)^2 = 4(a^2y^2 - a^2x^2 + a^4)$

$\Rightarrow (x^2 + y^2)^2 + 4a^2(x^2 - y^2) = 4a^2.$

34. c. Fourth vertex of parallelogram lies on circumcircle

$\Rightarrow$ parallelogram is cyclic.

⇒ parallelogram is a rectangle

⇒ tangents are perpendicular

⇒ locus of P is the director circle

35. a. Director circle of circle $x^2 + y^2 = a^2$ is $x^2 + y^2 = 2a^2$.

The semi-transverse axis is $\sqrt{3}a$.

Radius of the circle is $\sqrt{2}a$.

Hence, director circle and hyperbola do not intersect.

36. d. The equation of the normal to the hyperbola $\frac{x^2}{4} - \frac{y^2}{1} = 1$ at $(2\sec\theta, \tan\theta)$ is $2x\cos\theta + y\cot\theta = 5$

Slope of the normal is $-2\sin\theta = -1$

$$\Rightarrow \sin\theta = \frac{1}{2} \Rightarrow \theta = \frac{\pi}{6}$$

$$y\text{-intercept of the normal} = \frac{5}{\cot\theta} = \frac{5}{\sqrt{3}}$$

As it touches the ellipse $\frac{x^2}{a^2} + \frac{y^2}{b^2} = 1$, we have

$$a^2 + b^2 = \frac{25}{9}.$$

37. b. Equation of normals is given by $ty = t^3x - ct^4 + c = 0$. It passes through $(ct', c/t')$. Hence,

$$\frac{c}{t'} = t^3ct' - ct^4 + c = 0$$

$$t = t^3t'^2 - tt'^4 + t'$$

$$t^3t' = -1$$

38. a. Equation of normal at any point $\left(ct, \frac{c}{t}\right)$ is

$$ct^4 - xt^3 + ty - c = 0$$

$\Rightarrow$ Slope of normal $= t^2$

Let P be (h, k). Then,

$$ct^4 - ht^3 + tk - c = 0$$

$$\Rightarrow \Sigma t_i = \frac{h}{c} \text{ and } \Sigma t_i t_j = 0$$

$$\Rightarrow \Sigma t_i^2 = (\Sigma t_i)^2$$

$$\Rightarrow h^2 = c^2\lambda$$

Therefore, the required locus is $x^2 = \lambda c^2$.

39. d. The given hyperbola is

$$\frac{x^2}{16} - \frac{y^2}{9} = 1$$

$$\Rightarrow e = \frac{5}{4}$$

Its foci are $(\pm 5, 0)$.

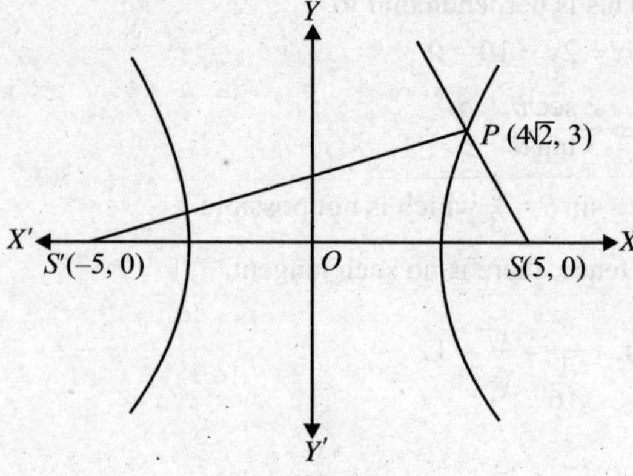

Fig. 5.9

The ray is incident at $P(4\sqrt{2}, 3)$.

The incident ray passes through (5, 0); so the reflected ray must pass through (−5, 0).

Its equation is $\frac{y - 0}{x + 5} = \frac{3}{4\sqrt{2} + 5}$

or $\quad 3x - y(4\sqrt{2} + 5) + 15 = 0$.

40. b.

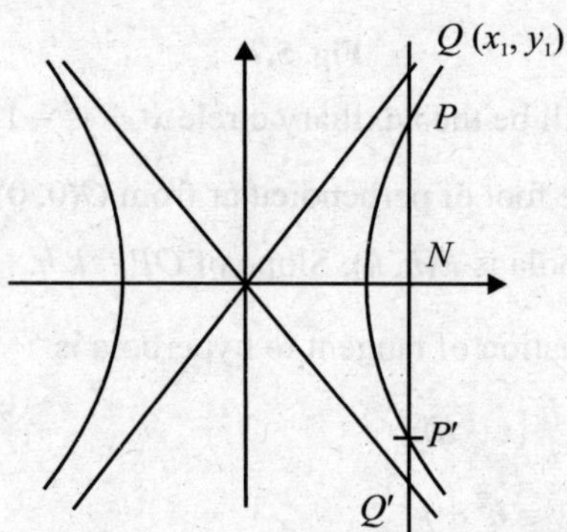

Fig. 5.10

$$NP = \frac{4}{5}\sqrt{x_1^2 - 25}$$

Q is on $y = \frac{4}{5}x$

$$NQ = \frac{4}{5}x_1$$

$$PQ = NQ - NP = \frac{4}{5}\left(x_1 - \sqrt{x_1^2 - 25}\right)$$

$$P'Q = \frac{4}{5}\left(x_1 + \sqrt{x_1^2 - 25}\right)$$

$$PQ \times P'Q = 16$$

41. b. Let the asymptotes be $2x + 3y + \lambda_1 = 0$ and $x + 2y + \lambda_2 = 0$

It will pass through centre (1, 2). Hence,

$\lambda_1 = -8, \lambda_2 = -5$

The equation of the hyperbola is

$(2x + 3y - 8)(x + 2y - 5) + \lambda = 0$

It passes through (2, 4),

$(4 + 12 - 8)(2 + 8 - 5) + \lambda = 0 \Rightarrow \lambda = -40$

Hence, equation of hyperbola is

$(2x + 3y - 8)(x + 2y - 5) = 40$

42. c. Let P be (h, k) be any point. The chord of contact of P w.r.t. the hyperbola is

$$\frac{hx}{a^2} - \frac{ky}{b^2} = 1 \quad (1)$$

The chord of contact of P w.r.t. the auxiliary circle is

$$hx + ky = a^2 - b^2 \quad (2)$$

Now, $\frac{h}{a^2} \times \frac{b^2}{k} \times \left(-\frac{h}{k}\right) = -1$

$\Rightarrow \frac{h^2}{a^2} - \frac{k^2}{b^2} = 0$

Therefore, P lies on one of the asymptotes.

43. c. Slopes of asymptotes are

$m_1 = \frac{b_1}{a_1}, m_2 = \frac{b_2}{a_2}$

According to the question,

$m_1 m_2 = -1$

$\Rightarrow a_1 a_2 + b_1 b_2 = 0$

44. c. For equation $S + K = 0$ to represent a pair of lines,

$$\begin{vmatrix} 1 & 2 & -2 \\ 2 & 3 & 1 \\ -2 & 1 & 1+k \end{vmatrix} = 0$$

$\Rightarrow 3(1 + k) - 1 - 2(2 + 2k + 2) - 2(2 + 6) = 0$

$\Rightarrow k = -22$

45. a.

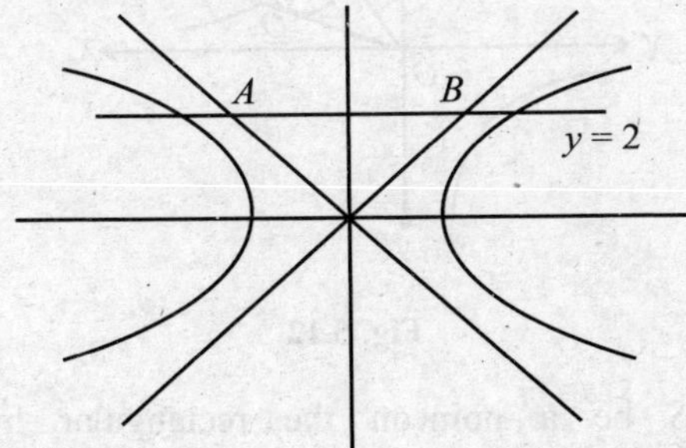

Fig. 5.11

For two distinct tangents on different branches the point should lie on the line $y = 2$ and between A and B (where A and B are the points on the asymptotes).

Equations of asymptotes are $4x = \pm 3y$

Solving with $y = 2$, we have

$x = \pm\frac{3}{2}$

$\therefore -\frac{3}{2} < \alpha < \frac{3}{2}$

46. d. Transverse axis is the equation of the angle bisector passing containing point (2, 3), which is given by

$$\frac{3x - 4y + 5}{5} = \frac{12x + 5y - 40}{13}$$

$\Rightarrow 21x + 77y = 265$

47. d. Let $P(x_1, y_1)$ be a point on the hyperbola $\frac{x^2}{a^2} - \frac{y^2}{b^2} = 1$.

The chord of contact of tangents from P to the hyperbola is given by

$$\frac{xx_1}{a^2} - \frac{yy_1}{b^2} = 1 \quad (1)$$

The equations of the asymptotes are

$\frac{x}{a} - \frac{y}{b} = 0$

and $\frac{x}{a} + \frac{y}{b} = 0$

The points of intersection of (1) with the two asymptotes are given by

$$x_1 = \frac{2a}{\frac{x_1}{a} - \frac{y_1}{b}}, \ y_1 = \frac{2b}{\frac{x_1}{a} - \frac{y_1}{b}}$$

$$x_2 = \frac{2a}{\frac{x_1}{a} + \frac{y_1}{b}}, \ y_2 = \frac{2b}{\frac{x_1}{a} + \frac{y_1}{b}}$$

Area of the said triangle $= \frac{1}{2}(x_1 y_2 - x_2 y_1)$

$$= \frac{1}{2}\left|\left(-\frac{4ab \times 2}{\frac{x_1^2}{a^2} - \frac{y_1^2}{b^2}}\right)\right| = 4ab$$

48. b. Since $c_1 c_2 (a_1 a_2 + b_1 b_2) < 0$, origin lies in acute angle.

$P(1, 2)$ lies in obtuse angle.

Acute angle between the asymptotes is $\frac{\pi}{3}$. Hence,

$e = \sec\frac{\theta}{2} = \sec\frac{\pi}{6} = \frac{2}{\sqrt{3}}$

49. a. Let the equation of asymptotes be

$$2x^2 + 5xy + 2y^2 + 4x + 5y + \lambda = 0 \quad (1)$$

This equation represents a pair of straight lines.

Therefore, $abc + 2fgh - af^2 - bg^2 - ch^2 = 0$.

We have

$4\lambda + 25 - \frac{25}{2} - 8 - \lambda\frac{25}{4} = 0$

$\Rightarrow -\frac{9\lambda}{4} + \frac{9}{2} = 0$

$\Rightarrow \lambda = 2$

Putting the value of λ in (1), we get

$2x^2 + 5xy + 2y^2 + 4x + 5y + 2 = 0$.

This is the equation of the asymptotes.

50. a. The given hyperbola is

$$xy - hx - ky = 0 \quad (1)$$

The equation of asymptotes is given by

$xy - hx - ky + c = 0$ (2)

Eq. (2) gives a pair of straight lines. So,

$$\begin{vmatrix} A & H & G \\ H & B & F \\ G & F & C \end{vmatrix} = 0 \Rightarrow \begin{vmatrix} 0 & \frac{1}{2} & -\frac{h}{2} \\ \frac{1}{2} & 0 & -\frac{k}{2} \\ -\frac{h}{2} & -\frac{k}{2} & c \end{vmatrix} = 0$$

$$\Rightarrow \frac{hk}{8} + \frac{hk}{8} - \frac{c}{4} = 0$$

$$\Rightarrow c = hk$$

Hence, asymptotes are

$xy - hx - ky + hk = 0$

or $(x - k)(y - h) = 0$

51. a. Any tangent to hyperbola forms a triangle with the asymptotes which has constant area ab.

Given $ab = a^2 \tan \lambda$, So

$$\frac{b}{a} = \tan \lambda$$

$$\Rightarrow e^2 - 1 = \tan^2 \lambda$$

$$\Rightarrow e^2 = 1 + \tan^2 \lambda = \sec^2 \lambda$$

$$\Rightarrow e = \sec \lambda$$

52. a. The asymptotes of a rectangular hyperbola are perpendicular to each other.

Given one asymptote, $x + y + c = 0$. Let the other asymptote be $x - y + \lambda = 0$.

We also know that the asymptotes pass through centre of the hyperbola. Therefore, the line $2x - y = 0$ and the asymptotes must be concurrent.

Thus, we have

$$\begin{vmatrix} 2 & -1 & 0 \\ 1 & 1 & c \\ 1 & -1 & \lambda \end{vmatrix} = 0$$

$$\Rightarrow \lambda = -3c$$

53. d. The equation of rectangular hyperbola is

$(x - 3)(y - 5) + \lambda = 0,$

which passes through (7, 8). Hence,

$4 \times 3 + \lambda \Rightarrow \lambda = -12$

$\therefore xy - 5x - 3y + 15 - 12 = 0$

$\Rightarrow xy - 3y - 5x + 3 = 0$

54. c. Foci of hyperbola lie on $y = x$. So, the major axis is $y = x$. Major axis of hyperbola bisects the asymptote. So equation of other asymptote is $x = 2y$.

Equation of hyperbola is $(y - 2x)(x - 2y) + k = 0$.

Given that it passes through $(3, 4) \Rightarrow k = 10$.

Hence, required equation is

$2x^2 + 2y^2 - 5xy + 10 = 0$

55. a. If (x_i, y_i) is the point of intersection of given curves, then

$$\frac{\sum_{j=1}^{4} x_i}{4} = \frac{1+1}{2} \text{ and } \frac{\sum_{j=1}^{4} y_i}{4} = 0$$

$$\text{Now } \frac{\sum_{i=1}^{3} x_i}{3} = \frac{4 - x_4}{3} \text{ and } \frac{\sum_{i=1}^{3} y_i}{3} = -\frac{y_4}{4}$$

Centroid $\left(\frac{\sum_{j=1}^{3} x_i}{3}, \frac{\sum_{i=1}^{3} y_i}{3}\right)$ lies on the line $y = 3x - 4$.

Hence,

$$\frac{-y_4}{3} = \frac{3(4 - x_4)}{3} - 4$$

$$\Rightarrow y_4 = 3x_4$$

Therefore, the locus of D is $y = 3x$.

56. b.

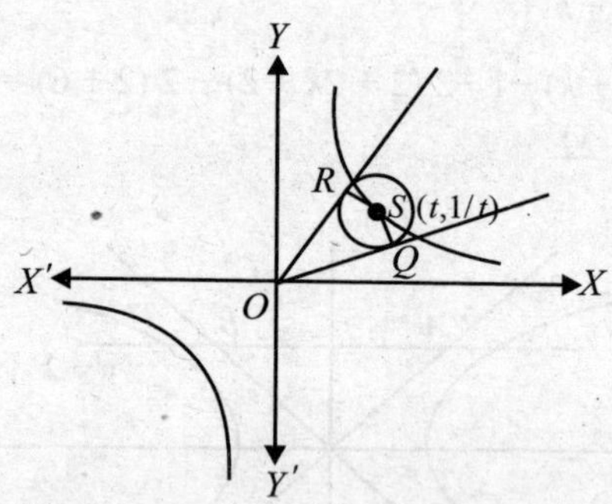

Fig. 5.12

Let S be a point on the rectangular hyperbola $\left[\text{say}\left(t, \frac{1}{t}\right)\right]$

Now, circumcircle of ΔOQR also passes through S. Therefore, circumcentre is the midpoint of OS. Hence,

$$x = \frac{t}{2}, y = \frac{1}{2t}$$

So, the locus of the circum centre is $xy = 1/4$.

57. a. The points are such that one of the points is the orthocentre of the triangle formed by other three points. When the vertices of a triangle lie on a rectangular hyperbola the orthocentre also lies on the same hyperbola.

58. b.

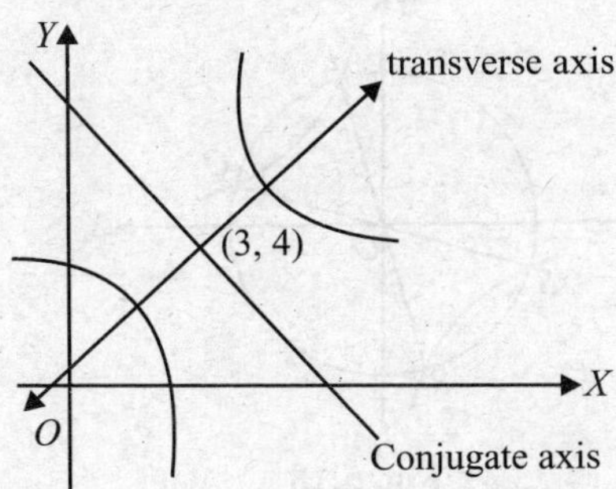

Fig. 5.13

$$(x-3)(y-4)=5$$

The axes of the hyperbola are $x = 3$ and $y = 4$.

Since the hyperbola is rectangular, axes are bisectors of these axes.

Hence, their slopes are ± 1, out of which conjugate axis has slope $m = -1$ and passes through (3, 4).

Hence, its equation is $y - 4 = -1(x - 3)$.

59. b.

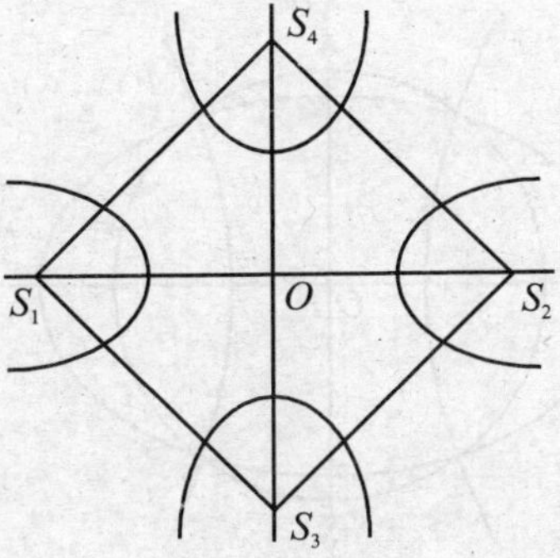

Fig. 5.14

Required area $= 4$ area $\Delta S_2OS_4 = 4 \times \frac{1}{2}\, ae \times 8\, be_1$

$-4 \times \frac{1}{2} \times 2 \times 3 \times ee_1$ (1)

$b^2 = a^2(e^2 - 1)$

$\Rightarrow e^2 = \frac{9}{4} + 1 = \frac{13}{4}$

Also $\frac{1}{e_1^2} = 1 - \frac{1}{e^2} = 1 - \frac{4}{13} = \frac{9}{13}$

$\Rightarrow e_1^2 = \frac{13}{9}$

Required area $= 12 \times \frac{\sqrt{13}}{2} \times \frac{\sqrt{13}}{3}$

$= 2 \times 13 = 26$

60. a. For $\lambda = -3$, the equation becomes $x^2 + y^2 - 3xy = 0$, which represents a pair of lines through origin.

61. d. $xy = c^2$

$\Rightarrow x\frac{dy}{dx} + y = 0$

Replacing $\frac{dy}{dx}$ by $-\frac{dx}{dy}$, we have

$-x\frac{dx}{dy} + y = 0$

$\Rightarrow ydy - xdx = 0$

Integrating, we have $x^2 - y^2 = k^2$, where k is the parameter which represents family of hyperbolas.

62. d.

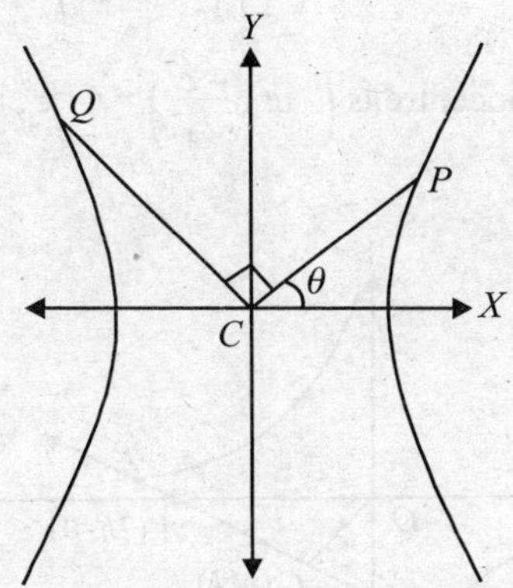

Fig. 5.15

Let $CP = r_1$ be inclined to transverse axis at an angle θ so that P is $(r_1 \cos\theta, r_1 \sin\theta)$ and P lies on the hyperbola. It gives

$$r_1^2\left(\frac{\cos^2\theta}{a^2} - \frac{\sin^2\theta}{b^2}\right) = 1$$

Replacing θ by $90° + \theta$, we have

$$r_2^2\left(\frac{\sin^2\theta}{a^2} - \frac{\cos^2\theta}{b^2}\right) = 1$$

$$\Rightarrow \frac{1}{r_1^2} + \frac{1}{r_2^2} = \frac{\cos^2\theta}{a^2} - \frac{\sin^2\theta}{b^2} + \frac{\sin^2\theta}{a^2} - \frac{\cos^2\theta}{b^2}$$

$$\Rightarrow \frac{1}{r_1^2} + \frac{1}{r_2^2} = \cos^2\theta\left(\frac{1}{a^2} - \frac{1}{b^2}\right) + \sin^2\theta \times \left(\frac{1}{a^2} - \frac{1}{b^2}\right)$$

$$\Rightarrow \frac{1}{r_1^2} + \frac{1}{r_2^2} = \frac{1}{a^2} - \frac{1}{b^2}$$

$$\Rightarrow \frac{1}{CP^2} + \frac{1}{CQ^2} = \frac{1}{a^2} - \frac{1}{b^2}$$

63. a. The midpoint of the chord is $\left(\frac{x_1 + x_2}{2}, \frac{y_1 + y_2}{2}\right)$.

The equation of the chord in terms of its midpoint is $T = S_1$.

$$\Rightarrow x\left(\frac{y_1+y_2}{2}\right)+y\left(\frac{x_1+x_2}{2}\right)-c^2=2\left(\frac{x_1+x_2}{2}\right)\left(\frac{y_1+y_2}{2}\right)-c^2$$

$$\Rightarrow x(y_1+y_2)+y(x_1+x_2)=(x_1+x_2)(y_1+y_2)$$

$$\Rightarrow \frac{x}{x_1+x_2}+\frac{y}{y_1+y_2}=1$$

64. c. A rectangular hyperbola circumscribing a triangle also passes through its orthocentre.
If $\left(ct_i, \frac{c}{t_i}\right)$, where $i = 1, 2, 3$, are the vertices of the triangle then the orthocentre is $\left(\frac{-c}{t_1t_2t_3}, -ct_1t_2t_3\right)$, where $t_1t_2t_3t_4 = 1$.

Hence, orthocentre is $\left(-ct_4, \frac{-c}{t_4}\right) = (-x_4, -y_4)$.

65. b.

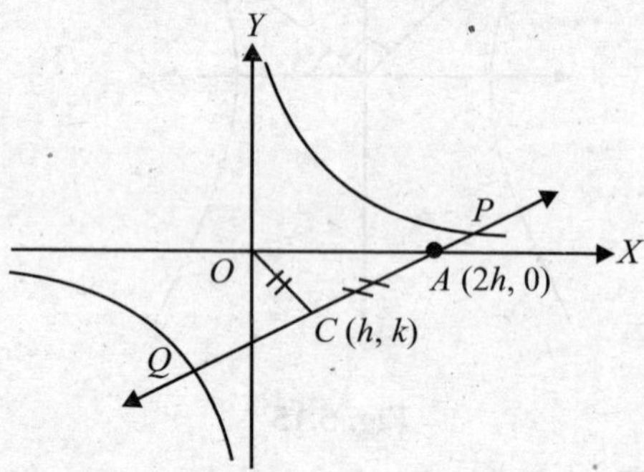

Fig. 5.16

Hyperbola is $xy = a^2$

or $2xy - 2a^2 = 0$

Chord with a given middle point is given by

$hy + kx - a^2 = 2hk - a^2$

$$\Rightarrow \frac{x}{h}+\frac{y}{k}=2$$

From the diagram, ΔOCA is isosceles with $OC = CA$.

66. b.

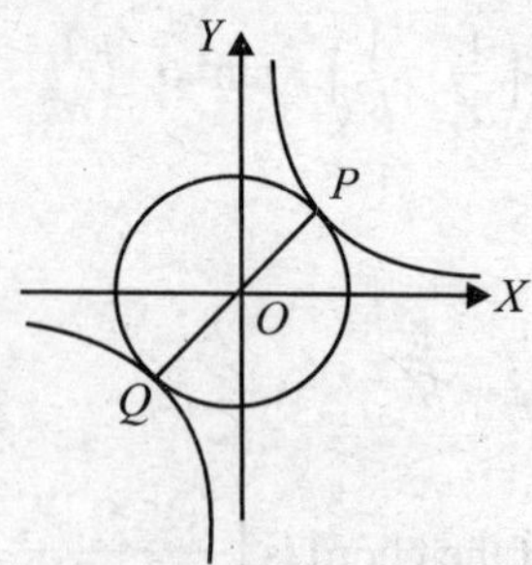

Fig. 5.17

From the diagram, PQ = diameter of the circle = 2

67. b. $x - 2 = m$

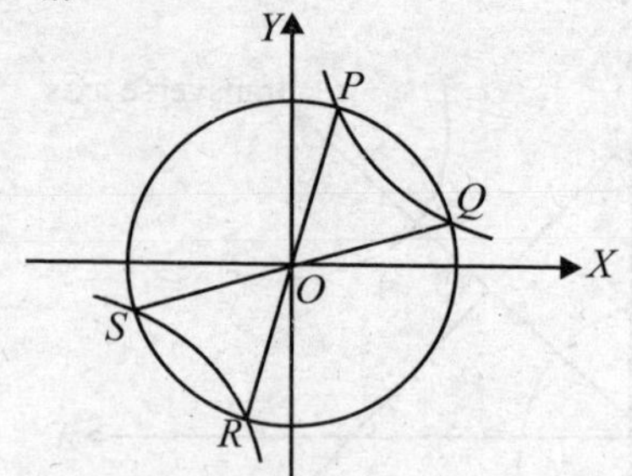

Fig. 5.18

$$y+1=\frac{4}{m}$$

$\therefore (x-2)(y+1) = 4$

$\Rightarrow XY = 4$, where $X = x - 2$, $Y = y + 1$

$S \equiv (x-2)^2 + (y+1)^2 = 25$

$\Rightarrow X^2 + Y^2 = 25$

Curve C and circle S both are concentric.

$\therefore OP^2 + OQ^2 + OR^2 + OS^2 = 4r^2 = 4 \times 25 = 100$

68. a.

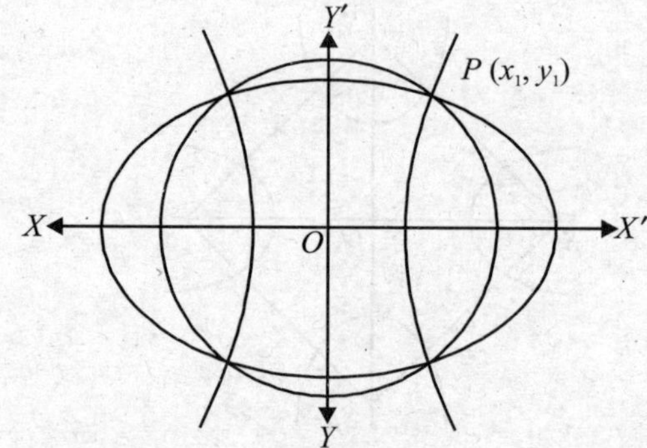

Fig. 5.19

Since ellipse and hyperbola intersect orthogonally, they are confocal. Hence, $a = 2$ (equating foci).

Let point of intersection in the first quadrant be $P(x_1, y_1)$. P lies on both the curves. Therefore,

$4x_1^2 + 9y_1^2 = 36$ and $4x_1^2 - y_1^2 = 4$.

Adding these two results, we get

$8(x_1^2 + y_1^2) = 40$

$\Rightarrow x_1^2 + y_1^2 = 5 \Rightarrow r = \sqrt{5}$

Hence, equation of the circle is $x^2 + y^2 = 5$.

69. b. The chord of contact of tangents from (x_1, y_1) is

$$\frac{xx_1}{a^2}+\frac{yy_1}{b^2}=1$$

It meets the axes at the points $\left(\frac{a^2}{x_1}, 0\right)$ and $\left(0, \frac{b^2}{y_1}\right)$.

Area of the triangle is $\frac{1}{2}\frac{a^2}{x_1}\frac{b^2}{y_1} = k$ (constant)

$\Rightarrow x_1 y_1 = \frac{a^2b^2}{2k} = c^2$ (c is constant)

$\Rightarrow xy = c^2$ is the required locus.

70. c.

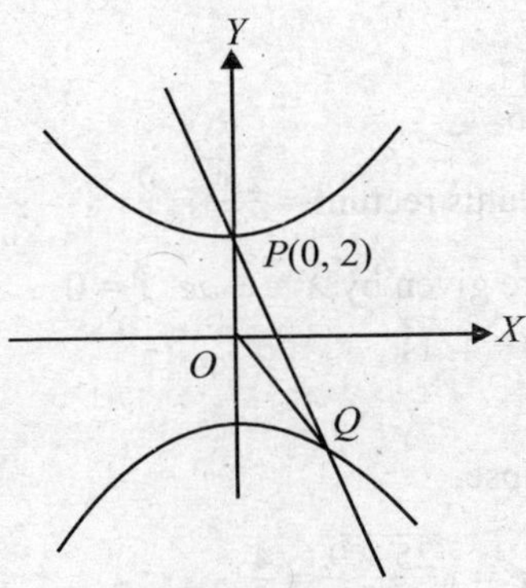

Fig. 5.20

Homogenizing the hyperbola using the straight line, we get pair of straight lines OP and OQ, which are given by

$$y^2 - x^2 = 4\left(\frac{\sqrt{3}x + y}{2}\right)^2$$

$$\Rightarrow y^2 - x^2 = 3x^2 + y^2 + 2\sqrt{3}xy$$

$$\Rightarrow 4x^2 + 2\sqrt{3}\ xy = 0$$

$$\Rightarrow x = 0 \text{ and } 2x + \sqrt{3}\ y = 0$$

Angle between the lines is

$$\frac{\pi}{2} - \tan^{-1}\left(\frac{2}{\sqrt{3}}\right) = \tan^{-1}\left(\frac{\sqrt{3}}{2}\right).$$

71. a. Let the variable chord be

$$x\cos\alpha + y\sin\alpha = p \qquad (1)$$

Let this chord intersect the hyperbola at A and B. Then the combined equation of OA and OB is given by

$$\frac{x^2}{a^2} - \frac{y^2}{b^2} = \left(\frac{x\cos\alpha + y\sin\alpha}{p}\right)^2$$

$$\Rightarrow x^2\left(\frac{1}{a^2} - \frac{\cos^2\alpha}{p^2}\right) - y^2\left(\frac{1}{b^2} + \frac{\sin^2\alpha}{p^2}\right) - \frac{2\sin\alpha\cos\alpha}{p}xy = 0$$

This chord subtends a right angle at the centre. Therefore,

Coefficient of x^2 + coefficient of $y^2 = 0$

$$\Rightarrow \frac{1}{a^2} - \frac{\cos^2\alpha}{p^2} - \frac{1}{b^2} - \frac{\sin^2\alpha}{p^2} = 0$$

$$\Rightarrow \frac{1}{a^2} - \frac{1}{b^2} = \frac{1}{p^2}$$

$$\Rightarrow p^2 = \frac{a^2b^2}{b^2 - a^2}$$

Hence, p is constant, i.e. it touches the fixed circle.

72. a. Equation of tangent at point $P(\alpha\cos\theta, \sin\theta)$ is

$$\frac{x}{\alpha}\cos\theta + \frac{y}{1}\sin\theta = 1 \qquad (1)$$

Let it cut the hyperbola at points P and Q.

Homogenizing the hyperbola $\alpha^2x^2 - y^2 = 1$ with the help of the above the equation, we get

$$\alpha^2x^2 - y^2 = \left(\frac{x}{\alpha}\cos\theta + y\sin\theta\right)^2.$$

This is a pair of straight lines OP and OQ.

Given $\angle POQ = \frac{\pi}{2}$. So

Coefficient of x^2 + coefficient of $y^2 = 0$

$$\Rightarrow \alpha^2 - \frac{\cos^2\theta}{\alpha^2} - 1 - \sin^2\theta = 0$$

$$\Rightarrow \alpha^2 - \frac{\cos^2\theta}{\alpha^2} - 1 - 1 + \cos^2\theta = 0$$

$$\Rightarrow \cos^2\theta = \frac{\alpha^2(2 - \alpha^2)}{\alpha^2 - 1}$$

Now, $0 \le \cos^2\theta \le 1$

$$\Rightarrow 0 \le \frac{\alpha^2(2 - \alpha^2)}{\alpha^2 - 1} \le 1$$

Solving, we get

$$\alpha^2 \in \left[\frac{\sqrt{5}+1}{2}, 2\right]$$

Multiple Correct Answers Type

1. a. We have,

$$\left|\sqrt{x^2 + (y-1)^2} - \sqrt{x^2 + (y+1)^2}\right| = K,$$

which is equivalent to $|S_1P - S_2P|$ = constant, where $S_1 \equiv (0, 1)$, $S_2 \equiv (0, -1)$ and $P \equiv (x, y)$.

The above equation represents a hyperbola. So, we have $2a = K$ and $2ae = S_1S_2 = 2$, where $2a$ is the transverse axis and e is the eccentricity.

Dividing, we have $e = 2/K$.

Since, $e > 1$ for a hyperbola, therefore $K < 2$.

Also, K must be a positive quantity.

So, we have $K \in (0, 2)$.

2. a., b., c.

$3x^4 - 2(19y + 8)x^2 + [(19y)^2 + (10)^2 + (10)^2 + y^4 + y^4 + 8^2] = 2(19 \times 10y + 10y^2 - 8y^2)$

$\Rightarrow 3x^4 - 2(19y + 8)x^2 + (19y - 10)^2 + (10 - y^2)^2 + (y^2 + 8)^2 = 0$

$\Rightarrow 3x^4 - 2(19y - 10 + 10 - y^2 + y^2 + 8)x^2 + (19y - 10)^2 + (10 - y^2)^2 + (y^2 + 8)^2 = 0$

$\Rightarrow [x^2 - (19y - 10)]^2 + [x^2 - (10 - y^2)]^2 + [x^2 - (y^2 + 8)]^2 = 0$

$\Rightarrow x^2 - 19y + 10 = 0, x^2 - 10 + y^2 = 0$ and $x^2 - y^2 - 8 = 0$

The three curves represented by the given equation are $x^2 = 19y - 10$ (parabola), $x^2 + y^2 = 10$ (circle) and $x^2 - y^2 = 8$ (hyperbola).

3. a., c. $|PS_1 - PS_2| = 2a$

$2a = K$

$\Rightarrow 2a = \sqrt{(24-0)^2 + (7-0)^2} - \sqrt{12^2 + 5^2} = 12$

$\therefore a = 6$

$2ae = \sqrt{(24-5)^2 + (12-7)^2}$

$= \sqrt{386}$

$\therefore e = \frac{\sqrt{386}}{12}$

Latus rectum $= \frac{2b^2}{a} = \frac{2a^2(e^2-1)}{a}$

$= 2 \times 6\left(\frac{386}{144} - 1\right) = \frac{121}{6}$

4. a., d.,

Let A (5, 12) and B (24, 7) be two fixed points. So $|OA - OB| = 12$ and $|OA + OB| = 38$.

If the conic is an ellipse, then

$e = \frac{\sqrt{386}}{38}$ $[\because 2ea = \sqrt{386}$ and $a = 19]$

If the conic is a hyperbola, then

$e = \frac{\sqrt{386}}{12}$ $[\because 2ae = \sqrt{386}$ and $a = 6]$

5. b. Locus of point of intersection of perpendicular tangents is director circle for $\frac{x^2}{a^2} - \frac{y^2}{b^2} = 1$. Equation of director circle is $x^2 + y^2 = a^2 - b^2$ which is real if $a > b$.

6. a., b., c., d.

Given hyperbola can be written as

$\frac{(x-1)^2}{16} - \frac{(y-1)^2}{9} = 1$

$\Rightarrow \frac{X^2}{16} - \frac{Y^2}{9} = 1$ [Where $X = x - 1$, $Y = y - 1$]

$e = \sqrt{1 + \frac{b^2}{a^2}} = \sqrt{1 + \frac{9}{16}} = \frac{5}{4}$

Directrices are $X = \pm \frac{a}{e}$

$\Rightarrow x - 1 = \pm \frac{16}{5} \Rightarrow x = \frac{21}{5}$

and $x = -\frac{11}{5}$.

Length of latus rectum $= \frac{2b^2}{a} = \frac{9}{2}$.

The foci are given by $X = \pm ae$, $Y = 0$

$\Rightarrow (6, 1), (-4, 1)$

7. a., b.,d.

For the ellipse,

$a = 5$ and $e = \sqrt{\frac{25-9}{25}} = \frac{4}{5}$

$\therefore ae = 4$

Hence, the foci are $(-4, 0)$ and $(4, 0)$.

For the hyperbola, $ae = 4$, $e = 2$

$\therefore a = 2$

$b^2 = 4(4 - 1) = 12$

$\Rightarrow b = \sqrt{12}$

8. a., b., c., d.

Solving, $xy = c^2$ and $x^2 + y^2 = a^2$, we have

$x^2 + \frac{c^4}{x^2} = a^2$

$\Rightarrow x^4 - a^2x^2 + c^4 = 0$

$\Rightarrow \Sigma x_i = 0$ and $x_1 x_2 x_3 x_4 = c^4$

Similarly, if we eliminate y, then $\Sigma y_i = 0$ and

$y_1 y_2 y_3 y_4 = c^4$

9. b., d.

$\frac{dx}{dy} = \frac{3y}{2x}$

$\Rightarrow \int 2x dx = \int 3y dy$

$\Rightarrow x^2 = \frac{3y^2}{2} + c$

or $\frac{x^2}{3} - \frac{y^2}{2} = \frac{c}{3}$

If c is positive, then

$$e=\sqrt{1+\frac{2}{3}}=\sqrt{\frac{5}{3}}$$

If c is negative, then

$$e=\sqrt{1+\frac{3}{2}}=\sqrt{\frac{5}{2}}$$

10. a., d.

Circle with points $P(2t_1, 2/t_1)$ and $Q(2t_2, 2/t_2)$ as diameter is given by

$$(x-2t_1)(x-2t_2)+\left(y-\frac{2}{t_1}\right)\left(y-\frac{2}{t_2}\right)=1 \qquad (1)$$

Also, slope of PQ is given by

$$-\frac{1}{t_1t_2}=1 \Rightarrow t_1t_2=-1$$

Hence, from (1), circle is

$(x^2+y^2-8)-(t_1+t_2)(x-y)=0$

which is of the form $S+\lambda L=0$.

Hence, circles passe through the points of intersection of the circle $x^2+y^2-8=0$ and the line $x=y$.

The points of intersection are (2, 2) and (–2, –2).

11. b., c., d.

$(x-\alpha)^2+(y-\beta)^2=k(lx+my+n)^2$

$$\Rightarrow \sqrt{(x-\alpha)^2+(y-\beta)^2}=\sqrt{k}\,\sqrt{l^2+m^2}\,\frac{(lx+my+n)}{\sqrt{l^2+m^2}}$$

$$\Rightarrow \frac{PS}{PM}=\sqrt{k}\,\sqrt{l^2+m^2},$$

where point $P(x, y)$ is any point on the curve.

Fixed point $S(\alpha, \beta)$ is focus and fixed line $lx+my+n=0$ is directrix.

If $k(l^2+m^2)=1$, P lies on a parabola.

If $k(l^2+m^2)<1$, P lies on an ellipse.

If $k(l^2+m^2)>1$, P lies on a hyperbola.

If $k=0$, P lies on a point circle.

12. a., b.

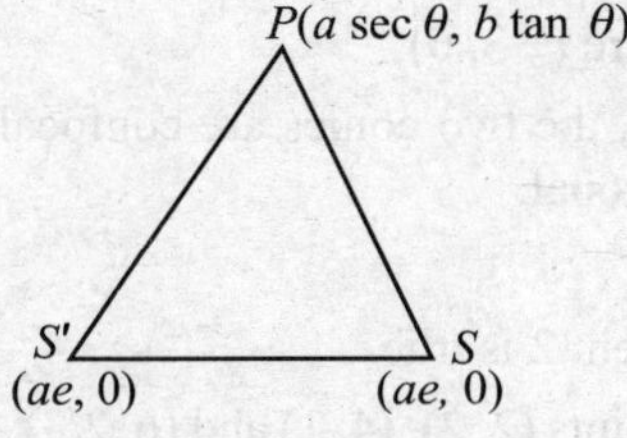

Fig. 5.21

Let (h, k) be excentre, then

$$h=\frac{a(ae\sec\theta+a)-ae(ae\sec\theta-a)-2ae(a\sec\theta)}{2ae(\sec\theta-1)}$$

$h=-a \Rightarrow x=-a$ (for $a\sec\theta>0$)

Similarly, $x=a$ for $a\sec\theta<0$

$\Rightarrow$ locus is $x^2=a^2$

Again let (h, k) be excentre opposite $\angle S'$,

$$\therefore h=\frac{2a^2e\sec\theta+a^2e^2\sec\theta+a^2e+a^2e^2\sec\theta-a^2e}{2a+2ae}$$

$$\Rightarrow h=ae\sec\theta,\ k=\frac{2aeb\tan\theta}{2a+2ae}$$

$\Rightarrow$ locus is hyperbola.

13. b., d. Differentiating $xy=1$ w.r.t. x, we have

$$\frac{dy}{dx}=-\frac{1}{x^2}<0$$

Hence, the slope of normal at any point $P(x_1, y_1)$ is $x_1^2>0$.

Therefore, slope of the normal must always be positive. Hence, possible lines are (b) and (d).

14. c., d.

Equations of asymptotes are $4y-3x=0$ and $4y+3x=0$

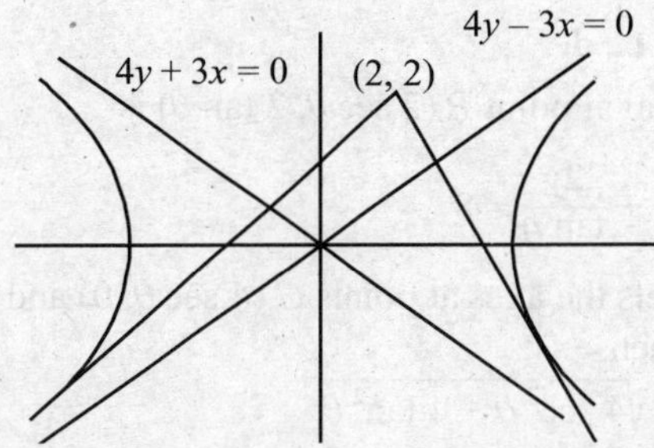

Fig. 5.22

As point (2, 2) lies above the asymptote. Points of contact of the tangents are in third and fourth quadrants.

15. b., c., d.

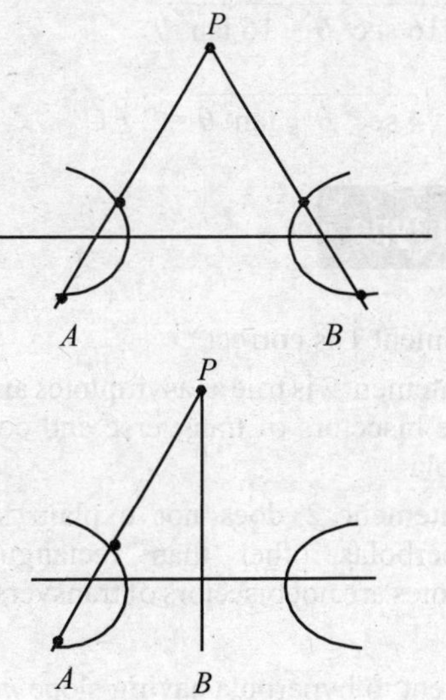

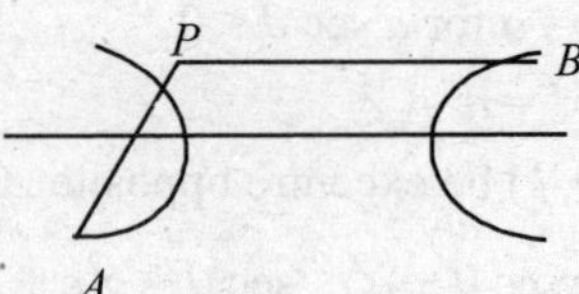

Fig. 5.23

Different cases have been shown in the above diagrams. Therefore, number of points can be 2, 3 or 4.

16. a., b., c.

Locus of point of intersection of perpendicular tangents is director circle $x^2 + y^2 = a^2 - b^2$. Now,

$$e^2 = 1 + \frac{b^2}{a^2}$$

If $a^2 > b^2$, then there are infinite (or more than 1) points on the circle $\Rightarrow e^2 < 2 \Rightarrow e < \sqrt{2}$.

If $a^2 < b^2$, there does not exist any point on the plane $\Rightarrow e^2 > 2 \Rightarrow e > \sqrt{2}$.

If $a^2 = b^2$, there is exactly one point (centre of the hyperbola) $\Rightarrow e = \sqrt{2}$.

17. a., b., c., d.

Normal at point P $(2 \sec \theta, 2 \tan \theta)$ is

$$\frac{2x}{\sec \theta} + \frac{2y}{\tan \theta} = 8$$

It meets the axes at points G $(4 \sec \theta, 0)$ and g $(0, 4 \tan \theta)$. Then

$$PG = \sqrt{4 \sec^2 \theta + 4 \tan^2 \theta}$$

$$Pg = \sqrt{4 \sec^2 \theta + 4 \tan^2 \theta}$$

$$PC = \sqrt{4 \sec^2 \theta + 4 \tan^2 \theta}$$

$$Gg = \sqrt{16 \sec^2 \theta + 16 \tan^2 \theta}$$

$$= 2\sqrt{4 \sec^2 \theta + \tan^2 \theta} = 2\, PC$$

Reasoning Type

1. b. Statement 1 is correct

Also statement 2 is true as asymptotes are perpendicular, they are bisectors of transverse and conjugate axes of hyperbola.

But statement 2 does not explain statement 1, as in hyperbolas other than rectangular hyperbolas asymptotes are not bisectors of transverse and conjugate axes.

2. a. Tangent to hyperbola having slope m is

$y = mx \pm \sqrt{4m^2 - 16}$,

which is real line if

$4m^2 - 16 > 0 \Rightarrow m^2 > 4 \Rightarrow m \in (-\infty, -2) \cup (2, \infty)$.

Hence, statement 2 is correct.

Also statement 1 is correct and statement 2 is correct explanation of statement 1.

3. a. Let P be the position of the gun and Q be the position of the target. Let u be the velocity of sound, v be the velocity of bullet and R be the position of the man. Then, we have

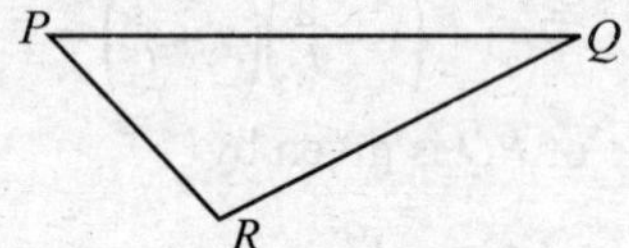

Fig. 5.24

$$\frac{PR}{u} = \frac{QR}{u} + \frac{PQ}{v}$$

$$\Rightarrow \frac{PR}{u} - \frac{QR}{u} = \frac{PQ}{v}$$

$$\Rightarrow PR - QR = \frac{u}{v} PQ = \text{constant}$$

and $\frac{u}{v} PQ < PQ$

Therefore, locus of R is a hyperbola.

4. a. For ellipse $\frac{x^2}{27/12} - \frac{y^2}{27/4} = 1$,

$$e = \sqrt{1 - \frac{16}{25}} = \frac{3}{5}$$

and $a = 5$

The foci are $(\pm 3, 0)$.

For hyperbola $\frac{x^2}{27/12} - \frac{y^2}{27/4} = 1$,

$$e = \sqrt{1 + \frac{12}{4}} = 2$$

and $a = \frac{3}{2}$

The foci are $(\pm 3, 0)$.

Therefore, the two conics are confocal. Hence, curves are orthogonal.

5. a.

6. d. Statement 2 is true.

For the points (2, 2), (4, 1) and (6, 2/3), $t_1 = 1$, $t_2 = 2$ and $t_3 = 3$, respectively.

For the point (1/4, 16), $t_4 = \frac{1}{8}$.

Now $t_1 t_2 t_3 t_4 = \frac{3}{4} \neq 1$.

Hence, statement 1 is false.

7. d.

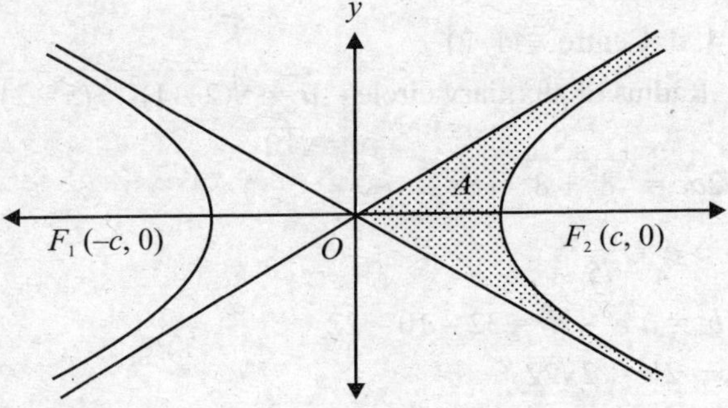

Fig. 5.25

Statement 1 is false as points in region A lie below the asymptote

$$y = \frac{b}{a}x \Rightarrow \frac{x_1^2}{a^2} - \frac{y_1^2}{b^2} > 0$$

Statement 2 is true (standard result). Indeed for points in region A

$$0 < \frac{x_1^2}{a^2} - \frac{y_1^2}{b^2} < 1$$

8. b. Given hyperbola is

$$\frac{x^2}{3} - \frac{y^2}{2} = 1$$

Now line having slope $m = 3$ is tangent to the hyperbola. So, its equation is

$$y = 3x \pm \sqrt{3(3)^2 - 2}$$

or $y = 3x \pm 5$

Hence, statement 1 is correct.

Also statement 2 is correct, but information is not enough to get the equations of tangents.

9. d. The locus of point of intersection of two mutually perpendicular tangents drawn on to hyperbola $\frac{x^2}{a^2} - \frac{y^2}{b^2} = 1$ is its director circle whose equation is $x^2 + y^2 = a^2 - b^2$.

For $\frac{x^2}{9} - \frac{y^2}{16} = 1$, $x^2 + y^2 = 9 - 16$.

So director circle does not exist.

10. b. Chord of contact z of $\frac{x^2}{a^2} - \frac{y^2}{b^2} = 1$ w.r.t. point P (x_1, y_1) is $\frac{xx_1}{a^2} - \frac{yy_1}{b^2} = 1$ (1)

Equation (1) can be written as

$$\frac{x(-x_1)}{a^2} - \frac{y(-y_1)}{b^2} = -1,$$

which is tangent to the hyperbola $\frac{x^2}{a^2} - \frac{y^2}{b^2} = -1$ at point $(-x_1, -y_1)$.

Obviously, points (x_1, y_1) and $(-x_1, -y_1)$ lie on the different branches of hyperbola

$$\frac{x^2}{a^2} - \frac{y^2}{b^2} = -1$$

Hence, statement 1 is correct.

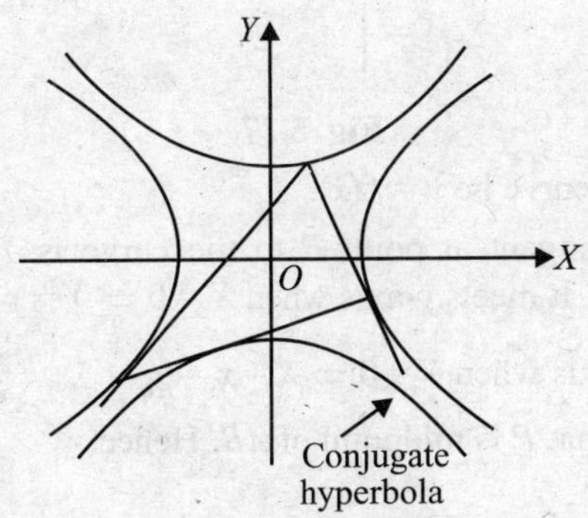

Fig. 5.26

Statement 2 is also correct but does not explain statement I.

11. d. We have $\sqrt{(\lambda - 3)^2 + 16} - 4 = 1 \Rightarrow \lambda = 0$ or 6.

12. a. Given that

$$\frac{r_2}{r_3} = k \text{ (constant)}$$

$$\Rightarrow \frac{\frac{\Delta}{s-b}}{\frac{\Delta}{s-c}} = k,$$

where Δ is area of triangle and s is semi-perimeter.

$$\Rightarrow \frac{s-c}{s-b} = k$$

$$\Rightarrow \frac{b-c}{2s-(b+c)} = \frac{k-1}{k+1}$$

$$\Rightarrow b - c = a\left(\frac{k-1}{k+1}\right) \text{ (constant) (as } BC = a \text{ is given)}$$

Therefore, the locus of vertex A is a hyperbola.

Linked Comprehension Type

For Problems 1–3

1. b., 2. c., 3. b.

Sol.

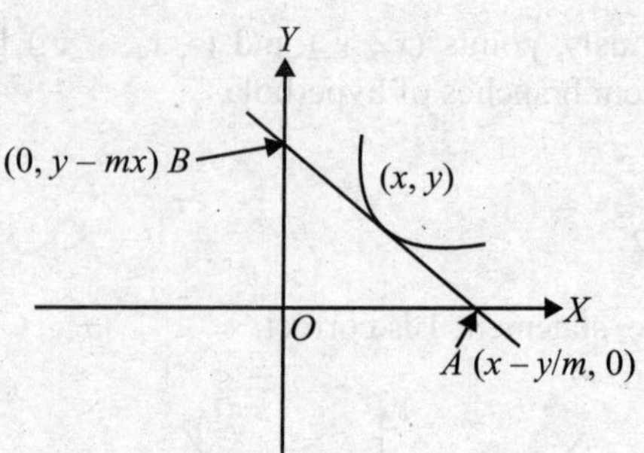

Fig. 5.27

Let the curve be $y = f(x)$.

Now tangent at point P to the curve is $Y - y = m(X - x)$. It meets y-axis when $X = 0 \Rightarrow Y = y - mx$

and x-axis when $Y = 0 \Rightarrow X = x - \frac{y}{m}$

Given that P is midpoint of AB. Hence,

$$x - \frac{y}{m} = 2x$$

$$\Rightarrow \frac{y}{m} = -x$$

$$\Rightarrow \frac{dy}{dx} = -\frac{y}{x}$$

$$\Rightarrow \frac{dy}{y} + \frac{dx}{x} = 0$$

$$\Rightarrow \log_e xy = c$$

$$\Rightarrow xy = c$$

As the curve passes through (2, 4), so $xy = 8$.

Solving with $y = x$, we get $x = 2\sqrt{2}$.

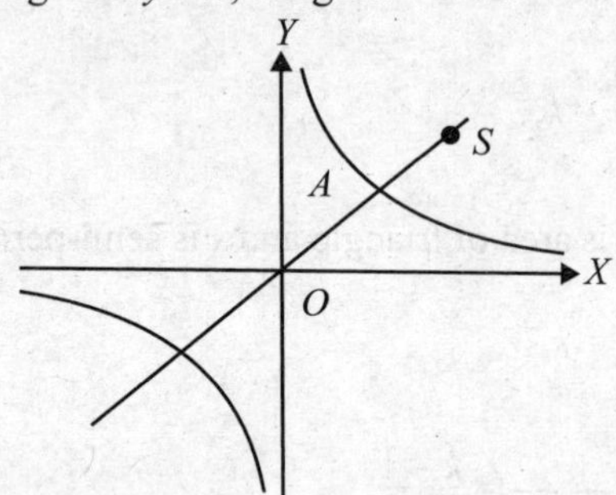

Fig. 5.28

$$\therefore OA = \sqrt{8 + 8} = 4$$

$$\Rightarrow OS = 4\sqrt{2}$$

Hence, coordinates of S are (4, 4) or (– 4, – 4).

Directrix is at distance $4/\sqrt{2}$ from origin.

Hence, its equations $x + y = \pm 4$.

For Problems 4–6

4. d., 5. b., 6. c.

Sol. 4. d. Centre ≡ (1, 2)

Radius of auxiliary circle $= a = \sqrt{(2-1)^2 + (5-2)^2} = \sqrt{10}$

$$2ae = \sqrt{8^2 + 8^2} = 8\sqrt{2}$$

$$\Rightarrow e = \frac{4}{\sqrt{5}}$$

$$b^2 = a^2e^2 - a^2 = 32 - 10 = 22$$

$$\Rightarrow 2b = 2\sqrt{22}$$

5. b. Directrix is perpendicular to the transverse axis. Let it be $x + y = k$.

Its distance from centre $\frac{a}{e}$

$$\Rightarrow \frac{|1 + 2 - k|}{\sqrt{2}} = \frac{5}{2\sqrt{2}} \Rightarrow k = 3 + \frac{5}{2} = \frac{11}{2}$$

6. c. The tangent is

$$y - 5 = -\frac{5-2}{6-5}(x - 2)$$

$$\Rightarrow 3x + y = 11$$

The hyperbola is

$$(x - 5)^2 + (y - 6)^2 = \frac{16}{5} \times \frac{(2x + 2y - 11)^2}{8}.$$

Solving, we get

$$(x - 5)^2 + (5 - 3x)^2 = \frac{2}{5}(2x + 22 - 6x - 11)^2$$

$$\Rightarrow 5[10x^2 - 40x + 50] = 2(11 - 4x)^2$$

$$\Rightarrow 9x^2 - 12x + 4 = 0$$

$$\Rightarrow (3x - 2)^2 = 0 \Rightarrow x = \frac{2}{3}, y = 9$$

For Problems 7–9

7. c., 8. d., 9. b.

Sol. 7. c. $2a = 3$

Distance between the foci (1, 2) and (5, 5) is 5.

$$\therefore 2ae = 5$$

$$\therefore e = \frac{5}{3}$$

Now if e' is eccentricity of the corresponding conjugate hyperbola, then

$$\frac{1}{e^2} + \frac{1}{e'^2} = 1$$

$$\Rightarrow e' = \frac{5}{4}$$

8. d. Director circle is given by

$$(x - h)^2 + (y - k)^2 = a^2 - b^2$$

where (h, k) is centre, i.e. the

midpoint of foci $\left(\frac{1+5}{2}, \frac{2+5}{2}\right) \equiv \left(3, \frac{7}{2}\right)$.

$$Y - y = m(X - x)$$

$b^2 = a^2(e^2 - 1)$

$= \left(\frac{3}{2}\right)^2 \left(\left(\frac{5}{3}\right)^2 - 1\right)$

$= 4$

Therefore, the director circle is

$(x - 3)^2 + \left(y - \frac{7}{2}\right)^2 = \frac{9}{4} - 4$

$\Rightarrow (x - 3)^2 + \left(\frac{y - 7}{2}\right)^2 = -\frac{7}{4}$

This does not represent any real point.

9. b. Slope of transverse axis is $\frac{3}{4}$.

Therefore, the angle of rotation is $\theta = \tan^{-1}(3/4)$.

For Problems 10–12

10. a., 11. c., 12. b.

Sol. 10. a. Equation of tangent in parametric form is given by

$\frac{x - 1}{-1/\sqrt{2}} = \frac{y - 1}{1/\sqrt{2}} = \pm 3\sqrt{2}$

$\Rightarrow A \equiv (4, -2),\ B \equiv (-2, 4)$

Equations of asymptotes (OA and OB) are given by

$y + 2 = \frac{-2}{4}(x - 4) \Rightarrow 2y + x = 0$

and $y - 4 = \frac{4}{-2}(x + 2) \Rightarrow 2x + y = 0$

Hence, the combined equation of asymptotes is

$(2x + y)(x + 2y) = 0$

$\Rightarrow 2x^2 + 2y^2 + 5xy = 0$

11. c. $m_{OA} = -\frac{1}{2},\ m_{OB} = -2$

$\tan\theta = \left|\frac{-1/2 + 2}{1 + 1}\right| = \frac{3}{4}$

$\Rightarrow \sin\theta = \frac{3}{5} \Rightarrow \theta = \sin^{-1}\left(\frac{3}{5}\right)$

12. b. Let the equation of the hyperbola be $2x^2 + 2y^2 + 5xy + \lambda = 0$.

It passes through (1, 1). Therefore,

$2 + 2 + 5 + \lambda = 0.$

$\Rightarrow \lambda = -9$

So, the hyperbola is $2x^2 + 2y^2 + 5xy = 9$.

Equation of the tangent at $\left(-1, \frac{7}{2}\right)$ is given by

$2x(-1) + 2y\left(\frac{7}{2}\right) + 5\,\frac{x(7/2) + (-1)y}{2} = 9$

$\Rightarrow 3x + 2y = 4$

For Problems 13–15

13. b., 14. c., 15 d.

13. b. Any point on the hyperbola $xy = 16$ is $\left(4t, \frac{4}{t}\right)$

Normal at this point is $y - 4/t = t^2(x - 4t)$.

If the normal passes through $P(h, k)$, then

$k - 4/t = t^2(h - 4t)$

$\Rightarrow 4t^4 - t^3h + tk - 4 = 0$

This equation has roots t_1, t_2, t_3, t_4 which are parameters of the four feet of normals on the hyperbola. Therefore,

$\Sigma t_1 = \frac{h}{4}$

$\Sigma t_1 t_2 = 0$

$\Sigma t_1 t_2 t_3 = -\frac{k}{4}$

and $t_1 t_2 t_3 t_4 = -1$

$\therefore \frac{1}{t_1} + \frac{1}{t_2} + \frac{1}{t_3} + \frac{1}{t_4} = \frac{k}{4}$

$\Rightarrow y_1 + y_2 + y_3 + y_4 = k$

According to the question,

$t_1^2 + t_2^2 + t_3^2 + t_4^2 = \frac{h^2}{16} = k.$

Hence, the locus of (h, k) is $x^2 = 16y$.

14. c. $x^2 = 16y$

Equation of tangent of P is

$x\,4t = \frac{16\,(y + t^2)}{2}.$

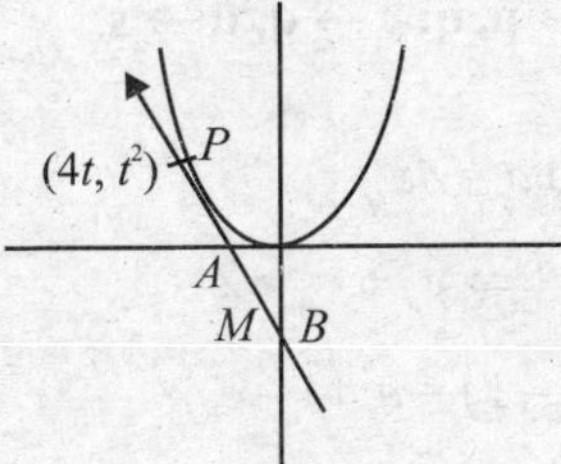

Fig. 5.29

$\Rightarrow 4tx = 8y + 8t^2$

$\Rightarrow tx = 2y + 2t^2$

$A = (2t, 0),\ B = (0, -t^2)$

$M(h, k)$ is the middle point of AB.

$h = t,\ k = -\frac{t^2}{2} \Rightarrow 2k = -h^2$

Therefore, the locus of $M(h, k)$ is $x^2 + 2y = 0$.

15. d. $\tan 30° = \frac{4t_1}{t_1^2} = \frac{4}{t_1}$

$\Rightarrow \frac{1}{\sqrt{3}} = \frac{4}{t_1} \Rightarrow t_1 = 4\sqrt{3}$

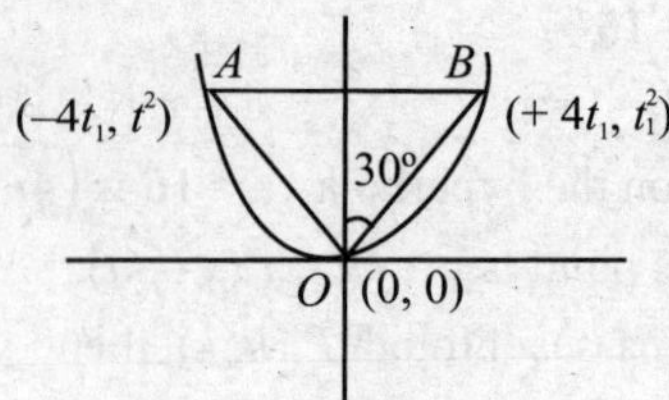

Fig. 5.30

$AB = 8t_1 = 32\sqrt{3}$

Area of $\Delta OAB = \frac{\sqrt{3}}{4} \times 32\sqrt{3} \times 32\sqrt{3}$

$= 768\sqrt{3}$ sq. units

For Problems 16–18

16. b., 17. c., 18. d.

Sol. 16. b. Perpendicular tangents intersect at the centre of rectangular hyperbola. Hence, centre of hyperbola is (1, 1) and equation of asymptotes are $x - 1 = 0$ and $y - 1 = 0$.

17. c. Let the equation of hyperbola is $xy - x - y + 1 + \lambda = 0$. It passes through (3, 2), so $\lambda = -2$. So the equation of hyperbola is $xy = x + y + 1$.

18. d. From the centre of hyperbola we can draw two real tangents to the rectangular hyperbola.

Matrix-Match Type

1. a → p; b → p, q; c → q; d → s.

We have

$A = ae_E$ and $a = Ae_H$

$\Rightarrow e_E\, e_H = 1 \Rightarrow e_E + e_H > 2$

$B^2 = A^2(e_H^2 - 1) = a^2(1 - e_E^2)$

$= b^2$

$\Rightarrow \frac{b}{B} = 1$

Also the angle between the asymptotes is

$2\tan^{-1}\frac{B}{A} = \frac{2\pi}{3}$

Also, $\frac{B}{A} = \sqrt{3} \Rightarrow \frac{b}{ae_E} = \sqrt{3} \Rightarrow e_E^2 = \frac{1}{4}$

Solving $\frac{x^2}{a^2} + \frac{y^2}{b^2} = 1$ and $\frac{x^2}{a^2e_E^2} - \frac{y^2}{b^2} = 1$,

we get

$x^2 = \frac{2a^2 e_E^2}{b^2(1 - e_E^2)} = 4$

2. a → p, q, r, s ; b → q, r ; c → p ; d → p, s.

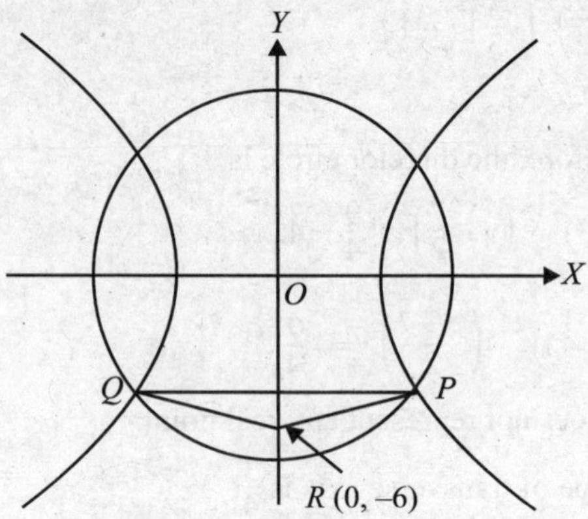

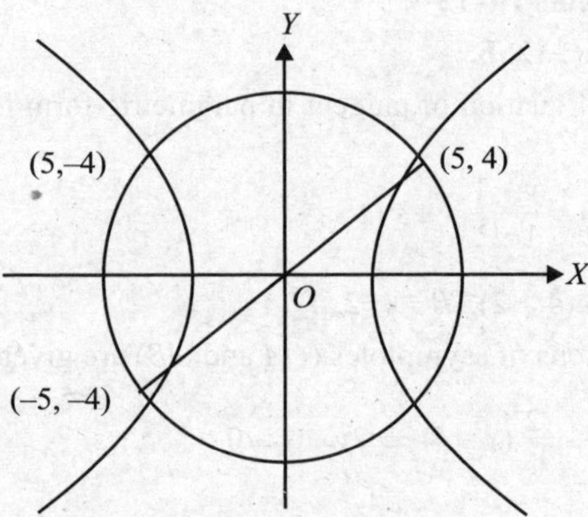

Fig. 5.31

a. Obviously all the points in column II are common to the hyperbola and circle.

b. Chord of contact of hyperbola w.r.t. $\left(0, -\frac{9}{4}\right)$ is

$\theta(x) - \left(-\frac{9}{4}\right)y = 9$ or $y = 4$

Solving this with hyperbola we have

$x^2 - 16 = 9 \Rightarrow x^2 = 25 \Rightarrow x = \pm 5.$

Hence, points of contact are (±5, 4).

c. Obviously the required point is (– 5, – 4).

d. Let the points on the hyperbola be $P(h, k)$ and $Q(-h, k)$.

Then area of triangle is $\frac{1}{2}|2h||-6-k| = 10$

$\Rightarrow |h||6 + k| = 10$ (1)

Also points P and Q lie on the hyperbola. Hence,

$h^2 - k^2 = 9$ (2)

Obviously points (±5, –4) satisfy both Eqs. (1) and (2).

3. **a → p; b → s, c → r; d → p.**

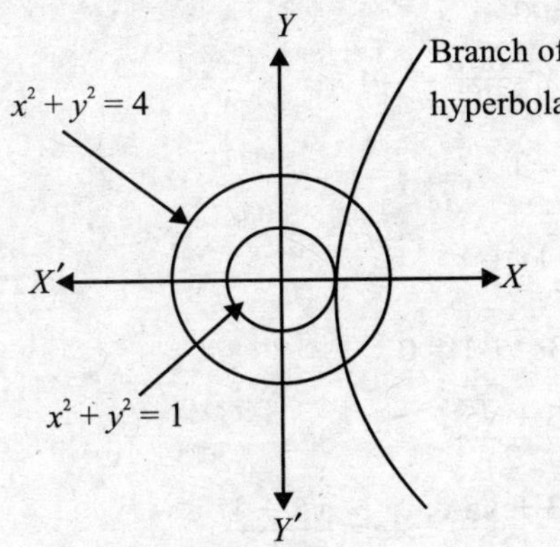

Fig. 5.32

Locus of point P satisfying $PA - PB = 2$ is a branch of the hyperbola $x^2 - y^2/3 = 1$.

For $r = 2$ the circle and the branch of the hyperbola intersect at two points. For $r = 1$ there is no point of intersection.

If m be the slope of the common tangent, then

$m^2 - 3 = r^2 (1 + m^2)$

$\Rightarrow m^2 = \dfrac{r^2 + 3}{1 - r^2}$

Hence, there are no common tangents for $r > 1$ and two common tangents for $r < 1$.

4. **a → r; b → p; c → s; d → q.**

a. $\text{Im}(z^2) = 3$

$\Rightarrow \text{Im}((x + iy)^2) = 3$

$\Rightarrow 2xy = 3$, which is a rectangular hyperbola having eccentricity $\sqrt{2}$.

b.

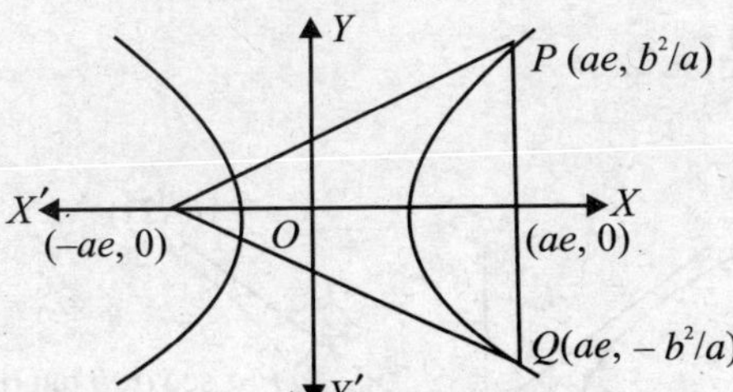

Fig. 5.33

$\tan 30° = \dfrac{b^2/a}{2ae}$

$\Rightarrow \dfrac{2}{\sqrt{3}} e = e^2 - 1$

$\Rightarrow \sqrt{3}e^2 - 2e - \sqrt{3} = 0$

$\Rightarrow e = \dfrac{2 \pm \sqrt{4 + 12}}{2\sqrt{3}} = \dfrac{2 \pm 4}{2\sqrt{3}}$

$\Rightarrow e = \dfrac{3}{\sqrt{3}} = \sqrt{3}$

c. Eccentricity of the hyperbola $= \dfrac{AB}{PA - PB} = \dfrac{6}{4} = \dfrac{3}{2}$

If eccentricity of conjugate hyperbola is e', then $\dfrac{1}{\left(\frac{3}{2}\right)^2} + \dfrac{1}{e'^2} = 1$

$\Rightarrow e' = \dfrac{3}{\sqrt{5}}$

d. Angle between the asymptotes is

$\tan^{-1}\left|\dfrac{2ab}{a^2 - b^2}\right| = \dfrac{\pi}{3}$

$\Rightarrow \left|\dfrac{2\frac{a}{b}}{\frac{a^2}{b^2} - 1}\right| = \sqrt{3}$

$\Rightarrow \dfrac{2\sqrt{e'^2 - 1}}{|e'^2 - 2|} = \sqrt{3}$ (where e' is eccentricity of conjugate hyperbola)

$\Rightarrow e' = 2$

5. **a → p., s.; b → q, r; c → r; d → p, s.**

a. We must have

$e_1 < 1 < e_2 \Rightarrow f(1) < 0 \Rightarrow 1 - a + 2 < 0 \Rightarrow a > 3$

b. We must have both the roots greater than 1.

i. $D > 0$ or $a^2 - 4 > 0$ or $a \in (-\infty, -2) \cup (2, \infty)$

ii. $1 \cdot f(1) > 0$ or $1 - a + 2 > 0$ or $a < 3$

iii. $\dfrac{a}{2} >= 1 \Rightarrow a > 2$

From Eqs. (1), (2) and (3) we have

$a \in (2, 3)$.

c. We must have

$\dfrac{1}{e_1^2} + \dfrac{1}{e_2^2} = 1$

$\Rightarrow \dfrac{(e_1 + e_2)^2 - 2e_1e_2}{e_1^2e_2^2} = 1$

$\Rightarrow \dfrac{a^2 - 4}{4} = 1$

$\Rightarrow a = \pm 2\sqrt{2}$

d. We must have $e_2 < \sqrt{2} < e_1$

$\Rightarrow f(\sqrt{2}) < 0$

$\Rightarrow 2 - a\sqrt{2} + 2 < 0$

$\Rightarrow a > 2\sqrt{2}$

Integer Type

1. (7) For a hyperbola,

$$e^2 = 1 + \frac{b^2}{a^2} = 1 + \frac{81}{144} = \frac{225}{144}$$

$$\Rightarrow e = \frac{15}{12} = \frac{15}{4}.$$

Also $a^2 = 144/25$.

Hence, the foci are

$$(\pm ae, 0) \equiv \left(\pm \frac{12}{5} \cdot \frac{5}{4}, 0\right) \equiv (\pm 3, 0)$$

Now for an ellipse, $ae = 3$ or $a^2e^2 = 9$.

Also $b^2 = a^2(1 - e^2)$. So

$b^2 = a^2 - a^2e^2 = 16 - 9 = 7$.

2. (2)

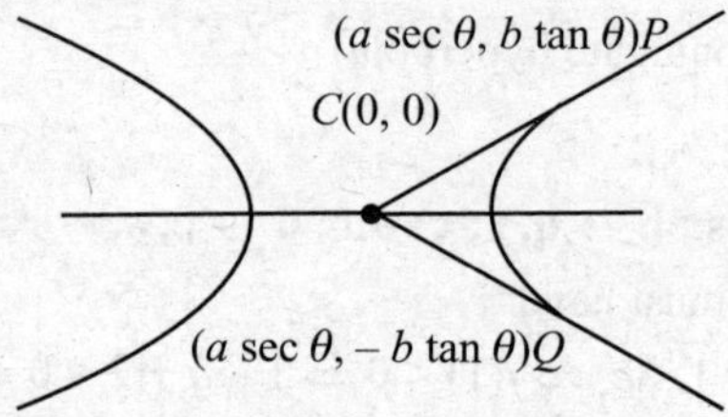

Fig. 5.34

Let the hyperbola be $(x^2/a^2) - (y^2/b^2) = 1$ and any double ordinate PQ be $(a \sec \theta, b \tan \theta)$, $(a \sec \theta, -b \tan \theta)$, and O be the centre $(0, 0)$.

ΔOPQ is equilateral. Therefore,

$$\tan 30° = \frac{b \tan \theta}{a \sec \theta}$$

$$\Rightarrow 3 \times \frac{b^2}{a^2} = \text{cosec}^2 \theta$$

$$\Rightarrow 3(e^2 - 1) = \text{cosec}^2 \theta$$

Now $\text{cosec}^2 \theta \geq 1$

$$\Rightarrow 3(e^2 - 1) \geq 1$$

$$\Rightarrow e^2 \geq \frac{4}{3}$$

$$\Rightarrow e > \frac{2}{\sqrt{3}}$$

3. (1)

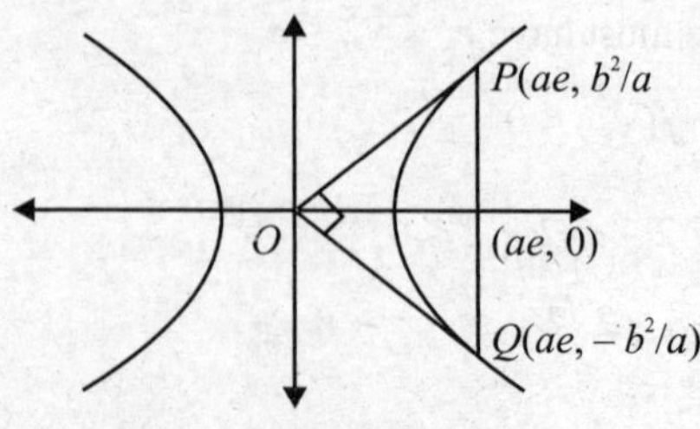

Fig. 5.35

$$m_{OP} \times m_{OQ} = -1$$

$$\Rightarrow \frac{b^2/a}{ae} \times \frac{-b^2/a}{ae} = -1$$

$$\Rightarrow \frac{b^4}{a^4e^2} = 1$$

$$\Rightarrow (e^2 - 1)^2 = e^2$$

$$\Rightarrow e^4 - 3e^2 + 1 = 0$$

$$\Rightarrow e^2 = \frac{3 + \sqrt{5}}{2}$$

$$\Rightarrow e^2 = \frac{3 + \sqrt{5}}{2} \Rightarrow e = \frac{\sqrt{5} + 1}{2}$$

4. (2)

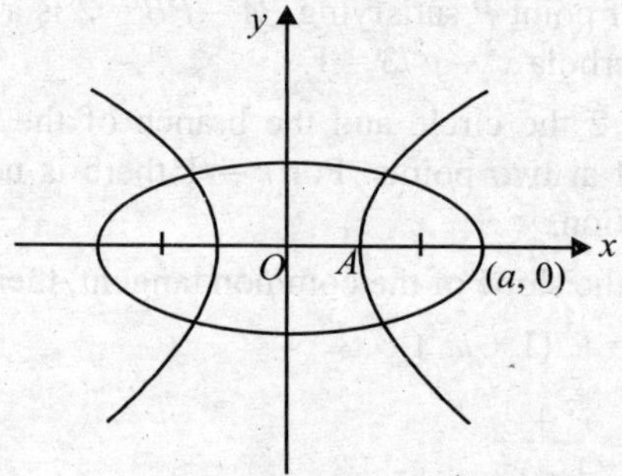

Fig. 5.36

Let the foci of the ellipse be $(\pm ae_1, 0)$ and those of the hyperbola be $(\pm Ae_2, 0)$.

According to the question, we have $ae_1 = Ae_2$ (1)

Also it is given that the conjugate axis of the hyperbola is equal to the minor axis of the ellipse. Therefore,

$a^2(1 - e_1^2) = A^2(e_2^2 - 1)$ (2)

From (1) and (2), we have

$$\frac{1}{e_1^2} + \frac{1}{e_2^2} = 2.$$

5. (0)

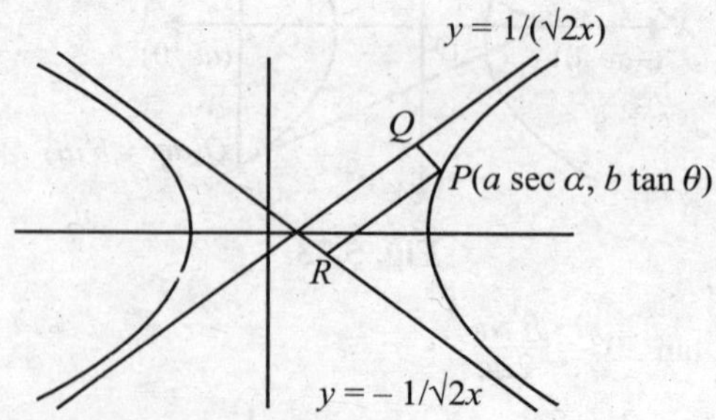

Fig. 5.37

The given hyperbola is $x^2 - 2y^2 - 2 = 0$

or $\frac{x^2}{2} - \frac{y^2}{1} = 1$

$$\Rightarrow PQ \times PR = \frac{|a \sec \alpha - \sqrt{2} b \tan \alpha|}{\sqrt{3}} \frac{|a \sec \alpha + \sqrt{2} b \tan \alpha|}{\sqrt{3}}$$

$$= \frac{a^2 \sec^2 \alpha - 2b^2 \tan^2 \alpha}{3}$$

$$= \frac{2(\sec^2 \alpha - \tan^2 \alpha)}{3}$$

[As $a = \sqrt{2}$ and $b = 1$]

$$= \frac{2}{3}$$

6. (1) Let the coordinates of P, Q, R, and S be $(ct_i, c/t_i)$, respectively, (where $t = 1, 2, 3$, and 4).

Now $PQ \perp RS$

$$\Rightarrow \frac{(c/t_2) - (c/t_1)}{ct_2 - ct_1} \times \frac{(c/t_1) - (c/t_2)}{ct_4 - ct_3} = -1$$

$$\Rightarrow t_1 t_2 t_3 t_4 = -1$$

The slope of CP is

$$\frac{c/t_1}{ct_1} = \frac{1}{t_1^2}.$$

Hence, the product of the slopes of CP, CQ, CR, and CS is

$$\frac{1}{t_1^2 t_2^2 t_3^2 t_4^2} = 1.$$

7. (1) The equation of the family of curves passing through the points of intersection of $x^2 - y^2 = a^2$ and $y = x^2$ is

$x^2 - y^2 - a^2 + \lambda (x^2 - y) = 0$

$\Rightarrow x^2(1 + \lambda) - y^2 - a^2 - \lambda y = 0$

It will be a circle if $\lambda = -2$. Therefore,

$-x^2 - y^2 - a^2 + 2y = 0$

$\Rightarrow x^2 + y^2 - 2y = -a^2$

$\Rightarrow x^2 + (y - 1)^2 = 1 - a^2$

$\Rightarrow 1 - a^2 > 0 \Rightarrow a^2 < 1$

$\Rightarrow a \in (-1, 1)$ (1)

Also both curves intersect at real points if $y^2 - y + a^2 = 0$ for some real y.

So if $-1/2 < a < 1/2$,

(1) and (2) gives $a \in (-1/2, 1/2)$.

8. (1)

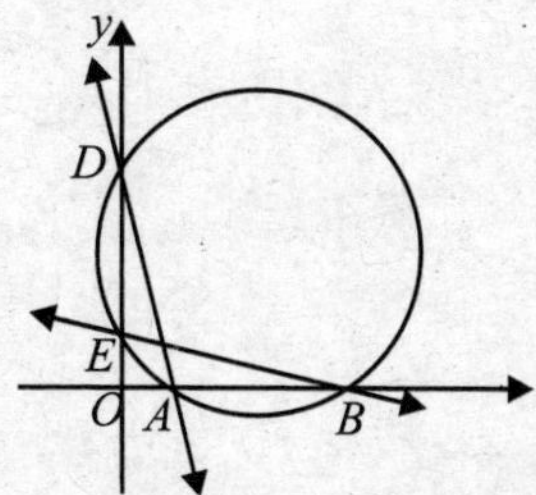

Fig. 5.38

Let the tangents be $y = m_1 x + \sqrt{a^2 m_1^2 - b^2}$ and $y = m_2 x + \sqrt{a^2 m_2^2 - b^2}$.

The points of intersection of these tangents with the axes are

$$A\left(\frac{-\sqrt{a^2 m_1^2 - b^2}}{m_1}, 0\right), B\left(\frac{-\sqrt{a^2 m_2^2 - b^2}}{m_2}, 0\right),$$

$$C\left(0, \sqrt{a^2 m_1^2 - b^2}\right), \text{ and } D\left(0, \sqrt{a^2 m_2^2 - b^2}\right).$$

Since the four points are concyclic,

$OA \times OB = OC \times OD$

$$\Rightarrow \left(\frac{-\sqrt{a^2 m_1^2 - b^2}}{m_1}, 0\right)\left(\frac{-\sqrt{a^2 m_2^2 - b^2}}{m_2}, 0\right)$$

$$= \sqrt{a^2 m_1^2 - b^2} \sqrt{a^2 m_2^2 - b^2}$$

$\Rightarrow m_1 m_2 = 1$

9. (1) A tangent to $(x^2/a^2) - (y^2/b^2) = 1$ is

$y = m_1 x \pm \sqrt{a^2 m_1^2 - b^2}$ (1)

The other hyperbola is $[x^2/(-b)^2] - [y^2/(-a^2)]$, then any tangent to it is

$y = m_2 x \pm \sqrt{(-b^2) m_2^2 - (-a^2)}$ (2)

If (1) and (2) are same, then $m_1 = m_2$ and $a^2 m_1^2 - b^2 = -b^2 m_1^2 + a^2$

$\Rightarrow a^2 m_1^2 - b^2 = a^2 - b^2 m_1^2$

$\Rightarrow (a^2 + b^2) m_1^2 = a^2 + b^2$

$\Rightarrow m_1^2 = 1$

$\Rightarrow m_1 = \pm 1$

Section D

Calculus

CHAPTER 1

Functions

Objective Type

Each question has four choices a, b, c, and d, out of which *only one* is correct.

1. The function $f: N \to N$ (N is the set of natural numbers) defined by $f(n) = 2n + 3$ is
   a. surjective only
   b. injective only
   c. bijective
   d. None of these
2. The function $f(x) = \sin\left(\log(x + \sqrt{1 + x^2})\right)$ is
   a. even function
   b. odd function
   c. neither even nor odd
   d. periodic function
3. If x is real, then the value of the expression $\dfrac{x^2 + 14x + 9}{x^2 + 2x + 3}$ lies between
   a. 5 and 4
   b. 5 and -4
   c. -5 and 4
   d. None of these
4. The function $f: R \to R$ is defined by $f(x) = \cos^2 x + \sin^4 x$ for $x \in R$, then the range of $f(x)$ is
   a. $\left(\dfrac{3}{4}, 1\right]$
   b. $\left[\dfrac{3}{4}, 1\right)$
   c. $\left[\dfrac{3}{4}, 1\right]$
   d. $\left(\dfrac{3}{4}, 1\right)$
5. The domain of the function $f(x) = \log_{3+x}(x^2 - 1)$ is
   a. $(-3, -1) \cup (1, \infty)$
   b. $[-3, -1) \cup [1, \infty)$
   c. $(-3, -2) \cup (-2, -1) \cup (1, \infty)$
   d. $[-3, -2) \cup (-2, -1) \cup [1, \infty)$
6. The domain of the function $f(x) = \left[\log_{10}\left(\dfrac{5x - x^2}{4}\right)\right]^{1/2}$ is
   a. $-\infty < x < \infty$
   b. $1 \le x \le 4$
   c. $4 \le x \le 16$
   d. $-1 \le x \le 1$
7. The domain of the function $f(x) = \dfrac{\sin^{-1}(3 - x)}{\ln(|x| - 2)}$ is
   a. $[2, 4]$
   b. $(2, 3) \cup (3, 4]$
   c. $[2, \infty)$
   d. $(-\infty, -3) \cup [2, \infty)$
8. The domain of $f(x) = \log|\log x|$ is
   a. $(0, \infty)$
   b. $(1, \infty)$
   c. $(0, 1) \cup (1, \infty)$
   d. $(-\infty, 1)$
9. The domain of $f(x) = \dfrac{\log_2(x + 3)}{x^2 + 3x + 2}$ is
   a. $R - \{-1, -2\}$
   b. $(-2, \infty)$
   c. $R - \{-1, -2, -3\}$
   d. $(-3, \infty) - \{-1, -2\}$
10. Let $f: \left[-\dfrac{\pi}{3}, \dfrac{2\pi}{3}\right]$ [0, 4] be a function defined as $f(x) = \sqrt{3}\sin x - \cos x + 2$. Then $f^{-1}(x)$ is given by
    a. $\sin^{-1}\left(\dfrac{x - 2}{2}\right) - \dfrac{\pi}{6}$
    b. $\sin^{-1}\left(\dfrac{x - 2}{2}\right) + \dfrac{\pi}{6}$
    c. $\dfrac{2\pi}{3} + \cos^{-1}\left(\dfrac{x - 2}{2}\right)$
    d. None of these
11. If $F(n + 1) = \dfrac{2F(n) + 1}{2}$ $n = 1, 2, \ldots$ and $F(1) = 2$, then $F(101)$ equals

a. 52 b. 49
c. 48 d. 51

12. The domain of the function

$f(x)=\dfrac{1}{\sqrt{{}^{10}C_{x-1}-3\times{}^{10}C_x}}$ contains the points

a. 9, 10, 11 b. 9, 10, 12
c. all natural numbers d. None of these

13. The domain of the function

$f(x)=\dfrac{x}{\sqrt{\sin(\ln x)-\cos(\ln x)}}$ is

a. $(e^{2n\pi}, e^{(3n+1/2)\pi})$ b. $(e^{(2n+1/4)\pi}, e^{(2n+5/4)\pi})$
c. $(e^{2n+1/4)\pi}, e^{(3n-3/4)\pi})$ d. None of these

14. If f is a function such that $f(0)=2$, $f(1)=3$ and $f(x+2)=2f(x)-f(x+1)$ for every real x, then $f(5)$ is

a. 7 b. 13
c. 1 d. 5

15. The range of $\sin^{-1}\left(\dfrac{x^2+1}{x^2+2}\right)$ is

a. $[0, \pi/2]$ b. $(0, \pi/6)$
c. $[\pi/6, \pi/2)$ d. None of these

16. The function $f(x)=\dfrac{\sec^{-1}x}{\sqrt{x-[x]}}$, where $[x]$ denotes the greatest integer less than or equal to x, is defined for all $x\in$

a. R
b. $R-\{(-1,1)\cup\{n\,|\,n\in Z\}\}$
c. $R^+-(0,1)$
d. $R^+-\{n\,|\,n\in N\}$

17. The domain of $f(x)=\cos^{-1}\left(\dfrac{2-|x|}{4}\right)+[\log(3-x)]^{-1}$ is

a. $[-2, 6]$ b. $[-6, 2)\cup(2, 3)$
c. $[-6, 2]$ d. $[-2, 2]\cup(2, 3)$

18. The domain of the function $f(x)=\sqrt{\log\left(\dfrac{1}{|\sin x|}\right)}$ is

a. $R-\{-\pi, \pi\}$ b. $R-\{n\pi\,|\,n\in Z\}$
c. $R-\{2n\pi\,|\,n\in z\}$ d. $(-\infty, \infty)$

19. The domain of the function

$f(x)=\log_2\left(-\log_{1/2}\left(1+\dfrac{1}{x^{1/4}}\right)-1\right)$ is

a. $(0, 1)$ b. $(0, 1]$
c. $[1, \infty)$ d. $(1, \infty)$

20. The range of $f(x)=\sin^{-1}(\sqrt{x^2+x+1})$ is

a. $\left(0, \dfrac{\pi}{2}\right]$ b. $\left(0, \dfrac{\pi}{3}\right]$
c. $\left[\dfrac{\pi}{3}, \dfrac{\pi}{2}\right]$ d. $\left[\dfrac{\pi}{6}, \dfrac{\pi}{3}\right]$

21. If $f(x)=\text{maximum}\left\{x^3, x^2, \dfrac{1}{64}\right\}\ \forall x\in[0,\infty)$, then

a. $f(x)=\begin{cases}x^2, & 0\le x\le 1\\ x^3, & x>1\end{cases}$ b. $f(x)=\begin{cases}\dfrac{1}{64}, & 0\le x\le\dfrac{1}{4}\\ x^2, & \dfrac{1}{4}<x\le 1\\ x^3, & x>1\end{cases}$

c. $f(x)=\begin{cases}\dfrac{1}{64}, & 0\le x\le\dfrac{1}{8}\\ x^2, & \dfrac{1}{8}<x\le 1\\ x^3, & x>1\end{cases}$ d. $f(x)=\begin{cases}\dfrac{1}{64}, & 0\le x\le\dfrac{1}{8}\\ x^3, & x>1/8\end{cases}$

22. If the period of $\dfrac{\cos(\sin(nx))}{\tan(x/n)}$, $n\in N$, is 6π then n is equal to

a. 3 b. 2
c. 6 d. 1

23. The number of real solutions of the equation $\log_{0.5}|x|=2|x|$ is

a. 1 b. 2
c. 0 d. None of these

24. The period of the function $\left|\sin^3\dfrac{x}{2}\right|+\left|\cos^5\dfrac{x}{5}\right|$ is

a. 2π b. 10π
c. 8π d. 5π

25. If $f(x)=\sqrt[n]{x^m}$, $n\in N$, is an even function, then m is

a. even integer
b. odd integer
c. any integer
d. $f(x)$-even is not possible

26. If f is periodic, g is non-periodic and $f(g(x))$ is periodic and $g(2)=3$, $g(4)=7$ then $g(6)$ is

a. 13 b. 15
c. 11 d. None of these

27. The period of function $2^{\{x\}}+\sin\pi x+3^{\{x/2\}}+\cos 2\pi x$ (where $\{x\}$ denotes the fractional part of x) is

a. 2 b. 1
c. 3 d. None of these

28. The equation $||x-2|+a|=4$ can have four distinct real solutions for x if a belongs to the interval

a. $(-\infty,-4)$ b. $(-\infty,0]$
c. $[4,\infty)$ d. None of these

29. Given the function $f(x)=\dfrac{a^x+a^{-x}}{2}$ (where $a>2$). Then $f(x+y)+f(x-y)=$

a. $2f(x)f(y)$ b. $f(x)f(y)$
c. $\dfrac{f(x)}{f(y)}$ d. None of these

30. If $\log_3(x^2-6x+11)\le 1$, then exhaustive range of values of x is

a. $(-\infty,2)\cup(4,\infty)$
b. $(2,4)$
c. $(-\infty,1)\cup(1,3)\cup(4,\infty)$
d. None of these

31. The domain of the function $f(x)=\sqrt{x^2-[x]^2}$, where $[x]$ = the greatest integer less than or equal to x, is

a. R b. $[0,+\infty)$
c. $(-\infty,0]$ d. None of these

32. The range of the function $f(x)=|x-1|+|x-2|$, $-1\le x\le 3$, is

a. $[1,3]$ b. $[1,5]$
c. $[3,5]$ d. None of these

33. Which of the following functions is inverse to itself?

a. $f(x)=\dfrac{1-x}{1+x}$ b. $f(x)=5^{\log x}$
c. $f(x)=2^{x(x-1)}$ d. None of these

34. A function $F(x)$ satisfies the functional equation $x^2F(x)+F(1-x)=2x-x^4$ for all real x. $F(x)$ must be

a. X^2 b. $1-x^2$
c. $1+x^2$ d. x^2+x+1

35. If $f(x)=\begin{cases} x^2\sin\dfrac{\pi x}{2}, & |x|<1 \\ x|x|, & |x|\ge 1\end{cases}$ then $f(x)$ is

a. an even function b. an odd function
c. a periodic function d. None of these

36. Function $f:(-\infty,-1)\to(0,e^5]$ defined by $f(x)=e^{x^3-3x+2}$ is not possible

a. many-one and onto b. many-one and into
c. one-one and onto d. one-one and into

37. If $f(x)=\dfrac{1}{x}$, $g(x)=\dfrac{1}{x^2}$ and $h(x)=x^2$

a. $fog(x)=x^2$ $x\ne 0$, $h(g(x))=\dfrac{1}{x^2}$

b. $h(g(x))=\dfrac{1}{x^2}$ $x\ne 0$, $fog(x)=x^2$

c. $fog(x)=\dfrac{1}{x^2}$ $x\ne 0$, $h(g(x))=(g(x))^2$, $x\ne 0$

d. None of these

38. If $[x]$ and $\{x\}$ represent the integral and fractional parts of x, respectively, then the value of $\sum_{r=1}^{2000}\dfrac{\{x+r\}}{2000}$ is

a. x b. $[x]$
c. $\{x\}$ d. $x+2001$

39. If $f(x)$ is a polynomial satisfying $f(x)f(1/x)=f(x)+f(1/x)$ and $f(3)=28$, then $f(4)$ is equal to

a. 63 b. 65
c. 17 d. None of these

40. The values of b and c for which the identity $f(x+1)-f(x)=8x+3$ is satisfied, where $f(x)=bx^2+cx+d$, are

a. $b=2, c=1$ b. $b=4, c=-1$
c. $b=-1, c=4$ d. $b=-1, c=1$

41. Let $f:R\to R$, $g:R\to R$ be two given functions such that f is injective and g is surjective, then which of the following is injective?

a. $g_0 f$ b. $f_0 g$
c. $g_0 g$ d. $f_0 f$

42. $f:N\to N$ where $f(x)=x-(-1)^x$ then f is

a. one-one and into b. many-one and into
c. one-one and onto d. many-one and onto

43. If $g(x)=x^2+x-2$ and $\dfrac{1}{2}gof(x)=2x^2-5x+2$, then $f(x)$ is

a. $2x-3$ b. $2x+3$
c. $2x^2+3x+1$ d. $2x^2-3x-1$

44. If $f:R\to R$ is an invertible function such that $f(x)$ and $f^{-1}(x)$ are symmetric about the line $y=-x$, then

a. $f(x)$ is odd
b. $f(x)$ and $f^{-1}(x)$ may not be symmetric about the line $y=x$
c. $f(x)$ may not be odd
d. None of these

45. Let $f:N\to N$ defined by $f(x)=x^2+x+1$, $x\in N$, then f is

a. One-one onto b. Many-one onto
c. One-one but not onto d. None of these

46. Let $f:X\to y$ $f(x)=\sin x+\cos x+2\sqrt{2}$ is invertible, then $X\to Y$ is/are

a. $\left[\dfrac{\pi}{4},\dfrac{5\pi}{4}\right]\to[\sqrt{2},3\sqrt{2}]$

b. $\left[-\dfrac{\pi}{4},\dfrac{3\pi}{4}\right]\to[\sqrt{2},3\sqrt{2}]$

c. $\left[-\frac{3\pi}{4}, \frac{3\pi}{4}\right] \to [\sqrt{2}, 3\sqrt{2}]$

d. $\left[-\frac{3\pi}{4}, -\frac{\pi}{4}\right] \to [\sqrt{2}, 3\sqrt{2}]$

47. If $f(x) = ax^7 + bx^3 + cx - 5$, a, b, c are real constants and $f(-7) = 7$, then the range of $f(7) + 17\cos x$ is

a. $[-34, 0]$ b. $[0, 34]$
c. $[-34, 34]$ d. None of these

48. If $f(x) = \frac{\sin([x]\pi)}{x^2 + x + 1}$, where [.] denotes the greatest integer function, then

a. f is one-one
b. f is not one-one and non-constant
c. f is a constant function
d. None of these

49. Let S be the set of all triangles and R^+ be the set of positive real numbers. Then the function $f: S \to R^+$, $f(\Delta) =$ area of Δ, where $\Delta \in S$ is

a. injective but not surjective
b. surjective but not injective
c. injective as well as surjective
d. neither injective nor surjective

50. A real-valued function $f(x)$ satisfies the function $f(x-y) = f(x)f(y) - f(a-x)f(a+y)$ where a is a given constant and $f(0) = 1$, then the graph of the function is symmetrical about

a. point $(2a, 0)$ b. point $(a, 0)$
c. line $x = 2a$ d. line $x = a$

51. If $g: [-2, 2] \to R$ where $f(x) = x^3 + \tan x + \left[\frac{x^2+1}{P}\right]$ is a odd function, then the value of parametric P where [.] denotes the greatest integer function is

a. $-5 < P < 5$ b. $P < 5$
c. $P > 5$ d. None of these

52. If $f\left(2x + \frac{y}{8}, 2x - \frac{y}{8}\right) = xy$, then $f(m, n) + f(n, m) = 0$

a. only when $m = n$ b. only when $m \neq n$
c. only when $m = -n$ d. for all m and n

53. If $f(x+y) = f(x) + f(y) - xy - 1\ \forall\ x, y \in R$ and $f(1) = 1$, then the number of solutions of $f(n) = n$, $n \in N$ is

a. 0 b. 1
c. 2 d. more than 2

54. The range of the function $f(x) = \frac{e^x - e^{|x|}}{e^x + e^{|x|}}$

a. $(-\infty, \infty)$ b. $[0, 1)$
c. $(-1, 0]$ d. $(-1, 1)$

55. If $f: R \to R$ is a function satisfying the property $f(2x+3) + f(2x+7) = 2, \forall\ x \in R$, then the fundamental period of $f(x)$ is

a. 2 b. 4
c. 8 d. 12

56. Let $f: R \to \left[0, \frac{\pi}{2}\right)$ defined by $f(x) = \tan^{-1}(x^2 + x + a)$, then the set of values of a for which f is onto is

a. $[0, \infty)$ b. $[2, 1]$
c. $\left[\frac{1}{4}, \infty\right)$ d. None of these

57. The domain of the function

$f(x) = \frac{1}{\sqrt{\{\sin x\} + \{\sin(\pi + x)\}}}$, where {.} denotes the fractional part, is

a. $[0, \pi]$ b. $(2n+1)\pi/2$
c. $(0, \pi)$ d. None of these

58. $f(x) = \frac{\cos x}{\left[\frac{2x}{\pi}\right] + \frac{1}{2}}$, where x is not an integral multiple of π and [.] denotes the greatest integer function is

a. An odd function b. Even function
c. Neither odd nor even d. None of these

59. If a function satisfies the condition $f\left(x + \frac{1}{x}\right) = x^2 + \frac{1}{x^2}$, $x \neq 0$, then the domain of $f(x)$ is

a. $[-2, 2]$ b. $(-\infty, -2] \cup [2, \infty)$
c. $(2, \infty)$ d. None of these

60. Let $f: [-10, 10] \to R$, where $f(x) = \sin x + [x^2/a]$ be an odd function. Then the set of values of parameter a is/are

a. $(-10, 10) \sim \{0\}$ b. $(0, 10)$
c. $[100, \infty)$ d. $(100, \infty)$

61. The function f satisfies the functional equation $3f(x) + 2f\left(\frac{x+59}{x-1}\right) = 10x + 30$ for all real $x \neq 1$. The value of $f(7)$ is

a. 8 b. 4
c. -8 d. 11
e. 44

62. The period of the function $f(x) = [6x + 7] + \cos \pi x - 6x$, where [.] denotes the greatest integer function, is

a. 3 b. 2π
c. 2 d. None of these

63. If the graph of the function $f(x) = \frac{a^x - 1}{x^n(a^x + 1)}$ is symmetrical about y-axis, then n equals

a. 2 b. $\frac{2}{3}$

c. $\frac{1}{4}$ d. $-\frac{1}{3}$

64. If $f(x)$ is an even function and satisfies the relation $x^2f(x)-2f\left(\frac{1}{x}\right)=g(x)$ where $g(x)$ is an odd function, then $f(5)$ equals

a. 0 b. $\frac{50}{75}$

c. $\frac{49}{75}$ d. None of these

65. If $f(x+y)=f(x).f(y)$ for all real x, y and $f(0)\neq 0$, then the function $g(x)=\frac{f(x)}{1+\{f(x)\}^2}$ is

a. Even function b. Odd function

c. Odd if $f(x)>0$ d. Neither even nor odd

66. The set of values of a, such that the equation $x^2+2ax+a=\sqrt{a^2+x-\frac{1}{16}-\frac{1}{16}}, x\geq -a$ has two distinct real rootsis given by

a. $[0, 1]$ b. $[-\infty, 0)$

c. $[0, \infty)$ d. $\left[\frac{3}{4}, \infty\right)$

67. Let $g(x)=f(x)-1$. If $f(x)+f(1-x)=2\ \forall\ x\in R$, then $g(x)$ is symmetrical about

a. Origin b. The line $x=\frac{1}{2}$

c. The point $(1, 0)$ d. The point $\left(\frac{1}{2}, 0\right)$

68. Domain (d) and range (R) of $f(x)=\sin^{-1}(\cos^{-1}[x])$ where [.] denotes the greatest integer function is

a. $D\equiv x\in[1, 2), R\in\{0\}$

b. $D\equiv x\in[0, 1], R\equiv\{-1, 0, 1\}$

c. $D\equiv x\in[-1, 1], R\equiv\left\{0, \sin^{-1}\left(\frac{\pi}{2}\right), \sin^{-1}(\pi)\right\}$

d. $D\equiv x\in[-1, 1], R\equiv\left\{-\frac{\pi}{2}, 0, \frac{\pi}{2}\right\}$

69. If $f(x+1)+f(x-1)=2f(x)$ and $f(0)=0$, then $f(n)$, $n\in N$, is

a. $nf(1)$ b. $\{f(1)\}^n$

c. 0 d. None of these

70. The range of the function f defined by $f(x)=\left[\frac{1}{\sin\{x\}}\right]$ [where [.] and {.}, respectively, denote the greatest integer and the fractional part functions] is

a. I, the set of integers

b. N, the set of natural numbers

c. W, the set of whole numbers

d. $\{2, 3, 4, \ldots\}$

71. If $[\cos^{-1}x]+[\cot^{-1}x]=0$, where [.] denotes the greatest integer function, then the complete set of values of x is

a. $(\cos 1, 1]$ b. $(\cos 1, \cot 1)$

c. $(\cot 1, 1]$ d. $[0, \cot 1)$

72. If $f(x)$ and $g(x)$ are periodic functions with period 7 and 11, respectively. Then the period of $F(x)=f(x)g\left(\frac{x}{5}\right)-g(x)f\left(\frac{x}{3}\right)$ is

a. 177 b. 222

c. 433 d. 1155

73. The period of the function

$f(x)=c^{\sin^2x+\sin^2\left(x+\frac{\pi}{3}\right)+\cos x\cos\left(x+\frac{\pi}{3}\right)}$

is (where c is constant)

a. 1 b. $\frac{\pi}{2}$

c. π d. Cannot be determined

74. If $f(x+f(y))=f(x)+y\ \forall\ x, y\in R$ and $f(0)=1$, then the value of $f(7)$ is

a. 1 b. 7

c. 6 d. 8

75. Let $f(x)=\sqrt{|x|-\{x\}}$ (where {.} denotes the fractional part of x) and X, Y are its domain and range, respectively, then

a. $x\in\left(-\infty, \frac{1}{2}\right]$ and $Y\in\left[\frac{1}{2}, \infty\right)$

b. $x\in\left(-\infty, -\frac{1}{2}\right]\cup[0, \infty)$ and $Y\in\left[\frac{1}{2}, \infty\right)$

c. $X\in\left(-\infty, -\frac{1}{2}\right]\cup[0, \infty)$ and $Y\in[0, \infty)$

d. None of these

76. Let f be a function satisfying of x then $f(xy)=\frac{f(x)}{y}$ for all positive real numbers x and y if $f(30)=20$, then the value of $f(40)$ is

a. 15 b. 20

c. 40 d. 60

77. The domain of the function

$f(x)=\sqrt{\ln_{(|x|-1)}(x^2+4x+4)}$ is

a. $[-3,-1]\cup[1,2]$
b. $(-2,-1)\cup[2,\infty)$
c. $(-\infty,-3]\cup(-2,-1)\cup(2,\infty)$
d. None of these

78. The range of $f(x) = [1+\sin x] + \left[2+\sin\frac{x}{2}\right] + \left[3+\sin\frac{x}{3}\right]+\cdots+\left[x+\sin\frac{\pi}{n}\right]$, $\forall x\in[0,p]$, where [.] denotes the greatest integer function, is

a. $\left\{\frac{n^2+n-2}{2},\frac{n(n+1)}{2}\right\}$
b. $\left\{\frac{n(n+1)}{2}\right\}$
c. $\left\{\frac{n^2+n-2}{2},\frac{n(n+1)}{2},\frac{n^2+n+2}{2}\right\}$
d. $\left\{\frac{n(n+1)}{2},\frac{n^2+n+2}{2}\right\}$

79. The total number of solutions of $[x]^2 = x + 2\{x\}$, where [.] and {.} denote the greatest integer function and fractional part, respectively, is

a. 2
b. 4
c. 6
d. None of these

80. The domain of $f(x)=\sqrt{2\{x\}^2-3\{x\}+1}$, where {.} denotes the fractional part in $[-1, 1]$, is

a. $[-1,1]\sim\left(\frac{1}{2},1\right)$
b. $\left[-1,-\frac{1}{2}\right]\cup\left[0,\frac{1}{2}\right]\cup\{1\}$
c. $\left[-1,\frac{1}{2}\right]$
d. $\left[-\frac{1}{2},1\right]$

81. The range of $\sin^{-1}\left[x^2+\frac{1}{2}\right]+\cos^{-1}\left[x^2-\frac{1}{2}\right]$, where [.] denotes the greatest integer function, is

a. $\left\{\frac{\pi}{2},\pi\right\}$
b. $\{\pi\}$
c. $\left\{\frac{\pi}{2}\right\}$
d. None of these

82. If the period of $\dfrac{\cos(\sin(nx))}{\tan\left(\frac{x}{n}\right)}$, $n\in N$ is 6π then

a. 3
b. 2
c. 6
d. 1

83. The domain of $f(x)=\ln(ax^3+(a+b)x^2+(b+c)x+c)$, where $a>0$, $b^2-4ac=0$, is

a. $(-1,\infty)\sim\left\{-\frac{b}{2a}\right\}$
b. $(1,\infty)\sim\left\{-\frac{b}{2a}\right\}$
c. $(-1,1)\sim\left\{-\frac{b}{2a}\right\}$
d. None of these

84. The period of $f(x)=[x]+[2x]+[3x]+[4x]+\cdots[nx] - \frac{n(n+1)}{2}x$, where $n\in N$, is

a. n
b. 1
c. $\frac{1}{n}$
d. None of these

85. If $f\left(x+\frac{1}{2}\right)+f\left(x-\frac{1}{2}\right)=f(x)$ for all $x\in R$, then the period of $f(x)$ is

a. 1
b. 2
c. 3
d. 4

86. If $f: R^+\to R$, $f(x)+3xf\left(\frac{1}{x}\right)=2(x+1)$, then $f(99)$ is equal to

a. 40
b. 30
c. 50
d. 60

87. If $f: X\to Y$, where X and Y are sets containing natural numbers, $f(x)=\frac{x+5}{x+2}$ then the number of elements in the domain and range of $f(x)$ are, respectively,

a. 1 and 1
b. 2 and 1
c. 2 and 2
d. 1 and 2

88. If $f(x)=\begin{cases}x^2 & \text{for } x\ge 0\\ x & \text{for } x<0\end{cases}$ then $f\mathrm{o}f(x)$ is given by

a. x^2 for $x\ge0$, x for $x<0$
b. x^4 for $x\ge0$, x^2 for $x<0$
c. x^4 for $x\ge0$, $-x^2$ for $x<0$
d. x^4 for $x\ge0$, x for $x<0$

89. If the graph of $y=f(x)$ is symmetrical about lines $x=1$ and $x=2$, then which of the following is true?

a. $f(x+1)=f(x)$
b. $f(x+3)=f(x)$
c. $f(x+2)=f(x)$
d. None of these

90. Let $f: R\to R$ and $f\left(x+\frac{5}{6}\right)+f(x)=f\left(x+\frac{1}{2}\right)+f\left(x+\frac{1}{3}\right)$, then

a. $f(x)$ is periodic
b. $f(x)$ is even
c. $f(x+2)-f(x+1)=f(x+1)-f(x)$
d. None of these

91. The domain of $f(x) = \sin^{-1}[2x^2 - 3]$, where [.] denotes the greatest integer function, is

a. $\left(-\sqrt{\frac{3}{2}}, \sqrt{\frac{3}{2}}\right)$

b. $\left(-\sqrt{\frac{3}{2}}, -1\right] \cup \left(-\sqrt{\frac{5}{2}}, \sqrt{\frac{5}{2}}\right)$

c. $\left(-\sqrt{\frac{5}{2}}, \sqrt{\frac{5}{2}}\right)$

d. $\left(-\sqrt{\frac{5}{2}}, -1\right] \cup \left[1, \sqrt{\frac{5}{2}}\right)$

92. The range of $f(x) = \cos^{-1}\left(\frac{1+x^2}{2x}\right) + \sqrt{2-x^2}$ is

a. $\left\{0, 1+\frac{\pi}{2}\right\}$ b. $\{0, 1+\pi\}$

c. $\left\{1, 1+\frac{\pi}{2}\right\}$ d. $\{1, 1+\pi\}$

93. If $f(x) = \begin{cases} x, & x \text{ is rational} \\ 1-x, & x \text{ is irrational} \end{cases}$ then $f(f(x))$ is

a. $x \; \forall \; x \in R$ b. $= \begin{cases} x, & x \text{ is irrational} \\ 1-x, & x \text{ is rational} \end{cases}$

c. $\begin{cases} x, & x \text{ is rational} \\ 1-x, & x \text{ is irrational} \end{cases}$ d. None of these

94. The range of $f(x) = [\,|\sin x| + |\cos x|\,]$, where [.] denotes the greatest integer function, is

a. $\{0\}$ b. $\{0, 1\}$

c. $\{1\}$ d. None of these

95. If $f(x) = \log_e\left(\frac{x^2+e}{x^2+1}\right)$, then the range of $f(x)$ is

a. $(0, 1)$ b. $[0, 1]$

c. $[0, 1)$ d. $(0, 1]$

96. The domain of the function

$f(x) = \frac{1}{\sqrt{4x - |x^2 - 10x + 9|}}$ is

a. $(7-\sqrt{40}, 7+\sqrt{40})$ b. $(0, 7+\sqrt{40})$

c. $(7-\sqrt{40}, \infty)$ d. None of these

97. If the function $f: [1, \infty) \to [1, \infty)$ is defined by $f(x) = 2^{x(x-1)}$, then $f^{-1}(x)$ is

a. $\left(\frac{1}{2}\right)^{x(x-1)}$ b. $\frac{1}{2}(1+\sqrt{1+4\log_2 x})$

c. $\frac{1}{2}(1-\sqrt{1+4\log_2 x})$ d. Not defined

98. The number of roots of the equation $x \sin x = 1$, $x \in [-2\pi, 0) \cup (0, 2\pi]$, is

a. 2 b. 3

c. 4 d. 0

99. The number of solutions of $2\cos x = |\sin x|$, $0 \le x \le 4\pi$, is

a. 0 b. 2

c. 4 d. Infinite

100. If $af(x+1) + bf\left(\frac{1}{x+1}\right) = x, x \ne -1, a \ne b$, then $f(2)$ is equal to

a. $\frac{2a+b}{2(a^2-b^2)}$ b. $\frac{a}{a^2-b^2}$

c. $\frac{a+2b}{a^2-b^2}$ d. None of these

101. The number of solutions of $\tan x - mx = 0$, $m > 1$ in $\left(-\frac{\pi}{2}, \frac{\pi}{2}\right)$ is

a. 1 b. 2

c. 3 d. m

102. The range of $f(x) = [\sin x + [\cos x + [\tan x + [\sec x]]]]$, $x \in (0, \pi/4)$, where [.] denotes the greatest integer function $\le x$, is

a. $\{0, 1\}$ b. $\{-1, 0, 1\}$

c. $\{1\}$ d. None of these

103. If $f(3x+2) + f(3x+29) = 0 \; \forall \; x \in R$, then the period of $f(x)$ is

a. 7 b. 8

c. 10 d. None of these

104. Let $f(x) = \begin{cases} \sin x + \cos x, & 0 \le x < \frac{\pi}{2} \\ a, & x = \pi/2 \\ \tan^2 x + \operatorname{cosec} x, & \pi/2 < x < \pi \end{cases}$

then its odd extension is

a. $\begin{cases} -\tan^2 x - \operatorname{cosec} x, & -\pi < x < -\frac{\pi}{2} \\ -a, & x = -\frac{\pi}{2} \\ -\sin x + \cos x, & -\frac{\pi}{2} < x < 0 \end{cases}$

b. $\begin{cases} -\tan^2 x + \operatorname{cosec} x, & -\pi < x < -\frac{\pi}{2} \\ -a, & x = -\frac{\pi}{2} \\ \sin x - \cos x, & -\frac{\pi}{2} < x < 0 \end{cases}$

c. $\begin{cases} -\tan^2 x + \operatorname{cosec} x, & -\pi < x < -\frac{\pi}{2} \\ a, & x = -\frac{\pi}{2} \\ \sin x - \cos x, & -\frac{\pi}{2} < x < 0 \end{cases}$

d. $\begin{cases} \tan^2 x + \cos x, & -\pi < x < -\frac{\pi}{2} \\ -a, & x = -\frac{\pi}{2} \\ \sin x + \cos x, & -\frac{\pi}{2} < x < 0 \end{cases}$

105. If f and g are one-one functions, then

a. $f+g$ is one-one
b. fg is one-one
c. fog is one-one
d. None of these

106. The domain of $f(x)$ is (0, 1), therefore, domain of $f(e^x)+f(\ln|x|)$ is

a. $(-1, e)$
b. $(1, e)$
c. $(-e, -1)$
d. $(-e, 1)$

107. The domain of $f(x) = \dfrac{1}{\sqrt{|\cos x| + \cos x}}$ is

a. $[-2n\pi, 2n\pi], n \in Z$
b. $(2n\pi, \overline{2n+1}\pi), n \in Z$
c. $\left(\dfrac{(4n+1)\pi}{2}, \dfrac{(4n+3)\pi}{2}\right), n \in Z$
d. $\left(\dfrac{(4n-1)\pi}{2}, \dfrac{(4n+1)\pi}{2}\right), n \in Z$

108. If $f(2x+3y, 2x-7y) = 20x$, then $f(x, y)$ equals

a. $7x-3y$
b. $7x+3y$
c. $3x-7y$
d. $x-ky$

109. Let $X = \{a_1, a_2, \dots, a_6\}$ and $Y = \{b_1, b_2, b_3\}$. The number of functions f from x to y such that it is onto and there are exactly three elements x in X such that $f(x) = b_1$ is

a. 75
b. 90
c. 100
d. 120

110. Let $f: R \to R$ and $g: R \to R$ be two one-one and onto functions such that they are the mirror images of each other about the line $y = a$. If $h(x) = f(x) + g(x)$, then $h(x)$ is

a. One-one and onto.
b. Only one-one and not onto.
c. Only onto but not one-one.
d. Neither one-one nor onto.

111. If $f(x) = (-1)^{[(2x)/\pi]}$, $g(x) = |\sin x| - |\cos x|$ and $\phi(x) = f(x)\,g(x)$ (where [.] denotes the greatest integer function) then the respective fundamental periods of $f(x), g(x)$ and $\phi(x)$ are

a. π, π, π
b. $\pi, 2\pi, \pi$
c. $\pi, \pi, \frac{\pi}{2}$
d. $\pi, \frac{\pi}{2}, \pi$

112. Let $f(n) = 1 + \frac{1}{2} + \frac{1}{3} + \dots + \frac{1}{n}$, then $f(1)+f(2)+f(3)+\dots+f(b)$ is equal to

a. $nf(b)-1$
b. $(n+1)f(n)-n$
c. $(n+1)f(n)+n$
d. $nf(n)+n$

113. Let $f(x) = e^{\{e^{|x|}\operatorname{sgn} x\}}$ and $g(x) = e^{[e^{|x|}\operatorname{sgn} x]}, x \in R$ where { } and [ ] denotes the fractional and integral part functions, respectively. Also $h(x) = \log(f(x)) + \log(g(x))$ then for real x, $h(x)$ is

a. An odd function.
b. An even function.
c. Neither an odd nor an even function.
d. Both odd as well as even function.

114. Let $f_1(x) = \begin{cases} x, & 0 \le x \le 1 \\ 1, & x > 1 \\ 0, & \text{otherwise} \end{cases}$

and $f_2(x) = f_1(-x)$ for all x
$f_3(x) = -f_2(x)$ for all x
$f_4(x) = f_3(-x)$ for all x

Which of the following is necessarily true?

a. $f_4(x) = f_1(x)$ for all x
b. $f_1(x) = -f_3(-x)$ for all x
c. $f_2(-x) = f_4(x)$ for all x
d. $f_1(x) + f_3(x) = 0$ for all x

115. The number of solutions of the equation $[y + [y]] = 2\cos x$, where $y = 1/3\,[\sin x + [\sin x + [\sin x]]]$ (where [.] denotes the greatest integer function) is

a. 4
b. 2
c. 3
d. 53

116. The sum of roots of the equation $\cos^{-1}(\cos x) = [x]$, where [.] denotes the greatest integer function, is

a. $2\pi+3$
b. $\pi+3$
c. $\pi-3$
d. $2\pi-3$

117. The range of

$f(x) = \sqrt{(1-\cos x)\sqrt{(1-\cos x)\sqrt{(1-\cos x)\sqrt{\dots\infty}}}}$ is

a. $[0, 1]$
b. $[0, 1/2]$
c. $[0, 2]$
d. None of these

118. Let $h(x) = |kx+5|$, the domain of $f(x)$ is $[-5, 7]$, the domain of $f(h(x))$ is $[-6, 1]$ and the range of $h(x)$ is the same as the domain of $f(x)$, then the value of k is

a. 1
b. 2
c. 3
d. 4

119. The range of $f(x)=(x+1)(x+2)(x+3)(x+4)+5$ for $x\in[-6,6]$ is

a. $[4,5045]$ b. $[0,5045]$

c. $[-20,5045]$ d. None of these

120. The exhaustive domain of

$f(x)=\sqrt{x^{12}-x^9+x^4-x+1}$ is

a. $[0,1]$ b. $[1,\infty)$

c. $(-\infty,1]$ d. R

121. The range of $f(x)=\sec^{-1}(\log_3\tan x+\log_{\tan x}3)$ is

a. $\left[\frac{\pi}{3},\frac{\pi}{2}\right)\cup\left(\frac{\pi}{2},\frac{2\pi}{3}\right]$ b. $\left[0,\frac{\pi}{2}\right)$

c. $\left(\frac{2\pi}{3},\pi\right]$ d. None of these

122. The range of the function $f(x)={}^{7-x}P_{x-3}$ is

a. $\{1,2,3\}$ b. $\{1,2,3,4,5,6\}$

c. $\{1,2,3,4\}$ d. $\{1,2,3,4,5\}$

123. A real-valued function $f(x)$ satisfies the functional equation $f(x-y)=f(x)f(y)-f(a-x)f(a+y)$, where a is a given constant and $f(0)=1$. $f(2a-x)$ is equal to

a. $f(x)$ b. $-f(x)$

c. $f(-x)$ d. $f(a)+f(a-x)$

Multiple Correct Answers Type

Each question has four choices a, b, c, and d, out of which *one or more* answers are correct.

1. Let $f(x)=\max\{1+\sin x,1,1-\cos x\}, x\in[0,2\pi]$ and $g(x)=\max\{1,|x-1|\}\ x\in R$, then

a. $g(f(0))=1$ b. $g(f(1))=1$

c. $f(f(1))=1$ d. $f(g(0))=1+\sin 1$

2. Which of the following functions are identical?

a. $f(x)=\ln x^2$ and $g(x)=2\ln x$

b. $f(x)=\log_x e$ and $g(x)=\dfrac{1}{\log_e x}$

c. $f(x)=\sin(\cos^{-1}x)$ and $g(x)=\cos(\sin^{-1}x)$

d. None of these

3. Which of the following function/functions have the graph symmetrical about the origin?

a. $f(x)$ given by $f(x)+f(y)=f\left(\dfrac{x+y}{1-xy}\right)$

b. $f(x)$ given by $f(x)+f(y)$ $=f\left(x\sqrt{1-y^2}+y\sqrt{1-x^2}\right)$

c. $f(x)$ given by $f(x+y)=f(x)+f(y)\ \forall\ x,y\in R$

d. None of these

4. If the function f satisfies the relation $f(x+y)+f(x-y)=2f(x)f(y)\ \forall\ x,y\in R$ and $f(0)\neq 0$, then

5. Consider the function $y=f(x)$ satisfying the condition $f\left(x+\dfrac{1}{x}\right)=x^2+\dfrac{1}{x^2}\ (x\neq 0)$, then

a. domain of $f(x)$ is R

b. domain of $f(x)$ is $R-(-2,2)$

c. range of $f(x)$ is $[-2,\infty)$

d. range of $f(x)$ is $[2,\infty)$

6. Let $f(x)+f(y)=f(x\sqrt{1-y^2}+y\sqrt{1-x^2})\times(f(x)$ is not identically zero). Then

a. $f(4x^3-3x)+3f(x)=0$

b. $f(4x^3-3x)=3f(x)$

c. $f(2x\sqrt{1-x^2})+2f(x)=0$

d. $f(2x\sqrt{1-x^2})=2f(x)$

7. Consider the real-valued function satisfying $2f(\sin x)+f(\cos x)=x$. Then

a. domain of $f(x)$ is R

b. domain of $f(x)$ is $[-1,1]$

c. range of $f(x)$ is $\left[-\dfrac{2\pi}{3},\dfrac{\pi}{3}\right]$

d. range of $f(x)$ is R

8. If $f(x)$ satisfies the relation $f(x+y)=f(x)+f(y)$ for all $x,y\in R$ and $f(1)=5$, then

a. $f(x)$ is an odd function b. $f(x)$ is an even function

c. $\sum_{r=1}^{m}f(r)=5\ {}^{m+1}C_2$ d. $\sum_{r=1}^{m}f(r)=\dfrac{5m(m+2)}{3}$

9. Let $f(x)=\begin{cases}x^2-4x+3, & x<3\\ x-4, & x\geq 3\end{cases}$ and $g(x)$ $=\begin{cases}x-3, & x<4\\ x^2+2x+2, & x\geq 4\end{cases}$ then, which of the the following is/are true?

a. $(f+g)(3.5)=0$ b. $f(g(3))=3$

c. $(fg)(2)=1$ d. $(f-g)(4)=0$

10. $f(x)=x^2-2ax+a(a+1), f:[a,\infty)\to[a,\infty)$. If one of the solutions of the equation $f(x)=f^{-1}(x)$ is 5049, then the other may be

a. 5051 b. 5048

c. 5052 d. 5050

11. Which of the following function is/are periodic

a. $f(x)=\begin{cases}1, & x \text{ is rational}\\ 0, & x \text{ is irrational}\end{cases}$

b. $f(x)=\begin{cases}x-[x]; & 2n\leq x<2n+1\\ \frac{1}{2}; & 2n+1\leq x<2n+2\end{cases}$, where [.] denotes the greatest integer function

c. $f(x) = (-1)^{[(2x)/\pi]}$, where [.] denotes the greatest integer function

d. $f(x) = ax - [ax + a] + \tan(\pi x/2)$, where [.] denotes the greatest integer function, and a is a rational number

12. If $f: R^+ \to R^+$ is a polynomial function satisfying the functional equation $f(f(x)) = 6x - f(x)$, then $f(17)$ is equal to

a. 17 b. −51

c. 34 d. −34

13. Let $f: R \to R$ be a function defined by $f(x+1) = \dfrac{f(x)-5}{f(x)-3}$ $\forall x \in R$. Then which of the following statement(s) is/are true

a. $f(2008)=f(2004)$ b. $f(2006)=f(2010)$

c. $f(2006)=f(2002)$ d. $f(2006)=f(2018)$

14. Let $f(x) = \sec^{-1}[1+\cos^2 x]$, where [.] denotes the greatest integer function. Then

a. the domain of f is R

b. the domain of f is $[1, 2]$

c. the domain of f is $[1, 2]$

d. the range of f is $[\sec^{-1} 1, \sec^{-1} 2]$

15. Which of the following pairs of functions is/are identical?

a. $f(x) = \tan(\tan^{-1} x)$ and $g(x) = \cot(\cot^{-1} x)$

b. $f(x) = \text{sgn}(x)$ and $g(x) = \text{sgn}(\text{sgn}(x))$

c. $f(x) = \cot^2 x \cdot \cos^2 x$ and $g(x) = \cot^2 x - \cos^2 x$

d. $f(x) = e^{\ln \sec^{-1} x}$ and $g(x) = \sec^{-1} x$

16. $f: R \to [-1, \infty)$ and $f(x) = \ln([|\sin 2x| + |\cos 2x|])$ (where [.] is the greatest integer function).

a. Z

b. $f(x)$ is periodic with fundamental period $\pi/4$

c. $f(x)$ is invertible in $\left[0, \dfrac{\pi}{4}\right]$

d. $f(x)$ is into function

17. Which of the following is not a function ([.] and {.} denotes the greatest integer and fractional part functions respectively)?

a. $\dfrac{1}{\ln[1-|x|]}$ b. $\dfrac{x!}{\{x\}}$

c. $x!\,\{x\}$ d. $\dfrac{\ln(x-1)}{\sqrt{(1-x^2)}}$

18. If the following functions are defined from $[-1, 1]$ to $[-1, 1]$, select those which are bijective

a. $\sin(\sin^{-1} x)$ b. $\dfrac{2}{\pi}\sin^{-1}(\sin x)$

c. $(\text{sgn}(x)) \ln(e^x)$ d. $x^3(\text{sgn}(x))$

19. If $f: R \to N \cup \{0\}$, where f(area of triangle joining points $P(5, 0)$, $Q(8, 4)$ and $R(x, y)$ such that the angle PRQ is a right) = number of triangle. Then, which of the following is true ?

a. $f(5)=4$ b. $f(7)=0$

c. $f(6.25)=2$ d. $f(x)$ is into

20. If $f(x)$ is a polynomial of degree n such that $f(0) = 0$, $f(1) = \dfrac{1}{2}, \ldots, f(n) = \dfrac{n}{n+1}$, then the value of $f(n+1)$ is

a. 1 when n is odd b. $\dfrac{n}{n+1}$ when n is even

c. $-\dfrac{n}{n+1}$ when n is odd d. −1 when n is even

21. Let $f(x) = \dfrac{3}{4}x + 1$, and $f^n(x)$ be defined as $f^2(x) = f(f(x))$, and for $n \geq 2$, $f^{n+1}(x) = f(f^n(x))$. If $\lambda = \lim\limits_{n\to\infty} f^n(x)$, then

a. λ is independent of x

b. λ is a linear polynomial in x

c. the line $y = \lambda$ has slope O

d. the line $4y = \lambda$ touches the unit circle with centre at the origin.

22. The domain of the function

$$f(x) = \log_e\left\{\log_{|\sin x|}(x^2 - 8x + 23) - \frac{3}{\log_2 |\sin x|}\right\},$$

contains which of the following interval / intervals.

a. $(3, \pi)$ b. $\left(\pi, \dfrac{3}{2}\right)$

c. $\left(\dfrac{3\pi}{2}, 5\right)$ d. None of these

Reasoning Type

Each question has four choices a, b, c, and d, out of which *only one* is correct. Each question contains Statement 1 and Statement 2.

a. if both the statements are True and Statement 2 is the correct explanation of Statement 1.

b. if both the statements are True but Statement 2 is Not the correct explanation of Statement 1.

c. if Statement 1 is True and Statement 2 is False.

d. if Statement 1 is False and Statement 2 is True.

1. Statement 1: $f(x) = \log_e x$ cannot be expressed as a sum of odd and even function.

Statement 2: $f(x) = \log_e x$ is neither odd nor even function.

2. Statement 1: If $g(x) = f(x) - 1$. If $f(x) + f(1-x) = 2$ $\forall x \in R$, then $g(x)$ is symmetrical about the point $(1/2, 0)$.

Statement 2: If $g(a-x)=-g(a+x)\ \forall\ x\in R$, then $g(x)$ is symmetrical about the point $(a, 0)$.

3. Consider the function satisfying the relation

$$\text{if } f\left(\frac{2\tan x}{1+\tan^2 x}\right)=\frac{(1+\cos 2x)(\sec^2 x+2\tan x)}{2}$$

Statement 1: Range of $y=f(x)$ is R.
Statement 2: Linear function has range R if domain is R.

4. Consider the function if $f(x)=\sin(kx)+\{x\}$, where $\{x\}$ represents the fractional part function.
Statement 1: $f(x)$ is periodic for $k=m\pi$ where m is a rational number.
Statement 2: The sum of two periodic functions is always periodic.

5. **Statement 1:** Function $f(x)=x^2+\tan^{-1}x$ is a non-periodic function.
Statement 2: The sum of two non-periodic function is always non-periodic.

6. **Statement 1:** If $x\in[1, \sqrt{3}]$, then the range of $f(x)=\tan^{-1}x$ is $[\pi/4, \pi/3]$.
Statement 2: If $x\in[a, b]$, then the range of $f(x)$ is $[f(a), f(b)]$.

7. **Statement 1:** $f: N\to R,\ f(x)=\sin x$ is a one-one function.
Statement 2: The period of $\sin x$ is 2π and 2π is an irrational number.

8. **Statement 1:** A continuous surjective function $f: R\to R, f(x)$ can never be a periodic function.
Statement 2: For a surjective function $f: R\to R, f(x)$ to be periodic, it should necessarily be a discontinuous function.

9. **Statement 1:** The solution of equation $||x^2-5x+4|-|2x-3||=|x^2-3x+1|$ is $x\in(-\infty, 1]\cup\left[\frac{3}{2}, 4\right]$.
Statement 2: If $|x+y|=|x|+|y|$, then $x.y\geq 0$.

10. Consider f and g be real-valued functions such that $f(x+y)+f(x-y)=2f(x).g(y)\ \forall\ x, y\in R$.
Statement 1: If $f(x)$ is not identically zero and $|f(x)|\leq 1\ \forall\ x\in R$, then $|g(y)|\leq 1\ \forall\ y\in R$.
Statement 2: For any two real numbers x and y, $|x+y|\leq|x|+|y|$.

11. **Statement 1:** $f(x)=\cos(x^2-\tan x)$ is a non-periodic function.
Statement 2: $x^2-\tan x$ is a non-periodic function.

12. **Statement 1:** The period of function $f(x)=\sin\{x\}$ is 1, where $\{.\}$ represents fractional part function.
Statement 2: $g(x)=\{x\}$ has period 1.

13. **Statement 1:** If $f: R\to R, y=f(x)$ is periodic and continuous function, then $y=f(x)$ cannot be onto.
Statement 2: A continuous periodic function is bounded.

14. Consider the functions $f(x)=\log_e x$ and $g(x)=2x+3$.
Statement 1: $f(g(x))$ is a one-one function.
Statement 2: $g(x)$ is a one-one function.

15. Consider the functions $f: R\to R,\ f(x)=x^3$ and $g: R\to R, g(x)=3x+4$.
Statement 1: $f(g(x))$ is an onto an function.
Statement 2: $g(x)$ is an onto function.

16. **Statement 1:** $f(x)=\sin x$ and $g(x)=\cos x$ are identical functions.
Statement 2: Both the functions have the same domain and range.

17. **Statement 1:** The period of $f(x)=\sin x$ is $2\pi\Rightarrow$ the period of $g(x)=|\sin x|$ is π.
Statement 2: The period of $f(x)=\cos x$ is $2\pi\Rightarrow$ the period of $g(x)=|\cos x|$ is π.

18. **Statement 1:** $f(x)=\sqrt{ax^2+bx+c}$ has a range $[0, \infty)$ if $b^2-4ac>0$.
Statement 2: $ax^2+bx+c=0$ has real roots if $b^2-4ac=0$.

19. **Statement 1:** If $f(x)=\cos x$ and $g(x)=x^2$, then $f(g(x))$ is an even function.
Statement 2: If $f(g(x))$ is an even function, then both $f(x)$ and $g(x)$ must be even function.

20. **Statement 1:** The graph of $y=\tan x$ is symmetrical about origin.
Statement 2: The graph of $y=\sec^2 x$ is symmetrical about y-axis.

Linked Comprehension Type

Based upon each paragraph, three multiple choice questions have to be answered. Each question has four choices a, b, c and d, out of which *only one* is correct.

For Problems 1–3

Consider the functions

$$f(x)=\begin{cases}x+1, & x\leq 1\\ 2x+1, & 1<x\leq 2\end{cases}\text{ and } g(x)=\begin{cases}x^2, & -1\leq x<2\\ x+2, & 2\leq x\leq 3\end{cases}.$$

1. The domain of the function $f(g(x))$ is
   a. $[0, \sqrt{2}]$ b. $[-1, 2]$
   c. $[-1, \sqrt{2}]$ d. None of these
2. The range of the function $f(g(x))$ is
   a. $[1, 5]$ b. $[2, 3]$
   c. $[1, 2]\cup(3, 5]$ d. None of these
3. The number of roots of the equation $f(g(x))=2$ is
   a. 1 b. 2
   c. 4 d. None of these

For Problems 4–6

Consider the function $f(x)$ satisfying the identity $f(x)+f\left(\frac{x-1}{x}\right)=1+x, \forall\ x\in R-\{0, 1\}$ and $g(x)=2f(x)-x+1$.

4. The domain of $y=\sqrt{g(x)}$ is

a. $\left(-\infty, \frac{1-\sqrt{5}}{2}\right] \cup \left[1, \frac{1+\sqrt{5}}{2}\right]$

b. $\left(-\infty, \frac{1-\sqrt{5}}{2}\right] \cup (0, 1) \cup \left[\frac{1+\sqrt{5}}{2}, \infty\right)$

c. $\left[\frac{-1-\sqrt{5}}{2}, 0\right] \cup \left[\frac{-1+\sqrt{5}}{2}, 1\right]$

d. None of these

5. The range of $y=g(x)$ is

a. $(-\infty, 5]$ b. $[1, \infty)$

c. $(-\infty, 1] \cup [5, \infty)$ d. None of these

6. The number of roots of the equation $g(x)=1$ is

a. 2 b. 1

c. 3 d. 0

For Problems 7–9

Let $f: N \to R$ be a function satisfying the following conditions, $f(1)=1$ and $f(1)+2f(2)+3f(3)+\cdots+nf(n) = n(n+1)f(n)$ for $n \ge 2$.

7. The value of $f(1003)=\frac{1}{K}$, where K equals

a. 1003 b. 2003

c. 2005 d. 2006

8. The value of $f(999)$ is $\frac{1}{K}$, where K equals

a. 999 b. 1000

c. 1998 d. 2000

9. $f(1), f(2), f(3), f(4), \ldots$ represents a series of

a. an A.P

b. a G.P.

c. a H.P.

d. An arithmetico-geometric

For Problems 10–12

If $(f(x))^2 \times f\left(\frac{1-x}{1+x}\right) = 64x, x \neq 0, 1$, then

10. $f(x)$ is equal to

a. $4x^{2/3}\left(\frac{1+x}{1-x}\right)^{1/3}$ b. $x^{1/3}\left(\frac{1-x}{1+x}\right)^{1/3}$

c. $x^{2/3}\left(\frac{1-x}{1+x}\right)^{1/3}$ d. $x\left(\frac{1+x}{1-x}\right)^{1/3}$

11. The domain of $f(x)$ is

a. $[0, \infty)$ b. $R-\{-1, 1\}$

c. $(-\infty, \infty)$ d. None of these

12. The value of $f(9/7)$ is

a. $8(7/9)^{2/3}$ b. $4(9/7)^{1/3}$

c. $-8(9/7)^{2/3}$ d. None of these

For Problems 13–15

$f(x) = \begin{cases} x-1, & -1 \le x \le 0 \\ x^2, & 0 \le x \le 1 \end{cases}$ and $g(x) = \sin x$. Consider the functions $h_1(x)=f(|g(x)|)$ and $h_2(x)=|f(g(x))|$.

13. Which of the following is not true about $h_1(x)$?

a. It is periodic function with period π

b. Range is [0, 1]

c. Domain is R

d. None of these

14. Which of the following is not true about $h_2(x)$?

a. Domain is R

b. It periodic function with period 2π

c. Range is [0, 1]

d. None of these

15. For $h_1(x)$ and $h_2(x)$ are identical function, then which of the following is not true?

a. Domain of $h_1(x)$ and $h_2(x)$, $x \in [2n\pi\ (2n+1)\pi]$, $n \in Z$

b. Range of $h_1(x)$ and $h_2(x)$ is [0, 1]

c. Period of $h_1(x)$ and $h_2(x)$ is 2π

d. None of these

For Problems 16–18

If $a_0=x$, $a_{n+1}=f(a_n)$, where $n=0, 1, 2, \ldots$, then answer the following questions.

16. If $f(x)=\sqrt[m]{(a-x^m)}, x>0, m \ge 2, m \in N$. Then

a. $a_n=x$, $n=2k+1$, where k is integer

b. $a_n=f(x)$ if $n=2k$, where k is integer

c. Inverse of a_n exists for any value of n and m

d. None of these

17. If $f(x)=\frac{1}{1-x}$, then which of the following is not true?

a. $a_n=\frac{1}{1-x}$ if $n=3k+1$

b. $a_n=\frac{1}{1-x}$ if $n=3k+2$

c. $a_n=x$ if $n=3k$

d. None of these

18. If $f: R \to R$ be given by $f(x)=3+4x$ and $a_n=A+Bx$, then which of the following is not true?

a. $A+B+1=2^{2n+1}$

b. $|A-B|=1$

c. $\lim_{n\to\infty} \frac{A}{B}=-1$

d. None of these

For Problems 19–21

Let $f(x)=f_1(x)-2f_2(x)$,

where $f_1(x)=\begin{cases}\min\{x^2,|x|\}, & |x|\le 1\\ \max\{x^2,|x|\}, & |x|>1\end{cases}$

and $f_2(x)=\begin{cases}\min\{x^2,|x|\}, & |x|>1\\ \max\{x^2,|x|\}, & |x|\le 1\end{cases}$

and $g(x)=\begin{cases}\min\{f(t): -3\le t\le x, & -3\le x<0\}\\ \max\{f(t): 0\le t\le x, & 0\le x\le 3\}\end{cases}$.

19. For $-3\le x\le -1$, the range of $g(x)$ is
   a. $[-1,3]$
   b. $[-1,-15]$
   c. $[-1,9]$
   d. None of these

20. For $x\in(-1,0)$, $f(x)+g(x)$ is
   a. x^2-2x+1
   b. x^2+2x-1
   c. x^2+2x+1
   d. x^2-2x-1

21. The graph of $y=g(x)$ in its domain is broken at
   a. 1 point
   b. 2 points
   c. 3 points
   d. None of these

For Problems 22–24

Let $f(x)=\begin{cases}2x+a, & x\ge -1\\ bx^2+3, & x<-1\end{cases}$

and $g(x)=\begin{cases}x+4, & 0\le x\le 4\\ -3x-2, & -2<x<0\end{cases}$

22. $g(f(x))$ is not defined if
   a. $a\in(10,\infty), b\in(5,\infty)$
   b. $a\in(4,10), b\in(5,\infty)$
   c. $a\in(10,\infty), b\in(0,1)$
   d. $a\in(4,10), b\in(1,5)$

23. If the domain of $g(f(x))$ is $[-1,4]$, then
   a. $a=1, b>5$
   b. $a=2, b>7$
   c. $a=2, b>10$
   d. $a=0, b\in R$

24. If $a=2$ and $b=3$, then the range of $g(f(x))$ is
   a. $(-2,8]$
   b. $(0,8]$
   c. $[4,8]$
   d. $[-1,8]$

For Problems 25–27

Let $f: R\to R$ is a function satisfying $f(2-x)=f(2+x)$ and $f(20-x)=f(x), \forall x\in R$. For this function f, answer the following.

25. If $f(0)=5$, then the minimum possible number of values of x satisfying $f(x)=5$, for $x\in[0,170]$, is
   a. 21
   b. 12
   c. 11
   d. 22

26. The graph of $y=f(x)$ is
   a. symmetrical about $x=16$
   b. symmetrical about $x=5$
   c. symmetrical about $x=8$
   d. symmetrical about $x=20$

27. If $f(2)\ne f(6)$, then the
   a. fundamental period of $f(x)$ is 1
   b. fundamental period of $f(x)$ may be 1
   c. period of $f(x)$ cannot be 1
   d. fundamental period of $f(x)$ is 8

For Problems 28–30

Consider two functions $f(x)=\begin{cases}[x], & -2\le x\le -1\\ |x|+1, & -1<x\le 2\end{cases}$ and $g(x)=\begin{cases}[x], & -\pi\le x<0\\ \sin x, & 0\le x\le\pi\end{cases}$, where [.] denotes the greatest integer function.

28. The exhaustive domain of $g(f(x))$ is
   a. $[0,2]$
   b. $[-2,0]$
   c. $[-2,2]$
   d. $[-1,2]$

29. The range of $g(f(x))$ is
   a. $[\sin 3, \sin 1]$
   b. $[\sin 3, 1]\cup\{-2,-1,0\}$
   c. $[\sin 1, 1]\cup\{-2,-1\}$
   d. $[\sin 1, 1]$

30. The number of integral points in the range of $g(f(x))$ is
   a. 2
   b. 4
   c. 3
   d. 5

Matrix-Match Type

Each question contains statements given in two columns which have to be matched. Statements a, b, c, d in column I have to be matched with statements p, q, r, s in column II. If the correct match is a→p, a→s, b→q, b→r, c→p, c→q and d→s, then the correctly bubbled 4×4 matrix should be as follows:

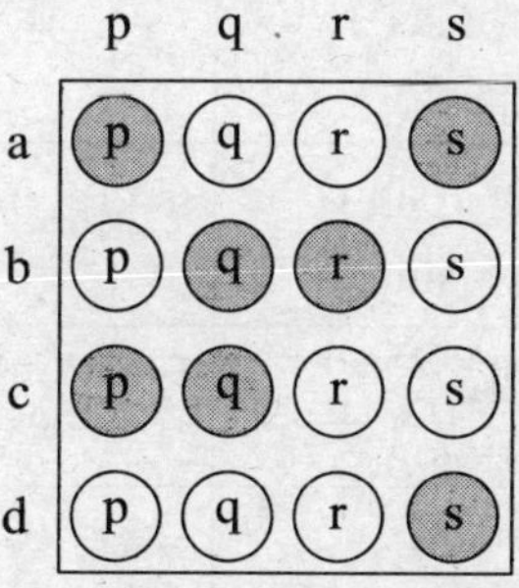

1. The function $f(x)$ is defined on the interval [0, 1]. Then match the following columns

Column I: Function	Column II: Domain
a. $f(\tan x)$	p. $\left[2n\pi, -\frac{\pi}{2}, 2n\pi+\frac{\pi}{2}\right]$, $n\in Z$
b. $f(\sin x)$	q. $\left[2n\pi, 2n\pi+\frac{\pi}{6}\right]$, $\cup\left[2n\pi+\frac{5\pi}{6}, (2n+1)\pi\right]$ $n\in Z$

c. $f(\cos x)$	r. $[2n\pi, (2n+1)\pi], n \in Z$
d. $f(2\sin x)$	s. $\left[n\pi, n\pi+\frac{\pi}{4}\right], n \in Z$

2.

Column I: Function	Column II: Type of function
a. $f(x)=\{(\operatorname{sgn} x)^{\operatorname{sgn} x}\}^n$; $x \neq 0$, n is an odd integer	p. odd function
b. $f(x)=\frac{x}{e^x-1}+\frac{x}{2}+1$	q. even function
c. $f(x)=\begin{cases}0, & \text{If } x \text{ is rational}\\ 1, & \text{If } x \text{ is irrational}\end{cases}$	r. neither odd nor even function
d. $f(x)=\max\{\tan x, \cot x\}$	s. periodic

3.

Column I: Functions	Column II: Values of x for which both the functions in any option of the column I are defined
a. $f(x)=\tan^{-1}\left(\frac{2x}{1-x^2}\right)$, $g(x)=2\tan^{-1}x$	p. $x \in \{-1, 1\}$
b. $f(x)=\sin^{-1}(\sin x)$ and $g(x)=\sin(\sin^{-1}x)$	q. $x \in [-1, 1]$
c. $f(x)=\log_{x^2} 25$ and $g(x)=\log_x 5$	r. $x \in (-1, 1)$
d. $f(x)=\sec^{-1}x + \operatorname{cosec}^{-1}x$, $g(x)=\sin^{-1}x+\cos^{-1}x$	s. $x \in (0, 1)$

4.

Column I	Column II
a. $f: R \rightarrow \left[\frac{3\pi}{4}, \pi\right)$ and $f(x)=\cot^{-1}(2x-x^2-2)$, then $f(x)$ is	p. one-one
b. $f: R \rightarrow R$ and $f(x)=e^{px}\sin qx$ where $p, q \in R^+$, then $f(x)$ is	q. into
c. $f: R^+ \rightarrow [4, \infty]$ and $f(x)=4+3x^2$, then $f(x)$ is	r. many-one
d. $f: X \rightarrow X$ and $f(f(x))=x$ $\forall x \in X$, then $f(x)$ is	s. onto

5. Let $f: R \rightarrow R$ and $g: R \rightarrow R$ be functions such that $f(g(x))$ is a one-one function.

Column I	Column II
a. Then $g(x)$	p. must be one-one
b. Then $f(x)$	q. may not be one-one
c. If $g(x)$ is onto then $f(x)$	r. may be many-one
d. If $g(x)$ is into then $f(x)$	s. must be many-one

6.

Column I: Function	Column II: Period
a. $f(x)=\cos(\lvert\sin x\rvert-\lvert\cos x\rvert)$	p. π
b. $f(x)=\cos(\tan x+\cot x)\cos(\tan x-\cot x)$	q. $\pi/2$
c. $f(x)=\sin^{-1}(\sin x)+e^{\tan x}$	r. 4π
d. $f(x)=\sin^3 x \sin 3x$	s. 2π

7. {.} denotes the fractional part function and [.] denotes the greatest integer function:

Column I: (Function)	Column II: (Period)
a. $f(x)=e^{\cos^4 \pi x+x-[x]+\cos^2 \pi x}$	p. 1/3
b. $f(x)=\cos 2\pi\{2x\}+\sin 2\pi\{2x\}$	q. 1/4
c. $f(x)=\sin 3\pi\{x\}+\tan \pi[x]$	r. 1/2
d. $f(x)=3x-[3x+a]-b$, where $a, b \in R^+$	s. 1

8.

Column I: (Function)	Column II: (Range)
a. $f(x)=\log_3(5+4x-x^2)$	p. function not defined
b. $f(x)=\log_3(x^2-4x-5)$	q. $[0, \infty)$
c. $f(x)=\log_3(x^2-4x+5)$	r. $(-\infty, 2]$
d. $f(x)=\log_3(4x-5-x^2)$	s. R

9.

Column I: Equation	Column II: Number of roots
a. $x^2 \tan x = 1, x \in [0, 2\pi]$	p. 5
b. $2^{\cos x} = \|\sin x\|, x \in [0, 2\pi]$	q. 2
c. If $f(x)$ is a polynomial of degree 5 with real coefficients such that $f(\|x\|) = 0$ has 8 real roots, then the number of roots of $f(x) = 0$	r. 3
d. $7^{\|x\|}(\|5 - \|x\|\|) = 1$	s. 4

Integer Type

1. The least value of a such that $(x^2 - x)/(1 - ax)$ attains all real values is ___.
2. If the domain of $f(x) = \sqrt{(0.625)^{4-3x} - (1.6)^{x(x+8)}}$ is $[a, b]$, then the value of $b - a$ is _____.
3. The greatest value of $f(x) = \log_3(5 - 4x - x^2)$ is_____.
4. The number of solutions of $2^x + 3^x + 4^x - 5^x = 0$ is _____.
5. The greatest value of x for which $f(x) = \sin^{-1}[\log_9(x^2/4)]$ is defined is _____.
6. The value of x for which the function $f(x) = \log_4\{\log_5[\log_3(18x - x^2 - 77)]\}$ is defined is _____.
7. The number of solutions of the equation $2[x] = x + \{x\}$, where [.] and {.} denote the greatest integer function and fractional part, respectively, is_____.
8. The number of solutions of the equation $x^2 - 4 - [x] = 0$ (where [.] denotes the greatest integer function) is _____.
9. The period of the function satisfying the relation $f(x) + f(x + 3) = 0$, $\forall x \in R$ is_____.
10. Number of roots of equation $x^2 - x + 1 = (1/2) + \sqrt{x - (3/4)}$, $x \geq 3/4$ is/are_____.
11. The number of integral values of n for which 3π is the period of the function $\cos(nx) \sin[(5x)/n]$ is _____.
12. Number of solutions of $\sin \pi x = |\ln |x||$ is/are_____.
13. If f is a polynomial function satisfying $2 + f(x) f(y) = f(x) + f(y) + f(xy)$, $\forall x, y \in R$ and if $f(2) = 5$, then the value of $f[f(1)]$ is _____.
14. If $f: R \to [0, \infty)$ is a function such that $f(x - 1) + f(x + 1) = \sqrt{3} f(x)$ and T is the period of $f(x)$, then the value of $T/2$ is _____.
15. Let $f(x, y)$ be a periodic function satisfying the condition $f(x, y) = f(2x + 2y, 2y - 2x)$, $\forall x, y \in R$ and let $g(x)$ be a function defined as $g(x) = f(2^x, 0)$. If T is the period of $g(x)$, then the value of $T/4$ is _____.
16. If $f(x) = \sin^2 x + \sin^2[x + (\pi/3)] + \cos x \cos[x + (\pi/3)]$ and $g(5/4) = 1$, then $(gof)(x)$ is equal to _____.
17. Let $E = \{1, 2, 3, 4\}$ and $F = \{1, 2\}$. If N is the number of onto functions from E to F, then the value of $N/2$ is _____.
18. The function f is continuous and has the property $f[f(x)] = 1 - x$. Then the value of $f(1/4) + f(3/4)$ is _____.
19. A continuous function $f(x)$ on $R \to R$ satisfies the relation $f(x) + f(2x + y) + 5xy = f(3x - y) + 2x^2 + 1, \forall x, y \in R$. Then the value of $|f(-4)|$ is _____.
20. The number of integral values of x satisfying the inequality $(3/4)^{6x+10-x^2} < (27/64)$ is _____.
21. A function f from integers to integers is defined as $f(x) = \begin{cases} n+3, & n \in \text{odd} \\ n/2, & n \in \text{even} \end{cases}$. Suppose $k \in$ odd and $f\{f[f(k)]\} = 27$, then the sum of the digits of k is _____.
22. If θ is the fundamental period of the function $f(x) = \sin^{99} x + \sin^{99}[x + (2\pi/3)] + \sin^{99}[x + (4\pi/3)]$, then the complex number $z = |z|(\cos\theta + i\sin\theta)$ lies in the quadrant _____.
23. If $x = 4/9$ satisfies the equation $\log_a(x^2 - x + 2) > \log_a(-x^2 + 2x + 3)$, then the sum of all possible distinct values of $[x]$ (where [.] represents the greatest integer function) is _____.
24. If $4^x - 2^{x+2} + 5 + ||b - 1| - 3| = |\sin y|$, $x, y, b \in R$, then the possible value of b is _____.
25. If $f: (0, \infty) \to (0, \infty)$ satisfies $f[xf(y)] = x^2 y^a$ $(a \in R)$, then a is equal to _____.
26. If $f: N \to N$, and $x_2 > x_1 \Rightarrow f(x_2) > f(x_1), \forall x_1, x_2 \in N$ and $f[f(n)] = 3n, \forall n \in N$, then $f(2)$ is equal to _____.
27. If the fundamental period of the function $f(x) = (\sin^2 ax)^{1/2} + (\cos^2 ax)^{1/2}$ is $\pi/8$, then the value of α is _____.
28. The number of integral values of a for which $f(x) = \log\{\log_{1/3}[\log_7(\sin x + a)]\}$ is defined for every real value of x is _____.
29. Let $f(x) = \sin^{23}x - \cos^{22}x$ and $g(x) = 1 + \frac{1}{2}\tan^{-1}|x|$, then the number of values of x in the interval $[-10\pi, 8\pi]$ satisfying the equation $f(x) = \text{sgn}[g(x)]$ is _____

30. Number of integral value of x satisfying $\log_2 (1 + x) = \log_3 (1 + 2^x)$ is________.

31. $f: R \to R, f(x^2 + x + 3) + 2f(x^2 - 3x + 5) = 6x^2 - 10x + 17\ \forall\ x \in R$, then the value of $f(5)$ is________.

32. The domain of definition of the function $f(x) = [2 \tan \pi x]^{\log [2 \tan \pi x]} \left(\frac{x^2 + 2x - 3}{4x^2 - 4x - 3}\right)$, where [ ] denotes the greatest integer function is given by the interval $\left[n + \frac{1}{n}, n + \frac{1}{2}\right)$, where $n \in I$ then absolute value of sum of all possible values of n is________.

33. Number of integers in the domain of function $f(x) + f(x^{-1}) = \frac{x^3 + 1}{x}$ is________.

34. Let $f: R^+ \to R$ be a function which satisfies $f(x) \cdot f(y) = f(xy) + 2\left(\frac{1}{x} + \frac{1}{y} + 1\right)$ for $x, y > 0$. Then sum of possible values of $f(1/10)$ is________.

35. $f: N \to N$ always satisfies $f(m)f(n) = f(m) + 3f(n)$, $\forall\ m, n \in N$, $(m \neq n)$, then $\underbrace{f(f(f(f...(2))))}_{10 \text{ times}}$ is equal to________.

36. Let a and b be real numbers and let $f(x) = a \sin x + b\sqrt[3]{x} + 4$, $\forall\ x \in R$. If $f(\log_{10} (\log_3 10)) = 5$, then the value of $f(\log_{10} (\log_{10} 3))$ is equal to________.

37. Let $f\left(x + \frac{1}{y}\right) + f\left(x - \frac{1}{y}\right) = 2f(x).f\left(\frac{1}{y}\right)\ \forall\ x, y \in R$ and $f(0) = 0$, then the value of $f(1) + f(2)$ is________.

ANSWERS AND SOLUTIONS

Objective Type

1. b. $f: N \to N, f(n) = 2n + 3.$

Here, the range of the function is {5, 6, 7, ...} or $N - \{1, 2, 3, 4\}$

which is a subset of N (co-domain).

Hence, function is into.

Also, it is clear that $f(n)$ is one-one or injective.

Hence, $f(n)$ is injective only.

2. b. $$f(x) = \sin\left(\log(x + \sqrt{1+x^2})\right)$$

$\Rightarrow f(-x) = \sin[\log(-x + \sqrt{1+x^2})]$

$\Rightarrow f(-x) = \sin\log\left((\sqrt{1+x^2} - x)\dfrac{(\sqrt{1+x^2}+x)}{(\sqrt{1+x^2}+x)}\right)$

$\Rightarrow f(-x) = \sin\log\left[\dfrac{1}{(x+\sqrt{1+x^2})}\right]$

$\Rightarrow f(-x) = \sin\left[-\log(x+\sqrt{1+x^2})\right]$

$\Rightarrow f(-x) = -\sin\left[\log(x+\sqrt{1+x^2})\right]$

$\Rightarrow f(-x) = -f(x)$

$\Rightarrow f(x)$ is an odd function.

3. c. $\dfrac{x^2+14x+9}{x^2+2x+3} = y$

$\Rightarrow x^2 + 14x + 9 = x^2y + 2xy + 3y$

$\Rightarrow x^2(y-1) + 2x(y-7) + (3y-9) = 0$

Since x is real,

$4(y-7)^2 - 4(3y-9)(y-1) > 0$

$\Rightarrow 4(y^2 + 49 - 14y) - 4(3y^2 + 9 - 12y) > 0$

$\Rightarrow 4y^2 + 196 - 56y - 12y^2 - 36 + 48y > 0$

$\Rightarrow 8y^2 + 8y - 160 < 0 \Rightarrow y^2 + y - 20 < 0$

$\Rightarrow (y+5)(y-4) < 0,$

Hence, y lies between -5 and 4.

4. c. $y = f(x) = \cos^2 x + \sin^4 x$

$\Rightarrow y = f(x) = \cos^2 x + \sin^2 x(1 - \cos^2 x)$

$\Rightarrow y = \cos^2 x + \sin^2 x - \sin^2 x\cos^2 x$

$\Rightarrow y = 1 - \sin^2 x\cos^2 x$

$\Rightarrow y = 1 - \dfrac{1}{4}\sin^2 2x$

$\therefore \dfrac{3}{4} \le f(x) \le 1 \qquad [\because 0 \le \sin^2 2x \le 1]$

$\Rightarrow f(R) \in [3/4, 1]$

5. c. $f(x)$ is to be defined when $x^2 - 1 > 0$ and $3 + x > 0$ and $3 + x \ne 1$

$\Rightarrow x^2 > 1$ and $x > -3$ and $x \ne -2$

$\Rightarrow x < -1$ or $x > 1$ and $x > -3$ and $x \ne -2$

$\therefore D_f = (-3, -2) \cup (-2, -1) \cup (1, \infty)$

6. b. We have $f(x) = \left[\log_{10}\left(\dfrac{5x-x^2}{4}\right)\right]^{1/2}$ (1)

From (1), clearly $f(x)$ is defined for those values of x for which $\log_{10}\left[\dfrac{5x-x^2}{4}\right] \ge 0$

$\Rightarrow \left(\dfrac{5x-x^2}{4}\right) \ge 10^0$

$\Rightarrow \left(\dfrac{5x-x^2}{4}\right) \ge 1$

$\Rightarrow x^2 - 5x + 4 \le 0$

$\Rightarrow (x-1)(x-4) \le 0$

Hence, the domain of the function is [1, 4].

7. b. $f(x) = \dfrac{\sin^{-1}(3-x)}{\log|x| - 2|}$

Let $g(x) = \sin^{-1}(3-x)$

$\Rightarrow -1 \le 3 - x \le 1$

The domain of $g(x)$ is [2, 4]

and let $h(x) = \log(|x| - 2|)$

$\Rightarrow |x| - 2 > 0$ or $|x| > 2$

$\Rightarrow x < -2$ or $x > 2$

$\Rightarrow (-\infty, -2) \cup (2, \infty)$

We know that

$(f/g)(x) = \dfrac{f(x)}{g(x)}\ \forall x \in D_1 \cap D_2 - \{x \in R : (g9x) = 0\}$

Therefore, the domain of $f(x) = (2, 4] - \{3\} = (2, 3) \cup (3, 4]$.

8. c. $f(x) = \log|\log x|$, $f(x)$ is defined if $|\log x| > 0$ and $x > 0$, i.e., if $x > 0$ and $x \ne 1$

$(\because |\log x| > 0$ if $x \ne 1)$

$\Rightarrow x \in (0, 1) \cup (1, \infty).$

9. d. Here $x + 3 > 0$ and $x^2 + 3x + 2 = 0$. Therefore,

$x > -3$ and $(x+1)(x+2) \ne 0$, i.e., $x \ne -1, -2$

Hence, the domain is $(-3, \infty) - \{-1, -2\}$.

10. b. $y = f(x) = \sqrt{3}\sin x - \cos x + 2 = 2\sin\left(x - \dfrac{\pi}{6}\right) + 2$

(1)

Since $f(x)$ is one-one and onto, f is invertible.

From (1) $\sin\left(x-\frac{\pi}{6}\right)=\frac{y-2}{2}$

$\Rightarrow x=\sin^{-1}\frac{y-2}{2}+\frac{\pi}{6}$

$\Rightarrow\ f^{-1}(x)=\sin^{-1}\left(\frac{x-2}{2}\right)+\frac{\pi}{6}$

11. a. $F(n+1)=\frac{2F(n)+1}{2}\Rightarrow F(n+1)-F(n)=\frac{1}{2}$

Put $n = 1, 2, 3, \ldots, 100$ and add, we get

$F(101)-F(1)=100\times\frac{1}{2}$

$\Rightarrow\quad F(101)=52 \qquad [\because F(1)=2]$

12. d. Given function is defined if ${}^{10}C_{x-1}>3\ {}^{10}C_x$

$\Rightarrow\quad \frac{1}{11-x}>\frac{3}{x}\quad\Rightarrow 4x>33$

$\Rightarrow\quad x\geq 9$ but $x\leq 10\quad\Rightarrow x=9, 10.$

13. b. For the domain $\sin(\ln x)>\cos(\ln x)$ and $x>0$

$2n\pi+\frac{\pi}{4}<\ln x<2n\pi+\frac{5}{4}.$

14. b. Put $x=0\ \Rightarrow\ f(2)=2f(0)-f(1)=2\times 2-3=1$

Put $x=1\ \Rightarrow\ f(3)=6-1=5$

Put $x=2\ \Rightarrow\ f(4)=2f(2)-f(3)=2\times 1-5=-3$

Put $x=3\ \Rightarrow\ f(5)=2f(3)-f(4)=2(5)-(-3)=13.$

15. c. Here, $\frac{x^2+1}{x^2+2}=1-\frac{1}{x^2+2}$

Now, $2\leq x^2+2<\infty$ for all $x\in R$

$\Rightarrow\quad \frac{1}{2}\geq\frac{1}{x^2+2}>0$

$\Rightarrow\quad -\frac{1}{2}\leq\frac{-1}{x^2+2}<0$

$\Rightarrow\quad \frac{1}{2}\leq 1-\frac{1}{x^2+2}<1$

$\Rightarrow\quad \frac{\pi}{6}\leq\sin^{-1}\left(1-\frac{1}{x^2+2}\right)<\frac{\pi}{2}$

16. b. The function $\sec^{-1}x$ is defined for all $x\in R-(-1, 1)$ and the function $\frac{1}{\sqrt{x-[x]}}$ is defined for all $x\in R–Z$

So the given function is defined for all $x\in R-\{(-1, 1)\cup\{n\mid n\in Z\}\}$.

17. b. $\cos^{-1}\left(\frac{2-|x|}{4}\right)$ exists if $-1\leq\frac{2-|x|}{4}\leq 1$

$\Rightarrow\quad -6\leq -|x|\leq 2$

$\Rightarrow\quad -2\leq |x|\leq 6$

$\Rightarrow\quad |x|\leq 6$

$\Rightarrow\quad -6\leq x\leq 6$

The function $[\log(3-x)]^{-1}=\frac{1}{\log(3-x)}$ is defined if $3-x>0$ and $x\neq 2$, i.e., if $x\neq 2$ and $x<3$. Thus, the domain of the given function is $\{x\mid -6\leq x\leq 6\}\cap\{x\mid x\neq 2, x<3\}=[-6, 2)\cup(2, 3).$

18. b. $f(x)$ is defined for $\log\left(\frac{1}{|\sin x|}\right)\geq 0$

$\Rightarrow\quad \frac{1}{|\sin x|}\geq 1$ and $|\sin x|\neq 0.$

$\Rightarrow\quad |\sin x|\neq 0$

$\left[\because \frac{1}{|\sin x|}\geq 1 \text{ for all } x\right]$

$\Rightarrow\quad x\neq n\pi, n\in Z$

Hence, the domain of $f(x)=R-\{n\pi : n\in Z\}$.

19. a. $f(x)$ is defined if $-\log_{1/2}\left(1+\frac{1}{x^{1/4}}\right)-1>0$

$\Rightarrow\quad \log_{1/2}\left(1+\frac{1}{x^{1/4}}\right)<-1$

$\Rightarrow\quad 1+\frac{1}{x^{1/4}}>\left(\frac{1}{2}\right)^{-1}$

$\Rightarrow\quad \frac{1}{x^{1/4}}>1$

$\Rightarrow\quad 0<x<1$

20. c. For the function to get defined $0\leq x^2+x+1\leq 1$,

but $x^2+x+1>\frac{3}{4}\Rightarrow\frac{\sqrt{3}}{2}\leq\sqrt{x^2+x+1}\leq 1$

$\Rightarrow\quad \frac{\pi}{3}\leq\sin^{-1}(\sqrt{x^2+x+1})\leq\frac{\pi}{2}.$

21. c.

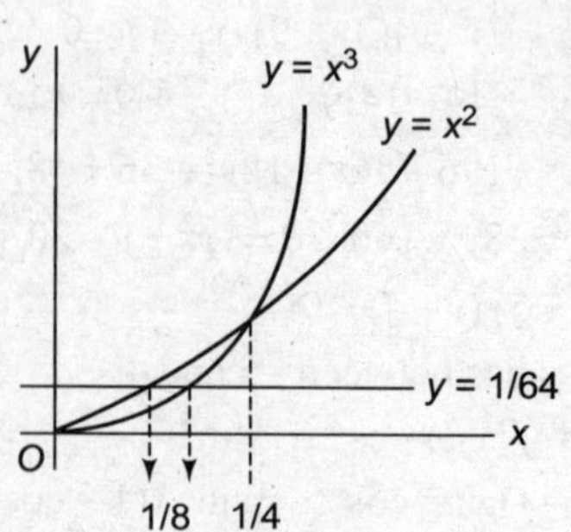

Fig. 1.1

Clearly, from the graph $f(x)=\begin{cases}\frac{1}{64}, & 0\leq x\leq\frac{1}{8}\\ x^2, & \frac{1}{8}<x\leq 1\\ x^3, & x>1\end{cases}$

22. c. The period of cos(sin nx) is π/n and the period of tan (x/n) is πn.

Thus, $6\pi = \text{LCM}\left(\frac{\pi}{n}, \pi n\right)$.

By checking for the different values of n, $n = 6$.

23. b. Draw the graph of $y = \log_{0.5}|x|$ and $y = 2|x|$

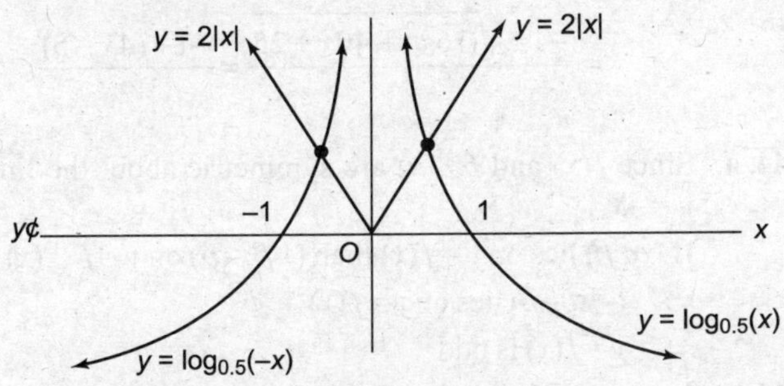

Fig. 1.2

Clearly, from the graph, there are two solutions.

24. b. $f(x) = \left|\sin^3\frac{x}{2}\right| + \left|\cos^5\frac{x}{5}\right|$

The period of $\sin^3 x$ is 2π

$\Rightarrow$ The period of $\sin^3\frac{x}{2}$ is $\frac{2\pi}{1/2} = 4\pi$.

$\Rightarrow$ The period of $\left|\sin^3\frac{x}{2}\right|$ is 2π

The period of $\cos^5 x$ is 2π

$\Rightarrow$ The period of $\cos^5\frac{x}{5}$ is $\frac{2\pi}{\left(\frac{1}{5}\right)} = 10\pi$

$\Rightarrow$ The period of $\left|\cos^5\frac{x}{2}\right|$ is 5π.

Now the period of $f(x) = \text{LCM of } \{2\pi, 10\pi\} = 10\pi$.

25. a. Given $f(x) = \sqrt[n]{x^m}$, $n \in N$ is an even function where $m \in I$.

$\Rightarrow f(x) = f(-x)$

$\Rightarrow \sqrt[n]{x^m} = \sqrt[n]{(-x)^m}$

$\Rightarrow x^m = (-x)^m$

$\Rightarrow$ m is an even integer

$\Rightarrow m = 2k, k \in I$

26. c. From the given data $g(x)$ must be linear function

Hence, $g(x) = ax + b$.

Also $g(2) = 2a + b = 3$ and $g(4) = 4a + b = 7$.

Solving, we get $a = 2$ and $b = -1$.

Hence, $g(x) = 2x - 1$.

Then, $g(6) = 11$.

27. a. The period of $\sin 2\pi x$ and $\cos 2\pi x$ is 1.

The period of $2^{\{x\}}$ is 1.

The period of $3^{\{x/2\}}$ is 2.

Hence, the period of $f(x)$ is LCM of 1 and 2 = 2.

28. a. $|x-2| + a = \pm 4$

$\Rightarrow |x-2| = \pm 4 - a$

For 4 real roots, $4 - a > 0$ and $-4 - a > 0$

$\Rightarrow a \in (-\infty, -4)$

29. a. We have $f(x+y) + f(x-y)$

$$= \frac{1}{2}[a^{x+y} + a^{-x-y} + a^{x-y} + a^{-x+y}]$$

$$= \frac{1}{2}[a^x(a^y + a^{-y}) + a^{-x}(a^y + a^{-y})]$$

$$= \frac{1}{2}(a^x + a^{-x})(a^y + a^{-y}) = 2f(x)f(y)$$

30. d. $\log_3(x^2 - 6x + 11) \le 1$

$\Rightarrow 0 < x^2 - 6x + 11 \le 3$

$\Rightarrow x \in [2, 4]$

31. d. $x^2 - [x]^2 \ge 0 \Rightarrow x^2 \ge [x]^2$

This is true for all positive values of x and all negative integer x.

32. b.

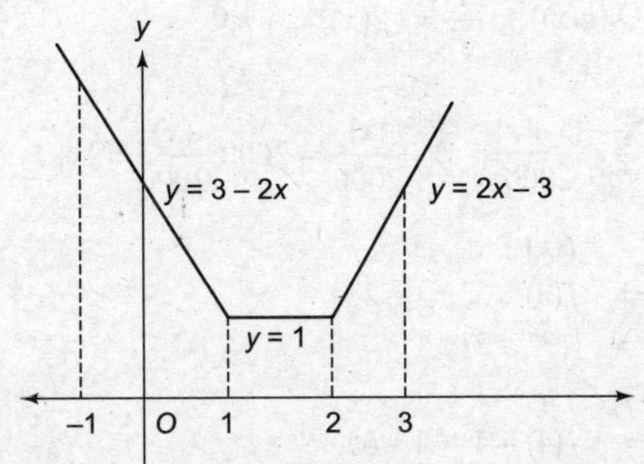

Fig. 1.3

In this interval, $f(x)$ is constant.

If $2 \le x \le 3$, $f(x) = x - 1 + x - 2 = 2x - 3$.

In this interval, $f(x)$ is increasing. Therefore, max $f(x)$ = the greatest among $f(-1)$ and $f(3) = 5$, min $f(x) = f(1) = 1$

So, the range is [1, 5].

33. a. By checking for different function, we find that for $f(x) = \frac{1-x}{1+x}$, $f^{-1}(x) = f(x)$.

34. b. $x^2F(x) + F(1-x) = 2x - x^4$ (1)

Replacing x by $1 - x$, we get

$\Rightarrow (1-x)^2F(1-x) + F(x) = 2(1-x) - (1-x)^4$ (2)

Eliminating $F(1-x)$ from (1) and (2), we get $F(x) = 1 - x^2$.

35. b. $f(-x)=\begin{cases}(-x)^2\sin\dfrac{\pi(-x)}{2}, & |-x|<1\\ (-x)|-x|, & |-x|\geq 1\end{cases}$

$=\begin{cases}-x^2\sin\dfrac{\pi x}{2}, & |x|<1\\ -x|x|, & |x|\geq 1\end{cases}$

$=-f(x).$

36. d. $f(x)=e^{x^3-3x+2}$

Let $g(x)=x^3-3x+2;\ g'(x)$

$=3x^2-3=3(x^2-1)$

$g'(x)\geq 0$ for $x\in(-\infty,-1]$

So $f(x)$ is increasing function and $f(x)$ is one-one.

Now, the range of $f(x)$ is $(0, e^4]$.

But co-domain is $(0, e^5]$. Therefore,

$f(x)$ is an into function.

37. c. $f(x)=\dfrac{1}{x},\ g(x)=\dfrac{1}{x^2}$ and $h(x)=x^2$

$f(g(x))=x^2, x\neq 0$

$h(g(x))=\dfrac{1}{x^4}=(g(x))^2,\ x\neq 0$

38. c. $\displaystyle\sum_{r=1}^{2000}\frac{\{x+r\}}{2000}=\sum_{r=1}^{2000}\frac{\{x\}}{2000}=2000\frac{\{x\}}{2000}=\{x\}.$

39. b. $f(x)=x^n+1$

$\Rightarrow\ f(3)=3^n+1=28$

$\Rightarrow\ 3^n=27$

$\therefore\ n=3$

$\Rightarrow\ f(4)=4^3+1=65.$

40. b. $\because\ f(x+1)-f(x)=8x+3$

$\Rightarrow\ \{b(x+1)^2+c(x+1)+d\}-\{bx^2+cx+d\}$

$=8x+3$

$\Rightarrow\ b\{(x+1)^2-x^2\}+c=8x+3$

$\Rightarrow\ b(2x+1)+c=8x+3$

On comparing co-efficients of x and constant term, we get

$2b=8$ and $b+c=3$

then $b=4$ and $c=-1$.

41. d. If f is injective and g is surjective

$\Rightarrow$ fog is injective

$\Rightarrow$ fof is injective.

42. c. $f(x)=\begin{cases}x-1, & x=\text{even}\\ x+1, & x=\text{odd}\end{cases}$, which is clearly are one-one and onto.

43. a. $\dfrac{1}{2}(gof)(x)=2x^2-5x+2$ or $\dfrac{1}{2}g[f(x)]$

$=2x^2-5x+2$

$\therefore\ [\{f(x)\}^2+\{f(x)\}-2]=2[2x^2-5x+2]$

$\Rightarrow\ f(x)^2+f(x)-(4x^2-10x+6)=0$

$\therefore f(x)=\dfrac{-1\pm\sqrt{1+4(4x^2-10x+6)}}{2}$

$=\dfrac{-1\pm\sqrt{(16x^2-40x+25}}{2}=\dfrac{-1+(4x-5)}{2}$

$=2x-3$

44. a. Since $f(x)$ and $f^{-1}(x)$ are symmetric about the line $y=-x$.

If (α,β) lies on $y=f(x)$ then $(-\beta,-\alpha)$ on $y=f^{-1}(x)$

$\Rightarrow\ (-\alpha,-\beta)$ lies on $y=f(x)$

$\Rightarrow\ y=f(x)$ is odd.

45. c. Let $x, y\in N$ such that $f(x)=f(y)$

Then $f(x)=f(y)$

$\Rightarrow\ x^2+x+1=y^2+y+1$

$\Rightarrow\ (x-y)(x+y+1)=0$

$\Rightarrow\ x=y$ or $x=(-y-1)\notin N$

Therefore, f is one-one.

Also, $f(x)$ does not take all positive integral values.

Hence f is into.

46.a. $f(x)=\sqrt{2}\sin\left(x+\dfrac{\pi}{4}\right)+2\sqrt{2}$

$\Rightarrow\ f(x)=\sqrt{2}\cos\left(x-\dfrac{\pi}{4}\right)+2\sqrt{2}$

$\Rightarrow\ Y=\left[\sqrt{2}, 3\sqrt{2}\right]$ and $X=\left[-\dfrac{3\pi}{4},\dfrac{\pi}{4}\right]$ or $\left[\dfrac{\pi}{4},\dfrac{5\pi}{4}\right]$

47. a. $f(7)+f(-7)=-10$

$\Rightarrow\ f(7)=-17$

$\Rightarrow\ f(7)+17\cos x=-17+17\cos x$ which has the range $[-34, 0]$.

48. c. $f(x)=\dfrac{\sin[x]\pi}{x^2+x+1}$

Let $[x]=n\in$ integer

$\Rightarrow\ \sin[x]n=0$

$\Rightarrow\ f(x)=0$

$\Rightarrow\ f(x)$ is constant functions.

49. b. Two triangles may have equal areas. Therefore, f is not one-one.

Since each positive real number can represent area of a triangle, f is onto.

50. c. Replace x by a and y by $x-a$

$\Rightarrow f(a-(x-a))=f(a)f(x-a)-f(0)f(x)$ (1)

Put $x=0, y=0; f(0)=(f(0))^2-[f(a)]^2$

$\Rightarrow f(a)=0$

$[\because f(0)=1]$

From (1), $f(2a-x)=-f(x)$.

51. c. $g(x)=x^3+\tan x+\left[\frac{x^2+1}{P}\right]$

$\Rightarrow g(-x)=(-x)^3+\tan(-x)+\left[\frac{(-x)^2+1}{P}\right]$

$\Rightarrow g(-x)=-x^3-\tan x+\left[\frac{x^2+1}{P}\right]$

$\Rightarrow g(x)+g(-x)=0$,

because $g(x)$ is a odd function. Therefore,

$$\left(-x^3-\tan x+\left[\frac{x^2+1}{P}\right]\right)+\left(\begin{matrix}-x^3-\tan x\\+\left[\frac{x^2+1}{P}\right]\end{matrix}\right)=0$$

$\Rightarrow 2\left[\frac{(x^2+1)}{P}\right]=0 \Rightarrow 0\le\frac{x^2+1}{P}<1$ because $x\in[-2,2]$

$\Rightarrow 0\le\frac{5}{P}<1\Rightarrow P>5$

52. d. Let $2x+\frac{y}{8}=\alpha$ and $2x-\frac{y}{8}=\beta$, then $x=\frac{\alpha+\beta}{4}$ and $y=4(\alpha-\beta)$.

Given, $f\left(2x+\frac{y}{8}, 2x-\frac{y}{8}\right)=xy$

$\Rightarrow f(\alpha,\beta)=\alpha^2-\beta^2$

$\Rightarrow f(m,n)+f(n,m)=m^2-n^2+n^2-m^2=0$ for all m, n

53. b. Given $f(x+y)=f(x)+f(y)-xy-1\ \forall\ x, y\in R$

$f(1)=1$

$f(2)=f(1+1)=f(1)+f(1)-1-1=0$

$f(3)=f(2+1)=f(2)+f(1)-2\times 1-1=-2$

$f(n+1)=f(n)+f(1)-n-1=f(n)-n<f(n)$

Thus, $f(1)>f(2)>f(3)>\ldots$ and $f(1)=1$.

Therefore, $f(1)=1$ and $f(n)<1$, for $n>1$

Hence, $f(n)=n, n\in N$ has only one solution $n=1$.

54. c. $f(x)=\frac{e^x-e^{|x|}}{e^x+e^{|x|}}=\begin{cases}0, & x\ge 0\\ \frac{e^x-e^{-x}}{e^x+e^{-x}}, & x<0\end{cases}$

Clearly, $f(x)$ is identically zero if $x\ge 0$ (1)

If $x<0$, let $y=f(x)=\frac{e^x-e^{-x}}{e^x+e^{-x}}\Rightarrow e^{2x}=\frac{1+y}{1-y}$

$\because x<0\Rightarrow e^{2x}<1\Rightarrow 0<e^{2x}<1$

$\therefore 0<\frac{1+y}{1-y}<1$

$\Rightarrow \frac{1+y}{1-y}>0$ and $\frac{1+y}{1-y}<1$

$\Rightarrow (y+1)(y-1)<0$ and $\frac{2y}{1-y}<0$

$\Rightarrow -1<y<1$ and $y<0$ or $y>1$

$\Rightarrow -1<y<0$ (2)

Combining (1) and (2), we get $-1<y\le 0$

$\Rightarrow$ Range $=(-1,0]$.

55. b. $f(2x+3)+f(2x+7)=2$ (1)

Replace x by $x+2$,

$f(2x+7)+f(2x+11)=2$ (2)

From (1) – (2) we get $f(2x+3)-f(2x+11)=0$

$\Rightarrow f(2x+3)=f(2x+11)$

$\Rightarrow f(2x+3)=f(2(x+4)+3)$

$\Rightarrow T=4$.

56. c. Since co-domain $=\left[0,\frac{\pi}{2}\right)$.

For f to be onto, the range $=\left[0,\frac{\pi}{2}\right)$

This is possible only when $x^2+x+a\ge 0$

$\therefore 1^2-4a\le 0\Rightarrow a\ge\frac{1}{4}$.

57. d. $f(x)=\frac{1}{\sqrt{\{\sin x\}+\{\sin(\pi+x)\}}}=\frac{1}{\sqrt{\{\sin x\}+\{-\sin x\}}}$

Now $\{\sin x\}+\{-\sin x\}=\begin{cases}0, & \sin x \text{ is an integer}\\ 1, & \sin x \text{ is not an integer}\end{cases}$

For $f(x)$ to get defined $\{\sin x\}+\{-\sin x\}\ne 0$

$\Rightarrow \sin x\ne$ integer

$\Rightarrow \sin x\ne\pm 1, 0$

$\Rightarrow x\ne\frac{n\pi}{2}, n\in I$

Hence, the domain is $R-\left\{\frac{n\pi}{2}/n\in I\right\}$.

58. a. $f(-x)=\frac{\cos(-x)}{\left[-\frac{2x}{\pi}\right]+\frac{1}{2}}$

$=\frac{\cos x}{-1-\left[\frac{2x}{\pi}\right]+\frac{1}{2}}$

[as x is not an integral multiple of π]

$\Rightarrow f(-x) = -\dfrac{\cos x}{\left[\dfrac{2x}{\pi}\right]+\dfrac{1}{2}} = -f(x)$

$\Rightarrow f(x)$ is an odd function.

59. b. $f\left(x+\dfrac{1}{x}\right) = \left(x+\dfrac{1}{x}\right)^2 - 2$

$\Rightarrow f(y) = y^2 - 2$, where $y = x + \dfrac{1}{x}$

for $x > 0, y = x + \dfrac{1}{x} \geq 2$ and for $x < 0, y = x + \dfrac{1}{x} \leq -2$

60. d. Since $f(x)$ is an odd function, $\left[\dfrac{x^2}{a}\right] = 0$ for all $x \in [-10, 10]$

$\Rightarrow 0 \leq \dfrac{x^2}{a} < 1$ for all $x \in [-10, 10] \Rightarrow a > 100$.

61. b. $3f(x) + 2f\left(\dfrac{x+59}{x-1}\right) = 10x + 30$

For $x = 7, 3f(7) + 2f(11) = 70 + 30 = 100$.
For $x = 11, 3f(11) + 2f(7) = 140$.

$\dfrac{f(7)}{-20} = \dfrac{f(11)}{-220} = \dfrac{-1}{9-4} \Rightarrow f(7) = 4$.

62. c. $f(x) = [6x + 7] + \cos \pi x - 6x$
$= [6x] + 7 + \cos \pi x - 6x$
$= 7 + \cos \pi x - \{6x\}$

$\{6x\}$ has the period 1/6 and $\cos \pi x$ has the period 2, then the period of $f(x) =$ LCM of 2 and 1/6 which is 2. Hence, the period is 2.

63. d. $f(x) = \dfrac{a^x - 1}{x^n(a^x + 1)}$

$f(x)$ is symmetrical about y-axis

$\Rightarrow f(x) = f(-x)$

$\Rightarrow \dfrac{a^x - 1}{x^n(a^x+1)} = \dfrac{a^{-x}-1}{(-x)^n(a^{-x}+1)}$

$\Rightarrow \dfrac{a^x - 1}{x^n(a^x+1)} = \dfrac{1-a^x}{(-x)^n(1+a^x)}$

$\Rightarrow x^n = -(-x)^n$

$\Rightarrow$ The value of n which satisfy this relation is $-\dfrac{1}{3}$.

64. a. $x^2 f(x) - 2f\left(\dfrac{1}{x}\right) = g(x)$ and $2f\left(\dfrac{1}{x}\right) - 4x^2 f(x)$

$= 2x^2 g(1/x)$

$\Rightarrow -3x^2 f(x) = g(x) + 2x^2 g\left(\dfrac{1}{x}\right)$

$\Rightarrow f(x) = -\left(\dfrac{g(x) + 2x^2 g\left(\dfrac{1}{x}\right)}{3x^2}\right)$

Since, $g(x)$ and x^2 are odd and even functions, respectively, $f(x)$ is an odd function. But $f(x)$ is given even

$\Rightarrow f(x) = 0 \; \forall x$. Hence, $f(5) = 0$.

65. a. Given $f(x + y) = f(x) f(y)$. Put $x = y = 0$, then $f(0) = 1$.

Put $y = -x$, then $f(0) = f(x) f(-x) \Rightarrow f(-x) = \dfrac{1}{f(x)}$

Now, $g(-x) = \dfrac{f(-x)}{1+\{f(x)\}^2}$

$g(-x) = \dfrac{f(-x)}{1+\{f(-x)\}^2} = \dfrac{\dfrac{1}{f(x)}}{1+\dfrac{1}{\{f(x)\}^2}}$

$= \dfrac{f(x)}{1+\{d(x)\}^2} = g(x)$

66. d. The equation is $x^2 + 2ax + \dfrac{1}{16} = -a + \sqrt{a^2 + x - \dfrac{1}{16}}$,

$\Rightarrow f(x) = f^{-1}(x)$,

which has a solution if $x^2 + 2ax + \dfrac{1}{16} = x$

$\Rightarrow x^2 + (2a-1)x + \dfrac{1}{16}$

For real and distinct roots $(2a-1)^2 - 4\dfrac{1}{16} \geq 0$

$\Rightarrow 2a - 1 \leq \dfrac{-1}{2}$ or $2a - 1 \geq \dfrac{1}{2} \Rightarrow a \leq \dfrac{1}{4}$ or $a \geq \dfrac{3}{4}$.

67. d. $f(x) - 1 + f(1-x) - 1 = 0$; so $g(x) + g(1-x) = 0$

Putting $x = x + \dfrac{1}{2}$, we get $g\left(\dfrac{1}{2}+x\right) + g\left(\dfrac{1}{2}-x\right) = 0$.

So, it is symmetrical about $\left(\dfrac{1}{2}, 0\right)$.

68. a. When $[x] = 0$ we have $\sin^{-1}(\cos^{-1}0) = \sin^{-1}(\pi/2)$, not defined.

When $[x] = -1$ we have $\sin^{-1}(\cos^{-1}-1) = \sin^{-1}(\pi)$, not defined.

When $[x] = 1$ we have $\sin^{-1}(\cos^{-1}1) = \sin^{-1}(0) = 0$.

Hence, $x \in [1, 2)$ and the range of function is $\{0\}$.

69. a. Putting $x=1, f(2)+f(0)=2f(1) \Rightarrow f(2)=2f(1)$

Putting $x=2, f(3)+f(1)=2f(2)$

$\Rightarrow \quad f(3)=2\times 2f(1)-f(1)=3f(1)$, and so on.

$\therefore \quad f(n)=nf(1)$, for $n=1, 2, \ldots, n$

$f(n+1)+f(n-1)=2f(n)$

$\Rightarrow \quad f(n+1)+(n-1)f(1)=2nf(1)$

$\Rightarrow \quad f(n+1)=(n+1)f(1)$

70. d. $\because \{x\} \in [0, 1)$

$\sin\{x\} \in (0, \sin 1)$ as $f(x)$ is defined if $\sin\{x\} \neq 0$

$$\Rightarrow \quad \frac{1}{\sin\{x\}} \in \left(\frac{1}{\sin 1}, \infty\right)$$

$$\Rightarrow \quad \left[\frac{1}{\sin\{x\}}\right] \in \{1, 2, 3, \ldots\}$$

Note that $1 < \frac{\pi}{3} \Rightarrow \sin 1 < \sin\frac{\pi}{3} = 0.866$

$$\Rightarrow \frac{1}{\sin 1} > 1.155.$$

71. c. We have $[\cos^{-1} x] \geq 0 \ \forall\, x \in [-1, 1]$

and $[\cot^{-1} x] \geq 0 \ \forall\, x \in R$.

Hence, $[\cot^{-1} x] + [\cot^{-1} x] = 0$

$\Rightarrow \quad [\cot^{-1} x] = [\cot^{-1} x] = 0$

If $[\cos^{-1} x] = 0 \Rightarrow x \in (\cos 1, 1]$

If $[\cot^{-1} x] = 0 \Rightarrow x \in (\cot 1, \infty)$

$\Rightarrow \quad x \in (\cot 1, 1]$

72. d. The period of $f(x)$ is 7

$\Rightarrow$ The period of $f\left(\frac{x}{3}\right)$ is $\frac{7}{1/3} = 21$

The period of $g(x)$ is 11

$\Rightarrow$ The period of $g\left(\frac{x}{5}\right)$ is $\frac{11}{1/5} = 55$

Hence, T_1 = period of $f(x)g\left(\frac{x}{5}\right) = 7\times 55 = 385$ and

T_2 = period of $g(x)\, f\left(\frac{x}{3}\right) = 11\times 21 = 231.$

$\therefore$ Period of $F(x) = \text{LCM}\,\{T_1, T_2\}$

$= \text{LCM}\,\{385, 231\}$

$= 7\times 11\times 3\times 5$

$= 1155.$

73. d. $\sin^2 x + \sin^2\left(x+\frac{\pi}{3}\right) + \cos x \cos\left(x+\frac{\pi}{3}\right)$

$$= \sin^2 x + \left(\frac{\sin x}{2} + \frac{\sqrt{3}\cos x}{2}\right)^2 + \cos x\left(\frac{\cos x}{2} - \frac{\sqrt{3}\sin x}{2}\right)$$

$$= \sin^2 x + \frac{\sin^2 x}{4} + \frac{3\cos^2 x}{4} + \frac{\cos^2 x}{2}$$

$$= \frac{5\sin^2 x}{4} + \frac{5\cos^2 x}{4} = 5/4.$$

Hence, $f(x) = c^{5/4}$ = constant, which is periodic whose period cannot be determined.

74. a. $f(x+f(y)) = f(x)+y, f(0)=1$

Putting $y=0$, we get $f(x+f(0)) = f(x)+0$

$\Rightarrow \quad f(x+1) = f(x) \ \forall\, x \in R$

Thus, $f(x)$ is the period with 1 as one of its period.

$\Rightarrow \quad f(7) = f(6) = f(5) = \cdots = f(1) = (0) = 1.$

75. c. $f(x) = \sqrt{|x| - \{x\}}$ is defined if $|x| \geq \{x\}$

$$\Rightarrow \quad x \in \left(-\infty - \frac{1}{2}\right] \cup [0, \infty) \Rightarrow Y \in [0, \infty).$$

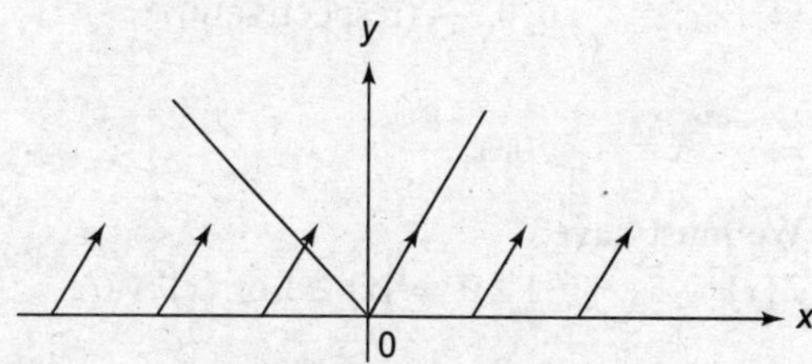

Fig.1.4

76. a. $f(xy) = \frac{f(x)}{y}$

$\Rightarrow \quad f(y) = \frac{f(1)}{y}$ [Putting $x=1$]

$\Rightarrow \quad f(30) = \frac{f(1)}{30}$ or $f(1) = 30\times f(30) = 30\times 20 = 600.$

Now, $f(40) = \frac{f(1)}{40} = \frac{600}{40} = 15.$

77.c. Case I

$0 < |x|-1 < 1 \Rightarrow 1 < |x| < 2$, then

$x^2+4x+4 \leq 1$

$\Rightarrow \quad x^2+4x+3 \leq 0$

$\Rightarrow \quad -3 \leq x \leq -1$

So $x \in (-2, -1)$ (1)

Case II

$|x|-1 > 1 \Rightarrow |x| > 2$, then $x^2+4x+4 \geq 1$

$\Rightarrow \quad x^2+4x+3 \geq 0$

$\Rightarrow \quad x \geq -1$ or $x \leq -3$

So, $x \in (-\infty, -3] \cup (2, \infty)$ (2)

From (1) and (2), $x \in (-\infty, -3] \cup (-2, -1) \cup (2, \infty)$.

78.d. $f(x)=\dfrac{n(n+1)}{2}+[\sin x]+\left[\sin\dfrac{x}{2}\right]+\cdots+\left[\sin\dfrac{x}{n}\right]$

Thus, the range of $f(x)=\left\{\dfrac{n(n+1)}{2},\dfrac{n(n+1)}{2}+1\right\}$ is $x\in[0,\pi]$.

79.b. $[x]^2=x+2\{x\}$

$\Rightarrow [x]^2=[x]+3\{x\}$

$\Rightarrow \{x\}=\dfrac{[x]^2-[x]}{3}$

$\Rightarrow 0\le\dfrac{[x]^2-[x]}{3}<1$

$\Rightarrow 0\le[x]^2-[x]<3$

$\Rightarrow [x]\in\left(\dfrac{1-\sqrt{3}}{2},0\right]\cup\left[1,\dfrac{1+\sqrt{3}}{2}\right)$

$\Rightarrow [x]=-1,0,1,2$

$\Rightarrow \{x\}=\dfrac{2}{3},0,0,\dfrac{2}{3}$, (respectively)

$\Rightarrow x=-\dfrac{1}{3},0,1,\dfrac{8}{3}$

80.b. We must have

$2\{x\}^2-3\{x\}+1\ge0\Rightarrow\{x\}\ge1$ or $\{x\}\le1/2$.

Thus, we have $0\le\{x\}\le1/2\Rightarrow x\in\left[n,n+\dfrac{1}{2}\right], n\in I.$

81.b. $\left[x^2+\dfrac{1}{2}\right]=\left[x^2-\dfrac{1}{2}+1\right]=1+\left[x^2-\dfrac{1}{2}\right].$

Thus, from domain point of view,

$\left[x^2-\dfrac{1}{2}\right]=0,-1\Rightarrow\left[x^2+\dfrac{1}{2}\right]=1,0$

$\Rightarrow f(x)=\sin^{-1}(1)+\cos^{-1}(0)$ or $\sin^{-1}(0)+\cos^{-1}(-1)$

$\Rightarrow f(x)=\{\pi\}$

82.c. The period of $\cos(\sin nx)$ is π/n and the period of $\tan(x/n)$ is πn.

Thus, $6\pi=\text{LCM}\left(\dfrac{\pi}{n},\pi n\right)$

$\Rightarrow 6\pi=\dfrac{\pi}{n}\lambda_1\Rightarrow n=\dfrac{\lambda_1}{6}$, and $6\pi=\lambda_2\pi n\Rightarrow n=\dfrac{6}{\lambda_2}$,

$\lambda_1,\lambda_2\in I^+$

From $n=\dfrac{6}{\lambda_2}\Rightarrow n=6,3,2,1.$

Clearly, for $n=6$, we get the period of $f(x)$ to be 6π.

83.a. We must have $ax^3+(a+b)x^2+(b+c)x+c>0$

$\Rightarrow ax^2(x+1)+bx(x+1)+c(x+1)>0$

$\Rightarrow (x+1)(ax^2+bx+c)>0$

$\Rightarrow a(x+1)\left(x+\dfrac{b}{2a}\right)^2>0$ as $b^2=4ac$

$\Rightarrow x>-1$ and $\ne-\dfrac{b}{2a}$

84.b. $f(x)=[x]+[2x]+[3x]+\cdots+[nx]-(x+2x+3x+\cdots+nx)$

$=-(\{x\}+\{2x\}+\{3x\}+\cdots+\{nx\})$

The period of $f(x)=\text{LCM}\left(1,\dfrac{1}{2},\dfrac{1}{3},\ldots,\dfrac{1}{n}\right)=1.$

85.c. $f\left(x+\dfrac{1}{2}\right)+f\left(x-\dfrac{1}{2}\right)=f(x)$

$\Rightarrow f(x+1)+f(x)=f\left(x+\dfrac{1}{2}\right)$

$\Rightarrow f(x+1)+f\left(x-\dfrac{1}{2}\right)=0$

$\Rightarrow f\left(x+\dfrac{3}{2}\right)=-f(x)$

$\Rightarrow f(x+3)=-f\left(x+\dfrac{3}{2}\right)=f(x)$

Therefore, $f(x)$ is periodic with period 3.

86.c. $f(x)+3xf\left(\dfrac{1}{x}\right)=2(x+1)$ (1)

Replacing x by $1/x$, we get

$f\left(\dfrac{1}{x}\right)+3\dfrac{1}{x}f(x)=2\left(\dfrac{1}{x}+1\right)$

$\Rightarrow xf\left(\dfrac{1}{x}\right)+3f(x)=2(x+1)$ (2)

From (1) and (2), we have $f(x)=\dfrac{x+1}{2}$

$\Rightarrow f(99)=50$

87.a. Let $y=\dfrac{x+5}{x+2}=1+\dfrac{3}{x+2}\Rightarrow x=1$

Also, $y-1=\dfrac{3}{x+2}$

$\Rightarrow x+2=\dfrac{3}{y-1}$

$\Rightarrow x=-2+\dfrac{3}{y-1}$

$\Rightarrow y=2$ only as x and y are natural numbers.

88.d. $f(f(x))=\begin{cases}(f(x))^2, & \text{for } f(x)\ge0\\ f(x), & \text{for } f(x)<0\end{cases}$

$$= \begin{cases} (x^2)^2, & x^2 \ge 0, x \ge 0 \\ x^2, & x \ge 0, x < 0 \\ x^2, & x^2 < 0, x \ge 0 \\ x, & x < 0, x < 0 \end{cases}$$

$$= \begin{cases} x^4, & x \ge 0 \\ x, & x < 0 \end{cases}$$

89.c. From the given data

$f(1-x) = f(1+x)$ (1)

and $f(2-x) = f(2+x)$ (2)

In (2) replacing x by $1+x$, we have

$f(1-x) = f(3+x)$

$\Rightarrow f(1+x) = f(3+x)$ [From (1)]

$\Rightarrow f(x) = f(2+x)$

90.c. We have observed that this function has a property that if A, B, C and D are four real numbers such that $A+B=C+D$.

Then $f(a)+f(b)=f(c)+f(d)$.

Now $x+2-(x+1)=x+1-x$.

91.d. We must have $-1 \le [2x^2-3] \le 1$

$$\Rightarrow -1 \le 2x^2 - 3 < 2 \Rightarrow 1 \le x^2 < \frac{5}{2}$$

$$\Rightarrow x \in \left(-\sqrt{\frac{5}{2}}, -1\right] \cup \left[1, \sqrt{\frac{5}{2}}\right)$$

92.c. $\cos^{-1}\left(\frac{1+x^2}{2x}\right)$ is defined if $\left|\frac{1+x^2}{2x}\right| \le 1$

$\Rightarrow x \ne 0$

$\Rightarrow 1+x^2-2|x| \le 0$

$\Rightarrow (|x|-1)^2 \le 0$

$\Rightarrow x = 1, -1$

Thus, the domain of $f(x)$ is $\{1, -1\}$. Hence, the range is $\{1, 1+\pi\}$.

93.a.
$$f(f(x)) = \begin{cases} f(x), & f(x) \text{ is rational} \\ 1-f(x), & f(x) \text{ is irrational} \end{cases}$$

$$\Rightarrow f(f(x)) = \begin{cases} x, & x \text{ is rational} \\ 1-(1-x) = x, & x \text{ is irrational}. \end{cases}$$

94.c. $y = |\sin x| + |\cos x|$

$\Rightarrow y^2 = 1 + |\sin 2x|$

$\Rightarrow 1 \le y^2 \le 2$

$\Rightarrow y \in [1, \sqrt{2}]$

$\Rightarrow f(x) = 1 \ \forall x \in R$

95.d. $f(x) = \ln\left(\frac{x^2+e}{x^2+1}\right) = \ln\left(\frac{x^2+1+e-1}{x^2+1}\right)$

$$= \ln\left(1 + \frac{e-1}{x^2+1}\right)$$

Now, $1 \le x^2+1 < \infty$

$$\Rightarrow 0 < \frac{1}{x^2+1} \le 1$$

$$\Rightarrow 0 < \frac{e-1}{x^2+1} \le e-1$$

$$\Rightarrow 1 < 1 + \frac{e-1}{x^2+1} \le e$$

$$\Rightarrow 0 < \ln\left(1 + \frac{e-1}{x^2+1}\right) \le 1$$

Hence, the range is (0, 1].

96.d. $f(x) = \frac{1}{\sqrt{4x - |x^2-10x+9|}}$

For $f(x)$ to be defined $|x^2-10x+9| < 4x$

$\Rightarrow x^2-10x+9 < 4x$ and $x^2-10x+9 > -4x$

$\Rightarrow x^2-14x+9 < 0$ and $x^2-6x+9 > 0$

$\Rightarrow x \in (7-\sqrt{40}, 7+\sqrt{40})$ and $x \in R - \{-3\}$

$\Rightarrow x \in (7-\sqrt{40}, -3) \cup (-3, 7+\sqrt{40})$

97.b. Given $y = 2^{x(x-1)}$

$\Rightarrow x(x-1) = \log_2 y$

$\Rightarrow x^2 - x - \log_2 y = 0$

$$\Rightarrow x = \frac{1 \pm \sqrt{1+4\log_2 y}}{2}$$

Only $x = \frac{1+\sqrt{1+4\log_2 y}}{2}$ lies in the domain.

$$\Rightarrow f^{-1}(x) = \frac{1}{2}[1+\sqrt{1+4\log_2 x}]a$$

98.c. $x \sin x = 1$ (1)

$$\Rightarrow y = \sin x = \frac{1}{x}$$

Root of equation (1) will be given by the point(s) of intersection of the graphs $y = \sin x$ and $y = \frac{1}{x}$.

Graphically, it is clear that we get four roots.

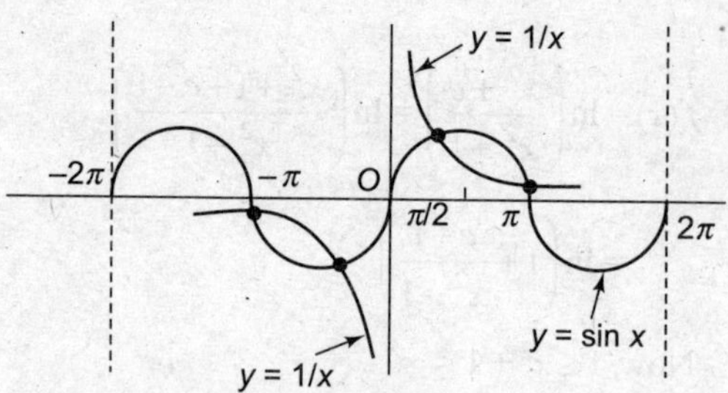

Fig.1.5

99.c. See the graph of $y = 2\cos x$ and $y = |\sin x|$. Their points of intersection represent the solution of the given equation.

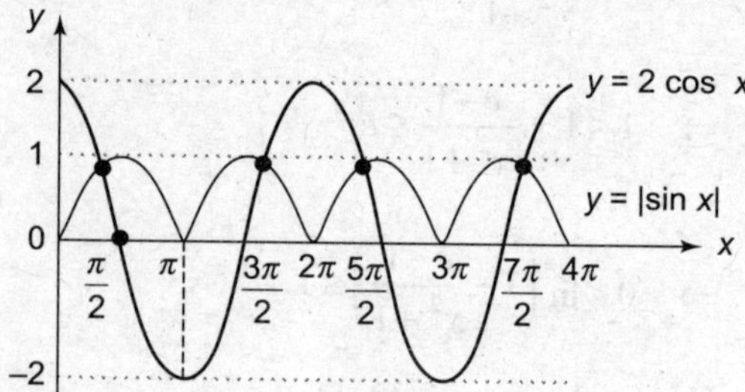

Fig.1.6

We find that the graphs intersect at four points. Hence, the equation has four solutions.

100.a. $af(x+1)+bf\left(\dfrac{1}{x+1}\right)=(x+1)-1$ (1)

Replacing $x+1$ by $\dfrac{1}{x+1}$, we get

$\therefore af\left(\dfrac{1}{x+1}\right)+bf(x+1)=\dfrac{1}{x+1}-1$ (2)

$(1)\times a-(2)\times b$

$\Rightarrow (a^2-b^2)f(x+1)=a(x+1)-a-\dfrac{b}{x+1}+b$

Putting $x=1$, $(a^2-b^2)f(2)=2a-a-\dfrac{b}{2}+b=a+\dfrac{b}{2}$

$=\dfrac{2a+b}{2}$

101.c.

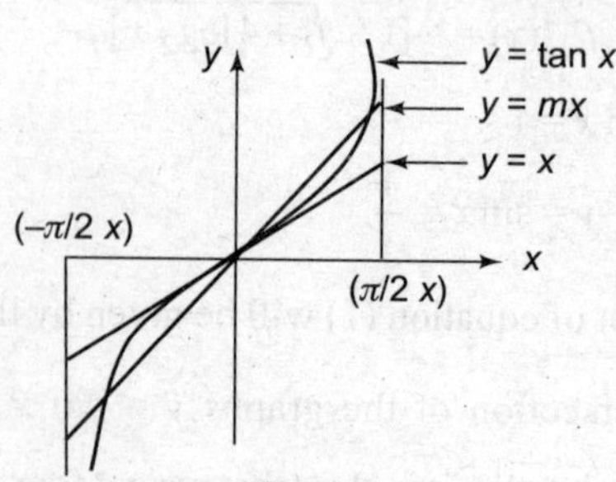

Fig.1.7

In $\left(-\dfrac{\pi}{2},0\right)$, the graph of $y=\tan x$ lies below $y=x$, which is a tangent at $x=0$ and in $(0,\pi/2)$ it lies above the line $y=x$.

For $m>1$, the line $y=mx$ lies below $y=x$ in $\left(-\dfrac{\pi}{2},0\right)$ and above $y=x$ in $(0,\pi/2)$. As $|\tan x|$ monotonically increases from O at $x=0$ to ∞ for $|x|=\pi/2$, the graphs of $y=\tan x$ and $y=mx$, $m>1$, meet at three points including

$x=0$ in $\left(-\dfrac{\pi}{2},\dfrac{\pi}{2}\right)$ independent of m.

102.c. Given $f(x)=[\sin x+[\cos x+[\tan x+[\sec x]]]]$

$=[\sin + p]$, where $p=[\cos x+[\tan x+[\sec x]]]$

$=[\sin x]+p$, (as p is an integer)

$=[\sin x]+[\cos x+[\tan x+[\sec x]]]$

$=[\sin x]+[\cos x]+[\tan x]+[\sec x]$

Now, for $x\in(0,\pi/4)$, $\sin x\in\left(0,\dfrac{1}{\sqrt{2}}\right)$,

$\cos x\in\left(\dfrac{1}{\sqrt{2}},1\right)$,

$\tan x\in(0,1)$, $\sec x\in(1,\sqrt{2})$

$\Rightarrow$ $[\sin x]=0$, $[\cos x]=0$, $[\tan x]=0$ and $[\sec x]=1$

$\Rightarrow$ The range of $f(x)$ is 1.

103.d. $f(3x+2)+f(3x+29)=0$ (1)

Replacing x by $x+9$, we get

$f(3(x+9)+2)+f(3(x+9)+29)=0$

$\Rightarrow f(3x+29)+f(3x+56)=0$ (2)

From (1) and (2), we get

$f(3x+2)=f(3x+56)$

$\Rightarrow f(3x+2)=f(3(x+18)+2)$

$\Rightarrow f(x)$ is periodic with period 18.

104.b. For odd function

$f(x)=-f(-x)$

$$=-\begin{cases}\sin(-x)+\cos(-x) & 0\le -x<\pi/2\\ a, & -x=\pi/2\\ \tan^{-2}(-x)+\operatorname{cosec}(-x), & \pi/2<-x<\pi\end{cases}$$

$$=\begin{cases}-\sin x+\cos x, & -\pi/2<x\le 0\\ -a, & x=-\pi/2\\ \tan^{-2}x-\operatorname{cosec}x, & -\pi<x<-\pi/2\end{cases}$$

105.c. (a) $f(x)=\sin x$ and $g(x)=\cos x$, $x\in[0,\pi/2]$

Here, both $f(x)$ and $g(x)$ are one-one functions,

but $h(x) = f(x) + g(x) = \sin x + \cos x$ is many-one as $h(0) = h(\pi/2) = 1$.

(b) $h(x) = f(x)\, g(x) = \sin x \cos x = \dfrac{\sin 2x}{2}$ is many-one, as $h(0) = h(\pi/2) = 0$.

(c) It is a fundamental property.

106.c. $f(x)$ is defined for $x \in (0, 1)$

$\Rightarrow$ $f(e^x) + f(\ln|x|)$ is defined for, $0 < e^x < 1$ and $0 < \ln|x| < 1$

$\Rightarrow$ $-\infty < x < 0$ and $1 < |x| < e$

$\Rightarrow$ $x \in (-\infty, 0)$ and $x \in (-e, -1) \cup (1, e)$

$\Rightarrow$ $x \in (-e, -1)$

107.d. $|\cos x| + \cos x = \begin{cases} 0, & \cos x \le 0 \\ 2\cos x, & \cos x > 0 \end{cases}$

For $f(x)$ to defined $\cos x > 0$

$\Rightarrow$ $x \in \left(\dfrac{(4n-1)\pi}{2}, \dfrac{(4n+1)\pi}{2}\right)$

(1st and 4th quadrant).

108.b. Let $2x + 3y = m$ and $2x - 7y = n$

$\Rightarrow$ $y = \dfrac{m-n}{10}$ and $x = \dfrac{7m+3n}{20}$

$\Rightarrow$ $f(m, n) = 7m + 3n$

$\Rightarrow$ $f(x, y) = 7x + 3y$

109.d. Image b_1 is assigned to any three of the six pre-images in ${}^6C_3$ ways.

Rest two images can be assigned to remaining three pre-images in $2^3 - 2$ ways (as function is onto).

Hence, number of functions are ${}^6C_3 \times (2^3 - 2) = 20 \times 6 = 120$

110.d. $y = f(x)$ and $y = g(x)$ are mirror image of each other about line $y = a$

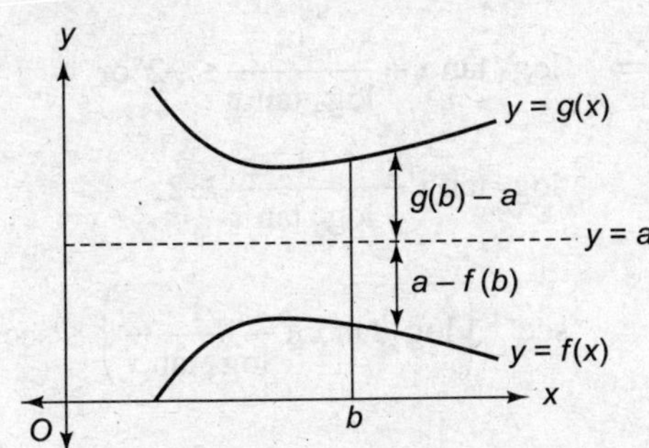

Fig.1.8

$\Rightarrow$ For some $x = b$, $g(b) - a = a - f(b)$

$\Rightarrow$ $f(b) + g(b) = 2a$

$\Rightarrow$ $h(b) = f(b) + g(b) = 2a$ (constant)

Hence, $h(x)$ is constant function. Thus it is neither one-one nor onto.

111.c. Clearly $f(x + \pi) = f(x)$, $g(x + \pi) = g(x)$ and

$$\phi\left(x + \frac{\pi}{2}\right) = \{(-1) f(x)\}\ \{(-1) g(x)\} = \phi(x).$$

112.b. In the sum, $f(1) + f(2) + f(3) + \cdots + f(n)$, 1 occurs n times, $\dfrac{1}{2}$ occurs $(n-1)$ times, $\dfrac{1}{3}$ occurs $(n-2)$ times and so on. Therefore,

$f(1) + f(2) + f(3) + \cdots + f(n)$

$$= n \times 1 + (n-1) \times \frac{1}{2} + (n-2) \times \frac{1}{3} + \cdots + 1 \times \frac{1}{n}$$

$$= n\left(1 + \frac{1}{2} + \frac{1}{3} + \cdots + \frac{1}{n}\right) - \left(\frac{1}{2} + \frac{2}{3} + \frac{3}{4} + \cdots + \frac{n-1}{n}\right)$$

$$= nf(n)$$

$$\left[\left(1 - \frac{1}{2}\right) + \left(1 - \frac{1}{3}\right) + \left(1 - \frac{1}{4}\right) + \cdots + \left(1 - \frac{1}{n}\right)\right]$$

$$= nf(n) - [n - f(n)]$$

$$= (n+1) f(n) - n$$

113.a. $h(x) = \log(f(x)\, g(x)) = \log e^{\{y\}+[y]} = \{y\} + [y]$

$= e^{|x|} \operatorname{sgn} x$

$$\therefore\ h(x) = e^{|x|} \operatorname{sgn} x = \begin{cases} e^x, & x > 0 \\ 0, & x = 0 \\ -e^{-x}, & x < 0 \end{cases}$$

$$\Rightarrow\ h(-x) = \begin{cases} e^{-x}, & x < 0 \\ 0, & x = 0 \\ -e^x, & x > 0 \end{cases}$$

$\Rightarrow$ $h(x) + h(-x) = 0$ for all x.

114.b.

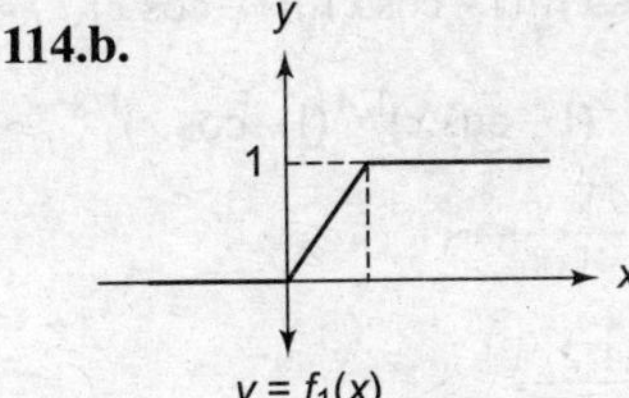

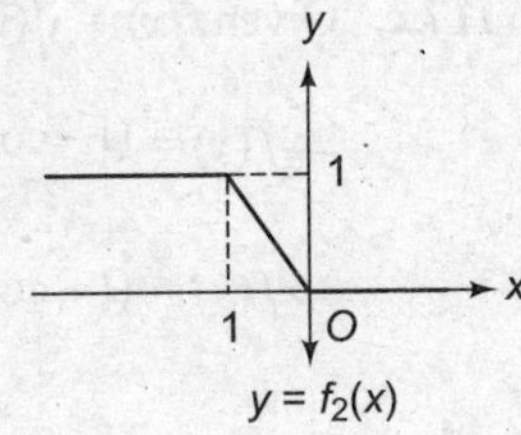

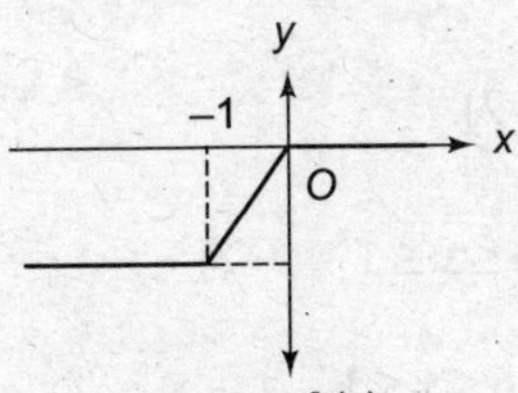

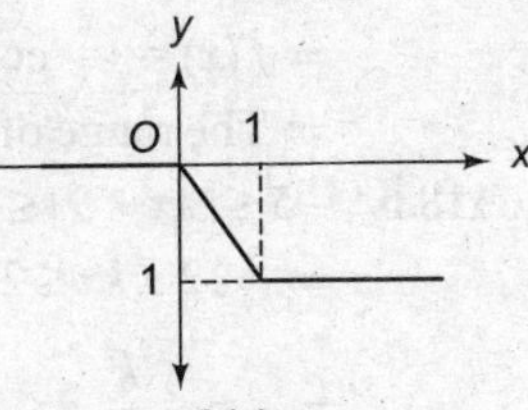

Fig.1.9

115.c. $[y + [y]] = 2\cos x$

$\Rightarrow$ $[y] + [y] = 2\cos x$ $\quad [\because [x + n] = [x] + n \text{ if } n \in I]$

$\Rightarrow 2[y] = 2\cos x \Rightarrow [y] = \cos x$ (1)

Also $y = \frac{1}{3}[\sin x + [\sin x + [\sin x]]]$

$= \frac{1}{3}(3[\sin x])$

$= [\sin x]$ (2)

From (1) and (2),

$[[\sin x]] = \cos x$

$\Rightarrow [\sin x] = \cos x$

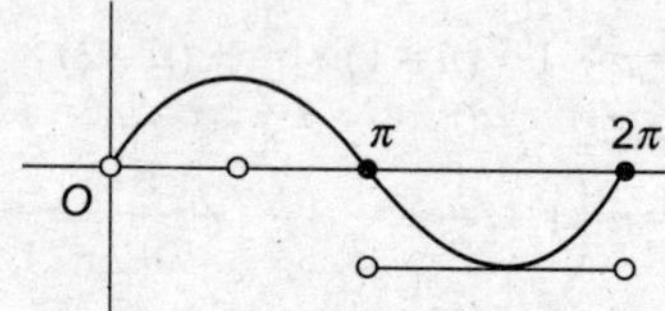

Fig.1.10

The number of solutions is 3.

116.a. $\cos^{-1}(\cos x) = [x]$

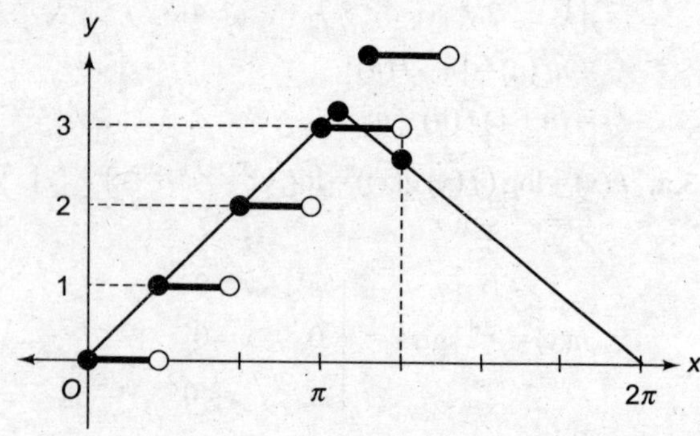

Fig.1.11

The solutions are clearly 0, 1, 2, 3 and $3 = 2\pi - x$ or $x = 2\pi - 3$.

117.c. Given $f(x) = \sqrt{(1-\cos x)\sqrt{(1-\cos x)\sqrt{(1-\cos x)\sqrt{\ldots\infty}}}}$

$\Rightarrow f(x) = (1-\cos x)^{1/2}(1-\cos x)^{1/4}(1-\cos x)^{1/8}\ldots\infty$

$\Rightarrow f(x) = (1-\cos x)^{\frac{1}{2}+\frac{1}{4}+\frac{1}{8}+\cdots\infty}$

$\Rightarrow f(x) = (1-\cos x)^{\frac{1/2}{1-(1/2)}}$

$\Rightarrow f(x) = 1 - \cos x$

$\Rightarrow$ The range of $f(x)$ is $[0, 2]$.

118.b. $-5 \le |kx+5| \le 7$

$\Rightarrow -12 \le kx \le 2$ where $-6 \le x \le 1$

$\Rightarrow -6 \le \frac{k}{2}x \le 1$ where $-6 \le x \le 1$

$\therefore k = 2.$

[$\because$ the range of $h(x)$ = the domain of $f(x)$]

119.a. Let $g(x) = (x+1)(x+2)(x+3)(x+4)$

Clearly, the rough graph of $g(x)$ is given as

Fig.1.12

$\therefore g(x) = (x+1)(x+2)(x+3)(x+4)$

$= (x+1)(x+4)(x+2)(x+3)$

$= (x^2+5x+4)(x^2+5x+6)$

$= t(t+2) = (t+1)^2 - 1,$

where $t = x^2 + 5x$

Now $g_{\min} = -1$, for which $x^2 + 5x = -1$ has real roots in $[-6, 6]$.

Also $g(6) = 7 \times 8 \times 9 \times 10 = 5040.$

Hence, the range of $g(x)$ is $[-1, 5040]$ for $x \in [-6, 6]$.

Then, the range of $f(x)$ is $[4, 5045]$.

120.d. $f(x) = \sqrt{x^{12} - x^9 + x^4 - x + 1}$

We must have $x^{12} - x^9 + x^4 - x + 1 \ge 0$ (1)

Obviously (1) is satisfied by $x \in (-\infty, 0]$.

Also $x^9(x^3-1) + x(x^3-1) + 1 \ge 0\ \forall\, x \in [1, \infty)$.

Further, $x^{12} - x^9 + x^4 - x + 1 = (1-x) + x^4(1-x^5) + x^{12}$ is also satisfied by $x \in (0, 1)$.

Hence, the domain is R.

121.a. $f(x) = \sec^{-1}(\log_3 \tan x + \log_{\tan x} 3)$.

$f(x) = \sec^{-1}\left(\log_3 \tan x + \frac{1}{\log_3 \tan x}\right)$

Now for $\log_3 \tan x$ to get defined, $\tan x \in (0, \infty)$

$\Rightarrow \log_3 \tan x \in (-\infty, \infty)$ or $\log_3 \tan x \in R$

Also $x + \frac{1}{x} \le -2$ or $x + \frac{1}{x} \ge 2$

$\Rightarrow \log_3 \tan x + \frac{1}{\log_3 \tan x} \le -2$ or

$\log_3 \tan x + \frac{1}{\log_3 \tan x} \ge 2$

$\Rightarrow \sec^{-1}\left(\log_3 \tan x + \frac{1}{\log_3 \tan x}\right) \le \sec^{-1}(-2)$ or

$\sec^{-1}\left(\log_3 \tan x + \frac{1}{\log_3 \tan x}\right) \ge \sec^{-1} 2$

$\Rightarrow f(x) \le \frac{2\pi}{3}$ or $f(x) \ge \frac{\pi}{3}$

$\Rightarrow \; f(x) \in \left[\frac{\pi}{3}, \frac{\pi}{2}\right) \cup \left(\frac{\pi}{2}, \frac{2\pi}{3}\right]$

122.a. We have $f(x) = {}^{7-x}P_{x-3} = \dfrac{(7-x)!}{(10-2x)!}$

We must have $7 - x > 0, x \ge 3$ and $7 - x \ge x - 3$

$\Rightarrow \; x < 7, x \ge 3$ and $x \le 5$

$\Rightarrow \; 3 \le x \le 5$

$\Rightarrow \; x = 3, 4, 5$

Now, $f(3) = \dfrac{4!}{4!} = 1, f(4) = \dfrac{3!}{2!} = 3, f(5) = \dfrac{2!}{0!} = 2$.

Hence, $R_f = \{1, 2, 3\}$.

123.b. We have $f(x-y) = f(x)f(y) - f(a-x)f(a+y)$

Putting $x = a$ and $y = a - x$, we get

$f(a-(x-a)) = f(a)f(x-a) - f(0)f(x)$ (1)

Putting $x = 0, y = 0$, we get

$f(0) = f(0)\,(f(0)) - f(a)f(a)$

$\Rightarrow \; f(0) = (f(0))^2 - (f(a))^2$

$\Rightarrow \; 1 = (1)^2 - (f(a))^2$

$\Rightarrow \; f(a) = 0$

$\Rightarrow \; f(2a - x) = -f(x)$

Multiple Correct Answers Type

1. a., b., d.

$f(0) = \max\{1 + \sin 0, 1, 1 - \cos 0\} = 1$

$g(0) = \max\{1, |0 - 1|\} = 1$

$f(1) = \max\{1 + \sin 1, 1, 1 - \cos 1\} = 1 + \sin 1$

$g(f(0)) = g(1) = \max\{1, |1 - 1|\} = 1$

$f(g(0)) = f(1) = 1 + \sin 1$

$g(f(1)) = g(1 + \sin 1) = \max\{1, |1 + \sin 1 - 1|\} = 1$

2. b., c.

(a) For $f(x) = \log x^2, x^2 > 0 \Rightarrow x \in R - \{0\}$

For $g(x) = 2 \log x, x > 0$

Hence, $f(x)$ and $g(x)$ are not identical.

(b) $f(x) = \log_x e = \dfrac{1}{\log_e x} = g(x)$

Hence, the functions are identical.

(c) $f(x) = \sin(\cos^{-1} x) = \sin\left(\dfrac{\pi}{2} - \sin^{-1} x\right)$

$= \cos\left(\sin^{-1} x\right)$

$= g(x)$

Hence, the functions are identical.

3. a., b., c.

$f(x) + f(y) = f\left(\dfrac{x+y}{1-xy}\right)$

Replace y by $-x \Rightarrow f(x) + f(-x) = f(0)$ (1)

Put $x = y = 0 \Rightarrow f(0) + f(0) = f(0) \Rightarrow f(0) = 0$

$\Rightarrow \; f(x) + f(-x) = 0$ (from (1))

Hence, $f(x)$ is an odd function.

$f(x) + f(y) = f\left(x\sqrt{1-y^2} + y\sqrt{1-x^2}\right)$

Replace y by $-x \Rightarrow f(x) + f(-x) = f(0)$ (2)

Put $x = y = 0 \Rightarrow f(0) + f(0) = f(0)$

$\Rightarrow \; f(0) = 0 \Rightarrow f(x) + f(-x) = 0$ (from (2))

Hence, $f(x)$ is an odd function.

$f(x+y) = f(x) + f(y)$

Replace y by $-x \Rightarrow f(0) = f(x) + f(-x)$ (3)

Put $x = y = 0 \Rightarrow f(0+0) = f(0) + f(0) \Rightarrow f(0) = 0$

$\Rightarrow f(x) + f(-x) = 0$ [From (3)]

Hence, $f(x)$ is an odd function.

4. a., c.

$f(x+y) + f(x-y) = 2f(x)\,.\,f(y)$ (1)

Put $x = 0 \Rightarrow f(y) + f(-y) = 2f(0)f(y)$ (2)

Put $x = y = 0 \Rightarrow f(0) + f(0) = 2f(0)f(0)$

$\Rightarrow \; f(0) = 1$ (as $f(0) \ne 0$)

$\Rightarrow \; f(-y) = f(y)$ (from (2))

Hence, the function is even. Then $f(-2) = f(2) = a$.

5. b., d.

$f\left(x + \dfrac{1}{x}\right) = x^2 + \dfrac{1}{x^2}$

$\Rightarrow f\left(x + \dfrac{1}{x}\right) = x^2 + \dfrac{1}{x^2} = \left(x + \dfrac{1}{x}\right)^2 - 2$

$\Rightarrow f(y) = \; y^2 - 2$

Now $y = x + \dfrac{1}{x} \ge 2$ or ≤ -2.

Hence, the domain of the function is $(-\infty, -2] \cup [2, \infty)$.

Also for these values of y, $y^2 \ge 4 \Rightarrow y^2 - 2 \ge 2$.

Hence, the range of the function is $[2, \infty)$.

6. a., d.

Given $f(x) + f(y) = \left(x\sqrt{1-y^2} + y\sqrt{1-x^2}\right)$ (1)

Replace y by $x \Rightarrow 2f(x) = f\left(2x\sqrt{1-x^2}\right)$

$3f(x) = f(x) + 2f(x)$

$= f(x) + f\left(2x\sqrt{1-x^2}\right)$

$= f\left(x\sqrt{1-4x^2(1-x^2)} + 2x\sqrt{1-x^2}\,\sqrt{1-x^2}\right)$

$= f\left(x\sqrt{(2x^2-1)^2} + 2x(1-x^2)\right)$

$= f\left(x|2x^2-1|+2x-2x^3\right)$

$= f\left(2x^3-x+2x-2x^3\right)$ or

$f\left(x-2x^3+2x-2x^3\right)$

$= f(x)$ or $f(3x-4x^3)$

$\Rightarrow f(x)=0$ or $3f(x)=f(3x-4x^3)$

Consider $3f(x)=f(3x-4x^3)$

Replace x by $-x$, we get

$3f(-x)=f(4x^3-3x)$ (2)

Also from (1), $f(x)+f(-x)=f(0)$

Put $x=y=0$ in (1), we have $f(0)=0$

$\Rightarrow f(x)+f(-x)=0$, thus $f(x)$ is an odd function.

Now from (2) $-3f(x)=f(4x^3-3x)$

$\Rightarrow\ f(4x^3-3x)+3f(x)=0$

7. b., c.

Given $2f(\sin x)+f(\cos x)=x$ (1)

Replace x by $\dfrac{\pi}{2}-x$

$\Rightarrow\ 2f(\cos x)+f(\sin x)=\dfrac{\pi}{2}-x$ (2)

Eliminating $f(\cos x)$ from (1) and (2), we get

$\Rightarrow\ 3f(\sin x)=3x-\dfrac{\pi}{2}$

$\Rightarrow\ f(\sin x)=x-\dfrac{\pi}{6}$

$\Rightarrow\ f(x)=\sin^{-1}x-\dfrac{\pi}{6}$

$f(x)$ has the domain $[-1, 1]$

Also, $\sin^{-1}x\in\left[-\dfrac{\pi}{2},\dfrac{\pi}{2}\right]\Rightarrow \sin^{-1}x-\dfrac{\pi}{6}$

$\in\left[-\dfrac{2\pi}{3},\dfrac{\pi}{3}\right]$.

8. a., c.

$f(2)=f(1+1)=2f(1)=10$

$f(3)=f(2+1)=f(2)+f(1)=10+5=15$

Then, $f(n)=5n$

$\Rightarrow\ \sum_{r=1}^{m} f(r)=5\sum_{r=1}^{m} r=\dfrac{5m(m+1)}{2}$

Replace y by $-x$, $\Rightarrow f(0)=f(x)+f(-x)$

Also put $x=y=0\Rightarrow f(0)=f(0)+f(0)\Rightarrow f(0)=0$.

$\Rightarrow\ f(x)+f(-x)=0$, hence, the function is odd.

9. a., b., c.

$(f+g)(3.5)=f(3.5)+g(3.5)=(-0.5)+0.5=0$

$f(g(3))=f(0)=3$

$(fg)(2)=f(2)g(2)=(-1)\times(-1)=1$

$(f-g)(4)=f(4)-g(4)=0-26=-26$

10. b., d.

$f(x)=x^2-2ax+a(a+1)$

$f(x)=(x-a)^2+a, x\in[a,\infty)$

Let $y=(x-a)^2+a$ clearly $y\geq a$

$\Rightarrow (x-a)^2=y-a$

$\Rightarrow x=a+\sqrt{y-a}$

$\therefore f^{-1}(x)=a+\sqrt{x-a}$

Now $f(x)=f^{-1}(x)$

$\Rightarrow (x-a)^2+a=a+\sqrt{x-a}$

$(x-a)^2=\sqrt{x-a}$

$\Rightarrow (x-a)^4=(x-a)$

$\Rightarrow x=a$ or $(x-a)^3=1$

$\Rightarrow x=a$ or $a+1$

If $a=5049$, then $a+1=5050$

If $a+1=5049$, then $a=5048$.

11. a., b., c., d.

$f(x)=\begin{cases}1, & x \text{ is rational}\\ 0, & x \text{ is irrational}\end{cases}$

$\Rightarrow f(x+k)=\begin{cases}1, & x+k \text{ is rational}\\ 0, & x+k \text{ is irrational}\end{cases}$

where k is any rational number

$\Rightarrow f(x+k)=\begin{cases}1, & x \text{ is rational}\\ 0, & x \text{ is irrational}\end{cases}$

$\Rightarrow f(x+k)=f(x)$

$\Rightarrow$ $f(x)$ is periodic function, but its fundamental period cannot be determined

$f(x)=\begin{cases}x-[x], & 2n\leq x<2n+1\\ 1/2, & 2n+1\leq x<2n+2\end{cases}$

Draw the graph from which it can be verified that period is 2.

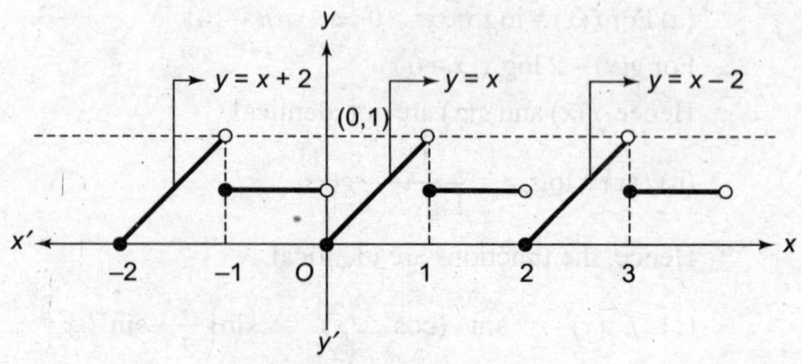

Fig.1.13

$(x)=(-1)^{[(2x)/\pi]}\Rightarrow f(x+\pi)$

$= (-1)\,(-1)^{[2(\pi+x)/\pi]} = (-1)^{[2+(2x/\pi)]} = (-1)^{[(2x/\pi)]+2}$

$= (-1)^{[(2x)/\pi]}$

Hence, the period is π.

$f(x) = ax + a - [ax + a] + \tan\left(\dfrac{\pi x}{2}\right)$

$= \{ax + a\} + \tan\left(\dfrac{\pi x}{2}\right)$

$\{ax + a\}$ is periodic with period $1/a$, $\tan\left(\dfrac{\pi x}{2}\right)p$ is periodic with period 2. Now, the LCM of $1/a$ and 2 is 2. Hence, the period of $f(x)$ is 2.

12. b., c.

$f(x)$ must be a linear function, let $f(x) = ax + b$

$\Rightarrow\ f(ax + b) = 6x - ax - b$

$\Rightarrow\ a(ax + b) + b = 6x - ax - b$

$\Rightarrow\ a^2 = 6 - a$ and $ab + b = -b$

$\Rightarrow\ a = 2$ or $-3 \Rightarrow b = 0$

$\Rightarrow\ f(x) = 2x$ or $-3x \Rightarrow f(17) = 34$ or -51

13. a., b., c., d.

$$f(x+1) = \frac{f(x)-5}{f(x)-3} \quad (1)$$

$f(x)f(x+1) - 3f(x+1) = f(x) - 5$

$$\Rightarrow\ f(x) = \frac{3f(x+1)-5}{f(x+1)-1}$$

Replacing x by $(x - 1)$, we get

$$f(x-1) = \frac{3f(x)-5}{f(x)-1} \quad (2)$$

$$\text{Using (1), } f(x+2) = \frac{f(x+1)-5}{f(x+1)-3} = \frac{\dfrac{f(x)-5}{f(x)-3}-5}{\dfrac{f(x)-5}{f(x)-3}-3}$$

$$= \frac{2f(x)-5}{f(x)-2} \quad (3)$$

$$\text{Using (2), } f(x-2) = \frac{3f(x-1)-5}{f(x-1)-1}$$

$$= \frac{3\left(\dfrac{3f(x)-5}{f(x)-1}\right)-5}{\dfrac{3f(x)-5}{f(x)-1}-1}$$

$$= \frac{2f(x)-5}{f(x)-2} \quad (4)$$

Using (3) and (4), we have $f(x+2) = f(x-2)$

$\Rightarrow\ f(x+4) = f(x) \Rightarrow f(x)$ is periodic with period 4.

14. a., d.

$f(x) = \sec^{-1}[1 + \cos^2 x]$

$f(x)$ is defined if $[1 + \cos^2 x] \le -1$ or $[1 + \cos^2 x] \ge 1$

$\Rightarrow\ [\cos^2 x] \le -2$ (not possible) or $[\cos^2 x] \ge 0$

$\Rightarrow\ \cos^2 \ge 0 \Rightarrow x \in R$

Now $0 \le \cos^2 x \le 1 \Rightarrow 1 \le 1 + \cos^2 x \le 2$

$\Rightarrow\ [1 + \cos^2 x] = 1, 2$

$\Rightarrow\ \sec^{-1}[1 + \cos^2 x] = \sec^{-1}1, \sec^{-1}2$

Hence, the range is $\{\sec^{-1}1, \sec^{-1}2\}$.

15. a., b., c.

$f(x) = \tan(\tan^{-1}x) = x$ for all x and $g(x) = \cot(\cot^{-1}x) = x$ for all x.

Hence, this pair is identical functions.

$f(x) = \text{sgn}(x)$ and $g(x) = \text{sgn}(\text{sgn}(x))$ have domain R

$f(x)$ has range $\{-1, 0, 1\}$ and $g(x) = \text{sgn}(\text{sgn}(x))$ has range $\{-1, 0, 1\}$

Also $f(x) = g(x)$ for any x, then this pair is identical functions

$g(x) = \cot^2 x - \cos^2 x = \cos^2 x(\text{cosec}^2 x - 1)$

$= \cos^2 x \cot^2 x = f(x)$

$f(x) = e^{\log_e \sec^{-1} x}$ has the domain $[1, \infty)$, whereas $g(x) = \sec^{-1}x$ has the domain $(-\infty, -1] \cup [1, \infty)$.

Hence, this pair is not identical functions.

16. b., d.

The period of $f(x) = |\sin 2x| + |\cos 2x|$ is $\pi/4$

$\Rightarrow\ [f(x)]$ is also periodic with period $\pi/4$.

Also $1 \le f(x) \le \sqrt{2}$

$\Rightarrow\ [f(x)] = 1\ f(x)$ is a many-one and into function.

17. a., b., d.

$f(x) = \dfrac{1}{\ln[1-|x|]}$ is defined if $[1-|x|] > 0$ and $1 - [x] \ne 1 \Rightarrow\ [1-|x|] \ge 2 \Rightarrow 1 - |x| \ge 2 \Rightarrow |x| \le -1$ which is not possible.

$f(x) = \dfrac{x!}{\{x\}}$. Here $x!$ is defined only when x is a natural number, but $\{x\}$ becomes zero for these values of x. Hence, $f(x)$ is not defined in this case.

$f(x) = x!\ \{x\}$ is defined for x being a natural number. Hence, $f(x)$ is a function whose domain $x \in N$.

$f(x) = \dfrac{\ln(x-1)}{\sqrt{(1-x^2)}}$. Here $\ln(x - 1)$ is defined only when $x - 1 > 0 \Rightarrow x > 1$. Also $1 - x^2 > 0$ for denominator, i.e., $-1 < x < 1$. Hence, $f(x)$ is not defined for any value of x.

18. a., b.

$f(x) = \sin(\sin^{-1} x) = x\ \forall\ x \in [-1, 1]$ which is one-one and onto.

$$f(x)=\frac{2}{\pi}\sin^{-1}(\sin x)=\frac{2}{\pi}x$$

The range of the function for $x\in[-1,1]$ is $\left[-\frac{2}{\pi},\frac{2}{\pi}\right]$, which is a subset of $[-1, 1]$.
Hence, the function is one-one but not onto, hence not bijective.

$$f(x)=(\text{sgn}(x))\ln(e^x)=(\text{sgn}(x))x=\begin{cases}x, & x>0\\ -x, & x<0\\ 0, & x=0\end{cases}.$$

This function has the range [0, 1] which is a subset of [–1, 1]. Hence, the function is into. Also, the function is many-one.

$$f(x)=x^3\,\text{sgn}(x)=\begin{cases}x^3, & x>0\\ -x^3, & x<0\\ 0, & x=0\end{cases}$$

which is many-one and into.

19. a., b., c., d.

Since $\angle PRQ=\pi/2$ and points P, Q, R lie on the circle with PQ as diameter.

Also $PQ=5$

Now, the maximum area of the triangle is

$$\Delta_{\max}=\frac{1}{2}(5)\left(\frac{5}{2}\right)=6.25$$

For area = 5, we have four symme trical positions of point P (shown as P_1, P_2, P_3, P_4)

For area = 6.25 we have exactly two points.

For area = 7, no such points exist.

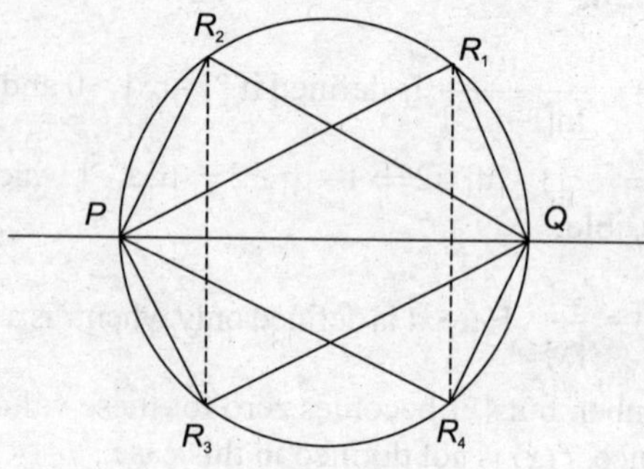

Fig.1.14

20. b., c.

$(x+1)f(x)-x$ is a polynomial degree $n+1$

$\Rightarrow\ (x+1)f(x)-x=k(x)[x-1][x-2]\ldots[x-n]$

$\Rightarrow\ [n+2]f(n+1)-(n+1)=k[(n+1)!]$

Also, $1=k(-1)(-2)\ldots((-n+1))$

$1=k(-1)^{n+1}(n+1)!$

$\Rightarrow\ (n+2)f(n+1)-(n+1)=(-1)^{n+1}$

$\Rightarrow\ f(n+1)=1$, if n is odd and $\frac{n}{n+1}$, if n is even.

21. a., c., d.

$$f^2(x)=f\left(\frac{3}{4}x+1\right)=\frac{3}{4}\left(\frac{3}{4}x+1\right)+1=\left(\frac{3}{4}\right)^2x+\frac{3}{4}+1 \quad (1)$$

$$f^3(x)=f\{f^2(x)\}=\frac{3}{4}\{f^2(x)+1\}$$

$$=\frac{3}{4}\left\{\left(\frac{3}{4}\right)^2x+\frac{3}{4}+1\right\}+1$$

$$=\left(\frac{3}{4}\right)^3x+\left(\frac{3}{4}\right)^2+\frac{3}{4}+1$$

$$\therefore\ f^n(x)=\left(\frac{3}{4}\right)^nx+\left(\frac{3}{4}\right)^{n-1}+\left(\frac{3}{4}\right)^{n-2}+\cdots+\left(\frac{3}{4}\right)+1$$

$$=\left(\frac{3}{4}\right)^nx+\frac{1-\left(\frac{3}{4}\right)^n}{1-\frac{3}{4}}$$

$$\therefore\ \lambda\lim_{n\to\infty}f^n(x)=0+4=4$$

22. a., b., c.

$f(x)$ is defined if $\log_{|\sin x|}(x^2-8x+23)-\frac{3}{\log_2|\sin x|}>0$

$$\Rightarrow\ \log_{|\sin x|}\left(\frac{x^2-8x+23}{8}\right)>0$$

This is true if $|\sin x|\ne 0, 1$ and $\frac{x^2-8x+23}{8}<1$

Now, $\frac{x^2-8x+23}{8}<1\Rightarrow x^2-8x+15<0$

$$\Rightarrow\ x\in(3,5)-\left\{\pi,\frac{3\pi}{2}\right\}$$

$$\text{Domain}=(3,\pi)\cup\left(\pi,\frac{3}{2}\right)\cup\left(\frac{3\pi}{2},5\right).$$

Reasoning Type

1.b. Any function can be expressed as a sum of odd and even function, no matter it is odd or even.
But $f(x)=\log e^x$ is not defined for $x<0$, hence statement 2 is true but not correct explanation of statement 1.

2.a. $f(x)-1+f(1-x)-1=0$; so $g(x)+g(1-x)=0$

Putting $x = x + \frac{1}{2}$, we get $g\left(\frac{1}{2}+x\right)+g\left(\frac{1}{2}-x\right)=0$.

So it is symmetrical about $\left(\frac{1}{2}, 0\right)$.

3.d. $f\left(\frac{2\tan x}{1+\tan^2 x}\right)=\frac{(1+\cos 2x)(\sec^2 x+2\tan x)}{2}$

$\Rightarrow\ f(\tan 2x)=\frac{2\cos^2 x(\sec^2 x+2\tan x)}{2}$

$= 1 + 2\sin x \cos x = 1 + \sin 2x$

$\Rightarrow\ f(y)=1+y$ where $y=\sin 2x$, now $\sin 2x \in [-1, 1]$

$\Rightarrow\ f(y)\in[0, 2]$

Hence, statement 1 is false but statement 2 is true.

4. c. $\sin(kx)$ has period $\frac{\pi}{k}$ and period of $\{x\}$ is 1

Now LCM of π/k and 1 exists only if k is a rational multiple of π (as LCM of rational and irrational number does not exist). Hence, statement 1 is true. But statement 2 is false as sum of two periodic function is not necessarily periodic. Consider $f(x) = \sin x + \{x\}$.

5. c. Obviously, $f(x)=x^2+\tan^{-1}x$ is non-periodic, but sum of two non-periodic function is not always non-periodic, as $f(x) = x$ and $g(x) = -[x]$, where [.] represents the greatest integer function.

$f(x)+g(x)=x-[x]=\{x\}$ is a periodic function ({.} represents the fractional part function).

6. c. $f(x)=\tan^{-1}x$ is an increasing function, then the range of function is $[\tan^{-1}1, \tan^{-1}\sqrt{3}\,]\equiv[\pi/4, \pi/3]$.

Hence, statement 1 is true. But statement 2 is not true in general. For non-monotonic function, statement 2 is false.

7. a. For any integer k, we have $f(k)=f(2n\pi+k)$ where $n \in Z$, but $2n\pi+k$ is not integer, hence $f(x)$ is one-one.

8. a. Consider $f(x) = \tan x$, which is surjective, periodic but discontinuous.

9. b. $||x^2-5x+4|-|2x-3||=|x^2-3x+1|$

$\Rightarrow\ ||x^2-5x+4|-|2x-3||=|(x^2-5x+4)+(2x-3)|$

$\Rightarrow\ (x^2-5x+4)(2x-3)\le 0$

$\Rightarrow\ (x-1)(2x-3)(x-4)\le 0$

$\Rightarrow\ x\in(-\infty, 1]\cup\left[\frac{3}{2}, 4\right]$

−	+	−	+
1	3/2	4	

Fig.1.15

Hence, statement 1 is true.

Statement 2 is true as it is the property of modulus function but is not a correct explanation of statement 1.

10. a. Let max $|f(x)| = M$ where $0 < M \le 1$ (since f is not identically zero and $|f(x)|\le 1\ \forall\, x\in R$)

Now, $f(x+y)+f(x-y)=2f(x)\,.\,g(y)$

$\Rightarrow\ |2f(x)\,g(y)|=|f(x+y)+f(x-y)|$

$\Rightarrow\ 2|f(x)|\,|g(y)|\le|f(x+y)|+|f(x-y)|\le M+M$

$\Rightarrow\ |g(y)|\le 1$ for $y\in R$.

11. b. Obviously, both the statements are true but statement 2 is not a correct explanation of statement 1, as function $f(x) = \cos(2x + 3)$ which is periodic though $g(x) = 2x + 3$ is non-periodic.

12. b. Obviously, both the statements are true but statement 2 is not a correct explanation of statement 1, as for $f(x) = \cos(\sin x)$ the period is π, where $\sin x$ has period 2π. Thus, the period of $f((g(x))$ is not always same as that of $g(x)$.

13. a. It is a fundamental concept

14. b. Both the statements are true, but statement 2 is not a correct explanation of statement 1 as $f(g(x))$ is one-one when $g(x)$ is one-one and $f(x)$ is many-one.

15. b. Both the statements are true, but statement 2 is not a correct explanation of statement 1, as for $f(g(x))$ is onto it is necessary that $f(x)$ is onto, but there is no restriction on $g(x)$.

16. d. Statement 1 is false, though $f(x) = \sin x$ and $g(x) = \cos x$ have same domain and range, $\cos x = \sin x$ does not hold for all $x \in R$.

However, the statement 2 is true.

17. a.

18. d. If $b^2 - 4ax > 0$ then $ax^2 + bx + c = 0$ has real distinct roots α, β.

If $a > 0$, then for $f(x)=\sqrt{ax^2+bx+c}$ to get defined, $ax^2 + bx + c \ge 0$, then the range of $f(x)$ is $[0, \infty)$

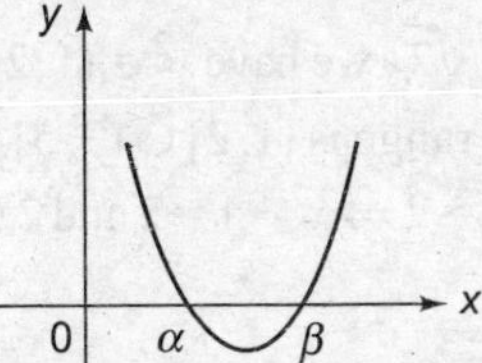

Fig.1.16

If $a < 0$, then for $f(x)$ to get defined, $ax^2 + bx + c \ge 0$, then the range of $f(x)$ is $\left[0, -\frac{b}{2a}\right]$.

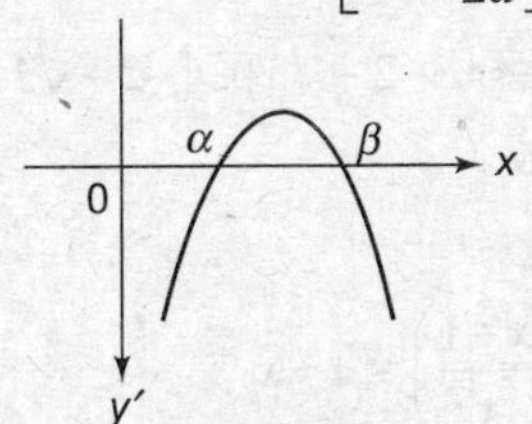

Fig.1.17

Hence, statement 1 is false, but statement 2 is true.

19 .c. $fog(x)$ can be even also when one of them is even and other is odd.

20. a. Obviously, the graph of $y=\tan x$ is symmetrical about origin, as it is an odd function.
Also derivative of an odd function is an even function, and $\sec^2 x$ is derivative of $\tan x$, hence both the statements are true, and statement 2 is a correct explanation of statement 1.

Linked Comprehension Type

For Problems 1–3

1. c., 2. c., 3.b

Sol. $f(x)=\begin{cases} x+1, & x\le 1 \\ 2x+1, & 1<x\le 2\end{cases}$

$g(x)=\begin{cases} x^2, & -1\le x<2 \\ x+2, & 2\le x\le 3\end{cases}$

$\Rightarrow f(x)=\begin{cases} g(x)+1, & g(x)\le 1 \\ 2g(x)+1, & 1<g(x)\le 2\end{cases}$

$\Rightarrow f(g(x))=\begin{cases} x^2+1, & x^2\le 1, \ -1\le x<2 \\ x+2+1, & x+2\le 1, \ 2\le x\le 3 \\ 2x^2+1, & 1<x^2\le 2, \ -1\le x<2 \\ 2(x+2)+1, & 1<x+2\le 2, \ 2\le x\le 3\end{cases}$

$\Rightarrow f(g(x))=\begin{cases} x^2+1, & -1\le x\le 1 \\ 2x^2+1, & 1<x\le \sqrt{2}\end{cases}$

1. c. Hence, the domain of $f(x)$ is $[-1, \sqrt{2}\,]$.

2. c. For $-1\le x\le 1$, we have $x^2\in[0,1]\Rightarrow x^2+1\in[1,2]$

For $1<x\le\sqrt{2}$, we have $x^2\in(1,2]\Rightarrow 2x^2+1\in(3,5]$
Hence, the range is $[1,2]\cup(3,5]$.

3. b. For $f(g(x))=2\Rightarrow x^2+1=2$ and $2x^2+1=2\Rightarrow x=\pm 1$ or $x=\pm\dfrac{1}{\sqrt{2}}$

$\Rightarrow \ x=\pm 1$ only. Hence, 2 roots

$\Rightarrow \ y^2+4y+1\ge 0$

$\Rightarrow \ [y-(-2-\sqrt{3})][y-(-2+\sqrt{3})]\ge 0$

$\Rightarrow \ y\in(-\infty,-2-\sqrt{3}]\cup[-2+\sqrt{3},\infty)$

For Problems 4–6

4. a., 5. b., 6. d.

Sol. $f(x)+f\left(\dfrac{x-1}{x}\right)=1+x \quad (1)$

In (1) replace x by $\dfrac{x-1}{x}$, we have $f\left(\dfrac{x-1}{x}\right)+f\left(\dfrac{\frac{x-1}{x}-1}{\frac{x-1}{x}}\right)=1+\dfrac{x-1}{x}$

$\Rightarrow \ f\left(\dfrac{x-1}{x}\right)+f\left(\dfrac{1}{1-x}\right)=1+\dfrac{x-1}{x} \quad (2)$

Now from (1) – (2), we have $f(x)-f\left(\dfrac{1}{1-x}\right)=x-\dfrac{x-1}{x} \quad (3)$

In (3) replacing x by $\dfrac{1}{1-x}$, we have

$f\left(\dfrac{1}{1-x}\right)-f\left(\dfrac{x-1}{x}\right)=\dfrac{1}{1-x}-\dfrac{\frac{1}{1-x}-1}{\frac{1}{1-x}}$

or $f\left(\dfrac{1}{1-x}\right)-f\left(\dfrac{x-1}{x}\right)=\dfrac{1}{1-x}-x \quad (4)$

Now from (1) + (3) + (4), we have

$2f(x)=1+x+x-\dfrac{x-1}{x}+\dfrac{1}{1-x}-x$

$\Rightarrow \ f(x)=\dfrac{x^3-x^2-1}{2x(x-1)}$

4. a. $\Rightarrow g(x)=\dfrac{x^3-x^2-1}{x(x-1)}-x+1$

$=\dfrac{-2+2x^2-2x}{2x(x-1)}=\dfrac{x^2-x-1}{x(x-1)}$

Now for $y=\sqrt{g(x)}$, we must have $\dfrac{x^2-x-1}{x(x-1)}\ge 0$

or $\dfrac{\left(x-\frac{1-\sqrt{5}}{2}\right)\left(x-\frac{1+\sqrt{5}}{2}\right)}{x(x-1)}\ge 0$

$\Rightarrow \ x\in\left(-\infty,\dfrac{1-\sqrt{5}}{2}\right]\cup(0,1)\cup\left[\dfrac{1+\sqrt{5}}{2},\infty\right)$

5.b. $y=g(x)=\dfrac{x^2-x-1}{x(x-1)}\Rightarrow(y-1)x^2+(1-y)x+1=0$

Now x is real, $\Rightarrow D\ge 0\Rightarrow(1-y)^2-4(y-1)\ge 0$

$\Rightarrow \ (y-1)(y-5)\ge 0$

$\Rightarrow \ y\in(-\infty,1]\cup[5,\infty)$

6.d. $g(x)=1 \Rightarrow \dfrac{x^2-x-1}{x(x-1)}=1 \Rightarrow -x-1=-x$, which has no solution.

For Problems 7–9

7. d., 8. c., 9. c.

Sol.

Here,

$f(1)+2f(2)+3f(3)+\cdots+nf(n)=n(n+1)f(n)$, for $n\ge 2$ (1)

Replacing n by $n+1$, we get

$f(1)+2f(2)+3f(3)+\cdots+(n+1)f(n+1)=(n+1)(n+2)f(n+1)$ (2)

From (2) – (1), we get

$(n+1)f(n+1)=(n+1)\{(n+2)f(n+1)-nf(n)\}$

$\Rightarrow f(n+1)=(n+2)f(n+1)-nf(n)$

$\Rightarrow nf(n)=(n+2)f(n+1)-f(n+1)$

$\Rightarrow nf(n)=(n+1)f(n+1)$

Putting $n=2, 3, 4, \ldots$, we get

$2f(2)=3f(3)=4f(4)=\ldots=nf(n)$

From (1), $f(1)+2f(2)+3f(3)+\cdots+nf(n)=n(n+1)f(n)$

$\Rightarrow f(1)+(n-1)\times nf(n)=n(n+1)f(n)$

$\Rightarrow f(1)=2nf(n)$

$\Rightarrow f(n)=\dfrac{f(1)}{2n}=\dfrac{1}{2n}$

7. d. $f(1003)=\dfrac{1}{2(1003)}=\dfrac{1}{2006}$

8. c. $f(999)=\dfrac{1}{2(999)}=\dfrac{1}{1998}$

9. c. $f(1), f(2), f(3), \ldots$ are in H.P.

$\dfrac{1}{2}, \dfrac{1}{4}, \dfrac{1}{6}, \dfrac{1}{8}, \ldots$ are in H.P.

Taking reciprocal, 2, 4, 6, 8, … are in A.P.

For Problems 10–11

10. a., 11. b., 12. c.

Sol. $(f(x))^2 f\left(\dfrac{1-x}{1+x}\right)=64x$ (1)

Putting $\dfrac{1-x}{1+x}=y$, and $x=\dfrac{1-y}{1+y}$, we get

$\left\{f\left(\dfrac{1-y}{1+y}\right)\right\}^2 \times f(y)=64\left(\dfrac{1-y}{1+y}\right)$

$\Rightarrow f(x)\times\left\{f\left(\dfrac{1-x}{1+x}\right)\right\}^2=64\left(\dfrac{1-x}{1+x}\right)$ (2)

From $(1)^2$ and (2), we get

$$\frac{f(x)^4\left\{f\left(\frac{1-x}{1+x}\right)\right\}^2}{f(x)\left\{f\left(\frac{1-x}{1+x}\right)\right\}^2}=\frac{(64x)^2}{64\left(\frac{1-x}{1+x}\right)}$$

$\Rightarrow \{f(x)\}^3=64x^2\left(\dfrac{1+x}{1-x}\right)$

$\Rightarrow f(x)=4x^{2/3}\left(\dfrac{1+x}{1-x}\right)^{1/3}$

$x=f(9/7)=-4(9/7)^{2/3}(2)$

For Problems 13–15

13. d., 14. c., 15. c.

Sol. $|g(x)|=|\sin x|, x\in R$

$$f(|g(x)|)=\begin{cases}|\sin x|-1, & -1\le|\sin x|<0\\ (|\sin x|)^2, & 0\le(|\sin x|)^2\le 1\end{cases}$$

$=\sin^2 x, x\in R$

$$f(g(x))=\begin{cases}\sin x-1, & -1\le\sin x<0\\ \sin^2 x, & 0\le\sin x\le 1\end{cases}$$

$$=\begin{cases}\sin x-1, & (2n+1)\pi<x<2n\pi\\ \sin^2 x, & 2n\pi\le x\le(2n+1)\pi\end{cases}, n\in Z.$$

$$\Rightarrow |f(g(x))|=\begin{cases}1-\sin x, & (2n+1)\pi<x<2n\pi\\ \sin^2 x, & 2n\pi\le x\le(2n+1)\pi\end{cases}, n\in Z$$

13. d. Clearly $h_1(x)=f(|g(x)|)=\sin^2 x$ has period π, range [0, 1] and domain R.

14. c. $h_2(x)=|f(g(x))|$ has domain R.

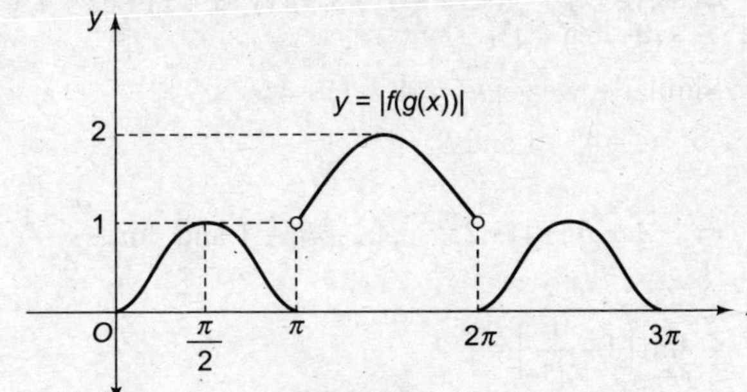

Fig.1.18

Also from the graph, it is periodic with period 2π and has range [0, 2].

15. c. For $h_1(x)\equiv h_2(x)=\sin^2 x$, $x\in[2n\pi,(2n+1)\pi]$, $n\in Z$ and has range [0, 1] for the common domain.

Also, the period is 2π (from the graph).

For Problems 16–18

16. d., 17. d., 18. c.

Sol. Given $a_{n+1} = f(a_n)$

Now $a_1 = f(a_0) = f(x)$

$\Rightarrow a_2 = f(a_1) = f(f(a_0)) = fof(x)$

$\Rightarrow a_n = \underbrace{fofof \ldots f(x)}_{n \text{ times}}$

16. d. $a_1 = f(x) = (a - x^m)^{1/m}$

$\Rightarrow a_2 = f(f(x)) = [a - \{(a - x^m)^{1/m}\}^m]^{1/m} = x$

$\Rightarrow a_3 = f(f(f(x))) = f(x)$

Obviously, the inverse does not exist when m is even and n is odd.

17. d.

Now if $f(x) = \dfrac{1}{1-x} \Rightarrow fof(x) = \dfrac{1}{1 - \dfrac{1}{1-x}} = \dfrac{x-1}{x}$

$\Rightarrow fofof(x) = \dfrac{\dfrac{1}{1-x} - 1}{\dfrac{1}{1-x}} = x$

$\Rightarrow a_n = \underbrace{fofof \ldots of(x)}_{n \text{ times}} = \dfrac{1}{1-x}$ if $n = 3k + 1$

$= \dfrac{x-1}{x}$ if $n = 3k + 2$

$= x$ if $n = 3k$

18. c. Since $a_1 = g(x) = 3 + 4x$.

$a_2 = g\{g(x)\} = g(3 + 4x) = 3 + 4(3 + 4x)$

$= (4^2 - 1) + 4^2 x$

$a_3 = g\{g^2(x)\} = g(15 + 4^2 x) = 3 + 4(15 + 4^2 x) = 63 + 4^3 x$

$= (4^3 - 1) + 4^3 x$

Similarly, we get $a_n = (4^n - 1) + 4^n x$

$\Rightarrow A = 4^n - 1$ and $B = 4^n$

$\Rightarrow A + B + 1 = 2^{2n+1}$, $|A - B| = 1$ and $\lim\limits_{n\to\infty} \dfrac{4^n - 1}{4^n}$

$= \lim\limits_{n\to\infty}\left(1 - \dfrac{1}{4^n}\right) = 1$

For Problems 19–21

19. a., 20. b., 21. a.

Sol. 19. a $f_1(x) = x^2$ and $f_2(x) = |x|$

$\Rightarrow f(x) = f_1(x) - 2f_2(x) = x^2 - 2|x|$

Graph of $f(x)$

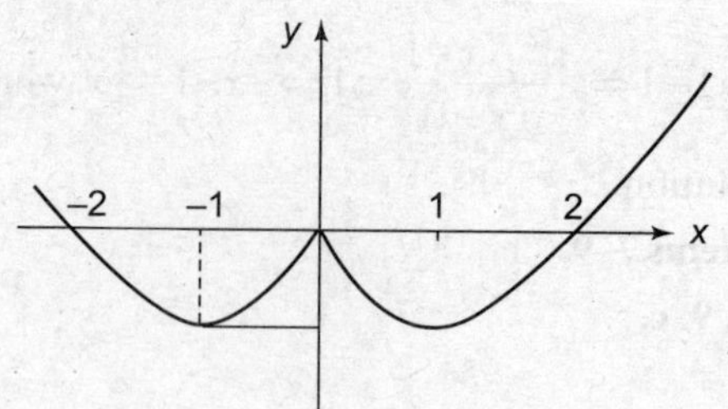

Fig. 1.19

$$g(x) = \begin{cases} f(x), & -3 \le x < -1 \\ -1, & -1 \le x < 0 \\ 0, & 0 \le x \le 2 \\ f(x), & 2 < x \le 3 \end{cases}$$

$$= \begin{cases} x^2 + 2x, & -3 \le x < -1 \\ -1, & -1 \le x < 0 \\ 0, & 0 \le x \le 2 \\ x^2 - 2x, & 2 < x \le 3 \end{cases}$$

The range of $g(x)$ for $[-3, -1]$ is $[-1, 3]$.

20. b. For $x \in (-1, 0)$, $f(x) + g(x) = x^2 + 2x - 1$.

21. a. Obviously, the graph is broken at $x = 0$.

For Problems 22–24

22. a., 23. d., 24. c.

Sol. 22. a.

$g(f(x))$ is not defined if

(i) $-2 + a > 8$ and (ii) $b + 3 > 8$

$a > 10$ and $b > 5$

23. d.

$x \in [-1, 2]$

$\Rightarrow -1 \le x \le 2$

$\Rightarrow -2 \le 2x \le 4$

$\Rightarrow -2 + a \le 2x + a \le 4 + a$

$\Rightarrow -2 + a \le -2$ and $4 + a \le 4$, i.e., $a = 0$

b can take any value.

24. c.

If $a = 2$, $b = 3$

$$f(x) = \begin{cases} 2x + 2 & : \ x \ge -1 \\ 3x^2 + 3 & : \ x < -1 \end{cases}$$

The range of $f(x)$ is $[0, \infty)$.

For Problems 25–27

25. a., 26. a., 27. c.

Sol. $f(2 - x) = f(2 + x)$ (1)

Replace x by $2 - x$, $\Rightarrow f(x) = f(4 - x)$ (2)

Also given $f(20 - x) = f(x)$ (3)

From (1) and (2), $f(4 - x) = f(20 - x)$

Replace x by $4-x$, $\Rightarrow f(x)=f(x+16)$

Hence, the period of $f(x)$ is 16.

25. a.

Given $f(0)=5$.

26. a.

Obvious by definition

27. c. If 1 is a period, then $f(x)=f(x+1), \forall\, x\in R$

$\Rightarrow f(2)=f(3)=f(4)=f(5)=f(6)$

which contradicts the given hypotheses that $f(2)\neq f(6)$

Hence, 1 cannot be period of $f(x)$.

For Problems 28–30

28. c., 29. c., 30. b.

Sol. $g(f(x))=\begin{cases}[f(x)] & -\pi\le f(x)<0\\ \sin f(x), & 0\le f(x)\le\pi\end{cases}$

$=\begin{cases}[[x]], & -\pi\le[x]<0, \quad -2\le x\le -1\\ [|x|+1], & -\pi\le|x|+1<0, \; -1<x\le 2\\ \sin[x], & 0\le[x]\le\pi, \quad -2\le x\le -1\\ \sin(|x|+1), & 0\le|x|+1\le\pi, \; -1<x\le 2\end{cases}$

$=\begin{cases}[x], & -2\le x\le -1\\ \sin(|x|+1), & -1<x\le 2\end{cases}$

Hence, the domain is $[-2, 2]$.

Also for $-2\le x\le -1$, $[x]=-2,-1$

and for $-1<x\le 2$, $|x|+1\in[1,3]$

$\Rightarrow \sin(|x|+1)\in[\sin 3, 1]$

Hence, the range is $\{-2,-1\}\cup[\sin 3, 1]$.

Also for $y\in[\sin 3, 1]$, $[y]=0, 1$

Hence, the number of integral points in the range is 4.

Matrix-Match Type

1. a → s; b → r; c → p; d → q.

$f(\tan x)$ is defined if $0\le\tan x\le 1$

$\Rightarrow \quad x\in\left[n\pi, n\pi+\frac{\pi}{4}\right], n\in I$

$f(\sin x)$ if defined if $0\le\sin x\le 1$

$\Rightarrow \quad x\in[2n\pi, (2n+1)\pi], n\in I$

$f(\cos x)$ is defined if $0\le\cos x\le 1$

$\Rightarrow \quad x\in\left[2n\pi-\frac{\pi}{2}, 2n\pi+\frac{\pi}{2}\right], n\in I$

$f(2\sin x)$ is defined if $0\le 2\sin x\le 1\Rightarrow 0\le\sin x\le 1/2$

$\Rightarrow \left[2n\pi, 2n\pi+\frac{\pi}{6}\right]\cup\left[2n\pi+\frac{5\pi}{6}, (2n+1)\pi\right], n\in I.$

2. a → p; b → q; c → q, s; d → p, s.

a. $f(x)=\{(\operatorname{sgn} x)^{\operatorname{sgn} x}\}^n=\begin{cases}\left[(1)^1\right]^n, & x>0\\ \left[(-1)^{-1}\right]^n, & x<0\end{cases}$

$=\begin{cases}1, & x>0\\ -1, & x<0\end{cases}.$

Hence, $f(x)$ is an odd function.

b. $f(x)=\dfrac{x}{e^x-1}+\dfrac{x}{2}+1$

$\Rightarrow f(-x)=\dfrac{-x}{e^{-x}-1}-\dfrac{x}{2}+1=\dfrac{xe^x}{e^x-1}-\dfrac{x}{2}+1$

$=\dfrac{xe^x-x+x}{e^x-1}-\dfrac{x}{2}+1$

$=x+\dfrac{x}{e^x-1}-\dfrac{x}{2}+1=\dfrac{x}{e^x-1}+\dfrac{x}{2}+1$

$=f(x)$

c. $f(x)=\begin{cases}0, & \text{If } x \text{ is rational}\\ 1, & \text{If } x \text{ is irrational}\end{cases}$

$\Rightarrow \; f(-x)=\begin{cases}0, & \text{If } -x \text{ is rational}\\ 1, & \text{If } -x \text{ is irrational}\end{cases}$

$=\begin{cases}0, & \text{If } x \text{ is rational}\\ 1, & \text{If } x \text{ is irrational}\end{cases}=f(x)$

d. $f(x)=\max\{\tan x, \cot x\}$

$\Rightarrow \quad f(-x)=\max\{\tan(-x), \cot(-x)\}$

$=\max\{-\tan(x), -\cot(x)\}$

$=-\max\{\tan(x), \cot(x)\}$

$=-f(x)$

Hence, $f(x)$ is an odd function.

Also $f(x+\pi)=\max\{\tan(x+\pi), \cot(x+\pi)\}$

$=\max\{\tan x, \cot x\}$

Hence, $f(x)$ is periodic with period π.

3. a → r, s; b → p, q, r, s; c → s. d → p.

a. $\tan^{-1}\left(\dfrac{2x}{1-x^2}\right)\in\left(-\dfrac{\pi}{2}, \dfrac{\pi}{2}\right)$

$\Rightarrow 2\tan^{-1}x\in\left(-\dfrac{\pi}{2}, \dfrac{\pi}{2}\right)$

$\Rightarrow \tan^{-1}x\in\left(-\dfrac{\pi}{4}, \dfrac{\pi}{4}\right) \Rightarrow \tan^{-1}x\in(-1, 1).$

b. $f(x)=\sin^{-1}(\sin x)$ and $g(x)=\sin(\sin^{-1}x)$

$f(x)$ is defined if $\sin x\in[-1, 1]$ which is true for all $x\in R$.

But $g(x)$ is defined for only $x \in [-1, 1]$.
Hence, $f(x)$ and $g(x)$ are identical if $x \in [-1, 1]$.

c. $f(x) = \log_{x^2} 25$ and $g(x) = \log_x 5$
$f(x)$ is defined for all $x \in R - \{0, 1\}$ and $g(x)$ is defined for $(0, \infty) - \{1\}$.
Hence, $f(x)$ and $g(x)$ are identical if $x \in (0, 1) \cup (1, \infty)$.

d. $f(x) = \sec^{-1}x + \text{cosec}^{-1}x$, $g(x) = \sin^{-1}x + \cos^{-1}x$
$f(x)$ has domain $R - (-1, 1)$ and $g(x)$ has domain $[-1, 1]$
Hence, both the functions are identical only if $x = -1, 1$.

4. a → r, s; b → r, s; c → p, q; d → p, s.

a. $f(x) = \cot^{-1}(2x - x^2 - 2)$
$= \cot^{-1}(-1 - (x-1)^2)$
$-1 - (x-1)^2 \le -1$
$\Rightarrow f(0) = f(2)$. Hence $f(x)$ is many-one.

$$\Rightarrow \cot^{-1}(2x - x^2 - 2) \in \left[\frac{3\pi}{4}, \pi\right)$$

Hence, $f(x)$ is onto.
Also $f(3) = f(-1)$, hence function is many-one.
$-1 - (x-1)^2 = -5$.

b.

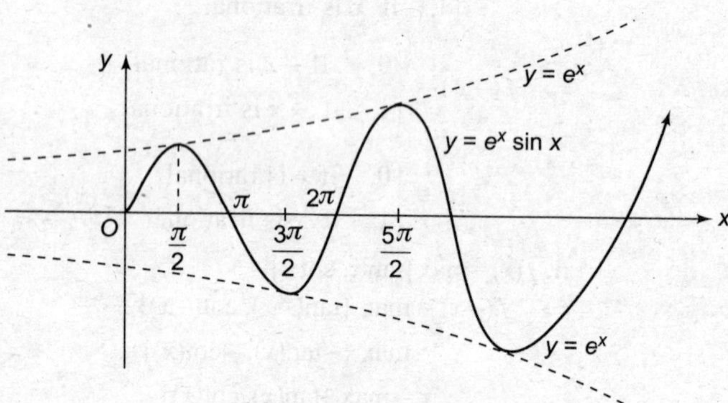

Fig.1.20

Clearly, from the graph that $f(x)$ is many-one and onto

c.

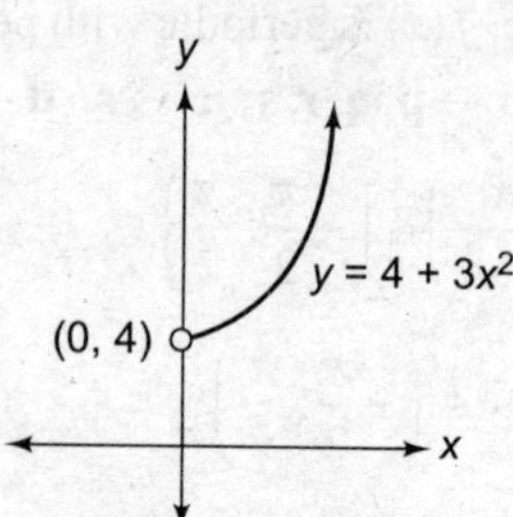

Fig.1.21

d. Let $X = \{x_1, x_2, \ldots, x_n\}$
Let $f(x_1) = x_2$
$\Rightarrow f(f(x_1)) = f(x_2) \Rightarrow x_1$
Thus $f(x)$ is one – one and onto.

5. a → p; b → q, r; c → p; d → q, r.

Since $f(g(x))$ is a one-one function
$\Rightarrow f(g(x_1)) \ne f(g(x_2))$ whenever $g(x_1) = g(x_2)$
$\Rightarrow (g(x_1)) \ne (g(x_2))$ whenever $x_1 \ne x_2$
$\Rightarrow g(x)$ is one-one.
If $f(x)$ is not one-one, then $f(x) = y$ is satisfied by $x = x_1, x_2$
$\Rightarrow f(x_1) = f(x_2) = y$ also if $g(x)$ is onto, then let $g(x_1) = x_1$ and $g(x_2) = x_2$
$\Rightarrow f(g(x_1)) = f(g(x_2))$
$\Rightarrow f(g(x))$ cannot be one-one.

6. a → q; b → q; c → s; d → p.

a. $f(x + \pi/2) = \cos(|\sin(x + \pi/2)| - |\cos(x + \pi/2)|)$
$= \cos(|\cos x| - |-\sin x|)$
$= \cos(|\cos x| - |\sin x|)$
$= \cos(|\sin x| - |\cos x|)$
$= f(x)$

b. $f(x + \pi/2) = \cos[\tan(x + \pi/2) + \cot(x + \pi/2)]$. $\cos[\tan(x + \pi/2) - \cot(x + \pi/2)]$
$= \cos[-\cot x - \tan x] \cos[-\cot x - \tan x]$
$= \cos(\tan x + \cot x) \cos(\tan x - \cot x)$
$= f(x)$

c. The period of $\sin^{-1}(\sin x)$ is 2π. The period of $e^{\tan x}$ is π. Thus, the period of $f(x) = \text{LCM}(2\pi, \pi) = 2\pi$

d. The given function is $f(x) = \sin^3 x \sin 3x$

$$\Rightarrow f(x) = \left(\frac{3\sin x - \sin 3x}{4}\right)\sin 3x$$

$$\Rightarrow f(x) = \frac{3}{8}(\cos 2x - \cos 4x) - \frac{1}{8}(1 - \cos 6x)$$

$\Rightarrow$ The period of $f(x)$ is π.

7. a → s; b → r; c → s; d → p.

a. $f(x) = e^{\cos^4 \pi x + x - [x] + \cos^2 \pi x}$
$\cos^2 \pi x$ and $\cos^4 \pi x$ has period 1
$x - [x] = \{x\}$ has period 1
Then the period of $f(x)$ is 1.

b. $f(x) = \cos 2\pi\{2x\} + \sin 2\pi\{2x\}$
The period $\{2x\}$ is 1/2, then the period of $f(x)$ is 1/2.

c. Clearly, $\tan \pi[x] = 0 \ \forall\ x \in R$ and the period of $\sin 3\pi\{x\}$ is equal to 1.

d. $f(x)=x-[x+a]-b=x+a-[x+a]-(a+b)$
$=\{x+a\}-(a+b)$

Thus, the period of $f(x)$ is 1.

8. a → r; b → s; c → q; d → p.

a. $f(x)=\log_3(5+4x-x^2)$
$=\log_3(9-(x-2)^2)$

Now $-\infty<9-(x-2)^2\le 9$

But for $f(x)$ to get defined, $0<9-(x-2)^2\le 9$

$\Rightarrow -\infty<\log_3(9-(x-2)^2)\le \log_3 9$

$\Rightarrow -\infty<\log_3(9-(x-2)^2)\le 2$

Hence, the range is $(-\infty, 2]$.

b. $f(x)=\log_3(x^2-4x-5)$
$=\log((x-2)^2-9)$

For $f(x)$ to get defined, $0<(x-2)^2-9<\infty$

$\Rightarrow \lim_{x\to 0}\log x<\log_e(x-2)^2-9<\lim_{x\to\infty}\log x$

$\Rightarrow -\infty<f(x)<\infty$

Hence, the range is R.

c. $f(x)=\log_3(x^2-4x+5)$
$=\log_3((x-2)^2+1)$
$(x-2)^2+1\in[1,\infty)$

$\Rightarrow \log_3(x^2-4x+5)\in[0,\infty)$.

d. $(x)=\log_3(4x-5-x^2)$
$=\log_3(-5-(x^2-4x))$
$=\log_3(-1-(x-2)^2)$

Now, $-1-(x-2)^2<0$ for all x

Hence, the function is not defined.

9. a → q. b → s. c → p. d → s.

p. $y=\tan x=\dfrac{1}{x^2}$

From the graph, it is clear that it will have two real roots

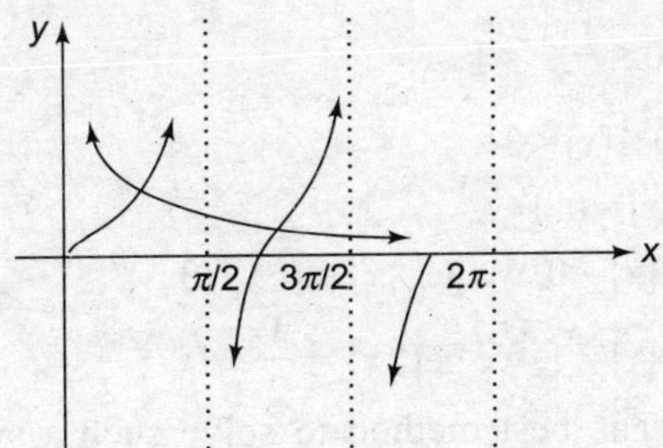

Fig.1.22

q. See the graphs of $y=2^{\cos x}$ and $y=|\sin x|$. Two curves meet at four points for $x\in[0,2\pi]$.

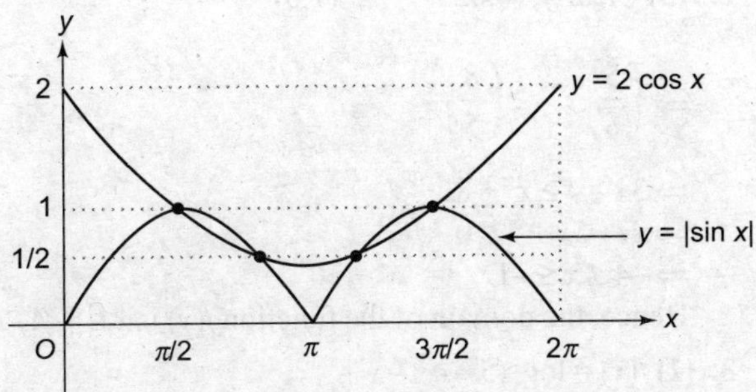

Fig.1.23

So the equation $2^{\cos x}=|\sin x|$ has four solutions.

r. Given that $f(|x|)=0$ has 8 real roots

$\Rightarrow f(x)=0$ has 4 positive roots.

Since $f(x)$ is a polynomial of degree 5, $f(x)$ cannot have even number of real roots.

$\Rightarrow f(x)$ has all the five roots real and one root is negative.

s. $7^{|x|}(|5-|x||)=1$.

$\Rightarrow |5-|x||=7^{-|x|}$

Draw the graph of $y=7^{-|x|}$ and $y=|5-|x||$.

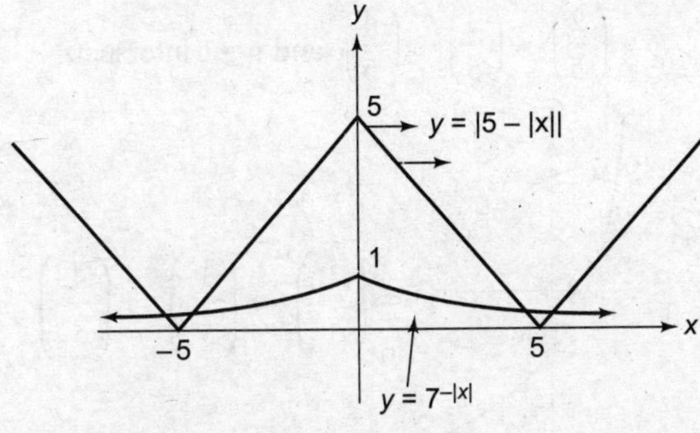

Fig.1.24

From the graph, the number of roots is 4.

Integer Type

1. (1) $y=\dfrac{x^2-x}{1-ax}$

$\Rightarrow x^2-x=y-axy$

$\Rightarrow x^2+x(ay-1)-y=0$

Since x is real,

$(ay-1)^2+4y^3\ge 0$

$\Rightarrow a^2y^2+2y(2-a)+1\ge 0,\ \forall y\in R$

$\Rightarrow a^2>0,\ 4(2-a)^2-4a^2\le 0$

$\Rightarrow 4-4a\le 0$

$\Rightarrow a\in[1,\infty)$

2. **(3)** Clearly, $(0.625)^{4-3x} \geq (1.6)^{x(x+8)}$

$$\Rightarrow \left(\frac{5}{8}\right)^{4-3x} \geq \left(\frac{8}{5}\right)^{x(x+8)}$$

$\Rightarrow 3x - 4 \geq x^2 + 8x$
$\Rightarrow x^2 + 5x + 4 \leq 0$
$\Rightarrow -4 \leq x \leq -1$

Hence, the domain of the function $f(x)$ is $x \in [-4, -1]$

3. **(2)** $f(x) = \log_3 (5 - 4x - x^2)$

$= \log_3 [9 - (x-2)^2]$

Now $f(x)$ is defined if $9 - (x-2)^2 > 0$.

But $9 - (x-2)^2 \leq 9$

$\Rightarrow 0 < 9 - (x-2)^2 \leq 9$

$\Rightarrow -\infty < \log_3 [9 - (x-2)^2] \leq \log_3 9$

Hence, the range is $(-\infty, 2]$.

4. **(1)** $2^x + 3^x + 4^x - 5^x = 0$

$\Rightarrow 2^x + 3^x + 4^x = 5^x$

$$\Rightarrow \left(\frac{2}{5}\right)^x + \left(\frac{3}{5}\right)^x + \left(\frac{4}{5}\right)^x = 1$$

Now the number of solutions of the equation is equal to the number of times

$$y = \left(\frac{2}{5}\right)^x + \left(\frac{3}{5}\right)^x + \left(\frac{4}{5}\right)^x \text{ and } y = 1 \text{ intersect.}$$

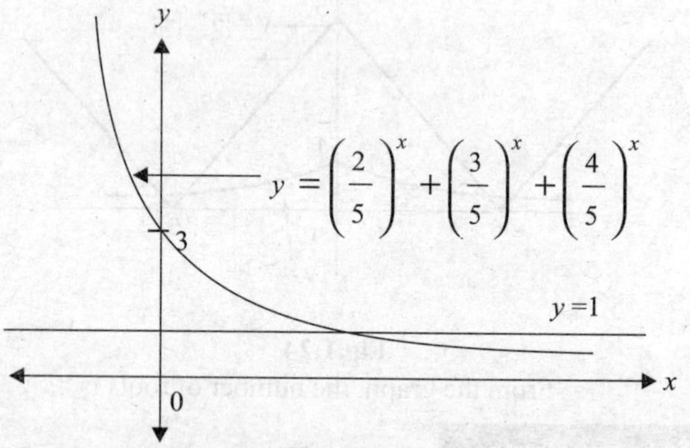

Fig.1.25

From the graph it can be seen that the equation has only one solution.

5. **(6)** We have $f(x) = \sin^{-1}\left[\log_9\left(\frac{x^2}{4}\right)\right]$

Since the domain of $\sin^{-1} x$ is $[-1\ 1]$,

$f(x) = \sin^{-1}\left[\log_9\left(\frac{x^2}{4}\right)\right]$ is defined if

$$-1 \leq \log_9\left(\frac{x^2}{4}\right) \leq 1$$

$$\Rightarrow 9^{-1} \leq \frac{x^2}{4} \leq 9^1$$

$$\Rightarrow \frac{4}{9} \leq x^2 \leq 36$$

$$\Rightarrow \frac{2}{3} \leq |x| \leq 6$$

$$\Rightarrow x \in \left[-6, -\frac{2}{3}\right] \cup \left[\frac{2}{3}, 6\right]$$

$(\because a \leq |x| \leq b \Leftrightarrow x \in [-b, -a] \cup [a, b])$

Hence, the domain of $f(x)$ is $\left[-6, -\frac{2}{3}\right] \cup \left[\frac{2}{3}, 6\right]$.

6. **(9)** We have $f(x) = \log_4 \{\log_5 [\log_3 (18x - x^2 - 77)]\}$

Since $\log_a x$ is defined for all $x > 0$. Therefore, $f(x)$ is defined if

$\log_5 [\log_3 (18x - x^2 - 77)] > 0$ and $18x - x^2 - 77 > 0$

$\Rightarrow \log_3 (18x - x^2 - 77) > 5^0$ and $x^2 - 18x + 77 < 0$

$\Rightarrow \log_3 (18x - x^2 - 77) > 1$ and $(x - 11)(x - 7) < 0$

$\Rightarrow 18x - x^2 - 77 > 3^1$ and $7 < x < 11$

$\Rightarrow 18x - x^2 - 80 > 0$ and $7 < x < 11$

$\Rightarrow x^2 - 18x + 80 < 0$ and $7 < x < 11$

$\Rightarrow (x - 10)(x - 8) < 0$ and $7 < x < 11$

$\Rightarrow 8 < x < 10$ and $7 < x < 11$

$\Rightarrow 8 < x < 10$

$\Rightarrow x \in (8, 10)$

Hence, the domain of $f(x)$ is $(8, 10)$.

7. **(2)** Given $2[x] = x + \{x\}$

$\Rightarrow 2[x] = [x] + 2\{x\}$

$$\Rightarrow \{x\} = \frac{[x]}{2}$$

$$\Rightarrow 0 \leq \frac{[x]}{2} < 1$$

$\Rightarrow 0 \leq [x] < 2$

$\Rightarrow [x] = 0, 1$

For $[x] = 0$, we get $\{x\} = 0\ x = 0$.

For $[x] = 1$, we get $\{x\} = \frac{1}{2} \Rightarrow x = \frac{3}{2}$.

8. **(2)** The best method to solve such a system is a graphical one.

The given equation is $x^2 - 4 = [x]$.

Then the solution of the equation are values of x where the graphs of $y = x^2 - 4$ and $y = [x]$ intersect.

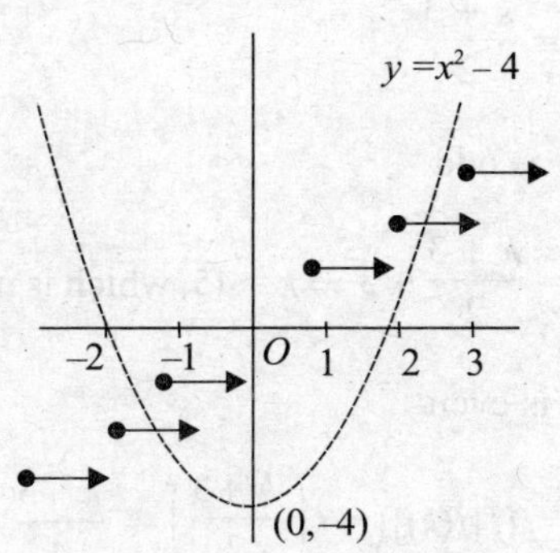

Fig.1.26

From the graph it is seen that the graph intersects when

$x^2 - 4 = 2$ and $x^2 - 4 = -2$

$\Rightarrow x^2 = 6$ or $x^2 = 2$

$\Rightarrow x = \sqrt{6}$ or $-\sqrt{2}$

9. **(6)** Given $f(x) + f(x+3) = 0$ (1)

Replacing x by $x + 3$, we have

$f(x+3) + f(x+6) = 0$ (2)

From (1) and (2), $f(x) = f(x+6)$.

Hence, the function has a period 6.

10. **(1)** $f(x) = x^2 - x + 1$ and $g(x) = (1/2) + \sqrt{x - (3/4)}$ are inverse of each other.

So $f(x) = g(x)$ when $f(x) = x$

$\Rightarrow x^2 - x + 1 = x$

$\Rightarrow x = 1$

11. **(8)** Let $f(x) = \cos(nx)\sin\left(\frac{5x}{n}\right)$

$f(x)$ is periodic

$\Rightarrow f(x + l) = f(x)$, where l is the period.

$$\Rightarrow \cos(nx + n\lambda)\sin\left(\frac{5x + 5\lambda}{n}\right) = \cos nx \sin\left(\frac{5x}{n}\right)$$

At $x = 0$, $\cos(n\lambda)\sin\left(\frac{5\lambda}{n}\right) = 0$

$\cos(n\lambda) = 0$

$\Rightarrow n\lambda = r\pi + \frac{\pi}{2}, r \in I$

$\Rightarrow n(3\pi) = r\pi + \frac{\pi}{2}$ $(\because \lambda = 3\pi)$

$\Rightarrow 3n - r = \frac{1}{2}$, which is impossible.

Again, let $\sin\left(\frac{5\lambda}{n}\right) = 0 \cdot$ Therefore,

$\frac{5\lambda}{n} = p\pi, p \in I$

$\Rightarrow \frac{5(3\pi)}{n} = p\pi$ $(\because \lambda = 3\pi)$

$\Rightarrow n = \frac{15}{p}$

For $p = \pm 1, \pm 3, \pm 5, \pm 15$

$n = \pm 15, \pm 5, \pm 3, \pm 1$ $(\because n \in I)$

Hence, $n = \pm 1, \pm 3, \pm 5, \pm 15$.

12. **(6)** There are exactly six solutions.

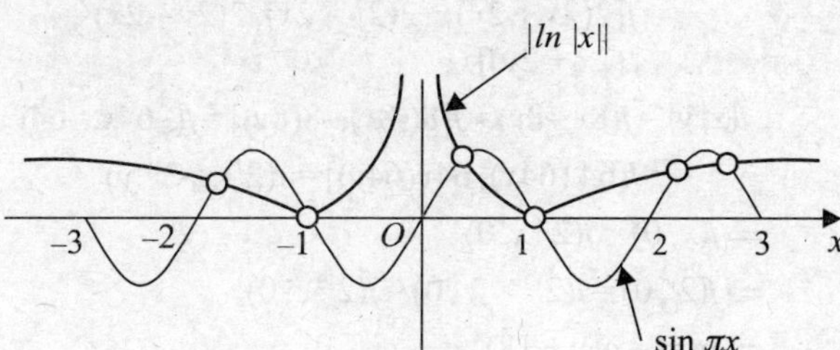

Fig.1.27

13. **(5)** Putting $y = \frac{1}{x}$, we get

$$2 + f(x)\, f\left(\frac{1}{x}\right) = f(x) + f\left(\frac{1}{x}\right) + f(1) \quad (1)$$

Now putting $x = 1$, we get

$2 + [f(1)]^2 = 3f(1)$

$\Rightarrow f(1) = 1$ or 2

But $f(1) \neq 1$; otherwise, from the given relation

$2 + f(x)f(1) = f(x) + f(1) + f(x)$ or $f(x) = 1$, which is not possible as we have $f(2) = 5$.

Hence, $f(1) = 2$.

From (1) we have

$$f(x)\, f\left(\frac{1}{x}\right) = f(x) + f\left(\frac{1}{x}\right)$$

$\Rightarrow f(x) = \pm x^n + 1$

$\Rightarrow f(2) = \pm 2^n + 1 = 5$

$\Rightarrow 2^n = 4 \Rightarrow n = 2$

$\Rightarrow f(x) = x^2 + 1$

$\Rightarrow f[f(1)] = f(2) = 5$

14. **(6)** $f(x-1) + f(x+1) = \sqrt{3} f(x)$ (1)

Putting $x + 2$ for x in relation (1) we get

$f(x+1) + f(x+3) = \sqrt{3}\, f(x+2)$ (2)

From (1) and (2), we get

$f(x-1) + 2f(x+1) + f(x+3) = \sqrt{3}\,[f(x) + f(x+2)]$

$= \sqrt{3}\left(\sqrt{3}\, f(x+1)\right) = 3f(x+1)$

$\Rightarrow f(x-1) + f(x+3) = f(x+1)$ (3)

Putting $x + 2$ for x in relation (3), we get

$f(x+1) + f(x+5) = f(x+3)$ (4)

Adding (3) and (4), we get $f(x-1) = -f(x+5)$

Now putting $x + 1$ for x, we get

$f(x) = -f(x+6)$ (5)

Putting $x+6$ in place of x in (5), we get

$f(x+6) = -f(x+12)$

From (5) again, we get $f(x) = -[-f(x+12)] = f(x+12)$.

Therefore, the period of $f(x)$ is 12.

15. **(3)** $f(x, y) = f(2x+2y, 2y-2x)$
$= f[2(2x+2y) + 2(2y-2x), 2(2y-2x) - 2(2x+2y)]$

$f(x, y) = f(8x, -8y) = f[8(-8x), -8(8y)] = f(-64x, -64y)$
$= f[64(64x), 64y(64y)] = f(2^{12}x, 2^{12}y)$

$\Rightarrow f(x, 0) = f(2^{12}x, 0)$

$\Rightarrow f(2^y, 0) = f(2^{12} \times 2^y, 0) = f(2^{12+y}, 0)$

$\Rightarrow g(y) = g(y+12)$

Hence, $g(x)$ is periodic and its period is 12.

16. **(1)** $f(x) = \sin^2 x + \sin^2\left(x + \frac{\pi}{3}\right) + \cos x \cos\left(x + \frac{\pi}{3}\right)$

$$= \sin^2 x + \frac{1}{4}\left(\sin x + \sqrt{3}\cos x\right)^2 + \frac{1}{2}\cos x\left(\cos x - \sqrt{3}\sin x\right)$$

$$= \frac{5}{4}\left(\sin^2 x + \cos^2 x\right) = \frac{5}{4}$$

Hence, $(gof)(x) = g[f(x)] = g(5/4) = 1$.

17. **(7)** From E to F, we can define $2 \times 2 \times 2 \times 2 = 16$ functions in all (two options for each elements of E), out of which two are into functions, when all the elements of E either map to 1 or to 2.

Therefore, the number of onto functions $= 16 - 2 = 14$.

18. **(1)** Given $f[f(x)] = -x + 1$.

Replacing $x \to f(x)$, we get

$f\{f[f(x)]\} = -f(x) + 1$

$f(1-x) = -f(x) + 1$

$f(x) + f(1-x) = 1$

19. **(7)** Let $2x + y = 3x - y \Rightarrow 2y = x \Rightarrow y = \frac{x}{2}$

Putting $y = \frac{x}{2}$, we get

$$f(x) + f\left(\frac{5x}{2}\right) + \frac{5x^2}{2} = f\left(\frac{5x}{2}\right) + 2x^2 + 1$$

$$\Rightarrow f(x) = 1 - \frac{x^2}{2}$$

20. **(7)** $\left(\frac{3}{4}\right)^{6x+10-x^2} < \frac{27}{64}$

$\Rightarrow 6x + 10 - x^2 > 3$

$\Rightarrow x^2 - 6x - 7 < 0$

$\Rightarrow (x+1)(x-7) < 0$

$\Rightarrow 0, 1, 2, 3, 4, 5, 6.$

21. **(6)** Since $k \in$ odd,

$f(k) = k + 3.$

$f[f(k)] = \frac{k+3}{2}$

$\frac{k+3}{2}$ is odd

$\Rightarrow 27 = \frac{k+3}{2} + 3 \Rightarrow k = 45$, which is not possible.

$\frac{k+3}{2}$ is even

$$\Rightarrow 27 = f\{f[f(k)]\} = f\left(\frac{k+3}{2}\right) = \frac{k+3}{4}$$

$\Rightarrow k = 105$

Verifying, we get

$f\{f[f(105)]\} = f[f(108)] = f(54) = 27.$

Hence, $k = 105$.

22. **(3)** Clearly, the fundamental period is $(4\pi)/3$. Hence, z lies in the third quadrant.

23. $\log_a(x^2 - x + 2) > \log_a(-x^2 + 2x + 3)$

Putting $x = 4/9$, we get

$$\log_a\left(\frac{142}{81}\right) > \log_a\left(\frac{299}{81}\right)$$

Now $\frac{142}{81} < \frac{299}{81}$

$\Rightarrow 0 < a < 1.$

Therefore, $\log_a(x^2 - x + 2) > \log_a(-x^2 + 2x + 3)$ gives $0 < x^2 - x + 2 < -x^2 + 2x + 3$.

Thus $x^2 - x + 2 > 0$ and $2x^2 - 3x - 1 < 0$ are true for all $x \in R$ and

$\frac{3-\sqrt{17}}{4} < x < \frac{3+\sqrt{17}}{4}$ respectively

$\Rightarrow \frac{3-\sqrt{17}}{4} < x < \frac{3+\sqrt{17}}{4}$

24. **(4)** $(2^{2x} - 4 \times 2^x + 4) + 1 + ||b-1| - 3| = |\sin y|$

$\Rightarrow (2^x - 2)^2 + 1 + ||b-1| - 3| = |\sin y|$

$\Rightarrow (2^x - 2)^2 + 1 + ||b-1| - 3| = |\sin y|$

LHS ≥ 1 and RHS ≤ 1. Therefore,

$2^x = 2, |b-1| - 3 = 0$

$\Rightarrow (b-1) = \pm 3$

$\Rightarrow b = 4, -2$

25. **(4)** $f: (0, \infty) \to (0, \infty)$

$f[xf(y)] = x^2 y^a \; (a \in R)$

Putting $x = 1$, we get $f[f(y)] = y^a$.

Putting $f(y) = \frac{1}{x}$, we get $f(1) = \frac{1}{[f(y)]^2} \times y^a$

and putting $y=1$, we get $(f(1))^3=1$. Therefore, $f(1)=1$.
For $y=1$, we have $f\{x[f(1)]\}=x^2$. Therefore, $f(x)=x^2$.
Thus $a=4$.

26. **(3)** $f(3n)=f\{f[f(n)]\}=3f(n),\ \cap\in N$
Putting $n=1$, we get $f(3)=3f(1)$.
If $f(1)=1$, then $f[f(1)]=f(1)=1$.
But $f[f(n)]=3n$
$\Rightarrow f[f(1)]=3$, which gives rise to $1=3$, which is not possible.
Therefore, $f(1)\neq 1$.
$3=f[f(1)]>f(1)>1$
Hence, $f(1)=2$
$f(2)=f[f(1)]=3.$

27. **(4)** $f(x)=|\sin \alpha x|+|\cos \alpha x|$
Period of $f(x)=\dfrac{\pi}{2|\alpha|}=\dfrac{\pi}{8}$ (given)
$\Rightarrow =\pm 4.$

28. $\log_{1/3}\log_7(\sin x+a)>0$
$\Rightarrow 0<\log_7(\sin x+\text{a})<1$
$\Rightarrow 1<(\sin x+a)<7,\quad x\in R$ ['a' should be less than the minimum value of $7-\sin x$ and 'a' must be greater than the maximum value of $1/0 \sin x$]
$\Rightarrow 1-\sin x<a<7-\sin x,\quad x\in R$
Hence, $2<a<6$

29. **(9)** $g(x)\ \dfrac{1}{2}\tan^{-1}|x|+1\Rightarrow \text{sgn}\,[g(x)]=1$

$\sin^{23}x-\cos^{22}x=1$

$\Rightarrow \sin^{23}x=1+\cos^{22}x$, which is possible if $\sin x=1$ and $\cos x=0$.

$\sin x=1, x=2n\pi+\dfrac{\pi}{2}$

Hence, $-10\pi\leq 2n\pi+\dfrac{\pi}{2}\leq 8\pi\Rightarrow -\dfrac{21}{4}\leq n\leq\dfrac{15}{4}$

$\Rightarrow -5\leq n\leq 3$

Hence, the number of values of x is 9.

30. **(2)** $x=1$ or $x=3\Rightarrow 2$ solutions

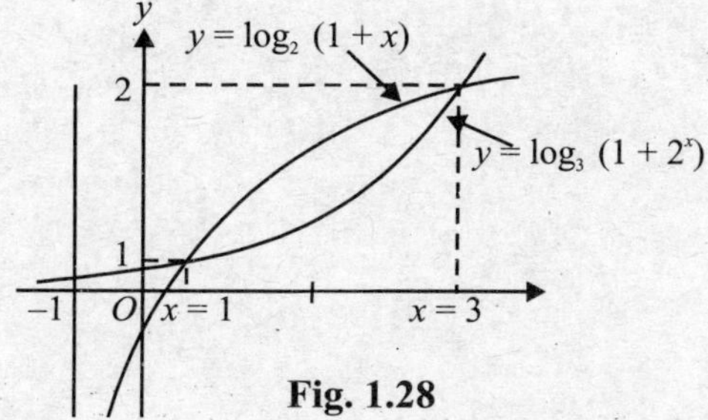

Fig. 1.28

31. **(7)** Obviously f is a linear polynomial
Let $f(x)=ax+b$, hence $f(x^2+x+3)+2f(x^2-3x+5)=6x^2-10x+17$ or $[a(x^2+x+3)+b]+2[a(x^2-3x+5)+b]=6x^2-10x+17$
$\Rightarrow a+2a=6$ (1)
and $a-6a=-10$ (2)
$\Rightarrow a=2$ (comparing coeff. of x^2 and coeff. of x on both sides)
Again, $3a+b+10a+2b=17$
$\Rightarrow b=-3$

32. **(5)** $f(x)=a^{\log_a N}=N\Rightarrow 0<[2\tan \pi x]<1$ or $[2\tan \pi x]>1$

and $\dfrac{x^2+2x-3}{4x^2-4x-3}>0$ i.e. $\dfrac{(x+3)(x-1)}{(2x-3)(2x+1)}>0$

$\Rightarrow x\in(-\infty,-3)\cup(-1/2,1)\cup(3/2,\infty)$ (1)
now $0<[2\tan \pi x]<1$ not possible $\therefore$ $[2\tan \pi x]>1$
$\Rightarrow 2\tan \pi x\geq 2\Rightarrow \tan \pi x\geq 1$

$\Rightarrow n\pi+\dfrac{\pi}{4}\leq \pi x<n\pi+\dfrac{\pi}{2}, n\in I$

$\Rightarrow n+\dfrac{1}{4}\leq x<n+\dfrac{1}{2}$ (2)

Common solution of (1) and (2) possible only if $n=0$, $n\geq 2$ or $n\leq -4$

33. **(4)** $f(x)+f\left(\dfrac{1}{x}\right)=\dfrac{x^3+1}{x}$

Replace x by $\dfrac{1}{x}$, $f\left(\dfrac{1}{x}\right)+f(x)=\dfrac{1}{x^2}+x$

$\Rightarrow x^2+\dfrac{1}{x}=\dfrac{1}{x^2}+x$

$\Rightarrow x-\dfrac{1}{x}=x^2-\dfrac{1}{x^2}$

$\Rightarrow \left(x-\dfrac{1}{x}\right)\left(x+\dfrac{1}{x}-1\right)=0$

$\Rightarrow x=\pm 1$

34. **(4)** Put $x=1$ and $y=1$
$f^2(1)-f(1)-6=0\quad\Rightarrow f(1)=3$ or $f(1)=-2$
Now put $y=1$

$f(x).f(1)=f(x)+2\left(\dfrac{2x+1}{x}\right)$

$\Rightarrow f(x)[f(1)-1]=\dfrac{2(2x+1)}{x}\Rightarrow f(x)=\dfrac{2(2x+1)}{x[f(1)-1]}$

Now for $f(1)=3$ or -2, we get two functions.

35. **(4)** Put $m=n=2\Rightarrow f(2)f(2)=f(2)+3f(2)\Rightarrow f(2)=4$
$\Rightarrow f(f(2))=f(4)$

Now $f(4)f(4) = f(4) + 3f(4) \Rightarrow f(4) = 4$

Thus $f(f(f(2)) = 4$ and so on.

Hence $\underbrace{f(f(f(f...(2))))}_{10 \text{ times}} = 4$

36. **(3)** $\log_{10}(\log_3 10) = \log_{10}\left(\dfrac{1}{\log_{10} 3}\right) = -\log_{10}(\log_{10} 3)$

Given $f(-\log_{10}(\log_{10} 3)) = 5$ (1)

now $f(x) = a\sin x + bx^{1/3} + 4$

$f(-x) = -a\sin x - bx^{1/3} + 4$

$\Rightarrow f(x) + f(-x) = 8$

$\Rightarrow f(\log_{10}(\log_{10} 3)) + f(-\log_{10}(\log_{10} 3)) = 8$

$\Rightarrow f(\log_{10}(\log_{10} 3)) + 5 = 8$

$\Rightarrow f(\log_{10}(\log_{10} 3) = 3$

37. **(0)** Putting $y = \dfrac{1}{x}$ and $x = 0$, we get $f(x) + f(-x) = 0$ (1)

Putting $x = 1, y = 1$, we get $f(2) = 2f(1)^2$ (2)

Putting $x = -1, y = -1$, we get

$f(-2) = 2f(-1) . f(-1)$

$\Rightarrow -f(2) = 2f(1)^2$ [from (1)] (3)

From (2) and (3), $f(2) = 0$ and $f(1) = 0$

$\therefore f(1) = f(2) = 0$

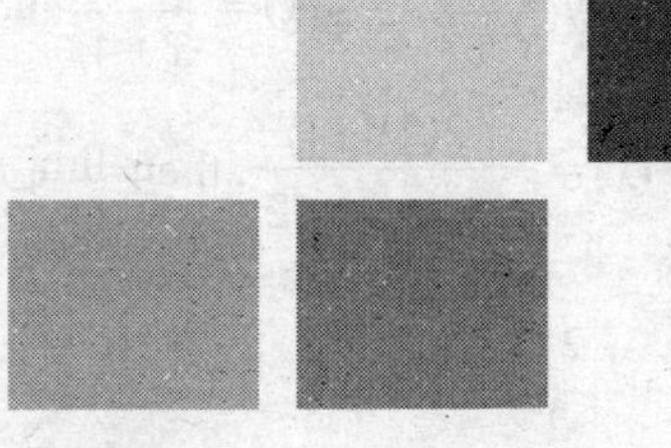

CHAPTER 2

Limits

Objective Type

Each question has four choices a, b, c, and d, out of which *only one* is correct.

1. If $f(x) = 0$ be a quadratic equation such that $f(-\pi) = f(\pi) = 0$ and $f\left(\frac{\pi}{2}\right) = -\frac{3\pi^2}{4}$, then $\lim_{x \to -\pi} \frac{f(x)}{\sin(\sin x)}$ is equal to

a. 0 **b.** π

c. 2π **d.** None of these

2. If $f(x) = \frac{\cos}{(1-\sin x)^{1/3}}$, then

a. $\lim_{x \to \pi/2^-} f(x) = -\infty$ **b.** $\lim_{x \to \pi^{+2}/2} f(x) = \infty$

c. $\lim_{x \to \pi/2} f(x) = \infty$ **d.** None of these

3. $\lim_{x \to \infty} \frac{x^2 \tan 1/x}{\sqrt{8x^2 + 7x + 1}}$ is equal to

a. $-\frac{1}{2\sqrt{2}}$ **b.** $\frac{1}{2\sqrt{2}}$

c. $\frac{1}{\sqrt{2}}$ **d.** Does not exist

4. $\lim_{x \to 0} \left[\frac{\sin(\text{sgn}(x))}{(\text{sgn}(x))}\right]$, where [.] denotes the greatest integer function, is equal to

a. 0 **b.** 1

c. -1 **d.** Does not exist

5. $\lim_{x \to \infty} \frac{2 + 2x + \sin 2x}{(2x + \sin 2x)e^{\sin x}}$ is equal to

a. 0 **b.** 1

c. -1 **d.** Does not exists

6. Let $\lim_{x \to 0} \frac{[x]^2}{x^2} = l$ and $\lim_{x \to 0} \frac{[x^2]}{x^2} = m$, where [ ] denotes greatest integer, then

a. l exists but m does not **b.** m exists but l does not

c. l and m both exist **d.** neither l nor m exists

7. $\text{Lim}_{x \to 1} \frac{x \sin(x - [x])}{x - 1}$, where [.] denotes the greatest integer function, is equal to

a. 0 **b.** -1

c. not existent **d.** None of these

8. $\lim_{x \to 0} \left(\frac{1 + \tan x}{1 + \sin x}\right)^{\text{cosec } x}$ is equal to

a. e **b.** $\frac{1}{e}$

c. 1 **d.** None of these

9. $\lim_{x \to \infty} \frac{\sin^4 x - \sin^2 x + 1}{\cos^4 x - \cos^2 x + 1}$ is equal to

a. 0 **b.** 1

c. $\frac{1}{3}$ **d.** $\frac{1}{2}$

10. $\lim_{x \to \infty} \left(\frac{x^3}{3x^2 - 4} - \frac{x^2}{3x + 2}\right)$ is equal to

a. Does not exist **b.** 1/3

c. 0 **d.** 2/9

11. If $f(x) = \dfrac{2}{x-3}$, $g(x) = \dfrac{x-3}{x+4}$ and $h(x) = -\dfrac{2(2x+1)}{x^2+x-12}$, then $\lim\limits_{x\to 3}[f(x)+g(x)+h(x)]$ is

a. -2
b. -1
c. $-\dfrac{2}{7}$
d. 0

12. $\lim\limits_{x\to 0}\dfrac{x(e^x-1)}{1-\cos x}$ is equal to

a. 0
b. ∞
c. -2
d. 2

13. $\lim\limits_{n\to\infty}\dfrac{n(2n+1)^2}{(n+2)(n^2+3n-1)}$ is equal to

a. 0
b. 2
c. 4
d. ∞

14. The value of $\lim\limits_{x\to\pi}\dfrac{1+\cos^3 x}{\sin^2 x}$ is

a. 1/3
b. 2/3
c. –1/4
d. 3/2

15. The value of $\lim\limits_{x\to\infty}\left(\sqrt{x^2+x+1}-\sqrt{x^2+1}\right)$ is

a. 2
b. –1/2
c. 1/2
d. 0

16. The value of $\lim\limits_{x\to 2}\dfrac{\sqrt{1+\sqrt{2+x}}-\sqrt{3}}{x-2}$ is

a. $\dfrac{1}{8\sqrt{3}}$
b. $\dfrac{1}{4\sqrt{3}}$
c. 0
d. None of these

17. $\lim\limits_{x\to\infty}\dfrac{(2x+1)^{40}(4x-1)^5}{(2x+3)^{45}}$ is equal to

a. 16
b. 24
c. 32
d. 8

18. $\lim\limits_{x\to\infty}\left[\sqrt{x+\sqrt{x+\sqrt{x}}}-\sqrt{x}\right]$ is equal to

a. 0
b. $\dfrac{1}{2}$
c. $\log 2$
d. e^4

19. $\lim\limits_{x\to\infty}\dfrac{(x+1)^{10}+(x+2)^{10}+\cdots+(x+100)^{10}}{x^{10}+10^{10}}$ is equal to

a. 0
b. 1
c. 10
d. 100

20. $\lim\limits_{x\to 0}\dfrac{x^a\sin^b x}{\sin(x^c)}$, where $a, b, c \in R\sim\{0\}$, exists and has non-zero value, then

a. $a+c=b$
b. $b+c=a$
c. $a+b=c$
d. None of these

21. $\lim\limits_{x\to\pi/2}\left[x\tan x-\left(\dfrac{\pi}{2}\right)\sec x\right]$ is equal to

a. 1
b. -1
c. 0
d. None of these

22. If $\lim\limits_{x\to\infty}\left(\dfrac{x^3+1}{x^2+1}-(ax+b)\right)=2$, then

a. $a=1, b=1$
b. $a=1, b=2$
c. $a=1, b=-2$
d. None of these

23. The value of $\lim\limits_{x\to 1}(2-x)^{\tan(\pi x)/2}$ is

a. $e^{-2/\pi}$
b. $e^{1/\pi}$
c. $e^{2/\pi}$
d. $e^{-1/\pi}$

24. $\lim\limits_{x\to 0}\dfrac{\sin x^n}{(\sin x)^m}$, $(m<n)$ is equal to

a. 1
b. 0
c. n/m
d. None of these

25. $\lim\limits_{x\to 0}\dfrac{x^4(\cot^4 x-\cot^2 x+1)}{(\tan^4 x-\tan^2 x+1)}$ is equal to

a. 1
b. 0
c. 2
d. None of these

26. $\lim\limits_{x\to\infty}\left(\dfrac{1}{e}-\dfrac{x}{1+x}\right)^x$ is equal to

a. $\dfrac{e}{1-e}$
b. 0
c. $e^{e/1-e}$
d. Does not exist

27. $\lim\limits_{x\to 1}\dfrac{1-x^2}{\sin 2\pi x}$ is equal to

a. $\dfrac{1}{2\pi}$
b. $\dfrac{-1}{\pi}$
c. $\dfrac{-2}{\pi}$
d. None of these

28. $\lim_{x\to 0}\frac{1}{x}\cos^{-1}\left(\frac{1-x^2}{1+x^2}\right)$ is equal to

a. 1 b. 0
c. 2 d. None of these

29. $\lim_{x\to\infty}\left(\frac{x^2+2x-1}{2x^2-3x-2}\right)^{\frac{2x+1}{2x-1}}$ is equal to

a. 0 b. ∞
c. 1/2 d. None of these

30. $\lim_{x\to\infty}\frac{2\sqrt{x}+3\sqrt[3]{x}+4\sqrt[4]{x}+\cdots+n\sqrt[n]{x}}{\sqrt{(2x-3)}+\sqrt[3]{(2x-3)}+\cdots+\sqrt[n]{(2x-3)}}$ is equal to

a. 1 b. ∞
c. $\sqrt{2}$ d. None of these

31. $\lim_{y\to 0}\frac{(x+y)\sec(x+y)-x\sec x}{y}$ is equal to

a. $\sec x\,(x\tan x+1)$ b. $x\tan x+\sec x$
c. $x\sec x+\tan x$ d. None of these

32. The value of $\lim_{m\to\infty}\left(\cos\frac{x}{m}\right)^m$ is

a. 1 b. e
c. e^{-1} d. None of these

33. $\lim_{x\to 1}\left[\operatorname{cosec}\frac{\pi x}{2}\right]^{1/1-x}$ (where [.] represents the greatest integer function) is equal to

a. 0 b. 1
c. ∞ d. Does not exist

34. $\lim_{n\to\infty}\left(\frac{n^2-n+1}{n^2-n-1}\right)^{n(n-1)}$ is equal to

a. e b. e^2
c. e^{-1} d. 1

35. If $f(x)=\lim_{n\to\infty}n(x^{1/n}-1)$, then for $x>0, y>0, f(xy)$ is equal to

a. $f(x)f(y)$ b. $f(x)+f(y)$
c. $f(x)-f(y)$ d. None of these

36. If $\lim_{x\to a}\left(\frac{f(x)}{g(x)}\right)$ exists, then

a. both $\lim_{x\to a}f(x)$ and $\lim_{x\to a}g(x)$ must exist
b. $\lim_{x\to a}f(x)$ need not exist but $\lim_{x\to a}g(x)$ exists
c. neither $\lim_{x\to a}f(x)$ nor $\lim_{x\to a}g(x)$ may exist
d. $\lim_{x\to a}f(x)$ exists but $\lim_{x\to a}g(x)$ need not exist

37. If $\lim_{n\to\infty}\frac{n\times 3^n}{n(x-2)^n+n\times 3^{n+1}-3^n}=\frac{1}{3}$, then the range of x is (where $n\in N$)

a. [2, 5) b. (1, 5)
c. (–1, 5) d. $(-\infty,\infty)$

38. The value of $\lim_{x\to 2}\frac{2^x+2^{3-x}-6}{\sqrt{2^{-x}}-2^{1-x}}$ is

a. 16 b. 8
c. 4 d. 2

39. $\lim_{n\to\infty}\left(\left(\frac{n}{n+1}\right)^{\alpha}+\sin\frac{1}{n}\right)^n$ (when $\alpha\in Q$) is equal to

a. $e^{-\alpha}$ b. $-\alpha$
c. $e^{1-\alpha}$ d. $e^{1+\alpha}$

40. $f(x)=\frac{\ln(x^2+e^x)}{\ln(x^4+e^{2x})}$. Then $\lim_{n\to\infty}f(x)$ is equal to

a. 1 b. 1/2
c. 2 d. None of these

41. $\lim_{x\to 1}\frac{1+\sin\pi\left(\frac{3x}{1+x^2}\right)}{1+\cos\pi x}$ is equal to

a. 0 b. 1
c. 2 d. 4

42. $\lim_{n\to\infty}\sum_{x=1}^{20}\cos^{2n}(x-10)$ is equal to

a. 0 b. 1
c. 19 d. 20

43. The value of $\lim_{x\to\infty}\frac{(2^{x^n})^{1/e^x}-(3^{x^n})^{1/e^x}}{x^n}$ (where $n\in N$) is

a. $\log n\left(\frac{2}{3}\right)$ b. 0
c. $n\log n\left(\frac{2}{3}\right)$ d. Not defined

44. If $f: (1, 2) \to R$ satisfies the inequality

$$\frac{\cos(2x-4)-33}{2} < f(x) < \frac{x^2|4x-8|}{x-2},\ \forall\ x \in (1, 2).$$

Then $\lim_{x\to 2^-} f(x)$ is

a. 16
b. Cannot be determined from the given information
c. -16
d. Does not exist

45. Let $f(x) = \lim_{n\to\infty} \frac{1}{\left(\frac{3}{\pi}\tan^{-1} 2x\right)^{2n} + 5}$. Then the set of values of x for which $f(x) = 0$ is

a. $|2x| > \sqrt{3}$ b. $|2x| < \sqrt{3}$
c. $|2x| \geq \sqrt{3}$ d. $|2x| \leq \sqrt{3}$

46. $\lim_{x\to 0}\left\{(1+x)^{2/x}\right\}$ (where $\{\cdot\}$ denotes the fractional part of x) is equal to

a. $e^2 - 7$ b. $e^2 - 8$
c. $e^2 - 6$ d. None of these

47. $\lim_{x\to 0} \frac{\sin(x^2)}{\ln(\cos(2x^2 - x))}$ is equal to

a. 2 b. -2
c. 1 d. -1

48. $\lim_{x\to -1} \frac{1}{\sqrt{|x|-\{-x\}}}$ (where $\{x\}$ denotes the fractional part of x) is equal to

a. Does not exist b. 1
c. ∞ d. $\frac{1}{2}$

49. If $f(x) = \begin{cases} x^n \sin(1/x^2), & x \neq 0 \\ 0, & x = 0 \end{cases}$, $(n \in I)$, then

a. $\lim_{x\to 0} f(x)$ exists for $n > 1$

b. $\lim_{x\to 0} f(x)$ exists for $n < 0$

c. $\lim_{x\to 0} f(x)$ does not exist for any value of n

d. $\lim_{x\to 0} f(x)$ cannot be determined

50. $\lim_{x\to 0} |x|^{[\cos x]}$ (where [.] denotes the greatest integer function) is equal to

a. 0 b. 1
c. Does not exist d. e

51. If $f(x) = e^x$, then $\lim_{x\to 0}(f(x))^{1/\{f(x)\}}$ (where { } denotes the fractional part of x) is equal to

a. $f(1)$ b. $f(0)$
c. $f(-\infty)$ d. Does not exist

52. The value of

$$\lim_{x\to 2}\left(\left(\frac{x^3-4x}{x^3-8}\right)^{-1} - \left(\frac{x+\sqrt{2x}}{x-2} - \frac{\sqrt{2}}{\sqrt{x}-\sqrt{2}}\right)^{-1}\right)$$ is

a. 1/2 b. 2
c. 1 d. None of these

53. $\lim_{x\to\infty} \frac{e^{1/x^2}-1}{2\tan^{-1}(x^2)-\pi}$ is equal to

a. 1 b. -1
c. $\frac{1}{2}$ d. $-\frac{1}{2}$

54. The value of $\lim_{x\to 0} \frac{1+\sin x - \cos x + \log(1-x)}{x^3}$ is

a. $\frac{1}{2}$ b. $-\frac{1}{2}$
c. 0 d. None of these

55. $\lim_{x\to 0} \frac{\cos(\tan x) - \cos x}{x^4}$ is equal to

a. 1/6 b. 1/3
c. 1/2 d. 1

56. If $x_1 = 3$ and $x_{n+1} = \sqrt{2 + x_n}$, $n \geq 1$, then $\lim_{n\to\infty} x_n$ is

a. -1 b. 2
c. $\sqrt{5}$ d. 3

57. $\lim_{x\to 0}\left(\frac{1^x + 2^x + 3^x + \cdots + n^x}{n}\right)^{1/x}$ is equal to

a. $(n!)^n$ b. $(n!)^{1/n}$
c. $n!$ d. $ln\,(n!)$

58. The value of the limit $\lim_{x\to 0} \frac{a^{\sqrt{x}} - a^{1/\sqrt{x}}}{a^{\sqrt{x}} + a^{1/\sqrt{x}}}$, $a > 1$ is

a. 4 b. 2
c. -1 d. 0

59. Among (i) $\lim_{x\to\infty} \sec^{-1}\left(\frac{x}{\sin x}\right)$ and

(ii) $\lim_{x\to\infty} \sec^{-1}\left(\frac{\sin x}{x}\right)$

a. (i) exists, (ii) does not exist
b. (i) does not exist, (ii) exists
c. both (i) and (ii) exist
d. neither (i) nor (ii) exists

60. If $\lim_{x\to 0} \frac{x^n - \sin x^n}{x - \sin^n x}$ is non-zero finite, then n must be equal

a. 4 b. 1
c. 2 d. 3

61. If $\lim_{x\to -2^-} \frac{ae^{1/|x+2|}-1}{2-e^{1/|x+2|}} = \lim_{x\to 2^+} \sin\left(\frac{x^4-16}{x^5+32}\right)$, then a is

a. $\sin\frac{3}{5}$ b. 2
c. $\sin\frac{2}{5}$ d. $\sin\frac{1}{5}$

62. $\lim_{x\to\infty}((x+5)\tan^{-1}(x+5)-(x+1)\tan^{-1}(x+1))$ is equal to

a. π b. 2π
c. $\pi/2$ d. None of these

63. $\lim_{x\to 1} \frac{(1-x)(1-x^2)\cdots(1-x^{2n})}{\{(1-x)(1-x^2)\cdots(1-x^n)\}^2}$, $n \in N$

a. ${}^{2n}P_n$ b. ${}^{2n}C_n$
c. $(2n)!$ d. None of these

64. The value of $\lim_{x\to 0}\left(\left[\frac{100x}{\sin x}\right]+\left[\frac{99\sin x}{x}\right]\right)$ (where [.] represents the greatest integral function) is

a. 199 b. 198
c. 0 d. None of these

65. The value of $\lim_{x\to 1/\sqrt{2}} \frac{x-\cos\left(\sin^{-1}x\right)}{1-\tan\left(\sin^{-1}x\right)}$ is

a. $-\frac{1}{\sqrt{2}}$ b. $\frac{1}{\sqrt{2}}$
c. $\sqrt{2}$ d. $-\sqrt{2}$

66. The value of $\lim_{x\to 1^-} \frac{1-\sqrt{x}}{\left(\cos^{-1}x\right)^2}$ is

a. 4 b. 1/2
c. 2 d. 1/4

67. $\lim_{x\to 0}\left[\min(y^2-4y+11)\frac{\sin x}{x}\right]$ (where [.] denotes the greatest integer function) is

a. 5 b. 6
c. 7 d. Does not exist

68. $\lim_{x\to \pi/2} \frac{\sin(x\cos x)}{\cos(x\sin x)}$ is equal to

a. 0 b. $\pi/2$
c. π d. 2π

69. If $\lim_{x\to 0}(x^{-3}\sin 3x + ax^{-2} + b)$ exists and is equal to 0, then

a. $a=-3$ and $b=9/2$ b. $a=3$ and $b=9/2$
c. $a=-3$ and $b=-9/2$ d. $a=3$ and $b=-9/2$

70. If $\lim_{x\to 0} \frac{x^n \sin^n x}{x^n - \sin^n x}$ is non-zero finite, then n is equal to

a. 1 b. 2
c. 3 d. None of these

71. $\lim_{x\to\infty} \frac{(1+x+x^2)}{x(\ln x)^3}$ is equal to

a. 2 b. e^2
c. e^{-2} d. None of these

Multiple Correct Answers Type

Each question has four choices a, b, c, and d, out of which *one or more* answers are correct.

1. Let $f(x) = \begin{cases} 1+\frac{2x}{a}, & 0\le x<1 \\ ax, & 1\le x<2 \end{cases}$. If $\lim_{x\to 1} f(x)$ exists, then a is

a. 1 b. –1
c. 2 d. –2

2. If $f(x) = |x-1| - [x]$, where $[x]$ = the greatest integer less than or equal to x, then

a. $f(1+0)=-1, f(1-0)=0$
b. $f(1+0)=0=f(1-0)$
c. $\lim_{x\to 1} f(x)$ exists
d. $\lim_{x\to 1} f(x)$ does not exist

3. If $\lim_{n\to\infty}\left(an-\frac{1+n^2}{1+n}\right)=b$, where a is a finite number, then

a. $a=1$ b. $a=0$
c. $b=1$ d. $b=-1$

4. If $m, n \in N$, $\lim_{x\to 0}\frac{\sin x^n}{(\sin x)^m}$ is

a. 1, if $n=m$ b. 0, if $n>m$
c. ∞, if $n<m$ d. n/m, if $n<m$

5. Which of the following is true ({.} denotes the fractional part of the function)?

a. $\lim_{x\to\infty}\frac{\log_e x}{\{x\}}=\infty$ b. $\lim_{x\to 2^+}\frac{x}{x^2-x-2}=\infty$
c. $\lim_{x\to -1^-}\frac{x}{x^2-x-2}=-\infty$ d. $\lim_{x\to\infty}\frac{\log_{0.5} x}{\{x\}}=\infty$

6. If $\lim_{x\to 1}(2-x+a[x-1]+b[1+x])$ exists, then a and b can take the values (where $[\cdot]$ denotes the greatest integer function)

a. $a=1/3, b=1$ b. $a=1, b=-1$
c. $a=9, b=-9$ d. $a=2, b=2/3$

7. $L=\lim_{x\to a}\frac{|2\sin x-1|}{2\sin x-1}$, then

a. limit does not exist when $a=\pi/6$
b. $L=-1$ when $a=\pi$
c. $L=1$ when $a=\pi/2$
d. $L=1$ when $a=0$

8. $f(x)=\lim_{n\to\infty}\frac{x}{x^{2n}+1}$, then

a. $f(1^+)+f(1^-)=0$ b. $f(1^+)+f(1^-)+f(1)=3/2$
c. $f(-1^+)+f(-1^-)=-1$ d. $f(1^+)+f(-1^-)=0$

9. $\lim_{n\to\infty}\frac{-3n+(-1)^n}{4n-(-1)^n}$ is equal to

a. $-\frac{3}{4}$ b. 0 if n is even
c. $-\frac{3}{4}$ if n is odd d. None of these

10. Given a real-valued function f such that

$$f(x)=\begin{cases}\frac{\tan^2\{x\}}{(x^2-[x]^2)} & \text{for } x>0\\ 1 & \text{for } x=0\\ \sqrt{\{x\}\cot\{x\}} & \text{for } x<0\end{cases}$$

where $[x]$ is the integral part and $\{x\}$ is the fractional part of x, then

a. $\lim_{x\to 0^+} f(x)=1$

b. $\lim_{x\to 0^-} f(x)=\cot 1$

c. $\cot^{-1}\left(\lim_{x\to 0^-} f(x)\right)^2=1$

d. $\tan^{-1}\left(\lim_{x\to 0^+} f(x)\right)=\frac{\pi}{4}$

Reasoning Type

Each question has four choices a, b, c, and d, out of which *only one* is correct. Each question contains Statement 1 and Statement 2.

a. if both the statements are True and Statement 2 is the correct explanation of Statement 1.
b. if both the statements are True but Statement 2 is Not the correct explanation of Statement 1.
c. if Statement 1 is True and Statement 2 is False.
d. if Statement 1 is False and Statement 2 is True.

1. **Statement 1:** $\left[\lim_{x\to 0}\frac{\sin x}{x}\right]=0$

Statement 2:
For $x\in(-\delta,\delta)$, where δ is positive and $\delta\to 0$, $\tan x>x$.

2. **Statement 1:** $\lim_{x\to\alpha}\frac{\sin(f(x))}{x-\alpha}$, where $f(x)=ax^2+bx+c$, is finite and non-zero, then $\lim_{x\to\alpha}\frac{e^{1/f(x)}-1}{e^{1/f(x)}+1}$ does not exist.

Statement 2: $\lim_{x\to\alpha}\frac{\sin(f(x))}{x-\alpha}$ can take finite value only when it takes $\frac{0}{0}$ form.

3. **Statement 1:** $\lim_{x\to 0}\sin^{-1}\{x\}$ does not exist (where {.} denotes fractional part function).
Statement 2: $\{x\}$ is discontinuous at $x=0$.

4. **Statement 1:** If a and b are positive and $[x]$ denotes greatest integer $\le x$, then $\lim_{x\to 0^+}\frac{x}{a}\left[\frac{b}{x}\right]=\frac{b}{a}$.

Statement 2: $\lim_{x\to\infty}\frac{\{x\}}{x}\to 0$, where $\{x\}$ denotes fractional part of x.

5. **Statement 1:** $\lim_{x\to\infty}\left(\frac{1^2}{x^3}+\frac{2^2}{x^3}+\frac{3^2}{x^3}+\dots+\frac{x^2}{x^3}\right)$

$= \lim_{x\to\infty} \frac{1^2}{x^3} + \lim_{x\to\infty} \frac{2^2}{x^3} + \dots + \lim_{x\to\infty} \frac{x^2}{x^3} = 0$

Statement 2: $\lim_{x\to a} (f_1(x) + f_2(x) + \dots + f_n(x))$

$= \lim_{x\to a} f_1(x) + \lim_{x\to a} f_2(x) + \dots + \lim_{x\to a} f_n(x)$, where $n \in N$.

6. Statement 1: $\lim_{x\to 0} \frac{\sqrt{1-\cos 2x}}{x}$ does not exist.

Statement 2: $f(x) = \frac{\sqrt{1-\cos 2x}}{x}$ is not defined at $x = 0$.

7. Statement 1: $\lim_{m\to\infty} \lim_{n\to\infty} \{\sin^{2m}(n!\pi x)\} = 0, m, n \in N$, when x is rational.

Statement 2: when $n \to \infty$ and x is rational, $n!\,x$ is integer.

8. Statement 1:

If $f(x) = \begin{cases} x, & \text{if } x \text{ is rational} \\ 1-x, & \text{if } x \text{ is irrational} \end{cases}$, then $\lim_{x\to 1/2} f(x)$ does not exist.

Statement 2 : $x \to 1/2$ can be rational or irrational value.

9. Statement 1: If $f(x) = \frac{(x-1)(x-2)}{(x-3)(x-4)}$, then $\lim_{x\to-\infty} \sin^{-1} f(x)$ exists, but $\lim_{x\to\infty} \cos^{-1} f(x)$ does not exist.

Statement 2: $\sin^{-1}x$ and $\cos^{-1}x$ are defined for $x \in [-1, 1]$.

10. Statement 1: $\lim_{x\to 0} [x]\left(\frac{e^{1/x}-1}{e^{1/x}+1}\right)$ (where [.] represents greatest integer function) does not exist.

Statement 2: $\lim_{x\to 0} \left(\frac{e^{1/x}-1}{e^{1/x}+1}\right)$ does not exist.

11. Statement 1: If $\lim_{x\to 0} \left(f(x) + \frac{\sin x}{x}\right)$ does not exist, then $\lim_{x\to 0} f(x)$ does not exist.

Statement 2: $\lim_{x\to 0} \frac{\sin x}{x}$ exists, and has value 1.

12. Statement 1: If $<a_n>$ be a sequence such that $a_1 = 1$ and $a_{n+1} = \sin a_n$, then $\lim_{n\to\infty} a_n = 0$.

Statement 2: Since $x > \sin x \ \forall \ x > 0$.

13. Statement 1: $\lim_{x\to 0} \log_e \left(\frac{\sin x}{x}\right) = 0$.

Statement 2: $\lim_{x\to a} f(g(x)) = f(\lim_{x\to 0} g(x))$.

Linked Comprehension Type

Based upon each paragraph, three multiple choice questions have to be answered. Each question has four choices a, b, c and d, out of which *only one* is correct.

For Problems 1–3

Let $f(x) = \frac{\sin^{-1}(1-\{x\}) \times \cos^{-1}(1-\{x\})}{\sqrt{2\{x\}} \times (1-\{x\})}$ where $\{x\}$ denotes the fractional part of x

1. $L = \lim_{x\to 0+} f(x)$ is equal to

a. $\pi/2$ b. $\frac{\pi}{2\sqrt{2}}$

c. $\frac{\pi}{\sqrt{2}}$ d. $\sqrt{2\pi}$

2. $R = \lim_{x\to 0-} f(x)$ is equal to

a. $\pi/2$ b. $\frac{\pi}{2\sqrt{2}}$

c. $\frac{\pi}{\sqrt{2}}$ d. $\sqrt{2\pi}$

3. Which of the following is not true?

a. $\cos L < \cos R$ b. $\tan(2L) > \tan R$

c. $\sin L > \sin R$ d. None of these

For Problems 4–6

$A_i = \frac{x - a_i}{|x - a_i|}$, $i = 1, 2, \dots, n$ and if $a_1 < a_2 < a_3 < \dots < a_n$.

4. If $1 \le m \le n$, $m \in N$, then the value of $L = \lim_{x\to a_m^-} (A_1 A_2 \cdots A_n)$ is

a. always 1 b. always – 1

c. $(-1)^{n-m+1}$ d. $(-1)^{n-m}$

5. If $1 \le m \le n$, $m \in N$, then the value of $R = \lim_{x\to a_m^+} (A_1 A_2 \cdots A_n)$ is

a. always 1 b. always – 1

c. $(-1)^{m+1}$ d. $(-1)^{n-m}$

6. If $a_m < a_1$, $m \in N$, then $\lim_{x \to a_m} (A_1 A_2 \cdots A_n)$

a. Is always equal to – 1. b. Is always equal to + 1.

c. Does not exist. d. Is equal to 1 or – 1.

For Problems 7–9

If $L = \lim_{x \to 0} \frac{\sin x + ae^x + be^{-x} + c\ln(1+x)}{x^3} = \infty$

7. The value of L is

a. 1/2 b. – 1/3

c. – 1/6 d. 3

8. Equation $ax^2 + bx + c = 0$ has

a. real and equal roots

b. complex roots

c. unequal positive real roots

d. unequal roots

9. The solution set of $||x + c| - 2a| < 4b$ is

a. $[-2, 2]$ b. $[0, 2]$

c. $[-1, 1]$ d. $[-2, 1]$

Matrix-Match Type

Each question contains statements given in two columns which have to be matched. Statements a, b, c, d in column I have to be matched with statements p, q, r, s in column II. If the correct match are a → p, a → s, b → r, c → p, c → q and d → s, then the correctly bubbled 4 × 4 matrix should be as follows:

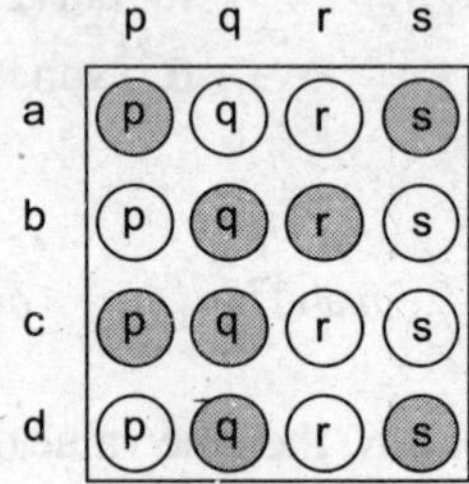

1.

Column I	Column II
a. If $L = \lim_{x \to -1} \frac{\sqrt[3]{(7-x)} - 2}{(x+1)}$, then $12L =$	p. –2
b. If $L = \lim_{x \to \pi/4} \frac{\tan^3 x - \tan x}{\cos\left(x + \frac{\pi}{4}\right)}$, then $-L/8 =$	q. –2
c. If $L = +\lim_{x \to 1} \frac{(2x-3)(\sqrt{x}-1)}{2x^2 + x - 3}$, then $20L =$	r. – 1
d. If $L = \lim_{x \to \infty} \frac{\log x^n - [x]}{[x]}$, where $n \in N$, ($[x]$ denotes greatest integer less than or equal to x), then $-2L =$	s. – 1

2.

Column I ([·] denotes the greatest integer function)	Column II
a. $\lim_{x \to 0}\left(\left[100\frac{\sin x}{x}\right] + \left[100\frac{\tan x}{x}\right]\right)$	p. 198
b. $\lim_{x \to 0}\left(\left[100\frac{x}{\sin x}\right] + \left[100\frac{\tan x}{x}\right]\right)$	q. 199
c. $\lim_{x \to 0}\left(\left[100\frac{\sin^{-1} x}{x}\right] + \left[100\frac{\tan^{-1} x}{x}\right]\right)$	r. 200
d. $\lim_{x \to 0}\left(\left[100\frac{x}{\sin^{-1} x}\right] + \left[100\frac{\tan^{-1} x}{x}\right]\right)$	s. 201

3.

Column I	Column II
a. If $\lim_{x \to \infty} (\sqrt{(x^2 - x - 1)} - ax - b) = 0$, where $a > 0$, then there exists at least one a and b for which point $(a, 2b)$ lies on the line.	p. $y = -3$
b. If $\lim_{x \to \infty} \frac{(1 + a^3) + 8e^{1/x}}{1 + (1 - b^3)\, e^{1/x}} = 2$, then there exists at least one a and b for which point (a, b^3) lies on the line.	q. $3x - 2y - 5 = 0$
c. If $\lim_{x \to \infty} (\sqrt{(x^4 - x^2 + 1)} - ax^2 - b) = 0$, then there exists at least one a and b for which point $(a, -2b)$ lies on the line.	r. $15x - 2y - 11 = 0$

d. If $\lim\limits_{x\to -a}\dfrac{x^7+a^7}{x+a}=7$, where $a<0$, then	**s.** $y=2$
there exists at least one a for which point $(a, 2)$ lies on the line.	

Integer Type

1. Let $f(x)=\begin{cases}x+1, & x>0\\ 2-x, & x\le 0\end{cases}$ and $g(x)=\begin{cases}x+3, & x<1\\ x^2-2x-2, & 1\le x<2\\ x-5, & x\ge 2\end{cases}$, then $\left|\lim\limits_{x\to 0} g(f(x))\right|$ is________.

2. The value of $\left|\lim\limits_{x\to 0}\dfrac{[(5\sin x-7\sin 2x+3\sin 3x)]}{x^2.\sin x}\right|$ is ________.

3. The value of $[\lim\limits_{x\to 2}\{(\sqrt{(x+7)}-3\sqrt{(2x-3)})/(\sqrt[3]{(x+6)}-2\sqrt[3]{(3x-5)})\}]$, where $[\,.\,]$ represents the greatest integer function is________.

4. The population of a country increases by 2% per year. If it increases k times in a century, then the value of $[k]$, where $[\cdot]$ represents the greatest integer function, is________.

5. Let $f(x)$ be a twice-differentiable function and $f''(0)=2$, then the value of $\lim\limits_{x\to 0}\{[2f(x)-3f(2x)+f(4x)]/x^2\}$ is________.

6. The value of $\lim\limits_{x\to 0}\log_{\tan^2 x}(\tan^2 2x)$ is________.

7. If $\lim\limits_{x\to a}[(a^x-x^a)/(x^x-a^a)]=-1$ and $a>0$, then the value of a is________.

8. If $L=\lim\limits_{x\to 0}[(\cos 4x+a\cos 2x+b)/x^4]$ is finite, then the value of L is________.

9. If $\lim\limits_{x\to 0}\{[x(1+a\cos x)-b\sin x]/x^3\}=1$, then the value of $|a+b|$ is________.

10. The integral value of n for which $\lim\limits_{x\to 0}\{[\cos^2 x-\cos x-e^x\cos x+e^x-(x^3/2)]/x^n\}$ is finite and non-zero is________.

11. If $f(n+1)=(1/2)\{f(n)+[9/f(n)]\}$, $n\in$ N and $f(n)>0$ for all $n\in$ N, then the value of $\lim\limits_{n\to\infty} f(n)$ is________.

12. The value of $\lim\limits_{x\to\infty}\left\{\dfrac{x}{x+\dfrac{\sqrt[3]{x}}{x+\dfrac{\sqrt[3]{x}}{x+\sqrt[3]{x}}\ldots\infty}}\right\}$ is________.

13. The value of $\lim\limits_{x\to\pi/2}\{[\sin x-(\sin x)^{\sin x}]/(1-\sin x+\ln\sin x)\}$ is ________.

14. Tangent lines are drawn at the end and middle points of a circular arc AB, and the points A and B are joined with a chord. Then the ratio of the areas of the triangles thus formed tends which value when the arc AB decreases infinitely?

15. If $f(x)=\begin{cases}\sin x, & x\ne n\pi,\ n\in \mathrm{I}\\ 2, & \textit{otherwise}\end{cases}$ and $g(x)=\begin{cases}x^2+1, & x\ne 0, 2\\ 4, & x=0\\ 5, & x=2\end{cases}$, then $\lim\limits_{x\to 0} g\{f(x)\}$ is________.

16. The integer n for which $\lim\limits_{x\to 0}\{[(\cos x-1)(\cos x-e^x)]/x^n\}$ is a finite non-zero number is________.

17. Let $\lim\limits_{x\to 1}[x^a-ax+a-1/(x-1)^2]=f(a)$. Then the value of $f(4)$ is________.

18. If $\lim\limits_{x\to 0}[(1-\sqrt{\cos 2x}\,\sqrt[3]{\cos 3x}\,\sqrt[4]{\cos 4x}\ldots\sqrt[n]{\cos nx})/x^2]$ is equal to 10, then the value of n is________.

19. If $f(x)=(3x^2+ax+a+1)/(x^2+x-2)$ and $\lim\limits_{x\to -2} f(x)$ exists, then the value of $(a-4)$ is___.

20. If $\lim\limits_{x\to\infty}\left(e^{2x}+e^x+x\right)^{\frac{1}{x}}=e^a$, then the value of a is________.

21. The value of $\lim\limits_{x\to\frac{\pi}{2}}\sqrt{\dfrac{\tan x-\sin(\tan^{-1}(\tan x))}{\tan x+\cos^2(\tan x)}}$ is_____.

22. The value of

$$\left[\lim_{n\to\infty}(2\times 3^2\times 2^3\times 3^4\ldots\times 2^{n-1}\times 3^n)^{\frac{1}{(n^2+1)}}\right]^4$$

is________.

23. If $L=\lim\limits_{x\to 0}\dfrac{\sin x^4-x^4\cos x^4+x^{20}}{x^4(e^{2x^4}-1-2x^4)}$, then the value of $1/L$ is________.

24. If $\lim\limits_{x\to 1}\dfrac{a\sin(x-1)+b\cos(x-1)+4}{x^2-1}=-2$ then $|a+b|$ is________.

25. If $\lim\limits_{x\to 0}\dfrac{\ln\cot\left(\dfrac{\pi}{4}-k_1x\right)}{\tan k_2x}=1$, then k_2/k_1 is________.

26. The value of $\lim\limits_{t\to 0}\dfrac{1-(1+t)^t}{\ln(1+t)-t}$ is equal to________.

27. If $\lim\limits_{x\to 0}\sin\left(\dfrac{\pi(1-\cos^m x)}{x^n}\right)$ exists, where $m, n\in N$, then the sum of all possible values of n is________.

ANSWERS AND SOLUTIONS

Objective Type

1. c. Given $f(x) = x^2 - \pi^2$

$$\lim_{x \to -\pi} \frac{x^2 - \pi^2}{\sin(\sin x)}$$

$$= \lim_{h \to 0} \frac{(-\pi + h)^2 - \pi^2}{\sin(\sin(-\pi + h))} = \lim_{h \to 0} \frac{-2h\pi + h^2}{-\sin(\sin h)}$$

$$= \lim_{h \to 0} \frac{h - 2\pi}{\dfrac{-\sin(\sin h)}{\sin h} \times \dfrac{\sin h}{h}} = 2\pi$$

2. d. $$\lim_{x \to \frac{\pi}{2}} \frac{\cos x}{(1 - \sin x)^{1/3}} = \lim_{t \to 0} \frac{-\sin t}{(1 - \cos t)^{1/3}}$$

$$= -\lim_{t \to 0} \frac{2\sin\frac{t}{2}\cos\frac{t}{2}}{\left(2\sin^2\frac{t}{2}\right)^{1/3}}$$

$$= -\lim_{t \to 0} 2^{2/3}\cot\frac{t}{2}\left(\sin\frac{t}{2}\right)^{1/3} = 0$$

3. a. $$\lim_{x \to -\infty} \frac{x^2 \tan\frac{1}{x}}{\sqrt{8x^2 + 7x + 1}} = \lim_{x \to -\infty} \frac{x^2 \tan\frac{1}{x}}{-x\sqrt{8 + \frac{7}{x} + \frac{1}{x^2}}}$$

$$= -\lim_{x \to -\infty} \frac{\tan\frac{1}{x}}{\frac{1}{x}\sqrt{8 + \frac{7}{x} + \frac{1}{x^2}}} = -\frac{1}{2\sqrt{2}}$$

4. a. $$\lim_{x \to 0+} \left[\frac{\sin(\operatorname{sgn} x)}{\operatorname{sgn}(x)}\right] = \lim_{x \to 0^+} \left[\frac{\sin 1}{1}\right]$$

$$= 0$$

$$= \lim_{x \to 0^-} \left[\frac{\sin(\operatorname{sgn} x)}{\operatorname{sgn}(x)}\right]$$

$$= \lim_{x \to 0^-} \left[\frac{\sin(-1)}{-1}\right]$$

$$= \lim_{x \to 0^-} [\sin 1]$$

$$= 0$$

Hence, the given limit is 0.

5. d. The given limit is

$$\lim_{x \to \infty} \frac{\frac{2}{x} + 2 + \frac{\sin 2x}{x}}{\left(2 + \frac{\sin 2x}{x}\right)e^{\sin x}}$$

$$= \frac{0 + 2 + 0}{(2 + 0) \times \infty} = 0 \quad \left[\because \lim_{x \to \infty} \sin x \in (-1, 1)\right]$$

6. b. $$\frac{[x]^2}{x^2} = \begin{bmatrix} 0 \text{ if } 0 < x < 1 \\ \frac{1}{x^2} \text{ if } -1 < x < 0 \end{bmatrix} \Rightarrow l \text{ does not exist}$$

$$\frac{[x^2]}{x^2} = \begin{bmatrix} 0 \text{ if } 0 < x < 1 \\ 0 \text{ if } -1 < x < 0 \end{bmatrix} \Rightarrow m \text{ exists and is equal to } 0]$$

7. c. $$\lim_{x \to 1} \frac{x \sin\{x - [x]\}}{x - 1}$$

Now L.H.L. $$= \lim_{h \to 0} \frac{(1 - h)\sin(1 - h - [1 - h])}{(1 - h) - 1}$$

$$= \lim_{h \to 0} \frac{(1 - h)\sin(1 - h)}{-h} = -\infty$$

R.H.L. $$= \lim_{h \to 0} \frac{(1 + h)\sin(1 + h - [1 + h])}{(1 + h) - 1}$$

$$= \lim_{h \to 0} \frac{(1 + h)\sin h}{h} = 1$$

Hence, the limit does not exist.

8. c. The given limit is

$$\lim_{x \to 0} [(1 + \tan x)^{\operatorname{cosec} x} x\, 1/(1 + \sin x)^{\operatorname{cosec} x}]$$

$$= \lim_{x \to 0} [(1 + \tan x)^{\cot x}\}^{\sec x} x\, \{1/(1 + \sin x)^{\operatorname{cosec} x}\}]$$

$$= e^{\sec 0}\frac{1}{e} = e\frac{1}{e} = 1$$

9. b. $$\lim_{x \to \infty} \frac{\sin^4 x - \sin^2 x + 1}{\cos^4 x - \cos^2 x + 1}$$

$$= \lim_{x \to \infty} \frac{(1 - \cos^2 x)^2 - (1 - \cos^2 x) + 1}{\cos^4 x - \cos^2 x + 1}$$

$$= \lim_{x \to \infty} \frac{\cos^4 x - \cos^2 x + 1}{\cos^4 x - \cos^2 x + 1}$$

$$= 1$$

10. d. $$\lim_{x \to \infty} \left(\frac{x^3}{3x^2 - 4} - \frac{x^2}{3x + 2}\right)$$

$$= \lim_{x\to\infty} \frac{x^3(3x+2)-x^2(3x^2-4)}{(3x^2-4)(3x+2)}$$

$$= \lim_{x\to\infty} \frac{2x^3+4x^2}{9x^3+6x^2-12x-8}$$

$$= \lim_{x\to\infty} \frac{2+\frac{4}{x}}{9+\frac{6}{x}-\frac{12}{x^2}-\frac{8}{x^3}}$$

$= 2/9$

11. c. We have $f(x)+g(x)+h(x)$

$$= \frac{x^2-4x+17-4x-2}{x^2+x-12}$$

$$= \frac{x^2-8x+15}{x^2+x-12} = \frac{(x-3)(x-5)}{(x-3)(x+4)}$$

$$\therefore \lim_{x\to 3}[f(x)+g(x)+h(x)] = \lim_{x\to 3}\frac{(x-3)(x-5)}{(x-3)(x+4)} = -\frac{2}{7}$$

12. d. $$\lim_{x\to 0} \frac{x(e^x-1)}{1-\cos x} = \lim_{x\to 0} \frac{2x(e^x-1)}{4\sin^2\frac{x}{2}}$$

$$= 2\lim_{x\to 0}\left[\frac{(x/2)^2}{\sin^2\frac{x}{2}}\right]\left(\frac{e^x-1}{x}\right) = 2$$

13. c. $$\lim_{n\to\infty} \frac{n(2n+1)^2}{(n+2)(n^2+3n-1)}$$

$$= \lim_{n\to\infty} \frac{\left(2+\frac{1}{n}\right)^2}{\left(1+\frac{2}{n}\right)\left(1+\frac{3}{n}-\frac{1}{n^2}\right)}$$

$$= \frac{(2+0)^2}{(1+0)(1+0+0)} = 4$$

14. d. We have $\lim_{x\to\pi} \frac{1+\cos^3 x}{\sin^2 x}$

$$= \lim_{x\to\pi} \frac{(1+\cos x)(1-\cos x+\cos^2 x)}{(1-\cos x)(1+\cos x)}$$

$$= \lim_{x\to\pi} \frac{1-\cos x+\cos^2 x}{1-\cos x} = \frac{1+1+1}{1+1} = \frac{3}{2}$$

15. c. Here the expression assumes the form $\infty - \infty$ as $x \to \infty$. So, we first reduce it to the rational form $\frac{f(x)}{g(x)}$.

We have $\lim_{x\to\infty}\left\{\sqrt{x^2+x+1}-\sqrt{x^2+1}\right\}$

$$= \lim_{x\to\infty} \frac{\left\{\sqrt{x^2+x+1}-\sqrt{x^2+1}\right\}}{\left\{\sqrt{x^2+x+1}+\sqrt{x^2+1}\right\}} \times \left\{\sqrt{x^2+x+1}+\sqrt{x^2+1}\right\}$$

$$= \lim_{x\to\infty} \frac{x^2+x+1-x^2-1}{\sqrt{x^2+x+1}+\sqrt{x^2+1}}$$

$$= \lim_{x\to\infty} \frac{x}{\sqrt{x^2+x+1}+\sqrt{x^2+1}}$$

$$= \lim_{x\to\infty} \frac{1}{\sqrt{1+\frac{1}{x}+\frac{1}{x^2}}+\sqrt{1+\frac{1}{x^2}}} = \frac{1}{2}$$

16. a. $$\lim_{x\to 2} \frac{\sqrt{1+\sqrt{2+x}}-\sqrt{3}}{x-2}$$

$$= \lim_{x\to 2} \frac{1+\sqrt{2+x}-3}{\left(\sqrt{1+\sqrt{2+x}}+\sqrt{3}\right)(x-2)} \quad \text{(Rationalizing)}$$

$$= \lim_{x\to 2} \frac{\sqrt{2+x}-2}{(\sqrt{1+\sqrt{2+x}}+\sqrt{3})(x-2)}$$

$$= \lim_{x\to 2} \frac{(x-2)}{(\sqrt{1+\sqrt{2+x}}+\sqrt{3})(\sqrt{2+x}+2)(x-2)}$$

(Rationalizing)

$$= \frac{1}{(2\sqrt{3})\,4} = \frac{1}{8\sqrt{3}}$$

17. c. $$\lim_{x\to\infty} \frac{(2x+1)^{40}(4x-1)^5}{(2x+3)^{45}}$$

$$= \lim_{x\to\infty} \frac{\left(2+\frac{1}{x}\right)^{40}\left(4-\frac{1}{x}\right)^5}{\left(2+\frac{3}{x}\right)^{45}}$$

[Dividing numerator and denominator by x^{45}]

$$= \frac{2^{40}4^5}{2^{45}} = 2^5 = 32$$

18. b. $\lim_{x\to\infty}\left[\sqrt{x+\sqrt{x+\sqrt{x}}}-\sqrt{x}\right]$

$$=\lim_{x\to\infty}\frac{x+\sqrt{x+\sqrt{x}}-x}{\sqrt{x+\sqrt{x+\sqrt{x}}}+\sqrt{x}}\text{ (Rationalizing)}$$

$$=\lim_{x\to\infty}\frac{\sqrt{1+x^{-1/2}}}{\sqrt{1+\sqrt{x^{-1}+x^{-3/2}}}+1}=\frac{1}{2}$$

19. d. $\lim_{x\to\infty}\frac{(x+1)^{10}+(x+2)^{10}+\cdots+(x+100)^{10}}{x^{10}+10^{10}}$

$$=\lim_{x\to\infty}\frac{x^{10}\left[\left(1+\frac{1}{x}\right)^{10}+\left(1+\frac{2}{x}\right)^{10}+\cdots+\left(1+\frac{100}{x}\right)^{10}\right]}{x^{10}\left[1+\frac{10^{10}}{x^{10}}\right]}$$

$=100$

20. c. $\lim_{x\to 0}\frac{x^a\sin^b x}{\sin x^c}=\lim_{x\to 0}x^a\left(\frac{\sin x}{x}\right)^b\left(\frac{x}{\sin x}\right)^c x^{b-c}$

$$=\lim_{x\to 0}x^{a+b-c}$$

This limit will have non-zero value if $a+b=c$.

21. b. $\lim_{x\to\pi/2}\left[x\tan x-\left(\frac{\pi}{2}\right)\sec x\right]$

$$=\lim_{x\to\pi/2}\frac{2x\sin x-\pi}{2\cos x}\quad\left(\frac{0}{0}\text{ form}\right)$$

$$=\lim_{x\to\pi/2}\frac{[2\sin x+2x\cos x]}{-2\sin x}$$

(Applying L' Hospital's Rule)

$=-1$

22. c. $\lim_{x\to\infty}\left(\frac{x^3+1}{x^2+1}-(ax+b)\right)=2$

$$\Rightarrow\lim_{x\to\infty}\frac{x^3(1-a)-bx^2-ax+(1-b)}{x^2+1}=2$$

$\Rightarrow 1-a=0$ and $-b=2$

$\Rightarrow a=1, b=-2$

23. c. $\lim_{x\to 1}(2-x)^{\tan(\pi x)/2}$

$$=\lim_{x\to 1}\{1+(1-x)\}^{\tan(\pi x/2)}$$

$$=e^{\lim_{x\to 1}(1-x)^{\tan(\pi x/2)}}$$

$$=e^{\lim_{x\to 1}(1-x)\cot\left(\frac{\pi}{2}-\frac{\pi x}{2}\right)}$$

$$=e^{\lim_{x\to 1}\frac{(1-x)}{\tan\left(\frac{\pi}{2}-\frac{\pi x}{2}\right)}}$$

$$=e^{\frac{2}{\pi}\lim_{x\to 1}\frac{\frac{\pi}{2}(1-x)}{\tan\left(\frac{\pi}{2}(1-x)\right)}}$$

$$=e^{2/\pi}$$

24. b. $\lim_{x\to 0}\frac{\sin x^n}{(\sin x)^m}=\lim_{x\to 0}\left(\frac{\sin x^n}{x^n}\right)\left(\frac{x^n}{x^m}\right)\left(\frac{x}{\sin x}\right)^m$

$$=\lim_{x\to 0}x^{n-m}=0$$

$[\because m<n]$

25. a. $\frac{x^4(\cot^4 x-\cot^2 x+1)}{(\tan^4 x-\tan^2 x+1)}$

$$=\frac{x^4(1-\tan^2 x+\tan^4 x)}{\tan^4 x(\tan^4 x-\tan^2 x+1)}=\frac{x^4}{\tan^4 x},\ x\neq 0$$

$$\Rightarrow\lim_{x\to 0}\frac{x^4(\cot^4 x-\cot^2 x+1)}{(\tan^4 x-\tan^2 x+1)}=\lim_{x\to 0}\frac{x^4}{\tan^4 x}=1$$

26. d. $\lim_{x\to\infty}\left(\frac{1}{e}-\frac{x}{1+x}\right)^x=\lim_{x\to\infty}\left(\frac{1}{e}-\frac{1}{\frac{1}{x}+1}\right)^x=\left(\frac{1}{e}-1\right)^\infty$

= (some negative value) ∞, which is not defined as base.

27. b. $\lim_{x\to 1}\frac{1-x^2}{\sin 2\pi x}$

$$=-\lim_{x\to 1}\frac{2\pi(1-x)(1+x)}{2\pi\sin(2\pi-2\pi x)}$$

$$=-\lim_{x\to 1}\frac{(2\pi-2\pi x)}{\sin(2\pi-2\pi x)}\frac{1+x}{2\pi}$$

$$=\frac{-1}{\pi}$$

28. d. We know that $\cos^{-1}\left(\frac{1-x^2}{1+x^2}\right)=\begin{cases}2\tan^{-1}x, & x\geq 0\\ -2\tan^{-1}x, & x\leq 0\end{cases}$

$$\Rightarrow\lim_{x\to 0^+}\frac{1}{x}\cos^{-1}\left(\frac{1-x^2}{1+x^2}\right)=\lim_{x\to 0^+}\frac{2\tan^{-1}x}{x}=2\text{, and}$$

$$\lim_{x\to 0^-}\frac{1}{x}\cos^{-1}\left(\frac{1-x^2}{1+x^2}\right)=\lim_{x\to 0^+}\left[-\frac{2\tan^{-1}x}{x}\right]=-2$$

29. c. $\lim_{x\to\infty}\left(\frac{x^2+2x-1}{2x^2-3x-2}\right)^{\frac{(2x+1)}{(2x-1)}} = \lim_{x\to\infty}\left(\frac{1+\frac{2}{x}-\frac{1}{x^2}}{2-\frac{3}{x}-\frac{2}{x^2}}\right)^{\frac{(2+1/x)}{(2-1/x)}}$

$= 1/2$

30. c. Since highest degree of x is 1/2, divide numerator and denominator by $\sqrt{x}$, then we have limit $2/\sqrt{2}$ or $\sqrt{2}$.

31. a. $\lim_{y\to 0}\left\{\frac{x\{\sec(x+y)-\sec x\}}{y}+\sec(x+y)\right\}$

$= \lim_{y\to 0}\left[\frac{x}{y}\left\{\frac{\cos x-\cos(x+y)}{\cos(x+y)\cos x}\right\}\right] + \lim_{y\to 0}\sec(x+y)$

$= \lim_{y\to 0}\left[\frac{x\,2\sin\left(x+\frac{y}{2}\right)\sin\left(\frac{y}{2}\right)}{y\cos(x+y)\cos x}\right]+\sec x$

$= \lim_{y\to 0}\left[\frac{x\sin\left(x+\frac{y}{2}\right)}{\cos(x+y)\cos x}\times\frac{\sin\left(\frac{y}{2}\right)}{\frac{y}{2}}\right]+\sec x$

$= x\tan x\sec x+\sec x$

$= \sec x\,(x\tan x+1)$

32. a. $\lim_{m\to\infty}\left(\cos\frac{x}{m}\right)^m = \lim_{m\to\infty}\left[1-\left(1-\cos\frac{x}{m}\right)\right]^m$

$= \lim_{m\to\infty}\left[1-2\sin^2\frac{x}{2m}\right]^m$

$= e^{\lim_{m\to\infty}\left(-2\sin^2\frac{x}{2m}\right)m}$

$= 1$

33. b. $\operatorname{cosec}\frac{\pi x}{2}\to 1$ when $x\to 1 \Rightarrow \left[\operatorname{cosec}\frac{\pi x}{2}\right]=1$

$\therefore$ limit $=1$

34. b. $\lim_{n\to\infty}\left(\frac{n^2-n+1}{n^2-n-1}\right)^{n(n-1)}$

$= \lim_{n\to\infty}\left(\frac{n(n-1)+1}{n(n-1)-1}\right)^{n(n-1)}$

$= \lim_{n\to\infty}\frac{\left(1+\frac{1}{n(n-1)}\right)^{n(n-1)}}{\left(1-\frac{1}{n(n-1)}\right)^{n(n-1)}} = \frac{e}{e^{-1}} = e^2$

35. b. $f(x) = \lim_{n\to\infty} n(x^{1/n}-1)$

$= \lim_{n\to\infty}\frac{x^{1/n}-1}{1/n}$

$= \lim_{m\to 0}\frac{x^m-1}{m}$ (where $\frac{1}{n}$ replaced by m)

$= \ln x$

$\Rightarrow f(xy) = \ln(xy) = \ln x+\ln y = f(x)+f(y)$

36. c. If $f(x) = \sin\left(\frac{1}{x}\right)$ and $g(x) = \frac{1}{x}$, then both $\lim_{x\to 0} f(x)$ and $\lim_{x\to 0} g(x)$ do not exist, but $\lim_{x\to 0}\frac{f(x)}{g(x)} = 0$ exists.

37. a. $\lim_{n\to\infty}\frac{n\times 3^n}{n(x-2)^n+n\times 3^{n+1}-3^n} = \frac{1}{3}$

$\Rightarrow \lim_{n\to\infty}\frac{1}{\frac{(x-2)^n}{3^n}+3-\frac{1}{n}}$ (Dividing N^r and D^r by $n\times 3^n$)

For $\lim_{n\to\infty}$ to be equal to 1/3

$\lim_{n\to\infty}\frac{1}{n}\to 0$ (which is true) and $\lim_{n\to\infty}\left(\frac{x-2}{3}\right)^n\to 0$

$\Rightarrow 2\le x<5$

38. b. $\lim_{x\to 2}\frac{2^x+2^{3-x}-6}{\sqrt{2^{-x}}-2^{1-x}} = \lim_{x\to 2}\frac{(2^x)^2-6\times 2^x+2^3}{\sqrt{2^x}-2}$

[Multiplying N^r and D^r by 2^x]

$= \lim_{x\to 2}\frac{(2^x-4)(2^x-2)(\sqrt{2^x}+2)}{(\sqrt{2^x}-2)(\sqrt{2^x}+2)}$

$= \lim_{x\to 2}\frac{(2^x-4)(2^x-2)(\sqrt{2^x}+2)}{(2^x-4)}$

$= \lim_{x\to 2}(2^x-2)(\sqrt{2^x}+2)$

$= (2^2-2)(2+2) = 8$

39. c. 1^∞ form

$$1 = e^{\lim_{n\to\infty} n\left(\left(\frac{n}{n+1}\right)^\alpha + \sin\frac{1}{n} - 1\right)} = e^{\lim_{n\to\infty} n\sin\frac{1}{n} + \lim_{n\to\infty} n\left(\left(\frac{n}{n+1}\right)^\alpha - 1\right)}$$

Consider, $\lim_{n\to\infty} n\left(\left(\frac{n}{n+1}\right)^\alpha - 1\right)$

$$= \lim_{n\to\infty} n\left(\left(\frac{1}{1+1/n}\right)^\alpha - 1\right)$$

Putting $n = 1/y$ we get

$$\lim_{y\to 0} \frac{1}{y}\left(\left(\frac{1}{1+y}\right)^\alpha - 1\right) = \lim_{y\to 0} \frac{1-(1+y)^\alpha}{y} = -a$$

(Using binomial)

$\therefore 1 = e^1 - a$

40. b. $L = \lim_{x\to\infty} \frac{\ln(x^2 + e^x)}{\ln(x^4 + e^{2x})} = \lim_{x\to\infty} \frac{\ln e^x\left(1+\frac{x^2}{e^x}\right)}{\ln e^{2x}\left(1+\frac{x^4}{e^{2x}}\right)}$

$$= \lim_{x\to\infty} \frac{x + \ln\left(1+\frac{x^2}{e^x}\right)}{2x + \ln\left(1+\frac{x^4}{e^{2x}}\right)}$$

$$= \lim_{x\to\infty} \frac{1 + \frac{1}{x}\ln\left(1+\frac{x^2}{e^x}\right)}{2 + \frac{1}{x}\ln\left(1+\frac{x^4}{e^{2x}}\right)}$$

Note that as $\frac{x^2}{e^x} \to 0$ and as $\frac{x^2}{e^{2x}} \to 0$ ($x\to\infty$)

(Using L' Hospital's Rule)

Hence, $L = 1/2$

41. a. $\lim_{x\to 1} \frac{1+\sin \pi\left(\frac{3x}{1+x^2}\right)}{1+\cos \pi x}$

$$= \lim_{x\to 1} \frac{1-\cos\left(\frac{3\pi}{2} - \frac{3\pi x}{1+x^2}\right)}{1-\cos(\pi - \pi x)}$$

$$= \lim_{x\to 1} \frac{2\sin^2\left(\frac{3\pi}{4} - \frac{3\pi x}{2(1+x^2)}\right)}{2\sin^2\left(\frac{\pi}{2} - \frac{\pi x}{2}\right)}$$

$$= \lim_{x\to 1} \left(\frac{\frac{3\pi}{4} - \frac{3\pi x}{2(1+x^2)}}{\frac{\pi}{2} - \frac{\pi x}{2}}\right)^2$$

$$= \lim_{x\to 1} 9\left(\frac{\frac{1}{2} - \frac{x}{1+x^2}}{1-x}\right)^2 = \lim_{x\to 1} 9\left(\frac{x-1}{2(1+x^2)}\right)^2 = 0$$

42. b. $\because \lim_{n\to\infty} \cos^{2n} x = \begin{cases} 1, & x = r\pi, r \in I \\ 0, & x \neq r\pi, r \in I \end{cases}$

Here, for $x = 10$, $\lim_{n\to\infty} \cos^{2n}(x-10) = 1$

and in all other cases it is zero.

$$\therefore \lim_{n\to\infty} \sum_{x=1}^{\infty} \cos^{2n}(x-10) = 1$$

43. b. $L = \lim_{x\to\infty} \frac{(2^{x^n})^{1/e^x} - (3^{x^n})^{1/e^x}}{x^n}$

$$= \lim_{x\to\infty} \frac{(3)^{x^n/e^x}\left(\left(\frac{2}{3}\right)^{x^n/e^x} - 1\right)}{x^n}$$

Now, $\lim_{x\to\infty} \frac{x^n}{e^x} = \lim_{x\to\infty} \frac{n!}{e^x} = 0$ (differentiating numerator and denominator n times for L' Hospital's Rule)

Hence $L = \lim_{x\to\infty}(3)^{x^n/e^x} \lim_{x\to\infty} \frac{\left(\frac{2}{3}\right)^{x^n/e^x} - 1}{\frac{x^n}{e^x}} \lim_{x\to\infty} \frac{1}{e^x}$

$$= 1 \times \log(2/3) \times 0 = 0$$

44. c. $\frac{\cos(2x-4)-33}{2} < f(x) < \frac{x^2|4x-8|}{x-2}$

$\Rightarrow$

$$\lim_{x\to 2^-} \frac{\cos(2x-4)-33}{2} < \lim_{x\to 2^-} f(x) < \lim_{x\to 2^-} \frac{x^2|4x-8|}{x-2}$$

$$\Rightarrow -16 < \lim_{x\to 2^-} f(x) < \lim_{x\to 2^-} \frac{x^2(8-4x)}{x-2}$$

$$\Rightarrow -16 < \lim_{x\to 2^-} f(x) < -16$$

$\Rightarrow \lim_{x\to 2^-} f(x) = -16$ (by Sandwich Theorem)

45. a. Given $g(x) = \lim_{n\to\infty} \dfrac{1}{\left(\dfrac{3}{\pi}\tan^{-1}2x\right)^{2n}+5} = 0$

$$\Rightarrow \left[\left(\frac{3}{\pi}\tan^{-1}2x\right)^2\right]^n \to \infty$$

$$\Rightarrow \left(\frac{3}{\pi}\tan^{-1}2x\right)^2 > 1$$

$$\Rightarrow |\tan^{-1}2x| > \frac{\pi}{3}$$

$$\Rightarrow \tan^{-1}2x < -\frac{\pi}{3} \text{ or } \tan^{-1}2x > \frac{\pi}{3}$$

$$\Rightarrow 2x < -\sqrt{3} \text{ or } 2x > \sqrt{3}$$

$$\Rightarrow |2x| > \sqrt{3}$$

46. a. $(1+x)^{2/x} = (1+x)^{2/x} - [(1+x)^{2/x}]$

Now, $\lim_{x\to 0}(1+x)^{2/x} = e^2$

$$\Rightarrow \lim_{x\to 0}\{(1+x)^{2/x}\} = e^2 - [e^2] = e^2 - 7$$

47. b. $\lim_{x\to 0} \dfrac{\sin(x^2)}{\ln(\cos(2x^2-x))}$

$$= \lim_{x\to 0} \frac{\sin(x^2)}{\log\left(1-2\sin^2\left(\dfrac{2x^2-x}{2}\right)\right)}$$

$$= \lim_{x\to 0} \frac{\dfrac{\sin(x^2)\,x^2}{x^2\log\left(1-2\sin^2\left(\dfrac{2x^2-x}{2}\right)\right)}}{-2\sin^2\left(\dfrac{2x^2-x}{2}\right)}\left[-2\sin^2\left(\frac{2x^2-x}{2}\right)\right]$$

$$= \lim_{x\to 0} - \frac{x^2}{\dfrac{2\sin^2\left(\dfrac{2x^2-x}{2}\right)}{\left(\dfrac{2x^2-x}{2}\right)^2}\left(\dfrac{2x^2-x}{2}\right)^2}$$

$$= \lim_{x\to 0} - \frac{2x^2}{(2x^2-x)^2} = \lim_{x\to 0} - \frac{2}{(2x-1)^2} = -2$$

48. a. L.H.L. $= \lim_{x\to -1^-} \dfrac{1}{\sqrt{|x|-\{-x\}}}$

$$= \lim_{x\to -1^-} \frac{1}{\sqrt{-x-(x+2)}}$$

$$= \lim_{x\to -1^-} \frac{1}{\sqrt{-2x-2}} = \infty$$

R.H.L. $= \lim_{x\to -1^+} \dfrac{1}{\sqrt{|x|-\{-x\}}} = \lim_{x\to -1^-} \dfrac{1}{\sqrt{-x-(x+1)}}$

$$= \lim_{x\to -1^-} \frac{1}{\sqrt{-2x-1}} = 1$$

Hence, the limit does not exist.

49. a. For $n > 1$,

$\lim_{x\to 0} x^n \sin(1/x^2) = 0 \times$ (any value between -1 to 1)

$= 0$

For $n < 0$,

$\lim_{x\to 0} x^n \sin(1/x^2) = \infty \times$ (any value between -1 to 1)

$= \infty$

50. b.

51. d. $\{e^x\} = \begin{cases} e^x, & x\to 0^- \\ e^x - 1, & x \to 0^+ \end{cases}$

$$\Rightarrow \lim_{x\to 0^-}(f(x))^{1/\{f(x)\}} = 1 \text{ and } \lim_{x\to 0^+}(f(x))^{1/\{f(x)\}} = e$$

52. a. $\lim_{x\to 2}\left[\left(\dfrac{x^3-4x}{x^3-8}\right)^{-1} - \left(\dfrac{\sqrt{x}(\sqrt{x}+\sqrt{2})}{(\sqrt{x}-\sqrt{2})(\sqrt{x}+\sqrt{2})} - \dfrac{\sqrt{2}}{\sqrt{x}-\sqrt{2}}\right)^{-1}\right]$

$$= \lim_{x\to 2}\left[\frac{x^2+2x+4}{x(x+2)} - \left(\frac{\sqrt{x}-\sqrt{2}}{\sqrt{x}-\sqrt{2}}\right)^{-1}\right]$$

$$= \lim_{x\to 2}\left[\frac{x^2+2x+4}{x(x+2)} - 1\right] = \frac{12}{8} - 1 = \frac{1}{2}$$

53. d. $\lim_{x\to\infty} \dfrac{e^{1/x^2}-1}{2\tan^{-1}(x^2)-\pi} = \lim_{t\to 0^+} \dfrac{e^{t^2}-1}{2\cot^{-1}t^2-\pi}$

$$= \lim_{t\to 0^+} \frac{e^{t^2}-1}{-2\tan t^2}$$

$$= \lim_{t\to 0^+} -\frac{1}{2}\,\frac{\dfrac{e^{t^2}-1}{t^2}}{\dfrac{\tan t^2}{t^2}} = -\frac{1}{2}$$

54. b. $\lim_{x\to 0} \dfrac{1+\sin x - \cos x + \log(1-x)}{x^3}$

$$= \lim_{x\to 0} \frac{1+\left(x-\dfrac{x^3}{3!}+\cdots\right)-\left(1-\dfrac{x^2}{2!}+\dfrac{x^4}{4!}-\cdots\right)+\left(-x-\dfrac{x^2}{2}-\dfrac{x^3}{3}-\cdots\right)}{x^3}$$

$$= -\frac{1}{3!} - \frac{1}{3} = -\frac{1}{2}$$

55. b. $\cos(\tan x) - \cos x = 2\sin\left(\frac{x+\tan x}{2}\right)\sin\left(\frac{x-\tan x}{2}\right)$

$$\Rightarrow \lim_{x\to 0}\frac{\cos(\tan x)-\cos x}{x^4} = \lim_{x\to 0}\frac{2\sin\left(\frac{x+\tan x}{2}\right)\sin\left(\frac{x-\tan x}{2}\right)}{x^4}$$

$$= \lim_{x\to 0}\frac{2\sin\left(\frac{x+\tan x}{2}\right)\sin\left(\frac{x-\tan x}{2}\right)}{x^4\left(\frac{x+\tan x}{2}\right)\left(\frac{x-\tan x}{2}\right)}\left(\frac{x^2-\tan^2 x}{4}\right)$$

$$= \frac{1}{2}\lim_{x\to 0}\frac{x^2-\tan^2 x}{x^4}$$

$$= \frac{1}{2}\lim_{x\to 0}\frac{x^2-\left(x+\frac{x^3}{3}+\frac{2}{15}x^5+\cdots\right)^2}{x^4}$$

$$= \frac{1}{2}\lim_{x\to 0}\frac{1}{x^2}\left(1-\left(1+\frac{x^2}{3}+\frac{2}{15}x^4+\cdots\right)^2\right) = \frac{1}{3}$$

56. b. $x_{n+1} = \sqrt{2+x_n}$

$\Rightarrow \lim_{n\to\infty} x_{n+1} = \sqrt{2+\lim_{n\to\infty} x_n}$

$\Rightarrow t = \sqrt{2+t}$

$\left(\because \lim_{x\to\infty} x_{n+1} = \lim_{x\to\infty} x_n = t\right)$

$\Rightarrow t^2 - t - 2 = 0$

$\Rightarrow (t-2)(t+1) = 0$

$\Rightarrow t = 2$

$(\because x_n > 0\ \forall\ n \therefore t > 0)$

57. b. $\lim_{x\to 0}\left(\frac{1^x+2^x+\cdots+n^x}{n}\right)^{1/x} = e^{\lim_{x\to 0}\left(\frac{1^x-1}{n}+\frac{2^x-1}{n}+\cdots+\frac{n^x-1}{n}\right)\frac{1}{x}}$

$= e^{\lim_{x\to 0}\frac{1}{n}\left\{\frac{1^x-1}{x}+\frac{2^x-1}{x}+\cdots+\frac{n^x-1}{x}\right\}}$

$= e^{\frac{1}{n}[\log 1+\log 2+\cdots+\log n]}$

$= e^{\frac{1}{n}(\log n!)}$

$= e^{\log(n!)^{\frac{1}{n}}} = (n!)^{\frac{1}{n}}$

58. c. $\lim_{x\to 0}\frac{a^{\sqrt{x}}-a^{1/\sqrt{x}}}{a^{\sqrt{x}}+a^{1/\sqrt{x}}}, a > 1$

Put $x = t^2$

$\therefore \lim_{t\to 0}\frac{a^t - a^{1/t}}{a^t + a^{1/t}}$

$\Rightarrow \lim_{t\to 0}\frac{a^{t-1/t}-1}{a^{t-1/t}+1} = \frac{a^{-\infty}-1}{a^{-\infty}+1} = \frac{0-1}{0+1} = -1$

59. a.i. $\lim_{x\to\infty}\sec^{-1}\left(\frac{x}{\sin x}\right) = \sec^{-1}\left(\frac{\infty}{\sin\infty}\right)$

$= \sec^{-1}\left(\frac{\infty}{\text{any value between } -1 \text{ to } 1}\right)$

$= \sec^{-1}(\pm\infty) = \frac{\pi}{2}$

ii. $\lim_{x\to\infty}\sec^{-1}\left(\frac{\sin x}{x}\right) = \sec^{-1}\left(\frac{\sin\infty}{\infty}\right)$

$\sec^{-1}\left(\frac{\text{any value between } -1 \text{ to } 1}{\infty}\right)$

$\sec^{-1} 0 =$ not defined

Hence (i) exists but (ii) does not exist.

60. b. For $n = 0$, we have $\lim_{x\to 0}\frac{1-\sin 1}{x-1} = \sin 1 - 1$

For $n = 1$, $\lim_{x\to 0}\frac{x-\sin x}{x-\sin x} = 1$

For $n = 2$, $\lim_{x\to 0}\frac{x^2-\sin^2 x}{x-\sin^2 x} = \lim_{x\to 0}\frac{1-\frac{\sin^2 x}{x^2}}{\frac{1}{x}-\frac{\sin^2 x}{x^2}}$

This does not exist.

For $n = 3$ also given limit does not exist.

Hence, $n = 0$ or 1.

61. c. $\lim_{x\to -2^-}\frac{ae^{1/|x+2|}-1}{2-e^{1/|x+2|}} = \lim_{x\to -2^-}\frac{a-e^{-1/|x+2|}}{2e^{-1/|x+2|}-1} = -a$

$\lim_{x\to -2^-}\sin\left(\frac{x^4-16}{x^5+32}\right) = \lim_{x\to -2^-}\sin\left(\frac{(x-2)(x^2+4)}{x^4-2x^3+4x^2-8x+16}\right)$

$= \sin\left(-\frac{2}{5}\right) \Rightarrow a = \sin\frac{2}{5}$

62. b. Given limit is $\lim_{x\to\infty}(x+1)[\tan^{-1}(x+5)-(x+1)] + 4\tan^{-1}(x+5)$

$= \lim_{x\to\infty}\left[(x+1)\tan^{-1}\frac{4}{1+(x+1)(x+5)} + 4\tan^{-1}(x+5)\right]$

$= \lim_{x\to\infty}\left[(x+1)\tan^{-1}\frac{\frac{4}{x^2+6x+6}}{\left(\frac{4}{x^2+6x+6}\right)}\times\frac{4}{x^2+6x+6} + 4\tan^{-1}(x+5)\right]$

$= 0 + 4\times\frac{\pi}{2} = 2\pi$

63. b. $\lim\limits_{x\to1}\dfrac{(1-x)(1-x^2)\cdots(1-x^{2n})}{\{(1-x)(1-x^2)\cdots(1-x^n)\}^2}$

$$=\lim_{x\to1}\frac{\left(\frac{1-x}{1-x}\right)\left(\frac{1-x^2}{1-x}\right)\cdots\left(\frac{1-x^{2n}}{1-x}\right)}{\left(\left(\frac{1-x}{1-x}\right)\left(\frac{1-x^2}{1-x}\right)\cdots\left(\frac{1-x^n}{1-x}\right)\right)^2}$$

$$=\frac{1\times2\times3\cdots(2n)}{(1\times2\times3\cdots n)^2}$$

$$=\frac{(2n)!}{n!n!}$$

$$={}^{2n}C_n$$

64. b. We know that $\lim\limits_{x\to0}\dfrac{\sin x}{x}\to1^-$ and $\lim\limits_{x\to0}\dfrac{x}{\sin x}\to1^+$

So, $\lim\limits_{x\to0}\left[100\dfrac{x}{\sin x}\right]+\lim\limits_{x\to0}\left[99\dfrac{\sin x}{x}\right]$

$=100+98=198$

65. a. Let $\sin^{-1}x=\theta$. Then, $x=\sin\theta$.

Now, $x\to\dfrac{1}{\sqrt2}\Rightarrow\sin\theta\to\dfrac{1}{\sqrt2}\Rightarrow\theta\to\dfrac{\pi}{4}$

$$\therefore\ \lim_{x\to1/\sqrt2}\frac{x-\cos\left(\sin^{-1}x\right)}{1-\tan\left(\sin^{-1}x\right)}=\lim_{\theta\to\pi/4}\frac{\sin\theta-\cos\theta}{1-\tan\theta}$$

$$=\lim_{\theta\to\pi/4}\frac{(\sin\theta-\cos\theta)}{(\cos\theta-\sin\theta)}\cos\theta$$

$$=\lim_{\theta\to\pi/4}-\cos\theta=-\frac{1}{\sqrt2}$$

66. d. We have $\lim\limits_{x\to1}\dfrac{1-\sqrt{x}}{\left(\cos^{-1}x\right)^2}=\lim\limits_{x\to1}\dfrac{\left(1-\sqrt{x}\right)\left(1+\sqrt{x}\right)}{\left(\cos^{-1}x\right)^2\left(1+\sqrt{x}\right)}$

$$=\lim_{x\to1}\frac{1-x}{\left(\cos^{-1}x\right)^2\left(1+\sqrt{x}\right)}$$

$$=\lim_{\theta\to0}\frac{1-\cos\theta}{\theta^2\left(1+\sqrt{\cos\theta}\right)}\text{1, where } x=\cos\theta$$

$[\because x\to1\Rightarrow\cos\theta\to1\Rightarrow\theta\to0]$

$$=\lim_{\theta\to0}\frac{1-\cos\theta}{\theta^2}\frac{1}{\left(1+\sqrt{\cos\theta}\right)}$$

$$=\lim_{\theta\to0}\frac{2\sin^2\frac{\theta}{2}}{4\frac{\theta^2}{4}}\left(\frac{1}{1+\sqrt{\cos\theta}}\right)$$

$$=\frac{1}{2}\lim_{\theta\to0}\left(\frac{\sin\frac{\theta}{2}}{\frac{\theta}{2}}\right)^2\frac{1}{\left(1+\sqrt{\cos\theta}\right)}=\frac{1}{2}(1)^2\frac{1}{(1+1)}=\frac{1}{4}$$

67. b. $\min(y^2-4y+11)=\min[(y-2)^2+7]=7$

$$\Rightarrow L=\lim_{x\to0}\left[\min(y^2-4y+11)\frac{\sin x}{x}\right]$$

$$=\lim_{x\to0}\left[\frac{7\sin x}{x}\right]$$

= a value slightly lesser than 7

$(|\sin x|<|x|$, when $x\to0)$

$$\Rightarrow L=\lim_{x\to0}\left[7\frac{\sin x}{x}\right]=6$$

68. b. $L=\lim\limits_{x\to\pi/2}\dfrac{\sin(x\cos x)}{\sin\left(\frac{\pi}{2}-x\sin x\right)}$

$$=\lim_{x\to\pi/2}\frac{\sin(x\cos x)}{(x\cos x)}\ \frac{x\cos x}{\sin\left(\frac{\pi}{2}-x\sin x\right)}\ \frac{\left(\frac{\pi}{2}-x\sin x\right)}{\left(\frac{\pi}{2}-x\sin x\right)}$$

$$=1\times1\lim_{x\to\pi/2}\frac{x\cos x}{\left(\frac{\pi}{2}-x\sin x\right)}$$

Put $x=\pi/2+h$

$$\text{Then, } L=\lim_{h\to0}\frac{\left(\frac{\pi}{2}+h\right)\cos\left(\frac{\pi}{2}+h\right)}{\frac{\pi}{2}-\left(\frac{\pi}{2}+h\right)\sin\left(\frac{\pi}{2}+h\right)}$$

$$=\lim_{h\to0}\frac{-\left(\frac{\pi}{2}+h\right)\sin h}{\frac{\pi}{2}(1-\cos h)-h\cos h}$$

$$=\lim_{h\to0}\frac{-\left(\frac{\pi}{2}+h\right)\left(\frac{\sin h}{h}\right)}{\frac{\pi}{2}\frac{(1-\cos h)}{h}-\cos h}$$

[Divide N^r and D^r by h]

$$=\frac{-\left(\frac{\pi}{2}+0\right)1}{0-1}=\frac{\pi}{2}$$

69. a. $\lim_{x\to 0} \frac{\sin 3x}{x^3} + \frac{a}{x^2} + b = \lim_{x\to 0} \frac{\sin 3x + ax + bx^3}{x^3}$

$$= \lim_{x\to 0} \frac{3\frac{\sin 3x}{3x} + a + bx^2}{x^2}$$

For existence, if limit $(3+a)=0$

$\Rightarrow a = -3$

$$\therefore\ 1 = \lim_{x\to 0} \frac{\sin 3x - 3x + bx^3}{x^3}$$

$$= 27 \lim_{t\to 0} \frac{\sin t - t}{t^3} + \text{b} = 0\ (3x = \text{t})$$

$$= -\frac{27}{6} + b = 0$$

$$\Rightarrow b = \frac{9}{2}$$

70. b. $\lim_{x\to 0} \frac{x^n \sin^n x}{x^n - \sin^n x}$

$$\Rightarrow \lim_{x\to 0} \frac{x^n\left(x - \frac{x^3}{3!} + \frac{x^5}{5!} - \cdots\right)^n}{x^n - \left(x - \frac{x^3}{3!} + \frac{x^5}{5!} - \cdots\right)^n}$$

$$= \lim_{x\to 0} \frac{\left(x - \frac{x^3}{3!} + \frac{x^5}{5!} - \cdots\right)^n}{1 - \left(1 - \frac{x^2}{3!} + \frac{x^4}{5!} - \cdots\right)^n}$$

$$= \lim_{x\to 0} \frac{x^n\left(1 - \frac{x^2}{3!} + \frac{x^4}{5!} - \cdots\right)^n}{1 - \left(1 - \frac{x^2}{3!} + \frac{x^4}{5!} - \cdots\right)^n}$$

For $n = 2$,

$$\lim_{x\to 0} \frac{x^2\left(1 - \frac{x^2}{3!} + \frac{x^4}{5!} - \cdots\right)^2}{1 - \left(1 - \frac{x^2}{3!} + \frac{x^4}{5!} - \cdots\right)^2}$$

$$\Rightarrow \lim_{x\to 0} \frac{x^2\left(1 - \frac{x^2}{3!} + \frac{x^4}{5!} - \cdots\right)^2}{\left(2 - \frac{x^2}{3!} + \frac{x^4}{5!} - \cdots\right)\left(\frac{x^2}{3!} - \frac{x^4}{5!} + \cdots\right)}$$

$$= \frac{1(1 - 0 + \cdots)^2}{(2 - 0 + 0)\left(\frac{1}{3!} - 0 + \cdots\right)}$$

$$= 3$$

71. d. $\lim_{x\to\infty} \frac{1 + x + x^2}{x(\ln x)^3} = \lim_{t\to 0^+} \frac{t^2 + t + 1}{t^2 \frac{1}{t}\left(\ln\left(\frac{1}{t}\right)\right)^3}$

$$= \lim_{t\to 0^+} \frac{1 + t + t^2}{-t(\ln t)^3} = +\infty$$

Multiple Correct Answers Type

1. b., c.

R.H.L. $= \lim_{h\to 0} f(1+h) = \lim_{h\to 0} a(1+h) = a$

L.H.L. $= \lim_{h\to 0} f(1+h)$

$$= \lim_{h\to 0} a\left\{1 + \frac{2}{a}(1+h)\right\} = 1 + \frac{2}{a}$$

$\lim_{x\to 1} f(x)$ exists $\Rightarrow$ R.H.L. = L.H.L. $\Rightarrow a = 1 + \frac{2}{a}$

$\Rightarrow a = 2, -1$

2. a., d.

$f(1+0) = \lim_{h\to 0}\{|1 + h - 1| - [1 + h]\} = \lim_{h\to 0}\{h - 1\}$

$= -1$

$f(1-0) = \lim_{h\to 0}\{|1 - h - 1| - [1 - h]\} = \lim_{h\to 0}\{h - 0\} = 0$

3. a., c.

$$\text{Limit} = \lim_{x\to\infty} \frac{an(1+n) - (1+n^2)}{1+n}$$

$$= \lim_{x\to\infty} \frac{(a-1)n^2 + an - 1}{n+1}$$

$$= \infty \text{ if } a - 1 \neq 0$$

If $a - 1 = 0$, limit $= \lim_{n\to\infty} \frac{an - 1}{n + 1} = a = b$

$\therefore\quad a = b = 1$

4. a., b., c.

$$L = \lim_{x\to 0} \frac{\sin x^n}{(\sin x)^m} = \lim_{x\to 0} \frac{\frac{\sin x^n}{x^n} x^n}{\frac{(\sin x)^m}{x^m} x^m} = \lim_{x\to 0} x^{n-m}$$

If $n = m$, then

$L = (\text{a very small value near to zero})^{\text{exactly zero}} = 1$

If $n > m$, then
$L = (\text{a very small value near to zero})^{\text{positive integer}} = 0$
If $n < m$, then
$L = (\text{a very small value near to zero})^{\text{negative integer}} = \infty$

5. a., b., c.

$$\lim_{x\to\infty} \frac{\log_e x}{\{x\}} = \frac{\text{Positive infinity}}{\text{A value between 0 and 1}} = \infty$$

$$\lim_{x\to 2^+} \frac{x}{x^2 - x - 2} = \lim_{x\to 2^+} \frac{x}{(x-2)(x+1)}$$

$$= \lim_{h\to 0} \frac{2+h}{h(3+h)} = \infty$$

$$\lim_{h\to -1} \frac{x}{x^2 - x - 2} = -\lim_{h\to -1^-} \frac{x}{(x-2)(x+1)}$$

$$= \lim_{h\to 0} \frac{1-h}{(-3-h)(-h)}$$

$$= \lim_{h\to 0} \frac{1+h}{(3+h)(h)} = -\infty$$

6. b., c.

Since the greatest integer function is discontinuous (sensitive) at integral values of x, then for a given limit to exist both left- and right-hand limit must be equal.

$$\text{L.H.L.} = \lim_{x\to 1^-} (2 - x + a[x-1] + b[1+x])$$

$$= 2 - 1 + a(-1) + b(1) = 1 - a + b$$

$$\text{R.H.L.} = \lim_{x\to 1^+} (2 - x + a[x-1] + b[1+x])$$

$$= 2 - 1 + a(0) + b(2) = 1 + 2b$$

On comparing we have $-a = b$

7. a., b., c.

$$L = \lim_{x\to a} \frac{|2\sin x - 1|}{2\sin x - 1}$$

For $a = \pi/6$, $\text{L.H.L.} = \lim_{x\to\pi^-/6} \frac{1 - 2\sin x}{2\sin x - 1} = -1$,

$$\text{R.H.L.} = \lim_{x\to\pi^+/6} \frac{2\sin x - 1}{2\sin x - 1} = 1$$

Hence the limit does not exist.

For $a = \pi$, $\lim_{x\to\pi^-} \frac{1 - 2\sin x}{2\sin x - 1} = -1$ (as in neighbourhood of π, $\sin x$ is less than ½).

For $a = \pi$, $\lim_{x\to\pi/2^-} \frac{2\sin x - 1}{2\sin x - 1}$ (as in neighbourhood of $\pi/2$, $\sin x$ approaches to 1).

8. b., c., d.

$$f(x) = \lim_{x\to\infty} \frac{x}{x^{2n} + 1}$$

$$= \begin{cases} x, & x^2 < 1 \\ 0, & x^2 > 1 \\ 1/2, & x = 1 \\ -1/2, & x = -1 \end{cases}$$

$$\Rightarrow f(1^+) = f(-1^-) = 0$$

$$f(1^-) = 1,\ f(-1^+) = -1$$

$$f(1) = 1/2$$

9. a., c.

Case I: If n is even, say $n = 2k$. Then the limit is

$$\lim_{k\to\infty} \frac{-6k + (-1)^{2k}}{8k - (-1)^{2k}} = \lim_{k\to\infty} \frac{-6k+1}{8k-1}$$

$$= \lim_{k\to\infty} \frac{-6 - \frac{2}{k}}{8 + \frac{1}{k}} = \frac{-6}{8} = \frac{-3}{4}$$

Case II: If n is odd, say $n = 2k + 1$. Then the limit is

$$\lim_{k\to\infty} \frac{-3(2k+1) - (-1)^{2k+1}}{4(2k+1) - (-1)^{2k+1}} = \lim_{k\to\infty} \frac{-6k - 3 + 1}{8k + 4 + 1}$$

$$= \lim_{k\to\infty} \frac{-6 - \frac{2}{k}}{8 + \frac{5}{k}} = \frac{-3}{4}$$

$$\therefore \lim_{k\to\infty} \frac{-3n + (-1)^n}{4n - (-1)^n} = \frac{-3}{4}, \ n \text{ even or odd.}$$

10. a., b., c., d.

We have $\lim_{k\to 0^+} f(x) = \lim_{k\to 0^+} \frac{\tan^2\{x\}}{\left(x^2 - [x]^2\right)}$

$$= \lim_{x\to 0^+} \frac{\tan^2 x}{x^2} = 1 \quad (1)$$

$[\because x \to 0^+; [x] = 0 \Rightarrow \{x\} = x]$

Also $\lim_{k\to 0^-} f(x) = \lim_{x\to 0^-} \sqrt{\{x\}\cot\{x\}} = \sqrt{\cot 1} \quad (2)$

$[\because x \to 0^-; [x] = -1 \Rightarrow \{x\} = x + 1 \Rightarrow \{x\} \to 1]$

Also, $\cot^{-1}\left(\lim_{x\to 0^-} f(x)\right)^2 = \cot^{-1}(\cot 1) = 1.$

Reasoning Type

1. b. For $x \in (-\delta, \delta)$, $\sin x < x \Rightarrow \lim_{x\to 0} \frac{\sin x}{x} = 1^-$

$$\Rightarrow \left[\lim_{x\to 0} \frac{\sin x}{x}\right] = 0$$

Also, $x \in (-\delta, \delta)$, $\tan x > x$, but from this nothing can be said about the relation between $\sin x$ and x. Hence, both the statements are true but statement 2 is not the correct explanation of statement 1.

2. a. For $\lim_{x \to \alpha} \frac{\sin(f(x))}{x-\alpha}$, denominator tends to 0, hence the numerator must also tend to 0 for limit to be finite. Then, a is a root of the equation $ax^2 + bx + c = 0$ or $f(\alpha) = 0$.
Also, consider $f(\alpha^+) \to 0^+$ and $f(\alpha^-) \to 0^-$

$$\Rightarrow \lim_{x \to \alpha^+} \frac{e^{1/f(x)} - 1}{e^{1/f(x)} + 1} = \lim_{x \to \alpha^+} \frac{1 - e^{-1/f(x)}}{1 + e^{-1/f(x)}} = 1$$

and $\lim_{x \to \alpha^-} \frac{e^{1/f(x)} - 1}{e^{1/f(x)} + 1} = -1$

Thus, both statements are true and statement 2 is the correct explanation of statement 1.

3. b. Limit of function $y = f(x)$ exists at $x = a$ though it is discontinuous at $x = a$, consider the function $f(x) = \frac{x^2 - 4}{x - 2}$. Here, $f(x)$ is not defined at $x = 2$, but limit of functions exists, as $\lim_{x \to 2^+} f(x) = \lim_{x \to 2^-} = 4$.

4. a. $L = \lim_{x \to 0^+} \frac{x}{a}\left[\frac{b}{x}\right]$

$$= \lim_{x \to 0^+} \frac{x}{a}\left(\frac{b}{x} - \left\{\frac{b}{x}\right\}\right)$$

$$= \lim_{x \to 0^+} \left(\frac{b}{a} - \frac{x}{a}\left\{\frac{b}{x}\right\}\right)$$

$$= \frac{b}{a} - \frac{b}{a} \lim_{x \to 0^+} \frac{\left\{\frac{b}{x}\right\}}{\frac{b}{x}}$$

$$= \frac{b}{a} - \frac{b}{a} \lim_{y \to \infty} \frac{\{y\}}{y} \quad \left[\text{where } y = \frac{b}{x} \text{ and } b > 0\right]$$

$$= \frac{b}{a}$$

Also, if $b < 0$, $L = \frac{b}{a} - \frac{b}{a} \lim_{y \to -\infty} \frac{\{y\}}{y} = \frac{b}{a}$

5. d. $\lim_{x \to \infty} \left(\frac{1^2}{x^3} + \frac{2^2}{x^3} + \frac{3^2}{x^3} + \dots + \frac{x^2}{x^3}\right)$

$$= \lim_{x \to \infty} \frac{x(x+1)(2x+1)}{6x^3} = \frac{1}{3}$$

6. b. $L = \lim_{x \to 0} \frac{\sqrt{1 - \cos 2x}}{x} = \lim_{x \to 0} \frac{\sqrt{2}\,|\sin x|}{x}$

$\Rightarrow$ L.H.L. $= -\sqrt{2}$ and R.H.L. $= \sqrt{2}$
Hence, limit of the function does not exist.
Also, statement 2 is true, but it is not the correct explanation of statement 1. As for limit to exist, it is not necessary that function is defined at that point.

7. a. When $n \to \infty$ and x is rational or $x = p/q$, where p and q are integers and $q \neq 0$.

$n!x = n! \times \frac{p}{q}$ is integer as $n!$ has factor q when $n \to \infty$.
Also, when $n!x$ is integer, $\sin(n!\pi x) = 0 \Rightarrow$ given limit is zero.

8. d. Obviously, statement 2 is true, as on the number line immediate neighbourhood of 1/2 is either rational or irrational, but this does not stop $f(x)$ to have limit at $x = 1/2$.
As $f(1/2) = 1/2, f(1/2^+) = \lim_{x \to 1/2^+} x = 1/2$ (if $1/2^+$ is rational)
or $\lim_{x \to 1/2^+} (1 - x) = 1 - 1/2 = 1/2$ (if $1/2^+$ is irrational)
Hence, $\lim_{x \to 1/2^+} f(x) = 1/2$.
With similar argument, we can prove that $\lim_{x \to 1/2^-} f(x) = 1/2$. Hence, limit of function exists at $x = 1/2$.

9. a. $\lim_{x \to \infty} \frac{(x-1)(x-2)}{(x-3)(x-4)} = \lim_{x \to \infty} \frac{x^2 - 3x + 2}{x^2 - 7x + 12}$

$$= \lim_{x \to \infty} \frac{1 - \frac{3}{x} + \frac{2}{x^2}}{1 - \frac{7}{x} + \frac{12}{x^2}} \to 1$$

[From right-hand side of 1]

Hence $\lim_{x \to \infty} \cos^{-1} f(x)$ does not exist as $\cos^{-1} x$ is defined for $x \in [-1, 1]$.

Also, $\lim_{x \to -\infty} \frac{1 - \frac{3}{x} + \frac{2}{x^2}}{1 - \frac{7}{x} + \frac{12}{x^2}} \to 1$ [From left-hand side of 1]

Hence, $\lim_{x \to \infty} \cos^{-1} f(x)$ exists.

10. b. $\lim_{x \to 0^+} [x]\left(\frac{e^{1/x} - 1}{e^{1/x} + 1}\right) = \lim_{h \to 0} [h]\left(\frac{1 - e^{-1/h}}{1 + e^{-1/h}}\right) = 0 \times 1 = 0$

$$\lim_{x \to 0^-} [x]\left(\frac{e^{1/x} - 1}{e^{1/x} + 1}\right) = \lim_{h \to 0} [-h]\left(\frac{e^{-1/h} - 1}{e^{-1/h} + 1}\right)$$

$$= -1 \times (-1) = 1$$

Thus, given limit does not exists. Also

$\lim_{x\to 0}\left(\frac{e^{1/x}-1}{e^{1/x}+1}\right)$ does not exist, but this cannot be taken as only reason for non-existence of

$\lim_{x\to 0}[x]\left(\frac{e^{1/x}-1}{e^{1/x}+1}\right)$.

11. a. If $\lim_{x\to 0} f(x)$ exists, then $\lim_{x\to 0}\left(f(x)+\frac{\sin x}{x}\right)$ always exists as $\lim_{x\to 0}\frac{\sin x}{x}$ exists finitely.

Hence $\lim_{x\to 0} f(x)$ must not exist.

12. a. $\lim_{n\to\infty} a_{n+1} = \lim_{n\to\infty} \sin a_n = \lim_{n\to\infty} a_n$

$\Rightarrow \lim_{n\to\infty}(a_n - \sin a_n) = 0$ which is possible only when $\lim_{n\to\infty} a_n = 0$.

13. c. Obviously, statement 1 is true, but statement 2 is not always true.
Consider, $f(x) = [x]$ and $g(x) = \sin x$.

Linked Comprehension Type

For Problems 1–3

1. a., 2. b., 3. d.

Sol.

We have $f(x) = \frac{\sin^{-1}(1-\{x\})\cos^{-1}(1-\{x\})}{\sqrt{2\{x\}}(1-\{x\})}$

$\therefore \lim_{x\to 0^+} f(x) = \lim_{h\to 0} f(0+h)$

$$= \lim_{h\to 0}\frac{\sin^{-1}(1-\{0+h\})\cos^{-1}(1-\{0+h\})}{\sqrt{2\{0+h\}}(1-\{0+h\})}$$

$$= \lim_{h\to 0}\frac{\sin^{-1}(1-h)\cos^{-1}(1-h)}{\sqrt{2h}(1-h)}$$

$$= \lim_{h\to 0}\frac{\sin^{-1}(1-h)}{(1-h)}\lim_{h\to 0}\frac{\cos^{-1}(1-h)}{\sqrt{2h}}$$

In second limit put $1-h=\cos\theta$

$$= \lim_{h\to 0}\frac{\sin^{-1}(1-h)}{(1-h)}\lim_{\theta\to 0}\frac{\cos^{-1}(\cos\theta)}{\sqrt{2(1-\cos\theta)}}$$

$$= \lim_{h\to 0}\frac{\sin^{-1}(1-h)}{(1-h)}\lim_{\theta\to 0}\frac{\theta}{2\sin(\theta/2)}$$

$(\because \theta > 0)$

$= \sin^{-1}1\times 1 = \pi/2$

and $\lim_{x\to 0^-} f(x) = \lim_{h\to 0} f(0-h)$

$$= \lim_{h\to 0}\frac{\sin^{-1}(1-\{0-h\})\cos^{-1}(1-\{0-h\})}{\sqrt{2\{0-h\}}\,(1-\{0-h\})}$$

$$= \lim_{h\to 0}\frac{\sin^{-1}(1+h-1)\cos^{-1}(1+h-1)}{\sqrt{2(-h+1)}\,(1+h-1)}$$

$$= \lim_{h\to 0}\frac{\sin^{-1}h}{h}\lim_{h\to 0}\frac{\cos^{-1}h}{\sqrt{2(1-h)}} = 1\frac{\pi/2}{\sqrt{2}} = \frac{\pi}{2\sqrt{2}}$$

For Problems 4–6

4. c., 5. d., 6. d.

Sol.

We have $A_i = \frac{x-a_i}{-(x-a_i)} = -1$, $i = 1, 2, \ldots, n$ and $a_1 < a_2 < \cdots < a_{n-1} < a_n$.

Let x be in the left neighbourhood of a_m.

Then $x-a_i < 0$ for $i = m, m+1, \ldots, n$ and $x-a_i > 0$ for $i = 1, 2, \cdots, m-1$. Therefore,

$A_i = \frac{x-a_i}{-(x-a_i)} = -1$ for $i = m, m+1, \ldots, n$ and

$A_i = \frac{x-a_i}{x-a_i} = 1$ for $i = 1, 2, \ldots, m-1$

Similarly, if x is in the right neighbourhood of a_m, then

$x-a_i < 0$ for $i = m+1, \ldots, n$ and $x-a_i > 0$ for $i = 1, 2, \ldots, m$.

$\therefore \quad A_i = \frac{x-a_i}{-(x-a_i)} = -1$ for $i = m+1, \ldots, n$ and

$A_i = \frac{x-a_i}{x-a_i} = 1$ for $i = 1, 2, \ldots, m$

Now, $\lim_{x\to a_m^-}(A_1 A_2 \cdots A_n) = (-1)^{n-m+1}$ and

$\lim_{x\to a_m^+}(A_1 A_2 \cdots A_n) = (-1)^{n-m}$

Hence, $\lim_{x\to a_m}(A_1 A_2 \cdots A_n)$ does not exist.

For Problems 7–9

7. b., 8. d., 9. c.

Sol.

$$L = \lim_{x\to 0}\frac{\sin x + ae^x + be^{-x} + c\ln(1+x)}{x^3}$$

$$= \lim_{x\to 0}\frac{\left(x-\frac{x^3}{3!}\right) + a\left(1+\frac{x}{1!}+\frac{x^2}{2!}+\frac{x^3}{3!}\right)}{x^3}$$

$$\frac{+b\left(1-\frac{x}{1!}+\frac{x^2}{2!}-\frac{x^3}{3!}\right)+c\left(x-\frac{x^2}{2}+\frac{x^3}{3}\right)}{x^3}$$

$$= \lim_{x\to 0}\frac{(a+b)+(1+a-b+c)x+\left(\frac{a}{2}+\frac{b}{2}-\frac{c}{2}\right)x^2}{x^3}$$

$$\frac{+\left(-\frac{1}{3!}+\frac{a}{3!}-\frac{b}{3!}+\frac{c}{3}\right)x^3}{x^3}$$

$$\Rightarrow a+b=0,\ 1+a-b+c=0,\ \frac{a}{2}+\frac{b}{2}-\frac{c}{2}=0$$

and $L=-\frac{1}{3!}+\frac{a}{3!}-\frac{b}{3!}+\frac{c}{3}$

Solving first three equations, we get $c = 0$, $a = -1/2$, $b = 1/2$.

Then, $L = -1/3$

Equation $ax^2 + bx + c = 0$ reduces to $x^2 - x = 0$ $\Rightarrow x = 0,$

$1||x+c|-2a| < 4b$ reduces to $||x|+1| < 2$.

$\Rightarrow -2 < |x|+1 < 2$

$\Rightarrow 0 \le |x| < 1$

$\Rightarrow x \in [-1, 1]$

Matrix-Match Type

1. a → s; b → r; c → p; d → q.

a. Let $x+1=h$

Then, $\lim_{x\to -1}\frac{\sqrt[3]{(7-x)}-2}{(x+1)} = \lim_{h\to 0}\frac{(8-h)^{1/3}-2}{h}$

$$= \lim_{h\to 0}\frac{2\left(1-\frac{h}{8}\right)^{1/3}-2}{h}$$

$$= 2\lim_{h\to 0}\frac{\left(1-\frac{1}{3}\frac{h}{8}\right)-1}{h}$$

$$= -\frac{1}{12}$$

b. We have $\lim_{x\to\pi/4}\frac{\tan^3 x-\tan x}{\cos(x+\pi/4)}$

$$= \lim_{x\to\pi/4}\frac{\tan x(\tan x-1)(\tan x+1)}{\cos(x+\pi/4)}$$

$$= \lim_{x\to\pi/4}\frac{\tan x(\sin x-\cos x)(\tan x+1)}{\cos x\cos(x+\pi/4)}$$

$$= -\lim_{x\to\pi/4}\frac{\tan x(\cos x-\sin x)(\tan x+1)}{\cos x\cos(x+\pi/4)}$$

$$= -\sqrt{2}\lim_{x\to\pi/2}\frac{\tan x\left(\frac{1}{\sqrt{2}}\cos x-\frac{1}{\sqrt{2}}\sin x\right)(\tan x+1)}{\cos x\cos(x+\pi/4)}$$

$$= -\sqrt{2}\lim_{x\to\pi/4}\frac{\tan x(\tan x+1)}{\cos x}$$

$$= -\sqrt{2}\times 2\times\sqrt{2} = -8$$

c. $\lim_{x\to 1}\frac{(2x-3)(\sqrt{x}-1)}{2x^2+x-3} = \lim_{x\to 1}\frac{(2x-3)(\sqrt{x}-1)}{(2x+3)(x-1)}$

$$= \lim_{x\to 1}\frac{(2x-3)(\sqrt{x}-1)}{(2x+3)(\sqrt{x}-1)(\sqrt{x}+1)}$$

$$= \lim_{x\to 1}\frac{(2x-3)}{(2x+3)(\sqrt{x}+1)}$$

$$\frac{2-3}{(2+3)(\sqrt{1}+1)}$$

$$= -1/10$$

d. $\lim_{x\to\infty}\frac{\log x^n-[x]}{[x]} = \lim_{x\to\infty}\frac{\log x^n}{[x]} - \lim_{x\to\infty}\frac{[x]}{[x]} = 0$

$$= \lim_{x\to\infty}\frac{\log x^n}{[x]} - \lim_{x\to\infty}\frac{[x]}{[x]}$$

$$= 0-1$$

$$= -1$$

2. a → q; b → r; c → q; d → p.

We know that $\lim_{x\to 0}\frac{\sin x}{x} = 1$ (but a value which is smaller than 1)

$\Rightarrow \left[\lim_{x\to 0}100\frac{\sin x}{x}\right] = 99$ (a value which is closer to 100 but smaller than 100)

$\left[\lim_{x\to 0}100\frac{x}{\sin x}\right] = 100$ (a value which is closer to 100 but bigger than 100)

$\left[\lim_{x\to 0}100\frac{\sin^{-1}x}{x}\right] = 100$ (a value which is closer to 100 but bigger than 100)

$\left[\lim_{x\to 0}100\frac{x}{\sin^{-1}x}\right] = 99$ (a value which is closer to 100 but smaller than 100)

$\lim_{x\to0}\frac{\tan x}{x}=1$ (but a value which is bigger than 1)

$\Rightarrow\left[\lim_{x\to0}100\frac{\tan x}{x}\right]=100$ (a value which is closer to 100 but bigger than 100)

$\left[\lim_{x\to0}100\frac{\tan^{-1}x}{x}\right]=99$ (a value which is closer to 100 but smaller than 100)

a. Hence $\lim_{x\to0}\left(\left[100\frac{\sin x}{x}\right]+\left[100\frac{\tan x}{x}\right]\right)=199$

b. $\lim_{x\to0}\left(\left[100\frac{x}{\sin x}\right]+\left[100\frac{\tan x}{x}\right]\right)=200$

c. $\lim_{x\to0}\left(\left[100\frac{\sin^{-1}x}{x}\right]+\left[100\frac{\tan^{-1}x}{x}\right]\right)=199$

d. $\lim_{x\to0}\left(\left[100\frac{x}{\sin^{-1}x}\right]+\left[100\frac{\tan^{-1}x}{x}\right]\right)=198$

3. a → r, q; b → p, q, r; c → r, s; d → r, s.

a. Here, $a>0$, if $a\le0$, then limit $=\infty$

$\therefore\lim_{x\to\infty}\frac{(\sqrt{(x^2-x+1)}-ax-b)(\sqrt{x^2-x+1)}+ax+b)}{(\sqrt{(x^2-x+1)}+ax+b)}$

$\Rightarrow\lim_{x\to\infty}\frac{(x^2-x+1)-(ax+b)^2}{\sqrt{(x^2-x+1)}+ax+b}$

$\Rightarrow\lim_{x\to\infty}\frac{(1-a^2)x^2-(1+2ab)x+(1-b^2)}{\sqrt{(x^2-x+1)}+ax+b}$

This is possible only when $1-a^2=0$ and $1+2ab=0$

$\therefore a=\pm1$

$\Rightarrow a=1\qquad(\because a>0)\qquad(1)$

$\Rightarrow b=-1/2$

$\Rightarrow(a,2b)=(1,-1)$

b. Divide numerator and denominator by $e^{1/x}$, then

$\lim_{x\to\infty}\frac{(1+a^3)e^{1/x}+8}{e^{1/x}+(1-b^3)}=2$

$\Rightarrow\frac{0+8}{0+1-b^3}=2$

$\Rightarrow1-b^3=4$

$\therefore b^3=-3\Rightarrow b=-3^{1/3}$

Then, $a\in R$

$\Rightarrow(a,b^3)=(a,-3)$

c. $\lim_{x\to\infty}(\sqrt{(x^4-x^2+1)}-ax^2-b)=0$

Put $x=\frac{1}{t}$ $\therefore\lim_{t\to0}\left(\sqrt{\left(\frac{1}{t^4}-\frac{1}{t^2}+1\right)}-\frac{a}{t^2}-b\right)=0$

$\Rightarrow\lim_{t\to0}\frac{\sqrt{(1-t^2+t^4)}-a-bt^2}{t^2}=0\qquad(1)$

Since R.H.S. is finite, numerator must be equal to 0 at $t\to0$.

$\therefore1-a=0\ \therefore a=1$

From equation (1), $\lim_{t\to0}\frac{\sqrt{(1-t^2+t^4)}-1-bt^2}{t^2}=0$

$\lim_{t\to0}(-1+t^2)\left(\frac{(1-t^2+t^4)^{1/2}-(1)^{1/2}}{(1-t^2+t^4)-1}\right)=b$

$\Rightarrow(-1)\left(\frac{1}{2}\right)=b\Rightarrow a=1,b=-\frac{1}{2}$

$\Rightarrow(a,-2b)=(1,2)$

d. $\lim_{x\to-a}\frac{a^7-(-x)^7}{a-(-x)}=7\Rightarrow7a^6=7$

$\Rightarrow a^6=1$

$\Rightarrow a=-1$

Integer Type

1. (3) As $x\to0^-\Rightarrow f(x)\to f(0^-)=2^+$

$\Rightarrow\quad\lim_{x\to0^-}g[f(x)]=g(2^+)=-3$

Also as $x\to0^+\Rightarrow f(x)\to f(0^+)=1^+$

$\Rightarrow\quad\lim_{x\to0^+}g[f(x)]=g(1^+)=-3$

Hence, $\lim_{x\to0}g[f(x)]$ exists and is equal to -3.

$\Rightarrow\quad\lim_{x\to0}g[f(x)]=-3$

2. (5) $\lim_{x\to0}\frac{5\sin x-7\sin2x+3\sin3x}{x^2\sin x}$

$=$

$$\lim_{x\to0}\frac{5\left(x-\frac{x^3}{3!}+\ldots\right)-7\left(2x-\frac{(2x)^3}{3!}+\ldots\right)+3\left(3x-\frac{(3x)^3}{3!}+\ldots\right)}{x^2\left(x-\frac{x^3}{3!}+\ldots\right)}$$

$$= \lim_{x\to 0}\frac{-\frac{5x^3}{3!}+\frac{56x^3}{3!}-\frac{81x^3}{3!}}{x^3\left(1-\frac{x^2}{3!}+...\right)}$$

$$= \frac{-5+56-81}{3!} = -5$$

3. (1) We have $L = \lim_{x\to 2}\frac{\sqrt{(x+7)}-3\sqrt{(2x-3)}}{\sqrt[3]{(x+6)}-2\sqrt[3]{(3x-5)}}\left(\frac{0}{0}\text{ form}\right)$

Let $x-2=t$ such that when $x\to 2, t\to 0$.

Then $L = \lim_{t\to 0}\frac{(t+9)^{1/2}-3(2t+1)^{1/2}}{(t+8)^{1/3}-2(3t+1)^{1/3}}\left(\frac{0}{0}\text{ form}\right)$

$$= \frac{3}{2}\lim_{t\to 0}\frac{\left(1+\frac{t}{9}\right)^{1/2}-(2t+1)^{1/2}}{\left(1+\frac{t}{8}\right)^{1/3}-(3t+1)^{1/3}}\left(\frac{0}{0}\text{ form}\right)$$

$$= \frac{3}{2}\lim_{t\to 0}\frac{\frac{1}{2}\frac{t}{9}-(2t)\frac{1}{2}}{\frac{t}{8}\frac{t}{3}-(3t)\frac{1}{3}} = \frac{3}{2}\frac{\frac{1}{18}-1}{\frac{1}{24}-1} = \frac{34}{23}$$

4. (7) If the initial number of inhabitant of a given country is A, then after a year the total population will be $A+\frac{A}{100}2 = \left(1+\frac{1}{50}\right)A$.

After two years, the population will be $[1+(1/50)]^2 A$. After 100 years, it will be $[1+(1/50)]^{100} A$, i.e., it will increase by $\{[1+(1/50)]^{50}\}^2$ times.

Taking into account that $\lim_{n\to\infty}[1+(1/n)]^n = e$, we can approximately consider that $[1+(1/50)]^{50}\approx e$.

Hence, after 100 years the population of the country will increase by $e^2\approx 7.39$ times.

Hence, $[k]=[7.39]=7$

5. (6) The given limit has 0/0 form.

Using L' Hospital Rule, we have

$$L = \lim_{x\to 0}\frac{2f'(x)-6f'(2x)+4f'(4x)}{2x}\left(\frac{0}{0}\text{ form}\right)$$

$$= \lim_{x\to 0}\frac{2f''(x)-12f''(2x)+16f''(4x)}{2}$$ (Using L' Hospital Rule)

$$= \frac{6f''(0)}{2} = 6$$

6. (1) $L = \lim_{x\to 0}\frac{\log(\tan^2 2x)}{\log(\tan^2 x)}\left(\frac{\infty}{\infty}\text{ form}\right)$

Using L' Hospital's Rule, we have

$$\lim_{x\to 0}\frac{\left(\frac{1}{\tan^2 2x}2\tan 2x\times\sec^2 2x\right)\times 2}{\frac{1}{\tan^2 x}2\tan x\times\sec^2 x}$$

$$= \lim_{x\to 0}\frac{2\left(\frac{1}{\sin 2x\cos 2x}\right)}{\left(\frac{1}{\sin x\cos x}\right)}$$

$$= \lim_{x\to 0}\frac{\left(\frac{1}{\sin 2x\cos 2x}\right)}{\left(\frac{1}{\sin 2x}\right)}$$

$$= \lim_{x\to 0}\frac{1}{\cos 2x} = 1$$

7. (1) Since $\lim_{x\to a}\frac{a^x-x^a}{x^x-a^a} = -1$

$$\Rightarrow \lim_{x\to a}\frac{a^x\log a - ax^{a-1}}{x^x(1+\log x)} = -1$$

$$\Rightarrow \frac{a^a(\log a-1)}{a^a(1+\log a)} = -1$$

$\Rightarrow \log a - 1 = -1-\log a$

$\Rightarrow 2\log a = 0$

$\Rightarrow \log a = 0$

$\Rightarrow a = 1$

8. (3) $\lim_{x\to 0}\frac{\cos 4x + a\cos 2x + b}{x^4}$ is finite.

Using expansion formula for cos 4x and cos 2x, we have

$$\lim_{x\to 0}\frac{\left(1-\frac{(4x)^2}{2!}+\frac{(4x)^4}{4!}\right)+a\left(1-\frac{(2x)^2}{2!}+\frac{(2x)^4}{4!}\right)+b}{x^4}$$

= finite

$$\Rightarrow \lim_{x\to 0}\frac{(1+a+b)+(-8-2a)x^2+\left(\frac{32}{3}+\frac{2}{3}a\right)x^4+...}{x^4}$$

$\Rightarrow 1+a+b=0$ (1)

and $-8-2a=0$ (2)

Solving (1) and (2) for a and b, we get $a=-4$ and $b=3$.

Also, $L = \frac{32}{3} + \frac{2}{3}a = \frac{32-8}{3} = 8$

9. (4) $\lim_{x\to 0} \frac{x(1+a\cos x) - b\sin x}{x^3} = 1 \left(\frac{0}{0}\text{form}\right)$

$\Rightarrow \lim_{x\to 0} \frac{1(1+a\cos x) + x(-a\sin x) - b\cos x}{3x^2} = 1$

(By L' Hospital's Rule)

Here numerator $\to 1 + a - b$ and denominator $\to 0$ and the limit is a finite number 1. Therefore,

$1 + a - b = 0,$ (1)

(If $1 + a - b \neq 0$, the limit will not be finite.)

$\Rightarrow \lim_{x\to 0} \frac{1 + a\cos x - ax\sin x - b\cos x}{3x^2} = 1 \left(\frac{0}{0}\text{form}\right)$

$\Rightarrow \lim_{x\to 0} \frac{0 - a\sin x - a\sin x - ax\cos x + b\sin x}{6x} = 1 \left(\frac{0}{0}\text{form}\right)$

$\Rightarrow$

$\lim_{x\to 0} \frac{-a\cos x - a\cos x - a\cos x + ax\sin x + b\cos x}{6}$

$= 1$

$\Rightarrow \frac{-3a + b}{6} = 1$

$\Rightarrow -3a + b = 6$ (2)

Solving (1) and (2), we get $a = -5/2$, $b = -3/2$.

10. (4) Given,

$\lim_{x\to 0} \frac{\cos^2 x - \cos x - e^x\cos x + e^x - \frac{x^3}{2}}{x^n}$

$= \lim_{x\to 0} \frac{(\cos x - 1)(\cos x - e^x) - \frac{x^3}{2}}{x^n}$

$= \lim_{x\to 0} \frac{\left(1 - \frac{x^2}{2!} + \frac{x^4}{4!} - \frac{x^6}{6!} + \ldots - 1\right)}{x^n}$

$= \frac{\left[\left(1 - \frac{x^2}{2!} + \frac{x^4}{4!} - \ldots\right) - \left(1 + x + \frac{x^2}{2!} + \frac{x^3}{3!} - \ldots\right)\right] - \frac{x^3}{2}}{x^n}$

$=$

$\lim_{x\to 0} \frac{\left(-\frac{x^2}{2!} + \frac{x^4}{4!} - \frac{x^6}{6!} + \ldots\right)\left[\left(-x - x^2 - \frac{x^3}{3!} - \frac{2x^5}{5!} - \ldots\right)\right] - \frac{x^3}{2}}{x^n}$

$= \lim_{x\to 0} \frac{\left(\frac{x^3}{2} + \frac{x^4}{2} + \frac{x^5}{12} - \frac{x^5}{24} + \ldots\right) - \frac{x^3}{2}}{x^n}$ = non-zero if n

$= 4.$

11. (3) As $n \to \infty$, $\lim_{n\to\infty} f(n) = \lim_{n\to\infty} f(n+1) = k$, say

We have $f(n+1) = \frac{1}{2}\left[f(n) + \frac{9}{f(n)}\right]$

$\Rightarrow \lim_{n\to\infty} f(n+1) = \lim_{n\to\infty} \frac{1}{2}\left[f(n) + \frac{9}{f(n)}\right]$

$\Rightarrow k = \frac{1}{2}\left(k + \frac{9}{k}\right)$

$\Rightarrow k^2 = 9$ or $k = 3$

$\Rightarrow \lim_{n\to\infty} f(n) = 3$

12. (1) Let $y = \cfrac{x}{x + \cfrac{\sqrt[3]{x}}{x + \sqrt[3]{x}\ldots\infty}}$

$= \cfrac{x}{x + \cfrac{1}{x^{2/3}} \times \cfrac{x}{x + \sqrt[3]{x}\ldots\infty}} = \cfrac{x}{x + \cfrac{y}{x^{2/3}}}$

$\Rightarrow y = \frac{x^{5/3}}{x^{5/3} + y}$

$\Rightarrow y^2 + (x^{5/3})y - x^{5/3} = 0$. Therefore,

$y = \frac{-x^{5/3} \pm \sqrt{x^{10/3} + 4x^{5/3}}}{2}$

$= \frac{-x^{5/3} + \sqrt{x^{10/3} + 4x^{5/3}}}{2}$ $(\because y > 0)$

$= \frac{4x^{5/3}}{2(\sqrt{(x^{10/3} + 4x^{5/3})} + x^{5/3})}$

$= \frac{2}{\sqrt{\left(1 + \frac{4}{x^{5/3}}\right)} + 1}$

Hence, $\lim_{x\to\infty} y = \frac{2}{\sqrt{1+0}+1} = \frac{2}{2} = 1$.

13. (2) Let $t = \sin x$

Then $P = \lim_{t\to 1} \frac{t - t^t}{1 - t + \ln t} \left(\frac{0}{0}\text{form}\right)$

Using L' Hospital's rule, we get

$$P=\frac{\lim\limits_{t\to1} 1-t^t(1+\ln t)}{0-1+\frac{1}{t}}\left(\frac{0}{0}\text{form}\right)$$

Again using L' Hospital's rule, we get

$$P=\lim_{t\to1}\frac{0-\left\{t^t\left(\frac{1}{t}\right)+(1+\ln t)\,t^t(1+\ln t)\right\}}{0-0-\frac{1}{t^2}}$$

$$=-\frac{(1+1)}{-1}=2$$

14. (4) Δ_1 = Area of ΔABC

$= R^2 \sin\theta\,(\sec\theta-\cos\theta)$

$= R^2\tan\theta\,(1-\cos^2\theta)$

Δ_2 = Area of ΔCDE

$$=\frac{R^2\,(1-\cos\theta)^2}{\cos^2\theta\tan\theta}\qquad \text{CM} = R\sec\theta - R$$

$$\text{DM} = \text{CM}\cos\theta$$

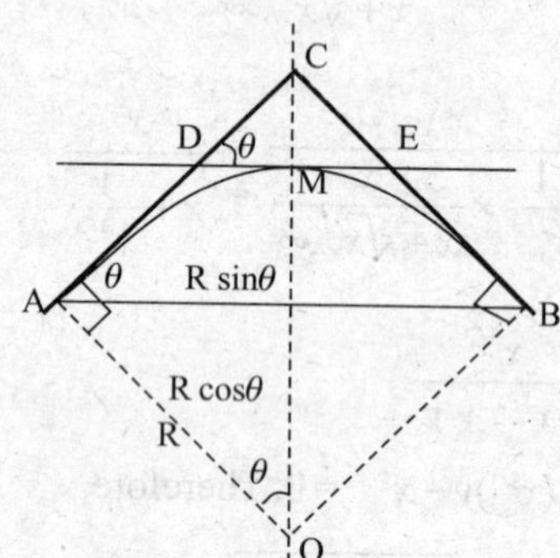

Fig. 2.1

$$\frac{\Delta_1}{\Delta_2}=\frac{\tan\theta\left(1-\cos^2\theta\right)\cos^2\theta\tan\theta}{(1-\cos\theta)^2}=L\text{, say}$$

$$\Rightarrow \lim_{\theta\to0}L=\lim_{\theta\to0}\frac{\left(\tan^2\theta\right)\cos^2\theta(1-\cos\theta)(1+\cos\theta)}{(1-\cos\theta)^2}$$

$$=1\times2\lim_{\theta\to0}\frac{\tan^2\theta}{\theta^2}\frac{\theta^2}{1-\cos\theta}=4$$

15. (1) $\lim\limits_{x\to0^+} g\{f(x)\} = g(f(0^+)) = g((\sin 0^+)) = g(0^+) = (0)^2 + 1 = 1$.

$\lim\limits_{x\to0^-} g\{f(x)\} = g[f(0^-)] = g[(\sin 0^-)] = g(0^-) = (0)^2 + 1 = 1$

Hence, $\lim\limits_{x\to0} g\{f(x)\} = 1$.

16. (3) $L=\lim\limits_{x\to0}\dfrac{(\cos x-1)(\cos x-e^x)}{x^n}$

$$=-\lim_{x\to0}\frac{(1-\cos x)(1+\cos x)(\cos x-e^x)}{(1+\cos x)x^n}$$

$$=\lim_{x\to0}\frac{\left(\frac{\sin x}{x}\right)^2\left(\frac{1-\cos x}{x}+\frac{e^x-1}{x}\right)}{x^{n-3}}\frac{1}{1+\cos x}$$

If L is finite and non–zero, then $n = 3$ (as for $n = 1, 2$, $L = 0$ and for $n = 4$, $L = \infty$).

17. (6) Putting $x = 1 + h$, we get

$$f(a)=\lim_{h\to0}\frac{(1+h)^a-a(1+h)+a-1}{h^2}$$

$$=\lim_{h\to0}\frac{\left(1+ah+\frac{a(a-1)}{2!}h^2+...\right)-a-ah+a-1}{h}$$

$$=\frac{a(a-1)}{2}$$

Therefore, $f(4) = 6$.

18. (6) $L=\lim\limits_{x\to0}=-\lim\limits_{x\to0}\dfrac{D\prod\limits_{r=2}^{n}(\cos rx)^{1/r}}{2x}$

Let $y=\prod\limits_{r=2}^{n}(\cos rx)^{1/r}$

$$\Rightarrow \ln y=\sum_{r=2}^{n}\left(\frac{1}{2}\ln(\cos rx)\right)$$

$$\Rightarrow \frac{1}{y}\frac{dy}{dx}=-\sum_{r=2}^{n}\tan(rx)$$

$$\Rightarrow -Dy=y\sum_{r=2}^{n}\tan(rx)$$

$$\Rightarrow D\prod_{r=1}^{n}(\cos rx)^{1/r}=-y\sum_{r=2}^{n}\tan(rx)$$

$$\Rightarrow L=\lim_{x\to0}\frac{y\sum\limits_{r=2}^{n}\tan(rx)}{2x}$$

$$=\frac{1}{2}(2+3+4+\cdots+n)$$

$$=\frac{1}{2}\left[\frac{n(n+1)}{2}-1\right]=\frac{n^2+n-2}{4}$$

Now $\dfrac{n^2+n-2}{4}=10$

$\Rightarrow n^2 + n - 42 = 0$

$\Rightarrow (n + 7)(n - 6) = 0$

$\Rightarrow n = 6$

19. (9) $f(x)=\dfrac{3x^2+ax+a+1}{(x+2)(x-1)}$

As $x \to -2$, $D^r \to 0$

So as $x \to -2$, $N^r \to 0$ Hence,

$12 - 2a + a + 1 = 0$

$\Rightarrow a = 13$

20. **(2)** $L = e^{\lim_{x\to\infty} \frac{\ln(e^{2x}+e^x+x)}{x}}$

Using L'Hospital's rule

$$L = e^{\lim_{x\to\infty} \frac{2e^{2x}+e^x+1}{e^{2x}+e^x+1}}$$

$$= e^{\lim_{x\to\infty} \frac{2+e^{-x}+e^{-2x}}{1+e^{-x}+e^{-2x}}} = e^2$$

21. **(1)** We have LHL

$$= \lim_{x\to\frac{\pi}{2}^-} \sqrt{\frac{\tan x - \sin\tan^{-1}(\tan x)}{\tan x + \cos^2(\tan x)}}$$

$$= \lim_{x\to\frac{\pi}{2}^-} \sqrt{\frac{\tan x - \sin x}{\tan x + \cos^2(\tan x)}}$$

$$= \lim_{x\to\frac{\pi}{2}^-} \sqrt{\frac{1-\frac{\sin x}{\tan x}}{1+\frac{\cos^2(\tan x)}{\tan x}}} = \sqrt{\frac{1-0}{1+0}} = 1$$

As $x \to \frac{\pi}{2}-$, $0 < x < \frac{\pi}{2}$, therefore $\tan^{-1}(\tan x) = x$.

Further as $x \to \frac{\pi}{2}-$, $\tan x \to \infty$ and $\cos^2(\tan x)$ is a real number between 0 and 1

$$\text{RHL} = \lim_{x\to\frac{\pi}{2}^+} \sqrt{\frac{\tan x - \sin\tan^{-1}(\tan x)}{\tan x + \cos^2(\tan x)}}$$

$$= \lim_{x\to\frac{\pi}{2}^+} \sqrt{\frac{\tan x + \sin x}{\tan x + \cos^2(\tan x)}}$$

$$= \lim_{x\to\frac{\pi}{2}^+} \sqrt{\frac{\frac{1+\sin x}{\tan x}}{\frac{1+\cos^2(\tan x)}{\tan x}}} = \sqrt{\frac{1+0}{1+0}} = 1$$

As $x \to \frac{\pi}{2}^+$, $x > \frac{\pi}{2} \Rightarrow \tan^{-1}\tan x = \tan^{-1}\tan(x-\pi)$

$= x - \pi$

Therefore, $\sin\tan^{-1}(\tan x) = \sin(x-\pi) = -\sin x$

Further as $x \to \frac{\pi}{2}+$, $\tan x \to -\infty$ and $\cos^2(\tan x)$ is a real number between 0 and 1

LHL = RHL = 1 Therefore required limit = 1

22. **(6)** It is obvious n is even, then

$$\lim_{n\to\infty}\left(2^{1+3+5+\ldots+n/2\text{ terms}} \times 3^{2+4+6+\ldots+n/2\text{ terms}}\right)^{\frac{1}{(n^2+1)}}$$

$$= \lim_{n\to\infty}\left(2^{\frac{n^2}{4}} \times 3^{\frac{n(n+2)}{4}}\right)^{\frac{1}{(n^2+1)}}$$

$$= \lim_{n\to\infty} 2^{\frac{n^2}{4(n^2+1)}} \times 3^{\frac{n(n+2)}{4(n^2+1)}}$$

$$= 2^{\lim_{n\to\infty}\frac{1}{4\left(1+\frac{1}{n^2}\right)}} \times 3^{\lim_{n\to\infty}\frac{\left(1+\frac{2}{n}\right)}{4\left(1+\frac{1}{n^2}\right)}}$$

$$= 2^{\frac{1}{4}}3^{\frac{1}{4}} = (6)^{\frac{1}{4}}$$

23. **(6)** $\lim_{x\to 0} \frac{\sin x^4 - x^4\cos x^4 + x^{20}}{x^4(e^{2x^4}-1-2x^4)}$

$$= \lim_{t\to 0} \frac{\sin t - t\cos t + t^5}{t(e^{2t}-1-2t)}$$

$$= \lim_{t\to 0} \frac{t - \frac{t^3}{3!} + \frac{t^5}{5!} \cdots - t\left(1 - \frac{t^2}{2!} + \frac{t^4}{4!} \cdots\right) + t^5}{t\left(1 + 2t + \frac{4t^2}{2!} + \frac{8t^3}{3!} + \frac{16t^4}{4!} + \cdots - 1 - 2t\right)}$$

$$= \lim_{t\to 0} \frac{-\frac{t^3}{6} + \frac{t^3}{2} + \frac{t^5}{5!} - \frac{t^5}{4!} + \cdots + t^5}{2t^3 + \frac{8t^4}{3!} + \cdots}$$

$$= \frac{-\frac{1}{6} + \frac{1}{2}}{2} = -\frac{-1+3}{12} = \frac{1}{6}$$

24. **(8)** Since RHS is a finite quantity, therefore

At $x \to 1$, numerator must be = 0,

$\therefore 0 + b + 4 = 0$

$b = -4$

Then $\lim_{x\to 1} \frac{a\sin(x-1) - 4\cos(x-1) + 4}{(x^2-1)} = -2$

Put $x = 1 + h$, then $\lim_{h\to 0} \frac{a\sin h + 4(1-\cos h)}{h(2+h)} = -2$

$$\Rightarrow \lim_{h\to 0} \frac{a\left(\frac{\sin h}{h}\right) + 4\left(\frac{1-\cos h}{h}\right)}{2+h} = -2$$

$$\Rightarrow \frac{a(1)+0}{2} = -2$$

$\Rightarrow a = -4$

$\Rightarrow (a, b) = (-4, -4)$

25. **(2)** $\lim_{x\to 0} \frac{\ln\cot\left(\frac{\pi}{4} - k_1x\right)}{\tan k_2x} = \lim_{x\to 0} \frac{\ln\left(\frac{1+\tan k_1x}{1-\tan k_1x}\right)}{\tan k_2x}$

$$= \lim_{x\to 0} \frac{\ln\left(1+\frac{2\tan k_1 x}{1-\tan k_1 x}\right)}{\frac{2\tan k_1 x}{1-\tan k_1 x}\left(\frac{\tan k_2 x}{k_2 x}\right)} \times \frac{2\tan k_1 x}{1-\tan k_1 x} \times \frac{1}{k_2 x}$$

$$= \lim_{x\to 0} \frac{2\tan k_1 x}{k_2 x(1-\tan k_1 x)}$$

$$\left[\because \lim_{x\to 0}\frac{\ln(1+x)}{x} = 1 \text{ and } \lim_{x\to 0}\frac{\tan x}{x} = 1\right]$$

Given $\lim_{x\to 0} \frac{\tan k_1 x}{\frac{k_2}{2}\times x} = 1 \Rightarrow k_1 = \frac{k_2}{2}$ or $2k_1 = k_2$

26. (2) $\lim_{t\to 0} \frac{1-(1+t)^t}{\ln(1+t)-t} = \lim_{t\to 0} \frac{\frac{1-(1+t)^t}{t^2}}{\frac{\ln(1+t)-t}{t^2}}$

$$= \lim_{t\to 0} \frac{\frac{1-(1+t\times t)}{t^2}}{\frac{t-\frac{t^2}{2}-t}{t^2}} = 2$$

(using expansion)

27. (3) $\lim_{x\to 0} \sin\left(\frac{\pi(1-\cos^m x)}{x^n}\right)$

$$= \sin\left(\lim_{x\to 0} \frac{\pi(1-\cos^m x)}{x^n}\right)$$

$$= \sin\left(\lim_{x\to 0} 2\pi m \frac{\sin^2 x/2}{x^n}\right)$$

$\Rightarrow m \in N$ and $n = 1$ or 2

CHAPTER 3

Continuity and Differentiability

Objective Type

Each question has four choices a, b, c, and d, out of which *only one* is correct.

1. Which of the following functions have finite number of points of discontinuity ([.] represents greatest integer function)?

   a. $\tan x$ b. $x[x]$
   c. $\frac{|x|}{x}$ d. $\sin[n\pi x]$

2. The function $f(x) = \frac{4-x^2}{4x-x^3}$ is

   a. discontinuous at only one point
   b. discontinuous exactly at two points
   c. discontinuous exactly at three points
   d. None of these

3. If $f(x) = \frac{\tan\left(\frac{\pi}{4}-x\right)}{\cot 2x}$, $(x \neq \pi/4)$ is continuous at $x = \pi/4$, then the value of $f(\pi/4)$ is

   a. 1 b. 1/2
   c. 1/3 d. −1

4. The function $f(x) = \frac{(3^x-1)^2}{\sin x \ln(1+x)}$, $x \neq 0$, is continuous at $x = 0$. Then the value of $f(0)$ is

   a. $2\log_e 3$ b. $(\log_e 3)^2$
   c. $\log_e 6$ d. None of these

5. If $f(x) = \begin{cases} \frac{1-|x|}{1+x}; & x \neq -1 \\ 1; & x = -1 \end{cases}$, then $f([2x])$ where [ ] represents the greatest integer function is

   a. discontinuous at $x = -1$ b. continuous at $x = 0$
   c. continuous at $x = 1/2$ d. continuous at $x = 1$

6. Let

   $$f(x) = \begin{cases} \frac{x-4}{|x-4|}+a, & x<4 \\ a+b, & x=4 \\ \frac{x-4}{|x-4|}+b, & x>4 \end{cases}$$ Then $f(x)$ is continous at $x = 4$ when,

   a. $a = 0, b = 0$ b. $a = 1, b = 1$
   c. $a = -1, b = 1$ d. $a = 1, b = -1$

7. If $f(x) = \frac{x-e^x+\cos 2x}{x^2}$, $x \neq 0$, is continuous at $x = 0$, then

   a. $f(0) = 5/2$ b. $[f(0)] = -2$
   c. $\{f(0)\} = -0.5$ d. $[f(0)]\{f(0)\} = -1.5$

   where $[x]$ and $\{x\}$ denotes the greatest integer and fractional part function.

8. Let $f(x) = \begin{cases} \frac{\left(x^3+x^2-16x+20\right)}{(x-2)^2}, & \text{if } x \neq 2 \\ k, & \text{if } x = 2 \end{cases}$

   if $f(x)$ is continuous for all x, then k is equal to

   a. 2 b. 3
   c. 6 d. 7

9. The value of $f(0)$, so that the function

$f(x) = \dfrac{2x - \sin^{-1} x}{2x + \tan^{-1} x}$ is continuous at each point in its domain, is equal to

a. 2 b. 1/3
c. 2/3 d. −1/3

10. Which of the following is true about

$$f(x)=\begin{cases}\dfrac{(x-2)}{|x-2|}\left(\dfrac{x^2-1}{x^2+1}\right); & x\neq 2\\ \dfrac{3}{5}; & x=2\end{cases}$$

a. $f(x)$ is continuous at $x = 2$.
b. $f(x)$ has removable discontinuity at $x = 2$.
c. $f(x)$ has non-removable discontinuity at $x = 2$.
d. Discontinuity at $x = 2$ can be removed by redefining function at $x = 2$.

11. $f(x) = \lim\limits_{n\to\infty} \dfrac{(x-1)^{2n}-1}{(x-1)^{2n}+1}$ is discontinuous at

a. $x = 0$ only b. $x = 2$ only
c. $x = 0$ and 2 d. None of these

12. If $f(x)=\begin{cases}\dfrac{8^x-4^x-2^x+1}{x^2}, & x>0\\ e^x\sin x+\pi x+\lambda\ln 4, & x\le 0\end{cases}$

is continuous at $x = 0$. Then the value of λ is

a. $4\log_e 2$ b. $2\log_e 2$
c. $\log_e 2$ d. None of these

13. If $f(x) = \dfrac{a\cos x - \cos bx}{x^2}$, $x \neq 0$ and $f(0) = 4$ is continuous at $x = 0$, then the ordered pair (a, b) is

a. $(\pm 1, 3)$ b. $(1, \pm 3)$
c. $(-1, -3)$ d. $(1, 3)$

14. The values of a and b so that the function

$$f(x)=\begin{cases}x+a\sqrt{2}\sin x, & 0\le x<\pi/4\\ 2x\cot x+b, & \pi/4\le x\le \pi/2\\ a\cos 2x-b\sin x, & \pi/2<x\le\pi\end{cases}$$ is

continuous for $0 \le x \le \pi$ are

a. $a = -\pi/12$ and $b = \pi/6$
b. $a = \pi/6$ and $b = -\pi/12$
c. $a = b = \pi/6$
d. None of these

15. Let $f: R \to R$ be given by $f(x) = 5x$, if $x \in Q$ and $f(x) = x^2 + 6$ if $x \in R \sim Q$, then

a. f is continuous at $x = 2$ and $x = 3$
b. f is not continuous at $x = 2$ and $x = 3$
c. f is continuous at $x = 2$ but not at $x = 3$
d. f is continuous at $x = 3$ but not at $x = 2$

16. The function $f(x) = |2 \operatorname{sgn} 2x| + 2$ has

a. jump discontinuity
b. removal discontinuity
c. infinite discontinuity
d. no discontinuity at $x = 0$

17. Let $f(x) = \lim\limits_{n\to\infty} (\sin x)^{2n}$, then which of the following is not true

a. Discontinuous at infinite number of points
b. Discontinuous at $x = \dfrac{\pi}{2}$
c. Discontinuous at $x = -\dfrac{\pi}{2}$
d. None of these

18. Let f be a continuous function on R such that $f(1/4n) = (\sin e^n)\, e^{-n^2} + \dfrac{n^2}{n^2+1}$. Then the value of $f(0)$ is

a. 1 b. 1/2
c. 0 d. None of these

19. If $f(x) = \dfrac{x^2 - bx + 25}{x^2 - 7x + 10}$ for $x \neq 5$ is continuous at $x = 5$, then the value of $f(5)$ is

a. 0 b. 5
c. 10 d. 25

20. Which of the following statements is always true? ([.] represents the greatest integer function)

a. If $f(x)$ is discontinuous, then $|f(x)|$ is discontinuous
b. If $f(x)$ is discontinuous, then $f(|x|)$ is discontinuous
c. $f(x) = [g(x)]$ is discontinuous when $g(x)$ is an integer
d. None of these

21. A function $f(x)$ is defined as

$f(x)=\begin{cases}\sin x, & x \text{ is rational}\\ \cos x, & x \text{ is irrational}\end{cases}$ is continuous

a. $x = n\pi + \pi/4, n \in I$ b. $x = n\pi + \pi/8, n \in I$
c. $x = n\pi + \pi/6, n \in I$ d. $x = n\pi + \pi/3, n \in I$

22. The number of points $f(x)=\begin{cases}[\cos \pi x], & x\le 1\\ |2x-3|[x-2], & x>1\end{cases}$

is discontinuous at ([.] denotes the greatest integral function)

a. two points b. three points
c. four points d. no points

23. A point where function $f(x)$ is not continuous where $f(x) = [\sin [x]]$ in $(0, 2\pi)$; [.] denotes the greatest integer $\le x$ is

a. (3, 0) b. (2, 0)
c. (1, 0) d. None of these

24. The function $f(x) = \sin(\log_e |x|), x \neq 0$, and 1 if $x = 0$
a. is continuous at $x = 0$
b. has removable discontinuity at $x = 0$
c. has jump of discontinuity at $x = 0$
d. has oscillating discontinuity at $x = 0$

25. The function defined by $f(x) = (-1)^{[x^3]}$ ([.] denotes the greatest integer function) satisfies
a. discontinuous for $x = n^{1/3}$, where n is any integer
b. $f(3/2) = 1$
c. $f'(x) = 1$ for $-1 < x < 1$
d. None of these

26. The function $f(x) = \{x\}\sin(\pi[x])$, where [.] denotes the greatest integer function and {.} is the fractional part function, is discontinuous at
a. all x
b. all integer points
c. no x
d. x which is not an integer

27. The function $f(x)$ defined by

$$f(x) = \begin{cases} \log_{(4x-3)}(x^2 - 2x + 5), & \frac{3}{4} < x < 1 \text{ and } x > 1 \\ 4, & x = 1 \end{cases}$$

a. is continuous at $x = 1$.
b. is discontinuous at $x = 1$ since $f(1^+)$ does not exist though $f(1^-)$ exists.
c. is discontinuous at $x = 1$ since $f(1^-)$ does not exist though $f(1^+)$ exists.
d. is discontinuous at $x = 1$ since neither $f(1^+)$ nor $f(1^-)$ exists.

28. Let $f(x) = [x]$ and $g(x) = \begin{cases} 0, & x \in Z \\ x^2, & x \in R - Z \end{cases}$. Then which of the following is not true ([.] represents greatest integer function)
a. $\lim_{x \to 1} g(x)$ exists but $g(x)$ is not continuous at $x = 1$.
b. $\lim_{x \to 1} f(x)$ does not exist and $f(x)$ is not continuous at $x = 1$.
c. gof is a discontinuous function.
d. fog is a discontinuous function.

29. $f(x) = \begin{cases} \dfrac{x}{2x^2 + |x|}, & x \neq 0 \\ 1, & x = 0 \end{cases}$ then $f(x)$ is
a. Continuous but non-differentiable at $x = 0$
b. Differentiable at $x = 0$
c. Discontinuous at $x = 0$
d. None of these

30. Let a function $f(x)$ be defined by $f(x) = \dfrac{x - |x - 1|}{x}$, then which of the following is not true
a. Discontinuous at $x = 0$
b. Discontinuous at $x = 1$
c. Not differentiable at $x = 0$
d. Not differentiable at $x = 1$

31. If $f(x) = x^3 \operatorname{sgn} x$, then
a. f is derivable at $x = 0$
b. f is continuous but not derivable at $x = 0$
c. L.H.D. at $x = 0$ is 1
d. R.H.D. at $x = 0$ is 1

32. Let $f(x) = \begin{cases} \min(\{x, x^2\}) x \geq 0 \\ \max\{2x, x - 1\} x < 0 \end{cases}$. Then which of the following is not true.
a. $f(x)$ is continuous at $x = 0$
b. $f(x)$ is differentiable at $x = 1$
c. $f(x)$ is not differnetiable at exactly three point
d. $f(x)$ is continuous every where

33. The function $f(x) = \sin^{-1}(\cos x)$ is
a. discontinuous at $x = 0$
b. continuous at $x = 0$
c. differentiable at $x = 0$
d. differentiable at $x = 2\pi$

34. Which of the following functions is non differentiable?
a. $f(x) = (e^x - 1)|e^{2x} - 1|$
b. $f(x) = \dfrac{x - 1}{x^2 + 1}$
c. $f(x) = \begin{cases} ||x - 3| - 1|, & x < 3 \\ \frac{x}{3}[x] - 2, & x \geq 3 \end{cases}$ at $x = 3$

where [.] represent the greatest integer function
d. $f(x) = 3(x - 2)^{3/4} + 3$

35. The number of values of $x \in [0, 2]$ at which $f(x) = \left|x - \frac{1}{2}\right| + |x - 1| + \tan x$ is not differentiable at
a. 0 b. 1
c. 2 d. None of these

36. The set of points where $x^2 |x|$ is thrice differentiable, is
a. R b. $R - \{0, \pm 1\}$
c. $R - \{0\}$ d. None of these

37. Which of the following function is not differentiable at $x = 1$?
a. $f(x) = (x^2 - 1)|(x - 1)(x - 2)|$
b. $f(x) = \sin(|x - 1|) - |x - 1|$
c. $f(x) = \tan(|x - 1|) + |x - 1|$
d. None of these

38. $f(x) = \begin{cases} xe^{-\left(\frac{1}{x}+\frac{1}{|x|}\right)}, & x \neq 0 \\ a, & x = 0 \end{cases}$. The value of a, such that $f(x)$ is differentiable at $x = 0$ at $x = 0$, is equal to
   a. 1 b. –1
   c. 0 d. None of these

39. If $f(x) = \begin{cases} ax^2+1, & x \leq 1 \\ x^2+ax+b, & x > 1 \end{cases}$ is differentiable at $x = 1$, then
   a. $a = 1, b = 1$ b. $a = 1, b = 0$
   c. $a = 2, b = 0$ d. $a = 2, b = 1$

40. If $f(x) = a|\sin x| + be^{|x|} + c|x|^3$ is differentiable at $x = 0$, then
   a. $a = b = c = 0$ b. $a = 0, b = 0, c \in R$
   c. $b = c = 0, a \in R$ d. $c = 0, a = 0, b \in R$

41. The number of points of non-differentiability for $f(x) = \max\{||x|-1|, 1/2\}$ is
   a. 4 b. 3
   c. 2 d. 5

42. Let $f(x) = \begin{cases} \sin 2x, & 0, x \leq \pi/6 \\ ax+b, & \pi/6 < x < 1 \end{cases}$. If $f(x)$ and $f'(x)$ are continuous, then
   a. $a = 1, b = \frac{1}{\sqrt{2}} + \frac{\pi}{6}$ b. $a = \frac{1}{\sqrt{2}}, b = \frac{1}{\sqrt{2}}$
   c. $a = 1, b = \frac{\sqrt{3}}{2} - \frac{\pi}{6}$ d. None of these

43. If $f(x) = \begin{cases} x^3, & x^2 < 1 \\ x, & x^2 \geq 1 \end{cases}$ then $f(x)$ is differentiable at
   a. $(-\infty, \infty) - \{1\}$ b. $(-\infty, \infty) \sim \{1 - 1\}$
   c. $(-\infty, \infty) \sim \{1 - 1, 0\}$ d. $(-\infty, \infty) \sim \{-1\}$

44. If $f(x) = (x^2 - 4)|x^3 - 6x^2 + 11x - 6| + \frac{x}{1+|x|}$, then the set of points at which the function $f(x)$ is not differentiable is
   a. $\{-2, 2, 1, 3\}$ b. $\{-2, 0, 3\}$
   c. $\{-2, 2, 0\}$ d. $\{1, 3\}$

45. If $f(x) = \cos \pi(|x| + [x])$, (where [.] denotes the greatest integral function), then which is not true?
   a. continuous at $x = 1/2$
   b. continuous at $x = 0$
   c. differentiable in $(-1, 0)$
   d. differentiable in $(0, 1)$

46. $f(x) = \begin{cases} e^{x^2} + x, & x > 0 \\ ax+b, & x \leq 0 \end{cases}$ is differentiable at $x = 0$ then
   a. $a = 1, b = -1$ b. $a = -1, b = 1$
   c. $a = 1, b = 1$ d. $a = -1, b = -1$

47. $f(x) = \begin{cases} e^{-1/x^2}, & x > 0 \\ 0, & x \leq 0 \end{cases}$, then $f(x)$ is
   a. differentiable at $x = 0$
   b. continuous but not differentiable at $x = 0$
   c. discontinuous at $x = 0$
   d. None of these

48. If $f(x) = \begin{cases} x-1, & x < 0 \\ x^2-2x, & x \geq 0 \end{cases}$ then
   a. $f(|x|)$ is discontinuous at $x = 0$
   b. $f(|x|)$ is differentiable at $x = 0$
   c. $|f(x)|$ is non-differentiable at $x = 0, 2$
   d. $|f(x)|$ is continuous at $x = 0$

49. If $f(x) = \begin{cases} |1-4x^2|, & 0 \leq x < 1 \\ [x^2-2x], & 1 \leq x < 2 \end{cases}$, where [.] denotes the greatest integer function. The $f(x)$ is
   a. Differentiable for all x
   b. Continuous at $x = 1$
   c. $f(x)$ is non-differentiable at $x = 1$
   d. None of these

   Discuss the continuity and differentiability of $f(x)$ in $[0, 2)$

50. Let $f(x) = \lim_{n \to \infty} \frac{\log(2+x) - x^{2n} \sin x}{1+x^{2n}}$. Then
   a. f is continuous at $x = 1$
   b. $\lim_{x \to 1^+} f(x) = \log 3$
   c. $\lim_{x \to 1^+} f(x) = -\sin 1$
   d. $\lim_{x \to 1^-} f(x)$ does not exist

51. If $f(x) = \begin{cases} x^a \sin\left(\frac{1}{x}\right), & x \neq 0 \\ 0, & x = 0 \end{cases}$ is continuous but non-differentiable at $x = 0$, then
   a. $a \in (-1, 0)$ b. $a \in (0, 2]$
   c. $a \in (0, 1]$ d. $a \in [1, 2)$

52. $f(x) = [\sin x] + [\cos x], x \in [0, 2\pi]$, where [.] denotes the greatest integer function. The total number of points, where $f(x)$ is non-differentiable, is equal to
   a. 2 b. 3
   c. 5 d. 4

53. If $x + 4|y| = 6y$, then y as a function of x is
   a. continuous at $x = 0$ b. derivable at $x = 0$
   c. $\frac{dy}{dx} = \frac{1}{2}$ for all x d. None of these

54. Let $g(x)$ be a polynomial of degree one and $f(x)$ be defined by $f(x) = \begin{cases} g(x), & x \le 0 \\ |x|^{\sin x}, & x > 0 \end{cases}$. If $f(x)$ is continuous satisfying $f'(1) = f(-1)$, then $g(x)$ is

a. $(1+\sin 1)x+1$ b. $(1-\sin 1)x+1$
c. $(1-\sin 1)x-1$ d. $(1+\sin 1)x-1$

55. If $f(x) = |1-x|$, then the points where $\sin^{-1}(f|x|)$ is non-differentiable are

a. $\{0, 1\}$ b. $\{0, -1\}$
c. $\{0, 1, -1\}$ d. None of these

56. Given that $f(x) = xg(x)/|x|$, $g(0) = g'(0) = 0$ and $f(x)$ is continuous at $x = 0$. Then the value of $f'(0)$

a. Does not exist b. is -1
c. is 1 d. is 0

57. The number of points, where the function $f(x) = \max(|\tan x| \cos |x|)$ is non-differentiable in the interval $(-\pi, \pi)$, is

a. 4 b. 6
c. 3 d. 2

58. If $f(x) = \begin{cases} \sin x, & x < 0 \\ \cos x - |x-1|, & x \ge 0 \end{cases}$ then $g(x) = f(|x|)$ is non-differentiable for

a. no value of x
b. exactly one value of x
c. exactly three values of x
d. None of these

59. If $f(x) = \begin{cases} 2x - [x] + x\sin(x-[x]); & x \ne 0 \\ 0; & x = 0 \end{cases}$, where [.] denotes the greatest integer function, then

a. $f(x)$ is differentiable at $x = 0$
b. $f(x)$ is differentiable at $x = 2$
c. $f(x)$ is continuous but not differentiable at $x = 0$
d. None of these

60. $f(x) = \max\{x/n, |\sin \pi x|\}, n \in N$ has maximum points of non-differentiability for $x \in (0, 4)$, then

a. maximum value of n is more than 4.5
b. least value of n is more that 3.5
c. maximum value of n is less than 4.5
d. least value of n is less than 3.5

61. $f(x) = [x^2] - \{x\}^2$, where [.] and {.} denote the greatest integer function and the fractional part respectively, is

a. continuous at $x = 1, -1$
b. continuous at $x = -1$ but not at $x = 1$
c. continuous at $x = -1$ but not at $x = -1$
d. discontinuous at $x = 1$ and $x = -1$

62. $f(x) = [\log_e x] + \sqrt{\{\log_e x\}}$, $x > 1$, where [.] and {.} denote the greatest integer function and the fractional part function respectively, then

a. $f(x)$ is continuous but non-differentiable at $x = e$
b. $f(x)$ is differentiable at $x = e$
c. $f(x)$ is discontinuous at $x = e$
d. None of these

63. $f(x) = \lim_{n\to\infty} \sin^{2n}(\pi x) + \left[x + \frac{1}{2}\right]$, where [.] denotes the greatest integer function is

a. continuous at $x = 1$ but discontinuous at $x = 3/2$
b. continuous at $x = 1$ and $x = 3/2$
c. discontinuous at $x = 1$ and $x = 3/2$
d. discontinuous at $x = 1$ but continuous at $x = 3/2$

64. If $f(x) = \text{sgn}(\sin^2 x - \sin x - 1)$ has exactly four points of discontinuity for $x \in (0, n\pi)$, $n \in N$, then

a. minimum value of n is 5
b. maximum value of n is 6
c. there are exactly two possible values of n
d. None of these

65. If $f(x) = \begin{cases} x^2 - ax + 3, & x \text{ is rational} \\ 2 - x, & x \text{ is irrational} \end{cases}$ is continuous at exactly two points, then the possible values of a are

a. $(2, \infty)$ b. $(-\infty, 3)$
c. $(-\infty, -1) \cup (3, \infty)$ d. None of these

66. $f(x) = \{x\}^2 - \{x^2\}$ ({.} denotes the fractional part function)

a. $f(x)$ is discontinuous at infinite number of integers but not all integers
b. $f(x)$ is discontinuous at finite number of integers
c. $f(x)$ is discontinuous at all integers
d. $f(x)$ is continuous at all integers

67. Let $f(x) = \begin{cases} g(x)\cos\frac{1}{x}, & x \ne 0 \\ 0, & x = 0 \end{cases}$, where $g(x)$ is an even function differentiable at $x = 0$, passing through the origin. The $f'(0)$

a. is equal to 1 b. is equal to 0
c. is equal to 2 d. does not exist

68. Let $f(x) = \begin{cases} 1 - \sqrt{1-x^2}, & \text{if } -1 \le x \le 1 \\ 1 + \log\frac{1}{x}, & \text{if } x > 1 \end{cases}$ is

a. continuous and differentiable at $x = 1$
b. continuous but not differentiable at $x = 1$
c. neither continuous nor differentiable at $x = 1$
d. None of these

69. If $f(x) = \sqrt{1 - \sqrt{1-x^2}}$, then $f(x)$ is

a. continuous on $[-1, 1]$ and differentiable on $(-1, 1)$

b. continuous [–1, 1] and differentiable on $(-1,0)\cup(0,1)$
c. continuous and differentiable on [–1, 1]
d. None of these

70. The set of all points, where $f(x)=\sqrt[3]{x^2|x|}-|x|-1$ is not differentiable, is
a. {0}
b. {–1, 0, 1}
c. {0, 1}
d. None of these

71. Let $f(x)$ be a function for all $x\in R$ and $f'(0)=1$. Then $g(x)=f(|x|)-\sqrt{\dfrac{1-\cos 2x}{2}}$, at $x=0$,
a. is differentiable at $x=0$ and its value is 1
b. is differentiable at $x=0$ and its value is 0
c. is non-differentiable at $x=0$ as its graph has sharp turn at $x=0$
d. is non-differentiable at $x=0$ as its graph has vertical tangent at $x=0$

72. A function $f(x)$ is defined as
$$f(x)=\begin{cases} x^m\sin\dfrac{1}{x}, & x\neq 0,\ m\in N \\ 0, & \text{if } x=0 \end{cases}$$
. The least value of m for which $f'(x)$ is continuous at $x=0$ is
a. 1
b. 2
c. 3
d. None

73. $f(x)=\begin{cases} x^2\left(\dfrac{e^{1/x}-e^{-1/x}}{e^{1/x}+e^{-1/x}}\right), & x\neq 0 \\ 0, & x=0 \end{cases}$. Then
a. $f(x)$ is discontinuous at $x=0$
b. $f(x)$ is continuous but non-differentiable at $x=0$
c. $f(x)$ is differentiable at $x=0$
d. $f'(0)=2$

Multiple Correct Answers Type

Each question has four choices a, b, c, and d out of which *one or more* answers are correct.

1. Which of the statement(s) is/are incorrect ?
a. If $f+g$ is continuous at $x=a$, then f and g are continuous at $x=a$.
b. If $\lim\limits_{x\to a}(fg)$ exists, then both $\lim\limits_{x\to a} f$ and $\lim\limits_{x\to a} g$ exist.
c. Discontinuity at $x=a \Rightarrow$ non-existence of limit.
d. All functions defined on a closed interval attain a maximum or a minimum value in that interval.

2. A function f is defined on an interval $[a, b]$. Which of the following statement(s) is/are incorrect ?
a. If $f(a)$ and $f(b)$ have opposite signs, then there must be a point $c\in(a, b)$ such that $f(c)=0$.
b. If f is continuous on $[a, b]$, $f(a)<0$ and $f(b)>0$, then there must be a point $c\in(a, b)$ such that $f(c)=0$.
c. If f is continuous on $[a, b]$ and there is a point c in (a, b) such that $f(c)=0$, then $f(a)$ and $f(b)$ have opposite signs.
d. If f has no zeros on $[a, b]$, then $f(a)$ and $f(b)$ have the same sign.

3. Which of the following function (s) f has/have a removable discontinuity at the indicated point ?
a. $f(x)=\dfrac{x^2-2x-8}{x+2}$ at $x=-2$
b. $f(x)=\dfrac{x-7}{|x-7|}$ at $x=7$
c. $f(x)=\dfrac{x^3+64}{x+4}$ at $x=-4$
d. $f(x)=\dfrac{3-\sqrt{x}}{9-x}$ at $x=0$

4. The function $f(x)=\begin{cases} 5x-4 & \text{for } 0<x\le 1 \\ 4x^2-3x & \text{for } 1<x<2 \\ 3x+4 & \text{for } x\ge 2 \end{cases}$
a. continuous at $x=1$ and $x=2$
b. continuous at $x=1$ but not derivable at $x=2$
c. continuous at $x=2$ but not derivable at $x=1$
d. continuous at $x=1$ and 2 but not derivable at $x=1$ and $x=2$

5. Which of the following is/are true for $f(x)=\text{sgn}(x)\times\sin x$
a. discontinuous no where
b. an even function
c. $f(x)$ is periodic
d. $f(x)$ is differentiable for all x

6. $f(x)=\lim\limits_{t\to\infty}\dfrac{(a+\sin\pi x)^t-1}{(a+\sin\pi x)^t+1}$, $x\in(0, 6)$ then
a. if $a=1$, then $f(x)$ has five points of discontinuity.
b. if $a=3$, then $f(x)$ has no point of discontinuity.
c. if $a=0.5$, then $f(x)$ has six points of discontinuity.
d. if $a=0$, then $f(x)$ has three points of discontinuity.

7. If $f(x)=\text{sgn}(x^2-ax+1)$ has maximum number of points of discontinuity, then
a. $a\in(2,\infty)$
b. $a\in(-\infty,-2)$
c. $a\in(-2, 2)$
d. None of these

8. If $f(x)=[|x|]$, where [.] denotes the greatest integer function, then which of the following is not true?

a. $f(x)$ is continuous $\forall\, x \in R$.

b. $f(x)$ is continuous from right and discontinuous from left $\forall\, x \in N \cup \{0\}$.

c. $f(x)$ is continuous from left and discontinuous from right $\forall\, x \in I$.

d. $f(x)$ is continuous at $x = 0$.

9. $f(x)$ is differentiable function and $(f(x)\; g(x))$ is differen-tiable at $x = a$, then

a. $g(x)$ must be differentiable at $x = a$.

b. If $g(x)$ is discontinuous, then $f(a) = 0$.

c. $f(a) \neq 0$, then $g(x)$ must be differentiable.

d. None of these.

10. The function defined as

$$f(x) = \lim_{n\to\infty}\begin{bmatrix} \cos^{2n} x \text{ if } x<0 \\ \sqrt[n]{1+x^n} \text{ if } 0 \le x \le 1, \\ \dfrac{1}{1+x^n} \text{ if } x>1 \end{bmatrix}$$

which of the following does not hold good?

a. continuous at $x = 0$ but discontinuous at $x = 1$.

b. continuous at $x = 1$ but discontinuous at $x = 0$.

c. continuous both at $x = 1$ and $x = 0$.

d. discontinuous both at $x = 1$ and $x = 0$.

11. Which of the following function(s) has/have removable discontinuity at $x = 1$?

a. $f(x) = \dfrac{1}{\ln|x|}$

b. $f(x) = \dfrac{x^2-1}{x^3-1}$

c. $f(x) = 2^{-2^{\frac{1}{1-x}}}$

d. $f(x) = \dfrac{\sqrt{x+1}-\sqrt{2x}}{x^2-x}$

12. $f(x) = \dfrac{[x]+1}{\{x\}+1}$ for $f:\left[0, \dfrac{5}{2}\right) \to \left(\dfrac{1}{2}, 3\right]$, where [.] represents the greatest integer function and {.} represents the fractional part of x, then which of the following is true.

a. $f(x)$ is injective discontinuous function.

b. $f(x)$ is surjective non-differentiable function.

c. $\min\left(\lim_{x\to 1^-} f(x),\ \lim_{x\to 1^+} f(x)\right) = f(1)$.

d. max (x values of point of discontinuity) $= f(1)$.

13. The function

$$f(x) = \begin{cases} 1, & |x| \ge 1 \\ \dfrac{1}{n^2}, & \dfrac{1}{n} < |x| < \dfrac{1}{n-1},\ n = 2, 3, \ldots \\ 0, & x = 0 \end{cases}$$

a. is discontinuous at finitely many points

b. is continuous everywhere

c. is discontinuous only at $x = \dfrac{1}{n}$, $n \in Z - \{0\}$

d. None of these

14. Let $f(x) = [x]$ and $g(x) = \begin{cases} 0, & x \in Z \\ x^2, & x \in R - Z \end{cases}$ ([.] represents greatest integer function). Then

a. $\lim_{x\to 1} g(x)$ exists but $g(x)$ is not continuous at $x = 1$.

b. $f(x)$ is not continuous at $x = 1$.

c. gof is continuous for all x.

d. fog is continuous for all x.

15. If $f(x) = \begin{cases} \dfrac{x \log \cos x}{\log(1+x^2)}, & x \neq 0 \\ 0, & x = 0 \end{cases}$ then

a. $f(x)$ is not continuous at $x = 0$.

b. $f(x)$ is continuous at $x = 0$.

c. $f(x)$ is continuous at $x = 0$ but not differentiable at $x = 0$.

d. $f(x)$ is differentiable at $x = 0$.

16. If $f(x) = x + |x| + \cos([\pi^2]x)$ and $g(x) = \sin x$, where [.] denotes the greatest integer function, then

a. $f(x) + g(x)$ is continuous everywhere.

b. $f(x) + g(x)$ is differentiable everywhere.

c. $f(x) \times g(x)$ is differentiable everywhere.

d. $f(x) \times g(x)$ is continuous but not differentiable at $x = 0$.

17. If $f(x) = \begin{cases} (\sin^{-1} x)^2 \cos(1/x), & x \neq 0 \\ 0, & x = 0 \end{cases}$ then

a. $f(x)$ is continuous everywhere in $x \in (-1, 1)$.

b. $f(x)$ is discontinuous in $x \in [-1, 1]$.

c. $f(x)$ is differentiable everywhere in $x \in (-1, 1)$.

d. $f(x)$ is non-differentiable nowhere in $x \in [-1, 1]$.

18. $f(x) = \begin{cases} x + a, & x \ge 0 \\ 2 - x, & x < 0 \end{cases}$ and $g(x) = \begin{cases} \{x\}, & x < 0 \\ \sin x + b, & x \ge 0 \end{cases}$ and if $f(g(x))$ is continuous at $x = 0$ then which of the following is/are true (where $\{x\}$ represents the fractional part function)

a. if $b = 1$, then a can take any real value.

b. if $b < 0$, then $a + b = 1$.

c. no values of a and b are possible.
d. there exist finites ordered pairs (a, b).

19. If $f(x) = \begin{cases} |x|-3, & x<1 \\ |x-2|+a, & x\geq 1 \end{cases}$ and $g(x) = \begin{cases} 2-|x|, & x<2 \\ \text{sgn}(x)-b, & x\geq 2 \end{cases}$ and $h(x) = f(x) + g(x)$ is discontinuous at exactly one point then which of the following values of a and b are possible

a. $a=-3, b=0$
b. $a=2, b=1$
c. $a=2, b=0$
d. $a=-3, b=1$

20. Let $f: R \to R$ be any function and $g(x) = \dfrac{1}{f(x)}$. Then which of following is/are not true

a. g is onto if f is onto.
b. g is one-one if f is one-one.
c. g is continuous if f is continuous.
d. g is differentiable if f is differentiable.

21. If $f(x) = \begin{cases} x^2(\text{sgn}[x]) + \{x\}, & 0\leq x<2 \\ \sin x + |x-3|, & 2\leq x<4 \end{cases}$,

where [ ] and { } represent the greatest integer and the fractional part function, respectively.

a. $f(x)$ is differentiable at $x=1$.
b. $f(x)$ is continuous but non-differentiable at $x=1$.
c. $f(x)$ is non-differentiable at $x=2$.
d. $f(x)$ is discontinuous at $x=2$.

22. $f(x) = \begin{cases} \left(\dfrac{3}{2}\right)^{(\cot 3x)/(\cot 2x)}; & 0<x<\dfrac{\pi}{2} \\ b+3; & x=\dfrac{\pi}{2} \\ (1+|\cot x|)^{(a|\tan x|)/b}; & \dfrac{\pi}{2}<x<\pi \end{cases}$ is

continuous at $x = \pi/2$, then

a. $a=0$
b. $a=2$
c. $b=-2$
d. $b=2$

Reasoning Type

Each question has four choices a, b, c, and d, out of which *only one* is correct. Each question contains Statement 1 and Statement 2.

a. if both the statements are True and Statement 2 is the correct explanation of Statement 1.
b. if both the statements are True but Statement 2 is Not the correct explanation of Statement 1.
c. if Statement 1 is True and Statement 2 is False.
d. if Statement 1 is False and Statement 2 is True.

1. **Statement 1:** $y = \sin x$ and $y = \sin^{-1}x$, both are differentiable functions.
Statement 2: Differentiability of $f(x)$ $\Rightarrow$ differentiability of $y = f^{-1}(x)$.

2. **Statement 1:** $f(x) = (2x-5)^{3/5}$ is non-differentiable at $x = 5/2$.
Statement 2: If the graph of $y = f(x)$ has sharp turn at $x = a$, then it is non-differentiable.

3. **Statement 1:** $f(x) = \text{sgn}(x^2 - 2x + 3)$ is continuous for all x.
Statement 2: $ax^2 + bx + c = 0$ has no real roots if $b^2 - 4ac < 0$.

4. **Statement 1:** $f(x) = \lim\limits_{x\to\infty} \dfrac{x^{2n}-1}{x^{2n}+1}$ is discontinuous at $x = 1$.
Statement 2: If limit of function exists at $x = a$ but not equal to $f(a)$, then $f(x)$ is discontinuous at $x = a$.

5. **Statement 1:** $f(x) = [\sin x] - [\cos x]$ is discontinuous at $x = \pi/2$, where [.] represent the greatest integer function.
Statement 2: If $f(x)$ and $g(x)$ are discontinuous at $x = a$, then $f(x) + g(x)$ is discontinuous at $x = a$.

6. **Statement 1:** $f(x) = \text{sgn}\, x$ is discontinuous at $x = 0$ $\Rightarrow f(x) = |\text{sgn}x|$ is discontinuous at $x = 0$.
Statement 2: Discontinuity of $f(x) \Rightarrow$ discontinuity of $|f(x)|$.

7. **Statement 1:** $f(x) = (\sin \pi x)(x-1)^{1/5}$ is differentiable at $x = 1$.
Statement 2: Product of two differentiable function is always differentiable.

8. **Statements 1:** The function $f(x) = \begin{cases} \dfrac{e^{1/x}-1}{e^{1/x}+1}, & x\neq 0 \\ \cos x & x=0 \end{cases}$ is discontinuous at $x = 0$.
Statements 2: $f(0) = 1$.

9. **Statement 1:** $f(x) = \sin x + [x]$ is discontinuous at $x = 0$, where [.] denotes the greatest integer function.
Statement 2: If $g(x)$ is continuous and $h(x)$ is discontinuous at $x = a$, then $g(x) + h(x)$ will necessarily be discontinuous at $x = a$.

10. **Statement 1:** $f(x) = |x| \sin x$ is non-differentiable at $x = 0$.
Statement 2: If $f(x)$ is not differentiate and $g(x)$ is differentiate at $x = a$, then $f(x)\, g(x)$ can still be differentiable at $x = a$.

11. **Statement 1:** If $f(x)$ is discontinuous at $x = e$ and $\lim\limits_{x\to a} g(x) = e$, then $\lim\limits_{x\to a} f(g(x))$ cannot be equal to $f\left(\lim\limits_{x\to a} g(x)\right)$.

Statement 2: If $f(x)$ is continuous at $x = e$ and $\lim_{x\to a} g(x) = e$, then $\lim_{x\to a} f(g(x)) = f\left(\lim_{x\to a} g(x)\right)$.

12. **Statement 1:** The function $|\ln x|$ and $\ln x$ are both continuous for all x.

Statement 2: Continuity of $|f(x)| \Rightarrow$ continuity of $f(x)$.

13. **Statement 1:** $f(x) = \tan^{-1} = \left(\dfrac{2x}{1-x^2}\right)$ is non-differentiable at $x = \pm 1$.

Statement 2: Principal value of $\tan^{-1} x$ are $\left[-\dfrac{\pi}{2}, \dfrac{\pi}{2}\right]$.

14. **Statement 1:** If $|f(x)| \le |x|$ for all $x \in R$, then $|f(x)|$ is continuous at 0.

Statement 2: If $f(x)$ is continuous, then $|f(x)|$ is also continuous.

15. **Statement 1:** $f(x) = ||x|^2 - 3|x| + 2|$ is not differentiable at 5 points.

Statement 2: If the graph of $f(x)$ crosses the x-axis at m distinct points, then $g(x) = |f(x)|$ is always non-differentiable at least at m distinct points.

16. **Statement 1:** The function $f(x) = a_1 e^{|x|} + a_2 |x|^5$, where a_1, a_2 are constants, is differentiable at $x = 0$ if $a_1 = 0$.

Statement 2: $e^{|x|}$ is a many-one function.

17. If $[\cdot]$ and $\{\cdot\}$ denote the greatest integer function and the fractional part function, respectively.

Let $f(x) = \{x\} + \sqrt{\{x\}}$.

Statement 1: f is not differentiable at integral values of x.

Statement 2: f is not continuous at integral points.

18. **Statement 1:** Let $f(x) = \lim_{m\to\infty} \left\{\lim_{n\to\infty} \cos^{2m}(n!\pi x)\right\}$, and

$g(x) = \begin{cases} 0, & \text{if } x \text{ is rational} \\ 1, & \text{if } x \text{ is irrational} \end{cases}$. Then $h(x) = f(x) + g(x)$ is continuous for all x.

Statement 2:
$f(x)$ and $g(x)$ are discontinuous for all $x \in R$.

19. **Statement 1:** If $f'(x)$ exists then $f'(x)$ is continuous.

Statement 2: Every differentiable function is continuous.

20. Consider the functions $f(x) = x^2 - 2x$ and $g(x) = -|x|$.

Statement 1: The composite function $F(x) = f(g(x))$ is not derivable at $x = 0$.

Statement 2: $F'(0^+) = 2$ and $F'(0^-) = -2$.

21. **Statement 1:** If $f(x)$ and $g(x)$ are two differentiable functions $\forall x \in R$, then $y = \max\{f(x), g(x)\}$ is always continuous but not differentiable at the point of intersection of graphs of $f(x)$ and $g(x)$.

Statement 2: $y = \max\{f(x), g(x)\}$ is always differentiable in between the two consecutive roots of $f(x) - g(x) = 0$ if both the functions $f(x)$ and $g(x)$ are differentiable $\forall x \in R$.

22. Consider the function

$f(x) = \cot^{-1}\left(\operatorname{sgn}\left(\dfrac{[x]}{2x - [x]}\right)\right)$, where [ ] denotes the greatest integer function.

Statement 1: $f(x)$ is discontinuous at $x = 1$.

Statement 2: $f(x)$ is non-differentiable at $x = 1$.

23. Consider the function $f(x) = \operatorname{sgn}(x - 1)$ and $g(x) = \cot^{-1}[x - 1]$, where [ ] denotes the greatest integer function.

Statement 1: The function $F(x) = f(x)g(x)$ is discontinuous at $x = 1$.

Statement 2: If $f(x)$ is discontinuous at $x = a$ and $g(x)$ is also discontinuous at $x = a$, then the product function $f(x)\,g(x)$ is discontinuous at $x = a$.

24. **Statement 1:** $f(x) = \min\{\sin x, \cos x\}$ is non-differentiable at $x = \pi/2$.

Statement 2: Non-differentiability of $\max\{g(x), h(x)\}$ $\Rightarrow$ non-differentiability of $\min\{g(x)\, h(x)\}$.

25. **Statement 1:** If $f(x)$ is a continuous function such that $f(0) = 1$ and $f(x) \ne x$, $\forall x \in R$, then $f(f(x)) > x$.

Statement 2: If $f: R \to R$, $f(x)$ is an onto function, then $f(x) = 0$ has at least one solution.

26. **Statement 1:** The function $f(x) = \left[\sqrt{x}\right]$ is discontinuous for all integral values of x in its domain (where $[x]$ is the greatest integer $\le x$).

Statement 2: $[g(x)]$ will be discontinuous for all x given by $g(x) = k$, where k is any integer.

Linked Comprehension Type

Based upon each paragraph, three multiple choice questions have to be answered. Each question has four choices a, b, c, and d, out of which *only one* is correct.

For Problems 1–3

$$\text{Let } f(x) = \begin{cases} \dfrac{a(1 - x\sin x) + b\cos x + 5}{x^2}, & x < 0 \\ 3, & x = 0, \\ \left\{1 + \left(\dfrac{P(x)}{x^2}\right)\right\}^{1/x}, & x > 0 \end{cases}$$

where $P(x)$ is a cubic function and f is continuous at $x = 0$.

1. The range of function $g(x) = 3a \sin x - b \cos x$ is
   a. $[-10, 10]$ b. $[-5, 5]$
   c. $[-12, 12]$ d. None of these
2. The value of $P''(0)$ is
   a. $\log_e 9$ b. $\log_e 2$
   c. 2 d. 1
3. If the leading co-efficient of $P(x)$ is positive, then the equation $P(x) = b$ has
   a. only one real, positive root
   b. only one real negative root
   c. three real roots
   d. None of these

For Problems 4–6

Let $f(x) = \begin{cases} x+2, & 0 \le x < 2 \\ 6-x, & x \ge 2 \end{cases}$,

$g(x) = \begin{cases} 1+\tan x, & 0 \le x < \dfrac{\pi}{4} \\ 3-\cot x, & \dfrac{\pi}{4} \le x < \pi \end{cases}$

4. $f(g(x))$ is
   a. discontinuous at $x = \pi/4$.
   b. differentiable at $x = \pi/4$.
   c. continuous but non-differentiable at $x = \pi/4$.
   d. differentiable at $x = \pi/4$, but derivative is not continuous.
5. The number of points of non-differentiability of $h(x) = |f(g(x))|$ is
   a. 1 b. 2
   c. 3 d. 4
6. The range of $h(x) = f(g(x))$ is
   a. $(-\infty, \infty)$ b. $(4, \infty)$
   c. $(-\infty, 4]$ d. None of these

For Problems 7–9

If $f(x) = x^2 + ax + 3$ and $g(x) = x + b$ and $F(x) = \lim_{n \to \infty} \dfrac{f(x) + x^{2n} g(x)}{1 + x^{2n}}$

7. If $F(x)$ is continuous at $x = 1$, then
   a. $b = a + 3$ b. $b = a - 1$
   c. $a = b - 2$ d. None of these
8. If $F(x)$ is continuous at $x = -1$, then
   a. $a + b = -2$ b. $a - b = 3$
   c. $a + b = 5$ d. None of these
9. If $F(x)$ is continuous at $x = \pm 1$, then $f(x) = g(x)$ has
   a. imaginary roots
   b. both the roots positive
   c. both the roots negative
   d. roots of opposite signs

For Problems 10 – 12

Let $f(x) = \begin{cases} [x], & -2 \le x \le -\dfrac{1}{2} \\ 2x^2 - 1, & -\dfrac{1}{2} < x \le 2 \end{cases}$ and $g(x) = f(|x|) + |f(x)|$, where [.] represents greatest integer function.

10. The number of points where $|f(x)|$ is non-differentiable is
    a. 3 b. 4
    c. 2 d. 5
11. The number of points where $g(x)$ is non-differentiable is
    a. 4 b. 5
    c. 2 d. 3
12. The number of points where $g(x)$ is discontinuous is
    a. 1 b. 2
    c. 3 d. None of these

Matrix-Match Type

Each question contains statements given in two columns which have to be matched. Statements a, b, c, d in column I have to be matched with statements p, q, r, s in column II. If the correct match is a → p, a → s, b → q, b → r, c → p, c → q and d → s, then the correctly bubbled 4 × 4 matrix should be as follows

	p	q	r	s
a	(p)	q	r	(s)
b	p	(q)	(r)	s
c	(p)	(q)	r	s
d	p	q	r	(s)

1.

Column 1	Column 2
a. $f(x) = \lvert x^3 \rvert$ is	p. continuous in $(-1, 1)$
b. $f(x) = \sqrt{\lvert x \rvert}$ is	q. differentiable in $(-1, 1)$
c. $f(x) = \lvert \sin^{-1} x \rvert$ is	r. differentiable in $(0, 1)$
d. $f(x) = \cos^{-1} \lvert x \rvert$ is	s. not differentiable at least at one point in $(-1, 1)$

2.

Column 1	Column 2
a. $f(x) = \begin{cases} \dfrac{1}{\lvert x \rvert} & \text{for } \lvert x \rvert \ge 1 \\ ax^2 + b & \text{for } \lvert x \rvert < 1 \end{cases}$ is differentiable	p. 2

everywhere and $\|k\| = a + b$, then the value of k is	
b. If $f(x) = \text{sgn}(x^2 - ax + 1)$ has exactly one point of discontinuity, then the value of a can be	**q.** -2
c. $f(x) = [2 + 3\|n\| \sin x], n \in N, x \in (0, \pi)$ has exactly 11 points of discontinuity, then the value of n is	**r.** 1
d. $f(x) = \|\|\|x\| - 2\| + a\|$ has exactly three points of non-differentiability, then the value of a is	**s.** -1

3. Consider the function $f(x) = x^2 + bx + c$, where $D = b^2 - 4c > 0$

Column 1	**Column 2**
Condition on b and c	Number of points of non-differentiability of $g(x) = \|f(\|x\|)\|$
a. $b < 0, c > 0$	**p.** 1
b. $c = 0, b < 0$	**q.** 2
c. $c = 0, b > 0$	**r.** 3
d. $b = 0, c < 0$	**s.** 5

4. Let $f(x) = \begin{cases} \dfrac{5e^{1/x} + 2}{3 - e^{1/x}}, & x \neq 0 \\ 0, & x = 0 \end{cases}$

Column 1	**Column 2**
a. $y = f(x)$ is	**p.** continuous at $x = 0$
b. $y = xf(x)$ is	**q.** discontinuous at $x = 0$
c. $y = x^2 f(x)$ is	**r.** differentiable at $x = 0$
d. $y = x^{-1} f(x)$ is	**s.** non-differentiable at $x = 0$

5.

Column 1	**Column 2**
a. $f(x) = \lim\limits_{n \to \infty} \cos^{2n}(2\pi x) + \left\{x + \dfrac{1}{2}\right\}$, where $\{.\}$ denotes the fractional part function at $x = \dfrac{1}{2}$	**p.** continuous
b. $f(x) = (\log_e x)(x - 1)^{1/5}$ at $x = 1$	**q.** discontinuous
c. $f(x) = [\cos 2\pi x] + \sqrt{\left\{\sin \pi \dfrac{x}{2}\right\}}$, where [.] and {.} denote the greatest integer and the fractional part function, respectively at $x = 1$	**r.** differentiable
d. $f(x) = \begin{cases} \cos 2x, & x \in Q \\ \sin x, & x \notin Q \end{cases}$ at $x = \dfrac{\pi}{6}$	**s.** non-differentiable

Integer Type

1. Let $g(x) = \begin{cases} a\sqrt{x+1}, & 0 < x < 3 \\ bx + 2, & 3 \leq x < 5 \end{cases}$. If $g(x)$ is differentiable on (0, 5), then $(a + b)$ is equal to ____.

2. If a is the number of points of continuity of $f(x) = \begin{cases} x - 1, & x \text{ is rational} \\ x^2 - x - 2, & x \text{ is irrational} \end{cases}$,

b is the number of points of discontinuity of $f(x) = \text{sgn}(x^3 - 3x + 1)$,

c is the number of points of non-differentiability of $f(x) = (\log x)\,|x^2 - 4x + 3| + 2(x - 2)^{1/3}$,

and d is the number of points where the graph of $f(x) = (\log x)\,|x^2 - 4x + 3| + 2(x - 2)^{1/3}$ has a sharp turn then the value of $a + b + c + d$ is ____.

3. Let $f(x) = \begin{cases} (\sin ax^2)/x^2, & x \neq 0 \\ (3/4) + [1/(4a)], & x = 0 \end{cases}$, $f(x)$ be continuous at $x = 0$, then the possible value of a is ____.

4. Let $f(x) = \begin{cases} (a + 3\cos x)/x^2, & x < 0 \\ b\tan\{\pi/[x + 3]\}, & x \geq 0 \end{cases}$, where [.] denotes the greatest integer function. Then the value of $|a|/b^2$ is ____.

5. If a and b are the numbers of points of non differentiability of $f(x) = [\sin^{-1} x]$ and $f(x) = \left[\dfrac{2}{1 + x^2}\right]$, $x \geq 0$, (where [.] represents greatest integer function) respectively, then the value of $a + b$ is ____.

6. The number of points of discontinuity of $f(x) = [2\cos x], x \in [0, 2\pi]$ ([·] represents the greatest integer function) are ____.
(where [.] represents greatest integer function)

7. The number of points of discontinuity of $f(x) = [\sin x + \cos x], x \in [0, 2\pi]$. ([·] represents the greatest integer function) is/are ____.

8. The least integral value of a for which the function $f(x) = [(x - 2)^3 / a] \sin(x - 2) + a\cos(x - 2)$, where [.] denotes the greatest integer function, is continuous in [0, 2] is ____.

9. If $f(x) = \begin{cases} \text{sgn}(x - 2) \times [\log_e x], & 1 \leq x \leq 3 \\ \{x^2\}, & 3 < x \leq 3.5 \end{cases}$, where [·] denotes the greatest integer function and {.} represents the fractional part function, then the number of points of discontinuity is________.

10. The number of points of discontinuity of $f(x) = \lim_{n\to\infty}[(x^{2n}-1)/(x^{2n}+1)]$ is ________.

11. The number of points where $f(x) = \begin{cases} x, & \text{if } x \text{ is rational} \\ 1-x, & \text{if } x \text{ is irrational} \end{cases}$ is continuous is ________.

12. If $f(x)$ is a continuous function $\forall\ x \in R$, and the range of $f(x)$ is $(2, \sqrt{26})$ and $g(x) = \left[\frac{f(x)}{c}\right]$ is continuous $\forall\ x \in R$, where [.] denotes the greatest integer function, then the least positive integral value of c is ________.

13. If $y = 1/(t^2+t-2)$, where $t = 1/(x-1)$, then the number of points where $f(x)$ is discontinuous is ____.

14. The number of points where the function $f(x) = [x] + |1-x|, x \in (-1, 3)$ where [.] represents greatest integer function is non-differentiable is ________.

15. The number of points where $f(x) = |\{x+(1/2)\}[x]|$, is discontinuons where [.] represents greatest integer function for $-2 \le x \le 2$ is ________.

16. The value of a for which $f(x) = \begin{cases} a+\sin^{-1}(x+b), & x \ge 1 \\ x, & x < 1 \end{cases}$ is differentiable at $x = 1$ is ________.

17. The number of points where $f(x) = \sin^{-1}[2x/(1+x^2)]$ is non-differentiable is ________.

18. A function $f(x)$ is defined as $f(x) = \begin{cases} x^2+ax+1, & x \text{ is rational} \\ ax^2+2x+b, & x \text{ is irrational} \end{cases}$ is continuous at $x = 1, 2$, then the value b is ________.

19. If $f(x)$ is a continuous function for all real values of x and satisfies $\{x^2+[f(x)-2]x+2\sqrt{3}-3-\sqrt{3}\}f(x) = 0, \forall\ x \in R$, then find the value of $\left(\{[f(\sqrt{3})]/2\}-1\right)^2$.

20. Let $f(x) =$

$$\begin{cases} (\cos x - \sin x)^{\operatorname{cosec} x}, & (-\pi)/2 < x < 0 \\ a, & x = 0 \\ (e^{1/x}+e^{2/x}+e^{3/x})/(ae^{2/x}+be^{3/x}), & 0 < x < \pi/2 \end{cases},$$

If $f(x)$ is continuous at $x = 0$, then the value of ab is________.

21. If $f(x) = \begin{cases} [ae^{1/(|x+2|)}-1]/[2-e^{1/(|x+2|)}], & -3 < x < -2 \\ b; & x = -2 \\ \sin[(x^4-16)/(x^5+32)], & -2 < x < 0 \end{cases}$ is continuous at $x = -2$, then the value of a/b is ________.

22. Function $f(x) = \begin{cases} [\sin(a+1)x+\sin x]/x, & x < 0 \\ c, & x = 0 \\ [(x+bx^2)^{1/2}-x^{1/2}]/(bx^{3/2}), & x > 0 \end{cases}$ is continuous at $x = 0$, then the value of $|a-b|$ is ____.

23. If a function f: $[-2a, 2a] \to$ R is an odd function such that $f(x) = f(2a-x)$ for $x \in [a, 2a]$ and the left-hand derivative at $x = a$ is 0, then the left-hand derivative at $x = -a$ is________.

24. Let $f(x)$ be a continuous function defined for $1 \le x \le 3$. If $f(x)$ takes rational values for all x and $f(2) = 6$, then $f(1.5)$ is equal to________.

25. The number of points where $f(x) = (x^2-1)|x^2-3x+2| + \cos(|x|)$ is non-differentiable is ________.

26. If $f(x) = \dfrac{\ln\left(e^{x^2}+2\sqrt{x}\right)}{\tan\sqrt{x}}$ is continuous at $x = 0$, then $f(0)$ must be equal to________.

27. $f(x) = \begin{cases} \left(\left|\sin\dfrac{2x^2}{a}+\cos\left(\dfrac{3x}{b}\right)\right|\right)^{ab/x^2}, & x \ne 0 \\ e^{x^2-2x+3}, & x = 0 \end{cases}$, is continuous at $x = 0\ \forall\ b \in R$, then $|1/a_{\min}|$ is_____.

28. Given the continuous function

$$f(x) = \begin{cases} 2x+3 & x \le 1 \\ -x^2+6 & x > 1, \end{cases}$$

then number of points where $|f(|x|)|$ is non-differentiable is________.

ANSWERS AND SOLUTIONS

Objective Type

1. c. $f(x) = \tan x$ is discontinuous when $x = (2n + 1)\pi/2$, $n \in Z$

$f(x) = x[x]$ is discontinuous when $x = k, k \in Z$

$f(x) = \sin [n\pi x]$ is discontinuous when $n\pi x = k, k \in Z$

Thus, all the above functions have infinite number of points of discontinuity.

But $f(x) = \dfrac{|x|}{x}$ is discontinuous when $x = 0$ only.

2. c. We have $f(x) = \dfrac{4-x^2}{x(4-x^2)}$

Clearly, there are three points of discontinuity, viz., 0, 2, –2, because $f(x)$ does not exist for these three values of c.

3. b. $f(x) = \dfrac{\tan\left(\frac{\pi}{4}-x\right)}{\cot 2x}$, $(x \neq \pi/4)$ is continuous at $x = \pi/4$

$$\Rightarrow f\left(\frac{\pi}{4}\right) = \lim_{x\to\pi/4} f(x)$$

$$= \lim_{x\to\pi/4} \frac{\tan\left(\frac{\pi}{4}-x\right)}{\cot 2x}$$

Now by applying L' Hospital's Rule,

$$= \lim_{x\to\frac{\pi}{4}} \frac{-\sec^2\left(\frac{\pi}{4}-x\right)}{-2\operatorname{cosec}^2(2x)} = \frac{1}{2}$$

4.b. Given $f(x)$ is continuous at $x = 0$

$$\Rightarrow \lim_{x\to 0} f(x) = f(0)$$

$$\Rightarrow \lim_{x\to 0} \frac{(3^x-1)^2}{\sin x \ln(1+x)} = f(0)$$

$$\Rightarrow f(0) = \lim_{x\to 0} \frac{\left(\frac{3^x-1}{x}\right)^2}{\left(\frac{\sin x}{x}\right)\left(\frac{\ln(1+x)}{x}\right)} = (\ln 3)^2$$

5. b. We have $f(x) = \begin{cases} \dfrac{1-|x|}{1+x}, & x \neq -1 \\ 1, & x = -1 \end{cases}$

$$= \begin{cases} 1, & x < 0, \\ \dfrac{1-x}{1+x}, & x \geq 0 \end{cases}$$

$[\because f(-1) = 1$ is given$]$

$$\Rightarrow f([2x]) = \begin{cases} 1, & [2x] < 0 \\ \dfrac{1-[2x]}{1+[2x]}, & [2x] \geq 0 \end{cases}$$

$$= \begin{cases} 1, & x < 0 \\ 1, & 0 \leq x < 1/2 \\ 0, & 1/2 \leq x < 1 \\ -1/3, & 1 \leq x < \dfrac{3}{2} \end{cases}$$

Clearly, $f(x)$ is continuous for all $x < 1/2$and discontinuous at $x = 1/2$, 1.

6. d. We have,

$$\text{L.H.L.} = \lim_{x\to 4^-} f(x)$$

$$= \lim_{h\to 0} f(4-h)$$

$$= \lim_{h\to 0} \frac{4-h-4}{|4-h-4|} + a$$

$$= \lim_{h\to 0} -\frac{h}{h} + a = a-1$$

$$\text{R.H.L.} = \lim_{x\to 4^+} f(x)$$

$$= \lim_{h\to 0} f(4+h)$$

$$= \lim_{h\to 0} \frac{4+h-4}{|4+h-4|} + b = b+1$$

$\Rightarrow f(4) = a+b$

Since $f(x)$ is continuous at $x = 4$,

$$\lim_{x\to 4^-} f(x) = f(4) = \lim_{x\to 4^+} f(x)$$

$\Rightarrow a-1 = a+b = b+1 \Rightarrow b = -1$ and $a = 1$.

7. d. $\displaystyle\lim_{x\to 0} \frac{x-e^x+1-(1-\cos 2x)}{x^2}$

$$= \lim_{x\to 0}\left[\frac{x-e^x+1}{x^2} - \frac{(1-\cos 2x)}{x^2}\right]$$

$$= \lim_{x\to 0}\left[\frac{x+1-\left(1+x+\frac{x^2}{2}\right)}{x^2} - \frac{2\sin^2 x}{x^2}\right]$$

(Using expansion of e^x)

$$= -\frac{1}{2} - 2 = -\frac{5}{2}$$

Hence, for continuity $f(0) = -5/2$

Now $[f(0)] = -3$; $\{f(0)\} = \left\{-\dfrac{5}{2}\right\} = \dfrac{1}{2}$.

Hence, $[f(0)]\ \{f(0)\} = -\dfrac{3}{2} = -1.5$.

8. d. We have $f(x)=\begin{cases}\dfrac{x^3+x^2-16x+20}{(x-2)^2}, & x\neq 2\\ k, & x=2\end{cases}$

Clearly, $f(x)$ is continuous for all values of x except possibly at $x=2$

It will be continuous at $x=2$ if $\lim\limits_{x\to 2} f(x)=f(2)$

$$\Rightarrow \lim_{x\to 2}\frac{x^3+x^2-16x+20}{(x-2)^2}=k$$

$$\Rightarrow k=\lim_{x\to 2}\frac{(x-2)^2(x+5)}{(x-2)^2}=\lim_{x\to 2}(x+5)=7$$

$\therefore k=7$

9. b. The function f is clearly continuous at each point in its domain except possibly at $x=0$. Given that $f(x)$ is continuous at $x=0$.

Therefore, $f(0)=\lim\limits_{x\to 0} f(x)$

$$=\lim_{x\to 0}\frac{2x-\sin^{-1}x}{2x+\tan^{-1}x}$$

$$=\lim_{x\to 0}\frac{2-(\sin^{-1}x)/x}{2+(\tan^{-1}x)\,x}=\frac{1}{3}$$

10. c. $\displaystyle\lim_{x\to 2^+}\frac{(x-2)}{|x-2|}\left(\frac{x^2-1}{x^2+1}\right)=\lim_{x\to 2^+}\frac{(x-2)}{(x-2)}\left(\frac{x^2-1}{x^2+1}\right)$

$$=\lim_{x\to 2^+}\left(\frac{x^2-1}{x^2+1}\right)=\frac{3}{5}$$

$$=\lim_{x\to 2^-}\frac{(x-2)}{|x-2|}\left(\frac{x^2-1}{x^2+1}\right)$$

$$=\lim_{x\to 2^-}\frac{(x-2)}{(2-x)}\left(\frac{x^2-1}{x^2+1}\right)=-\frac{3}{5}$$

Thus, L.H.L. $\neq$ R.H.L.

Hence, the function has non-removable discontinuous at $x=2$.

11. c. $f(x)=\displaystyle\lim_{n\to\infty}\frac{\left[(x-1)^2\right]^n-1}{\left[(x-1)^2\right]^n+1}$

$$=\lim_{n\to\infty}\frac{1-\dfrac{1}{\left[(x-1)^2\right]^n}}{1+\dfrac{1}{\left[(x-1)^2\right]^n}}$$

$$=\begin{cases}-1, & 0\le (x-1)^2<1\\ 0, & (x-1)^2=1\\ 1, & (x-1)^2>1\end{cases}$$

$$=\begin{cases}1, & x<0\\ 0, & x=0\\ -1, & 0<x<2\\ 0, & x=2\\ 1, & x>2\end{cases}$$

Thus, $f(x)$ is discontinuous at $x=0, 2$.

12. c. $f(0)=0+0+\lambda\ln 4=\lambda\ln 4$ (1)

$$\text{R.H.L.}=\lim_{x\to 0+} f(x)=\lim_{h\to 0} f(0+h)$$

$$=\lim_{h\to 0}\frac{8^h-4^h-2^h+1^h}{h^2}$$

$$=\lim_{h\to 0}\frac{(4^h-1)(2^h-1)}{h.h}$$

$$=\lim_{h\to 0}\left(\frac{4^h-1}{h}\right)\lim_{h\to 0}\left(\frac{2^h-1}{h}\right)$$

$=\ln 4\ln 2$ (2)

$\therefore f(0)=$ R.H.L.

$\Rightarrow \lambda=\ln 2$

13. b. We must have $\displaystyle\lim_{x\to 0}\frac{a\cos x-\cos bx}{x^2}=4$

$$\Rightarrow \lim_{x\to 0}\frac{a\left(1-\dfrac{x^2}{2!}\right)-\left(1-\dfrac{b^2x^2}{2!}\right)}{x^2}=4$$

$$\Rightarrow \lim_{x\to 0}\left[\frac{(a-1)}{x^2}-\left(\frac{a}{2}-\frac{b^2}{2}\right)\right]=4$$

$\Rightarrow a=1$ and $\dfrac{a}{2}-\dfrac{b^2}{2}=-4$

$\Rightarrow a=1$ and $b^2=9$

$\Rightarrow a=1$ and $b=\pm 3$

14. b. Given that

$f(x)=\begin{cases}x+a\sqrt{2}\sin x, & 0\le x<\pi/4\\ 2x\cot x+b, & \pi/4\le x\le \pi/2\\ a\cos 2x-b\sin x, & \pi/2<x\le\pi\end{cases}$ is

continuous for $0\le x\le\pi$

$$\therefore \lim_{x\to\pi/4^-} f(x)=\lim_{x\to\pi/4^+} f(x)$$

$$\Rightarrow \lim_{x\to\pi/4^-}\left(x+a\sqrt{2}\sin x\right)=\lim_{x\to\pi/4^+}(2x\cot x+b)$$

$\Rightarrow \dfrac{\pi}{4}+a=\dfrac{\pi}{2}+b$ (1)

Also, $\displaystyle\lim_{x\to\pi/2^-} f(x)=\lim_{x\to\pi/2^+} f(x)$

$\Rightarrow \lim\limits_{x\to\pi/2^-}(2x\cot x + b) = \lim\limits_{x\to\pi/2^+}(a\cos 2x - b\sin x)$

$\Rightarrow 0 + b = -a - b$ or $a + 2b = 0$ (2)

Solving equations (1) and (2), we have $a = \pi/6$ and $b = -\pi/12$.

15. a. $f(x)$ is continuous when $5x = x^2 + 6 \Rightarrow x = 2, 3$.

16. b. $f(x) = 2|\text{sgn}(2x)| + 2$

$$= 2\frac{2x}{|2x|} + 2 = 2\frac{x}{|x|} + 2$$

$\Rightarrow f(0) = 2$

$f(0^+) = 2\times 1 + 2 = 2$

$f(0^-) = 2\times 1 - 11 + 2 = 2$

Thus, $f(x)$ has removal discontinuity.

17.d. Since $\lim\limits_{n\to\infty} x^{2n} = \begin{cases} 0, & \text{if } |x|<1 \\ 1, & \text{if } |x|=1 \end{cases}$

$\therefore f(x) = \lim\limits_{x\to\infty}(\sin x)^{2n} = \begin{cases} 0, & \text{if } |\sin x|<1 \\ 1, & \text{if } |\sin x|=1 \end{cases}$

Thus, $f(x)$ is continuous at all x, except for those values of x for which $|\sin x| = 1$, i.e., $x = (2k+1)\dfrac{\pi}{2}$, $k \in Z$

18. a. As f is continuous so $f(0) = \lim\limits_{x\to 0} f(x)$

$f(0) = \lim\limits_{n\to\infty} f(1/4n)$

$$= \lim_{n\to\infty}\left((\sin e^n)\, e^{-n^2} + \frac{1}{1+1/n^2}\right) = 0 + 1 = 1.$$

19. a. $f(x) = \dfrac{x^2 - bx + 25}{x^2 - 7x + 10}, x \neq 5$

$f(x)$ is continuous at $x = 5$, only if $\lim\limits_{x\to 5}\dfrac{x^2 - bx + 25}{x^2 - 7x + 10}$ is finite.

Now $x^2 - 7x + 10 \to 0$ when $x \to 5$

then we must have $x^2 - bx + 25 \to 0$ for which $b = 10$

Hence, $\lim\limits_{x\to 5}\dfrac{x^2 - 10x + 25}{x^2 - 7x + 10} = \lim\limits_{x\to 5}\dfrac{x-5}{x-2} = 0.$

20. d.

21. a. $f(x)$ is continuous at some x where $\sin x = \cos x$ or $\tan x = 1$ or $x = n\pi + \pi/4$, $n \in I$.

22. b. Consider $x \in [0, 1]$.

From the graph given in Fig. 3.1, it is clear the $[\cos \pi x]$ is discontinuous at $x = 0, 1/2, 1$ (1)

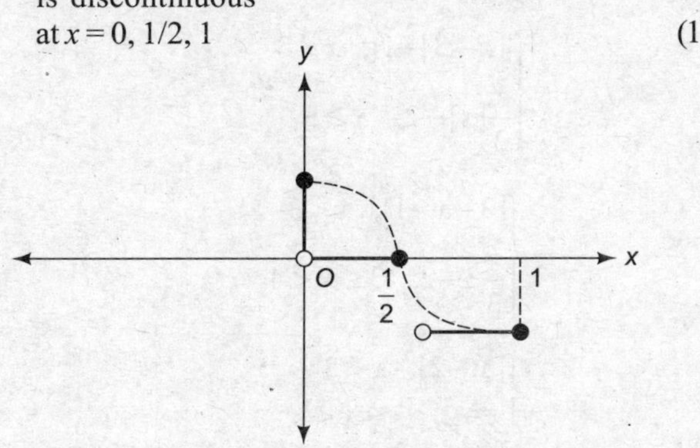

Fig. 3.1

Now consider $x \in (1, 2]$

$f(x) = [x-2]\,|2x-3|$

For $x \in (1, 2)$; $[x-2] = -1$ and for $x = 2$; $[x-2] = 0$

Also $|2x-3| = 0 \Rightarrow x = 3/2$

$\Rightarrow x = 3/2$ and 2 may be the points at which $f(x)$ may be discontinuous (2)

Combining equations (1) and (2), we have $x = 0, 1/2, 1, 3/2, 2$

Dividing $f(x)$ by about the 5 critical points,

we get $f(x) = \begin{cases} 1, & x = 0 \\ 0, & 0 < x \le \frac{1}{2} \\ -1, & \frac{1}{2} < x \le 1 \\ -(3-2x), & 1 < x \le 3/2 \\ -(2x-3), & 3/2 < x \le 2 \\ 0, & x = 2 \end{cases}$

Thus, $f(x)$ is continuous when $x \in [0, 2] - \{0, 1/2, 2\}$.

23. d. For $0 \le x < 1, f(x) = [\sin 0] = 0$,

$1 \le x < 2, f(x) = [\sin 1] = 0$

$2 \le x < 3, f(x) = [\sin 2] = 0$,

$3 \le x < 4, f(x) = [\sin 3] = 0$

$4 \le x < 5, f(x) = [\sin 4] = -1$

Hence, there is discontinuity at point $(4, -1)$

as $f(0+0) \neq f(0-0)$, thus $f(x)$ is discontinuous at $x = 0$.

24. d. We have $\lim\limits_{x\to 0^-} f(x) = \lim\limits_{h\to 0}\sin(\log_c|-h|)$

$= \lim\limits_{h\to 0}\sin(\log_e h)$, which does not exist but lies between -1 and 1.

Similarly, $\lim\limits_{x\to 0^+} f(x)$ lies between -1 and 1 but cannot be determined.

25. a. $f(x) = (-1)^{[x^3]}$ is discontinuous when $x^3 = n, n \in z \Rightarrow x = n^{1/3}$

$f\left(\dfrac{3}{2}\right) = (-1)^3 = -1$

For $x \in (-1, 0)$, $f(x) = (-1)^{-1} = -1$

$\Rightarrow f'(x) = 0$

For $x \in [0, 1)$, $f(x) = (-1)^{\circ} = 1$

$\Rightarrow f'(x) = 0$

26. c. $f(x) = \{x\}\sin(\pi[x])$

$= \{x\}\sin(\text{integral multiple of } \pi)$

$= 0$

Hence, $f(x)$ is continuous for all x.

27. d. We have $\lim\limits_{x\to 1^-} f(x) = \lim\limits_{h\to 0} f(1+h)$

$$= \lim_{h\to 0}\frac{\log(4+h^2)}{\log(1-4h)} = -\infty$$

and, $\lim\limits_{x\to 1^+} f(x) = \lim\limits_{h\to 0} f(1+h) = \lim\limits_{h\to 0}\dfrac{\log(4+h^2)}{\log(1+4h)} = \infty$

So, $f(1^-)$ and $f(1^+)$ do not exist.

28. c. Since $\lim_{x\to 1^-} g(x) = \lim_{x\to 1^+} g(x) = 1$ and $g(1) = 0$,

$g(x)$ is not continuous at $x = 1$ but $\lim_{x\to 1} g(x)$ exists.

We have $\lim_{x\to 1^-} f(x) = \lim_{h\to 0} f(1-h) = \lim_{h\to 0}[1-h] = 0$

and, $\lim_{x\to 1^+} f(x) = \lim_{h\to 0} f(1+h) = \lim_{h\to 0}[1+h] = 1$

So, $\lim_{x\to 1} f(x)$ does not exist and so $f(x)$ is not continuous at $x = 1$.

We have $gof(x) = g(f(x)) = g([x]) = 0,\ \forall\ x, y \in R$

So, gof is continuous for all x.

We have $fog(x) = f(g(x))$

$$= \begin{cases} f(0), & x \in Z \\ f(x^2), & x \in R - Z \end{cases}$$

$$= \begin{cases} 0, & x \in z \\ [x^2], & x \in R - Z \end{cases}$$

which is clearly not continuous.

29. c. $f(0+0) = \lim_{h\to 0} f(h)$

$$= \lim_{h\to 0} \frac{h}{2h^2 + h} = \lim_{h\to 0} \frac{1}{2h+1} = 1$$

and $f(0-0) = \lim_{h\to 0} f(-h)$

$$= \lim_{h\to 0} \frac{-h}{2h^2 + |-h|}$$

$$= \lim_{h\to 0} \frac{-h}{2h^2 + h} = \lim_{h\to 0} \frac{-1}{2h+1} = -1$$

30. b. We have

$$f(x) = \frac{x - |x-1|}{x} = \begin{cases} \dfrac{x + x - 1}{x}, & x < 1,\ x \ne 0 \\ \dfrac{x - (x-1)}{x}, & x \ge 1 \end{cases}$$

$$= \begin{cases} \dfrac{2x-1}{x}, & x < 1, x \ne 0 \\ \dfrac{1}{x}, & x \ge 1 \end{cases}$$

Clearly, $f(x)$ is discontinuous at $x = 0$ as it is not defined at $x = 0$. Since $f(x)$ is not defined at $x = 0$, therefore $f(x)$ cannot be differentiable at $x = 0$. Clearly $f(x)$ is continuous at $x = 1$, but it is not differentiable at $x = 1$, because $Lf'(1) = 1$ and $Rf'(1) = -1$.

31. a. We have $f(x) = \begin{cases} x^3, & x > 0 \\ 0, & x = 0 \\ -x^3, & x < 0 \end{cases}$

Clearly, $f(x)$ is continuous at $x = 0$

$$(\text{L.H.D. at } x = 0) = \left[\frac{d}{dx}(-x^3)\right]_{x=0} = [-3x^2]_{x=0} = 0$$

Similarly (R.H.D. at $x = 0$) $= 0$

So, $f(x)$ is differentiable at $x = 0$.

32.d.

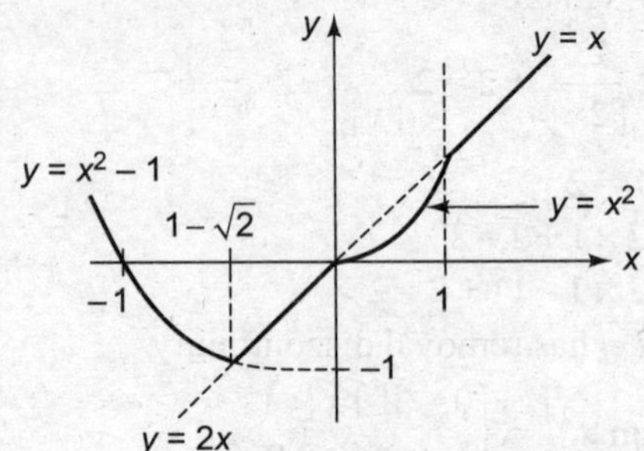

Fig. 3.2

From the graph it is clear that $f(x)$ is every where continuous but not differentiable at $x = 1 - \sqrt{2}$, 0, 1.

33.b. Since both $\cos x$ and $\sin^{-1} x$ are continuous function, $f(x) = \sin^{-1}(\cos x)$ is also a continuous function.

Now,

$$f'(x) = \frac{-\sin x}{\sqrt{1-\cos^2 x}} = \frac{\sin x}{|\sin x|} = \begin{cases} \dfrac{-\sin x}{-\sin x} = 1, & x < 0 \\ \dfrac{-\sin x}{\sin x} = -1, & x > 0 \end{cases}$$

$\therefore$ (L.H.D. at $x = 0$) $= 1$ and (R.H.D. at $x = 0$) $= -1$

Hence, $f(x)$ is not differentiable at $x = 0$.

34.d. $f(x) = (e^x - 1)|e^{2x} - 1|$

$= (e^x - 1)|e^x - 1||e^x + 1|$

$= (e^x + 1)(e^x - 1)|e^x - 1|$

Now, both $e^x + 1$ and $(e^x - 1)|e^x - 1|$ are differentiable

[as $g(x)|g(x)|$ is differentiable when $g(x) = 0$]

Hence, $f(x)$ is differentiable.

$f(x) = \dfrac{x-1}{x^2+1}$ is rational function in which denominator never becomes zero.

Hence, $f(x)$ is differentiable.

$$f(x) = \begin{cases} ||x-3|-1|, & x < 3 \\ \dfrac{x}{3}[x] - 2, & x \ge 3 \end{cases}$$

$$= \begin{cases} |3-x-1|, & x < 3 \\ \dfrac{x}{3}3 - 2, & 3 \le x < 4 \end{cases}$$

$$= \begin{cases} |x-2|, & x < 3 \\ x - 2, & 3 \le x < 4 \end{cases}$$

$= x - 2, x \in [2, 4)$

Hence, $f(x)$ is differentiable at $x = 3$.

$f(x) = 3(x-2)^{3/4} + 3 \Rightarrow f'(x) = \frac{9}{4}(x-2)^{-1/4}$

which is non-differentiable at $x = 2$.

Here $f(x)$ is continuous and the graph has vertical tangent at $x = 2$, however graph is smooth in neighbourhood of $x = 2$.

35.c. $\left|x - \frac{1}{2}\right|$ is continuous everywhere but not differentiable at $x = \frac{1}{2}$ $|x - 1|$ is continuous everywhere but not differentiable at $x = 1$ tan x is continuous in [0, 2] except at $x = \frac{\pi}{2}$.

Hence $f(x)$ is not differentiable at $x = \frac{1}{2}, 1, \frac{\pi}{2}$.

36.c. Let $f(x) = x^2|x|$ which could be expressed as

$$f(x) = \begin{cases} -x^3, & x < 0 \\ 0, & x = 0 \\ x^3, & x > 0 \end{cases} \Rightarrow f'(x) = \begin{cases} -3x^2, & x < 0 \\ 0, & x = 0 \\ 3x^2, & x > 0 \end{cases}$$

So, $f'(x)$ exists for all real x.

$$f''(x) = \begin{cases} -6x, & x < 0 \\ 0, & x = 0 \\ 6x, & x > 0 \end{cases}$$

So, $f''(x)$ exists for all real x.

$$f'''(x) = \begin{cases} -6, & x < 0 \\ 0, & x = 0 \\ 6, & x > 0 \end{cases}$$

However, $f'''(0)$ does not exist since $f'''(0^-) = -6$ and $f'''(0^+) = 6$ which are not equal. Thus, the set of points where $f(x)$ is thrice differentiable is $R - \{0\}$.

37.c $f(x) = (x^2 - 1)\,|(x-1)(x-2)|$

$$f'(1^+) = \lim_{h\to 0} \frac{((1+h)^2 - 1)\,|h(1+h-2)| - 0}{h} = 0,$$

$$f'(1^-) = \lim_{h\to 0} \frac{((1-h)^2 - 1)\,|-h(1-h-2)| - 0}{-h} = 0$$

Hence, it is differentiable at $x = 0$.

For $f(x) = \sin(|x-1|) - |x-1|$

$$f'(0^+) = \lim_{h\to 0} \frac{\sin h - h - 0}{h} = 0$$

$$f'(0^-) = \lim_{h\to 0} \frac{\sin|-h| - |-h|}{-h} = \lim_{h\to 0} \frac{\sin h - h}{-h} = 0$$

Hence, $f(x)$ is differentiable at $x = 0$.

For $f(x) = \tan(|x-1|) + |x-1|$

$$f'(0^+) = \lim_{h\to 0} \frac{\tan h + h - 0}{h} = 2$$

$$f'(0^-) = \lim_{h\to 0} \frac{\tan|-h| + |-h|}{-h} = \lim_{h\to 0} \frac{\tan h + h}{-h} = -2.$$

Hence, $f(x)$ is non-differentiable at $x = 0$.

38.d. $$f'(0+0) = \lim_{h\to 0} \frac{he^{-\left(\frac{1}{h}+\frac{1}{h}\right)} - a}{h}$$

$$= \lim_{h\to 0} \frac{he^{-2/h} - a}{h}$$

$$= 0 - \lim_{h\to 0} \frac{a}{h} = 0 \text{ if } a = 0$$

$$f'(0-0) = \lim_{h\to 0} \frac{-he^{-\left(-\frac{1}{h}+\frac{1}{h}\right)} - a}{-h}$$

$$= \lim_{h\to 0}\left(1 + \frac{a}{h}\right) = 1 \text{ if } a = 0$$

Thus, no values of a exists.

39.c. $f(x) = \begin{cases} ax^2 + 1, & x \le 1 \\ x^2 + ax + b, & x > 1 \end{cases}$ is differentiable at $x = 1$

Then $f(x)$ is continuous at $x = 1$

$\Rightarrow f(1^-) = f(1^+) \Rightarrow a + 1 = 1 + a + b \Rightarrow b = 0.$

Also $f'(x) = \begin{cases} 2ax, & x < 1 \\ 2x + a, & x > 1 \end{cases}$

We must have $f'(1^-) = f'(1^+) \Rightarrow 2a = 2 + a \Rightarrow a = 2.$

40.b. $|\sin x|$ and $e^{|x|}$ are not differentiable at $x = 0$ and $|x|^3$ is differentiable at $x = 0$.

Therefore, for $f(x)$ to be differentiable at $x = 0$, we must have $a = 0$, $b = 0$ and c can be any real number.

41.d.

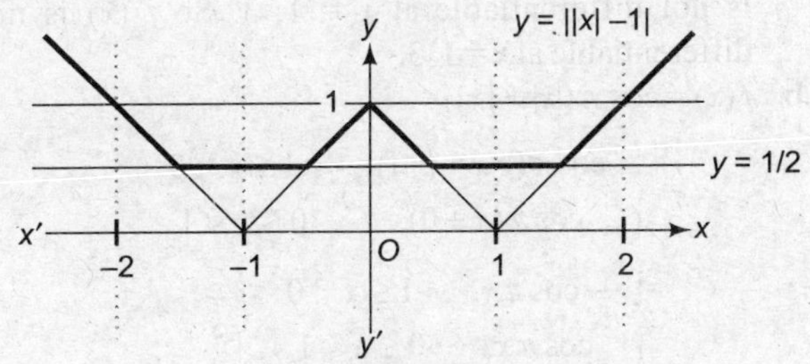

Fig. 3.3

Clearly, from the graph, $f(x)$ is non-differentiable at five points, $x = -2, -1, 0, 1, 2$.

42.c. Clearly, $f(x)$ is continuous for all x except possibly at $x = \pi/6$.

For $f(x)$ to be continuous at $x = \pi/6$, we must have

$$\lim_{x\to\pi/6^-} f(x) = \lim_{x\to\pi/6^+} f(x)$$

$$\Rightarrow \lim_{x\to\pi/6} \sin 2x = \lim_{x\to\pi/6} ax + b$$

$\Rightarrow \sin(\pi/3) = (\pi/6)\, a + b$

$\Rightarrow \frac{\sqrt{3}}{2} = \frac{\pi}{6} a + b$ (1)

For $f(x)$ to be differentiable at $x = \pi/6$, we must have L.H.D. at $x = \pi/6$ = R.H.D. at $x = \pi/6$

$\Rightarrow \lim_{x\to\pi/6} 2\cos 2x = \lim_{x\to\pi/6} a$

$\Rightarrow 2\cos \pi/3 = a \Rightarrow a = 1$

Putting $a = 1$ in equation (1), we get

$b = (\sqrt{3}/2) - \pi/6$.

43.b. $f(x)$ is clearly continuous for $x \in R$.

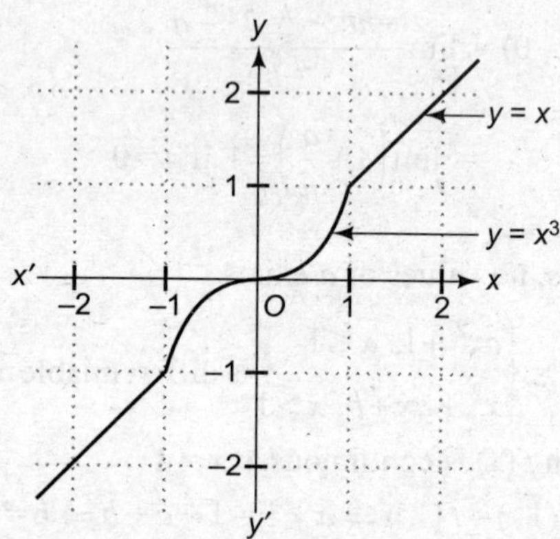

Fig. 3.4

$f'(x) = \begin{cases} 3x^2, & x^2 < 1 \\ 1, & x^2 > 1 \end{cases}$

thus $f(x)$ is non-differentiable at $x = 1, -1$.

44.d. $\frac{x}{1+|x|}$ is always differentiable (also at $x = 0$)

Also $(x-2)(x+2)\,|(x-1)(x-2)(x-3)|$ is not differentiable at $x = 1, 3$. So $f(x)$ is not differentiable at $x = 1, 3$.

45.b. $f(x) = \cos \pi(|x| + [x])$

$= \begin{cases} \cos\pi(-x+(-1)), & -1 \le x < 0 \\ \cos\pi(x+0), & 0 \le x < 1 \end{cases}$

$= \begin{cases} -\cos\pi x, & -1 \le x < 0 \\ \cos\pi x, & 0 \le x < 1 \end{cases}$

Obviously, $f(x)$ is discontinuous at $x = 0$ otherwise $f(x)$ is continuous and differentiable in (–1, 0) and (0, 1).

46.c. For $f(x)$ to be continuous at $x = 0$, we have

$f(0^-) = f(0^+) \Rightarrow 1 + 0 = a(0) + b \Rightarrow b = 1$

$f'(0^-) = \lim_{h\to 0} \frac{f(h) - f(0)}{h}$

$= \lim_{h\to 0} \frac{e^{h^2+h} - b}{h}$

$= \lim_{h\to 0} \frac{e^{h^2+h} - 1}{h}$

$= \lim_{h\to 0} \frac{e^{h^2+h} - 1}{h(h+1)}(h+1) = 1$

$\therefore f'(0^+) = a$

Hence, $a = 1$

47.a. Clearly $f(x)$ is continuous at $x = 0$

Now $f'(0^+) = \lim_{h\to 0} \frac{e^{-1/h^2} - 0}{h} = \lim_{h\to 0} \frac{1/h}{e^{1/h^2}}$

$= \lim_{h\to 0} \frac{-1/h^2}{-2/h^3\, e^{1/h^2}}$

(Applying L'Hospital's Rule)

$= \frac{1}{2} \lim_{h\to 0} \frac{h}{e^{1/h^2}} = 0$

Also $f(0^-) = 0$

Thus, $f(x)$ is differentiable at $x = 0$.

48.c. $f(x) = \begin{cases} |x| - 1, & |x| < 0 \\ |x|^2 - 2|x|, & |x| \ge 0 \end{cases}$

where $|x| < 0$ is not possible thus, neglecting we get,

$f(|x|) = |x|^2 - 2|x|, |x| \ge 0$

$f(|x|) = \begin{cases} x^2 + 2x, & x < 0 \\ x^2 - 2x, & x \ge 0 \end{cases}$ (1)

$\Rightarrow f'(|x|) = \begin{cases} 2x + 2, & x < 0 \\ 2x - 2, & x > 0 \end{cases}$

Clearly, $f(|x|)$ is continuous at $x = 0$ but non-differentiable at $x = 0$.

$f(|x|) = \begin{cases} |x| - 1, & |x| < 0 \\ |x|^2 - 2|x|, & |x| \ge 0 \end{cases}$

$|f(x)| = \begin{cases} 1 - x, & x < 0 \\ -x^2 + 2x, & 0 \le x < 2 \\ x^2 - 2x, & x \ge 2 \end{cases}$ (2)

Clearly $|f(x)|$ is discontinuous at $x = 0$, but continuous at $x = 0$

Also, $|f(x)| = \begin{cases} -1, & x < 0 \\ -2x + 2, & 0 < x < 2 \\ 2x - 2, & x > 2 \end{cases}$

$|f(x)$ is non-differentiable at $x = 0$ and $x = 2$.

49.c. Since $1 \le x < 2 \Rightarrow 0 \le x - 1 < 1$

$\Rightarrow [x^2 - 2x] = [(x-1)^2 - 1] = [(x-1)^2] - 1 = 0 - 1 = -1$

$$\therefore f(x) = \begin{cases} 1 - 4x^2, & 0 \le x < \frac{1}{2} \\ 4x^2 - 1, & \frac{1}{2} \le x < 1 \\ -1 & 1 \le x < 2 \end{cases}$$

Therefore, graph of $f(x)$ is as follows

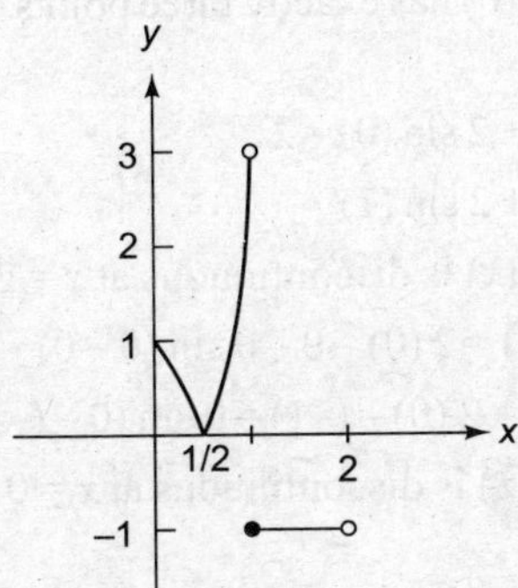

Fig. 3.5

It is clear from graph that $f(x)$ is discontinuous at $x = 1$ and not differentiable at $x = \frac{1}{2}$ and $x = 1$.

50.c. For $|x| < 1$, $x^{2n} \to 0$ as $n \to \infty$ and for $|x| > 1$, $1/x^{2n} \to 0$ as $n \to \infty$. So

$$f(x) = \begin{cases} \log(2+x), & |x| < 1 \\ \lim\limits_{n \to \infty} \dfrac{x^{-2n}\log(2+x) - \sin x}{x^{-2n} + 1} = -\sin x, & \text{if } |x| > 1 \\ \frac{1}{2}[\log(2+x) - \sin x], & |x| = 1 \end{cases}$$

Thus $\lim\limits_{x \to 1+} f(x) = \lim\limits_{x \to 1} (-\sin x) = -\sin 1$

and $\lim\limits_{x \to 1-} f(x) = \lim\limits_{x \to 1} \log(2+x) = \log 3.$

51.c. $f'(0) = \lim\limits_{h \to 0} \dfrac{f(h) - f(0)}{h}$

$$= \lim_{h \to 0} \frac{h^a \sin\left(\frac{1}{h}\right)}{h} = \lim_{h \to 0} h^{a-1} \sin\left(\frac{1}{h}\right)$$

This limit will not exist if $a - 1 \le 0 \Rightarrow a \le 1$.

Now $\lim\limits_{x \to 0} f(x) = \lim\limits_{x \to 0} x^a \sin\left(\frac{1}{x}\right) = 0$ if $a > 0$.

Thus, $a \in (0, 1]$.

52.c. $[\sin x]$ is non-differentiable at $x = \frac{\pi}{2}, \pi, 2\pi$

and $[\cos x]$ is non-differentiable at $x = 0, \frac{\pi}{2}, \frac{3\pi}{2}, 2\pi$.

Thus, $f(x)$ is definitely non-differentiable at $x = \pi, \frac{3\pi}{2}, 0$

Also, $f\left(\frac{\pi}{2}\right) = 1$, $f\left(\frac{\pi}{2} - 0\right) = 0$ and $f(2\pi) = 1$,

$f(2\pi - 0) = -1$

Thus, $f(x)$ is also non-differentiable at $x = \frac{\pi}{2}$ and 2π.

53.a We have $x + 4|y| = 6y$

$$\Rightarrow \begin{cases} x - 4y = 6y, & \text{if } y < 0 \\ x + 4y = 6y, & \text{if } y \ge 0 \end{cases}$$

$$\Rightarrow y = \begin{cases} \frac{1}{2}x, & \text{if } x \ge 0 \\ \frac{1}{10}x, & \text{if } x < 0 \end{cases}$$

$$\Rightarrow f'(x) = \begin{cases} \frac{1}{2}, & x \ge 0 \\ \frac{1}{10}, & x < 0 \end{cases}$$

Clearly, $f(x)$ is continuous at $x = 0$ but non-differentiable at $x = 0$.

54.b. $f(0^+) = \lim\limits_{x \to 0+} |x|^{\sin x} = e^{\lim\limits_{x \to 0} \sin x \log|x|}$

$$= e^{\lim\limits_{x \to 0} \frac{\log x}{\operatorname{cosec} x}} = e^0 = 1$$

(Using L' Hopsital's Rule)

$f(0^-) = g(0) = 1$

Let $g(x) = ax + b$

$\Rightarrow b = 1 \Rightarrow g(x) = ax + 1$

For $x > 0$, $f'(x) = e^{\sin x \ln(|x|)}\left[\cos x \ln(|x|) + \dfrac{\sin x}{x}\right]$

$f'(1) = 1[0 + \sin 1] = \sin 1$

$f(-1) = -a + 1 \Rightarrow a = 1 - \sin 1$

$g(x) = (1 - \sin 1)x + 1$

55.c. Given that $f(x) = |1 - x|$

$$\Rightarrow f(|x|) = \begin{cases} x - 1, & x > 1 \\ 1 - x, & 0 < x \le 1 \\ 1 + x, & -1 \le x \le 0 \\ -x - 1, & x < -1. \end{cases}$$

Clearly, the domain of $\sin^{-1}(f|x|)$ is $[-2, 2]$.

$\Rightarrow$ It is non-differentiable at the points $\{-1, 0, 1\}$.

56.d $f(x)$ is continuous at $x=0 \Rightarrow \lim_{x\to 0} f(x) = f(0)$

$$\Rightarrow f(0) = \lim_{x\to 0} f(0+h)$$

$$= \lim_{h\to 0} \frac{hg(h)}{|h|} = \lim_{h\to 0} g(h) = g(0) = 0$$

$$\text{Now } f'(0^+) = \lim_{h\to 0} \frac{f(0+h)-f(0)}{h}$$

$$= \lim_{h\to 0} \frac{\frac{hg(h)}{|h|}}{h}$$

$$= \lim_{h\to 0} \frac{g(h)}{h}$$

$$= \lim_{h\to 0} \frac{g(h)-g(0)}{h}$$

$$= g'(0) \text{ (as } g(0)=0) = 0$$

$$f'(0^-) = \lim_{h\to 0} \frac{f(0-h)-f(0)}{-h}$$

$$= \lim_{h\to 0} \frac{\frac{-hg(-h)}{|-h|}}{-h} = \lim_{h\to 0} \frac{g(-h)}{h}$$

$$= -\lim_{h\to 0} \frac{g(-h)-g(0)}{-h}$$

$$= -g'(0) = 0$$

Hence, $f'(0)$ exists and $f'(0) = 0$.

57.a.

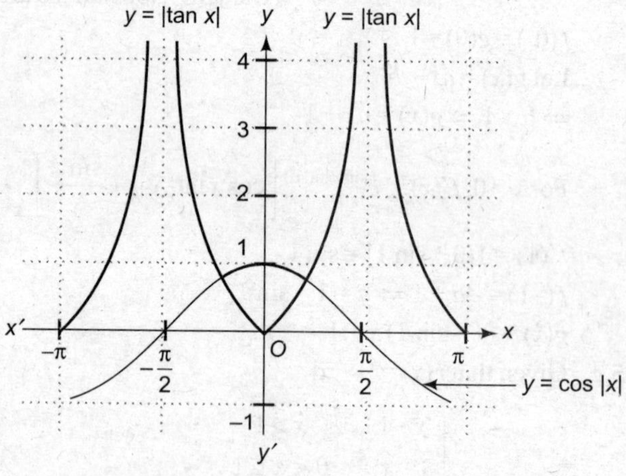

Fig. 3.6

The function is not differentiable and continuous at two points between $x = -\pi/2$ and $x = \pi/2$. Also the function is not continuous at $x = \frac{\pi}{2}$ and $x = -\frac{\pi}{2}$.

Hence, at four points, the function is not differentiable.

58.c. $f(|x|) = \begin{cases} \sin|x|, & |x|<0 \\ \cos(x) - ||x|-1|, & |x| \geq 0 \end{cases}$

$\Rightarrow f(|x|) = \cos(x) - ||x|-1|, x \in R$

[as $|x| < 0$ is not possible and $|x| \geq 0$ is true for all $x \in R$]

$\Rightarrow f(|x|) = \cos x - ||x|-1|, x \in R$

which is non-differentiable at $x = 0$ and when $|x|-1 = 0$ or $x = \pm 1$.

Hence, $f(|x|)$ has exactly three points of non-differentiability.

59.d. $f(2^+) = 2 + 2\sin(0) = 2$

$f(2^-) = 3 + 2\sin(1)$

Hence, $f(x)$ is discontinuous at $x = 2$.

Also $f(0^+) = 2(0) - 0 - 0\sin(0-0) = 0$

and $f(0^-) = 2(0) - (-1) - 0\sin(0-(-1)) = 1$

Hence, $f(x)$ is discontinuous at $x = 0$.

60.b. 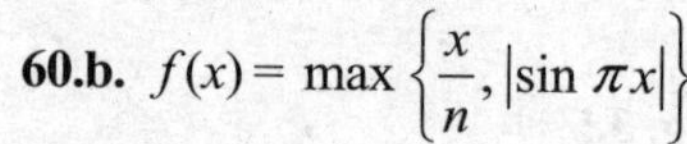
$f(x) = \max\left\{\frac{x}{n}, |\sin \pi x|\right\}$

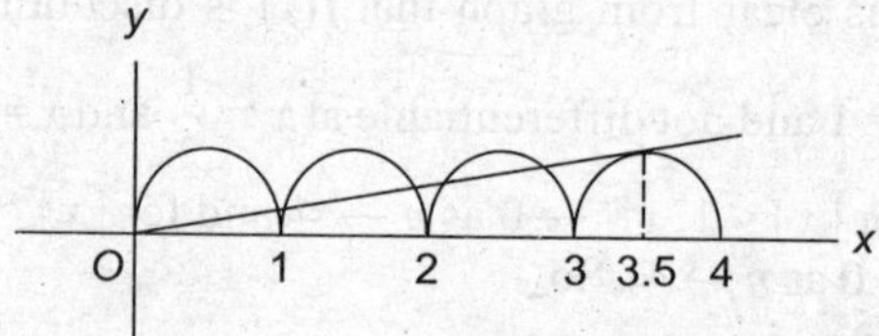

Fig. 3.7

Thus, for the maximum points of non differentiability, graphs of $y = x/7$ and $y = |\sin \pi x|$ must intersect at maximum number of points which occurs when $n > 3.5$.

Hence, the least value of n is 4.

61.d. $f(x) = [x^2] - \{x\}^2$

$f(-1) = 1, f(-1^-) = 1 - 1 = 0$

$f(1) = 1, f(1^+) = 1 - 0 = 1$

$f(1^-) = 0 - 1 = -1$

Thus, $f(x)$ is discontinuous at $x = 1, -1$.

62.a. $f(e) = [\log_e e] + \sqrt{\{\log_e e\}} = [1] + \sqrt{\{1\}} = 1 + 0 = 1$

$$f(e^+) = [\log_e e^+] + \sqrt{\{\log_e e^+\}}$$

$$= \lim_{h\to 0}[1+h] + \sqrt{\{1+h\}} = 1 + 0 = 1$$

$$f(e^-) = [\log_e e^-] + \sqrt{\{\log_e e^-\}}$$

$$= \lim_{h\to 0}[1-h] + \sqrt{\{1-h\}} = 0 + 1 = 1$$

Hence, $f(x)$ is continuous at $x = e$.

Now $f'(e^+) = \lim_{h\to 0} \frac{f(e+h)-f(e)}{h}$

$= \lim_{h\to 0} \frac{[1+h]+\sqrt{\{1+h\}}-1}{h}$

$= \lim_{h\to 0} \frac{1+\sqrt{h}-1}{h} = \lim_{h\to 0} \frac{1}{\sqrt{h}} \to \infty.$

Hence, $f(x)$ is non-differentiable at $x = 0$.

63.a. $f(x) = \lim_{n\to\infty} \left(\sin^2(\pi x)\right)^n + \left[x + \frac{1}{2}\right]$

Now $g(x) = \lim_{n\to\infty} \left(\sin^2(\pi x)\right)^n$ is discontinuous when

$\sin^2(\pi x) = 1$ or $\pi x = (2n+1)\frac{\pi}{2}$ or $x = \frac{(2n+1)}{2}$, $n \in z$

Thus, $g(x)$ is discontinuous at $x = 3/2$.

Also $h(x) = \left[x + \frac{1}{2}\right]$ is discontinuous at $x = 3/2$.

But $f(3/2) = \lim_{n\to\infty} \left(\sin^2(3\pi/2)\right)^n + \left[\frac{3}{2} + \frac{1}{2}\right] = 1 + 2 = 3.$

$f(3/2^+) = \lim_{n\to\infty} \left(\sin^2((3\pi/2)^+)\right)^n + \left[\left(\frac{3}{2}\right)^+ + \frac{1}{2}\right]$

$= 0 + 2 = 2.$

Hence, $f(x)$ is discontinuous at $x = 3/2$.

Both $g(x)$ and $h(x)$ are continuous at $x = 1$, hence $f(x)$ is continuous at $x = 1$.

64.c. $f(x) = \text{sgn}(\sin^2 x - \sin x - 1)$ is discontinuous when $\sin^2 x - \sin x - 1 = 0$

or $\sin x = \frac{1 \pm \sqrt{5}}{2}$ or $\sin x = \frac{1-\sqrt{5}}{2}$.

For exactly four points of discontinuity, n can take value 4 or 5 as shown in the diagram

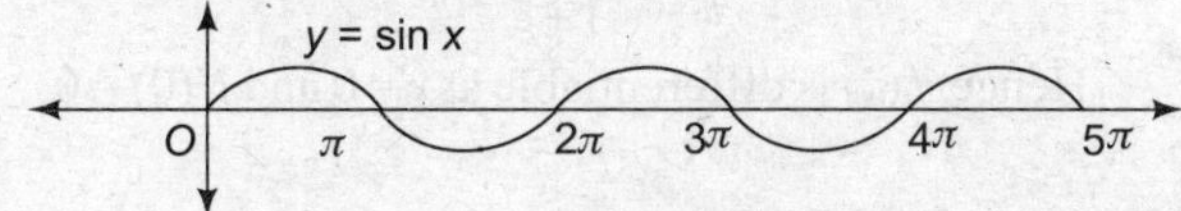

Fig. 3.8

65.c. $f(x) = \begin{cases} x^2 - ax + 3, & x \text{ is rational} \\ 2 - x, & x \text{ is irrational} \end{cases}$

is continuous when $x^2 - ax + 3 = 2 - x$ or $x^2 - (a-1)x + 1 = 0$, which must have two distinct roots for $(a-1)^2 - 4 > 0$

$\Rightarrow (a-1-2)(a-1+2) > 0$

$\Rightarrow a \in (-\infty, -1) \cup (3, \infty)$

66.a. Hence, check continuity at $x = k$, $k \in Z$

For positive integers.

$f(k) = \{k\}^2 - \{k^2\} = 0$

$f(k^+) = \{k^+\}^2 - \{(k^+)^2\} = 0 - 0$

$f(k^-) = \{k^-\}^2 - \{(k^-)^2\} = 1 - 1 = 0$

For negative integers,

$f(k) = \{k\}^2 - \{k^2\} = 0$

$f(k^+) = \{k^+\}^2 - \{(k^+)^2\} = 0 - 1 = -1$

$f(k^-) = \{k^-\}^2 - \{(k^-)^2\} = 1 - 0 = 1$

Hence, $f(x)$ is continuous at positive integers and discontinuous at negative integers.

67.b. $g(x)$ is an even function, then $g(x) = g(-x)$

$\Rightarrow g'(x) = -g'(-x) \Rightarrow g'(0) = -g'(0) \Rightarrow g'(0) = 0$

Now $f'(0) = \lim_{h\to 0} \frac{g(0+h)\cos(1/h) - 0}{h}$

$= \lim_{h\to 0} \frac{g(h)\cos(1/h)}{h} = \lim_{h\to 0} g'(0)\cos(1/h)$

$= 0$

68.b. $f(1) = 1 - \sqrt{1-1^2} = 1$

$f(1^-) = \lim_{x\to 1^+} \left(1 - \sqrt{1-x^2}\right) = 1$

$f(1^+) = \lim_{x\to 1^-} \left(1 + \log\frac{1}{x}\right) = 1 + \log\frac{1}{1} = 1$

Hence, $f(x)$ is continuous at $x = 1$.

$f'(1^+) = \lim_{h\to 0} \frac{f(1+h) - f(1)}{h}$

$= \lim_{h\to 0} \frac{1 + \log\frac{1}{1+h} - 1}{h}$

$= -\lim_{h\to 0} \frac{\log(1+h)}{h} = -1$

$f'(1^-) = \lim_{h\to 0} \frac{f(1-h) - f(1)}{-h}$

$= \lim_{h\to 0} \frac{1 - \sqrt{1-(1-h)^2} - 1}{-h} = \lim_{h\to 0} \frac{\sqrt{2-h}}{\sqrt{h}} = \infty$

Hence, $f(x)$ is non-differentiable at $x = 1$.

69.b. We have $f(x) = \sqrt{1 - \sqrt{1-x^2}}$.

The domain of definition of $f(x)$ is $[-1, 1]$.

For $x \neq 0$, $x \neq \pm 1$, we have

$f'(x) = \frac{1}{\sqrt{1-\sqrt{1-x^2}}} \times \frac{x}{\sqrt{1-x^2}}$

Since $f(x)$ is not defined on the right side of $x = 1$ and on the left side of $x = -1$.

Also, $f'(x) \to \infty$ when $x \to -1^+$ or $x \to 1^-$.

So, we check the differentiability at $x = 0$.

Now, L.H.D. at $x = 0$

$$= \lim_{x\to 0^-}\frac{f(x)-f(0)}{x-0}$$

$$= \lim_{h\to 0}\frac{f(0-h)-f(0)}{-h}$$

$$= \lim_{h\to 0}\frac{\sqrt{1-\sqrt{1-h^2}}-0}{-h}$$

$$= -\lim_{h\to 0}\frac{\sqrt{1-(1-(1/2)h^2+(3/8)h^4+\cdots)}}{h}$$

$$= -\lim_{h\to 0}\sqrt{\frac{1}{2}-\frac{3}{8}h^2+\cdots} = -\frac{1}{\sqrt{2}}$$

Similarly, R.H.D. at $x = 0$ is $\frac{1}{\sqrt{2}}$.

Hence, $f(x)$ is not differentiable at $x = 0$.

70.d. $f(x) = -x + x - 1, x < 0$

$\Rightarrow f(x) = -1 = a$ constant function.

Hence, differentiable for all x.

71.b. $g'(0^+) = \lim_{h\to 0}\frac{f(|h|)-|\sin h|-f(0)}{h}$

$$= \lim_{h\to 0}\frac{f(h)-f(0)}{h} - \lim_{h\to 0}\frac{\sin h}{h}$$

$$= 1 - 1 = 0$$

$$g'(0^-) = \lim_{h\to 0}\frac{f(|-h|)-|\sin(-h)|-f(0)}{-h}$$

$$= \lim_{h\to 0}\frac{f(h)-f(0)}{-h} + \lim_{h\to 0}\frac{\sin h}{h}$$

$$= -1 + 1 = 0$$

Thus $g(x)$ is differentiable and $g'(0) = 0$.

72.c. $f'(0^+) = \lim_{h\to 0}\frac{h^m \sin\frac{1}{h}}{h}$ must exist $\Rightarrow m > 1$

for $m > 1$, $h'(x)$

$$= \begin{cases} m x^{m-1}\sin\frac{1}{x} - x^{m-2}\cos\frac{1}{x}, & \text{if } x \neq 0 \\ 0, & \text{if } x = 0 \end{cases}$$

Now $\lim_{h\to 0} h(x) = \lim_{h\to 0}\left(m h^{m-1}\sin\frac{1}{h} - h^{m-2}\cos\frac{1}{h}\right)$

limit exists if $m > 2$

$\therefore m \in N \Rightarrow m = 3$

73.c. At $x = 0$,

L.H.L. $= \lim_{x\to 0-} f(x) = \lim_{h\to 0} f(0-h)$

$$= \lim_{h\to 0} h^2\left(\frac{e^{-1/h}-e^{1/h}}{e^{-1/h}+e^{1/h}}\right)$$

$$= \lim_{h\to 0} h^2\left(\frac{e^{-2/h}-1}{e^{-2/h}+1}\right)$$

$$= 0\left(\frac{0-1}{0+1}\right) = 0$$

R.H.L. $= \lim_{x\to 0+} f(x) = \lim_{h\to 0} f(0+h)$

$$= \lim_{h\to 0} h^2\left(\frac{e^{1/h}-e^{-1/h}}{e^{1/h}+e^{-1/h}}\right)$$

$$= \lim_{h\to 0} h^2\left(\frac{1-e^{-2/h}}{1+e^{-2/h}}\right)$$

$$= 0\left(\frac{1-0}{1+0}\right) = 0$$

and $f(0) = 0$

$\Rightarrow$ L.H.L. = R.H.L. = $f(0)$

Hence, $f(x)$ is continuous at $x = 0$.

Also L.H.D. $= \lim_{h\to 0}\frac{f(0-h)-f(0)}{-h}$

$$= \lim_{h\to 0}\frac{h^2\dfrac{e^{-1/h}-e^{1/h}}{e^{-1/h}+e^{1/h}}-0}{-h}$$

$$= -\lim_{h\to 0} h\frac{e^{-2/h}-1}{e^{-2/h}+1} = 0.$$

And R.H.D. $= \lim_{h\to 0}\frac{f(0+h)-f(0)}{h}$

$$= \lim_{h\to 0}\frac{h^2\dfrac{e^{1/h}-e^{-1/h}}{e^{1/h}+e^{-1/h}}-0}{-h}$$

$$= -\lim_{h\to 0} h\frac{1-e^{-2/h}}{1+e^{-2/h}} = 0.$$

Hence, $f(x)$ is differentiable at $x = 0$ and $f'(0) = 0$,

Multiple Correct Answers Type

1. **a., b, c., d.**

   a, b, and **c** are false. Refer to definitions.

   for **d** f must be continuous $\Rightarrow$ False

2. **a., c., d.**

   a. is wrong as continuity is a must for $f(x)$.

   b. is the correct form of intermediate value theorem.

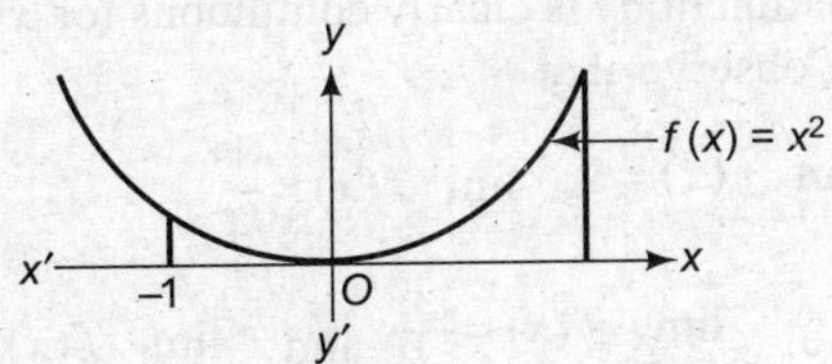

Fig. 3.9

c. As per the graph, the statement is incorrect.

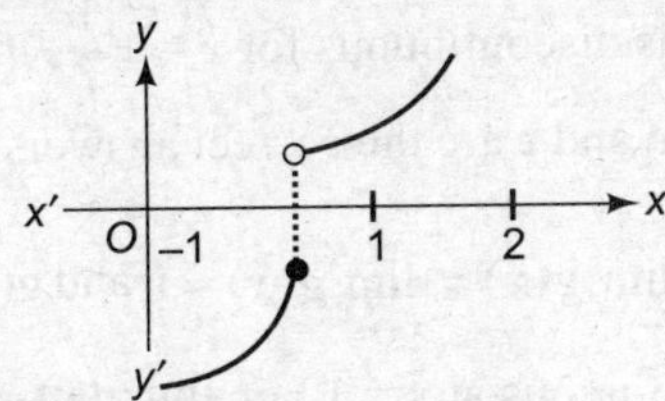

Fig. 3.10

d. is wrong if f is discontinuous.

3. a., c., d.

a. $\lim_{x\to-2} = -6$; **c.** $\lim_{x\to-4} = 48$; **d.** $\lim_{x\to-9} = \frac{1}{6}$

4. a., b.

$f(1^-)=1$; $f(1^+)=1$; $f(1)=1$
$f'(1^-)=5$; $f'(1^+)=5$
$f(2^+)=10$, $f(2^-)=10$
$f'(2^+)=3$; $f'(2^-)=13$.

5. a., b.

$f(x)=\text{sgn}(x)\sin x$
$f(0^+)=\text{sgn}(0^+)\sin(0^+)=1\times(0)=0$
$f(0^-)=\text{sgn}(0^-)\sin(0^-)=(-1)\times(0)=0$
Also $f(0)=0$
Hence, $f(x)$ is continuous everywhere.
Both sgn(x) and sin(x) are odd functions.
Hence, $f(x)$ is an even function.
Obviously, $f(x)$ is non-periodic.

$$\text{Now } f'(0^+)=\lim_{h\to0}\frac{f(h)-f(0)}{h}$$
$$=\lim_{h\to0}\frac{\text{sgn}(h)\sin h-0}{h}$$
$$=\lim_{h\to0}\frac{\sin h}{h}=1$$
$$\text{and } f'(0^+)=\lim_{h\to0}\frac{\text{sgn}(-h)\sin(-h)-0}{-h}$$
$$=\lim_{h\to0}\frac{-1\times(-\sin h)}{-h}=-1$$

Hence, $f(x)$ is non-differentiable at $x=0$.

6. a., b., c., d.

Given function is discontinuous when $a+\sin\pi x=1$
Now if $a=1\Rightarrow\sin\pi x=0\Rightarrow x=1,2,3,4,5$
If $a=3\Rightarrow\sin\pi x=-2$ not possible.
If $a=0.5\Rightarrow\sin\pi x=0.5$
$\Rightarrow x$ has 6 values, 2 each for one cycle of period 2.
If $a=0\Rightarrow\sin\pi x=1\Rightarrow x=\frac{1}{2},\frac{5}{2},\frac{9}{2}$
Hence, all the options are correct.

7. a., b.

For maximum points of discontinuity of
$f(x)=\text{sgn}(x^2-ax+1)$,
$x^2-ax+1=0$ must have two distinct roots,
for which $D=a^2-4>0$
$\Rightarrow a\in(-\infty,-2)\cup(2,\infty)$.

8. b., d.

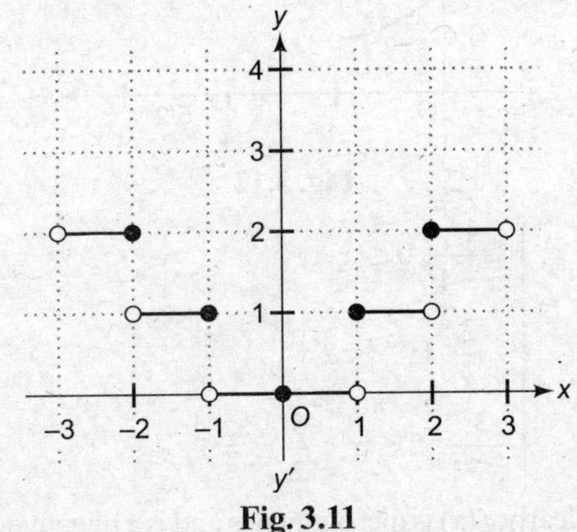

Fig. 3.11

9. b., c.

Option (a) is wrong as consider $f(x) = \sin x$ and $g(x)=|x|$,
$g(x)$ is non-differentiable at $x = 0$, but $f(x)g(x)$ is differentiable at $x=0$.

10. a., b., c.

$$f(0^-)=\lim_{n\to\infty}\left[\lim_{x\to0^-}\left(\cos^2 x\right)^n\right]$$
$$=(\text{a value lesser than }1)^\infty=0$$
$$f(0^+)=\lim_{n\to\infty}\left[\lim_{x\to0^+}\left(1+x^n\right)^{1/n}\right]$$
$$=\lim_{n\to\infty}\left[\lim_{x\to0^+}\left(1+x^n\right)^{1/n}\right]=1$$

Also $f(0)=1\Rightarrow$ discontinuous at $x=0$
Further, $f(1^-)=1$; $f(1^+)=0$; $f(1)=1$
$\Rightarrow$ discontinuous at $x=1$.

11. b., d.

a. $\lim_{x\to1^+}\frac{1}{\ln|x|}=\infty$ and $\lim_{x\to1^-}\frac{1}{\ln|x|}=-\infty$,
hence $f(x)$ has non-removable discontinuity.

b. $\lim_{x\to1}\frac{x^2-1}{x^3-1}=\frac{2}{3}$
$\therefore f(x)$ has removable discontinuity at $x=1$

c. $\lim_{x\to1^+}\left(2^{-2^{\frac{1}{1-x}}}\right)=1$ and $\lim_{x\to1^-}\left(2^{-2^{\frac{1}{1-x}}}\right)=0$.

Hence, the limit does not exist.

d. $\lim\limits_{x\to 1}\dfrac{\sqrt{x+1}-\sqrt{2x}}{x^2-x}=\dfrac{-1}{2\sqrt{2}}$ (Rationalizing)

Therefore, $f(x)$ has removable discontinuity at $x=1$.

12. a., b., d.

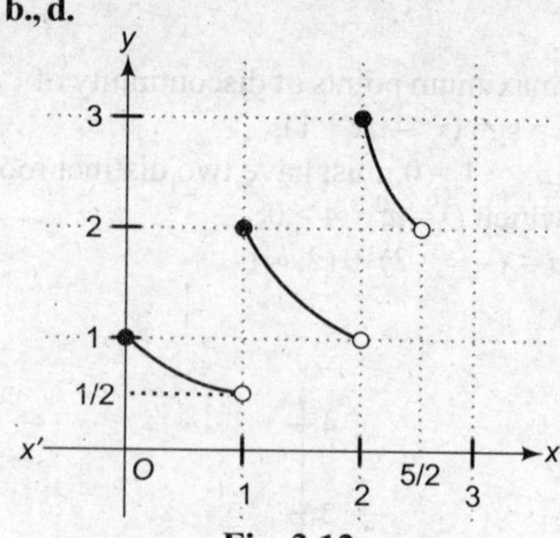

Fig. 3.12

$$f(x)\begin{cases}\dfrac{1}{x+1}, & 0\le x<1\\ \dfrac{2}{x}, & 1\le x<2\\ \dfrac{3}{x-1}, & 2\le x<\dfrac{5}{2}\end{cases}$$

Clearly, $f(x)$ is discontinuous and is a bijective function

$\lim\limits_{x\to 1^-} f(x)=\dfrac{1}{2}$, $\lim\limits_{x\to 1^+} f(x)=2$

$\min\left(\lim\limits_{x\to 1^-} f(x), \lim\limits_{x\to 1^+} f(x)\right)=\dfrac{1}{2}\ne f(1)$

$\max(1, 2)=2=f(1)$.

13. a., c.

$$f(x)=\begin{cases}1, & |x|\ge 1\\ \dfrac{1}{n^2}, & \dfrac{1}{n}<|x|<\dfrac{1}{n-1}, n=2,3,\ldots\\ 0, & x=0\end{cases}$$

$$=\begin{cases}1, & x\le \text{ or } x\ge 1\\ \dfrac{1}{4}, & x\in\left(-1,-\dfrac{1}{2}\right)\cup\left(\dfrac{1}{2},1\right)\\ \dfrac{1}{9}, & x\in\left(\dfrac{-1}{2},\dfrac{-1}{3}\right)\cup\left(\dfrac{1}{3},\dfrac{1}{2}\right)\\ \vdots\end{cases}$$

Fig. 3.13

The function f is clearly continuous for $|x|>1$.

We observe that

$\lim\limits_{x\to -1+} f(x)=1$, $\lim\limits_{x\to -1^-} f(x)=\dfrac{1}{4}$

Also, $\lim\limits_{x\to 1/n+} f(x)=\dfrac{1}{n^2}$ and $\lim\limits_{x\to 1/n-} f(x)$

$=\dfrac{1}{(n+1)^2}$

Thus f is discontinuous for $x=\pm\dfrac{1}{n}$, $n=1, 2, 3, \ldots$

Hence **a** and **c** are the correct answers.

14. a., b., c.

Since $\lim\limits_{x\to 1^-} g(x)=\lim\limits_{x\to 1^+} g(x)=1$ and $g(1)=0$, $g(x)$ is not continuous at $x=1$ but $\lim\limits_{x\to 1} g(x)$ exists.

We have $\lim\limits_{x\to 1^-} f(x)=\lim\limits_{h\to 0} f(1-h)=\lim\limits_{h\to 0}[1-h]=0$

and $\lim\limits_{x\to 1^+} f(x)=\lim\limits_{h\to 0} f(1+h)=\lim\limits_{h\to 0}[1+h]=1$

So $\lim\limits_{x\to 1} f(x)$ does not exist and so $f(x)$ is not continuous at $x=1$

We have $gof(x)=g(f(x))=g([x])=0$, for all $x, \in R$.

So, gof is continuous for all x.

We have $fog(x)=f(g(x))=\begin{cases}f(0), & x\in Z\\ f(x^2), & x\in R-Z\end{cases}$

$=\begin{cases}0, & x\in z\\ [x^2], & x\in R-Z\end{cases}$

which is clearly not continuous.

15. b., d.

We have $\lim\limits_{x\to 0}\dfrac{f(x)-f(0)}{x-0}=\lim\limits_{x\to 0}\dfrac{\log\cos x}{\log(1+x^2)}$

$=\lim\limits_{x\to 0}\dfrac{\log(1-1+\cos x)}{\log(1+x^2)}\dfrac{1-\cos x}{1-\cos x}$

$=\lim\limits_{x\to 0}\dfrac{\log\{1-(1-\cos x)\}}{1-\cos x}\dfrac{1-\cos x}{\log(1+x^2)}$

$=$

$-\lim\limits_{x\to 0}\dfrac{\log[1-(1-\cos x)]}{-(1-\cos x)}\dfrac{2\sin^2\dfrac{x}{2}}{4\left(\dfrac{x}{2}\right)^2}\dfrac{x^2}{\log(1+x^2)}=-\dfrac{1}{2}$

Hence, $f(x)$ is differentiable at $x=0$ and hence continuous at $x=0$.

Hence **b** and **d** are the correct answers.

16. a., c.

$f(x)=x+|x|+\cos 9x$, $g(x)=\sin x$

Since $f(x)$ and $g(x)$ are continuous everywhere, $f(x)+g(x)$ is also continuous everywhere. Therefore, $f(x)$ is non-differentiable and $x=0$.

Hence $f(x)+g(x)$ is non-differentiable at $x=0$

Now $h(x)=f(x)\times g(x)$

$$=\begin{cases}(\cos 9x)(\sin x), & x<0\\(2x+\cos 9x)(\sin x), & x\ge 0\end{cases}$$

Clearly, $h(x)$ is continuous at $x=0$

Also

$h'(x)$

$$=\begin{cases}\cos x\cos 9x-9\sin x\sin 9x, & x<0\\(2-9\sin 9x)\sin x+\cos x(2x+\cos 9x), & x>0\end{cases}$$

$h'(0^-)=1, h'(0^+)=1$

$\Rightarrow f(x)\times g(x)$ is differentiable everywhere.

17. a., c.

$$f(x)=\begin{cases}(\sin^{-1}x)^2\cos\left(\dfrac{1}{x}\right), & x\ne 0\\ 0, & x=0\end{cases}$$

$$\lim_{x\to 0}f(x)=\lim_{x\to 0}(\sin^{-1}x)^2\cos\left(\frac{1}{x}\right)$$

$=0\times$ (any value between -1 to 1) $=0$

Hence $f(x)$ is continuous at $x=0$

$$f'(0^+)=\lim_{h\to 0}\frac{(\sin^{-1}h)^2\cos\left(\dfrac{1}{h}\right)-0}{h}$$

$$=\left(\lim_{h\to 0}\frac{\sin^{-1}h}{h}\right)\left(\lim_{h\to 0}\sin^{-1}h\right)\left(\lim_{h\to 0}\cos\left(\frac{1}{h}\right)\right)$$

$=1\times(0)\times$ (any value between -1 to 1) $=0$

Similarly, $f'(0^-)=0$.

Hence, $f(x)$ is continuous and differentiable in $[-1, 1]$ and $(-1, 1)$, respectively.

18. a., b.

For $b=1$, we have $f(g(0))=f(\sin(0)+1)=f(1)=1+a$

Also $f(g(0^+))=\lim_{x\to 0^+}f(\sin x+1)=f(1)=1+a$

and $f(g(0^-))=\lim_{x\to 0^-}f(\{x\})=f(1^-)=2-1=1$

Hence, $f(g(x))$ is continuous for $a=1$.

For $b<0$,

$f(g(0))=f(\sin(0)+b)=f(b)=2-b$

$f(g(0^+))=\lim_{x\to 0^+}f(\sin x+b)=f(b)=2-b$

and $f(g(0^-))=\lim_{x\to 0^-}f(\{x\})=f(1)=1+a$

For continuity at $x=0$, we must have $2-b=1+a$ or $a+b=1$.

19. a., b., c.

$f(x)$ is continuous for all x if it is continuous at $x=1$ for which $|1|-3=|1-2|+a$ or $a=-3$

$g(x)$ is continuous for all x if it is continuous at $x=2$ for which $2-|2|=\text{sgn}(2)-b=1-b$ or $b=1$

Thus, $f(x)+g(x)$ is continuous for all x if $a=-3, b=1$.

Hence, $f(x)$ is discontinuous at exactly one point for options **a** and **b.**

20. a., c., d.

a is not correct as $f(x)=x$ from R to R is onto but its reciprocal function $g(x)=1/2$ is not onto on R.

b is obviously true.

Also $g(x)$ is not continuous, hence not differentiable though $f(x)$ is continuous and differentiable in the above case.

However **b** is correct.

21. a., c., d.

For continuity at $x=1$

$$\lim_{x\to 1^+}f(x)=\lim_{x\to 1^+}(x^2\,\text{sgn}[x]+\{x\})=1+0=1$$

$$\lim_{x\to 1^-}f(x)=\lim_{x\to 1^-}(x^2\,\text{sgn}[x]+\{x\})=1\,\text{sgn}(0)+1=1$$

Also, $f(1)=1$

$\therefore$ L.H.L. = R.H.L $=f(1)$. Hence, $f(x)$ is continuous at $x=1$.

Now for differentiability,

$$f'(1^+)=\lim_{h\to 0}\frac{f(1+h)-f(1)}{h}$$

$$=\lim_{h\to 0}\frac{(1+h)^2\,\text{sgn}[1+h]+\{1+h\}-1}{h}$$

$$=\lim_{h\to 0}\frac{(1+h)^2+h-1}{h}$$

$$=\lim_{h\to 0}\frac{h^2+3h}{h}$$

$=3$

$$\text{and } f'(1^-)=\lim_{h\to 0}\frac{f(1-h)-f(1)}{-h}$$

$$=\lim_{h\to 0}\frac{(1-h)^2\,\text{sgn}[1-h]+\{1-h\}-1}{-h}$$

$$=\lim_{h\to 0}\frac{(1-h)^2+1-h-1}{-h}$$

$$=\lim_{h\to 0}\frac{h^2-3h}{-h}$$

$=3$

$f'(1^+)=f'(1^-)$

Hence, $f(x)$ is differentiable at $x=1$.

Now at $x=2$,

$$\lim_{x\to 2^-}f(x)=\lim_{x\to 2^-}(x^2\,\text{sgn}[x]+\{x\})=4\times 0+1$$

$$\lim_{h\to 2^+}f(x)=\lim_{h\to 2^+}(\sin x+|x-3|)=1+\sin 2$$

Hence L.H.L $\ne$ R.H.L.

Hence, $f(x)$ is discontinuous at $x=2$ and then $f(x)$ is also non-differentiable at $x=2$.

22. a., c.

$$f\left(\frac{\pi^-}{2}\right)=\lim_{h\to 0}\left(\frac{3}{2}\right)^{\cot\left(3\left(\frac{\pi}{2}-h\right)\right)/\cot\left(2\left(\frac{\pi}{2}-h\right)\right)}$$

$$= \lim_{h\to 0}\left(\frac{3}{2}\right)^{\frac{\tan 3h}{-\cot 2h}}$$

$$= \lim_{h\to 0}\left(\frac{3}{2}\right)^{-(\tan 3h)(\tan 2h)} = 1$$

$$f\left(\frac{\pi}{2}^{+}\right) = \lim_{h\to 0}\left[1+\left|\cot\left(\frac{\pi}{2}+h\right)\right|\right]^{\left|a\right|\tan\left(\frac{\pi}{2}+h\right)\left|\right|/b}$$

$$= \lim_{h\to 0}(1+\tan h)^{\frac{a\cot h}{b}}$$

$$= e^{\lim_{h\to 0}(1+\tan h-1)\frac{a\cot h}{b}} = e^{a/b}$$

Also $f\left(\frac{\pi}{2}\right) = b+3$

$f(x)$ is continuous at $x = \pi/2$

$\Rightarrow 1 = b+3 = e^{a/b} \Rightarrow b = -2$ and $a = 0$.

Reasoning Type

1. b. Statement 1 is obviously true.

But statement 2 is false as $f(x) = x^3$ is differentiable, but $f^{-1}(x) = x^{1/3}$ is non-differentiable at $x = 0$.

$f^{-1}(x) = x^{1/3}$ has vertical tangent at $x = 0$.

2. b. $f(x) = (2x-5)^{3/5} \Rightarrow f'(x) = \dfrac{3}{5(2x-5)^{2/5}}$

Statement 2 as it is fundamental concept for non-differentiability.

But given function is non-differentiable at $x = 5/2$, as it has vertical tangent at $x = 5/2$, but not due to sharp turn.

The graph of the function is smooth in the neighbourhood of $x = 5/2$.

3. a. Statement 2 is true as it is fundamental concept.

Also $f(x) = \text{sgn}(g(x))$ is discontinuous when $g(x) = 0$.

Now given function $f(x) = \text{sgn}(x^2 - 2x + 3)$ may be discontinuous when $x^2 - 2x + 3 = 0$, which is not possible, it has imaginary roots as its discriminant is less than 0.

4. b. $f(x) = \lim\limits_{n\to\infty}\dfrac{x^{2n}-1}{x^{2n}+1}$ is discontinuous at $x = 1$

$$= \begin{cases} -1, & x^2 < 1 \\ 1, & x^2 > 1 \\ 0, & x^2 = 1 \end{cases}$$

$\Rightarrow f(1^+) = 1$ and $f(1^-) = -1$

Hence, $f(x)$ is discontinuous at $x = 1$ as limit of the function does not exist.

5. c. We know that both $[\sin x]$ and $[\cos x]$ are discontinuous at $x = \pi/2$.

Also $f(x) = [\sin x] - [\cos x]$ is discontinuous at $x = \pi/2$.

As $f(\pi/2) = 1 - 0 = 1$

And $f(\pi/2^+) = 0 - (-1) = 1$

$f(\pi/2^-) = 0 - 0 = 0$.

But difference of two discontinuous function is not necessarily discontinuous.

6. c. We know that sgn(x) is discontinuous at $x = 0$.

Also $f(x) = |\text{sgn}x| = \begin{cases} 1, & x \neq 0 \\ 0, & x = 0 \end{cases}$ which is discontinuous at $x = 0$.

Consider $g(x) = \begin{cases} -1, & x < 0 \\ 1, & x \geq 0 \end{cases}$. Here $g(x)$ is discontinuous at $x = 0$ but $|g(x)| = 1$ for all x is continuous at $x = 0$

Hence, answer is **c**.

7. b. $f(x) = (\sin \pi x)(x-1)^{1/5}$ is continuous function as both $(\sin \pi x)$ and $(x-1)^{1/5}$ are continuous

But $(x-1)^{1/5}$ is not differentiable at $x = 1$.

$$\text{However, } f'(1^-) = \lim_{h\to 0}\frac{f(1-h)-f(1)}{-h}$$

$$= \lim_{h\to 0}\frac{\sin[\pi(1-h)](1-h-1)^{1/5}-0}{-h}$$

$$= \lim_{h\to 0}\frac{\sin(\pi h)(-h)^{1/5}}{h} = 0$$

$$\text{And } f'(1^+) = \lim_{h\to 0}\frac{f(1+h)-f(1)}{h}$$

$$= \lim_{h\to 0}\frac{\sin[\pi(1+h)](1+h-1)^{1/5}-0}{h}$$

$$= \lim_{h\to 0}\frac{-\sin(\pi h)(h)^{1/5}}{h} = 0$$

Hence, $f(x)$ is differentiable at $x = 1$, though $(x-1)^{1/5}$ is not differentiable at $x = 1$.

However, statement 2 is correct but it is not a correct explanation of statement 1.

8. b. Statement 2 is true as $\cos 0 = 1$

$$\text{Now } \lim_{x\to 0+}\frac{e^{1/x}-1}{e^{1/x}+1} = \lim_{h\to 0}\frac{e^{1/h}-1}{e^{1/h}+1} = \lim_{h\to 0}\frac{1-e^{-1/h}}{1+e^{-1/h}} = 1$$

$$\text{and } \lim_{x\to 0-}\frac{e^{1/x}-1}{e^{1/x}+1} = \lim_{h\to 0}\frac{e^{-1/h}-1}{e^{-1/h}+1} = -1$$

Thus L.H.L. $\neq$ R.H.L. The function has non-removable discontinuity at $x = 0$. Hence, statement 2 is not a correct explanation of statement 1.

9. a. $\lim\limits_{x\to 0^+}(\sin x + [x]) = 0$, $\lim\limits_{x\to 0^-}(\sin x + [x]) = -1$

Thus, limit does not exist, hence $f(x)$ is discontinuous at $x = 0$.

Statement 2 is fundamental property and is a correct explanation of statement 1.

10. d. $f(x) = |x| \sin x$

$$\text{L.H.D.} = \lim_{h\to 0} \frac{|0-h|\sin(0-h)-0}{-h}$$

$$= \lim_{h\to 0} \frac{-h\sin h}{-h} = 0$$

$$\text{R.H.D} = \lim_{h\to 0} \frac{|0+h|\sin(0+h)-0}{h} = \lim_{h\to 0} \frac{h\sin h}{h} = 0$$

$\Rightarrow f(x)$ is differentiable at $x = 0$.

11. d. Statement 1 is incorrect because if $\lim_{x\to a} g(x)$ and $\lim_{x\to a} f(g(x))$ approach e from the same side of e (say right side), and $\lim_{x\to e^+} f(x) = f(e) \neq \lim_{x\to e^-} f(x)$, then $\lim_{x\to a} f(g(x)) = f(e^+) = f(e)$.

Statement 2 is correct.

12. c. Consider $f(x) = \begin{cases} 1, & \text{if } x \geq 0 \\ -1, & \text{if } x < 0 \end{cases}$

Hence, $|f(x)| = 1$ for all x is continuous at $x = 0$ but $f(x)$ is discontinuous at $x = 0$.

13. b. Statement 2 is obviously true.

But $f(x) = \tan^{-1}\left(\frac{2x}{1-x^2}\right)$ is non-differentiable at $x = \pm 1$ as $\frac{2x}{1-x^2}$ is not defined at $x = \pm 1$. Hence statement 1 is true but statement 2 is not correct explanation of statement 1.

14. b. $|f(x)| \leq |x|$

$\Rightarrow 0 \leq |f(x)| \leq |x|$

$\Rightarrow$ Graph of $y = |f(x)|$ lies between the graph of $y = 0$ and $y = |x|$

Also $|f(0)| \leq 0 \Rightarrow f(0) = 0$

Also from Sandwich Theorem,

$\lim_{x\to 0} 0 \leq \lim_{x\to 0} |f(x)| \leq \lim_{x\to 0} |x|$

$\Rightarrow \lim_{x\to 0} |f(x)| = 0$

$\Rightarrow y = f(x)$ is continuous at $x = 0$.

Also statement 2 is correct but it has no link with statement 1.

15. c. See the graph of $f(x) = ||x|^2 - 3|x| + 2|$,

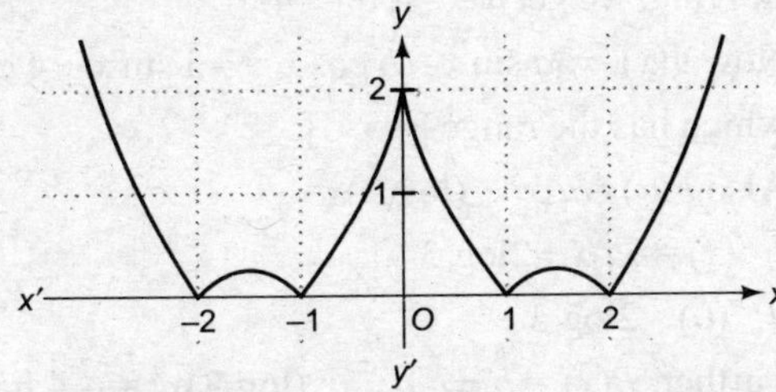

Fig. 3.14

which is non-differentiable at five points, $x = 0, \pm 1, \pm 2$.

However, statement 2 is false, as $f(x) = x^3$ crosses x-axis at $x = 0$, but $|f(x)| = |x^3|$ is differentiable at $x = 0$.

16. b. Statement 1 is correct as $e^{|x|}$ is non-differentiable at $x = 0$.

17. a. Let $x = k, k \in Z \Rightarrow f(k) = \{k\} + \sqrt{\{k\}} = 0$

$f(k^+) = 0 + 0 = 0, f(k^-) = 1 + 1 = 2$.

Hence, $f(x)$ is not continuous at integral points.

Hence, correct answer is **a.**

18. b. We know that $0 \leq \cos^2(n!\,\pi x) \leq 1$

Hence, $\lim_{m\to\infty} \cos^{2m}(n!\pi x) = 0$ or 1, as $0 \leq \cos^2(n!\,\pi x) < 1$ or $\cos^2(n!\,\pi x) = 1$

Also, since $n \to \infty$, then $n!\,x$ = integer if $x \in Q$ and $n!\,x \neq$ integer, if $x \in$ irrational.

Hence, $f(x) = \begin{cases} 1, & \text{if } x \text{ is rational} \\ 0, & \text{if } x \text{ is irrational} \end{cases}$

$\Rightarrow h(x) = 1$ when $\forall x \in R$ which is continuous for all x; however, statement 2 does not correctly explain statement 1 as addition of discontinuous functions may be continuous.

19. d. Consider $f(x) = \begin{cases} x^2 \sin\frac{1}{x}, & x \neq 0 \\ 0, & x = 0 \end{cases}$ which is differentiable at $x = 0$, but derivative is not continuous at $x = 0$.

However, statement 2 is correct.

20. a. $F(x) = f(g(x))$,.

$\Rightarrow F(x) = x^2 + 2|x|$

$\Rightarrow F'(x) = \begin{cases} 2x - 2, & x < 0 \\ 2x + 2, & x > 0 \end{cases}$

Hence, $F'(0^+) = 2$ and $F'(0^-) = -2$.

Hence, both statements are correct and statement 2 is a correct explanation of statement 1.

21. d. Statement 1 is false, as consider the function $f(x) = \max\{0, x^3\}$ which is equivalent to

$f(x) = \begin{cases} 0, & x < 0 \\ x^3, & x \geq 0 \end{cases}$.

Here $f(x)$ is continuous and differentiable at $x = 0$.

However, statement 2 is obviously true.

22. b. $f(x) = \begin{bmatrix} \pi/4, & x > 1 \\ \pi/4, & x = 1 \\ \pi/2, & x < 1 \end{bmatrix}$ [in the interval $(1-8, 1+8)$]

Hence, f is discontinuous and non-derivable, but non-derivability does not imply discontinuity.

23. c. $F(1) = 0, F(1^+) = \frac{\pi}{2}$ and $F(1^-) = -\frac{3\pi}{4}$

$\Rightarrow F$ is discontinuous

But for $f(x) = \begin{bmatrix} 1, & \text{if } x \geq 0 \\ -1, & \text{if } x < 0 \end{bmatrix}$ and

$g(x) = \begin{bmatrix} -1, & \text{if } x \geq 0 \\ 1, & \text{if } x < 0 \end{bmatrix}$,

$f(x)\,g(x)$ is continuous at $x = 0$.

24.c.

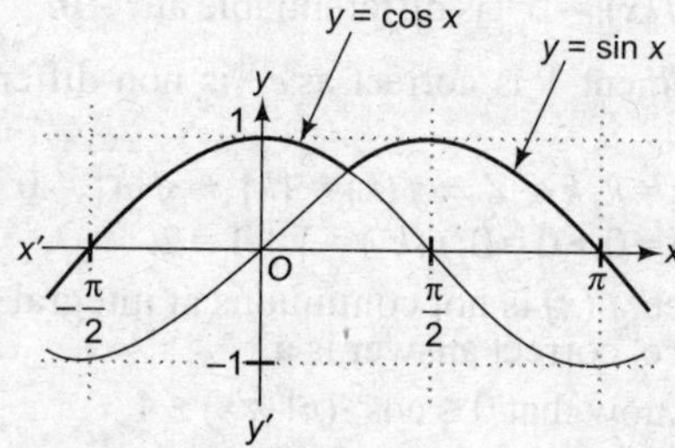

Fig. 3.15

From the graph, statement 1 is true.

Consider $f(x) = \min\{x, \sin|x|\}$ is differentiable at $x = 0$, though $g(x) = \max\{x, \sin|x|\}$ is non-differentiable at $x = 0$

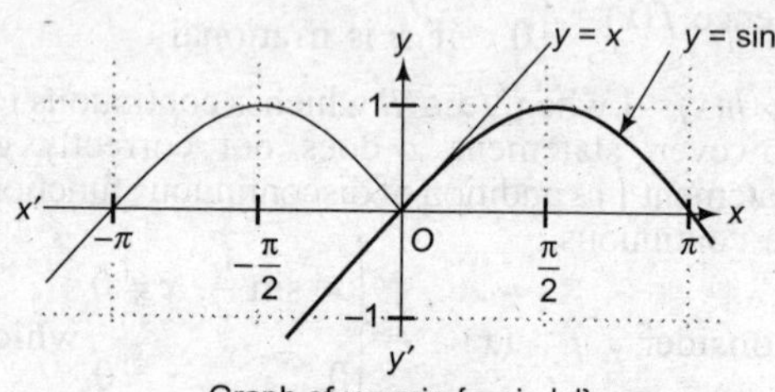

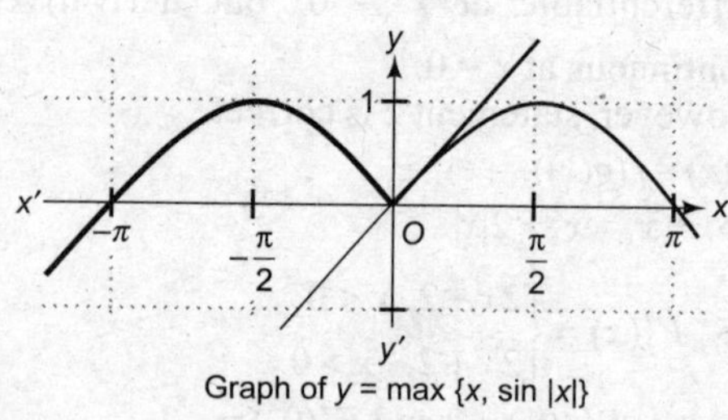

Fig. 3.16

25.b.

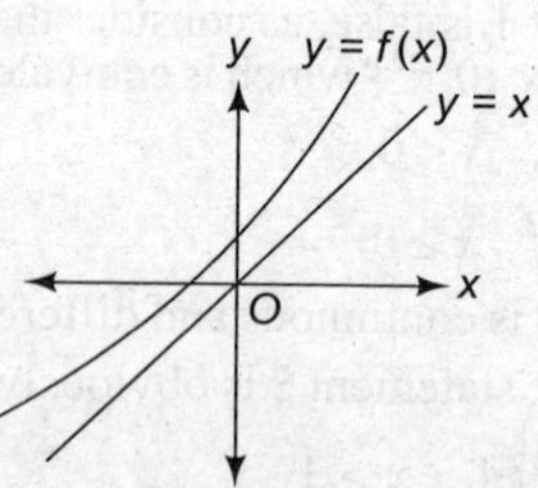

Fig. 3.17

Since $f(x)$ is a continuous function such that $f(0) = 1$ and $f(x) \neq x$, œ $x \in R$

The graph of $y = f(x)$ always lies above the graph of $y = x$. Hence $f(x) > x$.

Hence, $f(f(x)) > x$ (as $f(x)$ is onto function, $f(x)$ takes all real values which acts as x).

Statement 2 is fundamental property of continuous function, but does not explain statement 1.

26.c. Statement 1 is true as $\sqrt{x}$ is monotonic function. But statement 2 is false as $f(x) = [\sin x]$ is continuous at $x = 3\pi/2$, though $\sin(3\pi/2) = -1$ (integer).

Linked Comprehension Type

For Problems 1 – 3

1. b, 2. a, 3. b.

Sol. $$f(x) = \begin{cases} \dfrac{a(1 - x\sin x) + b\cos x + 5}{x^2}, & x < 0 \\ 3, & x = 0 \\ \left\{1 + \left(\dfrac{P(x)}{x}\right)\right\}^{1/x}, & x > 0 \end{cases}$$

where $P(x) = a_0 + a_1x + a_2x^2 + a_3x^3$

$f(0) = 3$

$$\text{R.H.L.} = \lim_{x\to 0+} f(x) = \lim_{h\to 0} f(0+h) = \lim_{h\to 0} f(h) = \lim_{h\to 0}\left\{1 + \left(\frac{P(h)}{h}\right)\right\}^{1/h}$$

Since f is continuous at $x = 0$, R.H.L. exists.

For the existence of R.H.L., $a_0, a_1 = 0$

$\Rightarrow$ R.H.L. $= \lim_{h\to 0}(1 + a_2h + a_3h^2)$ (1^∞ form)

$$= e^{\lim_{h\to 0}(1 + a_2h + a_3h^2 - 1)(1/h)} = e^{a_2}$$

$$\text{L.H.L.} = \lim_{x\to 0^-} f(x) = \lim_{h\to 0} f(0-h)$$

$$= \lim_{h\to 0}\frac{a(1 - (-h)\sin(-h)) + b\cos(-h) + 5}{(-h)^2}$$

$$= \lim_{h\to 0}\frac{a(1 - h(h)) + b\left(1 - \frac{h^2}{2!}\right) + 5}{h^2}$$

For finite value of L.H.L., $a + b + 5 = 0$ and $-a - \frac{b}{2} = 3$

Solving, we get $a = -1, b = -4$.

Now $g(x) = 3a \sin x - b\cos x = -3\sin x + 4\cos x$, which has the range $[-5, 5]$.

Also $P(x) = a_3x^3 + (\log_e 3)x^2$

$P''(x) = 6a_3x + 2\log_e 3$

$\Rightarrow P''(0) = 2\log_e 3$

Further, $P(x) = b \Rightarrow a_3x^3 + (\log_e 3)x^2 = -4$ has only one real root, as the graph of $P(x) = a_3x^3 + (\log_e 3)x^2$ meets $y = -4$ only once for negative value of x.

For Problems 4–6
4. c., 5. b., 6. c.

Sol. For $0 \le x < \frac{\pi}{4}$, $g(x) = 1 + \tan x$

$x \in \left[0, \frac{\pi}{4}\right) \Rightarrow 1 + \tan x \in [1, 2)$

so $f(g(x)) = f(1 + \tan x) = 1 + \tan x + 2$

and for $x \in \left[\frac{\pi}{4}, \pi\right)$, $g(x) = 3 - \cot x$

$x \in \left[\frac{\pi}{4}, \pi\right) \Rightarrow 3 - \cot x \in [2, \infty)$

so $f(g(x)) = f(3 - \cot x) = 6 - (3 - \cot x)$

Let $h(x) = f(g(x)) = \begin{cases} 3 + \tan x, & 0 \le x < \frac{\pi}{4} \\ 3 + \cot x, & \frac{\pi}{4} \le x < \pi \end{cases}$

Clearly, $f(g(x))$ is continuous in $[0, \pi)$

Now $h'\left(\frac{\pi^+}{4}\right) = \lim_{x \to \pi/4^+} (-\operatorname{cosec}^2 x) = -2$

$h'\left(\frac{\pi^-}{4}\right) = \lim_{x \to \pi/4^-} (\sec^2 x) = 2$

So $f(g(x))$ is differentiable everywhere in $[0, \pi)$ other than at $x = \pi/4$.

$|f(g(x))| = \begin{cases} |3 + \tan x|, & 0 \le x < \frac{\pi}{4} \\ |3 + \cot x|, & \frac{\pi}{4} \le x < \pi \end{cases}$

which is non-differentiable at $x = \pi/4$ and where $3 + \cot x = 0$ or $x = \cot^{-1}(-3)$

For $x \in \left[0, \frac{\pi}{4}\right)$, $3 + \tan x \in [3, 4)$

For $x \in \left[\frac{\pi}{4}, \pi\right)$, $3 + \cot x \in (-\infty, 4]$

Hence, the range is $(-\infty, 4]$.

For Problems 7–9
7. a., 8. c., 9. d.

Sol. $f(x) = \lim_{n \to \infty} \frac{f(x) + x^{2n} g(x)}{1 + x^{2n}}$

$$= \begin{cases} f(x), & 0 \le x^2 < 1 \\ \frac{f(x) + g(x)}{2}, & x^2 = 1 \\ g(x), & x^2 > 1 \end{cases}$$

$$= \begin{cases} g(x), & x < -1 \\ \frac{f(-1) + g(-1)}{2}, & x = -1 \\ f(x), & -1 < x < 1 \\ \frac{f(1) + g(1)}{2} & x = 1 \\ g(x) & x > 1 \end{cases}$$

If $F(x)$ is continuous $\forall\ x \in R$, $F(x)$ must be made continuous at $x = \pm 1$.

For continuity at $x = -1$, $f(-1) = g(-1) \Rightarrow 1 - a + 3 = b - 1 \Rightarrow a + b = 5$ (1)

For continuity at $x = 1$, $f(1) = g(1) \Rightarrow 1 + a + 3 = 1 + b \Rightarrow a - b = -3$ (2)

Solving equations (1) and (2), we get $a = 1$ and $b = 4$

$f(x) = g(x) \Rightarrow x^2 + x + 3 = x + 4 \Rightarrow x^2 = 1 \Rightarrow x = \pm 1$.

For Problems 10–12

10. a., 11. d., 12. b.

Sol.

$$f(x) = \begin{cases} [x], & -2 \le x \le -\frac{1}{2} \\ 2x^2 - 1, & -\frac{1}{2} < x \le 2 \end{cases}$$

$$= \begin{cases} -2, & -2 \le x < -1 \\ -1, & -1 \le x \le -\frac{1}{2} \\ 2x^2 - 1, & -\frac{1}{2} < x \le 2 \end{cases}$$

$$|f(x)| = \begin{cases} 2, & -2 \le x < -1 \\ 1, & -1 \le x \le -\frac{1}{2} \\ |2x^2 - 1|, & -\frac{1}{2} < x \le 2 \end{cases}$$

$$= \begin{cases} 2, & -2 \le x < -1 \\ 1, & -1 \le x \le -\frac{1}{2} \\ 1 - 2x^2, & -\frac{1}{2} < x \le \frac{1}{\sqrt{2}} \\ 2x^2 - 1, & \frac{1}{\sqrt{2}} < x \le 2 \end{cases}$$

$$f(|x|) = \begin{cases} -2, & -2 \le |x| < -1 \\ -1, & -1 \le |x| \le -\frac{1}{2} \\ 2|x|^2 - 1, & -\frac{1}{2} < |x| \le 2 \end{cases}$$

$= 2x^2 - 1,\ -2 \le x \le 2$

$$\Rightarrow g(x) = f(|x|) + |f(x)| = \begin{cases} 2x^2+1, & -2 \le x < -1 \\ 2x^2, & -1 \le x \le -\frac{1}{2} \\ 0, & -\frac{1}{2} < x < \frac{1}{\sqrt{2}} \\ 4x^2-2, & \frac{1}{\sqrt{2}} \le x \le 2 \end{cases}$$

$$g(-1^-) = \lim_{x\to -1}(2x^2+1) = 3,\ g(-1^+) = \lim_{x\to -1} 2x^2 = 2$$

$$g\left(-\frac{1}{2}^-\right) = \lim_{x\to -\frac{1}{2}} 2x^2 = \frac{1}{2},\ g\left(-\frac{1}{2}^+\right) = \lim_{x\to -\frac{1}{2}} 0 = 0$$

$$g\left(\frac{1}{\sqrt{2}}^-\right) = \lim_{x\to \frac{1}{\sqrt{2}}} 0 = 0 ,$$

$$g\left(\frac{1}{\sqrt{2}}^+\right) = \lim_{x\to \frac{1}{\sqrt{2}}} (4x^2-2) = 0.$$

Hence, $g(x)$ is discontinuous at $x = -1, -\frac{1}{2}$.

$g(x)$ is continuous at $x = \frac{1}{\sqrt{2}}$

Now, $g'\left(\frac{1}{\sqrt{2}}^-\right) = 0, g'\left(\frac{1}{\sqrt{2}}^+\right) = 8\left(\frac{1}{\sqrt{2}}\right) = \frac{8}{\sqrt{2}}$

Hence, $g(x)$ is non-differentiable at $x = \frac{1}{\sqrt{2}}$.

Matrix-Match Type

1. a→ p, q, r; b → p, r, s; c→ p, r, s; d → p, r, s.

a. $f(x) = |x^3| = x(x|x|)$ is continuous and differentiable.

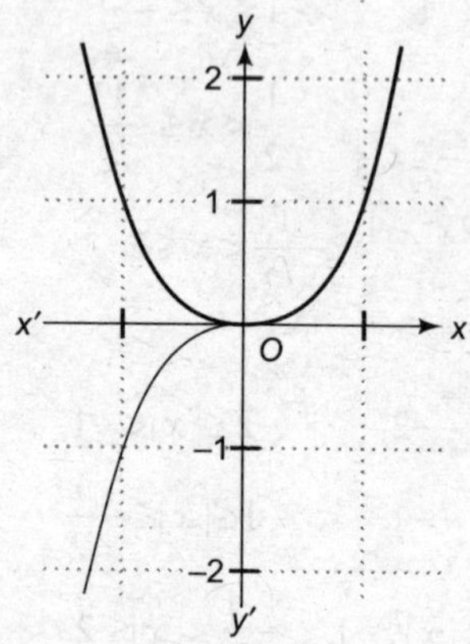

Fig. 3.18

b. $f(x) = \sqrt{|x|}$ is continuous

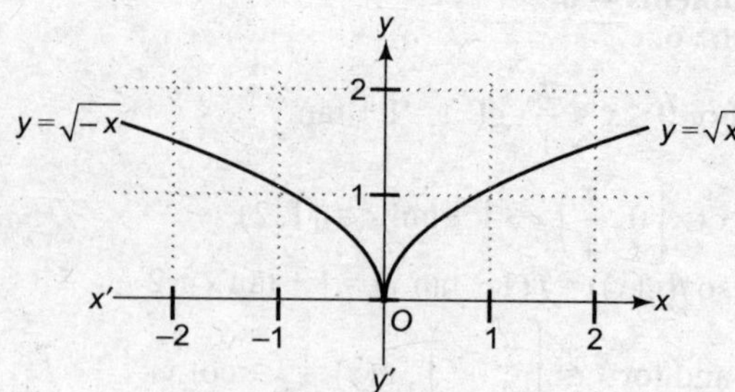

Fig. 3.19

Clearly, from the graph, $f(x)$ is non-differentiable at $x = 0$.

c. $f(x) = |\sin^{-1} x|$ is continuous

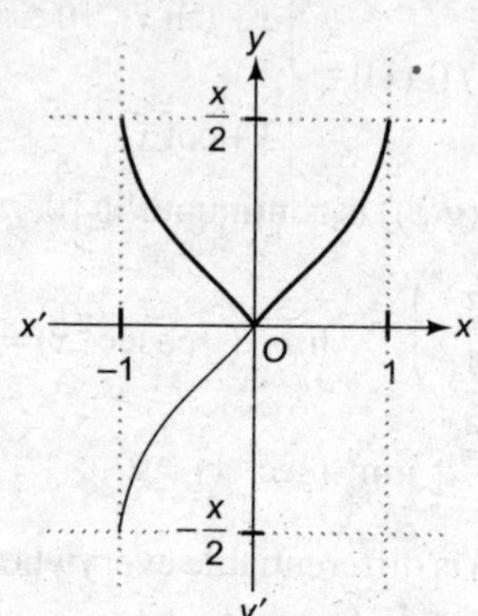

Fig. 3.20

Clearly, from the graph, $f(x)$ is non-differentiable at $x = 0$.

d. $f(x) = \cos^{-1}|x|$ is continuous

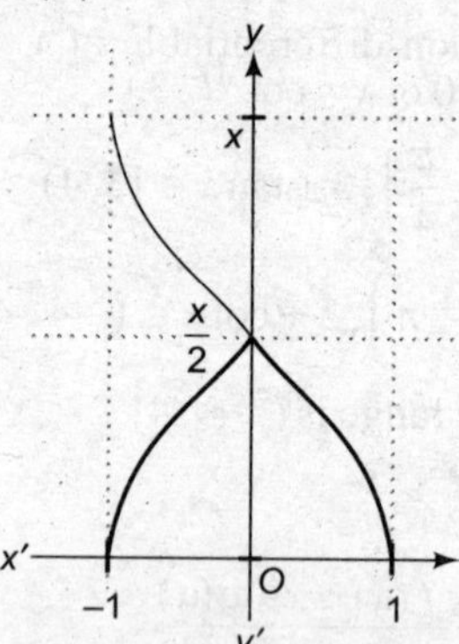

Fig. 3.21

Clearly, from the graph, $f(x)$ is non-differentiable at $x = 0$.

2. a → r, s; b → p, q; c→ p, q; d → p, r.

a. The given function is clearly continuous at all points except possibly at $x = \pm 1$.
As $f(x)$ is an even function, so we need to check its continuity only at $x = 1$.

$$\lim_{x\to 1^-} f(x) = \lim_{x\to 1^+} f(x) = f(1)$$

$$\Rightarrow \lim_{x\to 1^-}(ax^2+b) = \lim_{x\to 1^+} \frac{1}{|x|} \Rightarrow a = b = 1 \qquad (1)$$

Clearly, $f(x)$ is differentiable for all x, except possibly at $x=\pm 1$. As $f(x)$ is an even function, so we need to check its differentiability at $x=1$ only.

$$\lim_{x\to 1^-}\frac{f(x)-f(1)}{x-1}=\lim_{x\to 1^+}\frac{f(x)-f(1)}{x-1}$$

$$\Rightarrow \lim_{x\to 1}\frac{ax^2+b-1}{x-1}\ \lim_{x\to 1}\frac{ax^2+b-1}{x-1}=\lim_{x\to 1^+}\frac{\frac{1}{|x|}-1}{x-1}$$

$$\Rightarrow \lim_{x\to 1}\frac{ax^2-a}{x-1}=\lim_{x\to 1}\frac{-1}{x}\Rightarrow 2a=-1\Rightarrow a=1/2\ ½$$

Putting $a=-1/2$ in (1) we get $b=3/2\Rightarrow |k|=1\Rightarrow k=\pm 1$

b. $f(x)=\text{sgn}(x^2-ax+1)$ is discontinuous then $x^2-ax+1=0$ must have only one real root. Hence $a=\pm 2$.

c. $f(x)=[2+3|n|\sin x]$, $n\in N$ has exactly 10 points of discontinuity in $x\in(0,\pi)$.

The required number of points are $1+2(3|n|-1)=6|n|-1=11\Rightarrow n=\pm 2$.

d. If $f(x)=|||x|-2|+a|$ has exactly three points of non-differentiability.

$f(x)$ is non-differentiable at $x=0$, $|x|-2=0$ or $x=0,\pm 2$.

Hence, the value of a must be positive, as negative value of a allows $||x|-2|+a=0$ to have real roots, which gives two more points of non-differentiability.

3. a → s, b → r, c → p, d → q.

$g(x)=x^2+bx+c$	$g(\lvert x\rvert)=x^2+b\lvert x\rvert+c$	$f(x)=\lvert g(\lvert x\rvert)\rvert=\lvert x^2+b\lvert x\rvert+c\rvert$
$b<0, c>0$		
$c=0, b<0$		
$c=0, b>0$		
$b=0, c<0$		
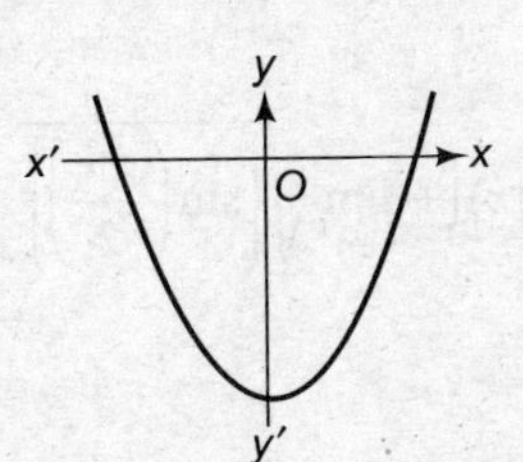	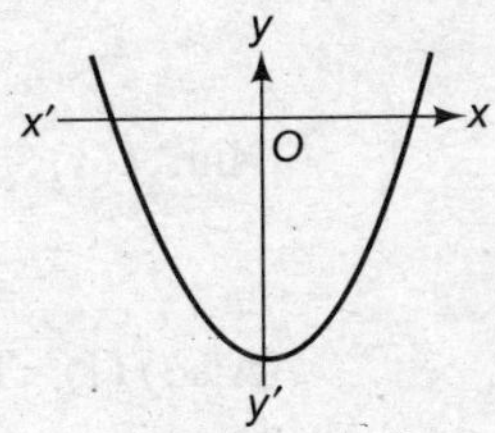	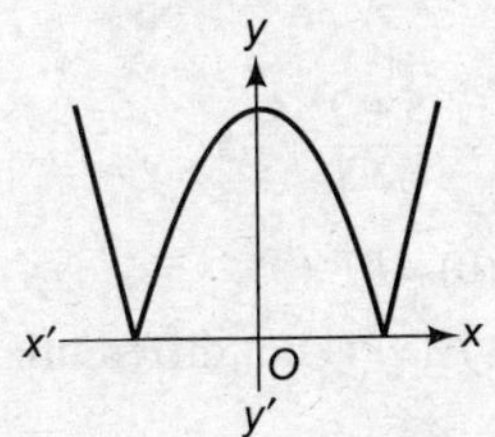

Fig. 3.22

4. a → q, s; b → p, s; c → p, r; d → q, s.

a. $f(x)=\begin{cases}\dfrac{5e^{1/x}+2}{3-e^{1/x}}, & x\neq 0\\ 0, & x=0\end{cases}$

$$f(0^+)=\lim_{h\to 0}\frac{5e^{1/h}+2}{3-e^{1/h}}=\lim_{h\to 0}\frac{5+2e^{-1/h}}{3e^{-1/h}-1}=-5.$$

Hence, $f(x)$ is discontinuous and non-differentiable at $x=0$.

b. $g(x)=xf(x)=\begin{cases}x\dfrac{5e^{1/x}+2}{3-e^{1/x}}, & x\neq 0\\ 0, & x=0\end{cases}$

$$f(0^+)=\lim_{h\to 0}h\frac{5e^{1/h}+2}{3-e^{1/h}}=\lim_{h\to 0}h\frac{5+2e^{-1/h}}{3e^{-1/h}-1}$$

$$=0\times(-5)$$

$$=0.$$

$$f(0^-)=\lim_{h\to 0}h\frac{5e^{-1/h}+2}{3-e^{-1/h}}=0\times(2/3)=0$$

Hence, $f(x)$ is continuous at $x=0$.

$$Lg'(0)=\lim_{h\to 0}\frac{g(0-h)-g(0)}{-h}$$

$$=\lim_{h\to 0}\frac{-hf(-h)-0}{-h}$$

$$=\lim_{h\to 0}f(-h)$$

$$=\lim_{h\to 0}\frac{5e^{-1/h}+2}{3-e^{-1/h}}=\frac{0+2}{3-0}=\frac{2}{3}$$

$$Rg'(0)=\lim_{h\to 0}\frac{g(0+h)-g(0)}{h}$$

$$=\lim_{h\to 0}\frac{g(h)-0}{h}$$

$$=\lim_{h\to 0}f(h)=\lim_{h\to 0}\frac{5e^{1/h}+2}{3-e^{1/h}}$$

$$=\lim_{h\to 0}\frac{5+2e^{-1/h}}{3e^{-1/h}-1}$$

$$=\frac{5+0}{0-1}=-5$$

$\because LF'(0)\neq RF'(0)$

Hence, $F(x)$ is not differentiable, but continuous at $x=0$.

c. For $x^2f(x)$,

Let $F(x)=x^2f(x)$

$$\therefore LF'(0)=\lim_{h\to 0}\frac{F(0-h)-F(0)}{-h}$$

$$=\lim_{h\to 0}\frac{h^2f(-h)-0}{-h}=0$$

$$RF'(0)=\lim_{h\to 0}\frac{F(0+h)-F(0)}{h}$$

$$=\lim_{h\to 0}\frac{h^2f(h)-0}{h}=0$$

$\therefore LF'(0)=RF'(0)$

Hence, $F(x)$ is differentiable at $x=0$, then it is always continuous at $x=0$.

d. Clearly from the above discussion $y=x^{-1}f(x)$ is discontinuous and hence non-differentiable at $x=0$.

5. a → q, s; b → p, r; c → p, r; d → p, s.

a. $f(x)=\lim_{n\to\infty}\left[\cos^2(2\pi x)\right]^n+\left\{x+\dfrac{1}{2}\right\}$

Obviously, $\lim_{x\to 1/2^+}f(x)=0+0=0$

And $\lim_{x\to 1/2^-}f(x)=0+1$

Hence, $f(x)$ is discontinuous at $x=1/2$.

b. $f(x)=(\log x)(x-1)^{1/5}$

Obviously, $f(x)$ is continuous at $x=1$

$$f'(1^+)=\lim_{h\to 0}\frac{f(1+h)-f(1)}{h}=\lim_{h\to 0}\frac{\log(1+h)h^{1/5}}{h}=0$$

$$f'(1^-)=\lim_{h\to 0}\frac{f(1-h)-f(1)}{h}$$

$$=\lim_{h\to 0}\frac{\log(1-h)(-h)^{1/5}}{-h}$$

$$=0$$

Hence, $f(x)$ is differentiable at $x=1$.

c. $f(x)=[\cos 2\pi x]+\sqrt{\left\{\sin\left(\dfrac{\pi x}{2}\right)\right\}}$

$$\lim_{x\to 1^-}f(x)=\lim_{x\to 1^-}[\cos 2\pi x]+\lim_{x\to 1^-}\sqrt{\left\{\sin\left(\frac{\pi x}{2}\right)\right\}}$$

$$=0+1=1$$

$$\lim_{x\to 1^+}f(x)=\lim_{x\to 1^+}[\cos(2\pi x)]+\lim_{x\to 1^+}\sqrt{\left\{\sin\left(\frac{\pi x}{2}\right)\right\}}$$

$$=0+1=1$$

Also $f(1)=1+0=1$.

$f(x)$ is continuous at $x=1$

$$f'(1^+) = \lim_{h\to 0}\frac{[\cos 2\pi(1+h)] + \sqrt{\left\{\sin\left(\frac{\pi(1+h)}{2}\right)\right\}} - 1}{h}$$

$$= \lim_{h\to 0}\frac{[\cos 2\pi h] + \sqrt{\left\{\cos\left(\frac{\pi h}{2}\right)\right\}} - 1}{h}$$

$$= \lim_{h\to 0}\frac{\sqrt{\cos\left(\frac{\pi h}{2}\right)} - 1}{h} = \lim_{h\to 0}\frac{-\frac{\pi}{2}\sin\left(\frac{\pi h}{2}\right)}{2\sqrt{\cos\left(\frac{\pi h}{2}\right)}} = 0$$

Similarly, $f'(1^-) = 0$.

d. $f(x) = \begin{cases}\cos 2x, & x\in Q\\ \sin x, & x\notin Q\end{cases}$ at $\frac{\pi}{6}$

$f(x)$ is continuous when $\cos 2x = \sin x$ which has $x = \frac{\pi}{6}$ as one of the solutions. Hence, it is continuous.

Also in the neighbourhood of $x = \frac{\pi}{6}$,

$$f'(x) = \begin{cases}-2\sin 2x, & \frac{\pi}{6} - \delta < x < \frac{\pi}{6}\\ \cos x, & \frac{\pi}{6} < x < \frac{\pi}{6} + \delta\end{cases}$$

Here, $f'\left(\frac{\pi}{6}^-\right) \neq f'\left(\frac{\pi}{6}^+\right)$.

$\Rightarrow f(x)$ is not differentiable at $x = \frac{\pi}{6}$

Integer Type

1. $g'(3^-)$

$$= \lim_{h\to 0}\frac{g(3-h) - g(h)}{-h} = \lim_{h\to 0}\frac{a\sqrt{4-h} - (3b+2)}{h} \quad (1)$$

For the existence of the limit, $\lim_{h\to 0} N^r = 0$. Therefore, $2a - 3b = 2$ (2)

$$\text{Now } g'(3^+) = \lim_{hk\to 0}\frac{b(3+h) + 2 - (3b+2)}{h} = b \quad (3)$$

Substituting $3b = 2 = 2a$ in equation (1), we get $g'(3^-) =$

$$\lim_{h\to 0}\frac{a\sqrt{4-h} - 2a}{-h} = \lim_{h\to 0}\left(\frac{(4-h)-4}{(-h)\left(\sqrt{4-h}+2\right)}\right) = \frac{a}{4}$$

Hence, $g'(3^-) = g'(3^+)$.

$$\frac{a}{4} = b \Rightarrow a = 4b \quad (4)$$

From (2) and (4), we get

$$8b - 3b = 2 \Rightarrow b = \frac{2}{5} \text{ and } a = \frac{8}{5} \Rightarrow a + b = 2.$$

2. (8) $f(x) = \begin{cases}x-1, & x \text{ is rational}\\ x^2 - x - 2, & x \text{ is irrational}\end{cases}$ is continuous when $x - 1 = x^2 - x - 2$

or $x^2 - 2x - 1 = 0$ or $x = \frac{2\pm\sqrt{8}}{2}$.

Hence, $f(x)$ is continuous at two points $\Rightarrow a = 2$.

$f(x) = \text{sgn}(x^3 - 3x + 1)$ is discontinuous when $x^3 - 3x + 1 = 0$.

Now $y = x^3 - 3x + 1$ has three distinct roots as $\frac{dy}{dx} = 3x^2 - 3 = 0 \Rightarrow x = \pm 1$.

Also $f(1) = 1 - 3 + 1 = -1$ and $f(-1) = -1 + 3 + 1 = 3$.

Hence, the graph of $y = x^3 - 3x + 1$ is

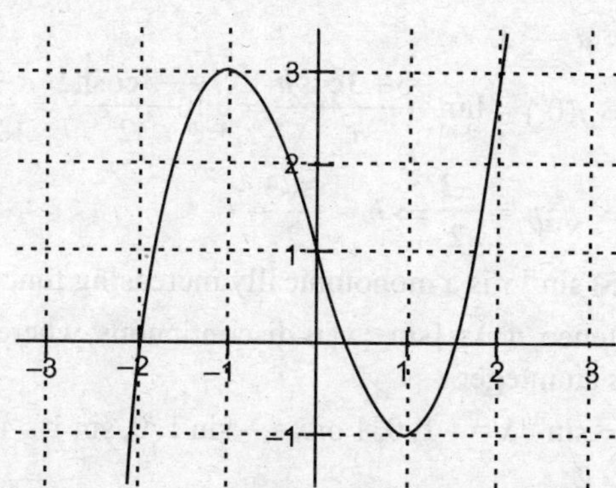

Fig. 3.23

Hence, $f(x) = \text{sgn}(x^3 - 3x + 1)$ is discontinuous at three points $\Rightarrow b = 3$.

$f(x) = (\log x)\,|x^2 - 4x + 3| + 2(x-2)^{1/3} = (\log x)\,|x-1||x-3| + 2(x-2)^{1/3}$,

which is non-differentiable at $x = 3$ and $x = 2$ as at $x = 3$, $f(x)$ has a sharp turn and at $x = 2$, $f(x)$ has a vertical tangent. At $x = 1$, $f(x)$ is differentiable. So $c = 2$ and $d = 1$.

3. (1) $f(x) = \begin{cases}\frac{\sin ax^2}{x^2}, & x\neq 0\\ \frac{3}{4} + \frac{1}{4a}, & x = 0\end{cases}$ is continuous at $x = 0$.

Therefore, $\lim_{x\to 0} f(x) = f(0)$

$$\Rightarrow \lim_{x\to 0}\frac{a\sin ax^2}{ax^2} = \frac{3}{4} + \frac{1}{4a}$$

$$\Rightarrow a = \frac{3}{4} + \frac{1}{4a}$$

$\Rightarrow 4a^2 - 3a - 1 = 0$
$\Rightarrow (4a+1)(a-1) = 0$
$\Rightarrow a = -1/4, 1$

4. (4) $f(x) = \begin{cases} \dfrac{a+3\cos x}{x^2}, & x<0 \\ b\tan\left(\dfrac{\pi}{[x+3]}\right), & x \geq 0 \end{cases}$

$f(0^+) = \lim_{h\to 0} b\tan\left(\dfrac{\pi}{[0+h+3]}\right) = b\tan\left(\dfrac{\pi}{3}\right) = \sqrt{3}b$

$f(0^-) = \lim_{h\to 0}\dfrac{a+3\cos(-h)}{(-h)^2}$

$= \lim_{h\to 0}\dfrac{a+3\cos h}{h^2}$

$= \lim_{h\to 0}\dfrac{a+3\cos h}{h^2}$

$\Rightarrow a+3 = 0$

As $f(x)$ is continuous at $x = 0$, $f(0^-)$ must be finite.

$\Rightarrow a = -3$

$\Rightarrow f(0^-) = \lim_{h\to 0}\dfrac{-3+3\cos h}{h^2} = \lim_{h\to 0}\dfrac{-3\cosh}{2} = \dfrac{-3}{2}$

$\Rightarrow \sqrt{3}b = \dfrac{-3}{2} \Rightarrow b = \dfrac{-\sqrt{3}}{2}$

5. (5) $\sin^{-1}x$ is a monotonically increasing function. Hence, $f(x) = [\sin^{-1}x]$ is discontinuous, where $\sin^{-1}x$ is an integer.

$\Rightarrow \sin^{-1}x = -1, 0, 1$ or $x = -\sin 1, 0, \sin 1$

$\dfrac{2}{1+x^2}$, $x \geq 0$, is a monotonically decreasing function.

Hence, $f(x) = \left[\dfrac{2}{1+x^2}\right]$, $x \geq 0$ is discontinuous when $\dfrac{2}{1+x^2}$ is an integer.

$\Rightarrow \dfrac{2}{1+x^2} = 1, 2$

$\Rightarrow x = 1, 0$

6. (8) $f(x) = [2\cos x]$

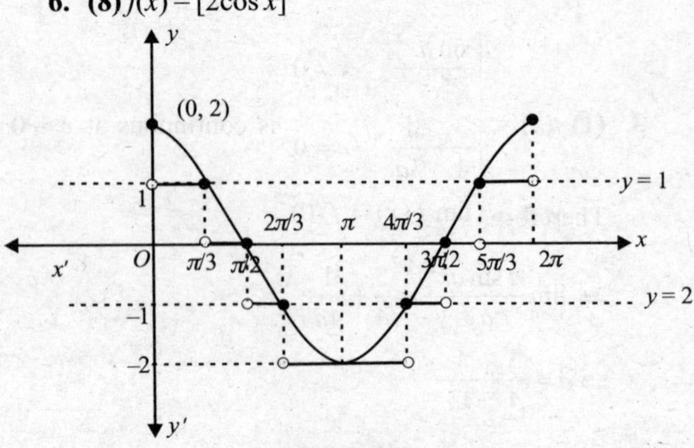

Fig. 3.24

Clearly, from the graph, it can be seen that $f(x)$ is discontinuous at $x = 0$,

$x = \dfrac{\pi}{3}, \dfrac{2\pi}{3}, \dfrac{4\pi}{3}, \dfrac{5\pi}{3}$

7. (6) $f(x) = [\sin x + \cos x] = [g(x)]$, where $g(x) = \sin x + \cos x$

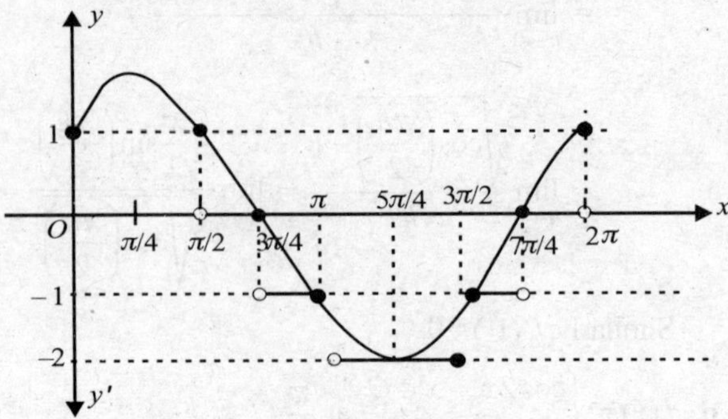

Fig. 3.25

$g(0) = 1,\ g\left(\dfrac{\pi}{4}\right) = \sqrt{2},\ g\left(\dfrac{\pi}{2}\right) = 1$

$g\left(\dfrac{3\pi}{4}\right) = 0,\ g(\pi) = -1,\ g\left(\dfrac{5\pi}{4}\right) = -\sqrt{2}$

$g\left(\dfrac{3\pi}{2}\right) = -1,\ g\left(\dfrac{7\pi}{4}\right) = 0,\ g(2\pi) = 1$

From the graph, it can be seen that there are six points of discontinuity.

8. (9) $\sin(x-2)$ and $\cos(x-2)$ are continuous for all x. Since $[x^3]$ is not continuous at integral values of x^3,

$f(x)$ is continuous in [0, 2] if $\left[\dfrac{(x-2)^3}{a}\right] = 0$, $\forall\, x \in [0, 2]$.

Now $(x-2)^3 \in [0, 8]$ for $x \in [4, 6]$

$\Rightarrow a > 8$ for $\left[\dfrac{(x-2)^3}{a}\right] = 0$

9. (4) i. Continuity should be checked at the end-points of intervals of each definition, i.e., $x = 1, 3, 3.5$.

ii. For $\{x^2\}$, continuity should be checked when $x^2 = 10, 11, 12$ or $x = \sqrt{10}, \sqrt{11}, \sqrt{12}$, $\{x^2\}$ is discontinuous for those values of x, where x^2 is an integer.
[Note: Here x^2 is monotonic for the given domain.]

iii. For $\text{sgn}(x-2)$, continuity should be checked when $x - 2 = 0$ or $x = 2$.

iv. For $[\log_e x]$, continuity should be checked when $\log_e x = 1$ or $x = e$ ($\in [1, 3]$).

Hence, the overall continuity must be checked at $x = 1, 2, e, 3, \sqrt{10}, \sqrt{11}, \sqrt{12}, 3.5$.

Further, $f(1) = 0$ and

$\lim_{x\to 1^+} f(x) = \lim_{x\to 1^+} \text{sgn}(x-2)\times[\log_e x] = 0$.

Hence, $f(x)$ is continuous at $x = 1$.

$\lim_{x\to 2^-} f(x) = \lim_{x\to 2^-} \text{sgn}(x-2)\times[\log_e x] = (-1)\times 0 = 0$

$\lim_{x\to 2^+} f(x) = \lim_{x\to 2^+} \text{sgn}(x-2)\times[\log_e x] = (1)\times 0 = 0$

Hence, $f(x)$ is continuous at $x = 2$.

$\lim_{x\to 3^-} f(x) = \lim_{x\to 3^-} \text{sgn}(x-2)\times[\log_e x] = 1$

$\Rightarrow \lim_{x\to 3^+} f(x) = \lim_{x\to 3^+}\{x^2\} = 0$

Hence, $f(x)$ is discontinuous at $x = 3$.

Also $\{x^2\}$ is discontinuous at $x = \sqrt{10}, \sqrt{11}, \sqrt{12}$.

Therefore,

$\lim_{x\to 3.5^-} f(x) = \lim_{x\to 3.5^-}\{x^2\} = 0.25 = f(3.5)$.

Hence, $f(x)$ is discontinuous at $x = 3, \sqrt{10}, \sqrt{11}, \sqrt{12}$.

10. (2) $f(x) = \lim_{n\to\infty} \frac{(x^2)^n - 1}{(x^2)^n + 1}$

$$= \lim_{n\to\infty} \frac{1 - \frac{1}{(x^2)^n}}{1 + \frac{1}{(x^2)^n}}$$

$$= \begin{cases} -1, & 0 \le x^2 < 1 \\ 0, & x^2 = 1 \\ 1, & x^2 > 1 \end{cases} = \begin{cases} 1, & x < -1 \\ 0, & x = -1 \\ -1, & -1 < x < 1 \\ 0, & x = 1 \\ 1, & x > 1 \end{cases}$$

Thus $f(x)$ is discontinuous at $x = \pm 1$.

11. (1) $f(x)$ is continuous at some $x = a$, where $x = 1 - x$ or $x = 1/2$.

Hence, $f(x)$ is continuous at $x = 1/2$.

Explanation:

We have $f(1/2) = 1/2$.

If $x \to 1/2^+$, then x may be rational or irrational

$\Rightarrow f(1/2^+) = 1/2$ or $1 - 1/2 = 1/2$.

If $x \to 1/2^-$, then x may be rational or irrational

$\Rightarrow f(1/2^-) = 1/2$ or $1 - 1/2 = 1/2$.

Hence, $f(x)$ is continuous at $x = 1/2$.

For some other point, say $x = 1, f(1) = 1$.

If $x \to 1^+$, then x may be rational or irrational

$\Rightarrow f(1^+) = 1$ or $1 - 1 = 0$.

Hence, $f(1^+)$ oscillates between 1 and 0, which causes discontinuity at $x = 1$.

Similarly, $f(x)$ oscillates between 0 and 1 for all $x \in R - \{1/2\}$.

12. (6) Obviously, if $g(x) = \left[\frac{f(x)}{c}\right]$ is continuous, then c must exceed the greatest value of $f(x)$ to make the ratio $f(x)/c$ lie between 0 and 1, for which the least positive integral value of c is 6.

13. (3) $t = \frac{1}{x-1}$ is discontinuous at $x = 1$.

Also $y = \frac{1}{t^2+t-2}$ is discontinuous at $t = -2$ and $t = 1$.

When $t = -2$,

$\frac{1}{x-1} = -2 \Rightarrow x = \frac{1}{2}$.

When $t = 1$,

$\frac{1}{x-1} = 1 \Rightarrow x = 2$.

So $y = f(x)$ is discontinuous at three points, $x = 1, 1/2$, and 2.

14. (3) $[x]$ is non-differentiable at $x = 0, 1, 2$, and $|1 - x|$ is non-differentiable at $x = 1$. Thus $f(x)$ is definitely non-differentiable at $x = 0, 2$. Moreover, $[x]$ is discontinuous at $x = 1$, whereas $|1 - x|$ is continuous at $x = 1$. Thus $f(x)$ is discontinuous and hence non-differentiable at $x = 1$.

$f(x) = \max\{\sec^{-1}x, \text{cosec}^{-1}x\}$

15. (4) Here $f(x) = \left|\left(x+\frac{1}{2}\right)[x]\right|, -2 \le x \le 2$

$$\Rightarrow f(x) = \begin{cases} \left|\left(x+\frac{1}{2}\right)(-2)\right|, & -2 \le x < -1 \\ \left|\left(x+\frac{1}{2}\right)(-1)\right|, & -1 \le x < 0 \\ \left|\left(x+\frac{1}{2}\right)(-0)\right|, & 0 \le x < 1 \\ \left|\left(x+\frac{1}{2}\right)(1)\right|, & 1 \le x < 2 \\ \left|\frac{3}{2}.2\right|, & x = 2 \end{cases}$$

$$\Rightarrow f(x) = \begin{cases} -(2x+1), & -2 \le x < -1 \\ -\left(x+\frac{1}{2}\right), & -1 \le x < -1/2 \\ (x+1/2), & -\frac{1}{2} \le x < 0 \\ 0, & 0 \le x < 1 \\ x+\frac{1}{2}, & 1 \le x < 2 \\ 3, & x = 2 \end{cases}$$

This can be plotted as below:

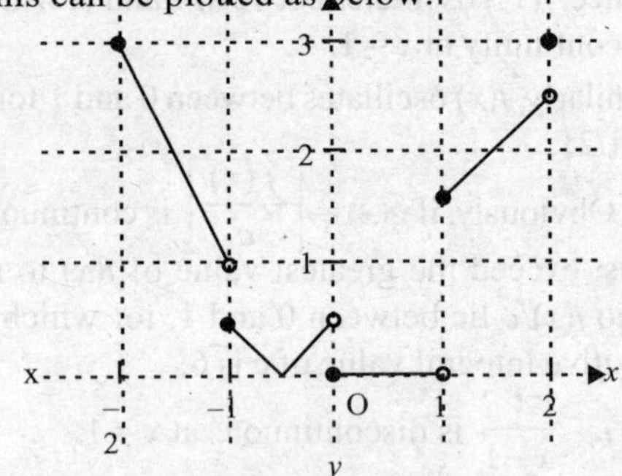

Fig. 3.26

The graph clearly shows that $f(x)$ is not continuous at $x = \{-1, 0, 1, 2\}$

16. (1) $f(x) = \begin{cases} a + \sin^{-1}(x+b), & x \ge 1 \\ x, & x < 1 \end{cases}$

$$\Rightarrow f'(x) = \begin{cases} \dfrac{1}{\sqrt{1-(x+b)^2}}, & x > 1 \\ 1, & x < 1 \end{cases}.$$

For $f(x)$ to be continuous at $x = 1$,

$f(1^+) = f(1^-) \Rightarrow a + \sin^{-1}(1+b) = 1$ (1)

Also $f'(1^+) = f'(1^-) \Rightarrow \dfrac{1}{\sqrt{1-(1+b)^2}} = 1 \Rightarrow b = -1.$

From (1), we get $a = 1$.

17. (2) $f(x) = \sin^{-1}\left(\dfrac{2x}{1+x^2}\right) = \begin{cases} 2\tan^{-1} x, & -1 \le x \le 1 \\ \pi - 2\tan^{-1} x, & \text{if } x > 1 \\ -\pi - 2\tan^{-1} x, & \text{if } x < -1 \end{cases}$

$$\Rightarrow f'(x) = \begin{cases} \dfrac{2}{1+x^2}, & -1 < x < 1 \\ -\dfrac{2}{1+x^2}, & \text{if } x > 1 \\ -\dfrac{2}{1+x^2}, & \text{if } x < -1 \end{cases} \qquad (1)$$

$\Rightarrow f'(-1^-) = -1, f'(-1^+) = 1$, and

$f'(1^-) = 1, f'(1^+) = -1.$

Hence, $f(x)$ is non–differentiable at $x = \pm 1$.

Graph of $f(x) = \sin^{-1}\left(\dfrac{2x}{1+x^2}\right)$

Fig. 3.27

Students find it difficult to remember all the cases of $\sin^{-1}[2x/(1+x^2)]$ in (1).

So the following short method should be used to check the differentiability.

Differentiating $f(x)$ w.r.t. x, we have

$$\frac{df(x)}{dx} = \frac{\dfrac{d\left(\dfrac{2x}{1+x^2}\right)}{dx}}{\sqrt{1-\left(\dfrac{2x}{1+x^2}\right)^2}}$$

$$= \frac{\dfrac{2(1+x^2) - 2x(2x)}{(1+x^2)}}{\sqrt{(1+x^2)^2 - 4x^2}}$$

$$= \frac{2(1-x^2)}{(1+x^2)|1-x^2|}$$

Clearly, $\dfrac{df(x)}{dx}$ is discontinuous at $x^2 = 1$ or $x = \pm 1$.

Hence, $f(x)$ is non-differentiable at $x = \pm 1$.

18. (0) $f(x) = \begin{cases} x^2 + ax + 1, & x \text{ is rational} \\ ax^2 + 2x + b, & x \text{ is irrational} \end{cases}$ is continuous at $x = 1$ and $x = 2$

$\Rightarrow x = 1$ and $x = 2$ are the roots of the equation

$x^2 + ax + 1 = ax^2 + 2x + b$ or

$(a-1)x^2 + (2-a)x + b - 1 = 0$

$\Rightarrow \dfrac{a-2}{a-1} = 3$ and $\dfrac{b-1}{a-1} = 2$

$\Rightarrow a = 1/2$ and $b = 0$

19. (3) $f(x)$ is continuous for all $x \in R$.

Thus $\lim\limits_{x \to \sqrt{3}} f(x) = f(\sqrt{3})$, where

$f(x) = \dfrac{x^2 - 2x + 2\sqrt{3} - 3}{\sqrt{3} - x}, x \ne \sqrt{3}$.

Now $\lim\limits_{x \to \sqrt{3}} f(x) = \lim\limits_{x \to \sqrt{3}} \dfrac{x^2 - 2x + 2\sqrt{3} - 3}{\sqrt{3} - x}$

$$= \lim_{x \to \sqrt{3}} \frac{(2-\sqrt{3}-x)(\sqrt{3}-x)}{(\sqrt{3}-x)}$$

$$= 2(1-\sqrt{3})$$

$\Rightarrow f(\sqrt{3}) = 2(1-\sqrt{3})$

20. (1) Here $f(x)$ is continuous at $x = 0$.

$\Rightarrow$ R.H.L. (at $x = 0$) = L.H.L. (at $x = 0$) = $f(0)$. Therefore,

R.H.L. (at $x = 0$) = $\lim\limits_{h \to 0} \dfrac{e^{1/h} + e^{2/h} + e^{3/h}}{ae^{2/h} + be^{3/h}}$ $\left(\dfrac{\infty}{\infty} \text{ form}\right)$

$$= \lim_{h\to 0}\frac{e^{3/h}\left\{\frac{1}{e^{2/h}}+\frac{1}{e^{1/h}}+1\right\}}{e^{3/h}\left\{\frac{a}{e^{1/h}}+b\right\}}$$

$$= \frac{1}{b} \quad (1)$$

Again, L.H.L. (at $x=0$) $= \lim_{h\to 0}(\cos h+\sin h)^{\text{cosec } h}$

$$= \lim_{h\to 0}[1+(\cos h+\sin h-1)]^{(-1/\sin h)} \quad [(1)^{\infty}\text{ form}]$$

$$= e^{\lim_{h\to 0}\{\cosh+\sinh-1\}\left(-\frac{1}{\sinh}\right)}$$

$$= e^{\lim_{h\to 0}\{-2\sin^2 h/2+2\sinh/2\cosh/2\}\left\{-\frac{1}{2\sinh/2\cosh/2}\right\}}$$

$$= e^{\lim_{h\to 0}\frac{\sinh/2-\cosh/2}{\cosh/2}} = e^{-1} \quad (2)$$

And $f(0)=a$

$\Rightarrow a=e^{-1}=\frac{1}{b}$ or ($a=e^{-1}$ and $b=e$)

21. **(1)** At $x=-2$,

$f(-2)=b$ (1)

R.H.L. $= \lim_{x\to -2+} f(x) = \lim_{h\to 0} f(-2+h)$

$$= \lim_{h\to 0}\sin\left(\frac{(-2+h)^4-16}{(-2+h)^5+32}\right)$$

$$= \sin\left\{\lim_{h\to 0}\frac{(h-2)^4-2^4}{2^5+(-2+h)^5}\right\} = \sin\left\{\lim_{h\to 0}\frac{(h-2)^4-(-2)^4}{(h-2)^5-(-2)^5}\right\}$$

$$= \sin\left\{\lim_{h\to 0}\frac{\frac{(h-2)^4-(-2)^4}{(h-2)-(-2)}}{\frac{(h-2)^5-(-2)^5}{(h-2)-(-2)}}\right\} = \sin\left\{\frac{4\times(-2)^{4-1}}{5\times(-2)^{5-1}}\right\}$$

$$= \sin\left\{\frac{4\times(-8)}{5\times(16)}\right\} = \sin\left(-\frac{2}{5}\right) \quad (2)$$

L.H.L. $= \lim_{x\to -2-} f(x) = \lim_{h\to 0} f(-2-h)$

$$= \lim_{h\to 0}\frac{ae^{1/(|-2-h+2|)}-1}{2-e^{1/(|-2-h+2|)}} = \lim_{h\to 0}\frac{ae^{1/h}-1}{2-e^{1/h}}$$

$$= \lim_{h\to 0}\frac{a-e^{-1/h}}{2e^{-1/h}-1} = \frac{a-0}{0-1} = -a \quad (3)$$

From (1), (2), and (3) we get

$-a=\sin\left(\frac{2}{5}\right)$ and $b=-\sin\left(\frac{2}{5}\right)$

22. **(2)** At $x=0, f(0)=c$ (1)

For $x<0$,

L.H.L. $= \lim_{x\to 0-} f(x) = \lim_{h\to 0} f(0-h)$

$$= \lim_{h\to 0}\frac{\sin(a+1)(0-h)+\sin(0-h)}{(0-h)}$$

$$= \lim_{h\to 0}\frac{-\sin(a+1)h-\sin h}{-h}$$

$$= \lim_{h\to 0}\left\{(a+1)\frac{\sin(a+1)h}{(a+1)h}+\frac{\sin h}{h}\right\}$$

$$= (a+1)\lim_{h\to 0}\frac{\sin(a+1)h}{(a+1)h}+\lim_{h\to 0}\frac{\sin h}{h}$$

$$= (a+1)\times 1+1 = a+2 \quad (2)$$

For $x>0$,

R.H.L. $= \lim_{x\to 0+} f(x) = \lim_{h\to 0} f(0+h)$

$$= \lim_{h\to 0}\frac{(h+bh^2)^{1/2}-h^{1/2}}{bh^{3/2}}$$

$$= \lim_{h\to 0}\frac{(1+bh)^{1/2}-1}{bh}$$

$$= \lim_{h\to 0}\frac{(1+bh)^{1/2}-(1)^{1/2}}{(1+bh)-1} = \frac{1}{2}(1)^{1/(2-1)} = \frac{1}{2} \quad (3)$$

$f(x)$ is continuous at $x=0$

$\Rightarrow$ L.H.L. = R.H.L. = $f(0)$ [from (1), (2), and (3)]

$\Rightarrow a+2=\frac{1}{2}=c$

$\Rightarrow a=-\frac{3}{2}, b\in R, c=\frac{1}{2}$

23. **(0)** Given $f(2a-x)=f(x),\ \forall\ x\in(a, 2a)$

and $f'(a^-)=0 \Rightarrow \lim_{h\to 0}\frac{f(a-h)-f(a)}{-h}=0$

Now $f'(-a^-) = \lim_{h\to 0}\frac{f(-a-h)-f(-a)}{-h}$

$$= \lim_{h\to 0}\frac{-f(a+h)+f(a)}{-h}$$

[$\because f(x)$ is an odd function]

$$= -\lim_{h\to 0}\frac{f(a+h)-f(a)}{h}$$

Now replacing x by $a-h$ in the original functional relation, we get

$\Rightarrow f(a-h)=f[2a-(a-h)]=f(a+h)$

$\Rightarrow f'(-a^-) = -\lim_{h\to 0}\frac{f(a-h)-f(a)}{h} = -f'(a^-)=0$

24. **(6)** As $f(x)$ is continuous in [1, 3], $f(x)$ will attain all values between $f(1)$ and $f(3)$. As $f(x)$ takes rational values for all x and there are innumerable irrational values between $f(1)$ and $f(3)$, this implies that $f(x)$ can take rational values for all x if $f(x)$ has a constant value at all points between $x=1$ and $x=3$.

Given that $f(2)=6$, then $f(1.5)=6$.

25. **(1)** $f(x) = (x^2 - 1)\,|x^2 - 3x + 2| + \cos(|x|)$

$= [(x-1)\,|x-1|]\,|x-2| + \cos x$

$(x-1)|x-1|$ and $\cos x$ are differentiable for all x.

But $|x-2|$ is non-differentiable at $x = 2$.

Hence, $f(x)$ is non-differentiable at $x = 2$.

26. **(2)** $\lim\limits_{x\to 0} \dfrac{\ln\left(e^{x^2} + 2\sqrt{x}\right)}{\tan\sqrt{x}}$

$$= \lim_{x\to 0} \frac{\ln\left(e^{x^2} + 2\sqrt{x}\right)}{\dfrac{\tan\sqrt{x}}{\sqrt{x}}\sqrt{x}}$$

$$= \lim_{x\to 0} \frac{\ln\left(e^{x^2} + 2\sqrt{x}\right)}{\sqrt{x}}$$

$$= \lim_{x\to 0} \frac{\dfrac{2xe^{x^2} + \dfrac{1}{\sqrt{x}}}{\left(e^{x^2} + 2\sqrt{x}\right)}}{\dfrac{1}{2\sqrt{x}}}$$

$$= 2\lim_{x\to 0} \frac{2x\sqrt{x}e^{x^2} + 1}{\left(e^{x^2} + 2\sqrt{x}\right)}$$

$= 2$

27. **(4)** $\lim\limits_{x\to 0} f(x) = e^{\frac{4b^2 - 9a}{2b}} = f(0) = e^3$

$\Rightarrow \dfrac{4b^2 - 9a}{2b} = 3$

$\Rightarrow 4b^2 - 6b - 9a = 0$

for real b, $D \geq 0 \Rightarrow a \geq -\dfrac{1}{4}$

28. **(5)**

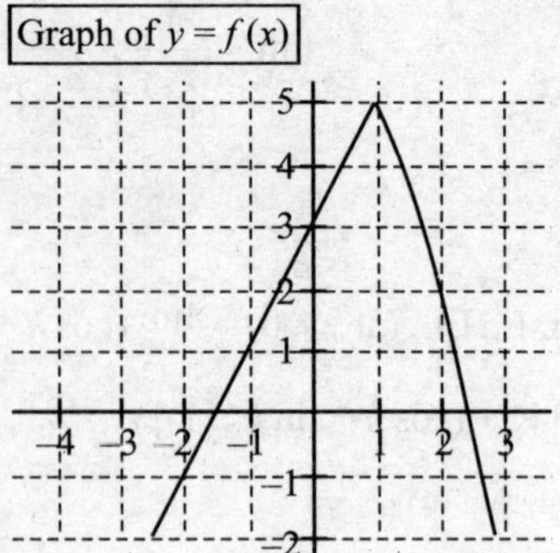

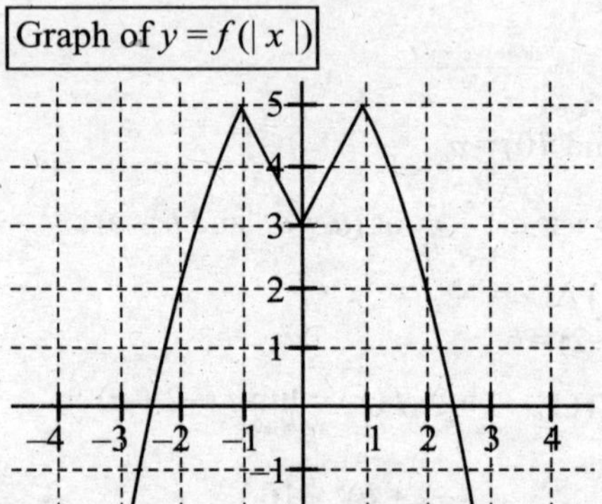

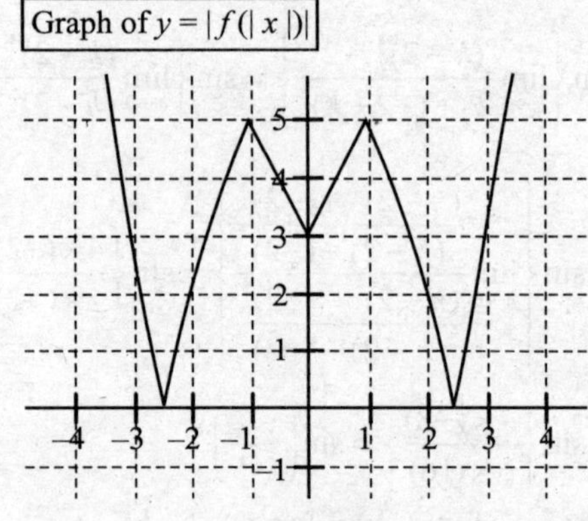

Fig. 3.28

CHAPTER 4

Methods of Differentiation

Objective Type

Each question has four choices a, b, c, and d, out of which *only one* is correct.

1. $\frac{dy}{dx}$ for $y = \tan^{-1}\left\{\sqrt{\frac{1+\cos x}{1-\cos x}}\right\}$, where $0 < x < 0\pi$, is

a. $-1/2$ b. 0
c. 1 d. -1

2. If $f(x) = |x^2 - 5x + 6|$, then $f'(x)$ equals

a. $2x - 5$ for $2 < x < 3$
b. $5 - 2x$ for $2 < x < 3$
c. $2x - 5$ for $2 \le x \le 3$
d. $5 - 2x$ for $2 \le x \le 3$

3. If $y = \tan^{-1}\left(\frac{\log(e/x^2)}{\log(ex^2)}\right) + \tan^{-1}\left(\frac{3+2\log x}{1-6\log x}\right)$, then $\frac{d^2y}{dx^2}$ is

a. 2 b. 1
c. 0 d. -1

4. If $f(0) = 0$, $f'(0) = 2$, then the derivative of $y = f(f(f(f(x)))$ at $x = 0$ is,

a. 2 b. 8
c. 16 d. 4

5. If $y = ax^{n+1} + bx^{-n}$, then $x^2\frac{d^2y}{dx^2}$ is equal to

a. $n(n-1)y$ b. $n(n+1)y$
c. ny d. n^2y

6. If $y = 1 + x + \frac{x^2}{2!} + \frac{x^3}{3!} + \cdots + \frac{x^n}{n!}$, then $\frac{dy}{dx}$ is equal to

a. y b. $y + \frac{x^n}{n!}$
c. $y - \frac{x^n}{n!}$ d. $y - 1 - \frac{x^n}{n!}$

7. If $y = a\sin x + b\cos x$, then $y^2 + \left(\frac{dy}{dx}\right)^2$ is a

a. function of x b. function of y
c. function of x and y d. constant

8. $\frac{d}{dx}\sqrt{\frac{1-\sin 2x}{1+\sin 2x}}$ is equal to

a. $\sec^2 x$ b. $-\sec^2\left(\frac{\pi}{4} - x\right)$
c. $\sec^2\left(\frac{\pi}{4} + x\right)$ d. $\sec^2\left(\frac{\pi}{4} - x\right)$

9. If $y = \left(x + \sqrt{x^2 + a^2}\right)^n$, then $\frac{dy}{dx}$ is

a. $\frac{ny}{\sqrt{x^2 + a^2}}$ b. $-\frac{ny}{\sqrt{x^2 + a^2}}$
c. $\frac{nx}{\sqrt{x^2 + a^2}}$ d. $-\frac{nx}{\sqrt{x^2 + a^2}}$

10. If $f(x)=\sqrt{1+\cos^2(x^2)}$, then $f'\left(\frac{\sqrt{\pi}}{2}\right)$ is

a. $\sqrt{\pi}/6$ b. $-\sqrt{(\pi/6)}$
c. $1/\sqrt{6}$ d. $\pi/\sqrt{6}$

11. $\frac{d}{dx}\cos^{-1}\sqrt{\cos x}$ is equal to

a. $\frac{1}{2}\sqrt{1+\sec x}$ b. $\sqrt{1+\sec x}$
c. $-\frac{1}{2}\sqrt{1+\sec x}$ d. $-\sqrt{1+\sec x}$

12. If $y=\log_{\sin x}(\tan x)$, then $\left(\frac{dy}{dx}\right)_{\pi/4}$ is equal to

a. $\frac{4}{\log 2}$ b. $-4\log 2$
c. $\frac{-4}{\log 2}$ d. None of these

13. If $y=\frac{\sin^{-1}x}{\sqrt{1-x^2}}$, then $(1-x^2)\frac{dy}{dx}$ is equal to

a. $x+y$ b. $1+xy$
c. $1-xy$ d. $xy-2$

14. If $y=\cot^{-1}\left[\frac{\sqrt{1+\sin x}+\sqrt{1-\sin x}}{\sqrt{1+\sin x}-\sqrt{1-\sin x}}\right]$, then $\frac{dy}{dx}=$

a. $\frac{1}{2}$ b. $\frac{2}{3}$
c. 3 d. 1

15. If $y=x(x^x)$, then $\frac{dy}{dx}$ is

a. $y[x^x(\log ex)\log x+x^x]$
b. $y[x^x(\log ex)\log x+x]$
c. $y[x^x(\log ex)\log x+x^{x-1}]$
d. $y[x^x(\log_e x)\log x+x^{x-1}]$

16. $\frac{d}{dx}\left[\sin^2\cot^{-1}\left\{\sqrt{\frac{1-x}{1+x}}\right\}\right]$ is equal to

a. -1 b. $\frac{1}{2}$
c. $-\frac{1}{2}$ d. 1

17. If $y=ae^{mx}+be^{-mx}$, then $\frac{d^2y}{dx^2}-m^2y$ is equal to

a. $m^2(ae^{mx}-be^{-mx})$ b. 1
c. 0 d. None of these

18. $\frac{d^n}{dx^n}(\log x)=$

a. $\frac{(n-1)!}{x^n}$ b. $\frac{n!}{x^n}$
c. $\frac{(n-2)!}{x^n}$ d. $(-1)^{n-1}\frac{(n-1)!}{x^n}$

19. If $y=\sqrt{\log x+\sqrt{\log x+\sqrt{\log x+\cdots\infty}}}$, then $\frac{dy}{dx}$ is

a. $\frac{x}{2y-1}$ b. $\frac{x}{2y+1}$
c. $\frac{1}{x(2y-1)}$ d. $\frac{1}{x(1-2y)}$

20. The differential coefficient of $f(\log_e x)$ with respect to x, where $f(x)=\log_e x$, is

a. $\frac{x}{\log_e x}$ b. $\frac{1}{x}\log_e x$
c. $\frac{1}{x\log_e x}$ d. None of these

21. If $y=\sec(\tan^{-1}x)$, then $\frac{dy}{dx}$ at $x=1$ is

a. $\frac{1}{\sqrt{2}}$ b. $-\frac{1}{\sqrt{2}}$
c. 1 d. None of these

22. If $f'(x)=\sqrt{2x^2-1}$ and $y=f(x^2)$, then $\frac{dy}{dx}$ at $x=1$ is

a. 2 b. 1
c. -2 d. None of these

23. If $u=f(x^3)$, $v=g(x^2)$, $f'(x)=\cos x$, and $g'(x)=\sin x$, then $\frac{du}{dv}$ is

a. $\frac{3}{2}x\cos x^3\operatorname{cosec} x^2$ b. $\frac{2}{3}\sin x^3\sec x^2$
c. $\tan x$ d. None of these

24. $x = t\cos t$, $y = t + \sin t$, then $\frac{d^2x}{dy^2}$ at $t = \frac{\pi}{2}$ is

a. $\frac{\pi+4}{2}$ b. $-\frac{\pi+4}{2}$

c. -2 d. None of these

25. If $f(x) = \sqrt{1-\sin 2x}$, then $f'(x)$ is equal to

a. $-(\cos x + \sin x)$, for $x \in (\pi/4, \pi/2)$

b. $\cos x + \sin x$, for $x \in (0, \pi/4)$

c. $-(\cos x + \sin x)$, for $x \in (0, \pi/4)$

d. $\cos x - \sin x$, for $x \in (\pi/4, \pi/2)$

26. If $y = x - x^2$, then the derivative of y^2 with respect to x^2 is

a. $1-2x$ b. $2-4x$

c. $3x-2x^2$ d. $1-3x+2x^2$

27. The first derivative of the function

$\left[\cos^{-1}\left(\sin\sqrt{\frac{1+x}{2}}\right)+x^x\right]$ with respect to x at $x = 1$ is

a. 3/4 b. 0

c. 1/2 d. $-1/2$

28. If $y = \sin px$ and y_n is the nth derivative of y, then

$\begin{vmatrix} y & y_1 & y_2 \\ y_3 & y_4 & y_5 \\ y_6 & y_7 & y_8 \end{vmatrix}$ is

a. 1 b. 0

c. -1 d. None of these

29. A function f, defined for all positive real numbers, satisfies the equation $f(x^2) = x^3$ for every $x > 0$. Then the value of $f'(4) =$

a. 12 b. 3

c. 3/2 d. Cannot be determined

30. Suppose $f(x) = e^{ax} + e^{bx}$, where $a \neq b$, and that $f''(x) - 2f'(x) - 15f(x) = 0$ for all x. Then the product ab is

a. 25 b. 9

c. -15 d. -9

31. If $y = \frac{(a-x)\sqrt{a-x}-(b-x)\sqrt{x-b}}{\sqrt{a-x}+\sqrt{x-b}}$, then $\frac{dy}{dx}$ wherever it is defined is

a. $\frac{x+(a+b)}{\sqrt{(a-x)(x-b)}}$ b. $\frac{2x-a-b}{2\sqrt{a-x}\sqrt{x-b}}$

c. $-\frac{(a+b)}{2\sqrt{(a-x)(x-b)}}$ d. $\frac{2x+(a+b)}{2\sqrt{(a-x)(x-b)}}$

32. The function $f(x) = e^x + x$, being differentiable and one to one, has a differentiable inverse $f^{-1}(x)$. The value of $\frac{d}{dx}(f^{-1})$ at the point $f(\log 2)$ is

a. $\frac{1}{\ln 2}$ b. $\frac{1}{3}$

c. $\frac{1}{4}$ d. None of these

33. Let $h(x)$ be differentiable for all x and let $f(x) = (kx + e^x)\,h(x)$, where k is some constant. If $h(0) = 5$, $h'(0) = -2$ and $f'(0) = 18$, then the value of k is

a. 5 b. 4

c. 3 d. 2.2

34. If $y = \tan^{-1}\left(\frac{2^x}{1+2^{2x+1}}\right)$, then $\frac{dy}{dx}$ at $x = 0$ is

a. 1 b. 2

c. ln 2 d. None of these

35. The nth derivative of the function $f(x) = \frac{1}{1-x^2}$ (where $x \in (-1, 1)$) at the point $x = 0$ where n is even is

a. 0 b. $n!$

c. ${}^nC_2$ d. 2^nC_2

36. $\frac{d^{20}y}{dx^{20}}(\cos 2x \cos 3x)$ is equal to

a. $2^{20}(\cos 2x - 2^{20}\cos 3x)$

b. $2^{20}(\cos 2x + 2^{20}\cos 4x)$

c. $2^{20}(\sin 2x + 2^{20}\sin 4x)$

d. $2^{20}(\sin 2x - 2^{20}\sin 4x)$

37. If $y = \sqrt{\frac{1-x}{1+x}}$, prove that $(1-x^2)\frac{dy}{dx}$ is equal to

a. y^2 b. $1/y$

c. $-y$ d. $-y/x$

38. The derivative of $y = (1-x)(2-x)\ldots(n-x)$ at $x = 1$ is

a. 0 b. $(-1)(n-1)!$

c. $n! - 1$ d. $(-1)^{n-1}(n-1)!$

39. If $y = \cos^{-1}\left(\frac{5\cos x - 12\sin x}{13}\right)$, where $x \in \left(0, \frac{\pi}{2}\right)$, then $\frac{dy}{dx}$ is

a. 1 b. -1

c. 0 d. None of these

40. If $y = \tan^{-1}\sqrt{\dfrac{x+1}{x-1}}$, then $\dfrac{dy}{dx}$ is

a. $\dfrac{-1}{2|x|\sqrt{x^2-1}}$ b. $\dfrac{-1}{2x\sqrt{x^2-1}}$

c. $\dfrac{1}{2x\sqrt{x^2-1}}$ d. None of these

41. If $\sin^{-1}\left(\dfrac{x^2-y^2}{x^2+y^2}\right) = \log a$, then $\dfrac{dy}{dx}$ is equal to

a. $\dfrac{x}{y}$ b. $\dfrac{y}{x^2}$

c. $\dfrac{x^2-y^2}{x^2+y^2}$ d. $\dfrac{y}{x}$

42. If $y = \cos^{-1}(\cos x)$, then $\dfrac{dy}{dx}$ at $x = \dfrac{5\pi}{4}$ is

a. 1 b. -1

c. $\dfrac{1}{\sqrt{2}}$ d. None of these

43. If $e^x = \dfrac{\sqrt{1+t}-\sqrt{1-t}}{\sqrt{1+t}+\sqrt{1-t}}$ and $\tan\dfrac{y}{2} = \sqrt{\dfrac{1-t}{1+t}}$, then $\dfrac{dy}{dx}$ at $t = \dfrac{1}{2}$ is

a. $-\dfrac{1}{2}$ b. $\dfrac{1}{2}$

c. 0 d. None of these

44. If $x^2 + y^2 = t - \dfrac{1}{t}$ and $x^4 + y^4 = t^2 + \dfrac{1}{t^2}$, then $x^3 y \dfrac{dy}{dx} =$

a. 0 b. 1

c. -1 d. None of these

45. If $y^{1/m} = (x + \sqrt{1+x^2})$, then $(1 + x^2)y_2 + xy_1$ is (where y_r represents rth derivative of y w.r.t. x)

a. m^2y b. my^2

c. m^2y^2 d. None of these

46. Suppose the function $f(x) - f(2x)$ has the derivative 5 at $x = 1$ and derivative 7 at $x = 2$. The derivative of the function $f(x) - f(4x)$ at $x = 1$ has the value equal to

a. 19 b. 9

c. 17 d. 14

47. If $f(x) = \sin^{-1}\cos x$, then the value of $f(10) + f'(10)$ is

a. $11 - \dfrac{7\pi}{2}$ b. $\dfrac{7\pi}{2} - 11$

c. $\dfrac{5\pi}{2} - 11$ d. None of these

48. If $(\sin x)(\cos y) = 1/2$, then d^2y/dx^2 at $(\pi/4, \pi/4)$ is

a. -4 b. -2

c. -6 d. 0

49. A function f satisfies the condition, $f(x) = f'(x) + f''(x) + f'''(x) + \cdots$ where $f(x)$ is a differentiable function indefinitely and dash denotes the order of derivative. If $f(0) = 1$, then $f(x)$ is

a. $e^{x/2}$ b. e^x

c. e^{2x} d. e^{4x}

50. If $\sqrt{1-x^2} + \sqrt{1-y^2} = a(x-y)$, then $\dfrac{dy}{dx}$ is equal to

a. $\sqrt{\dfrac{1-x^2}{1-y^2}}$ b. $\sqrt{\dfrac{1-y^2}{1-x^2}}$

c. $\sqrt{\dfrac{x^2-1}{1-y^2}}$ d. $\sqrt{\dfrac{y^2-1}{1-x^2}}$

51. If $y = x^2 + \dfrac{1}{x^2 + \dfrac{1}{x^2 + \dfrac{1}{x^2 + \cdots\infty}}}$, then $\dfrac{dy}{dx}$ is

a. $\dfrac{2xy}{2y - x^2}$ b. $\dfrac{xy}{y + x^2}$

c. $\dfrac{xy}{y - x^2}$ d. $\dfrac{2xy}{2 + \dfrac{x^2}{y}}$

52. $\dfrac{d}{dx}\left[\tan^{-1}\left(\dfrac{\sqrt{x}(3-x)}{1-3x}\right)\right] =$

a. $\dfrac{1}{2(1+x)\sqrt{x}}$ b. $\dfrac{3}{(1+x)\sqrt{x}}$

c. $\dfrac{2}{(1+x)\sqrt{x}}$ d. $\dfrac{3}{2(1+x)\sqrt{x}}$

53. Let $g(x)$ be the inverse of an invertible function $f(x)$ which is differentiable at $x = c$, then $g'(f(c))$ equals

a. $f'(c)$ b. $\dfrac{1}{f'(c)}$

c. $f(c)$ d. None of these

54. If $f(x) = x + \tan x$ and f is inverse of g, then $g'(x)$ equals

a. $\frac{1}{1+[g(x)-x]^2}$
b. $\frac{1}{2-[g(x)-x]^2}$
c. $\frac{1}{2+[g(x)-x]^2}$
d. None of these

55. If $y\sqrt{x^2+1}=\log(\sqrt{x^2+1}-x)$, then $(x^2+1)\frac{dy}{dx}+xy+1=$
a. 0
b. 1
c. 2
d. None of these

56. If $y=\frac{\sqrt{a+x}-\sqrt{a-x}}{\sqrt{a+x}+\sqrt{a-x}}$, then $\frac{dy}{dx}$ is equal to
a. $\frac{ay}{x\sqrt{a^2-x^2}}$
b. $\frac{ay}{\sqrt{a^2-x^2}}$
c. $\frac{ay}{x\sqrt{x^2-a^2}}$
d. None of these

57. If $f(x) = x^4\tan(x^3) - x\ln(1+x^2)$, then the value of $\frac{d^4(f(x))}{dx^4}$ at $x=0$ is
a. 0
b. 6
c. 12
d. 24

58. Let $g(x)$ be the inverse of an invertible function $f(x)$, which is differentiable for all real x, then $g''(f''(x))$ equals
a. $-\frac{f''(x)}{(f'(x))^3}$
b. $\frac{f'(x)f''(x)-(f'(x))^3}{f'(x)}$
c. $\frac{f'(x)f''(x)-(f'(x))^2}{(f'(x))^2}$
d. None of these

59. If $f(x)=|\log_e|x||$, then $f'(x)$ equals
a. $\frac{1}{|x|}$, where $x\neq 0$
b. $\frac{1}{x}$ for $|x|>1$ and $-\frac{1}{x}$ for $|x|<1$
c. $-\frac{1}{x}$ for $|x|>1$ and $\frac{1}{x}$ for $|x|<1$
d. $\frac{1}{x}$ for $x>0$ and $-\frac{1}{x}$ for $x<0$

60. If $y=|\cos x|+|\sin x|$, then $\frac{dy}{dx}$ at $x=\frac{2\pi}{3}$ is
a. $\frac{1-\sqrt{3}}{2}$
b. 0
c. $\frac{1}{2}(\sqrt{3}-1)$
d. None of these

61. If g is the inverse function of f and $f'(x)=\sin x$, then $g'(x)$ is
a. $\operatorname{cosec}\{g(x)\}$
b. $\sin\{g(x)\}$
c. $-\frac{1}{\sin\{g(x)\}}$
d. None of these

62. If $x=\phi(t), y=\psi(t)$, then $\frac{d^2y}{dx^2}$ is
a. $\frac{\phi'\psi''-\psi'\phi''}{(\phi')^2}$
b. $\frac{\phi'\psi''-\psi'\phi''}{(\phi')^3}$
c. $\frac{\phi''}{\psi''}$
d. $\frac{\psi''}{\phi''}$

63. $f(x)=e^x-e^{-x}-2\sin x-\frac{2}{3}x^3$, then the least value of n for which $\left.\frac{d^n}{dx^n}f(x)\right|_{x=0}$ is non-zero is
a. 5
b. 6
c. 7
d. 8

64. If $f(x)$ satisfies the relation $f\left(\frac{5x-3y}{2}\right)=\frac{5f(x)-3f(y)}{2}$ $\forall x, y \in R$, and $f(0)=3$ and $f'(0)=2$, then the period of $\sin(f(x))$ is
a. 2π
b. π
c. 3π
d. 4π

65. Instead of the usual definition of derivative $Df(x)$, if we define a new kind of derivative, $D^*F(x)$ by the formula $D^*(x) = \lim_{h\to 0}\frac{f^2(x+h)-f^2(x)}{h}$, where $f^2(x)$ means $[f(x)]^2$. If $f(x)=x\log x$, then $D^*f(x)|_{x=e}$ has the value
a. e
b. $2e$
c. $4e$
d. None of these

66. If $f(x)=2\sin^{-1}\sqrt{1-x}+\sin^{-1}\left(2\sqrt{x(1-x)}\right)$, where $x\in\left(0,\frac{1}{2}\right)$, then $f'(x)$ is
a. $\frac{2}{\sqrt{x(1-x)}}$
b. zero
c. $-\frac{2}{\sqrt{x(1-x)}}$
d. π

67. If $f''(x) = -f(x)$ and $g(x) = f'(x)$ and

$F(x) = \left(f\left(\frac{x}{2}\right)\right)^2 + \left(g\left(\frac{x}{2}\right)\right)^2$ and given that $F(5) = 5$, then $F(10)$ is

a. 5 b. 10
c. 0 d. 15

68. The derivative of $\tan^{-1}\left(\frac{\sqrt{1+x^2}-1}{x}\right)$ with respect to $\tan^{-1}\left(\frac{2x\sqrt{1-x^2}}{1-2x^2}\right)$ at $x = 0$ is

a. 1/8 b. 1/4
c. 1/2 d. 1

69. The nth derivative of xe^x vanishes when

a. $x = 0$ b. $x = -1$
c. $x = -n$ d. $x = n$

70. If $y^2 = ax^2 + bx + c$, then $y^3 \frac{d^2y}{dx^2}$ is

a. a constant b. a function of x only
c. a function of y only d. a function of x and y

71. If $y = \sin x + e^x$, then $\frac{d^2x}{dy^2} =$

a. $(-\sin x + e^x)^{-1}$ b. $\frac{\sin x - e^x}{(\cos x + e^x)^2}$

c. $\frac{\sin x - e^x}{(\cos x + e^x)^3}$ d. $\frac{\sin x + e^x}{(\cos x + e^x)^3}$

72. If $u = x^2 + y^2$ and $x = s + 3t$, $y = 2s - t$, then $\frac{d^2u}{ds^2}$ equals to

a. 12 b. 32
c. 36 d. 10

73. Let $y = t^{10} + 1$ and $x = t^8 + 1$, then $\frac{d^2y}{dx^2}$ is

a. $\frac{5}{2}t$ b. $20t^8$

c. $\frac{5}{16t^6}$ d. None of these

74. If $y = x\log\left(\frac{x}{a+bx}\right)$, then $x^3\frac{d^2y}{dx^2}$ equals to

a. $x\frac{dy}{dx} - y$ b. $\left(x\frac{dy}{dx} - y\right)^2$

c. $y\frac{dy}{dx} - x$ d. $\left(y\frac{dy}{dx} - x\right)^2$

75. Let $u(x)$ and $v(x)$ be differentiable functions such that $\frac{u(x)}{v(x)} = 7$. If $\frac{u'(x)}{v'(x)} = p$ and $\left(\frac{u(x)}{v(x)}\right)' = q$, then $\frac{p+q}{p-q}$ has the value equal to

a. 1 b. 0
c. 7 d. −7

76. If $ax^2 + 2hxy + by^2 = 1$, then $\frac{d^2y}{dx^2}$ is

a. $\frac{h^2 - ab}{(hx + by)^2}$ b. $\frac{ab - h^2}{(hx + by)^2}$

c. $\frac{h^2 + ab}{(hx + by)^2}$ d. None of these

Multiple Correct Answers Type

Each question has four choices a, b, c, and d, out of which *one* or *more* answers are correct.

1. If $y = e^{\sqrt{x}} + e^{-\sqrt{x}}$, then $\frac{dy}{dx}$ is equal to

a. $\frac{e^{\sqrt{x}} - e^{-\sqrt{x}}}{2\sqrt{x}}$ b. $\frac{e^{\sqrt{x}} - e^{-\sqrt{x}}}{2x}$

c. $\frac{1}{2\sqrt{x}}\sqrt{y^2 - 4}$ d. $\frac{1}{2\sqrt{x}}\sqrt{y^2 + 4}$

2. Let $y = \sqrt{x + \sqrt{x + \sqrt{x + \cdots\infty}}}$, then $\frac{dy}{dx}$ is equals to

a. $\frac{1}{2y - 1}$ b. $\frac{x}{x + 2y}$

c. $\frac{1}{\sqrt{1 + 4x}}$ d. $\frac{y}{2x + y}$

3. If 1 is a twice repeated root of the equation $ax^3 + bx^2 + bx + d = 0$, then

a. $a = b = d$ b. $a + b = 0$
c. $b + d = 0$ d. $a = d$

4. If $x^3 - 2x^2y^2 + 5x + y - 5 = 0$ and $y(1) = 1$, then

a. $y'(1) = 4/3$ b. $y''(1) = -4/3$

c. $y''(1) = -8\frac{22}{27}$ d. $y'(1) = 2/3$

5. $f(x) = |x^2 - 3|x| + 2|$, then which of the following is/are true

a. $f'(x) = 2x - 3$ for $x \in (0, 1) \cup (2, \infty)$
b. $f'(x) = 2x + 3$ for $x \in (-\infty, -2) \cup (-1, 0)$

c. $f'(x)=-2x-3$ for $x \in (-2,-1)$

d. None of these

6. If $y=\dfrac{x^4-x^2+1}{x^2+\sqrt{3}x+1}$ and $\dfrac{dy}{dx}=ax+b$, then the value of $a-b$ is

a. $\cot\dfrac{\pi}{8}$ **b.** $\cot\dfrac{5\pi}{12}$

c. $\tan\dfrac{5\pi}{12}$ **d.** $\tan\dfrac{5\pi}{8}$

7. Let $f(x)=\dfrac{\sqrt{x-2\sqrt{x-1}}}{\sqrt{x-1}-1}x$, then

a. $f'(10)=1$

b. $f'(3/2)=-1$

c. Domain of $f(x)$ is $x\ge 1$

d. Range of $f(x)$ is $(-2,-1]\cup(2,\infty)$

8. If $y=x^{(\log x)^{\log(\log x)}}$, then $\dfrac{dy}{dx}$ is

a. $\dfrac{y}{x}((\ln x^{\infty x-1})+2\ln x\ln(\ln x))$

b. $\dfrac{y}{x}(\log x)^{\log(\log x)}(2\log(\log x)+1)$

c. $\dfrac{y}{x\ln x}[(\ln x)^2+2\ln(\ln x)]$

d. $\dfrac{y}{x}\dfrac{\log y}{\log x}\left[2\log(\log x)+1\right]$

9. Which of the following is/are true?

a. $\dfrac{dy}{dx}$ for $y=\sin^{-1}(\cos x)$, where $x\in(0,\pi)$, is -1

b. $\dfrac{dy}{dx}$ for $y=\sin^{-1}(\cos x)$, where $x\in(\pi,2\pi)$, is 1

c. $\dfrac{dy}{dx}$ for $y=\cos^{-1}(\sin x)$, where $x\in\left(-\dfrac{\pi}{2},\dfrac{\pi}{2}\right)$, is -1

d. $\dfrac{dy}{dx}$ for $y=\cos^{-1}(\sin x)$, where $x\in\left(\dfrac{\pi}{2},\dfrac{3\pi}{2}\right)$, is -1

10. If $f(x-y), f(x)\ f(y)$ and $f(x+y)$ are in A.P. for all x, y, and $f(0)\neq 0$, then

a. $f(4)=f(-4)$ **b.** $f(2)+f(-2)=0$

c. $f'(4)+f'(4)=0$ **d.** $f'(2)=f'(-2)$

11. If $y=\cos^{-1}\left(\dfrac{2x}{1+x^2}\right)$, then $\dfrac{dy}{dx}$ is

a. $\dfrac{-2}{1+x^2}$ for all x **b.** $\dfrac{-2}{1+x^2}$ for all $|x|<1$

c. $\dfrac{2}{1+x^2}$ for $|x|>1$ **d.** None of these

12. $f: R^+\to R$ be a continuous function satisfying $f\left(\dfrac{x}{y}\right)=f(x)-f(y)\ \forall\ x,y\in R^+$. If $f'(1)=1$, then

a. 'f' is unbounded **b.** $\lim\limits_{x\to 0} f\left(\dfrac{1}{x}\right)=0$

c. $\lim\limits_{x\to 0}\dfrac{f(1+x)}{x}=1$ **d.** $\lim\limits_{x\to 0} x\,.\,f(x)=0$

13. If $f_n(x)=e^{f_{n-1}(x)}$ for all $n\in N$ and $f_0(x)=x$, then $\dfrac{d}{dx}\{f_n(x)\}$ is

a. $f_n(x)\dfrac{d}{dx}\{f_{n-1}(x)\}$

b. $f_n(x)\,f_{n-1}(x)$

c. $f_n(x)\,f_{n-1}(x)\ldots f_2(x).f_1(x)$

d. None of these

Reasoning Type

Each question has four choices a, b, c, and d, out of which *only one* is correct. Each question contains Statement 1 and Statement 2.

a. if both the statements are True and Statement 2 is the correct explanation of Statement 1

b. if both the statements are True but Statement 2 is Not the correct explanation of Statement 1

c. if Statement 1 is True and Statement 2 is False.

d. if Statement 1 is False and Statement 2 is True.

1. Statement 1: Let $f(x)=x[x]$ and [.] denotes greatest integral function, when x is not an integral, then rule for $f'(x)$ is given by $[x]$

Statement 2: $f'(x)$ does not exist for any $x\in$ integer.

2. Statement 1: If $f(x)$ is an odd function, then $f'(x)$ is an even function.

Statement 2: If $f'(x)$ is an even function, then $f(x)$ is an odd function.

3. Statement 1: Let $f: R\to R$ is a real-valued function $\forall\ x,y\in R$ such that $|f(x)-f(y)|\le|x-y|^3$, then $f(x)$ is a constant function.

Statement 2: If derivative of the function w.r.t. x is zero, then function is constant.

4. **Statement 1:** For $f(x) = \sin x, f'(\pi) = f'(3\pi)$.
   Statement 2: For $f(x) = \sin x, f(\pi) = f(3\pi)$.

5. **Statement 1:** If differentiable function $f(x)$ satisfies the relation $f(x) + f(x-2) = 0\ \forall\ x \in R$, and if $\left(\frac{d}{dx} f(x)\right)_{x=a} = b$, then $\left(\frac{d}{dx} f(x)\right)_{a+4000} = b$.
   Statement 2: $f(x)$ is a periodic function with period 4.

6. If for some differentiable function $f(\alpha) = 0$ and $f'(\alpha) = 0$,
   Statement 1: Then sign of $f(x)$ does not change in the neighborhood of $x = \alpha$.
   Statement 2: α is repeated root of $f(x) = 0$.

7. Consider function $f(x)$ satisfies the relation, $f(x + y^3) = f(x) + f(y^3),\ \forall\ x, y \in R$ and differentiable for all x.
   Statement 1: If $f'(2) = a$, then $f'(-2) = a$.
   Statement 2: $f(x)$ is an odd function.

Linked Comprehension Type

Based upon each paragraph, three multiple choice questions have to be answered. Each question has four choices a, b, c, and d, out of which *only one* is correct.

For Problems 1–3

$f(x)$ is a polynomial function $f: R \to R$ such that $f(2x) = f'(x)f''(x)$.

1. The value of $f(3)$ is
   a. 4 b. 12
   c. 15 d. None of these

2. $f(x)$ is
   a. one-one and onto
   b. one-one and into
   c. many-one and onto
   d. many-one and into

3. Equation $f(x) = x$ has
   a. three real and positive roots
   b. three real and negative roots
   c. one real roots
   d. three real roots such that sum of roots is zero

For Problems 4 – 6

$f: R \to R, f(x) = x^3 + x^2 f'(1) + x f''(2) + f'''(3)$ for all $x \in R$.

4. The value of $f(1)$ is
   a. 2 b. 3
   c. -1 d. 4

5. $f(x)$ is
   a. one-one and onto
   b. one-one and into
   c. many-one and onto
   d. many-one and into.

6. The value of $f'(1) + f''(2) + f'''(3)$ is
   a. 0 b. -1
   c. 2 d. 3

For Problems 7–9

Repeated roots : If equation $f(x) = 0$, where $f(x)$ is a polynomial function, and if it has roots $\alpha, \alpha, \beta, \ldots$ or α root is repeated root, then $f(x) = 0$ is equivalent to $(x-\alpha)^2(x-\beta)\ldots = 0$, from which we can conclude that $f'(x) = 0$ or $2(x-\alpha)[(x-\beta)\ldots] + (x-\alpha)^2[(x-\beta)\ldots]' = 0$ or $(x-\alpha)[2\{(x-\beta)\ldots\} + (x-\alpha)\{(x-\beta)\ldots\}'] = 0$ has root α.

Thus, if α root occurs twice in equation, then it is common in equations $f(x) = 0$ and $f'(x) = 0$.

Similarly, if α root occurs thrice in equation, then it is common in the equations $f(x) = 0, f'(x) = 0$ and $f''(x) = 0$.

7. If $x - c$ is a factor of order m of the polynomial $f(x)$ of degree $n (1 < m < n)$, then $x = c$ is a root of the polynomial (where $f^r(x)$ represent r^{th} derivative of $f(x)$ w.r.t. x)
   a. $f^m(x)$ b. $f^{m-1}(x)$
   c. $f^n(x)$ d. None of these

8. If $a_1x^3 + b_1x^2 + c_1x + d_1 = 0$ and $a_2x^3 + b_2x^2 + c_2x + d_2 = 0$ have a pair of repeated roots common, prove that
$$\begin{vmatrix} 3a_1 & 2b_1 & c_1 \\ 3a_2 & 2b_2 & c_2 \\ a_2b_1 - a_1b_2 & c_1a_2 - c_2a_1 & d_1a_2 - d_2a_1 \end{vmatrix} = 0.$$

9. If α root occurs p times and β root occurs q times in polynomial equation $f(x) = 0$ or n degree$(1 < p, q < n)$, then which of the following is not true. (where $f^r(x)$ represents r^{th} derivative of $f(x)$ w.r.t. x)
   a. if $p < q < n$, then α and β are two of the roots of the equation $f^{p-1}(x) = 0$
   b. if $q < p < n$, then α and β are two of the roots of the equation $f^{q-1}(x) = 0$
   c. If $p < q < n$, then equations $f(x) = 0$ and $f^p(x) = 0$ have exactly one root common
   d. If $q < p < n$, then equations $f^q(x) = 0$ and $f^p(x) = 0$ have exactly two roots common

For Problems 10–12

Equation $x^n - 1 = 0, n > 1, n \in N$, has roots $1, a_1, a_2, \ldots, a_n$

10. The value of $(1 - a_1)(1 - a_2)\ldots(1 - a_n)$ is

a. $n^2/2$ b. n
c. $(-1)^n n$ d. None of these

11. The value of $\sum_{r=2}^{n} \frac{1}{2-a_r}$ is

a. $\frac{2^{n-1}(n-2)+1}{2^n-1}$

b. $\frac{2^n(n-2)+1}{2^n-1}$

c. $\frac{2^{n-1}(n-1)-1}{2^n-1}$

d. None of these

12. The value of $\sum_{r=2}^{n} \frac{1}{1-a_r}$ is

a. $\frac{n}{4}$ b. $\frac{n(n-1)}{2}$

c. $\frac{n-1}{2}$ d. None of these

For Problems 13–15

$f(x)=x^2+xg'(1)+g''(2)$ and $g(x)=f(1)x^2+xf'(x)+f''(x)$

13. The value of $f(3)$ is

a. 1 b. 0
c. -1 d. -2

14. The value of $g(0)$ is

a. 0 b. -3
c. 2 d. None of these

15. The domain of the function $\sqrt{\frac{f(x)}{g(x)}}$ is

a. $(-\infty, 1] \cup (2, 3]$
b. $(-2, 0] \cup (1, \infty)$
c. $(-\infty, 0] \cup (2/3, 3]$
d. None of these

Matrix-Match Type

Each question contains statements given in two columns which have to be matched. Statements a, b, c, d, in column I have to be matched with statements p, q, r, s in column II. If the correct match are a → p, a → s, b → r, c → p, c → q and d → s, then the correctly bubbled 4 × 4 matrix should be as follows

	p	q	r	s
a	p	q	r	s
b	p	q	r	s
c	p	q	r	s
d	p	q	r	s

1.

Column I	Column II
a. Differentiable function $f(x)$ satisfies the relation $f(1-x)=f(1+x)$ for all $x \in R$	p. Graph of $f'(x)$ is symmetrical about point (1, 0)
b. Differentiable function $f(x)$ satisfies the relation $f(2-x)+f(x)=0$ for all $x \in R$	q. Graph of $f'(x)$ is symmetrical about line $x=1$
c. Differentiable function $f(x)$ satisfies the relation $f(x+2)+f(x)=0$ for all $x \in R$	r. $f'(-1)=f'(3)$
d. Differentiable function $f(x)$ satisfies the relation $f(x)+f(y)+f(x).f(y)=1$ for all x, y and $f(x)>0$	s. $f'(x)$ has period 4

2.

Column I	Column II
a. $y=f(x)$ be given by $x=t^5-5t^3-20t+7$ and $y=4t^3-3t^2-18t+3$, then $-5 \times \frac{dy}{dx}$ at $t=1$	p. 0
b. $P(x)$ be a polynomial of degree 4, with $P(2)=-1, P'(2)=0, P''(2)=2, P'''(2)=-12$ and $P^{iv}(2)=24$, then $P''(3)$	q. -2
c. $y=\frac{1}{x}$, then $\frac{\frac{dy}{\sqrt{1+y^4}}}{\frac{dx}{\sqrt{1+x^4}}}$	r. 2
d. $f\left(\frac{2x+3y}{5}\right)=\frac{2f(x)+3f(y)}{5}$ and $f'(0)=p$ and $f(0)=q$, then $f''(0)$	s. -1

3.

Column I	Column II
a. $y=\sin^{-1}\left(\frac{2x}{1+x^2}\right)$, then $\frac{dy}{dx}=-\frac{2}{1+x^2}$	p. for $x<0$
b. $y=\cos^{-1}\left(\frac{1}{\sqrt{1+x^2}}\right)$, then $\frac{dy}{dx}=\frac{1}{1+x^2}$	q. for $x>1$
c. $y=\lvert e^{\lvert x\rvert}-e\rvert$, then $\frac{dy}{dx}>0$	r. for $x<-1$
d. $u=\log\lvert 2x\rvert$, $v=\lvert\tan^{-1}x\rvert$, then $du/dv>2$	s. for $-1<x<0$

Integer Type

1. Let $f(x)=(x-1)(x-2)(x-3)\ldots(x-n)$, $n\in N$ and $f'(n)=5040$, then the value of n is________.
2. Let $y=f(x)$, where f satisfies the relation $f(x+y)=2f(x)+xf(y)+y\sqrt{f(x)}\ \forall\ x,y\in R$ and $f'(0)=0$. Then $f(6)$ is equal to________.
3. Let $f(x)=x+\cfrac{1}{2x+\cfrac{1}{2x+\cfrac{1}{2x+\ldots\infty}}}$. Then the value of $(\sqrt{f(50)f'(50)})/2$ is equal to________.
4. If $f(x)=\cos^{-1}((1/\sqrt{13})(2\cos x-3\sin x))+\sin^{-1}((1/\sqrt{13})(2\cos x+3\sin x))$ w.r.t $\sqrt{1+x^2}$, then $[(dy/dx)_{x=3/4}]$, where [.] represents the greatest integer function is________.
5. Let $f: R\to R$ be a function satisfying the condition $f(x+y^3)=f(x)+[f(y)]^3$ for all $x,y\in R$. If $f'(0)\geq 0$, then one of the possible values of $f(10)$ is________.
6. Let f be a twice-differentiable function such that $f''(x)=-f(x)$ and $f'(x)=g(x)$, $h(x)=[f(x)]^2+[g(x)]^2$. If $h(5)=4$, then $h(10)$ is________.
7. The value of $(dy)/(dx)$ at $x=-1$, when $(\sin y)^{\sin\left(\frac{\pi x}{2}\right)}(\sqrt{3}/2)\sec^{-1}(2x)+2^x\tan[\log(x+2)]=0$ is________.
8. Let $F(x)=f(x)g(x)h(x)$ for all real x, where $f(x)$, $g(x)$, and $h(x)$ are differentiable functions. At some point x_0, $F'(x_0)=21F(x_0)$, $f'(x_0)=4f(x_0)$, $g'(x_0)=-7g(x_0)$ and $h'(x_0)=kh(x_0)$. Then $k/4$ is equal to________.
9. If $\lvert a_1\sin x+a_2\sin 2x+\ldots+a_n\sin nx\rvert\leq\lvert\sin x\rvert$ for $x\in R$, then the maximum value of $\lvert a_1+2a_1+3a_3+\ldots+na_n\rvert$ is________.
10. If $f[(x+y)/3]=[2+f(x)+f(y)]/3$ for all real x and y and $f'(2)=2$, then $f(3)$ is equal to________.
11. Let $f(x)$ and $g(x)$ be two functions having finite non-zero third order derivatives $f'''(x)$ and $g''(x)$ for all $x\in R$. If

    $f(x)\,g(x)=1$ for all $x\in R$, then the value of $[(f'''/f')-(g'''/g')]/[(f''/f)-(g''/g)]$ is________.
12. The square of the reciprocal of the derivative of $f(\tan x)$ w.r.t. $g(\sec x)$ at $x=\pi/4$, where $f'(1)=2$ and $g'(\sqrt{2})=4$ is________.
13. If $y=(ax+b)/(x^2+c)$, then the value of $[(2xy'+y)y''']/[(xy''+y')y'']$, where a, b, c are constants and dashes denote differentiation w.r.t. x, is________.
14. Let $f: R\to R$ satisfying $\lvert f(x)\rvert\leq x^2$, $\forall\ x\in R$ be differentiable at $x=0$, then the value of $f'(0)$ is________.
15. If $f(x)+f(y)=f[(x+y)/(1-xy)]$ for all $x,y\in R\ (xy\neq 1)$ and $\lim_{x\to 0}[f(x)/x]=2$. Then $f'(1)$ is________.
16. Let $f(x+y)=f(x)f(y)$ for all x and y. Suppose $f(5)=2$ and $f'(0)=3$, then $f'(5)$ is________.
17. If $y=\begin{vmatrix}\sin x & \cos x & \sin x\\ \cos x & -\sin x & \cos x\\ x & 1 & 1\end{vmatrix}$, then $\frac{dy}{dx}$ is________.
18. If $f(x)=\begin{vmatrix}x^n & n! & 2\\ \cos x & \cos\frac{n\pi}{2} & 4\\ \sin x & \sin\frac{n\pi}{2} & 8\end{vmatrix}$, then the value of $\frac{d^n}{dx^n}$

    $[f(x)]_{x=0}$ is________.
19. Let $f(xy)=f(x)f(y)$, $\forall\ x,y\in R$ and f be differentiable at $x=1$ such that $f'(1)=1$ and $f(1)\neq 0$, $f(2)=3$, then find $f'(2)$.
20. If $x=\sqrt{a^{\sin^{-1}t}}$, $y=\sqrt{a^{\cos^{-1}t}}$, $a>0$ and $-1<t<1$, then the value of $xdy+ydx$ is________.
21. $(dy)/(dx)$ at $x=\pi/4$ for $x=a[\cos t+(1/2)\log\tan^2(t/2)]$ and $y=a\sin t$ is________.
22. The function f satisfies the relation $f(x)f'(-x)=f(-x)f'(x)$ for all x, and $f(0)=3$. Now if $f(3)=3$, then the value of $f(3)f(-3)$ is________.

23. If α is an integer satisfying $|\alpha| \le 4 - |[x]|$, where x is a real number for which $2x \tan^{-1} x$ is greater than or equal to $\ln(1 + x^2)$, then the number of maximum possible values of α (where [.] represents the greatest integer function) is ________.

24. Let $f(x)$ be a fourth differentiable function such that $f(2x^2 - 1) = 2xf(x)\ \forall\ x \in R$, then $f^{iv}(0)$ is equal to (where $f^{iv}(0)$ represents fourth derivative of $f(x)$ at $x = 0$)________.

25. If $g(x) = x^2 + \frac{1}{x^2}$ and $f(g(x)) = x^6 + \frac{1}{x^6}$, then $f'(2)$ equals________.

26. A function $f(x)$ satisfies the relation $f(x+y) = f(x) + f(y) + xy(x+y)\ \forall\ x, y \in R$. If $f'(0) = -1$, then $f'(3) = ?$

ANSWERS AND SOLUTIONS

Objective Type

1. a. $y = \tan^{-1}\left\{\sqrt{\dfrac{1+\cos x}{1-\cos x}}\right\}$

$= \tan^{-1}\left\{\sqrt{\dfrac{2\cos^2 x/2}{2\sin^2 x/2}}\right\}$

$= \tan^{-1}\left|\cot\dfrac{x}{2}\right| = \tan^{-1}\left(\cot\dfrac{x}{2}\right)$

$\Rightarrow y = \tan^{-1}\left\{\tan\left(\dfrac{\pi}{2}-\dfrac{x}{2}\right)\right\} = \dfrac{\pi}{2}-\dfrac{x}{2}$

$\therefore \dfrac{dy}{dx} = 0-\dfrac{1}{2} = -\dfrac{1}{2}$

2. b. $f(x) = |x^2-5x+6|$

$= \begin{cases} x^2-5x+6 & \text{if } x \ge 3 \text{ or } x \le 2 \\ -(x^2-5x+6), & \text{if } 2 < x < 3 \end{cases}$

$\Rightarrow f'(x) = \begin{cases} (2x-5), & \text{if } x \ge 3 \text{ or } x \le 2 \\ -(2x-5), & \text{if } 2 < x < 3 \end{cases}$

3. c. We have

$y = \tan^{-1}\left(\dfrac{\log e - \log x^2}{\log e + \log x^2}\right) + \tan^{-1}\left(\dfrac{3+2\log x}{1-6\log x}\right)$

$= \tan^{-1}\left(\dfrac{1-2\log x}{1+2\log x}\right) + \tan^{-1}\left(\dfrac{3+2\log x}{1-6\log x}\right)$

$= \tan^{-1} 1 - \tan^{-1}(2\log x) + \tan^{-1} 3$
$+ \tan^{-1}(2\log x)$
$= \tan^{-1} 1 + \tan^{-1} 3$

$\Rightarrow \dfrac{dy}{dx} = 0 \Rightarrow \dfrac{d^2y}{dx^2} = 0$

4.c. $y'(x) = f'(f(f(f(x))))f'(f(f(x)))f'(f(x))f'(x)$

$\Rightarrow y'(0) = f'(f(f(f(0))))f'(f(f(0)))f'(f(0))f'(0)$

$= f'(f(f(0)))f'(f(0))f'(0)f'(0)$

$= f'(f(0))f'(0)f'(0)f'(0)$

$= f'(0)f'(0)f'(0)f'(0)$

$= (f'(0))^4 = 2^4 = 16$

5.b. $y = ax^{n+1} + bx^{-n}$

$\Rightarrow \dfrac{dy}{dx} = (n+1)ax^n - nbx^{-n-1}$

$\Rightarrow \dfrac{d^2y}{dx^2} = n(n+1)ax^{n-1} + n(n+1)bx^{-n-2}$

$\Rightarrow x^2\dfrac{d^2y}{dx^2} = n(n+1)y$

6.b. $y = 1+x+\dfrac{x^2}{2!}+\dfrac{x^3}{3!}+\cdots+\dfrac{x^n}{n!}$

$\Rightarrow \dfrac{dy}{dx} = 0+1+x+\dfrac{x^2}{2!}+\cdots+\dfrac{x^{n-1}}{(n-1)!}$

$\Rightarrow \dfrac{dy}{dx}+\dfrac{x^n}{n!} = 1+x+\dfrac{x^2}{2!}+\cdots+\dfrac{x^n}{n!}$

$\Rightarrow \dfrac{dy}{dx} = y+\dfrac{x^n}{n!}$

7.d. $y = a\sin x + b\cos x$

Differentiating with respect to x, we get

$\dfrac{dy}{dx} = a\cos x - b\sin x$

Now, $\left(\dfrac{dy}{dx}\right)^2 = (a\cos x - b\sin x)^2$

$= a^2\cos^2 x + b^2\sin^2 x - 2ab\sin x\cos x$, and
$y^2 = (a\sin x + b\cos x)^2$
$= a^2\sin^2 x + b^2\cos^2 x + 2ab\sin x\cos x$

So,

$\left(\dfrac{dy}{dx}\right)^2 + y^2 = a^2(\sin^2 x+\cos^2 x)$

$+b^2(\sin^2 x+\cos^2 x)$

$\Rightarrow \left(\dfrac{dy}{dx}\right)^2 + y^2 = (a^2+b^2) = \text{constant.}$

8.b. $y = \sqrt{\dfrac{1-\sin 2x}{1+\sin 2x}} = \dfrac{\cos x - \sin x}{\cos x + \sin x}$

$= \dfrac{1-\tan x}{1+\tan x} = \tan\left(\dfrac{\pi}{4}-x\right)$

$\Rightarrow \dfrac{dy}{dx} = -\sec^2\left(\dfrac{\pi}{4}-x\right)$

9.a. Using chain rule, we have

$\dfrac{dy}{dx} = \dfrac{d}{dx}\left[\left(x+\sqrt{x^2+a^2}\right)^n\right]$

$= n\left(x+\sqrt{x^2+a^2}\right)^{n-1}\dfrac{d}{dx}\left(x+\sqrt{x^2+a^2}\right)$

$= n(x+\ x+\sqrt{x^2+a^2})^{n-1}\left(\dfrac{\sqrt{x^2+a^2}+x}{\sqrt{x^2+a^2}}\right)$

$$= \frac{n(x+\sqrt{x^2+a^2})^n}{\sqrt{x^2+a^2}}$$

$$= \frac{ny}{\sqrt{x^2+a^2}}$$

10.b. $f(x) = \sqrt{1+\cos^2(x^2)}$

$$\Rightarrow f'(x) = \frac{1}{2\sqrt{1+\cos^2(x)^2}}(2\cos x^2)(-\sin x^2)(2x)$$

$$\Rightarrow f'(x) = \frac{-x\sin 2x^2}{\sqrt{1+\cos^2(x^2)}}$$

$$\Rightarrow f'\left(\frac{\sqrt{\pi}}{2}\right) = \frac{-\frac{\sqrt{\pi}}{2}\sin\frac{2\pi}{4}}{\sqrt{1+\cos^2\frac{\pi}{4}}} = \frac{-\frac{\sqrt{\pi}}{2}1}{\sqrt{\frac{3}{2}}}$$

$$\therefore f'\left(\frac{\sqrt{\pi}}{2}\right) = -\sqrt{\frac{\pi}{6}}$$

11.a. $\frac{d}{dx}\cos^{-1}\sqrt{\cos x} = \frac{\sin x}{2\sqrt{\cos x}\sqrt{1-\cos x}}$

$$= \frac{\sqrt{1-\cos^2 x}}{2\sqrt{\cos x}\sqrt{1-\cos x}}$$

$$= \frac{1}{2}\sqrt{\frac{1+\cos x}{\cos x}}$$

12.c. $y = \frac{\log\tan x}{\log\sin x}$

$$\Rightarrow \frac{dy}{dx} = \frac{(\log\sin x)\left(\frac{\sec^2 x}{\tan x}\right) - (\log\tan x)(\cot x)}{(\log\sin x)^2}$$

$$\Rightarrow \left(\frac{dy}{dx}\right)_{\pi/4} = \frac{-4}{\log 2} \quad \text{[On simplification]}$$

13.b. $y = \frac{\sin^{-1} x}{\sqrt{1-x^2}}$

$$\frac{dy}{dx} = \frac{\sqrt{1-x^2}\frac{1}{\sqrt{1-x^2}} - (\sin^{-1} x)\frac{1}{2}\frac{(-2x)}{\sqrt{1-x^2}}}{1-x^2}$$

$$\Rightarrow (1-x^2)\frac{dy}{dx} = 1 + x\left(\frac{\sin^{-1} x}{\sqrt{1-x^2}}\right) = 1 + xy$$

14.a. $y = \cot^{-1}\left[\frac{\sqrt{1+\sin x}+\sqrt{1-\sin x}}{\sqrt{1+\sin x}-\sqrt{1-\sin x}}\right]$

$$= \cot^{-1}\left[\frac{2+2\cos x}{2\sin x}\right] = \cot^{-1}\left[\frac{1+\cos x}{\sin x}\right]$$

$$= \cot^{-1}\left[\cot\frac{x}{2}\right] = \frac{x}{2}$$

$$\therefore \frac{dy}{dx} = \frac{1}{2}$$

15.c. $y = x^{(x^x)}$

$\Rightarrow \log y = x^x \log x$

$$\Rightarrow \frac{1}{y}\frac{dy}{dx} = \frac{dz}{dx}\log x + \frac{1}{x}z \left(\text{where } x^x = z\right)$$

$$\Rightarrow \frac{dy}{dx} = x^{(x^x)}\left[x^x(\log_e x)\log x + x^{x-1}\right]$$

$$\left(\because \frac{dz}{dx} = x^x \log_e x\right)$$

16.b. Let $y = \sin^2\cot^{-1}\left\{\sqrt{\frac{1-x}{1+x}}\right\}$

Put $x = \cos\theta \Rightarrow \theta = \cos^{-1} x$

$$\Rightarrow y = \sin^2\cot^{-1}\left\{\sqrt{\frac{1-\cos\theta}{1+\cos\theta}}\right\} = \sin^2\cot^{-1}\left(\tan\frac{\theta}{2}\right)$$

$$\Rightarrow y = \sin^2\left(\frac{\pi}{2}-\frac{\theta}{2}\right)$$

$$= \cos^2\left(\frac{\theta}{2}\right) = \frac{1+\cos\theta}{2} = \frac{1+x}{2}$$

$$\Rightarrow \frac{dy}{dx} = \frac{1}{2}$$

17.c. $y = ae^{mx} + be^{-mx}$

$$\Rightarrow \frac{dy}{dx} = ame^{mx} - mbe^{-mx}$$

Again $\frac{d^2y}{dx^2} = am^2 e^{mx} + m^2 be^{-mx}$

$$\Rightarrow \frac{d^2y}{dx^2} = m^2\left(ae^{mx} + be^{-mx}\right) \Rightarrow \frac{d^2y}{dx^2} = m^2 y$$

$$\Rightarrow \frac{d^2y}{dx^2} - m^2 y = 0$$

18.d. Let $y = \log x$

$$\Rightarrow y_1 = \frac{1}{x}, y_2 = \frac{-1}{x^2}, y_3 = \frac{2}{x^3}, \cdots,$$

$$y_n = \frac{(-1)^{n-1}(n-1)!}{x^n}.$$

19.c. $y = \sqrt{\log x + y}$

$$\Rightarrow y^2 = \log x + y$$

$$\Rightarrow 2y\frac{dy}{dx} = \frac{1}{x} + \frac{dy}{dx} \Rightarrow \frac{dy}{dx} = \frac{1}{x(2y-1)}$$

20.c. $f(\log_e x) = \log_e(\log_e x)$

$$\therefore \frac{df(\log_e x)}{dx} = \frac{1}{\log_e x} \times \frac{1}{x}$$

21.a. $y = \sec(\tan^{-1} x) = \sec\left(\sec^{-1}\sqrt{1+x^2}\right) = \sqrt{1+x^2}$

Differentiating w.r.t. x, we have $\frac{dy}{dx} = \frac{x}{\sqrt{1+x^2}}$

$$\Rightarrow \left(\frac{dy}{dx}\right)_{x=1} = \frac{1}{\sqrt{2}}$$

22.a. $y = f(x^2) \Rightarrow \frac{dy}{dx} = f'(x^2)2x = 2x\sqrt{2(x^2)^2 - 1}$

At $x = 1$, $\frac{dy}{dx} = 2 \times 1 \times \sqrt{2-1}$

23.a. $$\frac{du}{dv} = \frac{\frac{du}{dx}}{\frac{dv}{dx}} = \frac{f'(x^3)3x^2}{g'(x^2)2x} = \frac{\cos x^3 3x^2}{\sin x^2 2x}$$

$$= \frac{3}{2} x \cos x^3 \operatorname{cosec} x^2$$

24.b. $$\frac{dx}{dy} = \frac{\frac{dx}{dt}}{\frac{dy}{dt}} = \frac{\cos t - t\sin t}{1+\cos t}$$

$$\therefore \frac{d^2x}{dy^2} = \frac{\frac{d}{dt}\left(\frac{dx}{dy}\right)}{\frac{dy}{dt}}$$

$$= \frac{\frac{(-2\sin t - t\cos t)(1+\cos t) - (\cos t - t\sin t)(-\sin t)}{(1+\cos t)^2}}{1+\cos t}$$

Now, put $t = \pi/2$

25.c. $f(x) = \sqrt{1-\sin 2x} = \sqrt{(\cos x - \sin x)^2}$

$$= |\cos x - \sin x|$$

$$= \begin{cases} \cos x - \sin x, & \text{for } 0 \le x \le \pi/4 \\ -(\cos x - \sin x), & \text{for } \pi/4 < x \le \pi/2 \end{cases}$$

$$\therefore f'(x) = \begin{cases} -(\cos x + \sin x), & \text{for } 0 < x < \pi/4 \\ (\cos x + \sin x), & \text{for } \pi/4 < x < \pi/2. \end{cases}$$

26.d. Let $u = y^2$ and $v = x^2$

$$\therefore \frac{du}{dx} = \frac{d}{dx}y^2 = \frac{d}{dy}y^2\frac{dy}{dx}$$

$$= 2y(1-2x) = 2(x-x^2)(1-2x) = 2x(1-x)(1-2x) \quad (1)$$

and $\frac{dv}{dx} = 2x$ (2)

Hence, $$\frac{du}{dv} = \frac{\left(\frac{du}{dx}\right)}{\left(\frac{dv}{dx}\right)} = \frac{2x(1-x)(1-2x)}{2x}$$ (from (1) and (2))

$$= (1-x)(1-2x) = 1 - 3x + 2x^2$$

27.a. $f(x) = \cos^{-1}\left[\cos\left(\frac{\pi}{2} - \sqrt{\frac{1+x}{2}}\right)\right] + x^x f(x)$

$$= \frac{\pi}{2} - \sqrt{\frac{1+x}{2}} + x^x$$

$$\Rightarrow f'(x) = -\frac{1}{\sqrt{2}} \times \frac{1}{2\sqrt{1+x}} + x^x(1 + \log x)$$

$$f'(1) = -\frac{1}{4} + 1 = \frac{3}{4}$$

28.b. $$D = \begin{vmatrix} \sin px & p\cos px & -p^2\sin px \\ -p^3\cos px & p^4\sin px & p^5\cos px \\ -p^6\sin px & -p^7\cos px & p^8\sin px \end{vmatrix}$$

$$= p^9\begin{vmatrix} \sin px & p\cos px & -p^2\sin px \\ -\cos px & p\sin px & p^2\cos px \\ -\sin px & -p\cos px & p^2\sin px \end{vmatrix}$$

$$= -p^9\begin{vmatrix} \sin px & p\cos px & -p^2\sin px \\ \cos px & p\sin px & p^2\cos px \\ \sin px & p\cos px & -p^2\sin px \end{vmatrix} = 0$$

29.b. $2xf'(x^2) = 3x^2$; $4f'(x) = 12 \Rightarrow f'(4) = 3$.

30.c. $(a^2 - 2a - 15)e^{ax} + (b^2 - 2b - 15)e^{bx} = 0$

$$\Rightarrow (a^2 - 2a - 15) = 0 \text{ and } b^2 - 2b - 15 = 0$$

$$\Rightarrow (a-5)(a+3) = 0 \text{ and } (b-5)(b+3) = 0$$

$$\Rightarrow a = 5 \text{ or } -3 \text{ and } b = 5 \text{ or } -3$$

$\therefore a \ne b$ hence $a = 5$ and $b = -3$

or $a = -3$ and $b = 5$

$\Rightarrow ab = -15$

31.b. $y = \dfrac{(a-x)^{3/2} + (x-b)^{3/2}}{\sqrt{a-x} + \sqrt{x-b}}$

$$= \frac{\left(\sqrt{a-x} + \sqrt{x-b}\right)\left(a-x-\sqrt{a-x}\sqrt{x-b}+x-b\right)}{\sqrt{a-x}+\sqrt{x-b}}$$

$$= a-b-\sqrt{a-x}\sqrt{x-b}$$

$$\Rightarrow \frac{dy}{dx} = \frac{1}{2\sqrt{a-x}}\sqrt{x-b} - \frac{1}{2\sqrt{x-b}}\sqrt{a-x}$$

$$= \frac{2x-a-b}{2\sqrt{a-x}\sqrt{x-b}}$$

32.b. $\dfrac{dy}{dx} = e^x + 1,\ \left.\dfrac{dy}{dx}\right|_{x=\log 2} = 3 \Rightarrow \dfrac{dy}{dx} = \dfrac{1}{3}$

33.c. $f'(x) = (kx + e^x)\, h'(x) + h(x)\,(k + e^x)$

$f'(0) = h'(0) + h(0)\,(k+1)$

$18 = -2 + 5(k+1) \Rightarrow k = 3$

34. d. $y = \tan^{-1}\left(\dfrac{2^{x+1} - 2^x}{1 + 2^x . 2^{x+1}}\right) = \tan^{-1} 2^{(x+1)} - \tan^{-1} 2^x$

$$\Rightarrow y' = \frac{2^{x+1}\ln 2}{1+\left(2^{x+1}\right)^2} - \frac{2^x \ln 2}{1+\left(2^x\right)^2}$$

$$\Rightarrow y'(0) = -\frac{1}{10}\ln 2$$

35.b. $f(x) = 1 + x^2 + x^4 + x^6 + \cdots \infty$, where $|x| \le 1$

$\Rightarrow f^n(0) = n!$, where n is even.

36.b. $y = 2\cos x \cos 3x = \cos 4x + \cos 2x$

$$\Rightarrow \frac{d^{20}y}{dx^{20}} = 4^{20}\cos 4x + 2^{20}\cos 2x$$

37.c. We have $y = \sqrt{\dfrac{1-x}{1+x}}$

Differentiating w. r. t. x, we get

$$\frac{dy}{dx} = \frac{1}{2}\left(\frac{1-x}{1+x}\right)^{(1/2-1)} \frac{d}{dx}\left(\frac{1-x}{1+x}\right)$$

$$= \frac{1}{2}\sqrt{\frac{1+x}{1-x}} \times \frac{(1+x)\frac{d}{dx}(1-x) - (1-x)\frac{d}{dx}(1+x)}{(1+x)^2}$$

$$= \frac{1}{2}\sqrt{\frac{1+x}{1-x}} \times \frac{(1+x)(-1)-(1-x)(1)}{(1+x)^2}$$

$$= -\sqrt{\frac{1+x}{1-x}}\,\frac{1}{(1+x)^2}$$

$$\Rightarrow (1-x^2)\frac{dy}{dx} = -\sqrt{\frac{1+x}{1-x}}\,\frac{1}{(1+x)^2}\left(1-x^2\right)$$

$$\Rightarrow (1-x)^2\frac{dy}{dx} = -\sqrt{\frac{1-x}{1+x}}$$

$$\Rightarrow (1-x^2)\frac{dy}{dx} = -y$$

$$\Rightarrow \left(1-x^2\right)\frac{dy}{dx} + y = 0$$

38.b. $\dfrac{dy}{dx} = -\,[(2-x)(3-x)\cdots(n-x) + (1-x)(3-x)$

$\cdots(n-x) + \cdots(1-x)(2-x)\cdots(n-1-x)]$

At $x = 1$

$\dfrac{dy}{dx} = -\,[(n-1)! + 0 + \cdots + 0] = (-1)(n-1)!$

39.a. Let $\cos\alpha = \dfrac{5}{13}$, then $\sin\alpha = \dfrac{12}{13}$.

So, $y = \cos^{-1}(\cos\alpha . \cos x - \sin\alpha . \sin x)$

$\Rightarrow y = \cos^{-1}\{\cos(x+\alpha)\} = x + \alpha$ [$\because x + \alpha$ is in the first or the second quadrant]

$$\Rightarrow \frac{dy}{dx} = 1$$

40.a. Let $x = \sec\theta$

Then $y = \tan^{-1}\sqrt{\dfrac{\sec\theta + 1}{\sec\theta - 1}}$

$$= \tan^{-1}\sqrt{\frac{1+\cos\theta}{1-\cos\theta}} = \tan^{-1}\left(\cot\frac{\theta}{2}\right)$$

$$\Rightarrow y = \tan^{-1}\left\{\tan\left(\frac{\pi}{2} - \frac{\theta}{2}\right)\right\} = \frac{\pi}{2} - \frac{1}{2}\sec^{-1}x$$

$$\Rightarrow \frac{dy}{dx} = -\frac{1}{2} \times \frac{1}{|x|\sqrt{x^2-1}}$$

41.d. We have $\sin^{-1}\left(\dfrac{x^2-y^2}{x^2+y^2}\right) = \log a$

$$\Rightarrow \frac{x^2-y^2}{x^2+y^2} = \sin(\log a)$$

$\Rightarrow \dfrac{1-\tan^2\theta}{1+\tan^2\theta} = \sin(\log a)$, on putting $y = x\tan\theta$

$\Rightarrow \cos 2\theta = \sin(\log a)$

$$\Rightarrow 2\theta = \cos^{-1}(\sin(\log a))$$

$$\Rightarrow \theta = \frac{1}{2}\cos^{-1}(\sin(\log a))$$

$$\Rightarrow \tan^{-1}\left(\frac{y}{x}\right) = \frac{1}{2}\cos^{-1}(\sin(\log a))$$

$$\Rightarrow \frac{y}{x} = \tan\left(\frac{1}{2}\cos^{-1}(\sin(\log a))\right)$$

Differentiating w.r.t. x

$$\Rightarrow \frac{x\frac{dy}{dx} - y}{x^2} = 0$$

$$\Rightarrow x\frac{dy}{dx} - y = 0$$

$$\Rightarrow \frac{dy}{dx} = \frac{y}{x}$$

42.b. $y = \cos^{-1}(\cos x) = \cos^{-1}\{\cos[2\pi - (2\pi - \pi)]\}$.
$= \cos^{-1}[\cos(2\pi - x)]$
$= 2\pi - x$

$$\therefore \frac{dy}{dx} = -1$$

43.a. Let $t = \cos 2\theta$

$$\text{Then } e^x = \frac{\sqrt{1+\cos 2\theta} - \sqrt{1-\cos 2\theta}}{\sqrt{1+\cos 2\theta} + \sqrt{1-\cos 2\theta}}$$

$$= \frac{\cos\theta - \sin\theta}{\cos\theta + \sin\theta} = \frac{1-\tan\theta}{1+\tan\theta} = \tan\left(\frac{\pi}{4} - \theta\right)$$

$$\tan\frac{y}{2} = \sqrt{\frac{1-\cos 2\theta}{1+\cos 2\theta}} = \tan\theta$$

At $t = \frac{1}{2}$, $\cos 2\theta = \frac{1}{2} \Rightarrow \theta = \frac{\pi}{6}$

Then $x = \log\tan\frac{\pi}{12}$, $y = \frac{\pi}{3}$

Differentiating w.r.t. θ, $e^x\frac{dx}{d\theta} = -\sec^2\left(\frac{\pi}{4} - \theta\right)$ and

$$\frac{1}{2}\sec^2\frac{y}{2}\frac{dy}{d\theta} = \sec^2\theta$$

$$\therefore \frac{dy}{dx} = \frac{\frac{dy}{d\theta}}{\frac{dx}{d\theta}} = \frac{2\sec^2\theta\cos^2\frac{y}{2}}{-e^{-x}\sec^2\left(\frac{\pi}{4} - \theta\right)}$$

At $t = \frac{1}{2}$, i.e., $\theta = \frac{\pi}{6}$, $\frac{dy}{dx} = \dfrac{2\sec^2\frac{\pi}{6}\cos^2\frac{\pi}{6}}{-e^{-\log\tan\pi/12}\sec^2\frac{\pi}{12}}$

$$\therefore \frac{dy}{dx} = \frac{2}{-\cot\frac{\pi}{12}\sec^2\frac{\pi}{12}}$$

$$= -2\tan\frac{\pi}{12}\cos^2\frac{\pi}{12} = -\sin\frac{\pi}{6} = -\frac{1}{2}$$

44.b. We have $x^2 + y^2 = t - \frac{1}{t}$ and $x^4 + y^4 = t^2 + \frac{1}{t^2}$

$$\Rightarrow (x^2+y^2)^2 = t^2 + \frac{1}{t^2} - 2$$

$$\Rightarrow (x^2+y^2)^2 = x^4 + y^4 - 2$$

$$\Rightarrow 2x^2y^2 = -2$$

$$\Rightarrow x^2y^2 = -1$$

$$\Rightarrow y^2 = -\frac{1}{x^2}$$

$$\Rightarrow 2y\frac{dy}{dx} = \frac{2}{x^3}$$

$$\Rightarrow x^3y\frac{dy}{dx} = 1$$

45.a. We have

$$y^{1/m} = \left(x + \sqrt{1+x^2}\right)$$

$$\Rightarrow y = \left(x + \sqrt{1+x^2}\right)^m$$

$$\Rightarrow \frac{dy}{dx} = m\left(x + \sqrt{1+x^2}\right)^{m-1}\left(1 + \frac{x}{\sqrt{x^2+1}}\right)$$

$$= m\frac{\left(x\sqrt{1+x^2}\right)^m}{\sqrt{1+x^2}}$$

$$\Rightarrow \frac{dy}{dx} = \frac{my}{\sqrt{1+x^2}}$$

$$\Rightarrow y_1^2(1+x^2) = m^2y^2$$

$$\Rightarrow 2y_1y_2(1+x^2) + 2xy_1^2 = 2m^2yy_1$$

$$\Rightarrow y_2(1+x^2) + xy_1 = m^2y$$

46.a. $y = f(x) - f(2x) \Rightarrow y' = f'(x) - 2f'(2x)$

$\Rightarrow y'(1) = f'(1) - 2f'(2) = 5$, and (1)

$y'(2) = f'(2) - 2f'(4) = 7$ (2)

Now let $y = f(x) - f(4x)$

$\Rightarrow y' = f'(x) - 4f'(4x)$

$\Rightarrow y'(1) = f'(1) - 4f'(4)$ (3)

Substituting the value of $f'(2) = 7 + 2f'(4)$ in (1), we get

$f'(1) - 2(7 + 2f'(4)) = 5$

$f'(1) - 4f'(4) = 19$

47.a. $f(10)=\sin^{-1}\cos 10=\sin^{-1}\sin\left(\frac{\pi}{2}-10\right)$

$$=-\sin^{-1}\sin\left(10-\frac{\pi}{2}\right)$$

$$=-\sin^{-1}\sin\left(3\pi-10+\frac{\pi}{2}\right)$$

$$=-\left(3\pi+\frac{\pi}{2}-10\right)=10-\frac{7\pi}{2}$$

$$f'(x)=\frac{-\sin x}{\sqrt{1-\cos^2 x}}=\frac{-\sin x}{|\sin x|}\Rightarrow f'(10)=\frac{-\sin 10}{|\sin 10|}=1.$$

So, $f(10)+f'(10)=11-\frac{7\pi}{2}$

48.a. $(\sin x)(\cos y)=\frac{1}{2}$

$\Rightarrow (\cos x)(\cos y)-\sin y\sin x\ \frac{dy}{dx}=0$

$\Rightarrow \frac{dy}{dx}=(\cot x)(\cot y)$

$\Rightarrow \frac{d^2y}{dx^2}=-\text{cosec}^2 x.\cot y-\text{cosec}^2 y\cot x.\frac{dy}{dx}$

Now $\left(\frac{dy}{dx}\right)_{(\pi/4,\ \pi/4)}=1$

$\Rightarrow \left(\frac{d^2y}{dx^2}\right)_{(\pi/4,\ \pi/4)}=-(2)(1)-(2)(1)(1)=-4$

49.a. Given $f=f^{\text{I}}+f^{\text{II}}+f^{\text{III}}+\cdots\infty$

$\Rightarrow f^{\text{I}}=f^{\text{II}}+f^{\text{II}}+f^{\text{IV}}+\cdots\infty$

$\Rightarrow f-f^{\text{I}}=f^{\text{I}}$

$\Rightarrow f=2f^{\text{I}}$

Hence, $\frac{f^{\text{I}}}{f}=1/2\Rightarrow\int\frac{f^{\text{I}}}{f}dx=\int\frac{1}{2}dx$

$\Rightarrow \log f(x)=x/2+c$

$\Rightarrow f(x)=e^{x/2+c}$

Also, $f(0)=1\Rightarrow c=0\Rightarrow f(x)=e^{x/2}$

50.b. Putting $x=\sin\theta$ and $y=\sin\phi$

$\cos\theta+\cos\phi=a(\sin\theta-\sin\phi)$

$\Rightarrow 2\cos\frac{\theta+\phi}{2}\cos\frac{\theta-\phi}{2}=a\left(2\cos\frac{\theta+\phi}{2}\sin\frac{\theta-\phi}{2}\right)$

$\Rightarrow \frac{\theta-\phi}{2}=\cot^{-1}a$

$\Rightarrow \theta-\phi=2\cot^{-1}a$

$\Rightarrow \sin^{-1}x-\sin^{-1}y=2\cot^{-1}a$

$\Rightarrow \frac{1}{\sqrt{1-x^2}}-\frac{1}{\sqrt{1-y^2}}\frac{dy}{dx}=0\Rightarrow\frac{dy}{dx}=\sqrt{\frac{1-y^2}{1-x^2}}$

51.a. $y=x^2+\frac{1}{y}$

$\Rightarrow y^2=x^2y+1$

$\Rightarrow 2y\frac{dy}{dx}=y\,2x+x^2\frac{dy}{dx}$

$\Rightarrow \frac{dy}{dx}=\frac{2xy}{2y-x^2}$

52.d. $\frac{d}{dx}\left(\tan^{-1}\frac{\left(\sqrt{x}(3-x)\right)}{1-3x}\right)$

Put $\sqrt{x}=\tan\theta\Rightarrow\theta=\tan^{-1}\sqrt{x}$

$\frac{d}{dx}\left(\tan^{-1}\frac{\left(\tan\theta(3-\tan^2\theta)\right)}{1-3\tan^2\theta}\right)$

$\frac{d}{dx}\left(\tan^{-1}\frac{\left(3\tan\theta-\tan^3\theta\right)}{1-3\tan^2\theta}\right)$

$\frac{d}{dx}\left(\tan^{-1}(\tan 3\theta)=\frac{d}{dx}(3\theta)\right)$

$\frac{d}{dx}\left(3\tan^{-1}\sqrt{x}\right)=\frac{3}{2\sqrt{x}\,(1+x)}$

53.b. Since $g(x)$ is the inverse of function $f(x)$, $gof(x)=I(x)$ for all x

Now $gof(x)=I(x),\ \forall x$

$\Rightarrow gof(x)=x,\ \forall\ x\Rightarrow(gof)'(x)=1,\ \forall\ x$

$\Rightarrow g'(f(x))f'(x)=1,\ \forall\ x$

$\Rightarrow g'(f(x))=\frac{1}{f'(x)},\ \forall\ x$

$\Rightarrow g'(f(c))=\frac{1}{f'(c)}$ (putting $x=c$)

54.c. $f(x)=x+\tan x$

$f(f^{-1}(y))=f^{-1}(y)+\tan f^{-1}(y)$

$y=g(y)+\tan g(y)$

$x=g(x)+\tan g(x)$

Differentiating, we get $1=g'(x)+\sec^2 g(x)\,g'(x)$

$\Rightarrow g'(x)=\frac{1}{1+\sec^2 g(x)}=\frac{1}{2+[g(x)-x]^2}$

55.a. $y\sqrt{x^2+1}=\log\left\{\sqrt{x^2+1}-x\right\}$

Differentiating both sides w.r.t. x., we get

$$\frac{dy}{dx}\sqrt{x^2+1}+y\frac{1}{2\sqrt{x^2+1}}2x=\frac{1}{\sqrt{x^2+1}-x}\times\left\{\frac{1}{2}\frac{2x}{\sqrt{x^2+1}}-1\right\}$$

$$\Rightarrow (x^2+1)\frac{dy}{dx}+xy=\sqrt{x^2+1}\ \frac{-1}{\sqrt{x^2+1}}.$$

$$\Rightarrow (x^2+1)\frac{dy}{dx}+xy+1=0$$

56.a. $y=\dfrac{\sqrt{a+x}-\sqrt{a-x}}{\sqrt{a+x}+\sqrt{a-x}}$

$$\Rightarrow y=\frac{\left(\sqrt{a+x}-\sqrt{a-x}\right)^2}{(a+x)-(a-x)}$$

$$\Rightarrow y=\frac{(a+x)+(a-x)-2\left(\sqrt{a^2-x^2}\right)}{2x}$$

$$=\frac{2a-2\sqrt{a^2-x^2}}{2x}=\frac{a-\sqrt{a^2-x^2}}{x} \qquad (1)$$

Differentiating w.r.t. x, we get

$$\frac{dy}{dx}=\frac{x\left[-\frac{1}{2\sqrt{a^2-x^2}}(-2x)\right]-\left(a-\sqrt{a^2-x^2}\right)}{x^2}$$

$$=\frac{x^2-a\sqrt{a^2-x^2}+a^2-x^2}{x^2\sqrt{a^2-x^2}}=\frac{a\left(a-\sqrt{a^2-x^2}\right)}{x^2\sqrt{a^2-x^2}}$$

$$=\frac{a}{x\sqrt{a^2-x^2}}\left[\frac{a-\sqrt{a^2-x^2}}{x}\right]=\frac{ay}{x\sqrt{a^2-x^2}}$$

[by (1)]

57.a. As $f(x)=x^4\tan(x^3)-x\ln(1+x^2)$ is odd, $\dfrac{d^4f(x)}{dx^4}=0$ at $x=0$.

58.a. Given that $g^{-1}(x)=f(x)\Rightarrow x=g(f(x))$ or $g'(f(x))f'(x)=1$

$$\Rightarrow g'(f(x))=\frac{1}{f'(x)}$$

$$\Rightarrow g''(f(x))f'(x)=\frac{f''(x)}{[f'(x)]^2}\Rightarrow g''(f(x))=\frac{f''(x)}{[f'(x)]^3}.$$

59.b. For $x>1$, we have $f(x)=|\log|x||=\log x$

$$\Rightarrow f'(x)=\frac{1}{x}$$

For $x<-1$, we have $f(x)=|\log|x||=\log(-x)$

$$\Rightarrow f'(x)=\frac{1}{x}$$

For $0<x<1$, we have $f(x)=|\log|x||=-\log x$

$$\Rightarrow f'(x)=\frac{-1}{x}$$

For $-1<x<0$, we have $f(x)=-\log(-x)$

$$\Rightarrow f'(x)=\frac{-1}{x}.$$

Hence, $f'(x)=\begin{cases}\dfrac{1}{x}, & |x|>1\\ -\dfrac{1}{x}, & |x|<1\end{cases}$

60.c. Around $x=\dfrac{2\pi}{3}$, $|\cos x|=-\cos x$ and $|\sin x|=\sin x$

$$\Rightarrow y=-\cos x+\sin x$$

$$\Rightarrow \frac{dy}{dx}=\sin x+\cos x$$

$$\Rightarrow \text{At } x=\frac{2\pi}{3},\ \frac{dy}{dx}=\sin\frac{2\pi}{3}+\cos\frac{2\pi}{3}.$$

61.a. Since g is the inverse function of f, we have $f\{g(x)\}=x$

$$\Rightarrow \frac{d}{dx}(f\{g(x)\})=1$$

$$\Rightarrow f'\{g(x)\}.g'(x)=1$$

$$\Rightarrow \sin\{g(x)\}\,g'(x)=1$$

$$\Rightarrow g'(x)=\frac{1}{\sin\{g(x)\}}$$

62.b. We have $x=f(t)$, $y=y(t)$. Therefore,

$$\frac{dy}{dx}=\frac{\frac{dy}{dt}}{\frac{dx}{dt}}=\frac{\psi'}{\phi'}$$

$$\Rightarrow \frac{d^2y}{dx^2}=\frac{d}{dx}\left(\frac{\psi'}{\phi'}\right)=\frac{d}{dt}\left(\frac{\psi'}{\phi'}\right)\frac{dt}{dx}$$

$$=\frac{\phi'\psi''-\psi'\phi''}{\phi'^2}\frac{1}{\phi'}=\frac{\phi'\psi''-\psi'\phi''}{\phi'^3}$$

63.c. $f(x)=e^x-e^{-x}-2\sin x-\dfrac{2}{3}x^3$

$f^{\mathrm{I}}(x)=e^x+e^{-x}-2\cos x-2x^2$

$f^{\mathrm{II}}(x)=e^x-e^{-x}+2\sin x-4x$

$f^{\mathrm{III}}(x)=e^x+e^{-x}+2\cos x-4$

$f^{\mathrm{IV}}(x)=e^x-e^{-x}-2\sin x$

$f^{\mathrm{V}}(x)=e^x+e^{-x}-2\cos x$

$f^{\mathrm{VI}}(x)=e^x-e^{-x}+2\sin x$

$f^{\mathrm{VII}}(x)=e^x+e^{-x}+2\cos x$

Clearly, $f^{\mathrm{VII}}(0)$ is non-zero.

64.b. Given $f\left(\dfrac{5x-3y}{2}\right)=\dfrac{5f(x)-3f(y)}{2}$

$\Rightarrow f\left(\frac{5x-3y}{5-3}\right)=\frac{5f(x)-3f(y)}{5-3}$, which satisfies section formula for abscissa on L.H.S. and ordinate on R.H.S. Hence, $f(x)$ must be the linear function (as only straight line satisfies such section formula).

Hence, $f(x)=ax+b$

But $f(0)=3\Rightarrow b=3, f'(0)=2\Rightarrow a=2$.

Thus, $f(x)=2x+3\Rightarrow$ Period of $\sin(f(x))=\sin(2x+3)$ is π.

65.c. $D^*(x)=\lim_{h\to 0}\frac{f^2(x+h)-f^2(x)}{h}$

$=\lim_{h\to 0}\frac{f(x+h)-f(x)}{h}(f(x+h)+f(x))$

$=2f(x)\lim_{h\to 0}\frac{f(x+h)-f(x)}{h}$

$=2f(x)\times f'(x)$

$\Rightarrow D^*(x\log x)=2x\log x(1+\log x)$

$\Rightarrow D^*f(x)|_{x=e}=4e$

66.b. $\sqrt{x}=\cos\theta$

$x\in\left(0,\frac{1}{2}\right)\Rightarrow\sqrt{x}=\cos\theta\in\left(0,\frac{1}{\sqrt{2}}\right)$

$\Rightarrow\theta\in\left(\frac{\pi}{4},\frac{\pi}{2}\right)$

$\Rightarrow 2\theta\in\left(\frac{\pi}{2},\pi\right)$

$\Rightarrow f(x)=2\sin^{-1}\sqrt{1-\cos^2\theta}$

$+\sin^{-1}\left(2\sqrt{\cos^2\theta\sin^2\theta}\right)$

$=2\sin^{-1}(\sin\theta)+\sin^{-1}(2\sin\theta\cos\theta)$

$=2\theta+\sin^{-1}(\sin 2\theta)$

$=2\theta+\pi-2\theta$

$=\pi$

$\Rightarrow f'(x)=0$

67.a. $F'(x)=\left[f\left(\frac{x}{2}\right)f'\left(\frac{x}{2}\right)+g\left(\frac{x}{2}\right)g'\left(\frac{x}{2}\right)\right]$

Here $g(x)=f'(x)$

and $g'(x)=f''(x)=-f(x)$. So

$F'(x)=f\left(\frac{x}{2}\right)g\left(\frac{x}{2}\right)-f\left(\frac{x}{2}\right)g\left(\frac{x}{2}\right)=0$

$\Rightarrow F(x)$ is a constant function.

$\Rightarrow F(10)=5$

68.b. Let $y=\tan^{-1}\left(\frac{\sqrt{1+x^2}-1}{x}\right)$

and $z=\tan^{-1}\left(\frac{2x\sqrt{1-x^2}}{1-2x^2}\right)$

Putting $x=\tan\theta$ in y, we get

$y=\tan^{-1}\left(\frac{\sec\theta-1}{\tan\theta}\right)=\tan^{-1}\left(\tan\frac{\theta}{2}\right)=\frac{1}{2}\tan^{-1}x$

$\Rightarrow\frac{dy}{dx}=\frac{1}{2(1+x^2)}$

Putting $x=\sin\theta$ in z, we get

$z=\tan^{-1}\left(\frac{2\sin\theta\cos\theta}{\cos 2\theta}\right)=\tan^{-1}(\tan 2\theta)$

$=2\theta=2\sin{-1}x$

$\Rightarrow\frac{dz}{dx}=\frac{2}{\sqrt{1-x^2}}$

Thus, $\frac{dy}{dz}=\frac{\frac{dy}{dx}}{\frac{dz}{dx}}=\frac{1}{4(1+x^2)}\sqrt{1-x^2}\Rightarrow\left(\frac{dy}{dz}\right)_{x=0}=\frac{1}{4}$

69.c. $f(x)=xe^x$

$f'(x)=e^x+xe^x$

$f''(x)=e^x+e^x+xe^x$

$f'''(x)=2e^x+e^x+xe^x=3e^x+xe^x$

...

...

$f^n(x)=ne^x+xe^x$.

Now, $f^n(x)=0$

$\Rightarrow ne^x+xe^x=0\Rightarrow x=-n$

70.a. $y^2=ax^2+bx+c$

$\Rightarrow 2y\frac{dy}{dx}=2ax+b$

$\Rightarrow 2\left(\frac{dy}{dx}\right)^2+2y\frac{d^2y}{dx^2}=2a$

$\Rightarrow y\frac{d^2y}{dx^2}=a-\left(\frac{dy}{dx}\right)^2$

$\Rightarrow y\frac{d^2y}{dx^2}=a-\left(\frac{2ax+b}{2y}\right)^2$

$$\Rightarrow y\frac{d^2y}{dx^2}=\frac{4ay^2-(2ax+b)^2}{4y^2}$$

$$\Rightarrow 4y^3\frac{d^2y}{dx^2}=4a\left(ax^2+bx+c\right)-\left(4a^2x^2+4abx+b^2\right)$$

$$\Rightarrow 4y^3\frac{d^2y}{dx^2}=4ac-b^2=\text{constant.}$$

71.c. $y=\sin x+e^x \Rightarrow \frac{dy}{dx}=\cos x+e^x$

$$\Rightarrow \frac{dx}{dy}=(\cos x+e^x)^{-1} \quad (1)$$

Again $\frac{d^2x}{dy^2}=-(\cos x+e^x)^{-2}(-\sin x+e^x)\frac{dx}{dy}$

Substituting the value of $\frac{dy}{dx}$ from (1)

$$\frac{d^2x}{dy^2}=\frac{\left(\sin x-e^x\right)}{\left(\cos x+e^x\right)^2}\left(\cos x+e^x\right)^{-1}=\frac{\sin x-e^x}{\left(\cos x+e^x\right)^3}$$

72.d. $u=x^2+y^2, x=s+3t, y=2s-t$

Now, $\frac{dx}{ds}=1, \frac{dy}{ds}=2$ (1)

$\frac{d^2x}{ds^2}=0, \frac{d^2y}{ds^2}=0$ (2)

Now $u=x^2+y^2, \frac{du}{ds}=2x\frac{dx}{ds}+2y\frac{dy}{ds}$

$$\frac{d^2u}{ds^2}=2\left(\frac{dx}{ds}\right)^2+2x\frac{d^2x}{ds^2}+2\left(\frac{dy}{ds}\right)^2+2y\left(\frac{d^2y}{ds^2}\right)$$

From (1) and (2), $\frac{d^2u}{ds^2}=2\times1+0+2\times4+0=10.$

73.c. Here, $y=t^{10}+1$ and $x=t^8+1$. Therefore,
$t^8=x-1 \Rightarrow t^2=(x-1)^{1/4}$
So $y=(x-1)^{5/4}+1$
Differentiat both sides w.r.t. x, we

get $\frac{dy}{dx}=\frac{5}{4}(x-1)^{1/4}$

Again, differentiating both sides w.r.t. x, we get

$$\frac{d^2y}{dx^2}=\frac{5}{16}(x-1)^{-3/4}$$

$$\frac{d^2y}{dx^2}=\frac{5}{16(x-1)^{3/4}}=\frac{5}{16\left(t^2\right)^3}=\frac{5}{16t^6}$$

74.b. From the given relation $\frac{x}{y}=\log x-\log(a+bx)$

Differentiating. w.r.t. x, we get $\frac{\left(x\frac{dy}{dx}\right)-y}{x^2}$

$$=\frac{1}{x}-\frac{1}{a+bx}b=\frac{a}{x(a+bx)}$$

$$\therefore x\frac{dy}{dx}-y=\frac{ax}{a+bx} \quad (1)$$

Differentiating. again w.r.t. x, we get

$$x\frac{d^2y}{dx^2}+\frac{dy}{dx}-\frac{dy}{dx}=\frac{(a+bx)a-ax.b}{(a+bx)^2}$$

$$\Rightarrow x\frac{d^2y}{dx^2}=\frac{a^2}{(a+bx)^2}$$

$$\Rightarrow x^3\frac{d^2y}{dx^2}=\frac{a^2x^2}{(a+bx)^2}=\left(x\frac{dy}{dx}-y\right)^2 \quad \text{[by (1)]}$$

75.a. $u(x)=7v(x)\Rightarrow u'(x)=7v'(x)\Rightarrow p=7$ (given)

Again $\frac{u(x)}{v(x)}=7\Rightarrow\left(\frac{u(x)}{v(x)}\right)'=0\Rightarrow q=0$

Now $\frac{p+q}{p-q}=\frac{7+0}{7-0}=1$

76.a. $ax^2+2hxy+by^2=1$

Differentiating. both sides w.r.t. x, we get

$$2ax+2hx\frac{dy}{dx}+2hy+2by\frac{dy}{dx}=0$$

$$\Rightarrow \frac{dy}{dx}=-\frac{ax+hy}{hx+by}$$

Again differentiating. w.r.t. x, we get

$$\Rightarrow \frac{d^2y}{dx^2}$$

$$=-\left[\frac{(hx+by)\left(a+h\frac{dy}{dx}\right)-(ax+hy)\left(h+b\frac{dy}{dx}\right)}{(hx+by)^2}\right]$$

$$=-\frac{\left[y\left(ab-h^2\right)+\frac{dy}{dx}\left(h^2x-abx\right)\right]}{(hx+by)^2}$$

$$= \frac{(h^2 - ab)\left(y - x\dfrac{dy}{dx}\right)}{(hx + by)^2}$$

$$= \frac{(h^2 - ab)}{(hx + by)^2}\left[y + x\frac{ax + hy}{hx + by}\right]$$

$$= \frac{h^2 - ab}{(hx + by)^3}$$

Multiple Correct Answers Type

1. a., c.

$$\frac{dy}{dx} = \frac{e^{\sqrt{x}}}{2\sqrt{x}} - \frac{e^{-\sqrt{x}}}{2\sqrt{x}}$$

$$= \frac{e^{\sqrt{x}} - e^{-\sqrt{x}}}{2\sqrt{x}}$$

$$= \frac{\sqrt{\left(e^{\sqrt{x}} + e^{-\sqrt{x}}\right)^2 - 4}}{2\sqrt{x}}$$

$$= \frac{\sqrt{y^2 - 4}}{2\sqrt{x}}$$

2. a., c., d.

$$y^2 = x + y \Rightarrow \frac{dy}{dx} = \frac{1}{2y - 1}$$

Also $y = \dfrac{x}{y} + 1 \Rightarrow \dfrac{dy}{dx} = \dfrac{y}{2x + y}$

Also $y^2 - y - x = 0 \Rightarrow y = \dfrac{1 \pm \sqrt{1 + 4x}}{2}$

$\Rightarrow y = \dfrac{1 + \sqrt{1 + 4x}}{2}$. (as $y > 0$)

3. b., c., d.

1 is a root of $f(x) = 0, f'(x) = 0$ and $f''(x) = 0$, or

1 is a root of $ax^3 + bx^2 + bx + d = 0$ (1)

$3ax^2 + 2bx + b = 0$ (2)

$\Rightarrow a + 2b + d = 0$

$a + b = 0$

$\Rightarrow b + d = 0$ and $a = d$.

4. a., c.

$x^3 - 2x^2y^2 + 5x + y - 5 = 0$

Differentiating. w.r.t. x, we get

$$\Rightarrow 3x^2 - 4xy^2 - 4x^2y\frac{dy}{dx} + 5 + \frac{dy}{dx} = 0$$

$$\Rightarrow y' = \frac{dy}{dx} = \frac{3x^2 - 4xy^2 + 5}{4x^2y - 1}$$

$$y'(1) = \frac{3 - 4 + 5}{4 - 1} = \frac{4}{3}$$

Also, y''

$$= \frac{(6x - 4y^2 - 8xyy')(4x^2y - 1) - (8xy + 4x^2y')(3x^2 - 4xy^2 + 5)}{(4x^2y - 1)^2}$$

$$\Rightarrow y''(1) = \frac{(6 - 4 - 8\times\frac{4}{3})(4 - 1) - (8 + 4\times\frac{4}{3})(3 - 4 + 5)}{(4 - 1)^2}$$

$$= -8 \times \frac{22}{27}$$

5. a., b., c.

$$f(x) = |x^2 - 3|x| + 2|$$

$$= \begin{cases} |x^2 - 3x + 2|, & x \ge 0 \\ |x^2 + 3x + 2|, & x < 0 \end{cases}$$

$$= \begin{cases} x^2 - 3x + 2, & x^2 - 3x + 2 \ge 0, \quad x \ge 0 \\ -x^2 + 3x - 2, & x^2 - 3x + 2 < 0, \quad x \ge 0 \\ x^2 + 3x + 2, & x^2 + 3x + 2 \ge 0, \quad x < 0 \\ -x^2 - 3x - 2, & x^2 + 3x + 2 < 0, \quad x < 0 \end{cases}$$

$$= \begin{cases} x^2 - 3x + 2, & x \in [0, 1] \cup [2, \infty) \\ -x^2 + 3x - 2, & x \in (1, 2) \\ x^2 + 3x + 2, & x \in (-\infty, -2] \cup [-1, 0) \\ -x^2 - 3x + 2, & x \in (-2, -1) \end{cases}$$

$$\Rightarrow f'(x) = \begin{cases} 2x - 3, & x \in (0, 1) \cup (2, \infty) \\ -2x + 3, & x \in (1, 2) \\ 2x + 3, & x \in (-\infty, -2) \cup (-1, 0) \\ -2x - 3, & x \in (-2, -1) \end{cases}$$

6. a., c.

$$y = \frac{\left(x^2 + 1\right)^2 - 3x^2}{x^2 + \sqrt{3}x + 1}$$

$$= \frac{\left(x^2 + 1 + \sqrt{3}x\right)\left(x^2 + 1 - \sqrt{3}\,x\right)}{x^2 + 1 + \sqrt{3}x}$$

$$\frac{dy}{dx} = 2x - \sqrt{3} \Rightarrow a = 2 \text{ and } b = -\sqrt{3}$$

$$a - b = 2 + \sqrt{3} = \tan\frac{5\pi}{12} = \cot\frac{\pi}{12}$$

7. a., b., d.

$$f(x)=\frac{\sqrt{\left(\sqrt{x-1}\right)^2+1-2\sqrt{x-1}}}{\sqrt{x-1}-1}x$$

$$=\frac{\left|\sqrt{x-1}-1\right|}{\sqrt{x-1}-1}x$$

$$=\begin{cases}-x \text{ if } x\in[1,\ 2)\\ x \quad \text{if } x\in(2,\ \infty)\end{cases}$$

8. b., d.

$y=x^{(\log x)^{\log(\log x)}}$

$\Rightarrow \log y=(\log x)\,(\log x)^{\log(\log x)}$ (1)

Taking log of both sides, we get

$\Rightarrow \log(\log y)=\log(\log x)+\log(\log x)\log(\log x)$

Differentiating. w.r.t. x, we get

$$\frac{1}{\log y}\times\frac{1}{y}\frac{dy}{dx}=\frac{1}{x\log x}+\frac{2\log(\log x)}{\log x}\frac{1}{x}$$

$$=\frac{2\log(\log x)+1}{x\log x}$$

$$\Rightarrow \frac{dy}{dx}=\frac{y}{x}\times\frac{\log y}{\log x}\left(2\log(\log x)+1\right)\Rightarrow D$$

Substituting the value of y from (1), we get

$$\frac{y}{x}(\log x)^{\log(\log x)}(2\log(\log x)+1)\Rightarrow B$$

9. a., b., c.

We have $\sin^{-1}(\cos x)=\dfrac{\pi}{2}-\cos^{-1}(\cos x)$

$$=\begin{cases}\dfrac{\pi}{2}-x, \text{if } 0<x\le\pi\\ \dfrac{\pi}{2}-(2\pi-x), \text{if } \pi<x<2\pi\end{cases}$$

$$=\begin{cases}\dfrac{\pi}{2}-x, & \text{if } 0<x\le\pi\\ x-\dfrac{3\pi}{2}, & \text{if } \pi<x<2\pi\end{cases}$$

$$\therefore \frac{d}{dx}\{\sin^{-1}(\cos x)\}=\begin{cases}-1, & \text{if } 0<x<\pi\\ 1, & \text{if } \pi<x<2\pi\end{cases}$$

We have $\cos^{-1}(\sin x)=\dfrac{\pi}{2}-\sin^{-1}(\sin x)$

$$=\begin{cases}\dfrac{\pi}{2}-x, \text{if } -\dfrac{\pi}{2}<x\le\dfrac{\pi}{2}\\ \dfrac{\pi}{2}-(\pi-x), \text{if } \dfrac{\pi}{2}<x<\dfrac{3\pi}{2}\end{cases}$$

$$=\begin{cases}\dfrac{\pi}{2}-x, & \text{if } -\dfrac{\pi}{2}<x\le\dfrac{\pi}{2}\\ x-\dfrac{\pi}{2}, & \text{if } \dfrac{\pi}{2}<x<\dfrac{3\pi}{2}\end{cases}$$

$$\therefore \frac{d}{dx}(\cos^{-1}(\sin x))=\begin{cases}-1, & \text{if } -\dfrac{\pi}{2}<x<\dfrac{\pi}{2}\\ 1, & \text{if } \dfrac{\pi}{2}<x<\dfrac{3\pi}{2}\end{cases}$$

10. a., c.

$f(x-y)$, $f(x)f(y)$, and $f(x+y)$ are in A.P.

$\Rightarrow f(x+y)+f(x-y)=2f(x)f(y)$ for all x, y

Putting $x=0, y=0$ in (1), we get

$f(0)+f(0)=2f(0)f(0)$

$\Rightarrow f(0)=1\ [\because f(0)\ne 0]$

Putting $x=0, y=x$, we get

$f(x)+f(-x)=2f(0)f(x)$

$\Rightarrow f(x)=f(-x)$ (1)

$\Rightarrow f(4)=f(-4), f(3)=f(-3)$

Differentiating (1) w.r.t. x, $f'(x)+f'(-x)=0$

$\Rightarrow f'(4)+f'(-4)=0$ and $f'(2)=f'(-2)$

11. b., c.

$$y=\cos^{-1}\left(\frac{2x}{1+x^2}\right)$$

$$\Rightarrow \frac{dy}{dx}=\frac{-1}{\sqrt{1-\dfrac{4x^2}{(1+x^2)^2}}}\frac{d}{dx}\left(\frac{2x}{1+x^2}\right)$$

$$\Rightarrow \frac{dy}{dx}=-\frac{1+x^2}{\sqrt{(1-x^2)^2}}\frac{2(1+x^2)-4x^2}{(1+x^2)^2}$$

$$\Rightarrow \frac{dy}{dx}=-2\frac{(1+x^2)}{|1-x^2|}\frac{1-x^2}{(1+x^2)^2}$$

$$\Rightarrow \frac{dy}{dx}=-2\left(\frac{1-x^2}{|1-x^2|}\right)\left(\frac{1}{1+x^2}\right)$$

$$\Rightarrow \frac{dy}{dx}=\begin{cases}\dfrac{2}{1+x^2}, & \text{if } |x|>1\\ \dfrac{-2}{1+x^2}, & \text{if } |x|<1\end{cases}$$

12. b., c., d.

$$f'(x)=\lim_{x\to 0}\frac{f(x+h)-f(x)}{h}$$

$$= \lim_{h\to 0} \frac{f\left(1+\frac{h}{x}\right)}{\frac{h}{x}x} = \frac{f'(1)}{x} = \frac{1}{x}$$

$\Rightarrow f(x) = \ln x$ as $f(1) = 0$

13. a., c.

$$\frac{d}{dx}\{f_n(x)\} = \frac{d}{dx}\{e^{f_{n-1}(x)}\}$$

$$= e^{f_{n-1}(x)}\frac{d}{dx}\{f_{n-1}(x)\} = f_n(x)\frac{d}{dx}\{f_{n-1}(x)\}$$

$$= f_n(x)\frac{d}{dx}\{e^{f_{n-2}(x)}\} = f_n(x)e^{f_{n-2}(x)}\frac{d}{dx}\{f_{n-2}(x)\}$$

$$= f_n(x)\, f_{n-1}(x)\frac{d}{dx}\{f_{n-2}(x)\}$$

...

$$= f_n(x) f_{n-1}(x)\cdots f_2(x)\frac{d}{dx}\{f_1(x)\}$$

$$= f_n(x)\, f_{n-1}(x)\cdots f_2(x)\frac{d}{dx}\{e^{f_0(x)}\}$$

$$= f_n(x)\, f_{n-1}(x)\cdots f_2(x)\, e^{f_0(x)}\frac{d}{dx}\{f_0(x)\}$$

Use $e^{f_0(x)} = f_1(x)$ and $f_0(x) = x$

Reasoning Type

1. a. $f(x) = x[x] = \begin{cases} -x, & -1 \le x < 0 \\ 0, & 0 \le x < 1 \\ x, & 1 \le x < 2 \\ 2x, & 2 \le x < 3 \\ \ldots & \ldots \end{cases}$

$\Rightarrow f'(x) = \begin{cases} -1, & -1 < x < 0 \\ 0, & 0 < x < 1 \\ 1, & 1 < x < 2 \\ 2, & 2 < x < 3 \\ \ldots & \ldots \end{cases} \Rightarrow f'(x) = [x]$

2. c. Statement 1 is always true, but statement 2 is not always true, as if $f'(x) = \cos x$, then $f(x)$ can be $\sin x$ which is odd function, but if $f(x) = -\sin x + 2$, then $f(x)$ is neither odd nor even.

3. a. Since $|f(x) - f(y)| \le |x - y|^3$, where $x \ne y$

$$\therefore \left|\frac{f(x)-f(y)}{x-y}\right| \le |x-y|^2$$

Taking lim as $y \to x$, we get

$$\lim_{y\to x}\left|\frac{f(x)-f(y)}{x-y}\right| \le \lim_{y\to x}|x-y|^2$$

$$\Rightarrow \left|\lim_{y\to x}\frac{f(x)-f(y)}{x-y}\right| \le \left|\lim_{y\to x}(x-y)^2\right|$$

$\Rightarrow |f'(x)| \le 0$

$\Rightarrow |f'(x)| = 0$ $(\because |f'(x)| \ge 0)$

$\therefore f'(x) = 0$

$\Rightarrow f(x) = c$ (constant)

4. b. Both statements are true, but statement 2 is not correct explanation of statement 1.
Statement 1 is true as period of $\sin x$ is 2π.
Or, in general if for $y = f(x), f(a) = f(b)$, we cannot say $f'(a) = f'(b)$.

5. a. $f(x) + f(x-2) = 0$ (1)
Replace x by $x - 2 \Rightarrow f(x-2) + f(x-4) = 0$ (2)
From (1) and (2), $f(x) - f(x-4) = 0$
Replace x by $x + 4 \Rightarrow f(x+4) = f(x)$.
$\Rightarrow f(x) = f(x+4) = f(x+8) = \cdots = f(x+4000)$
$\Rightarrow f'(x) = f'(x+4000)$.
Hence, both the statements are true and statement 2 is correct explanation of statement 1.
Hence, $f(x)$ is periodic with period 4.

6. d. Statement 2 is true as $f(\alpha) = 0$ and $f'(\alpha) = 0$, then definitely α is repeated root of $f(x) = 0$.
But from data, we are not sure how many times a root repeats.
Also $f(x) = (x - \alpha)^n \times g(x)$, which changes sign at $x = \alpha$, when n is odd and does not if n is even. Hence, statement 1 is false.

7. a. Given $f(x + y^3) = f(x) + f(y^3)\ \forall\ x, y \in R$
Put $x = y = 0$, we get $f(0+0) = f(0) + f(0) \Rightarrow f(0) = 0$.
Now, put $y = -x^{1/3}$, we get $f(0) = f(x) + f(-x)$
$\Rightarrow f(x) + f(-x) = 0$
$\Rightarrow f(x)$ is an odd function
$\Rightarrow f'(x)$ is an even function
$\Rightarrow f(-2) = a$

Linked Comprehension Type

For Problems 1–3

1. b., 2. a., 3. d.

Sol. Suppose degree of $f(x)$ is n, then degree of f' is $n - 1$ and degree of f'' is $n - 2$,
So $n = n - 1 + n - 2$
Hence, $n = 3$.
So put $f(x) = ax^3 + bx^2 + cx + d$. (where $a \ne 0$)
From $f(2x) = f'(x).f''(x)$, we have
$8ax^3 + 4bx^2 + 2cx + d$
$= (3ax^2 + 2bx + c)(6ax + 2b)$

$$= 18a^2x^3 + 18abx^2 + (6ac + 4b^2)x + 2bc$$

Comparing coefficients of terms, we have

$18a^2 = 8a \quad \Rightarrow a = 4/9$

$18ab = 4b \quad \Rightarrow b = 0$

$2c = 6ac + 4b^2 \quad \Rightarrow c = 0$

$d = 2bc \quad \Rightarrow d = 0$

$\Rightarrow f(x) = \dfrac{4x^3}{9}$, which is clearly one-one and onto.

$\Rightarrow f(3) = 12.$

Also, $\dfrac{4x^3}{9} = x \Rightarrow x = 0, x = \pm 3/2.$

Hence sum of roots of equation is zero.

For Problems 4–6

4. d., 5. c., 6. d.

Sol. Here, $f(x) = x^3 + x^2 f'(1) + xf''(2) + f'''(3)$

Put $f'(1) = a, f''(2) = b, f'''(3) = c$ (1)

$\therefore \quad f(x) = x^3 + ax^2 + bx + c$

$\Rightarrow f'(x) = 3x^2 + 2ax + b$ or

$f'(1) = 3 + 2a + b$ (2)

$\Rightarrow f''(x) = 6x + 2a$, or

$f''(2) = 12 + 2a$ (3)

$\Rightarrow f'''(x) = 6$, or

$f'''(3) = 6$ (4)

From (1) and (4), $c = 6$

From (1), (2), and (3), we have $a = -5, b = 2$. Therefore,

$f(x) = x^3 - 5x^2 + 2x + 6$

$f'(x) = 3x^2 - 10x + 2.$

For Problems 7–9

7. b., 9. d.

Sol. 7. b. From the given information, we have $f(x) = (x - c)^m g(x)$, where $g(x)$ is polynomial of degree $n - m$, Then $x = c$ is common root for the equations $f(x) = 0$, $f^1(x) = 0, f^2(x) = 0, \cdots, f^{m-1}(x) = 0$, where $f^r(x)$ represent r^{th} derivative of $f(x)$ w.r.t. x.

8. Let $f(x) = a_1x^3 + b_1x^2 + c_1x + d_1 = 0$ has roots α, α, β, then $g(x) = a_2x^3 + b_2x^2 + c_2x + d_2 = 0$ must have roots α, α, γ

$\Rightarrow a_1\alpha^3 + b_1\alpha^2 + c_1\alpha + d_1 = 0$, and (1)

$a_2\alpha^3 + b_2\alpha^2 + c_2\alpha + d_2 = 0$ (2)

α is also a root of equations $f'(x) = 3a_1x^2 + 2b_1x + c_1 = 0$ and $g'(x) = 3a_2x^2 + 2b_2x + c_2 = 0$

$\Rightarrow 3a_1\alpha^2 + 2b_1\alpha + c_1 = 0$, and (3)

$3a_2\alpha^2 + 2b_2\alpha + c_2 = 0$ (4)

Also from $a_2(1) - a_1(2)$, we have

$(a_2b_1 - a_1b_2)\alpha^2 + (c_1a_2 - c_2a_1)\alpha + d_1a_2 - d_2a_1 = 0$ (5)

Eliminating α from (3), (4) and (5) we have

$$\begin{vmatrix} 3a_1 & 2b_1 & c_1 \\ 3a_2 & 2b_2 & c_2 \\ a_2b_1 - a_1b_2 & c_1a_2 - c_2a_1 & d_1a_2 - d_2a_1 \end{vmatrix} = 0.$$

9. d.

For Problems 10–12

10. a., 11. a., 12. c.

Sol.

10.a. Since $1, a_1, a_2, \cdots, a_n$ are roots of $x^n - 1 = 0$, then

$x^n - 1 = (x - 1)(x - a_1)(x - a_2) \cdots (x - a_n)$ (1)

$$\Rightarrow \frac{x^n - 1}{x^n - 1} = (x - a_1)(x - a_2) \cdots (x - a_n)$$

$$\Rightarrow \lim_{x\to 1} \frac{x^n - 1}{x^n - 1} = \lim_{x\to 1} [(x - a_1)(x - a_2) \cdots (x - a_n)]$$

$$\Rightarrow (1 - a_1)(1 - a_2) \cdots (1 - a_n) = n.$$

11.a. From (1) $\log(x^n - 1) = \log(x - 1) + \log(x - a_1) + \cdots + \log(x - a_n)$

Differentiating. w.r.t. x, we get

$$\frac{nx^{n-1}}{x^n - 1} = \frac{1}{x - 1} + \frac{1}{x - a_1} + \frac{1}{x - a_2} + \cdots + \frac{1}{x - a_n} \quad (2)$$

Putting $x = 2$ in (2), we get

$$\frac{n2^{n-1}}{2^n - 1} = 1 + \frac{1}{2 - a_1} + \frac{1}{2 - a_2} + \cdots + \frac{1}{2 - a_n}$$

$$\Rightarrow \frac{1}{2 - a_1} + \frac{1}{2 - a_2} + \cdots + \frac{1}{2 - a_n}$$

$$= \frac{n2^{n-1}}{2^n - 1} - 1$$

$$= \frac{n2^{n-1} - 2^n + 1}{2^n - 1}$$

$$= \frac{2^{n-1}(n - 2) + 1}{2^n - 1}$$

12. c. From (2), $\dfrac{nx^{n-1}}{x^n - 1} - \dfrac{1}{x - 1}$

$$= \frac{1}{x - a_1} + \frac{1}{x - a_2} + \cdots + \frac{1}{x - a_n}$$

$$\Rightarrow \frac{nx^{n-1} - 1(1 + x + x^2 + \cdots + x^{n-1})}{x^n - 1}$$

$$= \frac{1}{x - a_1} + \frac{1}{x - a_2} + \cdots + \frac{1}{x - a_n}$$

$$\Rightarrow \lim_{x\to 1} \frac{nx^{n-1} - 1(1 + x + x^2 + \cdots + x^{n-1})}{x^n - 1}$$

$$= \lim_{x\to 1}\left(\frac{1}{x-a_1}+\frac{1}{x-a_2}+\cdots+\frac{1}{x-a_n}\right)$$

$$\Rightarrow \lim_{x\to 1}\frac{n(n-1)x^{n-2}-(1+2x+\cdots+(n-1)x^{n-2})}{nx^{n-1}}$$

$$=\frac{1}{1-a_1}+\frac{1}{1-a_2}+\cdots+\frac{1}{1-a_n}$$

[Applying L Hopsital's rule on L.H.S.]

$$\Rightarrow \frac{n(n-1)-(1+2+\cdots+(n-1))}{n}$$

$$=\frac{1}{1-a_1}+\frac{1}{1-a_2}+\cdots+\frac{1}{1-a_n}$$

$$\Rightarrow \frac{1}{1-a_1}+\frac{1}{1-a_2}+\cdots+\frac{1}{1-a_n}=\frac{n-1}{2}$$

For Problems 13–15

13.b., 14. c., 15. c.

Sol.

Here put $g'(1)=a, g''(2)=b$ (1)
Then $f(x) = x^2 + ax + b$, $f(1) = 1 + a + b$
$\Rightarrow f'(x) = 2x + a, f''(x) = 2.$
$\therefore g(x) = (1 + a + b)x^2 + (2x + a)x + 2$
$= x^2(3 + a + b) + ax + 2.$
$\Rightarrow g'(x) = 2x(3 + a + b) + a$ and $g''(x) = 2(3 + a + b).$
Hence, $g'(1) = 2(3 + a + b) + a$ (2)
$g''(2) = 2(3 + a + b)$ (3)
From (1), (2), and (3), we have $a = 2(3 + a + b) + a$ and $b = 2(3 + a + b)$
$\Rightarrow 3 + a + b = 0$ and $b + 2a + 6 = 0$
Hence, $b = 0$ and $a = -3$. So, $f(x) = x^2 - 3x$ and $g(x) = -3x + 2$.

$$\sqrt{\frac{f(x)}{g(x)}}=\sqrt{\frac{x^2-3x}{-3x+2}} \text{ is defined if } \frac{x^2-3x}{-3x+2}\ge 0$$

$$\Rightarrow \frac{x(x-3)}{(x-2/3)}\le 0 \Rightarrow x\in(-\infty,0]\cup(2/3,3]$$

Matrix-Match Type

1. **a.** $\to$ **p; b.** $\to$ **q, r; c.** $\to$ **s, r; d.** $\to$ **q, r.**

Sol.

a. $f(1-x)=f(1+x)$
$\Rightarrow -f'(1-x)=f'(1+x).$
Hence, graph of $f(x)$ is symmetrical about point (1, 0) (as if $f(x)=-f(-x)$, then $f(x)$ is odd and its graph is symmetrical about (0, 0). Now shift the graph at (1, 0)).

b. $f(2-x)+f(x)=0$
Replace x by $1+x$, then $f(2-(1+x))+f(1+x)=0$
$\Rightarrow f(1-x)+f(1+x)=0$
$\Rightarrow -f'(1-x)+f'(1+x)=0$
$\Rightarrow f'(1-x)=f'(1+x)$ (1)
$\Rightarrow$ Graph of $f'(x)$ is symmetrical about line $x=1$.
Also, put $x=2$ in (1), we get $f'(-1)=f'(3)$.

c. $f(x+2)+f(x)=0$ (1)
Replacing x by $x+2$, we get $f(x+4)+f(x+2)=0$ (2)
From (1) and (2), we have $f(x)=f(x+4)$
Hence, $f(x)$ is periodic with period 4.
Also,
$f'(x)=f'(x+4)$. Hence $f'(x)$ is periodic with period 4.
Put $x=-1$ in $f'(x)=f'(x+4)$, we get $f'(-1)=f'(3)$.

d. Putting $x=0, y=0$, we get $2f(0)+\{f(0)\}^2=1$
$\Rightarrow f(0)=\sqrt{2}-1$ $[\because f(0)>0]$
Putting $y=x$, $2f(x)+\{f(x)\}^2=1$
Differentiating. w.r.t. x, we get
$2f'(x)+2f(x).f'(x)=0$ or $f'(x)\{1+f(x)\}=0$
$\Rightarrow f'(x)=0$, because $f(x)>0$.

2. **a.** $\to$ **q; b.** $\to$ **r; c.** $\to$ **s; d.** $\to$ **p.**

Sol.

a. $$\frac{dy}{dx}=\frac{dy/dt}{dx/dt}=\frac{12t^2-6t-18}{5t^4-15t^2-20}$$

$$\Rightarrow \left.\frac{dy}{dx}\right|_{t=1}=\frac{12-6-18}{5-15-20}=\frac{2}{5}$$

$$\Rightarrow 5\left.\frac{dy}{dx}\right|_{t=1}=2 \text{ at } t=1.$$

b. Let us take $P(x)=a(x-2)^4+b(x-2)^3+c(x-2)^2+d(x-2)+e$

$-1=P(2)=e$

$0=P'(2)=d$

$2=P''(2)=2c\Rightarrow c=1$

$-12=P'''(2)=6b\Rightarrow b=-2$

Thus, $P''(x)=12(x-2)^2-12(x-2)+2$

$\Rightarrow P''(3)=12-12(1)+2=2$

c. Here $$\sqrt{(1+y^4)}=\sqrt{\left(1+\frac{1}{x^4}\right)}=\frac{\sqrt{1+x^4}}{x^2}$$ $\left(\because y=\frac{1}{x}\right)$

$$\Rightarrow \frac{\sqrt{1+y^4}}{\sqrt{1+x^4}}=\frac{1}{x^2} \quad (1)$$

But $y = \frac{1}{x}$

$\therefore \frac{dy}{dx} = -\frac{1}{x^2}$ (2)

From (1) and (2), $\frac{\sqrt{1+y^4}}{\sqrt{1+x^4}} = -\frac{dy}{dx}$

$\Rightarrow \frac{dy}{\sqrt{(1+y^4)}} + \frac{dx}{\sqrt{(1+x^4)}} = 0$

d. Obviously, $f(x)$ is a linear function.
Also from $f'(0) = p$ and $f(0) = q$, $f(x) = px + q$.
$\Rightarrow f''(0) = 0$

3. a. → q, r; b. → p, r; c. → q, s; d. → q, r.

Sol.

a. We know that

$$2\tan^{-1}x = \begin{cases} \sin^{-1}\left(\frac{2x}{1+x^2}\right), & \text{if } -1 \le x \le 1 \\ \pi - \sin^{-1}\left(\frac{2x}{1+x^2}\right), & \text{if } x > 1 \\ -\pi - \sin^{-1}\left(\frac{2x}{1+x^2}\right), & \text{if } x < -1 \end{cases}$$

$\Rightarrow \frac{dy}{dx} = -\frac{2}{1+x^2}$ if $x < -1$ or $x > 1$

b. $\cos^{-1}\left(\frac{1}{\sqrt{1+x^2}}\right) = \begin{cases} \tan^{-1}x, & x \ge 0 \\ -\tan^{-1}x, & x < 0 \end{cases}$

$\Rightarrow \frac{dy}{dx} = -\frac{1}{1+x^2}$ if $x < 0$

c. $y = |e^{|x|} - e| = \begin{cases} |e^x - e|, & x \ge 0 \\ |e^{-x} - e|, & x < 0 \end{cases} =$

$$\begin{cases} e^x - e, & x \ge 1 \\ e - e^x, & 0 \le x < 1 \\ e - e^{-x}, & -1 \le x < 0 \\ e^{-x} - e, & x < -1 \end{cases}$$

$\Rightarrow \frac{dy}{dx} > 0$ if $x > 1$ of $-1 < x < 0$.

d. $u = \log|2x|$, $v = |\tan^{-1}x|$

$\Rightarrow \frac{dy}{dx} = \frac{1}{x}$, and $\frac{dv}{dx} = \begin{cases} \frac{1}{1+x^2}, & x > 0 \\ -\frac{1}{1+x^2}, & x < 0 \end{cases}$

$\Rightarrow \frac{du}{dv} = \begin{cases} \frac{1+x^2}{x}, & x > 0 \\ -\frac{1+x^2}{x}, & x < 0 \end{cases}$

Now we know that $\frac{1+x^2}{x} = x + \frac{1}{x} > 2$ if $x > 1$ and < -2 if $x < -1$

$\Rightarrow \frac{du}{dv} > 2$ if $x < -1$ or $x > 1$

Integer Type

1. (8) $\ln[f(x)] = \ln(x-1) + \ln(x-2) + \ldots + \ln(x-n)$

$\Rightarrow f'(x) = f(x)\left[\frac{1}{x-1} + \frac{1}{x-2} + \ldots + \frac{1}{x-n}\right]$

$\Rightarrow f'(x) = (x-2)(x-3)(x-n) + (x-1)(x-3)\cdots(x-n) + \cdots + (x-1)(x-2)\cdots[x-(n-1)]$

$\Rightarrow f'(n) = (n-1)(n-2)(n-3) \times 3 \times 2 \times 1$ (All other factors except the last vanishes when $x = n$.)

$\Rightarrow 5040 = (n-1)!$

$\Rightarrow n = 8$

2. (9) $f'(x) = \lim_{h\to 0} \frac{f(x+h) - f(x+0)}{h}$

$\Rightarrow$

$\lim_{h\to 0} \frac{2f(x) + xf(h) + h\sqrt{f(x)} - 2f(x) - xf(0) - 0\sqrt{f(x)}}{h}$

$\Rightarrow \lim_{h\to 0} x\left(\frac{f(h) - f(0)}{h-0}\right) + \sqrt{f(x)}$ $(\because f(0) = 0)$

$\Rightarrow f'(x) = \sqrt{f(x)}$ $(\because f'(0) = 0)$

$\Rightarrow \int \frac{f'(x)}{\sqrt{f(x)}} dx = \int dx$

$\Rightarrow 2\sqrt{f(x)} = x + c$

$\Rightarrow f(x) = x^2/4$

3. (5) $f(x) = x + \cfrac{1}{x + x + \cfrac{1}{2x + \cfrac{1}{2x + \ldots\infty}}} = x + \frac{1}{x + f(x)}$

$\Rightarrow f(x) - x = \frac{1}{x + f(x)}$

$\Rightarrow f^2(x) - x^2 = 1$

Differentiating w.r.t. x, we get

$2f(x)f'(x) - 2x = 0$

$\Rightarrow f(x)f'(x) = x$

$\Rightarrow f(50)f'(50) = 100$

4. (3) $f(x) = \cos^{-1}\frac{1}{\sqrt{13}}(2\cos x - 3\sin x) + \sin^{-1}\frac{1}{\sqrt{13}}(2\cos x + 3\sin x)$

$$= \cos^{-1}\left[\frac{1}{\sqrt{13}}\sqrt{13}\cos\left(x+\tan^{-1}\frac{3}{2}\right)\right] + \sin^{-1}\left[\frac{1}{\sqrt{13}}\sqrt{13}\sin\left(x+\tan^{-1}\frac{2}{3}\right)\right]$$

$$= \cos^{-1}\left[\cos\left(x+\tan^{-1}\frac{3}{2}\right)\right] + \sin^{-1}\left[\sin\left(x+\tan^{-1}\frac{2}{3}\right)\right]$$

$$\Rightarrow f'(x) = \frac{\sin\left(x+\tan^{-1}\frac{3}{2}\right)}{\sqrt{1-\cos^2\left(x+\tan^{-1}\frac{3}{2}\right)}} + \frac{\cos\left(x+\tan^{-1}\frac{2}{3}\right)}{\sqrt{1-\sin^2\left(x+\tan^{-1}\frac{2}{3}\right)}}$$

$$= \frac{\sin\left(x+\tan^{-1}\frac{3}{2}\right)}{\left|\sin\left(x+\tan^{-1}\frac{3}{2}\right)\right|} + \frac{\cos\left(x+\tan^{-1}\frac{2}{3}\right)}{\left|\cos\left(x+\tan^{-1}\frac{2}{3}\right)\right|}$$

$\Rightarrow f'(3/4) = 1+1 = 2$

Now let $g(x) = \sqrt{1+x^2}$. Therefore,

$$g'(x) = \frac{x}{\sqrt{1+x^2}}$$

$$\Rightarrow g'\left(\frac{3}{4}\right) = \left(\frac{3}{5}\right)$$

$$\Rightarrow \frac{f'(3/4)}{g'(3/4)} = \frac{10}{3}$$

5. **(0)** Given $f(x+y^3) = f(x) + [f(y)]^3$ (1)
and $f'(0) \geq 0$ (2)
Replacing x, y by 0, we get
$f(0) = f(0) + f(0)^3 \Rightarrow f(0)$ (3)

Also $f'(0) = \lim\limits_{h\to 0}\dfrac{f(0+h)-f(0)}{h} = \lim\limits_{h\to 0}\dfrac{f(h)}{h}$ (4)

Let $I = f'(0) = \lim\limits_{h\to 0}\dfrac{f[0+(h^{1/3})^3]-f(0)}{(h^{1/3})^3}$

$$= \lim_{h\to 0}\frac{f[(h^{1/3})]^3}{(h^{1/3})^3} = \lim_{h\to 0}\left(\frac{f(h^{1/3})}{(h^{1/3})}\right)^3 = I^3$$

$\Rightarrow I = I^3$
or $I = 0, 1, -1$ as $f'(0) \geq 0$. Therefore,
$f'(0) = 0, 1$ (5)

Thus $f'(x)$

$$= \lim_{h\to 0}\frac{f(x+h)-f(x)}{h} = \lim_{h\to 0}\frac{f[x+(h^{1/3})^3]-f(x)}{(h^{1/3})^3}$$

$$\Rightarrow f'(x) = \lim_{h\to 0}\frac{f(x)+[f(h^{1/3})]^3-f(x)}{(h^{1/3})^3} \quad \text{[Using (1)]}$$

$$\Rightarrow f'(x) = \lim_{h\to 0}\left(\frac{f(h^{1/3})}{(h^{1/3})}\right)^3 = [f'(0)]^3$$

$\Rightarrow f'(x) = 0, 1,$ as $f'(0) = 0, 1$ [Using (5)]
Integrating both sides, we get
$f(x) = c$ or $x + c$, as $f(0) = 0$
$\Rightarrow f(x) = 0$ or x
Thus $f(10) = 0$ or 10.

6. **(4)** Given that f is a twice differentiable function such that $f''(x) = -f(x)$ and $f'(x) = g(x)$
$h(x) = [f(x)]^2 + [g(x)]^2$
$\Rightarrow h'(x) = 2f(x)f'(x) + 2g(x)g'(x)$
$= 2f(x)g(x) + 2g(x)f''(x)$
$[\because g(x) = f'(x) \Rightarrow g'(x) = f''(x)]$
$= 2f(x)g(x) + 2g(x)[-f(x)]$
$= 2f(x)g(x) - 2f(x)g(x)$
$= 0$
Thus $h'(x) = 0,\ \forall\ x$
$\Rightarrow h$ is a constant function. Therefore,
$h(5) = 4$
$\Rightarrow h(10) = 4.$

7. **(0)** Given $(\sin y)^{\sin\left(\frac{\pi x}{2}\right)} + \dfrac{\sqrt{3}}{2}\sec^{-1}(2x) + 2^x\tan[\log(x+2)] = 0$ (1)
For $x = -1$, we have

$$(\sin y)^{\sin\left(\frac{-\pi}{2}\right)} + \frac{\sqrt{3}}{2}\sec^{-1}(-2) + 2^{-1}\tan[\log(-1+2)] = 0$$

$$\Rightarrow (\sin y)^{-1} + \frac{\sqrt{3}}{2}\left(\frac{2\pi}{3}\right) + \frac{1}{2}\tan 0 = 0$$

$$\Rightarrow \frac{1}{\sin y} = -\frac{\pi}{\sqrt{3}} \Rightarrow \sin y = -\frac{\sqrt{3}}{\pi}, \text{ when } x = -1 \quad (2)$$

Now let $u = (\sin y)^{\sin\left(\frac{\pi x}{2}\right)}$

Taking log on both sides, we get

$$\log u = \sin\left(\frac{\pi x}{2}\right)\log \sin y.$$

Differentiating both sides w.r.t. x, we get

$$\frac{1}{u}\frac{du}{dx} = \frac{\pi}{2}\cos\left(\frac{\pi x}{2}\right)\log\sin y + \cot y\frac{dy}{dx}\sin\left(\frac{\pi x}{2}\right)$$

$$\Rightarrow \frac{du}{dx}$$

$$= (\sin y)^{\sin\left(\frac{\pi x}{2}\right)}$$

$$\times\left[\frac{\pi}{2}\cos\left(\frac{\pi x}{2}\right)\log\sin y + \sin\left(\frac{\pi x}{2}\right)\cot y\frac{dy}{dx}\right] \quad (3)$$

Now differentiating equation (1), we get

$$(\sin y)^{\sin\left(\frac{\pi x}{2}\right)}$$

$$\times\left[\frac{\pi}{2}\cos\left(\frac{\pi x}{2}\right)\log\sin y + \sin\left(\frac{\pi x}{2}\right)\cot y\frac{dy}{dx}\right]$$

$$+\frac{\sqrt{3}}{2x\sqrt{4x^2-1}} + 2^x(\log 2)\tan[\log(x+2)]$$

$$+\frac{2^x\sec^2[\log(x+2)]}{x+2} = 0$$

At $x = -1$ and $\sin y = -\frac{\sqrt{3}}{\pi}$, we get

$$\left(-\frac{\sqrt{3}}{\pi}\right)^{-1}\left[0+(-1)\sqrt{\frac{\pi^2}{3}-1}\left(\frac{dy}{dx}\right)_{x=-1}\right]$$

$$+\frac{\sqrt{3}}{-2\sqrt{3}} + 0 + 2^{-1} = 0$$

$$\Rightarrow \frac{\pi}{\sqrt{3}\sqrt{3}}\sqrt{\pi^2-3}\left(\frac{dy}{dx}\right)_{x=-1} - \frac{1}{2}+\frac{1}{2} = 0$$

$$\Rightarrow \frac{dy}{dx} = 0$$

8. **(6)** $F(x) = f(x)\,g(x)\,h(x),\ \forall\ x \in R$
$f(x), g(x), h(x)$ are differentiable functions,
$\Rightarrow F'(x) = f'(x)\,g(x)\,h(x) + f(x)\,g'(x)\,h(x) + f(x)\,g(x)\,h'(x)$
At $x = x_0$
$F'(x_0) = f'(x_0)\,g(x_0)\,h(x_0) + f(x_0)\,g'(x_0)\,h(x_0) + f(x_0)\,g(x_0)\,h'(x_0)$
Using the given values of $F'(x_0), f'(x_0), g'(x_0)$, and $h'(x_0)$, we get
$21F(x_0) = 4f(x_0)\,g(x_0)\,h(x_0) - 7f(x_0)\,g(x_0)\,h(x_0) + kh(x_0)f(x_0)\,g(x_0)$
$\Rightarrow 21 = 4 - 7 + k \qquad [\because F(x_0) = f(x_0)\,g(x_0)\,h(x_0)]$
$\Rightarrow k = 24$

9. **(1)** Clearly, we can get $a_1 + 2a_2 + 3a_3 + \ldots + na_n$ by differentiating $a_1\sin x + a_2\sin 2x + \ldots + a_n\sin nx$.
Let $f(x) = a_1\sin x + a_2\sin 2x + \cdots + a_n\sin nx$
$\Rightarrow f'(x) = a_1\cos x + 2a_2\cos 2x + \cdots + na_n\cos nx$
$\Rightarrow f'(0) = a_1 + 2a_2 + \cdots + na_n$
Also given $|f(x)| \le |\sin x|$ for $x \in R$ (1)
Putting $x = 0$, we get
$|f(0)| \le 0 \Rightarrow f(0) = 0.$

Now, $f'(0) = \lim_{h\to 0}\frac{f(h)-f(0)}{h}$

$\Rightarrow f'(0) = \lim_{h\to 0}\frac{f(h)}{h}$, as $f(0) = 0$

$\Rightarrow |f'(0)| = \lim_{h\to 0}\left|\frac{f(h)}{h}\right| \le \lim_{h\to 0}\left|\frac{\sin h}{h}\right| = 1$, as $|f(x)| \le |\sin x|$

Hence, $|f'(0)| \le 1$].

10. **(8)** Since $f\left(\frac{x+y}{3}\right) = \frac{2+f(x)+f(y)}{3}$ (1)

Replacing x by $3x$ and y by 0, we get

$f(x) = \frac{2+f(3x)+f(0)}{3}$

$\Rightarrow f(3x) - 3f(x) + 2 = -f(0)$ (2)

Putting $x = 0$ and $y = 0$ in (1), we get

$f(0) = 2$ (3)

Now $f'(x) = \lim_{h\to 0}\frac{f(x+h)-f(x)}{h}$

$= \lim_{h\to 0}\frac{f[(3x+3h)/3]-f(x)}{h}$

$= \lim_{h\to 0}\frac{\frac{2+f(3x)+f(3h)}{3}-f(x)}{h}$

$= \lim_{h\to 0}\frac{f(3x)-3f(x)+f(3h)+2}{3h}$

$= \lim_{h\to 0}\frac{f(3h)-f(0)}{3h}$ [From (2)]

$= f'(0) = c$, say. Therefore,
$f'(x) = c$
At $x = 2, f'(2) = c = 2$ (given)
Therefore, $f'(x) = 2$.
Integrating both sides, we get
$f(x) = 2x + d$ [From (2)]
Therefore, $d = 2$
Hence, $f(x) = 2x + 2$.

11. **(3)** We have $f(x)\,g(x) = 1$. Differentiating with respect to x, we get
$f'g + fg' = 0$ (1)
Differentiating (1) w.r.t. x, we get
$f''g + 2f'g' + fg'' = 0$ (2)
Differentiating (2) w.r.t. x, we get
$f'''g + g'''f + 3f''g' + 3g''f' = 0$
$\Rightarrow$

$$\frac{f'''}{f'}(f'g) + \frac{g'''}{g'}(fg') + \frac{3f''}{f}(fg') + \frac{3g''}{g}(gf') = 0$$

$$\Rightarrow \left(\frac{f'''}{f'}+\frac{3g''}{g}\right)(f'g) = -\left(\frac{g'''}{g'}+\frac{3f''}{f}\right)(f'g)$$

$$\Rightarrow -\left(\frac{f'''}{f'}+\frac{3g''}{g}\right)(fg') = -\left(\frac{g'''}{g'}+\frac{3f''}{g}\right)fg'$$

[Using (1)]

$$\Rightarrow \frac{f'''}{f'}+\frac{3g''}{g} = \frac{g'''}{g'}+\frac{3f''}{f}$$

$$\Rightarrow \frac{f'''}{f'}-\frac{g'''}{g'} = 3\left(\frac{f''}{f}-\frac{g''}{g}\right)$$

12. **(2)** Let $u = f(\tan x)$ and $v = g(\sec x)$

$$\Rightarrow \frac{du}{dx} = f'(\tan x)\sec^2 x$$

and $\dfrac{dv}{dx} = g'(\sec x)\sec x\tan x$

$$\Rightarrow \frac{du}{dv} = \frac{du}{dx}\Big/\frac{dv}{dx} = \frac{f'(\tan x)\sec^2 x}{g'(\sec x)\sec x\tan x}$$

$$\Rightarrow \left[\frac{du}{dv}\right]_{x=(\pi/4)} = \frac{f'\left(\tan\dfrac{\pi}{4}\right)}{g'\left(\sec\dfrac{\pi}{4}\right)\sin\dfrac{\pi}{4}} = \frac{f'(1)\times\sqrt{2}}{g'(\sqrt{2})}$$

$$= \frac{2\sqrt{2}}{4} = \frac{1}{\sqrt{2}}$$

13. **(2)** Given that $y(x^2+c) = ax+b$

Differentiating w.r.t. x, we get

$y'(x^2+c)+2xy = a$ (1)

Again differentiating w.r.t. x, we get

$y''(x^2+c)+y'\,2x+2xy'+2y = 0$

$\Rightarrow y''(x^2+c)+2(2xy'+y) = 0(2)$

Again differentiating w.r.t. x, we get

$y'''(x^2+c)+2xy''+2(2xy''+3y') = 0$

$\Rightarrow y'''(x^2+c)+6(xy''+y') = 0$ (3)

Multiplying (2) by y''' and (3) by y'' and then subtracting, we get

$2(2xy'+y)y'''-6(xy''+y')y'' = 0$

$\Rightarrow (2xy'+y)y''' = 3(xy''+y')y''$

14. **(0)** $|f(x)| \le x^2,\ \forall\ x\in R$ (1)

At $x = 0$, $|f(0)| \le 0 \Rightarrow f(0) = 0$ (2)

$$f'(0) = \lim_{h\to 0}\frac{f(h)-f(0)}{h} = \lim_{h\to 0}\frac{f(h)}{h} \quad (3)$$

Now, $\left|\dfrac{f(h)}{h}\right| \le |h|$ [From (1)]

$$\Rightarrow -|h| \le \frac{f(h)}{h} \le |h|$$

$$\Rightarrow \lim_{h\to 0}\frac{f(h)}{h} \to 0 \quad \text{(4) (Using sandwich theorem)}$$

Therefore, from (3) and (4), we get $f'(0) = 0$, i.e., $f(x)$ is differentiable at $x = 0$.

15. $f(x)+f(y) = f\left(\dfrac{x+y}{1-xy}\right)$ (1)

Putting $x = y = 0$, we get $f(0) = 0$.

Putting $y = -x$, we get $f(+x)+f(-x) = f(0)$

$\Rightarrow f(-x) = -f(x)$ (2)

Also $\displaystyle\lim_{x\to 0}\frac{f(x)}{x} = 2$

Now $f'(x) = \displaystyle\lim_{h\to 0}\frac{f(x+h)-f(x)}{h}$ (3)

$$= \lim_{h\to 0}\frac{f(x+h)+f(-x)}{h} \quad \text{[using (2) } -f(x) = f(-x)\text{]}$$

$$f'(x) = \lim_{h\to 0}\frac{f\left(\dfrac{x+h-x}{1-(x+h)(-x)}\right)}{h} \quad \text{[using (1)]}$$

$$\Rightarrow f'(x) = \lim_{h\to 0}\left[\frac{f\left(\dfrac{h}{1+x(x+h)}\right)}{h}\right]$$

$$\Rightarrow f'(x) = \lim_{h\to 0}\frac{f\left(\dfrac{h}{1+xh+x^2}\right)}{\left(\dfrac{h}{1+xh+x^2}\right)}\times\left(\frac{1}{1+xh+x^2}\right)$$

$$\Rightarrow f'(x) = \lim_{h\to 0}\frac{f\left(\dfrac{h}{1+xh+x^2}\right)}{\left(\dfrac{h}{1+xh+x^2}\right)}\times\lim_{h\to 0}\frac{1}{1+xh+x^2}$$

$$\left(\text{Using } \lim_{x\to 0}\frac{f(x)}{x} = 2\right)$$

$$\Rightarrow f'(x) = 2\times\frac{1}{1+x^2}$$

$$\Rightarrow f'(x) = \frac{2}{1+x^2}$$

Integrating both sides, we get $f(x) = 2\tan^{-1}(x)+c$. Therefore, $f(0) = 0 \quad \Rightarrow c = 0$.

Thus $f(x) = 2\tan^{-1} x$.

Hence, $f\left(\dfrac{1}{\sqrt{3}}\right) = 2\tan^{-1}\left(\dfrac{1}{\sqrt{3}}\right) = 2\times\dfrac{\pi}{6} = \dfrac{\pi}{3}$

and $f'(1) = \frac{2}{1+1^2} = \frac{2}{2} = 1.$

16. (6) $f(x+y) = f(x)f(y)$ (1)

$$f'(5) = \lim_{h\to 0} \frac{f(5+h)-f(5)}{h}$$

$$= \lim_{h\to 0} \frac{f(5)f(h)-f(5)}{h}$$

$$= f(5)\lim_{h\to 0} \frac{f(h)-1}{h}$$

$$= f(5)\lim_{h\to 0} \frac{f(h)-1}{h}$$

Replacing x by 5 and y by 0, we get

$f(5+0) = f(5)f(0)$

$\Rightarrow f(0) = 1$

$$\Rightarrow f'(5) = f(5)\lim_{h\to 0} \frac{f(h)-f0)}{h}$$

$= f(5)f'(0) = 2\times 3 = 6$

17. (1) $$\frac{dy}{dx} = \begin{vmatrix} \cos x & -\sin x & \cos x \\ \sin x & -\sin x & \cos x \\ x & 1 & 1 \end{vmatrix}$$

$$+\begin{vmatrix} \sin x & \cos x & \sin x \\ -\sin x & -\cos x & -\sin x \\ x & 1 & 1 \end{vmatrix}$$

$$+\begin{vmatrix} \sin x & \cos x & \sin x \\ \cos x & -\sin x & \cos x \\ 1 & 0 & 0 \end{vmatrix}$$

$$= 0 - \begin{vmatrix} \sin x & \cos x & \sin x \\ \sin x & \cos x & \sin x \\ x & 1 & 1 \end{vmatrix} + 1\begin{vmatrix} \cos x & \sin x \\ -\sin x & \cos x \end{vmatrix}$$

$= 0 + (\cos^2 x + \sin^2 x) = 1$

18. (0) $$\frac{d^n}{dx^n}[f(x)] = \begin{vmatrix} \frac{d^n}{dx^n}(x^n) & n! & 2 \\ \frac{d^n}{dx^n}(\cos x) & \cos\frac{n\pi}{2} & 4 \\ \frac{d^n}{dx^n}(\sin x) & \sin\frac{n\pi}{2} & 8 \end{vmatrix}$$

$$= \begin{vmatrix} n! & n! & 2 \\ \cos\left(x+\frac{n\pi}{2}\right) & \cos\frac{n\pi}{2} & 4 \\ \sin\left(x+\frac{n\pi}{2}\right) & \sin\frac{n\pi}{2} & 8 \end{vmatrix}$$

$$\Rightarrow \frac{d^n}{dx^n}[f(x)]_{x=0} = 0$$

19. (6) $f(xy) = f(x)f(y)$ (1)

$$f'(2) = \lim_{h\to 0} \frac{f(2+h)-f(2)}{h}$$

$$= \lim_{h\to 0} \frac{f\left[2\left(1+\frac{h}{2}\right)\right]-f(2)}{h}$$

$$= \lim_{h\to 0} \frac{f(2)f\left(1+\frac{h}{2}\right)-f(2)}{h}$$

$$= 2f(2)\lim_{h\to 0} \frac{f\left(1+\frac{h}{2}\right)-1}{\frac{h}{2}}$$

Replacing x and y by 0 in (1), we get

$f(0) = [f(0)]^2$

$\Rightarrow f(0) = 0$

$$\Rightarrow f'(2) = 2f(2)\lim_{h\to 0} \frac{f\left(1+\frac{h}{2}\right)-f(0)}{\frac{h}{2}} = 2f(2)f'(1)$$

$= 6$

20. (0) $x = \sqrt{a^{\sin^{-1}t}}$, $y = \sqrt{a^{\cos^{-1}t}}$

$\Rightarrow xy = \sqrt{a^{\sin^{-1}t+\cos^{-1}t}} = \sqrt{a^{\pi/2}}$

$\Rightarrow xdy + ydx = 0$

21. (1) $x = a\left[\cos t + \log\tan\frac{t}{2}\right]$ and $y = a\sin t$

Differentiating w.r.t. t, we get

$$\frac{dx}{dt} = a\left[-\sin t + \frac{1}{\tan t/2}\sec^2\frac{t}{2}\times\frac{1}{2}\right]$$

$$= a\left[-\sin t + \frac{1}{2\sin(t/2)\cos(t/2)}\right]$$

$$= a\left[-\sin t + \frac{1}{\sin t}\right]$$

and $\frac{dy}{dt} = a\cos t$

$$\Rightarrow \frac{dy}{dx} = \frac{dy/dt}{dx/dt} = \frac{a\cos t}{\frac{a\cos^2 t}{\sin t}} = \tan t$$

At $x = \pi/4$, $\frac{dy}{dx} = 1.$

22. **(1)** $f(x)f'(-x)=f(-x)f'(x)$

$\Rightarrow f'(x)f(-x)-f(x)f'(-x)=0$

$\Rightarrow \frac{d}{dx}[f(x)f(-x)]=0$

$\Rightarrow f(x)f(-x)=k$

Given $[f(0)]^2=k \Rightarrow k=9$.

Then $f(3)f(-3)=9$

23. **(9)** Let $y=2x\tan^{-1}x-\ln(1+x^2)$

$y'=2\tan^{-1}x+\frac{2x}{1+x^2}-\frac{2x}{1+x^2}$

$\Rightarrow y'>0,\ \forall\ x\in R^+, y'<0,\ \forall\ x\in R^-$

$\Rightarrow y\geq 0,\ \forall\ x\in R$

Therefore, $4-|[x]|$ takes the values 0, 1, 2, 3, 4.
$\{\because |\alpha|\leq 4-|[x]|\}$

$|\alpha|\leq 4-|[x]|$ is satisfied by $\alpha=0,\pm 1,\pm 2,\pm 3,\pm 4$.

Hence, the number of values of α is 9.

24. **(0)** Replace x by $-x$

$\Rightarrow x[f(x)-f(-x)]=0$

$\Rightarrow f(x)$ is odd function

$\Rightarrow f^{iv}(x)$ is also odd

$\Rightarrow f^{iv}(0)=0$

25. **(9)** $f(g(x))=\left(x^2+\frac{1}{x^2}\right)^3-3\left(x^2+\frac{1}{x^2}\right)$

$=(g(x))^3-3g(x)$

$\therefore f(x)=x^3-3x$

$f'(x)=3x^2-3$

26. **(8)** $f(x+y)=f(x)+f(y)+xy(x+y)$

$f(0)=0 \therefore \lim_{h\to 0}\frac{f(h)}{h}=-1$

$\therefore \lim_{h\to 0}\frac{f(x+h)-f(x)}{h}$

$=\lim_{h\to 0}\frac{f(x)+f(h)+xh(x+h)-f(x)}{h}$

$=\lim_{h\to 0}\frac{f(h)}{h}+\lim_{h\to 0}x(x+h)=-1+x^2$

$\therefore f'(x)=-1+x^2$

$\therefore f(x)=\frac{x^3}{3}-x+c$

$\therefore f(x)$ is a polynomial function, $f(x)$ is twice differentiable for all $x\in R$ and $f'(3)=3^2-1=8$

CHAPTER 5

Application of Derivatives: Tangents and Normals, Rate Measure

Objective Type

Each question has four choices a, b, c, and d, out of which *only one* answer is correct.

1. The number of tangents to the curve $x^{3/2} + y^{3/2} = a^{3/2}, a > 0$, which are equally inclined to the axes, is
   a. 2 b. 1
   c. 0 d. 4

2. The angle made by the tangent of the curve $x = a(t + \sin t \cos t); y = a(1 + \sin t)^2$ with the x-axis at any point on it is
   a. $\frac{1}{4}(\pi + 2t)$ b. $\frac{1 - \sin t}{\cos t}$
   c. $\frac{1}{4}(2t - \pi)$ d. $\frac{1 + \sin t}{\cos 2t}$

3. If m is the slope of a tangent to the curve $e^y = 1 + x^2$, then
   a. $|m| > 1$ b. $m > 1$
   c. $m > -1$ d. $|m| \le 1$

4. If at each point of the curve $y = x^3 - ax^2 + x + 1$, the tangent is inclined at an acute angle with the positive direction of the x-axis, then
   a. $a > 0$ b. $a \le \sqrt{3}$
   c. $-\sqrt{3} \le a \le \sqrt{3}$ d. None of these

5. The slope of the tangent to the curve $y = \sqrt{4 - x^2}$ at the point, where the ordinate and the abscissa are equal, is
   a. -1 b. 1
   c. 0 d. None of these

6. The curve given by $x + y = e^{xy}$ has a tangent parallel to the y-axis at the point
   a. (0, 1) b. (1, 0)
   c. (1, 1) d. None of these

7. If the line joining the points (0, 3) and (5, –2) is a tangent to the curve $y = \frac{c}{x+1}$, then the value of c is
   a. 1 b. -2
   c. 4 d. None of these

8. Let $f(x) = \begin{cases} -x^2, & \text{for } x < 0 \\ x^2 + 8, & \text{for } x \ge 0 \end{cases}$. Then x-intercept of the line that is tangent to the graph of $f(x)$ is
   a. zero b. -1
   c. -2 d. -4

9. The distance between the origin and the tangent to the curve $y = e^{2x} + x^2$ drawn at the point $x = 0$ is
   a. $\frac{1}{\sqrt{5}}$ b. $\frac{2}{\sqrt{5}}$
   c. $\frac{-1}{\sqrt{5}}$ d. $\frac{2}{\sqrt{3}}$

10. The point on the curve $3y = 6x - 5x^3$ the normal at which passes through the origin is
    a. (1, 1/3) b. (1/3, 1)
    c. (2, –28/3) d. (–1, – 1/3)

11. The normal to the curve $x = a(\cos\theta + \theta \sin\theta)$, $y = a(\sin\theta - \theta\cos\theta)$ at any point θ is such that

a. it makes angle $\frac{\pi}{2} + \theta$ with the x-axis

b. it passes through the origin

c. It is at a constant distance from the origin

d. it passes through $\left(\frac{a\pi}{2}, -a\right)$

12. At what points of curve $y = \frac{2}{3}x^3 + \frac{1}{2}x^2$, the tangent makes the equal angle with axis?

a. $\left(\frac{1}{2}, \frac{5}{24}\right)$ and $\left(-1, -\frac{1}{6}\right)$

b. $\left(\frac{1}{2}, \frac{4}{9}\right)$ and $(-1, 0)$

c. $\left(\frac{1}{3}, \frac{1}{7}\right)$ and $\left(-3, \frac{1}{2}\right)$

d. $\left(\frac{1}{3}, \frac{4}{47}\right)$ and $\left(-1, -\frac{1}{3}\right)$

13. The equation of the tangent to the curve $y = be^{-x/a}$ at the point where it crosses the y-axis is

a. $\frac{x}{a} - \frac{y}{b} = 1$ b. $ax + by = 1$

c. $ax - by = 1$ d. $\frac{x}{a} + \frac{y}{b} = 1$

14. The angle of intersection of the normals at the point $\left(-\frac{5}{\sqrt{2}}, \frac{3}{\sqrt{2}}\right)$ of the curves $x^2 - y^2 = 8$ and $9x^2 + 25y^2 = 225$ is

a. 0 b. $\frac{\pi}{2}$

c. $\frac{\pi}{3}$ d. $\frac{\pi}{4}$

15. A function $y = f(x)$ has a second order derivative $f''(x) = 6(x - 1)$. If its graph passes through the point (2, 1) and at that point tangent to the graph is $y = 3x - 5$, then the function is

a. $(x+1)^3$ b. $(x-1)^3$

c. $(x-1)^2$ d. $(x+1)^2$

16. If $x + 4y = 14$ is a normal to the curve $y^2 = \alpha x^3 - \beta$ at (2, 3), then the value of $\alpha + \beta$ is

a. 9 b. −5

c. 7 d. −7

17. The curve represented parametrically by the equations $x = 2 \ln \cot t + 1$ and $y = \tan t + \cot t$

a. tangent and normal intersect at the point (2, 1)

b. normal at $t = \pi/4$ is parallel to y-axis

c. tangent at $t = \pi/4$ is parallel to the line $y = x$

d. tangent at $t = \pi/4$ is parallel to x-axis

18. The abscissa of points P and Q on the curve $y = e^x + e^{-x}$ such that tangents at P and Q make 60° with x-axis

a. $\ln\left(\frac{\sqrt{3}+\sqrt{7}}{7}\right)$ and $\ln\left(\frac{\sqrt{3}+\sqrt{5}}{2}\right)$

b. $\ln\left(\frac{\sqrt{3}+\sqrt{7}}{2}\right)$

c. $\ln\left(\frac{\sqrt{7}-\sqrt{3}}{2}\right)$

d. $\pm\ln\left(\frac{\sqrt{3}+\sqrt{7}}{2}\right)$

19. If a variable tangent to the curve $x^2y = c^3$ makes intercepts a, b on x- and y-axis, respectively, then the value of a^2b is

a. $27c^3$ b. $\frac{4}{27}c^3$

c. $\frac{27}{4}c^3$ d. $\frac{4}{9}c^3$

20. Let C be the curve $y = x^3$ (where x takes all real values). The tangent at A meets the curve again at B. If the gradient at B is K times the gradient at A, then K is equal to

a. 4 b. 2

c. −2 d. $\frac{1}{4}$

21. A curve is represented by the equations $x = \sec^2 t$ and $y = \cot t$ where t is a parameter. If the tangent at the point P on the curve, where $t = \pi/4$, meets the curve again at the point Q, then $|PQ|$ is equal to

a. $\frac{5\sqrt{3}}{2}$ b. $\frac{5\sqrt{5}}{2}$

c. $\frac{2\sqrt{5}}{3}$ d. $\frac{3\sqrt{5}}{2}$

22. The x-intercept of the tangent at any arbitrary point of the curve $\frac{a}{x^2} + \frac{b}{y^2} = 1$ is proportional to

a. square of the abscissa of the point of tangency

b. square root of the abscissa of the point of tangency

c. cube of the abscissa of the point of tangency

d. cube root of the abscissa of the point of tangency

23. At any point on the curve $2x^2y^2 - x^4 = c$, the mean proportional between the abscissa and the

difference between the abscissa and the subnormal drawn to the curve at the same point is equal to

a. ordinate b. radius vector
c. x-intercept of tangent d. sub-tangent

24. If the length of sub-normal is equal to the length of sub-tangent at any point (3, 4) on the curve $y = f(x)$ and the tangent at (3, 4) to $y = f(x)$ meets the coordinate axes at A and B, then the maximum area of the triangle OAB, where O is origin, is

a. 45/2 b. 49/2
c. 25/2 d. 81/2

25. The number of points in the rectangle $\{(x, y)| -12 \le x \le 12$ and $-3 \le y \le 3\}$ which lie on the curve $y = x + \sin x$ and at which the tangent to the curve is parallel to the x-axis is

a. 0 b. 2
c. 4 d. 8

26. Tangent of acute angle between the curves $y = |x^2 - 1|$ and $y = \sqrt{7 - x^2}$ at their points of intersection is

a. $\frac{5\sqrt{3}}{2}$ b. $\frac{3\sqrt{5}}{2}$
c. $\frac{5\sqrt{3}}{4}$ d. $\frac{3\sqrt{5}}{4}$

27. The lines tangent to the curves $y^3 - x^2y + 5y - 2x = 0$ and $x^4 - x^3y^2 + 5x + 2y = 0$ at the origin intersect at an angle θ equal to

a. $\frac{\pi}{6}$ b. $\frac{\pi}{4}$
c. $\frac{\pi}{3}$ d. $\frac{\pi}{2}$

28. The two curves $x = y^2$, $xy = a^3$ cut orthogonally at a point, then a^2 is equal to

a. $\frac{1}{3}$ b. 3
c. 2 d. $\frac{1}{2}$

29. The curves $4x^2 + 9y^2 = 72$ and $x^2 - y^2 = 5$ at (3, 2)

a. touch each other b. cut orthogonally
c. intersect at 45° d. intersect at 60°

30. Let $f(1) = -2$ and $f'(x) \ge 4.2$ for $1 \le x \le 6$. The smallest possible value of $f(6)$ is

a. 9 b. 12
c. 15 d. 19

31. If $f(x) = x^3 + 7x - 1$, then $f(x)$ has a zero between $x = 0$ and $x = 1$. The theorem that best describes this is

a. Squeeze play theorem
b. Mean value theorem
c. Maximum-minimum value theorem
d. Intermediate value theorem

32. Consider the function $f(x) = \begin{cases} x\sin\frac{\pi}{x} & \text{for } x > 0 \\ 0, & \text{for } x = 0 \end{cases}$

then the number of points in (0, 1) where the derivative $f'(x)$ vanishes, is

a. 0 b. 1
c. 2 d. infinite

33. Let $f(x)$ and $g(x)$ be differentiable for $0 \le x \le 1$, such that $f(0) = 0, g(0) = 0, f(1) = 6$. Let there exists a real number c in (0, 1) such that $f'(c) = 2g'(c)$, then the value of $g(1)$ must be

a. 1 b. 3
c. −2 d. −1

34. If $3(a + 2c) = 4(b + 3d)$, then the equation $ax^3 + bx^2 + cx + d = 0$ will have

a. no real solution
b. at least one real root in (−1, 0)
c. at least one real root in (0, 1)
d. None of these

35. If the function $f(x) = ax^3 + bx^2 + 11x - 6$ satisfies conditions of Rolle's theorem in [1, 3] for $x = 2 + \frac{1}{\sqrt{3}}$, then value of a and b are respectively

a. −3, 2 b. 2, −4
c. 1, 6 d. None of these

36. A value of C for which the conclusion of mean value theorem holds for the function $f(x) = \log_e x$ on the interval [1, 3] is

a. $\frac{1}{2}\log_e 3$ b. $\log_3 e$
c. $\log_e 3$ d. $2\log_3 e$

37. Let $f(x)$ be a twice differentiable function for all real values of x and satisfies $f(1) = 1, f(2) = 4, f(3) = 9$. Then which of the following is definitely true?

a. $f''(x) = 2$, for $\forall x \in (1, 3)$
b. $f''(x) = f'(x) = 5$ for some $x \in (2, 3)$
c. $f''(x) = 3\ \forall x \in (2, 3)$
d. $f''(x) = 2$ for some $x \in (1, 3)$

38. The value of c in Lagrange's theorem for the function $f(x) = \log \sin x$ in the interval $[\pi/6, 5\pi/6]$

a. $\pi/4$ b. $\pi/2$
c. $2\pi/3$ d. None of these

39. In which of the following functions is Rolle's theorem applicable?

a. $f(x) = \begin{cases} x, & 0 \le x < 1 \\ 0, & x = 1 \end{cases}$ on [0, 1]

b. $f(x)=\begin{cases} \dfrac{\sin x}{x}, & -\pi \le x < 0 \\ 0, & x=0 \end{cases}$ on $[-\pi, 0]$

c. $f(x)=\dfrac{x^2-x-6}{x-1}$ on $[-2, 3]$

d. $f(x)=\begin{cases} \dfrac{x^3-2x^2-5x+6}{x-1}, & \text{if } x \ne 1, \text{ on } [-2,3] \\ -6, & \text{if } x=1 \end{cases}$

40. A point on the parabola $y^2 = 18x$ at which the ordinate increases at twice the rate of the abscissa is

a. $(2, 6)$ b. $(2, -6)$

c. $\left(\dfrac{9}{8}, -\dfrac{9}{2}\right)$ d. $\left(\dfrac{9}{8}, \dfrac{9}{2}\right)$

41. The rate of change of the volume of a sphere w.r.t. its surface area, when the radius is 2 cm, is

a. 1 b. 2

c. 3 d. 4

42. If there is an error of k % in measuring the edge of a cube, then the percent error in estimating its volume is

a. k b. $3k$

c. $\dfrac{k}{3}$ d. None of these

43. A lamp of negligible height is placed on the ground l_1 away from a wall. *A* man l_2 m tall is walking at a speed of $\dfrac{l_1}{10}$ *m*/s from the lamp to the nearest point on the wall. When he is midway between the lamp and the wall, the rate of change in the length of this shadow on the wall is

a. $-\dfrac{5l_2}{2}$ m/s b. $-\dfrac{2l_2}{5}$ m/s

c. $-\dfrac{l_2}{2}$ m/s d. $-\dfrac{l_2}{5}$ m/s

44. A man is moving away from a tower 41.6 m high at a rate of 2 m/s. If the eye level of the man is 1.6 m above the ground, then the rate at which the angle of elevation of the top of the tower changes, when he is at a distance of 30 m from the foot of the tower, is

a. $-\dfrac{4}{125}$ radian/s b. $-\dfrac{2}{25}$ radian/s

c. $-\dfrac{1}{625}$ radian/s d. None of these

45. The co-ordinates of a point on the parabola $y^2 = 8x$ whose distance from the circle $x^2 + (y + 6)^2 = 1$ is minimum is

a. $(2, 4)$ b. $(2, -4)$

c. $(18, -12)$ d. $(8, 8)$

46. At the point $P(a, a^n)$ on the graph of $y = x^n$ $(n \in N)$ in the first quadrant, a normal is drawn. The normal intersects the y-axis at the point $(0, b)$. If $\lim\limits_{a\to 0} b = 1/2$, then n equals

a. 1 b. 3

c. 2 d. 4

47. Suppose that f is differentiable for all x and that $f'(x) \le 2$ for all x. If $f(1) = 2$ and $f(4) = 8$, then $f(2)$ has the value equal to

a. 3 b. 4

c. 6 d. 8

48. The radius of a right circular cylinder increases at the rate of 0.1 cm/min, and the height decreases at the rate of 0.2 cm/min. The rate of change of the volume of the cylinder, in cm^3/min, when the radius is 2 cm and the height is 3 cm, is

a. -2π b. $-\dfrac{8\pi}{5}$

c. $-\dfrac{3\pi}{5}$ d. $\dfrac{2\pi}{5}$

49. A cube of ice melts without its changing shape at the uniform rate of 4 cm^3/min. The rate of change of the surface area of the cube, in cm^2/min, when the volume of the cube is 125 cm^3, is

a. -4 b. $-16/5$

c. $-16/6$ d. $-8/15$

50. The tangent to the curve $y = e^{kx}$ at a point $(0, 1)$ meets the x-axis at $(a, 0)$ where $a \in [-2, -1]$, then $k \in$

a. $[-1/2, 0]$ b. $[-1, -1/2]$

c. $[0, 1]$ d. $[1/2, 1]$

Multiple Correct Answers Type

Each question has four choices a, b, c, and d, out of which *one* or *more* answers are correct.

1. Points on the curve $f(x) = \dfrac{x}{1-x^2}$, where the tangent is inclined at an angle of $\dfrac{\pi}{4}$ to x-axis, are

a. $(0, 0)$ b. $\left(\sqrt{3}, \dfrac{\sqrt{3}}{2}\right)$

c. $\left(-2, \dfrac{2}{3}\right)$ d. $\left(-\sqrt{3}, \dfrac{\sqrt{3}}{2}\right)$

2. In the curve $y = ce^{x/a}$, the
   a. sub-tangent is constant
   b. sub-normal varies as the square of the ordinate
   c. tangent at (x_1, y_1) on the curve intersects the x-axis at a distance of $(x_1 - a)$ from the origin
   d. equation of the normal at the point where the curve cuts y-axis is $cy + ax = c^2$
3. Let $f(x) = a_5x^5 + a_4x^4 + a_3x^3 + a_2x^2 + a_1x$, where $a_i's$ are real and $f(x) = 0$ has a positive root α_0. Then
   a. $f'(x) = 0$ has a root α_1 such that $0 < \alpha_1 < \alpha_0$
   b. $f'(x) = 0$ has at least one real root
   c. $f''(x) = 0$ has at least one real root
   d. None of these
4. Let the parabolas $y = x(c - x)$ and $y = x^2 + ax + b$ touch each other at the point (1, 0), then
   a. $a + b + c = 0$ b. $a + b = 2$
   c. $b - c = 1$ d. $a + c = -2$
5. Which of the following pair(s) of curves is/are orthogonal
   a. $y^2 = 4ax; y = e^{-x/2a}$
   b. $y^2 = 4ax; x^2 = 4ay$ at (0, 0)
   c. $xy = a^2; x^2 - y^2 = b^2$
   d. $y = ax; x^2 + y^2 = c^2$
6. The co-ordinates of the point(s) on the graph of the function $f(x) = \frac{x^3}{3} - \frac{5x^2}{2} + 7x - 4$, where the tangent drawn cuts off intercepts from the co-ordinate axes which are equal in magnitude but opposite in sign, is
   a. (2, 8/3) b. (3, 7/2)
   c. (1, 5/6) d. None of these
7. The abscissa of a point on the curve $xy = (a + x)^2$, the normal at which cuts off numerically equal intercepts from the coordinate axes is
   a. $-\frac{a}{\sqrt{2}}$ b. $\sqrt{2}a$
   c. $\frac{a}{\sqrt{2}}$ d. $-\sqrt{2}a$
8. The angle formed by the positive y-axis and the tangent to $y = x^2 + 4x - 17$ at $(5/2, -3/4)$ is
   a. $\tan^{-1}(9)$ b. $\frac{\pi}{2} - \tan^{-1}(9)$
   c. $\frac{\pi}{2} + \tan^{-1}(9)$ d. None of these
9. If the tangent at any point $P(4m^2, 8m^3)$ of $x^3 - y^2 = 0$ is a normal also to the curve $x^3 - y^2 = 0$, then the value of m is
   a. $m = \frac{\sqrt{2}}{3}$ b. $m = -\frac{\sqrt{2}}{3}$
   c. $m = \frac{3}{\sqrt{2}}$ d. $m = -\frac{3}{\sqrt{2}}$
10. The angle between tangents to the curves $y = x^2$ and $x = y^2$ at (1, 1) is
    a. $\cos^{-1}\frac{4}{5}$ b. $\sin^{-1}\frac{3}{5}$
    c. $\tan^{-1}\frac{3}{4}$ d. $\tan^{-1}\frac{1}{3}$
11. The angle between the tangents at any point P and the line joining P to the origin 0, where P is a point on the curve $\ln(x^2 + y^2) = c\tan^{-1}\frac{y}{x}$, c is a constant, is
    a. independent of x
    b. independent of y
    c. independent of x but dependent on y
    d. independent of y but dependent on x
12. Given $f(x) = 4 - \left(\frac{1}{2} - x\right)^{2/3}$,
    $$g(x) = \begin{cases} \frac{\tan[x]}{x}, & x \neq 0 \\ 1, & x = 0 \end{cases},$$
    $h(x) = \{x\}$, $k(x) = 5^{\log_2(x+3)}$
    then in [0, 1] the Lagrange's mean value theorem is not applicable to the
    a. f b. g
    c. k d. h

Reasoning Type

Each question has four choices a, b, c and d, out of which *only one* is correct. Each question contains Statement 1 and Statement 2.

a. If both the statements are True and Statement 2 is the correct explanation of Statement 1
b. If both the statements are True but Statement 2 is Not the correct explanation of Statement 1
c. If Statement 1 is True and Statement 2 is False
d. If Statement 1 is False and Statement 2 is True

1. **Statements 1:** Conditions of Lagrange's mean value theorem fail in $f(x) = |x - 1|(x - 1)$.
   Statements 2: $|x - 1|$ is not differentiable at $x = 1$.
2. **Statements 1:** If $27a + 9b + 3c + d = 0$, then the equation $f(x) = 4ax^3 + 3bx^2 + 2cx + d = 0$ has at least one real root lying between (0, 3).

Statements 2: If $f(x)$ is continuous in $[a, b]$, derivable in (a, b) such that $f(a)=f(b)$, then at least one point $c \in (a, b)$ such that $f'(c)=0$.

3. **Statement 1:** If functions $f(t)$ and $g(t)$ are both continuous on the closed interval $[a,b]$, differentiable on the open interval (a,b), and $g'(t)$ is not zero on that open interval, then there exists some c in (a,b), such that

$$\frac{f'(c)}{g'(c)} = \frac{f(b)-f(a)}{g(b)-g(a)}.$$

Statement 2: If $f(t)$ and $g(t)$ are continuous and differentiable in $[a, b]$, then there exists some c in $(a.\ b)$ such that $f'(c) = \dfrac{f(b)-f(a)}{b-a}$ and $g'(c) = \dfrac{g(b)-g(a)}{b-a}$ from Lagrange's mean value theorem.

4. **Statement 1:** The maximum value of

$(\sqrt{-3+4x-x^2}+4)^2+(x-5)^2$ (where $1 \le x \le 3$) is 36.

Statement 2: The maximum distance between the point $(5, -4)$ and the point on the circle $(x-2)^2+y^2=1$ is 6.

5. **Statement 1:** If $g(x)$ is a differentiate function $g(2) \neq 0$, $g(-2) \neq 0$ and Rolle's theorem is not applicable to $f(x)=\dfrac{x^2-4}{g(x)}$ in $[-2, 2]$, then $g(x)$ has at least one root in $(-2, 2)$.

Statement 2: If $f(a)=f(b)$, then Rolle's theorem is applicable for $x \in (a, b)$.

6. **Statement 1:** The tangent at $x=1$ to the curve $y=x^3-x^2-x+2$ again meets the curve at $x=0$.

Statement 2: When the equation of a tangent solved with the curve, repeated roots are obtained at point of tangency.

Linked Comprehension Type

Based upon each paragraph, three multiple choice questions have to be answered. Each question has four choices a, b, c and d, out of which *only one* is correct.

For Problems 1–3

Tangent at a point P_1 (other than (0, 0)) on the curve $y=x^3$ meets the curve again at P_2. The tangent at P_2 meets the curve again at P_3 and so on.

1. If P_1 has co-ordinates (1, 1), then the sum $\lim\limits_{n\to\infty}\sum\limits_{r=1}^{n}\dfrac{1}{x_n}$ is (where $x_1, x_2, \ldots,$ are abscissas of $P_1, P_2, \ldots,$ respectively)

   a. 2/3 b. 1/3
   c. 1/2. d. 3/2

2. If P_1 has co-ordinates (1, 1), then the sum $\lim\limits_{n\to\infty}\sum\limits_{r=1}^{n}\dfrac{1}{y_n}$ is (where $y_1, y_2, \ldots,$ are ordinates of $P_1, P_2, \ldots,$ respectively)

   a. 1/8 b. 1/9
   c. 8/9 d. 9/8

3. Ratio of area of $\Delta P_1P_2P_3$ to that of $\Delta P_2P_3P_4$ is

   a. 1/4 b. 1/2
   c. 1/8 d. 1/16

For Problems 4–6

Consider the curve $x=1-3t^2, y=t-3t^3$. If a tangent at point $(1-3t^2, t-3t^3)$ inclined at an angle θ to positive x-axis and another tangent at point $P(-2, 2)$ cuts the curve again at Q.

4. The value of $\tan\theta + \sec\theta$ is equal to

   a. $3t$ b. t
   c. $t-t^2$ d. t^2-2t

5. The point Q will be

   a. $(1,-2)$ b. $\left(-\dfrac{1}{3}, -\dfrac{2}{3}\right)$
   c. $(-2, 1)$ d. None of these

6. The angle between the tangent at P and Q is

   a. $\dfrac{\pi}{4}$ b. $\dfrac{\pi}{6}$
   c. $\dfrac{\pi}{2}$ d. $\dfrac{\pi}{3}$

For Problems 7–8

A spherical balloon is being inflated so that its volume increases uniformly at the rate of 40 cm^3/min.

7. At $r=8$, its surface area increases at the rate

   a. 8 cm^2/min b. 10 cm^2/min
   c. 20 cm^2/min d. None of these

8. When $r=8$, then the increase in radius in the next 1/2 minute is

   a. 0.025 cm b. 0.050 cm
   c. 0.075 cm d. 0.01 cm

Matrix-Match Type

Each question contains statements given in two columns which have to be matched. Statements a, b, c, d in column I have to be matched with statements p, q, r, s in column II. If the correct match are a→p, a→s, b→r, c→p, c→q and d→s, then the correctly bubbled 4×4 matrix should be as follows:

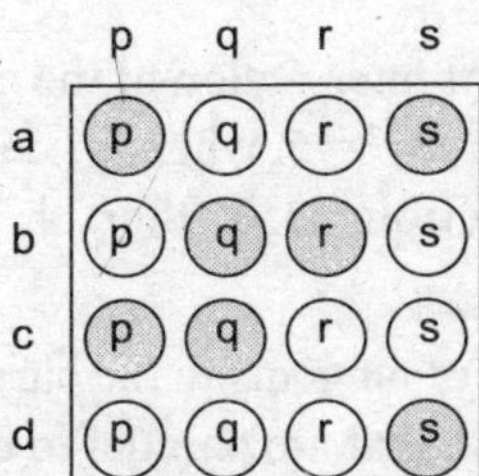

1.

Column I	Column II
a. The sides of a triangle vary slightly in such a way that its circum-radius remains constant, if $\frac{da}{\cos A} + \frac{db}{\cos B} + \frac{dc}{\cos C} + 1 = \lvert m \rvert$, then the values of m is	**p.** 1
b. The length of sub-tangent to the curve $x^2y^2 = 16$ at the point $(-2, 2)$ is $\lvert k \rvert$, then the value of k is	**q.** -1
c. The curve $y = 2e^{2x}$ intersects the y-axis at an angle $\cot^{-1} \lvert(8n-4)/3\rvert$, then the value of n is	**r.** 2
d. The area of a triangle formed by normal at the point $(1, 0)$ on the curve $x = e^{\sin y}$ with axes is $\lvert 2t + 1\rvert/6$ sq. units, then the value of t is	**s.** -2

2.

Column I	Column II
a. A circular plate is expanded by heat from radius 6 cm to 6.06 cm. Approximate increase in the area is	**p.** 5
b. If an edge of a cube increases by 2%, then percentage increase in the volume is	**q.** $0.72\,\pi$
c. If the rate of decrease of $\frac{x^2}{2} - 2x + 5$ is thrice the rate of decrease of x, then x is equal to (rate of decrease is non-zero)	**r.** 6
d. The rate of increase in the area of an equilateral triangle of side 30 cm, when each side increases at the rate of 0.1 cm/s is	**s.** $\frac{3\sqrt{3}}{2}$

3.

Column I: Curves	Column II: Angle between the curves
a. $y^2 = 4x$ and $x^2 = 4y$	**p.** 90°
b. $2y^2 = x^3$ and $y^2 = 32x$	**q.** any one of $\tan^{-1}\frac{3}{4}$ or $\tan^{-1}(16^{\frac{1}{3}})$
c. $xy = a^2$ and $x^2 + y^2 = 2a^2$	**r.** 0°
d. $y^2 = x$ and $x^3 + y^3 = 3xy$ at other than origin	**s.** $\tan^{-1}\frac{1}{2}$

Integer Type

1. If the tangent at any point $(4m^2, 8m^3)$ of $x^3 - y^2 = 0$ is a normal to the curve $x^3 - y^2 = 0$, then the value of $3m^2$ is__________.
2. If the distance between the points on the curve where tangents to the curve $y^2 - 2x^3 - 4y + 8 = 0$ pass through $(1, 2)$ is k, then the value of $[k]$, where [.] denotes the greatest integer function, is__________.
3. If the curve $y = ax^2 - 6x + b$ passes through $(0, 2)$ and has its tangent parallel to x-axis at $x = 3/2$, then the value of $\lvert a + b\rvert$ is__________.
4. If the sub-normal at any point on $y = a^{1-n}x^n$ is of constant length, then the value of $3/n$ is__________.
5. If the length of the tangent to the curve $y = x^3 + 3x^2 + 4x - 1$ at the point $x = 0$ is k, then the value of $[k]$, where [.] represents the greatest integer function, is__________.
6. If the angle between the curves $y^2 = 4x$ and $y = e^{-x/2}$ is θ, then the value of $\operatorname{cosec}^2(\theta/2)$ is__________.
7. The slope of the normal to the curve $x^3 + y^3 = 8xy$ at the point, other than origin, where it meets the curve $y^2 = 4x$ is__________.
8. The value of $3a^2$ if the curves $x^2/a^2 + y^2/4 = 1$ and $y^3 = 16x$ intersect orthogonally is__________.
9. If the curves $x^3 - 3xy^2 + 2 = 0$ and $3x^2y - y^3 - 2 = 0$ intersect at an angle θ, then the value of $\sin^2\theta$ is__________.
10. If the curves $ay + x^2 = 7$ and $x^3 = y$ intersect orthogonally at $(1, 1)$, then the value of a is__________.

11. If the shortest distance between the line $y = x - 2$ and the parabola $y = x^2 + 3x + 2$ is k, then the value of $2k^2$ is__________.

12. If P is the point on the curve $3x^2 - 4y^2 = 72$ which is nearest to the line $3x + 2y + 1 = 0$, then its distance from the point $(0, 3)$ is__________.

13. A man 1.6m high walks at the rate of 30m per minute away from a lamp which is 4m above ground. If k is the rate at which the man's shadow is lengthening, then the value of $k/4$ is__________.

14. A horse runs along a circle with a speed of 20 km/hr. A lantern is at the centre of the circle. A fence is along the tangent to the circle at the point at which the horse starts. If k is the speed with which the shadow of the horse moves along the fence at the moment when it covers 1/8 of the circle in km/hr, then the value of $k/5$ is__________.

15. If the function $f(x) = x^3 - 6x^2 + ax + b$ defined on $[1, 3]$ satisfies Rolle's theorem for $c = (2\sqrt{3}+1)/\sqrt{3}$, then the value of $(a - 4)$ is__________.

16. Consider the function $f(x) = 8x^2 - 7x + 5$ on the interval $[-6, 6]$. The value of c satisfying the conclusion of Lagrange's mean value theorem is__________.

17. Let f be differentiable for all x. If $f(1) = -2$ and $f'(x) \geq 2$ for all $x \in [1, 6]$, then the minimum value of $f(6)$ is__________.

18. Let $f(x)$ and $g(x)$ be differentiable for $0 \leq x \leq 2$ such that $f(0) = 2$, $g(0) = 1$, and $f(2) = 8$. Let there exist a real number c in $[0, 2]$ such that $f'(c) = 3g'(c)$, then the value of $g(2)$ is__________.

19. If the tangent at (x_1, y_1) to the curve $x^3 + y^3 = a^3$ meets the curve again at (x_2, y_2), then the value of $\left|\frac{x_2}{x_1} + \frac{x_1}{x_2}\right|$ is equal to__________.

20. If the angle of intersection of the curves $y = [|\sin x| + |\cos x|]$ and $x^2 + y^2 = 5$, where $[\cdot]$ denotes the greatest integer function, is θ, then the value of $\tan\theta$ is__________.

21. The number of tangents to the curve $y = \cos(x + y)$, $-2\pi \leq x \leq 2\pi$, that are parallel to the line $x + 2y = 0$ is/are__________.

22. The curve $y = ax^3 + bx^2 + cx + 5$ touches the x-axis at $P(-2, 0)$ and intersects the y-axis at a point Q where its gradient is 3. Then the value of $a + 2b + c$ is__________.

23. For a twice-differentiable function $f(x)$, $g(x)$ is defined as $g(x) = (f'(x)^2) + f''(x)f(x)$ on $[a, e]$. If for $a < b < c < d < e$, $f(a) = 0$, $f(b) = 2$, $f(c) = -1$, $f(d) = 2$, and $f(e) = 0$, then the minimum number of zero of $g(x)$ is__________.

24. The triangle formed by the tangent to the curve $f(x) = x^2 + bx - b$ at the point $(1, 1)$ and the coordinate axes lies in the first quadrant. If its area is 2, then the value of $|b|$ is__________.

25. If the portion of the tangent at any point on the curve $x = at^3$, $y = at^4$ between the axes is divided by the abscissa of the point of contact in the ratio of $m : n$ externally (where m and n are co-prime), then $|n + m|$ is equal to__________.

26. Let C be the curve $y = x^3$ (where x takes all real values). The tangent at A meets the curve again at B. If the gradient at B is K times the gradient at A, then $|K|$ is equal to__________.

27. Suppose a, b, c are such that the curve $y = ax^2 + bx + c$ is tangent to $y = 3x - 3$ at $(1, 0)$ and is also tangent to $y = x + 1$ at $(3, 4)$, then the value of $(2a - b - 4c)$ equals__________.

ANSWERS AND SOLUTIONS

Objective Type

1. b. Given curve is $x^{3/2}+y^{3/2}=2a^{3/2}$ (1)

$\therefore \frac{3}{2}\sqrt{x}+\frac{3}{2}\sqrt{y}\frac{dy}{dx}=0$

or $\frac{dy}{dx}=-\frac{\sqrt{x}}{\sqrt{y}}$

Since the tangent is equally inclined to the axes,

$\frac{dy}{dx}=\pm 1.$

$\therefore -\frac{\sqrt{x}}{\sqrt{y}}=\pm 1 \Rightarrow -\frac{\sqrt{x}}{\sqrt{y}}=-1 [\because \sqrt{x}>0, \sqrt{y}>0]$

$\Rightarrow \sqrt{x}=\sqrt{y}$

Putting $\sqrt{y}=\sqrt{x}$ in (1), we get

$2x^{3/2}=2a^{3/2} \Rightarrow x^3=a^3$

$\therefore x=a$ and so $y=a.$

2. a. $\frac{dx}{dt}=a+\frac{a}{2}2\cos 2t=a[1+\cos 2t]=2a\cos^2 t$

and $\frac{dy}{dt}=2a(1+\sin t)\cos t$

$\Rightarrow \frac{dy}{dx}=\frac{2a(1+\sin t)\cos t}{2a\cos^2 t}=\frac{(1+\sin t)}{\cos t}$

Then, the slope of the tangent

$\tan\theta=\frac{(\cos(t/2)+\sin(t/2))^2}{\cos^2(t/2)-\sin^2(t/2)}$

$=\frac{1+\tan\frac{t}{2}}{1-\tan\frac{t}{2}}=\tan\left(\frac{\pi}{4}+\frac{t}{2}\right)$

$\Rightarrow \theta=\frac{\pi+2t}{4}$

3. d. Differentiating w.r.t. x, we get $e^y\frac{dy}{dx}=2x$

$\Rightarrow \frac{dy}{dx}=\frac{2x}{1+x^2}$ $[\because e^y=1+x^2]$

$\Rightarrow m=\frac{2x}{1+x^2}$

or $|m|=\frac{2|x|}{1+|x|^2}$

But $1+|x|^2-2|x|=(1-|x|)^2\ge 0$

$\Rightarrow 1+|x|^2\ge 2|x|, \therefore |m|\le 1$

4. c. $\frac{dy}{dx}=3x^2-2ax+1$

Given that $\frac{dy}{dx}\ge 0$

$\Rightarrow 3x^2-2ax+1\ge 0$ for all x.

$\Rightarrow D\le 0$ or $4a^2-12\le 0$

$\Rightarrow -\sqrt{3}\le a\le\sqrt{3}$

5. a. Here $y>0$. Putting $y=x$ in $y=\sqrt{4-x^2}$, we get $x=\sqrt{2}, -\sqrt{2}$.

So the point is $(\sqrt{2}, \sqrt{2})$.

Differentiating $y^2+x^2=4$ w.r.t. x, $2y\frac{dy}{dx}+2x=0$

or $\frac{dy}{dx}=-\frac{x}{y}$

$\Rightarrow$ at $(\sqrt{2}, \sqrt{2})$, $\frac{dy}{dx}=-1$

6. b. Differentiating w.r.t. x, we get

$1+\frac{dy}{dx}=e^{xy}\left(y+x\frac{dy}{dx}\right)$

or $\frac{dy}{dx}=\frac{ye^{xy}-1}{1-xe^{xy}}$

$\frac{dy}{dx}=\infty \Rightarrow 1-xe^{xy}=0 \Rightarrow 1-x(x+y)=0.$

This holds for $x=1, y=0$.

7. c. The equation of the line is $y-3=\frac{3+2}{0-5}(x-0)$, i.e., $x+y-3=0$

$y=\frac{c}{x+1} \Rightarrow \frac{dy}{dx}=\frac{-c}{(x+1)^2}$

Let the line touches the curve at (α, β).

$\Rightarrow \alpha+\beta-3=0, \left(\frac{dy}{dx}\right)_{\alpha,\beta}=\frac{-c}{(\alpha+1)^2}=-1$

and $\beta=\frac{c}{\alpha+1}$

$\Rightarrow \frac{c}{(c/\beta)^2}=1$ or $\beta^2=c$ or $(3-\alpha)^2=c=(\alpha+1)^2$

$\Rightarrow 3-\alpha=\pm(\alpha+1)$ or $3-\alpha=\alpha+1$

$\Rightarrow \alpha=1$. So, $c=(1+1)^2=4$

8. b.

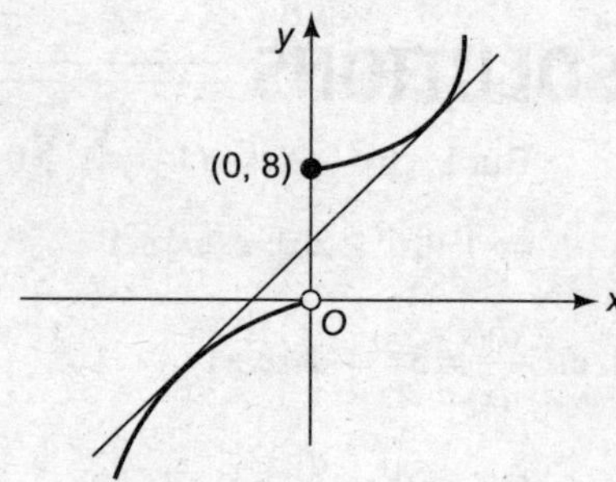

Fig. 5.1

Let $y = mx + c$ be a tangent to $f(x)$

$y = x^2 + 8$ for $x \geq 0$

$mx + c = x^2 + 8$

$x^2 - mx + 8 - c = 0$ (for the line to be tangent $D = 0$)

$\therefore m^2 = 4(8 - c)$ (1)

Again $y = -x^2$, for $x < 0$

$mx + c = -x^2$

$x^2 + mx + c = 0$

$D = 0 \Rightarrow m^2 = 4c$ (2)

From (1) and (2), we get

$c = 4, m = 4$

$\therefore y = 4x + 4$

Put $y = 0 \Rightarrow x = -1$.

9. a. Putting $x = 0$ in the given curve, we obtain $y = 1$. So the given point is (0, 1)

Now, $y = e^{2x} + x^2 \Rightarrow \frac{dy}{dx} = 2e^{2x} + 2x \Rightarrow \left(\frac{dy}{dx}\right)_{(0,1)} = 2$

The equation of the tangent at (0, 1) is

$y - 1 = 2(x - 0) \Rightarrow 2x - y + 1 = 0$ (1)

Required distance = length of the ⊥ from (0, 0) on (1)

$= \frac{1}{\sqrt{5}}$.

10. a. Let the required point be (x_1, y_1)

Now, $3y = 6x - 5x^3$

$\Rightarrow 3\frac{dy}{dx} = 6 - 15x^2$

$\Rightarrow \frac{dy}{dx} = 2 - 5x^2$

$\Rightarrow \left(\frac{dy}{dx}\right)_{(x_1, y_1)} = 2 - 5x_1^2$

The equation of the normal at (x_1, y_1) is

$y - y_1 = \frac{-1}{2 - 5x_1^2}(x - x_1)$

If it passes through the origin, then

$0 - y_1 = \frac{-1}{2 - 5x_1^2}(0 - x_1)$

$\Rightarrow y_1 = \frac{-x}{2 - 5x_1^2}$ (1)

Since (x_1, y_1) lies on the given curve.

Therefore, $3y_1 = 6x_1 - 5x_1^3$ (2)

Solving (1) and (2), we obtain $x_1 = 1$ and $y_1 = 1/3$

Hence, the required point is (1, 1/3).

11. a. We have $x = a(\cos\theta + \theta\sin\theta)$ and

$y = a(\sin\theta - \theta\cos\theta)$

$\therefore \frac{dx}{d\theta} = a(-\sin\theta + \theta\cos\theta + \sin\theta) = a\theta\cos\theta$

and $\frac{dy}{d\theta} = a(\cos\theta + \theta\sin\theta - \cos\theta) = a\theta\sin\theta$

$\Rightarrow \frac{dy}{dx} = \frac{dy/d\theta}{dx/d\theta} = \frac{a\theta\sin\theta}{a\theta\cos\theta} = \tan\theta$

$\Rightarrow$ Slope of the normal $= -\cot\theta = \tan\left(\frac{\pi}{2} + \theta\right)$

Hence, the normal makes an angle $\frac{\pi}{2} + \theta$ with the x-axis.

12. a. $y = \frac{2}{3}x^3 + \frac{1}{2}x^2$

$\therefore \frac{dy}{dx} = \frac{2}{3}3x^2 + \frac{1}{2}2x = 2x^2 + x$

Since the tangent makes equal angles with the axes.

$\Rightarrow \frac{dy}{dx} = \pm 1$

$\Rightarrow 2x^2 + x = \pm 1$

$\Rightarrow 2x^2 + x - 1 = 0$ ($2x^2 + x + 1 = 0$ has no real roots)

$\Rightarrow (2x - 1)(x + 1) = 0$

$\Rightarrow x = \frac{1}{2}$ or $x = -1$

13. d. $y = b\,e^{-x/a}$ meets the y-axis (0, b)

Again $\frac{dy}{dx} = b\,e^{-x/a}\left(-\frac{1}{a}\right)$

At (0, b), $\frac{dy}{dx} = be^0\left(-\frac{1}{a}\right) = -\frac{b}{a}$

Required tangent is $y - b = -\frac{b}{a}(x - 0)$

or $\frac{x}{a} + \frac{y}{b} = 1$.

14. b. $x^2 - y^2 = 8 \Rightarrow \dfrac{dy}{dx} = \dfrac{x}{y} \Rightarrow -\dfrac{1}{dy/dx} = -\dfrac{y}{x}$

At the point $\left(-\dfrac{5}{\sqrt{2}}, \dfrac{3}{\sqrt{2}}\right)$,

$$-\frac{1}{dy/dx} = \frac{-3/\sqrt{2}}{-5/\sqrt{2}} = \frac{3}{5}$$

Also $9x^2 + 25y^2 = 225$

$$\Rightarrow 18x + 50y\frac{dy}{dx} = 0$$

$$\Rightarrow \frac{dy}{dx} = -\frac{9x}{25y} \Rightarrow -\frac{dx}{dy} = \frac{25y}{9x}$$

At the point $\left(-\dfrac{5}{\sqrt{2}}, \dfrac{3}{\sqrt{2}}\right)$,

$$-\frac{dx}{dy} = \frac{25 \times 3/\sqrt{2}}{9(-5/\sqrt{2})} = -\frac{15}{9} = -\frac{5}{3}$$

Since the product of the slopes is -1,
the normals cut orthogonally, i.e., the required angle is equal to $\dfrac{\pi}{2}$.

15. b. We have $f''(x) = 6(x-1)$

Integrating, we get $f'(x) = 3(x-1)^2 + c$ (1)

At (2, 1), $y = 3x - 5$ is tangent to $y = f(x)$

$\therefore f'(2) = 3$

From (1), $3 = 3(2-1)^2 + c \Rightarrow 3 = 3 + c \Rightarrow c = 0$

$\therefore f'(x) = 3(x-1)^2$

Integrating, we get $f(x) = (x-1)^3 + c'$

Since the curve passes through (2, 1),

$1 = (2-1)^3 + c' \Rightarrow c' = 0$. Hence,

$\therefore f(x) = (x-1)^3$.

16. a. $y^2 = \alpha x^3 - \beta \Rightarrow \dfrac{dy}{dx} = \dfrac{3\alpha x^2}{2y}$

$\Rightarrow$ Slope of the normal at (2, 3) is

$$\left(-\frac{dx}{dy}\right)_{(2,3)} = -\frac{2\times 3}{3\,\alpha(2)^2} = -\frac{1}{2\alpha} = -\frac{1}{4}$$

$\Rightarrow \alpha = 2$

Also (2, 3) lies on the curve

$\Rightarrow 9 = 8\alpha - \beta \Rightarrow \beta = 16 - 9 = 7 \Rightarrow \alpha + \beta = 9$.

17. d. $x = 2\ln\cot t + 1$, $y = \tan t + \cot t$

Slope of the tangent

$$\left(\frac{dy}{dx}\right)_{t=\pi/4} = \left(\frac{\sec^2 t - \operatorname{cosec}^2 t}{-\dfrac{2}{\cot t}\operatorname{cosec}^2 t}\right)_{t=\pi/4} = 0$$

18. b. $y = e^x + e^{-x} \Rightarrow \dfrac{dy}{dx} = e^x - e^{-x} = \tan\theta$, where θ is the angle of the tangent with the x-axis

For $\theta = 60°$, we have $\tan 60° = e^x - e^{-x}$

$\Rightarrow e^{2x} - \sqrt{3}\,e^x - 1 = 0$

$$\Rightarrow e^x = \frac{\sqrt{3} \pm \sqrt{7}}{2}$$

$$\Rightarrow x = \log_e\left(\frac{\sqrt{3}+\sqrt{7}}{2}\right)$$

19. c. $x^2 y = c^3$

Differentiating w.r.t. x, we have

$$x^2\frac{dy}{dx} + 2xy = 0 \Rightarrow \frac{dy}{dx} = -\frac{2y}{x}$$

Equation of the tangent at (h, k) is

$$y - k = -\frac{2k}{h}(x - h).$$

$y = 0$ gives $x = \dfrac{3h}{2} = a$, and $x = 0$ gives $y = 3k = b$.

Now, $a^2 b = \dfrac{9h^2}{4}\,3k = \dfrac{27}{4}h^2 k = \dfrac{27}{4}c^3$

20. a.

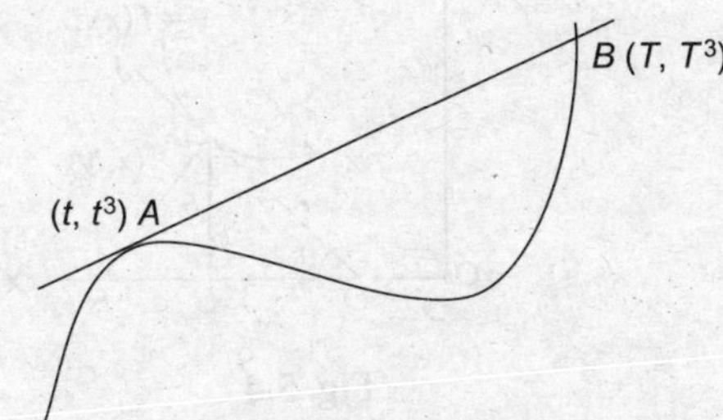

Fig. 5.2

$\dfrac{dy}{dx} = 3x^2 = 3t^2$ at A

$$\therefore 3t^2 = \frac{T^3 - t^3}{T - t} = T^2 + Tt + t^2$$

$\Rightarrow T^2 + Tt - 2t^2 = 0$

$\Rightarrow (T - t)(T + 2t) = 0$

$\Rightarrow T = t$ or $T = -2t$ [$T = t$ is not possible]

Now, $m_A = 3t^2$ and $m_B = 3T^2$

$$\Rightarrow \frac{m_B}{m_A} = \frac{T^2}{t^2} = \frac{4t^2}{t^2}$$ [Using $T = -2t$]

21. d.

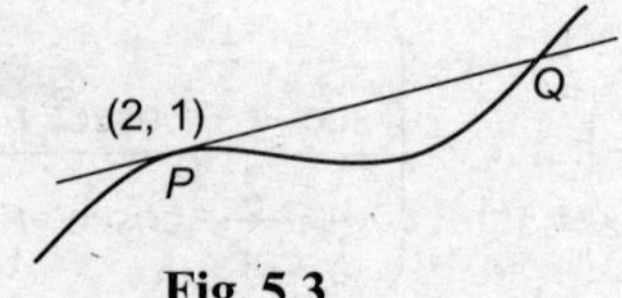

Fig. 5.3

Eliminating t gives $y^2(x-1)=1$.
Equation of the tangent at $P(2, 1)$ is $x+2y=4$.
Solving with curve $x=5$ and $y=-1/2$, we get

$Q(5, 1/2) \Rightarrow PQ=\dfrac{3\sqrt{5}}{2}$

22. c. $\dfrac{a}{x^2}+\dfrac{b}{y^2}=1 \Rightarrow ay^2+bx^2=x^2y^2$ (1)

$$-\frac{2a}{x^3}-\frac{2b}{y^3}\frac{dy}{dx}=0 \Rightarrow \frac{dy}{dx}=-\frac{ay^3}{bx^3}$$

Equation of the tangent at (h, k) is

$$y-k=-\frac{ak^3}{bh^3}(x-h)$$

for x-intercept, put $y=0$

$$\Rightarrow x=\frac{bh^3}{ak^2}+h$$

$$\Rightarrow x=h\left[\frac{bh^2+ak^2}{ak^2}\right]=h\left[\frac{h^2k^2}{ak^2}\right]=\frac{x^3}{a}$$

$\Rightarrow$ x-intercept is proportional to the cube of abscissa.

23. a.

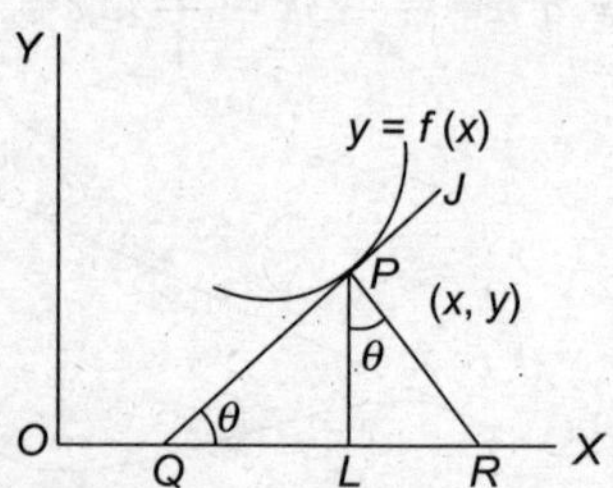

Fig. 5.4

Given curve is $2x^2y^2-x^4=c$ (1)

Sub-normal at $P(x, y)=yy_1=y\dfrac{dy}{dx}$ (2)

From (1), we get $2\left(x^2 2y\dfrac{dy}{dx}+2xy^2\right)-4x^3=0$

$$\Rightarrow \frac{dy}{dx}=\frac{x(x^2-y^2)}{x^2y} \quad (3)$$

Now, $x(x-yy_1)=x^2-xy\dfrac{dy}{dx}$

$=x^2-(x^2-y^2)$ [from (3)]

$=y^2$

$\Rightarrow$ Mean proportion $=\sqrt{x(x-yy_1)}=y$

24. b. Length of sub-normal = length of sub-tangent

$$\Rightarrow \frac{dy}{dx}=\pm 1$$

If $\dfrac{dy}{dx}=1$, equation of the tangent $y-4=x-3$

$\Rightarrow y-x=1$, area of $\Delta OAB=\dfrac{1}{2}\times 1\times 1=\dfrac{1}{2}$

If $\dfrac{dy}{dx}=-1$, equation of the tangent is $y-4=-x+3$

$\Rightarrow y+x=7$, area $=\dfrac{1}{2}\times 7\times 7=\dfrac{49}{2}$

25. a. $y=x+\sin x \Rightarrow$ If $\dfrac{dy}{dx}=1+\cos x=0$, then $\cos x=-1$

$\Rightarrow x=\pm\pi, \pm 3\pi \ldots$.
Also $y=\pm\pi, \pm 3\pi \ldots$. But for the given constrain on x and y, no such y exists.
Hence, no such tangent exists.

26. c. Solving $y=|x^2-1|$ and $y=\sqrt{7-x^2}$

we have $|x^2-1|=\sqrt{7-x^2}$

$\Rightarrow x^4-2x^2+1=7-x^2$

$\Rightarrow x^4-x^2-6=0$

$\Rightarrow (x^2-3)(x^2+2)=0$

$\Rightarrow x=\pm\sqrt{3}$

Points of intersection of the curves $y=|x^2-1|$ and $y=\sqrt{7-x^2}$ are $(\pm\sqrt{3}, 2)$.
Since both the curves are symmetrical about the y-axis, points of intersection are also symmetrical.

Now, $y=x^2-1 \Rightarrow \dfrac{dy}{dx}=2x$

$$\Rightarrow m_1=\left.\frac{dy}{dx}\right|_{(\sqrt{3},2)}=2\sqrt{3}$$

and $y=\sqrt{7-x^2} \quad \Rightarrow \dfrac{dy}{dx}=-\dfrac{x}{y}$

$$\Rightarrow m_2=\left.\frac{dy}{dx}\right|_{(\sqrt{3},2)}=-\frac{\sqrt{3}}{2}$$

$$\Rightarrow \tan\theta=\left|\frac{5\sqrt{3}}{4}\right|$$

27. d. Differentiating $y^3-x^2y+5y-2x=0$ w.r.t. x, we get
$3y^2y'-2xy-x^2y'+5y'-2=0$

$$\Rightarrow y'=\frac{2xy+2}{3y^2-x^2+5} \Rightarrow y'_{(0,0)}=2/5$$

Differentiating $x^4 - x^3y^2 + 5x + 2y = 0$ w.r.t. x, we have $4x^3 - 3x^2y^2 - 2x^3yy' + 5 + 2y' = 0$

$$\Rightarrow y' = \frac{3x^2y^2 - 4x^3 - 5}{2 - 2x^3y} \quad \Rightarrow y'_{(0,0)} = -5/2.$$

Thus, both the curves intersect at right angle.

28. d. Solving the curves, we get point of intersection (a^2, a)

For $x = y^2$, $\frac{dy}{dx} = \frac{1}{2y}$

At (a^2, a), $\frac{dy}{dx} = \frac{1}{2a}$

For $xy = a^3$, $\frac{dy}{dx} = -\frac{y}{x}$

At (a^2, a), $\frac{dy}{dx} = -\frac{a}{a^2} = -\frac{1}{a}$

Since the curves cut orthogonally,

$$\frac{1}{2a} \times -\frac{1}{a} = -1 \Rightarrow 2a^2 = 1 \Rightarrow a^2 = \frac{1}{2}$$

29. b. $4x^2 + 9y^2 = 72$

Differentiating w.r.t. x, we have

$$8x + 18y\frac{dy}{dx} = 0 \Rightarrow \frac{dy}{dx} = -\frac{4}{9}\frac{x}{y}$$

At $(3, 2)$, $\frac{dy}{dx} = -\frac{4}{9} \times \frac{3}{2} = -\frac{2}{3}$

Also $x^2 - y^2 = 5 \Rightarrow \frac{dy}{dx} = \frac{x}{y}$.

At $(3, 2)$, $\frac{dy}{dx} = \frac{3}{2}$

Hence, the curves cut orthogonally.

30. d. Using Lagrange's mean value theorem, for some $c \in (1, 6)$

such that $f'(c) = \frac{f(6) - f(1)}{5} = \frac{f(6)+2}{5} \geq 4.2$

$\Rightarrow f(6) + 2 \geq 21$

$\Rightarrow f(6) \geq 19$

31. d. $f(0) = -1$; $f(1) = 7$. So $f(0)$ and $f(1)$ have opposite sign.

32. d. $f(x)$ vanishes at points where

$$\sin\frac{\pi}{x} = 0, \text{ i.e., } \frac{\pi}{x} = k\pi, k = 1, 2, 3, 4, \ldots$$

Hence $x = \frac{1}{k}$.

Also $f'(x) = \sin\frac{\pi}{x} - \frac{\pi}{x}\cos\frac{\pi}{x}$, if $x \neq 0$.

Since the function has a derivative at any interior point of the interval $(0, 1)$, also continuous in $[0,1]$ and $f(0) = f(1)$. Hence, Rolle's theorem is applicable to any one of the interval $\left[\frac{1}{2}, 1\right], \left[\frac{1}{3}, \frac{1}{2}\right], \ldots, \left[\frac{1}{k+1}, \frac{1}{k}\right]$.

Hence, there exists at least one c in each of these intervals where $f'(c) = 0 \Rightarrow$ infinite points.

33. b. Applying Rolle's theorem to $F(x) = f(x) - 2g(x)$, we get

$F(0) = 0$,

$F(1) = f(1) - 2g(1)$

$\Rightarrow 0 = 6 - 2g(1)$

$\Rightarrow g(1) = 3$.

34. b. Let $f(x) = \frac{ax^4}{4} + \frac{bx^3}{3} + \frac{cx^2}{2} + dx$,

which is continuous and differentiable.

$$f(0) = 0, f(-1) = \frac{a}{4} - \frac{b}{3} + \frac{c}{2} - d$$

$$= \frac{1}{4}(a + 2c) - \frac{1}{3}(b + 3d) = 0$$

So according to Rolle's theorem, there exists at least one root of $f'(x) = 0$ in $(-1, 0)$.

35. a. $f(x) = ax^3 + bx^2 + 11x - 6$

satisfies conditions of Rolle's theorem in $[1, 3]$

$\Rightarrow f(1) = f(3)$

$\Rightarrow a + b + 11 - 6 = 27a + 9b + 33 - 6$

$\Rightarrow 13a + 4b = -11$ (1)

and $f'(x) = 3ax^2 + 2bx + 11$

$$\Rightarrow f'\left(2 + \frac{1}{\sqrt{3}}\right) = 3a\left(2 + \frac{1}{\sqrt{3}}\right)^2 + 2b\left(2 + \frac{1}{\sqrt{3}}\right) + 11$$

$= 0$

$$\Rightarrow 3a\left(4 + \frac{1}{3} + \frac{4}{\sqrt{3}}\right) + 2b\left(2 + \frac{1}{\sqrt{3}}\right) + 11 = 0 \quad (2)$$

From equations (1) and (2), we get $a = 1, b = -6$.

36. d. Here, $f(x) = \log_e x$. Therefore,

$$f'(c) = \frac{f(b) - f(a)}{b - a} \Rightarrow \frac{1}{c} = \frac{\log_e 3 - \log_e 1}{3 - 1}$$

$$\Rightarrow \frac{1}{c} = \frac{1}{2}\log_e 3 \Rightarrow c = 2\log_e 3.$$

37. d. Let $g(x) = f(x) - x^2$. We have

$g(1) = 0, g(2) = 0, g(3) = 0$

$[\because f(1) = 1, f(2) = 4, f(3) = 9]$

From Rolle's theorem on $g(x)$, $g'(x) = 0$ for at least $x \in (1, 2)$. Let $g'(c_1) = 0$ where $c_1, \in (1, 2)$.

Similarly, $g(x) = 0$ for at least one $x \in (2, 3)$. Let $g'(c_2) = 0$ where $c_2 \in (2, 3)$. Therefore,

$g'(c_1) = g'(c_2) = 0$

By Rolle's theorem, at least one $x \in (c_1, c_2)$ such that $g''(x) = 0 \Rightarrow f''(x) = 2$ for some $x \in (1, 3)$.

38. b. $f\left(\frac{5\pi}{6}\right) = \log \sin\left(\frac{5\pi}{6}\right) = \log \sin \frac{\pi}{6}$

$= \log \frac{1}{2} = -\log 2,$

$f\left(\frac{\pi}{6}\right) = \log \sin \frac{\pi}{6} = -\log 2.$

$f'(c) = \frac{1}{\sin x} \cos x = \cot x$

By Lagrange's mean value theorem,

$$\frac{f(5\pi/6) - f(\pi/6)}{(5\pi/6) - (\pi/6)} = \cot c$$

$\Rightarrow \cot c = 0 \Rightarrow c = \frac{\pi}{2}$

Thus, $c = \frac{\pi}{2} \in (\pi/6, 5\pi/6)$.

39. d. a. Discontinuous at $x = 1 \Rightarrow$ not applicable.

b. $F(x)$ is not continuous (jump discontinuity) at $x = 0$.

c. Discontinuity(missing point) at $x = 1 \Rightarrow$ not applicable.

d. Notice that $x^3 - 2x^2 - 5x + 6 = (x - 1)(x^2 - x - 6)$. Hence, $f(x) = x^2 - x - 6$ if $x \neq 1$ and $f(1) = -6$.

$\Rightarrow f$ is continuous at $x = 1$. So $f(x) = x^2 - x - 6$ throughout the interval $[-2, 3]$.

Also, note that $f(-2) = f(3) = 0$. Hence, Rolle's theorem applies. $f'(x) = 2x - 1$.

Setting $f'(x) = 0$, we obtain $x = 1/2$ which lies between -2 and 3.

40. d. We have $y^2 = 18x$. Therefore, (1)

$2y \frac{dy}{dx} = 18 \Rightarrow \frac{dy}{dx} = \frac{9}{y}$

Given that $\frac{dy}{dx} = 2 \Rightarrow \frac{9}{y} = 2 \Rightarrow y = \frac{9}{2}$

Putting in (1), we get $\frac{81}{4} = 18x \Rightarrow x = \frac{9}{8}$.

Hence, the point is $\left(\frac{9}{8}, \frac{9}{2}\right)$.

41. a. $V = \frac{4}{3}\pi r^3, S = 4\pi r^2$

$\frac{dV}{dr} = 4\pi r^2, \frac{dS}{dr} = 8\pi r$

$\Rightarrow \frac{dV}{dS} = \frac{dV/dr}{dS/dr} = \frac{4\pi r^2}{8\pi r} = \frac{r}{2}$

when $r = 2$, $\frac{dV}{ds} = \frac{2}{2} = 1$

42. b. $V = x^3$ and the percent error in measuring $x = \frac{dx}{x} \times 100 = k$

The percent error in measuring volume $= \frac{dV}{V} \times 100$

Now, $\frac{dV}{dx} = 3x^2$

$\Rightarrow dV = 3x^2 dx \Rightarrow \frac{dV}{V} = \frac{3x^2 dx}{x^3} = 3\frac{dx}{x}$

$\therefore \frac{dV}{V} \times 100 = 3\frac{dx}{x} \times 100 = 3k$

43. b.

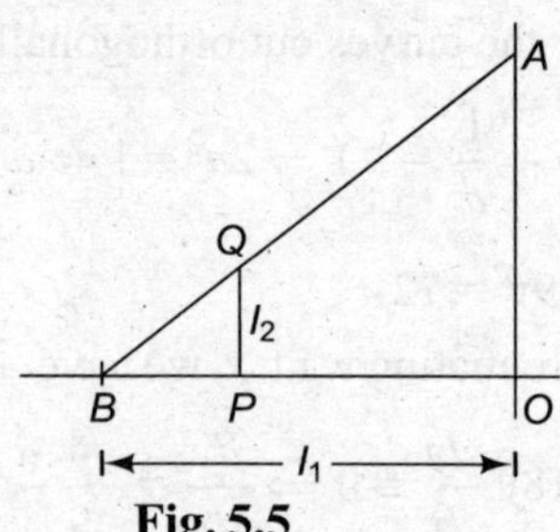

Fig. 5.5

Let $BP = x$. From similar Δ's property, we get

$\frac{AO}{l_1} = \frac{l_2}{x}$

$\Rightarrow AO = \frac{l_1 l_2}{x} \Rightarrow \frac{d(AO)}{dt} = \frac{-l_1 l_2}{x^2}\frac{dx}{dt}$, when

$x = \frac{l_1}{2}, \frac{d(AO)}{dt} = \frac{2l_2}{5}$ m/s.

44. a. Let CD be the position of man at any time t. Let $BD = x$, then $EC = x$. Let $\angle ACE = \theta$

Given, $AB = 41.6$ m, $CD = 1.6$ m and $\frac{dx}{dt} = 2$ m/s.

$AE = AB - EB = AB - CD = 41.6 - 1.6 = 40$ m

We have to find $\frac{d\theta}{dt}$ when $x = 30$ m

From ΔAEC, $\tan \theta = \frac{AE}{EC} = \frac{40}{x}$ (1)

Differentiating w.r.t. to t, $\sec^2 \theta \frac{d\theta}{dt} = \frac{-40}{x^2}\frac{dx}{dt}$

$\Rightarrow \sec^2 \theta \frac{d\theta}{dt} = \frac{-40}{x^2} \times 2$

$\Rightarrow \frac{d\theta}{dt} = \frac{-80}{x^2}\cos^2 \theta$

$$= -\frac{80}{x^2}\frac{x^2}{x^2+40^2} \quad \left[\because \cos\theta = \frac{x}{\sqrt{x^2+40^2}}\right]$$

$$\Rightarrow \frac{d\theta}{dt} = -\frac{80}{x^2+40^2} \quad (2)$$

when $x = 30$ m, $\frac{d\theta}{dt} = -\frac{80}{30^2+40^2} = -\frac{4}{125}$ radian/s.

45. b. Any point on the parabola $y^2 = 8x$ ($4a = 8$ or $a = 2$) is $(at^2, 2at)$ or $(2t^2, 4t)$.

For its minimum distance from the circle means its distance from the centre (0, – 6) of the circle.

Let D be the distance, then

$z = D^2 = (2t^2)^2 + (4t+6)^2 = 4(t^4 + 4t^2 + 12t + 9)$

$$\therefore \frac{dz}{dt} = 4(4t^3 + 8t + 12) = 0$$

$\Rightarrow 16(t^3 + 2t + 3) = 0$

$\Rightarrow 16(t+1)(t^2 - t + 3) = 0$

$\Rightarrow t = -1$

$$\frac{d^2z}{dt^2} = 16(3t^2+2) = \text{positive hence minimum.}$$

$\therefore$ point is (2, – 4).

46. c. $y = x^n$

$$\frac{dy}{dx} = nx^{n-1} = na^{n-1}$$

Slope of the normal is $-1/na^{n-1}$

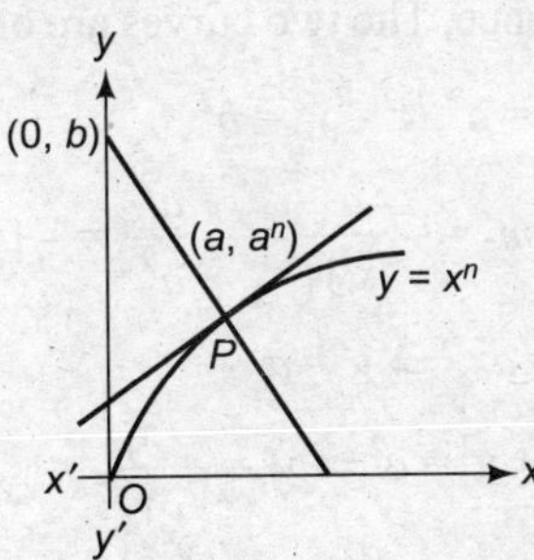

Fig. 5.6

Equation of the normal $y - a^n = -\frac{1}{na^{n-1}}(x - a)$

put $x = 0$ to get y-intercept

$$y = a^n + \frac{1}{na^{n-2}}; \text{ hence, } b = a^n + \frac{1}{na^{n-2}}$$

$$\lim_{a\to 0} b = \begin{cases} 0, & \text{if } n < 2 \\ \frac{1}{2}, & \text{if } n = 2 \\ \infty, & \text{if } n > 2 \end{cases}$$

47. b. Using Lagrange's mean value theorem for f in [1, 2]

for $c \in (1, 2)$, $\frac{f(2) - f(1)}{2-1} = f'(c) \le 2$

$\Rightarrow f(2) - f(1) \le 2$

$\Rightarrow f(2) \le 4$ (1)

Again using Lagrange's mean value theorem in [2, 4]

for $d \in (1, 2)$, $\frac{f(4) - f(2)}{4-2} = f'(d) \le 2$

$\Rightarrow f(4) - f(2) \le 4 \Rightarrow 8 - f(2) \le 4$

$\Rightarrow f(2) \ge 4$ (2)

From (1) and (2), $f(2) = 4$.

48. d. Given, $V = \pi r^2 h$

Differentiating both sides, we get

$$\frac{dV}{dt} = \pi\left(r^2\frac{dh}{dt} + 2r\frac{dr}{dt}h\right) = \pi r\left(r\frac{dh}{dt} + 2h\frac{dr}{dt}\right)$$

$$\frac{dr}{dt} = \frac{1}{10} \text{ and } \frac{dh}{dt} = -\frac{2}{10}$$

$$\frac{dV}{dt} = \pi r\left(r\left(-\frac{2}{10}\right) + 2h\left(\frac{1}{10}\right)\right) = \frac{\pi r}{5}(-r+h)$$

Thus, when $r = 2$ and $h = 3$,

$$\frac{dV}{dt} = \frac{\pi(2)}{5}(-2+3) = \frac{2\pi}{5}$$

49. b. $\frac{dV}{dt} = 4$ cm³/min; $\frac{dS}{dt} = ?$ where $V = 125$ cm²

$$V = x^3;\ S = 6x^2 \Rightarrow \frac{dV}{dt} = 3x^2\frac{dx}{dt}$$

$$-4 = 3x^2\frac{dx}{dt} \quad (1)$$

$$\text{Also } \frac{dS}{dt} = 12x\frac{dx}{dt}$$

$$-4 = 3x^2\frac{dS}{dt}\frac{1}{12x}$$

$$\frac{dS}{dt} = -\frac{16}{x}; \text{ where } V = 125 = x^3 \Rightarrow x = 5$$

$$\frac{dS}{dt} = -\frac{16}{x} \text{ cm}^2/\text{min} \Rightarrow \frac{dS}{dt} = -\frac{16}{5} \text{ cm}^2/\text{min.}$$

50. d. $\frac{dy}{dx} = ke^{kx} = k$ at (0, 1).

Equation of the tangent is $y - 1 = kx$.

Point of intersection with x-axis is $x = -\frac{1}{k}$, where

$$-2 \le -\frac{1}{k} \le -1 \Rightarrow k \in \left[\frac{1}{2}, 1\right].$$

Multiple Correct Answers Type

1. a., b., d.

$$f(x)=\frac{x}{1-x^2}$$

$$\therefore f'(x)=\frac{1+x^2}{(1-x^2)^2}=1, \text{ i.e., } x=0,-\sqrt{3},\sqrt{3}$$

$\Rightarrow$ The points are (0, 0), $\left(\pm\sqrt{3}, \frac{\sqrt{3}}{2}\right)$

2. a., b., c., d.

We have $y=ce^{x/a}$

$$\Rightarrow \frac{dy}{dx}=\frac{c}{a}e^{x/a} \Rightarrow \frac{dy}{dx}=\frac{1}{a}y$$

$$\Rightarrow \frac{y}{dy/dx}=a=\text{const.}$$

$\Rightarrow$ sub-tangent = const.

$$\Rightarrow \text{Length of the sub-normal} = y\frac{dy}{dx}$$

$$=y\frac{y}{a}=\frac{y^2}{a}\propto \text{(square of the ordinate)}$$

Equation of the tangent at (x_1, y_1) is

$$y-y_1=\frac{y_1}{a}(x-x_1)$$

This meets the x-axis at a point given by

$$-y_1=\frac{y_1}{a}(x-x_1) \Rightarrow x=x_1-a$$

The curve meets the y-axis at (0, c). Therefore,

$$\left(\frac{dy}{dx}\right)_{(0,\,c)}=c/a$$

So equation of the normal at (0, c) is

$$y-c=-\frac{1}{c/a}(x-0) \Rightarrow ax+cy=c^2$$

3. a., b., c.

Clearly, $f(0)=0$. So, $f(x)=0$ has two real roots 0, $\alpha_0(>0)$.

Therefore, $f'(x)=0$ has a real root α_1 lying between 0 and α_0. So $0<\alpha_1<\alpha_0$.

Again, $f'(x)=0$ is a fourth-degree equation. As imaginary roots occur in conjugate pairs, $f'(x)=0$ will have another real root α_2. Therefore, $f''(x)=0$ will have a real root lying between α_1 and α_2. As $f(x)=0$ is an equation of the fifth degree, it will have at least three real roots and so $f'(x)$ will have at least two real roots.

4. a., c., d.

$$y=x(c-x) \quad (1)$$

$$y=x^2+ax+b \quad (2)$$

Slope of (1) curve $=c-2x$

And at (1, 0), $c-2=m_1$ (say)

Slope of (2) curve $=2x+a$

at (1, 0), $2+a=m_1$ (say)

Curves are touching at (1, 0)

$\Rightarrow m_1=m_2$

$\Rightarrow 2+a=c-2$ (3)

Also (1, 0) lies on both the curves (4)

$\Rightarrow 0=c-1$ and $0=1+a+b$ and

Solving (3) and (4), we get

$a=-3, b=2, c=1$

5. a., b., c., d.

a. $y^2=4ax \Rightarrow m_1=y'=\frac{2a}{y}$

$$y=e^{-x}/2a \Rightarrow m_2=y'=-\frac{1}{2a}e^{-x/2a}=-\frac{1}{2a}y$$

$m_1m_2=-1$. Hence, orthogonal.

b. $y^2=4ax, x^2=4ay$

$$y'=\frac{4a}{2y_1}=\frac{2a}{y_1} \text{ Not defined at } x=0$$

$$y'=\frac{2x_1}{4a}=\frac{x_1}{2a}=0 \text{ at } x=0$$

Hence, The two curves are orthogonal at (0, 0).

c. $xy=a^2, x^2-y^2=b^2$

$$m_1m_2=-\frac{a^2}{x_1y_1}=-\frac{a^2}{a^2}=-1 \Rightarrow \text{orthogonal.}$$

d. $y=ax, \Rightarrow y'=a$

$$x^2+y^2=c^2 \Rightarrow y'=-\frac{x_1}{y_1}$$

$$m_1m_2=-\frac{ax_1}{y_1}=-\frac{y_1}{y_1}=-1 \text{ orthogonal.}$$

6. a., b.

Since the intercepts are equal in magnitude but opposite in sign

$$\Rightarrow \left.\frac{dy}{dx}\right|_P=1$$

$$\text{now } \frac{dy}{dx}=x^2-5x+7=1$$

$\Rightarrow x^2-5x+6=0$

$\Rightarrow x=2$ or 3.

7. a., c.

$xy = (a + x)^2$

$\Rightarrow y + xy' = 2(a + x)$

Now $y' = \pm 1$

$\Rightarrow y \pm x = 2(a + x)$

$\frac{(a+x)^2}{x} \pm x = 2(a + x)$

$\Rightarrow \pm x = 2(a + x) - \frac{(a+x)^2}{x}$

$\Rightarrow \pm x^2 = (a + x)(x - a)$

$\Rightarrow \pm x^2 = x^2 - a^2$

$\Rightarrow 2x^2 = a^2 \Rightarrow x = \pm \frac{a}{\sqrt{2}}$

8. b., c.

$y = x^2 + 4x - 17 \Rightarrow \frac{dy}{dx} = 2(x + 2)$

$\Rightarrow \left(\frac{dy}{dx}\right)_{x=5/2} = 9$

$\Rightarrow \tan\theta = 9$, where θ is the angle with the positive direction of x-axis.

$\Rightarrow$ Angle with y-axis is $\frac{\pi}{2} \pm \theta = \frac{\pi}{2} \pm \tan^{-1} 9$.

9. a., b.

$x^3 - y^2 = 0$ (1)

$\Rightarrow 2y \times \frac{dy}{dx} = 3x^2$

Slope of the tangent at P is

$\left.\frac{dy}{dx}\right|_P = \left.\frac{3x^2}{2y}\right|_{(4m^2, 8m^3)} = 3m$

Therefore, Equation of the tangent at P is

$y - 8m^3 = 3m(x - 4m^2)$ or $y = 3mx - 4m^3$ (2)

It cuts the curve again at point Q. Solving (1) and (2) we get $x = 4m^2, m^2$

Put $x = m^2$ in Eq. (2)

$\Rightarrow y = 3m(m^2) - 4m^3 = -m^3$ $\therefore Q$ is $(m^2, -m^3)$

Slope of the tangent at Q

$= \left.\frac{dy}{dx}\right|_{(m^2, -m^3)} = \frac{3(m^4)}{2 \times (-m^3)} = \frac{-3}{2} m$

Slope of the normal at $Q = \frac{1}{(-3/2)m} = \frac{2}{3m}$.

Since tangent at P is normal at $Q \Rightarrow \frac{2}{3m} = 3m$

$\Rightarrow 9m^2 = 2$

10. a., b., c.

$y = x^2 \Rightarrow \frac{dy}{dx} = 2x = 2$ at $(1, 1)$

$x = y^2 \Rightarrow y = \sqrt{x}$

$\Rightarrow \frac{dy}{dx} = \frac{1}{2\sqrt{x}} = \frac{1}{2}$ at $(1, 1)$

$\Rightarrow \tan\theta = \frac{2 - \frac{1}{2}}{1 + 2 - \frac{1}{2}} = \frac{\frac{3}{2}}{1 + 1} = \frac{3}{4}$

$\Rightarrow \theta = \tan^{-1}\frac{3}{4} = \cos^{-1}\frac{4}{5} = \sin^{-1}\frac{3}{5}$

11. a., b. Let $P(x, y)$ be a point on the curve

$\ln(x^2 + y^2) = c \tan^{-1}\frac{y}{x}$

Differentiating both sides with respect to x, we get

$\frac{2x + 2yy'}{(x^2 + y^2)} = \frac{c(xy' - y)}{(x^2 + y^2)} \Rightarrow y' = \frac{2x + cy}{cx - 2y} = m_1$ (say)

Slope of $OP = \frac{y}{x} = m_2$ (say)

Let the angle between the tangents at P and OP be θ

$\Rightarrow \tan\theta = \left|\frac{m_1 - m_2}{1 + m_1 m_2}\right| = \left|\frac{\frac{2x + cy}{cx - 2y} - \frac{y}{x}}{1 + \frac{2xy + cy^2}{cx^2 - 2xy}}\right| = \frac{2}{c}$

$\Rightarrow \theta = \tan^{-1}\left(\frac{2}{c}\right)$ which is independent of x and y.

12. a., b., d.

f is not differentiable at $x = 1/2$

g is not continuous in $[0, 1]$ at $x = 0$

h is not continuous in $[0, 1]$ at $x = 1$

$k(x) = (x + 3)^{\ln_2 5} = (x + 3)p$, where $2 < p < 3$, which is continuous and differentiable.

Reasoning Type

1. d. Though $|x - 1|$ is non-differentiable at $x = 1$, $(x - 1)|x - 1|$ is differentiable at $x = 1$, for which Lagrange's mean value theorem is applicable.

2. a. Consider $f'(x) = 4ax^3 + 3bx^2 + 2cx + d$

$\Rightarrow f(x) = ax^4 + bx^3 + cx^2 + dx + e$

$f(0) = e$ and $f(3) = 81a + 27b + 9c + 3d + e$

$= 3(27a + 9b + 3c + d) + e = 0$

Hence, Rolle's theorem is applicable for $f(x)$,

$\Rightarrow$ there exists at least one c in (a, b) such that $f'(c) = 0$.

3. c. Statement 1 is correct as it is the statement of Cauchy's mean value theorem. Statement 2 is false as it is necessary that c in both $f'(c) = \frac{f(b) - f(a)}{b - a}$ and $g'(c) = \frac{g(b) - g(a)}{b - a}$ is same.

4. a. Let $y = \sqrt{-3 + 4x - x^2}$

$\Rightarrow x^2 + y^2 - 4x + 3 = 0$ or point (x, y) lies on the this circle.

Then, the given expression is $(y+4)^2 + (x-5)^2$, which is the square of distance between point $P(5, -4)$ and any point on the circle $x^2 + y^2 - 4x + 3 = 0$ which has centre $C(2, 0)$ and radius 1.

Now $CP = 5$, then the maximum distance between the point P and any point on the circles is 6.

$\Rightarrow$ Maximum value of $(\sqrt{-3+4x-x^2} + 4)^2 + (x-5)^2$ is 36.

5. c. Statement 1 is correct as $f(-2) = f(2) = 0$ and Rolle's theorem is not applicable, then it implies that either $f(x)$ is discontinuous or $f'(x)$ does not exist at at least one point in $(-1, 1)$. Since it is given that $g(x)$ is differentiable, $g(x) = 0$ at at least one value of x in $(-1, 1)$.

Statement 2 is false as $f(x)$ must be differentiable in (a, b) is not given.

6. d. when $x = 1, y = 1$

$$\frac{dy}{dx} = 3x^2 - 2x - 1$$

$$\Rightarrow \left(\frac{dy}{dx}\right)_{x=1} = 0$$

$\Rightarrow$ Equation of the tangent is $y = 1$.

Solving with the curve, $x^3 - x^2 - x + 2 = 1$

$\Rightarrow x^3 - x^2 - x + 1 = 0 \Rightarrow x = -1, 1$ (1 is repeated root)

Therefore, the tangent meets the curve again at $x = -1$

Statement 1 is false and statement 2 is true.

Linked Comprehension Type

For Problems 1–3

1. a., 2.c., 3.d.

Sol. Let $P_1(t_1, t_1^3)$ be a point on the curve $y = x^3$. Therefore,

$$\left.\frac{dy}{dx}\right|_{(t_1, t_1^3)} = 3t_1^2$$

Tangent at P_1 is $y - t_1^3 = 3t_1^2 (x - t_1)$ (1)

The intersection of (1) and $y = x^3$

$\Rightarrow x^3 - t_1^3 = 3t_1^2(x - t_1)$

$\Rightarrow (x - t_1)(x^2 + xt_1 + t_1^2) - 3t_1^2(x - t_1) = 0$

$\Rightarrow (x - t_1)^2(x + 2t_1) = 0$

If $P_2(t_2, t_2^3)$, then $(t_2 - t_1)^2 (t_2 + 2t_1) = 0$. Therefore, $t_2 = -2t_1$ $(t_2 \neq t_1)$.

Similarly, the tangent at P_2 will meet the curve at the point $P_3(t_3\ t_3^3)$ when $t_3 = -2t_2 = 4t_1$ and so on.

The abscissae of $P_1, P_2, \ldots, P_n$ are

$t_1, -2t_1, 4t_1, \ldots, (-2)^{n-1} t_1$ in G.P.

$$\therefore \frac{t_2}{t_1} = \frac{t_3}{t_2} = \frac{t_4}{t_3} = \cdots = -2 \ (r \text{ say})$$

$\therefore t_2 = t_1 r, t_3 = t_2 r$ and $t_4 = t_3 r$

If $x_1 = 1$, then $x_2 = -2, t_3 = 4, \ldots$

Then $\lim\limits_{n\to\infty} \sum\limits_{r=1}^{n} \frac{1}{x_n}$ = sum of infinite G.P. with common ratio (–1/2) with first term 1

$$= \frac{1}{1 - \left(-\frac{1}{2}\right)} = \frac{2}{3}$$

Then $\lim\limits_{n\to\infty} \sum\limits_{r=1}^{n} \frac{1}{y_n}$ = sum of infinite G.P. with common ratio (–1/8) with first term 1

$$= \frac{1}{1 - \left(-\frac{1}{8}\right)} = \frac{8}{9}.$$

$$\text{Area of } \Delta P_1P_2P_3 = \frac{1}{2}\begin{vmatrix} t_1 & t_1^3 & 1 \\ t_3 & t_3^3 & 1 \\ t_4 & t_4^3 & 1 \end{vmatrix}$$

$$= \frac{1}{2}\begin{vmatrix} rt_1 & r^3t_1^3 & 1 \\ rt_2 & r^3t_2^3 & 1 \\ rt_3 & r^3t_3^3 & 1 \end{vmatrix}$$

$$= r^4 \text{ (Area of } (\Delta P_1P_2P_3))$$

$$\therefore \frac{\text{Area of } (\Delta P_1P_2P_3)}{\text{Area of } (\Delta P_2P_3P_4)} = \frac{1}{r^4} = \frac{1}{(-2)^4} = \frac{1}{16}$$

For Problems 4–6

4. b., 5.a., 6.c.

Sol.

4. a. $\frac{dy}{dx} = \frac{1-9t^2}{-6t} = \tan\theta$

$\Rightarrow 9t^2 - 6\tan\theta \times t - 1 = 0$

$\Rightarrow 3t = \tan\theta \pm \sec\theta$

$\Rightarrow \tan\theta + \sec\theta = 3t.$

5. c. $P(-2, 2) \Rightarrow t = -1 \Rightarrow \left.\frac{dy}{dx}\right|_{t=-1} = -\frac{4}{3}$

Equation of the tangent $y - 2 = -\frac{4}{3}(x+2)$

$\Rightarrow t - 3t^3 - 2 = -\frac{4}{3}(1 - 3t^2 + 2)$

$\Rightarrow 9t^3 + 12t^2 - 3t - 6 = 0$

$\Rightarrow 3t^3 + 4t^2 - t - 2 = 0$

$\Rightarrow (3t^2 + t - 2)(t+1) = 0$

$\Rightarrow (3t-2)(t+1)^2 = 0$

$\Rightarrow t = \frac{2}{3}$

$\Rightarrow Q = \left(-\frac{1}{3}, -\frac{2}{3}\right)$

6. c. $\left.\frac{dy}{dx}\right|_{t=2/3} = \frac{3}{4}$

$m_{po}m_Q = -1 \Rightarrow$ angle 90°

For Problems 7 – 8

7. b., 8. a.

Sol. Let V be the volume and r the radius of the balloon at any time, then

$V = \left(\frac{4}{3}\right)\pi r^3$

$\Rightarrow \frac{dV}{dt} = \left(\frac{4}{3}\right)(3\pi r^2)\frac{dr}{dt}$

$\Rightarrow \frac{dV}{dt} = 4\pi r^2 \frac{dr}{dt} = 40$ (given)

$\Rightarrow \frac{dr}{dt} = \frac{10}{\pi r^2}$ (1)

Now let S be the surface area of the balloon when its radius is r, then $S = 4\pi r^2$

$\Rightarrow \frac{dS}{dt} = 8\pi r\frac{dr}{dt}$ (2)

From (1) and (2), $\frac{dS}{dt} = 8\pi r \frac{10}{\pi r^2} = \frac{80}{r}$

When $r = 8$, the rate of increase of $S = \frac{80}{8}$ $= 10\,\text{cm}^2/\text{min}$.

$\Rightarrow$ Increase of S in $\frac{1}{2}$ minute $= 10 \times \left(\frac{1}{2}\right) = 5\,\text{cm}^2/\text{min}$.

If r_1 is the radius of the balloon after (1/2) min, then $4\pi r_1^2 = 4\pi(8)^2 + 5$

or $r_1^2 - 8^2 = \frac{5}{4\pi} = 0.397$ nearly or $r_1^2 = 64.397$

or $r_1 = 8.025$ nearly.

$\Rightarrow$ Required increase in the radius $= r_1 - 8$

$= 8.025 - 8$

$= 0.025$ cm.

Matrix-Match Type

1. a → p, q; b → r, s; c → r, q; d → p, s;

a. Given $\frac{a}{\sin A} = \frac{b}{\sin B} = \frac{c}{\sin C} = 2R$ (say)

$\therefore\ da = 2R\cos A\ dA$

$db = 2R\cos B\ dB$

$dc = 2R\cos C\ dC$

$\therefore\ \frac{da}{\cos A} + \frac{db}{\cos B} + \frac{dc}{\cos C}$

$= 2R(dA + dB + dC)$ (1)

Also $A + B + C = \pi$ So, $dA + dB + dC = 0$ (2)

From equations (1) and (2), we get

$\frac{da}{\cos A} + \frac{db}{\cos B} + \frac{dc}{\cos C} + 1 = 1$

$\Rightarrow m = \pm 1$

b. $x^2y^2 = 16 \Rightarrow xy = \pm 4$ (1)

$L_{ST} = \left|\frac{y}{dy/dx}\right|$

Differentiating (1) w.r.t. x, we get $y + xy' = 0$

$\Rightarrow y' = \frac{-y}{x}$

$L_{ST} = \left|\frac{y}{y/x}\right| = |x| \Rightarrow L_{ST} = 2$

$\Rightarrow k = \pm 2.$

c. $y = 2e^{2x}$ intersects y-axis at (0, 2)

$\frac{dy}{dx} = 4e^{2x}\ \therefore\ \left.\frac{dy}{dx}\right|_{\text{at } x=0} = 4.$

Angle of intersection with the y-axis is $\frac{\pi}{2} - \tan^{-1} 4 = \cot^{-1} 4$

$\Rightarrow n = 2$ or -1

d. $\frac{dy}{dx} = e^{\sin y}\cos y$: slope of the normal at (1, 0)

$= -1$

equation of the normal is $x + y = 1$

Area $= \frac{1}{2}$

$\Rightarrow t = 1, -2.$

2. a → q; b → r; c → p; d → s.

a. $r = 6$ cm $\delta r = 0.06$

$A = \pi r^2$ $\delta A = 2\pi r\delta r = 2\pi(6)(0.06) = 0.72\pi.$

b. $v = x^3$, $\delta v = 3x^2\,\delta x$

$$\frac{\delta v}{v} \times 100 = 3\frac{\delta x}{x} \times 100 = 3 \times 2 = 6.$$

c. $(x-2)\frac{dx}{dt} = 3\frac{dx}{dt}$

$\Rightarrow x = 5$

d. $A = \frac{\sqrt{3}}{4}x^2$

$$\Rightarrow \frac{dA}{dt} = \frac{\sqrt{3}}{2}\left(x\frac{dx}{dt}\right) = \frac{\sqrt{3}}{2} \times 30 \times \frac{1}{10} = \frac{3\sqrt{3}}{2}.$$

3. a → p, q; b → p, s; c → r; d → q;

a. $y^2 = 4x$ and $x^2 = 4y$ intersect at point (0, 0) and (4, 4)

$C_1 : y^2 = 4x \qquad C_2 : x^2 = 4y$

$$\frac{dy}{dx} = \frac{2}{y} \qquad \frac{dy}{dx} = \frac{x}{2}$$

$$\left.\frac{dy}{dx}\right|_{0,0} = \infty \qquad \left.\frac{dy}{dx}\right|_{0,0} = 0$$

Hence, $\tan\theta = 90°$ at point (0, 0)

$$\left.\frac{dy}{dx}\right|_{(4,4)} = \frac{1}{2} \quad \left.\frac{dy}{dx}\right|_{(4,4)} = 2$$

$$\tan\theta = \left|\frac{2 - \frac{1}{2}}{1 + 2\times\frac{1}{2}}\right| = \frac{3}{4}$$

b. Solving $2y^2 = x^3$ and $y^2 = 32x$, we get (0, 0), (8, 16) and (8, – 16)

$\sqrt{2}\,y = x\sqrt{x}$ or $\sqrt{2}\,y = -x\sqrt{x}$

at (0, 0) $\left.\frac{dy}{dx}\right|_{(0,0)} = 0$ for I

at (0, 0) $\left.\frac{dy}{dx}\right|_{(0,0)} = \infty$ for II

Hence, angle = 90°

now $\left.\frac{dy}{dx}\right|_{(8,16)} = \frac{3x^2}{4y} = \frac{3}{4}\frac{64}{16} = 3$ for I

$\left.\frac{dy}{dx}\right|_{(8,16)} = \frac{32}{2y} = \frac{16}{16} = 1$ for II

$$\therefore \tan\theta = \frac{3-1}{1+3} = \frac{2}{4} = \frac{1}{2}$$

Also tangent at the origin on $2y^2 = x^3$ is $y^2 = 0$, i.e., $y = 0$, i.e., x-axis and tangent to $y^2 = 32x$ at the origin is $x = 0$ ⇒ angle between the two curves at the origin is 90°.

c. The two curves are

$xy = a^2$ (1)

$x^2 + y^2 = 2a^2$ (2)

Solving (1) and (2), the points of intersection are (a, a) and $(-a, -a)$

Differentiating (1), $dy/dx = -y/x = m_1$ (say)

Differentiating (2), $dy/dx = -x/y = m_2$ (say)

At both points, $m_1 = -1 = m_2$

Hence, the two curves touch each other.

d. $y^2 = x$, $x^3 + y^3 = 3xy$

For 1st curve, $2y\frac{dy}{dx} = 1 \Rightarrow \left.\frac{dy}{dx}\right|_P = \frac{1}{2y_1}$

Again for the 2nd curve, $\left.\frac{dy}{dx}\right|_P = \frac{y_1 - x_1^2}{y_1^2 - x_1}$

solving $y^2 = x$ and $x^3 + y^3 = 3xy$;

$y^6 + y^3 = 3y^3 \Rightarrow y^3 + 1 = 3 \Rightarrow y^3 = 2$

$\therefore y_1 = 2^{1/3}$ and $x_1 = 2^{2/3}$.

Now $m_1 = \frac{1}{2\times 2^{1/3}} = \frac{1}{2^{4/3}}$; $m_2 = \frac{2^{1/3} - 2^{4/3}}{2^{2/3} - 2^{2/3}} \to \infty$

$$\tan\theta = \left|\frac{m_2 - m_1}{1 + m_1 m_2}\right| = \left|\frac{1 - \frac{m_1}{m_2}}{\frac{1}{m_2} + m_1}\right| = \left|\frac{1}{m_1}\right|$$

$= 2^{4/3} = 16^{1/3}$

Hence, $\theta = \tan^{-1}(16^{1/3})$

Integer Type

1. **(2)** Here, $y^2 = x^3$ (1)

$\Rightarrow 2y\dfrac{dy}{dx} = 3x^2$

Slope at $(4m^2, 8m^3)$ is

$\left(\dfrac{3x^2}{2y}\right)_{(4m^2,\, 8m^3)} = 3m.$

Equation of the tangent at $(4m^2, 8m^3)$ is

$\dfrac{y-8m^3}{x-4m^2} = 3m.$

$\Rightarrow y = 3mx - 4m^3$ (2)

For another point, solving (1) and (2), we get $x^3 = (3mx - 4m^3)^2$

$\Rightarrow x = 4m^2, m^2$

Thus $A(4m^2, 8m^3)$ and $B(m^2, -m^3)$

$\Rightarrow$ Slope of tangent at B,

$\left(\dfrac{dy}{dx}\right)_{(m^2,\,-m^3)} = \left(\dfrac{3x^2}{2y}\right)_{(m^2,\,-m^3)} = \dfrac{-3}{2}m$

$\Rightarrow$ Slope of normal at B is $2/(3m)$

Since the tangent and normal coincide,

$\dfrac{2}{3m} = 3m$

$\Rightarrow m^2 = \dfrac{2}{3} \Rightarrow m = \pm\sqrt{\dfrac{2}{3}}.$

2. **(6)** $y^2 - 2x^3 - 4y + 8 = 0$

Let a tangent be drawn to the curve at the point $Q(\alpha, \beta)$ passing through $P(1, 2)$.

Differentiating w.r.t. x, we get

$2y\dfrac{dy}{dx} - 6x^2 - 4\dfrac{dy}{dx} = 0$

$\Rightarrow \dfrac{dy}{dx} = \dfrac{3x^2}{y-2}$

Now the slope of line PQ is $\dfrac{dy}{dx}_{(\alpha,\,\beta)}$

$\Rightarrow \dfrac{\beta-2}{\alpha-1} = \dfrac{3\alpha^2}{\beta-2}$

$\Rightarrow (\beta-2)^2 = 3\alpha^2(\alpha-1)$ (1)

Also (α, β) satisfies the equation of the curve. Therefore,

$\beta^2 - 2\alpha^3 - 4\beta + 8 = 0$ or $(\beta-2)^2 = 2\alpha^3 - 4$ (2)

From (1) and (2), we get

$3\alpha^2(\alpha-1) = 2\alpha^3 - 4$

$\Rightarrow \alpha^3 - 3\alpha^2 + 4 = 0$ or $(\alpha-2)(\alpha^2-\alpha-2) = 0$ or $(\alpha-2)^2(\alpha+1) = 0$

When $\alpha = 2$, $(\beta-2)^2 = 12$ or $\beta = 2 \pm 2\sqrt{3}$.

When $\alpha = -1$, $(\beta-2)^2 = -6$ (which is not possible)

$\Rightarrow (\alpha, \beta) \equiv (2, 2 \pm 2\sqrt{3})$

3. **(4)** $y = ax^2 - 6x + b$ passes through $(0, 2)$

$\Rightarrow 2 = 0 - 0 + 6$

$\Rightarrow b = 2$

Again $\dfrac{dy}{dx} = 2ax - 6.$

At $x = \dfrac{3}{2}$, $\dfrac{dy}{dx} = 3a - 6.$

Since the tangent is parallel to x-axis, $\dfrac{dy}{dx} = 0$

$\Rightarrow 3a - 6 = 0 \Rightarrow a = 2$

Hence, $a = 2, b = 2.$

4. **(6)** $y = a^{1-n}x^n$

$\Rightarrow \dfrac{dy}{dx} = a^{1-n}nx^{n-1}$

$$\text{Sub-normal} = \left|y\dfrac{dy}{dx}\right|$$

$$= \left|ya^{1-n}nx^{n-1}\right|$$

$$= \left|a^{1-n}x^n a^{1-n}nx^{n-1}\right|$$

$$= \left|a^{2-2n}x^{2n-1}\right|,$$

which is constant if $2n - 1 = 0$ or $n = 1/2$.

5. **(1)** When $x = 0$, $y = -1$.

$\dfrac{dy}{dx} = 3x^2 + 6x + 4$

$\Rightarrow \left.\dfrac{dy}{dx}\right|_{x=0} = 4$

$$\Rightarrow \text{Length of tangent} = \left|y\sqrt{1+\left(\dfrac{dx}{dy}\right)^2}\right|$$

$$= \left|-1\sqrt{1+\dfrac{1}{16}}\right|$$

$$= \dfrac{\sqrt{17}}{4}.$$

6. **(2)**

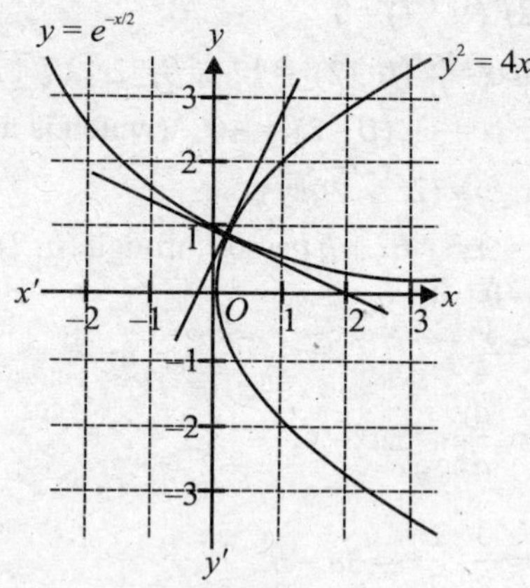

Fig. 5.7

Let the curves intersect at the point (x_1, y_1).

For $y^2 = 4x$, $\left.\dfrac{dy}{dx}\right|_{(x_1,\, y_1)} = \dfrac{2}{y_1}$

and for $y = e^{-x/2}$, $\left.\dfrac{dy}{dx}\right|_{(x_1,\, y_1)} = -\dfrac{1}{2}e^{-x_1/2} = -\dfrac{y_1}{2}$

$\Rightarrow m_1 m_2 = -1$

Hence, $\theta = 90°$.

Note: Here although we have not actually found the intersection point, we can geometrically see that the curves intersect.

7. **(1)** The curves are $x^3 + y^3 = 8xy$ (1)

and $y^2 = 4x$ (2)

Solving (1) and (2), we get

$x^3 + 4xy = 8xy$

$\Rightarrow x^3 = 4xy$

$\Rightarrow x^3 = 4x\,2\sqrt{x}$

$\Rightarrow x^{3/2}(x^{3/2} - 8) = 0$

$\Rightarrow x = 0$ or $x^{3/2} = 8 = 2^3$

$\Rightarrow x = 0$ or $x = 2^2 = 4$

Now when $x = 0$, we get $y = 0$

and when $x = 4$, we get $y^2 = 16$ or $y \pm 4$.

But $x = 4$ and $y = -4$ do not satisfy (1).

Thus (0, 0) and (4, 4) are the points of intersection of (1) and (2).

Differentiating (1), we get

$$\frac{dy}{dx} = \frac{8y - 3x^2}{3y^2 - 8x}.$$

At (4, 4), $\dfrac{dy}{dx} = -1$.

Hence, the equation of the normal to the curve (1) at (4, 4) is $(y - 4) = 1(x - 4)$ or $y - x = 0$.

8. **(4)** The two curves are

$$\frac{x^2}{a^2} + \frac{y^2}{4} = 1 \quad (1)$$

and $y^3 = 16x$ (2)

Differentiating (1), we get

$$\frac{dy}{dx} = -\frac{4x}{a^2 y} = m_1$$

Differentiating (2), we get

$$\frac{dy}{dx} = \frac{16}{3y^2} = m_2$$

Since the curves intersect orthogonally,

$m_1 m_2 = -1$

$$\Rightarrow \left(-\frac{4x}{a^2 y}\right)\left(\frac{16}{3y^2}\right) = -1$$

$\Rightarrow 64x = 3a^2 y^3$

$\Rightarrow 64x = 3a^2\, 16x,$ [Using (2)]

$\Rightarrow a^2 = \dfrac{4}{3}$

9. **(1)** We have $x^3 - 3xy^2 + 2 = 0$ (1)

and $3x^2y - y^3 - 2 = 0$ (2)

Differentiating (1) and (2) w.r.t. x, we get

$$\left(\frac{dy}{dx}\right)_{c_1} = \frac{x^2 - y^2}{2xy} \text{ and } \left(\frac{dy}{dx}\right)_{c_2} = \frac{-2xy}{x^2 - y^2}.$$

Therefore, $\left(\dfrac{dy}{dx}\right)_{c_1} \times \left(\dfrac{dy}{dx}\right)_{c_2} = -1.$

Hence, the two curves intersect at right angles.

10. **(6)** $ay + x^2 = 7$ and $x^3 = y$ intersect orthogonally.

Therefore, $\left(\dfrac{dy}{dx}\right) = -\dfrac{2x}{a}$ and $\left(\dfrac{dy}{dx}\right) = 3x^2$

$$\Rightarrow \left[\left(-\frac{2x}{a}\right)(3x^2)\right]_{(1,\,1)} = -1$$

$\Rightarrow -\dfrac{2}{a} \times 3 = -1 \Rightarrow a = 6$

11. **(3)** Let $P(x_1, y_1)$ be a point closest to the line $y = x - 2$.

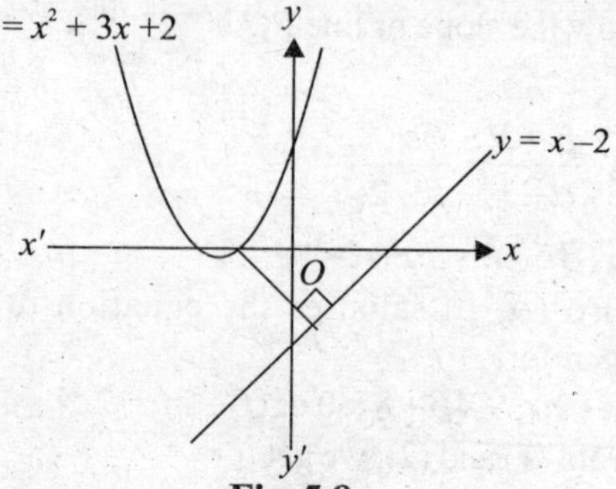

Fig. 5.8

Then $\left.\dfrac{dy}{dx}\right|_{(x_1, y_1)}$ = slope of the line

$\Rightarrow 2x_1 + 3 = 1$

$\Rightarrow x_1 = -1$

$\Rightarrow y_1 = 0$

The point (– 1, 0) is closest and its perpendicular distance from the line $y = x - 2$ will give the shortest distance.

Hence, the shortest distance is $\dfrac{3}{\sqrt{2}}$.

12. (6)

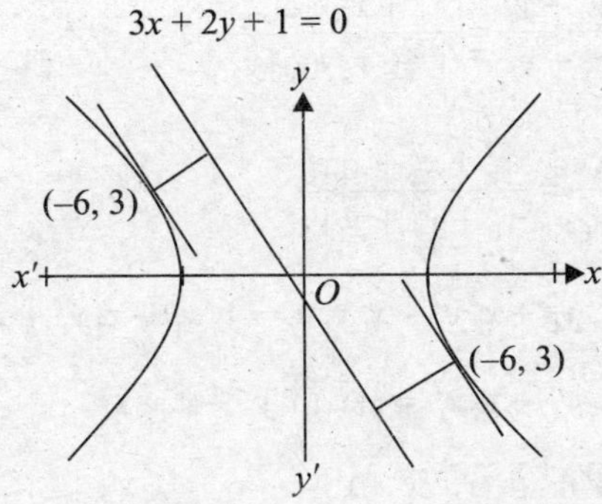

Fig. 5.9

Slope of the given line $3x + 2y + 1 = 0$ is $(-3/2)$.

Let us locate the point on the curve at which the tangent is parallel to the given line.

Differentiating both sides of the curve w.r.t. to x,

we get $6x - 8y\dfrac{dy}{dx} = 0$

$$\Rightarrow \left(\frac{dy}{dx}\right)_{(x_1, y_1)} = \frac{3x_1}{4y_1} = -\frac{3}{2}$$

(Since parallel to $3x + 2y = 1$)

$$\Rightarrow \frac{x_1}{y_1} = -2 \quad (1)$$

Also the point (x_1, y_1) lies on $3x^2 - 4y^2 = 72$. Therefore,

$$3x_1^2 - 4y_1^2 = 72$$

$$\Rightarrow 3\frac{x_1^2}{y_1^2} - 4 = \frac{72}{y_1^2} \quad (2)$$

$$\Rightarrow 3(4) - 4 = \frac{72}{y_1^2} \quad \text{[From (1)]}$$

$$\Rightarrow y_1^2 = 9 \Rightarrow y_1 = \pm 3$$

The required points are (–6, 3) and (6, –3).

Distance of (–6, 3) from the given line

$$= \frac{|-18+6+1|}{\sqrt{13}} = \frac{11}{\sqrt{13}}$$

and distance of (6, 3) from the given line

$$= \frac{|18-6+1|}{\sqrt{13}} = \frac{13}{\sqrt{13}} = \sqrt{13}.$$

Thus (–6, 3) is the required point.

13. (5)

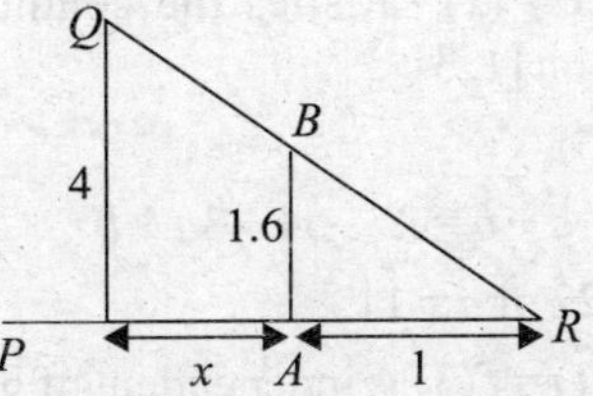

Fig. 5.10

Let $PQ = 4$ m be the height of the pole and $AB = 1.6$m be the height of the man.

Let the end of the shadow be R and let it be at a distance of l from A when the man is at a distance x from PQ at some instant.

Since ΔPQR and ΔABR are similar, we have

$$\frac{PQ}{AB} = \frac{PR}{AR}$$

$$\Rightarrow \frac{4}{1.6} = \frac{x+l}{l}$$

$$\Rightarrow 2x = 3l$$

$$\Rightarrow 2\frac{dx}{dt} = 3\frac{dl}{dt} \qquad \left(\text{Given } \frac{dx}{dt} = 30 \text{ m/min}\right)$$

$$\Rightarrow \frac{dl}{dt} = \frac{2}{3} \times 30 \text{ m/min} = 20 \text{ m/min}$$

14. (4)

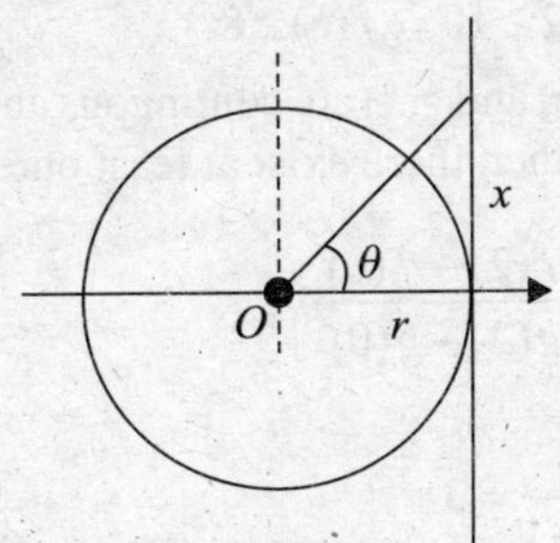

Fig. 5.11

$\tan\theta = \frac{x}{r} \Rightarrow x = r\tan\theta$

$\Rightarrow \frac{dx}{dt} = r\sec^2\theta\left(\frac{d\theta}{dt}\right)$

$= r\omega\sec^2\theta = v\sec^2\theta$

When $\theta = \pi/8$,

$\frac{dx}{dt} = v\sec^2\frac{\pi}{4} = 2v = 40$ km/hr.

Hence, $\theta = 45°$.

15. **(7)** Since $f(x)$ satisfies the conditions of Rolle's theorem on [1, 3],

$f(1) = f(3)$

$\Rightarrow 1 - 6 + a + b = 27 - 54 + 3a + b$

$\Rightarrow 2a = 22$ or $a = 11$.

Since $f(1) = f(3)$ is independent of b,

$a = 11$ and $b \in R$.

16. **(0)** $f'(c) = 16c - 7$

$= \frac{f(6) - f(-6)}{12}$

$= \frac{(8\times36 - 7\times6 + 5) - (8\times36 + 7\times6 + 5)}{12}$

$= -\frac{2\times7\times6}{12} = -7$

$\Rightarrow c = 0$

17. **(8)** By Lagrange's mean value theorem there exists $c \in (1, 6)$ such that

$f'(c) = \frac{f(6) - f(1)}{6 - 1}$

$\Rightarrow \frac{f(6) + 2}{5} = f'(c) \geq 2$

$\{\because f'(x) \geq 2 \text{ for all } x \in [1, 6]\}$

$\Rightarrow f(6) + 2 \geq 10 \Rightarrow f(6) \geq 8$

18. **(3)** As $f(x)$ and $g(x)$ are continuous and differentiable in [0, 2], then there exist at least one c such that

$\frac{f'(c)}{g'(c)} = \frac{f(2) - f(0)}{g(2) - g(0)}$

$\Rightarrow \frac{8 - 2}{g(2) - 1} = 3$

$\Rightarrow g(2) - 1 = 2$

$\Rightarrow g(2) = 3$

19. **(1)** Slope of the tangent at (x_1, y_1) is $(-x_1^2)/y_1^2$.

The tangent intersects the curve again at (x_2, y_2).

So slope of the tangent is $(y_2 - y_1)/(x_2 - x_1)$

$\Rightarrow -\frac{x_1^2}{y_1^2} = \frac{y_2 - y_1}{x_2 - x_1}$

Also $x_1^3 + y_1^3 = a^3$ and $x_2^3 + y_2^3 = a^3$. Therefore,

$x_1^3 + y_1^3 = x_2^3 + y_2^3$

$\frac{y_2^3 - y_1^3}{x_1^3 - x_2^3} = 1$

$\Rightarrow \frac{y_2 - y_1}{x_1 - x_2} = \frac{x_1^2 + x_2^2 + x_1x_2}{y_1^2 + y_2^2 + y_1y_2}$

$\Rightarrow \frac{x_1^2}{y_1^2} = \frac{x_1^2 + x_2^2 + x_1x_2}{y_1^2 + y_2^2 + y_1y_2}$

$\Rightarrow x_1^2y_1^2 + x_1^2y_2^2 + x_1^2y_1y_2 = y_1^2x_1^2 + x_2^2y_1^2 + y_1^2x_1x_2$

$\Rightarrow x_2^2y_1^2 - y_2^2x_1^2 = x_1y_1(x_1y_2 - x_2y_1)$

$\Rightarrow x_2y_1 + y_2x_1 = -x_1y_1$

$\Rightarrow \frac{x_2}{x_1} + \frac{y_2}{y_1} = -1$

20. **(2)** We know that

$1 \leq |\sin x| + |\cos x| \leq \sqrt{2}$

$\Rightarrow y = [|\sin x| + |\cos x|] = 1$

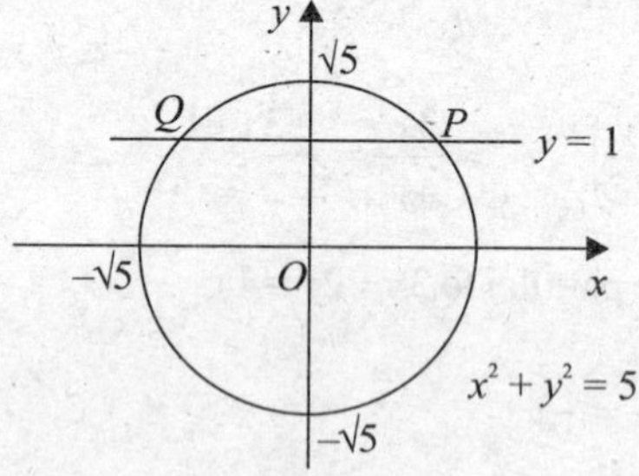

Fig. 5.12

Let P and Q be the points of intersection of the given curves.

Clearly, the given curves meet at points where $y = 1$. So we get $x^2 + 1 = 5 \Rightarrow x = \pm 2$.

Now $P(2, 1)$ and $Q(-2, 1)$.

Differentiating $x^2 + y^2 = 5$ w.r.t. x, we get $2x + 2y\frac{dy}{dx} = 0$

$\Rightarrow \frac{dy}{dx} = -\frac{x}{y}$

$\Rightarrow \left(\frac{dy}{dx}\right)_{(2,1)} = -2$ and $\left(\frac{dy}{dx}\right)_{(-2,1)} = 2$.

Clearly, the slopes of the line $y = 1$ is zero, and the slopes of the tangents at P and Q are (–2) and (2), respectively.

Thus the angle of intersection is $\tan^{-1}(2)$.

21. **(2)** Slope of the given line is $-1/2$ (1)

$\Rightarrow$ Slope of the tangent is $(dy)/(dx) = -1/2$.

Equation of the given curve is $y = \cos(x + y)$.

Differentiating the curve w.r.t. x, we get

$$\frac{dy}{dx} = -\sin(x+y)\left\{1+\frac{dy}{dx}\right\}$$

$$\Rightarrow \frac{dy}{dx} = -\frac{\sin(x+y)}{1+\sin(x+y)} = \text{slope of the tangent (2)}$$

From (1) and (2), we get

$$\frac{-\sin(x+y)}{1+\sin(x+y)} = -\frac{1}{2}$$

$\Rightarrow \sin(x+y) = 1$ (3)

$\Rightarrow \cos(x+y) = 0$

From the given curve $y = \cos(x+y)$, we get $y = 0$ (4)

From (3) and (4), we get

$\sin x = 1$

$$\Rightarrow x = \frac{\pi}{2}, -\frac{3\pi}{2}$$

Therefore, the points are $P(\pi/2, 0)$ and $Q(-3\pi/2, 0)$.

Hence, tangent at P is $x + 2y = \pi/2$ and tangent at Q is $x + 2y = -(3\pi)/2$.

22. **(1)** The curve $y = ax^3 + bx^2 + cx + 5$ touches the x-axis at $P(-2, 0)$. So the x-axis is a tangent at $(-2, 0)$.

The curve meets the y-axis at (0, 5). So we have

$$\frac{dy}{dx} = 3ax^2 + 2bx + c$$

$\Rightarrow \left.\frac{dy}{dx}\right|_{(0,5)} = 0 + 0 + c = 3$ (given)

$\Rightarrow c = 3$ (1)

and $\left.\frac{dy}{dx}\right|_{(-2,0)} = 0$

$\Rightarrow 12a - 4b + c = 0$

$\Rightarrow 12a - 4b + 3 = 0$ [From (1)] (2)

As $(-2, 0)$ lies on the curve,

$0 = -8a + 4b - 2c + 5$

$\Rightarrow 0 = -8a + 4b - 1$ ($\because c = 3$)

$\Rightarrow 8a - 4b + 1 = 0$ (3)

From (2) and (3), we get

$$a = -\frac{1}{2},\ b = -\frac{3}{4}$$

Hence, $a = -\frac{1}{2}$, $b = -\frac{3}{4}$, and $c = 3$.

23. **(6)** $g(x) = [f'(x)]^2 + f''(x)f(x) = \frac{d}{dx}[f(x)f'(x)]$

Let $h(x) = f(x)f'(x)$

Since $f(a) = 0, f(b) = 2, f(c) = -1, f(d) = 2, f(e) = 0$,

$f(x) = 0$ has four roots namely a, α, β, e, where $b < \alpha < c$ and $c < \beta < d$.

And $f'(x) = 0$ at three points k_1, k_2, k_3, where $a < k_1 < \alpha$, $\alpha < k_2 < \beta$, $\beta < k_3 < c$.

[$\because$ Between any two roots of a polynomial function $f(x) = 0$, there lies at least one root of $f'(x) = 0$]

Therefore, there are at least seven roots of $f(x)f'(x) = 0$.

$\Rightarrow$ There are at least six roots of $\frac{d}{dx}[f(x)f'(x)] = 0$, i.e., of $g(x) = 0$.

24. **(3)** $y = x^2 + bx - b$

$$\Rightarrow \frac{dy}{dx} = 2x + b$$

$\Rightarrow$ Equation of the tangent at (1, 1) is

$y - 1 = (2 + b)(x - 1)$

$\Rightarrow (b + 2)x - y = b + 1$

So x-intercept $= \frac{b+1}{b+2} = OA$ and y-intercept $= -(b + 1) = OB$.

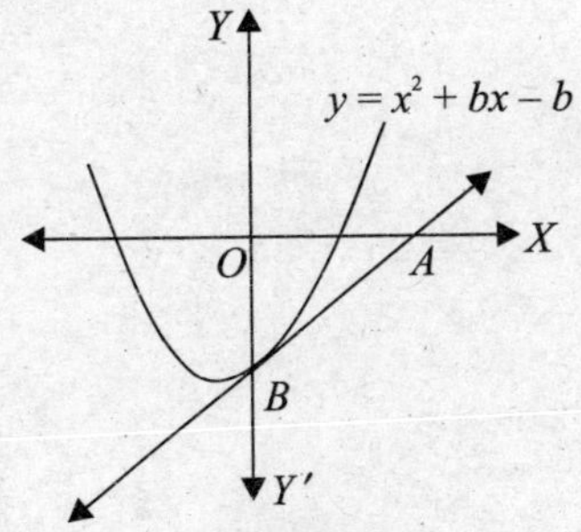

Fig. 5.13

Area of the triangle OAB is given as 2. Therefore,

$$\frac{1}{2}OA \times OB = 2$$

$$\Rightarrow \frac{1}{2}\left(\frac{b+1}{b+2}\right)[-(b+1)] = 2$$

$\Rightarrow b^2 + 2b + 1 = -4(b + 2)$

$\Rightarrow b^2 + 6b + 9 = 0$

$\Rightarrow (b + 2)^2 = 0 \Rightarrow b = -3$

25. (7) $\frac{dy}{dx} = \frac{4t}{3}$ Tangent is $y - at^4 = \frac{4t}{3}\left(x - at^3\right)$.

x-intercept $= \frac{at^3}{4}$ and y-intercept $= -\frac{at^4}{3}$

The points of intersection of the tangent with the axes are $[(at^3)/4, 0]$ and $[0, -(at^4)/3]$. Therefore,

$\frac{m}{n} = -\frac{3}{4}$

$\Rightarrow m = 3, n = 4$

$\Rightarrow m + n = 7$

26. (2)

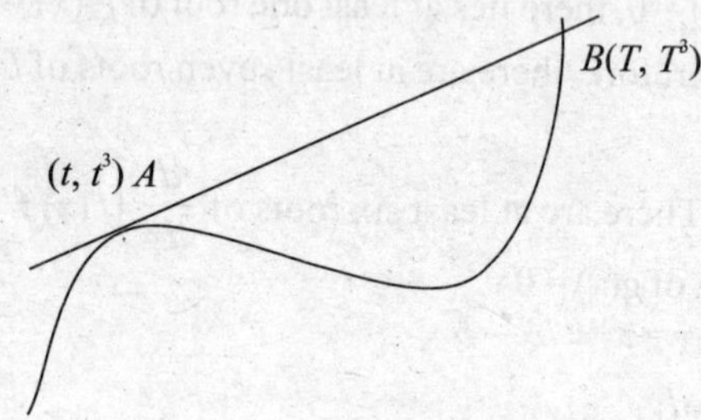

Fig. 5.14

$\frac{dy}{dx} = 3x^2 = 3t^2$ at A

$3t^2 = \frac{T^3 - t^3}{T - t} = T^2 + Tt + t^2$

$\Rightarrow T^2 + Tt - 2t^2 = 0$

$\Rightarrow (T - t)(T + 2t) = 0$

$\Rightarrow T = t$ or $T = -2t$ ($T = t$ is not possible)

27. (9) $y = ax^2 + bx + c$; $\frac{dy}{dx} = 2ax + b$

when $x = 1, y = 0 \Rightarrow a + b + c = 0$ (1)

$\left.\frac{dy}{dx}\right|_{x=1} = 3$ and $\left.\frac{dy}{dx}\right|_{x=3} = 1$

$2a + b = 3$ (2)

$6a + b = 1$ (3)

Solving (1), (2) and (3)

$a = -\frac{1}{2}$; $b = 4$, $c = -\frac{7}{2}$

$\therefore 2a - b - 4c = -1 - 4 + 14 = 9$

CHAPTER 6

Monotonocity and Maxima-Minima Functions

Objective Type

Each question has four choices a, b, c, and d, out of which *only one* is correct.

1. If $f(x) = kx^3 - 9x^2 + 9x + 3$ is monotonically increasing in each interval, then
   a. $k < 3$ b. $k \le 2$
   c. $k \ge 3$ d. None of these

2. If the function $f(x) = \dfrac{K\sin x + 2\cos x}{\sin x + \cos x}$ is strictly increasing for all values of x, then
   a. $K < 1$ b. $K > 1$
   c. $K < 2$ d. $K > 2$

3. Let $f: R \to R$ be a function such that $f(x) = ax + 3\sin x + 4\cos x$. Then $f(x)$ is invertible if
   a. $a \in (-5, 5)$ b. $a \in (-\infty, 5)$
   c. $a \in (-5, +\infty)$ d. None of these

4. Let $g(x) = 2f\left(\dfrac{x}{2}\right) + f(2-x)$ and $f''(x) < 0\ \forall x \in (0, 2)$. Then $g(x)$ increases in
   a. $(1/2, 2)$ b. $(4/3, 2)$
   c. $(0, 2)$ d. $(0, 4/3)$

5. On which of the following intervals is the function $x^{100} + \sin x - 1$ decreasing?
   a. $(0, \pi/2)$ b. $(0, 1)$
   c. $(\pi/2, \pi)$ d. None of these

6. A function is matched below against an interval where it is supposed to be increasing. Which of the following parts is incorrectly matched?

	Interval	Function
b.	$(-\infty, \infty)$	$x^3 - 3x^2 + 3x + 3$
c.	$(-\infty, -4]$	$x^3 + 6x^2 + 6$
d.	$\left(-\infty, \dfrac{1}{3}\right]$	$3x^2 - 2x + 1$

7. The function $f(x) = \tan^{-1}(\sin x + \cos x)$ is an increasing function in
   a. $\left(-\dfrac{\pi}{2}, \dfrac{\pi}{4}\right)$ b. $\left(0, \dfrac{\pi}{2}\right)$
   c. $\left(-\dfrac{\pi}{2}, \dfrac{\pi}{2}\right)$ d. $\left(\dfrac{\pi}{4}, \dfrac{\pi}{2}\right)$

8. The function x^x decreases in the interval
   a. $(0, e)$ b. $(0, 1)$
   c. $\left(0, \dfrac{1}{e}\right)$ d. None of these

9. The function $f(x) = \sum_{K=1}^{5}(x-K)^2$ assumes the minimum value of x given by
   a. 5 b. $\dfrac{5}{2}$
   c. 3 d. 2

10. Which of the following statements is always true?
    a. If $f(x)$ is increasing, then $f^{-1}(x)$ is decreasing
    b. If $f(x)$ is increasing, then $\dfrac{1}{f(x)}$ is also increasing

c. If f and g are positive function and f is increasing and g is decreasing, then f/g is a decreasing function

d. If f and g are positive function and f is decreasing and g is increasing, then f/g is a decreasing function

11. Let $f: R \to R$ be a differentiable function for all values of x and has the property that $f(x)$ and $f'(x)$ have opposite signs for all values of x. Then,

a. $f(x)$ is an increasing function

b. $f(x)$ is a decreasing function

c. $f^2(x)$ is a decreasing function

d. $|f(x)|$ is an increasing function

12. Let $f(x) = x\sqrt{4ax - x^2}$, $(a > 0)$. Then $f(x)$ is

a. increasing in $(0, 3a)$, decreasing in $(-\infty, 0) \cup (3a, \infty)$

b. increasing in $(a, 4a)$, decreasing in $(5a, \infty)$

c. increasing in $(0, 4a)$, decreasing in $(-\infty, 0)$

d. None of these

13. Let $f: R \to R$ be a differentiable function $\forall x \in k$. If the tangent drawn to the curve at any point $x \in (a, b)$ always lies below the curve, then

a. $f'(x) > 0, f''(x) < 0, \forall x \in (a, b)$

b. $f'(x) < 0, f''(x) < 0, \forall x \in (a, b)$

c. $f'(x) > 0, f''(x) > 0, \forall x \in (a, b)$

d. None of these

14. If $f'(x) = |x| - \{x\}$ where $\{x\}$ denotes the fractional part of x, then $f(x)$ is decreasing in

a. $\left(-\frac{1}{2}, 0\right)$ b. $\left(-\frac{1}{2}, 2\right)$

c. $\left(-\frac{1}{2}, 2\right)$ d. $\left(\frac{1}{2}, \infty\right)$

15. Function $f(x) = |x| - |x - 1|$ is monotonically increasing when

a. $x < 0$ b. $x > 1$

c. $x < 1$ d. $0 < x < 1$

16. Let f be continuous and differentiable function such that $f(x)$ and $f'(x)$ have opposite signs everywhere. Then

a. f is increasing

b. f is decreasing

c. $|f|$ is increasing and decreasing

d. $|f|$ is decreasing

17. If the function $f(x)$ increases in the interval (a, b), then the function $\phi(x) = [f(x)]^2$

a. increases in (a, b)

b. decreases in (a, b)

c. we cannot say that $\phi(x)$ increases or decreases in (a, b)

d. none of these

18. If $\phi(x)$ is a polynomial function and $\phi'(x) > \phi(x)$, $\forall x \geq 1$ and $f(1) = 0$, then

a. $\phi(x) \geq 0, \forall x \geq 1$ b. $\phi(x) < 0, \forall x \geq 1$

c. $\phi(x) = 0, \forall x \geq 1$ d. None of these

19. Which of the following statements is true for the

$$\text{function } f(x) = \begin{cases} \sqrt{x}, & x \geq 1 \\ x^3, & 0 \leq x \leq 1 \\ \frac{x^3}{3} - 4x, & x < 0 \end{cases}$$

a. It is monotonic increasing $\forall x \in R$.

b. $f'(x)$ fails to exist for 3 distinct real values of x

c. $f'(x)$ changes its sign twice as x varies from $-\infty$ to ∞

d. function attains its extreme values at x_1 and x_2, such that $x_1 x_2 > 0$

20. If $f''(x) > 0$, $\forall x \in R$, $f'(3) = 0$ and $g(x) = f(\tan^2 x - 2\tan x + 4)$, $0 < x < \pi/2$, then $g(x)$ is increasing in

a. $\left(0, \frac{\pi}{4}\right)$ b. $\left(\frac{\pi}{6}, \frac{\pi}{3}\right)$

c. $\left(0, \frac{\pi}{3}\right)$ d. $\left(\frac{\pi}{4}, \frac{\pi}{2}\right)$

21. Let $f(x)$ be a function such that $f'(x) = \log_{1/3}[\log_3(\sin x + a)]$. If $f(x)$ is decreasing for all real values of x, then

a. $a \in (1, 4)$ b. $a \in (4, \infty)$

c. $a \in (2, 3)$ d. $a \in (2, \infty)$

22. If $f(x) = x + \sin x$; $g(x) = e^{-x}$; $u = \sqrt{c+1} - \sqrt{c}$; $v = \sqrt{c} - \sqrt{c-1}$; $(c > 1)$ then

a. $fog(u) \geq fog(v)$ b. $gof(u) \leq gof(v)$

c. $gof(u) > gof(v)$ d. $fog(u) < fog(v)$

23. The length of the largest continuous interval in which the function $f(x) = 4x - \tan 2x$ is monotonic is

a. $\pi/2$ b. $\pi/4$

c. $\pi/8$ d. $\pi/16$

24. $f(x) = (x-1)|(x-2)(x-3)|$ then f decreases in

a. $\left(2 - \frac{1}{\sqrt{3}}, 2\right)$ b. $\left(2, 2 + \frac{1}{\sqrt{3}}\right)$

c. $\left(2 + \frac{1}{\sqrt{3}}, 4\right)$ d. $(3, \infty)$

25. The number of solutions of the equation $x^3 + 2x^2 + 5x + 2\cos x = 0$ in $[0, 2\pi]$ is

a. one b. two
c. three d. zero

26. $f(x) = (x-2)\,|x-3|$ is monotonically increasing in
a. $(-\infty, 5/2) \cup (3, \infty)$ b. $(5/2, \infty)$
d. $(2, \infty)$ d. $(-\infty, 3)$

27. $f(x) = (x-8)^4 (x-9)^5$, $0 \le x \le 10$, monotonically decreases in
a. $\left(\frac{76}{9}, 10\right]$ b. $\left(8, \frac{76}{9}\right)$
c. $[0, 8)$ d. $\left(\frac{76}{9}, 10\right]$

28. If $f(x) = x^3 + 4x^2 + \lambda x + 1$ is a monotonically decreasing function of x in the largest possible interval $(-2, -2/3)$ then
a. $\lambda = 4$ b. $\lambda = 2$
c. $\lambda = -1$ d. λ has no real value

29. $f(x) = |x \log_e x|$ monotonically decreases in
a. $(0, 1/e)$ b. $(1/e, 1)$
c. $(1, \infty)$ d. $(1/e, \infty)$

30. Given that $f'(x) > g'(x)$ for all real x, and $f(0) = g(0)$, then $f(x) < g(x)$ for all x belongs to
a. $(0, \infty)$ b. $(-\infty, 0)$
c. $(-\infty, \infty)$ d. None of these

31. A function $g(x)$ is defined as $g(x) = 1/4\, f(2x^2 - 1) + 1/2\, f(1 - x^2)$ and $f'(x)$ is an increasing function, then $g(x)$ is increasing in the interval
a. $(-1, 1)$
b. $\left(-\sqrt{\frac{2}{3}}, 0\right) \cup \left(\sqrt{\frac{2}{3}}, \infty\right)$
c. $\left(-\sqrt{\frac{2}{3}}, \sqrt{\frac{2}{3}}\right)$
d. None of these

32. Let $f(x)$ be a function defined as below:
$f(x) = \sin(x^2 - 3x), x \le 0$; and $6x + 5x^2, x > 0$
Then at $x = 0, f(x)$
a. has a local maximum b. has a local minimum
c. is discontinuous d. None of these

33. The greatest value of $f(x) = \cos(xe^{[x]} + 7x^2 - 3x)$, $x \in [-1, \infty)$ is
a. -1 b. 1
c. 0 d. None of these

34. If $f(x) = x^5 - 5x^4 + 5x^3 - 10$ has local max. and min. at $x = p$ and $x = q$, respectively, then $(p, q) =$
a. $(0, 1)$ b. $(1, 3)$
c. $(1, 0)$ d. None of these

35. The maximum value of $(\log x)/x$ is
a. 1 b. $2/e$
c. e d. $1/e$

36. If a function $f(x)$ has $f'(a) = 0$ and $f''(a) = 0$, then
a. $x = a$ is a maximum for $f(x)$
b. $x = a$ is a minimum for $f(x)$
c. it is difficult to say *a.* and *b*
d. $f(x)$ is necessarily a constant function

37. The minimum value of $2^{(x^2-3)^3+27}$ is
a. 2^{27} b. 2
c. 1 d. None of these

38. The maximum value of the function $\sin x\,(1 + \cos x)$ is
a. 3 b. $\frac{3\sqrt{3}}{4}$
c. 4 d. $\frac{3\sqrt{3}}{2}$

39. The greatest value of the function
$f(x) = \frac{\sin 2x}{\sin(x + \pi/4)}$ on the interval $\left[0, \frac{\pi}{2}\right]$ is
a. $\frac{1}{\sqrt{2}}$ b. $\sqrt{2}$
c. 1 d. $-\sqrt{2}$

40. The function $f(x) = (4\sin^2 x - 1)^n (x^2 - x + 1)$, $n \in N$, has a local minimum at $x = \pi/6$, then
a. n can be any even number
b. n can be an odd number
c. n can be odd prime number
d. n can be any natural number

41. All possible values of x for which the function $f(x) = x \ln x - x + 1$ is positive is
a. $(1, \infty)$ b. $(1/e, \infty)$
c. $[e, \infty)$ d. $(0, 1) \cup (1, \infty)$

42. The greatest value of $f(x) = (x+1)^{1/3} - (x-1)^{1/3}$ on $[0, 1]$ is
a. 1 b. 2
c. 3 d. $\frac{1}{3}$

43. If the function $f(x) = 2x^3 - 9ax^2 + 12a^2x + 1$, where $a > 0$, attains its maximum and minimum at p and q, respe- ctively such that $p^2 = q$, then a equals to
a. 1 b. 2
c. $\frac{1}{2}$ d. 3

44. The real number x when added to its inverse gives the minimum value of the sum at x equals to

a. 1 b. -1
c. -2 d. 2

45. The function $f(x) = \frac{x}{2} + \frac{2}{x}$ has a local minimum at

a. $x = 2$ b. $x = -2$
c. $x = 0$ d. $x = 1$

46. The maximum value of the function $f(x) = \sin\left(x + \frac{\pi}{6}\right) + \cos\left(x + \frac{\pi}{6}\right)$ in the interval $\left(0, \frac{\pi}{2}\right)$ occurs at

a. $\frac{\pi}{12}$ b. $\frac{\pi}{6}$
c. $\frac{\pi}{4}$ d. $\frac{\pi}{3}$

47. Let $f(x) = \begin{cases} x+2, & -1 \le x < 0 \\ 1, & x = 0 \\ \frac{x}{2}, & 0 < x \le 1 \end{cases}$

Then on $[-1, 1]$, this function has

a. a minimum
b. a maximum
c. either a maximum or a minimum
d. neither a maximum nor a minimum

48. The maximum value of $f(x) = \frac{x}{1 + 4x + x^2}$ is

a. $-\frac{1}{4}$ b. $-\frac{1}{3}$
c. $\frac{1}{6}$ d. $\frac{1}{5}$

49. The maximum slope of the curve $y = -x^3 + 3x^2 + 9x - 27$ is

a. 0 b. 12
c. 16 d. 32

50. Let $f(x) = \cos \pi x + 10x + 3x^2 + x^3, -2 \le x \le 3$. The absolute minimum value of $f(x)$ is

a. 0 b. -15
c. $3 - 2\pi$ d. None of these

51. The minimum value of $e^{(2x^2-2x+1)\sin^2 x}$ is

a. e b. $1/e$
c. 1 d. 0

52. The maximum value of $x^4 e^{-x^2}$ is

a. e^2 b. e^{-2}
c. $12e^{-2}$ d. $4e^{-2}$

53. If $a^2x^4 + b^2y^4 = c^6$, then the maximum value of xy is

a. $\frac{c^2}{\sqrt{ab}}$ b. $\frac{c^3}{ab}$
c. $\frac{c^3}{\sqrt{2ab}}$ d. $\frac{c^3}{2ab}$

54. The global maximum value of $f(x) = \log_{10}(4x^3 - 12x^2 + 11x - 3), x \in [2, 3]$ is

a. $-\frac{3}{2}\log_{10} 3$ b. $1 + \log_{10} 3$
c. $\log_{10} 3$ d. $\frac{3}{2}\log_{10} 3$

55. The least natural number a for which $x + ax^{-2} > 2$, $\forall \in (0, \infty)$ is

a. 1 b. 2
c. 5 d. None of these

56. A function f is defined by $f(x) = |x|^m |x-1|^n$, $\forall x \in R$. The maximum value of the function is $(m, n \in N)$

a. 1 b. $m^n n^m$
c. $\frac{m^m n^n}{(m+n)^{m+n}}$ d. $\frac{(mn)^{mn}}{(m+n)^{m+n}}$

57. $f(x) = \begin{cases} 4x - x^3 + \ln(a^2 - 3a + 3), & 0 \le x < 3 \\ x - 18, & x \ge 3 \end{cases}$

complete the set of values of a such that $f(x)$ as a local minima at $x = 3$ is

a. $[-1, 2]$ b. $(-\infty, 1) \cup (2, \infty)$
c. $[1, 2]$ d. $(-\infty, -1) \cup (2, \infty)$

58. Let the function $f(x)$ be defined as follows

$$f(x) = \begin{cases} x^3 + x^2 - 10x, & -1 \le x < 0 \\ \cos x, & 0 \le x < \pi/2 \\ 1 + \sin x, & \pi/2 \le x \le \pi \end{cases}$$

Then $f(x)$ has

a. a local minimum at $x = \pi/2$
b. a global maximum at $x = \pi/2$
c. an absolute minimum at $x = -1$
d. an absolute maximum at $x = \pi$

59. A differentiable function $f(x)$ has a relative minimum at $x = 0$, then the function $y = f(x) + ax + b$ has a relative minimum at $x = 0$ for

a. all a and all b b. all b if $a = 0$
c. all $b > 0$ d. all $a > 0$

60. If $f(x) = 4x^3 - x^2 - 2x + 1$ and

$$g(x) = \begin{cases} \min\{f(t) : 0 \le t \le x\} & ; \quad 0 \le x \le 1 \\ 3 - x & ; \quad 1 < x \le 2 \end{cases}$$ then

$g\left(\frac{1}{4}\right) + g\left(\frac{3}{4}\right) + g\left(\frac{5}{4}\right)$ has the value equal to

a. 7/4 b. 9/4

c. 13/4 d. 5/2

61. The set of value(s) of a for which the function $f(x) = ax^3/3 + (a+2)x^2 + (a-1)x + 2$ possesses a negative point of inflection

a. $(-\infty, -2) \cup (0, \infty)$ b. $\{-4/5\}$

c. $(-2, 0)$ d. empty set

62. Suppose that f is a polynomial of degree 3 and that $f''(x) \neq 0$ at any of the stationary point. Then

a. f has exactly one stationary point

b. f must have no stationary point

c. f must have exactly 2 stationary points

d. f has either 0 or 2 stationary point

63. The maximum value of the function $f(x) = \dfrac{(1+x)^{0.6}}{1+x^{0.6}}$ in the interval [0, 1] is

a. $2^{0.4}$ b. $2^{-0.4}$

c. 1 d. $2^{0.6}$

64. $f: R \to R, f(x)$ is differentiable such that $f(f(x)) = k(x^5 + x)$, $(k \neq 0)$, then $f(x)$ is always

a. increasing

b. decreasing

c. either increasing or decreasing

d. non-monotonic

65. If the function $f: R \to R$ and $g: R \to R$ are such that $f(x)$ is continuous at $x = \alpha$ and $f(\alpha) = a$ and $g(x)$ is discontinuous at $x = a$ but $g(f(x))$ is continuous at $x = \alpha$, then ($f(x)$ and $g(x)$ are non-constant functions)

a. $x = \alpha$ is an extremum of $f(x)$ and $x = a$ is an extremum of $g(x)$

b. $x = \alpha$ may not be an extremum of $f(x)$ and $x = a$ is an extremum of $g(x)$

c. $x = \alpha$ is an extremum of $f(x)$ and $x = a$ may not be an extremum of $g(x)$

d. None of these

66. If $f: R \to R$ and $g: R \to R$ are two functions such that $f(x) + f''(x) = -x\,g(x) f'(x)$ and $g(x) > 0\ \forall\, x \in R$, then the functions $f^2(x) + (f'(x))^2$ has

a. a maxima at $x = 0$

b. a minima at $x = 0$

c. a point of inflexion at $x = 0$

d. None of these

67. Let $h(x) = x^{m/n}$ for $x \in R$, where m and n are odd numbers and $0 < m < n$, then $y = h(x)$ has

a. no local extremums b. one local maximum

c. one local minimum d. None of these

68. $f(x) = 4\tan x - \tan^2 x + \tan^3 x, x \neq n\pi + \pi/2$

a. is monotonically increasing

b. is monotonically decreasing

c. has a point of maxima

d. has a point of minima

69. If for a function $f(x), f'(a) = 0, f''(a) = 0, f'''(a) > 0$, then at $x = a, f(x)$ is

a. minimum b. maximum

c. not an extreme point d. extreme point

70. The function $f(x) = x(x+4)e^{-x/2}$ has its local maxima at $x = a$, then

a. $a = 2\sqrt{2}$ b. $a = 1 - \sqrt{3}$

c. $a = -1 + \sqrt{3}$ d. $a = -4$

71. If $f(x) = \begin{cases} \sin^{-1}(\sin x), & x > 0 \\ \dfrac{\pi}{2}, & x = 0 \\ \cos^{-1}(\cos x), & x < 0 \end{cases}$, then

a. $x = 0$ is a point of maxima

b. $x = 0$ is a point of minima

c. $x = 0$ is a point of intersection

d. None of these

72. $f(x) = \begin{cases} 2 - |x^2 + 5x + 6|, & x \neq -2 \\ a^2 + 1, & x = -2 \end{cases}$, then the range of a, so that $f(x)$ has maxima at $x = -2$, is

a. $|a| \geq 1$ b. $|a| < 1$

c. $a > 1$ d. $a < 1$

73. If $A > 0, B > 0$ and $A + B = \pi/3$, then the maximum value of $\tan A \tan B$ is

a. $\dfrac{1}{\sqrt{3}}$ b. $\dfrac{1}{3}$

c. 3 d. $\sqrt{3}$

74. If the function $f(x) = \dfrac{t + 3x - x^2}{x - 4}$, where t is a parameter that has a minimum and maximum, then the range of values of t is

a. $(0, 4)$ b. $(0, \infty)$

c. $(-\infty, 4)$ d. $(4, \infty)$

75. The value of a for which the function $f(x) = a\sin x + (1/3)\sin 3x$ has an extremum at $x = \pi/3$ is

a. 1 b. −1

c. 0 d. 2

76. The least value of a for which the equation $\frac{4}{\sin x}+\frac{1}{1-\sin x}=a$ for at least one solution on the interval $(0, \pi/2)$ is

a. 9 b. 4
c. 8 d. 1

77. The largest term in the sequence $a_n=\frac{n^2}{n^3+200}$ is given by

a. $\frac{529}{49}$ b. $\frac{8}{89}$
c. $\frac{49}{543}$ d. None of these

78. The number of values of k for which the equation $x^3-3x+k=0$ has two distinct roots lying in the interval (0, 1) is

a. three
b. two
c. infinitely many
d. no value of k satisfies the requirement

79. Consider the function $f(x)=x\cos x-\sin x$, then identify the statement which is correct.

a. f is neither odd nor even
b. f is monotonic decreasing at $x=0$
c. f has a maxima at $x=\pi$
d. f has a minima at $x=-\pi$

80. Let $f(x)=ax^3+bx^2+cx+1$ have extrema at $x=\alpha, \beta$ such that $\alpha\beta<0$ and $f(\alpha)\,f(\beta)<0$. Then the equation $f(x)=0$ has

a. three equal real roots
b. one negative root if $f(\alpha)<0$ and $f(\beta)>0$
c. one positive root if $f(\alpha)>0$ and $f(\beta)<0$
d. None of these

81. A factory D is to be connected by a road with a straight railway line on which a town A is situated. The distance DB of the factory to the railway line is $5\sqrt{3}$ km. Length AB of the railway line is 20 km. Freight charges on the road are twice the charges on the railway. The point P $(AP<AB)$ on the railway line should the road DP be connected so as to ensure minimum freight charges from the factory to the town is

a. $BP=5$ km b. $AP=5$ km
c. $BP=7.5$ km d. None of these

82. The volume of the greatest cylinder which can be inscribed in a cone of height 30 cm and semi-vertical angle 30° is

a. $4000\,\pi/3$ cubic cm b. $400\,\pi/3$ cubic cm
c. $4000\,\pi/\sqrt{3}$ cubic cm d. None of these

83. A rectangle of the greatest area is inscribed in a trapezium $ABCD$. One of whose non-parallel sides AB is perpen dicular to the base, so that one of the rectangle's side lies on the larger base of the trapezium. The base of trapezium are 6 and 10 cm and AB is 8 cm long. Then the maximum area of the rectangle is

a. 24 sq. cm. b. 48 sq. cm.
c. 36 sq. cm. d. None of these

84. A bell tent consists of a conical portion above a cylindrical portion near the ground. For a given volume and a circular base of a given radius, the amount of the canvas used is a minimum when the semi-vertical angle of the cone is

a. $\cos^{-1}(2/3)$. b. $\sin^{-1}(2/3)$
c. $\cos^{-1}(1/3)$ d. None of these

85. A rectangle is inscribed in an equilateral triangle of side length $2a$ units. The maximum area of this rectangle can be

a. $\sqrt{3}a^2$ b. $\frac{\sqrt{3}a^2}{4}$
c. a^2 d. $\frac{\sqrt{3}a^2}{2}$

86. Tangents are drawn to $x^2+y^2=16$ from the point $P(0, h)$. These tangents meet the x–axis at A and B. If the area of triangle PAB is minimum, then

a. $h=12\sqrt{2}$ b. $h=6\sqrt{2}$
c. $h=8\sqrt{2}$ d. $h=4\sqrt{2}$

87. The largest area of a trapezium inscribed in a semi-circle of radius R, if the lower base is on the diameter, is

a. $\frac{3\sqrt{3}}{4}R^2$ b. $\frac{\sqrt{3}}{2}R^2$
c. $\frac{3\sqrt{3}}{8}R^2$ d. R^2

88. In a ΔABC, $\angle B=90°$ and $b+a=4$. The area of the triangle is the maximum when $\angle C$ is

a. $\pi/4$ b. $\pi/6$
c. $\pi/3$ d. None of these

89. The three sides of a trapezium are equal, each being 8 cm. The area of the trapezium, when it is maximum, is

a. $24\sqrt{3}$ sq. cm. b. $48\sqrt{3}$ sq. cm.
c. $72\sqrt{3}$ sq. cm. d. None of these

90. The fuel charges for running a train are proportional to the square of the speed generated in km per hour, and the cost is Rs. 48 at 16 km per hour. If the fixed charges amount to Rs. 300 per hour, the most economical speed is

a. 60 kmph b. 40 kmph
c. 48 kmph d. 36 kmph

91. A cylindrical gas container is closed at the top and open at the bottom, if the iron plate of the top is 5/4 times as thick as the plate forming the cylindrical sides, the ratio of the radius to the height of the cylinder using minimum material for the same capacity is

a. 3:4 b. 5:6
c. 4:5 d. None of these

92. The least perimeter of an isosceles triangle in which a circle of radius r can be inscribed is

a. $4\sqrt{3}r$ b. $2\sqrt{3}r$
c. $6\sqrt{3}r$ d. $8\sqrt{3}r$

93. A given right circular has a volume p, and the largest right circular cylinder that can be inscribed in the cone has a volume q. Then $p:q$ is

a. 9:4 b. 8:3
c. 7:2 d. None of these

94. A wire of length a is cut into two parts which are bent, respectively, in the form of a square and a circle. The least value of the sum of the areas so formed is

a. $\frac{a^2}{\pi+4}$ b. $\frac{a}{\pi+4}$
c. $\frac{a}{4(\pi+4)}$ d. $\frac{a^2}{4(\pi+4)}$

95. A box constructed from a rectangular metal sheet is 21 cm by 16 cm by cutting equal squares of sides x from the corners of the sheet and then turning up the projected portions. The value of x so that volume of the box is maximum is

a. 1 b. 2
c. 3 d. 4

96. The largest area if a trapezium is inscribed in a semi-circle of radius R, if the lower base is on the diameter, is

a. $\frac{3\sqrt{3}}{4}R^2$ b. $\frac{\sqrt{3}}{2}R^2$
c. $\frac{3\sqrt{3}}{8}R^2$ d. R^2

Multiple Correct Answers Type

Each question has four choices a, b, c, and d, out of which *one or more* answers are correct.

1. Let $f(x)=\begin{cases} x^2+3x, & -1\le x<0 \\ -\sin x, & 0\le x<\pi/2 \\ -1-\cos x, & \frac{\pi}{2}\le x\le \pi \end{cases}$. Then the global maxima of $f(x)$ and global minima of $f(x)$ are equal to

a. -1 b. 0
c. -3 d. -2

2. Let $f(x)=x^4-4x^3+6x^2-4x+1$, then

a. f increases on $[1,\infty)$
b. f decreases on $[1,\infty)$
c. f has a minimum at $x=1$
d. f has neither maximum nor minimum

3. Let $f(x)=2x-\sin x$ and $g(x)=\sqrt[3]{x}$, then

a. range of gof is R
b. gof is one-one
c. both f and g are one-one
d. both f and g are onto

4. If $f(x)=2x+\cot^{-1}x+\log(\sqrt{1+x^2}-x)$, then $f(x)$

a. increases in $[0,\infty)$
b. decreases in $[0,\infty)$
c. neither increases nor decreases in $[0,\infty)$
d. increases in $(-\infty,\infty)$

5. The function $f(x)=\frac{|x-1|}{x^2}$ is monotonically decreasing on

a. $(1,\infty)$ b. $(0,2)$
c. $(0,1)$ d. $(2,\infty)$

6. If $f(x)=\int_0^x \frac{\sin t}{t}\,dt$, $x>0$, then

a. $f(x)$ has a local maxima at $x=n\pi$ $(n=2k, k\in I^+)$
b. $f(x)$ has a local minima at $x=n\pi$ $(n=2k, k\in I^+)$
c. $f(x)$ has neither maxima nor minima at $x=n\pi$ $(n\in I^+)$
d. $f(x)$ has local maxima at $x=n\pi$ $(n=2k-1, k\in I^+)$

7. The values of parameter a for which the point of minimum of the function $f(x)=1+a^2x-x^3$ satisfies the inequality $\frac{x^2+x+2}{x^2+5x+6}<0$ are

a. $(2\sqrt{3}, 3\sqrt{3})$ b. $(-3\sqrt{3}, -2\sqrt{3})$
c. $(-2\sqrt{3}, 3\sqrt{3})$ d. $(-3\sqrt{2}, 2\sqrt{3})$

8. Let $f(x) = ax^2 - b|x|$, where a and b are constants. Then at $x = 0, f(x)$ has
   a. a maxima whenever $a>0, b>0$
   b. a maxima whenever $a>0, b<0$
   c. minima whenever $a>0, b<0$
   d. neither a maxima nor a minima whenever $a>0, b<0$

9. The function $y = \dfrac{2x-1}{x-2}$ $(x \neq 2)$
   a. is its own inverse
   b. decreases at all values of x in the domain
   c. has a graph entirely above x-axis
   d. is unbound for all x

10. Let $g'(x)>0$ and $f'(x)<0, \forall x \in R$, then
   a. $(f(x+1))>g(f(x-1))$
   b. $f(g(x-1))>f(g(x+1))$
   c. $g(f(x+1))<g(f(x-1))$
   d. $g(g(x+1))<g(g(x-1))$

11. If $f(x)=x^3-x^2+100x+2002$, then
   a. $f(1000)>f(1001)$
   b. $f\left(\dfrac{1}{2000}\right)>f\left(\dfrac{1}{2001}\right)$
   c. $f(x-1)>f(x-2)$
   d. $f(2x-3)>f(2x)$

12. If $f'(x) = g(x)(x-a)^2$ where $g(a) \neq 0$ and g is continuous at $x = a$, then
   a. f is increasing in the neighbourhood of a if $g(a)>0$
   b. f is increasing in the neighbourhood of a if $g(a)<0$
   c. f is decreasing in the neighbourhood of a if $g(a)>0$
   d. f is decreasing in the neighbourhood of a if $g(a)<0$

13. The value of a for which the function $f(x) = (4a-3)(x+\log 5) + 2(a-7)\cot\dfrac{x}{2}\sin^2\dfrac{x}{2}$ does not possess critical points is
   a. $(-\infty, -4/3)$ b. $(-\infty, -1)$
   c. $[1, \infty)$ d. $(2, \infty)$

14. Let $f(x)=(x-1)^4(x-2)^n, n \in N$. Then $f(x)$ has
   a. a maximum at $x=1$ if n is odd
   b. a maximum at $x=1$ if n is even
   c. a minimum at $x=1$ if n is even
   d. a minima at $x=2$ if n is even

15. Let $f(x)=\sin x+ax+b$, then which of the following is/are true.
   a. $f(x)=0$ has only one real root which is positive if $a>1, b<0$
   b. $f(x)=0$ has only one real root which is negative if $a>1, b>0$
   c. $f(x)=0$ has only one real root which is negative if $a<-1, b<0$
   d. None of these

16. The function $\dfrac{\sin(x+a)}{\sin(x+b)}$ has no maxima or minima if
   a. $b-a=n\pi, n \in I$ b. $b-a=(2n+1)\pi, n \in I$
   c. $b-a=2n\pi, n \in I$ d. None of these

17. If composite function $f_1(f_2(f_3(\ldots(f_n(x))))$ n times is an increasing function and if r of f_i s are decreasing function while rest are increasing, then maximum value of $r(n-r)$ is
   a. $\dfrac{n^2-1}{4}$, when n is an even number
   b. $\dfrac{n^2}{4}$, when n is an odd number
   c. $\dfrac{n^2-1}{4}$, when n is an odd number
   d. $\dfrac{n^2}{4}$, when n is even number

18. Let $f(x)=\begin{cases}\dfrac{(x-1)(6x-1)}{2x-1}, & \text{if } x \neq \dfrac{1}{2} \\ 0, & \text{if } x = \dfrac{1}{2}\end{cases}$ then at $x=\dfrac{1}{2}$, which of the following is/are not true.
   a. f has a local maxima
   b. f has a local minima
   c. f has an inflection point
   d. f has a removable discontinuity is

19. In which of the following graphs is $x = c$ the point of inflection?

a.
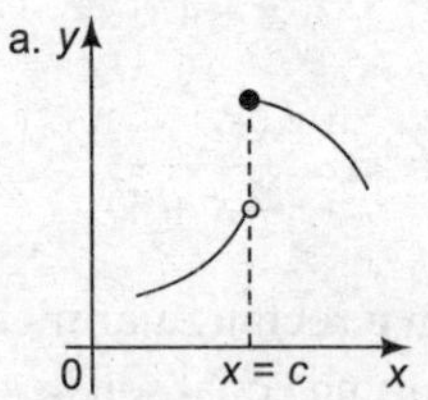

b.
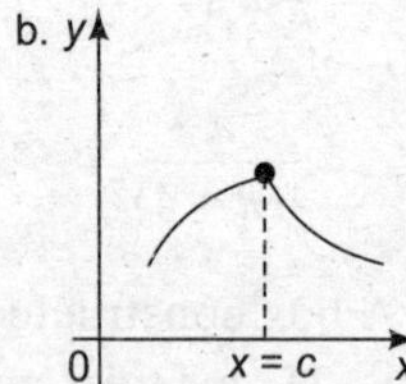

c.
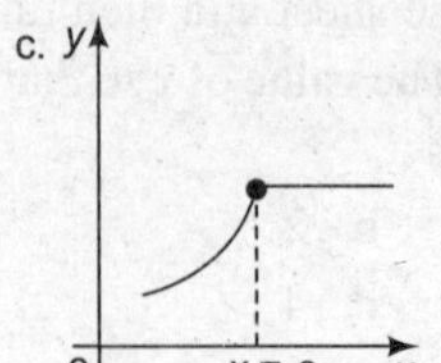

d.
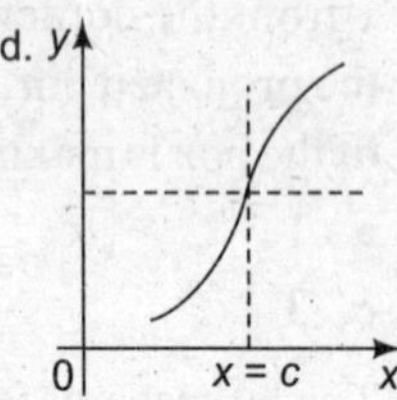

Fig. 6.1

20. Let $f(x)$ be an increasing function defined on $(0, \infty)$. If $f(2a^2+a+1)>f(3a^2-4a+1)$, then the possible integers in the range of a is/are
   a. 1 b. 2
   c. 3 d. 4

21. If $f(x)=(\sin^2 x-1)^n$, then $x=\pi/2$ is a point of

a. local maximum, if n is odd
b. local minimum, if n is odd
c. local maximum, if n is even
d. local minimum, if n is even

22. For the cubic function $f(x) = 2x^3 + 9x^2 + 12x + 1$ which one of the following statement/statements hold good?
a. $f(x)$ is non-monotonic
b. $f(x)$ increases in $(-\infty, -2) \cup (-1, \infty)$ and decreases in $(-2, -1)$
c. $f: R \to R$ is bijective
d. Inflection point occurs at $x = -3/2$

23. Let $f(x) = a_5x^5 + a_4x^4 + a_3x^3 + a_2x^2 + a_1x$, where a_i's are real and $f(x) = 0$ has a positive root α_0. Then
a. $f'(x) = 0$ has a root α_1 such that $0 < \alpha_1 < \alpha_0$
b. $f'(x) = 0$ has at least one real root
c. $f''(x) = 0$ has at least one real root.
d. None of these

24. If $f(x)$ and $g(x)$ are two positive and increasing functions, then which of the following is not always true?
a. $[f(x)]^{g(x)}$ is always increasing
b. if $[f(x)]^{g(x)}$ is decreasing, then $f(x) < 1$
c. if $[f(x)]^{g(x)}$ is increasing, then $f(x) > 1$
d. if $f(x) > 1$, then $[f(x)]^{g(x)}$ is increasing

25. An extremum of the function $f(x) = 2 - x/\pi \cos \pi(x + 3) + 1/\pi^2 \sin \pi(x + 3)$, $0 < x < 4$ occurs at
a. $x = 1$
b. $x = 2$
c. $x = 3$
d. $x = \pi$

Reasoning Type

Each question has four choices a, b, c and d, out of which *only one* is correct. Each question contains Statement 1 and Statement 2.

a. If both the statements are True and Statement 2 is the correct explanation of Statement 1
b. If both the statements are True but Statement 2 is Not the correct explanation of Statement 1
c. If Statement 1 is True and Statement 2 is False
d. If Statement 1 is False and Statement 2 is True

1. **Statement 1:** Both $\sin x$ and $\cos x$ are decreasing functions in $(\pi/2, \pi)$.

Statement 2: If a differentiable function decreases in an interval (a, b), then its derivative also decreases in (a, b).

2. **Statement 1:** $\alpha^\beta > \beta^\alpha$, for $2.91 < \alpha < \beta$.
Statement 2: $f(x) = \dfrac{\log_e x}{x}$ is a decreasing function for $x > e$.

3. **Statement 1:** $f(x) = |x - 1| + |x - 2| + |x - 3|$ has point of minima at $x = 3$.
Statement 2: $f(x)$ is non-differentiable at $x = 3$.

4. **Statement 1:** The function $f(x) = x \ln x$ is increasing in $(1/e, \infty)$.
Statement 2: If both $f(x)$ and $g(x)$ are increasing in (a, b) then $f(x)\, g(x)$ must be increasing in (a, b).

5. Let $f: R \to R$ is differentiable and strictly increasing function throughout its domain.
Statement 1: If $|f(x)|$ is also strictly increasing function, then $f(x) = 0$ has no real roots.
Statement 2: When $x \to \infty$ or $\to -\infty$, $f(x) \to 0$, but cannot be equal to zero.

6. **Statement 1:** Let $f(x) = 5 - 4(x - 2)^{2/3}$, then at $x = 2$ the function $f(x)$ attains neither the least value nor the greatest value.
Statement 2: At $x = 2$ first derivative does not exist.

7. **Statement 1:** $f(x) = x + \cos x$ is strictly increasing for $\forall x \in R$.
Statement 2: If $f(x)$ is strictly increasing, then $f'(x)$ may vanish at some finite number of points.

8. **Statement 1:** Both $f(x) = 2\cos x + 3\sin x$ and $g(x) = \sin^{-1} x/\sqrt{13} - \tan^{-1} 3/2$ are increasing, for $x \in (0, \pi/2)$.
Statement 2: If $f(x)$ is increasing then its inverse is also increasing.

9. **Statement 1:** $f(x) = \dfrac{x^3}{3} + \dfrac{ax^2}{2} + x + 5$ has positive point of maxima for $a < -2$.
Statement 2: $x^2 + ax + 1 = 0$ has both roots positive for $a < -2$.

10. **Statement 1:** For all $a, b \in R$ the function $f(x) = 3x^4 - 4x^3 + 6x^2 + ax + b$ has exactly one extremum.
Statement 2: If a cubic function is monotonic, then its graph cuts x-axis only once.

11. **Statement 1:** The value of $\left[\lim_{x \to 0+} \dfrac{\sin x \tan x}{x^2}\right]$ is 1, where [.] denotes the greatest integer function.
Statement 2: For $\left(0, \dfrac{\pi}{2}\right)$, $\sin x < x < \tan x$.

12. **Statement 1:** Let $f(x) = \sin(\cos x)$ in $\left[0, \dfrac{\pi}{2}\right]$, then $f(x)$ is decreasing in $\left[0, \dfrac{\pi}{2}\right]$.
Statement 2: $\cos x$ is a decreasing function $\forall x \in \left[0, \dfrac{\pi}{2}\right]$.

13. Let $f(x) = (x^3 - 6x^2 + 12x - 8)\,e^x$.
Statement 1: $f(x)$ is neither maximum nor minimum at $x = 2$.
Statement 2:
If a function $x = 2$ is a point of inflection, then it is not a point of extremum.

14. **Statement 1:** The function $f(x) = x^4 - 8x^3 + 22x^2 - 24x + 21$ is decreasing for every $x \in (2, 3) \cup (-\infty, 1)$.
Statement 2: $f(x)$ is increasing for $x \in (1, 2) \cup (3, \infty)$ and has no point of inflection.

15. **Statement 1:** If $f(0) = 0, f'(x) = \ln(x + \sqrt{1 + x^2})$, then $f(x)$ is positive for all $x \in R_0$.
Statements 2: $f(x)$ is increasing for $x > 0$ and decreasing for $x < 0$.

Linked Comprehension Type

Based upon each paragraph, three multiple choice questions have to be answered. Each question has four choices a, b, c, and d, out of which *only one* is correct.

For Problems 1–2

$$f(x) = \sin^{-1} x + x^2 - 3x + \frac{x^3}{3}$$

1. Which of the following is true about $f(x)$
   a. $f(x)$ has a point of a maxima
   b. $f(x)$ has a point of minima
   c. $f(x)$ is increasing
   d. $f(x)$ is decreasing

2. Which of the following is true for $x \in [0, 1]$?
   a. $\sin^{-1} x + x^2 - \frac{x(9 - x^2)}{3} \le 0$
   b. $\sin^{-1} x + x^2 - \frac{x(9 - x^2)}{3} \ge 0$
   c. $\sin^{-1} x + x^2 - \frac{x(9 - x^2)}{3} \le 1$
   d. $\sin^{-1} x + x^2 - \frac{x(9 - x^2)}{3} \ge 1$

For Problems 3–4

Let $f'(\sin x) < 0$ and $f''(\sin x) > 0\ \forall x \in \left(0, \frac{\pi}{2}\right)$ and $g(x) = f(\sin x) + f(\cos x)$.

3. Which of the following is true?
   a. g' is increasing
   b. g' is decreasing
   c. g' has a point of minima
   d. g' has a point of maxima

4. Which of the following is true
   a. $g(x)$ is decreasing in $\left(\frac{\pi}{4}, \frac{\pi}{2}\right)$
   b. $g(x)$ increasing in $\left(0, \frac{\pi}{4}\right)$
   c. $g(x)$ is monotonically increasing
   d. None of these

For Problems 5–8

Consider function $f(x) = \begin{cases} -x^2 + 4x + a, & x \le 3 \\ ax + b, & 3 < x < 4 \\ -\frac{b}{4}x + 6, & x \ge 4 \end{cases}$

(For questions 6 to 8 consider $f(x)$ is a continuous function).

5. Which of the following is true?
   a. $f(x)$ is discontinuous function for any value of a and b
   b. $f(x)$ is cont. for finite number of values of a and b.
   c. $f(x)$ cannot be differentiable for any value of a and b
   d. $f(x)$ is continuous for infinite values of a and b

6. If $x = 3$ is the only point of minima in its neighbourhood and $x = 4$ is neither a point of maxima nor a point of minima, then which of the following can be true?
   a. $a > 0, b < 0$
   b. $a < 0, b < 0$
   c. $a > 0, b \in R$
   d. None of these

7. If $x = 4$ is the only point of maxima in its neighbourhood but $x = 3$ is neither a point of maxima nor a point of minima, then which of the following can be true?
   a. $a < 0, b > 0$
   b. $a > 0, b < 0$
   c. $a > 0, b > 0$
   d. Not possible

8. If $x = 3$ is a point of minima and $x = 4$ is a point of maxima, then which of the following is true?
   a. $a < 0, b > 0$
   b. $a > 0, b < 0$
   c. $a > 0, b > 0$
   d. Not possible

For Problems 9–10

If $\phi(x)$ is a differentiable real-valued function satisfying $\phi'(x) + 2\phi(x) \le 1$, then it can be adjusted as $e^{2x}\phi'(x) + 2e^{2x}\phi(x) \le e^{2x}$ or $\frac{d}{dx}\left(e^{2x}\phi(x) - \frac{e^{2x}}{2}\right) \le 0$ or $\frac{d}{dx}e^{2x}\left(\phi(x) - \frac{1}{2}\right) \le 0$.

Here e^{2x} is called integrating factor which helps in creating single differential coefficient as shown above.

9. If $P(1) = 0$ and $\frac{dP(x)}{dx} > P(x)$ for all $x \ge 1$ then

a. $P(x) > 0 \ \forall x > 1$
b. $P(x)$ is a constant function
c. $P(x) < 0 \ \forall x > 1$
d. None of these

10. If $H(x_0) = 0$ for some $x = x_0$ and $\frac{d}{dx}H(x) > 2cxH(x)$ for all $x \geq x_0$, where $c > 0$, then
a. $H(x) = 0$ has root for $x > x_0$
b. $H(x) = 0$ has no roots for $x > x_0$
c. $H(x)$ is a constant function
d. None of these

For Problems 9–13

Let $h(x) = f(x) - a(f(x))^2 + a(f(x))^3$ for every real number x.

11. $h(x)$ increases as $f(x)$ increases for all real values of x if
a. $a \in [0, 3]$ b. $a \in [-2, 2]$
c. $[3, \infty)$ d. None of these

12. $h(x)$ increases as $f(x)$ decreases for all real values of x if
a. $a \in [0, 3]$ b. $a \in [-2, 2]$
c. $[3, \infty)$ d. None of these

13. $h(x)$ is non-monotonic function if
a. $a \in [0, 3]$ b. $a \in [-2, 2]$
c. $[3, \infty)$ d. $a \in (-\infty, 0] \cup [3, \infty)$

For Problems 14–16

$f(x) = x^3 - 9x^2 + 24x + c$ has three real and distinct roots α, β, and γ.

14. Possible values of c are
a. (–20, –16) b. (–20, –18)
c. (–18, –16) d. None of these

15. If $[\alpha] + [\beta] + [\gamma] = 8$, then the values of c, where [ ] represents the greatest integer function, are
a. (–20, –16) b. (–20, –18)
c. (–18, –16) d. None of these

16. If $[\alpha] + [\beta] + [\gamma] = 7$, then the values of c, where [ ] represents the greatest integer function, are
a. (–20, –16) b. (–20, –18)
c. (–18, –16) d. None of these

For Problems 17–21

Consider the graph of $y = g(x) = f'(x)$, given that $f(c) = 0$.

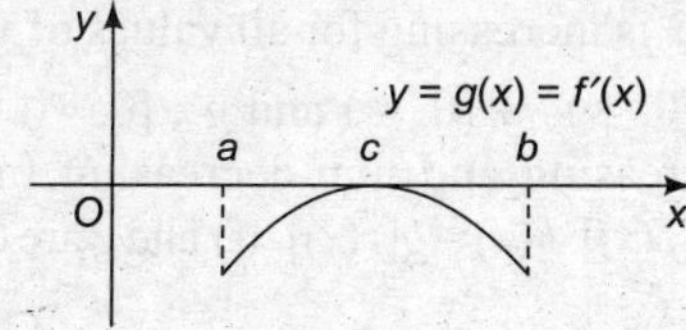

Fig. 6.2

17. The graph of $y = f(x)$ will intersect the x-axis
a. twice b. once
c. never d. None of these

18. The equation $f(x) = 0$, $a \leq x \leq b$ has
a. two repeated roots
b. no real roots
c. two distinct real roots.
d. at least three repeated roots

19. The graph of $y = f(x)$, $a \leq x \leq b$, has
a. two points of inflexion
b. one point of inflexion
c. no point of inflexion
d. None of these

20. The function $y = f(x)$, $a < x < b$, has
a. exactly one local maxima
b. one local minima and one maxima
c. exactly one local minima
d. None of these

21. The equation $f''(x) = 0$
a. cannot be decided x
b. has at least one real root
c. cannot be decided
d. has at least two distinct real roots

For Problems 22–24

Let $f(x) = 4x^2 - 4ax + a^2 - 2a + 2$ and the global minimum value of $f(x)$ for $x \in [0, 2]$ is equal to 3.

22. The number of values of a for which the global minimum value equal to 3 for $x \in [0, 2]$ occurs at the end point of interval [0, 2] is
a. 1 b. 2
c. 3 d. 0

23. The number of values of a for which the global minimum value equal to 3 occurs for $x \in [0, 2]$ occurs for the value of x lying in (0, 2) is
a. 1 b. 2
c. 3 d. 0

24. The values of a for which $f(x)$ is monotonic for $x \in [0, 2]$
a. $a \leq 0$ or $a \geq 4$ b. $0 \leq a \leq 4$
c. $a > 0$ d. None of these

For Problems 25–27

Let $f(x) = x^3 - 3(7 - a)x^2 - 3(9 - a^2)x + 2$.

25. The values of parameter a if $f(x)$ has a negative point of local minimum are
a. ϕ b. (–3, 3)
c. $\left(-\infty, \frac{58}{14}\right)$ d. None of these

26. The values of parameter a if $f(x)$ has a positive point of local maxima are

a. ϕ b. $(-\infty,-3)\cup(3,\infty)$

c. $\left(-\infty,\frac{58}{14}\right)$ d. None of these

27. The values of parameter a if $f(x)$ has points of extrema which are opposite in sign are

a. ϕ b. $(-3,3)$

c. $\left(-\infty,\frac{58}{14}\right)$ d. None of these

For Problems 28 – 30

Consider the function $f(x)=\left(1+\frac{1}{x}\right)^x$.

28. The domain of $f(x)$ is

a. $(-1,0)\cup(0,\infty)$ b. $R-\{0\}$

c. $(-\infty,-1)\cup(0,\infty)$ d. $(0,\infty)$

29. The function $f(x)$

a. has a maxima but no minima

b. has a minima but no maxima

c. has exactly one maxima and one minima

d. has neither a maxima nor a minima

30. The range of the function $f(x)$ is

a. $(0,\infty)$ b. $(-\infty,e)$

c. $(1,\infty)$ d. $(1,e)\cup(e,\infty)$

Matrix-Match Type

Each question contains statements given in two columns which have to be matched. Statements a, b, c, d in column I have to be matched with statements p, q, r, s in column II. If the correct match are a → p, a → s, b → q, b → r, c → p, c → q and d → s, then the correctly bubbled 4 × 4 matrix should be as follows:

	p	q	r	s
a	p	q	r	s
b	p	q	r	s
c	p	q	r	s
d	p	q	r	s

1. Consider function $f(x)=x^4-14x^2+24x-3$.

Column I : Number of roots Values of	Column II : $pf(x)+p=0$
a. Two negative real roots	p. $p>120$
b. Two real roots of opposite sign	q. $-8\le p\le -5$
c. Four real roots	r. $3<p\le 120$
d. No real roots	s. $p<-8$ or $-5<p<3$

2.

Column I	Column II
a. $f(x)=x^2\log x$	p. $f(x)$ has one point of minima
b. $f(x)=x\log_e x$	q. $f(x)$ has one point of maxima
c. $f(x)=\frac{\log x}{x}$	r. $f(x)$ increases in $(0, e)$
d. $f(x)=x^{-x}$	s. $f(x)$ decreases in $(0, 1/e)$

3. Let $f(x)=(x-1)^m(2-x)^n$; $m, n\in N$ and $m, n>2$

Column I	Column II
a. Both $x=1$ and $x=2$ are the points of minima if	p. m is even
b. $x=1$ is a point of minima and $x=2$ is a point of inflection if	q. m is odd
c. $x=2$ is a point of minima and $x=1$ is a point of inflection if	r. n is even
d. Both $x=1$ and $x=2$ are the points of inflection if	s. n is odd

4. The function $f(x)=\sqrt{(ax^3+bx^2+cx+d)}$ has its non-zero local minimum and maximum values at $x=-2$ and $x=2$, respectively. If a is a root of $x^2-x-6=0$.

Column I	Column II
a. The value/values of a	p. $=0$
b. The value/values of b	q. $=24$
c. The value/values of c	r. >32
d. The value/values of d	s. -2

Integer Type

1. Let $f(x)$ and $g(x)$ be two continuous functions defined from $R\to R$ such that $f(x_1)>f(x_2)$ and $g(x_1)<g(x_2)$ for all $x_1>x_2$. Then the number of integral values of a for which $f[g(\alpha^2-2\alpha)]>f[g(3\alpha-4)]$ is ____.

2. If k is the least value of k for which the function x^2+kx+1 is an increasing function in the interval $1<x<2$, then the value of $|k|$ is ____.

3. The least value of a for which $f(x)=2e^x-ae^{-x}+(2a+1)x-3$ is increasing for all values of x is ____.

4. Let $f:[0,\infty)\to[0,\infty)$ and $g:[0,\infty)\to[0,\infty)$ be non-increasing and non-decreasing functions and $h(x)=g[f(x)]$. $h(x)=g[f(x)]$. If f and g are differentiable for all points in their respective domains and $h(0) = 0$, then the value of $h(3)$ is ___.

5. If a is the maximum value of a for which the function $(a + 2)x^3 - 3ax^2 + 9ax - 1$ decreases monotonically throughout for all real x, then the value of $|a|$ is ___.

6. The least value of a for which $f(x) = \sqrt{3} \sin x - \cos x - 2ax + b$ decreases for all real values of x is ___.

7 The function $f(x) = (x^2 - 4)^n (x^2 - x + 1)$, $n \in N$ assumes a local minimum value at $x = 2$. Then the least positive integral value of n is ___.

8. The function $y = (ax + b)/[(x - 1)(x - 4)]$ has a turning point at $P(2, -1)$. Then the value of $|a + 6b|$ is ___.

9. The number of extrema of $f(x) = x/(1 + x \tan x)$, $x \in (0, \pi/2)$, is ___.

10. $f(x) = \begin{cases} \cos[(\pi x)/2], & x > 0 \\ x + a, & x \le 0 \end{cases}$. If $x = 0$ is the point of maxima, then the least value of $a + 6$ is ___.

11. The number of points of extrema of $f(x) = 2x + 3x^{2/3}$ is ___.

12. The number of integral values of k for which $x^3 - 3x + a = 0$ has three real distinct roots is ___.

13. A running track of 440 ft is to be laid out enclosing a football field, the shape of which is a rectangle with a semi-circle at each end. If L is the length of its sides for which the area of the rectangular portion is to be maximum, then the value of $L/10$ is ___.

14. A tangent to the parabola $y = x^2$ is drawn in such a way that the abscissa x_0 of the point of tangency belongs to the interval [1, 2]. The value of x_0 for which the triangle bounded by the tangent, the axis of ordinates, and the straight line $y = x_0^2$ has the greatest area is ___.

15. The number of points on the curve $5x^2 - 8xy + 5y^2 = 4$ whose distance from the origin is maximum or minimum is ___.

16. Let $f(x) = \sin^3 x + \lambda \sin^2 x$, $-(\pi/2) < x < (\pi/2)$. Then the number of integral values of λ in which λ lies in order that $f(x)$ has exactly one minimum and exactly one maximum are ___.

17. A cubic function $f(x)$ vanishes at $x = -2$ and has relative minimum/maximum at $x = -1$ and $x = 1/3$ if $\int_{-1}^{1} f(x)\, dx = 14/3$. The value of $f(0)$ is ___.

18. Let $f(x) =$

$$\begin{cases} -x^3 + [(b^3 - b^2 + b - 1)/(b^2 + 3b + 2)], & 0 \le x < 1 \\ 2x - 3, & 1 \le x \le 3 \end{cases}.$$

Then the least positive integral value of b such that $f(x)$ has the smallest value at $x = 1$ is ___.

19. Suppose $f(x)$ is a function satisfying the following conditions:

(a) $f(0) = 2, f(1) = 1$,

(b) f has a minimum value at $x = 5/2$, and

(c) for all x, $f'(x)$

$$= \begin{vmatrix} 2ax & 2ax - 1 & 2ax + b + 1 \\ b & b + 1 & -1 \\ 2(ax + b) & 2ax + 2b + 1 & 2ax + b \end{vmatrix},$$

where a, b are some constants.
Then the value of $f(0)$ is ___.

20. Let $p(x)$ be a polynomial of degree 3 satisfying $p(-1) = 10$, $p(1) = -6$, and $p(x)$ has a maximum at $x = -1$ and $p'(x)$ has a minimum at $x = 1$. If the distance between the local maxima and local minima of the curve is d, then the value of $[d/4]$, where [.] represents the greatest integer function, is ___.

21. The sum of the values of x for which the function $f(x) = \int_{-1}^{x} t(e^t - 1)(t - 1)(t - 2)^3 (t - 3)^5 dt$ has a local minimum is ___.

22. The minimum value of the function $f(x) = (x^p/p) + (x^{-q}/q)$, where $\frac{1}{p} + \frac{1}{q} = 1, p > 1$ is ___.

23. In a given solid cone of height H, another inverted cone having height h is carved in such a way that its volume is the maximum. Then the ratio $H : h$ is ___.

24. Let $f(x) =$

$$\begin{cases} |x^3 + x^2 + 3x + \sin x| [3 + \sin(1/x)], & x \ne 0 \\ 0, & x = 0 \end{cases},$$

then the number of points where $f(x)$ attains its minimum value is ___.

25. The number of integral values of $6b$ for which the equation $(x^3/3) - x = b$ has three distinct solutions is ___.

26. Let $f(x) = x^2 + ax + b$. If the maximum and the minimum values of $f(x)$ are 3 and 2, respectively for $0 \le x \le 2$, then the value of $a + b$ is ________.

27. The maximum value of

$f(x) = \left(3 - \sqrt{4 - x^2}\right)^2 + \left(1 + \sqrt{4 - x^2}\right)^3$ is A, then

$A/7 =$ ________.

ANSWERS AND SOLUTIONS

Objective Type

1. c. $f'(x) = 3kx^2 - 18x + 9 = 3[kx^2 - 6x + 3] \geq 0, \forall x \in R$

$\Rightarrow \quad D = b^2 - 4ac \leq 0, k > 0$, i.e., $36 - 12k \leq 0$

$\Rightarrow k \geq 3.$

2. d. Since $f(x) = \dfrac{K \sin x + 2\cos x}{\sin x + \cos x}$ is increasing for all x, therefore $f'(x) > 0$ for all x

$\Rightarrow \dfrac{K - 2}{(\sin x + \cos x)^2} > 0$ for all x

$\Rightarrow K - 2 > 0 \Rightarrow K > 2.$

3. d. $f'(x) = a + 3\cos x - 4\sin x$

$= a + 5\cos(x + \alpha)$, where $\cos\alpha = 3/5$

For invertible, $f(x)$ must be monotonic

$\Rightarrow f'(x) \geq 0\ \forall x$ or $f'(x) \leq 0\ \forall x$

$\Rightarrow a + 5\cos(x + \alpha) \geq 0$ or $a + 5\cos(x + \alpha) \leq 0$

$\Rightarrow a \geq -5\cos(x + \alpha)$ or $a \leq -5\cos(x + \alpha)$

$\Rightarrow a \geq 5$ or $a \leq -5.$

4. d. We have $g'(x) = f'\left(\dfrac{x}{2}\right) - f'(2 - x)$

Given $f''(x) < 0\ \forall x \in (0, 2).$

So $f'(x)$ is decreasing on $(0, 2)$

Let $\dfrac{x}{2} > 2 - x \Rightarrow f'\left(\dfrac{x}{2}\right) < f'(2 - x)$

Thus $\forall x > \dfrac{4}{3}, g'(x) < 0$

$\Rightarrow g(x)$ decreasing in $\left(\dfrac{4}{3}, 2\right)$

and increasing in $\left(0, \dfrac{4}{3}\right)$.

5. d. $f(x) = x^{100} + \sin x - 1$

$\Rightarrow f'(x) = 100x^{99} + \cos x$

If $0 < x < \pi/2$, then $f'(x) > 0$, Therefore $f(x)$ is increasing on $(0, \pi/2)$

If $0 < x < 1$, then

$100x^{99} > 0$ and $\cos x > 0$

[Since x lies between 0 and 1 radian]

$\Rightarrow f'(x) = 100x^{99} + \cos x > 0$

$\Rightarrow f(x)$ is increasing on $(0, 1)$.

If $\pi/2 < x < \pi$, then

$100x^{99} > 100 \quad [\because x > 1 \therefore x^{99} > 1]$

$\Rightarrow 100x^{99} + \cos x > 0$

$[\because \cos x \geq -1 \therefore 100x^{99} + \cos x > 99]$

$\Rightarrow f'(x) > 0 \Rightarrow f(x)$ is increasing in $(\pi/2, \pi)$.

6. d. When $f(x) = 3x^2 - 2x + 1$,

$f'(x) = 6x - 2.$

f is increasing $\Rightarrow f'(x) \geq 0$

$\Rightarrow 6x - 2 \geq 0 \Rightarrow x \geq 1/3$

7. a. Here $f(x) = \tan^{-1}(\sin x + \cos x)$. Therefore,

$f'(x) = \dfrac{1}{1 + (\sin x + \cos x)^2}(\cos x - \sin x)$

$= \dfrac{\cos x - \sin x}{2 + \sin 2x}$

For $-\dfrac{\pi}{2} < x < \dfrac{\pi}{4}$, $\cos x > \sin x$

Hence $y = f(x)$ is increasing in $\left(-\dfrac{\pi}{2}, \dfrac{\pi}{4}\right)$.

8. c. $\because f'(x) = x^x[1 + \log x] = x^x \log(ex)$

$0 < x < \dfrac{1}{e} \Rightarrow ex < 1 \Rightarrow \log ex < 0$

$\Rightarrow f'(x) < 0$

$\Rightarrow f(x)$ decreases

9. c. $f(x) = (x-1)^2 + (x-2)^2 + (x-3)^2 + (x-4)^2 + (x-5)^2$

$f'(x) = 2[x - 1 + x - 2 + x - 3 + x - 4 + x - 5] = 2[5x - 15]$

$f'(x) = 0$ gives $x = 3$ and $f''(x) > 0$ for all x

Hence, $f(x)$ is minimum for $x = 3$

10. d. If $f(x)$ increases then $f^{-1}(x)$ increases.

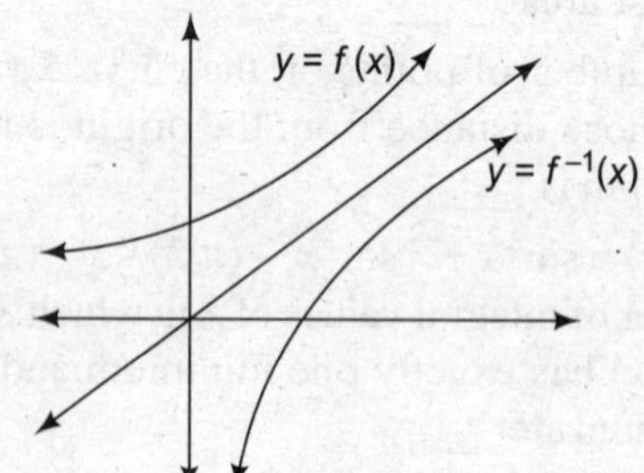

Fig. 6.3

If $f(x)$ increases, then $f'(x) > 0$

$\Rightarrow \dfrac{d}{dx}\left(\dfrac{1}{f(x)}\right) = -\dfrac{f'(x)}{f^2(x)} < 0$

$\Rightarrow \dfrac{1}{f(x)}$ decreases $\dfrac{d}{dx}\left(\dfrac{f}{g}\right) = \dfrac{f'g - fg'}{g^2}$

If f and g are positive functions and $f' < 0$ and $g' > 0$, then $\frac{d}{dx}\left(\frac{f}{g}\right) < 0$.

11. c. $f(x)\,f'(x) < 0\ \forall x \in R$

$\Rightarrow \frac{1}{2}\frac{d}{dx}(f^2(x)) < 0$

$\Rightarrow \frac{d}{dx}(f^2(x)) < 0$

$\Rightarrow f^2(x)$ is a decreasing function.

12. a. $f(x) = x\sqrt{4ax - x^2}$

$\Rightarrow\ f'(x) = \sqrt{4ax - x^2} + \frac{x(4a - 2x)}{2\sqrt{4ax - x^2}}$

$= \frac{2x(3a - x)}{\sqrt{4ax - x^2}}$

Now if $f'(x) > 0$

$\Rightarrow 2x\,(3a - x) > 0$

$\Rightarrow 2x\,(x - 3a) < 0$

$\Rightarrow x \in (0, 3a)$

Thus $f(x)$ increases in $(0, 3a)$ and decreases in $(3a, 4a)$

13. c. $f'(x) < 0, f''(x) < 0$

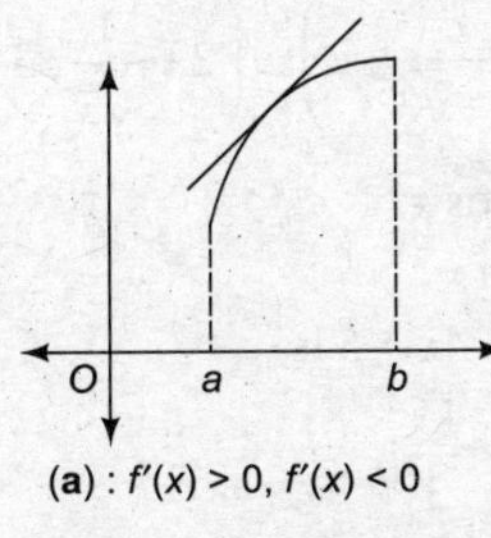

(a) : $f'(x) > 0,\ f'(x) < 0$

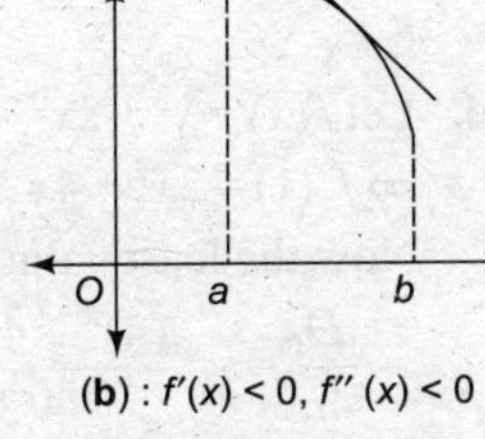

(b) : $f'(x) < 0,\ f''(x) < 0$

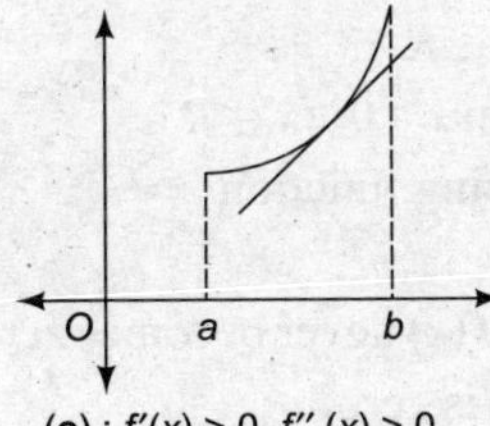

(c) : $f'(x) > 0,\ f''(x) > 0$

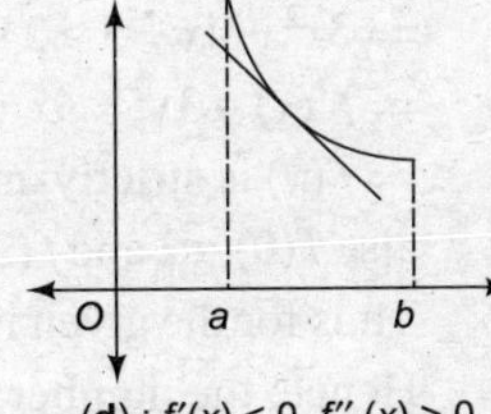

(d) : $f'(x) < 0,\ f''(x) > 0$

Fig. 6.4

Clearly for $f'(x) > 0,\ f''(x) > 0$ tangent always lies below the graph.

Or $f'(x) < 0, f''(x) > 0$. Tangent always lies below the graph.

14. a. $f(x) = |x| - \{x\} = |x| - (x - [x]) = |x| - x + [x]$

For $x \in (-1/2, 0)$,

$f'(x) = -x - x - 1 = -2x - 1$

Also for $-1/2 < x < 0 \Rightarrow 0 < -2x < 1$.

$\Rightarrow -1 < -2x - 1 < 0$

$\Rightarrow f'(x) < 0 \Rightarrow f(x)$ decreases in $(-1/2, 0)$.

Similarly, we can check for other given options say for $x \in (-1/2, 2)$,

$$f'(x) = \begin{cases} (-x) - x - 1, & -\frac{1}{2} < x < 0 \\ x - x + 0, & 0 \le x < 1 \\ x - x + 1, & 1 \le x < 2 \\ \vdots & \end{cases}$$

Here $f(x)$ decreases only in $(-1/2, 2)$, otherwise $f(x)$ in other intervals is constant. Hence (b), (c), (d) cannot be the correct options.

15. d. $f(x) = |x| - |x - 1|$

$$= \begin{cases} -x - (1 - x), & x < 0 \\ x - (1 - x), & 0 \le x < 1 \\ x - (x - 1), & x \ge 0 \end{cases}$$

$$= \begin{cases} -1, & x < 0 \\ 2x - 1, & 0 \le x < 1 \\ 1, & x \ge 1 \end{cases}$$

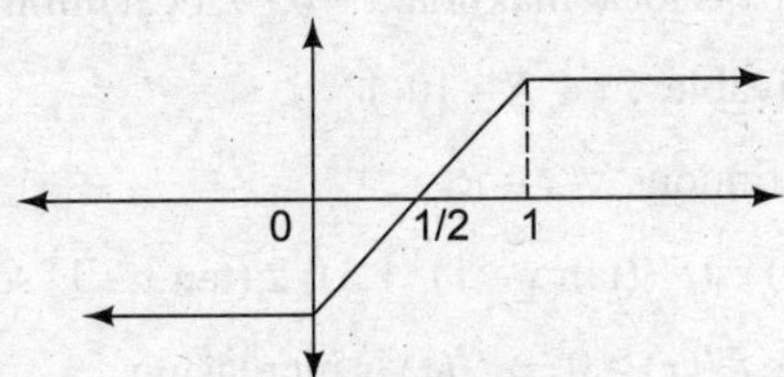

Fig. 6.5

Graph of the function is that $f(x)$ clearly increases in $(0, 1)$.

16. d. $|f(x)| = \begin{cases} f(x), & f(x) \ge 0 \\ -f(x), & f(x) < 0 \end{cases}$

$\Rightarrow \frac{d}{dx}|f(x)| = \begin{cases} f'(x), & f(x) \ge 0 \\ -f'(x), & f(x) < 0 \end{cases}$

Now as $f(x)$ and $f'(x)$ keep opposite sign, then $\frac{d}{dx}|f(x)| < 0$.

Hence $|f|$ is decreasing.

17. c. $\phi'(x) = 2f(x)\,f'(x)$

We do not know the sign of $f(x)$ in (a, b), so we cannot say about the sign of $f'(x)$.

18. a. Given $\phi'(x) - \phi(x) > 0\ \forall x \ge 1$

$\Rightarrow e^{-x}\{\phi'(x) - \phi(x)\} > 0\ \forall x \ge 1$

$\Rightarrow \frac{d}{dx}e^{-x}\phi(x) > 0\ \forall x \ge 1$

Therefore $e^{-x}\,\phi(x)$ is an increasing function $\forall x \ge 1$

Since $\phi(x)$ is a polynomial

$\Rightarrow e^{-x}\phi(x) > e^{-1}\phi(1) \Rightarrow e^{-x}\,\phi(x) > 0 \quad [\because \phi(1) = 0]$

$\Rightarrow \phi(x) > 0$. Also $\phi(1) = 0$

$\therefore \phi(x) \geq 0 \ \forall x \geq 1.$

19. c. Function is increasing in $(-\infty, -2) \cup (0, \infty)$, function is decreasing in $(-2, 0)$.

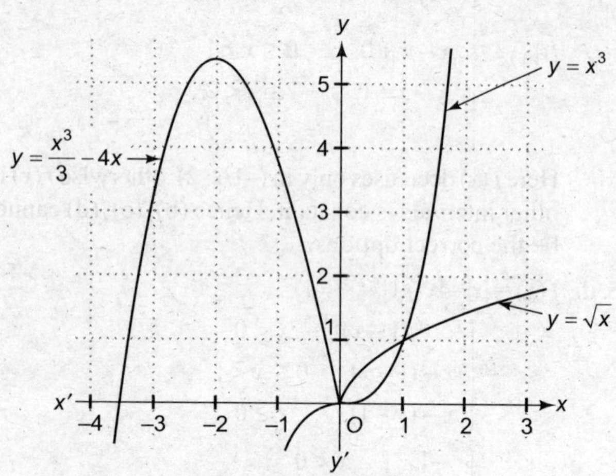

Fig. 6.6

$x = -2$ is local maxima, $x = 0 \rightarrow$ local minima

Derivable $\forall x \in R - \{0, 1\}$

Continuous $\forall x \in R$.

20. d. $g'(x) = (f'((\tan x - 1)^2 + 3))\, 2\, (\tan x - 1) \sec^2 x$

since $f''(x) > 0 \Rightarrow f'(x)$ is increasing

So, $f'((\tan x - 1)^2 + 3) > f'(3) = 0$,
$\forall x \in (0, \pi/4) \cup (\pi/4, \pi/2)$

Also $(\tan x - 1) > 0\ x \in (\pi/4, \pi/2)$.

So, $g(x)$ is increasing in $(\pi/4, \pi/2)$.

21. b. We must have $\log_{1/3} (\log_3(\sin x + a)) < 0 \ \forall x \in R$

$\Rightarrow \log_3(\sin x + a) > 1 \ \forall x \in R$

$\Rightarrow \sin x + a > 3 \ \forall x \in R$
$\Rightarrow a > 3 - \sin x \ \forall x \in R$
$\Rightarrow a > 4$

22. c. $u = \sqrt{c+1} - \sqrt{c}$

$u = \dfrac{1}{\sqrt{c+1}+\sqrt{c}}$ and $v = \dfrac{1}{\sqrt{c-1}+\sqrt{c}}$

Clearly $u < v$

Also f is increasing whereas g is decreasing

Thus $u < v$

$\Rightarrow f(u) < f(v)$

$\Rightarrow gof(u) > gof(v)$.

23. b. $f'(x) = 4 - 2\sec^2 2x = 2(1 - \tan^2 2x)$

For the continuous domain $\left(-\dfrac{\pi}{4}, \dfrac{\pi}{4}\right)$, $f'(x) \geq 0$ in $\left[-\dfrac{\pi}{8}, \dfrac{\pi}{8}\right]$ and $f'(x) \leq 0$ in $\left(-\dfrac{\pi}{4}, -\dfrac{\pi}{8}\right] \cup \left[\dfrac{\pi}{8}, \dfrac{\pi}{4}\right)$

So the required largest continuous interval is $\left[-\dfrac{\pi}{8}, \dfrac{\pi}{8}\right]$, length $= \dfrac{\pi}{4}$.

24. a.

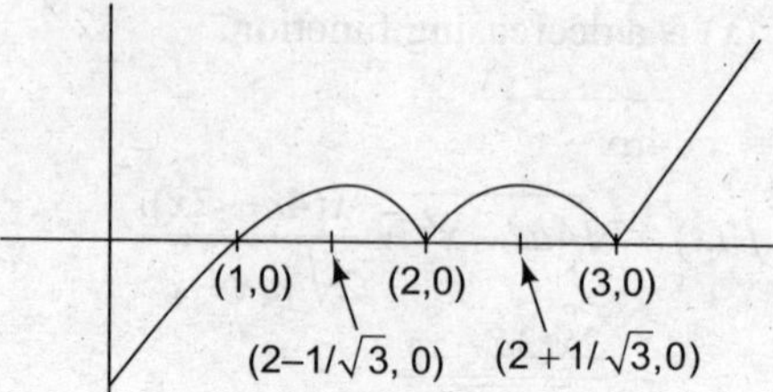

Fig. 6.7

$f(x) = (x-1)\,|(x-2)(x-3)|$
let $g(x) = (x-1)(x-2)(x-3)$
$= x^3 - 6x^2 + 11x - 6$
$\Rightarrow \ g'(x) = 3x^2 - 12x + 11$

$\Rightarrow \ x = \dfrac{12 \pm \sqrt{144-132}}{6} = \dfrac{12 \pm \sqrt{12}}{6} = 2 \pm \dfrac{1}{\sqrt{3}}$

Hence, $f(x)$ decreases in $\left(2 - \dfrac{1}{\sqrt{3}}, 2\right) \cup \left(2 + \dfrac{1}{\sqrt{3}}, 3\right)$.

25. d. Let $f(x) = x^3 + 2x^2 + 5x + 2\cos x$
$\Rightarrow f'(x) = 3x^2 + 4x + 5 - 2\sin x$
Now the least value of $3x^2 + 4x + 5$ is
$-\dfrac{D}{4a} = -\dfrac{(4)^2 - 4(3)(5)}{4(3)} = \dfrac{11}{3}$
and the greatest value of $2 \sin x = 2$
$\Rightarrow 3x^2 + 4x + 5 > 2 \sin x \ \forall x \in R$
$\Rightarrow f'(x) = 3x^2 + 4x + 5 - 2 \sin x > 0 \ \forall x \in R$
$\Rightarrow f(x)$ is strictly an increasing function
also $f(0) = 2$ and $f(2\pi) > 0$.
Thus for the given interval, $f(x)$ never becomes zero.
Hence, the number of roots is zero.

26. a. $f(x) = (x-2)\,|x-3|$
For, $f(x) = (x-2)(x-3) = x^2 - 5x + 6$
$f'(x) = 2x - 5 = 0 \Rightarrow x = 5/2$
Now, the graph of $f(x) = (x-2)\,|x-3|$ is

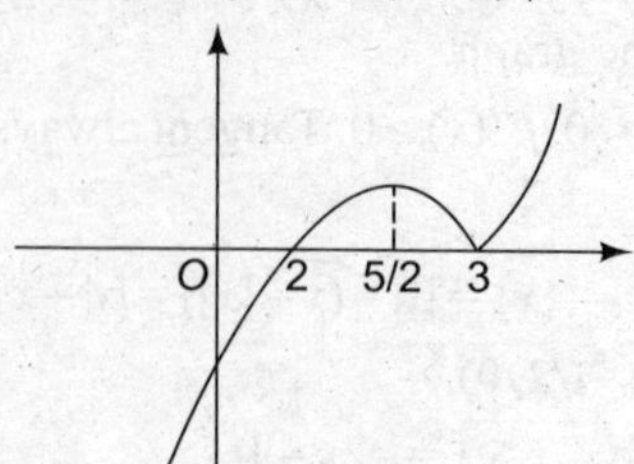

Fig. 6.8

Clearly from the graph, $f(x)$ increases in $(-\infty, 5/2) \cup (3, \infty)$.

27. b. $f(x) = (x-8)^4 (x-9)^5, 0 \le x \le 10$

$$\Rightarrow \quad f'(x) = 4(x-8)^3 (x-9)^5 + 5(x-9)^4 (x-8)^4$$
$$= (x-8)^3 (x-9)^4 [4(x-9) + 5(x-8)]$$
$$= 9(x-8)^3 (x-9)^4 \left(x - \frac{76}{9}\right).$$

Sign scheme of $f'(x)$

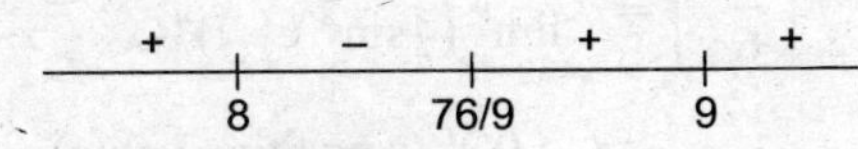

Fig. 6.9

$$f'(x) < 0, \text{ if } x \in \left(8, \frac{76}{9}\right)$$

$$\Rightarrow f(x) \text{ decreases if } x \in \left(8, \frac{76}{9}\right).$$

28. a. Here $f'(x) \le 0$

$$\Rightarrow 3x^2 + 8x + \lambda \le 0 \ \forall x \in \left(-2, -\frac{2}{3}\right)$$

Then situations for $f'(x)$ is as follow.

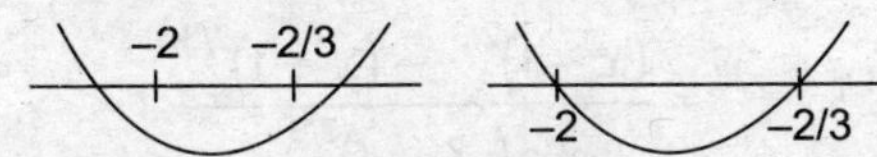

Fig. 6.10

Given that $f(x)$ decreases in the largest possible interval $(-2, -2/3)$, then $f'(x) = 0$ must have roots -2 and $-2/3$.

$$\Rightarrow \text{Product of roots is } (-2)\left(-\frac{2}{3}\right) = \frac{\lambda}{3} \Rightarrow \lambda = 4.$$

29. b. $f(x) = |x \log_e x|$

For $g(x) = x \log_e x$,

$$g'(x) = x \times \frac{1}{x} + \log_e x = 1 + \log_e x$$

$\Rightarrow$ $g(x)$ increases for $(1/e, \infty)$ and decreases for $(0, 1/e)$.

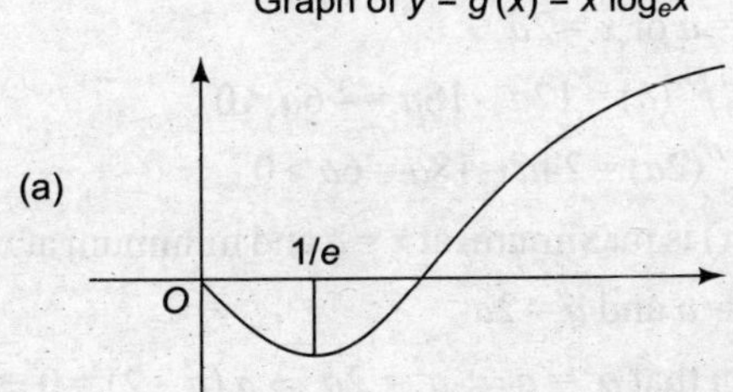

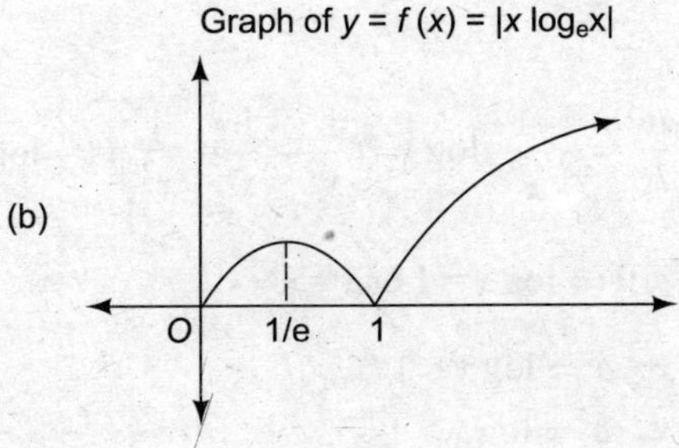

Fig. 6.11

From the graph, $f(x) = |x \log_e x|$ decreases in $(1/e, 1)$.

30. b. Let $h(x) = f(x) - g(x)$

$h'(x) = f'(x) - g'(x) > 0 \ \forall x \in R$

$\Rightarrow h(x)$ is an increasing function and

$h(0) = f(0) - g(0) = 0$

Therefore, $h(x) > 0 \ \forall \ x \in (0, \infty)$ and $h(x) < 0 \ \forall \ x \in (-\infty, 0)$.

31. b. $g'(x) = xf'(2x^2 - 1) - x f'(1 - x^2)$

if $x > 0$, $2x^2 - 1 > 1 - x^2$ (as f' is an increasing function)

$$\Rightarrow 3x^2 > 2 \Rightarrow x \in \left(-\infty, -\sqrt{\frac{2}{3}}\right) \cup \left(\sqrt{\frac{2}{3}}, \infty\right)$$

$$\Rightarrow x \in \left(\sqrt{\frac{2}{3}}, \infty\right)$$

If $x < 0$, $2x^2 - 1 < 1 - x^2$

$$3x^2 < 2 \Rightarrow x \in \left(-\sqrt{\frac{2}{3}}, \sqrt{\frac{2}{3}}\right) \Rightarrow x \in \left(-\sqrt{\frac{2}{3}}, 0\right).$$

32. a. $f(0) = \sin 0 = 0$

$f(0^+) \to 0^+$

$$f(0^-) = \lim_{x \to 0^-} \sin(x^2 - 3x) = \lim_{h \to 0} \sin(h^2 + 3h) \to 0^+$$

Thus, $f(0^+) > f(0)$ and $f(0^-) > f(0)$

Hence, $x = 0$ is a point of maxima.

33. b. Since $\cos \theta \le 1$ for all θ. Therefore, $f(x) \le 1$ for all x.

34. b. $\frac{dy}{dx} = 5x^2 (x-1)(x-3) = 0$

$\therefore x = 0, 1, 3$

Fig. 6.12

Clearly $x = 0$ is neither a point of maxima nor a point of minima as derivative does not change sign at $x = 0$.

$x = 1$ is a point of maxima and $x = 3$ is a point of minima.

35. d. $y = \dfrac{\log x}{x}$

$\Rightarrow \dfrac{dy}{dx} = -\dfrac{1}{x^2}\log x + \dfrac{1}{x} \cdot \dfrac{1}{x} = \dfrac{1}{x^2}(1-\log x) = 0$

$\dfrac{dy}{dx} = 0 \Rightarrow \log x = 1$ or $x = e$

For $x < e \Rightarrow \log x < 1$

and $x > e \Rightarrow \log x > 1$

At $x = e$, $\dfrac{dy}{dx}$ changes sign from + ive to –ive and hence y is max. at $x = e$ and its value is

$\dfrac{\log e}{e} = e^{-1}$.

36. c. When $f''(a) = 0$, then $f'''(a)$ must also be zero and sign of $f''''(a)$ will decide about max. or min.

37. c. Then given expression is minimum when

$y = (x^2 - 3)^3 + 27$ is minimum, which is so when $x = 0$

Hence $y_{min.} = 0$.

$\Rightarrow$ Minimum value of $2^{(x^2-3)^3+27}$ is $2^0 = 1$.

38. b. $y = \sin x(1 + \cos x) = \sin x + 1/2 \sin 2x$

$dy/dx = \cos x + \cos 2x = 0$

$\Rightarrow \cos 2x = -\cos x = \cos(\pi - x)$

$\Rightarrow 2x = \pi - x$ or $x = \pi/3$

$d^2y/dx^2 = -\sin x - 2\sin 2x$ which is clearly –ive at $x = \pi/3$ and hence y is max. at $x = \pi/3$ and its value is

$\dfrac{3\sqrt{3}}{4}$.

39. c. $f(x) = \dfrac{(\sin x + \cos x)^2 - 1}{1/\sqrt{2}(\sin x + \cos x)} = \sqrt{2}\dfrac{t^2-1}{t}$

or $f(x) = \phi(t) = \sqrt{2}\left(t - \dfrac{1}{t}\right)$

where $t = g(x) = \sin x + \cos x, x \in [0, \pi/2]$

$g'(x) = \cos x - \sin x = 0 \Rightarrow \tan x = 1$

$\Rightarrow x = \pi/4$ and $g''(x) = -$ ive

At $x = 0, t = 1 \therefore t \in [1, \sqrt{2}]$

Now $\phi(t) = \sqrt{2}\left(t - \dfrac{1}{t}\right)$ where $t \in [1, \sqrt{2}]$

$\phi'(t) = \sqrt{2}\left(1 + \dfrac{1}{t^2}\right) = +$ ive

Therefore, $\phi(t)$ is increasing.

Hence $\phi(t)$ is greatest at the endpoint of interval $[1, \sqrt{2}]$ i.e. $t = \sqrt{2}$

$\therefore f(x) = \phi(t) = \sqrt{2}\left[\sqrt{2} - \dfrac{1}{\sqrt{2}}\right] = 1$

40. a. $f(x) = (4\sin^2 x - 1)^n (x^2 - x + 1)$

$x^2 - x + 1 > 0 \; \forall x$

$f\left(\dfrac{\pi}{6}\right) = 0$

$f\left(\dfrac{\pi^+}{6}\right) = \lim\limits_{x \to \pi^+/6} (4\sin^2 x - 1)^n (x^2 - x + 1)$

$= \to 0^+$

$f\left(\dfrac{\pi^-}{6}\right) = \lim\limits_{x \to \pi^-/6} (4\sin^2 x - 1)^n (x^2 - x + 1)$

$= (\to 0^-)^n$ (a positive value)

$f\left(\dfrac{\pi^-}{6}\right) > 0$ if n is an even number.

41. d. $f(x) = x\ln x - x + 1$

$\therefore f(1) = 0$

$f'(x) = 1 + \ln x - 1 = \ln x$

$\therefore f'(x) < 0$ if $0 < x < 1$

and $f'(x) > 0$ if $x > 1$

42. b. We have $f(x) = (x+1)^{1/3} - (x-1)^{1/3}$

$\therefore \quad f'(x) = \dfrac{1}{3}(x+1)^{\frac{-2}{3}} - \dfrac{1}{3}(x-1)^{\frac{-2}{3}}$

$= \dfrac{(x-1)^{2/3} - (x+1)^{2/3}}{3\left(x^2-1\right)^{2/3}}$

Clearly $f'(x)$ does not exist at $x = \pm 1$

Now, $f'(x) = 0$

$\Rightarrow (x-1)^{2/3} = (x+1)^{2/3}$

$\Rightarrow (x-1)^2 = (x+1)^2$

$\Rightarrow -2x = 2x \Rightarrow 4x = 0 \Rightarrow x = 0$.

Clearly, $f'(x) \neq 0$ for any other values of $x \in [0, 1]$.

The value of $f(x)$ at $x = 0$ is 2.

Hence, the greatest value of $f(x) = 2$.

43. b. $f(x) = 2x^3 - 9ax^2 + 12a^2x + 1$

$\therefore f'(x) = 6x^2 - 18ax + 12a^2$ and $f''(x) = 12x - 18a$

For maximum/minimum, $6x^2 - 18ax + 12a^2 = 0$

$\Rightarrow x^2 - 3ax + 2a^2 = 0$

$\Rightarrow (x-a)(x-2a) = 0$

$\Rightarrow x = a$ or $x = 2a$

Now, $f''(a) = 12a - 18a = -6a < 0$

and $f''(2a) = 24a - 18a = 6a > 0$

$\therefore f(x)$ is maximum. at $x = a$ and minimum at $x = 2a$

$\Rightarrow p = a$ and $q = 2a$

Given that $p^2 = q \Rightarrow a^2 = 2a \Rightarrow a(a-2) = 0 \Rightarrow a = 2$.

Section D

44. a. Let $f(x)=x+\frac{1}{x}$

$\therefore f'(x)=1-\frac{1}{x^2}$ and $f''(x)=\frac{2}{x^3}$

For max./min., $f'(x)=0 \Rightarrow 1-\frac{1}{x^2}=0$

$\Rightarrow x^2=1 \Rightarrow x \pm 1$

$f(x)$ is minimum at $x=1$ $\quad [\because f''(x)=\frac{2}{1}=2>0]$

45. a. We have $f(x)=\frac{x}{2}+\frac{2}{x}$.

$f'(x)=\frac{1}{2}-\frac{2}{x^2}$ and $f''(x)=\frac{4}{x^3}$

Now $f'(x)=0 \Rightarrow x^2=4 \Rightarrow x=\pm 2$

$\therefore f''(x)>0$ for $x=2$

$\therefore f$ has local minima at $x=2$.

46. a. $f(x)=\sin\left(x+\frac{\pi}{6}\right)+\cos\left(x+\frac{\pi}{6}\right)$

$=\sqrt{2}\sin\left(x+\frac{\pi}{6}+\frac{\pi}{4}\right)$

$=\sqrt{2}\sin\left(x+\frac{5\pi}{12}\right)$

Its maximum value $=\sqrt{2}$ when $x+\frac{5\pi}{12}=\frac{\pi}{2}$

i.e., when $x=\frac{\pi}{2}-\frac{5\pi}{12}=\frac{6\pi-5\pi}{12}=\frac{\pi}{12}$.

47. d. $f(0)>f(0^+)$ and $f(0)<f(0^-)$, hence $x=0$ is neither a maximum nor a minimum.

48. c. $f'(x)=\frac{(1+4x+x^2)1-x(4+2x)}{(1+4x+x^2)^2}$

$=\frac{1-x^2}{(1+4x+x^2)^2}$

For maximum. or minimum. $f'(x)=0 \Rightarrow x=\pm 1$

For $x=1$, $f'(x)$ changes sign from positive to negative as x passes through 1.

$\therefore f(x)$ is maximum. for $x=1$, and maximum. value is

$\frac{1}{1+4+1}=\frac{1}{6}$.

49. b. $y=-x^3+3x^2+9x-27$

$\therefore \frac{dy}{dx}=-3x^2+6x+9$

Let the slope of tangent to the curve at any point be m (say)

$\Rightarrow m=-3x^2+6x+9$

$\Rightarrow \frac{dm}{dx}=-6x+6$

$\frac{d^2m}{dx^2}=-6<0$ for all x

Therefore, m is maximum. when $\frac{dm}{dx}=0$, i.e., when $x=1$

Hence maximum. slope $=-3+6+9=12$.

50. b. $f'(x)=-\pi\sin\pi x+10+6x+3x^2$

$=3(x+1)^2+7-\pi\sin\pi x>0$ for all x.

$\therefore f(x)$ is increasing in $-2\le x\le 3$.

So, the absolute minimum $=f(-2)=1-20+12-8$.

51. c. Given $y=e^{(2x^2-2x+1)\sin^2 x}=e^{2\left[\left(x-\frac{1}{2}\right)^2+\frac{1}{4}\right]\sin^2 x}$

Clearly, the minimum value occurs when $\sin^2 x=0$ as

$\left[\left(x-\frac{1}{2}\right)^2+\frac{1}{4}\right]\ge 1/4$.

52. d. $f(x)=x^4 e^{-x^2} \Rightarrow f'(x)=4x^3e^{-x^2}+x^4e^{-x^2}(-2x)$

For max. $f'(x)=0 \Rightarrow 4x^3e^{-x^2}-2x^5e^{-x^2}=0 \Rightarrow x^2=2$

$\Rightarrow x=\pm\sqrt{2}$

$f''(x)=12x^2e^{-x^2}+4x^3e^{-x^2}(-2x)-10x^4e^{-x^2}-2x^5e^{-x^2}(-2x)$

$\Rightarrow f''(\sqrt{2})=24e^{-2}-32e^{-2}-40e^{-2}+32e^{-2}=-\text{ve}$

Hence, $f(x)$ is max. at $x=\sqrt{2}$

$\Rightarrow$ Maximum value $=4e^{-2}$

53. c. $a^2x^4+b^2y^4=c^6$

$\Rightarrow y=\left(\frac{c^6-a^2x^4}{b^2}\right)^{1/4}$

$\Rightarrow f(x)=xy=x\left(\frac{c^6-a^2x^4}{b^2}\right)^{1/4}$

$\Rightarrow f(x)=\left(\frac{c^6x^4-a^2x^8}{b^2}\right)^{1/4}$

Differentiate $f(x)$ w.r.t x, then

$f'(x)=\frac{1}{4}\left(\frac{c^6x^4-a^2x^8}{b^2}\right)^{-3/4}\left(\frac{4x^3c^6}{b^2}-\frac{8x^7a^2}{b^2}\right)$

Put $f'(x)=0$, $\frac{4x^3c^6}{b^2}-\frac{8x^7a^2}{b^2}=0$

$\Rightarrow x^4=\frac{c^6}{2a^2} \Rightarrow x=\pm\frac{c^{3/2}}{2^{1/4}\sqrt{a}}$

At $x=\frac{c^{3/2}}{2^{1/4}\sqrt{a}}$, $f(x)$ will be maximum,

$$\text{so } f\left(\frac{c^{3/2}}{2^{1/4}\sqrt{a}}\right)=\left(\frac{c^{12}}{2a^2b^2}-\frac{c^{12}}{4a^2b^2}\right)^{1/4}=\left(\frac{c^{12}}{4a^2b^2}\right)^{1/4}$$

$$=\frac{c^3}{\sqrt{2ab}}$$

54. b. Let $g(x)=4x^3-12x^2+11x-3$

$\Rightarrow \ g'(x)=12x^2-24x+11$

$=12(x-1)^2-1$

$\Rightarrow \ g'(x)>0$ for $x\in[2,3]$

$\Rightarrow \ g(x)$ is increasing in $[2, 3]$

$f(x)_{\max}=f(3)=\log_{10}(4.27-12.9+11.3-3)$

$=\log_{10}(30)$

$=1+\log_{10}3.$

55. b. Let $f(x)=x+ax^{-2}-2$

$\Rightarrow \ f'(x)=1-2ax^{-3}=0\Rightarrow x=(2a)^{1/3}$

Also $f''(x)=6ax^{-4}\Rightarrow \ f''\left((2a)^{1/3}\right)>0$

$\Rightarrow \ x=(2a)^{1/3}$ is the point of minima.

For $x+ax^{-2}-2>0\ \forall x$ we must have $f\left((2a)^{1/3}\right)>0$

$\Rightarrow \ (2a)^{1/3}+a(2a)^{-2/3}-2>0$

$\Rightarrow \ 2a+a-2(2a)^{2/3}>0$

$\Rightarrow \ 3a>2(2a)^{2/3}$

$\Rightarrow \ 27a^3>32a^2$

$\Rightarrow \ a>32/27$

Hence, the least value of a is 2.

56. c. We have $f(x)=\begin{cases}(-1)^{m+n}x^m(x-1)^n, & \text{if } x<0\\ (-1)^n x^m(x-1)^n, & \text{if } 0\le x<1\\ x^m(x-1)^n, & \text{if } x\ge 1\end{cases}$

Let $g(x)=x^m(x-1)^n$, then

$g'(x)=mx^{m-1}(x-1)^n+nx^m(x-1)^{n-1}$

$=x^{m-1}(x-1)^{n-1}\{mx-m+nx\}$

Now $f'(x)=0\Rightarrow g'(x)=0\Rightarrow x=0,1$ or $\dfrac{m}{m+n}$

$f(0)=0, f(1)=0$ and

$$f\left(\frac{m}{m+n}\right)=(-1)^n\frac{m^m n^n(-1)^n}{(m+n)^{m+n}}$$

$$=\frac{m^m n^n}{(m+n)^{m+n}}>0$$

The maximum value is $\dfrac{m^m n^n}{(m+n)^{m+n}}$.

57. b. Clearly, $f(x)$ in decreasing just before $x=3$ and increasing after $x=3$. For $x=3$ to be the point of local minima, $f(3)\le f(3^-)$.

$\Rightarrow -15\le 12-27+\ln(a^2-3a+3)$

$\Rightarrow a^2-3a+3\ge 1$

$\Rightarrow a\in(-\infty,1)\cup(2,\infty).$

58. b.

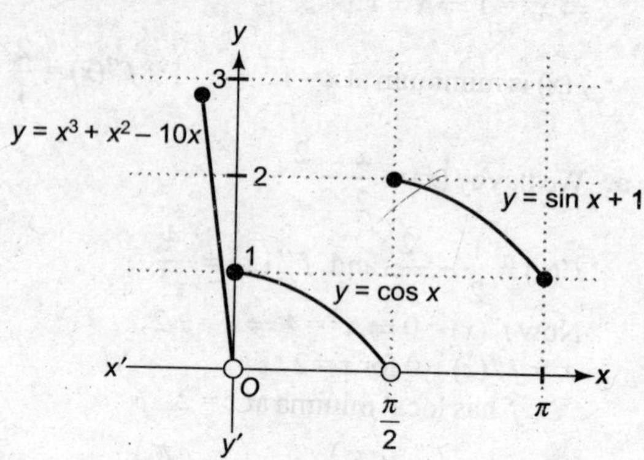

Fig. 6.13

59. b. Since $f(x)$ has a relative minimum at $x=0$, therefore $f'(0)=0$ and $f''(0)>0$

If the function $y=f(x)+ax+b$ has a relative minimum at $x=0$, then

$\dfrac{dy}{dx}=0$ at $x=0$

$\Rightarrow f'(x)+a=0$ for $x=0$

$\Rightarrow f'(0)+a=0\Rightarrow 0+a=0$ $\qquad [\because f'(0)=0]$

$\Rightarrow a=0$

Now, $\dfrac{d^2y}{dx^2}=f''(x)$

$\Rightarrow \left(\dfrac{d^2y}{dx^2}\right)_{x=0}=f''(0)>0$ $\qquad [\because f''(0)>0]$

Hence, y has a relative minimum at $x=0$ if $a=0$ and b can attain any real value.

60. d. $f'(x)=12x^2-2x-2=2[6x^2-x-1]$

$=2(3x+1)(2x-1)$

$$\text{Hence } g(x)=\begin{cases} f(x), & \text{if } 0\le x<\frac{1}{2}\\ f\left(\frac{1}{2}\right), & \text{if } \frac{1}{2}\le x\le 1\\ 3-x, & \text{if } 1<x\le 2\end{cases}$$

$\Rightarrow g\left(\dfrac{1}{4}\right)+g\left(\dfrac{3}{4}\right)+g\left(\dfrac{5}{4}\right)$

$=f\left(\dfrac{1}{4}\right)+f\left(\dfrac{1}{2}\right)+f\left(\dfrac{5}{4}\right)=\dfrac{5}{2}$

61. a. $f'(x)=ax^2+2(a+2)x+(a-1)$,

$f''(x)=2ax+2(a+2)=0$

$\Rightarrow x=-\dfrac{a+2}{a}$ which is the point of inflection

Given that, we must have $\dfrac{a+2}{a}<0$

$\Rightarrow(-\infty,-2)\cup(0,\infty)$.

62. d. The derivative of a degree 3 polynomial is quadratic. This must have either 0, 1 or 2 roots. If this has precisely one root, then this must be repeated. Hence we have $f'(x)=m(x-\alpha)^2$, where α is the repeated root and $m\in R$. So our original function f has a critical point at $x=\alpha$.

Also $f''(x)=2m(x-\alpha)$, in which case $f''(\alpha)=0$. But we are told that the 2nd derivative is non-zero at critical point. Hence there must be either 0 or 2 critical points.

63. c. $f'(x)=\dfrac{0.6(1+x)^{-0.4}(1+x^{0.6})-0.6x^{-0.4}(1+x)^{0.6}}{(1+x^{0.6})^2}$

$=0.6\dfrac{(1+x^{0.6})-x^{-0.4}(1+x)^1}{(1+x^{0.6})^2(1+x)^{0.4}}$

$=0.6\dfrac{(1+x^{0.6})x^{0.4}-(1+x)}{(1+x^{0.6})^2(1+x)^{0.4}x^{0.4}}$

$=0.6\dfrac{x^{0.4}-1}{(1+x^{0.6})^2(1+x)^{0.4}x^{0.4}}<0$

Hence $f(x)$ is decreasing

$\Rightarrow f(x)_{\max}=f(0)=1$

64. c. $f(f(x))=k(x^5+x)\Rightarrow f'(f(x))f'(x)=k(5x^4+1)$

$\Rightarrow f(x)$ is always increasing or decreasing as $k(5x^4+1)$ is either always negative or positive.

But if $f(x)$ is decreasing then $f'(f(x))f'(x)>0$.

65. c. $g[f(x)]$ is continuous at $x=\alpha\Rightarrow g[f(a)]=g(a)$

Also $\lim\limits_{x\to\alpha}g(f(x))=g(a)$

$\Rightarrow g[f(\alpha^-)]=g[f(\alpha^+)]=g(a)$

$\Rightarrow g(x)$ takes same limiting values at $f(\alpha^-)$, $f(\alpha^+)$ and $f(\alpha)$

$\Rightarrow f(\alpha^-)-f(\alpha^+)\Rightarrow h=a$ is an extremum of $f(x)$ and $x=\alpha$ may not be an extremum of $g(x)$.

66. a. $f(x)+f''(x)=-xg(x)f'(x)$

Let $h(x)=f^2(x)+(f'(x))^2$

$\Rightarrow h'(x)=2f(x)f'(x)+2f'(x)f''(x)$

$=2f'(x)[-x]g(x)f'(x)$

$=-2x(f'(x))^2g(x)$

$\Rightarrow x=0$ is a point of maxima for $h(x)$.

67. a. $h'(x)=\dfrac{m}{n}x^{\frac{m-n}{n}}=\dfrac{m}{n}x^{-\left(\frac{\text{even}}{\text{odd}}\right)}$

As $h'(x)$ is undefined at $x=0$ and $h'(x)$ does not change its sign in the neighbourhood. So, no extremums.

68. a. Here, $f(x)=4\tan x-\tan^2 x+\tan^3 x$

$\Rightarrow\quad f'(x)=4\sec^2 x-2\tan x\sec^2 x+3\tan^2 x\sec^2 x$

$=\sec^2 x(4-2\tan x+3\tan^2 x)$

$=3\sec^2 x\left\{\tan^2 x-\dfrac{2}{3}\tan x+\dfrac{4}{3}\right\}$

$=3\sec^2 x\left\{\left(\tan x-\dfrac{1}{3}\right)^2+\left(\dfrac{4}{3}-\dfrac{1}{9}\right)\right\}$

$=3\sec^2 x\left\{\left(\tan x-\dfrac{1}{3}\right)^2+\dfrac{11}{9}\right\}>0,\ \forall x$

Hence, $f(x)$ is increasing for all $x\in$ domain.

69. c. It is a fundamental property.

70. a. $f'(x)=-\dfrac{1}{2}e^{-\frac{x}{2}}(x^2-8)$.

Clearly, $x=2\sqrt{2}$ is the point of local maxima.

71. a. $f(0)=\pi/2$, $f(0^+)=0$, $f(0^-)=0$.

Hence, $x=0$ is the point of maxima.

72. a. $f(x)$ will have maxima only if $a^2+1\geq 2\Rightarrow|a|\geq 1$

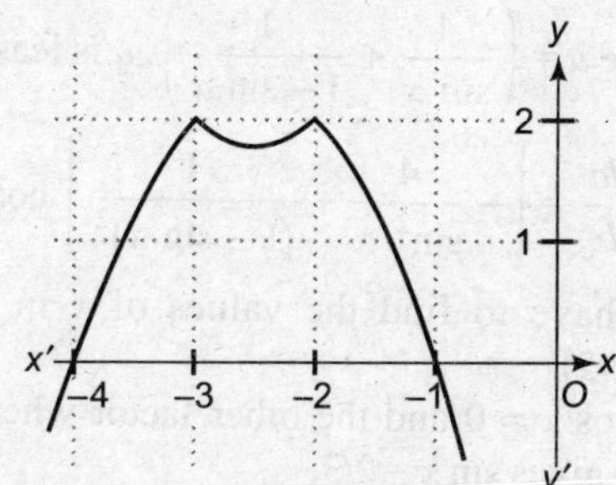

Fig. 6.14

73. b. Given $A+B=60°\Rightarrow B=60°-A$

$\Rightarrow\tan B=\tan(60°-A)=\dfrac{\sqrt{3}-\tan A}{1+\sqrt{3}\tan A}$

Now $z=\tan A\tan B$

or $z=\dfrac{t(\sqrt{3}-t)}{1+\sqrt{3}t}=\dfrac{\sqrt{3}t-t^2}{1+\sqrt{3}t}$

where $t=\tan A$

$\dfrac{dz}{dt}=-\dfrac{(t+\sqrt{3})(\sqrt{3}t-1)}{(1+\sqrt{3}t)^2}=0$

$\Rightarrow\quad t=1/\sqrt{3}$

$\Rightarrow\quad t=\tan A=\tan 30°$

The other value is rejected as both A and B are positive acute angles.

If $t < \frac{1}{\sqrt{3}}$, $\frac{dz}{dt} = +$ ive and if $t > \frac{1}{\sqrt{3}}$, $\frac{dz}{dt} = -$ ive.

Hence max. when $t = \frac{1}{\sqrt{3}}$ and max. value $= \frac{1}{3}$.

74. c. $f(x) = \frac{t+3x-x^2}{x-4}$;

$$f'(x) = \frac{(x-4)(3-2x)-(t+3x-x^2)}{(x-4)^2}$$

for maximum or minimum, $f'(x) = 0$

$-2x^2 + 11x - 12 - t - 3x + x^2 = 0$

$-x^2 + 8x - (12+t) = 0$

for one maxima and minima,

$D > 0$

$64 - 4(12+t) > 0$

$16 - 12 - t > 0 \Rightarrow 4 > t$ or $t < 4$.

75. d. If $f(x)$ has an extremum at $x = \pi/3$, then $f'(x) = 0$ at $x = \pi/3$

Now, $f(x) = a \sin x + \frac{1}{3} \sin 3x$

$\Rightarrow f'(x) = a \cos x + \cos 3x$

$\Rightarrow f'(\pi/3) = 0$

$\Rightarrow a \cos(\pi/3) + \cos \pi = 0$

$\Rightarrow a = 2$

76. a. Since $a = \left(\frac{4}{\sin x} + \frac{1}{1-\sin x}\right)$, a is least.

$$\Rightarrow \frac{da}{dx} = \left[-\frac{4}{\sin^2 x} + \frac{1}{(1-\sin x)^2}\right] \cos x = 0$$

We have to find the values of x in the interval $(0, \pi/2)$.

$\Rightarrow \cos x \neq 0$ and the other factor when equated to zero gives $\sin x = 2/3$.

$$\text{Now } \frac{d^2a}{dx^2} = \left[-\frac{4}{\sin^2 x} + \frac{1}{(1-\sin x)^2}\right](-\sin x)$$

$$+\left[\frac{8}{\sin^3 x} + \frac{2}{(1-\sin x)^3}\right] \cos^2 x$$

Put $\sin x = \frac{2}{3}$ or $\cos^2 x = 1 - \frac{4}{9} = \frac{5}{9}$

$$\therefore \frac{d^2a}{dx^2} = 0 + \left[\frac{8}{8/27} + 2 \times 27\right]\frac{5}{9} = 81 \times \frac{5}{9} = 45 > 0$$

$\Rightarrow a$ is minimum and its value is

$$\frac{4}{2/3} + \frac{1}{1-(2/3)} = 6 + 3 = 9.$$

77. c. Consider the function $f(x) = \frac{x^2}{(x^3+200)}$ (1)

$$f'(x) = x\frac{(400-x^3)}{(x^3+200)^2} = 0$$

When $x = (400)^{1/3}$, $(\because x \neq 0)$

$x = (400)^{1/3} - h \Rightarrow f'(x) > 0$

$x = (400)^{1/3} + h \Rightarrow f'(x) < 0$

Therefore, $f(x)$ has maxima at $x = (400)^{1/3}$

Since $7 < (400)^{1/3} < 8$, either a_7 or a_8 is the greatest term of the sequence.

$\because a_7 = \frac{49}{543}$ and $a_8 = \frac{8}{89}$ and $\frac{49}{543} > \frac{8}{89}$

$\Rightarrow a_7 = \frac{49}{543}$ is the greatest term.

78. d. Let there be a value of k for which $x^3 - 3x + k = 0$ has two distinct roots between 0 and 1.

Let a, b be two distinct roots of $x^3 - 3x + k = 0$ lying between 0 and 1 such that $a < b$. Let $f(x) = x^3 - 3x + k$. Then $f(a) = f(b) = 0$. Since between any two roots of a polynomial $f(x)$, there exists at least one root of its derivative $f'(x)$. Therefore, $f'(x) = 3x^2 - 3$ has at least one root between a and b. But $f'(x) = 0$ has two roots equal to ± 1 which do not lie between a and b. Hence $f(x) = 0$ has no real roots lying between 0 and 1 for any value of k.

79. b. $f'(x) = -x \sin x = 0$ when $x = 0$ or π

$$\left.\begin{aligned} f'(0^-) &= (-)(-)(-) < 0 \\ f'(0^+) &= (-)(+)(+) < 0 \end{aligned}\right\} \text{no sign change}$$

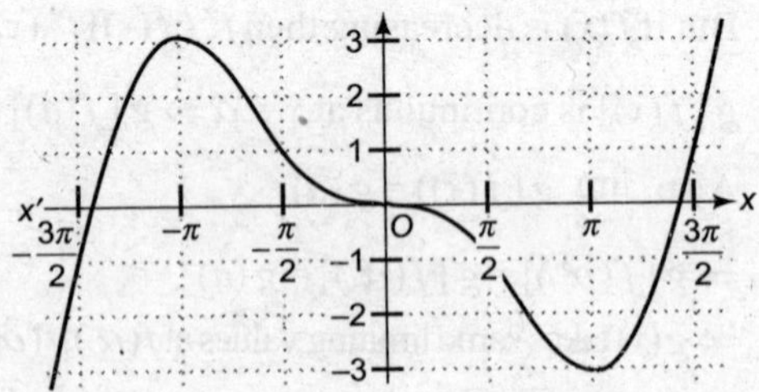

Fig. 6.15

This also implies that f is decreasing at $x = 0$

$\Rightarrow$ (b) is correct

$f''(x) = -(x \cos x + \sin x)$

$f''(\pi) = -(-\pi) > 0$ minima at $x = \pi$

$f''(-\pi) = -(\pi) < 0$ maxima at $x = \pi$

80. d. From the given data, graph of $f(x)$ can be shown as

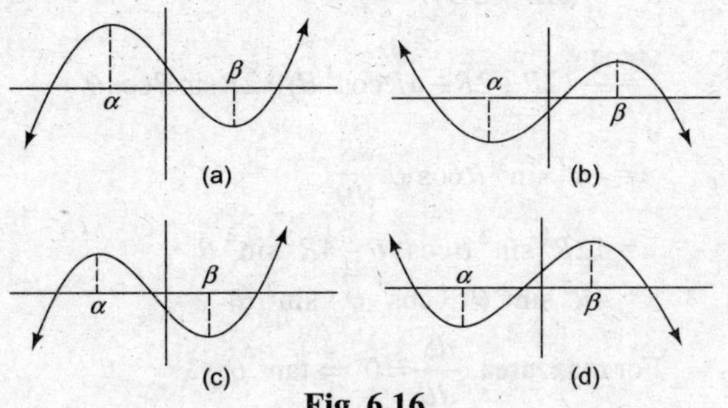

Fig. 6.16

Thus from graph, nothing can be said about roots when the sign of $f(\alpha)$ and $f(\beta)$ is given.

81. a.

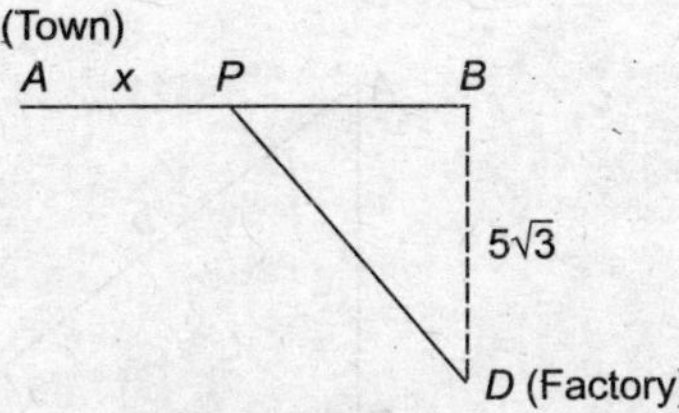

Fig. 6.17

Let the charges for railway line be k Rs./km.

Now the total freight charges,

$$\Rightarrow T = kx + 2k\sqrt{(20-x)^2+75}$$

Let $\frac{dT}{dx} = 0 \Rightarrow k + 2k\frac{2(x-20)}{2\sqrt{(x-20)^2+75}} = 0$

$\Rightarrow \quad 4(x-20)^2 = 75 + (x-20)^2$

$\Rightarrow \quad (x-20)^2 = 25 \Rightarrow x = 25, 15 \Rightarrow x = 15$ (as $AP < AB$).

$\Rightarrow \quad PB = AB - AP = 20 - 15 = 5$ km

82. a.

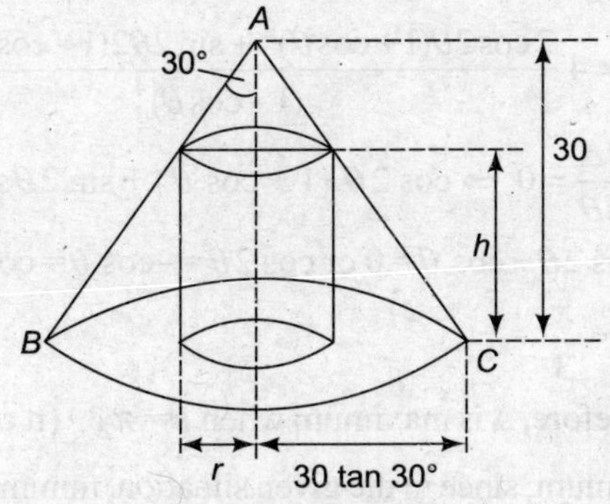

Fig. 6.18

From geometry, we have $\frac{r}{30\tan 30°} = \frac{30-h}{30}$

$\Rightarrow h = 30 - \sqrt{3}r$

Now, the volume of cylinder,

$V = \pi r^2 h = \pi r^2(30 - \sqrt{3}r)$

Now let $\frac{dv}{dr} = 0 \Rightarrow \pi(60r - 3\sqrt{3}r)^2 = 0 \Rightarrow r = \frac{20}{\sqrt{3}}$

Hence, $V_{\max} = \pi\left(\frac{20}{\sqrt{3}}\right)^2\left(30 - \sqrt{3}\frac{20}{\sqrt{3}}\right) = \pi\frac{400}{3} \times 10$

$= \frac{4000\pi}{3}$

83. b.

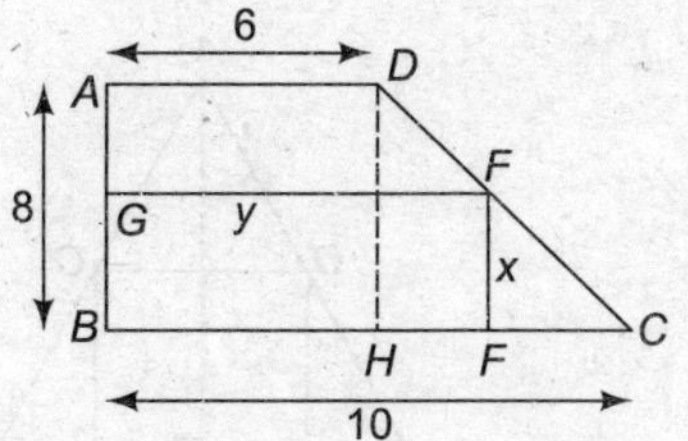

Fig. 6.19

Let rectangle $BEFG$ is inscribed.

Its area, $A = xy$

Now ΔFEC and ΔDHE are similar, i.e.,

$\Rightarrow \quad \frac{x}{8} = \frac{10-y}{4} \Rightarrow y = 10 - \frac{x}{2}$

$\Rightarrow A = x\left(10 - \frac{x}{2}\right)$ where $x \in (0, 8]$

Now $\frac{dA}{dx} = 10 - x$. Now for $x \in (0, 8)$

$\frac{dA}{dx} > 0 \Rightarrow A$ increases. Hence $A_{\max}$ occurs when $x = 8$. Hence. max area $= A_{\max} = 8(10 - \frac{8}{2}) = 48$ sq.cm.

84. a.

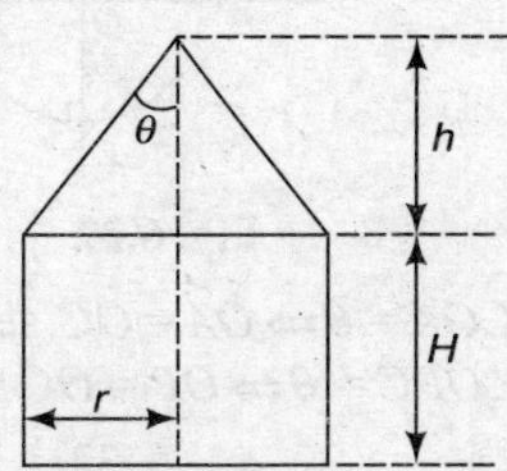

Fig. 6.20

Given volume and r

Now, V = volume of cone + volume of cylinder

$$= \frac{\pi}{3}r^2 h + \pi r^2 H$$

$V = \frac{\pi}{3}r^2(h + 3H) \Rightarrow H = \frac{\frac{3V}{\pi r^2} - h}{3}$

Now surface area, $S = \pi r l + 2\pi r H = \pi r$

$\sqrt{h^2+r^2} + 2\pi r \times \left(\frac{\frac{3V}{\pi r^2} - h}{3}\right)$

Now, let $\frac{dS}{dh} = 0 \Rightarrow \pi r\frac{h}{\sqrt{h^2+r^2}} - \frac{2\pi r}{3} = 0$

$$\Rightarrow \frac{h}{\sqrt{h^2+r^2}}=\frac{2}{3} \Rightarrow 5h^2=2r^2 \Rightarrow \frac{r}{h}=\frac{\sqrt{5}}{2}=\tan\theta$$

$$\Rightarrow \cos\theta=\frac{2}{3} \Rightarrow \theta=\cos^{-1}\frac{2}{3}$$

85. d.

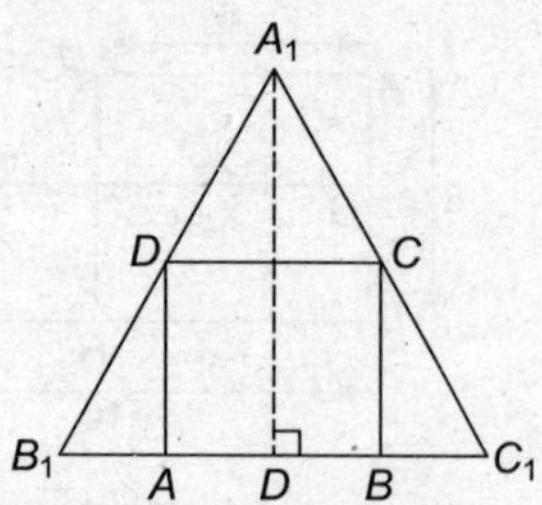

Fig. 6.21

Let $BD=x \Rightarrow BC_1=(a-x)$

$\Rightarrow BC=(a-x)\tan\frac{\pi}{3}=\sqrt{3}(a-x)$.

Now area of rectangle $ABCD$,

$\Delta=(AB)(BC)=2\sqrt{3}x(a-x)$.

$\Rightarrow \Delta \le 2\sqrt{3}\left(\frac{x+a-x}{2}\right)^2=\frac{\sqrt{3}a^2}{2}$ (using A.M. ≥ G.M.)

86. d.

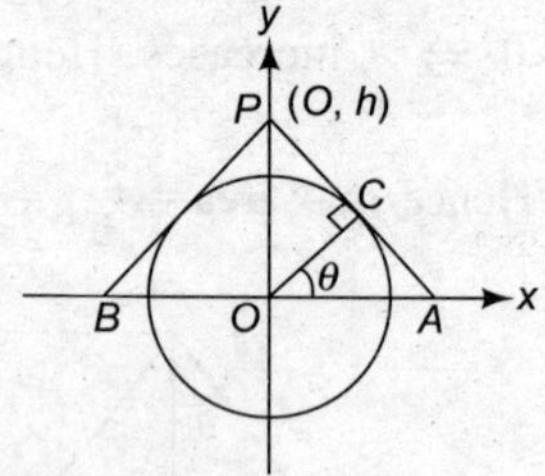

Fig. 6.22

Let $\angle COA=\theta \Rightarrow OA=OC\ \sec\theta=4\sec\theta$

Also $\angle OPC=\theta \Rightarrow OP=OC\ \operatorname{cosec}\theta=4\operatorname{cosec}\theta$

Now $\Delta_{PAB}=OA\ OP=\frac{32}{\sin 2\theta}$

For Δ_{PAB} to be minimum $\sin 2\theta=1 \Rightarrow \theta=\frac{\pi}{4}$

$\Rightarrow P=(0, 4\sqrt{2})$

87. a.

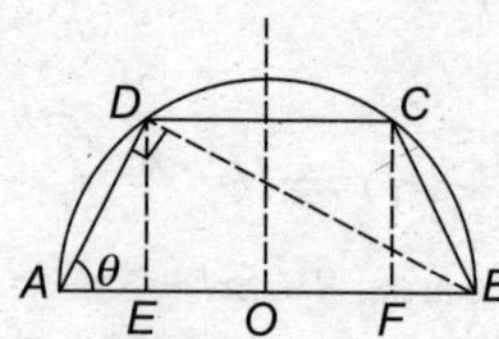

Fig. 6.23

$AD=AB\cos\theta=2R\cos\theta, AE=AD\cos\theta=2R\cos^2\theta$

$\Rightarrow EF=AB-2AE=2R-4R\cos^2\theta$

$DE=AD\sin\theta=2R\sin\theta\cos\theta$

$\Rightarrow$ Area of trapezium,

$$S=\frac{1}{2}(AB+CD)\times DE$$

$$=\frac{1}{2}(2R+2R-4R\cos^2\theta)\times 2R\sin\theta\cos\theta$$

$$=4R^2\sin^3\theta\cos\theta \quad \frac{dS}{d\theta}$$

$$=12R^2\sin^2\theta\cos^2\theta-4R^2\sin^4\theta$$

$$=4R^2\sin^2\theta(3\cos^2\theta-\sin^2\theta)$$

For max. area, $\frac{dS}{d\theta}=0 \Rightarrow \tan^2\theta=3$

$\Rightarrow \tan\theta=\sqrt{3}$ (θ is acute)

$\Rightarrow S_{\max}=\frac{3\sqrt{3}}{4}R^2$ [check for maxima].

88. c.

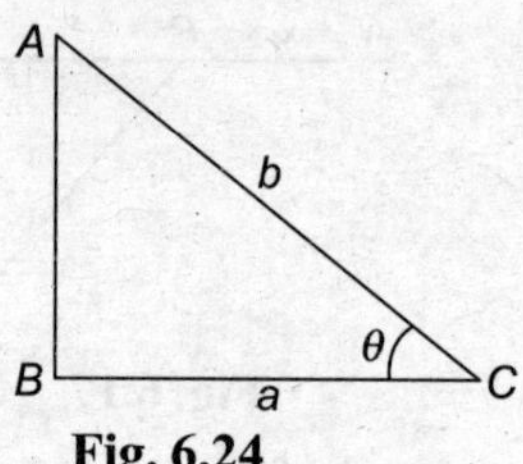

Fig. 6.24

$b\cos\theta=a \Rightarrow b\cos\theta+b=4$ or $b=\frac{4}{1+\cos\theta}$

$\Rightarrow a=\frac{4\cos\theta}{1+\cos\theta}$

$\Rightarrow$ area $=\Delta=\frac{1}{2}ba\sin\theta$

$$=\frac{1}{2}\frac{4}{1+\cos\theta}\frac{4\cos\theta}{1+\cos\theta}\times\sin\theta=\frac{4\sin 2\theta}{(1+\cos\theta)^2}$$

$$\Rightarrow \frac{d\Delta}{d\theta}$$

$$=4\frac{2\cos 2\theta(1+\cos\theta)^2+\sin 2\theta 2(1+\cos\theta)\sin\theta}{(1+\cos\theta)^4}$$

$\Rightarrow \frac{d\Delta}{d\theta}=0 \Rightarrow \cos 2\theta\,(1+\cos\theta)+\sin 2\theta\sin\theta=0$

or $\cos 2\theta+\cos\theta=0$ or $\cos 2\theta=-\cos\theta=\cos(\pi-\theta)$

or $\theta=\frac{\pi}{3}$

Therefore, Δ is maximum when $\theta=\pi/3$. (It cannot be minimum, since in the given situation, minimum is 0.)

89. b.

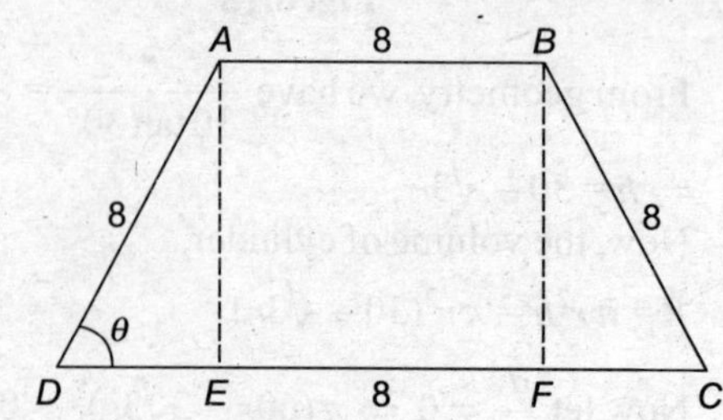

Fig. 6.25

$$A = (AB \times AE) + 2\left(\frac{1}{2} DE \times AE\right)$$
$$= (8 \times 8 \sin\theta) + 8 \sin\theta \times 8\cos\theta$$
$$= 64 \sin\theta + 32 \sin 2\theta$$

Let $\frac{dA}{d\theta} = 0 \Rightarrow 64\cos\theta + 64\cos 2\theta = 0$

$\Rightarrow 2\cos^2\theta + \cos\theta - 1 = 0$

$\Rightarrow (2\cos\theta - 1)(\cos\theta + 1) = 0$

$\Rightarrow \cos\theta = 1/2 \Rightarrow \theta = \pi/3$

$\Rightarrow A_{max} = 64\,\frac{\sqrt{3}}{2} + 32\frac{\sqrt{3}}{2}$

$= 32\sqrt{3} + 16\sqrt{3} = 48\sqrt{3}$.

90. b. $\because$ Fuel charges $\propto v^2$. Let F represents fuel charges

$\Rightarrow F \propto v^n \Rightarrow F = kv^2$ (1)

Given that F = Rs. 48 per hour, v = 16 km per hour

$\Rightarrow 48 = k(16)^2 \Rightarrow k = \frac{3}{16}$

From (1), $F = \frac{3v^2}{16}$

Let the train covers λ km in t hours

$\Rightarrow \lambda = vt$ or $t = \frac{\lambda}{v}$

$\Rightarrow$ Fuel charges in time $t = \frac{3}{16}v^2 \times \frac{\lambda}{v} = \frac{3v\lambda}{16}$

$\Rightarrow$ Total cost for running the train, C (say)

$= \frac{3v\lambda}{16} + 300 \times \frac{\lambda}{v}$

$\Rightarrow \frac{dC}{dv} = \frac{3\lambda}{16} - \frac{300\lambda}{v^2}$ and $\frac{d^2C}{dv^2} = \frac{600\lambda}{v^3}$

For the maximum or minimum value of C, $\frac{dC}{dv} = 0$

$\Rightarrow v = 40$ km/hr. Also, $\left.\frac{d^2C}{dv^2}\right|_{v=40} = \frac{60\lambda}{(40)^3} > 0\ (\because \lambda > 0)$

$\Rightarrow C$ is minimum when $v = 40$ km/hr.

91. c.

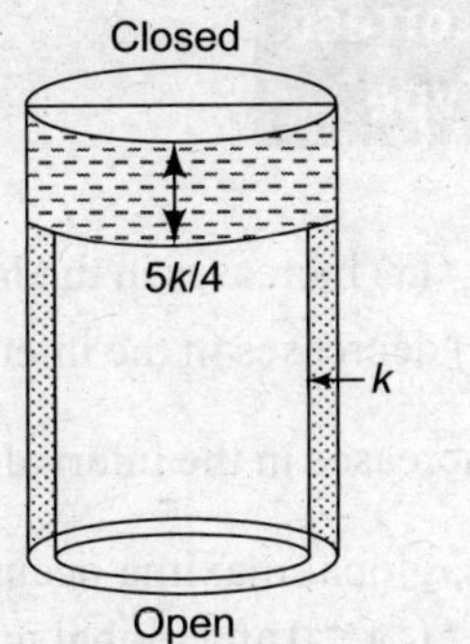

Fig. 6.26

Let x be the radius and y the height of the cylindrical gas container. Also let k be the thickness of the plates forming the cylindrical sides. Therefore, the thickness of the plate forming the top will be $5k/4$.

Capacity of the vessel = vol. of cylinder

$= \pi x^2 y = V$ (Given) $\Rightarrow y = V/(\pi x^2)$ (1)

Now the volume V_1 of the iron plate used for construction of the container is given by

$V_1 = \pi(x+k)^2(y + 5k/4) - \pi x^2 y$

$\Rightarrow \frac{dV_1}{dx} = 2Vk(x+k) \times \left(\frac{5\pi}{4V} - \frac{1}{x^3}\right)$

For max. or min. of V_1, $dV_1/dx = 0 \Rightarrow x = [4V/(5\pi)]^{1/3}$

For this value of x, d^2V_1/dx^2 is +ve

Hence, V_1 is min. when $x = [4V/(5\pi)]^{1/3}$

Now $x = [4V/(5\pi)]^{1/3}$

$\Rightarrow 5\pi x^3 = 4V = 4\pi x^2 y \Rightarrow x/y = 4/5$

Hence, the required ratio is 4 : 5.

92. c.

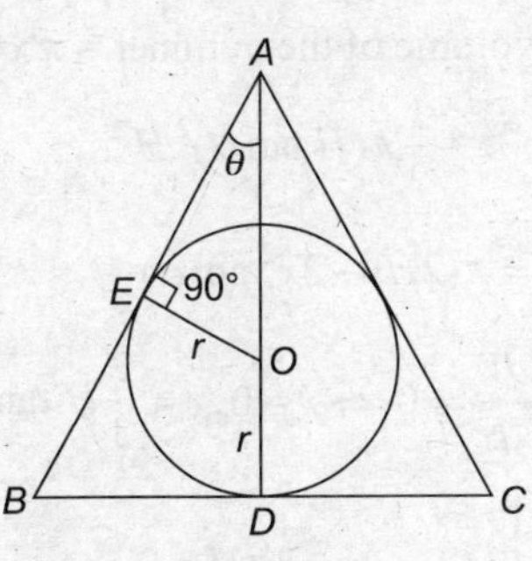

Fig. 6.27

Let ABC be an isosceles triangle in which a circle of radius r is inscribed.

Let $\angle BAD = \theta$ (semi-vertical angle)

In ΔOAE, $OA = OE$ cosec $\theta = r$ cosec θ, $AE = r\cot\theta$.

$\Rightarrow AD = OA + OD = r(\text{cosec}\,\theta + 1)$

Now, the perimeter of the ΔABC is

$S = AB + AC + BC = 2AB + 2BD$ $(\because AC = AB)$

$S = 2r(\text{cosec}\,\theta + 1)(\sec\theta + \tan\theta)$

or $S = \frac{4r(1+\sin\theta)^2}{\sin 2\theta}$

$\Rightarrow \frac{dS}{d\theta} = 4r\,[2(1+\sin\theta)\cos\theta\sin 2\theta - (1+\sin\theta)^2\, 2\cos 2\theta]/(\sin 2\theta)^2$

$= 8r(1+\sin\theta)[\sin 2\theta\cos\theta - \cos 2\theta\sin\theta - \cos 2\theta]$ $(\sin 2\theta)^2$

$= 8r(1+\sin\theta)(\sin\theta - 1 + 2\sin^2\theta)/(\sin 2\theta)^2$

$= 16r(1+\sin\theta)^2(\sin\theta - 1/2)/(\sin 2\theta)^2$

For max. or min. of S, $dS/d\theta = 0 \Rightarrow \sin\theta = 1/2$

$\therefore \theta = \pi/6$ $(\because \sin\theta \neq -1$ as θ is an acute angle)

Now if θ is little less and little greater than $\pi/6$, then sign of $dS/d\theta$ changes from –ve to +ve. Hence S is minimum when $\theta = \pi/6$, which is the only minima. Hence, the least perimeter of the

$\Delta = 4r\,[1 + \sin(\pi/6)]^2/\sin(\pi/3) = 6\sqrt{3}r.$

93. a.

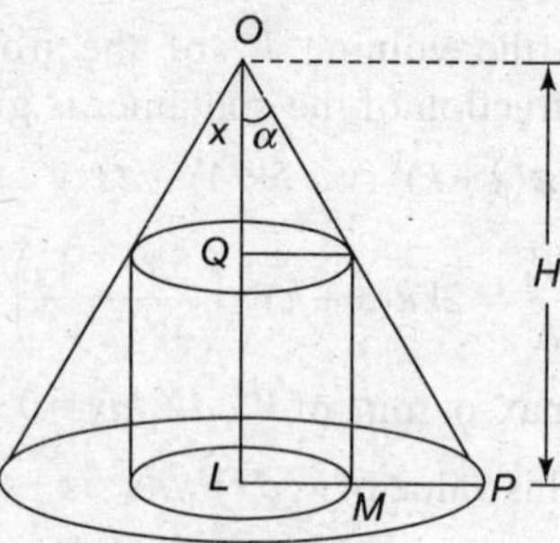

Fig. 6.28

Let H be the height of the cone and α be its semi-vertical angle. Suppose that x is the radius of the inscribed cylinder and h be its height $h = QL = OL - OQ = H - x\cot\alpha$,

V = volume of the cylinder $= \pi x^2(H - x\cot\alpha)$

Also $\pi = \frac{1}{3}\pi(H\tan\alpha)^2 H$ (1)

$$\frac{dV}{dx} = \pi(2Hx - 3x^2\cot\alpha)$$

So, $\frac{dV}{dx} = 0 \Rightarrow x = 0, x = \frac{2}{3}H\tan\alpha$;

$$\left.\frac{d^2V}{dx^2}\right|_{x=\frac{2}{3}H\tan\alpha} = -2\pi H < 0$$

So, V is max. when $x = \frac{2}{3}H\tan\alpha$, and $q = V_{\max} = \pi$

$$\frac{4}{9}H^2\tan^2\alpha\,\frac{1}{3}H = \frac{4}{27}\frac{\pi^3 p\tan^2\alpha}{\pi\tan^2\alpha} = \frac{4}{9}p. \text{ [from (1)]}$$

Hence $p : q = 9 : 4$.

94. d. Given $4x + 2\pi r = a$, where x is side length of the square and r is radius of the circle

$$A = x^2 + \pi r^2 = \frac{1}{16}(a - 2\pi r)^2 + \pi r^2$$

$\frac{dA}{dr} = 0$ gives $r = \frac{a}{2(\pi + 4)}$ for which $\frac{d^2A}{dr}$ is +ve and hence minimum.

$$\Rightarrow 4x = a - 2\pi r = a - \frac{a\pi}{\pi + 4} = \frac{4a}{\pi + 4}$$

$$\therefore x = \frac{a}{\pi + 4}$$

$$\therefore A = x^2 + r^2\pi = \frac{a^2}{4(\pi + 4)}.$$

95. c. The dimensions of the box after cutting equal squares of side x on the corner will be $21 - 2x$, $16 - 2x$ and height x.

$$V = x(21 - 2x)(16 - 2x)$$
$$= x(336 - 74x + 4x^2)$$

or $V = 4x^3 + 336x - 74x^2 \Rightarrow \frac{dv}{dx} = 12x^2 + 336 - 148x$

$\Rightarrow \frac{dV}{dx} = 0$ gives $x = 3$ for which $\frac{d^2V}{dx^2}$ is –ve and hence maximum.

96. a. See Fig. 6.29,

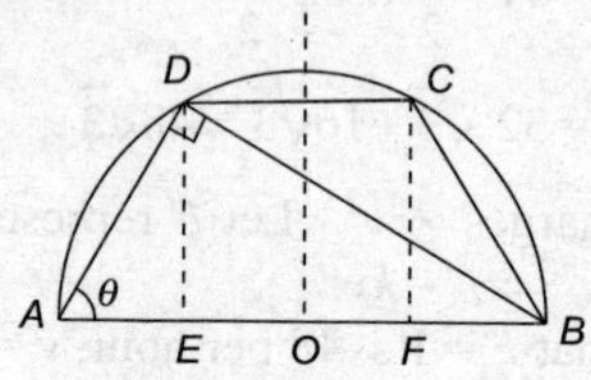

Fig. 6.29

$AD = AB\cos\theta = 2R\cos\theta$

$AE = AD\cos\theta = 2R\cos^2\theta$

$\therefore EF = AB - 2AE = 2R - 4R\cos^2\theta$

$DE = AD\sin\theta = 2R\sin\theta\cos\theta$

$\therefore$ Area of trapezium,

$$S = \frac{1}{2}(AB + CD) \times DE = \frac{1}{2}(2R + 2R - 4R\cos^2\theta) \times 2R\sin\theta\cos\theta$$
$$= 4R^3\sin^3\theta\cos\theta$$

$$\frac{dS}{d\theta} = 12R^2\sin^2\theta\cos^2\theta - 4R^2\sin^4\theta$$
$$= 4R^2\sin^2\theta(3\cos^2\theta - \sin^2\theta)$$

For maximum area, $\frac{dS}{d\theta} = 0 \Rightarrow \tan^2\theta = 3$

$\Rightarrow \tan\theta = \sqrt{3}$ (θ is acute)

Hence, maximum area $S_{\max} = \frac{3\sqrt{3}}{4}R^2$ (check for maxima).

Multiple Correct Answers Type

1. b., d.

Since, $f(x)$ increases in the interval $-1 \le x < 0$

$\Rightarrow f(x)$ decreases in the interval $0 \le x < \pi/2$

$f(x)$ increases in the interval $\frac{\pi}{2} \le x \le \pi$

Hence, global maxima occurs at $x = 0$ and $x = \pi$ ie, $f(0) = f(\pi) = 0$ and global minima occurs at $x = -1$, i.e., $f(-1) = -2$.

2. a., c.

$f'(x) = 4(x^3 - 3x^2 + 3x - 1) = 4(x-1)^3 > 0$ for $x > 1$.

Hence, f increases in $[1, \infty)$. Moreover, $f'(x) < 0$ for $x < 1$. Hence, f has a minimum at $x = 1$.

3. a., b., c., d.

$f(x) = 2x - \sin x \Rightarrow f'(x) = 2 - \cos x > 0\ \forall x$. Hence, $f(x)$ is strictly increasing, hence one-one and onto

$g(x) = x^{1/3}. \Rightarrow g'(x) = \frac{1}{3}x^{-2/3} > 0\ \forall x$, hence $g(x)$ is strictly increasing and hence one-one and onto.

Also gof is one-one.

$gof(x) = (2x - \sin x)^{1/3}$ has range R as the range of $2x - \sin x$ is R.

4. a., d.

We have

$$f(x) = 2x + \cot^{-1} x + \log(\sqrt{1+x^2} - x)$$

$$\therefore f'(x) = 2 - \frac{1}{1+x^2} + \frac{1}{\sqrt{1+x^2} - x}\left(\frac{x}{\sqrt{1+x^2}} - 1\right)$$

$$= \frac{1+2x^2}{1+x^2} - \frac{1}{\sqrt{1+x^2}} = \frac{1+2x^2}{1+x^2} - \frac{\sqrt{(1+x^2)}}{1+x^2}$$

$$= \frac{x^2 + \sqrt{1+x^2}(\sqrt{1+x^2} - 1)}{1+x^2} > 0 \text{ for all } x$$

Hence, $f(x)$ is an increasing function in $(-\infty, \infty)$ and in particular in $(0, \infty)$.

5. c., d.

We have

$$f(x) = \frac{|x-1|}{x^2} = \begin{cases} \frac{1-x}{x^2}, & x < 1, x \neq 0 \\ \frac{x-1}{x^2}, & x > 1 \end{cases}$$

$$\text{Now, } f'(x) < 0 \Rightarrow \begin{cases} \frac{x-2}{x^3} < 0, & \text{given that } x < 1 \\ \frac{2-x}{x^3} < 0, & \text{given that } x > 1 \end{cases}$$

$\Rightarrow 0 < x < 1$ or $x > 2$ for $f(x)$ to be decreasing

$\Rightarrow x \in (0, 1) \cup (2, \infty)$

6. b., d.

$$f'(x) = \frac{\sin x}{x}$$

For $f'(x) = 0$, $\frac{\sin x}{x} = 0 \Rightarrow x = n\pi (n \in I, n \neq 0)$

$$f''(x) = \frac{x\cos x - \sin x}{x^2}$$

$$f''(n\pi) = \frac{\cos n\pi}{n\pi} < 0 \text{ if } n = 2k-1 \text{ and } > 0 \text{ if } n = 2k,$$

$k \in I^+$.

Hence $f(x)$ has local maxima at $x = n\pi$, where $n = 2k - 1$ and local minima at $x = n\pi$, $n = 2k$, where $k \in I^+$.

7. a., b.

Given that $\frac{x^2 + x + 2}{x^2 + 5x + 6} < 0 \Rightarrow x \in (-3, -2)$

We have to find the extrema for the function

$f(x) = 1 + a^2x - x^3$

For maximum or minimum, $f'(x) = 0$

$\Rightarrow a^2 - 3x^2 = 0$ or $x = \pm\frac{a}{\sqrt{3}}$ and $f''(x) = -6x$ is positive when x is negative.

If a is positive, then the point of minima is $-\frac{a}{\sqrt{3}}$

i.e., $-3 < -\frac{a}{\sqrt{3}} < -2$ or $2\sqrt{3} < a < 3\sqrt{3}$.

And if a is negative, then the point of minima is $\frac{a}{\sqrt{3}}$

i.e., $-3 < \frac{a}{\sqrt{3}} < -2$ or $-3\sqrt{3} < a < -2\sqrt{3}$

Then, $a \in (-3\sqrt{3}, -2\sqrt{3}) \cup (2\sqrt{3}, 3\sqrt{3})$.

8. a., c.

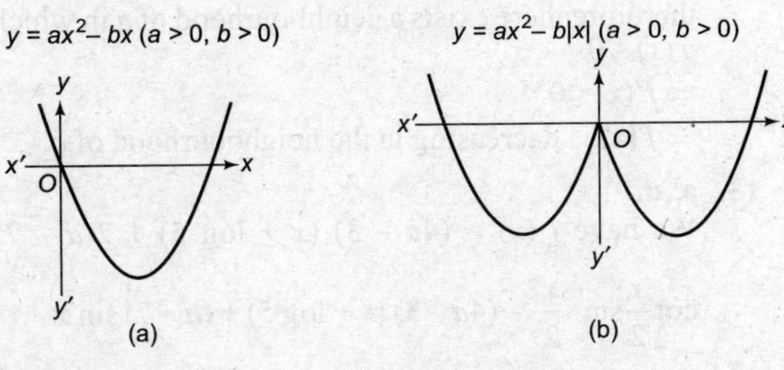

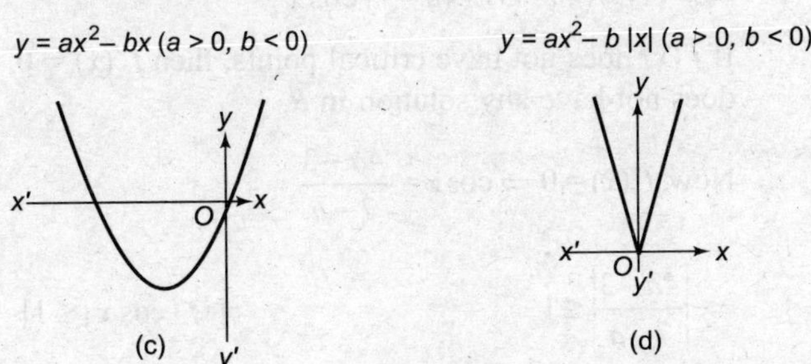

Fig. 6.30

9. a., b., d.

$$y = \frac{2x-1}{x-2}$$

$$\frac{dy}{dx} = \frac{2(x-2) - (2x-1)}{(x-2)^2} = \frac{-3}{(x-2)^2} < 0\ \forall x \neq 2$$

Therefore, y is decreasing in $(-\infty, 2)$ as well as in $(2, \infty)$

$$y = \frac{2x-1}{x-2} \Rightarrow x = \frac{2y-1}{y-2}$$

$$\therefore f^{-1}(x) = \frac{2x-1}{x-2} \quad \therefore f(x) \text{ is its own inverse.}$$

10. b., c.

Since. $g(x)$ is increasing and $f(x)$ is decreasing

$\Rightarrow g(x+1) > g(x-1)$ and $f(x+1) < f(x-1)$

$\Rightarrow f\{g(x+1)\} < f\{g(x-1)\}$

and $g\{f(x+1)\} < g\{f(x-1)\}$.

11. b., c.

$f(x) = x^3 - x^2 + 100x + 2002$

$f'(x) = 3x^2 - 2x + 100 > 2 \ \forall x \in R$

$\therefore f(x)$ is increasing (strictly)

Therefore, $f\left(\frac{1}{2000}\right) > f\left(\frac{1}{2001}\right)$

Also $f(x-1) > f(x-2)$ as $x-1 > x-2$ for $\forall x$.

12. a, d.

Since $g(a) \neq 0$, therefore either $g(a) > 0$ or $g(a) < 0$.

Let $g(a) > 0$. Since $g(x)$ is continuous at $x = a$, therefore there exists a neighbourhood of a in which $g(x) > 0$.

$\Rightarrow f'(x) > 0$

$\Rightarrow f(x)$ is increasing in the neighbourhood of a.

Let $g(a) < 0$. Since $g(x)$ is continuous at $x = a$, therefore there exists a neighbourhood of a in which $g(x) < 0$.

$\Rightarrow f'(x) < 0$

$\Rightarrow f(x)$ is decreasing in the neighbourhood of a.

13. a., d.

We have $f(x) = (4a-3)(x + \log 5) + 2(a-7)$

$$\cot\frac{x}{2}\sin^2\frac{x}{2} = (4a-3)(x+\log 5) + (a-7)\sin x$$

$$\Rightarrow f'(x) = (4a-3) + (a-7)\cos x$$

If $f(x)$ does not have critical points, then $f'(x) = 0$ does not have any solution in R.

Now, $f'(x) = 0 \Rightarrow \cos x = \dfrac{4a-3}{7-a}$

$$\Rightarrow \left|\frac{4a-3}{7-a}\right| \le 1 \qquad [\because |\cos x| \le 1]$$

$$\Rightarrow -1 \le \frac{4a-3}{7-a} \le 1 \Rightarrow a-7 \le 4a-3 \le 7-a$$

$\Rightarrow a \ge -4/3$ and $a \le 2$

Thus, $f'(x) = 0$ has solutions in R if $-4/3 \le a \le 2$

So, $f'(x) = 0$ is not solvable in R if $a < -4/3$ or $a > 2$, i.e., $a \in (-\infty, -4/3) \cup (2, \infty)$.

14. a, c, d.

Graph of $f(x)$

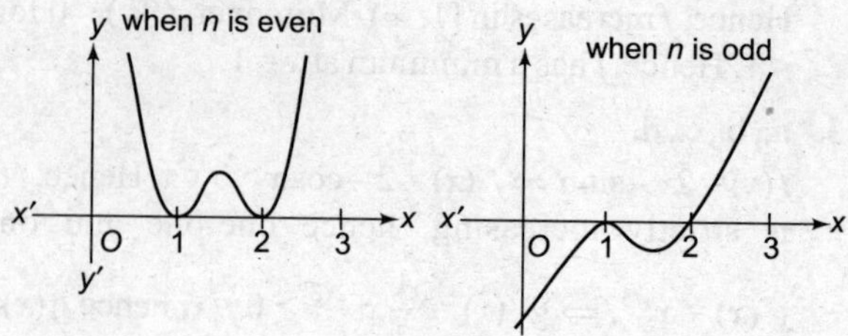

Fig. 6.31

15. a, b, c

$f'(x) = \cos x + a$

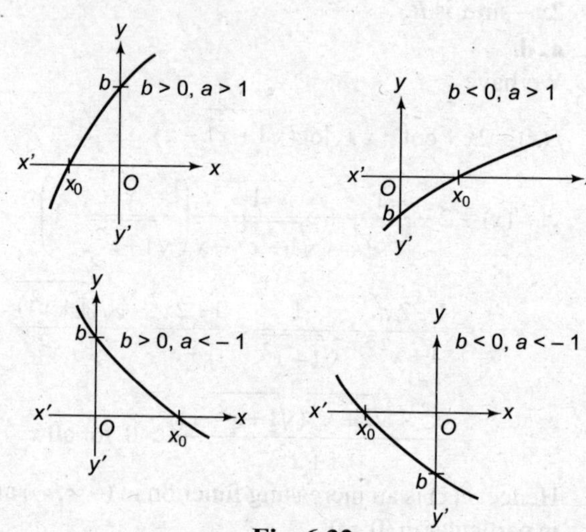

Fig. 6.32

If $a > 1$, then $f'(x) > 0$ or $f(x)$ is an increasing function, then $f(x) = 0$ has +ve root if $b < 0$ and –ve root if $b > 0$

$f'(x) = \cos x + a$

If $a < -1$, then $f'(x) < 0$ or $f(x)$ is a decreasing function, then $f(x) = 0$ has negative root if $b < 0$.

Hence, all the statements are correct.

16. a., b., c.

$$f(x) = \frac{\sin(x+a)}{\sin(x+b)}$$

$$f'(x) = \frac{\sin(x+b)\times\cos(x+a) - \sin(x+a)\cos(x+b)}{\sin^2(x+b)}$$

$$= \frac{\sin(b-a)}{\sin^2(x+b)}$$

If $\sin(b-a) = 0$, then $f'(x) = 0 \Rightarrow f(x)$ will be a constant,

i.e., $b - a = n\pi$ or $b - a = (2n+1)\pi$ or $b - a = 2n\pi$ then $f(x)$ has no minima.

17. c., d.

r must be an even integer because two decreasing functions are required to make it increasing function

Let $y = r(n-r)$

When n is odd, $r = \frac{n-1}{2}$ or $\frac{n+1}{2}$ for maximum values of y

when n is even, $r = \frac{n}{2}$ for maximum value of y.

Thererore, maximum $(y) = \frac{n^2-1}{4}$ when n is odd

and $\frac{n^2}{4}$ when n is even.

18. a., b., d.

$$f'(x) = \frac{12x^2 - 12x + 5}{(2x-1)^2} > 0 \ \forall\, x \in R$$

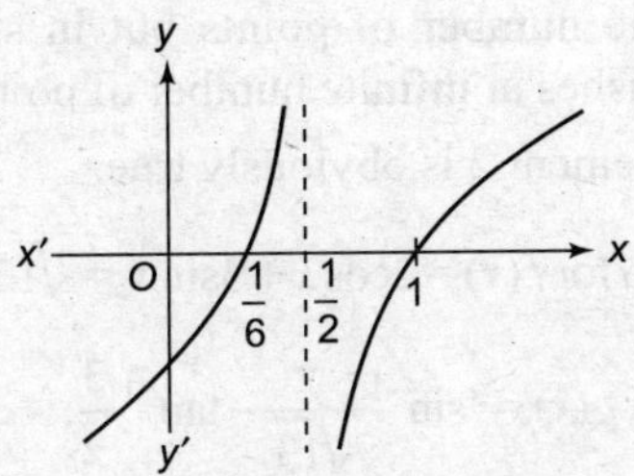

Fig. 6.33

Hence f is increasing $\forall x \in R$

$x = 1/2$ is the point of inflection as concavity changes at $x = 1/2$.

19. a., b., d.

At the point of inflection, concavity of the curve changes irrespective of any other factor.

20. b., c., d.

Since f is defined on $(0, \infty)$

$\therefore\ 2a^2 + a + 1 > 0$ which is true as $D < 0$

also $3a^2 - 4a + 1 > 0$

$(3a-1)(a-1) > 0 \Rightarrow a < 1/3$ or $a > 1$ (1)

as f is increasing hence

$f(2a^2 + a + 1) > f(3a^2 - 4a + 1)$

$\Rightarrow 2a^2 + a + 1 > 3a^2 - 4a + 1$

$\Rightarrow 0 > a^2 - 5a$

$\Rightarrow a(a-5) < 0 \Rightarrow (0, 5)$ (2)

From (1) and (2), we get

hence $a \in (0, 1/3) \cup (1, 5)$

Therefore, possible integers $\{2, 3, 4\}$.

21. b., d.

$f(x) = (\sin^2 x - 1)^n$

$$f\left(\frac{\pi}{2}\right) = 0$$

$$f\left(\frac{\pi^+}{2}\right) = (\to 0^-)^n \text{ and } f\left(\frac{\pi^-}{2}\right) = (\to 0^-)^n$$

If n is even $f\left(\frac{\pi^+}{2}\right)$ and $f\left(\frac{\pi^-}{2}\right) > 0$, then $x = \frac{\pi}{2}$ is the point of minima.

If n is odd $f\left(\frac{\pi^+}{2}\right)$ and $f\left(\frac{\pi^-}{2}\right) < 0$, then $x = \frac{\pi}{2}$ is the point of maxima.

22. a., b., d.

$f(x) = 2x^3 + 9x^2 + 12x + 1$

$\Rightarrow f'(x) = 6[x^2 + 3x + 2]$

$= 6(x+2)(x+1)$

$f'(x) < 0$ for $x \in (-2, -1)$, where $f(x)$ decreases.

$f'(x) > 0$ for $x \in (-\infty, -2) \cup (-1, \infty)$, where $f(x)$ increases.

$f''(x) = 2x + 3 = 0$

$\Rightarrow x = -3/2$ is the point of inflection.

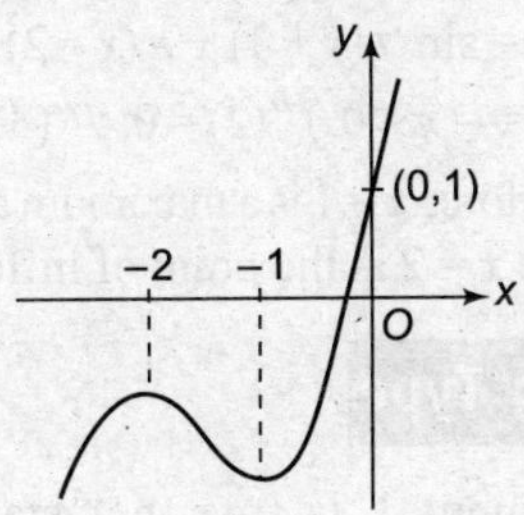

Fig. 6.34

From the graph, f is many-one, hence it is not bijective.

23. a., b., c.

$f(x) = a_5x^5 + a_4x^4 + a_3x^3 + a_2x^2 + a_1x$

$\Rightarrow f(x) = 0$ has one root $x = 0$

Also given that $f(x) = 0$ has positive root a_0.

Thus, the equation must have at least three roots real (as complex root occurs in conjugate pair). Thus $f'(x) = 0$ has at least two roots real as between two roots of $f(x) = 0$ there lies at least one root of $f'(x) = 0$.

Similarly, we can say that $f''(x) = 0$ has at least one real root.

Further, $f'(x) = 0$ has one root between roots $x = 0$ and $x = a_0$ of $f(x) = 0$.

24. a., b., c.

Let $y = f(x)^{g(x)}$

$$\Rightarrow \frac{dy}{dx} = f(x)^{g(x)}\left[g(x)\frac{f'(x)}{f(x)} + g'(x)\log f(x)\right]$$

$f(x)^{g(x)}$, $g(x)$, $f(x)$, $f'(x)$ and $g'(x)$ are positive, but $\log f(x)$ can be negative, which can cause $\frac{dy}{dx} < 0$, hence statement (a) is false.

If $f(x) < 1 \Rightarrow \log f(x) < 0$, which does not necessarily make $\frac{dy}{dx} < 0$, hence statement (b) is false.

$f(x) < 0$ can also cause $\frac{dy}{dx} > 0$, hence statement (c) is false.

But reverse of (c) is true.

25. a., c.

$$f(x) = \frac{2-x}{\pi}\cos\pi(x+3) + \frac{1}{\pi^2}\sin\pi(x+3)$$

$$f'(x) = -\frac{1}{\pi}\cos\pi(x+3) - (2-x)\sin\pi(x+3) + \frac{1}{\pi}\cos\pi(x+3)$$

$$= (x-2)\sin\pi(x+3) = 0$$

$x = 2, 1, 3$

$f''(x) = \sin\pi(x+3) + \pi(x-2)\cos\pi(x+3)$

$f''(1) = -\pi < 0, f''(2) = 0, f''(3) = \pi > 0$

Therefore, $x = 1$ is a maximum and $x = 3$ is a minimum, hence $x = 2$ is the point of inflection.

Reasoning Type

1. c. Statement 1 is true, but statement 2 is false as consider the functions in statement 1 in $\left(0, \frac{\pi}{2}\right)$.

2. a. $f(x) = \frac{\log_e x}{x} \Rightarrow f'(x) = \frac{1-\log_e x}{x^2}$

$f'(x) > 0$ for $1 - \log_e x > 0$ or $x < e \Rightarrow f(x)$ is increasing.

$f(x)$ is decreasing for $x > e$.

$e < 2.91 < \alpha < \beta$

$\Rightarrow f(\alpha) > f(\beta)$

$\Rightarrow \frac{\log_e \alpha}{\alpha} > \frac{\log_e \beta}{\beta}$

$\Rightarrow \beta \log_e \alpha > \alpha \log_e \beta$

$\Rightarrow \alpha^\beta > \beta^\alpha$.

3. d. Statement 2 is true as $f(x)$ is non-differentiable at $x = 1, 2, 3$.

But $f(x)$ has a point of minima at $x = 1$ and not at $x = 3$.

4. c. Both $f(x) = x$ and $g(x) = x^3$ are increasing in $(-1, 0)$ but $h(x) = x.x^3$ is decreasing.

5. a. Suppose $f(x) = 0$ has real root say $x = a$, then $f(x) < 0$ for all $x < a$.

Thus $|f(x)|$ becomes strictly decreasing in $(-\infty, a)$ which is a contradiction.

6. d. Statement 1 is false as $f(x) = 5 - 4(x-2)^{2/3}$ attains the greatest value at $x = 2$, though it is not differentiable at $x = 2$, and for extreme value it is not necessary that $f'(x)$ exists at that point.

Statement 2 is obviously true.

7. b. $f(x) = x + \cos x$

$\therefore f'(x) = 1 - \sin x > 0\ \forall x \in R$, except at $x = 2n\pi + \frac{\pi}{2}$

and $f'(x) = 0$ at $x = 2n\pi + \frac{\pi}{2}$

Therefore, $f(x)$ is strictly increasing.

Statement 2 is true but does not explain statement 1. Since according to Statement 2, $f'(x)$ may vanish at finite number of points but in statement 1 $f'(x)$ vanishes at infinite number of points.

8. a. Statement 2 is obviously true.

Also for $f(x) = 2\cos x + 3\sin x = \sqrt{13}\sin\left(x + \tan^{-1}\frac{3}{2}\right)$

$\Rightarrow \quad g(x) = \sin^{-1}\frac{x}{\sqrt{13}} - \tan^{-1}\frac{3}{2}$.

Hence statement 1 is true.

9. a. $f(x) = \frac{x^3}{3} + \frac{ax^2}{2} + x + 5$

$\Rightarrow f'(x) = x^2 + ax + 1$

If $f(x)$ has positive point of maxima, then point of minima is also positive. Hence both the roots of equation $x^2 + ax + 1 = 0$ must be positive.

$\Rightarrow$ sum or roots $-a > 0$, product of roots $1 > 0$ and discriminant $D = a^2 - 4 > 0$

$\Rightarrow a < -2$.

10. a. $\frac{dy}{dx} = 12x(x^2 - x + 1) - a$ and

$\frac{d^2y}{dx^2} = 12(3x^2 - 2x + 1) > 0$

$\Rightarrow \frac{dy}{dx}$ is an increasing function.

But $\frac{dy}{dx}$ is a polynomial of degree 3

$\Rightarrow$ it has exactly one real root root.

11. b. Let $f(x) = \sin x \tan x - x^2$

$\Rightarrow f'(x) = \sin x \sec^2 x + \sin x - 2x$

$\Rightarrow f''(x) = 2\sin x \sec^2 x \tan x + \sec x + \cos x - 2$

$= 2\sin x \tan x \sec^2 x + (\cos x + \sec x - 2)$

$> 0\ \forall x \in \left(0, \frac{\pi}{2}\right)$

$\Rightarrow f'(x)$ is an increasing function.

$\Rightarrow f'(x) > f'(0) \Rightarrow \sin x \sec^2 x + \sin x - 2x > 0$

$\Rightarrow f(x)$ is an increasing function

$\Rightarrow f(x) > f(0)$

$\Rightarrow \sin x \tan x - x^2 > 0$

$\Rightarrow \sin x \tan x > x^2$

Thus statement 1 is true, also statement 2 is true but it does not explain statement 1.

12. b. $f(x) = \sin(\cos x)$

$\Rightarrow f'(x) = -\sin x \cos(\cos x) < 0$ for $\forall x \in \left[0, \frac{\pi}{2}\right]$.

Statement 2 is also true, but it is not the only reason for statement 1 to be correct.

13. c. $f(x) = (x^3 - 6x^2 + 12x - 8)\, e^x$

$\Rightarrow f'(x) = e^x(x^3 - 6x^2 + 12x - 8) + e^x(3x^2 - 12x + 12)$

$= e^x(x^3 - 3x^2 + 4)$

$\Rightarrow f''(x) = e^x(x^3 - 3x^2 + 4) + e^x(3x^2 - 6x)$

$= e^x(x^3 - 6x + 4)$

$\Rightarrow f'''(x) = e^x(x^3 - 6x + 4) + e^x(3x^2 - 6)$

$= e^x(x^3 + 3x^2 - 6x - 2)$

Clearly, $f'(2) = f''(2) = 0$ and $f''' \neq 0$, hence $x = 2$ is the point of inflection and hence not a point of extrema. Thus statement 1 is true.

But statement 2 is false, as it is not necessary that at point of inflection, extrema does not occur. Consider the following graph.(Fig. 6.35).

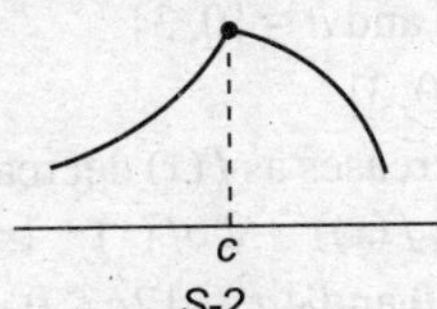

Fig. 6.35

14. a. $f(x) = x^4 - 8x^3 + 22x^2 - 24x + 21$

$f'(x) = 4x^3 - 24x^2 + 44x - 24$

$= 4(x^3 - 6x^2 + 11x - 6)$

$= 4(x-1)(x-2)(x-3)$

Sign scheme of $f'(x)$

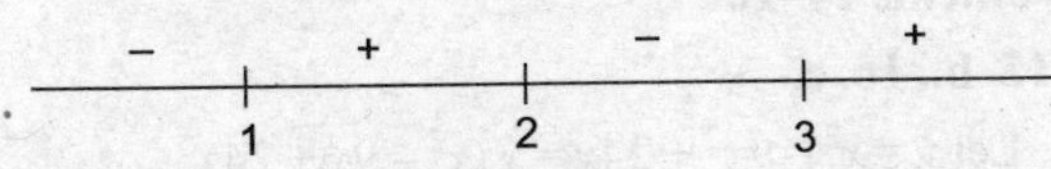

Fig. 6.36

From the sign scheme of $f'(x)$, $f(x)$ increases for $x \in (1, 2) \cup (3, \infty)$.

Since $f(x)$ is a polynomial function, which is continuous, and has no point of inflection, intervals of increase and decrease occur alternatively.

15. a. $f'(x) = \ln(x + \sqrt{1+x^2}) = -\ln(\sqrt{1+x^2} - x)$

$\Rightarrow f'(x) > 0$ for $x > 0$ and $f'(x) < 0$ for $x < 0$

$\Rightarrow f(x)$ is increasing when $x > 0$ and decreasing for $x < 0$

Hence for $x > 0, f(x) > f(0) \Rightarrow f(x) > 0$.

Again $f(x)$ is decreasing in $(-\infty, 0)$

Then for $x < 0, f(x) > f(0) \Rightarrow f(x) > 0$.

$\Rightarrow f(x)$ is positive for all $x \in R_0$

Thus, Statement 1 is true and follows from Statement 2.

Linked Comprehension Type

For Problems 1–2

1. b., 2. a.

Sol. Let $f(x) = \sin^{-1} x + x^2 - 3x + \frac{x^3}{3}$

$\Rightarrow f'(x) = \frac{1}{\sqrt{1-x^2}} + 2x - 3 + x^2$

$\Rightarrow f'(x) = 0$ for some $x = x_1 \in (0, 1)$

and $f''(x) = \frac{x}{(1-x^2)^{3/2}} + 2 + 2x > 0, \forall x \in (0, 1)$

$\Rightarrow x = x_1$ is the point of minimum

$f(x)$ is continuous $\forall x \in [0, 1]$.

Hence, the global maxima exist at $x = 0$ or $x = 1$

$f(0) = 0, f(1) = \pi/2 - 5/3 < 0$

$f(0)$ is global maxima $\forall x \in [0, 1]$

$\Rightarrow f(x) \leq f(0), x \in [0, 1]$

$\Rightarrow \sin^{-1} x + x^2 - 3x + x^3/3 \leq 0$

$\Rightarrow \sin^{-1} x + x^2 \leq \frac{x(9-x^2)}{3} \quad \forall x \in [0, 1]$

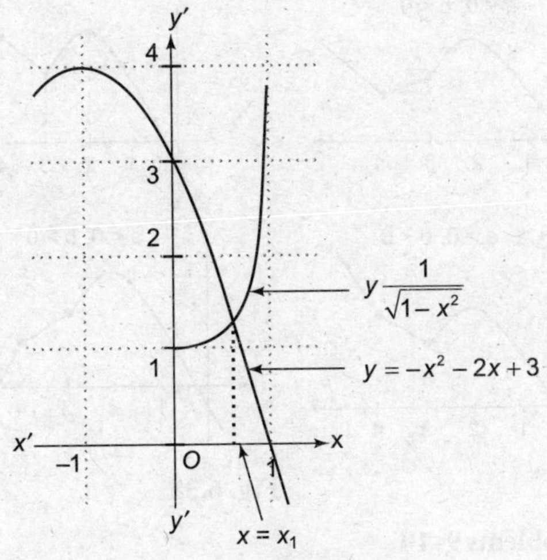

Fig. 6.37

Consider function $f(x) = \sin x + 2x - \frac{3x(x+1)}{\pi}$,

$x \in \left[0, \frac{\pi}{2}\right]$.

For Problems 3–4

3. a., 4. d.

Sol. $g'(x)=f'(\sin x)\cos x-f'(\cos x)\sin x$

$\Rightarrow g''(x) = -f'(\sin x)\sin x+\cos^2 x f''(\sin x) + f''(\cos x)\sin^2 x - f'(\cos x)\cos x > 0\ \forall x \in (0, \pi/2)$

(as it is given $f'(\cos x) = f'(\sin(\pi/2 - x)) < 0$ and $f''(\cos x) = f''(\sin(\pi/2 - x)) > 0$)

$\Rightarrow g'(x)$ is increasing in $(0, \pi/2)$. Also $g'(\pi/4) = 0$

$\Rightarrow g'(x) > 0\ \forall x \in \left(\frac{\pi}{4}, \frac{\pi}{2}\right)$ and $g'(x) < 0\ \forall x \in (0, \pi/4)$. Thus $g(x)$ is decreasing in $(0, \pi/4)$.

For Problems 5–8

5.d., 6.a., 7.d., 8. c.

Sol. If $f(x)$ is continuous then $f(3^-)=f(3^+)$

$\Rightarrow -9+12+a=3a+b \Rightarrow 2a+b=3$ (1)

Also $f(4^-)=f(4^+) \Rightarrow 4a+b=-b+6$

$\Rightarrow 2a+b=3$ (2)

$\Rightarrow f(x)$ is continuous for infinite values of a and b.

Also, $f'(x) = \begin{cases} -2x+4, & x<3 \\ a, & 3<x<4 \\ -\frac{b}{4}, & x>4 \end{cases}$. For $f(x)$ to be differentiable, $f'(3^-)=f(3^+)$

$\Rightarrow a=-2$ and $-\frac{b}{4}=a=-2 \Rightarrow b=8.$

Hence, $f(x)$ can be differentiable.

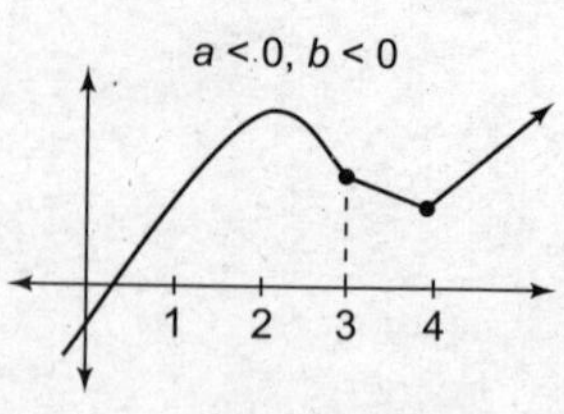

a > 0, b > 0

a > 0, b < 0

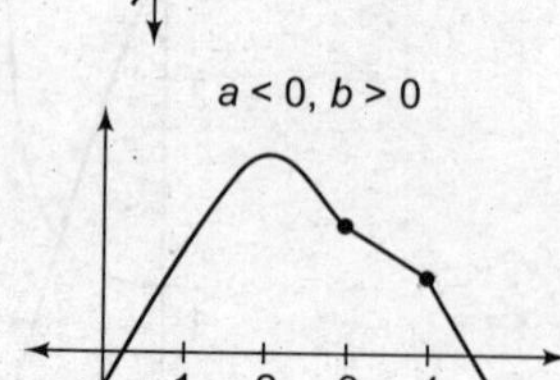

Fig. 6.38

For Problems 9–10

9.a., 10. b.

Sol.

9. a. $\frac{dP(x)}{dx} > P(x)$

$\Rightarrow \frac{dP(x)}{dx} - P(x) > 0$

$\Rightarrow \frac{d}{dx}\left(P(x)e^{-x}\right) > 0$

$\Rightarrow P(x)e^{-x}$ is an increasing function.

$\Rightarrow P(x)e^{-x} > P(1)e^{-1}\ \forall x \geq 1$

$\Rightarrow P(x)e^{-x} > 0\ \forall x > 1 \Rightarrow P(x) > 0\ \forall x > 1.$

10. b. Given that $\frac{d}{dx}H(x) > 2cxH(x)$

$\Rightarrow \frac{d}{dx}H(x) - 2cxH(x) > 0$

$\Rightarrow \frac{d}{dx}(H(x)e^{-cx^2}) > 0$

$\Rightarrow H(x)\ e^{-cx^2}$ is an increasing function.

But $H(x_0) = 0$ and e^{-cx^2} is always positive.

$\Rightarrow H(x) > 0$ for all $x > x_0$

$\Rightarrow H(x)$ cannot be zero for any $x > x_0$.

For Problems 11–13

11. a. $h(x) = f(x) - a(f(x))^2 + a(f(x))^3$

$\Rightarrow h'(x) = f'(x) - 2af(x)f'(x) + 3a(f(x))^2 f'(x)$

$\Rightarrow h'(x) = f'(x)[3a(f(x))^2 - 2af(x) + 1]$

Now $h(x)$ increases if $f(x)$ increases and $3a(f(x))^2 - 2af(x) + 1 > 0$ for all $x \in R$

$\Rightarrow 3a > 0$ and $4a^2 - 12a \leq 0$

$\Rightarrow a > 0$ and $a \in [0, 3]$

$\Rightarrow a \in [0, 3].$

12. d. $h(x)$ increases as $f(x)$ decreases for all real values of x if $3a(f(x))^2 - 2af(x) + 1 \leq 0$ for all $x \in R$

$\Rightarrow 3a < 0$ and $4a^2 - 12a \leq 0$

$\Rightarrow a < 0$ and $a \in [0, 3].$

$\Rightarrow$ no such a is possible.

13. d. $h(x)$ is non-monotonic functions if $3a(f(x))^2 - 2af(x) + 1$ changes sign for which $D \geq 0$ or $4a^2 - 12a \leq 0$

$\Rightarrow a \in (-\infty, 0] \cup [3, \infty).$

For Problems 14–16

14. a., 15. b., 16. c.

Sol. Let $y = x^3 - 9x^2 + 24x = x(x^2 - 9x + 24)$

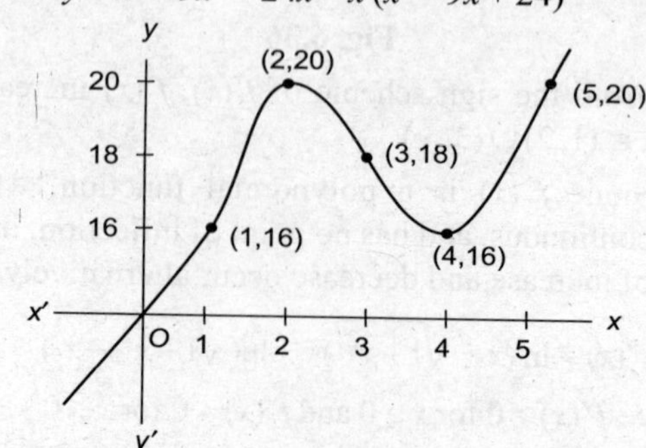

Fig. 6.39

Sign scheme:

$$\begin{array}{cccc} - & + & - & + \\ \hline & 2 & 0 & 4 \end{array}$$

Fig. 6.40

For three real roots of

$f(x) = x^3 - 9x^2 + 24x + c$, c must lie in the interval $(-20, -16)$

$f(0) = c < 0$

$f(1) = 1 - 9 + 24 + c = c + 16 < 0$ for $\forall c \in (-20, -16)$

$f(2) = 8 - 36 + 48 + c = c + 20 > 0$

$\alpha \in (1, 2) \Rightarrow [\alpha] = 1$

$f(3) = 27 - 81 + 72 = 18$

$\Rightarrow f(3) > 0$ if $c \in (-20, -18)$ or $f(3) < 0$ if $c \in (-18, -16)$

or $\beta \in (2, 3)$ if $c \in (-18, -16)$

and $\beta \in (3, 4)$ if $c \in (-20, -18)$

Now $f(4) = 64 - 144 + 96 + c = 16 + c < 0$ $\forall c \in (-20, -16)$

$f(5) = 125 - 225 + 120 + c = c + 20 > 0$ $\forall c \in (-20, -16)$

$\Rightarrow \gamma \in (4, 5) \Rightarrow [\gamma] = 4$

$$\text{Thus, } [\alpha] + [\beta] + [\gamma] = \begin{cases} 1+3+4, & -20 < c < -18 \\ 1+2+4, & -18 < c < -16 \end{cases}$$

Now if $c \in (-20, -18)$,

$\alpha \in (1, 2)$, $\beta \in (2, 3)$, $\gamma \in (4, 5)$

$\Rightarrow \quad [\alpha] + [\beta] + [\gamma] = 7$

if $c \in (-18, -16)$, $\alpha \in (1, 2)$, $\beta \in (3, 4)$, $\gamma \in (4, 5)$

$\Rightarrow \quad [\alpha] + [\beta] + [\gamma] = 8$.

For Problems 17–21

17. b., 18. d., 19. b., 20. d., 21. b.

Sol.

17. b. $f'(x) \le 0$ $\forall x \in [0, b]$, so $f(x)$ is a decreasing function and $f(c) = 0 \Rightarrow f(x)$ cuts x-axis once when $x = c$.

18. d. We note that $f(c) = 0, f'(c) = 0$. Also tangent to $f'(x)$ at $x = c$ is $y = 0$. So $f''(c) = 0$.
Therefore, $x = c$ is the repeated root of third order. That is, the equation $f(x) = 0$ has at least three repeated roots.

19. b. We have $f''(c) = 0$. So the graph of $y = f(x)$ has one point of inflexion at $x = c$.

20. d. As $f(x)$ is a decreasing functions for all $x \in (a, b)$, so $f(x)$ has no local maxima or minima.

21. b. $f''(c) = 0 \Rightarrow x = c$ is a root of $f''(x) = 0$.

For Problems 22–24

22. b., 23. d., 24. a.

Sol. $f(x) = 4x^2 - 4ax + a^2 - 2a + 2$. Vertex of this parabola is $\left(\frac{a}{2}, 2 - 2a\right)$.

Case 1: $0 < \frac{a}{2} < 2$

In this case, $f(x)$ will attain the minimum value at $x = \frac{a}{2}$. Thus $f\left(\frac{a}{2}\right) = 3$.

$\Rightarrow 3 = -2a + 2 \Rightarrow a = -\frac{1}{2}$ (Rejected).

Case 2: $\frac{a}{2} \ge 2$

In this, $f(x)$ attains the global minimum value at $x = 2$. Thus $f(2) = 3$

$\Rightarrow 3 = 16 - 8a + a^2 - 2a + 2 \Rightarrow a = 5 \pm \sqrt{10}$. Thus $a = 5 + \sqrt{10}$.

Case 3: $\frac{a}{2} \le 0$

In this case, $f(x)$ attains the global minimum value at $x = 0$. Thus $f(0) = 3$

$\Rightarrow 3 = a^2 - 2a + 2 \Rightarrow a = 1 \pm \sqrt{2}$. Thus, $a = 1 - \sqrt{2}$.

Hence, the permissible values of a are $1 - \sqrt{2}$ and $5 + \sqrt{10}$.

$f(x) = 4x^2 - 49x + a^2 - 2a + 2$ is monotonic in $[0, 2]$.
Hence, the point of minima of function should not lie in $[0,2]$.

Now $f'(x) = 0 \Rightarrow 8x - 4a = 0 \Rightarrow x = a/2$. If $\frac{a}{2} \in [0, 2]$

$\Rightarrow a \in [0, 4]$.

For $f(x)$ to be monotonic in $[0, 2]$, $a \notin [0, 4]$

$\Rightarrow a \le 0$ or $a \ge 4$.

For Problems 25–27

25. a., 26. c., 27. d.

Sol. $f(x) = x^3 - 3(7 - a)x^2 - 3(9 - a^2)x + 2$

$f'(x) = 3x^2 - 6(7 - a)x - 3(9 - a^2)$

For real root, $D \ge 0$,

$\Rightarrow 49 + a^2 - 14a + 9 - a^2 \ge 0 \Rightarrow a \le \frac{58}{14}$ (1)

For local minimum $f''(x) = 6x - 6(7 - a) > 0$

$\Rightarrow 7 - a < 0$ has x must be negative

$\Rightarrow 7 - a < 0 \Rightarrow a > 7$ (2)

From (1) and (2), we get no possible values of a.

For local maxima $f''(x) = 6x - 6(7 - \text{a}) < 0$

$\Rightarrow 7 - a > x > 0 \Rightarrow a < 7$ (3)

From (1) and (3), $a \le \frac{58}{14}$

For points of extrema of opposite sign, we must have roots of $f'(x)$ of opposite sign.

For which $-\frac{3(9 - a^2)}{3} < 0 \Rightarrow a^2 < 9 \Rightarrow a \in (-3, 3)$.

For Problems 28 – 30

28. c., 29. d., 30. d.

Sol. $f(x)=\left(1+\frac{1}{x}\right)^x$, $f(x)$ is defined if $1+\frac{1}{x}>0$

$\Rightarrow \frac{x+1}{x}>0 \Rightarrow (-\infty,-1)\cup(0,\infty)$

Now $f'(x)=\left(1+\frac{1}{x}\right)^x\left[\ln\left(1+\frac{1}{x}\right)-\frac{x}{1+\frac{1}{x}}\frac{-1}{x^2}\right]$

$=\left(1+\frac{1}{x}\right)^x\left[\ln\left(1+\frac{1}{x}\right)-\frac{1}{x+1}\right]$

Now $\left(1+\frac{1}{x}\right)^x$ is always positive, hence the sign of $f'(x)$ depends on sign of $\ln\left(1+\frac{1}{x}\right)-\frac{1}{1+x}$

Let $g(x)=\ln\left(1+\frac{1}{x}\right)-\frac{1}{x+1}$

$g'(x)=\frac{1}{1+\frac{1}{x}}\frac{-1}{x^2}+\frac{1}{(x+1)^2}=\frac{-1}{x(x+1)^2}$ (1)

for $x\in(0,\infty)$, $g'(x)<0$

$\Rightarrow g(x)$ is monotonically decreasing for $x\in(0,\infty)$

$\Rightarrow g(x)>\lim_{x\to\infty} g(x) \Rightarrow g(x)>0$

and since $g(x)>0 \Rightarrow f'(x)>0$ (2)

for $x\in(-\infty,-1)$, $g'(x)>0$

$\Rightarrow g(x)$ is monotonically increasing for $x\in(-\infty,-1)$

$\Rightarrow g(x)>\lim_{x\to-\infty} g(x)$

$\Rightarrow g(x)>0 \Rightarrow f'(x)>0$

Hence from (1) and (2) we get $f'(x)>0$ for all $x\in(-\infty,-1)\cup(0,\infty)$

$\Rightarrow f(x)$ is monotonically increasing in its domain

Also $\lim_{x\to\pm\infty}\left(1+\frac{1}{x}\right)^x=e$

$\lim_{x\to 0^+}\left(1+\frac{1}{x}\right)^x=1$ and $\lim_{x\to -1}\left(1+\frac{1}{x}\right)^x=\infty$

The graph of $f(x)$ is sown in Fig. 6.41.

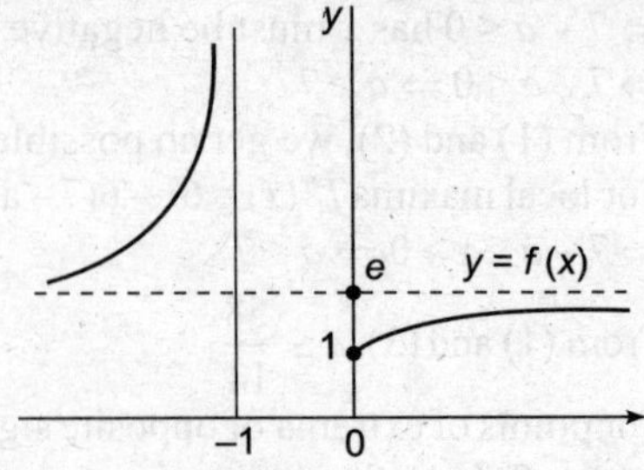

Fig. 6.41

Range is $y\in(1,\infty)-\{e\}$.

Matrix-Match Type

1. a → r; b → s; c → q; d → p

$f'(x)=4x^3-28x+24$
$=4(x^3-7x+6)$
$=4(x^3-x^2+x^2-x-6x+6)$
$=4(x-1)(x^2+x-6)$
$=4(x-1)(x+3)(x-2)$

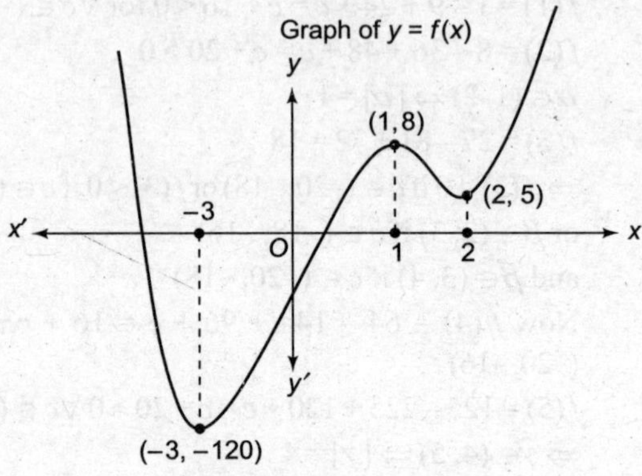

Fig. 6.42

Now, nature of roots of $f(x)+p=0$ can be obtained by shifting the graph of $y=f(x)$ by p units upward or downward depending on whether p is positive or negative.

2. a → p; b → p, s; c → q, r; d → q.

a. $f(x)=x^2\log x$

for $f'(x)=x(2\log x+1)=0$, $x=1/\sqrt{e}$ which is the point of minima as derivative changes sign from negative to positive.

Also the function decreases in $\left(1,1/\sqrt{e}\right)$.

b. $y=x\log x$

$\Rightarrow \frac{dy}{dx}=x\times\frac{1}{x}+\log x\times 1=1+\log x$

and $\frac{d^2y}{dx^2}=\frac{1}{x}$

For $\frac{dy}{dx}=0$

$\Rightarrow \log x=-1 \Rightarrow x=\frac{1}{e}$

$\frac{d^2y}{dx^2}=\frac{1}{1/e}=e>0$ at $x=\frac{1}{e}$

$\Rightarrow y$ is min for $x=\frac{1}{e}$

$\Rightarrow$ min. $y=\frac{1}{e}\log\frac{1}{e}=-\frac{1}{e}$

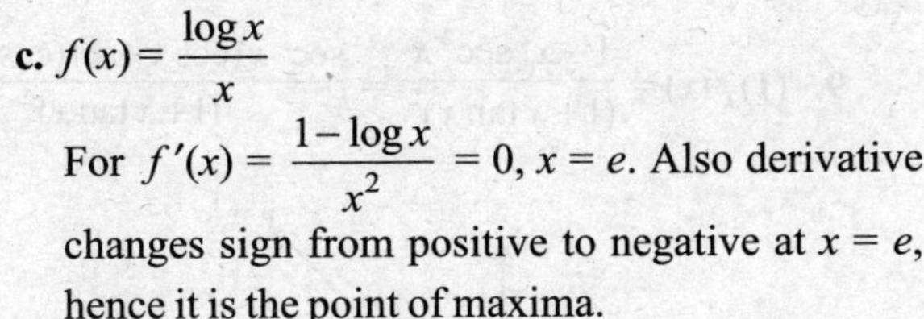

c. $f(x)=\dfrac{\log x}{x}$

For $f'(x)=\dfrac{1-\log x}{x^2}=0$, $x=e$. Also derivative changes sign from positive to negative at $x=e$, hence it is the point of maxima.

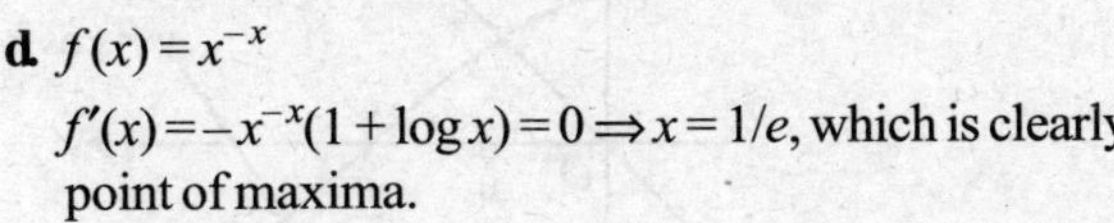

d. $f(x)=x^{-x}$

$f'(x)=-x^{-x}(1+\log x)=0 \Rightarrow x=1/e$, which is clearly point of maxima.

Also $f(x)$ increases from $(0, e)$.

3. **a → p, r; b → p, s; c → q, r; d → q, s.**

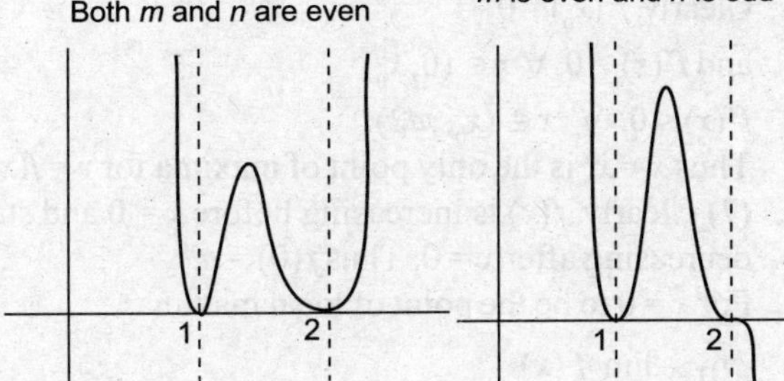

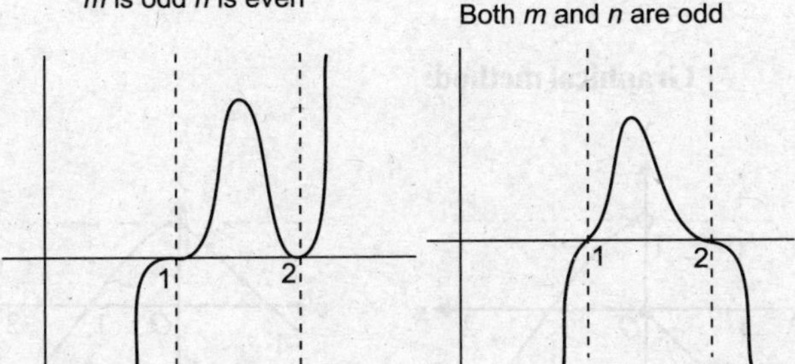

Fig. 6.43

4. **a → s; b → p; c → q; d → r.**

Since $f(x)$ is minimum at $x=-2$ and maximum at $x=2$, let $g(x)=ax^3+bx^2+cx+d$

Therefore, $g(x)$ is also minimum at $x=-2$ and maximum at $x=2$

$\therefore a<0$

Since, a is a root of $x^2-x-6=0$, i.e., $x=3, -2$

$\therefore a=-2$

Then, $g(x)=-2x^3+bx^2+cx+d$.

$\therefore g'(x)=-6x^2+2bx+c=-6(x+2)(x-2)$

(Since $g(x)$ is minimum at $x=-2$ and maximum at $x=2$)

On comparing, we get

$b=0$ and $c=24$

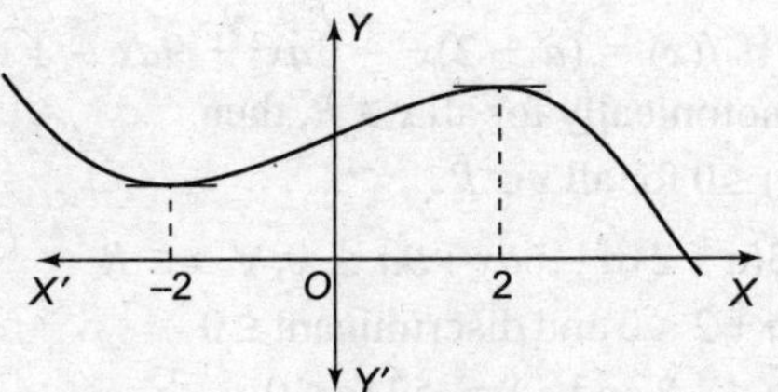

Fig. 6.44

Since minimum and maximum values are positive

$\therefore g(-2)>0 \Rightarrow 16-48+d>0 \Rightarrow d>32$

and $g(2)>0 \Rightarrow -16+48+d>0 \Rightarrow d>-32$

It is clear $d>32$

Hence, $a=-2, b=0, c=24, d>32$.

Integer Type

1. **(2)** Obviously, f is increasing and g is decreasing in (x_1, x_2).

Hence, $f[g(\alpha^2-2\alpha)]>f[g(3\alpha-4)]$ as f is increasing.

$\Rightarrow g(a^2-2\alpha)>g(3\alpha-4)$

$\Rightarrow \alpha^2-2\alpha<3\alpha-4$ as g is decreasing

$\Rightarrow \alpha^2-5\alpha+4<0$

$\Rightarrow (\alpha-1)(\alpha-4)<0$

$\Rightarrow a\in(1,4)$

2. **(2)** $f(x)=x^2+kx+1$

For $f(x)$ to be increasing,

$f'(x)>0$

$\Rightarrow \dfrac{d}{dx}\left(x^2+kx+1\right)>0$

$\Rightarrow 2x+k>0$

$\Rightarrow k>-2x$

For $x\in(1,2)$, the least value of k is -2.

3. **(0)** $f'(x)=2e^x+ae^{-x}+2a+1$

$=e^{-x}[2e^{2x}+(2a+1)e^x+a]$

$=2e^{-x}\left[e^{2x}+\left(a+\dfrac{1}{2}\right)e^x+\dfrac{a}{2}\right]$

$=2e^{-x}(e^x+a)\left(e^x+\dfrac{1}{2}\right)$

For $f(x)$ to be increasing,

$f'(x)\geq 0, \forall x\in R$

$\Rightarrow e^x+a\geq 0, \forall x\in R$

$\Rightarrow a\geq 0$

4. **(0)** Here $h(x)=g[f(x)]$, since $g(x)\in[0,\infty)$.

$h(x)\geq 0, \forall x\in$ domain

Also $h'(x)=g'[f(x)]f'(x)\leq 0$ as $g'(x)\geq 0$

and $h(x)\leq 0, \forall x\in$ domain as $h(0)=0$.

Hence, $h(x)=0, \forall x\in$ domain.

5. **(3)** If $f(x) = (a + 2)x^3 - 3ax^2 + 9ax - 1$ decreases monotonically for all $x \in R$, then
$f'(x) \le 0$ for all $x \in R$.
$\Rightarrow 3(a+2)x^2 - 6ax + 9a \le 0, \forall\ x \in R$
$\Rightarrow a + 2 < 0$ and discriminant ≤ 0
$\Rightarrow a < -2$ and $-8a^2 - 24a \le 0$
$\Rightarrow a < -2$ and $a(a+3) \ge 0$
$\Rightarrow a < -2$ and $a \le -3$ or $a \ge 0$
$\Rightarrow a \le -3$

6. **(1)** Since $f(x) = \sqrt{3}\sin x - \cos x - 2ax + b$ is decreasing for all real values of x,
$f'(x) < 0, \forall\ x$
$\Rightarrow \sqrt{3}\cos x + \sin x - 2a \le 0, \forall\ x$
$\Rightarrow \frac{\sqrt{3}}{2}\cos x + \frac{1}{2}\sin x \le a, \forall\ x$
$\Rightarrow \sin\left(x + \frac{\pi}{3}\right) \le a, \forall\ x$
$\Rightarrow a \ge 1$ $\left[\because \sin\left(x+\frac{\pi}{3}\right) \le 1\right]$

7. **(2)** $f(x) = (x^2 - 4)^n (x^2 - x + 1)$ and
$f(2) = 0$
Now $x^2 - x + 1 > 0, \forall\ x$
$f(2^+) = \lim_{x \to 2^+} (x^2 - 4)^n (x^2 - x + 1)$
$= 3 \lim_{h \to 0} [(h+2)^2 - 4]^n$
$= 3 \lim_{h \to 0} (4h + h^2)^n > 0$
$f(2^-) = \lim_{x \to 2^-} (x^2 - 4)^n (x^2 - x + 1)$
$= 3 \lim_{h \to 0} [(h-2)^2 - 4]^n$
$= 3 \lim_{h \to 0} (h^2 - 4h)^n$
$= 3 \times (\text{very small negative value})^n$
For $x = 0$ to be the point of minima, we must have $f(2^-) > 0$ for which n must be an even integer.

8. **(6)** $y = \frac{ax+b}{(x-1)(x-4)} = \frac{ax+b}{x^2 - 5x + 4}$
has a turning point at $P(2, -1)$.
$\Rightarrow P(2, -1)$ lies on the curve
$\Rightarrow 2a + b = 2$ (1)
Also $\frac{dy}{dx} = 0$ at $P(2, -1)$
Now $\frac{dy}{dx} = \frac{a(x^2 - 5x + 4) - (2x - 5)(ax + b)}{(x^2 - 5x + 4)^2}$
At $P(2, -1)$, $\frac{dy}{dx} = \frac{-2a + 2a + b}{4} = 0$
$\Rightarrow b = 0 \Rightarrow a = 1$ [From (1)]

9. **(1)** $f'(x) = \frac{1 - x^2\sec^2 x}{(1 + x\tan x)^2} = \frac{\sec^2 x(\cos x + x)(\cos x - x)}{(1 + x\tan x)^2}$

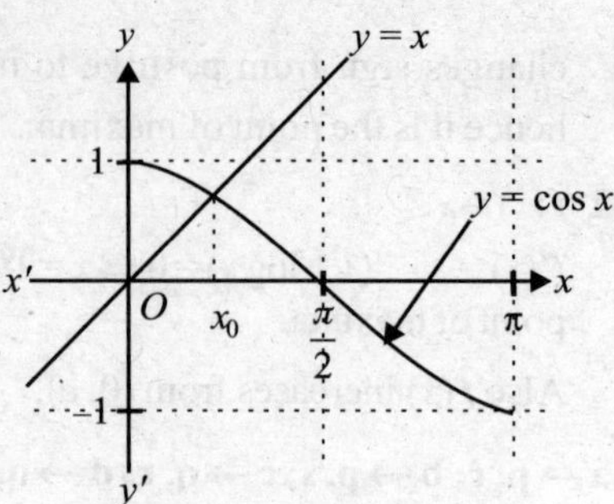

Fig. 6.45

Clearly, $f'(x_0) = 0$
and $f'(x) > 0, \forall\ x \in (0, x_0)$
$f'(x) < 0, \forall\ x \in (x_0, \pi/2)$.
Thus $x = x_0$ is the only point of maxima for $y = f(x)$.

10. **(7)** Clearly, $f(x)$ is increasing before $x = 0$ and starts decreasing after $x = 0$. Thus $f(0) = a$.
For $x = 0$ to be the point of local maxima,
$f(0) \ge \lim_{x \to 0+} f(x)$
$\Rightarrow f(0) \ge \lim_{x \to 0+} \cos\left(\frac{\pi x}{2}\right)$
$\Rightarrow a \ge 1$

Graphical method:

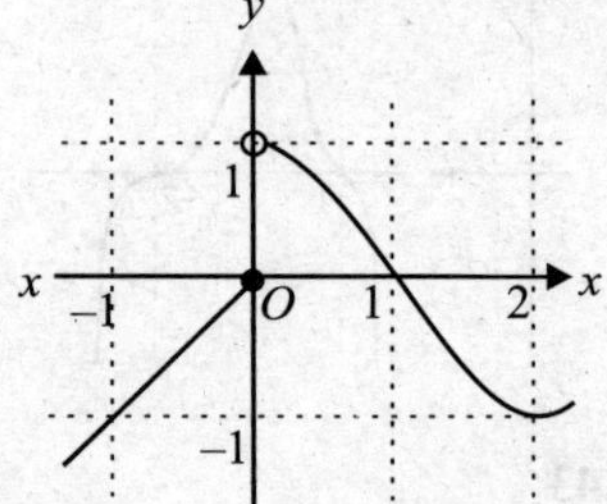

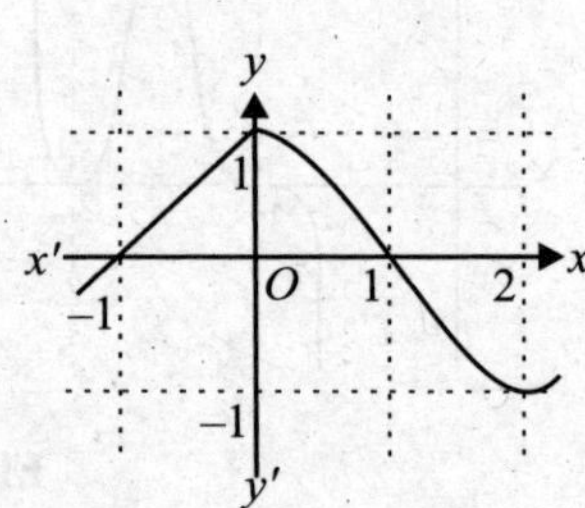

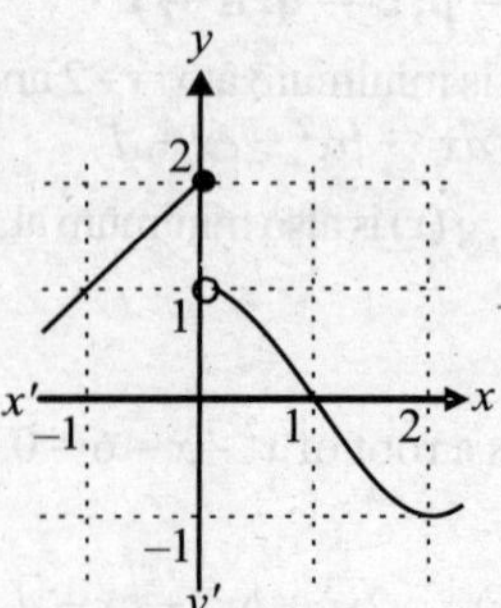

Fig. 6.46

For $a = 0$, $x = 0$ is not the point of extrema.

The graph of $y = x + a$ must shift at least one unit upward for $x = 0$ to be the point of maxima.

Hence, $a \ge 1$

11. **(2)** $f(x) = 2x + 3x^{2/3}$

$f'(x) = 2 + 3 \times \frac{2}{3}x^{-1/3} = 2(1 + x^{-1/3})$

Let $f'(x) = 0$. Therefore,
$x^{1/3} + 1 = 0 \Rightarrow x = -1$

$\Rightarrow f''(x) = -\frac{2}{3}x^{-4/3}$

and $f''(-1) = -\frac{2}{3}(-1)^{-4/3} = -\frac{2}{3} < 0$

$\Rightarrow x = -1$ is the point of maxima.
Also $f(x)$ is non-differentiable at $x = 0$.

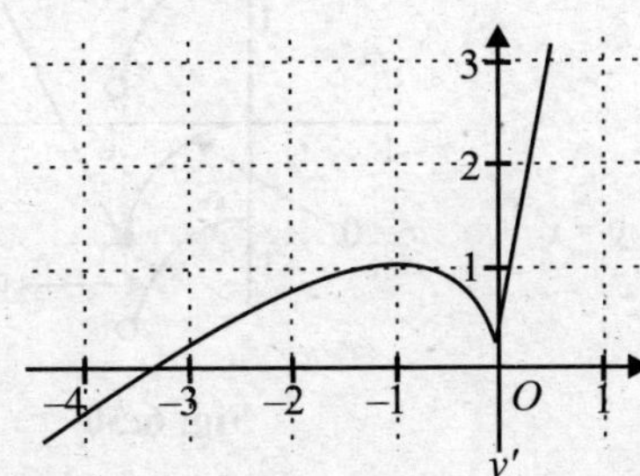

Fig. 6.47

From the graph it can be seen that $x = 0$ is the point of local minima.

12. **(3)** Let $f(x) = x^3 - 3x + a$
Let $f'(x) = 0$. Therefore,
$3x^2 - 3 = 0 \Rightarrow x = \pm 1$.
For three distinct roots,
$f(1) f(-1) < 0$
$\Rightarrow (1 - 3 + a)(-1 + 3 + a) < 0$
$\Rightarrow (a + 2)(a - 2) < 0$
$\Rightarrow -2 < a < 2$

13. **(7)**

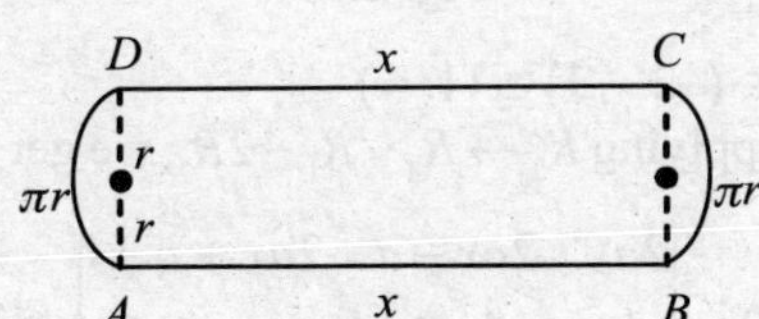

Fig. 6.48

Perimeter = 440 *ft*
$2x + \pi r + \pi r = 440$ or $2x + 2\pi r = 440$
A = Area of rectangular portion = $x\,2r$

$A = x\frac{(440 - 2x)}{\pi} = \frac{1}{\pi}(440x - 2x^2)$

Let $\frac{dA}{dx} = \frac{1}{\pi}(440 - 4x) = 0$

$\Rightarrow x = 110$ for which $(d^2A)/(dx^2)$ is negative
$\Rightarrow A$ is maximum when $x = 110$

$\Rightarrow 2r = \frac{440 - 2x}{\pi} = \frac{440 - 220}{22/7} = 70$

14. **(2)**

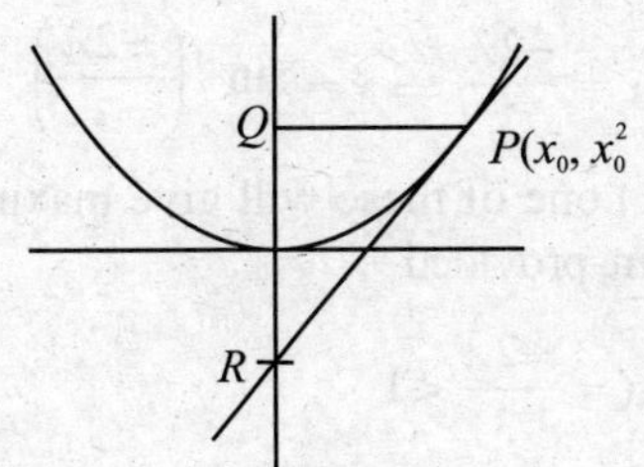

Fig. 6.49

$y = x^2$

$\Rightarrow \frac{dy}{dx} = 2x$

$\Rightarrow$ Equation of the tangent at (x_0, x_0^2) is $y - x_0^2 = 2x_0(x - x_0)$.
It meets the y-axis at $R(0, -x_0^2)$. Q is $(0, x_0^2)$.
$\Rightarrow Z$ = area of the triangle PQR

$= \frac{1}{2} \times 2x_0^2 \times x_0 = x_0^3,\ 1 \le x_0 \le 2$

$\frac{dZ}{dx_0} = 3x_0^2 > 0$ in $1 \le x_0 \le 2$.

$\Rightarrow Z$ is an increasing function in [1, 2].
Hence Z, i.e., the area of ΔPQR is greatest at $x_0 = 2$.

15. **(4)** Let (r, θ) be the polar co-ordinates of any point P on the curve where r is the distance of the point from origin
$\Rightarrow r^2\,[5(\cos^2\theta + \sin^2\theta) - 8\sin\theta\cos\theta] = 4$

$\Rightarrow r^2 = \frac{4}{5 - 4\sin 2\theta}$

r^2 is maximum when $5 - 4\sin 2\theta$ is minimum $= 5 - 4 = 1$ when
$\sin 2\theta = 1$
$\Rightarrow 2\theta = 90° \Rightarrow \theta = 45° \Rightarrow r = \pm 2, \theta = 45°$ (1)
Again r^2 is minimum when $5 - 4\sin 2\theta$ is maximum $= 5 + 4 = 9$ when

$\sin 2\theta = -1 \Rightarrow 2\theta = \frac{3\pi}{2} \Rightarrow \theta = \frac{3\pi}{4}$

$\Rightarrow r = \pm\frac{2}{3}, \theta = \frac{3\pi}{4}$ (2)

Hence, the points are $(r\cos\theta, r\sin\theta)$ where r, θ are given by (1) and (2).

Thus we get the four points as $(\sqrt{2}, \sqrt{2})$, $(-\sqrt{2}, -\sqrt{2})$, $(\sqrt{2}/3, -\sqrt{2}/3)$, and $(-\sqrt{2}/3, \sqrt{2}/3)$.

16. **(2)** $f(x) = \sin^3 x + \lambda \sin^2 x$
$f'x) = 3\sin^2 x(\cos x) + \lambda\, 2\sin x(\cos x)$
$= \sin x \cos x(3\sin x + 2\lambda)$
For extremum, let $f'(x) = 0$. Therefore,
$\sin x = 0$, $\cos x = 0$, $\sin x = (-2\lambda)/3$.
Since $-\pi/2 < x < \pi/2$, $\cos x \ne 0$

$\sin x = 0 \Rightarrow x = 0$

and $\sin x = \dfrac{-2\lambda}{3} \Rightarrow x = \sin^{-1}\left(\dfrac{-2\lambda}{3}\right)$ (1)

From (1) one of these will give maximum and one minimum, provided

$-1 < \sin x = \dfrac{-2\lambda}{3} < 1$

i.e., $-1 < \dfrac{-2\lambda}{3} < 1$

$\Rightarrow -3 < -2\lambda < 3$

$\Rightarrow -3 < 2\lambda < 3$

i.e., $-3/2 < \lambda < 3/2$

But if $\lambda = 0$, then $\sin x = 0$ has only one solution. Therefore,

$\lambda \in (-3/2, 3/2) - \{0\}$

$\Rightarrow \lambda \in (-3/2, 0) \cup (0, 3/2)$

For this value of λ, there are two distinct solutions. Since $f(x)$ is continuous, these solutions give one maximum and one minimum because for a continuous function, there must be one minimum between two maxima and vice versa.

17. **(2)** Let $f(x) = ax^3 + bx^2 + cx + d$

$f(x)$ vanishes at $x = -2$

$\Rightarrow -8a + 4b - 2c + d = 0$ (1)

and $f'(x) = 3ax^2 + 2bx + c$

Also $f(x)$ has relative maximum/minimum at $x = -1$ and $x = 1/3$.

$\Rightarrow f'(-1) = 0 = f'(1/3)$

$\Rightarrow a + 2b + 3c = 0$ (2)

and $a + 2b + 3c = 0$ (3)

Also $\displaystyle\int_{-1}^{1} f(x)dx = \frac{14}{3}$

$\Rightarrow \dfrac{14}{3}\left(\dfrac{ax^4}{4} + \dfrac{bx^3}{3} + \dfrac{cx^2}{2} + dx\right)_{-1}^{1} = \dfrac{14}{3}$

$\Rightarrow \left(\dfrac{a}{4} + \dfrac{b}{3} + \dfrac{c}{2} + d\right) - \left(\dfrac{a}{4} - \dfrac{b}{3} + \dfrac{c}{2} - d\right) = \dfrac{14}{3}$

$\Rightarrow \dfrac{b}{3} + d = \dfrac{7}{3}$

$\Rightarrow b + 3d = 7$ (4)

From (1), (2), (3), and (4), we get on solving $a = 1, b = 1, c = -1, d = 2$.

Therefore, the required cubic is $x^3 + x^2 - x + 2$.

18. **(2)** We have

$$f(x) = \begin{cases} -x^3 + \dfrac{(b^3 - b^2 + b - 1)}{(b^2 + 3b + 2)}, & 0 \le x < 1 \\ 2x - 3, & 1 \le x \le 3 \end{cases}$$

Let $\dfrac{(b^3 - b^2 + b - 1)}{(b^2 + 3b + 2)} = a$ (constant).

Let $a = 0$, then $f(x) = \begin{cases} -x^3, & 0 \le x < 1 \\ 2x - 3, & 1 \le x \le 3 \end{cases}$.

The graph is as follows

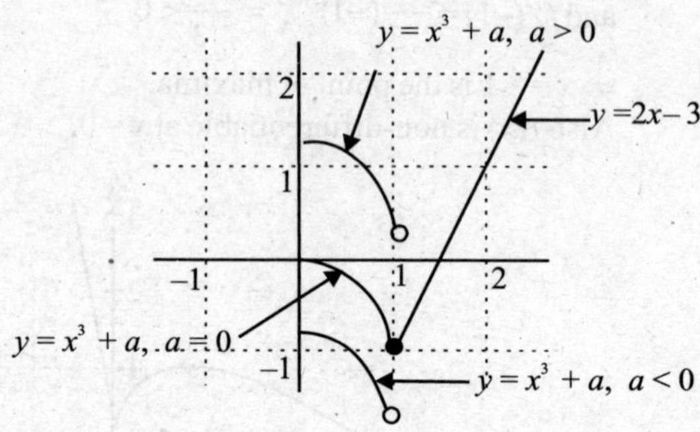

Fig. 6.50

If $a > 0$, the graph of $(-x^3 + a)$ shifts upward.

If $a < 0$, the graph of $(-x^3 + a)$ shifts downward.

For the point of minima, $a > 0$

$\Rightarrow \dfrac{b^3 - b^2 + b - 1}{b^2 + 3b + 2} \ge 0$

$\Rightarrow \dfrac{(b^2 + 1)(b - 1)}{(b + 2)(b + 1)} \ge 0$

$\Rightarrow (b - 1)(b + 1)(b + 2) \ge 0$

$\Rightarrow$ (sign scheme: − on $(-\infty,-2)$, + on $(-2,-1)$, − on $(-1,1)$, + on $(1,\infty)$; points −2, −1, 1)

Fig. 6.51

$\Rightarrow b \in (-2, -1) \cup (1, \infty)$

19. **(2)** Applying $R_3 \to R_3 - R_1 - 2R_2$, we get

$$f'(x) = \begin{vmatrix} 2ax & 2ax - a & 2ax + b + 1 \\ b & b + 1 & -1 \\ 0 & 0 & 1 \end{vmatrix}$$

$= \begin{vmatrix} 2ax & 2ax - 1 \\ b & b + 1 \end{vmatrix} = \begin{vmatrix} 2ax & -1 \\ b & 1 \end{vmatrix}$ (Using $C_2 \to C_2 - C_1$)

$\Rightarrow f'(x) = 2ax + b$

Integrating, we get

$f(x) = ax^2 + bx + c$, where c is an arbitrary constant.

Since f has a maximum at $x = 5/2$,

$f'(5/2) = 0 \Rightarrow 5a + b = 0$ (1)

Also $f(0) = 2 \Rightarrow c = 2$

And $f(1) = 1 \Rightarrow a + b + c = 1$. Therefore,

$a + b = -1$ (2)

Solving (1) and (2) for a, b, we get
$a = 1/4, b = -5/4$

Thus $f(x) = \frac{1}{4}x^2 - \frac{5}{4}x + 2$.

20. (8) Let $p(x) = ax^3 + bx^2 + cx + d$

$p(-1) = 10 \Rightarrow -a + b - c + d = 10$ (1)

$p(1) = -6 \Rightarrow a + b + c + d = -6$ (2)

$p(x)$ has a maximum at $x = -1$

$\Rightarrow p'(-1) = 0$

$\Rightarrow 3a - 2b + c = 0$ (3)

$p'(x)$ has a minimum at $x = 1$

$\Rightarrow p''(1) = 0$

$\Rightarrow 6a + 2b = 0$ (4)

Solving (1), (2), (3), and (4), we get

from (4), $b = -3a$

from (3), $3a + 6a + c = 0 \Rightarrow c = -9a$

from (2) $a - 3a - 9a + d = -6 \Rightarrow d = 11a - 6$

from (1), $-a - 3a + 9a + 11a - 6 = 10$

$\Rightarrow 16a = 16 \Rightarrow a = 1$

$\Rightarrow b = -3, c = -9, d = 5$

$\Rightarrow p(x) = x^3 - 3x^2 - 9x + 5$

So $p'(x) = 0$

$\Rightarrow 3x^2 - 6x - 9 = 0$

$\Rightarrow 3(x + 1)(x - 3) = 0$

$\Rightarrow x = -1$ is a point of maxima (given)

and $x = 3$ is a point of minima.

($\because$ maximum and minimum occur alternatively)

Points of local maximum is $(-1, 10)$ and local minimum is $(3, -22)$.

And the distance between them is equal to

$$\sqrt{[3 - (1)]^2 + (-22 - 10)^2}$$

$$= \sqrt{16 + 1024}$$

$$= \sqrt{1040} = 4\sqrt{65}$$

21. (4) $f(x) = \int_{-1}^{x} t(e^t - 1)(t - 1)(e - 2)^3(e - 3)^5\, dt$

$\Rightarrow f'(x) = x(e^x - 1)(x - 1)(x - 2)^3(x - 3)^5$

Critical points are 0, 1, 2, 3.

Sign scheme of $f'(x)$ is

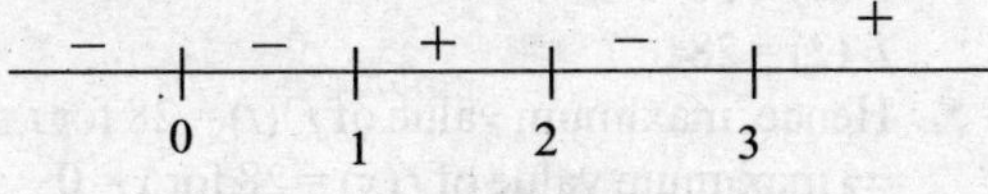

Fig. 6.52

Clearly, $x = 1$ and $x = 3$ are the points of minima.

22. (1) We have

$$f(x) = \frac{x^p}{p} + \frac{x^{-q}}{q}$$

$$\Rightarrow f'(x) = x^{p-1} - x^{-q-1} = x^{p-1} - \frac{1}{x^{q+1}} = \frac{x^{p+q} - 1}{x^{q+1}}$$

$f'(x) = 0$ gives $x^{p+q} = 1$

$\Rightarrow x = 1$

When $x < 1, f'(x) < 0$, and when $x > 1, f'(x) > 0$.

Hence, $f'(x)$ changes sign from negative to positive as x passes through the value $x = 1$ and hence, $f(x)$ has a minimum value at $x = 1$.

23. (3)

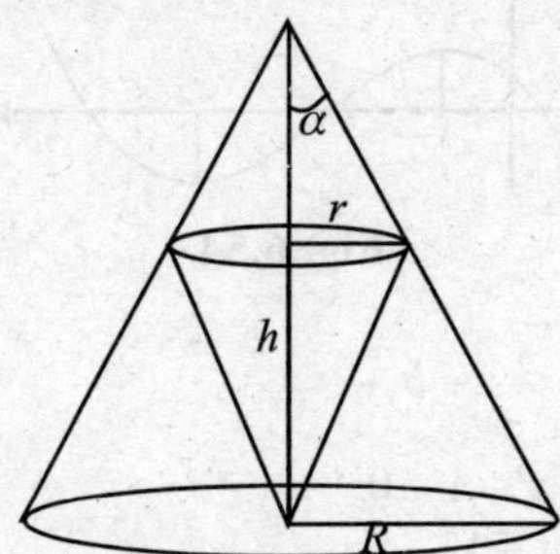

Fig. 6.53

$$\frac{r}{R} = \frac{H - h}{H}$$

$$r = \frac{R(H - h)}{H}$$

$$\text{Volume} = \frac{1}{3}\pi \frac{R^2(H - h)^2}{H^2} \times h$$

$V = \frac{\pi R^2}{3H^2}(H - h)^2 h$. Therefore,

$$\frac{dv}{dh} = \frac{\pi R^2}{3H^2}\left[(H - h)^2 - 2h(H - h)\right]$$

$$= \frac{\pi R^2}{3H^2}(H - h)(H - h - 2h)$$

So $\frac{dV}{dh} = 0$ if $h = \frac{H}{3}$ and

$h = \frac{H}{3}$ is a point of maximum.

Volume is maximum when $h = \frac{H}{3}$.

24. (1) $f(x)$

$$= \begin{cases} |x^3 + x^2 + 3x + \sin x|\,[3 + \sin(1/x)], & x \neq 0 \\ 0, & x = 0 \end{cases}$$

Let $g(x) = x^3 + x^2 + 3x + \sin x$. Therefore,

$f'(x) = 3x^2 + 2x + 3 + \cos x$

$$= 3\left(x^2 + \frac{2x}{3} + 1\right) + \cos x$$

$$= 3\left[\left(x + \frac{1}{3}\right)^2 + \frac{8}{9}\right] + \cos x > 0$$

and $2 < 3 + \sin\left(\frac{1}{x}\right) < 4$.

Hence, the minimum value of $f(x)$ is 0 at $x = 0$.
Hence, the number of points is 1.

25. (7)

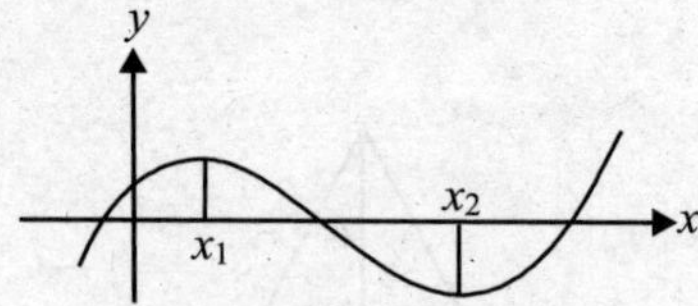

Fig. 6.54

$f(x) = \frac{x^3}{3} - x - b$

$\Rightarrow f'(x) = x^2 - 1 = 0$

$\Rightarrow x = 1$ or -1

For three distinct roots,
$f(x_1) f(x_2) < 0$, where x_1 and x_2 are the roots of $f'(x) = 0$

$\Rightarrow \left(\frac{1}{3} - 1 - b\right)\left(-\frac{1}{3} + 1 - b\right) < 0$

$\Rightarrow \left(b + \frac{2}{3}\right)\left(b - \frac{2}{3}\right) < 0$

$\Rightarrow b \in \left(-\frac{2}{3}, \frac{2}{3}\right)$

26. (1) $f(x) = x^2 + ax + b$

Vertex of $f(x)$ is at $-\frac{a}{2}$

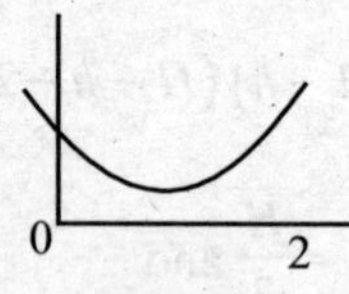

Fig. 6.55

If $0 \le -\frac{a}{2} \le 2$

i.e. $0 \le -a \le 4$; $-4 \le a \le 0$

$f(x)]_{\min} = f\left(-\frac{a}{2}\right) = \frac{a^2}{4} - \frac{a^2}{2} + b = b - \frac{a^2}{4} = 2$ (1)

and $f(x)]_{\max} = f(0)$ or $f(2)$
but $f(0) = b$ and $f(2) = 4 + 2a + b$
Let $b = 3$

From (1), $3 - \frac{a^2}{4} = 2 \Rightarrow \frac{a^2}{4} = 1 \Rightarrow a = 2$ or -2

but $a = 2$ is rejected, hence $a = -2$
If $f(2)$ is maximum
$4 + 2a + b = 3 \Rightarrow 2a + b = -1$ (2)

(2) – (1) gives $2a + \frac{a^2}{4} = -3 \Rightarrow a^2 + 8a + 12 = 0$,

$(a + 6)(a + 2) = 0$
$a = -2$ or $a = -6$ (rejected)
For vertex < 0 or vertex >2, there are no accepted ordered pairs.

27. (4) $f(x) = \left(3 - \sqrt{4 - x^2}\right)^2 + \left(1 + \sqrt{4 - x^2}\right)^2$

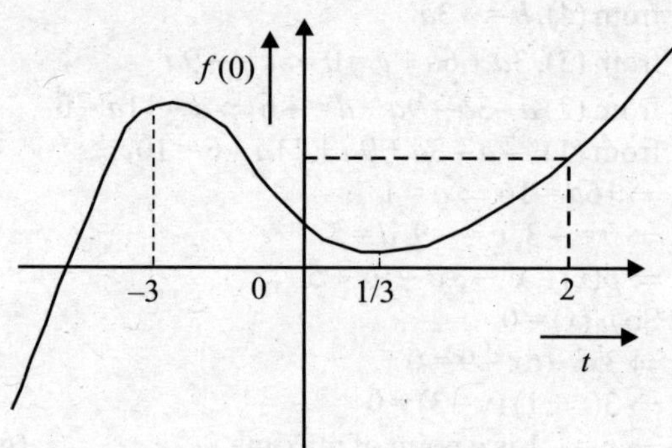

Fig. 6.56

Let $t = \sqrt{4 - x^2}$, clearly $0 \le t \le 2$
$\Rightarrow F(t) = (3 - t)^2 + (1 + t)^3$
For maxima and minima,
$\Rightarrow F'(t) = 0$
$\Rightarrow -2(3 - t) + 3(1 + t)^2 = 0$
$\Rightarrow 3t^2 + 8t - 3 = 0$
$\Rightarrow (3t - 1)(t + 3) = 0$

$\Rightarrow t = -3, \frac{1}{3}$

Also $F''(t)|_{t=-3} = -10 \Rightarrow$ maxima
And $F''(t)|_{t=1/3} = -10 \Rightarrow$ minima
As $t \ne -3$, hence maximum value of $F(t)$ will occur at the end points for which
$F(0) = 10$
$F(2) = 28$
Hence, maximum value of $F(t) = 28$ for $t = 2$
$\Rightarrow$ maximum value of $f(x) = 28$ for $x = 0$

CHAPTER 7

Indefinite Integration

Objective Type

Each question has four choices a, b, c, and d, out of which *only one* is correct.

1. $\int \frac{\sin 2x}{\sin 5x \sin 3x} dx$ is equal to

 a. $\log \sin 3x - \log \sin 5x + c$

 b. $\frac{1}{3}\log \sin 3x + \frac{1}{5}\log \sin 5x + c$

 c. $\frac{1}{3}\log \sin 3x - \frac{1}{5}\log \sin 5x + c$

 d. $3 \log \sin 3x - 5 \log \sin 5x + c$

2. If $I = \int \sqrt{1 + \sin x}\, dx$, then I is equal to

 a. $-2\sqrt{1 - \sin x} + C$

 b. $\sin(x/2) + \cos(x/2) + C$

 c. $\cos(x/2) - \sin(x/2) + C$

 d. $2\sqrt{1 - \sin x} + C$

3. $\int \frac{\sin^8 x - \cos^8 x}{1 - 2\sin^2 x \cos^2 x} dx$ is equal to

 a. $\frac{1}{2}\sin 2x$ b. $-\frac{1}{2}\sin 2x$

 c. $-\frac{1}{2}\sin x$ d. $-\sin^2 x$

4. If $\int \frac{\cos 4x + 1}{\cot x - \tan x} dx = A\cos 4x + B$, then

 a. $A = -1/2$ b. $A = -1/8$

 c. $A = -1/4$ d. None of these

5. The primitive of the function $x\,|\cos x|$ when $\frac{\pi}{2} < x < \pi$ is given by

 a. $\cos x + x \sin x$ b. $-\cos x - x \sin x$

 c. $x \sin x - \cos x$ d. None of these

6. $\int \frac{dx}{x\left(x^n + 1\right)}$ is equal to

 a. $\frac{1}{n}\log\left(\frac{x^n}{x^n + 1}\right) + c$ b. $\frac{1}{n}\log\left(\frac{x^n + 1}{x^n}\right) + c$

 c. $\log\left(\frac{x^n}{x^n + 1}\right) + c$ d. None of these

7. $\int 4\sin x \cos\frac{x}{2} \cos\frac{3x}{2} dx$ is equal to

 a. $\cos x + \frac{1}{2}\cos 2x - \frac{1}{3}\cos 3x + C$

 b. $\cos x - \frac{1}{2}\cos 2x - \frac{1}{3}\cos 3x + C$

 c. $\cos x + \frac{1}{2}\cos 2x + \frac{1}{3}\cos 3x + C$

 d. $\cos x - \frac{1}{2}\cos 2x + \frac{1}{3}\cos 3x + C$

8. Let $x = f''(t)\cos t + f'(t)\sin t$ and $y = -f''(t)\sin t + f'(t)\cos t$. Then $\int \left[\left(\frac{dx}{dt}\right)^2 + \left(\frac{dy}{dt}\right)^2\right]^{1/2} dt$ equals

a. $f'(t)+f''(t)+c$ b. $f''(t)+f'''(t)+c$
c. $f(t)+f''(t)+c$ d. $f'(t)-f''(t)+c$

9. $\int \frac{1}{\sqrt{\sin^3 x \sin(x+\alpha)}} dx, \alpha \neq n\pi, n \in Z$ is equal to

a. $-2 \operatorname{cosec} \alpha (\cos \alpha - \tan x \sin \alpha)^{1/2} + C$
b. $-2(\cos \alpha + \cot x \sin \alpha)^{1/2} + C$
c. $-2 \operatorname{cosec} \alpha (\cos \alpha + \cot x \sin \alpha)^{1/2} + C$
d. $-2 \operatorname{cosec} \alpha (\sin \alpha + \cot x \cos \alpha)^{1/2} + C$

10. $\int \frac{px^{p+2q-1} - qx^{q-1}}{x^{2p+2q} + 2x^{p+q} + 1} dx$ is equal to

a. $-\frac{x^p}{x^{p+q}+1} + C$ b. $\frac{x^q}{x^{p+q}+1} + C$
c. $-\frac{x^q}{x^{p+q}+1} + C$ d. $\frac{x^p}{x^{p+q}+1} + C$

11. If $I_n = \int (\ln x)^n \, dx$, then $I_n + nI_{n-1}$

a. $\frac{(\ln x)^n}{x}$ b. $x(\ln x)^{n-1}$
c. $x(\ln x)^n$ d. None of these

12. $\int \frac{x+2}{(x^2+3x+3)\sqrt{x+1}} dx$ is equal to

a. $\frac{1}{\sqrt{3}} \tan^{-1}\left(\frac{x}{\sqrt{3(x+1)}}\right)$

b. $\frac{2}{\sqrt{3}} \tan^{-1}\left(\frac{x}{\sqrt{3(x+1)}}\right)$

c. $\frac{2}{\sqrt{3}} \tan^{-1}\left(\frac{x}{\sqrt{x+1}}\right)$

d. None of these

13. $\int \frac{\sin 2x}{\sin^4 x + \cos^4 x} dx$ is equal to

a. $\cot^{-1}(\tan^2 x) + c$ b. $\tan^{-1}(\tan^2 x) + c$
c. $\cot^{-1}(\cot^2 x) + c$ d. $\tan^{-1}(\cot^2 x) + c$

14. $\int \frac{\sec x \, dx}{\sqrt{\sin(2x+A) + \sin A}}$ is equal to

a. $\frac{\sec A}{\sqrt{2}} \sqrt{\tan x \cos A - \sin A} + c$

b. $\sqrt{2} \sec A \sqrt{\tan x \cos A - \sin A} + c$

c. $\sqrt{2} \sec A \sqrt{\tan x \cos A + \sin A} + c$

d. None of these

15. If $\int \sqrt{1+\sin x} \, f(x) \, dx = \frac{2}{3}(1+\sin x)^{3/2} + c$, then $f(x)$ equals

a. $\cos x$ b. $\sin x$
c. $\tan x$ d. 1

16. Let $\int e^x \{f(x) - f'(x)\} dx = \phi(x)$. Then $\int e^x f(x) dx$ is

a. $\phi(x) = e^x f(x)$
b. $\phi(x) - e^x f(x)$
c. $\frac{1}{2}\{\phi(x) + e^x f(x)\}$
d. $\frac{1}{2}\{\phi(x) + e^x f'(x)\}$

17. Let $f(x) = \int \frac{x^2 dx}{(1+x^2)(1+\sqrt{1+x^2})}$ and $f(0) = 0$, then the value of $f(1)$ be

a. $\log(1+\sqrt{2})$ b. $\log(1+\sqrt{2}) - \frac{\pi}{4}$
c. $\log(1+\sqrt{2}) + \frac{\pi}{2}$ d. None of these

18. If $y = \int \frac{dx}{(1+x^2)^{3/2}}$ and $y = 0$ when $x = 0$, find the value of y when $x = 1$ is

a. $\frac{1}{\sqrt{2}}$ b. $\sqrt{2}$
c. $2\sqrt{2}$ d. None of these

19. $\int \sqrt{x}\,(1+x^{1/3})^4 \, dx$ is equal to

a. $2\left\{x^{2/3} + \frac{4}{11}x^{11/6} + \frac{6}{13}x^{13/6} + \frac{4}{15}x^{5/2} + \frac{1}{17}x^{17/6}\right\} + c$

b. $6\left\{x^{2/3} - \frac{4}{11}x^{11/6} + \frac{6}{13}x^{13/6} - \frac{4}{15}x^{5/2} + \frac{1}{17}x^{17/6}\right\} + c$

c. $6\left\{x^{2/3} + \frac{4}{11}x^{11/6} + \frac{6}{13}x^{13/6} + \frac{4}{15}x^{5/2} + \frac{1}{17}x^{17/6}\right\} + c$

d. None of these

20. If $\int x^5 (1+x^3)^{2/3} dx = A(1+x^3)^{8/3} + B(1+x^3)^{5/3} + c$, then

a. $A=\frac{1}{4}, B=\frac{1}{5}$ b. $A=\frac{1}{8}, B=-\frac{1}{5}$

c. $A=-\frac{1}{8}, B=\frac{1}{5}$ d. None of these

21. The value of the integral $\int \frac{(1-\cos\theta)^{2/7}}{(1+\cos\theta)^{9/7}} d\theta$ is

a. $\frac{7}{11}\left(\tan\frac{\theta}{2}\right)^{11/7} + C$

b. $\frac{7}{11}\left(\cos\frac{\theta}{2}\right)^{11/7} + C$

c. $\frac{7}{11}\left(\sin\frac{\theta}{2}\right)^{11/7} + C$

d. None of these

22. If $\int \frac{1-x^7}{x(1+x^7)} dx = a\ln|x| + b\ln|x^7+1| + c$, then

a. $a=1, b=\frac{2}{7}$ b. $a=-1, b=\frac{2}{7}$

c. $a=1, b=-\frac{2}{7}$ d. $a=-1, b=-\frac{2}{7}$

23. $\int \sin^{-1}\left(\frac{2x}{1+x^2}\right) dx$ is equal to

a. $x\tan^{-1}x - \ln|\sec(\tan^{-1}x)| + c$

b. $x\tan^{-1}x + \ln|\sec(\tan^{-1}x)| + c$

c. $x\tan^{-1}x - \ln|\cos(\tan^{-1}x)| + c$

d. None of these

24. $\int \frac{\ln(\tan x)}{\sin x\cos x} dx$ is equal to

a. $\frac{1}{2}\ln(\tan x) + c$ b. $\frac{1}{2}\ln(\tan^2 x) + c$

c. $\frac{1}{2}(\ln(\tan x))^2 + c$ d. None of these

25. $\int \frac{2\sin x}{(3+\sin 2x)} dx$ is equal to

a. $\frac{1}{2}\ln\left|\frac{2+\sin x-\cos x}{2-\sin x+\cos x}\right| - \frac{1}{\sqrt{2}}\tan^{-1}\left(\frac{\sin x+\cos x}{\sqrt{2}}\right) + c$

b. $\frac{1}{2}\ln\left|\frac{2+\sin x-\cos x}{2-\sin x+\cos x}\right| - \frac{1}{2\sqrt{2}}\tan^{-1}\left(\frac{\sin x+\cos x}{\sqrt{2}}\right) + c$

c. $\frac{1}{4}\ln\left|\frac{2+\sin x-\cos x}{2-\sin x+\cos x}\right| - \frac{1}{\sqrt{2}}\tan^{-1}\left(\frac{\sin x+\cos x}{\sqrt{2}}\right) + c$

d. None of these

26 $\int \frac{x^9\, dx}{(x^2+4)^6}$ is equal to

a. $\frac{1}{5x}\left(4+\frac{1}{x^2}\right)^{-5} + c$

b. $\frac{1}{5}\left(4+\frac{1}{x^2}\right)^{-5} + c$

c. $\frac{1}{10x}\left(1+4x^2\right)^{-5} + c$

d. $\frac{1}{10x}\left(4+\frac{1}{x^2}\right)^{-5} + c$

27. $\int e^{\tan^{-1}x}(1+x+x^2)\, d(\cot^{-1}x)$ is equal to

a. $-e^{\tan^{-1}x} + c$ b. $e^{\tan^{-1}x} + c$

c. $-x\, e^{\tan^{-1}x} + c$ d. $x\, e^{\tan^{-1}x} + c$

28. If $\int \frac{dx}{\sqrt{\sin^3 x\cos^5 x}} = a\sqrt{\cot x} + b\sqrt{\tan^3 x} + c$, then

a. $a=-1, b=1/3$ b. $a=-3, b=2/3$

c. $a=-2, b=4/3$ d. None of these

29. $\int \frac{\cos 4x - 1}{\cot x - \tan x} dx$ is equal to

a. $\frac{1}{2}\ln|\sec 2x| - \frac{1}{4}\cos^2 2x + c$

b. $\frac{1}{2}\ln|\sec 2x| + \frac{1}{4}\cos^2 x + c$

c. $\frac{1}{2}\ln|\cos 2x| - \frac{1}{4}\cos^2 2x + c$

d. $\frac{1}{2}\ln|\cos 2x| + \frac{1}{4}\cos^2 x + c$

30. If $\int \frac{1}{x\sqrt{1-x^3}} dx = a\log\left|\frac{\sqrt{1-x^2}-1}{\sqrt{1-x^2}+1}\right| + b$, then a is equal to

a. 1/3 b. 2/3

c. –1/3 d. –2/3

31. If $\int \frac{dx}{x^2(x^n+1)^{(n-1)/n}} = -[f(x)]^{1/n} + c$, then $f(x)$ is

a. $(1+x^n)$ b. $1+x^{-n}$

c. x^n+x^{-n} d. None of these

32. $\int \frac{\sqrt{x-1}}{x\sqrt{x+1}}\,dx$ is equal to

a. $\ln|x-\sqrt{x^2-1}|-\tan^{-1}x+c$

b. $\ln|x+\sqrt{x^2-1}|-\tan^{-1}x+c$

c. $\ln|x-\sqrt{x^2-1}|-\sec^{-1}x+c$

d. $\ln|x+\sqrt{x^2-1}|-\sec^{-1}x+c$

33. If $I=\int \frac{dx}{(2ax+x^2)^{3/2}}$, then I is equal to

a. $-\frac{x+a}{\sqrt{2ax+x^2}}+c$

b. $-\frac{1}{a}\frac{x+a}{\sqrt{2ax+x^2}}+c$

c. $-\frac{1}{a^2}\frac{x+a}{\sqrt{2ax+x^2}}+c$

d. $-\frac{1}{a^3}\frac{x+a}{\sqrt{2ax+x^3}}+c$

34. If $f'(x)=\frac{1}{-x+\sqrt{x^2+1}}$ and $f(0)=-\frac{1+\sqrt{2}}{2}$, then $f(1)$ is equal to

a. $-\log(\sqrt{2}+1)$ b. 1

c. $1+\sqrt{2}$ d. $\log(1+\sqrt{2})$

35. $\int e^x\left(\frac{2\tan x}{1+\tan x}+\cot^2\left(x+\frac{\pi}{4}\right)\right)dx$ is equal to

a. $e^x\tan\left(\frac{\pi}{4}-x\right)+c$ b. $e^x\tan\left(x-\frac{\pi}{4}\right)+c$

c. $e^x\tan\left(\frac{3\pi}{4}-x\right)+c$ d. None of these

36. The value of the integral $\int (x^2+x)(x^{-8}+2x^{-9})^{1/10}\,dx$ is

a. $\frac{5}{11}(x^2+2x)^{11/10}+c$ b. $\frac{5}{6}(x+1)^{11/10}+c$

c. $\frac{6}{7}(x+1)^{11/10}+c$ d. None of these

37. If $\int \frac{dx}{(x+2)(x^2+1)}=a\ln(1+x^2)+b\tan^{-1}x+\frac{1}{5}\ln|x+2|+C$, then

a. $a=-\frac{1}{10},\ b=-\frac{2}{5}$ b. $a=\frac{1}{10},\ b=-\frac{2}{5}$

c. $a=-\frac{1}{10},\ b=\frac{2}{5}$ d. $a=\frac{1}{10},\ b=\frac{2}{5}$

38. If $\int \frac{3\sin x+2\cos x}{3\cos x+2\sin x}dx=ax+b\ln|2\sin x+3\cos x|+C$, then

a. $a=-\frac{12}{13},\ b=\frac{15}{39}$

b. $a=-\frac{7}{13},\ b=\frac{6}{13}$

c. $a=\frac{12}{13},\ b=-\frac{15}{39}$

d. $a=-\frac{7}{13},\ b=-\frac{6}{13}$

39. $\int \frac{3e^x-5e^{-x}}{4e^x+5e^{-x}}dx=ax+b\ln(4e^x+5e^{-x})+C$, then

a. $a=-\frac{1}{8},\ b=\frac{7}{8}$ b. $a=\frac{1}{8},\ b=\frac{7}{8}$

c. $a=-\frac{1}{8},\ b=-\frac{7}{8}$ d. $a=\frac{1}{8},\ b=-\frac{7}{8}$

40. $\int \sqrt{\frac{\cos x-\cos^3 x}{1-\cos^3 x}}\,dx$ is equal to

a. $\frac{2}{3}\sin^{-1}(\cos^{3/2}x)+C$

b. $\frac{3}{2}\sin^{-1}(\cos^{3/2}x)+C$

c. $\frac{2}{3}\cos^{-1}(\cos^{3/2}x)+C$

d. None of these

41. If $l^r(x)$ means log log log … x, the log being repeated r times, then $\int [x\,l(x)l^2(x)l^3(x)\ldots l^r(x)]^{-1}dx$ is equal to

a. $l^{r+1}(x)+C$ b. $\frac{l^{r+1}(x)}{r+1}+C$

c. $l^r(x)+C$ d. None of these

42. If $I=\int\left(\sqrt{\cot x}-\sqrt{\tan x}\right)dx$, then I equals

a. $\sqrt{2}\log\left(\sqrt{\tan x}-\sqrt{\cot x}\right)+C$

b. $\sqrt{2}\log|\sin x+\cos x+\sqrt{\sin 2x}|+C$

b. $\sqrt{2}\log|\sin x-\cos x+\sqrt{2}\sin x\cos x|+C$

d. $\sqrt{2}\log|\sin(x+\pi/4)+\sqrt{2}\sin x\cos x|+C$

43. If $I=\int\frac{dx}{\left(a^2-b^2x^2\right)^{3/2}}$, then I equals

a. $\frac{x}{\sqrt{a^2-b^2x^2}}+C$

b. $\frac{x}{a^2\sqrt{a^2-b^2x^2}}+C$

c. $\frac{ax}{\sqrt{a^2-b^2x^2}}+C$

d. None of these

44. $\int e^{x^4}(x+x^3+2x^5)e^{x^2}dx$ is equal to

a. $\frac{1}{2}xe^{x^2}e^{x^4}+c$ b. $\frac{1}{2}x^2e^{x^4}+c$

c. $\frac{1}{2}e^{x^2}e^{x^4}+c$ d. $\frac{1}{2}x^2e^{x^2}e^{x^4}+c$

45. $\int x\left(\frac{\ln a^{a^{x/2}}}{3a^{5x/2}b^{3x}}+\frac{\ln b^{b^x}}{2a^{2x}b^{4x}}\right)dx$ (where $a, b \in R^+$) is equal to

a. $\frac{1}{6\ln a^2b^3}a^{2x}b^{3x}\ln\frac{a^{2x}b^{3x}}{e}+k$

b. $\frac{1}{6\ln a^2b^3}\frac{1}{a^{2x}b^{3x}}\ln\frac{1}{ea^{2x}b^{3x}}+k$

c. $\frac{1}{6\ln a^2b^3}\frac{1}{a^{2x}b^{3x}}\ln(a^{2x}b^{3x})+k$

d. $-\frac{1}{6\ln a^2b^3}\frac{1}{a^{2x}b^{3x}}\ln(a^{2x}b^{3x})+k$

46. If $\int x\frac{\ln(x+\sqrt{1+x^2})}{\sqrt{1+x^2}}dx$

$=a\sqrt{1+x^2}\ln(x+\sqrt{1+x^2})+bx+c$, then

a. $a=1, b=-1$ b. $a=1, b=1$

c. $a=-1, b=1$ d. $a=-1, b=-1$

47. $\int\frac{\operatorname{cosec}^2 x-2005}{\cos^{2005}x}dx$ is equal to

a. $\frac{\cot x}{(\cos x)^{2005}}+c$ b. $\frac{\tan x}{(\cos x)^{2005}}+c$

c. $\frac{-(\tan x)}{(\cos x)^{2005}}+c$ d. None of these

48. If $xf(x)=3f^2(x)+2$, then

$\int\frac{2x^2-12xf(x)+f(x)}{(6f(x)-x)(x^2-f(x))^2}dx$ equals

a. $\frac{1}{x^2-f(x)}+c$ b. $\frac{1}{x^2+f(x)}+c$

c. $\frac{1}{x-f(x)}+c$ d. $\frac{1}{x+f(x)}+c$

49. If $\int f(x)\sin x\cos x\,dx=\frac{1}{2\left(b^2-a^2\right)}\ln f(x)+c$, then $f(x)$ is equal to

a. $\frac{1}{a^2\sin^2x+b^2\cos^2x}$

b. $\frac{1}{a^2\sin^2x-b^2\cos^2}$

c. $\frac{1}{a^2\cos^2x+b^2\sin^2x}$

d. $\frac{1}{a^2\cos^2x-b^2\sin^2x}$

50. The value of integral $\int e^x\left(\frac{1}{\sqrt{1+x^2}}+\frac{1-2x^2}{\sqrt{(1+x^2)^5}}\right)dx$ is equal to

a. $e^x\left(\frac{1}{\sqrt{1+x^2}}+\frac{x}{\sqrt{(1+x^2)^3}}\right)+c$

b. $e^x\left(\frac{1}{\sqrt{1+x^2}}-\frac{x}{\sqrt{(1+x^2)^3}}\right)+c$

c. $e^x\left(\frac{1}{\sqrt{1+x^2}}+\frac{x}{\sqrt{(1+x^2)^5}}\right)+c$

d. None of these

51. $\int\frac{dx}{\left(1+\sqrt{x}\right)\sqrt{\left(x-x^2\right)}}$ is equal to

a. $\frac{1+\sqrt{x}}{(1-x)^2}+c$ b. $\frac{1+\sqrt{x}}{(1+x)^2}+c$

c. $\frac{1-\sqrt{x}}{(1-x)^2}+c$ d. $\frac{2\left(\sqrt{x}-1\right)}{\sqrt{(1-x)}}+c$

52. The value of $\int \frac{(ax^2 - b)\,dx}{x\sqrt{c^2 x^2 - (ax^2 + b)^2}}$ is equal to

a. $\frac{1}{c}\sin^{-1}\left(ax + \frac{b}{x}\right) + k$

b. $c\sin^{-1}\left(a + \frac{b}{x}\right) + c$

c. $\sin^{-1}\left(\frac{ax + \frac{b}{x}}{c}\right) + k$

d. None of these

53. If $\int \frac{dx}{\cos^3 x\sqrt{\sin 2x}} = a(\tan^2 x + b)\sqrt{\tan x} + c$, then

a. $a = \frac{\sqrt{2}}{5}, b = \frac{1}{\sqrt{5}}$ b. $a = \frac{\sqrt{2}}{5}, b = 5$

c. $a = \frac{\sqrt{2}}{5}, b = -\frac{1}{\sqrt{5}}$ d. $a = \frac{\sqrt{2}}{5}, b = \sqrt{5}$

54. If $\int x\log(1 + 1/x)dx = f(x)\log(x+1) + g(x)x^2 + Ax + C$, then

a. $f(x) = \frac{1}{2}x^2$ b. $g(x) = \log x$

c. $A = 1$ d. None of these

55. $\int\left[\sqrt{\tan x} + \sqrt{\cot x}\right] dx$ is equal to

a. $\sqrt{2}\tan^{-1}\left(\frac{\tan x}{\sqrt{2\tan x}}\right) + c$

b. $\sqrt{2}\tan^{-1}\left(\frac{\tan x - 1}{\sqrt{2\tan x}}\right) + c$

c. $\frac{\tan x}{\sqrt{2}}\tan^{-1}\left(\frac{\cot x + 1}{\sqrt{2\tan x}}\right) + c$

d. $\frac{\tan x}{\sqrt{2}}\tan^{-1}\left(\frac{\cot x - 1}{\sqrt{2\cot x}}\right) + c$

56. If $I_{m,n} = \int \cos^m x \sin nx\, dx$, then $7I_{4,3} - 4I_{3,2}$ is equal to

a. constant b. $-\cos^2 x + C$

c. $-\cos^4 x\cos 3x + C$ d. $\cos 7x - \cos 4x + C$

57. If $\frac{dx}{x^2(x^n + 1)^{(n-1)/n}} = -[f(x)]^{1/n} + C$, then $f(x)$ is

a. $(1 + x^n)$ b. $1 + x^{-n}$

c. $x^n + x^{-n}$ d. None of these

58. $4\int \frac{\sqrt{a^6 + x^8}}{x} dx$ is equal to

a. $\sqrt{a^6 + x^8} + \frac{a^3}{2}\ln\left|\frac{\sqrt{a^6 + x^8} + a^3}{\sqrt{a^6 + x^8} - a^3}\right| + c$

b. $a^6\ln\left|\frac{\sqrt{a^6 + x^8} - a^3}{\sqrt{a^6 + x^8} + a^3}\right| + c$

c. $\sqrt{a^6 + x^8} + \frac{a^3}{2}\ln\left|\frac{\sqrt{a^6 + x^8} - a^3}{\sqrt{a^6 + x^8} + a^3}\right| + c$

d. $a^6\ln\left|\frac{\sqrt{a^6 + x^8} + a^3}{\sqrt{a^6 + x^8} - a^3}\right| + c$

59. If $I = \int e^{-x}\log(e^x + 1)\,dx$, then I equals

a. $x + (e^{-x} + 1)\log(e^x + 1) + C$

b. $x + (e^x + 1)\log(e^x + 1) + C$

c. $x - (e^{-x} + 1)\log(e^x + 1) + C$

d. None of these

60. If $\int xe^x\cos x dx = ae^x(b(1 - x)\sin x + cx\cos x) + d$, then

a. $a = 1, b = 1, c = -1$ b. $a = \frac{1}{2}, b = -1, c = 1$

c. $a = 1, b = -1, c = 1$ d. $a = \frac{1}{2}, b = 1, c = -1$

61. If $I = \int\sqrt{\frac{5 - x}{2 + x}}\,dx$, then I equals

a. $\sqrt{x + 2}\sqrt{5 - x} + 3\sin^{-1}\sqrt{\frac{x + 2}{3}} + C$

b. $\sqrt{x + 2}\sqrt{5 - x} + 7\sin^{-1}\frac{\sqrt{x + 2}}{7} + C$

c. $\sqrt{x + 2}\sqrt{5 - x} + 5\sin^{-1}\sqrt{\frac{x + 2}{5}} + C$

d. None of these

62. $\int e^{\tan x}(\sec x - \sin x)dx$, is equal to

a. $e^{\tan x}\cos x+C$ b. $e^{\tan x}\sin x+C$

c. $-e^{\tan x}\cos x+C$ d. $e^{\tan x}\sec x+C$

63. $\int\frac{x^3dx}{\sqrt{1+x^2}}$ is equal to

a. $\frac{1}{3}\sqrt{1+x^2}(2+x^2)+C$

b. $\frac{1}{3}\sqrt{1+x^2}(x^2-1)+C$

c. $\frac{1}{3}(1+x^2)^{3/2}+C$

d. $\frac{1}{3}\sqrt{1+x^2}(x^2-2)+C$

64. If $I=\int\frac{dx}{\sec x+\operatorname{cosec} x}$, then I equals

a. $\frac{1}{2}\left(\cos x+\sin x-\frac{1}{\sqrt{2}}\log(\operatorname{cosec} x-\cos x)\right)+C$

b. $\frac{1}{2}\left(\sin x-\cos x-\frac{1}{\sqrt{2}}\log|\operatorname{cosec} x+\cot x|\right)+C$

c. $\frac{1}{\sqrt{2}}\left(\sin x+\cos x+\frac{1}{2}\log|\operatorname{cosec} x-\cos x|\right)+C$

d. $\frac{1}{2}[\sin x-\cos x]-\frac{1}{\sqrt{2}}\log|\operatorname{cosec}(x+\pi/4)-\cot(x+\pi/4)|+C$

65. If $I=\int\frac{\sin 2x}{(3+4\cos x)^3}dx$, then I equals

a. $\frac{3\cos x+8}{(3+4\cos x)^2}+C$

b. $\frac{3+8\cos x}{16(3+4\cos x)^2}+C$

c. $\frac{3+\cos x}{(3+4\cos x)^2}+C$

d. $\frac{3-8\cos x}{16(3+4\cos x)^2}+C$

66. $\int\frac{\ln\left(\frac{x-1}{x+1}\right)}{x^2-1}dx$ is equal to

a. $\frac{1}{2}\left(\ln\left(\frac{x-1}{x+1}\right)\right)^2+C$

b. $\frac{1}{2}\left(\ln\left(\frac{x+1}{x-1}\right)\right)^2+C$

c. $\frac{1}{4}\left(\ln\left(\frac{x-1}{x+1}\right)\right)^2+C$

d. $\frac{1}{4}\left(\ln\left(\frac{x+1}{x-1}\right)\right)$

67. $\int\sqrt{e^x-1}\,dx$ is equal to

a. $2\left[\sqrt{e^x-1}-\tan^{-1}\sqrt{e^x-1}\right]+c$

b. $\sqrt{e^x-1}-\tan^{-1}\sqrt{e^x-1}+c$

c. $\sqrt{e^x-1}+\tan^{-1}\sqrt{e^x-1}+c$

d. $2\left[\sqrt{e^x-1}+\tan^{-1}\sqrt{e^x-1}\right]+c$

68. $\int x\sin x\sec^3 x\,dx$ is equal to

a. $\frac{1}{2}\left[\sec^2 x-\tan x\right]+c$

b. $\frac{1}{2}\left[x\sec^2 x-\tan x\right]+c$

c. $\frac{1}{2}[x\sec^2 x+\tan x]+c$

d. $\frac{1}{2}[\sec^2 x+\tan x]+c$

69. $\int e^x\frac{(x^2+1)}{(x+1)^2}dx$ is equal to

a. $\left(\frac{x-1}{x+1}\right)e^x+c$ b. $e^x\left(\frac{x+1}{x-1}\right)+c$

c. $e^x(x+1)(x-1)+c$ d. None of these

70. $\int\left(\frac{x+2}{x+4}\right)^2 e^x\,dx$ is equal to

a. $e^x\left(\frac{x}{x+4}\right)+c$ b. $e^x\left(\frac{x+2}{x+4}\right)+c$

c. $e^x\left(\frac{x-2}{x+4}\right)+c$ d. $\left(\frac{2xe^2}{x+4}\right)+c$

71. $\int \frac{3+2\cos x}{(2+3\cos x)^2}dx$ is equal to

a. $\left(\frac{\sin x}{3\cos x+2}\right)+c$ b. $\left(\frac{2\cos x}{3\sin x+2}\right)+c$

c. $\left(\frac{2\cos x}{3\cos x+2}\right)+c$ d. $\left(\frac{2\sin x}{3\sin x+2}\right)+c$

Multiple Correct Answers Type

Each question has four choices a, b, c, and d, out of which *one or more* answers are correct.

1. $\int \frac{x^2+\cos^2 x}{x^2+1}\operatorname{cosec}^2 x dx$ is equal to

a. $\cot x - \cot^{-1}x + c$

b. $c - \cot x + \cot^{-1}x$

c. $-\tan^{-1}x - \frac{\operatorname{cosec} x}{\sec x}+c$

d. $-e^{\log \tan^{-1}x} - \cot x + c$

2. If $\int \sin x d(\sec x) = f(x) - g(x) + c$, then

a. $f(x) = \sec x$ b. $f(x) = \tan x$

c. $g(x) = 2x$ d. $g(x) = x$

3. $\int \sqrt{1+\operatorname{cosec} x}\, dx$ equals

a. $2\sin^{-1}\sqrt{\sin x}+c$

b. $\sqrt{2}\cos^{-1}\sqrt{\cos x}+c$

c. $c - 2\sin^{-1}(1-2\sin x)$

d. $\cos^{-1}(1-2\sin x)+c$

4. If $I = \int \sec^2 x \operatorname{cosec}^4 x dx = A\cot^3 x + B\tan x + C\cot x + D$, then

a. $A = -\frac{1}{3}$ b. $B = 2$

c. $C = -2$ d. None of these

5. A curve $g(x) = \int x^{27}(1+x+x^2)^6(6x^2+5x+4)dx$ is passing through origin, then

a. $g(1) = \frac{3^7}{7}$ b. $g(1) = \frac{2^7}{7}$

c. $g(-1) = \frac{1}{7}$ d. $g(-1) = \frac{3^7}{14}$

6. If $\int \sqrt{\operatorname{cosec} x + 1}\, dx = k fog(x) + c$, where k is a real constant, then

a. $k=-2, f(x) = \cot^{-1}x, g(x) = \sqrt{\operatorname{cosec} x - 1}$

b. $k=-2, f(x) = \tan^{-1}x, g(x) = \sqrt{\operatorname{cosec} x - 1}$

c. $k=2, f(x) = \tan^{-1}x, g(x) = \frac{\cot x}{\sqrt{\operatorname{cosec} x - 1}}$

d. $k=2, f(x) = \cot^{-1}x, g(x) = \frac{\cot x}{\sqrt{\operatorname{cosec} x + 1}}$

7. If $I = \int \frac{\sin x + \sin^3 x}{\cos 2x}dx = P\cos x + Q\log|f(x)| + R$ then

a. $P = 1/2, Q = -\frac{3}{4\sqrt{2}}$

b. $P = 1/4, Q = -\frac{1}{\sqrt{2}}, f(x) = \frac{\sqrt{2}\cos x - 1}{\sqrt{2}\cos x + 1}$

c. $P = 1/2, Q = -\frac{3}{4\sqrt{2}}, f(x) = \frac{\sqrt{2}\cos x + 1}{\sqrt{2}\cos x - 1}$

d. $P = -1/2, Q = -\frac{3}{4\sqrt{2}}, f(x) = \frac{\sqrt{2}\cos x + 1}{\sqrt{2}\cos x - 1}$

8. If $\int \frac{e^{x-1}}{(x^2-5x+4)}2x\,dx = AF(x-1) + BF(x-4) + C$ and $F(x) = \int \frac{e^x}{x}dx$, then

a. $A = -2/3$ b. $B = 4/3\, e^3$

c. $A = 2/3$ d. $B = 8/3\, e^3$

9. If $\int x^2 e^{-2x}dx = e^{-2x}(ax^2+bx+c)+d$, then

a. $a = 1$ b. $b = 2$

c. $c = \frac{1}{2}$ d. $d \in R$

10. If $\int \frac{x^4+1}{x^6+1}dx = \tan^{-1}f(x) - \frac{2}{3}\tan^{-1}g(x) + C$, then

a. both $f(x)$ and $g(x)$ are odd functions

b. $f(x)$ is monotonic function

c. $f(x) = g(x)$ has no real roots

d. $\int \frac{f(x)}{g(x)}dx = -\frac{1}{x} + \frac{3}{x^3} + c$

11. If $\int \frac{x^2-x+1}{(x^2+1)^{\frac{3}{2}}} e^x dx = e^x f(x)+c$, then

a. $f(x)$ is an even function
b. $f(x)$ is a bounded function
c. The range of $f(x)$ is (0, 1]
d. $f(x)$ has two points of extrema

12. If $\int \frac{\cos 4x+1}{\cot x-\tan x} dx = Af(x)+B$, then

a. $A=-\frac{1}{8}$
b. $B=\frac{1}{2}$
c. $f(x)$ has fundamental period $\frac{\pi}{2}$
d. $f(x)$ is an odd function

13. If $\int \sin^{-1} x \cos^{-1} x\, dx = f^{-1}(x)$

$\left[Ax - x f^{-1}(x) - 2\sqrt{1-x^2}\right] + 2x + C$, then

a. $f(x)=\sin x$
b. $f(x)=\cos x$
c. $A=\frac{\pi}{4}$
d. $A=\frac{\pi}{2}$

14. If $f(x)=\int \frac{x^8+4}{x^4-2x^2+2} dx$ and $f(0)=0$, then

a. $f(x)$ is an odd function
b. $f(x)$ has range R
c. $f(x)$ has at least one real root
d. $f(x)$ is a monotonic function

15. $\int \frac{dx}{x^2+ax+1} = f(g(x))+c$, then

a. $f(x)$ is inverse trigonometric function for $|a|>2$
b. $f(x)$ is logarithmic function for $|a|<2$
c. $g(x)$ is quadratic function for $|a|>2$
d. $g(x)$ is rational function for $|a|<2$

Reasoning Type

Each question has four choices a, b, c, and d, out of which *only one* is correct. Each question contains Statement 1 and Statement 2.

a. If both the statements are True and Statement 2 is the correct explanation of Statement 1.
b. If both the statements are True but Statement 2 is Not the correct explanation of Statement 1.
c. if Statement 1 is True and Statement 2 is False.
d. if Statement 1 is False and Statement 2 is True.

1. Statement 1: $\int e^x \sin x dx = \frac{e^x}{2}(\sin x - \cos x)+c.$

Statement 2: $\int e^x (f(x)+f'(x))\, dx = e^x f(x)+c.$

2. Statement 1: For $-1<a<4$, $\int \frac{dx}{x^2+2(a-1)x+a+5}$ $=\lambda \log|g(x)|+c$, where λ and c are constants.

Statement 2: For $-1<a<4$, $\frac{1}{x^2+2(a-1)x+a+5}$ is a continuous function.

3 Statement 1: $\int \frac{\sin x\, dx}{x}$ $(x>0)$ cannot be evaluated.

Statement 2: Only differentiable functions can be integrated.

4. Statement 1: $\int \frac{dx}{x^3\sqrt{1+x^4}} = -\frac{1}{2}\sqrt{1+\frac{1}{x^4}}+C.$

Statements 2: For integration by parts we have to follow IIATE rule.

5. Statement 1: If the primitive of $f(x) = \pi \sin \pi x + 2x - 4$, has the value 3 for $x = 1$, then there are exactly two value of x for which primitive of $f(x)$ vanishes is.

Statement 2: $\cos \pi x$ has period 2.

6. Statement 1: $\int \frac{\{f(x)\,\phi'(x)-f'(x)\,\phi(x)\}}{f(x)\,\phi(x)} \{\log \phi(x)$ $-\log f(x)\}\, dx = \frac{1}{2}\left\{\log \frac{\phi(x)}{f(x)}\right\}^2 + c.$

Statement 2: $\int (h(x))^n h'(x) dx = \frac{(h(x))^{n+1}}{n+1}+c.$

Linked Comprehension Type

Based upon each paragraph, three multiple choice questions have to be answered. Each question has four choices a, b, c, and d, out of which *only one* is correct.

For Problems 1–3

$y=f(x)$ is a polynomial function passing through point (0, 1) and which increases in the intervals (1, 2) and $(3, \infty)$ and decreases in the intervals $(-\infty, 1)$ and (2, 3).

1. If $f(1)=-8$, then the value of $f(2)$ is

a. 1–3
b. –6
c. –20
d. –7

2. If $f(1) = -8$, then the range of $f(x)$ is

a. $[3, \infty)$ b. $[-8, \infty)$

c. $[-7, \infty)$ d. $(-\infty, 6]$

3. If $f(x) = 0$ has four real roots, then the range of values of leading co-efficient of polynomial is

a. $[4/9, 1/2]$ b. $[4/9, 1]$

c. $[1/3, 1/2]$ d. None of these

For Problems 4–6

If A is square matrix and e^A is defined as $e^A = I + A + \frac{A^2}{2!} + \frac{A^3}{3!} + \ldots = \frac{1}{2}\begin{bmatrix} f(x) & g(x) \\ g(x) & f(x) \end{bmatrix}$, where $A = \begin{bmatrix} x & x \\ x & x \end{bmatrix}$ and $0 < x < 1$, I is an identity matrix.

4. $\int \frac{g(x)}{f(x)} dx$ is equal to

a. $\log(e^x + e^{-x}) + c$ b. $\log(e^x - e^{-x}) + c$

c. $\log(e^{2x} - 1) + c$ d. None of these

5. $\int (g(x) + 1) \sin x dx$ is equal to

a. $\frac{e^x}{2}(\sin x - \cos x)$

b. $\frac{e^{2x}}{5}(2\sin x - \cos x)$

c. $\frac{e^x}{5}(\sin 2x - \cos 2x)$

d. None of these

6. $\int \frac{f(x)}{\sqrt{g(x)}} dx$ is equal to

a. $\frac{1}{2\sqrt{e^x - 1}} - \operatorname{cosec}^{-1}(e^x) + c$

b. $\frac{2}{\sqrt{e^x - e^{-x}}} - \sec^{-1}(e^x) + c$

c. $\frac{1}{2\sqrt{e^{2x} - 1}} + \sec^{-1}(e^x) + c$

d. None of these

For Problems 7–9

Euler's substitution

Integrals of the form $\int R(x, \sqrt{ax^2 + bx + c})\, dx$ are calculated with the aid of one of the three Euler substitutions

1. $\sqrt{ax^2 + bx + c} = t \pm x\sqrt{a}$ if $a > 0$;

2. $\sqrt{ax^2 + bx + c} = tx \pm \sqrt{c}$ if $c > 0$;

3. $\sqrt{ax^2 + bx + c} = (x - a)\, t$ if $ax^2 + bx + c = a(x - a)(x - b)$ i.e., if α is a real root of $ax^2 + bx + c = 0$.

7. Which of the following functions does not appear in the primitive of $\frac{1}{1 + \sqrt{x^2 + 2x + 2}}$ if t is a function of x ?

a. $\log_e|t + 1|$ b. $\log_e|t + 2|$

c. $\frac{1}{t + 1}$ d. None of these

8. Which of the following functions does not appear in the primitive of $\frac{dx}{x + \sqrt{x^3 - x + 1}}$ if t is a function of x?

a. $\log_e|t|$ b. $\log_e|t - 2|$

c. $\log_e|t - 1|$ d. $\log|t + 1|$

9. $\int \frac{x dx}{\left(\sqrt{7x - 10 - x^2}\right)^3}$ can be evaluated by substituting for x as

a. $x = \frac{5 + 2t^2}{t^2 + 1}$ b. $x = \frac{5 - t^2}{t^2 + 2}$

c. $x = \frac{2t^2 - 5}{3t^2 - 1}$ d. None of these

Matrix-Match Type

Each question contains statements given in two columns which have to be matched. Statements (a, b, c, d) in column 1 have to be matched with statements (p, q, r, s) in column 2. If the correct match are a → p, s, b → r, c → p, q and d → s, then the correctly bubbled 4 × 4 matrix should be as follows:

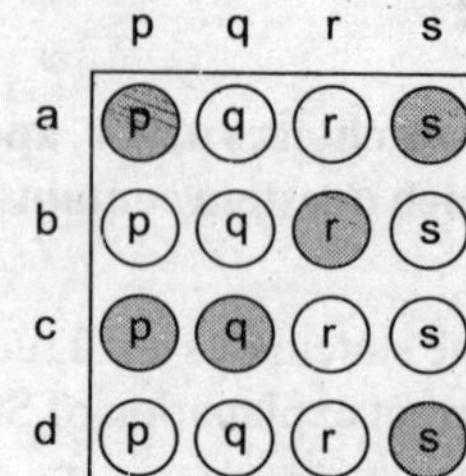

1.

Column 1	Column 2
a. If $\int \frac{2^x}{\sqrt{1-4^x}} dx = k\sin^{-1}(f(x)) + C$, then k is greater than	**p.** 0
b. If $\int \frac{(\sqrt{x})^5}{(\sqrt{x})^7 + x^6} dx = a\ln\frac{x^k}{x^k+1} + c$, then ak is less than	**q.** 1
c. $\int \frac{x^4+1}{x(x^2+1)^2} dx = k\ln\|x\| + \frac{m}{1+x^2} + n$, where n is the constant of integration, then mk is greater than	**r.** 3
d. $\int \frac{dx}{5+4\cos x} = k\tan^{-1}\left(m\tan\frac{x}{2}\right) + C$, then k/m is greater than	**s.** 4

2.

Column 1	Column 2
a. $\int \frac{e^{2x}-1}{e^{2x}+1} dx$ is equal to	**p.** $x - \log\left[1+\sqrt{1-e^{2x}}\right] + c$
b. $\int \frac{1}{(e^x+e^{-x})^2} dx$ is equal to	**q.** $\log(e^x+1) - x - e^{-x} + c$,
c. $\int \frac{e^{-x}}{1+e^x} dx$ is equal to	**r.** $\log(e^{2x}+1) - x + c$
d. $\int \frac{1}{\sqrt{1-e^{2x}}} dx$ is equal to	**s.** $-\frac{1}{2(e^{2x}+1)} + c$

3.

Column 1	Column 2 (which of the following functions appear in Column 1)
a. $I = \int \frac{x^2-x+1}{x^3-4x^2+4x} dx$	**p.** $\log\|x\|$
b. $\int \frac{x^2-1}{x(x-2)^3} dx$	**q.** $\log\|x-2\|$
c. $\int \frac{x^3+1}{x(x-2)^2} dx$	**r.** x
d. $\int \frac{x^5+1}{x(x-2)^3} dx$	**s.** x
	t. x^2

Integer Type

1. If $f(x) = \int \frac{3x^4-1}{(x^4+x+1)^2} dx$ and $f(0) = 0$, then $f(-1) =$

2. If $\int \frac{4e^x+6e^{-x}}{9e^x-4e^{-x}} dx = Ax + B\log(9e^{2x}-4) + C$, then $|4A| =$

3. If $\int \frac{x^2-1}{(x^2+1)\sqrt{1+x^4}} dx = \frac{1}{a}\tan^{-1}\frac{\sqrt{x^2+1/x^2}}{\sqrt{2}} + c$, then the value of a is ______.

4. Let $f(x) = \int \left(\frac{\cos x}{x} - \log x^{\sin x}\right) dx$ and $f(1) = 0$, then the value of $f(\pi/2)$ is ______.

5. If $f(x) = \int \frac{1}{(x+1)\sqrt{x^2-1}} dx$ and $f(1) = 0$, then the value of $5(f(3/2))^2$ is ______.

6. If $\int \frac{(x-x^3)^{1/3}}{x^4} dx = a\left(\frac{1}{x^2}-1\right)^b + c$, then the value of $1/|ab|$ is ______.

7. If $\int \frac{1}{[(x-1)^3(x+2)^5]^{1/4}} dx = a\left(\frac{x-1}{x+2}\right)^b + c$, then the value of $1/ab$ is ______.

8. If $f(x) = \int \frac{\sqrt{\tan x}}{\sin x\cos x} dx$, and $f(0) = 3$, then the value of $f(\pi/4)$ is______.

9. If $\int \frac{5x^4+4x^5}{(x^5+x+1)^2} dx = f(x) + C$, then the value of $1/f(1)$ is ______.

10. If $\int \frac{\cos^2 x + \sin 2x}{(2\cos x - \sin x)^2} dx = \frac{\cos x}{2\cos x - \sin x} + ax + b\ln|2\cos x - \sin x| + c$, then $|a+2b|$ is ______.

11. If $f(x) = \int \frac{2+\sqrt{x}}{(x+1+\sqrt{x})^2} dx$ and $f(0) = 0$, then the value of $[f(4)]$ is ______ (where [.] represents the greatest integer function).

ANSWERS AND SOLUTIONS

Objective Type

1. c. $\int \frac{\sin 2x}{\sin 5x \sin 3x} dx$

$= \int \frac{\sin (5x - 3x)}{\sin 5x \sin 3x}$

$= \int \frac{\sin 5x \cos 3x - \cos 5x \sin 3x}{\sin 5x \sin 3x} dx$

$= \frac{1}{3} \log \sin 3x - \frac{1}{5} \log \sin 5x + c$

2. a. $I = \int \frac{\sqrt{1+\sin x}\sqrt{1-\sin x}}{\sqrt{1-\sin x}} dx$

$= \int \frac{\cos x}{\sqrt{1-\sin x}} dx = -2\sqrt{1-\sin x} + c$

3. b. $\int \frac{\sin^8 x - \cos^8 x}{1-2\sin^2 x \cos^2 x} dx$

$= \int \frac{(\sin^2 x - \cos^2 x)(\sin^4 x + \cos^4 x)}{1-2\sin^2 x\cos^2 x}$

$= \int -\cos 2x dx = -\frac{1}{2}\sin 2x + c$

4. b. $\int \frac{\cos 4x + 1}{\cot x - \tan x} dx$

$= \int \frac{2\cos^2 2x}{\cos^2 x - \sin^2 x} \sin x \cos x \, dx$

$= \int \cos 2x \sin 2x dx$

$= \frac{1}{4}\int \sin 4x dx = -\frac{1}{8}\cos 4x + c$

5. b. $f(x) = x|\cos x|, \frac{\pi}{2} < x < \pi = -x \cos x$, because $\cos x$ is negative in $\left(\frac{\pi}{2}, \pi\right)$.

Therefore, the required primitive function

$= \int -x \cos x dx$

Now, use integration by parts

6. a. $I = \int \frac{dx}{x(x^n+1)} = \int \frac{x^{n-1}}{x^n(x^n+1)} dx$

Putting $x^n = t$ so that $n x^{n-1} dx = dt$

$\Rightarrow x^{n-1} dx = \frac{1}{n} dt$

$\therefore \quad I = \int \frac{\frac{1}{n}dt}{t(t+1)} = \frac{1}{n}\int\left(\frac{1}{t} - \frac{1}{t+1}\right) dt$

$= \frac{1}{n}(\log t - \log(t+1)) + c$

$= \frac{1}{n}\log\left(\frac{x^n}{x^n+1}\right) + c$

7. b. $I = \int 4 \sin x \cos\frac{x}{2} \cos\frac{3x}{2} dx$

$= \int 2\sin x(\cos 2x + \cos x)dx$

$= \int (\sin 3x - \sin x + \sin 2x)dx$

$= \cos x - \frac{1}{3}\cos 3x - \frac{1}{2}\cos 2x + c$

8. c. $\frac{dx}{dt} = f'''(t)\cos t - f''(t)\sin t + f''(t)\sin t + f'(t)\cos t$

$= [f'''(t) + f'(t)] \cos t$

$\frac{dy}{dt} = -f'''(t)\sin t - f''(t)\cos t + f''(t)\cos t - f'(t)$
$\sin t = -[f'''(t) + f'(t)]\sin t$

$\Rightarrow \quad \left[\left(\frac{dx}{dt}\right)^2 + \left(\frac{dy}{dt}\right)^2\right]^{1/2}$

$= [(f'''(t) + f'(t))^2 (\cos^2 t + \sin^2 t)]^{1/2}$

$= f'''(t) + f'(t)$

$\Rightarrow \quad \int\left[\left(\frac{dx}{dt}\right)^2 + \left(\frac{dy}{dt}\right)^2\right]^{1/2} dt = f''(t) + f(t) + c$

9. c. $\sin^3 x \sin(x+\alpha)$
$= \sin^3 x(\sin x \cos\alpha + \cos x \sin\alpha)$
$= \sin^4 x(\cos\alpha + \cot x \sin\alpha)$

$I = \int \frac{1}{\sqrt{\sin^3 x \sin(x+\alpha)}} dx$

$= \int \frac{1}{\sin^2 x\sqrt{\cos\alpha + \cot x \sin\alpha}} dx$

$= \int \frac{\operatorname{cosec}^2 x}{\sqrt{\cos\alpha + \cot x \sin\alpha}} dx$

Putting $\cos\alpha + \cot x \sin\alpha = t$ and $-\operatorname{cosec}^2 x \sin\alpha dx = dt$, we have

$I = \int -\frac{1}{\sin\alpha\sqrt{t}} dt = -\frac{1}{\sin\alpha}\int t^{-1/2} dt$

$$= \frac{1}{\sin\alpha}\left(\frac{t^{1/2}}{1/2}\right)+C$$

$\Rightarrow \quad I = -2 \operatorname{cosec}\alpha\sqrt{t} + C = -2\operatorname{cosec}\alpha(\cos\alpha + \cot x \sin\alpha)^{1/2} + c$

10. c. $\int \frac{px^{p+2q-1}-qx^{q-1}}{(x^{p+q}+1)^2}dx = \int \frac{px^{p-1}-qx^{-q-1}}{(x^p+x^{-q})^2}dx$

[Taking x^q as x^{2q} common from D^r and take it in N^r]

$$= \int \frac{dt}{t^2} = -\frac{1}{t}+c = -\frac{1}{x^p+x^{-q}}+C$$

$$= -\frac{x^q}{x^{p+q}+1}+C$$

11. c. $I_n = x(\ln x)^n - \int \frac{x(n)(\ln x)^{n-1}}{x}dx$

$$= x(\ln x)^n - n\,I_{(n-1)}$$

$$\Rightarrow I_n + n\,I_{n-1} = x(\ln x)^n$$

12. b. Let $I = \int \frac{x+2}{(x^2+3x+3)\sqrt{x+1}}dx$

Putting $x+1 = t^2$, $dx = 2t\,dt$, we get

$$I = 2\int \frac{t^2+1}{t^4+t^2+1}dt$$

$$= 2\int \frac{1+(1/t)^2}{\left(t-\frac{1}{t}\right)^2+3}dt$$

$$= \frac{2}{\sqrt{3}}\tan^{-1}\left(\frac{t-\frac{1}{t}}{\sqrt{3}}\right)+C$$

$$= \frac{2}{\sqrt{3}}\tan^{-1}\left(\frac{x}{\sqrt{3(x+1)}}\right)+C$$

13. b. $I = \int \frac{\sin 2x}{\sin^4 x+\cos^4 x}dx$

$$= \int \frac{2\sin x\cos x}{\sin^4 x+\cos^4 x}dx$$

$$= \int \frac{2\tan x\sec^2 x}{1+\tan^4 x}dx$$

Let $\tan^2 x = t \Rightarrow 2\tan x\sec^2 x\,dx = dt$,

$$\Rightarrow \quad I = \int \frac{dt}{1+t^2} = \tan^{-1} t + c = \tan^{-1}(\tan^2 x)+c$$

14. c. $I = \int \frac{\sec x\,dx}{\sqrt{2\sin(x+A)\cos x}}$

$$= \int \frac{\sec^2 x\,dx}{\sqrt{\frac{2\sin(x+A)}{\cos x}}}$$

$$= \frac{1}{\sqrt{2}}\int \frac{\sec^2 x\,dx}{\sqrt{\tan x\cos A+\sin A}}$$

$$= \frac{\sec A}{\sqrt{2}}\int \frac{2p\,dp}{p}$$

If $\tan x\cos A+\sin A = p^2$, then $\cos A\sec^2 x\,dx = 2p\,dp$

$$I = \sqrt{2}\sec A\int dp$$

$$= \sqrt{2}\sec A\sqrt{\tan x\cos A+\sin A}+c$$

15. a. Differentiating both sides, we get

$$\sqrt{1+\sin x}\;f(x) = \frac{2}{3}\frac{3}{2}(1+\sin x)^{1/2}\cos x$$

$$\Rightarrow f(x) = \cos x.$$

16. c. Here, $\int e^x\{f(x)-f'(x)\}dx = \phi(x)$

and $\int e^x\{f(x)+f'(x)\}dx = e^x f(x)$

On adding, we get $2\int e^x f(x)dx = \phi(x)+e^x f(x)$

17. b. $f(x) = \int \frac{x^2dx}{(1+x^2)(1+\sqrt{1+x^2})}$

let $x = \tan\theta \Rightarrow dx = \sec^2\theta\,d\theta = (1+x^2)\,d\theta$

$$\Rightarrow \quad f(x) = \int \frac{x^2dx}{(1+x^2)(1+\sqrt{1+x^2})}$$

$$= \int \frac{\tan^2\theta\sec^2\theta\,d\theta}{\sec^2\theta(1+\sec\theta)}$$

$$= \int \frac{\tan^2\theta\,d\theta}{1+\sec\theta}$$

$$= \int \frac{\sin^2\theta\,d\theta}{\cos\theta(1+\cos\theta)}$$

$$= \int \frac{1-\cos^2\theta\,d\theta}{\cos\theta(1+\cos\theta)}$$

$$= \int \frac{(1-\cos\theta)\,d\theta}{\cos\theta}$$

$$= \int \sec\theta\,d\theta - \int d\theta$$

$$= \log(x+\sqrt{1+x^2}) - \tan^{-1}x + C$$

Given $f(0) = 0$

$\Rightarrow \quad 0 = \log 1 - 0 + C$

$\Rightarrow \quad C = 0$

$\Rightarrow \; f(1) = \log(1 + \sqrt{1 + x^2}) - \tan^{-1}(1)$

$= \log(1 + \sqrt{2}) - \dfrac{\pi}{4}$

18. a. Let $x = \tan\theta$, then $dx = \sec^2\theta d\theta$

Now $y = \int \dfrac{dx}{(1+x^2)^{\frac{3}{2}}} = \int \dfrac{\sec^2\theta}{(1+\tan^2\theta)^{\frac{3}{2}}} d\theta$

$= \int \dfrac{\sec^2\theta}{(\sec^2\theta)^{\frac{3}{2}}} d\theta$

$= \int \dfrac{\sec^2\theta}{\sec^3\theta} d\theta = \int \dfrac{d\theta}{\sec\theta} = \int \cos\theta d\theta$

Hence, $y = \sin\theta + c = \dfrac{x}{\sqrt{1+x^2}} + c \quad (1)$

$\left[\because \tan\theta = x = \dfrac{x}{1} \therefore \sin\theta = \dfrac{x}{\sqrt{1^2+x^2}}\right]$

Given when $x = 0, y = 0 \Rightarrow$ from equation(1), $0 = 0 + c$

$\Rightarrow c = 0$

$\Rightarrow$ from equation(1), $y = \dfrac{x}{\sqrt{1+x^2}}$

$\Rightarrow$ when $x = 1, y = \dfrac{1}{\sqrt{2}}$

19. c. Let $x = t^6 \Rightarrow dx = 6t^5\,dt$

$\Rightarrow I = \int t^3 (1+t^2)^4 6t^5\,dt$

$\Rightarrow I = 6\int t^8(1 + 4t^2 + 6t^4 + 4t^6 + t^8)\,dt$

$= 6\int (t^8 + 4t^{10} + 6t^{12} + 4t^{14} + t^{16})\,dt$

$= 6\left\{\dfrac{t^9}{9} + \dfrac{4t^{11}}{11} + \dfrac{6t^{13}}{13} + \dfrac{4t^{15}}{15} + \dfrac{t^{17}}{17}\right\} + C$

$= 6\left\{x^{2/3} + \dfrac{4}{11}x^{11/6} + \dfrac{6}{13}x^{13/6} + \dfrac{4}{15}x^{5/2} + \dfrac{1}{17}x^{17/6}\right\}$

$+ C$

20. b. Here, $\int x^5(1+x^3)^{2/3}\,dx$

Let $1 + x^3 = t^2$ and $3x^2\,dx = 2t\,dt$

$\therefore \int x^5(1+x^3)^{2/3}\,dx = \int x^3(1+x^3)^{2/3}x^2\,dx$

$= \int (t^2 - 1)(t^2)^{2/3}x^2\,dx$

$= \dfrac{2}{3}\int (t^2 - 1)\,t^{7/3}\,dt$

$= \dfrac{2}{3}\int (t^{13/3} - t^{7/3})\,dt$

$= \dfrac{2}{3}\left\{\dfrac{3}{16}t^{16/3} - \dfrac{3}{10}t^{10/3}\right\} + C$

$= \dfrac{1}{8}(1+x^3)^{8/3} - \dfrac{1}{5}(1+x^3)^{5/3} + C$

21. a. Let $I = \int \dfrac{(1-\cos\theta)^{2/7}}{(1+\cos\theta)^{9/7}} d\theta$

$I = \int \dfrac{(2\sin^2\theta/2)^{2/7}}{(2\cos^2\theta/2)^{9/2}} d\theta = \dfrac{1}{2}\int \dfrac{(\sin\theta/2)^{4/7}}{(\cos\theta/2)^{18/7}} d\theta$

Put $\dfrac{\theta}{2} = t \therefore \dfrac{d\theta}{2} = dt$

$\Rightarrow I = \int \dfrac{(\sin t)^{4/7}}{(\cos t)^{18/7}} dt$ (Here $m + n = -2$)

$= \int (\tan t)^{4/7}\sec^2 t dt$

Put $\tan t = u \therefore \sec^2 t\,dt = du$

$\Rightarrow I = \int u^{4/7} du = \dfrac{u^{11/7}}{11/7} + c = \dfrac{7}{11}(\tan t)^{11/7} + C$

$= \dfrac{7}{11}\left(\tan\dfrac{\theta}{2}\right)^{11/7} + C$

22. c. $I = \int \dfrac{1-x^7}{x(1+x^7)} dx = a\ln|x| + b\ln|1+x^7| + C$

Diff. both sides, we get $\dfrac{1-x^7}{x(1+x^7)} = \dfrac{a}{x} + b\dfrac{7x^6}{1+x^7}$

$\Rightarrow 1 - x^7 = a(1+x^7) + 7bx^7$

$\Rightarrow a = 1, a + 7b = -1$

$\Rightarrow b = -2/7$

23. d. $I = \int \sin^{-1}\left(\dfrac{2x}{1+x^2}\right)dx$, let $x = \tan\theta$

$\Rightarrow dx = \sec^2\theta\,d\theta$

$\Rightarrow I = \int \sin^{-1}\left(\dfrac{2\tan\theta}{1+\tan^2\theta}\right)\sec^2\theta d\theta = 2\int \theta\sec^2\theta\,d\theta$

$= 2(\theta\tan\theta - \ln|\sec\theta|) + c$

$= 2(x\tan^{-1}x - \ln|\sec(\tan^{-1}x)|) + c$

24. c. $I = \int \dfrac{\ln(\tan x)}{\sin x\cos x} dx$, let $t = \ln(\tan x)$

$\Rightarrow \dfrac{dt}{dx} = \dfrac{\sec^2 x}{\tan x}$

$$\Rightarrow dt = \frac{dx}{\sin x \cos x}$$

$$\Rightarrow I = \int t\,dt = \frac{1}{2}t^2 + c = \frac{1}{2}(\ln(\tan x))^2 + c$$

25. c. $I = \int \frac{2\sin x}{(3+\sin 2x)}dx$

$$= \int \frac{\sin x + \cos x + \sin x - \cos x}{(3+\sin 2x)}dx$$

$$= \underset{I_1}{\underbrace{\int \frac{\sin x + \cos x}{3+\sin 2x}dx}} - \underset{I_2}{\underbrace{\int \frac{-\sin x + \cos x}{(3+\sin 2x)}dx}}$$

Putting $t_1 = \sin x - \cos x$ in I_1 and $t_2 = \sin x + \cos x$ in I_2, we get

$$I = \int \frac{dt_1}{[3+(1-t_1^2)]} \int \frac{dt_2}{[3+(t_2^2-1)]}$$

$$= \int \frac{dt_1}{4-t_1^2} - \int \frac{dt_2}{2+t_2^2}$$

$$= \frac{1}{4}\ln\left|\frac{2+t_1}{2-t_1}\right| - \frac{1}{\sqrt{2}}\tan^{-1}\left(\frac{t_2}{\sqrt{2}}\right) + c$$

$$= \frac{1}{2}\ln\left|\frac{2+\sin x - \cos x}{2-\sin x + \cos x}\right| - \frac{1}{\sqrt{2}}\tan^{-1}\left(\frac{\sin x + \cos x}{\sqrt{2}}\right) + c$$

26. d. $I = \int \frac{x^9\,dx}{(x^2+4)^6}$

$$= \int \frac{dx}{x^3\left(4+\frac{1}{x^2}\right)^6}$$

$$= -\frac{1}{2}\int \frac{d\left(4+\frac{1}{x^2}\right)}{\left(4+\frac{1}{x^2}\right)^6}$$

$$= -\frac{1}{2}\frac{\left(4+\frac{1}{x^2}\right)^{-5}}{-5} + c = \frac{1}{10}\left(4+\frac{1}{x^2}\right)^{-5} + c$$

27. c. $I = \int e^{\tan^{-1}x}(1+x+x^2) - \left(\frac{1}{1+x^2}\right)dx$

$$= -\int e^{\tan^{-1}x}\left(1+\frac{x}{1+x^2}\right)dx$$

$$= -\int e^{\tan^{-1}x}dx - \int x\frac{e^{\tan^{-1}x}}{1+x^2}dx$$

$$= -\int e^{\tan^{-1}x}dx - x\,e^{\tan^{-1}x} + \int e^{\tan^{-1}x}dx + c$$

$$= x\,e^{\tan^{-1}x} + c$$

28. d. $I = \int \frac{dx}{\sqrt{\sin^3 x \cos^5 x}}$

$$= \int \frac{dx}{\sqrt{\frac{\sin^3 x}{\cos^3 x}\cos^8 x}}$$

$$= \int \frac{\sec^4 x}{\sqrt{\tan^3 x}}dx$$

$$= \int \frac{(1+\tan^2 x)\sec^2 x}{\sqrt{\tan^3 x}}dx$$

Let $t = \tan x \Rightarrow dt = \sec^2 x\,dx$

$$\Rightarrow \quad I = \int \frac{1+t^2}{t^{3/2}}dt$$

$$= \int (t^{-3/2} + t^{1/2})\,dt$$

$$= -2t^{-\frac{1}{2}} + \frac{2}{3}t^{3/2} + c$$

$$= -2\sqrt{\cot x} + \frac{2}{3}\sqrt{\tan^3 x} + c$$

$$\Rightarrow \quad a = -2,\, b = \frac{2}{3}$$

29. c. $I = \int \frac{\cos 4x - 1}{\cot x - \tan x}dx$

$$= \int \frac{-2\sin^2 2x(\sin x\cos x)}{(\cos^2 x - \sin^2 x)}dx$$

$$= -\int \frac{\sin^2 2x \sin 2x}{\cos 2x}x$$

$$= \int \frac{(\cos^2 2x - 1)\sin 2x}{\cos 2x}dx$$

Let $t = \cos{}^2x \Rightarrow dt = -2\sin{}^2x\,dx$

$$\Rightarrow \quad I = \frac{1}{2}\int \frac{(1-t^2)}{t}dt = \frac{1}{2}\ln|t| - \frac{t^2}{4} + c$$

$$= \frac{1}{2}\ln|\cos 2x| - \frac{1}{4}\cos^2 2x + c$$

30. a. Putting $1 - x^3 = y^2, -3x^2dx = 2y\,dy$, we get

$$\int \frac{1}{x\sqrt{1-x^3}}dx = -\frac{2}{3}\int \frac{1}{1-y^2}dy$$

$$= \frac{1}{3}\log\left|\frac{y-1}{y+1}\right| + c$$

$$= \frac{1}{3}\log\left|\frac{\sqrt{1-x^3}-1}{\sqrt{1-x^3}+1}\right| + c$$

$$\Rightarrow a = \frac{1}{3}$$

31. b. We have $\int \frac{dx}{x^2(x^n+1)^{(n-1)/n}}$

$$= \int \frac{dx}{x^2\, x^{n-1}\left(1+\frac{1}{x^n}\right)^{(n-1)/n}}$$

$$= \int \frac{dx}{x^{n+1}(1+x^{-n})^{(n-1)/n}}$$

Put $1 + x - n = t$

$\therefore \quad -nx^{-n-1}\, dx = dt \Rightarrow \frac{dx}{x^{n+1}} = -\frac{dt}{n}$

$$\Rightarrow \int \frac{dx}{x^2(x^n+1)^{(n-1)/n}} = -\frac{1}{n}\int \frac{dt}{t^{(n-1)/n}}$$

$$= -\frac{1}{n}\int t^{1/n-1} dt = -\frac{1}{n}\frac{t^{1/n-1+1}}{1/n-1+1} + c$$

$$= -t^{1/n} + c = -(1+x^{-n})^{1/n} + c$$

32. d. $I = \int \frac{\sqrt{x-1}}{x\sqrt{x+1}} dx$

$$= \int \frac{x-1}{x\sqrt{x^2-1}} dx$$

$$= \int \frac{dx}{\sqrt{x^2-1}} - \int \frac{dx}{x\sqrt{x^2-1}}$$

$$= \ln|x+\sqrt{x^2+1}| - \sec^{-1} x + c$$

33. c. Write $2ax + x^2 = (x+a)^2 - a^2$, and put $x + a = a \sec\theta$, so that $dx = a \sec\theta \tan\theta\, d\theta$

$$\therefore \quad I = \int \frac{a \sec\theta \tan\theta}{a^3 \tan^3\theta} d\theta$$

$$= \frac{1}{a^2}\int \frac{\cos\theta}{\sin^2} d\theta$$

$$= -\frac{1}{a^2 \sin\theta} + C$$

$$= -\frac{1}{a^2}\frac{\sec\theta}{\tan\theta} + C = -\frac{1}{a^2}\frac{x+a}{\sqrt{2ax+x^2}} + c$$

34. d. By rationalizing the integrand, the given integral can be written as

$$f(x) = \int \left(x + \sqrt{x^2+1}\right) dx$$

$$= \frac{x^2}{2} + \frac{x}{2}\sqrt{x^2+1} + \log\left|x+\sqrt{x^2+1}\right| + C$$

Putting $x = 0$, we have $f(0) = C$ so $C = -1/2 - 1/\sqrt{2}$

and $f(1) = \frac{1}{2} + \frac{1}{2}\sqrt{2} + \log|1+\sqrt{2}| + \left(-\frac{1}{2} - \frac{1}{\sqrt{2}}\right)$

$$= \log(1+\sqrt{2}) = -\log(\sqrt{2}-1)$$

35. b. $\int e^x \left(\frac{2\tan x}{1+\tan x} + \tan^2\left(x - \frac{\pi}{4}\right)\right) dx$

$$= \int e^x \left(\tan\left(x-\frac{\pi}{4}\right) + \sec^2\left(x-\frac{\pi}{4}\right)\right) dx$$

$$= e^x \tan\left(x-\frac{\pi}{4}\right) + c$$

36. a. Given that $I = \int (x^2+x)(x^{-8}+2x^{-9})^{1/10} dx$

or $\quad I = \int (x+1)(x^2+2x)^{1/10} dx$

Now put $x^2 + 2x = 1 \Rightarrow (x+1)dx = \frac{dt}{2}$

$$\Rightarrow \quad I = \int t^{1/10}\frac{dt}{2} = \frac{1}{2}\times\frac{10}{11}t^{11/10} = \frac{5}{11}t^{11/10} + c$$

$$= \frac{5}{11}(x^2+2x)^{11/10} + c$$

37. c. $\int \frac{dx}{(x+2)(x^2+1)} = a \ln(1+x^2) + b \tan^{-1} x + \frac{1}{5}\ln|x+2| + C$

Differentiating both sides, we get

$$\frac{1}{(x+2)(x^2+1)} = \frac{2ax}{(1+x^2)} + \frac{b}{(1+x^2)} + \frac{1}{5(x+2)}$$

$$\Rightarrow \quad \frac{1}{(x+2)(x^2+1)} = \frac{(x+2)(5b+10ax)+1+x^2}{5(1+x^2)(x+2)}$$

$$\Rightarrow \quad 5 = (1+x^2) + 5(b+2ax)(x+2)$$

Comparing the like powers of x in both sides, we get

$1 + 10a = 0,\ b + 4a = 0,\ 10b + 1 = 5$

$$\Rightarrow \quad a = -\frac{1}{10},\ b = \frac{2}{5}$$

38. c. Differentia both sides, we get

$$\frac{3\sin x + 2\cos x}{3\cos x + 2\sin x} = a + \frac{b(2\cos x - 3\sin x)}{(2\sin x + 3\cos x)}$$

$$= \frac{\sin x\,(2a-3b) + \cos x\,(3a+2b)}{(3\cos x + 2\sin x)}$$

Comparing like terms on both sides, we get

$$3 = 2a - 3b,\ 2 = 3a + 2b \Rightarrow a = \frac{12}{13},\ b = -\frac{15}{39}$$

39. a. $\int \frac{3e^x - 5e^{-x}}{4e^x + 5e^{-x}} = ax + b\ln(4e^x + 5e^{-x}) + C$

Differentiating both sides, we get

$$\frac{3e^x - 5e^{-x}}{4e^x + 5e^{-x}} = a + b\frac{(4e^x - 5e^{-x})}{4e^x + 5e^{-x}}$$

$$\Rightarrow \quad 3e^x - 5e^{-x} = a(4e^x + 5e^{-x}) + b(4e^x - 5e^{-x})$$

Comparing the coefficient of like terms on both sides, we get

$3 = 4(a+b), -5 = 5a - 5b \Rightarrow a = -\frac{1}{8},\ b = \frac{7}{8}$

40. c. $\int \sqrt{\frac{\cos x - \cos^3 x}{1-\cos^3 x}}dx = \int \sqrt{\frac{\cos x}{1-\cos^3 x}}\sin x dx$

$= \int \sqrt{\frac{t}{1-t^3}}dt = -\int \frac{\sqrt{t}}{\sqrt{1-(t^{3/2})^2}}dt,$

where $t = \cos x$

$= -\frac{2}{3}\int \frac{\frac{3}{2}\sqrt{t}}{\sqrt{1-(t^{3/2})^2}}dt = \frac{2}{3}\cos^{-1}(t^{3/2}) + C$

41. a. Putting,

$l^{r+1}(x) = t$ and $\frac{1}{xl(x)l^2(x)\ldots l^r(x)}dx = dt$, we get

$\int \frac{1}{xl^2(x)l^3(x)\ldots l^r(x)} = \int 1\, dt = t + C = l^{r+1}(x) + C$

42. b. $I = \int \frac{\cos x - \sin x}{\sqrt{\cos x \sin x}}dx$

Put $\sin x + \cos x = t$, so that $2\sin x\cos x = t^2 - 1$

$\therefore \quad I = \sqrt{2}\int \frac{dt}{\sqrt{t^2-1}} = \sqrt{2}\log\left|t + \sqrt{t^2-1}\right| + C$

$= \sqrt{2}\log|\sin x + \cos x + \sqrt{\sin 2x}| + C$

43. b. Write $I = \int \frac{dx}{x^3\left(a^2/x^2 - b^2\right)^{3/2}}$

and put $a^2/x^2 = t + b^2$, so that $(-2a^2/x^3)\,dx = dt$

$\therefore \quad I = \int \frac{\left(-1/2a^2\right)dt}{t^{3/2}}$

$= -\frac{1}{2a^2}\int t^{-3/2}\,dt = \frac{1}{a^2\sqrt{t}} + C$

$= \frac{1}{a^2\left(a^2/x^2 - b^2\right)^{1/2}} + C$

$= \frac{x}{a^2\left(a^2 - b^2x^2\right)^{1/2}} + C$

44. d. Putting $x^2 = t$,

$I = \frac{1}{2}\int e^{t^2}(1 + t + 2t^2)e^t\,dt$

$= \frac{1}{2}\int e^t[te^{t^2} + (e^{t^2} + 2t^2e^{t^2})]dt$

$= \frac{1}{2}\int e^t[f(t) + f'(t)]dt = \frac{1}{2}e^t(te^{t^2}) + C$

where $t = x^2$

45. b. $I = \int x\left(\frac{\ln a^{a^{x/2}}}{3a^{5x/2}b^{3x}} + \frac{\ln b^{b^x}}{2a^{2x}b^{4x}}\right)dx = \int \frac{\ln a^{2x}b^{3x}}{6a^{2x}b^{3x}}dx$

let $a^{2x}b^{3x} = t$, then $t\ln a^2b^3\,dx = dt$

$I = \int \frac{1}{6\ln a^2b^3}\frac{\ln t}{t^2}dt$

$= \frac{1}{6\ln a^2b^3}\left(\frac{-\ln t}{t} - \int \frac{-1}{t^2}dt\right)$

$= -\frac{1}{6\ln a^2b^3}\left(\frac{\ln et}{t}\right) + k$

$= -\frac{1}{6\ln a^2b^3}\left(\frac{\ln a^{2x}b^{3x}e}{a^{2x}b^{3x}}\right) + k$

46. a. $I = \int x\frac{\ln(x + \sqrt{x^2+1})}{\sqrt{x^2+1}}dx$, let $t = \sqrt{x^2+1}$

$\Rightarrow \quad \frac{dt}{dx} = \frac{x}{\sqrt{x^2+1}}$

$\Rightarrow \quad I = \int \ln(t + \sqrt{t^2-1})dt$

$= \ln(t + \sqrt{t^2-1})\,t - \int \frac{1 + \frac{t}{\sqrt{t^2-1}}}{t + \sqrt{t^2-1}}\,t\,dt$

$= t\ln(t + \sqrt{t^2-1}) - \frac{1}{2}\int \frac{2t}{\sqrt{t^2-1}}dt$

$= t\ln(t + \sqrt{t^2-1}) - \sqrt{t^2-1} + C$

$= \sqrt{1+x^2}\ln(x + \sqrt{1+x^2}) - x + C$

$\Rightarrow \quad a = 1, b = -1$

47. d. $\int \frac{\operatorname{cosec}^2 x - 2005}{\cos^{2005} x}dx$

$= \int (\cos x)^{-2005}\operatorname{cosec}^2 x dx - 2005\int \frac{dx}{\cos^{2005} x}$

$= (\cos x)^{-2005}(-\cot x)$

$- \int (-2005)(\cos x)^{-2006}(-\sin x)(-\cot x)dx - 2005$

$\int \frac{dx}{\cos^{2005} x} = -\frac{\cot x}{(\cos x)^{2005}} + C$

48. a. $f'(x) = \frac{f(x)}{6f(x) - x}$

Now $I = \int \frac{2x(x - 6f(x)) + f(x)}{(6f(x) - x)(x^2 - f(x))^2}\,dx$

$\Rightarrow \quad I = -\int \frac{2x - f'(x)}{(x^2 - f(x))^2}\,dx = \frac{1}{x^2 - f(x)} + C$

49. a. Differentiating, we get

$$\frac{f'(x)}{f(x)^2} = 2(b^2 - a^2)\sin x\cos x$$

$$\Rightarrow -\frac{1}{f(x)} = -b^2\cos^2 x - a^2\sin^2 x$$

$$\Rightarrow \quad f(x) = \frac{1}{a^2\sin^2 x + b^2\cos^2 x}$$

50. a. $\int e^x\left(\frac{1}{\sqrt{1+x^2}} - \frac{x}{\sqrt{(1+x^2)^3}} + \frac{x}{\sqrt{(1+x^2)^3}} + \frac{1-2x^2}{\sqrt{(1+x^2)^5}}\right)$

$$= e^x\frac{1}{\sqrt{1+x^2}} + e^x\frac{x}{\sqrt{(1+x^2)^3}}$$

$$= e^x\left(\frac{1}{1+x^2} + \frac{x}{\sqrt{(1+x^2)^3}}\right) + C$$

Using $\int e^x(f(x) + f'(x))dx$, we get

$$= e^x f(x) + c$$

51. d. Let $I = \int\frac{dx}{\left(1+\sqrt{x}\right)\sqrt{\left(x-x^2\right)}}$

If $\sqrt{x} = \sin p$, then $\frac{1}{2\sqrt{x}}\,dx = \cos p\, dp$

$$I = \int\frac{2\sin p\cos p\, dp}{(1+\sin p)\sin p\cos p}$$

$$= 2\int\frac{dp}{(1+\sin p)}$$

$$= 2\int\frac{(1-\sin p)\,dp}{\cos^2 p}$$

$$= 2\left\{\int\sec^2 p\, dp - \int(\tan p\sec p)\, dp\right\}$$

$$= 2(\tan p - \sec p)$$

$$= 2\left(\sqrt{\frac{x}{(1-x)}} - \frac{1}{\sqrt{(1-x)}}\right) + C = \frac{2\left(\sqrt{x}-1\right)}{\sqrt{(1-x)}} + C$$

52. c. Let $I = \int\frac{\left(ax^2 - b\right)dx}{x\sqrt{c^2x^2 - \left(ax^2+b\right)^2}}$

$$= \int\frac{\left(a - \frac{b}{x^2}\right)dx}{\sqrt{c^2 - \left(ax + \frac{b}{x}\right)^2}}, \quad \left\{\begin{array}{l}\text{put } ax + \frac{b}{x} = t \\ \therefore \left(a - \frac{b}{x^2}\right)dx = dt\end{array}\right.$$

$$= \int\frac{dt}{\sqrt{c^2 - t^2}} = \sin^{-1}\left(\frac{t}{c}\right) + k$$

$$= \sin^{-1}\left(\frac{ax + \frac{b}{x}}{c}\right) + k$$

53. b. $I = \int\frac{dx}{\cos^3 x\sqrt{\sin 2x}}$

$$= \int\frac{dx}{\cos^3 x\sqrt{\frac{2\sin x\cos x}{\cos^2 x}\cos^2 x}}$$

$$= \int\frac{\sec^4 x\, dx}{\sqrt{2\tan x}} = \frac{1}{\sqrt{2}}\int\frac{\sec^2 x(1+\tan^2 x)}{\sqrt{\tan x}}dx$$

let $t = \sqrt{\tan x}$

$$\Rightarrow \quad dt = \frac{\sec^2 x dx}{2\sqrt{\tan x}}$$

$$\Rightarrow \quad I = \frac{2}{\sqrt{2}}\int(1+t^4)dt$$

$$= \sqrt{2}\left(t + \frac{t^5}{5}\right) + C$$

$$= \frac{\sqrt{2}}{5}t(t^4+5) + C$$

$$= \frac{\sqrt{2}}{5}\sqrt{\tan x}\,(\tan^2 x + 5) + C$$

$$\Rightarrow \quad a = \frac{\sqrt{2}}{5}, b = 5$$

54. d. $\int x\log\left(1+\frac{1}{x}\right)dx$

$$= \int x\log(x+1)dx - \int x\log x dx$$

$$= \frac{x^2}{2}\log(x+1) - \frac{1}{2}\int\frac{x^2}{x+1}dx - \frac{x^2}{2}\log x + \frac{1}{2}\int\frac{x^2}{x}dx$$

$$= \frac{x^2}{2}\log(x+1) - \frac{1}{2}\int\left(x - 1 + \frac{1}{x+1}\right)dx - \frac{x^2}{2}\log x + \frac{1}{4}x^2$$

$$= \frac{x^2}{2}\log(x+1) - \frac{x^2}{2}\log x - \frac{1}{2}\left(\frac{x^2}{2} - x\right) - \frac{1}{2}\log(x+1) + \frac{1}{4}$$

$$= \frac{x^2}{2}\log(x+1) - \frac{x^2}{2}\log x - \frac{1}{2}\log(x+1) + \frac{1}{2}x + C$$

Hence, $f(x)=\dfrac{x^2}{2}-\dfrac{1}{2}, g(x)=-\dfrac{1}{2}\log x$ and $A=\dfrac{1}{2}$

55. b. Let $I=\int\left(\sqrt{\tan x}+\sqrt{\cot x}\right)dx$

$$=\int\frac{\sqrt{2}\,(\sin x+\cos x)}{\sqrt{\sin 2x}}dx$$

Put $\sin x-\cos x=z$

$\therefore(\cos x+\sin x)\,dx=dz$

Also $\sin^2 x+\cos^2 x-2\sin x\cos x=z^2$

$\Rightarrow \sin 2x=1-z^2$

$$\Rightarrow \quad I=\int\frac{\sqrt{2}\,dz}{\sqrt{1-z^2}}=\sqrt{2}\sin^{-1}z+C$$

$$=\sqrt{2}\sin^{-1}(\sin x-\cos x)+C$$

$$=\sqrt{2}\tan^{-1}\frac{\tan x-1}{\sqrt{2\tan x}}+C$$

56. c. $I_{4,3}=\int\cos^4 x\sin 3x dx$

Integrating by parts, we have

$$I_{4,3}=-\frac{\cos 3x\cos^4 x}{3}-\frac{4}{3}\int\cos^3 x\sin x\cos 3x dx$$

But $\sin x\cos 3x=-\sin 2x+\sin 3x\cos x$, so

$$I_{4,3}=-\frac{\cos x\cos^4 x}{3}+\frac{4}{3}\int\cos^3 x\sin 2x dx$$

$$-\frac{4}{3}\int\cos^4 x\sin 3x dx+C$$

$$=-\frac{\cos 3x\cos^4 x}{3}+\frac{4}{3}I_{3,2}-\frac{4}{3}I_{4,3}+C$$

Therefore, $\dfrac{7}{3}I_{4,3}-\dfrac{4}{3}I_{3,2}=-\dfrac{\cos 3x\cos^3 x}{3}+C$

or $7I_{4,3}-4I_{3,2}=-\cos 3x\cos^4 x+$ constant.

57. b. We have $\displaystyle\int\frac{dx}{x^2(x^n+1)^{(n-1)/n}}$

$$=\int\frac{dx}{x^2x^{n-1}\left(1+\dfrac{1}{x^n}\right)^{(n-1)/n}}$$

$$=\int\frac{dx}{x^{n+1}(1+x^{-n})^{(n-1)/n}}$$

Put $1+x^{-n}=t \therefore -nx^{-n-1}dx=dt\Rightarrow \dfrac{dx}{x^{n+1}}=-\dfrac{dt}{n}$

$$\Rightarrow \quad \int\frac{dx}{x^2(x^n+1)^{(n-1)/n}}=-\frac{1}{n}\int\frac{dt}{t^{(n-1)/n}}$$

$$=-\frac{1}{n}\int t^{1/n-1}dt=-\frac{1}{n}\frac{t^{1/n-1+1}}{1/n-1+1}+C$$

$$=-t^{1/n}+c=-(1+x^{-n})^{1/n}+C$$

58. c. Putting $a^6+x^8=t^2$, we get

$$\Rightarrow \quad I=\int\frac{t^2}{t^2-a^6}dt=t+\frac{a^3}{2}\ln\left|\frac{t-a^3}{t+a^3}\right|+c$$

59. c. $I=-e^{-x}\log\left(e^x+1\right)+\displaystyle\int\frac{e^{-x}e^x}{e^x+1}dx$

$$=-e^{-x}\log(e^x+1)+\int\frac{e^{-x}}{e^{-x}+1}dx.$$

$$=-e^{-x}\log(e^x+1)-\log(e^{-x}+1)+C$$

$$=-e^{-x}\log(e^x+1)-\log(1+e^x)+x+C$$

$$=-(e^{-x}+1)\log(e^x+1)+x+C$$

60. b. $I=\int xe^x\cos x dx=xe^x\sin x-\int(xe^x+e^x)\sin x dx$

$$=xe^x\sin x-xe^x(-\cos x)+\int(xe^x+e^x)\cos x\,dx$$

$$-\int e^x\sin x dx$$

$$=xe^x\sin x+xe^x\cos x-\int xe^x\cos x dx$$

$$-\int e^x(\cos x+\sin x)dx$$

$$\Rightarrow \quad 2I=xe^x(\sin x+\cos x)-e^x\sin x+d$$

$$\Rightarrow \quad 2I=e^x((x-1)\sin x+x\cos x)+d$$

$$\Rightarrow \quad I=\frac{1}{2}e^x((x-1)\sin x+x\cos x)+d$$

$$\Rightarrow \quad a=\frac{1}{2},\ b=-1,\ c=1$$

61. b. Put $2+x=t^2$, so that $dx=2t$ and

$$I=\int\frac{\sqrt{7-t^2}}{t}(2t)\,dt\ 2\int\sqrt{7-t^2}\,dt$$

$$=t\sqrt{7-t^2}+7\sin^{-1}\left(\frac{t}{\sqrt{7}}\right)+C$$

$$=\sqrt{x+2}\sqrt{5-x}+7\sin^{-1}\left(\frac{\sqrt{x+2}}{\sqrt{7}}\right)+C$$

62. c. $I=\int e^{\tan x}(\sec x-\sin x)dx$

$$=\int\sin x\,e^{\tan x}dx-\int\sec x\,e^{\tan x}dx$$

$$=-e^{\tan x}\cos x+\int\cos x\,e^{\tan x}\sec^2 x dx-\int\sec x\,e^{\tan x}dx$$

$$=-\cos x\,e^{\tan x}+C$$

63. d. $I=\displaystyle\int\frac{x^3dx}{\sqrt{1+x^2}}=\int\frac{x\times x^2dx}{\sqrt{1+x^2}}$, let $t=\sqrt{1+x^2}$

$$\Rightarrow \quad \frac{dt}{dx}=\frac{x}{\sqrt{1+x^2}}$$

$$\Rightarrow \quad I = \int (t^2 - 1)dt$$

$$= \frac{t^3}{3} - t + C = \frac{t}{3}(t^2 - 3) + C$$

$$= \frac{1}{3}\sqrt{1+x^2}(x^2 - 2) + C$$

64. d. $I = \int \frac{\sin x \cos x}{\sin x + \cos x} dx$

$$= \frac{1}{2}\int \frac{(\sin x + \cos x)^2 - 1}{\sin x + \cos x} dx$$

$$= \frac{1}{2}\int \left[\sin x + \cos x - \frac{1}{\sqrt{2}\sin(x+\pi/4)}\right] dx$$

$$= \frac{1}{2}[\sin x - \cos x] - \frac{1}{\sqrt{2}}\log\left|\operatorname{cosec}(x+\pi/4) - \cot(x+\pi/4)\right| + C$$

65. b. $I = \int \frac{\sin 2x}{(3 + 4\cos x)^3} dx$

and put $3 + 4\cos x = t$, so that $-4\sin x\, dx = dt$

$$I = \frac{-1}{8}\int \frac{(t-3)}{t^3} dt = \frac{1}{8}\left(\frac{1}{t} - \frac{3}{2}\frac{1}{t^2}\right) + C$$

$$= \frac{2t - 3}{16t^2} = \frac{8\cos x + 3}{16(3 + 4\cos x)^2} + C$$

66. c. $I = \int \frac{\ln\left(\frac{x-1}{x+1}\right)}{x^2 - 1} dx$, let $t = \ln\left(\frac{x-1}{x+1}\right)$

$$\Rightarrow \quad \frac{dt}{dx} = \frac{x+1}{x-1}\left\{\frac{x+1-(x-1)}{(x+1)^2}\right\} = \frac{2}{(x^2-1)}$$

$$\Rightarrow \quad \frac{dx}{x^2 - 1} = \frac{dt}{2}$$

$$\Rightarrow \quad I = \frac{1}{2}\int t\,dt = \frac{1}{4}t^2 + C = \frac{1}{4}\left(\ln\left(\frac{x-1}{x+1}\right)^2\right) + C$$

67. a. $A = \int \sqrt{e^x - 1}\, dx$

Let $e^x - 1 = t^2 \Rightarrow e^x dx = 2t\, dt \Rightarrow dx = \frac{2t}{t^2 + 1} dt$

$$\Rightarrow \quad I = \int t \frac{2t}{t^2 + 1} dt = \int \frac{2t^2}{t^2 + 1} dt$$

$$= \int \frac{2(t^2 + 1) - 2}{t^2 + 1} dt = \int 2dt - \int \frac{2dt}{t^2 + 1}$$

$$= 2t - 2\tan^{-1} t + c$$

$$= 2\sqrt{e^x - 1} - 2\tan^{-1}\sqrt{e^x - 1} + C$$

68. b. $\int x \sin x \sec^3 x\, dx$

$$= \int x \sin x \frac{1}{\cos^3 x} dx$$

$$= \int x \tan x \sec^2 x\, dx$$

$$= x\int \sec x(\sec x \tan x)dx - \int[\sec x(\sec x \tan x)dx]dx + C$$

$$= x\frac{\sec^2 x}{2} - \int \frac{\sec^2 x}{2} dx + C$$

$$= x\frac{\sec^2 x}{2} - \frac{\tan x}{2} + C$$

69. a. $\int \frac{e^x (x^2 + 1)}{(x+1)^2} dx$

$$= \int \frac{e^x (x^2 - 1 + 2)}{(x+1)^2} dx$$

$$= \int e^x \left[\frac{x-1}{x+1} + \frac{2}{(x+1)^2}\right] dx$$

$= \int e^x [f(x) + f'(x)]\, dx$, where $f(x) = \frac{x-1}{x+1}$ and

$$f'(x) = \frac{2}{(x+1)^2}$$

$$= e^x\left(\frac{x-1}{x+1}\right) + c$$

70. a. $I = \int \left(\frac{x+2}{x+4}\right)^2 e^x\, dx = \int e^x \left[\frac{x^2 + 4x + 4}{(x+4)^2}\right] dx$

$$\Rightarrow \quad I = \int e^x \left[\frac{x(x+4)}{(x+4)^2} + \frac{4}{(x+4)^2}\right] dx$$

$$= e^x \left[\frac{x}{x+4} + \frac{4}{(x+4)^2}\right] dx$$

$$= e^x\left(\frac{x}{x+4}\right) + c$$

71. a. Let $I = \int \frac{3 + 2\cos x}{(2 + 3\cos x)^2} dx$. Multiplying N^r and D^r by

$\operatorname{cosec}^2 x$, we get

$$\Rightarrow I = \int \frac{(3\operatorname{cosec}^2 x + 2\cot x \operatorname{cosec} x)}{(2\operatorname{cosec} x + 3\cot x)^2} dx$$

$$=-\int \frac{-3\operatorname{cosec}^2 x-2\cot x\operatorname{cosec} x}{(2\operatorname{cosec} x+3\cot x)^2}dx$$

$$=\frac{1}{2\operatorname{cosec} x+3\cot x}+C=\left(\frac{\sin x}{2+3\cos x}\right)+C$$

Multiple Correct Answers Type

1. b., c., d.

$$I=\int \frac{x^2+\cos^2 x}{x^2+1}\operatorname{cosec}^2 x dx$$

$$=\int \frac{x^2+1+\cos^2 x-1}{x^2+1}\operatorname{cosec}^2 x dx$$

$$=\int\left(1-\frac{\sin^2 x}{x^2+1}\right)\operatorname{cosec}^2 x dx$$

$$=\int\left(\operatorname{cosec}^2 x-\frac{1}{x^2+1}\right)dx$$

$$=-\cot x-\tan^{-1}x+c$$

$$=-\cot x+\cot^{-1}x-\frac{\pi}{2}+c$$

$$=-\cot x+\cot^{-1}x+c$$

2. b., d.

$$\int \sin x d(\sec x)$$

$$=\int \sin x\frac{d(\sec x)}{dx}dx=\int \sin x\sec x\tan x dx$$

$$=\int \tan^2 x dx=\int(\sec^2 x-1)dx=\tan x-x+c$$

$$\Rightarrow f(x)=\tan x,\ g(x)=x$$

3. a., d.

$$I=\int \frac{\sqrt{(1+\sin x)(1-\sin x)}}{\sqrt{\sin x(1-\sin x)}}dx$$

$$=\int \frac{\cos x}{\sqrt{\sin x(1-\sin x)}}dx$$

$$=\int \frac{\cos x}{\sqrt{\frac{1}{4}-\left(\frac{1}{2}-\sin x\right)^2}}dx$$

$$=\int \frac{-dt}{\sqrt{\left(\frac{1}{2}\right)^2-t^2}}\quad\left(\text{Putting } \frac{1}{2}-\sin x=t\right)$$

$$=-\sin^{-1}\left(\frac{t}{1/2}\right)+C=-\sin^{-1}(1-2\sin x)+C$$

$$=\cos^{-1}(1-2\sin x)+C-\frac{\pi}{2}$$

$$=\cos^{-1}(1-2\sin x)+C$$

$$=\cos^{-1}\left(1-2\left(\sqrt{\sin x}\right)^2\right)+C$$

$$=\cos^{-1}\left(1-2\sin^2 t\right)+C\quad\left(\text{Putting } \sqrt{\sin x}=\sin t\right)$$

$$=\cos^{-1}(\cos 2t)+C$$

$$=2t+C\left(\because\ \sqrt{\sin x}>0\Rightarrow \sin t>0\Rightarrow t\in\left(0\ \frac{\pi}{2}\right)\right)$$

$$=2\sin^{-1}\left(\sqrt{\sin x}\right)+C$$

4. a., c. $I=\int \sec^2 x\operatorname{cosec}^4 x dx$

$$=\int \frac{(\sin^2 x+\cos^2 x)^2}{\cos^2 x\sin^4 x}dx$$

$$=\int \frac{\sin^4 x+\cos^4 x+2\sin^2 x\cos^2 x}{\cos^2 x\sin^4 x}dx$$

$$=\int\left(\sec^2 x+2\operatorname{cosec}^2 x+\frac{\cos^2 x}{\sin^4 x}\right)dx$$

$$=\tan x-2\cot x+\int \cot^2 x\operatorname{cosec}^2 x dx$$

$$=\tan x-2\cot x-\frac{\cot^3 x}{3}+D$$

5. a., c. $g(x)=\int x^{27}(1+x+x^2)^6(6x^2+5x+4)\,dx$

$$=\int (x^4+x^5+x^6)^6(6x^5+5x^4+4x^3)\,dx$$

Let $x^6+x^5+x^4=t\Rightarrow(6x^5+5x^4+4x^3)dx=dt$

$$\therefore\quad g(x)=\int t^6 dt=\frac{t^7}{7}+c=\frac{1}{7}(x^4+x^5+x^6)^7+C$$

$$g(0)=0\Rightarrow x=0\ \Rightarrow g(1)=\frac{3^7}{7}\ \text{ also } g(-1)=\frac{1}{7}$$

6. b., d. $I=\int\sqrt{\operatorname{cosec} x+1}\,dx=\int\frac{\cot x}{\sqrt{\operatorname{cosec} x-1}}dx$

Put $\operatorname{cosec} x-1=t^2\Rightarrow-\operatorname{cosec} x\cot x\,dx=2tdt$

$$\Rightarrow\quad I=-\int\frac{-\cot x\operatorname{cosec} x}{\operatorname{cosec} x\sqrt{\operatorname{cosec} x-1}}dx=-\int\frac{2dt}{1+t^2}$$

$$=-2\tan^{-1}t+c=-2\tan^{-1}\sqrt{\operatorname{cosec} x-1}+c$$

$$=-2\left[\frac{\pi}{2}-\cot^{-1}\sqrt{\operatorname{cosec} x-1}\right]+c$$

$$=2\cot^{-1}\sqrt{\operatorname{cosec} x-1}+c$$

$$=2\cot^{-1}\frac{\cot x}{\sqrt{\operatorname{cosec} x+1}}+c$$

7. a., c. Let $\cos x = t$, $\Rightarrow \cos x = t$

$\Rightarrow \cos 2x = 2t^2 - 1$ and $dt = -\sin x\, dx$.

$$\text{Thus,}\quad I = \int \frac{t^2-2}{2t^2-1}dt = \frac{1}{2}\int \frac{2t^2-4}{2t^2-1}dt$$

$$= \frac{1}{2}\int dt - \frac{3}{2}\int \frac{dt}{2t^2-1}$$

$$= \frac{1}{2}t - \frac{3}{2\sqrt{2}} \times \frac{1}{2}\log\left|\frac{\sqrt{2}\,t-1}{\sqrt{2}\,t+1}\right| + C$$

$$= \frac{1}{2}\cos x - \frac{3}{4\sqrt{2}}\log\left|\frac{\sqrt{2}\cos x - 1}{\sqrt{2}\cos x + 1}\right| + C$$

So, $P = 1/2, Q = -\dfrac{3}{4\sqrt{2}}, f(x) = \dfrac{\sqrt{2}\cos x - 1}{\sqrt{2}\cos x + 1}$

or $P = 1/2, Q = \dfrac{3}{4\sqrt{2}}, f(x) = \dfrac{\sqrt{2}\cos x + 1}{\sqrt{2}\cos x - 1}$

8. a., d.

$$\frac{2x}{(x-1)(x-4)} = \frac{C}{x-1} + \frac{D}{x-4}$$

$$2x = C(x-4) + D(x-1)$$

$\therefore \quad C = -2/3, D = 8/3$

$$\therefore \quad \int \frac{e^{x-1}}{(x-1)(x-4)}2x\,dx$$

$$= \int e^{x-1}\left(\frac{-2/3}{x-1} + \frac{8/3}{x-4}\right)dx$$

$$= -\frac{2}{3}F(x-1) + \frac{8}{3}e^3 F(x-4) + C$$

$\therefore \quad A = -2/3, B = 8/3\, e^3$

9. a., c., d.

$$\int x^2 e^{-2x}dx = e^{-2x}(ax^2+bx+c)+d$$

Differentiating both sides, we get

$$x^2 e^{-2x} = e^{-2x}(2ax+b) + (ax^2+bx+c)(-2e^{-2x})$$

$$= e^{-2x}(-2ax^2 + 2(a-b)x + b - 2c)$$

$\Rightarrow \quad a = 1, 2(a-b) = 0, b - 2c = 0$

$\Rightarrow \quad b = 1, c = \dfrac{1}{2}$

10. a., c., d.

$$\text{Let}\quad I = \int \frac{(x^4+1)}{(x^6+1)}dt$$

$$= \frac{(x^2+1)^2 - 2x^2}{(x^2+1)(x^4-x^2+1)}dx$$

$$= \int \frac{(x^2+1)dx}{(x^4-x^2+1)} - 2\int \frac{x^2dx}{(x^6+1)}$$

$$= \int \frac{\left(1+\frac{1}{x^2}\right)dx}{\left(x^2-1+\frac{1}{x^2}\right)} - 2\int \frac{x^2dx}{(x^3)^2+1}$$

In first integral, put $x - \dfrac{1}{x} = t$

$\therefore \left(1+\dfrac{1}{x^2}\right)dx = dt$ and in second integral put $x^3 = u$

$\therefore \quad x^2dx = \dfrac{du}{3}$

$$\text{then}\quad I = \int \frac{dt}{1+t^2} - \frac{2}{3}\int \frac{du}{1+u^2}$$

$$= \tan^{-1} t - \frac{2}{3}\tan^{-1} u + C$$

$$= \tan^{-1}\left(x - \frac{1}{x}\right) - \frac{2}{3}\tan^{-1}(x^3) + C$$

Here, $f(x) = x - \dfrac{1}{x}$ and $g(x) = x^3$

Both the functions are one-one.

Also $f'(x) = 1 + \dfrac{1}{x^2} \neq 0$. Hence, $f(x)$ is monotonic.

$$\text{Also}\int \frac{f(x)}{g(x)}dx = \int \frac{x - \frac{1}{x}}{x^3}dx = \int\left(\frac{1}{x^2} - \frac{1}{x^4}\right)dx$$

$$= -\frac{1}{x} + \frac{3}{x^3} + C$$

11. a., b., c.

$$I = \int \frac{x^2-x+1}{(x^2+1)^{3/2}}e^x dx$$

$$= \int e^x\left[\frac{x^2+1}{(x^2+1)^{3/2}} - \frac{x}{(x^2+1)^{3/2}}\right]dx$$

$$= \int e^x\left[\frac{1}{\sqrt{x^2+1}} + \left\{\frac{-x}{(x^2+1)^{3/2}}\right\}\right]dx$$

$$= \int e^x[f(x) + f'(x)]dx, \text{ where } f(x) = \frac{1}{\sqrt{x^2+1}}$$

$$= e^x f(x) + c = \frac{e^x}{\sqrt{x^2+1}} + c$$

The graph of $f(x)$ is given in Fig. 7.1

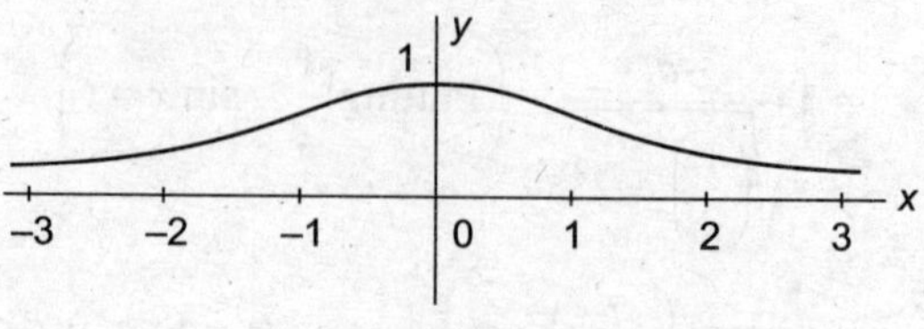

Fig. 7.1

From the graph, $f(x)$ is even, bounded function and has the range (0, 1].

12. a., c.

$$\int \frac{\cos^2 2x \sin 2x \, dx}{\cos 2x} = \frac{1}{2}\int \sin 4x \, dx$$

$$= -\frac{1}{8}\cos 4x + B$$

13. a., d.

$$\int \sin^{-1} x \cos^{-1} x \, dx = \left[\frac{\pi}{2}\sin^{-1} x - \left(\sin^{-1} x\right)^2\right] dx$$

$$= \frac{\pi}{2}\left(x\sin^{-1} x + \sqrt{1-x^2}\right)$$

$$- \left(x\left(\sin^{-1} x\right)^2 + \sin^{-1} x\sqrt{1-x^2} - x\right) + C \text{ (integrating by parts)}$$

$$= \sin^{-1} x\left[\frac{\pi}{2}x - x\sin^{-1} x - 2\sqrt{1-x^2}\right] + \frac{\pi}{2}\sqrt{1-x^2}$$

$$+ 2x + c$$

$$\therefore f^{-1}(x) = \sin^{-1} x, f(x) = \sin x$$

14. a., b., c., d.

$$\int \frac{\left(x^8 + 4 + 4x^4\right) - 4x^4}{x^4 - 2x^2 + 2} dx$$

$$= \int \frac{\left(x^4 + 2\right)^2 - \left(2x^2\right)^2}{\left(x^4 - 2x^2 + 2\right)} dx$$

$$= \int \frac{\left(x^4 + 2 - 2x^2\right)\left(x^4 + 2 + 2x^2\right)}{\left(x^4 - 2x^2 + 2\right)} dx$$

$$= \frac{x^5}{5} + \frac{2x^3}{3} + 2x + C$$

15. a., b., d.

$$\int \frac{dx}{x^2 + ax + 1} = \int \frac{dx}{\left(x + \frac{a}{2}\right)^2 + \left(1 - \frac{a^2}{4}\right)}$$

Reasoning Type

1. a. $\int e^x \sin x dx$

$$= \frac{1}{2}\int e^x (\sin x + \cos x + \sin x - \cos x) dx$$

$$= \frac{1}{2}\left(\int e^x (\sin x + \cos x) dx - \int e^x (\cos x - \sin x) dx\right)$$

$$= \frac{1}{2}\left(e^x \sin x - e^x \cos x\right) + c = \frac{1}{2}e^x(\sin x - \cos x) + c$$

2. d. For $x^2 + 2(a-1)x + a + 5 = 0$

if $D < 0 \Rightarrow 4(a-1)^2 - 4(a+5) < 0$

$\Rightarrow a^2 - 3a - 4 < 0$ or $(a-4)(a+1) < 0$ or $-1 < a < 4$

Thus for these value of $a, x^2 + 2(a-1)x + a + 5$ cannot be factorized, hence

$$\int \frac{dx}{x^2 + 2(a-1)x + a + 5} = \lambda \tan^{-1}|g(x)| + c$$

Hence, statement 1 is false and statement 2 is true.

3. b. $\int \frac{\sin x dx}{x}$ cannot be evaluated as there does not exist any method for evaluating this (integration by parts also does not work); however, $\frac{\sin x}{x}$ $(x > 0)$ is a differentiable function.

Hence, both the statements are true but statement 2 is not a correct explanation of statement 1.

4. b. $I = \int \frac{dx}{x^3\sqrt{1+x^4}} = \int \frac{dx}{x^5\sqrt{\frac{1}{x^4} + 1}}$

Let $\frac{1}{x^4} + 1 = t \Rightarrow dt = \frac{-4}{x^5}dx$

$$\Rightarrow I = -\frac{1}{4}\int \frac{dt}{\sqrt{t}} = -\frac{1}{2}\sqrt{t} = -\frac{1}{2}\sqrt{1 + \frac{1}{x^4}} + C$$

Thus, both the statements are true but statement 2 is not correct explanation of statement 1.

5. b. $f(x) = \pi \sin \pi x + 2x - 4$

$\Rightarrow g(x) = \int (\pi \sin \pi x + 2x - 4) dx$

$= -\cos \pi x + x^2 - 4x + c$

Also $f(1) = 3 \Rightarrow 1 + 1 - 4 + c = 3 \Rightarrow c = 0$

$\Rightarrow g(x) = -\cos \pi x + x^2 - 4x$

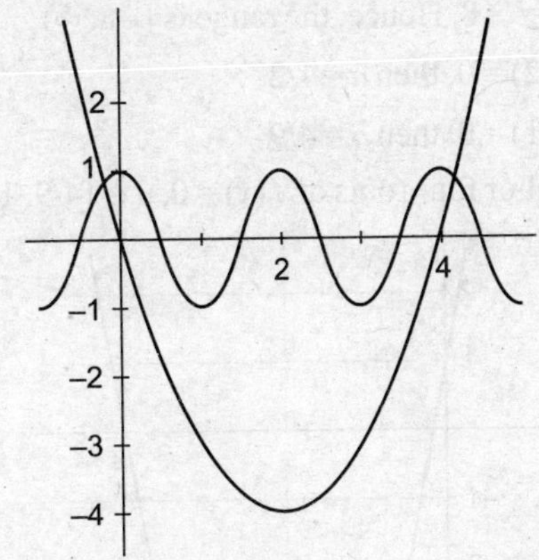

Fig. 7.2

Hence, both the statements are true but statement 2 is not a correct explanation of statement 1.

6. a. $I = \int \frac{\{f(x)\,\phi'(x) - f'(x)\,\phi(x)\}}{f(x)\,\phi(x)} \{\log \phi(x) - \log f(x)\}\,dx$

$$= \int \log \frac{\phi(x)}{f(x)} d\left\{\log \frac{\phi(x)}{f(x)}\right\} = \frac{1}{2}\left\{\log \frac{\phi(x)}{f(x)}\right\}^2 + c$$

Linked Comprehension Type

For Problems 1–3

1. d., 2. b., 3. a.

Sol. From the given data, we can conclude that $\frac{dy}{dx} = 0$, at $x = 1, 2, 3$.

Hence, $f'(x) = a(x-1)(x-2)(x-3), a > 0$

$$\Rightarrow f(x) = \int a(x^3 - 6x^2 + 11x - 6)dx$$

$$= a\int (x^3 - 6x^2 + 11x - 6)dx$$

$$= a\left(\frac{x^4}{4} - 2x^3 + \frac{11x^2}{2} - 6x\right) + C$$

Also $f(0) = 1 \Rightarrow c = 1$

$$\Rightarrow f(x) = a\left(\frac{x^4}{4} - 2x^3 + \frac{11x^2}{2} - 6x\right) + 1 \quad (1)$$

$$f(1) = a\left(-\frac{9}{4}\right) + 1, f(2) = -2a + 1,$$

$$f(3) = a\left(-\frac{9}{4}\right) + 1 \quad (2)$$

$\Rightarrow$ The graph is symmetrical about line $x = 2$ and the range is $[f(1), \infty)$ or $[f(3), \infty)$.

1. d. $f(1) = -8 \Rightarrow a = 4$ (from (2))

$\Rightarrow f(2) = -7$

2. b. $f(3) = -8$. Hence, the range is $[-8, \infty)$

3. a. If $f(2) = 0$, then $a = 1/2$

If $f(1) = 0$, then $a = 4/9$

$\Rightarrow$ For four roots of $f(x) = 0$, $a \in [4/9, 1/2]$

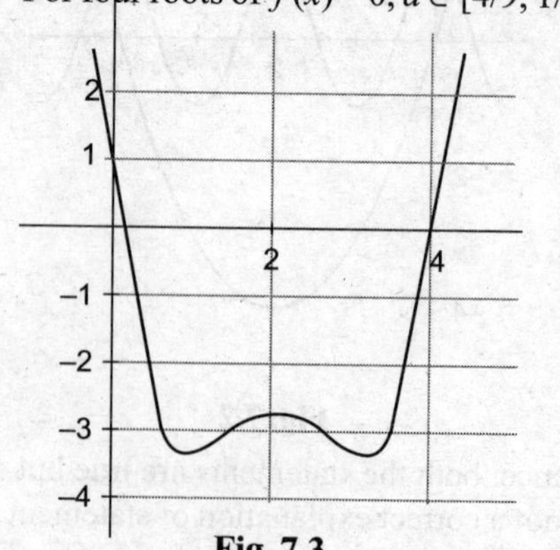

Fig. 7.3

For Problems 4–6

4. a., 5. b., 6. c.

Sol. $A = \begin{bmatrix} x & x \\ x & x \end{bmatrix} \Rightarrow A^2 = \begin{bmatrix} 2x^2 & 2x^2 \\ 2x^2 & 2x^2 \end{bmatrix}$,

$A^3 = \begin{bmatrix} 2^2x^2 & 2^2x^2 \\ 2^2x^2 & 2^2x^2 \end{bmatrix}$ and so on

Then $e^A = I + A + \frac{A^2}{2!} + \frac{A^3}{3!} + \dots +$

$$= \begin{bmatrix} 1 + x + \frac{2x^2}{2!} + \frac{2^2x^3}{3!} + \dots & x + \frac{2x^2}{2!} + \frac{2^2x^3}{3!} + \dots \\ x + \frac{2x^2}{2!} + \frac{2^2x^3}{3!} + \dots & 1 + x + \frac{2x^2}{2!} + \frac{2^2x^3}{3!} + \dots \end{bmatrix}$$

$$= \begin{bmatrix} \frac{1}{2}\left(1 + 2x + \frac{2^2x^2}{2!} + \frac{2^3x^3}{3!} + \dots\right) + \frac{1}{2} & \frac{1}{2}\left(1 + 2x + \frac{2^2x^2}{2!} + \dots\right) - \frac{1}{2} \\ \frac{1}{2}\left(1 + 2x + \frac{2^2x^2}{2!} + \frac{2^3x^3}{3!} + \dots\right) - \frac{1}{2} & \frac{1}{2}\left(1 + 2x + \frac{2^2x^2}{2!} + \dots\right) + \frac{1}{2} \end{bmatrix}$$

$$= \frac{1}{2}\begin{bmatrix} e^{2x} + 1 & e^{2x} - 1 \\ e^{2x} - 1 & e^{2x} + 1 \end{bmatrix}$$

$$\Rightarrow f(x) = e^{2x} + 1 \text{ and } g(x) = e^{2x} - 1$$

4. a. $\int \frac{e^{2x} - 1}{e^{2x} + 1} dx = \int \frac{e^x - e^{-x}}{e^x + e^{-x}} dx$

5. b. $\int (g(x) + 1)\sin x\,dx = \int e^{2x} \sin x\,dx$

$$= \frac{e^{2x}}{5}(2\sin x - \cos x).$$

6. c. $\int \frac{e^{2x} + 1}{\sqrt{e^{2x} - 1}} dx = \int \frac{e^{2x}}{\sqrt{e^{2x} - 1}} dx + \int \frac{1}{\sqrt{e^{2x} - 1}} dx$

$$= \int \frac{e^{2x}}{\sqrt{e^{2x} - 1}} dx + \int \frac{e^x}{e^x\sqrt{e^{2x} - 1}} dx$$

$$= \frac{1}{2\sqrt{e^{2x} - 1}} + \sec^{-1}(e^x) + C.$$

For Problems 7–9
7. d., 8. b., 9. a
Sol.

7. d. Here $a = 1 > 0$, therefore we make the substitution $\sqrt{x^2 + 2x + 2} = t - x$. Squaring both sides of this equality and reducing the similar terms, we get $2x + 2tx = t^2 - 2 \Rightarrow x$

$$= \frac{t^2 - 2}{2(1+t)}\, dx = \frac{t^2 + 2t + 2}{2(1+t)^2} dt;$$

$$1 + \sqrt{x^2 + 2x + 2} = 1 + t - \frac{t^2 - 2}{2(1+t)} = \frac{t^2 + 4t + 4}{2(1+t)}.$$

Substituting into the integral, we get

$$I = \int \frac{2(1+t)(t^2 + 2t + 2)}{(t^2 + 4t + 4)\, 2(1+t)^2} dt$$

$$= \int \frac{(t^2 + 2t + 2)\, dt}{(1+t)(t+2)^2}$$

Now let us expand the obtained proper rational fraction into per tail fractions:

$$\frac{t^2 + 2t + 2}{(t+1)(t+2)^3} = \frac{A}{t+1} + \frac{B}{t+2}\, \frac{D}{(t+2)^2}.$$

8. b. $I = \int \frac{dx}{x + \sqrt{x^3 - x + 1}}$

Since here $c = 1 > 0$, we can apply the second Euler substitution

$\sqrt{x^2 - x + 1} = tx - 1$

$\Rightarrow (2t - 1)x = (t^2 - 1)x^2; x = \frac{2t - 1}{t^2 - 1}$

Substituting into I, we get an integral of a rational fraction:

$$\int \frac{dx}{x + \sqrt{x^2 - x + 1}} = \int \frac{-2t^2 + 2t - 2}{t(t-1)(t+1)^2} dt,$$

$$\text{Now } \frac{-2t^2 + 2t - 2}{t(t-1)(t+1)} = \frac{A}{t} + \frac{B}{t-1} + \frac{D}{(t+1)^2} + \frac{E}{t+1}$$

9. a. In this case $a < 0$ and $c < 0$, therefore neither the first, nor the second Euler substitution is applicable. But the quadratic $7x - 10 - x^2$ has real roots $\alpha = 2, \beta = 5$, therefore we use the third Euler substitution:

$\sqrt{7x - 10 - x^2} = \sqrt{(x-2)(5-x)} = (x-2)t$

$\Rightarrow 5 - x = (x-2)t^2;$

$\Rightarrow \quad x = \frac{5 + 2t^2}{t^2 + 1}$

Matrix-Match Type

1. a.→p, q; b.→p; c.→r, s; d.→p, q.

a. Let $I = \int \frac{2^x}{\sqrt{1-4^x}} dx = \frac{1}{\log 2} \int \frac{1}{\sqrt{1-t^2}} dt$

Putting $2^x = t$, $2^x \log 2\, dx = dt$

$$I = \frac{1}{\log 2} \sin^{-1}\left(\frac{t}{1}\right) + C = \frac{1}{\log 2} \sin^{-1}(2^x) + C$$

$$\therefore K = \frac{1}{\log 2}$$

b. $\int \frac{dx}{(\sqrt{x})^2 + (\sqrt{x})^7} = \int \frac{dx}{(\sqrt{x})^7 \left(1 + \frac{1}{(\sqrt{x})^5}\right)}$

Put $\frac{1}{(\sqrt{x})^5} = y, \frac{dy}{dx} = -\frac{5}{2(\sqrt{x})^7}$

$$\therefore I = \int \frac{-2dy}{5(1+y)} = -\frac{2}{5} \ln|1+y| + c$$

$$= \frac{2}{5} \ln \left(\frac{1}{1 + \frac{1}{(\sqrt{x})^5}} \right)$$

$$\Rightarrow a = \frac{2}{5}, k = \frac{5}{2}$$

c. Add and subtract $2x^2$ in the numerator, then $k = 1$ and $m = 1$.

d. $I = \int \frac{dx}{5 + 4\cos x}$

$$= \int \frac{dx}{5\left(\sin^2 \frac{x}{2} + \cos^2 \frac{x}{2}\right) + 4\left(\cos^2 \frac{x}{2} - \sin^2 \frac{x}{2}\right)}$$

$$= \int \frac{dx}{9\cos^2 \frac{x}{2} + \sin^2 \frac{x}{2}} = \int \frac{\sec^2 \frac{x}{2}}{9 + \tan^2 \frac{x}{2}} dx$$

Let $t = \tan \frac{x}{2} \Rightarrow 2\, dt = \sec^2 \frac{x}{2} dx$

$$\Rightarrow \quad I = \int \frac{2dt}{9 + t^2} = \frac{2}{3} \tan^{-1}\left(\frac{t}{3}\right) + C$$

$$= \frac{2}{3} \tan^{-1} \left(\frac{\tan\left(\frac{x}{2}\right)}{3} \right) + C$$

$$\Rightarrow \quad k = \frac{2}{3}, m = \frac{1}{3}$$

2. a.→r, b.→s; c→q; d→p.

a. $\int \frac{e^{2x}-1}{e^{2x}+1}dx = \int \frac{e^x - e^{-x}}{e^x + e^{-x}}dx = \int \frac{(e^x + e^{-x})'}{e^x + e^{-x}}dx$

$= \log(e^x + e^{-x}) = \log(e^{2x}+1) - x + C$

b. $I = \int \frac{1}{(e^x + e^{-x})^2}dx = \int \frac{e^{2x}}{(e^{2x}+1)^2}dx$

Put $e^{2x}+1 = t \Rightarrow 2e^{2x}\,dx = dt$, we get

$\Rightarrow I = \frac{1}{2}\int \frac{1}{t^2}dt = -\frac{1}{2}\frac{1}{t} + C = -\frac{1}{2(e^{2x}+1)} + C.$

c. $I = \int \frac{e^{-x}}{1+e^x}dx = \int \frac{e^{-x}\,e^{-x}}{e^{-x}+1}dx$

Put $e^{-x}+1 = t \Rightarrow -e^{-x}\,dx = dt$

$\Rightarrow \quad I = -\int \frac{(t-1)}{t}dt = \int \left(\frac{1}{t}-1\right)dt$

$= \log t - t + C$

$= \log(e^{-x}+1) - (e^{-x}+1) + C$

$= \log(e^x+1) - x - e^{-x} - 1 + C$

$= \log(e^x+1) - x - e^{-x} + C$

d. $I = \int \frac{1}{\sqrt{1-e^{2x}}}dx = \int \frac{e^{-x}}{\sqrt{e^{-2x}-1}}dx$

Put $e^{-x} = t \Rightarrow -e^{-x}\,dx = dt$,

$\Rightarrow \quad I = -\int \frac{1}{\sqrt{t^2-1}}dt$

$= -\log\left[t + \sqrt{t^2-1}\right] + C$

$= -\log\left[e^{-x} + \sqrt{e^{-2x}-1}\right] + C$

$= -\log\left[\frac{1}{e^x} + \frac{\sqrt{1-e^{2x}}}{e^x}\right] + C$

$= -\log\left[1 + \sqrt{1-e^{2x}}\right] + \log e^x + C$

$= x - \log\left[1 + \sqrt{1-e^{2x}}\right] + C$

3. a.→p, q, r; b.→p, q, r; c.→p, q, r, s; d.→p, q, r, s; t.

a. $\int \frac{x^2-x+1}{x^3-4x^2+4x}dx = \int\left[\frac{A}{x} + \frac{B}{x-2} + \frac{C}{(x-2)^2}\right]dx.$

b. $\int \frac{x^2-1}{x(x-2)^3}dx$

$= \int\left[\frac{A}{x} + \frac{B}{x-2} + \frac{C}{(x-2)^2} + \frac{D}{(x-2)^3}\right]dx.$

c. $\int \frac{x^3+1}{x(x-2)^2}dx = \int\left[\left(\frac{x^3+1}{x(x-2)^2} - 1\right) + 1\right]dx$

$= \int\left[\left(\frac{x^3+1-x(x-2)^2}{x(x-2)^2}\right) + 1\right]dx$

$= \int\left[\left(\frac{A}{x} + \frac{B}{x-2} + \frac{C}{(x-2)^2}\right) + 1\right]dx$

d. $\int \frac{x^5+1}{x(x-2)^3}dx = \int\left[x + kx + \frac{g(x)}{x(x-2)^3}\right]dx,$

where k is constant $\neq 0$ and $g(x)$ is polynomial of degree less than four.

Integer Type

1. (1) $\int \frac{3x^4-1}{x^2(x^3+1+x^{-1})^2}dx = \int \frac{3x^2-x^{-2}}{(x^3+1+x^{-1})^2}dx$

$= -\frac{x}{x^4+x+1} + c$

$f(0) = 0 \Rightarrow c = 0 \Rightarrow f(-1) = 1$

2. (6) We have $\int \frac{4e^x+6e^{-x}}{9e^x-4e^{-x}}dx = Ax + B\ln(9e^{2x}-4) + C$

Differentiating both sides w.r.t. x, we get

$\frac{4e^x+6e^{-x}}{9e^x-4e^{-x}} = A + \frac{18\,B\,e^x}{9e^x-4e^{-x}}$

$\Rightarrow \frac{4e^x+6e^{-x}}{9e^x-4e^{-x}} = \frac{(9A+18\,B)e^x - 4Ae^{-x}}{9e^x-4e^{-x}}$

$\Rightarrow 9A + 18B = 4; -4A = 6$

$\Rightarrow A = -3/2,\ B = \left(4+\frac{27}{2}\right)\frac{1}{18} = \frac{35}{36}$, C can have any real value.

3. (2) $I = \int \frac{x^2-1}{(x^2+1)\sqrt{x^4+1}}dx$

$= \int \frac{x^2(1-1/x^2)}{x^2(x+1/x)\sqrt{x^2+1/x^2}}dx$

$= \int \frac{(1-1/x^2)dx}{(x+1/x)\sqrt{(x+1/x)^2-2}}$

Putting $x + (1/x) = t$, we have $I = \int \frac{dt}{t\sqrt{t^2-2}}$.

Again putting $t^2 - 2 = y^2$, $2t\,dt = 2y\,dy$, we get

$$I = \int \frac{y\,dy}{(y^2+2)y} = \frac{1}{\sqrt{2}}\tan^{-1}\frac{y}{\sqrt{2}}$$

$$= \frac{1}{2}\tan^{-1}\frac{\sqrt{x^2+1/x^2}}{\sqrt{2}} + c$$

4. (0) $\int \left(\frac{\cos x}{x} - \log x^{\sin x}\right) dx$

$$= \int \frac{\cos x}{x}dx - \int \sin x \log x dx$$

$$= \cos x \log x - \int -\sin x \log x dx - \int \sin x \log x dx$$

(integration 1st integral by parts)

$= \cos x \log x + c$

$f(1) = 0 + c = 0 \Rightarrow c = 0$

$\Rightarrow f(\pi/2) = 0$

5. (1) Let $I = \int \frac{1}{(x+1)\sqrt{x^2-1}}dx$

Putting $x + 1 = \frac{1}{t}$ and $dx = -\frac{1}{t^2}dt$, we get

$$I = \int \frac{1}{\frac{1}{t}\sqrt{\left(\frac{1}{t}-1\right)^2 - 1}}\left(-\frac{1}{t^2}\right) dt$$

$$= -\int \frac{dt}{\sqrt{1-2t}} = -\int (1-2t)^{-1/2}\, dt$$

$$= -\frac{(1-2t)^{1/2}}{(-2)\left(\frac{1}{2}\right)} + C = \sqrt{1-2t} + C$$

$$= \sqrt{1 - \frac{2}{x+1}} + C = \sqrt{\frac{x-1}{x+1}} + C$$

$f(1) = 0 \Rightarrow c = 0$. Then

$$f\left(\frac{3}{2}\right) = \sqrt{\frac{1}{5}} \Rightarrow 5\left(f\left(\frac{3}{2}\right)\right)^2 = 1.$$

6. (2) $I = \int \frac{(x-x^3)^{1/3}}{x^4}dx = \int \frac{\left(\frac{1}{x^2}-1\right)^{1/3}}{x^3}dx$

Putting $\frac{1}{x^2} = t$, $\frac{1}{x^3}dx = -\frac{dt}{2}$, we get

$$I = -\frac{1}{2}\int t^{1/3}dt = -\frac{3}{8}t^{4/3} + C = -\frac{3}{8}\left(\frac{1}{x^2}-1\right)^{4/3} + C$$

$\Rightarrow a = -3/8$, $b = 4/3 \Rightarrow ab = -1/2 \Rightarrow 1/|ab| = 2$

7. (3) $I = \int \frac{1}{[(x-1)^3(x+2)^5]^{1/4}}dx$

$$= \int \frac{1}{\left(\frac{x-1}{x+2}\right)^{3/4}(x+2)^2}dx$$

Let $\frac{x-1}{x+2} = t \Rightarrow \frac{3dx}{(x+2)^2} = dt$

$$\Rightarrow I = \frac{1}{3}\int \frac{1}{t^{3/4}}dt$$

$$= \frac{1}{3}\left(\frac{t^{1/4}}{1/4}\right) + C = \frac{4}{3}t^{1/4} + C = \frac{4}{3}\left(\frac{x-1}{x+2}\right)^{1/4} + C$$

$\Rightarrow a = 4/3$ and $b = 1/4 \Rightarrow ab = 1/3$

8. (5) $I = \int \frac{\sqrt{\tan x}}{\sin x \cos x}dx = \int \frac{\sqrt{\tan x}}{\tan x}\sec^2 x\, dx$

$$= \int \frac{1}{\sqrt{t}}dt, \text{ where } t = \tan x$$

$I = 2t^{1/2} + C = 2\sqrt{\tan x} + C$

9. (3) Let $I = \int \frac{(5x^4 + 4x^5)}{(x^5+x+1)^2}dx$

Dividing above and below by x^{10}, we get

$$I = \int \frac{\left(\frac{5}{x^6}+\frac{4}{x^5}\right)dx}{\left(1+\frac{1}{x^4}+\frac{1}{x^5}\right)^2}$$

Puting $1 + \frac{1}{x^4} + \frac{1}{x^5} = t$,

$\left(-\frac{4}{x^5} - \frac{5}{x^6}\right)dx = dt$ or $\left(\frac{4}{x^5} + \frac{5}{x^6}\right)dx = -dt$, we get

$$I = -\int \frac{dt}{t^2} = \frac{1}{t} + c = \frac{1}{\left(1+\frac{1}{x^4}+\frac{1}{x^5}\right)} + c$$

$$= \frac{x^5}{(x^5+x+1)} + c$$

$f(1) = 1/3$

10. (1) Let $I = \int \frac{\cos^2 x + \sin 2x)}{(2\cos x - \sin^2 x)}dx$

$$= \int \frac{\cos x + 2\sin x \cos x}{(2\cos x - \sin x)^2} dx$$

Integrating by parts, taking cos x as the first and $\frac{(\cos x + 2\sin x)}{(2\cos x - \sin x)^2}$ as the second function, we have

$$I = \cos x \left\{\frac{1}{2\cos x - \sin x}\right\} - \int \frac{-\sin x dx}{(2\cos x - \sin x)}$$

$$= \cos x \left\{\frac{1}{2\cos x - \sin x}\right\} + \int \frac{\sin x dx}{(2\cos x - \sin x)}$$

$$= \frac{\cos x}{(2\cos x - \sin x)}$$

$$+ \int \frac{-\frac{1}{5}(2\cos x - \sin x) - \frac{2}{5}(-2\sin x - \cos x)}{(2\cos x - \sin x)} dx$$

$$\left[N' = \lambda D' + \mu \frac{d}{dx} D'\right]$$

$$= \frac{\cos x}{(2\cos x - \sin x)}$$

$$-\frac{1}{5}\int dx - \frac{2}{5}\int \frac{(-2\sin x - \cos x)}{(2\cos x - \sin x)} dx$$

$$= \frac{\cos x}{2\cos x - \sin x} - \frac{1}{5}x - \frac{2}{5}\ln|2\cos x - \sin x| + c$$

$$\Rightarrow a = -\frac{1}{5}, b = -\frac{2}{5} \Rightarrow |a + 2b| = 1.$$

11. (1) $I = \int \frac{\left(\frac{2}{x^2} + \frac{1}{x\sqrt{x}}\right) dx}{\left(1 + \frac{1}{\sqrt{x}} + \frac{1}{x}\right)^2}$

Let $1 + \frac{1}{\sqrt{x}} + \frac{1}{x} = t$

$$\Rightarrow I = -2\int \frac{dt}{t^2} = \frac{2}{t} + c$$

$$= \frac{2x}{x + \sqrt{x} + 1} + c$$

CHAPTER 8

Definite Integration

Objective Type

Each question has four choices a, b, c, and d, out of which *only one* is correct.

1. If $S_n = \left[\frac{1}{1+\sqrt{n}} + \frac{1}{2+\sqrt{2n}} + \cdots + \frac{1}{n+\sqrt{n^2}}\right]$, then $\lim_{n\to\infty} S_n$ is equal to

a. $\log 2$ b. $\log 4$
c. $\log 8$ d. None of these

2. The value of $\lim_{n\to\infty} \sum_{r=1}^{4n} \frac{\sqrt{n}}{\sqrt{r}\left(3\sqrt{r}+4\sqrt{n}\right)^2}$ is equal to

a. $\frac{1}{35}$ b. $\frac{1}{14}$
c. $\frac{1}{10}$ d. $\frac{1}{5}$

3. Which of the following is incorrect?

a. $\int_{a+c}^{b+c} f(x)\,dx = \int_a^b f(x+c)dx$

b. $\int_{ac}^{bc} f(x)dx = c\int_a^b f(cx)dx$

c. $\int_{-a}^{a} f(x)dx = \frac{1}{2}\int_{-a}^{a} (f(x)+f(-x)\,dx$

d. None of these

4. The solution for x of the equation $\int_{\sqrt{2}}^{x} \frac{dt}{t\sqrt{t^2-1}} = \frac{\pi}{2}$ is

a. π b. $\frac{\sqrt{3}}{2}$
c. $2\sqrt{2}$ d. None of these

5. $\int_{-1}^{1/2} \frac{e^x\left(2-x^2\right)dx}{(1-x)\sqrt{1-x^2}}$ is equal to

a. $\frac{\sqrt{e}}{2}\left(\sqrt{3}+1\right)$ b. $\frac{\sqrt{3e}}{2}$
c. $\sqrt{3e}$ d. $\sqrt{\frac{e}{3}}$

6. The value of the integral $\int_{-\pi}^{\pi} \sin mx \sin nx\,dx$ for $m \neq n\ (m, n \in I)$ is

a. 0 b. π
c. $\pi/2$ d. 2π

7. The value of the integral $\int_0^\infty \frac{x\log x}{(1+x^2)^2}dx$ is

a. 0 b. $\log 7$
c. $5\log 13$ d. None of these

8. $\int_0^a \frac{dx}{x+\sqrt{a^2-x^2}}$ is

a. $\frac{a^2}{4}$ b. $\frac{\pi}{2}$
c. $\frac{\pi}{4}$ d. π

9. $\int_0^{\pi/2} |\sin x - \cos x|\,dx$ is equal to

a. 0
b. $2\left(\sqrt{2}-1\right)$
c. $\sqrt{2}-1$
d. $2\left(\sqrt{2}+1\right)$

10. $\int_0^{\pi} \frac{x \tan x}{\sec x + \cos x} dx$ is

a. $\frac{\pi^2}{4}$
b. $\frac{\pi^2}{2}$
c. $\frac{3\pi^2}{2}$
d. $\frac{\pi^2}{3}$

11. If $\int_{-1}^{4} f(x)dx = 4$ and $\int_{2}^{4} (3 - f(x))dx = 7$, then the value of $\int_{2}^{-1} f(x)dx$ is

a. 2
b. –3
c. –5
d. None of these

12. If $\int_0^1 e^{x^2}(x-\alpha)dx = 0$, then

a. $1<\alpha<2$
b. $\alpha<0$
c. $0<\alpha<1$
d. $\alpha=0$

13. If $\int_0^{\infty} \frac{x^2\, dx}{\left(x^2+a^2\right)\left(x^2+b^2\right)\left(x^2+c^2\right)}$

$= \frac{\pi}{2(a+b)(b+c)(c+a)}$, then the value of

$\int_0^{\infty} \frac{dx}{\left(x^2+4\right)\left(x^2+9\right)}$ is

a. $\frac{\pi}{60}$
b. $\frac{\pi}{20}$
c. $\frac{\pi}{40}$
d. $\frac{\pi}{80}$
e. $\frac{\pi}{10}$

14. The value of the integral $\int_0^1 \frac{dx}{x^2 + 2x\cos\alpha + 1}$ is equal to

a. $\sin\alpha$
b. $\alpha \sin\alpha$
c. $\frac{\alpha}{2\sin\alpha}$
d. $\frac{\alpha}{2}\sin\alpha$

15. The value of $\int_1^e \frac{1 + x^2 \ln x}{x + x^2 \ln x} dx$ is

a. e
b. $\ln(1+e)$
c. $e + \ln(1+e)$
d. $e - \ln(1+e)$

16. The value of the integral $\int_0^{1/\sqrt{3}} \frac{dx}{(1+x^2)\sqrt{1-x^2}}$ must be

a. $\frac{\pi}{2\sqrt{2}}$
b. $\frac{\pi}{4\sqrt{2}}$
c. $\frac{\pi}{8\sqrt{2}}$
d. None of these

17. The value of the integral $\int_0^{\log 5} \frac{e^x\sqrt{e^x - 1}}{e^x + 3} dx$ is

a. $3+2\pi$
b. $4-\pi$
c. $2+\pi$
d. None of these

18. $\int_0^{\infty} \frac{xdx}{(1+x)(1+x^2)}$ is equal to

a. $\frac{\pi}{4}$
b. $\frac{\pi}{2}$
c. π
d. None of these

19. $\int_0^{\infty} \frac{dx}{[x+\sqrt{x^2+1}]^3}$ is equal to

a. $\frac{3}{8}$
b. $\frac{1}{8}$
c. $-\frac{3}{8}$
d. None of these

20. Given $\int_0^{\pi/2} \frac{dx}{1+\sin x+\cos x} = \log 2$, then the value of the definite integral $\int_0^{\pi/2} \frac{\sin x}{1+\sin x+\cos x} dx$ is equal to

a. $\frac{1}{2}\log 2$
b. $\frac{\pi}{2} - \log 2$
c. $\frac{\pi}{4} - \frac{1}{2}\log 2$
d. $\frac{\pi}{2} + \log 2$

21. If $I_1 = \int_{-100}^{101} \frac{dx}{\left(5+2x-2x^2\right)\left(1+e^{2-4x}\right)}$ and

$I_2 = \int_{-100}^{101} \frac{dx}{5+2x-2x^2}$ then $\frac{I_1}{I_2}$ is

a. 2
b. $\frac{1}{2}$
c. 1
d. $-\frac{1}{2}$

22. If $f(x) = \frac{e^x}{1+e^x}$, $I_1 = \int_{f(-a)}^{f(a)} xg\left(x(1-x)\right) dx$ and

$I_2 = \int_{f(-a)}^{f(a)} g\left(x(1-x)\right) dx$, then the value of $\frac{I_2}{I_1}$ is

a. -1 b. -2
c. 2 d. 1

23. If $f(y)=e^y, g(y)=y, y>0$ and $F(t)=\int_0^t f(t-y)g(y)\,dt$, then
a. $F(t)=e^t-(1+t)$ b. $F(t)=te^t$
c. $F(t)=te^{-t}$ d. $F(t)=1-e^t(1+t)$

24. The value of the definite integral $\int_0^{\sqrt{\ln(\pi/4)}} \cos(e^{x^2})\, 2xe^{x^2}\,dx$ is
a. 1 b. $1+(\sin 1)$
c. $1-(\sin 1)$ d. $(\sin 1)-1$

25. The value of $\int_1^{(1+\sqrt{5})/2} \frac{x^2+1}{x^4-x^2+1}\log\left(1+x-\frac{1}{x}\right)dx$ is
a. $\frac{\pi}{8}\log_e 2$ b. $\frac{\pi}{2}\log_e 2$
c. $-\frac{\pi}{2}\log_e 2$ d. None of these

26. If $f(x)$ satisfies the condition of Rolle's theorem in $[1, 2]$, then $\int_1^2 f'(x)\,dx$ is equal to
a. 1 b. 3
c. 0 d. None of these

27. The value of the integral
$$\int_{-1}^{3}\left(\tan^{-1}\frac{x}{x^2+1}+\tan^{-1}\frac{x^2+1}{x}\right)dx \text{ is equal to}$$
a. π b. 2π
c. 4π d. None of these

28. If $P(x)$ is a polynomial of the least degree that has a maximum equal to 6 at $x=1$, and a minimum equal to 2 at $x=3$, then $\int_0^1 P(x)dx$ equals
a. $\frac{17}{4}$ b. $\frac{13}{4}$
c. $\frac{19}{4}$ d. $\frac{5}{4}$

29. The numbers of possible continuous $f(x)$ defined in $[0, 1]$ for which $I_1=\int_0^1 f(x)\,dx=1, I_2=\int_0^1 x\,f(x)\,dx=a, I_3=\int_0^1 x^2\,f(x)\,dx=a^2$ is/are
a. 1 b. ∞
c. 2 d. 0

30. The value of the definite integral $\int_0^{\pi/2}\sqrt{\tan x}\,dx$ is
a. $\sqrt{2}\pi$ b. $\frac{\pi}{\sqrt{2}}$
c. $2\sqrt{2}\pi$ d. $\frac{\pi}{2\sqrt{2}}$

31. Suppose that $F(x)$ is an antiderivative of $f(x)=\frac{\sin x}{x}$, where $x>0$, then $\int_1^3 \frac{\sin 2x}{x}dx$ can be expressed as
a. $F(6)-F(2)$ b. $\frac{1}{2}(F(6)-F(2))$
c. $\frac{1}{2}(F(3)-F(1))$ d. $2(F(6)-F(2))$

32. If $\int_0^1 \cot^{-1}(1-x+x^2)dx=\lambda\int_0^1 \tan^{-1} x dx$, then λ is equal to
a. 1 b. 2
c. 3 d. 4

33. The value of the integral $\int_{-3\pi/4}^{5\pi/4}\frac{(\sin x+\cos x)}{e^{x-\pi/4}+1}dx$ is
a. 0 b. 1
c. 2 d. None of these

34. $\int_{2-a}^{2+a} f(x)dx$ is equal to (where $f(2-\alpha)=f(2+\alpha)\ \forall\ \alpha\in R$)
a. $2\int_2^{2+a} f(x)dx$ b. $2\int_0^a f(x)dx$
c. $2\int_2^2 f(x)dx$ d. None of these

35. The value of the integral $\int_0^1 e^{x^2}dx$ lies in the interval
a. $(0,1)$ b. $(-1,0)$
c. $(1,e)$ d. None of these

36. $I_1=\int_0^{\pi/2}\ln(\sin x)dx$, $I_2=\int_{-\pi/4}^{\pi/4}\ln(\sin x+\cos x)dx$, then
a. $I_1=2I_2$ b. $I_2=2I_1$
c. $I_1=4I_2$ d. $I_2=4I_1$

37. If $I_1=\int_0^{\pi/2}\frac{\cos^2 x}{1+\cos^2 x}dx$, $I_2=\int_0^{\pi/2}\frac{\sin^2 x}{1+\sin^2 x}dx$,
$I_3=\int_0^{\pi/2}\frac{1+2\cos^2 x\sin^2 x}{4+2\cos^2 x\sin^2 x}dx$, then

a. $I_1 = I_2 > I_3$
b. $I_3 > I_1 = I_2$
c. $I_1 = I_2 = I_3$
d. None of these

38. If $f(x)$ is continuous for all real values of x, then $\sum_{r=1}^{n} \int_0^1 f(r-1+x)\,dx$ is equal to

a. $\int_0^n f(x)dx$
b. $\int_0^1 f(x)dx$
c. $n\int_0^1 f(x)dx$
d. $(n-1)\int_0^1 f(x)dx$

39. $I_1 = \int_0^{\frac{\pi}{2}} \frac{\sin x - \cos x}{1+\sin x \cos x}\,dx,\ I_2 = \int_0^{2\pi} \cos^6 x\,dx,$

$I_3 = \int_{-\frac{\pi}{2}}^{\frac{\pi}{2}} \sin^3 x\,dx,\ I_4 = \int_0^1 \ln\left(\frac{1}{x}-1\right)dx$, then

a. $I_2 = I_3 = I_4 = 0, I_1 \neq 0$
b. $I_1 = I_2 = I_3 = 0, I_4 \neq 0$
c. $I_1 = I_3 = I_4 = 0, I_2 \neq 0$
d. $I_1 = I_2 = I_3 = 0, I_4 \neq 0$

40. If $f(x)$ and $g(x)$ are continuous functions, the $\int_{\ln \lambda}^{\ln 1/\lambda} \frac{f(x^2/4)[f(x)-f(-x)]}{g(x^2/4)[g(x)+g(-x)]}dx$ is

a. depend on λ
b. a non-zero constant
c. zero
d. None of these

41. $\int_{-\pi}^{\pi} \frac{2x(1+\sin x)}{1+\cos^2 x}dx$ is equal to

a. π
b. π^2
c. 0
d. None of these

42. $f(x) > 0 \ \forall\ x \in R$ and is bounded. If

$$\lim_{n\to\infty}\left[\int_0^a \frac{f(x)dx}{f(x)+f(a-x)} + a\int_a^{2a} \frac{f(x)dx}{f(x)+f(3a-x)} + a^2\int_{2a}^{3a} \frac{f(x)dx}{f(x)+f(5a-x)} + \dots + a^{n-1}\int_{(n-1)a}^{na} \frac{f(x)dx}{f(x)+f[(2n-1)a-x]}\right] = 7/5$$

(where $a < 1$), then a is equal to

a. $\frac{2}{7}$
b. $\frac{1}{7}$
c. $\frac{14}{19}$
d. $\frac{9}{14}$

43. If $f(x) = \int_0^{\pi} \frac{t \sin t\,dt}{\sqrt{1+\tan^2 x \sin^2 t}}$ for $0 < x < \frac{\pi}{2}$, then

a. $f(0^+) = -\pi$
b. $f\left(\frac{\pi}{4}\right) = \frac{\pi^2}{8}$
c. f is continuous and differentiable in $\left(0, \frac{\pi}{2}\right)$
d. f is continuous but not differentiable in $\left(0, \frac{\pi}{2}\right)$

44. If $\int_{-\pi/4}^{3\pi/4} \frac{e^{\pi/4}dx}{(e^x + e^{\pi/4})(\sin x + \cos x)} = k\int_{-\pi/2}^{\pi/2} \sec x dx$, then the value of k is

a. $\frac{1}{2}$
b. $\frac{1}{\sqrt{2}}$
c. $\frac{1}{2\sqrt{2}}$
d. $-\frac{1}{\sqrt{2}}$

45. $\int_{-\pi/3}^{0} \left[\cot^{-1}\left(\frac{2}{2\cos x - 1}\right) + \cot^{-1}\left(\cos x - \frac{1}{2}\right)\right] dx$ is equal to

a. $\frac{\pi^2}{6}$
b. $\frac{\pi^2}{3}$
c. $\frac{\pi^2}{8}$
d. $\frac{3\pi^2}{8}$

46. $\int_0^{\infty} \left(\frac{\pi}{1+\pi^2 x^2} - \frac{1}{1+x^2}\right) \log x\,dx$ is equal to

a. $-\frac{\pi}{2}\ln \pi$
b. 0
c. $\frac{\pi}{2}\ln 2$
d. None of these

47. If $f(x) = \cos(\tan^{-1}x)$, then the value of the integral $\int_0^1 x f''(x)\,dx$ is

a. $\frac{3-\sqrt{2}}{2}$
b. $\frac{3+\sqrt{2}}{2}$
c. 1
d. $1 - \frac{3}{2\sqrt{2}}$

48. The equation of the curve is $y = f(x)$. The tangents at $[1, f(1)]$ $[2, f(2)]$ and $[3, f(3)]$ make angle $\frac{\pi}{6}, \frac{\pi}{3}, \frac{\pi}{4}$, respectively, with the positive direction of x-axis, then the value of $\int_2^3 f'(x) f''(x)\,dx + \int_1^3 f''(x)dx$ is equal to

a. $-1/\sqrt{3}$
b. $1/\sqrt{3}$
c. 0
d. None of these

49. The value of $\int_1^e \left(\frac{\tan^{-1} x}{x}+\frac{\log x}{1+x^2}\right)dx$ is

a. $\tan e$ b. $\tan^{-1} e$
c. $\tan^{-1}(1/e)$ d. None of these

50. If $f(\pi)=2$ and $\int_0^{\pi}(f(x)+f''(x))\sin x\,dx=5$, then $f(0)$ is equal to (it is given that $f(x)$ is continuous in $[0,\pi]$)

a. 7 b. 3
c. 5 d. 1

51. If $\int_1^2 e^{x^2}dx=a$, then $\int_e^{e^4}\sqrt{\ln x}\,dx$ is equal to

a. $2e^4-2e-a$ b. $2e^4-e-a$
c. $2e^4-e-2a$ d. e^4-e-a

52. $\int_{-\pi/2}^{\pi/2}\frac{e^{|\sin x|}\cos x}{(1+e^{\tan x})}dx$ is equal to

a. $e+1$ b. $1-e$
c. $e-1$ d. None of these

53. The value of the expression $\frac{\int_0^a x^4\sqrt{a^2-x^2}\,dx}{\int_0^a x^2\sqrt{a^2-x^2}\,dx}$ is equal to

a. $\frac{a^2}{6}$ b. $\frac{3a^2}{2}$
c. $\frac{3a^2}{4}$ d. $\frac{a^2}{2}$

54. If $A=\int_0^{\pi}\frac{\cos x}{(x+2)^2}dx$, then $\int_0^{\pi/2}\frac{\sin 2x}{x+1}dx$ is equal to

a. $\frac{1}{2}+\frac{1}{\pi+2}-A$ b. $\frac{1}{\pi+2}-A$
c. $1+\frac{1}{\pi+2}-A$ d. $A-\frac{1}{2}-\frac{1}{\pi+2}$

55. $\int_0^4\frac{(y^2-4y+5)\sin(y-2)dy}{[2y^2-8y+11]}$ is equal to

a. 0 b. 2
c. -2 d. None of these

56. $\int_{\sin\theta}^{\cos\theta} f(x\tan\theta)dx$ $\left(\text{where } \theta\neq\frac{n\pi}{2}, n\in I\right)$ is equal to

a. $-\cos\theta\int_1^{\tan\theta} f(x\sin\theta)dx$
b. $-\tan\theta\int_{\cos\theta}^{\sin\theta} f(x)dx$
c. $\sin\theta\int_1^{\tan\theta} f(x\cos\theta)dx$
d. $\frac{1}{\tan\theta}\int_{\sin\theta}^{\sin\theta\tan\theta} f(x)dx$

57. Let $I_1=\int_0^1\frac{e^x dx}{1+x}$ and $I_2=\int_0^1\frac{x^2 dx}{e^{x^3}(2-x^3)}$, then $\frac{I_1}{I_2}$ is equal to

a. $3/e$ b. $e/3$
c. $3e$ d. $1/3e$

58. If $\int_0^1\frac{\sin t}{1+t}dt=\alpha$, then the value of the integral $\int_{4\pi-2}^{4\pi}\frac{\sin\frac{t}{2}}{4\pi+2-t}dt$ in terms of α is given by

a. 2α b. -2α
c. α d. $-\alpha$

59. $\int_0^1\frac{\tan^{-1}x}{x}dx$ is equal to

a. $\int_0^{\frac{\pi}{2}}\frac{\sin x}{x}dx$ b. $\int_0^{\frac{\pi}{2}}\frac{x}{\sin x}dx$
c. $\frac{1}{2}\int_0^{\frac{\pi}{2}}\frac{\sin x}{x}dx$ d. $\frac{1}{2}\int_0^{\frac{\pi}{2}}\frac{x}{\sin x}dx$

60. If $I_K=\int_1^e(\ln x)^k dx$ (where $k\in I^+$) then I_4 equals

a. $9e-24$ b. $12-2e$
c. $24-9e$ d. $6e-12$

61. If $I(m,n)=\int_0^1 x^{m-1}(1-x)^{n-1}dx$, then $(m,n\in I, m, n\geq 0)$

a. $I(m,n)=\int_0^{\infty}\frac{x^{m-1}}{(1+x)^{m+n}}dx$
b. $I(m,n)=\int_0^{\infty}\frac{x^m}{(1+x)^{m+n}}dx$
c. $I(m,n)=\int_0^{\infty}\frac{x^{n-1}}{(1+x)^{m+n}}dx$
d. $I(m,n)=\int_0^{\infty}\frac{x^n}{(1+x)^{m+n}}dx$

62. The value of $\int_0^{\pi} \dfrac{\sin\left(n+\frac{1}{2}\right)x}{\sin\left(\frac{x}{2}\right)} dx$ is, $n \in I, n \geq 0$

a. $\frac{\pi}{2}$ b. 0
c. π d. 2π

63. The value of the definite integral $\int_0^{\pi/2} \frac{\sin 5x}{\sin x} dx$ is

a. 0 b. $\frac{\pi}{2}$
c. π d. 2π

64. If $I_n = \int_0^{\pi} e^x (\sin x)^n \, dx$, then $\frac{I_3}{I_1}$ is equal to

a. 3/5 b. 1/5
c. 1 d. 2/5

65. Given $I_m = \int_1^e (\log x)^m \, dx$. If $\frac{I_m}{K} + \frac{I_{m-2}}{L} = e$, then the values of K and L are

a. $\frac{1}{1-m}, \frac{1}{m}$ b. $(1-m), \frac{1}{m}$
c. $\frac{1}{1-m}, \frac{m(m-2)}{m-1}$ d. $\frac{m}{m-1}, m-2$

66. Let $f(x) = \min(\{x\}, \{-x\}) \; \forall \; n \in R$, where $\{n\}$ denotes the fractional part of x, then $\int_{-100}^{100} f(x)\,dx$ is equal to

a. 50 b. 100
c. 200 d. None of these

67. $\int_1^4 \{x - 0.4\} dx$ equals (where $\{x\}$ is a fractional part of x)

a. 13 b. 6.3
c. 1.5 d. 7.5

68. The value of $\int_1^a [x] \, f'(x) dx$, where $a > 1$, where $[x]$ denotes the greatest integer not exceeding x is

a. $afA - \{f(1) + f(2) + \cdots + f([a])\}$
b. $[a]fA - \{f(1) + f(2) + \cdots + f([a]\}$
c. $[a]f([a]) - \{f(1) + f(2) + \cdots + fA\}$
d. $af([a]) - \{f(1) + f(2) + \cdots + fA\}$

69. The value of $\int_0^x [\cos t] dt$, $x \in \left[(4n+1)\frac{\pi}{2}, (4n+3)\frac{\pi}{2}\right]$ and $n \in N$ is equal to

a. $\frac{\pi}{2}(2n-1) - 2x$ b. $\frac{\pi}{2}(2n-1) + x$
c. $\frac{\pi}{2}(2n+1) - x$ d. $\frac{\pi}{2}(2n+1) + x$

70. If $f(x) = \int_0^1 \frac{dt}{1+|x-t|}$, then $f'\left(\frac{1}{2}\right)$ is equal to

a. 0 b. $\frac{1}{2}$
c. 1 d. None of these

71. Let $f(x) = \int_2^x \frac{dt}{\sqrt{1+t^4}}$ and g be the inverse of f. Then the value of $g'(0)$ is

a. 1 b. 17
c. $\sqrt{17}$ d. None of these

72. Let $f(x) = \int_1^x \sqrt{2-t^2} \, dt$. Then the real roots of the equation $x^2 - f'(x) = 0$ are

a. ± 1 b. $\pm 1/\sqrt{2}$
c. $\pm 1/2$ d. 0 and 1

73. If $x = \int_c^{\sin t} \sin^{-1} z \, dz$, $y = \int_k^{\sqrt{t}} \frac{\sin z^2}{z} dz$, then $\frac{dy}{dx}$ is equal to

a. $\frac{\tan t}{2t}$ b. $\frac{\tan t}{t^2}$
c. $\frac{\tan t}{2t^2}$ d. $\frac{\tan t^2}{2t^2}$

74. If $f(x) = \cos x - \int_0^x (x-t) f(t) \, dt$, then $f''(x) + f(x)$ is equal to

a. $-\cos x$ b. $-\sin x$
c. $\int_0^x (x-t) \, f(t) \, dt$ d. 0

75. A function f is continuous for all x (and not every where zero) such that $f^2(x) = \int_0^x f(t) \frac{\cos t}{2+\sin t} dt$, then $f(x)$ is

a. $\frac{1}{2} \ln\left(\frac{x+\cos x}{2}\right)$; $x \neq 0$
b. $\frac{1}{2} \ln\left(\frac{3}{2+\cos x}\right)$; $x \neq 0$
c. $\frac{1}{2} \ln\left(\frac{2+\sin x}{2}\right)$; $x \neq n\pi, n \in I$
d. $\frac{\cos x + \sin x}{2+\sin x}$; $x \neq n\pi + \frac{3\pi}{4}$, $n \in I$

76. $\lim\limits_{x\to 0}\frac{1}{x}\left[\int_y^a e^{\sin^2 t}dt-\int_{x+y}^a e^{\sin^2 t}dt\right]$ is equal to

a. $e^{\sin^2 y}$ b. $\sin 2y e^{\sin^2 y}$
c. 0 d. None of these

77. $f(x)=\int_1^x \frac{e^t}{t}dt$, where $x\in R^+$. Then the complete set of values of x for which $f(x)\le \ln x$ is

a. $(0,1]$ b. $[1,\infty)$
c. $(0,\infty)$ d. None of these

78. If $\int_0^x f(t)dt = x+\int_x^1 tf(t)dt$, then the value of $f(1)$ is

a. 1/2 b. 0
c. 1 d. −1/2

79. If $\int_{\cos x}^1 t^2 f(t)dt = 1-\cos x \;\forall\; x\in\left(0,\frac{\pi}{2}\right)$, then the value of $\left[f\left(\frac{\sqrt{3}}{4}\right)\right]$ is ([.] denotes the greatest integer function)

a. 4 b. 5
c. 6 d. −7

80. If $\int\limits_0^{f(x)} t^2\,dt = x\cos\pi x$, then $f'(9)$ is

a. $-\frac{1}{9}$ b. $-\frac{1}{3}$
c. $\frac{1}{3}$ d. non-existent

81. If $f(x)=1+\frac{1}{x}\int_1^x f(t)dt$, then the value of $f(e^{-1})$ is

a. 1 b. 0
c. −1 d. None of these

82. If $A=\int\limits_0^1 x^{50}(2-x)^{50}\,dx$; $B=\int\limits_0^1 x^{50}(1-x)^{50}\,dx$, which of the following is true ?

a. $A=2^{50}B$ b. $A=2^{-50}B$
c. $A=2^{100}B$ d. $A=2^{-100}B$

83. The value of $\int\limits_0^1\left(\prod\limits_{r=1}^{n}(x+r)\right)\left(\sum\limits_{k=1}^{n}\frac{1}{x+k}\right)dx$ equals

a. n b. $n!$
c. $(n+1)!$ d. $n\cdot n!$

84. If $I=\int\limits_{-20\pi}^{20\pi}|\sin x|[\sin x]dx$ (where [.] denotes the greatest integer function), then the value of I is

a. −40 b. 40
c. 20 d. −20

85. Given that f satisfies $|f(u)-f(v)|\le|u-v|$ for u and v in $[a,b]$, then $\left|\int\limits_a^b f(x)\,dx-(b-a)f(a)\right|\le$

a. $\frac{(b-a)}{2}$ b. $\frac{(b-a)^2}{2}$
c. $(b-a)^2$ d. None of these

86. $\int_0^\infty \frac{\sin^2 x}{x^2}dx$ must be same as

a. $\int_0^\infty \frac{\sin x}{x}dx$ b. $\left(\int_0^\infty \frac{\sin x}{x}dx\right)^2$
c. $\int_0^\infty \frac{\cos^2 x}{x^2}dx$ d. None of these

87. If $\int_0^\infty \frac{\sin x}{x}dx=\frac{\pi}{2}$, then $\int_0^\infty \frac{\sin^3 x}{x}dx$ is equal to

a. $\pi/2$ b. $\pi/4$
c. $\pi/6$ d. $3\pi/2$

88. $\int_0^x [\sin t]\,dt$, where $x\in(2n\pi,(2n+1)\pi)$, $n\in N$ and [.] denotes the greatest integer function, is equal to

a. $-n\pi$ b. $-(n+1)\pi$
c. $-2n\pi$ d. $-(2n+1)\pi$

89. $f(x)$ is a continuous function for all real values of x and satisfies $\int_0^x f(t)dt=\int_x^1 t^2 f(t)dt+\frac{x^{16}}{8}+\frac{x^6}{3}+a$, then the value of a is equal to

a. $-\frac{1}{24}$ b. $\frac{17}{168}$
c. $\frac{1}{7}$ d. $-\frac{167}{840}$

90. $\int_0^x \frac{2^t}{2^{[t]}}dt$, where [.] denotes the greatest integer function, and $x\in R^+$, is equal to

a. $\frac{1}{\ln 2}([x]+2^{\{x\}}-1)$ b. $\frac{1}{\ln 2}([x]+2^{\{x\}})$
c. $\frac{1}{\ln 2}([x]-2^{\{x\}})$ d. $\frac{1}{\ln 2}([x]+2^{\{x\}}+1)$

91. $f(x)$ is a continuous for all real values of x and satisfies $\int_n^{n+1} f(x)dx=\frac{n^2}{2}\,\forall\, n\in I$, then $\int_{-3}^{5} f(|x|)$ dx is equal to

a. 19/2 b. 35/2
c. 17/2 d. None of these

92. The value of $\int_{1/e}^{\tan x} \frac{t\,dt}{1+t^2} + \int_{1/e}^{\cot x} \frac{dt}{t(1+t^2)}$, where $x \in \left(\frac{\pi}{6}, \frac{\pi}{3}\right)$, is equal to

a. 0 b. 2
c. 1 d. None of these

93. Let $I_1 = \int_{-2}^{2} \frac{x^6+3x^5+7x^4}{x^4+2} dx$ and $I_2 = \int_{-3}^{1} \frac{2(x+1)^2+11(x+1)+14}{(x+1)^4+2} dx$, then the value of $I_1 + I_2$ is

a. 8 b. 200/3
c. 100/3 d. None of these

94. For $x \in R$ and a continuous function f, let $I_1 = \int_{\sin^2 t}^{1+\cos^2 t} x f\{x(2-x)\} dx$ and $I_2 = \int_{\sin^2 t}^{1+\cos^2 t} f\{x(2-x)\} dx$. Then $\frac{I_1}{I_2}$ is

a. -1 b. 1
c. 2 d. 3

95. Given a function $f: [0, 4] \to R$ is differentiable, then for some $\alpha, \beta \in (0, 2)$, $\int_0^4 f(t)dt$ equals to

a. $f(\alpha^2)+f(\beta^2)$ b. $2\alpha f(\alpha^2)+2\beta f(\beta^2)$
c. $\alpha f(b^2)+\beta f(\alpha^2)$ d. $f(\alpha)f(\beta)[f(\alpha)+f(\beta)]$

96. $\int_{-3}^{3} x^8\{x^{11}\} dx$ is equal to (where {.} is the fractional part of x)

a. 3^8 b. 3^7
c. 3^9 d. None of these

97. If $S = \left(\frac{1}{2}\right)^2 + \left(\frac{1}{2}\right)^2 \frac{1}{3} + \left(\frac{1}{2}\right)^3 \frac{1}{4} + \left(\frac{1}{2}\right)^4 \frac{1}{5} + \cdots$, then

a. $S = \ln 8 - 2$ b. $S = \ln \frac{4}{e}$
c. $S = \ln 4 + 1$ d. None of these

98. Let $f: R \to R$ be a continuous function and $f(x) = f(2x)$ is true $\forall\ x \in R$. If $f(1) = 3$, then the value of $\int_{-1}^{1} f(f(x))dx$ is equal to

a. 6 b. 0
c. $3f(3)$ d. $2f(0)$

99. $\int_{-1}^{2} \left[\frac{[x]}{1+x^2}\right] dx$, where [ ] denotes the greatest integer function, is equal to

a. -2 b. -1
c. zero d. None of these

100. f is an odd function. It is also known that $f(x)$ is continuous for all values of x and is periodic with period 2. If $g(x) = \int_0^x f(t)\,dt$ then

a. $g(x)$ is odd
b. $g(n) = 0, n \in N$
c. $g(2n) = 0, n \in N$
d. $g(x)$ is non-periodic

101. $\int_0^x |\sin t|\,dt$, where $x \in (2n\pi, (2n+1)\pi)$, where $n \in N$, is equal to

a. $4n - \cos x$ b. $4n - \sin x$
c. $4n + 1 - \cos x$ d. $4n - 1 - \cos x$

102. If $f(x) = \int_1^x |t|\,dt$, then for any $x \geq 0$, $f(x)$ equals

a. $\frac{1}{2}(1-x^2)$ b. $\frac{1}{2}x^2$
c. $\frac{1}{2}(1+x^2)$ d. None of these

103. If $g(x) = \int_0^x (|\sin t| + |\cos t|)\,dt$, then $g\left(x + \frac{\pi n}{2}\right)$ is equal to, where $n \in N$

a. $g(x)+g(\pi)$ b. $g(x)+g\left(\frac{n\pi}{2}\right)$
c. $g(x)+g\left(\frac{\pi}{2}\right)$ d. None of these

104. The value of $\int_{-2}^{1}\left[x\left[1+\cos\left(\frac{\pi x}{2}\right)\right]+1\right]dx$, where [.] denotes the greatest integer function, is

a. 1 b. 1/2
c. 2 d. None of these

105. If $a > 0$ and $A = \int_0^a \cos^{-1} x\,dx$, then $\int_{-a}^{a} (\cos^{-1} x - \sin^{-1} \sqrt{1-x^2})dx = \pi a - \lambda A$, then λ is

a. 0 b. 2
c. 3 d. None of these

106. The value of $\int_a^b (x-a)^3(b-x)^4\,dx$ is

a. $\frac{(b-a)^4}{6^4}$ b. $\frac{(b-a)^8}{280}$
c. $\frac{(b-a)^7}{7^3}$ d. None of these

107. If $\int_0^t \frac{bx\cos 4x - a\sin 4x}{x^2}dx = \frac{a\sin 4t}{t} - 1$, where $0 < t < \frac{\pi}{4}$, then the values of a, b are equal to

a. $\frac{1}{4}, 1$ b. $-1, 4$
c. $2, 2$ d. $2, 4$

108. If $\lambda = \int_0^1 \frac{e^t}{1+t}$, then $\int_5^1 e^t \log_e(1+t)dt$ is equal to

a. 2λ b. $e\log_e 2 - \lambda$
c. λ d. $e\log_e 2 + \lambda$

109. Let f be integrable over $[0, a]$ for any real value of a. If $I_1 = \int_0^{\pi/2} \cos\theta\, f(\sin\theta + \cos^2\theta)d\theta$ and $I_2 = \int_0^{\pi/2} \sin 2\theta\, f(\sin\theta + \cos^2\theta)\,d\theta$, then

a. $I_1 = -2I_2$ b. $I_1 = I_2$
c. $2I_1 = I_2$ d. $I_1 = -I_2$

110. The range of the function $f(x) = \int_{-1}^{1} \frac{\sin x\, dx}{(1 - 2t\cos x + t^2)}$ is

a. $\left[-\frac{\pi}{2}, \frac{\pi}{2}\right]$ b. $[0, \pi]$
c. $\{0, \pi\}$ d. $\left\{-\frac{\pi}{2}, \frac{\pi}{2}\right\}$

111. The value of $\lim_{n\to\infty}\left[\tan\frac{\pi}{2n}\tan\frac{2\pi}{2n}\cdots\tan\frac{n\pi}{2n}\right]^{1/n}$ is

a. e b. e^2
c. 1 d. e^3

112. If $f'(x) = f(x) + \int_0^1 f(x)dx$, given $f(0) = 1$, then the value of $f(\log_e 2)$ is

a. $\frac{1}{3+e}$ b. $\frac{5-e}{3-e}$
c. $\frac{2+e}{e-2}$ d. None of these

113. If $f(x)$ is monotonic differentiable function on $[a, b]$, then $\int_a^b f(x)dx + \int_{f(a)}^{f(b)} f^{-1}(x)dx =$

a. $bf(A) - af(B)$ b. $bf(B) - af(A)$
c. $f(A) + f(B)$ d. cannot be found

114. If $\alpha, \beta\ (\beta > \alpha)$ are the roots of $g(x) = ax^2 + bx + c = 0$ and $f(x)$ is an even function, then $\int_\alpha^\beta \frac{e^{f\left(\frac{g(x)}{x-\alpha}\right)}dx}{e^{f\left(\frac{g(x)}{x-\alpha}\right)} + e^{f\left(\frac{g(x)}{x-\beta}\right)}}$ is equal to

a. $\left|\frac{b}{2a}\right|$ b. $\frac{\sqrt{b^2-4ac}}{|2a|}$
c. $\left|\frac{b}{a}\right|$ d. None of these

115. If $y^r = \frac{n!\,{}^{n+r-1}C_{r-1}}{r^n}$, where $n = kr$ (k is constant), then $\lim_{x\to\infty} y$ is equal to

a. $(k-1)\log_e(1+k) - k$
b. $(k+1)\log_e(k-1) + k$
c. $(k+1)\log_e(k-1) - k$
d. $(k-1)\log_e(k-1) + k$

116. $\int_3^{10} [\log[x]]dx$ is equal to (where [.] represents the greatest integer function)

a. 9 b. $16 - e$
c. 10 d. $10 + e$

117. If the function $f: [0, 8] \to R$ is differentiable, then for $0 < a, b < 2$, $\int_0^8 f(t)dt$ is equal to

a. $3[\alpha^3 f(\alpha^2) + \beta^2 f(\beta^2)]$ b. $3[\alpha^3 f(\alpha) + \beta^3 f(\beta)]$
c. $3[\alpha^2 f(\alpha^3) + \beta^2 f(\beta^3)]$ d. $3[\alpha^2 f(\alpha^2) + \beta^2 f(\beta^2)]$

118. The function f and g are positive and continuous, f is increasing and g is decreasing, then $\int_0^1 f(x)\,[g(x) - g(1-x)]\,dx$

a. is always non-positive
b. is always non-negative
c. can take positive and negative values
d. None of these

Multiple Correct Answers Type

Each question has four choices a, b, c, and d, out of which *one or more* answers are correct.

1. A function $f(x)$ which satisfies the relation $f(x) = e^x + \int_0^1 e^x f(t)\,dt$, then

a. $f(0)<0$

b. $f(x)$ is a decreasing function

c. $f(x)$ is an increasing function

d. $\int_5^1 f(x)\,dx>0$

2. Let $f(x)=\int_1^x \frac{3^t}{1+t^2}dt$, where $x>0$, then

a. for $0<\alpha<\beta, f(\alpha)<f(\beta)$

b. for $0<\alpha<\beta, f(\alpha)>f(\beta)$

c. $f(x)+\pi/4<\tan^{-1}x,\ \forall\ x\geq 1$

d. $f(x)+\pi/4>\tan^{-1}x,\ \forall\ x\geq 1$

3. The values of a for which the integral $\int_5^2 |x-a|dx\geq 1$ is satisfied are

a. $[2,\infty)$ b. $(-\infty,0]$

c. $(0,2)$ d. None of these

4. If $\int_a^b |\sin x|\,dx=8$ and $\int_0^{a+b}|\cos x|\,dx=9$, then

a. $a+b=\frac{9\pi}{2}$ b. $|a-b|=4\pi$

c. $\frac{a}{b}=15$ d. $\int_a^b \sec^2 x dx=0$

5. Let $I=\int_1^3 \sqrt{3+x^3}\,dx$, then the values of I will lie in the interval

a. $[4,6]$ b. $[1,3]$.

c. $[4, 2\sqrt{30}\,]$ d. $\left[\sqrt{15},\sqrt{30}\right]$

6. If $g(x)=\int_0^x 2|t|\,dt$, then

a. $g(x)=x|x|$

b. $g(x)$ is monotonic

c. $g(x)$ is differentiable at $x=0$

d. $g'(x)$ is differentiable at $x=0$

7. Let $f:[1,\infty)\to R$ and $f(x)=x\int_1^x \frac{e^t}{t}dt-e^x$, then

a. $f(x)$ is an increasing function

b. $\lim_{x\to\infty} f(x)\to\infty$

c. $f'(x)$ has a maxima at $x=e$

d. $f(x)$ is a decreasing function

8. The value of $\int_0^1 \frac{2x^2+3x+3}{(x+1)(x^2+2x+2)}dx$ is

a. $\frac{\pi}{4}+2\log 2-\tan^{-1}2$

b. $\frac{\pi}{4}+2\log 2-\tan^{-1}\frac{1}{3}$

c. $2\log 2-\cot^{-1}3$

d. $-\frac{\pi}{4}+\log 4+\cot^{-1}2$

9. If $A_n=\int_0^{\pi/2}\frac{\sin(2n-1)x}{\sin x}dx$; $B_n=\int_0^{\pi/2}\left(\frac{\sin nx}{\sin x}\right)^2 dx$, for $n\in N$, then

a. $A_{n+1}=A_n$ b. $B_{n+1}=B_n$

c. $A_{n+1}-A_n=B_{n+1}$ d. $B_{n+1}-B_n=A_{n+1}$

10. If $f(x)=\int_a^x [f(x)]^{-1}dx$ and $\int_a^1 [f(x)]^{-1}dx=\sqrt{2}$, then

a. $f(2)=2$ b. $f'(2)=1/2$

c. $f^{-1}(2)=2$ d. $\int_0^1 f(x)dx=\sqrt{2}$

11. The value of $\int_0^\infty \frac{dx}{1+x^4}$ is

a. same as that of $\int_0^\infty \frac{x^2+1dx}{1+x^4}$

b. $\frac{\pi}{2\sqrt{2}}$

c. same as that of $\int_0^\infty \frac{x^2dx}{1+x^4}$

d. $\frac{\pi}{\sqrt{2}}$

12. If $f(x)=\int_0^x |t-1|\,dt$, where $0\leq x\leq 2$, then

a. range of $f(x)$ is $[0,1]$

b. $f(x)$ is differentiable at $x=1$

c. $f(x)=\cos^{-1}x$ has two real roots

d. $f'(1/2)=1/2$

13. If $I_n=\int_0^{\pi/4}\tan^n x\,dx$ ($n>1$ and is an integer), then

a. $I_n+I_{n-2}=\frac{1}{n+1}$

b. $I_n+I_{n-2}=\frac{1}{n-1}$

c. $I_2+I_4, I_4+I_6, \ldots$, are in H.P.

d. $\frac{1}{2(n+1)}<I_n<\frac{1}{2(n-1)}$

14. If $\int_a^b \frac{f(x)}{f(a)+f(a+b-x)}dx = 10$, then

a. $b=22, a=2$ b. $b=15, a=-5$

c. $b=10, a=-10$ d. $b=10, a=-2$

15. If $I_n = \int_0^1 \frac{dx}{(1+x^2)^n}$, where $n \in N$, which of the following statements holds good ?

a. $2n\,I_{n+1} = 2^{-n} + (2n-1)I_n$

b. $I_2 = \frac{\pi}{8} + \frac{1}{4}$

c. $I_2 = \frac{\pi}{8} - \frac{1}{4}$

d. $I_3 = \frac{3\pi}{32} + \frac{1}{4}$

16. If $f(x)$ is integrable over [1, 2], then $\int_1^2 f(x)\,dx$ is equal to

a. $\lim_{n\to\infty} \frac{1}{n}\sum_{r=1}^{2n} f\left(\frac{r}{n}\right)$ b. $\lim_{n\to\infty} \frac{1}{n}\sum_{r=n+1}^{2n} f\left(\frac{r}{n}\right)$

c. $\lim_{n\to\infty} \frac{1}{n}\sum_{r=1}^{n} f\left(\frac{r+n}{n}\right)$ d. $\lim_{n\to\infty} \frac{1}{n}\sum_{r=1}^{2n} f\left(\frac{r}{n}\right)$

17. If $f(2-x)=f(2+x)$ and $f(4-x)=f(4+x)$ for all x and $f(x)$ is a function for which $\int_0^2 f(x)\,dx = 5$, then $\int_0^{50} f(x)\,dx$ is equal to

a. 125 b. $\int_{-4}^{46} f(x)\,dx$

c. $\int_1^{51} f(x)\,dx$ d. $\int_2^{52} f(x)\,dx$

18. $\int_0^x \left\{\int_0^u f(t)\,dt\right\} du$ is equal to

a. $\int_0^x (x-u)f(u)du$ b. $\int_0^x uf(x-u)du$

c. $x\int_0^x f(u)du$ d. $x\int_0^x uf(u-x)du$

19. If $f(x) = \int_1^x \frac{e^t}{1+t^2}dt$, $\forall\ x>0$, then

a. $g(x) = \tan^{-1}x - f(x)$ is a decreasing function for $x>0$

b. $f(x) \ge \tan^{-1}x - \frac{\pi}{4}$, $\forall\ x \ge 1$

c. $f(x) \le \tan^{-1}x - \frac{\pi}{4}$, $\forall\ x \le 1$

d. $g(x) = \tan^{-1}x - f(x)$ is an increasing function for $x<1$

20. The value of $\int_0^1 e^{x^2-x}\,dx$ is

a. <1 b. >1

c. $> e^{-\frac{1}{4}}$ d. $< e^{-\frac{1}{4}}$

21. If $f(x) = \int_0^x (\cos(\sin t) + \cos(\cos t))dt$, then $f(x+\pi)$ is

a. $f(x)+f(\pi)$ b. $f(x)+2f(\pi)$

c. $f(x)+f\left(\frac{\pi}{2}\right)$ d. $f(x)+2f\left(\frac{\pi}{2}\right)$

Reasoning Type

Each question has four choices a, b, c and d, out of which *only one* is correct. Each question contains Statement 1 and Statement 2.

a. If both the statements are True and Statement 2 is the correct explanation of Statement 1

b. If both the statements are True but Statement 2 is Not the correct explanation of Statement 1

c. If Statement 1 is True and Statement 2 is False

d. If Statement 1 is False and Statement 2 is True

1. Let $f(x)$ is continuous and positive for $x \in [a, b]$, $g(x)$ is continuous for $x \in [a,b]$ and $\int_a^b |g(x)|\,dx > \left|\int_a^b g(x)dx\right|$, then

Statement 1: The value of $\int_a^b f(x)g(x)dx$ can be zero.

Statement 2: Equation $g(x)=0$ has at least one root in $x \in (a,b)$.

2. Statement 1: The value of $\int_{-4}^{-5} \sin(x^2-3)dx + \int_{-2}^{-1} \sin(x^2+12x+33)$ is zero.

Statement 2: $\int_{-a}^{a} f(x)dx = 0$ if $f(x)$ is an odd function.

3. Statement 1: The value of $\int_0^1 \tan^{-1}\frac{2x-1}{(1+x-x^2)}dx = 0$.

Statement 2: $\int_a^b f(x)dx = \int_a^b f(a+b-x)dx$.

4. Statement 1: On the interval $\left[\frac{5\pi}{4}, \frac{4\pi}{3}\right]$, the least value of the function $f(x) = \int_{5\pi/4}^{x} (3\sin t + 4\cos t)\,dt$ is 0.

Statement 2: If $f(x)$ is a decreasing function on the interval $[a,b]$, then the least value of $f(x)$ is $f(b)$.

5. Consider the function $f(x)$ satisfying the relation $f(x+1)+f(x+7)=0,\ \forall\ x\in R$,
Statement 1: The possible least value of t for which $\int_a^{a+t} f(x)dx$ is independent of a is 12.
Statement 2: $f(x)$ is a periodic function.

6. Consider $I_1 = \int_0^{\pi/4} e^{x^2}dx,\ I_2 = \int_0^{\pi/4} e^{x}dx,$
$I_3 = \int_0^{\pi/4} e^{x^2}\cos x\, dx,\ I_4 = \int_0^{\pi/4} e^{x^2}\sin x\, dx.$
Statement 1: $I_2 > I_1 > I_3 > I_4.$
Statement 2: for $x\in(0,1)$, $x>x^2$ and $\sin x>\cos x$.

7. Statement 1: Let m be any integer. Then the value of $I_m = \int_0^{\pi} \frac{\sin 2mx}{\sin x}dx$ is zero.
Statement 2: $I_1 = I_2 = I_3 = \cdots = I_m.$

8. Statement 1: $\int_0^{\pi}\sqrt{1-\sin^2 x}dx = 0.$
Statement 2: $\int_0^{\pi}\cos x dx = 0.$

9. Statement 1: The value of $\int_0^{2\pi}\cos^{99} x dx$ is 0.
Statement 2: $\int_0^{2a} f(x)dx = 2\int_0^{a} f(x)dx$, if $f(2a-x) = f(x)$.

10. Statement 1: $\int_a^x f(t)dt$ is an even function if $f(x)$ is an odd function.
Statement 2: $\int_a^x f(t)dt$ is an odd function if $f(x)$ is an even function.

11. Statement 1: $f(x)$ is symmetrical about $x=2$, then $\int_{2-a}^{2+a} f(x)dx$ is equal to $2\int_2^{2+a} f(x)dx.$
Statement 2: If $f(x)$ is symmetrical about $x=b$, then $f(b-\alpha)=f(b+\alpha)\ \forall\ (\alpha\in R)$.

12. Statement 1: The value of $\int_0^{\pi/4}\log(1+\tan\theta)d\theta = \frac{\pi}{8}\log 2.$
Statement 2: The value of $\int_0^{\pi/2}\log\sin\theta d\theta = -\pi \log 2.$

13. Statement 1: The value of $\int_0^{\pi}\sin^{100} x\cos^{99} x\, dx$ is zero.
Statement 2: $\int_a^b f(x)\,dx = \int_{a+c}^{b+c} f(x-c)\,dx.$

14. Statement 1: $\int_0^{\pi} x\sin x\cos^2 x\, dx = \frac{\pi}{2}\int_0^{\pi}\sin x\cos^2 x\, dx$
Statement 2: $\int_a^b x f(x)\, dx = \frac{a+b}{2}\int_a^b f(x)dx.$

15. Let f be a polynomial function of degree n.
Statement 1: There exist a number $x\in[a,b]$ such that $\int_a^x f(t)\,dt = \int_x^b f(t)\,dt.$
Statement 2: $f(x)$ is a continuous function.

16. Statement 1: $\int_0^x |\sin t|\,dt$, for $x\in[0,2\pi]$ is a non-differentiable function.
Statement 2: $|\sin t|$ is non-differentiable at $x=\pi$.

17. Statement 1: If $f(x)$ is continuous on $[a,b]$, then there exists a point $c\in(a,b)$ such that $\int_a^b f(x)dx = f(c)(b-a)$
Statement 2: For $a<b$, if m and M are, respectively, the smallest and greatest values of $f(x)$ on $[a,b]$, then $m(b-a)\le\int_a^b f(x)dx\le(b-a)M.$

Linked Comprehension Type

Based upon each paragraph, three multiple choice questions have to be answered. Each question has four choices a, b, c, and d, out of which *only one* is correct.

For Problems 1–3

$y=f(x)$ satisfies the relation $\int_2^x f(t)dt = \frac{x^2}{2} + \int_x^2 t^2 f(t)dt - 2$

1. The range of $y=f(x)$ is
   a. $[0,\infty)$ b. R
   c. $(-\infty,0]$ d. $\left(-\frac{1}{2},\frac{1}{2}\right)$

2. The value of $\int_{-2}^{2} f(x)dx$ is
   a. 0 b. -2
   c. $2\log_e 2$ d. None of these

3. The value of x for which $f(x)$ is increasing is
   a. $(-\infty,1]$ b. $[-1,\infty)$
   c. $[-1,1]$ d. None of these

For Problems 4–6

Let $f: R\to R$ be a differentiable function such that

$f(x) = x^2 + \int_0^x e^{-t} f(x-t)dt.$

4. $f(x)$ increases for

a. $x>1$ b. $x<-2$

c. $x>2$ d. None of these

5. $y=f(x)$ is

a. injective but not surjective

b. surjective but not injective

c. bijective

d. neither injective nor surjective

6. The value $\int_0^1 f(x)dx$ is

a. $\frac{1}{4}$ b. $-\frac{1}{12}$

c. $\frac{5}{12}$ d. $\frac{12}{7}$

For Problems 7–9

$f(x)$ satisfies the relation $f(x)-\lambda\int_0^{\pi/2}\sin x\cos t f(t)dt=\sin x$

7. If $\lambda>2$, then $f(x)$ decreases in which of the following interval

a. $(0, \pi)$ b. $(\pi/2, 3\pi/2)$

c. $(-\pi/2, \pi/2)$ d. None of these

8. If $f(x)=2$ has at least one real root, then

a. $\lambda\in[1,4]$ b. $\lambda\in[-1,2]$

c. $\lambda\in[0,1]$ d. $\lambda\in[1,3]$

9. If $\int_0^{\pi/2} f(x)dx=3$, then the value of λ is

a. 1 b. 3/2

c. 4/3 d. None of these

For Problems 10–13

Let $f(x)$ and $\phi(x)$ are two continuous functions on R satisfying $\phi(x)=\int_a^x f(t)dt$, $a\neq 0$ and another continuous function $g(x)$ satisfying $g(x+\alpha)+g(x)=0\ \forall\ x\in R, \alpha>0$ and $\int_b^{2k} g(t)dt$ is independent of b.

10. If $f(x)$ is an odd function, then

a. $\phi(x)$ is also an odd function

b. $\phi(x)$ is an even function

c. $\phi(x)$ is neither as even nor an odd function

d. For $\phi(x)$ to be an even function, it must satisfy $\int_0^a f(x)\,dx=0$

11. If $f(x)$ is an even function, then

a. $\phi(x)$ is also an even function

b. $\phi(x)$ is an odd function

c. If $f(a-x)=-f(x)$, then $\phi(x)$ is an even function

d. If $f(a-x)=-f(x)$, then $\phi(x)$ is an odd function

12. Least positive value of c if c, k, b are in A.P. is

a. 0 b. 1

c. α d. 2α

13. If m, n are even integers and $p, q\in R$, then $\int_{p+m\alpha}^{q+n\alpha} g(t)dt$ is equal to

a. $\int_p^q g(x)dx$

b. $(n-m)\int_0^\alpha g(x)dx$

c. $\int_p^q g(x)dx+(n-m)\int_0^\alpha g(2x)dx$

d. $\int_p^q g(x)dx+(n-m)\int_0^\alpha g(x)dx$

For Problems 14–17

Evaluating Integrals Dependent on a Parameter

Differentiate I w.r.t. the parameter within the sign of integrals taking variable of the integrand as constant. Now, evaluate the integral so obtained as usual as a function of the parameter and then integrate the result to get I. Constant of integration can be computed by giving some arbitrary values to the parameter and the corresponding value of I.

14. The value of $\int_0^1 \frac{x^a-1}{\log x}dx$ is

a. $\log(a-1)$ b. $\log(a+1)$

c. $a\log(a+1)$ d. None of these

15. The value of $\int_0^{\pi/2}\log(\sin^2\theta+k^2\cos^2\theta)d\theta$, where $k\geq 0$ is

a. $\pi\log(1+k)+\pi\log 2$ b. $\pi\log(1+k)$

c. $\pi\log(1+k)-\pi\log 2$ d. $\log(1+k)-\log 2$

16. The value of $\frac{dI}{da}$ when

$I=\int_0^{\pi/2}\log\left(\frac{1+a\sin x}{1-a\sin x}\right)\frac{dx}{\sin x}$ (where $|a|<1$) is

a. $\frac{\pi}{\sqrt{1-a^2}}$ b. $-\pi\sqrt{1-a^2}$

c. $\sqrt{1-a^2}$ d. $\frac{\sqrt{1-a^2}}{\pi}$

17. If $\int_0^{\pi} \frac{dx}{(a-\cos x)} = \frac{\pi}{\sqrt{a^2-1}}$, then the value of $\int_0^{\pi} \frac{dx}{(\sqrt{10}-\cos x)^3}$ is

a. $\frac{\pi}{81}$ b. $\frac{7\pi}{162}$

c. $\frac{7\pi}{81}$ d. None of these

For Problems 18–20

$$f(x) = \sin x + \int_{-\pi/2}^{\pi/2} (\sin x + t\cos x) f(t)dt$$

18. The range of $f(x)$ is

a. $\left[-\frac{\sqrt{3}}{2}, \frac{\sqrt{3}}{2}\right]$ b. $\left[-\frac{\sqrt{5}}{3}, \frac{\sqrt{5}}{3}\right]$

c. $\left[-\frac{\sqrt{5}}{2}, \frac{\sqrt{5}}{2}\right]$ d. None of these

19. $f(x)$ is not invertible for

a. $x \in \left[-\frac{\pi}{2} - \tan^{-1} 2, \frac{\pi}{2} - \tan^{-1} 2\right]$

b. $x \in \left[\tan^{-1}\frac{1}{2}, \pi + \tan^{-1}\frac{1}{2}\right]$

c. $x \in \left[\pi + \cot^{-1} 2, 2\pi + \cot^{-1} 2\right]$

d. None of these

20. The value of $\int_0^{\pi/2} f(x)dx$ is

a. 1 b. -2

c. -1 d. 2

Matrix-Match Type

Each question contains statements given in two columns which have to be matched.

Statements a, b, c, d in column 1 have to be matched with statements p, q, r, s in column 2. If the correct match is a → p, a → s, b → r, c → p, and d → s, then the correctly bubbled 4 × 4 matrix should be as follows:

	p	q	r	s
a	(p)	(q)	(r)	(s)
b	(p)	(q)	(r)	(s)
c	(p)	(q)	(r)	(s)
d	(p)	(q)	(r)	(s)

1. If [.] denotes the greatest integer function, then match the following columns:

Column 1	Column 2
a. $\int_{-1}^{1} [x+[x+[x]]]\, dx$	p. 3
b. $\int_{2}^{5} ([x]+[-x])\, dx$	q. 1/5
c. $\int_{-1}^{3} \text{sgn}(x-[x])dx$	r. 4
d. $25\int_0^{\pi/4}\left(\tan^6(x-[x])+\tan^4(x-[x])\right)dx$	s. -3

2.

Column 1	Column 2
a. $\lim_{n\to\infty}\left[\int_0^2 \frac{\left(1+\frac{t}{n+1}\right)^n}{n+1} dt\right]$ is equal to	p. $e - \frac{1}{2}e^2 - \frac{3}{2}$
b. Let $f(x)$ be a function satisfying $f'(x) = f(x)$ with $f(0) = 1$ and g be the function satisfying $f(x) + g(x) = x^2$. Then the value of the integral $\int_0^1 f(x)g(x)\,dx$ is	q. e^2
c. $\int_0^1 e^{e^x}(1+xe^x)dx$ is equal to	r. $e^2 - 1$
d. $\lim_{k\to 0}\frac{1}{k}\int_0^k (1+\sin 2x)^{\frac{1}{x}} dx$ is equal to	s. e^e

3.

Column 1	Column 2
a. If $f(x)$ is an integrable function for $x \in \left[\frac{\pi}{6}, \frac{\pi}{3}\right]$ and $I_1 = \int_{\pi/6}^{\pi/3} \sec^2\theta f(2\sin 2\theta)\, d\theta$ and $I_2 = \int_{\pi/6}^{\pi/3} \text{cosec}^2\theta f(2\sin 2\theta)d\theta$, then I_1/I_2	p. $\frac{3}{2}$

b. If $f(x+1)=f(3+x)$ for $\forall\ x$, and the value of $\int_a^{a+b} f(x)dx$ is independent of a then the value of b can be	**q.** 1
c. The value of $\int_1^4 \frac{\tan^{-1}[x^2]}{\tan^{-1}[x^2]+\tan^{-1}[25+x^2-10x]}$ (where [.]denotes the greatest integer function) is	**r.** 2
d. If $I=\int_0^2 \sqrt{x+\sqrt{x+\sqrt{x+\cdots\infty}}}\,dx$ (where $x>0$), then $[I]$ is equal to (where [.] denotes the greatest integer function)	**s.** 4

4.

Column 1	Column 2
a. If $I=\int_{-2}^{2}(\alpha x^3+\beta x+\gamma)dx$, then I is	**p.** independent of α
b. Let α, β be the distinct positive roots of the equation $\tan x = 2x$, then $\gamma\int_0^1(\sin\alpha x\cdot\sin\beta x)dx$ (where $\gamma\neq0$) is	**q.** independent of β
c. If $f(x+\alpha)+f(x)=0$, where $\alpha>0$, then $\int_\beta^{\beta+2\gamma\alpha} f(x)dx$, where $\gamma\in N$, is	**r.** independent of γ
d. $\gamma\int_0^\alpha[\sin x]dx$ is, where $\gamma\neq0$, $\alpha\in[(2\beta+1)\pi,(2\beta+2)\pi]$ $n\in N$, and where [.] denotes the greatest integer function	**s.** depends on α

Integer Type

1. If $L=\lim_{n\to\infty}\{[(n+1)(n+2)\cdots(n+n)]^{1/n}/n\}$, then the value of eL is ________.
2. If
$$L=\lim_{n\to\infty}\frac{(1^2+2^2+3^3+\cdots+n^2)(1^3+2^3+3^3+\cdots+n^3)}{1^6+2^6+3^6+\cdots+n^6},$$
then the value of $12L$ is ________.
3. Let $[dF(x)]/(dx)=\left(e^{\sin x}/x\right)$, $x>0$. If $\int_1^4(3/x)e^{\sin x^3}\,dx=F(k)-F(1)$, then the possible value of $\sqrt{k}$ is ___.
4. If $f(0)=1$, $f(2)=3$, and $f'(2)=3$, then the value of $\int_0^1 x f''(2x)dx$ is________.
5. If $A=\int_{-\pi/2}^{2\pi}\sin^{-1}(\sin x)\,dx$, then the value of $\left|(\pi^2/A)\right|$ is ___.
6. The value of $3\int_{-\pi/2}^{\pi/2}\sqrt{\cos x-\cos^3 x}\,dx$ is ___.
7. If $P=\int_{-\pi/4}^{\pi/4}\log(\sin x+\cos x)dx$ and $Q=\int_0^\infty\log(x+1/x)\times[(dx)/(1+x^2)]$, then the value of $|Q/P|$ is ___.
8. The value of $\int_{-\pi/4}^{\pi/4}[(x^9-3x^5+7x^3-x+1)/\cos^2 x]dx$ is ___.
9. If $\int_{-1/2}^{1/2}\left\{[(x+1)/(x-1)]^2+[(x-1)/(x+1)]^2-2\right\}^{1/2}dx$ $=k\log(4/3)$, then the value of k is ___.
10. If $\left(\int_0^n[x]dx\right)/\left(\int_0^n\{x\}dx\right)=7$, where $[x]$ and $\{x\}$ are the integral and fractional parts of x, respectively, and $n\in N$, then the value of n is ___.
11. Let $f: R\to R$ be a differentiable function having $f(2)=6$, $f'(2)=1/48$. Then the value of $\lim_{x\to2}(1/2)\int_6^{f(x)}[(4t^3)/(x-2)]dt$ is ___.
12. The value of $\lim_{x\to\infty}\left[\left(\int_0^x e^{x^2}dx\right)^2/\left(\int_0^x e^{2x^2}dx\right)\right]$ is ___.
13. If $x=\int_0^y[(dt)/\sqrt{1+9t^2}]$ and $(d^2y)/(dx^2)=ay$, then the value of a is________.
14. If $I_n=\int_0^\pi x^n\sin x\,dx$ and $I_5+20\,I_3=\pi^k$, then the value of k is ___.
15. The value of $\int_{-4}^{-5}e^{(x+5)^2}dx+3\int_{1/3}^{2/3}e^{9[x-(2/3)]^2}dx$ is ___.
16. If $\int_0^{\pi/2}\log\sin\theta\,d\theta=k$, then the value of $\left|\left[\int_0^{\pi/2}(\theta/\sin\theta)^2d\theta\right]/\left(\int_0^{\pi/2}\log\sin\theta\,d\theta\right)\right|$ is________.
17. If $f(x)=\begin{cases}1-|x|, & |x|\le1\\ 0, & |x|>1\end{cases}$ and $g(x)=f(x-1)+f(x+1)$, then the value of $\int_{-3}^{5}g(x)\,dx$ is ___.
18. If $\int_0^{\pi/2}x^n\sin x\,dx=(3/4)(\pi^2-8)$, then the value of n is ___.

19. Determine a positive integer $n \le 5$ such that $\int_0^1 e^x (x-1)^n \, dx = 16 - 6e$.

20. For $x > 0$, let $f(x) = \int_e^x [(\log t)/(1+t)]dt$. Then the value of $3/f(e)$ is ___.

21. If $f(x) = \begin{cases} e^{\cos x} \sin x, & \text{for} |x| \le 2 \\ 2, & \text{otherwise} \end{cases}$, then $4\int_{-2}^{3} f(x)\, dx$ is equal to ___.

22. $\int_{-2}^{0} [x^3 + 3x^2 + 3x + 3 + (x+1)\cos(x+1)]\, dx$ is equal to ___.

23. $\int_1^e [(1/\sqrt{x \ln x}) + \sqrt{(\ln x)/x}\,]dx = L$, then the value of $3L/\sqrt{e}$ is ___.

24. Let $A = \int_0^\infty [(\ln x)/(1+x^3)]dx$ and $B = \int_0^\infty [(x \ln x)/(1+x^3)]dx$, then the value of $|A/B|$ is equal to ___.

25. Let $u = \int_0^1 \{[\ln(x+1)]/(x^2+1)\}dx$ and $v = \int_0^{\pi/2} \ln(\sin 2x)\, dx$, then the value of $-v/u$ is ___.

26. If $\int_a^b \frac{x^n}{x^n + (16-x)^n}\, dx = 6$, then b/a is ___.

27. $\lim_{x \to 0} \frac{1}{\sin x} \int_0^{\ln(1+x)} (1 - \tan 2y)^{1/y} dy = 1/e^a$, then a is equal to ___.

28. If $\int_{-1}^{1} (1+x)^{1/2} (1-x)^{3/2}\, dx = I$, then the value of $\sec^3 (I/2)$ is ___.

29. The value of $\left[\lim_{n \to \infty} \left(n^{-3/2}\right) \sum_{j=1}^{6n} \sqrt{j} \right]$ is (where [.] represents greatest integer function) ___.

30. Number of solutions of the equation $\int_0^x t^2 . \sin(x-t)\, dt = x^2$ for $x \in [0, 8\pi]$ are ___.

31. Let $f(x)$ is a derivable function satisfying $f(x) = \int_0^x e^t \sin(x-t)\, dt$ and $g(x) = f''(x) - f(x)$, then the number of integer in the range of $g(x)$ are ___.

32. If $f(x) = \int_0^1 (x f(t) + 1)dt$, then $\frac{1}{4}\int_0^3 f(x)$ is equal to ___.

33. Let $I = \int_0^\infty \frac{x^{2008}}{1+x^2} dx$ and $J = \int_0^\infty \frac{x^{2008} + x^{-2008}}{1+x^2} dx$, then J/I is ___.

34. If $I = \int_0^{3\pi/4} ((1+x)\sin x + (1-x)\cos x)\, dx$, then the value of [$I$], where [.] represents greatest integer function, is ___.

35. A function f is defined by $f(x) = \int_0^\pi \cos t \cos(x-t)\, dt$, $0 \le x \le 2\pi$ then maximum value of $\frac{14 f(0)}{\pi}$ is ___.

36. The function f is continuous and has the property $f(f(x)) = 1 - x$ for all $x \in [0, 1]$ and $J = \int_0^1 f(x)\, dx$, then the value of $1/J$ is ___.

37. Let $L = \lim_{n \to \infty} \int_a^\infty \frac{n\, dx}{1 + n^2 x^2}$, where $a \in R$ then the sum of all possible values of $\cos L$ is ___.

ANSWERS AND SOLUTIONS

Objective Type

1. b. $\lim_{n\to\infty} S_n$

$= \lim_{n\to\infty}\left[\frac{1}{1+\sqrt{n}}+\frac{1}{2+\sqrt{2n}}+\cdots+\frac{1}{n+\sqrt{n^2}}\right]$

$= \lim_{n\to\infty}\frac{1}{n}\left[\frac{1}{\frac{1}{n}+\frac{1}{\sqrt{n}}}+\frac{1}{\frac{2}{n}+\sqrt{\frac{2}{n}}}+\cdots+\frac{1}{\frac{n}{n}+\sqrt{\frac{n}{n}}}\right]$

$= \int_0^1 \frac{dx}{\sqrt{x}\left(\sqrt{x}+1\right)}$

Put $\sqrt{x}=z$, $\therefore \frac{1}{2\sqrt{x}}dx = dz$

$\Rightarrow \lim_{n\to\infty} S_n = \int_0^1 \frac{2\,dz}{z+1} = 2\left|\log(z+1)\right|_0^1$

$= 2(\log 2 - \log 1) = 2\log 2 = \log 4$

2. c. $\lim_{n\to\infty}\sum_{r=1}^{4n}\frac{\sqrt{n}}{\sqrt{r}\left(3\sqrt{r}+4\sqrt{n}\right)^2}$

$T_r = \frac{1}{\sqrt{\frac{r}{n}}\,n\left(3\sqrt{\frac{r}{n}}+4\right)^2}$

$S = \frac{1}{n}\sum_1^{4n}\frac{1}{\left(3\sqrt{\frac{r}{n}}+4\right)^2\sqrt{\frac{r}{n}}}$

$= \int_0^4 \frac{dx}{\sqrt{x}(3\sqrt{x}+4)^2}$

Put $3\sqrt{x}+4=t \Rightarrow \frac{3}{2}\frac{1}{\sqrt{x}}dx = dt$

$= \frac{2}{3}\int_4^{10}\frac{dt}{t^2} = \frac{2}{3}\left[\frac{1}{t}\right]_{10}^4 = \frac{1}{10}$

3. d. $I = \int_{a+c}^{b+c} f(x)\,dx$, putting $x = t+c$

$\Rightarrow dx = dt$, we get $I = \int_a^b f(t+c)\,dt = \int_a^b f(x+c)\,dx$

$I = \int_{ac}^{bc} f(x)\,dx$

Putting $x = tc \Rightarrow dx = c\,dt$,

we get $I = c\int_a^b f(ct)\,dt = c\int_a^b f(cx)\,dx$

$f(x) = \frac{1}{2}(f(x)+f(-x)+f(x)-f(-x))$

$\Rightarrow \int_{-a}^{a} f(x)\,dx$

$= \frac{1}{2}\int_{-a}^{a}(f(x)+f(-x)+f(x)-f(-x))\,dx$

$= \frac{1}{2}\int_{-a}^{a}(f(x)+f(-x))\,dx + \frac{1}{2}\int_{-a}^{a}(f(x)-f(-x))\,dx$

$= \frac{1}{2}\int_{-a}^{a}(f(x)+f(-x)\,dx$

as $f(x)+f(-x)$ is even and $f(x)-f(-x)$ is odd.

4. d. $\int_{\sqrt{2}}^{x}\frac{dt}{t\sqrt{t^2-1}} = \frac{\pi}{2}$

$\Rightarrow [\sec^{-1}t]_{\sqrt{2}}^{x} = \frac{\pi}{2}$

$\Rightarrow \sec^{-1}x - \sec^{-1}\sqrt{2} = \frac{\pi}{2}$

$\Rightarrow \sec^{-1}x - \frac{\pi}{4} = \frac{\pi}{2}$

$\Rightarrow \sec^{-1}x = \frac{\pi}{2}+\frac{\pi}{4} = \frac{3\pi}{4}, \Rightarrow x = -\sqrt{2}$

5. c. $\int_{-1}^{1/2}\frac{e^x\left(2-x^2\right)dx}{(1-x)\sqrt{1-x^2}}$

$= \int_{-1}^{1/2}\frac{e^x\left(1-x^2+1\right)}{(1-x)\sqrt{1-x^2}}$

$= \int_{-1}^{1/2} e^x\left[\sqrt{\frac{1+x}{1-x}}+\frac{1}{(1-x)\sqrt{1-x^2}}\right]dx$

$= e^x\sqrt{\frac{1+x}{1-x}}\Bigg|_{-1}^{1/2}$

$= \sqrt{3e}$

6. a. $\int_{-\pi}^{\pi}\sin nx\sin mx\,dx$

$= \int_0^{\pi} 2\sin mx\sin nx\,dx$

$= \int_0^{\pi}[\cos(m-n)x - \cos(m+n)x]\,dx$

$= \left|\frac{\sin(m-n)x}{m-n} - \frac{\sin(m+n)x}{m+n}\right|_0^{\pi} = 0$

7. a. $I = \int_0^{\infty}\frac{x\log x\,dx}{(1+x^2)^2}$

Let $x = \frac{1}{t}$

$$\Rightarrow I=\int_{\infty}^{0}\frac{\left(\frac{1}{t}\right)\log\left(\frac{1}{t}\right)\left(-\frac{1}{t^2}\right)dt}{\left(1+\frac{1}{t^2}\right)^2}$$

$$=-\int_0^{\infty}\frac{t\log t}{(1+t^2)^2}dt$$

$$=-I$$

$$\Rightarrow I=0$$

8.c. Put $x=a\sin\theta$ $\therefore$ $dx=a\cos d\theta$

When $x=0$, $\theta=0$; $x=a$, $\theta=\frac{\pi}{2}$

$$\therefore \text{ given integral } I=\int_0^{\pi/2}\frac{a\cos\theta\, d\theta}{a\sin\theta+a\cos\theta}$$

$$=\int_0^{\pi/2}\frac{\cos\theta d\theta}{\sin\theta+\cos\theta}$$

$$\text{Also } I=\int_0^{\pi/2}\frac{\cos\left(\frac{\pi}{2}-\theta\right)d\theta}{\sin\left(\frac{\pi}{2}-\theta\right)+\cos\left(\frac{\pi}{2}-\theta\right)}$$

$$=\int_0^{\pi/2}\frac{\sin\theta\, d\theta}{\cos\theta+\sin\theta}$$

$$\therefore 2I=\int_0^{\pi/2}\frac{\cos\theta+\sin\theta}{\cos\theta+\sin\theta}d\theta=\int_0^{\pi/2}d\theta=\frac{\pi}{2}$$

$$\therefore I=\frac{\pi}{4}$$

9. b. $\int_0^{\pi/2}|\sin x-\cos x|\,dx$

$$=\int_0^{\pi/4}-(\sin x-\cos x)dx+\int_{\pi/4}^{\pi/2}(\sin x-\cos x)dx$$

$$=|\cos x+\sin x|_0^{\pi/4}+|-\cos x-\sin x|_{\pi/4}^{\pi/2}$$

$$=\left(\frac{1}{\sqrt2}+\frac{1}{\sqrt2}-1-0\right)+\left(-0-1+\frac{1}{\sqrt2}+\frac{1}{\sqrt2}\right)$$

$$=\frac{4}{\sqrt2}-2=2\sqrt2-2=2\left(\sqrt2-1\right)$$

10. a. Let $I=\int_0^{\pi}\frac{x\tan x}{\sec x+\cos x}dx$ (1)

$$=\int_0^{\pi}\frac{(\pi-x)\tan(\pi-x)}{\sec(\pi-x)+\cos(\pi-x)}dx$$

$$=\int_0^{\pi}\frac{(\pi-x)\tan x}{\sec x+\cos x}dx \quad (2)$$

Adding equations (1) and (2) gives

$$2I=\pi\int_0^{\pi}\frac{\tan x}{\sec x+\cos x}dx$$

$$=\pi\int_0^{\pi}\frac{\frac{\sin x}{\cos x}}{\frac{1}{\cos x}+\cos x}dx=\pi\int_0^{\pi}\frac{\sin x}{1+\cos^2 x}dx$$

Put $\cos x=z$ $\therefore$ $-\sin x\,dx=dz$

when $x=0, z=1, x=\pi, z=-1$

$$\therefore 2I=\pi\int_1^{-1}\frac{-dz}{1+z^2}=\pi\int_{-1}^{1}\frac{dz}{1+z^2}$$

$$=\pi|\tan^{-1}z|_{-1}^{1}$$

$$=\pi[\tan^{-1}1-\tan^{-1}(-1)]$$

$$=\pi\left(\frac{\pi}{4}+\frac{\pi}{4}\right)=\frac{2\pi^2}{4}$$

$$\Rightarrow I=\frac{\pi^2}{4}$$

11. c. We have $\int_2^4(3-f(x))dx=7$

$$\Rightarrow 6-\int_2^4 f(x)dx=7\Rightarrow\int_2^4 f(x)dx=-1$$

Now,

$$\int_2^{-1}f(x)dx=-\int_{-1}^{2}f(x)dx=-\left[\int_{-1}^{4}f(x)dx+\int_4^2 f(x)dx\right]$$

$$=-\left[\int_{-1}^{4}f(x)dx-\int_2^4 f(x)dx\right]=-[4+1]=-5$$

12. e. We have $\int_0^1 e^{x^2}(x-\alpha)\,dx=0$

$$\Rightarrow\int_0^1 e^{x^2}x\,dx=\int_0^1 e^{x^2}\alpha\,dx$$

$$\Rightarrow\frac{1}{2}\int_0^1 e^t dt=\alpha\int_0^1 e^{x^2}dx\text{, where } t=x^2$$

$$\Rightarrow\frac{1}{2}(e-1)=\alpha\int_0^1 e^{x^2}dx \quad (1)$$

Since e^{x^2} is an increasing function for $0\le x\le 1$, therefore, $1\le e^{x^2}\le e$ when $0\le x\le 1$

$$\Rightarrow 1(1-0)\le\int_0^1 e^{x^2}dx\le e(1-0)$$

$$\Rightarrow 1\le\int_0^1 e^{x^2}dx\le e \quad (2)$$

From equations (1) and (2), we find that L.H.S. of equation (1) is positive and $\int_0^1 e^{x^2}dx$ lies between 1 and e. Therefore, α is a positive real number.

Now, from equation (1) $\Rightarrow\alpha=\dfrac{\frac{1}{2}(e-1)}{\int_0^1 e^{x^2}dx}$ (3)

The denominator of equation (3) is greater than unity and the numerator lies between 0 and 1.Therefore, $0<\alpha<1$.

13.a. Putting $a=2, b=3, c=0$, we get

$$\int_0^\infty \frac{dx}{(x^2+4)(x^2+9)}=\frac{\pi}{2(2+3)(3+0)(0+2)}$$

$$=\frac{\pi}{60}$$

14.c. Given integral

$$=\int_0^1 \frac{dx}{(x+\cos\alpha)^2+(1-\cos^2\alpha)}$$

$$=\int_0^1 \frac{dx}{(x+\cos\alpha)^2+\sin^2\alpha}$$

$$=\frac{1}{\sin\alpha}\left|\tan^{-1}\frac{x+\cos\alpha}{\sin\alpha}\right|_0^1$$

$$=\frac{1}{\sin\alpha}\left[\tan^{-1}\frac{1+\cos\alpha}{\sin\alpha}-\tan^{-1}\frac{\cos\alpha}{\sin\alpha}\right]$$

$$=\frac{1}{\sin\alpha}\left[\tan^{-1}\cot\frac{\alpha}{2}-\tan^{-1}(\cot\alpha)\right]$$

$$=\frac{1}{\sin\alpha}\left[\tan^{-1}\tan\left(\frac{\pi}{2}-\frac{\alpha}{2}\right)-\tan^{-1}\tan\left(\frac{\pi}{2}-\alpha\right)\right]$$

$$=\frac{1}{\sin\alpha}\left[\left(\frac{\pi}{2}-\frac{\alpha}{2}\right)-\left(\frac{\pi}{2}-\alpha\right)\right]=\frac{\alpha}{2\sin\alpha}$$

15.d. $I=\int_1^e\left(\frac{1}{x}+1\right)dx-\int_1^e\frac{1+\ln x}{1+x\ln x}dx$

$$=[\ln x+x]_1^e-[\ln(1+x\ln x)]_1^e$$

$$=e-\ln(1+e)$$

16.b. On putting $x=\sin\theta$, we get $dx=\cos\theta\,d\theta$

Integral (without limits) $=\int\frac{\cos\theta d\theta}{(1+\sin^2\theta)(\cos\theta)}$

$$=\int\frac{d\theta}{1+\sin^2\theta}=\int\frac{\text{cosec}^2\theta d\theta}{2+\cot^2\theta}$$

$$=\int\frac{-dt}{2+t^2}\text{ where } t=\cot\theta$$

$$=-\frac{1}{\sqrt{2}}\tan^{-1}\frac{t}{\sqrt{2}}=-\frac{1}{\sqrt{2}}\tan^{-1}\frac{\cot\theta}{\sqrt{2}}$$

$$=-\frac{1}{\sqrt{2}}\tan^{-1}\frac{1}{\sqrt{2}}\left(\frac{\sqrt{1-x^2}}{x}\right)$$

$$\Rightarrow \text{Definite integral}=-\frac{1}{\sqrt{2}}\tan^{-1}1+\frac{1}{\sqrt{2}}\tan^{-1}\infty$$

$$=-\frac{\pi}{4\sqrt{2}}+\frac{\pi}{2\sqrt{2}}=\frac{\pi}{4\sqrt{2}}$$

17.b. Putting $e^x-1=t^2$ in the given integral, we have

$$\int_0^{\log 5}\frac{e^x\sqrt{e^x-1}}{e^x+3}dx=2\int_0^2\frac{t^2}{t^2+4}dt=2\left(\int_0^2 1dt-4\int_0^2\frac{dt}{t^2+4}\right)$$

$$=2\left[\left(t-2\tan^{-1}\left(\frac{t}{2}\right)\right)_0^2\right]$$

$$=2[(2-2\times\pi/4)]=4-\pi$$

18. a. Put $x=\tan\theta \therefore dx=\sec^2\theta d\theta$

When $x=\infty$, $\tan\theta=\infty$, $\therefore \theta=\pi/2$

$$\therefore I=\int_0^{\pi/2}\frac{\tan\theta\sec^2\theta}{(1+\tan\theta)(\sec^2\theta)}d\theta \quad (1)$$

Now changing equation (1) into $\sin\theta$ and $\cos\theta$

$$\therefore I=\int_0^{\pi/2}\frac{\sin\theta\,d\theta}{\cos\theta+\sin\theta}=\frac{\pi}{4}$$

19.a. Putting $x=\tan\theta$, we get

$$\int_0^{\pi/2}\frac{dx}{[x+\sqrt{x^2+1}]^3}=\int_0^\infty\frac{\sec^2\theta d\theta}{(\tan\theta+\sec\theta)^3}$$

$$=\int_0^{\pi/2}\frac{\cos\theta}{(1+\sin\theta)^3}d\theta$$

$$=\left[-\frac{1}{2(1+\sin\theta)^2}\right]_0^{\pi/2}$$

$$=-\frac{1}{8}+\frac{1}{2}=\frac{3}{8}$$

20. c. $I=\int_0^{\pi/2}\frac{\sin xdx}{1+\sin x+\cos x}=\int_0^{\pi/2}\frac{\cos xdx}{1+\sin x+\cos x}$

$$\Rightarrow 2I=\int_0^{\pi/2}\frac{\sin x+\cos x+1-1}{\sin x+\cos x+1}dx$$

$$\Rightarrow 2I=\frac{\pi}{2}-\log 2$$

$$\Rightarrow I=\frac{\pi}{4}-\frac{1}{2}\log 2$$

21.b. $I_1=\int_{-100}^{101}\frac{dx}{(5+2x-2x^2)(1+e^{2-4x})}$

$$=\int_{-100}^{101}\frac{dx}{(5+2(1-x)-2(1-x)^2)(1+e^{2-4(1-x)})}$$

$$\Rightarrow 2I_1=\int_{-100}^{101}\frac{dx}{5+2x-2x^2}=I_2$$

$$\Rightarrow \frac{I_1}{I_2}=\frac{1}{2}$$

22.c. $f(x)=\dfrac{e^x}{1+e^x}$ $\therefore f(a)=\dfrac{e^a}{1+e^a}$

and $f(-a)=\dfrac{e^{-a}}{1+e^{-a}}$

$=\dfrac{e^{-a}}{1+\dfrac{1}{e^a}}=\dfrac{1}{1+e^a}$

$\Rightarrow f(a)+f(-a)=\dfrac{e^a+1}{1+e^a}=1$

Let $f(-a)=\alpha \therefore f(a)=1-a$

Now, $I_1=\displaystyle\int_{\alpha}^{1-\alpha} xg(x(1-x))\,dx$

$=\displaystyle\int_{\alpha}^{1-\alpha}(1-x)g((1-x)(1-(1-x))\,dx$

$=\displaystyle\int_{\alpha}^{1-\alpha}(1-x)g(x(1-x))\,dx$

$\therefore 2I_1=\displaystyle\int_{\alpha}^{1-\alpha} g(x(1-x))\,dx=I_2$ $\therefore \dfrac{I_2}{I_1}=2$

23.a. We have $f(y)=e^y, g(y)=y : y>0$

$F(t)=\int_0^t f(t-y)g(y)dy$

$=\int_0^t e^{t-y}y\,dy$

$=e^t\int_0^t e^{-y}y\,dy$

$=e^t\left(\left[-ye^{-y}\right]_0^t+\int_0^t e^{-y}\,dy\right)$

$=e^t\left(-te^{-t}-\left[e^{-y}\right]_0^t\right)$

$=e^t\left(-te^{-t}-e^{-t}+1\right)$

$=e^t-(1+t)$

24.c. $I=\displaystyle\int_0^{\sqrt{\ln\left(\frac{\pi}{2}\right)}}\cos\left(e^{x^2}\right)2xe^{x^2}\,dx$

Put $e^{x^2}=t\Rightarrow e^{x^2}2x\,dx=dt$

$\Rightarrow I=\displaystyle\int_1^{\pi/2}\cos t\,dt=[\sin t]_1^{\pi/2}=1-(\sin 1)$

25. a. $\displaystyle\int_1^{\frac{1+\sqrt{5}}{2}}\frac{1+\frac{1}{x^2}}{x^2-1+\frac{1}{x^2}}\log\left(1+x-\frac{1}{x}\right)dx$

$=\displaystyle\int_1^{\frac{1+\sqrt{5}}{2}}\frac{1+\frac{1}{x^2}}{\left(x-\frac{1}{x}\right)^2+1}\log\left(1+x-\frac{1}{x}\right)dx$

Put $x-\dfrac{1}{x}=t$

$\left(1+\dfrac{1}{x^2}\right)dx=dt$

If $x=1, t=0$, and

$x=\dfrac{\sqrt{5}+1}{2}, t=1$

$\Rightarrow I=\displaystyle\int_0^1\frac{\ln(1+t)\,dt}{1+t^2}$ Put $t=\tan\theta$ $\therefore dt=\sec^2\theta d\theta$

$I=\displaystyle\int_0^{\pi/4}\ln(1+\tan\theta)\,d\theta=\frac{\pi}{8}\log_e 2$

26. c. As $f(x)$ satisfies the conditions of Rolle's theorem in [1, 2],
$f(x)$ is continuous in the interval and $f(1)=f(2)$.

$\therefore \int_1^2 f'(x)dx=[f(x)]_1^2=f(2)-f(1)=0$

27. a. $\int_{-1}^{3}\left(\tan^{-1}\dfrac{x}{x^2+1}+\tan^{-1}\dfrac{x^2+1}{x}\right)dx$

$=\int_{-1}^{0}\left(\tan^{-1}\dfrac{x}{x^2+1}+\tan^{-1}\dfrac{x^2+1}{x}\right)dx$

$+\int_0^3\left(\tan^{-1}\dfrac{x}{x^2+1}+\tan^{-1}\dfrac{x^2+1}{x}\right)dx$

$=\int_{-1}^0 -\dfrac{\pi}{2}dx+\int_0^3\dfrac{\pi}{2}dx$

$=\left[-\dfrac{\pi}{2}x\right]_{-1}^0+\left[\dfrac{\pi}{2}x\right]_0^3$

$=\pi$

28.c. The polynomial is an every where differentiable function. Therefore, the points of extremum can only be the roots of the derivative. Furthermore, the derivative of a polynomial is a polynomial. The polynomial of theleast degree with roots $x=1$ and $x=3$ has the form $a(x-1)(x-3)$.
Hence, $P(x)=a(x-1)(x-3)$.
Since at $x=1$, we must have $P(1)=6$, we have

$P(x)=\int_1^x P'(x)dx+6=a\int_1^x(x^2-4x+3)dx+6$

$$= a\left(\frac{x^3}{3} - 2x^2 + 3x - \frac{4}{3}\right) + 6$$

Also, $P(3) = 2$ so $a = 3$. Hence, $P(x) = x^3 - 6x^2 + 9x + 2$.

Thus, $\int_0^1 P(x)dx = \frac{1}{4} - 2 + \frac{9}{2} + 2 = \frac{19}{4}$

29.d. Since $a^2 I_1 - 2aI_2 + I_3 = 0$

$$\Rightarrow \int_0^1 (a-x)^2 f(x)dx = 0$$

Hence, no such positive function $f(x)$.

30.b. $I = \int_0^{\pi/2} \sqrt{\tan x}\, dx$ (1)

$\Rightarrow I = \int_0^{\pi/2} \sqrt{\cot x}\, dx$ (2)

Adding equations (1) and (2), we get

$$2I = \int_0^{\pi/2} \left(\sqrt{\tan x} + \sqrt{\cot x}\right) dx$$

$$= \sqrt{2}\int_0^{\pi/2} \frac{\sin x + \cos x}{\sqrt{\sin 2x}} dx$$

$$= \sqrt{2}\int_0^{\pi/2} \frac{\sin x + \cos x}{\sqrt{1-(\sin x - \cos x)^2}} dx$$

$= \sqrt{2}\int_{-1}^{1} \frac{dt}{\sqrt{1-t^2}}$ (where $\sin x - \cos x = t$)

$$= 2\sqrt{2}\int_0^1 \frac{dt}{\sqrt{1-t^2}}$$

$$= \sqrt{2}\,\pi$$

$$\Rightarrow I = \frac{\pi}{\sqrt{2}}$$

31.a. Let $I = \int_1^3 \frac{\sin 2x}{x} dx$

Put $2x = t$,

$$\Rightarrow dx = \frac{dt}{2}$$

$$\Rightarrow I = \frac{2}{2}\int_2^6 \frac{\sin t}{t} dt = \int_2^6 \frac{\sin t}{t} dt$$

But given $\int \frac{\sin x}{x} dx = F(x)$

$$\Rightarrow \int_2^6 \frac{\sin t}{t} dt = F(6) - F(2)$$

32.b. $\int_0^1 \cot^{-1}(1 - x + x^2)\, dx$

$$= \int_0^1 \tan^{-1}\left(\frac{1}{1-x+x^2}\right) dx$$

$$= \int_0^1 \tan^{-1}\left(\frac{x+(1-x)}{1-x(1-x)}\right) dx$$

$$= \int_0^1 \tan^{-1} x dx + \int_0^1 \tan^{-1}(1-x) dx$$

$$= \int_0^1 \tan^{-1} x dx + \int_0^1 \tan^{-1}[1-(1-x)] dx$$

$$= 2\int_0^1 \tan^{-1} x dx$$

$$\Rightarrow \lambda = 2$$

33.a. Let $I = \int_{-3\pi/4}^{5\pi/4} \frac{(\sin x + \cos x)}{e^{x-\pi/4} + 1} dx$

$$\Rightarrow I = \int_{-3\pi/4}^{5\pi/4} \frac{\sqrt{2}\cos\left(x - \frac{\pi}{4}\right)}{e^{x-\pi/4} + 1} dx$$

Putting $x - \frac{\pi}{4} = t$

$$\Rightarrow dx = dt$$

$\Rightarrow I = \int_{-\pi}^{\pi} \frac{\sqrt{2}\cos t}{e^t + 1} dt$ (1)

Replacing t by $\pi + (-\pi) - t$ or $-t$, we get

$I = \int_{-\pi}^{\pi} \frac{\sqrt{2}\cos(-t)}{e^{-t} + 1} dt = \int_{-\pi}^{\pi} \frac{e^t \sqrt{2}\cos t}{e^t + 1} dt$ (2)

Adding equations (1) and (2), we get

$$2I = \sqrt{2}\int_{-\pi}^{\pi} \cos t dt \Rightarrow I = 0.$$

34. a. $f(2-\alpha) = f(2+\alpha)$

$\Rightarrow$ function is symmetric about the line $x = 2$.

$$\int_{2-a}^{2+a} f(x)dx = 2\int_2^{2+a} f(x)dx.$$

35. c. Since e^{x^2} is an increasing function on (0, 1), therefore $m = e^0 = 1$, $M = e^1 = e$ (m and M are minimum and maximum values of $f(x) = e^{x^2}$ in the interval (0, 1))

$\Rightarrow 1 < e^{x^2} < e$, for all $x \in (0, 1)$

$$\Rightarrow 1(1-0) < \int_0^1 e^{x^2} dx < e(1-0)$$

$$\Rightarrow 1 < \int_0^1 e^{x^2} dx < e$$

36.a. $I_2=\int_{-\pi/4}^{\pi/4}\ln(\sin x+\cos x)\,dx$

$=\int_0^{\pi/4}(\ln(\sin x+\cos x)+\ln(\sin(-x)+\cos(-x)))\,dx$

$=\int_0^{\pi/4}(\ln(\sin x+\cos x)+\ln(\cos x-\sin x))\,dx$

$=\int_0^{\pi/4}\ln(\cos^2 x-\sin^2 x)\,dx$

$=\int_0^{\pi/4}\ln(\cos 2x)\,dx$

Putting $2x=t$, i.e., $\frac{dt}{2}=dx$, we get

$I_2=\frac{1}{2}\int_0^{\pi/2}\ln(\cos t)\,dt=\frac{1}{2}\int_0^{\pi/2}\ln\left(\cos\left(\frac{\pi}{2}-t\right)\right)dt$

$=\frac{1}{2}\int_0^{\pi/2}\ln(\sin t)dt=\frac{1}{2}I_1$

$\Rightarrow I_1=2I_2$

37.c. $I_1=\int_0^{\pi/2}\frac{\cos^2 x}{1+\cos^2 x}dx$

$=\int_0^{\pi/2}\frac{\cos^2(\pi/2-x)}{1+\cos^2(\pi/2-x)}dx$

$=\int_0^{\pi/2}\frac{\sin^2 x}{1+\sin^2 x}dx=I_2$

Also $I_1+I_2=\int_0^{\pi/2}\left(\frac{\sin^2 x}{1+\sin^2 x}+\frac{\cos^2 x}{1+\cos^2 x}\right)dx$

$=\int_0^{\pi/2}\frac{\sin^2 x+\sin^2 x\cos^2 x+\cos^2 x+\sin^2 x\cos^2 x}{1+\sin^2 x+\cos^2 x+\sin^2 x\cos^2 x}dx$

$=\int\frac{1+2\sin^2 x\cos^2 x}{2+\sin^2 x\cos^2 x}dx=2I_3$

$2I_1=2I_3\Rightarrow I_1=I_3\ \Rightarrow I_1=I_2=I_3$

38.a. $\sum_{r=1}^{n}\int_0^1 f(r-1+x)\,dx$

$=\int_0^1 f(x)dx+\int_0^1 f(1+x)dx+\int_0^1 f(2+x)dx+\cdots$

$+\int_0^1 f(n-1+x)dx$

$=\int_0^1 f(x)dx+\int_1^2 f(x)dx+\int_2^3 f(x)dx+\int_{r-1}^{2} f(x)dx+\cdots$

$+\int_{n-1}^{n} f(x)dx.$

$=\int_0^n f(x)\,dx$

39. c. $I_1=\int_0^{\pi/2}\frac{\sin x-\cos x}{1+\sin x\cos x}dx$

$=\int_0^{\pi/2}\frac{\sin\left(\frac{\pi}{2}-x\right)-\cos\left(\frac{\pi}{2}-x\right)}{1+\sin\left(\frac{\pi}{2}-x\right)\cos\left(\frac{\pi}{2}-x\right)}dx$

$=\int_0^{\pi/2}\frac{\cos x-\sin x}{1+\sin x\cos x}dx=-I_1$

$\Rightarrow I_1=0$

$I_3=0$ as $\sin^3 x$ is odd.

$I_4=\int_0^1\ln\left(\frac{1-x}{x}\right)dx$

$=\int_0^1\ln\left(\frac{1-(1-x)}{1-x}\right)dx$

$=\int_0^1\ln\frac{x}{1-x}dx=-I_4$

$\Rightarrow I_4=0$

$I_2=\int_0^{2\pi}\cos^6 x dx=2\int_0^{\pi}\cos^6 x\,dx\neq 0$

40. c. $I=\int_{\log\lambda}^{\log\frac{1}{\lambda}}\frac{f(x^2/4)[f(x)-f(-x)]}{g(x^2/4)[g(x)+g(-x)]}dx$

$=\int_{\log\lambda}^{-\log\lambda}\frac{f(x^2/4)[f(x)-f(-x)]}{g(x^2/4)[g(x)+g(-x)]}=0$

(as function inside the integration is odd)

41. b. $I=0+2\int_0^{\pi}\frac{2x\sin x}{1+\cos^2 x}=4\int_0^{\pi}\frac{x\sin x}{1+\cos^2 x}dx$

$=4\frac{\pi^2}{4}=\pi^2$

42. c. $\int_0^a\frac{f(x)}{f(x)+f(a-x)}=\frac{a}{2}$

$\Rightarrow\lim_{n\to\infty}\left[\frac{a}{2}+\frac{a^2}{2}+\frac{a^3}{2}+\cdots+\frac{a^n}{2}\right]=\frac{7}{5}$

$\Rightarrow\lim_{n\to\infty}\frac{a}{1-a}=\frac{14}{5}$

$\Rightarrow 5a=14-14a$

$\Rightarrow a=\frac{14}{19}$

43.c. $f(x)=\int_0^{\pi}\frac{t\sin t}{\sqrt{1+\tan^2 x\sin^2 t}}dt$ (1)

Replacing t by $\pi-t$ and then adding $f(x)$ with equation (1).

$$f(x)=\frac{\pi}{2}\int_0^{\pi}\frac{\sin t}{\sqrt{1+\tan^2 x\sin^2 t}}dt$$

$$=\pi\int_0^{\pi/2}\frac{\sin t}{\sqrt{1+\tan^2 x(1-\cos^2 t)}}dt$$

$$=\pi\int_0^{\pi/2}\frac{\sin t}{\sqrt{\sec^2 x-\tan^2 x\cos^2 t}}dt$$

Let $y=\cos t$

$\therefore dy=-\sin t\,dt$

$$=\pi\int_0^1\frac{dy}{\sqrt{\sec^2 x-(\tan^2 x)y^2}}$$

$$=\frac{\pi}{\tan x}\int_0^1\frac{dy}{\sqrt{\operatorname{cosec}^2 x-y^2}}$$

$$=\frac{\pi}{\tan x}\left\{\sin^{-1}\frac{y}{\operatorname{cosec}x}\right\}_0^1$$

$$=\frac{\pi}{\tan x}\sin^{-1}(\sin x)=\frac{\pi x}{\tan x}$$

44. c. $I=\int_{-\pi/4}^{3\pi/4}\frac{dx}{\sqrt{2}(e^{x-\pi/4}+1)\cos\left(x-\frac{\pi}{4}\right)}$

Putting $x-\frac{\pi}{4}=t$, we get

$$\Rightarrow I=\frac{1}{\sqrt{2}}\int_{-\pi/2}^{\pi/2}\frac{dt}{(e^t+1)\cos t}$$

$$=\frac{1}{\sqrt{2}}\int_{-\pi/2}^{\pi/2}\frac{e^t dt}{(e^t+1)\cos t}$$

Adding, we get $2I=\frac{1}{\sqrt{2}}\int_{-\pi/2}^{\pi/2}\sec t\,dt$

$\therefore I=\frac{1}{2\sqrt{2}}\int_{-\pi/2}^{\pi/2}\sec x\,dx \quad \therefore k=\frac{1}{2\sqrt{2}}$

45. a. For $x\in\left(-\frac{\pi}{3},0\right)$, $2\cos x-1>0$

$$\Rightarrow I=\int_{-\pi/3}^{0}\frac{\pi}{2}dx=\frac{\pi^2}{6}$$

46. a. $\int_0^{\infty}\left(\frac{\pi}{1+\pi^2x^2}-\frac{1}{1+x^2}\right)\log x\,dx$

$$=\int_0^{\infty}\frac{\log\left(\frac{y}{\pi}\right)dy}{1+y^2}-\int_0^{\infty}\frac{\log x}{1+x^2}dx$$

$$=-\int_0^{\infty}\frac{\log\pi}{1+y^2}dy=-\frac{\pi}{2}\ln\pi$$

47. d. $f(x)=\cos(\tan^{-1}x)$

$$\Rightarrow f'(x)=-\frac{\sin(\tan^{-1}x)}{1+x^2}$$

$$\Rightarrow I=\int_0^1 x\,f''(x)\,dx$$

$$=[x\,f'(x)]_0^1-\int_0^1 f'(x)\,dx \text{ (Integrating by parts)}$$

$$=[f'(1)]-[f(x)]_0^1$$

$$=f'(1)-f(1)+f(0)$$

Now $f(0)=1;\ f'(1)=-\frac{1}{2\sqrt{2}};\ f(1)=\frac{1}{\sqrt{2}}$

$$\Rightarrow I=1-\frac{3}{2\sqrt{2}}$$

48. a. Given $f'(1)=\tan\pi/6, f'(2)=\tan\pi/3, f'(3)=\tan\pi/4$

Now $\int_2^3 f'(x)f''(x)dx+\int_1^3 f''(x)\,dx$

$$=\left[\frac{(f'(x))^2}{2}\right]_2^3+[f'(x)]_1^3$$

$$=\frac{(f'(3))^2-(f'(2))^2}{2}+f'(3)-f'(1)$$

$$=\frac{(1)^2-(\sqrt{3})^2}{2}+\left(1-\frac{1}{\sqrt{3}}\right)$$

$$=\frac{1-3}{2}+1-\frac{1}{\sqrt{3}}=\frac{-1}{\sqrt{3}}$$

49. b. $\int_1^e\left(\frac{\tan^{-1}x}{x}+\frac{\log x}{1+x^2}\right)dx$

$$=\int_1^e\frac{\tan^{-1}x}{x}dx+\int_1^e\frac{\log x}{1+x^2}dx$$

$$=\int_1^e\frac{\tan^{-1}x}{x}dx+\left(\log x\tan^{-1}x\right)_1^e-\int_1^e\frac{\tan^{-1}x}{x}dx$$

$$=\tan^{-1}e$$

50. b. $\int_0^{\pi}[f(x)+f''(x)]\sin x\,dx$

$$=\int_0^{\pi}f(x)\sin x\,dx+\int_0^{\pi}f''(x)\sin x\,dx$$

$$=(f(x)-\cos x)_0^{\pi}+\int_0^{\pi}f'(x)\cos x\,dx$$

$$+\sin x\,f'(x)|_0^{\pi}-\int_0^{\pi}\cos x\,f'(x)dx$$

$$=f(\pi)+f(0)=5 \text{ (given)}$$

$$\Rightarrow f(0)=5-f(\pi)=5-2=3$$

51. b. $I_1=\int_e^{e^4}\sqrt{\ln x}dx$, putting $t=\sqrt{\ln x}$, i.e.,

$$dt=\frac{dx}{2x\sqrt{\ln x}}$$

$$\Rightarrow dx=2t\,e^{t^2}\,dt$$

$$\Rightarrow \int_e^{e^4}\sqrt{\ln x}dx=\int_1^2 2t^2e^{t^2}\,dt$$

$$=t\,e^{t^2}\Big|_1^2-\int_1^2 e^{t^2}\,dt=2e^4-e-a$$

52. c. $\int_{-\frac{\pi}{2}}^{\frac{\pi}{2}}\frac{e^{|\sin x|}\cos x}{(1+e^{\tan x})}dx$

$$=\int_0^{\frac{\pi}{2}}\left(\frac{e^{|\sin x|}\cos x}{1+e^{\tan x}}+\frac{e^{|\sin x|}\cos x}{1+e^{-\tan x}}\right)dx$$

$$=\int_0^{\frac{\pi}{2}}e^{|\sin x|}\cos x dx$$

$$=\int_0^{\frac{\pi}{2}}e^{\sin x}\cos x dx$$

$$=e^{\sin x}\Big|_0^{\frac{\pi}{2}}=e-1$$

53. d. $\int_0^a x^4\sqrt{a^2-x^2}\,dx$

$$=\left[\frac{-x^3\left(a^2-x^2\right)^{3/2}}{3}\right]_0^a+a^2\,\frac{3}{6}\int_0^a x^2\sqrt{a^2-x^2}\,dx$$

Integrating by parts with x^3 as first function and $x\sqrt{a^2-x^2}$ as second function.

$$=\frac{a^2}{2}\int_0^a x^2\sqrt{a^2-x^2}\,dx$$

$$\Rightarrow\frac{\int_0^a x^4\sqrt{a^2-x^2}\,dx}{\int_0^a x^2\sqrt{a^2-x^2}\,dx}=\frac{a^2}{2}$$

54. a. $I=\int_0^{\pi/2}\frac{\sin 2x}{x+1}dx$. Put $x=y/2$

$$\Rightarrow I=\int_0^{\pi}\frac{\sin y}{y+2}dy$$

$$=\left(\frac{-\cos y}{y+2}\right)_0^{\pi}-\int_0^{\pi}\frac{\cos y}{(y+2)^2}dy \text{ (integrating by parts)}$$

$$\Rightarrow I=\frac{1}{\pi+2}+\frac{1}{2}-A.$$

55. a. $I=\int_0^4\frac{(y^2-4y+5)\sin(y-2)}{(2y^2-8y+1)}dy$, put $y-2=z$

$$\Rightarrow I=\int_{-2}^{2}\frac{z^2+1}{2z^2-7}\sin(z)dz=0$$

56. a. Putting $x\tan\theta=z\sin\theta\Rightarrow dx=\cos\theta\,dz$

$$\Rightarrow I=\cos\theta\int_{\tan\theta}^{1}f(z\sin\theta)\,dz$$

$$=-\cos\theta\int_1^{\tan\theta}f(x\sin\theta)\,dx$$

57. c. $I_1=\int_0^1\frac{e^x dx}{1+x}$, $I_2=\int_0^1\frac{x^2 dx}{e^{x^3}(2-x^3)}$

In I_2, put $1-x^3=t$

$$\Rightarrow I_2=\frac{1}{3}\int_1^0\frac{-dt}{e^{1-t}(1+t)}$$

$$=\frac{1}{3e}\int_0^1\frac{e^t dt}{1+t}=\frac{1}{3e}I_1$$

$$\Rightarrow\frac{I_1}{I_2}=3e$$

58. d. $I=\int_{4\pi-2}^{4\pi}\frac{\sin\frac{t}{2}}{4\pi+2-t}\,dt=\frac{1}{2}\int_{4\pi-2}^{4\pi}\frac{\sin\frac{t}{2}}{1+\left(2\pi-\frac{t}{2}\right)}dt$

Put $2\pi-\frac{t}{2}=z$

$$\therefore -\frac{1}{2}dt=dz, \text{ i.e., } dt=-2\,dz$$

When $t=4\pi-2, z=2\pi-2\pi+1=1$

When $t=4\pi, z=2\pi-2\pi=0$

$$\Rightarrow I=\frac{1}{2}\int_1^0\frac{\sin(2\pi-z)(-2\,dz)}{1+z}$$

$$=\int_0^1\frac{-\sin z\,dz}{z+1}=-\int\frac{\sin t}{1+t}dt=-\alpha$$

Hence, the given integral $=-\alpha$

59. d. $I=\int_0^1\frac{\tan^{-1}x}{x}dx$

Putting $x=\tan\theta\Rightarrow dx=\sec^2\theta\,d\theta$

$$\Rightarrow I=\int_0^{\pi/4}\frac{\theta}{\tan\theta}\sec^2\theta d\theta$$

$$=\int_0^{\pi/4}\frac{2\theta}{\sin 2\theta}d\theta$$

Putting $2\theta=t$, i.e., $2d\theta=dt$,

we get $I=2\int_0^{\pi/2}\frac{t/2}{\sin t}\frac{dt}{2}$

$$=\frac{1}{2}\int_0^{\pi/2}\frac{t}{\sin t}dt$$

$$=\frac{1}{2}\int_0^{\pi/2}\frac{x}{\sin x}dx.$$

60. a. $I_K = \int_1^e (\ln x)^k\,dx = |x(\ln x)^k|_1^e - k\int_1^e (\ln x)^{k-1}dx$

$\Rightarrow I_k = e - kI_{k-1}$

$\Rightarrow I_4 = e - 4I_3$

$= e - 4[e - 3(e - 2I_1)]$

$= 9e - 24 \quad (\because I_1 = 1)$

61. c. Putting $x = \dfrac{1}{1+y}$, $dx = -\dfrac{1}{(1+y)^2}dy$,

we get $I_{(m,n)} = \int_0^1 x^{m-1}(1-x)^{n-1}dx$

$$= \int_\infty^0 \frac{1}{(1+y)^{m-1}}\left(1-\frac{1}{1+y}\right)^{n-1}\frac{(-1)}{(1+y)^2}dy$$

$$= \int_0^\infty \frac{y^{n-1}}{(1+y)^{m+n}}dy = \int_0^\infty \frac{x^{n-1}}{(1+x)^{m+n}}dx$$

Since $I(m,n) = I(n,m)$

Therefore, $I(m,n) = \int_0^\infty \dfrac{x^{m-1}}{(1+x)^{m+n}}dx$

$$= \int_0^\infty \frac{x^{n-1}}{(1+x)^{m+n}}dx$$

62. c. We have $I_{n+1} - I_n = 2\int_0^\pi \cos(n+1)x\,dx = 0$

$\therefore I_{n+1} = I_n \Rightarrow I_{n+1} = I_n = \cdots = I_0$

$\Rightarrow I_n = \pi$ for all $n \ge 0$

63. b. $\sin nx - \sin(n-2)x = 2\cos(n-1)x \sin x$

$$\Rightarrow \int \frac{\sin nx}{\sin x}dx = \int 2\cos(n-1)\,dx + \int \frac{\sin(n-2)x}{\sin x}dx$$

$$\therefore \int_0^{\pi/2} \frac{\sin 5x}{\sin x}dx = \int_0^{\pi/2} 2\cos 4x\,dx + \int_0^{\pi/2} \frac{\sin 3x}{\sin x}dx$$

$$= 0 + \int_0^{\pi/2} \frac{\sin 3x}{\sin x}dx = \int_0^{\pi/2} dx = \frac{\pi}{2}$$

64. a. $I_3 = \int_0^\pi e^x(\sin x)^3\,dx$

$$= e^x(\sin x)^3\Big|_0^\pi - 3\int_0^\pi (\sin x)^2\cos x\, e^x dx$$

$$= 0 - 3(\sin x)^2\cos x\, e^x\Big|_0^\pi + 3\int_0^\pi (2\sin x\cos x\cos x - \sin x\sin^2 x)\,e^x dx$$

$$= 0 + 6\int_0^\pi \sin x\cos^2 x e^x dx - 3\int_0^\pi \sin^3 x\, e^x dx$$

$$= 6\int_0^\pi \sin x(1-\sin^2 x)e^x dx - 3\int_0^\pi \sin^3 x\, e^x dx$$

$$= 6\int_0^\pi \sin x e^x dx - 9\int_0^\pi \sin x^3 x e^x dx$$

$= 6I_1 - 9I_3$

$\Rightarrow 10I_3 = 6I_1$

$\Rightarrow \dfrac{I_3}{I_1} = \dfrac{3}{5}$

65. b. $I_m = \int_1^e (\log x)^m\,dx = I_m$

$$= \left[x(\log x)^m\right]_1^e - \int_1^e x\frac{m(\log x)^{m-1}}{x}dx$$

(integrating by parts)

$\Rightarrow I_m = e - m\int_1^e (\log x)^{m-1}dx = e - mI_{m-1}$ (1)

Replacing m by $m-1$

$I_{m-1} = e - (m-1)I_{m-2}$ (2)

From equations (1) and (2), we have

$I_m = e - m[e - (m-1)I_{m-2}]$

$\Rightarrow I_m - m(m-1)I_{m-1} = e(1-m) \Rightarrow \dfrac{I_m}{1-m} + mI_{m-2} = e$

$\Rightarrow K = 1 - m$ and $L = \dfrac{1}{m}$

66. a.

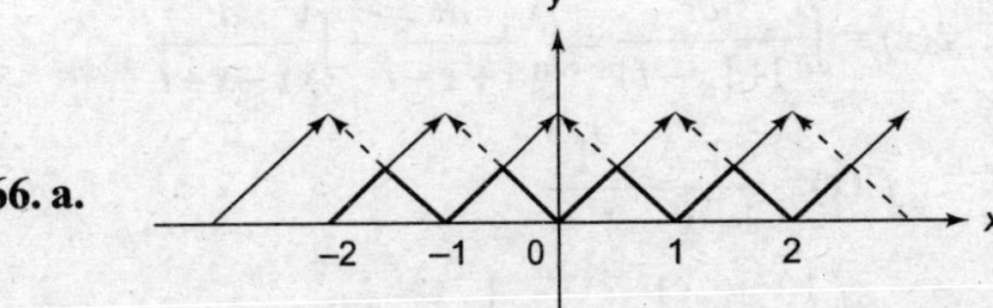

Fig. 8.1

The graph with solid line is the graph of $f(x) = \{x\}$ and the graph with dotted lines is the graph of $f(x) = \{-x\}$. Now the graph of $\min(\{x\}, \{-x\})$ is the graph with dark solid lines.

$\int_{-100}^{100} f(x)\,dx$ = area of 200 triangles shown as solid dark lines in the diagram $= 200\dfrac{1}{2}(1)(\dfrac{1}{2}) = 50$.

67. c. Put $x - 0.4 = t$

$$\Rightarrow \int_{0.6}^{3.6}\{t\}\,dt = \int_{0.6}^{0.6+3}\{t\}\,dt$$

$$= \int_0^1 (t-[t])dt = 3\left(\frac{t^2}{2}\right)_0^1 = \frac{3}{2} = 1.5$$

68.b. Let $I = \int_1^a [x]\, f'(x)\, dx,\ a > 1$

Let $a = k + h$, where $[a] = k$, and $0 \le h < 1$

$$\therefore \int_1^a [x]\, f'(x)\, dx = \int_1^2 1\, f'(x) dx + \int_2^3 2\, f'(x)\, dx + \cdots + \int_{k-1}^{k} (k-1) f'(x)\, dx + \int_k^{k+h} k\, f'(x)\, dx$$

$$= [f(2) - f(1)] + 2[f(3) - f(2)] + \cdots + (k-1)[f(k) - f(k-1)] + k[f(k+h) - f(k)]$$

$$= -f(1) - f(2) - f(3) \cdots - f(k) + kf(k+h)$$

$$= [a] f(A) - [f(1) + f(2) + \cdots + f([a])]$$

69.c. $I = \int\limits_0^x [\cos t] dt = \int\limits_0^{2n\pi} [\cos t] dt + \int\limits_{2n\pi}^{x} [\cos t] dt$

$$= n\int\limits_0^{2\pi} [\cos t] dt + \int\limits_{2n\pi}^{2n\pi+\pi/2} [\cos t] dt + \int\limits_{2n\pi+\frac{\pi}{2}}^{x} [\cos t] dt$$

$$= -n\pi + 0 + (x - (2n\pi + \pi/2))(-1)$$

$$= -n\pi + 2n\pi + \pi/2 - x$$

$$= (2n+1)\pi/2 - x$$

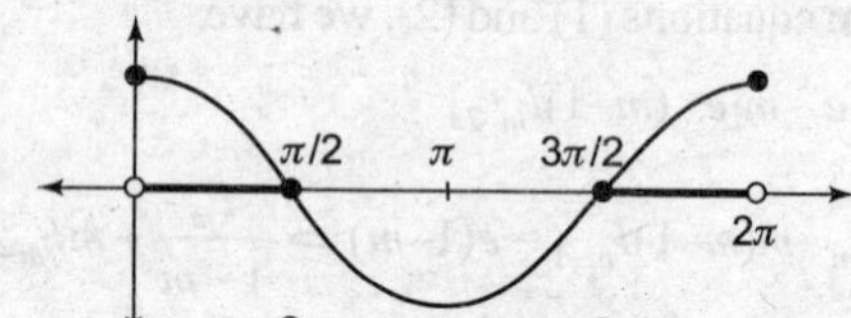

Fig. 8.2

70. b. $f(x) = \int_0^1 \frac{dt}{1 + |x - t|} = \int_0^x \frac{dt}{1 + x - t} + \int_x^1 \frac{dt}{1 - x + t}$

$$\therefore f'(x) = \frac{1}{1+x} - \frac{1}{2-x}$$

$$\Rightarrow f'\left(\frac{1}{2}\right) = \frac{1}{1+1/2} - \frac{1}{2-1/2} = 0$$

71. c. $f(x) = \int\limits_2^x \frac{dt}{\sqrt{1+t^4}}$

$$\Rightarrow f'(x) = \frac{1}{\sqrt{1+x^4}} = \frac{dy}{dx}$$

Now $g'(x) = \frac{dx}{dy} = \sqrt{1+x^4}$

f

x

f(x) = 0

g

Fig. 8.3

when $y = 0$, i.e., $\int\limits_2^x \frac{dt}{\sqrt{1+t^4}} = 0$ then $x = 2$

Hence, $g'(0) = \sqrt{1+16} = \sqrt{17}$

72. a. $f'(x) = \sqrt{2 - x^2} \Rightarrow x^2 - f'(x) = 0 \Rightarrow x^2 = \sqrt{2 - x^2}$

$$\Rightarrow x^4 + x^2 - 2 = 0$$

$$\Rightarrow (x^2 - 1)(x^2 + 2) = 0$$

$$\Rightarrow x^2 - 1 = 0\ (x^2 + 2 \ne 0)$$

$$\Rightarrow x = \pm 1.$$

73. c. $\frac{dx}{dt} = \sin^{-1}(\sin t) \cos t = t \cos t$ and $\frac{dy}{dt} = \frac{\sin t}{\sqrt{t}} \cdot \frac{1}{2\sqrt{t}}$

$$= \frac{\sin t}{2t} \Rightarrow \frac{dy}{dx} = \frac{\sin t}{2t.t \cos t} = \frac{\tan t}{2t^2}$$

74. a. $f(x) = \cos x - \int_0^x (x - t) f(t) dt$

$$\Rightarrow f(x) = \cos x - x\int_0^x f(t) dt + \int_0^x t f(t) dt$$

(differentiating w.r.t. x by Leibnitz's Rule)

$$\Rightarrow f'(x) = -\sin x - xf(x) - \int_0^x f(t) dt + xf(x)$$

$$\Rightarrow f'(x) = -\sin x - \int_0^x f(t) dt$$

(again differentiating w.r.t. x)

$$\Rightarrow f''(x) = -\cos x - f(x)$$

$$\Rightarrow f''(x) + f(x) = -\cos x$$

75. c. $f^2(x) = \int_0^x f(t) \frac{\cos t}{2 + \sin t} dt$

$$\Rightarrow 2f(x) f'(x) = f(x) \frac{\cos x}{2 + \sin x}$$

(differentiating w.r.t. x using Leibnitz rule)

$$\Rightarrow 2 f'(x) = \frac{\cos x}{2 + \sin x}\ [\text{as } f(x) \text{ is not zero everywhere}]$$

$$\Rightarrow 2\int f'(x) dx = \int \frac{\cos x}{2 + \sin x} dx$$

$$\Rightarrow 2f(x) = \log_e(2 + \sin x) + \log C$$

Put $x = 0$ we have $2f(0) = \log 2 + \log C$,

or $\log C = -\log 2$

$$\Rightarrow f(x) = \frac{1}{2} \ln\left(\frac{2 + \sin x}{2}\right);\ x \ne n\pi,\ n \in I.$$

76. a. $\lim\limits_{x \to 0} \frac{1}{x}\left[\int_y^a e^{\sin^2 t} dt + \int_a^{x+y} e^{\sin^2 t} dt\right]$

$$= \lim_{x \to 0} \frac{1}{x} \int_y^{x+y} e^{\sin^2 t} dt \left(\frac{0}{0} \text{ form}\right)$$

Apply L'Hopital Rule

$$= \lim_{x\to 0} \frac{e^{\sin^2(x+y)}\left(1+\frac{dy}{dx}\right) - e^{\sin^2 y}\frac{dy}{dx}}{1}$$

$$= e^{\sin^2 y}\left[1+\frac{dy}{dx}-\frac{dy}{dx}\right] = e^{\sin^2 y}$$

77. a. $f(x) = \int_1^x \frac{e^t}{t} dt \Rightarrow f(1) = 0$ and $f'(x) = \frac{e^x}{x}$

Let $g(x) = f(x) - \ln(x), x \in R^+$

$\Rightarrow g'(x) = f'(x) - \frac{1}{x} = \frac{e^x - 1}{x} > 0 \ \forall \ x \in R^+$

$\Rightarrow g(x)$ is increasing for $x \in R^+$, $g(1) = f(1) - \ln 1$
$= 0 - 0 = 0$

$\Rightarrow g(x) > 0 \ \forall \ x > 1$ and $g(x) \le 0 \ \forall \ x \in (0, 1]$

$\Rightarrow \ln x \ge f(x) \ \forall \ x \in (0, 1]$

78. a. $\int_0^x f(t)dt = x + \int_x^1 tf(t)\,dt$

$$\Rightarrow \frac{d}{dx}\left(\int_0^x f(t)dt\right) = \frac{d}{dx}\left(x + \int_x^1 tf(t)dt\right)$$

$\Rightarrow f(x) = 1 + 0 - xf(x)$ [using Leibnitz's Rule]

$\Rightarrow f(x) = 1 - xf(x)$

$\Rightarrow f(x) = \frac{1}{x+1} \Rightarrow f(1) = \frac{1}{2}$

79. b. $\int_{\cos x}^1 t^2 f(t)dt = 1 - \cos x$

Diff. both sides w.r.t. x

$$\frac{d}{dx}\int_{\cos x}^1 t^2 f(t)\,dt = \frac{d}{dx}(1 - \cos x)$$

$\Rightarrow 1f(1)0 - \cos^2 x f(\cos x)(-\sin x) = \sin x$

$\Rightarrow \cos^2 x f(\cos x) \sin x = \sin x$

$\Rightarrow f(\cos x) = \frac{1}{\cos^2 x}$

Now $f\left(\frac{\sqrt{3}}{4}\right)$ is attained when $\cos x = \frac{\sqrt{3}}{4}$

$f\left(\frac{\sqrt{3}}{4}\right) = \frac{16}{3} = 5.33$

$\left[f\left(\frac{\sqrt{3}}{4}\right)\right] = 5$

80. a. $\int_0^{f(x)} t^2\,dt = x\cos \pi x$ (1)

$\Rightarrow \left.\frac{t^3}{3}\right|_0^{f(x)} = x\cos\pi x$

$\Rightarrow [f(x)]^3 = 3x\cos\pi x$ (2)

$\Rightarrow [f(9)]^3 = -27 \ \Rightarrow f(9) = -3$

Also diff. equation (1) w.r.t. x, we get

$[f(x)]^2 f'(x) = \cos\pi x - x\pi\sin\pi x$

$\Rightarrow [f(9)]^2 f'(9) = -1$

$\Rightarrow f'(9) = -\frac{1}{(f(9))^2} = -\frac{1}{9}$

81. b. Given $xf(x) = x + \int_1^x f(t)dt$

$f(x) + xf'(x) = 1 + f(x)$

$\Rightarrow f(x) = \log|x| + c$

$f(1) = 1$

$\Rightarrow f(x) = \log|x| + 1$

$\Rightarrow f(e^{-1}) = 0$

82. c. Given $A = \int_5^1 x^{50}(2-x)^{50}\,dx$; $B = \int_0^1 x^{50}(1-x)^{50}\,dx$

In A, put $x = 2t \Rightarrow dx = 2dt$

$\Rightarrow A = 2\int_0^{1/2} 2^{50}.t^{50}2^{50}(1-t)^{50}\,dt$ (1)

Now, $B = 2\int_0^{1/2} x^{50}(1-x)^{50}\,dx$ (2)

[using $\int_0^{2a} f(x)dx = 2\int_0^a f(x)dx$ if $f(2a-x) = f(x)$]

From equations (1) and (2), we get

$A = 2^{100}B$

83. d. The given integran]d is a perfect differential coeff. of $\prod_{r=1}^n (x+r)$

$\Rightarrow I = \left[\prod_{r=1}^n (x+r)\right]_0^1 = (n+1)! - n! = n \cdot n!.$

84. a. $\int_{-20\pi}^{20\pi} |\sin x|[\sin x]dx$

$= \int_0^{20\pi} |\sin x|([\sin x] + [-\sin x])dx$

$= -20\int_0^{\pi}(\sin x)dx = -20(-\cos x)_0^{\pi} = 20(-2) = -40$

85. b. $\left|\int_a^b f(x)\,dx - (b-a)f(a)\right|$

$= \left|\int_a^b f(x)\,dx - \int_a^b f(a)dx\right|$

$$= \left|\int_a^b (f(x) - f(a))\,dx\right|$$

$$\le \int_a^b |f(x) - f(a)|\,dx$$

$$\le \int_a^b |x-a|\,dx = \int_a^b (x-a)\,dx = \frac{(b-a)^2}{2}.$$

86. a. On integrating by parts taking $\sin^2 x$ as first function and $\frac{1}{x^2}$ as second function, we get

$$\int_0^\infty \frac{\sin^2 x}{x^2}dx$$

$$= \left|\sin^2 x\left(-\frac{1}{x}\right)\right|_0^\infty - \int_0^\infty 2\sin x\cos x\left(-\frac{1}{x}\right)dx$$

Now $\lim\limits_{x\to\infty} \sin^2 x\left(-\frac{1}{x}\right) = 0$, and $\left(\because \frac{\text{finite}}{\text{infinity}} \text{ form}\right)$

$$\lim_{x\to 0}\frac{\sin^2 x}{x} = \lim_{x\to 0}(\sin x)\frac{\sin x}{x} = 0$$

Thus, $\int_0^\infty \frac{\sin^2 x}{x^2}dx = 0 + \int_0^\infty \frac{\sin 2x}{x}dx$

Now put $2x = t$, then $dx = dt/2$ and $\int_0^\infty \frac{\sin 2x}{x}dx$

$$= \int_0^\infty \frac{\sin t}{t/2}\frac{dt}{2} = \int_0^\infty \frac{\sin t}{t}dt = \int_0^\infty \frac{\sin x}{x}dx$$

87. b. $\sin^3 x = \frac{3}{4}\sin x - \frac{1}{4}\sin 3x$

$$\Rightarrow \int_0^\infty \frac{\sin^3 x}{x}dx$$

$$= \frac{3}{4}\int_0^\infty \frac{\sin x}{x}dx - \frac{1}{4}\int_0^\infty \frac{\sin 3x}{x}dx$$

$$= \frac{3}{4}\int_0^\infty \frac{\sin x}{x}dx - \frac{1}{4}\int_0^\infty \frac{\sin u}{u}du \quad (u = 3x)$$

$$= \frac{3}{4}\frac{\pi}{2} - \frac{1}{4}\frac{\pi}{2} = \frac{\pi}{4}$$

88. a. $I = \int_0^x [\sin t]dt = \int_0^{2n\pi}[\sin t]dt + \int_{2n\pi}^x [\sin t]dt$

$= n\int_0^{2\pi}[\sin t]dt + \int_{2n\pi}^x [\sin t]dt$ (as $[\sin x]$ is periodic with period 2π)

$$= n\left(\int_0^{\pi/2}[\sin t]dt + \int_{\pi/2}^{\pi}[\sin t]dt + \int_{\pi}^{2\pi}[\sin t]dt\right) + \int_{2n\pi}^x [\sin t]dt$$

$$\Rightarrow I = n\left(0 + 0 - \int_\pi^{2\pi} dt\right) + 0 = -n\pi$$

89. d. $\int_0^x f(t)dt = \int_x^1 t^2 f(t)dt + \frac{x^{16}}{8} + \frac{x^6}{3} + a$ (1)

For $x = 1$, $\int_0^1 f(t)dt = 0 + \frac{1}{8} + \frac{1}{3} + a = \frac{11}{24} + a$

Diff. both sides of equation (1) w.r.t. x we get,

$$f(x) = 0 - x^2 f(x) + 2x^{15} + 2x^5$$

$$\Rightarrow f(x) = \frac{2(x^{15}+x^5)}{1+x^2}$$

$$\Rightarrow 2\int_0^1 \frac{x^{15}+x^5}{1+x^2}dx = \frac{11}{24} + a$$

$$\Rightarrow 2\int_0^1 (x^{13} - x^{11} + x^9 - x^7 + x^5)dx = \frac{11}{24} + a$$

$$\Rightarrow 2\left(\frac{1}{14} - \frac{1}{12} + \frac{1}{10} - \frac{1}{8} + \frac{1}{6}\right) = \frac{11}{24} + a$$

$$\Rightarrow a = -\frac{167}{840}$$

90.a. Let $n \le x < n+1$ where $n \in I, I \ge 0$

$$I = \int_0^x \frac{2^t}{2^{[t]}}dt = \int_0^n 2^{\{t\}}dt + \int_n^x 2^{\{t\}}dt$$

$$= n\int_0^1 2^{\{t\}}dt + \int_n^x 2^{\{t\}}dt$$

$$= n\int_0^1 2^t dt + \int_n^x 2^{t-n}dt$$

$$= n\left.\frac{2^t}{\ln 2}\right|_0^1 + \frac{1}{2^n}\left.\frac{2^t}{\ln 2}\right|_n^x$$

$$= \frac{n}{\ln 2}(2-1) + \frac{1}{2^n \ln 2}(2^x - 2^n)$$

$$= \frac{n}{\ln 2} + \frac{1}{\ln 2}(2^{x-n} - 1)$$

$$= \frac{[x] + 2^{\{x\}} - 1}{\ln 2}$$

91.b. $\int_{-3}^5 f(|x|)dx = \int_{-3}^3 (f|x|)dx + \int_3^5 f(|x|)dx$

$$= 2\int_0^3 f(x)dx + \int_3^5 f(x)dx$$

$$= 2\left(\int_0^1 f(x)dx + \int_1^2 f(x)dx + \int_2^3 f(x)dx\right) + \left(\int_3^4 f(x)dx + \int_4^5 f(x)dx\right)$$

$$= 2\left(0 + \frac{1}{2} + \frac{2^2}{2}\right) + \left(\frac{9}{2} + \frac{16}{2}\right) = \frac{35}{2}$$

92.c. $f(x)=\int_{\frac{1}{e}}^{\tan x}\frac{tdt}{(1+t^2)}+\int_{\frac{1}{e}}^{\cot x}\frac{dt}{t(1+t^2)}$

$$\Rightarrow f'(x)=\frac{\tan x}{1+\tan^2 x}\sec^2 x+\frac{1}{\cot x(1+\cot^2 x)}(-\operatorname{cosec}^2 x)$$

$$=\tan x-\tan x=0$$

$\Rightarrow f(x)$ is a constant function.

$$f\left(\frac{\pi}{4}\right)=\int_{\frac{1}{e}}^{1}\frac{tdt}{(1+t^2)}+\int_{\frac{1}{e}}^{1}\frac{dt}{t(1+t^2)}$$

$$=\int_{\frac{1}{e}}^{1}\frac{1}{t}dt=\ln t\Big|_{1/e}^{1}=1$$

93.c. In I_2, put $x+1=t$, then

$$I_2=\int_{-2}^{2}\frac{2t^2+11t+14}{t^4+2}dt=\int_{-2}^{2}\frac{2x^2+11x+14}{x^4+2}dx$$

$$\therefore I_1+I_2=\int_{-2}^{2}\frac{x^6+3x^5+7x^4+2x^2+11x+14}{x^4+2}dx$$

$$=\int_{-2}^{2}\frac{(x^2+3x+7)(x^4+2)+5x}{x^4+2}dx$$

$$=\int_{-2}^{2}(x^2+3x+7)dx+5\int_{-2}^{2}\frac{x}{x^4+2}dx$$

$$=2\int_{0}^{2}(x^2+7)dx=\frac{100}{3}$$

(The other integrals are zero, being integrals of odd functions.)

94. b. $I_1=\int_{\sin^2 t}^{1+\cos^2 t}xf(x(2-x))dx$

$$=\int_{\sin^2 t}^{1+\cos^2 t}(2-x)f(x(2-x))dx=2I_2-I_1$$

$$2I_1=2I_2\Rightarrow\frac{I_1}{I_2}=1$$

95. b. $I=\int_0^4 f(t)dt$, put $t=x^2$

$\Rightarrow dt=2xdx$, then

$$I=2\int_0^2 xf(x^2)\,dx$$

From Lagrange's Mean Value Theorem

$$\frac{\int_0^2 2xf(x^2)dx-\int_0^0 2xf(x^2)dx}{2-0}=2yf(y^2)$$

for some $y\in(0,2)$

$$\Rightarrow\int_0^2 2xf(x^2)\,dx=2\times 2yf(y^2)$$

$$=2\left\{\frac{2\alpha f(\alpha^2)+2\beta f(\beta^2)}{2}\right\}$$

(where $0<\beta<y<\alpha<2$, and using intermediate Mean Value Theorem).

96. b. $I=\int_{-3}^{3}x^8\{x^{11}\}dx$ (1)

Replacing x by $-x$, we have $I=\int_{-3}^{3}x^8\{-x^{11}\}dx$ (2)

Adding equations (1) and (2), we get

$$2I=\int_{-3}^{3}x^8\left(\{x^{11}\}+\{-x^{11}\}\right)dx=2\int_{0}^{3}x^8dx$$

$$=2\left(\frac{x^9}{9}\right)_0^3=2.37$$

$\Rightarrow I=3^7$ [as $\{x\}+\{-x\}=1$ for x is not an integer]

97. b. Let $S'=1+x+x^2+x^3+\dots=\frac{1}{1-x}$

Integrating w.r.t. x, we get $\left|\left(x+\frac{x^2}{2}+\frac{x^3}{3}+\cdots\right)\right|_0^{1/2}$

$$=-|\ln(1-x)|_0^{1/2}\Rightarrow=\frac{1}{2}+\frac{1}{2}(S)=\ln 2\quad\Rightarrow S=\ln\frac{4}{e}$$

98. a. $f(2x)=f(x)=f\left(\frac{x}{2}\right)=f\left(\frac{x}{2^2}\right)=\dots=f\left(\frac{x}{2^n}\right)$

So, when $n\to\infty\Rightarrow f(2x)=f(0)$

($f(x)$ is continuous), i.e., $f(x)$ is a constant function.

$$\Rightarrow f(x)=f(1)=3,\ \int_{-1}^{1}f(f(x))dx=\int_{-1}^{1}3dx=6.$$

99. b. $[x]=0,\ \forall\ x\in[0,1)$

For $x\in[1,2)$, $[x]=1$

$$\Rightarrow\frac{[x]}{1+x^2}=\frac{1}{1+x^2}<1,\forall\ x\in[1,2)\Rightarrow\left[\frac{[x]}{1+x^2}\right]=0$$

For $x\in[-1,0)$, $[x]=-1\Rightarrow\frac{[x]}{1+x^2}=-\frac{1}{1+x^2}$

Clearly, $2\ge 1+x^2>1,\ \forall\ x\in[-1,0)$

$$\frac{1}{2}\le\frac{1}{1+x^2}<1\Rightarrow-\frac{1}{2}\ge-\frac{1}{1+x^2}>-1$$

$$\Rightarrow\left[\frac{[x]}{1+x^2}\right]=-1\ \forall\ x\in[-1,0)$$

Thus, the given integral $=-\int_{-1}^{0}dx=-1$.

100. c. $g(x)=\int_0^x f(t)dt$

$$g(-x)=\int_0^{-x}f(t)dt=-\int_0^x f(-t)dt$$

$$=\int_0^x f(t)dt\text{ as }f(-t)=-f(t)$$

$\Rightarrow g(-x) = g(x)$, thus $g(x)$ is even.

Also $g(x+2) = \int_0^{x+2} f(t)dt$

$$= \int_0^2 f(t)dt + \int_2^{2+x} f(t)dt$$

$$= g(2) + \int_0^x f(t+2)dt$$

$$= g(2) + \int_0^x f(t)dt$$

$$= g(2) + g(x)$$

Now, $g(2) = \int_0^2 f(t)dt = \int_0^1 f(t)dt + \int_1^2 f(t)dt$

$$= \int_0^1 f(t)dt + \int_{-1}^0 f(t+2)dt$$

$$= \int_0^1 f(t)dt + \int_{-1}^0 f(t)dt$$

$$= \int_{-1}^1 f(t)dt = 0 \text{ as } f(t) \text{ is odd}$$

$\Rightarrow g(2) = 0 \Rightarrow g(x+2) = g(x)$

$\Rightarrow g(x)$ is periodic with period 2.

$\Rightarrow g(4) = 0 \Rightarrow f(6) = 0, g(2n) = 0, n \in N.$

101. c. $\int_0^x |\sin t|\, dt = \int_0^{2n\pi} |\sin t|\, dt + \int_{2n\pi}^x |\sin t|\, dt$

$$= 2n\int_0^{\pi} |\sin t|\, dt + \int_{2n\pi}^x \sin t dt \text{ (as } x \text{ lies in either 1st or 2nd quadrant)}$$

$$= 2n(-\cos t)_0^{\pi} + (-\cos t)_{2n\pi}^x$$

$$= 4n - \cos x + 1$$

102.c. $f(x) = \begin{cases} \int_{-1}^x -tdt & -1 \le x \le 0 \\ \int_{-1}^0 -tdt + \int_0^x tdt & x \ge 0 \end{cases}$

$$= \begin{cases} \frac{1}{2}(1-x^2), & -1 \le x \le 0 \\ \frac{1}{2}(1+x^2), & x \ge 0 \end{cases}$$

103.b. $g\left(x + \frac{\pi n}{2}\right) = \int_0^{x+\frac{n\pi}{2}} (|\sin t| + |\cos t|)\, dt$

$$= \int_0^x (|\sin t| + |\cos t|)\, dt + \int_x^{x+\frac{n\pi}{2}} (|\sin t| + |\cos t|)\, dt$$

$$= g(x) + \int_0^{\frac{n\pi}{2}} (|\sin t| + |\cos t|)\, dt$$

(as $|\sin t| + |\cos t|$ has a period $\pi/2$)

$$= g(x) + g\left(\frac{n\pi}{2}\right)$$

104.c.

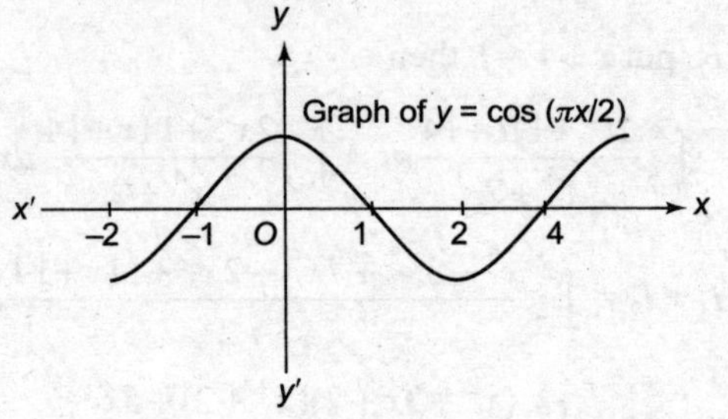

Graph of $y = \cos(\pi x/2)$

Fig. 8.4

From graph, $\int_{-2}^1 \left[x\left[1+\cos\frac{\pi x}{2}\right]+1\right]dx$

$$= \int_{-2}^{-1} [x[1+(-1)]+1]dx + \int_{-1}^{1} [x[1+0]+1]dx$$

$$= (x)_{-2}^{-1} + \int_{-1}^1 [x+1]dx$$

$$= (-1-(-2)) + \int_{-1}^0 0dx + \int_0^1 1dx = 2$$

105.b. $I = \int_{-a}^a \left(\cos^{-1} x - \sin^{-1}\sqrt{1-x^2}\right) dx$

$$= \int_{-a}^0 \cos^{-1} x\, dx + A - 2\int_0^a \sin^{-1}\sqrt{1-x^2}\, dx$$

$$= \int_0^a \left(\pi - \cos^{-1} x\right) dx + A - 2A$$

$$= a\pi - 2A \Rightarrow \lambda = 2$$

106.b. Put $x = a\cos^2\theta + b\sin^2\theta$, so that $dx = 2(b-a)\sin\theta \cos\theta\, d\theta$, then

$$\int_a^b (x-a)^3(b-x)^4 dx = 2(b-a)\int_0^{\pi/2} (a\cos^2\theta + b\sin^2\theta - a)^3(b - a\cos^2\theta - b\sin^2\theta)^4 \sin\theta\cos\theta d\theta$$

$$= 2(b-a)^8 \int_0^{\pi/2} \sin^7\theta\cos^9\theta d\theta$$

$$= 2(b-a)^8 \int_0^{\pi/2} \sin^7\theta(1-\sin^2\theta)^4 \cos\theta d\theta$$

$$= 2(b-a)^8 \int_0^1 x^7(1-x^2)^4 dx$$

$$=2(b-a)^8\int_0^1 x^7(1-x^2)^4dx$$

$$=2(b-a)^8\int_0^1 x^7(1-4x^2+6x^4-4x^6+x^8)dx$$

$$=2(b-a)^8\left[\frac{1}{8}-\frac{4}{10}+\frac{6}{12}-\frac{4}{14}+\frac{1}{16}\right]=\frac{(b-a)^8}{280}$$

107. a. $I=b\int_0^1 \frac{1}{x}\cos 4x\,dx-a\int_0^1 \frac{1}{x^2}\sin 4x\,dx=bI_1-aI_2$

$$I_2=\int_0^1 \frac{1}{x^2}\sin 4x\,dx$$

$$=\left\{\left[-\frac{1}{x}\sin 4x\right]_0^t+4\int_0^t \frac{\cos 4x}{x}dx\right\}$$

$$I_2=\left[-\frac{\sin 4t}{t}+4+4I_1\right],\left\{\lim_{x\to 0}\frac{\sin 4x}{x}=4\right\}$$

$$\therefore I=bI_1-a\left\{-\frac{\sin 4t}{t}+4+4I_1\right\}$$

$$=(b-4a)\int_0^1 \frac{1}{x}\cos 4x\,dx+\frac{a\sin 4t}{t}-4a$$

$$=\frac{a\sin 4t}{t}-1$$

$$\therefore (b-4a)\int_0^1 \frac{1}{x}\cos 4x\,dx=4a-1$$

L.H.S. is a function of t, whereas R.H.S. is a constant.
Hence, we must have $b-4a=0$ and $4a-1=0$.

$$\therefore a=\frac{1}{4}, b=1$$

108. b. Given $\lambda=\int_0^1 \frac{e^t}{1+t}dt$

$$\int_0^1 e^t\log_e(1+t)dt=\left[\log_e(1+t)e^t\right]_0^1-\int_0^1 \frac{e^t}{1+t}$$

$$=e\log_e 2-\lambda$$

109. b. $I_1-I_2=\int_0^{\pi/2}(\cos\theta-\sin 2\theta)f(\sin\theta+\cos^2\theta)d\theta$

Put $t=\sin\theta+\cos^2\theta\Rightarrow dt=(\cos\theta-\sin 2\theta)\,d\theta$

$$\Rightarrow I_1-I_2=\int_1^1 f(t)dt=0$$

110. d. We have $f(x)=\int_{-1}^1 \frac{\sin x\,dt}{\sin^2 x+(t-\cos x)^2}$

$$=\frac{\sin x}{\sin x}\tan^{-1}\left(\frac{t-\cos x}{\sin x}\right)\Bigg|_{-1}^1$$

$$=\tan^{-1}\left(\frac{1-\cos x}{\sin x}\right)-\tan^{-1}\left(\frac{-1-\cos x}{\sin x}\right)$$

$$=\tan^{-1}(\tan x/2)+\tan^{-1}(\cot x/2)$$

Now, we know that $\tan^{-1}x+\tan^{-1}\frac{1}{x}=\begin{cases}\frac{\pi}{2}, & x>0\\ -\frac{\pi}{2}, & x<0\end{cases}$

$$\Rightarrow \tan^{-1}\left(\tan\frac{x}{2}\right)+\tan^{-1}\left(\frac{1}{\tan\frac{x}{2}}\right)=\begin{cases}\frac{\pi}{2}, & \tan\frac{x}{2}>0\\ -\frac{\pi}{2}, & \tan\frac{x}{2}<0\end{cases}$$

Hence, range of $f(x)$ is $\left\{-\frac{\pi}{2},\frac{\pi}{2}\right\}$

111. c. Let $A=\lim_{n\to\infty}\left[\tan\frac{\pi}{2n}\tan\frac{2\pi}{2n}\cdots\tan\frac{n\pi}{2n}\right]^{1/n}$

$$\therefore \log A=\lim_{n\to\infty}\frac{1}{n}\left[\log\tan\frac{\pi}{2n}+\log\tan\frac{2\pi}{2n}+\cdots+\log\tan\frac{n\pi}{2n}\right]$$

$$=\lim_{n\to\infty}\sum_{r=1}^n \frac{1}{n}\log\tan\frac{\pi r}{2n}=\int_0^1 \log\tan\left(\frac{\pi}{2}x\right)dx$$

$$=\frac{2}{\pi}\int_0^{\pi/2}\log\tan y\,dy \quad (1)$$

$$\left[\text{Putting } \frac{1}{2}\pi x=y \therefore dx=(2/\pi)\,dy\right]$$

Now let $I=\int_0^{\pi/2}\log\tan y\,dy$

$$I=\int_0^{\pi/2}\log\tan\left(\frac{1}{2}\pi-y\right)dy \quad \text{(by Property IV)}$$

$$=\int_0^{\pi/2}\log\cot y\,dy$$

$$=-\int_0^{\pi/2}\log\tan y\,dy=-I$$

or $I+I=0$ or $2I=0$ or $I=0$

$\therefore$ from equation (1), $\log A=0 \therefore A=e^0=1$

112. b. Differentiating, we get $f''(x)=f'(x)$

$$\int\frac{df'(x)}{f'(x)}=\int dx\Rightarrow \ln f'(x)=x+c\Rightarrow f'(x)=Ae^x \quad (1)$$

$$\Rightarrow \int f'(x)dx=\int Ae^x dx\Rightarrow f(x)=Ae^x+B \quad (2)$$

Now, $f(0)=1\Rightarrow A+B=1$

$$\therefore f'(x)=f(x)+\int_0^1(Ae^x+1-A)dx$$

$$Ae^x=(Ae^x+1-A)+|(Ae^x+(1-A)x)|_0^1$$

$$1-A+(Ae+1-A-A)=0$$

$$\Rightarrow A(e-3)=-2$$

$$\Rightarrow A=\frac{2}{3-e}\text{ and } B=1-\frac{2}{3-e}=\frac{1-e}{3-e}$$

$$\Rightarrow f(\log_e 2)=\frac{4}{3-e}+\frac{1-e}{3-e}=\frac{5-e}{3-e}$$

113. b. $\int_a^b f(x)dx = [xf(x)]_a^b - \int_a^b xf'(x)dx$ (1)

Now put $f(x) = t \therefore x = f^{-1}(t)$

and $f'(x)\,dx = dt$ and adjust the limits

$\therefore \int_a^b f(x)\,dx = [bf(B) - af(A)] - \int_{f(a)}^{f(b)} f^{-1}(t)\,dt$ by (1)

$\therefore \int_a^b f(x) + \int_{f(a)}^{f(b)} f^{-1}(x)\,dx = bf(B) - af(A)$

114. b. $2I = \int_5^\beta \frac{e^{f\left(\frac{g(x)}{x-\alpha}\right)}dx}{e^{f\left(\frac{g(x)}{x-\alpha}\right)} + e^{f\left(\frac{g(x)}{x-\beta}\right)}}$

$+ \int_\alpha^\beta \frac{e^{f\left(\frac{g(\alpha+\beta-x)}{\beta-x}\right)}dx}{e^{f\left(\frac{g(\alpha+\beta-x)}{\beta-x}\right)} + e^{f\left(\frac{g(\alpha+\beta-x)}{\alpha-x}\right)}}$

$\Rightarrow I = \frac{1}{2}(\beta - \alpha) = \frac{\sqrt{b^2 - 4ac}}{2a}.$

115. a. $y^r = \left(1+\frac{1}{r}\right)\left(1+\frac{2}{r}\right)\left(1+\frac{3}{r}\right)\cdots\left(1+\frac{n-1}{r}\right)$

$\Rightarrow \log y = \frac{1}{r}\sum_{p=1}^{n-1} \log\left(1+\frac{p}{r}\right)$

$\lim_{n\to\infty} y = \lim_{r\to\infty} y = \int_0^k \log(1+x)dx$

$= (k-1)\log_e(1+k) - k$

116. a. When $e \le [x] \le e^2$ $1 < \log[x] < 2$

when $e^2 \le [x] \le e^3$ $2 < \log[x] < 3$

$\therefore \int_3^8 1\,dx + \int_8^{10} 2\,dx = 9$

117. c. Let $g(x) = \int_0^{x^3} f(t)dt$

Now $\int_0^8 f(t)dt = g(2) = \frac{g(2)-g(1)}{2-1} + \frac{g(1)-g(0)}{1-0}$

$= g'(\alpha) + g'(\beta)$

$= 3[\alpha^2 f(\alpha^3) + \beta^2 f(\beta^3)]$

118. a. $I = \int_0^1 f(x)[g(x) - g(1-x)]\,dx$

$= -\int_0^1 f(1-x)[g(x) - g(1-x)]\,dx$

$\Rightarrow 2I = \int_0^1 [f(x) - f(1-x)][g(x) - g(1-x)]\,dx \le 0$

Multiple Correct Answers Type

1. a., b.

$f(x) = e^x + \int_0^1 e^x f(t)\,dt = e^x + k e^x$ where $k = \int_0^1 f(t)\,dt$

$\therefore k = \int_0^1 \left(e^t + ke^t\right)dt = e + ke - 1 - k$

$\therefore k = \frac{e-1}{2-e}$, thus $f(x) = e^x\left(1 + \frac{e-1}{2-e}\right) = \frac{e^x}{2-e}$

Obviously, $f(0) = \frac{1}{2-e} < 0$

Also $f'(x) = \frac{e^x}{2-e} < 0$ for $\forall\ x \in R$.

Hence, $f(x)$ is a decreasing function.

Also $\int_0^1 f(x)\,dx = \int_0^1 \frac{e^x}{2-e}\,dx$

$= \left[\frac{e^x}{2-e}\right]_0^1$

$= \frac{e-1}{2-e} < 0$

2. a., d.

$f'(x) = \frac{3^x}{1+x^2} > 0\ \forall\ x > 0$

$\Rightarrow f'(x) = \frac{3^x}{1+x^2} > \frac{1}{1+x^2},\ \forall\ x \ge 1$

$\Rightarrow \int_1^x f'(x)\,dx > \int_1^x \frac{1}{1+x^2}dx$

$\Rightarrow f(x) > \tan^{-1}x - \tan^{-1}1 \Rightarrow f(x) + \pi/4 > \tan^{-1}x$

3. a., b., c.

For $a \le 0$,

given equation becomes

$\int_0^2 (x-a)dx \ge 1 \Rightarrow a \le \frac{1}{2} \Rightarrow a \le 0$

For $0 < a < 2$,

$\int_0^2 |x-a|\,dx \ge 1 \Rightarrow \int_0^a (a-x)dx + \int_a^2 (x-a)dx \ge 1$

$\Rightarrow \frac{a^2}{2} + 2 - 2a + \frac{a^2}{2} \ge 1 \Rightarrow a^2 - 2a + 1 \ge 0$

$\Rightarrow (a-1)^2 \ge 0$

For $a \ge 2$,

$$\int_0^2 |x-a|\,dx \ge 1 \Rightarrow \int_0^2 (a-x)dx \ge 1$$

$\Rightarrow 2a-2 \ge 1$

$\Rightarrow a \ge \frac{3}{2} \Rightarrow a \ge 2$

4. a., b., d.

We know $\int_a^b |\sin x|\,dx$ represents the area under the curve from $x = a$ to $x = b$. We also know that area from $x = a$ to $x = a + \pi$ is 2.

$\therefore \int_a^b |\sin x|\,dx = 8 \Rightarrow b-a = \frac{8\pi}{2}$ (1)

Similarly, $\int_0^{a+b} |\cos x|\,dx = 9 \Rightarrow a+b-0 = \frac{9\pi}{2}$ (2)

From (1) and (2), $a = \frac{\pi}{4}$ and $b = \frac{17\pi}{4}$

$\Rightarrow |a+b| = \frac{9\pi}{2}$, $|a-b| = 4\pi$, $\frac{a}{b} = 17$ and

$\int_a^b \sec^2 x dx = [\tan x]_{\pi/4}^{17\pi/4} = 0$

5. a., c.

Let $f(x) = \sqrt{3+x^3}$

Clearly, $f(x)$ is increasing in [1, 3]

$\Rightarrow$ The least value of the function,

$m = f(1) = \sqrt{3+1^3} = 2$

and the greatest value of the function,

$M = f(3) = \sqrt{3+3^3} = \sqrt{30}$

Therefore, $(3-1)\,2 \le \int_1^3 \sqrt{3+x^3}\,dx \le (3-1)\sqrt{30}$

Here, $4 \le \int_1^3 \sqrt{3+x^3}\,dx \le 2\sqrt{30}$

6. a., b., c.

$$g(x) = \int_0^x 2|t|\,dt$$

$$= \begin{cases} \int_0^x -2t dt, & x<0 \\ \int_0^x 2t dt, & x \ge 0 \end{cases}$$

$$= \begin{cases} \left[-t^2\right]_0^x, & x<0 \\ \left[t^2\right]_0^x, & x \ge 0 \end{cases}$$

$$= \begin{cases} -x^2, & x<0 \\ x^2, & x \ge 0 \end{cases}$$

$= x|x|$

Clearly, continuous and differentiable at $x = 0$

Also, $g'(x) = \begin{cases} -2x, & x<0 \\ 2x, & x \ge 0 \end{cases}$ which is non-differentiable at $x = 0$.

7. a., b.

$$f(x) = x\int_1^x \frac{e^t}{t}dt - e^x$$

$$\Rightarrow f'(x) = x\frac{e^x}{x} + \int_1^x \frac{e^t}{t}dt - e^x$$

$$\Rightarrow f'(x) = \int_1^x \frac{e^t}{t}dt > 0\ [\because x \in [1, \infty)]$$

$\Rightarrow f(x)$ is an increasing function.

8. a., c., d.

$$I = \int_0^1 \frac{2x^2+3x+3}{(x+1)\left(x^2+2x+2\right)}$$

$$= \int_0^1 \frac{2(x^2+2x+2)-(x+1)}{(x+1)\left(x^2+2x+2\right)}dx$$

$$= \int_0^1 \left(\frac{2}{x+1} \frac{1}{x^2+2x+2}\right)dx$$

$$= \left[2\log(x+1) - \tan^{-1}(x+1)\right]_0^1$$

$$= 2\log 2 - \tan^{-1}2 + \tan^{-1}1 \qquad (1)$$

$$= 2\log 2 - \tan^{-1}2 + \frac{\pi}{4}$$

$$= \log 2 - \left(\frac{\pi}{2} - \cot^{-1}2\right) + \frac{\pi}{4}$$

$$= -\frac{\pi}{4} + \log 4 + \cot^{-1}2$$

From equation (1), $I = 2\log 2 - \tan^{-1}\left(\frac{2-1}{1+2\times 1}\right)$

$$= 2\log 2 - \tan^{-1}\frac{1}{3}$$

$$= 2\log 2 - \cot^{-1}3$$

9. a., d.

$$A_{n+1} - A_n = \int_0^{\pi/2} \frac{\sin(2n+1)x - \sin(2n-1)x}{\sin x}$$

$$= \int_0^{\pi/2} 2\cos 2nx\,dx = 0$$

$\Rightarrow A_{n+1} = A_n$

$$B_{n+1} - B_n = \int_0^{\pi/2} \frac{\sin^2(n+1)x - \sin^2 x}{\sin^2 x}dx$$

$$= \int_0^{\pi/2} \frac{\sin(2n+1)x}{\sin x}dx = A_{n+1}$$

10. a., b., c.

$f(x)=\int_a^x \frac{1}{f(x)}dx \Rightarrow f'(x)=\frac{1}{f(x)}\times 1-0$

$\Rightarrow f(x)f'(x)=1$

$\Rightarrow \int f(x)f'(x)dx=\int 1dx$

$\Rightarrow \frac{1}{2}[f(x)]^2=x+c$ (1)

Now given that $\int_a^1 [f(x)]^{-1}dx=\sqrt{2}$

$\Rightarrow f(1)=\sqrt{2}$

$\Rightarrow$ From (1), $\frac{1}{2}[f(1)]^2=1+c\Rightarrow c=0$

$\Rightarrow f(x)=\pm\sqrt{2x}$

But $f(1)=\sqrt{2}\Rightarrow f(x)=\sqrt{2x}\Rightarrow f(2)=2$

Also $f'(x)=\frac{1}{\sqrt{2x}}\Rightarrow f'(2)=1/2$

$\int_0^1 f(x)dx=\int_0^1\sqrt{2x}dx=\left[\frac{(2x)^{3/2}}{3}\right]_0^1=\frac{(2)^{3/2}}{3}$

Also, $f^{-1}(x)=\frac{x^2}{2}\Rightarrow f^{-1}(2)=2$

11. b., c.

$I=\int_0^\infty \frac{dx}{1+x^4}$ (1)

$=\int_0^\infty \frac{x^2+1-x^2}{1+x^4}dx$

$=\int_0^\infty \frac{x^2}{1+x^4}dx+\int_0^\infty\frac{1-x^2}{1+x^4}dx$

$=I_1+I_2$

$I_2=\int_0^\infty \frac{\frac{1}{x^2}-1}{\frac{1}{x^2}+x^2}dx,$

Put $x+\frac{1}{x}=y$

$\Rightarrow I_2=\int_\infty^\infty \frac{-1}{y^2-2}dy=0$

$\Rightarrow I=\int_0^\infty\frac{dx}{1+x^4}=\int_0^\infty\frac{x^2dx}{1+x^4}$ (2)

Adding equations (1) and (2), we get

$\Rightarrow 2I=\int_0^\infty\frac{1+x^2dx}{1+x^4}=\int_0^\infty\frac{\frac{1}{x^2}+1}{\frac{1}{x^2}+x^2}dx$, put $x-\frac{1}{x}=y$

$\Rightarrow I=\int_{-\infty}^\infty\frac{dy}{y^2+2}=\left[\frac{1}{\sqrt{2}}\tan^{-1}\frac{y}{\sqrt{2}}\right]_{-\infty}^\infty=\frac{\pi}{\sqrt{2}}$

$\Rightarrow I=\frac{\pi}{2\sqrt{2}}$

12. a., b., d.

Given that $f(x)=\int_0^x|t-1|\,dt$

$\Rightarrow f(x)=\int_0^x(1-t)\,dt,\ 0\le x\le 1$

$=x-\frac{x^2}{2}$

Also $f(x)=\int_0^1(1-t)dt+\int_1^x(t-1)\,dt$, where $1\le x\le 2$

$=\frac{1}{2}+\frac{x^2}{2}-x+\frac{1}{2}=\frac{x^2}{2}-x+1$

Thus, $f(x)=\begin{cases}x-\frac{x^2}{2}, & 0\le x\le 1\\ \frac{x^2}{2}-x+1, & 1<x\le 2\end{cases}$

$\Rightarrow f'(x)=\begin{cases}1-x, & 0\le x<1\\ x-1, & 1<x<2\end{cases}$

Thus, $f(x)$ is continuous as well as differentiable at $x=1$.

Also $f(x)=\cos^{-1}x$ has one real root, draw the graph and verify.

For range of $f(x)$:

$f(x)=\int_0^x|t-1|\,dt$ is the value of area bounded by the curve $y=|t-1|$ and x-axis between the limits $t=0$ and $t=x$.

Obviously, minimum area is obtained when $t=0$ and $t=x$ coincide or $x=0$.

Maximum value of area occurs when $t=2$, hence $f(2)=$ area of shaded region $=1$.

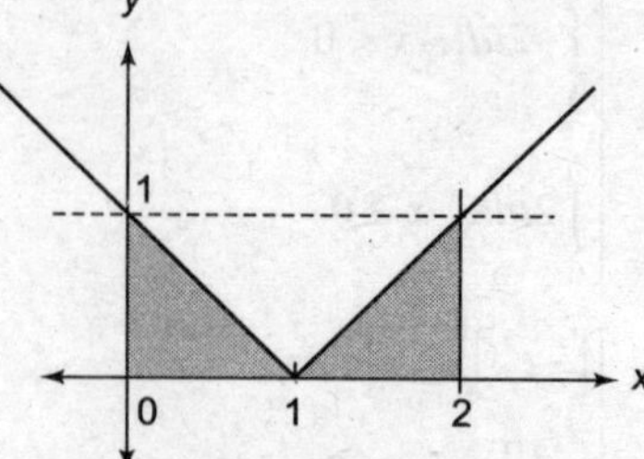

Fig. 8.5

13. b., c., d.

$I_n=\int_0^{\pi/4}\tan^n x\,dx$

$$= \int_0^{\pi/4} \tan^{n-2} x \tan^2 x\, dx$$

$$= \int_0^{\pi/4} \sec^2 x \tan^{n-2} x\, dx - \int_0^{\pi/4} \tan^{n-2} x\, dx$$

$$= \int_0^1 t^{n-2}\, dt - I_{n-2} \quad \text{where } t = \tan x$$

$$I_n + I_{n-2} = \left(\frac{t^{n-1}}{n-1}\right)_0^1$$

$$\Rightarrow I_n + I_{n-2} = \frac{1}{n-1}$$

$\Rightarrow I_2 + I_4, I_4 + I_6, \ldots$ are in H.P.

For $0 < x < \pi/4$, we have $0 < \tan^n x < \tan^{n-2} x$

So that $0 < I_n < I_{n-2} \Rightarrow I_n + I_{n+2} < 2I_n < I_n + I_{n-2}$

$$\Rightarrow \frac{1}{n+1} < 2I_n < \frac{1}{n-1} \Rightarrow \frac{1}{2(n+1)} < I_n < \frac{1}{2(n-1)}$$

14. a., b., c.

Let $I = \int_a^b \frac{f(x)}{f(x) + f(a+b-x)} dx$ (1)

$$= \int_\alpha^\beta \frac{f(a+b-x)}{f(a+b-x) + f(x)} dx \quad (2)$$

Adding equations (1) and (2), we get

$$\Rightarrow 2I = \int_a^b 1\, dx = b - a$$

$$\Rightarrow I = \left(\frac{b-a}{2}\right) = 10 \quad \text{(given)}$$

$\therefore b - a = 20$

15. a., b., d.

$$I_n = \int_0^1 \frac{dx}{\left(1+x^2\right)^n} = \int_0^1 \left(1+x^2\right)^{-n} dx$$

$$= \left.\frac{x}{\left(1+x^2\right)^n}\right|_0^1 - \int_0^1 (-n)\left(1+x^2\right)^{-n-1} 2x \times x dx$$

$$= \frac{1}{2^n} + 2n \int_0^1 \frac{x^2 dx}{\left(1+x^2\right)^{n+1}}$$

$$= \frac{1}{2^n} + 2n \int_0^1 \frac{1 + x^2 - 1}{\left(1+x^2\right)^{n+1}} dx$$

$$= \frac{1}{2^n} + 2n I_n - 2n I_{n+1}$$

$$\Rightarrow 2n I_{n+1} = 2^{-n} + (2n-1) I_n$$

$$\Rightarrow 2I_2 = \frac{1}{2} + I_1 = \frac{1}{2} + \tan^{-1} x|_0^1$$

$$\Rightarrow I_2 = \frac{1}{4} + \frac{\pi}{8}$$

Also $4I_3 = 2^{-2} + 3I_2$

$$= \frac{1}{4} + 3\left(\frac{1}{4} + \frac{\pi}{8}\right) = \frac{1}{4} + \frac{3\pi}{32}$$

16. b., c.

$$\lim_{n\to\infty} \frac{1}{n} \sum_{r=n+1}^{2n} f\left(\frac{r}{n}\right) = \int_1^2 f(x)\, dx$$

$$\lim_{n\to\infty} \frac{1}{n} \sum_{r=1}^{n} f\left(\frac{r+n}{n}\right) = \int_0^1 f(1+x)\, dx$$

$$= \int_1^2 f(t)\, dt = \int_1^2 f(x) dx$$

$$\lim_{n\to\infty} \frac{1}{n} \sum_{r=1}^{n} f\left(\frac{r}{n}\right) = \int_0^1 f(x) dx$$

$$\lim_{n\to\infty} \frac{1}{n} \sum_{r=1}^{2n} f\left(\frac{r}{n}\right) = \int_0^2 f(x)\, dx$$

17. a., b., d.

$f(2-x) = f(2+x), f(4-x) = f(4+x)$

$\Rightarrow f(4+x) = f(4-x) = f(2+2-x) = f(2-(2-x)) = f(x)$

$\Rightarrow 4$ is a period of $f(x)$

$$\int_0^{50} f(x)\, dx = \int_0^{48} f(x) dx + \int_{48}^{50} f(x)\, dx$$

$$= 12\int_0^4 f(x) dx + \int_0^2 f(x)\, dx$$

(in second integral replacing x by $x + 48$ and then using $f(x) = f(x + 48)$)

$$= 12\left(\int_0^2 f(x)\, dx + \int_0^2 f(4-x)\, dx\right) + 5$$

$$= 12\left(\int_0^2 f(x)\, dx + \int_0^2 f(4+x) dx\right) + 5$$

$$= 24\int_0^2 f(x)\, dx + 5 = 125$$

$$\int_{-4}^{46} f(x) dx = \int_{-4}^{-2} f(x) dx + \int_{-2}^{-2+48} f(x) dx$$

$$= \int_0^2 f(x+4) dx + 12\int_0^4 f(x) dx$$

$$= \int_0^2 f(x) dx + 24\int_0^2 f(x) dx$$

$= 125$

Also $\int_2^{52} f(x)dx = \int_2^4 f(x)dx + \int_4^{4+48} f(x)dx$

$= \int_0^2 f(4-x)dx + 12\int_0^4 f(x)dx$

$= \int_0^2 f(4+x)dx + 24\int_0^2 f(x)dx$

$= \int_0^2 f(x)dx + 24\int_0^2 f(x)dx$

$= 125$

$\int_1^{51} f(x)dx = \int_1^3 f(x)dx + \int_3^{3+48} f(x)dx$

$= \int_1^3 f(x)dx + 12\int_0^4 f(x)dx$

$= \int_0^2 f(x+1)dx + 24\int_0^2 f(x)dx$

$\neq 125$

18. a., b.

L.H.S. $= \int_0^x \left\{\int_0^u f(t)\, dt\right\} du$

Integrating by parts choose '1' as the second funciton

$= \left\{\int_0^u f(t)\, dt\, u\right\}_0^x - \int_0^x f(u)\, u\, du$

$= x\int_0^x f(t)\, dt - \int_0^x f(u)\, u\, du$

$= x\int_0^x f(u)\, du - \int_0^x f(u)\, u\, du = \int_0^x f(u)\,(x-u) du$

= R.H.S.

19. a., b.

Let $g(x) = \tan^{-1} x - f(x)$ (1)

Diff. (1) w.r.t. x, we get $g'(x) = \dfrac{1}{1+x^2} - f'(x)$

Also $f'(x) = \dfrac{e^x}{1+x^2} \Rightarrow g'(x) = \dfrac{1-e^x}{1+x^2} < 0\ \forall\ x > 0$

$\Rightarrow g(x)$ is a decreasing function for $x > 0$

$\Rightarrow g(x) \le g(1)\ \forall\ x \ge 1$

$g(1) = \dfrac{\pi}{4}$ $[\because f(1) \equiv 0]$

$\Rightarrow g(x) \le \pi/4\ \forall\ x \ge 1$ or $\tan^{-1} x - f(x) \le \pi/4$

$\Rightarrow f(x) \ge \tan^{-1} x - \dfrac{\pi}{4}\ \forall\ x \ge 1$

20. a., c.

$\int_0^1 e^{x^2-x}\, dx$

For $x \in (0, 1)$, $x^2 - x \in (-1/4, 0)$

$\Rightarrow e^{-1/4} < e^{x^2-x} < e^0$

$\Rightarrow e^{-\frac{1}{4}} < \int_0^1 e^{x^2-x}\, dx < 1$

21. a., d.

$f(x+\pi) = \int_0^{x+\pi} (\cos(\sin t) + \cos(\cos t))\, dt$

$= \int_0^{\pi} (\cos(\sin t) + \cos(\cos t))\, dt$

$+ \int_{\pi}^{x+\pi} (\cos(\sin t) + \cos(\cos t))\, dt$

$= f(\pi) + \int_0^x (\cos(\sin t) + \cos(\cos t))\, dt$

($\because$ for $g(x) = \cos(\sin t) + \cos(\cos t)$, $f(x+\pi)$ $= f(x)$)

$= f(\pi) + f(x)$

$= f(\pi) + 2f\left(\dfrac{\pi}{2}\right)$ ($\because g(x)$ has period $\pi/2$)

Reasoning Type

1. a. Given that $\int_a^b |g(x)|\, dx > \left|\int_a^b g(x)\, dx\right| \Rightarrow y = g(x)$ cuts the graph at least once, then $y = f(x)\, g(x)$ changes sign at least once in (a, b), hence $\int_a^b f(x)g(x)\, dx$ can be zero.

2. b. $I = \int_{-4}^{-5} \sin(x^2 - 3)dx + \int_{-2}^{-1} \sin(x^2 + 12x + 33)dx$

$= I_1 + I_2$

$I_2 = \int_{-2}^{-1} \sin(x^2 + 12x + 33)\, dx$

$= \int_{-2}^{-1} \sin((x+6)^2 - 3)\, dx,$

put $x + 6 = -y$

$\Rightarrow I_2 = -\int_{-4}^{-5} \sin(y^2 - 3)\, dy = -I_1$

$\Rightarrow I_1 + I_2 = 0 \Rightarrow I = 0$

3. a. $I=\int_0^1 \tan^{-1}\dfrac{2(1-x)-1}{1+(1-x)-(1-x)^2}dx$

$=\int_0^1 \tan^{-1}\dfrac{1-2x}{1+x-x^2}dx$

$=-I$

$\Rightarrow I=0$

4. d. $f(x)=\int_{5\pi/4}^{x}(3\sin t+4\cos t)\,dt$

$\Rightarrow f'(x)=3\sin x+4\cos x,\ x\in\left[\dfrac{5\pi}{4},\dfrac{4\pi}{3}\right]$

These values of x are in third quadrant where both $\sin x$ and $\cos x$ are negative.

Then $f'(x)<0$ for $x\in\left[\dfrac{5\pi}{4},\dfrac{4\pi}{3}\right]$

Hence, $f(x)$ is decreasing for these values of x. Then, the least value of function occurs at $x=4\pi/3$.

$\Rightarrow f_{\text{minimum}}=\int_{5\pi/4}^{4\pi/3}(3\sin t+4\cos t)\,dt$

$=\dfrac{3}{2}+\dfrac{1}{\sqrt{2}}-2\sqrt{3}.$

5. a. Given $f(x+1)+f(x+7)=0,\ \forall\ x\in R$

Replace x by $x-1$, we have $f(x)+f(x+6)=0$ (1)

Now replace x by $x+6$, we have

$f(x+6)+f(x+12)=0$ (2)

From equations (1) and (2), we have

$f(x)=f(x+12)$ (3)

Hence, $f(x)$ is periodic with period 12.

$\Rightarrow \int_a^{a+t} f(x)\,dx$ is independent of a if t is positive integral multiple of 12 then possible value of t is 12.

6.c. $x>x^2\ \forall\ x\in\left(0,\dfrac{\pi}{4}\right)\Rightarrow e^x>e^{x^2}\ \forall\ x\in\left(0,\dfrac{\pi}{4}\right)$

$\cos x>\sin x\ \forall\ x\in\left(0,\dfrac{\pi}{4}\right)$

$\Rightarrow e^{x^2}\cos x>e^{x^2}\sin x$

$\Rightarrow e^x>e^{x^2}>e^{x^2}\cos x>e^{x^2}\sin x\ \forall\ x\in\left(0,\dfrac{\pi}{4}\right)$

$\Rightarrow I_2>I_1>I_3>I_4$

7.a. Let $I_m=\int_0^{\pi}\dfrac{\sin 2mx}{\sin x}dx$. Then,

$I_m-I_{m-1}=\int_0^{\pi}\dfrac{\sin 2mx-\sin 2(m-1)x}{\sin x}dx$

$=\int_0^{\pi}2\cos(2m-1)x\,dx$

$=\dfrac{2}{2m-1}[\sin(2m-1)x]_0^{\pi}=0$

$I_m=I_{m-1}$ for all $m\in N$

$\Rightarrow I_m=I_{m-1}=I_{m-2}=\cdots=I_1$

But, $I_1=\int_0^{\pi}\dfrac{\sin 2x}{\sin x}dx=2\int_0^{\pi}\cos x\,dx=0.$

$\therefore I_m=0$ for all $m\in N$

8.d. $\int_0^{\pi}\sqrt{1-\sin^2 x}\,dx=\int_0^{\pi}|\cos x|\,dx$

$=\int_0^{\pi/2}\cos x\,dx+\int_{\pi/2}^{\pi}-\cos x\,dx$

$=1+1=2$

Hence, statement 1 is false.

However, statement 2 is true.

9.a. Let $I=\int_0^{2\pi}\cos^{99}x\,dx$.

Then,

$I=2\int_0^{\pi}\cos^{99}x\,dx\qquad[\because \cos^{99}(2\pi-x)=\cos^{99}x]$

Now, $\int_0^{\pi}\cos^{99}x\,dx=0\qquad[\because \cos^{99}(\pi-x)=-\cos^{99}x]$

$\Rightarrow I=2\times 0=0$

10.c. Statement 1 is true as it is a fundamental property.

Let $g(x)=\int_a^x f(t)\,dt$

If $f(x)$ is an even function

Then $g(-x)=\int_a^{-x} f(t)\,dt$

$=-\int_{-a}^{x} f(-y)\,dy$

$=-\int_{-a}^{x} f(y)\,dy$

$=-\int_{-a}^{a} f(y)\,dy-\int_a^x f(y)\,dy$

$\neq -g(x)$

Hence, statement 2 is false.

11. a. Statement 2 is a fundamental concept, also we have $f(2-a)=f(2+a)$

$\int_{2-a}^{2+a} f(x)\,dx=2\int_2^{2+a} f(x)\,dx$

12. c. Both the statements are true independently, but statement 2 is not a correct explanation of statement 1.

13. a. To prove $\int_a^b f(x)\,dx = \int_{a+c}^{b+c} f(x-c)\,dx$

Put $z = x - c$, then $dz = dx$

When $x = a + c, z = a$ and when $x = b + c, z = b$

$\therefore \int_{a+c}^{b+c} f(x-c)\,dx = \int_a^b f(z)\,dz = \int_a^b f(x)\,dx$

Thus, statement 2 is true

$\int_a^b f(x)\,dx = \int_{a+c}^{b+c} f(x-c)\,dx$

Putting $f(x) = \sin^{100} x \cos^{99} x$, $a = 0$, $b = \pi$, and $c = -\frac{\pi}{2}$, we get

$\int_0^{\pi} \sin^{100} x \cos^{99} x\,dx$

$= \int_{-\pi/2}^{\pi/2} \sin^{100}\left(x + \frac{\pi}{2}\right)\cos^{99}\left(x + \frac{\pi}{2}\right)dx$

$= -\int_{-\pi/2}^{\pi/2} \cos^{100} x \sin^{99} x\,dx$

$= 0$ $\quad [\because \cos^{100} x \sin^{99} x$ is an odd function$]$

14.c. $\int_a^b x f(x)\,dx = \int_a^b (a+b-x)\,f(a+b-x)\,dx$

$= (a+b)\int_a^b f(a+b-x)\,dx - \int_a^b x\,f(a+b-x)\,dx$

Therefore, Statement 2 is true only when $f(a+b-x) = f(x)$ which holds in statement 1.

Therefore, Statement 2 is false and statement 1 is true.

15.a. Let $g(x) = \int_a^x f(t)\,dt - \int_x^b f(t)\,dt$, where $x \in [a,b]$

We have $g(a) = -\int_a^b f(t)\,dt$ and $g(b) = \int_a^b f(t)\,dt$

$\Rightarrow g(a)g(b) = -\left(\int_a^b f(t)\,dt\right)^2 \le 0$

Clearly, $g(x)$ is continuous in $[a, b]$ and $g(a)g(b) \le 0$

It implies that $g(x)$ will becomes zero at least once in $[a, b]$. Hence, $\int_a^x f(t)\,dt = \int_x^b f(t)\,dt$ for at least one value of $x \in [a, b]$.

Hence, both the statements are true and statement 2 is a correct explanation of statement 1.

16. d. Obviously, $|\sin t|$ is non-differentiable at $x = \pi$.

But $\int_0^x |\sin t|\,dt$

$$= \begin{cases} \int_0^x \sin t, & 0 \le x < \pi \\ \int_0^{\pi} \sin t\,dt + \int_{\pi}^x -\sin t\,dt, & \pi \le x \le 2\pi \end{cases}$$

$$= \begin{cases} -\cos x + 1, & 0 \le x < \pi \\ 3 + \cos x, & \pi \le x \le 2\pi \end{cases}$$

which is continuous as well as differentiable at $x = \pi$.

Hence, statement 1 is false.

17.a. For $a < b$. If m and M are the smallest and greatest values of $f(x)$ on $[a, b]$

then $m(b-a) \le \int_a^b f(x)dx \le (b-a)M$

or $m \le \frac{1}{(b-a)}\int_a^b f(x)dx \le M$

Since $f(x)$ is continuous on $[a, b]$, it takes on all intermediate values between m and M.

Therefore, some values $f(c)$ $(a \le f(c) \le b)$, we will have $\frac{1}{(b-c)}\int_a^b f(x)\,dx = f(c)$

or $\int_a^b f(x)\,dx = f(c)(b-a)$.

Hence, both the statements are true and statement 2 is a correct explanation of statement 1.

Linked Comprehension Type

For Problems 1–3

1. d., 2. a., 3. c.

Sol.

$\int_2^x f(t)\,dt = \frac{x^2}{2} + \int_x^2 t^2 f(t)\,dt$

Differentiating w.r.t. x, we get

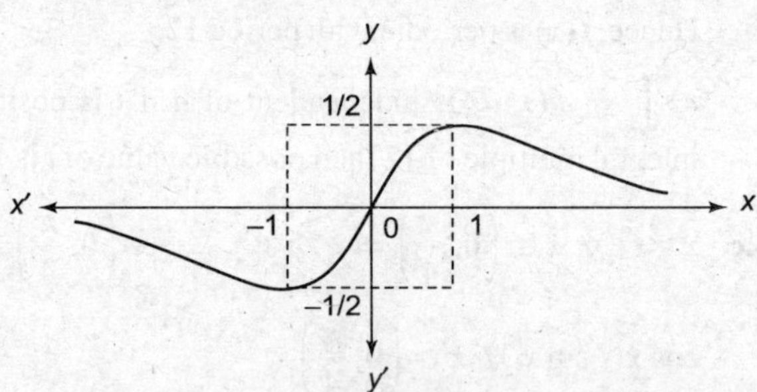

Fig. 8.6

$f(x) = x + (-x^2 f(x))$

$\Rightarrow f(x)\,[1 + x^2] = x$

$\Rightarrow y = f(x) = \frac{x}{1+x^2}$

$\Rightarrow yx^2 - x + y = 0$

Since x is real, $D \ge 0$

$\Rightarrow 1 - 4y^2 \ge 0$

$\Rightarrow y \in \left(-\frac{1}{2}, \frac{1}{2}\right)$

Also $f(x)$ is an odd function, hence $\int_{-2}^{2} f(x)\,dx$

$$f'(x) = \frac{1+x^2-2x^2}{1+x^2} = \frac{1-x^2}{1+x^2} \geq 0$$

$\Rightarrow x^2 - 1 \leq 0$

$\Rightarrow x \in [-1, 1]$

For Problems 4–6

4.b., 5. b., 6. c.

Sol.

$$f(x) = x^2 + \int_0^x e^{-t} f(x-t)\,dt \quad (1)$$

$$= x^2 + \int_0^x e^{-(x-t)} f(x-(x-t))\,dt$$

$$\left[\text{Using } \int_a^b f(x)\,dx = \int_a^b f(a+b-x)\,dx\right]$$

$$= x^2 + e^{-x}\int_0^x e^t f(t)\,dt \quad (2)$$

Diff. w.r.t. x, we get

$$\Rightarrow f'(x) = 2x - e^{-x}\int_0^x e^t f(t)\,dt + e^{-x}e^x f(x)$$

$$= 2x - e^{-x}\int_0^x e^t f(t)\,dt + f(x)$$

$\Rightarrow f'(x) = 2x + x^2$ [using equation (2)]

$$\Rightarrow f(x) = \frac{x^3}{3} + x^2 + c$$

Also $f(0) = 0$ [from equation (1)]

$$\Rightarrow f(x) = \frac{x^3}{3} + x^2$$

$\Rightarrow f'(x) = x^2 + 2x$

$\Rightarrow f'(x) = 0$ has real roots, hence $f(x)$ is non-monotonic.

Hence, $f(x)$ is many-one, but range is R, hence surjective.

$$\int_0^1 f(x)\,dx = \int_0^1 \left(\frac{x^3}{3} + x^2\right) dx$$

$$= \left[\frac{x^4}{12} + \frac{x^3}{3}\right]_0^1$$

$$= \frac{1}{12} + \frac{1}{3} = \frac{5}{12}$$

For Problems 7–9

7. c., 8. d., 9. c.

Sol.

$$f(x) - \lambda \int_0^{\pi/2} \sin x \cos t\, f(t)\,dt = \sin x$$

$$\Rightarrow f(x) - \lambda \sin x \int_0^{\pi/2} \cos t\, f(t)\,dt = \sin x$$

$\Rightarrow f(x) - A\sin x = \sin x$ or

$f(x) = (A+1)\sin x$, where $A = \lambda \int_0^{\pi/2} \cos t\, f(t)\,dt$

$$\Rightarrow A = \lambda \int_0^{\pi/2} \cos t (A+1) \sin t\,dt$$

$$= \frac{\lambda(A+1)}{2}\int_0^{\pi/2} \sin 2t\,dt$$

$$= \frac{\lambda(A+1)}{2}\left[\frac{-\cos 2t}{2}\right]_0^{\pi/2}$$

$$= \frac{\lambda(A+1)}{2}$$

$$\Rightarrow A = \frac{\lambda}{2-\lambda}$$

$$\Rightarrow f(x) = \left(\frac{\lambda}{2-\lambda} + 1\right)\sin x$$

$$\Rightarrow f(x) = \left(\frac{2}{2-\lambda}\right)\sin x$$

$$\left(\frac{2}{2-\lambda}\right)\sin x = 2$$

$\Rightarrow \sin x = (2-\lambda)$

$\Rightarrow |2-\lambda| \leq 1$

$\Rightarrow -1 \leq \lambda - 2 \leq 1$

$\Rightarrow 1 \leq \lambda \leq 3$

$$\int_0^{\pi/2} f(x)\,dx = 3$$

$$\Rightarrow \int_0^{\pi/2} \frac{2}{2-\lambda}\sin x\,dx = 3$$

$$\Rightarrow -\left[\frac{2}{2-\lambda}\cos x\right]_0^{\pi/2} = 3$$

$$\Rightarrow \frac{2}{2-\lambda} = 3$$

$\Rightarrow \lambda = 4/3$

For Problems 10–13

10. b., 11. d., 12. d., 13. d.

Sol.

10. b. $f(x)$ is an odd function

$\Rightarrow f(x) = -f(-x)$

$$\phi(-x) = \int_a^{-x} f(t)\,dt, \text{ put } t = -y$$

$$\Rightarrow \phi(-x) = \int_{-a}^{x} f(-t)(-dt) = \int_{-a}^{x} f(t)\,dt$$

$$= \int_{-a}^{a} f(t)\,dt + \int_a^x f(t)\,dt = 0 + \int_a^x f(t)\,dt = \phi(x).$$

11.d. If $f(x)$ is an even function, then

$$\phi(-x) = -\int_{-a}^{x} f(t)\,dt$$

$$= -\int_{-a}^{a} f(t)\,dt - \int_{a}^{x} f(t)\,dt$$

$$= -2\int_{0}^{a} f(t)\,dt - \int_{a}^{x} f(t)\,dt \quad \text{(as } f(x) \text{ is an even function)}$$

Now, $\int_0^a f(t)\,dt = \int_0^a f(a-t)\,dt$

$$= \int_0^a -f(t)\,dt \text{ [using } f(a-x) = -f(x)\text{)g]}$$

$$\Rightarrow \int_0^a f(t)\,dt = 0$$

$$\Rightarrow \phi(-x) = -\int_a^x f(t)\,dt = -f(x)$$

$\Rightarrow \phi(x)$ is an odd function.

12.d. $g(x+\alpha) + g(x) = 0$

$\Rightarrow g(x+2\alpha) + g(x+\alpha) = 0$

$\Rightarrow g(x+2\alpha) = g(x)$

$\Rightarrow g(x)$ is periodic with period 2α

$\Rightarrow \int_b^{2k} g(t)\,dt = \int_b^{b+c} g(x)\,dx \quad (\because b, k, c \text{ are in A.P.})$

This is independent of b then c has least value 2α.

13. d. $\int_{p+m\alpha}^{q+n\alpha} g(t)\,dt$

$$= \int_{p+m\alpha}^{p} g(x)\,dx + \int_p^q g(x)\,dx + \int_q^{q+n\alpha} g(x)\,dx$$

$$= -m\int_0^{\alpha} g(x)\,dx + \int_p^q g(x)\,dx + n\int_0^{\alpha} g(x)\,dx$$

$$= \int_p^q g(x)\,dx + (n-m)\int_0^{\alpha} g(x)\,dx$$

For Problems 14–17

14. b., 15. c., 16. a., 17. c.

Sol.

14. b. Let $I(a) = \int_0^1 \frac{x^a - 1}{\log x}\,dx \quad (1)$

Differentiating w.r.t. a keeping x as constant

$$\therefore \frac{dI(a)}{da} = \int_0^1 \frac{d}{da}\left(\frac{x^a - 1}{\log x}\right) dx$$

$$= \int_0^1 \frac{x^a \ln x}{\log x}\,dx$$

$$= \int_0^1 x^a\,dx$$

$$= \left.\frac{x^{a+1}}{a+1}\right|_0^1$$

$$= \frac{1}{(a+1)}$$

Integrating both sides w.r.t. a, we get

$I(a) = \log(a+1) + c$

for $a = 0, I(0) = \log 1 + c$ [from equation (1)]

$0 = 0 + c$

$\therefore I = \log(a+1)$

15.c. Let $F(k) = \int_0^{\pi/2} \ln(\sin^2\theta + k^2\cos^2\theta)\,d\theta$

$$F'(k) = \int_0^{\pi/2} \frac{1}{\sin^2\theta + k^2\cos^2\theta} 2k\cos^2\theta\,d\theta$$

$$= 2k\int_0^{\pi/2} \frac{\cos^2\theta}{\sin^2\theta + k^2\cos^2\theta}\,d\theta$$

$$= 2k\int_0^{\pi/2} \frac{d\theta}{\tan^2\theta + k^2}$$

$$= 2k\int_0^{\pi/2} \frac{\sec^2\theta - \tan^2\theta}{\tan^2\theta + k^2}\,d\theta$$

$$= 2k\int_0^{\infty} \frac{dt}{t^2+k^2} - 2k\int_0^{\pi/2} d\theta + 2k^3\int_0^{\pi/2} \frac{d\theta}{\tan^2\theta + k^2} \quad \text{(Putting } t = \tan\theta\text{)}$$

$$= 2k\left.\frac{1}{k}\tan^{-1}\frac{1}{k}\right|_0^{\infty} - 2k\frac{\pi}{2} + k^2F'(k)$$

$\Rightarrow (1-k^2)F'(k) = \pi - k\pi = \pi(1-k)$

$\Rightarrow F'(k) = \frac{\pi}{1+k}$

$\Rightarrow F(k) = \pi\log(1+k) + c$

For $k = 1, F(1) = 0 \Rightarrow c = -\pi\log 2$

$\Rightarrow F(k) = \pi\log(1+k) - \pi\log 2$

16.a. Let $I(a) = \int_0^{\pi/2} \log\left(\frac{1+a\sin x}{1-a\sin x}\right)\frac{dx}{\sin x}$

$$\frac{dI}{da} = \int_0^{\pi/2} \frac{2\sin x}{1-a^2\sin^2 x}\frac{dx}{\sin x}$$

$$= \int_0^{\pi/2} \frac{2\sec^2 x dx}{1+\tan^2 x - a^2 \tan^2 x}$$

$$= \int_0^{\pi/2} \frac{2\sec^2 x dx}{1+(1-a^2)\tan^2 x}$$

$$= \int_0^{\infty} \frac{2dt}{1+(1-a^2)t^2} \text{ (put } \tan x = t)$$

$$= \frac{2}{\sqrt{1-a^2}}\left[\tan^{-1}\left(t\sqrt{1-a^2}\right)\right]_0^{\infty}$$

$$= \frac{\pi}{\sqrt{1-a^2}}$$

$\Rightarrow I = \pi \sin^{-1} a$ [as when $I(0) = 0$]

17. c. $\displaystyle\int_0^{\pi} \frac{dx}{(a-\cos x)} = \frac{\pi}{\sqrt{a^2-1}}$

Diff. both sides with respect to a, we get

$$-\int_0^{\pi} \frac{dx}{(a-\cos x)^2} = \frac{-\pi a}{(a^2-1)^{3/2}}$$

Again diff. with respect to a, we get

$$2\int_0^{\pi} \frac{dx}{(a-\cos x)^3} = \frac{\pi(1+2a^2)}{(a^2-1)^{5/2}}$$

Put $a = \sqrt{10}$, we get $\displaystyle\int_0^{\pi} \frac{dx}{(\sqrt{10}-\cos x)^3} = \frac{7\pi}{81}$

For Problems 18–20

18.b., 19. d., 20. c.

Sol.

$$f(x) = \sin x + \sin x \int_{-\pi/2}^{\pi/2} f(t)\,dt + \cos x \int_{-\pi/2}^{\pi/2} tf(t)\,dt$$

$$= \sin x\left(1 + \int_{-\pi/2}^{\pi/2} f(t)\,dt\right) + \cos x \int_{-\pi/2}^{\pi/2} tf(t)\,dt$$

$$= A \sin x + B \cos x$$

Thus, $A = 1 + \displaystyle\int_{-\pi/2}^{\pi/2} f(t)\,dt$

$$= 1 + \int_{-\pi/2}^{\pi/2} (A\sin t + B\cos t)\,dt$$

$$= 1 + 2B \int_0^{\pi/2} \cos t\,dt$$

$$\Rightarrow \quad A = 1 + 2B \qquad (1)$$

$$B = \int_{-\pi/2}^{\pi/2} tf(t)\,dt$$

$$= \int_{-\pi/2}^{\pi/2} t(A\sin t + B\cos t)\,dt$$

$$= 2A \int_0^{\pi/2} t\sin t\,dt$$

$$= 2A[-t\cos t + \sin t]_0^{\pi/2}$$

$$\Rightarrow \quad B = 2A \qquad (2)$$

From equations (1) and (2), we get

$A = -1/3$, $B = -2/3$

$$\Rightarrow f(x) = -\frac{1}{3}(\sin x + 2\cos x)$$

Thus, the range of $f(x)$ is $\left[-\frac{\sqrt{5}}{3}, \frac{\sqrt{5}}{3}\right]$

$$f(x) = -\frac{1}{3}(\sin x + 2\cos x)$$

$$= -\frac{\sqrt{5}}{3}\sin\left(x + \tan^{-1} 2\right)$$

$$= -\frac{\sqrt{5}}{3}\cos\left(x - \tan^{-1}\frac{1}{2}\right)$$

$f(x)$ is invertible if $-\frac{\pi}{2} \le x + \tan^{-1} 2 \le \frac{\pi}{2}$

$$\Rightarrow -\frac{\pi}{2} - \tan^{-1} 2 \le x \le \frac{\pi}{2} - \tan^{-1} 2$$

or $0 \le x - \tan^{-1}\frac{1}{2} \le \pi$

$$\Rightarrow \tan^{-1}\frac{1}{2} \le x \le \pi + \tan^{-1}\frac{1}{2}$$

or $\pi \le x - \tan^{-1}\frac{1}{2} \le 2\pi$

$$\Rightarrow x \in \left[\pi + \cot^{-1} 2,\ 2\pi + \cot^{-1} 2\right]$$

$$\int_0^{\pi/2} f(x)\,dx = -\frac{1}{3}\int_0^{\pi/2} (\sin x + 2\cos x)\,dx$$

$$= -\frac{1}{3}[-\cos x + 2\sin x]_0^{\pi/2}$$

$$= -1$$

Matrix-Match Type

1. a → s; b → s; c → r; d → q.

a. $\displaystyle\int_{-1}^{1} [x + [x + [x]]]\,dx$

(use property $[x+n] = [x] + n$ if n is integer)

$= \int_{-1}^{1} 3[x]\,dx = 3\int_{-1}^{1}[x]\,dx = 3\int_{0}^{1}([x]+[-x])\,dx$

$= -3$ (as $[x]+[-x]=-1$)

b. $\int_2^5 ([x]+[-x])\,dx = \int_2^5 -1\,dx = -3$

c. $\text{sgn}(x-[x]) = \begin{cases} 1, \text{ if } x \text{ is not an integer} \\ 0, \text{ if } x \text{ is an integer} \end{cases}$

Hence, $\int_{-1}^{3} \text{sgn}(x-[x])\,dx = 4(1-0) = 4.$

d. Let $I = \int_0^{\pi/4} \left(\tan^6 (x-[x]) + \tan^4 (x-[x])\right) dx$

$\left\{\because 0 < x \le \frac{\pi}{4} \Rightarrow [x] = 0\right\}$

$\therefore\ I = \int_0^{\pi/4} \left(\tan^6 x + \tan^4 x\right) dx$

$= \int_0^{\pi/4} \tan^4 x\left(\tan^2 x + 1\right) dx$

$= \int_0^{\pi/4} \tan^4 x \sec^2 x\,dx$

$= \left(\frac{\tan^5 x}{5}\right)_0^{\pi/4}$

$= \frac{1}{5}$

2. a → r; b → p; c → s; d → q.

a. $\lim_{n\to\infty} \left[\frac{\int_0^2 \left(1+\frac{t}{n+1}\right)^n dt}{n+1}\right]$

$= \lim_{n\to\infty} \left[\left(1+\frac{t}{n+1}\right)^{n+1}\right]_0^2$

$= \lim_{n\to\infty} \left(1+\frac{2}{n+1}\right)^{n+1} - 1$

$= e^2 - 1$

b. $f'(x) = f(x) \Rightarrow f(x) = C\,e^x$ and since $f(0) = 1$

$\therefore 1 = f(0) = C$

$\therefore f(x) = e^x$ and hence $g(x) = x^2 - e^x$.

Thus, $\int_0^1 f(x)g(x)\,dx$

$= \int_0^1 (x^2 e^x - e^{2x})\,dx$

$= x^2 e^x\Big|_0^1 - 2\int_0^1 x e^x\,dx - \frac{e^{2x}}{2}\Big|_0^1$

$= (e-0) - 2\,xe^x\Big|_0^1 + 2\,e^x\Big|_0^1 - \frac{1}{2}(e^2-1)$

$= (e-0) - 2\,e + 2\,e - 2 - \frac{1}{2}(e^2-1)$

$= e - \frac{1}{2}e^2 - \frac{3}{2}$

c. $I = \int_0^1 e^{e^x}(1 + x e^x)\,dx$

Let $e^x = t$

$\Rightarrow \int_1^e e^t(1 + t\log t)\frac{dt}{t} = \int_1^e e^t\left(\frac{1}{t} + \log t\right) dt$

$= \left[e^t \log t\right]_1^e$

$= e^e$

d. $L = \lim_{k\to 0} \frac{\int_0^k (1+\sin 2x)^{\frac{1}{x}}\,dx}{k} \left(\text{form } \frac{0}{0}\right)$

$\Rightarrow L = \lim_{k\to 0} (1+\sin 2k)^{\frac{1}{k}}$

$= e^{\lim_{k\to 0}\frac{1}{k}(\sin 2k)} = e^2$

3. a → q; b → r, s; c → p; d → p.

a. $I_1 = \int_{\pi/6}^{\pi/3} \sec^2\theta\, f(2\sin 2\theta)\,d\theta$

Applying property $\int_a^b f(a+b-x)\,dx = \int_a^b f(x)\,dx$

$I_1 = \int_{\pi/6}^{\pi/3} \sec^2\left(\frac{\pi}{2}-\theta\right) f\left(2\sin 2\left(\frac{\pi}{2}-\theta\right)\right) d\theta$

$I_1 = \int_{\pi/6}^{\pi/3} \text{cosec}^2\,\theta\, f(2\sin 2\theta)\,d\theta = I_2$

b. $f(x+1) = f(x+3) \Rightarrow f(x) = f(x+2)$

$\Rightarrow f(x)$ is periodic with period 2.

Then $\int_a^{a+b} f(x)\,dx$ is independent of a, for which b is multiple of 2.

$\Rightarrow b = 2, 4, 6 \ldots$

c. Let $I = \int_1^4 \frac{\tan^{-1}[x^2]}{\tan^{-1}[x^2] + \tan^{-1}[25 + x^2 - 10x]}$ (1)

Applying $\int_a^b f(x)\,dx = \int_a^b f(a+b-x)\,dx$, we get

$$I=\int_1^4 \frac{\tan[(5-x)^2]}{\tan^{-1}[(5-x)^2]+\tan^{-1}[x^2]}dx \quad (2)$$

Adding equations (1) and (2), we get

$$2I=\int_1^4 dx \Rightarrow 2I=3 \Rightarrow I=3/2$$

d. Let $y=\sqrt{x+y}$

$y^2-y-x=0$

$y=\dfrac{1\pm\sqrt{1+4x}}{2\times 1}\; y>1$

$y=\dfrac{1+\sqrt{1+4x}}{2}$

$$I=\int_0^2 \frac{1+\sqrt{1+4x}}{2}dx=\left[\frac{x}{2}+\frac{(1+4x)^{3/2}}{\frac{3}{2}\times 2\times 4}\right]_0^2$$

$$=\left[\left(1+\frac{27}{12}\right)-\left(0+\frac{1}{12}\right)\right]=1+\frac{26}{12}=\frac{19}{6}$$

$\Rightarrow [I]=3$

4. a → p, q; b → p, q, r; c → q, s; d → s.

a. $I=\int_{-2}^{2}(\alpha x^3+\beta x+\gamma)dx$

$\alpha x^3+\beta x$ is an odd function

$I=0+2\int_0^2 \gamma dx=2.2\gamma=4\gamma$

b. $I=\dfrac{1}{2}\int_0^1 2\sin\alpha x\sin\beta x\,dx$

$$=\frac{1}{2}\int_0^1(\cos(\alpha-\beta)x-\cos(\alpha+\beta)x)dx$$

$$=\frac{1}{2}\left[\frac{\sin(\alpha-\beta)x}{\alpha-\beta}-\frac{\sin(\alpha+\beta)x}{\alpha+\beta}\right]_0^1$$

$$=\frac{1}{2}\left[\frac{\sin(\alpha-\beta)}{\alpha-\beta}-\frac{\sin(\alpha+\beta)}{\alpha+\beta}\right] \quad (1)$$

Also $2\alpha=\tan\alpha$ and $2\beta=\tan\beta$

$\Rightarrow 2(\alpha-\beta)=\tan\alpha-\tan\beta$ and

$2(\alpha+\beta)=\tan\alpha+\tan\beta$

$2(\alpha-\beta)=\dfrac{\sin(\alpha-\beta)}{\cos\alpha\cos\beta}$ and $2(\alpha+\beta)=\dfrac{\sin(\alpha+\beta)}{\cos\alpha\cos\beta}$

Substituting these values, we get

$I=(\cos\alpha\cos\beta)-(\cos\alpha\cos\beta)=0.$

c. $f(x+\alpha)+f(x)=0$

$\Rightarrow f(x+2\alpha)+f(x+\alpha)=0$

$\Rightarrow f(x+2\alpha)=f(x)$

$\Rightarrow f(x)$ is periodic with period 2α

$$\Rightarrow \int_\beta^{\beta+2\gamma\alpha}(\alpha x^3+\beta x+\gamma)dx=\gamma\int_0^{2\alpha}f(x)dx.$$

d. Let $I=\int_0^\alpha[\sin x]dx$, $\alpha\in[(2\beta+1)\pi,(2\beta+2)\pi]$, $\beta\in N$, [where $[\cdot]$ denotes the greatest integer function.]

$$I=\int_0^{2\beta\pi}[\sin x]dx+\int_{2\beta\pi}^{(2\beta+1)\pi}[\sin x]dx+\int_{(2\beta+1)\pi}^{\alpha}[\sin x]dx$$

$$=\beta\int_0^{2\pi}[\sin x]dx+0+\int_{(2\beta+1)\pi}^{\alpha}(-1)dx$$

$=-\beta\pi+(2\beta+1)\pi-\alpha$

$=(\beta+1)\pi-\alpha$

$\Rightarrow \gamma\int_0^\alpha[\sin x]dx$ depends on α, β and γ

Integer Type

1. (4) Let $L=\lim\limits_{n\to\infty}\dfrac{[(n+1)(n+2)\cdots(n+n)]^{1/n}}{n}$

$$=\lim_{n\to\infty}\left[\frac{(n+1)(n+2)\cdots(n+n)}{n^n}\right]^{1/n}$$

$$=\lim_{n\to\infty}\left[\frac{n+1}{n}\,\frac{n+2}{n}\cdots\frac{n+n}{n}\right]^{1/n}$$

$$=\lim_{n\to\infty}\left[\left(1+\frac{1}{n}\right)\left(1+\frac{2}{n}\right)\cdots\left(1+\frac{n}{n}\right)\right]^{1/n}$$

$$\log L=\lim_{n\to\infty}\frac{1}{n}\left[\log\left(1+\frac{1}{n}\right)+\log\left(1+\frac{2}{n}\right)+\cdots+\log\left(1+\frac{n}{n}\right)\right]$$

$$=\lim_{n\to\infty}\sum_{r=1}^{n}\frac{1}{n}\log\left(1+\frac{r}{n}\right)$$

$$=\int_0^1\log(1+x)dx$$

$$=\left[x\log(1+x)\right]_0^1-\int_0^1\frac{x}{1+x}dx$$

$$=\log 2-\int_0^1\left(1-\frac{1}{1+x}\right)dx$$

$$=\log 2-[x-\log(1+x)]_0^1$$

$$=\log 2-[(1-\log 2)-(0-\log 1)]$$

$$=2\log 2-1=\log(2^2/e)$$

Therefore, $L=2^2/e=4/e$.

2. (7) The given limit is

$$\lim_{n\to\infty}\frac{\sum_{r=1}^{n} r^2\times\sum_{r=1}^{n} r^3}{\sum_{r=1}^{n} r^6} = \lim_{n\to\infty}\frac{\frac{1}{n}\sum_{r=1}^{n}\left(\frac{r}{n}\right)^2\times\frac{1}{n}\sum_{r=1}^{n}\left(\frac{r}{n}\right)^3}{\frac{1}{n}\sum_{r=1}^{n}\left(\frac{r}{n}\right)^6}$$

$$= \frac{\int_0^1 x^2dx\times\int_0^1 x^3dx}{\int_0^1 x^6dx}$$

$$= \frac{7}{12}$$

3. (8) In $\int_1^4 \frac{3}{x}e^{\sin x^3}dx$, put $x^3 = t$ so that $3x^2dx = dt$.

Also when $x = 1, t = 1$ and when $x = 4, t = 64$. Therefore,

$$\int_1^4 \frac{3}{x}\ e^{\sin x^3}\ dx = \int_1^{64}\frac{3}{x}\ e^{\sin t}\ \frac{dt}{3x^2}$$

$$= \int_1^{64}\frac{1}{t}\ e^{\sin t}\ dt$$

$$= \int_1^{64}\frac{d}{dt}\ F(t)dt = F(64) - F(1)$$

Hence, $k = 64$.

4. (2) $I_1 = \int_0^1 x f''(2x)dx$

Putting $t = 2x$, i.e., $dx = \frac{dt}{2}$, we get

$$I_1 = \frac{1}{4}\int_0^2 t\, f''(t)dt$$

$$= \frac{1}{4}\left[t\ f'(t)\big|_0^2 - \int_0^2 f'(t)dt\right] \quad \text{(Integrating by parts)}$$

$$= \frac{1}{4}\left[t\ f'(t)\big|_0^2 - f(t)\big|_0^2\right]$$

$$\Rightarrow I_1 = \frac{1}{4}[2f'(2) - f(2) + f(0)] = \frac{1}{4}(10 - 3 + 1) = 2$$

5. (8) The graph of $f(x) = \sin^{-1}(\sin x)$ is given below.

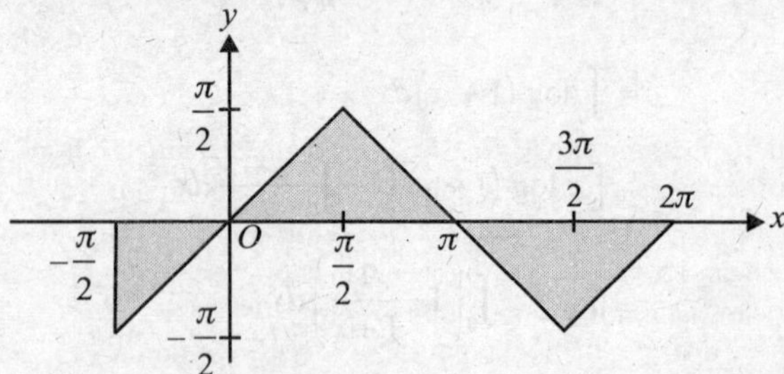

Fig. 8.7

$$\Rightarrow \int_{-\pi/2}^{2\pi}\sin^{-1}(\sin x)\,dx$$

$$= \int_{-\pi/2}^{0}\sin^{-1}(\sin x)\,dx + \int_0^{\pi}\sin^{-1}(\sin x)\,dx + \int_{\pi}^{2\pi}\sin^{-1}(\sin x)\,dx$$

= Area of shaded region

$$= -\left(\frac{1}{2}\times\frac{\pi}{2}\times\frac{\pi}{2}\right) + \left(\frac{1}{2}\times\pi\times\frac{\pi}{2}\right) - \left(\frac{1}{2}\times\pi\times\frac{\pi}{2}\right)$$

$$= -\frac{\pi^2}{8}$$

6. (4) Given integral

$$I = \int_{-\pi/2}^{\pi/2}\sqrt{\left[\cos x\left(1-\cos^2 x\right)\right]}\ dx$$

$$= \int_{-\pi/2}^{\pi/2}\sqrt{\left(\cos x\ \sin^2 x\right)}\ dx$$

$$= \int_{-\pi/2}^{\pi/2}\sqrt{(\cos x)}\ |\sin x|\ dx \qquad (1)$$

$$\text{Now } |\sin x| = \begin{cases}-\sin x, \ if\ -\pi/2 \le x < 0\\ \sin x, \ if\ 0 < x \le \pi/2\end{cases}$$

From (1), we have

$$I = \int_{-\pi/2}^{0}\sqrt{(\cos x)}\ (-\sin x)dx + \int_0^{\pi/2}\sqrt{(\cos x)}\ \sin x\ dx$$

Putting $\cos x = t$, $-\sin x\ dx = dt$, we get

$$I = \int_0^1 t^{1/2}\ dt - \int_1^0 t^{1/2}\ dt = 2\int_0^1 t^{1/2}\ dt$$

$$= 2\times\left(\frac{2}{3}\right)\left[t^{3/2}\right]_0^1$$

$$= \frac{4}{3}$$

7. (4) $P = \int_{-\pi/4}^{\pi/4}\log\left[\sqrt{2}\sin\left(x+\frac{\pi}{4}\right)\right]dx$

Putting $x + \frac{\pi}{4} = \theta$, $dx = d\theta$,

$$= \int_0^{\pi/2}\log\left(\sqrt{2}\sin\theta\right)d\theta$$

$$= \frac{1}{2}\int_0^{\pi/2}\log 2\ d\theta + \int_0^{\pi/2}\log\sin\theta\ d\theta$$

$$= \left(\frac{1}{4}\pi\log 2\right) - \frac{1}{2}\pi\log 2$$

$$= -\frac{1}{4}\pi\log 2$$

$$Q = \int_0^{\infty}\log(x+1/x)\ \frac{dx}{1+x^2}$$

Putting $x = \tan\theta$, $dx = \sec^2\theta\ d\theta$, the given integral

$$Q = \int_0^{\pi/2}\frac{\log\left(\tan\theta+\cot\theta\right)}{1+\tan^2\theta}\sec^2\theta\ d\theta$$

$$= \int_0^{\pi/2}\log\left(\frac{\sin\theta}{\cos\theta}+\frac{\cos\theta}{\sin\theta}\right)d\theta$$

$$= \int_0^{\pi/2} \log\left(\frac{1}{\sin\theta \cos\theta}\right) d\theta$$

$$= -\int_0^{\pi/2} \log\sin\theta \, d\theta - \int_0^{\pi/2} \log\cos\theta \, d\theta$$

$$= -2\left(-\frac{1}{2}\pi \log 2\right) = \pi \log 2$$

8. **(2)** $f(x) = \dfrac{x^9 - 3x^5 + 7x^3 - x}{\cos^2 x} + \sec^2 x$

$$= \sec^2 x\,(x^9 - 3x^5 + 7x^3 - x) + \sec^2 x$$

$$\Rightarrow \int_{-\pi/4}^{\pi/4} f(x)\,dx = \int_{-\pi/4}^{\pi/4} \sec^2 x\,dx$$

$$= 2\int_0^{\pi/4} \sec^2 x\,dx$$

$$= 2\tan x\Big|_0^{\pi/4} = 2$$

9. Given integral

$$= \int_{-1/2}^{1/2} \left[\left(\frac{x+1}{x-1} - \frac{x-1}{x+1}\right)^2\right]^{1/2} dx$$

$$= \int_{-1/2}^{1/2} \left|\frac{x+1}{x-1} - \frac{x-1}{x+1}\right| dx$$

$$= \int_{-1/2}^{1/2} \left|\frac{4x}{x^2-1}\right| dx = 2\int_0^{1/2} \left|\frac{4x}{x^2-1}\right| dx$$

$$= 2\int_0^{1/2} \frac{4x}{1-x^2}\,dx \quad \because \left|\frac{4x}{x^2-1}\right| = -\frac{4x}{x^2-1}, \text{ when } 0 < x \le \frac{1}{2}$$

$$= -4\,[\log(1-x^2)]_0^{1/2}$$

$$= -4\log(3/4) = 4\log(4/3)$$

10. **(8)** $I = \dfrac{\int_0^n [x]\,dx}{\int_0^n \{x\}\,dx}$

$$= \frac{\int_0^n (x - \{x\})\,dx}{\int_0^n \{x\}\,dx}$$

$$= \frac{\int_0^n x\,dx}{\int_0^n \{x\}\,dx} - 1$$

$$= \frac{\left.\frac{x^2}{2}\right|_0^n}{n\int_0^1 \{x\}\,dx} - 1$$

$$= \frac{\frac{n^2}{2}}{n\int_0^1 \{x\}\,dx} - 1$$

$$= \frac{\frac{n^2}{2}}{n\frac{1}{2}} - 1 = n - 1$$

11. **(9)** $\displaystyle\lim_{x\to 2} \int_6^{f(x)} \frac{4t^3}{x-2}\,dt = \lim_{x\to 2} \frac{\int_6^{f(x)} 4t^3\,dt}{x-2} \quad \left(\frac{0}{0} \text{ form}\right)$

$$= \lim_{x\to 2} \frac{4\,[f(x)]^3 f'(x)}{1}$$

(Applying L' Hospital's rule)

$$= 4[f(2)]^3 \times f'(2)$$

$$= 4(6)^3 \times \frac{1}{48}$$

$$= 18$$

12. **(0)** Since $e^{x^2} > 0$, $e^{2x^2} > 0$ in $[0, x], x > 0$,

$\int_0^x e^{x^2}\,dx$ and $\int_0^x e^{2x^2}\,dx \to \infty$ as $x \to \infty$.

$L = \displaystyle\lim_{x\to\infty} \frac{\left(\int_0^x e^{x^2}\right)^2}{\int_0^x e^{2x^2}}$ is of the form $\dfrac{\infty}{\infty}$.

Using L' Hospital's rule, we get

$$L = \lim_{x\to\infty} \frac{2e^{x^2}\int_0^x e^{x^2}\,dx}{e^{2x^2}}$$

$$= 2\lim_{x\to\infty} \frac{\int_0^x e^{x^2}\,dx}{e^{2x^2}} \quad \left(\frac{\infty}{\infty} \text{ form}\right)$$

$$= 2\lim_{x\to\infty} \frac{e^{x^2}}{2xe^{x^2}}$$

$$= \lim_{x\to\infty} \frac{1}{x} = 0$$

13. **(9)** $x = \displaystyle\int_0^y \frac{dt}{\sqrt{1+9t^2}}$

Differentiating w.r.t. y, we get

$$\frac{dx}{dy} = \frac{1}{\sqrt{1+9y^2}}$$

$$\Rightarrow \frac{dy}{dx} = \sqrt{1+9y^2}$$

$$\Rightarrow \frac{d}{dx}\left(\frac{dy}{dx}\right) = \frac{d}{dy}\left(\sqrt{1+9y^2}\right)\frac{dy}{dx}$$

$$\Rightarrow \frac{d^2y}{dx^2} = \frac{18y}{2\sqrt{1+9y^2}}\sqrt{1+9y^2} = 9y$$

$$\Rightarrow a = 9$$

14. **(5)** $I_n = \int_0^{\pi} x^n \sin x \, dx$

$$= \left[-x^n \cos x\right]_0^{\pi} + n\int_0^{\pi} x^{n-1} \cos x \, dx$$

$$= \pi^n + n\,[x^{n-1} \sin x]_0^{\pi} - n(n-1)\int_0^{\pi} x^{n-2} \sin x dx$$

$$\Rightarrow I_n = \pi^n + n \times 0 - n(n-1)\, I_{n-2}$$

Putting $n = 5$, we get

$I_5 = \pi^5 - 20I_3$

$\Rightarrow I_5 + 20I_3 = \pi^5$

15. **(0)** $I = \int_{-4}^{-5} e^{(x+5)^2} dx + 3\int_{1/3}^{2/3} e^{9[x-(2/3)]^2} dx$

$$= \int_{-4}^{-5} e^{(x+5)^2} dx + 3\int_{1/3}^{2/3} e^{(3x-2)^2} dx$$

$$= I_1 + I_2$$

Note that in both I_1 and I_2, the function has the same format, i.e., e^{t^2}.

Also e^{t^2} is not integrable.

Now in I_1, let $x + 5 = y$ and in I_2, $3x - 2 = -t$. Therefore,

$$I = \int_1^0 e^{y^2} dy + \int_1^0 e^{t^2} (-dt) = 0$$

16. **(2)** $I = \int_0^{\pi/2} \left(\frac{\theta}{\sin\theta}\right)^2 d\theta$

$$= \int_0^{\pi/2} \theta^2 \text{cosec}^2\theta \, d\theta$$

$$= [\theta^2(-\cot\theta)]_0^{\pi/2} - \int_0^{\pi/2} 2\theta\,(-\cot\theta)\, d\theta$$

(Integrating by parts)

$$= [-\lim_{\theta\to\pi/2} \theta^2 \cot\theta] + 2\int_0^{\pi/2} \theta \cot\theta \, d\theta$$

$$= 0 + 2\left[[\theta \log\sin\theta]_0^{\pi/2} - \int_0^{\pi/2} \log\sin\theta \, d\theta\right]$$

(Integrating by parts)

$$= 2\left[-\lim_{\theta\to 0} \theta \ln\sin\theta - k\right]$$

$$= -2k$$

17. **(2)**

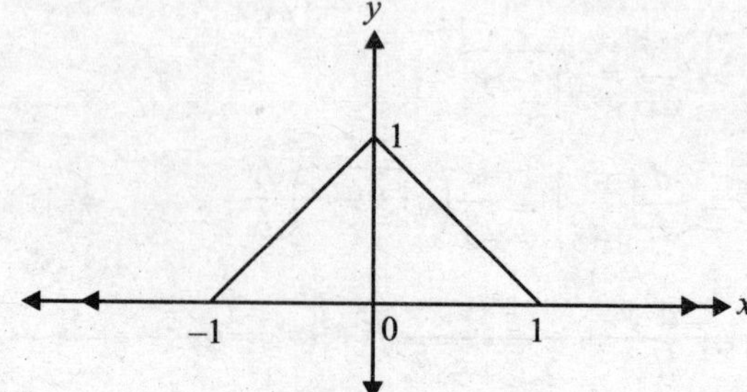

Graph of $y = f(x)$

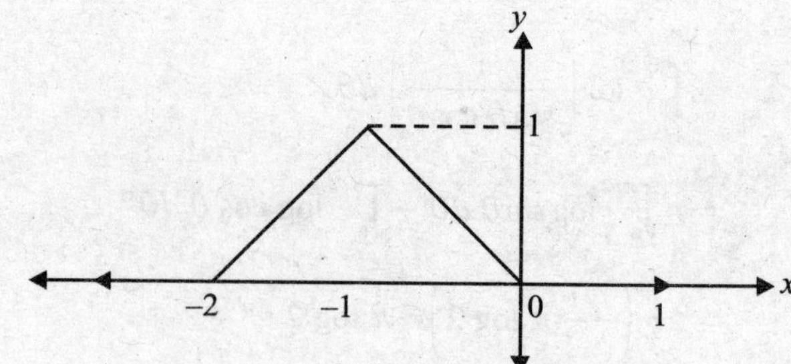

Graph of $y = f(x + 1)$

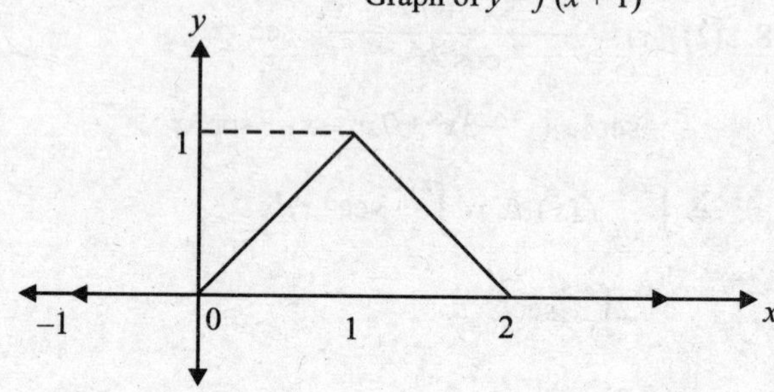

Graph of $y = f(x - 1)$

Fig. 8.8

$$\int_{-3}^{5} g(x)dx = \int_{-3}^{5} f(x-1)\,dx + \int_{-3}^{5} f(x-1)\,dx$$

= Area of triangle in the graph of $y = f(x + 1)$ + Area of triangle in the graph of $y = f(x + 1)$

$$= 2 \times \frac{1}{2} \times 2 \times 1$$

$$= 2$$

18. **(3)** Let $I_n = \int_0^{\pi/2} x^n \sin x \, dx$

Integrating by parts choosing sin x as the second function, we get

$$I_n = \left[x^n\,(-\cos x)\right]_0^{\pi/2} - \int_0^{\pi/2} nx^{n-1}\,(-\cos x)\, dx$$

$$= 0 - n\int_0^{\pi/2} x^{n-1} \cos x \, dx$$

Again integrating by parts,

$$I_n = n\,[x^{n-1} \sin x]_0^{\pi/2} - n(n-1)\int_0^{\pi/2} x^{n-2} \sin x \, dx$$

$$\Rightarrow I_n = n\left(\frac{\pi}{2}\right)^{n-1} - n(n-1)\, I_{n-2}$$

R.H.S. contains π^2. So putting $n = 3$, we get

$$I_3 = 3\left(\frac{\pi}{2}\right)^2 - 3 \times 2I_1$$

$$= \frac{3\pi^2}{4} - 6\int_0^{\pi/2} x \sin x \, dx$$

$$= \frac{3\pi^2}{4} - 6[x\,(-\cos x) + \sin\, x]_0^{\pi/2}$$

$$= \frac{3\pi^2}{4} - 6\,(1) = \frac{3}{4}\left(\pi^2 - 8\right),$$ which is true.

Hence, $n = 3$.

19. **(3)** Given $\int_0^1 e^x (x-1)^n \, dx = 16 - 6e$, where $n \in N$ and $n \le 5$.

To find the value of n,

let $I_n = \int_0^1 e^x (x-1)^n \, dx$

$= [(x-1)^n e^x]_0^1 - \int_0^1 n(x-1)^{n-1} e^x \, dx$

$= -(-1)^n - \int_0^1 n(x-1)^{n-1} e^x \, dx$

$\Rightarrow I_n = (-1)^{n+1} - nI_{n-1}$ (1)

Also $I_1 = \int_0^1 e^x (x-1) \, dx$

$= \left[e^x (x-1)\right]_0^1 - \int_0^1 e^x \, dx$

$= -(-1) - [e^x]_0^1$

$= 1 - (e-1) = 2 - e$

Using equation (1), we get

$I_2 = (-1)^3 - 2I_1$

$= -1 - 2(2-e) = 2e - 5$

Similarly, $I_3 = (-1)^4 - 3I_2$

$= 1 - 3(2e-5) = 16 - 6e$

Hence, $n = 3$.

20. **(6)** Let $F(x) = f(x) + f\left(\frac{1}{x}\right)$

$= \int_1^x \frac{\log t}{1+t} dt + \int_1^{1/x} \frac{\log t}{1+t} dt$

In 2nd integral, let

$t = \frac{1}{y}$

$\Rightarrow dt = -\frac{1}{y^2} dy$

$\Rightarrow F(x) = \int_1^x \frac{\log t}{1+t} dt + \int_1^x \frac{-\log y}{1+\frac{1}{y}}\left(-\frac{dy}{y^2}\right)$

$= \int_1^x \frac{\log t}{1+t} dt + \int_1^x \frac{\log y}{y(1+y)} dy$

$= \int_1^x \frac{\log t}{1+t} dt + \int_1^x \frac{\log t}{t(1+t)} dt$

$= \int_1^x \frac{\log t}{t} dt = \frac{1}{2} (\log x)^2$

Hence, $F(e) = \frac{1}{2}$.

21. **(8)** $f(x) = \begin{cases} e^{\cos x} \sin x, & \text{for } |x| \le 2 \\ 2, & \text{otherwise} \end{cases}$

$\Rightarrow \int_{-2}^{2} f(x) \, dx = \int_{-2}^{2} f(x) \, dx + \int_2^3 f(x) \, dx$

$= \int_{-2}^{2} e^{\cos x} \sin x \, dx + \int_2^3 2 \, dx$

$= 0 + 2[x]_2^3$

($\because e^{\cos x} \sin x$ is an odd function)

$= 2[3-2]$

$= 2$

22. **(4)** $I = \int_{-2}^{0} [x^3 + 3x^2 + 3x + 3 + (x+1)\cos(x+1)] dx$

$= \left[\frac{x^4}{4} + x^3 + \frac{3x^2}{2} + 3x + (x+1)\sin (x+1) + \cos(x+1)\right]_{-2}^{0}$

$= (\sin 1 + \cos 1) - (4 - 8 + 6 - 6 + \sin 1 + \cos 1)$

$= 4$

23. **(6)** $I = \int_1^e \frac{1 + \ln x}{\sqrt{x \ln x}} dx$

Putting $(x \ln x) = t^2$, $(\ln x + 1) \, dx = 2t \, dt$, we get

$I = \int_0^{\sqrt{e}} \frac{2t \, dt}{t} = 2\sqrt{e}$

24. **(1)** $A + B = \int_0^{\infty} \frac{(x+1)\ln x}{1+x^3} dx = \int_0^{\infty} \frac{\ln x}{x^2 - x + 1} dx$

Let $x = \frac{1}{t}$

$\Rightarrow A + B = -(A + B)$

$\Rightarrow A + B = 0$

25. **(4)** $u = \int_0^1 \frac{\ln(x+1)}{x^2+1} dx$, (Putting $x = \tan\theta$)

$= \int_0^{\pi/4} \ln(1 + \tan\theta) \, d\theta$

$= \int_0^{\pi/4} \ln\left(1 + \frac{1-\tan\theta}{1+\tan\theta}\right) d\theta$

$= \int_0^{\pi/4} \ln \frac{2}{1+\tan\theta} d\theta = \frac{\pi}{4} \ln 2 - u$. Therefore,

$u = \frac{\pi}{8} \ln 2$

$\Rightarrow 4u = \frac{\pi}{2} \ln 2$ (1)

Again $v = \int_0^{\pi/2} \ln(\sin 2x) \, dx$, (Putting $2x = t$)

$\Rightarrow v = \frac{1}{2}\int_0^{\pi} \ln(\sin t) \, dt = \int_0^{\pi/2} \ln(\sin t) \, dt$

$\Rightarrow v = -\frac{\pi}{2} \ln 2$ (2)

$(1)+(2) \Rightarrow 4u+v=0$

$\Rightarrow -v/u = 4.$

26. (7) $\int_a^b \frac{x^n}{x^n+(16-x)^n}dx = 6$ (1)

Let $a+b=16$, then by property $\int_a^b \frac{(16-x^n)}{(16-x^n)+x^n}dx$

$=6$ (2)

Adding Equations (1) and (2), $\int_a^b 1\times dx = 12$ or

$b-a=12$

Solving $a+b=16$ and $b-a=12$, we get $a=2, b=14$ and $n\in R$

27. (2) $L = \lim_{x\to 0}\frac{\int_0^{\ln(1+x)}(1-\tan 2y)^{1/y}\,dy}{\frac{\sin x}{x}\times x}$

Using L'Hospital rule,

$L = \lim_{x\to 0}\frac{[1-\tan 2(\ln(1+x))]^{\frac{1}{\ln(1+x)}}}{(1+x)}\ (1^\infty)$

$= e^{-\lim_{x\to 0}\frac{1}{\ln(1+x)}\tan 2(\ln(1+x))}$

$= e^{-\lim_{x\to 0}\frac{\tan(2\ln(1+x))}{2\ln(1+x)}\times 2}$

$= e^{-2}$ (using $\lim_{x\to 0}\frac{\tan\theta}{\theta}=1$)

28. (8) Using property $\int_a^b f(x)\,dx = \int_a^b f(a+b-x)\,dx$

$I = \int_{-1}^{1}(1-x)^{1/2}(1+x)^{3/2}\,dx$

$\Rightarrow 2I = \int_{-1}^{1}(1+x)^{1/2}(1-x)^{1/2}[(1-x)+(1+x)]dx$

$\Rightarrow 2I = 2\int_{-1}^{1}\sqrt{1-x^2}\,dx$

$\Rightarrow I = 2\int_0^1\sqrt{1-x^2}\,dx\ (x=\sin\theta)\Rightarrow dx=\cos d\theta$

$\Rightarrow I = 2\int_0^{\pi/2}\cos^2\theta\,d\theta = \frac{\pi}{2}$

29. (9) $1 = \lim_{n\to\infty}\frac{\sqrt{1}+\sqrt{2}+\sqrt{3}+...+\sqrt{6n}}{n\sqrt{n}}$

$= \lim_{n\to\infty}\frac{1}{n}\sum_{r=1}^{6n}\sqrt{\frac{r}{n}} = \int_0^6\sqrt{x}\,dx = \left[\frac{2}{3}x^{3/2}\right]_0^6 = \frac{2}{3}\times 6\sqrt{6}$

$= \sqrt{96}$

30. (5) $I = \int_0^x t^2\sin(x-t))\,dt = x^2$ (integrating by parts)

$= t^2\cos(x-t)]_0^x - 2\int t,\ \cos(x-t)\,dt = x^2$

$= (x^2-0) - 2\left[-t\sin(x-t)]_0^x + \int_0^x \sin(x-t)\,dt\right]$

$= x^2$

$= -2\,[0+\cos(x-t)]_0^x] = 0;\ 1-\cos x = 0$

$\Rightarrow x = 2n\pi,\ n\in I$

31. (3) $f(x) = \int_0^x e^t\sin(x-t)dt = \int_0^x e^{x-t}\sin(t)dt$

$f(x) = e^x\int_0^x e^{-t}\sin t dt$

$f'(x) = e^x.e^{-x}\sin x + \left(\int_0^x e^{-t}\sin t\,dt\right)e^x$

$f'(x) = \sin x + f(x)$ (1)

$f''(x) = \cos x + f'(x) = \cos x + \sin x + f(x)$

[Using (1)]

$f''(x) - f(x) = \sin x + \cos x$ (2)

$g(x) = \sin x + \cos x \Rightarrow g(x)\in[-\sqrt{2},\sqrt{2}]$

32. (3) Let $\int_0^1 f(t)dt = k$, so $f(x) = kx+1$

Now $\int_0^1(kt+1)dt = k$

$\Rightarrow \frac{k}{2}+1 = k$, so $k=2$

$\therefore f(x) = 2x+1$

Also $\int_0^3 f(x)dx = 12$

33. (2) In I put $x = \frac{1}{t}$

$\therefore I = \int_0^\infty \frac{t^{-2008}}{1+t^2}dt = \int_0^\infty\frac{x^{-2008}}{1+x^2}dx$

$\therefore 2I = \int_0^\infty\frac{x^{2008}+x^{-2008}}{1+x^2}dx = J$

34. (4) $I = \int_0^{3\pi/4}(\sin x+\cos x)\,dx + \int_0^{3\pi/4}\underset{\text{I}}{x}\underbrace{(\sin x-\cos x)}_{\text{II}}\,dx$

$= \int_0^{3\pi/4}(\sin x+\cos x)\,dx + \underbrace{x(-\cos x-\sin x)|_0^{3\pi/4}}_{\text{zero}}$

$$+\int_0^{3\pi/4}(\sin x+\cos x)\,dx$$

$$=2\int_0^{3\pi/4}(\sin x+\cos x)\,dx=2(\sqrt{2}+1)$$

35. **(7)** $f(x)=\int_0^{\pi}\cos t\cos(x-t)\,dt$ (1)

$$=\int_0^{\pi}-\cos t\ \cos(x-\pi+t)\,dt \quad \text{(using King)}$$

$f(x)=\int_0^{\pi}\cos t.\cos(x+t)\,dt$ (2)

(1) + (2) gives

$$2f(x)=\int_0^{\pi}\cos t\ (2\cos x\cos t)\,dt$$

$$\therefore f(x)=\cos x\int_0^{\pi}\cos^2 t\,dt=2\cos x\int_0^{\pi/2}\cos^2 t\,dt$$

$$f(x)=\frac{\pi\cos x}{2}$$

36. **(2)** Given $f(f(x))=-x+1$

Replacing $x\to f(x)$

$f(f(f(x)))=-f(x)+1$

$f(1-x)=-f(x)+1$

$f(x)+f(1-x)=1$ (1)

Now $J=\int_0^1 f(x)\,dx=\int_0^1 f(1-x)\,dx$

$$2J=\int\left(f(x)+f(1-x)\right)dx;\ 2J=\int_0^1 dx=1$$

$$\Rightarrow J=\frac{1}{2}$$

37. **(0)** Consider $I=\int_a^{\infty}\frac{n\,dx}{n^2\left(x^2+\frac{1}{n^2}\right)}=\frac{1}{n}\times n\left(\tan^{-1}nx\right)_a^{\infty}$

$$=\left(\frac{\pi}{2}-\tan^{-1}an\right)$$

$$\therefore L=\lim_{n\to\infty}\left(\frac{\pi}{2}-\tan^{-1}an\right)=\begin{bmatrix}\pi \text{ if } a<0\\ \pi/2 \text{ if } a=0\\ 0 \text{ if } a>0\end{bmatrix}$$

CHAPTER 9

Area

Objective Type

Each question has four choices a, b, c, and d, out of which *only one* is correct.

1. Area enclosed by the curve $y = f(x)$ defined parametrically as $x = \frac{1-t^2}{1+t^2}$, $y = \frac{2t}{1+t^2}$ is equal to

   a. π sq. units b. $\pi/2$ sq. units

   c. $\frac{3\pi}{4}$ sq. units d. $\frac{3\pi}{2}$ sq. units

2. The area inside the parabola $5x^2 - y = 0$ but outside the parabola $2x^2 - y + 9 = 0$ is

   a. $12\sqrt{3}$ sq. units b. $6\sqrt{3}$ sq. units

   c. $8\sqrt{3}$ sq. units d. $4\sqrt{3}$ sq. units

3. Let $f(x) = \text{minimum}\ (x+1, \sqrt{1-x})$ for all $x \le 1$. Then the area bounded by $y = f(x)$ and the x-axis is

   a. $\frac{7}{3}$ sq. units b. $\frac{1}{6}$ sq. units

   c. $\frac{11}{6}$ sq. units d. $\frac{7}{6}$ sq. units

4. Area enclosed between the curves $|y| = 1 - x^2$ and $x^2 + y^2 = 1$ is

   a. $\frac{3\pi - 8}{3}$ sq. units b. $\frac{\pi - 8}{3}$ sq. units

   c. $\frac{2\pi - 8}{3}$ sq. units d. None of these

5. The area of the region enclosed by the curves $y = x \log x$ and $y = 2x - 2x^2$ is

   a. $\frac{7}{12}$ sq. units b. $\frac{1}{2}$ sq. units

   c. $\frac{5}{12}$ sq. units d. None of these

6. The area enclosed between the curves $y = \log_e (x + e)$, $x = \log_e \left(\frac{1}{y}\right)$, and the x-axis is

   a. 2 sq. units b. 1 sq. units

   c. 4 sq. units d. None of these

7. Area bounded by $y = \frac{1}{x^2 - 2x + 2}$ and the x-axis is

   a. 2π sq. units b. $\frac{\pi}{2}$ sq. units

   c. 2 sq. units d. π sq. units

8. Area bounded by the curve $xy^2 = a^2(a - x)$ and the y-axis is

   a. $\pi a^2/2$ sq. units b. πa^2 sq. units

   c. $3\pi a^2$ sq. units d. None of these

9. The area of the closed figure bounded by $x = -1$, $y = 0$, $y = x^2 + x + 1$ and the tangent to the curve $y = x^2 + x + 1$ at $A(1, 3)$ is

   a. 4/3 sq. units b. 7/3 sq. units

   c. 7/6 sq. units d. None of these

10. The area bounded by $y = \sec^{-1} x$, $y = \text{cosec}^{-1} x$, and line $x - 1 = 0$ is

    a. $\log(3 + 2\sqrt{2}) - \frac{\pi}{2}$ sq. units

    b. $\frac{\pi}{2} - \log(3 + 2\sqrt{2})$ sq. units

c. $\pi - \log_e 3$ sq. units
d. None of these

11. The area of the region whose boundaries are defined by the curves $y = 2\cos x$, $y = 3\tan x$, and the y-axis is

a. $1 + 3\ln\left(\frac{2}{\sqrt{3}}\right)$ sq. units

b. $1 + \frac{3}{2}\ln 3 - 3\ln 2$ sq. units

c. $1 + \frac{3}{2}\ln 3 - \ln 2$ sq. units

d. $\ln 3 - \ln 2$ sq. units

12. The area between the curve $y = 2x^4 - x^2$, the x-axis, and the ordinates of the two minima of the curve is

a. 11/60 sq. units b. 7/120 sq. units
c. 1/30 sq. units d. 7/90 sq. units

13. The area bounded by the curve $a^2y = x^2(x + a)$ and the x-axis is

a. $a^2/3$ sq. units b. $a^2/4$ sq. units
c. $3a^2/4$ sq. units d. $a^2/12$ sq. units

14. The area of the region in 1st quadrant bounded by the y-axis, $y = \frac{x}{4}$, $y = 1 + \sqrt{x}$ and $y = \frac{2}{\sqrt{x}}$ is

a. 2/3 sq. units b. 8/3 sq. units
c. 11/3 sq. units d. 13/6 sq. units

15. The area of the closed figure bounded by $y = \frac{x^2}{2} - 2x + 2$ and the tangents to it at (1, 1/2) and (4, 2) is

a. 9/8 sq. units b. 3/8 sq. units
c. 3/2 sq. units d. 9/4 sq. units

16. The area of the closed figure bounded by $x = -1$, $x = 2$, and $y = \begin{cases} -x^2 + 2, & x \le 1 \\ 2x - 1, & x > 1 \end{cases}$ and the abscissa axis is

a. 16/3 sq. units b. 10/3 sq. units
c. 13/3 sq. units d. 7/3 sq. units

17. The area of the region bounded by $x^2 + y^2 - 2x - 3 = 0$ and $y = |x| + 1$ is

a. $\frac{\pi}{2} - 1$ sq. units b. 2π sq. units
c. 4π sq. units d. $\pi/2$ sq. units

18. The value of the parameter a such that the area bounded by $y = a^2x^2 + ax + 1$, coordinate axes and the line $x = 1$ attains its least value, is equal to

a. $-\frac{1}{4}$ sq. units b. $-\frac{1}{2}$ sq. units
c. $-\frac{3}{4}$ sq. units d. -1 sq. units

19. The area enclosed by the curve $y = \sqrt{4 - x^2}$, $y \ge \sqrt{2}\sin\left(\frac{x\pi}{2\sqrt{2}}\right)$, and the x-axis is divided by the y-axis in the ratio

a. $\frac{\pi^2 - 8}{\pi^2 + 8}$ b. $\frac{\pi^2 - 4}{\pi^2 + 4}$

c. $\frac{\pi - 4}{\pi - 4}$ d. $\frac{2\pi^2}{2\pi + \pi^2 - 8}$

20. If $f(x) = \sin x$, $\forall\, x \in \left[0, \frac{\pi}{2}\right]$, $f(x) + f(\pi - x) = 2$, $\forall\, x \in \left(\frac{\pi}{2}, \pi\right]$, and $f(x) = f(2\pi - x)$, $\forall\, x \in (\pi, 2\pi)$, then the area enclosed by $y = f(x)$ and x-axis is

a. π sq. units b. 2π sq. units
c. 2 sq. units d. 4 sq. units

21. The area of the region bounded by $x = 0$, $y = 0$, $x = 2$, $y = 2$, $y \le e^x$ and $y \ge \ln x$, is

a. $6 - 4\ln 2$ sq. units b. $4\ln 2 - 2$ sq. units
c. $2\ln 2 - 4$ sq. units d. $6 - 2\ln 2$ sq. units

22. The area of the loop of the curve, $ay^2 = x^2(a - x)$ is

a. $4a^2$ sq. units b. $\frac{8a^2}{15}$ sq. units

c. $\frac{16a^2}{9}$ sq. units d. None of these

23. The area of the region enclosed between the curves $x = y^2 - 1$ and $x = |y|\sqrt{1 - y^2}$ is

a. 1 sq. units b. 4/3 sq. units
c. 2/3 sq. units d. 2 sq. units

24. The area bounded by the loop of the curve $4y^2 = x^2(4 - x^2)$ is

a. 7/3 sq. units b. 8/3 sq. units
c. 11/3 sq. units d. 16/3 sq. units

25. The area enclosed by the curves $xy^2 = a^2(a - x)$ and $(a - x)y^2 = a^2x$ is

a. $(\pi - 2)a^2$ sq. units b. $(4 - \pi)a^2$ sq. units
c. $\pi a^2/3$ sq. units d. None of these

26. The area bounded by the curves $y = xe^x$, $y = xe^{-x}$ and the line $x = 1$ is

a. $\frac{2}{e}$ sq. units b. $1 - \frac{2}{e}$ sq. units

c. $\frac{1}{e}$ sq. units d. $1 - \frac{1}{e}$ sq. units

27. The area of the figure bounded by the parabola $(y - 2)^2 = x - 1$, the tangent to it at the point with the ordinate $x = 3$ and the x-axis is

a. 7 sq. units b. 6 sq. units
c. 9 sq. units d. None of these

28. The area bounded by $y = 3 - |3 - x|$ and $y = \frac{6}{|x + 1|}$ is

a. $\frac{15}{2} - 6\ln 2$ sq. units b. $\frac{13}{2} - 3\ln 2$ sq. units

c. $\frac{13}{2} - 6\ln 2$ sq. units d. None of these

29. The area of the region of the plane bounded by $\max(|x|, |y|) \le 1$ and $xy \le \frac{1}{2}$ is

a. $1/2 + \ln 2$ sq. units b. $3 + \ln 2$ sq. units
c. $31/4$ sq. units d. $1 + 2\ln 2$ sq. units

30. The area bounded by the two branches of curve $(y-x)^2 = x^3$ and the straight line $x = 1$ is
a. $1/5$ sq. units b. $3/5$ sq. units
c. $4/5$ sq. units d. $8/4$ sq. units

31. Area bounded by the curves $y = \log_e x$ and $y = (\log_e x)^2$ is
a. $e - 2$ sq. units b. $3 - e$ sq. units
c. e sq. units d. $e - 1$ sq. units

32. The area of the region containing the points (x, y) satisfying $4 \le x^2 + y^2 \le 2(|x| + |y|)$ is
a. 8 sq. units b. 2 sq. units
c. 4π sq. units d. 2π sq. units

33. Let $f(x) = x^3 + 3x + 2$ and $g(x)$ is the inverse of it. Then the area bounded by $g(x)$, the x-axis and the ordinate at $x = -2$ and $x = 6$ is
a. $1/4$ sq. units b. $4/3$ sq. units
c. $5/4$ sq. units d. $7/3$ sq. units

34. The area bounded by the curve $f(x) = x + \sin x$ and its inverse function between the ordinates $x = 0$ and $x = 2\pi$ is
a. 4π square units b. 8π square units
c. 4 square units d. 8 square units

35. The area bounded by the x-axis, the curve $y = f(x)$ and the lines $x = 1, x = b$ is equal to $\sqrt{b^2+1} - \sqrt{2}$ for all $b > 1$, then $f(x)$ is
a. $\sqrt{x-1}$ b. $\sqrt{x+1}$
c. $\sqrt{x^2+1}$ d. $\dfrac{x}{\sqrt{1+x^2}}$

36. Let $f(x)$ be a non-negative continuous function such that the area bounded by the curve $y = f(x)$, the x-axis, and the ordinates $x = \dfrac{\pi}{4}$ and $x = \beta > \dfrac{\pi}{4}$ is $\beta \sin\beta + \dfrac{\pi}{4}\cos\beta + \sqrt{2}\beta$. Then $f'\left(\dfrac{\pi}{2}\right)$ is
a. $\left(\dfrac{\pi}{2} - \sqrt{2} - 1\right)$ b. $\left(\dfrac{\pi}{4} + \sqrt{2} - 1\right)$
c. $-\dfrac{\pi}{2}$ d. $\left(1 - \dfrac{\pi}{4} - \sqrt{2}\right)$

37. The area bounded by the curves $y = \sin^{-1}|\sin x|$ and $y = (\sin^{-1}|\sin x|)^2$, where $0 \le x \le 2\pi$, is
a. $\dfrac{1}{3} + \dfrac{\pi^2}{4}$ sq. units b. $\dfrac{1}{6} + \dfrac{\pi^3}{8}$ sq. units
c. 2 sq. units d. None of these

38. Consider two curves $C_1: y^2 = 4[\sqrt{y}]x$ and $C_2: x^2 = 4[\sqrt{x}]y$, where [.] denotes the greatest integer function. Then the area of region enclosed by these two curves within the square formed by the lines $x=1, y=1, x=4, y=4$ is
a. $8/3$ sq. units b. $10/3$ sq. units
c. $11/3$ sq. units d. $11/4$ sq. units

39. The area enclosed between the curve $y^2(2a - x) = x^3$ and the line $x = 2$ above the x-axis is
a. πa^2 sq. units b. $\dfrac{3\pi a^2}{2}$ sq. units
c. $2\pi a^2$ sq. units d. $3\pi a^2$ sq. units

40. The area bounded by the curve $y^2 = 1 - x$ and the lines $y = \dfrac{|x|}{x}$, $x = -1$, and $x = \dfrac{1}{2}$ is
a. $\dfrac{3}{\sqrt{2}} - \dfrac{11}{6}$ sq. units b. $3\sqrt{2} - \dfrac{11}{4}$ sq. units
c. $\dfrac{6}{\sqrt{2}} - \dfrac{11}{5}$ sq. units d. None of these

Multiple Correct Answers Type

Each question has four choices a, b, c, and d out of which *one or more* answers are correct.

1. Let $A(k)$ be the area bounded by the curves $y = x^2 - 3$ and $y = kx + 2$.
a. The range of $A(k)$ is $\left[\dfrac{10\sqrt{5}}{3}, \infty\right)$.
b. The range of $A(k)$ is $\left[\dfrac{20\sqrt{5}}{3}, \infty\right)$.
c. If function $k \to A(k)$ is defined for $k \in [-2, \infty)$, then $A(k)$ is many-one function.
d. The value of k for which area is minimum is 1.

2. The parabolas $y^2 = 4x$ and $x^2 = 4y$ divide the square region bounded by the lines $x = 4, y = 4$ and the co-ordinate axes. If S_1, S_2, and S_3 are the areas of these parts numbered from top to bottom, respectively, then
a. $S_1 : S_2 \equiv 1:1$ b. $S_2 : S_3 \equiv 1:2$
c. $S_1 : S_3 \equiv 1:1$ d. $S_1 : (S_1 + S_2) = 1:2$

3. Which of the following have the same bounded area
a. $f(x) = \sin x, g(x) = \sin^2 x$, where $0 \le x \le 10\pi$
b. $f(x) = \sin x, g(x) = |\sin x|$, where $0 \le x \le 20\pi$
c. $f(x) = |\sin x|, g(x) = \sin^3 x$, where $0 \le x \le 10\pi$
d. $f(x) = \sin x, g(x) = \sin^4 x$, where $0 \le x \le 10\pi$

4. If the curve $y = ax^{1/2} + bx$ passes through the point $(1, 2)$ and lies above the x-axis for $0 \le x \le 9$ and the area enclosed by the curve, the x-axis and the line $x = 4$ is 8 sq. units. Then
a. $a = 1$ b. $b = 1$
c. $a = 3$ d. $b = -1$

5. The area enclosed by the curves $x = a\sin^3 t$ and $y = a\cos^3 t$ is equal to

a. $12a\int_0^{\pi/2}\cos^4 t\sin^2 t\,dt$

b. $12a\int_0^{\pi/2}\cos^2 t\sin^4 t\,dt$

c. $2\int_{-a}^{a}\left(a^{2/3}-x^{2/3}\right)^{3/2}dx$

d. $4\int_0^{a}\left(a^{2/3}-x^{2/3}\right)^{3/2}dx$

6. If A_i is the area bounded by $|x-a_i|+|y|=b_i$, $i\in N$, where $a_{i+1}=a_i+\frac{3}{2}b_i$ and $b_{i+1}=\frac{b_i}{2}$, $a_1=0$, $b_1=32$, then

a. $A_3=128$

b. $A_3=256$

c. $\lim_{n\to\infty}\sum_{i=1}^{n}A_i=\frac{8}{3}(32)^2$

d. $\lim_{n\to\infty}\sum_{i=1}^{n}A_i=\frac{4}{3}(16)^2$

Reasoning Type

Each question has four choices a, b, c, and d out of which *only one* is correct. Each question contains Statement 1 and Statement 2.

a. If both the statements are True and Statement 2 is the correct explanation of Statement 1

b. If both the statements are True but Statement 2 is Not the correct explanation of Statement 1

c. If Statement 1 is True and Statement 2 is False.

d. If Statement 1 is False and Statement 2 is True.

1. **Statement 1:** The area bounded by $y=e^x$, $y=0$ and $x=0$ is 1 sq. units.
Statement 2: The area bounded by $y=\log_e x$, $x=0$ and $y=0$ is 1 sq. units.

2. $f(x)$ is a polynomial of degree 3 passing through origin having local extrema at $x=\pm 2$.
Statement 1: Ratio of areas in which $f(x)$ cuts the circle $x^2+y^2=36$ is 1:1.
Statement 2: Both $y=f(x)$ and the circle are symmetric about the origin.

3. **Statement 1:** The area bounded by parabola $y=x^2-4x+3$ and $y=0$ is 4/3 sq. units
Statement 2: The area bounded by curve $y=f(x)\geq 0$ and $y=0$ between ordinates $x=a$ and $x=b$ (where $b>a$) is $\int_a^b f(x)\,dx$.

4. **Statement 1:** The area enclosed between the parabolas $y^2-2y+4x+5=0$ and $x^2+2x-y+2=0$ is same as that of bounded by curves $y^2=-4x$ and $x^2=y$.
Statement 2: Shifting of origin to point (h, k) does not change the bounded area.

5. **Statement 1:** The area of the region bounded by the curve $2y=\log_e x$, $y=e^{2x}$ and the pair of lines $(x+y-1)\times(x+y-3)=0$ is $2k$ sq. units.
Statement 2: The area of the region bounded by the curves $y=e^{2x}$, $y=x$ and the pair of lines $x^2+y^2+2xy-4x-4y+3=0$ is k units.

6. **Consider two regions**
R_1 : Point P is nearer to (1, 0) than to $x=-1$.
R_2 : Point P is nearer to (0, 0) than to (8, 0).
Statement 1: Area of the region common to R_1 and R_2 is $\frac{128}{3}$ sq. units.
Statement 2: Area bounded by $x=4\sqrt{y}$ and $y=4$ is $\frac{32}{3}$ sq. units.

7. **Statement 1:** Area bounded by $2\geq\max.\{|x-y|, |x+y|\}$ is 8 sq. units.
Statement 2: Area of the square of side length 4 is 16 sq. units.

Linked Comprehension Type

Based upon each paragraph, three multiple choice questions have to be answered. Each question has four choices a, b, c, and d out of which *only one* is correct.

For Problems 1–2

Let A_r be the where area of the region bounded between the curves $y^2=(e^{-kr})x$ (where $k>0$, $r\in N$) and the line $y=mx$ (where $m\neq 0$), k and m are some constants.

1. $A_1, A_2, A_3, \ldots$ are in G.P. with common ratio

a. e^{-k}
b. e^{-2k}
c. e^{-4k}
d. None of these

2. $\lim_{n\to\infty}\sum_{i=1}^{n}A_i=\frac{1}{48(e^{2k}-1)}$, then the value of m is

a. 3
b. 1
c. 2
d. 4

For Problems 3–5

If $y=f(x)$ is a monotonic function in (a, b), then the area bounded by the ordinates at $x=a, x=b, y=f(x)$ and $y=f(c)$ (where $c\in(a, b)$) is minimum when $c=\frac{a+b}{2}$.

Proof: $A=\int_a^c (f(c)-f(x))\,dx+\int_c^b (f(x)-f(c))\,dx$

$=f(c)(c-a)-\int_a^c (f(x))\,dx+\int_c^b (f(x))\,dx-f(c)(b-c)$

$$\Rightarrow A = [2c-(a+b)]f(c) + \int_c^b (f(x))\,dx - \int_a^c (f(x))\,dx$$

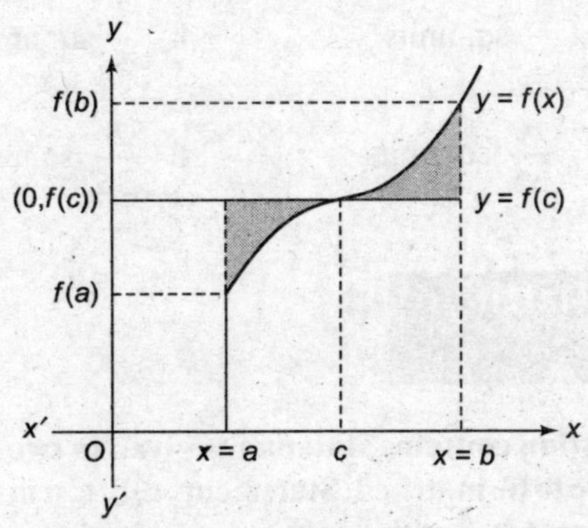

Fig. 9.1

Differentiating w.r.t. c, we get

$$\frac{dA}{dc} = [2c-(a+b)]f'(c) + 2f(c) + 0 - f(c) - (f(c) - 0)$$

For maxima and minima $\frac{dA}{dc} = 0$

$\Rightarrow f'(c)\,[2c-(a+b)] = 0$ (as $f'(c) \neq 0$)

Hence, $c = \frac{a+b}{2}$

also $c < \frac{a+b}{2}$, $\frac{dA}{dc} < 0$, and $c > \frac{a+b}{2}$, $\frac{dA}{dc} > 0$.

Hence, A is minimum when $c = \frac{a+b}{2}$.

3. If the area bounded by $f(x) = \frac{x^3}{3} - x^2 + a$ and the straight lines $x = 0$, $x = 2$, and the x-axis is minimum, then the value of a is
   a. 1/3 b. 2
   c. 1 d. 2/3
4. The value of the parameter a for which the area of the figure bounded by the abscissa axis, the graph of the function $y = x^3 + 3x^2 + x + a$ and the straight lines, which are parallel to the axis of ordinates and cut the abscissa axis at the point of extremum of the function, is the least, is
   a. 2 b. 0
   c. –1 d. 1
5. If the area enclosed by $f(x) = \sin x + \cos x$, $y = a$ between two consecutive points of extremum is minimum, then the value of a is
   a. 0 b. –1
   c. 1 d. 2

For Problems 6–8

Consider the areas S_0, S_1, S_2, ... bounded by the x-axis and half-waves of the curve $y = e^{-x}\sin x$, where $x \geq 0$.

6. The value of S_0 is
   a. $\frac{1}{2}(1+e^{\pi})$ sq. units
   b. $\frac{1}{2}(1+e^{-\pi})$ sq. units
   c. $\frac{1}{2}(1-e^{-\pi})$ sq. units
   d. $\frac{1}{2}(e^{\pi}-1)$ sq. units
7. The sequence $S_0, S_1, S_2, \ldots$, forms a G.P. with common ratio
   a. $\frac{e^{\pi}}{2}$ b. $e^{-\pi}$
   c. e^{π} d. $\frac{e^{-\pi}}{2}$
8. $\sum_{n=0}^{\infty} S_n$ is equal to
   a. $\frac{1+e^{\pi}}{1-e^{-\pi}}$ b. $\frac{\frac{1}{2}(1+e^{\pi})}{1-e^{-\pi}}$
   c. $\frac{1}{2(1-e^{-\pi})}$ d. None of these

For Problems 9–11

Two curves $C_1 \equiv [f(y)]^{2/3} + [f(x)]^{1/3} = 0$ and $C_2 \equiv [f(y)]^{2/3} + [f(x)]^{2/3} = 12$, satisfying the relation $f(x-y)f(x+y) - (x+y)f(x-y) = 4xy\,(x^2-y^2)$.

9. The area bounded by C_1 and C_2 is
   a. $2\pi - \sqrt{3}$ sq. units b. $2\pi + \sqrt{3}$ sq. units
   c. $\pi + \sqrt{6}$ sq. units d. $2\sqrt{3} - \pi$ sq. units
10. The area bounded by the curve C_2 and $|x| + |y| = \sqrt{12}$ is
    a. $12\pi - 2\sqrt{12}$ sq. units b. $6 - \sqrt{12}$ sq. units
    c. $2\sqrt{12} - 6$ sq. units d. None of these
11. The area bounded by C_1 and $x + y + 2 = 0$ is
    a. 5/2 sq. units b. 7/2 sq. units
    c. 9/2 sq. units d. None of these

For Problems 12–13

Consider the two curves C_1: $y = 1 + \cos x$ and C_2 : $y = 1 + \cos(x - \alpha)$ for $\alpha \in \left(0, \frac{\pi}{2}\right)$; where $x \in [0, \pi]$. Also the area of the figure bounded by the curves C_1, C_2, and $x = 0$ is same as that of the figure bounded by C_2, $y = 1$ and $x = \pi$.

12. The value of α is
    a. $\frac{\pi}{4}$ b. $\frac{\pi}{3}$
    c. $\frac{\pi}{6}$ d. $\frac{\pi}{8}$

13. For the values of α, area bounded by C_1, C_2, $x = 0$. and $x = \pi$ is

a. 1 sq. units
b. 2 sq. units
c. $2 + \sqrt{3}$ sq. units
d. None of these

For Problems 14–16

Consider the function defined implicitly by the equation $y^2 - 2ye^{\sin^{-1}x} + x^2 - 1 + [x] + e^{2\sin^{-1}x} = 0$ (where $[x]$ denotes the greatest integer function).

14. The area of the region bounded by the curve and the line $x = -1$ is

a. $\pi + 1$ sq. units
b. $\pi - 1$ sq. units
c. $\frac{\pi}{2} + 1$ sq. units
d. $\frac{\pi}{2} - 1$ sq. units

15. Line $x = 0$ divide the region mention above in two parts. The ratio of area of left-hand side of line to that of right-hand side of line is

a. $2 + \pi : \pi$
b. $2 - \pi : \pi$
c. 1:1
d. $\pi + 2 : \pi$

16. The area of the region of curve and line $x = 0$ and $x = 1/2$ is

a. $\frac{\sqrt{3}}{4} + \frac{\pi}{6}$ sq. units
b. $\frac{\sqrt{3}}{2} + \frac{\pi}{6}$ sq. units
c. $\frac{\sqrt{3}}{4} - \frac{\pi}{6}$ sq. units
d. $\frac{\sqrt{3}}{2} - \frac{\pi}{6}$ sq. units

For Problems 17–19

Computing area with parametrically represented boundaries: If the boundary of a figure is represented by parametric equation, i.e., $x = x(t)$, $y = y(t)$, then the area of the figure is evaluated by one of the three formulas

$$S = -\int_{\alpha}^{\beta} y(t)x'(t)\,dt,\quad S = \int_{\alpha}^{\beta} x(t)y'(t)\,dt,\quad S = \frac{1}{2}\int_{\alpha}^{\beta} (xy' - yx')\,dt,$$

where α and β are the values of the parameter t corresponding respectively to the beginning and the end of the traversal of the curve corresponding to increasing t.

17. The area of the region bounded by an arc of the cycloid $x = a(t - \sin t)$, $y = a(1 - \cos t)$ and the x-axis is

a. $6\pi a^2$ sq. units
b. $3\pi a^2$ sq. units
c. $4\pi a^2$ sq. units
d. None of these

18. The area of the loop described as $x = \frac{t}{3}(6 - t)$, $y = \frac{t^2}{8}(6 - t)$ is

a. $\frac{27}{5}$ sq. units
b. $\frac{24}{5}$ sq. units
c. $\frac{27}{6}$ sq. units
d. $\frac{21}{5}$ sq. units

19. If the curve given by parametric equation $x = t - t^3$, $y = 1 - t^4$ forms a loop for all values of $t \in [-1, 1]$, then the area of the loop is

a. $\frac{1}{7}$ sq. units
b. $\frac{3}{5}$ sq. units
c. $\frac{16}{35}$ sq. units
d. $\frac{8}{35}$ sq. units

Matrix-Match Type

Each question contains statements given in two columns which have to be matched. Statements a, b, c, d in column I have to be matched with statements p, q, r, s in column II. If the correct match are a→p, a→s, b→r, c→p, c→q and d→s, then the correctly bubbled 4 × 4 matrix should be as follows:

	p	q	r	s
a	(p)	q	r	(s)
b	p	q	(r)	s
c	(p)	(q)	r	s
d	p	q	r	(s)

1.

Column 1	Column 2
a. The area bounded by the curve $y = x\,\lvert x\rvert$, x-axis, and the ordinates $x = 1$, $x = -1$	p. 10/3 sq. units
b. The area of the region lying between the line $x - y + 2 = 0$ and the curve $x = \sqrt{y}$	q. 64/3 sq. units
c. The area enclosed between the curves $y^2 = x$ and $y = \lvert x\rvert$	r. 2/3 sq. units
d. The area bounded by parabola $y^2 = x$, straight line $y = 4$, and y-axis	s. 1/6 sq. units

2.

Column 1	Column 2
a. Area enclosed by $y = [x]$ and $y = \{x\}$, where $[\cdot]$ and $\{\cdot\}$ represent greatest integer and fractional part functions.	p. 16/5 sq. units

b. The area bounded by the curves $y^2 = x^3$ and $\lfloor y \rfloor = 2x$.	**q.** 1 sq. units
c. The smaller area included between the curves $\sqrt{x} + \sqrt{\lvert y \rvert} = 1$ and $\lvert x \rvert + \lvert y \rvert = 1$.	**r.** 4 sq. units
d. Area bounded by the curves $y = \left[\dfrac{x^2}{64} + 2\right]$ (where [.] denotes the greatest integer function), $y = x - 1$ and $x = 0$ above the x-axis.	**s.** 2/3 sq. units

3.

Column 1: [·] represents greatest integer function.	Column 2
a. Area enclosed by $[x]^2 = [y]^2$ for $1 \le x \le 4$	**p.** 8 sq. units
b. Area enclosed by $[\lvert x \rvert] + [\lvert y \rvert] = 2$	**q.** 6 sq. units
c. Area enclosed by $[\lvert x \rvert]\,[\lvert y \rvert] = 2$	**r.** 4 sq. units
d. Area enclosed by $\dfrac{[\lvert x \rvert]}{[\lvert y \rvert]} = 2$, $-5 \le x \le 5$	**s.** 12 sq. units

Integer Type

1. Consider the region formed by the lines $x = 0, y = 0, x = 2, y = 2$. The area enclosed by the curves $y = e^x$ and $y = \ln x$ within this region has been removed. If the area of the remaining region is $k(\ln 2 - 1)$, then the value of k is__________.
2. The area bounded by $y = \log_e x$, $y = -\log_e x$, $y = \log_e(-x)$, and $y = -\log_e(-x)$ is__________.
3. If the area bounded by $y = \sin^{-1}x$, $y = \cos^{-1} x$, and x-axis is k sq. units, the value of $[k]$, where [·] represents the greatest integer function, is__________.
4. If the area bounded by $y = \tan^{-1} x$, $y = \cot^{-1} x$, and y-axis in the first quadrant is $(1/2)\log k$ sq. units, then the value of k is__________.
5. The possible value of $b > 0$ so that the area of the bounded region enclosed between the parabolas $y = x - bx^2$ and $y = x^2/b$ is maximum is__________.
6. Let C_1 and C_2 be the graphs of the functions $y = x^2$ and $y = 2x$, $0 \le x \le 1$, respectively. Let C_3 be the graph of a function $y = f(x)$, $0 \le x \le 1$, $f(0) = 0$. For a point P on C_1, let the lines through P, parallel to the axes, meet C_2 and C_3 at Q and R, respectively (see figure). If for every position of P (on C_1), the areas of the shaded regions OPQ and ORP are equal, then the value of $f(2)$ is__________.

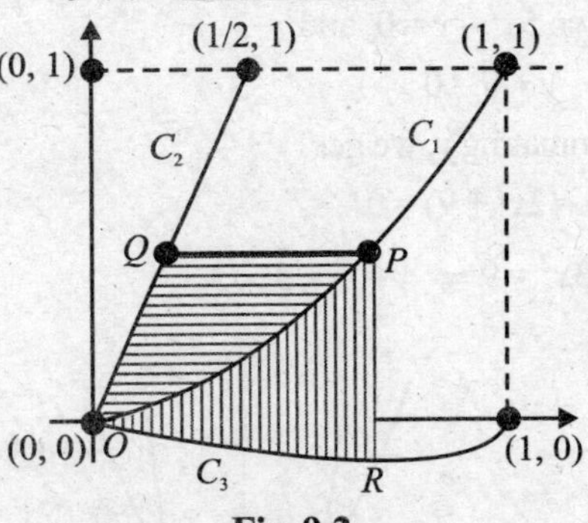

Fig. 9.2

7. Let $O(0, 0)$, $A(2, 0)$, and $B(1, 1/\sqrt{3})$ be the vertices of a triangle. Let R be the region consisting of all those points P inside ΔOAB that satisfy $d(P, OA) \le \min[d(P, OB), d(P, AB)]$, where d denotes the distance from the point to the corresponding line. If the area of region R is k sq. units, then the value of $(k - 2)^2$ is__________.
8. The area bounded by the curves $y = \sqrt{x}$, $2y + 3 = x$, and x-axis in the first quadrant is__________.
9. Find the area bounded by the x-axis, part of the curve $y = 1 + (8/x^2)$, and the ordinates at $x = 2$ and $x = 4$. If the ordinates at $x = a$ divide the area into two equal parts, then the value of a^2 is__________.
10. If the area bounded by the curves $y = x^2$ and $y = 2/(1 + x^2)$ is k sq. units, then the value of $[k]$ is__________.

ANSWERS AND SOLUTIONS

Objective Type

1. a. Clearly t can be any real number

Let $t = \tan\theta \Rightarrow x = \dfrac{1-\tan^2\theta}{1+\tan^2\theta}$

$\Rightarrow x = \cos 2\theta$, and $y = \dfrac{2\tan\theta}{1+\tan^2\theta} = \sin 2\theta$

$\Rightarrow x^2 + y^2 = 1$

Thus, required area $= \pi$ sq. units.

2. a. Given $5x^2 - y = 0$, and (1)

$2x^2 - y + 9 = 0$ (2)

Eliminating y, we get

$5x^2 - (2x^2 + 9) = 0$

$\Rightarrow 3x^2 = 9 \Rightarrow x = -\sqrt{3}, \sqrt{3}$

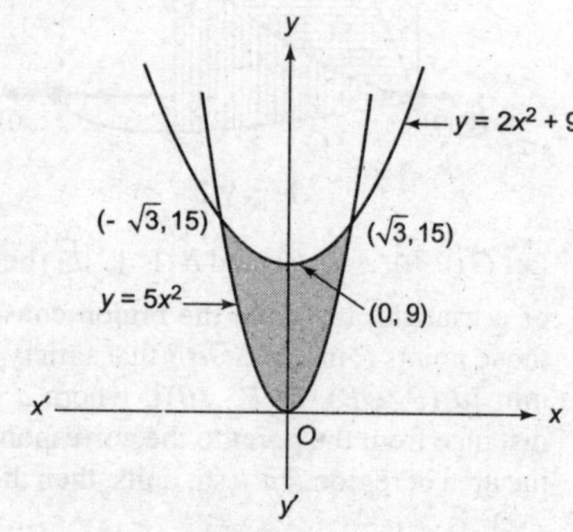

Fig. 9.3

Therefore, required area

$= 2\int_0^{\sqrt{3}} \left((2x^2+9) - 5x^2\right) dx$

$= 2\int_0^{\sqrt{3}} \left(9 - 3x^2\right) dx$

$= 2\left[9x - x^3\right]_0^{\sqrt{3}}$

$= 2\left[9\sqrt{3} - 3\sqrt{3}\right] = 12\sqrt{3}$ sq. units

3. d.

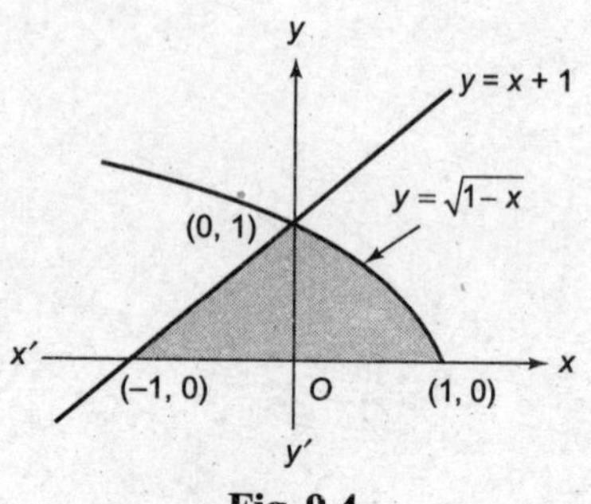

Fig. 9.4

Required area = shaded region

$$= \int_0^1 (x_2 - x_1)dy$$

(integrating along y-axis)

$$= \int_0^1 [(1-y^2) - (y-1)]dy$$

$$= \frac{7}{6} \text{ sq. unit}$$

4. a.

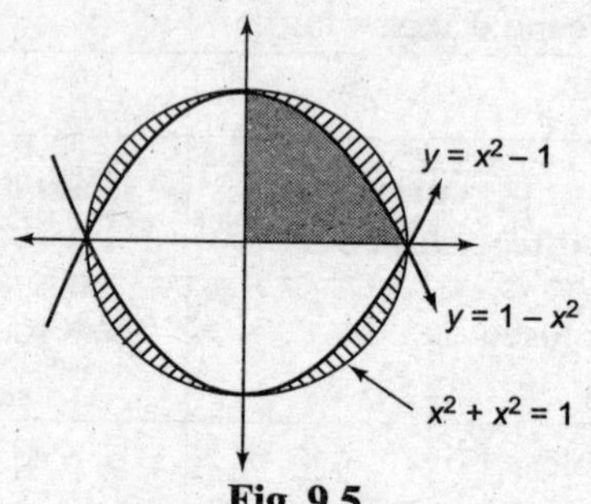

Fig. 9.5

The dotted area is

$$A = \int_0^1 (1-x^2)dx = \left(x - \frac{x^3}{3}\right)_0^1 = 1 - \frac{1}{3} = \frac{2}{3}.$$

Hence, area bounded by circle $x^2 + y^2 = 1$ and $|y| = 1 - x^2$

= lined area

= Area of circle – area bounded by $|y| = 1 - x^2$

$= \pi - 4 \times \left(\dfrac{2}{3}\right) = \dfrac{3\pi - 8}{3}$ sq. units.

5. a. Curve tracing : $y = x\log_e x$

Clearly, $x > 0$,

For $0 < x < 1$, $x\log_e x < 0$, and for $x > 1$, $x\log_e x > 0$

Also $x\log_e x = 0 \Rightarrow x = 1$.

Further, $\dfrac{dy}{dx} = 0 \Rightarrow 1 + \log_e x = 0 \Rightarrow x = 1/e$, which is a point of minima.

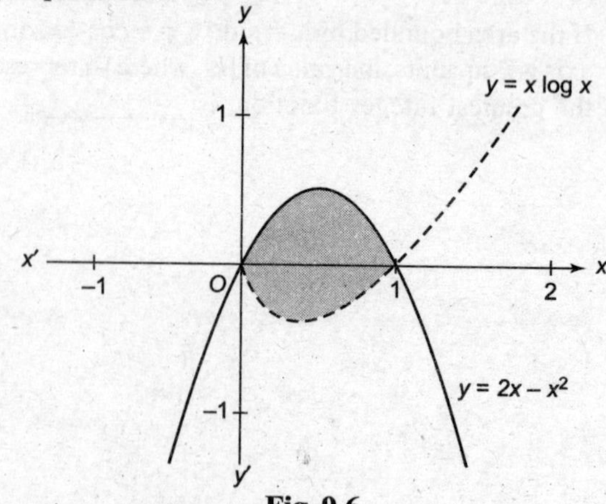

Fig. 9.6

Required area

$$= \int_0^1 (2x - 2x^2)dx - \int_0^1 x \log x \, dx$$

$$= \left[x^2 - \frac{2x^3}{3}\right]_0^1 - \left[\frac{x^2}{2}\log x - \frac{x^2}{4}\right]_0^1$$

$$= \left(1 - \frac{2}{3}\right) - \left[0 - \frac{1}{4} - \frac{1}{2}\lim_{x\to 0} x^2 \log x\right]$$

$$= \frac{1}{3} + \frac{1}{4} = \frac{7}{12}$$

6. a. $y = \log_e(x + e), x = \log_e\left(\frac{1}{y}\right) \Rightarrow y = e^{-x}$.

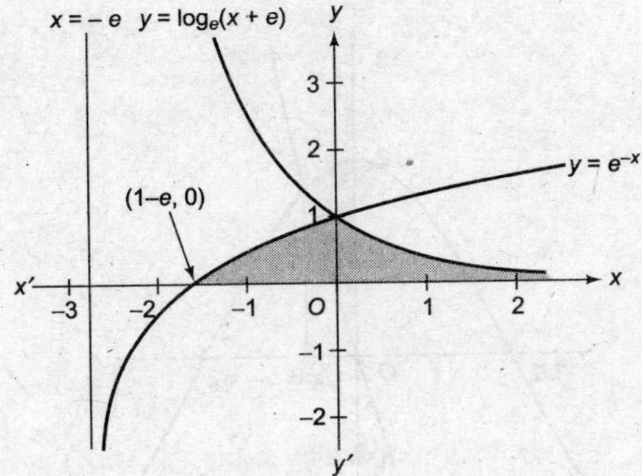

Fig. 9.7

$$\text{Required area} = \int_{1-e}^{0} \log_e(x+e)dx + \int_0^{\infty} e^{-x}\, dx$$

$$= |x \log_e (x-e)|_{1-e}^{0} - \int_{1-e}^{0} \frac{x}{x+e}dx - |e^{-x}|_0^{\infty}$$

$$= \int_0^{1-e}\left(1 - \frac{e}{x+e}\right)dx - e^{-\infty} + e^0$$

$$= |x - e\log(x+e)|_0^{1-e} - 0 + 1$$

$$= 1 - e + e \log e + 1 = 2 \text{ sq. units.}$$

7. d $y = \dfrac{1}{(x-1)^2 + 1}$

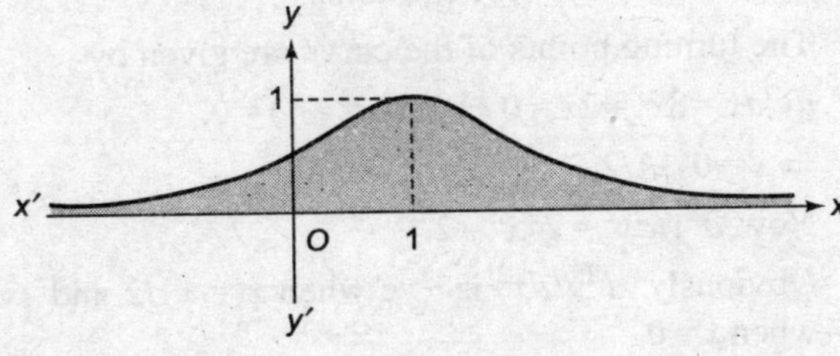

Fig. 9.8

$$\text{Area} = 2\int_1^{\infty} \frac{1}{(x-1)^2+1}dx = 2[\tan^{-1}(x-1)]_1^{\infty}$$

$$= \pi \text{ sq. units.}$$

8. b $xy^2 = a^2(a - x)$

$$\Rightarrow x = \frac{a^3}{y^2 + a^2}$$

The given curve is symmetrical about the x-axis, and meets it at $(a, 0)$.

The line $x = 0$, i.e., y-axis is an asymptote.

$$\text{Area} = \int_0^{\infty} x\, dy = 2\int_0^{\infty} \frac{a^3}{y^2 + a^2}dx$$

$$= 2a^2 \frac{1}{a}\left[\tan^{-1}\frac{y}{a}\right]_0^{\infty} = 2a^2 \frac{\pi}{2} = \pi a^2 \text{ sq. units.}$$

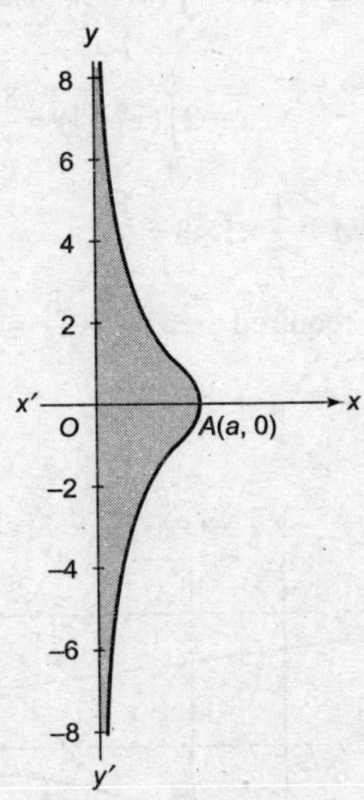

Fig. 9.9

9. c. Given $y = x^2 + x + 1 = \left(x + \frac{1}{2}\right)^2 + \frac{3}{4}$

$$\Rightarrow y - \frac{3}{4} = \left(x + \frac{1}{2}\right)^2.$$

This is a parabola with vertex at $\left(-\frac{1}{2}, \frac{3}{4}\right)$ and the curve is concave upwards.

$$y = x^2 + x + 1 \Rightarrow \frac{dy}{dx} = 2x + 1 \Rightarrow \left(\frac{dy}{dx}\right)_{(1,3)} = 3$$

Equation of the tangent at A (1, 3) is $y = 3x$.

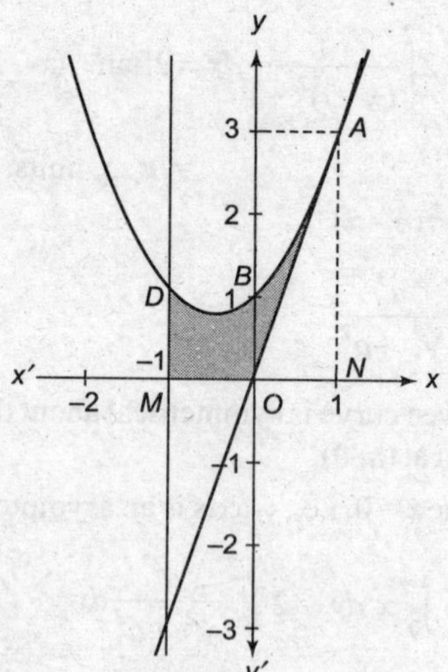

Fig. 9.10

Required (shaded) area = area $ABDMN$ – area ONA

Now, area $ABDMN = \int_{-1}^{1} (x^2 + x + 1)dx$

$= 2\int_{0}^{1} (x^2 + 1) = \frac{8}{3}$

Area of $ONA = \frac{1}{2} \times 1 \times 3 = \frac{3}{2}$.

Therefore, required area $= \frac{8}{3} - \frac{3}{2} = \frac{16-9}{6} = \frac{7}{6}$ sq. units.

10. a.

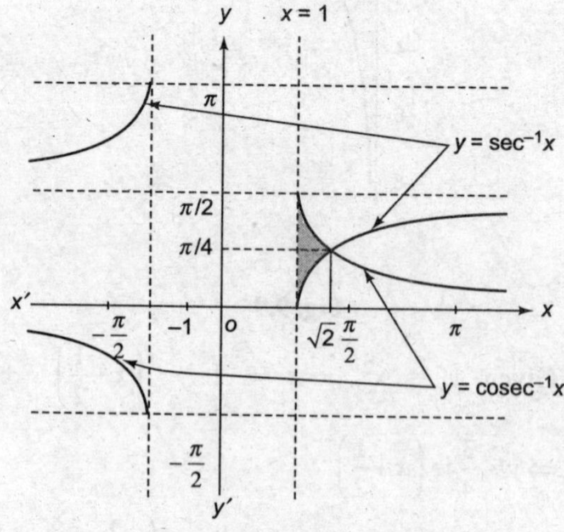

Fig. 9.11

Integrating along the x-axis, we get

$$A = \int_{1}^{\sqrt{2}} \left(\text{cosec}^{-1} x - \sec^{-1} x\right) dx$$

Integrating along the y-axis, we get

$$A = 2\int_{0}^{\pi/4} (\sec y - 1)\, dy$$

$$= 2\left[\log|\sec y + \tan y| - y\right]_0^{\pi/4}$$

$$= 2\left[\log\left|\sqrt{2}+1\right| - \frac{\pi}{4}\right]$$

$$= \log(3 + 2\sqrt{2}) - \frac{\pi}{2} \text{ sq. units.}$$

11. b. Solving $2\cos x = 3\tan x$, we get

$$2 - 2\sin^2 x = 3\sin x \Rightarrow \sin x = \frac{1}{2} \Rightarrow x = \frac{\pi}{6}.$$

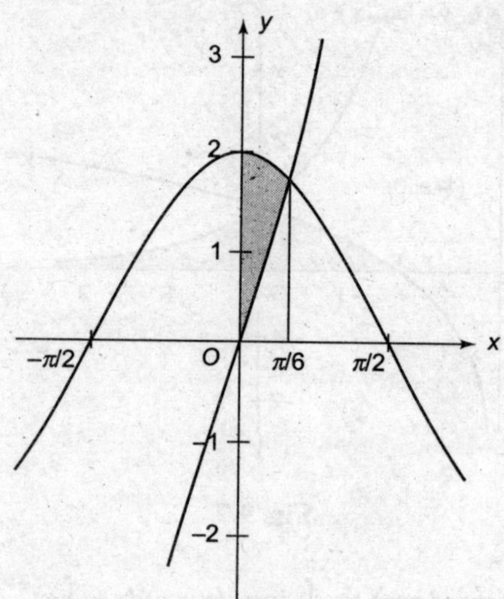

Fig. 9.12

$$\text{Required area} = \int_{0}^{\pi/6} (2\cos x - 3\tan x)dx$$

$$= 2\sin x - 3\log\sec x\Big|_0^{\pi/6}$$

$$= 1 - 3\ln 2 + \frac{3}{2}\ln 3 \text{ sq. unit}$$

12. b. The curve is $y = 2x^4 - x^2 = x^2(2x^2 - 1)$

The curve is symmetrical about the axis of y.

The curve passes through the origin and the tangent at the origin is $y = 0$, i.e., x-axis.

The turning points of the curve are given by $dy/dx = 8x^3 - 2x = 0 \Rightarrow 2x(4x^2 - 1) = 0$.

$\Rightarrow x = 0, \pm 1/2$

Now, $d^2y/dx^2 = 24x^2 - 2$.

Obviously, d^2y/dx^2 is +ve when $x = \pm 1/2$ and –ve when $x = 0$.

Thus, the ordinates for the two minima of the curve are $x = -1/2$ and $x = 1/2$.

At $x = -1/2$ or $1/2$, min $y = -1/8$.

The curve intersects the axes at $O(0, 0)$, $A(-1/\sqrt{2}, 0)$, and $B(1/\sqrt{2}, 0)$.

Thus, the graph of the curve is shown in the figure.

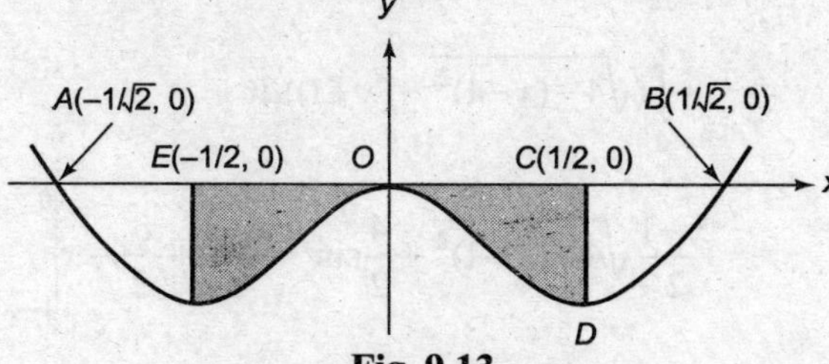

Fig. 9.13

Here, $y \le 0$, as x varies from $x = -1/2$ to $x = 1/2$

$\therefore$ The required area = 2 Area $OCDO$

$$= 2\left|\int_0^{1/2} y\,dx\right|$$

$$= 2\left|\int_0^{1/2} (2x^4 - x^2)\,dx\right|$$

$= 7/120$ sq. units

13. d. The curve is $y = \dfrac{x^2(x+a)}{a^2}$, which is a cubic polynomial.

Since $\dfrac{x^2(x+a)}{a^2} = 0$ has repeated root $x = 0$, it touches the x-axis at $(0, 0)$ and intersects at $(1, 0)$.

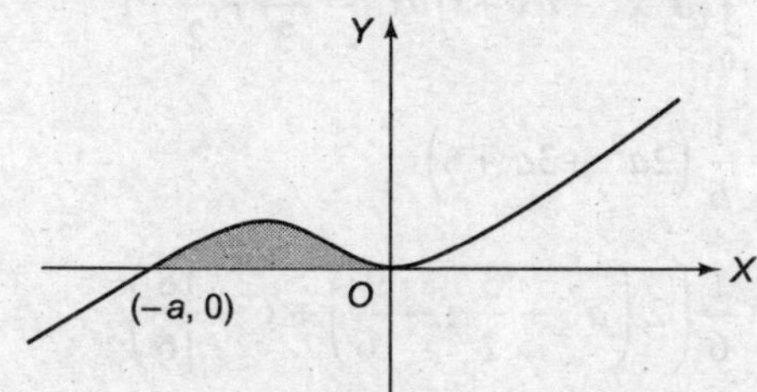

Fig. 9.14

Required area $= \displaystyle\int_{-a}^{0} y\,dx = \int_{-a}^{0}\left[\frac{x^2(x+a)}{a^2}\right]dx$

$= a^2/12$ sq. units.

14. c.

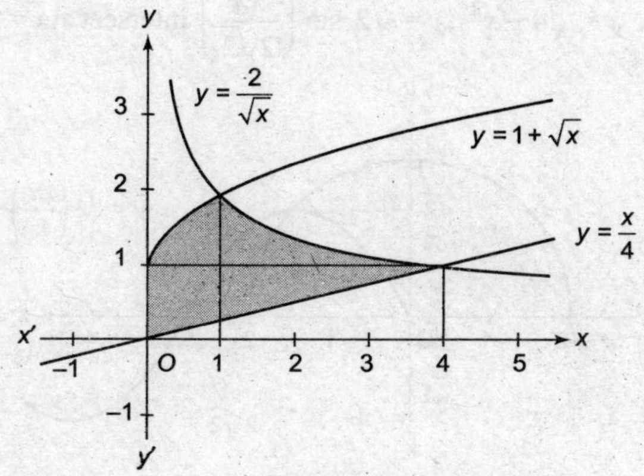

Fig. 9.15

$$A_1 = \int_0^1 \left(1 + \sqrt{x} - \frac{x}{4}\right)dx$$

$$= \left[x + \frac{2x^{3/2}}{3} - \frac{x^2}{8}\right]_0^1 = 1 + \frac{2}{3} - \frac{1}{8} = \frac{37}{24}.$$

$$A_2 = \int_1^4 \left(\frac{2}{\sqrt{x}} - \frac{x}{4}\right)dx$$

$$= \left[4\sqrt{x} - \frac{x^2}{8}\right]_1^4$$

$$= \left[8 - 2 - 4 + \frac{1}{8}\right] = \frac{17}{8}$$

$$\Rightarrow A = A_1 + A_2 = = \frac{88}{24} = \frac{11}{3} \text{ sq. units}$$

15. a. $y = \dfrac{x^2}{2} - 2x + 2 = \dfrac{(x-2)^2}{2}$,

$$\frac{dy}{dx} = x - 2,\ \left(\frac{dy}{dx}\right)_{x=1} = -1,\ \left(\frac{dy}{dx}\right)_{x=4} = 2$$

$\Rightarrow$ Tangent at $(1, 1/2)$ is $y - 1/2 = -1(x-1)$ or $2x + 2y - 3 = 0$

Tangent at $(4, 2)$ is $y - 2 = 2(x-4)$ or $2x - y - 6 = 0$

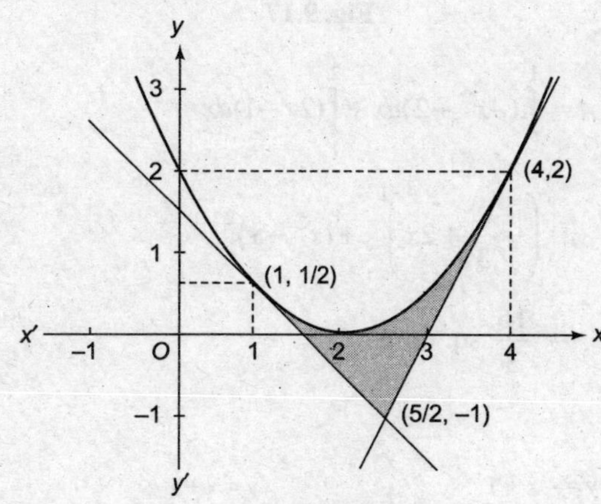

Fig. 9.16

$$\text{Hence, } A = \int_1^{5/2}\left(\frac{x^2}{2} - 2x + 2 - \frac{3-2x}{2}\right)dx$$

$$+ \int_{5/2}^{4}\left(\frac{x^2}{2} - 2x + 2 - (2x-6)\right)dx$$

$$= \int_1^4\left(\frac{x^2}{2} - 2x + 2\right)dx - \int_1^{5/2}\left(\frac{3-2x}{2}\right)dx$$

$$- \int_{5/2}^{4} (2x-6)\,dx$$

$$= \left(\frac{x^3}{6} - x^2 + 2x\right)_1^4 - \frac{1}{2}\left(3x - x^2\right)_1^{5/2} - \left(x^2 - 6x\right)_{5/2}^4$$

$$= \left(\frac{63}{6} - 15 + 6\right) - \frac{1}{2}\left(3 \times \frac{3}{2} - \left(\frac{25}{4} - 1\right)\right)$$

$$- \left(\left(16 - \frac{25}{4}\right) - 6\left(4 - \frac{5}{2}\right)\right)$$

$$= \frac{3}{2} - \frac{1}{2}\left(\frac{9}{2} - \frac{21}{4}\right) - \left(\frac{39}{4} - 6\left(\frac{3}{2}\right)\right) = \frac{9}{8} \text{ sq. units}$$

16. a.

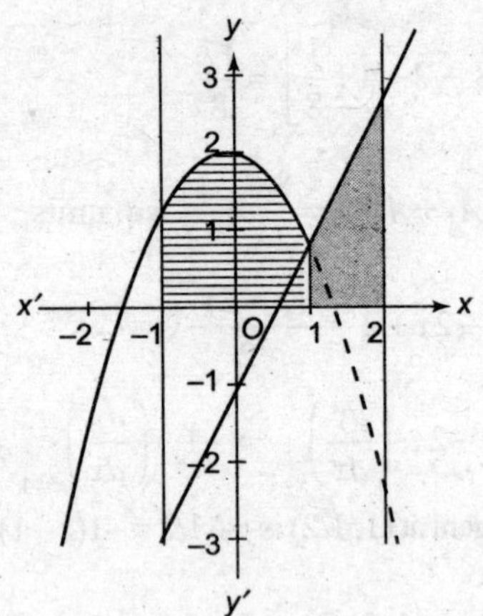

Fig. 9.17

$$A = \int_{-1}^{1} (-x^2 + 2)dx + \int_{1}^{2} (2x - 1)dx$$

$$= \left(-\frac{x^3}{3} + 2x\right)_{-1}^{1} + (x^2 - x)_1^2$$

$$= \frac{16}{3} \text{ sq. units}$$

17. a.

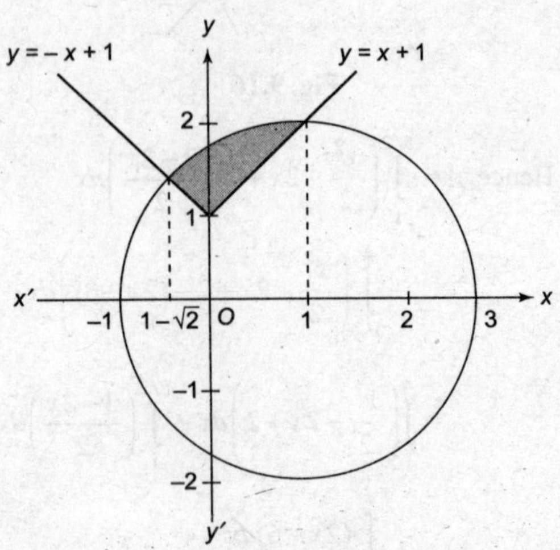

Fig. 9.18

$$x^2 + y^2 - 2x - 3 = 0$$

$$\Rightarrow (x-1)^2 + y^2 = 4$$

$$A = \int_{1-\sqrt{2}}^{0} (\sqrt{4-(x-1)^2} - (-x+1))dx$$

$$+ \int_{0}^{1} (\sqrt{4-(x-1)^2} - (x+1))dx$$

$$= \frac{x-1}{2}\sqrt{4-(x-1)^2} + \frac{4}{2}\sin^{-1}\frac{x-1}{2} + \frac{x^2}{2} - x\Bigg|_{1-\sqrt{2}}^{0}$$

$$+ \frac{x-1}{2}\sqrt{4-(x-1)^2} + \frac{4}{2}\sin^{-1}\frac{x-1}{2} - \frac{x^2}{2} - x\Bigg|_{0}^{1}$$

$$= \left(-\frac{\sqrt{3}}{2} - \frac{\pi}{3}\right) - \left(\frac{-\sqrt{2}}{2}\sqrt{2} - \frac{\pi}{2} + \frac{3-2\sqrt{2}}{2} - 1 + \sqrt{2}\right)$$

$$+ \left(-\frac{1}{2} - 1\right) - \left(-\frac{\sqrt{3}}{2} - \frac{\pi}{3}\right)$$

$$= -\left(-1 - \frac{\pi}{2} + \frac{3}{2} - \sqrt{2} - 1 + \sqrt{2}\right) - \frac{3}{2} = \frac{\pi}{2} - 1 \text{ sq. unit.}$$

18. c. $a^2x^2 + ax + 1$ is clearly positive for all real values of x.
Area under consideration

$$A = \int_{0}^{1} \left(a^2x^2 + ax + 1\right)dx = \frac{a^2}{3} + \frac{a}{2} + 1$$

$$= \frac{1}{6}\left(2a^2 + 3a + 6\right)$$

$$= \frac{1}{6}\left(2\left(a^2 + \frac{3}{2}a + \frac{9}{16}\right) + 6 - \frac{18}{16}\right)$$

$$= \frac{1}{6}\left(2\left(a + \frac{3}{4}\right)^2 + \frac{39}{8}\right),$$

which is clearly minimum for $a = -\dfrac{3}{4}$.

19. d. $y = \sqrt{4 - x^2}$, $y = \sqrt{2}\sin\left(\dfrac{x\pi}{2\sqrt{2}}\right)$ intersect at $x = \sqrt{2}$

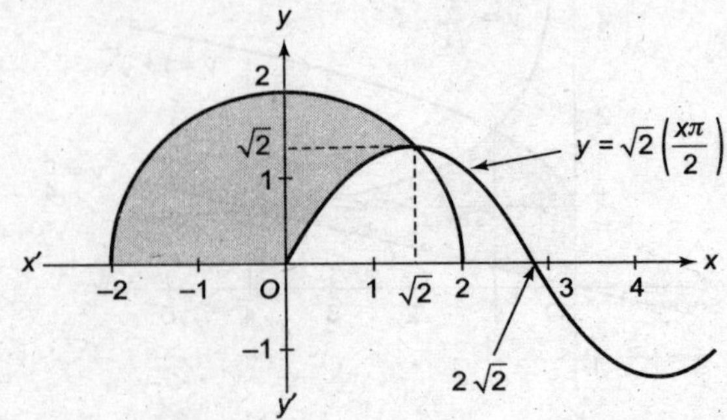

Fig. 9.19

Area of the left of y-axis is π.

Area of the right of y-axis

$$= \int_0^{\sqrt{2}} \left(\sqrt{4-x^2} - \sqrt{2} \sin \frac{x\pi}{2\sqrt{2}} \right) dx$$

$$= \left(\frac{x\sqrt{4-x^2}}{2} + \frac{4}{2} \sin^{-1} \frac{x}{2} \right)_0^{\sqrt{2}} + \left(\frac{4}{\pi} \cos \frac{x\pi}{2\sqrt{2}} \right)_0^{\sqrt{2}}$$

$$= \left(1 + 2\times\frac{\pi}{4} \right) + \frac{4}{\pi}(0-1)$$

$$= 1 + \frac{\pi}{2} - \frac{4}{\pi}$$

$$= \frac{2\pi - \pi^2 - 8}{2\pi} \text{ sq. units.}$$

$$\therefore \text{ratio} = \frac{2\pi^2}{2\pi + \pi^2 - 8}.$$

20. b. $f(x) = \sin x$

$f(x) + f(\pi - x) = 2$

$f(x) = 2 - f(\pi - x) = 2 - \sin(\pi - x) = 2 - \sin x$, where $x \in \left(\frac{\pi}{2}, \pi\right]$

$f(x) = f(2\pi - x) = 2 - \sin(2\pi - x)$, where $x \in \left(\pi, \frac{3\pi}{2}\right]$

$f(x) = f(2\pi - x) = -\sin x$, where $x \in \left(\frac{3\pi}{2}, 2\pi\right]$

$$f(x) = \begin{cases} \sin x, x \in \left[0, \frac{\pi}{2}\right] \\ 2 - \sin x, x \in \left(\frac{\pi}{2}, \pi\right] \\ 2 + \sin x, x \in \left(\pi, \frac{3\pi}{2}\right] \\ -\sin x, x \in \left(\frac{3\pi}{2}, 2\pi\right] \end{cases}$$

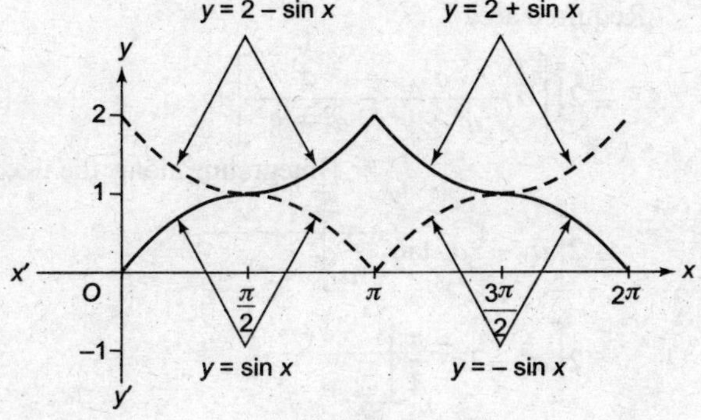

Fig. 9.20

$$\text{Area} = \int_0^{\pi/2} \sin x \sin dx + \int_{\pi/2}^{\pi} (2 - \sin x)\, dx + \int_{\pi}^{3\pi/2} (2 + \sin x)\, dx + \int_{3\pi/2}^{2\pi} (-\sin x)\, dx$$

$$= 1 + 2\times\frac{\pi}{2} - 1 + 2.\frac{\pi}{2} - 1 + 1 = 2\pi \text{ sq. units.}$$

21. a.

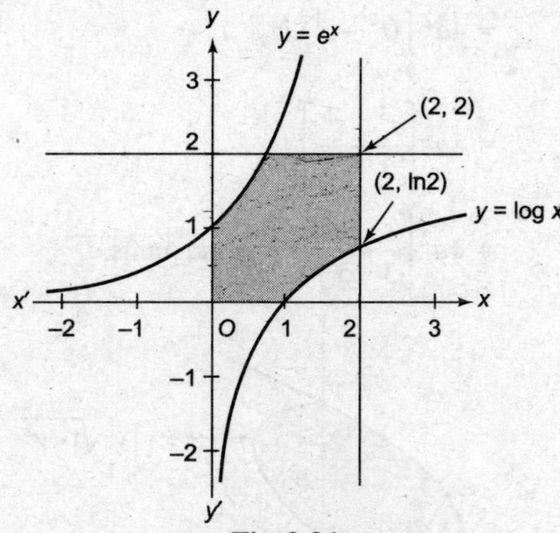

Fig. 9.21

$$A = \int_1^2 \ln x\, dx$$

$$= [x \log x - x]_1^2$$

$$= 2 \log 2 - 1$$

$\Rightarrow$ Required area $= 4 - 2(2 \ln 2 - 1)$

$\Rightarrow 6 - 4 \ln 2$ sq. units.

22. b. $ay^2 = x^2(a - x) \Rightarrow y = \pm x \sqrt{\frac{a-x}{a}}$

Curve tracing : $y = x \sqrt{\frac{a-x}{a}}$

We must have $x \le a$

For $0 < x \le a, y > 0$ and for $x < 0, y < 0$

Also $y = 0 \Rightarrow x = 0, a$

Curve is symmetrical about x-axis.

When $x \to -\infty, y \to -\infty$

Also, it can be verified that y has only one point of maxima for $0 < x < a$.

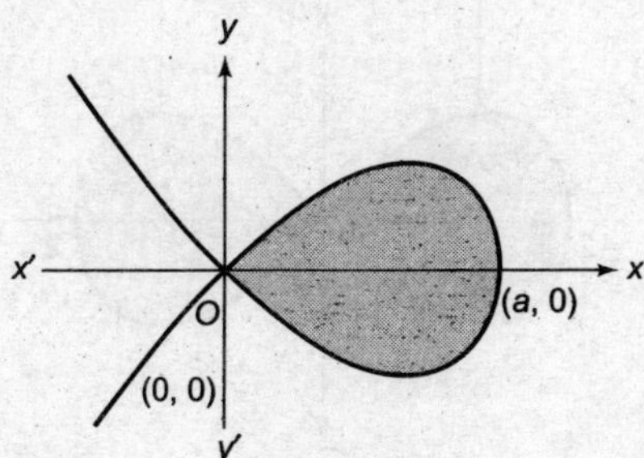

Fig. 9.22

$$\text{Area} = 2\int_0^a x\sqrt{\frac{a-x}{a}}\,dx$$

$$\sqrt{\frac{a-x}{a}} = t \Rightarrow 1-\frac{x}{a} = t^2 \Rightarrow x = a(1-t^2)$$

$$\Rightarrow A = 2\int_1^0 a(1-t^2)t(-2at)dt$$

$$= 4a^2\int_0^1 (t^2-t^4)dt$$

$$= 4a^2\left[\frac{t^3}{3}-\frac{t^5}{5}\right]_0^1$$

$$= 4a^2\left[\frac{1}{3}-\frac{1}{5}\right] = \frac{8a^2}{15} \text{ sq. units.}$$

23. d.

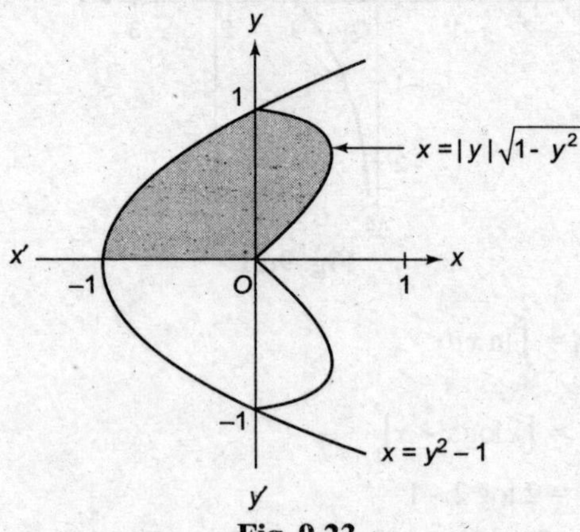

Fig. 9.23

$$A = 2\int_0^1\left[y\sqrt{1-y^2} - (y^2-1)\right]dy = 2 \text{ sq. units}$$

24. d. $4y^2 = x^2(4-x^2)$ (1)

$$\Rightarrow y = \pm\frac{1}{2}\sqrt{x^2(4-x^2)}$$

$$\Rightarrow y = \pm\frac{x}{2}\sqrt{(4-x^2)}$$

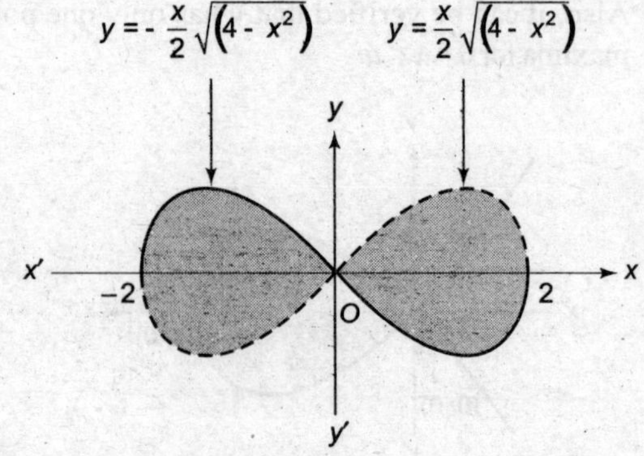

Fig. 9.24

$$\therefore \text{Area } (A) = 4\times\int_0^2 \frac{2x}{2\times 2}\sqrt{(4-x^2)}\,dx$$

Let $4-x^2 = t \Rightarrow -2x\,dx = dt$

$$\Rightarrow A = \frac{-4}{4}\int_4^0 \sqrt{t}\,dt = \int_0^4 \sqrt{t}\,dt$$

$$= \left[\frac{t^{3/2}}{3/2}\right]_0^4 = \frac{2}{3}\times[\sqrt{64}-0]$$

$$\Rightarrow A = \frac{16}{3} \text{ sq.units}$$

25. a. The two curves are

$$xy^2 = a^2(a-x) \Rightarrow x = \frac{a^3}{a^2+y^2} \quad (1)$$

and $(a-x)y^2 = a^2x$

$$\Rightarrow x = \frac{ay^2}{a^2+y^2} = \frac{ay^2+a^3-a^3}{a^2+y^2} = a - \frac{a^3}{a^2+y^2} \quad (2)$$

Curve (1) is symmetrical about the x-axis, and have y-axis as the asymptote.
Curve (2) is symmetrical about x-axis, tangent at origin as y-axis and the asymptote $x = a$.
The two curves intersect at the point $P(a/2, a)$ and $Q(a/2, -a)$.

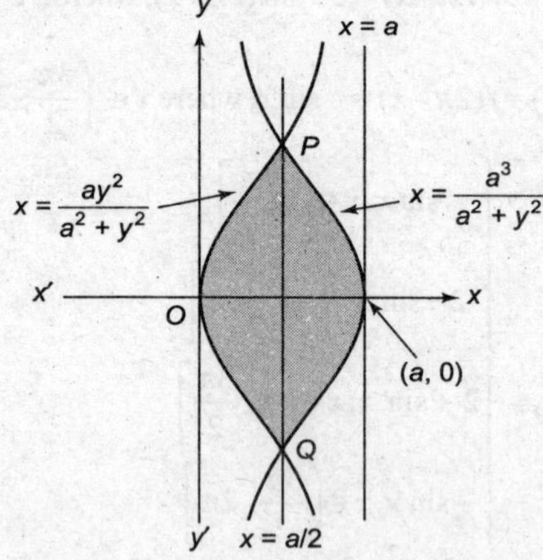

Fig. 9.25

Required area

$$= 2\int_0^a\left[a - \frac{a^3}{a^2+y^2} - \frac{a^3}{a^2+y^2}\right]dy$$

[Integrating along the y-axis]

$$= 2\left[ay - 2a^2\tan^{-1}\frac{y}{a}\right]_0^a$$

$$= 2\left[a^2 - 2a^2\frac{\pi}{4}\right]$$

$$= (\pi-2)a^2 \text{ sq. units.}$$

26. a. Curve tracing : $y = x e^x$

Let $\frac{dy}{dx} = 0 \Rightarrow e^x + xe^x = 0 \Rightarrow x = -1$.

Also, at $x = -1$, $\frac{dy}{dx}$ changes sign from negative to positive,

hence, $x = -1$ is a point of minima.

When $x \to \infty, y \to \infty$

Also $\lim_{x \to -\infty} xe^x = \lim_{x \to -\infty} \frac{x}{e^{-x}} = \lim_{x \to -\infty} \frac{1}{-e^{-x}} = 0$

With similar types of arguments, we can draw the graph of $y = x e^{-x}$.

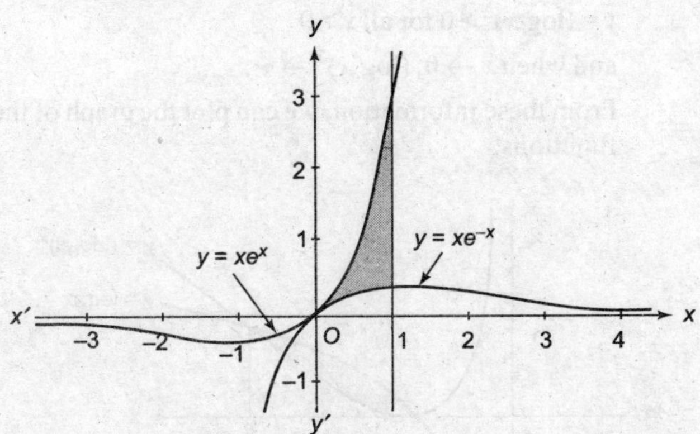

Fig. 9.26

Required area

$$= \int_0^1 xe^x dx - \int_0^1 xe^{-x} dx$$

$$= \left[xe^x\right]_0^1 - \int_0^1 e^x dx - \left(\left[-xe^{-x}\right]_0^1 + \int_0^1 e^{-x} dx\right)$$

$$= e - (e-1) - \left(-e^{-1} - (e^{-1} - 1)\right) = \frac{2}{e} \text{ sq. units.}$$

27. c. Given parabola is $(y-2)^2 = x - 1$

$$\Rightarrow \frac{dy}{dx} = \frac{1}{2(y-2)}$$

When $y = 3, x = 2$

$$\frac{dy}{dx} = \frac{1}{2(3-2)} = \frac{1}{2}. \text{ Therefore,}$$

Tangent at (2, 3) is $y - 3 = \frac{1}{2}(x-2) \Rightarrow x - 2y + 4 = 0$

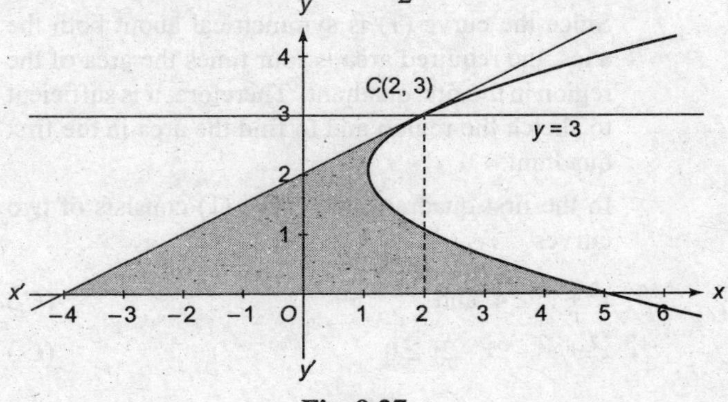

Fig. 9.27

Required area

$$= \int_0^3 \left((y-2)^2 + 1\right) dy - \int_0^3 (2y-4)\, dy$$

$$= \left|\frac{(y-2)^3}{3} + y\right|_0^3 - \left|y^2 - 4y\right|_0^3$$

$$= \frac{1}{3} + 3 + \frac{8}{3} - (9 - 12) = 9 \text{ sq. units.}$$

28. c.

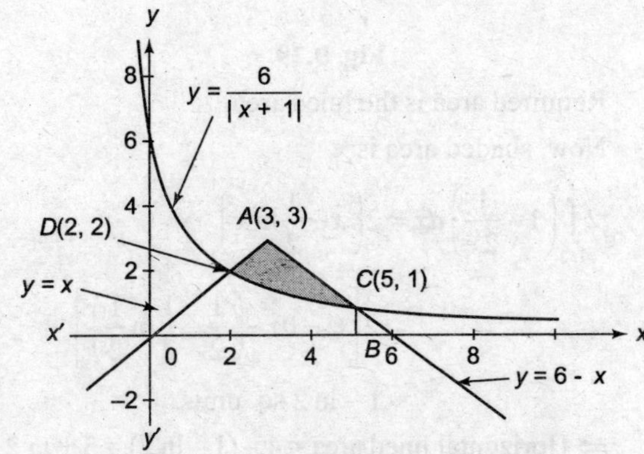

Fig. 9.28

First consider $y = 3 - |3 - x|$

For $x < 3$; $y = 3 - (3 - x) = x$

For $x \geq 3$; $y = 3 - (x - 3) = 6 - x$

Consider $y = \frac{6}{|x+1|}$

For $x < -1$; $y = \frac{6}{-1-x}$

$\Rightarrow (1 + x)y = -6$

For $x > -1$; $y = \frac{6}{x+1}$

Required area

$$= \left[\int_2^3 \left(x - \frac{6}{x+1}\right) dx + \int_3^5 \left((6-x) - \frac{6}{x+1}\right) dx\right]$$

$$= \left[\left(\frac{x^2}{2}\right)_2^3 + \left(6x - \frac{x^2}{2}\right)_3^5 - \left(6\log(x+1)\right)_2^5\right]$$

$$= \left[\frac{5}{2} + 4 - 6\log 2\right]$$

$$= \frac{13}{2} - 6 \ln 2 \text{ sq. units.}$$

29. b. $\max(|x|, |y|) \leq 1 \Rightarrow |x| \leq 1$, and $|y| \leq 1$

which represent a square bounded by $x = \pm 1$ and $y = \pm 1$

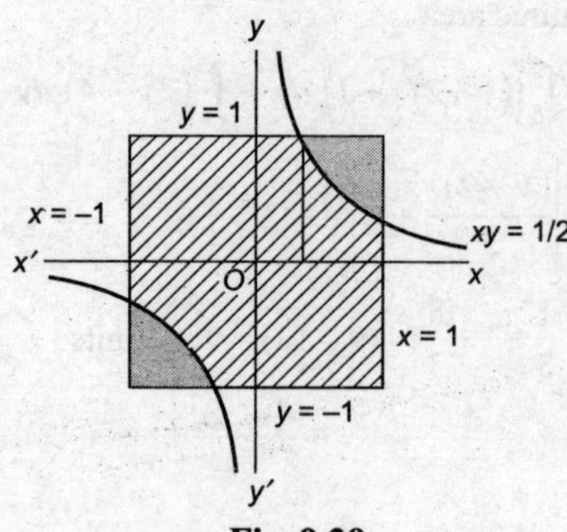

Fig. 9.29

Required area is the lined area.

Now, shaded area is

$$2\int_{1/2}^{1}\left(1-\frac{1}{2x}\right)dx = 2\left(x-\frac{1}{2}\ln x\right)_{1/2}^{1}$$

$$= 2\left[(1-0)-\left(\frac{1}{2}-\frac{1}{2}\ln\frac{1}{2}\right)\right]$$

$$= 1 - \ln 2 \text{ sq. units.}$$

$\Rightarrow$ Horizontal lined area $= 4 - (1 - \ln 2) = 3 + \ln 2$ sq. units.

30. c. $(y-x)^2 = x^3$, where $x \geq 0$

$\Rightarrow \quad y - x = \pm x^{3/2}$

$\Rightarrow \quad y = x + x^{3/2}$ (1)

$y = x - x^{3/2}$ (2)

Function (1) is an increasing function.

Function (2) meets x-axis, when $x - x^{3/2} = 0$ or $x = 0, 1$.

Also, for $0 < x < 1$, $x - x^{3/2} > 0$ and for $x > 1$, $x - x^{3/2} < 0$.

When $x \to \infty$, $x - x^{3/2} \to -\infty$.

From these information, we can plot the graph as below:

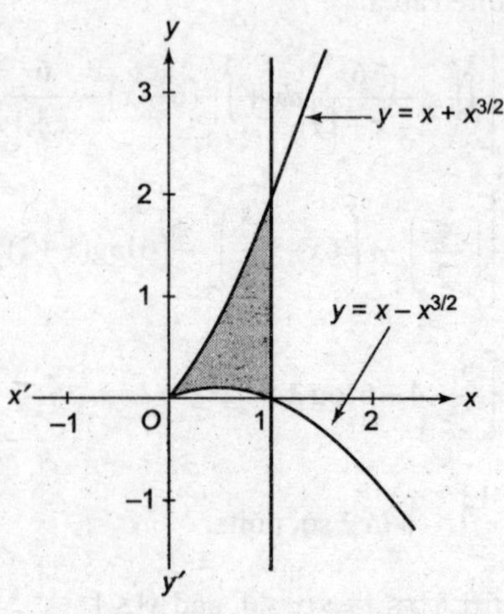

Fig. 9.30

Required area

$$= \int_0^1\left[\left(x+x^{3/2}\right)-\left(x+x^{3/2}\right)\right]dx = 2\int_0^1 x^{3/2}dx$$

$$= 2\left[\frac{x^{5/2}}{5/2}\right]_0^1 = \frac{4}{5} \text{ sq. units.}$$

31. b. Given curves are $y = \log_e x$ and $y = (\log_e x)^2$

Solving $\log_e x = (\log_e x)^2 \Rightarrow \log_e x = 0, 1 \Rightarrow x = 1$ and $x = e$

Also, for $1 < x < e$, $0 < \log_e x < 1 \Rightarrow \log_e x > (\log_e x)^2$

For $x > e$, $\log_e x < (\log_e x)^2$

$y = (\log_e x)^2 > 0$ for all $x > 0$

and when $x \to 0$, $(\log_e x)^2 \to \infty$.

From these information, we can plot the graph of the functions.

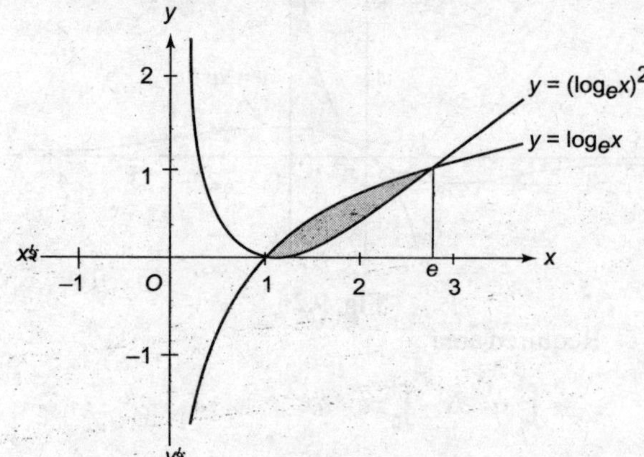

Fig. 9.31

Then the required area $= \int_1^e (\log x - (\log_e x)^2)dx$

$$= \int_1^e \log x dx - \int_1^e (\log_e x)^2 dx$$

$$= \left[x\log_e x - x\right]_1^e - \left[x(\log_e x)^2\right]_1^e + \int_1^e \frac{2\log_e x}{x} x dx$$

$$= 1 - e + 2\left[x\log_e x - x\right]_1^e$$

$$= 3 - e \text{ sq. units.}$$

32. a. The points in the required region satisfy

$4 \leq x^2 + y^2 \leq 2(|x| + |y|)$ (1)

Since the curve (1) is symmetrical about both the axes, the required area is four times the area of the region in the first quadrant. Therefore, it is sufficient to sketch the region and to find the area in the first quadrant.

In the first quadrant, the curve (1) consists of two curves

$x^2 + y^2 \geq 4$ and (C_1)

$x^2 + y^2 - 2x - 2y \geq 0$ (C_2)

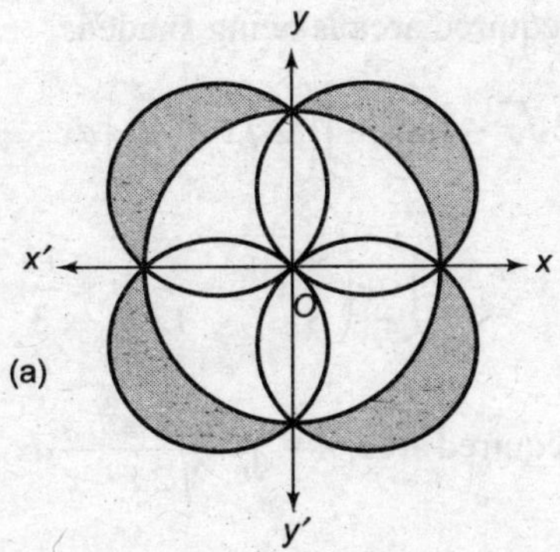

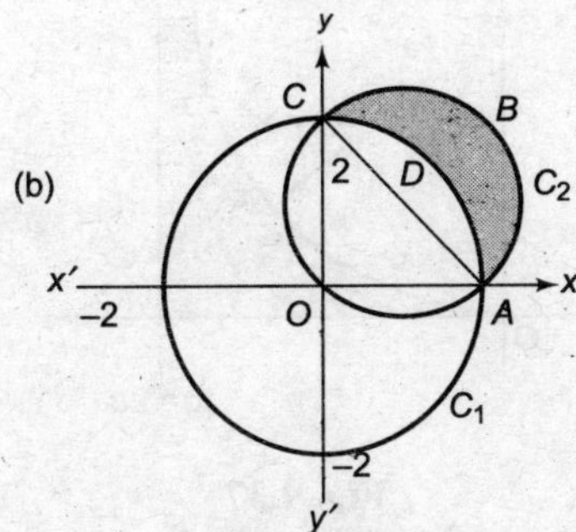

Fig. 9.32

Required area = 4 area $ABCDA$

= 4(area of semi-circle $ABCA$) − (area of sector $ADCA$)

= 4(area of semi-circle $ABCA$) − (area of sector $OADCO$ − area of triangle OAC)

$= 4\{\pi - (\pi - 2\} = 8$ sq. units.

33. c. The required area will be equal to the area enclosed by $y = f(x)$, y-axis between the abscissa at $y = -2$ and $y = 6$

Hence, $A = \int_0^1 (6 - f(x))\,dx + \int_{-1}^0 (f(x) - (-2))\,dx$

$= \int_0^1 (4 - x^3 - x)\,dx + \int_{-1}^0 (x^3 + 3x + 4)\,dx = \frac{5}{4}$ sq. units.

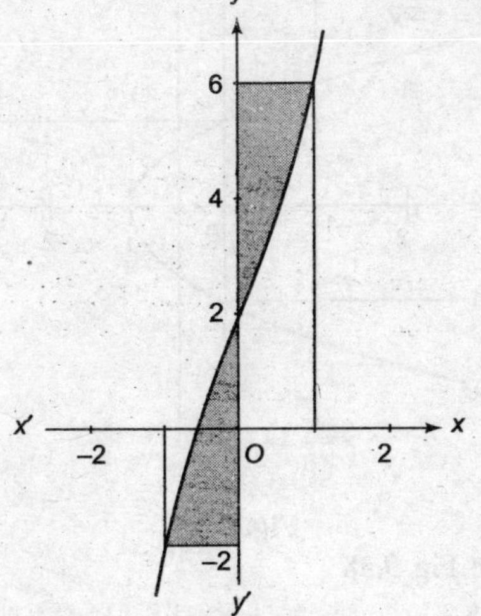

Fig. 9.33

34. d. Curve tracing: $y = x + \sin x$

$$\frac{dy}{dx} = 1 + \cos x \geq 0 \;\forall\; x$$

Also $\frac{d^2y}{dx^2} = -\sin x = 0$ when $x = n\pi, n \in Z$

Hence, $x = n\pi$ are points of inflection, where curve changes its concavity.

Also for $x \in (0, \pi)$, $\sin x > 0 \Rightarrow x + \sin x > x$.

And for $x \in (\pi, 2\pi)$, $\sin x < 0 \Rightarrow x + \sin x < x$.

From these information, we can plot the graph of $y = f(x)$ and its inverse.

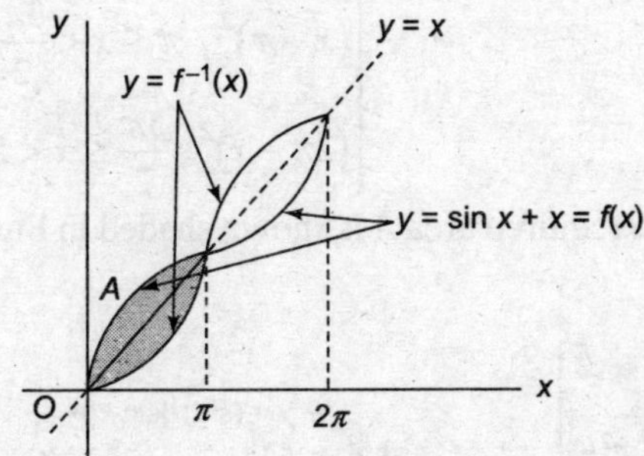

Fig. 9.34

Required area = $4A$, where

$$A = \int_0^\pi (x + \sin x)dx - \int_0^\pi x dx$$

$$= \frac{\pi^2}{2} - \cos\pi + \cos 0 - \frac{\pi^2}{2} = 2 \text{ square units.}$$

35. d. Area $= \int_1^b f(x)\,dx = \sqrt{b^2+1} - \sqrt{2}$

$$= \sqrt{b^2+1} - \sqrt{1+1}$$

$$= \left|\sqrt{x^2+1}\right|_1^b$$

$$\therefore f(x) = \frac{d}{dx}\left(\sqrt{x^2+1}\right) = \frac{1}{2}\frac{2x}{\sqrt{x^2+1}} = \frac{x}{\sqrt{x^2+1}}.$$

36. c. $\int_{\pi/4}^{\beta} f(x)dx = \beta\sin\beta + \frac{\pi}{4}\cos\beta + \sqrt{2}\,\beta$

Differentiating both sides w.r.t. β, we get

$$f(\beta) = \beta\cos\beta + \sin\beta - \frac{\pi}{4}\sin\beta + \sqrt{2}$$

$$\Rightarrow f'(\beta) = -\beta\sin\beta + \cos\beta + \cos\beta - \frac{\pi}{4}\cos\beta$$

$$\Rightarrow f'\left(\frac{\pi}{2}\right) = -\frac{\pi}{2}$$

37. d. $y=\sin^{-1}|\sin x|=\begin{cases} x, & 0\le x<\dfrac{\pi}{2} \\ \pi-x, & \dfrac{\pi}{2}\le x<\pi \\ x-\pi, & \pi\le x<\dfrac{3\pi}{2} \\ 2\pi-x, & \dfrac{3\pi}{2}\le x<2\pi \end{cases}$

$y=(\sin^{-1}|\sin x|)^2=\begin{cases} x^2, & 0\le x<\dfrac{\pi}{2} \\ (\pi-x)^2, & \dfrac{\pi}{2}\le x<\pi \\ (x-\pi)^2, & \pi\le x<\dfrac{3\pi}{2} \\ (2\pi-x)^2, & \dfrac{3\pi}{2}\le x<2\pi. \end{cases}$

The required area A is shown shaded in Fig.9.35

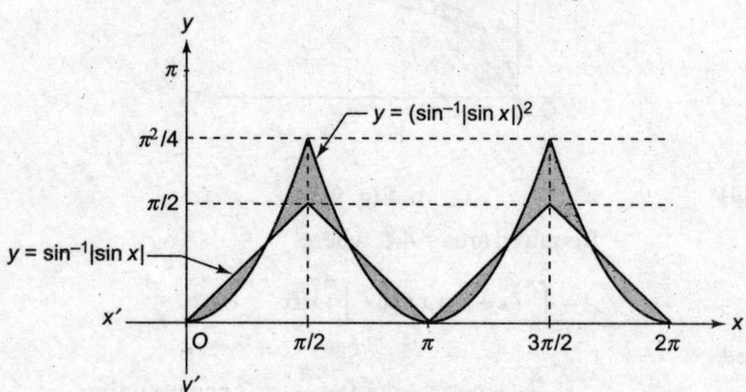

Fig. 9.35

$$\Rightarrow 4\int_0^1 (x-x^2)\,dx+4\int_1^{\pi/2}(x^2-x)\,dx$$

$$=\frac{4}{3}+\pi^2\left[\frac{2\pi-3}{6}\right] \text{ sq. units.}$$

38. c. $y^2=4\left[\sqrt{y}\right]x$

For $y\in[1,4)$, $\left[\sqrt{y}\right]=1\Rightarrow y^2=4x$.

Similarly, for $x\in[1,4)$, $\left\lfloor\sqrt{x}\right\rfloor=1$ and

$x^2=4\left\lfloor\sqrt{x}\right\rfloor y$ would transform into $x^2=4y$.

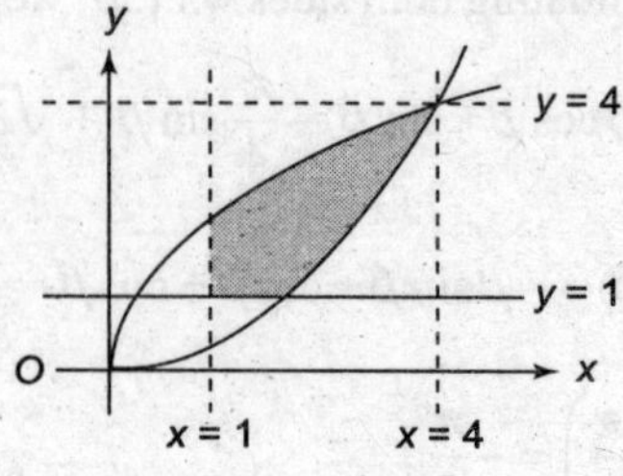

Fig. 9.36

The required area is being shaded.

$$A=\int_1^2(2\sqrt{x}-1)dx+\int_2^4\left(2\sqrt{x}-\frac{x^2}{4}\right)dx$$

$$=\left(\frac{4}{3}x^{3/2}-x\right)_1^2+\left(\frac{4}{3}x^{3/2}-\frac{x^3}{12}\right)_2^4=\frac{11}{3} \text{ sq. units.}$$

39. b. The required area, $A=\int_0^{2a}\sqrt{\dfrac{x^3}{2a-x}}\,dx$

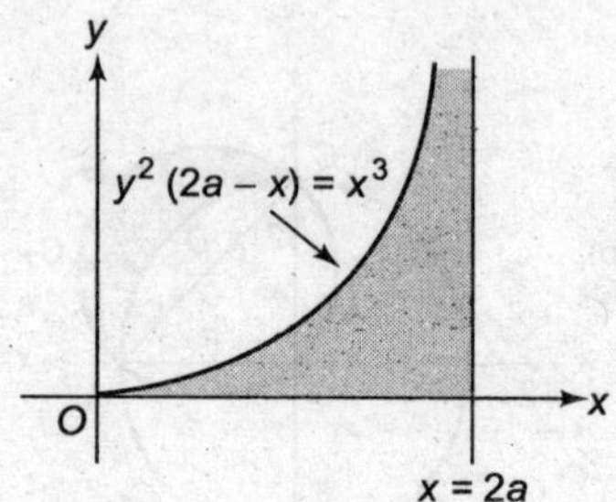

Fig. 9.37

Put $x=2a\sin^2\theta$

$\Rightarrow dx=2a\,2\sin\theta\cos\theta\,d\theta$

$$\Rightarrow A=\int_0^{\pi/2}\frac{\sqrt{8a^3}\,\sin^6\theta\,4a\sin\theta\cos\theta\,d\theta}{\sqrt{2a-2a\sin^2\theta}}$$

$$=8a^2\int_0^{\pi/2}\sin^4\theta\,d\theta=8a^2\,\frac{3}{4}\,\frac{1}{2}\,\frac{\pi}{2}$$

$$=\frac{3\pi a^2}{2} \text{ sq. units.}$$

40. a.

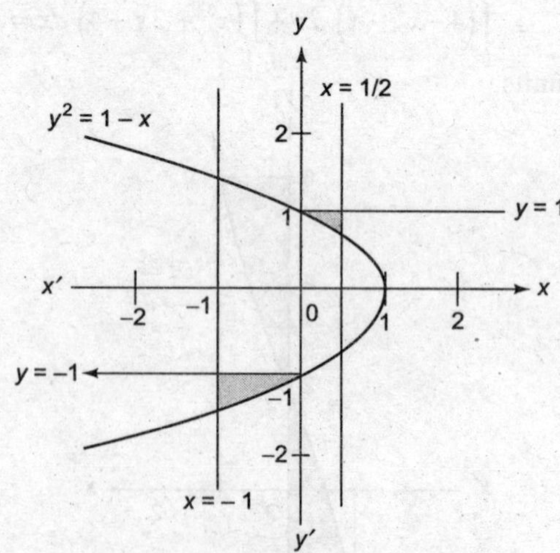

Fig. 9.38

From Fig. 9.38

$$A=\int_{-1}^{0}\left(-1-\left(-\sqrt{1-x}\right)\right)dx+\int_0^{1/2}(1-\sqrt{1-x})dx$$

$$= \left[-x - \frac{(1-x)^{3/2}}{3/2}\right]_{-1}^{0} + \left[x + \frac{(1-x)^{3/2}}{3/2}\right]_{0}^{1/2}$$

$$= \left[-\frac{2}{3} - \left(1 - \frac{2 \times 2^{3/2}}{3}\right)\right] + \left[\frac{1}{2} + \frac{2}{3 \times 2^{3/2}} - \frac{2}{3}\right]$$

$$= \frac{2}{3 \times 2^{3/2}} + \frac{2 \times 2^{3/2}}{3} - \frac{4}{3} - \frac{1}{2}$$

$$= \frac{3}{\sqrt{2}} - \frac{4}{3} - \frac{1}{2}$$

$$= \frac{3}{\sqrt{2}} - \frac{11}{6} \text{ sq. units}$$

Multiple Correct Answers Type

1. b., c.

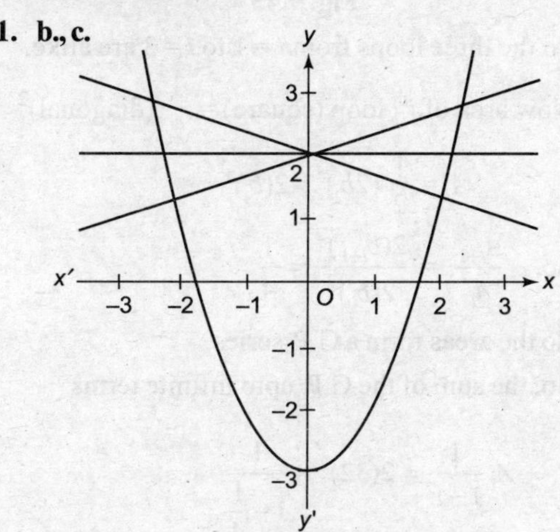

Fig. 9.39

Line $y = kx + 3$ passes through fixed point (0, 2) for different value of k.

Also, it is obvious that minimum $A(k)$ occurs when $k = 0$, as when line is rotated from this position about point (0, 2) the increased part of area is more than the decreased part of area.

$$\text{Area} = 2\int_0^{\sqrt{5}} (2-(x^2-3))dx$$

$$= 2\int_0^{\sqrt{5}} (5-x^2)dx$$

$$= 2\left[5x - \frac{x^3}{3}\right]_0^{\sqrt{5}}$$

$$= 2\left[5\sqrt{5} - \frac{5\sqrt{5}}{3}\right] = \frac{20\sqrt{5}}{3} \text{ sq. units}$$

2. a., b., d.

$y^2 = 4x$ and $x^2 = 4y$ meet at $O(0, 0)$ and a (4, 4).

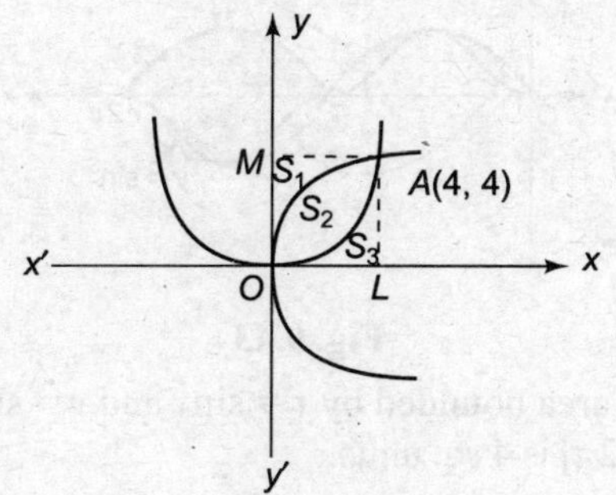

Fig. 9.40

$$\text{Now } S_3 = \int_0^4 \frac{x^2}{4} dx$$

$$= \frac{1}{4}\left[\frac{x^3}{3}\right]_0^4 = \frac{1}{12}[64 - 0] = \frac{16}{3}.$$

$$S_2 = \int_0^4 2\sqrt{x}\, dx - S_3 = 2\left[\frac{x^{3/2}}{3/2}\right]_0^4 - \frac{16}{3}$$

$$= \frac{4}{3}[8-0] - \frac{16}{3} = \frac{16}{3}.$$

And $\quad S_1 = 4 \times 4 - (S_2 + S_3) = 16 - \left(\frac{16}{3} + \frac{16}{3}\right) = \frac{16}{3}.$

Hence, $S_1 : S_2 : S_3 = 1:1:1$

3. a., c., d.

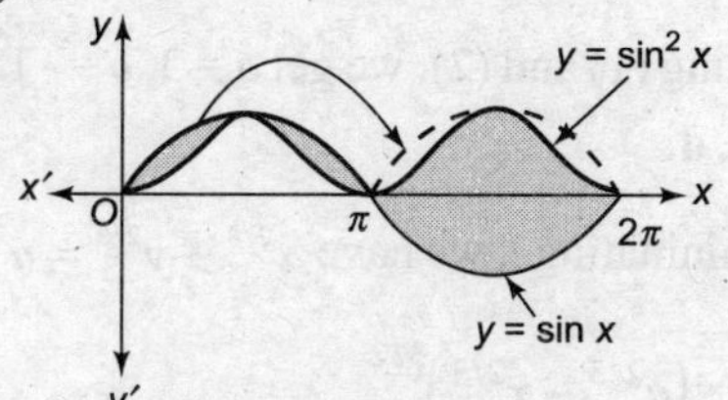

Fig. 9.41

We know that area bounded by $y = \sin x$ and x-axis for $x \in [0, \pi]$ is 2 sq. units.

Then area bounded by $y = \sin x$ and $y = \sin^2 x$ is 4 sq. units for $x \in [0, 2\pi]$.

Then for $x \in [0, 10\pi]$, the area bounded is 20 sq. units.

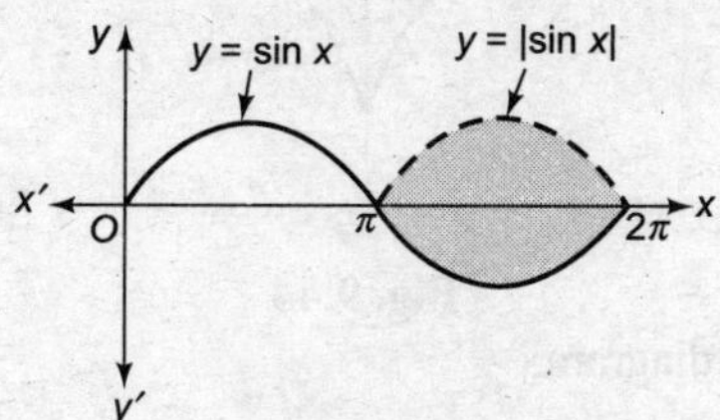

Fig. 9.42

The area bounded by $y = \sin x$ and $y = |\sin x|$ for $x \in [0, 2\pi]$ is 4 sq. units.

Then for $x \in [0, 20\pi]$, the area bounded is 40 sq. units.

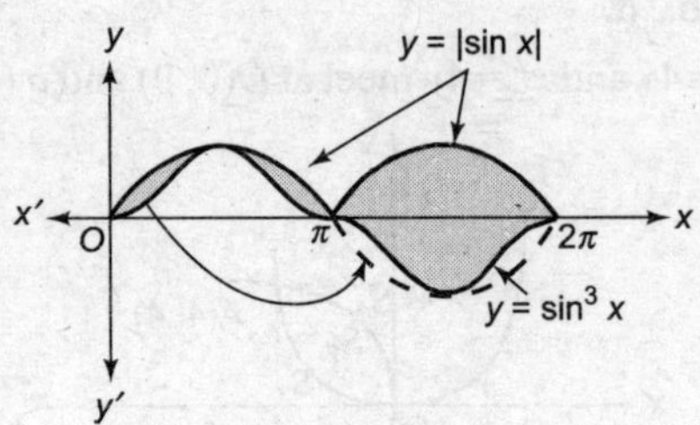

Fig. 9.43

The area bounded by $y = \sin x$ and $y = \sin^3 x$ for $x \in [0, 2\pi]$ is 4 sq. units.

Then for $x \in [0, 10\pi]$, the area bounded is 20 sq. units.

Similarly, the area bounded by $y = \sin x$ and $y = \sin^4 x$ for $x \in [0, 10\pi]$ is 20 sq. units.

4. c., d.

Since the curve $y = ax^{1/2} + bx$ passes through the point (1, 2). Therefore,

$2 = a + b(1)$

By observation the curve also passes through (0, 0). Therefore, the area enclosed by the curve, x-axis and $x = 4$ is given by

$$A = \int_0^4 (ax^{1/2} + bx)dx = 8 \Rightarrow \frac{2a}{3} \times 8 + \frac{b}{2} \times 16 = 8$$

$$\Rightarrow \frac{2a}{3} + b + 1. \qquad (2)$$

Solving (1) and (2), we get $a = 3, b = -1$.

5. a., c., d.

Elimininating t, we have $x^{2/3} + y^{2/3} = a^{2/3}$

$$\Rightarrow y = \left(a^{2/3} - x^{2/3}\right)^{3/2}.$$

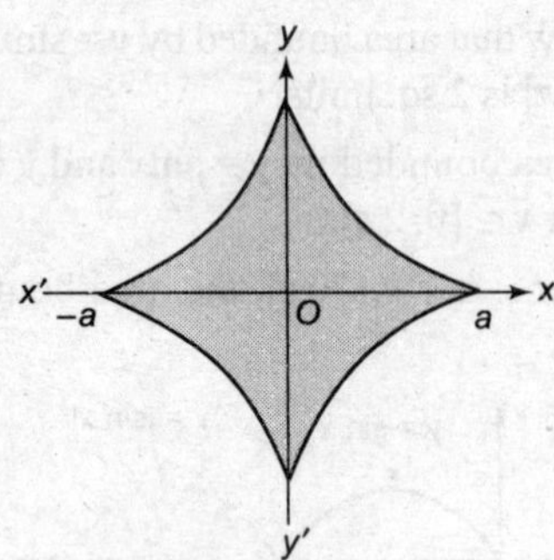

Fig. 9.44

From diagram,

$$\Rightarrow \quad A = 2\int_{-a}^{a} \left(a^{2/3} - x^{2/3}\right)^{3/2} dx$$

$$= 4\int_0^a \left(a^{2/3} - x^{2/3}\right)^{3/2} dx$$

$$A = 4\int_0^a y\,dx$$

$$= 4a^2 \int_0^{\pi/2} 3\cos^3 t \sin^2 t \cos t \, dt.$$

6. a., c.

$$a_1 = 0, b_1 = 32, a_2 = a_1 + \frac{3}{2} b_1 = 48, b_2 = \frac{b_1}{2} = 16$$

$$a_3 = 48 + \frac{3}{2} \times 16 = 72, b_3 = \frac{16}{2} = 8$$

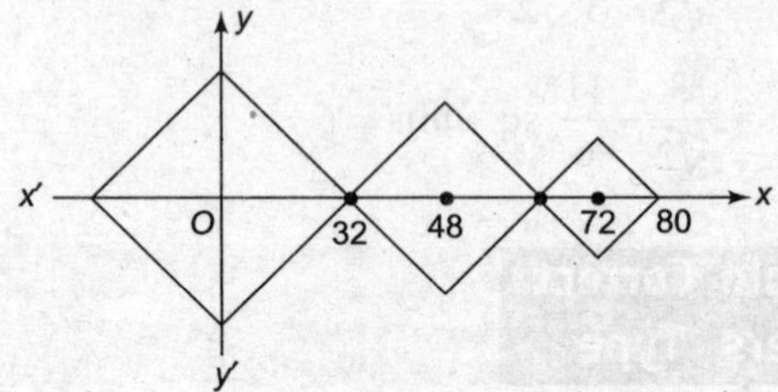

Fig. 9.45

So the three loops from $i = 1$ to $i = 3$ are alike.

Now area of ith loop (square) $= \frac{1}{2}$ (diagonal)^2

$$A_i = \frac{1}{2}(2b_i)^2 = 2(b_i)^2$$

So, $$\frac{A_{i+1}}{A_i} = \frac{2(b_{i+1})^2}{2(b_i)^2} = \frac{1}{4}.$$

So the areas form a G.P. series

So, the sum of the G.P. upto infinite terms

$$= A_1 \frac{1}{1-r} = 2(32)^2 \times \frac{1}{1-\frac{1}{4}}$$

$$= 2 \times (32)^2 \times \frac{4}{3} = \frac{8}{3}(32)^2 \text{ square units.}$$

Reasoning Type

1. a. Since $y = e^x$ and $y = \log_e x$ are inverse to each other.

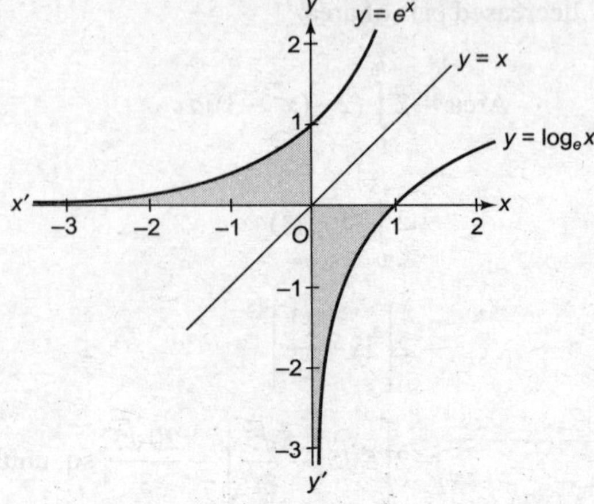

Fig. 9.46

2. a. Statement 2 is correct as $y = f(x)$ is odd and hence statement 1 is correct.

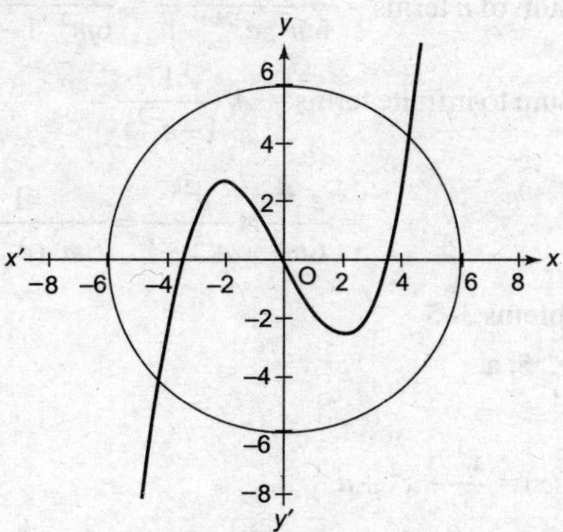

Fig. 9.47

3. b. Area $= \int_1^3 -(x^2 - 4x + 3)\, dx$

$$= -\left(\frac{x^3}{3} - \frac{4x^2}{2} + 3x\right)\Bigg|_1^3 = \frac{4}{3} \text{ sq. units.}$$

Therefore, statement 1 is true.

Obviously, statement 2 is true, but does not explain statement 1.

4. a. Given curves are $y^2 - 2y + 4x + 5 = 0$ and $x^2 + 2x - y + 2 = 0$ or $(y-1)^2 = -4(x+1)$ and $(x+1)^2 = y - 1$.

Shifting origin to (–1, 1), equation of given curves changes to $Y^2 = -4X$ and $X^2 = Y$.

Hence, statement 1 is true and statement 2 is correct explanation of statement 1.

5. a.

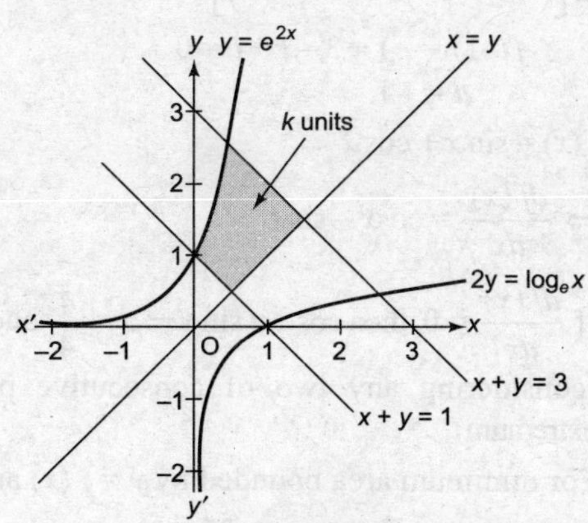

Fig. 9.48

$y = e^{2x}$ and $2y = \log_e x$ are inverse of each other

The shaded area is given as k sq. units.

⇒ The required area is $2k$ sq. units.

6. d. R_1 : points $P(x, y)$ is nearer to (1, 0) than to $x = -1$

$\Rightarrow \sqrt{(x-1)^2 + y^2} < |x-1|$

$\Rightarrow y^2 < 4x$

⇒ Point P lies inside parabola $y^2 = 4x$.

R_2 : Point $P(x, y)$ is nearer to (0, 0) than to (8, 0)

$\Rightarrow |x| < |x-8|$

$\Rightarrow x^2 < x^2 - 16x + 64$

$\Rightarrow x < 4$

⇒ Point P is towards left side of line $x = 4$.

The area of common region of R_1 and R_2 is the area bounded by $x = 4$ and $y^2 = 4x$.

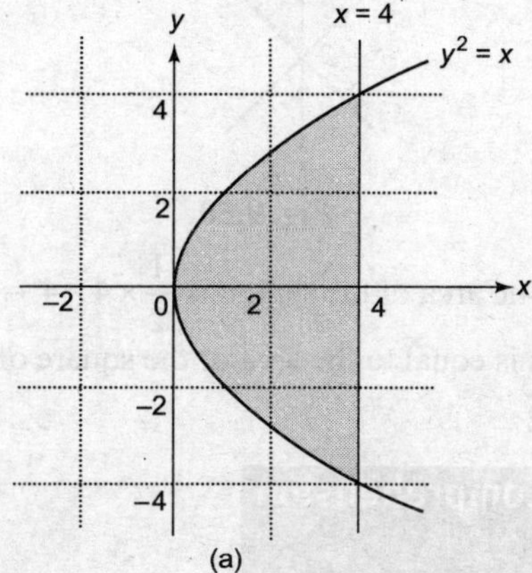

(a)

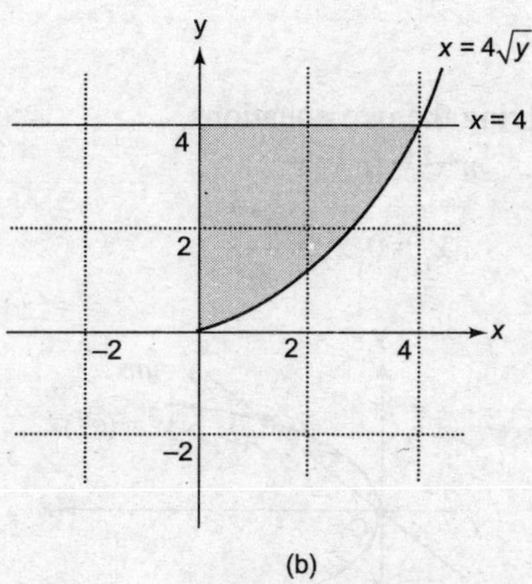

(b)

Fig. 9.49

This area is twice the area bounded by $x = 4\sqrt{y}$ and $y = 4$. Now, the area bounded by $x = 4\sqrt{y}$ and $y = 4$ is

$$A = \int_0^4 \left(4 - \frac{x^2}{4}\right) dx = \left[4x - \frac{x^3}{12}\right]_0^4 = \left[16 - \frac{64}{12}\right] = \frac{32}{3} \text{ sq. units.}$$

Hence, the area bounded by R_1 and R_2 is $\frac{64}{3}$ sq. units.

Thus, statement 1 is false but statement 2 is true.

7. b. $2 \geq \max.\{|x-y|, |x+y|\}$
$\Rightarrow |x-y| \leq 2$ and $|x+y| \leq 2$, which forms a square of diagonal length 4 units.

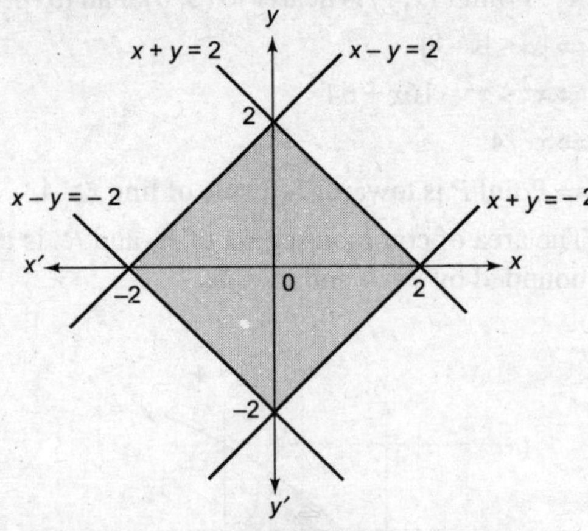

Fig. 9.50

$\Rightarrow$ The area of the region is $\frac{1}{2} \times 4 \times 4 = 8$ sq. units.

This is equal to the area of the square of side length $2\sqrt{2}$.

Linked Comprehension Type

For Problems 1–2

1. b., 2. c

Sol.

Solving the two equations,

$$m^2x^2 = (e^{-kr})\,x$$

$$x_1 = 0, x_2 = \frac{e^{-kr}}{m^2}.$$

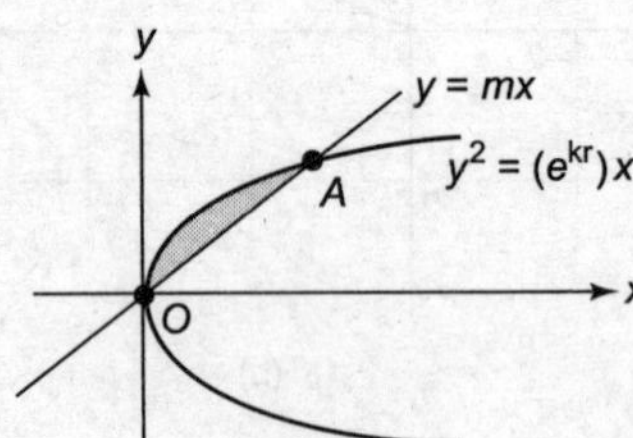

Fig. 9.51

$$\text{So, } A_r = \int_0^{x_2} \left(e^{-\frac{kr}{2}} \sqrt{x} - mx \right) dx$$

$$= \frac{2}{3} e^{-kr/2} x_2^{3/2} - m\frac{x_2^2}{2}$$

$$= \frac{2}{3} e^{-kr/2} \frac{e^{-3kr/2}}{m^3} - \frac{m}{2}\frac{e^{-2kr}}{m^4} = \frac{e^{-2kr}}{6m^3}.$$

$$\text{Now, } \frac{A_{r+1}}{A_r} = \frac{e^{-2k(r+1)}}{e^{-2kr}} = e^{-2k} = \text{constant.}$$

So, the sequence $A_1, A_2, A_3, \ldots$ is in G.P.

$$\text{Sum of } n \text{ terms} = \frac{e^{-2k}}{6m^3} \frac{e^{-2nk}-1}{e^{-2k}-1} = \frac{1}{6m^3} \frac{e^{-2nk}-1}{1-e^{2k}}$$

$$\text{Sum to infinite terms} = A_1 \frac{1}{1-e^{-2k}}$$

$$= \frac{e^{-2k}}{6m^3} \times \frac{e^{2k}}{e^{2k}-1} = \frac{1}{6m^3(e^{2k}-1)}.$$

For Problems 3–5

3. d., 4. c., 5. a.

Sol.

3. d. $f(x) = \frac{x^3}{3} - x^2 + a$

$f'(x) = x^2 - 2x = x(x-2) < 0$ (note that $f(x)$ is monotonic in $(0, 2)$)

Hence for the minimum and $f(x)$ must cross the x-axis at $\frac{0+2}{2} = 1$.

Hence, $f(1) = \frac{1}{3} - 1 + a = 0$

$\Rightarrow \quad a = \frac{2}{3}$

4. c. $f(x) = x^3 + 3x^2 + x + a$

$f'(x) = 3x^2 + 6x + 1 = 0$

$\Rightarrow \quad x = -1 \pm \frac{\sqrt{6}}{3}.$

Hence, $f(x)$ cuts the x-axis at

$$\frac{1}{2}\left[\left(-1+\frac{\sqrt{6}}{3}\right)+\left(-1-\frac{\sqrt{6}}{3}\right)\right] = -1.$$

$$f(-1) = -1 + 3 - 1 + a = 0$$

$$a = -1.$$

5. a. $f(x) = \sin x + \cos x$

$\Rightarrow \quad \frac{df(x)}{dx} = \cos x - \sin x$

If $\frac{df(x)}{dx} = 0$, then $\cos x = \sin x \Rightarrow x = \frac{\pi}{4}$ and $x = \frac{5\pi}{4}$ (considering any two of consecutive points of extremum).

For minimum area bounded by $y = f(x)$ and $y = a$, between $x = \frac{\pi}{4}$ and $x = \frac{5\pi}{4}$, graphs of $g(x)$ must cut $y = a$ at $c = \frac{\frac{\pi}{4}+\frac{5\pi}{4}}{2} = \frac{3\pi}{4}$.

$$a = f\left(\frac{3\pi}{4}\right) \Rightarrow a = \sin\left(\frac{3\pi}{4}\right) + \cos\left(\frac{3\pi}{4}\right) = 0.$$

For Problems 6 – 8

6. a., 7. c., 8. b.

Sol. Since $-1 \le \sin x \le 1$, the curve $y = e^{-x} \sin x$ is bounded by the curves $y = e^{-x}$ and $y = e^{-x}$.

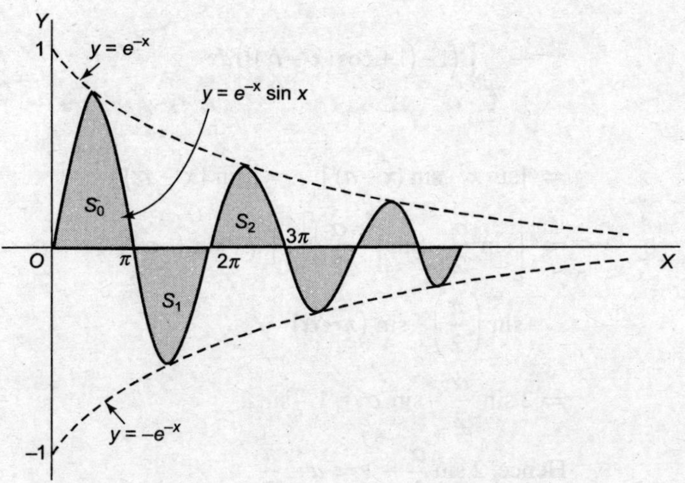

Fig. 9.52

Also, the curve $y = e^{-x} \sin x$ intersects the positive semi-axis OX at the points where $\sin x = 0$, where $x_n = n\pi, n \in Z$.

Also $|y_n| = |y$ coordinate in the half-wave $S_n|$

$= (-1)^n e^{-x} \sin x$, and

In S_n, $n\pi \le x \le (n+1)\pi$

$$\therefore S_n = (-1)^n \int_{n\pi}^{(n+1)\pi} e^{-x} \sin x dx$$

$$= \frac{(-1)^{n+1}}{2}\left[e^{-x}(-\sin x + \cos x)\right]_{n\pi}^{(n+1)\pi}$$

$$= \frac{(-1)^{n+1}}{2}\left[e^{-(n+1)\pi}(-1)^{n+1} - e^{n\pi}\beta(-1)^n\right]$$

$$= \frac{e^{-n\pi}}{2}\left(1+e^{\pi}\right)$$

$$\Rightarrow \frac{S_{n+1}}{S_n} = e^{-\pi} \quad \text{and } S_0 = \frac{1}{2}\left(1+e^{\pi}\right).$$

Therefore, the sequence S_0, S_1, S_2, ... forms an infinite G.P. with common ratio $e^{-\alpha\pi/\beta}$.

$$\therefore \sum_{n=0}^{\infty} S_n = \frac{\frac{1}{2}(1+e^{\pi})}{1-e^{-\pi}}$$

For Problems 9–11

9. b.,10. a., 11. c.

Sol.

9. b. Given

$(x-y)f(x+y) - (x+y)f(x-y) = 4xy(x^2-y^2)$

$= (x^2-y^2)\,[(x+y)^2-(x-y)^2]$

$= (x-y)(x+y)^3 - (x+y)(x-y)^3$

$\Rightarrow f(x+y) = (x+y)^3 \Rightarrow f(x) = x^3, f(y) = y^3$

Now equations of given curves are

$$y^2 + x = 0 \quad (1)$$

$$x^2 + y^2 = 12 \quad (2)$$

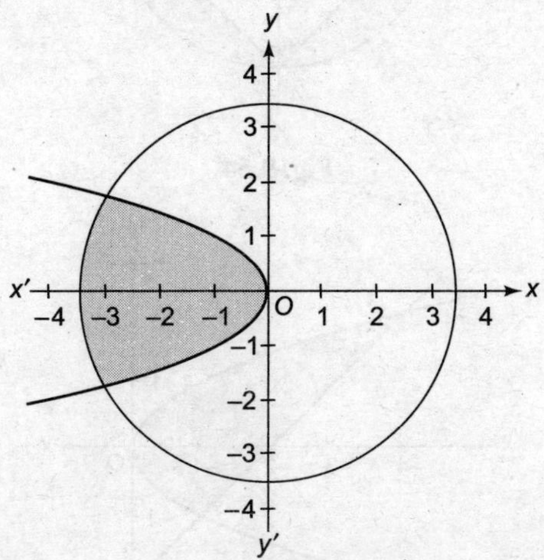

Fig. 9.53

Solving Eq's (1) and (2), we get $x = -3, y = \pm\sqrt{3}$

The area bounded by curves

$$A = 2\left[\left|\int_{-2\sqrt{3}}^{-3}\sqrt{12-x^2}\,dx\right| + \left|\int_{-3}^{0}\sqrt{-x}dx\right|\right]$$

$$I_1 = 2\int_{-2\sqrt{3}}^{-3}\sqrt{12-x^2}\,dx = 2\int_{-\pi/2}^{-\pi/3} 12\cos^2\theta d\theta$$

$$= 12\left[\int_{-\pi/2}^{-\pi/3}(1+\cos 2\theta)d\theta\right]$$

$$= 12\left[\theta + \frac{\sin\theta}{2}\right]_{-\pi/2}^{-\pi/3} = 12\left[-\frac{\pi}{3} - \frac{\sqrt{3}}{4} + \frac{\pi}{2}\right]$$

$$= 12\left[\frac{\pi}{6} - \frac{\sqrt{3}}{4}\right] = 2\pi - 3\sqrt{3}.$$

$$I_2 = 2\int_{-3}^{0}\sqrt{-x}dx = \frac{2[(-x)^{3/2}]_{-3}^{0}}{-3/2} = -\frac{4}{3}[0-3^{3/2}]$$

$$= 4\sqrt{3}.$$

$A = 2\pi - 3\sqrt{3} + 4\sqrt{3} = 2\pi + \sqrt{3}$ sq. units.

10. a. The required area is = area of circle – area of square

$= 12\pi - 24$ sq. units.

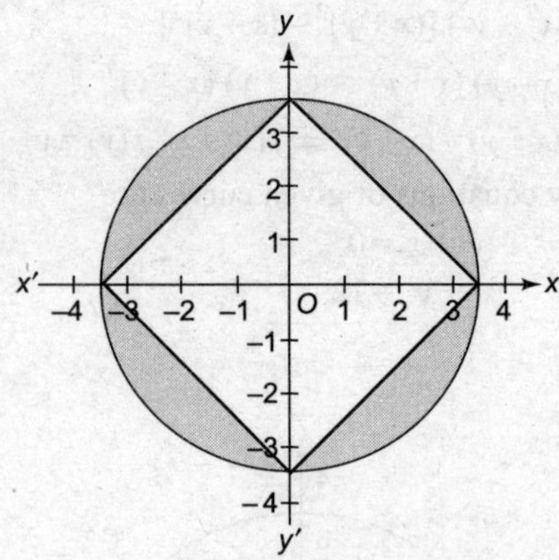

Fig. 9.54

11. c.

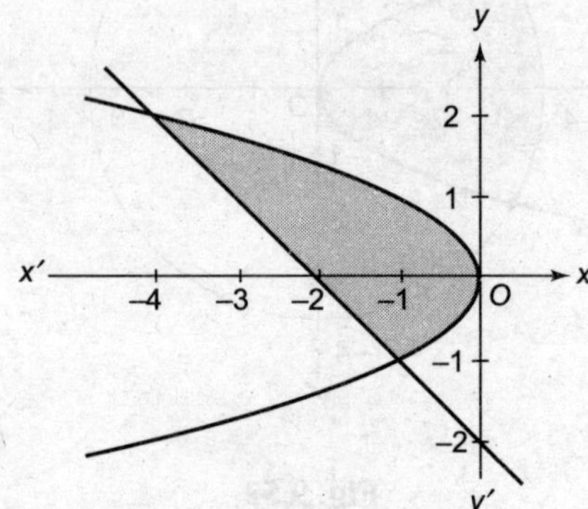

Fig. 9.55

The required area

$= \int_{-1}^{2} (-y^2 - (-y-2))dy$

$= \left[\frac{y^2}{2} + 2y - \frac{y^3}{3}\right]_{-1}^{2}$

$= \left[\frac{4}{2} + 4 - \frac{8}{3} - \left(\frac{1}{2} - 2 + \frac{1}{3}\right)\right]$

$= 9/2$ sq. units.

For Problems 12–13

12. c.,13. b.

Sol.

12. c.

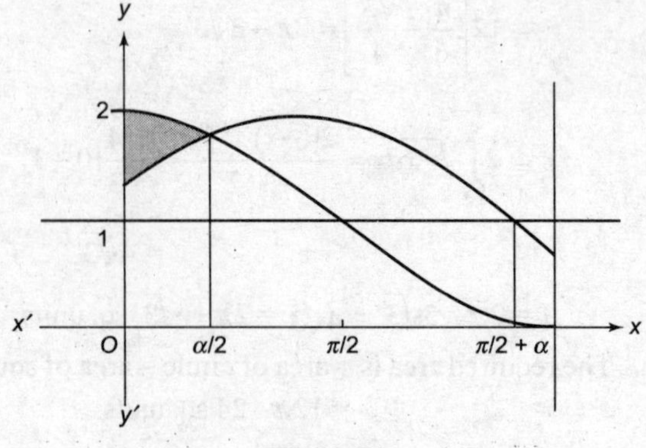

Fig. 9.56

$1 + \cos x = 1 + \cos(x - \alpha)$

$x = \alpha - x \Rightarrow x = \frac{\alpha}{2}$

Now $\int_0^{\alpha/2} \left((1+\cos x) - (1+\cos(x-\alpha))\right) dx$

$= -\int_{\frac{\pi}{2}+\alpha}^{\pi} \left(1 - (1+\cos(x-\alpha))\right) dx$

$\Rightarrow [\sin x - \sin(x-a)]_0^{\alpha/2} = [\sin(x-\alpha)]_{\pi}^{\frac{\pi}{2}+\alpha}$

$\Rightarrow \left[\sin\frac{\alpha}{2} - \sin\left(-\frac{\alpha}{2}\right)\right] - [0 - \sin(-\alpha)]$

$= \sin\left(\frac{\pi}{2}\right) - \sin(\pi - \alpha)$

$\Rightarrow 2\sin\frac{\alpha}{2} - \sin\alpha = 1 - \sin a$

Hence, $2\sin\frac{\alpha}{2} = 1 \Rightarrow a = \frac{\pi}{3}$.

13. b. $\int_0^{\pi/6} \left((1+\cos x) - \left(1+\cos\left(x - \frac{\pi}{3}\right)\right)\right) dx$

$+ \int_{\pi/6}^{\pi} \left(\left(1+\cos\left(x-\frac{\pi}{3}\right)\right) - (1+\cos x)\right) dx$

$= \left[\sin x - \sin\left(x - \frac{\pi}{3}\right)\right]_0^{\pi/6}$

$+ \left[\sin\left(x - \frac{\pi}{3}\right) - \sin x\right]_{\pi/6}^{\pi}$

$= \left[\left(\frac{1}{2} + \frac{1}{2}\right) - \frac{\sqrt{3}}{2}\right] + \left[\frac{\sqrt{3}}{2} - \left(-\frac{1}{2} - \frac{1}{2}\right)\right]$

$= 2$ sq. units.

For Problems 14–16

14. a., 15. d., 16. a.

Sol.

14. a. For $-1 \le x < 0$

$(y - e^{\sin^{-1} x})^2 = 2 - x^2$

$y = e^{\sin^{-1} x} \pm \sqrt{2 - x^2}$

$A = \int_{-1}^{0} \left(e^{\sin^{-1} x} + \sqrt{2-x^2}\right) - \left(e^{\sin^{-1} x} - \sqrt{2-x^2}\right) dx$

$= 2\int_{-1}^{0} \sqrt{2-x^2}\, dx$

$= 2\left(\frac{1}{2}x\sqrt{2-x^2}\Big|_{-1}^{0} + \frac{2}{2}\sin^{-1}\frac{x}{\sqrt{2}}\Big|_{-1}^{0}\right)$

$= \left[1 + 2\left(0 - \left(-\frac{\pi}{4}\right)\right)\right]$

$= \dfrac{\pi}{2} + 1$ sq. units.

For $0 \le x < 1, y = \sin^{-1} x \pm \sqrt{1 - x^2}$

$$A = 2\int_0^1 \sqrt{1 - x^2}\, dx$$

$$= 2\left[\frac{x}{2}\sqrt{1 - x^2}\Big|_0^1 + \frac{1}{2}\sin^{-1}\frac{x}{1}\Big|_0^1\right]$$

$$= 0 + \sin^{-1}(1) = \frac{\pi}{2} \text{ sq. units.}$$

$$\text{Total area} = \left(\frac{\pi}{2} + 1\right) + \frac{\pi}{2} = \pi + 1.$$

15. d. $\text{Ratio} = \dfrac{\frac{\pi}{2} + 1}{\frac{\pi}{2}} = \dfrac{\pi + 2}{\pi}.$

16. a. $A = 2\displaystyle\int_0^{1/2} \sqrt{1 - x^2}\, dx$

$$= 2\left[\frac{x}{2}\sqrt{1 - x^2}\Big|_0^{\frac{1}{2}} - \frac{1}{2}\sin^{-1} x\Big|_0^{\frac{1}{2}}\right]$$

$$= \frac{\sqrt{3}}{4} + \frac{\pi}{6} \text{ sq. unit.}$$

For Problems 17–19

17. b., 18. a., 19. c.

Sol.

17. b. $S = -\displaystyle\int_0^{2\pi} a(1 - \cos t)a(1 - \cos t)dt$

$$= -a^2\int_0^{2\pi}(1 - 2\cos t + \cos^2 t)dt$$

$$= -a^2\int_0^{2\pi}\left(1 - 2\cos t + \left(\frac{1 + \cos 2t}{2}\right)\right)dt$$

$$= -\frac{a^2}{2}\int_0^{2\pi}(3 - 4\cos t + \cos 2t)dt$$

$$= -\frac{a^2}{2}[3t - 4\cos t + \cos 2t]_0^{2\pi}$$

$= -3\pi a^2 = 3\pi a^2$ sq. units (absolute value).

18. a. $\displaystyle\int_0^6\left(\frac{3}{2}t^2 - \frac{1}{2}t^3 + \frac{1}{24}t^4\right)dt$

$$= \frac{3}{2}\frac{6^3}{3} - \frac{1}{2}\frac{6^4}{4} + \frac{1}{24}\frac{6^5}{5} = \frac{6^3}{2} - \frac{6^4}{8} + \frac{6^4}{20}$$

$$= 6^4\left(\frac{1}{12} - \frac{1}{8} + \frac{1}{20}\right) = \frac{54}{5}.$$

$$\therefore \frac{1}{2}\int_0^6 (xy' - yx')dx = \frac{1}{2} \times \frac{54}{5} = \frac{27}{5} \text{ sq. units.}$$

19. c. $\dfrac{dx}{dt} = 1 - 3t^2$ and $\dfrac{dy}{dt} = 1 - 4t^3$.

So, $x\dfrac{dy}{dt} - y\dfrac{dx}{dt}$

$$= (t - t^3)(1 - 4t^3) - (1 - t^4)(1 - 3t^2)$$

$$= t^6 - 3t^4 - t^3 + 3t^2 + t - 1$$

$$\text{Required area} = \frac{1}{2}\int_{-1}^{1}(t^6 - 3t^4 - t^3 + 3t^2 + t - 1)dt$$

$= \dfrac{16}{35}$ sq. units [Taking absolute value]

Matrix-Match Type

1. a. → r; c. → s; b. → p; d. → q.

Sol.

a.

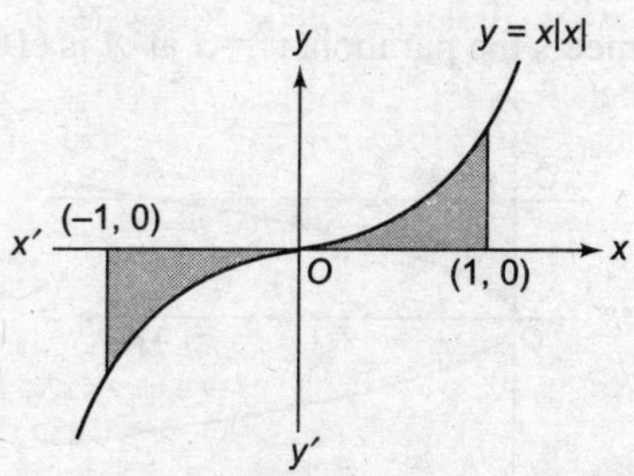

Fig. 9.57

$$\text{Required area} = \left|\int_{-1}^{1} x\,|x|\,dx\right|$$

$$= \left|\int_{-1}^{0} x\,|x|\,dx\right| + \int_0^1 x\,|x|\,dx = \left|\int_{-1}^{0} -x^2 dx\right| + \int_0^1 x^2 dx$$

$$= \left|\left[-\frac{x^3}{3}\right]_{-1}^{0}\right| - \left[\frac{x^3}{3}\right]_0^1 = \frac{1}{3} + \frac{1}{3} = \frac{2}{3} \text{ sq. units.}$$

b.

Fig. 9.58

Required area $= \int_0^2 (y_1 - y_2)dx$

$= \int_0^2 [(x+2)-(x^2)]dx = \left[\frac{x^2}{2}+2x-\frac{x^3}{3}\right]_0^2$

$= = 2+4-\frac{8}{3} = \frac{10}{3}$ sq. units.

c. Required. area $= \int_0^1 \left(\sqrt{x} - x\right) dx = \left[\frac{x^{3/2}}{3/2} - \frac{x^2}{2}\right]_0^1$

$= \left(\frac{1}{3/2} - \frac{1}{2}\right) = \frac{2}{3} - \frac{1}{2} = \frac{1}{6}$ sq. units.

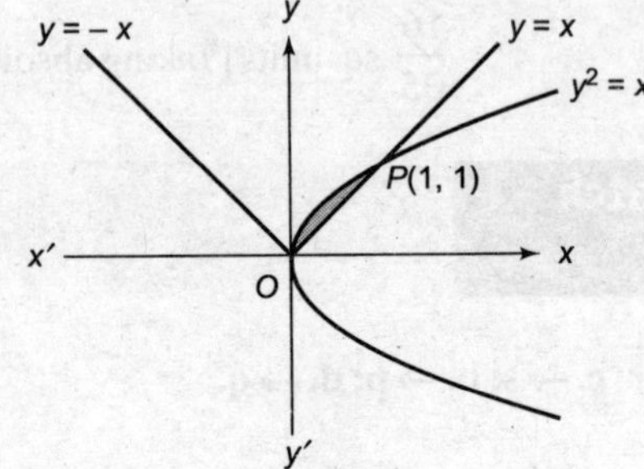

Fig. 9.59

d. $y = 4$ meets the parabola $y^2 = x$ at A is (16, 4)

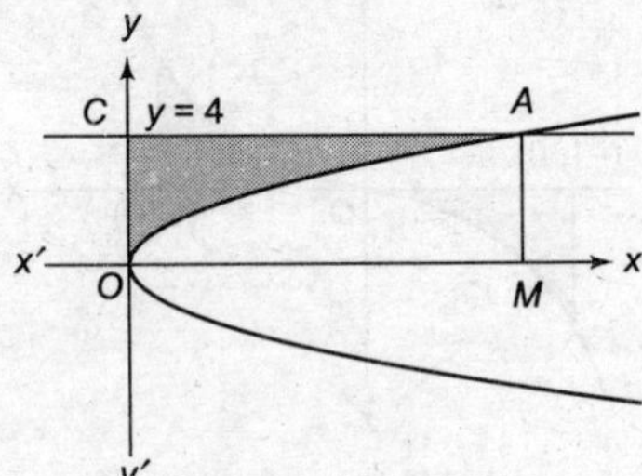

Fig. 9.60

Required area = Area of rectangle $OMAC$ – Area OMA

$= 4 \times 16 - \int_0^{16} \sqrt{x}\, dx = 64 - \left|\frac{x^{3/2}}{3/2}\right|_0^{16}$

$= 64 - \frac{2}{3}(4)^3 = 64 - \frac{128}{3} = \frac{64}{3}$ sq. units.

2. a. → q; c. → s; b. → p; d. → r.

Sol.

a. Area $= 2\left(\frac{1}{2}1\times 1\right) = 1$ sq. units.

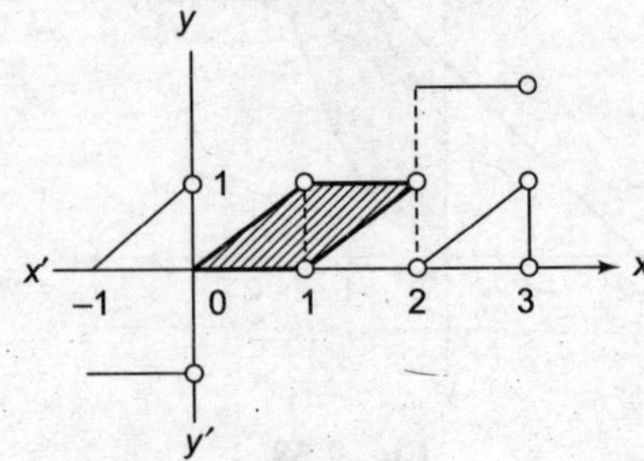

Fig. 9.61

b. $y^2 = x^3$ and $|y| = 2x$, both curves are symmetric about the y-axis

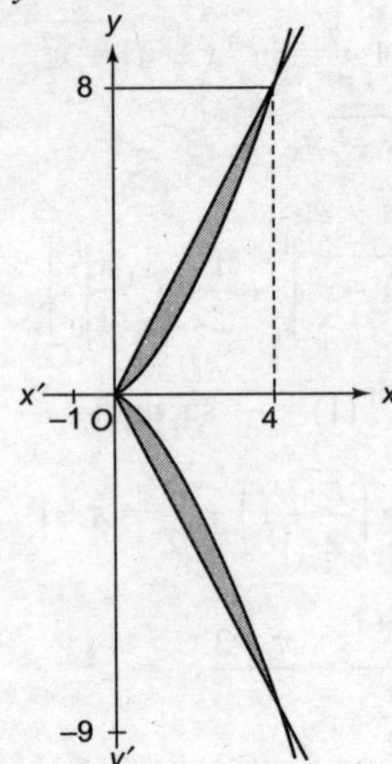

Fig. 9.62

$4x^2 = x^3 \Rightarrow x = 0, 4.$

The required area $= 2\int_0^4 \left(2x - x^{3/2}\right)dx = \frac{16}{5}$ sq. units.

c. $\sqrt{x} + \sqrt{|y|} = 1$

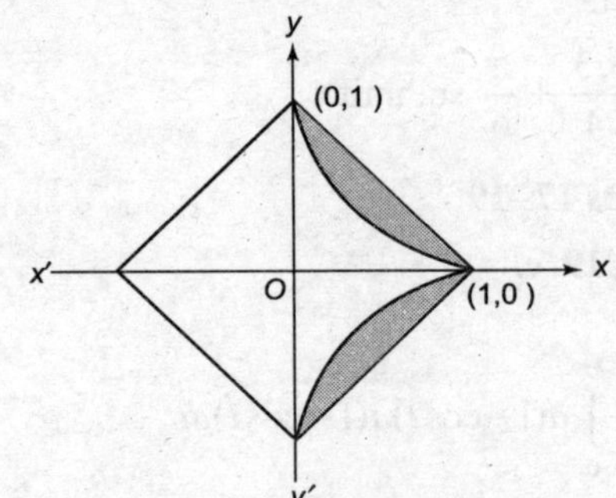

Fig. 9.63

The curve is symmetrical about the x-axis

$\sqrt{|y|} = 1 - \sqrt{x}$ and $\sqrt{x} = 1 - \sqrt{|y|}$

$\Rightarrow$ for $x > 0, y > 0$ $\sqrt{y} = 1 - \sqrt{x}$

$\frac{1}{2\sqrt{y}}\frac{dy}{dx} = -\frac{1}{2\sqrt{x}}$

$\frac{dy}{dx} = -\sqrt{\frac{x}{y}}$

$\frac{dy}{dx} < 0$, function is decreasing, the required area

$= 2\int_0^1 \left((1-x) - (1 - 2\sqrt{x} + x\right)dx$

$= 4\int_0^1 \left(\sqrt{x} - x\right)dx$

$$= 4\left[\frac{x^{3/2}}{3/2} - \frac{x^2}{2}\right]_0^1$$

$$= 4\left[\frac{2}{3} - \frac{1}{2}\right]$$

$$= \frac{2}{3} \text{ sq. units.}$$

d. If $-8 < x < 8$, then $y = 2$.

If $x \in (-8\sqrt{2}, -8] \cup [8, \sqrt{2})$, then $y = 3$, and so on

Intersection of $y = x - 1$ and $y = 2$. We get $x = 3 \in (-8, 8)$.

Intersection of $y = x - 1$ and $y = 3$,

we get $x = 4 \notin (-8\sqrt{2}, -8] \cup [8, 8\sqrt{2})$.

Similarly, $y = x - 1$ will not intersect $y = \left[\frac{x^2}{64} + 2\right]$ at any other integral, except in the interval $x \in (-8, 8)$.

The required area (shaded region) $= 2 \times 3 - \frac{1}{2} \times 2 \times 2 = 4$ sq. units.

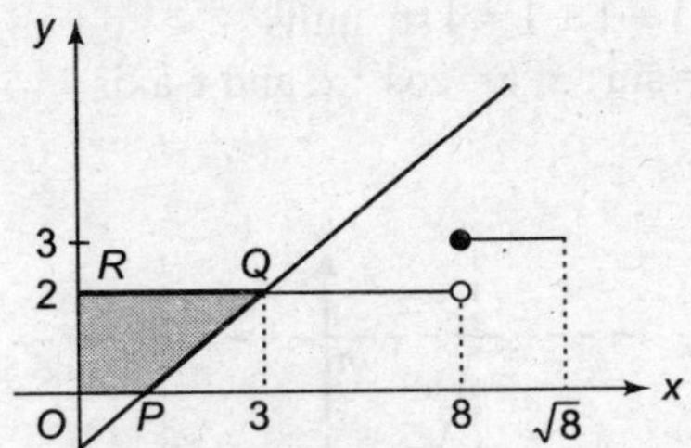

Fig. 9.64

3. a. → q; c. → p; b. → s; d. → p.

Sol.

a. $[x]^2 = [y]^2$, where $1 \le x \le 4$

$\Rightarrow [x] = \pm [y]$

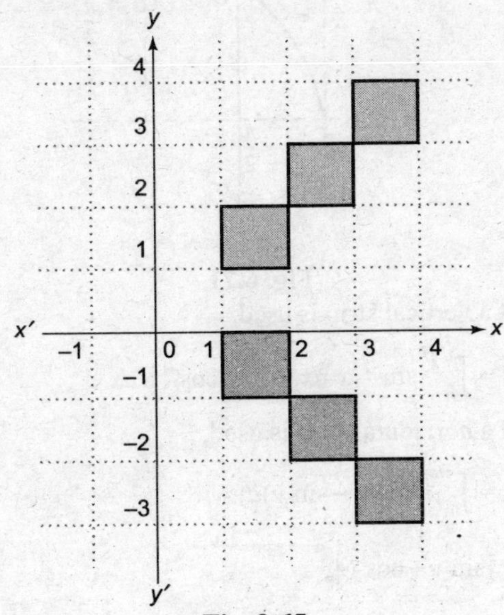

Fig. 9.65

b. $[|x|] \, | \, [|y|] = 2$

The graph is symmetrical about both the x-axis and y-axis.

For $x, y > 0$; $[x] + [y] = 2$.

$\Rightarrow [x] = 0$ and $[y] = 2$, $[x] = 1$ and $[y] = 1$ or $[x] = 2$ and $[y] = 0$.

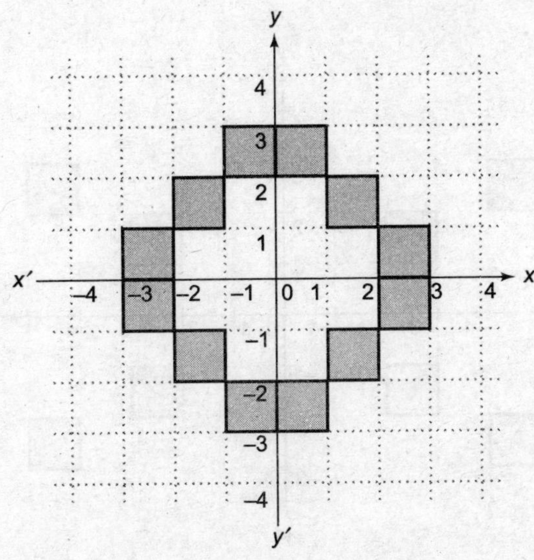

Fig. 9.66

c. $[|x|] \, [|y|] = 2$

The graph is symmetrical about both the x-axis and y-axis.

For $x, y > 0$; $[x][y] = 2 \Rightarrow [x] = 1$ and $[y] = 2$ or $[x] = 2$ and $[y] = 1$.

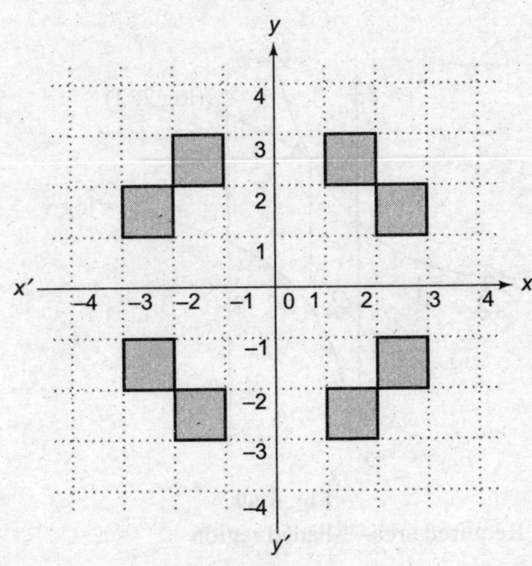

Fig. 9.67

d. $\frac{[|x|]}{[|y|]} = 2$, where $-5 \le x \le 5$.

The graph is symmetrical about both the axes.

For $x, y > 0$, $[x] = 2[y]$, $[y] \ne 0$.

$\Rightarrow [x] = 2$ and $[y] = 1$ or $[x] = 4$ and $[y] = 2$.

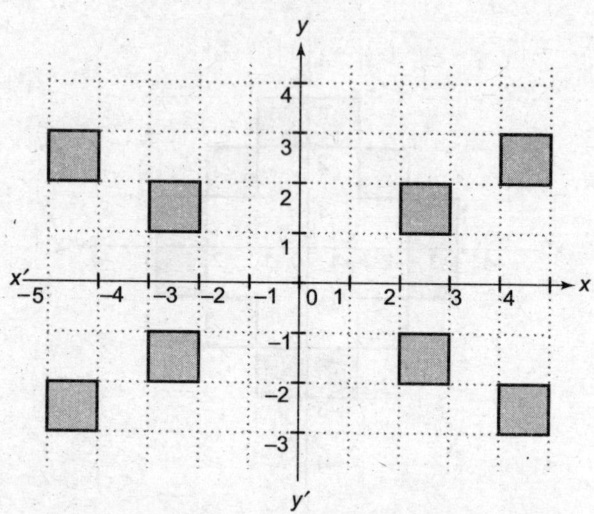

Fig. 9.68

Integer Type

1. (4)

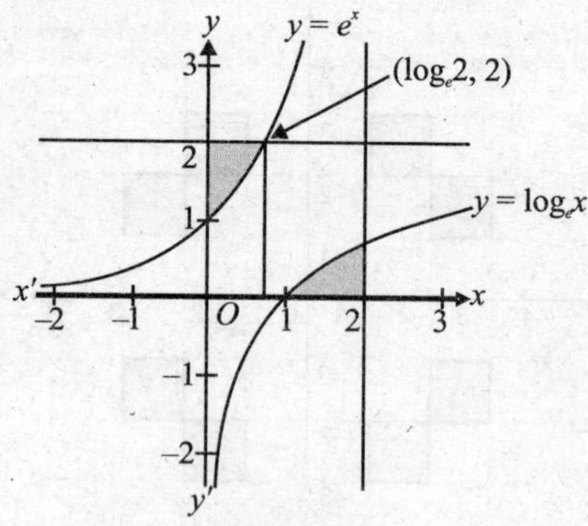

Fig. 6.69

Required area = shaded region

$= 2\int_0^{\ln 2} (2 - e^x)dx$

$= 2\,(2\ln 2 - 2)$ sq. units

$= 4(\ln 2 - 1)$ sq. units

2. (4)

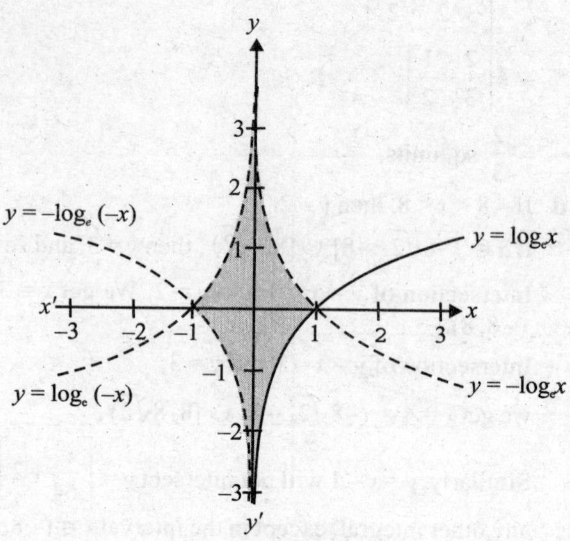

Fig. 6.70

From the figure, required area = area of shaded region $= 1 + 1 + 1 + 1 = 4$ sq. units

3. (0) $y = \sin^{-1} x$, $y = \cos^{-1} x$, and x-axis

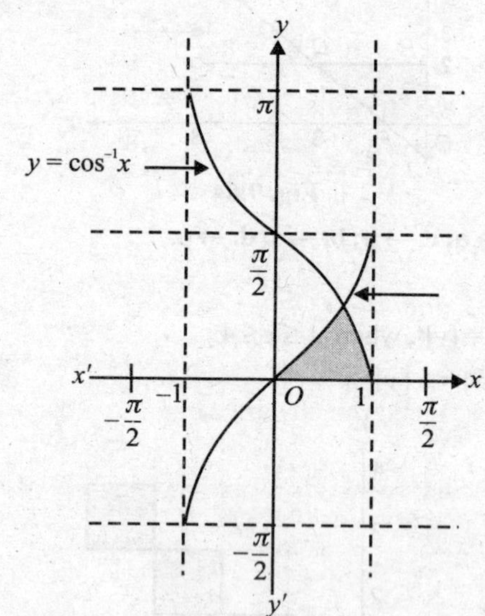

Fig. 6.71

If a vertical strip is used,

$$A = \int_0^{1/\sqrt{2}} \sin^{-1} x\,dx + \int_{1/\sqrt{2}}^{1} \cos^{-1} x\,dx$$

If a horizontal strip is used,

$$A = \int_0^{\pi/4} (\cos y - \sin y)dy$$

$$= [\sin y + \cos y]_0^{\pi/4}$$

$$= \left[\frac{1}{\sqrt{2}} + \frac{1}{\sqrt{2}} - 1\right]$$

$$= \sqrt{2} - 1$$

4. (2)

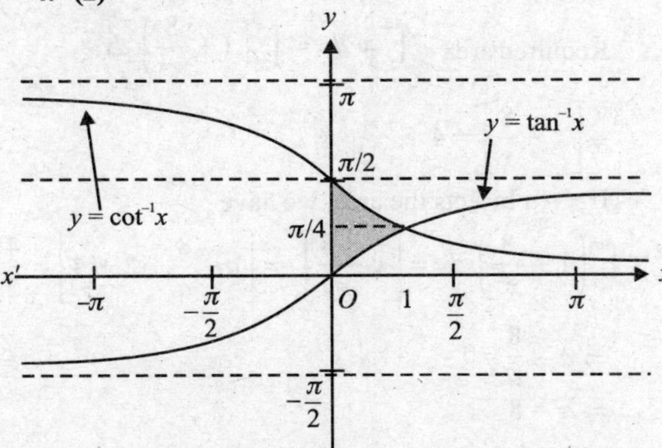

Fig. 6.72

Integrating along x-axis, we get

$$A = \int_0^1 (\cot^{-1} x - \tan^{-1} x)\, dx$$

$$= \int_0^1 \left(\frac{\pi}{2} - 2\tan^{-1} x\right) dx$$

Integrating along y-axis, we get

$$A = 2\int_0^{\pi/4} x\, dy = 2\int_0^{\pi/4} \tan y\, dy = [\log(\sec y)]_0^{\pi/4}$$

$$= \log\sqrt{2} \text{ sq. units}$$

5. (1) The given curves are

$y = x - bx^2$ (1)

and $y = x^2/b$ (2)

$$\Rightarrow \left(y - \frac{1}{4b}\right) = -b\left(x - \frac{1}{2b}\right)^2 \text{ and } x^2 = by$$

Clearly, the first curve is a downward parabola which meets the x-axis at (0, 0) and $(1/b, 0)$.

The second is an upward parabola with vertex at (0, 0).

Solving (1) and (2), we get the intersection points of the two curves as (0, 0) and

$$\left[\frac{b}{1+b^2}, \frac{b}{(1+b^2)^2}\right].$$

Hence, the graph of the given curves is as below.

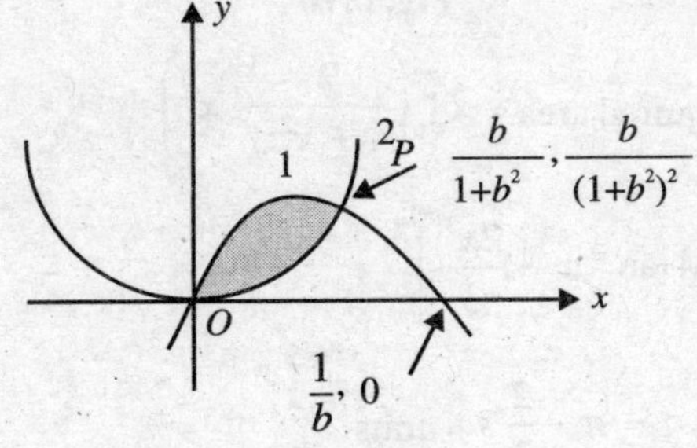

Fig. 6.73

Shaded portion represents the required area given by

$$A = \int_0^{b/(1+b^2)} \left(x - bx^2 - \frac{x^2}{b}\right) dx$$

$$= \left[\frac{x^2}{2} - \frac{bx^3}{3} - \frac{x^3}{3b}\right]_0^{b/(1+b^2)}$$

$$= \frac{b^2}{2(1+b^2)^2} - \frac{b^4}{3(1+b^2)^3} - \frac{b^2}{3(1+b^2)^3}$$

$$= \frac{b^4 + b^2}{6(1+b^2)^3}$$

$$= \frac{b^2}{6(1+b^2)^2}$$

$$= \frac{1}{6\left(\frac{1}{b} + b\right)^2} \text{ sq. units}$$

Now $1/b + b \geq 2$ or ≤ -2. So

$$\left(\frac{1}{b} + b\right)^2 \geq 4$$

So the area is maximum when $[(1/b) + b]^2_{\min} = 4$, for which $b = \pm 1$.

But since $b > 0$, we have $b = 1$.

6. (4) Let P on $C_1, y = x^2$, be (t, t^2). So the y-coordinate of Q is also t^2. Since Q is on $y = 2x$ and $y = t^2$, $x = t^2/2$.

Thus $Q\left(\frac{t^2}{2}, t^2\right)$

For point R, $x = t$, and it is on $y = f(x)$. Therefore, $R(t, f(t))$.

Given that Area OPQ = Area OPR

$$\Rightarrow \int_0^{t^2} \left(\sqrt{y} - \frac{y}{2}\right) dy = \int_0^t \left(x^2 - f(x)\right) dx$$

Differentiating both sides w.r.t. t, we get

$$\left(\sqrt{t^2} - \frac{t^2}{2}\right)(2t) = t^2 - f(t)$$

$$\Rightarrow f(t) = t^3 - t^2$$

$$= x^3 - x^2$$

7. (3)

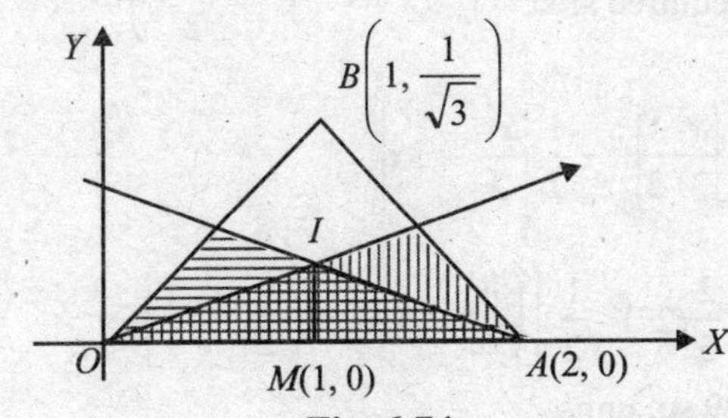

Fig. 6.74

$d(P, OA) \le \min[d(P, OB), d(P, AB)]$
$\Rightarrow d(P, OA) \le d(P, OB)$ and $d(P, OA) \le d(P, AB)$
When $d(P, OA) = d(P, OB)$, P is equidistant from OA and OB, or P lies on the angle bisector of the lines OA and OB. Hence, when $d(P, OA) \le d(P, OB)$, point P is nearer to OA than OB or lies below the bisector of OA and OB.
Similarly, when $d(P, OA) \le d(P, AB)$, P is nearer to OA than OB, or lies below the bisector of OA and AB.
Required area = area or ΔOIA

Now, $\tan \angle BOA = \frac{1/\sqrt{3}}{1} = \frac{1}{\sqrt{3}}$

$\Rightarrow \angle BOA = 30°$
$\Rightarrow \angle IOA = 15°$
$\Rightarrow IM = \tan 15° = 2 - \sqrt{3}$

Hence, area of $\Delta OIA = \frac{1}{2} OA \times IM$

$= \frac{1}{2} \times 2 \times (2 - \sqrt{3})$

$= 2 - \sqrt{3}$ sq. units

8. **(9)** To find the area between the curves $y = \sqrt{x}$, $2y + 3 = x$, and x-axis in the first quadrant.

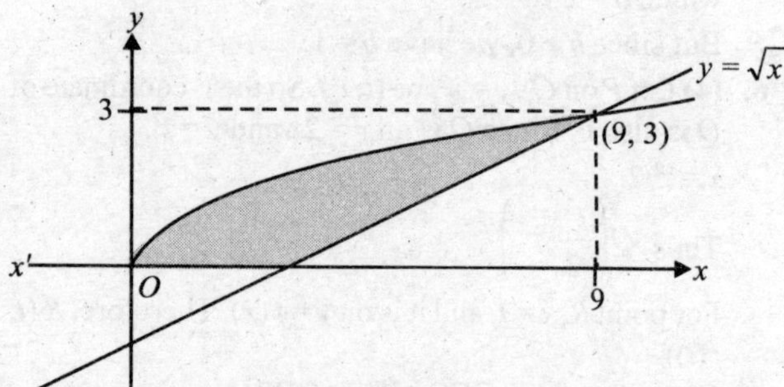

Fig. 6.75

The given curves intersect when
$y^2 = 2y + 3$
$\Rightarrow y^2 - 2y - 3 = 0$
$\Rightarrow (y - 3)(y + 1) = 0$
$\Rightarrow y = 3, -1$
When $y = 3, x = 9$. (First quadrant)

Required area $= \int_0^9 \sqrt{x}\, dx - \int_3^9 \left(\frac{x-3}{2}\right) dx$

$= \left|\frac{x^{3/2}}{3/2}\right|_0^9 - \frac{1}{2}\left|\frac{x^2}{2} - 3x\right|_3^9$

$= \frac{2}{3}(27) - \frac{1}{2}\left[\left(\frac{81}{2} - 27\right) - \left(\frac{9}{2} - 9\right)\right]$

$= 9$ sq. units.

9. **(8)** Given $y = 1 + \frac{8}{x^2}$.

Here y is always positive. So the curve lies above the x-axis.

Required area $= \int_2^4 y\, dx = \int_2^4 \left(1 + \frac{8}{x^2}\right) dx$

$= \left[x - \frac{8}{x}\right]_2^4 = 4$

If $x = a$ bisects the area, we have

$\int_2^a \left(1 + \frac{8}{x^2}\right) dx = \left[x - \frac{8}{x}\right]_2^a = \left[a - \frac{8}{a} - 2 + 4\right] = \frac{4}{2}$

$\Rightarrow a - \frac{8}{a} = 0$
$\Rightarrow a^2 = 8$
$\Rightarrow a = \pm 2\sqrt{2}$
Since $a > 2$, $a = 2\sqrt{2}$.

10. **(2)** The given curves are $y = x^2$ (1)

and $y = \frac{2}{1 + x^2}$ (2)

Solving (1) and (2), we get

$x^2 = \frac{2}{1 + x^2}$

$\Rightarrow x^4 + x^2 - 2 = 0$
$\Rightarrow (x^2 - 1)(x^2 + 2) = 0$
$\Rightarrow x = \pm 1$
Also $y = 2/(1 + x^2)$ is an even function. Hence, its graph is symmetrical about the y-axis.
At $x = 0, y = 2$.
When the value of x is increased, y decreases, and when $x \to \infty, y \to 0$.
Thus $y = 0$ is an asymptote of the given curve.
Thus the graph of the two curves is as follows.

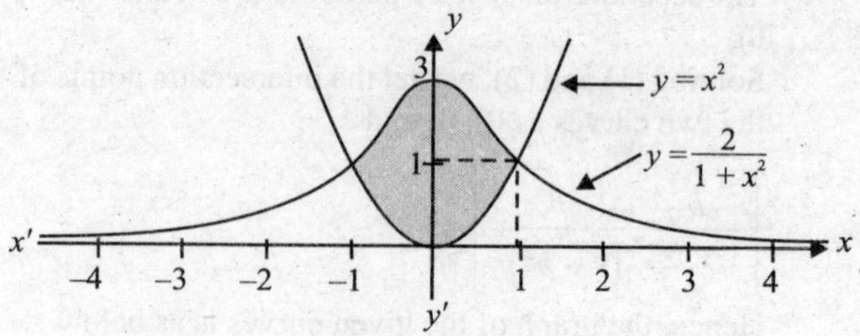

Fig. 6.76

Required area $= 2\int_0^1 \left(\frac{2}{1 + x^2} - x^2\right) dx$

$= \left[4 \tan^{-1} x - \frac{2x^3}{3}\right]_0^1$

$= \pi - \frac{2}{3}$ sq. units

CHAPTER 10

Differential Equations

Objective Type

Each question has four choices a, b, c, and d, out of which *only one* is correct.

1. The degree of the differential equation satisfying $\sqrt{1-x^2}+\sqrt{1-y^2}=a(x-y)$ is

 a. 1 b. 2

 c. 3 d. None of these

2. The differential equation whose solution is $Ax^2 + By^2 = 1$, where A and B are arbitrary constants is of

 a. second order and second degree

 b. first order and second degree

 c. first order and first degree

 d. second order and first degree

3. The differential equation of the family of curves $y = e^x(A\cos x + B\sin x)$, where A and B are arbitrary constants, is

 a. $\dfrac{d^2y}{dx^2}-2\dfrac{dy}{dx}+2y=0$

 b. $\dfrac{d^2y}{dx^2}+2\dfrac{dy}{dx}-2y=0$

 c. $\dfrac{d^2y}{dx^2}+\left(\dfrac{dy}{dx}\right)^2+y=0$

 d. $\dfrac{d^2y}{dx^2}-7\dfrac{dy}{dx}+2y=0$

4. Differential equation of the family of circles touching the line $y = 2$ at $(0, 2)$ is

 a. $x^2+(y-2)^2+\dfrac{dy}{dx}(y-2)=0$

 b. $x^2+(y-2)\left(2-2x\dfrac{dx}{dy}-y\right)=0$

 c. $x^2+(y-2)^2+\left(\dfrac{dx}{dy}+y-2\right)(y-2)=0$

 d. None of these

5. The differential equation of all parabolas whose axis are parallel to the y-axis is

 a. $\dfrac{d^3y}{dx^3}=0$ b. $\dfrac{d^2x}{dy^2}=C$

 c. $\dfrac{d^3y}{dx^3}+\dfrac{d^2x}{dy^2}=0$ d. $\dfrac{d^2y}{dx^2}+2\dfrac{dy}{dx}=C$

6. A differential equation associated to the primitive $y = a + be^{5x} + ce^{-7x}$ is

 a. $y_3+2y_2-y_1=0$ b. $4y_3+5y_2-20y_1=0$

 c. $y_3+2y_2-35y_1=0$ d. None of these

 where y_n represents n^{th} order derivative.

7. The differential equation of all circles which pass through the origin and whose centres lie on the y-axis is

 a. $(x^2-y^2)\dfrac{dy}{dx}-2xy=0$

 b. $(x^2-y^2)\dfrac{dy}{dx}+2xy=0$

c. $(x^2-y^2)\dfrac{dy}{dx}-xy=0$

d. $(x^2-y^2)\dfrac{dy}{dx}+xy=0$

8. The form of the differential equation of the central conics, is

a. $x=y\dfrac{dy}{dx}$ b. $x+y\dfrac{dy}{dx}=0$

c. $x\left(\dfrac{dy}{dx}\right)^2+xy\dfrac{d^2y}{dx^2}=y\dfrac{dy}{dx}$ d. None of these

9. The differential equation for the family of curve $x^2+y^2-2ay=0$, where a is an arbitrary constant, is

a. $2(x^2-y^2)y'=xy$ b. $2(x^2+y^2)y'=xy$

c. $(x^2-y^2)y'=2xy$ d. $(x^2+y^2)y'=2xy$

10. The differential equation of all circle passing through the origin and having their centres on the x-axis is

a. $x^2=y^2+3xy\dfrac{dy}{dx}$ b. $y^2=x^2+2xy\dfrac{dy}{dx}$

c. $y^2=x^2-2xy\dfrac{dy}{d}$ d. $x^2=y^2+xy\dfrac{dy}{dx}$

11. If $y=\dfrac{x}{\log|cx|}$ (where c is an arbitrary constant) is the general solution of the differential equation $dy/dx=y/x+\phi(x/y)$ then the function $\phi(x/y)$ is

a. x^2/y^2 b. $-x^2/y^2$

c. y^2/x^2 d. $-y^2/x^2$

12. The differential equation whose general solution is given by, $y=(c_1\cos(x+c_2)-(c_3e^{(-x+c_4)})+(c_5\sin x)$, where c_1, c_2, c_3, c_4, c_5 are arbitrary constants, is

a. $\dfrac{d^4y}{dx^4}-\dfrac{d^2y}{dx^2}+y=0$

b. $\dfrac{d^3y}{dx^3}+\dfrac{d^2y}{dx^2}+\dfrac{dy}{dx}+y=0$

c. $\dfrac{d^5y}{dx^5}+y=0$

d. $\dfrac{d^3y}{dx^3}-\dfrac{d^2y}{dx^2}+\dfrac{dy}{dx}-y=0$

13. The solution to the differential equation $y\log y+xy'=0$, where $y(1)=e$, is

a. $x(\log y)=1$ b. $xy(\log y)=1$

c. $(\log y)^2=2$ d. $\log y+\left(\dfrac{x^2}{2}\right)y=1$

14. If $y=y(x)$ and $\dfrac{2+\sin x}{y+1}\left(\dfrac{dy}{dx}\right)=-\cos x$, $y(0)=1$, then $y(\pi/2)$ equals

a. 1/3 b. 2/3

c. −1/3 d. 1

15. The equation of the curves through the point (1, 0) and whose slope is $\dfrac{y-1}{x^2+x}$ is

a. $(y-1)(x+1)+2x=0$

b. $2x(y-1)+x+1=0$

c. $x(y-1)(x+1)+2=0$

d. None of these

16. The solution of the equation $\dfrac{dy}{dx}=\dfrac{x(2\log x+1)}{\sin y+y\cos y}$ is

a. $y\sin y=x^2\log x+\dfrac{x^2}{2}+c$

b. $y\cos y=x^2(\log x+1)+c$

c. $y\cos y=x^2\log x+\dfrac{x^2}{2}+c$

d. $y\sin y=x^2\log x+c$

17. The solution of the equation $\log(dy/dx)=ax+by$ is

a. $\dfrac{e^{by}}{b}=\dfrac{e^{ax}}{a}+c$ b. $\dfrac{e^{-by}}{-b}=\dfrac{e^{ax}}{a}+c$

c. $\dfrac{e^{-by}}{a}=\dfrac{e^{ax}}{b}+c$ d. None of these

18. The solution of the equation $(x^2y+x^2)dx+y^2(x-1)\,dy=0$ is given by

a. $x^2+y^2+2(x-y)+2\ln\dfrac{(x-1)(y+1)}{c}=0$

b. $x^2+y^2+2(x-y)+\ln\dfrac{(x-1)(y+1)}{c}=0$

c. $x^2+y^2+2(x-y)-2\ln\dfrac{(x-1)(y+1)}{c}=0$

d. None of these

19. Solution of differential equation $dy-\sin x\sin y\,dx=0$ is

a. $e^{\cos x}\tan\dfrac{y}{2}=c$ b. $e^{\cos x}\tan y=c$

c. $\cos x\tan y=c$ d. $\cos x\sin y=c$

20. The solution of $\frac{dv}{dt}+\frac{k}{m}v=-g$ is

a. $v=ce^{-\frac{k}{m}t}-\frac{mg}{k}$ b. $v=c-\frac{mg}{k}e^{-\frac{k}{m}t}$

c. $ve^{-\frac{k}{m}t}=c-\frac{mg}{k}$ d. $ve^{\frac{k}{m}t}=c-\frac{mg}{k}$

21. The solution of the equation $dy/dx=\cos(x-y)$ is

a. $y+\cot\left(\frac{x-y}{2}\right)=C$

b. $x+\cot\left(\frac{x-y}{2}\right)=C$

c. $x+\tan\left(\frac{x-y}{2}\right)=C$

d. None of these

22. Solution of $\frac{dy}{dx}+2xy=y$ is

a. $y=ce^{x-x^2}$ b. $y=ce^{x^2-x}$

c. $y=ce^{x}$ d. $y=ce^{-x^2}$

23. The general solution of the differential equation $\frac{dy}{dx}+\sin\frac{x+y}{2}=\sin\frac{x-y}{2}$ is

a. $\log\tan\left(\frac{y}{2}\right)=c-2\sin x$

b. $\log\tan\left(\frac{y}{4}\right)=c-2\sin\left(\frac{x}{2}\right)$

c. $\log\tan\left(\frac{y}{2}+\frac{\pi}{4}\right)=c-2\sin x$

d. $\log\tan\left(\frac{y}{4}+\frac{\pi}{4}\right)=c-2\sin\left(\frac{x}{2}\right)$

24. The solutions of $(x+y+1)\,dy=dx$ is

a. $x+y+2=Ce^{y}$

b. $x+y+4=C\log y$

c. $\log(x+y+2)=Cy$

d. $\log(x+y+2)=C-y$

25. If for the differential equation $y'=\frac{y}{x}+\phi\left(\frac{x}{y}\right)$ the general solution is $y=\frac{x}{\log|Cx|}$ then $\phi(x/y)$ is given by

a. $-x^2/y^2$ b. y^2/x^2

c. x^2/y^2 d. $-y^2/x^2$

26. The slope of the tangent at (x, y) to a curve passing through $\left(1, \frac{\pi}{4}\right)$ is given by $\frac{y}{x}-\cos^2\left(\frac{y}{x}\right)$, then the equation of the curve is

a. $y=\tan^{-1}\left(\log\left(\frac{e}{x}\right)\right)$

b. $y=x\tan^{-1}\left(\log\left(\frac{x}{e}\right)\right)$

c. $y=x\tan^{-1}\left(\log\left(\frac{e}{x}\right)\right)$

d. None of these

27. If $x\frac{dy}{dx}=y(\log y-\log x+1)$ then the solution of the equation is

a. $\log\frac{x}{y}=cy$ b. $\log\frac{y}{x}=cy$

c. $\log\frac{x}{y}=cx$ d. None of these

28. The solution of differential equation $yy'=x\left(\frac{y^2}{x^2}+\frac{f(y^2/x^2)}{f'(y^2/x^2)}\right)$ is

a. $f(y^2/x^2)=cx^2$ b. $x^2f(y^2/x^2)=c^2y^2$

c. $x^2f(y^2/x^2)=c$ d. $f(y^2/x^2)=cy/x$

29. The solution of $(x^2+xy)\,dy=(x^2+y^2)\,dx$ is

a. $\log x=\log(x-y)+\frac{y}{x}+c$

b. $\log x=2\log(x-y)+\frac{y}{x}+c$

c. $\log x=\log(x-y)+\frac{x}{y}+c$

d. None of these

30. The solution of $(y+x+5)\,dy=(y-x+1)\,dx$ is

a. $\log((y+3)^2+(x+2)^2)+\tan^{-1}\frac{y+3}{y+2}+C$

b. $\log((y+3)^2+(x-2)^2)+\tan^{-1}\frac{y-3}{x-2}=C$

c. $\log((y+3)^2+(x+2)^2)+2\tan^{-1}\frac{y+3}{x+2}=C$

d. $\log((y+3)^2+(x+2)^2)-2\tan^{-1}\frac{y+3}{x+2}=C$

31. The slope of the tangent at (x, y) to a curve passing through a point $(2, 1)$ is $\frac{x^2+y^2}{2xy}$, then the equation of the curve is

a. $2(x^2-y^2)=3x$ b. $2(x^2-y^2)=6y$
c. $x(x^2-y^2)=6$ d. $x(x^2+y^2)=10$

32. If $x\dfrac{dy}{dx}=y(\log y-\log x+1)$, then the solution of the equation is

a. $x\log\left(\dfrac{y}{x}\right)=cy$ b. $y\log\left(\dfrac{x}{y}\right)=cx$
c. $\log\left(\dfrac{x}{y}\right)=cy$ d. $\log\left(\dfrac{y}{x}\right)=cx$

33. The general solution of the differential equation, $y'+y\phi'(x)-\phi(x).\phi'(x)=0$ where $\phi(x)$ is a known function is

a. $y=ce^{-\phi(x)}+\phi(x)-1$
b. $y=ce^{+\phi(x)}+\phi(x)-1$
c. $y=ce^{-\phi(x)}-\phi(x)+1$
d. $y=ce^{-\phi(x)}+\phi(x)+1$

where c is an arbitrary constant.

34. The solution of $\dfrac{dy}{dx}=\dfrac{x^2+y^2+1}{2xy}$ satisfying $y(1)=1$ is given by

a. a system of parabolas
b. a system of circles
c. $y^2=x(1+x)-1$
d. $(x-2)^2+(y-3)^2=5$

35. The integrating factor of the differential equation $\dfrac{dy}{dx}(x\log_e x)+y=2\log_e x$ is given by

a. x b. e^x
c. $\log_e x$ d. $\log_e(\log_e x)$

36. The solution of the differential equation $x(x^2+1)(dy/dx)=y(1-x^2)+x^3\log x$ is

a. $y(x^2+1)/x=\dfrac{1}{4}x^2\log x+\dfrac{1}{2}x^2+c$
b. $y^2(x^2-1)/x=\dfrac{1}{2}x^2\log x-\dfrac{1}{4}x^2+c$
c. $y(x^2+1)/x=\dfrac{1}{2}x^2\log x-\dfrac{1}{4}x^2+c$
d. None of these

37. Integrating factor of differential equation $\cos x\dfrac{dy}{dx}+y\sin x=1$ is

a. $\cos x$ b. $\tan x$
c. $\sec x$ d. $\sin x$

38. Solution of the equation $\cos^2 x\dfrac{dy}{dx}-(\tan 2x)y=\cos^4 x$, $|x|<\dfrac{\pi}{4}$, when $y\left(\dfrac{\pi}{6}\right)=\dfrac{3\sqrt{3}}{8}$ is

a. $y=\tan 2x\cos^2 x$ b. $y=\cot 2x\cos^2 x$
c. $y=\dfrac{1}{2}\tan 2x\cos^2 x$ d. $y=\dfrac{1}{2}\cot 2x\cos^2 x$

39. If integrating factor of $x(1-x^2)\,dy+(2x^2y-y-ax^3)\,dx=0$ is $e^{\int p dx}$, then P is equal to

a. $\dfrac{2x^2-ax^3}{x(1-x^2)}$ b. $2x^3-1$
c. $\dfrac{2x^2-a}{ax^3}$ d. $\dfrac{2x^2-1}{x(1-x^2)}$

40. A function $y=f(x)$ satisfies $(x+1)f'(x)-2(x^2+x)f(x)=\dfrac{e^{x^2}}{(x+1)}$, $\forall\ x>-1$.

If $f(0)=5$, then $f(x)$ is

a. $\left(\dfrac{3x+5}{x+1}\right)e^{x^2}$ b. $\left(\dfrac{6x+5}{x+1}\right)e^{x^2}$
c. $\left(\dfrac{6x+5}{(x+1)^2}\right)e^{x^2}$ d. $\left(\dfrac{5-6x}{x+1}\right)e^{x^2}$

41. The solution of the differential equation $\dfrac{dy}{dx}=\dfrac{1}{xy[x^2\sin y^2+1]}$ is

a. $x^2(\cos y^2-\sin y^2-2Ce^{-y^2})=2$
b. $y^2(\cos x^2-\sin y^2-2Ce^{-y^2})=2$
c. $x^2(\cos y^2-\sin y^2-e^{-y^2})=4C$
d. None of these

42. The general solution of the equation $\dfrac{dy}{dx}=1+xy$ is

a. $y=ce^{-x^2/2}$ b. $y=ce^{x^2/2}$
c. $y=(x+c),e^{-x^2/2}$ d. None of these

43. The solution of the differential equation, $(x+2y^3)\dfrac{dy}{dx}=y$ is

a. $\dfrac{x}{y^2}=y+c$ b. $\dfrac{x}{y}=y^2+c$
c. $\dfrac{x^2}{y}=y^2+c$ d. $\dfrac{y}{x}=x^2+c$

44. The solution of the differential equation, $x^2\dfrac{dy}{dx}\cos\dfrac{1}{x}-y\sin\dfrac{1}{x}=-1$, where $y\to-1$ as $x\to\infty$ is

a. $y=\sin\dfrac{1}{x}-\cos\dfrac{1}{x}$ b. $y=\dfrac{x+1}{x\sin\frac{1}{x}}$
c. $y=\cos\dfrac{1}{x}+\sin\dfrac{1}{x}$ d. $y=\dfrac{x+1}{x\cos 1/x}$

45. The solution of the differential equation, $2x^2y\dfrac{dy}{dx} = \tan(x^2y^2) - 2xy^2$ given $y(1) = \sqrt{\dfrac{\pi}{2}}$ is

a. $\sin x^2y^2 = e^{x-1}$ b. $\sin(x^2y^2) = x$

c. $\cos x^2y^2 + x = 0$ d. $\sin(x^2y^2) = e\,e^x$

46. Solution of the differential equation

$$\left\{\frac{1}{x} - \frac{y^2}{(x-y)^2}\right\}dx + \left\{\frac{x^2}{(x-y)^2} - \frac{1}{y}\right\}dy = 0 \text{ is}$$

a. $\ln\left|\dfrac{x}{y}\right| + \dfrac{xy}{x-y} = c$ b. $\dfrac{xy}{x-y} = ce^{x/y}$

c. $\ln|xy| = c + \dfrac{xy}{x-y}$ d. None of these

47. If $y + x\dfrac{dy}{dx} = x\dfrac{\phi(xy)}{\phi'(xy)}$ then $\phi(xy)$ is equation to

a. $ke^{x^2/2}$ b. $ke^{y^2/2}$

c. $ke^{xy/2}$ d. ke^{xy}

48. The solution of differential equation $(2y + xy^3)\,dx + (x + x^2y^2)\,dy = 0$ is

a. $x^2y + \dfrac{x^3y^3}{3} = c$ b. $xy^2 + \dfrac{x^3y^3}{3} = c$

c. $x^2y + \dfrac{x^4y^4}{4} = c$ d. None of these

49. The solution of $ye^{-x/y}dx - (xe^{(-x/y)} + y^3)dy = 0$ is

a. $e^{-x/y} + y^2 = C$ b. $xe^{-x/y} + y = C$

c. $2e^{-x/y} + y^2 = C$ d. $e^{-x/y} + 2y^2 = C$

50. The curve satisfying the equation $\dfrac{dy}{dx} = \dfrac{y\,(x+y^3)}{x\,(y^3 - x)}$ and passing through the point $(4, -2)$ is

a. $y^2 = -2x$ b. $y = -2x$

c. $y^3 = -2x$ d. None of these

51. The solution of differential equation

$$\frac{x + y\dfrac{dy}{dx}}{y - x\dfrac{dy}{dx}} = \frac{x\cos^2(x^2+y^2)}{y^3} \text{ is}$$

a. $\tan(x^2+y^2) = \dfrac{x^2}{y^2} + c$

b. $\cot(x^2+y^2) = \dfrac{x^2}{y^2} + c$

c. $\tan(x^2+y^2) = \dfrac{y^2}{x^2} + c$

d. $\cot(x^2+y^2) = \dfrac{y^2}{x^2} + c$

52. The solution of the differential equation

$$\frac{dy}{dx} = \frac{3x^2y^4 + 2xy}{x^2 - 2x^3y^3} \text{ is}$$

a. $\dfrac{y^2}{x} - x^3y^2 = c$ b. $\dfrac{x^2}{y^2} + x^3y^3 = c$

c. $\dfrac{x^2}{y} + x^3y^2 = c$ d. $\dfrac{x^2}{3y} - 2x^3y^2 = c$

53. The solution of the differential equation

$$\{1 + x\sqrt{(x^2+y^2)}\}dx + \{\sqrt{(x^2+y^2)} - 1\}\,y\,dy = 0 \text{ is}$$

a. $x^2 + \dfrac{y^2}{2} + \dfrac{1}{3}(x^2+y^2)^{3/2} = c$

b. $x - \dfrac{y^3}{3} + \dfrac{1}{2}(x^2+y^2)^{1/2} = c$

c. $x - \dfrac{y^2}{2} + \dfrac{1}{3}(x^2+y^2)^{3/2} = c$

d. None of these

54. The differential equation of family of curves whose tangent form an angle of $\pi/4$ with the hyperbola $xy = C^2$ is

a. $\dfrac{dy}{dx} = \dfrac{x^2 + C^2}{x^2 - C^2}$ b. $\dfrac{dy}{dx} = \dfrac{x^2 - C^2}{x^2 + C^2}$

c. $\dfrac{dy}{dx} = -\dfrac{C^2}{x^2}$ d. None of these

55. Tangent to a curve intercepts the y-axis at a point P. A line perpendicular to this tangent through P passes through another point $(1, 0)$. The differential equation of the curve is

a. $y\dfrac{dy}{dx} - x\left(\dfrac{dy}{dx}\right)^2 = 1$

b. $\dfrac{xd^2y}{dx^2} + \left(\dfrac{dy}{dx}\right)^2 = 0$

c. $y\dfrac{dx}{dy} + x = 1$

d. None of these

56. The differential equation of the curve for which the initial ordinate of any tangent is equal to the corresponding subnormal

a. is linear

b. is homogeneous of 2nd degree

c. has separable variables

d. is of 2nd order

57. Orthogonal trajectories of family of the curve $x^{2/3}+y^{2/3}=a^{2/3}$, where a is any arbitrary constant, is

a. $x^{2/3}-y^{2/3}=c$ b. $x^{4/3}-y^{4/3}=c$

c. $x^{4/3}+y^{4/3}=c$ d. $x^{1/3}-y^{1/3}=c$

58. The differential equation of all non-horizontal lines in a plane is

a. $\frac{d^2y}{dx^2}$ b. $\frac{d^2x}{dy^2}=0$

c. $\frac{dy}{dx}=0$ d. $\frac{dx}{dy}=0$

59. The curve for which the normal at any point (x, y) and the line joining origin to that point form an isosceles triangle with the x-axis as base is

a. an ellipse

b. a rectangular hyperbola

c. a circle

d. None of these

60. The equation of the curve which is such that the portion of the axis of x cut off between the origin and tangent at any point is proportional to the ordinate of that point is

a. $x=y(a-b\log x)$ b. $\log x=by^2+a$

c. $x^2=y(a-b\log y)$ d. None of these

(b is a constant of proportionality)

61. The family of curves represented by $\frac{dy}{dx}=\frac{x^2+x+1}{y^2+y+1}$ and $\frac{dy}{dx}+\frac{y^2+y+1}{x^2+x+1}=0$

a. Touch each other b. Are orthogonal

c. Are one and the same d. None of these

62. A normal at $P(x, y)$ on a curve meets the x-axis at Q and N is the foot of the ordinate at P. If $NQ=\frac{x(1+y^2)}{1+x^2}$, then the equation of curve given that it passes through the point (3, 1) is

a. $x^2-y^2=8$ b. $x^2+2y^2=11$

c. $x^2-5y^2=4$ d. None of these

63. A curve is such that the mid point of the portion of the tangent intercepted between the point where the tangent is drawn and the point where the tangent meets y-axis, lies on the line $y=x$. If the curve passes through (1, 0), then the curve is

a. $2y=x^2-x$ b. $y=x^2-x$

c. $y=x-x^2$ d. $y=2(x-x^2)$

64. The equation of a curve passing through (2, 7/2) and having gradient $1-\frac{1}{x^2}$ at (x, y) is

a. $y=x^2+x+1$ b. $xy=x^2+x+1$

c. $xy=x+1$ d. None of these

65. A normal at any point (x, y) to the curve $y=f(x)$ cuts triangle of unit area with the axis, the equation of the curve is

a. $y^2-x^2\left(\frac{dy}{dx}\right)^2=4\frac{dy}{dx}$

b. $x^2-y^2\left(\frac{dy}{dx}\right)^2=\frac{dy}{dx}$

c. $x+y\frac{dy}{dx}=y$

d. None of these

66. The normal to a curve at $P(x, y)$ meets the x–axis at G. If the distance of G from the origin is twice the abscissa of P, then the curve is a

a. parabola b. circle

c. hyperbola d. ellipse

67. The x-intercept of the tangent to a curve is equal to the ordinate of the point of contact. The equation of the curve through the point (1, 1) is

a. $ye^{x/y}=e$ b. $xe^{x/y}=e$

c. $xe^{y/x}=e$ d. $ye^{y/x}=e$

68. The equation of a curve passing through (1, 0) for which the product of the abscissa of a point P and the intercept made by a normal at P on the x-axis equals twice the square of the radius vector of the point P, is

a. $x^2+y^2=x^4$ b. $x^2+y^2=2x^4$

c. $x^2+y^2=4x^4$ d. None of these

69. The differential equation of all parabolas each of which has a latus rectum $4a$ and whose axis are parallel to the x-axis is

a. of order 1 and degree 2

b. of order 2 and degree 3

c. of order 2 and degree 1

d. of order 2 and degree 2

70. The curve, with the property that the projection of the ordinate on the normal is constant and has a length equal to a, is

a. $a \ln\left(\sqrt{y^2 - a^2} + y\right) = x + c$

b. $x + \sqrt{a^2 - y^2} = c$

c. $(y - a)^2 = cx$

d. $ay = \tan^{-1}(x + c)$

71. The solution of the differential equation $y(2x^4 + y)\dfrac{dy}{dx} = (1 - 4xy^2)x^2$ is given by

a. $3(x^2y)^2 + y^3 - x^3 = c$

b. $xy^2 + \dfrac{y^3}{3} - \dfrac{x^3}{3} + c = 0$

c. $\dfrac{2}{5}yx^5 + \dfrac{y^3}{3} = \dfrac{x^3}{3} - \dfrac{4xy^3}{3} + c$

d. None of these

72. The solution of the differential equation $(x \cot y + \log \cos x)\, dy + (\log \sin y - y \tan x)\, dx = 0$

a. $(\sin x)^y (\cos y)^x = c$ b. $(\sin y)^x (\cos x)^y = c$

c. $(\sin x)^x (\cos y)^y = c$ d. None of these

73. Spherical rain drop evaporates at a rate proportional to its surface area. The differential equation corresponding to the rate of change of the radius of the rain drop if the constant of proportionality is $K > 0$, is

a. $\dfrac{dr}{dt} + K = 0$ b. $\dfrac{dr}{dt} - K = 0$

c. $\dfrac{dr}{dt} = Kr$ d. None of these

74. Water is drained from a vertical cylindrical tank by opening a valve at the base of the tank. It is known that the rate at which the water level drops is proportional to the square root of water depth y, where the constant of proportionality $k > 0$ depends on the acceleration due to gravity and the geometry of the hole. If t is measured in minutes and $k = 1/15$ then the time to drain the tank if the water is 4 m deep to start with is

a. 30 min b. 45 min

c. 60 min d. 80 min

75. The population of a country increases at a rate proportional to the number of inhabitants. f is the population which doubles in 30 years, then the population will triple in approximately

a. 30 years b. 45 years

c. 48 years d. 54 years

76. An object falling from rest in air is subject not only to the gravitational force but also to air resistnace. Assume that the air resistance is proportional to the velocity with constant of proportionality as $k > 0$, and acts in a direction opposite to motion ($g = 9.8$ m/s^2). Then velocity cannot exceed

a. $9.8/k$ m/s b. $98/k$ m/s

c. $\dfrac{k}{9.8}$ m/s d. None of these

77. The solution to differential equation

$$x^2 = 1 + \left(\frac{x}{y}\right)^{-1}\frac{dy}{dx} + \frac{\left(\frac{x}{y}\right)^{-2}\left(\frac{dy}{dx}\right)^2}{2!} + \frac{\left(\frac{x}{y}\right)^{-3}\left(\frac{dy}{dx}\right)^3}{3!} + \cdots \text{ is}$$

a. $y^2 = x^2(\ln x - 1) + c$ b. $y = x^2(\ln x - 1) + c$

c. $y^2 = x(\ln x - 1) + c$ d. $y = x^2 e^{x^2} + c$

78. The solution to the differential equation $y'y''' = 3(y'')^2$ is

a. $x = A_1y^2 + A_2y + A_3$ b. $x = A_1y + A_2$

c. $x = A_1y^2 + A_2y$ d. None of these

79. Number of values of $m \in N$ for which $y = e^{mx}$ is a solution of the differential equation

$$\frac{d^3y}{dx^3} - 3\frac{d^2y}{dx^2} - 4\frac{dy}{dx} + 12y = 0$$

a. 0 b. 1

c. 2 d. More than 2

80. A curve passing through (2, 3) and satisfying the differential equation $\int_0^x t\,y(t)\,dt = x^2y(x)$, $(x > 0)$ is

a. $x^2 + y^2 = 13$ b. $y^2 = \dfrac{9}{2}x$

c. $\dfrac{x^2}{8} + \dfrac{y^2}{18} = 1$ d. $xy = 6$

81. The solution to the differential equation $\dfrac{d^2y}{dx^2} = \sin 3x + e^x + x^2$ when $y_1(0) = 1$ and $y(0) = 0$ is

a. $\dfrac{-\sin 3x}{9} + e^x + \dfrac{x^4}{12} + \dfrac{1}{3}x - 1$

b. $\dfrac{-\sin 3x}{9} + e^x + \dfrac{x^4}{12} + \dfrac{1}{3}x$

c. $\dfrac{-\cos 3x}{3} + e^x + \dfrac{x^4}{12} + \dfrac{1}{3}x + 1$

d. None of these

82. The solution to the differential equation

$$\frac{x + \frac{x^3}{3!} + \frac{x^5}{5!} + \cdots}{1 + \frac{x^2}{2!} + \frac{x^4}{4!} + \cdots} = \frac{dx - dy}{dx + dy} \text{ is}$$

a. $2y\,e^{2x} = C\,e^{2x} + 1$ b. $2y\,e^{2x} = C\,e^{2x} - 1$

c. $y\,e^{2x} = C\,e^{2x} + 2$ d. None of these

83. The solution to the differential equation

$$x=1+xy\frac{dy}{dx}+\frac{x^2y^2}{2!}\left(\frac{dy}{dx}\right)^2+\frac{x^3y^3}{3!}\left(\frac{dy}{dx}\right)^3+\cdots \text{ is}$$

a. $y=\ln(x)+c$ b. $y^2=(\ln x)^2+c$

c. $y=\log x+xy$ d. $xy=x^y+c$

84. The differential equation of the curve

$\frac{x}{c-1}+\frac{y}{c+1}=1$ is given by

a. $\left(\frac{dy}{dx}-1\right)\left(y+x\frac{dy}{dx}\right)=2\frac{dy}{dx}$

b. $\left(\frac{dy}{dx}+1\right)\left(y-x\frac{dy}{dx}\right)=\frac{dy}{dx}$

c. $\left(\frac{dy}{dx}+1\right)\left(y-x\frac{dy}{dx}\right)=2\frac{dy}{dx}$

d. None of these

85. The function $f(\theta)=\frac{d}{d\theta}\int_0^{\theta}\frac{dx}{1-\cos\theta\cos x}$ satisfies the differential equation

a. $\frac{df}{d\theta}+2f(\theta)\cot\theta=0$

b. $\frac{df}{d\theta}-2f(\theta)\cot\theta=0$

c. $\frac{df}{d\theta}+2f(\theta)=0$

d. $\frac{df}{d\theta}-2f(\theta)=0$

86. Differential equation of the family of curves $v=A/r+B$, where A and B are arbitrary constants, is

a. $\frac{d^2v}{dr^2}+\frac{1}{r}\frac{dv}{dr}=0$ b. $\frac{d^2v}{dr^2}-\frac{2}{r}\frac{dv}{dr}=0$

c. $\frac{d^2v}{dr^2}+\frac{2}{r}\frac{dv}{dr}=0$ d. None of these

87. The solution to the differential equation $y'''-8y''=0$ where $y(0)=\frac{1}{8}, y_1(0)=0, y_2(0)=1$ is

a. $y=\frac{1}{8}\left(\frac{e^{8x}}{8}+x-\frac{7}{9}\right)$

b. $y=\frac{1}{8}\left(\frac{e^{8x}}{8}+x+\frac{7}{8}\right)$

c. $y=\frac{1}{8}\left(\frac{e^{8x}}{8}-x+\frac{7}{8}\right)$

d. None of these

88. Solution of the differential equation

$$\left(e^{x^2}+e^{y^2}\right)y\frac{dy}{dx}+e^{x^2}(xy^2-x)=0 \text{ is}$$

a. $e^{x^2}(y^2-1)+e^{y^2}=C$

b. $e^{y^2}(x^2-1)+e^{x^2}=C$

c. $e^{y^2}(y^2-1)+e^{x^2}=C$

d. $e^{x^2}(y-1)+e^{y^2}=C$

Multiple Correct Answers Type

Each question has four choices a, b, c, and d, out of which *one or more* answers are correct.

1. Which one of the following functions is/are homogeneous?

a. $f(x,y)=\frac{x-y}{x^2+y^2}$

b. $f(x,y)=x^{1/3}y^{-2/3}\tan^{-1}\frac{x}{y}$

c. $f(x,y)=x(\ln\sqrt{x^2+y^2}-\ln y)+ye^{x/y}$

d. $f(x,y)=x\left[\ln\frac{2x^2+y^2}{x}-\ln(x+y)\right]+y^2\tan\frac{x+2y}{3x-y}$

2. For the differential equation whose solution is $(x-h)^2+(y-k)^2=a^2$ is (a is a constant)

a. order is 2 b. order is 3

c. degree 2 d. degree 3

3. The equation of the curve satisfying the differential equation $y\left(\frac{dy}{dx}\right)^2+(x-y)\frac{dy}{dx}-x=0$ can be a

a. Circle b. Straight line

c. Parabola d. Ellipse

4. Which of the following equation(s) is/are linear?

a. $\frac{dy}{dx}+\frac{y}{x}=\log x$ b. $y\left(\frac{dy}{dx}\right)+4x=0$

c. $(2x+y^3)\left(\frac{dy}{dx}\right)=3y$ d. None of these

5. The solution of $\frac{dy}{dx} = \frac{ax+h}{by+k}$ represents a parabola when

a. $a=0, b\neq 0$ b. $a\neq 0, b\neq 0$

c. $b=0, a\neq 0$ d. $a=0, b\in R$

6. The equation of the curve satisfying the differential equation $y_2(x^2+1)=2xy_1$ passing through the point (0, 1) and having slope of tangent at $x=0$ as 3 (where y_2 and y_1 represents 2nd and 1st order derivative), then

a. $y=f(x)$ is strictly increasing function

b. $y=f(x)$ is non-monotonic function

c. $y=f(x)$ has three distinct real roots

d. $y=f(x)$ has only one negative root

7. Identify the statement(s) which is/are true.

a. $f(x,y)=e^{y/x}+\tan(y/x)$ is homogeneous of degree zero.

b. $x\ln\frac{y}{x}dx+\frac{y^2}{x}\sin^{-1}\frac{y}{x}dy=0$ is homogeneous differential equation.

c. $f(x,y)=x^2+\sin x\cos y$ is not homogeneous.

d. $(x^2+y^2)dx-(xy^2-y^3)dy=0$ is a homegeneous differential equation.

8. The graph of the function $y=f(x)$ passing through the point (0, 1) and satisfying the differential equation $\frac{dy}{dx}+y\cos x=\cos x$ is such that

a. It is a constant function

b. It is periodic

c. It is neither an even nor an odd function

d. It is continuous and differentiable for all x

9. If $f(x)$, $g(x)$ be twice differential functions on [0, 2] satisfying $f''(x)=g''(x)$, $f'(1)=2g'(1)=4$ and $f(2)=3g(2)=9$, then

a. $f(4)-g(4)=10$

b. $|f(x)-g(x)|<2\Rightarrow -2<x<0$

c. $f(2)=g(2)\Rightarrow x=-1$

d. $f(x)-g(x)=2x$ has real root

10. A solution of the differential equation $(x^2y^2-1)\,dy+2x\,y^3\,dx=0$ is

a. $1+x^2y^2=cx$ b. $1+x^2y^2=cy$

c. $y=0$ d. $y=-\frac{1}{x^2}$

11. $y=ae^{-1/x}+b$ is a solution of $\frac{dy}{dx}=\frac{y}{x^2}$ then

a. $a\in R$

b. $b=0$

c. $b=1$

d. a takes finite number of values

12. For equation of the curve whose subnormal is constant, then

a. Its eccentricity is 1

b. Its eccentricity is $\sqrt{2}$

c. Its axis is x-axis

d. Its axis is y-axis.

13. The solution of $\frac{xdx+ydy}{xdy-ydx}=\sqrt{\frac{1-x^2-y^2}{x^2+y^2}}$ is

a. $\sqrt{x^2+y^2}=\sin\{\tan^{-1}(y/x)+C\}$

b. $\sqrt{x^2+y^2}=\cos\{(\tan^{-1}y/x)+C\}$

c. $\sqrt{x^2+y^2}=(\tan(\sin^{-1}y/x)+C)$

d. $y=x\tan\left(c+\sin^{-1}\sqrt{x^2+y^2}\right)$

14. The curve for which the length of the normal is equal to the length of the radius vector is/are

a. circles

b. rectangular hyperbola

c. ellipses

d. straight lines

15. In which of the following differential equation degree is not defined

a. $\frac{d^2y}{dx^2}+3\left(\frac{dy}{dx}\right)^2=x\log\frac{d^2y}{dx^2}$

b. $\left(\frac{d^2y}{dx^2}\right)^2+\left(\frac{dy}{dx}\right)^2=x\sin\left(\frac{d^2y}{dx^2}\right)$

c. $x=\sin\left(\frac{dy}{dx}-2y\right), |x|<1$

d. $x-2y=\log\left(\frac{dy}{dx}\right)$

Reasoning Type

Each question has four choices a, b, c, and d out of which *only one* is correct. Each question contains Statement 1 and Statement 2.

a. If both the statements are True and Statement 2 is the correct explanation of Statement 1

b. If both the statements are True but Statement 2 is Not the correct explanation of Statement 1

c. If Statement 1 is True and Statement 2 is False.

d. If Statement 1 is False and Statement 2 is True.

1. **Statement 1:** The differential equation of all circles in a plane must be of order 3.

   Statement 2: There is only one circle passing through three non-collinear points.

2. **Statement 1:** The differential equation of the family of curves represented by $y = Ae^x$ is given by $\frac{dy}{dx} = y$.

   Statement 2: $\frac{dy}{dx} = y$ is valid for every member of the given family.

3. **Statement 1:** Degree of differential equation $2x - 3y + 2 = \log\left(\frac{dy}{dx}\right)$ is not defined.

   Statement 2: In the given differential equation the power of highest order derivative when expressed as a polynomials of derivatives is called degree.

4. **Statement 1:** Order of differential equation represents number of arbitrary constants in the general solution.

   Statement 2: Degree of differential equation represents number of family of curves.

5. **Statement 1:** The order of the differential equation whose general solution is $y = c_1\cos 2x + \cos_2\sin^2 x + c_3\cos^2 x + c_4 e^{2x} + c_5\, e^{2x+c_6}$ is 3.

   Statement 2: Total number of arbitrary parameters in the given general solution in the statement (1) is 3.

Linked Comprehension Type

Based upon each paragraph, three multiple choice questions have to be answered. Each question has four choices a, b, c, and d out of which *only one* is correct.

For Problems 1–3

Let $f(x)$ be non-positive continuous function and $F(x) = \int_0^x f(t)dt \;\forall\, x \geq 0$ and $f(x) \geq cF(x)$ where $c > 0$ and let $g: [0, \infty) \to R$ be a function such that $\frac{dg(x)}{dx} < g(x) \;\forall\, x > 0$ and $g(0) = 0$.

1. The total number of roots of the equation $f(x) = g(x)$ is/are

   a. ∞ b. 1
   c. 2 d. 0

2. The number of solutions of the equation $|x^2 + x - 6| = f(x) + g(x)$ is/are

   a. 2 b. 1
   c. 0 d. 3

3. The solution set of inequation $g(x)(\cos^{-1}x - \sin^{-1}x) \leq 0$

   a. $\left[-1, \frac{1}{\sqrt{2}}\right]$ b. $\left[\frac{1}{\sqrt{2}}, 1\right]$
   c. $\left[0, \frac{1}{\sqrt{2}}\right]$ d. $\left(0, \frac{1}{\sqrt{2}}\right]$

For Problems 4–6

The differential equation $y = px + f(p)$, (1)

where $p = \frac{dy}{dx}$, is known as Clairout's Equation. To solve equation (1), differentiate it w.r.t. x, which gives either $\frac{dp}{dx} = 0$

$\Rightarrow p = c$ (2)

or $x + f'(p) = 0$ (3)

Note:

a. *If p is eliminated between equations* (1) *and* (2), *the solution obtained is a general solution of equation* (1).

b. *If p is eliminated between equation* (1) *and* (3), *then solution obtained does not contain any arbitrary constant and is not particular solution of equation* (1). *This solution is called singular solution of equation* (1).

4. Which of the following is about solutions of differential equation $y = xy' + \sqrt{1 + y'^2}$?

   a. The general solution of equation is family of parabolas
   b. The general solution of equation is family of circles
   c. The singular solution of equation is circle
   d. The singular solution of equation is ellipse

5. The number of solution of the equation $f(x) = -1$ and the singular solution of the equation $y = x\frac{dy}{dx} + \left(\frac{dy}{dx}\right)^2$ is

   a. 1 b. 2
   c. 4 d. 0

6. The singular solution of the differential equation $y = mx + m - m^3$ where, $m = \frac{dy}{dx}$ passes through the point

   a. (0, 0) b. (0, 1)
   c. (1, 0) d. (−1, 0)

For Problems 7–9

For certain curves $y = f(x)$ satisfying $\dfrac{d^2y}{dx^2} = 6x - 4, f(x)$ has local minimum value 5 when $x = 1$.

7. Number of critical point for $y = f(x)$ for $x \in [0, 2]$

a. 0 b. 1
c. 2 d. 3

8. Global minimum value of $y = f(x)$ for $x \in [0, 2]$ is

a. 5 b. 7
c. 8 d. 9

9. Global maximum value of $y = f(x)$ for $x \in [0, 2]$ is

a. 5 b. 7
c. 8 d. 9

For Problems 10–12

A certain radioactive material is known to decay at a rate proportional to the amount present. Initially there is 50 kg of the material present and after two hours it is observed that the material has lost 10 percent of its original mass. Based on these data answer the following questions.

10. The expression for the mass of the material remaining at any time t

a. $N = 50e^{-(1/2)(\ln 0.9)\,t}$ b. $50e^{-(1/4)(\ln 9)\,t}$
c. $N = 50e^{-(\ln 0.9)\,t}$ d. None of these

11. The mass of the material after four hours

a. $50^{-0.5\ln 9}$ b. $50e^{-2\ln 9}$
c. $50e^{-2\ln 0.9}$ d. None of these

12. The time at which the material has decayed to one half of its initial mass.

a. (ln 1/2) / (1/2 ln 9) hr
b. (ln 2) / (–1/2 ln 0.9) hr
c. (ln 1/2) / (–1/2 ln 0.9) hr
d. None of these

For Problems 13–15

Consider a tank which initially holds V_0 litre. of brine that contains a pound of salt. Another brine solution, containing b pound of salt/litre., is poured into the tank at the rate of e litre./minute while, simultaneously, the well-stirred solution leaves the tank at the rate of f litre./minute. The problem is to find the amount of salt in the tank at any time t.

Let Q denote the amount of salt in the tank at any time. The time rate of change of Q, dQ/dt, equals the rate at which salt enters the tank minus the rate at which salt leaves the tank. Salt enters the tank at the rate of be pound/minute. To determine the rate at which salt leaves the tank, we first calculate the volume of brine in the tank at any time t, which is the initial volume V_0 plus the volume of brine added et minus the volume of brine removed ft. Thus, the volume of brine at any time is

$$V_0 + et - ft \quad \text{(a)}$$

The concentration of salt in the tank at any time is $Q/(V_0 + et - ft)$, from which it follows that salt leaves the tank at the rate of $f\left(\dfrac{Q}{V_0 + et - ft}\right)$ lb/min.

Thus, $$\frac{dQ}{dt} = be - f\left(\frac{Q}{V_0 + et - ft}\right) \quad \text{(b)}$$

or $$\frac{dQ}{dt} + \frac{f}{V_0 + et - ft}Q = be$$

13. A tank initially holds 100 l. of a brine solution containing 20 lb of salt. At $t = 0$, fresh water is poured into the tank at the rate of 5 l./min, while the well-stirred mixture leaves the tank at the same rate. Then the amount of salt in the tank after 20 min.

a. $20/e$ b. $10/e$
c. $40/e^2$ d. $5/e$

14. A 50-1l. tank initially contains 10 l. of fresh water. At $t = 0$, a brine solution containing 1 lb of salt per gallon is poured into the tank at the rate of 4 l./min, while the well-stirred mixture leaves the tank at the rate of 2 l./min. Then the amount of time required for overflow to occur is

a. 30 min b. 20 min
c. 10 min d. 40 min

15. In above question the amount of salt in the tank at the moment of overflow is

a. 20 lb b. 50 lb
c. 30 lb d. None of these

Matrix-Match Type

Each question contains statements given in two columns which have to be matched. Statements a, b, c, d in column I have to be matched with statements p, q, r, s in column II. If the correct match are a → p, a → s, b → q, r, c → p, q and d → s, then the correctly bubbled 4 × 4 matrix should be as follows:

	p	q	r	s
a	p	q	r	s
b	p	q	r	s
c	p	q	r	s
d	p	q	r	s

1.

Column I	Column II : Differential equation
a. order 1	p. of all parabolas whose axis is the x-axis
b. order 2	q. of family of curves $y = a(x + a)^2$, where a is an arbitrary constant

c. degree 1	r. $\left(1+3\dfrac{dy}{dx}\right)^{2/3} = \dfrac{4d^3y}{dx^3}$
d. degree 3	s. of family of curve $y^2 = 2c\left(x+\sqrt{c}\right)$, where $c > 0$

2.

Column I	Column II
a. If the function $y = e^{4x} + 2e^{-x}$ is a solution of the differential equation $\dfrac{\dfrac{d^3y}{dx^3} - 13\dfrac{dy}{dx}}{y} = K$ then the value of $K/3$ is	p. 3
b. Number of straight lines which satisfy the differential equation $\dfrac{dy}{dx} + x\left(\dfrac{dy}{dx}\right)^2 - y = 0$ is	q. 4
c. If real value of m for which the substitution, $y = u^m$ will transform the differential equation, $2x^4y\dfrac{dy}{dx} + y^4 = 4x^6$ into a homogeneous equation then the value of $2m$ is	r. 2
d. If the solution of differential equation $x^2\dfrac{d^2y}{dx^2} + 2x\dfrac{dy}{dx} = 12y$ is $y = Ax^m + Bx^{-n}$ then $\|m+n\|$ is	s. 1

Integer Type

1. The sum of the order and degree of the differential equation $\dfrac{d^2y}{dx^2} = \left[y + \left(\dfrac{dy}{dx}\right)^6\right]^{1/4}$ is________.

2. If the differential equation having $y = (\sin^{-1}x)^2 + A\cos^{-1}x + B$, where A and B are arbitrary constants, is $(1-x^2)\dfrac{d^2y}{dx^2} - x\dfrac{dy}{dx} = k$, then the value of k is________.

3. The slope of a curve passing through (3, 4) at any point is the reciprocal of twice the ordinate of that point. If m is the slope of the curve when the abscissa is 3, then the value of $|1/m|$ is________.

4. The curve such that the intercept on the x-axis cut off between the origin and the tangent at a point is twice the abscissa and which passes through the point (1, 2). If the ordinate of the point on the curve is 1/3, then the value of the abscissa is________.

5. The integral part of half of the constant interest rate required if an initial deposit placed into an account accrues interest compounded continuously to double its value in six years ($\ln|2| = 0.6930$) is________.

6. A pair of curves such that
   a. the tangents drawn at points with equal abscissas intersect on the y-axis,
   b. the normal drawn at points with equal abscissas intersect on the x-axis,
   c. one curve passes through (1, 1) and other passes through (2, 3),

   Then the number of points of intersection of the curves is________.

7. A cyclist moving on a level road at 4 m/s stops pedalling and lets the wheels come to rest. The retardation of the cycle has two components: a constant 0.08 m/s^2 due to friction in the working parts and resistance of $0.02\,v^2/m$ where v is speed in meters per second. If d is the distance traversed by the cycle before it comes to rest, the value of $[d/10]$, where $[\cdot]$ represents the greatest integer function, (consider $\ln 5 = 1.61$) is________.

8. If $y = y(x)$ and $\dfrac{2+\sin x}{y+1}\left(\dfrac{dy}{dx}\right) = -\cos x$, $y(0) = 1$, then $9y\left(\dfrac{\pi}{2}\right)$ equals________.

9. For the primitive integral equation $ydx + y^2dy = x\,dy$, $x \in R$, $y > 0$, $y(1) = 1$, then $y(-3)$ is________.

10. The sum of the order and degree of the differential equation representing the family of curves $y^2 = 2c\left(x+\sqrt{c}\right)$, where c is a positive parameter, is________.

11. The order of the differential equation whose general solution is given by $y = (c_1 + c_2)\cos(x + c_3) - c_4e^{x+c_5}$, where c_1, c_2, c_3, c_4, and c_5 are orbitrary constants, is________.

ANSWERS AND SOLUTIONS

Objective Type

1. a. Putting $x = \sin A$ and $y = \sin B$ in the given relation, we get

$\cos A + \cos B = a(\sin A - \sin B)$

$\Rightarrow A - B = 2\cot^{-1} a$

$\Rightarrow \sin^{-1} x - \sin^{-1} y = 2\cot^{-1} a$

Differentiating w.r.t. x, we get

$$\frac{1}{\sqrt{1-x^2}} - \frac{1}{\sqrt{1-y^2}}\frac{dy}{dx} = 0$$

Clearly, it is differential equation of degree one.

2. d. $Ax^2 + By^2 = 1$ (1)

Differentiating w.r.t. x, we get

$$2Ax + 2By\frac{dy}{dx} = 0 \Rightarrow Ax + By\frac{dy}{dx} = 0 \quad (2)$$

Again differentiating. $A + By\dfrac{d^2y}{dx^2} + B\left(\dfrac{dy}{dx}\right)^2 = 0$ (3)

From Eqs. (2) and (3), we get

$$x\left[-By\frac{d^2y}{dx^2} - B\left(\frac{dy}{dx}\right)^2\right] + By\frac{dy}{dx} = 0$$

$$\Rightarrow xy\frac{d^2y}{dx^2} + x\left(\frac{dy}{dx}\right)^2 - y\frac{dy}{dx} = 0$$

$\therefore$ order = 2 and degree = 1

3. a. $y = e^x(A\cos x + B\sin x)$

$$\frac{dy}{dx} = e^x[-A\sin x + B\cos x] + e^x[A\cos x + B\sin x]$$

$$\frac{dy}{dx} = e^x[-A\sin x + B\cos x] + y \quad (1)$$

Again differentiating w.r.t. x, we get

$$\frac{d^2y}{dx^2} = e^x[-A\sin x + B\cos x] + e^x[-A\cos x - B\sin x] + \frac{dy}{dx}$$

$$\frac{d^2y}{dx^2} = \left(\frac{dy}{dx} - y\right) - y + \frac{dy}{dx} \quad \text{[using (1)]}$$

$$\frac{d^2y}{dx^2} - 2\frac{dy}{dx} + 2y = 0$$

4. d. Equation of circle will be $x^2 + (y-2)^2 + \lambda(y-2) = 0$

Differentiating, we get

$$2x + 2(y-2)\frac{dy}{dx} + \lambda\frac{dy}{dx} = 0$$

Therefore, the equation is

$$x^2 + (y-2)^2 - (y-2)\left(2x\frac{dx}{dy} + 2y - 4\right) = 0$$

5. a. The equation of a member of the family of parabolas having axis parallel to y-axis is

$y = Ax^2 + Bx + C$ (1)

where A, B, and C are arbitrary constants

Differentiating Eq. (1) w.r.t. x, we get

$$\frac{dy}{dx} = 2Ax + B \quad (2)$$

which on again differentiating w.r.t. x gives $\dfrac{d^2y}{dx^2} = 2A$ (3)

Differentiating (3) w.r.t. x, we get $\dfrac{d^3y}{dx^3} = 0$

6. c. Differentiating the given equation successively, we get

$y_1 = 5be^{5x} - 7ce^{-7x}$ (1)

$y_2 = 25be^{5x} + 49ce^{-7x}$ (2)

$y_3 = 125be^{5x} - 343ce^{-7x}$ (3)

Multiplying Eq. (1) by 7 and then adding to equation (2),

we get $y_2 + 7y_1 = 60be^{5x}$ (4)

Multiplying Eq. (1) by 5 and then subtracting it from equation (2),

we get $y_2 - 5y_1 = 84ce^{-7x}$ (5)

Putting the values of b and c, obtained from equation (4) and (5), respectively, in equation (1), we get

$y_3 + 2y_2 - 35y_1 = 0$

7. a. If $(0, k)$ be the centre on y-axis then its radius will be k as it passes through origin. Hence its equation is

$x^2 + (y-k)^2 = k^2$

or $x^2 + y^2 = 2ky$ (1)

$$\therefore\ 2x + 2y\frac{dy}{dx} = 2k\frac{dy}{dx}$$

$$= \frac{x^2 + y^2}{y}\frac{dy}{dx} \quad \text{[By (1)]}$$

$$\therefore\ 2xy = (x^2 + y^2 - 2y^2)\frac{dy}{dx}$$

or $(x^2 - y^2)\dfrac{dy}{dx} - 2xy = 0$

8. c. Central conics are given by

$ax^2 + by^2 = 1 \Rightarrow 2ax + 2byy_1 = 0$

Differentiating. w.r.t. x, we get

$\Rightarrow ax+byy_1=0 \Rightarrow \frac{-a}{b}=\frac{yy_1}{x}$ (1)

Again differentiating. w.r.t. x, we get

$\Rightarrow a+by_1^2+byy_2=0 \Rightarrow \frac{-a}{b}=y_1^2+yy_2$ (2)

From Eq. (1) and (2), we get

$\frac{yy_1}{x}=y_1^2+yy_2$

$\Rightarrow yy_1=xy_1^2+xyy_2$

9. c. The given family of curve is $x^2+y^2-2ay=0$ (1)

Differentiating. w.r.t. x, we get $2x+2y\frac{dy}{dx}-2a\frac{dy}{dx}=0$

$\Rightarrow 2x+2y\frac{dy}{dx}-\frac{x^2+y^2}{y}\frac{dy}{dx}=0$ [Using equation (1)]

$\Rightarrow 2xy+(2y^2-x^2-y^2)y'=0$

$\Rightarrow (y^2-x^2)y'+2xy=0$

$\Rightarrow (x^2-y^2)y'=2xy$

10. b. Here equation of circle is

$(x-h)^2+y^2=h^2$

$\Rightarrow x^2+y^2-2hx=0$ (1)

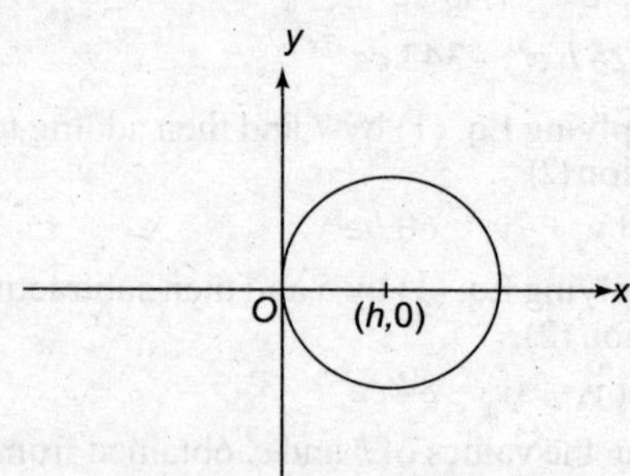

Fig. 10.1

Differentiating. w.r.t. x, $2x+2y\frac{dy}{dx}-2h=0$

$\Rightarrow 2h=2x+2y\frac{dy}{dx}$

Putting in Eq. (1), we get $x^2+y^2-(2x+2y\frac{dy}{dx})x=0$

$\Rightarrow y^2=x^2+2xy\frac{dy}{dx}$

11. d. $\log c+\log|x|=\frac{x}{y}$

Differentiating. w.r.t. x, $\frac{1}{x}=\frac{y-x\frac{dy}{dx}}{y^2}$

$\Rightarrow \frac{y^2}{x}=y-x\frac{dy}{dx}$

$\Rightarrow \frac{dy}{dx}=\frac{y}{x}-\frac{y^2}{x^2}$

$\Rightarrow \phi\left(\frac{x}{y}\right)=-\frac{y^2}{x^2}$

12. b. $y=c_1\cos(x+c_2)-(c_3e^{-x+c_4})+(c_5\sin x)$

$\Rightarrow y=c_1(\cos x\cos c_2-\sin x\sin c_2)$

$-(c_3e^{c_4}e^{-x})+(c_5\sin x)$

$\Rightarrow y=(c_1\cos c_2)\cos x-(c_1\sin c_2-c_5)\sin x$

$-(c_3e^{c_4})e^{-x}$

$\Rightarrow y=l\cos x+m\sin x-ne^{-x}$ (1)

where l, m, n are arbitrary constant

$\Rightarrow \frac{dy}{dx}=-l\sin x+m\cos x+ne^{-x}$ (2)

$\Rightarrow \frac{d^2y}{dx^2}=-l\cos x-m\sin x-ne^{-x}$ (3)

$\Rightarrow \frac{d^3y}{dx^3}=l\sin x-m\cos x+ne^{-x}$ (4)

From Eq. (1) + (3), $\frac{d^2y}{dx^2}+y=-2ne^{-x}$ (5)

From Eq. (2) + (4), $\frac{d^3y}{dx^3}+\frac{dy}{dx}=2ne^{-x}$ (6)

From Eq. (5) + (6), we get

$\frac{d^3y}{dx^3}+\frac{d^2y}{dx^2}+\frac{dy}{dx}+y=0$

13. a. $x\frac{dy}{dx}+y(\log y)=0$

$\Rightarrow \int\frac{dx}{x}+\int\frac{dy}{y(\ln y)}=c$

$\Rightarrow \log x+\log(\log y)=\log c$

$\Rightarrow x\log y=c$

$y(1)=e \Rightarrow c=1$

Hence, the equation of the curve is $x\log y=1$

14. a. Variables separable $\frac{1}{y+1}dy=-\frac{\cos x}{2+\sin x}dx$

Integrating, we get

$\log(y+1)+\log k+\log(2+\sin x)=0$

$\therefore$ $k(y+1)(2+\sin x)=1$ when $x=0, y=1$ where k is constant.

$\therefore$ $4k=1$ or $k=1/4$

$\therefore$ $(y+1)(2+\sin x)=4$

Now put $x=\pi/2$ $\therefore (y+1)3=4$

$\therefore$ $y=\frac{1}{3}$

15. a. Slope $= \frac{dy}{dx} \Rightarrow \frac{dy}{dx} = \frac{y-1}{x^2+x}$

$$\Rightarrow \frac{dy}{y-1} = \frac{dx}{x^2+x}$$

$$\Rightarrow \int \frac{1}{y-1}dy = \int \left(\frac{1}{x} - \frac{1}{x+1}\right)dx + C$$

$$\Rightarrow \frac{(y-1)(x+1)}{x} = k$$

Putting $x = 1, y = 0$, we get $k = -2$

The equation is $(y-1)(x+1) + 2x = 0$

16. d. $(y \cos y + \sin y)\, dy = (2x \log x + x)\, dx$

$$\Rightarrow y \sin y - \int \sin y\, dy + \int \sin y\, dx$$

$$= x^2 \log x - \int x^2 \frac{1}{x} dx + \int x\, dx + c$$

$\therefore \quad y \sin y = x^2 \log x + c$

17. b. $\frac{dy}{dx} = e^{ax+by} = e^{ax} e^{by}$

or $e^{-by} dy = e^{ax} dx$

$\therefore \quad -\frac{1}{b} e^{-by} = \frac{1}{a} e^{ax} + c$

18. a. $x^2(y+1)\, dx + y^2(x-1) dy = 0$

$$\Rightarrow \frac{x^2 dx}{x-1} = -\frac{y^2 dy}{y+1}$$

$$\Rightarrow \int \left[x+1+\frac{1}{x-1}\right]dx = -\int \left[y-1+\frac{1}{y+1}\right]dy$$

$$\Rightarrow \frac{x^2}{2} + x + \ln(x-1) = -\left[\frac{y^2}{2} - y + \ln(y+1)\right] + \ln c$$

$$\Rightarrow \frac{x^2+y^2}{2} + (x-y) + \ln\left(\frac{(x-1)(y+1)}{c}\right) = 0$$

19. a. $dy - \sin x \sin y\, dx = 0$

$\Rightarrow dy = \sin x \sin y\, dx$

$$\Rightarrow \int \operatorname{cosec} y dy = \int \sin x dx$$

$$\Rightarrow \log \tan \frac{y}{2} = -\cos x + \log c$$

$$\Rightarrow \log \frac{\tan \frac{y}{2}}{c} = -\cos x$$

$$\Rightarrow \frac{\tan \frac{y}{2}}{c} = e^{-\cos x}$$

$$\Rightarrow e^{\cos x} \tan \frac{y}{2} = c$$

20. a. $\frac{dv}{dt} + \frac{k}{m} v = -g$

$$\Rightarrow \frac{dv}{dt} = -\frac{k}{m}\left(v + \frac{mg}{k}\right)$$

$$\Rightarrow \frac{dv}{v + mg/k} = -\frac{k}{m} dt$$

$$\Rightarrow \log\left(v + \frac{mg}{k}\right) = -\frac{k}{m} t + \log C$$

$$\Rightarrow u + \frac{mg}{k} = Ce^{-k/mt}$$

$$\Rightarrow u = ce^{-\frac{k}{m}t} - \frac{mg}{k}$$

21. b. Putting $u = x - y$, we get $du/dx = 1 - dy/dx$. The given equation can be written as $1 - du/dx = \cos u$

$\Rightarrow (1 - \cos u) = du/dx$

$$\Rightarrow \int \frac{du}{1-\cos u} = \int dx + C$$

$$\Rightarrow \frac{1}{2} \int \operatorname{cosec}^2 (u/2)\, du = \int dx + C$$

$\Rightarrow x + \cot (u/2) = c$

$$\Rightarrow x + \cot \frac{x-y}{2} = C$$

22. a. $\frac{dy}{dx} + 2xy = y$

$$\Rightarrow \frac{dy}{dx} = y(1-2x)$$

$$\Rightarrow \frac{dy}{y} = (1-2x)\, dx$$

$\Rightarrow \log y = x - x^2 + c_1$

$\Rightarrow y = e^{x-x^2} e^{c_1} = c e^{x-x^2}$ where $c = e^{c_1}$

$\Rightarrow y = ce^{x-x^2}$ is the required solution.

23. b. We have $\frac{dy}{dx} = \sin \frac{x-y}{2} - \sin \frac{x+y}{2}$

$$= -2\cos \frac{x}{2} \sin \frac{y}{2}$$

$$\Rightarrow \log \tan \frac{y}{4} = -\frac{\sin \frac{x}{2}}{\frac{1}{2}} + c$$

$$\Rightarrow \log \tan\left(\frac{y}{4}\right) = c - 2\sin \frac{x}{2}$$

24. a. Putting $x + y + 1 = u$, we have $du = dx + dy$ and the given equations reduces to

$u(du - dx) = dx$

$\Rightarrow \dfrac{u\ du}{u+1} = dx$

$\Rightarrow u - \log(u+1) = x$

$\Rightarrow \log(x+y+2) = y + \text{constant}$

$\Rightarrow x+y+2 = Ce^y$

25. d. Putting $v = y/x$ so that $x\dfrac{dv}{dx} + v = \dfrac{dy}{dx}$.

we have $x\dfrac{dv}{dx} + v = v + \phi(1/v)$

$\Rightarrow \dfrac{dv}{\phi(1/v)} = \dfrac{dx}{x}$

$\Rightarrow \log|Cx| = \int \dfrac{dv}{\phi(1/v)}$

[C being constant of integration]

But $y = \dfrac{x}{\log|Cx|}$ is the general solution

so $\dfrac{x}{y} = \dfrac{1}{v} = \log|Cx| = \int \dfrac{dv}{\phi(1/v)}$

$\Rightarrow \phi(1/v) = -1/v^2$ [Differentiating w.r.t. v both sides]

$\Rightarrow \phi(x/y) = -y^2/x^2$

26. c. We have, $\dfrac{dy}{dx} = \dfrac{y}{x} - \cos^2\left(\dfrac{y}{x}\right)$

Putting $y = ux$, so that $\dfrac{dy}{dx} = \upsilon + x\dfrac{d\upsilon}{dx}$, we get

$\upsilon + x\dfrac{d\upsilon}{dx} = \upsilon - \cos^2 \upsilon$

$\Rightarrow \dfrac{d\upsilon}{\cos^2 \upsilon} = -\dfrac{dx}{x}$

$\Rightarrow \sec^2 u\ du = -\dfrac{1}{x}dx$

On integration, we get

$\tan u = -\log x + \log C$

$\Rightarrow \tan\left(\dfrac{y}{x}\right) = -\log x + \log C$

This passes through $(1, \pi/4)$, therefore $1 = \log C$.

So, $\tan\left(\dfrac{y}{x}\right) = -\log x + 1$

$\Rightarrow \tan\left(\dfrac{y}{x}\right) = -\log x + \log e$

$\Rightarrow y = x\tan^{-1}\left(\log\left(\dfrac{e}{x}\right)\right)$

27. d. $\dfrac{dy}{dx} = \dfrac{y}{x}\left[\log\dfrac{y}{x} + 1\right]$

Put $y = vx$

$v + x\dfrac{dy}{dx} = v\log v + v$

$\therefore \dfrac{dv}{v\log v} = \dfrac{dx}{x}$

$\therefore \log(\log v) = \log x + \log k = \log kx$

$\therefore \log\dfrac{y}{x} = kx$

28. a. The given equation can be written as

$\dfrac{y}{x}\dfrac{dy}{dx} = \left\{\dfrac{y^2}{x^2} + \dfrac{f(y^2/x^2)}{f'(y^2/x^2)}\right\}$

Above equation is a homogeneous equation.

Putting $y = vx$, we get

$v\left[v + x\dfrac{dv}{dx}\right] = v^2 + \dfrac{f(v^2)}{f'(v^2)}$

$\Rightarrow vx\dfrac{dv}{dx} = \dfrac{f(v^2)}{f'(v^2)}$ variable separable

$\Rightarrow \dfrac{2vf'(v^2)}{f(v^2)}dv = 2\dfrac{dv}{x}$

Now integrating both sides, we get

$\log f(v^2) = \log x^2 + \log c$ [$\log c$ = constant]

or $\log f(v^2) = \log cx^2$

or $f(v^2) = cx^2$

or $f(y^2/x^2) = cx^2$

29. b. $(x^2 + xy)\,dy = (x^2 + y^2)dx \Rightarrow \dfrac{dy}{dx} = \dfrac{x^2 + y^2}{x^2 + xy}$

Let $\dfrac{y}{x} = v \Rightarrow \dfrac{dy}{dx} = v + x\dfrac{d\upsilon}{dx}$

Therefore, equation reduces to

$x\dfrac{dv}{dx} = \dfrac{1+v^2-v}{1+v}$

$= \dfrac{1+v^2-v-v^2}{1+v}$

$= \dfrac{1-\upsilon}{1+\upsilon}$

$\Rightarrow \int \dfrac{1+\upsilon}{1-\upsilon}d\upsilon = \int \dfrac{dx}{x}$

$\Rightarrow -\int\left(1 - \dfrac{2}{1-\upsilon}\right)d\upsilon = \int \dfrac{dx}{x}$

$\Rightarrow -\upsilon - 2\log(1-\upsilon) = \log x + \log c$

$\Rightarrow -\dfrac{y}{x} - 2\log\left(\dfrac{x-y}{x}\right) = \log x + \log c$

$\Rightarrow \dfrac{-y}{x} - 2\log(x-y) + 2\log x = \log x + \log c$

$\Rightarrow \log x = 2\log(x-y) + \dfrac{y}{x} + k$ where $k = \log c$

30. c. The intersection of $y - x + 1 = 0$ and $y + x + 5 = 0$ is $(-2, -3)$. Put $x = X - 2$, $y = Y - 3$.

The given equation reduces to $\dfrac{dY}{dX} = \dfrac{Y - X}{Y + X}$.

Putting $Y = vX$, we get

$$X\frac{dv}{dX} = -\frac{v^2+1}{v+1}$$

$$\Rightarrow \left(-\frac{v}{v^2+1} - \frac{1}{v^2+1}\right)dv = \frac{dX}{X}$$

$$\Rightarrow -\frac{1}{2}\log\left(v^2+1\right) - \tan^{-1} v = \log|X| + \text{constant}$$

$$\Rightarrow \log(Y^2 + X^2) + 2\tan^{-1}\frac{Y}{X} = \text{constant}$$

$$\Rightarrow \log((y+3)^2 + (x+2)^2) + 2\tan^{-1}\frac{y+3}{x+2} = C$$

31. a. $\dfrac{dy}{dx} = \dfrac{x^2+y^2}{2xy}$ (1)

Put $y = vx$ $\therefore$ $\dfrac{dy}{dx} = v + x\dfrac{dv}{dx}$

Therefore, Eq. (1) transforms to

$$v + x\frac{dv}{dx} = \frac{x^2+v^2x^2}{2x\,vx} = \frac{1+v^2}{2v}$$

$$\Rightarrow x\frac{dv}{dx} = \frac{1+v^2}{2v} - v = \frac{1-v^2}{2v}$$

$$\Rightarrow \frac{2v\,dv}{1-v^2} = \frac{dx}{x}$$

$$\Rightarrow \log x + \log(1-v^2) = \log C$$

$$\Rightarrow x(1-v^2) = C$$

$$\Rightarrow x\left(1 - \frac{y^2}{x^2}\right) = C$$

$$\Rightarrow x^2 - y^2 = Cx$$

It passes through (2, 1).

$\therefore$ $4 - 1 = 2C \Rightarrow C = \dfrac{3}{2}$

$\therefore$ $x^2 - y^2 = \dfrac{3}{2}x \Rightarrow 2(x^2 - y^2) = 3x$

32. d. The given equation is $\dfrac{dy}{dx} = \dfrac{y}{x}\left(\log\dfrac{y}{x} + 1\right)$ (1)

Put $\dfrac{y}{x} = v$, i.e., $y = vx$

so that $\dfrac{dy}{dx} = v + x\dfrac{dv}{dx}$

Therefore, Eq. (1) becomes $v + x\dfrac{dv}{dx} = v(\log v + 1)$

$$\Rightarrow x\frac{dv}{dx} = v\log v$$

$$\Rightarrow \frac{dv}{v\log v} = \frac{dx}{x}$$

Integrating, we get $\log(\log v) = \log x + \log c$

$$\Rightarrow \log v = cx \Rightarrow \log\frac{y}{x} = cx$$

33. a. $\dfrac{dy}{dx} + y\,\phi'(x) = \phi(x)\,\phi'(x)$

I.F. $= e^{\int \phi'(x)dx} = e^{\phi(x)}$

Hence, the solution is

$$ye^{\phi(x)} = \int e^{\phi(x)}\,\phi(x)\,\phi'(x)\,dx$$

$= \int e^t\,t\,dt$, where $\phi(x) = t$

$= te^t - e^t + C$

$= \phi(x)\,e^{\phi(x)} - e^{\phi(x)} + C$

$\therefore$ $y = ce^{-\phi(x)} + \phi(x) - 1$

34. c. Rewriting the given equation as

$$2xy\frac{dy}{dx} - y^2 = 1 + x^2$$

$$\Rightarrow 2y\frac{dy}{dx} - \frac{1}{x}y^2 = \frac{1}{x} + x$$

Putting $y^2 = u$, we have

$$\frac{du}{dx} - \frac{1}{x}u = \frac{1}{x} + x$$

$$\text{I.F.} = e^{-\int \frac{1}{x}dx} = \frac{1}{x}$$

Solution is $u\dfrac{1}{x} = \displaystyle\int\left(\frac{1}{x^2} + 1\right)dx = -\frac{1}{x} + x + C$

$\Rightarrow y^2 = (x^2 - 1) + Cx$

Since $y(1) = 1$ so $C = 1$.

Hence $y^2 = x(1 + x) - 1$, which represents a system of hyperbola.

35. c. $\because$ $\dfrac{dy}{dx} + \dfrac{y}{x\log_e x} = \dfrac{2}{x}$

$\therefore$ I.F. $= e^{\int \frac{1}{x\log_e x}dx}$

$= e^{\log_e \log_e x}$

$= \log_e x$

36. c. The given equation can be rewritten as

$$\frac{dy}{dx} + \frac{x^2-1}{x\left(x^2+1\right)}y = \frac{x^2\log x}{\left(x^2+1\right)} \quad (1)$$

which is linear. Also

$$P = \frac{x^2-1}{x\left(x^2+1\right)} \text{ and } Q = \frac{x^2\log x}{\left(x^2+1\right)}$$

$$\int P\,dx = \int\left[\frac{2x}{x^2+1} - \frac{1}{x}\right]dx$$

[resolving into partial fractions]

$= \log (x^2+1) - \log x$

$\therefore$ I.F. $= e^{\log[(x^2+1)/x]} = \dfrac{x^2+1}{x}$

Hence the required solution of equation (1) is

$$\frac{y(x^2+1)}{x} = \int \frac{(x^2+1)}{x} \frac{x^2 \log x}{(x^2+1)} dx + c$$

$$= \int x \log x \, dx + c$$

$$= \frac{1}{2} x^2 \log x - \int \frac{1}{x} \frac{x^2}{2} dx + c$$

$$\therefore \; y(x^2+1)/x = \frac{1}{2} x^2 \log x - \frac{1}{4} x^2 + c$$

37. c. $\cos x \dfrac{dy}{dx} + y \sin x = 1$

$$\Rightarrow \frac{dy}{dx} + y \frac{\sin x}{\cos x} = \sec x$$

$$\therefore \int P\, dx = \int \frac{\sin x}{\cos x} dx$$

$$= -\log \cos x$$

$$= \log \sec x$$

$$\therefore \text{ I.F.} = e^{\int P dx} = e^{\log \sec x} = \sec x$$

38. c. The given differential equation can be written as

$\dfrac{dy}{dx} - \dfrac{\tan 2x}{\cos^2 x} y = \cos^2 x$ which is linear differential equation of first order.

$$\int P\, dx = \int \frac{-\sin 2x}{\cos 2x \cos^2 x} dx$$

$$= -\int \frac{2 \sin 2x\, dx}{\cos 2x (1+\cos 2x)}$$

$$= \int \frac{dt}{t(1+t)}$$

$$= \int \left(\frac{1}{t} - \frac{1}{1+t} \right) dt$$

$$= \log \frac{t}{1+t} \text{ where } t = \cos 2x$$

$$= \log \frac{\cos 2x}{1+\cos 2x} \qquad \left[\because -\frac{\pi}{2} < 2x < \frac{\pi}{2} \right]$$

$$\therefore \; e^{\int P dx} = e^{\log \frac{\cos 2x}{1+\cos 2x}}$$

$$= \frac{\cos 2x}{1+\cos 2x} = \frac{\cos 2x}{2\cos^2 x}$$

Therefore, the solution is,

$$y \frac{\cos 2x}{2\cos^2 x} = \int \frac{\cos^2 x \cos 2x}{2 \cos^2 x} dx + C$$

$$= \frac{1}{4} \sin 2x + C$$

When $x = \dfrac{\pi}{6}, y = \dfrac{3\sqrt{3}}{8}$

$$\therefore \; \frac{3\sqrt{3}}{8} \cdot \frac{4}{2 \times 2 \times 3} = \frac{1}{4} \frac{\sqrt{3}}{2} + C \Rightarrow C = 0$$

$$\therefore \; y = \frac{1}{2} \tan 2x \cos^2 x$$

39. d. $x(1-x^2)dy + (2x^2y - y - ax^3)dx = 0$

$$\Rightarrow x(1-x^2)\frac{dy}{dx} + 2x^2y - y - ax^3 = 0$$

$$\Rightarrow x(1-x^2)\frac{dy}{dx} + y(2x^2-1) = ax^3$$

$$\Rightarrow \frac{dy}{dx} + \frac{2x^2-1}{x(1-x^2)} y = \frac{ax^3}{x(1-x^2)}$$

which is of the form $\dfrac{dy}{dx} + Py = Q$.

Its integrating factor is $e^{\int P dx}$.

Here $P = \dfrac{2x^2-1}{x(1-x^2)}$

40. b. $f'(x) - \dfrac{2x(x+1)}{x+1} f(x) = \dfrac{e^{x^2}}{(x+1)^2}$

I. F. $= e^{\int -2x\, dx} = e^{-x^2}$

Therefore, solution is $f(x)$

$$e^{-x^2} = \int \frac{dx}{(x+1)^2}$$

$$\Rightarrow f(x) e^{-x^2} = -\frac{1}{x+1} + C$$

Given $f(0) = 5 \Rightarrow C = 6$

$$\therefore \; f(x) = \left(\frac{6x+5}{x+1} \right) e^{x^2}$$

41. a. $\dfrac{dy}{dx} = \dfrac{1}{xy[x^2 \sin y^2 + 1]}$

$$\Rightarrow \frac{dx}{dy} = xy[x^2 \sin y^2 + 1]$$

$$\Rightarrow \frac{1}{x^3} \frac{dx}{dy} - \frac{1}{x^2} y = y \sin y^2$$

Putting $-1/x^2 = u$, the last equation can be written as

$$\frac{du}{dy} + 2uy = 2y \sin y^2.$$

I.F. $= e^{y^2}$. Therefore, solution is

$$u e^{y^2} = \int 2y \sin y^2 e^{y^2} dy + C$$

$= \int (\sin t)e^t\, dt + C$

$= \frac{1}{2}e^{y^2}(\sin y^2 - \cos y^2) + c'$

$\Rightarrow 2u = (\sin y^2 - \cos y^2) + 2Ce^{-y^2}$

$\Rightarrow 2 = x^2 [\cos y^2 - \sin y^2 - 2Ce^{-y^2}]$

42. d. $\frac{dy}{dx} = 1 + xy$

$\Rightarrow \frac{dy}{dx} - xy = 1$

I.F. $= e^{\int -x dx} = e^{-x^2/2}$

Hence solution is $y.e^{-x^2/2} = \int e^{-x^2/2} dx + c.$

$\int e^{-x^2/2} dx$ is not further integrable.

43. b. $\frac{dx}{dy} = \frac{x + 2y^3}{y}$

$\Rightarrow \frac{dx}{dy} - \frac{1}{y}x = 2y^2$ which is linear

I.F. $= e^{\int -\frac{1}{y} dy} = e^{-\log y} = 1/y$. Therefore, solution is

$\frac{1}{y}x = \int \frac{1}{y} 2y^2\, dy = y^2 + c$

$\Rightarrow \frac{x}{y} = y^2 + c$

44. a. $x^2 \frac{dy}{dx} \cos\frac{1}{x} - y \sin\frac{1}{x} = -1$

$\Rightarrow \frac{dy}{dx} - \frac{y}{x^2}\tan\frac{1}{x} = -\sec\frac{1}{x}\frac{1}{x^2}$ (linear)

I.F. $= e^{-\int \frac{1}{x^2}\tan\frac{1}{x} dx} = \sec\frac{1}{x}$

$\Rightarrow$ solution is $y \sec\frac{1}{x} = -\int \sec^2\left(\frac{1}{x}\right)\frac{1}{x^2}dx$

$= \tan\frac{1}{x} + c$

Given $y \to -1, x \to \infty \Rightarrow c = -1$

Hence equation of curve is $y = \sin\frac{1}{x} - \cos\frac{1}{x}$.

45. d. $2x^2y\frac{dy}{dx} = \tan(x^2y^2) - 2xy^2$

$\Rightarrow x^2\, 2y\frac{dy}{dx} + y^2\, 2x = \tan(x^2 y^2)$

$\Rightarrow \frac{d}{dx}(x^2 y^2) = \tan(x^2 y^2)$

$\Rightarrow \int \cot(x^2 y^2) d(x^2 y^2) = \int dx$

$\Rightarrow \log(\sin(x^2y^2)) = x + c$

when $x = 1$, $y = \sqrt{\frac{\pi}{2}} \Rightarrow c = 1$

$\Rightarrow$ Equation of curve is $x = \log\sin(x^2y^2) - 1$

$\Rightarrow \log\sin(x^2y^2) = x - 1$

$\Rightarrow \sin(x^2 y^2) = e^{x-1}$

46. a. $\left\{\frac{1}{x} - \frac{y^2}{(x-y)^2}\right\}dx + \left\{\frac{x^2}{(x-y)^2} - \frac{1}{y}\right\}dy = 0$

$\Rightarrow \left(\frac{dx}{x} - \frac{dy}{y}\right) + \left(\frac{x^2dy - y^2dx}{(x-y)^2}\right) = 0$

$\Rightarrow \left(\frac{dx}{x} - \frac{dy}{y}\right) + \left(\frac{dy/y^2 - dx/x^2}{(1/y - 1/x)^2}\right) = 0$

Integrating, we get $\ln|x| - \ln|y| - \frac{1}{(1/x - 1/y)} = c$

$\Rightarrow \ln\left|\frac{x}{y}\right| + \frac{xy}{x-y} = c$

47. a. Put $xy = v \therefore y + x\frac{dy}{dx} = \frac{dv}{dx}$

$\Rightarrow \frac{dv}{dx} = x\frac{\phi(v)}{\phi'(v)}$

$\therefore \frac{\phi'(v)}{\phi(v)}dv = x\,dx$. Integrating, we get

$\log\phi(v) = \frac{x^2}{2} + \log k$

$\Rightarrow \log\frac{\phi(v)}{k} = \frac{x^2}{2}$

or $\phi(v) = ke^{x^2/2} \Rightarrow \phi(xy) = ke^{x^2/2}$

48. a. $(2y + xy^3)dx + (x + x^2y^2)dy = 0$

$\Rightarrow (2y\,dx + x dy) + (xy^3dx + x^2y^2dy) = 0$

Multiplying by x, we get

$(2xy\,dx + x^2\,dy) + (x^2y^3\,dx + x^3y^2\,dy) = 0$

$\Rightarrow d(x^2y) + \frac{1}{3}d(x^3y^3) = 0$

Integrating, we get $x^2y + \frac{x^3y^3}{3} = c$

49. c. $ye^{-x/y}dx - (xe^{(-x/y)} + y^3)dy = 0$

$\Rightarrow (ydx - xdy)\, e^{-x/y} - y^3dy = 0$

$\Rightarrow \frac{ydx - xdy}{y^2}e^{-x/y} = ydy$

$\Rightarrow d(x/y)e^{-x/y} = ydy$

$\Rightarrow -e^{-x/y} = \frac{y^2}{2} + C$

$\Rightarrow 2e^{-x/y} + y^2 = C$

50. c. $(xy^3 - x^2)\,dy - (xy + y^4)\,dx = 0$

$\Rightarrow y^3(x\,dy - y\,dx) - x(x\,dy + y\,dx) = 0$

$\Rightarrow x^2y^3\dfrac{(x\,dy - y\,dx)}{x^2} - x(x\,dy + y\,dx) = 0$

$\Rightarrow x^2y^3\,d\left(\dfrac{y}{x}\right) - x\,d(xy) = 0$

$\Rightarrow$ Dividing by x^3y^2, we get

$\Rightarrow \dfrac{y}{x}\,d\left(\dfrac{y}{x}\right) - \dfrac{d(xy)}{x^2y^2} = 0$

Now integrating $\dfrac{1}{2}\left(\dfrac{y}{x}\right)^2 + \dfrac{1}{xy} = c$

It passes through the point (4, – 2).

$\Rightarrow \dfrac{1}{8} - \dfrac{1}{8} = c \Rightarrow c = 0$

$\therefore \quad y^3 = -2x$

51. a. The given equation can be written as

$$\frac{x\,dx + y\,dy}{(y\,dx - x\,dy)/y^2} = y^2\,\frac{x}{y^3}\cos^2(x^2 + y^2)$$

$$\Rightarrow \frac{xdx + ydy}{\cos^2(x^2 + y^2)} = \frac{x}{y}\left(\frac{ydx - xdy}{y^2}\right)$$

$$\Rightarrow \frac{1}{2}\sec^2(x^2 + y^2)\,d(x^2 + y^2) = \frac{x}{y}\,d\left(\frac{x}{y}\right)$$

On integrating, we get

$$\frac{1}{2}\tan(x^2 + y^2) = \frac{1}{2}\left(\frac{x}{y}\right)^2 + \frac{c}{2}$$

or $\tan(x^2 + y^2) = \dfrac{x^2}{y^2} + c$

52. c. Re-writing the differential equation as

$(2xy\,dx - x^2\,dy) + y^2(3x^2y^2\,dx + 2x^3y\,dy) = 0$

Dividing by y^2, we get

$$\frac{y\,2x\,dx - x^2\,dy}{y^2} + y^2\,3x^2\,dx + x^3\,2y\,dy = 0$$

or $d\left(\dfrac{x^2}{y}\right) + d\left(x^3y^2\right) = 0$

Integrating, we get the solution

$$\frac{x^2}{y} + x^3y^2 = c$$

53. c. $\left\{1 + x\sqrt{(x^2 + y^2)}\right\}dx + \left\{\sqrt{(x^2 + y^2)} - 1\right\}y\,dy = 0$

$\Rightarrow dx - y\,dy + \sqrt{(x^2 + y^2)}\,(x\,dx + y\,dy) = 0$

$\Rightarrow dx - y\,dy + \dfrac{1}{2}\sqrt{(x^2 + y^2)}\,d(x^2 + y^2) = 0$

Integrating, we have

$$x - \frac{y^2}{2} + \frac{1}{2}\int\sqrt{t}\,dt = c, \left\{t = \sqrt{(x^2 + y^2)}\right\}$$

or $x - \dfrac{y^2}{2} + \dfrac{1}{3}\left(x^2 + y^2\right)^{3/2} = c$

54. b. Let the slope of tangent of required family be $dx/dx = m_1$. Also $y = \dfrac{c^2}{x}$. Therefore we have

$$\frac{dy}{dx} = -\frac{c^2}{x^2} = m_2\,(\text{say})$$

By the given condition,

we have $\tan\dfrac{\pi}{4} = \dfrac{m_1 - m_2}{1 + m_1m_2}$

$\Rightarrow 1 + m_1m_2 = m_1 - m_2$

$\Rightarrow \dfrac{dy}{dx} + \dfrac{c^2}{x^2} = 1 - \dfrac{c^2}{x^2}\dfrac{dy}{dx}$

$\Rightarrow \dfrac{dy}{dx}\left(1 + \dfrac{c^2}{x^2}\right) = 1 - \dfrac{c^2}{x^2}$

$\Rightarrow \dfrac{dy}{dx} = \dfrac{x^2 - c^2}{x^2 + c^2}$

55. a.

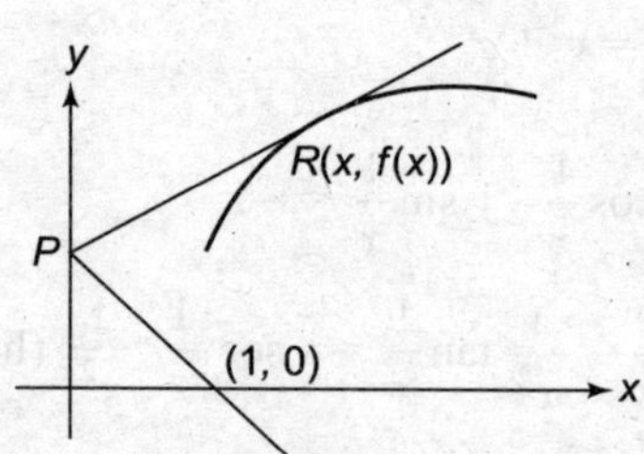

Fig. 10.2

The equation of the tangent at the point $R(x, f(x))$ is $Y - f(x) = f'(x)(X - x)$

The coordinates of the point P are $(0, f(x) - xf'(x))$

The slope of the perpendicular line through P is $\dfrac{f(x) - xf'(x)}{-1} = -\dfrac{1}{f'(x)}$

$\Rightarrow f(x)f'(x) - x(f'(x))^2 = 1$

$\Rightarrow \dfrac{ydy}{dx} - x\left(\dfrac{dy}{dx}\right)^2 = 1$ which is the required differential equation to the curve at $y = f(x)$.

56. a. If $y = f(x)$ is the curve,

$Y - y = \dfrac{dy}{dx}(X - x)$ is the equation of the tangent at (x, y)

Putting $X = 0$, the initial ordinate of the tangent is therefore $y - xf'(x)$.

The subnormal at this point is given by $y\dfrac{dy}{dx}$, so we have $y\dfrac{dy}{dx}=y-x\dfrac{dy}{dx} \Rightarrow \dfrac{y}{x+y}=\dfrac{dy}{dx}$.

This is a homogeneous equation and, by rewriting it as $\dfrac{dx}{dy}=\dfrac{x+y}{y}=\dfrac{x}{y}+1 \Rightarrow \dfrac{dx}{dy}-\dfrac{x}{y}=1$ we see that it is also a linear equation.

57. b. $x^{2/3}+y^{2/3}=a^{2/3}$

$\Rightarrow \dfrac{2}{3}x^{-1/3}+\dfrac{2}{3}y^{-1/3}\dfrac{dy}{dx}=0$

$\Rightarrow \dfrac{dy}{dx}=-\dfrac{x^{-1/3}}{y^{-1/3}}$ (1)

Replacing $\dfrac{dy}{dx}$ by $-\dfrac{dx}{dy}$

$\Rightarrow \dfrac{dx}{dy}=\dfrac{x^{-1/3}}{y^{-1/3}}$

$\Rightarrow \int x^{1/3}dx=\int y^{1/3}dy$

$\Rightarrow x^{4/3}-y^{4/3}=c$

58. b. The general equation of all non-horizontal lines in xy-plane is $ax+by=1$, where $a\neq 0$.

Now, $ax+by=1$

$\Rightarrow a\dfrac{dx}{dy}+b=0$ [Diff. w.r.t. y]

$\Rightarrow a\dfrac{d^2x}{dy^2}=0$ [Diff. w.r.t. y]

$\Rightarrow \dfrac{d^2x}{dy^2}=0$ [$\because a\neq 0$]

Hence, the required differential equation is $\dfrac{d^2x}{dy^2}=0$

59. b. It is given that the triangle OPG is an isosceles triangle.

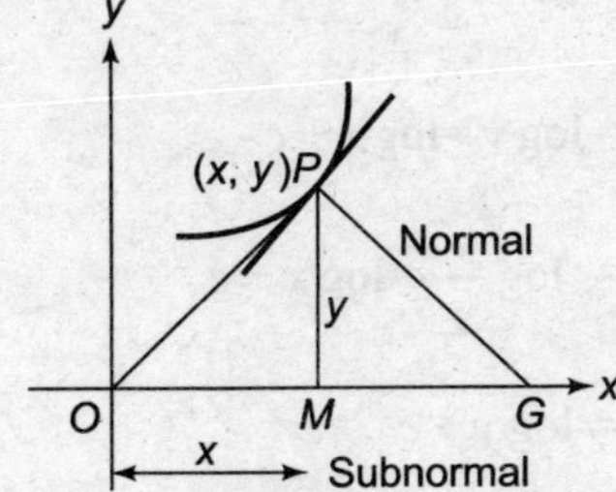

Fig. 10.3

Therefore, $OM=MG=$ sub-normal

$\Rightarrow x=y\dfrac{dy}{dx}\Rightarrow x\,dx=y\,dx$

On integration, we get $x^2-y^2=C$, which is a rectangular hyperbola.

60. a. Let the equation of the curve be $y=f(x)$.

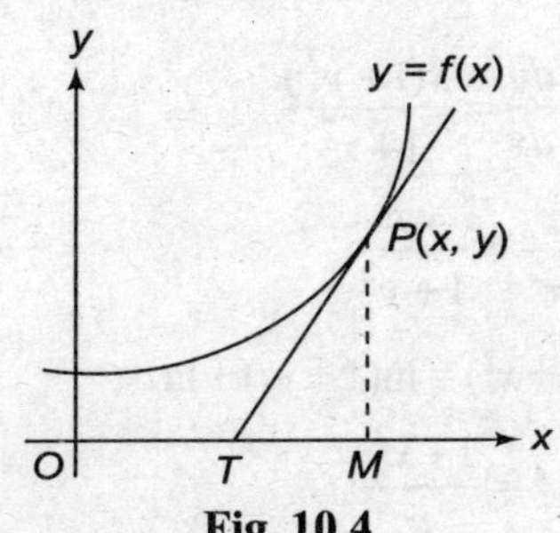

Fig. 10.4

It is given that $OT\propto y$

$\Rightarrow OT=by$

$\Rightarrow OM-TM=by$

$\Rightarrow x-\dfrac{y}{dy/dx}=by$

[$\because TM=$ Length of the subtangent]

$\Rightarrow x-y\dfrac{dx}{dy}=by$

$\Rightarrow \dfrac{dx}{dy}-\dfrac{x}{y}=-b$

It is linear differential equation.

Its solution is $\dfrac{x}{y}=-b\log y+a$.

$\Rightarrow x=y(a-b\log y)$

61. b. For the family of curves represented by the first differential equation the slope of the tangent at any point (x, y) is given by

$\left(\dfrac{dy}{dx}\right)_{c_1}=\dfrac{x^2+x+1}{y^2+y+1}$

For the family of curves represented by the second differential the slope of the tangent at any point is given by

$\left(\dfrac{dy}{dx}\right)_{c_2}=-\dfrac{y^2+y+1}{x^2+x+1}$

Clearly, $\left(\dfrac{dy}{dx}\right)_{c_1}\times\left(\dfrac{dy}{dx}\right)_{c_2}=-1$

Hence, the two curves are orthogonal.

62. c. Equation of normal at point $P(x, y)$, $Y-y=-\dfrac{dx}{dy}(X-x)$

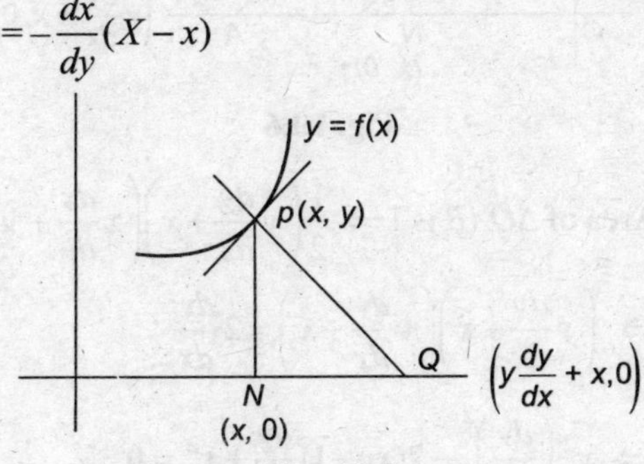

Fig. 10.5

$NQ = y\dfrac{dy}{dx} = \dfrac{x(1+y^2)}{1+x^2}$

$\Rightarrow \dfrac{xdx}{1+x^2} = \dfrac{ydy}{1+y^2}$

$\Rightarrow \ln(1+x^2) = \ln(1+y^2) + \ln c$

$\Rightarrow 1+y^2 = \dfrac{1+x^2}{c}$

It passes through (3, 1) $\Rightarrow 1+1 = \dfrac{1+(3)^2}{c} \Rightarrow c = 5$

$\Rightarrow$ curve is $5+5y^2 = 1+x^2$ or $x^2 - 5y^2 = 4$

63. c. The point on the y-axis is $\left(0,\ y - x\dfrac{dy}{dx}\right)$.

According to given condition,

$\dfrac{x}{2} = y - \dfrac{x}{2}\dfrac{dy}{dx} \Rightarrow \dfrac{dy}{dx} = 2\dfrac{y}{x} - 1$

Putting $\dfrac{y}{x} = v$, we get $x\dfrac{dv}{dx} = v - 1$

$\Rightarrow \ln\left|\dfrac{y}{x} - 1\right| = \ln|x| + c$

$\Rightarrow 1 - \dfrac{y}{x} = x$ [As $y(1) = 0$]

64. b. We have, $\dfrac{dy}{dx} = 1 - \dfrac{1}{x^2} \Rightarrow y = x + \dfrac{1}{x} + C$

This passes through (2, 7/2). Therefore,

$\dfrac{7}{2} = 2 + \dfrac{1}{2} + C \Rightarrow C = 1$

Thus the equation of the curve is

$y = x + \dfrac{1}{x} + 1 \Rightarrow xy = x^2 + x + 1$

65. d. Equation of normal at point p is $Y - y = -\dfrac{dx}{dy}(X - x)$

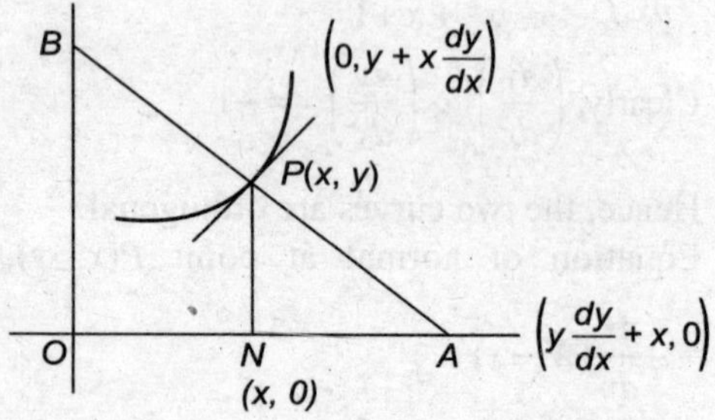

Fig. 10.6

Area of ΔOAB is $1 \Rightarrow \dfrac{1}{2}\left(y\dfrac{dy}{dx} + x\right)\left(x\dfrac{dx}{dy} + y\right) = 1$

$\Rightarrow \left(y\dfrac{dy}{dx} + x\right)\left(y\dfrac{dy}{dx} + x\right) = 2\dfrac{dy}{dx}$

$\Rightarrow y^2\left(\dfrac{dy}{dx}\right)^2 + 2(xy - 1)\dfrac{dy}{dx} + x^2 = 0$

66. c. Slope of tangent $= \dfrac{dy}{dx}$

$\therefore$ slope of normal $= -\dfrac{dx}{dy}$

Therefore, the equation of normal is

$Y - y = -\dfrac{dx}{dy}(X - x)$

This meets the x-axis ($y = 0$), where

$-y = -\dfrac{dx}{dy}(X - x) \Rightarrow X = x + y\dfrac{dy}{dx}$

$\therefore$ G is $\left(x + y\dfrac{dy}{dx},\ 0\right)$

$\therefore$ $OG = 2x \Rightarrow x + y\dfrac{dy}{dx} = 2x$

$\Rightarrow y\dfrac{dy}{dx} = x \Rightarrow y\,dy = xdx$

Integrating, we get $\dfrac{y^2}{2} = \dfrac{x^2}{2} + \dfrac{C}{2}$

$\Rightarrow y^2 - x^2 = c$, which is a hyperbola.

67. a. Equation of tangent is $Y - y = \dfrac{dy}{dx}(X - x)$

For X-intercept $Y = 0 \Rightarrow X = x - y\dfrac{dx}{dy}$

According to question $x - y\dfrac{dx}{dy} = y$

$\Rightarrow \dfrac{dy}{dx} = \dfrac{y}{x - y}$

putting $y = vx$, we get

$\Rightarrow v + x\dfrac{dv}{dx} = \dfrac{v}{1 - v}$

$\Rightarrow x\dfrac{dv}{dx} = \dfrac{v}{1 - v} - v = \dfrac{v - v + v^2}{1 - v}$

$\Rightarrow \displaystyle\int \dfrac{1 - v}{v^2}dv = \int \dfrac{dx}{x}$

$\Rightarrow -\dfrac{1}{v} - \log v = \log x + c$

$\Rightarrow -\dfrac{x}{y} - \log\dfrac{y}{x} = \log x + c$

$\Rightarrow -\dfrac{x}{y} = \log y + c$

Given when $x = 1, y = 1 \Rightarrow c = -1$

Hence, equation of curve is $1 - \dfrac{x}{y} = \log y$

$\Rightarrow y = e\,e^{-x/y} \Rightarrow e^{x/y} = \dfrac{e}{y}$

$\Rightarrow ye^{x/y} = e$

$$\Rightarrow \frac{y'-1}{y'+1}=c$$

Putting value of c in Eq. (1), we get

$$\Rightarrow \frac{x}{\frac{y'-1}{y'+1}-1}+\frac{y}{\frac{y'-1}{y'+1}+1}=1$$

$$\Rightarrow \frac{x(y'+1)}{-2}+\frac{y(y'+1)}{2y'}=1$$

$$\Rightarrow \frac{(y'+1)}{2}\left(\frac{y}{y'}-x\right)=1$$

$$\Rightarrow \left(1+\frac{dy}{dx}\right)\left(y-x\frac{dy}{dx}\right)=2\frac{dy}{dx}$$

85. a. We have

$$f(\theta)=\frac{d}{d\theta}\int_0^{\theta}\frac{dx}{1-\cos\theta\cos x}=\frac{1}{1-\cos^2\theta}=\operatorname{cosec}^2\theta$$

[using Leibnitz's Rule]

$$\Rightarrow \frac{df(\theta)}{d\theta}=-2\operatorname{cosec}^2\theta\cot\theta$$

$$\Rightarrow \frac{df(\theta)}{d\theta}+2f(\theta)\cot\theta=0 \qquad [\because f(\theta)=\operatorname{cosec}^2\theta]$$

86. c. $v=\frac{A}{r}+B$ (1)

$\frac{dv}{dr}=-\frac{A}{r^2}$ (2)

$\frac{d^2v}{dr^2}=\frac{2A}{r^3}$ (3)

Eliminating A between equations (2) and (3), we get

$$r\frac{d^2v}{dr^2}=\frac{2A}{r^2}=2\left(-\frac{dv}{dr}\right)$$

$$\therefore \frac{d^2v}{dr^2}+\frac{2}{r}\frac{dv}{dr}=0$$

87. c. $\frac{y'''}{y''}=8\Rightarrow \log y_2=8x+c$

When $x=0, y_2=1$ and $\log 1=0$ $\therefore c=0$

$\therefore y^2=e^{8x}$. Integrate

$y'=\frac{e^{8x}}{8}+\lambda x=0, y'(0)=0$

$\therefore \lambda=-1/8$

$\therefore y'=\frac{e^{8x}}{8}-\frac{1}{8}$. Integrate again

$y=\frac{e^{8x}}{64}-\frac{x}{8}+k$

Also when $x=0, y=\frac{1}{8}$ $\therefore k=\frac{7}{64}$

$$\therefore y=\frac{1}{8}\left(\frac{e^{8x}}{8}-x+\frac{7}{8}\right)$$

88. a. $y^2=t$; $2y\frac{dy}{dx}=\frac{dt}{dx}$; hence the differential equation becomes

$$\left(e^{x^2}+e^t\right)\frac{dt}{dx}+2e^{x^2}(xt-x)=0$$

$$e^{x^2}+e^t+2e^{x^2}x(t-1)\frac{dx}{dt}=0$$

put $e^{x^2}=z$; $e^{x^2}2x\frac{dx}{dt}=\frac{dz}{dt}$

$$\Rightarrow z+e^t+\frac{dz}{dt}(t-1)=0$$

$$\Rightarrow \frac{dz}{dt}+\frac{z}{(t-1)}=-\frac{e^t}{(t-1)}; \text{ I.F.}=e^{\int\frac{dt}{t-1}}=e^{\ln(t-1)}=t-1$$

$$\Rightarrow z(t-1)=-\int(e^t)\,dt$$

$$\Rightarrow z(t-1)=-e^t+C$$

$$\Rightarrow e^{x^2}(y^2-1)=-e^{y^2}+C$$

$$\Rightarrow e^{x^2}(y^2-1)+e^{y^2}=C$$

Multiple Correct Answers Type

1. a., b., c.

a. $f(\lambda x,\lambda y)=\frac{\lambda(x-y)}{\lambda^2(x^2+y^2)}=\lambda^{-1}f(x,y)$

$\Rightarrow$ Homogeneous of degree (–1).

b. $f(\lambda x,\lambda y)=(\lambda x)^{1/3}(\lambda y)^{-2/3}\tan^{-1}\frac{x}{y}$

$$=\lambda^{-1/3}x^{1/3}y^{-2/3}\tan^{-1}\frac{x}{y}$$

$$=\lambda^{-1/3}f(x,y)$$

$\Rightarrow$ Homogeneous

c. $f(\lambda x,\lambda y)=\lambda x\left(\ln\sqrt{\lambda^2(x^2+y^2)}-\ln\lambda y\right)+\lambda ye^{x/y}$

$$=\lambda x\left[\ln\left(\frac{\lambda\sqrt{(x^2+y^2)}}{\lambda y}\right)\right]+\lambda ye^{x/y}$$

$$=\lambda\left[x\left(\ln\sqrt{x^2+y^2}-\ln y\right)+ye^{x/y}\right]$$

$$=\lambda f(x,y)$$

$\Rightarrow$ Homogeneous

d. $f(\lambda x,\lambda y)=\lambda x\left[\ln\frac{2\lambda^2x^2+\lambda^2y^2}{\lambda x\lambda(x+y)}\right]+\lambda^2x^2\tan\frac{x+2y}{3x-y}$

$$=\lambda x\left[\ln\frac{2x^2+y^2}{x(x+y)}\right]+\lambda^2x^2\tan\frac{x+2y}{3x-y}$$

$\Rightarrow$ non-homogeneous

2. a., c. We have $(x-h)^2+(y-k)^2=a^2$ (1)

Differentiating w.r.t. x, we get

$$2(x-h)+2(y-k)\frac{dy}{dx}=0$$

$$\Rightarrow (x-h)+(y-k)\frac{dy}{dx}=0 \quad (2)$$

Differentiating w.r.t. x, we get

$$1+\left(\frac{dy}{dx}\right)^2+(y-k)\frac{d^2y}{dx^2}=0 \quad (3)$$

From Eq. (3), $y-k=-\left(\frac{1+p^2}{q}\right)$, where $p=\frac{dy}{dx}$, $q=\frac{d^2y}{dx^2}$

Putting the value of $y-k$ in Eq. (2), we get

$$x-h=\frac{(1+p^2)p}{q}$$

Substituting the values of $x-h$ and $y-k$ in Eq. (1), we get

$$\left(\frac{1+p^2}{q}\right)^2(1+p^2)=a^2$$

$$\Rightarrow \left[1+\left(\frac{dy}{dx}\right)^2\right]^3=a^2\left(\frac{d^2y}{dx^2}\right)^2,$$

which is the required differential equation.

3. a., b. $y\left(\frac{dy}{dx}\right)^2+(x-y)\frac{dy}{dx}-x=0$

$$\Rightarrow \frac{dy}{dx}=\frac{(y-x)\pm\sqrt{(x-y)^2+4xy}}{2y}$$

$\Rightarrow \frac{dy}{dx}=1$, which gives a straight line

or $\frac{dy}{dx}=-\frac{x}{y}$, which gives a circle.

4. a., b, c.

Obviously **(a)** is linear differential equations with $P=\frac{1}{x}$ and $Q=\log x$

$y\left(\frac{dy}{dx}\right)+4x=0 \Rightarrow \frac{dy}{dx}+\frac{4x}{y}=0$. Hence, not linear.

$$(2x+y^3)\left(\frac{dy}{dx}\right)=3y$$

$$\Rightarrow \frac{dx}{dy}=\frac{2x}{3y}+\frac{y^2}{3}$$

$\Rightarrow \frac{dx}{dy}-\frac{2x}{3y}=\frac{y^2}{3}$, which is linear with $P=\frac{2}{3y}$ and $Q=\frac{y^2}{3}$

$ydx+dy=0 \Rightarrow \frac{dy}{dx}+y=0$ which is linear differential equations with $P=1$ and $Q=0$

5. a., c. $\frac{dy}{dx}=\frac{ax+h}{by+k} \Rightarrow (by+k)\,dy=(ax+h)\,dx$

$$\Rightarrow b\frac{y^2}{2}+ky=\frac{a}{2}x^2+hx+C$$

For this to represent a parabola, one of the two terms x^2 or y^2 is zero.

Therefore, either $a=0, b\neq 0$ or $a\neq 0, b=0$

6. a., d. The given differential equation is

$$y_2(x^2+1)=2xy_1 \Rightarrow \frac{y_2}{y_1}=\frac{2x}{x^2+1}$$

Integrating both sides, we get

$\log y_1=\log(x^2+1)+\log C$

$\Rightarrow y_1=C(x^2+1)$ (1)

It is given that $y_1=3$ at $x=0$

Putting $x=0$, $y_1=3$ in equation (1), we get $3=C$

Substituting the value of C in (1), we obtain

$y_1=3(x^2+1)$ (2)

Integrating both sides w.r.t. to x, we get

$y=x^3+3x+C_2$

This passes through the point (0, 1). Therefore, $1=C_2$

Hence, the required equation of the curve is $y=x^3+3x+1$.

Obviously it is strictly increasing from equation (2)

Also $f(0)=1>0$, then the only root is negative.

7. a., b., c.

8. a., b., d.

$\frac{dy}{dx}+y\cos x=\cos x$ (linear)

$\text{I.F.}=e^{\int \cos x\,dx}=e^{\sin x}$

Therefore, solution is $y\,e^{\sin x}=\int e^{\sin x}\cos x\,dx=e^{\sin x}+c$

when $x=0, y=1$ then $c=0$

$\Rightarrow y=1$. Hence, options **(a), (b), and (d)** are true.

9. a., b., c.

We have, $f''(x)=g''(x)$. On integration, We get

$f'(x)=g'(x)+C$ (1)

Putting $x=1$, we get

$f'(1)=g'(1)+C \Rightarrow 4=2+C \Rightarrow C=2$

$\therefore f'(x)=g'(x)+2$

Integrating w.r.t. x, we get $f(x)=g(x)+2x+c_1$ (2)

Putting $x=2$, we get

$f(2)=g(2)+4+c_1 \Rightarrow 9=3+4+c_1 \Rightarrow c_1=2$

Therefore, $f(x)=g(x)+2x+2$. Putting $x=4$, we get $f(4)-g(4)=10$

$|f(x)-g(x)|<2 \Rightarrow |2x+2|<2 \Rightarrow |x+1|<1 \Rightarrow -2<x<0$

Also $f(2)=g(2) \Rightarrow x=-1$

$f(x)-g(x)=2x$ has no solution.

10. b.

$$(x^2y^2-1)\,dy+2xy^3\,dx=0$$

$$\Rightarrow x^2y^2dy+2xy^3\,dx=dy$$

$$\Rightarrow x^2\,dy+2xy\,dx=\frac{dy}{y^2}$$

$$\Rightarrow \int d\left(x^2y\right)=\int\frac{dy}{y^2}+c$$

$$\Rightarrow x^2y=\frac{y^{-1}}{-1}+c$$

$$\Rightarrow x^2y^2=-1+cy$$

i.e., $1+x^2y^2=cy$

11. a., b.

$$\frac{dy}{dx}=\frac{y}{x^2}\Rightarrow\frac{dy}{y}=\frac{dx}{x^2}\Rightarrow \ln y=-\frac{1}{x}+\ln c\Rightarrow\frac{y}{c}=e^{-1/x}$$

$$\Rightarrow y=ce^{-1/x}$$

Comparing with $y=a\,e^{-1/x}+b,\ a\in R,\ b=0$

12. b. We have $y\dfrac{dy}{dx}=k$ (constant)

$$\Rightarrow y\,dy=k\,dx\Rightarrow\frac{y^2}{2}=kx+C\Rightarrow y^2=2kx+2C$$

$$\Rightarrow y^2=2ax+b,\text{ where } a=k,\ b=2C$$

13. a., d.

The differential equation can be re-written as

$$\frac{x\,dx+y\,dy}{\sqrt{1-(x^2+y^2)}}=\frac{x\,dy-y\,dx}{\sqrt{x^2+y^2}}$$

Since $d\tan^{-1}(y/x)=\dfrac{x\,dy-y\,dx}{x^2+y^2}$, and

$d(x^2+y^2)=2(x\,dx+y\,dy)$,

Therefore, we have $\dfrac{\frac{1}{2}d\left(x^2+y^2\right)}{\sqrt{x^2+y^2}\sqrt{1-\left(x^2+y^2\right)}}$

$$=\frac{x\,dy-y\,dx}{x^2+y^2}$$

$$=d\{\tan^{-1}(y/x)\}$$

Put $x^2+y^2=t^2$ in the L.H.S and get

$$\frac{t\,dt}{t\sqrt{1-t^2}}=d\{\tan^{-1}(y/x)\}$$

Integrating both sides, we get

$\sin^{-1}t=\tan^{-1}(y/x)+c$

i.e., $\sin^{-1}\sqrt{\left(x^2+y^2\right)}=\tan^{-1}(y/x)+c$

14. a., b.

We have length of the normal = radius vector

$$\Rightarrow y\sqrt{1+\left(\frac{dy}{dx}\right)^2}=\sqrt{x^2+y^2}$$

$$\Rightarrow y^2\left(1+\left(\frac{dy}{dx}\right)^2\right)=x^2+y^2$$

$$\Rightarrow y^2\left(\frac{dy}{dx}\right)^2=x^2$$

$$\Rightarrow x=\pm y\frac{dy}{dx}$$

$$\Rightarrow x=y\frac{dy}{dx}\text{ or } x=-y\frac{dy}{dx}$$

$$\Rightarrow x\,dx-y\,dy=0\text{ or } x\,dx+y\,dy=0$$

$$\Rightarrow x^2-y^2=c_1\text{ or } x^2+y^2=c_2$$

Clearly, $x^2-y^2=c_1$ represents a rectangular hyperbola and $x^2+y^2=c_2$ represents circles.

15. a., b.

$$x=\sin\left(\frac{dy}{dx}-2y\right)\Rightarrow\frac{dy}{dx}-2y=\sin^{-1}x$$

$$x-2y=\log\left(\frac{dy}{dx}\right)\Rightarrow\frac{dy}{dx}=e^{x-2y}$$

Reasoning Type

1. a. The equation of circle contains. Three independent constants if it passes through three non-collinear points, therefore a is true and follows from *R*.

2. a. $y=Ae^x$

On differentiation we get $\dfrac{dy}{dx}=Ae^x$

3. d. Statement 2 is obviously true. But statement 1 is false as $2x-3y+2=\log\left(\dfrac{dy}{dx}\right)$

$$\Rightarrow\left(\frac{dy}{dx}\right)=e^{2x-3y+2}\text{ which has degree 1.}$$

4. b. Statement 1 is obviously true. Even statement 2 is also true but it does not explain statement 1.

5. a. $y=c_1\cos 2x+c_2\sin^2x+c_3\cos^2x+c_4e^{2x}+c_5e^{2x+c_6}$

$$=c_1\cos2x+c_2\left(\frac{1-\cos 2x}{2}\right)+c_3\left(\frac{\cos 2x-1}{2}\right)$$

$$+c_4e^{2x}+c_5e^{c_6}e^{2x}$$

$$=\left(c_1-\frac{c_2}{2}+\frac{c_3}{2}\right)\cos 2x+\left(\frac{c_2}{2}-\frac{c_3}{2}\right)+(c_4+c_5')e^{2x}$$

$$=\lambda_1\cos 2x+\lambda_2e^{2x}+\lambda_3$$

$\Rightarrow$ Total number of independent parameters in the given general solution is 3.

Hence, statement 1 is true, also statement 2 is true which explains statement 1.

Linked Comprehension Type

For Problems 1–3

1. b., 2., c., 3., a.

Sol.

1. b. $f(x) \le 0$ and $F'(x) = f(x)$

$\Rightarrow f(x) \ge cF(x)$

$\Rightarrow F'(x) - cF(x) \ge 0$

$\Rightarrow e^{-cx} F'(x) - c\,e^{-cx} F(x) \ge 0$

$\Rightarrow \frac{d}{dx}\left(e^{-cx} F(x)\right) \ge 0$

$\Rightarrow e^{-cx} F(x)$ is an increasing function

$\Rightarrow e^{-cx} F(x) \ge e^{-c(0)} F(0)$

$\Rightarrow e^{-cx} F(x) \ge 0$

$\Rightarrow F(x) \ge 0$

$\Rightarrow f(x) \ge 0$

[As $f(x) \ge cF(x)$ and c is positive]

$\Rightarrow f(x) = 0$

Also $\left(\frac{d\,g(x)}{dx}\right) < g(x)\ \forall\ x > 0$

$\Rightarrow e^{-x}\frac{d(g(x))}{dx} - e^{-x} g(x) < 0$

$\Rightarrow \frac{d}{dx}(e^{-x} g(x)) < 0$

$\Rightarrow e^{-x} g(x)$ is a decreasing function

$\Rightarrow e^{-x} g(x) < e^{-(0)} g(0)$

$\Rightarrow g(x) < 0$ (as $g(0) = 0$)

Thus $f(x) = g(x)$ has one solution $x = 0$

2. c. $|x^2 + x - 6| = f(x) + g(x) \Rightarrow |x^2 + x - 6| = g(x)$

$\Rightarrow$ no solution

3. a. $g(x)(\cos^{-1} x - \sin^{-1} x) \le 0$

$\Rightarrow (\cos^{-1} x - \sin^{-1} x) \ge 0 \Rightarrow x \in \left[-1, \frac{1}{\sqrt{2}}\right]$

For Problems 4–6

4. c., 5., b., 6., d.

Sol.

4. c. Given equation can be rewritten as

$$y = xp + \sqrt{(1+p^2)},\ p = \frac{dy}{dx} \quad (1)$$

Differentiating w.r.t. x, we get

$$p = p + x\frac{dp}{dx} + \frac{1}{2\sqrt{1+p^2}}\,2p\,\frac{dp}{dx}$$

$$\Rightarrow \frac{dp}{dx} = 0 \text{ or } \frac{p}{\sqrt{1+p^2}} = -x$$

$$\Rightarrow p = c \text{ or } p = \frac{x}{\sqrt{1-x^2}}$$

$\Rightarrow y = cx + \sqrt{(1+c^2)}$ gives the general solution and $x^2 + y^2 = 1$ as singular solution

5. b. $y = xp + p^2 \left(p = \frac{dy}{dx}\right)$ (1)

Differentiating Eq. (1) w.r.t. x, we get

$$p = p + x\frac{dp}{dx} + 2p\frac{dp}{dx}$$

$$\Rightarrow \frac{dp}{dx}(x + 2p) = 0$$

$$\Rightarrow \frac{dp}{dx} = 0 \text{ or } p = -\frac{x}{2}$$

Eliminating p from Eq. (1), we get

$$y = -\frac{x^2}{2} + \frac{x^2}{4} = -\frac{x^2}{4} \quad (4)$$

Clearly $f(x) = -\frac{x^2}{4} = -1$ has two solutions

6. d. $y = mx + m - m^3$ (1)

Differentiating w.r.t. x, we get

$$\frac{dy}{dx} = m + x\frac{dm}{dx} + \frac{dm}{dx} - 3m^2\frac{dm}{dx}$$

$$m = m + x\frac{dm}{dx} + \frac{dm}{dx} - 3m^2\frac{dm}{dx}$$

$$\frac{dm}{dx}(x + 1 - 3m^2) = 0$$

$$\frac{dm}{dx} = 0 \Rightarrow m = c \quad (2)$$

$$\text{or } x + 1 - 3m^2 = 0 \Rightarrow m^2 = \frac{x+1}{3} \quad (3)$$

Eliminating m between Eqs. (1) and (3), we get

$y = m(x + 1 - m^2)$

$$\Rightarrow y = \left(\frac{x+1}{3}\right)^{1/2}\left(x + 1 - \frac{x+1}{3}\right)$$

$$\Rightarrow y = \left(\frac{x+1}{3}\right)^{1/2}\frac{2}{3}(x+1)$$

$$\Rightarrow y = 2\left(\frac{x+1}{3}\right)^{3/2}$$

$$\Rightarrow y^2 = \frac{4}{27}(x+1)^3$$

$$\Rightarrow 27y^2 = 4(x+1)^3$$

For Problems 7–9

7. c., 8., a., 9., b.

Sol. Integrating $\frac{d^2y}{dx^2} = 6x - 4$, we get $\frac{dy}{dx} = 3x^2 - 4x + A$

When $x = 1$, $\frac{dy}{dx} = 0$ so that $A = 1$. Hence

$\frac{dy}{dx} = 3x^2 - 4x + 1$ (1)

Integrating, we get $y = x^3 - 2x^2 + x + B$.

When $x = 1, y = 5$, so that $B = 5$.

Thus, we have $y = x^3 - 2x^2 + x + 5$

From Eq. (1), we get the critical points $x = 1/3$, $x = 1$

At the critical point $x = 1/3$, d^2y/dx^2 is negative.

Therefore, at $x = 1/3$, y has a local maximum.

At $x = 1$, $\frac{d^2y}{dx^2}$ is positive.

Therefore, at $x = 1$ y has a local minimum.

Also $f(1) = 5, f\left(\frac{1}{3}\right) = \frac{157}{27}$ $f(0) = 5, f(2) = 7$

Hence the global maximum value = 7 and the global minimum value = 5.

For Problems 10–12

10. a., 11., c., 12., c.

10. a. Let N denote the amount of material present at time t. Then,

$\frac{dN}{dt} - kN = 0$

This differential equation is separable and linear, its solution is $N = ce^{kt}$ (1)

At $t = 0$, we are given that $N = 50$. Therefore, from equation (1), $50 = ce^{k(0)}$, or $c = 50$.

Thus, $N = 50\,e^{kt}$ (2)

At $t = 2$, 10 percent of the original mass of 50 mg or 5 mg, has decayed.

Hence, at $t = 2$, $N = 50 - 5 = 45$.

Substituting these values into Eq. (2) and solving for k, we have $45 = 50e^{2k}$ or $k = \frac{1}{2}\log\frac{45}{50}$

Substituting this value into (2), we obtain the amount of mass present at any time t as

$N = 50e^{-(1/2)(\ln 0.9)\,t}$ (3)

where t is measured in hours.

11. c. We require N at $t = 4$. Substituting $t = 4$ into (3) and then solving for N, we find

$N = 50e^{-2\ln 0.9}$

12. c. We require t when $N = 50/2 = 25$. Substituting $N = 25$ into equation (3) and solving for t, we find $25 = 50\,e^{-(1/2)(\ln 0.9)\,t}$ or $t = (\ln 1/2)/(-1/2 \ln 0.9)$ hr.

For Problems 13–15

13. c., 14., b., 15., d.

13. a. Here, $V_0 = 100$, $a = 20$, $b = 0$, and $e = f = 5$. Hence

$\frac{dQ}{dt} + \frac{1}{20}Q = 0$

The solution of this linear equation is $Q = ce^{-t/20}$ (1)

At $t = 0$, we are given that $Q = a = 20$.

Substituting these values into Eq. (1), we find that $c = 20$, so that Eq. (1) can be rewritten as $Q = 20\,e^{-t/20}$.

For $t = 20$, $Q = 20/e$

14. b. Here $a = 0, b = 1, e = 4, f = 2$, and $V_0 = 10$.

The volume of brine in the tank at any time t is given as $V_0 + et - ft = 10 + 2t$.

We require t when $10 + 2t = 50$, hence, $t = 20$ min.

15. d. For the equation $\frac{dQ}{dt} + \frac{2}{10+2t}Q = 4$

This is a linear equation. Its solution is

$Q = \frac{40t + 4t^2 + c}{10 + 2t}$ (1)

At $t = 0$, $Q = a = 0$. Substituting these values in Eq. (1), we find that $c = 0$. We require Q at the moment of overflow, which from part (a) is $t = 20$. Thus

$Q = \frac{40(20) + 4(20)^2}{10 + 2(20)} = 40$ lb

Matrix-Match Type

1. a → q, s; b → p; c → p; d → q, r, s.

p. The equation of the required parabola is of the form $y^2 = 4a(x - h)$. Differentiating, we have

$2y\frac{dy}{dx} = 4a \Rightarrow y\frac{dy}{dx} = 2a \Rightarrow \left(\frac{dy}{dx}\right)^2 + y\frac{d^2y}{dx^2} = 0$

The degree of this differential equation is 1 and the order is 2.

q. We have $y = a(x + a)^2$ (1)

$\Rightarrow \frac{dy}{dx} = 2a(x + a)$ (2)

Dividing Eqs. (1) by (2), we get $\frac{y}{\frac{dy}{dx}} = \frac{x+a}{2}$

$\Rightarrow x + a = \frac{2y}{y_1}$, where $y_1 = \frac{dy}{dx}$

Substituting $a = \frac{2y}{y_1} - x$ in Eq. (1),

we get $y = \left(\frac{2y}{y_1} - x\right)\left(\frac{2y}{y_1}\right)^2 \Rightarrow y_1^3 y = 4(2y - xy_1)y^2$

Clearly, it is a differential equation of degree 3.

r. The given equation is $\left(1+3\frac{dy}{dx}\right)^{2/3} = 4\frac{d^3y}{dx^3}$

Cubing, we get $\left(1+3\frac{dy}{dx}\right)^2 = 64\left(\frac{d^3y}{dx^2}\right)^3$

Hence, order = degree = 3.

s. We have $y^2 = 2c\left(x + \sqrt{c}\right)$ (1)

Differentiating. w.r.t. x, we get $2y\frac{dy}{dx} = 2c$

$\Rightarrow c = y\frac{dy}{dx}$

Putting in Eq. (1), we get

$$y^2 = 2\left(y\frac{dy}{dx}\right)x + 2\left(y\frac{dy}{dx}\right)^{3/2}$$

$$\Rightarrow \left(y^2 - 2xy\frac{dy}{dx}\right)^2 = 4y^3\left(\frac{dy}{dx}\right)^3$$

Its order is 1 and degree is 3

2. **a → q; b → r; c → p; d → s.**

a. $y = e^{4x} + 2e^{-x}$; $y_1 = 4e^{4x} - 2e^{-x}$; $y_2 = 16e^{4x} + 2e^{-x}$;

$y_3 = 64e^{4x} - 2e^{-x}$

Now, $y_3 - 13y_1 = (64e^{4x} - 2e^{-x}) - 13(4e^{4x} - 2e^{-x})$

$= 12e^{4x} + 24e^{-x}$

$y_3 - 13y = 12(e^{4x} + 2e^{-x}) = 12y$

Therefore, $K = 12$ and $K/3 = 4$.

b. Let $y = mx + c \Rightarrow \frac{dy}{dx} = m$

$\Rightarrow mx + c\ m + xm^2$

$\Rightarrow m = m^2$ and $c = m$

$\Rightarrow m = 0$ or $m = 1$

c. $y = u^m \Rightarrow \frac{dy}{dx} = m\,u^{m-1}\frac{du}{dx}$

Substituting the value of y and $\frac{dy}{dx}$ in

$$2x^4y\frac{dy}{dx} + y^4 = 4x^6$$

we have $2x^4\,u^m\,m\,u^{m-1}\frac{du}{dx} + u^{4m} = 4\,x^6$.

$$\Rightarrow \frac{du}{dx} = \frac{4x^6 - u^{4m}}{2m\,x^4\,u^{2m-1}}$$

For homogeneous $4m = 6 \Rightarrow m = 3/2$.

and $2m - 1 = 2 \Rightarrow m = \frac{3}{2}$

d. $y = Ax^m + Bx^{-n}$

$\Rightarrow \frac{dy}{dx} = Amx^{m-1} - nBx^{-n-1}$

$\Rightarrow \frac{d^2y}{dx^2} = Am(m-1)x^{m-2} + n(n+1)Bx^{-n-2}$

Putting these values in $x^2\frac{d^2y}{dx^2} + 2x\frac{dy}{dx} = 12y$

we have $= m(m+1)Ax^m + n(n-1)Bx^{-n} = 12(Ax^m + Bx^{-n})$

$\Rightarrow m(m+1) = 12$ or $n(n-1) = 12$

$\Rightarrow m = 3, -4$ or $n = 4, -3$

Integer Type

1. **(6)** $\frac{d^2y}{dx^2} = \left[y + \left(\frac{dy}{dx}\right)^6\right]^{1/4}$

$$\Rightarrow \left(\frac{d^2y}{dx^2}\right)^4 = \left[y + \left(\frac{dy}{dx}\right)^6\right]$$

Hence, the order is 2 and degree is 4.

2. **(2)** $y = (\sin^{-1}x)^2 + A\cos^{-1}x + B$

$= (\sin^{-1}x)^2 - A\sin^{-1}x + \frac{\pi A}{2} + B$

Differentiating w.r.t. x, we get

$$\frac{dy}{dx} = \frac{2\sin^{-1}x}{\sqrt{1-x^2}} - \frac{A}{\sqrt{1-x^2}}$$

$$\Rightarrow (1-x^2)\left(\frac{dy}{dx}\right)^2 = 4(\sin^{-1}x)^2 - 4A\sin^{-1}x + A^2$$

$$= 4y - 4B + A^2 - 2\pi A$$

Differentiating again w.r.t. x, we have

$$2(1-x^2)\left(\frac{dy}{dx}\right)\left(\frac{d^2y}{dx^2}\right) - 2x\left(\frac{dy}{dx}\right)^2 = 4\frac{dy}{dx}$$

$\Rightarrow (1-x^2)\frac{d^2y}{dx^2} - x\frac{dy}{dx} = 2$, which is the required differential equation.

3. **(8)** It is given that $\frac{dy}{dx} = \frac{1}{2y}$. So $2y\,dy = dx$

Integrating, we get $y^2 = x + c$.

Now when $x = 3, y = 4$ gives $c = 13$.

Hence the equation of the required curve is $y^2 = x + 13$, which is a parabola

When $x = 3, y = \pm 4$.

4. **(6)** The equation of the tangent at any point $P(x, y)$ is

$$Y - y = \frac{dy}{dx}(X - x) \quad (1)$$

Given that

intercept on the x–axis (putting $Y = 0$) = 2 (x – coordinates of P)

$$\Rightarrow x - y\frac{dx}{dy} = 2x$$

$$\Rightarrow -\frac{dy}{y} = \frac{dx}{x}$$

Integrating, we get $xy = c$.

Since the curve passes through (1, 2), $c = 2$.

Hence, the equation of the required curve is $xy = 2$.

5. **(5)** The balance $N(t)$ in the account at any time t is

$$\frac{dN}{dt} - kN = 0. \text{ Its solution is } N(t) = ce^{kt} \quad (1)$$

Let the initial deposit be N_0.

At $t = 0$, $N(0) = N_0$, which when substituted into (1) yields $N_0 = ce^{k(0)} = c$ and (1) becomes $N(t) = N_0e^{kt}$.

We seek the value of k for which $N = 2N_0$ when $t = 6$. Substituting these values into (1)and solving for k

we find $2N_0 = N_0e^{k(6)} \Rightarrow e^{6k} = 2 \Rightarrow k = \frac{1}{6}\ln|2| = 0.1155$.

An interest rate of 11.55 percent is required.

6. **(0)** Let the curves be $y = f_1(x)$ and $y = f_2(x)$.

Equation of tangents with equal abscissa x are $Y - f_1(x) = f'(x)(X - x)$ and $Y - f_2(x) = f'(x) = (X - x)$.

These tangent intersect at the y-axis. So their y-intercepts are same. Therefore,

$$-xf'_1(x) + f_1(x) = -xf'_2(x) + f_2(x)$$

$$\Rightarrow f_1(x) - f_2(x) = x(f'_1(x) - f'_2(x))$$

$$\Rightarrow \int \frac{f_1'(x) - f_2'(x)}{f_1(x) - f_2(x)} = \int \frac{dx}{x}$$

$$\Rightarrow \ln|f_1(x) - f_2(x)| = \ln|x| + \log C_1$$

$$\Rightarrow f_1(x) - f_2(x) = \pm C_1 x \quad (1)$$

Now the equation of the normal with equal abscissa x are

$$(Y - f_1(x)) = -\frac{1}{f_1'(x)}(X - x)$$

$$\text{and } (Y - f_2(x)) = -\frac{1}{f_2'(x)}(X - x).$$

As these normals intersect on the x-axis, we have

$$x + f_1(x)\, f_1'(x) = x + f_2(x)\, f_2'(x)$$

$$\Rightarrow f_1(x)\, f_1'(x) - f_2(x)\, f_2'(x) = 0$$

Integrating, we get $f_1^2(x) - f_1^2(x) = C_2$

$$\Rightarrow f_1(x) + f_2(x) = \frac{C_2}{f_1(x) - f_2(x)}$$

$$= \pm\frac{C_2}{C_1 x} = \frac{\pm\lambda_2}{x} \quad (2)$$

From (1) and (2), we get,

$$2f_1(x) = \pm\left(\frac{\lambda_2}{x} + C_1 x\right) \text{ and } 2f_2(x) = \pm\left(\frac{\lambda_2}{x} - C_1 x\right).$$

we have $f_1(1) = 1$ and $f(2) = 3$

$$\Rightarrow f_1(x) = \frac{2}{x} - x,\ f_2(x) = \frac{2}{x} + x$$

7. **(4)** Let the cyclist starts to moving from the point O and moving along OX, attains a velocity v at the point P in time t such that $OP = x$. Let the acceleration of the moving cycle at P be a. Then we know that

$$v = \frac{dx}{dt} \text{ and a} = \frac{dv}{dt} = \frac{d^2x}{dt^2} = v\frac{dv}{dx} \quad (1)$$

By hypothesis, retardation = 0.08 + 0.02 v^2 = 0.02 $(4 + v^2)$. Therefore,

$$v\frac{dv}{dx} = -0.02(4 + v^2) \text{ or}$$

$$dx = -\frac{1}{0.02}\frac{d\ dv}{4 + v^2} \quad (2)$$

Integrating (2) between the limits $x = 0$, $v = 4\ m/s$, and $x = x'$ meters, $v = 0$, we get

$$\int_0^{x'} dx = -\frac{1}{0.04}\int_4^0 \frac{2vdv}{4 + v^2}$$

$$\text{or } x' = -\frac{1}{0.04}[\ln(4 + v^2)]_4^0$$

$$x' = -\frac{1}{0.04}[\ln 4 - \ln 20]$$

$$= \frac{\ln 5}{0.04} = \frac{1.61}{0.04} = \frac{161}{4} \text{ meters.}$$

8. **(3)** $\frac{dy}{dx}\left(\frac{2 + \sin x}{1 + y}\right) = -\cos x,\ y(0) = 1$

$$\Rightarrow \frac{dy}{1 + y} = \frac{-\cos x}{2 + \sin x}dx$$

Integrating both sides, we get
$\ln(1+y) = -\ln(2+\sin x) + C$
Putting $x = 0$ and $y = 1$, we get
$\ln 2 = -\ln 2 + C$
$\Rightarrow C = \ln 4.$
Putting $x = \pi/2$, we get
$\ln(1+y) = -\ln 3 + \ln 4 = \ln 4/3$
$\Rightarrow y = 1/3.$

9. **(3)** The given equation is
$ydx + y^2 dy = x dy;\ \ x \in R, y > 0, y(1) = 1$

$$\Rightarrow \frac{ydx - xdy}{y^2} + dy = 0$$

$$\Rightarrow \frac{d}{dx}\cdot\left(\frac{x}{y}\right) + dy = 0$$

On integrating, we get

$$\frac{x}{y} + y = C$$

$y(1) = 1 \Rightarrow 1 + 1 = C \Rightarrow C = 2$

$$\therefore\ \frac{x}{y} + y = 2$$

Now to find $y-3)$, putting $x = -3$ in the above equation we get,

$$-\frac{3}{y} + y = 2$$

$\Rightarrow y^2 - 2y - 3 = 0 \Rightarrow y = 3, -1$
But given that $y > 0$ Therefore, $y = 3$.

10. **(4)** $y^2 = 2c\left(x + \sqrt{c}\right)$

Differentiating w.r.t. x, we get
$2yy' = 2c \Rightarrow c = yy'$.
Eliminating c, we get,

$y^2 = 2yy_1\,(x + \sqrt{yy_1})$ or $(y^2 - 2x\,yy_1)^2 = 4y^3\,y_1^3$.

It involves only first order derivative, so its order is 1 but its degree is 3 as y_1^3 is there.

11. **(3)** $y = (c_1 + c_2)\cos(x + c_3) - c_4\,e^{x+c_5}$
$= (c_1 + c_2)\cos(x + c_3) - c_4\,e^{c_5}e^x$
$= A\cos(x + c_3) - Be^x$
[Taking $c_1 + c_2 = A$, $c_4\,e^{c_5} = B$]
Thus, there are actually three arbitrary constants and hence this differential equation should be of order 3.

Section E

Vectors & 3D Geometry

CHAPTER 1

Introduction to Vectors

Objective Type

Each question has four choices a, b, c and d, out of which *only one* answer is correct. Find the correct answer.

1. Four non-zero vectors will always be
   a. linearly dependent
   b. linearly independent
   c. either **a.** or **b.**
   d. none of these

2. Let $\vec{a}, \vec{b}$, and $\vec{c}$ be three units vectors such that $3\vec{a}+4\vec{b}+5\vec{c}=0$. Then which of the following statements is true?
   a. $\vec{a}$ is parallel to $\vec{b}$
   b. $\vec{a}$ is perpendicular to $\vec{b}$
   c. $\vec{a}$ is neither parallel nor perpendicular to $\vec{b}$
   d. none of these

3. Let ABC be a triangle, the position vectors of whose vertices are, respectively, $\hat{i}+2\hat{j}+4\hat{k}$, $2\hat{i}+2\hat{j}+\hat{k}$, and $2\hat{i}+4\hat{j}-3\hat{k}$. Then ΔABC is
   a. isosceles b. equilateral
   c. right angled d. none of these

4. If $|\vec{a}+\vec{b}|<|\vec{a}-\vec{b}|$, then the angle between $\vec{a}$ and $\vec{b}$ can lie in the interval
   a. $(-\pi/2, \pi/2)$ b. $(0, \pi)$
   c. $(\pi/2, 3\pi/2)$ d. $(0, 2\pi)$

5. A point O is the centre of a circle circumscribed about a triangle ABC. Then $\overrightarrow{OA}\sin 2A+\overrightarrow{OB}\sin 2B+\overrightarrow{OC}\sin 2C$ is equal to
   a. $(\overrightarrow{OA}+\overrightarrow{OB}+\overrightarrow{OC})\sin 2A$
   b. $3\,.\,\overrightarrow{OG}$, where G is the centroid of triangle ABC
   c. $\vec{0}$
   d. none of these

6. If G is the centroid of a triangle ABC, then $\overrightarrow{GA}+\overrightarrow{GB}+\overrightarrow{GC}$ is equal to
   a. $\vec{0}$ b. $3\overrightarrow{GA}$
   c. $3\overrightarrow{GB}$ d. $3\overrightarrow{GC}$

7. If $\vec{a}$ is a non-zero vector of modulus a and m is a non-zero scalar, then $m\vec{a}$ is a unit vector if
   a. $m=\pm 1$ b. $a=|m|$
   c. $a=1/|m|$ d. $a=1/m$

8. $ABCD$ a parallelogram, and A_1 and B_1 are the midpoints of sides BC and CD, respectively. If $\overrightarrow{AA_1}+\overrightarrow{AB_1}=\lambda\,\overrightarrow{AC}$, then λ is equal to
   a. $\frac{1}{2}$ b. 1
   c. $\frac{3}{2}$ d. 2

9. The position vectors of the points P and Q with respect to the origin O are $\vec{a}=\hat{i}+3\hat{j}-2\hat{k}$ and $\vec{b}=3\hat{i}-\hat{j}-2\hat{k}$ equal to, respectively. If M is a point on PQ, such that OM is the bisector of POQ, then $\overrightarrow{OM}$ is
   a. $2(\hat{i}-\hat{j}+\hat{k})$ b. $2\hat{i}+\hat{j}-2\hat{k}$
   c. $2(-\hat{i}+\hat{j}-\hat{k})$ d. $2(\hat{i}+\hat{j}+\hat{k})$

10. $ABCD$ is a quadrilateral. E is the point of intersection of the line joining the midpoints of the opposite

sides. If O is any point and $\overrightarrow{OA}+\overrightarrow{OB}+\overrightarrow{OC}+\overrightarrow{OD} = x\overrightarrow{OE}$, then x is equal to

a. 3 b. 9
c. 7 d. 4

11. If vectors $\overrightarrow{AB} = -3\hat{i}+4\hat{k}$ and $\overrightarrow{AC}=5\hat{i}-2\hat{j}+4\hat{k}$ are the sides of a ΔABC, then the length of the median through A is

a. $\sqrt{14}$ b. $\sqrt{18}$
c. $\sqrt{29}$ d. 5

12. A, B, C and D have position vectors $\vec{a},\vec{b},\vec{c}$ and $\vec{d}$ respectively, such that $\vec{a}-\vec{b}=2(\vec{d}-\vec{c})$. Then

a. AB and CD bisect each other
b. BD and AC bisect each other
c. AB and CD trisect each other
d. BD and AC trisect each other

13. If $\vec{a}$ and $\vec{b}$ are two unit vectors and θ is the angle between them, then the unit vector along the angular bisector of $\vec{a}$ and $\vec{b}$ will be given by

a. $\dfrac{\vec{a}-\vec{b}}{2\cos(\theta/2)}$ b. $\dfrac{\vec{a}+\vec{b}}{2\cos(\theta/2)}$
c. $\dfrac{\vec{a}-\vec{b}}{\cos(\theta/2)}$ d. none of these

14. Let us define the length of a vector $a\hat{i}+b\hat{j}+c\hat{k}$ as $|a|+|b|+|c|$. This definition coincides with the usual definition of length of a vector $a\hat{i}+b\hat{j}+c\hat{k}$ if and only if

a. $a=b=c=0$
b. any two of a, b and c are zero
c. any one of a, b and c is zero
d. $a+b+c=0$

15. Given three vectors $\vec{a}=6\hat{i}-3\hat{j}, \vec{b}=2\hat{i}-6\hat{j}$ and $\vec{c}=-2\hat{i}+21\hat{j}$ such that $\vec{\alpha}=\vec{a}+\vec{b}+\vec{c}$. Then the resolution of the vector $\vec{\alpha}$ into components with respect to $\vec{a}$ and $\vec{b}$ is given by

a. $3\vec{a}-2\vec{b}$ b. $3\vec{b}-2\vec{a}$
c. $2\vec{a}-3\vec{b}$ d. $\vec{a}-2\vec{b}$

16. If $\vec{\alpha}+\vec{\beta}+\vec{\gamma}=a\vec{\delta}$ and $\vec{\beta}+\vec{\gamma}+\vec{\delta}=b\vec{\alpha}$, $\vec{\alpha}$ and $\vec{\delta}$ are non-collinear, then $\vec{\alpha}+\vec{\beta}+\vec{\gamma}+\vec{\delta}$ equals

a. $a\vec{\alpha}$ b. $b\vec{\delta}$
c. 0 d. $(a+b)\,\vec{\gamma}$

17. In triangle ABC, $\angle A = 30°$, H is the orthocentre and D is the midpoint of BC. Segment HD is produced to T such that $HD = DT$. The length AT is equal to

a. $2BC$ b. $3BC$
c. $\frac{4}{3}BC$ d. none of these

18. Let $\vec{r}_1, \vec{r}_2, \vec{r}_3, \ldots, \vec{r}_n$ be the position vectors of points $P_1, P_2, P_3, \ldots, P_n$ relative to the origin O. If the vector equation $a_1\vec{r}_1+a_2\vec{r}_2+\cdots+a_n\vec{r}_n=0$ holds, then a similar equation will also hold w.r.t. to any other origin provided

a. $a_1+a_2+\cdots+a_n=n$
b. $a_1+a_2+\cdots+a_n=1$
c. $a_1+a_2+\cdots+a_n=0$
d. $a_1=a_2=a_3=\cdots=a_n=0$

19. Given three non-zero, non-coplanar vectors $\vec{a},\vec{b}$ and $\vec{c}$, $\vec{r}_1 = p\vec{a}+q\vec{b}+\vec{c}$ and $\vec{r}_2=\vec{a}+p\vec{b}+q\vec{c}$. The vectors $\vec{r}_1+2\vec{r}_2$ and $2\vec{r}_1+\vec{r}_2$ are collinear, then (p, q) is

a. (0,0) b. (1,–1)
c. (–1, 1) d. (1, 1)

20. If the vectors $\vec{a}$ and $\vec{b}$ are linearly independent satisfying $(\sqrt{3}\tan\theta+1)\,\vec{a}+(\sqrt{3}\sec\theta-2)\,\vec{b}=0$, then the most general values of θ are

a. $n\pi-\frac{\pi}{6}, n\in Z$ b. $2n\pi\pm\frac{11\pi}{6}, n\in Z$
c. $n\pi\pm\frac{\pi}{6}, n\in Z$ d. $2n\pi+\frac{11\pi}{6}, n\in Z$

21. In a trapezium, vector $\overrightarrow{BC}=\alpha\,\overrightarrow{AD}$. We will then find that $\vec{p}=\overrightarrow{AC}+\overrightarrow{BD}$ is collinear with $\overrightarrow{AD}$. If $\vec{p}=\mu\,\overrightarrow{AD}$, then which one is true?

a. $\mu=\alpha+2$ b. $\mu+\alpha=1$
c. $\alpha=\mu+1$ d. $\mu=\alpha+1$

22. Vectors $\vec{a}=\hat{i}+2\hat{j}+3\hat{k}$; $\vec{b}=2\hat{i}-\hat{j}+\hat{k}$ and $\vec{c}=3\hat{i}+\hat{j}+4\hat{k}$ are so placed that the end point of one vector is the starting point of the next vector. Then the vectors are

a. not coplanar
b. coplanar but cannot form a triangle
c. coplanar and form a triangle
d. coplanar and can form a right-angled triangle

23. Vectors $\vec{a}=-4\hat{i}+3\hat{k}$; $\vec{b}=14\hat{i}+2\hat{j}-5\hat{k}$ are laid off from one point. Vector $\vec{d}$, which is being laid off from the same point dividing the angle between vectors $\vec{a}$ and $\vec{b}$ in equal halves and having the magnitude $\sqrt{6}$, is

a. $\hat{i}+\hat{j}+2\hat{k}$ b. $\hat{i}-\hat{j}+2\hat{k}$

c. $\hat{i}+\hat{j}-2\hat{k}$ d. $2\hat{i}-\hat{j}-2\hat{k}$

24. If $\hat{i}-3\hat{j}+5\hat{k}$ bisects the angle between $\hat{a}$ and $-\hat{i}+2\hat{j}+2\hat{k}$, where $\hat{a}$ is a unit vector, then

a. $\hat{a}=\frac{1}{105}(41\hat{i}+88\hat{j}-40\hat{k})$

b. $\hat{a}=\frac{1}{105}(41\hat{i}+88\hat{j}+40\hat{k})$

c. $\hat{a}=\frac{1}{105}(-41\hat{i}+88\hat{j}-40\hat{k})$

d. $\hat{a}=\frac{1}{105}(41\hat{i}-88\hat{j}-40\hat{k})$

25. If $4\hat{i}+7\hat{j}+8\hat{k}$, $2\hat{i}+3\hat{j}+4\hat{k}$, and $2\hat{i}+5\hat{j}+7\hat{k}$ are the position vectors of the vertices A, B and C, respectively, of triangle ABC, the position vector of the point where the bisector of angle A meets BC, is

a. $\frac{2}{3}(-6\hat{i}-8\hat{j}-6\hat{k})$ b. $\frac{2}{3}(6\hat{i}+8\hat{j}+6\hat{k})$

c. $\frac{1}{3}(6\hat{i}+13\hat{j}+18\hat{k})$ d. $\frac{1}{3}(5\hat{j}+12\hat{k})$

26. If $\vec{b}$ is vector whose initial point divides the join of $5\hat{i}$ and $5\hat{j}$ in the ratio $k:1$ and terminal pint is origin and $|\vec{b}|\le\sqrt{37}$, then k lies in the interval

a. $[-6,-1/6]$

b. $(-\infty,-6]\cup[-1/6,\infty)$

c. $[0,6]$

d. none of these

27. Find the value of λ so that the points P, Q, R, and S on the sides OA, OB, OC and AB, respectively, of a regular tetrahedron $OABC$ are coplanar. It is given that $\frac{OP}{OA}=\frac{1}{3}$, $\frac{OQ}{OB}=\frac{1}{2}$, $\frac{OR}{OC}=\frac{1}{3}$ and $\frac{OS}{AB}=\lambda$.

a. $\lambda=\frac{1}{2}$ b. $\lambda=-1$

c. $\lambda=0$ d. for no value of λ

28. I is the incentre of triangle ABC, whose corresponding sides are a, b, c, respectively. $a\,\overrightarrow{IA}+b\,\overrightarrow{IB}+c\,\overrightarrow{IC}$ is always equal to

a. $\vec{0}$ b. $(a+b+c)\,\overrightarrow{BC}$

c. $(\vec{a}+\vec{b}+\vec{c})\,\overrightarrow{AC}$ d. $(a+b+c)\,\overrightarrow{AB}$

29. Let $x^2+3y^2=3$ be the equation of an ellipse in the x-y plane. A and B are two points whose position vectors are $-\sqrt{3}\,\hat{i}$ and $-\sqrt{3}\,\hat{i}+2\hat{k}$. Then the position vector of a point P on the ellipse such that $\angle APB=\pi/4$ is

a. $\pm\hat{j}$ b. $\pm(\hat{i}+\hat{j})$

c. $\pm\hat{i}$ d. none of these

30. Locus of the point P, for which $\overrightarrow{OP}$ represents a vector with direction cosine $\cos\alpha=\frac{1}{2}$ ('O' is the origin) is

a. a circle parallel to the y-z plane with centre on the x-axis

b. a cone concentric with the positive x-axis having vertex at the origin and the slant height equal to the magnitude of the vector

c. a ray emanating from the origin and making an angle of 60° with the x-axis

d. a disc parallel to the y-z plane with centre on the x-axis and radius equal to $|\overrightarrow{OP}|\sin 60°$

31. If $\vec{x}$ and $\vec{y}$ are two non-collinear vectors and ABC is a triangle with side lengths a, b and c satisfying $(20a-15b)\,\vec{x}+(15b-12c)\,\vec{y}+(12c-20a)\,(\vec{x}\times\vec{y})=\vec{0}$, then triangle ABC is

a. an acute-angled triangle

b. an obtuse-angled triangle

c. a right-angled triangle

d. an isosceles triangle

32. A uni-modular tangent vector on the curve $x=t^2+2$, $y=4t-5$, $z=2t^2-6t$ at $t=2$ is

a. $\frac{1}{3}(2\hat{i}+2\hat{j}+\hat{k})$ b. $\frac{1}{3}(\hat{i}-\hat{j}-\hat{k})$

c. $\frac{1}{6}(2\hat{i}+\hat{j}+\hat{k})$ d. $\frac{2}{3}(\hat{i}+\hat{j}+\hat{k})$

33. If $\vec{x}$ and $\vec{y}$ are two non-collinear vectors and a, b and c represent the sides of a ΔABC satisfying $(a-b)\,\vec{x}+(b-c)\,\vec{y}+(c-a)\,(\vec{x}\times\vec{y})=0$, then ΔABC is

a. an acute-angled triangle

b. an obtuse-angled triangle

c. a right-angled triangle

d. a scalene triangle

34. $\vec{A}$ is a vector with direction cosines, $\cos\alpha$, $\cos\beta$, and $\cos\gamma$. Assuming the y-z plane as a mirror, the direction cosines of the reflected image of $\vec{A}$ in the y-z plane are

a. $\cos\alpha$, $\cos\beta$, $\cos\gamma$

b. $\cos\alpha$, $-\cos\beta$, $\cos\gamma$

c. $-\cos\alpha$, $\cos\beta$, $\cos\gamma$

d. $-\cos\alpha$, $-\cos\beta$, $-\cos\gamma$

Multiple Correct Answers Type

Each question has four choices a, b, c, and d, out of which *one or more* are correct.

1. If $(x, y, z) \neq (0, 0, 0)$ and $(\hat{i}+\hat{j}+3\hat{k})x+(3\hat{i}-3\hat{j}+\hat{k})y+(-4\hat{i}+5\hat{j})z=a(x\hat{i}+y\hat{j}+z\hat{k})$, then the value of a is
   a. 1 b. −1
   c. 2 d. 0

2. The sides of a parallelogram are $2\hat{i}+4\hat{j}-5\hat{k}$ and $\hat{i}+2\hat{j}+3\hat{k}$. Then the unit vector parallel to one of the diagonals is
   a. $\frac{1}{7}(3\hat{i}+6\hat{j}-2\hat{k})$
   b. $\frac{1}{7}(3\hat{i}-6\hat{j}-2\hat{k})$
   c. $\frac{1}{\sqrt{69}}(\hat{i}+2\hat{j}+8\hat{k})$
   d. $\frac{1}{\sqrt{69}}(-\hat{i}-2\hat{j}+8\hat{k})$

3. A vector $\vec{a}$ has the components $2p$ and 1 w.r.t. a rectangular Cartesian system. This system is rotated through a certain angle about the origin in the counterclockwise sense. If with respect to a new system, $\vec{a}$ has components $(p+1)$ and 1, then p is equal to
   a. −1 b. −1/3
   c. 1 d. 2

4. $\vec{a}=\hat{i}+\hat{j}+\hat{k}$, $\vec{b}=4\hat{i}+3\hat{j}+4\hat{k}$ and $\vec{c}=\hat{i}+x_1\hat{j}+x_2\hat{k}$ are linearly dependent vectors and $|\vec{c}|=\sqrt{3}$. Then
   a. $x_2=\pm 1$ b. $x_2=1$
   c. $x_1=-1$ d. $x_1=\pm 1$

5. If $\vec{a}$, $\vec{b}$ and $\vec{c}$ are coplanar vectors and λ is a real number, then the vectors $\vec{a}+2\vec{b}+3\vec{c}$, $\lambda\vec{b}+\mu\vec{c}$ and $(2\lambda-1)\vec{c}$ are coplanar for
   a. all values of μ b. $\lambda=\frac{1}{2}$
   c. $\lambda=0$ d. no value of λ

6. If the resultant of three forces $\vec{F}_1=p\hat{i}+3\hat{j}-\hat{k}$, $\vec{F}_2=6\hat{i}-\hat{k}$ and $\vec{F}_3=-5\hat{i}+\hat{j}+2\hat{k}$ acting on a particle has a magnitude equal to 5 units, then the value of p is
   a. −6 b. −4
   c. 2 d. 4

7. If the vectors $\hat{i}-\hat{j}$, $\hat{j}+\hat{k}$ and $\vec{a}$ form a triangle, then $\vec{a}$ may be
   a. $-\hat{i}-\hat{k}$ b. $\hat{i}-2\hat{j}-\hat{k}$
   c. $2\hat{i}+\hat{j}+\hat{k}$ d. $\hat{i}+\hat{k}$

8. The vector $\hat{i}+x\hat{j}+3\hat{k}$ is rotated through an angle θ and doubled in magnitude and becomes $4\hat{i}+(4x-2)\hat{j}+2\hat{k}$. The values of x are
   a. 1 b. −2/3
   c. 2 d. 4/3

9. $\vec{a}$, $\vec{b}$, and $\vec{c}$ are three coplanar unit vectors such that $\vec{a}+\vec{b}+\vec{c}=0$. If three vectors $\vec{p}$, $\vec{q}$, and $\vec{r}$ are parallel to $\vec{a}$, $\vec{b}$, and $\vec{c}$, respectively, and have integral but different magnitudes, then among the following options, $|\vec{p}+\vec{q}+\vec{r}|$ can take a value equal to
   a. 1 b. 0
   c. $\sqrt{3}$ d. 2

10. If non-zero vectors $\vec{a}$ and $\vec{b}$ are equally inclined to vector $\vec{c}$, then $\vec{c}$ is
    a. $\frac{|\vec{a}|}{|\vec{a}|+2|\vec{b}|}\vec{a}+\frac{|\vec{b}|}{|\vec{a}|+|\vec{b}|}\vec{b}$
    b. $\frac{|\vec{b}|}{|\vec{a}|+|\vec{b}|}\vec{a}+\frac{|\vec{a}|}{|\vec{a}|+|\vec{b}|}\vec{b}$
    c. $\frac{|\vec{a}|}{|\vec{a}|+2|\vec{b}|}\vec{a}+\frac{|\vec{b}|}{|\vec{a}|+2|\vec{b}|}\vec{b}$
    d. $\frac{|\vec{b}|}{2|\vec{a}|+|\vec{b}|}\vec{a}+\frac{|\vec{a}|}{2|\vec{a}|+|\vec{b}|}\vec{b}$

11. If $A(-4, 0, 3)$ and $B(14, 2, -5)$, then which one of the following points lie on the bisector of the angle between $\overrightarrow{OA}$ and $\overrightarrow{OB}$ (O is the origin of reference)?
    a. (2, 2, 4) b. (2, 11, 5)
    c. (−3, −3, −6) d. (1, 1, 2)

12. In a four-dimensional space where unit vectors along the axes are $\hat{i}, \hat{j}, \hat{k}$ and $\hat{l}$ and $\vec{a}_1, \vec{a}_2, \vec{a}_3, \vec{a}_4$ are four non-zero vectors such that no vector can be expressed as linear combination of others and $(\lambda-1)(\vec{a}_1-\vec{a}_2)+\mu(\vec{a}_2+\vec{a}_3)+\gamma(\vec{a}_3+\vec{a}_4-2\vec{a}_2)+\vec{a}_3+\delta\vec{a}_4=\vec{0}$ then
    a. $\lambda=1$ b. $\mu=-2/3$
    c. $\gamma=2/3$ d. $\delta=1/3$

13. Let ABC be a triangle, the position vectors of whose vertices are $7\hat{j}+10\hat{k}$, $-\hat{i}+6\hat{j}+6\hat{k}$, and $-4\hat{i}+9\hat{j}+6\hat{k}$. Then ΔABC is

a. isosceles
b. equilateral
c. right angled
d. none of these

Reasoning Type

Each question has four choices a, b, c, and d, out of which *only one* is correct. Each question contains Statement 1 and Statement 2.

a. Both the statements are True, and Statement 2 is correct explanation for Statement 1.

b. Both the statements are True, but Statement 2 is Not the correct explanation for Statement 1.

c. Statement 1 is True and Statement 2 is False.

d. Statment 1 is False and Statement 2 is True.

1. A vector has components p and 1 with respect to a rectangular Cartesian system. The axes are rotated through an angle α about the origin in the anticlockwise sense.

Statement 1: If the vector has component $p+2$ and 1 with respect to the new system, then $p=-1$

Statement 2: Magnitude of the original vector and the new system remains the same.

2. **Statement 1:** If three points P, Q and R have position vectors $\vec{a}$, $\vec{b}$, and $\vec{c}$, respectively, and $2\vec{a}+3\vec{b}-5\vec{c}=0$, then the points P, Q, and R must be collinear.

Statement 2: If for three points A, B, and C; $\overrightarrow{AB}=\lambda\,\overrightarrow{AC}$, then the points A, B, and C must be collinear.

3. **Statement 1:** If $\vec{u}$ and $\vec{v}$ are unit vectors inclined at an angle α and $\vec{x}$ is a unit vector bisecting the angle between them, then $\vec{x}=(\vec{u}+\vec{v})/(2\sin(\alpha/2))$.

Statement 2: If ΔABC is an isosceles triangle with $AB=AC=1$, then the vector representing the bisector of angle A is given by $\overrightarrow{AD}=(\overrightarrow{AB}+\overrightarrow{AC})/2$.

4. **Statement 1:** If $\cos\alpha$, $\cos\beta$, and $\cos\gamma$ are the direction cosines of any line segment, $\cos^2\alpha+\cos^2\beta+\cos^2\gamma=1$.

Statement 2: If $\cos\alpha$, $\cos\beta$, and $\cos\gamma$ are the direction cosines of a line segment, $\cos 2\alpha+\cos 2\beta+\cos 2\gamma=-1$.

5. **Statement 1:** The direction cosines of one of the angular bisectors of two intersecting lines having direction cosines as l_1, m_1, n_1, and l_2, m_2, n_2 are proportional to $l_1+l_2, m_1+m_2, n_1+n_2$.

Statement 2: The angle between the two intersecting lines having direction cosines as l_1, m_1, n_1 and l_2, m_2, n_2 is given by $\cos\theta=l_1l_2+m_1m_2+n_1n_2$.

6. **Statement 1:** In ΔABC, $\overrightarrow{AB}+\overrightarrow{BC}+\overrightarrow{CA}=0$

Statement 2: If $\overrightarrow{OA}=\vec{a}$, $\overrightarrow{OB}=\vec{b}$, then $\overrightarrow{AB}$ $\vec{a}+\vec{b}$

7. **Statement 1:** $\vec{a}=3\vec{i}+p\vec{j}+3\vec{k}$ and $\vec{b}=2\vec{i}+3\vec{j}+q\vec{k}$ are parallel vectors if $p=9/2$ and $q=2$.

Statement 2: If $\vec{a}=a_1\vec{i}+a_2\vec{j}+a_3\vec{k}$ and $\vec{b}=b_1\vec{i}+b_2\vec{j}+b_3\vec{k}$ are parallel, $\frac{a_1}{b_1}=\frac{a_2}{b_2}=\frac{a_3}{b_3}$

8. **Statement 1:** If $|\vec{a}+\vec{b}|=|\vec{a}-\vec{b}|$, then $\vec{a}$ and $\vec{b}$ are perpendicular to each other.

Statement 2 : If the diagonals of a parallelogram are equal in magnitude, then the parallelogram is a rectangle.

9. **Statement 1:** Let $A(\vec{a})$, $B(\vec{b})$, and $C(\vec{c})$ be three points such that $\vec{a}=2\hat{i}+\hat{k}$, $\vec{b}=3\hat{i}-\hat{j}+3\hat{k}$, and $\vec{c}=-\hat{i}+7\hat{j}-5\hat{k}$. Then $OABC$ is a tetrahedron.

Statement 2: Let $A(\vec{a})$, $B(\vec{b})$, and $C(\vec{c})$ be three points such that $\vec{a}$, $\vec{b}$, and $\vec{c}$ are non-coplanar. Then $OABC$ is a tetrahedron, where O is the origin.

10. **Statement 1:** Let $\vec{a}$, $\vec{b}$, $\vec{c}$, and $\vec{d}$ be the position vectors of four points A, B, C, and D and $3\vec{a}-2\vec{b}+5\vec{c}-6\vec{d}=\vec{0}$. Then points A, B, C, and D are coplanar.

Statement 2: Three non-zero, linearly dependent coinitial vectors ($\overrightarrow{PQ}, \overrightarrow{PR}$ and $\overrightarrow{PS}$) are coplanar.

11. **Statement 1:** If $|\vec{a}|=3$, $|\vec{b}|=4$ and $|\vec{a}+\vec{b}|=5$, then $|\vec{a}-\vec{b}|=5$.

Statement 2: The length of the diagonals of a rectangle is the same.

Linked Comprehension Type

Based on each paragraph, some multiple choice questions have to be answered. Each question has four choices a, b, c, and d, out of which *only one* is correct.

For Problems 1–3

ABCD is a parallelogram. *L* is a point on *BC* which divides *BC* in the ratio 1 : 2. *AL* intersects *BD* at *P*. *M* is a point on *DC* which divides *DC* in the ratio 1 : 2 and *AM* intersects *BD* in *Q*.

1. Point *P* divides *AL* in the ratio

   a. 1 : 2   b. 1 : 3
   c. 3 : 1   d. 2 : 1

2. Point *Q* divides *DB* in the ratio

   a. 1 : 2   b. 1 : 3
   c. 3 : 1   d. 2 : 1

3. $PQ : DB$ is equal to

   a. 2/3   b. 1/3
   c. 1/2   d. 3/4

For Problems 4–5

Let *OABCD* be a pentagon in which the sides *OA* and *CB* are parallel and the sides *OD* and *AB* are parallel. Also $OA : CB = 2 : 1$ and $OD : AB = 1 : 3$.

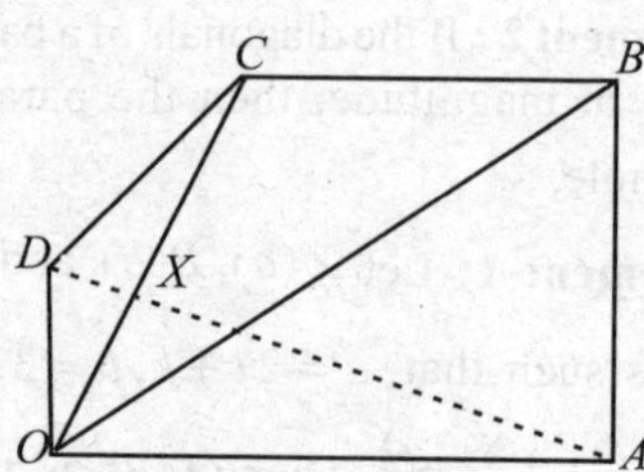

Fig. 1.1

4. The ratio $\dfrac{OX}{XC}$ is

   a. 3/4   b. 1/3
   c. 2/5   d. 1/2

5. The ratio $\dfrac{AX}{XD}$ is

   a. 5/2   b. 6
   c. 7/3   d. 4

For Problems 6–7

Consider the regular hexagon *ABCDEF* with centre at *O* (origin).

6. $\overrightarrow{AD} + \overrightarrow{EB} + \overrightarrow{FC}$ is equal to

   a. $2\overrightarrow{AB}$ b. $3\overrightarrow{AB}$
   c. $4\overrightarrow{AB}$ d. none of these

7. Five forces $\overrightarrow{AB}, \overrightarrow{AC}, \overrightarrow{AD}, \overrightarrow{AE}, \overrightarrow{AF}$ act at the vertex *A* of a regular hexagon *ABCDEF*. Then their resultant is

   a. $6\overrightarrow{AO}$ b. $2\overrightarrow{AO}$
   c. $4\overrightarrow{AO}$ d. $6\overrightarrow{AO}$

Matrix-Match Type

Each question contains statements given in two columns which have to be matched. Statements (a, b, c, d) in Column I have to be matched with statements (p, q, r, s) in Column II. If the correct matches are a → p, s; b → q, r; c → p, q and d → s, then the correctly bubbled 4 × 4 matrix should be as follows:

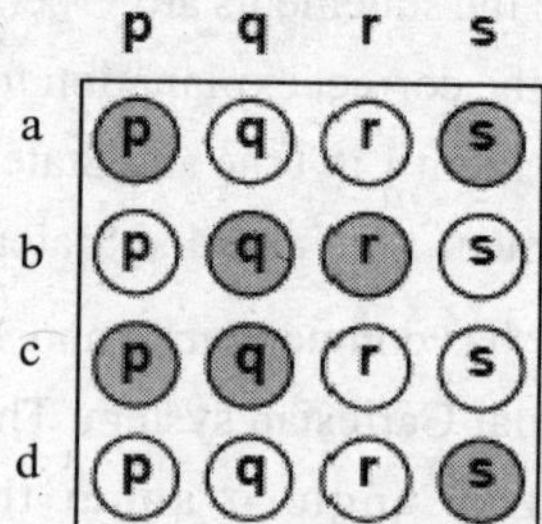

1. Refer to the following diagram:

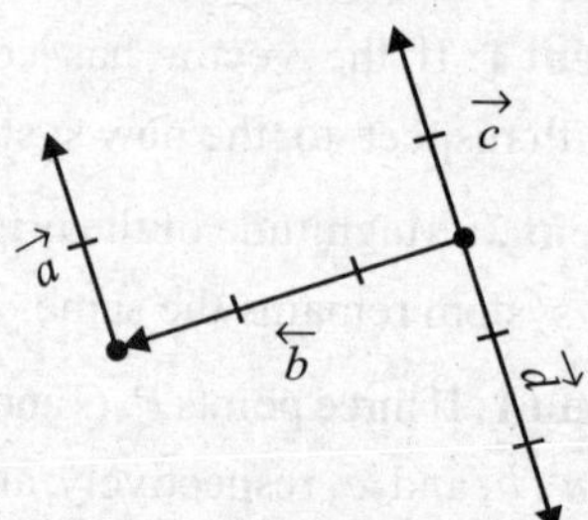

Fig. 1.2

Column I	Column II
a. Collinear vectors	p. $\vec{a}$
b. Coinitial vectors	q. $\vec{b}$
c. Equal vectors	r. $\vec{c}$
d. Unlike vectors (same initial point)	s. $\vec{d}$

2. $\vec{a}$ and $\vec{b}$ form the consecutive sides of a regular hexagon *ABCDEF*.

Column I	Column II
a. If $\overrightarrow{CD} = x\vec{a} + y\vec{b}$, then	p. $x = -2$
b. If $\overrightarrow{CE} = x\vec{a} + y\vec{b}$, then	q. $x = -1$
c. If $\overrightarrow{AE} = x\vec{a} + y\vec{b}$, then	r. $y = 1$
d. $\overrightarrow{AD} = -x\vec{a}$, then	s. $y = 2$

Integer Type

1. If $ABCDEF$ is a regular hexagon and $\overrightarrow{AD}+\overrightarrow{EB}+\overrightarrow{FC}=k\overrightarrow{AB}$, then the value of k is_______.
2. $ABCD$ is a quadrilateral and E is the point of intersection of the lines joining the middle points of the opposite sides. If the resultant of $\overrightarrow{OA}$, $\overrightarrow{OB}$, $\overrightarrow{OC}$, and $\overrightarrow{OD}$ is equal to $k\ \overrightarrow{OE}$, where O is any point, then the value of k is______.
3. The vectors $2\hat{i}+3\hat{j}$, $5\hat{i}+6\hat{j}$, and $8\hat{i}+\lambda\hat{j}$ have their initial point at (1, 1). The value of λ so that the vectors terminate on one straight line is______.
4. If $\vec{a}$, $\vec{b}$, and $\vec{c}$ are three non-zero vectors, no two of which are collinear, $\vec{a}+2\vec{b}$ is collinear with $\vec{c}$, and $\vec{b}+3\vec{c}$ is collinear with $\vec{a}$, then the value of $|\vec{a}+2\vec{b}+6\vec{c}|$ is ______.
5. The vectors $\vec{a}$ and $\vec{b}$ are non-collinear. If the vectors $\vec{c}=(x-2)\vec{a}+\vec{b}$ and $\vec{d}=(2x+1)\vec{a}-\vec{b}$ are collinear, then the value of $1/x$ is ________.
6. The points with position vectors $10\hat{i}+3\hat{j}$, $12\hat{i}-5\hat{j}$, and $a\hat{i}+11\hat{j}$ are collinear if a equals ________.
7. $ABCD$ is a tetrahedron and O is any point. If the lines joining O to the vertices meet the opposite faces at P, Q, R, and S, respectively, then the value of $(OP/AP)+(OQ/BQ)+(OR/CR)+(OS/DS)$ is______.
8. Consider the vectors $\hat{i}+\cos(\beta-\alpha)\hat{j}+\cos(\gamma-\alpha)\hat{k}$, $\cos(\alpha-\beta)\hat{i}+\hat{j}+\cos(\gamma-\beta)\hat{k}$, and $\cos(\alpha-\gamma)\hat{i}+\cos(\beta-\gamma)\hat{j}+a\hat{k}$, where α, β, and γ are different angles. If these vectors are coplanar, then the value of a is ______.
9. In a triangle PQR, S and T are points on QR and PR, respectively, such that $QS=3SR$ and $PT=4TR$. Let M be the point of intersection of PS and QT. Then the value of $[QM/MT]$, where [.] represents the greatest integer function, is ______.
10. If the vectors $a\hat{i}+\hat{j}+\hat{k}$, $\hat{i}+b\hat{j}+\hat{k}$, and $\hat{i}+\hat{j}+c\hat{k}$ $(a\neq b\neq c\neq 1)$ are coplanar, then the value of $\frac{1}{1-a}+\frac{1}{1-b}+\frac{1}{1-c}$ is ________.

ANSWERS AND SOLUTIONS

Objective Type

1. **a.** Four or more than four non-zero vectors are always linearly dependent.

2. **d.** $3\vec{a}+4\vec{b}+5\vec{c}=0$

$\Rightarrow \vec{a}, \vec{b}$ and $\vec{c}$ are coplanar.

No other conclusion can be derived from it.

3. **c.** $\overrightarrow{BC}=\overrightarrow{OC}-\overrightarrow{OB}=4\hat{i}+2\hat{j}-4\hat{k}$

$AB=-3\hat{i}-3\hat{k},\ AC=\hat{i}+2\hat{j}-7\hat{k}$

$BC^2=36, AB^2=18, AC^2=54$

Clearly, $AC^2=BC^2+AB^2$

$\therefore \angle B=90°$

4. **c.** $|\vec{a}+\vec{b}|<|\vec{a}-\vec{b}|2$

$\Rightarrow$ Longer diagonal of the parallelogram is smaller than the smaller diagonal.

$\Rightarrow \frac{\pi}{2}<\theta<\frac{3\pi}{2}$

5. **c.** The position vector of the point O with respect to itself is

$$\frac{\overrightarrow{OA}\sin 2A+\overrightarrow{OB}\sin 2B+\overrightarrow{OC}\sin 2C}{\sin 2A+\sin 2B+\sin 2C}$$

$$\Rightarrow \frac{\overrightarrow{OA}\sin 2A+\overrightarrow{OB}\sin 2B+\overrightarrow{OC}\sin 2C}{\sin 2A+\sin 2B+\sin 2C}=\vec{0}$$

$$\Rightarrow \overrightarrow{OA}\sin 2A+\overrightarrow{OB}\sin 2B+\overrightarrow{OC}\sin 2C=\vec{0}$$

6. **a.** We have $\overrightarrow{GB}+\overrightarrow{GC}=(1+1)\overrightarrow{GD}=2\overrightarrow{GD}$, where D is the midpoint of BC.

$\therefore \overrightarrow{GA}+\overrightarrow{GB}+\overrightarrow{GC}=\overrightarrow{GA}+2\overrightarrow{GD}=\overrightarrow{GA}-\overrightarrow{GA}=0$

(Since G divides AC in the ratio 2 : 1, $2\overrightarrow{GD}=-\overrightarrow{GA}$)

7. **c.** $m\vec{a}$ is a unit vector if and only if

$|m\vec{a}|=1\Rightarrow|m|\,|\vec{a}|=1\Rightarrow|m|a=1\Rightarrow a=\frac{1}{|m|}$

8. **c.**

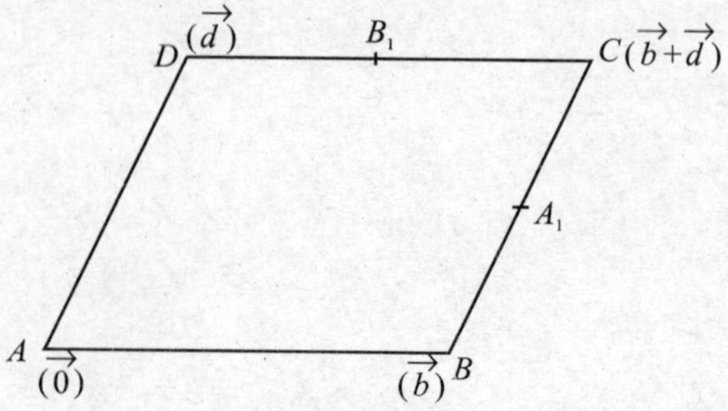

Fig. 1.3

Let P.V. of A, B and D be $\vec{0}$, $\vec{b}$, and $\vec{d}$, respectively.

Then P.V. of C, $\vec{c}=\vec{b}+\vec{d}$

Also P.V. of $A_1=\vec{b}+\frac{\vec{d}}{2}$

and P.V. of $B_1=\vec{d}+\frac{\vec{b}}{2}$

$\Rightarrow \overrightarrow{AA_1}+\overrightarrow{AB_1}=\frac{3}{2}(\vec{b}+\vec{d})=\frac{3}{2}\overrightarrow{AC}$

9. **b.** Since $|\overrightarrow{OP}|=|\overrightarrow{OQ}|=\sqrt{14}$, ΔOPQ is isosceles.

Hence the internal bisector OM is perpendicular to PQ and M is the midpoint of P and Q.

$\therefore \overrightarrow{OM}=\frac{1}{2}(\overrightarrow{OP}+\overrightarrow{OQ})=2\hat{i}+\hat{j}-2\hat{k}$

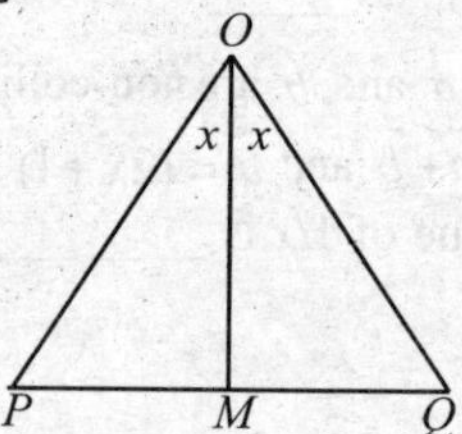

Fig. 1.4

10. **d.**

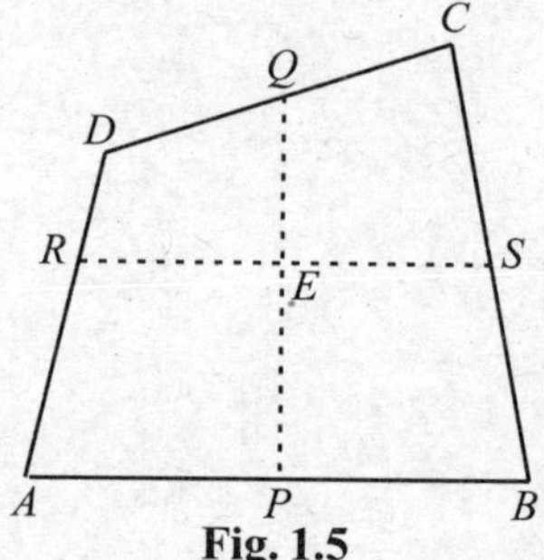

Fig. 1.5

Let $\overrightarrow{OA}=\vec{a}$, $\overrightarrow{OB}=\vec{b}$, $\overrightarrow{OC}=\vec{c}$, and $\overrightarrow{OD}=\vec{d}$.

Therefore,

$\overrightarrow{OA}+\overrightarrow{OB}+\overrightarrow{OC}+\overrightarrow{OD}=\vec{a}+\vec{b}+\vec{c}+\vec{d}$

P, the midpoint of AB, is $\frac{\vec{a}+\vec{b}}{2}$

Q, the midpoint of $\overrightarrow{CD}$, is $\frac{\vec{c}+\vec{d}}{2}$

Therefore, the midpoint of $\overrightarrow{PQ}$ is $\frac{\vec{a}+\vec{b}+\vec{c}+\vec{d}}{4}$.

Similarly the midpoint of RS is $\frac{\vec{a}+\vec{b}+\vec{c}+\vec{d}}{4}$, i.e.,

$\overrightarrow{OE}=\frac{\vec{a}+\vec{b}+\vec{c}+\vec{d}}{4}\Rightarrow x=4$

11. b.

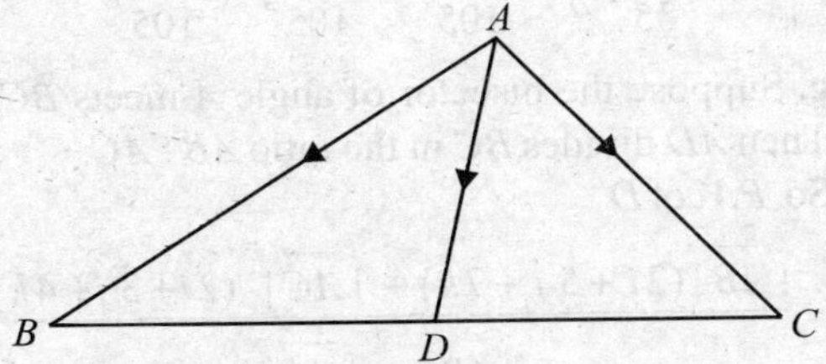

Fig. 1.6

$\overrightarrow{AB}+\overrightarrow{AC}+2\overrightarrow{AD}$

$\therefore \overrightarrow{AD}=\frac{1}{2}\{(-3\hat{i}+4\hat{k})+(5\hat{i}-2\hat{j}+4\hat{k})\}$

$=\hat{i}-\hat{j}+4\hat{k}$

Length of $AD=\sqrt{1+1+16}=\sqrt{18}$

12. d. $\vec{a}-\vec{b}=2(\vec{d}-\vec{c})$

$\therefore \frac{\vec{a}+2\vec{c}}{2+1}=\frac{\vec{b}+2\vec{d}}{2+1}$

$\Rightarrow$ AC and BD trisect each other as L.H.S. is the position vector of a point trisecting A an C, and R.H.S. that of B and D.

13. b. Vector in the direction of angular bisector of $\vec{a}$ and $\vec{b}=\frac{\vec{a}+\vec{b}}{2}$.

$\frac{\vec{a}+\vec{b}}{2}$ has magnitude $\cos(\theta/2)$

So the unit vector in this direction will have magnitude $\frac{\vec{a}+\vec{b}}{2\cos\frac{\theta}{2}}$.

14. b. $|a|+|b|+|c|=\sqrt{a^2+b^2+c^2}$

$\Leftrightarrow 2|ab|+2|bc|+2|ca|=0$

$\Leftrightarrow ab=bc=ca=0 \Leftrightarrow$ any two of a, b, and c are zero

15. c. $\vec{\alpha}=\vec{a}+\vec{b}+\vec{c}=6\hat{i}+12\hat{j}$

Let $\vec{\alpha}=x\vec{a}+y\vec{b}\Rightarrow 6x+2y=6$

$-3x-6y=12$

$\therefore x=2, y=-3$

$\therefore \vec{\alpha}=2\vec{a}-3\vec{b}$

16. c. Given $\vec{\alpha}+\vec{\beta}+\vec{\gamma}=a\vec{\delta}$ (1)

$\vec{\beta}+\vec{\gamma}+\vec{\delta}=b\vec{\alpha}$ (2)

From (1), $\vec{\alpha}+\vec{\beta}+\vec{\gamma}+\vec{\delta}=(a+1)\vec{\delta}$ (3)

From (2), $\vec{\alpha}+\vec{\beta}+\vec{\gamma}+\vec{\delta}=(b+1)\vec{\alpha}$ (4)

From (3) and (4),

$(a+1)\vec{\delta}=(b+1)\vec{\alpha}$ (5)

Since $\vec{\alpha}$ is not parallel to $\vec{\delta}$,

from (5), $a+1=0$ and $b+1=0$

From (3), $\vec{\alpha}+\vec{\beta}+\vec{\gamma}+\vec{\delta}=0$.

17. a. Let the origin of reference be the circumcentre of the triangle.

Let $\overrightarrow{OA}=\vec{a}, \overrightarrow{OB}=\vec{b}, \overrightarrow{OC}=\vec{c}$, and $\overrightarrow{OT}=\vec{t}$

Then $|\vec{a}|=|\vec{b}|=|\vec{c}|=R$ (circumradius)

Again $\overrightarrow{OA}+\overrightarrow{OB}+\overrightarrow{OC}=\overrightarrow{OA}+2\overrightarrow{OD}=\overrightarrow{OA}+\overrightarrow{AH}$

$=\overrightarrow{OH}$

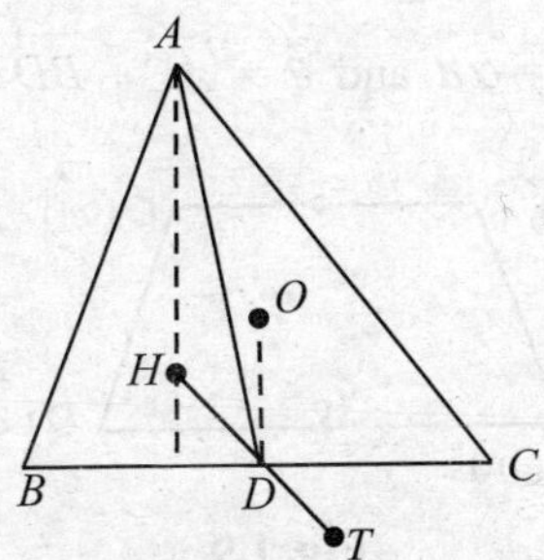

Fig. 1.7

Therefore, the P.V. of H is $\vec{a}+\vec{b}+\vec{c}$. Since D is the midpoint of HT, we have

$\frac{\vec{a}+\vec{b}+\vec{c}+\vec{t}}{2}=\frac{\vec{b}+\vec{c}}{2}\Rightarrow \vec{t}=-\vec{a}$

$\therefore \overrightarrow{AT}=-2\vec{a}\Rightarrow \overrightarrow{AT}=|-2\vec{a}|=2|\vec{a}|=2R.$

But $BC=2R\sin A=R$. So,

$AT=2BC$

18. c. Given $a_1\vec{r_1}+a_2\vec{r_2}+\cdots+a_n\vec{r_n}=0$

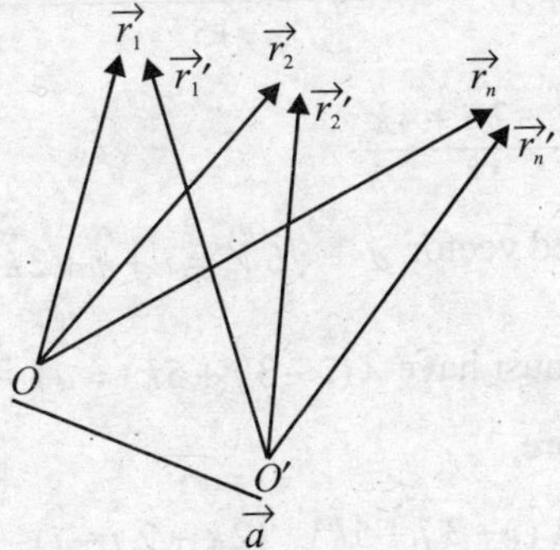

Fig. 1.8

Now $\vec{a}+\vec{r_1}'=\vec{r_1}$ and so on

Hence $a_1(\vec{a}+\vec{r_1}')+a_2(\vec{a}+\vec{r_2}')+\cdots+a_n(\vec{a}+\vec{r_n}')=0$

$a_1\vec{r_1}'+a_2\vec{r_2}'+\cdots+a_n\vec{r_n}'+\vec{a}(a_1+a_2+\cdots+a_n)=0$

Hence $a_1\vec{r_1}'+a_2\vec{r_2}'+\cdots+a_n\vec{r_n}'=0$ if $a_1+a_2+\cdots+a_n=0$.

19. **d.** $\vec{r_1} + 2\vec{r_2} = (p\vec{a} + q\vec{b} + \vec{c}) + 2(\vec{a} + p\vec{b} + q\vec{c})$

$$= (p+2)\vec{a} + (q+2p)\vec{b} + (1+2q)\vec{c}$$

$$2\vec{r_1} + \vec{r_2} = (2p+1)\vec{a} + (2q+p)\vec{b} + (2+q)\vec{c}$$

$$\frac{p+2}{2p+1} = \frac{q+2p}{2q+p} = \frac{1+2q}{2+q} = \frac{p+q+2p+2q+3}{p+q+2p+2q+3} = 1$$

$\Rightarrow p = 1$ and $q = 1$

20. **d.** $\sqrt{3}\tan\theta + 1 = 0$ and $\sqrt{3}\sec\theta - 2 = 0$

$$\Rightarrow \theta = \frac{11\pi}{6}$$

$$\Rightarrow \theta = 2n\pi + \frac{11\pi}{6},\ n \in Z$$

21. **d.** $\vec{c} - \vec{b} = \alpha\vec{d}$ and $\vec{P} = \overrightarrow{AC} + \overrightarrow{BD} = \mu\,\overrightarrow{AD}$

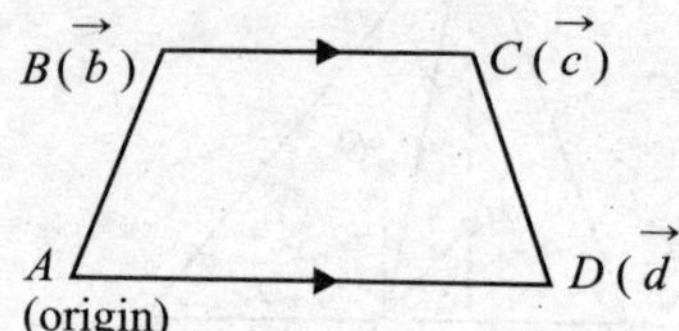

Fig. 1.9

Hence $\vec{p} = \vec{c} + \vec{d} - \vec{b} = \mu\,\vec{d}$ (using $\vec{c} - \vec{b} = \alpha\vec{d}$)

or $\alpha + 1 = \mu$

22. **b.** Note that $\vec{a} + \vec{b} = \vec{c}$

23. **a.** $\hat{a} = \dfrac{-4\hat{i} + 3\hat{k}}{5}$; $\hat{b} = \dfrac{14\hat{i} + 2\hat{j} - 5\hat{k}}{15}$

A vector $\vec{V}$ bisecting the angle between $\vec{a}$ and $\vec{b}$ is

$$\vec{V} = \hat{a} + \hat{b} = \frac{-12\hat{i} + 9\hat{k} + 14\hat{i} + 2\hat{j} - 5\hat{k}}{15}$$

$$= \frac{2\hat{i} + 2\hat{j} + 4\hat{k}}{15}.$$

Required vector $\vec{d} = \sqrt{6}\,\hat{V} = \hat{i} + \hat{j} + 2\hat{k}$

24. **d.** We must have $\lambda(\hat{i} - 3\hat{j} + 5\hat{k}) = \vec{a} + \dfrac{2\hat{k} + 2\hat{j} - \hat{i}}{3}$.

Therefore,

$$3\vec{a} = 3\lambda(\hat{i} - 3\hat{j} + 5\hat{k}) - (2\hat{k} + 2\hat{j} - \hat{i})$$

$$= \hat{i}(3\lambda + 1) - \hat{j}(2 + 9\lambda) + \hat{k}(15\lambda - 2)$$

$$\Rightarrow 3|\vec{a}| = \sqrt{(3\lambda+1)^2 + (2+9\lambda)^2 + (15\lambda-2)^2}$$

$$\Rightarrow 9 = (3\lambda+1)^2 + (2+9\lambda)^2 + (15\lambda-2)^2$$

$$\Rightarrow 315\lambda^2 - 18\lambda = 0 \Rightarrow \lambda = 0, \frac{2}{35}.$$

If $\lambda = 0$, $\vec{a} = \hat{i} - 2\hat{j} - 2\hat{k}$ (not acceptable)

For $\lambda = \dfrac{2}{35}$, $\vec{a} = \dfrac{41}{105}\hat{i} - \dfrac{88}{105}\hat{j} - \dfrac{40}{105}\hat{k}$

25. **c.** Suppose the bisector of angle A meets BC at D. Then AD divides BC in the ratio $AB : AC$. So, $P.V.$ of D

$$= \frac{|\overrightarrow{AB}|(2\hat{i} + 5\hat{j} + 7\hat{k}) + |\overrightarrow{AC}|\,(2\hat{i} + 3\hat{j} + 4\hat{k})}{|\overrightarrow{AB}| + |\overrightarrow{AC}|}$$

But $\overrightarrow{AB} = -2\hat{i} - 4\hat{j} - 4\hat{k}$ and $\overrightarrow{AC} = -2\hat{i} - 2\hat{j} - \hat{k}$.

$\Rightarrow |\overrightarrow{AB}| = 6$ and $|\overrightarrow{AC}| = 3$. Therefore,

$$\text{P.V. of } D = \frac{6(2\hat{i} + 5\hat{j} + 7\hat{k}) + 3(2\hat{i} + 3\hat{j} + 4\hat{k})}{6+3}$$

$$= \frac{1}{3}(6\hat{i} + 13\hat{j} + 18\hat{k})$$

26. **b.** The point that divides $5\hat{i}$ and $5\hat{j}$ in the ratio of

$k : 1$ is $\dfrac{(5\hat{j})k + (5\hat{i}).1}{k+1}$

$$\therefore \vec{b} = \frac{5\hat{i} + 5k\hat{j}}{k+1}$$

Also, $|\vec{b}| \le \sqrt{37}$

Fig. 1.10

$$\Rightarrow \frac{1}{k+1}\sqrt{25 + 25k^2} \le \sqrt{37}$$

$$\Rightarrow 5\sqrt{1+k^2} \le \sqrt{37}(k+1)$$

Squaring both sides

$25(1 + k^2) \le 37(k^2 + 2k + 1)$

or $6k^2 + 37k + 6 \ge 0 \Rightarrow (6k+1)(k+6) \ge 0$

$$k \in (-\infty, -6] \cup \left[-\frac{1}{6}, \infty\right)$$

27. **b.** Let $\overrightarrow{OA} = \vec{a}$, $\overrightarrow{OB} = \vec{b}$, and $\overrightarrow{OC} = \vec{c}$, then $\overrightarrow{AB}$

$= \vec{b} - \vec{a}$ and $\overrightarrow{OP} = \dfrac{1}{3}\vec{a}$, $\overrightarrow{OQ} = \dfrac{1}{2}\vec{b}$, $\overrightarrow{OR} = \dfrac{1}{3}\vec{c}$.

Since P, Q, R, and S are coplanar,

$\overrightarrow{PS} = \alpha\,\overrightarrow{PQ} + \beta\,\overrightarrow{PR}$ ($\overrightarrow{PS}$ can be written as a linear combination of $\overrightarrow{PQ}$ and $\overrightarrow{PR}$)

$$= \alpha(\overrightarrow{OQ} - \overrightarrow{OP}) + \beta(\overrightarrow{OR} - \overrightarrow{OP})$$

i.e., $\overrightarrow{OS} - \overrightarrow{OP} = -(\alpha + \beta)\dfrac{\vec{a}}{3} + \dfrac{\alpha}{2}\vec{b} + \dfrac{\beta}{3}\vec{c}$

$\Rightarrow \overrightarrow{OS} = (1-\alpha-\beta)\frac{\vec{a}}{3} + \frac{\alpha}{2}\vec{b} + \frac{\beta}{3}\vec{c}$ (1)

Given $\overrightarrow{OS} = \lambda\,\overrightarrow{AB} = \lambda\,(\vec{b}-\vec{a})$ (2)

From (1) and (2), $\beta = 0, \frac{1-\alpha}{3} = -\lambda$ and $\frac{\alpha}{2} = \lambda$

$\Rightarrow 2\lambda = 1 + 3\lambda$

$\Rightarrow \lambda = -1$

28. **a.** Let the incentre be at the origin and be $A(\vec{p}), B(\vec{q})$ and $C(\vec{r})$. Then

$\overrightarrow{IA} = \vec{p},\ \overrightarrow{IB} = \vec{q}$ and $\overrightarrow{IC} = \vec{r}$.

Incentre I is $\frac{a\vec{p}+b\vec{q}+c\vec{r}}{a+b+c}$, where $p = BC, q = AC$ and $r = AB$.

Incentre is at the origin. Therefore,

$\frac{a\vec{p}+b\vec{q}+c\vec{r}}{a+b+c} = \vec{0}$, or $a\vec{p}+b\vec{q}+c\vec{r} = \vec{0}$

$\Rightarrow a\,\overrightarrow{IA} + b\,\overrightarrow{IB} + c\,\overrightarrow{IC} = \vec{0}$

29. **a.**

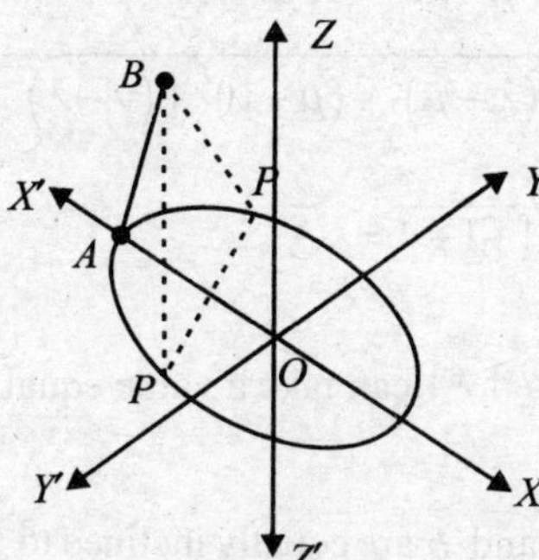

Fig. 1.11

Point P lies on $x^2 + 3y^2 = 3$. (1)

Now from the diagram, according to given conditions, $AP = AB$

or $(x+\sqrt{3})^2 + (y-0)^2 = 4$ or $(x+\sqrt{3})^2 + y^2 = 4$ (2)

Solving (1) and (2), we get $x = 0$ and $y = \pm 1$

Hence, point P has position vector $\pm\hat{j}$.

30. **b.**

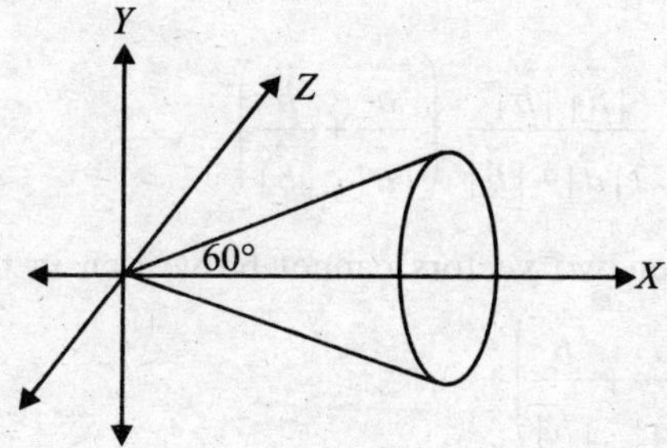

Fig. 1.12

From the diagram, it is obvious that locus is cone concentric with the positive x-axis having vertex at the origin and the slant height equal to the magnitude of the vector.

31. **c.** Since $\vec{x}, \vec{y}$ and $\vec{x}\times\vec{y}$ are linearly independent,

$20a - 15b = 15b + 12c - 20a = 0$

$\Rightarrow \frac{a}{3} = \frac{b}{4} = \frac{c}{5}$

$\Rightarrow c^2 = a^2 + b^2$

Hence, ΔABC is right angled.

32. **a.** The position vector of any point at t is

$\vec{r} = (2+t^2)\,\hat{i} + (4t-5)\,\hat{j} + (2t^2-6)\,\hat{k}$

$\Rightarrow \frac{d\vec{r}}{dt} = 2t\,\hat{i} + 4\,\hat{j} + (4t-6)\,\hat{k}$

$\Rightarrow \left.\frac{d\vec{r}}{dt}\right|_{t=2} = 4\hat{i} + 4\hat{j} + 2\hat{k}$ and $\left|\left.\frac{d\vec{r}}{dt}\right|_{t=2}\right|$

$= \sqrt{16+16+4} = 4$

Hence, the required unit tangent vector at $t = 2$ is $\frac{1}{3}(2\hat{i} + 2\hat{j} + \hat{k})$.

33. **a.** As $\vec{x}, \vec{y}$ and $\vec{x}\times\vec{y}$ are non-collinear vectors,

$\Rightarrow$ vectors are linearly independent.

$\Rightarrow a - b = 0 = b - c = c - a$

$\Rightarrow a = b = c$

Therefore, the triangle is equilateral.

34. **c.**

Multiple Correct Answers Type

1. **b., d.**

The given vector equation can be written as

$[(1-a)x + 3y - 4z]\,\hat{i} + [x - (3+a)\,y + 5z]\,\hat{j} + [3x + y - az]\,\hat{k} = \vec{0}$

$\Rightarrow (1-a)x + 3y - 4z = 0$

$x - (3+a)\,y + 5z = 0$ and $3x + y - az = 0$

The above system of equation has a non-trivial solution.

$$\Rightarrow \begin{vmatrix} 1-a & 3 & -4 \\ 1 & -(3+a) & 5 \\ 3 & 1 & -a \end{vmatrix} = 0$$

$\Rightarrow a = 0, -1$

2. **a., d.**

Let $\vec{a} = 2\hat{i} + 4\hat{j} - 5\hat{k}$ and $\vec{b} = \hat{i} + 2\hat{j} + 3\hat{k}$.

Then the diagonals of the parallelogram are $\vec{p} = \vec{a} + \vec{b}$ and $\vec{q} = \vec{b} - \vec{a}$, i.e., $\vec{p}=3\hat{i}+6\hat{j}-2\hat{k}$, $\vec{q}=-\hat{i}-2\hat{j}+8\hat{k}$

So unit vectors along the diagonals are $\frac{1}{7}(3\hat{i}+6\hat{j}-2\hat{k})$ and $\frac{1}{\sqrt{69}}(-\hat{i}-2\hat{j}+8\hat{k})$.

3. b., c.

We have, $\vec{a}=2p\,\hat{i}+\hat{j}$

On rotation, let $\vec{b}$, be the vector with components $(p+1)$ and 1 so that $\vec{b}=(p+1)\hat{i}+\hat{j}$.

Now, $|\vec{a}|=|\vec{b}|\Rightarrow a^2=b^2$

$\Rightarrow 4p^2+1=(p+1)^2+1$

$\Rightarrow 4p^2=(p+1)^2$

$\Rightarrow 2p=\pm(p+1)$

$\Rightarrow 3p=-1$ or $p=1$

$\therefore p=-1/3$ or $p=1$

4. b., d.

Clearly $\vec{a}$, $\vec{b}$, and $\vec{c}$ must be coplanar.

$$\begin{vmatrix}1 & 1 & 1\\ 4 & 3 & 4\\ 1 & x_1 & x_2\end{vmatrix}=0\Rightarrow x_2=1$$

Also $|\vec{c}|=\sqrt{3}\Rightarrow 3=1+x_1^2+x_2^2\Rightarrow x_1^2=1\Rightarrow x_1=\pm 1$

5. a., b., c.

For coplanar vectors, $\begin{vmatrix}1 & 2 & 3\\ 0 & \lambda & \mu\\ 0 & 0 & 2\lambda-1\end{vmatrix}=0$

$\Rightarrow (2\lambda-1)\lambda=0\Rightarrow \lambda=0,\ \frac{1}{2}$

6. b., c.

Let $\vec{R}$ be the resultant.

Then $\vec{R}=\vec{F}_1+\vec{F}_2+\vec{F}_3=(p+1)\hat{i}+4\hat{j}$

Given, $|\vec{R}|=5$. Therefore,

$(p+1)^2+16=25$

$\Rightarrow p+1=\pm 3$

$\therefore p=2,-4$

7. a., b., d.

$$\vec{a}=\left[\pm\left(\hat{i}-\hat{j}\right)\pm\left(\hat{j}+\hat{k}\right)\right]$$

$$=\pm\left(\hat{i}+\hat{k}\right),\pm\left(\hat{i}-2\hat{j}-\hat{k}\right)$$

8. b., c.

Let $\vec{\alpha}=\hat{i}+x\hat{j}+3\hat{k}$, $\vec{\beta}=4\hat{i}+(4x-2)\hat{j}+2\hat{k}$

Given, $2|\vec{\alpha}|=|\vec{\beta}|$

$\Rightarrow 2\sqrt{10+x^2}=\sqrt{20+4(2x-1)^2}$

$\Rightarrow 10+x^2=5+(4x^2-4x+1)$

$\Rightarrow 3x^2-4x-4=0$

$\Rightarrow x=2,-\frac{2}{3}$

9. c., d.

Let $\vec{a}$, $\vec{b}$ and $\vec{c}$ lie in the *x-y* plane.

Let $\vec{a}=\hat{i}$, $\vec{b}=-\frac{1}{2}\hat{i}+\frac{\sqrt{3}}{2}\hat{i}$, and $\vec{c}=-\frac{1}{2}\hat{i}-\frac{\sqrt{3}}{2}\hat{j}$.

Therefore, $|\vec{p}+\vec{q}+\vec{r}|=|\lambda\vec{a}+\mu\vec{b}+\nu\vec{c}|$

$$=\left|\lambda\hat{i}+\mu\left(-\frac{1}{2}\hat{i}+\frac{\sqrt{3}}{2}\hat{j}\right)+\nu\left(-\frac{1}{2}\hat{i}-\frac{\sqrt{3}}{2}\hat{j}\right)\right|$$

$$=\left|\left(\lambda-\frac{\mu}{2}-\frac{\nu}{2}\right)\hat{i}+\frac{\sqrt{3}}{2}(\mu-\nu)\hat{j}\right|$$

$$=\sqrt{\left(\lambda-\frac{\mu}{2}-\frac{\nu}{2}\right)^2+\frac{3}{4}(\mu-\nu)^2}$$

$$=\sqrt{\lambda^2+\mu^2+\nu^2-\lambda\mu-\lambda\nu-\mu\nu}$$

$$=\frac{1}{\sqrt{2}}\sqrt{(\lambda-\mu)^2+(\mu-\nu)^2+(\nu-\lambda)^2}$$

$$\geq\frac{1}{\sqrt{2}}\sqrt{1+1+4}=\sqrt{3}$$

$\Rightarrow |\vec{p}+\vec{q}+\vec{r}|$ can take a value equal to $\sqrt{3}$ and 2.

10. b., d.

Since $\vec{a}$ and $\vec{b}$ are equally inclined to $\vec{c}$, $\vec{c}$ must be of the form $t\left(\frac{\vec{a}}{|\vec{a}|}+\frac{\vec{b}}{|\vec{b}|}\right)$. Now,

$$\frac{|\vec{b}|}{|\vec{a}|+|\vec{b}|}\vec{a}+\frac{|\vec{a}|}{|\vec{a}|+|\vec{b}|}\vec{b}=\frac{|\vec{a}||\vec{b}|}{|\vec{a}|+|\vec{b}|}\left(\frac{\vec{a}}{|\vec{a}|}+\frac{\vec{b}}{|\vec{b}|}\right)$$

Also, $\frac{|\vec{b}|}{2|\vec{a}|+|\vec{b}|}\vec{a}+\frac{|\vec{a}|}{2|\vec{a}|+|\vec{b}|}\vec{b}$

$$=\frac{|\vec{a}||\vec{b}|}{2|\vec{a}|+|\vec{b}|}\left(\frac{\vec{a}}{|\vec{a}|}+\frac{\vec{b}}{|\vec{b}|}\right)$$

Other two vectors cannot be written in the form $t\left(\frac{\vec{a}}{|\vec{a}|}+\frac{\vec{b}}{|\vec{b}|}\right)$.

11. a., c., d.

$\overrightarrow{OA} = -4\hat{i} + 3\hat{k}$; $\overrightarrow{OB} = 14\hat{i} + 2\hat{j} - 5\hat{k}$

$\hat{a} = \dfrac{-4\hat{i} + 3\hat{k}}{5}$; $\hat{b} = \dfrac{14\hat{i} + 2\hat{j} - 5\hat{k}}{15}$

$\vec{r} = \dfrac{\lambda}{15}[-12\hat{i} + 9\hat{j} + 14\hat{i} + 2\hat{j} - 5\hat{k}]$

$\vec{r} = \dfrac{\lambda}{15}[2\hat{i} + 2\hat{j} + 4\hat{k}]$

$\vec{r} = \dfrac{2\lambda}{15}[\hat{i} + \hat{j} + 2\hat{k}]$

12. a., b., d.

$(\lambda - 1)(\vec{a}_1 - \vec{a}_2) + \mu(\vec{a}_2 + \vec{a}_3) + \gamma(\vec{a}_3 + \vec{a}_4 - 2\vec{a}_2) + \vec{a}_3 + \delta\vec{a}_4 = \vec{0}$

i.e., $(\lambda - 1)\vec{a}_1 + (1 - \lambda + \mu - 2\gamma)\vec{a}_2 + (\mu + \gamma + 1)\vec{a}_3 + (\gamma + \delta)\vec{a}_4 = \vec{0}$

Since $\vec{a}_1, \vec{a}_2, \vec{a}_3$ and $\vec{a}_4$ are linearly independent

$\lambda - 1 = 0,\ 1 - \lambda + \mu - 2\gamma = 0,\ \gamma + y + 1 = 0$, and $\gamma + \delta = 0$

i.e., $\lambda = 1, \mu = 2\gamma, \mu + \gamma + 1 = 0, \gamma + \delta = 0$

i.e., $\lambda = 1, \mu = -\dfrac{2}{3}, \gamma = -\dfrac{1}{3}, \delta = \dfrac{1}{3}$

13. a. c.

We have, $\overrightarrow{AB} = -\hat{i} - \hat{j} - 4\hat{k}$, $\overrightarrow{BC} = -3\hat{i} + 3\hat{j}$, and, $\overrightarrow{CA} = 4\hat{i} - 2\hat{j} + 4\hat{k}$. Therefore,

$|\overrightarrow{AB}| = |\overrightarrow{BC}| = 3\sqrt{2}$ and $|\overrightarrow{CA}| = 6$

Clearly, $|\overrightarrow{AB}|^2 + |\overrightarrow{BC}|^2 = |\overrightarrow{AC}|^2$

Hence, the triangle is right-angled isosceles triangle.

Reasoning Type

1. a. $\sqrt{(p+2)^2 + 1} = \sqrt{p^2 + 1}$

$\Rightarrow p^2 + 4 + 4p + 1 = p^2 + 1$

$\Rightarrow 4p = -4$

$\Rightarrow p = -1$

Hence ***a.*** is the correct option.

2. a. $2\vec{a} + 3\vec{b} - 5\vec{c} = 0$

$\Rightarrow 3(\vec{b} - \vec{a}) = 5(\vec{c} - \vec{a}) \Rightarrow \overrightarrow{AB} = \dfrac{5}{3}\overrightarrow{Ac}$

$\Rightarrow \overrightarrow{AB}$ and $\overrightarrow{AC}$ must be parallel since there is a common point A. The points A, B, and C must be collinear.

3. d. In an isosceles triangle ABC is which $AB = AC$, the median and the bisector from A must be the same line $\Rightarrow$ Statement 2 is true.

Now $\overrightarrow{AD} = \dfrac{\vec{u} + \vec{v}}{2}$ and $|\overrightarrow{AD}|^2 = \dfrac{1}{2} 2\cos^2\dfrac{\alpha}{2}$; so

$|\overrightarrow{AD}| = \cos\dfrac{\alpha}{2}$

Therefore, unit vector along AD, i.e., x, is given by

$\vec{x} = \dfrac{\overrightarrow{AD}}{|AD|} = \dfrac{\vec{u} + \vec{v}}{2\cos\dfrac{\alpha}{2}}$

Hence, statement 2 is false.

4. b. Obviously, statement 1 is true.

$\cos 2\alpha + \cos 2\beta + \cos 2\gamma$

$= 2\cos^2\alpha - 1 + 2\cos^2\beta - 1 + 2\cos^2\gamma - 1$

$= 2(\cos^2\alpha + \cos^2\beta + \cos^2\gamma) - 3$

$= 2 - 3 = -1.$

Hence, statement 2 is true but does not explain statement 1 as it is a result derived using result in the statement.

5. b. Unit vector along $AB = l_1 i + m_1 j + n_1 k$

Unit vector along $AC = l_2 i + m_2 j + n_2 k$

So angular bisector is $l_1 i + m_1 j + n_1 k + l_2 i + n_2 j + n_2 k = (l_1 + l_2)i + (m_1 + m_2)j + (n_1 + n_2)k.$

Unit vector along angular bisector is

$\dfrac{(l_1 + l_2)i + (m_1 + m_2)j + (n_1 + n_2)k}{\sqrt{(l_1 + l_2)^2 + (m_1 + m_2)^2 + (n_1 + n_2)^2}}$

$\Rightarrow \dfrac{(l_1 + l_2)}{2\cos\theta/2} i + \dfrac{(m_1 + m_1)}{2\cos\theta/2} j + \dfrac{(n_1 + n_2)}{2\cos\theta/2} k$ (θ is the angle between two lines)

So direction cosines are $\dfrac{(l_1 + l_2)}{2\cos\theta/2}, \dfrac{(m_1 + m_2)}{2\cos\theta/2}$, and $\dfrac{(n_1 + n_2)}{2\cos\theta/2}$.

6. c. In ΔABC, $\overrightarrow{AB} + \overrightarrow{BC} = \overrightarrow{AC} = -\overrightarrow{CA} \Rightarrow \overrightarrow{AB} + \overrightarrow{BC} + \overrightarrow{CA} = \vec{O}$ $\overrightarrow{OA} + \overrightarrow{AB} = \overrightarrow{OB}$ is the triangle law of addition.

Hence, statements 1 is true and statement 2 is false.

7. a.

$\dfrac{3}{2} = \dfrac{p}{3} = \dfrac{3}{q} \Rightarrow p = \dfrac{9}{2}$ and $q = 2$

Thus, both statement are true and statement 2 is the correct explanation for statement 1.

8. a. $\vec{a} + \vec{b} = \vec{a} - \vec{b}$ are the diagonals of a parallelogram whose sides are $\vec{a}$ and $\vec{b}$.

$|\vec{a} + \vec{b}| = |\vec{a} - \vec{b}|$

$\Rightarrow$ Diagonals of the parallelogram have the same length.

$\Rightarrow$ The parallelogram is a rectangle $\Rightarrow \vec{a} \perp \vec{b}$

9. **d.** Statement 1is false as points *O*, *A*, *B* and *C* are coplanar.

10. **a.** $3\vec{a} - 2\vec{b} + 5\vec{c} - 6\vec{d}$

$= (2\vec{a} - 2\vec{b}) + (-5\vec{a} + 5\vec{c}) + (6\vec{a} - 6\vec{d})$

$= -2\overrightarrow{AB} + 5\overrightarrow{AC} - 6\overrightarrow{AD} = \vec{0}$

Therefore, $\overrightarrow{AB}$, $\overrightarrow{AC}$ and $\overrightarrow{AD}$ are linearly dependent. Hence by Statement 2, Statement 1is true.

11. **a.** We have adjacent sides of triangle $|\vec{a}| = 3$, $|\vec{b}| = 4$.

The length of the diagonal is $|\vec{a} + \vec{b}| = 5$.

Since it satisfies the Pythagoras theorem, $\vec{a} \perp \vec{b}$. Hence the parallelogram is a rectangle.

Linked Comprehension Type

For Problems 1–3

1. c., 2. b., 3. c.

Sol.

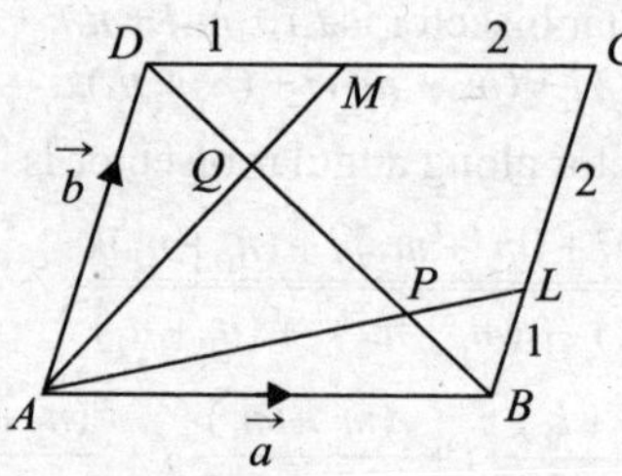

Fig. 1.13

$\overrightarrow{BL} = \frac{1}{3}\vec{b}$

$\therefore \overrightarrow{AL} = \vec{a} + \frac{1}{3}\vec{b}$

Let $\overrightarrow{AP} = \lambda \overrightarrow{AL}$ and *P* divides *DB* in the ratio $\mu : 1 - \mu$.

Then $\overrightarrow{AP} = \lambda \vec{a} + \frac{\lambda}{3}\vec{b}$ (1)

Also $\overrightarrow{AP} = \mu \vec{a} + (1-\mu)\vec{b}$ (2)

From (1) and (2), $\lambda \vec{a} + \frac{\lambda}{3}\vec{b} = \mu\vec{a} + (1-\mu)\vec{b}$

$\therefore \lambda = \mu$ and $\frac{\lambda}{3} = 1 - \mu$

$\therefore \lambda = \frac{3}{4}$

Hence, *P* divides *AL* in the ratio 3 : 1 and *P* divides *DB* in the ratio 3 : 1.

Similarly *Q* divides *DB* in the ratio 1 : 3.

Thus $DQ = \frac{1}{4}DB$ and $PB = \frac{1}{4}DB$

$\therefore PQ = \frac{1}{2}DB$ i.e. $PQ : DB = 1 : 2$

For Problems 4–5

4. c., 5. b.

Sol.

Let the position vectors of *A*, *B*, *C* and *D* be $\vec{a}$, $\vec{b}$, $\vec{c}$ and $\vec{d}$, respectively.

Then, $OA : CB = 2 : 1$

$\Rightarrow \overrightarrow{OA} = 2\overrightarrow{CB}$

$\Rightarrow \vec{a} = 2(\vec{b} - \vec{c})$ (1)

and $OD : AB = 1 : 3$

$3\overrightarrow{OD} = \overrightarrow{AB}$

$\Rightarrow 3\vec{d} = (\vec{b} - \vec{a}) = \vec{b} - 2(\vec{b} - \vec{c})$ [Using (1)]

$\Rightarrow 3\vec{d} = -\vec{b} + 2\vec{c}$ (2)

Let $OX : XC = \lambda : 1$ and $AX : XD = \mu : 1$

Now, *X* divides *OC* in the ratio $\lambda : 1$. Therefore,

P.V. of $X = \frac{\lambda \vec{c}}{\lambda + 1}$ (3)

X also divides *AD* in the ratio $\mu : 1$

P.V. of $X = \frac{\mu \vec{d} + \vec{a}}{\mu + 1}$ (4)

From (3) and (4), we get

$$\frac{\lambda \vec{c}}{\lambda + 1} = \frac{\mu \vec{d} + \vec{a}}{\mu + 1}$$

$$\Rightarrow \left(\frac{\lambda}{\lambda+1}\right)\vec{c} = \left(\frac{\mu}{\mu+1}\right)\vec{d} + \left(\frac{1}{\mu+1}\right)\vec{a}$$

$$\Rightarrow \left(\frac{\lambda}{\lambda+1}\right)\vec{c} = \left(\frac{\mu}{\mu+1}\right)\left(\frac{-\vec{b}+2\vec{c}}{3}\right) + \left(\frac{1}{\mu+1}\right)2\left(\vec{b}-\vec{c}\right) \quad \text{[Using (1) and (4)]}$$

$$\Rightarrow \left(\frac{\lambda}{\lambda+1}\right)\vec{c} = \left(\frac{6-\mu}{3(\mu+1)}\right)\vec{b} + \left(\frac{2\mu}{3(\mu+1)} - \frac{2}{\mu+1}\right)\vec{c}$$

$$\Rightarrow \left(\frac{\lambda}{\lambda+1}\right)\vec{c} = \left(\frac{6-\mu}{3(\mu+1)}\right)\vec{b} + \left(\frac{2\mu-6}{3(\mu+1)}\right)\vec{c}$$

$$\Rightarrow \left(\frac{6-\mu}{3(\mu+1)}\right)\vec{b} + \left(\frac{2\mu-6}{3(\mu+1)} - \frac{\lambda}{\lambda+1}\right)\vec{c} = \vec{0}$$

$$\Rightarrow \frac{6-\mu}{3(\mu+1)} = 0 \text{ and } \frac{2\mu-6}{3(\mu+1)} - \frac{\lambda}{\lambda+1} = 0$$

[As $\vec{b}$ and $\vec{c}$ are non-collinear]

$\Rightarrow \mu = 6, \lambda = \dfrac{2}{5}$

Hence, $OX : XC = 2 : 5$

For Problems 6–7

6. c., 7. d.

Sol.

Consider the regular hexagon $ABCDEF$ with centre at O (origin).

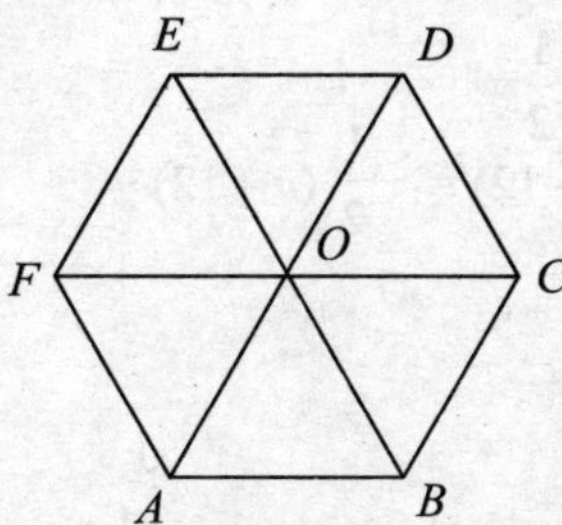

Fig. 1.14

$$\overrightarrow{AD} + \overrightarrow{EB} + \overrightarrow{FC} = 2\overrightarrow{AO} + 2\overrightarrow{OB} + 2\overrightarrow{OC}$$

$$= 2(\overrightarrow{AO} + \overrightarrow{OB}) + 2\overrightarrow{OC}$$

$$= 2\overrightarrow{AB} + 2\overrightarrow{AB} \text{ [Since } \overrightarrow{OC} = \overrightarrow{AB}]$$

$$= 4\overrightarrow{AB}$$

$$\vec{R} = \overrightarrow{AB} + \overrightarrow{AC} + \overrightarrow{AD} + \overrightarrow{AE} + \overrightarrow{AF}$$

$$= \overrightarrow{ED} + \overrightarrow{AC} + \overrightarrow{AD} + \overrightarrow{AE} + \overrightarrow{CD}$$

$$(\because \overrightarrow{AB} = \overrightarrow{ED} \text{ and } \overrightarrow{AF} = \overrightarrow{CD})$$

$$= (\overrightarrow{AC} + \overrightarrow{ED}) + (\overrightarrow{AE} + \overrightarrow{ED}) + \overrightarrow{AD}$$

$$= \overrightarrow{AD} + \overrightarrow{AD} + \overrightarrow{AD} = 3\overrightarrow{AD} = 6\overrightarrow{AO}$$

Matrix-Match Type

2. a → p, r, s; b → q, r, s; c → p, r; d → r, s.

Sol.

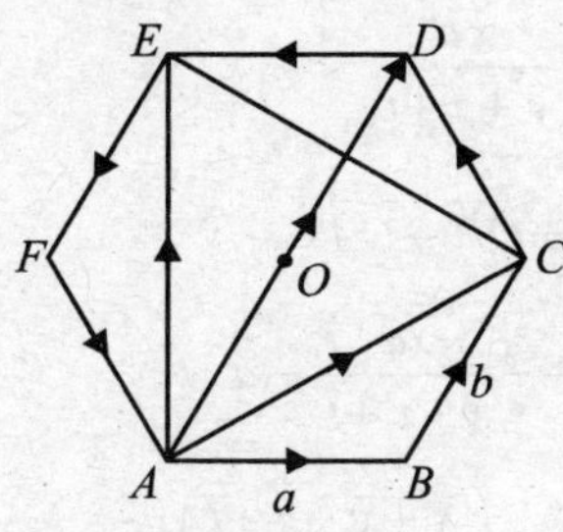

Fig. 1.15

$\overrightarrow{AB} = \vec{a}, \overrightarrow{BC} = \vec{b}$

$$\therefore \overrightarrow{AC} = \overrightarrow{AB} + \overrightarrow{BC} = \vec{a} + \vec{a} \quad (1)$$

$$\overrightarrow{AD} = 2\overrightarrow{BC} = 2\vec{b} \quad (2)$$

[Because AD is parallel to BC and is twice its length].

$$\overrightarrow{CD} = \overrightarrow{AD} - \overrightarrow{AC} = 2\vec{b} - (\vec{a} + \vec{b})$$

$$= \vec{b} - \vec{a}$$

$$\overrightarrow{FA} = -\overrightarrow{CD} = \vec{a} - \vec{b} \quad (3)$$

$$\overrightarrow{DE} = -\overrightarrow{AB} = -\vec{a} \quad (4)$$

$$\overrightarrow{EF} = -\overrightarrow{BC} = -\vec{b} \quad (5)$$

$$\overrightarrow{AE} = \overrightarrow{AD} + \overrightarrow{DE} = 2\vec{b} - \vec{a} \quad (6)$$

$$\overrightarrow{CE} = \overrightarrow{CD} + \overrightarrow{DE} = \vec{b} - \vec{a} - \vec{a} = \vec{b} - 2\vec{a} \quad (7)$$

Integer Type

1. (4)

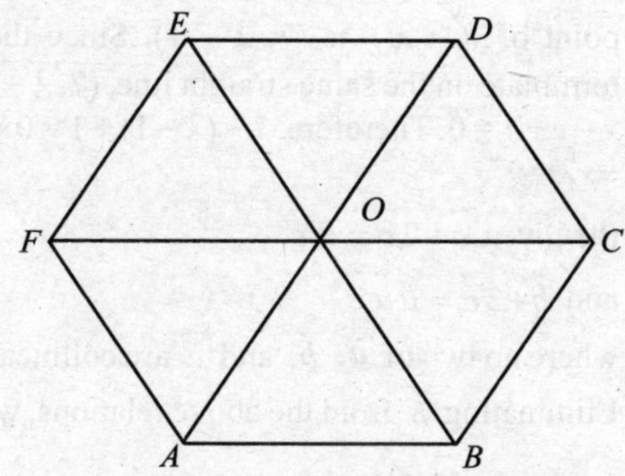

Fig. 1.16

$$\overrightarrow{AD} + \overrightarrow{EB} + \overrightarrow{FC} = 2\overrightarrow{AO} + 2\overrightarrow{OB} + 2\overrightarrow{OC}$$

$$= 2(\overrightarrow{AO} + \overrightarrow{OB}) + 2\overrightarrow{OC}$$

$$= 2\overrightarrow{AB} + 2\overrightarrow{AB} \; (\because \overrightarrow{OC} = \overrightarrow{AB})$$

$$= 4\overrightarrow{AB}$$

2. (4) We know that the figure formed by the lines joining the midpoints of the sides of a quadrilateral is a parallelogram. Hence $MPNQ$ is a parallelogram whose diagonals are MN and PQ, which intersect at E, the midpoint of both MN and PQ.

For any origin O, we have

$\overrightarrow{OA} + \overrightarrow{OB} = 2(\overrightarrow{OM})$, where M is the midpoint of AB.

Also $\overrightarrow{OC} + \overrightarrow{OB} = 2(\overrightarrow{ON})$, where N is the midpoint of CD. Therefore,

$$\overrightarrow{OA} + \overrightarrow{OB} + \overrightarrow{OC} + \overrightarrow{OD} = 2(\overrightarrow{OM} + \overrightarrow{ON})$$

$$= 2[2\overrightarrow{OE}] = 4\overrightarrow{OE},$$

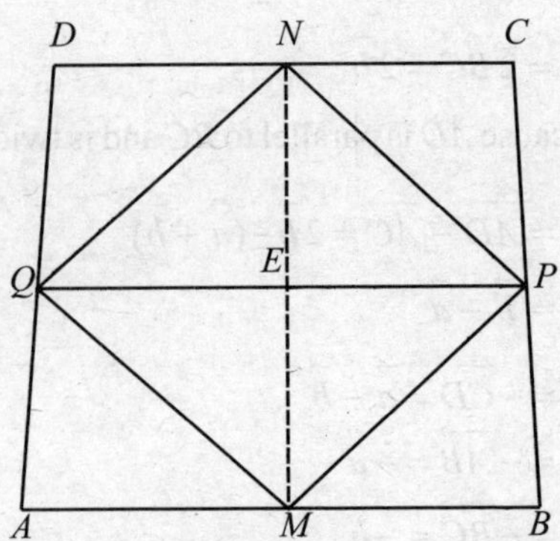

Fig. 1.17

where E is the midpoint of MN as it is the point of intersection of the diagonals of a parallelogram.

3. **(9)** Since the vectors $2\hat{i}+3\hat{j}$ and $5\hat{i}+6\hat{j}$ have (1, 1) as the initial point, their terminal points are (1, 2) and (4, 5), respectively. The equation of the line joining these two points is $x-y+1=0$. The terminal point of $8\hat{i}+\lambda\hat{j}$ is $(7, \lambda-1)$. Since the vectors terminate on the same straight line, $(7, \lambda-1)$ lies on $x-y+1=0$. Therefore, $7-(\lambda-1)+1=0$

$\Rightarrow \lambda=9.$

4. **(0)** Given $\vec{a}+2\vec{b}=\lambda\vec{c}$ (1)

and $\vec{b}+3\vec{c}=\mu\vec{a}$, (2)

where no two of $\vec{a}$, $\vec{b}$, and $\vec{c}$ are collinear vectors.

Eliminating $\vec{b}$ from the above relations, we get

$\vec{a}-6\vec{c}=\lambda\vec{c}-2\mu\vec{a}$

$\Rightarrow \vec{a}(1+2\mu)=(\lambda+6)\vec{c}$

$\Rightarrow \mu=-1/2$ and $\lambda=-6$ as $\vec{a}$ and $\vec{c}$ are non-collinear.

$\Rightarrow |\vec{a}+2\vec{b}+2\vec{c}|=0$

5. **(3)** Both vectors $\vec{c}$ and $\vec{d}$ are non-zero as the coefficients of $\vec{b}$ in both are non-zero.

The vectors $\vec{c}$ and $\vec{d}$ are collinear if one of them is a linear multiple of the other. Therefore,

$\vec{d}=\lambda\vec{c}$

or $(2x+1)\vec{a}-\vec{b}=\lambda[(x-2)\vec{a}+\vec{b}]$ (1)

or $[(2x+1)-\lambda(x-2)]\vec{a}-(1+\lambda)\vec{b}=0$

The above equation is of the form $p\vec{a}+q\vec{b}=0$, where $\vec{a}$ and $\vec{b}$ are non-collinear. Hence, $p=0$ and $q=0$. Therefore,

$2x+1-\lambda(x-2)=0$ (2)

$1+\lambda=0$ (3)

From (2), we get $\lambda=-1$ and putting it in (1), we get $x=1/3$.

Alternative method:

$\vec{c}=(x-2)\vec{a}+\vec{b}$ and $\vec{d}=(2x+1)\vec{a}-\vec{b}$ are collinear if

$$\frac{x-2}{2x+1}=\frac{1}{-1}$$

$$\Rightarrow x=\frac{1}{3}$$

6. **(8)** If the given points are A, B, and C, then

$\overrightarrow{AB}=k(\overrightarrow{BC})$

$\Rightarrow 2\hat{i}-8\hat{j}=k[(a-12)\hat{i}+16\hat{j}]$

Comparing the coefficients of $\hat{i}$ and $\hat{j}$, we get

$-8=16k$

$$\Rightarrow k=-\frac{1}{2}$$

$$2=k(a-12)=-\frac{1}{2}(a-12)$$

$\Rightarrow a=8$

7. **(1)**

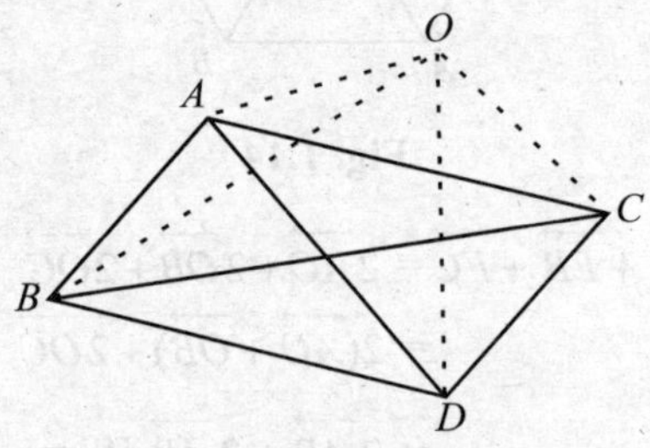

Fig. 1.18

Here $ABCD$ is a tetrahedron.

Let O be the origin and let the position vectors of A, B, C, and D be $\vec{a}$, $\vec{b}$, $\vec{c}$, and $\vec{d}$, respectively.

We know that four linearly dependent vectors can be expressed as

$x\vec{a}+y\vec{b}+z\vec{c}+t\vec{d}=0$, where x, y, z, and t are scalars

or $y\vec{b}+z\vec{c}+t\vec{d}=-x\vec{a}$

$$\Rightarrow \frac{y\vec{b}+z\vec{c}+t\vec{d}}{y+z+t}=-\frac{x\vec{a}}{y+z+t},$$

where L.H.S. is the position vector of a point on the plane BCD and R.H.S. is a point on the line $\overrightarrow{AO}$. Therefore, there must be a point common to the plane and the straight line, i.e.,

$$\overrightarrow{OP}=-\frac{x\vec{a}}{y+z+t}$$

But $\overrightarrow{AP}=\overrightarrow{OP}-\overrightarrow{OA}$

$$=-\frac{x\vec{a}}{y+z+t}-\vec{a}$$

$$=-\left(\frac{x+y+z+t}{y+z+t}\right)\vec{a}$$

$$\Rightarrow \overrightarrow{OP}=\frac{x}{y+z+t}\left(\frac{y+z+t}{x+y+z+t}\right)\overrightarrow{AP}$$

$$=\left(\frac{x}{x+y+z+t}\right)\overrightarrow{AP}$$

$$\Rightarrow \frac{OP}{AP}=\frac{x}{x+y+z+t}$$

Similarly,

$$\frac{OQ}{BQ}=\frac{y}{x+y+z+t},$$

$$\frac{OR}{CR}=\frac{z}{x+y+z+t} \text{ and}$$

$$\frac{OS}{DS}=\frac{t}{x+y+z+t}$$

Hence,

$$\frac{OP}{AP}+\frac{OQ}{BQ}+\frac{OR}{CR}+\frac{OS}{DS}=1$$

8. **(1)** Since the vectors are coplanar,

$$\begin{vmatrix} 1 & \cos(\beta-\alpha) & \cos(\gamma-\alpha) \\ \cos(\alpha-\beta) & 1 & \cos(\gamma-\beta) \\ \cos(\alpha-\gamma) & \cos(\beta-\gamma) & a \end{vmatrix}=0$$

$$\Rightarrow \begin{vmatrix} \cos\alpha & \sin\alpha & 0 \\ \cos\beta & \sin\beta & 0 \\ \cos\gamma & \sin\gamma & a-1 \end{vmatrix}\begin{vmatrix} \cos\alpha & \sin\alpha & 0 \\ \cos\beta & \sin\beta & 0 \\ \cos\gamma & \sin\gamma & 1 \end{vmatrix}=0$$

$$\Rightarrow a=1$$

9. **(3)** Let $QM:MT=\lambda:1$ and $PM:MS=\mu:1$

and $\overrightarrow{QP}=\vec{a}$, $\overrightarrow{QR}=\vec{b}$

$$\Rightarrow \overrightarrow{QT}=\frac{4\vec{b}+\vec{a}}{5}$$

$$\text{and } \overrightarrow{QM}=\frac{\lambda}{\lambda+1}\left(\frac{4\vec{b}+\vec{a}}{5}\right) \qquad (1)$$

$$\overrightarrow{QS}=\frac{3}{4}\vec{b},\ \overrightarrow{QM}=\frac{\mu\left(\frac{3}{4}\vec{b}\right)+\vec{a}}{\mu+1} \qquad (2)$$

From (1) and (2), we get

$$\frac{1}{\mu+1}=\frac{\gamma}{5(\lambda+1)} \text{ and } \frac{4\lambda}{5(\lambda+1)}=\frac{3\mu}{4(\mu+1)}$$

$\Rightarrow \lambda=15/4$ and $\mu=16/3$

Hence, $QM:MT=15:4$.

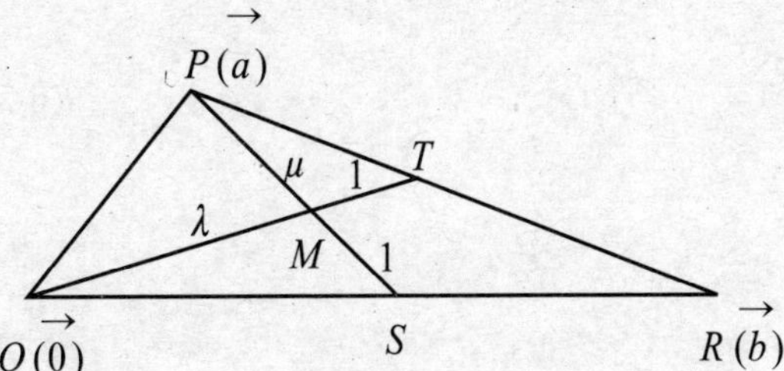

Fig. 1.19

10. **(1)** The given vectors $\vec{u}=a\hat{i}+\hat{j}+\hat{k}$, $\vec{v}=\hat{i}+b\hat{j}+\hat{k}$, and $\vec{w}=\hat{i}+\hat{j}+c\hat{k}$, where $a\neq b\neq c\neq 1$, are coplanar. Therefore,

$$[\vec{u}\,\vec{v}\,\vec{w}]=0$$

$$\Rightarrow \begin{vmatrix} a & 1 & 1 \\ 1 & b & 1 \\ 1 & 1 & c \end{vmatrix}=0$$

Operating $C_1\to C_1-C_2$ and $C_2\to C_2-C_3$, we get

$$\begin{vmatrix} a-1 & 0 & 1 \\ 1-b & b-1 & 1 \\ 0 & 1-c & c \end{vmatrix}=0$$

Expanding, we get

$$c(a-1)(b-1)+(1-b)(1-c)-(1-c)(a-1)=0$$

$$\Rightarrow \frac{c}{1-c}+\frac{1}{1-a}+\frac{1}{1-b}=0$$

$$\Rightarrow \frac{c}{1-c}+1+\frac{1}{1-a}+\frac{1}{1-b}=1$$

$$\Rightarrow \frac{1}{1-c}+\frac{1}{1-a}+\frac{1}{1-b}=1$$

CHAPTER 2

Different Products of Vectors and Their Geometrical Applications

Objective Type

Each question has four choices a, b, c, and d, out of which *only one* answer is correct. Find the correct answer.

1. Two vectors are equal only if they have equal component in
   a. a given direction
   b. two given directions
   c. three given directions
   d. in any arbitrary direction

2. Let $\vec{a}, \vec{b}$, and $\vec{c}$ be the three vectors having magnitudes 1, 5, and 3, respectively, such that the angle between $\vec{a}$ and $\vec{b}$ is θ and $\vec{a} \times (\vec{a} \times \vec{b}) = \vec{c}$, then $\tan\theta$ is equal to
   a. 0 b. 2/3
   c. 3/5 d. 3/4

3. $\vec{a}, \vec{b}$, and $\vec{c}$ are three vectors of equal magnitude. The angle between each pair of vectors is $\pi/3$ such that $|\vec{a}+\vec{b}+\vec{c}| = \sqrt{6}$. Then $|\vec{a}|$ is equal to
   a. 2 b. -1
   c. 1 d. $\sqrt{6}/3$

4. If $\vec{a}, \vec{b}$, and $\vec{c}$ are three mutually perpendicular vectors, then the vector which is equally inclined to these vectors is
   a. $\vec{a}+\vec{b}+\vec{c}$ b. $\dfrac{\vec{a}}{|\vec{a}|}+\dfrac{\vec{b}}{|\vec{b}|}+\dfrac{\vec{c}}{|\vec{c}|}$
   c. $\dfrac{\vec{a}}{|\vec{a}|^2}+\dfrac{\vec{b}}{|\vec{b}|^2}+\dfrac{\vec{c}}{|\vec{c}|^2}$ d. $|\vec{a}|\vec{a}-|\vec{b}|\vec{b}+|\vec{c}|\vec{c}$

5. Let $\vec{a} = \hat{i}+\hat{j}; \vec{b} = 2\hat{i}-\hat{k}$. Then vector $\vec{r}$ satisfying the equations $\vec{r}\times\vec{a} = \vec{b}\times\vec{a}$ and $\vec{r}\times\vec{b} = \vec{a}\times\vec{b}$ is
   a. $\hat{i}-\hat{j}+\hat{k}$ b. $3\hat{i}-\hat{j}+\hat{k}$
   c. $3\hat{i}+\hat{j}-\hat{k}$ d. $\hat{i}-\hat{j}-\hat{k}$

6. If $\vec{a}$ and $\vec{b}$ are two vectors, such that $\vec{a}\cdot\vec{b} < 0$ and $|\vec{a}\cdot\vec{b}| = |\vec{a}\times\vec{b}|$, then the angle between vectors $\vec{a}$ and $\vec{b}$ is
   a. π b. $7\pi/4$
   c. $\pi/4$ d. $3\pi/4$

7. If $\hat{a}, \hat{b}$, and $\hat{c}$ are three unit vectors, such that $\hat{a}+\hat{b}+\hat{c}$ is also a unit vector and θ_1, θ_2, and θ_3 are angles between the vectors $\hat{a}, \hat{b}; \hat{b}, \hat{c}$ and $\hat{c}, \hat{a}$, respectively, then among θ_1, θ_2, and θ_3
   a. all are acute angles
   b. all are right angles
   c. at least one is obtuse angle
   d. none of these

8. If $\vec{a}, \vec{b}, \vec{c}$ are unit vectors such that $\vec{a}\cdot\vec{b} = 0 = \vec{a}\cdot\vec{c}$ and the angle between $\vec{b}$ and $\vec{c}$ is $\pi/3$. Then the value of $|\vec{a}\times\vec{b} - \vec{a}\times\vec{c}|$ is
   a. 1/2 b. 1
   c. 2 d. none of these

9. $P(\vec{p})$ and $Q(\vec{q})$ are the position vectors of two fixed points and $R(\vec{r})$ is the position vector of a variable point. If R moves such that $(\vec{r}-\vec{p})\times(\vec{r}-\vec{q}) = \vec{0}$, then the locus of R is

a. a plane containing the origin O and parallel to two non-collinear vectors $\overrightarrow{OP}$ and $\overrightarrow{OQ}$

b. the surface of a sphere described on PQ as its diameter

c. a line passing through points P and Q

d. a set of lines parallel to line PQ

10. Two adjacent sides of a parallelogram $ABCD$ are $2\hat{i}+4\hat{j}-5\hat{k}$ and $\hat{i}+2\hat{j}+3\hat{k}$. Then the value of $|\overrightarrow{AC}\times\overrightarrow{BD}|$ is

a. $20\sqrt{5}$ b. $22\sqrt{5}$

c. $24\sqrt{5}$ d. $26\sqrt{5}$

11. $\vec{r}\times\vec{a}=\vec{b}\times\vec{a};\vec{r}\times\vec{b}=\vec{a}\times\vec{b};\vec{a}\neq 0;\vec{b}\neq 0;\ \vec{a}\neq\lambda\vec{b}$; $\vec{a}$ is not perpendicular to $\vec{b}$. Then $\vec{r}$ is equal to

a. $\vec{a}-\vec{b}$ b. $\vec{a}+\vec{b}$

c. $\vec{a}\times\vec{b}+\vec{a}$ d. $\vec{a}\times\vec{b}+\vec{b}$

12. Let the pairs $\vec{a},\vec{b}$ and $\vec{c}$, $\vec{d}$ each determine a plane. Then the planes are parallel if

a. $(\vec{a}\times\vec{c})\times(\vec{b}\times\vec{d})=\vec{0}$

b. $(\vec{a}\times\vec{c})\cdot(\vec{b}\times\vec{d})=\vec{0}$

c. $(\vec{a}\times\vec{b})\times(\vec{c}\times\vec{d})=\vec{0}$

d. $(\vec{a}\times\vec{b})\cdot(\vec{c}\times\vec{d})=\vec{0}$

13. If $\vec{r}\cdot\vec{a}=\vec{r}\cdot\vec{b}=\vec{r}\cdot\vec{c}=0$, where $\vec{a}$, $\vec{b}$ and $\vec{c}$ are non-coplanar, then

a. $\vec{r}\perp(\vec{c}\times\vec{a})$ b. $\vec{r}\perp(\vec{a}\times\vec{b})$

c. $\vec{r}\perp(\vec{b}\times\vec{c})$ d. $\vec{r}=\vec{0}$

14. If $\vec{a}$ satisfies $\vec{a}\times(\hat{i}+2\hat{j}+\hat{k})=\hat{i}-\hat{k}$, then $\vec{a}$ is equal to

a. $\lambda\hat{i}+(2\lambda-1)\hat{j}+\lambda\hat{k},\lambda\in R$

b. $\lambda\hat{i}+(1-2\lambda)\hat{j}+\lambda\hat{k},\lambda\in R$

c. $\lambda\hat{i}+(2\lambda+1)\hat{j}+\lambda\hat{k},\lambda\in R$

d. $\lambda\hat{i}-(1+2\lambda)\hat{j}+\lambda\hat{k},\lambda\in R$

15. Vectors $3\vec{a}-5\vec{b}$ and $2\vec{a}+\vec{b}$ are mutually perpendicular. If $\vec{a}+4\vec{b}$ and $\vec{b}-\vec{a}$ are also mutually perpendicular, then the cosine of the angle between $\vec{a}$ and $\vec{b}$ is

a. $\frac{19}{5\sqrt{43}}$ b. $\frac{19}{3\sqrt{43}}$

c. $\frac{19}{2\sqrt{45}}$ d. $\frac{19}{6\sqrt{43}}$

16. The unit vector orthogonal to vector $-\hat{i}+2\hat{j}+2\hat{k}$ and making equal angles with the x- and y-axes is

a. $\pm\frac{1}{3}(2\hat{i}+2\hat{j}-\hat{k})$ b. $\pm\frac{1}{3}(\hat{i}+\hat{j}-\hat{k})$

c. $\pm\frac{1}{3}(2\hat{i}-2\hat{j}-\hat{k})$ d. None of these

17. The value of x for which the angle between $\vec{a}=2x^2\hat{i}+4x\hat{j}+\hat{k}$ and $\vec{b}=7\hat{i}-2\hat{j}+x\hat{k}$ is obtuse and the angle between $\vec{b}$ and the z-axis and less than $\pi/6$, are

a. $a<x<1/2$ b. $1/2<x<15$

c. $x>1/2$ or $x<0$ d. none of these

18. If vectors $\vec{a}$ and $\vec{b}$ are two adjacent sides of a parallelogram, then the vectors representing the altitude of the parallelogram which is perpendicular to $\vec{a}$ is

a. $\vec{b}+\frac{\vec{b}\times\vec{a}}{|\vec{a}|^2}$ b. $\frac{\vec{a}\cdot\vec{b}}{|\vec{b}|^2}$

c. $\vec{b}-\frac{\vec{b}\cdot\vec{a}}{|\vec{a}|^2}\vec{a}$ d. $\frac{\vec{a}\times(\vec{b}\times\vec{a})}{|\vec{b}|^2}$

19. A parallelogram is constructed on $3\vec{a}+\vec{b}$ and $\vec{a}-4\vec{b}$, where $|\vec{a}|=6$ and $|\vec{b}|=8$ and $\vec{a}$ and $\vec{b}$ are anti-parallel. Then the length of the longer diagonal is

a. 40 b. 64

c. 32 d. 48

20. If $\vec{a}\cdot\vec{b}=0$, where $\vec{a}$ and $\vec{b}$ are unit vectors and the unit vector $\vec{c}$ is inclined at an angle θ to both $\vec{a}$ and $\vec{b}$. If $\vec{c}=m\vec{a}+n\vec{b}+p(\vec{a}\times\vec{b})$, $(m, n, p\in R)$, then

a. $-\frac{\pi}{4}\le\theta\le\frac{\pi}{4}$ b. $\frac{\pi}{4}\le\theta\le\frac{3\pi}{4}$

c. $0\le\theta\le\frac{\pi}{4}$ d. $0\le\theta\le\frac{3\pi}{4}$

21. $\vec{a}$ and $\vec{c}$ are unit vectors and $|\vec{b}|=4$. The angle between $\vec{a}$ and $\vec{c}$ is $\cos^{-1}(1/4)$. Then $\vec{b}-2\vec{c}=\lambda\vec{a}$. The value of λ is

a. 3, – 4 b. 1/4, 3/4

c. –3, 4 d. –1/4, 3/4

22. Let the position vectors of the points P and Q be $4\hat{i}+\hat{j}+\lambda\hat{k}$ and $2\hat{i}-\hat{j}+\lambda\hat{k}$, respectively. Vector $\hat{i}-\hat{j}+6\hat{k}$ is perpendicular to the plane containing the origin and the points P and Q. Then λ equals

a. $-1/2$ b. $1/2$
c. 1 d. none of these

23. Let $\hat{a}, \hat{b}$, and $\hat{c}$ be unit vectors and α, β, and γ the angles between the vectors $\hat{a}, \hat{b}; \hat{b}, \hat{c}$, and $\hat{c}, \hat{a}$, respectively. If $\hat{a}+\hat{b}+\hat{c}$ is also a unit vector, then $\cos\alpha+\cos\beta+\cos\gamma$ is equal to

a. -1 b. 3
c. -3 d. 1

24. If unit vector $\vec{c}$ makes an angle $\pi/3$ with $\hat{i}+\hat{j}$, then minimum and maximum values of $(\hat{r}\times\hat{j})\cdot\vec{c}$, respectively, are

a. $0, \sqrt{3}/2$ b. $-1/2\sqrt{3}/2$
c. $-\sqrt{3}/2, \sqrt{3}/2$ d. none of these

25. G is the centroid of triangle ABC and A_1 and B_1 are the midpoints of sides AB and AC, respectively. If Δ_1 is the area of quadrilateral GA_1AB_1 and Δ is the area of triangle ABC, then Δ/Δ_1 is equal to

a. $\frac{3}{2}$ b. 3
c. $\frac{1}{3}$ d. none of these

26. Vectors $\vec{a}, \vec{b}, \vec{c}$ and $\vec{d}$ are coplanar and $(\sin\alpha)\vec{a}+(2\sin 2\beta)\vec{b}+(3\sin 3\gamma)\vec{c}-\vec{d}=\vec{0}$. Then the least value of $\sin^2\alpha+\sin^2 2\beta+\sin^2 3\gamma$ is

a. 1/14 b. 14
c. 6 d. $1/\sqrt{6}$

27. If $\vec{a}$ and $\vec{b}$ are any two vectors of magnitudes 1 and 2, respectively, and $(1-3\vec{a}\cdot\vec{b})^2+|2\vec{a}+\vec{b}+3(\vec{a}\times\vec{b})|^2=47$, then the angle between $\vec{a}$ and $\vec{b}$ is

a. $\pi/3$ b. $\pi-\cos^{-1}(1/4)$
c. $\frac{2\pi}{3}$ d. $\cos^{-1}(1/4)$

28. If $\vec{a}$ and $\vec{b}$ are any two vectors of magnitudes 2 and 3, respectively, such that $|2(\vec{a}\times\vec{b})|+|3(\vec{a}\cdot\vec{b})|=k$, then the maximum value of k is

a. $\sqrt{13}$ b. $2\sqrt{13}$
c. $6\sqrt{13}$ d. $10\sqrt{13}$

29. $\vec{a}$, $\vec{b}$, and $\vec{c}$ are unit vectors such that $|\vec{a}+\vec{b}+3\vec{c}|=4$. Angle between $\vec{a}$ and $\vec{b}$ is θ_1, between $\vec{b}$ and $\vec{c}$ is θ_2 and between $\vec{a}$ and $\vec{c}$ varies $[\pi/6, 2\pi/3]$. Then the maximum value of $\cos\theta_1+3\cos\theta_2$ is

a. 3 b. 4
c. $2\sqrt{2}$ d. 6

30. If the vector product of a constant vector $\overrightarrow{OA}$ with a variable vector $\overrightarrow{OB}$ in a fixed plane OAB be a constant vector, then the locus of B is

a. a straight line perpendicular to $\overrightarrow{OA}$
b. a circle with centre O radius equal to $|\overrightarrow{OA}|$
c. a straight line parallel to $\overrightarrow{OA}$
d. none of these

31. Let $\vec{u}, \vec{v}$ and $\vec{w}$ be such that $|\vec{u}|=1$, $|\vec{v}|=2$ and $|\vec{w}|=3$. If the projection of $\vec{v}$ along $\vec{u}$ is equal to that of $\vec{w}$ along $\vec{u}$ and vectors $\vec{v}$ and $\vec{w}$ are perpendicular to each other, then $|\vec{u}-\vec{v}+\vec{w}|$ equals

a. 2 b. $\sqrt{7}$
c. $\sqrt{14}$ d. 14

32. If the two adjacent sides of two rectangles are represented by vectors $\vec{p}=5\vec{a}-3\vec{b}$; $\vec{q}=-\vec{a}-2\vec{b}$ and $\vec{r}=-4\vec{a}-\vec{b}$; $\vec{s}=-\vec{a}+\vec{b}$, respectively, then the angle between the vectors $\vec{x}=\frac{1}{3}(\vec{p}+\vec{r}+\vec{s})$ and $\vec{y}=\frac{1}{5}(\vec{r}+\vec{s})$ is

a. $-\cos^{-1}\left(\frac{19}{5\sqrt{43}}\right)$ b. $\cos^{-1}\left(\frac{19}{5\sqrt{43}}\right)$
c. $\pi\cos^{-1}\left(\frac{19}{5\sqrt{43}}\right)$ d. cannot be evaluated

33. If $\vec{A}=(1,1,1)$ and $\vec{C}=(0,1,1)$ are given vectors, then a vector $\vec{B}$ satisfying the equation $\vec{A}\times\vec{B}=\vec{C}$ and $\vec{A}\cdot\vec{B}=3$ is

a. $(5,2,2)$ b. $(5/3, 2/3, 2/3)$
c. $(2/3, 5/3, 2/3)$ d. $(2/3, 2/3, 5/3)$

34. The position vectors of the points A, B, and C are $\hat{i}+\hat{j}+\hat{k}, \hat{i}+5\hat{j}-\hat{k}$ and $2\hat{i}+3\hat{j}+5\hat{k}$, respectively. The greatest angle of triangle ABC is

a. 120° b. 90°
c. $\cos^{-1}(3/4)$ d. none of these

35. Given three vectors $\vec{a}$, $\vec{b}$, and $\vec{c}$, two of which are non-collinear. Further if $(\vec{a}+\vec{b})$ is collinear with $\vec{c}$, $(\vec{b}+\vec{c})$ is collinear with $\vec{a}$, $|\vec{a}| = |\vec{b}| = |\vec{c}| = \sqrt{2}$, then the value of $\vec{a}\cdot\vec{b}+\vec{b}\cdot\vec{c}+\vec{c}\cdot\vec{a}$ is

a. 3 b. −3
c. 0 d. cannot be evaluated

36. If $\vec{a}$ and $\vec{b}$ are unit vectors such that $(\vec{a}+\vec{b}).\ (2\vec{a}+3\vec{b})\times(3\vec{a}-2\vec{b}) = \vec{0}$, then angle between $\vec{a}$ and $\vec{b}$ is

a. 0 b. $\pi/2$
c. π d. indeterminate

37. If in a right-angled triangle ABC, the hypotenuse $AB = p$, then $\overrightarrow{AB}\cdot\overrightarrow{AC}+\overrightarrow{BC}\cdot\overrightarrow{BA}+\overrightarrow{CA}\cdot\overrightarrow{CB}$ is equal to

a. $2p^2$ b. $\frac{p^2}{2}$
c. p^2 d. none of these

38. Resolved part of vector $\vec{a}$ and along vector $\vec{b}$ is $\vec{a_1}$ and that perpendicular to $\vec{b}$ is $\vec{a_2}$ then $\vec{a_1}\times\vec{a_2}$ is equal to

a. $\frac{(\vec{a}\times\vec{b})\cdot\vec{b}}{|\vec{b}|^2}$ b. $\frac{(\vec{a}\cdot\vec{b})\,\vec{a}}{|\vec{a}|^2}$

c. $\frac{(\vec{a}\cdot\vec{b})(\vec{b}\times\vec{a})}{|\vec{b}|^2}$ d. $\frac{(\vec{a}\cdot\vec{b})(\vec{b}\times\vec{a})}{|\vec{b}\times\vec{a}|}$

39. $\vec{a} = 2\hat{i}-\hat{j}+\hat{k}, \vec{b} = \hat{i}+2\hat{j}-\hat{k}, \vec{c} = \hat{i}+\hat{j}-2\hat{k}$. A vector coplanar with $\vec{b}$ and $\vec{c}$ whose projection on $\vec{a}$ is magnitude $\sqrt{\frac{2}{3}}$ is

a. $2\hat{i}+3\hat{j}-3\hat{k}$ b. $-2\hat{i}-\hat{j}+5\hat{k}$
c. $2\hat{i}+3\hat{j}+3\hat{k}$ d. $2\hat{i}+\hat{j}+5\hat{k}$

40. P is any arbitrary point on the circumcircle of the equilateral triangle of side length l units then $|\overrightarrow{PA}|^2+|\overrightarrow{PB}|^2+|\overrightarrow{PC}|^2$ is always equal to

a. $2l^2$ b. $2\sqrt{3}\,l^2$
c. l^2 d. $3l^2$

41. $\vec{a}$ and $\vec{c}$ are unit vectors and $|\vec{b}| = 4$. If angle between $\vec{b}$ and $\vec{c}$ is $\cos^{-1}\left(\frac{1}{4}\right)$ and $\vec{a}\times\vec{b} = 2\vec{a}\times\vec{c}$, then $\vec{b} = \lambda\vec{a}+2\vec{c}$, where λ is equal to

a. $\pm\frac{1}{4}$ b. $\pm\frac{1}{2}$
c. ±1 d. ±4

42. $\vec{a}$ and $\vec{b}$ are two unit vectors that are mutually perpendicular. A unit vector that is equally inclined to $\vec{a}$, $\vec{b}$ and $\vec{a}\times\vec{b}$ is

a. $\frac{1}{\sqrt{2}}(\vec{a}+\vec{b}+\vec{a}\times\vec{b})$ b. $\frac{1}{2}(\vec{a}\times\vec{b}+\vec{a}+\vec{b})$

c. $\frac{1}{\sqrt{3}}(\vec{a}+\vec{b}+\vec{a}\times\vec{b})$ d. $\frac{1}{3}(\vec{a}+\vec{b}+\vec{a}\times\vec{b})$

43. $\vec{a}$ and $\vec{b}$ are two mutually perpendicular unit vectors and $\vec{c}$ is a unit vector inclined at an angle θ to both $\vec{a}$ and $\vec{b}$ if $\vec{c} = x\,\vec{a}+x\,\vec{b}+y(\vec{a}\times\vec{b})$, where $x, y \in R$, exhaustive range of θ is

a. $[\pi/2, 3\pi/2]$ b. $[\pi/4, 3\pi/4]$
c. $[\pi/2, \pi/2]$ d. none of these

44. The position vectors of the vertices A, B and C of a triangle are three unit vectors $\hat{a}$, $\hat{b}$, and $\hat{c}$, respectively. A vector $\vec{d}$ is such that $\vec{d}\cdot\hat{a} = \vec{d}\cdot\hat{b} = \vec{d}\cdot\hat{c}$ and $\vec{d} = \lambda(\hat{b}+\hat{c})$. Then triangle ABC is

a. acute angled b. obtuse angled
c. right angled d. none of these

45. If a is a real constant and A, B, and C are variable angles and $\sqrt{a^2-4}\tan A + a\tan B + \sqrt{a^2+4}\tan c = 6a$, then the least value of $\tan^2 A + \tan^2 B + \tan^2 C$ is

a. 6 b. 10
c. 12 d. 3

46. The vertex A of triangle ABC is on the line $\vec{r} = \hat{i}+\hat{j}+\lambda\hat{k}$ and the vertices B and C have respective position vectors $\hat{i}$ and $\hat{j}$. Let Δ be the area of the triangle and $\Delta \in [3/2, \sqrt{33}/2]$. Then the range of values of λ corresponding to A is

a. $[-8,-4]\cup[4,8]$ b. $[-4,4]$
c. $[-2,2]$ d. $[-4,-2]\cup[2,4]$

47. If $\vec{p}$ and $\vec{s}$ are not perpendicular to each other and $\vec{r}\times\vec{p} = \vec{q}\times\vec{p}$ and $\vec{r}\cdot\vec{s} = 0$, then $\vec{r}$ is equal to

a. $\vec{p}\cdot\vec{s}$

b. $\vec{q}-\left(\frac{\vec{q}\cdot\vec{s}}{\vec{p}\cdot\vec{s}}\right)\vec{p}$

c. $\vec{q}+\left(\frac{\vec{q}\cdot\vec{p}}{\vec{p}\cdot\vec{s}}\right)\vec{p}$

d. $\vec{q}+\mu\,\vec{p}$ for all scalars μ

48. Position vector $\hat{k}$ is rotated about origin by angle 135° in such a way that the plane made by it bisects the angle between $\hat{i}$ and $\hat{j}$. Then its new position is

a. $\pm\frac{\hat{i}}{\sqrt{2}}\pm\frac{\hat{j}}{\sqrt{2}}$ b. $\pm\frac{\hat{i}}{\sqrt{2}}\pm\frac{\hat{j}}{\sqrt{2}}-\frac{\hat{k}}{\sqrt{2}}$

c. $\frac{\hat{i}}{\sqrt{2}}-\frac{\hat{k}}{\sqrt{2}}$ d. none of these

49. In a quadrilateral $ABCD$, $\overrightarrow{AC}$ is the bisector of $\overrightarrow{AB}$ and $\overrightarrow{AD}$, angle between $\overrightarrow{AB}$ and $\overrightarrow{AD}$ is $2\pi/3$, $15|\overrightarrow{AC}|=3|\overrightarrow{AB}|=5|\overrightarrow{AD}|$. Then the angle between $\overrightarrow{BA}$ and $\overrightarrow{CD}$ is

a. $\cos^{-1}\frac{\sqrt{14}}{7\sqrt{2}}$ b. $\cos^{-1}\frac{\sqrt{21}}{7\sqrt{3}}$

c. $\cos^{-1}\frac{2}{\sqrt{7}}$ d. $\cos^{-1}\frac{2\sqrt{7}}{14}$

50. In the following figure, AB, DE, and GF are parallel to each other and AD, BG, and EF are parallel to each other. If $CD : CE = CG : CB = 2:1$, then the value of area (ΔAEG) : area (ΔABD) is equal to

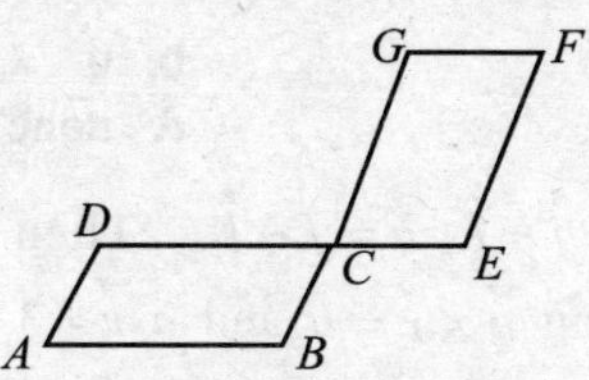

Fig. 2.1

a. 7/2 b. 3

c. 4 d. 9/2

51. Vector $\hat{a}$ in the plane of $\vec{b} = 2\hat{i}+\hat{j}$ and $\vec{c}=\hat{i}-\hat{j}+\hat{k}$ is such that it is equally inclined to $\vec{b}$ and $\vec{d}$ where $\vec{d}=\hat{j}+2\hat{k}$ is

a. $\frac{\hat{i}+\hat{j}+\hat{k}}{\sqrt{3}}$ b. $\frac{\hat{i}-\hat{j}+\hat{k}}{\sqrt{3}}$

c. $\frac{2\hat{i}+\hat{j}}{\sqrt{5}}$ d. $\frac{2\hat{i}+\hat{j}}{\sqrt{5}}$

52. Let $ABCD$ be a tetrahedron such that the edges AB, AC, and AD are mutually perpendicular. Let the area of triangles ABC, ACD and ADB be 3, 4, and 5 sq. units, respectively. Then the area of triangle BCD is

a. $5\sqrt{2}$ b. 5

c. $\frac{\sqrt{5}}{2}$ d. $\frac{5}{2}$

53. Let $\vec{f}(t)=[t]\hat{i}+(t-[t])\hat{j}+[t+1]\hat{k}$, where [.] denotes the greatest integer function. Then the vectors $\vec{f}\left(\frac{5}{4}\right)$ and $\vec{f}(t)$, $0<t<1$, are

a. parallel to each other

b. perpendicular to each other

c. inclined at an angle $\cos^{-1}\frac{2}{\sqrt{7(1-t^2)}}$

d. inclined at $\cos^{-1}\frac{8+t}{9\sqrt{1+t^2}}$

54. If $\vec{a}$ is parallel to $\vec{b}\times\vec{c}$, then $(\vec{a}\times\vec{b})\cdot(\vec{a}\times\vec{c})$ is equal to

a. $|\vec{a}|^2(\vec{b}\cdot\vec{c})$ b. $|\vec{b}|^2(\vec{a}\cdot\vec{c})$

c. $|\vec{c}|^2(\vec{a}\cdot\vec{b})$ d. none of these

55. Three vectors $\hat{i}+\hat{j}$, $\hat{j}+\hat{k}$ and $\hat{k}+\hat{i}$ taken two at a time form three planes. The three unit vectors drawn perpendicular to these three planes form a parallelepiped of volume

a. 1/3 b. 4

c. $(3\sqrt{3})/4$ d. $4/(3\sqrt{3})$

56. If $\vec{d}=\vec{a}\times\vec{b}+\vec{b}\times\vec{c}+\vec{c}\times\vec{a}$ is a non-zero vector and $|(\vec{d}\cdot\vec{c})(\vec{a}\times\vec{b})+(\vec{d}\cdot\vec{a})(\vec{b}\times\vec{c})+(\vec{d}\cdot\vec{b})(\vec{c}\times\vec{a})|=0$, then

a. $|\vec{a}|=|\vec{b}|=|\vec{c}|$

b. $|\vec{a}|+|\vec{b}|+|\vec{c}|=|\vec{d}|$

c. $\vec{a},\vec{b}$ and $\vec{c}$ are coplanar

d. none of these

57. If $|\vec{a}|=2$ and $|\vec{b}|=3$ and $\vec{a}\cdot\vec{b}$ =, then $(\vec{a}\times(\vec{a}\times(\vec{a}\times(\vec{a}\times\vec{b}))))$ is equal to

a. $48\hat{b}$ b. $-48\hat{b}$

c. $48\hat{a}$ d. $-48\hat{a}$

58. If the two diagonals of one of its faces are $6\hat{i}+6\hat{k}$ and $4\hat{j}+2\hat{k}$ and of the edges not containing the given diagonal is $\vec{c}=4\hat{j}-8\hat{k}$ then the volume of parallelepiped is

a. 60 b. 80

c. 100 d. 120

59. The volume of a tetrahedron formed by the coterminus edges $\vec{a}$, $\vec{b}$, and $\vec{c}$ is 3. Then the volume of the parallelepiped formed by the coterminus edges $\vec{a}+\vec{b}$, $\vec{b}+\vec{c}$, and $\vec{c}+\vec{a}$ is

a. 6 b. 18

c. 36 d. 9

60. If $\vec{a}, \vec{b}$, and $\vec{c}$ are three mutually orthogonal unit vectors, then the triple product $[\vec{a}+\vec{b}+\vec{c}\;\; \vec{a}+\vec{b}\;\; \vec{b}+\vec{c}]$ equals

a. 0 b. 1 or -1
c. 1 d. 3

61. Vector $\vec{c}$ is perpendicular to vectors $\vec{a} = (2, -3, 1)$ and $\vec{b} = (1, -2, 3)$ and satisfies the condition $\vec{c}\cdot(\hat{i}+2\hat{j}-7\hat{k}) = 10$. Then vector $\vec{c}$ is equal to

a. $(7, 5, 1)$ b. $(-7, -5, -1)$
c. $(1, 1, -1)$ d. none of these

62. Given $\vec{a} = x\hat{i} + y\hat{j} + 2\hat{k}$, $\vec{b} = \hat{i} - \hat{j} + \hat{k}$, $\vec{c} = \hat{i} + 2\hat{j}$; $\vec{a} \perp \vec{b}$, $\vec{a}\cdot\vec{c} = 4$. Then

a. $[\vec{a}\,\vec{b}\,\vec{c}]^2 = |\vec{a}|$ b. $[\vec{a}\,\vec{b}\,\vec{c}] = |\vec{a}|$
c. $[\vec{a}\,\vec{b}\,\vec{c}] = 0$ d. $[\vec{a}\,\vec{b}\,\vec{c}] = |\vec{a}|^2$

63. Let $\vec{a} = a_1\hat{i} + a_2\hat{j} + a_3\hat{k}$, $\vec{b} = b_1\hat{i} + b_2\hat{j} + b_3\hat{k}$, and $\vec{c} = c_1\hat{i} + c_2\hat{j} + c_3\hat{k}$ be three non-zero vectors such that $\vec{c}$ is a unit vector perpendicular to both $\vec{a}$ and $\vec{b}$. If the angle between $\vec{a}$ and $\vec{b}$ is $\pi/6$, then the value of $\begin{vmatrix} a_1 & b_1 & c_1 \\ a_2 & b_2 & c_2 \\ a_3 & b_3 & c_3 \end{vmatrix}^2$ is

a. 0
b. 1
c. $\frac{1}{4}(a_1^2+a_2^2+a_3^2)(b_1^2+b_2^2+b_3^2)$
d. $\frac{3}{4}(a_1^2+a_2^2+a_3^2)(b_1^2+b_2^2+b_3^2)$

64. Let $\vec{r}, \vec{a}, \vec{b}$, and $\vec{c}$ be four non zero vectors such that $\vec{r}\cdot\vec{a} = 0$, $|\vec{r}\times\vec{b}| = |\vec{r}||\vec{b}|$ and $|\vec{r}\times\vec{c}| = |\vec{r}||\vec{c}|$. Then $[a\,b\,c]$ is equal to

a. $|a||b||c|$ b. $-|a||b||c|$
c. 0 d. none of these

65. If $\vec{a}, \vec{b}$, and $\vec{c}$ are such that $[\vec{a}\,\vec{b}\,\vec{c}] = 1$, $\vec{c} = \lambda\vec{a}\times\vec{b}$, angle between $\vec{a}$ and $\vec{b}$ is $2\pi/3$, $|\vec{a}| = \sqrt{2}$, $|\vec{b}| = \sqrt{3}$ and $|\vec{c}| = \frac{1}{\sqrt{3}}$, then the angle between $\vec{a}$ and $\vec{b}$ is

a. $\frac{\pi}{6}$ b. $\frac{\pi}{4}$
c. $\frac{\pi}{3}$ d. $\frac{\pi}{2}$

66. If $4\vec{a} + 5\vec{b} + 9\vec{c} = 0$, then $(\vec{a}\times\vec{b})\times[(\vec{b}\times\vec{c})\times(\vec{c}\times\vec{a})]$ is equal to

a. a vector perpendicular to the plane of $\vec{a}, \vec{b}$, and $\vec{c}$
b. a scalar quantity
c. $\vec{0}$
d. none of these

67. Value of $[\vec{a}\times\vec{b}\;\; \vec{a}\times\vec{c}\;\; \vec{d}]$ is always equal to

a. $(\vec{a}\cdot\vec{d})[\vec{a}\,\vec{b}\,\vec{c}]$ b. $(\vec{a}\cdot\vec{c})[\vec{a}\,\vec{b}\,\vec{c}]$
c. $(\vec{a}\cdot\vec{b})[\vec{a}\,\vec{b}\,\vec{c}]$ d. none of these

68. Let $\hat{a}$ and $\hat{b}$ be mutually perpendicular unit vectors. Then for any arbitrary $\vec{r}$,

a. $\vec{r} = (\vec{r}\cdot\hat{a})\hat{a} + (\vec{r}\cdot\hat{b})\hat{b} + (\vec{r}\cdot(\hat{a}\times\hat{b}))(\hat{a}\times\hat{b})$
b. $\vec{r} = (\vec{r}\cdot\hat{a})\hat{a} - (\vec{r}\cdot\hat{b})\hat{b} - (\vec{r}\cdot(\hat{a}\times\hat{b}))(\hat{a}\times\hat{b})$
c. $\vec{r} = (\vec{r}\cdot\hat{a})\hat{a} - (\vec{r}\cdot\hat{b})\hat{b} + (\vec{r}\cdot(\hat{a}\times\hat{b}))(\hat{a}\times\hat{b})$
d. none of these

69. Let $\vec{a}$ and $\vec{b}$ be unit vectors that are perpendicular to each other. Then $[\vec{a} + (\vec{a}\times\vec{b})\;\; \vec{b} + (\vec{a}\times\vec{b})\;\; \vec{a}\times\vec{b}]$ will always be equal to

a. 1 b. 0
c. -1 d. none of these

70. $\vec{a} = \hat{i} + \hat{j} + \hat{k}$, $\vec{b} = \hat{j} - \hat{k}$. Then a vector $\vec{r}$ satisfying $\vec{a}\times\vec{r} = \vec{b}$ and $\vec{a}\cdot\vec{r} = 3$ is

a. $\frac{5}{3}\hat{i} - \frac{2}{3}\hat{j} + \frac{2}{3}\hat{k}$ b. $\frac{5}{3}\hat{i} + \frac{2}{3}\hat{j} - \frac{2}{3}\hat{k}$
c. $\frac{5}{3}\hat{i} + \frac{2}{3}\hat{j} + \frac{2}{3}\hat{k}$ d. $\frac{5}{3}\hat{i} - \frac{2}{3}\hat{j} - \frac{2}{3}\hat{k}$

71. $\vec{b}$ and $\vec{c}$ are unit vectors. Then for any arbitrary vector $\vec{a}$, $(((\vec{a}\times\vec{b}) + (\vec{a}\times\vec{c}))\times(\vec{b}\times\vec{c}))\cdot(\vec{b}-\vec{c})$ is always equal to

a. $|\vec{a}|$ b. $\frac{1}{2}|\vec{a}|$
c. $\frac{1}{3}|\vec{a}|$ d. none of these

72. If $\vec{a}\cdot\vec{b} = \beta$ and $\vec{a}\times\vec{b} = \vec{c}$, then $\vec{b}$ is

a. $\frac{(\beta\vec{a} - \vec{a}\times\vec{c})}{|\vec{a}|^2}$ b. $\frac{(\beta\vec{a} + \vec{a}\times\vec{c})}{|\vec{a}|^2}$
c. $\frac{(\beta\vec{c} - \vec{a}\times\vec{c})}{|\vec{a}|^2}$ d. $\frac{(\beta\vec{a} + \vec{a}\times\vec{c})}{|\vec{a}|^2}$

73. If $a(\vec{\alpha}\times\vec{\beta}) + b(\vec{\beta}\times\vec{\gamma}) + c(\vec{\gamma}\times\vec{\alpha}) = 0$ and at least one of a, b and c is non-zero, then vectors $\vec{\alpha}, \vec{\beta}$, and $\vec{\gamma}$ are

a. parallel
b. coplanar
c. mutually perpendicular
d. none of these

74. If $(\vec{a}\times\vec{b})\times(\vec{b}\times\vec{c}) = \vec{b}$, where $\vec{a}, \vec{b}$ and $\vec{c}$ are non-zero vectors, then

a. a, b and c can be coplanar
b. a, b and c must be coplanar
c. $\vec{a}, \vec{b}$ and $\vec{c}$ cannot be coplanar
d. none of these

75. If $\vec{r}\cdot\vec{a} = \vec{r}\cdot\vec{b} = \vec{r}\cdot\vec{c} = \frac{1}{2}$ for some non-zero vector $\vec{r}$, then the area of the triangle whose vertices are $A(\vec{a})$, $B(\vec{b})$ and $C(\vec{c})$ is ($\vec{a},\vec{b},\vec{c}$ are non-coplanar)

a. $|[\vec{a}\,\vec{b}\,\vec{c}]|$
b. $|\vec{r}|$
c. $|[\vec{a}\,\vec{b}\,\vec{c}]\vec{r}|$
d. none of these

76. If non-zero vectors $\vec{a}$ and $\vec{b}$ are perpendicular to each other, then the solution to the equation $\vec{r}\times\vec{a} = \vec{b}$ is

a. $\vec{r} = \alpha\vec{b} - \frac{1}{|\vec{b}|^2}(\vec{a}\times\vec{b})$
b. $\vec{r} = \alpha\vec{a} + \frac{1}{|\vec{a}|^2}(\vec{a}\times\vec{b})$
c. $\vec{r} = \alpha\vec{a} + \frac{1}{|\vec{b}|^2}(\vec{a}\times\vec{b})$
d. none of these

77. If $\vec{a}$ and $\vec{b}$ are two unit vectors inclined at an angle $\pi/3$, then $\{\vec{a}\times(\vec{b}+\vec{a}\times\vec{b})\}\cdot\vec{b}$ is equal to

a. $\frac{-3}{4}$
b. $\frac{1}{4}$
c. $\frac{3}{4}$
d. $\frac{1}{2}$

78. If $\vec{a}$ and $\vec{b}$ are orthogonal unit vectors, then for a vector $\vec{r}$ noncoplanar with $\vec{a}$ and $\vec{b}$, vector $\vec{r}\times\vec{a}$ is equal to

a. $[\vec{r}\,\vec{a}\,\vec{b}]\vec{b} - (\vec{r}\cdot\vec{b})(\vec{b}\times\vec{a})$
b. $[\vec{r}\,\vec{a}\,\vec{b}](\vec{a}+\vec{b})$
c. $[\vec{r}\,\vec{a}\,\vec{b}]\vec{a} + (\vec{r}\cdot\vec{a})\vec{a}\times\vec{b}$
d. none of these

79. If $\vec{a}, \vec{b}$, and $\vec{c}$ are any three non-coplanar vectors then the equation $[\vec{b}\times\vec{c}\;\; \vec{c}\times\vec{a}\;\; \vec{a}\times\vec{b}]x^2 + [\vec{a}+\vec{b}\;\; \vec{b}+\vec{c}\;\; \vec{c}+\vec{a}]x + 1 + [\vec{b}-\vec{c}\;\; \vec{c}-\vec{a}\;\; \vec{a}-\vec{b}] = 0$ has roots

a. real and distinct
b. real
c. equal
d. imaginary

80. For any four vectors, $(\vec{b}\times\vec{c})\cdot(\vec{a}\times\vec{d}) + (\vec{c}\times\vec{a})\cdot(\vec{b}\times\vec{d}) + (\vec{a}\times\vec{b})\cdot(\vec{c}\times\vec{d})$ is equal to

a. $[\vec{a}\,\vec{b}\,\vec{c}]$
b. $[\vec{a}\,\vec{c}\,\vec{d}]$
c. $[\vec{b}\,\vec{c}\,\vec{d}]$
d. none of these

81. The condition for the equation $\vec{r}\times\vec{a} = \vec{b}$ and $\vec{r}\times\vec{c} = \vec{d}$ to be consistent is

a. $\vec{b}\cdot\vec{c} = \vec{a}\cdot\vec{d}$
b. $\vec{a}\cdot\vec{b} = \vec{c}\cdot\vec{d}$
c. $\vec{b}\cdot\vec{c} + \vec{a}\cdot\vec{d} = 0$
d. $\vec{a}\cdot\vec{b} + \vec{c}\cdot\vec{d} = 0$

82. If $\vec{a}$ and $\vec{b}$ are non-zero non-collinear vectors, then $[\vec{a}\,\vec{b}\,\hat{i}]\hat{i} + [\vec{a}\,\vec{b}\,\hat{j}]\hat{j} + [\vec{a}\,\vec{b}\,\hat{k}]\hat{k}$ is equal to

a. $\vec{a}+\vec{b}$
b. $\vec{a}\times\vec{b}$
c. $\vec{a}-\vec{b}$
d. $\vec{b}\times\vec{a}$

83. If $\vec{a} = 2\hat{i}+\hat{j}+\hat{k}$, $\vec{b} = \hat{i}+2\hat{j}+2\hat{k}$, $\vec{c} = \hat{i}+\hat{j}+2\hat{k}$, and $(1+\alpha)\hat{i} + \beta(1+\alpha)\hat{j} + \gamma(1+\alpha)(1+\beta)\hat{k} = \vec{a}\times(\vec{b}\times\vec{c})$, then α, β, and γ are

a. $-2, -4, -\frac{2}{3}$
b. $2, -4, \frac{2}{3}$
c. $-2, 4, \frac{2}{3}$
d. $2, 4, -\frac{2}{3}$

84. If $\vec{a} = \vec{b}+\vec{c}$, $\vec{b}\times\vec{d} = \vec{0}$ and $\vec{c}\cdot\vec{d} = 0$, then $\frac{\vec{d}\times(\vec{a}\times\vec{d})}{|\vec{d}|^2}$ is always equal to

a. $\vec{a}$
b. $\vec{d}$
c. $\vec{b}$
d. $\vec{c}$

85. For any two vectors $\vec{a}$ and $\vec{b}$, $(\vec{a}\times\hat{i})\cdot(\vec{b}\times\hat{i}) + (\vec{a}\times\hat{j})\cdot(\vec{b}\times\hat{j}) + (\vec{a}\times\hat{k})\cdot(\vec{b}\times\hat{k})$ is always equal to

a. $\vec{a}\cdot\vec{b}$
b. $2\vec{a}\cdot\vec{b}$
c. zero
d. none of these

86. Let $\vec{a}, \vec{b}$ and $\vec{c}$ be three non-coplanar vectors and $\vec{r}$ be any arbitrary vector. Then $(\vec{a}\times\vec{b})\times(\vec{r}\times\vec{c}) + (\vec{b}\times\vec{c})\times(\vec{r}\times\vec{a}) + (\vec{c}\times\vec{a})\times(\vec{r}\times\vec{b})$ is always equal to

a. $[\vec{a}\,\vec{b}\,\vec{c}]\vec{r}$ b. $2[\vec{a}\,\vec{b}\,\vec{c}]\vec{r}$

c. $3[\vec{a}\,\vec{b}\,\vec{c}]\vec{r}$ d. none of these

87. If $\vec{p}=\frac{\vec{b}\times\vec{c}}{[\vec{a}\,\vec{b}\,\vec{c}]}$, $\vec{q}=\frac{\vec{c}\times\vec{a}}{[\vec{a}\,\vec{b}\,\vec{c}]}$, and $\vec{r}=\frac{\vec{a}\times\vec{b}}{[\vec{a}\,\vec{b}\,\vec{c}]}$, where $\vec{a}$, $\vec{b}$, and $\vec{c}$ are three non-coplanar vectors, then the value of the expression $(\vec{a}+\vec{b}+\vec{c})\cdot(\vec{p}+\vec{q}+\vec{r})$ is

a. 3 b. 2

c. 1 d. 0

88. $A(\vec{a})$, $B(\vec{b})$, and $C(\vec{c})$ are the vertices of triangle ABC and $R(\vec{r})$ is any point in the plane of triangle ABC, then $\vec{r}\cdot(\vec{a}\times\vec{b}+\vec{b}\times\vec{c}+\vec{c}\times\vec{a})$ is always equal to

a. zero b. $[\vec{a}\,\vec{b}\,\vec{c}]$

c. $-[\vec{a}\,\vec{b}\,\vec{c}]$ d. none of these

89. If $\vec{a}$, $\vec{b}$ and $\vec{c}$ are non-coplanar vectors and $\vec{a}\times\vec{c}$ is perpendicular to $\vec{a}\times(\vec{b}\times\vec{c})$, then the value of $[\vec{a}\times(\vec{b}\times\vec{c})]\times\vec{c}$ is equal to

a. $[\vec{a}\,\vec{b}\,\vec{c}]\,\vec{c}$ b. $[\vec{a}\,\vec{b}\,\vec{c}]\vec{b}$

c. $\vec{0}$ d. $[\vec{a}\,\vec{b}\,\vec{c}]\vec{a}$

90. If V be the volume of a tetrahedron and V' be the volume of another tetrahedran formed by the centroids of the previous tetrahedron, if $V = KV'$; then K is equal to

a. 9 b. 12

c. 27 d. 81

91. c. $[(\vec{a}\times\vec{b})\times(\vec{b}\times\vec{c})(\vec{b}\times\vec{c})\times(\vec{c}\times\vec{a})(\vec{c}\times\vec{a})\times(\vec{a}\times\vec{b})]$ is equal to (where $\vec{a}$, $\vec{b}$ and $\vec{c}$ are non-zero non-coplanar vectors)

a. $[\vec{a}\,\vec{b}\,\vec{c}]^2$ b. $[\vec{a}\,\vec{b}\,\vec{c}]^3$

c. $[\vec{a}\,\vec{b}\,\vec{c}]^4$ d. $[\vec{a}\,\vec{b}\,\vec{c}]$

92. If $\vec{r}=x_1(\vec{a}\times\vec{b})+x_2(\vec{b}\times\vec{c})+x_3(\vec{c}\times\vec{d})$ and $4[\vec{a}\,\vec{b}\,\vec{c}]=1$ then $x_1+x_2+x_3$ is equal to

a. $\frac{1}{2}\vec{r}\cdot(\vec{a}+\vec{b}+\vec{c})$ b. $\frac{1}{4}\vec{r}\cdot(\vec{a}+\vec{b}+\vec{c})$

c. $2\vec{r}\cdot(\vec{a}+\vec{b}+\vec{c})$ d. $4\vec{r}\cdot(\vec{a}+\vec{b}+\vec{c})$

93. If $\vec{a}\perp\vec{b}$, then vector $\vec{v}$ in terms of $\vec{a}$ and $\vec{b}$ satisfying the equations $\vec{v}\cdot\vec{a}=0$ and $\vec{v}\cdot\vec{b}=1$ and $[\vec{v}\,\vec{a}\,\vec{b}]=1$ is

a. $\frac{\vec{b}}{|\vec{b}|^2}+\frac{\vec{a}\times\vec{b}}{|\vec{a}\times\vec{b}|^2}$ b. $\frac{\vec{b}}{|\vec{b}|}+\frac{\vec{a}\times\vec{b}}{|\vec{a}\times\vec{b}|^2}$

c. $\frac{\vec{b}}{|\vec{b}|^2}+\frac{\vec{a}\times\vec{b}}{|\vec{a}\times\vec{b}|}$ d. none of these

94. If $\vec{a}'=\hat{i}+\hat{j}$, $\vec{b}'=\hat{i}-\hat{j}+2\hat{k}$ and $\vec{c}'=2\hat{i}+\hat{j}-\hat{k}$, then the altitude of the parallelepiped formed by the vectors $\vec{a}$, $\vec{b}$, and $\vec{c}$ having base formed by $\vec{b}$ and $\vec{c}$ is where (a, is reciprocal vector a, etc.)

a. 1 b. $3\sqrt{2}/2$

c. $1/\sqrt{6}$ d. $1/\sqrt{2}$

95. If $\vec{a}=\hat{i}+\hat{j}$, $\vec{b}=\hat{j}+\hat{k}$, and $\vec{c}=\hat{k}+\hat{i}$, then in the reciprocal system of vectors $\vec{a},\vec{b},\vec{c}$ reciprocal $\vec{a}$ of vector $\vec{a}$ is

a. $\frac{\hat{i}+\hat{j}+\hat{k}}{2}$ b. $\frac{\hat{i}-\hat{j}+\hat{k}}{2}$

c. $\frac{-\hat{i}-\hat{j}+\hat{k}}{2}$ d. $\frac{\hat{i}+\hat{j}-\hat{k}}{2}$

Multiple Correct Answers Type

Each question has four choices a, b, c, and d, out of which *one or more* answers are correct.

1. If unit vectors $\vec{a}$ and $\vec{b}$ are inclined at an angle 2θ such that $|\vec{a}-\vec{b}|<1$ and $0\le\theta\le\pi$, then θ lies in the interval

a. $[0, \pi/6)$ b. $(5\pi/6, \pi]$

c. $[\pi/6, \pi/2]$ d. $[\pi/2, 5\pi/6]$

2. $\vec{b}$ and $\vec{c}$ are non-collinear if $\vec{a}\times(\vec{b}\times\vec{c})+(\vec{a}\cdot\vec{b})\vec{b}=(4-2x-\sin y)\,\vec{b}+(x^2-1)\,\vec{c}$ and $(\vec{c}\cdot\vec{c})\vec{a}=\vec{c}$. Then

a. $x=1$ b. $x=-1$

c. $y=(4n+1)\frac{\pi}{2}, n\in I$ d. $y=(2n+1)\frac{\pi}{2}, n\in I$

3. Unit vectors $\vec{a}$ and $\vec{b}$ are perpendicular, and unit vector $\vec{c}$ is inclined at an angle θ to both $\vec{a}$ and $\vec{b}$. If $\vec{c}=\alpha\vec{a}+\beta\vec{b}+\gamma(\vec{a}\times\vec{b})$, then

a. $a=\beta$ b. $\gamma^2=1-2\alpha^2$

c. $\gamma^2=-\cos 2\theta$ d. $\beta^2=\frac{1+\cos 2\theta}{2}$

4. $\vec{a}$ and $\vec{b}$ are two given vectors. With these vectors as adjacent sides, a parallelogram is constructed. The vector which is the altitude of the parallelogram and which is perpendicular to $\vec{a}$ is

a. $\frac{(\vec{a}\cdot\vec{b})}{|\vec{a}|^2}\vec{a}-\vec{b}$

b. $\frac{1}{|\vec{a}|^2}\{|\vec{a}|^2\vec{b}-(\vec{a}\cdot\vec{b})\vec{a}\}$

c. $\frac{\vec{a}\times(\vec{a}\times\vec{b})}{|\vec{a}|^2}$

d. $\frac{\vec{a}\times(\vec{b}\times\vec{a})}{|\vec{b}|^2}$

5. If $\vec{a}\times(\vec{b}\times\vec{c})$ is perpendicular to $(\vec{a}\times\vec{b})\times\vec{c}$, we may have

a. $(\vec{a}\cdot\vec{c})|\vec{b}|^2=(\vec{a}\cdot\vec{b})(\vec{b}\cdot\vec{c})$

b. $\vec{a}\cdot\vec{b}=0$

c. $\vec{a}\cdot\vec{c}=0$

d. $\vec{b}\cdot\vec{c}=0$

6. Let $\vec{a}$, $\vec{b}$, and $\vec{c}$ be vectors forming right-hand triad. Let $\vec{p}=\frac{\vec{b}\times\vec{c}}{[\vec{a}\,\vec{b}\,\vec{c}]}$, $\vec{q}=\frac{\vec{c}\times\vec{a}}{[\vec{a}\,\vec{b}\,\vec{c}]}$ and $\vec{r}=\frac{\vec{a}\times\vec{b}}{[\vec{a}\,\vec{b}\,\vec{c}]}$.

If $x\in R^+$, then

a. $x[\vec{a}\,\vec{b}\,\vec{c}]+\frac{[\vec{p}\,\vec{q}\,\vec{r}]}{x}$ has least value 2

b. $x^4[\vec{a}\,\vec{b}\,\vec{c}]^2+\frac{[\vec{p}\,\vec{q}\,\vec{r}]}{x^2}$ has least value $(3/2^{2/3})$

c. $[\vec{p}\,\vec{q}\,\vec{r}]>0$

d. none of these

7. $a_1, a_2, a_3\in R-\{0\}$ and $a_1+a_2\cos 2x+a_3\sin^2 x=0$ for all $x\in R$, then

a. vectors $\vec{a}=a_1\hat{i}+a_2\hat{j}+a_3\hat{k}$ and $\vec{b}=4\hat{i}+2\hat{j}+\hat{k}$ are perpendicular to each other

b. vectors $\vec{a}=a_1\hat{i}+a_2\hat{j}+a_3\hat{k}$ and $\vec{b}=-\hat{i}+\hat{j}+2\hat{k}$ are parallel to each other

c. if vector $\vec{a}=a_1\hat{i}+a_2\hat{j}+a_3\hat{k}$ is of length $\sqrt{6}$ units, then one of the ordered tripplet $(a_1, a_2, a_3)=(1,-1,-2)$

d. if $2a_1+3a_2+6a_3=26$, then $|a_1\hat{i}+a_2\hat{j}+a_3\hat{k}|$ is $2\sqrt{6}$

8. If $\vec{a}$ and $\vec{b}$ are two vectors and angle between them is θ, then

a. $|\vec{a}\times\vec{b}|^2+(\vec{a}\cdot\vec{b})^2=|\vec{a}|^2|\vec{b}|^2$

b. $|\vec{a}\times\vec{b}|=(\vec{a}\cdot\vec{b})$, if $\theta=\pi/4$

c. $\vec{a}\times\vec{b}=(\vec{a}\cdot\vec{b})\hat{n}$, ($\hat{n}$ is normal unit vector), if $\theta=\pi/4$

d. $(\vec{a}\times\vec{b})\cdot(\vec{a}+\vec{b})=0$

9. Let $\vec{a}$ and $\vec{b}$ be two non-zero perpendicular vectors. A vector $\vec{r}$ satisfying the equation $\vec{r}\times\vec{b}=\vec{a}$ can be

a. $\vec{b}-\frac{\vec{a}\times\vec{b}}{|\vec{b}|^2}$ b. $2\vec{b}-\frac{\vec{a}\times\vec{b}}{|\vec{b}|^2}$

c. $|\vec{a}|\vec{b}-\frac{\vec{a}\times\vec{b}}{|\vec{b}|^2}$ d. $|\vec{b}|\vec{b}-\frac{\vec{a}\times\vec{b}}{|\vec{b}|^2}$

10. If vectors $\vec{b}=(\tan\alpha, -1, 2\sqrt{\sin\alpha/2})$ and $\vec{c}=\left(\tan\alpha, \tan\alpha, -\frac{3}{\sqrt{\sin\alpha/2}}\right)$ are orthogonal and vector $\vec{a}=((1, 3, \sin 2\alpha)$ makes an obtuse angle with the z-axis, then the value of α is

a. $\alpha=(4n+1)\pi+\tan^{-1}2$

b. $\alpha=(4n+1)\pi-\tan^{-1}2$

c. $\alpha=(4n+2)\pi+\tan^{-1}2$

d. $\alpha=(4n+2)\pi-\tan^{-1}2$

11. Let $\vec{r}$ be a unit vector satisfying $\vec{r}\times\vec{a}=\vec{b}$, where $|\vec{a}|=\sqrt{3}$ and $|\vec{b}|=\sqrt{2}$. Then

a. $\vec{r}=\frac{2}{3}(\vec{a}+\vec{a}\times\vec{b})$ b. $\vec{r}=\frac{1}{3}(\vec{a}+\vec{a}\times\vec{b})$

c. $\vec{r}=\frac{2}{3}(\vec{a}-\vec{a}\times\vec{b})$ d. $\vec{r}=\frac{1}{3}(-\vec{a}+\vec{a}\times\vec{b})$

12. If $\vec{a}$ and $\vec{b}$ are unequal unit vectors such that $(\vec{a}-\vec{b})\times[(\vec{b}+\vec{a})\times(2\vec{a}+\vec{b})]=\vec{a}+\vec{b}$, then angle θ between $\vec{a}$ and $\vec{b}$ is

a. 0 b. $\pi/2$

c. $\pi/4$ d. π

13. If $\vec{a}$ and $\vec{b}$ are two unit vectors perpendicular to each other and $\vec{c}=\lambda_1\vec{a}+\lambda_2\vec{b}+\lambda_3(\vec{a}\times\vec{b})$, then which of the following is (are) true?

a. $\lambda_1=\vec{a}\cdot\vec{c}$

b. $\lambda_2=|\vec{b}\times\vec{c}|$

c. $\lambda_3=|(\vec{a}\times\vec{b})\times\vec{c}|$

d. $\lambda_1+\lambda_2+\lambda_3=(\vec{a}+\vec{b}+\vec{a}\times\vec{b})\cdot\vec{c}$

14. If vectors $\vec{a}$ and $\vec{b}$ are non-collinear, then $\frac{\vec{a}}{|\vec{a}|}+\frac{\vec{b}}{|\vec{b}|}$ is

a. a unit vector

b. in the plane of $\vec{a}$ and $\vec{b}$

c. equally inclined to $\vec{a}$ and $\vec{b}$

d. perpendicular to $\vec{a}\times\vec{b}$

15. If $\vec{a}$ and $\vec{b}$ are non zero vectors such that $|\vec{a}+\vec{b}|=|\vec{a}-2\vec{b}|$, then

a. $2\vec{a}\cdot\vec{b}=|\vec{b}|^2$

b. $\vec{a}\cdot\vec{b}=|\vec{b}|^2$

c. least value of $\vec{a}\cdot\vec{b}+\frac{1}{|\vec{b}|^2+2}$ is $\sqrt{2}$

d. least value of $\vec{a}\cdot\vec{b}+\frac{1}{|\vec{b}|+2}$ is $\sqrt{2}-1$

16. Let $\vec{a}, \vec{b}$, and $\vec{c}$ be non-zero vectors and $\vec{V_1}=\vec{a}\times(\vec{b}\times\vec{c})$ and $\vec{V_2}=(\vec{a}\times\vec{b})\times\vec{c}$. Vectors $\vec{V_1}$ and $\vec{V_2}$ are equal. Then

a. a and $\vec{b}$ are orthogonal

b. a and $\vec{c}$ are collinear

c. $\vec{b}$ and $\vec{c}$ are orthogonal

d. $\vec{b}=\lambda(\vec{a}\times\vec{c})$ when λ is a scalar

17. Vectors $\vec{A}$ and $\vec{B}$ satisfying the vector equation $\vec{A}+\vec{B}=\vec{a}$, $\vec{A}\times\vec{B}=\vec{b}$ and $\vec{A}\cdot\vec{a}=1$, where $\vec{a}$ and $\vec{b}$ are given vectors, are

a. $\vec{A}=\frac{(\vec{a}\times\vec{b})-\vec{a}}{a^2}$

b. $\vec{B}=\frac{(\vec{b}\times\vec{a})+\vec{a}(a^2-1)}{a^2}$

c. $\vec{A}=\frac{(\vec{a}\times\vec{b})+\vec{a}}{a^2}$

d. $\vec{B}=\frac{(\vec{b}\times\vec{a})-\vec{a}(a^2-1)}{a^2}$

18. A vector $\vec{d}$ is equally inclined to three vectors $\vec{a}=\hat{i}-\hat{j}+\hat{k}$, $\vec{b}=2\hat{i}+\hat{j}$, and $\vec{c}=3\hat{j}-2\hat{k}$. Let $\vec{x}, \vec{y}$ and $\vec{z}$ be three vectors in the plane of $\vec{a}, \vec{b}$; $\vec{b}, \vec{c}$; $\vec{c}, \vec{a}$, respectively. Then

a. $\vec{x}\cdot\vec{d}=-1$

b. $\vec{y}\cdot\vec{d}=1$

c. $\vec{z}\cdot\vec{d}=0$

d. $\vec{r}\cdot\vec{d}=0$, where $\vec{r}=\lambda\vec{x}+\mu\vec{y}+\delta\vec{z}$

19. Vectors perpendicular to $\hat{i}-\hat{j}-\hat{k}$ and in the plane of $\hat{i}+\hat{j}+\hat{k}$ and $-\hat{i}+\hat{j}+\hat{k}$ are

a. $\hat{i}+\hat{k}$
b. $2\hat{i}+\hat{j}+\hat{k}$
c. $3\hat{i}+2\hat{j}+\hat{k}$
d. $-4\hat{i}-2\hat{j}-2\hat{k}$

20. If the side $\overrightarrow{AB}$ of an equilateral triangle ABC lying in the x-y plane is $3\hat{i}$, then the side $\overrightarrow{CB}$ can be

a. $-\frac{3}{2}(\hat{i}-\sqrt{3}\hat{j})$
b. $\frac{3}{2}(\hat{i}-\sqrt{3}\hat{j})$
c. $-\frac{3}{2}(\hat{i}+\sqrt{3}\hat{j})$
d. $\frac{3}{2}(\hat{i}+\sqrt{3}\hat{j})$

21. The angles of a triangle, two of whose sides are represented by vectors $\sqrt{3}(\hat{a}\times\vec{b})$ and $\vec{b}-(\hat{a}\cdot\vec{b})\hat{a}$, where $\vec{b}$ is a non-zero vector and $\hat{a}$ is a unit vector in the direction of $\vec{b}$, are

a. $\tan^{-1}(\sqrt{3})$
b. $\tan^{-1}(1/\sqrt{3})$
c. $\cot^{-1}(0)$
d. $\tan^{-1}(1)$

22. $\vec{a}, \vec{b}$, and $\vec{c}$ are unimodular and coplanar. A unit vector $\vec{d}$ is perpendicular to them. If $(\vec{a}\times\vec{b})\times(\vec{c}\times\vec{d})=\frac{1}{6}\hat{i}-\frac{1}{3}\hat{j}+\frac{1}{3}\hat{k}$, and the angle between $\vec{a}$ and $\vec{b}$ is 30°, then $\vec{c}$ is

a. $(\hat{i}-2\hat{j}+2\hat{k})/3$
b. $(-\hat{i}+2\hat{j}-2\hat{k})/3$
c. $(2\hat{i}+2\hat{j}-\hat{k})/3$
d. $(-2\hat{i}-2\hat{j}+\hat{k})/3$

23. If $\vec{a}+2\vec{b}+3\vec{c}=\vec{0}$, then $\vec{a}\times\vec{b}+\vec{b}\times\vec{c}+\vec{c}\times\vec{a}=$

a. $2(\vec{a}\times\vec{b})$
b. $6(\vec{b}\times\vec{c})$
c. $3(\vec{c}\times\vec{a})$
d. $\vec{0}$

24. If $\vec{a}$ and $\vec{b}$ are two non-collinear unit vectors and $\vec{u}=\vec{a}-(\vec{a}\cdot\vec{b})\vec{b}$ and $\vec{v}=\vec{a}\times\vec{b}$. Then $|\vec{v}|$ is

a. $|\vec{u}|$
b. $|\vec{u}|+|\vec{u}\cdot\vec{b}|$
c. $|\vec{u}|+|\vec{u}\cdot\vec{a}|$
d. none of these

25. If $\vec{a}\times\vec{b}=\vec{c}$, $\vec{b}\times\vec{c}=\vec{a}$, where $\vec{c}\neq\vec{0}$, then

a. $|\vec{a}|=|\vec{c}|$
b. $|\vec{a}|=|\vec{b}|$
c. $|\vec{b}|=1$
d. $|\vec{a}|=|\vec{b}|=|\vec{c}|=1$

26. Let $\vec{a}, \vec{b}$, and $\vec{c}$ be three non-coplanar vectors and $\vec{d}$ be a non-zero vector perpendicular to $(\vec{a}+\vec{b}+\vec{c})$. Now $\vec{d} = (\vec{a}\times\vec{b})\sin x + (\vec{b}\times\vec{c})\cos y + 2(\vec{c}\times\vec{a})$. Then

a. $\dfrac{\vec{d}\cdot(\vec{a}+\vec{c})}{[\vec{a}\,\vec{b}\,\vec{c}]} = 2$

b. $\dfrac{\vec{d}\cdot(\vec{a}+\vec{c})}{[\vec{a}\,\vec{b}\,\vec{c}]} = -2$

c. minimum value of x^2+y^2 is $\pi^2/4$

d. minimum value of x^2+y^2 is $5\pi^2/4$

27. If $\vec{a}, \vec{b}$, and $\vec{c}$ are three unit vectors such that $\vec{a}\times(\vec{b}\times\vec{c}) = \frac{1}{2}\vec{b}$, then ($\vec{b}$ and $\vec{c}$ being non-parallel)

a. angle between $\vec{a}$ and $\vec{b}$ is $\pi/3$

b. angle between $\vec{a}$ and $\vec{c}$ is $\pi/3$

c. angle between $\vec{a}$ and $\vec{b}$ is $\pi/2$

d. angle between $\vec{a}$ and $\vec{c}$ is $\pi/2$

28. If in triangle ABC, $\overrightarrow{BC} = \dfrac{\vec{u}}{|\vec{u}|} - \dfrac{\vec{v}}{|\vec{v}|}$ and $\overrightarrow{AC} = \dfrac{2\vec{u}}{|\vec{u}|}$, where $|\vec{u}| \neq |\vec{v}|$, then

a. $1+\cos 2A+\cos 2B+\cos 2C = 0$

b. $\sin A = \cos C$

c. projection of AB on BC is equal to BC

d. projection of AB on BC is equal to AB

29. $[\vec{a}\times\vec{b}\;\;\vec{c}\times\vec{d}\;\;\vec{e}\times\vec{f}]$ is equal to

a. $[\vec{a}\,\vec{b}\,\vec{d}][\vec{c}\,\vec{e}\,\vec{f}] - [\vec{a}\,\vec{b}\,\vec{c}][\vec{d}\,\vec{e}\,\vec{f}]$

b. $[\vec{a}\,\vec{b}\,\vec{e}][\vec{f}\,\vec{c}\,\vec{d}] - [\vec{a}\,\vec{b}\,\vec{f}][\vec{e}\,\vec{c}\,\vec{d}]$

c. $[\vec{c}\,\vec{d}\,\vec{a}][\vec{b}\,\vec{e}\,\vec{f}] - [\vec{a}\,\vec{d}\,\vec{b}][\vec{a}\,\vec{e}\,\vec{f}]$

d. $[\vec{a}\,\vec{c}\,\vec{e}][\vec{b}\,\vec{d}\,\vec{f}]$

30. The scalars l and m such that $l\vec{a}+m\vec{b} = \vec{c}$, where $\vec{a}, \vec{b}$, and $\vec{c}$ are given vectors, are equal to

a. $l = \dfrac{(\vec{c}\times\vec{b})\cdot(\vec{a}\times\vec{b})}{(\vec{a}\times\vec{b})^2}$ b. $l = \dfrac{(\vec{c}\times\vec{a})\cdot(\vec{b}\times\vec{a})}{(\vec{b}\times\vec{a})}$

c. $m = \dfrac{(\vec{c}\times\vec{a})\cdot(\vec{b}\times\vec{a})}{(\vec{b}\times\vec{a})^2}$ d. $m = \dfrac{(\vec{c}\times\vec{a})\cdot(\vec{b}\times\vec{a})}{(\vec{b}\times\vec{a})}$

31. If $(\vec{a}\times\vec{b})\times(\vec{c}\times\vec{d})\cdot(\vec{a}\times\vec{d}) = 0$, then which of the following may be true?

a. $\vec{a}, \vec{b}, \vec{c}$ and $\vec{d}$ are necessarily coplanar

b. $\vec{a}$ lies in the plane of $\vec{c}$ and $\vec{d}$

c. $\vec{b}$ lies in the plane of $\vec{a}$ and $\vec{d}$

d. $\vec{c}$ lies in the plane of $\vec{a}$ and $\vec{d}$

32. A, B, C, and D are four points such that $\overrightarrow{AB} = m(2\hat{i}-6\hat{j}+2\hat{k})$, $\overrightarrow{BC} = (\hat{i}-2\hat{j})$ and $\overrightarrow{CD} = n(-6\hat{i}+15\hat{j}-3\hat{k})$. Find the conditions on the scalars m and n so that CD intersects AB at some point E.

a. $m \geq 1/2$,

b. $n \geq 1/3$

c. the area of the triangle BCE is $\sqrt{6}/2$

d. $m < n$

33. If vectors $\vec{a}, \vec{b}$, and $\vec{c}$ are non-coplanar and l, m, and n are distinct scalars, then $[(l\vec{a}+m\vec{b}+n\vec{c})\,(l\vec{b}+m\vec{c}+n\vec{a})(l\vec{c}+m\vec{a}+n\vec{b})] = 0$ implies

a. $l+m+n = 0$

b. roots of the equation $lx^2+mx+n = 0$ are real

c. $l^2+m^2+n^2 = 0$

d. $l^3+m^3+n^3 = 3lmn$

34. Let $\vec{\alpha} = a\hat{i}+b\hat{j}+c\hat{k}$, $\vec{\beta} = b\hat{i}+c\hat{j}+a\hat{k}$ and $\vec{\gamma} = c\hat{i}+a\hat{j}+b\hat{k}$ be three coplanar vectors with $a \neq b$, and $\vec{v} = \hat{i}+\hat{j}+\hat{k}$. Then $\vec{v}$ is perpendicular to

a. $\vec{\alpha}$ b. $\vec{\beta}$

c. $\vec{\gamma}$ d. none of these

35. If vectors $\vec{A} = 2\hat{i}+3\hat{j}+4\hat{k}$, $\vec{B} = \hat{i}+\hat{j}+5\hat{k}$ and $\vec{C}$ form a left-handed system, then $\vec{C}$ is

a. $11\hat{i}-6\hat{j}-\hat{k}$ b. $-11\hat{i}+6\hat{j}+\hat{k}$

c. $11\hat{i}-6\hat{j}+\hat{k}$ d. $-11\hat{i}+6\hat{j}-\hat{k}$

Reasoning Type

Each question has four choices a, b, c, and d, out of which *only one* is correct. Each question contains Statement 1 and Statement 2.

a. If both the statements are True and Statement 2 is the correct explanation of Statement 1

b. If both the statements are True but Statement 2 is Not the correct explanation of Statement 1

c. If Statement 1 is True and Statement 2 is False.

d. If Statement 1 is False and Statement 2 is True.

1. **Statement 1:** Vector $\vec{c} = -5\hat{i}+7\hat{j}+2\hat{k}$ is along the bisector of angle between $\vec{a} = \hat{i}+2\hat{j}+2\hat{k}$, $\vec{b} = -8\hat{i}+\hat{j}-4\hat{k}$.

   Statement 2: $\vec{c}$ is equally inclined to $\vec{a}$ and $\vec{b}$.

2. **Statement 1:** A component of vector $\vec{b} = 4\hat{i}+2\hat{j}+3\hat{k}$ in the direction perpendicular to the direction of the vector $\vec{a} = \hat{i}+\hat{j}+\hat{k}$ is $\hat{i}-\hat{j}$.

   Statement 2: A component of vector in the direction of $\vec{a} = \hat{i}+\hat{j}+\hat{k}$ is $2\hat{i}+2\hat{j}+2\hat{k}$.

3. **Statement 1:** Distance of point $D(1, 0, -1)$ from the plane of points $A(1, -2, 0)$, $B(3, 1, 2)$, and $C(-1, 1, -1)$ is $\frac{8}{\sqrt{129}}$.

   Statement 2: Volume of tetrahedron formed by the points A, B, C, and D is $\frac{\sqrt{129}}{2}$.

4. Let $\vec{r}$ be a non-zero vector satisfying $\vec{r}\cdot\vec{a} = \vec{r}\cdot\vec{b} = \vec{r}\cdot\vec{c} = 0$ for given non-zero vectors $\vec{a}, \vec{b}$ and $\vec{c}$.

   Statement 1: $[\vec{a}-\vec{b}\ \ \vec{b}-\vec{c}\ \ \vec{c}-\vec{a}] = 0$

   Statement 2: $[\vec{a}\,\vec{b}\,\vec{c}] = 0$

5. **Statement 1:** If $a_1\hat{i}+a_2\hat{j}+a_3\hat{k}$, $b_1\hat{i}+b_2\hat{j}+b_3\hat{k}$ and $c_1\hat{i}+c_2\hat{j}+c_3\hat{k}$ are three mutually perpendicular unit vectors, then $a_1\hat{i}+b_1\hat{j}+c_1\hat{k}$, $a_2\hat{i}+b_2\hat{j}+c_2\hat{k}$, and $a_3\hat{i}+b_3\hat{j}+c_3\hat{k}$ may be mutually perpendicular unit vectors.

   Statement 2: Value of determinant and its transpose are the same.

6. **Statement 1:** If $\vec{A} = 2\hat{i}+3\hat{j}+6\hat{k}$, $\vec{B} = \hat{i}+\hat{j}-2\hat{k}$, and $\vec{C} = \hat{i}+2\hat{j}+\hat{k}$, then $|\vec{A}\times(\vec{A}\times(\vec{A}\times\vec{B}))\cdot\vec{C}| = 243$.

   Statement 2: $|\vec{A}\times(\vec{A}\times(\vec{A}\times\vec{B}))\cdot\vec{C}| = |\vec{A}|^2|[\vec{A}\vec{B}\vec{C}]|$

7. **Statement 1:** If $\vec{a}, \vec{b}$, and $\vec{c}$ are three mutually perpendicular unit vectors and $\vec{d}$ is a vector such that $\vec{a}, \vec{b}, \vec{c}$, and $\vec{d}$ are non-coplanar. If $[\vec{d}\,\vec{b}\,\vec{c}] = [\vec{d}\,\vec{a}\,\vec{b}] = [\vec{d}\,\vec{c}\,\vec{a}] = 1$, then $\vec{d} = \vec{a}+\vec{b}+\vec{c}$.

   Statement 2: $[\vec{d}\,\vec{b}\,\vec{c}] = [\vec{d}\,\vec{a}\,\vec{b}] = [\vec{d}\,\vec{c}\,\vec{a}] \Rightarrow \vec{d}$ is equally inclined to $\vec{a}, \vec{b}$, and $\vec{c}$.

8. Consider three vectors $\vec{a}, \vec{b}$, and $\vec{c}$.

   Statement 1: $\vec{a}\times\vec{b} = ((\hat{i}\times\vec{a})\cdot\vec{b})\hat{i} + ((\hat{j}\times\vec{a})\cdot\vec{b})\hat{j} + ((\hat{k}\times\vec{a})\cdot\vec{b})\hat{k}$

   Statement 2: $\vec{c} = (\hat{i}\cdot\vec{c})\hat{i} + (\hat{j}\cdot\vec{c})\hat{j} + (\hat{k}\cdot\vec{c})\hat{k}$

Linked Comprehension Type

Based upon each paragraph, three multiple choice questions have to be answered. Each question has four choices a, b, c, and d, out of which *only one* is correct.

For Problems 1–3

Let $\vec{u}, \vec{v}$, and $\vec{w}$ be three unit vectors such that $\vec{u}+\vec{v}+\vec{w} = \vec{a}$, $\vec{u}\times(\vec{v}\times\vec{w}) = \vec{b}$, $(\vec{u}\times\vec{v})\times\vec{w} = \vec{c}$, $\vec{a}\cdot\vec{u} = 3/2$, $\vec{a}\cdot\vec{v} = 7/4$ and $|\vec{a}| = 2$.

1. Vector $\vec{u}$ is

   a. $\vec{a} - \frac{2}{3}\vec{b} + \vec{c}$ b. $\vec{a} + \frac{4}{3}\vec{b} + \frac{8}{3}\vec{c}$

   c. $2\vec{a} - \vec{b} + \frac{1}{3}\vec{c}$ d. $\frac{4}{3}\vec{a} - \vec{b} + \frac{2}{3}\vec{c}$

2. Vector $\vec{v}$ is

   a. $2\vec{a} - 3\vec{c}$ b. $3\vec{b} - 4c$

   c. $-4\vec{c}$ d. $\vec{a} + \vec{b} + 2\vec{c}$

3. Vector $\vec{w}$ is

   a. $\frac{2}{3}(2\vec{c} - \vec{b})$ b. $\frac{1}{3}(\vec{a} - \vec{b} - \vec{c})$

   c. $\frac{1}{3}\vec{a} - \frac{2}{3}\vec{b} - 2\vec{c}$ d. $\frac{4}{3}(\vec{c} - \vec{b})$

For Problems 4–6

Vectors $\vec{x}, \vec{y}$, and $\vec{z}$ each of magnitude $\sqrt{2}$ make angle of 60° with each other. $\vec{x}\times(\vec{y}\times\vec{z}) = \vec{a}$, $\vec{y}\times(\vec{z}\times\vec{x}) = \vec{b}$, and $\vec{x}\times\vec{y} = \vec{c}$.

4. Vector $\vec{x}$ is

   a. $\frac{1}{2}[(\vec{a}-\vec{b})\times\vec{c} + (\vec{a}+\vec{b})]$

   b. $\frac{1}{2}[(\vec{a}+\vec{b})\times\vec{c} + (\vec{a}-\vec{b})]$

   c. $\frac{1}{2}[-(\vec{a}+\vec{b})\times\vec{c} + (\vec{a}+\vec{b})]$

   d. $\frac{1}{2}[(\vec{a}+\vec{b})\times\vec{c} - (\vec{a}+\vec{b})]$

5. Vector $\vec{y}$ is

   a. $\frac{1}{2}[(\vec{a}+\vec{c})\times\vec{b} - \vec{b} - \vec{a}]$

   b. $\frac{1}{2}[(\vec{a}-\vec{c})\times\vec{c} + \vec{b} + \vec{a}]$

c. $\frac{1}{2}[(\vec{a}+\vec{b})\times\vec{c}+\vec{b}+\vec{a}]$

d. $\frac{1}{2}[(\vec{a}-\vec{c})\times\vec{a}+\vec{b}-\vec{a}]$

6. Vector $\vec{z}$ is

a. $\frac{1}{2}[(\vec{a}-\vec{c})\times\vec{c}-\vec{b}+\vec{a}]$

b. $\frac{1}{2}[(\vec{a}+\vec{b})\times\vec{c}+\vec{b}-\vec{a}]$

c. $\frac{1}{2}[\vec{c}\times(\vec{a}-\vec{b})+\vec{b}+\vec{a}]$

d. none of these

For Problems 7–9

If $\vec{x}\times\vec{y}=\vec{a},\ \vec{y}\times\vec{z}=\vec{b},\ \vec{x}\cdot\vec{b}=\gamma,\ \vec{x}\cdot\vec{y}=1$, and $\vec{y}\cdot\vec{z}=1$

7. Vector $\vec{x}$ is

a. $\frac{1}{|\vec{a}\times\vec{b}|^2}[\vec{a}\times(\vec{a}\times\vec{b})]$

b. $\frac{\gamma}{|\vec{a}\times\vec{b}|^2}[\vec{a}\times\vec{b}-\vec{a}\times(\vec{a}\times\vec{b})]$

c. $\frac{\gamma}{|\vec{a}\times\vec{b}|^2}[\vec{a}\times\vec{b}+\vec{b}\times(\vec{a}\times\vec{b})]$

d. none of these

8. Vector $\vec{y}$ is

a. $\frac{\vec{a}\times\vec{b}}{\gamma}$

b. $\vec{a}+\frac{\vec{a}\times\vec{b}}{\gamma}$

c. $\vec{a}+\vec{b}+\frac{\vec{a}\times\vec{b}}{\gamma}$

d. none of these

9. Vector $\vec{z}$ is

a. $\frac{\gamma}{|\vec{a}\times\vec{b}|^2}[\vec{a}+\vec{b}\times(\vec{a}\times\vec{b})]$

b. $\frac{\gamma}{|\vec{a}\times\vec{b}|^2}[\vec{a}+\vec{b}-\vec{a}\times(\vec{a}\times\vec{b})]$

c. $\frac{\gamma}{|\vec{a}\times\vec{b}|^2}[\vec{a}\times\vec{b}+\vec{b}\times(\vec{a}\times\vec{b})]$

d. none of these

For Problems 10–12

Given two orthogonal vectors $\vec{A}$ and $\vec{B}$ each of length unity. Let $\vec{P}$ be the vector satisfying the equation $\vec{P}\times\vec{B}=\vec{A}-\vec{P}$. Then

10. $(\vec{P}\times\vec{B})\times\vec{B}$ is equal to

a. $\vec{P}$

b. $-\vec{P}$

c. $2\vec{B}$

d. $\vec{A}$

11. $\vec{P}$ is equal to

a. $\frac{\vec{A}}{2}+\frac{\vec{A}\times\vec{B}}{2}$

b. $\frac{\vec{A}}{2}+\frac{\vec{B}\times\vec{A}}{2}$

c. $\frac{\vec{A}\times\vec{B}}{2}-\frac{\vec{A}}{2}$

d. $\vec{A}\times\vec{B}$

12. Which of the following statements is false?

a. Vectors $\vec{P}$, $\vec{A}$, and $\vec{P}\times\vec{B}$ are linearly dependent.

b. Vectors $\vec{P}$, $\vec{B}$, and $\vec{P}\times\vec{B}$ are linearly independent.

c. $\vec{P}$ is orthogonal to $\vec{B}$ and has length $1/\sqrt{2}$.

d. None of the above.

For Problems 13–15

Let $\vec{a}=2\hat{i}+3\hat{j}-6\hat{k}$, $\vec{b}=2\hat{i}-3\hat{j}+6\hat{k}$, and $\vec{c}=-2\hat{i}+3\hat{j}+6\hat{k}$. Let $\vec{a}_1$ be the projection of $\vec{a}$ on $\vec{b}$ and $\vec{a}_2$ be the projection of $\vec{a}_1$ on $\vec{c}$. Then

13. $\vec{a}_2$ is equal to

a. $\frac{943}{49}(2\hat{i}-3\hat{j}-6\hat{k})$

b. $\frac{943}{49^2}(2\hat{i}-3\hat{j}-6\hat{k})$

c. $\frac{943}{49}(-2\hat{i}+3\hat{j}+6\hat{k})$

d. $\frac{943}{49^2}(-2\hat{i}+3\hat{j}+6\hat{k})$

14. $\vec{a}_1\cdot\vec{b}$ is equal to

a. -41

b. $-41/7$

c. 41

d. 287

15. Which of the following is true?

a. $\vec{a}$ and $\vec{a}_2$ are collinear.

b. $\vec{a}_1$ and $\vec{c}$ are collinear.

c. $\vec{a}$, $\vec{a}_1$, and $\vec{b}$ are coplanar.

d. $\vec{a}$, $\vec{a}_1$, and $\vec{a}_2$ are coplanar.

For Problems 16–18

Consider a triangular pyramid *ABCD* the position vectors of whose angular points are *A*(3, 0, 1), *B*(–1, 4, 1), *C*(5, 2, 3), and *D*(0, –5, 4). Let *G* be the point of intersection of the medians of triangle *BCD*.

16. The length of vector $\overrightarrow{AG}$ is

a. $\sqrt{17}$

b. $\sqrt{51}/3$

c. $3/\sqrt{6}$

d. $\sqrt{59}/4$

17. Area of triangle ABC in square units is

a. 24 b. $8\sqrt{6}$

c. $4\sqrt{6}$ d. none of these

18. The length of the perpendicular from the vertex D on the opposite face is

a. $14/\sqrt{6}$ b. $2/\sqrt{6}$

c. $3/\sqrt{6}$ d. none of these

For Problems 19–21

Vertices of a parallelogram taken in order are $A(2, -1, 4)$, $B(1, 0 -1)$, $C(1, 2, 3)$, and D.

19. The distance between the parallel lines AB and CD is

a. $\sqrt{6}$ b. $3\sqrt{6/5}$

c. $2\sqrt{2}$ d. 3

20. Distance of the point $P(8, 2, -12)$ from the plane of the parallelogram is

a. $\frac{4\sqrt{6}}{9}$ b. $\frac{32\sqrt{6}}{9}$

c. $\frac{16\sqrt{6}}{9}$ d. none

21. The orthogonal projections of the parallelogram on the three coordinate planes xy, yz, and zx, respectively, are

a. 14, 4, 2 b. 2, 4, 14

c. 4, 2, 14 d. 2, 14, 4

For Problems 22–24

Let $\vec{r}$ be a position vector of a variable point in Cartesian OXY plane such that $\vec{r}\cdot(10\hat{j}-8\hat{i}-\vec{r}) = 40$ and $p_1 = \max\{|\vec{r}+2\hat{i}-3\hat{j}|^2\}$, $p_2 = \min\{|\vec{r}+2\hat{i}-3\hat{j}|^2\}$. A tangent line is drawn to the curve $y = 8/x^2$ at the point A with abscissa 2. The drawn line cuts the x-axis at a point B.

22. p_2 is equal to

a. 9 b. $2\sqrt{2}-1$

c. $6\sqrt{2}+3$ d. $9-4\sqrt{2}$

23. $p_1 + p_2$ is equal to

a. 2 b. 10

c. 18 d. 5

24. $\overrightarrow{AB}\cdot\overrightarrow{OB}$ is equal to

a. 1 b. 2

c. 3 d. 4

Matrix-Match Type

Each question contains statements given in two columns which have to be matched. Statements (a, b, c, d) in column I have to be matched with statements (p, q, r, s) in column II. If the correct match are a → p, a → s, b → q, c → p, c → q, and d → s, then the correctly bubbled 4 × 4 matrix should be as follows:

	p	q	r	s
a	(p)	(q)	(r)	(s)
b	(p)	(q)	(r)	(s)
c	(p)	(q)	(r)	(s)
d	(p)	(q)	(r)	(s)

1.

Column I	Column II
a. The possible value of a if $\vec{r} = (\hat{i} + \hat{j}) + \lambda(\hat{i} + 2\hat{j} - \hat{k})$ and $\vec{r} = (\hat{i} + 2\hat{j}) + \mu(-\hat{i} + \hat{j} + a\hat{k})$ are not consistent, where l and m are scalars, is	p. −4
b. The angle between vectors $\vec{a} = \lambda\hat{i} - 3\hat{j} - \hat{k}$ and $\vec{b} = 2\lambda\hat{i} + \lambda\hat{j} - \hat{k}$ is acute, whereas vector $\vec{b}$ makes with the axes of coordinates an obtuse angle. Then l may be	q. −2
c. The possible value of a such that $2\hat{i} - \hat{j} + \hat{k}$, $\hat{i} + 2\hat{j} + (1+a)\hat{k}$ and $3\hat{i} + a\hat{j} + 5\hat{k}$ are coplanar is	r. 2
d. If $\vec{A} = 2\hat{i} + \lambda\hat{j} + 3\hat{k}$, $\vec{B} = 2\hat{i} + \lambda\hat{j} + \hat{k}$, $\vec{C} = 3\hat{i} + \hat{j}$ and $\vec{A} + \lambda\vec{B}$ is perpendicular to $\vec{C}$, then $\|2\lambda\|$ is	s. 3

2.

Column I	Column II
a. If $\vec{a}, \vec{b}$, and $\vec{c}$ are three mutually perpendicular vectors where $\vert\vec{a}\vert = \vert\vec{b}\vert = 2$, $\vert\vec{c}\vert = 1$, then $[\vec{a}\times\vec{b}\ \ \vec{b}\times\vec{c}\ \ \vec{c}\times\vec{a}]$ is	**p.** -12
b. If $\vec{a}$ and $\vec{b}$ are two unit vectors inclined at $\pi/3$, then $16[\vec{a}\ \vec{b} + \vec{a}\times\vec{b}\ \vec{b}]$ is	**q.** 0
c. If $\vec{b}$ and $\vec{c}$ are orthogonal unit vectors and $\vec{b}\times\vec{c} = \vec{a}$, then $[\vec{a}+\vec{b}+\vec{c}\ \ \vec{a}+\vec{b}\ \ \vec{b}+\vec{c}]$ is	**r.** 16
d. If $[\vec{x}\ \vec{y}\ \vec{a}] = [\vec{x}\ \vec{y}\ \vec{b}] = [\vec{a}\ \vec{b}\ \vec{c}] = 0$, each vector being a non-zero vectors, then $[\vec{x}\ \vec{y}\ \vec{c}]$ is	**s.** 1

3.

Column I	Column II
a. If $\vert\vec{a}\vert = \vert\vec{b}\vert = \vert\vec{c}\vert$, angle between each pair of vectors is $\frac{\pi}{3}$ and $\vert\vec{a}+\vec{b}+\vec{c}\vert = \sqrt{6}$, then $2\vert\vec{a}\vert$ is equal to	**p.** 3
b. If $\vec{a}$ is a perpendicular to $\vec{b}+\vec{c}$, $\vec{b}$ is perpendicular to $\vec{c}+\vec{a}$, $\vec{c}$ is perpendicular to $\vec{a}+\vec{b}$, $\vert\vec{a}\vert = 2$, $\vert\vec{b}\vert = 3$ and $\vert\vec{c}\vert = 6$. Then $\vert\vec{a}+\vec{b}+\vec{c}\vert - 2$ is equal to	**q.** 2
c. $\vec{a} = 2\hat{i} + 3\hat{j} - \hat{k}$, $\vec{b} = -\hat{i} + 2\hat{j} - 4\hat{k}$, $\vec{c} = \hat{i} + \hat{j} + \hat{k}$ and $\vec{d} = 3\hat{i} + 2\hat{j} + \hat{k}$ then $\frac{1}{7}(\vec{a}\times\vec{b})\cdot(\vec{c}\times\vec{d})$ is equal to	**r.** 4
d. If $\vert\vec{a}\vert = \vert\vec{b}\vert = \vert\vec{c}\vert = 2$ and $\vec{a}\cdot\vec{b} = \vec{b}\cdot\vec{c} = \vec{c}\cdot\vec{a} = 2$, then $[\vec{a}\ \vec{b}\ \vec{c}]\cos 45^\circ$ is equal to	**s.** 5

4. Given two vectors $\vec{a} = -\hat{i} + \hat{j} + 2\hat{k}$ and $\vec{b} = -\hat{i} - 2\hat{j} - \hat{k}$.

Column I	Column II
a. Area of triangle formed by $\vec{a}$, and $\vec{b}$	**p.** $3\sqrt{6}$
b. Area of parallelogram having sides $\vec{a}$ and $\vec{b}$	**q.** $12\sqrt{3}$
c. Area of parallelogram having diagonals $2\vec{a}$ and $4\vec{b}$	**r.** $3\sqrt{3}$
d. Volume of parallelepiped formed by $\vec{a}$ and $\vec{b}$ and $\vec{c} = \hat{i} + \hat{j} + \hat{k}$	**s.** $\frac{3\sqrt{3}}{2}$

5. Given two vectors $\vec{a} = -\hat{i} + 2\hat{j} + 2\hat{k}$ and $\vec{b} = -2\hat{i} + \hat{j} + 2\hat{k}$.

Column I	Column II
a. A vector coplanar with $\vec{a}$ and $\vec{b}$	**p.** $-3\hat{i} + 3\hat{j} + 4\hat{k}$
b. A vector which is perpendicular to both $\vec{a}$ and $\vec{b}$	**q.** $2\hat{i} - 2\hat{j} + 3\hat{k}$
c. A vector which is equally inclined to $\vec{a}$ and $\vec{b}$	**r.** $\hat{i} + \hat{j}$
d. A vector which is forms a triangle with $\vec{a}$ and $\vec{b}$	**s.** $\hat{i} - \hat{j} + 5\hat{k}$

6.

Column I	Column II
a. If $\lvert \vec{a}+\vec{b}\rvert = \lvert \vec{a}+2\vec{b}\rvert$, then angle between $\vec{a}$ and $\vec{b}$ is	p. 90°
b. If $\lvert \vec{a}+\vec{b}\rvert = \lvert \vec{a}-2\vec{b}\rvert$, then angle between $\vec{a}$ and $\vec{b}$ is	q. obtuse
c. If $\lvert \vec{a}+\vec{b}\rvert = \lvert \vec{a}-\vec{b}\rvert$, then angle between $\vec{a}$ and $\vec{b}$ is	r. 0°
d. Angle between $\vec{a}\times\vec{b}$ and a vector perpendicular to the vector $\vec{c}\times(\vec{a}\times\vec{b})$ is	s. acute

7. Volume of parallelepiped formed by vectors $\vec{a}\times\vec{b},\ \vec{b}\times\vec{c}$, and $\vec{c}\times\vec{a}$ is 36 sq. units.

Column I	Column II
a. Volume of parallelepiped formed by vectors $\vec{a}, \vec{b}$, and $\vec{c}$ is	p. 0 sq. units
b. Volume of tetrahedron formed by vectors $\vec{a}, \vec{b}$, and $\vec{c}$ is	q. 12 sq. units
c. Volume of parallelepiped formed by vectors $\vec{a}+\vec{b},\ \vec{b}+\vec{c}$, and $\vec{c}+\vec{a}$	r. 6 sq. units
d. Volume of parallelepiped formed by vectors $\vec{a}-\vec{b},\ \vec{b}-\vec{c}$, and $\vec{c}-\vec{a}$	s. 1 sq. units

Integer Type

1. If $\vec{a}=x\hat{i}+(x-1)\hat{j}+\hat{k}$ and $\vec{b}=(x+1)\hat{i}+\hat{j}+a\hat{k}$ always make an acute angle for all $x \in R$, then the least integral value of a is__________.

2. If $\vec{a}\cdot\hat{i}=\vec{a}\cdot(\hat{i}+\hat{j})=\vec{a}\cdot(\hat{i}+\hat{j}+\hat{k})$, then the value of $\lvert\vec{a}\rvert$ is__________.

3. A particle acted by constant forces $4\hat{i}+\hat{j}-3\hat{k}$ and $3\hat{i}+\hat{j}-\hat{k}$ is displaced from the point $\hat{i}+2\hat{j}+3\hat{k}$ to the point $5\hat{i}+4\hat{j}+\hat{k}$. If w is the total work done by the force units, then the value of $w/8$ is__________.

4. If $\lvert\vec{a}\rvert=\lvert\vec{b}\rvert=\lvert\vec{a}+\vec{b}\rvert=1$, then the value of $\lvert\vec{a}-\vec{b}\rvert^2$ is__________.

5. If $\vec{a}, \vec{b}$ and $\vec{c}$ are vectors such that $\lvert\vec{a}\rvert=3$, $\lvert\vec{b}\rvert=4$, and $\lvert\vec{c}\rvert=5$, and $(\vec{a}+\vec{b})$ is perpendicular to $\vec{c}$, $(\vec{b}+\vec{c})$ is perpendicular to $\vec{a}$, and $(\vec{c}+\vec{a})$ is perpendicular to $\vec{b}$, then the value of $[\lvert\vec{a}+\vec{b}+\vec{c}\rvert]$, where $[\cdot]$ represents the greatest integer function, is__________.

6. Let $\vec{u}=\hat{i}+\hat{j}$, $\vec{v}=\hat{i}-\hat{j}$, and $\vec{w}=\hat{i}+2\hat{j}+3\hat{k}$. If $\hat{n}$ is a unit vector such that $\vec{u}\cdot\hat{n}=0$ and $\vec{v}\cdot\hat{n}=0$, then the value of $\lvert\vec{w}\cdot\hat{n}\rvert$ is__________.

7. Let $\vec{A}$ and $\vec{B}$ be two non-parallel unit vectors in a plane. If $\left(\alpha\vec{A}+\vec{B}\right)$ bisects the internal angle between $\vec{A}$ and $\vec{B}$, then the value of α is__________.

8. If $\vec{a}$ and $\vec{b}$ are unit vectors, then the greatest value of $(\lvert\vec{a}+\vec{b}\rvert+\lvert\vec{a}-\vec{b}\rvert)^2$ is__________.

9. If $\lvert\vec{a}\rvert=2$, then the value of $\lvert\vec{a}\times\hat{i}\rvert^2+\lvert\vec{a}\times\hat{j}\rvert^2+\lvert\vec{a}\times\hat{k}\rvert^2$ is__________.

10. A, B, C, and D are any four points in space. If $\lvert(\overrightarrow{AB}\times\overrightarrow{CD})+(\overrightarrow{BC}\times\overrightarrow{AD})+(\overrightarrow{CA}\times\overrightarrow{BD})\rvert=k$(area of ΔABC), then the value of k is__________.

11. Let $\vec{a}, \vec{b}$, and $\vec{c}$ be three vectors such that $\vec{a}\neq 0$ and $\vec{a}\times\vec{b}=2\vec{a}\times\vec{c}$, $\lvert\vec{a}\rvert=\lvert\vec{c}\rvert=1$, $\lvert\vec{b}\rvert=4$, and $\lvert\vec{b}\times\vec{c}\rvert=\sqrt{15}$. If $\vec{b}-2\vec{c}=\lambda\vec{a}$, then the value of $\lvert\lambda\rvert$ is__________.

12. If $\lvert\vec{a}\rvert=2$, $\lvert\vec{b}\rvert=5$, and $\lvert\vec{a}\times\vec{b}\rvert=8$, then the value of $\vec{a}\cdot\vec{b}$ is__________.

13. Let $\vec{a}, \vec{b}, \vec{c}$ be three non-zero vectors such that $\vec{a}+\vec{b}+\vec{c}=\vec{0}$ and $\lambda\,\vec{b}\times\vec{a}+\vec{b}\times\vec{c}+\vec{c}\times\vec{a}=\vec{0}$, then the value of λ is__________.

14. If $(\vec{a}\times\vec{b})^2+(\vec{a}\cdot\vec{b})^2=144$ and $|\vec{a}|=4$, then the value of $|\vec{b}|$ is__________.

15. Given $|\vec{a}|=|\vec{b}|=1$ and $|\vec{a}+\vec{b}|=\sqrt{3}$. If $\vec{c}$ is a vector such that $\vec{c}-\vec{a}-2\vec{b}=3(\vec{a}\times\vec{b})$, then the value of $[\vec{c}\cdot\vec{b}]$, where $[\cdot]$ represents the greatest integer function, is__________.

16. If $\vec{a}$ and $\vec{b}$ are two vectors such that $|\vec{a}\times\vec{b}|=2$, then the value of $[\vec{a}\ \vec{b}\ \vec{a}\times\vec{b}]$ is

17. Let G_1, G_2, and G_3 be the centroids of the trianglular faces *OBC*, *OCA*, and *OAB* of a tetrahedron *OABC*. If V_1 denotes the volume of the tetrahedron *OABC* and V_2 that of the parallelepiped with OG_1, OG_2, and OG_3 as three concurrent edges, then the value of $4V_1/V_2$ is__________.

18. Let *O* be an interior point of ΔABC such that $\overrightarrow{OA}+2\overrightarrow{OB}+3\overrightarrow{OC}=\vec{0}$, then the ratio of the area of ΔABC to the area of ΔAOC is__________.

19. If $\vec{e}_1, \vec{e}_2, \vec{e}_3$ and $\vec{E}_1, \vec{E}_2, \vec{E}_3$ are two set of vectors such that $\vec{e}_i\cdot\vec{E}_j=1$, if $i=j$ and $\vec{e}_i\cdot\vec{E}_j=0$, if $i\neq j$, then the value of $[\vec{e}_1\ \vec{e}_2\ \vec{e}_3]\ [\vec{E}_1\ \vec{E}_2\ \vec{E}_3]$ is__________.

20. Let $\overrightarrow{OA}=\vec{a}$, $\overrightarrow{OB}=10\vec{a}+2\vec{b}$, and $\overrightarrow{OC}=\vec{b}$, where *O*, *A*, and *C* are non-collinear points. Let *P* denote the area of the quadrilateral *OABC*, and let *q* denote the area of the parallelogram with *OA* and *OC* as adjacent sides. If $p=kq$, then the value of *k* is__________.

21. Let $\vec{a}, \vec{b}$, and $\vec{c}$ be three non-coplanar vectors, and $\vec{p}, \vec{q}$, and $\vec{r}$ be vectors defined by the relations $\vec{p}=(\vec{b}\times\vec{c})/[\vec{a}\,\vec{b}\,\vec{c}]$, $\vec{q}=(\vec{c}\times\vec{a})/[\vec{a}\,\vec{b}\,\vec{c}]$, and $\vec{r}=(\vec{a}\times\vec{b})/[\vec{a}\,\vec{b}\,\vec{c}]$. Then the value of the expression $(\vec{a}+\vec{b})\cdot\vec{p}+(\vec{b}+\vec{c})\cdot\vec{q}+(\vec{c}+\vec{a})\cdot\vec{r}$ is__________.

22. Let $\vec{a}=2i+j-2k$ and $b=i+j$. If *c* is a vector such that $\vec{a}\cdot\vec{c}=|\vec{c}|$, $|\vec{c}-\vec{a}|=2\sqrt{2}$, and the angle between $\vec{a}\times\vec{b}$ and $\vec{c}$ is 30°, then the value of $|2(\vec{a}\times\vec{b})\times\vec{c}|$ is__________.

23. If $\hat{a}, \hat{b}$, and $\hat{c}$ are unit vectors, then $|\hat{a}-\hat{b}|^2+|\hat{b}-\hat{c}|^2+|\hat{c}-\hat{a}|^2$ does NOT exceed__________.

24. Let $\vec{V}=2\vec{i}+\vec{j}-\vec{k}$ and $\vec{W}=\vec{i}+3\vec{k}$. If $\vec{U}$ is a unit vector, then the maximum integral value of the scalar triple product $\left[\vec{U}\ \vec{V}\ \vec{W}\right]$ is__________.

ANSWERS AND SOLUTIONS

Objective Type

1. c. If $\vec{x}=\vec{y} \Rightarrow \hat{a}\cdot\vec{x}=\hat{a}\cdot\vec{y}$. This equality must hold for any arbitrary $\hat{a}$

2. d. $\vec{a}\times(\vec{a}\times\vec{b})=\vec{c} \Rightarrow |\vec{a}||\vec{a}\times\vec{b}|=|\vec{c}|$

$[\because \vec{a}\perp(\vec{a}\times\vec{b})]$

$1\,(1\times 5)\sin\theta=3 \Rightarrow \sin\theta=\frac{3}{5} \Rightarrow \tan\theta=\frac{3}{4}$.

3. c. $|\vec{a}+\vec{b}+\vec{c}|^2=6$

$\Rightarrow |\vec{a}|^2+|\vec{b}|^2+|\vec{c}|^2+2(\vec{a}\cdot\vec{b}+\vec{b}\cdot\vec{c}+\vec{c}\cdot\vec{a})=6$

$\Rightarrow |\vec{a}|=|\vec{b}|=|\vec{c}|$ and $\vec{a}\cdot\vec{b}=|\vec{a}||\vec{b}|\cos\frac{\pi}{3}$

i.e., $\vec{a}\cdot\vec{b}=\frac{1}{2}|\vec{a}|^2$

$\therefore 3|\vec{a}|^2+3|\vec{a}|^2=6$

$\Rightarrow |\vec{a}|^2 \Rightarrow |\vec{a}|=1$

4. b. Let $\vec{\alpha}=\frac{\vec{a}}{|\vec{a}|}+\frac{\vec{b}}{|\vec{b}|}+\frac{\vec{c}}{|\vec{c}|}$

Since $\vec{a},\vec{b},$ and $\vec{c}$ are mutually perpendicular vectors, if α makes angles θ, ϕ ψ with $\vec{a},\vec{b},$ and $\vec{c}$, respectively, then

$\cos\theta=\frac{\vec{\alpha}\cdot\vec{a}}{|\vec{\alpha}||\vec{a}|}=\frac{1}{|\vec{a}|}$

$\cos\phi=\frac{1}{|\vec{\alpha}|}$, $\cos\psi=\frac{1}{|\vec{\alpha}|}$

$\therefore \theta=\phi=\psi$

5. c. $\vec{r}\times\vec{a}=\vec{b}\times\vec{a} \Rightarrow (\vec{r}-\vec{b})\times\vec{a}=0$

$\vec{r}\times\vec{b}=\vec{a}\times\vec{b} \Rightarrow (\vec{r}-\vec{a})\times\vec{b}=0$

If $\vec{r}=x\hat{i}+y\hat{j}+z\hat{k}$, then

$$\begin{vmatrix}\hat{i} & \hat{j} & \hat{k}\\ x-2 & y & z+1\\ 1 & 1 & 0\end{vmatrix}=0 \text{ and } \begin{vmatrix}\hat{i} & \hat{j} & \hat{k}\\ x-1 & y-1 & z\\ 2 & 0 & -1\end{vmatrix}=0$$

$\Rightarrow z+1=0, x-y=2$ and $y-1=0, x-1+2z=0$

$\Rightarrow x=3,\ y=1, z=-1$

6. d. $|\vec{a}\cdot\vec{b}|=|\vec{a}\times\vec{b}|$

$\Rightarrow |\vec{a}||\vec{b}|\,|\cos\theta|=|\vec{a}||\vec{b}|\,|\sin\theta|$

[where θ is the angle between $\vec{a}$ and $\vec{b}$]

$\Rightarrow |\cos\theta|=|\sin\theta|$

$\Rightarrow \theta=\frac{\pi}{4}$ or $\frac{3\pi}{4}$ (as $0\le\theta\le\pi$).

But $\vec{a}\cdot\vec{b}<0 \therefore \theta=\frac{3\pi}{4}$

7. c. $|\vec{a}+\vec{b}+\vec{c}|^2=1$

$\Rightarrow |\vec{a}|^2+|\vec{b}|^2+|\vec{c}|^2+2|\vec{a}||\vec{b}|\cos\theta_1$

$+2|\vec{b}||\vec{c}|\cos\theta_2+2|\vec{c}||\vec{a}|\cos\theta_3=1$

$\Rightarrow \cos\theta_1+\cos\theta_2+\cos\theta_3=-1$

$\Rightarrow$ One of θ_1, θ_2 and θ_3 should be obtuse angle.

8. b. $|\vec{a}\times\vec{b}-\vec{a}\times\vec{c}|^2\ |\vec{a}\times(\vec{b}-\vec{c})|^2$

$=|\vec{a}|^2|\vec{b}-\vec{c}|^2-(\vec{a}\cdot(\vec{b}-\vec{c}))^2=|\vec{b}-\vec{c}|^2$

$=|\vec{b}|^2+|\vec{c}|^2-2|\vec{b}||\vec{c}|\cos\frac{\pi}{3}=1$

9. c. $R(\vec{r})$ moves on PQ.

$P(\vec{p})$ —— $R(\vec{r})$ —— $Q(\vec{q})$

10. b. $|\vec{AC}\times\vec{BD}|=2|\vec{AB}\times\vec{AD}|$

$$=2\begin{vmatrix}\hat{i} & \hat{j} & \hat{k}\\ 2 & 4 & -5\\ 1 & 2 & 3\end{vmatrix}$$

$=|2[\hat{i}(12+10)-\hat{j}(6+5)+\hat{k}$

$(4-4)]|$

$=|2[22\hat{i}-11\hat{j}]|$

$=22|[2\hat{i}-\hat{j}]|$

$=22\times\sqrt{5}$

11. b. $\vec{r}\times\vec{a}-\vec{b}\times\vec{a}=0$ and $\vec{r}\times\vec{b}+\vec{b}\times\vec{a}=0$

Adding, we get $\vec{r}\times(\vec{a}+\vec{b})=0$

But as we are given $\vec{a}\neq\lambda\vec{b}$, $\vec{r}=\vec{a}+\vec{b}$

12. c. $\vec{a}\times\vec{b}$ is a vector perpendicular to the plane containing $\vec{a}$ and $\vec{b}$. Similarly, $\vec{c}\times\vec{d}$ is a vector perpendicular to the plane containing $\vec{c}$ and $\vec{d}$. Thus, the two planes will be parallel if their normals, i.e., $\vec{a}\times\vec{b}$ and $\vec{c}\times\vec{d}$ are parallel.

$\Rightarrow (\vec{a}\times\vec{b})\times(\vec{c}\times\vec{d})=\vec{0}$

13. d. Let $\vec{r} \neq \vec{0}$. Then $\vec{r}\cdot\vec{a} = \vec{r}\cdot\vec{b} = \vec{r}\cdot\vec{c} = 0$

$\Rightarrow \vec{a}, \vec{b}$, and $\vec{c}$ are coplanar, which is a contradiction.

Therefore, $\vec{r} = \vec{0}$.

14. c. $\vec{a}\times(\hat{i}\times 2\hat{j}+\hat{k}) = \hat{i}-\hat{k} = [\hat{j}\times(\hat{i}+2\hat{j}+\hat{k})]$

$\Rightarrow (\vec{a}-\hat{j})\times(\hat{i}+2\hat{j}+\hat{k}) = \vec{0}$

$\Rightarrow \vec{a}-\hat{j} = \lambda(\hat{i}+2\hat{j}+\hat{k})$

$\Rightarrow \vec{a} = \lambda\hat{i}+(2\lambda+1)\hat{j}+\lambda\hat{k}, \lambda\in R$

15. a. $(3\vec{a}-5\vec{b})\cdot(2\vec{a}+\vec{b}) = 0$

$\Rightarrow 6|\vec{a}|^2 - 5|\vec{b}|^2 = 7\vec{a}\cdot\vec{b}$

Also, $(\vec{a}+4\vec{b})\cdot(\vec{b}-\vec{a}) = 0$

$\Rightarrow -|\vec{a}|^2 + 4|\vec{b}|^2 = 3\vec{a}\cdot\vec{b}$

$\Rightarrow \frac{6}{7}|\vec{a}|^2 - \frac{5}{7}|\vec{b}|^2 = -\frac{1}{3}|\vec{a}|^2 + \frac{4}{3}|\vec{b}|^2$

$\Rightarrow 25|\vec{a}|^2 = 43|\vec{b}|^2$

$\Rightarrow 3\vec{a}\cdot\vec{b} = -|\vec{a}|^2 + 4|\vec{b}|^2 = \frac{57}{25}|\vec{b}|^2$

$\Rightarrow 3|\vec{a}||\vec{b}|\cos\theta = \frac{57}{25}|\vec{b}|^2$

$\Rightarrow 3\sqrt{\frac{43}{25}}|\vec{b}|^2\cos\theta = \frac{57}{25}|\vec{b}|^2$

$\Rightarrow \cos\theta = \frac{19}{5\sqrt{43}}$

16. a. Let l, m, and n be the direction cosines of the required vector.
Then, $l = m$ (given). Therefore

Required vector $\vec{r} = l\hat{i}+m\hat{j}+n\hat{k} = l\hat{i}+l\hat{j}+n\hat{k}$

Now, $l^2+m^2+n^2 = 1 \Rightarrow 2l^2+n^2 = 1$ (1)

Since, $\hat{r}$ is perpendicular to $-\hat{i}+2\hat{j}+2\hat{k}$,

$\vec{r}\cdot(-\hat{i}+2\hat{j}+2\hat{k}) = 0 \Rightarrow -l+2l+2n = 0$

$\Rightarrow l+2n = 0$ (2)

From (1) and (2), we get $n = \mp\frac{1}{3}, l = \pm\frac{2}{3}$

Hence, required vector $\vec{r} = \frac{1}{3}(\pm 2\hat{i}\pm 2\hat{j}\mp\hat{k})$

$= \pm\frac{1}{3}(2\hat{i}+2\hat{j}-\hat{k})$

17. d. The angle between $\vec{a}$ and $\vec{b}$ is obtuse. Therefore, $\vec{a}\cdot\vec{b} < 0$

$\Rightarrow 14x^2 - 8x + x < 0$

$\Rightarrow 7x(2x-1) < 0$

$\Rightarrow 0 < x < 1/2$ (1)

The angle between $\vec{b}$ and the z-axis is acute and less than $\pi/6$. Therefore,

$\frac{\vec{b}\cdot\vec{k}}{|\vec{b}||\vec{k}|} > \cos\pi/6$ ($\because \theta < \pi/6 \Rightarrow \cos\theta > \cos\pi/6$)

$\Rightarrow \frac{x}{\sqrt{x^2+53}} > \frac{\sqrt{3}}{2}$

$\Rightarrow 4x^2 > 3x^2 + 159$

$\Rightarrow x^2 > 159$

$\Rightarrow x > \sqrt{159}$ or $x < -\sqrt{159}$ (2)

Clearly, (1) and (2) cannot hold together.

18. c.

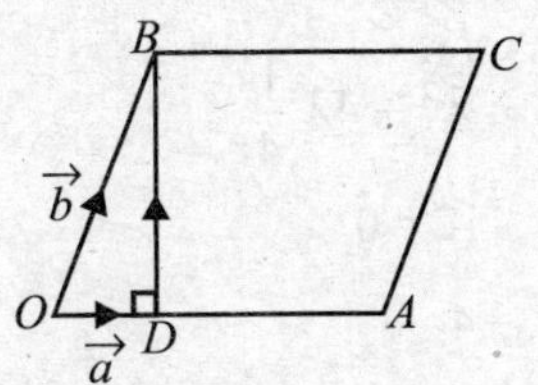

Fig. 2.2

Let $\overrightarrow{OD} = t\vec{a}$

$\therefore \overrightarrow{DB} = \vec{b} - t\vec{a}$ [$\because \overrightarrow{DB}\perp\overrightarrow{OA}$]

$\therefore (\vec{b}-t\vec{a})\cdot\vec{a} = 0$

$\Rightarrow t = \frac{\vec{b}\cdot\vec{a}}{|\vec{a}|^2}$

$\therefore \overrightarrow{DB} = \vec{b} - \frac{(\vec{b}.\vec{a})\vec{a}}{|\vec{a}|^2}$

19. d. $(3\vec{a}+\vec{b})\cdot(\vec{a}-4\vec{b})$

$= 3|\vec{a}|^2 - 11\vec{a}\cdot\vec{b} - 4|\vec{b}|^2$

$= 3\times 36 - 11\times 6\times 8\cos\pi - 4\times 64 > 0$

Therefore, the angle between $\vec{a}$ and $\vec{b}$ is acute.

The longer diagonal is given by

$\vec{\alpha} = (3\vec{a}+\vec{a}) + (\vec{a}-4\vec{b}) = 4\vec{a}-3\vec{b}$

Now, $|\vec{\alpha}|^2 = |4\vec{a}-3\vec{b}|^2 = 16|\vec{a}|^2 + 9|\vec{b}|^2 - 24\vec{a}\cdot\vec{b}$

$= 16\times 36 + 9\times 64 - 24\times 6\times 8\cos\pi$

$= 16\times 144$

$\Rightarrow |4\vec{a}-3\vec{b}| = 48$

20. b. $\vec{c} = m\vec{a} + n\vec{b} + p(\vec{a}\times\vec{b})$

Taking dot product with $\vec{a}$ and $\vec{b}$, we have

$m = n = \cos\theta$

$\Rightarrow |\vec{c}| = |\cos\theta\,\vec{a} + \cos\theta\,\vec{b} + p\,(\vec{a}\times\vec{b})| = 1$

Squaring both sides, we get

$\cos^2\theta + \cos^2\theta + p^2 = 1$

$\Rightarrow \cos\theta = \pm\dfrac{\sqrt{1-p^2}}{\sqrt{2}}$

Now $-\dfrac{1}{\sqrt{2}} \le \cos\theta \le \dfrac{1}{\sqrt{2}}$ [For real value of θ]

$\therefore \dfrac{\pi}{4} \le \cos\theta \le \dfrac{3\pi}{4}$

21. a. $\vec{b} - 2\vec{c} = \lambda\vec{a}$

$\Rightarrow \vec{b} = 2\vec{c} + \lambda\vec{a}$

$\Rightarrow |\vec{b}|^2 = |2\vec{c} + \lambda\vec{a}|^2$

$\Rightarrow 16 = 4|\vec{c}|^2 + \lambda^2|\vec{a}|^2 + 4\lambda\,\vec{a}\cdot\vec{c}$

$\Rightarrow 16 = 4 + \lambda^2 + 4\lambda\dfrac{1}{4}$

$\Rightarrow \lambda^2 + \lambda - 12 = 0$

$\Rightarrow \lambda = 3, -4$

22. a. A vector perpendicular to the plane of O, P, and Q is $\overrightarrow{OP} \times \overrightarrow{OQ}$.

Now, $\overrightarrow{OP}\times\overrightarrow{OQ} = \begin{vmatrix} \hat{i} & \hat{j} & \hat{k} \\ 4 & 1 & \lambda \\ 2 & -1 & \lambda \end{vmatrix} = 2\lambda\hat{i} - 2\lambda\hat{j} - 6\hat{k}.$

Therefore, $\hat{i} - \hat{j} + 6\hat{k}$ is parallel to $2\lambda\hat{i} - 2\lambda\hat{j} - 6\hat{k}$. Hence,

$\dfrac{1}{2\lambda} = \dfrac{-1}{-2\lambda} = \dfrac{6}{-6}$

$\lambda = -\dfrac{1}{2}$

23. a. $\hat{a}\cdot\hat{b} = \cos\alpha,\ \hat{b}\cdot\hat{c} = \cos\beta,\ \hat{c}\cdot\hat{a} = \cos\gamma$

Now $|\hat{a} + \hat{b} + \hat{c}| = 1$

$\therefore (\hat{a} + \hat{b} + \hat{c})^2 = 1$

$\Rightarrow |\hat{a}|^2 + |\hat{b}|^2 + |\hat{c}|^2 + 2(\hat{a}\cdot\hat{b} + \hat{b}\cdot\hat{c} + \hat{c}\cdot\hat{a}) = 1$

$\Rightarrow 1 + 1 + 1 + 2(\cos\alpha + \cos\beta + \cos\gamma) = 1$

$\Rightarrow \cos\alpha + \cos\beta + \cos\gamma = -1$

24. c. $(\hat{i}\times\hat{j})\cdot\vec{c} \le |\hat{i}\times\hat{j}|\,|\vec{c}|\cos\dfrac{\pi}{6}$

$\Rightarrow \dfrac{\sqrt{3}}{2} \le (\hat{i}\times\hat{j})\cdot\vec{c} \le \dfrac{\sqrt{3}}{2}$

25. b.

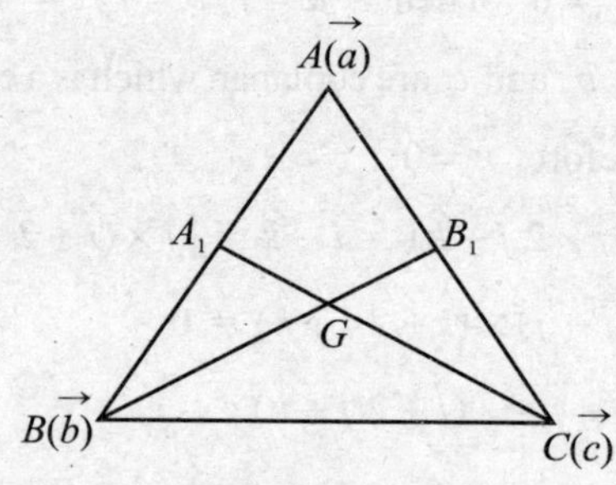

Fig. 2.3

Let P.V. of A, B, and C be $\vec{0}$, $\vec{b}$, and $\vec{c}$, respectively. Therefore,

$\vec{G} = \dfrac{\vec{b} + \vec{c}}{3}$

$\vec{A_1} = \dfrac{\vec{b}}{2},\ \vec{B_1} = \dfrac{\vec{c}}{2}$

$\Delta_{AB_1G} = \dfrac{1}{2}|\overrightarrow{AG}\times\overrightarrow{AB_1}| = \dfrac{1}{2}\left|\dfrac{\vec{b}+\vec{c}}{3}\times\left(\dfrac{\vec{c}}{2}\right)\right|$

$= \dfrac{1}{12}|\vec{b}\times\vec{c}|$

$\Delta_{AA_1G} = \dfrac{1}{2}|\overrightarrow{AG}\times\overrightarrow{AA_1}| = \dfrac{1}{2}\left|\dfrac{\vec{b}+\vec{c}}{3}\times\left(\dfrac{\vec{b}}{2}\right)\right|$

$= \dfrac{1}{12}|\vec{b}\times\vec{c}|$

$\Rightarrow \Delta_{GA_1AB_1} = \dfrac{1}{6}|\vec{b}\times\vec{c}| = \dfrac{1}{3}\times\dfrac{1}{2}|\vec{b}\times\vec{c}| = \dfrac{1}{3}\Delta_{ABC}$

$\Rightarrow \dfrac{\Delta}{\Delta_1} = 3$

26. a. Vectors $\vec{a}, \vec{b}, \vec{c}$, and $\vec{d}$ are coplanar. Therefore,

$\sin\alpha + 2\sin 2\beta + 3\sin 3\gamma = 1.$

Now $|\sin\alpha + 2\sin 2\beta + 3\sin 3\gamma|$

$\le \sqrt{1+4+9}\times\sqrt{\sin^2\alpha + \sin^2 2\beta + \sin^2 3\gamma}$

$\Rightarrow \sin^2\alpha + \sin^2 2\beta + \sin^2 3\gamma \ge \dfrac{1}{14}$

27. c. $1 + 9(\vec{a}\cdot\vec{b})^2 - 6(\vec{a}\cdot\vec{b}) + 4|\vec{a}|^2 + |\vec{b}|^2 + 9|\vec{a}\times\vec{b}|^2 + 4\vec{a}\cdot\vec{b} = 47$

$\Rightarrow 1 + 4 + 4 + 36 - 4\cos\theta = 47$

$\Rightarrow \cos\theta = -\dfrac{1}{2}$

$\Rightarrow$ Angle between $\vec{a}$ and $\vec{b}$ is $\dfrac{2\pi}{3}$.

28. c. $|2(\vec{a}\times\vec{b})|+3|\vec{a}\cdot\vec{b}|\le\sqrt{|\vec{a}\times\vec{b}|^2+|\vec{a}\cdot\vec{b}|^2}\times\sqrt{4+9}$

$\le 6\sqrt{13}$

29. b. $|\vec{a}+\vec{b}+3\vec{c}|^2=16$

$\Rightarrow|\vec{a}|^2+|\vec{b}|^2+9|\vec{c}|^2+2\cos\theta_1+6\cos\theta_2+6\cos\theta_3=16,\ \theta_3\in[\pi/6,2\pi/3]$

$\Rightarrow 2\cos\theta_1+6\cos\theta_2=5-6\cos\theta_3$

$\Rightarrow(\cos\theta_1+3\cos\theta_2)_{\max}=4$

30. c. $|\vec{a}\times\vec{r}|=|\vec{c}|$

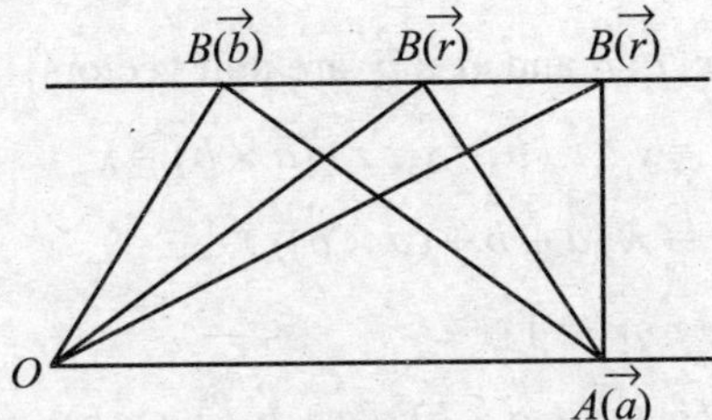

Fig. 2.4

Triangles on the same base and between the same parallel will have the same area.

31. c. Given $\vec{v}\cdot\vec{u}=\vec{w}\cdot\vec{u}$

and $\vec{v}\perp\vec{w}\Rightarrow\vec{v}\cdot\vec{w}=0$

Now, $|\vec{u}-\vec{v}+\vec{w}|^2$

$=|\vec{u}|^2+|\vec{v}|^2+|\vec{w}|^2-2\vec{u}\cdot\vec{v}-2\vec{w}\cdot\vec{v}+2\vec{u}\cdot\vec{w}$

$=1+4+9$

So $|-\vec{u}-\vec{v}+\vec{w}|=\sqrt{14}$

32. b. We have

$\vec{p}\cdot\vec{q}=0$

$\Rightarrow(5\vec{a}-3\vec{b})\cdot(-\vec{a}-2\vec{b})=0$

$\Rightarrow 6|\vec{b}|^2-5|\vec{a}|^2-7\vec{a}\cdot\vec{b}=0$ (1)

Also $\vec{r}\cdot\vec{s}=0$

$\Rightarrow(-4\vec{a}-\vec{b})\ (-\vec{a}+\vec{b})$

$\Rightarrow 4|\vec{a}|^2-|\vec{b}|^2-3\vec{a}\cdot\vec{b}=0$ (2)

Now $\vec{x}=\frac{1}{3}(\vec{p}+\vec{r}+\vec{s})$

$=\frac{1}{3}(5\vec{a}-3\vec{b}-4\vec{a}-\vec{b}-\vec{a}+\vec{b})=-\vec{b}$

and $\vec{y}=\frac{1}{5}(\vec{r}+\vec{s})=\frac{1}{5}(-5\vec{a})=-\vec{a}$

Angle between $\vec{x}$ and y, i.e.,

$\cos\theta=\frac{\vec{x}\cdot\vec{y}}{|\vec{x}||\vec{y}|}=\frac{\vec{a}\cdot\vec{b}}{|\vec{a}||\vec{b}|}$ (3)

From (1) and (2), $|\vec{a}|=\sqrt{\frac{25}{19}}\sqrt{\vec{a}\cdot\vec{b}}$ and $|\vec{b}|=\sqrt{\frac{43}{19}}\sqrt{\vec{a}\cdot\vec{b}}$. Therefore

$|\vec{a}||\vec{b}|=\frac{\sqrt{25\times43}}{19}\vec{a}\cdot\vec{b}$

$\theta=\cos^{-1}\left(\frac{19}{5\sqrt{43}}\right)$

33. c. Let $\vec{B}=b_1\hat{i}+b_2\hat{j}+b_3\hat{k}$

$\vec{A}\cdot\vec{B}=3\Rightarrow b_1+b_2+b_3=3$ (1)

$\vec{A}\times\vec{B}=\vec{C}\Rightarrow\begin{vmatrix}\hat{i}&\hat{j}&\hat{k}\\1&1&1\\b_1&b_2&b_3\end{vmatrix}=0\hat{i}+\hat{j}+\hat{k}$

$\Rightarrow b_3=b_2;\ b_2-b_1=-1;\ b_1-b_3=1$ (2)

Solving (1) and (2), we have $\vec{B}\equiv\left(\frac{5}{3},\frac{2}{3},\frac{2}{3}\right)$

34. b. Since, $\overrightarrow{OA}=\hat{i}+\hat{j}+\hat{k}$

$\overrightarrow{OB}=\hat{i}+5\hat{j}-\hat{k}$

$\overrightarrow{OC}=2\hat{i}+3\hat{j}+5\hat{k}$

$a=BC=|\overrightarrow{BC}|=|\overrightarrow{OC}-\overrightarrow{OB}|=|\hat{i}-2\hat{j}+6\hat{k}|=\sqrt{41}$

$b=CA=|\overrightarrow{CA}|=|\overrightarrow{OA}-\overrightarrow{OC}|=|-\hat{i}-2\hat{j}-4\hat{k}|=\sqrt{21}$

and $c=AB=|\overrightarrow{AB}|=|\overrightarrow{OB}-\overrightarrow{OA}|=|0\hat{i}+4\hat{j}-2\hat{k}|$

$=\sqrt{20}$

Since $a>b>c$, A is the greatest angle,

$\cos A=\frac{b^2+c^2-a^2}{2bc}=\frac{21+20-41}{2\times\sqrt{21}\times\sqrt{20}}=0$

$\therefore\angle A=90^\circ$

35. b. $\vec{a}+\vec{b}=\lambda\vec{c}$ (1)

and $\vec{b}+\vec{c}=\mu\vec{a}$ (2)

$\therefore(\lambda\vec{c}-\vec{a})+\vec{c}=\mu\vec{a}$ [Putting $\vec{b}=\lambda\vec{c}-\vec{a}$]

$\Rightarrow(\lambda+1)\vec{c}=(\mu+1)\vec{a}$

$\Rightarrow\lambda=\mu=-1$

$\Rightarrow\vec{a}+\vec{b}+\vec{c}=0$

$\Rightarrow|\vec{a}|^2+|\vec{b}|^2+|\vec{c}|^2+2(\vec{a}\cdot\vec{b}+\vec{b}\cdot\vec{c}+\vec{c}\cdot\vec{a})=0$

$\Rightarrow\vec{a}\cdot\vec{b}+\vec{b}\cdot\vec{c}+\vec{c}\cdot\vec{a}=-3$

36. d. $\vec{0}=(\vec{a}+\vec{b})\cdot(2\vec{a}+3\vec{b})\times(3\vec{a}-2\vec{b})$

$=(\vec{a}+\vec{b})\cdot(-4\vec{a}\times\vec{b}-9\vec{a}\times\vec{b})$

$=-13(\vec{a}+\vec{b})\cdot(\vec{a}\times\vec{b})$,

which is true for all values of $\vec{a}$ and $\vec{b}$.

37. c. We have

$\overrightarrow{AB}\cdot\overrightarrow{AC}+\overrightarrow{BC}\cdot\overrightarrow{BA}+\overrightarrow{CA}\cdot\overrightarrow{CB}$

$=(AB)(AC)\cos\theta+(BC)(BA)\sin\theta+0$

$=AB(AC\cos\theta+BC\sin\theta)$

$=AB\left(\frac{(AC)^2}{AB}+\frac{(BC)^2}{AB}\right)$

$=AC^2+BC^2=AB^2=p^2$

38. c. $\vec{a}_1=(\vec{a}\cdot\hat{b})\hat{b}=\frac{(\vec{a}\cdot\vec{b})\vec{b}}{|\vec{b}|^2}$.

$\Rightarrow \vec{a}_2=\vec{a}-\vec{a}_1=\vec{a}-\frac{(\vec{a}\cdot\vec{b})\vec{b}}{|\vec{b}|^2}$

Thus, $\vec{a}_1\times\vec{a}_2=\frac{(\vec{a}\cdot\vec{b})\vec{b}}{|\vec{b}|^2}\times\left(\vec{a}-\frac{(\vec{a}\cdot\vec{b})\vec{b}}{|\vec{b}|^2}\right)$

$=\frac{(\vec{a}\cdot\vec{b})(\vec{b}\times\vec{a})}{|\vec{b}|^2}$

39. b. Let the required vector be $\vec{r}$. Then $\vec{r}=x_1\vec{b}+x_2\vec{c}$

and $\vec{r}\cdot\vec{a}=\sqrt{\frac{2}{3}}\ (|\vec{a}|)=2$

Now, $\vec{r}\cdot\vec{a}=x_1\vec{a}\cdot\vec{b}+x_2\vec{a}\cdot\vec{c}$

$\Rightarrow 2=x_1(2-2-1)+x_2(2-1-2)\Rightarrow x_1+x_2=-2$

$\Rightarrow \vec{r}=x_1(\hat{i}+2\hat{j}-\hat{k})+x_2(\hat{i}+\hat{j}-2\hat{k})$

$=\hat{i}(x_1+x_2)+\hat{j}(2x_1+x_2)-\hat{k}(2x_2+x_1)$

$=-2\hat{i}+\hat{j}(x_1-2)-\hat{k}(-4-x_1)$, where $x_1\in R$.

40. a. Let P.V. of P, A, B, and C be $\vec{p}$, $\vec{a}$, $\vec{b}$, and $\vec{c}$, respectively, and $O(\vec{0})$ be the circumcentre of equilateral triangle ABC. Then

$|\vec{p}|=|\vec{b}|=|\vec{a}|=|\vec{c}|=\frac{l}{\sqrt{3}}$

Now $|\overrightarrow{PA}|^2=|\vec{a}-\vec{p}|^2=|\vec{a}|^2+|\vec{p}|^2-2\vec{p}\cdot\vec{a}$

Similarly, $|\overrightarrow{PB}|^2=|\vec{b}|^2+|\vec{p}|^2-2\vec{p}\cdot\vec{b}$

and $|\overrightarrow{PC}|^2=|\vec{c}|^2+|\vec{p}|^2-2\vec{p}\cdot\vec{c}$

$\Rightarrow \Sigma|\overrightarrow{PA}|^2=6\times\frac{l^2}{3}-2\vec{p}\cdot(\vec{a}+\vec{b}+\vec{c})=2l^2$

as $(\vec{a}+\vec{b}+\vec{c}/3=\vec{0})$

41. d. $\vec{a}\times\vec{b}=2\vec{a}\times\vec{c}$

$\vec{a}\times\vec{b}-2\vec{a}\times\vec{c}=\vec{0}$

$\vec{a}\times(\vec{b}-2\vec{c})=\vec{0}$

$\vec{a}$ and $\vec{b}-2\vec{c}$ are collinear.

$\Rightarrow \vec{b}-2\vec{c}=\lambda\vec{a}$

$|\vec{b}-2\vec{c}|^2=\lambda^2|\vec{a}|^2$

$|\vec{b}|^2+4|\vec{c}|^2-4\vec{b}\cdot\vec{c}=\lambda^2|\vec{a}|^2$

$|\vec{b}|^2+4|\vec{c}|^2-4\vec{b}\cdot\vec{c}=\lambda^2|\vec{a}|^2$

$16+4-4(4)(1)\frac{1}{4}=\lambda^2(1)$

42. c. Let the required vector $\vec{r}$ be such that

$\vec{r}=x_1\vec{a}+x_2\vec{b}+x_3\vec{a}\times\vec{b}$

We must have $\vec{r}\cdot\vec{a}=\vec{r}\cdot\vec{b}=\vec{r}\cdot(\vec{a}\times\vec{b})$

[As $\vec{r},\vec{a},\vec{b}$ and $\vec{a}\times\vec{b}$ are unit vectors]

$\vec{r}\cdot\vec{a}=x_1$, $\vec{r}\cdot\vec{b}=x_2$, $\vec{r}\cdot(\vec{a}\times\vec{b})=x_3$

$\Rightarrow \vec{r}=\lambda(\vec{a}+\vec{b}+(\vec{a}\times\vec{b}))$

Also, $\vec{r}\cdot\vec{r}=1$

$\Rightarrow \lambda^2(\vec{a}+\vec{b}+\vec{a}\times\vec{b})\cdot(\vec{a}+\vec{b}+(\vec{a}\times\vec{b}))=1$

$\Rightarrow \lambda^2(|\vec{a}|^2+|\vec{b}|^2+|\vec{a}\times\vec{b}|^2)=1$

$\Rightarrow \lambda^2=\frac{1}{3}\Rightarrow\lambda=\pm\frac{1}{\sqrt{3}}$

$\Rightarrow \vec{r}=\pm\frac{1}{\sqrt{3}}(\vec{a}+\vec{b}+\vec{a}\times\vec{b})$

43. b. $\vec{c}=x\vec{a}+x\vec{b}+y(\vec{a}\times\vec{b})$

$\vec{a}\cdot\vec{c}=\vec{b}\cdot\vec{c}=\cos\theta$

[as $\vec{a}$, $\vec{b}$, and $\vec{c}$ are unit vectors]

$\Rightarrow \vec{a}\cdot\vec{c}=x|\vec{a}|^2\Rightarrow x=\cos\theta$

Also, $\vec{b}\cdot\vec{c}=x|\vec{b}|^2=\cos\theta$

$\Rightarrow \vec{c}=\cos\theta(\vec{a}+\vec{b})+y(\vec{a}\times\vec{b})$

Now, $|\vec{c}|^2=1=(\cos\theta(\vec{a}+\vec{b})+y(\vec{a}\times\vec{b}))\cdot(\cos\theta(\vec{a}+\vec{b})+y(\vec{a}\times\vec{b}))$

$\Rightarrow 1=\cos^2\theta(\vec{a}+\vec{b})\cdot(\vec{a}+\vec{b})+y^2(\vec{a}\times\vec{b})\cdot(\vec{a}\times\vec{b})$

$\Rightarrow 1=\cos^2\theta(|\vec{a}|^2+|\vec{b}|^2)+y^2|\vec{a}|^2|\vec{b}|^2$

$\Rightarrow 1=2\cos^2\theta+y^2$

$\Rightarrow \cos^2\theta=\frac{1}{2}-\frac{y^2}{2}$

$\Rightarrow 0\le\cos^2\theta\le\frac{1}{2}$

$\Rightarrow -\frac{1}{\sqrt{2}}\le\cos\theta\le\frac{1}{\sqrt{2}}$

$\Rightarrow \theta\in\left[\frac{\pi}{4},\frac{3\pi}{4}\right]$

44. c. $\vec{d}\cdot\hat{a} = \vec{d}\cdot\hat{b} = \vec{d}\cdot\hat{c}$

$\Rightarrow \lambda(1+\hat{b}\cdot\hat{c}) = \lambda(1+\hat{b}\cdot\hat{c}) = \lambda(\hat{a}\cdot\hat{b}+\hat{a}\cdot\hat{c})$

$\Rightarrow 1+\hat{b}\cdot\hat{c} = \hat{a}\cdot\hat{b}+\hat{a}\cdot\hat{c}$

$\Rightarrow 1-\hat{a}\cdot\hat{b}+\hat{b}\cdot\hat{c}-\hat{a}\cdot\hat{c}=0$

$\Rightarrow 1-\hat{a}\cdot\hat{b}+(\hat{b}-\hat{a})\cdot\hat{c}=0$

$\Rightarrow \hat{a}\cdot(\hat{a}-\hat{b})+(\hat{b}-\hat{a})\cdot\hat{c}=0$

$\Rightarrow (\hat{a}-\hat{c})\cdot(\hat{a}-\hat{b})=0$

$\Rightarrow \hat{a}-\hat{c}$ is perpendicular to $(\hat{a}-\hat{b})$

$\Rightarrow$ The triangle is right angled.

45. c. The given relation can be rewritten as the vector expression

$(\sqrt{a^2-4}\,\hat{i}+a\,\hat{j}+\sqrt{a^2+4}\,\hat{k})\cdot(\tan A\,\hat{i}\ += 6a$

$\Rightarrow \sqrt{a^2-4+a^2+a^2+4}$

$\sqrt{\tan^2 A+\tan^2 B+\tan^2 C}\times(\cos\theta)=6a$

$[\because \vec{a}\cdot\vec{b}=|\vec{a}||\vec{b}|\cos\theta]$

$\Rightarrow \sqrt{3}\,a\sqrt{\tan^2 A+\tan^2 B+\tan^2 C}\times(\cos\theta)=6a$

$\Rightarrow \tan^2 A+\tan^2 B+\tan^2 C=12\sec^2\theta \geq 12$

$[\because \sec^2\theta \geq 1]$

The least value of $\tan^2 A+\tan^2 B+\tan^2 C$ is 12.

46. d. $\Delta=\frac{1}{2}|(\hat{j}+\lambda\hat{k})\times(\hat{i}+\lambda\hat{k})|=\frac{1}{2}|-\hat{k}+\lambda\hat{i}+\lambda\hat{j}|$

$=\frac{1}{2}\sqrt{2\lambda^2+1}$

$\Rightarrow \frac{9}{4}\leq\frac{1}{4}(2\lambda^2+1)\leq\frac{33}{4}$

$\Rightarrow 4\leq\lambda^2\leq 16$

$\Rightarrow 2\leq|\lambda|\leq 4$

47. b. $(\vec{r}-\vec{q})\times\vec{p}=\vec{0}$

$\Rightarrow \vec{r}-\vec{q}=\lambda\vec{p}$

$\Rightarrow \vec{r}=\lambda\vec{p}+\vec{q}$ (1)

Now $\vec{r}\cdot\vec{s}=0$

$\Rightarrow (\lambda\vec{p}+\vec{q})\cdot\vec{s}=0$

$\Rightarrow \lambda=-\dfrac{\vec{q}\cdot\vec{s}}{\vec{p}\cdot\vec{s}}$ (2)

Puting λ from (2) in (1) $\Rightarrow \vec{r}=\vec{q}-\left(\dfrac{\vec{q}\cdot\vec{s}}{\vec{p}\cdot\vec{s}}\right)\vec{p}$

48. b. Let $\vec{r}$ be the new position. Then $\vec{r}=\lambda\hat{k}+\mu(\hat{i}+\hat{j})$

Also $\vec{r}\cdot\hat{k}=-\frac{1}{\sqrt{2}} \Rightarrow \lambda=-\frac{1}{\sqrt{2}}$

Also, $\lambda^2+2\mu^2=1 \Rightarrow 2\mu^2=\frac{1}{2} \Rightarrow \mu=\pm\frac{1}{2}$

$\therefore \vec{r}=\pm\frac{1}{2}(\hat{i}+\hat{j})-\frac{\hat{k}}{\sqrt{2}}$

49. c.

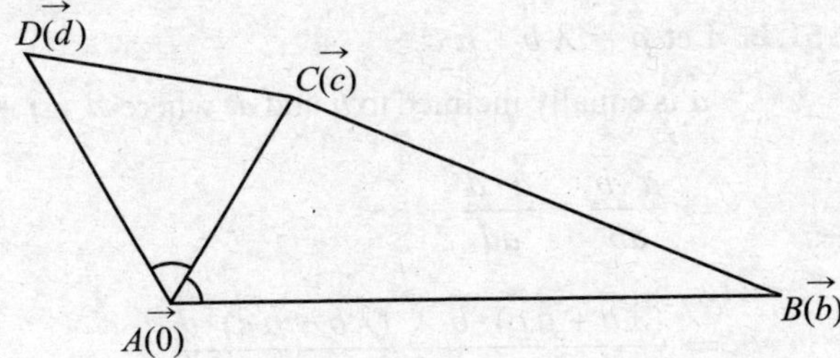

Fig. 2.5

Let $|\overrightarrow{AC}|=\lambda>0$

Then from $15|\overrightarrow{AC}|=3|\overrightarrow{AB}|=5|\overrightarrow{AD}|$

$|\overrightarrow{AB}|=5\lambda$

Let θ be the angle between $\overrightarrow{BA}$ and $\overrightarrow{CD}$.

$\Rightarrow \cos\theta=\dfrac{\overrightarrow{BA}\cdot\overrightarrow{CD}}{|\overrightarrow{BA}||\overrightarrow{CD}|}=\dfrac{-\vec{b}\cdot(\vec{d}-\vec{c})}{|\vec{b}||\vec{d}-\vec{c}|}$ (1)

Now $-\vec{b}\cdot(\vec{d}-\vec{c})=\vec{b}\cdot\vec{c}-\vec{b}\cdot\vec{d}$

$=|\vec{b}||\vec{c}|\cos\frac{\pi}{3}-|\vec{b}||\vec{d}|\cos\frac{2\pi}{3}$

$=(5\lambda)(\lambda)\frac{1}{2}+(5\lambda)(3\lambda)\frac{1}{2}$

$=\dfrac{5\lambda^2+15\lambda^2}{2}$

$=10\lambda^2$

Denominator of (1) is $|\vec{b}||\vec{d}-\vec{c}|$

Now $|\vec{d}-\vec{c}|^2=|\vec{d}|^2+|\vec{c}|^2-2\,\vec{c}\cdot\vec{d}$

$=9\lambda^2+\lambda^2-2(\lambda)(3\lambda)(1/2)$

$=10\lambda^2-3\lambda^2$

$=7\lambda^2$

Denominator of (1) is $(5)(\sqrt{7}\lambda)=5\sqrt{7}\lambda^2$

$\therefore \cos\theta=\dfrac{10\lambda^2}{5\sqrt{7}\lambda^2}=\dfrac{2}{\sqrt{7}}$.

50. a. Let A be origin $\overrightarrow{AB}=\vec{a}$, $\overrightarrow{AD}=\vec{b}$

so, $\overrightarrow{AE}=\vec{b}+\frac{3}{2}\vec{a}$, $\overrightarrow{AG}=\vec{a}+3\vec{b}$.

So the required ratio $= \dfrac{\frac{1}{2}\left|(\vec{a}+3\vec{b})\times\left(\vec{b}+\frac{3}{2}\vec{a}\right)\right|}{\frac{1}{2}|\vec{a}\times\vec{b}|}$

$= \dfrac{7}{2}$

51. b. Let $\vec{a} = \lambda\vec{b}+\mu\vec{c}$.

$\vec{a}$ is equally inclined to $\vec{b}$ and $\vec{d}$ where $\vec{d}=\hat{j}+2\hat{k}$.

$\Rightarrow \dfrac{\vec{a}\cdot\vec{b}}{ab} = \dfrac{\vec{a}\cdot\vec{d}}{ad}$

$\Rightarrow \dfrac{(\lambda\vec{b}+\mu\vec{c})\cdot\vec{b}}{b} = \dfrac{(\lambda\vec{b}+\mu\vec{c})\cdot\vec{d}}{d}$

$\Rightarrow \dfrac{[\lambda(2\hat{i}+\hat{j})+\mu(\hat{i}-\hat{j}+\hat{k})]\cdot(2\hat{i}+\hat{j})}{\sqrt{5}}$

$= \dfrac{[\lambda(2\hat{i}+\hat{j})+\mu(\hat{i}-\hat{j}+\hat{k})]\cdot(\hat{j}+2\hat{k})}{\sqrt{5}}$

$\Rightarrow \lambda(4+1)+\mu(2-1)=\lambda(1)+\mu(-1+2)$

$\Rightarrow 4\lambda=0$, i.e., $\lambda=0$

$\therefore\ \hat{a} = \dfrac{\hat{i}-\hat{j}+\hat{k}}{\sqrt{3}}$

52. a. Area of $\Delta BCD = \frac{1}{2}|\overrightarrow{BC}\times\overrightarrow{BD}|$

$= \frac{1}{2}|(b\hat{i}-c\hat{j})\times(b\hat{i}-d\hat{k})|$

$= \frac{1}{2}|bd\,\hat{j}+bc\,\hat{k}+dc\,\hat{i}|$

$= \frac{1}{2}\sqrt{b^2c^2+c^2d^2+d^2b^2}$ (1)

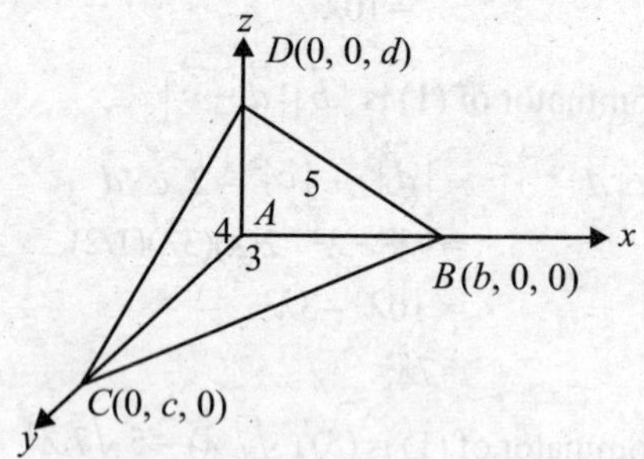

Fig. 2.6

Now $6 = bc$; $8 = cd$; $10 = bd$

$b^2c^2+c^2d^2+d^2b^2 = 200$

Substituting the value in (1)

$A = \frac{1}{2}\sqrt{200} = 5\sqrt{2}$

53. d. $\vec{f}\left(\frac{5}{4}\right) = \left[\frac{5}{4}\right]\hat{i}+\left(\frac{5}{4}-\left[\frac{5}{4}\right]\right)\hat{j}+\left[\frac{5}{4}+1\right]\hat{k}$

$= \hat{i}+\left(\frac{5}{4}-1\right)\hat{j}+2\hat{k} = \hat{i}+\frac{1}{4}\hat{j}+2\hat{k}$

When $0<t<1$, $\vec{f}(t) = 0\,\hat{i}+\{t-0\}\,\hat{j}+\hat{k} = t\,\hat{j}+\hat{k}$

$\vec{f}\left(\frac{5}{4}\right)\cdot\vec{f}(t) = 2+\frac{t}{4}$

So $\cos\theta = \dfrac{2+\frac{t}{4}}{\left|\hat{i}+\frac{1}{4}\hat{j}+2\hat{k}\right|\,|t\,\hat{j}+\hat{k}|}$

$= \dfrac{2+\frac{t}{4}}{\sqrt{1+\frac{1}{16}+4}\,\sqrt{1+t^2}}$

$= \dfrac{8+t}{9\sqrt{1+t^2}}$

54. a. $(\vec{a}\times\vec{b})\cdot(\vec{a}\times\vec{c}) = (\vec{a}\times\vec{b})\cdot\vec{u}$, where $\vec{u} = \vec{a}\times\vec{c}$

$\Rightarrow \vec{a}\cdot(\vec{b}\times\vec{u}) = \vec{a}\cdot[\vec{b}\times(\vec{a}\times\vec{c})]$.

$= \vec{a}\cdot[(\vec{b}\cdot\vec{c})\,\vec{a}-(\vec{a}\cdot\vec{b})\,\vec{c}]$

$= \vec{a}\cdot(\vec{b}\cdot\vec{c})\,\vec{a}\ \ (\vec{a}\cdot\vec{b}=0)$

$= |\vec{a}|^2(\vec{b}\cdot\vec{c})$

55. d. $(\hat{i}+\hat{j})\times(\hat{j}+\hat{k}) = \hat{i}-\hat{j}+\hat{k}$ so that unit vector perpendicular to the plane of $\hat{i}+\hat{j}$ and $\hat{j}+\hat{k}$ is $\frac{1}{\sqrt{3}}(\hat{i}-\hat{j}+\hat{k})$.

Similarly, the other two unit vectors are $\frac{1}{\sqrt{3}}(\hat{i}+\hat{j}-\hat{k})$ and $\frac{1}{\sqrt{3}}(-\hat{i}+\hat{j}+\hat{k})$.

The required volume $= \dfrac{1}{3\sqrt{3}}\begin{vmatrix}1 & -1 & 1\\ 1 & 1 & -1\\ -1 & 1 & 1\end{vmatrix} = \dfrac{4}{3\sqrt{3}}$

56. a. $\frac{1}{4} = (\hat{a}\times\hat{b})\cdot(\hat{a}\times\hat{b}) \Rightarrow ab\sin\theta = 1/2$

$\Rightarrow \theta = \pi/6$, as $a = b = 1$

57. a. $(\vec{a}\times(\vec{a}\times(\vec{a}\times(\vec{a}\times\vec{b}))))$

$= (\vec{a}\times(\vec{a}\times((\vec{a}\cdot\vec{b})\vec{a}-(\vec{a}\cdot\vec{a})\vec{b})))$

$= (\vec{a}\times(\vec{a}\times(-4\vec{b}))) = -4(\vec{a}\times(\vec{a}\times\vec{b}))$

$= -4((\vec{a}\cdot\vec{b})\vec{a}-(\vec{a}\cdot\vec{a})\vec{b}) = -4(-4\vec{b}) = -4(-4\vec{b})$

$= 16\,\vec{b} = 48\,\hat{b}$

58. d. Let $\vec{a} = 6\hat{i} + 6\hat{k}, \vec{b} = 4\hat{j} + 2\hat{k}, \vec{c} = 4\hat{j} - 8k$

then $\vec{a} \times \vec{b} = -24\hat{i} - 12\hat{j} + 24\hat{k}$

$= 12(-2\hat{i} - \hat{j} + 2\hat{k})$

Therefore, area of the base of the parallelepiped

$= \frac{1}{2} |\vec{a} \times \vec{b}|$

$= \frac{1}{2} (12 \times 3) = 18$

Height of the parallelepiped

= length of projection of $\vec{c}$ on $\vec{a} \times \vec{b}$

$= \frac{|\vec{c} \cdot \vec{a} \times \vec{b}|}{|\vec{a} \times \vec{b}|}$

$= \frac{|12(-4-16)|}{36}$

$= \frac{20}{3}$

Therefore volume of the parallelepiped is $18 \times \frac{20}{3}$ $= 120.$

59. c. $3 = \frac{1}{6}[\vec{a}\,\vec{b}\,\vec{c}]$

$\Rightarrow [\vec{a}\,\vec{b}\,\vec{c}] = 18$

Volume of the required parallelepiped is

$[\vec{a} + \vec{b}\;\; \vec{b} + \vec{c}\;\; \vec{c} + \vec{a}] = 2[\vec{a}\,\vec{b}\,\vec{c}] = 36.$

60. b. Here $[\vec{a}\,\vec{b}\,\vec{c}] = \pm 1$

$[\vec{a} + \vec{b} + \vec{c}\;\; \vec{a} + \vec{b}\;\; \vec{b} + \vec{c}]$

$= (\vec{a} + \vec{b} + \vec{c}) \times (\vec{a} + \vec{b}) \cdot (\vec{b} + \vec{c})$

$= \vec{c} \times (\vec{a} + \vec{b}) \cdot (\vec{b} + \vec{c})$

$= (\vec{c} \times \vec{a} + \vec{c} \times \vec{b}) \cdot (\vec{b} + \vec{c})$

$= \vec{c} \times \vec{a} \cdot \vec{b} = [\vec{a}\,\vec{b}\,\vec{c}] = \pm 1$

61. a. Let $\vec{c} = \lambda(\vec{a} \times \vec{b})$.

Hence $\lambda(\vec{a} \times \vec{b}) \cdot (\hat{i} + 2\hat{j} - 7\hat{k}) = 10$

$\Rightarrow \lambda \begin{vmatrix} 2 & -3 & 1 \\ 1 & -2 & 3 \\ 1 & 2 & -7 \end{vmatrix} = 10$

$\Rightarrow \lambda = -1$

$\Rightarrow \vec{c} = -(\vec{a} \times \vec{b})$

62. d. $\vec{a} \perp \vec{b} \Rightarrow x - y + 2 = 0$

$\vec{a} \cdot \vec{c} = 4 \Rightarrow x + 2y = 4$

Solving we get $x = 0$; $y = 2$

$\Rightarrow \vec{a} = 2\hat{j} + 2\hat{k}$

$\Rightarrow [\vec{a}\,\vec{b}\,\vec{c}] = \begin{vmatrix} 0 & 2 & 2 \\ 1 & -1 & 1 \\ 1 & 2 & 0 \end{vmatrix} = 8 = |\vec{a}|^2$

63. c. $(\vec{a} \times \vec{b} \cdot \vec{c})^2 = |\vec{a}|^2 |\vec{b}|^2 |\vec{c}|^2 \sin^2\theta \cos^2\phi$ [θ is the angle between $\vec{a}$ and $\vec{b}$, $\phi = 0$]

$= \frac{1}{4}(a_1^2 + a_2^2 + a_3^2)(b_1^2 + b_2^2 + b_3^2)$

64. c. $\vec{r} \cdot \vec{a} = 0, |\vec{r} \times \vec{b}| = |\vec{r}||\vec{b}|$, and $|\vec{r} \times \vec{c}| = |\vec{r}||\vec{c}|$

$\Rightarrow \vec{r} \perp \vec{a}, \vec{b}, \vec{c}$

$\therefore [\vec{a}\,\vec{b}\,\vec{c}] = 0$

65. b. $\vec{c} = \lambda(\vec{a} \times \vec{b})$

$\Rightarrow \vec{c} \cdot \vec{c} = \lambda(\vec{a} \times \vec{b}) \cdot \vec{c}$

$\Rightarrow \frac{1}{3} = \lambda$

Also $|\vec{c}|^2 = \lambda^2 |\vec{a} \times \vec{b}|^2$

$\Rightarrow \frac{1}{3} = \frac{1}{9}(a^2b^2 \sin^2\theta) = \frac{1}{9} \times 2 \times 3 \sin^2\theta$

$\Rightarrow \sin^2\theta = \frac{1}{2}$

$\Rightarrow \theta = \frac{\pi}{4}$

66. c. $4\vec{a} + 5\vec{b} + 9\vec{c} = 0$

$\Rightarrow$ Vectors $\vec{a}, \vec{b}$ and $\vec{c}$ are coplanar.

$\Rightarrow \vec{b} \times \vec{c}$ and $\vec{c} \times \vec{a}$ are collinear

$\Rightarrow (\vec{b} \times \vec{c}) \times (\vec{c} \times \vec{a}) = \vec{0}.$

67. a. $[\vec{a} \times \vec{b}\;\; \vec{a} \times \vec{c}\;\; \vec{d}] = (\vec{a} \times \vec{b}) \cdot ((\vec{a} \times \vec{c}) \times \vec{d})$

$= (\vec{a} \times \vec{b}) \cdot ((\vec{a} \cdot \vec{d})\vec{c} - (\vec{c} \cdot \vec{d})\vec{a})$

$= (\vec{a} \cdot \vec{d})[\vec{a}\,\vec{b}\,\vec{c}]$

68. a. Let $\vec{r} = x_1\hat{a} + x_2\hat{b} + x_3(\hat{a} \times \hat{b})$

$\Rightarrow \vec{r} \cdot \hat{a} = x_1 + x_2\hat{a} \cdot \hat{b} + x_3\hat{a} \cdot (\hat{a} \times \hat{b}) = x_1$

Also, $\vec{r} \cdot \hat{b} = x_1\hat{a} \cdot \hat{b} + x_2 + x_3\hat{b} \cdot (\hat{a} + \hat{b}) = x_2$

and $\vec{r} \cdot (\hat{a} \times \hat{b}) = x_1\hat{a} \cdot (\hat{a} \times \hat{b}) + x_2\hat{b} \cdot (\hat{a} \times \hat{b})$
$+ x_3(\hat{a} \times \hat{b}) \cdot (\hat{a} \times \hat{b}) = x_3$

$\Rightarrow \vec{r} = (\vec{r} \cdot \hat{a})\hat{a} + (\vec{r} \cdot \hat{b})\hat{b} + (\vec{r} \cdot (\hat{a} \times \hat{b}))(\hat{a} \times \hat{b})$

69. a. $[\vec{a}+(\vec{a}\times\vec{b})\ \vec{b}+(\vec{a}\times\vec{b})\ \vec{a}\times\vec{b}]$

$= (\vec{a}+(\vec{a}\times\vec{b}))\cdot((\vec{b}+(\vec{a}\times\vec{b}))\times(\vec{a}\times\vec{b}))$

$= (\vec{a}+(\vec{a}\times\vec{b}))\cdot(\vec{b}\times(\vec{a}\times\vec{b}))$

$= (\vec{a}+(\vec{a}\times\vec{b}))\cdot(\vec{a}-(\vec{a}\cdot\vec{b})\vec{b})$

$= \vec{a}\cdot\vec{a}=1$ (as $\vec{a}\cdot\vec{b}=0,\ \vec{a}\cdot(\vec{a}\times\vec{b})=0$)

70. c. $\vec{a}\times\vec{r}=\vec{b}$

$\Rightarrow \vec{a}\times(\vec{a}\times\vec{r}) = \vec{a}\times\vec{b}$

$\Rightarrow (\vec{a}\cdot\vec{r})\vec{a}-(\vec{a}\cdot\vec{a})\vec{r} = \vec{a}\times\vec{b}$

$\Rightarrow (\vec{a}\cdot\vec{a})\vec{r} = 3\vec{a}-(\vec{a}\times\vec{b})$

$\Rightarrow 3\vec{r} = 3(\hat{i}+\hat{j}+\hat{k})-(-2\hat{i}+\hat{j}+\hat{k})$

$\Rightarrow \vec{r}=\frac{1}{3}(5\hat{i}+2\hat{j}+2\hat{k})$

71. d. $((\vec{a}\times\vec{b})+(\vec{a}\times\vec{c}))\times(\vec{b}\times\vec{c})$

$= (\vec{a}\times\vec{b})\times(\vec{b}\times\vec{c})+(\vec{a}\times\vec{c})\times(\vec{b}\times\vec{c})$

$= ((\vec{a}\times\vec{b})\cdot\vec{c})\vec{b}-((\vec{a}\times\vec{b})\cdot\vec{b})\vec{c}+((\vec{a}\times\vec{c})\cdot\vec{c})\vec{b}$

$-((\vec{a}\times\vec{c})\cdot\vec{b})\vec{c} = [\vec{a}\,\vec{b}\,\vec{c}](\vec{b}+\vec{c})$

$\Rightarrow ((\vec{a}\times\vec{b})+(\vec{a}\times\vec{c}))\times(\vec{b}\times\vec{c}))\cdot(\vec{b}-\vec{c})$

$= [\vec{a}\,\vec{b}\,\vec{c}](\vec{b}+\vec{c}).(\vec{b}-\vec{c})$

$= [\vec{a}\,\vec{b}\,\vec{c}](|\vec{b}|^2-|\vec{c}|^2)=0$

72. a. $\vec{a}\times\vec{b}=\vec{c}$

$\Rightarrow \vec{a}\times(\vec{a}\times\vec{b})=\vec{a}\times\vec{c}$

$\Rightarrow (\vec{a}\cdot\vec{b})\vec{a}-|\vec{a}|^2\vec{b}=\vec{a}\times\vec{c}$

$\Rightarrow \vec{b}=\dfrac{\beta\vec{a}-\vec{a}\times\vec{c}}{|\vec{a}|^2}$ $[\because\ \vec{a}\cdot\vec{b}=\beta]$

73. b. Taking dot product of

$a(\vec{\alpha}\times\vec{\beta})+b(\vec{\beta}\times\vec{\gamma})+c(\vec{\gamma}\times\vec{\alpha})=0$ with

$\vec{\gamma}$, $\vec{\alpha}$, and $\vec{\beta}$, respectively, we have

$a[\vec{\alpha}\,\vec{\beta}\,\vec{\gamma}]=0$

$b[\vec{\alpha}\,\vec{\beta}\,\vec{\gamma}]=0$

$c[\vec{\alpha}\,\vec{\beta}\,\vec{\gamma}]=0$

Since at least one of a, b, and $c\neq 0$,

$[\vec{\alpha}\,\vec{\beta}\,\vec{\gamma}]=0$

Hence, $\vec{\alpha}$, $\vec{\beta}$, and $\vec{\gamma}$ are coplanar.

74. c. $(\vec{a}\times\vec{b})\times(\vec{b}\times\vec{c})=\vec{b}$

$\Rightarrow [\vec{a}\,\vec{b}\,\vec{c}]\vec{b}=\vec{b}$

$\Rightarrow [\vec{a}\,\vec{b}\,\vec{c}]=1$

$\therefore$ $\vec{a}$, $\vec{b}$, and $\vec{c}$ cannot be coplanar.

75. c. Any vector $\vec{r}$ can be represented in terms of three non-coplanar vectors $\vec{a}$, $\vec{b}$, and $\vec{c}$ as

$\vec{r}=x(\vec{a}\times\vec{b})+y(\vec{b}\times\vec{c})+z(\vec{c}\times\vec{a})$ (1)

Taking dot product with $\vec{a}$, $\vec{b}$, and $\vec{c}$, respectively, we have,

$x=\dfrac{\vec{r}\cdot\vec{c}}{[\vec{a}\,\vec{b}\,\vec{c}]}$, $y=\dfrac{\vec{r}\cdot\vec{a}}{[\vec{a}\,\vec{b}\,\vec{c}]}$, and $z=\dfrac{\vec{r}\cdot\vec{b}}{[\vec{a}\,\vec{b}\,\vec{c}]}$

From (1)

$[\vec{a}\,\vec{b}\,\vec{c}]\vec{r}=\frac{1}{2}(\vec{a}\times\vec{b}+\vec{b}\times\vec{c}+\vec{c}\times\vec{a})$

Therefore, area of $\Delta ABC=\frac{1}{2}|\vec{a}\times\vec{b}+\vec{b}\times\vec{c}+\vec{c}\times\vec{a}|$

$=|[\vec{a}\,\vec{b}\,\vec{c}]\vec{r}|$

76. b. Let $\vec{r}=\alpha\vec{a}+\beta\vec{b}+\gamma(\vec{a}\times\vec{b})$

Taking cross product with $\vec{a}$, we get

$\vec{b}=\beta(\vec{b}\times\vec{a})+\gamma|\vec{a}|^2\vec{b}$

Equating the coefficients, we get

$\beta=0,\ \gamma=\dfrac{1}{|\vec{a}|^2}$

$\therefore\ \vec{r}=\alpha\vec{a}+\dfrac{1}{|\vec{a}|^2}(\vec{a}\times\vec{b})$

77. a. $\{\vec{a}\times(\vec{b}+\vec{a}\times\vec{b})\}\cdot\vec{b}$

$=\{\vec{a}\times\vec{b}+\vec{a}\times(\vec{a}\times\vec{b})\}\cdot\vec{b}$

$=[\vec{a}\,\vec{b}\,\vec{b}]+\{(\vec{a}\cdot\vec{b})\vec{a}-(\vec{a}\cdot\vec{a})\vec{b}\}\cdot\vec{b}$

$=0+(\vec{a}\cdot\vec{b})^2-(\vec{a}\cdot\vec{a})(\vec{b}\cdot\vec{b})$

$=\cos^2\frac{\pi}{3}-1=-\frac{3}{4}$

78. a. $\vec{r}\times\vec{a}=\lambda\vec{a}+\mu\vec{b}+\gamma\vec{a}\times\vec{b}$

$\therefore\ [\vec{r}\,\vec{a}\,\vec{a}]=\lambda\vec{a}\cdot\vec{a}+\mu\vec{b}\cdot\vec{a}+\gamma[\vec{a}\,\vec{b}\,\vec{a}]$

$\therefore\ \lambda=0$ $[\because\ \vec{a}\perp\vec{a}]$

Also $[\vec{r}\,\vec{a}\,\vec{b}]=\lambda\vec{a}\cdot\vec{b}+\mu\vec{b}\cdot\vec{b}+\gamma[\vec{a}\,\vec{b}\,\vec{b}]=\mu$

Also $(\vec{r}\times\vec{a})\times\vec{b}=\gamma(\vec{a}\times\vec{b})\times\vec{b}$

$\Rightarrow (\vec{r}\cdot\vec{b})\vec{a}-(\vec{a}\cdot\vec{b})\vec{r}=\gamma\{(\vec{a}\cdot\vec{b})\vec{b}-(\vec{b}\cdot\vec{b})\vec{a}\}$

$\Rightarrow (\vec{r}\cdot\vec{b})\vec{a}=-\gamma\vec{a}\ \therefore\ \gamma=-(\vec{r}\cdot\vec{b})$

79. c. The given equation reduces to

$[\vec{a}\,\vec{b}\,\vec{c}]^2 x^2 + 2[\vec{a}\,\vec{b}\,\vec{c}]x + 1 = 0 \Rightarrow D = 0$

80. d. $(\vec{b}\times\vec{c})\cdot(\vec{a}\times\vec{d}) = (\vec{b}\cdot\vec{a})(\vec{c}\cdot\vec{d}) - (\vec{b}\cdot\vec{d})\cdot(\vec{c}\cdot\vec{a})$

$(\vec{c}\times\vec{a})\cdot(\vec{b}\times\vec{d}) = (\vec{c}\cdot\vec{b})(\vec{a}\cdot\vec{d}) - (\vec{c}\cdot\vec{d})(\vec{a}\cdot\vec{b})$

$(\vec{a}\times\vec{b})\cdot(\vec{c}\times\vec{d}) = (\vec{a}\cdot\vec{c})(\vec{b}\cdot\vec{d}) - (\vec{a}\cdot\vec{d})(\vec{b}\cdot\vec{c})$

$\Rightarrow (\vec{b}\times\vec{c})\cdot(\vec{a}\times\vec{d}) + (\vec{c}\times\vec{a})\cdot(\vec{b}\times\vec{d}) + (\vec{a}\times\vec{b})\cdot(\vec{c}\times\vec{d}) = 0$

81. c. $\vec{r}\times\vec{a} = \vec{b}$

$\Rightarrow \vec{d}\times(\vec{r}\times\vec{a}) = \vec{d}\times\vec{b}$

$\Rightarrow (\vec{a}\cdot\vec{d})\vec{r} - (\vec{d}\cdot\vec{r})\vec{a} = \vec{d}\times\vec{b}$ (1)

$\vec{r}\times\vec{c} = \vec{d}$

$\Rightarrow \vec{b}\times(\vec{r}\times\vec{c}) = \vec{b}\times\vec{d}$

$\Rightarrow (\vec{b}\cdot\vec{c})\vec{r} - (\vec{b}\cdot\vec{r})\vec{c} = \vec{b}\times\vec{d}$ (2)

Adding (1) and (2), we get

$(\vec{a}\cdot\vec{d} + \vec{b}\cdot\vec{c})\vec{r} - (\vec{d}\cdot\vec{r})\vec{a} - (\vec{b}\cdot\vec{r})\vec{c} = \vec{0}$

Now $\vec{r}\cdot\vec{d} = 0$ and $\vec{b}\cdot\vec{r} = 0$ as $\vec{d}$ and $\vec{r}$ as well as $\vec{b}$ and $\vec{r}$ are mutually perpendicular.

Hence, $(\vec{b}\cdot\vec{c} + \vec{a}\cdot\vec{d})\vec{r} = \vec{0}$.

82. b. Let $\vec{a}\times\vec{b} = x\hat{i} + y\hat{j} + z\hat{k}$. Therefore,

$[\vec{a}\,\vec{b}\,\hat{i}] = (\vec{a}\times\vec{b})\cdot\hat{i} = x$

$[\vec{a}\,\vec{b}\,\hat{j}] = (\vec{a}\times\vec{b})\cdot\hat{j} = y$

$[\vec{a}\,\vec{b}\,\hat{k}] = (\vec{a}\times\vec{b})\cdot\hat{k} = z$

Hence, $[\vec{a}\,\vec{b}\,\hat{i}]\hat{i} + [\vec{a}\,\vec{b}\,\hat{j}]\hat{j} = [\vec{a}\,\vec{b}\,\hat{k}]\hat{k}$

$= x\hat{i} + y\hat{j} + z\hat{k}$

$= \vec{a}\times\vec{b}$

83. a. $\vec{a}\times(\vec{b}\times\vec{c}) = (\vec{a}\cdot\vec{c})\vec{b} - (\vec{a}\cdot\vec{b})\vec{c}$

$= 5(\hat{i} + 2\hat{j} + 2\hat{k}) - 6(\hat{i} + \hat{j} + 2\hat{k})$

$\Rightarrow (1+\alpha)\hat{i} + \beta(1+\alpha)\hat{j} + \gamma(1+\alpha)(1+\beta)\hat{k}$

$= -\hat{i} + 4\hat{j} - 2\hat{k}$

$\Rightarrow 1+\alpha = -1, \beta = -4$, and $\gamma(-1)(-3) = -2$

$\Rightarrow \gamma = -\frac{2}{3}$

84. d. $\vec{a} = \vec{b} + \vec{c},\ \vec{b}\times\vec{d} = \vec{0},\ \vec{c}\cdot\vec{d} = 0$

$\Rightarrow \vec{a}\times\vec{d} = \vec{b}\times\vec{d} + \vec{c}\times\vec{d} = \vec{c}\times\vec{d}$

$\Rightarrow \vec{d}\times(\vec{a}\times\vec{d}) = \vec{d}\times(\vec{c}\times\vec{d})$

$= (\vec{d}\cdot\vec{d})\vec{c} - (\vec{c}\cdot\vec{d})\vec{d} = |\vec{d}|^2\vec{c}$

$\Rightarrow \vec{c} = \dfrac{\vec{d}\times(\vec{a}\times\vec{d})}{|\vec{d}|^2}$

85. b. $(\vec{a}\times\hat{i})\cdot(\vec{b}\times\hat{i}) = \begin{vmatrix}\vec{a}\cdot\vec{b} & \vec{a}\cdot\hat{i}\\ \vec{b}\cdot\hat{i} & \hat{i}\cdot\hat{i}\end{vmatrix} = (\vec{a}\cdot\vec{b}) - (\vec{a}\cdot\hat{i})(\vec{b}\cdot\hat{i})$

Similarly, $(\vec{a}\times\hat{j})\cdot(\vec{b}\times\hat{j}) = (\vec{a}\cdot\vec{b}) - (\vec{a}\cdot\hat{j})(\vec{b}\cdot\hat{j})$

and $(\vec{a}\times\hat{k})\cdot(\vec{b}\times\hat{k}) = \vec{a}\cdot\vec{b} - (\vec{a}\cdot\hat{k})(\vec{b}\cdot\hat{k})$

Let $\vec{a} = a_1\hat{i} + a_2\hat{j} + a_3\hat{k}$, $\vec{b} = b_1\hat{i} + b_2\hat{j} + b_3\hat{k}$.

Therefore, $(\vec{a}\cdot\hat{i}) = a_1$, $\vec{a}\cdot\hat{j} = a_2$, $\vec{a}\cdot\hat{k} = a_3$, $\vec{b}\cdot\hat{i} = b_1$, $\vec{b}\cdot\hat{j} = b_2$, $(\vec{b}\cdot\hat{k}) = b_3$

$\Rightarrow (\vec{a}\times\hat{i})\cdot(\vec{b}\times\hat{i}) + (\vec{a}\times\hat{j})\cdot(\vec{b}\times\hat{j}) + (\vec{a}\times\hat{k})\cdot(\vec{b}\times\hat{k})$

$= 3\vec{a}\cdot\vec{b} - (a_1b_1 + a_2b_2 + a_3b_3)$

$= 3\vec{a}\cdot\vec{b} - \vec{a}\cdot\vec{b} = 2\vec{a}\cdot\vec{b}$

86. b. $(\vec{a}\times\vec{b})\times(\vec{r}\times\vec{c})$

$= ((\vec{a}\times\vec{b})\cdot\vec{c})\vec{r} - ((\vec{a}\times\vec{b})\cdot\vec{r})\cdot\vec{c}$

$= [\vec{a}\,\vec{b}\,\vec{c}]\vec{r} - [\vec{a}\,\vec{b}\,\vec{r}]\vec{c}$

Similarly, $(\vec{b}\times\vec{c})\times(\vec{r}\times\vec{a}) = [\vec{b}\,\vec{c}\,\vec{a}]\vec{r} - [\vec{b}\,\vec{c}\,\vec{r}]\vec{a}$

and, $(\vec{c}\times\vec{a})\times(\vec{r}\times\vec{b}) = [\vec{c}\,\vec{a}\,\vec{b}]\vec{r} - [\vec{c}\,\vec{a}\,\vec{r}]\vec{b}$

$\Rightarrow (\vec{a}\times\vec{b})\times(\vec{r}\times\vec{c}) + (\vec{b}\times\vec{c})\times(\vec{r}\times\vec{a}) + (\vec{c}\times\vec{a})\times(\vec{r}\times\vec{b})$

$= 3[\vec{a}\,\vec{b}\,\vec{c}]\vec{r} - ([\vec{b}\,\vec{c}\,\vec{r}]\vec{a} + [\vec{c}\,\vec{a}\,\vec{r}]\vec{b} + [\vec{a}\,\vec{b}\,\vec{r}]\vec{c})$

$= 3[\vec{a}\,\vec{b}\,\vec{c}]\vec{r} - [\vec{a}\,\vec{b}\,\vec{c}]\vec{r} = 2[\vec{a}\,\vec{b}\,\vec{c}]\vec{r}$

87. a. We have,

$\vec{a}\cdot\vec{p} = \vec{a}\cdot\dfrac{(\vec{b}\times\vec{c})}{[\vec{a}\,\vec{b}\,\vec{c}]} = \dfrac{\vec{a}\cdot(\vec{b}\times\vec{c})}{[\vec{a}\,\vec{b}\,\vec{c}]} = \dfrac{[\vec{a}\,\vec{b}\,\vec{c}]}{[\vec{a}\,\vec{b}\,\vec{c}]} = 1$

$\vec{a}\cdot\vec{q} = \vec{a}\cdot\dfrac{\vec{c}\times\vec{a}}{[\vec{a}\,\vec{b}\,\vec{c}]} = \dfrac{[\vec{a}\,\vec{c}\,\vec{a}]}{[\vec{a}\,\vec{b}\,\vec{c}]} = 0$

Similarly, $\vec{a}\cdot\vec{r} = 0$, $\vec{b}\cdot\vec{p} = 0$, $\vec{b}\cdot\vec{q} = 1$, $\vec{b}\cdot\vec{r} = 0$, $\vec{c}\cdot\vec{p} = 0$, $\vec{c}\cdot\vec{q} = 0$, and $\vec{c}\cdot\vec{r} = 1$

$\therefore (\vec{a} + \vec{b} + \vec{c})\cdot(\vec{p} + \vec{q} + \vec{r}) = \vec{a}\cdot\vec{p} + \vec{a}\cdot\vec{q} + \vec{a}\cdot\vec{r} + \vec{b}\cdot\vec{p} + \vec{b}\cdot\vec{q} + \vec{b}\cdot\vec{r} + \vec{c}\cdot\vec{p} + \vec{c}\cdot\vec{q} + \vec{c}\cdot\vec{r}$

$= 1 + 1 + 1 = 3$

88. b. A vector perpendicular to the plane of $A(\vec{a})$, $B(\vec{b})$, and $C(\vec{c})$ is

$(\vec{b}-\vec{a})\times(\vec{c}-\vec{a}) = \vec{a}\times\vec{b}+\vec{b}\times\vec{c}+\vec{c}\times\vec{a}$.

Now for any point $R(\vec{r})$ in the plane of A, B, and C is $(\vec{r}-\vec{a})\cdot(\vec{a}\times\vec{b}+\vec{b}\times\vec{c}+\vec{c}\times\vec{a})=0$. Therefore,

$\vec{r}\cdot(\vec{a}\times\vec{b}+\vec{b}\times\vec{c}+\vec{c}\times\vec{a})-\vec{a}\cdot(\vec{a}\times\vec{b}+\vec{b}\times\vec{c}$
$+\vec{c}\times\vec{a})=0$

$\vec{r}\cdot(\vec{a}\times\vec{b}+\vec{b}\times\vec{c}+\vec{c}\times\vec{a})=0+\vec{a}\cdot\vec{b}\times\vec{c}+0$

$\vec{r}\cdot(\vec{a}\times\vec{b}+\vec{b}\times\vec{c}+\vec{c}\times\vec{a})=[\vec{a}\,\vec{b}\,\vec{c}]$

89. c. Given that $\vec{a}$, $\vec{b}$, and $\vec{c}$ are non-coplanar.

$\Rightarrow [\vec{a}\,\vec{b}\,\vec{c}]\neq 0$ (1)

Again $\vec{a}\times(\vec{b}\times\vec{c})\cdot(\vec{a}\times\vec{c})=0$

$\Rightarrow [(\vec{a}\cdot\vec{c})\vec{b}-(\vec{a}\cdot\vec{b})\vec{c}]\cdot(\vec{a}\times\vec{c})=0$

$\Rightarrow (\vec{a}\cdot\vec{c})[\vec{b}\,\vec{a}\,\vec{c}]=0$

$\Rightarrow (\vec{a}\cdot\vec{c})=0$

$\Rightarrow \vec{a}$ and $\vec{c}$ are perpendicular. (2)

$\vec{a}\times(\vec{b}\times\vec{c})=(\vec{a}\cdot\vec{c})\vec{b}-(\vec{a}\cdot\vec{b})\vec{c}$

$\Rightarrow [\vec{a}\times(\vec{b}\times\vec{c})]\times\vec{c}=\vec{0}$

90. c. Consider a tetrahedron with vertices $O(0, 0, 0)$, $A(a, 0, 0)$, $B(0, b, 0)$, and $C(0, 0, c)$.

Its volume $V=\frac{1}{6}[\vec{a}\,\vec{b}\,\vec{c}]$

Now centroids of the faces OAB, OAC, OBC, and ABC are $G_1(a/3, b/3, 0)$, $G_2(a/3, 0, c/3)$, $G_3(0, b/3, c/3)$, and $G_4(a/3, b/3, c/3)$, respectively.

$\overrightarrow{G_4G_1}=\vec{a}/3$, $\overrightarrow{G_4G_2}=\vec{b}/3$, $\overrightarrow{G_4G_3}=\vec{a}/3$.

Volume of tetrahedron by centroids

$$V'=\frac{1}{6}\left[\frac{\vec{a}}{3}\ \frac{\vec{b}}{3}\ \frac{\vec{c}}{3}\right]=\frac{1}{27}V$$

$\Rightarrow K=27$

91. c. $[(\vec{a}\times\vec{b})\times(\vec{b}\times\vec{c})(\vec{b}\times\vec{c})\times(\vec{c}\times\vec{a})(\vec{c}\times\vec{a})$
$\times(\vec{a}\times\vec{b})]=[[\vec{a}\,\vec{b}\,\vec{c}]\,\vec{b}\,[\vec{a}\,\vec{b}\,\vec{c}]\,\vec{c}\,[\vec{a}\,\vec{b}\,\vec{c}]\cdot\vec{a}]$
$=[\vec{a}\,\vec{b}\,\vec{c}]^3\,[\vec{b}\,\vec{c}\,\vec{a}]=[\vec{a}\,\vec{b}\,\vec{c}]^4$

92. d. $\vec{r}=x_1(\vec{a}\times\vec{b})+x_2(\vec{b}\times\vec{c})+x_3(\vec{c}\times\vec{a})$

$\Rightarrow \vec{r}\cdot\vec{a}=x_2[\vec{a}\,\vec{b}\,\vec{c}]$, $\vec{r}\cdot\vec{b}=x_3[\vec{b}\,\vec{c}\,\vec{a}]$

and $\vec{r}\cdot\vec{c}=x_1[\vec{c}\,\vec{a}\,\vec{b}]=x_1[\vec{a}\,\vec{b}\,\vec{c}]$

$\Rightarrow x_1+x_2+x_3=4\vec{r}\cdot(\vec{a}+\vec{b}+\vec{c})$

93. a. Let $\vec{v}=x\vec{a}+y\vec{b}+z\vec{a}\times\vec{b}$

Given: $\vec{a}\cdot\vec{b}=0$, $\vec{v}\cdot\vec{a}=0$, $\vec{v}\cdot\vec{b}=1$, $[\vec{v}\,\vec{a}\,\vec{b}]=1$

$\Rightarrow \vec{v}\cdot\vec{a}=x\,\vec{a}\cdot\vec{a}=x|\vec{a}|^2$ $[\because \vec{a}\cdot\vec{b}=0, \vec{a}\cdot\vec{a}\times\vec{b}=0]$

$\Rightarrow x=0$

Again, $\vec{v}\cdot\vec{b}=y|\vec{b}|^2\Rightarrow 1=yb^2$

$\therefore y=\frac{1}{b^2}$

Again, $\vec{v}\cdot(\vec{a}\times\vec{b})=z(\vec{a}\times\vec{b})^2$

$\Rightarrow 1=z(\vec{a}\times\vec{b})^2 \Rightarrow z=\frac{1}{|\vec{a}\times\vec{b}|^2}$

Hence, $\vec{v}=\frac{1}{|\vec{b}|^2}\vec{b}+\frac{1}{|\vec{a}\times\vec{b}|^2}\vec{a}\times\vec{b}$.

94. d. Volume of the parallelepiped formed by $\vec{a}'$, $\vec{b}'$, and $\vec{c}'$ is 4.

Therefore, the volume of the parallelepiped formed by $\vec{a}$, $\vec{b}$ and $\vec{c}$ is $\frac{1}{4}$.

$\vec{b}\times\vec{c}=\frac{(\vec{c}'\times\vec{a}')\times\vec{c}}{4}=\frac{1}{4}\vec{a}'$,

$|\vec{b}\times\vec{c}|=\frac{\sqrt{2}}{4}=\frac{1}{2\sqrt{2}}\ 2$

Length of altitude $=\frac{1}{4}\times 2\sqrt{2}=\frac{1}{\sqrt{2}}$

95. d. $\vec{a}'=\frac{\vec{b}\times\vec{c}}{[\vec{a}\,\vec{b}\,\vec{c}]}=\frac{\hat{i}+\hat{j}-\hat{k}}{2}$

Multiple Correct Answers Type

1. a., b. We have, $|\vec{a}-\vec{b}|^2=|\vec{a}|^2+|\vec{b}|^2-2(\vec{a}\cdot\vec{b})$

$\Rightarrow |\vec{a}-\vec{b}|^2=|\vec{a}|^2+|\vec{b}|^2-2|\vec{a}||\vec{b}|\cos 2\theta$

$\Rightarrow |\vec{a}-\vec{b}|^2=2-2\cos 2\theta$ $(\because |\vec{a}|=|\vec{b}|=1)$

$\Rightarrow |\vec{a}-\vec{b}|^2=4\sin^2\theta$

$\Rightarrow |\vec{a}-\vec{b}|=2|\sin\theta|$

Now, $|\vec{a}-\vec{b}|<1$

$\Rightarrow 2|\sin\theta|<1$

$\Rightarrow |\sin\theta|<\frac{1}{2}$

$\Rightarrow \theta\in[0, \pi/6)$ or $\theta\in(5\pi/6, \pi]$

2. a., c.

$\vec{a}\times(\vec{b}\times\vec{c})+(\vec{a}\cdot\vec{b})\vec{b}$

$= (4-2x-\sin y)\vec{b}+(x^2-1)\vec{c}$

$\Rightarrow (\vec{a}\cdot\vec{c})\vec{b}-(\vec{a}\cdot\vec{b})\vec{c}+(\vec{a}\cdot\vec{b})\vec{b}$

$= (4-2x-\sin y)\vec{b}+(x^2-1)\vec{c}$

Now, $(\vec{c}\cdot\vec{c})\vec{a}=\vec{c}$. Therefore,

$(\vec{c}\cdot\vec{c})(\vec{c}\cdot\vec{c})=(\vec{c}\cdot\vec{c})\Rightarrow \vec{a}\cdot\vec{c}=1$

$\Rightarrow 1+\vec{a}\cdot\vec{b}=4-2x-\sin y,\ x^2-1=-(\vec{a}\cdot\vec{b})$

$\Rightarrow 1=4-2x-\sin y+x^2-1$

$\Rightarrow \sin y=x^2-2x+2=(x-1)^2+1$

But $\sin y\le 1\Rightarrow x=1,\ \sin y=1$

$\Rightarrow y=(4n+1)\dfrac{\pi}{2},\quad n\in I$

3. a., b., c., d.

Since $\vec{a}$, $\vec{b}$, and $\vec{c}$ are unit vectors inclined at an angle θ.

$|\vec{a}|=|\vec{b}|=1$ and $\cos\theta=\vec{a}\cdot\vec{c}=\vec{b}\cdot\vec{c}$

Now, $\vec{c}=\alpha\vec{a}+\beta\vec{b}+\gamma(\vec{a}\times\vec{b})$ (1)

$\Rightarrow \vec{a}\cdot\vec{c}=\alpha(\vec{a}\cdot\vec{a})+\beta(\vec{a}\cdot\vec{b})+\gamma\{\vec{a}\cdot(\vec{a}\times\vec{b})\}$

$\Rightarrow \cos\theta=\alpha|\vec{a}|^2 \qquad [\because\ \vec{a}\cdot\vec{b}=0,\ \vec{a}\cdot(\vec{a}\times\vec{b})=0]$

$\Rightarrow \cos\theta=\alpha$

Similarly, by taking dot product on both sides of (1) by $\vec{b}$, we get $\beta=\cos\theta$.

$\therefore \alpha=\beta$

Again, $\vec{c}=\alpha\vec{a}+\beta\vec{b}+\gamma(\vec{a}\times\vec{b})$

$\Rightarrow |\vec{c}|^2=|\alpha\vec{a}+\beta\vec{b}+\gamma(\vec{a}\times\vec{b})|^2$

$=\alpha^2|\vec{a}|^2+\beta^2|\vec{b}|^2+\gamma^2|\vec{a}\times\vec{b}|^2+2\alpha\beta(\vec{a}\cdot\vec{b})$

$+2\alpha\gamma\{\vec{a}\cdot(\vec{a}\times\vec{b})\}+2\beta\gamma(\vec{b}\cdot\{\vec{a}\times\vec{b}\})$

$\Rightarrow 1=\alpha^2+\beta^2+\gamma^2|\vec{a}\times\vec{b}|^2$

$\Rightarrow 1=2\alpha^2+\gamma^2\{|\vec{a}|^2|\vec{b}|^2\sin^2\pi/2\}$

$\Rightarrow 1=2\alpha^2+\gamma^2\Rightarrow \alpha^2=\dfrac{1-\gamma^2}{2}$

But $\alpha=\beta=\cos\theta$.

$1=2\alpha^2+\gamma^2\Rightarrow\gamma^2=1-2\cos^2\theta=-\cos 2\theta$

$\therefore \beta^2=\dfrac{1-\gamma^2}{2}=\dfrac{1+\cos 2\theta}{2}$

4. a., b., c.

We have,

$AM=$ projection of $\vec{b}$ on $\vec{a}=\dfrac{\vec{a}.\vec{b}}{|\vec{a}|}$

$\therefore\ \overrightarrow{AM}=\left(\dfrac{\vec{a}\cdot\vec{b}}{|\vec{a}|^2}\right)\vec{a}$

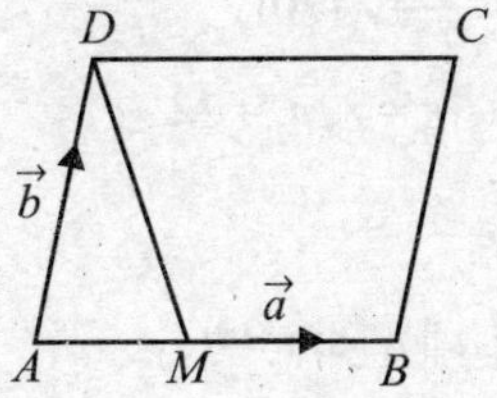

Fig. 2.7

Now, in ΔADM

$\overrightarrow{AD}=\overrightarrow{AM}+\overrightarrow{MD}\Rightarrow\overrightarrow{DM}=\overrightarrow{AM}-\overrightarrow{AD}$

$\Rightarrow \overrightarrow{DM}=\dfrac{(\vec{a}\cdot\vec{b})\vec{a}}{|\vec{a}|^2}-\vec{b}$

Also, $\overrightarrow{DM}=\dfrac{1}{|\vec{a}|^2}[(\vec{a}\cdot\vec{b})\vec{a}-|\vec{a}|^2\vec{b}]$

$\Rightarrow \overrightarrow{MD}=\dfrac{1}{|\vec{a}|^2}[|\vec{a}|^2\vec{b}-(\vec{a}\cdot\vec{b})\vec{a}]$

Now, $\dfrac{\vec{a}\times(\vec{a}\times\vec{b})}{|\vec{a}|^2}=\dfrac{1}{|\vec{a}|^2}[(\vec{a}\cdot\vec{b})\vec{a}-(\vec{a}\cdot\vec{a})\vec{b}]=\overrightarrow{DM}$

5. a., c. $\vec{a}\times(\vec{b}\times\vec{c})=(\vec{a}\cdot\vec{c})\vec{b}-(\vec{a}\cdot\vec{b})\vec{c}$ and

$(\vec{a}\times\vec{b})\times\vec{c}=-(\vec{c}\cdot\vec{b})\vec{a}+(\vec{a}\cdot\vec{c})\vec{b}$

We have been given $(\vec{a}\times(\vec{b}\times\vec{c}))\cdot((\vec{a}\times\vec{b})\times\vec{c})=0$. Therefore,

$((\vec{a}\cdot\vec{c})\vec{b}-(\vec{a}\cdot\vec{b})\vec{c})\cdot((\vec{a}\cdot\vec{c})\vec{b}-(\vec{c}\cdot\vec{b})\vec{a})=0$

$(\vec{a}\cdot\vec{c})^2|\vec{b}|^2-(\vec{a}\cdot\vec{c})(\vec{b}\cdot\vec{c})(\vec{a}\cdot\vec{b})-(\vec{a}\cdot\vec{b})(\vec{a}\cdot\vec{c})(\vec{b}\cdot\vec{c})$
$+(\vec{a}\cdot\vec{b})(\vec{b}\cdot\vec{c})(\vec{c}\cdot\vec{a})=0$

$(\vec{a}\cdot\vec{c})^2|\vec{b}|^2$

$=(\vec{a}\cdot\vec{c})(\vec{a}\cdot\vec{b})(\vec{b}\cdot\vec{c})\ (\vec{a}\cdot\vec{c})((\vec{a}\cdot\vec{c})(\vec{b}\cdot\vec{b})-(\vec{a}\cdot\vec{b})(\vec{b}\cdot\vec{c}))$

$=0$

$\vec{a}\cdot\vec{c}=0$ or $(\vec{a}\cdot\vec{c})\ |\vec{b}|^2=(\vec{a}\cdot\vec{b})(\vec{b}\cdot\vec{c})$

6. a., c.

We have $[\vec{p}\ \vec{q}\ \vec{r}]=\dfrac{1}{[\vec{a}\ \vec{b}\ \vec{c}]}$. Therefore, $[\vec{p}\ \vec{q}\ \vec{r}]>0$

a. $x>0, x[\vec{a}\ \vec{b}\ \vec{c}]+\dfrac{[\vec{p}\ \vec{q}\ \vec{r}]}{x}\ge 2$ [Using A.M. $\ge$ G.M]

b. Similarly, use A.M. $\ge$ G.M.

7. a., b., c.

$a_1 + a_2 \cos 2x + a_3 \sin^2 x = 0, \forall x \in R$

$\Rightarrow (a_1 + a_2) + \sin^2 x\,(a_3 - 2a_2) = 0$

$\Rightarrow a_1 + a_2 = 0 \;\&\; a_3 - 2a_2 = 0$

$\frac{a_1}{-1} = \frac{a_2}{1} = \frac{a_3}{2} = \lambda (\neq 0)$

$\Rightarrow a_1 = -\lambda, a_2 = \lambda, a_3 = 2\lambda$

8. a., b., c., d.

$\vec{a} \times \vec{b} = |\vec{a}||\vec{b}| \sin\theta\, \hat{n}$

$\Rightarrow |\vec{a} \times \vec{b}| = |\vec{a}||\vec{b}| \sin\theta \quad \Rightarrow \sin\theta = \frac{|\vec{a}\times\vec{b}|}{|\vec{a}||\vec{b}|}$ (1)

$\vec{a}\cdot\vec{b} = |\vec{a}||\vec{b}| \cos\theta$

$\Rightarrow \cos\theta = \frac{|\vec{a}\cdot\vec{b}|}{|\vec{a}||\vec{b}|}$ (2)

From (1) and (2),

$\sin^2\theta + \cos^2\theta = 1$

$\Rightarrow |\vec{a}\times\vec{b}|^2 + (\vec{a}\cdot\vec{b})^2 = |\vec{a}|^2|\vec{b}|^2$

If $\theta = \pi/4$, then $\sin\theta = \cos\theta = 1/\sqrt{2}$. Therefore,

$|\vec{a}\times\vec{b}| = \frac{|\vec{a}||\vec{b}|}{\sqrt{2}}$ and $\vec{a}\cdot\vec{b} = \frac{|\vec{a}||\vec{b}|}{\sqrt{2}}$

$|\vec{a}\times\vec{b}| = \vec{a}\cdot\vec{b}$

$\vec{a}\times\vec{b} = |\vec{a}||\vec{b}| \sin\theta\,\hat{n} = \frac{|\vec{a}||\vec{b}|}{\sqrt{2}}\hat{n}$

$\vec{a}\times\vec{b} = (\vec{a}\cdot\vec{b})\hat{n}$

9. a., b., c., d.

Since $\vec{a}, \vec{b}$ and $\vec{a}\times\vec{b}$ are non-coplanar,

$\vec{r} = x\vec{a} + y\vec{b} + z(\vec{a}\times\vec{b})$

$\therefore \vec{r}\times\vec{b} = \vec{a}$

$\Rightarrow x\vec{a}\times\vec{b} + z\{(\vec{a}\cdot\vec{b})\vec{b} - (\vec{b}\cdot\vec{b})\vec{a}\} = \vec{a}$

$\Rightarrow -(1 + z|\vec{b}|^2)\vec{a} + x\vec{a}\times\vec{b} = 0$ [Since $\vec{a}\cdot\vec{b} = 0$]

$\therefore x = 0$ and $z = -\frac{1}{|\vec{b}|^2}$

Thus, $\vec{r} = y\vec{b} - \frac{\vec{a}\times\vec{b}}{|\vec{b}|^2}$, where y is the parameter.

10. b., d. Since $\vec{a} = (1, 3, \sin 2\alpha)$ makes an obtuse angle with the z-axis, its z-component is negative. Therefore, $2\alpha < 0$

$\Rightarrow -1 \le \sin 2\alpha \le 0$ (1)

But $\vec{b}\cdot\vec{c} = 0$ [$\because$ orthogonal]

$\tan^2\alpha - \tan\alpha - 6 = 0$

$\therefore (\tan\alpha - 3)(\tan\alpha + 2) = 0$

$\Rightarrow \tan\alpha = 3, -2$

Now, $\tan\alpha = 3$. Therefore,

$\sin 2\alpha = \frac{2\tan\alpha}{1+\tan^2\alpha} = \frac{6}{1+9} = \frac{3}{5} = 0$

Not possible as $\sin 2\alpha < 0$

Now, if $\tan\alpha = -2$,

$\Rightarrow \sin 2\alpha = \frac{2\tan\alpha}{1+\tan^2\alpha} = \frac{-4}{1+4} = \frac{4}{5} = 0$

$\Rightarrow \tan 2\alpha > 0$

$\Rightarrow 2\alpha$ is the third quadrant. Also, $\sqrt{\sin\alpha/2}$ is meaningful. If $0 < \sin\alpha/2 < 1$, then $\alpha = (4n+1)\pi - \tan^{-1} 2$ and $\alpha = (4n+2)\pi - \tan^{-1} 2$.

11. b., d.

$\vec{a}\times(\vec{r}\times\vec{a}) = \vec{a}\times\vec{b}$

$3\vec{r} - (\vec{a}\cdot\vec{r})\,\vec{a} = \vec{a}\times\vec{b}$

Also $|\vec{r}\times\vec{a}| = |\vec{b}|$

$\Rightarrow \sin^2\theta = 2$

$\Rightarrow (1 - \cos^2\theta) = \frac{2}{3}$

$\Rightarrow \frac{1}{3} = \cos^2\theta$

$\Rightarrow \vec{a}.\vec{r} = \pm 1$

$\Rightarrow \vec{r} \pm \vec{a} = \vec{a}\times\vec{b}$

$\Rightarrow \vec{r} = \frac{1}{3}(\vec{a}\times\vec{b} \pm \vec{a})$

12. b., d.

$(\vec{a} - \vec{b})\times[(\vec{b} + \vec{a})\times(2\vec{a} + \vec{b})] = \vec{b} + \vec{a}$

$\Rightarrow \{(\vec{a} - \vec{b})\cdot(2\vec{a} + \vec{b})\}(\vec{b} + \vec{a}) - \{(\vec{a} - \vec{b})\cdot(\vec{b} + \vec{a})\}$

$(2\vec{a} + \vec{b}) = \vec{b} + \vec{a}$

$\Rightarrow (2 - \vec{a}\cdot\vec{b} - 1)(\vec{b} + \vec{a}) = \vec{b} + \vec{a}$

$\Rightarrow$ either $\vec{b} + \vec{a} = \vec{0}$ or $1 - \vec{a}\cdot\vec{b} = 1$

$\Rightarrow$ either $\vec{b} = -\vec{a}$ or $\vec{a}\cdot\vec{b} = 0$

$\Rightarrow$ either $\theta = \pi$ or $\theta = \pi/2$

13. a., d.

Given $\vec{c} = \lambda_1\vec{a} + \lambda_2\vec{b} + \lambda_3(\vec{a}\times\vec{b})$ (1)

and $\vec{a}\cdot\vec{b} = 0, |\vec{a}| = 1, |\vec{b}| = 1$

From (1), $\vec{a}\cdot\vec{c} = \lambda_1, \vec{c}\cdot\vec{b} = \lambda_2$

and $\vec{c}\cdot(\vec{a}\times\vec{b}) = |\vec{a}\times\vec{b}|^2\lambda_3$

$= (1.1 \sin 90°)^2 \lambda_3 = \lambda_3$

Hence $\lambda_1 + \lambda_2 + \lambda_3 = (\vec{a} + \vec{b} + \vec{a}\times\vec{b})\cdot\vec{c}$

14. b., c., d.

Obviously, $\frac{\vec{a}}{|\vec{a}|}+\frac{\vec{b}}{|\vec{b}|}$ is a vector in the plane of $\vec{a}$ and $\vec{b}$ and hence perpendicular to $\vec{a}\times\vec{b}$. It is also equally inclined to $\vec{a}$ and $\vec{b}$ as it is along the angle bisector.

15. a., d.

$|\vec{a}+\vec{b}|=|\vec{a}-2\vec{b}|$

$\Rightarrow \vec{a}\cdot\vec{b}=\frac{|\vec{b}|^2}{2}$

Also $\vec{a}\cdot\vec{b}+\frac{1}{|\vec{b}|^2+2}=\frac{|\vec{b}|^2+2}{2}+\frac{1}{|\vec{b}|^2+2}-1$

$\geq\sqrt{2}-1$ [Using A.M. $\geq$ G.M.]

16. b., d.

$\vec{V}_1=\vec{V}_2$

$\vec{a}\times(\vec{b}\times\vec{c})=(\vec{a}\times\vec{b})\times\vec{c}$

$\Rightarrow (\vec{a}\cdot\vec{c})\vec{b}-(\vec{a}\cdot\vec{b})\vec{c}=(\vec{a}\cdot\vec{c})\vec{b}-(\vec{b}\cdot\vec{c})\vec{a}$

$\Rightarrow (\vec{a}\cdot\vec{b})\vec{c}=(\vec{b}\cdot\vec{c})\vec{a}$

$\Rightarrow$ either $\vec{c}$ and $\vec{a}$ are collinear or $\vec{b}$ is perpendicular to both $\vec{a}$ and $\vec{c}\Rightarrow\vec{b}=\lambda(\vec{a}\times\vec{c})$.

17. b., c.

We have $\vec{A}+\vec{B}=\vec{a}$

$\Rightarrow \vec{A}\cdot\vec{a}+\vec{B}\cdot\vec{a}=\vec{a}\cdot\vec{a}$

$\Rightarrow 1+\vec{B}\cdot\vec{a}=a^2$ (given $\vec{A}\cdot\vec{a}=1$)

$\Rightarrow \vec{B}\cdot\vec{a}=a^2-1$ (1)

Also $\vec{A}\times\vec{B}=\vec{b}$

$\Rightarrow \vec{a}\times(\vec{A}\times\vec{B})=\vec{a}\times\vec{b}$

$\Rightarrow (\vec{a}\cdot\vec{B})\vec{A}-(\vec{a}\cdot\vec{A})\vec{B}=\vec{a}\times\vec{b}$

$\Rightarrow (a^2-1)\vec{A}-\vec{B}=\vec{a}\times\vec{b}$ (using (1) and $\vec{a}\cdot\vec{A}=1$) (2)

and $\vec{A}+\vec{B}=\vec{a}$ (3)

From (2) and (3)

$\vec{A}=\frac{(\vec{a}\times\vec{b})+\vec{a}}{a^2}$ and

$\vec{B}=\vec{a}-\left\{\frac{(\vec{a}\times\vec{b})+\vec{a}}{a^2}\right\}$

or $\vec{B}=\frac{(\vec{b}\times\vec{a})+\vec{a}(a^2-1)}{a^2}$

Thus $\vec{A}=\frac{(\vec{a}\times\vec{b})+\vec{a}}{a^2}$ and $\vec{B}=\frac{(\vec{b}\times\vec{a})+\vec{a}(a^2-1)}{a^2}$

18. c., d.

Since $[\vec{a}\,\vec{b}\,\vec{c}\,]=0$, $\vec{a}$, $\vec{b}$, and $\vec{c}$ are coplanar vectors.

Further, since $\vec{d}$ is equally inclined to $\vec{a}$, $\vec{b}$, and $\vec{c}$,

$\vec{d}\cdot\vec{a}=\vec{d}\cdot\vec{b}=\vec{d}\cdot\vec{c}=0$

$\vec{d}\cdot\vec{x}=\vec{d}\cdot\vec{y}=\vec{d}\cdot\vec{z}=0$

$\vec{d}\cdot\vec{r}=0$

19. b., d. Let $\vec{\alpha}=\hat{i}-\hat{j}-\hat{k}$, $\vec{\beta}=\hat{i}+\hat{j}+\hat{k}$ and $\vec{\gamma}=-\hat{i}+\hat{j}+\hat{k}$.

Let $\vec{a}=\hat{i}+\hat{k}$, $\vec{b}=2\hat{i}+\hat{j}+\hat{k}$,

$\vec{c}=3\hat{i}+2\hat{j}+\hat{k}$, $\vec{d}=-4\hat{i}-2\hat{j}-2\hat{k}$

Clearly, $\vec{\alpha}$ is perpendicular to $\vec{a}$, $\vec{b}$, $\vec{c}$, and $\vec{d}$.

Also, $[\vec{a}\,\vec{\beta}\,\vec{\gamma}]=\begin{vmatrix}1 & 0 & 1\\ 1 & 1 & 1\\ -1 & 1 & 1\end{vmatrix}=2\neq 0$

$[\vec{b}\,\vec{\beta}\,\vec{\gamma}]=\begin{vmatrix}2 & 1 & 1\\ 1 & 1 & 1\\ -1 & 1 & 1\end{vmatrix}=0$

$[\vec{c}\,\vec{\beta}\,\vec{\gamma}]=\begin{vmatrix}3 & 2 & 1\\ 1 & 1 & 1\\ -1 & 1 & 1\end{vmatrix}=2\neq 0$

$[\vec{d}\,\vec{\beta}\,\vec{\gamma}]=\begin{vmatrix}-4 & -2 & -2\\ 1 & 1 & 1\\ -1 & 1 & 1\end{vmatrix}=0$

20. b., d. Let $\vec{a}=-\frac{3}{2}(\hat{i}-\sqrt{3}\hat{j})$, $\vec{b}=\frac{3}{2}(\hat{i}-\sqrt{3}\hat{j})$,

$\vec{c}=-\frac{3}{2}(\hat{i}+\sqrt{3}\hat{j})$, $\vec{d}=\frac{3}{2}(\hat{i}+\sqrt{3}\hat{j})$

$\overrightarrow{AB}=3\hat{i}=\vec{\alpha}$ (say)

Clearly, $|\vec{a}|=|\vec{b}|=|\vec{c}|=|\vec{d}|=3$

If $\vec{\alpha}$ makes angle α, β, and γ and δ with $\vec{a}$, $\vec{b}$, $\vec{c}$, and $\vec{d}$, respectively, then

$\cos\alpha=\frac{\vec{\alpha}\cdot\vec{a}}{|\vec{\alpha}||\vec{a}|}=\frac{-9/2}{3\cdot 3}=-\frac{1}{2}$

$$\cos\beta=\frac{\vec{\alpha}\cdot\vec{b}}{|\vec{\alpha}||\vec{b}|}=\frac{9/2}{3\times 3}=\frac{1}{2}\Rightarrow\beta=60°$$

$$\cos\gamma=\frac{\vec{\alpha}\cdot\vec{c}}{|\vec{\alpha}||\vec{c}|}=\frac{-9/2}{3\times 3}\Rightarrow\gamma=-\frac{1}{2}$$

$$\cos\delta=\frac{\vec{\alpha}\cdot\vec{d}}{|\vec{\alpha}||\vec{d}|}=\frac{9/2}{3\times 3}=\frac{1}{2}\Rightarrow\delta=60°$$

21. a., b., c.

Consider $\vec{V}_1\cdot\vec{V}_2=0\Rightarrow A=90°$

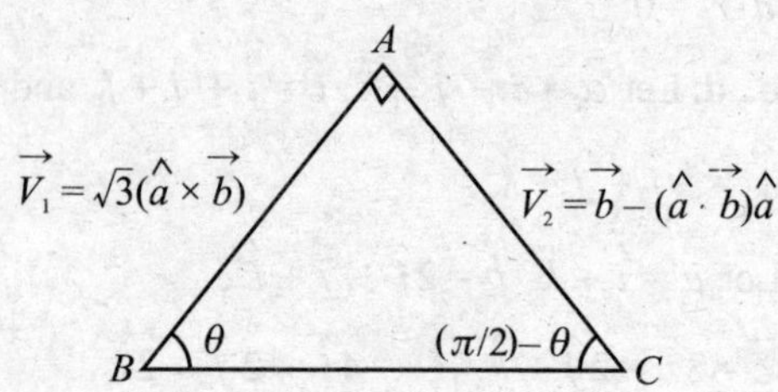

Fig. 2.8

Using the sine law, $\left|\frac{\vec{b}-(\hat{a}\cdot\vec{b})a}{\sin\theta}\right|=\frac{\sqrt{3}|\hat{a}\times\vec{b}|}{\cos\theta}$

$$\Rightarrow\tan\theta=\frac{1}{\sqrt{3}}\frac{|\vec{b}-(\vec{a}\cdot\vec{b})\vec{a}|}{|\vec{a}\times\vec{b}|}$$

$$=\frac{1}{\sqrt{3}}\frac{|(\vec{a}\times\vec{b})\times\vec{a}|}{|\vec{a}\times\vec{b}|}$$

$$=\frac{1}{\sqrt{3}}\frac{|\vec{a}\times\vec{b}||\vec{a}|\sin 90°}{|\vec{a}\times\vec{b}|}=\frac{1}{\sqrt{3}}$$

$$\Rightarrow\theta=\frac{\pi}{6}$$

22. a., b. Given, $\frac{1}{6}\hat{i}-\frac{1}{3}\hat{j}+\frac{1}{3}\hat{k}=(\vec{a}\times\vec{b})\times(\vec{c}\times\vec{d})$

$$=[\vec{a}\,\vec{b}\,\vec{d}]\vec{c}-[\vec{a}\,\vec{b}\,\vec{c}]\vec{d}$$

$$=[\vec{a}\,\vec{b}\,\vec{d}]\vec{c} \qquad (1)$$

[$\because$ $\vec{a},\vec{b}$ and $\vec{c}$ are coplanar]

$$[\vec{a}\,\vec{b}\,\vec{d}]=(\vec{a}\times\vec{b})\cdot\vec{d}$$

$$=|\vec{a}\times\vec{b}||\vec{d}|\cos\theta$$

$$[\because \vec{d}\perp\vec{a},\vec{d}\perp\vec{b}\ \therefore\vec{d}\parallel\vec{a}\times\vec{b}]$$

$$=ab\sin 30°.1.(\pm 1) \qquad [\because\theta=0\text{ or }\pi]$$

$$=1\times 1\times\frac{1}{2}\times 1(\pm 1)=\pm\frac{1}{2}$$

From (1),

$$\vec{c}=\pm\left(\frac{1}{3}\hat{i}-\frac{2}{3}\hat{j}+\frac{2}{3}\hat{k}\right)=\pm\frac{\hat{i}-2\hat{j}+2\hat{k}}{3}$$

23. a., b., c.

We know that $\vec{a}+\vec{b}+\vec{c}=\vec{0}$, then $\vec{a}\times\vec{b}=\vec{b}\times\vec{c}=\vec{c}\times\vec{a}$.

Given $\vec{a}+2\vec{b}+3\vec{c}=\vec{0}$

$$\Rightarrow 2\vec{a}\times\vec{b}=6\vec{b}\times\vec{c}=3\vec{c}\times\vec{a}.$$

Hence, $\vec{a}\times\vec{b}+\vec{b}\times\vec{c}+\vec{c}\times\vec{a}=2(\vec{a}\times\vec{b})$ or $6(\vec{b}\times\vec{c})$ or $3(\vec{c}\times\vec{a})$.

24. a., b. $\vec{u}=\vec{a}-(\vec{a}\cdot\vec{b})\vec{b}$

$$=\vec{a}(\vec{b}\cdot\vec{b})-(\vec{a}\cdot\vec{b})\vec{b}$$

$$=\vec{b}\times(\vec{a}\times\vec{b})$$

$$\Rightarrow|\vec{u}|=|\vec{b}\times(\vec{a}\times\vec{b})|$$

$$=|\vec{b}||\vec{a}\times\vec{b}|\sin 90°$$

$$=|\vec{b}||\vec{a}\times\vec{b}|$$

$$=|\vec{v}|$$

Also $\vec{u}\cdot\vec{b}=\vec{b}\cdot\vec{b}\times(\vec{a}\times\vec{b})$

$$=[\vec{b}\ \vec{b}\ \vec{a}\times\vec{b}]$$

$$=0$$

$$\Rightarrow|\vec{v}|=|\vec{u}|+|\vec{u}\cdot\vec{b}|$$

25. a., c.

$\vec{a}\times\vec{b}=\vec{c},\ \vec{b}\times\vec{c}=\vec{a}$

Taking cross with $\vec{b}$ in first equation, we get

$$\vec{b}\times(\vec{a}\times\vec{b})=\vec{b}\times\vec{c}=\vec{a}$$

$$\Rightarrow|\vec{b}|^2\vec{a}-(\vec{a}\cdot\vec{b})\vec{b}=\vec{a}\Rightarrow|\vec{b}|=1\text{ and }\vec{a}\cdot\vec{b}=0$$

Also $|\vec{a}\times\vec{b}|=|\vec{c}|\Rightarrow|\vec{a}||\vec{b}|\sin\frac{\pi}{2}=|\vec{c}|$

$$\Rightarrow|\vec{a}|=|\vec{c}|$$

26. b., d. $\vec{d}\cdot\vec{a}=[\vec{a}\ \vec{b}\ \vec{c}]\cos y=-\vec{d}\cdot(\vec{b}+\vec{c})$

$$\Rightarrow\cos y=-\frac{\vec{d}\cdot(\vec{b}+\vec{c})}{[\vec{a}\cdot\vec{b}\cdot\vec{c}]}$$

Similarly, $\sin x = -\dfrac{\vec{d}\cdot(\vec{a}+\vec{b})}{[\vec{a}\ \vec{b}\ \vec{c}]}$ and $\dfrac{\vec{d}\cdot(\vec{a}+\vec{c})}{[\vec{a}\ \vec{b}\ \vec{c}]} = -2$

$\therefore\ \sin x + \cos y + 2 = 0$

$\Rightarrow \sin x + \cos y = -2$

$\Rightarrow \sin x = -1, \cos y = -1$

Since we want the minimum value of $x^2+y^2, x = -\pi/2$, $y = \pi$

$\therefore$ The minimum value of x^2+y^2 is $5\pi^2/4$.

27. b., c.

$\vec{a}\times(\vec{b}\times\vec{c}) = \frac{1}{2}\vec{b}$

$\Rightarrow (\vec{a}\cdot\vec{c})\vec{b} - (\vec{a}\cdot\vec{b})\vec{c} = \frac{1}{2}\vec{b}$

$\Rightarrow \vec{a}\cdot\vec{c} = \frac{1}{2}$ and $\vec{a}\cdot\vec{b} = 0$

$\Rightarrow 1\times 1\cos\alpha = \frac{1}{2}$ and $\vec{a}\perp\vec{b}$

$\Rightarrow \alpha = \frac{\pi}{3}$ and $\vec{a}\perp\vec{b}$

28. a., b., c.

$\overrightarrow{AB} + \overrightarrow{BC} = \overrightarrow{AC}$

$\overrightarrow{BC} = \dfrac{2\vec{u}}{|\vec{u}|} - \dfrac{\vec{u}}{|\vec{u}|} + \dfrac{\vec{v}}{|\vec{v}|} = \dfrac{\vec{u}}{|\vec{u}|} + \dfrac{\vec{v}}{|\vec{v}|}$

$\overrightarrow{AB}.\overrightarrow{BC} = \left(\dfrac{\vec{u}}{|\vec{u}|} - \dfrac{\vec{v}}{|\vec{v}|}\right)\left(\dfrac{\vec{u}}{|\vec{u}|} + \dfrac{\vec{v}}{|\vec{v}|}\right)$

$= (\hat{u}-\hat{v}).(\hat{u}+\hat{v}) = 1 - 1 = 0.$

$\Rightarrow \angle B = 90°$

$\Rightarrow 1 + \cos 2A + \cos 2B + \cos 2C = 0.$

29. a., b., c.

Let $\vec{A} = \vec{a}\times\vec{b}$, $\vec{B} = \vec{c}\times\vec{d}$ and $\vec{C} = \vec{e}\times\vec{f}$.

We know that

$\vec{A}\cdot(\vec{B}\times\vec{C}) = \vec{B}\cdot(\vec{C}\times\vec{A}) = \vec{C}\cdot(\vec{A}\times\vec{B})$

$= (\vec{a}\times\vec{b})\cdot[(\vec{c}\times\vec{d})\times(\vec{e}\times\vec{f})]$

$= (\vec{a}\times\vec{b})\cdot[\{(\vec{c}\times\vec{d})\cdot\vec{f}\}\vec{e} - \{(\vec{c}\times\vec{d})\cdot\vec{e}\}\vec{f}\}$

$= [\vec{c}\ \vec{d}\ \vec{f}][\vec{a}\ \vec{b}\ \vec{e}] - [\vec{c}\ \vec{d}\ \vec{e}][\vec{a}\ \vec{b}\ \vec{f}]$

Similarly, other parts can be obtained.

30. a., c.

Here $(l\vec{a} + m\vec{b})\times\vec{b} = \vec{c}\times\vec{b} \Rightarrow l\vec{a}\times\vec{b} = \vec{c}\times\vec{b}$

$\Rightarrow l(\vec{a}\times\vec{b})^2 = (\vec{c}\times\vec{b})\cdot(\vec{a}\times\vec{b})$

$\Rightarrow l = \dfrac{(\vec{c}\times\vec{b})\cdot(\vec{a}\times\vec{b})}{(\vec{a}\times\vec{b})^2}$

Similarly, $m = \dfrac{(\vec{c}\times\vec{a})\cdot(\vec{b}\times\vec{a})}{(\vec{b}\times\vec{a})^2}$

31. b., c., d.

$(\vec{a}\times\vec{b})\times(\vec{c}\times\vec{d})\cdot(\vec{a}\times\vec{d}) = 0$

$\Rightarrow ([\vec{a}\ \vec{c}\ \vec{d}]\vec{b} - [\vec{b}\ \vec{c}\ \vec{d}]\vec{a})\cdot(\vec{a}\times\vec{d}) = 0$

$\Rightarrow [\vec{a}\ \vec{c}\ \vec{d}][\vec{b}\ \vec{a}\ \vec{d}] = 0$

$\Rightarrow$ Either $\vec{c}$ or $\vec{b}$ must lie in the plane of $\vec{a}$ and $\vec{d}$.

32. a., b., c.

Let $\overrightarrow{EB} = p\,\overrightarrow{AB}$ and $\overrightarrow{CE} = q\,\overrightarrow{CD}$.

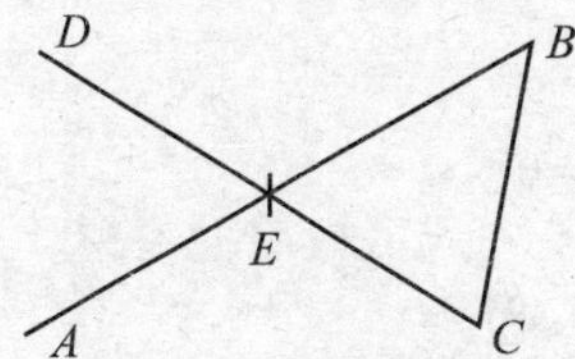

Fig. 2.9

Then $0 < p$ and $q \le 1$

Since $\overrightarrow{EB} + \overrightarrow{BC} + \overrightarrow{CE} = 0$

$\therefore pm(2\hat{i} - 6\hat{j} + 2\hat{k}) + (-2) + qn(-6 + 15 - 3) = 0$

$\Rightarrow (2pm + 1 - 6qn)\hat{i} + (-6pm - 2 + 15qn)\hat{j}$

$+ (2pm - 6qn)\hat{k} = 0$

$\Rightarrow 2pm - 6qn + 1 = 0, -6pm - 2 + 15qn = 0,$

$2pm - 6qn = 0$

Solving these, we get

$p = 1/(2m)$ and $q = 1/(3n)$

$\therefore 0 < 1/(2m) \le 1$ and $0 < 1/(3n) \le 1$

$\Rightarrow m \ge 1/2$ and $n \ge 1/3$

The area of $\Delta BCE = \frac{1}{2}|\overrightarrow{EB}\times\overrightarrow{BC}| = \frac{1}{2}\sqrt{6}$.

33. a., b., d.

$$\left.\begin{aligned} \vec{V}_1 &= l\vec{a} + m\vec{b} + n\vec{c} \\ \vec{V}_2 &= n\vec{a} + l\vec{b} + m\vec{c} \\ \vec{V}_3 &= m\vec{a} + n\vec{b} + l\vec{c} \end{aligned}\right\}$$ when $\vec{a}, \vec{b}$ and $\vec{c}$ are non-coplanar.

Therefore,

$$[\vec{V}_1\ \vec{V}_2\ \vec{V}_3] = \begin{vmatrix} l & m & n \\ n & l & m \\ m & n & l \end{vmatrix} [\vec{a}\ \vec{b}\ \vec{c}] = 0$$

But $[\vec{a}\,\vec{b}\,\vec{c}] \neq 0 \Rightarrow \begin{vmatrix} l & m & n \\ n & l & m \\ m & n & l \end{vmatrix} = 0$. Therefore,

$(l+m+n)\,[(l-m)^2+(m-n)^2+(n-l)^2]=0$

$l+m+n=0$ (1)

Obviously, $lx^2+mx+n=0$ is satisfied by $x=1$ due to (1).

$l^3+m^3+n^3=3lmn$

$\Rightarrow (l+m+n)(l^2+m^2+n^2-lm-mn-ln)=0$, which is true.

34. a., b., c.

It is given that $\vec{\alpha}$, $\vec{b}$, and $\vec{\gamma}$ are coplanar vectors.

Therefore,

$[\vec{\alpha}\,\vec{\beta}\,\vec{\gamma}]=0$

$$\Rightarrow \begin{vmatrix} a & b & c \\ b & c & a \\ c & a & b \end{vmatrix} = 0$$

$\Rightarrow 3abc-a^3-b^3-c^3=0$

$\Rightarrow a^3+b^3+c^3-3abc=0$

$\Rightarrow (a+b+c)(a^2+b^2+c^2-ab-bc-ca)=0$

$\Rightarrow a+b+c=0\ \ [\because a^2+b^2+c^2-ab-bc-ca \neq 0]$

$\Rightarrow \vec{v}\cdot\vec{\alpha} = \vec{v}\cdot\vec{\beta} = \vec{v}\cdot\vec{\gamma} = 0$

$\Rightarrow \vec{v}$ is perpendicular to $\vec{\alpha}$, $\vec{\beta}$, and $\vec{\gamma}$.

35. b., d.

For $\vec{A}$, $\vec{B}$ and $\vec{C}$ to form a left-handed system $[\vec{A}\vec{B}\vec{C}]<0$

$$\vec{A}\times\vec{B} \begin{vmatrix} \hat{i} & \hat{j} & \hat{k} \\ 2 & 3 & 4 \\ 1 & 1 & 5 \end{vmatrix} = 11\hat{i}-6\hat{j}-\hat{k} \quad (1)$$

(1) is satisfied by options (b) and (d)

Reasoning Type

1. b. A vector along the bisector is

$$\frac{\vec{a}}{|\vec{a}|}+\frac{\vec{b}}{|\vec{b}|}=\frac{-5\hat{i}+7\hat{j}+2\hat{k}}{9}$$

Hence, $\vec{c}=-5\hat{i}+7\hat{j}+2\hat{k}$ is along the bisector. It is obvious that $\vec{c}$ makes an equal angle with $\vec{a}$ and $\vec{b}$.

However, statement 2 does not explain statement 1, as a vector inclined to given two vectors is not necessarily coplanar.

2. c. Component of vector $\vec{b}=4\hat{i}+2\hat{j}+3\hat{k}$ in the direction of $\vec{a}=\hat{i}+\hat{j}+\hat{k}$ is $\frac{\vec{a}\cdot\vec{b}\ \vec{a}}{|\vec{a}|\,|\vec{a}|}$ or $3\hat{i}+3\hat{j}+3\hat{k}$. Then component in the direction perpendicular to the direction of $\vec{a}=\hat{i}+\hat{j}+\hat{k}$ is $\vec{b}-3\hat{i}+3\hat{j}+3\hat{k}=\hat{i}-\hat{j}$

3. d. $\overrightarrow{AD}=2\hat{j}-\hat{k}$, $\overrightarrow{BD}=-2\hat{i}-\hat{j}-3\hat{k}$ and $\overrightarrow{CD}=2\hat{i}-\hat{j}$

Volume of tetrahedron is

$$\frac{1}{6}[\overrightarrow{AD}\ \overrightarrow{BD}\ \overrightarrow{CD}]=\frac{1}{6}\left\|\begin{matrix} 0 & 2 & -1 \\ -2 & -1 & -3 \\ 2 & -1 & 0 \end{matrix}\right\|=\frac{8}{3}$$

Also, the area of the triangle ABC is

$$\frac{1}{2}|\overrightarrow{AB}\times\overrightarrow{AC}|=\frac{1}{2}\left\|\begin{matrix} \hat{i} & \hat{j} & \hat{k} \\ 2 & 3 & 2 \\ -2 & 3 & -1 \end{matrix}\right\|$$

$$=\frac{1}{2}|-9\hat{i}-2\hat{j}+12\hat{k}|$$

$$=\frac{\sqrt{129}}{2}$$

Then $\frac{8}{3}=\frac{1}{3}\times$ (distance of D from base ABC) $\times$ (area of triangle ABC)

Distance of D from base $ABC = 16/\sqrt{129}$

4. b. $\vec{r}\cdot\vec{a}=\vec{r}\cdot\vec{b}=\vec{r}\cdot\vec{c}=0$ only if $\vec{a}$, $\vec{b}$, and $\vec{c}$ are coplanar.

$\Rightarrow [\vec{a}\,\vec{b}\,\vec{c}]=0$

Hence, statement 2 is true.

Also, $[\vec{a}-\vec{b}\ \ \vec{b}-\vec{c}\ \ \vec{c}-\vec{a}]=0$ even if $[\vec{a}\ \vec{b}\ \vec{c}]\neq 0$.

Hence, statement 2 is not the correct explanation for statement 1.

5. a. Let the three given unit vectors be $\hat{a}$, $\hat{b}$, and $\hat{c}$. Since they are mutually perpendicular, $\hat{a}\cdot(\hat{b}\times\hat{c})=1$. Therefore,

$$\begin{vmatrix} a_1 & a_2 & a_3 \\ b_1 & b_2 & b_3 \\ c_1 & c_2 & c_3 \end{vmatrix} = 1$$

$$\begin{vmatrix} a_1 & b_1 & c_1 \\ a_2 & b_2 & c_2 \\ a_3 & b_3 & c_3 \end{vmatrix} = 1$$

Hence, $a_1\hat{i} + b_1\hat{j} + c_1\hat{k}$, $a_2\hat{i} + b_2\hat{j} + c_2\hat{k}$, and $a_3\hat{i} + b_3\hat{j} + c_3\hat{k}$ may be mutually perpendicular.

6. d. $\vec{V} = \vec{A}\times((\vec{A}\cdot\vec{B})\vec{A} - (\vec{A}\cdot\vec{A})\vec{B})\cdot\vec{C}$

$$= \left(\underbrace{\vec{A}\times(\vec{A}\cdot\vec{B})\vec{A}}_{\text{zero}} - (\vec{A}\cdot\vec{A})\vec{A}\times\vec{B}\right)\cdot\vec{C}$$

$$= -|\vec{A}|^2[\vec{A}\ \vec{B}\ \vec{C}]$$

Now, $|\vec{A}|^2 = 4+9+36 = 49$

$$[\vec{A}\vec{B}\vec{C}] = \begin{vmatrix} 2 & 3 & 6 \\ 1 & 1 & -2 \\ 1 & 2 & 1 \end{vmatrix}$$

$$= 2(1+4) - 1(3-12) + 1(-6-6)$$

$$= 10 + 9 - 12 = 7$$

$\therefore |-|\vec{A}|^2[\vec{A}\vec{B}\vec{C}]| = 49\times 7 = 343$

7. b. Let $\vec{d} = \lambda_1\vec{a} + \lambda_2\vec{b} + \lambda_3\vec{c}$

$\Rightarrow [\vec{d}\,\vec{a}\,\vec{b}] = \lambda_3[\vec{c}\,\vec{a}\,\vec{b}] = \lambda_3 = 1$

$[\vec{c}\,\vec{a}\,\vec{b}] = 1$ (because $\vec{a}$, $\vec{b}$, and $\vec{c}$ are three mutually perpendicular unit vectors)

Similarly, $\lambda_1 = \lambda_2 = 1$

$\Rightarrow \vec{d} = \vec{a} + \vec{b} + \vec{c}$

Hence, statement 1 and statement 2 are correct, but statement 2 does not explain statement 1 as it does not give the value of dot products.

8. a. Statement 2 is true (see properties of dot product)

Also, $(\hat{i}\times\vec{a})\cdot\vec{b} = \hat{i}\cdot(\vec{a}\times\vec{b})$

$\Rightarrow \vec{a}\times\vec{b} = (\hat{i}\cdot(\vec{a}\times\vec{b}))\hat{i} + (\hat{j}\cdot(\vec{a}\times\vec{b}))\hat{j}$

$+ (\hat{k}\cdot(\vec{a}\times\vec{b}))\hat{k}$

Linked Comprehension Type

For Problems 1–3

1.b., 2. c., 3. d.

Sol. Taking dot product of $\vec{u}+\vec{v}+\vec{w} = \vec{a}$ with $\vec{u}$, we have

$$1 + \vec{u}\cdot\vec{v} + \vec{u}\cdot\vec{w} = \vec{a}\cdot\vec{u} = \frac{3}{2} \Rightarrow \vec{u}\cdot\vec{v} + \vec{u}\cdot\vec{w} = \frac{1}{2} \quad (1)$$

Similarly, taking dot product with $\vec{v}$, we have

$$\vec{u}\cdot\vec{v} + \vec{w}\cdot\vec{v} = \frac{3}{4} \quad (2)$$

Also, $\vec{a}\cdot\vec{u} + \vec{a}\cdot\vec{v} + \vec{a}\cdot\vec{w} = \vec{a}\cdot\vec{a} = 4$

$$\Rightarrow \vec{a}\cdot\vec{w} = 4 - \left(\frac{3}{2} + \frac{7}{4}\right) = \frac{3}{4}$$

Again, taking dot product with $\vec{w}$, we have

$$\vec{u}\cdot\vec{w} + \vec{v}\cdot\vec{w} = \frac{3}{4} - 1 = -\frac{1}{4} \quad (3)$$

Adding (1), (2), and (3), we have

$$2(\vec{u}\cdot\vec{v} + \vec{u}\cdot\vec{w} + \vec{v}\cdot\vec{w}) = 1$$

$$\vec{u}\cdot\vec{v} + \vec{u}\cdot\vec{w} + \vec{v}\cdot\vec{w} = \frac{1}{2} \quad (4)$$

Subtracting (1), (2), and (3), from (4), we have

$$\vec{v}\cdot\vec{w} = 0,\ \vec{u}\cdot\vec{w} = -\frac{1}{4} \text{ and } \vec{u}\cdot\vec{v} = \frac{3}{4}$$

Now, the equations $\vec{u}\times(\vec{v}\times\vec{w}) = \vec{b}$ and $(\vec{u}\times\vec{v})\times\vec{w} = \vec{c}$ can be written as

$$(\vec{u}\cdot\vec{w})\vec{v} - (\vec{u}\cdot\vec{v})\vec{w} = \vec{b}$$

and $(\vec{u}\cdot\vec{w})\vec{v} - (\vec{v}\cdot\vec{w})\vec{u} = \vec{c}$

$$\Rightarrow -\frac{1}{4}\vec{v} - \frac{3}{4}\vec{w} = \vec{b},\ -\frac{1}{4}\vec{v} = \vec{c}, \text{ i.e., } \vec{v} = -4\vec{c}$$

$$\Rightarrow \vec{c} - \frac{3}{4}\vec{w} = \vec{b} \Rightarrow \vec{w} = \frac{4}{3}(\vec{c} - \vec{b}) \text{ and}$$

$$\vec{u} = \vec{a} - \vec{v} - \vec{w} = \vec{a} + 4\vec{c} - \frac{4}{3}\vec{c} + \frac{4}{3}\vec{b} = \vec{a} + \frac{4}{3}\vec{b} + \frac{8}{3}\vec{c}$$

For Problems 4–6

4.d., 5. c., 6. b.

Sol. Given that $|\vec{x}| = |\vec{y}| = |\vec{z}| = \sqrt{2}$ and they are inclined at an angle of 60° with each other.

$$\therefore \vec{x}\cdot\vec{y} = \vec{y}\cdot\vec{z} = \vec{z}\cdot\vec{x} = \sqrt{2}\times\sqrt{2}\cos 60° = 1$$

$$\vec{x}\times(\vec{y}\times\vec{z}) = \vec{a} \Rightarrow (\vec{x}\cdot\vec{z})\vec{y} - (\vec{x}\cdot\vec{y})\vec{z} = \vec{a}$$

$$\Rightarrow \vec{y} - \vec{z} = \vec{a} \quad (1)$$

Similarly, $\vec{y}\times(\vec{z}\times\vec{x})=\vec{b}\Rightarrow \vec{z}-\vec{x}=\vec{b}$ (2)

$\vec{y}=\vec{a}+\vec{z},\ \vec{x}=\vec{z}-\vec{b}$ [From (1) and (2)] (3)

Now, $\vec{x}\times\vec{y}=\vec{c}$

$\Rightarrow (\vec{z}-\vec{b})\times(\vec{z}+\vec{a})=\vec{c}$

$\Rightarrow \vec{z}\times\vec{a}-\vec{b}\times\vec{z}-\vec{b}\times\vec{a}=\vec{c}$

$\Rightarrow \vec{z}\times(\vec{a}+\vec{b})=\vec{c}+(\vec{b}\times\vec{a})$ (4)

$\Rightarrow (\vec{a}+\vec{b})\times\{\vec{z}\times(\vec{a}+\vec{b})\}$

$=(\vec{a}+\vec{b})\times\vec{c}+(\vec{a}+\vec{b})\times(\vec{b}\times\vec{a})$

$\Rightarrow (\vec{a}+\vec{b})^2\vec{z}-\{(\vec{a}+\vec{b})\cdot\vec{z}\}(\vec{a}+\vec{b})$

$=(\vec{a}+\vec{b})\times\vec{c}+|\vec{a}|^2\vec{b}-|\vec{b}|^2\vec{a}+(\vec{a}\cdot\vec{b})(\vec{b}-\vec{a})$ (5)

Now, (1) $\Rightarrow |\vec{a}|^2=|\vec{y}-\vec{z}|^2=2+2-2=2$

Similarly, (2) $\Rightarrow |\vec{b}|^2=2$

Also (1) and (2)

$\Rightarrow \vec{a}+\vec{b}=\vec{y}-\vec{x}\quad\Rightarrow |\vec{a}+\vec{b}|^2=2$ (6)

Also $(\vec{a}+\vec{b})\cdot\vec{z}=(\vec{y}-\vec{x})\cdot\vec{z}=\vec{y}\cdot\vec{z}-\vec{x}\cdot\vec{z}$

$=1-1=0$

and $\vec{a}\cdot\vec{b}=\vec{y}\cdot\vec{x}-|\vec{z}|^2+\vec{z}\cdot\vec{x}=-1$

Thus from (5), we have

$2\vec{z}=(\vec{a}+\vec{b})\times\vec{c}+2(\vec{b}-\vec{a})-(\vec{b}-\vec{a})$ or

$\vec{z}=(1/2)[(\vec{a}+\vec{b})\times\vec{c}+\vec{b}-\vec{a}]$

$\therefore\ \vec{y}=\vec{a}+\vec{z}=(1/2)[(\vec{a}+\vec{b})\times\vec{c}+\vec{b}+\vec{a}]$ and

$\vec{x}=\vec{z}-\vec{b}=(1/2)[(\vec{a}+\vec{b})\times\vec{c}-(\vec{a}+\vec{b})]$

For Problems 7–9

7.d., 8. a., 9. c.

Sol. Given

$\vec{x}\times\vec{y}=\vec{a}$ (1)

$\vec{y}\times\vec{z}=\vec{b}$ (2)

$\vec{x}\cdot\vec{b}=\gamma$ (3)

$\vec{x}\cdot\vec{y}=1$ (4)

$\vec{y}\cdot\vec{z}=1$ (5)

From (2), $\vec{x}\cdot(\vec{y}\times\vec{z})=\vec{x}\cdot\vec{b}=\gamma\Rightarrow[\vec{x}\ \vec{y}\ \vec{z}]=\gamma$

From (1) and (2), $(\vec{x}\times\vec{y})\times(\vec{y}\times\vec{z})=\vec{a}\times\vec{b}$

$\therefore\ [\vec{x}\ \vec{y}\ \vec{z}]\vec{y}-[\vec{y}\ \vec{y}\ \vec{z}]\vec{x}=\vec{a}\times\vec{b}\Rightarrow \vec{y}=\dfrac{\vec{a}\times\vec{b}}{\gamma}$ (7)

Also from (1) and (2), we get $(\vec{x}\times\vec{y})\times\vec{y}=\vec{a}\times\vec{y}$

$\Rightarrow (\vec{x}\cdot\vec{y})\vec{y}-(\vec{y}\cdot\vec{y})\vec{x}=\vec{a}\times\vec{y}$

$\Rightarrow \vec{x}=(1/|\vec{y}|^2)(\vec{y}-\vec{a}\times\vec{y})$

$$=\frac{\gamma^2}{|\vec{a}\times\vec{b}|^2}\left[\frac{\vec{a}\times\vec{b}}{\gamma}-\frac{\vec{a}\times(\vec{a}\times\vec{b})}{\gamma}\right]$$

$$\Rightarrow \vec{x}=\frac{\gamma}{|\vec{a}\times\vec{b}|^2}[\vec{a}\times\vec{b}-\vec{a}\times(\vec{a}\times\vec{b})]$$

Also from (2), $(\vec{y}\times\vec{z})\times\vec{y}=\vec{b}\times\vec{y}$

$\Rightarrow |\vec{y}|^2\vec{z}-(\vec{z}\cdot\vec{y})\vec{y}=\vec{b}\times\vec{y}$

$\Rightarrow \vec{z}=(1/|\vec{y}|^2)[\vec{y}+\vec{b}\times\vec{y}]$

$$=\frac{\gamma}{|\vec{a}\times\vec{b}|^2}[\vec{a}\times\vec{b}+\vec{b}\times(\vec{a}\times\vec{b})]$$

For Problems 10–12

10.d., 11. b., 12. b.

Sol.

$\vec{P}\times\vec{B}=\vec{A}-\vec{P}$ and $|\vec{A}|=|\vec{B}|=1$ and $\vec{A}\cdot\vec{B}=0$ is given

Now $\vec{P}\times\vec{B}=\vec{A}-\vec{P}$ (1)

$(\vec{P}\times\vec{B})\times\vec{B}=(\vec{A}-\vec{P})\times\vec{B}$ [Taking cross product with $\vec{B}$ on both sides]

$(\vec{P}\cdot\vec{B})\vec{B}-(\vec{B}\cdot\vec{B})\vec{P}=\vec{A}\times\vec{B}-\vec{P}\times\vec{B}$

$(\vec{P}\cdot\vec{B})\vec{B}-\vec{P}=\vec{A}\times\vec{B}-\vec{A}+\vec{P}$

$\Rightarrow 2\vec{P}=\vec{A}-\vec{A}\times\vec{B}-(\vec{P}\cdot\vec{B})\vec{B}$

$$\Rightarrow \vec{P}=\frac{\vec{A}-\vec{A}\times\vec{B}-(\vec{P}\cdot\vec{B})\vec{B}}{2} \quad (2)$$

Taking dot product with $\vec{B}$ on both sides of (1) we get

$[\vec{P}\ \vec{B}\ \vec{B}]=\vec{A}\cdot\vec{B}-\vec{P}\cdot\vec{B}$

$\Rightarrow \vec{P}\cdot\vec{B}=0$ (3)

$$\vec{P}=\frac{\vec{A}+\vec{B}\times\vec{A}}{2}$$

Now $(\vec{P}\times\vec{B})\times\vec{B}=(\vec{P}\cdot\vec{B})\vec{B}-(\vec{B}\cdot\vec{B})\vec{P}=-\vec{P}$.

For Problems 13–15

13.b., 14. a., 15. c.

Sol.

13. b.

$$\vec{a}_1=\left[(2\hat{i}+3\hat{j}-6\hat{k})\cdot\frac{(2\hat{i}-3\hat{j}+6\hat{k})}{7}\right]\frac{2\hat{i}-3\hat{j}+6\hat{k}}{7}$$

$$=\frac{-41}{49}(2\hat{i}-3\hat{j}+6\hat{k})$$

$$\vec{a}_2=\frac{-41}{49}\left((2\hat{i}-3\hat{j}+6\hat{k})\cdot\frac{(-2\hat{i}+3\hat{j}+6\hat{k})}{7}\right)$$

$$\frac{(-2\hat{i}+3\hat{j}+6\hat{k})}{7}$$

$$=\frac{-41}{(49)^2}(-4-9+36)(-2\hat{i}+3\hat{j}+6\hat{k})$$

$$=\frac{943}{49^2}(2\hat{i}-3\hat{j}-6\hat{k})$$

14. a. $\vec{a}_1\cdot\vec{b}=\frac{-41}{49}(2\hat{i}-3\hat{j}+6\hat{k})\cdot(2\hat{i}-3\hat{j}+6\hat{k})=-41$

15. c. $\vec{a}$, $\vec{a}_1$, and $\vec{b}$ are coplanar because $\vec{a}_1$ and $\vec{b}$ are collinear.

For Problems 16–18

16.b., 17. c., 18. a.

Sol. Point G is $\left(\frac{4}{3},\frac{1}{3},\frac{8}{3}\right)$. Therefore,

$$\Rightarrow\left|\overrightarrow{AG}\right|^2=\left(\frac{5}{3}\right)^2+\frac{1}{9}+\left(\frac{5}{3}\right)^2=\frac{51}{9}$$

$$\Rightarrow\left|\overrightarrow{AG}\right|=\frac{\sqrt{51}}{3}$$

$$\overrightarrow{AB}=-4\hat{i}+4\hat{j}+0\hat{k}$$

$$\overrightarrow{AC}=2\hat{i}+2\hat{j}+2\hat{k}$$

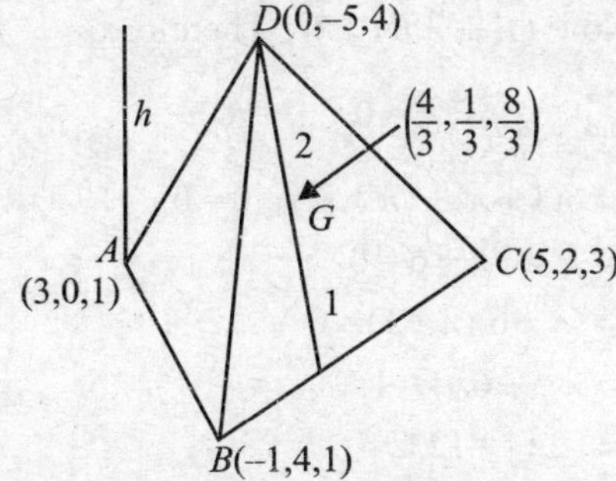

Fig. 2.10

$$\therefore\ \overrightarrow{AB}\times\overrightarrow{AC}=-8\begin{vmatrix}\hat{i}&\hat{j}&\hat{k}\\1&-1&0\\1&1&1\end{vmatrix}$$

$$=8(\hat{i}+\hat{j}-2\hat{k})$$

$$\text{Area of }\Delta ABC=\frac{1}{2}|\overrightarrow{AB}\times\overrightarrow{AC}|=4\sqrt{6}$$

$$\overrightarrow{AD}=-3\hat{i}-5\hat{j}+3\hat{k}$$

The length of the perpendicular from the vertex D on the opposite face

$$=|\text{projection of }\overrightarrow{AD}\text{ on }\overrightarrow{AB}\times\overrightarrow{AC}|$$

$$=\left|\frac{(-3\hat{i}-5\hat{j}+3\hat{k})(\hat{i}+\hat{j}-2\hat{k})}{\sqrt{6}}\right|$$

$$=\left|\frac{-3-5-6}{\sqrt{6}}\right|=\frac{14}{\sqrt{6}}$$

For Problems 19–21

19.c., 20. b., 21. d.

Sol.

19. c. Let the point D be (a_1, a_2, a_3)

$a_1+1=3\Rightarrow a_1=2$

$a_2+0=1\Rightarrow a_2=1$

$a_3-1=7\Rightarrow a_3=8$

$\therefore D(2, 1, 8)$

$$\vec{d}=\left|\frac{(\overrightarrow{AB})\times(\overrightarrow{AD})}{|\overrightarrow{AB}|}\right|$$

$$\overrightarrow{AB}=\hat{i}-\hat{j}+5\hat{k}$$

$$\overrightarrow{AD}=0\hat{i}+2\hat{j}+4\hat{k}$$

$$\overrightarrow{AB}\times\overrightarrow{AD}=\begin{vmatrix}\hat{i}&\hat{j}&\hat{k}\\1&-1&5\\0&2&4\end{vmatrix}$$

$$=-14\hat{i}-4\hat{j}+2\hat{k}$$

$$=-2(7\hat{i}+2\hat{j}-\hat{k})$$

$$\Rightarrow|\overrightarrow{AB}\times\overrightarrow{AD}|=2\sqrt{2}$$

20. b.

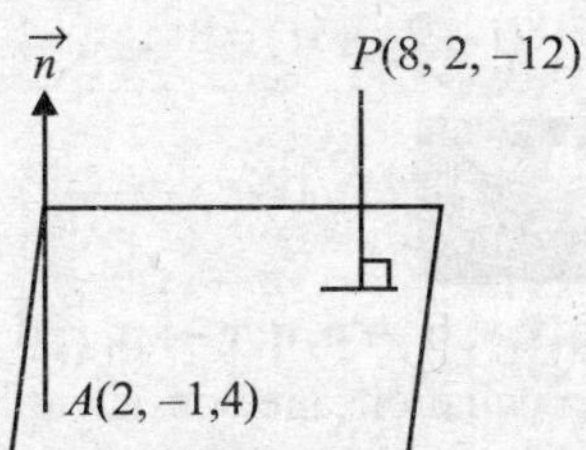

Fig. 2.11

$\vec{n}=7\hat{i}+2\hat{j}-\hat{k}$ is normal to the plane $P\equiv(8, 2, -12)$.

$$|\overrightarrow{AP}|=6\hat{i}+3\hat{j}-16\hat{k}$$

$\therefore$ distance $d = \left|\dfrac{\overrightarrow{AP}\cdot\vec{n}}{|\vec{n}|}\right|$

$= \left|\dfrac{42+6+16}{\sqrt{49+4+1}}\right|$

$= \dfrac{64}{\sqrt{54}}$

$= \dfrac{64}{3\sqrt{6}} = \dfrac{64\sqrt{6}}{18} = \dfrac{32\sqrt{6}}{9}$

21. d. Vector normal to the plane

$\overrightarrow{AD} \times \overrightarrow{AB} = +2(7\hat{i} + 2\hat{j} - \hat{k})$

Projection of $xy = 2$;

$yz = 14$

$zx = 4$

For Problems 22–24

22.d., 23. c., 24. c.

Sol. Let $\vec{r} = x\hat{i} + y\hat{j}$

$x^2 + y^2 + 8x - 10y + 40 = 0$

centre $C(-4, 5)$, radius $r = 1$

$p_1 = \max\{(x+2)^2 + (y-3)^2\}$

$p_2 = \min\{(x+2)^2 + (y-3)^2\}$

Let P be $(-2, 3)$. Then

$CP = 2\sqrt{2}, r = 1$

$p_1 = (2\sqrt{2} - 1)^2$

$p_2 = (2\sqrt{2} + 1)^2$

$p_1 + p_2 = 18$

Slope $= AB = \left(\dfrac{dy}{dx}\right)_{(2,2)} = -2.$

Equation of AB, $2x + y = 6$

$\overrightarrow{OA} = 2\hat{i} + 2\hat{j},\ \overrightarrow{OB} = 3\hat{i}$

$\overrightarrow{AB} = \hat{i} - 2\hat{j}$

$\overrightarrow{AB}\cdot\overrightarrow{OB} = (\hat{i} - 2\hat{j})(3\hat{i}) = 3$

Matrix-Match Type

1. a. → p, q, r, s; b → p, q; c → p, r; d → r

a. Given equations are consistent if

$(\hat{i} + \hat{j}) + \lambda(\hat{i} + 2\hat{j} - \hat{k})$

$= (\hat{i} + 2\hat{j}) + \mu(-\hat{i} + \hat{j} + a\hat{k})$

$\Rightarrow 1 + \lambda = 1 - \mu, 1 + 2\lambda = 2 + \mu, -\lambda = a\mu$

$\Rightarrow \lambda = -1/3$ and $\mu = 1/3$

$\Rightarrow a = 1$

b. $\vec{a} = \lambda\hat{i} - 3\hat{j} - \hat{k}$

$\vec{b} = 2\lambda\hat{i} + \lambda\hat{j} - \hat{k}$

Angle between $\vec{a}$ and $\vec{b}$ is acute. Therefore,

$\vec{a}\cdot\vec{b} > 0$

$\Rightarrow 2\lambda^2 - 3\lambda + 1 > 0$

$\Rightarrow (2\lambda - 1)(\lambda - 1) > 0$

$\Rightarrow \lambda \in \left(-\infty, \dfrac{1}{2}\right) \cup (1, \infty)$

Also $\vec{b}$ makes an obtuse angle with the axes. Therefore,

$\vec{b}\cdot\hat{i} < 0 \Rightarrow \lambda < 0$

$\vec{b}\cdot\hat{j} < 0 \Rightarrow \lambda < 0$ (2)

Combining these two, we get $\lambda = -4, -2$.

c. If vectors $2\hat{i} - \hat{j} + \hat{k}$, $\hat{i} + 2\hat{j} + (1 + a)\hat{k}$, and $3\hat{i} + a\hat{j} + 5\hat{k}$ are coplanar, then

$\begin{vmatrix} 2 & -1 & 1 \\ 1 & 2 & 1+a \\ 3 & a & 5 \end{vmatrix} = 0$

$\Rightarrow a^2 + 2a - 8 = 0$

$\Rightarrow (a+4)(a-2) = 0$

$\Rightarrow a = -4, 2$

d. $\vec{A} = 2\hat{i} + \lambda\hat{j} + 3\hat{k}$

$B = 2\hat{i} + \lambda\hat{j} + \hat{k}$

$C = 3\hat{i} + \hat{j} + 0.\hat{k}$

$\therefore \vec{A} + \lambda\vec{B} = 2(1 + \lambda)\hat{i} + (\lambda + \lambda^2)\hat{j} + (3 + \lambda)\hat{k}$

Now $(\vec{A} + \lambda\vec{B}) \perp \vec{C}$. Therefore,

$(\vec{A} + \lambda\vec{B})\cdot\vec{C} = 0$

$\Rightarrow 6(1 + \lambda) + (\lambda + \lambda^2) + 0 = 0$

$\Rightarrow \lambda^2 + 7\lambda + 6 = 0$

$\Rightarrow (\lambda + 6)(\lambda + 1) = 0$

$\Rightarrow \lambda = -6, = -1$

$\Rightarrow |2\lambda| = 12, 2$

2. a. → r; b → p; c → s; d → q

a. If $\vec{a}$, $\vec{b}$, and $\vec{c}$ are mutually perendicular

$[\vec{a}\times\vec{b}\ \ \vec{b}\times\vec{c}\ \ \vec{c}\times\vec{a}] = [\vec{a}\,\vec{b}\,\vec{c}]^2$

$= (|\vec{a}||\vec{b}||\vec{c}|)^2 = 16$

b. Given $\vec{a}$ and $\vec{b}$ are two unit vectors, i.e., $|\vec{a}| = |\vec{b}| = 1$ and angle between them is $\pi/3$.

$$\sin\theta = \frac{|\vec{a}\times\vec{b}|}{|\vec{a}||\vec{b}|} \Rightarrow \sin\frac{\pi}{3} = |\vec{a}\times\vec{b}|$$

$$\Rightarrow \frac{\sqrt{3}}{2} = |\vec{a}\times\vec{b}|$$

Now

$$[\vec{a}\ \ \vec{b}+\vec{a}\times\vec{b}\ \ \vec{b}] = [\vec{a}\ \vec{b}\ \vec{b}] + [\vec{a}\ \ \vec{a}\times\vec{b}\ \ \vec{b}]$$
$$= 0 + [\vec{a}\ \ \vec{a}\times\vec{b}\ \ \vec{b}]$$
$$= (\vec{a}\times\vec{b})\cdot(\vec{b}\times\vec{a})$$
$$= -(\vec{a}\times\vec{b})\cdot(\vec{a}\times\vec{b})$$
$$= -|\vec{a}\times\vec{b}|^2$$
$$= -\frac{3}{4}$$

c. If $\vec{b}$ and $\vec{c}$ are orthogonal, $\vec{b}\cdot\vec{c} = 0$.

Also, it is given that $\vec{b}\times\vec{c} = \vec{a}$. Now

$$[\vec{a}+\vec{b}+\vec{c}\ \ \vec{a}+\vec{b}\ \ \vec{b}+\vec{c}]$$
$$= [\vec{a}\ \ \vec{a}+\vec{b}\ \ \vec{b}+\vec{c}] + [\vec{b}+\vec{c}\ \ \vec{a}+\vec{b}\ \ \vec{b}+\vec{c}]$$
$$= [\vec{a}\ \vec{b}\ \vec{c}]$$
$$= \vec{a}\cdot(\vec{b}\times\vec{c})$$
$$= \vec{a}\cdot\vec{a} = |\vec{a}|^2 = 1 \text{ (because } \vec{a} \text{ is unit vector)}$$

d. $[\vec{x}\ \vec{y}\ \vec{a}] = 0$

Therefore, $\vec{x}$, $\vec{y}$ and $\vec{a}$ are coplanar. (1)

$[\vec{x}\ \vec{y}\ \vec{b}] = 0$

Therefore, $\vec{x}$, $\vec{y}$ and $\vec{b}$ are coplanar. (2)

Also, $[\vec{a}\ \vec{b}\ \vec{c}] = 0$

Therefore, $\vec{a}$, $\vec{b}$ and $\vec{c}$ are coplanar (3)

From (1), (2) and (3),

$\vec{x}$, $\vec{y}$ and $\vec{c}$ are coplanar. Therefore,

$[\vec{x}\ \vec{y}\ \vec{c}] = 0$

3. **a → q; b → s; c → p; d → r.**

a. $|\vec{a}+\vec{b}+\vec{c}| = \sqrt{6}$

$$\Rightarrow \vec{a}^2 + \vec{b}^2 + \vec{c}^2 + 2\vec{a}\cdot\vec{b} + 2\vec{b}\cdot\vec{c} + 2\vec{c}\cdot\vec{a} = 6$$

$\therefore |\vec{a}| = 1$

b. $\vec{a}$ is perpendicular $\vec{b}+\vec{c} \Rightarrow \vec{a}\cdot\vec{b} + \vec{a}\cdot\vec{c} = 0$ (1)

$\vec{b}$ is perpendicular $\vec{a}+\vec{c} \Rightarrow \vec{b}\cdot\vec{c} + \vec{b}\cdot\vec{c} = 0$ (2)

$\vec{c}$ is perpendicular $\vec{a}+\vec{b}$

$\Rightarrow \vec{c}\cdot\vec{a} + \vec{a}\cdot\vec{c}\cdot\vec{b} = 0$ (3)

From (1), (2) and (3), we get

$\vec{a}\cdot\vec{b} = \vec{b}\cdot\vec{c} = \vec{c}\cdot\vec{a} = 0$

$\therefore |\vec{a}+\vec{b}+\vec{c}| = 7$

c. $(\vec{a}\cdot\vec{c})(\vec{b}\cdot\vec{d}) - (\vec{b}\cdot\vec{c})(\vec{a}\cdot\vec{d}) = 21$

d. We know that $[\vec{a}\times\vec{b}\ \ \vec{b}\times\vec{c}\ \ \vec{c}\times\vec{a}] = [\vec{a}\ \vec{b}\ \vec{c}]^2$

$$\text{And } [\vec{a}\ \vec{b}\ \vec{c}]^2 = \begin{vmatrix} \vec{a}\cdot\vec{a} & \vec{a}\cdot\vec{b} & \vec{a}\cdot\vec{c} \\ \vec{b}\cdot\vec{a} & \vec{b}\cdot\vec{b} & \vec{b}\cdot\vec{c} \\ \vec{c}\cdot\vec{a} & \vec{c}\cdot\vec{b} & \vec{c}\cdot\vec{c} \end{vmatrix}$$

$$= \begin{vmatrix} 4 & 2 & 2 \\ 2 & 4 & 2 \\ 2 & 2 & 4 \end{vmatrix}$$

$$= 32$$

$\therefore [\vec{a}\ \vec{b}\ \vec{c}] = 4\sqrt{2}$

4. **a. → s; b → r; c → q; d → p**

a. $$\vec{a}\times\vec{b} = \begin{vmatrix} \hat{i} & \hat{j} & \hat{k} \\ -1 & 1 & 2 \\ -1 & -2 & -1 \end{vmatrix} = 3\hat{i} - 3\hat{j} + 3\hat{k}$$

Hence, the area of the triangle is $\frac{3\sqrt{3}}{2}$.

b. The area of the parallelogram is $3\sqrt{3}$.

c. The area of a parallelogram whose diagonals area $2\vec{a}$ and $4\vec{b}$ is $\frac{1}{2}|2\vec{a}\times 4\vec{b}| = 12\sqrt{3}$.

d. The volume of the parallelepiped is $|(\vec{a}\times\vec{b})\cdot\vec{c}| = \sqrt{9+36+9} = 3\sqrt{6}$.

5. **a → p, r; b → q; c → s; d → p**

a. Vector $\hat{i}+\hat{j}$ is coplanar with $\vec{a}$ and $\vec{b}$.

b. $\vec{a}\times\vec{b} = \begin{vmatrix} \hat{i} & \hat{j} & \hat{k} \\ -1 & 2 & 2 \\ -2 & 1 & 2 \end{vmatrix}$

$= 2\hat{i}-2\hat{j}+3\hat{k}$

c. If $\vec{c}$ is equally inclined to $\vec{a}$ and $\vec{b}$, then we must have $\vec{a}\cdot\vec{c} = \vec{b}\cdot\vec{c}$, which is true for $\vec{c} = \hat{i}-\hat{j}+5\hat{k}$.

d. Vector is forming triangle with $\vec{a}$ and $\vec{b}$. Then $\vec{c} = \vec{a}+\vec{b} = -3\hat{i}+3\hat{j}+4\hat{k}$.

6. **a.→q; b→s; c→p; d→r**

a. $|\vec{a}+\vec{b}| = |\vec{a}+2\vec{b}|$

$\Rightarrow a^2+b^2+2\vec{a}\cdot\vec{b} = a^2+4b^2+4\vec{a}\cdot\vec{b}$

$\Rightarrow 2\vec{a}\cdot\vec{b} = -3b^2 < 0$

Hence, angle between $\vec{a}$ and $\vec{b}$ is obtuse.

b. $|\vec{a}+\vec{b}| = |\vec{a}-2\vec{b}|$

$\Rightarrow a^2+b^2+2\vec{a}\cdot\vec{b} = a^2+4b^2-4\vec{a}\cdot\vec{b}$

$\Rightarrow 6\vec{a}\cdot\vec{b} = 3b^2$

Hence, angle between $\vec{a}$ and $\vec{b}$ is acute.

c. $|\vec{a}+\vec{b}| = |\vec{a}-\vec{b}|$

$\Rightarrow \vec{a}\cdot\vec{b}$

$\Rightarrow$ $\vec{a}$ is perpendicular to $\vec{b}$.

d. $\vec{c}\times(\vec{a}\times\vec{b})$ lies in the plane of vectors $\vec{a}$ and $\vec{b}$. A vector perpendicular to this plane is parallel to $\vec{a}\times\vec{b}$. Hence, angle is 0°.

7. **a→r; b→s; c→q; d→p**

$[\vec{a}\times\vec{b}\ \ \vec{b}\times\vec{c}\ \ \vec{c}\times\vec{a}] = 36$

$\Rightarrow [\vec{a}\ \vec{b}\ \vec{c}] = 6$

$\Rightarrow$ Volume of tetrahedron formed by vectors $\vec{a}, \vec{b}$ and $\vec{c}$ is $\frac{1}{6}[\vec{a}\ \vec{b}\ \vec{c}] = 1$.

$[\vec{a}+\vec{b}\ \ \vec{b}+\vec{c}\ \ \vec{c}+\vec{a}] = 2[\vec{a}\ \vec{b}\ \vec{c}] = 12$

$\vec{a}-\vec{b},\ \vec{b}-\vec{c}$, and $\vec{c}-\vec{a}$ are coplanar, So

$[\vec{a}-\vec{b}\ \ \vec{b}-\vec{c}\ \ \vec{c}-\vec{a}] = 0.$

Integer Type

1. (3) $\vec{a}\cdot\vec{b} = [x\hat{i}+(x-1)\hat{j}+\hat{k}]\cdot[(x+1)\hat{i}+\hat{j}+a\hat{k}]$

$= x(x+1)+x-1+a$

$= x^2+2x+a-1$

We must have $\vec{a}\cdot\vec{b} > 0, \forall x\in R$

$\Rightarrow x^2+2x+a-1 > 0, \forall x\in R$

$\Rightarrow 4-4(a-1) < 0$

$\Rightarrow a > 2$

2. (1) Let $\vec{a} = x\hat{i}+y\hat{j}+z\hat{k}$

Then $\vec{a}\cdot\hat{i} = (x\hat{i}+y\hat{j}+z\hat{k})\cdot\hat{i} = x$

and $\vec{a}\cdot(\hat{i}+\hat{j}) = x+y$

and $\vec{a}\cdot(\hat{i}+\hat{j}+\hat{k}) = x+y+z$ ($\because$ given that $x = x+y = x+y+z$)

Now $x = x+y \Rightarrow y = 0$ and $x+y = x+y+z \Rightarrow z = 0$

Therefore, $x = 1$.

Hence, $\vec{a} = \hat{i}$

3. (5) Here $\vec{F} = \vec{F_1}+\vec{F_2} = (4\hat{i}+\hat{j}-3\hat{k})+(3\hat{i}+\hat{j}-\hat{k})$

$= 7\hat{i}+2\hat{j}-4\hat{k}$ and $\vec{d} = \vec{d_2}-\vec{d_1}$

$= (5\hat{i}+4\hat{j}+\hat{k})-(\hat{i}+2\hat{j}+3\hat{k})$

$= 4\hat{i}+2\hat{j}-2\hat{k}$

Hence, work done $= \vec{F}\cdot\vec{d}$

$= (7\hat{i}+2\hat{j}-4\hat{k})\cdot(4\hat{i}+2\hat{j}-2\hat{k})$

$= (7)(4)+(2)(2)+(-4)(-2)$

$= 28+4+8 = 40$ units

4. (3) We have

$|\vec{a}+\vec{b}|^2+|\vec{a}-\vec{b}|^2 = 2(|\vec{a}|^2+|\vec{b}|^2)$

$\Rightarrow 1+|\vec{a}-\vec{b}|^2 = 4 \Rightarrow |\vec{a}-\vec{b}| = \sqrt{3}$

5. (7) Given $\left(\vec{a}+\vec{b}\right)\cdot\vec{c} = 0 \Rightarrow \vec{a}\cdot\vec{c}+\vec{b}\cdot\vec{c} = 0$

$\left(\vec{b}+\vec{c}\right)\cdot\vec{a} = 0 \Rightarrow \vec{a}\cdot\vec{b}+\vec{c}\cdot\vec{a} = 0$

$\left(\vec{c}+\vec{a}\right)\cdot\vec{b} = 0 \Rightarrow \vec{b}\cdot\vec{c}+\vec{a}\cdot\vec{b} = 0$. Therefore,

$2\left(\vec{a}\cdot\vec{b}+\vec{b}\cdot\vec{c}+\vec{c}\cdot\vec{a}\right) = 0$. Now,

$|\vec{a}+\vec{b}+\vec{c}|^2 = |\vec{a}|^2+|\vec{b}|^2+|\vec{c}|^2$

$+2\left(\vec{a}\cdot\vec{b}+\vec{b}\cdot\vec{c}+\vec{c}\cdot\vec{a}\right) = 50$

$\Rightarrow |\vec{a}+\vec{b}+\vec{c}| = 5\sqrt{2}$

6. (3) $\hat{n} = a_1\hat{i} + a_2\hat{j} + a_3\hat{k}$, where $a_1^2 + a_2^2 + a_3^2 = 1$

Given that $\hat{u}\cdot\hat{n} = 0 \Rightarrow a_1 + a_2 = 0$

Also $\hat{v}\cdot\hat{n} = 0 \Rightarrow a_1 - a_2 = 0$. Therefore,

$a_1 = a_2 = 0$,
and $a_3 = 1$ or -1

$\hat{n} = \hat{k}$ or $-\hat{k}$

Hence, $|\hat{w}\cdot\hat{n}| = 3$

7. (1) $\vec{A}\cdot\left(\alpha\vec{A}+\vec{B}\right) = \vec{B}\cdot\left(\alpha\vec{A}+\vec{B}\right)$

$\Rightarrow \alpha + \vec{A}\cdot\vec{B} = \alpha\,\vec{A}\cdot\vec{B} + 1$

$\Rightarrow (\vec{A}\cdot\vec{B})(1-\alpha) = (1-\alpha)$

Since $\vec{A}\cdot\vec{B} \neq 0, \alpha = 1$

8. (8) Let θ be the angle between the unit vectors $\vec{a}$ and $\vec{b}$.

Then $\vec{a}\cdot\vec{b} = \cos\theta$

$\Rightarrow |\vec{a}|^2 + |\vec{b}|^2 + 2\vec{a}\cdot\vec{b} = 2 + 2\cos\theta = 4\cos^2\frac{\theta}{2}$

$\Rightarrow |\vec{a}+\vec{b}| = 2\cos\frac{\theta}{2}, |\vec{a}-\vec{b}| = 2\sin\frac{\theta}{2}$

$\Rightarrow |\vec{a}+\vec{b}| + |\vec{a}-\vec{b}| = 2\left(\cos\frac{\theta}{2} + \sin\frac{\theta}{2}\right)$
$\leq 2\sqrt{2}$.

9. (8) $|\vec{a}\times\hat{i}|^2 = \begin{vmatrix} \hat{i} & \hat{j} & \hat{k} \\ a_1 & a_2 & a_3 \\ 1 & 0 & 0 \end{vmatrix}^2$,

(Since $\vec{a} = a_1\hat{i} + a_2\hat{j} + a_3\hat{k}$)

$= |a_3\hat{j} - a_2\hat{k}|^2 = a_3^2 + a_2^2$

Similarly, $|\vec{a}\times\hat{j}|^2 = a_1^2 + a_3^2$

and $|\vec{a}\times\hat{k}|^2 = a_1^2 + a_2^2$

Hence, the required result can be given as

$2(a_1^2 + a_2^2 + a_3^2) = 2|\vec{a}|^2 = 8.$

10. (4) Let the position vectors of A, B, C, and D be $\vec{a}, \vec{b}, \vec{c}$, and $\vec{0}$. Therefore,

$\overrightarrow{AB}\times\overrightarrow{CD} = (\vec{b}-\vec{a})\times(-\vec{c})$,

$\overrightarrow{BC}\times\overrightarrow{AD} = (\vec{c}-\vec{b})\times -\vec{a}$,

and $\overrightarrow{CA}\times\overrightarrow{BD} = (\vec{a}-\vec{c})\times -\vec{b}$

$\Rightarrow (\overrightarrow{AB}\times\overrightarrow{CD}) + (\overrightarrow{BC}\times\overrightarrow{AD}) + (\overrightarrow{CA}\times\overrightarrow{BD})$

$= \vec{c}\times\vec{b} + \vec{a}\times\vec{c} + \vec{a}\times\vec{c} + \vec{b}\times\vec{a} - \vec{a}\times\vec{b} + \vec{c}\times\vec{b}$

$= 2(\vec{c}\times\vec{b} + \vec{b}\times\vec{a} + \vec{a}\times\vec{c})$

$= 2(\vec{c}\times(\vec{b}-\vec{a}) - \vec{a}\times(\vec{b}-\vec{a}))$

$= 2((\vec{c}-\vec{a})\times(\vec{b}-\vec{a}))$

$= 2(\overrightarrow{AC}\times\overrightarrow{AB})$

$\Rightarrow |\overrightarrow{AB}\times\overrightarrow{CD} + \overrightarrow{BC}\times\overrightarrow{AD} + \overrightarrow{CA}\times\overrightarrow{BD}|$

$= 4\left|\frac{1}{2}(\overrightarrow{AC}\times\overrightarrow{AB})\right| = 4\Delta ABC$

11. (4) If the angle between $\vec{b}$ and $\vec{c}$ is α and $|\vec{b}\times\vec{c}| = \sqrt{15}$, then

$|\vec{b}||\vec{c}|\sin\alpha = \sqrt{15}$

$\Rightarrow \sin\alpha = \frac{\sqrt{15}}{4} \Rightarrow \cos\alpha = \frac{1}{4}$

$\vec{b} - 2\vec{c} = \lambda\vec{a}$

$\Rightarrow |\vec{b}-2\vec{c}|^2 = \lambda^2|\vec{a}|^2$

$\Rightarrow |\vec{b}|^2 + 4|\vec{c}|^2 - 4\vec{b}\cdot\vec{c} = \lambda^2|\vec{a}|^2$

$\Rightarrow 16 + 4 - 4[|\vec{b}||\vec{c}|\cos\alpha] = \lambda^2$

$\Rightarrow 16 + 4 - 4(4\times1\times\frac{1}{4}) = \lambda^2$

$\Rightarrow \lambda^2 = 16 \Rightarrow \lambda = \pm 4$

12. (6) $\vec{a}\cdot\vec{b} = |\vec{a}||\vec{b}|\cos\theta$

But $|\vec{a}\times\vec{b}| = |\vec{a}||\vec{b}|\sin\theta$

$\Rightarrow \sin\theta = \frac{|\vec{a}\times\vec{b}|}{|\vec{a}||\vec{b}|} = \frac{4}{5} \Rightarrow \cos\theta = \frac{3}{5}$

Therefore, $\vec{a}\cdot\vec{b} = 2\times5\times\frac{3}{5} = 6$

13. (2)

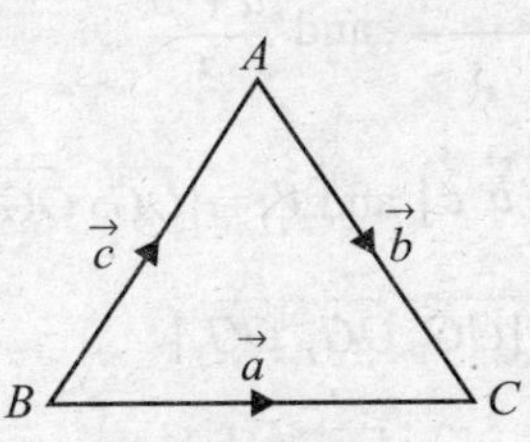

Fig.2.12

Clearly, $\vec{a}, \vec{b}$, and $\vec{c}$ represent the sides of a triangle. Therefore,

$\vec{a}\times\vec{b} = \vec{b}\times\vec{c} = \vec{c}\times\vec{a}$

$$\Rightarrow \vec{a}\times\vec{b}+\vec{b}\times\vec{c}+\vec{c}\times\vec{a}=3\vec{a}\times\vec{b}$$

$$\Rightarrow 2\vec{b}\times\vec{a}+\vec{b}\times\vec{c}+\vec{c}\times\vec{a}=0$$

14. (3) We have $(\vec{a}\times\vec{b})^2+(\vec{a}\cdot\vec{b})^2=144$

If the angle between $\vec{a}$ and $\vec{b}$ is θ, then

$$|\vec{a}|^2\,|\vec{b}|^2\sin^2\theta+|\vec{a}|^2\,|\vec{b}|^2\cos^2\theta=144$$

$$\Rightarrow |\vec{a}|^2\,|\vec{b}|^2=144$$

$$\Rightarrow |\vec{a}|\,|\vec{b}|=12$$

$$\Rightarrow 4|\vec{b}|=12$$

$$\Rightarrow |\vec{b}|=3$$

15. (2) We have $|\vec{a}+\vec{b}|=\sqrt{3}$

$$\Rightarrow |\vec{a}+\vec{b}|^2=3$$

$$\Rightarrow |\vec{a}|^2+|\vec{b}|^2+2(\vec{a}\cdot\vec{b})=3$$

$$\Rightarrow 1+1+2(\vec{a}\cdot\vec{b})=3$$

$$\Rightarrow \vec{a}\cdot\vec{b}=\frac{1}{2} \qquad (1)$$

Now $\vec{c}-\vec{a}-2\vec{b}=3(\vec{a}\times\vec{b})$

$$\Rightarrow (\vec{c}-\vec{a}-2\vec{b})\cdot\vec{b}=3\,[(\vec{a}\times\vec{b})\cdot\vec{b}]$$

$$\Rightarrow \vec{c}\cdot\vec{b}-\vec{a}\cdot\vec{b}-2(\vec{b}\cdot\vec{b})=0 \qquad (\because [\vec{a}\,\vec{b}\,\vec{b}]=0)$$

$$\Rightarrow \vec{c}\cdot\vec{b}-\frac{1}{2}-2\times1=0 \qquad \text{[Using (1)]}$$

$$\Rightarrow \vec{c}\cdot\vec{b}=\frac{5}{2}$$

16. (4) $\left[\bar{a}\;\bar{b}\;\bar{a}\times\bar{b}\right]=(\bar{a}\times\bar{b})\times(\bar{a}\times\bar{b})$

$$=|\bar{a}\times\bar{b}|^2=4$$

17. (9) Taking O as the origin, let the position vectors of A, B, and C be $\vec{a}$, $\vec{b}$, and $\vec{c}$, respectively. Then the position vectors of G_1, G_2 and G_3 are

$$\frac{\vec{b}+\vec{c}}{3},\ \frac{\vec{c}+\vec{a}}{3},\ \text{and}\ \frac{\vec{a}+\vec{b}}{3}.$$

$V_1=\frac{1}{6}[\vec{a}\,\vec{b}\,\vec{c}]$ and $V_2=[\overrightarrow{OG_1}\;\overrightarrow{OG_2}\;\overrightarrow{OG_3}]$

Now $V_2=[\overrightarrow{OG_1}\;\overrightarrow{OG_2}\;\overrightarrow{OG_3}]$

$$\Rightarrow V_2=\frac{1}{27}[\vec{b}+\vec{c}\;\;\vec{c}+\vec{a}\;\;\vec{a}+\vec{b}]$$

$$\Rightarrow V_2=\frac{2}{27}[\vec{a}\,\vec{b}\,\vec{c}]$$

$$\Rightarrow V_2=\frac{2}{27}\times 6V_1 \Rightarrow 9V_2=4V_1$$

18. (3) $\dfrac{\text{Area of }\Delta ABC}{\text{Area of }\Delta AOC}=\dfrac{(1/2)\,|\vec{a}\times\vec{b}+\vec{b}\times\vec{c}+\vec{c}\times\vec{a}|}{(1/2)\,|\vec{a}\times\vec{c}\,|}$

Now $\vec{a}+2\vec{b}+3\vec{c}=0$

Cross with $\vec{b}$, $\vec{a}\times\vec{b}+3\vec{c}\times\vec{b}=0$

$$\Rightarrow \vec{a}\times\vec{b}=3(\vec{b}\times\vec{c})$$

Cross with $\vec{a}$, $2\vec{a}\times\vec{b}+3\vec{a}\times\vec{c}=0$

$$\Rightarrow \vec{a}\times\vec{b}=\frac{3}{2}(\vec{b}\times\vec{c})$$

Hence, $\vec{a}\times\vec{b}=\frac{3}{2}(\vec{c}\times\vec{a})=3(\vec{b}\times\vec{c})$

Let $(\vec{c}\times\vec{a})=\vec{p}$. Then

$$\vec{a}\times\vec{b}=\frac{3\vec{p}}{2},\ \vec{b}\times\vec{c}=\frac{\vec{p}}{2}$$

Hence, the ratio is

$$\frac{|\vec{a}\times\vec{b}+\vec{b}\times\vec{c}+\vec{c}\times\vec{a}|}{|\vec{c}\times\vec{a}|}$$

$$=\frac{\left|(3\vec{p}/2)+(\vec{p}/2)+\vec{p}\right|}{|\vec{p}|}$$

$$=\frac{3|\vec{p}|}{|\vec{p}|}=3$$

19. (1) We know that $[\vec{e}_1\,\vec{e}_2\,\vec{e}_3][\vec{E}_1\,\vec{E}_2\,\vec{E}_3]$

$$=\begin{vmatrix}\vec{e}_1\cdot\vec{E}_1 & \vec{e}_1\cdot\vec{E}_2 & \vec{e}_1\cdot\vec{E}_3\\ \vec{e}_2\cdot\vec{E}_1 & \vec{e}_2\cdot\vec{E}_2 & \vec{e}_2\cdot\vec{E}_2\\ \vec{e}_3\cdot\vec{E}_1 & \vec{e}_3\cdot\vec{E}_2 & \vec{e}_3\cdot\vec{E}_3\end{vmatrix}$$

$$=\begin{vmatrix}1 & 0 & 0\\ 0 & 1 & 0\\ 0 & 0 & 1\end{vmatrix}=1$$

20. (6) q = area of the parallelogram with $\overrightarrow{OA}$ and $\overrightarrow{OC}$ as adjacent sides

$$=|\overrightarrow{OA}\times\overrightarrow{OC}|$$

$$=|\vec{a}\times\vec{b}|$$

and p = area of quadrilateral $OABC$

$$=\frac{1}{2}|\overrightarrow{OA}\times\overrightarrow{OB}|+\frac{1}{2}|\overrightarrow{OB}\times\overrightarrow{OC}|$$

$$=\frac{1}{2}|\vec{a}\times(10\vec{a}+2\vec{b})|+\frac{1}{2}|(10\vec{a}+2\vec{b})\times\vec{b}|$$

$= |\vec{a}\times\vec{b}| + 5|\vec{a}\times\vec{b}| = 6|\vec{a}\times\vec{b}|$

Hence $p = 6q \Rightarrow k = 6$

21. (3) Given that $\vec{a}$, $\vec{b}$, and $\vec{c}$ are non-coplanar. Therefore, $[\vec{a}\,\vec{b}\,\vec{c}] \neq 0$. Also

$$\vec{p} = \frac{\vec{b}\times\vec{c}}{[\vec{a}\,\vec{b}\,\vec{c}]},\ \vec{q} = \frac{\vec{c}\times\vec{a}}{[\vec{a}\,\vec{b}\,\vec{c}]},\ \vec{r} = \frac{\vec{a}\times\vec{b}}{[\vec{a}\,\vec{b}\,\vec{c}]} \quad (1)$$

Now $\left(\vec{a}+\vec{b}\right)\cdot\vec{p} + \left(\vec{b}+\vec{c}\right)\cdot\vec{q} + \left(\vec{c}+\vec{a}\right)\cdot\vec{r}$

$$= \left(\vec{a}+\vec{b}\right)\cdot\frac{\vec{b}\times\vec{c}}{[\vec{a}\,\vec{b}\,\vec{c}]} + \left(\vec{b}+\vec{c}\right)\cdot\frac{\vec{c}\times\vec{a}}{[\vec{a}\,\vec{b}\,\vec{c}]} + \left(\vec{c}+\vec{a}\right)\cdot\frac{\vec{a}\times\vec{b}}{[\vec{a}\,\vec{b}\,\vec{c}]}$$

$$= \frac{\vec{a}\cdot\vec{b}\times\vec{c}}{[\vec{a}\,\vec{b}\,\vec{c}]} + \frac{\vec{b}\cdot\vec{c}\times\vec{a}}{[\vec{a}\,\vec{b}\,\vec{c}]} + \frac{\vec{c}\cdot\vec{a}\times\vec{b}}{[\vec{a}\,\vec{b}\,\vec{c}]}$$

$(\because \vec{b}\cdot\vec{b}\times\vec{c} = \vec{c}\cdot\vec{c}\times\vec{a} = \vec{a}\cdot\vec{a}\times\vec{b} = 0)$

$$= \frac{[\vec{a}\,\vec{b}\,\vec{c}]}{[\vec{a}\,\vec{b}\,\vec{c}]} + \frac{[\vec{a}\,\vec{b}\,\vec{c}]}{[\vec{a}\,\vec{b}\,\vec{c}]} + \frac{[\vec{a}\,\vec{b}\,\vec{c}]}{[\vec{a}\,\vec{b}\,\vec{c}]}$$

$= 1 + 1 + 1$

$= 3$

22. (3) $|(\vec{a}\times\vec{b})\times\vec{c}| = |\vec{a}\times\vec{b}|\,|\vec{c}|\,\sin 30°$

$$= \frac{1}{2}|\vec{a}\times\vec{b}||\vec{c}| \quad (1)$$

We have $\vec{a} = 2\hat{i} + \hat{j} - 2\hat{k}$ and $\vec{b} = \hat{i} + \hat{j}$

$\Rightarrow \vec{a}\times\vec{b} = 2\hat{i} - 2\hat{j} + \hat{k}$

$\Rightarrow |\vec{a}\times\vec{b}| = \sqrt{9} = 3$

Also given $|\vec{c}-\vec{a}| = 2\sqrt{2}$

$\Rightarrow |\vec{c}-\vec{a}|^2 = 8$

$\Rightarrow |\vec{c}|^2 + |\vec{a}|^2 - 2\vec{a}\cdot\vec{c} = 8$

Given $|\vec{a}| = 3$ and $\vec{a}\cdot\vec{c} = |\vec{c}|$. Therefore,

$|\vec{c}|^2 - 2|\vec{c}| + 1 = 0$

$\Rightarrow (|\vec{c}| - 1)^2 = 0$

$\Rightarrow |\vec{c}| = 1$

Substituting the values of $|\vec{a}\times\vec{b}|$ and $|\vec{c}|$ in (1), we get

$(|\vec{a}\times\vec{b})\times\vec{c}| = \frac{1}{2}\times 3\times 1 = 3/2$

23. (9) Given $\hat{a}$, $\hat{b}$, and $\hat{c}$ are unit vectors.

Now $x = |\hat{a}-\hat{b}|^2 + |\hat{b}-\hat{c}|^2 + |\hat{c}-\hat{a}|^2$

$= \hat{a}\cdot\hat{a} + \hat{b}\cdot\hat{b} + \hat{c}\cdot\hat{c} - 2\hat{a}\cdot\hat{b} - 2\hat{b}\cdot\hat{c} - 2\hat{c}\cdot\hat{a}$

$$\Rightarrow 6 - 2(\hat{a}\cdot\hat{b} + \hat{b}\cdot\hat{c} + \hat{c}\cdot\hat{a}) \quad (1)$$

Also

$|\hat{a}+\hat{b}+\hat{c}| \geq 0$

$\Rightarrow \hat{a}\cdot\hat{a} + \hat{b}\cdot\hat{b} + \hat{c}\cdot\hat{c} + 2(\hat{a}\cdot\hat{b} + \hat{b}\cdot\hat{c} + \hat{c}\cdot\hat{a}) \geq 0$

$\Rightarrow 3 + 2(\hat{a}\cdot\hat{b} + \hat{b}\cdot\hat{c} + \hat{c}\cdot\hat{a}) \geq 0$

$\Rightarrow 2(\hat{a}\cdot\hat{b} + \hat{b}\cdot\hat{c} + \hat{c}\cdot\hat{a}) \geq -3$

$\Rightarrow -2(\hat{a}\cdot\hat{b} + \hat{b}\cdot\hat{c} + \hat{c}\cdot\hat{a}) \leq 3$

$$\Rightarrow 6 - 2(\hat{a}\cdot\hat{b} + \hat{b}\cdot\hat{c} + \hat{c}\cdot\hat{a}) \leq 9 \quad (2)$$

From (1) and (2), we get $x \leq 9$

Hence, x does not exceed 9.

24. (7) Given that $\vec{V} = 2\vec{i} + \vec{j} - \vec{k}$ and $\vec{W} = \vec{i} + 3\vec{k}$ and $\vec{u}$ is a unit vector. Thus $|\vec{U}| = 1$.

Now $\left[\vec{U}\ \vec{V}\ \vec{W}\right]$

$= \vec{U}\cdot(\vec{V}\times\vec{W})$

$= \vec{U}\cdot(2\hat{i} + \hat{j} - \hat{k})\times(\hat{i} + 3\hat{k})$

$= \vec{U}\cdot(3\hat{i} - 7\hat{j} - \hat{k})$

$= \sqrt{3^2 + 7^2 + 1^2}\cos\theta$, which is maximum when $\cos\theta = 1$.

Hence, the maximum value of $\left[\vec{U}\ \vec{V}\ \vec{W}\right]$ is $\sqrt{59}$.

CHAPTER 3

Three-Dimensional Geometry

Objective Type

Each question has four choices a, b, c and d, out of which *only* *one* answer is correct. Find the correct answer.

1. In a three-dimensional xyz space, the equation $x^2 - 5x + 6 = 0$ represents
   a. points
   b. planes
   c. curves
   d. pair of straight lines

2. The line $\dfrac{x-2}{3} = \dfrac{y+1}{2} = \dfrac{z-1}{-1}$ intersects the curve $xy = c^2, z = 0$ if c is equal to
   a. ± 1
   b. $\pm 1/3$
   c. $\pm \sqrt{5}$
   d. none of these

3. If the equation of a line and a plane be $\dfrac{x+3}{2} = \dfrac{y-4}{3} = \dfrac{z+5}{2}$ and $4x - 2y - z = 1$, respectively, then
   a. the line is parallel to the plane
   b. the line is perpendicular to the plane
   c. the line lies in the plane
   d. none of these

4. The length of the perpendicular from the origin to the plane passing through the point $\vec{a}$ and containing the line $\vec{r} = \vec{b} + \lambda \vec{c}$ is
   a. $\dfrac{[\vec{a}\,\vec{b}\,\vec{c}]}{|\vec{a}\times\vec{b} + \vec{b}\times\vec{c} + \vec{c}\times\vec{a}|}$
   b. $\dfrac{[\vec{a}\,\vec{b}\,\vec{c}]}{|\vec{a}\times\vec{b} + \vec{b}\times\vec{c}|}$
   c. $\dfrac{[\vec{a}\,\vec{b}\,\vec{c}]}{|\vec{b}\times\vec{c} + \vec{c}\times\vec{a}|}$
   d. $\dfrac{[\vec{a}\,\vec{b}\,\vec{c}]}{|\vec{c}\times\vec{a} + \vec{a}\times\vec{b}|}$

5. The distance of point A (–2, 3, 1) from the line PQ through P (–3, 5, 2), which make equal angles with the axes is
   a. $2/\sqrt{3}$
   b. $\sqrt{14/3}$
   c. $16/\sqrt{3}$
   d. $5/\sqrt{3}$

6. Vector equation of the plane $\vec{r} = \hat{i} - \hat{j} + \lambda(\hat{i} + \hat{j} + \hat{k}) + \mu(\hat{i} - 2\hat{j} + 3\hat{k})$ in the scalar dot product form is
   a. $\vec{r}\cdot(5\hat{i} - 2\hat{j} + 3\hat{k}) = 7$
   b. $\vec{r}\cdot(5\hat{i} + 2\hat{j} - 3\hat{k}) = 7$
   c. $\vec{r}\cdot(5\hat{i} - 2\hat{j} - 3\hat{k}) = 7$
   d. $\vec{r}\cdot(5\hat{i} + 2\hat{j} + 3\hat{k}) = 7$

7. A unit vector parallel to the intersection of the planes, $\vec{r}\cdot\left(\hat{i} - \hat{j} + \hat{k}\right) = 5$ and $\vec{r}\cdot\left(2\hat{i} + \hat{j} - 3\hat{k}\right) = 4$ is

a. $\frac{2\hat{i}+5\hat{j}-3\hat{k}}{\sqrt{38}}$ b. $\frac{2\hat{i}-5\hat{j}+3\hat{k}}{\sqrt{38}}$

c. $\frac{-2\hat{i}-5\hat{j}-3\hat{k}}{\sqrt{38}}$ d. $\frac{-2\hat{i}+5\hat{j}-3\hat{k}}{\sqrt{38}}$

8. Let L_1 be the line $\vec{r_1}=2\hat{i}+\hat{j}-\hat{k}+\lambda(\hat{i}+2\hat{k})$ and let L_2 be the line $\vec{r_2}=3\hat{i}+\hat{j}+\mu(\hat{i}+\hat{j}-\hat{k})$.
Let Π be the plane which contains the line L_1 and is parallel to L_2. The distance of the plane Π from the origin is

a. $\sqrt{2/7}$ b. 1/7
c. $\sqrt{6}$ d. none

9. For the line $\frac{x-1}{1}=\frac{y-2}{2}=\frac{z-3}{3}$, which one of the following is incorrect?
a. it lies in the plane $x-2y+z=0$
b. it is same as line $\frac{x}{1}=\frac{y}{2}=\frac{z}{3}$
c. it passes through (2, 3, 5)
d. it is parallel to the plane $x-2y+z-6=0$

10. The value of m for which straight line $3x-2y+z+3=0=4x-3y+4z+1$ is parallel to the plane $2x-y+mz-2=0$ is
a. –2 b. 8
c. –18 d. 11

11. The intercept made by the plane $\vec{r}\cdot\vec{n}=q$ on the x-axis is

a. $\frac{q}{\hat{i}\cdot\vec{n}}$ b. $\frac{\hat{i}\cdot\vec{n}}{q}$

c. $\frac{\hat{i}\cdot\vec{n}}{q}$ d. $\frac{q}{|\vec{n}|}$

12. Equation of lines in the plane $\pi\equiv 2x-y+z-4=0$ which is perpendicular to the line l whose equation is $\frac{x-2}{1}=\frac{y-2}{-1}=\frac{z-3}{-2}$ and which passes through the point intersection of l and π is

a. $\frac{x-2}{1}=\frac{y-1}{5}=\frac{z-1}{-1}$

b. $\frac{x}{3}=\frac{y-3}{5}=\frac{z-5}{-1}$

c. $\frac{x+2}{2}=\frac{y+1}{-1}=\frac{z+1}{1}$

d. $\frac{x-2}{2}=\frac{y-1}{-1}=\frac{z-1}{1}$

13. If the foot of the perpendicular from the origin to a plane is $P(a, b, c)$ the equation of the plane is

a. $\frac{x}{a}+\frac{y}{b}+\frac{z}{c}=3$
b. $ax+by+cz=3$
c. $ax+by+cz=a^2+b^2+c^2$
d. $ax+by+cz=a+b+c$

14. The equation of plane which passes through the point of intersection of lines $\frac{x-1}{3}=\frac{y-2}{1}=\frac{z-3}{2}$ and $\frac{x-3}{1}=\frac{y-1}{2}=\frac{z-2}{3}$ and at greatest distance from the point (0, 0, 0) is
a. $4x+3y+5z=25$ b. $4x+3y=5z=50$
c. $3x+4y+5z=49$ d. $x+7y-5z=2$

15. Let $A(\vec{a})$ and $B(\vec{b})$ be points on two skew line $\vec{r}=\vec{a}+\lambda\vec{p}$ and $\vec{r}=\vec{b}+u\vec{q}$ and the shortest distance between the skew line is 1, where $\vec{p}$ and $\vec{q}$ are unit vectors forming adjacent sides of a parallelogram enclosing an area of 1/2 units. If an angle between AB and the line of shortest distance is 60°, then $AB=$

a. $\frac{1}{2}$ b. 2
c. 1 d. $\lambda\in R-\{0\}$

16. Let A (1, 1, 1), B (2, 3, 5) and C (–1, 0, 2) be three points, then equation of a plane parallel to the plane ABC which is a distance 2 is
a. $2x-3y+z+2\sqrt{14}=0$
b. $2x-3y+z-\sqrt{14}=0$
c. $2x-3y+z+2=0$
d. $2x-3y+z-2=0$

17. The point on the line $\frac{x-2}{1}=\frac{y+3}{-2}=\frac{z+5}{-2}$ at a distance of 6 from the point (2, – 3, – 5) is
a. $(3,-5,-3)$ b. $(4,-7,-9)$
c. $(0,2,-1)$ d. $(-3,5,3)$

18. The coordinates of the foot of the perpendicular drawn from the origin to the line joining the points (–9, 4, 5) and (10, 0, –1) will be
a. (–3, 2, 1) b. (1, 2, 2)
c. (4, 5, 3) d. None of these

19. If P_1: $\vec{r}\cdot\vec{n_1}-d_1=0$, P_2: $\vec{r}\cdot\vec{n_2}-d_2=0$ and P_3: $\vec{r}\cdot\vec{n_3}-d_3=0$ are three planes and $\vec{n_1},\vec{n_2}$ and $\vec{n_3}$ are three non–coplanar vectors, then three lines $P_1=0$, $P_2=0$; $P_2=0$, $P_3=0$, and $P_3=0$, $P_1=0$ are
a. parallel lines b. coplanar lines
c. coincident lines d. concurrent lines

20. The length of projection of the line segment the points (1, 0, – 1) and (–1, 2, 2) to the plane $x + 3y - 5z = 6$, is equal to

a. 2 b. $\sqrt{\frac{271}{53}}$

c. $\sqrt{\frac{472}{31}}$ d. $\sqrt{\frac{474}{35}}$

21. The number of planes that are equidistant from four non-coplanar points is

a. 3 b. 4

c. 7 d. 9

22. In a three dimensional co-ordinate system *P*, *Q*, and *R* are images of a point *A* (*a*, *b*, *c*) in the *xy*, the *yz*, and *zx* planes, respectively. If *G* is the centroid of triangle *PQR* then area of triangle *AOG* is (*O* is the origin)

a. 0 b. $a^2 + b^2 + c^2$

c. $\frac{2}{3}(a^2 + b^2 + c^2)$ d. None of these

23. A plane passing though (1, 1, 1) cuts positive direction of co-ordinates axes at *A*, *B*, and *C* the Volume of tetrahedron *OABC* satisfies

a. $V \le \frac{9}{2}$ b. $V \ge \frac{9}{2}$

c. $V = \frac{9}{2}$ d. none of these

24. If lines $x = y = z$ and $x = y/2 = z/3$ and third line passing through (1, 1, 1) form a triangle of area $\sqrt{6}$ units, then point of intersection of third line with second line will be

a. $(1, 2, 3)$ b. $(2, 4, 6)$

c. $\left(\frac{4}{3}, \frac{8}{3}, \frac{12}{3}\right)$ d. None of these

25. The point of intersection of the line, passing through (0, 0, 1) and intersecting the lines $x + 2y + z = 1$, $-x + y - 2z = 2$, and $x + y = 2$, $x + z = 2$ with *xy* plane is

a. $\left(\frac{5}{3}, -\frac{1}{3}, 0\right)$ b. $(1, 1, 0)$

c. $\left(\frac{2}{3}, -\frac{1}{3}, 0\right)$ d. $\left(-\frac{5}{3}, \frac{1}{3}, 0\right)$

26. Shortest distance between the lines $\frac{x-1}{1} = \frac{y-1}{1} = \frac{z-1}{1}$ and $\frac{x-2}{1} = \frac{y-3}{1} = \frac{z-4}{1}$ is equal to

a. $\sqrt{14}$ b. $\sqrt{7}$

c. $\sqrt{2}$ d. None of these

27. Distance of $P\ (\vec{p})$ from the plane $\vec{r} \cdot \vec{n} = 0$ is

a. $|\vec{p} \cdot \vec{n}|$ b. $\frac{|\vec{p} \times \vec{n}|}{|\vec{n}|}$

c. $\frac{|\vec{p} \cdot \vec{n}|}{|\vec{n}|}$ d. None of these

28. The reflection of the point $\vec{a}$ in the plane $\vec{r} \cdot \vec{n} = q$ is

a. $\vec{a} + \frac{(\vec{q} - \vec{a} \cdot \vec{n})}{|\vec{n}|}$

b. $\vec{a} + 2\left(\frac{(\vec{q} - \vec{a} \cdot \vec{n})}{|\vec{n}|^2}\right)\vec{n}$

c. $\vec{a} + \frac{2(\vec{q} + \vec{a} \cdot \vec{n})}{|\vec{n}|}\vec{n}$

d. None of these

29. The line $\vec{r} = \vec{a} + \lambda\vec{b}$ will not meet the plane $\vec{r} \cdot \vec{n} = q$, if

a. $\vec{b} \cdot \vec{n} = 0, \vec{a} \cdot \vec{n} = q$

b. $\vec{b} \cdot \vec{n} \ne 0, \vec{a} \cdot \vec{n} \ne q$

c. $\vec{b} \cdot \vec{n} = 0, \vec{a} \cdot \vec{n} \ne q$

d. $\vec{b} \cdot \vec{n} \ne 0, \vec{a} \cdot \vec{n} = q$

30. If a line makes an angle of $\pi/4$ with the positive direction of each of *x*-axis and *y*-axis, then the angle that the line makes with the positive direction of the *z*-axis is

a. $\frac{\pi}{3}$ b. $\frac{\pi}{4}$

c. $\frac{\pi}{2}$ d. $\frac{\pi}{6}$

31. The ratio in which the plane $\vec{r} \cdot (\vec{i} - 2\vec{j} + 3\vec{k}) = 17$ divides the line joining the points $-2\vec{i} + 4\vec{j} + 7\vec{k}$ and $3\vec{i} - 5\vec{j} + 8\vec{k}$ is

a. 1:5 b. 1:10

c. 3:5 d. 3:10

32. The image of the point (–1, 3, 4) in the plane $x - 2y = 0$ is

a. $\left(-\frac{17}{3}, -\frac{19}{3}, 4\right)$ b. $(15, 11, 4)$

c. $\left(-\frac{17}{3}, -\frac{19}{3}, 1\right)$ d. $\left(\frac{9}{5}, -\frac{13}{5}, 4\right)$

33. The distance between the line $\vec{r} = 2\hat{i} - 2\hat{j} + 3\hat{k} + \lambda(\hat{i} - \hat{j} + 4\hat{k})$ and the plane $\vec{r} \cdot (\hat{i} + 5\hat{j} + \hat{k}) = 5$ is

a. $\frac{10}{3\sqrt{3}}$ b. $\frac{10}{9}$

c. $\frac{10}{3}$ d. $\frac{3}{10}$

34. Let L be the line of intersection of the planes $2x + 3y + z = 1$ and $x + 3y + 2z = 2$

If L makes an angle α with the positive x-axis, then $\cos \alpha$ equals;

a. $\frac{1}{2}$ b. 1

c. $\frac{1}{\sqrt{2}}$ d. $\frac{1}{\sqrt{3}}$

35. The length of the perpendicular drawn form (1, 2, 3) to the line $\frac{x-6}{3} = \frac{y-7}{2} = \frac{z-7}{-2}$ is

a. 4 b. 5

c. 6 d. 7

36. If the angle θ between the line $\frac{x+1}{1} = \frac{y-1}{2} = \frac{z-2}{2}$ and the plane $2x - y + \sqrt{\lambda}\, z + 4 = 0$ is such that $\sin \theta = \frac{1}{3}$, the value of λ is

a. $\frac{-3}{5}$ b. $\frac{5}{3}$

c. $\frac{-4}{3}$ d. $\frac{3}{4}$

37. The intersection of the spheres $x^2 + y^2 + z^2 + 7x - 2y - z = 13$ and $x^2 + y^2 + z^2 - 3x + 3y + 4z = 8$ is the same as the intersection of one of the spheres and the plane

a. $x - y - z = 1$ b. $x - 2y - z = 1$

c. $x - y - 2z = 1$ d. $2x - y - z = 1$

38. A plane makes intercepts OA, OB, and OC whose measurements are a, b and c on the OX, OY, and OZ axes. The area of triangle ABC is

a. $\frac{1}{2}(ab + bc + ca)$

b. $\frac{1}{2}abc\,(a + b + c)$

c. $\frac{1}{2}(a^2b^2 + b^2c^2 + c^2a^2)^{1/2}$

d. $\frac{1}{2}(a + b + c)^2$

39. A line makes some angle θ, with each of the x, and z-axis. If the angle β, which it makes with y-axis, is such that $\sin^2 \beta = 3 \sin^2 \theta$, the $\cos^2 \theta$ equals

a. $\frac{2}{3}$ b. $\frac{1}{5}$

c. $\frac{3}{5}$ d. $\frac{2}{5}$

40. The shortest distance from the plane $12x + y + 3z = 327$ to the sphere $x^2 + y^2 + z^2 + 4x - 2y - 6z = 155$ is

a. 39 b. 26

c. $41\frac{4}{13}$ d. 13

41. A tetrahedron has vertices O (0, 0, 0), A (1, 2, 1), B (2, 1, 3), and C (–1, 1, 2), then angle between OAB and ABC will be:

a. $\cos^{-1}\left(\frac{17}{31}\right)$ b. 30°

c. 90° d. $\cos^{-1}\left(\frac{19}{35}\right)$

42. The radius of the circle in which the sphere $x^2 + y^2 + z^2 + 2z - 2y - 4z - 19 = 0$ is cut by the plane $x + 2y + 2z + 7 = 0$ is

a. 2 b. 3

c. 4 d. 1

43. The lines $\frac{x-2}{1} = \frac{y-3}{1} = \frac{z-4}{-k}$ and $\frac{x-1}{k} = \frac{y-4}{2} = \frac{z-5}{1}$ are co-planar if:

a. $k = 1$ or -1 b. $k = 0$ or 3

c. $k = 3$ or -3 d. $k = 0$ or -1

44. The point of intersection of the lines $\frac{x-5}{3} = \frac{y-7}{-1} = \frac{z+2}{1}$ and $\frac{x+3}{-36} = \frac{y-3}{2} = \frac{z-6}{4}$ is

a. $\left(21, \frac{5}{3}, \frac{10}{3}\right)$ b. (2, 10, 4)

c. (–3, 3, 6) d. (5, 7, –2)

45. Two systems of rectangular axes have the same origin. If a plane cuts them at distance a, b, c, and a', b', c' from the origin, then:

a. $\frac{1}{a^2} + \frac{1}{b^2} + \frac{1}{c^2} + \frac{1}{a'^2} + \frac{1}{b'^2} + \frac{1}{c'^2} = 0$

b. $\frac{1}{a^2} - \frac{1}{b^2} - \frac{1}{c^2} + \frac{1}{a'^2} - \frac{1}{b'^2} - \frac{1}{c'^2} = 0$

c. $\frac{1}{a^2} + \frac{1}{b^2} + \frac{1}{c^2} - \frac{1}{a'^2} - \frac{1}{b'^2} - \frac{1}{c'^2} = 0$

d. $\frac{1}{a^2} + \frac{1}{b^2} + \frac{1}{c^2} + \frac{1}{a'^2} + \frac{1}{b'^2} + \frac{1}{c'^2} = 0$

46. The plane, which passes through the point (3, 2, 0) and the line $\frac{x-3}{1} = \frac{y-6}{5} = \frac{z-4}{4}$ is

a. $x - y + z = 1$ b. $x + y + z = 5$

c. $x + 2y - z = 1$ d. $2x - y + z = 5$

47. The direction ratios of a normal to the plane through (1, 0, 0) and (0, 1, 0), which makes an angle of $\pi/4$ with the plane $x + y = 3$ are

a. $<1, \sqrt{2}, 1>$ b. $<1, 1, \sqrt{2}>$
c. $<1, 1, 2>$ d. $<\sqrt{2}, 1, 1>$

48. The centre of the circle given by $\vec{r} \cdot (\hat{i} + 2\hat{j} + 2\hat{k}) = 15$ and $|\vec{r} - (\hat{j} + 2\hat{k})| = 4$ is

a. $(0, 1, 2)$ b. $(1, 3, 4)$
c. $(-1, 3, 4)$ d. None of these

49. The lines which intersect the skew lines $y = mx$, $z = c$; $y = -mx$, $z = -c$ and the x-axis lie on the surface

a. $cz = mxy$ b. $xy = cmz$
c. $cy = mxz$ d. none of these

50. Distance of the point $P(\vec{p})$ from the line $\vec{r} = \vec{a} + \lambda\vec{b}$ is

a. $\left|(\vec{a} - \vec{p}) + \frac{((\vec{p} - \vec{a}) \cdot \vec{b})\vec{b}}{|\vec{b}|^2}\right|$

b. $\left|(\vec{b} - \vec{p}) + \frac{((\vec{p} - \vec{a}) \cdot \vec{b})\vec{b}}{|\vec{b}|^2}\right|$

c. $\left|(\vec{a} - \vec{p}) + \frac{((\vec{p} - \vec{b}) \cdot \vec{b})\vec{b}}{|\vec{b}|^2}\right|$

d. None of these

51. From the point P(a, b, c), let perpendiculars PL and PM be drawn to YOZ and ZOX planes, respectively. Then the equation of the plane OLM is

a. $\frac{x}{a} + \frac{y}{b} + \frac{z}{c} = 0$ b. $\frac{x}{a} + \frac{y}{b} - \frac{z}{c} = 0$
c. $\frac{x}{a} - \frac{y}{b} - \frac{z}{c} = 0$ d. $\frac{x}{a} - \frac{y}{b} + \frac{z}{c} = 0$

52. The plane $\vec{r} \cdot \vec{n} = q$ will contain the line $\vec{r} = \vec{a} + \lambda\vec{b}$ if

a. $\vec{b} \cdot \vec{n} \neq 0, \vec{a} \cdot \vec{n} \neq q$
b. $\vec{b} \cdot \vec{n} = 0, \vec{a} \cdot \vec{n} \neq q$
c. $\vec{b} \cdot \vec{n} = 0, \vec{a} \cdot \vec{n} = q$
d. $\vec{b} \cdot \vec{n} \neq 0, \vec{a} \cdot \vec{n} = q$

53. The projection of point $P(\vec{p})$ on the plane $\vec{r} \cdot \vec{n} = q$ is s $(\vec{s})$ then

a. $\vec{s} = \frac{(q - \vec{p} \cdot \vec{n})\vec{n}}{|\vec{n}|^2}$

b. $\vec{s} = \vec{p} + \frac{(q - \vec{p} \cdot \vec{n})\vec{n}}{|\vec{n}|^2}$

c. $\vec{s} = \vec{p} - \frac{(\vec{p} \cdot \vec{n})\vec{n}}{|\vec{n}|^2}$

d. $\vec{s} = \vec{p} - \frac{(q - \vec{p} \cdot \vec{n})\vec{n}}{|\vec{n}|^2}$

54. The angle between $\hat{i}$ line of the intersection of the plane $\vec{r} \cdot (\hat{i} + 2\hat{j} + 3\hat{k}) = 0$ and $\vec{r} \cdot (3\hat{i} + 3\hat{j} + \hat{k}) = 0$, is

a. $\cos^{-1}\left(\frac{1}{3}\right)$ b. $\cos^{-1}\left(\frac{1}{\sqrt{3}}\right)$
c. $\cos^{-1}\left(\frac{2}{\sqrt{3}}\right)$ d. None of these

55. The median AD of the triangle ABC is bisected at E, BE meets AC in F, then $AF : AC =$

a. $3:4$ b. $1:3$
c. $1:2$ d. $1:4$

56. The equation of the plane which passes through the line of intersection of the planes $\vec{r} \cdot \vec{n_1} = q_1$, $\vec{r} \cdot \vec{n_2} = q_2$ and is parallel to the line of instruction of the planes $\vec{r} \cdot \vec{n_3} = q_3$ and $\vec{r} \cdot \vec{n_4} = q_4$, is

a. $[\vec{n_2}\ \vec{n_3}\ \vec{n_4}](\vec{r} \cdot \vec{n_1} - q_1) = [\vec{n_1}\ \vec{n_3}\ \vec{n_4}](\vec{r} \cdot \vec{n_2} - q_2)$

b. $[\vec{n_1}\ \vec{n_2}\ \vec{n_3}](\vec{r} \cdot \vec{n_4} - q_4) = [\vec{n_4}\ \vec{n_3}\ \vec{n_1}](\vec{r} \cdot \vec{n_2} - q_2)$

c. $[\vec{n_4}\ \vec{n_3}\ \vec{n_1}](\vec{r} \cdot \vec{n_4} - q_4) = [\vec{n_1}\ \vec{n_2}\ \vec{n_3}](\vec{r} \cdot \vec{n_2} - q_2)$

d. none of these

57. Consider the triangle AOB in the x-y plane where $A \equiv (1, 0, 0)$; $B \equiv (0, 2, 0)$; and $O \equiv (0, 0, 0)$. The new position of O, when triangle is rotated about side AB by 90° can be

a. $\left(\frac{4}{5}, \frac{3}{5}, \frac{2}{\sqrt{5}}\right)$ b. $\left(\frac{-3}{5}, \frac{\sqrt{2}}{5}, \frac{2}{\sqrt{5}}\right)$
c. $\left(\frac{4}{5}, \frac{2}{5}, \frac{2}{\sqrt{5}}\right)$ d. $\left(\frac{4}{5}, \frac{2}{5}, \frac{1}{\sqrt{5}}\right)$

58. Let $\vec{a} = \hat{i} + \hat{j}$ and $\vec{b} = 2\hat{i} - \hat{k}$ then the point of intersection of the the lines $\vec{r} \times \vec{a} = \vec{b} \times \vec{a}$ and $\vec{r} \times \vec{b} = \vec{a} \times \vec{b}$ is

a. $(3, -1, 1)$ b. $(3, 1, -1)$
c. $(-3, 1, 1)$ d. $(-3, -1, -1)$

59. The coordinates of the point P on the line $\vec{r} = (\hat{i} + \hat{j} + \hat{k}) + \lambda(-\hat{i} + \hat{j} - \hat{k})$ which is nearest to the origin is

a. $\left(\frac{2}{3}, \frac{4}{3}, \frac{2}{3}\right)$ b. $\left(-\frac{2}{3}, -\frac{4}{3}, \frac{2}{3}\right)$

c. $\left(\frac{2}{3}, \frac{4}{3}, -\frac{2}{3}\right)$ d. None of these

60. The ratio in which the line segment joining the points whose position vectors are $2\hat{i}-4\hat{j}-7\hat{k}$ and $-3\hat{i}+5\hat{j}-8\hat{k}$ is divided by the plane whose equations is $\hat{r}(\hat{i}-2\hat{j}+3\hat{k})=13$, is

a. 13 : 12 internally b. 12 : 25 externally

c. 13 : 25 internally d. 37 : 25 internally

61. Which of the following are equations for the plane passing through the points $P(1, 1, -1)$, $Q(3, 0, 2)$, and $R(-2, 1, 0)$?

a. $(2\hat{i}-3\hat{j}+3\hat{k})\cdot((x+2)\hat{i}+(y-1)\hat{j}+z\hat{k})=0$

b. $x=3-t, y=-11t, z=2-3t$

c. $(x+2)+11(y-1)=3z$

d. $(2\hat{i}-\hat{j}+3\hat{k})\times(-3\hat{i}+\hat{k})\cdot((x+2)\hat{i}\,(y-1)\hat{j}+z\hat{k})=0$

62. Given $\vec{\alpha}=3\hat{i}+\hat{j}+2\hat{k}$ and $\vec{\beta}=\hat{i}-2\hat{j}-4\hat{k}$ are the position vectors of the point A and B. Then the distance of the point $-\hat{i}+\hat{j}+\hat{k}$ from the plane passing through B and perpendicular to AB is

a. 5 b. 10

c. 15 d. 20

63. L_1 and L_2 are two lines whose vector equations are

$L_1: \vec{r}=\lambda((\cos\theta+\sqrt{3})\hat{i}+(\sqrt{2}\sin\theta)\hat{j}+(\cos\theta-\sqrt{3})\hat{k})$

$L_2: \vec{r}=\mu\left(a\hat{i}+b\hat{j}+c\hat{k}\right)$, where λ and μ are scalars and α is the acute angle between L_1 and L_2. If the angle α is independent of θ then the value of α is

a. $\frac{\pi}{6}$ b. $\frac{\pi}{4}$

c. $\frac{\pi}{3}$ d. $\frac{\pi}{2}$

64. The shortest distance between the lines $\frac{x-3}{3}=\frac{y-8}{-1}=\frac{z-3}{-1}$ and $\frac{x+3}{-3}=\frac{y+7}{2}=\frac{z-6}{4}$ is

a. $\sqrt{30}$ b. $2\sqrt{30}$

c. $5\sqrt{30}$ d. $3\sqrt{30}$

65. The line through $\hat{i}+3\hat{j}+2\hat{k}$ and perpendicular to the line $\vec{r}=(\hat{i}+2\hat{j}-\hat{k})+\lambda(2\hat{i}+\hat{j}+\hat{k})$ and $\vec{r}=(2\hat{i}+6\hat{j}+\hat{k})+\mu(\hat{i}+2\hat{j}+3\hat{k})$ is

a. $\vec{r}=(\hat{i}+2\hat{j}-\hat{k})+\lambda(-\hat{i}+5\hat{j}-3\hat{k})$

b. $\vec{r}=\hat{i}+3\hat{j}+2\hat{k}+\lambda(\hat{i}-5\hat{j}+3\hat{k})$

c. $\vec{r}=\hat{i}+3\hat{j}+2\hat{k}+\lambda(\hat{i}+5\hat{j}+3\hat{k})$

d. $\vec{r}=\hat{i}+3\hat{j}+2\hat{k}+\lambda(-\hat{i}-5\hat{j}-3\hat{k})$

66. The equation of the plane passing through the lines $\frac{x-4}{1}=\frac{y-3}{1}=\frac{z-2}{2}$ and $\frac{x-3}{1}=\frac{y-2}{-4}=\frac{z}{5}$ is

a. $11x-y-3z=35$ b. $11x+y-3z=35$

c. $11x-y+3z=35$ d. None of these

67. The three planes $4y+6z=5$; $2x+3y+5z=5$ and $6x+5y+9z=10$.

a. meet in a point

b. have a line in common

c. form a triangular prism

d. none of these

68. The equation of the plane through the line of intersection of the planes $ax+by+cz+d=0$ and $a'x+b'y+c'z+d'=0$ and parallel to the line $y=0$ and $z=0$ is

a. $(ab'-a'b)x+(bc'-b'c)y+(ad'-a'd)=0$

b. $(ab'-a'b)x+(bc'-b'c)y+(ad'-a'd)z=0$

c. $(ab'-a'b)y+(ac'-a'c)z+(ad'-a'd)=0$

d. None of these

69. Equation of the plane passing through the points (2, 2, 1) (9, 3, 6) and perpendicular to the plane $2x+6y+6z-1=0$ is

a. $3x+4y+5z=9$ b. $3x+4y-5z=9$

c. $3x+4y-5z=9$ d. None of the above

70. Value of K such that the line $\frac{x-1}{2}=\frac{y-1}{3}=\frac{z-1}{\lambda}$ is perpendicular to normal to the plane $\vec{r}\cdot(2\vec{i}+3\vec{j}+4\vec{k})=0$ is

a. $-\frac{13}{4}$ b. $-\frac{17}{4}$

c. 4 d. none of these

71. The equation of the plane through the intersection of the planes $x+2y+3z-4=0$ and $4x+3y+2z+1=0$ and passing through the origin is

a. $17x+14y+11z=0$

b. $7x+4y+z=0$

c. $x+14y+11z=0$

d. $17x+y+z=0$

72. The plane $4x+7y+4z+81=0$ is rotated through a right angle about its line of intersection with the plane $5x+3y+10z=25$. The equation of the plane in its new position is

a. $x-4y+6z=106$ b. $x-8y+13z=103$

c. $x-4y+6z=110$ d. $x-8y+13z=105$

73. The vector equation of the plane passing through the origin and the line of intersection of the plane $\vec{r}\cdot\vec{a}=\lambda$ and $\vec{r}\cdot\vec{b}=\mu$ is

a. $\vec{r}\cdot(\lambda\vec{a}-\mu\vec{b})=0$ b. $\vec{r}\cdot(\lambda\vec{b}-\mu\vec{a})=0$

c. $\vec{r}\cdot(\lambda\vec{a}+\mu\vec{b})=0$ d. $\vec{r}\cdot(\lambda\vec{b}+\mu\vec{a})=0$

74. The line $\vec{r}=\vec{a}+\lambda(\vec{b}\times\vec{c})$ and $\vec{r}=\vec{b}+\mu(\vec{c}\times\vec{a})$ will intersect if

a. $\vec{a}\times\vec{c}=\vec{b}\times\vec{c}$ b. $\vec{a}\cdot\vec{c}=\vec{b}\cdot\vec{c}$

c. $\vec{b}\times\vec{a}=\vec{c}\times\vec{a}$ d. None of these

75. The projection of the line $\frac{x+1}{-1}=\frac{y}{2}=\frac{z-1}{3}$ on the plane $x-2y+z=6$ is the line of intersection of this plane with the plane

a. $2x+y+2=0$ b. $3x+y-z=2$

c. $2x-3y+8z=3$ d. none of these

76. The direction cosines of a line satisfy the relations $\lambda(l+m)=n$ and $mn+nl+lm=0$. The value of λ for which the two lines are perpendicular to each other is

a. 1 b. 2

c. 1/2 d. none of these

77. The intercepts made on the axes by the plane which bisects the line joining the points (1, 2, 3) and (–3, 4, 5) at right angles are

a. $\left(-\frac{9}{2},9,9\right)$ b. $\left(\frac{9}{2},9,9\right)$

c. $\left(9,-\frac{9}{2},9\right)$ d. $\left(9,\frac{9}{2},9\right)$

78. The pair of lines whose direction cosines are given by the equations $3l+m+5n=0$ and $6mn-2nl+5lm=0$, are

a. parallel b. perpendicular

c. inclined at $\cos^{-1}\left(\frac{1}{6}\right)$ d. none of these

79. A sphere of constant radius $2k$ passes through the origin and meets the axes in A, B and C. The locus of a centroid of the tetrahedron $OABC$ is

a. $x^2+y^2+z^2=4k^2$ b. $x^2+y^2+z^2=k^2$

c. $2(k^2+y^2+z)^2=k^2$ d. none of these

80. A plane passes through a fixed point (a, b, c). The locus of the foot of the perpendicular to it from the origin is a sphere of radius

a. $\frac{1}{2}\sqrt{a^2+b^2+c^2}$ b. $\sqrt{a^2+b^2+c^2}$

c. $a^2+b^2+c^2$ d. $\frac{1}{2}(a^2+b^2+c^2)$

81. Which is the nature of the intersection of the set of planes $x+ay+(b+c)z+d=0$, $x+by+(c+a)z+d=0$ and $x+cy(a+b)z+d=0$?

a. They meet at a point.

b. They form a triangular prism.

c. They pass through a line.

d. They are at equal distance from the origin.

82. What is the equation of the plane which bisects the line joining the points (3, –2, 1) and (1, 4, –3) at right angles ?

a. $x-3y+2z+3=0$

b. $3x-2y+z+3=0$

c. $x+4y-3z+2=0$

d. $x-3y+2z+2=0$

83. What is the equation of the plane which passes through the z-axis and is perpendicular to the line $\frac{x-a}{\cos\theta}=\frac{y+2}{\sin\theta}=\frac{z-3}{0}$?

a. $x+y\tan\theta=0$ b. $y+x\tan\theta=0$

c. $x\cos\theta-y\sin\theta=0$ d. $x\sin\theta-y\cos\theta=0$

84. A straight line L on the XY-plane bisects the angle between OX and OY. What are the direction cosines of L ?

a. $<1/\sqrt{2},1/\sqrt{2},0>$ b. $<1/2,\sqrt{3}/2,0>$

c. $<0, 0, 1>$ d. $<2/3, 2/3, 1/3>$

85. For what value(s) of a, will the two points $(1, a, 1)$ and $(-3, 0, a)$ lie on opposite sides of the plane $3x+4y-12z+13=0$?

a. $a<-1$ or $a>1/3$ b. $a=0$ only

c. $0<a<1$ d. $-1<a<1$

86. If the plane $\frac{x}{2}+\frac{y}{3}+\frac{z}{6}=1$ cuts the axes of coordinates at points A, B, and C, then the area of the triangle ABC.

a. 18 sq unit b. 36 sq unit

c. $3\sqrt{14}$ sq unit d. $2\sqrt{14}$ sq unit

Multiple Correct Answers Type

Each question has four choices a, b, c, and d, out of which *one or more* are correct.

1. Let PM be the perpendicular from the point $P(1, 2, 3)$ to the x-y plane. If $\overrightarrow{OP}$ makes an angle θ with the positive direction of the z-axis and $\overrightarrow{OM}$ makes an angle ϕ with the positive direction of x-axis, where O is the origin and θ and ϕ are acute angles, then

a. $\cos\theta\cos\phi = 1/\sqrt{14}$

b. $\sin\theta\sin\phi = 2/\sqrt{14}$

c. $\tan\phi = 2$

d. $\tan\theta = \sqrt{5}/3$

2. Let P_1 denote the equation of the planes to which the vector $(\hat{i}+\hat{j})$ is normal and which contains the line whose equation is $\vec{r} = \hat{i}+\hat{j}+\hat{k}+\lambda(\hat{i}-\hat{j}-\hat{k})$ and P_2 denote he equation of the plane containing the line L and a point with position vector $\hat{j}$. Which of the following holds good?

a. The equation of P_1 is $x+y=2$.

b. The equation of P_2 is $\vec{r}\cdot(\hat{i}-2\hat{j}+\hat{k})=2$.

c. The acute angle between P_1 and P_2 is $\cot^{-1}(\sqrt{3})$.

d. The angle between the plane P_2 and the line L is $\tan^{-1}\sqrt{3}$.

3. If the planes $\vec{r}\cdot(\hat{i}+\hat{j}+\hat{k}) = q_1$, $\vec{r}\cdot(\hat{i}+2a\hat{j}+\hat{k}) = q_2$, and $\vec{r}\cdot(a\hat{i}+a^2\hat{j}+\hat{k}) = q_3$ intersect in a line, then the value of a is

a. 1 b. 1/2

c. 2 d. 0

4. A line with direction cosines proportional to $1, -5$, and -2 meets lines $x = y+5 = z+1$ and $x+5 = 3y = 2z$. The coordinates of each of the points of the intersection are given by

a. $(2, -3, 1)$ b. $(1, 2, 3)$

c. $(0, 5/3, 5/2)$ d. $(3, -2, 2)$

5. Let $P=0$ be the equation of a plane passing through the line of intersection of the planes $2x-y=0$ and $3z-y=0$ and perpendicular to the plane $4x+5y-3z=8$. Then the points which lie on the plane $P=0$ is/are

a. $(0, 9, 17)$ b. $(1/7, 2, 1/9)$

c. $(1, 3, -4)$ d. $(1/2, 1, 1/3)$

6. The equation of the lines $x+y+z-1=0$ and $4x+y-2z+2=0$ written in the symmetrical form is

a. $\frac{x-1}{2} = \frac{y+2}{-1} = \frac{z-2}{2}$

b. $\frac{x+1/2}{1} = \frac{y-1}{-2} = \frac{z-1/2}{1}$

c. $\frac{x}{1} = \frac{y}{-2} = \frac{z-1}{1}$

d. $\frac{x+1}{1} = \frac{y-2}{-2} = \frac{z-0}{1}$

7. Consider the planes $3x-6y+2z+5=0$ and $4x-12y+3z=3$. The plane $67x-162y+47z+44=0$ bisects that angle between the given planes which

a. contains the origin b. is acute

c. is obtuse d. none of these

8. If the lines $\frac{x-2}{1} = \frac{y-3}{1} = \frac{z-4}{\lambda}$ and $\frac{x-1}{\lambda} = \frac{y-4}{2} = \frac{z-5}{1}$ intersect, then

a. $\lambda = -1$ b. $\lambda = 2$

c. $\lambda = -3$ d. $\lambda = 0$

9. The equations of the plane which passes through (0, 0, 0) and which is equally inclined to the planes $x-y+z-3=0$ and $x+y+z+4=0$ is/are

a. $y=0$ b. $x=0$

c. $x+y=0$ d. $x+z=0$

10. The xy plane is rotated about its line of intersection with the yz plane by 45°, then the equation of the new plane is/are

a. $z+x=0$ b. $z-y=0$

c. $x+y+z=0$ d. $z-x=0$

11. The equation of the plane which is equally inclined to the lines $\frac{x-1}{2} = \frac{y}{-2} = \frac{z+2}{-1}$ and $\frac{x+3}{8} = \frac{y-4}{1} = \frac{z}{-4}$ and passing through the origin is/are

a. $14x-5y-7z=0$ b. $2x+7y-z=0$

c. $3x-4y-z=0$ d. $x+2y-5z=0$

12. Which of the following lines lie on the plane $x+2y-z+4=0$?

a. $\frac{x-1}{1} = \frac{y}{-1} = \frac{z-5}{-1}$

b. $x-y+z = 2x+y-z=0$

c. $\vec{r} = 2\hat{i}-\hat{j}+4\hat{k}+\lambda(3\hat{i}+\hat{j}+5\hat{k})$

d. none of these

13. If the volume of tetrahedron $ABCD$ is 1 cubic units, where $A(0, 1, 2)$ and $B(-1, 2, 1)$, $C(1, 2, 1)$, then the locus of point D is

a. $x+y-z=3$ b. $y+z=6$

c. $y+z=0$ d. $y+z=-3$

14. A rod of length 2 units whose one end is $(1, 0, -1)$ and other end touches the plane $x-2y+2z+4=0$, then

a. The rod sweeps the figure whose volume is 3π cubic units.

b. The area of the region which the rod traces on the plane is 2π.

c. The length of the projection of the rod on the plane is $\sqrt{3}$ units.

d. The centre of the region which the rod traces on the plane is (4/3, –2/3, 1/3).

15. Consider a set of points in the space which is at a distance of 2 units from the line $\frac{x}{1}=\frac{y-1}{-1}=\frac{z+2}{2}$ between the planes $x-y-z+3=0$ and $x-y-z-2=0$.

a. The volume of the bounded figure by points R and the planes is $(10/3\sqrt{3})\pi$ cube units.

b. The area of the curved surface formed by the set of points R is $(4/\sqrt{3})\pi$ sq. units.

c. The volume of the bounded figure by the set of points R and the planes is $(20/\sqrt{3})(\pi$ cubic units.

d. The area of the curved surface formed by the set of points R is $(10/\sqrt{3})\pi$ sq. units

16. The equation of a line passing through the point $\vec{a}$ parallel to the plane $\vec{r}.\vec{n}=q$ and perpendicular to the line $\vec{r}=\vec{b}+t\vec{c}$ is

a. $\vec{r}=\vec{a}+\lambda(\vec{n}\times\vec{c})$

b. $(\vec{r}-\vec{a})\times(\vec{n}\times\vec{c})=0$

c. $\vec{r}=\vec{b}+\lambda(\vec{n}\times\vec{c})$

d. none of these

Reasoning Type

Each question has four choices a, b, c, and d, out of which *only one* is correct. Each question contains Statement 1 and Statement 2.

a. Both the statements areTrue, and Statement 2 is the correct explanation For Statement 1.

b. Both the statements are True, but Statement 2 is Not the correct explanation for Statement 1.

c. Statement 1 is True and Statement 2 is False.

d. Statment 1 is False and Statement 2 is True.

1. **Statement 1:** Lines $\vec{r}=\hat{i}-\hat{j}+\lambda(\hat{i}+\hat{j}-\hat{k})$ and $\vec{r}=2\hat{i}-\hat{j}+\mu(\hat{i}+\hat{j}-\hat{k})$ do not intersect.

Statement 2: Skew lines never intersect.

2. **Statement 1:** Lines $\vec{r}=\hat{i}+\hat{j}-\hat{k}+\lambda(3\hat{i}-\hat{j})$ and $\vec{r}=4\hat{i}-\hat{k}+\mu(2\hat{i}+3\hat{k})$ intersect.

Statement 2: If $\vec{b}\times\vec{d}=0$, then lines $\vec{r}=\vec{a}+\lambda\vec{b}$ and $\vec{r}=\vec{c}+\lambda\vec{d}$ do not intersect.

3. The equation of two straight lines are $\frac{x-1}{2}=\frac{y+3}{1}=\frac{z-2}{-3}$ and $\frac{x-2}{1}=\frac{y-1}{-3}=\frac{z+3}{2}$.

Statement 1: The given lines are coplanar.

Statement 2: The equations $2x_1-y_1=1, x_1+3y_1=4$ and $3x_1+2y_1=5$ are consistent.

4. **Statement 1:** A plane passes through the point $A(2, 1, -3)$. If distance of this plane from origin is maximum, then its equation is $2x+y-3z=14$.

Statement 2: If the plane passing though the point $A(\vec{a})$ is at maximum distance from origin than normal to plane is vector $\vec{a}$.

5. **Statement 1 :** Line $\frac{x-1}{1}=\frac{y-0}{2}=\frac{z+2}{-1}$ lies in the plane $2x-3y-4z-10=0$.

Statement 2 : If line $\vec{r}=\vec{a}+\lambda\vec{b}$ lies in the plane $\vec{r}\cdot\vec{c}=n$ (where n is scalar), then $\vec{b}\cdot\vec{c}=0$.

6 **Statement 1:** Let θ be the angle between the line $\frac{x-2}{2}=\frac{y-1}{-3}=\frac{z+2}{-2}$ and the plane $x+y-z=5$. Then $\theta=\sin^{-1}(1/\sqrt{51})$.

Statement 2: The angle between a straight line and a plane is the complement of the angle between the line and the normal to the plane.

7. **Statement 1:** Let A $(\vec{i}+\vec{j}+\vec{k})$ and B $(\vec{i}-\vec{j}+\vec{k})$ be two points. Then point P $(2\vec{i}+3\vec{j}+\vec{k})$ lies exterior to the sphere with AB as its diameter.

Statement 2: If A and B are any two points and P is a point in space such that $\overrightarrow{PA}\cdot\overrightarrow{PB}>0$, then the point P lies exterior to the sphere with AB as its diameter.

8. **Statement 1:** There exists a unique sphere which passes through the three points and which has the least radius.

Statement 2: The centre of such a sphere lies on the plane determined by the given three points.

9. **Statement 1:** There exist two points on the line $\frac{x-1}{1}=\frac{y}{-1}=\frac{z+2}{2}$ which are at a distance of 2 units from point (1, 2, – 1).

Statement 2: Perpendicular distance of point (1, 2, –1) from the line $\frac{x-1}{1}=\frac{y}{-1}=\frac{z+2}{2}$ is 1 unit.

10. **Statement 1:** The shortest distance between the lines $\frac{x}{-3}=\frac{y-1}{1}=\frac{z+1}{-1}$ and $\frac{x-2}{1}=\frac{y-3}{2}=\left(\frac{z+(13/7)}{-1}\right)$ is zero.

Statement 2: The given lines are perpendicular.

Linked Comprehension Type

Based on each paragraph, three multiple-choice questions have to be answered. Each question has four choices *a, b, c,* and *d*, out of which *only one* is correct.

For Problems 1–3

Given four points $A(2, 1, 0)$, $B(1, 0, 1)$, $C(3, 0, 1)$, and $D(0, 0, 2)$. Point D lies on a line L orthogonal to the plane determined by the points A, B, and C.

1. The equation of the plane ABC is

   a. $x+y+z-3=0$ b. $y+z-1=0$

   c. $x+z-1=0$ d. $2y+z-1=0$

2. The equation of the line L is

   a. $\vec{r}=2\hat{k}+\lambda(\hat{i}+\hat{k})$

   b. $\vec{r}=2\hat{k}+\lambda(2\hat{j}+\hat{k})$

   c. $\vec{r}=2\hat{k}+\lambda(\hat{j}+\hat{k})$

   d. none

3. The perpendicular distance of D from the plane ABC is

   a. $\sqrt{2}$ b. $1/2$

   c. 2 d. $1/\sqrt{2}$

For Problems 4–6

A ray of light comes along the line $L=0$ and strikes the plane mirror kept along the plane $P=0$ at B. $A(2, 1, 6)$ is a point on the line $L=0$ whose image about $P=0$ is A'. It is given that $L=0$ is $\frac{x-2}{3}=\frac{y-1}{4}=\frac{z-6}{5}$ and $P=0$ is $x+y-2z=3$.

4. The coordinates of A' are

   a. $(6, 5, 2)$ b. $(6, 5, -2)$

   c. $(6, -5, 2)$ d. none of these

5. The coordinates of B are

   a. $(5, 10, 6)$ b. $(10, 15, 11)$

   c. $(-10, -15, -14)$ d. none of these

6. If $L_1=0$ is the reflected ray, then its equation is

   a. $\frac{x+10}{4}=\frac{y-5}{4}=\frac{z+2}{3}$

   b. $\frac{x+10}{3}=\frac{y+15}{5}=\frac{z+14}{5}$

   c. $\frac{x+10}{4}=\frac{y+15}{5}=\frac{z+14}{3}$

   d. none of these

For Problems 7–9

Consider three planes $2x+py+6z=8$, $x+2y+qz=5$ and $x+y+3z=4$.

7. Three planes intersect at a point if

   a. $p=2, q\neq 3$ b. $p\neq 2, q\neq 3$

   c. $p\neq 2, q=3$ d. $p=2, q=3$

8. Three planes do not have any common point of intersection if

   a. $p=2, q\neq 3$ b. $p\neq 2, q\neq 3$

   c. $p\neq 2, q=3$ d. $p=2, q=3$

9. The planes have infinite points common among them if

   a. $p=2, q\in 3$ b. $p\in 2, q\in 3$

   c. $p\neq 2, q=3$ d. $p=2, q=3$

For Problems 10–12

Consider a plane $x+y-z=1$ and the point $A(1, 2, -3)$. A line L has the equation $x=1+3r$, $y=2-r$ and $z=3+4r$.

10. The coordinate of a point B of line L such that AB is parallel to the plane is

    a. $(10, -1, 15)$ b. $(-5, 4, -5)$

    c. $(4, 1, 7)$ d. $(-8, 5, -9)$

11. The equation of the plane containing the line L and the point A has the equation

    a. $x-3y+5=0$ b. $x+3y-7=0$

    c. $3x-y-1=0$ d. $3x+y-5=0$

12. The distance between the points on the line which are at a distance of $4/\sqrt{3}$ from the plane is

    a. $4\sqrt{26}$ b. 20

    c. $10\sqrt{13}$ d. none of these

Matrix-Match Type

Each question contains statements given in two columns which have to be matched. Statements (a, b, c, d) in Column I have to be matched with statements (p, q, r, s) in Column II. If the correct matches are a → p, s; b → q, r; c → p, q and d → s, then the correctly bubbled 4 × 4 matrix should be as follows:

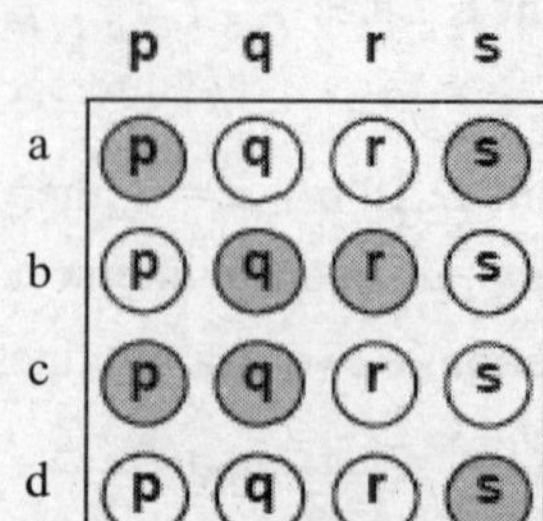

1.

Column I	Column II
a. A vector perpendicular to the line $x=2t+1$, $y=t+2$ and $z=-t-3$	**p.** $7\hat{i}+3\hat{j}+5\hat{k}$
b. A vector parallel to the planes $x+y+z-3=0$ and $2x-y+3z=0$	**q.** $4\hat{i}-\hat{j}-3\hat{k}$
c. A vector along which the distance between the lines $\frac{x}{2}=\frac{y}{-3}=\frac{z}{-1}$ and $\vec{r}=(3\hat{i}-\hat{j}+\hat{k})+t(\hat{i}+\hat{j}-2\hat{k})$ is the shortest	**r.** $-11\hat{i}+7\hat{j}+5\hat{k}$
d. A vector normal to the plane $\vec{r}=-\hat{i}+4\hat{j}-6\hat{k}+\lambda(\hat{i}+3\hat{j}-2\hat{k})+\mu(-\hat{i}+2\hat{j}-5\hat{k})$	**s.** $\hat{i}+3\hat{j}+\hat{k}$

2.

Column I	Column II
a. Lines $\frac{x-1}{-2}=\frac{y+2}{3}=\frac{z}{-1}$ and $\vec{r}=(3\hat{i}-\hat{j}+\hat{k})+t(\hat{i}+\hat{j}+\hat{k})$ are	**p.** intersecting
b. Lines $\frac{x+5}{1}=\frac{y-3}{7}=\frac{z+3}{3}$ and $x-y+2z-4=0$ $=2x+y-3z+5=0$ are	**q.** perpendicular
c. Lines ($x=t-3, y=-2t+1, z=-3t-2$) and $\vec{r}=(t+1)\hat{i}+(2t+3)\hat{j}+(-t-9)\hat{k}$	**r.** parallel
d. Lines $\vec{r}=(\hat{i}+3\hat{j}-\hat{k})+t(2\hat{i}-\hat{j}-\hat{k})$ and $\vec{r}=(-\hat{i}-2\hat{j}+5\hat{k})+s(\hat{i}-2\hat{j}+\hat{k})$	**s.** skew

3.

Column I	Column II
a. The coordinates of a point on the line $x=4y+5, z=3y-6$ at a distance 3 from the point $(5, 3, -6)$ is/are	**p.** $(-1, -2, 0)$
b. The plane containing the lines $\frac{x-2}{3}=\frac{y+3}{5}=\frac{z+5}{7}$ and $\frac{x+2}{1}=\frac{y-4}{4}=\frac{z-6}{7}$ passes through	**q.** $(5, 0, -6)$
c. A lines passes through two points $A(2, -3, -1)$ and $B(8, -1, 2)$. The coordinates of a point on this line nearer to the origin at a distance of 14 units from A is/are	**r.** $(2, 5, 7)$
d. The coordinates of the foot of the perpendicular from the point $(3, -1, 11)$ on the line $\frac{x}{2}=\frac{y-2}{3}=\frac{z-3}{4}$ is/are	**s.** $(14, 1, 5)$

4.

Column I	Column II
a. The distance between the line $\vec{r}=(2i-2j+3k)+\lambda(i-j+4k)$ and plane $\vec{r}\cdot(\hat{i}+5\hat{j}+\hat{k})=5$	**p.** $\frac{25}{3\sqrt{14}}$
b. Distance between parallel planes $\vec{r}\cdot(2\hat{i}-\hat{j}+3\hat{k})=4$ and $\vec{r}\cdot(6i-3j+9k)+13=0$ is	**q.** 13/7
c. The distance of a point $(2, 5, -3)$ from the plane $\vec{r}\cdot(6i-3j+2k)=4$ is	**r.** $\frac{10}{3\sqrt{3}}$
d. The distance of the point $(1, 0, -3)$ from the plane $x-y-z-9=0$ measured of parallel to line $\frac{x-2}{2}=\frac{y+2}{3}=\frac{z-6}{-6}$	**s.** 7

5.

	Column I	Column II
a.	Image of the point (3, 5, 7) in the plane $2x+y+z=-18$ is	p. $(-1,-1,-1)$
b.	The point of intersection of the line $\frac{x-2}{-3}=\frac{y-1}{-2}=\frac{z-3}{2}$ and the plane $2x+y-z=3$ is	q. $(-21,-7,-5)$
c.	The foot of the perpendicular from the point (1, 1 2) to the plane $2x-2y+4z+5=0$ is	r. $\left(\frac{5}{2},\frac{2}{3},\frac{8}{3}\right)$
d.	The intersection point of the lines $\frac{x-1}{2}=\frac{y-2}{3}=\frac{z-3}{4}$ and $\frac{x-4}{5}=\frac{y-1}{2}=z$ is	s. $\left(-\frac{1}{12},\frac{25}{12},\frac{-2}{12}\right)$

Integer Type

1. If the angle between the lines whose direction cosines are connected by the relations $l+m+n=0$ and $2lm+2nl-mn=0$ is θ, then the value of $4\sin^2\theta$ is____.
2. If the length of perpendicular drawn from the point (2, 3, 4) to the line $\frac{4-x}{2}=\frac{y}{6}=\frac{1-z}{3}$ is L, then the value of $[L]$ is, where $[\cdot]$ represents the greatest integer function, ________.
3. If the shortest distance between the lines $\frac{x-1}{2}=\frac{y-2}{3}=\frac{z-3}{4};\frac{x-2}{3}=\frac{y-4}{4}=\frac{z-5}{5}$ is d, then the value of $6d^2$ is __________.
4. If the straight line $x=1+s, y=3-\lambda s, z=1+\lambda s$ and $x=\frac{t}{2}, y=1+t, z=2-t$ with parameters s and t respectively, are coplanar, then the value of $|\lambda|$ is ____.
5. If (p, q, r) is the image of point (1, 2, 3) in the line $\frac{x-6}{3}=\frac{y-7}{2}=\frac{z-7}{-2}$, then the value of $(p-q+r)$ is __________.
6. If the lines $\frac{x-1}{2}=\frac{y+1}{3}=\frac{z-1}{4}$ and $\frac{x-3}{1}=\frac{y-k}{2}=\frac{z}{1}$ intersect, then find the value of $2k$.
7. The distance of the point (1, 0, – 3) from the plane $x-y-z=9$ measured parallel to the line $\frac{x-2}{2}=\frac{y+2}{3}=\frac{z-6}{-6}$ is__________.
8. If d is the distance between the line $\frac{x+1}{-3}=\frac{y-3}{2}=\frac{z-2}{1}$ and the plane $x+y+z+3=0$, then the value of $\sqrt{3}d$ is_________.
9. The radius of the circular section of the sphere $|\vec{r}|=5$ by the plane $\vec{r}\cdot(\vec{i}+2\vec{j}-\vec{k})=4\sqrt{3}$ is_____.
10. A sphere of constant radius k passes through the origin and meets the axes in A, B and C. If the centroid of the triangle ABC lies on the sphere $m(x^2+y^2+z^2)=4k^2$, then the value of m is________.
11. If the perpendicular distance of a corner of a unit cube from a diagonal not passing through it is d, then the value of $3d^2$ is ________.
12. If the planes $x-cy-bz=0$, $cx-y+az=0$ and $bx+ay-z=0$ pass through a straight line, then the value of $a^2+b^2+c^2+2abc$ is___________.
13. The value of k such that $\frac{x-4}{1}=\frac{y-2}{1}=\frac{z-k}{2}$ lies in the plane $2x-4y+z=7$ is________.
14. A variable plane, at a distance of the one unit from the origin, cuts the coordinates axes at A, B and C. If the centroid $D(x, y, z)$ of triangle ABC satisfies the relation $\frac{1}{x^2}+\frac{1}{y^2}+\frac{1}{z^2}=k$, then the value k is_____.
15. A plane which is perpendicular to two planes $2x-2y+z=0$ and $x-y+2z=4$, passes through (1, –2, 1). The square of the distance of the plane from the point (1, 2, 2) is___________.
16. Plane passing through (1, 1, 1) and parallel to the lines L_1, L_2 have direction ratios (1, 0, –1), (1, – 1, 0). If V is the volume of tetrahedron formed by origin and the points where these planes intersect the coordinates axes, then the value of $2V$ is________.
17. A line with positive direction cosines passes through the point P (2, – 1, 2) and makes equal angles with the coordinate axes. The line meets the plane $2x+y+z=9$ at point Q. Then the square of the length of the line segment PQ equals____________.
18. Given $\vec{\alpha}=3i+j+2k$, $\vec{\beta}=i-2j-4k$ are the position vectors of points A and B. Then the distance of point $-i+j+k$ from the plane passing through B and perpendicular to AB is ____________.
19. The distance of the point P (3, 8, 2) from the line $\frac{1}{2}(x-1)=\frac{1}{4}(y-3)=\frac{1}{3}(z-2)$ measured parallel to the plane $3x+2y-2z+15=0$ is____________.
20. Through the point P (1, 2,2), a plane drawn at right angle to OP meets the axes in A,B,C. If the area of the triangle is $243\lambda/72$, then the value of $|\lambda|$ is .

ANSWERS AND SOLUTIONS

Objective Type

1. **b.** $x^2 - 5x + 6 = 0$
$\Rightarrow x - 2 = 0, x - 3 = 0,$
which represents planes.

2. **c.** We have $z = 0$ for the point, where the line intersects the curve. Therefore,
$$\frac{x-2}{3} = \frac{y+1}{2} = \frac{0-1}{-1}$$
$$\Rightarrow \frac{x-2}{3} = 1 \text{ and } \frac{y+1}{2} = 1$$
$\Rightarrow x = 5$ and $y = 1$
Putting these values in $xy = c^2$, we get
$5 = c^2 \Rightarrow c = \pm\sqrt{5}$

3. **a.** $4(2) - 2(3) - 1(2) = 0$
Also, point (–3, 4, –5) does not lie on the plane.
Therefore, the line is parallel to the plane.

4. **c.** The given plane passes through $\vec{a}$ and is parallel to the vectors $\vec{b} - \vec{a}$ and $\vec{c}$. So it is normal to $(\vec{b} - \vec{a}) \times \vec{c}$. Hence, its equation is
$(\vec{r} - \vec{a}) \cdot ((\vec{b} - \vec{a}) \times \vec{c}) = 0$
or $\vec{r} \cdot (\vec{b} \times \vec{c} + \vec{c} \times \vec{a}) = [\vec{a}\,\vec{b}\,\vec{c}]$
The length of the perpendicular from the origin to this plane is
$$\frac{[\vec{a}\,\vec{b}\,\vec{c}]}{|\vec{b} \times \vec{c} + \vec{c} \times \vec{a}|}$$

5. **b.** Here, $\alpha = \beta = \gamma$.
Since $\cos^2\alpha + \cos^2\beta + \cos^2\gamma = 1$,
$$\cos\alpha = \frac{1}{\sqrt{3}}.$$
DC's of PQ are $\left(\frac{1}{\sqrt{3}}, \frac{1}{\sqrt{3}}, \frac{1}{\sqrt{3}}\right)$.

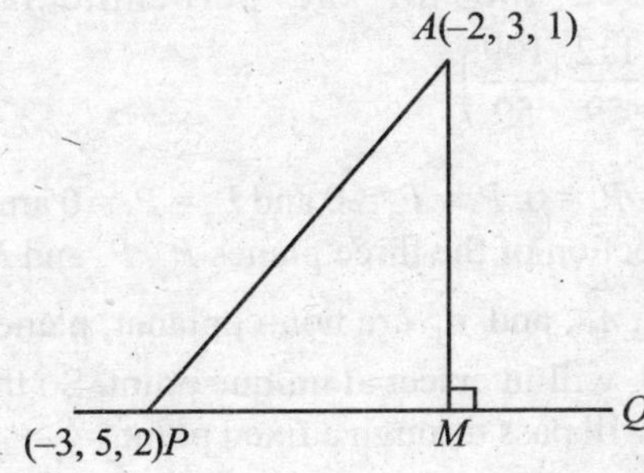

Fig. 3.1

PM = Projection of AP on PQ
$$= \left|(-2+3)\frac{1}{\sqrt{3}} + (3-5)\frac{1}{\sqrt{3}} + (1-2)\frac{1}{\sqrt{3}}\right|$$
$$= \frac{2}{\sqrt{3}}$$
and $AP = \sqrt{(-2+3)^2 + (3-5)^2 + (1-2)^2} = \sqrt{6}$
$$AM = \sqrt{(AP)^2 - (PM)^2} = \sqrt{6 - \frac{4}{3}} = \sqrt{\frac{14}{3}}$$

6. **c.**

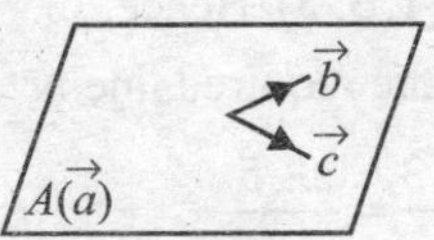

Fig. 3.2

Plane through $\vec{a}$ and parallel to two non-collinear vector is $(\vec{r} - \vec{a}) \cdot (\vec{b} \times \vec{c}) = 0$.
Here, $\vec{a} = \hat{i} - \hat{j}$, $\vec{b} = \hat{i} + \hat{j} + \hat{k}$, $\vec{c} = \hat{i} - 2\hat{j} + 3\hat{k}$

7. **c.** $\hat{a} = \pm\dfrac{\vec{n}_1 \times \vec{n}_2}{|\vec{n}_1 \times \vec{n}_2|} = \pm\dfrac{2\hat{i} + 5\hat{j} + 3\hat{k}}{\sqrt{38}}$ [where $\vec{n}_1$ and $\vec{n}_2$ are normal to planes]

8. **a.** Equation of the plane containing L_1 $A(x-2) + B(y-1) + C(z+1) = 0$
where $A + 2C = 0; A + B - C = 0$
$\Rightarrow A = -2C,\ B = 3C,\ C = C$
$\Rightarrow$ Plane is $-2(x-2) + 3(y-1) + z + 1 = 0$ or $2x - 3y - z - 2 = 0$
Hence, $p = \left|\dfrac{-2}{\sqrt{14}}\right| = \sqrt{\dfrac{2}{7}}$

9. **c.** On (1, 2, 3) satisfies the plane $x - 2y + z = 0$ and also $(\hat{i} + 2\hat{j} + 3\hat{k}) \cdot (\hat{i} - 2\hat{j} + \hat{k}) = 0$
Since the lines $\dfrac{x-1}{1} = \dfrac{y-2}{2} = \dfrac{z-3}{3}$ and $\dfrac{x}{1} = \dfrac{y}{2} = \dfrac{z}{3}$ both satisfy (0, 0, 0) and (1, 2, 3), both are same. Given line is obviously parallel to the plane $x - 2y + z = 6$.

10. **a.** Vector $\left((3\hat{i} - 2\hat{j} + \hat{k}) \times (4\hat{i} - 3\hat{j} + 4\hat{k})\right)$ is perpendicular to $2\hat{i} - \hat{j} + m\hat{k}$
$$\Rightarrow \begin{vmatrix} 3 & -2 & 1 \\ 4 & -3 & 4 \\ 2 & -1 & m \end{vmatrix} = 0 \qquad \Rightarrow m = -2$$

11. **a.** x intercept is say x_1

$\Rightarrow$ Plane passes through it

$\therefore\ x_1 \hat{i}\cdot\vec{n}=q \Rightarrow x_1=\dfrac{q}{\hat{i}\cdot\vec{n}}$

12. **b.** Let direction ratios of the line be (a, b, c), then

$2a-b+c=0$ and $a-b-2c=0$, i.e., $\dfrac{a}{3}=\dfrac{b}{5}=\dfrac{c}{-1}$

Therefore, direction ratios of the line are $(3, 5, -1)$.

Any point on the line is $(2+\lambda, 2-\lambda, 3-2\lambda)$, it lies on the plane π if $2(2+\lambda)-(2-\lambda)+(3-2\lambda)=4$

$\Rightarrow 4+2\lambda-2+\lambda+3-2\lambda=4 \Rightarrow \lambda=-1$

Therefore, the point of intersection of the line and the plane is $(1, 3, 5)$. Hence, equation of the required line is

$$\frac{x-1}{3}=\frac{y-3}{5}=\frac{z-5}{-1}$$

13. **c.** Direction ratios of OP are (a, b, c) Therefore, equation of the plane is

$a(x-a)+b(y-b)+c(z-c)=0$

i.e. $xa+yb+zc=a^2+b^2+c^2$

14. **b.** Let a point $(3\lambda+1, \lambda+2, 2\lambda+3)$ of the first line also lies on the second line

Then $\dfrac{3\lambda+1-3}{1}=\dfrac{\lambda+2-1}{2}=\dfrac{2\lambda+3-2}{3} \Rightarrow \lambda=1$

Hence, the point of intersection P of the two lines is $(4, 3, 5)$.

Equation of plane perpendicular to OP where O is $(0, 0, 0)$ and passing through P is $4x+3y+5z=50$.

15 **b.** $1=\left|(\vec{b}-\vec{a})\cdot\dfrac{(\vec{p}\times\vec{q})}{|\vec{p}\times\vec{q}|}\right|$

$\Rightarrow |\vec{b}-\vec{a}|\cos 60^\circ=1\ AB=2$

16. **a.** $A(1, 1, 1), B(2, 3, 5), C(-1, 0, 2)$ direction ratios of AB are $<1, 2, 4>$.

Direction ratios of AC are $<-2, -1, 1>$.

Therefore, direction ratios of normal to plane ABC are $<2, -3, 1>$

As a result, equation of the plane ABC is $2x-3y+z=0$.

Let the equation of the required plane is $2x-3y+z$

$=k$, then $\left|\dfrac{k}{\sqrt{4+9+1}}\right|=2k=\pm 2\sqrt{14}$

Hence, equation of the required plane is $2x-3y+z+2\sqrt{14}=0$

17. **b.** Direction cosines of the given are $\dfrac{1}{3}, -\dfrac{2}{3}, -\dfrac{2}{3}$

Hence, the equation of line can be point in the form

$$\frac{x-2}{1/3}=\frac{y+3}{-2/3}=\frac{z+5}{-2/3}=r$$

Therefore, any point on the line is

$\left(2+\dfrac{r}{3}, -3-\dfrac{2r}{3}, -5-\dfrac{2r}{3}\right)$, where $r=\pm 6$.

Points are $(4, -7, -9)$ and $(0, 1, -1)$.

18. **d.** Let AD be perpendicular and D be foot of the perpendicular which divides BC in the ratio $\lambda : 1$, then

$$D\left(\frac{10\lambda-9}{\lambda+1}, \frac{4}{\lambda+1}, \frac{-\lambda+5}{\lambda+1}\right). \quad (1)$$

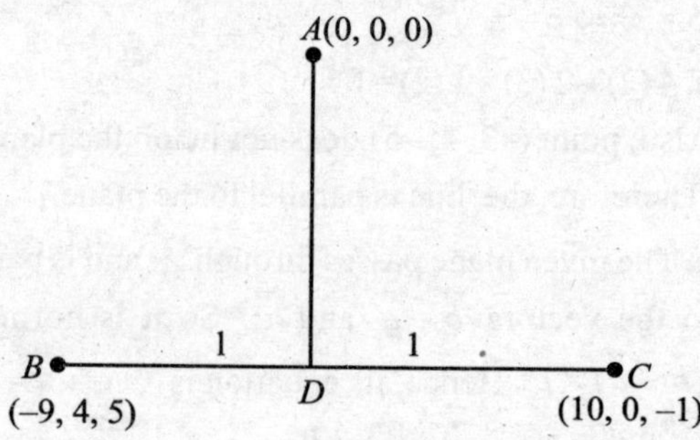

Fig. 3.3

The direction ratios of AD are $\dfrac{10\lambda-9}{\lambda+1}, \dfrac{4}{\lambda+1}$, and $\dfrac{-\lambda+5}{\lambda+1}$ and direction ratios of BC are $19, -4$ and -6.

Since $AD \perp BC$, we get

$$19\left(\frac{10\lambda-9}{\lambda+1}\right)-4\left(\frac{4}{\lambda+1}\right)-6\left(\frac{-\lambda+5}{\lambda+1}\right)=0$$

$\Rightarrow \lambda=\dfrac{31}{28}$

Hence, on putting the value of λ in (1), we get required foot of the perpendicular, i.e., $\left(\dfrac{58}{59}, \dfrac{112}{59}, \dfrac{109}{59}\right)$.

19. **d.** $P_1=P_2=0, P_2=P_3=0$ and $P_3=P_1=0$ are lines of intersection of the three planes P_1, P_2 and P_3.

As $\vec{n}_1, \vec{n}_2$, and $\vec{n}_3$ are non-coplanar, planes P_1, P_2 and P_3 will intersect at unique point. So the given lines will pass through a fixed point.

20. **d.** Let $A(1, 0, -1), B(-1, 2, 2)$

Direction ratios of segment AB are $<2, -2, -3>$.

$$\cos\theta = \frac{|2\times1+3(-2)-5(-3)|}{\sqrt{1+9+25}\sqrt{4+4+9}}$$

$$= \frac{11}{\sqrt{17}\sqrt{35}} = \frac{11}{\sqrt{595}}$$

Length of projection is

$(AB)\sin\theta$

$$= \sqrt{(2)^2+(2)^2+(3)^2} \times \sqrt{1-\frac{121}{595}}$$

$$= \sqrt{17}\cdot\frac{\sqrt{474}}{\sqrt{17}\sqrt{35}} = \sqrt{\frac{474}{35}} \text{ units}$$

21. **c.** Let the point be A, B, C, and D.
The number of planes which have three points on one side and the fourth point on the other side is 4.
The number of planes which have two points on each side of the plane is 3.
$\Rightarrow$ Number of planes is 7.

22. **a.** Point A is $(a, b, c) \Rightarrow$ Points P, Q, R are $(a, b, -c)$, $(-a, b, c)$ and $(a, -b, c)$ respectively

$\Rightarrow$ Centroid of triangle PQR is $\left(\frac{a}{3}, \frac{b}{3}, \frac{c}{3}\right)$

$\Rightarrow G \equiv \left(\frac{a}{3}, \frac{b}{3}, \frac{c}{3}\right)$

$\Rightarrow A, O, G$ are collinear $\Rightarrow$ area of triangle AOG is zero.

23. **b.** Let the equation of the plane be $\frac{x}{a}+\frac{y}{b}+\frac{z}{c}=1$

$$\Rightarrow \frac{1}{a}+\frac{1}{b}+\frac{1}{c}=1$$

$\Rightarrow$ Volume of tetrahedron $OABC$ is $V=\frac{1}{6}(a\,b\,c)$

Now $(abc)^{1/3} \geq \dfrac{3}{\frac{1}{a}+\frac{1}{b}+\frac{1}{c}} \geq 3$ (G.M. $\geq$ H.M.)

$$\Rightarrow a\,b\,c \geq 27 \Rightarrow V \geq \frac{9}{2}$$

24. **b.** Let any point on second line be $(\lambda, 2\lambda, 3\lambda)$

$$\cos\theta = \frac{6}{\sqrt{42}},\ \sin\theta = \frac{\sqrt{6}}{\sqrt{42}}$$

$$\Delta_{OAB} = \frac{1}{2}(OA)\ OB\sin\theta$$

$$= \frac{1}{2}\sqrt{3}\,\lambda\,\sqrt{14} \times \frac{6}{\sqrt{42}} = \sqrt{6}$$

$\Rightarrow \lambda = 2$

So B is $(2, 4, 6)$.

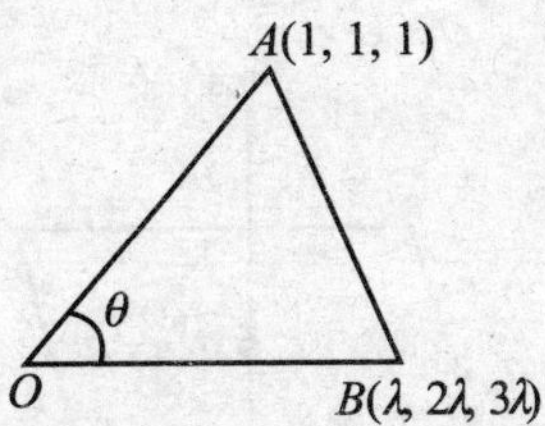

Fig. 3.4

25. **a.** Equation of line $x+2y+z-1+\lambda(-x+y-2z-2)=0$ (1)

$x+y-2+\mu(x+z-2)=0$ (2)

$(0, 0, 1)$ lies on it $\Rightarrow \lambda=-1/2, \mu=-2$

For point of intersection point $z = 0$ and solve (1) and (2).

26. **c.** Since the given lines are parallel.

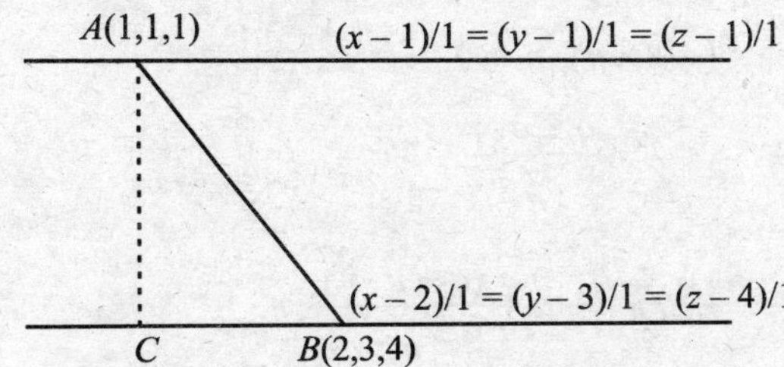

Fig. 3.5

From the figure, we get

$$BC = \frac{(2-1)1}{\sqrt{3}} + \frac{(3-1)1}{\sqrt{3}} + \frac{(4-1)1}{\sqrt{3}}$$

$$= \frac{1+2+3}{\sqrt{3}} = 2\sqrt{3}\quad AB = \sqrt{1+4+9} = \sqrt{14}$$

Shortest distance $= AC = \sqrt{14-12} = \sqrt{2}$.

27. **c.** Let $Q\ (\vec{q})$ be the foot of altitude drawn from P to the plane $\vec{r}\cdot\vec{n}=0$.

$$\Rightarrow \vec{q}-\vec{p}=\lambda\vec{n} \quad \Rightarrow \vec{q}=\vec{p}+\lambda\vec{n}$$

Also $\hat{q}\cdot\vec{n}=0 \quad \Rightarrow (\vec{p}+\lambda\vec{n})\cdot\vec{n}=0$

$$\Rightarrow \lambda = -\frac{\vec{p}\cdot\vec{n}}{|\vec{n}|^2} \Rightarrow \vec{q}-\vec{p} = -\frac{(\vec{p}\cdot\vec{n})}{|\vec{n}|^2}\vec{n}$$

Thus, required distance $= |\vec{q}-\vec{p}|$

$$= \frac{|\vec{p}\cdot\vec{n}|}{|\vec{n}|} = |\vec{p}\cdot\hat{n}|$$

28. **b.** Given plane is $\vec{r}\cdot\vec{n}=q$ (1)

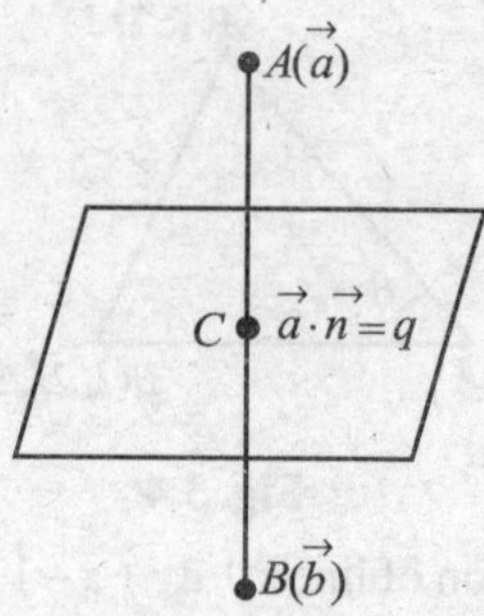

Fig. 3.6

Let the image of $A\ (\vec{a})$ in plane be $B\ (\vec{b})$.

Equation of AC is $\vec{r} = \vec{a} + \lambda\vec{n}$

[$\because$ AC is normal to plane] (2)

Solving (1) and (2), we get

$$(\vec{a} + \lambda\vec{n})\cdot\vec{n} = q$$

$$\Rightarrow \lambda = \frac{q - \vec{a}\cdot\vec{n}}{|\vec{n}|^2}$$

$$\therefore\ \overrightarrow{OC} = \vec{a} + \frac{(q - \vec{a}\cdot\vec{n})}{|\vec{n}|^2}\cdot\vec{n}$$

But $\overrightarrow{OC} = \dfrac{\vec{a}+\vec{b}}{2}$

$$\therefore\ \vec{a} + \frac{(q - \vec{a}\cdot\vec{n})\vec{n}}{|\vec{n}|^2} = \frac{\vec{a}+\vec{b}}{2}$$

$$\Rightarrow \vec{b} = \vec{a} + 2\left(\frac{q - \vec{a}\cdot\vec{n}}{|\vec{n}|^2}\right)\vec{n}$$

29. **c.** We must have $\vec{b}\cdot\vec{n} = 0$ (because the line and the plane must be parallel) and $\vec{a}\cdot\vec{n} \neq q$. (as point $\vec{a}$ on the line should not lie on the plane)

30. **c.** Here $l = \cos\dfrac{\pi}{4}$, $m = \cos\dfrac{\pi}{4}$

Let the line make an angle γ with z-axis. Therefore,

$l^2 + m^2 + n^2 = 1$

$$\Rightarrow \cos^2\frac{\pi}{4} + \cos^2\frac{\pi}{4} + \cos^2\gamma = 1$$

$$\Rightarrow \frac{1}{2} + \frac{1}{2} + \cos^2\gamma = 1$$

$$\Rightarrow 2\cos^2\gamma = 0 \Rightarrow \cos\gamma = 0 \Rightarrow \gamma = \frac{\pi}{2}$$

31. **d.** Let the plane $\vec{r}\cdot(\vec{i} - 2\vec{j} + 3\vec{k}) = 17$ divide the line joining the points $-2\vec{i} + 4\vec{j} + 7\vec{k}$ and $3\vec{i} - 5\vec{j} + 8\vec{k}$ in the ratio $t:1$ at the point P.

Therefore, point P is

$$\frac{3t-2}{t+1}\vec{i} + \frac{-5t+4}{t+1}\vec{j} + \frac{8t+7}{t+1}\vec{k}$$

This lies on the given plane. Therefore,

$$\frac{3t-2}{t+1}(1) + \frac{-5t+4}{t+1}(-2) + \frac{8t+7}{t+1}(3) = 17$$

Solving, we get

$$t = \frac{3}{10}$$

32. **d.** Let $P(\alpha, \beta, \gamma)$ be the image of the point $Q(-1, 3, 4)$.

Midpoint of PQ lies on $x - 2y = 0$. Then,

$$\frac{\alpha-1}{2} - 2\left(\frac{\beta+3}{2}\right) = 0$$

$$\Rightarrow \alpha - 1 - 2\beta - 6 = 0 \Rightarrow \alpha - 3\beta = 7 \qquad (1)$$

Also PQ is perpendicular to the normal to the plane. Then,

$$\frac{\alpha+1}{1} = \frac{\beta-3}{-2} = \frac{\gamma-4}{0} \qquad (2)$$

Solving (1) and (2), we get

$$\alpha = \frac{9}{5}, \beta = -\frac{13}{5}, \gamma = 4$$

Therefore, image is

$$\left(\frac{9}{5}, -\frac{13}{5}, 4\right)$$

33. **a.** It is obvious that given line and plane are parallel. Given point on the line is $A(2, -2, 3)$. $B(0, 0, 5)$ is a point on the plane. Therefore,

$$\overrightarrow{AB} = (2-0)\hat{i} + (-2-0)\hat{j} + (3-5)\hat{k}$$

Then distance of B from the plane = projection of $\overrightarrow{AB}$ on vector $\hat{i} + 5\hat{j} + \hat{k}$. Therefore,

$$p = \left|\frac{(2\hat{i} - 2\hat{j} - 2\hat{k})\cdot(\hat{i} + 5\hat{j} + \hat{k})}{\sqrt{1+25+1}}\right|$$

$$= \left|\frac{2-10-2}{\sqrt{27}}\right| = \frac{10}{3\sqrt{3}}$$

34. **d.** Direction ratios of the line of intersection is

$$\begin{vmatrix} \hat{i} & \hat{j} & \hat{k} \\ 2 & 3 & 1 \\ 1 & 3 & 2 \end{vmatrix} = 3\hat{i} - 3\hat{j} + 3\hat{k}$$

Hence, $\cos\alpha = \dfrac{3}{\sqrt{9+9+9}} = \dfrac{1}{\sqrt{3}}$.

35. **d.** Let P be the point (1, 2, 3) and PN be the length of the perpendicular from P on the given line.

Coordinates of point N are $(3\lambda+6, 2\lambda+7, -2\lambda+7)$.

Now PN is perpendicular to given line or vector $3\vec{i}+2\vec{j}-2\vec{k}$

$\Rightarrow 3(3\lambda+6-1)+2(2\lambda+7-2)-2(-2\lambda+7-3)=0$

$\Rightarrow \lambda=-1$

Then, point N is (3, 5, 9).

$\Rightarrow PN=7$

36. **b.** The line is $\dfrac{x+1}{1}=\dfrac{y-1}{2}=\dfrac{z-2}{2}$ and the plane is $2x-y+\sqrt{\lambda}\,z+4=0.$

If θ be the angle between the line and the plane, then $90°-\theta$ is the angle between the line and normal to the plane

$$\Rightarrow \cos(90°-\theta)=\frac{(1)(2)+(2)(-1)+(2)(\sqrt{\lambda})}{\sqrt{1+4+4}\sqrt{4+1+\lambda}}$$

$$\Rightarrow \sin\theta=\frac{2-2+2\sqrt{\lambda}}{3\sqrt{5+\lambda}}\Rightarrow \frac{1}{3}=\frac{2\sqrt{\lambda}}{3\sqrt{5+\lambda}}$$

$\Rightarrow \sqrt{5+\lambda}=2\sqrt{\lambda}$

$\Rightarrow 5+\lambda=4\lambda$

$\Rightarrow 3\lambda=5$

$\Rightarrow \lambda=\dfrac{5}{3}$

37. **d.** The given spheres are

$x^2+y^2+z^2+7x-2y-z-13=0$ (1)

and $x^2+y^2+z^2-3x+3y+4z-8=0$ (2)

Subtracting (2) from (1), we get

$10x-5y-5z-5=0$

$\Rightarrow 2x-y-z=1$

38. **c.** Required area $=\sqrt{(\Delta^2 xy+\Delta^2 yz+\Delta^2 zx)}$, (1)

$$\text{where } \Delta xy=\frac{1}{2}\left|\begin{vmatrix} a & 0 & 1\\ 0 & b & 1\\ 0 & 0 & 1\end{vmatrix}\right|=\frac{1}{2}ab$$

$$\Delta yz=\frac{1}{2}\left|\begin{vmatrix} 0 & 0 & 1\\ b & 0 & 1\\ 0 & c & 1\end{vmatrix}\right|=\frac{1}{2}bc$$

$$\text{and } \Delta zx=\frac{1}{2}\left|\begin{vmatrix} 0 & c & 1\\ a & 0 & 1\\ 0 & 0 & 1\end{vmatrix}\right|=\frac{1}{2}ac$$

From (1), area is $\dfrac{1}{2}\sqrt{(a^2b^2+b^2c^2+c^2a^2)}$.

39. **c.** Here $\sin^2\beta=3\sin^2\theta$ (1)

By the question $\cos^2\theta+\cos^2\theta+\cos^2\beta=1$ (2)

$\Rightarrow \cos^2\beta=1-2\cos^2\theta$ (3)

Adding (1) and (3), we get

$1=1+3\sin^2\theta-2\cos^2\theta$

$\Rightarrow 1=1+3(1-\cos^2\theta)-2\cos^2\theta$

$\Rightarrow 5\cos^2\theta=3$

$\Rightarrow \cos^2\theta=\dfrac{3}{5}$

40. **d.** The given sphere is

$x^2+y^2+z^2+4x-2y-6z-155=0$

Its centre is $(-2, 1, 3)$

and radius $=\sqrt{4+1+9+155}=\sqrt{169}=13.$

Therefore, distance of centre $(-2, 1, 3)$ from the plane $12x+4y+3z=327$

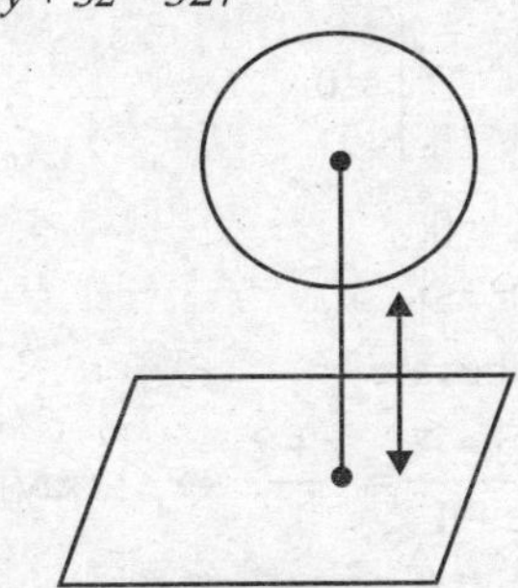

Fig. 3.7

$$=\frac{|12(-2)+4(1)+3(3)-327|}{\sqrt{144+16+9}}=26$$

Hence, shortest distance is 13.

41. **d.** Vector perpendicular to the face OAB is

$$\overrightarrow{OA}\times\overrightarrow{OB}=(\hat{i}+2\hat{j}+\hat{k})\times(2\hat{i}+\hat{j}+3\hat{k})$$
$$=5\hat{i}-\hat{j}-3\hat{k}$$

Vector perpendicular to face ABC is

$$\overrightarrow{AB}\times\overrightarrow{AC}=(\hat{i}-\hat{j}+2\hat{k})\times(-2\hat{i}-\hat{j}+\hat{k})$$
$$=\hat{i}-5\hat{j}-3\hat{k}$$

Since the angle between the face = angle between their normal,

$$\cos\theta=\frac{5+5+9}{\sqrt{35}\sqrt{35}}=\frac{19}{35}\Rightarrow\theta=\cos^{-1}\left(\frac{19}{35}\right)$$

42. **b.** Center of the sphere is $(-1, 1, 2)$ and it radius is $\sqrt{1+1+4+19}=5$

And CL perpendicular distance of C from plane is

$$\left|\frac{-1+2+4+7}{\sqrt{1+4+4}}\right|=4$$

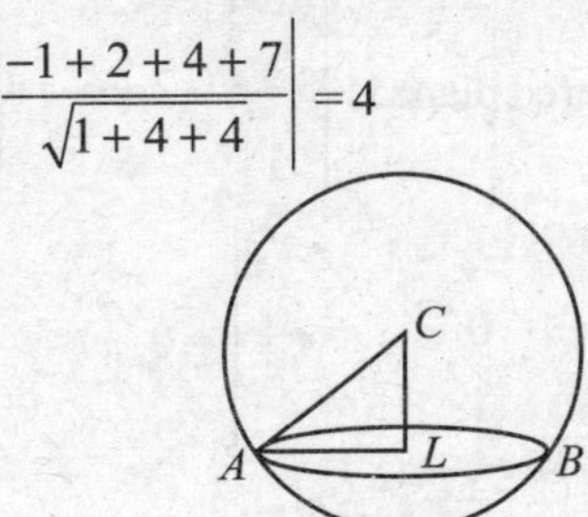

Fig. 3.8

Now $AL^2 = CA^2 - CL^2 = 25 - 16 = 9$
Hence, radius of the centre = $\sqrt{9} = 3$.

43. **b.** The lines $\dfrac{x-2}{1} = \dfrac{y-3}{1} = \dfrac{z-4}{-k}$ (1)

and $\dfrac{x-1}{k} = \dfrac{y-4}{2} = \dfrac{z-5}{1}$ (2)

are coplanar if $\begin{vmatrix} x_2 - x_1 & y_2 - y_1 & z_2 - z_1 \\ l_1 & m_1 & n_1 \\ l_2 & m_1 & n_2 \end{vmatrix} = 0$

or $\begin{vmatrix} 1 & -1 & -1 \\ 1 & 1 & -k \\ k & 2 & 1 \end{vmatrix} = 0$

$\Rightarrow k^2 + 3k = 0$
$\Rightarrow k = 0$ or -3

44. **a.** Given lines are

$\dfrac{x-5}{3} = \dfrac{y-7}{-1} = \dfrac{z+2}{1} = r_1$ (say)

and $\dfrac{x+3}{-36} = \dfrac{y-3}{2} = \dfrac{z-6}{4} = r_2$ (say)

$\therefore x = 3r_1 + 5 = -36r_2 - 3$,
$y = -r_1 + 7 = 3 + 2r_2$
and $z = r_1 - 2 = 4r_2 + 6$
On solving, we get

$x = 21, y = \dfrac{5}{3}, z = \dfrac{10}{3}$

45. **c.** The planes are $\dfrac{x}{a} + \dfrac{y}{b} + \dfrac{z}{c} = 1$ and $\dfrac{x}{a'} + \dfrac{y}{b'} + \dfrac{z}{c'} = 1$

Since the perpendicular distance of origin on the planes is same,

$$\left| \frac{-1}{\sqrt{\frac{1}{a^2} + \frac{1}{b^2} + \frac{1}{c^2}}} \right| = \left| \frac{-1}{\sqrt{\frac{1}{a'^2} + \frac{1}{b'^2} + \frac{1}{c'^2}}} \right|$$

$\Rightarrow \dfrac{1}{a^2} + \dfrac{1}{b^2} + \dfrac{1}{c^2} - \dfrac{1}{a'^2} - \dfrac{1}{b'^2} - \dfrac{1}{c'^2} = 0$

46. **a.** The required plane is $\begin{vmatrix} x-3 & y-6 & z-4 \\ 3-3 & 2-6 & 0-4 \\ 1 & 5 & 4 \end{vmatrix} = 0$

$\Rightarrow \begin{vmatrix} x-3 & y-z-2 & z-4 \\ 0 & 0 & -4 \\ 1 & 1 & 4 \end{vmatrix} = 0$

[Operating $C_2 \to C_2 - C_3$]
$\Rightarrow 4(x - 3 - y + z + 2) = 0$
$\Rightarrow x - y + z = 1$

47. **b.** Any plane through (1, 0, 0) is
$a(x-1) + by + cz = 0.$ (1)
It passes through (0, 1, 0).
$\therefore a(0-1) + b(1) + c(0) = 0$
$\Rightarrow -a + b = 0$ (2)
(1) makes an angle of $\pi/4$ with $x + y = 3$. Therefore,

$\cos\dfrac{\pi}{4} = \dfrac{a(1) + b(1) + c(0)}{\sqrt{a^2 + b^2 + c^2}\sqrt{1+1+0}}$

$\Rightarrow \dfrac{1}{\sqrt{2}} = \dfrac{a+b}{\sqrt{2}\sqrt{a^2 + b^2 + c^2}}$

$\Rightarrow a + b = \sqrt{a^2 + b^2 + c^2}$

Squaring, we get
$a^2 + b^2 + 2ab = a^2 + b^2 + c^2$
$\Rightarrow 2ab = c^2 \Rightarrow 2a^2 = c^2$ (using (2))
$\Rightarrow c = \sqrt{2}\,a$
Hence, $a : b : c = a : a : \sqrt{2}\,a$
$= 1 : 1 : \sqrt{2}$

48. **b.** The equation of the line through the centre $\hat{j} + 2\hat{k}$ and normal to the given plane is
$\vec{r} = \hat{j} + 2\hat{k} + \lambda(\hat{i} + 2\hat{j} + 2\hat{k})$ (1)
This meets the plane for which
$[\hat{j} + 2\hat{k} + \lambda(\hat{i} + 2\hat{j} + 2\hat{k})] \cdot (\hat{i} + 2\hat{j} + 2\hat{k}) = 15$
$\Rightarrow 6 + 9\lambda = 15 \Rightarrow \lambda = 1$
Putting in (1), we get
$\vec{r} = \hat{j} + 2\hat{k} + (\hat{i} + 2\hat{j} + 2\hat{k}) = \hat{i} + 3\hat{j} + 4\hat{k}$
Hence, centre is (1, 3, 4).

49. **c.** Equation of the planes through $y = mx, z = c$ and $y = -mx, z = -c$ are, respectively,
$(y - mx) + \lambda_1(z - c) = 0$ (1)
and $(y + mx) + \lambda_2(z + c) = 0$ (2)
It meets at x-axis, i.e., $y = 0 = z$. Therefore, $\lambda_2 = \lambda_1$.

From (1) and (2), $\dfrac{y - mx}{z - c} = \dfrac{y + mx}{z + c}$
$\therefore cy = mzx$

50. **c.** Let $Q(\vec{q})$ be the foot of altitude drawn from $P(\vec{p})$ to the line $\vec{r} = \vec{a} + \lambda\vec{b}$
$\Rightarrow (\vec{q} - \vec{p}) \cdot \vec{b} = 0$ and $\vec{q} = \vec{a} + \lambda\vec{b}$
$\Rightarrow (\vec{a} + \lambda\vec{b} - \vec{p}) \cdot \vec{b} = 0$
$\Rightarrow (\vec{a} - \vec{p}) \cdot \vec{b} + \lambda|\vec{b}|^2 = 0$

$\lambda = \dfrac{(\vec{p} - \vec{a}) \cdot \vec{b}}{|\vec{b}|^2}$

$$\Rightarrow \vec{q}-\vec{p}=\vec{a}+\frac{((\vec{p}-\vec{a})\cdot\vec{b})\,\vec{b}}{|\vec{b}|^2}-\vec{p}$$

$$\Rightarrow |\vec{q}-\vec{p}|=\left|(\vec{a}-\vec{p})+\frac{((\vec{p}-\vec{a})\cdot\vec{b})\,\vec{b}}{|\vec{b}|^2}\right|$$

51. **b.** Coordinates of L and M are $(0, b, c)$ and $(a, 0, c)$, respectively. Therefore, the equation of the plane passing through $(0, 0, 0)$, $(0, b, c)$ and $(a, 0, c)$ is

$$\begin{vmatrix} x-0 & y-0 & z-0 \\ 0 & b & c \\ a & 0 & c \end{vmatrix}=0,\text{ or } \frac{x}{a}+\frac{y}{b}-\frac{z}{c}=0.$$

52. **c.** We must have $\vec{b}\cdot\vec{n}=0$ and $\vec{a}\cdot\vec{n}=q$.

53. **b.** We have $\vec{s}-\vec{p}=\lambda\vec{n}$ and $\vec{s}\cdot\vec{n}=q$.

$$\Rightarrow (\lambda\vec{n}+\vec{p})\cdot\vec{n}=q$$

$$\Rightarrow \lambda=\frac{q-\vec{p}\cdot\vec{n}}{|\vec{n}|^2}$$

$$\Rightarrow \vec{s}=\vec{p}+\frac{(q-\vec{p}\cdot\vec{n})\,\vec{n}}{|\vec{n}|^2}$$

54. **d.** Line of intersection of $\vec{r}\cdot(\hat{i}+2\hat{j}+3\hat{k})=0$ and $\vec{r}\cdot(3\hat{i}+3\hat{j}+\hat{k})=0$ will be parallel to $(3\hat{i}+3\hat{j}+\hat{k})\times(\hat{i}+2\hat{j}+3\hat{k})$, i.e., $7\hat{i}-8\hat{j}+3\hat{k}$.

If the required angle is θ, then

$$\cos\theta=\frac{7}{\sqrt{49+64+9}}=\frac{7}{\sqrt{122}}$$

55. **b.**

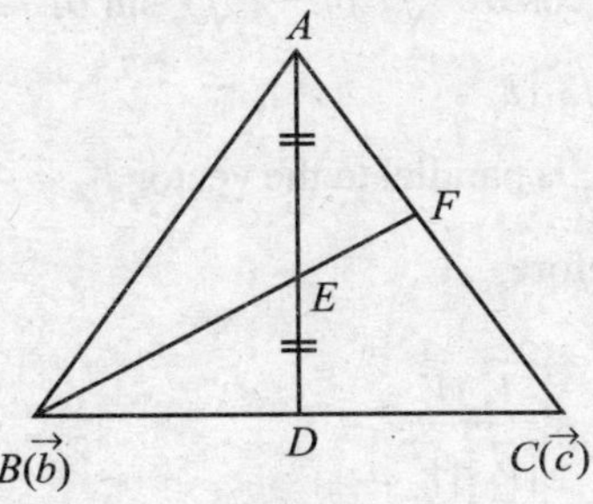

Fig. 3.9

Taking A is the origin, let P.V. of B and C be $\vec{b}$ and $\vec{c}$, respectively.

P.V. of D is $\frac{\vec{b}+\vec{c}}{2}$ and P.V. of E is $\frac{\vec{b}+\vec{c}}{4}$.

Equation of line BF is $\vec{r}=\vec{b}+\lambda\left(\frac{\vec{b}+\vec{c}}{4}-\vec{b}\right)$.

Equation of line AC is $\vec{r}=0+\mu\vec{c}$.

For the point of intersection, $\vec{b}+\lambda\left(\frac{\vec{b}+\vec{c}}{4}-\vec{b}\right)=\mu\vec{c}$

$$\Rightarrow 1-\frac{3\lambda}{4}=0 \text{ and } \frac{\lambda}{4}=\mu\left(\frac{\vec{b}+\vec{c}}{2}\right)$$

$\Rightarrow \lambda=\frac{4}{3}$ and $\mu=\frac{1}{3}$. Therefore, P.V. of F is $\vec{r}=\frac{1}{3}\vec{c}$

$$\Rightarrow \overrightarrow{AF}=\frac{1}{3}\overrightarrow{AC}\Rightarrow AF:AC=1:3$$

56. **a.** $\vec{r}\cdot\vec{n}_1+l\,\vec{r}\cdot\vec{n}_2=q_1+lq_2$ (1),

where λ is a parameter.

So, $\vec{n}_1+\lambda\,\vec{n}_2$ is normal to the plane (1). Now, any plane parallel to the line of intersection of the planes $\vec{r}\cdot\vec{n}_3=q_3$ and $\vec{r}\cdot\vec{n}_4=q_4$ is of the form $\vec{r}\cdot(\vec{n}_3\times\vec{n}_4)=d$. Hence, we must have

$$\Rightarrow [\vec{n}_1+\lambda\vec{n}_2]\cdot[\vec{n}_3\times\vec{n}_4]=0$$

$$\Rightarrow [\vec{n}_1\,\vec{n}_3\,\vec{n}_4]+\lambda[\vec{n}_2\,\vec{n}_3\,\vec{n}_4]=0$$

$$\Rightarrow \lambda=\frac{-[\vec{n}_1\,\vec{n}_3\,\vec{n}_4]}{[\vec{n}_2\,\vec{n}_3\,\vec{n}_4]}$$

$\Rightarrow$ On putting this value in Eq. (1), we have the equation of the required plane as

$$\vec{r}\cdot\vec{n}_1-q_1=\frac{[\vec{n}_1\,\vec{n}_3\,\vec{n}_4]}{[\vec{n}_2\,\vec{n}_3\,\vec{n}_4]}(\vec{r}\cdot\vec{n}_2-q_2)$$

$$\Rightarrow [\vec{n}_2\,\vec{n}_3\,\vec{n}_4](\vec{r}\cdot\vec{n}_1-q_1)=[\vec{n}_1\,\vec{n}_3\,\vec{n}_4](\vec{r}\cdot\vec{n}_2-q_2).$$

57. **c.**

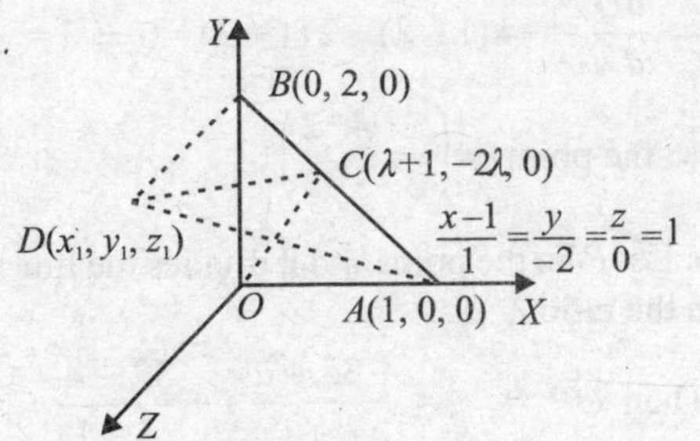

Fig. 3.10

Equation of line AB is $\frac{x-1}{1}=\frac{y}{-2}=\frac{z}{0}=\lambda$.

Now $AB\perp OC\Rightarrow 1(\lambda+1)+(-2\lambda)(-2)=0$

$$\Rightarrow 5\lambda=-1\Rightarrow\lambda=-\frac{1}{5}$$

$\Rightarrow C$ is $\left(\frac{4}{5},\frac{2}{5},0\right)$. Now

$$x_1^2+(y_1-2)^2+z_1^2=4 \qquad (1)$$

and $(x_1 - 1)^2 + y_1^2 + z_1^2 = 1$ (2)

Now $OC \perp CD$

$$\Rightarrow \left(x_1 - \frac{4}{5}\right)\frac{4}{5} + \left(y_1 - \frac{2}{5}\right)\frac{2}{5} + (z_1 - 0)\,0 = 0 \quad (3)$$

From (1) and (2), we get

$-4y_1 + 2x_1 = 0 \Rightarrow x_1 = 2y_1$

From (3), putting $x_1 = 2y_1 \Rightarrow 2y_1 = \frac{4}{5} \Rightarrow y_1 = \frac{2}{5}$

$\Rightarrow x_1 = \frac{4}{5}$. Putting this value of x_1 and y_1 in (1), we get

$z_1 = \pm\frac{2}{\sqrt{5}}$.

58. **b.** Let $\vec{r} \times \vec{a} = \vec{b} \times \vec{a}$

$\Rightarrow (\vec{r} - \vec{b}) \times \vec{a} = \vec{0} \Rightarrow \vec{r} = \vec{b} + t\vec{a}$

Similarly, $\vec{r} = \vec{a} + k\vec{b}$, where t and k are scalars.

Now,

$\vec{a} + k\vec{b} = \vec{b} + t\vec{a}$

$\Rightarrow t = 1, k = 1$

[Equating the coefficients of $\vec{a}$ and $\vec{b}$]

$\therefore \vec{r} = \vec{a} + \vec{b} = \hat{i} + \hat{j} + 2\hat{i} - \hat{k} = 3\hat{i} + \hat{j} - \hat{k}$

i.e., (3, 1, –1)

59. **a.** Let the point P be (x, y, z), then the vector $x\hat{i} + y\hat{j} + z\hat{k}$ will lie on the line so

$\Rightarrow (x-1)\hat{i} + (y-1)\hat{j} + (z-1)\hat{k} = -\lambda\hat{i} + \lambda\hat{j} - \lambda\hat{k}$

$\Rightarrow x = 1 - \lambda, y + 1 + \lambda$, and $z = 1 - \lambda$

Now the point P is nearest to the origin

$\Rightarrow D = (1-\lambda)^2 + (1+\lambda)^2 + (1-\lambda)^2$

$\Rightarrow \frac{dD}{d\lambda} = -4(1-\lambda) + 2(1+\lambda) = 0 \Rightarrow \lambda = \frac{1}{3}$

$\Rightarrow$ the point is $\left(\frac{2}{3}, \frac{4}{3}, \frac{2}{3}\right)$.

60. **b.** Let P be the point and it divides the line segment in the ratio $\lambda : 1$.

Then $\overrightarrow{OP} = \vec{r} = \frac{-3\lambda + 2}{\lambda + 1}\hat{i} + \frac{5\lambda - 4}{\lambda + 1}\hat{j} + \frac{-8\lambda - 7}{\lambda + 1}\hat{k}$.

It satisfies $\vec{r} \cdot (\hat{i} - 2\hat{j} + 3\hat{k})\, 13$. So

$$\frac{-3\lambda + 2}{\lambda + 1} - 2\frac{5\lambda - 4}{\lambda + 1} + 3\frac{-8\lambda - 7}{\lambda + 1} = 13$$

or $-3\lambda + 2 - 2(5\lambda - 4) + 3(-8\lambda - 7) = 13(\lambda + 1)$

or $-37\lambda - 11 = 13\lambda + 13$ or $50\lambda = -24$ or $\lambda = -\frac{12}{25}$

61. **d.** $\vec{V_1}, \vec{V_2}, \overrightarrow{PS}$ are in the same plane

$\therefore (2\hat{i} - \hat{j} + 3\hat{k}) \times (-3\hat{i} + \hat{k})\ ((x+2)\hat{i} + (y-1)\hat{j} + z\hat{k}) \cdot ((x+2)\hat{i} + (y-1)\hat{j} + z\hat{k}) = 0$

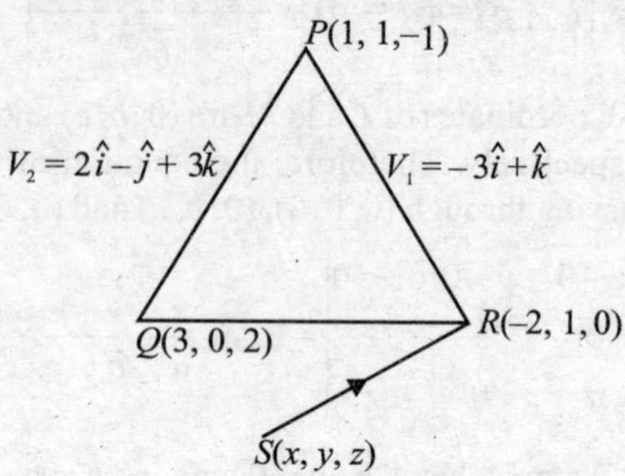

Fig. 3.11

62. **a.** $\overrightarrow{AB} = \vec{\beta} - \vec{\alpha} = -2\hat{i} - 3\hat{j} - 6\hat{k}$

Equation of plane passing through B and perpendicular to AB is

$(\vec{r} - \overrightarrow{OB}) \cdot \overrightarrow{AB} = 0$

$\vec{r} \cdot (2\hat{i} + 3\hat{j} + 6\hat{k}) + 28 = 0$

$\Rightarrow$ Hence, required distance from $\vec{r} = -\hat{i} + \hat{j} + \hat{k}$ is

$$\left|\frac{(-\hat{i} + \hat{j} + \hat{k}) \cdot (2i + 3j + 6k) + 28}{|2i + 3j + 6k|}\right|$$

$$= \left|\frac{-2 + 3 + 6 + 28}{7}\right| = 5 \text{ units.}$$

63. **a.** Both the lines pass through origin. Line L_1 is parallel to the vector

$\vec{V_1} = (\cos\theta + \sqrt{3})\hat{i} + (\sqrt{2}\sin\theta)\hat{j} + (\cos\theta - \sqrt{3})\hat{k}$

and L_2 is parallel to the vector $\vec{V_2} = a\hat{i} + b\hat{j} + c\hat{k}$.

Therefore,

$$\cos\alpha = \frac{\vec{V_1} \cdot \vec{V_2}}{|\vec{V_1}||\vec{V_2}|}$$

$$= \frac{a(\cos\theta + \sqrt{3}) + (b\sqrt{2})\sin\theta + c(\cos\theta - \sqrt{3})}{\sqrt{a^2 + b^2 + c^2}\sqrt{(\cos\theta + \sqrt{3})^2 + 2\sin^2\theta + (\cos\theta - \sqrt{3})^2}}$$

$$= \frac{(a + c)\cos\theta + b\sqrt{3}\sin\theta + (a - c)\sqrt{3}}{\sqrt{a^2 + b^2 + c^2}\sqrt{2 + 6}}$$

In order that $\cos\alpha$ is independent of θ, we get $a + c = 0$ and $b = 0$. Therefore,

$$\cos\alpha = \frac{2a\sqrt{3}}{a\sqrt{2}\,2\sqrt{2}} = \frac{\sqrt{3}}{2}$$

$\Rightarrow \alpha = \frac{\pi}{6}$

64. **d.** Given lines are $\vec{r} = (3, 8, 3) + l\,(3, -1, 1)$ and $\vec{r} = (-3, -7, 6) + m\,(-3, 2, 4)$.

Required *S.D.* is

$$\frac{(-3-3, -7-8, 6-3)\,[(3,-1,1)\times(-3,2,4)]}{|(3,-1,1)\times(-3,2,4)|}$$

$$= \frac{(-6,-15,3)\,(-6,-15,3)}{|(-6,-15,3)|}$$

$$= \frac{36+225+9}{\sqrt{36+225+9}} = \frac{270}{\sqrt{270}} = \sqrt{270} = 3\sqrt{30}$$

65. **b.** The required line passes through the point $\hat{i}+3\hat{j}+2\hat{k}$ and is perpendicular to the lines $\vec{r} = (\hat{i}+2\hat{j}-\hat{k}) + \lambda\,(2\hat{i}+\hat{j}+\hat{k})$, $\vec{r} = (2\hat{i}+6\hat{j}+\hat{k}) + \mu\,(\hat{i}+2\hat{j}+3\hat{k})$. Therefore it is parallel to the vector

$\vec{b} = (2\hat{i}+\hat{j}+\hat{k})\times(\hat{i}+2\hat{j}+3\hat{k}) = (\hat{i}-5\hat{j}+3\hat{k})$.

Hence, the equation of the required line is

$\vec{r} = (\hat{i}+3\hat{j}+2\hat{k}) + \lambda\,(\hat{i}-5\hat{j}+3\hat{k})$

66. **d.** Here, the required plane is

$a\,(x-4) + b\,(y-3) + c\,(z-2) = 0$.

Also $a + b + 2c = 0$ and $a - 4b + 5c = 0$.

Solving, we have

$$\frac{a}{5+8} = \frac{b}{2-5} = \frac{c}{-4-1} = k$$

$$\frac{a}{13} = \frac{b}{-3} = \frac{c}{-5} = k$$

Therefore, the required equation of plane is $-13x + 3y + 5z + 33 = 0$.

67. **b.** Plane passing through the line of intersection of planes $4y + 6z = 5$ and $2x + 3y + 5z = 5$ is

$(4y + 6z - 5) + \lambda\,(2x + 3y + 5z - 5) = 0$, or

$2\lambda x + (3\lambda + 4)\,y + (5\lambda + 6)\,z - 5\lambda - 5 = 0$

Clearly, for $\lambda = -3$, we get the plane $6x + 5y + 9z = 10$.

Hence, given three planes have common line of intersection.

68. **c.** The equation of a plane through the line of intersection of the planes $ax + by + cz + d = 0$ and $a'x + b'y + c'z + d' = 0$ is

$(ax + by + cz + d) + \lambda\,(a'x + b'y + c'z + d') = 0$

or $x\,(a + \lambda a') + y\,(b + \lambda b') + z\,(c + \lambda c') + d + \lambda d' = 0$ (1)

This is parallel to x-axis, i.e., $y = 0, z = 0$. Therefore,

$1\,(a + \lambda a') + 0(b + \lambda b') + 0(c + \lambda c') = 0$

$\Rightarrow \lambda = -\dfrac{a}{a'}$

Putting the value of λ in (1), the required plane is

$y\,(a'b - ab') + z\,(a'c - ac') + a'd - ad' = 0$

or $(ab' - a'b)\,y + (ac' - a'c)z + ad' - a'd = 0$.

69. **b.** Any plane through (2, 2, 1) is

$a\,(x-2) + b\,(y-2) + c\,(z-1) = 0$ (1)

It passes through (9, 3, 6) if $7a + b + 5c = 0$. (2)

Also (1) is perpendicualr to $2x + 6y + 6z - 1 = 0$, we have $2a + 6b + 6c = 0$. Therefore,

$a + 3b + 3c = 0$ (3)

$\therefore \dfrac{a}{-12} = \dfrac{b}{-16} = \dfrac{c}{20}$ or $\dfrac{a}{3} = \dfrac{b}{4} = \dfrac{c}{-5}$

[from (2) and (3)]

Therefore, the required plane is $3(x-2) + 4\,(y-2) - 5\,(z-1) = 0$ or $3x + 4y - 5z - 9 = 0$.

70. **a.** Since line is parallel to the plane vector, $2\vec{i} + 3\vec{j} + \lambda\vec{k}$ is perpendicular to the normal to the plane

$2\vec{i} + 3\vec{j} + 4\vec{k}$

$\Rightarrow 2\times 2 + 3\times 3 + 4\lambda = 0$

$\Rightarrow \lambda = -\dfrac{13}{4}$

71. **a.** Any plane through the given planes is $x + 2y + 3z - 4 + \lambda\,(4x + 3y + 2z + 1) = 0$

It passes through (0, 0, 0). Therefore,

$-4 + \lambda = 0$ or $\lambda = 4$.

Therefore, the required plane is $x + 2y + 3z + 4\,(4x + 3y + 2z) = 0$ or $17x + 14y + 11z = 0$.

72. **a.** The equation of the plane through the line of intersection of the planes $4x + 7y + 4z + 81 = 0$ and $5x + 3y + 10z = 25$ if

$(4x + 7y + 4z + 81) + \lambda(5x + 3y + 10z - 25) = 0$

$\Rightarrow (4 + 5\lambda)x + (7 + 3\lambda)y + (4 + 10\lambda)z + 81 - 25\lambda = 0$, (1)

which is perpendicular to $4x + 7y + 4z + 81 = 0$.

$\Rightarrow 4(4 + 5\lambda) + 7(7 + 3\lambda) + 4(4 + 10\lambda) = 0$

$\Rightarrow 81\lambda + 81 = 0$

$\Rightarrow \lambda = -1$

Hence, the plane is $x - 4y + 6z = 106$.

73. **b.** The equation of a plane through the line of intersection of the planes $\vec{r}\cdot\vec{a} = \lambda$ and $\vec{r}\cdot\vec{b} = \mu$ is

$(\vec{r}\cdot\vec{a} - \lambda) + k\,(\vec{r}\cdot\vec{b} - \mu) = 0$

or $\vec{r}\cdot(\vec{a} + k\vec{b}) = \lambda + k\mu$ (1)

This passes through the origin. Therefore

$\vec{0}\,(\vec{a} + k\vec{b}) = \lambda + \mu k \Rightarrow k = \dfrac{-\lambda}{\mu}$

Putting the value of k in (1), we get the equation of the required plane as

$\vec{r}\cdot(\mu\vec{a}-\lambda\vec{b})=0 \Rightarrow \vec{r}\cdot(\lambda\vec{b}-\mu\vec{a})=0$

74. **b.** The lines $\vec{r}=\vec{a}+\lambda(\vec{b}\times\vec{c})$ and $\vec{r}=\vec{b}+\mu(\vec{c}\times\vec{a})$ pass through points $\vec{a}$ and $\vec{b}$, respectively, and are parallel to the vectors $\vec{b}\times\vec{c}$ and $\vec{c}\times\vec{a}$ respectively. Therefore, they intersect if $\vec{a}-\vec{b}, \vec{b}\times\vec{c}$ and $\vec{c}\times\vec{a}$ are coplanar and so

$(\vec{a}-\vec{b})\cdot\{(\vec{b}\times\vec{c})\times(\vec{c}\times\vec{a})\}=0$

$\Rightarrow (\vec{a}-\vec{b})\cdot([\vec{b}\ \vec{c}\ \vec{a}]\vec{c}-[\vec{b}\ \vec{c}\ \vec{c}]\vec{a})=0$

$\Rightarrow ((\vec{a}-\vec{b})\cdot\vec{c})[\vec{b}\ \vec{c}\ \vec{a}]=0$

$\Rightarrow \vec{a}\cdot\vec{c}-\vec{b}\cdot\vec{c}=0 \Rightarrow \vec{a}\cdot\vec{c}=\vec{b}\cdot\vec{c}$

75. **a.** Equation of plane through (–1, 0, 1) is

$a(x+1)+b(y-0)+c(z-1)=0$ (1)

which is parallel to the given line and perpendicular to the given plane

$-a+2b+3c=0$ (2) and $a-2b+c=0.$ (3)

From Eqs. (2) and (3), we get $c=0, a=2b$.

From Equation (1), $2b(x+1)+by=0$

$\Rightarrow 2x+y+2=0.$

76. **b.** Eliminating n, we get

$\lambda(l+m)^2+lm=0$

$\Rightarrow \frac{\lambda l^2}{m^2}+(2\lambda+1)\frac{l}{m}+\lambda=0$

$\Rightarrow \frac{l_1l_2}{m_1m_2}=1$ (product of roots $\frac{l_1}{m_1}$ and $\frac{l_2}{m_2}$),

where l_1/m_1 and l_2/m_2 are the roots of this equation, further eliminating m, we get

$\lambda l^2-ln-n^2=0$

$\Rightarrow \frac{l_1l_2}{n_1n_2}=-\frac{1}{\lambda}$

Since the lines with direction cosines (l_1, m_1, n_1) and (l_2, m_2, n_2) are perpendicular, we have

$l_1l_2+m_1m_2+n_1n_2=0$

$\Rightarrow 1+1-\lambda=0$

$\Rightarrow \lambda=2$

77. **a.** Direction ratios of the line joining points $P(1, 2, 3)$ and $Q(-3, 4, 5)$ are –4, 2, 2 which are direction ratios of the normal to the plane.

Then, equation of plane is $-4x+2y+2z=k$.

Also this plane passes through midpoint of PQ, (–1, 3, 4)

$\Rightarrow -4(-1)+2(3)+2(4)=k$

$\Rightarrow k=18$

$\Rightarrow$ Equation of plane is $2x-y-z=-9$.

Then, intercepts are –9/2, 9 and 9.

78. **c.** $3l+m+5n=0$ (1)

$6mn-2nl+5ml=0$ (2)

Substituting the value of n from Eq. (1) in Eq. (2), we get

$6l^2+9lm-6m^2=0$

$\Rightarrow 6\left(\frac{l}{m}\right)^2+9\left(\frac{l}{m}\right)-6=0$

$\therefore \frac{l_1}{m_1}=\frac{1}{2}$ and $\frac{l_2}{m_2}=-2$

From Eq. (1), we get

$\frac{l_1}{n_1}=-1$ and $\frac{l_2}{n_2}=-2$

$\therefore \frac{l_1}{1}=\frac{m_1}{2}=\frac{n_1}{-1}=\sqrt{\frac{l_1^2+m_1^2+n_1^2}{1+4+1}}=\frac{1}{\sqrt{6}}$

and $\frac{l_2}{2}=\frac{m_2}{-1}=\frac{n_2}{-1}=\frac{\sqrt{l_2^2+m_2^2+n_2^2}}{\sqrt{4+1+1}}=\frac{1}{\sqrt{6}}$

If θ is the angle between the lines, then

$$\cos\theta=\left(\frac{1}{\sqrt{6}}\right)\left(\frac{2}{\sqrt{6}}\right)+\left(\frac{2}{\sqrt{6}}\right)\left(-\frac{1}{\sqrt{6}}\right)+\left(-\frac{1}{\sqrt{6}}\right)\left(-\frac{1}{\sqrt{6}}\right)=\frac{1}{6}.$$

Hence, $\theta=\cos^{-1}\left(\frac{1}{6}\right)$

79. **b.** Let the equation of the sphere be $x^2+y^2+z^2-ax-by-cz=0$. This meets the axes at $A(a, 0, 0)$, $B(0, b, 0)$, and $C(0, 0, c)$.

Let (α, β, γ) be the coordinates of the centroid of the tetrahedron $OABC$. Then

$\frac{a}{4}=\alpha, \frac{b}{4}=\beta, \frac{c}{4}=\gamma$

$\Rightarrow a=4\alpha, b=4\beta, c=4\gamma$

Now, radius of the sphere = $2k$

$\Rightarrow \frac{1}{2}\sqrt{a^2+b^2+c^2}=2k \Rightarrow a^2+b^2+c^2=16k^2$

$\Rightarrow 16(\alpha^2+\beta^2+\gamma^2)=16k^2$

Hence, the locus of (α, β, γ) is $(x^2+y^2+z^2)=k^2$.

80. **a.** Let the foot of the perpendicular from the origin on the given plane be $P(\alpha, \beta, \gamma)$. Since the plane passes through $A(a, b, c)$,

$AP \perp OP \Rightarrow \overrightarrow{AP}\cdot\overrightarrow{OP}=0$

$\Rightarrow [(\alpha-a)\hat{i}+(\beta-b)\hat{j}+(\gamma-c)\hat{k}]$

$\cdot(\alpha\hat{i}+\beta\hat{j}+\gamma\hat{k})=0$

$\Rightarrow \alpha(\alpha-a)+\beta(\beta-b)+\gamma(\gamma-c)=0$

Hence, the locus of (α, β, γ) is

$x(x-a)+y(y-b)+z(z-c)=0$

$x^2+y^2+z^2-ax-by-cz=0,$

which is a sphere of radius $\frac{1}{2}\sqrt{a^2+b^2+c^2}$.

81. c. $\begin{vmatrix} 1 & a & b+c \\ 1 & b & c+a \\ 1 & c & a+b \end{vmatrix} = \begin{vmatrix} 1 & a & a+b+c \\ 1 & b & a+b+c \\ 1 & c & a+b+c \end{vmatrix} = 0$

82. a. The direction ratios of line AB are

$(3-1); (-2-4); [(1-(-3)]$

$=2, -6, 4$

If c is middle point of AB, then the coordinates of c are given as

$\frac{3+1}{2}, \frac{-2+4}{2}$ and $\frac{1-3}{2}$, or 2, 1 and -1

Hence, the equation of plane that bisects the given points at right angle is given as

$2(x-2)+(-6)(y-1)+4(z+1)=0$

$\Rightarrow 2x-4-6y+6+4z+4=0$

$\Rightarrow 2x-6y+4z+6=0$

$\Rightarrow x-3y+2z+3=0$

83. a. The plane is perpendicular to the line $\frac{x-a}{\cos\theta}=\frac{y+2}{\sin\theta}=\frac{z-3}{0}$.

Hence, the direction ratios of the normal of the plane are $\cos\theta$, $\sin\theta$ and 0. (1)

Now, the required plane passes through the z-axis. Hence the point (0, 0, 0) lies on the plane.

From Eqs. (1) and (2), we get equation of the plane as

$\cos\theta(x-0)+\sin\theta(y-0)+0(z-0)=0$

$\cos\theta x+\sin\theta y=0$

$x+y\tan\theta=0$

84. a. The given line makes angles of $\pi/4$, $\pi/4$, and $\pi/2$ with the x-, y- and z-axes, respectively.

$\Rightarrow$ Direction cosines of the given line are $\cos(\pi/4)$, $\cos(\pi/4)$ and $\cos(\pi/2)$, or $1/\sqrt{2}, 1/\sqrt{2}$ and 0.

85. a. We must have $(3+4a-12+13)(-9-12a+13)<0$.

$\Rightarrow (a+1)(12a-4)>0$

$\Rightarrow a<-1$ or $a>1/3$

86. c. Here the given plane is $\frac{x}{2}+\frac{y}{3}+\frac{z}{6}=1$ the cuts the x, y and z-axis on (2,0,0) (0,3,0) and (0,0,6) respectively.

Therefore, the projections of the triangle ABC on the y-z, z-x and x-y planes are Δx, Δy, and Δz, respectively. Then

$\Delta x=\frac{1}{2}\times3\times6=9$

$\Delta y=\frac{1}{2}\times6\times2=6$

$\Delta z=\frac{1}{2}\times2\times3=3$

Therefore, the area of triangle ABC is

$\sqrt{\Delta x^2+\Delta y^2+\Delta z^2}=\sqrt{9^2+6^2+3^2}$.

Multiple Correct Answers Type

1. b., c., d.

If P be (x, y, z), then from the figure,

$x=r\sin\theta\cos\phi$, $y=r\sin\theta\sin\phi$ and $z=r\cos\theta$.

$1=r\sin\theta\cos\phi$, $2=r\sin\theta\sin\phi$ and $3=r\cos\theta$.

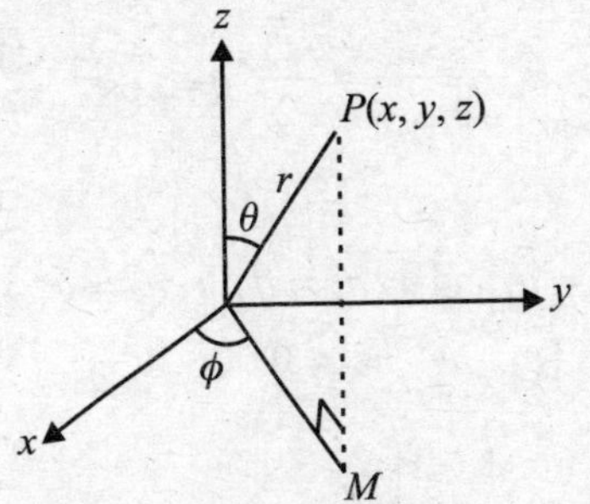

Fig. 3.12

$\Rightarrow 1^2+2^2+3^2=r^2 \Rightarrow r=\pm\sqrt{14}$.

$\therefore \sin\theta\cos\phi=\frac{1}{\sqrt{14}}$, $\sin\theta\sin\phi=\frac{2}{\sqrt{14}}$, and

$\cos\theta=\frac{3}{\sqrt{14}}$

[Neglecting negative sign as θ and ϕ are acute]

$\therefore \frac{\sin\theta\sin\phi}{\sin\theta\cos\phi}=\frac{2}{1}\Rightarrow \tan\phi=2.$

Also, $\tan\theta=\sqrt{5}/3$.

2. a., c.

$(\vec{r}-\vec{a})\cdot\vec{n}_1=0$

$\vec{r}\cdot\vec{n}_1=\vec{a}\cdot\vec{n}$

$(x\hat{i}+y\hat{j}+z\hat{k})\cdot(\hat{i}+\hat{j})=2$

$x+y=2$

now $\overrightarrow{AB}=\hat{i}+\hat{k}$

$\overrightarrow{AB}\times\vec{V}=\vec{n}_2=\begin{vmatrix} \hat{i} & \hat{j} & \hat{k} \\ 1 & 0 & 1 \\ 1 & -1 & -1 \end{vmatrix}$

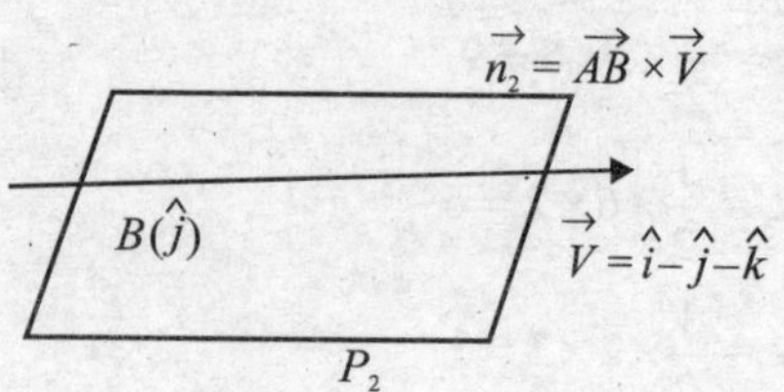

Fig. 3.13

$\vec{n_2} = \hat{i}(0+1) - \hat{j}(-1 \ \ -1) + \hat{k}(-1) = \vec{n_2} = \hat{i} + 2\hat{j} - \hat{k}$

Hence, the equation of P_2 is

$(\vec{r} - \hat{j}) \cdot (\hat{i} + 2\hat{j} - \hat{k}) = 0.$

$\vec{r} \cdot (\hat{i} + 2\hat{j} - \hat{k}) = 2$

If θ is the acute angle between P_1 and P_2, then

$$\cos\theta = \frac{\vec{n_1} \cdot \vec{n_2}}{|\vec{n_1}||\vec{n_2}|} = \left| \frac{(\hat{i} + \hat{j}) \cdot (\hat{i} + 2\hat{j} - \hat{k})}{\sqrt{2} \cdot \sqrt{6}} \right|$$

$$= \frac{3}{\sqrt{2} \cdot \sqrt{6}} = \frac{\sqrt{3}}{2} \quad \theta = \cos^{-1}\frac{\sqrt{3}}{2} = \frac{\pi}{6}$$

3. a., b.

$\vec{r} \cdot \vec{n_1} = q_1$ and $\vec{r} \cdot \vec{n_2} = q_2$, $\vec{r} \cdot \vec{n_3} = q_3$ intersect in a line if $[\vec{n_1} \ \vec{n_2} \ \vec{n_3}] = 0$

$$\text{So,} \begin{vmatrix} 1 & 1 & 1 \\ 1 & 2a & 1 \\ a & a^2 & 1 \end{vmatrix} = 0$$

$$\Rightarrow \begin{vmatrix} 0 & 0 & 1 \\ 1-2a & 2a-1 & 1 \\ a-a^2 & a^2-1 & 1 \end{vmatrix} = 0$$

$(1-2a)(a^2-1) - (a-a^2)(2a-1) = 0$

$a^2 - 1 - 2a^3 + 2a - 2a^2 + a + 2a^3 - a^2 = 0$

$-1 + 3a - 2a^2 = 0$

$a = 1/2, 1.$

4. a., b.

Let the coordinates of the point(s) be a, b, and c.

Therefore, the equation of the line passing through (a, b, c) and whose direction ratios are 1, -5 and -2 is

$$\frac{x-a}{1} = \frac{y-b}{-5} = \frac{z-c}{-2} \quad (1)$$

Line (1) intersects the line $x = y + 5 = z + 1$.

Therefore,

$$\frac{x}{1} = \frac{y+5}{1} = \frac{z+1}{1} \quad (2)$$

Therefore, these are coplanar.

$$\begin{vmatrix} 1 & -5 & -2 \\ 1 & 1 & 1 \\ a & b+5 & c+1 \end{vmatrix} = 0$$

or $a + b - 2c + 3 = 0$

Also, by using same procedure with the second equation, we get the condition

$11a + 15b - 32c + 55 = 0.$

5. a., d.

The equation of the plane passing through the intersection of the planes $2x - y = 0$ and $3z - y = 0$ is

$2x - y + \lambda(3z - y) = 0 \quad (1)$

or $2x - y(\lambda + 1) + 3\lambda z = 0$

Plane (1) is perpendicular to $4x + 5y - 3z = 8$.

Therefore,

$4 \times 2 - 5(\lambda + 1) - 9\lambda = 0$

$8 - 5\lambda - 5 - 9\lambda = 0$

$\Rightarrow 3 - 14\lambda = 0$

$\lambda = 3/14$

$\therefore 2x + y + \frac{3}{14}(3z - y) = 0$

$28x - 17y + 9z = 0.$

6. b., c., d.

$x + y + z - 1 = 0$

$4x + y - 2z + 2 = 0$

Therefore, the line is along the vector

$(\hat{i} + \hat{j} + \hat{k}) \times (4\hat{i} + \hat{j} - 2\hat{k}) = 3\hat{i} + 6\hat{j} + 3\hat{k}$

Let $z = k$. Then $x = k - 1$ and $y = 2 - 2k$

Therefore, $(k - 1, 2 - 2k, k)$ is any point on the line.

Hence, $(-1, 2, 0)$, $(0, 0, 1)$ and $(-1/2, 1, 1/2)$ are the points on the line.

7. a., b.

$3x - 6y + 2z + 5 = 0 \quad (1)$

$-4x + 12y - 3z + 3 = 0 \quad (2)$

$$\text{Bisectors are } \frac{3x - 6y + 2z + 5}{\sqrt{9 + 36 + 4}}$$

$$= \pm \frac{-4x + 12y - 3z + 3}{\sqrt{16 + 144 + 9}}.$$

The plane which bisects the angle between the planes that contains the origin.

$13(3x - 6y + 2z + 5) = 7(-4x + 12y - 3z + 3)$

$67x - 162y + 47z + 44 = 0 \quad (3)$

Further, $3 \times (-4) + (-6)(12) + 2 \times (-3) < 0$

Hence, the origin lies in the acute angle.

8. **a., d.**

The given lines intersect each other if

$$\begin{vmatrix} 2-1 & 3-4 & 4-5 \\ 1 & 1 & \lambda \\ \lambda & 2 & 1 \end{vmatrix} = 0 \Rightarrow \lambda = 0, -1.$$

9. **a., c.**

The required plane is parallel to the bisector of the given planes.

Bisectors are $\frac{x-y+z-3}{\sqrt{3}} = \pm\frac{x+y+z+4}{\sqrt{3}}$

or $y+7=0$ and $2x+2y+1=0$. Hence, the planes are $y=0$ and $x+y=0$.

10. **a., d.**

Ther equation of a plane passsing through the line of intersection of the *xy* and *yz* planes is

$z+\lambda x=0, \quad \lambda \in R$

This plane makes an angle 45° with the *xy* plane $(z=0)$.

$\Rightarrow \cos 45^\circ = \frac{1}{\sqrt{1}\sqrt{\lambda^2+1}}$

$\Rightarrow \lambda = \pm 1$

11. **a., b.**

The plane is equally inclined to the lines. Hence, it is perpendicular to the angle bisector of the vectors $2\hat{i}-2\hat{j}-\hat{k}$ and $8\hat{i}+\hat{j}-4\hat{k}$.

Vector along the angle bisectors of the vectors are

$\frac{2\hat{i}-2\hat{j}-\hat{k}}{3} \pm \frac{8\hat{i}+\hat{j}-4\hat{k}}{9}$, or

$\frac{14\hat{i}-5\hat{j}-7\hat{k}}{9}$ and $\frac{-2\hat{i}-7\hat{j}+\hat{k}}{9}$.

Hence, the equation of the planes is $14x-5y-7z=0$ or $2x+7y-z=0$.

12. **a., c.**

For line $\frac{x-1}{1} = \frac{y}{-1} = \frac{z-5}{1}$, point (1, 0, 5) lies on the plane. Also, the vector along the line $\hat{i}-\hat{j}-\hat{k}$ is perpendicular to the normal $\hat{i}+2\hat{j}-\hat{k}$ to the plane.

For line $\vec{r} = 2\hat{i}-\hat{j}+4\hat{k}+\lambda(3\hat{i}+\hat{j}+5\hat{k})$, point (2, −1, 4) lies on the plane and vector $3\hat{i}+\hat{j}+5\hat{k}$ is perpendicular to the normal $\hat{i}+2\hat{j}-\hat{k}$.

Line $x-y+z=2x+y-z=0$ passes through the origin, which is not on the given plane.

13. **b., c.**

Volume of tetrahedron *ABCD* is

$\frac{1}{6}|[\overrightarrow{AB}\ \ \overrightarrow{AC}\ \ \overrightarrow{AD}]| = 1$ cubic units.

$$\Rightarrow \begin{vmatrix} -1 & 1 & -1 \\ 1 & 1 & -1 \\ x-0 & y-1 & z-2 \end{vmatrix} = \pm 6$$

$\Rightarrow -2(y-1)-2(z-2)=\pm 6$

$\Rightarrow y-1+z-2=\pm 3$

$\Rightarrow y+z=6$ or $y+z=0$.

14. **a., c., d**

The rod sweeps out the figure which is a cone.

The distance of point $A(1, 0, -1)$ from the plane is

$\frac{|1-2+4|}{\sqrt{9}} = 1$ unit.

The slant height l of the cone is 2 units.

Then the radius of the base of the cone is

$\sqrt{l^2-1} = \sqrt{4-1} = \sqrt{3}$.

Hence, the volume of the cone is $\pi(\sqrt{3})^2(1) = 3\pi$ cubic units.

Area of the circle on the plane which the rod traces is 3π.

Also, the centre of the circle is $Q(x, y, z)$. Then

$\frac{x-1}{1} = \frac{y-0}{-2} = \frac{z+1}{2} = \frac{-(1-0-2+4)}{1^2+(-2)^2+2^2}$, or

$Q(x, y, z) \equiv \left(\frac{4}{3}, \frac{-2}{3}, -\frac{1}{3}\right)$.

15. **b., c.**

Distance between the planes is $h = 5/\sqrt{3}$.

Also the figure formed is cylinder, whose radius is $r = 2$ units.

Hence, the volume of the cylinder is $\pi r^2 h$ $= (20/\sqrt{3})\pi$ cubic units.

Also the curved surface area is $2\pi rh = 4\pi\sqrt{3}$

16. **a., b.**

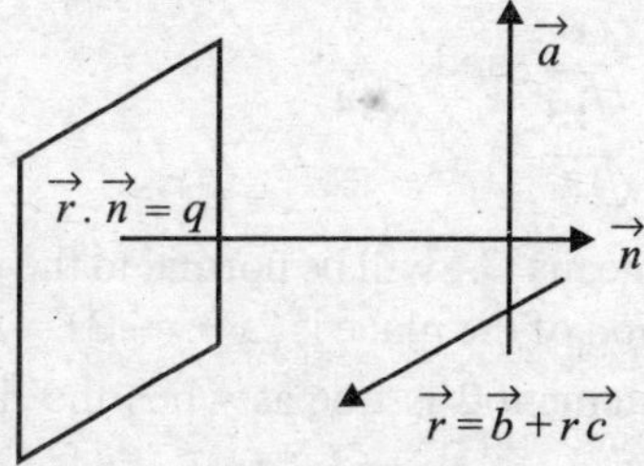

Fig. 3.14

Required line is parallel to $\vec{n} \times \vec{c}$.

The equation of line is $\vec{r} = \vec{a} + \lambda(\vec{n} \times \vec{c})$

$\Rightarrow (\vec{r} - \vec{a}) = \lambda(\vec{n} \times \vec{c})$

$\therefore (\vec{r} - \vec{a}) \times (\vec{n} \times \vec{c}) = 0$

Reasoning Type

1. **b.** Given lines are parallel as both are directed along same vector $(\hat{i}+\hat{j}-\hat{k})$; so they do not intersect. Also Statement 2 is correct by definition of skew lines, but skew lines are those which are neither parallel nor intersecting. Hence, both the statements are true, but statement 2 is not the correct explanation for statement 1.

2. **b.** For the given lines let $\vec{a}_1 = \hat{i} + \hat{j} - \hat{k}$, $\vec{a}_2 = 4\hat{i} - \hat{k}$, $\vec{b}_1 = 3\hat{i} - \hat{j}$ and $\vec{b}_2 = 2\hat{i} + 3\hat{k}$. Therefore,

$$[\vec{a}_2 - \vec{a}_1 \;\; \vec{b}_1 - \vec{b}_2] = \begin{vmatrix} 4-1 & 0-1 & -1+1 \\ 3 & -1 & 0 \\ 2 & 0 & 3 \end{vmatrix}$$

$$= \begin{vmatrix} 3 & -1 & 0 \\ 3 & -1 & 0 \\ 2 & 0 & 3 \end{vmatrix} = 0$$

Hence, the lines are coplanar. Also vectors $\vec{b}_1$ and $\vec{b}_2$ along which the lines are directed are not collinear. Hence, the lines intersect. When $\vec{b} \times \vec{d} = 0$, vectors $\vec{b}$ and $\vec{d}$ are collinear; therefore, lines $\vec{r} = \vec{a} + \lambda \vec{b}$ and $\vec{r} = \vec{c} + \lambda \vec{d}$ are parallel and do not the intersect. But this statement is not the correct explanation for Statement 1.

3. **a.** Any point on the first line is $(2x_1 + 1, x_1 - 3, -3x_1 + 2)$.
Any point on the second line is $(y_1 + 2, -3y_1 + 1, 2y_1 - 3)$.
If two lines are coplanar, then $2x_1 - y_1 = 1, x_1 + 3y_1 = 4$, and $3x_1 + 2y_1 = 5$ are consistent.

4. **a.** The direction cosines of segment OA are $\frac{2}{\sqrt{14}}, \frac{1}{\sqrt{14}}$, and $\frac{-3}{\sqrt{14}}$.
$OA = \sqrt{14}$
This means OA will be normal to the plane and the equation of the plane is $2x + y - 3z = 14$.

5. **b.** Statement 2 is true as when the line lies in the plane, vector $\vec{b}$ along which the line is directed is perpendicular to the normal $\vec{c}$ of the plane, but it does not explain statement 1 as for $\vec{b} \cdot \vec{c} = 0$, the line may be parallel to the plane. However, Statement 1 is correct as any point on the line $(t + 1, 2t, -t - 2)$ lies on the plane for $t \in R$.

6. **a.** $\sin \theta = \left| \frac{2-3+2}{\sqrt{4+9+4\sqrt{3}}} \right| = \frac{1}{\sqrt{51}}$.
Therefore, statement 1 is true and statement 2 is also true by definition.

7. **a.** $\overrightarrow{PA} \cdot \overrightarrow{PB} = 9 > 0$. Therefore, P is exterior to the sphere. statement 2 is also true (standard result).

8. **a.** Obviously answer is (*a*).

9. **c.** Any point on the line $\frac{x-1}{1} = \frac{y}{-1} = \frac{z+2}{2}$ is B $(t+1, -t, 2t-2), t \in R$.
Also, AB is perpendicular to the line, where A is $(1, 2, -4)$.
$\Rightarrow 1(t) - (-t-2) + 2(2t+2) = 0$
$\Rightarrow 6t + 6 = 0$
$\Rightarrow t = -1$
point B is $(0, 1, -4)$.
Hence, $AB = \sqrt{1+1+0} = \sqrt{2}$.

10. **d.** Direction ratios of the given lines are $(-3, 1, -1)$ and $(1, 2, -1)$, Hence, the lines are perpendicular as $(-3)(1) + (1)(2) + (-1)(-1) = 0$.
Also lines are coplanar as

$$\begin{vmatrix} 0-2 & 1-3 & -1+13/7 \\ -3 & 1 & -1 \\ 1 & 2 & -1 \end{vmatrix} = 0.$$

But statement 2 is not enough reason for the shortest distance to be zero, as two skew lines can be perpendicular also.

Linked Comprehension Type

For Problems 1–3

1. b., 2. c. 3. d.

Sol.

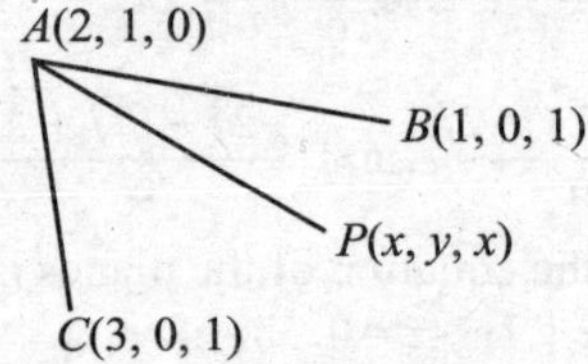

Fig. 3.15

$$\begin{vmatrix} x-2 & y-1 & z \\ -1 & -1 & 1 \\ 1 & -1 & 1 \end{vmatrix} = 0$$

$(x-2)[(-1)-(-1)] - (y-1)[(-1)-1] + z[1+1] = 0$
$2(y-1) + 2z = 0$
$\Rightarrow y + z - 1 = 0$

The vector normal to the plane is $\vec{r} = 0\hat{i} + \hat{j} + \hat{k}$.

The equation of the line through $(0, 0, 2)$ and parallel to $\vec{n}$ is $\vec{r} = 2\hat{k} + \lambda(\hat{j} + \hat{k})$.

The perpendicular distance of $D(0, 0, 2)$ from plane ABC is $\left| \frac{2-1}{\sqrt{1^2 + 1^2}} \right| = \frac{1}{\sqrt{2}}$.

For Problems 4–6

4. b., 5. c., 6. c.

Sol.

4. **b.** Let $Q(x_2, y_2, z_2)$ be the image of $A(2, 1, 6)$ about mirror $x+y-2z=3$. Then,

$$\frac{x_2-2}{1}=\frac{y_2-1}{1}=\frac{z_2-6}{-2}=\frac{-2(2+1-12-3)}{1^2+1^2+2^2}=4$$

$\Rightarrow (x_2, y_2, z_2) \equiv (6, 5, -2)$.

5. **c.** Let $\frac{x-2}{3}=\frac{y-1}{4}=\frac{z-6}{5}=\lambda$.

$x=2+3\lambda, y=1+4\lambda, z=6+5\lambda$ lies on plane $x+y-2z=3$.

$\Rightarrow 2+3\lambda+1+4\lambda-2(6+5\lambda)=3$

$\Rightarrow 3+7\lambda-12-10\lambda=3 \Rightarrow -3\lambda=12 \Rightarrow \lambda=-4$

Point $B \equiv (-10, -15, -14)$.

6. **c.** The equation of the reflected ray $L_1=0$ is the line joining $Q(x_2, y_2, z_2)$ and $B(-10, -15, -14)$.

$$\frac{x+10}{16}=\frac{y+15}{20}=\frac{z+14}{12}$$

$$\frac{x+10}{4}=\frac{y+15}{5}=\frac{z+14}{3}$$

For Problems 7–9

7. b., 8. c., 9.

Sol.

The given system of equation is

$2x+py+6z=8$

$x+2y+qz=5$

$x+y+3z=4$

$$\Delta=\begin{vmatrix}2 & p & 6\\1 & 2 & q\\1 & 1 & 3\end{vmatrix}=(2-p)(3-q)$$

By Cramer's rule, if $\Delta \neq 0$, i.e., $p \neq 2$ and $q \neq 3$, the system has a unique solution.

If $p=2$ or $q=3$, $\Delta=0$, then if $\Delta_x=\Delta_y=\Delta_z=0$, the system has infinite solutions and if any one of Δ_x, Δ_y, and $\Delta_z \neq 0$, the system has no solution.

$$\text{Now } \Delta_x=\begin{vmatrix}8 & p & 6\\5 & 2 & q\\4 & 1 & 3\end{vmatrix}$$

$=30-8q-15p+4pq$

$$\Delta_y=\begin{vmatrix}1 & 8 & 6\\1 & 5 & q\\1 & 4 & 3\end{vmatrix}$$

$=-8q+8q=0$

$$\Delta_z=\begin{vmatrix}2 & p & 8\\1 & 2 & 5\\1 & 1 & 4\end{vmatrix}$$

$=p-2$

Thus, if $p=2$, $\Delta_x=\Delta_y=\Delta_z=0$ for all $q \in R$, the system has infinite solutions.

If $p \neq 2$, $q=3$ and $\Delta_z \neq 0$, then the system has no solution.

Hence, the system has (1) no solution if $p \neq 2$ and $q=3$, (2) a unique solution if $p \neq 2$ and $q \neq 3$ and (3) infinite solutions if $p=2$ and $q \in R$.

For Problems 10–12

10. d., 11. b., 12. a.

Sol.

10. **d.** The line $\frac{x-1}{3}=\frac{y-2}{-1}=\frac{z-3}{4}=r$

Any point say $B \equiv 3r+1, 2-r, 3+4r$ (on the line L)

$\overrightarrow{AB}=3r, -r, 4r+6$

Hence,

$\overrightarrow{AB}$ is parallel to $x+y-z=1$.

$\Rightarrow 3r-r-4r-6=0$ or $r=-3$.

B is $(-8, 5, -9)$.

11. **b.** The equation of plane containing the line L is $A(x-1)+B(y-2)+C(z-3)=0$, where $3A-B+4C=0$ (1)

Therefore, (1) also contains the point $A(1, 2, -3)$.

Hence, $C=0$ and $3A=B$.

The equation of the plane is $x-1+3(y-2)=0$ or $x+3y-7=0$.

12. **a.** The distance of point $(1+3r, 2-r, 3+4r)$ from the plane is $\frac{|1+3r+2-r-3-4r|}{\sqrt{1+1+1}}=\frac{|2r|}{\sqrt{3}}=\frac{4}{\sqrt{3}}$

$\Rightarrow r=\pm 2$

Hence, the points are $A(7, 0, 11)$ and $B(-5, 4, -5)$.

$\Rightarrow AB=\sqrt{416}=4\sqrt{26}$.

Matrix-Match Type

1. **a → s; b → q; c → p; d → r.**

a. Line $x=2t+1, y=t+2, z=-t-3$ or $\frac{x-1}{2}=\frac{y-2}{1}=\frac{z+3}{-1}$, which is along the vector $2\hat{i}+\hat{j}-\hat{k}$.

Vector $\hat{i}+3\hat{j}+\hat{k}$ is perpendicular to the line.

b. Normals to the planes $x+y+z-3=0$ and $2x-y+3z=0$ are $\vec{n}_1=\hat{i}+\hat{j}+\hat{k}$ and $\vec{n}_2=2\hat{i}-\hat{j}+3\hat{k}$. Then the vector along the line of intersection of planes is $\vec{n}_1\times\vec{n}_2 = \begin{vmatrix}\hat{i} & \hat{j} & \hat{k}\\ 1 & 1 & 1\\ 2 & -1 & 3\end{vmatrix} = 4\hat{i}-\hat{j}-3\hat{k}$.

c. The shortest distance between the lines $\frac{x}{2}=\frac{y}{-3}=\frac{z}{-1}$ and $\vec{r}=(3\hat{i}-\hat{j}+\hat{k})+t\ (\hat{i}+\hat{j}-2\hat{k})$ occurs along the vector

$$(2\hat{i}-3\hat{j}-\hat{k})\times(\hat{i}+\hat{j}-2\hat{k}) = \begin{vmatrix}\hat{i} & \hat{j} & \hat{k}\\ 2 & -3 & -1\\ 1 & 1 & -2\end{vmatrix} = 7\hat{i}+3\hat{j}+5\hat{k}.$$

d. Normal to the plane $\vec{r}=-\hat{i}+4\hat{j}-6\hat{k}+\lambda(\hat{i}+3\hat{j}-2\hat{k})+\mu(-\hat{i}+2\hat{j}-5\hat{k})$ is

$$\begin{vmatrix}\hat{i} & \hat{j} & \hat{k}\\ 1 & 3 & -2\\ -1 & 2 & -5\end{vmatrix} = -11\hat{i}+7\hat{j}+5\hat{k}$$

2. **a → q, s; b → r; c → p, q; d → p.**

a. Line $\frac{x-1}{-2}=\frac{y+2}{3}=\frac{z}{-1}$ is along the vector $\vec{a}=-2\hat{i}+3\hat{j}-\hat{k}$ and the line $\vec{r}=(3\hat{i}-\hat{j}+\hat{k})+t(\hat{i}+\hat{j}+\hat{k})$ is along the vector $\vec{b}=\hat{i}+\hat{j}+\hat{k}$.

Here $\vec{a}\perp\vec{b}$.

$$\text{Also}\begin{vmatrix}3-1 & -1-(-2) & 1-0\\ -2 & 3 & -1\\ 1 & 1 & 1\end{vmatrix}\neq 0.$$

b. The direction ratios of the line $x-y+2z-4=0=2x+y-3z+5$ are $\begin{vmatrix}\hat{i} & \hat{j} & \hat{k}\\ 1 & -1 & 2\\ 2 & 1 & -3\end{vmatrix}$

$=\hat{i}+7\hat{j}+3\hat{k}$. Hence, the given two lines are parallel.

c. The given lines are $(x=t-3,\ y=-2t+1,\ z=-3t-2)$ and $\vec{r}=(t+1)\hat{i}+(2t+3)\hat{j}+(-t-9)\hat{k}$, or $\frac{x+3}{1}=\frac{y-1}{-2}=\frac{z+2}{-3}$ and $\frac{x-1}{1}=\frac{y-3}{2}=\frac{z+9}{-1}$.

The lines are perpendicular as $(1)(1)+(-2)(2)+(-3)(-1)=0$.

$$\text{Also}\begin{vmatrix}-3-1 & 1-3 & -2-(-9)\\ 1 & -2 & -3\\ 1 & 2 & -1\end{vmatrix}=0$$

Hence, the lines are intersecting.

d. The given lines are $\vec{r}=(\hat{i}+3\hat{j}-\hat{k})+t(2\hat{i}-\hat{j}-\hat{k})$ and $\vec{r}=(-\hat{i}-2\hat{j}+5\hat{k})+s(\hat{i}-2\hat{j}+\hat{k})$.

$$\begin{vmatrix}1-(-1) & 3-(-2) & -1-5\\ 2 & -1 & -1\\ 1 & -2 & 1\end{vmatrix}=0$$

Hence, the lines are coplanar and hence, intersecting (as the lines are not parallel).

3. **a → q; b → p; c. → s; d → r.**

a. The given line is $x=4y+5, z=3y-6$, or $\frac{x-5}{4}=y,\ \frac{z+6}{3}=y$

or $\frac{x-5}{4}=\frac{y}{1}=\frac{z+6}{3}=\lambda$ (say)

Any point on the line is of the form $(4\lambda+5, \lambda, 3\lambda-6)$.

The distance between $(4\lambda+5, \lambda, 3\lambda-6)$ and $(5, 3, -6)$ is 3 units (given). Therefore,

$(4\lambda+5-5)^2+(\lambda-3)^2+(3\lambda-6+6)^2=9$

$\Rightarrow 16\lambda^2+\lambda^2+9-6\lambda+9\lambda^2=9$

$\Rightarrow 26\lambda^2-6\lambda=0$

$\Rightarrow \lambda=0, 3/13$

The point is $(5, 0, -6)$.

b. The equation of the plane containing the lines $\frac{x-2}{3}=\frac{y+3}{5}=\frac{z+5}{7}$ and $\frac{x+2}{1}=\frac{y-4}{4}=\frac{z-6}{7}$ is

$$\begin{vmatrix}x-2 & y+3 & z+5\\ 1 & 4 & 7\\ 3 & 5 & 7\end{vmatrix}=0$$

$\Rightarrow x-2y+z-3=0$

Point $(-1, -2, 0)$ lies on this plane.

c. The line passing through points $A(2, -3, -1)$ and $B(8, -1, 2)$ is $\frac{x-2}{8-2}=\frac{y+3}{-1+3}=\frac{z+1}{2+1}$, or

$\frac{x-2}{6}=\frac{y+3}{2}=\frac{z+1}{3}=\lambda$ (say).

Any point on this line is of the form $P(6\lambda+2, 2\lambda-3, 3\lambda-1)$, whose distance from point $A(2, -3, -1)$ is 14 units. Therefore,

$PA=14$

$PA^2=(14)^2$
$(6\lambda)^2+(2\lambda)^2+(3\lambda)^2=196$
$49\lambda^2=196$
$\lambda^2=4$
$\lambda=\pm 2$
The required points are (14, 1, 5) and (−10, −7, −7). The point nearer to the origin is (14, 1, 5).

d. Any point on line AB, $\frac{x}{2}=\frac{y-2}{3}=\frac{z-3}{4}=\lambda$ is $M(2\lambda, 3\lambda+2, 4\lambda+3)$. Therefore, the direction ratios of PM are $2\lambda-3$, $3\lambda+3$ and $4\lambda-8$.

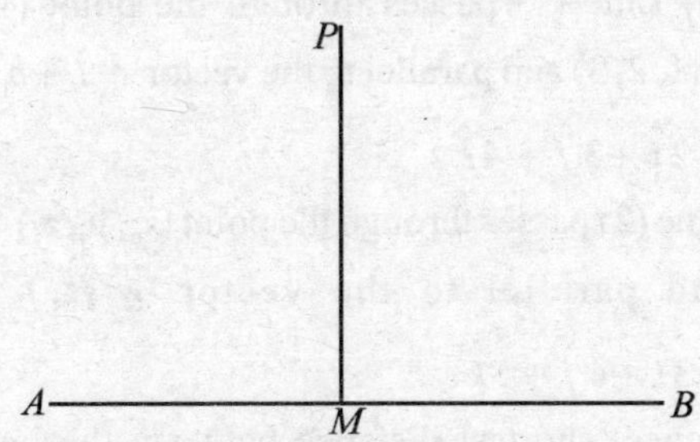

Fig. 3.16

But $PM \perp AB$
$\frac{x}{2}=\frac{y-2}{3}=\frac{z-3}{4}$
$\therefore 2(2\lambda-3)+3(3\lambda+3)+4(4\lambda-8)=0$
$4\lambda-6+9\lambda+9+16\lambda-32=0$
$29\lambda-29=0$
$\lambda=1$
Therefore, foot of the perpendicular is $M(2, 5, 7)$.

4. a → r; b → p; c → q; d → s.

a. The given line and plane are $\vec{r}=(2\hat{i}-2\hat{j}+3\hat{k})+\lambda(\hat{i}-\hat{j}+4\hat{k})$ and $\vec{r}\cdot(\hat{i}+5\hat{j}+\hat{k})=5$, respectively. Since $(\hat{i}-\hat{j}+4\hat{k})\cdot(\hat{i}+5\hat{j}+\hat{k})=0$, line and plane are parallel.
Hence, the required distance = distance of point (2, −2, 3) from the plane $x+5y+z-5=0$, which is $\frac{|2-10+3-5|}{\sqrt{1+25+1}}=\frac{10}{3\sqrt{3}}$.

b. The distance between two parallel planes $\vec{r}\cdot(2i-j+3k)=4$ and $\vec{r}\cdot(6i-3j+9k)+13=0$ is

$$d=\frac{|4-(-13/3)|}{\sqrt{(2)^2+(-1)^2+(3)^2}}=\frac{(25/3)}{\sqrt{14}}=\frac{25}{3\sqrt{14}}$$

c. The perpendicular distance of the point (2, 5, −3) from the plane $\vec{r}\cdot(6i-3j+2k)=4$ or $6x-3y+2z-4=0$ is

$$d=\frac{|12-15-6-4|}{\sqrt{36+9+4}}$$
$$=13/\sqrt{49}=13/7.$$

d. The equation of the line AB is
$$\frac{x-2}{2}=\frac{y+2}{3}=\frac{z-6}{-6}$$
The equation of line passing through (1, 0, −3) and parallel to AB is
$$\frac{x-1}{2}=\frac{y}{3}=\frac{z+3}{-6}=r \text{ (say)}$$

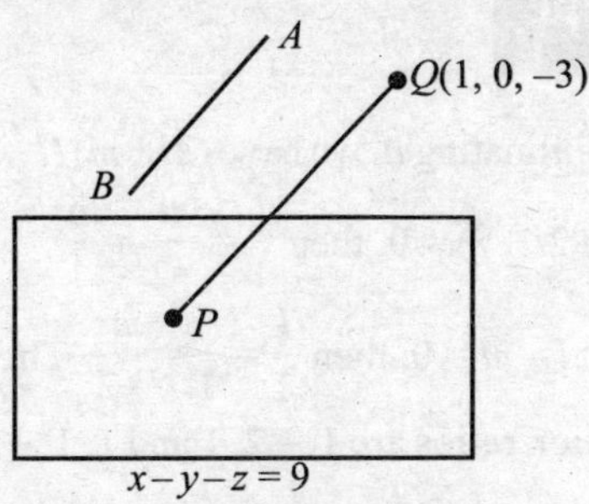

Fig. 3.17

The coordinates of any point on line $P(2r+1, 3r, -6r-3)$ which lie on plane
$(2r+1)-(3r)-(-6r-3)=9$ $r=1$
Point $P\equiv(3, 3, -9)$
Required distance PQ
$=\sqrt{(3-1)^2+(3-0)^2+(-9+3)^2}$
$=\sqrt{4+9+36}=7.$

5. a → q; b → r; c → s; d → p.

a. If the required image is (x, y, z), then $\frac{x-3}{2}=\frac{y-5}{1}=\frac{z-7}{1}=-\frac{2(6+5+7+18)}{2^2+1^2+1^2}=-12$, or (−21, −7, −5).

b. Any point on the line $\frac{x-2}{-3}=\frac{y-1}{2}=\frac{z-3}{2}=\lambda$ is $(-3\lambda+2, 2\lambda+1, 2\lambda+3)$, which lies on plane $2x+y-z=3$. Therefore,
$-6\lambda+4+2\lambda+1-2\lambda-3=3$
$-6\lambda=1$
$\lambda=-1/6$
Therefore, the point is $\left(\frac{5}{2},\frac{2}{3},\frac{8}{3}\right)$.

c. If (x, y, z) is required foot of the perpendicular, then
$$\frac{x-1}{2}=\frac{y-1}{-2}=\frac{z-2}{4}=-\frac{(2-2+8+5)}{2^2+(-2)^2+4^2},$$
or $(x, y, z)\equiv\left(\frac{-1}{12},\frac{25}{12},\frac{-2}{12}\right)$.

d. Any point on the line $\frac{x-1}{2}=\frac{y-2}{3}=\frac{z-3}{4}=\lambda$ is $P(2\lambda+1, 3\lambda+2, 4\lambda+3)$, which satisfies the line $\frac{x-4}{5}=\frac{y-1}{2}=\frac{z}{1}$ or $\frac{2\lambda+1-4}{5}$
$=\frac{3\lambda+2-1}{2}=\frac{4\lambda+3}{1}$
$\Rightarrow \lambda=-1$
The required point is $(-1,-1,-1)$.

Integer Type

1. (3) Eliminating n, we have $(2l+m)(l-m)=0$

When $2l+m=0$, then $\frac{l}{1}=\frac{m}{-2}=\frac{n}{1}$

When $l-m=0$, then $\frac{l}{1}=\frac{m}{1}=\frac{n}{-2}$. Therefore, direction ratios are $1,-2,1$ and $1,1,-2$.

$$\cos\theta=\frac{\Sigma l_1 l_2}{\sqrt{(\Sigma l_1^2)}\sqrt{(\Sigma l_2^2)}}=-\frac{1}{2}$$

$\Rightarrow \theta=120°=\frac{2\pi}{3}$

2. Let P be the foot of the perpendicular from A (2, 3, 4) to the given line l whose equation is $\frac{4-x}{2}=\frac{y}{6}=\frac{1-z}{3}$ or $\frac{x-4}{-2}=\frac{y}{6}=\frac{z-1}{-3}=k$ (say)

$\Rightarrow x=4-2k, y=6k, z=1-3k$

As P lies on (1), the coordinates of P are $(4-2k, 6k, 1-3k)$ for some value of k.

The direction ratios of AP are $4-2k-2, 6k-3, 1-3k-4$ or $2-2k, 6k-3, -3-3k$.

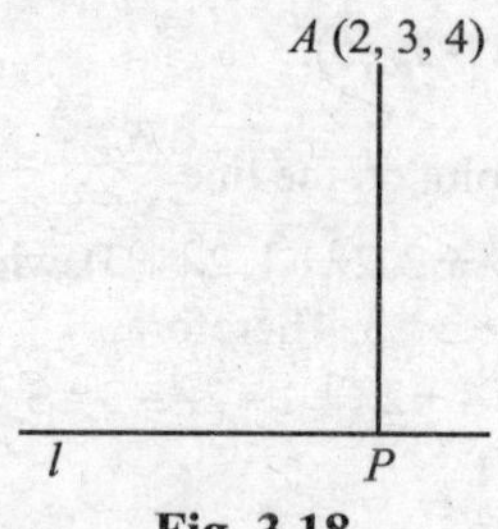

Fig. 3.18

Also the direction ratios of l are $-2, -6, -3$.

Since AP is perpendicular to l, $a_1a_2+b_1b_2+c_1c_2=0$

$\Rightarrow -2(2-2k)+6(6k-3)-3(-3-3k)=0$

$\Rightarrow -4+4k+36k-18+9+9k=0$ or $49-13=0$

or $k=13/49$.

We have

$AP^2=(4-2k-2)^2+(6k-3)^2+(1-3k-4)^2$
$=(2-2k)^2+(6k-3)^2+(-3-3k)^2$
$=4-8k+4k^2+36k^2-36k+9+9+18k+9k^2$
$=22-26k+49k^2$

$$=22-26\left(\frac{13}{49}\right)+49\left(\frac{13}{49}\right)^2$$

$$=\frac{22\times 49-26\times 13+13^2}{49}=\frac{909}{49}$$

$\Rightarrow AP=\frac{3}{7}\sqrt{101}$

3. (1) Line (1) passes through the point $(x_1, y_1, z_1) \equiv (1, 2, 3)$ and parallel to the vector $a_1\hat{i}+b_1\hat{j}+c_1\hat{k} \equiv 2\hat{i}+3\hat{j}+4\hat{k}$

Line (2) passes through the point $(x_2, y_2, z_2) \equiv (2, 4, 5)$ and parallel to the vector $a_2\hat{i}+b_2\hat{j}+c_2\hat{k} \equiv 3\hat{i}+4\hat{j}+5\hat{k}$.

Hence, shortest distance between the lines using the formula

$$\frac{\begin{vmatrix} x_2-x_1 & y_2-y_1 & z_2-z_1 \\ a_1 & b_1 & c_1 \\ a_2 & b_2 & c_2 \end{vmatrix}}{\left\|\begin{matrix} \hat{i} & \hat{j} & \hat{k} \\ a_1 & b_1 & c_1 \\ a_2 & b_2 & c_2 \end{matrix}\right\|}$$

is

$$\frac{\begin{vmatrix} 2-1 & 4-2 & 5-3 \\ 2 & 3 & 4 \\ 3 & 4 & 5 \end{vmatrix}}{\left\|\begin{matrix} \hat{i} & \hat{j} & \hat{k} \\ 2 & 3 & 4 \\ 3 & 4 & 5 \end{matrix}\right\|}=\frac{1}{\sqrt{6}}.$$

4. (2) The given lines are

$\frac{x-1}{1}=\frac{y-3}{-\lambda}=\frac{z-1}{\lambda}(=s)$

and $\frac{x-0}{1/2}=\frac{y-1}{1}=\frac{y-2}{-1}(=t)$

are coplanar if $\begin{vmatrix} 0+1 & 1-3 & 2-1 \\ 1 & -\lambda & \lambda \\ 1/2 & 1 & -1 \end{vmatrix}=0$

$\Rightarrow \begin{vmatrix} 1 & -2 & 1 \\ 1 & -\lambda & \lambda \\ 1/2 & 1 & -1 \end{vmatrix}=0$

$\Rightarrow 1(\lambda-\lambda)+2\left(-1-\frac{\lambda}{2}\right)+1\left(1+\frac{\lambda}{2}\right)=0$

$\Rightarrow \lambda=-2$

5. **(8)** Let the image of the point A (1, 2, 3) on the line l whose equations are

$$\frac{x-6}{3}=\frac{y-7}{2}=\frac{z-7}{-2}=k \text{ (say)} \quad (1)$$

be A'. Then AA' is perpendicular to l and the point of intersection of AA' and l is the midpoint of AA'. Note that M is the foot of the perpendicular from A to l.

The coordinates of any point on (1) are of the form $(3k+6, 2k+7, -2k+7)$. Therefore, coordinates of M are $(3k+6, 2k+7, -2k+7)$, for some value of k. The direction ratios of AM are $3k+6-1, 2k+7-2, -2k+7-3$ or $3k+5, 2k+5, -2k+4$.

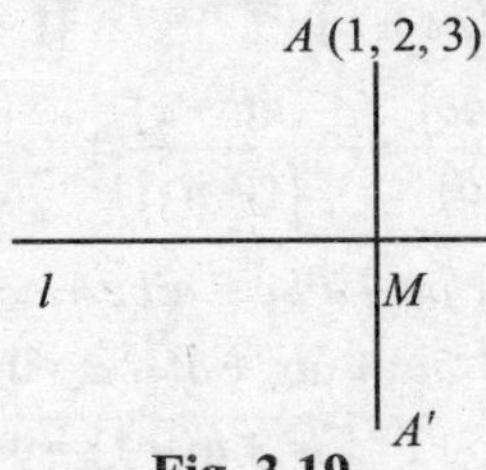

Fig. 3.19

Also the direction ratios of l are 3, 2, –2.

Since AM is perpendicular to l, $a_1a_2+b_1b_2+c_1c_2=0$

$\Rightarrow 3(3k+5)+2(2k+5)-2(2k+4)=0$

$\Rightarrow 17k+17=0$ or $k=-1$

Thus the coordinates of M are (3, 5, 9).

Suppose the coordinates of A' be (x, y, z).

The coordinates of the midpoint of AA' are

$$\left(\frac{x+1}{2}, \frac{y+2}{2}, \frac{z+3}{2}\right).$$

But the midpoint of AA' is (5, 3, 9). Therefore,

$$\frac{x+1}{2}=5, \frac{y+2}{2}=3, \frac{z+3}{2}=9 \Rightarrow x=9, y=4, z=15$$

Thus the image of A in l is (17, 16, 7).

6. **(9)** $\frac{x-1}{2}=\frac{y+1}{3}=\frac{z-1}{4}=\lambda$

$\Rightarrow x=2\lambda+1, y=3\lambda-1$, and $z=4\lambda+1$

$\frac{x-3}{1}=\frac{y-k}{2}=\frac{z}{1}=\mu$

$\Rightarrow x=3+\mu, y=k+2\mu$, and $z=\mu$

Since the above lines intersect,

$2\lambda+1=3+\mu \quad (1)$

$3\lambda-1=2\mu+k \quad (2)$

$\mu=4\lambda+1 \quad (3)$

Solving (1) and (3) and putting the value of λ and μ in (2) we get $k=9/2$.

7. **(7)**

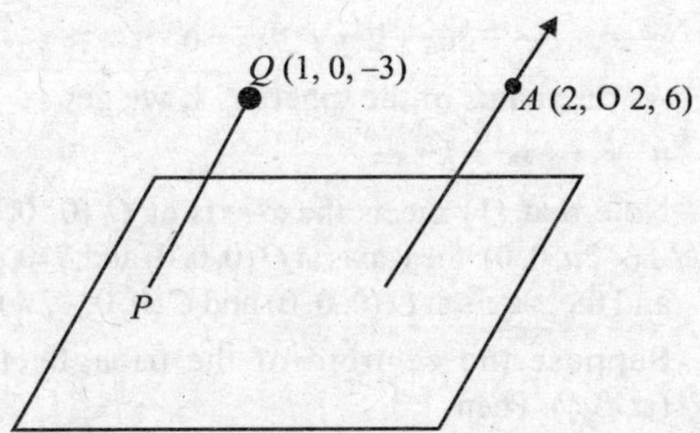

Fig. 3.20

The given plane is $x-y-z=9 \quad (1)$

The given line AB is $\frac{x-2}{2}=\frac{y+2}{3}=\frac{z-6}{-6} \quad (2)$

Equation of the line passing through (1, 0, – 3) and parallel to

$\frac{x-2}{2}=\frac{y+2}{3}=\frac{z-6}{-6}$ is

$\frac{x-1}{2}=\frac{y-0}{3}=\frac{z+3}{-6}=r.$

The coordinate of any point on (3) may be given as $P(2r+1, 3r, -6r-3)$

If P is the intersection point of (1) and (3), then it must lie on (1). Therefore,

$(2r+1)-(3r)-(-6r-3)=9$

$\Rightarrow 2r+1-3r+6r+3=9 \Rightarrow r=1.$

Therefore coordinate of $P(3, 3, -9)$

$\Rightarrow$ Distance between $Q(1, 0, -3)$ and $P(3, 3, -9)$

$=\sqrt{(3-1)^2+(3-0)^2+(-9+3)^2}$

$=\sqrt{4+9+36}=7$

8. **(4)** The line is along the vector $\vec{a}=-3\hat{i}+2\hat{j}+\hat{k}$ and the plane is normal to the vector $\vec{b}=\hat{i}+\hat{j}+\hat{k}$.

Since $\vec{a}\cdot\vec{b}=0$, the line is parallel to the plane.

Hence the distance between the line and the plane is the distance of the point (–1, 3, 2) from the plane

or $\frac{|-1+3+2|}{\sqrt{1+1+1}}=\frac{4}{\sqrt{3}}.$

9. **(3)** The radius of the sphere is 5.

The given plane is $x+y-z=4\sqrt{3}$.

The length of the perpendicular from the centre (0, 0, 0) of the sphere on the plane is $\frac{4\sqrt{3}}{\sqrt{1+1+1}}=4.$

Hence, the radius of the circular section is $\sqrt{25-16}=\sqrt{9}=3.$

10. **(9)** Let the equation of any sphere passing through the origin and having radius k be

$$x^2+y^2+z^2+2ux+2vy+2wz=0 \quad (1)$$

As the radius of the sphere is k, we get

$$u^2+v^2+w^2=k^2 \quad (2)$$

Note that (1) meets the x-axis at O (0, 0, 0) and $A(-2u, 0, 0)$; the y-axis at $O(0, 0, 0)$ and $B(0, -2v, 0)$; and the z-axis at $O(0, 0, 0)$ and $C(0, 0, -2w)$.

Suppose the centroid of the triangle ABC be (α, β, γ). Then

$$\alpha=-\frac{2u}{3}, \beta=-\frac{2v}{3}, \text{ and } \gamma=-\frac{2w}{3}$$

$$\Rightarrow u=-\frac{3\alpha}{2}, v=-\frac{3\beta}{2}, \text{ and } w=\frac{3v}{2}$$

Putting this in (2), we get

$$\left(\frac{-3}{2}\alpha\right)^2+\left(\frac{-3}{2}\beta\right)^2+\left(\frac{-3}{2}\gamma\right)^2=k^2$$

$$\Rightarrow \alpha^2+\beta^2+\gamma^2=\frac{4}{9}k^2.$$

This show that the centroid of the triangle ABC lies on $x^2+y^2+z^2=\frac{4}{9}k^2$.

11. **(2)**

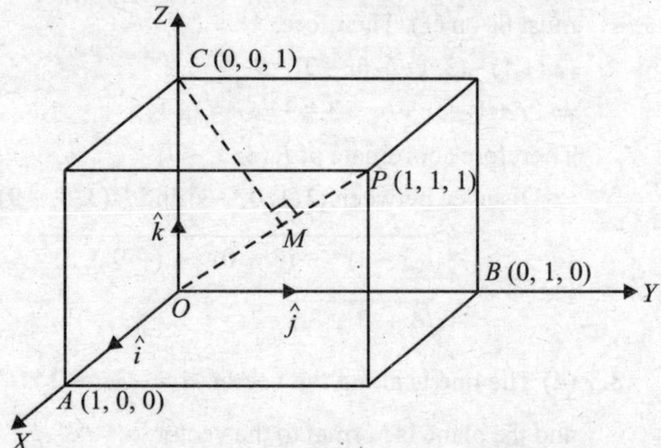

Fig. 3.21

Let the edges OA, OB, and OC of the unit cube be along OX, OY, and OZ, respectively. Since $OA=OB=OC=1$ unit, $\overrightarrow{OA}=\hat{i}, \overrightarrow{OB}=\hat{j}, \overrightarrow{OC}=\hat{k}$.

Let CM be the perpendicular from the corner C on the diagonal OP. The vector equation of OP is $\vec{r}=\lambda\left(\hat{i}+\hat{j}+\hat{k}\right)$. Therefore,

$OM=$ projection of $\overrightarrow{OC}$ on $\overrightarrow{OP}=\overrightarrow{OC}\cdot\overrightarrow{OP}$

$$=\hat{k}\cdot\frac{\left(\hat{i}+\hat{j}+\hat{k}\right)}{\sqrt{3}}=\frac{1}{\sqrt{3}}$$

Now $OC^2=OM^2+CM^2$

$$\Rightarrow CM^2=|\overrightarrow{OC}|^2-OM^2=1-\frac{1}{3}=\frac{2}{3}\Rightarrow CM=\sqrt{\frac{2}{3}}$$

12. **(1)** Given planes are

$$x-cy-bz=0 \quad (1)$$

$$cx-y+az=0 \quad (2)$$

$$bx+ay-z=0 \quad (3)$$

The equation of planes passing through the line of intersection of planes (1) and (2) may be taken as

$(x-cy-bz)+\lambda(cx-y+az)=0$

or $x(1+\lambda c)-y(c+\lambda)+z(-b+a\lambda)=0 \quad (4)$

If planes (3) and (4) are same, then Eq. (3) and (4) will be identical.

$$\frac{1+x\lambda}{b}=\frac{-(c+\lambda)}{a}=\frac{-b+a\lambda}{-1}$$

$$\text{or } \lambda=-\frac{(a+bc)}{(ac+b)} \text{ and } \lambda=-\frac{(ab+c)}{\left(1-a^2\right)}$$

$$\therefore \frac{-(a+bc)}{(ac+b)}=-\frac{(ab+c)}{\left(1-a^2\right)}$$

$$\Rightarrow a-a^3+bc-a^2bc=a^2bc+ac^2+ab^2+bc$$

$$\Rightarrow 2a^2bc+ac^2+ab^2+a^3-a=0$$

$$\Rightarrow a(2abc+c^2+b^2+a^2-1)=0$$

$$\Rightarrow a^2+b^2+c^2+2abc=1$$

13. **(7)** As the line $\frac{x-4}{1}=\frac{y-2}{1}=\frac{z-k}{2}$ lies in the plane $2x-4y+z=7$, the point $(4, 2, k)$ through which the line passes must also lie on the given plane and hence $2\times4-4\times2+k=7$

$\Rightarrow k=7.$

14. **(9)** Let the equation of the variable plane be $\frac{x}{a}+\frac{y}{b}+\frac{z}{c}=1$, which meets the axes at A $(a, 0, 0)$, B $(0, b, 0)$, and $C(0, 0, c)$. Therefore, centroid of ΔABC is $(a/3, b/3, c/3)$ and it satisfies the relation

$$\frac{1}{x^2}+\frac{1}{y^2}+\frac{1}{z^2}=k.$$

$$\Rightarrow \frac{9}{a^2}+\frac{9}{b^2}+\frac{9}{c^2}=k$$

$$\Rightarrow \frac{1}{a^2}+\frac{1}{b^2}+\frac{1}{c^2}=\frac{k}{9} \quad (1)$$

Also given that the distance of plane $\frac{x}{a}+\frac{y}{b}+\frac{z}{c}=1$ from (0, 0, 0) is 1 unit

$$\Rightarrow \frac{1}{\sqrt{\frac{1}{a^2}+\frac{1}{b^2}+\frac{1}{c^2}}}=1\Rightarrow \frac{1}{a^2}+\frac{1}{b^2}+\frac{1}{c^2}=1 \quad (2)$$

From (1) and (2), we get $k/9=1$ i.e. $k=9$.

15. **(8)** The equation of the plane through the point $(1, -2, 1)$ and perpendicular to the planes $2x - 2y + z = 0$ and $x - y + 2z = 4$ is given by

$$\begin{vmatrix} x-1 & y+2 & z-1 \\ 2 & -2 & 1 \\ 1 & -1 & 2 \end{vmatrix} = 0$$

$\Rightarrow x + y + 1 = 0$

Its distance from the point (1, 2, 2) is $\left|\dfrac{1+2+1}{\sqrt{2}}\right|$

$= 2\sqrt{2}.$

16. **(9)** Since, the plane is parallel to the lines L_1 and L_2 with direction ratios $(1, 0, -1)$ and $(1, -1, 0)$, a vector perpendicular to L_1 and L_2 will be parallel to the normal $\vec{n}$ to the plane.

$$\therefore \vec{n} = \begin{vmatrix} \hat{i} & \hat{j} & \hat{k} \\ 1 & 0 & -1 \\ 1 & -1 & 0 \end{vmatrix} = -\hat{i} - \hat{j} - \hat{k}$$

Therefore, equation of the plane through (1, 1, 1) and having normal vector $\vec{n} = -\hat{i} - \hat{j} - \hat{k}$ is given by

$(\vec{r} - \vec{a}) \cdot \vec{n} = 0$

$\Rightarrow -1(x-1) - 1(y-1) - 1(z-1) = 0$

$\Rightarrow x + y + z = 3$

$\Rightarrow \dfrac{x}{3} + \dfrac{y}{3} + \dfrac{z}{3} = 1$ (1)

The plane meets the axes at A (3, 0, 0), B (0, 3, 0), and C (0, 0, 3) or $A\,(3\hat{i})$, $B\,(3\hat{j})$, and $C\,(3\hat{k})$.

Hence the volume of tetrahedron $OABC$ is

$\dfrac{1}{6}[3\hat{i}\ 3\hat{j}\ 3\hat{k}] = \dfrac{27}{6} = \dfrac{9}{2}$ cubic units.

17. The direction cosines of the line are $1/\sqrt{3}, 1/\sqrt{3}, 1/\sqrt{3}$.

Any point on the line at a distance t from $P(2, -1, 2)$ is $\left(2 - \dfrac{1}{\sqrt{3}}, -1 + \dfrac{t}{\sqrt{3}}, 2 + \dfrac{1}{\sqrt{3}}\right)$,

which lies on $2x + y + z - 9 \Rightarrow t = \sqrt{3}$.

18. **(5)** $\overrightarrow{AB} = \vec{\beta} - \vec{\alpha} = -2i - 3j - 6k$

The equation of the plane passing through B and perpendicular to AB is $(\vec{r} - \overrightarrow{OB}) \cdot \overrightarrow{AB} = 0$

$\Rightarrow \vec{r} \cdot (2i + 3j + 6k) + 28 = 0.$

Hence, the required distance from $\vec{c} = -i + j + k$ is

$$\left|\frac{\vec{c} \cdot (2i + 3j + 6k) + 28}{|2i + 3j + 6k|}\right| = \left|\frac{-2 + 3 + 6 + 28}{7}\right| = 5 \text{ units.}$$

19. **(7)**

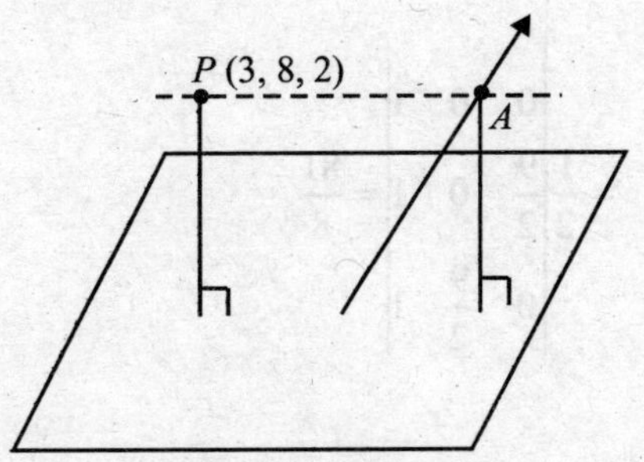

Fig. 3.22

Let a general point on the line be $A(2\lambda + 1, 4\lambda + 3, 3\lambda + 2)$.

Let this point lie at the same distance as the point P (3, 8, 2) from the plane $3x + 2y - 2z + 15 = 0$. Therefore,

$$\frac{3 \times 3 + 2 \times 8 - 2 \times 2 + 15}{\sqrt{17}}$$

$$= \frac{3(2\lambda + 1) + 2(4\lambda + 3) - 2(3\lambda + 2) + 15}{\sqrt{17}}$$

$\Rightarrow 36 = 8\lambda + 20$

$\Rightarrow \lambda = 2$

Therefore, A is (5, 11, 8).

$PA = \sqrt{(5-3)^2 + (11-8)^2 + (8-2)^2}$

$= \sqrt{4 + 9 + 36} = 7$

20. **(9)** The direction ratios of the line OP are 1,2,2.
Also, the equation of any plane through $P(1,2,2)$ is
$A(x-1) + B(y-2) + C(z-2) = 0$ (1)
If it is perpendicular to OP, then the normal to this plane is parallel to OP and so $\dfrac{A}{1} = \dfrac{B}{2} = \dfrac{C}{2}$

From (1), the equation of the plane through $P(1,2,2)$ and perpendicular to OP is given by
$1(x-1) + 2(y-2) + 2(z-2) = 0$

$\Rightarrow x+2y+2z=9$

The intercepts made by this plane on coordinate axes $= 9, \frac{9}{2}, \frac{9}{2}$

The coordinates of A (9, 0, 0), $B\left(0, \frac{9}{2}, 0\right)$, $C\left(0, 0 \frac{9}{2}\right)$.

The coordinates of the projections of ΔABCon the yz-plane are (0, 0, 0), $\left(0, \frac{9}{2}, 0\right), \left(0, 0, \frac{9}{2}\right)$, respectively

Δ_x = area projection of ΔABC on yz-plane

$$= \frac{1}{2}\begin{vmatrix} 0 & 0 & 1 \\ \frac{9}{2} & 0 & 1 \\ 0 & \frac{9}{2} & 1 \end{vmatrix} = \frac{81}{8}$$

Δ_y = area of projection of ΔABC on zx-plane $= \frac{81}{4}$

Δ_z = area of projection if ΔABC on xy-plane $= \frac{81}{4}$

The required area $= \sqrt{\Delta_x^2 + \Delta_y^2 + \Delta_z^2}$

$$= \sqrt{\left(\frac{81}{8}\right)^2 + \left(\frac{81}{4}\right)^2 + \left(\frac{81}{4}\right)^2}$$

$$= \left(\frac{81}{4}\right)\sqrt{\frac{1}{4} + 1 + 1}$$

$$= \frac{81}{4} \times \frac{3}{2} = \frac{243}{8}$$

According to the equations,

$$\text{Area} = \frac{243\lambda}{72} = \frac{243}{8}.$$

$\lambda = 9$